W9-ASV-001

# The first person to invent a car that runs on water…

… may be sitting right in your classroom! Every one of your students has the potential to make a difference. And realizing that potential starts right here, in your course.

When students succeed in your course—when they stay on-task and make the breakthrough that turns confusion into confidence—they are empowered to realize the possibilities for greatness that lie within each of them. We know your goal is to create an environment where students reach their full potential and experience the exhilaration of academic success that will last them a lifetime. *WileyPLUS* can help you reach that goal.

Wiley**PLUS** is an online suite of resources—including the complete text—that will help your students:

- come to class better prepared for your lectures
- get immediate feedback and context-sensitive help on assignments and quizzes
- track their progress throughout the course

"I just wanted to say how much this program helped me in studying… I was able to actually see my mistakes and correct them. … I really think that other students should have the chance to use *WileyPLUS*."

Ashlee Krisko, *Oakland University*

www.wiley.com/college/wileyplus

**80%** of students surveyed said it improved their understanding of the material. *

# WileyPLUS is built around the activities you perform

## Prepare & Present

Create outstanding class presentations using a wealth of resources, such as PowerPoint™ slides, image galleries, interactive simulations, demonstration videos, and more. Plus you can easily upload any materials you have created into your course, and combine them with the resources Wiley provides you with.

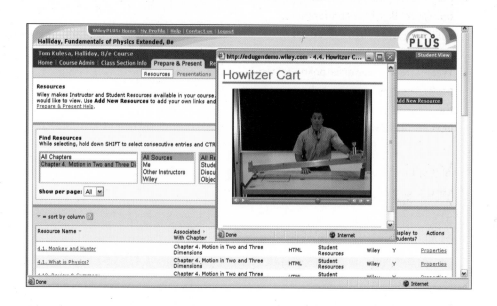

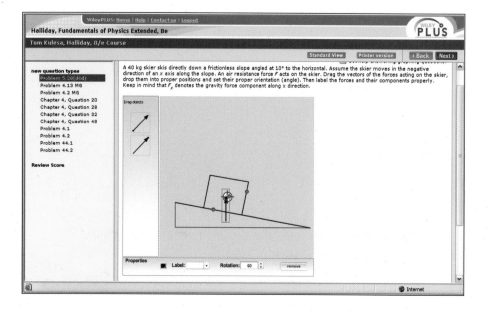

## Create Assignments

All of the end-of-chapter problems from the 8th edition of *Fundamentals of Physics* are available for assignment in an algorithmic format. Each problem has a link to the relevant portion of the text and a Hint or other problem solving help that can be made available to the student at the instructor's discretion. New to this edition are GO (Guided Online) problems that provide problem solving guidance in an interactive tutorial format. Also new are vector drawing and vector diagram problems and problems built around interactive simulations.

*Based on a spring 2005 survey of 972 student users of *WileyPLUS*

## TO THE STUDENT

# You have the potential to make a difference!

Will you be the first person to land on Mars? Will you invent a car that runs on water? But, first and foremost, will you get through this course?

*WileyPLUS* is a powerful online system packed with features to help you make the most of your potential, and get the best grade you can!

## With WileyPLUS you get:

### A complete online version of your text and other study resources

Study more effectively and get instant feedback when you practice on your own. Resources like self-assessment quizzes, concept simulations, and Interactive LearningWare bring the subject matter to life, and help you master the material.

### Problem-solving help, instant grading, and feedback on your homework and quizzes

You can keep all of your assigned work in one location, making it easy for you to stay on task. Plus, many homework problems contain direct links to the relevant portion of your text to help you deal with problem-solving obstacles at the moment they come up.

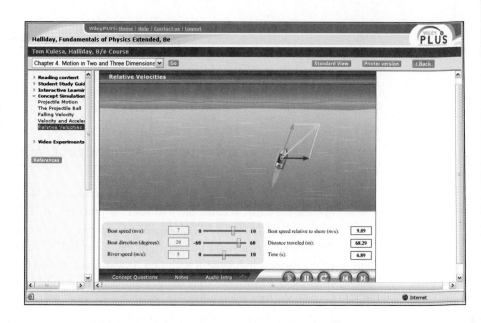

### The ability to track your progress and grades throughout the term.

A personal gradebook allows you to monitor your results from past assignments at any time. You'll always know exactly where you stand.

If your instructor uses *WileyPLUS*, you will receive a URL for your class. If not, your instructor can get more information about *WileyPLUS* by visiting www.wiley.com/college/wileyplus

"It has been a great help, and I believe it has helped me to achieve a better grade."
Michael Morris, *Columbia Basin College*

# 69% of students surveyed said it helped them get a better grade. *

## THE WILEY BICENTENNIAL–KNOWLEDGE FOR GENERATIONS

*E*ach generation has its unique needs and aspirations. When Charles Wiley first opened his small printing shop in lower Manhattan in 1807, it was a generation of boundless potential searching for an identity. And we were there, helping to define a new American literary tradition. Over half a century later, in the midst of the Second Industrial Revolution, it was a generation focused on building the future. Once again, we were there, supplying the critical scientific, technical, and engineering knowledge that helped frame the world. Throughout the 20th Century, and into the new millennium, nations began to reach out beyond their own borders and a new international community was born. Wiley was there, expanding its operations around the world to enable a global exchange of ideas, opinions, and know-how.

For 200 years, Wiley has been an integral part of each generation's journey, enabling the flow of information and understanding necessary to meet their needs and fulfill their aspirations. Today, bold new technologies are changing the way we live and learn. Wiley will be there, providing you the must-have knowledge you need to imagine new worlds, new possibilities, and new opportunities.

Generations come and go, but you can always count on Wiley to provide you the knowledge you need, when and where you need it!

**WILLIAM J. PESCE**
PRESIDENT AND CHIEF EXECUTIVE OFFICER

**PETER BOOTH WILEY**
CHAIRMAN OF THE BOARD

# Fundamentals of Physics

**8**E

HALLIDAY / RESNICK

# Fundamentals of Physics

**8E**

# Jearl Walker

Cleveland State University

John Wiley & Sons, Inc.

BICENTENNIAL
1807
WILEY
2007
BICENTENNIAL

ACQUISITIONS EDITOR  Stuart Johnson
PROJECT EDITOR  Geraldine Osnato
EDITORIAL ASSISTANT  Aly Rentrop
SENIOR MARKETING MANAGER  Amanda Wygal
SENIOR PRODUCTION EDITOR  Elizabeth Swain
TEXT DESIGNER  Madelyn Lesure
COVER DESIGNER  Norm Christiansen
DUMMY DESIGNER  Lee Goldstein
PHOTO EDITOR  Hilary Newman
ILLUSTRATION EDITOR  Anna Melhorn
ILLUSTRATION STUDIO  Radiant Illustrations Inc.
SENIOR MEDIA EDITOR  Thomas Kulesa
MEDIA PROJECT MANAGER  Bridget O'Lavin
COVER IMAGE  ©Eric Heller/Photo Researchers
BICENTENNIAL LOGO DESIGN  Richard J. Pacifico

This book was set in 10/12 Times Ten by Progressive Information Technologies and printed and bound by Von Hoffmann Press. The cover was printed by Von Hoffmann Press.

This book is printed on acid free paper.

Copyright © 2008 John Wiley & Sons, Inc. All rights reserved. No part of this publication may be reproduced, stored in a retrieval system or transmitted in any form or by any means, electronic, mechanical, photocopying, recording, scanning or otherwise, except as permitted under Sections 107 or 108 of the 1976 United States Copyright Act, without either the prior written permission of the Publisher, or authorization through payment of the appropriate per-copy fee to the Copyright Clearance Center, Inc. 222 Rosewood Drive, Danvers, MA 01923, website www.copyright.com. Requests to the Publisher for permission should be addressed to the Permissions Department, John Wiley & Sons, Inc., 111 River Street, Hoboken, NJ 07030-5774, (201)748-6011, fax (201)748-6008, or online at http://www.wiley.com/go/permissions.

To order books or for customer service please call 1-800-CALL WILEY (225-5945).

*Library of Congress Cataloging-in-Publication Data*
Halliday, David
  Fundamentals of physics.—8th ed./David Halliday, Robert Resnick, Jearl Walker.
    p. cm.
  Includes index.
  ISBN 978-0-470-04472-8 (acid-free paper)
  *Also catalogued as*
  Extended version:   ISBN 978-0-471-75801-3  (acid-free paper)
  1. Physics—Textbooks. I. Resnick, Robert II. Walker, Jearl III. Title.
  QC21.3.H35 2008
  530—dc22
                                                            2006041375

Printed in the United States of America

10  9  8  7  6  5

# Brief Contents

# Contents

## 19 The Kinetic Theory of Gases  507

What causes the fog that appears when a carbonated drink is opened?

## 20 Entropy and the Second Law of Thermodynamics  536

What is the connection between a rubber band's stretch and the direction of time?

## 21 Electric Charge  561

How can a video monitor in a surgical room increase the risk of bacterial contamination?

## 22 Electric Fields  580

How does a bee use electrostatics to collect and then distribute pollen grains?

## 23 Gauss' Law  605

How can lightning harm you even if it does not strike you?

## 24 Electric Potential  628

What danger does a sweater pose to a computer?

## 31 Electromagnetic Oscillations and Alternating Current  826

*How did a solar eruption knock out the power-grid system of Quebec?*

## 32 Maxwell's Equations; Magnetism of Matter  861

*How can a mural painting record the direction of Earth's magnetic field?*

## 33 Electromagnetic Waves  889

*What causes a sundog, the bright, colorful spot that can appear left or right of the Sun?*

## 34 Images  924

*How can a fish see clearly in both air and water simultaneously?*

## 35 Interference  958

*How do color-shifting inks on paper currency shift colors?*

## Appendices

Fun with a big challenge. That is how I have regarded physics since the day when Sharon, one of the students in a class I taught as a graduate student, suddenly demanded of me, "What has any of this got to do with my life?" Of course I immediately responded, "Sharon, this has everything to do with your life—this is physics."

She asked me for an example. I thought and thought but could not come up with a single one. That night I created *The Flying Circus of Physics* for Sharon but also for me because I realized her complaint was mine. I had spent six years slugging my way through many dozens of physics textbooks that were carefully written with the best of pedagogical plans, but there was something missing. Physics is the most interesting subject in the world because it is about how the world works, and yet the textbooks had been thoroughly wrung of any connection with the real world. The fun was missing.

I have packed a lot of real-world physics into this HRW book, connecting it with the new edition of *The Flying Circus of Physics*. Much of the material comes from the HRW classes I teach, where I can judge from the faces and blunt comments what material and presentations work and what do not. The notes I make on my successes and failures there help form the basis of this book. My message here is the same as I had with every student I've met since Sharon so long ago: "Yes, you *can* reason from basic physics concepts all the way to valid conclusions about the real world, and that understanding of the real world is where the fun is."

I have many goals in writing this book but the overriding one is to provide instructors with a tool by which they can teach students how to effectively read scientific material, identify fundamental concepts, reason through scientific ques-

tions, and solve quantitative problems. This process is not easy for either students or instructors. Indeed, the course associated with this book may be one of the most challenging of all the courses taken by a student. However, it can also be one of the most rewarding because it reveals the world's fundamental clockwork from which all scientific and engineering applications spring.

Many users of the seventh edition (both instructors and students) sent in comments and suggestions to improve the book. These improvements are now incorporated into the narrative and problems throughout the book. The publisher John Wiley & Sons and I regard the book as an ongoing project and encourage more input from users. You can send suggestions, corrections, and positive or negative comments to John Wiley & Sons (http:www.wiley.com/college/halliday) or Jearl Walker (mail address: Physics Department, Cleveland State University, Cleveland, OH 44115 USA; fax number: (USA) 216 687 2424; or email address: physics@wiley.com; or the blog site at www.flyingcircus-ofphysics.com). We may not be able to respond to all suggestions, but we keep and study each of them.

## Major Content Changes

• Flying Circus material has been incorporated into the text in several ways: chapter opening Puzzlers, Sample Problems, text examples, and end-of-chapter Problems. The purpose of this is two-fold: (1) make the subject more interesting and engaging, (2) show the student that the world around them can be examined and understood using the fundamental principles of physics.

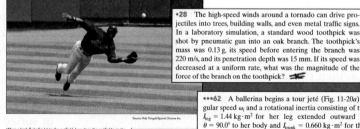

**4    Motion in Two and Three Dimensions**

When a high fly ball is hit to the outfield, how does the outfielder in the area know where to be in order to catch it? Often the outfielder will jog or run at a measured pace to the catch site, arriving just as the ball does. Playing experience surely helps, but some other factor seems to be involved.

*What clue is hidden in the ball's motion?*

The answer is in this chapter.

Source: Rob Tringali/Sports Chrome Inc.

58

---

**Sample Problem    4-8**

Suppose a baseball batter $B$ hits a high fly ball to the outfield, directly toward an outfielder $F$ and with a launch speed of $v_0 = 40$ m/s and a launch angle of $\theta_0 = 35°$. During the flight, a line from the outfielder to the ball makes an angle $\phi$ with the ground. Plot elevation angle $\phi$ versus time $t$, assuming that the outfielder is already positioned to catch the ball, is 6.0 m too close to the batter, and is 6.0 m too far away.

**KEY IDEAS**   (1) If we neglect air drag, the ball is a projectile for which the vertical motion and the horizontal motion can be analyzed separately. (2) Assuming the ball is caught at approximately the height it is hit, the horizontal distance traveled by the ball is the range $R$, given by Eq. 4-26 ($R = (v_0^2/g) \sin 2\theta_0$).

**Calculations:** The ball can be caught if the outfielder's distance from the batter is equal to the range $R$ of the ball. Using Eq. 4-26, we find

$$R = \frac{v_0^2}{g} \sin 2\theta_0 = \frac{(40 \text{ m/s})^2}{9.8 \text{ m/s}^2} \sin (70°) = 153.42 \text{ m}.$$

**FIG. 4-17**   The elevation a fielder is (a) defined and (b)

Figure 4-17a shows a snapshot of the ball in flight when the ball is at height $y$ and horizontal distance $x$ from the batter (who is at the origin). The horizontal distance of the ball from the outfielder is $R - x$, and the elevation angle $\phi$ of the ball in the outfielder's view is given by $\tan \phi = y/(R - x)$. For the height $y$, we use Eq. 4-22

•28   The high-speed winds around a tornado can drive projectiles into trees, building walls, and even metal traffic signs. In a laboratory simulation, a standard wood toothpick was shot by pneumatic gun into an oak branch. The toothpick's mass was 0.13 g, its speed before entering the branch was 220 m/s, and its penetration depth was 15 mm. If its speed was decreased at a uniform rate, what was the magnitude of the force of the branch on the toothpick?

•••62   A ballerina begins a tour jeté (Fig. 11-20a) with angular speed $\omega_i$ and a rotational inertia consisting of two parts: $I_{leg} = 1.44 \text{ kg} \cdot \text{m}^2$ for her leg extended outward at angle $\theta = 90.0°$ to her body and $I_{trunk} = 0.660 \text{ kg} \cdot \text{m}^2$ for the rest of her body (primarily her trunk). Near her maximum height she holds both legs at angle $\theta = 30.0°$ to her body and has angular speed $\omega_f$ (Fig. 11-20b). Assuming that $I_{trunk}$ has not changed, what is the ratio $\omega_f/\omega_i$?

3. *Long jump*   When an athlete takes off from the ground in a running long jump, the forces on the launching foot give the athlete an angular momentum with a forward rotation around a horizontal axis. Such rotation would not allow the jumper to land properly: In the landing, the legs should be together and extended forward at an angle so that the heels mark the sand at the greatest distance. Once airborne, the angular momentum cannot change (it is conserved) because no external torque acts to change it. However, the jumper can shift most of the angular momentum to the arms by rotating them in windmill fashion (Fig. 11-19). Then the body remains upright and in the proper orientation for landing.

• Links to *The Flying Circus of Physics* are shown throughout the text material and end-of-chapter problems with a biplane icon. In the electronic version of this book, clicking on the icon takes you to the corresponding item in *Flying Circus*. The bibliography of *Flying Circus* (over 10 000 references to scientific and engineering journals) is located at www.flyingcircusofphysics.com.

• The Newtonian gravitational law, the Coulomb law, and the Biot-Savart law are now introduced in unit-vector notation.

• Most of the chapter-opening puzzlers (the examples of applied physics designed to entice a reader into each chapter) are new and come straight from research journals in many different fields.

• Several thousand of the end-of-chapter problems have been rewritten to streamline both the presentation and the answer. Many new problems of the moderate and difficult categories have been included.

## Chapter Features

**Opening puzzlers.** A curious puzzling situation opens each chapter and is explained somewhere within the chapter, to entice a student to read the chapter. These features, which are a hallmark of *Fundamentals of Physics*, are based on current research as reported in scientific, engineering, medical, and legal journals.

**What is physics?** The narrative of every chapter now begins with this question, and with an answer that pertains to the subject of the chapter. (A plumber once asked me, "What do you do for a living?" I replied, "I teach physics." He thought for several minutes and then asked, "What is physics?" The plumber's career was entirely based on physics, yet he did not even know what physics is. Many students in introductory physics do not know what physics is but assume that it is irrelevant to their chosen career.)

**Checkpoints** are stopping points that effectively ask the student, "Can you answer this question with some reasoning based on the narrative or sample problem that you just read?" If not, then the student should go back over that previous material before traveling deeper into the chapter. For example, see Checkpoint 1 on page 62 and Checkpoint 2 on page 280. *Answers to all checkpoints are in the back of the book.*

**Sample problems** are chosen to demonstrate how problems can be solved with reasoned solutions rather than quick and simplistic plugging of numbers into an equation with no regard for what the equation means.

The sample problems with the label "Build your skill" are typically longer, with more guidance.

**Key Ideas** in the sample problems focus a student on the basic concepts at the root of the solution to a problem. In effect, these key ideas say, "We start our solution by using this basic concept, a procedure that prepares us for solving many other problems. We don't start by grabbing an equation for a quick plug-and-chug, a procedure that prepares us for nothing."

**Problem-solving tactics** contain helpful instructions to guide the beginning physics student as to how to solve problems and avoid common errors.

**Review & Summary** is a brief outline of the chapter contents that contains the essential concepts but which is not a substitute for reading the chapter.

**Questions** are like the checkpoints and require reasoning and understanding rather than calculations. *Answers to the odd-numbered questions are in the back of the book.*

**Problems** are grouped under section titles and are labeled according to difficulty. *Answers to the odd-numbered problems are in the back of the book.*

**Icons for additional help.** When worked-out solutions are provided either in print or electronically for certain of the odd-numbered problems, the statements for those problems include a trailing icon to alert both student and instructor as to where the solutions are located. An icon guide is provided here and at the beginning of each set of problems:

| | |
|---|---|
| **GO** | Tutoring problem available (at instructor's discretion) in *WileyPLUS* and WebAssign |
| **SSM** | Worked-out solution available in Student Solutions Manual |
| **WWW** | Worked-out solution is at |
| • – ••• | Number of dots indicates level of problem difficulty |
| **ILW** | Interactive solution is at |
| | Additional information available in *The Flying Circus of Physics* and at flyingcircusofphysics.com |

http://www.wiley.com/college/halliday

SSM Solution is in the Student Solutions Manual.
WWW Solution is at
 http://www.wiley.com/college/halliday
ILW Interactive LearningWare solution is at
 http://www.wiley.com/college/halliday

**Additional problems.** These problems are not ordered or sorted in any way so that a student must determine which parts of the chapter apply to any given problem.

## Additional Features

**Reasoning versus plug-and-chug.** A primary goal of this book is to teach students to reason through challenging situations, from basic principles to a solution. Although some plug-and-chug homework problems remain in the book (on purpose), most homework problems emphasize reasoning.

**Chapters of reasonable length.** To avoid producing a book thick enough to stop a bullet (and thus also a student), I have made the chapters of reasonable length. I explain enough to get a student going but not so much that a student no longer must analyze and fuse ideas. After all, a student will need the skill of analyzing and fusing ideas long after this book is read and the course is completed.

**Use of vector-capable calculators.** When vector calculations in a sample problem can be performed directly on-screen with a vector-capable calculator, the solution of the sample problem indicates that fact but still carries through the traditional component analysis. When vector calculations cannot be performed directly on-screen, the solution explains why.

**Graphs as puzzles.** These are problems that give a graph and ask for a result that requires much more than just reading off a data point from the graph. Rather, the solution requires an understanding of the physical arrangement in a problem and the principles behind the associated equations. These problems are more like Sherlock Holmes puzzles because a student must decide what data are important. For examples, see problem 50 on page 80, problem 12 on page 108, and problem 22 on page 231.

**Problems with applied physics,** based on published research, appear in many places, either as the opening puzzler of a chapter, a sample problem, or a homework problem. For example, see the opening puzzler for Chapter 4 on page 58, Sample Problem 4-8 on pages 69-70, and homework problem 62 on page 302. For an example of homework problems that build on a continuing story, see problems 2, 39, and 61 on pages 131, 134, and 136.

**Problems with novel situations.** Here is one of several hundred such problems: Problem 69 on page 113 relates a true story of how Air Canada flight 143 ran out of fuel at an altitude of 7.9 km because the crew and airport personnel did not consider the units for the fuel (an important lesson for students who tend to "blow off" units).

## Versions of the Text

To accommodate the individual needs of instructors and students, the eighth edition of *Fundamentals of Physics* is available in a number of different versions.

The **Regular Edition** consists of Chapters 1 through 37 (ISBN 978-0-470-04472-8).

The **Extended Edition** contains six additional chapters on quantum physics and cosmology, Chapters 1–44 (ISBN 978-0-471-75801-3).

Both editions are available as single, hard-cover books, or in the following alternative versions:

- **Volume 1** - Chapters 1–20 (Mechanics and Thermodynamics), hardcover, ISBN 978-0-47004473-5

- **Volume 2** - Chapters 21–44 (E&M, Optics, and Quantum Physics), hardcover, ISBN 978-0-470-04474-2

- **Part 1** - Chapters 1–11, paperback, ISBN 978-0-470-04475-9

- **Part 2** - Chapters 12–20, paperback, ISBN 978-0-470-04476-6

- **Part 3** - Chapters 21–32, paperback, ISBN 978-0-470-04477-3

- **Part 4** - Chapters 33–37, paperback, ISBN 978-0-470-04478-0

- **Part 5** - Chapters 38–44, paperback, ISBN 978-0-470-04479-7

## WileyPLUS

There have been several significant additions to the WileyPLUS course that accompanies *Fundamentals of Physics:*

- All of the end-of-chapter problems have been coded and are now available for assignment.

- Every problem has an associated Hint that can made available to the students at the instructor's discretion.

- There are approximately 400 additional Sample Problems available to the student **at the instructor's discretion.** The Sample Problems are written in the same style and format as those in the text, i.e., they are intended to give the student transferable problem-solving skills rather than specific recipes.

- For every chapter, approximately 6 problems are available in a tutorial format that provides step-by-step, interactive problem-solving guidance. Most are marked here in this book with the icon ![GO] ("Guided Online") but more are being added.

- For every chapter, approximately 6 problems are available in a version that requires the student to enter an algebraic answer.

- There are vector drawing and vector diagram problems that use "drag 'n drop" functionality to assess the students' ability to draw vectors and vector diagrams.

- There are simulation problems that require the student to work with a java applet.

The overall purpose of this new material is to move the on-line homework experience beyond simple Right/Wrong grading and provide meaningful problem solving guidance and support.

## Instructor's Supplements

**Instructor's Solutions Manual** by Sen-Ben Liao, Lawrence Livermore National Laboratory. This manual provides worked-out solutions for all problems found at the end of each chapter.

### Instructor Companion Site

http://www.wiley.com/college/halliday

• **Instructor's Manual** by J. Richard Christman, U.S. Coast Guard Academy. This resource contains lecture notes outlining the most important topics of each chapter; demonstration experiments; laboratory and computer projects; film and video sources; answers to all Questions, Exercises, Problems, and Checkpoints; and a correlation guide to the Questions, Exercises, and Problems in the previous edition. It also contains a complete list of all problems for which solutions are available to students (SSM, WWW, and ILW).

• **Lecture PowerPoint Slides** by Athos Petrou and John Cerne of the University of Buffalo. These PowerPoints cover the entire book and are heavily illustrated with figures from the text.

• **Classroom Response Systems ("Clicker") Questions** by David Marx, Illinois State University. There are two sets of questions available: Reading Quiz and Interactive Lecture. The Reading Quiz questions are intended to be relatively straightforward for any student who read the assigned material. The Interactive Lecture questions are intended to for use in an interactive lecture setting.

• **Wiley Physics Simulations** by Andrew Duffy, Boston University. 50 interactive simulations (Java applets) that can be used for classroom demonstrations.

• **Wiley Physics Demonstrations** by David Maiullo, Rutgers University. This is a collection of digital videos of 80 standard physics demonstrations. They can be shown in class or accessed from the student companion site. There is an accompanying Instructor's Guide that includes "clicker" questions.

• **Test Bank** by J. Richard Christman, U.S. Coast Guard Academy. The Test Bank includes more than 2200 multiple-choice questions. These items are also available in the Computerized Test Bank which provides full editing features to help you customize tests (available in both IBM and Macintosh versions).

• All of the *Instructor's Solutions Manual* in MSWord, and pdf files

• All text illustrations, suitable for both classroom projection and printing.

**On-line homework and quizzing.** In addition to Wiley*PLUS, Fundamentals of Physics*, eighth edition also supports WebAssignPLUS and CAPA, which are other programs that give instructors the ability to deliver and grade homework and quizzes on-line.

**WebCT and Blackboard.** A variety of materials have been prepared for easy incorporation in either WebCT or Blackboard. WebCT and Blackboard are powerful and easy-to-use web-based course-management systems that allow instructors to set up complete on-line courses with chat rooms, bulletin boards, quizzing, student tracking, etc.

## Student's Supplements

**Student Companion site.** This web site

http://www.wiley.com/college/halliday

was developed specifically for *Fundamentals of Physics*, eighth edition, and is designed to further assist students in the study of physics. The site includes solutions to selected end-of-chapter problems (which are identified with a www icon in the text); self-quizzes; simulation exercises; tips on how to make best use of a programmable calculator; and the Interactive LearningWare tutorials that are described below.

**Student Study Guide.** The student study guide consists of an overview of the chapter's important concepts, hints for solving end-of-chapter questions/problems, and practice quizzes.

**Student's Solutions Manual** by J. Richard Christman, U.S. Coast Guard Academy and Edward Derringh, Wentworth Institute. This manual provides student with complete worked-out solutions to 15 percent of the problems found at the end of each chapter within the text. These problems are indicated with an ssm icon.

**Interactive LearningWare.** This software guides students through solutions to 200 of the end-of-chapter problems. These problems are indicated with an ilw icon. The solutions process is developed interactively, with appropriate feedback and access to error-specific help for the most common mistakes.

**Wiley Desktop Edition.** An electronic version of *Fundamentals of Physics*, eighth edition containing the complete, extended version of the text is available for download at:

www.wiley.com/college/desktop

Wiley Desktop Editions are a cost effective alternative to the printed text.

**Physics as a Second Language:** *Mastering Problem Solving* by Thomas Barrett, of Ohio State University. This brief paperback teaches the student how to approach problems more efficiently and effectively. The student will learn how to recognize common patterns in physics problems, break problems down into manageable steps, and apply appropriate techniques. The book takes the student step-by-step through the solutions to numerous examples.

# Acknowledgments

A great many people have contributed to this book. J. Richard Christman, of the U.S. Coast Guard Academy, has once again created many fine supplements; his recommendations to this book have been invaluable. Sen-Ben Liao of Lawrence Livermore National Laboratory, James Whitenton of Southern Polytechnic State University, and Jerry Shi, of Pasadena City College, performed the Herculean task of working out solutions for every one of the homework problems in the book. At John Wiley publishers, the book received support from Stuart Johnson, the editor who oversaw the entire project, Tom Kulesa, who coordinated the state-of-the-art media package, and Geraldine Osnato, who managed a super team to create an impressive supplements package. We thank Elizabeth Swain, the production editor, for pulling all the pieces together during the complex production process. We also thank Maddy Lesure, for her design of both the text and the book cover; Lee Goldstein for her page make-up; Helen Walden for her copyediting; Anna Melhorn for managing the illustration program; and Lilian Brady for her proofreading. Hilary Newman was inspired in the search for unusual and interesting photographs. Both the publisher John Wiley & Sons, Inc. and Jearl Walker would like to thank the following for comments and ideas about the 7th edition: Richard Woodard, University of Florida; David Wick, Clarkson University; Patrick Rapp, University of Puerto Rico at Mayagüez; Nora Thornber, Raritan Valley Community College; Laurence I. Gould, University of Hartford; Greg Childers, California State University at Fullerton; Asha Khakpour of California State University at Fullerton; Joe F. McCullough, Cabrillo College. Finally, our external reviewers have been outstanding and we acknowledge here our debt to each member of that team.

Maris A. Abolins
*Michigan State University*

Edward Adelson
*Ohio State University*

Nural Akchurin
*Texas Tech*

Barbara Andereck
*Ohio Wesleyan University*

Mark Arnett
*Kirkwood Community College*

Arun Bansil
*Northeastern University*

Richard Barber
*Santa Clara University*

Neil Basecu
*Westchester Community College*

Anand Batra
*Howard University*

Richard Bone
*Florida International University*

Michael E. Browne
*University of Idaho*

Timothy J. Burns
*Leeward Community College*

Joseph Buschi
*Manhattan College*

Philip A. Casabella
*Rensselaer Polytechnic Institute*

Randall Caton
*Christopher Newport College*

Roger Clapp
*University of South Florida*

W. R. Conkie
*Queen's University*

Renate Crawford
*University of Massachusetts-Dartmouth*

Mike Crivello
*San Diego State University*

Robert N. Davie, Jr.
*St. Petersburg Junior College*

Cheryl K. Dellai
*Glendale Community College*

Eric R. Dietz
*California State University at Chico*

N. John DiNardo
*Drexel University*

Eugene Dunnam
*University of Florida*

Robert Endorf
*University of Cincinnati*

F. Paul Esposito
*University of Cincinnati*

Jerry Finkelstein
*San Jose State University*

Robert H. Good
*California State University-Hayward*

John B. Gruber
*San Jose State University*

Ann Hanks
*American River College*

Randy Harris
*University of California-Davis*

Samuel Harris
*Purdue University*

Harold B. Hart
*Western Illinois University*

Rebecca Hartzler
*Seattle Central Community College*

John Hubisz
*North Carolina State University*

Joey Huston
*Michigan State University*

David Ingram
*Ohio University*

Shawn Jackson
*University of Tulsa*

Hector Jimenez
*University of Puerto Rico*

Sudhakar B. Joshi
*York University*

Leonard M. Kahn
*University of Rhode Island*

Leonard Kleinman
*University of Texas at Austin*

Craig Kletzing
*University of Iowa*

Arthur Z. Kovacs
*Rochester Institute of Technology*

Kenneth Krane
*Oregon State University*

Priscilla Laws
*Dickinson College*

Edbertho Leal
*Polytechnic University of Puerto Rico*

Vern Lindberg
*Rochester Institute of Technology*

Peter Loly
*University of Manitoba*

Andreas Mandelis
*University of Toronto*

Robert R. Marchini
*Memphis State University*

Paul Marquard
*Caspar College*

David Marx
*Illinois State University*

James H. McGuire
*Tulane University*

David M. McKinstry
*Eastern Washington University*

Eugene Mosca
*United States Naval Academy*

James Napolitano
*Rensselaer Polytechnic Institute*

Michael O'Shea
*Kansas State University*

Patrick Papin
*San Diego State University*

Kiumars Parvin
*San Jose State University*

Robert Pelcovits
*Brown University*

Oren P. Quist
*South Dakota State University*

Joe Redish
*University of Maryland*

Timothy M. Ritter
*University of North Carolina at Pembroke*

Gerardo A. Rodriguez
*Skidmore College*

John Rosendahl
*University of California at Irvine*

Todd Ruskell
*Colorado School of Mines*

Michael Schatz
*Georgia Institute of Technology*

Darrell Seeley
*Milwaukee School of Engineering*

Bruce Arne Sherwood
*North Carolina State University*

Ross L. Spencer
*Brigham Young University*

Paul Stanley
*Beloit College*

Harold Stokes
*Brigham Young University*

Michael G. Strauss
*University of Oklahoma*

Jay D. Strieb
*Villanova University*

Dan Styer
*Oberlin College*

Michael Tammaro
*University of Rhode Island*

Marshall Thomsen
*Eastern Michigan University*

David Toot
*Alfred University*

Tung Tsang
*Howard University*

J. S. Turner
*University of Texas at Austin*

T. S. Venkataraman
*Drexel University*

Gianfranco Vidali
*Syracuse University*

Fred Wang
*Prairie View A & M*

Robert C. Webb
*Texas A & M University*

William M. Whelan
*Ryerson Polytechnic University*

George Williams
*University of Utah*

David Wolfe
*University of New Mexico*

# Fundamentals of Physics

**8**E

# Measurement

*When an earthquake strikes a populated region, it can shake apart buildings and other structures or cause them to topple over. However, in some regions it can cause structures to sink into the ground until they are significantly submerged, as if the structures were on a dense fluid instead of solid ground.*

## How can a building sink into the ground?

The answer is in this chapter.

©AP/Wide World Photos

## 1-1 | WHAT IS PHYSICS?

Science and engineering are based on measurements and comparisons. Thus, we need rules about how things are measured and compared, and we need experiments to establish the units for those measurements and comparisons. One purpose of physics (and engineering) is to design and conduct those experiments.

For example, physicists strive to develop clocks of extreme accuracy so that any time or time interval can be precisely determined and compared. You may wonder whether such accuracy is actually needed or worth the effort. Here is one example of the worth: Without clocks of extreme accuracy, the Global Positioning System (GPS) that is now vital to worldwide navigation would be useless.

## 1-2 | Measuring Things

We discover physics by learning how to measure the quantities involved in physics. Among these quantities are length, time, mass, temperature, pressure, and electric current.

We measure each physical quantity in its own units, by comparison with a **standard.** The **unit** is a unique name we assign to measures of that quantity—for example, meter (m) for the quantity length. The standard corresponds to exactly 1.0 unit of the quantity. As you will see, the standard for length, which corresponds to exactly 1.0 m, is the distance traveled by light in a vacuum during a certain fraction of a second. We can define a unit and its standard in any way we care to. However, the important thing is to do so in such a way that scientists around the world will agree that our definitions are both sensible and practical.

Once we have set up a standard—say, for length—we must work out procedures by which any length whatever, be it the radius of a hydrogen atom, the wheelbase of a skateboard, or the distance to a star, can be expressed in terms of the standard. Rulers, which approximate our length standard, give us one such procedure for measuring length. However, many of our comparisons must be indirect. You cannot use a ruler, for example, to measure the radius of an atom or the distance to a star.

There are so many physical quantities that it is a problem to organize them. Fortunately, they are not all independent; for example, speed is the ratio of a length to a time. Thus, what we do is pick out—by international agreement—a small number of physical quantities, such as length and time, and assign standards to them alone. We then define all other physical quantities in terms of these *base quantities* and their standards (called *base standards*). Speed, for example, is defined in terms of the base quantities length and time and their base standards.

Base standards must be both accessible and invariable. If we define the length standard as the distance between one's nose and the index finger on an outstretched arm, we certainly have an accessible standard—but it will, of course, vary from person to person. The demand for precision in science and engineering pushes us to aim first for invariability. We then exert great effort to make duplicates of the base standards that are accessible to those who need them.

## 1-3 | The International System of Units

In 1971, the 14th General Conference on Weights and Measures picked seven quantities as base quantities, thereby forming the basis of the International System of Units, abbreviated SI from its French name and popularly known as the *metric system*. Table 1-1 shows the units for the three base quantities—length, mass, and time—that we use in the early chapters of this book. These units were defined to be on a "human scale."

**TABLE 1-1**

**Units for Three SI Base Quantities**

| Quantity | Unit Name | Unit Symbol |
|----------|-----------|-------------|
| Length   | meter     | m           |
| Time     | second    | s           |
| Mass     | kilogram  | kg          |

Many SI *derived units* are defined in terms of these base units. For example, the SI unit for power, called the **watt** (W), is defined in terms of the base units for mass, length, and time. Thus, as you will see in Chapter 7,

$$1 \text{ watt} = 1 \text{ W} = 1 \text{ kg} \cdot \text{m}^2/\text{s}^3, \tag{1-1}$$

where the last collection of unit symbols is read as kilogram-meter squared per second cubed.

To express the very large and very small quantities we often run into in physics, we use *scientific notation*, which employs powers of 10. In this notation,

$$3\,560\,000\,000 \text{ m} = 3.56 \times 10^9 \text{ m} \tag{1-2}$$

and

$$0.000\,000\,492 \text{ s} = 4.92 \times 10^{-7} \text{ s}. \tag{1-3}$$

Scientific notation on computers sometimes takes on an even briefer look, as in 3.56 E9 and 4.92 E−7, where E stands for "exponent of ten." It is briefer still on some calculators, where E is replaced with an empty space.

As a further convenience when dealing with very large or very small measurements, we use the prefixes listed in Table 1-2. As you can see, each prefix represents a certain power of 10, to be used as a multiplication factor. Attaching a prefix to an SI unit has the effect of multiplying by the associated factor. Thus, we can express a particular electric power as

$$1.27 \times 10^9 \text{ watts} = 1.27 \text{ gigawatts} = 1.27 \text{ GW} \tag{1-4}$$

or a particular time interval as

$$2.35 \times 10^{-9} \text{ s} = 2.35 \text{ nanoseconds} = 2.35 \text{ ns}. \tag{1-5}$$

Some prefixes, as used in milliliter, centimeter, kilogram, and megabyte, are probably familiar to you.

**TABLE 1-2**

**Prefixes for SI Units**

| Factor | Prefix[a] | Symbol |
|---|---|---|
| $10^{24}$ | yotta- | Y |
| $10^{21}$ | zetta- | Z |
| $10^{18}$ | exa- | E |
| $10^{15}$ | peta- | P |
| $10^{12}$ | tera- | T |
| **$10^{9}$** | **giga-** | **G** |
| **$10^{6}$** | **mega-** | **M** |
| **$10^{3}$** | **kilo-** | **k** |
| $10^{2}$ | hecto- | h |
| $10^{1}$ | deka- | da |
| $10^{-1}$ | deci- | d |
| **$10^{-2}$** | **centi-** | **c** |
| **$10^{-3}$** | **milli-** | **m** |
| **$10^{-6}$** | **micro-** | **$\mu$** |
| **$10^{-9}$** | **nano-** | **n** |
| **$10^{-12}$** | **pico-** | **p** |
| $10^{-15}$ | femto- | f |
| $10^{-18}$ | atto- | a |
| $10^{-21}$ | zepto- | z |
| $10^{-24}$ | yocto- | y |

[a]The most frequently used prefixes are shown in bold type.

## 1-4 | Changing Units

We often need to change the units in which a physical quantity is expressed. We do so by a method called *chain-link conversion*. In this method, we multiply the original measurement by a **conversion factor** (a ratio of units that is equal to unity). For example, because 1 min and 60 s are identical time intervals, we have

$$\frac{1 \text{ min}}{60 \text{ s}} = 1 \quad \text{and} \quad \frac{60 \text{ s}}{1 \text{ min}} = 1.$$

Thus, the ratios (1 min)/(60 s) and (60 s)/(1 min) can be used as conversion factors. This is *not* the same as writing $\frac{1}{60} = 1$ or $60 = 1$; each *number* and its *unit* must be treated together.

Because multiplying any quantity by unity leaves the quantity unchanged, we can introduce conversion factors wherever we find them useful. In chain-link conversion, we use the factors to cancel unwanted units. For example, to convert 2 min to seconds, we have

$$2 \text{ min} = (2 \text{ min})(1) = (2 \text{ min})\left(\frac{60 \text{ s}}{1 \text{ min}}\right) = 120 \text{ s}. \tag{1-6}$$

If you introduce a conversion factor in such a way that unwanted units do *not* cancel, invert the factor and try again. In conversions, the units obey the same algebraic rules as variables and numbers.

Appendix D gives conversion factors between SI and other systems of units, including non-SI units still used in the United States. However, the conversion factors are written in the style of "1 min = 60 s" rather than as a ratio. The following sample problem gives an example of how to set up such ratios.

## Sample Problem 1-1

When, according to legend, Pheidippides ran from Marathon to Athens in 490 B.C. to bring word of the Greek victory over the Persians, he probably ran at a speed of about 23 rides per hour (rides/h). The ride is an ancient Greek unit for length, as are the stadium and the plethron: 1 ride was defined to be 4 stadia, 1 stadium was defined to be 6 plethra, and, in terms of a modern unit, 1 plethron is 30.8 m. How fast did Pheidippides run in kilometers per second (km/s)?

**KEY IDEA**    In chain-link conversions, we write the conversion factors as ratios that will eliminate unwanted units.

**Calculation:** Here we write

$$23 \text{ rides/h} = \left(23 \frac{\text{rides}}{\text{h}}\right)\left(\frac{4 \text{ stadia}}{1 \text{ ride}}\right)\left(\frac{6 \text{ plethra}}{1 \text{ stadium}}\right)$$

$$\times \left(\frac{30.8 \text{ m}}{1 \text{ plethron}}\right)\left(\frac{1 \text{ km}}{1000 \text{ m}}\right)\left(\frac{1 \text{ h}}{3600 \text{ s}}\right)$$

$$= 4.7227 \times 10^{-3} \text{ km/s} \approx 4.7 \times 10^{-3} \text{ km/s.}$$
(Answer)

## Sample Problem 1-2

The cran is a British volume unit for freshly caught herrings: 1 cran = 170.474 liters (L) of fish, about 750 herrings. Suppose that, to be cleared through customs in Saudi Arabia, a shipment of 1255 crans must be declared in terms of cubic covidos, where the covido is an Arabic unit of length: 1 covido = 48.26 cm. What is the required declaration?

**KEY IDEA**    From Appendix D we see that 1 L is equivalent to 1000 cm³. To convert from *cubic* centimeters to *cubic* covidos, we must *cube* the conversion ratio between centimeters and covidos.

**Calculation:** We write the following chain-link conversion:

1255 crans

$$= (1255 \text{ crans})\left(\frac{170.474 \text{ L}}{1 \text{ cran}}\right)\left(\frac{1000 \text{ cm}^3}{1 \text{ L}}\right)\left(\frac{1 \text{ covido}}{48.26 \text{ cm}}\right)^3$$

$$= 1.903 \times 10^3 \text{ covidos}^3.$$
(Answer)

---

### PROBLEM-SOLVING TACTICS

*Tactic 1: Significant Figures and Decimal Places* If you calculated the answer to Sample Problem 1-1 without your calculator automatically rounding it off, the number 4.722 666 666 67 × 10⁻³ might have appeared in the display. The precision implied by this number is meaningless. We rounded the answer to 4.7 × 10⁻³ km/s so as not to imply that it is more precise than the given data. The given speed of 23 rides/h consists of two digits, called **significant figures.** Thus, we rounded the answer to two significant figures. In this book, final results of calculations are often rounded to match the least number of significant figures in the given data. (However, sometimes an extra significant figure is kept.) When the leftmost of the digits to be discarded is 5 or more, the last remaining digit is rounded up; otherwise it is retained as is. For example, 11.3516 is rounded to three significant

figures as 11.4 and 11.3279 is rounded to three significant figures as 11.3. (The answers to sample problems in this book are usually presented with the symbol = instead of ≈ even if rounding is involved.)

When a number such as 3.15 or 3.15 × 10³ is provided in a problem, the number of significant figures is apparent, but how about the number 3000? Is it known to only one significant figure (3 × 10³)? Or is it known to as many as four significant figures (3.000 × 10³)? In this book, we assume that all the zeros in such given numbers as 3000 are significant, but you had better not make that assumption elsewhere.

Don't confuse *significant figures* with *decimal places*. Consider the lengths 35.6 mm, 3.56 m, and 0.00356 m. They all have three significant figures but they have one, two, and five decimal places, respectively.

# 1-5 | Length

In 1792, the newborn Republic of France established a new system of weights and measures. Its cornerstone was the meter, defined to be one ten-millionth of the distance from the north pole to the equator. Later, for practical reasons, this Earth standard was abandoned and the meter came to be defined as the distance between two fine lines engraved near the ends of a platinum–iridium bar, the **standard meter bar,** which was kept at the International Bureau of Weights and Measures near Paris. Accurate copies of the bar were sent to standardizing laboratories throughout the world. These **secondary standards** were used to produce other, still more accessible standards, so that ultimately every measuring device derived its authority from the standard meter bar through a complicated chain of comparisons.

Eventually, a standard more precise than the distance between two fine scratches on a metal bar was required. In 1960, a new standard for the meter, based on the wavelength of light, was adopted. Specifically, the standard for the meter was redefined to be 1 650 763.73 wavelengths of a particular orange-red light emitted by atoms of krypton-86 (a particular isotope, or type, of krypton) in a gas discharge tube. This awkward number of wavelengths was chosen so that the new standard would be close to the old meter-bar standard.

By 1983, however, the demand for higher precision had reached such a point that even the krypton-86 standard could not meet it, and in that year a bold step was taken. The meter was redefined as the distance traveled by light in a specified time interval. In the words of the 17th General Conference on Weights and Measures:

> The meter is the length of the path traveled by light in a vacuum during a time interval of 1/299 792 458 of a second.

This time interval was chosen so that the speed of light $c$ is exactly

$$c = 299\ 792\ 458 \text{ m/s.}$$

Measurements of the speed of light had become extremely precise, so it made sense to adopt the speed of light as a defined quantity and to use it to redefine the meter.

Table 1-3 shows a wide range of lengths, from that of the universe (top line) to those of some very small objects.

**TABLE 1-3**

**Some Approximate Lengths**

| Measurement | Length in Meters |
|---|---|
| Distance to the first galaxies formed | $2 \times 10^{26}$ |
| Distance to the Andromeda galaxy | $2 \times 10^{22}$ |
| Distance to the nearby star Proxima Centauri | $4 \times 10^{16}$ |
| Distance to Pluto | $6 \times 10^{12}$ |
| Radius of Earth | $6 \times 10^{6}$ |
| Height of Mt. Everest | $9 \times 10^{3}$ |
| Thickness of this page | $1 \times 10^{-4}$ |
| Length of a typical virus | $1 \times 10^{-8}$ |
| Radius of a hydrogen atom | $5 \times 10^{-11}$ |
| Radius of a proton | $1 \times 10^{-15}$ |

## PROBLEM-SOLVING TACTICS

*Tactic 2:* **Order of Magnitude** The *order of magnitude* of a number is the power of ten when the number is expressed in scientific notation. For example, if $A = 2.3 \times 10^4$ and $B = 7.8 \times 10^4$, then the orders of magnitude of both $A$ and $B$ are 4.

Often, engineering and science professionals will estimate the result of a calculation to the *nearest* order of magnitude. For our example, the nearest order of magnitude is 4 for $A$ and 5 for $B$. Such estimation is common when detailed or precise data required in the calculation are not known or easily found. Sample Problem 1-3 gives an example.

## Sample Problem | 1-3 | Build your skill

The world's largest ball of string is about 2 m in radius. To the nearest order of magnitude, what is the total length $L$ of the string in the ball?

**KEY IDEA** We could, of course, take the ball apart and measure the total length $L$, but that would take great effort and make the ball's builder most unhappy. Instead, because we want only the nearest order of magnitude, we can estimate any quantities required in the calculation.

*Calculations:* Let us assume the ball is spherical with radius $R = 2$ m. The string in the ball is not closely packed (there are uncountable gaps between adjacent sections of string). To allow for these gaps, let us somewhat overestimate the cross-sectional area of the string by assuming the cross section is square, with an edge

length $d = 4$ mm. Then, with a cross-sectional area of $d^2$ and a length $L$, the string occupies a total volume of

$$V = (\text{cross-sectional area})(\text{length}) = d^2 L.$$

This is approximately equal to the volume of the ball, given by $\frac{4}{3}\pi R^3$, which is about $4R^3$ because $\pi$ is about 3. Thus, we have

$$d^2 L = 4R^3,$$

or

$$L = \frac{4R^3}{d^2} = \frac{4(2 \text{ m})^3}{(4 \times 10^{-3} \text{ m})^2}$$

$$= 2 \times 10^6 \text{ m} \approx 10^6 \text{ m} = 10^3 \text{ km.}$$

(Answer)

(Note that you do not need a calculator for such a simplified calculation.) To the nearest order of magnitude, the ball contains about 1000 km of string!

## 1-6 | Time

Time has two aspects. For civil and some scientific purposes, we want to know the time of day so that we can order events in sequence. In much scientific work, we want to know how long an event lasts. Thus, any time standard must be able to answer two questions: "*When* did it happen?" and "What is its *duration*?" Table 1-4 shows some time intervals.

**TABLE 1-4**

**Some Approximate Time Intervals**

| Measurement | Time Interval in Seconds |
| --- | --- |
| Lifetime of the proton (predicted) | $3 \times 10^{40}$ |
| Age of the universe | $5 \times 10^{17}$ |
| Age of the pyramid of Cheops | $1 \times 10^{11}$ |
| Human life expectancy | $2 \times 10^{9}$ |
| Length of a day | $9 \times 10^{4}$ |
| Time between human heartbeats | $8 \times 10^{-1}$ |
| Lifetime of the muon | $2 \times 10^{-6}$ |
| Shortest lab light pulse | $1 \times 10^{-16}$ |
| Lifetime of the most unstable particle | $1 \times 10^{-23}$ |
| The Planck time[a] | $1 \times 10^{-43}$ |

[a]This is the earliest time after the big bang at which the laws of physics as we know them can be applied.

FIG. 1-1 When the metric system was proposed in 1792, the hour was redefined to provide a 10-hour day. The idea did not catch on. The maker of this 10-hour watch wisely provided a small dial that kept conventional 12-hour time. Do the two dials indicate the same time? *(Steven Pitkin)*

Any phenomenon that repeats itself is a possible time standard. Earth's rotation, which determines the length of the day, has been used in this way for centuries; Fig. 1-1 shows one novel example of a watch based on that rotation. A quartz clock, in which a quartz ring is made to vibrate continuously, can be calibrated against Earth's rotation via astronomical observations and used to measure time intervals in the laboratory. However, the calibration cannot be carried out with the accuracy called for by modern scientific and engineering technology.

To meet the need for a better time standard, atomic clocks have been developed. An atomic clock at the National Institute of Standards and Technology (NIST) in Boulder, Colorado, is the standard for Coordinated Universal Time (UTC) in the United States. Its time signals are available by shortwave radio (stations WWV and WWVH) and by telephone (303-499-7111). Time signals (and related information) are also available from the United States Naval Observatory at website http://tycho.usno.navy.mil/time.html. (To set a clock extremely accurately at your particular location, you would have to account for the travel time required for these signals to reach you.)

Figure 1-2 shows variations in the length of one day on Earth over a 4-year period, as determined by comparison with a cesium (atomic) clock. Because the variation displayed by Fig. 1-2 is seasonal and repetitious, we suspect the rotating Earth when there is a difference between Earth and atom as timekeepers. The variation is due to tidal effects caused by the Moon and to large-scale winds.

The 13th General Conference on Weights and Measures in 1967 adopted a standard second based on the cesium clock:

> One second is the time taken by 9 192 631 770 oscillations of the light (of a specified wavelength) emitted by a cesium-133 atom.

Atomic clocks are so consistent that, in principle, two cesium clocks would have to run for 6000 years before their readings would differ by more than 1 s. Even such accuracy pales in comparison with that of clocks currently being developed; their precision may be 1 part in $10^{18}$—that is, 1 s in $1 \times 10^{18}$ s (which is about $3 \times 10^{10}$ y).

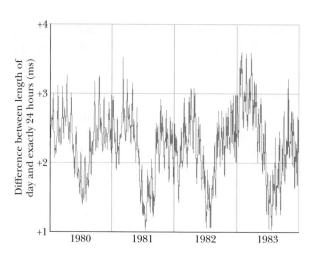

FIG. 1-2 Variations in the length of the day over a 4-year period. Note that the entire vertical scale amounts to only 3 ms (= 0.003 s).

# 1-7 | Mass

## The Standard Kilogram

The SI standard of mass is a platinum–iridium cylinder (Fig. 1-3) kept at the International Bureau of Weights and Measures near Paris and assigned, by international agreement, a mass of 1 kilogram. Accurate copies have been sent to standardizing laboratories in other countries, and the masses of other bodies can be determined by balancing them against a copy. Table 1-5 shows some masses expressed in kilograms, ranging over about 83 orders of magnitude.

The U.S. copy of the standard kilogram is housed in a vault at NIST. It is removed, no more than once a year, for the purpose of checking duplicate copies that are used elsewhere. Since 1889, it has been taken to France twice for recomparison with the primary standard.

## A Second Mass Standard

The masses of atoms can be compared with one another more precisely than they can be compared with the standard kilogram. For this reason, we have a second mass standard. It is the carbon-12 atom, which, by international agreement, has been assigned a mass of 12 **atomic mass units** (u). The relation between the two units is

$$1 \text{ u} = 1.660\,538\,86 \times 10^{-27} \text{ kg}, \tag{1-7}$$

with an uncertainty of $\pm 10$ in the last two decimal places. Scientists can, with reasonable precision, experimentally determine the masses of other atoms relative to the mass of carbon-12. What we presently lack is a reliable means of extending that precision to more common units of mass, such as a kilogram.

## Density

As we shall discuss further in Chapter 14, the **density** $\rho$ of a material is the mass per unit volume:

$$\rho = \frac{m}{V}. \tag{1-8}$$

Densities are typically listed in kilograms per cubic meter or grams per cubic centimeter. The density of water (1.00 gram per cubic centimeter) is often used as a comparison. Fresh snow has about 10% of that density; platinum has a density that is about 21 times that of water.

FIG. 1-3 The international 1 kg standard of mass, a platinum–iridium cylinder 3.9 cm in height and in diameter. *(Courtesy Bureau International des Poids et Mesures, France)*

**TABLE 1-5**

**Some Approximate Masses**

| Object | Mass in Kilograms |
|---|---|
| Known universe | $1 \times 10^{53}$ |
| Our galaxy | $2 \times 10^{41}$ |
| Sun | $2 \times 10^{30}$ |
| Moon | $7 \times 10^{22}$ |
| Asteroid Eros | $5 \times 10^{15}$ |
| Small mountain | $1 \times 10^{12}$ |
| Ocean liner | $7 \times 10^{7}$ |
| Elephant | $5 \times 10^{3}$ |
| Grape | $3 \times 10^{-3}$ |
| Speck of dust | $7 \times 10^{-10}$ |
| Penicillin molecule | $5 \times 10^{-17}$ |
| Uranium atom | $4 \times 10^{-25}$ |
| Proton | $2 \times 10^{-27}$ |
| Electron | $9 \times 10^{-31}$ |

## Sample Problem 1-4

A heavy object can sink into the ground during an earthquake if the shaking causes the ground to undergo *liquefaction,* in which the soil grains experience little friction as they slide over one another. The ground is then effectively quicksand. The possibility of liquefaction in sandy ground can be predicted in terms of the *void ratio e* for a sample of the ground:

$$e = \frac{V_{\text{voids}}}{V_{\text{grains}}}. \tag{1-9}$$

Here, $V_{\text{grains}}$ is the total volume of the sand grains in the sample and $V_{\text{voids}}$ is the total volume between the grains (in the *voids*). If $e$ exceeds a critical value of 0.80,

liquefaction can occur during an earthquake. What is the corresponding sand density $\rho_{\text{sand}}$? Solid silicon dioxide (the primary component of sand) has a density of $\rho_{\text{SiO}_2} = 2.600 \times 10^3$ kg/m³.

**KEY IDEA** The density of the sand $\rho_{\text{sand}}$ in a sample is the mass per unit volume—that is, the ratio of the total mass $m_{\text{sand}}$ of the sand grains to the total volume $V_{\text{total}}$ of the sample:

$$\rho_{\text{sand}} = \frac{m_{\text{sand}}}{V_{\text{total}}}. \tag{1-10}$$

**Calculations:** The total volume $V_{total}$ of a sample is

$$V_{total} = V_{grains} + V_{voids}.$$

Substituting for $V_{voids}$ from Eq. 1-9 and solving for $V_{grains}$ lead to

$$V_{grains} = \frac{V_{total}}{1 + e}. \qquad (1\text{-}11)$$

From Eq. 1-8, the total mass $m_{sand}$ of the sand grains is the product of the density of silicon dioxide and the total volume of the sand grains:

$$m_{sand} = \rho_{SiO_2} V_{grains}. \qquad (1\text{-}12)$$

Substituting this expression into Eq. 1-10 and then substituting for $V_{grains}$ from Eq. 1-11 lead to

$$\rho_{sand} = \frac{\rho_{SiO_2}}{V_{total}} \frac{V_{total}}{1 + e} = \frac{\rho_{SiO_2}}{1 + e}. \qquad (1\text{-}13)$$

Substituting $\rho_{SiO_2} = 2.600 \times 10^3 \text{ kg/m}^3$ and the critical value of $e = 0.80$, we find that liquefaction occurs when the sand density exceeds

$$\rho_{sand} = \frac{2.600 \times 10^3 \text{ kg/m}^3}{1.80} = 1.4 \times 10^3 \text{ kg/m}^3.$$

(Answer)

---

## REVIEW & SUMMARY

**Measurement in Physics**  Physics is based on measurement of physical quantities. Certain physical quantities have been chosen as **base quantities** (such as length, time, and mass); each has been defined in terms of a **standard** and given a **unit** of measure (such as meter, second, and kilogram). Other physical quantities are defined in terms of the base quantities and their standards and units.

**SI Units**  The unit system emphasized in this book is the International System of Units (SI). The three physical quantities displayed in Table 1-1 are used in the early chapters. Standards, which must be both accessible and invariable, have been established for these base quantities by international agreement. These standards are used in all physical measurement, for both the base quantities and the quantities derived from them. Scientific notation and the prefixes of Table 1-2 are used to simplify measurement notation.

**Changing Units**  Conversion of units may be performed by using *chain-link conversions* in which the original data are multiplied successively by conversion factors written as unity

and the units are manipulated like algebraic quantities until only the desired units remain.

**Length**  The meter is defined as the distance traveled by light during a precisely specified time interval.

**Time**  The second is defined in terms of the oscillations of light emitted by an atomic (cesium-133) source. Accurate time signals are sent worldwide by radio signals keyed to atomic clocks in standardizing laboratories.

**Mass**  The kilogram is defined in terms of a platinum–iridium standard mass kept near Paris. For measurements on an atomic scale, the atomic mass unit, defined in terms of the atom carbon-12, is usually used.

**Density**  The density $\rho$ of a material is the mass per unit volume:

$$\rho = \frac{m}{V}. \qquad (1\text{-}8)$$

---

## PROBLEMS

 Tutoring problem available (at instructor's discretion) in *WileyPLUS* and WebAssign
**SSM**  Worked-out solution available in Student Solutions Manual
• – •••  Number of dots indicates level of problem difficulty
**WWW**  Worked-out solution is at
**ILW**  Interactive solution is at ─ http://www.wiley.com/college/halliday
Additional information available in *The Flying Circus of Physics* and at flyingcircusofphysics.com

### sec. 1-5  Length

•1  The micrometer (1 $\mu$m) is often called the *micron*. (a) How many microns make up 1.0 km? (b) What fraction of a centimeter equals 1.0 $\mu$m? (c) How many microns are in 1.0 yd?

•2  Spacing in this book was generally done in units of points and picas: 12 points = 1 pica, and 6 picas = 1 inch. If a figure was misplaced in the page proofs by 0.80 cm, what was the misplacement in (a) picas and (b) points?

•3  Horses are to race over a certain English meadow for a distance of 4.0 furlongs. What is the race distance in (a) rods and (b) chains? (1 furlong = 201.168 m, 1 rod = 5.0292 m, and 1 chain = 20.117 m.)  **SSM WWW**

•4  A *gry* is an old English measure for length, defined as 1/10 of a line, where *line* is another old English measure for length, defined as 1/12 inch. A common measure for length in the publishing business is a *point,* defined as 1/72 inch. What is an area of 0.50 gry$^2$ in points squared (points$^2$)?

•5  Earth is approximately a sphere of radius $6.37 \times 10^6$ m. What are (a) its circumference in kilometers, (b) its surface area in square kilometers, and (c) its volume in cubic kilometers?  **SSM**

••6   Harvard Bridge, which connects MIT with its fraternities across the Charles River, has a length of 364.4 Smoots plus one ear. The unit of one Smoot is based on the length of Oliver Reed Smoot, Jr., class of 1962, who was carried or dragged length by length across the bridge so that other pledge members of the Lambda Chi Alpha fraternity could mark off (with paint) 1-Smoot lengths along the bridge. The marks have been repainted biannually by fraternity pledges since the initial measurement, usually during times of traffic congestion so that the police cannot easily interfere. (Presumably, the police were originally upset because the Smoot is not an SI base unit, but these days they seem to have accepted the unit.) Figure 1-4 shows three parallel paths, measured in Smoots (S), Willies (W), and Zeldas (Z). What is the length of 50.0 Smoots in (a) Willies and (b) Zeldas?

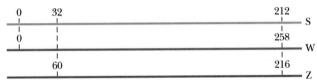

FIG. 1-4   Problem 6.

••7   Antarctica is roughly semicircular, with a radius of 2000 km (Fig. 1-5). The average thickness of its ice cover is 3000 m. How many cubic centimeters of ice does Antarctica contain? (Ignore the curvature of Earth.)

FIG. 1-5   Problem 7.

••8   You can easily convert common units and measures electronically, but you still should be able to use a conversion table, such as those in Appendix D. Table 1-6 is part of a conversion table for a system of volume measures once common in Spain; a volume of 1 fanega is equivalent to 55.501 dm³ (cubic decimeters). To complete the table, what numbers (to three significant figures) should be entered in (a) the cahiz column, (b) the fanega column, (c) the cuartilla column, and (d) the almude column, starting with the top blank? Express 7.00 almudes in (e) medios, (f) cahizes, and (g) cubic centimeters (cm³).

**TABLE 1-6**

**Problem 8**

|            | cahiz | fanega | cuartilla | almude | medio |
|------------|-------|--------|-----------|--------|-------|
| 1 cahiz =  | 1     | 12     | 48        | 144    | 288   |
| 1 fanega = |       | 1      | 4         | 12     | 24    |
| 1 cuartilla = |    |        | 1         | 3      | 6     |
| 1 almude = |       |        |           | 1      | 2     |
| 1 medio =  |       |        |           |        | 1     |

••9   Hydraulic engineers in the United States often use, as a unit of volume of water, the *acre-foot*, defined as the volume of water that will cover 1 acre of land to a depth of 1 ft. A severe thunderstorm dumped 2.0 in. of rain in 30 min on a town of area 26 km². What volume of water, in acre-feet, fell on the town?   **ILW**

## sec. 1-6   Time

•10   The fastest growing plant on record is a *Hesperoyucca whipplei* that grew 3.7 m in 14 days. What was its growth rate in micrometers per second?

•11   A fortnight is a charming English measure of time equal to 2.0 weeks (the word is a contraction of "fourteen nights"). That is a nice amount of time in pleasant company but perhaps a painful string of microseconds in unpleasant company. How many microseconds are in a fortnight?

•12   A lecture period (50 min) is close to 1 microcentury. (a) How long is a microcentury in minutes? (b) Using

$$\text{percentage difference} = \left( \frac{\text{actual} - \text{approximation}}{\text{actual}} \right) 100,$$

find the percentage difference from the approximation.

•13   For about 10 years after the French Revolution, the French government attempted to base measures of time on multiples of ten: One week consisted of 10 days, one day consisted of 10 hours, one hour consisted of 100 minutes, and one minute consisted of 100 seconds. What are the ratios of (a) the French decimal week to the standard week and (b) the French decimal second to the standard second?

•14   Time standards are now based on atomic clocks. A promising second standard is based on *pulsars*, which are rotating neutron stars (highly compact stars consisting only of neutrons). Some rotate at a rate that is highly stable, sending out a radio beacon that sweeps briefly across Earth once with each rotation, like a lighthouse beacon. Pulsar PSR 1937+21 is an example; it rotates once every 1.557 806 448 872 75 ± 3 ms, where the trailing ±3 indicates the uncertainty in the last decimal place (it does *not* mean ±3 ms). (a) How many rotations does PSR 1937+21 make in 7.00 days? (b) How much time does the pulsar take to rotate exactly one million times and (c) what is the associated uncertainty?

•15   Three digital clocks A, B, and C run at different rates and do not have simultaneous readings of zero. Figure 1-6 shows simultaneous readings on pairs of the clocks for four occasions. (At the earliest occasion, for example, B reads 25.0 s and C reads 92.0 s.) If two events are 600 s apart on clock A, how far apart are they on (a) clock B and (b) clock C? (c) When clock A reads 400 s, what does clock B read? (d) When clock C reads 15.0 s, what does clock B read? (Assume negative readings for prezero times.)

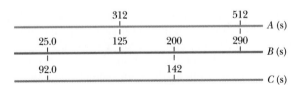

FIG. 1-6   Problem 15.

•16   Until 1883, every city and town in the United States kept its own local time. Today, travelers reset their watches only when the time change equals 1.0 h. How far, on the average, must you travel in degrees of longitude between the time-zone boundaries at which your watch must be reset by 1.0 h? (*Hint:* Earth rotates 360° in about 24 h.)

•17   Five clocks are being tested in a laboratory. Exactly at noon, as determined by the WWV time signal, on successive days of a week the clocks read as in the following table. Rank the five clocks according to their relative value as good time-keepers, best to worst. Justify your choice. **SSM**

| Clock | Sun. | Mon. | Tues. | Wed. | Thurs. | Fri. | Sat. |
|---|---|---|---|---|---|---|---|
| A | 12:36:40 | 12:36:56 | 12:37:12 | 12:37:27 | 12:37:44 | 12:37:59 | 12:38:14 |
| B | 11:59:59 | 12:00:02 | 11:59:57 | 12:00:07 | 12:00:02 | 11:59:56 | 12:00:03 |
| C | 15:50:45 | 15:51:43 | 15:52:41 | 15:53:39 | 15:54:37 | 15:55:35 | 15:56:33 |
| D | 12:03:59 | 12:02:52 | 12:01:45 | 12:00:38 | 11:59:31 | 11:58:24 | 11:57:17 |
| E | 12:03:59 | 12:02:49 | 12:01:54 | 12:01:52 | 12:01:32 | 12:01:22 | 12:01:12 |

**••18** Because Earth's rotation is gradually slowing, the length of each day increases: The day at the end of 1.0 century is 1.0 ms longer than the day at the start of the century. In 20 centuries, what is the total of the daily increases in time?

**•••19** Suppose that, while lying on a beach near the equator watching the Sun set over a calm ocean, you start a stopwatch just as the top of the Sun disappears. You then stand, elevating your eyes by a height $H = 1.70$ m, and stop the watch when the top of the Sun again disappears. If the elapsed time is $t = 11.1$ s, what is the radius $r$ of Earth?

**sec. 1-7   Mass**

**•20** Gold, which has a density of 19.32 g/cm$^3$, is the most ductile metal and can be pressed into a thin leaf or drawn out into a long fiber. (a) If a sample of gold, with a mass of 27.63 g, is pressed into a leaf of 1.000 $\mu$m thickness, what is the area of the leaf? (b) If, instead, the gold is drawn out into a cylindrical fiber of radius 2.500 $\mu$m, what is the length of the fiber?

**•21** (a) Assuming that water has a density of exactly 1 g/cm$^3$, find the mass of one cubic meter of water in kilograms. (b) Suppose that it takes 10.0 h to drain a container of 5700 m$^3$ of water. What is the "mass flow rate," in kilograms per second, of water from the container? **SSM**

**•22** The record for the largest glass bottle was set in 1992 by a team in Millville, New Jersey—they blew a bottle with a volume of 193 U.S. fluid gallons. (a) How much short of 1.0 million cubic centimeters is that? (b) If the bottle were filled with water at the leisurely rate of 1.8 g/min, how long would the filling take? Water has a density of 1000 kg/m$^3$. **GO**

**•23** Earth has a mass of $5.98 \times 10^{24}$ kg. The average mass of the atoms that make up Earth is 40 u. How many atoms are there in Earth?

**••24** One cubic centimeter of a typical cumulus cloud contains 50 to 500 water drops, which have a typical radius of 10 $\mu$m. For that range, give the lower value and the higher value, respectively, for the following. (a) How many cubic meters of water are in a cylindrical cumulus cloud of height 3.0 km and radius 1.0 km? (b) How many 1-liter pop bottles would that water fill? (c) Water has a density of 1000 kg/m$^3$. How much mass does the water in the cloud have?

**••25** Iron has a density of 7.87 g/cm$^3$, and the mass of an iron atom is $9.27 \times 10^{-26}$ kg. If the atoms are spherical and tightly packed, (a) what is the volume of an iron atom and (b) what is the distance between the centers of adjacent atoms?

**••26** A mole of atoms is $6.02 \times 10^{23}$ atoms. To the nearest order of magnitude, how many moles of atoms are in a large domestic cat? The masses of a hydrogen atom, an oxygen

atom, and a carbon atom are 1.0 u, 16 u, and 12 u, respectively. (*Hint:* Cats are sometimes known to kill a mole.)

**••27** On a spending spree in Malaysia, you buy an ox with a weight of 28.9 piculs in the local unit of weights: 1 picul = 100 gins, 1 gin = 16 tahils, 1 tahil = 10 chees, and 1 chee = 10 hoons. The weight of 1 hoon corresponds to a mass of 0.3779 g. When you arrange to ship the ox home to your astonished family, how much mass in kilograms must you declare on the shipping manifest? (*Hint:* Set up multiple chain-link conversions.)

**••28** Grains of fine California beach sand are approximately spheres with an average radius of 50 $\mu$m and are made of silicon dioxide, which has a density of 2600 kg/m$^3$. What mass of sand grains would have a total surface area (the total area of all the individual spheres) equal to the surface area of a cube 1.00 m on an edge? **GO**

**••29** During heavy rain, a section of a mountainside measuring 2.5 km horizontally, 0.80 km up along the slope, and 2.0 m deep slips into a valley in a mud slide. Assume that the mud ends up uniformly distributed over a surface area of the valley measuring 0.40 km $\times$ 0.40 km and that mud has a density of 1900 kg/m$^3$. What is the mass of the mud sitting above a 4.0 m$^2$ area of the valley floor?

**••30** Water is poured into a container that has a leak. The mass $m$ of the water is given as a function of time $t$ by $m = 5.00t^{0.8} - 3.00t + 20.00$, with $t \geq 0$, $m$ in grams, and $t$ in seconds. (a) At what time is the water mass greatest, and (b) what is that greatest mass? In kilograms per minute, what is the rate of mass change at (c) $t = 2.00$ s and (d) $t = 5.00$ s?

**•••31** A vertical container with base area measuring 14.0 cm by 17.0 cm is being filled with identical pieces of candy, each with a volume of 50.0 mm$^3$ and a mass of 0.0200 g. Assume that the volume of the empty spaces between the candies is negligible. If the height of the candies in the container increases at the rate of 0.250 cm/s, at what rate (kilograms per minute) does the mass of the candies in the container increase?

**Additional Problems**

**32** Table 1-7 shows some old measures of liquid volume. To complete the table, what numbers (to three significant figures) should be entered in (a) the wey column, (b) the chaldron column, (c) the bag column, (d) the pottle column, and (e) the gill column, starting with the top blank? (f) The volume of 1 bag is equal to 0.1091 m$^3$. If an old story has a witch cooking up some vile liquid in a cauldron of volume 1.5 chaldrons, what is the volume in cubic meters?

**TABLE 1-7**

**Problem 32**

| | wey | chaldron | bag | pottle | gill |
|---|---|---|---|---|---|
| 1 wey = | 1 | 10/9 | 40/3 | 640 | 120 240 |
| 1 chaldron = | | | | | |
| 1 bag = | | | | | |
| 1 pottle = | | | | | |
| 1 gill = | | | | | |

**33** An old English children's rhyme states, "Little Miss Muffet sat on a tuffet, eating her curds and whey, when along came a spider who sat down beside her. ..." The spider sat down not because of the curds and whey but because Miss Muffet had a stash of 11 tuffets of dried flies. The volume measure of a tuffet is given by 1 tuffet = 2 pecks = 0.50 Imperial bushel, where 1 Imperial bushel = 36.3687 liters (L). What was Miss Muffet's stash in (a) pecks, (b) Imperial bushels, and (c) liters?

**34** An old manuscript reveals that a landowner in the time of King Arthur held 3.00 acres of plowed land plus a livestock area of 25.0 perches by 4.00 perches. What was the total area in (a) the old unit of roods and (b) the more modern unit of square meters? Here, 1 acre is an area of 40 perches by 4 perches, 1 rood is an area of 40 perches by 1 perch, and 1 perch is the length 16.5 ft.

**35** A tourist purchases a car in England and ships it home to the United States. The car sticker advertised that the car's fuel consumption was at the rate of 40 miles per gallon on the open road. The tourist does not realize that the U.K. gallon differs from the U.S. gallon:

$$1 \text{ U.K. gallon} = 4.545\ 963\ 1 \text{ liters}$$
$$1 \text{ U.S. gallon} = 3.785\ 306\ 0 \text{ liters.}$$

For a trip of 750 miles (in the United States), how many gallons of fuel does (a) the mistaken tourist believe she needs and (b) the car actually require? **SSM**

**36** Two types of *barrel* units were in use in the 1920s in the United States. The apple barrel had a legally set volume of 7056 cubic inches; the cranberry barrel, 5826 cubic inches. If a merchant sells 20 cranberry barrels of goods to a customer who thinks he is receiving apple barrels, what is the discrepancy in the shipment volume in liters?

**37** The description for a certain brand of house paint claims a coverage of 460 ft$^2$/gal. (a) Express this quantity in square meters per liter. (b) Express this quantity in an SI unit (see Appendices A and D). (c) What is the inverse of the original quantity, and (d) what is its physical significance?

**38** In the United States, a doll house has the scale of 1:12 of a real house (that is, each length of the doll house is $\frac{1}{12}$ that of the real house) and a miniature house (a doll house to fit within a doll house) has the scale of 1:144 of a real house. Suppose a real house (Fig. 1-7) has a front length of 20 m, a depth of 12 m, a height of 6.0 m, and a standard sloped roof (vertical triangular faces on the ends) of height 3.0 m. In cubic meters, what are the volumes of the corresponding (a) doll house and (b) miniature house?

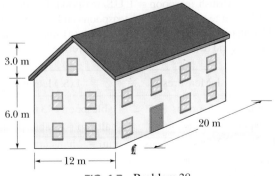

**FIG. 1-7**  Problem 38.

**39** A *cord* is a volume of cut wood equal to a stack 8 ft long, 4 ft wide, and 4 ft high. How many cords are in 1.0 m$^3$? **SSM**

**40** One molecule of water (H$_2$O) contains two atoms of hydrogen and one atom of oxygen. A hydrogen atom has a mass of 1.0 u and an atom of oxygen has a mass of 16 u, approximately. (a) What is the mass in kilograms of one molecule of water? (b) How many molecules of water are in the world's oceans, which have an estimated total mass of $1.4 \times 10^{21}$ kg?

**41** A ton is a measure of volume frequently used in shipping, but that use requires some care because there are at least three types of tons: A *displacement ton* is equal to 7 barrels bulk, a *freight ton* is equal to 8 barrels bulk, and a *register ton* is equal to 20 barrels bulk. A *barrel bulk* is another measure of volume: 1 barrel bulk = 0.1415 m$^3$. Suppose you spot a shipping order for "73 tons" of M&M candies, and you are certain that the client who sent the order intended "ton" to refer to volume (instead of weight or mass, as discussed in Chapter 5). If the client actually meant displacement tons, how many extra U.S. bushels of the candies will you erroneously ship if you interpret the order as (a) 73 freight tons and (b) 73 register tons? (1 m$^3$ = 28.378 U.S. bushels.) **SSM**

**42** Strangely, the wine for a large wedding reception is to be served in a stunning cut-glass receptacle with the interior dimensions of 40 cm × 40 cm × 30 cm (height). The receptacle is to be initially filled to the top. The wine can be purchased in bottles of the sizes given in the following table. Purchasing a larger bottle instead of multiple smaller bottles decreases the overall cost of the wine. To minimize the cost, (a) which bottle sizes should be purchased and how many of each should be purchased and, once the receptacle is filled, how much wine is left over in terms of (b) standard bottles and (c) liters?

1 standard bottle

1 magnum = 2 standard bottles

1 jeroboam = 4 standard bottles

1 rehoboam = 6 standard bottles

1 methuselah = 8 standard bottles

1 salmanazar = 12 standard bottles

1 balthazar = 16 standard bottles = 11.356 L

1 nebuchadnezzar = 20 standard bottles

**43** A typical sugar cube has an edge length of 1 cm. If you had a cubical box that contained a mole of sugar cubes, what would its edge length be? (One mole = $6.02 \times 10^{23}$ units.)

**44** Using conversions and data in the chapter, determine the number of hydrogen atoms required to obtain 1.0 kg of hydrogen. A hydrogen atom has a mass of 1.0 u.

**45** An astronomical unit (AU) is the average distance between Earth and the Sun, approximately $1.50 \times 10^8$ km. The speed of light is about $3.0 \times 10^8$ m/s. Express the speed of light in astronomical units per minute. **SSM**

**46** What mass of water fell on the town in Problem 9? Water has a density of $1.0 \times 10^3$ kg/m$^3$.

**47** A person on a diet might lose 2.3 kg per week. Express the mass loss rate in milligrams per second, as if the dieter could sense the second-by-second loss.

**48** The *corn–hog ratio* is a financial term used in the pig market and presumably is related to the cost of feeding a pig until it is large enough for market. It is defined as the ratio of the market price of a pig with a mass of 3.108 slugs to the market price of a U.S. bushel of corn. (The word "slug" is derived from an old German word that means "to hit"; we have the same meaning for "slug" as a verb in modern English.) A U.S. bushel is equal to 35.238 L. If the corn–hog ratio is listed as 5.7 on the market exchange, what is it in the metric units of

$$\frac{\text{price of 1 kilogram of pig}}{\text{price of 1 liter of corn}} ?$$

(*Hint:* See the Mass table in Appendix D.)

**49** You are to fix dinners for 400 people at a convention of Mexican food fans. Your recipe calls for 2 jalapeño peppers per serving (one serving per person). However, you have only habanero peppers on hand. The spiciness of peppers is measured in terms of the *scoville heat unit* (SHU). On average, one jalapeño pepper has a spiciness of 4000 SHU and one habanero pepper has a spiciness of 300 000 SHU. To get the desired spiciness, how many habanero peppers should you substitute for the jalapeño peppers in the recipe for the 400 dinners?

**50** A unit of area often used in measuring land areas is the *hectare,* defined as $10^4$ m$^2$. An open-pit coal mine consumes 75 hectares of land, down to a depth of 26 m, each year. What volume of earth, in cubic kilometers, is removed in this time?

**51** (a) A unit of time sometimes used in microscopic physics is the *shake*. One shake equals $10^{-8}$ s. Are there more shakes in a second than there are seconds in a year? (b) Humans have existed for about $10^6$ years, whereas the universe is about $10^{10}$ years old. If the age of the universe is defined as 1 "universe day," where a universe day consists of "universe seconds" as a normal day consists of normal seconds, how many universe seconds have humans existed?

**52** As a contrast between the old and the modern and between the large and the small, consider the following: In old rural England 1 hide (between 100 and 120 acres) was the area of land needed to sustain one family with a single plough for one year. (An area of 1 acre is equal to 4047 m$^2$.) Also, 1 wapentake was the area of land needed by 100 such families. In quantum physics, the cross-sectional area of a nucleus (defined in terms of the chance of a particle hitting and being absorbed by it) is measured in units of barns, where 1 barn is $1 \times 10^{-28}$ m$^2$. (In nuclear physics jargon, if a nucleus is "large," then shooting a particle at it is like shooting a bullet at a barn door, which can hardly be missed.) What is the ratio of 25 wapentakes to 11 barns?

**53** A traditional unit of length in Japan is the ken (1 ken = 1.97 m). What are the ratios of (a) square kens to square meters and (b) cubic kens to cubic meters? What is the volume of a cylindrical water tank of height 5.50 kens and radius 3.00 kens in (c) cubic kens and (d) cubic meters?

**54** You receive orders to sail due east for 24.5 mi to put your salvage ship directly over a sunken pirate ship. However, when your divers probe the ocean floor at that location and find no evidence of a ship, you radio back to your source of information, only to discover that the sailing distance was supposed to be 24.5 *nautical miles,* not regular miles. Use the Length table in Appendix D.

**55** A standard interior staircase has steps each with a rise (height) of 19 cm and a run (horizontal depth) of 23 cm. Research suggests that the stairs would be safer for descent if the run were, instead, 28 cm. For a particular staircase of total height 4.57 m, how much farther into the room would the staircase extend if this change in run were made?

**56** The common Eastern mole, a mammal, typically has a mass of 75 g, which corresponds to about 7.5 moles of atoms. (A mole of atoms is $6.02 \times 10^{23}$ atoms.) In atomic mass units (u), what is the average mass of the atoms in the common Eastern mole?

**57** An *astronomical unit* (AU) is equal to the average distance from Earth to the Sun, about $92.9 \times 10^6$ mi. A *parsec* (pc) is the distance at which a length of 1 AU would subtend an angle of exactly 1 second of arc (Fig. 1-8). A *light-year* (ly) is the distance that light, traveling through a vacuum with a speed of 186 000 mi/s, would cover in 1.0 year. Express the Earth–Sun distance in (a) parsecs and (b) light-years. **SSM**

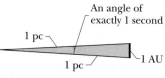
**FIG. 1-8** Problem 57.

**58** In purchasing food for a political rally, you erroneously order shucked medium-size Pacific oysters (which come 8 to 12 per U.S. pint) instead of shucked medium-size Atlantic oysters (which come 26 to 38 per U.S. pint). The filled oyster container shipped to you has the interior measure of 1.0 m × 12 cm × 20 cm, and a U.S. pint is equivalent to 0.4732 liter. By how many oysters is the order short of your anticipated count?

**59** The cubit is an ancient unit of length based on the distance between the elbow and the tip of the middle finger of the measurer. Assume that the distance ranged from 43 to 53 cm, and suppose that ancient drawings indicate that a cylindrical pillar was to have a length of 9 cubits and a diameter of 2 cubits. For the stated range, what are the lower value and the upper value, respectively, for (a) the cylinder's length in meters, (b) the cylinder's length in millimeters, and (c) the cylinder's volume in cubic meters?

**60** An old English cookbook carries this recipe for cream of nettle soup: "Boil stock of the following amount: 1 breakfastcup plus 1 teacup plus 6 tablespoons plus 1 dessertspoon. Using gloves, separate nettle tops until you have 0.5 quart; add the tops to the boiling stock. Add 1 tablespoon of cooked rice and 1 saltspoon of salt. Simmer for 15 min." The following table gives some of the conversions among old (premetric) British measures and among common (still premetric) U.S. measures. (These measures scream for metrication.) For liquid measures, 1 British teaspoon = 1 U.S. teaspoon. For dry measures, 1 British teaspoon = 2 U.S. teaspoons and 1 British quart = 1 U.S. quart. In U.S. measures, how much (a) stock, (b) nettle tops, (c) rice, and (d) salt are required in the recipe?

| Old British Measures | U.S. Measures |
| --- | --- |
| teaspoon = 2 saltspoons | tablespoon = 3 teaspoons |
| dessertspoon = 2 teaspoons | half cup = 8 tablespoons |
| tablespoon = 2 dessertspoons | cup = 2 half cups |
| teacup = 8 tablespoons | |
| breakfastcup = 2 teacups | |

# Motion Along a Straight Line

Jeremy Woodhouse/Masterfile

A woodpecker hammers its beak into the limb of a tree to search for insects to eat, to create storage space, or to audibly advertise for a mate. The motion toward the limb may be very rapid, but the stopping once the limb is reached is extremely rapid and would be fatal to a human. Thus, a woodpecker should seemingly fall from the tree either dead or unconscious every time it slams its beak into the tree. Not only does it survive, but it rapidly repeats the motion, sending out a rat-tat-tat signal through the air.

*Why can a woodpecker survive the severe impacts with a tree limb?*

The answer is in this chapter.

One purpose of physics is to study the motion of objects—how fast they move, for example, and how far they move in a given amount of time. NASCAR engineers are fanatical about this aspect of physics as they determine the performance of their cars before and during a race. Geologists use this physics to measure tectonic-plate motion as they attempt to predict earthquakes. Medical researchers need this physics to map the blood flow through a patient when diagnosing a partially closed artery, and motorists use it to determine how they might slow sufficiently when their radar detector sounds a warning. There are countless other examples. In this chapter, we study the basic physics of motion where the object (race car, tectonic plate, blood cell, or any other object) moves along a single axis. Such motion is called *one-dimensional motion.*

## 2-2 | Motion

The world, and everything in it, moves. Even seemingly stationary things, such as a roadway, move with Earth's rotation, Earth's orbit around the Sun, the Sun's orbit around the center of the Milky Way galaxy, and that galaxy's migration relative to other galaxies. The classification and comparison of motions (called **kinematics**) is often challenging. What exactly do you measure, and how do you compare?

Before we attempt an answer, we shall examine some general properties of motion that is restricted in three ways.

1. The motion is along a straight line only. The line may be vertical, horizontal, or slanted, but it must be straight.

2. Forces (pushes and pulls) cause motion but will not be discussed until Chapter 5. In this chapter we discuss only the motion itself and changes in the motion. Does the moving object speed up, slow down, stop, or reverse direction? If the motion does change, how is time involved in the change?

3. The moving object is either a **particle** (by which we mean a point-like object such as an electron) or an object that moves like a particle (such that every portion moves in the same direction and at the same rate). A stiff pig slipping down a straight playground slide might be considered to be moving like a particle; however, a tumbling tumbleweed would not.

## 2-3 | Position and Displacement

To locate an object means to find its position relative to some reference point, often the **origin** (or zero point) of an axis such as the $x$ axis in Fig. 2-1. The **positive direction** of the axis is in the direction of increasing numbers (coordinates), which is to the right in Fig. 2-1. The opposite is the **negative direction.**

For example, a particle might be located at $x = 5$ m, which means it is 5 m in the positive direction from the origin. If it were at $x = -5$ m, it would be just as far from the origin but in the opposite direction. On the axis, a coordinate of $-5$ m is less than a coordinate of $-1$ m, and both coordinates are less than a coordinate of $+5$ m. A plus sign for a coordinate need not be shown, but a minus sign must always be shown.

A change from position $x_1$ to position $x_2$ is called a **displacement** $\Delta x$, where

$$\Delta x = x_2 - x_1. \tag{2-1}$$

(The symbol $\Delta$, the Greek uppercase delta, represents a change in a quantity, and it means the final value of that quantity minus the initial value.) When numbers are inserted for the position values $x_1$ and $x_2$ in Eq. 2-1, a displacement in the positive direction (to the right in Fig. 2-1) always comes out positive, and a dis-

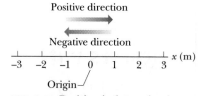

**FIG. 2-1** Position is determined on an axis that is marked in units of length (here meters) and that extends indefinitely in opposite directions. The axis name, here $x$, is always on the positive side of the origin.

placement in the opposite direction (left in the figure) always comes out negative. For example, if the particle moves from $x_1 = 5$ m to $x_2 = 12$ m, then $\Delta x = (12$ m$)$ $- (5$ m$) = +7$ m. The positive result indicates that the motion is in the positive direction. If, instead, the particle moves from $x_1 = 5$ m to $x_2 = 1$ m, then $\Delta x = (1$ m$) - (5$ m$) = -4$ m. The negative result indicates that the motion is in the negative direction.

The actual number of meters covered for a trip is irrelevant; displacement involves only the original and final positions. For example, if the particle moves from $x = 5$ m out to $x = 200$ m and then back to $x = 5$ m, the displacement from start to finish is $\Delta x = (5$ m$) - (5$ m$) = 0$.

A plus sign for a displacement need not be shown, but a minus sign must always be shown. If we ignore the sign (and thus the direction) of a displacement, we are left with the **magnitude** (or absolute value) of the displacement. For example, a displacement of $\Delta x = -4$ m has a magnitude of 4 m.

Displacement is an example of a **vector quantity,** which is a quantity that has both a direction and a magnitude. We explore vectors more fully in Chapter 3 (in fact, some of you may have already read that chapter), but here all we need is the idea that displacement has two features: (1) Its *magnitude* is the distance (such as the number of meters) between the original and final positions. (2) Its *direction,* from an original position to a final position, can be represented by a plus sign or a minus sign if the motion is along a single axis.

*What follows is the first of many checkpoints you will see in this book. Each consists of one or more questions whose answers require some reasoning or a mental calculation, and each gives you a quick check of your understanding of a point just discussed. The answers are listed in the back of the book.*

✓ **CHECKPOINT 1**    Here are three pairs of initial and final positions, respectively, along an $x$ axis. Which pairs give a negative displacement: (a) $-3$ m, $+5$ m; (b) $-3$ m, $-7$ m; (c) $7$ m, $-3$ m?

## 2-4 I Average Velocity and Average Speed

A compact way to describe position is with a graph of position $x$ plotted as a function of time $t$—a graph of $x(t)$. (The notation $x(t)$ represents a function $x$ of $t$, not the product $x$ times $t$.) As a simple example, Fig. 2-2 shows the position function $x(t)$ for a stationary armadillo (which we treat as a particle) over a 7 s time interval. The animal's position stays at $x = -2$ m.

Figure 2-3a is more interesting, because it involves motion. The armadillo is apparently first noticed at $t = 0$ when it is at the position $x = -5$ m. It moves toward $x = 0$, passes through that point at $t = 3$ s, and then moves on to increasingly larger positive values of $x$. Figure 2-3b depicts the straight-line motion of the armadillo and is something like what you would see. The graph in Fig. 2-3a is more abstract and quite unlike what you would see, but it is richer in information. It also reveals how fast the armadillo moves.

Actually, several quantities are associated with the phrase "how fast." One of them is the **average velocity** $v_{avg}$, which is the ratio of the displacement $\Delta x$ that occurs during a particular time interval $\Delta t$ to that interval:

$$v_{avg} = \frac{\Delta x}{\Delta t} = \frac{x_2 - x_1}{t_2 - t_1}. \tag{2-2}$$

The notation means that the position is $x_1$ at time $t_1$ and then $x_2$ at time $t_2$. A common unit for $v_{avg}$ is the meter per second (m/s). You may see other units in the problems, but they are always in the form of length/time.

On a graph of $x$ versus $t$, $v_{avg}$ is the **slope** of the straight line that connects two particular points on the $x(t)$ curve: one is the point that corresponds to $x_2$ and $t_2$,

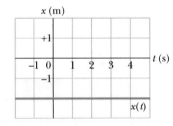

FIG. 2-2   The graph of $x(t)$ for an armadillo that is stationary at $x = -2$ m. The value of $x$ is $-2$ m for all times $t$.

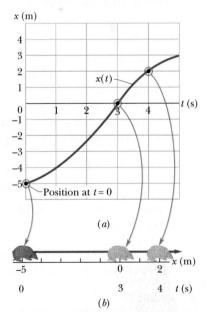

FIG. 2-3   (a) The graph of $x(t)$ for a moving armadillo. (b) The path associated with the graph. The scale below the $x$ axis shows the times at which the armadillo reaches various $x$ values.

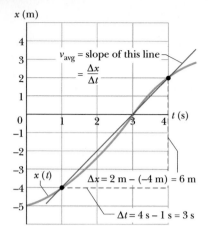

**FIG. 2-4** Calculation of the average velocity between $t = 1$ s and $t = 4$ s as the slope of the line that connects the points on the $x(t)$ curve representing those times.

and the other is the point that corresponds to $x_1$ and $t_1$. Like displacement, $v_{avg}$ has both magnitude and direction (it is another vector quantity). Its magnitude is the magnitude of the line's slope. A positive $v_{avg}$ (and slope) tells us that the line slants upward to the right; a negative $v_{avg}$ (and slope) tells us that the line slants downward to the right. The average velocity $v_{avg}$ always has the same sign as the displacement $\Delta x$ because $\Delta t$ in Eq. 2-2 is always positive.

Figure 2-4 shows how to find $v_{avg}$ in Fig. 2-3 for the time interval $t = 1$ s to $t = 4$ s. We draw the straight line that connects the point on the position curve at the beginning of the interval and the point on the curve at the end of the interval. Then we find the slope $\Delta x/\Delta t$ of the straight line. For the given time interval, the average velocity is

$$v_{avg} = \frac{6 \text{ m}}{3 \text{ s}} = 2 \text{ m/s}.$$

**Average speed** $s_{avg}$ is a different way of describing "how fast" a particle moves. Whereas the average velocity involves the particle's displacement $\Delta x$, the average speed involves the total distance covered (for example, the number of meters moved), independent of direction; that is,

$$s_{avg} = \frac{\text{total distance}}{\Delta t}. \tag{2-3}$$

Because average speed does *not* include direction, it lacks any algebraic sign. Sometimes $s_{avg}$ is the same (except for the absence of a sign) as $v_{avg}$. However, as is demonstrated in Sample Problem 2-1, the two can be quite different.

## Sample Problem 2-1

You drive a beat-up pickup truck along a straight road for 8.4 km at 70 km/h, at which point the truck runs out of gasoline and stops. Over the next 30 min, you walk another 2.0 km farther along the road to a gasoline station.

**(a)** What is your overall displacement from the beginning of your drive to your arrival at the station?

**KEY IDEA** Assume, for convenience, that you move in the positive direction of an $x$ axis, from a first position of $x_1 = 0$ to a second position of $x_2$ at the station. That second position must be at $x_2 = 8.4$ km + 2.0 km = 10.4 km. Then your displacement $\Delta x$ along the $x$ axis is the second position minus the first position.

*Calculation:* From Eq. 2-1, we have

$$\Delta x = x_2 - x_1 = 10.4 \text{ km} - 0 = 10.4 \text{ km}. \quad \text{(Answer)}$$

Thus, your overall displacement is 10.4 km in the positive direction of the $x$ axis.

**(b)** What is the time interval $\Delta t$ from the beginning of your drive to your arrival at the station?

**KEY IDEA** We already know the walking time interval $\Delta t_{wlk}$ ($= 0.50$ h), but we lack the driving time interval $\Delta t_{dr}$. However, we know that for the drive the displacement $\Delta x_{dr}$ is 8.4 km and the average velocity $v_{avg,dr}$ is 70 km/h. Thus, this average velocity is the ratio of the displacement for the drive to the time interval for the drive.

*Calculations:* We first write

$$v_{avg,dr} = \frac{\Delta x_{dr}}{\Delta t_{dr}}.$$

Rearranging and substituting data then give us

$$\Delta t_{dr} = \frac{\Delta x_{dr}}{v_{avg,dr}} = \frac{8.4 \text{ km}}{70 \text{ km/h}} = 0.12 \text{ h}.$$

So,
$$\Delta t = \Delta t_{dr} + \Delta t_{wlk}$$
$$= 0.12 \text{ h} + 0.50 \text{ h} = 0.62 \text{ h}. \quad \text{(Answer)}$$

**(c)** What is your average velocity $v_{avg}$ from the beginning of your drive to your arrival at the station? Find it both numerically and graphically.

**KEY IDEA** From Eq. 2-2 we know that $v_{avg}$ *for the entire trip* is the ratio of the displacement of 10.4 km *for the entire trip* to the time interval of 0.62 h *for the entire trip*.

*Calculation:* Here we find

$$v_{avg} = \frac{\Delta x}{\Delta t} = \frac{10.4 \text{ km}}{0.62 \text{ h}}$$
$$= 16.8 \text{ km/h} \approx 17 \text{ km/h}. \quad \text{(Answer)}$$

To find $v_{avg}$ graphically, first we graph the function $x(t)$ as shown in Fig. 2-5, where the beginning and arrival points on the graph are the origin and the point labeled as "Station." Your average velocity is the slope of the straight line connecting those points; that is, $v_{avg}$ is the

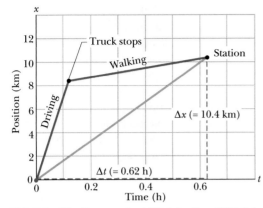

**FIG. 2-5** The lines marked "Driving" and "Walking" are the position–time plots for the driving and walking stages. (The plot for the walking stage assumes a constant rate of walking.) The slope of the straight line joining the origin and the point labeled "Station" is the average velocity for the trip, from the beginning to the station.

ratio of the *rise* ($\Delta x = 10.4$ km) to the *run* ($\Delta t = 0.62$ h), which gives us $v_{avg} = 16.8$ km/h.

**(d)** Suppose that to pump the gasoline, pay for it, and walk back to the truck takes you another 45 min. What is your average speed from the beginning of your drive to your return to the truck with the gasoline?

**KEY IDEA** Your average speed is the ratio of the total distance you move to the total time interval you take to make that move.

**Calculation:** The total distance is 8.4 km + 2.0 km + 2.0 km = 12.4 km. The total time interval is 0.12 h + 0.50 h + 0.75 h = 1.37 h. Thus, Eq. 2-3 gives us

$$s_{avg} = \frac{12.4 \text{ km}}{1.37 \text{ h}} = 9.1 \text{ km/h}. \quad \text{(Answer)}$$

---

## PROBLEM-SOLVING TACTICS

**Tactic 1:   Do You Understand the Problem?**   The common difficulty is simply not understanding the problem. The best test of understanding is this: Can *you* explain the problem?

Write down the given data, with units, using the symbols of the chapter. (In Sample Problem 2-1, the given data allow you to find your net displacement $\Delta x$ in part (a) and the corresponding time interval $\Delta t$ in part (b).) Identify the unknown and its symbol. (In the sample problem, the unknown in part (c) is your average velocity $v_{avg}$.) Then find the connection between the unknown and the data. (The connection is provided by Eq. 2-2, the definition of average velocity.)

**Tactic 2:   Are the Units OK?**   Be sure to use a consistent set of units when putting numbers into the equations. In Sample Problem 2-1, the logical units in terms of the given data are kilometers for distances, hours for time intervals, and kilometers per hour for velocities. You may sometimes need to convert units.

**Tactic 3:   Is Your Answer Reasonable?**   Does your answer make sense, or is it far too large or far too small? Is the sign correct? Are the units appropriate? In part (c) of Sample Problem 2-1, for example, the correct answer is 17 km/h. If you find 0.00017 km/h, −17 km/h, 17 km/s, or 17 000 km/h, you should realize at once that you have done something wrong. The error may lie in your method, in your algebra, or in your keystroking of numbers on a calculator.

**Tactic 4:   Reading a Graph**   Figures 2-2, 2-3a, 2-4, and 2-5 are graphs you should be able to read easily. In each graph, the variable on the horizontal axis is the time $t$, with the direction of increasing time to the right. In each, the variable on the vertical axis is the position $x$ of the moving particle with respect to the origin, with the positive direction of $x$ upward. Always note the units (seconds or minutes; meters or kilometers) in which the variables are expressed.

# 2-5 | Instantaneous Velocity and Speed

You have now seen two ways to describe how fast something moves: average velocity and average speed, both of which are measured over a time interval $\Delta t$. However, the phrase "how fast" more commonly refers to how fast a particle is moving at a given instant—its **instantaneous velocity** (or simply **velocity**) $v$.

The velocity at any instant is obtained from the average velocity by shrinking the time interval $\Delta t$ closer and closer to 0. As $\Delta t$ dwindles, the average velocity approaches a limiting value, which is the velocity at that instant:

$$v = \lim_{\Delta t \to 0} \frac{\Delta x}{\Delta t} = \frac{dx}{dt}. \quad (2\text{-}4)$$

Note that $v$ is the rate at which position $x$ is changing with time at a given instant; that is, $v$ is the derivative of $x$ with respect to $t$. Also note that $v$ at any instant is the slope of the position–time curve at the point representing that instant. Velocity is another vector quantity and thus has an associated direction.

**Speed** is the magnitude of velocity; that is, speed is velocity that has been stripped of any indication of direction, either in words or via an algebraic sign. (*Caution:* Speed and average speed can be quite different.) A velocity of $+5$ m/s and one of $-5$ m/s both have an associated speed of 5 m/s. The speedometer in a car measures speed, not velocity (it cannot determine the direction).

> ✓ **CHECKPOINT 2**    The following equations give the position $x(t)$ of a particle in four situations (in each equation, $x$ is in meters, $t$ is in seconds, and $t > 0$): (1) $x = 3t - 2$; (2) $x = -4t^2 - 2$; (3) $x = 2/t^2$; and (4) $x = -2$. (a) In which situation is the velocity $v$ of the particle constant? (b) In which is $v$ in the negative $x$ direction?

### Sample Problem 2-2

Figure 2-6a is an $x(t)$ plot for an elevator cab that is initially stationary, then moves upward (which we take to be the positive direction of $x$), and then stops. Plot $v(t)$.

**KEY IDEA**    We can find the velocity at any time from the slope of the $x(t)$ curve at that time.

**Calculations:** The slope of $x(t)$, and so also the velocity, is zero in the intervals from 0 to 1 s and from 9 s on, so then the cab is stationary. During the interval $bc$, the slope is constant and nonzero, so then the cab moves with constant velocity. We calculate the slope of $x(t)$ then as

$$\frac{\Delta x}{\Delta t} = v = \frac{24 \text{ m} - 4.0 \text{ m}}{8.0 \text{ s} - 3.0 \text{ s}} = +4.0 \text{ m/s}.$$

The plus sign indicates that the cab is moving in the positive $x$ direction. These intervals (where $v = 0$ and $v = 4$ m/s) are plotted in Fig. 2-6b. In addition, as the cab initially begins to move and then later slows to a stop, $v$ varies as indicated in the intervals 1 s to 3 s and 8 s to 9 s. Thus, Fig. 2-6b is the required plot. (Figure 2-6c is considered in Section 2-6.)

Given a $v(t)$ graph such as Fig. 2-6b, we could "work backward" to produce the shape of the associated $x(t)$ graph (Fig. 2-6a). However, we would not know the actual values for $x$ at various times, because the $v(t)$ graph indicates only *changes* in $x$. To find such a change in $x$ during any interval, we must, in the language of calculus, calculate the area "under the curve" on the $v(t)$ graph for that interval. For example, during the interval 3 s to 8 s in which the cab has a velocity of 4.0 m/s, the change in $x$ is

$$\Delta x = (4.0 \text{ m/s})(8.0 \text{ s} - 3.0 \text{ s}) = +20 \text{ m}.$$

(This area is positive because the $v(t)$ curve is above the $t$ axis.) Figure 2-6a shows that $x$ does indeed increase by 20 m in that interval. However, Fig. 2-6b does not tell us the *values* of $x$ at the beginning and end of the interval. For that, we need additional information, such as the value of $x$ at some instant.

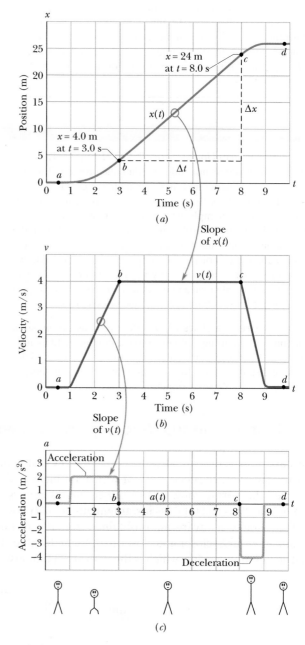

**FIG. 2-6** (a) The $x(t)$ curve for an elevator cab that moves upward along an $x$ axis. (b) The $v(t)$ curve for the cab. Note that it is the derivative of the $x(t)$ curve ($v = dx/dt$). (c) The $a(t)$ curve for the cab. It is the derivative of the $v(t)$ curve ($a = dv/dt$). The stick figures along the bottom suggest how a passenger's body might feel during the accelerations.

**Sample Problem** **2-3**

The position of a particle moving on an $x$ axis is given by

$$x = 7.8 + 9.2t - 2.1t^3, \qquad (2\text{-}5)$$

with $x$ in meters and $t$ in seconds. What is its velocity at $t = 3.5$ s? Is the velocity constant, or is it continuously changing?

**KEY IDEA** Velocity is the first derivative (with respect to time) of the position function $x(t)$.

**Calculations:** For simplicity, the units have been omitted from Eq. 2-5, but you can insert them if you like by changing the coefficients to 7.8 m, 9.2 m/s, and $-2.1$ m/s$^3$. Taking the derivative of Eq. 2-5, we write

$$v = \frac{dx}{dt} = \frac{d}{dt}(7.8 + 9.2t - 2.1t^3),$$

which becomes

$$v = 0 + 9.2 - (3)(2.1)t^2 = 9.2 - 6.3t^2. \qquad (2\text{-}6)$$

At $t = 3.5$ s,

$$v = 9.2 - (6.3)(3.5)^2 = -68 \text{ m/s.} \quad \text{(Answer)}$$

At $t = 3.5$ s, the particle is moving in the negative direction of $x$ (note the minus sign) with a speed of 68 m/s. Since the quantity $t$ appears in Eq. 2-6, the velocity $v$ depends on $t$ and so is continuously changing.

## 2-6 | Acceleration

When a particle's velocity changes, the particle is said to undergo **acceleration** (or to accelerate). For motion along an axis, the **average acceleration** $a_{\text{avg}}$ over a time interval $\Delta t$ is

$$a_{\text{avg}} = \frac{v_2 - v_1}{t_2 - t_1} = \frac{\Delta v}{\Delta t}, \qquad (2\text{-}7)$$

where the particle has velocity $v_1$ at time $t_1$ and then velocity $v_2$ at time $t_2$. The **instantaneous acceleration** (or simply **acceleration**) is

$$a = \frac{dv}{dt}. \qquad (2\text{-}8)$$

In words, the acceleration of a particle at any instant is the rate at which its velocity is changing at that instant. Graphically, the acceleration at any point is the slope of the curve of $v(t)$ at that point. We can combine Eq. 2-8 with Eq. 2-4 to write

$$a = \frac{dv}{dt} = \frac{d}{dt}\left(\frac{dx}{dt}\right) = \frac{d^2x}{dt^2}. \qquad (2\text{-}9)$$

In words, the acceleration of a particle at any instant is the second derivative of its position $x(t)$ with respect to time.

A common unit of acceleration is the meter per second per second: m/(s·s) or m/s$^2$. Other units are in the form of length/(time·time) or length/time$^2$. Acceleration has both magnitude and direction (it is yet another vector quantity). Its algebraic sign represents its direction on an axis just as for displacement and velocity; that is, acceleration with a positive value is in the positive direction of an axis, and acceleration with a negative value is in the negative direction.

Figure 2-6c is a plot of the acceleration of the elevator cab discussed in Sample Problem 2-2. Compare this $a(t)$ curve with the $v(t)$ curve—each point on the $a(t)$ curve shows the derivative (slope) of the $v(t)$ curve at the corresponding time. When $v$ is constant (at either 0 or 4 m/s), the derivative is zero and so also is the acceleration. When the cab first begins to move, the $v(t)$ curve has a positive derivative (the slope is positive), which means that $a(t)$ is positive. When the cab slows to a stop, the derivative and slope of the $v(t)$ curve are negative; that is, $a(t)$ is negative.

Next compare the slopes of the $v(t)$ curve during the two acceleration periods. The slope associated with the cab's slowing down (commonly called "deceleration") is steeper because the cab stops in half the time it took to get up to speed. The steeper slope means that the magnitude of the deceleration is larger than that of the acceleration, as indicated in Fig. 2-6c.

**FIG. 2-7** Colonel J. P. Stapp in a rocket sled as it is brought up to high speed (acceleration out of the page) and then very rapidly braked (acceleration into the page). *(Courtesy U.S. Air Force)*

The sensations you would feel while riding in the cab of Fig. 2-6 are indicated by the sketched figures at the bottom. When the cab first accelerates, you feel as though you are pressed downward; when later the cab is braked to a stop, you seem to be stretched upward. In between, you feel nothing special. In other words, your body reacts to accelerations (it is an accelerometer) but not to velocities (it is not a speedometer). When you are in a car traveling at 90 km/h or an airplane traveling at 900 km/h, you have no bodily awareness of the motion. However, if the car or plane quickly changes velocity, you may become keenly aware of the change, perhaps even frightened by it. Part of the thrill of an amusement park ride is due to the quick changes of velocity that you undergo (you pay for the accelerations, not for the speed). A more extreme example is shown in the photographs of Fig. 2-7, which were taken while a rocket sled was rapidly accelerated along a track and then rapidly braked to a stop.

Large accelerations are sometimes expressed in terms of $g$ units, with

$$1g = 9.8 \text{ m/s}^2 \qquad (g \text{ unit}). \qquad (2\text{-}10)$$

(As we shall discuss in Section 2-9, $g$ is the magnitude of the acceleration of a falling object near Earth's surface.) On a roller coaster, you may experience brief accelerations up to $3g$, which is $(3)(9.8 \text{ m/s}^2)$, or about $29 \text{ m/s}^2$, more than enough to justify the cost of the ride.

---

**PROBLEM-SOLVING TACTICS**

*Tactic 5: An Acceleration's Sign* In common language, the sign of an acceleration has a nonscientific meaning: positive acceleration means that the speed of an object is increasing, and negative acceleration means that the speed is decreasing (the object is decelerating). In this book, however, the sign of an acceleration indicates a direction, not whether an object's speed is increasing or decreasing.

For example, if a car with an initial velocity $v = -25$ m/s is braked to a stop in 5.0 s, then $a_{\text{avg}} = +5.0$ m/s$^2$. The acceleration is *positive*, but the car's speed has decreased. The reason is the difference in signs: the direction of the acceleration is opposite that of the velocity.

Here then is the proper way to interpret the signs:

If the signs of the velocity and acceleration of a particle are the same, the speed of the particle increases. If the signs are opposite, the speed decreases.

**CHECKPOINT 3** A wombat moves along an $x$ axis. What is the sign of its acceleration if it is moving (a) in the positive direction with increasing speed, (b) in the positive direction with decreasing speed, (c) in the negative direction with increasing speed, and (d) in the negative direction with decreasing speed?

---

**Sample Problem** | **2-4** | **Build your skill**

A particle's position on the $x$ axis of Fig. 2-1 is given by

$$x = 4 - 27t + t^3,$$

with $x$ in meters and $t$ in seconds.

(a) Because position $x$ depends on time $t$, the particle must be moving. Find the particle's velocity function $v(t)$ and acceleration function $a(t)$.

**KEY IDEAS** (1) To get the velocity function $v(t)$, we differentiate the position function $x(t)$ with respect to time. (2) To get the acceleration function $a(t)$, we differentiate the velocity function $v(t)$ with respect to time.

*Calculations:* Differentiating the position function, we find

$$v = -27 + 3t^2, \qquad \text{(Answer)}$$

**FIG. 2-7**  *Continued*

with $v$ in meters per second. Differentiating the velocity function then gives us

$$a = +6t, \qquad \text{(Answer)}$$

with $a$ in meters per second squared.

**(b)** Is there ever a time when $v = 0$?

***Calculation:*** Setting $v(t) = 0$ yields

$$0 = -27 + 3t^2,$$

which has the solution

$$t = \pm 3 \text{ s}. \qquad \text{(Answer)}$$

Thus, the velocity is zero both 3 s before and 3 s after the clock reads 0.

**(c)** Describe the particle's motion for $t \geq 0$.

***Reasoning:*** We need to examine the expressions for $x(t)$, $v(t)$, and $a(t)$.

At $t = 0$, the particle is at $x(0) = +4$ m and is moving with a velocity of $v(0) = -27$ m/s—that is, in the negative direction of the $x$ axis. Its acceleration is $a(0) = 0$ because just then the particle's velocity is not changing.

For $0 < t < 3$ s, the particle still has a negative velocity, so it continues to move in the negative direction. However, its acceleration is no longer 0 but is increasing and positive. Because the signs of the velocity and the acceleration are opposite, the particle must be slowing.

Indeed, we already know that it stops momentarily at $t = 3$ s. Just then the particle is as far to the left of the origin in Fig. 2-1 as it will ever get. Substituting $t = 3$ s into the expression for $x(t)$, we find that the particle's position just then is $x = -50$ m. Its acceleration is still positive.

For $t > 3$ s, the particle moves to the right on the axis. Its acceleration remains positive and grows progressively larger in magnitude. The velocity is now positive, and it too grows progressively larger in magnitude.

## 2-7 | Constant Acceleration: A Special Case

In many types of motion, the acceleration is either constant or approximately so. For example, you might accelerate a car at an approximately constant rate when a traffic light turns from red to green. Then graphs of your position, velocity, and acceleration would resemble those in Fig. 2-8. (Note that $a(t)$ in Fig. 2-8c is constant, which requires that $v(t)$ in Fig. 2-8b have a constant slope.) Later when you brake the car to a stop, the acceleration (or deceleration in common language) might also be approximately constant.

Such cases are so common that a special set of equations has been derived for dealing with them. One approach to the derivation of these equations is given in this section. A second approach is given in the next section. Throughout both sections and later when you work on the homework problems, keep in mind that *these equations are valid only for constant acceleration (or situations in which you can approximate the acceleration as being constant).*

When the acceleration is constant, the average acceleration and instantaneous acceleration are equal and we can write Eq. 2-7, with some changes in notation, as

$$a = a_{\text{avg}} = \frac{v - v_0}{t - 0}.$$

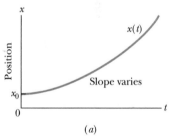

(a)

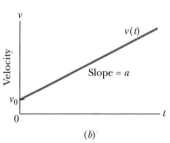

(b)

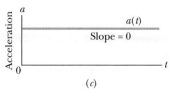

(c)

**FIG. 2-8**  (*a*) The position $x(t)$ of a particle moving with constant acceleration. (*b*) Its velocity $v(t)$, given at each point by the slope of the curve of $x(t)$. (*c*) Its (constant) acceleration, equal to the (constant) slope of the curve of $v(t)$.

Here $v_0$ is the velocity at time $t = 0$ and $v$ is the velocity at any later time $t$. We can recast this equation as

$$v = v_0 + at. \qquad (2\text{-}11)$$

As a check, note that this equation reduces to $v = v_0$ for $t = 0$, as it must. As a further check, take the derivative of Eq. 2-11. Doing so yields $dv/dt = a$, which is the definition of $a$. Figure 2-8b shows a plot of Eq. 2-11, the $v(t)$ function; the function is linear and thus the plot is a straight line.

In a similar manner, we can rewrite Eq. 2-2 (with a few changes in notation) as

$$v_{\text{avg}} = \frac{x - x_0}{t - 0}$$

and then as

$$x = x_0 + v_{\text{avg}}t, \qquad (2\text{-}12)$$

in which $x_0$ is the position of the particle at $t = 0$ and $v_{\text{avg}}$ is the average velocity between $t = 0$ and a later time $t$.

For the linear velocity function in Eq. 2-11, the *average* velocity over any time interval (say, from $t = 0$ to a later time $t$) is the average of the velocity at the beginning of the interval ($= v_0$) and the velocity at the end of the interval ($= v$). For the interval from $t = 0$ to the later time $t$ then, the average velocity is

$$v_{\text{avg}} = \tfrac{1}{2}(v_0 + v). \qquad (2\text{-}13)$$

Substituting the right side of Eq. 2-11 for $v$ yields, after a little rearrangement,

$$v_{\text{avg}} = v_0 + \tfrac{1}{2}at. \qquad (2\text{-}14)$$

Finally, substituting Eq. 2-14 into Eq. 2-12 yields

$$x - x_0 = v_0 t + \tfrac{1}{2}at^2. \qquad (2\text{-}15)$$

As a check, note that putting $t = 0$ yields $x = x_0$, as it must. As a further check, taking the derivative of Eq. 2-15 yields Eq. 2-11, again as it must. Figure 2-8a shows a plot of Eq. 2-15; the function is quadratic and thus the plot is curved.

Equations 2-11 and 2-15 are the *basic equations for constant acceleration;* they can be used to solve any constant acceleration problem in this book. However, we can derive other equations that might prove useful in certain specific situations. First, note that as many as five quantities can possibly be involved in any problem about constant acceleration—namely, $x - x_0, v, t, a$, and $v_0$. Usually, one of these quantities is *not* involved in the problem, *either as a given or as an unknown.* We are then presented with three of the remaining quantities and asked to find the fourth.

Equations 2-11 and 2-15 each contain four of these quantities, but not the same four. In Eq. 2-11, the "missing ingredient" is the displacement $x - x_0$. In Eq. 2-15, it is the velocity $v$. These two equations can also be combined in three ways to yield three additional equations, each of which involves a different "missing variable." First, we can eliminate $t$ to obtain

$$v^2 = v_0^2 + 2a(x - x_0). \qquad (2\text{-}16)$$

This equation is useful if we do not know $t$ and are not required to find it. Second, we can eliminate the acceleration $a$ between Eqs. 2-11 and 2-15 to produce an equation in which $a$ does not appear:

$$x - x_0 = \tfrac{1}{2}(v_0 + v)t. \qquad (2\text{-}17)$$

Finally, we can eliminate $v_0$, obtaining

$$x - x_0 = vt - \tfrac{1}{2}at^2. \qquad (2\text{-}18)$$

Note the subtle difference between this equation and Eq. 2-15. One involves the initial velocity $v_0$; the other involves the velocity $v$ at time $t$.

Table 2-1 lists the basic constant acceleration equations (Eqs. 2-11 and 2-15) as well as the specialized equations that we have derived. To solve a simple constant acceleration problem, you can usually use an equation from this list (*if you have the list with you*). Choose an equation for which the only unknown variable is the variable requested in the problem. A simpler plan is to remember only Eqs. 2-11 and 2-15, and then solve them as simultaneous equations whenever needed.

✓ **CHECKPOINT 4**     The following equations give the position $x(t)$ of a particle in four situations: (1) $x = 3t - 4$; (2) $x = -5t^3 + 4t^2 + 6$; (3) $x = 2/t^2 - 4/t$; (4) $x = 5t^2 - 3$. To which of these situations do the equations of Table 2-1 apply?

**TABLE 2-1**

**Equations for Motion with Constant Acceleration[a]**

| Equation Number | Equation | Missing Quantity |
|---|---|---|
| 2-11 | $v = v_0 + at$ | $x - x_0$ |
| 2-15 | $x - x_0 = v_0 t + \frac{1}{2}at^2$ | $v$ |
| 2-16 | $v^2 = v_0^2 + 2a(x - x_0)$ | $t$ |
| 2-17 | $x - x_0 = \frac{1}{2}(v_0 + v)t$ | $a$ |
| 2-18 | $x - x_0 = vt - \frac{1}{2}at^2$ | $v_0$ |

[a]Make sure that the acceleration is indeed constant before using the equations in this table.

## Sample Problem    2-5

The head of a woodpecker is moving forward at a speed of 7.49 m/s when the beak makes first contact with a tree limb. The beak stops after penetrating the limb by 1.87 mm. Assuming the acceleration to be constant, find the acceleration magnitude in terms of g.

**KEY IDEA**    We can use the constant-acceleration equations; in particular, we can use Eq. 2-16 ($v^2 = v_0^2 + 2a(x - x_0)$), which relates velocity and displacement.

**Calculations:** Because the woodpecker's head stops, the final velocity is $v = 0$. The initial velocity is $v_0 = 7.49$ m/s, and the displacement during the constant acceleration is $x - x_0 = 1.87 \times 10^{-3}$ m. Substituting these values into Eq. 2-16, we have

$$0^2 = (7.49 \text{ m/s})^2 + 2a(1.87 \times 10^{-3} \text{ m}),$$

or         $a = -1.500 \times 10^4$ m/s$^2$.

Dividing by $g = 9.8$ m/s$^2$ and taking the absolute value, we find that the magnitude of the head's acceleration is

$$a = (1.53 \times 10^3)g. \qquad \text{(Answer)}$$

**Comment:** This typical acceleration magnitude for a woodpecker is about 70 times the acceleration magnitude of Colonel Stapp in Fig. 2-7 and certainly would have been lethal to him. The ability of a woodpecker to withstand such huge acceleration magnitudes is not well understood, but there are two main arguments. (1) The woodpecker's motion is almost along a straight line. Some researchers believe that concussion can occur in humans and animals when the head is rapidly rotated around the neck (and brain stem), but that it is less likely in straight-line motion. (2) The woodpecker's brain is attached so well to the skull that there is little residual movement or oscillation of the brain just after the impact and no chance for the tissue connecting the skull and brain to tear.

## Sample Problem    2-6    Build your skill

Figure 2-9 gives a particle's velocity $v$ versus its position as it moves along an $x$ axis with constant acceleration. What is its velocity at position $x = 0$?

**KEY IDEA**    We can use the constant-acceleration equations; in particular, we can use Eq. 2-16 ($v^2 = v_0^2 + 2a(x - x_0)$), which relates velocity and position.

**First try:** Normally we want to use an equation that includes the requested variable. In Eq. 2-16, we can identify $x_0$ as 0 and $v_0$ as being the requested variable. Then we can identify a second pair of values as being $v$ and $x$. From the graph, we have two such pairs: (1) $v = 8$ m/s and $x = 20$ m, and (2) $v = 0$ and $x = 70$ m. For example, we can write Eq. 2-16 as

$$(8 \text{ m/s})^2 = v_0^2 + 2a(20 \text{ m} - 0). \qquad (2\text{-}19)$$

However, we know neither $v_0$ nor $a$.

**Second try:** Instead of directly involving the requested

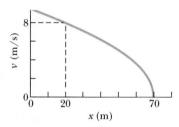

**FIG. 2-9** Velocity versus position.

variable, let's use Eq. 2-16 with the two pairs of known data, identifying $v_0 = 8$ m/s and $x_0 = 20$ m as the first pair and $v = 0$ m/s and $x = 70$ m as the second pair. Then we can write

$$(0 \text{ m/s})^2 = (8 \text{ m/s})^2 + 2a(70 \text{ m} - 20 \text{ m}),$$

which gives us $a = -0.64$ m/s$^2$. Substituting this value into Eq. 2-19 and solving for $v_0$ (the velocity associated with the position of $x = 0$), we find

$$v_0 = 9.5 \text{ m/s}. \qquad \text{(Answer)}$$

**Comment:** Some problems involve an equation that includes the requested variable. A more challenging problem requires you to first use an equation that does *not* include the requested variable but that gives you a value needed to find it. Sometimes that procedure takes *physics courage* because it is so indirect. However, if you build your solving skills by solving lots of problems, the procedure gradually requires less courage and may even become obvious. Solving problems of any kind, whether physics or social, requires practice.

## 2-8 | Another Look at Constant Acceleration*

The first two equations in Table 2-1 are the basic equations from which the others are derived. Those two can be obtained by integration of the acceleration with the condition that $a$ is constant. To find Eq. 2-11, we rewrite the definition of acceleration (Eq. 2-8) as

$$dv = a\, dt.$$

We next write the *indefinite integral* (or *antiderivative*) of both sides:

$$\int dv = \int a\, dt.$$

Since acceleration $a$ is a constant, it can be taken outside the integration. We obtain

$$\int dv = a \int dt$$

or

$$v = at + C. \tag{2-20}$$

To evaluate the constant of integration $C$, we let $t = 0$, at which time $v = v_0$. Substituting these values into Eq. 2-20 (which must hold for all values of $t$, including $t = 0$) yields

$$v_0 = (a)(0) + C = C.$$

Substituting this into Eq. 2-20 gives us Eq. 2-11.

To derive Eq. 2-15, we rewrite the definition of velocity (Eq. 2-4) as

$$dx = v\, dt$$

and then take the indefinite integral of both sides to obtain

$$\int dx = \int v\, dt.$$

Next, we substitute for $v$ with Eq. 2-11:

$$\int dx = \int (v_0 + at)\, dt.$$

Since $v_0$ is a constant, as is the acceleration $a$, this can be rewritten as

$$\int dx = v_0 \int dt + a \int t\, dt.$$

Integration now yields

$$x = v_0 t + \tfrac{1}{2} at^2 + C', \tag{2-21}$$

where $C'$ is another constant of integration. At time $t = 0$, we have $x = x_0$. Substituting these values in Eq. 2-21 yields $x_0 = C'$. Replacing $C'$ with $x_0$ in Eq. 2-21 gives us Eq. 2-15.

## 2-9 | Free-Fall Acceleration

If you tossed an object either up or down and could somehow eliminate the effects of air on its flight, you would find that the object accelerates downward at a certain constant rate. That rate is called the **free-fall acceleration,** and its magnitude is represented by $g$. The acceleration is independent of the object's characteristics, such as mass, density, or shape; it is the same for all objects.

*This section is intended for students who have had integral calculus.

Two examples of free-fall acceleration are shown in Fig. 2-10, which is a series of stroboscopic photos of a feather and an apple. As these objects fall, they accelerate downward—both at the same rate g. Thus, their speeds increase at the same rate, and they fall together.

The value of g varies slightly with latitude and with elevation. At sea level in Earth's midlatitudes the value is 9.8 m/s² (or 32 ft/s²), which is what you should use as an exact number for the problems in this book unless otherwise noted.

The equations of motion in Table 2-1 for constant acceleration also apply to free fall near Earth's surface; that is, they apply to an object in vertical flight, either up or down, when the effects of the air can be neglected. However, note that for free fall: (1) The directions of motion are now along a vertical $y$ axis instead of the $x$ axis, with the positive direction of $y$ upward. (This is important for later chapters when combined horizontal and vertical motions are examined.) (2) The free-fall acceleration is negative—that is, downward on the $y$ axis, toward Earth's center—and so it has the value $-g$ in the equations.

> The free-fall acceleration near Earth's surface is $a = -g = -9.8$ m/s², and the *magnitude* of the acceleration is $g = 9.8$ m/s². Do not substitute $-9.8$ m/s² for $g$.

Suppose you toss a tomato directly upward with an initial (positive) velocity $v_0$ and then catch it when it returns to the release level. During its *free-fall flight* (from just after its release to just before it is caught), the equations of Table 2-1 apply to its motion. The acceleration is always $a = -g = -9.8$ m/s², negative and thus downward. The velocity, however, changes, as indicated by Eqs. 2-11 and 2-16: during the ascent, the magnitude of the positive velocity decreases, until it momentarily becomes zero. Because the tomato has then stopped, it is at its maximum height. During the descent, the magnitude of the (now negative) velocity increases.

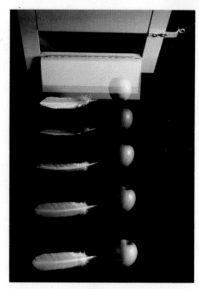

**FIG. 2-10** A feather and an apple free fall in vacuum at the same magnitude of acceleration g. The acceleration increases the distance between successive images. In the absence of air, the feather and apple fall together. *(Jim Sugar/Corbis Images)*

✓ **CHECKPOINT 5**    (a) If you toss a ball straight up, what is the sign of the ball's displacement for the ascent, from the release point to the highest point? (b) What is it for the descent, from the highest point back to the release point? (c) What is the ball's acceleration at its highest point?

**Sample Problem    2-7**

On September 26, 1993, Dave Munday went over the Canadian edge of Niagara Falls in a steel ball equipped with an air hole and then fell 48 m to the water (and rocks). Assume his initial velocity was zero, and neglect the effect of the air on the ball during the fall.

(a) How long did Munday fall to reach the water surface?

**KEY IDEA**    Because Munday's fall was a free fall, the constant-acceleration equations of Table 2-1 apply.

**Calculations:** Let us place a $y$ axis along the path of his fall, with $y = 0$ at his starting point and the positive direction up the axis (Fig. 2-11). Then the acceleration is $a = -g$ along that axis, and the water level is at $y = -48$ m (negative because it is below $y = 0$). Let the fall begin at time $t = 0$, with initial velocity $v_0 = 0$.

From Table 2-1 we choose Eq. 2-15 (but in $y$ notation) because it contains the requested time $t$. We find

$$y - y_0 = v_0 t - \tfrac{1}{2}gt^2,$$
$$-48 \text{ m} - 0 = 0t - \tfrac{1}{2}(9.8 \text{ m/s}^2)t^2,$$
$$t^2 = 48/4.9,$$

and                            $t = 3.1$ s.                    (Answer)

Note that Munday's displacement $y - y_0$ is a negative quantity—Munday fell down, in the *negative direction* of the $y$ axis (he did not fall up!). Also note that 48/4.9 has two square roots: 3.1 and −3.1. Here we choose the positive root because Munday obviously reaches the water surface *after* he begins to fall at $t = 0$.

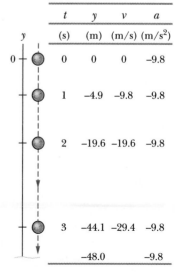

| | $t$ | $y$ | $v$ | $a$ |
|---|---|---|---|---|
| $y$ | (s) | (m) | (m/s) | (m/s²) |
| 0 | 0 | 0 | 0 | −9.8 |
| | 1 | −4.9 | −9.8 | −9.8 |
| | 2 | −19.6 | −19.6 | −9.8 |
| | 3 | −44.1 | −29.4 | −9.8 |
| | | −48.0 | | −9.8 |

**FIG. 2-11** The position, velocity, and acceleration of a freely falling object, here the steel ball ridden by Dave Munday over Niagara Falls.

(b) Munday could count off the three seconds of free fall but could not see how far he had fallen with each count. Determine his position at each full second.

*Calculations:* We again use Eq. 2-15 but now we substitute, in turn, the values $t = 1.0$ s, $2.0$ s, and $3.0$ s, and solve for Munday's position $y$. The results are shown in Fig. 2-11.

(c) What was Munday's velocity as he reached the water surface?

*Calculation:* To find the velocity from the original data without using the time of fall from (a), we rewrite Eq. 2-16 in $y$ notation and then substitute known data:

$$v^2 = v_0^2 - 2g(y - y_0) = 0 - (2)(9.8 \text{ m/s}^2)(-48 \text{ m}),$$
so $\quad v = -30.67 \text{ m/s} \approx -31 \text{ m/s} = -110 \text{ km/h}.$
$$\text{(Answer)}$$

We chose the negative root here because the velocity was in the negative direction.

(d) What was Munday's velocity at each count of one full second? Was he aware of his increasing speed?

*Calculations:* To find the velocities from the original data without using the positions from (b), we let $a = -g$ in Eq. 2-11 and then substitute, in turn, the values $t = 1.0$ s, $2.0$ s, and $3.0$ s. Here is an example:

$$v = v_0 - gt$$
$$= 0 - (9.8 \text{ m/s}^2)(1.0 \text{ s}) = -9.8 \text{ m/s}. \quad \text{(Answer)}$$

The other results are shown in Fig. 2-11.

Once he was in free fall, Munday was unaware of the increasing speed because the acceleration during the fall was always $-9.8$ m/s$^2$, as noted in the last column of Fig. 2-11. He was, of course, sharply aware of hitting the water because then the acceleration abruptly changed. (Munday survived the fall but then faced stiff legal fines for his daredevil action.)

## Sample Problem 2-8

In Fig. 2-12, a pitcher tosses a baseball up along a $y$ axis, with an initial speed of 12 m/s.

(a) How long does the ball take to reach its maximum height?

**KEY IDEAS** (1) Once the ball leaves the pitcher and before it returns to his hand, its acceleration is the free-fall acceleration $a = -g$. Because this is constant, Table 2-1 applies to the motion. (2) The velocity $v$ at the maximum height must be 0.

*Calculation:* Knowing $v$, $a$, and the initial velocity $v_0 = 12$ m/s, and seeking $t$, we solve Eq. 2-11, which con-

tains those four variables. This yields

$$t = \frac{v - v_0}{a} = \frac{0 - 12 \text{ m/s}}{-9.8 \text{ m/s}^2} = 1.2 \text{ s}. \quad \text{(Answer)}$$

(b) What is the ball's maximum height above its release point?

*Calculation:* We can take the ball's release point to be $y_0 = 0$. We can then write Eq. 2-16 in $y$ notation, set $y - y_0 = y$ and $v = 0$ (at the maximum height), and solve for $y$. We get

$$y = \frac{v^2 - v_0^2}{2a} = \frac{0 - (12 \text{ m/s})^2}{2(-9.8 \text{ m/s}^2)} = 7.3 \text{ m}. \quad \text{(Answer)}$$

(c) How long does the ball take to reach a point 5.0 m above its release point?

*Calculations:* We know $v_0$, $a = -g$, and displacement $y - y_0 = 5.0$ m, and we want $t$, so we choose Eq. 2-15. Rewriting it for $y$ and setting $y_0 = 0$ give us

$$y = v_0 t - \tfrac{1}{2}gt^2,$$

or $\quad 5.0 \text{ m} = (12 \text{ m/s})t - (\tfrac{1}{2})(9.8 \text{ m/s}^2)t^2.$

If we temporarily omit the units (having noted that they are consistent), we can rewrite this as

$$4.9t^2 - 12t + 5.0 = 0.$$

Solving this quadratic equation for $t$ yields

$$t = 0.53 \text{ s} \quad \text{and} \quad t = 1.9 \text{ s}. \quad \text{(Answer)}$$

There are two such times! This is not really surprising because the ball passes twice through $y = 5.0$ m, once on the way up and once on the way down.

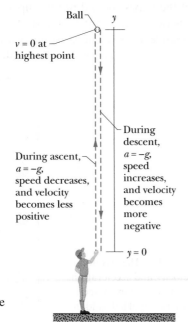

Ball

$v = 0$ at highest point

During ascent, $a = -g$, speed decreases, and velocity becomes less positive

During descent, $a = -g$, speed increases, and velocity becomes more negative

$y = 0$

**FIG. 2-12** A pitcher tosses a baseball straight up into the air. The equations of free fall apply for rising as well as for falling objects, provided any effects from the air can be neglected.

*Tactic 6: Meanings of Minus Signs* In Sample Problems 2-7 and 2-8, we established a vertical axis (the *y* axis) and we chose—quite arbitrarily—its upward direction to be positive. We then chose the origin of the *y* axis (that is, the $y = 0$ position) to suit the problem. In Sample Problem 2-7, the origin was at the top of the falls, and in Sample Problem 2-8 it was at the pitcher's hand. A negative value of *y* then means that the body is below the chosen origin. A negative velocity means that the body is moving in the negative direction of the *y* axis—that is, downward. This is true no matter where the body is located.

We take the acceleration to be negative ($-9.8$ m/s$^2$) in all problems dealing with falling bodies. A negative acceleration means that, as time goes on, the velocity of the body becomes either less positive or more negative. This is true no matter where the body is located and no matter how fast or in what direction it is moving. In Sample Problem 2-8, the acceleration of the ball is negative (downward) throughout its flight, whether the ball is rising or falling.

*Tactic 7: Unexpected Answers* Mathematics often generates answers that you might not have thought of as possibilities, as in Sample Problem 2-8c. If you get more answers than you expect, do not automatically discard the ones that do not seem to fit. Examine them carefully for physical meaning. If time is your variable, even a negative value can mean something; negative time simply refers to time before $t = 0$, the (arbitrary) time at which you decided to start your stopwatch.

## 2-10 | Graphical Integration in Motion Analysis

When we have a graph of an object's acceleration versus time, we can integrate on the graph to find the object's velocity at any given time. Because acceleration *a* is defined in terms of velocity as $a = dv/dt$, the Fundamental Theorem of Calculus tells us that

$$v_1 - v_0 = \int_{t_0}^{t_1} a\, dt. \tag{2-22}$$

The right side of the equation is a definite integral (it gives a numerical result rather than a function), $v_0$ is the velocity at time $t_0$, and $v_1$ is the velocity at later time $t_1$. The definite integral can be evaluated from an $a(t)$ graph, such as in Fig. 2-13*a*. In particular,

$$\int_{t_0}^{t_1} a\, dt = \left( \begin{array}{c} \text{area between acceleration curve} \\ \text{and time axis, from } t_0 \text{ to } t_1 \end{array} \right). \tag{2-23}$$

If a unit of acceleration is 1 m/s$^2$ and a unit of time is 1 s, then the corresponding unit of area on the graph is

$$(1\ \text{m/s}^2)(1\ \text{s}) = 1\ \text{m/s},$$

which is (properly) a unit of velocity. When the acceleration curve is above the time axis, the area is positive; when the curve is below the time axis, the area is negative.

Similarly, because velocity *v* is defined in terms of the position *x* as $v = dx/dt$, then

$$x_1 - x_0 = \int_{t_0}^{t_1} v\, dt, \tag{2-24}$$

where $x_0$ is the position at time $t_0$ and $x_1$ is the position at time $t_1$. The definite integral on the right side of Eq. 2-24 can be evaluated from a $v(t)$ graph, like that shown in Fig. 2-13*b*. In particular,

$$\int_{t_0}^{t_1} v\, dt = \left( \begin{array}{c} \text{area between velocity curve} \\ \text{and time axis, from } t_0 \text{ to } t_1 \end{array} \right). \tag{2-25}$$

If the unit of velocity is 1 m/s and the unit of time is 1 s, then the corresponding unit of area on the graph is

$$(1\ \text{m/s})(1\ \text{s}) = 1\ \text{m},$$

which is (properly) a unit of position and displacement. Whether this area is positive or negative is determined as described for the $a(t)$ curve of Fig. 2-13*a*.

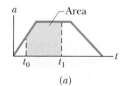

(a)

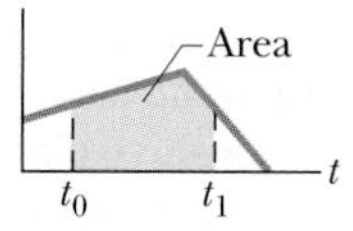

(b)

FIG. 2-13 The area between a plotted curve and the horizontal time axis, from time $t_0$ to time $t_1$, is indicated for (*a*) a graph of acceleration *a* versus *t* and (*b*) a graph of velocity *v* versus *t*.

**Sample Problem 2-9**

"Whiplash injury" commonly occurs in a rear-end collision where a front car is hit from behind by a second car. In the 1970s, researchers concluded that the injury was due to the occupant's head being whipped back over the top of the seat as the car was slammed forward. As a result of this finding, head restraints were built into cars, yet neck injuries in rear-end collisions continued to occur.

In a recent test to study neck injury in rear-end collisions, a volunteer was strapped to a seat that was then moved abruptly to simulate a collision by a rear car moving at 10.5 km/h. Figure 2-14a gives the accelerations of the volunteer's torso and head during the collision, which began at time $t = 0$. The torso acceleration was delayed by 40 ms because during that time interval the seat back had to compress against the volunteer. The head acceleration was delayed by an additional 70 ms. What was the torso speed when the head began to accelerate?

**KEY IDEA** We can calculate the torso speed at any time by finding an area on the torso $a(t)$ graph.

**Calculations:** We know that the initial torso speed is $v_0 = 0$ at time $t_0 = 0$, at the start of the "collision." We want the torso speed $v_1$ at time $t_1 = 110$ ms, which is when the head begins to accelerate.

Combining Eqs. 2-22 and 2-23, we can write

$$v_1 - v_0 = \begin{pmatrix} \text{area between acceleration curve} \\ \text{and time axis, from } t_0 \text{ to } t_1 \end{pmatrix}. \quad (2\text{-}26)$$

For convenience, let us separate the area into three regions (Fig. 2-14b). From 0 to 40 ms, region $A$ has no area:

$$\text{area}_A = 0.$$

From 40 ms to 100 ms, region $B$ has the shape of a triangle, with area

$$\text{area}_B = \tfrac{1}{2}(0.060 \text{ s})(50 \text{ m/s}^2) = 1.5 \text{ m/s}.$$

From 100 ms to 110 ms, region $C$ has the shape of a rectangle, with area

$$\text{area}_C = (0.010 \text{ s})(50 \text{ m/s}^2) = 0.50 \text{ m/s}.$$

Substituting these values and $v_0 = 0$ into Eq. 2-26 gives us

$$v_1 - 0 = 0 + 1.5 \text{ m/s} + 0.50 \text{ m/s},$$

or

$$v_1 = 2.0 \text{ m/s} = 7.2 \text{ km/h}. \quad \text{(Answer)}$$

**Comments:** When the head is just starting to move forward, the torso already has a speed of 7.2 km/h. Researchers argue that it is this difference in speeds during the early stage of a rear-end collision that injures the neck. The backward whipping of the head happens later and could, especially if there is no head restraint, increase the injury.

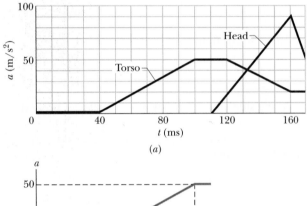

(a)

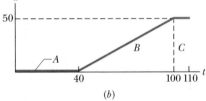

(b)

**FIG. 2-14** (a) The $a(t)$ curve of the torso and head of a volunteer in a simulation of a rear-end collision. (b) Breaking up the region between the plotted curve and the time axis to calculate the area.

# REVIEW & SUMMARY

**Position** The *position x* of a particle on an *x* axis locates the particle with respect to the **origin**, or zero point, of the axis. The position is either positive or negative, according to which side of the origin the particle is on, or zero if the particle is at the origin. The **positive direction** on an axis is the direction of increasing positive numbers; the opposite direction is the **negative direction.**

**Displacement** The *displacement* $\Delta x$ of a particle is the change in its position:

$$\Delta x = x_2 - x_1. \quad (2\text{-}1)$$

Displacement is a vector quantity. It is positive if the particle

has moved in the positive direction of the *x* axis and negative if the particle has moved in the negative direction.

**Average Velocity** When a particle has moved from position $x_1$ to position $x_2$ during a time interval $\Delta t = t_2 - t_1$, its *average velocity* during that interval is

$$v_{\text{avg}} = \frac{\Delta x}{\Delta t} = \frac{x_2 - x_1}{t_2 - t_1}. \quad (2\text{-}2)$$

The algebraic sign of $v_{\text{avg}}$ indicates the direction of motion ($v_{\text{avg}}$ is a vector quantity). Average velocity does not depend on the actual distance a particle moves, but instead depends on its original and final positions.

On a graph of $x$ versus $t$, the average velocity for a time interval $\Delta t$ is the slope of the straight line connecting the points on the curve that represent the two ends of the interval.

**Average Speed** The *average speed* $s_{avg}$ of a particle during a time interval $\Delta t$ depends on the total distance the particle moves in that time interval:

$$s_{avg} = \frac{\text{total distance}}{\Delta t}. \qquad (2\text{-}3)$$

**Instantaneous Velocity** The *instantaneous velocity* (or simply **velocity**) $v$ of a moving particle is

$$v = \lim_{\Delta t \to 0} \frac{\Delta x}{\Delta t} = \frac{dx}{dt}, \qquad (2\text{-}4)$$

where $\Delta x$ and $\Delta t$ are defined by Eq. 2-2. The instantaneous velocity (at a particular time) may be found as the slope (at that particular time) of the graph of $x$ versus $t$. **Speed** is the magnitude of instantaneous velocity.

**Average Acceleration** *Average acceleration* is the ratio of a change in velocity $\Delta v$ to the time interval $\Delta t$ in which the change occurs:

$$a_{avg} = \frac{\Delta v}{\Delta t}. \qquad (2\text{-}7)$$

The algebraic sign indicates the direction of $a_{avg}$.

**Instantaneous Acceleration** *Instantaneous acceleration* (or simply **acceleration**) $a$ is the first time derivative of velocity $v(t)$ and the second time derivative of position $x(t)$:

$$a = \frac{dv}{dt} = \frac{d^2x}{dt^2}. \qquad (2\text{-}8, 2\text{-}9)$$

On a graph of $v$ versus $t$, the acceleration $a$ at any time $t$ is the slope of the curve at the point that represents $t$.

**Constant Acceleration** The five equations in Table 2-1 describe the motion of a particle with constant acceleration:

$$v = v_0 + at, \qquad (2\text{-}11)$$

$$x - x_0 = v_0 t + \tfrac{1}{2}at^2, \qquad (2\text{-}15)$$

$$v^2 = v_0^2 + 2a(x - x_0), \qquad (2\text{-}16)$$

$$x - x_0 = \tfrac{1}{2}(v_0 + v)t, \qquad (2\text{-}17)$$

$$x - x_0 = vt - \tfrac{1}{2}at^2. \qquad (2\text{-}18)$$

These are *not* valid when the acceleration is not constant.

**Free-Fall Acceleration** An important example of straight-line motion with constant acceleration is that of an object rising or falling freely near Earth's surface. The constant acceleration equations describe this motion, but we make two changes in notation: (1) we refer the motion to the vertical $y$ axis with $+y$ vertically *up*; (2) we replace $a$ with $-g$, where $g$ is the magnitude of the free-fall acceleration. Near Earth's surface, $g = 9.8 \text{ m/s}^2 \, (= 32 \text{ ft/s}^2)$.

## QUESTIONS

**1** Figure 2-15 shows four paths along which objects move from a starting point to a final point, all in the same time interval. The paths pass over a grid of equally spaced straight lines. Rank the paths according to (a) the average velocity of the objects and (b) the average speed of the objects, greatest first.

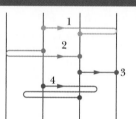

FIG. 2-15 Question 1.

**2** Figure 2-16 is a graph of a particle's position along an $x$ axis versus time. (a) At time $t = 0$, what is the sign of the particle's position? Is the particle's velocity positive, negative, or 0 at (b) $t = 1$ s, (c) $t = 2$ s, and (d) $t = 3$ s? (e) How many times does the particle go through the point $x = 0$?

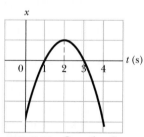

FIG. 2-16 Question 2.

**3** Figure 2-17 gives the velocity of a particle moving on an $x$ axis. What are (a) the initial and (b) the final directions of travel? (c) Does the particle stop momentarily? (d) Is the acceleration positive or negative? (e) Is it constant or varying?

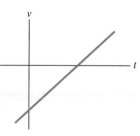

FIG. 2-17 Question 3.

**4** Figure 2-18 gives the acceleration $a(t)$ of a Chihuahua as it chases a German shepherd along an axis. In which of the time periods indicated does the Chihuahua move at constant speed?

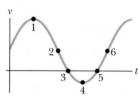

FIG. 2-18 Question 4.

**5** Figure 2-19 gives the velocity of a particle moving along an axis. Point 1 is at the highest point on the curve; point 4 is at the lowest point; and points 2 and 6 are at the same height. What is the direction of travel at (a) time $t = 0$ and (b) point 4? (c) At which of the six numbered points does the particle reverse its direction of travel? (d) Rank the six points according to the magnitude of the acceleration, greatest first.

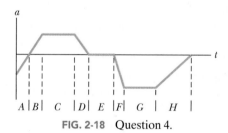

FIG. 2-19 Question 5.

**6** The following equations give the velocity $v(t)$ of a particle in four situations: (a) $v = 3$; (b) $v = 4t^2 + 2t - 6$; (c) $v = 3t - 4$; (d) $v = 5t^2 - 3$. To which of these situations do the equations of Table 2-1 apply?

**7** In Fig. 2-20, a cream tangerine is thrown directly upward past three evenly spaced windows of equal heights. Rank the windows according to (a) the average speed of the cream tangerine while passing them, (b) the time the cream tangerine takes to pass them, (c) the magnitude of the acceleration of the cream tangerine while passing them, and (d) the change $\Delta v$ in the speed of the cream tangerine during the passage, greatest first.

**8** At $t = 0$, a particle moving along an $x$ axis is at position

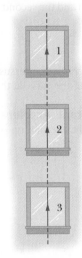

FIG. 2-20   Question 7.

$x_0 = -20$ m. The signs of the particle's initial velocity $v_0$ (at time $t_0$) and constant acceleration $a$ are, respectively, for four situations: (1) +, +; (2) +, −; (3) −, +; (4) −, −. In which situations will the particle (a) stop momentarily, (b) pass through the origin, and (c) never pass through the origin?

**9** Hanging over the railing of a bridge, you drop an egg (no initial velocity) as you throw a second egg downward. Which curves in Fig. 2-21 give the velocity $v(t)$ for (a) the dropped egg and (b) the thrown egg? (Curves $A$ and $B$ are parallel; so are $C$, $D$, and $E$; so are $F$ and $G$.)

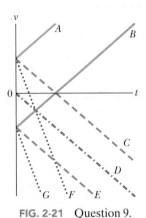

FIG. 2-21   Question 9.

# PROBLEMS

GO   Tutoring problem available (at instructor's discretion) in *WileyPLUS* and WebAssign

SSM   Worked-out solution available in Student Solutions Manual   WWW   Worked-out solution is at ___

•–•••   Number of dots indicates level of problem difficulty   ILW   Interactive solution is at ___   http://www.wiley.com/college/halliday

Additional information available in *The Flying Circus of Physics* and at flyingcircusofphysics.com

## sec. 2-4   Average Velocity and Average Speed

**•1** An automobile travels on a straight road for 40 km at 30 km/h. It then continues in the same direction for another 40 km at 60 km/h. (a) What is the average velocity of the car during this 80 km trip? (Assume that it moves in the positive $x$ direction.) (b) What is the average speed? (c) Graph $x$ versus $t$ and indicate how the average velocity is found on the graph. **SSM WWW**

**•2** A car travels up a hill at a constant speed of 40 km/h and returns down the hill at a constant speed of 60 km/h. Calculate the average speed for the round trip.

**•3** During a hard sneeze, your eyes might shut for 0.50 s. If you are driving a car at 90 km/h during such a sneeze, how far does the car move during that time?

**•4** The 1992 world speed record for a bicycle (human powered vehicle) was set by Chris Huber. His time through the measured 200 m stretch was a sizzling 6.509 s, at which he commented, "Cogito ergo zoom!" (I think, therefore I go fast!). In 2001, Sam Whittingham beat Huber's record by 19.0 km/h. What was Whittingham's time through the 200 m?

**•5** The position of an object moving along an $x$ axis is given by $x = 3t - 4t^2 + t^3$, where $x$ is in meters and $t$ in seconds. Find the position of the object at the following values of $t$: (a) 1 s, (b) 2 s, (c) 3 s, and (d) 4 s. (e) What is the object's displacement between $t = 0$ and $t = 4$ s? (f) What is its average velocity for the time interval from $t = 2$ s to $t = 4$ s? (g) Graph $x$ versus $t$ for $0 \leq t \leq 4$ s and indicate how the answer for (f) can be found on the graph. **SSM**

**•6** Compute your average velocity in the following two cases: (a) You walk 73.2 m at a speed of 1.22 m/s and then run 73.2 m at a speed of 3.05 m/s along a straight track. (b)

You walk for 1.00 min at a speed of 1.22 m/s and then run for 1.00 min at 3.05 m/s along a straight track. (c) Graph $x$ versus $t$ for both cases and indicate how the average velocity is found on the graph.

**••7** In 1 km races, runner 1 on track 1 (with time 2 min, 27.95 s) appears to be faster than runner 2 on track 2 (2 min, 28.15 s). However, length $L_2$ of track 2 might be slightly greater than length $L_1$ of track 1. How large can $L_2 - L_1$ be for us still to conclude that runner 1 is faster? **ILW**

**••8** To set a speed record in a measured (straight-line) distance $d$, a race car must be driven first in one direction (in time $t_1$) and then in the opposite direction (in time $t_2$). (a) To eliminate the effects of the wind and obtain the car's speed $v_c$ in a windless situation, should we find the average of $d/t_1$ and $d/t_2$ (method 1) or should we divide $d$ by the average of $t_1$ and $t_2$? (b) What is the fractional difference in the two methods when a steady wind blows along the car's route and the ratio of the wind speed $v_w$ to the car's speed $v_c$ is 0.0240?

**••9** You are to drive to an interview in another town, at a distance of 300 km on an expressway. The interview is at 11:15 A.M. You plan to drive at 100 km/h, so you leave at 8:00 A.M. to allow some extra time. You drive at that speed for the first 100 km, but then construction work forces you to slow to 40 km/h for 40 km. What would be the least speed needed for the rest of the trip to arrive in time for the interview?

**••10** *Panic escape.* Figure 2-22 shows a general situation in which a stream of people attempt to escape through an exit door that turns out to be locked. The people move toward the door at speed $v_s = 3.50$ m/s, are

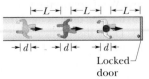

FIG. 2-22   Problem 10.

each $d = 0.25$ m in depth, and are separated by $L = 1.75$ m. The arrangement in Fig. 2-22 occurs at time $t = 0$. (a) At what average rate does the layer of people at the door increase? (b) At what time does the layer's depth reach 5.0 m? (The answers reveal how quickly such a situation becomes dangerous.) ✈

**••11** Two trains, each having a speed of 30 km/h, are headed at each other on the same straight track. A bird that can fly 60 km/h flies off the front of one train when they are 60 km apart and heads directly for the other train. On reaching the other train, the bird flies directly back to the first train, and so forth. (We have no idea *why* a bird would behave in this way.) What is the total distance the bird travels before the trains collide?

**•••12** *Traffic shock wave.* An abrupt slowdown in concentrated traffic can travel as a pulse, termed a *shock wave*, along the line of cars, either downstream (in the traffic direction) or upstream, or it can be stationary. Figure 2-23 shows a uniformly spaced line of cars moving at speed $v = 25.0$ m/s toward a uniformly spaced line of slow cars moving at speed $v_s = 5.00$ m/s. Assume that each faster car adds length $L = 12.0$ m (car length plus buffer zone) to the line of slow cars when it joins the line, and assume it slows abruptly at the last instant. (a) For what separation distance $d$ between the faster cars does the shock wave remain stationary? If the separation is twice that amount, what are the (b) speed and (c) direction (upstream or downstream) of the shock wave? ✈

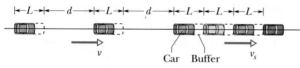

**FIG. 2-23** Problem 12.

**•••13** You drive on Interstate 10 from San Antonio to Houston, half the *time* at 55 km/h and the other half at 90 km/h. On the way back you travel half the *distance* at 55 km/h and the other half at 90 km/h. What is your average speed (a) from San Antonio to Houston, (b) from Houston back to San Antonio, and (c) for the entire trip? (d) What is your average velocity for the entire trip? (e) Sketch $x$ versus $t$ for (a), assuming the motion is all in the positive $x$ direction. Indicate how the average velocity can be found on the sketch. **ILW**

**sec. 2-5 Instantaneous Velocity and Speed**
**•14** The position function $x(t)$ of a particle moving along an $x$ axis is $x = 4.0 - 6.0t^2$, with $x$ in meters and $t$ in seconds. (a) At what time and (b) where does the particle (momentarily) stop? At what (c) negative time and (d) positive time does the particle pass through the origin? (e) Graph $x$ versus $t$ for the range $-5$ s to $+5$ s. (f) To shift the curve rightward on the graph, should we include the term $+20t$ or the term $-20t$ in $x(t)$? (g) Does that inclusion increase or decrease the value of $x$ at which the particle momentarily stops?

**•15** (a) If a particle's position is given by $x = 4 - 12t + 3t^2$ (where $t$ is in seconds and $x$ is in meters), what is its velocity at $t = 1$ s? (b) Is it moving in the positive or negative direction of $x$ just then? (c) What is its speed just then? (d) Is the speed increasing or decreasing just then? (Try answering the next two questions without further calculation.) (e) Is there ever an

instant when the velocity is zero? If so, give the time $t$; if not, answer no. (f) Is there a time after $t = 3$ s when the particle is moving in the negative direction of $x$? If so, give the time $t$; if not, answer no. **GO**

**•16** An electron moving along the $x$ axis has a position given by $x = 16te^{-t}$ m, where $t$ is in seconds. How far is the electron from the origin when it momentarily stops?

**••17** The position of a particle moving along the $x$ axis is given in centimeters by $x = 9.75 + 1.50t^3$, where $t$ is in seconds. Calculate (a) the average velocity during the time interval $t = 2.00$ s to $t = 3.00$ s; (b) the instantaneous velocity at $t = 2.00$ s; (c) the instantaneous velocity at $t = 3.00$ s; (d) the instantaneous velocity at $t = 2.50$ s; and (e) the instantaneous velocity when the particle is midway between its positions at $t = 2.00$ s and $t = 3.00$ s. (f) Graph $x$ versus $t$ and indicate your answers graphically.

**sec. 2-6 Acceleration**
**•18** (a) If the position of a particle is given by $x = 20t - 5t^3$, where $x$ is in meters and $t$ is in seconds, when, if ever, is the particle's velocity zero? (b) When is its acceleration $a$ zero? (c) For what time range (positive or negative) is $a$ negative? (d) Positive? (e) Graph $x(t)$, $v(t)$, and $a(t)$.

**•19** At a certain time a particle had a speed of 18 m/s in the positive $x$ direction, and 2.4 s later its speed was 30 m/s in the opposite direction. What is the average acceleration of the particle during this 2.4 s interval? **SSM**

**•20** The position of a particle moving along an $x$ axis is given by $x = 12t^2 - 2t^3$, where $x$ is in meters and $t$ is in seconds. Determine (a) the position, (b) the velocity, and (c) the acceleration of the particle at $t = 3.0$ s. (d) What is the maximum positive coordinate reached by the particle and (e) at what time is it reached? (f) What is the maximum positive velocity reached by the particle and (g) at what time is it reached? (h) What is the acceleration of the particle at the instant the particle is not moving (other than at $t = 0$)? (i) Determine the average velocity of the particle between $t = 0$ and $t = 3$ s.

**••21** The position of a particle moving along the $x$ axis depends on the time according to the equation $x = ct^2 - bt^3$, where $x$ is in meters and $t$ in seconds. What are the units of (a) constant $c$ and (b) constant $b$? Let their numerical values be 3.0 and 2.0, respectively. (c) At what time does the particle reach its maximum positive $x$ position? From $t = 0.0$ s to $t = 4.0$ s, (d) what distance does the particle move and (e) what is its displacement? Find its velocity at times (f) 1.0 s, (g) 2.0 s, (h) 3.0 s, and (i) 4.0 s. Find its acceleration at times (j) 1.0 s, (k) 2.0 s, (l) 3.0 s, and (m) 4.0 s.

**••22** From $t = 0$ to $t = 5.00$ min, a man stands still, and from $t = 5.00$ min to $t = 10.0$ min, he walks briskly in a straight line at a constant speed of 2.20 m/s. What are (a) his average velocity $v_{avg}$ and (b) his average acceleration $a_{avg}$ in the time interval 2.00 min to 8.00 min? What are (c) $v_{avg}$ and (d) $a_{avg}$ in the time interval 3.00 min to 9.00 min? (e) Sketch $x$ versus $t$ and $v$ versus $t$, and indicate how the answers to (a) through (d) can be obtained from the graphs.

**sec. 2-7 Constant Acceleration: A Special Case**
**•23** An electron has a constant acceleration of $+3.2$ m/s$^2$. At a certain instant its velocity is $+9.6$ m/s. What is its velocity (a) 2.5 s earlier and (b) 2.5 s later?

•24 A muon (an elementary particle) enters a region with a speed of $5.00 \times 10^6$ m/s and then is slowed at the rate of $1.25 \times 10^{14}$ m/s². (a) How far does the muon take to stop? (b) Graph $x$ versus $t$ and $v$ versus $t$ for the muon.

•25 Suppose a rocket ship in deep space moves with constant acceleration equal to 9.8 m/s², which gives the illusion of normal gravity during the flight. (a) If it starts from rest, how long will it take to acquire a speed one-tenth that of light, which travels at $3.0 \times 10^8$ m/s? (b) How far will it travel in so doing? **SSM**

•26 On a dry road, a car with good tires may be able to brake with a constant deceleration of 4.92 m/s². (a) How long does such a car, initially traveling at 24.6 m/s, take to stop? (b) How far does it travel in this time? (c) Graph $x$ versus $t$ and $v$ versus $t$ for the deceleration.

•27 An electron with an initial velocity $v_0 = 1.50 \times 10^5$ m/s enters a region of length $L$ = 1.00 cm where it is electrically accelerated (Fig. 2-24). It emerges with $v = 5.70 \times 10^6$ m/s. What is its acceleration, assumed constant? **SSM**

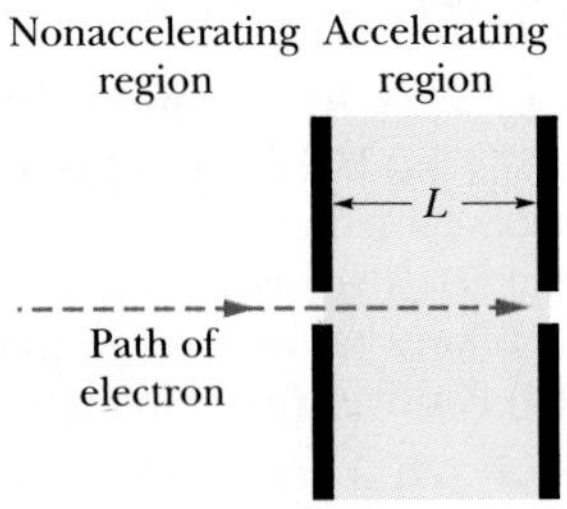

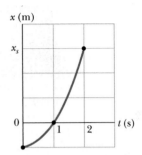

Nonaccelerating region    Accelerating region

← L →

Path of electron

FIG. 2-24 Problem 27.

•28 *Catapulting mushrooms.* Certain mushrooms launch their spores by a catapult mechanism. As water condenses from the air onto a spore that is attached to the mushroom, a drop grows on one side of the spore and a film grows on the other side. The spore is bent over by the drop's weight, but when the film reaches the drop, the drop's water suddenly spreads into the film and the spore springs upward so rapidly that it is slung off into the air. Typically, the spore reaches a speed of 1.6 m/s in a 5.0 $\mu$m launch; its speed is then reduced to zero in 1.0 mm by the air. Using that data and assuming constant accelerations, find the acceleration in terms of $g$ during (a) the launch and (b) the speed reduction.

•29 An electric vehicle starts from rest and accelerates at a rate of 2.0 m/s² in a straight line until it reaches a speed of 20 m/s. The vehicle then slows at a constant rate of 1.0 m/s² until it stops. (a) How much time elapses from start to stop? (b) How far does the vehicle travel from start to stop?

•30 A world's land speed record was set by Colonel John P. Stapp when in March 1954 he rode a rocket-propelled sled that moved along a track at 1020 km/h. He and the sled were brought to a stop in 1.4 s. (See Fig. 2-7.) In terms of $g$, what acceleration did he experience while stopping?

•31 A certain elevator cab has a total run of 190 m and a maximum speed of 305 m/min, and it accelerates from rest and then back to rest at 1.22 m/s². (a) How far does the cab move while accelerating to full speed from rest? (b) How long does it take to make the nonstop 190 m run, starting and ending at rest? **ILW**

•32 The brakes on your car can slow you at a rate of 5.2 m/s². (a) If you are going 137 km/h and suddenly see a state trooper, what is the minimum time in which you can get your car under the 90 km/h speed limit? (The answer reveals the futility of braking to keep your high speed from being detected with a radar or laser gun.) (b) Graph $x$ versus $t$ and $v$ versus $t$ for such a slowing.

•33 A car traveling 56.0 km/h is 24.0 m from a barrier when the driver slams on the brakes. The car hits the barrier 2.00 s later. (a) What is the magnitude of the car's constant acceleration before impact? (b) How fast is the car traveling at impact? **SSM ILW**

••34 A car moves along an $x$ axis through a distance of 900 m, starting at rest (at $x = 0$) and ending at rest (at $x = 900$ m). Through the first $\frac{1}{4}$ of that distance, its acceleration is +2.25 m/s². Through the next $\frac{3}{4}$ of that distance, its acceleration is −0.750 m/s². What are (a) its travel time through the 900 m and (b) its maximum speed? (c) Graph position $x$, velocity $v$, and acceleration $a$ versus time $t$ for the trip.

••35 Figure 2-25 depicts the motion of a particle moving along an $x$ axis with a constant acceleration. The figure's vertical scaling is set by $x_s = 6.0$ m. What are the (a) magnitude and (b) direction of the particle's acceleration?

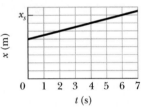

FIG. 2-25 Problem 35.

••36 (a) If the maximum acceleration that is tolerable for passengers in a subway train is 1.34 m/s² and subway stations are located 806 m apart, what is the maximum speed a subway train can attain between stations? (b) What is the travel time between stations? (c) If a subway train stops for 20 s at each station, what is the maximum average speed of the train, from one start-up to the next? (d) Graph $x$, $v$, and $a$ versus $t$ for the interval from one start-up to the next.

••37 Cars $A$ and $B$ move in the same direction in adjacent lanes. The position $x$ of car $A$ is given in Fig. 2-26, from time $t = 0$ to $t = 7.0$ s. The figure's vertical scaling is set by $x_s = 32.0$ m. At $t = 0$, car $B$ is at $x = 0$, with a velocity of 12 m/s and a negative constant acceleration $a_B$. (a) What must $a_B$ be such that the cars are (momentarily) side by side (momentarily at the same value of $x$) at $t = 4.0$ s? (b) For that value of $a_B$, how many times are the cars side by side? (c) Sketch the position $x$ of car $B$ versus time $t$ on Fig. 2-26. How many times will the cars be side by side if the magnitude of acceleration $a_B$ is (d) more than and (e) less than the answer to part (a)?

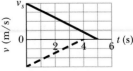

FIG. 2-26 Problem 37.

••38 You are driving toward a traffic signal when it turns yellow. Your speed is the legal speed limit of $v_0 = 55$ km/h; your best deceleration rate has the magnitude $a = 5.18$ m/s². Your best reaction time to begin braking is $T = 0.75$ s. To avoid having the front of your car enter the intersection after the light turns red, should you brake to a stop or continue to move at 55 km/h if the distance to the intersection and the duration of the yellow light are (a) 40 m and 2.8 s, and (b) 32 m and 1.8 s? Give an answer of brake, continue, either (if either strategy works), or neither (if neither strategy works and the yellow duration is inappropriate).

••39 As two trains move along a track, their conductors suddenly notice that they are headed toward each other. Figure 2-27

FIG. 2-27 Problem 39.

gives their velocities $v$ as functions of time $t$ as the conductors slow the trains. The figure's vertical scaling is set by $v_s = 40.0$ m/s. The slowing processes begin when the trains are 200 m apart. What is their separation when both trains have stopped?

**••40** In Fig. 2-28, a red car and a green car, identical except for the color, move toward each other in adjacent lanes and parallel to an $x$ axis. At time $t = 0$, the red car is at $x_r = 0$ and the green car is at $x_g = 220$ m. If the red car has a constant velocity of 20 km/h, the cars pass each other at $x = 44.5$ m, and if it has a constant velocity of 40 km/h, they pass each other at $x = 76.6$ m. What are (a) the initial velocity and (b) the acceleration of the green car?

FIG. 2-28    Problems 40 and 41.

**••41** Figure 2-28 shows a red car and a green car that move toward each other. Figure 2-29 is a graph of their motion, showing the positions $x_{g0} = 270$ m and $x_{r0} = -35.0$ m at time $t = 0$. The green car has a constant speed of 20.0 m/s and the red car begins from rest. What is the acceleration magnitude of the red car?

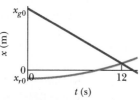

FIG. 2-29    Problem 41.

**•••42** When a high-speed passenger train traveling at 161 km/h rounds a bend, the engineer is shocked to see that a locomotive has improperly entered onto the track from a siding and is a distance $D = 676$ m ahead (Fig. 2-30). The locomotive is moving at 29.0 km/h. The engineer of the high-speed train immediately applies the brakes. (a) What must be the magnitude of the resulting constant deceleration if a collision is to be just avoided? (b) Assume that the engineer is at $x = 0$ when, at $t = 0$, he first spots the locomotive. Sketch $x(t)$ curves for the locomotive and high-speed train for the cases in which a collision is just avoided and is not quite avoided. GO

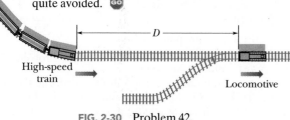

FIG. 2-30    Problem 42.

**•••43** You are arguing over a cell phone while trailing an unmarked police car by 25 m; both your car and the police car are traveling at 110 km/h. Your argument diverts your attention from the police car for 2.0 s (long enough for you to look at the phone and yell, "I won't do that!"). At the beginning of that 2.0 s, the police officer begins braking suddenly at 5.0 m/s². (a) What is the separation between the two cars when your attention finally returns? Suppose that you take another 0.40 s to realize your danger and begin braking. (b) If you too brake at 5.0 m/s², what is your speed when you hit the police car?

### sec. 2-9   Free-Fall Acceleration

**•44** Raindrops fall 1700 m from a cloud to the ground. (a) If they were not slowed by air resistance, how fast would the

drops be moving when they struck the ground? (b) Would it be safe to walk outside during a rainstorm?

**•45** At a construction site a pipe wrench struck the ground with a speed of 24 m/s. (a) From what height was it inadvertently dropped? (b) How long was it falling? (c) Sketch graphs of $y$, $v$, and $a$ versus $t$ for the wrench.   SSM

**•46** A hoodlum throws a stone vertically downward with an initial speed of 12.0 m/s from the roof of a building, 30.0 m above the ground. (a) How long does it take the stone to reach the ground? (b) What is the speed of the stone at impact?

**•47** (a) With what speed must a ball be thrown vertically from ground level to rise to a maximum height of 50 m? (b) How long will it be in the air? (c) Sketch graphs of $y$, $v$, and $a$ versus $t$ for the ball. On the first two graphs, indicate the time at which 50 m is reached.   SSM WWW

**•48** When startled, an armadillo will leap upward. Suppose it rises 0.544 m in the first 0.200 s. (a) What is its initial speed as it leaves the ground? (b) What is its speed at the height of 0.544 m? (c) How much higher does it go?

**•49** A hot-air balloon is ascending at the rate of 12 m/s and is 80 m above the ground when a package is dropped over the side. (a) How long does the package take to reach the ground? (b) With what speed does it hit the ground?   SSM

**••50** A bolt is dropped from a bridge under construction, falling 90 m to the valley below the bridge. (a) In how much time does it pass through the last 20% of its fall? What is its speed (b) when it begins that last 20% of its fall and (c) when it reaches the valley beneath the bridge?

**••51** A key falls from a bridge that is 45 m above the water. It falls directly into a model boat, moving with constant velocity, that is 12 m from the point of impact when the key is released. What is the speed of the boat?   SSM ILW

**••52** At time $t = 0$, apple 1 is dropped from a bridge onto a roadway beneath the bridge; somewhat later, apple 2 is thrown down from the same height. Figure 2-31 gives the vertical positions $y$ of the apples versus $t$ during the falling, until both apples have hit the roadway. With approximately what speed is apple 2 thrown down?

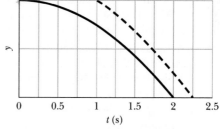

FIG. 2-31    Problem 52.

**••53** As a runaway scientific balloon ascends at 19.6 m/s, one of its instrument packages breaks free of a harness and free-falls. Figure 2-32 gives the vertical velocity of the package versus time, from before it breaks free to when it reaches the ground. (a) What maximum

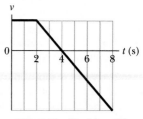

FIG. 2-32    Problem 53.

height above the break-free point does it rise? (b) How high is the break-free point above the ground?

••**54** Figure 2-33 shows the speed $v$ versus height $y$ of a ball tossed directly upward, along a $y$ axis. Distance $d$ is 0.40 m. The speed at height $y_A$ is $v_A$. The speed at height $y_B$ is $\frac{1}{3}v_A$. What is speed $v_A$?

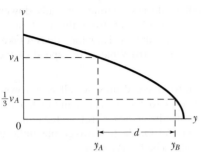

**FIG. 2-33** Problem 54.

••**55** A ball of moist clay falls 15.0 m to the ground. It is in contact with the ground for 20.0 ms before stopping. (a) What is the magnitude of the average acceleration of the ball during the time it is in contact with the ground? (Treat the ball as a particle.) (b) Is the average acceleration up or down? **SSM**

••**56** A stone is dropped into a river from a bridge 43.9 m above the water. Another stone is thrown vertically down 1.00 s after the first is dropped. The stones strike the water at the same time. (a) What is the initial speed of the second stone? (b) Plot velocity versus time on a graph for each stone, taking zero time as the instant the first stone is released.

••**57** To test the quality of a tennis ball, you drop it onto the floor from a height of 4.00 m. It rebounds to a height of 2.00 m. If the ball is in contact with the floor for 12.0 ms, (a) what is the magnitude of its average acceleration during that contact and (b) is the average acceleration up or down?

••**58** A rock is thrown vertically upward from ground level at time $t = 0$. At $t = 1.5$ s it passes the top of a tall tower, and 1.0 s later it reaches its maximum height. What is the height of the tower?

••**59** Water drips from the nozzle of a shower onto the floor 200 cm below. The drops fall at regular (equal) intervals of time, the first drop striking the floor at the instant the fourth drop begins to fall. When the first drop strikes the floor, how far below the nozzle are the (a) second and (b) third drops?

••**60** An object falls a distance $h$ from rest. If it travels 0.50$h$ in the last 1.00 s, find (a) the time and (b) the height of its fall. (c) Explain the physically unacceptable solution of the quadratic equation in $t$ that you obtain.

••**61** A drowsy cat spots a flowerpot that sails first up and then down past an open window. The pot is in view for a total of 0.50 s, and the top-to-bottom height of the window is 2.00 m. How high above the window top does the flowerpot go?

•••**62** A ball is shot vertically upward from the surface of another planet. A plot of $y$ versus $t$ for the ball is shown in Fig. 2-34,

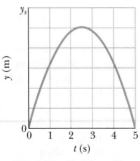

**FIG. 2-34** Problem 62.

where $y$ is the height of the ball above its starting point and $t = 0$ at the instant the ball is shot. The figure's vertical scaling is set by $y_s = 30.0$ m. What are the magnitudes of (a) the free-fall acceleration on the planet and (b) the initial velocity of the ball?

•••**63** A steel ball is dropped from a building's roof and passes a window, taking 0.125 s to fall from the top to the bottom of the window, a distance of 1.20 m. It then falls to a sidewalk and bounces back past the window, moving from bottom to top in 0.125 s. Assume that the upward flight is an exact reverse of the fall. The time the ball spends below the bottom of the window is 2.00 s. How tall is the building? **GO**

•••**64** A basketball player grabbing a rebound jumps 76.0 cm vertically. How much total time (ascent and descent) does the player spend (a) in the top 15.0 cm of this jump and (b) in the bottom 15.0 cm? Do your results explain why such players seem to hang in the air at the top of a jump?

### sec. 2-10 Graphical Integration in Motion Analysis

•**65** In Sample Problem 2-9, at maximum head acceleration, what is the speed of (a) the head and (b) the torso?

•**66** A salamander of the genus *Hydromantes* captures prey by launching its tongue as a projectile: The skeletal part of the tongue is shot forward, unfolding the rest of the tongue, until the outer portion lands on the prey, sticking to it. Figure 2-35 shows the accelera-

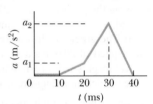

**FIG. 2-35** Problem 66.

tion magnitude $a$ versus time $t$ for the acceleration phase of the launch in a typical situation. The indicated accelerations are $a_2 = 400$ m/s² and $a_1 = 100$ m/s². What is the outward speed of the tongue at the end of the acceleration phase?

••**67** How far does the runner whose velocity–time graph is shown in Fig. 2-36 travel in 16 s? The figure's vertical scaling is set by $v_s = 8.0$ m/s. **ILW**

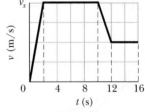

**FIG. 2-36** Problem 67.

••**68** In a forward punch in karate, the fist begins at rest at the waist and is brought rapidly forward until the arm is fully extended. The speed $v(t)$ of the fist is given in Fig. 2-37 for someone skilled in karate. How far has the fist moved at (a) time $t = 50$ ms and (b) when the speed of the fist is maximum?

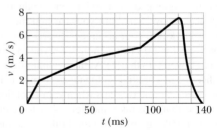

**FIG. 2-37** Problem 68.

••**69** When a soccer ball is kicked toward a player and the player deflects the ball by "heading" it, the acceleration of the head during the collision can be significant. Figure 2-38 gives the measured acceleration $a(t)$ of a soccer player's head for a bare head and a helmeted head, starting from rest. At time $t = 7.0$ ms, what is the difference in the speed acquired by the bare head and the speed acquired by the helmeted head?

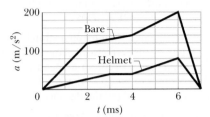

FIG. 2-38   Problem 69.

•••**70**   Two particles move along an $x$ axis. The position of particle 1 is given by $x = 6.00t^2 + 3.00t + 2.00$ (in meters and seconds); the acceleration of particle 2 is given by $a = -8.00t$ (in meters per seconds squared and seconds) and, at $t = 0$, its velocity is 20 m/s. When the velocities of the particles match, what is their velocity?

### Additional Problems

**71**   At the instant the traffic light turns green, an automobile starts with a constant acceleration $a$ of 2.2 m/s². At the same instant a truck, traveling with a constant speed of 9.5 m/s, overtakes and passes the automobile. (a) How far beyond the traffic signal will the automobile overtake the truck? (b) How fast will the automobile be traveling at that instant?   **GO**

**72**   Figure 2-39 shows part of a street where traffic flow is to be controlled to allow a *platoon* of cars to move smoothly along the street. Suppose that the platoon leaders have just reached intersection 2, where the green appeared when they were distance $d$ from the intersection. They continue to travel at a certain speed $v_p$ (the speed limit) to reach intersection 3, where the green appears when they are distance $d$ from it. The intersections are separated by distances $D_{23}$ and $D_{12}$. (a) What should be the time delay of the onset of green at intersection 3 relative to that at intersection 2 to keep the platoon moving smoothly?

Suppose, instead, that the platoon had been stopped by a red light at intersection 1. When the green comes on there, the leaders require a certain time $t_r$ to respond to the change and an additional time to accelerate at some rate $a$ to the cruising speed $v_p$. (b) If the green at intersection 2 is to appear when the leaders are distance $d$ from that intersection, how long after the light at intersection 1 turns green should the light at intersection 2 turn green?

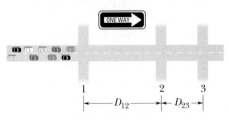

FIG. 2-39   Problem 72.

**73**   In an arcade video game, a spot is programmed to move across the screen according to $x = 9.00t - 0.750t^3$, where $x$ is distance in centimeters measured from the left edge of the screen and $t$ is time in seconds. When the spot reaches a screen edge, at either $x = 0$ or $x = 15.0$ cm, $t$ is reset to 0 and the spot starts moving again according to $x(t)$. (a) At what time after starting is the spot instantaneously at rest? (b) At what value of $x$ does this occur? (c) What is the spot's acceleration (including sign) when this occurs? (d) Is it moving right or left just prior to coming to rest? (e) Just after? (f) At what time $t > 0$ does it first reach an edge of the screen?

**74**   A lead ball is dropped in a lake from a diving board 5.20 m above the water. It hits the water with a certain velocity and then sinks to the bottom with this same constant velocity. It reaches the bottom 4.80 s after it is dropped. (a) How deep is the lake? What are the (b) magnitude and (c) direction (up or down) of the average velocity of the ball for the entire fall? Suppose that all the water is drained from the lake. The ball is now thrown from the diving board so that it again reaches the bottom in 4.80 s. What are the (d) magnitude and (e) direction of the initial velocity of the ball?

**75**   The single cable supporting an unoccupied construction elevator breaks when the elevator is at rest at the top of a 120-m-high building. (a) With what speed does the elevator strike the ground? (b) How long is it falling? (c) What is its speed when it passes the halfway point on the way down? (d) How long has it been falling when it passes the halfway point?

**76**   Two diamonds begin a free fall from rest from the same height, 1.0 s apart. How long after the first diamond begins to fall will the two diamonds be 10 m apart?

**77**   If a baseball pitcher throws a fastball at a horizontal speed of 160 km/h, how long does the ball take to reach home plate 18.4 m away?

**78**   A proton moves along the $x$ axis according to the equation $x = 50t + 10t^2$, where $x$ is in meters and $t$ is in seconds. Calculate (a) the average velocity of the proton during the first 3.0 s of its motion, (b) the instantaneous velocity of the proton at $t = 3.0$ s, and (c) the instantaneous acceleration of the proton at $t = 3.0$ s. (d) Graph $x$ versus $t$ and indicate how the answer to (a) can be obtained from the plot. (e) Indicate the answer to (b) on the graph. (f) Plot $v$ versus $t$ and indicate on it the answer to (c).

**79**   A motorcycle is moving at 30 m/s when the rider applies the brakes, giving the motorcycle a constant deceleration. During the 3.0 s interval immediately after braking begins, the speed decreases to 15 m/s. What distance does the motorcycle travel from the instant braking begins until the motorcycle stops?

**80**   A pilot flies horizontally at 1300 km/h, at height $h = 35$ m above initially level ground. However, at time $t = 0$, the pilot begins to fly over ground sloping upward at angle $\theta = 4.3°$ (Fig. 2-40). If the pilot does not change the airplane's heading, at what time $t$ does the plane strike the ground?

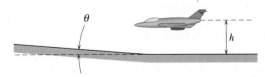

FIG. 2-40   Problem 80.

**81**   A shuffleboard disk is accelerated at a constant rate from rest to a speed of 6.0 m/s over a 1.8 m distance by a player using a cue. At this point the disk loses contact with the cue and slows at a constant rate of 2.5 m/s² until it stops. (a) How much time elapses from when the disk begins to accelerate until it stops? (b) What total distance does the disk travel?

**82** The head of a rattlesnake can accelerate at 50 m/s² in striking a victim. If a car could do as well, how long would it take to reach a speed of 100 km/h from rest?

**83** A jumbo jet must reach a speed of 360 km/h on the runway for takeoff. What is the lowest constant acceleration needed for takeoff from a 1.80 km runway?

**84** An automobile driver increases the speed at a constant rate from 25 km/h to 55 km/h in 0.50 min. A bicycle rider speeds up at a constant rate from rest to 30 km/h in 0.50 min. What are the magnitudes of (a) the driver's acceleration and (b) the rider's acceleration?

**85** To stop a car, first you require a certain reaction time to begin braking; then the car slows at a constant rate. Suppose that the total distance moved by your car during these two phases is 56.7 m when its initial speed is 80.5 km/h, and 24.4 m when its initial speed is 48.3 km/h. What are (a) your reaction time and (b) the magnitude of the acceleration?

**86** A red train traveling at 72 km/h and a green train traveling at 144 km/h are headed toward each other along a straight, level track. When they are 950 m apart, each engineer sees the other's train and applies the brakes. The brakes slow each train at the rate of 1.0 m/s². Is there a collision? If so, answer yes and give the speed of the red train and the speed of the green train at impact, respectively. If not, answer no and give the separation between the trains when they stop.

**87** At time $t = 0$, a rock climber accidentally allows a piton to fall freely from a high point on the rock wall to the valley below him. Then, after a short delay, his climbing partner, who is 10 m higher on the wall, throws a piton downward. The positions $y$ of the pitons versus $t$ during the falling are given in Fig. 2-41. With what speed is the second piton thrown?

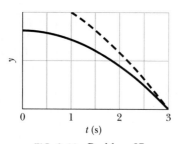

**FIG. 2-41** Problem 87.

**88** A rock is shot vertically upward from the edge of the top of a tall building. The rock reaches its maximum height above the top of the building 1.60 s after being shot. Then, after barely missing the edge of the building as it falls downward, the rock strikes the ground 6.00 s after it is launched. In SI units: (a) with what upward velocity is the rock shot, (b) what maximum height above the top of the building is reached by the rock, and (c) how tall is the building?

**89** A particle's acceleration along an $x$ axis is $a = 5.0t$, with $t$ in seconds and $a$ in meters per second squared. At $t = 2.0$ s, its velocity is $+17$ m/s. What is its velocity at $t = 4.0$ s? **SSM**

**90** A train started from rest and moved with constant acceleration. At one time it was traveling 30 m/s, and 160 m farther on it was traveling 50 m/s. Calculate (a) the acceleration, (b) the time required to travel the 160 m mentioned, (c) the time required to attain the speed of 30 m/s, and (d) the distance moved from rest to the time the train had a speed of 30 m/s. (e) Graph $x$ versus $t$ and $v$ versus $t$ for the train, from rest.

**91** A hot rod can accelerate from 0 to 60 km/h in 5.4 s. (a) What is its average acceleration, in m/s², during this time? (b) How far will it travel during the 5.4 s, assuming its acceleration is constant? (c) From rest, how much time would it require to go a distance of 0.25 km if its acceleration could be maintained at the value in (a)? **SSM**

**92** A rocket-driven sled running on a straight, level track is used to investigate the effects of large accelerations on humans. One such sled can attain a speed of 1600 km/h in 1.8 s, starting from rest. Find (a) the acceleration (assumed constant) in terms of $g$ and (b) the distance traveled.

**93** Figure 2-42 shows a simple device for measuring your reaction time. It consists of a cardboard strip marked with a scale and two large dots. A friend holds the strip *vertically*, with thumb and forefinger at the dot on the right in Fig. 2-42. You then position your thumb and forefinger at the other dot (on the left in Fig. 2-42), being careful not to touch the strip. Your friend releases the strip, and you try to pinch it as soon as possible after you see it begin to fall. The mark at the place where you pinch the strip gives your reaction time. (a) How far from the lower dot should you place the 50.0 ms mark? How much higher should you place the marks for (b) 100, (c) 150, (d) 200, and (e) 250 ms? (For example, should the 100 ms marker be 2 times as far from the dot as the 50 ms marker? If so, give an answer of 2 times. Can you find any pattern in the answers?)

Reaction time (ms)

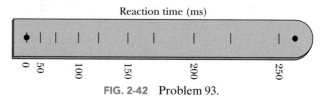

**FIG. 2-42** Problem 93.

**94** Figure 2-43 gives the acceleration $a$ versus time $t$ for a particle moving along an $x$ axis. The $a$-axis scale is set by $a_s = 12.0$ m/s². At $t = -2.0$ s, the particle's velocity is 7.0 m/s. What is its velocity at $t = 6.0$ s?

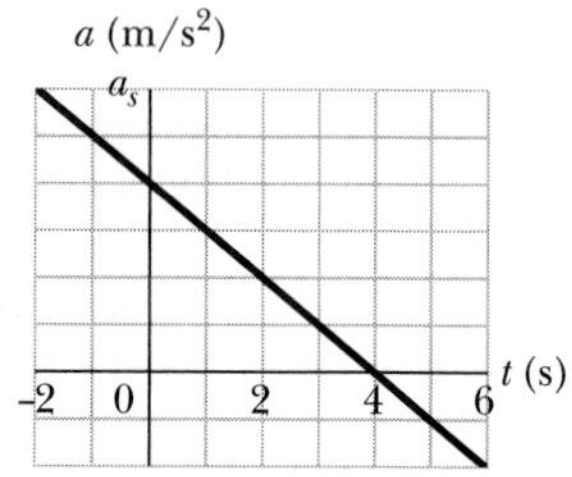

**FIG. 2-43** Problem 94.

**95** A mining cart is pulled up a hill at 20 km/h and then pulled back down the hill at 35 km/h through its original level. (The time required for the cart's reversal at the top of its climb is negligible.) What is the average speed of the cart for its round trip, from its original level back to its original level?

**96** On average, an eye blink lasts about 100 ms. How far does a MiG-25 "Foxbat" fighter travel during a pilot's blink if the plane's average velocity is 3400 km/h?

**97** When the legal speed limit for the New York Thruway was increased from 55 mi/h to 65 mi/h, how much time was saved by a motorist who drove the 700 km between the Buffalo entrance and the New York City exit at the legal speed limit? **SSM**

**98** A motorcyclist who is moving along an $x$ axis directed toward the east has an acceleration given by $a = (6.1 - 1.2t)$ m/s² for $0 \le t \le 6.0$ s. At $t = 0$, the velocity and position of the cyclist are 2.7 m/s and 7.3 m. (a) What is the maximum speed achieved by the cyclist? (b) What total distance does the cyclist travel between $t = 0$ and 6.0 s?

**99** A certain juggler usually tosses balls vertically to a height *H*. To what height must they be tossed if they are to spend twice as much time in the air? **SSM**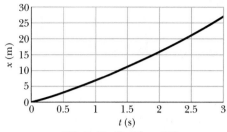

**100** A car moving with constant acceleration covered the distance between two points 60.0 m apart in 6.00 s. Its speed as it passed the second point was 15.0 m/s. (a) What was the speed at the first point? (b) What was the magnitude of the acceleration? (c) At what prior distance from the first point was the car at rest? (d) Graph *x* versus *t* and *v* versus *t* for the car, from rest ($t = 0$).

**101** A rock is dropped from a 100-m-high cliff. How long does it take to fall (a) the first 50 m and (b) the second 50 m?

**102** Two subway stops are separated by 1100 m. If a subway train accelerates at +1.2 m/s$^2$ from rest through the first half of the distance and decelerates at −1.2 m/s$^2$ through the second half, what are (a) its travel time and (b) its maximum speed? (c) Graph *x*, *v*, and *a* versus *t* for the trip.

**103** A certain sprinter has a top speed of 11.0 m/s. If the sprinter starts from rest and accelerates at a constant rate, he is able to reach his top speed in a distance of 12.0 m. He is then able to maintain this top speed for the remainder of a 100 m race. (a) What is his time for the 100 m race? (b) In order to improve his time, the sprinter tries to decrease the distance required for him to reach his top speed. What must this distance be if he is to achieve a time of 10.0 s for the race?

**104** A particle starts from the origin at $t = 0$ and moves along the positive *x* axis. A graph of the velocity of the particle as a function of the time is shown in Fig. 2-44; the *v*-axis scale is set by $v_s = 4.0$ m/s. (a) What is the coordinate of the particle at $t = 5.0$ s? (b) What is the velocity of the particle at $t = 5.0$ s? (c) What is the acceleration of the particle at $t = 5.0$ s? (d) What is the average velocity of the particle between $t = 1.0$ s and $t = 5.0$ s? (e) What is the average acceleration of the particle between $t = 1.0$ s and $t = 5.0$ s?

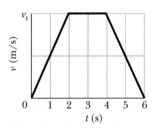

**FIG. 2-44** Problem 104.

**105** A stone is thrown vertically upward. On its way up it passes point *A* with speed *v*, and point *B*, 3.00 m higher than *A*, with speed $\frac{1}{2}v$. Calculate (a) the speed *v* and (b) the maximum height reached by the stone above point *B*.

**106** A rock is dropped (from rest) from the top of a 60-m-tall building. How far above the ground is the rock 1.2 s before it reaches the ground?

**107** An iceboat has a constant velocity toward the east when a sudden gust of wind causes the iceboat to have a constant acceleration toward the east for a period of 3.0 s. A plot of *x* versus *t* is shown in Fig. 2-45, where $t = 0$ is taken to be the instant the wind starts to blow and the positive *x* axis is toward the east. (a) What is the acceleration of the iceboat during the 3.0 s interval? (b) What is the velocity of the iceboat at the end of the 3.0 s interval? (c) If the acceleration remains constant for an additional 3.0 s, how far does the iceboat travel during this second 3.0 s interval? **SSM**

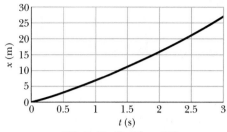

**FIG. 2-45** Problem 107.

**108** A ball is thrown vertically downward from the top of a 36.6-m-tall building. The ball passes the top of a window that is 12.2 m above the ground 2.00 s after being thrown. What is the speed of the ball as it passes the top of the window?

**109** The speed of a bullet is measured to be 640 m/s as the bullet emerges from a barrel of length 1.20 m. Assuming constant acceleration, find the time that the bullet spends in the barrel after it is fired.

**110** A parachutist bails out and freely falls 50 m. Then the parachute opens, and thereafter she decelerates at 2.0 m/s$^2$. She reaches the ground with a speed of 3.0 m/s. (a) How long is the parachutist in the air? (b) At what height does the fall begin?

**111** The Zero Gravity Research Facility at the NASA Glenn Research Center includes a 145 m drop tower. This is an evacuated vertical tower through which, among other possibilities, a 1 m diameter sphere containing an experimental package can be dropped. (a) How long is the sphere in free fall? (b) What is its speed just as it reaches a catching device at the bottom of the tower? (c) When caught, the sphere experiences an average deceleration of 25*g* as its speed is reduced to zero. Through what distance does it travel during the deceleration?

**112** A ball is thrown *down* vertically with an initial *speed* of $v_0$ from a height of *h*. (a) What is its speed just before it strikes the ground? (b) How long does the ball take to reach the ground? What would be the answers to (c) part a and (d) part b if the ball were thrown *upward* from the same height and with the same initial speed? Before solving any equations, decide whether the answers to (c) and (d) should be greater than, less than, or the same as in (a) and (b).

**113** A car can be braked to a stop from the autobahn-like speed of 200 km/h in 170 m. Assuming the acceleration is constant, find its magnitude in (a) SI units and (b) in terms of *g*. (c) How much time $T_b$ is required for the braking? Your *reaction time* $T_r$ is the time you require to perceive an emergency, move your foot to the brake, and begin the braking. If $T_r = 400$ ms, then (d) what is $T_b$ in terms of $T_r$, and (e) is most of the full time required to stop spent in reacting or braking? Dark sunglasses delay the visual signals sent from the eyes to the visual cortex in the brain, increasing $T_r$. (f) In the extreme case in which $T_r$ is increased by 100 ms, how much farther does the car travel during your reaction time?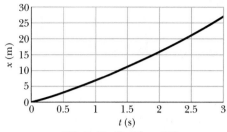

**114** The sport with the fastest moving ball is jai alai, where measured speeds have reached 303 km/h. If a professional jai alai player faces a ball at that speed and involuntarily blinks, he blacks out the scene for 100 ms. How far does the ball move during the blackout?

# 3 Vectors

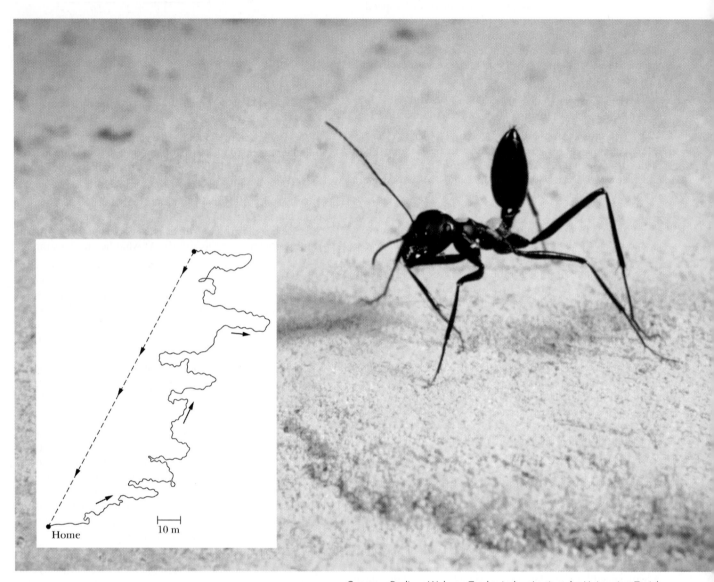

Courtesy Rudiger Wehner, Zoologisches Institut der Universitat Zurich

*The desert ant* Cataglyphis fortis *lives in the plains of the Sahara desert. When one of the ants forages for food, it travels from its home nest along a haphazard search path like the one shown here. The ant may travel more than 500 m along such a complicated path over flat, featureless sand that contains no landmarks. Yet, when the ant decides to return home, it turns and then runs directly home.*

*How does the ant know the way home with no guiding clues on the desert plain?*

The answer is in this chapter.

## 3-1 | WHAT IS PHYSICS?

Physics deals with a great many quantities that have both size and direction, and it needs a special mathematical language—the language of vectors—to describe those quantities. This language is also used in engineering, the other sciences, and even in common speech. If you have ever given directions such as "Go five blocks down this street and then hang a left," you have used the language of vectors. In fact, navigation of any sort is based on vectors, but physics and engineering also need vectors in special ways to explain phenomena involving rotation and magnetic forces, which we get to in later chapters. In this chapter, we focus on the basic language of vectors.

## 3-2 | Vectors and Scalars

A particle moving along a straight line can move in only two directions. We can take its motion to be positive in one of these directions and negative in the other. For a particle moving in three dimensions, however, a plus sign or minus sign is no longer enough to indicate a direction. Instead, we must use a *vector*.

A **vector** has magnitude as well as direction, and vectors follow certain (vector) rules of combination, which we examine in this chapter. A **vector quantity** is a quantity that has both a magnitude and a direction and thus can be represented with a vector. Some physical quantities that are vector quantities are displacement, velocity, and acceleration. You will see many more throughout this book, so learning the rules of vector combination now will help you greatly in later chapters.

Not all physical quantities involve a direction. Temperature, pressure, energy, mass, and time, for example, do not "point" in the spatial sense. We call such quantities **scalars**, and we deal with them by the rules of ordinary algebra. A single value, with a sign (as in a temperature of $-40°F$), specifies a scalar.

The simplest vector quantity is displacement, or change of position. A vector that represents a displacement is called, reasonably, a **displacement vector**. (Similarly, we have velocity vectors and acceleration vectors.) If a particle changes its position by moving from $A$ to $B$ in Fig. 3-1a, we say that it undergoes a displacement from $A$ to $B$, which we represent with an arrow pointing from $A$ to $B$. The arrow specifies the vector graphically. To distinguish vector symbols from other kinds of arrows in this book, we use the outline of a triangle as the arrowhead.

In Fig. 3-1a, the arrows from $A$ to $B$, from $A'$ to $B'$, and from $A''$ to $B''$ have the same magnitude and direction. Thus, they specify identical displacement vectors and represent the same *change of position* for the particle. A vector can be shifted without changing its value *if* its length and direction are not changed.

The displacement vector tells us nothing about the actual path that the particle takes. In Fig. 3-1b, for example, all three paths connecting points $A$ and $B$ correspond to the same displacement vector, that of Fig. 3-1a. Displacement vectors represent only the overall effect of the motion, not the motion itself.

## 3-3 | Adding Vectors Geometrically

Suppose that, as in the vector diagram of Fig. 3-2a, a particle moves from $A$ to $B$ and then later from $B$ to $C$. We can represent its overall displacement (no matter what its actual path) with two successive displacement vectors, $AB$ and $BC$. The *net* displacement of these two displacements is a single displacement from $A$ to $C$. We call $AC$ the **vector sum** (or **resultant**) of the vectors $AB$ and $BC$. This sum is not the usual algebraic sum.

In Fig. 3-2b, we redraw the vectors of Fig. 3-2a and relabel them in the way that we shall use from now on, namely, with an arrow over an italic symbol, as

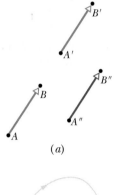

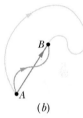

**FIG. 3-1** (a) All three arrows have the same magnitude and direction and thus represent the same displacement. (b) All three paths connecting the two points correspond to the same displacement vector.

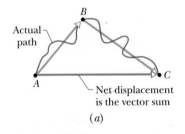

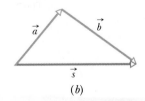

**FIG. 3-2** (a) $AC$ is the vector sum of the vectors $AB$ and $BC$. (b) The same vectors relabeled.

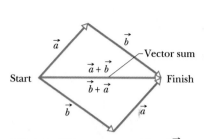

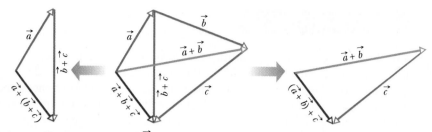

**FIG. 3-4** The three vectors $\vec{a}, \vec{b},$ and $\vec{c}$ can be grouped in any way as they are added; see Eq. 3-3.

**FIG. 3-3** The two vectors $\vec{a}$ and $\vec{b}$ can be added in either order; see Eq. 3-2.

in $\vec{a}$. If we want to indicate only the magnitude of the vector (a quantity that lacks a sign or direction), we shall use the italic symbol, as in $a$, $b$, and $s$. (You can use just a handwritten symbol.) A symbol with an overhead arrow always implies both properties of a vector, magnitude and direction.

We can represent the relation among the three vectors in Fig. 3-2$b$ with the *vector equation*

$$\vec{s} = \vec{a} + \vec{b}, \qquad (3\text{-}1)$$

which says that the vector $\vec{s}$ is the vector sum of vectors $\vec{a}$ and $\vec{b}$. The symbol $+$ in Eq. 3-1 and the words "sum" and "add" have different meanings for vectors than they do in the usual algebra because they involve both magnitude *and* direction.

Figure 3-2 suggests a procedure for adding two-dimensional vectors $\vec{a}$ and $\vec{b}$ geometrically. (1) On paper, sketch vector $\vec{a}$ to some convenient scale and at the proper angle. (2) Sketch vector $\vec{b}$ to the same scale, with its tail at the head of vector $\vec{a}$, again at the proper angle. (3) The vector sum $\vec{s}$ is the vector that extends from the tail of $\vec{a}$ to the head of $\vec{b}$.

Vector addition, defined in this way, has two important properties. First, the order of addition does not matter. Adding $\vec{a}$ to $\vec{b}$ gives the same result as adding $\vec{b}$ to $\vec{a}$ (Fig. 3-3); that is,

$$\vec{a} + \vec{b} = \vec{b} + \vec{a} \quad \text{(commutative law)}. \qquad (3\text{-}2)$$

**FIG. 3-5** The vectors $\vec{b}$ and $-\vec{b}$ have the same magnitude and opposite directions.

Second, when there are more than two vectors, we can group them in any order as we add them. Thus, if we want to add vectors $\vec{a}$, $\vec{b}$, and $\vec{c}$, we can add $\vec{a}$ and $\vec{b}$ first and then add their vector sum to $\vec{c}$. We can also add $\vec{b}$ and $\vec{c}$ first and then add *that* sum to $\vec{a}$. We get the same result either way, as shown in Fig. 3-4. That is,

$$(\vec{a} + \vec{b}) + \vec{c} = \vec{a} + (\vec{b} + \vec{c}) \quad \text{(associative law)}. \qquad (3\text{-}3)$$

The vector $-\vec{b}$ is a vector with the same magnitude as $\vec{b}$ but the opposite direction (see Fig. 3-5). Adding the two vectors in Fig. 3-5 would yield

$$\vec{b} + (-\vec{b}) = 0.$$

Thus, adding $-\vec{b}$ has the effect of subtracting $\vec{b}$. We use this property to define the difference between two vectors: let $\vec{d} = \vec{a} - \vec{b}$. Then

$$\vec{d} = \vec{a} - \vec{b} = \vec{a} + (-\vec{b}) \quad \text{(vector subtraction)}; \qquad (3\text{-}4)$$

that is, we find the difference vector $\vec{d}$ by adding the vector $-\vec{b}$ to the vector $\vec{a}$. Figure 3-6 shows how this is done geometrically.

As in the usual algebra, we can move a term that includes a vector symbol from one side of a vector equation to the other, but we must change its sign. For example, if we are given Eq. 3-4 and need to solve for $\vec{a}$, we can rearrange the equation as

$$\vec{d} + \vec{b} = \vec{a} \quad \text{or} \quad \vec{a} = \vec{d} + \vec{b}.$$

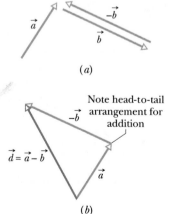

(a)

Note head-to-tail arrangement for addition

(b)

**FIG. 3-6** (a) Vectors $\vec{a}, \vec{b},$ and $-\vec{b}$. (b) To subtract vector $\vec{b}$ from vector $\vec{a}$, add vector $-\vec{b}$ to vector $\vec{a}$.

Remember that, although we have used displacement vectors here, the rules for addition and subtraction hold for vectors of all kinds, whether they represent velocities, accelerations, or any other vector quantity. However, we can add only vectors of the same kind. For example, we can add two displacements, or two velocities, but adding a displacement and a velocity makes no sense. In the arithmetic of scalars, that would be like trying to add 21 s and 12 m.

> **✓ CHECKPOINT 1**   The magnitudes of displacements $\vec{a}$ and $\vec{b}$ are 3 m and 4 m, respectively, and $\vec{c} = \vec{a} + \vec{b}$. Considering various orientations of $\vec{a}$ and $\vec{b}$, what is (a) the maximum possible magnitude for $\vec{c}$ and (b) the minimum possible magnitude?

**Sample Problem  3-1**

In an orienteering class, you have the goal of moving as far (straight-line distance) from base camp as possible by making three straight-line moves. You may use the following displacements in any order: (a) $\vec{a}$, 2.0 km due east (directly toward the east); (b) $\vec{b}$, 2.0 km 30° north of east (at an angle of 30° toward the north from due east); (c) $\vec{c}$, 1.0 km due west. Alternatively, you may substitute either $-\vec{b}$ for $\vec{b}$ or $-\vec{c}$ for $\vec{c}$. What is the greatest distance you can be from base camp at the end of the third displacement?

**Reasoning:** Using a convenient scale, we draw vectors $\vec{a}, \vec{b}, \vec{c}, -\vec{b}$, and $-\vec{c}$ as in Fig. 3-7a. We then mentally slide the vectors over the page, connecting three of them at a time in head-to-tail arrangements to find their vector sum $\vec{d}$. The tail of the first vector represents base camp. The head of the third vector represents the point at which you stop. The vector sum $\vec{d}$ extends from the tail of the first vector to the head of the third vector. Its magnitude $d$ is your distance from base camp.

We find that distance $d$ is greatest for a head-to-tail

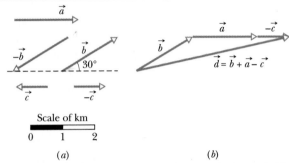

**FIG. 3-7**  (a) Displacement vectors; three are to be used. (b) Your distance from base camp is greatest if you undergo displacements $\vec{a}, \vec{b}$, and $-\vec{c}$, in any order.

arrangement of vectors $\vec{a}, \vec{b}$, and $-\vec{c}$. They can be in any order, because their vector sum is the same for any order. The order shown in Fig. 3-7b is for the vector sum

$$\vec{d} = \vec{b} + \vec{a} + (-\vec{c}).$$

Using the scale given in Fig. 3-7a, we measure the length $d$ of this vector sum, finding

$$d = 4.8 \text{ m.} \qquad \text{(Answer)}$$

## 3-4 | Components of Vectors

Adding vectors geometrically can be tedious. A neater and easier technique involves algebra but requires that the vectors be placed on a rectangular coordinate system. The $x$ and $y$ axes are usually drawn in the plane of the page, as shown in Fig. 3-8a. The $z$ axis comes directly out of the page at the origin; we ignore it for now and deal only with two-dimensional vectors.

A **component** of a vector is the projection of the vector on an axis. In Fig. 3-8a, for example, $a_x$ is the component of vector $\vec{a}$ on (or along) the $x$ axis and $a_y$ is the component along the $y$ axis. To find the projection of a vector along an axis, we draw perpendicular lines from the two ends of the vector to the axis, as shown. The projection of a vector on an $x$ axis is its $x$ *component*, and similarly the projection on the $y$ axis is the $y$ *component*. The process of finding the components of a vector is called **resolving the vector.**

A component of a vector has the same direction (along an axis) as the vector. In Fig. 3-8, $a_x$ and $a_y$ are both positive because $\vec{a}$ extends in the positive direction of both axes. (Note the small arrowheads on the components, to indicate their direction.) If we were to reverse vector $\vec{a}$, then both components would be negative and their arrowheads would point toward negative $x$ and $y$. Resolving vector $\vec{b}$ in Fig. 3-9 yields a positive component $b_x$ and a negative component $b_y$.

In general, a vector has three components, although for the case of Fig. 3-8a

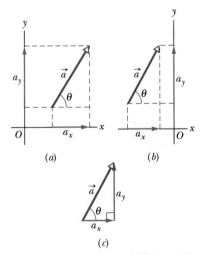

**FIG. 3-8**  (a) The components $a_x$ and $a_y$ of vector $\vec{a}$. (b) The components are unchanged if the vector is shifted, as long as the magnitude and orientation are maintained. (c) The components form the legs of a right triangle whose hypotenuse is the magnitude of the vector.

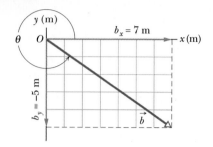

**FIG. 3-9** The component of $\vec{b}$ on the x axis is positive, and that on the y axis is negative.

the component along the z axis is zero. As Figs. 3-8a and b show, if you shift a vector without changing its direction, its components do not change.

We can find the components of $\vec{a}$ in Fig. 3-8a geometrically from the right triangle there:

$$a_x = a \cos \theta \quad \text{and} \quad a_y = a \sin \theta, \tag{3-5}$$

where $\theta$ is the angle that the vector $\vec{a}$ makes with the positive direction of the x axis, and a is the magnitude of $\vec{a}$. Figure 3-8c shows that $\vec{a}$ and its x and y components form a right triangle. It also shows how we can reconstruct a vector from its components: we arrange those components *head to tail*. Then we complete a right triangle with the vector forming the hypotenuse, from the tail of one component to the head of the other component.

Once a vector has been resolved into its components along a set of axes, the components themselves can be used in place of the vector. For example, $\vec{a}$ in Fig. 3-8a is given (completely determined) by a and $\theta$. It can also be given by its components $a_x$ and $a_y$. Both pairs of values contain the same information. If we know a vector in *component notation* ($a_x$ and $a_y$) and want it in *magnitude-angle notation* (a and $\theta$), we can use the equations

$$a = \sqrt{a_x^2 + a_y^2} \quad \text{and} \quad \tan \theta = \frac{a_y}{a_x} \tag{3-6}$$

to transform it.

In the more general three-dimensional case, we need a magnitude and two angles (say, a, $\theta$, and $\phi$) or three components ($a_x$, $a_y$, and $a_z$) to specify a vector.

✓ **CHECKPOINT 2**
In the figure, which of the indicated methods for combining the x and y components of vector $\vec{a}$ are proper to determine that vector?

(a)   (b)   (c)

(d)   (e)   (f)

---

**Sample Problem** | **3-2**

A small airplane leaves an airport on an overcast day and is later sighted 215 km away, in a direction making an angle of 22° east of due north. How far east and north is the airplane from the airport when sighted?

**KEY IDEA** We are given the magnitude (215 km) and the angle (22° east of due north) of a vector and need to find the components of the vector.

**Calculations:** We draw an xy coordinate system with the positive direction of x due east and that of y due north (Fig. 3-10). For convenience, the origin is placed at

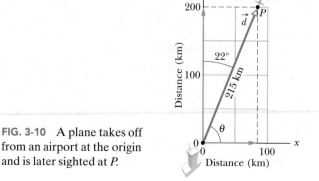

**FIG. 3-10** A plane takes off from an airport at the origin and is later sighted at P.

the airport. The airplane's displacement $\vec{d}$ points from the origin to where the airplane is sighted.

To find the components of $\vec{d}$, we use Eq. 3-5 with $\theta = 68°$ ($= 90° - 22°$):

$$d_x = d \cos\theta = (215 \text{ km})(\cos 68°)$$
$$= 81 \text{ km} \qquad \text{(Answer)}$$
$$d_y = d \sin\theta = (215 \text{ km})(\sin 68°)$$
$$= 199 \text{ km} \approx 2.0 \times 10^2 \text{ km}. \qquad \text{(Answer)}$$

Thus, the airplane is 81 km east and $2.0 \times 10^2$ km north of the airport.

### Sample Problem 3-3

For two decades, spelunking teams sought a connection between the Flint Ridge cave system and Mammoth Cave, which are in Kentucky. When the connection was finally discovered, the combined system was declared the world's longest cave (more than 200 km long). The team that found the connection had to crawl, climb, and squirm through countless passages, traveling a net 2.6 km westward, 3.9 km southward, and 25 m upward. What was their displacement from start to finish?

**KEY IDEA**  We have the components of a three-dimensional vector, and we need to find the vector's magnitude and two angles to specify the vector's direction.

***Horizontal Components:*** We first draw the components as in Fig. 3-11a. The horizontal components (2.6 km west and 3.9 km south) form the legs of a horizontal right triangle. The team's horizontal displacement forms the hypotenuse of the triangle, and its

magnitude $d_h$ is given by the Pythagorean theorem:

$$d_h = \sqrt{(2.6 \text{ km})^2 + (3.9 \text{ km})^2} = 4.69 \text{ km}.$$

Also from the horizontal triangle in Fig. 3-11a, we see that this horizontal displacement is directed south of due west by an angle $\theta_h$ given by

$$\tan\theta_h = \frac{3.9 \text{ km}}{2.6 \text{ km}},$$

so

$$\theta_h = \tan^{-1}\frac{3.9 \text{ km}}{2.6 \text{ km}} = 56°, \qquad \text{(Answer)}$$

which is one of the two angles we need to specify the direction of the overall displacement.

***Overall Displacement:*** To include the vertical component (25 m = 0.025 km), we now take a side view of Fig. 3-11a, looking northwest. We get Fig. 3-11b, where the vertical component and the horizontal displacement $d_h$ form the legs of another right triangle. Now the team's overall displacement forms the hypotenuse of that triangle, with a magnitude $d$ given by

$$d = \sqrt{(4.69 \text{ km})^2 + (0.025 \text{ km})^2} = 4.69 \text{ km}$$
$$\approx 4.7 \text{ km}. \qquad \text{(Answer)}$$

This displacement is directed upward from the horizontal displacement by the angle

$$\theta_v = \tan^{-1}\frac{0.025 \text{ km}}{4.69 \text{ km}} = 0.3°. \qquad \text{(Answer)}$$

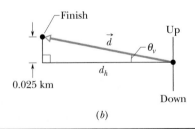

**FIG. 3-11** (*a*) The components of the spelunking team's overall displacement and their horizontal displacement $d_h$. (*b*) A side view showing $d_h$ and the team's overall displacement vector $\vec{d}$.

Thus, the team's displacement vector had a magnitude of 4.7 km and was at an angle of 56° south of west and at an angle of 0.3° upward. The net vertical motion was, of course, insignificant compared with the horizontal motion. However, that fact would have been of no comfort to the team, which had to climb up and down countless times to get through the cave. The route they actually covered was quite different from the displacement vector.

### PROBLEM-SOLVING TACTICS

***Tactic 1:  Angles—Degrees and Radians***  Angles that are measured relative to the positive direction of the *x* axis are positive if they are measured in the counterclockwise direction and negative if measured clockwise. For example, 210° and −150° are the same angle.

Angles may be measured in degrees or radians (rad). To relate the two measures, recall that a full circle is 360° and $2\pi$ rad. To convert, say, 40° to radians, write

$$40°\,\frac{2\pi \text{ rad}}{360°} = 0.70 \text{ rad}.$$

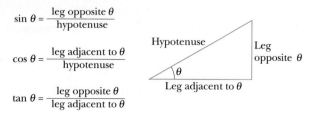

$$\sin \theta = \frac{\text{leg opposite } \theta}{\text{hypotenuse}}$$

$$\cos \theta = \frac{\text{leg adjacent to } \theta}{\text{hypotenuse}}$$

$$\tan \theta = \frac{\text{leg opposite } \theta}{\text{leg adjacent to } \theta}$$

**FIG. 3-12** A triangle used to define the trigonometric functions. See also Appendix E.

**Tactic 2: Trig Functions** You need to know the definitions of the common trigonometric functions—sine, cosine, and tangent—because they are part of the language of science and engineering. They are given in Fig. 3-12 in a form that does not depend on how the triangle is labeled.

You should also be able to sketch how the trig functions vary with angle, as in Fig. 3-13, in order to be able to judge whether a calculator result is reasonable. Even knowing the signs of the functions in the various quadrants can be of help.

**Tactic 3: Inverse Trig Functions** When the inverse trig functions $\sin^{-1}$, $\cos^{-1}$, and $\tan^{-1}$ are taken on a calculator, you must consider the reasonableness of the answer you get, because there is usually another possible answer that the calculator does not give. The range of operation for a calculator in taking each inverse trig function is indicated in Fig. 3-13. As an example, $\sin^{-1} 0.5$ has associated angles of 30° (which is displayed by the calculator, since 30° falls within its range of operation) and 150°. To see both values, draw a horizontal line through 0.5 in Fig. 3-13a and note where it cuts the sine curve.

How do you distinguish a correct answer? It is the one that seems more reasonable for the given situation. As an example, reconsider the calculation of $\theta_h$ in Sample Problem 3-3, where $\tan \theta_h = 3.9/2.6 = 1.5$. Taking $\tan^{-1} 1.5$ on your calculator tells you that $\theta_h = 56°$, but $\theta_h = 236°$ ($= 180° + 56°$) also has a tangent of 1.5. Which is correct? From the physical situation (Fig. 3-11a), 56° is reasonable and 236° is clearly not.

**Tactic 4: Measuring Vector Angles** The equations for $\cos \theta$ and $\sin \theta$ in Eq. 3-5 and for $\tan \theta$ in Eq. 3-6 are valid only if the angle is measured from the positive direction of the $x$ axis. If it is measured relative to some other direction, then the trig functions in Eq. 3-5 may have to be interchanged and the

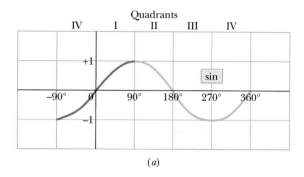

(a)

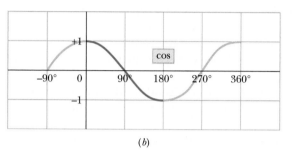

(b)

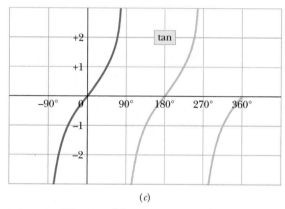

(c)

**FIG. 3-13** Three useful curves to remember. A calculator's range of operation for taking *inverse* trig functions is indicated by the darker portions of the colored curves.

ratio in Eq. 3-6 may have to be inverted. A safer method is to convert the angle to one measured from the positive direction of the $x$ axis.

## 3-5 | Unit Vectors

A **unit vector** is a vector that has a magnitude of exactly 1 and points in a particular direction. It lacks both dimension and unit. Its sole purpose is to point—that is, to specify a direction. The unit vectors in the positive directions of the $x$, $y$, and $z$ axes are labeled $\hat{i}$, $\hat{j}$, and $\hat{k}$, where the hat ˆ is used instead of an overhead arrow as for other vectors (Fig. 3-14). The arrangement of axes in Fig. 3-14 is said to be a **right-handed coordinate system.** The system remains right-handed if it is rotated rigidly. We use such coordinate systems exclusively in this book.

Unit vectors are very useful for expressing other vectors; for example, we can express $\vec{a}$ and $\vec{b}$ of Figs. 3-8 and 3-9 as

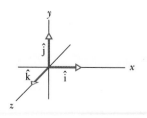

**FIG. 3-14** Unit vectors $\hat{i}$, $\hat{j}$, and $\hat{k}$ define the directions of a right-handed coordinate system.

and

$$\vec{a} = a_x \hat{i} + a_y \hat{j} \tag{3-7}$$

$$\vec{b} = b_x \hat{i} + b_y \hat{j}. \tag{3-8}$$

These two equations are illustrated in Fig. 3-15. The quantities $a_x\hat{i}$ and $a_y\hat{j}$ are vectors, called the **vector components** of $\vec{a}$. The quantities $a_x$ and $a_y$ are scalars, called the **scalar components** of $\vec{a}$ (or, as before, simply its **components**).

As an example, let us write the displacement $\vec{d}$ of the spelunking team of Sample Problem 3-3 in terms of unit vectors. First, superimpose the coordinate system of Fig. 3-14 on the one shown in Fig. 3-11a. Then the directions of $\hat{i}$, $\hat{j}$, and $\hat{k}$ are toward the east, up, and toward the south, respectively. Thus, displacement $\vec{d}$ from start to finish is neatly expressed in unit-vector notation as

$$\vec{d} = -(2.6\text{ km})\hat{i} + (0.025\text{ km})\hat{j} + (3.9\text{ km})\hat{k}. \qquad (3\text{-}9)$$

Here $-(2.6\text{ km})\hat{i}$ is the vector component $d_x\hat{i}$ along the $x$ axis, and $-(2.6\text{ km})$ is the $x$ component $d_x$.

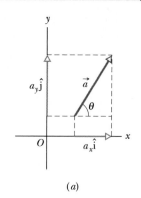

(a)

## 3-6 | Adding Vectors by Components

Using a sketch, we can add vectors geometrically. On a vector-capable calculator, we can add them directly on the screen. A third way to add vectors is to combine their components axis by axis, which is the way we examine here.

To start, consider the statement

$$\vec{r} = \vec{a} + \vec{b}, \qquad (3\text{-}10)$$

which says that the vector $\vec{r}$ is the same as the vector $(\vec{a} + \vec{b})$. Thus, each component of $\vec{r}$ must be the same as the corresponding component of $(\vec{a} + \vec{b})$:

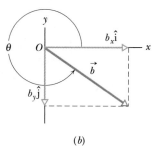

(b)

**FIG. 3-15** (a) The vector components of vector $\vec{a}$. (b) The vector components of vector $\vec{b}$.

$$r_x = a_x + b_x \qquad (3\text{-}11)$$
$$r_y = a_y + b_y \qquad (3\text{-}12)$$
$$r_z = a_z + b_z. \qquad (3\text{-}13)$$

In other words, two vectors must be equal if their corresponding components are equal. Equations 3-10 to 3-13 tell us that to add vectors $\vec{a}$ and $\vec{b}$, we must (1) resolve the vectors into their scalar components; (2) combine these scalar components, axis by axis, to get the components of the sum $\vec{r}$; and (3) combine the components of $\vec{r}$ to get $\vec{r}$ itself. We have a choice in step 3. We can express $\vec{r}$ in unit-vector notation (as in Eq. 3-9) or in magnitude-angle notation (as in the answer to Sample Problem 3-3).

This procedure for adding vectors by components also applies to vector subtractions. Recall that a subtraction such as $\vec{d} = \vec{a} - \vec{b}$ can be rewritten as an addition $\vec{d} = \vec{a} + (-\vec{b})$. To subtract, we add $\vec{a}$ and $-\vec{b}$ by components, to get

$$d_x = a_x - b_x, \quad d_y = a_y - b_y, \quad \text{and} \quad d_z = a_z - b_z,$$

where

$$\vec{d} = d_x\hat{i} + d_y\hat{j} + d_z\hat{k}.$$

✓ **CHECKPOINT 3**   (a) In the figure here, what are the signs of the $x$ components of $\vec{d_1}$ and $\vec{d_2}$? (b) What are the signs of the $y$ components of $\vec{d_1}$ and $\vec{d_2}$? (c) What are the signs of the $x$ and $y$ components of $\vec{d_1} + \vec{d_2}$?

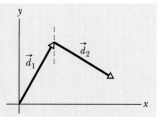

**Sample Problem** | **3-4**

Figure 3-16a shows the following three vectors:

$$\vec{a} = (4.2\text{ m})\hat{i} - (1.5\text{ m})\hat{j},$$
$$\vec{b} = (-1.6\text{ m})\hat{i} + (2.9\text{ m})\hat{j},$$

and
$$\vec{c} = (-3.7\text{ m})\hat{j}.$$

What is their vector sum $\vec{r}$ which is also shown?

**KEY IDEA**  We can add the three vectors by components, axis by axis, and then combine the components to write the vector sum $\vec{r}$.

**Calculations:** For the $x$ axis, we add the $x$ components of $\vec{a}$, $\vec{b}$, and $\vec{c}$, to get the $x$ component of the vector sum $\vec{r}$:

$$r_x = a_x + b_x + c_x$$
$$= 4.2 \text{ m} - 1.6 \text{ m} + 0 = 2.6 \text{ m}.$$

Similarly, for the $y$ axis,

$$r_y = a_y + b_y + c_y$$
$$= -1.5 \text{ m} + 2.9 \text{ m} - 3.7 \text{ m} = -2.3 \text{ m}.$$

We then combine these components of $\vec{r}$ to write the vector in unit-vector notation:

$$\vec{r} = (2.6 \text{ m})\hat{i} - (2.3 \text{ m})\hat{j}, \qquad \text{(Answer)}$$

where $(2.6 \text{ m})\hat{i}$ is the vector component of $\vec{r}$ along the $x$ axis and $-(2.3 \text{ m})\hat{j}$ is that along the $y$ axis. Figure 3-16b shows one way to arrange these vector components to form $\vec{r}$. (Can you sketch the other way?)

We can also answer the question by giving the magnitude and an angle for $\vec{r}$. From Eq. 3-6, the magnitude is

$$r = \sqrt{(2.6 \text{ m})^2 + (-2.3 \text{ m})^2} \approx 3.5 \text{ m} \quad \text{(Answer)}$$

and the angle (measured from the $+x$ direction) is

$$\theta = \tan^{-1}\left(\frac{-2.3 \text{ m}}{2.6 \text{ m}}\right) = -41°, \quad \text{(Answer)}$$

where the minus sign means clockwise.

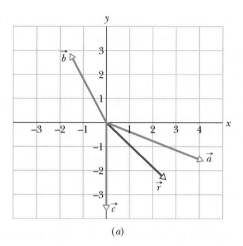

(a)

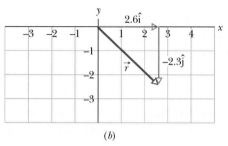

(b)

**FIG. 3-16** Vector $\vec{r}$ is the vector sum of the other three vectors.

---

**Sample Problem** | **3-5**

According to experiments, the desert ant shown in the chapter opening photograph keeps track of its movements along a mental coordinate system. When it wants to return to its home nest, it effectively sums its displacements along the axes of the system to calculate a vector that points directly home. As an example of the calculation, let's consider an ant making five runs of 6.0 cm each on an $xy$ coordinate system, in the directions shown in Fig. 3-17a, starting from home. At the end of the fifth run, what are the magnitude and angle of the ant's net displacement vector $\vec{d}_{net}$, and what are those of the homeward vector $\vec{d}_{home}$ that extends from the ant's final position back to home?

**KEY IDEAS** (1) To find the net displacement $\vec{d}_{net}$, we need to sum the five individual displacement vectors:

$$\vec{d}_{net} = \vec{d}_1 + \vec{d}_2 + \vec{d}_3 + \vec{d}_4 + \vec{d}_5.$$

(2) We evaluate this sum for the $x$ components alone,

$$d_{net,x} = d_{1x} + d_{2x} + d_{3x} + d_{4x} + d_{5x}, \qquad (3\text{-}14)$$

and for the $y$ components alone,

$$d_{net,y} = d_{1y} + d_{2y} + d_{3y} + d_{4y} + d_{5y}. \qquad (3\text{-}15)$$

(3) We construct $\vec{d}_{net}$ from its $x$ and $y$ components.

**Calculations:** To evaluate Eq. 3-14, we apply the $x$ part of Eq. 3-5 to each run:

$$d_{1x} = (6.0 \text{ cm}) \cos 0° = +6.0 \text{ cm}$$
$$d_{2x} = (6.0 \text{ cm}) \cos 150° = -5.2 \text{ cm}$$
$$d_{3x} = (6.0 \text{ cm}) \cos 180° = -6.0 \text{ cm}$$
$$d_{4x} = (6.0 \text{ cm}) \cos(-120°) = -3.0 \text{ cm}$$
$$d_{5x} = (6.0 \text{ cm}) \cos 90° = 0.$$

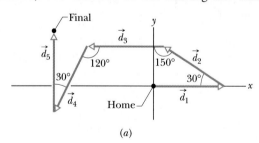

(a)

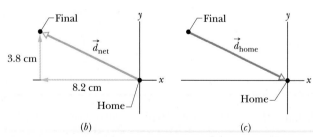

(b)          (c)

**FIG. 3-17** (a) A search path of five runs. (b) The $x$ and $y$ components of $\vec{d}_{net}$. (c) Vector $\vec{d}_{home}$ points the way to the home nest.

Equation 3-14 then gives us

$$d_{net,x} = +6.0 \text{ cm} + (-5.2 \text{ cm}) + (-6.0 \text{ cm})$$
$$+ (-3.0 \text{ cm}) + 0$$
$$= -8.2 \text{ cm}.$$

Similarly, we evaluate the individual $y$ components of the five runs using the $y$ part of Eq. 3-5. The results are shown in Table 3-1. Substituting the results into Eq. 3-15 then gives us

$$d_{net,y} = +3.8 \text{ cm}.$$

Vector $\vec{d}_{net}$ and its $x$ and $y$ components are shown in Fig. 3-17b. To find the magnitude and angle of $\vec{d}_{net}$ from its components, we use Eq. 3-6. The magnitude is

$$d_{net} = \sqrt{d_{net,x}^2 + d_{net,y}^2}$$
$$= \sqrt{(-8.2 \text{ cm})^2 + (3.8 \text{ cm})^2} = 9.0 \text{ cm}.$$

To find the angle (measured from the positive direction of $x$), we take an inverse tangent:

$$\theta = \tan^{-1}\left(\frac{d_{net,y}}{d_{net,x}}\right)$$
$$= \tan^{-1}\left(\frac{3.8 \text{ cm}}{-8.2 \text{ cm}}\right) = -24.86°.$$

*Caution:* Recall from Problem-Solving Tactic 3 that taking an inverse tangent on a calculator may not give the correct answer. The answer $-24.86°$ indicates that

the direction of $\vec{d}_{net}$ is in the fourth quadrant of our $xy$ coordinate system. However, when we construct the vector from its components (Fig. 3-17b), we see that the direction of $\vec{d}_{net}$ is in the second quadrant. Thus, we must "fix" the calculator's answer by adding 180°:

$$\theta = -24.86° + 180° = 155.14° \approx 155°.$$

Thus, the ant's displacement $\vec{d}_{net}$ has magnitude and angle

$$d_{net} = 9.0 \text{ cm at } 155°. \qquad \text{(Answer)}$$

Vector $\vec{d}_{home}$ directed from the ant to its home has the same magnitude as $\vec{d}_{net}$ but the opposite direction (Fig. 3-17c). We already have the angle ($-24.86° \approx -25°$) for the direction opposite $\vec{d}_{net}$. Thus, $\vec{d}_{home}$ has magnitude and angle

$$d_{home} = 9.0 \text{ cm at } -25°. \qquad \text{(Answer)}$$

A desert ant traveling more than 500 m from its home will actually make thousands of individual runs. Yet, it somehow knows how to calculate $\vec{d}_{home}$ (without studying this chapter).

**TABLE 3-1**

| Run | $d_x$ (cm) | $d_y$ (cm) |
|-----|-----------|-----------|
| 1 | +6.0 | 0 |
| 2 | −5.2 | +3.0 |
| 3 | −6.0 | 0 |
| 4 | −3.0 | −5.2 |
| 5 | 0 | +6.0 |
| net | −8.2 | +3.8 |

---

### Sample Problem | 3-6 | Build your skill

Here is a problem involving vector addition that *cannot* be solved directly on a vector-capable calculator, using the vector notation of the calculator. A fellow camper is to walk away from you in a straight line (vector $\vec{A}$), turn, walk in a second straight line (vector $\vec{B}$) and then stop. How far must you walk in a straight line (vector $\vec{C}$) to reach her?

The three vectors (shown in Fig. 3-18) are related by

$$\vec{C} = \vec{A} + \vec{B}. \qquad (3\text{-}16)$$

$\vec{A}$ has a magnitude of 22.0 m and is directed at an angle of $-47.0°$ (clockwise) from the positive direction of an $x$ axis. $\vec{B}$ has a magnitude of 17.0 m and is directed counterclockwise from the positive direction of the $x$ axis by angle $\phi$. $\vec{C}$ is in the positive direction of the $x$ axis. What is the magnitude of $\vec{C}$?

**KEY IDEA** We cannot answer the question by adding $\vec{A}$ and $\vec{B}$ directly on a vector-capable calculator, say, in the generic form of

[magnitude $A \angle$ angle $A$] + [magnitude $B \angle$ angle $B$]

because we do not know the value for the angle $\phi$ of $\vec{B}$. However, we *can* express Eq. 3-16 in terms of components for either the $x$ axis or the $y$ axis.

**Calculations:** Since $\vec{C}$ is directed along the $x$ axis, we choose that axis and write

$$C_x = A_x + B_x.$$

We next express each $x$ component in the form of the $x$ part of Eq. 3-5 and substitute known data. We then have

$$C \cos 0° = 22.0 \cos(-47.0°) + 17.0 \cos \phi. \qquad (3\text{-}17)$$

However, this hardly seems to help, because we still cannot solve for $C$ without knowing $\phi$.

Let us now express Eq. 3-16 in terms of components along the $y$ axis:

$$C_y = A_y + B_y.$$

We then cast these $y$ components in the form of the $y$ part of Eq. 3-5 and substitute known data, to write

$$C \sin 0° = 22.0 \sin(-47.0°) + 17.0 \sin \phi,$$

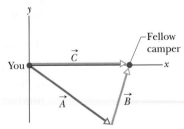

**FIG. 3-18** $\vec{C}$ equals the sum $\vec{A} + \vec{B}$.

which yields

$$0 = 22.0 \sin(-47.0°) + 17.0 \sin \phi.$$

Solving for $\phi$ then gives us

$$\phi = \sin^{-1} -\frac{22.0 \sin(-47.0°)}{17.0} = 71.17°.$$

Substituting this result into Eq. 3-17 leads us to

$$C = 20.5 \text{ m.} \qquad \text{(Answer)}$$

Note the technique of solution: When we got stuck with components on the $x$ axis, we worked with components on the $y$ axis, to evaluate $\phi$. We next moved back to the $x$ axis, to evaluate $C$.

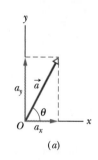

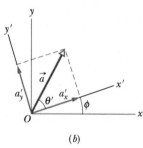

**FIG. 3-19** (*a*) The vector $\vec{a}$ and its components. (*b*) The same vector, with the axes of the coordinate system rotated through an angle $\phi$.

## 3-7 | Vectors and the Laws of Physics

So far, in every figure that includes a coordinate system, the $x$ and $y$ axes are parallel to the edges of the book page. Thus, when a vector $\vec{a}$ is included, its components $a_x$ and $a_y$ are also parallel to the edges (as in Fig. 3-19*a*). The only reason for that orientation of the axes is that it looks "proper"; there is no deeper reason. We could, instead, rotate the axes (but not the vector $\vec{a}$) through an angle $\phi$ as in Fig. 3-19*b*, in which case the components would have new values, call them $a'_x$ and $a'_y$. Since there are an infinite number of choices of $\phi$, there are an infinite number of different pairs of components for $\vec{a}$.

Which then is the "right" pair of components? The answer is that they are all equally valid because each pair (with its axes) just gives us a different way of describing the same vector $\vec{a}$; all produce the same magnitude and direction for the vector. In Fig. 3-19 we have

$$a = \sqrt{a_x^2 + a_y^2} = \sqrt{a'^2_x + a'^2_y} \qquad (3\text{-}18)$$

and

$$\theta = \theta' + \phi. \qquad (3\text{-}19)$$

The point is that we have great freedom in choosing a coordinate system, because the relations among vectors do not depend on the location of the origin or on the orientation of the axes. This is also true of the relations of physics; they are all independent of the choice of coordinate system. Add to that the simplicity and richness of the language of vectors and you can see why the laws of physics are almost always presented in that language: one equation, like Eq. 3-10, can represent three (or even more) relations, like Eqs. 3-11, 3-12, and 3-13.

## 3-8 | Multiplying Vectors*

There are three ways in which vectors can be multiplied, but none is exactly like the usual algebraic multiplication. As you read this section, keep in mind that a vector-capable calculator will help you multiply vectors only if you understand the basic rules of that multiplication.

### Multiplying a Vector by a Scalar

If we multiply a vector $\vec{a}$ by a scalar $s$, we get a new vector. Its magnitude is the product of the magnitude of $\vec{a}$ and the absolute value of $s$. Its direction is the direction of $\vec{a}$ if $s$ is positive but the opposite direction if $s$ is negative. To divide $\vec{a}$ by $s$, we multiply $\vec{a}$ by $1/s$.

### Multiplying a Vector by a Vector

There are two ways to multiply a vector by a vector: one way produces a scalar (called the *scalar product*), and the other produces a new vector (called the *vector product*). (Students commonly confuse the two ways.)

*This material will not be employed until later (Chapter 7 for scalar products and Chapter 11 for vector products), and so your instructor may wish to postpone assignment of this section.

## The Scalar Product

The **scalar product** of the vectors $\vec{a}$ and $\vec{b}$ in Fig. 3-20a is written as $\vec{a} \cdot \vec{b}$ and defined to be

$$\vec{a} \cdot \vec{b} = ab \cos \phi, \qquad (3\text{-}20)$$

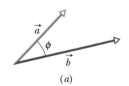

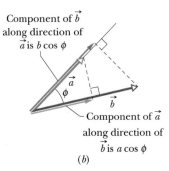

where $a$ is the magnitude of $\vec{a}$, $b$ is the magnitude of $\vec{b}$, and $\phi$ is the angle between $\vec{a}$ and $\vec{b}$ (or, more properly, between the directions of $\vec{a}$ and $\vec{b}$). There are actually two such angles: $\phi$ and $360° - \phi$. Either can be used in Eq. 3-20, because their cosines are the same.

Note that there are only scalars on the right side of Eq. 3-20 (including the value of $\cos \phi$). Thus $\vec{a} \cdot \vec{b}$ on the left side represents a *scalar* quantity. Because of the notation, $\vec{a} \cdot \vec{b}$ is also known as the **dot product** and is spoken as "a dot b."

A dot product can be regarded as the product of two quantities: (1) the magnitude of one of the vectors and (2) the scalar component of the second vector along the direction of the first vector. For example, in Fig. 3-20b, $\vec{a}$ has a scalar component $a \cos \phi$ along the direction of $\vec{b}$; note that a perpendicular dropped from the head of $\vec{a}$ onto $\vec{b}$ determines that component. Similarly, $\vec{b}$ has a scalar component $b \cos \phi$ along the direction of $\vec{a}$.

**FIG. 3-20** (*a*) Two vectors $\vec{a}$ and $\vec{b}$, with an angle $\phi$ between them. (*b*) Each vector has a component along the direction of the other vector.

If the angle $\phi$ between two vectors is 0°, the component of one vector along the other is maximum, and so also is the dot product of the vectors. If, instead, $\phi$ is 90°, the component of one vector along the other is zero, and so is the dot product.

Equation 3-20 can be rewritten as follows to emphasize the components:

$$\vec{a} \cdot \vec{b} = (a \cos \phi)(b) = (a)(b \cos \phi). \qquad (3\text{-}21)$$

The commutative law applies to a scalar product, so we can write

$$\vec{a} \cdot \vec{b} = \vec{b} \cdot \vec{a}.$$

When two vectors are in unit-vector notation, we write their dot product as

$$\vec{a} \cdot \vec{b} = (a_x \hat{i} + a_y \hat{j} + a_z \hat{k}) \cdot (b_x \hat{i} + b_y \hat{j} + b_z \hat{k}), \qquad (3\text{-}22)$$

which we can expand according to the distributive law: Each vector component of the first vector is to be dotted with each vector component of the second vector. By doing so, we can show that

$$\vec{a} \cdot \vec{b} = a_x b_x + a_y b_y + a_z b_z. \qquad (3\text{-}23)$$

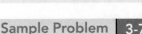 **CHECKPOINT 4** Vectors $\vec{C}$ and $\vec{D}$ have magnitudes of 3 units and 4 units, respectively. What is the angle between the directions of $\vec{C}$ and $\vec{D}$ if $\vec{C} \cdot \vec{D}$ equals (a) zero, (b) 12 units, and (c) −12 units?

---

**Sample Problem** 3-7

What is the angle $\phi$ between $\vec{a} = 3.0\hat{i} - 4.0\hat{j}$ and $\vec{b} = -2.0\hat{i} + 3.0\hat{k}$? (*Caution:* Although many of the following steps can be bypassed with a vector-capable calculator, you will learn more about scalar products if, at least here, you use these steps.)

**KEY IDEA** The angle between the directions of two vectors is included in the definition of their scalar product (Eq. 3-20):

$$\vec{a} \cdot \vec{b} = ab \cos \phi. \qquad (3\text{-}24)$$

*Calculations:* In Eq. 3-24, $a$ is the magnitude of $\vec{a}$, or

$$a = \sqrt{3.0^2 + (-4.0)^2} = 5.00, \qquad (3\text{-}25)$$

and $b$ is the magnitude of $\vec{b}$, or

$$b = \sqrt{(-2.0)^2 + 3.0^2} = 3.61. \qquad (3\text{-}26)$$

We can separately evaluate the left side of Eq. 3-24 by writing the vectors in unit-vector notation and using the distributive law:

$$\begin{aligned}
\vec{a} \cdot \vec{b} &= (3.0\hat{i} - 4.0\hat{j}) \cdot (-2.0\hat{i} + 3.0\hat{k}) \\
&= (3.0\hat{i}) \cdot (-2.0\hat{i}) + (3.0\hat{i}) \cdot (3.0\hat{k}) \\
&\quad + (-4.0\hat{j}) \cdot (-2.0\hat{i}) + (-4.0\hat{j}) \cdot (3.0\hat{k}).
\end{aligned}$$

We next apply Eq. 3-20 to each term in this last expression. The angle between the unit vectors in the first term ($\hat{\i}$ and $\hat{\i}$) is 0°, and in the other terms it is 90°. We then have

$$\vec{a} \cdot \vec{b} = -(6.0)(1) + (9.0)(0) + (8.0)(0) - (12)(0)$$
$$= -6.0.$$

Substituting this result and the results of Eqs. 3-25 and 3-26 into Eq. 3-24 yields

$$-6.0 = (5.00)(3.61) \cos\phi,$$

so $\quad \phi = \cos^{-1}\dfrac{-6.0}{(5.00)(3.61)} = 109° \approx 110°.$ (Answer)

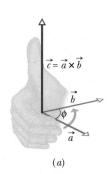

(a)

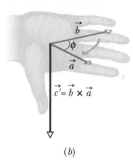

(b)

**FIG. 3-21** Illustration of the right-hand rule for vector products. (a) Sweep vector $\vec{a}$ into vector $\vec{b}$ with the fingers of your right hand. Your outstretched thumb shows the direction of vector $\vec{c} = \vec{a} \times \vec{b}$. (b) Showing that $\vec{b} \times \vec{a}$ is the reverse of $\vec{a} \times \vec{b}$.

## The Vector Product

The **vector product** of $\vec{a}$ and $\vec{b}$, written $\vec{a} \times \vec{b}$, produces a third vector $\vec{c}$ whose magnitude is

$$c = ab \sin\phi, \tag{3-27}$$

where $\phi$ is the *smaller* of the two angles between $\vec{a}$ and $\vec{b}$. (You must use the smaller of the two angles between the vectors because $\sin\phi$ and $\sin(360° - \phi)$ differ in algebraic sign.) Because of the notation, $\vec{a} \times \vec{b}$ is also known as the **cross product,** and in speech it is "a cross b."

> If $\vec{a}$ and $\vec{b}$ are parallel or antiparallel, $\vec{a} \times \vec{b} = 0$. The magnitude of $\vec{a} \times \vec{b}$, which can be written as $|\vec{a} \times \vec{b}|$, is maximum when $\vec{a}$ and $\vec{b}$ are perpendicular to each other.

The direction of $\vec{c}$ is perpendicular to the plane that contains $\vec{a}$ and $\vec{b}$. Figure 3-21a shows how to determine the direction of $\vec{c} = \vec{a} \times \vec{b}$ with what is known as a **right-hand rule.** Place the vectors $\vec{a}$ and $\vec{b}$ tail to tail without altering their orientations, and imagine a line that is perpendicular to their plane where they meet. Pretend to place your *right* hand around that line in such a way that your fingers would sweep $\vec{a}$ into $\vec{b}$ through the smaller angle between them. Your outstretched thumb points in the direction of $\vec{c}$.

The order of the vector multiplication is important. In Fig. 3-21b, we are determining the direction of $\vec{c}' = \vec{b} \times \vec{a}$, so the fingers are placed to sweep $\vec{b}$ into $\vec{a}$ through the smaller angle. The thumb ends up in the opposite direction from previously, and so it must be that $\vec{c}' = -\vec{c}$; that is,

$$\vec{b} \times \vec{a} = -(\vec{a} \times \vec{b}). \tag{3-28}$$

In other words, the commutative law does not apply to a vector product.

In unit-vector notation, we write

$$\vec{a} \times \vec{b} = (a_x\hat{\i} + a_y\hat{\j} + a_z\hat{k}) \times (b_x\hat{\i} + b_y\hat{\j} + b_z\hat{k}), \tag{3-29}$$

which can be expanded according to the distributive law; that is, each component of the first vector is to be crossed with each component of the second vector. The cross products of unit vectors are given in Appendix E (see "Products of Vectors"). For example, in the expansion of Eq. 3-29, we have

$$a_x\hat{\i} \times b_x\hat{\i} = a_x b_x(\hat{\i} \times \hat{\i}) = 0,$$

because the two unit vectors $\hat{\i}$ and $\hat{\i}$ are parallel and thus have a zero cross product. Similarly, we have

$$a_x\hat{\i} \times b_y\hat{\j} = a_x b_y(\hat{\i} \times \hat{\j}) = a_x b_y \hat{k}.$$

In the last step we used Eq. 3-27 to evaluate the magnitude of $\hat{\i} \times \hat{\j}$ as unity. (These vectors $\hat{\i}$ and $\hat{\j}$ each have a magnitude of unity, and the angle between them is 90°.) Also, we used the right-hand rule to get the direction of $\hat{\i} \times \hat{\j}$ as being in the positive direction of the $z$ axis (thus in the direction of $\hat{k}$).

Continuing to expand Eq. 3-29, you can show that

$$\vec{a} \times \vec{b} = (a_y b_z - b_y a_z)\hat{\i} + (a_z b_x - b_z a_x)\hat{\j} + (a_x b_y - b_x a_y)\hat{k}. \tag{3-30}$$

A determinant (Appendix E) or a vector-capable calculator can also be used.

To check whether any *xyz* coordinate system is a right-handed coordinate system, use the right-hand rule for the cross product $\hat{i} \times \hat{j} = \hat{k}$ with that system. If your fingers sweep $\hat{i}$ (positive direction of *x*) into $\hat{j}$ (positive direction of *y*) with the outstretched thumb pointing in the positive direction of *z*, then the system is right-handed.

✓ **CHECKPOINT 5**    Vectors $\vec{C}$ and $\vec{D}$ have magnitudes of 3 units and 4 units, respectively. What is the angle between the directions of $\vec{C}$ and $\vec{D}$ if the magnitude of the vector product $\vec{C} \times \vec{D}$ is (a) zero and (b) 12 units?

## Sample Problem    3-8

In Fig. 3-22, vector $\vec{a}$ lies in the *xy* plane, has a magnitude of 18 units and points in a direction 250° from the +*x* direction. Also, vector $\vec{b}$ has a magnitude of 12 units and points in the +*z* direction. What is the vector product $\vec{c} = \vec{a} \times \vec{b}$?

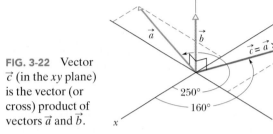

**FIG. 3-22**  Vector $\vec{c}$ (in the *xy* plane) is the vector (or cross) product of vectors $\vec{a}$ and $\vec{b}$.

**KEY IDEA**    When we have two vectors in magnitude-angle notation, we find the magnitude of their cross product with Eq. 3-27 and the direction of their cross product with the right-hand rule of Fig. 3-21.

**Calculations:** For the magnitude we write

$$c = ab \sin \phi = (18)(12)(\sin 90°) = 216. \quad \text{(Answer)}$$

To determine the direction in Fig. 3-22, imagine placing the fingers of your right hand around a line perpendicular to the plane of $\vec{a}$ and $\vec{b}$ (the line on which $\vec{c}$ is shown) such that your fingers sweep $\vec{a}$ into $\vec{b}$. Your out-

stretched thumb then gives the direction of $\vec{c}$. Thus, as shown in the figure, $\vec{c}$ lies in the *xy* plane. Because its direction is perpendicular to the direction of $\vec{a}$, it is at an angle of

$$250° - 90° = 160° \quad \text{(Answer)}$$

from the positive direction of the *x* axis.

## Sample Problem    3-9

If $\vec{a} = 3\hat{i} - 4\hat{j}$ and $\vec{b} = -2\hat{i} + 3\hat{k}$, what is $\vec{c} = \vec{a} \times \vec{b}$?

**KEY IDEA**    When two vectors are in unit-vector notation, we can find their cross product by using the distributive law.

**Calculations:** Here we write

$$\vec{c} = (3\hat{i} - 4\hat{j}) \times (-2\hat{i} + 3\hat{k})$$
$$= 3\hat{i} \times (-2\hat{i}) + 3\hat{i} \times 3\hat{k} + (-4\hat{j}) \times (-2\hat{i})$$
$$+ (-4\hat{j}) \times 3\hat{k}.$$

We next evaluate each term with Eq. 3-27, finding the direction with the right-hand rule. For the first term here, the angle $\phi$ between the two vectors being crossed is 0. For the other terms, $\phi$ is 90°. We find

$$\vec{c} = -6(0) + 9(-\hat{j}) + 8(-\hat{k}) - 12\hat{i}$$
$$= -12\hat{i} - 9\hat{j} - 8\hat{k}. \quad \text{(Answer)}$$

This vector $\vec{c}$ is perpendicular to both $\vec{a}$ and $\vec{b}$, a fact you can check by showing that $\vec{c} \cdot \vec{a} = 0$ and $\vec{c} \cdot \vec{b} = 0$; that is, there is no component of $\vec{c}$ along the direction of either $\vec{a}$ or $\vec{b}$.

### PROBLEM-SOLVING TACTICS

**Tactic 5:    *Common Errors with Cross Products***    Several errors are common in finding a cross product. (1) Failure to arrange vectors tail to tail is tempting when an illustration presents them head to tail; you must mentally shift (or better, redraw) one vector to the proper arrangement without changing its orientation. (2) Failing to use the right hand in applying the right-hand rule is easy when the right hand is occupied with a calculator or pencil. (3) Failure to sweep the first vector

of the product into the second vector can occur when the orientations of the vectors require an awkward twisting of your hand to apply the right-hand rule. Sometimes that happens when you try to make the sweep mentally rather than actually using your hand. (4) Failure to work with a right-handed coordinate system results when you forget how to draw such a system. See Fig. 3-14 for one perspective. Practice drawing other perspectives, such as the (correct ones) shown in Fig. 3-25 on page 53.

## REVIEW & SUMMARY

**Scalars and Vectors** *Scalars*, such as temperature, have magnitude only. They are specified by a number with a unit (10°C) and obey the rules of arithmetic and ordinary algebra. *Vectors*, such as displacement, have both magnitude and direction (5 m, north) and obey the rules of vector algebra.

**Adding Vectors Geometrically** Two vectors $\vec{a}$ and $\vec{b}$ may be added geometrically by drawing them to a common scale and placing them head to tail. The vector connecting the tail of the first to the head of the second is the vector sum $\vec{s}$. To subtract $\vec{b}$ from $\vec{a}$, reverse the direction of $\vec{b}$ to get $-\vec{b}$; then add $-\vec{b}$ to $\vec{a}$. Vector addition is commutative and obeys the associative law.

**Components of a Vector** The (scalar) *components* $a_x$ and $a_y$ of any two-dimensional vector $\vec{a}$ along the coordinate axes are found by dropping perpendicular lines from the ends of $\vec{a}$ onto the coordinate axes. The components are given by

$$a_x = a \cos \theta \quad \text{and} \quad a_y = a \sin \theta, \tag{3-5}$$

where $\theta$ is the angle between the positive direction of the $x$ axis and the direction of $\vec{a}$. The algebraic sign of a component indicates its direction along the associated axis. Given its components, we can find the magnitude and orientation of the vector $\vec{a}$ with

$$a = \sqrt{a_x^2 + a_y^2} \quad \text{and} \quad \tan \theta = \frac{a_y}{a_x}. \tag{3-6}$$

**Unit-Vector Notation** *Unit vectors* $\hat{i}, \hat{j}$, and $\hat{k}$ have magnitudes of unity and are directed in the positive directions of the $x$, $y$, and $z$ axes, respectively, in a right-handed coordinate system. We can write a vector $\vec{a}$ in terms of unit vectors as

$$\vec{a} = a_x \hat{i} + a_y \hat{j} + a_z \hat{k}, \tag{3-7}$$

in which $a_x\hat{i}$, $a_y\hat{j}$, and $a_z\hat{k}$ are the **vector components** of $\vec{a}$ and $a_x, a_y$, and $a_z$ are its **scalar components.**

**Adding Vectors in Component Form** To add vectors

in component form, we use the rules

$$r_x = a_x + b_x \quad r_y = a_y + b_y \quad r_z = a_z + b_z. \tag{3-11 to 3-13}$$

Here $\vec{a}$ and $\vec{b}$ are the vectors to be added, and $\vec{r}$ is the vector sum.

**Product of a Scalar and a Vector** The product of a scalar $s$ and a vector $\vec{v}$ is a new vector whose magnitude is $sv$ and whose direction is the same as that of $\vec{v}$ if $s$ is positive, and opposite that of $\vec{v}$ if $s$ is negative. To divide $\vec{v}$ by $s$, multiply $\vec{v}$ by $1/s$.

**The Scalar Product** The **scalar** (or **dot**) **product** of two vectors $\vec{a}$ and $\vec{b}$ is written $\vec{a} \cdot \vec{b}$ and is the *scalar* quantity given by

$$\vec{a} \cdot \vec{b} = ab \cos \phi, \tag{3-20}$$

in which $\phi$ is the angle between the directions of $\vec{a}$ and $\vec{b}$. A scalar product is the product of the magnitude of one vector and the scalar component of the second vector along the direction of the first vector. In unit-vector notation,

$$\vec{a} \cdot \vec{b} = (a_x\hat{i} + a_y\hat{j} + a_z\hat{k}) \cdot (b_x\hat{i} + b_y\hat{j} + b_z\hat{k}), \tag{3-22}$$

which may be expanded according to the distributive law. Note that $\vec{a} \cdot \vec{b} = \vec{b} \cdot \vec{a}$.

**The Vector Product** The **vector** (or **cross**) **product** of two vectors $\vec{a}$ and $\vec{b}$ is written $\vec{a} \times \vec{b}$ and is a *vector* $\vec{c}$ whose magnitude $c$ is given by

$$c = ab \sin \phi, \tag{3-27}$$

in which $\phi$ is the smaller of the angles between the directions of $\vec{a}$ and $\vec{b}$. The direction of $\vec{c}$ is perpendicular to the plane defined by $\vec{a}$ and $\vec{b}$ and is given by a right-hand rule, as shown in Fig. 3-21. Note that $\vec{a} \times \vec{b} = -(\vec{b} \times \vec{a})$. In unit-vector notation,

$$\vec{a} \times \vec{b} = (a_x\hat{i} + a_y\hat{j} + a_z\hat{k}) \times (b_x\hat{i} + b_y\hat{j} + b_z\hat{k}), \tag{3-29}$$

which we may expand with the distributive law.

## QUESTIONS

**1** Being part of the "Gators," the University of Florida golfing team must play on a putting green with an alligator pit. Figure 3-23 shows an overhead view of one putting challenge of the team; an $xy$ coordinate system is superimposed. Team members must putt from the origin to the hole, which is at $xy$ coordinates (8 m, 12 m), but they can putt the golf ball using only one or more of the following displacements, one or more times:

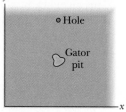
FIG. 3-23 Question 1.

$$\vec{d_1} = (8 \text{ m})\hat{i} + (6 \text{ m})\hat{j}, \qquad \vec{d_2} = (6 \text{ m})\hat{j}, \qquad \vec{d_3} = (8 \text{ m})\hat{i}.$$

The pit is at coordinates (8 m, 6 m). If a team member putts the ball into or through the pit, the member is automatically trans-

ferred to Florida State University, the arch rival. What sequence of displacements should a team member use to avoid the pit?

**2** Equation 3-2 shows that the addition of two vectors $\vec{a}$ and $\vec{b}$ is commutative. Does that mean subtraction is commutative, so that $\vec{a} - \vec{b} = \vec{b} - \vec{a}$?

**3** Can the sum of the magnitudes of two vectors ever be equal to the magnitude of the sum of the same two vectors? If no, why not? If yes, when?

**4** The two vectors shown in Fig. 3-24 lie in an $xy$ plane. What are the signs of the $x$ and $y$ components, respectively, of (a) $\vec{d_1} + \vec{d_2}$, (b) $\vec{d_1} - \vec{d_2}$, and (c) $\vec{d_2} - \vec{d_1}$?

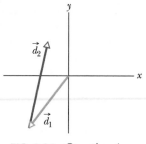
FIG. 3-24 Question 4.

**5** If $\vec{d} = \vec{a} + \vec{b} + (-\vec{c})$, does (a) $\vec{a} + (-\vec{d}) = \vec{c} + (-\vec{b})$, (b) $\vec{a} = (-\vec{b}) + \vec{d} + \vec{c}$, and (c) $\vec{c} + (-\vec{d}) = \vec{a} + \vec{b}$?

**6** Describe two vectors $\vec{a}$ and $\vec{b}$ such that

(a) $\vec{a} + \vec{b} = \vec{c}$   and   $a + b = c$;

(b) $\vec{a} + \vec{b} = \vec{a} - \vec{b}$;

(c) $\vec{a} + \vec{b} = \vec{c}$   and   $a^2 + b^2 = c^2$.

**7** Which of the arrangements of axes in Fig. 3-25 can be

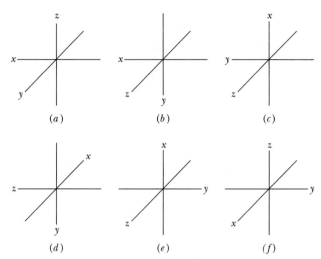

(a)        (b)        (c)

(d)        (e)        (f)

FIG. 3-25   Question 7.

labeled "right-handed coordinate system"? As usual, each axis label indicates the positive side of the axis.

**8** Figure 3-26 shows vector $\vec{A}$ and four other vectors that have the same magnitude but differ in orientation. (a) Which of those other four vectors have the same dot product with $\vec{A}$? (b) Which have a negative dot product with $\vec{A}$?

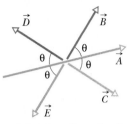

FIG. 3-26   Question 8.

**9** If $\vec{F} = q(\vec{v} \times \vec{B})$ and $\vec{v}$ is perpendicular to $\vec{B}$, then what is the direction of $\vec{B}$ in the three situations shown in Fig. 3-27 when constant $q$ is (a) positive and (b) negative?

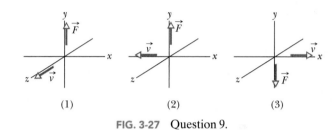

(1)        (2)        (3)

FIG. 3-27   Question 9.

**10** If $\vec{a} \cdot \vec{b} = \vec{a} \cdot \vec{c}$, must $\vec{b}$ equal $\vec{c}$?

## PROBLEMS

| GO | Tutoring problem available (at instructor's discretion) in *WileyPLUS* and WebAssign |
| SSM | Worked-out solution available in Student Solutions Manual  **WWW** Worked-out solution is at |
| • – ••• | Number of dots indicates level of problem difficulty  **ILW** Interactive solution is at  http://www.wiley.com/college/halliday |
| | Additional information available in *The Flying Circus of Physics* and at flyingcircusofphysics.com |

### sec. 3-4   Components of Vectors

•**1** The $x$ component of vector $\vec{A}$ is $-25.0$ m and the $y$ component is $+40.0$ m. (a) What is the magnitude of $\vec{A}$? (b) What is the angle between the direction of $\vec{A}$ and the positive direction of $x$?  **SSM**

•**2** Express the following angles in radians: (a) $20.0°$, (b) $50.0°$, (c) $100°$. Convert the following angles to degrees: (d) $0.330$ rad, (e) $2.10$ rad, (f) $7.70$ rad.

•**3** What are (a) the $x$ component and (b) the $y$ component of a vector $\vec{a}$ in the $xy$ plane if its direction is $250°$ counterclockwise from the positive direction of the $x$ axis and its magnitude is $7.3$ m?  **SSM**

•**4** In Fig. 3-28, a heavy piece of machinery is raised by sliding it a distance $d = 12.5$ m along a plank oriented at angle $\theta = 20.0°$ to the horizontal. How far is it moved (a) vertically and (b) horizontally?

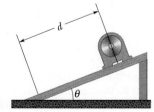

FIG. 3-28   Problem 4.

•**5** A ship sets out to sail to a point 120 km due north. An unexpected storm blows the ship to a point 100 km due east of its starting point. (a) How far and (b) in what direction must it now sail to reach its original destination?

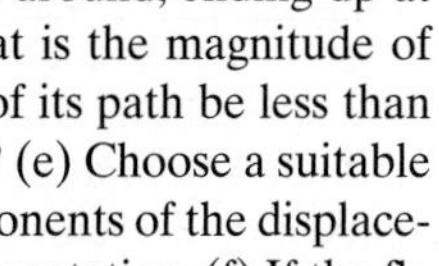

•**6** A displacement vector $\vec{r}$ in the $xy$ plane is 15 m long and directed at angle $\theta = 30°$ in Fig. 3-29. Determine (a) the $x$ component and (b) the $y$ component of the vector.

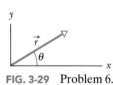

FIG. 3-29   Problem 6.

••**7** A room has dimensions 3.00 m (height) × 3.70 m × 4.30 m. A fly starting at one corner flies around, ending up at the diagonally opposite corner. (a) What is the magnitude of its displacement? (b) Could the length of its path be less than this magnitude? (c) Greater? (d) Equal? (e) Choose a suitable coordinate system and express the components of the displacement vector in that system in unit-vector notation. (f) If the fly walks, what is the length of the shortest path? (*Hint:* This can be answered without calculus. The room is like a box. Unfold its walls to flatten them into a plane.)  **SSM WWW**

### sec. 3-6   Adding Vectors by Components

•**8** A car is driven east for a distance of 50 km, then north for 30 km, and then in a direction 30° east of north for 25 km. Sketch the vector diagram and determine (a) the magnitude and (b) the angle of the car's total displacement from its starting point.

•9 (a) In unit-vector notation, what is the sum $\vec{a} + \vec{b}$ if $\vec{a} = (4.0 \text{ m})\hat{i} + (3.0 \text{ m})\hat{j}$ and $\vec{b} = (-13.0 \text{ m})\hat{i} + (7.0 \text{ m})\hat{j}$? What are the (b) magnitude and (c) direction of $\vec{a} + \vec{b}$? **SSM**

•10 A person walks in the following pattern: 3.1 km north, then 2.4 km west, and finally 5.2 km south. (a) Sketch the vector diagram that represents this motion. (b) How far and (c) in what direction would a bird fly in a straight line from the same starting point to the same final point?

•11 A person desires to reach a point that is 3.40 km from her present location and in a direction that is 35.0° north of east. However, she must travel along streets that are oriented either north–south or east–west. What is the minimum distance she could travel to reach her destination?

•12 For the vectors $\vec{a} = (3.0 \text{ m})\hat{i} + (4.0 \text{ m})\hat{j}$ and $\vec{b} = (5.0 \text{ m})\hat{i} + (-2.0 \text{ m})\hat{j}$, give $\vec{a} + \vec{b}$ in (a) unit-vector notation, and as (b) a magnitude and (c) an angle (relative to $\hat{i}$). Now give $\vec{b} - \vec{a}$ in (d) unit-vector notation, and as (e) a magnitude and (f) an angle.

•13 Two vectors are given by

$$\vec{a} = (4.0 \text{ m})\hat{i} - (3.0 \text{ m})\hat{j} + (1.0 \text{ m})\hat{k}$$

and $$\vec{b} = (-1.0 \text{ m})\hat{i} + (1.0 \text{ m})\hat{j} + (4.0 \text{ m})\hat{k}.$$

In unit-vector notation, find (a) $\vec{a} + \vec{b}$, (b) $\vec{a} - \vec{b}$, and (c) a third vector $\vec{c}$ such that $\vec{a} - \vec{b} + \vec{c} = 0$.

•14 Find the (a) $x$, (b) $y$, and (c) $z$ components of the sum $\vec{r}$ of the displacements $\vec{c}$ and $\vec{d}$ whose components in meters along the three axes are $c_x = 7.4$, $c_y = -3.8$, $c_z = -6.1$; $d_x = 4.4$, $d_y = -2.0$, $d_z = 3.3$.

•15 An ant, crazed by the Sun on a hot Texas afternoon, darts over an $xy$ plane scratched in the dirt. The $x$ and $y$ components of four consecutive darts are the following, all in centimeters: (30.0, 40.0), ($b_x$, −70.0), (−20.0, $c_y$), (−80.0, −70.0). The overall displacement of the four darts has the $xy$ components (−140, −20.0). What are (a) $b_x$ and (b) $c_y$? What are the (c) magnitude and (d) angle (relative to the positive direction of the $x$ axis) of the overall displacement? **GO**

•16 In the sum $\vec{A} + \vec{B} = \vec{C}$, vector $\vec{A}$ has a magnitude of 12.0 m and is angled 40.0° counterclockwise from the $+x$ direction, and vector $\vec{C}$ has a magnitude of 15.0 m and is angled 20.0° counterclockwise from the $-x$ direction. What are (a) the magnitude and (b) the angle (relative to $+x$) of $\vec{B}$?

•17 The two vectors $\vec{a}$ and $\vec{b}$ in Fig. 3-30 have equal magnitudes of 10.0 m and the angles are $\theta_1 = 30°$ and $\theta_2 = 105°$. Find the (a) $x$ and (b) $y$ components of their vector sum $\vec{r}$, (c) the magnitude of $\vec{r}$, and (d) the angle $\vec{r}$ makes with the positive direction of the $x$ axis. **SSM ILW WWW**

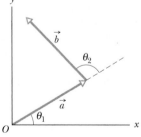

**FIG. 3-30** Problem 17.

•18 You are to make four straight-line moves over a flat desert floor, starting at the origin of an $xy$ coordinate system and ending at the $xy$ coordinates (−140 m, 30 m). The $x$ component and $y$ component of your moves are the following, respectively, in meters: (20 and 60), then ($b_x$ and −70), then (−20 and $c_y$), then (−60 and −70). What are (a) component $b_x$ and (b) component $c_y$? What are (c) the magnitude and (d) the angle (relative to the positive direction of the $x$ axis) of the overall displacement?

•19 Three vectors $\vec{a}$, $\vec{b}$, and $\vec{c}$ each have a magnitude of 50 m and lie in an $xy$ plane. Their directions relative to the positive direction of the $x$ axis are 30°, 195°, and 315°, respectively. What are (a) the magnitude and (b) the angle of the vector $\vec{a} + \vec{b} + \vec{c}$, and (c) the magnitude and (d) the angle of $\vec{a} - \vec{b} + \vec{c}$? What are the (e) magnitude and (f) angle of a fourth vector $\vec{d}$ such that $(\vec{a} + \vec{b}) - (\vec{c} + \vec{d}) = 0$? **ILW**

•20 (a) What is the sum of the following four vectors in unit-vector notation? For that sum, what are (b) the magnitude, (c) the angle in degrees, and (d) the angle in radians?

$\vec{E}$: 6.00 m at +0.900 rad   $\vec{F}$: 5.00 m at −75.0°

$\vec{G}$: 4.00 m at +1.20 rad   $\vec{H}$: 6.00 m at −210°

•21 In a game of lawn chess, where pieces are moved between the centers of squares that are each 1.00 m on edge, a knight is moved in the following way: (1) two squares forward, one square rightward; (2) two squares leftward, one square forward; (3) two squares forward, one square leftward. What are (a) the magnitude and (b) the angle (relative to "forward") of the knight's overall displacement for the series of three moves?

••22 An explorer is caught in a whiteout (in which the snowfall is so thick that the ground cannot be distinguished from the sky) while returning to base camp. He was supposed to travel due north for 5.6 km, but when the snow clears, he discovers that he actually traveled 7.8 km at 50° north of due east. (a) How far and (b) in what direction must he now travel to reach base camp?

••23 Oasis $B$ is 25 km due east of oasis $A$. Starting from oasis $A$, a camel walks 24 km in a direction 15° south of east and then walks 8.0 km due north. How far is the camel then from oasis $B$? **GO**

••24 Two beetles run across flat sand, starting at the same point. Beetle 1 runs 0.50 m due east, then 0.80 m at 30° north of due east. Beetle 2 also makes two runs; the first is 1.6 m at 40° east of due north. What must be (a) the magnitude and (b) the direction of its second run if it is to end up at the new location of beetle 1?

••25 If $\vec{B}$ is added to $\vec{C} = 3.0\hat{i} + 4.0\hat{j}$, the result is a vector in the positive direction of the $y$ axis, with a magnitude equal to that of $\vec{C}$. What is the magnitude of $\vec{B}$?

••26 Vector $\vec{A}$, which is directed along an $x$ axis, is to be added to vector $\vec{B}$, which has a magnitude of 7.0 m. The sum is a third vector that is directed along the $y$ axis, with a magnitude that is 3.0 times that of $\vec{A}$. What is that magnitude of $\vec{A}$?

••27 Typical backyard ants often create a network of chemical trails for guidance. Extending outward from the nest, a trail branches (*bifurcates*) repeatedly, with 60° between the branches. If a roaming ant chances upon a trail, it can tell the way to the nest at any branch point: If it is moving away from the nest, it has two choices of path requiring a small turn in its travel direction, either 30° leftward or 30° rightward. If it is moving toward the nest, it has only one such choice. Figure 3-31 shows a typical ant trail, with lettered straight sec-

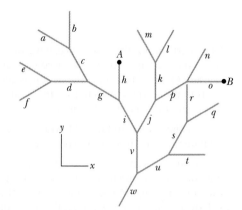

**FIG. 3-31**
Problem 27.

tions of 2.0 cm length and symmetric bifurcation of 60°. What are the (a) magnitude and (b) angle (relative to the positive direction of the superimposed $x$ axis) of an ant's displacement from the nest (find it in the figure) if the ant enters the trail at point $A$? What are the (c) magnitude and (d) angle if it enters at point $B$?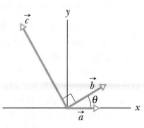

•• **28** Here are two vectors:

$$\vec{a} = (4.0\text{ m})\hat{i} - (3.0\text{ m})\hat{j} \quad \text{and} \quad \vec{b} = (6.0\text{ m})\hat{i} + (8.0\text{ m})\hat{j}.$$

What are (a) the magnitude and (b) the angle (relative to $\hat{i}$) of $\vec{a}$? What are (c) the magnitude and (d) the angle of $\vec{b}$? What are (e) the magnitude and (f) the angle of $\vec{a} + \vec{b}$; (g) the magnitude and (h) the angle of $\vec{b} - \vec{a}$; and (i) the magnitude and (j) the angle of $\vec{a} - \vec{b}$? (k) What is the angle between the directions of $\vec{b} - \vec{a}$ and $\vec{a} - \vec{b}$?

•• **29** If $\vec{d}_1 + \vec{d}_2 = 5\vec{d}_3$, $\vec{d}_1 - \vec{d}_2 = 3\vec{d}_3$, and $\vec{d}_3 = 2\hat{i} + 4\hat{j}$, then what are, in unit-vector notation, (a) $\vec{d}_1$ and (b) $\vec{d}_2$?

•• **30** What is the sum of the following four vectors in (a) unit-vector notation, and as (b) a magnitude and (c) an angle?

$\vec{A} = (2.00\text{ m})\hat{i} + (3.00\text{ m})\hat{j}$     $\vec{B}$: 4.00 m, at $+65.0°$
$\vec{C} = (-4.00\text{ m})\hat{i} + (-6.00\text{ m})\hat{j}$     $\vec{D}$: 5.00 m, at $-235°$

••• **31** In Fig. 3-32, a cube of edge length $a$ sits with one corner at the origin of an $xyz$ coordinate system. A *body diagonal* is a line that extends from one corner to another through the center. In unit-vector notation, what is the body diagonal that extends from the corner at (a) coordinates $(0, 0, 0)$, (b) coordinates $(a, 0, 0)$, (c) coordinates $(0, a, 0)$, and (d) coordinates $(a, a, 0)$? (e) Determine the angles that the body diagonals make with the adjacent edges. (f) Determine the length of the body diagonals in terms of $a$.

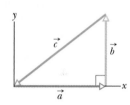

**FIG. 3-32** Problem 31.

### sec. 3-7 **Vectors and the Laws of Physics**

• **32** In Fig. 3-33, a vector $\vec{a}$ with a magnitude of 17.0 m is directed at angle $\theta = 56.0°$ counterclockwise from the $+x$ axis. What are the components (a) $a_x$ and (b) $a_y$ of the vector? A second coordinate system is inclined by angle $\theta' = 18.0°$

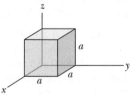

**FIG. 3-33** Problem 32.

with respect to the first. What are the components (c) $a'_x$ and (d) $a'_y$ in this primed coordinate system?

### sec. 3-8 **Multiplying Vectors**

• **33** Two vectors, $\vec{r}$ and $\vec{s}$, lie in the $xy$ plane. Their magnitudes are 4.50 and 7.30 units, respectively, and their directions are 320° and 85.0°, respectively, as measured counterclockwise from the positive $x$ axis. What are the values of (a) $\vec{r} \cdot \vec{s}$ and (b) $\vec{r} \times \vec{s}$?

• **34** If $\vec{d}_1 = 3\hat{i} - 2\hat{j} + 4\hat{k}$ and $\vec{d}_2 = -5\hat{i} + 2\hat{j} - \hat{k}$, then what is $(\vec{d}_1 + \vec{d}_2) \cdot (\vec{d}_1 \times 4\vec{d}_2)$?

• **35** Three vectors are given by $\vec{a} = 3.0\hat{i} + 3.0\hat{j} - 2.0\hat{k}$, $\vec{b} = -1.0\hat{i} - 4.0\hat{j} + 2.0\hat{k}$, and $\vec{c} = 2.0\hat{i} + 2.0\hat{j} + 1.0\hat{k}$. Find (a) $\vec{a} \cdot (\vec{b} \times \vec{c})$, (b) $\vec{a} \cdot (\vec{b} + \vec{c})$, and (c) $\vec{a} \times (\vec{b} + \vec{c})$.

• **36** Two vectors are given by $\vec{a} = 3.0\hat{i} + 5.0\hat{j}$ and $\vec{b} = 2.0\hat{i} + 4.0\hat{j}$. Find (a) $\vec{a} \times \vec{b}$, (b) $\vec{a} \cdot \vec{b}$, (c) $(\vec{a} + \vec{b}) \cdot \vec{b}$, and (d) the component of $\vec{a}$ along the direction of $\vec{b}$. (*Hint:* For (d), consider Eq. 3-20 and Fig. 3-20.)

• **37** For the vectors in Fig. 3-34, with $a = 4$, $b = 3$, and $c = 5$, what are (a) the magnitude and (b) the direction of $\vec{a} \times \vec{b}$, (c) the magnitude and (d) the direction of $\vec{a} \times \vec{c}$, and (e) the magnitude and (f) the direction of $\vec{b} \times \vec{c}$? (The $z$ axis is not shown.)

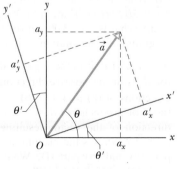

**FIG. 3-34** Problems 37 and 50.

•• **38** Displacement $\vec{d}_1$ is in the $yz$ plane 63.0° from the positive direction of the $y$ axis, has a positive $z$ component, and has a magnitude of 4.50 m. Displacement $\vec{d}_2$ is in the $xz$ plane 30.0° from the positive direction of the $x$ axis, has a positive $z$ component, and has magnitude 1.40 m. What are (a) $\vec{d}_1 \cdot \vec{d}_2$, (b) $\vec{d}_1 \times \vec{d}_2$, and (c) the angle between $\vec{d}_1$ and $\vec{d}_2$?

•• **39** Use the definition of scalar product, $\vec{a} \cdot \vec{b} = ab \cos \theta$, and the fact that $\vec{a} \cdot \vec{b} = a_x b_x + a_y b_y + a_z b_z$ to calculate the angle between the two vectors given by $\vec{a} = 3.0\hat{i} + 3.0\hat{j} + 3.0\hat{k}$ and $\vec{b} = 2.0\hat{i} + 1.0\hat{j} + 3.0\hat{k}$. **SSM ILW WWW**

•• **40** For the following three vectors, what is $3\vec{C} \cdot (2\vec{A} \times \vec{B})$?

$\vec{A} = 2.00\hat{i} + 3.00\hat{j} - 4.00\hat{k}$
$\vec{B} = -3.00\hat{i} + 4.00\hat{j} + 2.00\hat{k}$     $\vec{C} = 7.00\hat{i} - 8.00\hat{j}$

•• **41** Vector $\vec{A}$ has a magnitude of 6.00 units, vector $\vec{B}$ has a magnitude of 7.00 units, and $\vec{A} \cdot \vec{B}$ has a value of 14.0. What is the angle between the directions of $\vec{A}$ and $\vec{B}$?

•• **42** In the product $\vec{F} = q\vec{v} \times \vec{B}$, take $q = 2$,

$$\vec{v} = 2.0\hat{i} + 4.0\hat{j} + 6.0\hat{k} \quad \text{and} \quad \vec{F} = 4.0\hat{i} - 20\hat{j} + 12\hat{k}.$$

What then is $\vec{B}$ in unit-vector notation if $B_x = B_y$? **GO**

•• **43** The three vectors in Fig. 3-35 have magnitudes $a = 3.00$ m, $b = 4.00$ m, and $c = 10.0$ m and angle $\theta = 30.0°$. What are (a) the $x$ component and (b) the $y$ component of $\vec{a}$; (c) the $x$ component and (d) the $y$ com-

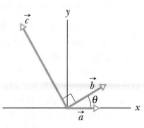

**FIG. 3-35** Problem 43.

ponent of $\vec{b}$; and (e) the $x$ component and (f) the $y$ component of $\vec{c}$? If $\vec{c} = p\vec{a} + q\vec{b}$, what are the values of (g) $p$ and (h) $q$?

**SSM ILW**

••44  In a meeting of mimes, mime 1 goes through a displacement $\vec{d}_1 = (4.0\text{ m})\hat{i} + (5.0\text{ m})\hat{j}$ and mime 2 goes through a displacement $\vec{d}_2 = (-3.0\text{ m})\hat{i} + (4.0\text{ m})\hat{j}$. What are (a) $\vec{d}_1 \times \vec{d}_2$, (b) $\vec{d}_1 \cdot \vec{d}_2$, (c) $(\vec{d}_1 + \vec{d}_2) \cdot \vec{d}_2$, and (d) the component of $\vec{d}_1$ along the direction of $\vec{d}_2$? (*Hint:* For (d), see Eq. 3-20 and Fig. 3-20.)

**Additional Problems**

45  Rock *faults* are ruptures along which opposite faces of rock have slid past each other. In Fig. 3-36, points $A$ and $B$ coincided before the rock in the foreground slid down to the right. The net displacement $\overrightarrow{AB}$ is along the plane of the fault. The horizontal component of $\overrightarrow{AB}$ is the *strike-slip AC*. The component of $\overrightarrow{AB}$ that is directed down the plane of the fault is the *dip-slip AD*. (a) What is the magnitude of the net displacement $\overrightarrow{AB}$ if the strike-slip is 22.0 m and the dip-slip is 17.0 m? (b) If the plane of the fault is inclined at angle $\phi = 52.0°$ to the horizontal, what is the vertical component of $\overrightarrow{AB}$?

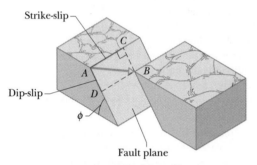

Strike-slip

Dip-slip

$\phi$

Fault plane

**FIG. 3-36**  Problem 45.

46  Two vectors $\vec{a}$ and $\vec{b}$ have the components, in meters, $a_x = 3.2$, $a_y = 1.6$, $b_x = 0.50$, $b_y = 4.5$. (a) Find the angle between the directions of $\vec{a}$ and $\vec{b}$. There are two vectors in the $xy$ plane that are perpendicular to $\vec{a}$ and have a magnitude of 5.0 m. One, vector $\vec{c}$, has a positive $x$ component and the other, vector $\vec{d}$, a negative $x$ component. What are (b) the $x$ component and (c) the $y$ component of $\vec{c}$, and (d) the $x$ component and (e) the $y$ component of vector $\vec{d}$?

47  A vector $\vec{a}$ of magnitude 10 units and another vector $\vec{b}$ of magnitude 6.0 units differ in directions by 60°. Find (a) the scalar product of the two vectors and (b) the magnitude of the vector product $\vec{a} \times \vec{b}$.  **SSM**

48  Vector $\vec{a}$ has a magnitude of 5.0 m and is directed east. Vector $\vec{b}$ has a magnitude of 4.0 m and is directed 35° west of due north. What are (a) the magnitude and (b) the direction of $\vec{a} + \vec{b}$? What are (c) the magnitude and (d) the direction of $\vec{b} - \vec{a}$? (e) Draw a vector diagram for each combination.

49  A particle undergoes three successive displacements in a plane, as follows: $\vec{d}_1$, 4.00 m southwest; then $\vec{d}_2$, 5.00 m east; and finally $\vec{d}_3$, 6.00 m in a direction 60.0° north of east. Choose a coordinate system with the $y$ axis pointing north and the $x$ axis pointing east. What are (a) the $x$ component and (b) the $y$ component of $\vec{d}_1$? What are (c) the $x$ component and (d) the $y$ component of $\vec{d}_2$? What are (e) the $x$ component and (f) the $y$ component of $\vec{d}_3$? Next, consider the *net* displacement

of the particle for the three successive displacements. What are (g) the $x$ component, (h) the $y$ component, (i) the magnitude, and (j) the direction of the net displacement? If the particle is to return directly to the starting point, (k) how far and (l) in what direction should it move?

50  For the vectors in Fig. 3-34, with $a = 4$, $b = 3$, and $c = 5$, calculate (a) $\vec{a} \cdot \vec{b}$, (b) $\vec{a} \cdot \vec{c}$, and (c) $\vec{b} \cdot \vec{c}$.

51  A sailboat sets out from the U.S. side of Lake Erie for a point on the Canadian side, 90.0 km due north. The sailor, however, ends up 50.0 km due east of the starting point. (a) How far and (b) in what direction must the sailor now sail to reach the original destination?  **SSM**

52  Find the sum of the following four vectors in (a) unit-vector notation, and as (b) a magnitude and (c) an angle relative to $+x$.

$\vec{P}$:  10.0 m, at 25.0° counterclockwise from $+x$

$\vec{Q}$:  12.0 m, at 10.0° counterclockwise from $+y$

$\vec{R}$:  8.00 m, at 20.0° clockwise from $-y$

$\vec{S}$:  9.00 m, at 40.0° counterclockwise from $-y$

53  Vectors $\vec{A}$ and $\vec{B}$ lie in an $xy$ plane. $\vec{A}$ has magnitude 8.00 and angle 130°; $\vec{B}$ has components $B_x = -7.72$ and $B_y = -9.20$. What are the angles between the negative direction of the $y$ axis and (a) the direction of $\vec{A}$, (b) the direction of the product $\vec{A} \times \vec{B}$, and (c) the direction of $\vec{A} \times (\vec{B} + 3.00\hat{k})$?

54  Here are three displacements, each in meters: $\vec{d}_1 = 4.0\hat{i} + 5.0\hat{j} - 6.0\hat{k}$, $\vec{d}_2 = -1.0\hat{i} + 2.0\hat{j} + 3.0\hat{k}$, and $\vec{d}_3 = 4.0\hat{i} + 3.0\hat{j} + 2.0\hat{k}$. (a) What is $\vec{r} = \vec{d}_1 - \vec{d}_2 + \vec{d}_3$? (b) What is the angle between $\vec{r}$ and the positive $z$ axis? (c) What is the component of $\vec{d}_1$ along the direction of $\vec{d}_2$? (d) What is the component of $\vec{d}_1$ that is perpendicular to the direction of $\vec{d}_2$ and in the plane of $\vec{d}_1$ and $\vec{d}_2$? (*Hint:* For (c), consider Eq. 3-20 and Fig. 3-20; for (d), consider Eq. 3-27.)

55  Vectors $\vec{A}$ and $\vec{B}$ lie in an $xy$ plane. $\vec{A}$ has magnitude 8.00 and angle 130°; $\vec{B}$ has components $B_x = -7.72$ and $B_y = -9.20$. (a) What is $5\vec{A} \cdot \vec{B}$? What is $4\vec{A} \times 3\vec{B}$ in (b) unit-vector notation and (c) magnitude-angle notation with spherical coordinates (see Fig. 3-37)? (d) What is the angle between the directions of $\vec{A}$ and $4\vec{A} \times 3\vec{B}$? (*Hint:* Think a bit before you resort to a calculation.) What is $\vec{A} + 3.00\hat{k}$ in (e) unit-vector notation and (f) magnitude-angle notation with spherical coordinates?

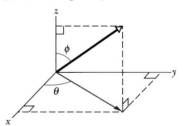

**FIG. 3-37**  Problem 55.

56  Vector $\vec{d}_1$ is in the negative direction of a $y$ axis, and vector $\vec{d}_2$ is in the positive direction of an $x$ axis. What are the directions of (a) $\vec{d}_2/4$ and (b) $\vec{d}_1/(-4)$? What are the magnitudes of products (c) $\vec{d}_1 \cdot \vec{d}_2$ and (d) $\vec{d}_1 \cdot (\vec{d}_2/4)$? What is the direction of the vector resulting from (e) $\vec{d}_1 \times \vec{d}_2$ and (f) $\vec{d}_2 \times \vec{d}_1$? What is the magnitude of the vector product in (g) part (e) and (h) part (f)? What are the (i) magnitude and (j) direction of $\vec{d}_1 \times (\vec{d}_2/4)$?

**57** Here are three vectors in meters:

$$\vec{d_1} = -3.0\hat{i} + 3.0\hat{j} + 2.0\hat{k}$$
$$\vec{d_2} = -2.0\hat{i} - 4.0\hat{j} + 2.0\hat{k}$$
$$\vec{d_3} = 2.0\hat{i} + 3.0\hat{j} + 1.0\hat{k}.$$

What results from (a) $\vec{d_1} \cdot (\vec{d_2} + \vec{d_3})$, (b) $\vec{d_1} \cdot (\vec{d_2} \times \vec{d_3})$, and (c) $\vec{d_1} \times (\vec{d_2} + \vec{d_3})$?

**58** A golfer takes three putts to get the ball into the hole. The first putt displaces the ball 3.66 m north, the second 1.83 m southeast, and the third 0.91 m southwest. What are (a) the magnitude and (b) the direction of the displacement needed to get the ball into the hole on the first putt?

**59** Consider $\vec{a}$ in the positive direction of $x$, $\vec{b}$ in the positive direction of $y$, and a scalar $d$. What is the direction of $\vec{b}/d$ if $d$ is (a) positive and (b) negative? What is the magnitude of (c) $\vec{a} \cdot \vec{b}$ and (d) $\vec{a} \cdot \vec{b}/d$? What is the direction of the vector resulting from (e) $\vec{a} \times \vec{b}$ and (f) $\vec{b} \times \vec{a}$? (g) What is the magnitude of the vector product in (e)? (h) What is the magnitude of the vector product in (f)? What are (i) the magnitude and (j) the direction of $\vec{a} \times \vec{b}/d$ if $d$ is positive?

**60** A vector $\vec{d}$ has a magnitude of 2.5 m and points north. What are (a) the magnitude and (b) the direction of $4.0\vec{d}$? What are (c) the magnitude and (d) the direction of $-3.0\vec{d}$?

**61** Let $\hat{i}$ be directed to the east, $\hat{j}$ be directed to the north, and $\hat{k}$ be directed upward. What are the values of products (a) $\hat{i} \cdot \hat{k}$, (b) $(-\hat{k}) \cdot (-\hat{j})$, and (c) $\hat{j} \cdot (-\hat{j})$? What are the directions (such as east or down) of products (d) $\hat{k} \times \hat{j}$, (e) $(-\hat{i}) \times (-\hat{j})$, and (f) $(-\hat{k}) \times (-\hat{j})$?

**62** Consider two displacements, one of magnitude 3 m and another of magnitude 4 m. Show how the displacement vectors may be combined to get a resultant displacement of magnitude (a) 7 m, (b) 1 m, and (c) 5 m.

**63** A bank in downtown Boston is robbed (see the map in Fig. 3-38). To elude police, the robbers escape by helicopter, making three successive flights described by the following displacements: 32 km, 45° south of east; 53 km, 26° north of west; 26 km, 18° east of south. At the end of the third flight they are captured. In what town are they apprehended?

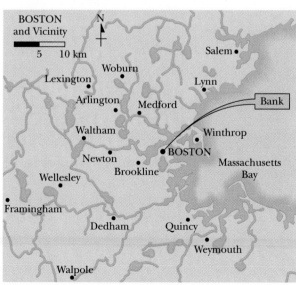

FIG. 3-38 Problem 63.

**64** A wheel with a radius of 45.0 cm rolls without slipping along a horizontal floor (Fig. 3-39). At time $t_1$, the dot $P$ painted on the rim of the wheel is at the point of contact between the wheel and the floor. At a later time $t_2$, the wheel has rolled through one-half of a revolution. What are (a) the magnitude and (b) the angle (relative to the floor) of the displacement of $P$?

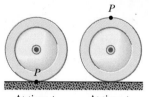

FIG. 3-39 Problem 64.

**65** $\vec{A}$ has the magnitude 12.0 m and is angled 60.0° counter-clockwise from the positive direction of the $x$ axis of an $xy$ coordinate system. Also, $\vec{B} = (12.0 \text{ m})\hat{i} + (8.00 \text{ m})\hat{j}$ on that same coordinate system. We now rotate the system counterclockwise about the origin by 20.0° to form an $x'y'$ system. On this new system, what are (a) $\vec{A}$ and (b) $\vec{B}$, both in unit-vector notation?

**66** A woman walks 250 m in the direction 30° east of north, then 175 m directly east. Find (a) the magnitude and (b) the angle of her final displacement from the starting point. (c) Find the distance she walks. (d) Which is greater, that distance or the magnitude of her displacement?

**67** (a) In unit-vector notation, what is $\vec{r} = \vec{a} - \vec{b} + \vec{c}$ if $\vec{a} = 5.0\hat{i} + 4.0\hat{j} - 6.0\hat{k}$, $\vec{b} = -2.0\hat{i} + 2.0\hat{j} + 3.0\hat{k}$, and $\vec{c} = 4.0\hat{i} + 3.0\hat{j} + 2.0\hat{k}$? (b) Calculate the angle between $\vec{r}$ and the positive $z$ axis. (c) What is the component of $\vec{a}$ along the direction of $\vec{b}$? (d) What is the component of $\vec{a}$ perpendicular to the direction of $\vec{b}$ but in the plane of $\vec{a}$ and $\vec{b}$? (*Hint:* For (c), see Eq. 3-20 and Fig. 3-20; for (d), see Eq. 3-27.)

**68** If $\vec{a} - \vec{b} = 2\vec{c}$, $\vec{a} + \vec{b} = 4\vec{c}$, and $\vec{c} = 3\hat{i} + 4\hat{j}$, then what are (a) $\vec{a}$ and (b) $\vec{b}$?

**69** A protester carries his sign of protest, starting from the origin of an $xyz$ coordinate system, with the $xy$ plane horizontal. He moves 40 m in the negative direction of the $x$ axis, then 20 m along a perpendicular path to his left, and then 25 m up a water tower. (a) In unit-vector notation, what is the displacement of the sign from start to end? (b) The sign then falls to the foot of the tower. What is the magnitude of the displacement of the sign from start to this new end?

**70** A vector $\vec{d}$ has a magnitude 3.0 m and is directed south. What are (a) the magnitude and (b) the direction of the vector $5.0\vec{d}$? What are (c) the magnitude and (d) the direction of the vector $-2.0\vec{d}$?

**71** If $\vec{B}$ is added to $\vec{A}$, the result is $6.0\hat{i} + 1.0\hat{j}$. If $\vec{B}$ is subtracted from $\vec{A}$, the result is $-4.0\hat{i} + 7.0\hat{j}$. What is the magnitude of $\vec{A}$? **SSM**

**72** A fire ant, searching for hot sauce in a picnic area, goes through three displacements along level ground: $\vec{d_1}$ for 0.40 m southwest (that is, at 45° from directly south and from directly west), $\vec{d_2}$ for 0.50 m due east, $\vec{d_3}$ for 0.60 m at 60° north of east. Let the positive $x$ direction be east and the positive $y$ direction be north. What are (a) the $x$ component and (b) the $y$ component of $\vec{d_1}$? What are (c) the $x$ component and (d) the $y$ component of $\vec{d_2}$? What are (e) the $x$ component and (f) the $y$ component of $\vec{d_3}$?

What are (g) the $x$ component, (h) the $y$ component, (i) the magnitude, and (j) the direction of the ant's net displacement? If the ant is to return directly to the starting point, (k) how far and (l) in what direction should it move?

# 4

# Motion in Two and Three Dimensions

*Source:* Rob Tringali/Sports Chrome Inc.

When a high fly ball is hit to the outfield, how does the outfielder in the area know where to be in order to catch it? Often the outfielder will jog or run at a measured pace to the catch site, arriving just as the ball does. Playing experience surely helps, but some other factor seems to be involved.

## What clue is hidden in the ball's motion?

The answer is in this chapter.

## 4-1 | WHAT IS PHYSICS?

In this chapter we continue looking at the aspect of physics that analyzes motion, but now the motion can be in two or three dimensions. For example, medical researchers and aeronautical engineers might concentrate on the physics of the two- and three-dimensional turns taken by fighter pilots in dogfights because a modern high-performance jet can take a tight turn so quickly that the pilot immediately loses consciousness. A sports engineer might focus on the physics of basketball. For example, in a *free throw* (where a player gets an uncontested shot at the basket from about 4.3 m), a player might employ the *overhand push shot,* in which the ball is pushed away from about shoulder height and then released. Or the player might use an *underhand loop shot,* in which the ball is brought upward from about the belt-line level and released. The first technique is the overwhelming choice among professional players, but the legendary Rick Barry set the record for free-throw shooting with the underhand technique.

Motion in three dimensions is not easy to understand. For example, you are probably good at driving a car along a freeway (one-dimensional motion) but would probably have a difficult time in landing an airplane on a runway (three-dimensional motion) without a lot of training.

In our study of two- and three-dimensional motion, we start with position and displacement.

## 4-2 | Position and Displacement

One general way of locating a particle (or particle-like object) is with a **position vector** $\vec{r}$, which is a vector that extends from a reference point (usually the origin) to the particle. In the unit-vector notation of Section 3-5, $\vec{r}$ can be written

$$\vec{r} = x\hat{i} + y\hat{j} + z\hat{k}, \tag{4-1}$$

where $x\hat{i}, y\hat{j},$ and $z\hat{k}$ are the vector components of $\vec{r}$ and the coefficients $x, y,$ and $z$ are its scalar components.

The coefficients $x, y,$ and $z$ give the particle's location along the coordinate axes and relative to the origin; that is, the particle has the rectangular coordinates $(x, y, z)$. For instance, Fig. 4-1 shows a particle with position vector

$$\vec{r} = (-3\text{ m})\hat{i} + (2\text{ m})\hat{j} + (5\text{ m})\hat{k}$$

and rectangular coordinates $(-3\text{ m}, 2\text{ m}, 5\text{ m})$. Along the $x$ axis the particle is 3 m from the origin, in the $-\hat{i}$ direction. Along the $y$ axis it is 2 m from the origin, in the $+\hat{j}$ direction. Along the $z$ axis it is 5 m from the origin, in the $+\hat{k}$ direction.

As a particle moves, its position vector changes in such a way that the vector always extends to the particle from the reference point (the origin). If the position vector changes—say, from $\vec{r}_1$ to $\vec{r}_2$ during a certain time interval—then the particle's **displacement** $\Delta\vec{r}$ during that time interval is

$$\Delta\vec{r} = \vec{r}_2 - \vec{r}_1. \tag{4-2}$$

Using the unit-vector notation of Eq. 4-1, we can rewrite this displacement as

$$\Delta\vec{r} = (x_2\hat{i} + y_2\hat{j} + z_2\hat{k}) - (x_1\hat{i} + y_1\hat{j} + z_1\hat{k})$$

or as

$$\Delta\vec{r} = (x_2 - x_1)\hat{i} + (y_2 - y_1)\hat{j} + (z_2 - z_1)\hat{k}, \tag{4-3}$$

where coordinates $(x_1, y_1, z_1)$ correspond to position vector $\vec{r}_1$ and coordinates $(x_2, y_2, z_2)$ correspond to position vector $\vec{r}_2$. We can also rewrite the displacement by substituting $\Delta x$ for $(x_2 - x_1), \Delta y$ for $(y_2 - y_1),$ and $\Delta z$ for $(z_2 - z_1)$:

$$\Delta\vec{r} = \Delta x\hat{i} + \Delta y\hat{j} + \Delta z\hat{k}. \tag{4-4}$$

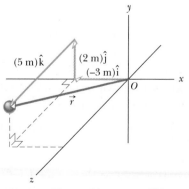

**FIG. 4-1** The position vector $\vec{r}$ for a particle is the vector sum of its vector components.

**Sample Problem** 4-1

In Fig. 4-2, the position vector for a particle initially is

$$\vec{r}_1 = (-3.0\text{ m})\hat{i} + (2.0\text{ m})\hat{j} + (5.0\text{ m})\hat{k}$$

and then later is

$$\vec{r}_2 = (9.0\text{ m})\hat{i} + (2.0\text{ m})\hat{j} + (8.0\text{ m})\hat{k}.$$

What is the particle's displacement $\Delta\vec{r}$ from $\vec{r}_1$ to $\vec{r}_2$?

**KEY IDEA** The displacement $\Delta\vec{r}$ is obtained by subtracting the initial $\vec{r}_1$ from the later $\vec{r}_2$.

**Calculation:** The subtraction gives us

$$\Delta\vec{r} = \vec{r}_2 - \vec{r}_1$$
$$= [9.0 - (-3.0)]\hat{i} + [2.0 - 2.0]\hat{j} + [8.0 - 5.0]\hat{k}$$
$$= (12\text{ m})\hat{i} + (3.0\text{ m})\hat{k}. \qquad \text{(Answer)}$$

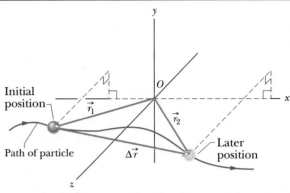

FIG. 4-2 The displacement $\Delta\vec{r} = \vec{r}_2 - \vec{r}_1$ extends from the head of the initial position vector $\vec{r}_1$ to the head of the later position vector $\vec{r}_2$.

This displacement vector is parallel to the $xz$ plane because it lacks a $y$ component.

**Sample Problem** 4-2

A rabbit runs across a parking lot on which a set of coordinate axes has, strangely enough, been drawn. The coordinates (meters) of the rabbit's position as functions of time $t$ (seconds) are given by

$$x = -0.31t^2 + 7.2t + 28 \qquad (4\text{-}5)$$

and $\qquad y = 0.22t^2 - 9.1t + 30. \qquad (4\text{-}6)$

(a) At $t = 15$ s, what is the rabbit's position vector $\vec{r}$ in unit-vector notation and in magnitude-angle notation?

**KEY IDEA** The $x$ and $y$ coordinates of the rabbit's position, as given by Eqs. 4-5 and 4-6, are the scalar components of the rabbit's position vector $\vec{r}$.

**Calculations:** We can write

$$\vec{r}(t) = x(t)\hat{i} + y(t)\hat{j}. \qquad (4\text{-}7)$$

(We write $\vec{r}(t)$ rather than $\vec{r}$ because the components are functions of $t$, and thus $\vec{r}$ is also.)

At $t = 15$ s, the scalar components are

$$x = (-0.31)(15)^2 + (7.2)(15) + 28 = 66\text{ m}$$

and $\quad y = (0.22)(15)^2 - (9.1)(15) + 30 = -57\text{ m},$

so $\qquad \vec{r} = (66\text{ m})\hat{i} - (57\text{ m})\hat{j}, \qquad \text{(Answer)}$

which is drawn in Fig. 4-3a. To get the magnitude and angle of $\vec{r}$, we use Eq. 3-6:

$$r = \sqrt{x^2 + y^2} = \sqrt{(66\text{ m})^2 + (-57\text{ m})^2}$$
$$= 87\text{ m}, \qquad \text{(Answer)}$$

and $\quad \theta = \tan^{-1}\dfrac{y}{x} = \tan^{-1}\left(\dfrac{-57\text{ m}}{66\text{ m}}\right) = -41°.$

(Answer)

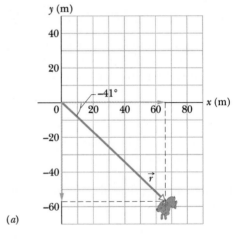

(a)

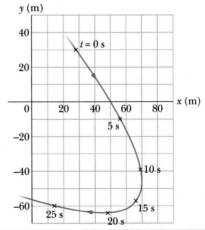

(b)

FIG. 4-3 (a) A rabbit's position vector $\vec{r}$ at time $t = 15$ s. The scalar components of $\vec{r}$ are shown along the axes. (b) The rabbit's path and its position at five values of $t$.

**Check:** Although $\theta = 139°$ has the same tangent as $-41°$, the components of $\vec{r}$ indicate that the desired angle is $139° - 180° = -41°$.

(b) Graph the rabbit's path for $t = 0$ to $t = 25$ s.

**Graphing:** We can repeat part (a) for several values of $t$ and then plot the results. Figure 4-3b shows the plots for five values of $t$ and the path connecting them. We can also plot Eqs. 4-5 and 4-6 on a calculator.

## 4-3 | Average Velocity and Instantaneous Velocity

If a particle moves from one point to another, we might need to know how fast it moves. Just as in Chapter 2, we can define two quantities that deal with "how fast": *average velocity* and *instantaneous velocity*. However, here we must consider these quantities as vectors and use vector notation.

If a particle moves through a displacement $\Delta\vec{r}$ in a time interval $\Delta t$, then its **average velocity** $\vec{v}_{avg}$ is

$$\text{average velocity} = \frac{\text{displacement}}{\text{time interval}},$$

or

$$\vec{v}_{avg} = \frac{\Delta\vec{r}}{\Delta t}. \tag{4-8}$$

This tells us that the direction of $\vec{v}_{avg}$ (the vector on the left side of Eq. 4-8) must be the same as that of the displacement $\Delta\vec{r}$ (the vector on the right side). Using Eq. 4-4, we can write Eq. 4-8 in vector components as

$$\vec{v}_{avg} = \frac{\Delta x\hat{i} + \Delta y\hat{j} + \Delta z\hat{k}}{\Delta t} = \frac{\Delta x}{\Delta t}\hat{i} + \frac{\Delta y}{\Delta t}\hat{j} + \frac{\Delta z}{\Delta t}\hat{k}. \tag{4-9}$$

For example, if the particle in Sample Problem 4-1 moves from its initial position to its later position in 2.0 s, then its average velocity during that move is

$$\vec{v}_{avg} = \frac{\Delta\vec{r}}{\Delta t} = \frac{(12\text{ m})\hat{i} + (3.0\text{ m})\hat{k}}{2.0\text{ s}} = (6.0\text{ m/s})\hat{i} + (1.5\text{ m/s})\hat{k}.$$

That is, the average velocity (a vector quantity) has a component of 6.0 m/s along the x axis and a component of 1.5 m/s along the z axis.

When we speak of the **velocity** of a particle, we usually mean the particle's **instantaneous velocity** $\vec{v}$ at some instant. This $\vec{v}$ is the value that $\vec{v}_{avg}$ approaches in the limit as we shrink the time interval $\Delta t$ to 0 about that instant. Using the language of calculus, we may write $\vec{v}$ as the derivative

$$\vec{v} = \frac{d\vec{r}}{dt}. \tag{4-10}$$

Figure 4-4 shows the path of a particle that is restricted to the $xy$ plane. As the particle travels to the right along the curve, its position vector sweeps to the right. During time interval $\Delta t$, the position vector changes from $\vec{r}_1$ to $\vec{r}_2$ and the particle's displacement is $\Delta\vec{r}$.

To find the instantaneous velocity of the particle at, say, instant $t_1$ (when the particle is at position 1), we shrink interval $\Delta t$ to 0 about $t_1$. Three things happen as we do so. (1) Position vector $\vec{r}_2$ in Fig. 4-4 moves toward $\vec{r}_1$ so that $\Delta\vec{r}$ shrinks toward zero. (2) The direction of $\Delta\vec{r}/\Delta t$ (and thus of $\vec{v}_{avg}$) approaches the direction of the line tangent to the particle's path at position 1. (3) The average velocity $\vec{v}_{avg}$ approaches the instantaneous velocity $\vec{v}$ at $t_1$.

In the limit as $\Delta t \to 0$, we have $\vec{v}_{avg} \to \vec{v}$ and, most important here, $\vec{v}_{avg}$ takes on the direction of the tangent line. Thus, $\vec{v}$ has that direction as well:

> The direction of the instantaneous velocity $\vec{v}$ of a particle is always tangent to the particle's path at the particle's position.

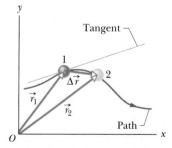

**FIG. 4-4** The displacement $\Delta\vec{r}$ of a particle during a time interval $\Delta t$, from position 1 with position vector $\vec{r}_1$ at time $t_1$ to position 2 with position vector $\vec{r}_2$ at time $t_2$. The tangent to the particle's path at position 1 is shown.

The result is the same in three dimensions: $\vec{v}$ is always tangent to the particle's path. To write Eq. 4-10 in unit-vector form, we substitute for $\vec{r}$ from Eq. 4-1:

$$\vec{v} = \frac{d}{dt}(x\hat{i} + y\hat{j} + z\hat{k}) = \frac{dx}{dt}\hat{i} + \frac{dy}{dt}\hat{j} + \frac{dz}{dt}\hat{k}.$$

This equation can be simplified somewhat by writing it as

$$\vec{v} = v_x\hat{i} + v_y\hat{j} + v_z\hat{k}, \tag{4-11}$$

where the scalar components of $\vec{v}$ are

$$v_x = \frac{dx}{dt}, \quad v_y = \frac{dy}{dt}, \quad \text{and} \quad v_z = \frac{dz}{dt}. \tag{4-12}$$

For example, $dx/dt$ is the scalar component of $\vec{v}$ along the $x$ axis. Thus, we can find the scalar components of $\vec{v}$ by differentiating the scalar components of $\vec{r}$.

Figure 4-5 shows a velocity vector $\vec{v}$ and its scalar $x$ and $y$ components. Note that $\vec{v}$ is tangent to the particle's path at the particle's position. *Caution:* When a position vector is drawn, as in Figs. 4-1 through 4-4, it is an arrow that extends from one point (a "here") to another point (a "there"). However, when a velocity vector is drawn, as in Fig. 4-5, it does *not* extend from one point to another. Rather, it shows the instantaneous direction of travel of a particle at the tail, and its length (representing the velocity magnitude) can be drawn to any scale.

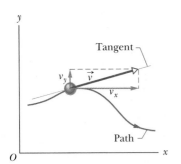

**FIG. 4-5** The velocity $\vec{v}$ of a particle, along with the scalar components of $\vec{v}$.

✓**CHECKPOINT 1**    The figure shows a circular path taken by a particle. If the instantaneous velocity of the particle is $\vec{v} = (2 \text{ m/s})\hat{i} - (2 \text{ m/s})\hat{j}$, through which quadrant is the particle moving at that instant if it is traveling (a) clockwise and (b) counterclockwise around the circle? For both cases, draw $\vec{v}$ on the figure.

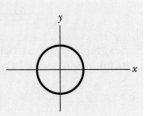

---

**Sample Problem**    **4-3**

For the rabbit in Sample Problem 4-2 find the velocity $\vec{v}$ at time $t = 15$ s.

**KEY IDEA**    We can find $\vec{v}$ by taking derivatives of the components of the rabbit's position vector.

**Calculations:** Applying the $v_x$ part of Eq. 4-12 to Eq. 4-5, we find the $x$ component of $\vec{v}$ to be

$$v_x = \frac{dx}{dt} = \frac{d}{dt}(-0.31t^2 + 7.2t + 28)$$

$$= -0.62t + 7.2. \tag{4-13}$$

At $t = 15$ s, this gives $v_x = -2.1$ m/s. Similarly, applying the $v_y$ part of Eq. 4-12 to Eq. 4-6, we find

$$v_y = \frac{dy}{dt} = \frac{d}{dt}(0.22t^2 - 9.1t + 30)$$

$$= 0.44t - 9.1. \tag{4-14}$$

At $t = 15$ s, this gives $v_y = -2.5$ m/s. Equation 4-11 then yields

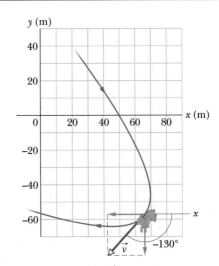

**FIG. 4-6** The rabbit's velocity $\vec{v}$ at $t = 15$ s.

$$\vec{v} = (-2.1 \text{ m/s})\hat{i} + (-2.5 \text{ m/s})\hat{j}, \quad \text{(Answer)}$$

which is shown in Fig. 4-6, tangent to the rabbit's path and in the direction the rabbit is running at $t = 15$ s.

To get the magnitude and angle of $\vec{v}$, either we use a vector-capable calculator or we follow Eq. 3-6 to write

$$v = \sqrt{v_x^2 + v_y^2} = \sqrt{(-2.1 \text{ m/s})^2 + (-2.5 \text{ m/s})^2}$$
$$= 3.3 \text{ m/s} \hspace{3cm} \text{(Answer)}$$

and $\quad \theta = \tan^{-1}\dfrac{v_y}{v_x} = \tan^{-1}\left(\dfrac{-2.5 \text{ m/s}}{-2.1 \text{ m/s}}\right)$

$$= \tan^{-1} 1.19 = -130°. \hspace{1cm} \text{(Answer)}$$

**Check:** Is the angle $-130°$ or $-130° + 180° = 50°$?

## 4-4 | Average Acceleration and Instantaneous Acceleration

When a particle's velocity changes from $\vec{v}_1$ to $\vec{v}_2$ in a time interval $\Delta t$, its **average acceleration** $\vec{a}_{avg}$ during $\Delta t$ is

$$\frac{\text{average}}{\text{acceleration}} = \frac{\text{change in velocity}}{\text{time interval}},$$

or

$$\vec{a}_{avg} = \frac{\vec{v}_2 - \vec{v}_1}{\Delta t} = \frac{\Delta \vec{v}}{\Delta t}. \hspace{1cm} (4\text{-}15)$$

If we shrink $\Delta t$ to zero about some instant, then in the limit $\vec{a}_{avg}$ approaches the **instantaneous acceleration** (or **acceleration**) $\vec{a}$ at that instant; that is,

$$\vec{a} = \frac{d\vec{v}}{dt}. \hspace{1cm} (4\text{-}16)$$

If the velocity changes in *either* magnitude *or* direction (or both), the particle must have an acceleration.

We can write Eq. 4-16 in unit-vector form by substituting Eq. 4-11 for $\vec{v}$ to obtain

$$\vec{a} = \frac{d}{dt}(v_x \hat{i} + v_y \hat{j} + v_z \hat{k})$$

$$= \frac{dv_x}{dt}\hat{i} + \frac{dv_y}{dt}\hat{j} + \frac{dv_z}{dt}\hat{k}.$$

We can rewrite this as

$$\vec{a} = a_x \hat{i} + a_y \hat{j} + a_z \hat{k}, \hspace{1cm} (4\text{-}17)$$

where the scalar components of $\vec{a}$ are

$$a_x = \frac{dv_x}{dt}, \quad a_y = \frac{dv_y}{dt}, \quad \text{and} \quad a_z = \frac{dv_z}{dt}. \hspace{1cm} (4\text{-}18)$$

To find the scalar components of $\vec{a}$, we differentiate the scalar components of $\vec{v}$.

Figure 4-7 shows an acceleration vector $\vec{a}$ and its scalar components for a particle moving in two dimensions. *Caution:* When an acceleration vector is drawn, as in Fig. 4-7, it does *not* extend from one position to another. Rather, it shows the direction of acceleration for a particle located at its tail, and its length (representing the acceleration magnitude) can be drawn to any scale.

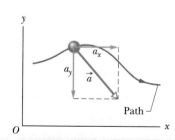

**FIG. 4-7** The acceleration $\vec{a}$ of a particle and the scalar components of $\vec{a}$.

✓**CHECKPOINT 2**   Here are four descriptions of the position (in meters) of a puck as it moves in an $xy$ plane:

(1) $x = -3t^2 + 4t - 2$   and   $y = 6t^2 - 4t$     (3) $\vec{r} = 2t^2\hat{i} - (4t + 3)\hat{j}$
(2) $x = -3t^3 - 4t$   and   $y = -5t^2 + 6$     (4) $\vec{r} = (4t^3 - 2t)\hat{i} + 3\hat{j}$

Are the $x$ and $y$ acceleration components constant? Is acceleration $\vec{a}$ constant?

**Sample Problem** **4-4**

For the rabbit in Sample Problems 4-2 and 4-3, find the acceleration $\vec{a}$ at time $t = 15$ s.

**KEY IDEA** We can find $\vec{a}$ by taking derivatives of the rabbit's velocity components.

**Calculations:** Applying the $a_x$ part of Eq. 4-18 to Eq. 4-13, we find the $x$ component of $\vec{a}$ to be

$$a_x = \frac{dv_x}{dt} = \frac{d}{dt}(-0.62t + 7.2) = -0.62 \text{ m/s}^2.$$

Similarly, applying the $a_y$ part of Eq. 4-18 to Eq. 4-14 yields the $y$ component as

$$a_y = \frac{dv_y}{dt} = \frac{d}{dt}(0.44t - 9.1) = 0.44 \text{ m/s}^2.$$

We see that the acceleration does not vary with time (it is a constant) because the time variable $t$ does not appear in the expression for either acceleration component. Equation 4-17 then yields

$$\vec{a} = (-0.62 \text{ m/s}^2)\hat{i} + (0.44 \text{ m/s}^2)\hat{j}, \quad \text{(Answer)}$$

which is superimposed on the rabbit's path in Fig. 4-8.

To get the magnitude and angle of $\vec{a}$, either we use a vector-capable calculator or we follow Eq. 3-6. For the magnitude we have

$$a = \sqrt{a_x^2 + a_y^2} = \sqrt{(-0.62 \text{ m/s}^2)^2 + (0.44 \text{ m/s}^2)^2}$$
$$= 0.76 \text{ m/s}^2. \quad \text{(Answer)}$$

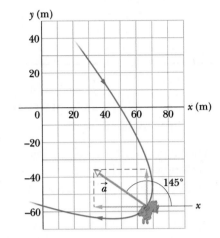

**FIG. 4-8** The acceleration $\vec{a}$ of the rabbit at $t = 15$ s. The rabbit happens to have this same acceleration at all points on its path.

For the angle we have

$$\theta = \tan^{-1} \frac{a_y}{a_x} = \tan^{-1}\left(\frac{0.44 \text{ m/s}^2}{-0.62 \text{ m/s}^2}\right) = -35°.$$

However, this angle, which is the one displayed on a calculator, indicates that $\vec{a}$ is directed to the right and downward in Fig. 4-8. Yet, we know from the components that $\vec{a}$ must be directed to the left and upward. To find the other angle that has the same tangent as $-35°$ but is not displayed on a calculator, we add $180°$:

$$-35° + 180° = 145°. \quad \text{(Answer)}$$

This *is* consistent with the components of $\vec{a}$. Note that $\vec{a}$ has the same magnitude and direction throughout the rabbit's run because the acceleration is constant.

**Sample Problem** **4-5**

A particle with velocity $\vec{v}_0 = -2.0\hat{i} + 4.0\hat{j}$ (in meters per second) at $t = 0$ undergoes a constant acceleration $\vec{a}$ of magnitude $a = 3.0$ m/s$^2$ at an angle $\theta = 130°$ from the positive direction of the $x$ axis. What is the particle's velocity $\vec{v}$ at $t = 5.0$ s?

**KEY IDEA** Because the acceleration is constant, Eq. 2-11 ($v = v_0 + at$) applies, but we must use it separately for motion parallel to the $x$ axis and motion parallel to the $y$ axis.

**Calculations:** We find the velocity components $v_x$ and $v_y$ from the equations

$$v_x = v_{0x} + a_x t \quad \text{and} \quad v_y = v_{0y} + a_y t.$$

In these equations, $v_{0x}$ ($= -2.0$ m/s) and $v_{0y}$ ($= 4.0$ m/s) are the $x$ and $y$ components of $\vec{v}_0$, and $a_x$ and $a_y$ are the $x$ and $y$ components of $\vec{a}$. To find $a_x$ and $a_y$, we resolve $\vec{a}$ either with a vector-capable calculator or with Eq. 3-5:

$$a_x = a \cos \theta = (3.0 \text{ m/s}^2)(\cos 130°) = -1.93 \text{ m/s}^2,$$
$$a_y = a \sin \theta = (3.0 \text{ m/s}^2)(\sin 130°) = +2.30 \text{ m/s}^2.$$

When these values are inserted into the equations for $v_x$ and $v_y$, we find that, at time $t = 5.0$ s,

$$v_x = -2.0 \text{ m/s} + (-1.93 \text{ m/s}^2)(5.0 \text{ s}) = -11.65 \text{ m/s},$$
$$v_y = 4.0 \text{ m/s} + (2.30 \text{ m/s}^2)(5.0 \text{ s}) = 15.50 \text{ m/s}.$$

Thus, at $t = 5.0$ s, we have, after rounding,

$$\vec{v} = (-12 \text{ m/s})\hat{i} + (16 \text{ m/s})\hat{j}. \quad \text{(Answer)}$$

Either using a vector-capable calculator or following Eq. 3-6, we find that the magnitude and angle of $\vec{v}$ are

$$v = \sqrt{v_x^2 + v_y^2} = 19.4 \approx 19 \text{ m/s} \quad \text{(Answer)}$$

and

$$\theta = \tan^{-1} \frac{v_y}{v_x} = 127° \approx 130°. \quad \text{(Answer)}$$

**Check:** Does 127° appear on your calculator's display, or does $-53°$ appear? Now sketch the vector $\vec{v}$ with its components to see which angle is reasonable.

# 4-5 | Projectile Motion

We next consider a special case of two-dimensional motion: A particle moves in a vertical plane with some initial velocity $\vec{v}_0$ but its acceleration is always the free-fall acceleration $\vec{g}$, which is downward. Such a particle is called a **projectile** (meaning that it is projected or launched), and its motion is called **projectile motion.** A projectile might be a tennis ball (Fig. 4-9) or baseball in flight, but it is not an airplane or a duck in flight. Many sports (from golf and football to lacrosse and racquetball) involve the projectile motion of a ball, and much effort is spent in trying to control that motion for an advantage. For example, the racquetball player who discovered the Z-shot in the 1970s easily won his games because the ball's peculiar flight to the rear of the court always perplexed his opponents.

Our goal here is to analyze projectile motion using the tools for two-dimensional motion described in Sections 4-2 through 4-4 and making the assumption that air has no effect on the projectile. Figure 4-10, which is analyzed in the next section, shows the path followed by a projectile when the air has no effect. The projectile is launched with an initial velocity $\vec{v}_0$ that can be written as

$$\vec{v}_0 = v_{0x}\hat{i} + v_{0y}\hat{j}. \tag{4-19}$$

The components $v_{0x}$ and $v_{0y}$ can then be found if we know the angle $\theta_0$ between $\vec{v}_0$ and the positive $x$ direction:

$$v_{0x} = v_0 \cos\theta_0 \quad \text{and} \quad v_{0y} = v_0 \sin\theta_0. \tag{4-20}$$

During its two-dimensional motion, the projectile's position vector $\vec{r}$ and velocity vector $\vec{v}$ change continuously, but its acceleration vector $\vec{a}$ is constant and *always* directed vertically downward. The projectile has *no* horizontal acceleration.

Projectile motion, like that in Figs. 4-9 and 4-10, looks complicated, but we have the following simplifying feature (known from experiment):

> In projectile motion, the horizontal motion and the vertical motion are independent of each other; that is, neither motion affects the other.

This feature allows us to break up a problem involving two-dimensional motion into two separate and easier one-dimensional problems, one for the horizontal motion (with *zero acceleration*) and one for the vertical motion (with *constant downward acceleration*). Here are two experiments that show that the horizontal motion and the vertical motion are independent.

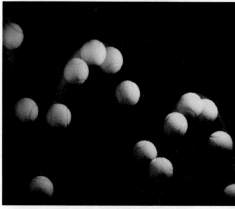

**FIG. 4-9** A stroboscopic photograph of a yellow tennis ball bouncing off a hard surface. Between impacts, the ball has projectile motion. *Source: Richard Megna/Fundamental Photographs.*

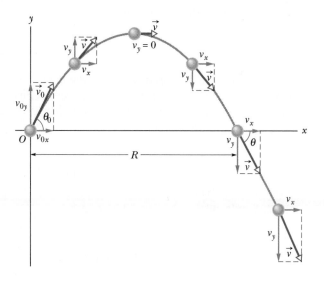

**FIG. 4-10** The path of a projectile that is launched at $x_0 = 0$ and $y_0 = 0$, with an initial velocity $\vec{v}_0$. The initial velocity and the velocities at various points along its path are shown, along with their components. Note that the horizontal velocity component remains constant but the vertical velocity component changes continuously. The *range R* is the horizontal distance the projectile has traveled *when it returns to its launch height.*

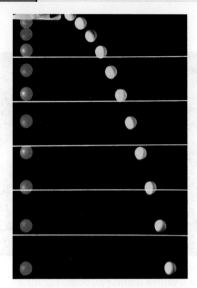

FIG. 4-11 One ball is released from rest at the same instant that another ball is shot horizontally to the right. Their vertical motions are identical. *Source:* Richard Megna/ Fundamental Photographs.

### Two Golf Balls

Figure 4-11 is a stroboscopic photograph of two golf balls, one simply released and the other shot horizontally by a spring. The golf balls have the same vertical motion, both falling through the same vertical distance in the same interval of time. *The fact that one ball is moving horizontally while it is falling has no effect on its vertical motion; that is, the horizontal and vertical motions are independent of each other.*

### A Great Student Rouser

Figure 4-12 shows a demonstration that has enlivened many a physics lecture. It involves a blowgun G, using a ball as a projectile. The target is a can suspended from a magnet M, and the tube of the blowgun is aimed directly at the can. The experiment is arranged so that the magnet releases the can just as the ball leaves the blowgun.

If g (the magnitude of the free-fall acceleration) were zero, the ball would follow the straight-line path shown in Fig. 4-12 and the can would float in place after the magnet released it. The ball would certainly hit the can.

However, g is *not* zero, but the ball *still* hits the can! As Fig. 4-12 shows, during the time of flight of the ball, both ball and can fall the same distance h from their zero-g locations. The harder the demonstrator blows, the greater is the ball's initial speed, the shorter the flight time, and the smaller the value of h.

✓**CHECKPOINT 3** At a certain instant, a fly ball has velocity $\vec{v} = 25\hat{i} - 4.9\hat{j}$ (the x axis is horizontal, the y axis is upward, and $\vec{v}$ is in meters per second). Has the ball passed its highest point?

## 4-6 | Projectile Motion Analyzed

Now we are ready to analyze projectile motion, horizontally and vertically.

### The Horizontal Motion

Because there is *no acceleration* in the horizontal direction, the horizontal component $v_x$ of the projectile's velocity remains unchanged from its initial value $v_{0x}$ throughout the motion, as demonstrated in Fig. 4-13. At any time t, the projectile's horizontal displacement $x - x_0$ from an initial position $x_0$ is given by Eq. 2-15 with a = 0, which we write as

$$x - x_0 = v_{0x}t.$$

Because $v_{0x} = v_0 \cos \theta_0$, this becomes

$$x - x_0 = (v_0 \cos \theta_0)t. \tag{4-21}$$

### The Vertical Motion

The vertical motion is the motion we discussed in Section 2-9 for a particle in free fall. Most important is that the acceleration is constant. Thus, the equations of Table 2-1 apply, provided we substitute $-g$ for a and switch to y notation. Then, for example, Eq. 2-15 becomes

$$y - y_0 = v_{0y}t - \tfrac{1}{2}gt^2$$
$$= (v_0 \sin \theta_0)t - \tfrac{1}{2}gt^2, \tag{4-22}$$

where the initial vertical velocity component $v_{0y}$ is replaced with the equivalent $v_0 \sin \theta_0$. Similarly, Eqs. 2-11 and 2-16 become

$$v_y = v_0 \sin \theta_0 - gt \tag{4-23}$$

and

$$v_y^2 = (v_0 \sin \theta_0)^2 - 2g(y - y_0). \tag{4-24}$$

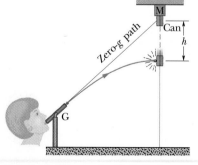

FIG. 4-12 The projectile ball always hits the falling can. Each falls a distance h from where it would be were there no free-fall acceleration.

As is illustrated in Fig. 4-10 and Eq. 4-23, the vertical velocity component behaves just as for a ball thrown vertically upward. It is directed upward initially, and its magnitude steadily decreases to zero, *which marks the maximum height of the path.* The vertical velocity component then reverses direction, and its magnitude becomes larger with time.

## The Equation of the Path

We can find the equation of the projectile's path (its **trajectory**) by eliminating time $t$ between Eqs. 4-21 and 4-22. Solving Eq. 4-21 for $t$ and substituting into Eq. 4-22, we obtain, after a little rearrangement,

$$y = (\tan \theta_0)x - \frac{gx^2}{2(v_0 \cos \theta_0)^2} \quad \text{(trajectory)}. \quad (4\text{-}25)$$

This is the equation of the path shown in Fig. 4-10. In deriving it, for simplicity we let $x_0 = 0$ and $y_0 = 0$ in Eqs. 4-21 and 4-22, respectively. Because $g$, $\theta_0$, and $v_0$ are constants, Eq. 4-25 is of the form $y = ax + bx^2$, in which $a$ and $b$ are constants. This is the equation of a parabola, so the path is *parabolic.*

## The Horizontal Range

The *horizontal range* $R$ of the projectile, as Fig. 4-10 shows, is the *horizontal distance the projectile has traveled when it returns to its initial (launch) height.* To find range $R$, let us put $x - x_0 = R$ in Eq. 4-21 and $y - y_0 = 0$ in Eq. 4-22, obtaining

$$R = (v_0 \cos \theta_0)t$$

and

$$0 = (v_0 \sin \theta_0)t - \tfrac{1}{2}gt^2.$$

Eliminating $t$ between these two equations yields

$$R = \frac{2v_0^2}{g} \sin \theta_0 \cos \theta_0.$$

Using the identity $\sin 2\theta_0 = 2 \sin \theta_0 \cos \theta_0$ (see Appendix E), we obtain

$$R = \frac{v_0^2}{g} \sin 2\theta_0. \quad (4\text{-}26)$$

*Caution:* This equation does *not* give the horizontal distance traveled by a projectile when the final height is not the launch height.

Note that $R$ in Eq. 4-26 has its maximum value when $\sin 2\theta_0 = 1$, which corresponds to $2\theta_0 = 90°$ or $\theta_0 = 45°$.

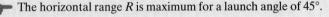 The horizontal range $R$ is maximum for a launch angle of 45°.

However, when the launch and landing heights differ, as in shot put, hammer throw, and basketball, a launch angle of 45° does not yield the maximum horizontal distance.

## The Effects of the Air

We have assumed that the air through which the projectile moves has no effect on its motion. However, in many situations, the disagreement between our calculations and the actual motion of the projectile can be large because the air resists (opposes) the motion. Figure 4-14, for example, shows two paths for a fly ball that leaves the bat at an angle of 60° with the horizontal and an initial speed of 44.7 m/s. Path I (the baseball player's fly ball) is a calculated path that approximates normal conditions of play, in air. Path II (the physics professor's fly ball) is the path the ball would follow in a vacuum.

**FIG. 4-13** The vertical component of this skateboarder's velocity is changing but not the horizontal component, which matches the skateboard's velocity. As a result, the skateboard stays underneath him, allowing him to land on it. *Source:* Jamie Budge/Liaison/ Getty Images, Inc.

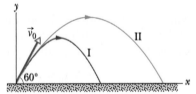

**FIG. 4-14** (I) The path of a fly ball calculated by taking air resistance into account. (II) The path the ball would follow in a vacuum, calculated by the methods of this chapter. See Table 4-1 for corresponding data. (Adapted from "The Trajectory of a Fly Ball," by Peter J. Brancazio, *The Physics Teacher,* January 1985.)

**TABLE 4-1**

**Two Fly Balls**[a]

|  | Path I (Air) | Path II (Vacuum) |
|---|---|---|
| Range | 98.5 m | 177 m |
| Maximum height | 53.0 m | 76.8 m |
| Time of flight | 6.6 s | 7.9 s |

[a]See Fig. 4-14. The launch angle is 60° and the launch speed is 44.7 m/s.

✓ CHECKPOINT 4    A fly ball is hit to the outfield. During its flight (ignore the effects of the air), what happens to its (a) horizontal and (b) vertical components of velocity? What are the (c) horizontal and (d) vertical components of its acceleration during ascent, during descent, and at the topmost point of its flight?

## Sample Problem | 4-6 | Build your skill

In Fig. 4-15, a rescue plane flies at 198 km/h (= 55.0 m/s) and constant height $h = 500$ m toward a point directly over a victim, where a rescue capsule is to land.

(a) What should be the angle $\phi$ of the pilot's line of sight to the victim when the capsule release is made?

**KEY IDEAS**    Once released, the capsule is a projectile, so its horizontal and vertical motions can be considered separately (we need not consider the actual curved path of the capsule).

**Calculations:** In Fig. 4-15, we see that $\phi$ is given by

$$\phi = \tan^{-1} \frac{x}{h}, \qquad (4\text{-}27)$$

where $x$ is the horizontal coordinate of the victim (and of the capsule when it hits the water) and $h = 500$ m. We should be able to find $x$ with Eq. 4-21:

$$x - x_0 = (v_0 \cos \theta_0)t. \qquad (4\text{-}28)$$

Here we know that $x_0 = 0$ because the origin is placed at the point of release. Because the capsule is *released* and not shot from the plane, its initial velocity $\vec{v}_0$ is equal to the plane's velocity. Thus, we know also that the initial velocity has magnitude $v_0 = 55.0$ m/s and angle $\theta_0 = 0°$ (measured relative to the positive direction of the $x$ axis). However, we do not know the time $t$ the capsule takes to move from the plane to the victim.

To find $t$, we next consider the *vertical* motion and specifically Eq. 4-22:

$$y - y_0 = (v_0 \sin \theta_0)t - \tfrac{1}{2}gt^2. \qquad (4\text{-}29)$$

Here the vertical displacement $y - y_0$ of the capsule is $-500$ m (the negative value indicates that the capsule moves *downward*). So,

$$-500 \text{ m} = (55.0 \text{ m/s})(\sin 0°)t - \tfrac{1}{2}(9.8 \text{ m/s}^2)t^2.$$

Solving for $t$, we find $t = 10.1$ s. Using that value in Eq. 4-28 yields

$$x - 0 = (55.0 \text{ m/s})(\cos 0°)(10.1 \text{ s}),$$

or    $x = 555.5$ m.

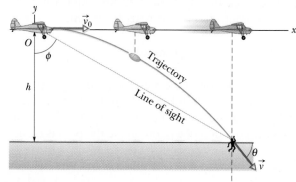

**FIG. 4-15**  A plane drops a rescue capsule while moving at constant velocity in level flight. While falling, the capsule remains under the plane.

Then Eq. 4-27 gives us

$$\phi = \tan^{-1} \frac{555.5 \text{ m}}{500 \text{ m}} = 48.0°. \qquad \text{(Answer)}$$

(b) As the capsule reaches the water, what is its velocity $\vec{v}$ in unit-vector notation and in magnitude-angle notation?

**KEY IDEAS**    (1) The horizontal and vertical components of the capsule's velocity are independent. (2) Component $v_x$ does not change from its initial value $v_{0x} = v_0 \cos \theta_0$ because there is no horizontal acceleration. (3) Component $v_y$ changes from its initial value $v_{0y} = v_0 \sin \theta_0$ because there is a vertical acceleration.

**Calculations:** When the capsule reaches the water,

$$v_x = v_0 \cos \theta_0 = (55.0 \text{ m/s})(\cos 0°) = 55.0 \text{ m/s}.$$

Using Eq. 4-23 and the capsule's time of fall $t = 10.1$ s, we also find that when the capsule reaches the water,

$$v_y = v_0 \sin \theta_0 - gt \qquad (4\text{-}30)$$
$$= (55.0 \text{ m/s})(\sin 0°) - (9.8 \text{ m/s}^2)(10.1 \text{ s})$$
$$= -99.0 \text{ m/s}.$$

Thus, at the water

$$\vec{v} = (55.0 \text{ m/s})\hat{i} - (99.0 \text{ m/s})\hat{j}. \qquad \text{(Answer)}$$

Using Eq. 3-6 as a guide, we find that the magnitude and the angle of $\vec{v}$ are

$$v = 113 \text{ m/s} \quad \text{and} \quad \theta = -60.9°. \qquad \text{(Answer)}$$

## Sample Problem    4-7

Figure 4-16 shows a pirate ship 560 m from a fort defending a harbor entrance. A defense cannon, located at sea level, fires balls at initial speed $v_0 = 82$ m/s.

(a) At what angle $\theta_0$ from the horizontal must a ball be fired to hit the ship?

**KEY IDEAS** (1) A fired cannonball is a projectile. We want an equation that relates the launch angle $\theta_0$ to the ball's horizontal displacement as it moves from cannon to ship. (2) Because the cannon and the ship are at the same height, the horizontal displacement is the range.

**Calculations:** We can relate the launch angle $\theta_0$ to the range $R$ with Eq. 4-26 ($R = (v_0^2/g) \sin 2\theta_0$), which, after rearrangement, gives

$$\theta_0 = \frac{1}{2} \sin^{-1} \frac{gR}{v_0^2} = \frac{1}{2} \sin^{-1} \frac{(9.8 \text{ m/s}^2)(560 \text{ m})}{(82 \text{ m/s})^2}$$

$$= \frac{1}{2} \sin^{-1} 0.816. \qquad (4\text{-}31)$$

One solution of $\sin^{-1}$ (54.7°) is displayed by a calculator; we subtract it from 180° to get the other solution (125.3°). Thus, Eq. 4-31 gives us

$$\theta_0 = 27° \quad \text{and} \quad \theta_0 = 63°. \quad \text{(Answer)}$$

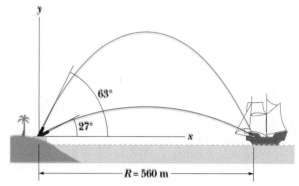

**FIG. 4-16**    A pirate ship under fire.

(b) What is the maximum range of the cannonballs?

**Calculations:** We have seen that maximum range corresponds to an elevation angle $\theta_0$ of 45°. Thus,

$$R = \frac{v_0^2}{g} \sin 2\theta_0 = \frac{(82 \text{ m/s})^2}{9.8 \text{ m/s}^2} \sin (2 \times 45°)$$

$$= 686 \text{ m} \approx 690 \text{ m.} \qquad \text{(Answer)}$$

As the pirate ship sails away, the two elevation angles at which the ship can be hit draw together, eventually merging at $\theta_0 = 45°$ when the ship is 690 m away. Beyond that distance the ship is safe.

## Sample Problem    4-8

Suppose a baseball batter $B$ hits a high fly ball to the outfield, directly toward an outfielder $F$ and with a launch speed of $v_0 = 40$ m/s and a launch angle of $\theta_0 = 35°$. During the flight, a line from the outfielder to the ball makes an angle $\phi$ with the ground. Plot elevation angle $\phi$ versus time $t$, assuming that the outfielder is already positioned to catch the ball, is 6.0 m too close to the batter, and is 6.0 m too far away.

**KEY IDEAS** (1) If we neglect air drag, the ball is a projectile for which the vertical motion and the horizontal motion can be analyzed separately. (2) Assuming the ball is caught at approximately the height it is hit, the horizontal distance traveled by the ball is the range $R$, given by Eq. 4-26 ($R = (v_0^2/g) \sin 2\theta_0$).

**Calculations:** The ball can be caught if the outfielder's distance from the batter is equal to the range $R$ of the ball. Using Eq. 4-26, we find

$$R = \frac{v_0^2}{g} \sin 2\theta_0 = \frac{(40 \text{ m/s})^2}{9.8 \text{ m/s}^2} \sin (70°) = 153.42 \text{ m.}$$

$$(4\text{-}32)$$

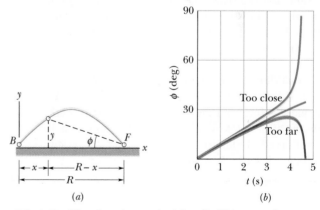

**FIG. 4-17**    The elevation angle $\phi$ for a ball hit toward an outfielder is (a) defined and (b) plotted versus time $t$.

Figure 4-17a shows a snapshot of the ball in flight when the ball is at height $y$ and horizontal distance $x$ from the batter (who is at the origin). The horizontal distance of the ball from the outfielder is $R - x$, and the elevation angle $\phi$ of the ball in the outfielder's view is given by $\tan \phi = y/(R - x)$. For the height $y$, we use Eq. 4-22 ($y - y_0 = (v_0 \sin \theta_0)t - \frac{1}{2}gt^2$), setting $y_0 = 0$. For the horizontal distance $x$, we substitute with Eq. 4-21 ($x - x_0 = (v_0 \cos \theta_0)t$), setting $x_0 = 0$. Thus, using $v_0 = 40$ m/s and $\theta_0 = 35°$, we have

$$\phi = \tan^{-1} \frac{(40 \sin 35°)t - 4.9t^2}{153.42 - (40 \cos 35°)t}. \qquad (4\text{-}33)$$

Graphing this function versus $t$ gives the middle plot in Fig. 4-17b. We see that the ball's angle in the outfielder's view increases at an almost steady rate throughout the flight.

If the outfielder is 6.0 m too close to the batter, we replace the distance of 153.42 m in Eq. 4-33 with

153.42 m − 6.0 m = 147.42 m. Regraphing the function gives the "Too close" plot in Fig. 4-17b. Now the elevation angle of the ball rapidly increases toward the end of the flight as the ball soars over the outfielder's head. If the outfielder is 6.0 m too far away from the batter, we replace the distance of 153.42 m in Eq. 4-33 with 159.42 m. The resulting plot is labeled "Too far" in the figure: The angle first increases and then rapidly decreases. Thus, if a ball is hit directly toward an outfielder, the player can tell from the change in the ball's elevation angle $\phi$ whether to stay put, run toward the batter, or back away from the batter.

---

## Sample Problem    4-9    Build your skill

At time $t = 0$, a golf ball is shot from ground level into the air, as indicated in Fig. 4-18a. The angle $\theta$ between the ball's direction of travel and the positive direction of the x axis is given in Fig. 4-18b as a function of time $t$. The ball lands at $t = 6.00$ s. What is the magnitude $v_0$ of the ball's launch velocity, at what height ($y - y_0$) above the launch level does the ball land, and what is the ball's direction of travel just as it lands?

**KEY IDEAS**    (1) The ball is a projectile, and so its horizontal and vertical motions can be considered separately. (2) The horizontal component $v_x$ ($= v_0 \cos \theta_0$) of the ball's velocity does not change during the flight. (3) The vertical component $v_y$ of its velocity *does* change and is zero when the ball reaches maximum height. (4) The ball's direction of travel at any time during the flight is at the angle of its velocity vector $\vec{v}$ just then. That angle is given by $\tan \theta = v_y/v_x$, with the velocity components evaluated at that time.

*Calculations:* When the ball reaches its maximum height, $v_y = 0$. So, the direction of the velocity $\vec{v}$ is horizontal, at angle $\theta = 0°$. From the graph, we see that this condition occurs at $t = 4.0$ s. We also see that the launch angle $\theta_0$ (at $t = 0$) is 80°. Using Eq. 4-23 ($v_y = v_0 \sin \theta_0 - gt$), with $t = 4.0$ s, $g = 9.8$ m/s$^2$, $\theta_0 = 80°$, and $v_y = 0$, we find

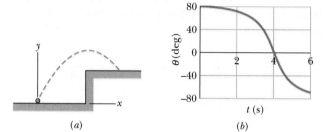

FIG. 4-18    (a) Path of a golf ball shot onto a plateau. (b) The angle $\theta$ that gives the ball's direction of motion during the flight is plotted versus time $t$.

$$v_0 = 39.80 \approx 40 \text{ m/s.} \qquad \text{(Answer)}$$

The ball lands at $t = 6.00$ s. Using Eq. 4-22 ($y - y_0 = (v_0 \sin \theta_0)t - \frac{1}{2}gt^2$) with $t = 6.00$ s, we obtain

$$y - y_0 = 58.77 \text{ m} \approx 59 \text{ m.} \qquad \text{(Answer)}$$

Just as the ball lands, its horizontal velocity $v_x$ is still $v_0 \cos \theta_0$; substituting for $v_0$ and $\theta_0$ gives us $v_x = 6.911$ m/s. We find its vertical velocity just then by using Eq. 4-23 ($v_y = v_0 \sin \theta_0 - gt$) with $t = 6.00$ s, which yields $v_y = -19.60$ m/s. Thus, the angle of the ball's direction of travel at landing is

$$\theta = \tan^{-1} \frac{v_y}{v_x} = \tan^{-1} \frac{-19.60 \text{ m/s}}{6.911 \text{ m/s}} \approx -71°. \qquad \text{(Answer)}$$

---

## 4-7 | Uniform Circular Motion

A particle is in **uniform circular motion** if it travels around a circle or a circular arc at constant (*uniform*) speed. Although the speed does not vary, *the particle is accelerating* because the velocity changes in direction.

Figure 4-19 shows the relationship between the velocity and acceleration vectors at various stages during uniform circular motion. Both vectors have constant magnitude, but their directions change continuously. The velocity is always directed tangent to the circle in the direction of motion. The acceleration is always directed *radially inward*. Because of this, the acceleration associated with uniform circular motion is called a **centripetal** (meaning "center seeking")

**acceleration.** As we prove next, the magnitude of this acceleration $\vec{a}$ is

$$a = \frac{v^2}{r} \quad \text{(centripetal acceleration)}, \quad (4\text{-}34)$$

where $r$ is the radius of the circle and $v$ is the speed of the particle.

In addition, during this acceleration at constant speed, the particle travels the circumference of the circle (a distance of $2\pi r$) in time

$$T = \frac{2\pi r}{v} \quad \text{(period)}. \quad (4\text{-}35)$$

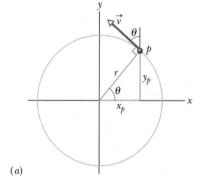

**FIG. 4-19** Velocity and acceleration vectors for uniform circular motion.

$T$ is called the *period of revolution*, or simply the *period*, of the motion. It is, in general, the time for a particle to go around a closed path exactly once.

## Proof of Eq. 4-34

To find the magnitude and direction of the acceleration for uniform circular motion, we consider Fig. 4-20. In Fig. 4-20a, particle $p$ moves at constant speed $v$ around a circle of radius $r$. At the instant shown, $p$ has coordinates $x_p$ and $y_p$.

Recall from Section 4-3 that the velocity $\vec{v}$ of a moving particle is always tangent to the particle's path at the particle's position. In Fig. 4-20a, that means $\vec{v}$ is perpendicular to a radius $r$ drawn to the particle's position. Then the angle $\theta$ that $\vec{v}$ makes with a vertical at $p$ equals the angle $\theta$ that radius $r$ makes with the $x$ axis.

The scalar components of $\vec{v}$ are shown in Fig. 4-20b. With them, we can write the velocity $\vec{v}$ as

$$\vec{v} = v_x\hat{i} + v_y\hat{j} = (-v\sin\theta)\hat{i} + (v\cos\theta)\hat{j}. \quad (4\text{-}36)$$

Now, using the right triangle in Fig. 4-20a, we can replace $\sin\theta$ with $y_p/r$ and $\cos\theta$ with $x_p/r$ to write

$$\vec{v} = \left(-\frac{vy_p}{r}\right)\hat{i} + \left(\frac{vx_p}{r}\right)\hat{j}. \quad (4\text{-}37)$$

To find the acceleration $\vec{a}$ of particle $p$, we must take the time derivative of this equation. Noting that speed $v$ and radius $r$ do not change with time, we obtain

$$\vec{a} = \frac{d\vec{v}}{dt} = \left(-\frac{v}{r}\frac{dy_p}{dt}\right)\hat{i} + \left(\frac{v}{r}\frac{dx_p}{dt}\right)\hat{j}. \quad (4\text{-}38)$$

Now note that the rate $dy_p/dt$ at which $y_p$ changes is equal to the velocity component $v_y$. Similarly, $dx_p/dt = v_x$, and, again from Fig. 4-20b, we see that $v_x = -v\sin\theta$ and $v_y = v\cos\theta$. Making these substitutions in Eq. 4-38, we find

$$\vec{a} = \left(-\frac{v^2}{r}\cos\theta\right)\hat{i} + \left(-\frac{v^2}{r}\sin\theta\right)\hat{j}. \quad (4\text{-}39)$$

This vector and its components are shown in Fig. 4-20c. Following Eq. 3-6, we find

$$a = \sqrt{a_x^2 + a_y^2} = \frac{v^2}{r}\sqrt{(\cos\theta)^2 + (\sin\theta)^2} = \frac{v^2}{r}\sqrt{1} = \frac{v^2}{r},$$

as we wanted to prove. To orient $\vec{a}$, we find the angle $\phi$ shown in Fig. 4-20c:

$$\tan\phi = \frac{a_y}{a_x} = \frac{-(v^2/r)\sin\theta}{-(v^2/r)\cos\theta} = \tan\theta.$$

Thus, $\phi = \theta$, which means that $\vec{a}$ is directed along the radius $r$ of Fig. 4-20a, toward the circle's center, as we wanted to prove.

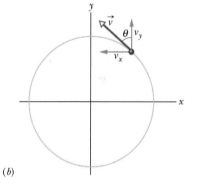

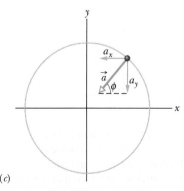

**FIG. 4-20** Particle $p$ moves in counterclockwise uniform circular motion. (a) Its position and velocity $\vec{v}$ at a certain instant. (b) Velocity $\vec{v}$. (c) Acceleration $\vec{a}$.

✓**CHECKPOINT 5**    An object moves at constant speed along a circular path in a horizontal $xy$ plane, with the center at the origin. When the object is at $x = -2$ m, its velocity is $-(4\text{ m/s})\hat{j}$. Give the object's (a) velocity and (b) acceleration at $y = 2$ m.

**Sample Problem** | **4-10**

"Top gun" pilots have long worried about taking a turn too tightly. As a pilot's body undergoes centripetal acceleration, with the head toward the center of curvature, the blood pressure in the brain decreases, leading to loss of brain function.

There are several warning signs. When the centripetal acceleration is 2g or 3g, the pilot feels heavy. At about 4g, the pilot's vision switches to black and white and narrows to "tunnel vision." If that acceleration is sustained or increased, vision ceases and, soon after, the pilot is unconscious—a condition known as g-LOC for "g-induced loss of consciousness."

What is the magnitude of the acceleration, in g units, of a pilot whose aircraft enters a horizontal circular turn with a velocity of $\vec{v}_i = (400\hat{i} + 500\hat{j})$ m/s and 24.0 s later leaves the turn with a velocity of $\vec{v}_f = (-400\hat{i} - 500\hat{j})$ m/s?

**KEY IDEAS** We assume the turn is made with uniform circular motion. Then the pilot's acceleration is centripetal and has magnitude $a$ given by Eq. 4-34 ($a = v^2/R$), where $R$ is the circle's radius. Also, the time

required to complete a full circle is the period given by Eq. 4-35 ($T = 2\pi R/v$).

**Calculations:** Because we do not know radius $R$, let's solve Eq. 4-35 for $R$ and substitute into Eq. 4-34. We find

$$a = \frac{2\pi v}{T}.$$

Speed $v$ here is the (constant) magnitude of the velocity during the turning. Let's substitute the components of the initial velocity into Eq. 3-6:

$$v = \sqrt{(400 \text{ m/s})^2 + (500 \text{ m/s})^2} = 640.31 \text{ m/s}.$$

To find the period $T$ of the motion, first note that the final velocity is the reverse of the initial velocity. This means the aircraft leaves on the opposite side of the circle from the initial point and must have completed half a circle in the given 24.0 s. Thus a full circle would have taken $T = 48.0$ s. Substituting these values into our equation for $a$, we find

$$a = \frac{2\pi(640.31 \text{ m/s})}{48.0 \text{ s}} = 83.81 \text{ m/s}^2 \approx 8.6g. \quad \text{(Answer)}$$

# 4-8 | Relative Motion in One Dimension

Suppose you see a duck flying north at 30 km/h. To another duck flying alongside, the first duck seems to be stationary. In other words, the velocity of a particle depends on the **reference frame** of whoever is observing or measuring the velocity. For our purposes, a reference frame is the physical object to which we attach our coordinate system. In everyday life, that object is the ground. For example, the speed listed on a speeding ticket is always measured relative to the ground. The speed relative to the police officer would be different if the officer were moving while making the speed measurement.

Suppose that Alex (at the origin of frame $A$ in Fig. 4-21) is parked by the side of a highway, watching car $P$ (the "particle") speed past. Barbara (at the origin of frame $B$) is driving along the highway at constant speed and is also watching car $P$. Suppose that they both measure the position of the car at a given moment. From Fig. 4-21 we see that

$$x_{PA} = x_{PB} + x_{BA}. \quad (4\text{-}40)$$

The equation is read: "The coordinate $x_{PA}$ of $P$ as measured by $A$ *is equal to* the coordinate $x_{PB}$ of $P$ as measured by $B$ *plus* the coordinate $x_{BA}$ of $B$ as measured by $A$." Note how this reading is supported by the sequence of the subscripts.

Taking the time derivative of Eq. 4-40, we obtain

$$\frac{d}{dt}(x_{PA}) = \frac{d}{dt}(x_{PB}) + \frac{d}{dt}(x_{BA}).$$

Thus, the velocity components are related by

$$v_{PA} = v_{PB} + v_{BA}. \quad (4\text{-}41)$$

This equation is read: "The velocity $v_{PA}$ of $P$ as measured by $A$ *is equal to* the velocity $v_{PB}$ of $P$ as measured by $B$ *plus* the velocity $v_{BA}$ of $B$ as measured by $A$." The term $v_{BA}$ is the velocity of frame $B$ relative to frame $A$.

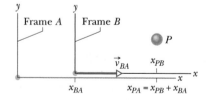

**FIG. 4-21** Alex (frame $A$) and Barbara (frame $B$) watch car $P$, as both $B$ and $P$ move at different velocities along the common $x$ axis of the two frames. At the instant shown, $x_{BA}$ is the coordinate of $B$ in the $A$ frame. Also, $P$ is at coordinate $x_{PB}$ in the $B$ frame and coordinate $x_{PA} = x_{PB} + x_{BA}$ in the $A$ frame.

Here we consider only frames that move at constant velocity relative to each other. In our example, this means that Barbara (frame $B$) drives always at constant velocity $v_{BA}$ relative to Alex (frame $A$). Car $P$ (the moving particle), however, can change speed and direction (that is, it can accelerate).

To relate an acceleration of $P$ as measured by Barbara and by Alex, we take the time derivative of Eq. 4-41:

$$\frac{d}{dt}(v_{PA}) = \frac{d}{dt}(v_{PB}) + \frac{d}{dt}(v_{BA}).$$

Because $v_{BA}$ is constant, the last term is zero and we have

$$a_{PA} = a_{PB}. \tag{4-42}$$

In other words,

> Observers on different frames of reference that move at constant velocity relative to each other will measure the same acceleration for a moving particle.

## Sample Problem | 4-11

In Fig. 4-21, suppose that Barbara's velocity relative to Alex is a constant $v_{BA} = 52$ km/h and car $P$ is moving in the negative direction of the $x$ axis.

(a) If Alex measures a constant $v_{PA} = -78$ km/h for car $P$, what velocity $v_{PB}$ will Barbara measure?

**KEY IDEAS** We can attach a frame of reference $A$ to Alex and a frame of reference $B$ to Barbara. Because the frames move at constant velocity relative to each other along one axis, we can use Eq. 4-41 ($v_{PA} = v_{PB} + v_{BA}$) to relate $v_{PB}$ to $v_{PA}$ and $v_{BA}$.

**Calculation:** We find

$$-78 \text{ km/h} = v_{PB} + 52 \text{ km/h}.$$

Thus, $\qquad v_{PB} = -130$ km/h. (Answer)

**Comment:** If car $P$ were connected to Barbara's car by a cord wound on a spool, the cord would be unwinding at a speed of 130 km/h as the two cars separated.

(b) If car $P$ brakes to a stop relative to Alex (and thus relative to the ground) in time $t = 10$ s at constant acceleration, what is its acceleration $a_{PA}$ relative to Alex?

**KEY IDEAS** To calculate the acceleration of car $P$ relative to Alex, we must use the car's velocities relative to Alex. Because the acceleration is constant, we can use

Eq. 2-11 ($v = v_0 + at$) to relate the acceleration to the initial and final velocities of $P$.

**Calculation:** The initial velocity of $P$ relative to Alex is $v_{PA} = -78$ km/h and the final velocity is 0. Thus,

$$a_{PA} = \frac{v - v_0}{t} = \frac{0 - (-78 \text{ km/h})}{10 \text{ s}} \frac{1 \text{ m/s}}{3.6 \text{ km/h}}$$
$$= 2.2 \text{ m/s}^2. \qquad \text{(Answer)}$$

(c) What is the acceleration $a_{PB}$ of car $P$ relative to Barbara during the braking?

**KEY IDEA** To calculate the acceleration of car $P$ relative to Barbara, we must use the car's velocities relative to Barbara.

**Calculation:** We know the initial velocity of $P$ relative to Barbara from part (a) ($v_{PB} = -130$ km/h). The final velocity of $P$ relative to Barbara is $-52$ km/h (this is the velocity of the stopped car relative to the moving Barbara). Thus,

$$a_{PB} = \frac{v - v_0}{t} = \frac{-52 \text{ km/h} - (-130 \text{ km/h})}{10 \text{ s}} \frac{1 \text{ m/s}}{3.6 \text{ km/h}}$$
$$= 2.2 \text{ m/s}^2. \qquad \text{(Answer)}$$

**Comment:** We should have foreseen this result: Because Alex and Barbara have a constant relative velocity, they must measure the same acceleration for the car.

## 4-9 | Relative Motion in Two Dimensions

Our two observers are again watching a moving particle $P$ from the origins of reference frames $A$ and $B$, while $B$ moves at a constant velocity $\vec{v}_{BA}$ relative to $A$. (The corresponding axes of these two frames remain parallel.) Figure 4-22 shows a certain instant during the motion. At that instant, the position vector of the origin of $B$

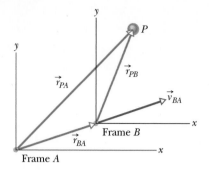

**FIG. 4-22** Frame $B$ has the constant two-dimensional velocity $\vec{v}_{BA}$ relative to frame $A$. The position vector of $B$ relative to $A$ is $\vec{r}_{BA}$. The position vectors of particle $P$ are $\vec{r}_{PA}$ relative to $A$ and $\vec{r}_{PB}$ relative to $B$.

relative to the origin of $A$ is $\vec{r}_{BA}$. Also, the position vectors of particle $P$ are $\vec{r}_{PA}$ relative to the origin of $A$ and $\vec{r}_{PB}$ relative to the origin of $B$. From the arrangement of heads and tails of those three position vectors, we can relate the vectors with

$$\vec{r}_{PA} = \vec{r}_{PB} + \vec{r}_{BA}. \tag{4-43}$$

By taking the time derivative of this equation, we can relate the velocities $\vec{v}_{PA}$ and $\vec{v}_{PB}$ of particle $P$ relative to our observers:

$$\vec{v}_{PA} = \vec{v}_{PB} + \vec{v}_{BA}. \tag{4-44}$$

By taking the time derivative of this relation, we can relate the accelerations $\vec{a}_{PA}$ and $\vec{a}_{PB}$ of the particle $P$ relative to our observers. However, note that because $\vec{v}_{BA}$ is constant, its time derivative is zero. Thus, we get

$$\vec{a}_{PA} = \vec{a}_{PB}. \tag{4-45}$$

As for one-dimensional motion, we have the following rule: Observers on different frames of reference that move at constant velocity relative to each other will measure the *same* acceleration for a moving particle.

## Sample Problem 4-12

In Fig. 4-23a, a plane moves due east while the pilot points the plane somewhat south of east, toward a steady wind that blows to the northeast. The plane has velocity $\vec{v}_{PW}$ relative to the wind, with an airspeed (speed relative to the wind) of 215 km/h, directed at angle $\theta$ south of east. The wind has velocity $\vec{v}_{WG}$ relative to the ground with speed 65.0 km/h, directed 20.0° east of north. What is the magnitude of the velocity $\vec{v}_{PG}$ of the plane relative to the ground, and what is $\theta$?

**KEY IDEAS** The situation is like the one in Fig. 4-22. Here the moving particle $P$ is the plane, frame $A$ is attached to the ground (call it $G$), and frame $B$ is "attached" to the wind (call it $W$). We need a vector diagram like Fig. 4-22 but with three velocity vectors.

**Calculations:** First we construct a sentence that relates the three vectors shown in Fig. 4-23b:

$$\begin{array}{c}\text{velocity of plane}\\\text{relative to ground}\\(PG)\end{array} = \begin{array}{c}\text{velocity of plane}\\\text{relative to wind}\\(PW)\end{array} + \begin{array}{c}\text{velocity of wind}\\\text{relative to ground.}\\(WG)\end{array}$$

This relation is written in vector notation as

$$\vec{v}_{PG} = \vec{v}_{PW} + \vec{v}_{WG}. \tag{4-46}$$

We need to resolve the vectors into components on the coordinate system of Fig. 4-23b and then solve Eq. 4-46 axis by axis. For the $y$ components, we find

$$v_{PG,y} = v_{PW,y} + v_{WG,y}$$

or $\quad 0 = -(215 \text{ km/h}) \sin \theta + (65.0 \text{ km/h})(\cos 20.0°).$

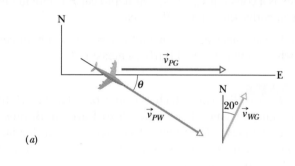

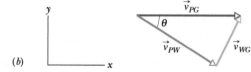

(a)

(b)

**FIG. 4-23** A plane flying in a wind.

Solving for $\theta$ gives us

$$\theta = \sin^{-1} \frac{(65.0 \text{ km/h})(\cos 20.0°)}{215 \text{ km/h}} = 16.5°. \quad \text{(Answer)}$$

Similarly, for the $x$ components we find

$$v_{PG,x} = v_{PW,x} + v_{WG,x}.$$

Here, because $\vec{v}_{PG}$ is parallel to the $x$ axis, the component $v_{PG,x}$ is equal to the magnitude $v_{PG}$. Substituting this notation and the value $\theta = 16.5°$, we find

$$v_{PG} = (215 \text{ km/h})(\cos 16.5°) + (65.0 \text{ km/h})(\sin 20.0°)$$
$$= 228 \text{ km/h}. \quad \text{(Answer)}$$

# REVIEW & SUMMARY

**Position Vector**   The location of a particle relative to the origin of a coordinate system is given by a *position vector* $\vec{r}$, which in unit-vector notation is

$$\vec{r} = x\hat{i} + y\hat{j} + z\hat{k}. \qquad (4\text{-}1)$$

Here $x\hat{i}$, $y\hat{j}$, and $z\hat{k}$ are the vector components of position vector $\vec{r}$, and $x$, $y$, and $z$ are its scalar components (as well as the coordinates of the particle). A position vector is described either by a magnitude and one or two angles for orientation, or by its vector or scalar components.

**Displacement**   If a particle moves so that its position vector changes from $\vec{r}_1$ to $\vec{r}_2$, the particle's *displacement* $\Delta\vec{r}$ is

$$\Delta\vec{r} = \vec{r}_2 - \vec{r}_1. \qquad (4\text{-}2)$$

The displacement can also be written as

$$\Delta\vec{r} = (x_2 - x_1)\hat{i} + (y_2 - y_1)\hat{j} + (z_2 - z_1)\hat{k} \qquad (4\text{-}3)$$

$$= \Delta x\hat{i} + \Delta y\hat{j} + \Delta z\hat{k}. \qquad (4\text{-}4)$$

**Average Velocity and Instantaneous Velocity**   If a particle undergoes a displacement $\Delta\vec{r}$ in time interval $\Delta t$, its *average velocity* $\vec{v}_{\text{avg}}$ for that time interval is

$$\vec{v}_{\text{avg}} = \frac{\Delta\vec{r}}{\Delta t}. \qquad (4\text{-}8)$$

As $\Delta t$ in Eq. 4-8 is shrunk to 0, $\vec{v}_{\text{avg}}$ reaches a limit called either the *velocity* or the *instantaneous velocity* $\vec{v}$:

$$\vec{v} = \frac{d\vec{r}}{dt}, \qquad (4\text{-}10)$$

which can be rewritten in unit-vector notation as

$$\vec{v} = v_x\hat{i} + v_y\hat{j} + v_z\hat{k}, \qquad (4\text{-}11)$$

where $v_x = dx/dt$, $v_y = dy/dt$, and $v_z = dz/dt$. The instantaneous velocity $\vec{v}$ of a particle is always directed along the tangent to the particle's path at the particle's position.

**Average Acceleration and Instantaneous Acceleration**   If a particle's velocity changes from $\vec{v}_1$ to $\vec{v}_2$ in time interval $\Delta t$, its *average acceleration* during $\Delta t$ is

$$\vec{a}_{\text{avg}} = \frac{\vec{v}_2 - \vec{v}_1}{\Delta t} = \frac{\Delta\vec{v}}{\Delta t}. \qquad (4\text{-}15)$$

As $\Delta t$ in Eq. 4-15 is shrunk to 0, $\vec{a}_{\text{avg}}$ reaches a limiting value called either the *acceleration* or the *instantaneous acceleration* $\vec{a}$:

$$\vec{a} = \frac{d\vec{v}}{dt}. \qquad (4\text{-}16)$$

In unit-vector notation,

$$\vec{a} = a_x\hat{i} + a_y\hat{j} + a_z\hat{k}, \qquad (4\text{-}17)$$

where $a_x = dv_x/dt$, $a_y = dv_y/dt$, and $a_z = dv_z/dt$.

**Projectile Motion**   *Projectile motion* is the motion of a particle that is launched with an initial velocity $\vec{v}_0$. During its flight, the particle's horizontal acceleration is zero and its vertical acceleration is the free-fall acceleration $-g$. (Upward is taken to be a positive direction.) If $\vec{v}_0$ is expressed as a magnitude (the speed $v_0$) and an angle $\theta_0$ (measured from the horizontal), the particle's equations of motion along the horizontal $x$ axis and vertical $y$ axis are

$$x - x_0 = (v_0 \cos\theta_0)t, \qquad (4\text{-}21)$$

$$y - y_0 = (v_0 \sin\theta_0)t - \tfrac{1}{2}gt^2, \qquad (4\text{-}22)$$

$$v_y = v_0 \sin\theta_0 - gt, \qquad (4\text{-}23)$$

$$v_y^2 = (v_0 \sin\theta_0)^2 - 2g(y - y_0). \qquad (4\text{-}24)$$

The **trajectory** (path) of a particle in projectile motion is parabolic and is given by

$$y = (\tan\theta_0)x - \frac{gx^2}{2(v_0 \cos\theta_0)^2}, \qquad (4\text{-}25)$$

if $x_0$ and $y_0$ of Eqs. 4-21 to 4-24 are zero. The particle's **horizontal range** $R$, which is the horizontal distance from the launch point to the point at which the particle returns to the launch height, is

$$R = \frac{v_0^2}{g} \sin 2\theta_0. \qquad (4\text{-}26)$$

**Uniform Circular Motion**   If a particle travels along a circle or circular arc of radius $r$ at constant speed $v$, it is said to be in *uniform circular motion* and has an acceleration $\vec{a}$ of constant magnitude

$$a = \frac{v^2}{r}. \qquad (4\text{-}34)$$

The direction of $\vec{a}$ is toward the center of the circle or circular arc, and $\vec{a}$ is said to be *centripetal*. The time for the particle to complete a circle is

$$T = \frac{2\pi r}{v}. \qquad (4\text{-}35)$$

$T$ is called the *period of revolution*, or simply the *period*, of the motion.

**Relative Motion**   When two frames of reference $A$ and $B$ are moving relative to each other at constant velocity, the velocity of a particle $P$ as measured by an observer in frame $A$ usually differs from that measured from frame $B$. The two measured velocities are related by

$$\vec{v}_{PA} = \vec{v}_{PB} + \vec{v}_{BA}, \qquad (4\text{-}44)$$

where $\vec{v}_{BA}$ is the velocity of $B$ with respect to $A$. Both observers measure the same acceleration for the particle:

$$\vec{a}_{PA} = \vec{a}_{PB}. \qquad (4\text{-}45)$$

## QUESTIONS

**1** Figure 4-24 shows the initial position $i$ and the final position $f$ of a particle. What are the (a) initial position vector $\vec{r}_i$ and (b) final position vector $\vec{r}_f$, both in unit-vector notation? (c) What is the $x$ component of displacement $\Delta\vec{r}$?

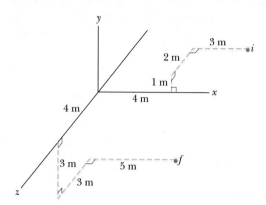

FIG. 4-24   Question 1.

**2** Figure 4-25 shows the path taken by a skunk foraging for trash food, from initial point $i$. The skunk took the same time $T$ to go from each labeled point to the next along its path. Rank points $a$, $b$, and $c$ according to the magnitude of the average velocity of the skunk to reach them from initial point $i$, greatest first.

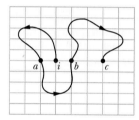

FIG. 4-25   Question 2.

**3** You are to launch a rocket, from just above the ground, with one of the following initial velocity vectors: (1) $\vec{v}_0 = 20\hat{i} + 70\hat{j}$, (2) $\vec{v}_0 = -20\hat{i} + 70\hat{j}$, (3) $\vec{v}_0 = 20\hat{i} - 70\hat{j}$, (4) $\vec{v}_0 = -20\hat{i} - 70\hat{j}$. In your coordinate system, $x$ runs along level ground and $y$ increases upward. (a) Rank the vectors according to the launch speed of the projectile, greatest first. (b) Rank the vectors according to the time of flight of the projectile, greatest first.

**4** Figure 4-26 shows three situations in which identical projectiles are launched (at the same level) at identical initial speeds and angles. The projectiles do not land on the same terrain, however. Rank the situations according to the final speeds of the projectiles just before they land, greatest first.

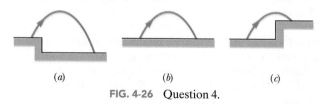

(a)                    (b)                    (c)

FIG. 4-26   Question 4.

**5** When Paris was shelled from 100 km away with the WWI long-range artillery piece "Big Bertha," the shells were fired at an angle greater than 45° to give them a greater range, possibly even twice as long as at 45°. Does that result mean that the air density at high altitudes increases with altitude or decreases?

**6** In Fig. 4-27, a cream tangerine is thrown up past windows 1, 2, and 3, which are identical in size and regularly spaced

vertically. Rank those three windows according to (a) the time the cream tangerine takes to pass them and (b) the average speed of the cream tangerine during the passage, greatest first.

The cream tangerine then moves down past windows 4, 5, and 6, which are identical in size and irregularly spaced horizontally. Rank those three windows according to (c) the time the cream tangerine takes to pass them and (d) the average speed of the cream tangerine during the passage, greatest first.

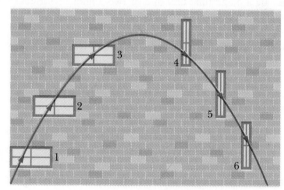

FIG. 4-27   Question 6.

**7** Figure 4-28 shows three paths for a football kicked from ground level. Ignoring the effects of air, rank the paths according to (a) time of flight, (b) initial vertical velocity component, (c) initial horizontal velocity component, and (d) initial speed, greatest first.

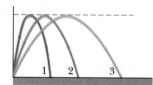

FIG. 4-28   Question 7.

**8** The only good use of a fruitcake is in catapult practice. Curve 1 in Fig. 4-29 gives the height $y$ of a catapulted fruitcake versus the angle $\theta$ between its velocity vector and its acceleration vector during flight. (a) Which of the lettered points on that curve corresponds to the landing of the fruitcake on the ground? (b) Curve 2 is a similar plot for the same launch speed but for a different launch angle. Does the fruitcake now land farther away or closer to the launch point?

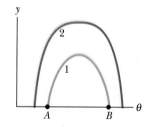

FIG. 4-29   Question 8.

**9** An airplane flying horizontally at a constant speed of 350 km/h over level ground releases a bundle of food supplies. Ignore the effect of the air on the bundle. What are the bundle's initial (a) vertical and (b) horizontal components of velocity? (c) What is its horizontal component of velocity just before hitting the ground? (d) If the airplane's speed were, instead, 450 km/h, would the time of fall be longer, shorter, or the same?

**10** A ball is shot from ground level over level ground at a certain initial speed. Figure 4-30 gives the range $R$

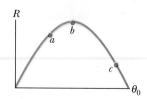

FIG. 4-30   Question 10.

of the ball versus its launch angle $\theta_0$. Rank the three lettered points on the plot according to (a) the total flight time of the ball and (b) the ball's speed at maximum height, greatest first.

**11** In Fig. 4-31, particle $P$ is in uniform circular motion, centered on the origin of an $xy$ coordinate system. (a) At what values of $\theta$ is the vertical component $r_y$ of the position vector greatest in magnitude? (b) At what values of $\theta$ is the vertical component $v_y$ of the particle's velocity greatest in magnitude? (c) At what values of $\theta$ is the vertical component

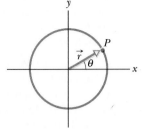

**FIG. 4-31** Question 11.

$a_y$ of the particle's acceleration greatest in magnitude?

**12** (a) Is it possible to be accelerating while traveling at constant speed? Is it possible to round a curve with (b) zero acceleration and (c) a constant magnitude of acceleration?

**13** Figure 4-32 shows four tracks (either half- or quarter-circles) that can be taken by a train, which moves at a constant speed. Rank the tracks according to the magnitude of a train's acceleration on the curved portion, greatest first.

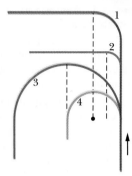

**FIG. 4-32** Question 13.

# PROBLEMS

 **GO** Tutoring problem available (at instructor's discretion) in *WileyPLUS* and WebAssign

**SSM** Worked-out solution available in Student Solutions Manual

• – ••• Number of dots indicates level of problem difficulty

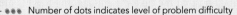 Additional information available in *The Flying Circus of Physics* and at flyingcircusofphysics.com

**WWW** Worked-out solution is at

**ILW** Interactive solution is at — http://www.wiley.com/college/halliday

### sec. 4-2  Position and Displacement

•**1** A positron undergoes a displacement $\Delta \vec{r} = 2.0\hat{i} - 3.0\hat{j} + 6.0\hat{k}$, ending with the position vector $\vec{r} = 3.0\hat{j} - 4.0\hat{k}$, in meters. What was the positron's initial position vector?

•**2** A watermelon seed has the following coordinates: $x = -5.0$ m, $y = 8.0$ m, and $z = 0$ m. Find its position vector (a) in unit-vector notation and as (b) a magnitude and (c) an angle relative to the positive direction of the $x$ axis. (d) Sketch the vector on a right-handed coordinate system. If the seed is moved to the $xyz$ coordinates (3.00 m, 0 m, 0 m), what is its displacement (e) in unit-vector notation and as (f) a magnitude and (g) an angle relative to the positive $x$ direction?

•**3** The position vector for an electron is $\vec{r} = (5.0\text{ m})\hat{i} - (3.0\text{ m})\hat{j} + (2.0\text{ m})\hat{k}$. (a) Find the magnitude of $\vec{r}$. (b) Sketch the vector on a right-handed coordinate system.

•••**4** The minute hand of a wall clock measures 10 cm from its tip to the axis about which it rotates. The magnitude and angle of the displacement vector of the tip are to be determined for three time intervals. What are the (a) magnitude and (b) angle from a quarter after the hour to half past, the (c) magnitude and (d) angle for the next half hour, and the (e) magnitude and (f) angle for the hour after that?

### sec. 4-3  Average Velocity and Instantaneous Velocity

•**5** An ion's position vector is initially $\vec{r} = 5.0\hat{i} - 6.0\hat{j} + 2.0\hat{k}$, and 10 s later it is $\vec{r} = -2.0\hat{i} + 8.0\hat{j} - 2.0\hat{k}$, all in meters. In unit-vector notation, what is its $\vec{v}_{\text{avg}}$ during the 10 s?

•**6** An electron's position is given by $\vec{r} = 3.00t\hat{i} - 4.00t^2\hat{j} + 2.00\hat{k}$, with $t$ in seconds and $\vec{r}$ in meters. (a) In unit-vector notation, what is the electron's velocity $\vec{v}(t)$? At $t = 2.00$ s, what is $\vec{v}$ (b) in unit-vector notation and as (c) a magnitude and (d) an angle relative to the positive direction of the $x$ axis?

•**7** A train at a constant 60.0 km/h moves east for 40.0 min, then in a direction 50.0° east of due north for 20.0 min, and

then west for 50.0 min. What are the (a) magnitude and (b) angle of its average velocity during this trip? **SSM**

••**8** A plane flies 483 km east from city $A$ to city $B$ in 45.0 min and then 966 km south from city $B$ to city $C$ in 1.50 h. For the total trip, what are the (a) magnitude and (b) direction of the plane's displacement, the (c) magnitude and (d) direction of its average velocity, and (e) its average speed?

••**9** Figure 4-33 gives the path of a squirrel moving about on level ground, from point $A$ (at time $t = 0$), to points $B$ (at $t = 5.00$ min), $C$ (at $t = 10.0$ min), and finally $D$ (at $t = 15.0$ min). Consider the average velocities of the squirrel from point $A$ to each of the other three points. Of them, what are the (a) magnitude and (b) angle of the one with the least magnitude and the (c) magnitude and (d) angle of the one with the greatest magnitude?

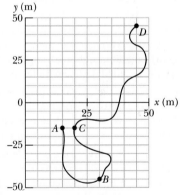

**FIG. 4-33** Problem 9.

•••**10** The position vector $\vec{r} = 5.00t\hat{i} + (et + ft^2)\hat{j}$ locates a particle as a function of time $t$. Vector $\vec{r}$ is in meters, $t$ is in seconds, and factors $e$ and $f$ are constants. Figure 4-34 gives the angle $\theta$ of the particle's direction of travel as a function of $t$ ($\theta$ is measured from the positive $x$ direction). What are (a) $e$ and (b) $f$, including units?

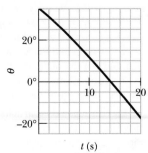

**FIG. 4-34** Problem 10.

### sec. 4-4 Average Acceleration and Instantaneous Acceleration

•11 A particle moves so that its position (in meters) as a function of time (in seconds) is $\vec{r} = \hat{i} + 4t^2\hat{j} + t\hat{k}$. Write expressions for (a) its velocity and (b) its acceleration as functions of time. SSM

•12 A proton initially has $\vec{v} = 4.0\hat{i} - 2.0\hat{j} + 3.0\hat{k}$ and then 4.0 s later has $\vec{v} = -2.0\hat{i} - 2.0\hat{j} + 5.0\hat{k}$ (in meters per second). For that 4.0 s, what are (a) the proton's average acceleration $\vec{a}_{avg}$ in unit-vector notation, (b) the magnitude of $\vec{a}_{avg}$, and (c) the angle between $\vec{a}_{avg}$ and the positive direction of the x axis?

•13 The position $\vec{r}$ of a particle moving in an xy plane is given by $\vec{r} = (2.00t^3 - 5.00t)\hat{i} + (6.00 - 7.00t^4)\hat{j}$, with $\vec{r}$ in meters and t in seconds. In unit-vector notation, calculate (a) $\vec{r}$, (b) $\vec{v}$, and (c) $\vec{a}$ for t = 2.00 s. (d) What is the angle between the positive direction of the x axis and a line tangent to the particle's path at t = 2.00 s? GO

•14 At one instant a bicyclist is 40.0 m due east of a park's flagpole, going due south with a speed of 10.0 m/s. Then 30.0 s later, the cyclist is 40.0 m due north of the flagpole, going due east with a speed of 10.0 m/s. For the cyclist in this 30.0 s interval, what are the (a) magnitude and (b) direction of the displacement, the (c) magnitude and (d) direction of the average velocity, and the (e) magnitude and (f) direction of the average acceleration?

••15 A cart is propelled over an xy plane with acceleration components $a_x = 4.0$ m/s² and $a_y = -2.0$ m/s². Its initial velocity has components $v_{0x} = 8.0$ m/s and $v_{0y} = 12$ m/s. In unit-vector notation, what is the velocity of the cart when it reaches its greatest y coordinate?

••16 A moderate wind accelerates a pebble over a horizontal xy plane with a constant acceleration $\vec{a} = (5.00$ m/s²$)\hat{i} + (7.00$ m/s²$)\hat{j}$. At time t = 0, the velocity is (4.00 m/s)$\hat{i}$. What are the (a) magnitude and (b) angle of its velocity when it has been displaced by 12.0 m parallel to the x axis?

••17 A particle leaves the origin with an initial velocity $\vec{v} = (3.00\hat{i})$ m/s and a constant acceleration $\vec{a} = (-1.00\hat{i} - 0.500\hat{j})$ m/s². When it reaches its maximum x coordinate, what are its (a) velocity and (b) position vector? SSM ILW

••18 The velocity $\vec{v}$ of a particle moving in the xy plane is given by $\vec{v} = (6.0t - 4.0t^2)\hat{i} + 8.0\hat{j}$, with $\vec{v}$ in meters per second and t (> 0) in seconds. (a) What is the acceleration when t = 3.0 s? (b) When (if ever) is the acceleration zero? (c) When (if ever) is the velocity zero? (d) When (if ever) does the speed equal 10 m/s?

•••19 The acceleration of a particle moving only on a horizontal xy plane is given by $\vec{a} = 3t\hat{i} + 4t\hat{j}$, where $\vec{a}$ is in meters per second-squared and t is in seconds. At t = 0, the position vector $\vec{r} = (20.0$ m$)\hat{i} + (40.0$ m$)\hat{j}$ locates the particle, which then has the velocity vector $\vec{v} = (5.00$ m/s$)\hat{i} + (2.00$ m/s$)\hat{j}$. At t = 4.00 s, what are (a) its position vector in unit-vector notation and (b) the angle between its direction of travel and the positive direction of the x axis?

•••20 In Fig. 4-35, particle A moves along the line y = 30 m with a constant velocity $\vec{v}$ of magnitude 3.0 m/s and parallel

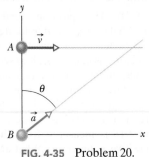

FIG. 4-35 Problem 20.

to the x axis. At the instant particle A passes the y axis, particle B leaves the origin with zero initial speed and constant acceleration $\vec{a}$ of magnitude 0.40 m/s². What angle θ between $\vec{a}$ and the positive direction of the y axis would result in a collision?

### sec. 4-6 Projectile Motion Analyzed

•21 A projectile is fired horizontally from a gun that is 45.0 m above flat ground, emerging from the gun with a speed of 250 m/s. (a) How long does the projectile remain in the air? (b) At what horizontal distance from the firing point does it strike the ground? (c) What is the magnitude of the vertical component of its velocity as it strikes the ground?

•22 In the 1991 World Track and Field Championships in Tokyo, Mike Powell jumped 8.95 m, breaking by a full 5 cm the 23-year long-jump record set by Bob Beamon. Assume that Powell's speed on takeoff was 9.5 m/s (about equal to that of a sprinter) and that g = 9.80 m/s² in Tokyo. How much less was Powell's range than the maximum possible range for a particle launched at the same speed?

•23 The current world-record motorcycle jump is 77.0 m, set by Jason Renie. Assume that he left the take-off ramp at 12.0° to the horizontal and that the take-off and landing heights are the same. Neglecting air drag, determine his take-off speed.

•24 A small ball rolls horizontally off the edge of a tabletop that is 1.20 m high. It strikes the floor at a point 1.52 m horizontally from the table edge. (a) How long is the ball in the air? (b) What is its speed at the instant it leaves the table?

•25 A dart is thrown horizontally with an initial speed of 10 m/s toward point P, the bull's-eye on a dart board. It hits at point Q on the rim, vertically below P, 0.19 s later. (a) What is the distance PQ? (b) How far away from the dart board is the dart released?

•26 In Fig. 4-36, a stone is projected at a cliff of height h with an initial speed of 42.0 m/s directed at angle $\theta_0 = 60.0°$ above the horizontal. The stone strikes at A, 5.50 s after launching. Find (a) the height h of the cliff, (b) the speed of the stone just before impact at A, and (c) the maximum height H reached above the ground.

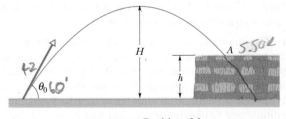

FIG. 4-36 Problem 26.

•27 A certain airplane has a speed of 290.0 km/h and is diving at an angle of θ = 30.0° below the horizontal when the pilot releases a radar decoy (Fig. 4-37). The horizontal distance between the release point and the point where the decoy strikes the ground is d = 700 m. (a) How long is the decoy in the air? (b) How high was the release point? ILW

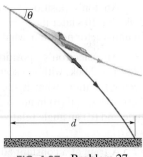

FIG. 4-37 Problem 27.

**•28** A stone is catapulted at time $t = 0$, with an initial velocity of magnitude 20.0 m/s and at an angle of 40.0° above the horizontal. What are the magnitudes of the (a) horizontal and (b) vertical components of its displacement from the catapult site at $t = 1.10$ s? Repeat for the (c) horizontal and (d) vertical components at $t = 1.80$ s, and for the (e) horizontal and (f) vertical components at $t = 5.00$ s.

**••29** A lowly high diver pushes off horizontally with a speed of 2.00 m/s from the platform edge 10.0 m above the surface of the water. (a) At what horizontal distance from the edge is the diver 0.800 s after pushing off? (b) At what vertical distance above the surface of the water is the diver just then? (c) At what horizontal distance from the edge does the diver strike the water?  **SSM WWW**

**••30** A trebuchet was a hurling machine built to attack the walls of a castle under siege. A large stone could be hurled against a wall to break apart the wall. The machine was not placed near the wall because then arrows could reach it from the castle wall. Instead, it was positioned so that the stone hit the wall during the second half of its flight. Suppose a stone is launched with a speed of $v_0 = 28.0$ m/s and at an angle of $\theta_0 = 40.0°$. What is the speed of the stone if it hits the wall (a) just as it reaches the top of its parabolic path and (b) when it has descended to half that height? (c) As a percentage, how much faster is it moving in part (b) than in part (a)?

**••31** A plane, diving with constant speed at an angle of 53.0° with the vertical, releases a projectile at an altitude of 730 m. The projectile hits the ground 5.00 s after release. (a) What is the speed of the plane? (b) How far does the projectile travel horizontally during its flight? What are the (c) horizontal and (d) vertical components of its velocity just before striking the ground?  **SSM**

**••32** During a tennis match, a player serves the ball at 23.6 m/s, with the center of the ball leaving the racquet horizontally 2.37 m above the court surface. The net is 12 m away and 0.90 m high. When the ball reaches the net, (a) does the ball clear it and (b) what is the distance between the center of the ball and the top of the net? Suppose that, instead, the ball is served as before but now it leaves the racquet at 5.00° below the horizontal. When the ball reaches the net, (c) does the ball clear it and (d) what now is the distance between the center of the ball and the top of the net?

**••33** In a jump spike, a volleyball player slams the ball from overhead and toward the opposite floor. Controlling the angle of the spike is difficult. Suppose a ball is spiked from a height of 2.30 m with an initial speed of 20.0 m/s at a downward angle of 18.00°. How much farther on the opposite floor would it have landed if the downward angle were, instead, 8.00°?

**••34** A soccer ball is kicked from the ground with an initial speed of 19.5 m/s at an upward angle of 45°. A player 55 m away in the direction of the kick starts running to meet the ball at that instant. What must be his average speed if he is to meet the ball just before it hits the ground?

**••35** A projectile's launch speed is five times its speed at maximum height. Find launch angle $\theta_0$.

**••36** Suppose that a shot putter can put a shot at the world-class speed $v_0 = 15.00$ m/s and at a height of 2.160 m. What horizontal distance would the shot travel if the launch angle $\theta_0$ is (a) 45.00° and (b) 42.00°? The answers indicate that the

angle of 45°, which maximizes the range of projectile motion, does not maximize the horizontal distance when the launch and landing are at different heights.

**••37** A ball is shot from the ground into the air. At a height of 9.1 m, its velocity is $\vec{v} = (7.6\hat{i} + 6.1\hat{j})$ m/s, with $\hat{i}$ horizontal and $\hat{j}$ upward. (a) To what maximum height does the ball rise? (b) What total horizontal distance does the ball travel? What are the (c) magnitude and (d) angle (below the horizontal) of the ball's velocity just before it hits the ground?  **ILW**

**••38** You throw a ball toward a wall at speed 25.0 m/s and at angle $\theta_0 = 40.0°$ above the horizontal (Fig. 4-38). The wall is distance $d = 22.0$ m from the release point of the ball. (a) How far above the release point does the ball hit the wall? What are the (b) horizontal and (c) vertical components of its velocity as it hits the wall? (d) When it hits, has it passed the highest point on its trajectory?

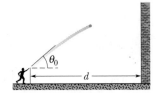

**FIG. 4-38** Problem 38.

**••39** A rifle that shoots bullets at 460 m/s is to be aimed at a target 45.7 m away. If the center of the target is level with the rifle, how high above the target must the rifle barrel be pointed so that the bullet hits dead center?  **SSM**

**••40** A baseball leaves a pitcher's hand horizontally at a speed of 161 km/h. The distance to the batter is 18.3 m. (a) How long does the ball take to travel the first half of that distance? (b) The second half? (c) How far does the ball fall freely during the first half? (d) During the second half? (e) Why aren't the quantities in (c) and (d) equal?

**••41** In Fig. 4-39, a ball is thrown leftward from the left edge of the roof, at height $h$ above the ground. The ball hits the ground 1.50 s later, at distance $d = 25.0$ m from the building and at angle $\theta = 60.0°$ with the horizontal. (a) Find $h$. (*Hint:* One way is to reverse the motion, as if on videotape.) What are the (b) magnitude and (c) angle relative to the horizontal of the velocity at which the ball is thrown? (d) Is the angle above or below the horizontal?

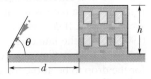

**FIG. 4-39** Problem 41.

**••42** A golf ball is struck at ground level. The speed of the golf ball as a function of the time is shown in Fig. 4-40, where $t = 0$ at the instant the ball is struck. (a) How far does the golf ball travel horizontally before returning to ground level? (b) What is the maximum height above ground level attained by the ball?

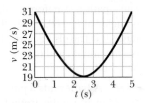

**FIG. 4-40** Problem 42.

**••43** In Fig. 4-41, a ball is launched with a velocity of magnitude 10.0 m/s, at an angle of 50.0° to the horizontal. The launch point is at the base of a ramp of horizontal length $d_1 = 6.00$ m and height $d_2 = 3.60$ m. A plateau is located at the top of the ramp. (a) Does the ball land on the ramp or

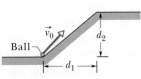

**FIG. 4-41** Problem 43.

$y_0 = \frac{1}{2}at^2 + v_it$

$v_f = at + v_i$

the plateau? When it lands, what are the (b) magnitude and (c) angle of its displacement from the launch point?

**••44**  In 1939 or 1940, Emanuel Zacchini took his human-cannonball act to an extreme: After being shot from a cannon, he soared over three Ferris wheels and into a net (Fig. 4-42). (a) Treating him as a particle, calculate his clearance over the first wheel. (b) If he reached maximum height over the middle wheel, by how much did he clear it? (c) How far from the cannon should the net's center have been positioned (neglect air drag)?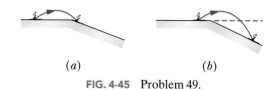

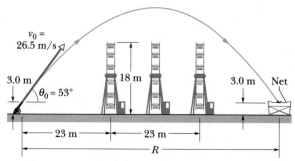

FIG. 4-42   Problem 44.

**••45**  Upon spotting an insect on a twig overhanging water, an archer fish squirts water drops at the insect to knock it into the water (Fig. 4-43). Although the fish sees the insect along a straight-line path at angle $\phi$ and distance $d$, a drop must be launched at a different angle $\theta_0$ if its parabolic path is to intersect the insect. If $\phi = 36.0°$, $d = 0.900$ m, and the launch speed is 3.56 m/s, what $\theta_0$ is required for the drop to be at the top of the parabolic path when it reaches the insect?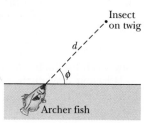

FIG. 4-43   Problem 45.

**••46**  In Fig. 4-44, a ball is thrown up onto a roof, landing 4.00 s later at height $h = 20.0$ m above the release level. The ball's path just before landing is angled at $\theta = 60.0°$ with the roof. (a) Find the horizontal distance $d$ it travels. (See the hint to Problem 41.) What are the (b) magnitude and (c) angle (relative to the horizontal) of the ball's initial velocity?

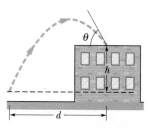

FIG. 4-44   Problem 46.

**••47**  A batter hits a pitched ball when the center of the ball is 1.22 m above the ground. The ball leaves the bat at an angle of 45° with the ground. With that launch, the ball should have a horizontal range (returning to the *launch* level) of 107 m. (a) Does the ball clear a 7.32-m-high fence that is 97.5 m horizontally from the launch point? (b) At the fence, what is the distance between the fence top and the ball center? **SSM WWW**

**••48**  In basketball, *hang* is an illusion in which a player seems to weaken the gravitational acceleration while in midair. The illusion depends much on a skilled player's ability to rapidly shift the ball between hands during the flight, but it might also be supported by the longer horizontal distance the

player travels in the upper part of the jump than in the lower part. If a player jumps with an initial speed of $v_0 = 7.00$ m/s at an angle of $\theta_0 = 35.0°$, what percent of the jump's range does the player spend in the upper half of the jump (between maximum height and half maximum height)?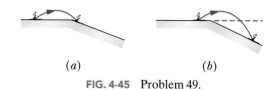

**•••49**  A skilled skier knows to jump upward before reaching a downward slope. Consider a jump in which the launch speed is $v_0 = 10$ m/s, the launch angle is $\theta_0 = 9.0°$, the initial course is approximately flat, and the steeper track has a slope of 11.3°. Figure 4-45a shows a *prejump* that allows the skier to land on the top portion of the steeper track. Figure 4-45b shows a jump at the edge of the steeper track. In Fig. 4-45a, the skier lands at approximately the launch level. (a) In the landing, what is the angle $\phi$ between the skier's path and the slope? In Fig. 4-45b, (b) how far below the launch level does the skier land and (c) what is $\phi$? (The greater fall and greater $\phi$ can result in loss of control in the landing.)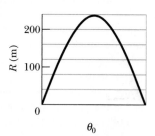

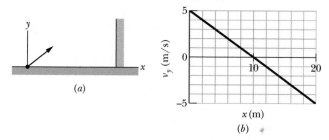

(a)                              (b)

FIG. 4-45   Problem 49.

**•••50**  A ball is to be shot from level ground toward a wall at distance $x$ (Fig. 4-46a). Figure 4-46b shows the $y$ component $v_y$ of the ball's velocity just as it would reach the wall, as a function of that distance $x$. What is the launch angle?

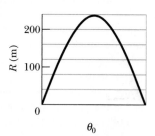

FIG. 4-46   Problem 50.

**•••51**  A football kicker can give the ball an initial speed of 25 m/s. What are the (a) least and (b) greatest elevation angles at which he can kick the ball to score a field goal from a point 50 m in front of goalposts whose horizontal bar is 3.44 m above the ground? **SSM**

**•••52**  A ball is to be shot from level ground with a certain speed. Figure 4-47 shows the range $R$ it will have versus the launch angle $\theta_0$. The value of $\theta_0$ determines the flight time; let $t_{max}$ represent the maximum flight time. What is the least speed the ball will have during its flight if $\theta_0$ is chosen such that the flight time is $0.500t_{max}$?

FIG. 4-47   Problem 52.

**•••53**  A ball rolls horizontally off the top of a stairway with a speed of 1.52 m/s. The steps are 20.3 cm high and 20.3 cm wide. Which step does the ball hit first? **SSM**

•••**54** Two seconds after being projected from ground level, a projectile is displaced 40 m horizontally and 53 m vertically above its launch point. What are the (a) horizontal and (b) vertical components of the initial velocity of the projectile? (c) At the instant the projectile achieves its maximum height above ground level, how far is it displaced horizontally from the launch point? **GO**

•••**55** In Fig. 4-48, a baseball is hit at a height $h = 1.00$ m and then caught at the same height. It travels alongside a wall, moving up past the top of the wall 1.00 s after it is hit and then down past the top of the wall 4.00 s later, at distance $D = 50.0$ m farther along the wall. (a) What horizontal distance is traveled by the ball from hit to catch? What are the (b) magnitude and (c) angle (relative to the horizontal) of the ball's velocity just after being hit? (d) How high is the wall?

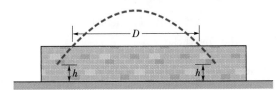

FIG. 4-48   Problem 55.

## sec. 4-7   Uniform Circular Motion

•**56** A centripetal-acceleration addict rides in uniform circular motion with period $T = 2.0$ s and radius $r = 3.00$ m. At $t_1$ his acceleration is $\vec{a} = (6.00 \text{ m/s}^2)\hat{i} + (-4.00 \text{ m/s}^2)\hat{j}$. At that instant, what are the values of (a) $\vec{v} \cdot \vec{a}$ and (b) $\vec{r} \times \vec{a}$?

•**57** A woman rides a carnival Ferris wheel at radius 15 m, completing five turns about its horizontal axis every minute. What are (a) the period of the motion, the (b) magnitude and (c) direction of her centripetal acceleration at the highest point, and the (d) magnitude and (e) direction of her centripetal acceleration at the lowest point? **ILW**

•**58** What is the magnitude of the acceleration of a sprinter running at 10 m/s when rounding a turn of a radius 25 m?

•**59** When a large star becomes a *supernova,* its core may be compressed so tightly that it becomes a *neutron star,* with a radius of about 20 km (about the size of the San Francisco area). If a neutron star rotates once every second, (a) what is the speed of a particle on the star's equator and (b) what is the magnitude of the particle's centripetal acceleration? (c) If the neutron star rotates faster, do the answers to (a) and (b) increase, decrease, or remain the same?

•**60** An Earth satellite moves in a circular orbit 640 km above Earth's surface with a period of 98.0 min. What are the (a) speed and (b) magnitude of the centripetal acceleration of the satellite?

•**61** A carnival merry-go-round rotates about a vertical axis at a constant rate. A man standing on the edge has a constant speed of 3.66 m/s and a centripetal acceleration $\vec{a}$ of magnitude 1.83 m/s$^2$. Position vector $\vec{r}$ locates him relative to the rotation axis. (a) What is the magnitude of $\vec{r}$? What is the direction of $\vec{r}$ when $\vec{a}$ is directed (b) due east and (c) due south?

•**62** A rotating fan completes 1200 revolutions every minute. Consider the tip of a blade, at a radius of 0.15 m.

(a) Through what distance does the tip move in one revolution? What are (b) the tip's speed and (c) the magnitude of its acceleration? (d) What is the period of the motion?

••**63** A purse at radius 2.00 m and a wallet at radius 3.00 m travel in uniform circular motion on the floor of a merry-go-round as the ride turns. They are on the same radial line. At one instant, the acceleration of the purse is $(2.00 \text{ m/s}^2)\hat{i} + (4.00 \text{ m/s}^2)\hat{j}$. At that instant and in unit-vector notation, what is the acceleration of the wallet?

••**64** A particle moves along a circular path over a horizontal $xy$ coordinate system, at constant speed. At time $t_1 = 4.00$ s, it is at point $(5.00$ m, $6.00$ m$)$ with velocity $(3.00 \text{ m/s})\hat{j}$ and acceleration in the positive $x$ direction. At time $t_2 = 10.0$ s, it has velocity $(-3.00 \text{ m/s})\hat{i}$ and acceleration in the positive $y$ direction. What are the (a) $x$ and (b) $y$ coordinates of the center of the circular path if $t_2 - t_1$ is less than one period?

••**65** At $t_1 = 2.00$ s, the acceleration of a particle in counterclockwise circular motion is $(6.00 \text{ m/s}^2)\hat{i} + (4.00 \text{ m/s}^2)\hat{j}$. It moves at constant speed. At time $t_2 = 5.00$ s, its acceleration is $(4.00 \text{ m/s}^2)\hat{i} + (-6.00 \text{ m/s}^2)\hat{j}$. What is the radius of the path taken by the particle if $t_2 - t_1$ is less than one period? **GO**

••**66** A particle moves horizontally in uniform circular motion, over a horizontal $xy$ plane. At one instant, it moves through the point at coordinates $(4.00$ m, $4.00$ m$)$ with a velocity of $-5.00\hat{i}$ m/s and an acceleration of $+12.5\hat{j}$ m/s$^2$. What are the (a) $x$ and (b) $y$ coordinates of the center of the circular path?

•••**67** A boy whirls a stone in a horizontal circle of radius 1.5 m and at height 2.0 m above level ground. The string breaks, and the stone flies off horizontally and strikes the ground after traveling a horizontal distance of 10 m. What is the magnitude of the centripetal acceleration of the stone during the circular motion? **SSM WWW**

•••**68** A cat rides a merry-go-round turning with uniform circular motion. At time $t_1 = 2.00$ s, the cat's velocity is $\vec{v}_1 = (3.00 \text{ m/s})\hat{i} + (4.00 \text{ m/s})\hat{j}$, measured on a horizontal $xy$ coordinate system. At $t_2 = 5.00$ s, its velocity is $\vec{v}_2 = (-3.00 \text{ m/s})\hat{i} + (-4.00 \text{ m/s})\hat{j}$. What are (a) the magnitude of the cat's centripetal acceleration and (b) the cat's average acceleration during the time interval $t_2 - t_1$, which is less than one period?

## sec. 4-8   Relative Motion in One Dimension

•**69** A cameraman on a pickup truck is traveling westward at 20 km/h while he videotapes a cheetah that is moving westward 30 km/h faster than the truck. Suddenly, the cheetah stops, turns, and then runs at 45 km/h eastward, as measured by a suddenly nervous crew member who stands alongside the cheetah's path. The change in the animal's velocity takes 2.0 s. What are the (a) magnitude and (b) direction of the animal's acceleration according to the cameraman and the (c) magnitude and (d) direction according to the nervous crew member?

•**70** A boat is traveling upstream in the positive direction of an $x$ axis at 14 km/h with respect to the water of a river. The water is flowing at 9.0 km/h with respect to the ground. What are the (a) magnitude and (b) direction of the boat's velocity with respect to the ground? A child on the boat walks from front to rear at 6.0 km/h with respect to the boat. What are the (c) magnitude and (d) direction of the child's velocity with respect to the ground?

**••71** A suspicious-looking man runs as fast as he can along a moving sidewalk from one end to the other, taking 2.50 s. Then security agents appear, and the man runs as fast as he can back along the sidewalk to his starting point, taking 10.0 s. What is the ratio of the man's running speed to the sidewalk's speed?

### sec. 4-9 Relative Motion in Two Dimensions

**•72** A rugby player runs with the ball directly toward his opponent's goal, along the positive direction of an *x* axis. He can legally pass the ball to a teammate as long as the ball's velocity relative to the field does not have a positive *x* component. Suppose the player runs at speed 4.0 m/s relative to the field while he passes the ball with velocity $\vec{v}_{BP}$ relative to himself. If $\vec{v}_{BP}$ has magnitude 6.0 m/s, what is the smallest angle it can have for the pass to be legal?

**••73** Two ships, *A* and *B*, leave port at the same time. Ship *A* travels northwest at 24 knots, and ship *B* travels at 28 knots in a direction 40° west of south. (1 knot = 1 nautical mile per hour; see Appendix D.) What are the (a) magnitude and (b) direction of the velocity of ship *A* relative to *B*? (c) After what time will the ships be 160 nautical miles apart? (d) What will be the bearing of *B* (the direction of *B*'s position) relative to *A* at that time? **SSM ILW**

**••74** A light plane attains an airspeed of 500 km/h. The pilot sets out for a destination 800 km due north but discovers that the plane must be headed 20.0° east of due north to fly there directly. The plane arrives in 2.00 h. What were the (a) magnitude and (b) direction of the wind velocity?

**••75** Snow is falling vertically at a constant speed of 8.0 m/s. At what angle from the vertical do the snowflakes appear to be falling as viewed by the driver of a car traveling on a straight, level road with a speed of 50 km/h? **SSM**

**••76** After flying for 15 min in a wind blowing 42 km/h at an angle of 20° south of east, an airplane pilot is over a town that is 55 km due north of the starting point. What is the speed of the airplane relative to the air?

**••77** A train travels due south at 30 m/s (relative to the ground) in a rain that is blown toward the south by the wind. The path of each raindrop makes an angle of 70° with the vertical, as measured by an observer stationary on the ground. An observer on the train, however, sees the drops fall perfectly vertically. Determine the speed of the raindrops relative to the ground. **SSM**

**••78** A 200-m-wide river flows due east at a uniform speed of 2.0 m/s. A boat with a speed of 8.0 m/s relative to the water leaves the south bank pointed in a direction 30° west of north. What are the (a) magnitude and (b) direction of the boat's velocity relative to the ground? (c) How long does the boat take to cross the river? **GO**

**••79** Two highways intersect as shown in Fig. 4-49. At the instant shown, a police car *P* is distance $d_P = 800$ m from the intersection and moving at speed $v_P = 80$ km/h. Motorist *M* is distance $d_M = 600$ m from the intersection and moving at speed $v_M = 60$ km/h. (a) In unit-vector notation, what is the velocity of the motorist with respect to the police car? (b) For the instant shown in Fig. 4-49, what is the angle between the velocity found in (a) and the line of sight between the two

cars? (c) If the cars maintain their velocities, do the answers to (a) and (b) change as the cars move nearer the intersection?

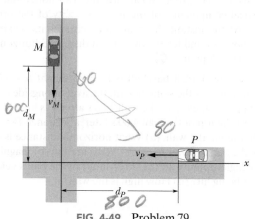

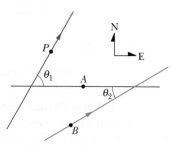

**FIG. 4-49** Problem 79.

**••80** In the overhead view of Fig. 4-50, Jeeps *P* and *B* race along straight lines, across flat terrain, and past stationary border guard *A*. Relative to the guard, *B* travels at a constant speed of 20.0 m/s, at the angle $\theta_2 = 30.0°$. Relative to the guard, *P* has accelerated from rest at a constant rate of 0.400 m/s² at the angle $\theta_1 = 60.0°$. At a certain time during the acceleration, *P* has a speed of 40.0 m/s. At that time, what are the (a) magnitude and (b) direction of the velocity of *P* relative to *B* and the (c) magnitude and (d) direction of the acceleration of *P* relative to *B*?

**FIG. 4-50** Problem 80.

**•••81** Ship *A* is located 4.0 km north and 2.5 km east of ship *B*. Ship *A* has a velocity of 22 km/h toward the south, and ship *B* has a velocity of 40 km/h in a direction 37° north of east. (a) What is the velocity of *A* relative to *B* in unit-vector notation with $\hat{i}$ toward the east? (b) Write an expression (in terms of $\hat{i}$ and $\hat{j}$) for the position of *A* relative to *B* as a function of *t*, where *t* = 0 when the ships are in the positions described above. (c) At what time is the separation between the ships least? (d) What is that least separation?

**•••82** A 200-m-wide river has a uniform flow speed of 1.1 m/s through a jungle and toward the east. An explorer wishes to leave a small clearing on the south bank and cross the river in a powerboat that moves at a constant speed of 4.0 m/s with respect to the water. There is a clearing on the north bank 82 m upstream from a point directly opposite the clearing on the south bank. (a) In what direction must the boat be pointed in order to travel in a straight line and land in the clearing on the north bank? (b) How long will the boat take to cross the river and land in the clearing?

### Additional Problems

**83** You are kidnapped by political-science majors (who are upset because you told them political science is not a real science). Although blindfolded, you can tell the speed of their car (by the whine of the engine), the time of travel (by mentally counting off seconds), and the direction of travel (by turns along the rectangular street system). From these clues, you know that you are taken along the following course: 50 km/h for 2.0 min, turn 90° to the right, 20 km/h for 4.0 min, turn 90° to the right, 20 km/h for 60 s, turn 90° to the left, 50 km/h for 60 s, turn 90° to the right, 20 km/h for 2.0 min, turn 90° to the left, 50 km/h for 30 s. At that point, (a) how far are you from your starting point, and (b) in what direction relative to your initial direction of travel are you?

**84** *Curtain of death.* A large metallic asteroid strikes Earth and quickly digs a crater into the rocky material below ground level by launching rocks upward and outward. The following table gives five pairs of launch speeds and angles (from the horizontal) for such rocks, based on a model of crater formation. (Other rocks, with intermediate speeds and angles, are also launched.) Suppose that you are at $x = 20$ km when the asteroid strikes the ground at time $t = 0$ and position $x = 0$ (Fig. 4-51). (a) At $t = 20$ s, what are the x and y coordinates of the rocks headed in your direction from launches $A$ through $E$? (b) Plot these coordinates and then sketch a curve through the points to include rocks with intermediate launch speeds and angles. The curve should indicate what you would see as you look up into the approaching rocks and what dinosaurs must have seen during asteroid strikes long ago. 🛬

| Launch | Speed (m/s) | Angle (degrees) |
|--------|-------------|-----------------|
| A | 520 | 14.0 |
| B | 630 | 16.0 |
| C | 750 | 18.0 |
| D | 870 | 20.0 |
| E | 1000 | 22.0 |

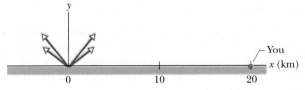

**FIG. 4-51** Problem 84.

**85** In Fig. 4-52, a lump of wet putty moves in uniform circular motion as it rides at a radius of 20.0 cm on the rim of a wheel rotating counterclockwise with a period of 5.00 ms. The lump then happens to fly off the rim at the 5 o'clock position (as if on a clock face). It leaves the rim at a height of $h = 1.20$ m from the floor and at a distance $d = 2.50$ m from a wall. At what height on the wall does the lump hit?

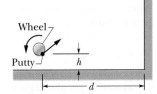

**FIG. 4-52** Problem 85.

**86** A particle is in uniform circular motion about the origin of an xy coordinate system, moving clockwise with a period of 7.00 s. At one instant, its position vector (from the origin) is $\vec{r} = (2.00 \text{ m})\hat{i} - (3.00 \text{ m})\hat{j}$. At that instant, what is its velocity in unit-vector notation?

**87** In Fig. 4-53, a ball is shot directly upward from the ground with an initial speed of $v_0 = 7.00$ m/s. Simultaneously, a construction elevator cab begins to move upward from the ground with a constant speed of $v_c = 3.00$ m/s. What maximum height does the ball reach relative to (a) the ground and (b) the cab floor? At what rate does the speed of the ball change relative to (c) the ground and (d) the cab floor?

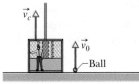

**FIG. 4-53** Problem 87.

**88** In Fig. 4-54a, a sled moves in the negative x direction at constant speed $v_s$ while a ball of ice is shot from the sled with a velocity $\vec{v}_0 = v_{0x}\hat{i} + v_{0y}\hat{j}$ relative to the sled. When the ball lands, its horizontal displacement $\Delta x_{bg}$ relative to the ground (from its launch position to its landing position) is measured. Figure 4-54b gives $\Delta x_{bg}$ as a function of $v_s$. Assume the ball lands at approximately its launch height. What are the values of (a) $v_{0x}$ and (b) $v_{0y}$? The ball's displacement $\Delta x_{bs}$ relative to the sled can also be measured. Assume that the sled's velocity is not changed when the ball is shot. What is $\Delta x_{bs}$ when $v_s$ is (c) 5.0 m/s and (d) 15 m/s?

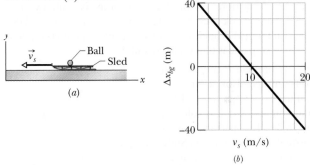

**FIG. 4-54** Problem 88.

**89** A woman who can row a boat at 6.4 km/h in still water faces a long, straight river with a width of 6.4 km and a current of 3.2 km/h. Let $\hat{i}$ point directly across the river and $\hat{j}$ point directly downstream. If she rows in a straight line to a point directly opposite her starting position, (a) at what angle to $\hat{i}$ must she point the boat and (b) how long will she take? (c) How long will she take if, instead, she rows 3.2 km *down* the river and then back to her starting point? (d) How long if she rows 3.2 km *up* the river and then back to her starting point? (e) At what angle to $\hat{i}$ should she point the boat if she wants to cross the river in the shortest possible time? (f) How long is that shortest time?

**90** In Fig. 4-55, a radar station detects an airplane approaching directly from the east. At first observation, the airplane is at distance $d_1 = 360$ m from the station and at angle $\theta_1 = 40°$ above the horizon. The airplane is tracked through an angular change $\Delta\theta = 123°$ in the vertical east–west plane; its distance is then $d_2 = 790$ m. Find the (a) magnitude and (b) direction of the airplane's displacement during this period.

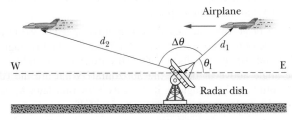

**FIG. 4-55** Problem 90.

**91** A rifle is aimed horizontally at a target 30 m away. The bullet hits the target 1.9 cm below the aiming point. What are (a) the bullet's time of flight and (b) its speed as it emerges from the rifle? **SSM**

**92** The fast French train known as the TGV (Train à Grande Vitesse) has a scheduled average speed of 216 km/h. (a) If the train goes around a curve at that speed and the magnitude of the acceleration experienced by the passengers is to be limited to 0.050$g$, what is the smallest radius of curvature for the track that can be tolerated? (b) At what speed must the train go around a curve with a 1.00 km radius to be at the acceleration limit?

**93** A magnetic field can force a charged particle to move in a circular path. Suppose that an electron moving in a circle experiences a radial acceleration of magnitude $3.0 \times 10^{14}$ m/s$^2$ in a particular magnetic field. (a) What is the speed of the electron if the radius of its circular path is 15 cm? (b) What is the period of the motion?

**94** The position vector for a proton is initially $\vec{r} = 5.0\hat{i} - 6.0\hat{j} + 2.0\hat{k}$ and then later is $\vec{r} = -2.0\hat{i} + 6.0\hat{j} + 2.0\hat{k}$, all in meters. (a) What is the proton's displacement vector, and (b) to what plane is that vector parallel?

**95** A particle $P$ travels with constant speed on a circle of radius $r = 3.00$ m (Fig. 4-56) and completes one revolution in 20.0 s. The particle passes through $O$ at time $t = 0$. State the following vectors in magnitude-angle notation (angle relative to the positive direction of $x$). With respect to $O$, find the particle's position vector at the times $t$ of (a) 5.00 s, (b) 7.50 s, and (c) 10.0 s.

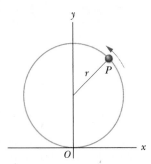

**FIG. 4-56** Problem 95.

(d) For the 5.00 s interval from the end of the fifth second to the end of the tenth second, find the particle's displacement. For that interval, find (e) its average velocity and its velocity at the (f) beginning and (g) end. Next, find the acceleration at the (h) beginning and (i) end of that interval.

**96** An iceboat sails across the surface of a frozen lake with constant acceleration produced by the wind. At a certain instant the boat's velocity is $(6.30\hat{i} - 8.42\hat{j})$ m/s. Three seconds later, because of a wind shift, the boat is instantaneously at rest. What is its average acceleration for this 3 s interval?

**97** In 3.50 h, a balloon drifts 21.5 km north, 9.70 km east, and 2.88 km upward from its release point on the ground. Find (a) the magnitude of its average velocity and (b) the angle its average velocity makes with the horizontal.

**98** A ball is thrown horizontally from a height of 20 m and hits the ground with a speed that is three times its initial speed. What is the initial speed?

**99** A projectile is launched with an initial speed of 30 m/s at an angle of 60° above the horizontal. What are the (a) magnitude and (b) angle of its velocity 2.0 s after launch, and (c) is the angle above or below the horizontal? What are the (d) magnitude and (e) angle of its velocity 5.0 s after launch, and (f) is the angle above or below the horizontal?

**100** An airport terminal has a moving sidewalk to speed passengers through a long corridor. Larry does not use the moving sidewalk; he takes 150 s to walk through the corridor. Curly, who simply stands on the moving sidewalk, covers the same distance in 70 s. Moe boards the sidewalk and walks along it. How long does Moe take to move through the corridor? Assume that Larry and Moe walk at the same speed.

**101** A football player punts the football so that it will have a "hang time" (time of flight) of 4.5 s and land 46 m away. If the ball leaves the player's foot 150 cm above the ground, what must be the (a) magnitude and (b) angle (relative to the horizontal) of the ball's initial velocity?

**102** For women's volleyball the top of the net is 2.24 m above the floor and the court measures 9.0 m by 9.0 m on each side of the net. Using a jump serve, a player strikes the ball at a point that is 3.0 m above the floor and a horizontal distance of 8.0 m from the net. If the initial velocity of the ball is horizontal, (a) what minimum magnitude must it have if the ball is to clear the net and (b) what maximum magnitude can it have if the ball is to strike the floor inside the back line on the other side of the net?

**103** Figure 4-57 shows the straight path of a particle across an $xy$ coordinate system as the particle is accelerated from rest during time interval $\Delta t_1$. The acceleration is constant. The $xy$ coordinates for point $A$ are (4.00 m, 6.00 m); those for point $B$ are (12.0 m, 18.0 m). (a) What is the ratio $a_y/a_x$ of the acceleration components? (b) What are the coordinates of the particle if the motion is continued for another interval equal to $\Delta t_1$?

**FIG. 4-57**
Problem 103.

**104** An astronaut is rotated in a horizontal centrifuge at a radius of 5.0 m. (a) What is the astronaut's speed if the centripetal acceleration has a magnitude of 7.0$g$? (b) How many revolutions per minute are required to produce this acceleration? (c) What is the period of the motion?

**105** (a) What is the magnitude of the centripetal acceleration of an object on Earth's equator due to the rotation of Earth? (b) What would Earth's rotation period have to be for objects on the equator to have a centripetal acceleration of magnitude 9.8 m/s$^2$?

**106** A person walks up a stalled 15-m-long escalator in 90 s. When standing on the same escalator, now moving, the person is carried up in 60 s. How much time would it take that person to walk up the moving escalator? Does the answer depend on the length of the escalator?

**107** A baseball is hit at ground level. The ball reaches its maximum height above ground level 3.0 s after being hit. Then 2.5 s after reaching its maximum height, the ball barely clears a fence that is 97.5 m from where it was hit. Assume the ground is level. (a) What maximum height above ground level is reached by the ball? (b) How high is the fence? (c) How far beyond the fence does the ball strike the ground? **SSM**

**108** The range of a projectile depends not only on $v_0$ and $\theta_0$ but also on the value $g$ of the free-fall acceleration, which varies from place to place. In 1936, Jesse Owens established a world's running broad jump record of 8.09 m at the Olympic Games at Berlin (where $g = 9.8128$ m/s$^2$). Assuming the same values of $v_0$ and $\theta_0$, by how much would his record have differed if he had competed instead in 1956 at Melbourne (where $g = 9.7999$ m/s$^2$)?

**109** During volcanic eruptions, chunks of solid rock can be blasted out of the volcano; these projectiles are called *volcanic bombs.* Figure 4-58 shows a cross section of Mt. Fuji, in Japan. (a) At what initial speed would a bomb have to be ejected, at angle $\theta_0 = 35°$ to the horizontal, from the vent at $A$ in order to fall at the foot of the volcano at $B$, at vertical distance $h = 3.30$ km and horizontal distance $d = 9.40$ km? Ignore, for the moment, the effects of air on the bomb's travel. (b) What would be the time of flight? (c) Would the effect of the air increase or decrease your answer in (a)?

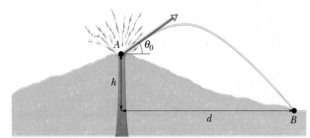

FIG. 4-58 Problem 109.

**110** Long flights at midlatitudes in the Northern Hemisphere encounter the jet stream, an eastward airflow that can affect a plane's speed relative to Earth's surface. If a pilot maintains a certain speed relative to the air (the plane's *airspeed*), the speed relative to the surface (the plane's *ground speed*) is more when the flight is in the direction of the jet stream and less when the flight is opposite the jet stream. Suppose a round-trip flight is scheduled between two cities separated by 4000 km, with the outgoing flight in the direction of the jet stream and the return flight opposite it. The airline computer advises an airspeed of 1000 km/h, for which the difference in flight times for the outgoing and return flights is 70.0 min. What jet-stream speed is the computer using?

**111** A particle starts from the origin at $t = 0$ with a velocity of $8.0\hat{j}$ m/s and moves in the $xy$ plane with constant acceleration $(4.0\hat{i} + 2.0\hat{j})$ m/s². When the particle's $x$ coordinate is 29 m, what are its (a) $y$ coordinate and (b) speed? **SSM**

**112** A sprinter running on a circular track has a velocity of constant magnitude 9.2 m/s and a centripetal acceleration of magnitude 3.8 m/s². What are (a) the track radius and (b) the period of the circular motion?

**113** An electron having an initial horizontal velocity of magnitude $1.00 \times 10^9$ cm/s travels into the region between two horizontal metal plates that are electrically charged. In that region, the electron travels a horizontal distance of 2.00 cm and has a constant downward acceleration of magnitude $1.00 \times 10^{17}$ cm/s² due to the charged plates. Find (a) the time the electron takes to travel the 2.00 cm, (b) the vertical distance it travels during that time, and the magnitudes of its (c) horizontal and (d) vertical velocity components as it emerges from the region.

**114** An elevator without a ceiling is ascending with a constant speed of 10 m/s. A boy on the elevator shoots a ball directly upward, from a height of 2.0 m above the elevator floor, just as the elevator floor is 28 m above the ground. The initial speed of the ball with respect to the elevator is 20 m/s. (a) What maximum height above the ground does the ball reach? (b) How long does the ball take to return to the elevator floor?

**115** Suppose that a space probe can withstand the stresses of a 20$g$ acceleration. (a) What is the minimum turning radius of such a craft moving at a speed of one-tenth the speed of light? (b) How long would it take to complete a 90° turn at this speed?

**116** At what initial speed must the basketball player in Fig. 4-59 throw the ball, at angle $\theta_0 = 55°$ above the horizontal, to make the foul shot? The horizontal distances are $d_1 = 1.0$ ft and $d_2 = 14$ ft, and the heights are $h_1 = 7.0$ ft and $h_2 = 10$ ft.

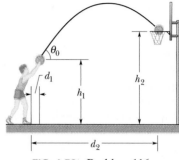

FIG. 4-59 Problem 116.

**117** A wooden boxcar is moving along a straight railroad track at speed $v_1$. A sniper fires a bullet (initial speed $v_2$) at it from a high-powered rifle. The bullet passes through both lengthwise walls of the car, its entrance and exit holes being exactly opposite each other as viewed from within the car. From what direction, relative to the track, is the bullet fired? Assume that the bullet is not deflected upon entering the car, but that its speed decreases by 20%. Take $v_1 = 85$ km/h and $v_2 = 650$ m/s. (Why don't you need to know the width of the boxcar?)

**118** You are to throw a ball with a speed of 12.0 m/s at a target that is height $h = 5.00$ m above the level at which you release the ball (Fig. 4-60). You want the ball's velocity to be horizontal at the instant it reaches the target. (a) At what angle $\theta$ above the horizontal must you throw the ball? (b) What is the horizontal distance from the release point to the target? (c) What is the speed of the ball just as it reaches the target?

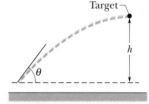

FIG. 4-60 Problem 118.

**119** Figure 4-61 shows the path taken by a drunk skunk over level ground, from initial point $i$ to final point $f$. The angles are $\theta_1 = 30.0°$, $\theta_2 = 50.0°$, and $\theta_3 = 80.0°$, and the distances are $d_1 = 5.00$ m, $d_2 = 8.00$ m, and $d_3 = 12.0$ m. What are the (a) magnitude and (b) angle of the skunk's displacement from $i$ to $f$?

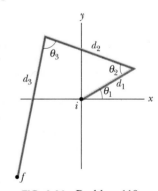

FIG. 4-61 Problem 119.

**120** A projectile is fired with an initial speed $v_0 = 30.0$ m/s from level ground at a target that is on the ground, at distance $R = 20.0$ m, as shown in Fig. 4-62. What are the (a) least and (b) greatest launch angles that will allow the projectile to hit the target?

**121** Oasis $A$ is 90 km due west of oasis $B$. A desert camel

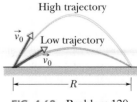

FIG. 4-62 Problem 120.

leaves *A* and takes 50 h to walk 75 km at 37° north of due east. Next it takes 35 h to walk 65 km due south. Then it rests for 5.0 h. What are the (a) magnitude and (b) direction of the camel's displacement relative to *A* at the resting point? From the time the camel leaves *A* until the end of the rest period, what are the (c) magnitude and (d) direction of its average velocity and (e) its average speed? The camel's last drink was at *A*; it must be at *B* no more than 120 h later for its next drink. If it is to reach *B* just in time, what must be the (f) magnitude and (g) direction of its average velocity after the rest period? **SSM**

**122** *A graphing surprise.* At time $t = 0$, a burrito is launched from level ground, with an initial speed of 16.0 m/s and launch angle $\theta_0$. Imagine a position vector $\vec{r}$ continuously directed from the launching point to the burrito during the flight. Graph the magnitude $r$ of the position vector for (a) $\theta_0 = 40.0°$ and (b) $\theta_0 = 80.0°$. For $\theta_0 = 40.0°$, (c) when does $r$ reach its maximum value, (d) what is that value, and how far (e) horizontally and (f) vertically is the burrito from the launch point? For $\theta_0 = 80.0°$, (g) when does $r$ reach its maximum value, (h) what is that value, and how far (i) horizontally and (j) vertically is the burrito from the launch point?

**123** In Sample Problem 4-7b, a ball is shot through a horizontal distance of 686 m by a cannon located at sea level and angled at 45° from the horizontal. How much greater would the horizontal distance have been had the cannon been 30 m higher?

**124** (a) If an electron is projected horizontally with a speed of $3.0 \times 10^6$ m/s, how far will it fall in traversing 1.0 m of horizontal distance? (b) Does the answer increase or decrease if the initial speed is increased?

**125** The magnitude of the velocity of a projectile when it is at its maximum height above ground level is 10 m/s. (a) What is the magnitude of the velocity of the projectile 1.0 s before it achieves its maximum height? (b) What is the magnitude of the velocity of the projectile 1.0 s after it achieves its maximum height? If we take $x = 0$ and $y = 0$ to be at the point of maximum height and positive $x$ to be in the direction of the velocity there, what are the (c) $x$ coordinate and (d) $y$ coordinate of the projectile 1.0 s before it reaches its maximum height and the (e) $x$ coordinate and (f) $y$ coordinate 1.0 s after it reaches its maximum height?

**126** A frightened rabbit moving at 6.0 m/s due east runs onto a large area of level ice of negligible friction. As the rabbit slides across the ice, the force of the wind causes it to have a constant acceleration of 1.4 m/s², due north. Choose a coordinate system with the origin at the rabbit's initial position on the ice and the positive $x$ axis directed toward the east. In unit-vector notation, what are the rabbit's (a) velocity and (b) position when it has slid for 3.0 s?

**127** The pilot of an aircraft flies due east relative to the ground in a wind blowing 20 km/h toward the south. If the speed of the aircraft in the absence of wind is 70 km/h, what is the speed of the aircraft relative to the ground?

**128** The pitcher in a slow-pitch softball game releases the ball at a point 3.0 ft above ground level. A stroboscopic plot of the position of the ball is shown in Fig. 4-63, where the readings are 0.25 s apart and the ball is released at $t = 0$. (a) What is the initial speed of the ball? (b) What is the speed of the ball at the instant it reaches its maximum height above ground level? (c) What is that maximum height?

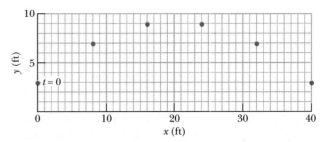

**FIG. 4-63** Problem 128.

**129** The New Hampshire State Police use aircraft to enforce highway speed limits. Suppose that one of the airplanes has a speed of 135 mi/h in still air. It is flying straight north so that it is at all times directly above a north–south highway. A ground observer tells the pilot by radio that a 70.0 mi/h wind is blowing but neglects to give the wind direction. The pilot observes that in spite of the wind the plane can travel 135 mi along the highway in 1.00 h. In other words, the ground speed is the same as if there were no wind. (a) From what direction is the wind blowing? (b) What is the heading of the plane; that is, in what direction does it point?

**130** The position $\vec{r}$ of a particle moving in the $xy$ plane is given by $\vec{r} = 2t\hat{i} + 2\sin[(\pi/4 \text{ rad/s})t]\hat{j}$, where $\vec{r}$ is in meters and $t$ is in seconds. (a) Calculate the $x$ and $y$ components of the particle's position at $t = 0, 1.0, 2.0, 3.0$, and 4.0 s and sketch the particle's path in the $xy$ plane for the interval $0 \le t \le 4.0$ s. (b) Calculate the components of the particle's velocity at $t = 1.0$, 2.0, and 3.0 s. Show that the velocity is tangent to the path of the particle and in the direction the particle is moving at each time by drawing the velocity vectors on the plot of the particle's path in part (a). (c) Calculate the components of the particle's acceleration at $t = 1.0, 2.0$, and 3.0 s.

**131** A golfer tees off from the top of a rise, giving the golf ball an initial velocity of 43 m/s at an angle of 30° above the horizontal. The ball strikes the fairway a horizontal distance of 180 m from the tee. Assume the fairway is level. (a) How high is the rise above the fairway? (b) What is the speed of the ball as it strikes the fairway?

**132** A track meet is held on a planet in a distant solar system. A shot-putter releases a shot at a point 2.0 m above ground level. A stroboscopic plot of the position of the shot is shown in Fig. 4-64, where the readings are 0.50 s apart and the shot is released at time $t = 0$. (a) What is the initial velocity of the shot in unit-vector notation? (b) What is the magnitude of the free-fall acceleration on the planet? (c) How long after it is released does the shot reach the ground? (d) If an identical throw of the shot is made on the surface of Earth, how long after it is released does it reach the ground?

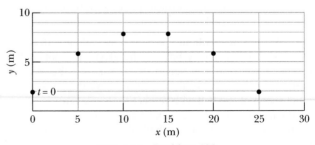

**FIG. 4-64** Problem 132.

# Force and Motion—I

Blue Water Photo/Index Stock

Many roller-coaster enthusiasts prefer riding in the first car because they enjoy being the first to go over an "edge" and onto a downward slope. However, many other enthusiasts prefer the rear car— they claim that going over the edge is far more frightening there. The roller coaster is certainly moving faster when the last car is dragged over the edge by the rest of the roller coaster. But there seems to be some other, more subtle element that brings out the fear as that last car comes to the edge.

**What is the subtle fear factor in riding the last car in a roller coaster?**

The answer is in this chapter.

## 5-1 WHAT IS PHYSICS?

We have seen that part of physics is a study of motion, including accelerations, which are changes in velocities. Physics is also a study of what can *cause* an object to accelerate. That cause is a **force,** which is, loosely speaking, a push or pull on the object. The force is said to *act* on the object to change its velocity. For example, when a dragster accelerates, a force from the track acts on the rear tires to cause the dragster's acceleration. When a defensive guard knocks down a quarterback, a force from the guard acts on the quarterback to cause the quarterback's backward acceleration. When a car slams into a telephone pole, a force on the car from the pole causes the car to stop. Science, engineering, legal, and medical journals are filled with articles about forces on objects, including people.

## 5-2 | Newtonian Mechanics

The relation between a force and the acceleration it causes was first understood by Isaac Newton (1642–1727) and is the subject of this chapter. The study of that relation, as Newton presented it, is called *Newtonian mechanics.* We shall focus on its three primary laws of motion.

Newtonian mechanics does not apply to all situations. If the speeds of the interacting bodies are very large—an appreciable fraction of the speed of light—we must replace Newtonian mechanics with Einstein's special theory of relativity, which holds at any speed, including those near the speed of light. If the interacting bodies are on the scale of atomic structure (for example, they might be electrons in an atom), we must replace Newtonian mechanics with quantum mechanics. Physicists now view Newtonian mechanics as a special case of these two more comprehensive theories. Still, it is a very important special case because it applies to the motion of objects ranging in size from the very small (almost on the scale of atomic structure) to astronomical (galaxies and clusters of galaxies).

## 5-3 | Newton's First Law

Before Newton formulated his mechanics, it was thought that some influence, a "force," was needed to keep a body moving at constant velocity. Similarly, a body was thought to be in its "natural state" when it was at rest. For a body to move with constant velocity, it seemingly had to be propelled in some way, by a push or a pull. Otherwise, it would "naturally" stop moving.

These ideas were reasonable. If you send a puck sliding across a wooden floor, it does indeed slow and then stop. If you want to make it move across the floor with constant velocity, you have to continuously pull or push it.

Send a puck sliding over the ice of a skating rink, however, and it goes a lot farther. You can imagine longer and more slippery surfaces, over which the puck would slide farther and farther. In the limit you can think of a long, extremely slippery surface (said to be a **frictionless surface**), over which the puck would hardly slow. (We can in fact come close to this situation by sending a puck sliding over a horizontal air table, across which it moves on a film of air.)

From these observations, we can conclude that a body will keep moving with constant velocity if no force acts on it. That leads us to the first of Newton's three laws of motion:

**Newton's First Law:** If no force acts on a body, the body's velocity cannot change; that is, the body cannot accelerate.

In other words, if the body is at rest, it stays at rest. If it is moving, it continues to move with the same velocity (same magnitude *and* same direction).

## 5-4 | Force

We now wish to define the unit of force. We know that a force can cause the acceleration of a body. Thus, we shall define the unit of force in terms of the acceleration that a force gives to a standard reference body, which we take to be the standard kilogram of Fig. 1-3. This body has been assigned, exactly and by definition, a mass of 1 kg.

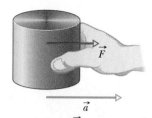

FIG. 5-1   A force $\vec{F}$ on the standard kilogram gives that body an acceleration $\vec{a}$.

We put the standard body on a horizontal frictionless table and pull the body to the right (Fig. 5-1) so that, by trial and error, it eventually experiences a measured acceleration of 1 m/s². We then declare, as a matter of definition, that the force we are exerting on the standard body has a magnitude of 1 newton (abbreviated N).

We can exert a 2 N force on our standard body by pulling it so that its measured acceleration is 2 m/s², and so on. Thus in general, if our standard body of 1 kg mass has an acceleration of magnitude $a$, we know that a force $F$ must be acting on it and that the magnitude of the force (in newtons) is equal to the magnitude of the acceleration (in meters per second per second).

Thus, a force is measured by the acceleration it produces. However, acceleration is a vector quantity, with both magnitude and direction. Is force also a vector quantity? We can easily assign a direction to a force (just assign the direction of the acceleration), but that is not sufficient. We must prove by experiment that forces are vector quantities. Actually, that has been done: forces are indeed vector quantities; they have magnitudes and directions, and they combine according to the vector rules of Chapter 3.

This means that when two or more forces act on a body, we can find their **net force,** or **resultant force,** by adding the individual forces vectorially. A single force that has the magnitude and direction of the net force has the same effect on the body as all the individual forces together. This fact is called the **principle of superposition for forces.** The world would be quite strange if, for example, you and a friend were to pull on the standard body in the same direction, each with a force of 1 N, and yet somehow the net pull was 14 N.

In this book, forces are most often represented with a vector symbol such as $\vec{F}$, and a net force is represented with the vector symbol $\vec{F}_{net}$. As with other vectors, a force or a net force can have components along coordinate axes. When forces act only along a single axis, they are single-component forces. Then we can drop the overhead arrows on the force symbols and just use signs to indicate the directions of the forces along that axis.

Instead of the wording used in Section 5-3, the more proper statement of Newton's First Law is in terms of a *net* force:

> **Newton's First Law:** If no *net* force acts on a body ($\vec{F}_{net} = 0$), the body's velocity cannot change; that is, the body cannot accelerate.

There may be multiple forces acting on a body, but if their net force is zero, the body cannot accelerate.

### Inertial Reference Frames

Newton's first law is not true in all reference frames, but we can always find reference frames in which it (as well as the rest of Newtonian mechanics) is true. Such frames are called **inertial reference frames,** or simply **inertial frames.**

> An inertial reference frame is one in which Newton's laws hold.

For example, we can assume that the ground is an inertial frame provided we can neglect Earth's astronomical motions (such as its rotation).

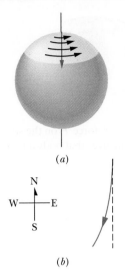

(a)

N
W——+——E
S

(b)

**FIG. 5-2** (a) The path of a puck sliding from the north pole as seen from a stationary point in space. Earth rotates to the east. (b) The path of the puck as seen from the ground.

That assumption works well if, say, a puck is sent sliding along a *short* strip of frictionless ice—we would find that the puck's motion obeys Newton's laws. However, suppose the puck is sent sliding along a *long* ice strip extending from the north pole (Fig. 5-2a). If we view the puck from a stationary frame in space, the puck moves south along a simple straight line because Earth's rotation around the north pole merely slides the ice beneath the puck. However, if we view the puck from a point on the ground so that we rotate with Earth, the puck's path is not a simple straight line. Because the eastward speed of the ground beneath the puck is greater the farther south the puck slides, from our ground-based view the puck appears to be deflected westward (Fig. 5-2b). However, this apparent deflection is caused not by a force as required by Newton's laws but by the fact that we see the puck from a rotating frame. In this situation, the ground is a **noninertial frame.**

In this book we usually assume that the ground is an inertial frame and that measured forces and accelerations are from this frame. If measurements are made in, say, an elevator that is accelerating relative to the ground, then the measurements are being made in a noninertial frame and the results can be surprising. We see an example of this in Sample Problem 5-8.

✓ **CHECKPOINT 1** Which of the figure's six arrangements correctly show the vector addition of forces $\vec{F}_1$ and $\vec{F}_2$ to yield the third vector, which is meant to represent their net force $\vec{F}_{net}$?

(a) $\vec{F}_1$ $\vec{F}_2$   (b) $\vec{F}_1$ $\vec{F}_2$   (c) $\vec{F}_1$ $\vec{F}_2$

(d) $\vec{F}_2$ $\vec{F}_1$   (e) $\vec{F}_1$ $\vec{F}_2$   (f) $\vec{F}_1$ $\vec{F}_2$

## 5-5 | Mass

Everyday experience tells us that a given force produces different magnitudes of acceleration for different bodies. Put a baseball and a bowling ball on the floor and give both the same sharp kick. Even if you don't actually do this, you know the result: The baseball receives a noticeably larger acceleration than the bowling ball. The two accelerations differ because the mass of the baseball differs from the mass of the bowling ball—but what, exactly, is mass?

We can explain how to measure mass by imagining a series of experiments in an inertial frame. In the first experiment we exert a force on a standard body, whose mass $m_0$ is defined to be 1.0 kg. Suppose that the standard body accelerates at 1.0 m/s². We can then say the force on that body is 1.0 N.

We next apply that same force (we would need some way of being certain it is the same force) to a second body, body $X$, whose mass is not known. Suppose we find that this body $X$ accelerates at 0.25 m/s². We know that a *less massive* baseball receives a *greater acceleration* than a more massive bowling ball when the same force (kick) is applied to both. Let us then make the following conjecture: The ratio of the masses of two bodies is equal to the inverse of the ratio of their accelerations when the same force is applied to both. For body $X$ and the standard body, this tells us that

$$\frac{m_X}{m_0} = \frac{a_0}{a_X}.$$

Solving for $m_X$ yields

$$m_X = m_0 \frac{a_0}{a_X} = (1.0 \text{ kg}) \frac{1.0 \text{ m/s}^2}{0.25 \text{ m/s}^2} = 4.0 \text{ kg}.$$

Our conjecture will be useful, of course, only if it continues to hold when we change the applied force to other values. For example, if we apply an 8.0 N force to the standard body, we obtain an acceleration of 8.0 m/s². When the 8.0 N force is applied to body $X$, we obtain an acceleration of 2.0 m/s². Our conjecture then gives us

$$m_X = m_0 \frac{a_0}{a_X} = (1.0 \text{ kg}) \frac{8.0 \text{ m/s}^2}{2.0 \text{ m/s}^2} = 4.0 \text{ kg},$$

consistent with our first experiment. Many experiments yielding similar results indicate that our conjecture provides a consistent and reliable means of assigning a mass to any given body.

Our measurement experiments indicate that mass is an *intrinsic* character-istic of a body—that is, a characteristic that automatically comes with the existence of the body. They also indicate that mass is a scalar quantity. However, the nagging question remains: What, exactly, is mass?

Since the word *mass* is used in everyday English, we should have some intu-itive understanding of it, maybe something that we can physically sense. Is it a body's size, weight, or density? The answer is no, although those characteristics are sometimes confused with mass. We can say only that *the mass of a body is the characteristic that relates a force on the body to the resulting acceleration*. Mass has no more familiar definition; you can have a physical sensation of mass only when you try to accelerate a body, as in the kicking of a baseball or a bowling ball.

## 5-6 | Newton's Second Law

All the definitions, experiments, and observations we have discussed so far can be summarized in one neat statement:

**Newton's Second Law:** The net force on a body is equal to the product of the body's mass and its acceleration.

In equation form,

$$\vec{F}_{net} = m\vec{a} \qquad \text{(Newton's second law)}. \qquad (5\text{-}1)$$

This equation is simple, but we must use it cautiously. First, we must be certain about which body we are applying it to. Then $\vec{F}_{net}$ must be the vector sum of *all* the forces that act on *that* body. Only forces that act on *that* body are to be included in the vector sum, not forces acting on other bodies that might be involved in the given situation. For example, if you are in a rugby scrum, the net force on *you* is the vector sum of all the pushes and pulls on *your* body. It does not include any push or pull on another player from you.

Like other vector equations, Eq. 5-1 is equivalent to three component equa-tions, one for each axis of an $xyz$ coordinate system:

$$F_{net,x} = ma_x, \quad F_{net,y} = ma_y, \quad \text{and} \quad F_{net,z} = ma_z. \qquad (5\text{-}2)$$

Each of these equations relates the net force component along an axis to the acceleration along that same axis. For example, the first equation tells us that the

**TABLE 5-1**

**Units in Newton's Second Law (Eqs. 5-1 and 5-2)**

| System | Force | Mass | Acceleration |
|---|---|---|---|
| SI | newton (N) | kilogram (kg) | $m/s^2$ |
| CGS[a] | dyne | gram (g) | $cm/s^2$ |
| British[b] | pound (lb) | slug | $ft/s^2$ |

[a] 1 dyne = 1 g·cm/s².
[b] 1 lb = 1 slug·ft/s².

sum of all the force components along the x axis causes the x component $a_x$ of the body's acceleration, but causes no acceleration in the y and z directions. Turned around, the acceleration component $a_x$ is caused only by the sum of the force components along the x axis. In general,

> The acceleration component along a given axis is caused *only* by the sum of the force components along that *same* axis, and not by force components along any other axis.

Equation 5-1 tells us that if the net force on a body is zero, the body's acceleration $\vec{a} = 0$. If the body is at rest, it stays at rest; if it is moving, it continues to move at constant velocity. In such cases, any forces on the body *balance* one another, and both the forces and the body are said to be in *equilibrium*. Commonly, the forces are also said to *cancel* one another, but the term "cancel" is tricky. It does *not* mean that the forces cease to exist (canceling forces is not like canceling dinner reservations). The forces still act on the body.

For SI units, Eq. 5-1 tells us that

$$1 \text{ N} = (1 \text{ kg})(1 \text{ m/s}^2) = 1 \text{ kg} \cdot \text{m/s}^2. \tag{5-3}$$

Some force units in other systems of units are given in Table 5-1 and Appendix D.

To solve problems with Newton's second law, we often draw a **free-body diagram** in which the only body shown is the one for which we are summing forces. A sketch of the body itself is preferred by some teachers but, to save space in these chapters, we shall usually represent the body with a dot. Also, each force on the body is drawn as a vector arrow with its tail on the body. A coordinate system is usually included, and the acceleration of the body is sometimes shown with a vector arrow (labeled as an acceleration).

A **system** consists of one or more bodies, and any force on the bodies inside the system from bodies outside the system is called an **external force.** If the bodies making up a system are rigidly connected to one another, we can treat the system as one composite body, and the net force $\vec{F}_{net}$ on it is the vector sum of all external forces. (We do not include **internal forces**—that is, forces between two bodies inside the system.) For example, a connected railroad engine and car form a system. If, say, a tow line pulls on the front of the engine, the force due to the tow line acts on the whole engine–car system. Just as for a single body, we can relate the net external force on a system to its acceleration with Newton's second law, $\vec{F}_{net} = m\vec{a}$, where m is the total mass of the system.

✓ **CHECKPOINT 2**   The figure here shows two horizontal forces acting on a block on a frictionless floor. If a third horizontal force $\vec{F}_3$ also acts on the block, what are the magnitude and direction of $\vec{F}_3$ when the block is (a) stationary and (b) moving to the left with a constant speed of 5 m/s?

## Sample Problem 5-1

Figures 5-3a to c show three situations in which one or two forces act on a puck that moves over frictionless ice along an x axis, in one-dimensional motion. The puck's mass is $m = 0.20$ kg. Forces $\vec{F}_1$ and $\vec{F}_2$ are directed along the axis and have magnitudes $F_1 = 4.0$ N and $F_2 = 2.0$ N. Force $\vec{F}_3$ is directed at angle $\theta = 30°$ and has magnitude $F_3 = 1.0$ N. In each situation, what is the acceleration of the puck?

**KEY IDEA**  In each situation we can relate the acceleration $\vec{a}$ to the net force $\vec{F}_{net}$ acting on the puck with Newton's second law, $\vec{F}_{net} = m\vec{a}$. However, because the motion is along only the x axis, we can simplify each situation by writing the second law for x components only:

$$F_{net,x} = ma_x. \tag{5-4}$$

The free-body diagrams for the three situations are given in Figs. 5-3d to f with the puck represented by a dot.

**Situation A:** For Fig. 5-3d, where only one horizontal force acts, Eq. 5-4 gives us

$$F_1 = ma_x,$$

which, with given data, yields

$$a_x = \frac{F_1}{m} = \frac{4.0 \text{ N}}{0.20 \text{ kg}} = 20 \text{ m/s}^2. \quad \text{(Answer)}$$

The positive answer indicates that the acceleration is in the positive direction of the x axis.

**Situation B:** In Fig. 5-3e, two horizontal forces act on the puck, $\vec{F}_1$ in the positive direction of x and $\vec{F}_2$ in the negative direction. Now Eq. 5-4 gives us

$$F_1 - F_2 = ma_x,$$

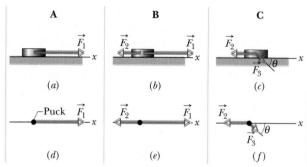

FIG. 5-3 (a)–(c) In three situations, forces act on a puck that moves along an x axis. (d)–(f) Free-body diagrams.

which, with given data, yields

$$a_x = \frac{F_1 - F_2}{m} = \frac{4.0 \text{ N} - 2.0 \text{ N}}{0.20 \text{ kg}} = 10 \text{ m/s}^2.$$

(Answer)

Thus, the net force accelerates the puck in the positive direction of the x axis.

**Situation C:** In Fig. 5-3f, force $\vec{F}_3$ is not directed along the direction of the puck's acceleration; only x component $F_{3,x}$ is. (Force $\vec{F}_3$ is two-dimensional but the motion is only one-dimensional.) Thus, we write Eq. 5-4 as

$$F_{3,x} - F_2 = ma_x. \tag{5-5}$$

From the figure, we see that $F_{3,x} = F_3 \cos \theta$. Solving for the acceleration and substituting for $F_{3,x}$ yield

$$a_x = \frac{F_{3,x} - F_2}{m} = \frac{F_3 \cos \theta - F_2}{m}$$

$$= \frac{(1.0 \text{ N})(\cos 30°) - 2.0 \text{ N}}{0.20 \text{ kg}} = -5.7 \text{ m/s}^2.$$

(Answer)

Thus, the net force accelerates the puck in the negative direction of the x axis.

## Sample Problem 5-2

In the overhead view of Fig. 5-4a, a 2.0 kg cookie tin is accelerated at 3.0 m/s² in the direction shown by $\vec{a}$, over a frictionless horizontal surface. The acceleration is caused by three horizontal forces, only two of which are shown: $\vec{F}_1$ of magnitude 10 N and $\vec{F}_2$ of magnitude 20 N. What is the third force $\vec{F}_3$ in unit-vector notation and in magnitude-angle notation?

**KEY IDEA**  The net force $\vec{F}_{net}$ on the tin is the sum of the three forces and is related to the acceleration $\vec{a}$ via Newton's second law ($\vec{F}_{net} = m\vec{a}$). Thus,

$$\vec{F}_1 + \vec{F}_2 + \vec{F}_3 = m\vec{a}, \tag{5-6}$$

which gives us

$$\vec{F}_3 = m\vec{a} - \vec{F}_1 - \vec{F}_2. \tag{5-7}$$

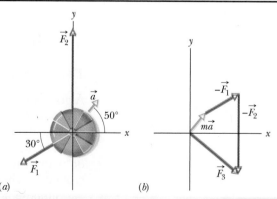

FIG. 5-4 (a) An overhead view of two of three horizontal forces that act on a cookie tin, resulting in acceleration $\vec{a}$. $\vec{F}_3$ is not shown. (b) An arrangement of vectors $m\vec{a}$, $-\vec{F}_1$, and $-\vec{F}_2$ to find force $\vec{F}_3$.

**Calculations:** Because this is a two-dimensional problem, we *cannot* find $\vec{F}_3$ merely by substituting the magnitudes for the vector quantities on the right side of Eq. 5-7. Instead, we must vectorially add $m\vec{a}$, $-\vec{F}_1$ (the reverse of $\vec{F}_1$), and $-\vec{F}_2$ (the reverse of $\vec{F}_2$), as shown in Fig. 5-4*b*. This addition can be done directly on a vector-capable calculator because we know both magnitude and angle for all three vectors. However, here we shall evaluate the right side of Eq. 5-7 in terms of components, first along the *x* axis and then along the *y* axis.

**x components:** Along the *x* axis we have

$$F_{3,x} = ma_x - F_{1,x} - F_{2,x}$$
$$= m(a\cos 50°) - F_1\cos(-150°) - F_2\cos 90°.$$

Then, substituting known data, we find

$$F_{3,x} = (2.0\text{ kg})(3.0\text{ m/s}^2)\cos 50° - (10\text{ N})\cos(-150°)$$
$$- (20\text{ N})\cos 90°$$
$$= 12.5\text{ N}.$$

**y components:** Similarly, along the *y* axis we find

$$F_{3,y} = ma_y - F_{1,y} - F_{2,y}$$
$$= m(a\sin 50°) - F_1\sin(-150°) - F_2\sin 90°$$
$$= (2.0\text{ kg})(3.0\text{ m/s}^2)\sin 50° - (10\text{ N})\sin(-150°)$$
$$- (20\text{ N})\sin 90°$$
$$= -10.4\text{ N}.$$

**Vector:** In unit-vector notation, we can write

$$\vec{F}_3 = F_{3,x}\hat{i} + F_{3,y}\hat{j} = (12.5\text{ N})\hat{i} - (10.4\text{ N})\hat{j}$$
$$\approx (13\text{ N})\hat{i} - (10\text{ N})\hat{j}. \qquad \text{(Answer)}$$

We can now use a vector-capable calculator to get the magnitude and the angle of $\vec{F}_3$. We can also use Eq. 3-6 to obtain the magnitude and the angle (from the positive direction of the *x* axis) as

$$F_3 = \sqrt{F_{3,x}^2 + F_{3,y}^2} = 16\text{ N}$$

and

$$\theta = \tan^{-1}\frac{F_{3,y}}{F_{3,x}} = -40°. \qquad \text{(Answer)}$$

---

### PROBLEM-SOLVING TACTICS

***Tactic 1: Dimensions and Vectors*** When you are dealing with forces, you cannot just add or subtract their magnitudes to find their net force unless they happen to be directed *along the same axis*. If they are not, you must use vector addition, either by means of a vector-capable calculator or by finding components along axes, one axis at a time, as is done in Sample Problem 5-2.

***Tactic 2: Reading Force Problems*** Read the problem statement several times until you have a good mental picture of what the situation is, what data are given, and what is requested. If you know what the problem is about but don't know what to do next, put the problem aside and reread the text. If you are hazy about Newton's second law, reread that section. Study the sample problems. And remember that solving physics problems (like repairing cars and designing computer chips) takes training.

***Tactic 3: Draw Two Types of Figures*** You may need two figures. One is a rough sketch of the actual situation. When you draw the forces, place the tail of each force vector either on the boundary of or within the body on which that force acts. The other figure is a free-body diagram: the forces on a *single* body are drawn, with the body represented by a dot or a sketch. Place the tail of each force vector on the dot or sketch.

***Tactic 4: What Is Your System?*** If you are using Newton's second law, you must know what body or system you are applying it to. In Sample Problem 5-1 it is the puck (not the ice). In Sample Problem 5-2, it is the cookie tin.

***Tactic 5: Choose Your Axes Wisely*** Often, we can save a lot of work by choosing one of our coordinate axes to coincide with one of the forces.

## 5-7 | Some Particular Forces

### The Gravitational Force

A **gravitational force** $\vec{F}_g$ on a body is a certain type of pull that is directed toward a second body. In these early chapters, we do not discuss the nature of this force and usually consider situations in which the second body is Earth. Thus, when we speak of *the* gravitational force $\vec{F}_g$ on a body, we usually mean a force that pulls on it directly toward the center of Earth—that is, directly down toward the ground. We shall assume that the ground is an inertial frame.

Suppose a body of mass *m* is in free fall with the free-fall acceleration of magnitude *g*. Then, if we neglect the effects of the air, the only force acting on the body is the gravitational force $\vec{F}_g$. We can relate this downward force and downward acceleration with Newton's second law ($\vec{F} = m\vec{a}$). We place a vertical $y$ axis along the body's path, with the positive direction upward. For this axis, Newton's second law can be written in the form $F_{net,y} = ma_y$, which, in our situation, becomes

$$-F_g = m(-g)$$

or $$F_g = mg. \tag{5-8}$$

In words, the magnitude of the gravitational force is equal to the product $mg$.

This same gravitational force, with the same magnitude, still acts on the body even when the body is not in free fall but is, say, at rest on a pool table or moving across the table. (For the gravitational force to disappear, Earth would have to disappear.)

We can write Newton's second law for the gravitational force in these vector forms:

$$\vec{F}_g = -F_g\hat{\jmath} = -mg\hat{\jmath} = m\vec{g}, \tag{5-9}$$

where $\hat{\jmath}$ is the unit vector that points upward along a $y$ axis, directly away from the ground, and $\vec{g}$ is the free-fall acceleration (written as a vector), directed downward.

## Weight

The **weight** $W$ of a body is the magnitude of the net force required to prevent the body from falling freely, as measured by someone on the ground. For example, to keep a ball at rest in your hand while you stand on the ground, you must provide an upward force to balance the gravitational force on the ball from Earth. Suppose the magnitude of the gravitational force is 2.0 N. Then the magnitude of your upward force must be 2.0 N, and thus the weight $W$ of the ball is 2.0 N. We also say that the ball *weighs* 2.0 N and speak about the ball *weighing* 2.0 N.

A ball with a weight of 3.0 N would require a greater force from you—namely, a 3.0 N force—to keep it at rest. The reason is that the gravitational force you must balance has a greater magnitude—namely, 3.0 N. We say that this second ball is *heavier* than the first ball.

Now let us generalize the situation. Consider a body that has an acceleration $\vec{a}$ of zero relative to the ground, which we again assume to be an inertial frame. Two forces act on the body: a downward gravitational force $\vec{F}_g$ and a balancing upward force of magnitude $W$. We can write Newton's second law for a vertical $y$ axis, with the positive direction upward, as

$$F_{net,y} = ma_y.$$

In our situation, this becomes

$$W - F_g = m(0) \tag{5-10}$$

or $$W = F_g \quad \text{(weight, with ground as inertial frame).} \tag{5-11}$$

This equation tells us (assuming the ground is an inertial frame) that

The weight $W$ of a body is equal to the magnitude $F_g$ of the gravitational force on the body.

Substituting $mg$ for $F_g$ from Eq. 5-8, we find

$$W = mg \quad \text{(weight),} \tag{5-12}$$

which relates a body's weight to its mass.

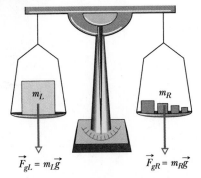

**FIG. 5-5** An equal-arm balance. When the device is in balance, the gravitational force $\vec{F}_{gL}$ on the body being weighed (on the left pan) and the total gravitational force $\vec{F}_{gR}$ on the reference bodies (on the right pan) are equal. Thus, the mass $m_L$ of the body being weighed is equal to the total mass $m_R$ of the reference bodies.

To *weigh* a body means to measure its weight. One way to do this is to place the body on one of the pans of an equal-arm balance (Fig. 5-5) and then place reference bodies (whose masses are known) on the other pan until we strike a balance (so that the gravitational forces on the two sides match). The masses on the pans then match, and we know the mass of the body. If we know the value of $g$ for the location of the balance, we can also find the weight of the body with Eq. 5-12.

We can also weigh a body with a spring scale (Fig. 5-6). The body stretches a spring, moving a pointer along a scale that has been calibrated and marked in either mass or weight units. (Most bathroom scales in the United States work this way and are marked in the force unit pounds.) If the scale is marked in mass units, it is accurate only where the value of $g$ is the same as where the scale was calibrated.

The weight of a body must be measured when the body is not accelerating vertically relative to the ground. For example, you can measure your weight on a scale in your bathroom or on a fast train. However, if you repeat the measurement with the scale in an accelerating elevator, the reading differs from your weight because of the acceleration. Such a measurement is called an *apparent weight.*

*Caution:* A body's weight is not its mass. Weight is the magnitude of a force and is related to mass by Eq. 5-12. If you move a body to a point where the value of $g$ is different, the body's mass (an intrinsic property) is not different but the weight is. For example, the weight of a bowling ball having a mass of 7.2 kg is 71 N on Earth but only 12 N on the Moon. The mass is the same on Earth and Moon, but the free-fall acceleration on the Moon is only 1.6 m/s$^2$.

### The Normal Force

If you stand on a mattress, Earth pulls you downward, but you remain stationary. The reason is that the mattress, because it deforms downward due to you, pushes up on you. Similarly, if you stand on a floor, it deforms (it is compressed, bent, or buckled ever so slightly) and pushes up on you. Even a seemingly rigid concrete floor does this (if it is not sitting directly on the ground, enough people on the floor could break it).

The push on you from the mattress or floor is a **normal force** $\vec{F}_N$. The name comes from the mathematical term *normal,* meaning perpendicular: The force on you from, say, the floor is perpendicular to the floor.

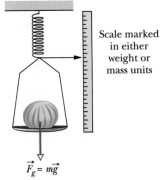

**FIG. 5-6** A spring scale. The reading is proportional to the *weight* of the object on the pan, and the scale gives that weight if marked in weight units. If, instead, it is marked in mass units, the reading is the object's weight only if the value of $g$ at the location where the scale is being used is the same as the value of $g$ at the location where the scale was calibrated.

Scale marked in either weight or mass units

$\vec{F}_g = m\vec{g}$

When a body presses against a surface, the surface (even a seemingly rigid one) deforms and pushes on the body with a normal force $\vec{F}_N$ that is perpendicular to the surface.

Figure 5-7$a$ shows an example. A block of mass $m$ presses down on a table, deforming it somewhat because of the gravitational force $\vec{F}_g$ on the block. The table pushes up on the block with normal force $\vec{F}_N$. The free-body diagram for the block is given in Fig. 5-7$b$. Forces $\vec{F}_g$ and $\vec{F}_N$ are the only two forces on the block and they are both vertical. Thus, for the block we can write Newton's second law for a positive-upward $y$ axis ($F_{\text{net},y} = ma_y$) as

$$F_N - F_g = ma_y.$$

From Eq. 5-8, we substitute $mg$ for $F_g$, finding

$$F_N - mg = ma_y.$$

Then the magnitude of the normal force is

$$F_N = mg + ma_y = m(g + a_y) \qquad (5\text{-}13)$$

for any vertical acceleration $a_y$ of the table and block (they might be in an accelerating elevator). If the table and block are not accelerating relative to the ground, then $a_y = 0$ and Eq. 5-13 yields

$$F_N = mg. \qquad (5\text{-}14)$$

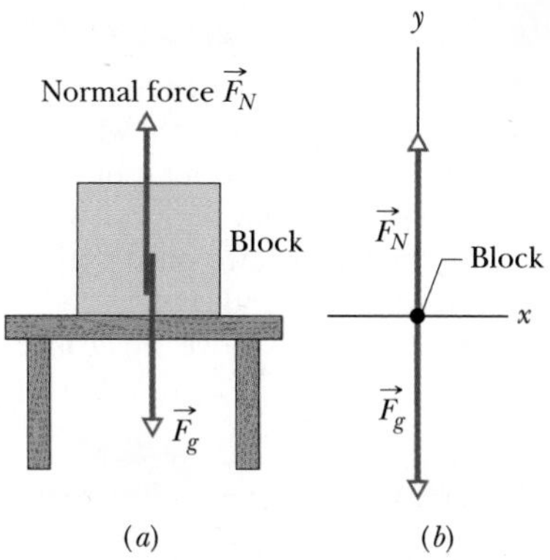

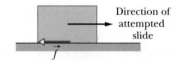

✓ **CHECKPOINT 3**  In Fig. 5-7, is the magnitude of the normal force $\vec{F}_N$ greater than, less than, or equal to $mg$ if the block and table are in an elevator moving upward (a) at constant speed and (b) at increasing speed?

**FIG. 5-7**  (*a*) A block resting on a table experiences a normal force $\vec{F}_N$ perpendicular to the tabletop. (*b*) The free-body diagram for the block.

## Friction

If we either slide or attempt to slide a body over a surface, the motion is resisted by a bonding between the body and the surface. (We discuss this bonding more in the next chapter.) The resistance is considered to be a single force $\vec{f}$, called either the **frictional force** or simply **friction.** This force is directed along the surface, opposite the direction of the intended motion (Fig. 5-8). Sometimes, to simplify a situation, friction is assumed to be negligible (the surface is *frictionless*).

## Tension

When a cord (or a rope, cable, or other such object) is attached to a body and pulled taut, the cord pulls on the body with a force $\vec{T}$ directed away from the body and along the cord (Fig. 5-9*a*). The force is often called a *tension force* because the cord is said to be in a state of *tension* (or to be *under tension*), which means that it is being pulled taut. The *tension in the cord* is the magnitude $T$ of the force on the body. For example, if the force on the body from the cord has magnitude $T = 50$ N, the tension in the cord is 50 N.

A cord is often said to be *massless* (meaning its mass is negligible compared to the body's mass) and *unstretchable*. The cord then exists only as a connection between two bodies. It pulls on both bodies with the same force magnitude $T$, even if the bodies and the cord are accelerating and even if the cord runs around a *massless, frictionless pulley* (Figs. 5-9*b* and *c*). Such a pulley has negligible mass compared to the bodies and negligible friction on its axle opposing its rotation. If the cord wraps halfway around a pulley, as in Fig. 5-9*c*, the net force on the pulley from the cord has the magnitude $2T$.

**FIG. 5-8**  A frictional force $\vec{f}$ opposes the attempted slide of a body over a surface.

✓ **CHECKPOINT 4**  The suspended body in Fig. 5-9*c* weighs 75 N. Is $T$ equal to, greater than, or less than 75 N when the body is moving upward (a) at constant speed, (b) at increasing speed, and (c) at decreasing speed?

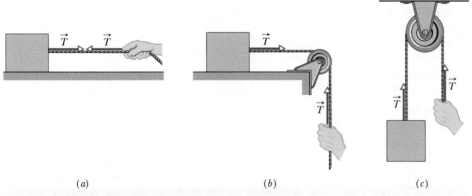

**FIG. 5-9**  (*a*) The cord, pulled taut, is under tension. If its mass is negligible, the cord pulls on the body and the hand with force $\vec{T}$, even if the cord runs around a massless, frictionless pulley as in (*b*) and (*c*).

*Tactic 6: Normal Force* Equation 5-14 for the normal force on a body holds only when $\vec{F}_N$ is directed upward and the body's vertical acceleration is zero; so we do *not* apply it for other orientations of $\vec{F}_N$ or when the vertical acceleration is not zero. Instead, we must derive a new expression for $\vec{F}_N$ from Newton's second law.

We are free to move $\vec{F}_N$ around in a figure as long as we maintain its orientation. For example, in Fig. 5-7a we can slide it downward so that its head is at the boundary between block and tabletop. However, $\vec{F}_N$ is least likely to be misinterpreted when its tail is either at that boundary or somewhere within the block (as shown). An even better technique is to draw a free-body diagram as in Fig. 5-7b, with the tail of $\vec{F}_N$ directly on the dot or sketch representing the block.

## Sample Problem 5-3

*Takeoff illusion.* A jet plane taking off from an aircraft carrier is propelled by its powerful engines while being thrown forward by a catapult mechanism installed in the carrier deck. The resulting high acceleration allows the plane to reach takeoff speed in a short distance on the deck. However, that high acceleration also compels the pilot to angle the plane sharply nose-down as it leaves the deck. Pilots are trained to ignore this compulsion, but occasionally a plane is flown straight into the ocean. Let's explore the physics behind the compulsion.

Your sense of vertical depends on visual clues and on the vestibular system located in your inner ear. That system contains tiny hair cells in a fluid. When you hold your head upright, the hairs are vertically in line with the gravitational force $\vec{F}_g$ on you and the system signals your brain that your head is upright. When you tilt your head backward by some angle $\phi$, the hairs are bent and the system signals your brain about the tilt. The hairs are also bent when you are accelerated forward by an applied horizontal force $\vec{F}_{app}$. The signal sent to your brain then indicates, erroneously, that your head is tilted back, to be in line with an extension through the vector sum $\vec{F}_{sum} = \vec{F}_g + \vec{F}_{app}$ (Fig. 5-10a). However, the erroneous signal is ignored when visual clues clearly indicate no tilt, such as when you are accelerated in a car.

A pilot being hurled along the deck of an aircraft carrier at night has almost no visual clues. The illusion of tilt is strong and very convincing, with the result that the pilot feels as though the plane leaves the deck headed sharply upward. Without proper training, a pilot will attempt to level the plane by bringing its nose sharply down, sending the plane into the ocean.

Suppose that, starting from rest, a pilot undergoes constant horizontal acceleration to reach a takeoff speed of 85 m/s in 90 m. What is the angle $\phi$ of the illusionary tilt experienced by the pilot?

**KEY IDEAS** (1) We can use Newton's second law to relate the magnitude $F_{app}$ of the force on the pilot (from the seatback) to the resulting acceleration $a_x$: $F_{app} = ma_x$, where $m$ is the mass of the pilot. (2) Because the acceleration is constant, we can use the equations of Table 2-1 to find $a_x$.

**Calculations:** We need to find the tilt angle $\phi$ of the line that extends through $\vec{F}_{sum}$, the vector sum of the vertical gravitational force $\vec{F}_g$ acting on the pilot and the horizontal applied force $\vec{F}_{app}$. We can find $\phi$ by rearranging the force vectors as in Fig. 5-10b and then writing

$$\tan \phi = \frac{F_{app}}{F_g},$$

or

$$\phi = \tan^{-1}\left(\frac{F_{app}}{F_g}\right). \qquad (5\text{-}15)$$

Since we know the initial speed ($v_0 = 0$), the final speed ($v_x = 85$ m/s), and the displacement ($x - x_0 = 90$ m), we use Eq. 2-16 ($v^2 = v_0^2 + 2a(x - x_0)$) to write

$$(85 \text{ m/s})^2 = 0^2 + 2a_x(90 \text{ m}),$$

or

$$a_x = 40.1 \text{ m/s}^2.$$

Then, by Newton's second law, $F_{app} = m(40.1 \text{ m/s}^2)$. Substituting this result and the result $F_g = m(9.8 \text{ m/s}^2)$ in Eq. 5-15 gives us

$$\phi = \tan^{-1}\left(\frac{m(40.1 \text{ m/s}^2)}{m(9.8 \text{ m/s}^2)}\right) = 76°. \quad \text{(Answer)}$$

Thus, as the plane is accelerated along the carrier deck, the pilot feels an illusion of a backward tilt of 76°, as though the plane is angled nose-up by 76°. The illusion may compel the pilot to put the plane nose-down by 76° just after takeoff.

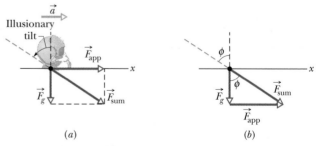

**FIG. 5-10** (a) Force $\vec{F}_{app}$, directed to the right, is applied to the pilot during takeoff. The pilot's head feels as though it is tilted back along the red dashed line. (b) The vector sum $\vec{F}_{sum}$ ($= \vec{F}_g + \vec{F}_{app}$) is at angle $\phi$ from the vertical.

# 5-8 | Newton's Third Law

Two bodies are said to *interact* when they push or pull on each other—that is, when a force acts on each body due to the other body. For example, suppose you position a book $B$ so it leans against a crate $C$ (Fig. 5-11a). Then the book and crate interact: There is a horizontal force $\vec{F}_{BC}$ on the book from the crate (or due to the crate) and a horizontal force $\vec{F}_{CB}$ on the crate from the book (or due to the book). This pair of forces is shown in Fig. 5-11b. Newton's third law states that

> **Newton's Third Law:** When two bodies interact, the forces on the bodies from each other are always equal in magnitude and opposite in direction.

For the book and crate, we can write this law as the scalar relation

$$F_{BC} = F_{CB} \qquad \text{(equal magnitudes)}$$

or as the vector relation

$$\vec{F}_{BC} = -\vec{F}_{CB} \qquad \text{(equal magnitudes and opposite directions),}$$

where the minus sign means that these two forces are in opposite directions. We can call the forces between two interacting bodies a **third-law force pair.** When any two bodies interact in any situation, a third-law force pair is present. The book and crate in Fig. 5-11a are stationary, but the third law would still hold if they were moving and even if they were accelerating.

As another example, let us find the third-law force pairs involving the cantaloupe in Fig. 5-12a, which lies on a table that stands on Earth. The cantaloupe interacts with the table and with Earth (this time, there are three bodies whose interactions we must sort out).

Let's first focus on the forces acting on the cantaloupe (Fig. 5-12b). Force $\vec{F}_{CT}$ is the normal force on the cantaloupe from the table, and force $\vec{F}_{CE}$ is the gravitational force on the cantaloupe due to Earth. Are they a third-law force pair? No, because they are forces on a single body, the cantaloupe, and not on two interacting bodies.

To find a third-law pair, we must focus not on the cantaloupe but on the interaction between the cantaloupe and one other body. In the cantaloupe–Earth interaction (Fig. 5-12c), Earth pulls on the cantaloupe with a gravitational force $\vec{F}_{CE}$ and the cantaloupe pulls on Earth with a gravitational force $\vec{F}_{EC}$. Are these forces a third-law force pair? Yes, because they are forces on two interacting bodies, the force on each due to the other. Thus, by Newton's third law,

$$\vec{F}_{CE} = -\vec{F}_{EC} \qquad \text{(cantaloupe – Earth interaction).}$$

Next, in the cantaloupe–table interaction, the force on the cantaloupe from the table is $\vec{F}_{CT}$ and, conversely, the force on the table from the cantaloupe is $\vec{F}_{TC}$ (Fig. 5-12d). These forces are also a third-law force pair, and so

$$\vec{F}_{CT} = -\vec{F}_{TC} \qquad \text{(cantaloupe–table interaction).}$$

✓ **CHECKPOINT 5** Suppose that the cantaloupe and table of Fig. 5-12 are in an elevator cab that begins to accelerate upward. (a) Do the magnitudes of $\vec{F}_{TC}$ and $\vec{F}_{CT}$ increase, decrease, or stay the same? (b) Are those two forces still equal in magnitude and opposite in direction? (c) Do the magnitudes of $\vec{F}_{CE}$ and $\vec{F}_{EC}$ increase, decrease, or stay the same? (d) Are those two forces still equal in magnitude and opposite in direction?

# 5-9 | Applying Newton's Laws

The rest of this chapter consists of sample problems. You should pore over them, learning their procedures for attacking a problem. Especially important is knowing how to translate a sketch of a situation into a free-body diagram with appropriate axes, so that Newton's laws can be applied.

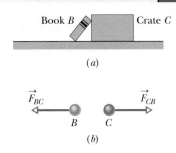

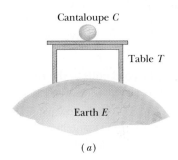

**FIG. 5-11** (a) Book $B$ leans against crate $C$. (b) Forces $\vec{F}_{BC}$ (the force on the book from the crate) and $\vec{F}_{CB}$ (the force on the crate from the book) have the same magnitude and are opposite in direction.

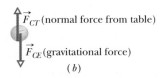

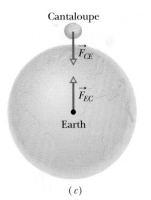

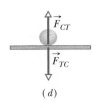

**FIG. 5-12** (a) A cantaloupe lies on a table that stands on Earth. (b) The forces *on the cantaloupe* are $\vec{F}_{CT}$ and $\vec{F}_{CE}$. (c) The third-law force pair for the cantaloupe–Earth interaction. (d) The third-law force pair for the cantaloupe–table interaction.

**Sample Problem** | **5-4** | **Build your skill**

Figure 5-13 shows a block $S$ (the *sliding block*) with mass $M = 3.3$ kg. The block is free to move along a horizontal frictionless surface and connected, by a cord that wraps over a frictionless pulley, to a second block $H$ (the *hanging block*), with mass $m = 2.1$ kg. The cord and pulley have negligible masses compared to the blocks (they are "massless"). The hanging block $H$ falls as the sliding block $S$ accelerates to the right. Find (a) the acceleration of block $S$, (b) the acceleration of block $H$, and (c) the tension in the cord.

**Q**  *What is this problem all about?*

You are given two bodies—sliding block and hanging block—but must also consider *Earth,* which pulls on both bodies. (Without Earth, nothing would happen here.) A total of five forces act on the blocks, as shown in Fig. 5-14:

1.  The cord pulls to the right on sliding block $S$ with a force of magnitude $T$.

2.  The cord pulls upward on hanging block $H$ with a force of the same magnitude $T$. This upward force keeps block $H$ from falling freely.

3.  Earth pulls down on block $S$ with the gravitational force $\vec{F}_{gS}$, which has a magnitude equal to $Mg$.

4.  Earth pulls down on block $H$ with the gravitational force $\vec{F}_{gH}$, which has a magnitude equal to $mg$.

5.  The table pushes up on block $S$ with a normal force $\vec{F}_N$.

There is another thing you should note. We assume that the cord does not stretch, so that if block $H$ falls 1 mm in a certain time, block $S$ moves 1 mm to the right in that same time. This means that the blocks move together and their accelerations have the same magnitude $a$.

**Q**  *How do I classify this problem? Should it suggest a particular law of physics to me?*

Yes. Forces, masses, and accelerations are involved, and they should suggest Newton's second law of motion, $\vec{F}_{net} = m\vec{a}$. That is our starting **Key Idea**.

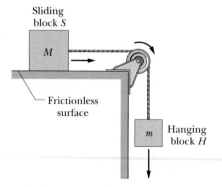

FIG. 5-13  A block $S$ of mass $M$ is connected to a block $H$ of mass $m$ by a cord that wraps over a pulley.

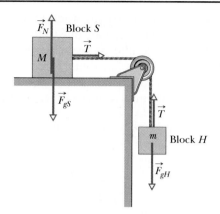

FIG. 5-14  The forces acting on the two blocks of Fig. 5-13.

**Q**  *If I apply Newton's second law to this problem, to which body should I apply it?*

We focus on two bodies, the sliding block and the hanging block. Although they are *extended objects* (they are not points), we can still treat each block as a particle because every part of it moves in exactly the same way. A second **Key Idea** is to apply Newton's second law separately to each block.

**Q**  *What about the pulley?*

We cannot represent the pulley as a particle because different parts of it move in different ways. When we discuss rotation, we shall deal with pulleys in detail. Meanwhile, we eliminate the pulley from consideration by assuming its mass to be negligible compared with the masses of the two blocks. Its only function is to change the cord's orientation.

**Q**  *OK. Now how do I apply $\vec{F}_{net} = m\vec{a}$ to the sliding block?*

Represent block $S$ as a particle of mass $M$ and draw *all* the forces that act *on* it, as in Fig. 5-15a. This is the block's free-body diagram. Next, draw a set of axes. It makes sense to draw the $x$ axis parallel to the table, in the direction in which the block moves.

**Q**  *Thanks, but you still haven't told me how to apply $\vec{F}_{net} = m\vec{a}$ to the sliding block. All you've done is explain how to draw a free-body diagram.*

You are right, and here's the third **Key Idea**: The expression $\vec{F}_{net} = M\vec{a}$ is a vector equation, so we can write it as three component equations:

$$F_{net,x} = Ma_x \quad F_{net,y} = Ma_y \quad F_{net,z} = Ma_z \quad (5\text{-}16)$$

in which $F_{net,x}$, $F_{net,y}$, and $F_{net,z}$ are the components of the net force along the three axes. Now we apply each component equation to its corresponding direction. Because block $S$ does not accelerate vertically, $F_{net,y} = Ma_y$ becomes

$$F_N - F_{gS} = 0 \quad \text{or} \quad F_N = F_{gS}.$$

Thus in the $y$ direction, the magnitude of the normal force is equal to the magnitude of the gravitational force.

No force acts in the $z$ direction, which is perpendicular to the page.

In the $x$ direction, there is only one force component, which is $T$. Thus, $F_{net,x} = Ma_x$ becomes

$$T = Ma. \tag{5-17}$$

This equation contains two unknowns, $T$ and $a$; so we cannot yet solve it. Recall, however, that we have not said anything about the hanging block.

**Q**  *I agree. How do I apply $\vec{F}_{net} = m\vec{a}$ to the hanging block?*

We apply it just as we did for block $S$: Draw a free-body diagram for block $H$, as in Fig. 5-15$b$. Then apply $\vec{F}_{net} = m\vec{a}$ in component form. This time, because the acceleration is along the $y$ axis, we use the $y$ part of Eq. 5-16 ($F_{net,y} = ma_y$) to write

$$T - F_{gH} = ma_y.$$

We can now substitute $mg$ for $F_{gH}$ and $-a$ for $a_y$ (negative because block $H$ accelerates in the negative direction of the $y$ axis). We find

$$T - mg = -ma. \tag{5-18}$$

Now note that Eqs. 5-17 and 5-18 are simultaneous equations with the same two unknowns, $T$ and $a$. Subtracting these equations eliminates $T$. Then solving for $a$ yields

$$a = \frac{m}{M + m} g. \tag{5-19}$$

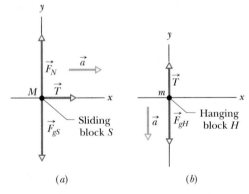

(a)                         (b)

**FIG. 5-15**  ($a$) A free-body diagram for block $S$ of Fig. 5-13. ($b$) A free-body diagram for block $H$ of Fig. 5-13.

Substituting this result into Eq. 5-17 yields

$$T = \frac{Mm}{M + m} g. \tag{5-20}$$

Putting in the numbers gives, for these two quantities,

$$a = \frac{m}{M + m} g = \frac{2.1 \text{ kg}}{3.3 \text{ kg} + 2.1 \text{ kg}} (9.8 \text{ m/s}^2)$$

$$= 3.8 \text{ m/s}^2 \qquad \text{(Answer)}$$

and  $$T = \frac{Mm}{M + m} g = \frac{(3.3 \text{ kg})(2.1 \text{ kg})}{3.3 \text{ kg} + 2.1 \text{ kg}} (9.8 \text{ m/s}^2)$$

$$= 13 \text{ N}. \qquad \text{(Answer)}$$

**Q**  *The problem is now solved, right?*

That's a fair question, but the problem is not really finished until we have examined the results to see whether they make sense. (If you made these calculations on the job, wouldn't you want to see whether they made sense before you turned them in?)

Look first at Eq. 5-19. Note that it is dimensionally correct and that the acceleration $a$ will always be less than $g$. This is as it must be, because the hanging block is not in free fall. The cord pulls upward on it.

Look now at Eq. 5-20, which we can rewrite in the form

$$T = \frac{M}{M + m} mg. \tag{5-21}$$

In this form, it is easier to see that this equation is also dimensionally correct, because both $T$ and $mg$ have dimensions of forces. Equation 5-21 also lets us see that the tension in the cord is always less than $mg$, and thus is always less than the gravitational force on the hanging block. That is a comforting thought because, if $T$ were *greater* than $mg$, the hanging block would accelerate upward.

We can also check the results by studying special cases, in which we can guess what the answers must be. A simple example is to put $g = 0$, as if the experiment were carried out in interstellar space. We know that in that case, the blocks would not move from rest, there would be no forces on the ends of the cord, and so there would be no tension in the cord. Do the formulas predict this? Yes, they do. If you put $g = 0$ in Eqs. 5-19 and 5-20, you find $a = 0$ and $T = 0$. Two more special cases you might try are $M = 0$ and $m \to \infty$.

---

**Sample Problem**  **5-5**

In Fig. 5-16$a$, a cord pulls on a box of sea biscuits up along a frictionless plane inclined at $\theta = 30°$. The box has mass $m = 5.00$ kg, and the force from the cord has magnitude $T = 25.0$ N. What is the box's acceleration component $a$ along the inclined plane?

**KEY IDEA**  The acceleration along the plane is set by the force components along the plane (not by force components perpendicular to the plane), as expressed by Newton's second law (Eq. 5-1).

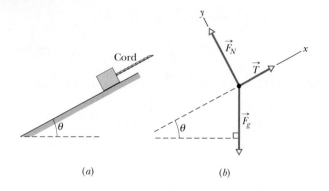

Cord

$(a)$

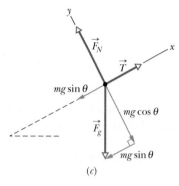

$(b)$

**FIG. 5-16** $(a)$ A box is pulled up a plane by a cord. $(b)$ The three forces acting on the box: the cord's force $\vec{T}$, the gravittional force $\vec{F}_g$, and the normal force $\vec{F}_N$. $(c)$ The components of $\vec{F}_g$ along the plane and pependicular to it.

$mg \sin \theta$

$mg \cos \theta$

$mg \sin \theta$

$(c)$

***Calculation:*** For convenience, we draw a coordinate system and a free-body diagram as shown in Fig. 5-16$b$. The positive direction of the $x$ axis is up the plane. Force $\vec{T}$ from the cord is up the plane and has magnitude $T = 25.0$ N. The gravitational force $\vec{F}_g$ is downward and has magnitude $mg = (5.00$ kg$)(9.8$ m/s$^2) = 49.0$ N. More important, its component along the plane is down the plane and has magnitude $mg \sin \theta$ as indicated in Fig. 5-16$c$. (To see why that trig function is involved, compare the right triangles in Figs. 5-16$b$ and $c$.) To indicate the direction, we can write the component as $-mg \sin \theta$. The normal force $\vec{F}_N$ is perpendicular to the plane and thus does not determine acceleration along the plane.

We write Newton's second law ($\vec{F}_{net} = m\vec{a}$) for motion along the $x$ axis as

$$T - mg \sin \theta = ma. \tag{5-22}$$

Substituting data and solving for $a$, we find

$$a = 0.100 \text{ m/s}^2, \tag{Answer}$$

where the positive result indicates that the box accelerates up the plane.

---

## Sample Problem | 5-6

Let's return to the chapter opening question: What produces the fear factor in the last car of a traditional gravity-driven roller coaster? Let's consider a coaster having 10 identical cars with total mass $M$ and massless interconnections. Figure 5-17$a$ shows the coaster just after the first car has begun its descent along a frictionless slope with an angle $\theta$. Figure 5-17$b$ shows the coaster just before the last car begins its descent. What is the acceleration of the coaster in these two situations?

**KEY IDEAS** (1) The net force on an object causes the object's acceleration, as related by Newton's second law (Eq. 5-1, $\vec{F}_{net} = m\vec{a}$). (2) When the motion is along a single axis, we write that law in component form (such as $F_{net,x} = ma_x$) and we use only force components along that axis. (3) When several objects move together at the same velocity and with the same acceleration, they can be regarded as a single composite object. *Internal forces* act between the individual objects, but only *external forces* can cause the composite object to accelerate.

***Calculations for Fig. 5-17a:*** Figure 5-17$c$ shows free-body diagrams associated with Fig. 5-17$a$, with convenient axes superimposed. The tilted $x'$ axis has its positive direction up the slope. $T$ is the magnitude of the interconnection force between the car on the slope and the cars still on the plateau. Because the coaster consists of 10 identical cars with total mass $M$, the mass of the

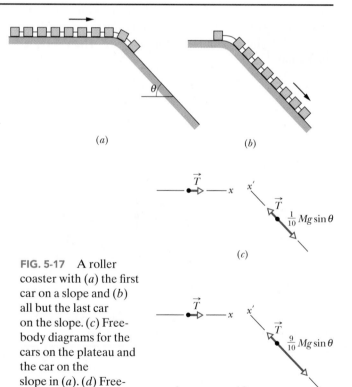

$(a)$

$(b)$

$\vec{T}$ — $x$

$x'$

$\vec{T}$

$\frac{1}{10} Mg \sin \theta$

$(c)$

**FIG. 5-17** A roller coaster with $(a)$ the first car on a slope and $(b)$ all but the last car on the slope. $(c)$ Free-body diagrams for the cars on the plateau and the car on the slope in $(a)$. $(d)$ Free-body diagrams for $(b)$.

$\vec{T}$ — $x$

$x'$

$\vec{T}$

$\frac{9}{10} Mg \sin \theta$

$(d)$

car on the slope is $\frac{1}{10}M$ and the mass of the cars on the plateau is $\frac{9}{10}M$. Only a single *external* force acts along the $x$ axis on the nine-car composite—namely, the interconnection force with magnitude $T$. (The forces between the nine cars are internal forces.) Thus, Newton's second law for motion along the $x$ axis ($F_{net,\,x} = ma_x$) becomes

$$T = \tfrac{9}{10}Ma, \tag{5-23}$$

where $a$ is the magnitude of the acceleration $a_x$ along the $x$ axis.

Along the tilted $x'$ axis, two forces act on the car on the slope: the interconnection force with magnitude $T$ (in the positive direction of the axis) and the $x'$ component of the gravitational force (in the negative direction of the axis). From Sample Problem 5-5, we know to write that gravitational component as $-mg \sin \theta$, where $m$ is the mass. Because we know that the car accelerates *down* the slope in the negative $x'$ direction with magnitude $a$, we can write the acceleration as $-a$. Thus, for this car, with mass $\tfrac{1}{10}M$ we write Newton's second law for motion along the $x'$ axis as

$$T - \tfrac{1}{10}Mg \sin \theta = \tfrac{1}{10}M(-a). \tag{5-24}$$

Substituting for $T$ from Eq. 5-23 and solving for $a$, we have

$$a = \tfrac{1}{10}g \sin \theta. \qquad \text{(Answer)}$$

**Calculations for Fig. 5-17b:** Figure 5-17d shows free-body diagrams associated with Fig. 5-17b. For the car

still on the plateau, we rewrite Eq. 5-23 as

$$T = \tfrac{1}{10}Ma.$$

For the nine cars on the slope, we rewrite Eq. 5-24 as

$$T - \tfrac{9}{10}Mg \sin \theta = \tfrac{9}{10}M(-a).$$

Again solving for $a$, we now find

$$a = \tfrac{9}{10}g \sin \theta. \qquad \text{(Answer)}$$

**The fear factor:** This last answer is 9 times the first answer. Thus, in general, the acceleration of the cars greatly increases as more of them go over the edge and onto the slope. That increase in acceleration occurs regardless of your car choice, but your interpretation of the acceleration depends on the choice. In the first car, most of the acceleration occurs on the slope and is due to the component of the gravitational force along the slope, which is reasonable. In the last car, most of the acceleration occurs on the plateau and is due to the push on you from the back of your seat. That push rapidly increases as you approach the edge, giving you the frightening sensation that you are about to be hurled off the plateau and into the air.

---

**Sample Problem** | **5-7** | **Build your skill**

Figure 5-18a shows the general arrangement in which two forces are applied to a 4.00 kg block on a frictionless floor, but only force $\vec{F}_1$ is indicated. That force has a fixed magnitude but can be applied at angle $\theta$ to the positive direction of the $x$ axis. Force $\vec{F}_2$ is horizontal and fixed in both magnitude and angle. Figure 5-18b gives the horizontal acceleration $a_x$ of the block for any given value of $\theta$ from 0° to 90°. What is the value of $a_x$ for $\theta = 180°$?

**KEY IDEAS** (1) The horizontal acceleration $a_x$ depends on the net horizontal force $F_{net,\,x}$, as given by Newton's second law. (2) The net horizontal force is the sum of the horizontal components of forces $\vec{F}_1$ and $\vec{F}_2$.

**Calculations:** The $x$ component of $\vec{F}_2$ is $F_2$ because the vector is horizontal. The $x$ component of $\vec{F}_1$ is $F_1 \cos \theta$. Using these expressions and a mass $m$ of 4.00 kg, we can write Newton's second law ($\vec{F}_{net} = m\vec{a}$) for motion along the $x$ axis as

$$F_1 \cos \theta + F_2 = 4.00a_x. \tag{5-25}$$

From this equation we see that when $\theta = 90°$, $F_1 \cos \theta$ is zero and $F_2 = 4.00a_x$. From the graph we see that the corresponding acceleration is 0.50 m/s². Thus,

$F_2 = 2.00$ N and $\vec{F}_2$ must be in the positive direction of the $x$ axis.

From Eq. 5-25, we find that when $\theta = 0°$,

$$F_1 \cos 0° + 2.00 = 4.00a_x. \tag{5-26}$$

From the graph we see that the corresponding acceleration is 3.0 m/s². From Eq. 5-26, we then find that $F_1 = 10$ N.

Substituting $F_1 = 10$ N, $F_2 = 2.00$ N, and $\theta = 180°$ into Eq. 5-25 leads to

$$a_x = -2.00 \text{ m/s}^2. \qquad \text{(Answer)}$$

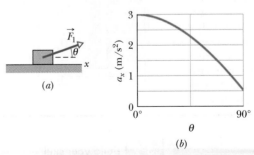

(a)

(b)

**FIG. 5-18** (a) One of the two forces applied to a block is shown. Its angle $\theta$ can be varied. (b) The block's acceleration component $a_x$ versus $\theta$.

**Sample Problem** 5-8 Build your skill

In Fig. 5-19a, a passenger of mass $m = 72.2$ kg stands on a platform scale in an elevator cab. We are concerned with the scale readings when the cab is stationary and when it is moving up or down.

(a) Find a general solution for the scale reading, whatever the vertical motion of the cab.

**KEY IDEAS** (1) The reading is equal to the magnitude of the normal force $\vec{F}_N$ on the passenger from the scale. The only other force acting on the passenger is the gravitational force $\vec{F}_g$, as shown in the free-body diagram of Fig. 5-19b. (2) We can relate the forces on the passenger to his acceleration $\vec{a}$ by using Newton's second law ($\vec{F}_{net} = m\vec{a}$). However, recall that we can use this law only in an inertial frame. If the cab accelerates, then it is *not* an inertial frame. So we choose the ground to be our inertial frame and make any measure of the passenger's acceleration relative to it.

**Calculations:** Because the two forces on the passenger and his acceleration are all directed vertically, along the $y$ axis in Fig. 5-19b, we can use Newton's second law written for $y$ components ($F_{net,y} = ma_y$) to get

$$F_N - F_g = ma$$

or
$$F_N = F_g + ma. \quad (5\text{-}27)$$

This tells us that the scale reading, which is equal to $F_N$, depends on the vertical acceleration. Substituting $mg$ for $F_g$ gives us

$$F_N = m(g + a) \quad \text{(Answer)} \quad (5\text{-}28)$$

for any choice of acceleration $a$.

(b) What does the scale read if the cab is stationary or moving upward at a constant 0.50 m/s?

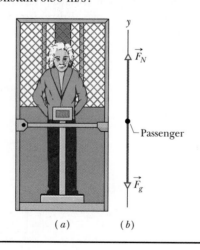

**FIG. 5-19** (a) A passenger stands on a platform scale that indicates either his weight or his apparent weight. (b) The free-body diagram for the passenger, showing the normal force $\vec{F}_N$ on him from the scale and the gravitational force $\vec{F}_g$.

**KEY IDEA** For any constant velocity (zero or otherwise), the acceleration $a$ of the passenger is zero.

**Calculation:** Substituting this and other known values into Eq. 5-28, we find

$$F_N = (72.2 \text{ kg})(9.8 \text{ m/s}^2 + 0) = 708 \text{ N}. \quad \text{(Answer)}$$

This is the weight of the passenger and is equal to the magnitude $F_g$ of the gravitational force on him.

(c) What does the scale read if the cab accelerates upward at 3.20 m/s² and downward at 3.20 m/s²?

**Calculations:** For $a = 3.20$ m/s², Eq. 5-28 gives

$$F_N = (72.2 \text{ kg})(9.8 \text{ m/s}^2 + 3.20 \text{ m/s}^2)$$
$$= 939 \text{ N}, \quad \text{(Answer)}$$

and for $a = -3.20$ m/s², it gives

$$F_N = (72.2 \text{ kg})(9.8 \text{ m/s}^2 - 3.20 \text{ m/s}^2)$$
$$= 477 \text{ N}. \quad \text{(Answer)}$$

For an upward acceleration (either the cab's upward speed is increasing or its downward speed is decreasing), the scale reading is greater than the passenger's weight. That reading is a measurement of an apparent weight, because it is made in a noninertial frame. For a downward acceleration (either decreasing upward speed or increasing downward speed), the scale reading is less than the passenger's weight.

(d) During the upward acceleration in part (c), what is the magnitude $F_{net}$ of the net force on the passenger, and what is the magnitude $a_{p,cab}$ of his acceleration as measured in the frame of the cab? Does $\vec{F}_{net} = m\vec{a}_{p,cab}$?

**Calculation:** The magnitude $F_g$ of the gravitational force on the passenger does not depend on the motion of the passenger or the cab; so, from part (b), $F_g$ is 708 N. From part (c), the magnitude $F_N$ of the normal force on the passenger during the upward acceleration is the 939 N reading on the scale. Thus, the net force on the passenger is

$$F_{net} = F_N - F_g = 939 \text{ N} - 708 \text{ N} = 231 \text{ N}, \quad \text{(Answer)}$$

during the upward acceleration. However, his acceleration $a_{p,cab}$ relative to the frame of the cab is zero. Thus, in the noninertial frame of the accelerating cab, $F_{net}$ is not equal to $ma_{p,cab}$, and Newton's second law does not hold.

**Sample Problem** 5-9 Build your skill

In Fig. 5-20a, a constant horizontal force $\vec{F}_{app}$ of magnitude 20 N is applied to block A of mass $m_A = 4.0$ kg, which pushes against block B of mass $m_B = 6.0$ kg. The blocks slide over a frictionless surface, along an x axis.

(a) What is the acceleration of the blocks?

**Serious Error:** Because force $\vec{F}_{app}$ is applied directly to block $A$, we use Newton's second law to relate that force to the acceleration $\vec{a}$ of block $A$. Because the motion is along the $x$ axis, we use that law for $x$ components ($F_{net,x} = ma_x$), writing it as

$$F_{app} = m_A a.$$

However, this is seriously wrong because $\vec{F}_{app}$ is not the only horizontal force acting on block $A$. There is also the force $\vec{F}_{AB}$ from block $B$ (Fig. 5-20b).

**Dead-End Solution:** Let us now include force $\vec{F}_{AB}$ by writing, again for the $x$ axis,

$$F_{app} - F_{AB} = m_A a.$$

(We use the minus sign to include the direction of $\vec{F}_{AB}$.) Because $F_{AB}$ is a second unknown, we cannot solve this equation for $a$.

**Successful Solution:** Because of the direction in which force $\vec{F}_{app}$ is applied, the two blocks form a rigidly connected system. We can relate the net force *on the system* to the acceleration *of the system* with Newton's second law. Here, once again for the $x$ axis, we can write that law as

$$F_{app} = (m_A + m_B)a,$$

where now we properly apply $\vec{F}_{app}$ to the system with total mass $m_A + m_B$. Solving for $a$ and substituting known values, we find

$$a = \frac{F_{app}}{m_A + m_B} = \frac{20\ \text{N}}{4.0\ \text{kg} + 6.0\ \text{kg}} = 2.0\ \text{m/s}^2.$$
(Answer)

Thus, the acceleration of the system and of each block is in the positive direction of the $x$ axis and has the magnitude 2.0 m/s².

(b) What is the (horizontal) force $\vec{F}_{BA}$ on block $B$ from block $A$ (Fig. 5-20c)?

**KEY IDEA** We can relate the net force on block $B$ to the block's acceleration with Newton's second law.

**Calculation:** Here we can write that law, still for components along the $x$ axis, as

$$F_{BA} = m_B a,$$

which, with known values, gives

$$F_{BA} = (6.0\ \text{kg})(2.0\ \text{m/s}^2) = 12\ \text{N}. \quad \text{(Answer)}$$

Thus, force $\vec{F}_{BA}$ is in the positive direction of the $x$ axis and has a magnitude of 12 N.

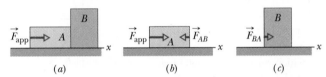

(a)  (b)  (c)

**FIG. 5-20** (a) A constant horizontal force $\vec{F}_{app}$ is applied to block $A$, which pushes against block $B$. (b) Two horizontal forces act on block $A$. (c) Only one horizontal force acts on block $B$.

## REVIEW & SUMMARY

**Newtonian Mechanics** The velocity of an object can change (the object can accelerate) when the object is acted on by one or more **forces** (pushes or pulls) from other objects. *Newtonian mechanics* relates accelerations and forces.

**Force** Forces are vector quantities. Their magnitudes are defined in terms of the acceleration they would give the standard kilogram. A force that accelerates that standard body by exactly 1 m/s² is defined to have a magnitude of 1 N. The direction of a force is the direction of the acceleration it causes. Forces are combined according to the rules of vector algebra. The **net force** on a body is the vector sum of all the forces acting on the body.

**Newton's First Law** If there is no net force on a body, the body remains at rest if it is initially at rest or moves in a straight line at constant speed if it is in motion.

**Inertial Reference Frames** Reference frames in which Newtonian mechanics holds are called *inertial reference frames* or *inertial frames*. Reference frames in which Newtonian mechanics does not hold are called *noninertial reference frames* or *noninertial frames*.

**Mass** The **mass** of a body is the characteristic of that body that relates the body's acceleration to the net force causing the acceleration. Masses are scalar quantities.

**Newton's Second Law** The net force $\vec{F}_{net}$ on a body with mass $m$ is related to the body's acceleration $\vec{a}$ by

$$\vec{F}_{net} = m\vec{a}, \tag{5-1}$$

which may be written in the component versions

$$F_{net,x} = ma_x \quad F_{net,y} = ma_y \quad \text{and} \quad F_{net,z} = ma_z. \tag{5-2}$$

The second law indicates that in SI units

$$1\ \text{N} = 1\ \text{kg} \cdot \text{m/s}^2. \tag{5-3}$$

A **free-body diagram** is a stripped-down diagram in which only *one* body is considered. That body is represented by either a sketch or a dot. The external forces on the body are drawn, and a coordinate system is superimposed, oriented so as to simplify the solution.

**Some Particular Forces** A **gravitational force** $\vec{F}_g$ on a body is a pull by another body. In most situations in this book, the other body is Earth or some other astronomical body. For Earth, the force is directed down toward the ground, which is assumed to be an inertial frame. With that assumption, the magnitude of $\vec{F}_g$ is

$$F_g = mg, \qquad (5\text{-}8)$$

where $m$ is the body's mass and $g$ is the magnitude of the free-fall acceleration.

The **weight** $W$ of a body is the magnitude of the upward force needed to balance the gravitational force on the body. A body's weight is related to the body's mass by

$$W = mg. \qquad (5\text{-}12)$$

A **normal force** $\vec{F}_N$ is the force on a body from a surface against which the body presses. The normal force is always perpendicular to the surface.

A **frictional force** $\vec{f}$ is the force on a body when the body slides or attempts to slide along a surface. The force is always parallel to the surface and directed so as to oppose the sliding. On a *frictionless surface,* the frictional force is negligible.

When a cord is under **tension,** each end of the cord pulls on a body. The pull is directed along the cord, away from the point of attachment to the body. For a *massless* cord (a cord with negligible mass), the pulls at both ends of the cord have the same magnitude $T$, even if the cord runs around a *massless, frictionless pulley* (a pulley with negligible mass and negligible friction on its axle to oppose its rotation).

**Newton's Third Law**    If a force $\vec{F}_{BC}$ acts on body $B$ due to body $C$, then there is a force $\vec{F}_{CB}$ on body $C$ due to body $B$:

$$\vec{F}_{BC} = -\vec{F}_{CB}.$$

# QUESTIONS

**1**  In Fig. 5-21, forces $\vec{F}_1$ and $\vec{F}_2$ are applied to a lunchbox as it slides at constant velocity over a frictionless floor. We are to decrease angle $\theta$ without changing the magnitude of $\vec{F}_1$. For constant velocity, should we increase, decrease, or maintain the magnitude of $\vec{F}_2$?

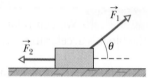

**FIG. 5-21**    Question 1.

**2**  At time $t = 0$, constant $\vec{F}$ begins to act on a rock moving through deep space in the $+x$ direction. (a) For time $t > 0$, which are possible functions $x(t)$ for the rock's position: (1) $x = 4t - 3$, (2) $x = -4t^2 + 6t - 3$, (3) $x = 4t^2 + 6t - 3$? (b) For which function is $\vec{F}$ directed opposite the rock's initial direction of motion?

**3**  Figure 5-22 shows overhead views of four situations in which forces act on a block that lies on a frictionless floor. If the force magnitudes are chosen properly, in which situations is it possible that the block is (a) stationary and (b) moving with a constant velocity?

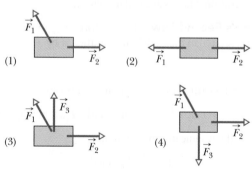

**FIG. 5-22**    Question 3.

**4**  Two horizontal forces,

$$\vec{F}_1 = (3 \text{ N})\hat{i} - (4 \text{ N})\hat{j} \quad \text{and} \quad \vec{F}_2 = -(1 \text{ N})\hat{i} - (2 \text{ N})\hat{j}$$

pull a banana split across a frictionless lunch counter. Without using a calculator, determine which of the vectors in the free-body diagram of Fig. 5-23 best represent (a) $\vec{F}_1$ and (b) $\vec{F}_2$.

What is the net-force component along (c) the $x$ axis and (d) the $y$ axis? Into which quadrants do (e) the net-force vector and (f) the split's acceleration vector point?

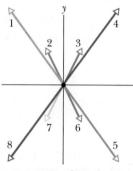

**FIG. 5-23**    Question 4.

**5**  Figure 5-24 gives the free-body diagram for four situations in which an object is pulled by several forces across a frictionless floor, as seen from overhead. In which situations does the object's acceleration $\vec{a}$ have (a) an $x$ component and (b) a $y$ component? (c) In each situation, give the direction of $\vec{a}$ by naming either a quadrant or a direction along an axis. (This can be done with a few mental calculations.)

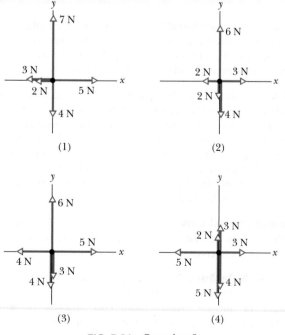

**FIG. 5-24**    Question 5.

**6** Figure 5-25 gives three graphs of velocity component $v_x(t)$ and three graphs of velocity component $v_y(t)$. The graphs are not to scale. Which $v_x(t)$ graph and which $v_y(t)$ graph best correspond to each of the four situations in Question 5 and Fig. 5-24?

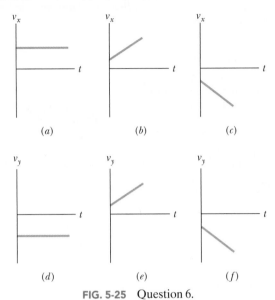

FIG. 5-25  Question 6.

**7** Figure 5-26 shows a train of four blocks being pulled across a frictionless floor by force $\vec{F}$. What total mass is accelerated to the right by (a) force $\vec{F}$, (b) cord 3, and (c) cord 1? (d) Rank the blocks according to their accelerations, greatest first. (e) Rank the cords according to their tension, greatest first. (Warm-up for Problems 50 and 51)

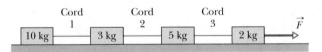

FIG. 5-26  Question 7.

**8** Figure 5-27 shows the same breadbox in four situations where horizontal forces are applied. Rank the situations according to the magnitude of the box's acceleration, greatest first.

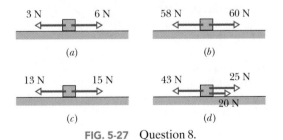

FIG. 5-27  Question 8.

**9** A vertical force $\vec{F}$ is applied to a block of mass $m$ that lies on a floor. What happens to the magnitude of the normal force $\vec{F}_N$ on the block from the floor as magnitude $F$ is increased from zero if force $\vec{F}$ is (a) downward and (b) upward?

**10** Figure 5-28 shows four choices for the direction of a force of magnitude $F$ to be applied to a block on an inclined plane. The di-

rections are either horizontal or vertical. (For choices $a$ and $b$, the force is not enough to lift the block off the plane.) Rank the choices according to the magnitude of the normal force on the block from the plane, greatest first.

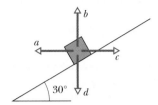

FIG. 5-28  Question 10.

**11** July 17, 1981, Kansas City: The newly opened Hyatt Regency is packed with people listening and dancing to a band playing favorites from the 1940s. Many of the people are crowded onto the walkways that hang like bridges across the wide atrium. Suddenly two of the walkways collapse, falling onto the merrymakers on the main floor.

The walkways were suspended one above another on vertical rods and held in place by nuts threaded onto the rods. In the original design, only two long rods were to be used, each extending through all three walkways (Fig. 5-29$a$). If each walkway and the merrymakers on it have a combined mass of $M$, what is the total mass supported by the threads and two nuts on (a) the lowest walkway and (b) the highest walkway?

Threading nuts on a rod is impossible except at the ends, so the design was changed: Instead, six rods were used, each connecting two walkways (Fig. 5-29$b$). What now is the total mass supported by the threads and two nuts on (c) the lowest walkway, (d) the upper side of the highest walkway, and (e) the lower side of the highest walkway? It was this design that failed.

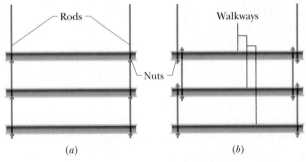

FIG. 5-29  Question 11.

**12** Figure 5-30 shows three blocks being pushed across a frictionless floor by horizontal force $\vec{F}$. What total mass is accelerated to the right by (a) force $\vec{F}$, (b) force $\vec{F}_{21}$ on block 2 from block 1, and (c) force $\vec{F}_{32}$ on block 3 from block 2? (d) Rank the blocks according to their acceleration magnitudes, greatest first. (e) Rank forces $\vec{F}$, $\vec{F}_{21}$, and $\vec{F}_{32}$ according to magnitude, greatest first. (Warm-up for Problem 53)

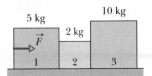

FIG. 5-30  Question 12.

## PROBLEMS

**GO**    Tutoring problem available (at instructor's discretion) in *WileyPLUS* and WebAssign

**SSM**    Worked-out solution available in Student Solutions Manual    **WWW**    Worked-out solution is at ──

• – •••    Number of dots indicates level of problem difficulty    **ILW**    Interactive solution is at ──    http://www.wiley.com/college/halliday

✈    Additional information available in *The Flying Circus of Physics* and at flyingcircusofphysics.com

### sec. 5-6    Newton's Second Law

**•1**    If the 1 kg standard body has an acceleration of 2.00 m/s² at 20.0° to the positive direction of an x axis, what are (a) the x component and (b) the y component of the net force acting on the body, and (c) what is the net force in unit-vector notation?

**•2**    Two horizontal forces act on a 2.0 kg chopping block that can slide over a frictionless kitchen counter, which lies in an xy plane. One force is $\vec{F}_1 = (3.0 \text{ N})\hat{i} + (4.0 \text{ N})\hat{j}$. Find the acceleration of the chopping block in unit-vector notation when the other force is (a) $\vec{F}_2 = (-3.0 \text{ N})\hat{i} + (-4.0 \text{ N})\hat{j}$, (b) $\vec{F}_2 = (-3.0 \text{ N})\hat{i} + (4.0 \text{ N})\hat{j}$, and (c) $\vec{F}_2 = (3.0 \text{ N})\hat{i} + (-4.0 \text{ N})\hat{j}$.

**•3**    Only two horizontal forces act on a 3.0 kg body that can move over a frictionless floor. One force is 9.0 N, acting due east, and the other is 8.0 N, acting 62° north of west. What is the magnitude of the body's acceleration?

**••4**    A 2.00 kg object is subjected to three forces that give it an acceleration $\vec{a} = -(8.00 \text{ m/s}^2)\hat{i} + (6.00 \text{ m/s}^2)\hat{j}$. If two of the three forces are $\vec{F}_1 = (30.0 \text{ N})\hat{i} + (16.0 \text{ N})\hat{j}$ and $\vec{F}_2 = -(12.0 \text{ N})\hat{i} + (8.00 \text{ N})\hat{j}$, find the third force.

**••5**    There are two forces on the 2.00 kg box in the overhead view of Fig. 5-31, but only one is shown. For $F_1 = 20.0$ N, $a = 12.0$ m/s², and $\theta = 30.0°$, find the second force (a) in unit-vector notation and as (b) a magnitude and (c) an angle relative to the positive direction of the x axis. **SSM**

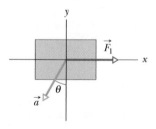

**FIG. 5-31**    Problem 5.

**••6**    While two forces act on it, a particle is to move at the constant velocity $\vec{v} = (3 \text{ m/s})\hat{i} - (4 \text{ m/s})\hat{j}$. One of the forces is $\vec{F}_1 = (2 \text{ N})\hat{i} + (-6 \text{ N})\hat{j}$. What is the other force?

**••7**    Three astronauts, propelled by jet backpacks, push and guide a 120 kg asteroid toward a processing dock, exerting the forces shown in Fig. 5-32, with $F_1 = 32$ N, $F_2 = 55$ N, $F_3 = 41$ N, $\theta_1 = 30°$, and $\theta_3 = 60°$. What is the asteroid's acceleration (a) in unit-vector notation and as (b) a magnitude and (c) a direction relative to the positive direction of the x axis? **GO**

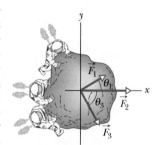

**FIG. 5-32**    Problem 7.

**••8**    In a two-dimensional tug-of-war, Alex, Betty, and Charles pull horizontally on an automobile tire at the angles shown in the overhead view of Fig. 5-33. The tire remains stationary in spite of the three pulls. Alex pulls with force $\vec{F}_A$ of magnitude 220 N, and Charles pulls with force $\vec{F}_C$ of magnitude 170 N. Note that the direction of $\vec{F}_C$ is not given. What is the magnitude of Betty's force $\vec{F}_B$?

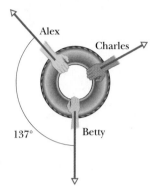

**FIG. 5-33**    Problem 8.

**••9**    A 2.0 kg particle moves along an x axis, being propelled by a variable force directed along that axis. Its position is given by $x = 3.0 \text{ m} + (4.0 \text{ m/s})t + ct^2 - (2.0 \text{ m/s}^3)t^3$, with x in meters and t in seconds. The factor c is a constant. At $t = 3.0$ s, the force on the particle has a magnitude of 36 N and is in the negative direction of the axis. What is c?

**••10**    A 0.150 kg particle moves along an x axis according to $x(t) = -13.00 + 2.00t + 4.00t^2 - 3.00t^3$, with x in meters and t in seconds. In unit-vector notation, what is the net force acting on the particle at $t = 3.40$ s?

**••11**    A 0.340 kg particle moves in an xy plane according to $x(t) = -15.00 + 2.00t - 4.00t^3$ and $y(t) = 25.00 + 7.00t - 9.00t^2$, with x and y in meters and t in seconds. At $t = 0.700$ s, what are (a) the magnitude and (b) the angle (relative to the positive direction of the x axis) of the net force on the particle, and (c) what is the angle of the particle's direction of travel?

**•••12**    Two horizontal forces $\vec{F}_1$ and $\vec{F}_2$ act on a 4.0 kg disk that slides over frictionless ice, on which an xy coordinate system is laid out. Force $\vec{F}_1$ is in the positive direction of the x axis and has a magnitude of 7.0 N. Force $\vec{F}_2$ has a magnitude of 9.0 N. Figure 5-34 gives the x component $v_x$ of the velocity of the disk as a function of time t during the sliding. What is the angle between the constant directions of forces $\vec{F}_1$ and $\vec{F}_2$?

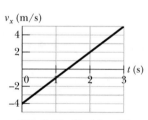

**FIG. 5-34**    Problem 12.

### sec. 5-7    Some Particular Forces

**•13**    (a) An 11.0 kg salami is supported by a cord that runs to a spring scale, which is supported by a cord hung from the ceiling (Fig. 5-35a). What is the reading on the scale, which is marked in weight units? (b) In Fig. 5-35b the salami is supported by a cord that runs around a pulley and to a scale. The opposite end of the scale is attached by a cord to a wall. What is the reading on the scale? (c) In Fig. 5-35c the wall has been replaced with a second 11.0 kg salami, and the assembly is stationary. What is the reading on the scale? **SSM**

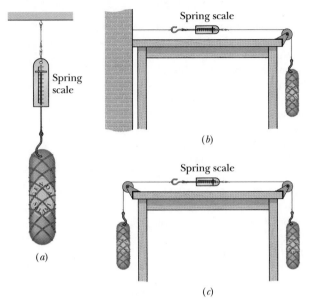

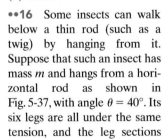

FIG. 5-35   Problem 13.

•**14**   A block with a weight of 3.0 N is at rest on a horizontal surface. A 1.0 N upward force is applied to the block by means of an attached vertical string. What are the (a) magnitude and (b) direction of the force of the block on the horizontal surface?

•**15**   Figure 5-36 shows an arrangement in which four disks are suspended by cords. The longer, top cord loops over a frictionless pulley and pulls with a force of magnitude 98 N on the wall to which it is attached. The tensions in the shorter cords are $T_1 = 58.8$ N, $T_2 = 49.0$ N, and $T_3 = 9.8$ N. What are the masses of (a) disk $A$, (b) disk $B$, (c) disk $C$, and (d) disk $D$?

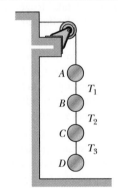

FIG. 5-36   Problem 15.

••**16**   Some insects can walk below a thin rod (such as a twig) by hanging from it. Suppose that such an insect has mass $m$ and hangs from a horizontal rod as shown in Fig. 5-37, with angle $\theta = 40°$. Its six legs are all under the same tension, and the leg sections

FIG. 5-37   Problem 16.

nearest the body are horizontal. (a) What is the ratio of the tension in each tibia (forepart of a leg) to the insect's weight? (b) If the insect straightens out its legs somewhat, does the tension in each tibia increase, decrease, or stay the same?

### sec. 5-9   Applying Newton's Laws

•**17**   A customer sits in an amusement park ride in which the compartment is to be pulled downward in the negative direction of a $y$ axis with an acceleration magnitude of $1.24g$, with $g = 9.80$ m/s². A 0.567 g coin rests on the customer's knee.

Once the motion begins and in unit-vector notation, what is the coin's acceleration relative to (a) the ground and (b) the customer? (c) How long does the coin take to reach the compartment ceiling, 2.20 m above the knee? In unit-vector notation, what are (d) the actual force on the coin and (e) the apparent force according to the customer's measure of the coin's acceleration?

•**18**   Tarzan, who weighs 820 N, swings from a cliff at the end of a 20.0 m vine that hangs from a high tree limb and initially makes an angle of 22.0° with the vertical. Assume that an $x$ axis extends horizontally away from the cliff edge and a $y$ axis extends upward. Immediately after Tarzan steps off the cliff, the tension in the vine is 760 N. Just then, what are (a) the force on him from the vine in unit-vector notation and the net force on him (b) in unit-vector notation and as (c) a magnitude and (d) an angle relative to the positive direction of the $x$ axis? What are the (e) magnitude and (f) angle of Tarzan's acceleration just then?

•**19**   In Fig. 5-38, let the mass of the block be 8.5 kg and the angle $\theta$ be 30°. Find (a) the tension in the cord and (b) the normal force acting on the block. (c) If the cord is cut, find the magnitude of the resulting acceleration of the block. **SSM WWW**

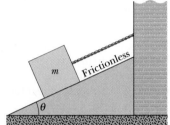

FIG. 5-38   Problem 19.

•**20**   There are two horizontal forces on the 2.0 kg box in the overhead view of Fig. 5-39 but only one (of magnitude $F_1 = 20$ N) is shown. The box moves

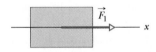

FIG. 5-39   Problem 20.

along the $x$ axis. For each of the following values for the acceleration $a_x$ of the box, find the second force in unit-vector notation: (a) 10 m/s², (b) 20 m/s², (c) 0, (d) $-10$ m/s², and (e) $-20$ m/s².

•**21**   A constant horizontal force $\vec{F}_a$ pushes a 2.00 kg FedEx package across a frictionless floor on which an $xy$ coordinate system has been drawn. Figure 5-40 gives the package's $x$ and $y$ velocity components versus time $t$. What are the (a) magnitude and (b) direction of $\vec{F}_a$?

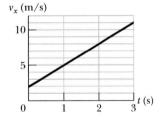
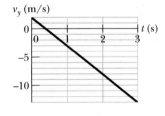

FIG. 5-40   Problem 21.

•**22**   In April 1974, John Massis of Belgium managed to move two passenger railroad cars. He did so by clamping his teeth down on a bit that was attached to the cars with a rope and then leaning backward while pressing his feet against the railway ties. The cars together weighed 700 kN (about 80 tons). Assume that he pulled with a constant force that was 2.5 times his body weight, at an upward angle $\theta$ of 30° from the horizontal. His mass was 80 kg, and he moved the cars by 1.0 m. Neglecting any retarding force from the wheel rotation, find the speed of the cars at the end of the pull.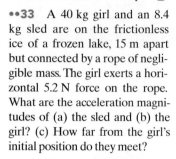

•23 *Sunjamming.* A "sun yacht" is a spacecraft with a large sail that is pushed by sunlight. Although such a push is tiny in everyday circumstances, it can be large enough to send the spacecraft outward from the Sun on a cost-free but slow trip. Suppose that the spacecraft has a mass of 900 kg and receives a push of 20 N. (a) What is the magnitude of the resulting acceleration? If the craft starts from rest, (b) how far will it travel in 1 day and (c) how fast will it then be moving?

•24 The tension at which a fishing line snaps is commonly called the line's "strength." What minimum strength is needed for a line that is to stop a salmon of weight 85 N in 11 cm if the fish is initially drifting at 2.8 m/s? Assume a constant deceleration.

•25 A 500 kg rocket sled can be accelerated at a constant rate from rest to 1600 km/h in 1.8 s. What is the magnitude of the required net force? **SSM**

•26 A car traveling at 53 km/h hits a bridge abutment. A passenger in the car moves forward a distance of 65 cm (with respect to the road) while being brought to rest by an inflated air bag. What magnitude of force (assumed constant) acts on the passenger's upper torso, which has a mass of 41 kg?

•27 A firefighter who weighs 712 N slides down a vertical pole with an acceleration of 3.00 m/s$^2$, directed downward. What are the (a) magnitude and (b) direction (up or down) of the vertical force on the firefighter from the pole and the (c) magnitude and (d) direction of the vertical force of the pole on the firefighter?

•28 The high-speed winds around a tornado can drive projectiles into trees, building walls, and even metal traffic signs. In a laboratory simulation, a standard wood toothpick was shot by pneumatic gun into an oak branch. The toothpick's mass was 0.13 g, its speed before entering the branch was 220 m/s, and its penetration depth was 15 mm. If its speed was decreased at a uniform rate, what was the magnitude of the force of the branch on the toothpick?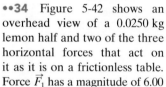

•29 An electron with a speed of $1.2 \times 10^7$ m/s moves horizontally into a region where a constant vertical force of $4.5 \times 10^{-16}$ N acts on it. The mass of the electron is $9.11 \times 10^{-31}$ kg. Determine the vertical distance the electron is deflected during the time it has moved 30 mm horizontally. **SSM**

•30 A car that weighs $1.30 \times 10^4$ N is initially moving at 40 km/h when the brakes are applied and the car is brought to a stop in 15 m. Assuming the force that stops the car is constant, find (a) the magnitude of that force and (b) the time required for the change in speed. If the initial speed is doubled, and the car experiences the same force during the braking, by what factors are (c) the stopping distance and (d) the stopping time multiplied? (There could be a lesson here about the danger of driving at high speeds.)

••31 The velocity of a 3.00 kg particle is given by $\vec{v} = (8.00t\hat{i} + 3.00t^2\hat{j})$ m/s, with time $t$ in seconds. At the instant the net force on the particle has a magnitude of 35.0 N, what are the direction (relative to the positive direction of the $x$ axis) of (a) the net force and (b) the particle's direction of travel?

••32 In Fig. 5-41, a crate of mass $m = 100$ kg is pushed at constant speed up a frictionless ramp ($\theta = 30.0°$) by a horizontal force $\vec{F}$. What are the magnitudes of (a) $\vec{F}$ and (b) the force on the crate from the ramp? **GO**

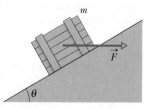

FIG. 5-41 Problem 32.

••33 A 40 kg girl and an 8.4 kg sled are on the frictionless ice of a frozen lake, 15 m apart but connected by a rope of negligible mass. The girl exerts a horizontal 5.2 N force on the rope. What are the acceleration magnitudes of (a) the sled and (b) the girl? (c) How far from the girl's initial position do they meet?

••34 Figure 5-42 shows an overhead view of a 0.0250 kg lemon half and two of the three horizontal forces that act on it as it is on a frictionless table. Force $\vec{F_1}$ has a magnitude of 6.00 N and is at $\theta_1 = 30.0°$. Force $\vec{F_2}$ has a magnitude of 7.00 N and is at $\theta_2 = 30.0°$. In unit-vector notation, what is the third force if the lemon half (a) is stationary, (b) has constant velocity $\vec{v} = (13.0\hat{i} - 14.0\hat{j})$ m/s, and (c) has varying velocity $\vec{v} = (13.0t\hat{i} - 14.0t\hat{j})$ m/s$^2$, where $t$ is time?

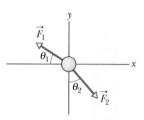

FIG. 5-42 Problem 34.

••35 A block is projected up a frictionless inclined plane with initial speed $v_0 = 3.50$ m/s. The angle of incline is $\theta = 32.0°$. (a) How far up the plane does the block go? (b) How long does it take to get there? (c) What is its speed when it gets back to the bottom? **SSM WWW**

••36 A 40 kg skier skis directly down a frictionless slope angled at 10° to the horizontal. Assume the skier moves in the negative direction of an $x$ axis along the slope. A wind force with component $F_x$ acts on the skier. What is $F_x$ if the magnitude of the skier's velocity is (a) constant, (b) increasing at a rate of 1.0 m/s$^2$, and (c) increasing at a rate of 2.0 m/s$^2$?

••37 A sphere of mass $3.0 \times 10^{-4}$ kg is suspended from a cord. A steady horizontal breeze pushes the sphere so that the cord makes a constant angle of 37° with the vertical. Find (a) the push magnitude and (b) the tension in the cord. **ILW**

••38 A dated box of dates, of mass 5.00 kg, is sent sliding up a frictionless ramp at an angle of $\theta$ to the horizontal. Figure 5-43 gives, as a function of time $t$, the component $v_x$ of the box's velocity along an $x$ axis that extends directly up the ramp. What is the magnitude of the normal force on the box from the ramp?

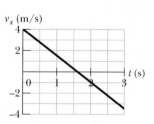

FIG. 5-43 Problem 38.

••39 An elevator cab and its load have a combined mass of 1600 kg. Find the tension in the supporting cable when the cab, originally moving downward at 12 m/s, is brought to rest with constant acceleration in a distance of 42 m.

••40 Holding on to a towrope moving parallel to a frictionless ski slope, a 50 kg skier is pulled up the slope, which is at an angle of 8.0° with the horizontal. What is the magnitude $F_{rope}$ of the force on the skier from the rope when (a) the magnitude $v$ of the skier's velocity is constant at 2.0 m/s and (b) $v = 2.0$ m/s as $v$ increases at a rate of 0.10 m/s$^2$?

**••41** An elevator cab that weighs 27.8 kN moves upward. What is the tension in the cable if the cab's speed is (a) increasing at a rate of 1.22 m/s² and (b) decreasing at a rate of 1.22 m/s²?

**••42** A lamp hangs vertically from a cord in a descending elevator that decelerates at 2.4 m/s². (a) If the tension in the cord is 89 N, what is the lamp's mass? (b) What is the cord's tension when the elevator ascends with an upward acceleration of 2.4 m/s²?

**••43** Using a rope that will snap if the tension in it exceeds 387 N, you need to lower a bundle of old roofing material weighing 449 N from a point 6.1 m above the ground. (a) What magnitude of the bundle's acceleration will put the rope on the verge of snapping? (b) At that acceleration, with what speed would the bundle hit the ground?

**••44** An elevator cab is pulled upward by a cable. The cab and its single occupant have a combined mass of 2000 kg. When that occupant drops a coin, its acceleration relative to the cab is 8.00 m/s² downward. What is the tension in the cable?

**••45** In Fig. 5-44, a chain consisting of five links, each of mass 0.100 kg, is lifted vertically with constant acceleration of magnitude $a = 2.50$ m/s². Find the magnitudes of (a) the force on link 1 from link 2, (b) the force on link 2 from link 3, (c) the force on link 3 from link 4, and (d) the force on link 4 from link 5. Then find the magnitudes of (e) the force $\vec{F}$ on the top link from the person lifting the chain and (f) the *net* force accelerating each link. **SSM**

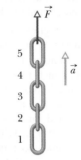
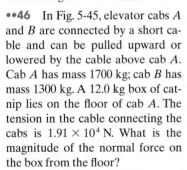

FIG. 5-44 Problem 45.

**••46** In Fig. 5-45, elevator cabs $A$ and $B$ are connected by a short cable and can be pulled upward or lowered by the cable above cab $A$. Cab $A$ has mass 1700 kg; cab $B$ has mass 1300 kg. A 12.0 kg box of catnip lies on the floor of cab $A$. The tension in the cable connecting the cabs is $1.91 \times 10^4$ N. What is the magnitude of the normal force on the box from the floor?

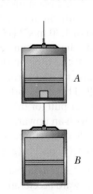

FIG. 5-45 Problem 46.

**••47** In Fig. 5-46, a block of mass $m = 5.00$ kg is pulled along a horizontal frictionless floor by a cord that exerts a force of magnitude $F = 12.0$ N at an angle $\theta = 25.0°$. (a) What is the magnitude of the block's acceleration? (b) The force magnitude $F$ is slowly increased. What is its value just before the block is lifted (completely) off the floor? (c) What is the magnitude of the block's acceleration just before it is lifted (completely) off the floor?

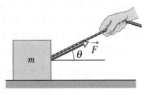

FIG. 5-46 Problems 47 and 62.

**••48** In earlier days, horses pulled barges down canals in the manner shown in Fig. 5-47. Suppose the horse pulls on the rope with a force of 7900 N at an angle of $\theta = 18°$ to the direction of motion of the barge, which is headed straight along the positive direction of an $x$ axis. The mass of the barge is 9500 kg, and the magnitude of its acceleration is 0.12 m/s². What are the (a) magnitude and (b) direction (relative to positive $x$) of the force on the barge from the water? **GO**

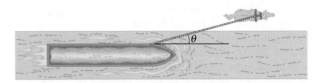

FIG. 5-47 Problem 48.

**••49** The Zacchini family was renowned for their human-cannonball act in which a family member was shot from a cannon using either elastic bands or compressed air. In one version of the act, Emanuel Zacchini was shot over three Ferris wheels to land in a net at the same height as the open end of the cannon and at a range of 69 m. He was propelled inside the barrel for 5.2 m and launched at an angle of 53°. If his mass was 85 kg and he underwent constant acceleration inside the barrel, what was the magnitude of the force propelling him? (*Hint:* Treat the launch as though it were along a ramp at 53°. Neglect air drag.)

**••50** Figure 5-48 shows four penguins that are being playfully pulled along very slippery (frictionless) ice by a curator. The masses of three penguins and the tension in two of the cords are $m_1 = 12$ kg, $m_3 = 15$ kg, $m_4 = 20$ kg, $T_2 = 111$ N, and $T_4 = 222$ N. Find the penguin mass $m_2$ that is not given.

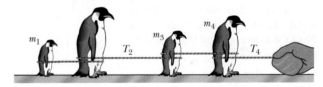

FIG. 5-48 Problem 50.

**••51** In Fig. 5-49, three connected blocks are pulled to the right on a horizontal frictionless table by a force of magnitude $T_3 = 65.0$ N. If $m_1 = 12.0$ kg, $m_2 = 24.0$ kg, and $m_3 = 31.0$ kg, calculate (a) the magnitude of the system's acceleration, (b) the tension $T_1$, and (c) the tension $T_2$.

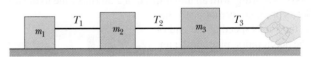

FIG. 5-49 Problem 51.

**••52** In Fig. 5-50a, a constant horizontal force $\vec{F}_a$ is applied to block $A$, which pushes against block $B$ with a 20.0 N force directed horizontally to the right. In Fig. 5-50b, the same force $\vec{F}_a$ is applied to block $B$; now block $A$ pushes on block $B$ with a

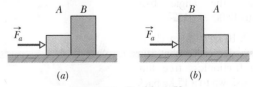

FIG. 5-50 Problem 52.

10.0 N force directed horizontally to the left. The blocks have a combined mass of 12.0 kg. What are the magnitudes of (a) their acceleration in Fig. 5-50a and (b) force $\vec{F}_a$?

**••53** Two blocks are in contact on a frictionless table. A horizontal force is applied to the larger block, as shown in Fig. 5-51. (a) If $m_1 = 2.3$ kg, $m_2 = 1.2$ kg, and $F = 3.2$ N, find the magnitude of the force between the two blocks. (b) Show that if a force of the same magnitude $F$ is applied to the smaller block but in the opposite direction, the magnitude of the force between the blocks is 2.1 N, which is not the same value calculated in (a). (c) Explain the difference. SSM ILW WWW

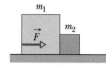

FIG. 5-51   Problem 53.

**••54** In Fig. 5-52, three ballot boxes are connected by cords, one of which wraps over a pulley having negligible friction on its axle and negligible mass. The three masses are $m_A = 30.0$ kg, $m_B = 40.0$ kg, and $m_C = 10.0$ kg. When the assembly is released from rest, (a) what is the tension in the cord connecting $B$ and $C$, and (b) how far does $A$ move in the first 0.250 s (assuming it does not reach the pulley)?

FIG. 5-52   Problem 54.

**••55** Figure 5-53 shows two blocks connected by a cord (of negligible mass) that passes over a frictionless pulley (also of negligible mass). The arrangement is known as *Atwood's machine.* One block has mass $m_1 = 1.30$ kg; the other has mass $m_2 = 2.80$ kg. What are (a) the magnitude of the blocks' acceleration and (b) the tension in the cord? GO

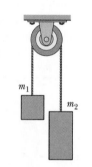

FIG. 5-53   Problems 55 and 63.

**••56** In shot putting, many athletes elect to launch the shot at an angle that is smaller than the theoretical one (about 42°) at which the distance of a projected ball at the same speed and height is greatest. One reason has to do with the speed the athlete can give the shot during the acceleration phase of the throw. Assume that a 7.260 kg shot is accelerated along a straight path of length 1.650 m by a constant applied force of magnitude 380.0 N, starting with an initial speed of 2.500 m/s (due to the athlete's preliminary motion). What is the shot's speed at the end of the acceleration phase if the angle between the path and the horizontal is (a) 30.00° and (b) 42.00°? (*Hint:* Treat the motion as though it were along a ramp at the given angle.) (c) By what percent is the launch speed decreased if the athlete increases the angle from 30.00° to 42.00°?

**••57** A 10 kg monkey climbs up a massless rope that runs over a frictionless tree limb and back down to a 15 kg package on the ground (Fig. 5-54).

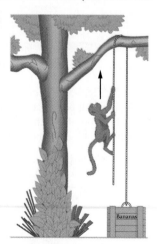

FIG. 5-54   Problem 57.

(a) What is the magnitude of the least acceleration the monkey must have if it is to lift the package off the ground? If, after the package has been lifted, the monkey stops its climb and holds onto the rope, what are the (b) magnitude and (c) direction of the monkey's acceleration and (d) the tension in the rope? SSM

**••58** An 85 kg man lowers himself to the ground from a height of 10.0 m by holding onto a rope that runs over a frictionless pulley to a 65 kg sandbag. With what speed does the man hit the ground if he started from rest?

**••59** A block of mass $m_1 = 3.70$ kg on a frictionless plane inclined at angle $\theta = 30.0°$ is connected by a cord over a massless, frictionless pulley to a second block of mass $m_2 = 2.30$ kg (Fig. 5-55). What are (a) the magnitude of the acceleration of each block, (b) the direction of the acceleration of the hanging block, and (c) the tension in the cord? ILW

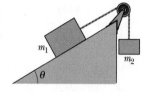

FIG. 5-55   Problem 59.

**••60** Figure 5-56 shows a man sitting in a bosun's chair that dangles from a massless rope, which runs over a massless, frictionless pulley and back down to the man's hand. The combined mass of man and chair is 95.0 kg. With what force magnitude must the man pull on the rope if he is to rise (a) with a constant velocity and (b) with an upward acceleration of 1.30 m/s²? (*Hint:* A free-body diagram can really help.)

If the rope on the right extends to the ground and is pulled by a co-worker, with what force magnitude must the co-worker pull for the man to rise (c) with a constant velocity and (d) with an upward acceleration of 1.30 m/s²? What is the magnitude of the force on the ceiling from the pulley system in (e) part a, (f) part b, (g) part c, and (h) part d?

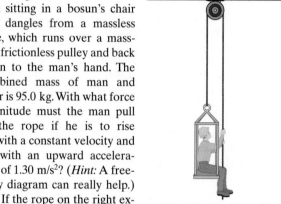

FIG. 5-56   Problem 60.

**••61** A hot-air balloon of mass $M$ is descending vertically with downward acceleration of magnitude $a$. How much mass (ballast) must be thrown out to give the balloon an upward acceleration of magnitude $a$? Assume that the upward force from the air (the lift) does not change because of the decrease in mass. SSM ILW

**••62** Figure 5-46 shows a 5.00 kg block being pulled along a frictionless floor by a cord that applies a force of constant magnitude 20.0 N but with an angle $\theta(t)$ that varies with time. When angle $\theta = 25.0°$, at what rate is the block's acceleration changing if (a) $\theta(t) = (2.00 \times 10^{-2}$ deg/s$)t$ and (b) $\theta(t) = -(2.00 \times 10^{-2}$ deg/s$)t$? (*Hint:* Switch to radians.)

**•••63** Figure 5-53 shows *Atwood's machine,* in which two containers are connected by a cord (of negligible mass) passing over a frictionless pulley (also of negligible mass). At time $t = 0$, container 1 has mass 1.30 kg and container 2 has mass 2.80 kg, but container 1 is losing mass (through a leak) at the constant rate of 0.200 kg/s. At what rate is the acceleration magnitude of the containers changing at (a) $t = 0$ and (b) $t = 3.00$ s? (c) When does the acceleration reach its maximum value?

•••**64** A shot putter launches a 7.260 kg shot by pushing it along a straight line of length 1.650 m and at an angle of 34.10° from the horizontal, accelerating the shot to the launch speed from its initial speed of 2.500 m/s (which is due to the athlete's preliminary motion). The shot leaves the hand at a height of 2.110 m and at an angle of 34.10°, and it lands at a horizontal distance of 15.90 m. What is the magnitude of the athlete's average force on the shot during the acceleration phase? (*Hint:* Treat the motion during the acceleration phase as though it were along a ramp at the given angle.)

•••**65** Figure 5-57 shows three blocks attached by cords that loop over frictionless pulleys. Block *B* lies on a frictionless table; the masses are $m_A =$ 6.00 kg, $m_B = 8.00$ kg, and $m_C =$ 10.0 kg. When the blocks are released, what is the tension in the cord at the right?

**FIG. 5-57** Problem 65.

•••**66** Figure 5-58 shows a box of mass $m_2 = 1.0$ kg on a frictionless plane inclined at angle $\theta = 30°$. It is connected by a cord of negligible mass to a box of mass $m_1 = 3.0$ kg on a horizontal frictionless surface. The pulley is frictionless and massless. (a) If the magnitude of horizontal force $\vec{F}$ is 2.3 N, what is the tension in the connecting cord? (b) What is the largest value the magnitude of $\vec{F}$ may have without the cord becoming slack?

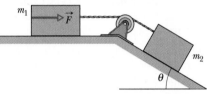

**FIG. 5-58** Problem 66.

•••**67** Figure 5-59 gives, as a function of time *t*, the force component $F_x$ that acts on a 3.00 kg ice block that can move only along the *x* axis. At $t = 0$, the block is moving in the positive direction of the axis, with a speed of 3.0 m/s. What are its (a) speed and (b) direction of travel at $t = 11$ s?

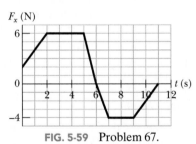

**FIG. 5-59** Problem 67.

•••**68** Figure 5-60 shows a section of a cable-car system. The maximum permissible mass of each car with occupants is 2800 kg. The cars, riding on a support cable, are pulled by a second cable attached to the support tower on each car. Assume that the cables are taut and inclined at angle $\theta = 35°$. What is the difference in tension between adjacent sections of pull cable if the cars are at the maximum permissible mass and are being accelerated up the incline at 0.81 m/s²?

Support cable
Pull cable

**FIG. 5-60** Problem 68.

### Additional Problems

**69** *Blowing off units.* Throughout your physics course, your instructor will expect you to be careful with units in your calculations. Yet some students tend to neglect them and just trust that they always work out properly. Maybe this real-world example will keep you from such a sloppy habit.

On July 23, 1983, Air Canada Flight 143 was being readied for its long trip from Montreal to Edmonton when the flight crew asked the ground crew to determine how much fuel was already on board. The flight crew knew they needed to begin the trip with 22 300 kg of fuel. They knew that amount in kilograms because Canada had recently switched to the metric system; previously fuel had been measured in pounds. The ground crew could measure the onboard fuel only in liters, which they reported as 7682 L. Thus, to determine how much fuel was on board and how much additional fuel was needed, the flight crew asked the ground crew for the conversion factor from liters to kilograms of fuel. The response was 1.77, which the flight crew used (1.77 kg corresponds to 1 L). (a) How many kilograms of fuel did the flight crew think they had? (In this problem, take all given data as being exact.) (b) How many liters did they ask to be added?

Unfortunately, the response from the ground crew was based on pre-metric habits—1.77 was the conversion factor not from liters to kilograms but rather from liters to *pounds* of fuel (1.77 lb corresponds to 1 L). (c) How many kilograms of fuel were actually on board? (Except for the given 1.77, use four significant figures for other conversion factors.) (d) How many liters of additional fuel were actually needed? (e) When the airplane left Montreal, what percentage of the required fuel did it have?

En route to Edmonton, at an altitude of 7.9 km, the airplane ran out of fuel and began to fall. Although the airplane had no power, the pilot managed to put it into a downward glide. Because the nearest working airport was too far to reach by gliding only, the pilot angled the glide toward an old, nonworking airport.

Unfortunately, the runway at that airport had been converted to a track for race cars, and a steel barrier had been constructed across it. Fortunately, as the airplane hit the runway, the front landing gear collapsed, dropping the nose of the airplane onto the runway. The skidding slowed the airplane so that it stopped just short of the steel barrier, with stunned race drivers and fans looking on. All on board the airplane emerged safely. The point here is this: Take care of the units.

**70** The only two forces acting on a body have magnitudes of 20 N and 35 N and directions that differ by 80°. The resulting acceleration has a magnitude of 20 m/s². What is the mass of the body?

**71** Figure 5-61 is an overhead view of a 12 kg tire that is to be pulled by three horizontal ropes. One rope's force ($F_1 =$ 50 N) is indicated. The forces from the other ropes are to be oriented such that the tire's acceleration magnitude *a* is least.

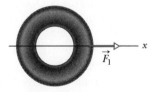

**FIG. 5-61** Problem 71.

What is that least *a* if (a) $F_2 = 30$ N, $F_3 = 20$ N; (b) $F_2 = 30$ N, $F_3 = 10$ N; and (c) $F_2 = F_3 = 30$ N?

**72** A block of mass *M* is pulled along a horizontal frictionless surface by a rope of mass *m*, as shown in Fig. 5-62. A horizontal force $\vec{F}$ acts on one end of the rope. (a) Show that the rope *must* sag, even if only by an imperceptible amount. Then, assuming that the sag is negligible, find (b) the acceleration of rope and block, (c) the force on the block from the rope, and (d) the tension in the rope at its midpoint.

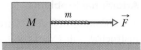

FIG. 5-62 Problem 72.

**73** A worker drags a crate across a factory floor by pulling on a rope tied to the crate. The worker exerts a force of magnitude $F = 450$ N on the rope, which is inclined at an upward angle $\theta = 38°$ to the horizontal, and the floor exerts a horizontal force of magnitude $f = 125$ N that opposes the motion. Calculate the magnitude of the acceleration of the crate if (a) its mass is 310 kg and (b) its weight is 310 N. **SSM**

**74** Three forces act on a particle that moves with unchanging velocity $\vec{v} = (2 \text{ m/s})\hat{i} - (7 \text{ m/s})\hat{j}$. Two of the forces are $\vec{F}_1 = (2 \text{ N})\hat{i} + (3 \text{ N})\hat{j} + (-2 \text{ N})\hat{k}$ and $\vec{F}_2 = (-5 \text{ N})\hat{i} + (8 \text{ N})\hat{j} + (-2 \text{ N})\hat{k}$. What is the third force?

**75** A 52 kg circus performer is to slide down a rope that will break if the tension exceeds 425 N. (a) What happens if the performer hangs stationary on the rope? (b) At what magnitude of acceleration does the performer just avoid breaking the rope?

**76** An 80 kg man drops to a concrete patio from a window 0.50 m above the patio. He neglects to bend his knees on landing, taking 2.0 cm to stop. (a) What is his average acceleration from when his feet first touch the patio to when he stops? (b) What is the magnitude of the average stopping force exerted on him by the patio? ✈

**77** In Fig. 5-63, 4.0 kg block $A$ and 6.0 kg block $B$ are connected by a string of negligible mass. Force $\vec{F}_A = (12 \text{ N})\hat{i}$ acts on block $A$; force $\vec{F}_B = (24 \text{ N})\hat{i}$ acts on block $B$. What is the tension in the string?

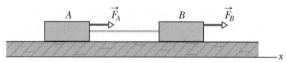

FIG. 5-63 Problem 77.

**78** In the overhead view of Fig. 5-64, five forces pull on a box of mass $m = 4.0$ kg. The force magnitudes are $F_1 = 11$ N, $F_2 = 17$ N, $F_3 = 3.0$ N, $F_4 = 14$ N, and $F_5 = 5.0$ N, and angle $\theta_4$ is 30°. Find the box's acceleration (a) in unit-vector notation and as (b) a magnitude and (c) an angle relative to the positive direction of the $x$ axis.

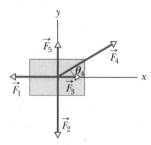

FIG. 5-64 Problem 78.

**79** A certain force gives an object of mass $m_1$ an acceleration of 12.0 m/s² and an object of mass $m_2$ an acceleration of 3.30 m/s². What acceleration would the force give to an object of mass (a) $m_2 - m_1$ and (b) $m_2 + m_1$? **SSM**

**80** Imagine a landing craft approaching the surface of Callisto, one of Jupiter's moons. If the engine provides an upward force (thrust) of 3260 N, the craft descends at constant

speed; if the engine provides only 2200 N, the craft accelerates downward at 0.39 m/s². (a) What is the weight of the landing craft in the vicinity of Callisto's surface? (b) What is the mass of the craft? (c) What is the magnitude of the free-fall acceleration near the surface of Callisto?

**81** An object is hung from a spring balance attached to the ceiling of an elevator cab. The balance reads 65 N when the cab is standing still. What is the reading when the cab is moving upward (a) with a constant speed of 7.6 m/s and (b) with a speed of 7.6 m/s while decelerating at a rate of 2.4 m/s²?

**82** In Fig. 5-65, a force $\vec{F}$ of magnitude 12 N is applied to a FedEx box of mass $m_2 = 1.0$ kg. The force is directed up a plane tilted by $\theta = 37°$. The box is connected by a cord to a UPS box of mass $m_1 = 3.0$ kg on the floor. The floor, plane, and pulley are frictionless, and the masses of the pulley and cord are negligible. What is the tension in the cord?

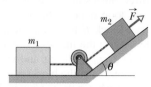

FIG. 5-65 Problem 82.

**83** A certain particle has a weight of 22 N at a point where $g = 9.8$ m/s². What are its (a) weight and (b) mass at a point where $g = 4.9$ m/s²? What are its (c) weight and (d) mass if it is moved to a point in space where $g = 0$?

**84** Compute the weight of a 75 kg space ranger (a) on Earth, (b) on Mars, where $g = 3.7$ m/s², and (c) in interplanetary space, where $g = 0$. (d) What is the ranger's mass at each location?

**85** A 1400 kg jet engine is fastened to the fuselage of a passenger jet by just three bolts (this is the usual practice). Assume that each bolt supports one-third of the load. (a) Calculate the force on each bolt as the plane waits in line for clearance to take off. (b) During flight, the plane encounters turbulence, which suddenly imparts an upward vertical acceleration of 2.6 m/s² to the plane. Calculate the force on each bolt now.

**86** An 80 kg person is parachuting and experiencing a downward acceleration of 2.5 m/s². The mass of the parachute is 5.0 kg. (a) What is the upward force on the open parachute from the air? (b) What is the downward force on the parachute from the person?

**87** Suppose that in Fig. 5-13, the masses of the blocks are 2.0 kg and 4.0 kg. (a) Which mass should the hanging block have if the magnitude of the acceleration is to be as large as possible? What then are (b) the magnitude of the acceleration and (c) the tension in the cord?

**88** You pull a short refrigerator with a constant force $\vec{F}$ across a greased (frictionless) floor, either with $\vec{F}$ horizontal (case 1) or with $\vec{F}$ tilted upward at an angle $\theta$ (case 2). (a) What is the ratio of the refrigerator's speed in case 2 to its speed in case 1 if you pull for a certain time $t$? (b) What is this ratio if you pull for a certain distance $d$?

**89** A spaceship lifts off vertically from the Moon, where $g = 1.6$ m/s². If the ship has an upward acceleration of 1.0 m/s² as it lifts off, what is the magnitude of the force exerted by the ship on its pilot, who weighs 735 N on Earth?

**90** Compute the initial upward acceleration of a rocket of mass $1.3 \times 10^4$ kg if the initial upward force produced by its engine (the thrust) is $2.6 \times 10^5$ N. Do not neglect the gravitational force on the rocket.

**91** Figure 5-66a shows a mobile hanging from a ceiling; it consists of two metal pieces ($m_1 = 3.5$ kg and $m_2 = 4.5$ kg) that are strung together by cords of negligible mass. What is the tension in (a) the bottom cord and (b) the top cord? Figure 5-66b shows a mobile consisting of three metal pieces. Two of the masses are $m_3 = 4.8$ kg and $m_5 = 5.5$ kg. The tension in the top cord is 199 N. What is the tension in (c) the lowest cord and (d) the middle cord? **SSM**

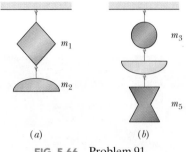

(a)               (b)

**FIG. 5-66**    Problem 91.

**92** If the 1 kg standard body is accelerated by only $\vec{F}_1 = (3.0$ N$)\hat{i} + (4.0$ N$)\hat{j}$ and $\vec{F}_2 = (-2.0$ N$)\hat{i} + (-6.0$ N$)\hat{j}$, then what is $\vec{F}_{net}$ (a) in unit-vector notation and as (b) a magnitude and (c) an angle relative to the positive $x$ direction? What are the (d) magnitude and (e) angle of $\vec{a}$?

**93** A nucleus that captures a stray neutron must bring the neutron to a stop within the diameter of the nucleus by means of the *strong force*. That force, which "glues" the nucleus together, is approximately zero outside the nucleus. Suppose that a stray neutron with an initial speed of $1.4 \times 10^7$ m/s is just barely captured by a nucleus with diameter $d = 1.0 \times 10^{-14}$ m. Assuming the strong force on the neutron is constant, find the magnitude of that force. The neutron's mass is $1.67 \times 10^{-27}$ kg.

**94** A 15 000 kg helicopter lifts a 4500 kg truck with an upward acceleration of 1.4 m/s$^2$. Calculate (a) the net upward force on the helicopter blades from the air and (b) the tension in the cable between helicopter and truck.

**95** A motorcycle and 60.0 kg rider accelerate at 3.0 m/s$^2$ up a ramp inclined 10° above the horizontal. What are the magnitudes of (a) the net force on the rider and (b) the force on the rider from the motorcycle? **SSM**

**96** An interstellar ship has a mass of $1.20 \times 10^6$ kg and is initially at rest relative to a star system. (a) What constant acceleration is needed to bring the ship up to a speed of 0.10$c$ (where $c$ is the speed of light, $3.0 \times 10^8$ m/s) relative to the star system in 3.0 days? (b) What is that acceleration in $g$ units? (c) What force is required for the acceleration? (d) If the engines are shut down when 0.10$c$ is reached (the speed then remains constant), how long does the ship take (start to finish) to journey 5.0 light-months, the distance that light travels in 5.0 months?

**97** For sport, a 12 kg armadillo runs onto a large pond of level, frictionless ice. The armadillo's initial velocity is 5.0 m/s along the positive direction of an $x$ axis. Take its initial position on the ice as being the origin. It slips over the ice while being pushed by a wind with a force of 17 N in the positive direction of the $y$ axis. In unit-vector notation, what are the animal's (a) velocity and (b) position vector when it has slid for 3.0 s?

**98** A 50 kg passenger rides in an elevator cab that starts from rest on the ground floor of a building at $t = 0$ and rises to the top floor during a 10 s interval. The cab's acceleration as a function of the time is shown in Fig. 5-67, where positive val-

ues of the acceleration mean that it is directed upward. What are the (a) magnitude and (b) direction (up or down) of the maximum force on the passenger from the floor, the (c) magnitude and (d) direction of the minimum force on the passenger from the floor, and the (e) magnitude and (f) direction of the maximum force on the floor from the passenger?

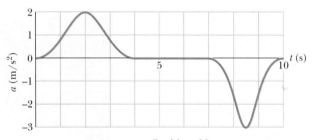

**FIG. 5-67**    Problem 98.

**99** Figure 5-68 shows a box of dirty money (mass $m_1 = 3.0$ kg) on a frictionless plane inclined at angle $\theta_1 = 30°$. The box is connected via a cord of negligible mass to a box of laundered money (mass $m_2 = 2.0$ kg) on a friction-

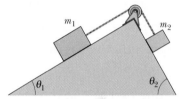

**FIG. 5-68**    Problem 99.

less plane inclined at angle $\theta_2 = 60°$. The pulley is frictionless and has negligible mass. What is the tension in the cord? **SSM**

**100** Suppose the 1 kg standard body accelerates at 4.00 m/s$^2$ at 160° from the positive direction of an $x$ axis due to two forces; one is $\vec{F}_1 = (2.50$ N$)\hat{i} + (4.60$ N$)\hat{j}$. What is the other force (a) in unit-vector notation and as (b) a magnitude and (c) an angle?

**101** In Fig. 5-69, a tin of antioxidants ($m_1 = 1.0$ kg) on a frictionless inclined surface is connected to a tin of corned beef ($m_2 = 2.0$ kg). The pulley is massless and frictionless. An upward force of magnitude $F = 6.0$ N acts on the corned beef tin, which has a downward acceleration of 5.5 m/s$^2$. What are (a) the tension in the connecting cord and (b) angle $\beta$? **SSM**

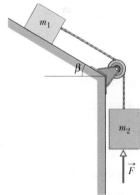

**FIG. 5-69**    Problem 101.

**102** A rocket and its payload have a total mass of $5.0 \times 10^4$ kg. How large is the force produced by the engine (the thrust) when the rocket is (a) "hovering" over the launchpad just after ignition and (b) accelerating upward at 20 m/s$^2$?

**103** A motorcycle of weight 2.0 kN accelerates from 0 to 88.5 km/h in 6.0 s. What are the magnitudes of (a) the constant acceleration and (b) the net force causing the acceleration?

**104** An initially stationary electron (mass $= 9.11 \times 10^{-31}$ kg) undergoes a constant acceleration through 1.5 cm, reaching $6.0 \times 10^6$ m/s. What are (a) the magnitude of the force accelerating the electron and (b) the electron's weight?

The Great Pyramid, built about 4500 years ago, consists of about 2 300 000 stone blocks, most with a mass of 2000 to 3000 kg. How did the engineers and workers manage to lift the stones into place to construct this pyramid, which is over 140 m high? Some researchers argue that during the construction a large team of men would pull a block up a giant earthen ramp that ran at a modest angle up one side of the pyramid. However, no evidence (such as rubble or painted pictures) exists to support this theory. Other researchers argue that a spiral ramp ran around the pyramid. However, such a ramp would have been highly unstable and, besides, maneuvering a 2000 kg stone around the 90° corners along the ramp would have been daunting, if not impossible.

Will & Deni McIntyre/Stone/Getty Images

*How did the ancient people move the blocks up and into position?*

The answer is in this chapter.

## 6-1 | WHAT IS PHYSICS?

In this chapter we focus on the physics of three common types of force: frictional force, drag force, and centripetal force. An engineer preparing a car for the Indianapolis 500 must consider all three types. Frictional forces acting on the tires are crucial to the car's acceleration out of the pit and out of a curve (if the car hits an oil slick, the friction is lost and so is the car). Drag forces acting on the car from the passing air must be minimized or else the car will consume too much fuel and have to pit too early (even one 14 s pit stop can cost a driver the race). Centripetal forces are crucial in the turns (if there is insufficient centripetal force, the car slides into the wall). We start our discussion with frictional forces.

## 6-2 | Friction

Frictional forces are unavoidable in our daily lives. If we were not able to counteract them, they would stop every moving object and bring to a halt every rotating shaft. About 20% of the gasoline used in an automobile is needed to counteract friction in the engine and in the drive train. On the other hand, if friction were totally absent, we could not get an automobile to go anywhere, and we could not walk or ride a bicycle. We could not hold a pencil, and, if we could, it would not write. Nails and screws would be useless, woven cloth would fall apart, and knots would untie.

Here we deal with the frictional forces that exist between dry solid surfaces, either stationary relative to each other or moving across each other at slow speeds. Consider three simple thought experiments:

1. Send a book sliding across a long horizontal counter. As expected, the book slows and then stops. This means the book must have an acceleration parallel to the counter surface, in the direction opposite the book's velocity. From Newton's second law, then, a force must act on the book parallel to the counter surface, in the direction opposite its velocity. That force is a frictional force.

2. Push horizontally on the book to make it travel at constant velocity along the counter. Can the force from you be the only horizontal force on the book? No, because then the book would accelerate. From Newton's second law, there must be a second force, directed opposite your force but with the same magnitude, so that the two forces balance. That second force is a frictional force, directed parallel to the counter.

3. Push horizontally on a heavy crate. The crate does not move. From Newton's second law, a second force must also be acting on the crate to counteract your force. Moreover, this second force must be directed opposite your force and have the same magnitude as your force, so that the two forces balance. That second force is a frictional force. Push even harder. The crate still does not move. Apparently the frictional force can change in magnitude so that the two forces still balance. Now push with all your strength. The crate begins to slide. Evidently, there is a maximum magnitude of the frictional force. When you exceed that maximum magnitude, the crate slides.

Figure 6-1 shows a similar situation. In Fig. 6-1a, a block rests on a tabletop, with the gravitational force $\vec{F}_g$ balanced by a normal force $\vec{F}_N$. In Fig. 6-1b, you exert a force $\vec{F}$ on the block, attempting to pull it to the left. In response, a frictional force $\vec{f}_s$ is directed to the right, exactly balancing your force. The force $\vec{f}_s$ is called the **static frictional force.** The block does not move.

Figures 6-1c and 6-1d show that as you increase the magnitude of your applied force, the magnitude of the static frictional force $\vec{f}_s$ also increases and the block remains at rest. When the applied force reaches a certain magnitude, however, the block "breaks away" from its intimate contact with the tabletop and

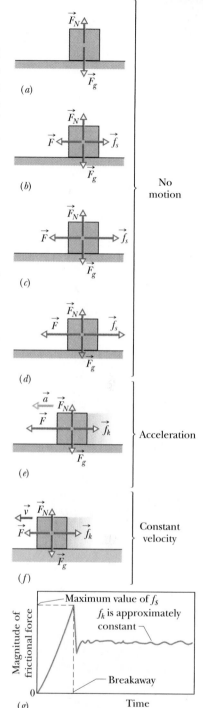

**FIG. 6-1** (a) The forces on a stationary block. (b–d) An external force $\vec{F}$, applied to the block, is balanced by a static frictional force $\vec{f}_s$. As $F$ is increased, $f_s$ also increases, until $f_s$ reaches a certain maximum value. (e) The block then "breaks away," accelerating suddenly in the direction of $\vec{F}$. (f) If the block is now to move with constant velocity, $F$ must be reduced from the maximum value it had just before the block broke away. (g) Some experimental results for the sequence (a) through (f).

accelerates leftward (Fig. 6-1*e*). The frictional force that then opposes the motion is called the **kinetic frictional force** $\vec{f}_k$.

Usually, the magnitude of the kinetic frictional force, which acts when there is motion, is less than the maximum magnitude of the static frictional force, which acts when there is no motion. Thus, if you wish the block to move across the surface with a constant speed, you must usually decrease the magnitude of the applied force once the block begins to move, as in Fig. 6-1*f*. As an example, Fig. 6-1*g* shows the results of an experiment in which the force on a block was slowly increased until breakaway occurred. Note the reduced force needed to keep the block moving at constant speed after breakaway.

A frictional force is, in essence, the vector sum of many forces acting between the surface atoms of one body and those of another body. If two highly polished and carefully cleaned metal surfaces are brought together in a very good vacuum (to keep them clean), they cannot be made to slide over each other. Because the surfaces are so smooth, many atoms of one surface contact many atoms of the other surface, and the surfaces *cold-weld* together instantly, forming a single piece of metal. If a machinist's specially polished gage blocks are brought together in air, there is less atom-to-atom contact, but the blocks stick firmly to each other and can be separated only by means of a wrenching motion. Usually, however, this much atom-to-atom contact is not possible. Even a highly polished metal surface is far from being flat on the atomic scale. Moreover, the surfaces of everyday objects have layers of oxides and other contaminants that reduce cold-welding.

When two ordinary surfaces are placed together, only the high points touch each other. (It is like having the Alps of Switzerland turned over and placed down on the Alps of Austria.) The actual *micro*scopic area of contact is much less than the apparent *macro*scopic contact area, perhaps by a factor of $10^4$. Nonetheless, many contact points do cold-weld together. These welds produce static friction when an applied force attempts to slide the surfaces relative to each other.

If the applied force is great enough to pull one surface across the other, there is first a tearing of welds (at breakaway) and then a continuous re-forming and tearing of welds as movement occurs and chance contacts are made (Fig. 6-2). The kinetic frictional force $\vec{f}_k$ that opposes the motion is the vector sum of the forces at those many chance contacts.

If the two surfaces are pressed together harder, many more points cold-weld. Now getting the surfaces to slide relative to each other requires a greater applied force: The static frictional force $\vec{f}_s$ has a greater maximum value. Once the surfaces are sliding, there are many more points of momentary cold-welding, so the kinetic frictional force $\vec{f}_k$ also has a greater magnitude.

Often, the sliding motion of one surface over another is "jerky" because the two surfaces alternately stick together and then slip. Such repetitive *stick-and-slip* can produce squeaking or squealing, as when tires skid on dry pavement, fingernails scratch along a chalkboard, or a rusty hinge is opened. It can also produce beautiful sounds, as when a bow is drawn properly across a violin string.

## 6-3 | Properties of Friction

Experiment shows that when a dry and unlubricated body presses against a surface in the same condition and a force $\vec{F}$ attempts to slide the body along the surface, the resulting frictional force has three properties:

**Property 1.** If the body does not move, then the static frictional force $\vec{f}_s$ and the component of $\vec{F}$ that is parallel to the surface balance each other. They are equal in magnitude, and $\vec{f}_s$ is directed opposite that component of $\vec{F}$.

**Property 2.** The magnitude of $\vec{f}_s$ has a maximum value $f_{s,\text{max}}$ that is given by

$$f_{s,\text{max}} = \mu_s F_N, \tag{6-1}$$

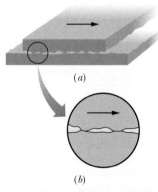

*(a)*

*(b)*

**FIG. 6-2** The mechanism of sliding friction. (*a*) The upper surface is sliding to the right over the lower surface in this enlarged view. (*b*) A detail, showing two spots where cold-welding has occurred. Force is required to break the welds and maintain the motion.

where $\mu_s$ is the **coefficient of static friction** and $F_N$ is the magnitude of the normal force on the body from the surface. If the magnitude of the component of $\vec{F}$ that is parallel to the surface exceeds $f_{s,max}$, then the body begins to slide along the surface.

Property 3.    If the body begins to slide along the surface, the magnitude of the frictional force rapidly decreases to a value $f_k$ given by

$$f_k = \mu_k F_N, \tag{6-2}$$

where $\mu_k$ is the **coefficient of kinetic friction.** Thereafter, during the sliding, a kinetic frictional force $\vec{f}_k$ with magnitude given by Eq. 6-2 opposes the motion.

The magnitude $F_N$ of the normal force appears in properties 2 and 3 as a measure of how firmly the body presses against the surface. If the body presses harder, then, by Newton's third law, $F_N$ is greater. Properties 1 and 2 are worded in terms of a single applied force $\vec{F}$, but they also hold for the net force of several applied forces acting on the body. Equations 6-1 and 6-2 are *not* vector equations; the direction of $\vec{f}_s$ or $\vec{f}_k$ is always parallel to the surface and opposed to the attempted sliding, and the normal force $\vec{F}_N$ is perpendicular to the surface.

The coefficients $\mu_s$ and $\mu_k$ are dimensionless and must be determined experimentally. Their values depend on certain properties of both the body and the surface; hence, they are usually referred to with the preposition "between," as in "the value of $\mu_s$ *between* an egg and a Teflon-coated skillet is 0.04, but that *between* rock-climbing shoes and rock is as much as 1.2." We assume that the value of $\mu_k$ does not depend on the speed at which the body slides along the surface.

✓ **CHECKPOINT 1**    A block lies on a floor. (a) What is the magnitude of the frictional force on it from the floor? (b) If a horizontal force of 5 N is now applied to the block, but the block does not move, what is the magnitude of the frictional force on it? (c) If the maximum value $f_{s,max}$ of the static frictional force on the block is 10 N, will the block move if the magnitude of the horizontally applied force is 8 N? (d) If it is 12 N? (e) What is the magnitude of the frictional force in part (c)?

## Sample Problem   6-1

If a car's wheels are "locked" (kept from rolling) during emergency braking, the car slides along the road. Ripped-off bits of tire and small melted sections of road form the "skid marks" that reveal that cold-welding occurred during the slide. The record for the longest skid marks on a public road was reportedly set in 1960 by a Jaguar on the M1 highway in England (Fig. 6-3a)— the marks were 290 m long! Assuming that $\mu_k = 0.60$ and the car's acceleration was constant during the braking, how fast was the car going when the wheels became locked?

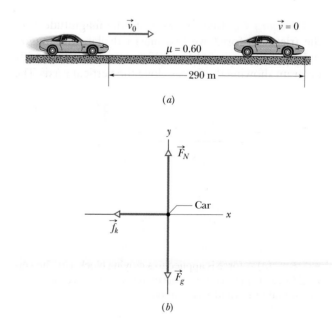

(a)

(b)

**KEY IDEAS**    (1) Because the acceleration $a$ is assumed constant, we can use the constant-acceleration equa-

**FIG. 6-3**   (a) A car sliding to the right and finally stopping after a displacement of 290 m. (b) A free-body diagram for the car.

tions of Table 2-1 to find the car's initial speed $v_0$. (2) If we neglect the effects of the air on the car, acceleration $a$ was due only to a kinetic frictional force $\vec{f}_k$ on the car from the road, directed opposite the direction of the car's motion, assumed to be in the positive direction of an $x$ axis (Fig. 6-3$b$). We can relate this force to the acceleration by writing Newton's second law for $x$ components ($F_{\text{net},x} = ma_x$) as

$$-f_k = ma, \qquad (6\text{-}3)$$

where $m$ is the car's mass. The minus sign indicates the direction of the kinetic frictional force.

**Calculations:** From Eq. 6-2, the frictional force has the magnitude $f_k = \mu_k F_N$, where $F_N$ is the magnitude of the normal force on the car from the road. Because the car is not accelerating vertically, we know from Fig. 6-3$b$ and Newton's second law that the magnitude of $\vec{F}_N$ is equal to the magnitude of the gravitational force $\vec{F}_g$ on the car, which is $mg$. Thus, $F_N = mg$.

Now solving Eq. 6-3 for $a$ and substituting $f_k = \mu_k F_N = \mu_k mg$ for $f_k$ yield

$$a = -\frac{f_k}{m} = -\frac{\mu_k mg}{m} = -\mu_k g, \qquad (6\text{-}4)$$

where the minus sign indicates that the acceleration is in the negative direction of the $x$ axis, opposite the direction of the velocity. Next, let's use Eq. 2-16,

$$v^2 = v_0^2 + 2a(x - x_0),$$

from the constant-acceleration equations of Chapter 2. We know that the displacement $x - x_0$ was 290 m and assume that the final speed $v$ was 0. Substituting for $a$ from Eq. 6-4 and solving for $v_0$ give

$$v_0 = \sqrt{2\mu_k g(x - x_0)} = \sqrt{(2)(0.60)(9.8 \text{ m/s}^2)(290 \text{ m})}$$
$$= 58 \text{ m/s} = 210 \text{ km/h.} \quad \text{(Answer)}$$

We assumed that $v = 0$ at the far end of the skid marks. Actually, the marks ended only because the Jaguar left the road after 290 m. So $v_0$ was at *least* 210 km/h.

**Sample Problem** 6-2

In Fig. 6-4$a$, a block of mass $m = 3.0$ kg slides along a floor while a force $\vec{F}$ of magnitude 12.0 N is applied to it at an upward angle $\theta$. The coefficient of kinetic friction between the block and the floor is $\mu_k = 0.40$. We can vary $\theta$ from 0 to 90° (the block remains on the floor). What $\theta$ gives the maximum value of the block's acceleration magnitude $a$?

**KEY IDEAS** Because the block is moving, a *kinetic* frictional force acts on it. The magnitude is given by Eq. 6-2 ($f_k = \mu_k F_N$, where $F_N$ is the normal force). The direction is opposite the motion (the friction opposes the sliding).

**Calculating $F_N$:** Because we need the magnitude $f_k$ of the frictional force, we first must calculate the magnitude $F_N$ of the normal force. Figure 6-4$b$ is a free-body diagram showing the forces along the vertical $y$ axis. The

normal force is upward, the gravitational force $\vec{F}_g$ with magnitude $mg$ is downward, and (note) the vertical component $F_y$ of the applied force is upward. That component is shown in Fig. 6-4$c$, where we can see that $F_y = F \sin \theta$. We can write Newton's second law ($\vec{F}_{\text{net}} = m\vec{a}$) for those forces along the $y$ axis as

$$F_N + F \sin \theta - mg = m(0), \qquad (6\text{-}5)$$

where we substituted zero for the acceleration along the $y$ axis (the block does not even move along that axis). Thus,

$$F_N = mg - F \sin \theta. \qquad (6\text{-}6)$$

**Calculating acceleration $a$:** Figure 6-4$d$ is a free-body diagram for motion along the $x$ axis. The horizontal component $F_x$ of the applied force is rightward; from Fig. 6-4$c$, we see that $F_x = F \cos \theta$. The frictional force has magnitude $f_k$ ($= \mu_k F_N$) and is leftward. Writing Newton's second law for motion along the $x$ axis gives us

$$F \cos \theta - \mu_k F_N = ma. \qquad (6\text{-}7)$$

Substituting for $F_N$ from Eq. 6-6 and solving for $a$ lead to

$$a = \frac{F}{m} \cos \theta - \mu_k \left( g - \frac{F}{m} \sin \theta \right). \qquad (6\text{-}8)$$

**Finding a maximum:** To find the value of $\theta$ that maximizes $a$, we take the derivative of $a$ with respect to $\theta$ and set the result equal to zero:

$$\frac{da}{d\theta} = -\frac{F}{m} \sin \theta + \mu_k \frac{F}{m} \cos \theta = 0.$$

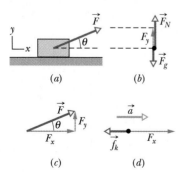

(a)    (b)

(c)    (d)

**FIG. 6-4** (*a*) A force is applied to a moving block. (*b*) The vertical forces. (*c*) The components of the applied force. (*d*) The horizontal forces and acceleration.

Rearranging and using the identity $(\sin \theta)/(\cos \theta) = \tan \theta$ give us

$$\tan \theta = \mu_k.$$

Solving for $\theta$ and substituting the given $\mu_k = 0.40$, we find that the acceleration will be maximum if

$$\theta = \tan^{-1} \mu_k$$
$$= 21.8° \approx 22°. \qquad \text{(Answer)}$$

**Comment:** As we increase $\theta$ from 0, more of the applied force $\vec{F}$ is upward, relieving the normal force. The decrease in the normal force causes a decrease in the frictional force, which opposes the block's motion. Thus, the block's acceleration tends to increase. However, the increase in $\theta$ also decreases the horizontal component of $\vec{F}$, and so the block's acceleration tends to decrease. These opposing tendencies produce a maximum acceleration at $\theta = 22°$.

---

## Sample Problem  6-3

Although many ingenious schemes have been attributed to the building of the Great Pyramid, the stone blocks were probably hauled up the side of the pyramid by men pulling on ropes. Figure 6-5a represents a 2000 kg stone block in the process of being pulled up the finished (smooth) side of the Great Pyramid, which forms a plane inclined at angle $\theta = 52°$. The block is secured to a wood sled and is pulled by multiple ropes (only one is shown). The sled's track is lubricated with water to decrease the coefficient of static friction to 0.40. Assume negligible friction at the (lubricated) point where the ropes pass over the edge at the top of the side. If each man on top of the pyramid pulls with a (reasonable) force of 686 N, how many men are needed to put the block on the verge of moving?

**KEY IDEAS** (1) Because the block is on the *verge* of moving, the static frictional force must be at its maximum possible value; that is, $f_s = f_{s,max}$. (2) Because the block is on the verge of moving *up* the plane, the frictional force must be *down* the plane (to oppose the pending motion). (3) From Sample Problem 5-5, we know that the component of the gravitational force down the plane is $mg \sin \theta$ and the component perpendicular to (and inward from) the plane is $mg \cos \theta$ (Fig. 6-5b).

**Calculations:** Figure 6-5c is a free-body diagram for the block, showing the force $\vec{F}$ applied by the ropes, the static frictional force $\vec{f}_s$, and the two components of the gravitational force. We can write Newton's second law ($\vec{F}_{net} = m\vec{a}$) for forces along the x axis as

$$F - mg \sin \theta - f_s = m(0). \qquad (6\text{-}9)$$

Because the block is on the verge of sliding and the frictional force is at the maximum possible value $f_{s,\ max}$, we use Eq. 6-1 to replace $f_s$ with $\mu_s F_N$:

$$f_s = f_{s,max}$$
$$= \mu_s F_N. \qquad (6\text{-}10)$$

From Figure 6-5c, we see that along the y axis Newton's

second law becomes

$$F_N - mg \cos \theta = m(0). \qquad (6\text{-}11)$$

Solving Eq. 6-11 for $F_N$ and substituting the result into Eq. 6-10, we have

$$f_s = \mu_s mg \cos \theta. \qquad (6\text{-}12)$$

Substituting this expression into Eq. 6-9 and solving for $F$ lead to

$$F = \mu_s mg \cos \theta + mg \sin \theta. \qquad (6\text{-}13)$$

Substituting $m = 2000$ kg, $\theta = 52°$, and $\mu_s = 0.40$, we find that the force required to put the stone block on the verge of moving is $2.027 \times 10^4$ N. Dividing this by the assumed pulling force of 686 N from each man, we find that the required number of men is

$$N = \frac{2.027 \times 10^4 \text{ N}}{686 \text{ N}} = 29.5 \approx 30 \text{ men}. \qquad \text{(Answer)}$$

**Comment:** Once the stone block began to move, the friction was kinetic friction and the coefficient was about 0.20. You can show that the required number of men was then 26 or 27. Thus, the huge stone blocks of the Great Pyramid could be pulled up into position by reasonably small teams of men.

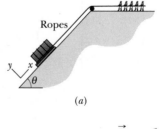

(a)

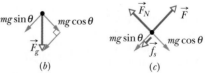

(b)          (c)

**FIG. 6-5** (a) A stone block on the verge of being pulled up the side of the Great Pyramid. (b) The components of the gravitational force. (c) A free-body diagram for the block.

## 6-4 | The Drag Force and Terminal Speed

A **fluid** is anything that can flow—generally either a gas or a liquid. When there is a relative velocity between a fluid and a body (either because the body moves through the fluid or because the fluid moves past the body), the body experiences a **drag force** $\vec{D}$ that opposes the relative motion and points in the direction in which the fluid flows relative to the body.

Here we examine only cases in which air is the fluid, the body is blunt (like a baseball) rather than slender (like a javelin), and the relative motion is fast enough so that the air becomes turbulent (breaks up into swirls) behind the body. In such cases, the magnitude of the drag force $\vec{D}$ is related to the relative speed $v$ by an experimentally determined **drag coefficient** $C$ according to

$$D = \tfrac{1}{2}C\rho A v^2, \tag{6-14}$$

where $\rho$ is the air density (mass per volume) and $A$ is the **effective cross-sectional area** of the body (the area of a cross section taken perpendicular to the velocity $\vec{v}$). The drag coefficient $C$ (typical values range from 0.4 to 1.0) is not truly a constant for a given body because if $v$ varies significantly, the value of $C$ can vary as well. Here, we ignore such complications.

Downhill speed skiers know well that drag depends on $A$ and $v^2$. To reach high speeds a skier must reduce $D$ as much as possible by, for example, riding the skis in the "egg position" (Fig. 6-6) to minimize $A$.

When a blunt body falls from rest through air, the drag force $\vec{D}$ is directed upward; its magnitude gradually increases from zero as the speed of the body increases. This upward force $\vec{D}$ opposes the downward gravitational force $\vec{F}_g$ on the body. We can relate these forces to the body's acceleration by writing Newton's second law for a vertical $y$ axis ($F_{\text{net},y} = ma_y$) as

$$D - F_g = ma, \tag{6-15}$$

where $m$ is the mass of the body. As suggested in Fig. 6-7, if the body falls long enough, $D$ eventually equals $F_g$. From Eq. 6-15, this means that $a = 0$, and so the body's speed no longer increases. The body then falls at a constant speed, called the **terminal speed** $v_t$.

To find $v_t$, we set $a = 0$ in Eq. 6-15 and substitute for $D$ from Eq. 6-14, obtaining

$$\tfrac{1}{2}C\rho A v_t^2 - F_g = 0,$$

which gives

$$v_t = \sqrt{\frac{2F_g}{C\rho A}}. \tag{6-16}$$

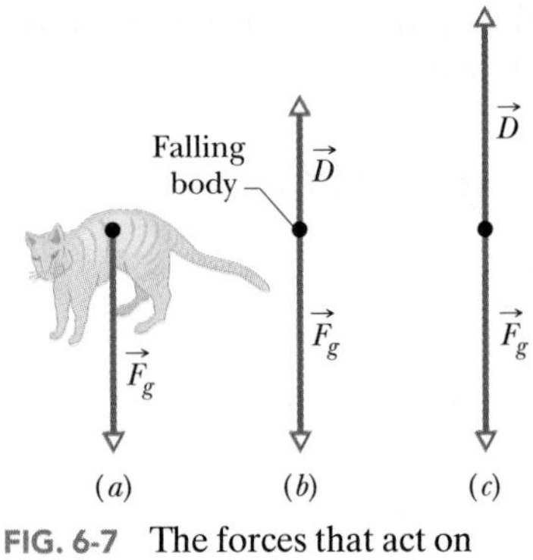

Table 6-1 gives values of $v_t$ for some common objects.

According to calculations* based on Eq. 6-14, a cat must fall about six floors to reach terminal speed. Until it does so, $F_g > D$ and the cat accelerates downward because of the net downward force. Recall from Chapter 2 that your body is an accelerometer, not a speedometer. Because the cat also senses the acceleration, it is frightened and keeps its feet underneath its body, its head tucked in, and its spine bent upward, making $A$ small, $v_t$ large, and injury likely.

However, if the cat does reach $v_t$ during a longer fall, the acceleration vanishes and the cat relaxes somewhat, stretching its legs and neck horizontally outward and straightening its spine (it then resembles a flying squirrel). These actions increase area $A$ and thus also, by Eq. 6-14, the drag $D$. The cat begins to slow because now $D > F_g$ (the net force is upward), until a new, smaller $v_t$ is reached. The decrease

*W. O. Whitney and C. J. Mehlhaff, "High-Rise Syndrome in Cats." *The Journal of the American Veterinary Medical Association,* 1987.

**FIG. 6-6** This skier crouches in an "egg position" so as to minimize her effective cross-sectional area and thus minimize the air drag acting on her. *(Karl-Josef Hildenbrand/dpa/Landov LLC)*

Falling body

$\vec{D}$    $\vec{D}$

$\vec{F}_g$    $\vec{F}_g$    $\vec{F}_g$

(a)    (b)    (c)

**FIG. 6-7** The forces that act on a body falling through air: (a) the body when it has just begun to fall and (b) the free-body diagram a little later, after a drag force has developed. (c) The drag force has increased until it balances the gravitational force on the body. The body now falls at its constant terminal speed.

**TABLE 6-1**

**Some Terminal Speeds in Air**

| Object | Terminal Speed (m/s) | 95% Distance$^a$ (m) |
|---|---|---|
| Shot (from shot put) | 145 | 2500 |
| Sky diver (typical) | 60 | 430 |
| Baseball | 42 | 210 |
| Tennis ball | 31 | 115 |
| Basketball | 20 | 47 |
| Ping-Pong ball | 9 | 10 |
| Raindrop (radius = 1.5 mm) | 7 | 6 |
| Parachutist (typical) | 5 | 3 |

$^a$This is the distance through which the body must fall from rest to reach 95% of its terminal speed.

*Source:* Adapted from Peter J. Brancazio, *Sport Science,* 1984, Simon & Schuster, New York.

in $v_t$ reduces the possibility of serious injury on landing. Just before the end of the fall, when it sees it is nearing the ground, the cat pulls its legs back beneath its body to prepare for the landing.

Humans often fall from great heights for the fun of skydiving. However, in April 1987, during a jump, sky diver Gregory Robertson noticed that fellow sky diver Debbie Williams had been knocked unconscious in a collision with a third sky diver and was unable to open her parachute. Robertson, who was well above Williams at the time and who had not yet opened his parachute for the 4 km plunge, reoriented his body head-down so as to minimize $A$ and maximize his downward speed. Reaching an estimated $v_t$ of 320 km/h, he caught up with Williams and then went into a horizontal "spread eagle" (as in Fig. 6-8) to increase $D$ so that he could grab her. He opened her parachute and then, after releasing her, his own, a scant 10 s before impact. Williams received extensive internal injuries due to her lack of control on landing but survived.

**FIG. 6-8** Sky divers in a horizontal "spread eagle" maximize air drag. (*Steve Fitchett/Taxi/Getty Images*)

## Sample Problem    6-4

If a falling cat reaches a first terminal speed of 97 km/h while it is tucked in and then stretches out, doubling $A$, how fast is it falling when it reaches a new terminal speed?

**KEY IDEA**  The terminal speeds of the cat depend on (among other things) the effective cross-sectional areas $A$ of the cat, according to Eq. 6-16. Thus, we can use that

equation to set up a ratio of speeds. We let $v_{to}$ and $v_{tn}$ represent the original and new terminal speeds, and $A_o$ and $A_n$ the original and new areas. Then by Eq. 6-16,

$$\frac{v_{tn}}{v_{to}} = \frac{\sqrt{2F_g/C\rho A_n}}{\sqrt{2F_g/C\rho A_o}} = \sqrt{\frac{A_o}{A_n}} = \sqrt{\frac{A_o}{2A_o}} = \sqrt{0.5} \approx 0.7,$$

which means that $v_{tn} \approx 0.7 v_{to}$, or about 68 km/h.

## Sample Problem    6-5

A raindrop with radius $R = 1.5$ mm falls from a cloud that is at height $h = 1200$ m above the ground. The drag coefficient $C$ for the drop is 0.60. Assume that the drop is spherical throughout its fall. The density of water $\rho_w$ is 1000 kg/m$^3$, and the density of air $\rho_a$ is 1.2 kg/m$^3$.

(a) What is the terminal speed of the drop?

**KEY IDEA**  The drop reaches a terminal speed $v_t$ when the gravitational force on it is balanced by the air drag force on it, so its acceleration is zero. We could then

apply Newton's second law and the drag force equation to find $v_t$, but Eq. 6-16 does all that for us.

***Calculations:*** To use Eq. 6-16, we need the drop's effective cross-sectional area $A$ and the magnitude $F_g$ of the gravitational force. Because the drop is spherical, $A$ is the area of a circle ($\pi R^2$) that has the same radius as the sphere. To find $F_g$, we use three facts: (1) $F_g = mg$, where $m$ is the drop's mass; (2) the (spherical) drop's volume is $V = \frac{4}{3}\pi R^3$; and (3) the density of the water in the drop is the mass per volume, or $\rho_w = m/V$. Thus, we find

$$F_g = V\rho_w g = \tfrac{4}{3}\pi R^3 \rho_w g.$$

We next substitute this, the expression for $A$, and the given data into Eq. 6-16. Being careful to distinguish between the air density $\rho_a$ and the water density $\rho_w$, we obtain

$$v_t = \sqrt{\frac{2F_g}{C\rho_a A}} = \sqrt{\frac{8\pi R^3 \rho_w g}{3C\rho_a \pi R^2}} = \sqrt{\frac{8R\rho_w g}{3C\rho_a}}$$

$$= \sqrt{\frac{(8)(1.5 \times 10^{-3}\,\text{m})(1000\,\text{kg/m}^3)(9.8\,\text{m/s}^2)}{(3)(0.60)(1.2\,\text{kg/m}^3)}}$$

$$= 7.4\,\text{m/s} \approx 27\,\text{km/h}. \quad \text{(Answer)}$$

Note that the height of the cloud does not enter into the calculation. As Table 6-1 indicates, the raindrop reaches terminal speed after falling just a few meters.

**(b)** What would be the drop's speed just before impact if there were no drag force?

**KEY IDEA** With no drag force to reduce the drop's speed during the fall, the drop would fall with the constant free-fall acceleration $g$, so the constant-acceleration equations of Table 2-1 apply.

**Calculation:** Because we know the acceleration is $g$, the initial velocity $v_0$ is 0, and the displacement $x - x_0$ is $-h$, we use Eq. 2-16 to find $v$:

$$v = \sqrt{2gh} = \sqrt{(2)(9.8\,\text{m/s}^2)(1200\,\text{m})}$$

$$= 153\,\text{m/s} \approx 550\,\text{km/h}. \quad \text{(Answer)}$$

Had he known this, Shakespeare would scarcely have written, "it droppeth as the gentle rain from heaven, upon the place beneath." In fact, the speed is close to that of a bullet from a large-caliber handgun!

## 6-5 | Uniform Circular Motion

From Section 4-7, recall that when a body moves in a circle (or a circular arc) at constant speed $v$, it is said to be in uniform circular motion. Also recall that the body has a centripetal acceleration (directed toward the center of the circle) of constant magnitude given by

$$a = \frac{v^2}{R} \quad \text{(centripetal acceleration)}, \quad (6\text{-}17)$$

where $R$ is the radius of the circle.

Let us examine two examples of uniform circular motion:

1. *Rounding a curve in a car.* You are sitting in the center of the rear seat of a car moving at a constant high speed along a flat road. When the driver suddenly turns left, rounding a corner in a circular arc, you slide across the seat toward the right and then jam against the car wall for the rest of the turn. What is going on?

   While the car moves in the circular arc, it is in uniform circular motion; that is, it has an acceleration that is directed toward the center of the circle. By Newton's second law, a force must cause this acceleration. Moreover, the force must also be directed toward the center of the circle. Thus, it is a **centripetal force,** where the adjective indicates the direction. In this example, the centripetal force is a frictional force on the tires from the road; it makes the turn possible.

   If you are to move in uniform circular motion along with the car, there must also be a centripetal force on you. However, apparently the frictional force on you from the seat was not great enough to make you go in a circle with the car. Thus, the seat slid beneath you, until the right wall of the car jammed into you. Then its push on you provided the needed centripetal force on you, and you joined the car's uniform circular motion.

2. *Orbiting Earth.* This time you are a passenger in the space shuttle *Atlantis*. As it and you orbit Earth, you float through your cabin. What is going on?

   Both you and the shuttle are in uniform circular motion and have accelerations directed toward the center of the circle. Again by Newton's second law, centripetal forces must cause these accelerations. This time the centripetal forces are gravitational pulls (the pull on you and the pull on the shuttle) exerted by Earth and directed radially inward, toward the center of Earth.

In both car and shuttle you are in uniform circular motion, acted on by a centripetal force—yet your sensations in the two situations are quite different. In the car, jammed up against the wall, you are aware of being compressed by the wall. In the orbiting shuttle, however, you are floating around with no sensation of any force acting on you. Why this difference?

The difference is due to the nature of the two centripetal forces. In the car, the centripetal force is the push on the part of your body touching the car wall. You can sense the compression on that part of your body. In the shuttle, the centripetal force is Earth's gravitational pull on every atom of your body. Thus, there is no compression (or pull) on any one part of your body and no sensation of a force acting on you. (The sensation is said to be one of "weightlessness," but that description is tricky. The pull on you by Earth has certainly not disappeared and, in fact, is only a little less than it would be with you on the ground.)

Another example of a centripetal force is shown in Fig. 6-9. There a hockey puck moves around in a circle at constant speed $v$ while tied to a string looped around a central peg. This time the centripetal force is the radially inward pull on the puck from the string. Without that force, the puck would slide off in a straight line instead of moving in a circle.

Note again that a centripetal force is not a new kind of force. The name merely indicates the direction of the force. It can, in fact, be a frictional force, a gravitational force, the force from a car wall or a string, or any other force. For any situation:

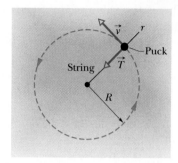

**FIG. 6-9** An overhead view of a hockey puck moving with constant speed $v$ in a circular path of radius $R$ on a horizontal frictionless surface. The centripetal force on the puck is $\vec{T}$, the pull from the string, directed inward along the radial axis $r$ extending through the puck.

> A centripetal force accelerates a body by changing the direction of the body's velocity without changing the body's speed.

From Newton's second law and Eq. 6-17 ($a = v^2/R$), we can write the magnitude $F$ of a centripetal force (or a net centripetal force) as

$$F = m\,\frac{v^2}{R} \qquad \text{(magnitude of centripetal force)}. \qquad (6\text{-}18)$$

Because the speed $v$ here is constant, the magnitudes of the acceleration and the force are also constant.

However, the directions of the centripetal acceleration and force are not constant; they vary continuously so as to always point toward the center of the circle. For this reason, the force and acceleration vectors are sometimes drawn along a radial axis $r$ that moves with the body and always extends from the center of the circle to the body, as in Fig. 6-9. The positive direction of the axis is radially outward, but the acceleration and force vectors point radially inward.

✓**CHECKPOINT 2** When you ride in a Ferris wheel at constant speed, what are the directions of your acceleration $\vec{a}$ and the normal force $\vec{F}_N$ on you (from the always upright seat) as you pass through (a) the highest point and (b) the lowest point of the ride?

---

**Sample Problem** 6-6

Igor is a cosmonaut on the International Space Station, in a circular orbit around Earth, at an altitude $h$ of 520 km and with a constant speed $v$ of 7.6 km/s. Igor's mass $m$ is 79 kg.

(a) What is his acceleration?

**KEY IDEA** Igor is in uniform circular motion and thus has a centripetal acceleration of magnitude given by Eq. 6-17 ($a = v^2/R$).

**Calculation:** The radius $R$ of Igor's motion is $R_E + h$, where $R_E$ is Earth's radius ($6.37 \times 10^6$ m, from Appendix C). Thus,

$$a = \frac{v^2}{R} = \frac{v^2}{R_E + h}$$

$$= \frac{(7.6 \times 10^3 \text{ m/s})^2}{6.37 \times 10^6 \text{ m} + 0.52 \times 10^6 \text{ m}}$$

$$= 8.38 \text{ m/s}^2 \approx 8.4 \text{ m/s}^2. \qquad \text{(Answer)}$$

This is the value of the free-fall acceleration at Igor's altitude. If he were lifted to that altitude and released, instead of being put into orbit there, he would fall toward Earth's center, starting out with that value for his acceleration. The difference in the two situations is that when he orbits Earth, he always has a "sideways" motion as well: As he falls, he also moves to the side, so that he ends up moving along a curved path around Earth.

(b) What force does Earth exert on Igor?

**KEY IDEAS** (1) There must be a centripetal force on Igor if he is to be in uniform circular motion. (2) That force is the gravitational force $\vec{F}_g$ on him from Earth, directed toward his center of rotation (at the center of Earth).

**Calculation:** From Newton's second law, written along the radial axis $r$, this force has the magnitude

$$F_g = ma = (79 \text{ kg})(8.38 \text{ m/s}^2)$$
$$= 662 \text{ N} \approx 660 \text{ N}. \qquad \text{(Answer)}$$

If Igor were to stand on a scale placed on the top of a tower of height $h = 520$ km, the scale would read 660 N. In orbit, the scale (if Igor could "stand" on it) would read zero because he and the scale are in free fall together, and therefore his feet do not actually press against it.

---

## Sample Problem 6-7

In a 1901 circus performance, Allo "Dare Devil" Diavolo introduced the stunt of riding a bicycle in a loop-the-loop (Fig. 6-10a). Assuming that the loop is a circle with radius $R = 2.7$ m, what is the least speed $v$ Diavolo could have at the top of the loop to remain in contact with it there?

**KEY IDEA** We can assume that Diavolo and his bicycle travel through the top of the loop as a single particle in uniform circular motion. Thus, at the top, the acceleration $\vec{a}$ of this particle must have the magnitude $a = v^2/R$ given by Eq. 6-17 and be directed downward, toward the center of the circular loop.

**Calculations:** The forces on the particle when it is at the top of the loop are shown in the free-body diagram of Fig 6-10b. The gravitational force $\vec{F}_g$ is directed downward along a $y$ axis; so is the normal force $\vec{F}_N$ on the particle from the loop; so also is the centripetal acceleration of the particle. Thus, Newton's second law for $y$ components ($F_{net,y} = ma_y$) gives us

$$-F_N - F_g = m(-a)$$

and

$$-F_N - mg = m\left(-\frac{v^2}{R}\right). \qquad (6\text{-}19)$$

If the particle has the *least speed* $v$ needed to remain in contact, then it is on the *verge of losing contact* with the loop (falling away from the loop), which means that $F_N = 0$ at the top of the loop (the particle and loop touch but without any normal force). Substituting 0 for $F_N$ in Eq. 6-19, solving for $v$, and then substituting known values give us

$$v = \sqrt{gR} = \sqrt{(9.8 \text{ m/s}^2)(2.7 \text{ m})}$$
$$= 5.1 \text{ m/s}. \qquad \text{(Answer)}$$

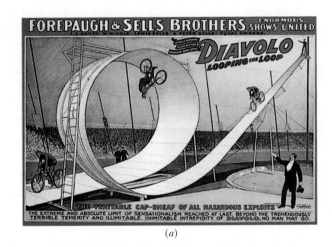

(a)

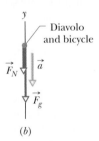

(b)

**FIG. 6-10** (*a*) Contemporary advertisement for Diavolo and (*b*) free-body diagram for the performer at the top of the loop. (*Photograph in part a reproduced with permission of Circus World Museum*)

**Comments:** Diavolo made certain that his speed at the top of the loop was greater than 5.1 m/s so that he did not lose contact with the loop and fall away from it. Note that this speed requirement is independent of the mass of Diavolo and his bicycle. Had he feasted on, say, pierogies before his performance, he still would have had to exceed only 5.1 m/s to maintain contact as he passed through the top of the loop.

Even some seasoned roller-coaster riders blanch at the thought of riding the Rotor, which is essentially a large, hollow cylinder that is rotated rapidly around its central axis (Fig. 6-11). Before the ride begins, a rider enters the cylinder through a door on the side and stands on a floor, up against a canvas-covered wall. The door is closed, and as the cylinder begins to turn, the rider, wall, and floor move in unison. When the rider's speed reaches some predetermined value, the floor abruptly and alarmingly falls away. The rider does not fall with it but instead is pinned to the wall while the cylinder rotates, as if an unseen (and somewhat unfriendly) agent is pressing the body to the wall. Later, the floor is eased back to the rider's feet, the cylinder slows, and the rider sinks a few centimeters to regain footing on the floor. (Some riders consider all this to be fun.)

Suppose that the coefficient of static friction $\mu_s$ between the rider's clothing and the canvas is 0.40 and that the cylinder's radius $R$ is 2.1 m.

(a) What minimum speed $v$ must the cylinder and rider have if the rider is not to fall when the floor drops?

**KEY IDEAS**

1. The gravitational force $\vec{F}_g$ on the rider tends to slide her down the wall, but she does not move because a frictional force from the wall acts upward on her (Fig. 6-11).

2. If she is to be on the verge of sliding down, that upward force must be a *static* frictional force $\vec{f}_s$ at its maximum value $\mu_s F_N$, where $F_N$ is the magnitude of the normal force $\vec{F}_N$ on her from the cylinder (Fig. 6-11).

3. This normal force is directed horizontally toward the central axis and is the centripetal force that causes the rider to move in a circular

path, with centripetal acceleration of magnitude $a = v^2/R$ and directed toward the center of the circle.

We want speed $v$ in that last expression, for the condition that the rider is on the verge of sliding.

**Vertical calculations:** We first place a vertical $y$ axis through the rider, with the positive direction upward. We can then apply Newton's second law to the rider, writing it for $y$ components ($F_{\text{net},y} = ma_y$) as

$$f_s - mg = m(0),$$

where $m$ is the rider's mass and $mg$ is the magnitude of $\vec{F}_g$. Because the rider is on the verge of sliding, we substitute the maximum value $\mu_s F_N$ for $f_s$ in this equation, getting

$$\mu_s F_N - mg = 0,$$

or

$$F_N = \frac{mg}{\mu_s}. \tag{6-20}$$

**Radial calculations:** Next we place a radial $r$ axis through the rider, with the positive direction outward. We can then write Newton's second law for components along that axis as

$$-F_N = m\left(-\frac{v^2}{R}\right). \tag{6-21}$$

Substituting Eq. 6-20 for $F_N$ and then solving for $v$, we find

$$v = \sqrt{\frac{gR}{\mu_s}} = \sqrt{\frac{(9.8 \text{ m/s}^2)(2.1 \text{ m})}{0.40}}$$

$$= 7.17 \text{ m/s} \approx 7.2 \text{ m/s}. \quad \text{(Answer)}$$

Note that the result is independent of the rider's mass; it holds for anyone riding the Rotor, from a child to a sumo wrestler, which is why no one has to "weigh in" to ride the Rotor.

(b) If the rider's mass is 49 kg, what is the magnitude of the centripetal force on her?

**Calculation:** According to Eq. 6-21,

$$F_N = m\frac{v^2}{R} = (49 \text{ kg})\frac{(7.17 \text{ m/s})^2}{2.1 \text{ m}}$$

$$\approx 1200 \text{ N}. \quad \text{(Answer)}$$

Although this force is directed toward the central axis, the rider has an overwhelming sensation that the force pinning her against the wall is directed radially outward. Her sensation stems from the fact that she is in a noninertial frame (she and it are accelerating). As measured from such frames, forces can be illusionary. The illusion is part of the Rotor's attraction.

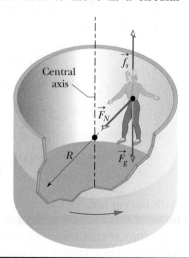

**FIG. 6-11** A Rotor in an amusement park, showing the forces on a rider. The centripetal force is the normal force $\vec{F}_N$ with which the wall pushes inward on the rider.

**Sample Problem** | **6-9** | **Build your skill**

*Upside-down racing:* A modern race car is designed so that the passing air pushes down on it, allowing the car to travel much faster through a flat turn in a Grand Prix without friction failing. This downward push is called *negative lift.* Can a race car have so much negative lift that it could be driven upside down on a long ceiling, as done fictionally by a sedan in the first *Men in Black* movie?

Figure 6-12a represents a Grand Prix race car of mass $m = 600$ kg as it travels on a flat track in a circular arc of radius $R = 100$ m. Because of the shape of the car and the wings on it, the passing air exerts a negative lift $\vec{F}_L$ downward on the car. The coefficient of static friction between the tires and the track is 0.75. (Assume that the forces on the four tires are identical.)

(a) If the car is on the verge of sliding out of the turn when its speed is 28.6 m/s, what is the magnitude of $\vec{F}_L$?

**KEY IDEAS**

1. A centripetal force must act on the car because the car is moving around a circular arc; that force must be directed toward the center of curvature of the arc (here, that is horizontally).

2. The only horizontal force acting on the car is a frictional force on the tires from the road. So the required centripetal force is a frictional force.

3. Because the car is not sliding, the frictional force must be a *static* frictional force $\vec{f}_s$ (Fig. 6-12a).

4. Because the car is on the verge of sliding, the magnitude $f_s$ is equal to the maximum value $f_{s,\text{max}} = \mu_s F_N$, where $F_N$ is the magnitude of the normal force $\vec{F}_N$ acting on the car from the track.

*Radial calculations:* The frictional force $\vec{f}_s$ is shown in the free-body diagram of Fig. 6-12b. It is in the negative direction of a radial axis $r$ that always extends from the center of curvature through the car as the car moves. The force produces a centripetal acceleration of magnitude $v^2/R$. We can relate the force and acceleration by writing Newton's second law for components along the $r$ axis ($F_{\text{net},r} = ma_r$) as

$$-f_s = m\left(-\frac{v^2}{R}\right). \qquad (6\text{-}22)$$

Substituting $f_{s,\text{max}} = \mu_s F_N$ for $f_s$ leads us to

$$\mu_s F_N = m\left(\frac{v^2}{R}\right). \qquad (6\text{-}23)$$

*Vertical calculations:* Next, let's consider the vertical forces on the car. The normal force $\vec{F}_N$ is directed up, in the positive direction of the $y$ axis in Fig. 6-12b. The gravitational force $\vec{F}_g = m\vec{g}$ and the negative lift $\vec{F}_L$ are directed down. The acceleration of the car along the $y$ axis is zero. Thus we can write Newton's second law

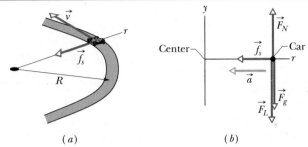

(a) (b)

**FIG. 6-12** (a) A race car moves around a flat curved track at constant speed $v$. The frictional force $\vec{f}_s$ provides the necessary centripetal force along a radial axis $r$. (b) A free-body diagram (not to scale) for the car, in the vertical plane containing $r$.

for components along the $y$ axis ($F_{\text{net},y} = ma_y$) as

$$F_N - mg - F_L = 0,$$

or

$$F_N = mg + F_L. \qquad (6\text{-}24)$$

*Combining results:* Now we can combine our results along the two axes by substituting Eq. 6-24 for $F_N$ in Eq. 6-23. Doing so and then solving for $F_L$ lead to

$$F_L = m\left(\frac{v^2}{\mu_s R} - g\right)$$

$$= (600\text{ kg})\left(\frac{(28.6\text{ m/s})^2}{(0.75)(100\text{ m})} - 9.8\text{ m/s}^2\right)$$

$$= 663.7\text{ N} \approx 660\text{ N}. \qquad \text{(Answer)}$$

(b) The magnitude $F_L$ of the negative lift on a car depends on the square of the car's speed $v^2$, just as the drag force does (Eq. 6-14). Thus, the negative lift on the car here is greater when the car travels faster, as it does on a straight section of track. What is the magnitude of the negative lift for a speed of 90 m/s?

**KEY IDEA** $F_L$ is proportional to $v^2$.

*Calculations:* Thus we can write a ratio of the negative lift $F_{L,90}$ at $v = 90$ m/s to our result for the negative lift $F_L$ at $v = 28.6$ m/s as

$$\frac{F_{L,90}}{F_L} = \frac{(90\text{ m/s})^2}{(28.6\text{ m/s})^2}.$$

Substituting $F_L = 663.7$ N and solving for $F_{L,90}$ give us

$$F_{L,90} = 6572\text{ N} \approx 6600\text{ N}. \qquad \text{(Answer)}$$

*Upside-down racing:* The gravitational force is

$$F_g = mg = (600\text{ kg})(9.8\text{ m/s}^2)$$

$$= 5880\text{ N}.$$

With the car upside down, the negative lift is an *upward* force of 6600 N, which exceeds the downward 5880 N. Thus, the car could run on a long ceiling *provided* that it moves at about 90 m/s (= 324 km/h = 201 mi/h).

Curved portions of highways are always banked (tilted) to prevent cars from sliding off the highway. When a highway is dry, the frictional force between the tires and the road surface may be enough to prevent sliding. When the highway is wet, however, the frictional force may be negligible, and banking is then essential. Figure 6-13a represents a car of mass $m$ as it moves at a constant speed $v$ of 20 m/s around a banked circular track of radius $R = 190$ m. (It is a normal car, rather than a race car, which means any vertical force from the passing air is negligible.) If the frictional force from the track is negligible, what bank angle $\theta$ prevents sliding?

**KEY IDEA**  Unlike Sample Problem 6-9, the track is banked so as to tilt the normal force $\vec{F}_N$ on the car to-

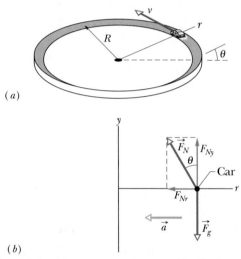

(a)

(b)

**FIG. 6-13**  (a) A car moves around a curved banked road at constant speed $v$. The bank angle is exaggerated for clarity. (b) A free-body diagram for the car, assuming that friction between tires and road is zero and that the car lacks negative lift. The radially inward component $F_{Nr}$ of the normal force (along radial axis $r$) provides the necessary centripetal force and radial acceleration.

ward the center of the circle (Fig. 6-13b). Thus, $\vec{F}_N$ now has a centripetal component of magnitude $F_{Nr}$, directed inward along a radial axis $r$. We want to find the value of the bank angle $\theta$ such that this centripetal component keeps the car on the circular track without need of friction.

**Radial calculation:**  As Fig. 6-13b shows (and as you should verify), the angle that force $\vec{F}_N$ makes with the vertical is equal to the bank angle $\theta$ of the track. Thus, the radial component $F_{Nr}$ is equal to $F_N \sin \theta$. We can now write Newton's second law for components along the $r$ axis ($F_{net,r} = ma_r$) as

$$-F_N \sin \theta = m\left(-\frac{v^2}{R}\right). \qquad (6\text{-}25)$$

We cannot solve this equation for the value of $\theta$ because it also contains the unknowns $F_N$ and $m$.

**Vertical calculations:**  We next consider the forces and acceleration along the $y$ axis in Fig. 6-13b. The vertical component of the normal force is $F_{Ny} = F_N \cos \theta$, the gravitational force $\vec{F}_g$ on the car has the magnitude $mg$, and the acceleration of the car along the $y$ axis is zero. Thus we can write Newton's second law for components along the $y$ axis ($F_{net,y} = ma_y$) as

$$F_N \cos \theta - mg = m(0),$$

from which

$$F_N \cos \theta = mg. \qquad (6\text{-}26)$$

**Combining results:**  Equation 6-26 also contains the unknowns $F_N$ and $m$, but note that dividing Eq. 6-25 by Eq. 6-26 neatly eliminates both those unknowns. Doing so, replacing $(\sin \theta)/(\cos \theta)$ with $\tan \theta$, and solving for $\theta$ then yield

$$\theta = \tan^{-1}\frac{v^2}{gR}$$

$$= \tan^{-1}\frac{(20 \text{ m/s})^2}{(9.8 \text{ m/s}^2)(190 \text{ m})} = 12°. \quad \text{(Answer)}$$

## REVIEW & SUMMARY

**Friction**  When a force $\vec{F}$ tends to slide a body along a surface, a **frictional force** from the surface acts on the body. The frictional force is parallel to the surface and directed so as to oppose the sliding. It is due to bonding between the body and the surface.

If the body does not slide, the frictional force is a **static frictional force** $\vec{f}_s$. If there is sliding, the frictional force is a **kinetic frictional force** $\vec{f}_k$.

1. If a body does not move, the static frictional force $\vec{f}_s$ and the component of $\vec{F}$ parallel to the surface are equal in magnitude, and $\vec{f}_s$ is directed opposite that component. If the component increases, $f_s$ also increases.

2. The magnitude of $\vec{f}_s$ has a maximum value $f_{s,max}$ given by

$$f_{s,max} = \mu_s F_N, \qquad (6\text{-}1)$$

where $\mu_s$ is the **coefficient of static friction** and $F_N$ is the magnitude of the normal force. If the component of $\vec{F}$ parallel to the surface exceeds $f_{s,max}$, the body slides on the surface.

3. If the body begins to slide on the surface, the magnitude of the frictional force rapidly decreases to a constant value $f_k$ given by

$$f_k = \mu_k F_N, \qquad (6\text{-}2)$$

where $\mu_k$ is the **coefficient of kinetic friction.**

**Drag Force**  When there is relative motion between air (or some other fluid) and a body, the body experiences a **drag force** $\vec{D}$ that opposes the relative motion and points in the direction in which the fluid flows relative to the body. The magnitude of $\vec{D}$ is related to the relative speed $v$ by an experimentally determined **drag coefficient** $C$ according to

$$D = \tfrac{1}{2}C\rho Av^2, \qquad (6\text{-}14)$$

where $\rho$ is the fluid density (mass per unit volume) and $A$ is the **effective cross-sectional area** of the body (the area of a cross section taken perpendicular to the relative velocity $\vec{v}$).

**Terminal Speed**    When a blunt object has fallen far enough through air, the magnitudes of the drag force $\vec{D}$ and the gravitational force $\vec{F}_g$ on the body become equal. The body then falls at a constant **terminal speed** $v_t$ given by

$$v_t = \sqrt{\frac{2F_g}{C\rho A}}. \qquad (6\text{-}16)$$

**Uniform Circular Motion**    If a particle moves in a circle or a circular arc of radius $R$ at constant speed $v$, the particle is said to be in **uniform circular motion.** It then has a **centripetal acceleration** $\vec{a}$ with magnitude given by

$$a = \frac{v^2}{R}. \qquad (6\text{-}17)$$

This acceleration is due to a net **centripetal force** on the particle, with magnitude given by

$$F = \frac{mv^2}{R}, \qquad (6\text{-}18)$$

where $m$ is the particle's mass. The vector quantities $\vec{a}$ and $\vec{F}$ are directed toward the center of curvature of the particle's path.

## QUESTIONS

**1**    In Fig. 6-14, horizontal force $\vec{F}_1$ of magnitude 10 N is applied to a box on a floor, but the box does not slide. Then, as the magnitude of vertical force $\vec{F}_2$ is increased from zero, do the following quantities increase, decrease, or stay the same: (a) the magnitude of the frictional force $\vec{f}_s$ on the box; (b) the magnitude of the normal force $\vec{F}_N$ on the box from the floor; (c) the maximum value $f_{s,\text{max}}$ of the magnitude of the static frictional force on the box? (d) Does the box eventually slide?

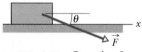

FIG. 6-14    Question 1.

**2**    In three experiments, three different horizontal forces are applied to the same block lying on the same countertop. The force magnitudes are $F_1 = 12$ N, $F_2 = 8$ N, and $F_3 = 4$ N. In each experiment, the block remains stationary in spite of the applied force. Rank the forces according to (a) the magnitude $f_s$ of the static frictional force on the block from the countertop and (b) the maximum value $f_{s,\text{max}}$ of that force, greatest first.

**3**    In Fig. 6-15, if the box is stationary and the angle $\theta$ between the horizontal and force $\vec{F}$ is increased somewhat, do the following quantities increase, decrease, or remain the same: (a) $F_x$; (b) $f_s$; (c) $F_N$; (d) $f_{s,\text{max}}$? (e) If, instead, the box is sliding and $\theta$ is increased, does the magnitude of the frictional force on the box increase, decrease, or remain the same?

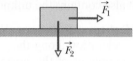

FIG. 6-15    Question 3.

**4**    Repeat Question 3 for force $\vec{F}$ angled upward instead of downward as drawn.

**5**    If you press an apple crate against a wall so hard that the crate cannot slide down the wall, what is the direction of (a) the static frictional force $\vec{f}_s$ on the crate from the wall and (b) the normal force $\vec{F}_N$ on the crate from the wall? If you increase your push, what happens to (c) $f_s$, (d) $F_N$, and (e) $f_{s,\text{max}}$?

**6**    In Fig. 6-16, a block of mass $m$ is held stationary on a ramp by the frictional force on it from the ramp. A force $\vec{F}$, directed up the ramp, is then applied to the block and gradually increased in magnitude from

FIG. 6-16    Question 6.

zero. During the increase, what happens to the direction and magnitude of the frictional force on the block?

**7**    Reconsider Question 6 but with the force $\vec{F}$ now directed down the ramp. As the magnitude of $\vec{F}$ is increased from zero, what happens to the direction and magnitude of the frictional force on the block?

**8**    In Fig. 6-17, a horizontal force of 100 N is to be applied to a 10 kg slab that is initially stationary on a frictionless floor, to accelerate the slab. A 10 kg block lies on top of the slab; the coefficient of friction $\mu$ between the block and the slab is not known, and the block might slip. (a) Considering that possibility, what is the possible range of values for the magnitude of the slab's acceleration $a_{\text{slab}}$? (*Hint:* You don't need written calculations; just consider extreme values for $\mu$.) (b) What is the possible range for the magnitude $a_{\text{block}}$ of the block's acceleration?

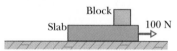

FIG. 6-17    Question 8.

**9**    A person riding a Ferris wheel moves through positions at (1) the top, (2) the bottom, and (3) midheight. If the wheel rotates at a constant rate, rank these three positions according to (a) the magnitude of the person's centripetal acceleration, (b) the magnitude of the net centripetal force on the person, and (c) the magnitude of the normal force on the person, greatest first.

**10**    In 1987, as a Halloween stunt, two sky divers passed a pumpkin back and forth between them while they were in free fall just west of Chicago. The stunt was great fun until the last sky diver with the pumpkin opened his parachute. The pumpkin broke free from his grip, plummeted about 0.5 km, ripped through the roof of a house, slammed into the kitchen floor, and splattered all over the newly remodeled kitchen. From the sky diver's viewpoint and from the pumpkin's viewpoint, why did the sky diver lose control of the pumpkin?

**11**    Figure 6-18 shows the path of a park ride that travels at constant speed through five circular arcs of radii $R_0$, $2R_0$, and $3R_0$. Rank the arcs according to the magnitude of the centripetal force on a rider traveling in the arcs, greatest first.

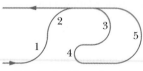

FIG. 6-18    Question 11.

# PROBLEMS

| **GO** | Tutoring problem available (at instructor's discretion) in *WileyPLUS* and WebAssign | | |
|---|---|---|---|
| **SSM** | Worked-out solution available in Student Solutions Manual | **WWW** | Worked-out solution is at ——— |
| **• – •••** | Number of dots indicates level of problem difficulty | **ILW** | Interactive solution is at ——— |
| | Additional information available in *The Flying Circus of Physics* and at flyingcircusofphysics.com | | |

http://www.wiley.com/college/halliday

## sec. 6-3    Properties of Friction

**•1** A bedroom bureau with a mass of 45 kg, including drawers and clothing, rests on the floor. (a) If the coefficient of static friction between the bureau and the floor is 0.45, what is the magnitude of the minimum horizontal force that a person must apply to start the bureau moving? (b) If the drawers and clothing, with 17 kg mass, are removed before the bureau is pushed, what is the new minimum magnitude?   **SSM WWW**

**•2** *The mysterious sliding stones.* Along the remote Racetrack Playa in Death Valley, California, stones sometimes gouge out prominent trails in the desert floor, as if the stones had been migrating (Fig. 6-19). For years curiosity mounted about why the stones moved. One explanation was that strong winds during occasional rainstorms would drag the rough stones over ground softened by rain. When the desert dried out, the trails behind the stones were hard-baked in place. According to measurements, the coefficient of kinetic friction between the stones and the wet playa ground is about 0.80. What horizontal force must act on a 20 kg stone (a typical mass) to maintain the stone's motion once a gust has started it moving? (Story continues with Problem 39.)

**FIG. 6-19** Problem 2. What moved the stone? *(Jerry Schad/ Photo Researchers)*

**•3** A person pushes horizontally with a force of 220 N on a 55 kg crate to move it across a level floor. The coefficient of kinetic friction is 0.35. What is the magnitude of (a) the frictional force and (b) the crate's acceleration?   **SSM ILW**

**•4** A baseball player with mass $m = 79$ kg, sliding into second base, is retarded by a frictional force of magnitude 470 N. What is the coefficient of kinetic friction $\mu_k$ between the player and the ground?

**•5** The floor of a railroad flatcar is loaded with loose crates having a coefficient of static friction of 0.25 with the floor. If the train is initially moving at a speed of 48 km/h, in how short a distance can the train be stopped at constant acceleration without causing the crates to slide over the floor?

**•6** A slide-loving pig slides down a certain 35° slide in twice the time it would take to slide down a frictionless 35° slide. What is the coefficient of kinetic friction between the pig and the slide?

**•7** A 3.5 kg block is pushed along a horizontal floor by a force $\vec{F}$ of magnitude 15 N at an angle $\theta = 40°$ with the horizontal (Fig. 6-20). The coefficient of kinetic friction between the block and the floor is 0.25. Calculate the magnitudes of (a) the frictional force on the block from the floor and (b) the block's acceleration.   **GO**

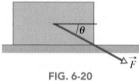

**FIG. 6-20**
Problems 7 and 24.

**•8** In a pickup game of dorm shuffleboard, students crazed by final exams use a broom to propel a calculus book along the dorm hallway. If the 3.5 kg book is pushed from rest through a distance of 0.90 m by the horizontal 25 N force from the broom and then has a speed of 1.60 m/s, what is the coefficient of kinetic friction between the book and floor?

**•9** A 2.5 kg block is initially at rest on a horizontal surface. A horizontal force $\vec{F}$ of magnitude 6.0 N and a vertical force $\vec{P}$ are then applied to the block (Fig. 6-21). The coefficients of friction for the block and surface are $\mu_s = 0.40$ and $\mu_k = 0.25$. Determine the magnitude of the frictional force acting on the block if the magnitude of $\vec{P}$ is (a) 8.0 N, (b) 10 N, and (c) 12 N.   **GO**

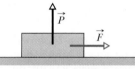

**FIG. 6-21**  Problem 9.

**•10** In about 1915, Henry Sincosky of Philadelphia suspended himself from a rafter by gripping the rafter with the thumb of each hand on one side and the fingers on the opposite side (Fig. 6-22). Sincosky's mass was 79 kg. If the coefficient of static friction between hand and rafter was 0.70, what was the least magnitude of the normal force on the rafter from each thumb or opposite fingers? (After suspending himself, Sincosky chinned himself on the rafter and then moved hand-over-hand along the rafter. If you do not think Sincosky's grip was remarkable, try to repeat his stunt.)

**FIG. 6-22**  Problem 10.

**•11** A worker pushes horizontally on a 35 kg crate with a force of magnitude 110 N. The coefficient of static friction between the crate and the floor is 0.37. (a) What is the value of $f_{s,max}$ under the circumstances? (b) Does the crate move? (c) What is the frictional force on the crate from the floor? (d) Suppose, next, that a second worker pulls directly upward on the crate to help out. What is the least vertical pull that will allow the first worker's 110 N push to move the crate? (e) If, instead, the second worker pulls horizontally to help out, what is the least pull that will get the crate moving?

•**12** Figure 6-23 shows the cross section of a road cut into the side of a mountain. The solid line $AA'$ represents a weak bedding plane along which sliding is possible. Block $B$ directly above the highway is

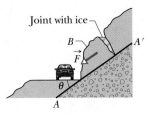

**FIG. 6-23** Problem 12.

separated from uphill rock by a large crack (called a *joint*), so that only friction between the block and the bedding plane prevents sliding. The mass of the block is $1.8 \times 10^7$ kg, the *dip angle* $\theta$ of the bedding plane is 24°, and the coefficient of static friction between block and plane is 0.63. (a) Show that the block will not slide. (b) Water seeps into the joint and expands upon freezing, exerting on the block a force $\vec{F}$ parallel to $AA'$. What minimum value of force magnitude $F$ will trigger a slide down the plane?

•**13** A 68 kg crate is dragged across a floor by pulling on a rope attached to the crate and inclined 15° above the horizontal. (a) If the coefficient of static friction is 0.50, what minimum force magnitude is required from the rope to start the crate moving? (b) If $\mu_k = 0.35$, what is the magnitude of the initial acceleration of the crate? **SSM**

•**14** Figure 6-24 shows an initially stationary block of mass $m$ on a floor. A force of magnitude $0.500mg$ is then applied at upward angle $\theta = 20°$. What is the magnitude of the acceleration of the block across the floor if (a) $\mu_s = 0.600$ and $\mu_k = 0.500$ and (b) $\mu_s = 0.400$ and $\mu_k = 0.300$?

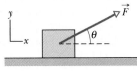

**FIG. 6-24** Problem 14.

•**15** The coefficient of static friction between Teflon and scrambled eggs is about 0.04. What is the smallest angle from the horizontal that will cause the eggs to slide across the bottom of a Teflon-coated skillet?

••**16** You testify as an *expert witness* in a case involving an accident in which car $A$ slid into the rear of car $B$, which was stopped at a red light along a road headed down a hill (Fig. 6-25). You find that the slope of the hill is $\theta = 12.0°$, that the cars were separated by distance $d = 24.0$ m when the driver of car $A$ put the car into a slide (it lacked any automatic anti-brake-lock system), and that the speed of car $A$ at the onset of braking was $v_0 = 18.0$ m/s. With what speed did car $A$ hit car $B$ if the coefficient of kinetic friction was (a) 0.60 (dry road surface) and (b) 0.10 (road surface covered with wet leaves)?

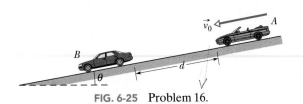

**FIG. 6-25** Problem 16.

••**17** A 12 N horizontal force $\vec{F}$ pushes a block weighing 5.0 N against a vertical wall (Fig. 6-26). The coefficient of static friction between the wall and the block is 0.60, and the coefficient of kinetic friction is 0.40. Assume that the

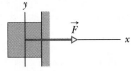

**FIG. 6-26** Problem 17.

block is not moving initially. (a) Will the block move? (b) In unit-vector notation, what is the force on the block from the wall?

••**18** A 4.10 kg block is pushed along a floor by a constant applied force that is horizontal and has a magnitude of 40.0 N. Figure 6-27 gives the block's speed $v$ versus time $t$ as the block moves along an $x$ axis on the floor. The scale of the figure's vertical axis is set by $v_s = 5.0$

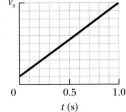

**FIG. 6-27** Problem 18.

m/s. What is the coefficient of kinetic friction between the block and the floor?

••**19** An initially stationary box of sand is to be pulled across a floor by means of a cable in which the tension should not exceed 1100 N. The coefficient of static friction between the box and the floor is 0.35. (a) What should be the angle between the cable and the horizontal in order to pull the greatest possible amount of sand, and (b) what is the weight of the sand and box in that situation?

••**20** A loaded penguin sled weighing 80 N rests on a plane inclined at angle $\theta = 20°$ to the horizontal (Fig. 6-28). Between the sled and the plane, the coefficient of static friction is 0.25, and the coefficient of kinetic friction is 0.15. (a) What is the least magni-

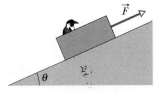

**FIG. 6-28** Problems 20 and 26.

tude of the force $\vec{F}$, parallel to the plane, that will prevent the sled from slipping down the plane? (b) What is the minimum magnitude $F$ that will start the sled moving up the plane? (c) What value of $F$ is required to move the sled up the plane at constant velocity?

••**21** In Fig. 6-29, a force $\vec{P}$ acts on a block weighing 45 N. The block is initially at rest on a plane inclined at angle $\theta = 15°$ to the horizontal. The positive di-

**FIG. 6-29** Problem 21.

rection of the $x$ axis is up the plane. The coefficients of friction between block and plane are $\mu_s = 0.50$ and $\mu_k = 0.34$. In unit-vector notation, what is the frictional force on the block from the plane when $\vec{P}$ is (a) $(-5.0 \text{ N})\hat{i}$, (b) $(-8.0 \text{ N})\hat{i}$, and (c) $(-15 \text{ N})\hat{i}$?

••**22** In Fig. 6-30, a box of Cheerios (mass $m_C = 1.0$ kg) and a box of Wheaties (mass $m_W = 3.0$ kg) are accelerated across a horizontal surface by a horizontal force $\vec{F}$ applied to the Cheerios box. The magnitude of the frictional force on the Cheerios box is 2.0 N, and the magnitude of the frictional force on the Wheaties box is 4.0 N. If the magnitude of $\vec{F}$ is

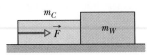

**FIG. 6-30** Problem 22.

12 N, what is the magnitude of the force on the Wheaties box from the Cheerios box?

**••23** Block *B* in Fig. 6-31 weighs 711 N. The coefficient of static friction between block and table is 0.25; angle $\theta$ is 30°; assume that the cord between *B* and the knot is horizontal. Find the maximum weight of block *A* for which the system will be stationary. **SSM WWW**

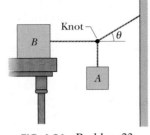

**FIG. 6-31** Problem 23.

**••24** A block is pushed across a floor by a constant force that is applied at downward angle $\theta$ (Fig. 6-20). Figure 6-32 gives the acceleration magnitude *a* versus a range of values for the coefficient of kinetic friction $\mu_k$ between block and floor: $a_1$ = 3.0 m/s², $\mu_{k2}$ = 0.20, and $\mu_{k3}$ = 0.40. What is the value of $\theta$?

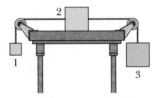

**FIG. 6-32** Problem 24.

**••25** When the three blocks in Fig. 6-33 are released from rest, they accelerate with a magnitude of 0.500 m/s². Block 1 has mass *M*, block 2 has 2*M*, and block 3 has 2*M*. What is the coefficient of kinetic friction between block 2 and the table?

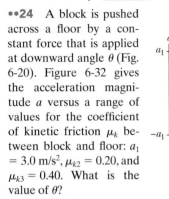

**FIG. 6-33** Problem 25

**••26** In Fig. 6-28, a sled is held on an inclined plane by a cord pulling directly up the plane. The sled is to be on the verge of moving up the plane. In Fig. 6-34, the magnitude *F* required of the cord's force on the sled is plotted versus a range of values for the coefficient of static friction $\mu_s$ between sled and plane: $F_1$ = 2.0 N, $F_2$ = 5.0 N, and $\mu_2$ = 0.50. At what angle $\theta$ is the plane inclined?

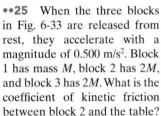

**FIG. 6-34** Problem 26.

**••27** Two blocks, of weights 3.6 N and 7.2 N, are connected by a massless string and slide down a 30° inclined plane. The coefficient of kinetic friction between the lighter block and the plane is 0.10; that between the heavier block and the plane is 0.20. Assuming that the lighter block leads, find (a) the magnitude of the acceleration of the blocks and (b) the tension in the string. **SSM**

**••28** Figure 6-35 shows three crates being pushed over a concrete floor by a horizontal force $\vec{F}$ of magnitude 440 N. The masses of the crates are $m_1$ = 30.0 kg, $m_2$ = 10.0 kg, and $m_3$ = 20.0 kg. The coefficient of kinetic friction between the floor and each of the crates is 0.700. (a) What is the magni-

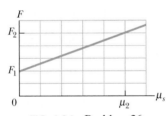

**FIG. 6-35** Problem 28.

tude $F_{32}$ of the force on crate 3 from crate 2? (b) If the crates then slide onto a polished floor, where the coefficient of kinetic friction is less than 0.700, is magnitude $F_{32}$ more than, less than, or the same as it was when the coefficient was 0.700? **GO**

**••29** Body *A* in Fig. 6-36 weighs 102 N, and body *B* weighs 32 N. The coefficients of friction between *A* and the incline are $\mu_s$ = 0.56 and $\mu_k$ = 0.25. Angle $\theta$ is 40°. Let the positive direction of an *x* axis be up the incline. In unit-vector notation, what is the acceleration of *A* if *A* is initially (a) at rest, (b) moving up the incline, and (c) moving down the incline?

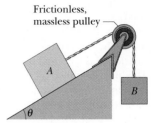

**FIG. 6-36** Problems 29 and 30.

**••30** In Fig. 6-36, two blocks are connected over a pulley. The mass of block *A* is 10 kg, and the coefficient of kinetic friction between *A* and the incline is 0.20. Angle $\theta$ of the incline is 30°. Block *A* slides down the incline at constant speed. What is the mass of block *B*?

**••31** In Fig. 6-37, blocks *A* and *B* have weights of 44 N and 22 N, respectively. (a) Determine the minimum weight of block *C* to keep *A* from sliding if $\mu_s$ between *A* and the table is 0.20. (b) Block *C* suddenly is lifted off *A*. What is the acceleration of block *A* if $\mu_k$ between *A* and the table is 0.15?

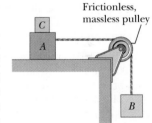

**FIG. 6-37** Problem 31.

**••32** A toy chest and its contents have a combined weight of 180 N. The coefficient of static friction between toy chest and floor is 0.42. The child in Fig. 6-38 attempts to move the chest across the floor by pulling on an attached rope. (a) If $\theta$ is 42°, what is the magnitude of the force $\vec{F}$ that the child must exert on the rope to put the chest on the verge of moving? (b) Write an expression for the magnitude *F* required to put the chest on the verge of moving as a function of the angle $\theta$. Determine (c) the value of $\theta$ for which *F* is a minimum and (d) that minimum magnitude.

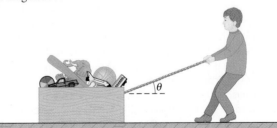

**FIG. 6-38** Problem 32.

**••33** The two blocks (*m* = 16 kg and *M* = 88 kg) in Fig. 6-39 are not attached to each other. The coefficient of static friction between the blocks is $\mu_s$ = 0.38, but the surface beneath the larger block is frictionless.

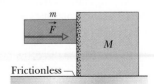

**FIG. 6-39** Problem 33.

What is the minimum magnitude of the horizontal force $\vec{F}$ required to keep the smaller block from slipping down the larger block? **ILW**

•••**34** In Fig. 6-40, a slab of mass $m_1 = 40$ kg rests on a frictionless floor, and a block of mass $m_2 = 10$ kg rests on top of the slab. Between block and slab, the coefficient of static friction is 0.60, and the coefficient of kinetic friction is 0.40. The block is pulled by a horizontal force $\vec{F}$ of magnitude 100 N. In unit-vector notation, what are the resulting accelerations of (a) the block and (b) the slab? **GO**

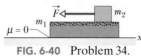

FIG. 6-40 Problem 34.

•••**35** A 1000 kg boat is traveling at 90 km/h when its engine is shut off. The magnitude of the frictional force $\vec{f}_k$ between boat and water is proportional to the speed $v$ of the boat: $f_k = 70v$, where $v$ is in meters per second and $f_k$ is in newtons. Find the time required for the boat to slow to 45 km/h. **SSM**

### sec. 6-4 The Drag Force and Terminal Speed

•**36** The terminal speed of a sky diver is 160 km/h in the spread-eagle position and 310 km/h in the nosedive position. Assuming that the diver's drag coefficient $C$ does not change from one position to the other, find the ratio of the effective cross-sectional area $A$ in the slower position to that in the faster position.

••**37** Calculate the ratio of the drag force on a jet flying at 1000 km/h at an altitude of 10 km to the drag force on a prop-driven transport flying at half that speed and altitude. The density of air is 0.38 kg/m³ at 10 km and 0.67 kg/m³ at 5.0 km. Assume that the airplanes have the same effective cross-sectional area and drag coefficient $C$.

••**38** In downhill speed skiing a skier is retarded by both the air drag force on the body and the kinetic frictional force on the skis. (a) Suppose the slope angle is $\theta = 40.0°$, the snow is dry snow with a coefficient of kinetic friction $\mu_k = 0.0400$, the mass of the skier and equipment is $m = 85.0$ kg, the cross-sectional area of the (tucked) skier is $A = 1.30$ m², the drag coefficient is $C = 0.150$, and the air density is 1.20 kg/m³. (a) What is the terminal speed? (b) If a skier can vary $C$ by a slight amount $dC$ by adjusting, say, the hand positions, what is the corresponding variation in the terminal speed?

••**39** *Continuation of Problem 2.* Now assume that Eq. 6-14 gives the magnitude of the air drag force on the typical 20 kg stone, which presents to the wind a vertical cross-sectional area of 0.040 m² and has a drag coefficient $C$ of 0.80. Take the air density to be 1.21 kg/m³, and the coefficient of kinetic friction to be 0.80. (a) In kilometers per hour, what wind speed $V$ along the ground is needed to maintain the stone's motion once it has started moving? Because winds along the ground are retarded by the ground, the wind speeds reported for storms are often measured at a height of 10 m. Assume wind speeds are 2.00 times those along the ground. (b) For your answer to (a), what wind speed would be reported for the storm? (c) Is that value reasonable for a high-speed wind in a storm? (Story continues with Problem 61.)

••**40** Assume Eq. 6-14 gives the drag force on a pilot plus ejection seat just after they are ejected from a plane traveling horizontally at 1300 km/h. Assume also that the mass of the seat is equal to the mass of the pilot and that the drag coefficient is that of a sky diver. Making a reasonable guess of the pilot's mass and using the appropriate $v_t$ value from Table 6-1, estimate the magnitudes of (a) the drag force on the *pilot + seat* and (b) their horizontal deceleration (in terms of $g$), both just after ejection. (The result of (a) should indicate an engineering requirement: The seat must include a protective barrier to deflect the initial wind blast away from the pilot's head.)

### sec. 6-5 Uniform Circular Motion

•**41** What is the smallest radius of an unbanked (flat) track around which a bicyclist can travel if her speed is 29 km/h and the $\mu_s$ between tires and track is 0.32? **ILW**

•**42** During an Olympic bobsled run, the Jamaican team makes a turn of radius 7.6 m at a speed of 96.6 km/h. What is their acceleration in terms of $g$?

•**43** A cat dozes on a stationary merry-go-round, at a radius of 5.4 m from the center of the ride. Then the operator turns on the ride and brings it up to its proper turning rate of one complete rotation every 6.0 s. What is the least coefficient of static friction between the cat and the merry-go-round that will allow the cat to stay in place, without sliding?

•**44** Suppose the coefficient of static friction between the road and the tires on a car is 0.60 and the car has no negative lift. What speed will put the car on the verge of sliding as it rounds a level curve of 30.5 m radius?

••**45** A circular-motion addict of mass 80 kg rides a Ferris wheel around in a vertical circle of radius 10 m at a constant speed of 6.1 m/s. (a) What is the period of the motion? What is the magnitude of the normal force on the addict from the seat when both go through (b) the highest point of the circular path and (c) the lowest point?

••**46** A roller-coaster car has a mass of 1200 kg when fully loaded with passengers. As the car passes over the top of a circular hill of radius 18 m, its speed is not changing. At the top of the hill, what are the (a) magnitude $F_N$ and (b) direction (up or down) of the normal force on the car from the track if the car's speed is $v = 11$ m/s? What are (c) $F_N$ and (d) the direction if $v = 14$ m/s?

••**47** In Fig. 6-41, a car is driven at constant speed over a circular hill and then into a circular valley with the same radius. At the top of the hill, the normal force on the driver from the car seat is 0. The driver's mass is 70.0 kg. What is the magnitude of the normal force on the driver from the seat when the car passes through the bottom of the valley?

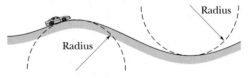

FIG. 6-41 Problem 47.

••**48** A police officer in hot pursuit drives her car through a circular turn of radius 300 m with a constant speed of 80.0 km/h. Her mass is 55.0 kg. What are (a) the magnitude and (b) the angle (relative to vertical) of the *net* force of the officer on the car seat? (*Hint:* Consider both horizontal and vertical forces.)

••**49** A student of weight 667 N rides a steadily rotating Ferris wheel (the student sits upright). At the highest point, the magnitude of the normal force $\vec{F}_N$ on the student from the seat is 556 N. (a) Does the student feel "light" or "heavy" there? (b) What is the magnitude of $\vec{F}_N$ at the lowest point? If the wheel's speed is doubled, what is the magnitude $F_N$ at the (c) highest and (d) lowest point? **SSM ILW**

**••50** An amusement park ride consists of a car moving in a vertical circle on the end of a rigid boom of negligible mass. The combined weight of the car and riders is 5.0 kN, and the circle's radius is 10 m. At the top of the circle, what are the (a) magnitude $F_B$ and (b) direction (up or down) of the force on the car from the boom if the car's speed is $v = 5.0$ m/s? What are (c) $F_B$ and (d) the direction if $v = 12$ m/s?

**••51** An old streetcar rounds a flat corner of radius 9.1 m, at 16 km/h. What angle with the vertical will be made by the loosely hanging hand straps?

**••52** In designing circular rides for amusement parks, mechanical engineers must consider how small variations in certain parameters can alter the net force on a passenger. Consider a passenger of mass $m$ riding around a horizontal circle of radius $r$ at speed $v$. What is the variation $dF$ in the net force magnitude for (a) a variation $dr$ in the radius with $v$ held constant, (b) a variation $dv$ in the speed with $r$ held constant, and (c) a variation $dT$ in the period with $r$ held constant?

**••53** An airplane is flying in a horizontal circle at a speed of 480 km/h (Fig. 6-42). If its wings are tilted at angle $\theta = 40°$ to the horizontal, what is the radius of the circle in which the plane is flying? Assume that the required force is provided entirely by an "aerodynamic lift" that is perpendicular to the wing surface. **SSM WWW**

**FIG. 6-42** Problem 53.

**••54** An 85.0 kg passenger is made to move along a circular path of radius $r = 3.50$ m in uniform circular motion. (a) Figure 6-43a is a plot of the required magnitude $F$ of the net centripetal force for a range of possible values of the passenger's speed $v$. What is the plot's slope at $v = 8.30$ m/s? (b) Figure 6-43b is a plot of $F$ for a range of possible values of $T$, the period of the motion. What is the plot's slope at $T = 2.50$ s?

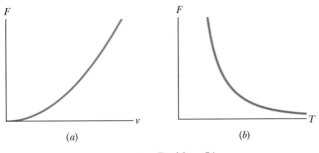

**FIG. 6-43** Problem 54.

**••55** A puck of mass $m = 1.50$ kg slides in a circle of radius $r = 20.0$ cm on a frictionless table while attached to a hanging cylinder of mass $M = 2.50$ kg by a cord through a hole in the table (Fig. 6-44). What speed keeps the cylinder at rest? **GO**

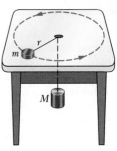

**FIG. 6-44** Problem 55.

**••56** *Brake or turn?* Figure 6-45 depicts an overhead view of a car's path as the car travels toward a wall. Assume that the driver begins to brake the car when the distance to the wall is $d = 107$ m, and take the car's mass as $m = 1400$ kg, its initial speed as $v_0 = 35$ m/s, and the coefficient of static friction as $\mu_s = 0.50$. Assume that the car's weight is distributed evenly on the four wheels, even during braking. (a) What magnitude of static friction is needed (between tires and road) to stop the car just as it reaches the wall? (b) What is the maximum possible static friction $f_{s,\,max}$? (c) If the coefficient of kinetic friction between the (sliding) tires and the road is $\mu_k = 0.40$, at what speed will the car hit the wall? To avoid the crash, a driver could elect to turn the car so that it just barely misses the wall, as shown in the figure. (d) What magnitude of frictional force would be required to keep the car in a circular path of radius $d$ and at the given speed $v_0$? (e) Is the required force less than $f_{s,\,max}$ so that a circular path is possible?

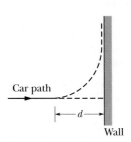

**FIG. 6-45** Problem 56.

**••57** A bolt is threaded onto one end of a thin horizontal rod, and the rod is then rotated horizontally about its other end. An engineer monitors the motion by flashing a strobe lamp onto the rod and bolt, adjusting the strobe rate until the bolt appears to be in the same eight places during each full rotation of the rod (Fig. 6-46). The strobe rate is 2000 flashes per second; the bolt has mass 30 g and is at radius 3.5 cm. What is the magnitude of the force on the bolt from the rod?

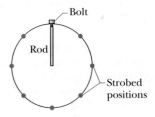

**FIG. 6-46** Problem 57.

**••58** A banked circular highway curve is designed for traffic moving at 60 km/h. The radius of the curve is 200 m. Traffic is moving along the highway at 40 km/h on a rainy day. What is the minimum coefficient of friction between tires and road that will allow cars to take the turn without sliding off the road? (Assume the cars do not have negative lift.)

**•••59** In Fig. 6-47, a 1.34 kg ball is connected by means of two massless strings, each of length $L = 1.70$ m, to a vertical, rotating rod. The strings are tied to the rod with separation $d = 1.70$ m and are taut. The tension in the upper string is 35 N. What are the (a) tension in the lower string, (b) magnitude of the net force $\vec{F}_{net}$ on the ball, and (c) speed of the ball? (d) What is the direction of $\vec{F}_{net}$? **SSM ILW**

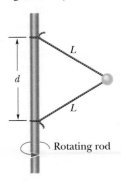

**FIG. 6-47** Problem 59.

**Additional Problems**

**60** Figure 6-48 shows a *conical pendulum*, in which the bob (the small object at the lower end of the cord) moves in a horizontal circle at constant speed. (The cord sweeps out a cone as the bob rotates.) The bob has a mass of 0.040 kg, the string has length $L = 0.90$ m and negligible mass, and the bob follows a circular path of circumference 0.94 m. What are (a) the tension in the string and (b) the period of the motion?

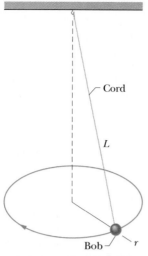

FIG. 6-48 Problem 60.

**61** *Continuation of Problems 2 and 39.* Another explanation is that the stones move only when the water dumped on the playa during a storm freezes into a large, thin sheet of ice. The stones are trapped in place in the ice. Then, as air flows across the ice during a wind, the air-drag forces on the ice and stones move them both, with the stones gouging out the trails. The magnitude of the air-drag force on this horizontal "ice sail" is given by $D_{ice} = 4C_{ice}\rho A_{ice}v^2$, where $C_{ice}$ is the drag coefficient ($2.0 \times 10^{-3}$), $\rho$ is the air density (1.21 kg/m³), $A_{ice}$ is the horizontal area of the ice, and $v$ is the wind speed along the ice.

Assume the following: The ice sheet measures 400 m by 500 m by 4.0 mm and has a coefficient of kinetic friction of 0.10 with the ground and a density of 917 kg/m³. Also assume that 100 stones identical to the one in Problem 2 are trapped in the ice. To maintain the motion of the sheet, what are the required wind speeds (a) near the sheet and (b) at a height of 10 m? (c) Are these reasonable values for high-speed winds in a storm?

**62** *Engineering a highway curve.* If a car goes through a curve too fast, the car tends to slide out of the curve. For a banked curve with friction, a frictional force acts on a fast car to oppose the tendency to slide out of the curve; the force is directed down the bank (in the direction water would drain). Consider a circular curve of radius $R = 200$ m and bank angle $\theta$, where the coefficient of static friction between tires and pavement is $\mu_s$. A car (without negative lift) is driven around the curve as shown in Fig. 6-13. (a) Find an expression for the car speed $v_{max}$ that puts the car on the verge of sliding out. (b) On the same graph, plot $v_{max}$ versus angle $\theta$ for the range 0° to 50°, first for $\mu_s = 0.60$ (dry pavement) and then for $\mu_s = 0.050$ (wet or icy pavement). In kilometers per hour, evaluate $v_{max}$ for a bank angle of $\theta = 10°$ and for (c) $\mu_s = 0.60$ and (d) $\mu_s = 0.050$. (Now you can see why accidents occur in highway curves when icy conditions are not obvious to drivers, who tend to drive at normal speeds.)

**63** In Fig. 6-49, the coefficient of kinetic friction between the block and inclined plane is 0.20, and angle $\theta$ is 60°. What are the (a) magnitude $a$ and (b) direction (up or down the plane) of the block's acceleration if the block is sliding

FIG. 6-49 Problem 63.

down the plane? What are (c) $a$ and (d) the direction if the block is sent sliding up the plane?

**64** In Fig. 6-50, block 1 of mass $m_1 = 2.0$ kg and block 2 of mass $m_2 = 3.0$ kg are connected by a string of negligible mass and are initially held in place. Block 2 is on a frictionless surface tilted at $\theta = 30°$.

FIG. 6-50 Problem 64.

The coefficient of kinetic friction between block 1 and the horizontal surface is 0.25. The pulley has negligible mass and friction. Once they are released, the blocks move. What then is the tension in the string?

**65** A block of mass $m_t = 4.0$ kg is put on top of a block of mass $m_b = 5.0$ kg. To cause the top block to slip on the bottom one while the bottom one is held fixed, a horizontal force of at least 12 N must be applied to the top block. The assembly of blocks is now placed on a horizontal, frictionless table (Fig. 6-51). Find the magnitudes of (a) the maximum horizontal force $\vec{F}$ that can be applied to the lower block so that the blocks will move together and (b) the resulting acceleration of the blocks. **SSM**

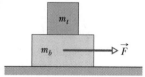

FIG. 6-51 Problem 65.

**66** A box of canned goods slides down a ramp from street level into the basement of a grocery store with acceleration 0.75 m/s² directed down the ramp. The ramp makes an angle of 40° with the horizontal. What is the coefficient of kinetic friction between the box and the ramp?

**67** An 8.00 kg block of steel is at rest on a horizontal table. The coefficient of static friction between the block and the table is 0.450. A force is to be applied to the block. To three significant figures, what is the magnitude of that applied force if it puts the block on the verge of sliding when the force is directed (a) horizontally, (b) upward at 60.0° from the horizontal, and (c) downward at 60.0° from the horizontal?

**68** In Fig. 6-52, a box of ant aunts (total mass $m_1 = 1.65$ kg) and a box of ant uncles (total mass $m_2 = 3.30$ kg) slide down an inclined plane while attached by a massless rod parallel to the plane. The angle of incline is $\theta = 30.0°$. The coefficient of kinetic friction between the aunt box and the incline is $\mu_1 = 0.226$; that between the uncle box and the incline is $\mu_2 = 0.113$. Compute (a) the tension in the rod and (b) the magnitude of the common acceleration of the two boxes. (c) How would the answers to (a) and (b) change if the uncles trailed the aunts?

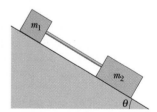

FIG. 6-52 Problem 68.

**69** In Fig. 6-53, a crate slides down an inclined right-angled

FIG. 6-53 Problem 69.

trough. The coefficient of kinetic friction between the crate and the trough is $\mu_k$. What is the acceleration of the crate in terms of $\mu_k$, $\theta$, and $g$?

**70** A student wants to determine the coefficients of static friction and kinetic friction between a box and a plank. She places the box on the plank and gradually raises one end of the plank. When the angle of inclination with the horizontal reaches 30°, the box starts to slip, and it then slides 2.5 m down the plank in 4.0 s at constant acceleration. What are (a) the coefficient of static friction and (b) the coefficient of kinetic friction between the box and the plank?

**71** A locomotive accelerates a 25-car train along a level track. Every car has a mass of $5.0 \times 10^4$ kg and is subject to a frictional force $f = 250v$, where the speed $v$ is in meters per second and the force $f$ is in newtons. At the instant when the speed of the train is 30 km/h, the magnitude of its acceleration is 0.20 m/s². (a) What is the tension in the coupling between the first car and the locomotive? (b) If this tension is equal to the maximum force the locomotive can exert on the train, what is the steepest grade up which the locomotive can pull the train at 30 km/h?

**72** A house is built on the top of a hill with a nearby slope at angle $\theta = 45°$ (Fig. 6-54). An engineering study indicates that the slope angle should be reduced because the top layers of soil along the slope might slip past the lower layers. If the coefficient of static friction between two such layers is 0.5, what is the least angle $\phi$ through which the present slope should be reduced to prevent slippage?

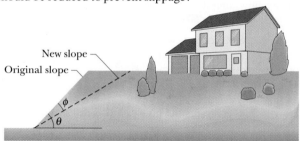

**FIG. 6-54** Problem 72.

**73** What is the terminal speed of a 6.00 kg spherical ball that has a radius of 3.00 cm and a drag coefficient of 1.60? The density of the air through which the ball falls is 1.20 kg/m³.

**74** A high-speed railway car goes around a flat, horizontal circle of radius 470 m at a constant speed. The magnitudes of the horizontal and vertical components of the force of the car on a 51.0 kg passenger are 210 N and 500 N, respectively. (a) What is the magnitude of the net force (of *all* the forces) on the passenger? (b) What is the speed of the car?

**75** An 11 kg block of steel is at rest on a horizontal table. The coefficient of static friction between block and table is 0.52. (a) What is the magnitude of the horizontal force that will put the block on the verge of moving? (b) What is the magnitude of a force acting upward 60° from the horizontal that will put the block on the verge of moving? (c) If the force acts downward at 60° from the horizontal, how large can its magnitude be without causing the block to move?

**76** Calculate the magnitude of the drag force on a missile 53 cm in diameter cruising at 250 m/s at low altitude, where the density of air is 1.2 kg/m³. Assume $C = 0.75$.

**77** A bicyclist travels in a circle of radius 25.0 m at a constant speed of 9.00 m/s. The bicycle–rider mass is 85.0 kg. Calculate the magnitudes of (a) the force of friction on the bicycle from the road and (b) the *net* force on the bicycle from the road. SSM

**78** A 110 g hockey puck sent sliding over ice is stopped in 15 m by the frictional force on it from the ice. (a) If its initial speed is 6.0 m/s, what is the magnitude of the frictional force? (b) What is the coefficient of friction between the puck and the ice?

**79** In Fig. 6-55, a 49 kg rock climber is climbing a "chimney." The coefficient of static friction between her shoes and the rock is 1.2; between her back and the rock is 0.80. She has reduced her push against the rock until her back and her shoes are on the verge of slipping. (a) Draw a free-body diagram of her. (b) What is the magnitude of her push against the rock? (c) What fraction of her weight is supported by the frictional force on her shoes?

**FIG. 6-55** Problem 79.

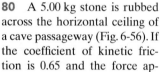

**FIG. 6-56** Problem 80.

**80** A 5.00 kg stone is rubbed across the horizontal ceiling of a cave passageway (Fig. 6-56). If the coefficient of kinetic friction is 0.65 and the force applied to the stone is angled at $\theta = 70.0°$, what must the magnitude of the force be for the stone to move at constant velocity?

**81** Block A in Fig. 6-57 has mass $m_A = 4.0$ kg, and block B has mass $m_B = 2.0$ kg. The coefficient of kinetic friction between block B and the horizontal plane is $\mu_k = 0.50$. The inclined plane is frictionless and at angle $\theta = 30°$. The pulley serves only to change the direction of the cord connecting the blocks. The cord has negligible mass. Find (a) the tension in the cord and (b) the magnitude of the acceleration of the blocks. SSM

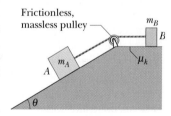

**FIG. 6-57** Problem 81.

**82** A ski that is placed on snow will stick to the snow. However, when the ski is moved along the snow, the rubbing warms and partially melts the snow, reducing the coefficient of kinetic friction and promoting sliding. Waxing the ski makes it water repellent and reduces friction with the resulting layer of water. A magazine reports that a new type of plastic ski is especially water repellent and that, on a gentle 200 m slope in the Alps, a skier reduced his top-to-bottom time from 61 s with standard skis to 42 s with the new skis. Determine the magnitude of his average acceleration with (a) the standard skis and (b) the new skis. Assuming a 3.0° slope, compute the coefficient of kinetic friction for (c) the standard skis and (d) the new skis.

**83** Playing near a road construction site, a child falls over a barrier and down onto a dirt slope that is angled downward at 35° to the horizontal. As the child slides *down* the slope, he has an acceleration that has a magnitude of 0.50 m/s² and that is directed *up* the slope. What is the coefficient of kinetic friction between the child and the slope?

**84** In Fig. 6-58, a stuntman drives a car (without negative lift) over the top of a hill, the cross section of which can be approximated by a circle of radius $R = 250$ m. What is the greatest speed at which he can drive without the car leaving the road at the top of the hill?

**FIG. 6-58** Problem 84.

**85** A car weighing 10.7 kN and traveling at 13.4 m/s without negative lift attempts to round an unbanked curve with a radius of 61.0 m. (a) What magnitude of the frictional force on the tires is required to keep the car on its circular path? (b) If the coefficient of static friction between the tires and the road is 0.350, is the attempt at taking the curve successful? **SSM**

**86** A 100 N force, directed at an angle $\theta$ above a horizontal floor, is applied to a 25.0 kg chair sitting on the floor. If $\theta = 0°$, what are (a) the horizontal component $F_h$ of the applied force and (b) the magnitude $F_N$ of the normal force of the floor on the chair? If $\theta = 30.0°$, what are (c) $F_h$ and (d) $F_N$? If $\theta = 60.0°$, what are (e) $F_h$ and (f) $F_N$? Now assume that the coefficient of static friction between chair and floor is 0.420. Does the chair slide or remain at rest if $\theta$ is (g) 0°, (h) 30.0°, and (i) 60.0°?

**87** A student, crazed by final exams, uses a force $\vec{P}$ of magnitude 80 N and angle $\theta = 70°$ to push a 5.0 kg block across the ceiling of his room (Fig. 6-59). If the coefficient of kinetic friction between the block and the ceiling is 0.40, what is the magnitude of the block's acceleration?

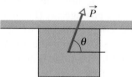

**FIG. 6-59** Problem 87.

**88** A certain string can withstand a maximum tension of 40 N without breaking. A child ties a 0.37 kg stone to one end and, holding the other end, whirls the stone in a vertical circle of radius 0.91 m, slowly increasing the speed until the string breaks. (a) Where is the stone on its path when the string breaks? (b) What is the speed of the stone as the string breaks?

**89** You must push a crate across a floor to a docking bay. The crate weighs 165 N. The coefficient of static friction between crate and floor is 0.510, and the coefficient of kinetic friction is 0.32. Your force on the crate is directed horizontally. (a) What magnitude of your push puts the crate on the verge of sliding? (b) With what magnitude must you then push to keep the crate moving at a constant velocity? (c) If, instead, you then push with the same magnitude as the answer to (a), what is the magnitude of the crate's acceleration?

**90** A child weighing 140 N sits at rest at the top of a playground slide that makes an angle of 25° with the horizontal. The child keeps from sliding by holding onto the sides of the slide. After letting go of the sides, the child has a constant acceleration of 0.86 m/s² (down the slide, of course). (a) What is the coefficient of kinetic friction between the child and the slide? (b) What maximum and minimum values for the coefficient of static friction between the child and the slide are consistent with the information given here?

**91** A filing cabinet weighing 556 N rests on the floor. The coefficient of static friction between it and the floor is 0.68, and the coefficient of kinetic friction is 0.56. In four different attempts to move it, it is pushed with horizontal forces of magnitudes (a) 222 N, (b) 334 N, (c) 445 N, and (d) 556 N. For each attempt, calculate the magnitude of the frictional force on it from the floor. (The cabinet is initially at rest.) (e) In which of the attempts does the cabinet move? **SSM**

**92** A sling-thrower puts a stone (0.250 kg) in the sling's pouch (0.010 kg) and then begins to make the stone and pouch move in a vertical circle of radius 0.650 m. The cord between the pouch and the person's hand has negligible mass and will break when the tension in the cord is 33.0 N or more. Suppose the sling-thrower could gradually increase the speed of the stone. (a) Will the breaking occur at the lowest point of the circle or at the highest point? (b) At what speed of the stone will that breaking occur?

**93** A four-person bobsled (total mass = 630 kg) comes down a straightaway at the start of a bobsled run. The straightaway is 80.0 m long and is inclined at a constant angle of 10.2° with the horizontal. Assume that the combined effects of friction and air drag produce on the bobsled a constant force of 62.0 N that acts parallel to the incline and up the incline. Answer the following questions to three significant digits. (a) If the speed of the bobsled at the start of the run is 6.20 m/s, how long does the bobsled take to come down the straightaway? (b) Suppose the crew is able to reduce the effects of friction and air drag to 42.0 N. For the same initial velocity, how long does the bobsled now take to come down the straightaway?

**94** In Fig. 6-60, force $\vec{F}$ is applied to a crate of mass $m$ on a floor where the coefficient of static friction between crate and floor is $\mu_s$. Angle $\theta$ is initially 0° but is gradually increased so that the force vector rotates clockwise in the figure. During the rotation, the magnitude $F$ of the force is continuously adjusted so that the crate is always on the verge of sliding. For $\mu_s = 0.70$, (a) plot the ratio $F/mg$ versus $\theta$ and (b) determine the angle $\theta_{inf}$ at which the ratio approaches an infinite value. (c) Does lubricating the floor increase or decrease $\theta_{inf}$, or is the value unchanged? (d) What is $\theta_{inf}$ for $\mu_s = 0.60$?

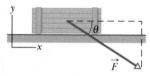

**FIG. 6-60** Problem 94.

**95** In the early afternoon, a car is parked on a street that runs down a steep hill, at an angle of 35.0° relative to the horizontal. Just then the coefficient of static friction between the tires and the street surface is 0.725. Later, after nightfall, a sleet storm hits the area, and the coefficient decreases due to both the ice and a chemical change in the road surface because of the temperature decrease. By what percentage must the coefficient decrease if the car is to be in danger of sliding down the street?

**96** In Fig. 6-61, block 1 of mass $m_1 = 2.0$ kg and block 2 of mass $m_2 = 1.0$ kg are connected by a string of negligible mass. Block 2 is pushed by force $\vec{F}$ of magnitude 20 N and angle $\theta = 35°$. The coefficient of kinetic friction between

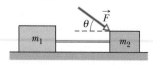

**FIG. 6-61** Problem 96.

each block and the horizontal surface is 0.20. What is the tension in the string?

**97** In Fig. 6-62 a fastidious worker pushes directly along the handle of a mop with a force $\vec{F}$. The handle is at an angle $\theta$ with the vertical, and $\mu_s$ and $\mu_k$ are the coefficients of static and kinetic friction between the head of the mop and the floor. Ignore the mass of the handle and assume that all the mop's mass $m$ is in its head. (a)

**FIG. 6-62** Problem 97.

If the mop head moves along the floor with a constant velocity, then what is $F$? (b) Show that if $\theta$ is less than a certain value $\theta_0$, then $\vec{F}$ (still directed along the handle) is unable to move the mop head. Find $\theta_0$.

**98** A circular curve of highway is designed for traffic moving at 60 km/h. Assume the traffic consists of cars without negative lift. (a) If the radius of the curve is 150 m, what is the correct angle of banking of the road? (b) If the curve were not banked, what would be the minimum coefficient of friction between tires and road that would keep traffic from skidding out of the turn when traveling at 60 km/h?

**99** A block slides with constant velocity down an inclined plane that has slope angle $\theta$. The block is then projected up the same plane with an initial speed $v_0$. (a) How far up the plane will it move before coming to rest? (b) After the block comes to rest, will it slide down the plane again? Give an argument to back your answer. **SSM**

**100** In Fig. 6-63, a block weighing 22 N is held at rest against a vertical wall by a horizontal force $\vec{F}$ of magnitude 60 N. The coefficient of static friction between the wall and the block is 0.55, and the coefficient of kinetic friction between them is 0.38. In six experiments, a second force $\vec{P}$ is applied to the block and directed parallel

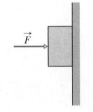

**FIG. 6-63** Problem 100.

to the wall with these magnitudes and directions: (a) 34 N, up, (b) 12 N, up, (c) 48 N, up, (d) 62 N, up, (e) 10 N, down, and (f) 18 N, down. In each experiment, what is the magnitude of the frictional force on the block? In which does the block move (g) up the wall and (h) down the wall? (i) In which is the frictional force directed down the wall?

**101** When a small 2.0 g coin is placed at a radius of 5.0 cm on a horizontal turntable that makes three full revolutions in 3.14 s, the coin does not slip. What are (a) the coin's speed, the (b) magnitude and (c) direction (radially inward or outward) of the coin's acceleration, and the (d) magnitude and (e) direction (inward or outward) of the frictional force on the coin? The coin is on the verge of slipping if it is placed at a radius of 10 cm. (f) What is the coefficient of static friction between coin and turntable?

**102** A child places a picnic basket on the outer rim of a merry-go-round that has a radius of 4.6 m and revolves once every 30 s. (a) What is the speed of a point on that rim? (b) What is the lowest value of the coefficient of static friction between basket and merry-go-round that allows the basket to stay on the ride?

**103** A 1.5 kg box is initially at rest on a horizontal surface when at $t = 0$ a horizontal force $\vec{F} = (1.8t)\hat{i}$ N (with $t$ in seconds) is applied to the box. The acceleration of the box as a function of time $t$ is given by $\vec{a} = 0$ for $0 \le t \le 2.8$ s and $\vec{a} = (1.2t - 2.4)\hat{i}$ m/s$^2$ for $t > 2.8$ s. (a) What is the coefficient of static friction between the box and the surface? (b) What is the coefficient of kinetic friction between the box and the surface?

**104** A trunk with a weight of 220 N rests on the floor. The coefficient of static friction between the trunk and the floor is 0.41, and the coefficient of kinetic friction is 0.32. (a) What is the magnitude of the minimum horizontal force with which a person must push on the trunk to start it moving? (b) Once the trunk is moving, what magnitude of horizontal force must the person apply to keep it moving with constant velocity? (c) If the person continued to push with the force used to start the motion, what would be the magnitude of the trunk's acceleration?

**105** A warehouse worker exerts a constant horizontal force of magnitude 85 N on a 40 kg box that is initially at rest on the horizontal floor of the warehouse. When the box has moved a distance of 1.4 m, its speed is 1.0 m/s. What is the coefficient of kinetic friction between the box and the floor? **SSM**

**106** Imagine that the standard kilogram is placed on Earth's equator, where it moves in a circle of radius $6.40 \times 10^6$ m (Earth's radius) at a constant speed of 465 m/s due to Earth's rotation. (a) What is the magnitude of the centripetal force on the standard kilogram during the rotation? Imagine that the standard kilogram hangs from a spring balance at that location and assume that it would weigh exactly 9.80 N if Earth did not rotate. (b) What is the reading on the spring balance; that is, what is the magnitude of the force on the spring balance from the standard kilogram?

**107** As a 40 N block slides down a plane that is inclined at 25° to the horizontal, its acceleration is 0.80 m/s$^2$, directed up the plane. What is the coefficient of kinetic friction between the block and the plane?

**108** Luggage is transported from one location to another in an airport by a conveyor belt. At a certain location, the belt moves down an incline that makes an angle of 2.5° with the horizontal. Assume that with such a slight angle there is no slipping of the luggage. Determine the magnitude of the frictional force by the belt on a box weighing 69 N when the box is on the inclined portion of the belt and the belt speed is (a) 0 and constant, (b) 0.65 m/s and constant, (c) 0.65 m/s and increasing at a rate of 0.20 m/s$^2$, (d) 0.65 m/s and decreasing at a rate of 0.20 m/s$^2$, and (e) 0.65 m/s and increasing at a rate of 0.57 m/s$^2$. (f) For which of these five situations is the frictional force directed down the incline?

**109** In Fig. 6-64, a 5.0 kg block is sent sliding up a plane inclined at $\theta = 37°$ while a horizontal force $\vec{F}$ of magnitude 50 N acts on it. The coefficient of kinetic friction between block and plane is 0.30. What are the (a) magnitude and (b) direction (up or down the plane) of the block's

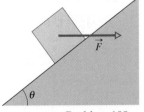

**FIG. 6-64** Problem 109.

acceleration? The block's initial speed is 4.0 m/s. (c) How far up the plane does the block go? (d) When it reaches its highest point, does it remain at rest or slide back down the plane?

# 7 Kinetic Energy and Work

UPI Photo/Dilip Vishwanat/Landov LLC

*The driver of a funny car prepares for a timed run along a quarter-mile track by spinning the wheels, to make the tires and track sticky so that traction is high. Then the driver waits at the starting line until the countdown on the Christmas tree lights* reaches green. *The car's forward surge is so powerful that the car is effectively launched like a horizontal rocket. The science and engineering of funny cars is now so advanced that winning and losing is often determined by elapsed times differing by only 1 ms.*

## What property of a car determines the winning time?

The answer is in this chapter.

One of the fundamental goals of physics is to investigate something that everyone talks about: energy. The topic is obviously important. Indeed, our civilization is based on acquiring and effectively using energy.

For example, everyone knows that any type of motion requires energy: Flying across the Pacific Ocean requires it. Lifting material to the top floor of an office building or to an orbiting space station requires it. Throwing a fastball requires it. We spend a tremendous amount of money to acquire and use energy. Wars have been started because of energy resources. Wars have been ended because of a sudden, overpowering use of energy by one side. Everyone knows many examples of energy and its use, but what does the term *energy* really mean?

## 7-2 | What Is Energy?

The term *energy* is so broad that a clear definition is difficult to write. Technically, energy is a scalar quantity associated with the state (or condition) of one or more objects. However, this definition is too vague to be of help to us now.

A looser definition might at least get us started. Energy is a number that we associate with a system of one or more objects. If a force changes one of the objects by, say, making it move, then the energy number changes. After countless experiments, scientists and engineers realized that if the scheme by which we assign energy numbers is planned carefully, the numbers can be used to predict the outcomes of experiments and, even more important, to build machines, such as flying machines. This success is based on a wonderful property of our universe: Energy can be transformed from one type to another and transferred from one object to another, but the total amount is always the same (energy is *conserved*). No exception to this *principle of energy conservation* has ever been found.

Think of the many types of energy as being numbers representing money in many types of bank accounts. Rules have been made about what such money numbers mean and how they can be changed. You can transfer money numbers from one account to another or from one system to another, perhaps electronically with nothing material actually moving. However, the total amount (the total of all the money numbers) can always be accounted for: It is always conserved.

In this chapter we focus on only one type of energy (*kinetic energy*) and on only one way in which energy can be transferred (*work*). In the next chapter we examine a few other types of energy and how the principle of energy conservation can be written as equations to be solved.

## 7-3 | Kinetic Energy

**Kinetic energy** $K$ is energy associated with the *state of motion* of an object. The faster the object moves, the greater is its kinetic energy. When the object is stationary, its kinetic energy is zero.

For an object of mass $m$ whose speed $v$ is well below the speed of light,

$$K = \tfrac{1}{2}mv^2 \qquad \text{(kinetic energy).} \qquad (7\text{-}1)$$

For example, a 3.0 kg duck flying past us at 2.0 m/s has a kinetic energy of $6.0 \text{ kg} \cdot \text{m}^2/\text{s}^2$; that is, we associate that number with the duck's motion.

The SI unit of kinetic energy (and every other type of energy) is the **joule** (J), named for James Prescott Joule, an English scientist of the 1800s. It is defined directly from Eq. 7-1 in terms of the units for mass and velocity:

$$1 \text{ joule} = 1 \text{ J} = 1 \text{ kg} \cdot \text{m}^2/\text{s}^2. \qquad (7\text{-}2)$$

Thus, the flying duck has a kinetic energy of 6.0 J.

## Sample Problem 7-1

In 1896 in Waco, Texas, William Crush parked two locomotives at opposite ends of a 6.4-km-long track, fired them up, tied their throttles open, and then allowed them to crash head-on at full speed (Fig. 7-1) in front of 30,000 spectators. Hundreds of people were hurt by flying debris; several were killed. Assuming each locomotive weighed $1.2 \times 10^6$ N and its acceleration was a constant 0.26 m/s$^2$, what was the total kinetic energy of the two locomotives just before the collision?

**KEY IDEAS** (1) We need to find the kinetic energy of each locomotive with Eq. 7-1, but that means we need each locomotive's speed just before the collision and its mass. (2) Because we can assume each locomotive had constant acceleration, we can use the equations in Table 2-1 to find its speed $v$ just before the collision.

*Calculations:* We choose Eq. 2-16 because we know values for all the variables except $v$:

$$v^2 = v_0^2 + 2a(x - x_0).$$

With $v_0 = 0$ and $x - x_0 = 3.2 \times 10^3$ m (half the initial separation), this yields

$$v^2 = 0 + 2(0.26 \text{ m/s}^2)(3.2 \times 10^3 \text{ m}),$$

or

$$v = 40.8 \text{ m/s}$$

(about 150 km/h).

We can find the mass of each locomotive by divid-

**FIG. 7-1** The aftermath of an 1896 crash of two locomotives. *(Courtesy Library of Congress)*

ing its given weight by $g$:

$$m = \frac{1.2 \times 10^6 \text{ N}}{9.8 \text{ m/s}^2} = 1.22 \times 10^5 \text{ kg}.$$

Now, using Eq. 7-1, we find the total kinetic energy of the two locomotives just before the collision as

$$K = 2(\tfrac{1}{2}mv^2) = (1.22 \times 10^5 \text{ kg})(40.8 \text{ m/s})^2$$
$$= 2.0 \times 10^8 \text{ J}. \qquad \text{(Answer)}$$

This collision was like an exploding bomb.

## 7-4 | Work

If you accelerate an object to a greater speed by applying a force to the object, you increase the kinetic energy $K$ ($= \tfrac{1}{2}mv^2$) of the object. Similarly, if you decelerate the object to a lesser speed by applying a force, you decrease the kinetic energy of the object. We account for these changes in kinetic energy by saying that your force has transferred energy *to* the object from yourself or *from* the object to yourself. In such a transfer of energy via a force, **work** $W$ is said to be *done on the object by the force*. More formally, we define work as follows:

> Work $W$ is energy transferred to or from an object by means of a force acting on the object. Energy transferred to the object is positive work, and energy transferred from the object is negative work.

"Work," then, is transferred energy; "doing work" is the act of transferring the energy. Work has the same units as energy and is a scalar quantity.

The term *transfer* can be misleading. It does not mean that anything material flows into or out of the object; that is, the transfer is not like a flow of water. Rather, it is like the electronic transfer of money between two bank accounts: The number in one account goes up while the number in the other account goes down, with nothing material passing between the two accounts.

Note that we are not concerned here with the common meaning of the word "work," which implies that *any* physical or mental labor is work. For example, if you push hard against a wall, you tire because of the continuously repeated muscle contractions that are required, and you are, in the common sense, working.

However, such effort does not cause an energy transfer to or from the wall and thus is not work done on the wall as defined here.

To avoid confusion in this chapter, we shall use the symbol $W$ only for work and shall represent a weight with its equivalent $mg$.

## 7-5 | Work and Kinetic Energy

### Finding an Expression for Work

Let us find an expression for work by considering a bead that can slide along a frictionless wire that is stretched along a horizontal $x$ axis (Fig. 7-2). A constant force $\vec{F}$, directed at an angle $\phi$ to the wire, accelerates the bead along the wire. We can relate the force and the acceleration with Newton's second law, written for components along the $x$ axis:

$$F_x = ma_x, \tag{7-3}$$

where $m$ is the bead's mass. As the bead moves through a displacement $\vec{d}$, the force changes the bead's velocity from an initial value $\vec{v}_0$ to some other value $\vec{v}$. Because the force is constant, we know that the acceleration is also constant. Thus, we can use Eq. 2-16 to write, for components along the $x$ axis,

$$v^2 = v_0^2 + 2a_x d. \tag{7-4}$$

Solving this equation for $a_x$, substituting into Eq. 7-3, and rearranging then give us

$$\tfrac{1}{2}mv^2 - \tfrac{1}{2}mv_0^2 = F_x d. \tag{7-5}$$

The first term on the left side of the equation is the kinetic energy $K_f$ of the bead at the end of the displacement $d$, and the second term is the kinetic energy $K_i$ of the bead at the start of the displacement. Thus, the left side of Eq. 7-5 tells us the kinetic energy has been changed by the force, and the right side tells us the change is equal to $F_x d$. Therefore, the work $W$ done on the bead by the force (the energy transfer due to the force) is

$$W = F_x d. \tag{7-6}$$

If we know values for $F_x$ and $d$, we can use this equation to calculate the work $W$ done on the bead by the force.

> To calculate the work a force does on an object as the object moves through some displacement, we use only the force component along the object's displacement. The force component perpendicular to the displacement does zero work.

From Fig. 7-2, we see that we can write $F_x$ as $F \cos \phi$, where $\phi$ is the angle between the directions of the displacement $\vec{d}$ and the force $\vec{F}$. Thus,

$$W = Fd \cos \phi \qquad \text{(work done by a constant force).} \tag{7-7}$$

Because the right side of this equation is equivalent to the scalar (dot) product $\vec{F} \cdot \vec{d}$, we can also write

$$W = \vec{F} \cdot \vec{d} \qquad \text{(work done by a constant force),} \tag{7-8}$$

where $F$ is the magnitude of $\vec{F}$. (You may wish to review the discussion of scalar products in Section 3-8.) Equation 7-8 is especially useful for calculating the work when $\vec{F}$ and $\vec{d}$ are given in unit-vector notation.

*Cautions:* There are two restrictions to using Eqs. 7-6 through 7-8 to calculate work done on an object by a force. First, the force must be a *constant force;* that is, it must not change in magnitude or direction as the object moves. (Later, we shall discuss what to do with a *variable force* that changes in magnitude.) Second, the object must be *particle-like.* This means that the object must be *rigid;* all parts

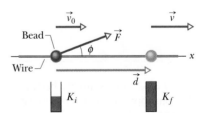

**FIG. 7-2** A constant force $\vec{F}$ directed at angle $\phi$ to the displacement $\vec{d}$ of a bead on a wire accelerates the bead along the wire, changing the velocity of the bead from $\vec{v}_0$ to $\vec{v}$. A "kinetic energy gauge" indicates the resulting change in the kinetic energy of the bead, from the value $K_i$ to the value $K_f$.

of it must move together, in the same direction. In this chapter we consider only particle-like objects, such as the bed and its occupant being pushed in Fig. 7-3.

*Signs for work.* The work done on an object by a force can be either positive work or negative work. For example, if the angle $\phi$ in Eq. 7-7 is less than 90°, then cos $\phi$ is positive and thus so is the work. If $\phi$ is greater than 90° (up to 180°), then cos $\phi$ is negative and thus so is the work. (Can you see that the work is zero when $\phi = 90$°?) These results lead to a simple rule. To find the sign of the work done by a force, consider the force vector component that is parallel to the displacement:

> ☞ A force does positive work when it has a vector component in the same direction as the displacement, and it does negative work when it has a vector component in the opposite direction. It does zero work when it has no such vector component.

*Units for work.* Work has the SI unit of the joule, the same as kinetic energy. However, from Eqs. 7-6 and 7-7 we can see that an equivalent unit is the newton-meter (N · m). The corresponding unit in the British system is the foot-pound (ft · lb). Extending Eq. 7-2, we have

$$1 \text{ J} = 1 \text{ kg} \cdot \text{m}^2/\text{s}^2 = 1 \text{ N} \cdot \text{m} = 0.738 \text{ ft} \cdot \text{lb}. \tag{7-9}$$

*Net work done by several forces.* When two or more forces act on an object, the **net work** done on the object is the sum of the works done by the individual forces. We can calculate the net work in two ways. (1) We can find the work done by each force and then sum those works. (2) Alternatively, we can first find the net force $\vec{F}_{net}$ of those forces. Then we can use Eq. 7-7, substituting the magnitude $F_{net}$ for $F$ and also the angle between the directions of $\vec{F}_{net}$ and $\vec{d}$ for $\phi$. Similarly, we can use Eq. 7-8 with $\vec{F}_{net}$ substituted for $\vec{F}$.

## Work–Kinetic Energy Theorem

Equation 7-5 relates the change in kinetic energy of the bead (from an initial $K_i = \frac{1}{2}mv_0^2$ to a later $K_f = \frac{1}{2}mv^2$) to the work $W$ (= $F_x d$) done on the bead. For such particle-like objects, we can generalize that equation. Let $\Delta K$ be the change in the kinetic energy of the object, and let $W$ be the net work done on it. Then

$$\Delta K = K_f - K_i = W, \tag{7-10}$$

which says that

$$\left( \begin{array}{c} \text{change in the kinetic} \\ \text{energy of a particle} \end{array} \right) = \left( \begin{array}{c} \text{net work done on} \\ \text{the particle} \end{array} \right).$$

We can also write

$$K_f = K_i + W, \tag{7-11}$$

which says that

$$\left( \begin{array}{c} \text{kinetic energy after} \\ \text{the net work is done} \end{array} \right) = \left( \begin{array}{c} \text{kinetic energy} \\ \text{before the net work} \end{array} \right) + \left( \begin{array}{c} \text{the net} \\ \text{work done} \end{array} \right).$$

These statements are known traditionally as the **work–kinetic energy theorem** for particles. They hold for both positive and negative work: If the net work done on a particle is positive, then the particle's kinetic energy increases by the amount of the work. If the net work done is negative, then the particle's kinetic energy decreases by the amount of the work.

For example, if the kinetic energy of a particle is initially 5 J and there is a net transfer of 2 J to the particle (positive net work), the final kinetic energy is 7 J. If, instead, there is a net transfer of 2 J from the particle (negative net work), the final kinetic energy is 3 J.

**FIG. 7-3** A contestant in a bed race. We can approximate the bed and its occupant as being a particle for the purpose of calculating the work done on them by the force applied by the student.

> ✔ **CHECKPOINT 1**     A particle moves along an $x$ axis. Does the kinetic energy of the particle increase, decrease, or remain the same if the particle's velocity changes (a) from −3 m/s to −2 m/s and (b) from −2 m/s to 2 m/s? (c) In each situation, is the work done on the particle positive, negative, or zero?

## Sample Problem 7-2

Figure 7-4a shows two industrial spies sliding an initially stationary 225 kg floor safe a displacement $\vec{d}$ of magnitude 8.50 m, straight toward their truck. The push $\vec{F_1}$ of spy 001 is 12.0 N, directed at an angle of 30.0° downward from the horizontal; the pull $\vec{F_2}$ of spy 002 is 10.0 N, directed at 40.0° above the horizontal. The magnitudes and directions of these forces do not change as the safe moves, and the floor and safe make frictionless contact.

(a) What is the net work done on the safe by forces $\vec{F_1}$ and $\vec{F_2}$ during the displacement $\vec{d}$?

**KEY IDEAS** (1) The net work $W$ done on the safe by the two forces is the sum of the works they do individually. (2) Because we can treat the safe as a particle and the forces are constant in both magnitude and direction, we can use either Eq. 7-7 ($W = Fd \cos \phi$) or Eq. 7-8 ($W = \vec{F} \cdot \vec{d}$) to calculate those works. Since we know the magnitudes and directions of the forces, we choose Eq. 7-7.

**Calculations:** From Eq. 7-7 and the free-body diagram for the safe in Fig. 7-4b, the work done by $\vec{F_1}$ is

$$W_1 = F_1 d \cos \phi_1 = (12.0 \text{ N})(8.50 \text{ m})(\cos 30.0°)$$
$$= 88.33 \text{ J},$$

and the work done by $\vec{F_2}$ is

$$W_2 = F_2 d \cos \phi_2 = (10.0 \text{ N})(8.50 \text{ m})(\cos 40.0°)$$
$$= 65.11 \text{ J}.$$

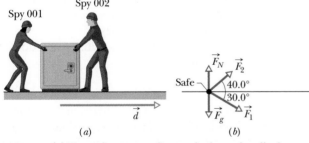

Spy 002
Spy 001

Safe

40.0°
30.0°

$\vec{F_N}$  $\vec{F_2}$
$\vec{F_g}$  $\vec{F_1}$

$\vec{d}$

(a)          (b)

**FIG. 7-4** (a) Two spies move a floor safe through a displacement $\vec{d}$. (b) A free-body diagram for the safe.

Thus, the net work $W$ is

$$W = W_1 + W_2 = 88.33 \text{ J} + 65.11 \text{ J}$$
$$= 153.4 \text{ J} \approx 153 \text{ J}. \qquad \text{(Answer)}$$

During the 8.50 m displacement, therefore, the spies transfer 153 J of energy to the kinetic energy of the safe.

(b) During the displacement, what is the work $W_g$ done on the safe by the gravitational force $\vec{F_g}$ and what is the work $W_N$ done on the safe by the normal force $\vec{F_N}$ from the floor?

**KEY IDEA** Because these forces are constant in both magnitude and direction, we can find the work they do with Eq. 7-7.

**Calculations:** Thus, with $mg$ as the magnitude of the gravitational force, we write

$$W_g = mgd \cos 90° = mgd(0) = 0 \quad \text{(Answer)}$$

and $\qquad W_N = F_N d \cos 90° = F_N d(0) = 0. \quad \text{(Answer)}$

We should have known this result. Because these forces are perpendicular to the displacement of the safe, they do zero work on the safe and do not transfer any energy to or from it.

(c) The safe is initially stationary. What is its speed $v_f$ at the end of the 8.50 m displacement?

**KEY IDEA** The speed of the safe changes because its kinetic energy is changed when energy is transferred to it by $\vec{F_1}$ and $\vec{F_2}$.

**Calculations:** We relate the speed to the work done by combining Eqs. 7-10 and 7-1:

$$W = K_f - K_i = \tfrac{1}{2}mv_f^2 - \tfrac{1}{2}mv_i^2.$$

The initial speed $v_i$ is zero, and we now know that the work done is 153.4 J. Solving for $v_f$ and then substituting known data, we find that

$$v_f = \sqrt{\frac{2W}{m}} = \sqrt{\frac{2(153.4 \text{ J})}{225 \text{ kg}}}$$
$$= 1.17 \text{ m/s}. \qquad \text{(Answer)}$$

## Sample Problem 7-3

During a storm, a crate of crepe is sliding across a slick, oily parking lot through a displacement $\vec{d} = (-3.0 \text{ m})\hat{i}$ while a steady wind pushes against the crate with a force $\vec{F} = (2.0 \text{ N})\hat{i} + (-6.0 \text{ N})\hat{j}$. The situation and coordinate axes are shown in Fig. 7-5.

(a) How much work does this force do on the crate during the displacement?

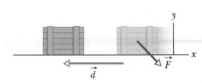

$y$
$x$
$\vec{F}$
$\vec{d}$

**FIG. 7-5** Force $\vec{F}$ slows a crate during displacement $\vec{d}$.

**KEY IDEA** Because we can treat the crate as a particle and because the wind force is constant ("steady") in both magnitude and direction during the displacement, we can use either Eq. 7-7 ($W = Fd \cos \phi$) or Eq. 7-8 ($W = \vec{F} \cdot \vec{d}$) to calculate the work. Since we know $\vec{F}$ and $\vec{d}$ in unit-vector notation, we choose Eq. 7-8.

**Calculations:** We write

$$W = \vec{F} \cdot \vec{d} = [(2.0 \text{ N})\hat{i} + (-6.0 \text{ N})\hat{j}] \cdot [(-3.0 \text{ m})\hat{i}].$$

Of the possible unit-vector dot products, only $\hat{i} \cdot \hat{i}$, $\hat{j} \cdot \hat{j}$, and $\hat{k} \cdot \hat{k}$ are nonzero (see Appendix E). Here we obtain

$$W = (2.0 \text{ N})(-3.0 \text{ m})\hat{i} \cdot \hat{i} + (-6.0 \text{ N})(-3.0 \text{ m})\hat{j} \cdot \hat{i}$$
$$= (-6.0 \text{ J})(1) + 0 = -6.0 \text{ J}. \qquad \text{(Answer)}$$

Thus, the force does a negative 6.0 J of work on the crate, transferring 6.0 J of energy from the kinetic energy of the crate.

**(b)** If the crate has a kinetic energy of 10 J at the beginning of displacement $\vec{d}$, what is its kinetic energy at the end of $\vec{d}$?

**KEY IDEA** Because the force does negative work on the crate, it reduces the crate's kinetic energy.

**Calculation:** Using the work–kinetic energy theorem in the form of Eq. 7-11, we have

$$K_f = K_i + W = 10 \text{ J} + (-6.0 \text{ J}) = 4.0 \text{ J}. \qquad \text{(Answer)}$$

Less kinetic energy means that the crate has been slowed.

## 7-6 | Work Done by the Gravitational Force

We next examine the work done on an object by the gravitational force acting on it. Figure 7-6 shows a particle-like tomato of mass $m$ that is thrown upward with initial speed $v_0$ and thus with initial kinetic energy $K_i = \frac{1}{2}mv_0^2$. As the tomato rises, it is slowed by a gravitational force $\vec{F}_g$; that is, the tomato's kinetic energy decreases because $\vec{F}_g$ does work on the tomato as it rises. Because we can treat the tomato as a particle, we can use Eq. 7-7 ($W = Fd \cos \phi$) to express the work done during a displacement $\vec{d}$. For the force magnitude $F$, we use $mg$ as the magnitude of $\vec{F}_g$. Thus, the work $W_g$ done by the gravitational force $\vec{F}_g$ is

$$W_g = mgd \cos \phi \qquad \text{(work done by gravitational force).} \qquad (7\text{-}12)$$

For a rising object, force $\vec{F}_g$ is directed opposite the displacement $\vec{d}$, as indicated in Fig. 7-6. Thus, $\phi = 180°$ and

$$W_g = mgd \cos 180° = mgd(-1) = -mgd. \qquad (7\text{-}13)$$

The minus sign tells us that during the object's rise, the gravitational force acting on the object transfers energy in the amount $mgd$ from the kinetic energy of the object. This is consistent with the slowing of the object as it rises.

After the object has reached its maximum height and is falling back down, the angle $\phi$ between force $\vec{F}_g$ and displacement $\vec{d}$ is zero. Thus,

$$W_g = mgd \cos 0° = mgd(+1) = +mgd. \qquad (7\text{-}14)$$

The plus sign tells us that the gravitational force now transfers energy in the amount $mgd$ to the kinetic energy of the object. This is consistent with the speeding up of the object as it falls. (Actually, as we shall see in Chapter 8, energy transfers associated with lifting and lowering an object involve the full object–Earth system.)

### Work Done in Lifting and Lowering an Object

Now suppose we lift a particle-like object by applying a vertical force $\vec{F}$ to it. During the upward displacement, our applied force does positive work $W_a$ on the object while the gravitational force does negative work $W_g$ on it. Our applied force tends to transfer energy to the object while the gravitational force tends to transfer energy from it. By Eq. 7-10, the change $\Delta K$ in the kinetic energy of the object due to these two energy transfers is

$$\Delta K = K_f - K_i = W_a + W_g, \qquad (7\text{-}15)$$

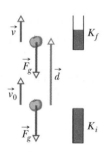

**FIG. 7-6** Because the gravitational force $\vec{F}_g$ acts on it, a particle-like tomato of mass $m$ thrown upward slows from velocity $\vec{v}_0$ to velocity $\vec{v}$ during displacement $\vec{d}$. A kinetic energy gauge indicates the resulting change in the kinetic energy of the tomato, from $K_i$ ($= \frac{1}{2}mv_0^2$) to $K_f$ ($= \frac{1}{2}mv^2$).

in which $K_f$ is the kinetic energy at the end of the displacement and $K_i$ is that at the start of the displacement. This equation also applies if we lower the object, but then the gravitational force tends to transfer energy *to* the object while our force tends to transfer energy *from* it.

In one common situation, the object is stationary before and after the lift—for example, when you lift a book from the floor to a shelf. Then $K_f$ and $K_i$ are both zero, and Eq. 7-15 reduces to

$$W_a + W_g = 0$$

or
$$W_a = -W_g. \qquad (7\text{-}16)$$

Note that we get the same result if $K_f$ and $K_i$ are not zero but are still equal. Either way, the result means that the work done by the applied force is the negative of the work done by the gravitational force; that is, the applied force transfers the same amount of energy to the object as the gravitational force transfers from the object. Using Eq. 7-12, we can rewrite Eq. 7-16 as

$$W_a = -mgd \cos \phi \qquad \text{(work done in lifting and lowering; } K_f = K_i\text{),} \qquad (7\text{-}17)$$

with $\phi$ being the angle between $\vec{F}_g$ and $\vec{d}$. If the displacement is vertically upward (Fig. 7-7a), then $\phi = 180°$ and the work done by the applied force equals $mgd$. If the displacement is vertically downward (Fig. 7-7b), then $\phi = 0°$ and the work done by the applied force equals $-mgd$.

Equations 7-16 and 7-17 apply to any situation in which an object is lifted or lowered, with the object stationary before and after the lift. They are independent of the magnitude of the force used. For example, if you lift a mug from the floor to over your head, your force on the mug varies considerably during the lift. Still, because the mug is stationary before and after the lift, the work your force does on the mug is given by Eqs. 7-16 and 7-17, where, in Eq. 7-17, $mg$ is the weight of the mug and $d$ is the distance you lift it.

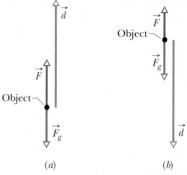

FIG. 7-7 (a) An applied force $\vec{F}$ lifts an object. The object's displacement $\vec{d}$ makes an angle $\phi = 180°$ with the gravitational force $\vec{F}_g$ on the object. The applied force does positive work on the object. (b) An applied force $\vec{F}$ lowers an object. The displacement $\vec{d}$ of the object makes an angle $\phi = 0°$ with the gravitational force $\vec{F}_g$. The applied force does negative work on the object.

## Sample Problem 7-4

One of the lifts of Paul Anderson (Fig. 7-8) in the 1950s remains a record: Anderson stooped beneath a reinforced wood platform, placed his hands on a short stool to brace himself, and then pushed upward on the platform with his back, lifting the platform straight up by 1.0 cm. The platform held automobile parts and a safe filled with lead, with a total weight of 27 900 N (6270 lb).

(a) As Anderson lifted the load, how much work was done on it by the gravitational force $\vec{F}_g$?

**KEY IDEA** We can treat the load as a single particle because the components moved rigidly together. Thus we can use Eq. 7-12 ($W_g = mgd \cos \phi$) to find the work $W_g$ done on the load by $\vec{F}_g$.

**Calculation:** The angle $\phi$ between the directions of the downward gravitational force and the upward displacement was 180°. Substituting this and the given data into Eq. 7-12, we find

$$W_g = mgd \cos \phi = (27\ 900 \text{ N})(0.010 \text{ m})(\cos 180°)$$
$$= -280 \text{ J.} \qquad \text{(Answer)}$$

(b) How much work was done by the force Anderson applied to make the lift?

FIG. 7-8 Using a harness across his back, Paul Anderson lifted a platform and a scout troop off the ground. (©AP/WideWorld Photos)

**KEY IDEAS** Anderson's force was certainly not constant. Thus, we *cannot* just substitute a force magnitude into Eq. 7-7 to find the work done. However, we know that the load was stationary both at the start and at the end of the lift. Therefore, we know that the work $W_A$ done by Anderson's applied force was the negative of the work $W_g$ done by the gravitational force $\vec{F}_g$.

*Calculation:* Equation 7-16 gives us

$$W_A = -W_g = +280 \text{ J.} \qquad \text{(Answer)}$$

*Comments:* This is hardly more than the work needed to lift a stuffed school backpack from the floor to shoulder level. So, why was Anderson's lift so amazing? Work (energy transfer) and force are different quantities; although Anderson's lift required an unremarkable energy transfer, it required a truly remarkable force.

---

**Sample Problem | 7-5**

An initially stationary 15.0 kg crate of cheese wheels is pulled, via a cable, a distance $d = 5.70$ m up a frictionless ramp to a height $h$ of 2.50 m, where it stops (Fig. 7-9a).

(a) How much work $W_g$ is done on the crate by the gravitational force $\vec{F}_g$ during the lift?

**KEY IDEA** We treat the crate as a particle and use Eq. 7-12 ($W_g = mgd \cos \phi$) to find the work $W_g$ done by $\vec{F}_g$.

*Calculations:* We do not know the angle $\phi$ between the directions of $\vec{F}_g$ and displacement $\vec{d}$. However, from the crate's free-body diagram in Fig. 7-9b, we find that $\phi$ is $\theta + 90°$, where $\theta$ is the (unknown) angle of the ramp. Equation 7-12 then gives us

$$W_g = mgd \cos(\theta + 90°) = -mgd \sin \theta, \quad (7\text{-}18)$$

where we have used a trigonometic identity to simplify the expression. The result seems to be useless because $\theta$ is unknown. But (continuing with physics courage) we see from Fig. 7-9a that $d \sin \theta = h$, where $h$ is a known quantity. With this substitution, Eq. 7-18 gives us

$$\begin{aligned} W_g &= -mgh \qquad\qquad\qquad (7\text{-}19) \\ &= -(15.0 \text{ kg})(9.8 \text{ m/s}^2)(2.50 \text{ m}) \\ &= -368 \text{ J.} \qquad\qquad\qquad \text{(Answer)} \end{aligned}$$

Note that Eq. 7-19 tells us that the work $W_g$ done by the gravitational force depends on the vertical displacement but (surprisingly) not on the horizontal displacement. (We return to this point in Chapter 8.)

(b) How much work $W_T$ is done on the crate by the force $\vec{T}$ from the cable during the lift?

**KEY IDEA** We cannot just substitute the force magnitude $T$ for $F$ in Eq. 7-7 ($W = Fd \cos \phi$) because we do not know the value of $T$. However, to get us going we can treat the crate as a particle and then apply the work–kinetic energy theorem ($\Delta K = W$) to it.

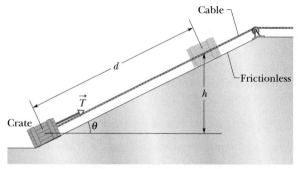

(a)

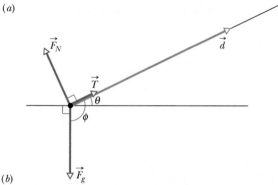

(b)

**FIG. 7-9** (*a*) A crate is pulled up a frictionless ramp by a force $\vec{T}$ parallel to the ramp. (*b*) A free-body diagram for the crate, showing also the displacement $\vec{d}$.

*Calculations:* Because the crate is stationary before and after the lift, the change $\Delta K$ in its kinetic energy is zero. For the net work $W$ done on the crate, we must sum the works done by all three forces acting on the crate. From (a), the work $W_g$ done by the gravitational force $\vec{F}_g$ is $-368$ J. The work $W_N$ done by the normal force $\vec{F}_N$ on the crate from the ramp is zero because $\vec{F}_N$ is perpendicular to the displacement. We want the work $W_T$ done by $\vec{T}$. Thus, the work–kinetic energy theorem gives us

$$\Delta K = W_T + W_g + W_N$$

or

$$0 = W_T - 368 \text{ J} + 0,$$

and so

$$W_T = 368 \text{ J.} \qquad \text{(Answer)}$$

---

**Sample Problem | 7-6 | Build your skill**

An elevator cab of mass $m = 500$ kg is descending with speed $v_i = 4.0$ m/s when its supporting cable begins to slip, allowing it to fall with constant acceleration $\vec{a} = \vec{g}/5$ (Fig. 7-10a).

(a) During the fall through a distance $d = 12$ m, what

is the work $W_g$ done on the cab by the gravitational force $\vec{F}_g$?

**KEY IDEA** We can treat the cab as a particle and thus use Eq. 7-12 ($W_g = mgd \cos \phi$) to find the work $W_g$.

*Calculation:* From Fig. 7-10*b*, we see that the angle between the directions of $\vec{F}_g$ and the cab's displacement $\vec{d}$ is 0°. Then, from Eq. 7-12, we find

$$W_g = mgd \cos 0° = (500 \text{ kg})(9.8 \text{ m/s}^2)(12 \text{ m})(1)$$
$$= 5.88 \times 10^4 \text{ J} \approx 59 \text{ kJ.} \qquad \text{(Answer)}$$

**(b)** During the 12 m fall, what is the work $W_T$ done on the cab by the upward pull $\vec{T}$ of the elevator cable?

**KEY IDEAS** (1) We can calculate the work $W_T$ with Eq. 7-7 ($W = Fd \cos \phi$) if we first find an expression for the magnitude $T$ of the cable's pull. (2) We can find that expression by writing Newton's second law for components along the $y$ axis in Fig. 7-10*b* ($F_{\text{net,}y} = ma_y$).

*Calculations:* We get

$$T - F_g = ma.$$

Solving for $T$, substituting $mg$ for $F_g$, and then substituting the result in Eq. 7-7, we obtain

$$W_T = Td \cos \phi = m(a + g)d \cos \phi.$$

Next, substituting $-g/5$ for the (downward) acceleration $a$ and then 180° for the angle $\phi$ between the directions of forces $\vec{T}$ and $m\vec{g}$, we find

$$W_T = m\left(-\frac{g}{5} + g\right)d \cos \phi = \frac{4}{5} mgd \cos \phi$$

$$= \frac{4}{5} (500 \text{ kg})(9.8 \text{ m/s}^2)(12 \text{ m}) \cos 180°$$

$$= -4.70 \times 10^4 \text{ J} \approx -47 \text{ kJ.} \qquad \text{(Answer)}$$

*Caution:* Note that $W_T$ is not simply the negative of $W_g$. The reason is that, because the cab accelerates during the fall, its speed changes during the fall, and thus its kinetic energy also changes. Therefore, Eq. 7-16 (which assumes that the initial and final kinetic energies are equal) does *not* apply here.

**(c)** What is the net work $W$ done on the cab during the fall?

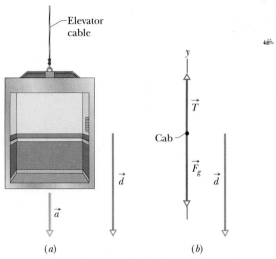

**FIG. 7-10** An elevator cab, descending with speed $v_i$, suddenly begins to accelerate downward. (*a*) It moves through a displacement $\vec{d}$ with constant acceleration $\vec{a} = \vec{g}/5$. (*b*) A free-body diagram for the cab, displacement included.

*Calculation:* The net work is the sum of the works done by the forces acting on the cab:

$$W = W_g + W_T = 5.88 \times 10^4 \text{ J} - 4.70 \times 10^4 \text{ J}$$
$$= 1.18 \times 10^4 \text{ J} \approx 12 \text{ kJ.} \qquad \text{(Answer)}$$

**(d)** What is the cab's kinetic energy at the end of the 12 m fall?

**KEY IDEA** The kinetic energy changes *because* of the net work done on the cab, according to Eq. 7-11 ($K_f = K_i + W$).

*Calculation:* From Eq. 7-1, we can write the kinetic energy at the start of the fall as $K_i = \frac{1}{2}mv_i^2$. We can then write Eq. 7-11 as

$$K_f = K_i + W = \frac{1}{2}mv_i^2 + W$$
$$= \frac{1}{2}(500 \text{ kg})(4.0 \text{ m/s})^2 + 1.18 \times 10^4 \text{ J}$$
$$= 1.58 \times 10^4 \text{ J} \approx 16 \text{ kJ.} \qquad \text{(Answer)}$$

## 7-7 | Work Done by a Spring Force

We next want to examine the work done on a particle-like object by a particular type of *variable force*—namely, a **spring force**, the force from a spring. Many forces in nature have the same mathematical form as the spring force. Thus, by examining this one force, you can gain an understanding of many others.

### The Spring Force

Figure 7-11*a* shows a spring in its **relaxed state**—that is, neither compressed nor extended. One end is fixed, and a particle-like object—a block, say—is attached to the other, free end. If we stretch the spring by pulling the block to the right as in Fig. 7-11*b*, the spring pulls on the block toward the left. (Because a spring

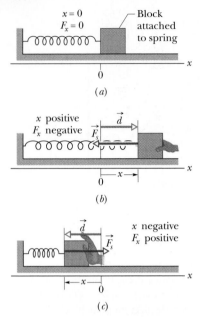

**FIG. 7-11** (a) A spring in its relaxed state. The origin of an $x$ axis has been placed at the end of the spring that is attached to a block. (b) The block is displaced by $\vec{d}$, and the spring is stretched by a positive amount $x$. Note the restoring force $\vec{F}_s$ exerted by the spring. (c) The spring is compressed by a negative amount $x$. Again, note the restoring force.

force acts to restore the relaxed state, it is sometimes said to be a *restoring force.*) If we compress the spring by pushing the block to the left as in Fig. 7-11c, the spring now pushes on the block toward the right.

To a good approximation for many springs, the force $\vec{F}_s$ from a spring is proportional to the displacement $\vec{d}$ of the free end from its position when the spring is in the relaxed state. The *spring force* is given by

$$\vec{F}_s = -k\vec{d} \qquad \text{(Hooke's law)}, \qquad (7\text{-}20)$$

which is known as **Hooke's law** after Robert Hooke, an English scientist of the late 1600s. The minus sign in Eq. 7-20 indicates that the direction of the spring force is always opposite the direction of the displacement of the spring's free end. The constant $k$ is called the **spring constant** (or **force constant**) and is a measure of the stiffness of the spring. The larger $k$ is, the stiffer the spring; that is, the larger $k$ is, the stronger the spring's pull or push for a given displacement. The SI unit for $k$ is the newton per meter.

In Fig. 7-11 an $x$ axis has been placed parallel to the length of the spring, with the origin ($x = 0$) at the position of the free end when the spring is in its relaxed state. For this common arrangement, we can write Eq. 7-20 as

$$F_x = -kx \qquad \text{(Hooke's law)}, \qquad (7\text{-}21)$$

where we have changed the subscript. If $x$ is positive (the spring is stretched toward the right on the $x$ axis), then $F_x$ is negative (it is a pull toward the left). If $x$ is negative (the spring is compressed toward the left), then $F_x$ is positive (it is a push toward the right). Note that a spring force is a *variable force* because it is a function of $x$, the position of the free end. Thus $F_x$ can be symbolized as $F(x)$. Also note that Hooke's law is a *linear* relationship between $F_x$ and $x$.

## The Work Done by a Spring Force

To find the work done by the spring force as the block in Fig. 7-11a moves, let us make two simplifying assumptions about the spring. (1) It is *massless;* that is, its mass is negligible relative to the block's mass. (2) It is an *ideal spring;* that is, it obeys Hooke's law exactly. Let us also assume that the contact between the block and the floor is frictionless and that the block is particle-like.

We give the block a rightward jerk to get it moving and then leave it alone. As the block moves rightward, the spring force $F_x$ does work on the block, decreasing the kinetic energy and slowing the block. However, we *cannot* find this work by using Eq. 7-7 ($W = Fd \cos \phi$) because that equation assumes a constant force. The spring force is a variable force.

To find the work done by the spring, we use calculus. Let the block's initial position be $x_i$ and its later position $x_f$. Then divide the distance between those two positions into many segments, each of tiny length $\Delta x$. Label these segments, starting from $x_i$, as segments 1, 2, and so on. As the block moves through a segment, the spring force hardly varies because the segment is so short that $x$ hardly varies. Thus, we can approximate the force magnitude as being constant within the segment. Label these magnitudes as $F_{x1}$ in segment 1, $F_{x2}$ in segment 2, and so on.

With the force now constant in each segment, we *can* find the work done within each segment by using Eq. 7-7. Here $\phi = 180°$, and so $\cos \phi = -1$. Then the work done is $-F_{x1} \Delta x$ in segment 1, $-F_{x2} \Delta x$ in segment 2, and so on. The net work $W_s$ done by the spring, from $x_i$ to $x_f$, is the sum of all these works:

$$W_s = \sum -F_{xj} \Delta x, \qquad (7\text{-}22)$$

where $j$ labels the segments. In the limit as $\Delta x$ goes to zero, Eq. 7-22 becomes

$$W_s = \int_{x_i}^{x_f} -F_x \, dx. \qquad (7\text{-}23)$$

From Eq. 7-21, the force magnitude $F_x$ is $kx$. Thus, substitution leads to

$$W_s = \int_{x_i}^{x_f} -kx \, dx = -k \int_{x_i}^{x_f} x \, dx$$

$$= (-\tfrac{1}{2}k)[x^2]_{x_i}^{x_f} = (-\tfrac{1}{2}k)(x_f^2 - x_i^2). \qquad (7\text{-}24)$$

Multiplied out, this yields

$$W_s = \tfrac{1}{2}kx_i^2 - \tfrac{1}{2}kx_f^2 \qquad \text{(work by a spring force)}. \qquad (7\text{-}25)$$

This work $W_s$ done by the spring force can have a positive or negative value, depending on whether the *net* transfer of energy is to or from the block as the block moves from $x_i$ to $x_f$. *Caution:* The final position $x_f$ appears in the *second* term on the right side of Eq. 7-25. Therefore, Eq. 7-25 tells us:

> ☛ Work $W_s$ is positive if the block ends up closer to the relaxed position ($x = 0$) than it was initially. It is negative if the block ends up farther away from $x = 0$. It is zero if the block ends up at the same distance from $x = 0$.

If $x_i = 0$ and if we call the final position $x$, then Eq. 7-25 becomes

$$W_s = -\tfrac{1}{2}kx^2 \qquad \text{(work by a spring force)}. \qquad (7\text{-}26)$$

### The Work Done by an Applied Force

Now suppose that we displace the block along the $x$ axis while continuing to apply a force $\vec{F}_a$ to it. During the displacement, our applied force does work $W_a$ on the block while the spring force does work $W_s$. By Eq. 7-10, the change $\Delta K$ in the kinetic energy of the block due to these two energy transfers is

$$\Delta K = K_f - K_i = W_a + W_s, \qquad (7\text{-}27)$$

in which $K_f$ is the kinetic energy at the end of the displacement and $K_i$ is that at the start of the displacement. If the block is stationary before and after the displacement, then $K_f$ and $K_i$ are both zero and Eq. 7-27 reduces to

$$W_a = -W_s. \qquad (7\text{-}28)$$

> ☛ If a block that is attached to a spring is stationary before and after a displacement, then the work done on it by the applied force displacing it is the negative of the work done on it by the spring force.

*Caution:* If the block is not stationary before and after the displacement, then this statement is *not* true.

✓ **CHECKPOINT 2**    For three situations, the initial and final positions, respectively, along the $x$ axis for the block in Fig. 7-11 are (a) $-3$ cm, 2 cm; (b) 2 cm, 3 cm; and (c) $-2$ cm, 2 cm. In each situation, is the work done by the spring force on the block positive, negative, or zero?

Sample Problem    7-7

A package of spicy Cajun pralines lies on a frictionless floor, attached to the free end of a spring in the arrangement of Fig. 7-11a. A rightward applied force of magnitude $F_a = 4.9$ N would be needed to hold the package at $x_1 = 12$ mm.

(a) How much work does the spring force do on the package if the package is pulled rightward from $x_0 = 0$ to $x_2 = 17$ mm?

**KEY IDEA**    As the package moves from one position to another, the spring force does work on it as given by Eq. 7-25 or Eq. 7-26.

*Calculations:* We know that the initial position $x_i$ is 0 and the final position $x_f$ is 17 mm, but we do not know the spring constant $k$. We can probably find $k$ with Eq. 7-21 (Hooke's law), but we need this fact to use it: Were the package held stationary at $x_1 = 12$ mm, the spring force would have to balance the applied force (according to Newton's second law). Thus, the spring force $F_x$ would have to be $-4.9$ N (toward the left in Fig. 7-11b); so Eq. 7-21 ($F_x = -kx$) gives us

$$k = -\frac{F_x}{x_1} = -\frac{-4.9 \text{ N}}{12 \times 10^{-3} \text{ m}} = 408 \text{ N/m}.$$

Now, with the package at $x_2 = 17$ mm, Eq. 7-26 yields

$$W_s = -\tfrac{1}{2}kx_2^2 = -\tfrac{1}{2}(408 \text{ N/m})(17 \times 10^{-3} \text{ m})^2$$
$$= -0.059 \text{ J.} \qquad \text{(Answer)}$$

(b) Next, the package is moved leftward to $x_3 = -12$ mm. How much work does the spring force do on

the package during this displacement? Explain the sign of this work.

**Calculation:** Now $x_i = +17$ mm and $x_f = -12$ mm, and Eq. 7-25 yields

$$W_s = \tfrac{1}{2}kx_i^2 - \tfrac{1}{2}kx_f^2 = \tfrac{1}{2}k(x_i^2 - x_f^2)$$
$$= \tfrac{1}{2}(408 \text{ N/m})[(17 \times 10^{-3} \text{ m})^2 - (-12 \times 10^{-3} \text{ m})^2]$$
$$= 0.030 \text{ J} = 30 \text{ mJ.} \qquad \text{(Answer)}$$

This work done on the block by the spring force is positive because the spring force does more positive work as the block moves from $x_i = +17$ mm to the spring's relaxed position than it does negative work as the block moves from the spring's relaxed position to $x_f = -12$ mm.

---

## Sample Problem  7-8

In Fig. 7-12, a cumin canister of mass $m = 0.40$ kg slides across a horizontal frictionless counter with speed $v = 0.50$ m/s. It then runs into and compresses a spring of spring constant $k = 750$ N/m. When the canister is momentarily stopped by the spring, by what distance $d$ is the spring compressed?

**FIG. 7-12** A canister of mass $m$ moves at velocity $\vec{v}$ toward a spring that has spring constant $k$.

the canister as

$$K_f - K_i = -\tfrac{1}{2}kd^2.$$

Substituting according to the third idea makes this expression

$$0 - \tfrac{1}{2}mv^2 = -\tfrac{1}{2}kd^2.$$

Simplifying, solving for $d$, and substituting known data then give us

$$d = v\sqrt{\frac{m}{k}} = (0.50 \text{ m/s})\sqrt{\frac{0.40 \text{ kg}}{750 \text{ N/m}}}$$
$$= 1.2 \times 10^{-2} \text{ m} = 1.2 \text{ cm.} \qquad \text{(Answer)}$$

**KEY IDEAS**

1. The work $W_s$ done on the canister by the spring force is related to the requested distance $d$ by Eq. 7-26 ($W_s = -\tfrac{1}{2}kx^2$), with $d$ replacing $x$.

2. The work $W_s$ is also related to the kinetic energy of the canister by Eq. 7-10 ($K_f - K_i = W$).

3. The canister's kinetic energy has an initial value of $K = \tfrac{1}{2}mv^2$ and a value of zero when the canister is momentarily at rest.

**Calculations:** Putting the first two of these ideas together, we write the work–kinetic energy theorem for

---

## 7-8 | Work Done by a General Variable Force

### One-Dimensional Analysis

Let us return to the situation of Fig. 7-2 but now consider the force to be in the positive direction of the $x$ axis and the force magnitude to vary with position $x$. Thus, as the bead (particle) moves, the magnitude $F(x)$ of the force doing work on it changes. Only the magnitude of this variable force changes, not its direction, and the magnitude at any position does not change with time.

Figure 7-13a shows a plot of such a *one-dimensional variable force*. We want an expression for the work done on the particle by this force as the particle moves from an initial point $x_i$ to a final point $x_f$. However, we *cannot* use Eq. 7-7 ($W = Fd \cos \phi$) because it applies only for a constant force $\vec{F}$. Here, again, we shall use calculus. We divide the area under the curve of Fig. 7-13a into a number of narrow strips of width $\Delta x$ (Fig. 7-13b). We choose $\Delta x$ small enough to permit us to take the force $F(x)$ as being reasonably constant over that interval. We let $F_{j,\text{avg}}$ be the average value of $F(x)$ within the $j$th interval. Then in Fig. 7-13b, $F_{j,\text{avg}}$ is the height of the $j$th strip.

With $F_{j,\text{avg}}$ considered constant, the increment (small amount) of work $\Delta W_j$ done by the force in the $j$th interval is now approximately given by Eq. 7-7 and is

$$\Delta W_j = F_{j,\text{avg}} \, \Delta x. \tag{7-29}$$

In Fig. 7-13$b$, $\Delta W_j$ is then equal to the area of the $j$th rectangular, shaded strip.

To approximate the total work $W$ done by the force as the particle moves from $x_i$ to $x_f$, we add the areas of all the strips between $x_i$ and $x_f$ in Fig. 7-13$b$:

$$W = \sum \Delta W_j = \sum F_{j,\text{avg}} \, \Delta x. \tag{7-30}$$

Equation 7-30 is an approximation because the broken "skyline" formed by the tops of the rectangular strips in Fig. 7-13$b$ only approximates the actual curve of $F(x)$.

We can make the approximation better by reducing the strip width $\Delta x$ and using more strips (Fig. 7-13$c$). In the limit, we let the strip width approach zero; the number of strips then becomes infinitely large and we have, as an exact result,

$$W = \lim_{\Delta x \to 0} \sum F_{j,\text{avg}} \, \Delta x. \tag{7-31}$$

This limit is exactly what we mean by the integral of the function $F(x)$ between the limits $x_i$ and $x_f$. Thus, Eq. 7-31 becomes

$$W = \int_{x_i}^{x_f} F(x) \, dx \qquad \text{(work: variable force).} \tag{7-32}$$

If we know the function $F(x)$, we can substitute it into Eq. 7-32, introduce the proper limits of integration, carry out the integration, and thus find the work. (Appendix E contains a list of common integrals.) Geometrically, the work is equal to the area between the $F(x)$ curve and the $x$ axis, between the limits $x_i$ and $x_f$ (shaded in Fig. 7-13$d$).

### Three-Dimensional Analysis

Consider now a particle that is acted on by a three-dimensional force

$$\vec{F} = F_x\hat{i} + F_y\hat{j} + F_z\hat{k}, \tag{7-33}$$

in which the components $F_x$, $F_y$, and $F_z$ can depend on the position of the particle; that is, they can be functions of that position. However, we make three simplifications: $F_x$ may depend on $x$ but not on $y$ or $z$, $F_y$ may depend on $y$ but not on $x$ or $z$, and $F_z$ may depend on $z$ but not on $x$ or $y$. Now let the particle move through an incremental displacement

$$d\vec{r} = dx\hat{i} + dy\hat{j} + dz\hat{k}. \tag{7-34}$$

The increment of work $dW$ done on the particle by $\vec{F}$ during the displacement $d\vec{r}$ is, by Eq. 7-8,

$$dW = \vec{F} \cdot d\vec{r} = F_x \, dx + F_y \, dy + F_z \, dz. \tag{7-35}$$

The work $W$ done by $\vec{F}$ while the particle moves from an initial position $r_i$ having coordinates $(x_i, y_i, z_i)$ to a final position $r_f$ having coordinates $(x_f, y_f, z_f)$ is then

$$W = \int_{r_i}^{r_f} dW = \int_{x_i}^{x_f} F_x \, dx + \int_{y_i}^{y_f} F_y \, dy + \int_{z_i}^{z_f} F_z \, dz. \tag{7-36}$$

If $\vec{F}$ has only an $x$ component, then the $y$ and $z$ terms in Eq. 7-36 are zero and the equation reduces to Eq. 7-32.

### Work–Kinetic Energy Theorem with a Variable Force

Equation 7-32 gives the work done by a variable force on a particle in a one-dimensional situation. Let us now make certain that the calculated work is

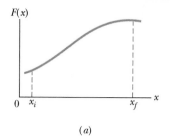

(a)

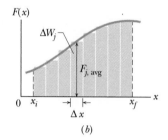

(b)

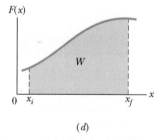

(c)

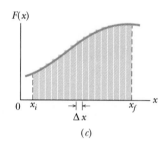

(d)

**FIG. 7-13** (a) A one-dimensional force $\vec{F}(x)$ plotted against the displacement $x$ of a particle on which it acts. The particle moves from $x_i$ to $x_f$. (b) Same as (a) but with the area under the curve divided into narrow strips. (c) Same as (b) but with the area divided into narrower strips. (d) The limiting case. The work done by the force is given by Eq. 7-32 and is represented by the shaded area between the curve and the $x$ axis and between $x_i$ and $x_f$.

indeed equal to the change in kinetic energy of the particle, as the work–kinetic energy theorem states.

Consider a particle of mass $m$, moving along an $x$ axis and acted on by a net force $F(x)$ that is directed along that axis. The work done on the particle by this force as the particle moves from position $x_i$ to position $x_f$ is given by Eq. 7-32 as

$$W = \int_{x_i}^{x_f} F(x)\, dx = \int_{x_i}^{x_f} ma\, dx, \tag{7-37}$$

in which we use Newton's second law to replace $F(x)$ with $ma$. We can write the quantity $ma\, dx$ in Eq. 7-37 as

$$ma\, dx = m \frac{dv}{dt}\, dx. \tag{7-38}$$

From the chain rule of calculus, we have

$$\frac{dv}{dt} = \frac{dv}{dx} \frac{dx}{dt} = \frac{dv}{dx} v, \tag{7-39}$$

and Eq. 7-38 becomes

$$ma\, dx = m \frac{dv}{dx} v\, dx = mv\, dv. \tag{7-40}$$

Substituting Eq. 7-40 into Eq. 7-37 yields

$$W = \int_{v_i}^{v_f} mv\, dv = m \int_{v_i}^{v_f} v\, dv$$

$$= \tfrac{1}{2} mv_f^2 - \tfrac{1}{2} mv_i^2. \tag{7-41}$$

Note that when we change the variable from $x$ to $v$ we are required to express the limits on the integral in terms of the new variable. Note also that because the mass $m$ is a constant, we are able to move it outside the integral.

Recognizing the terms on the right side of Eq. 7-41 as kinetic energies allows us to write this equation as

$$W = K_f - K_i = \Delta K,$$

which is the work–kinetic energy theorem.

## Sample Problem 7-9

In an epidural procedure, as used in childbirth, a surgeon or an anesthetist must run a needle through the skin on the patient's back, through various tissue layers and into a narrow region called the epidural space that lies within the spinal canal surrounding the spinal cord. The needle is intended to deliver an anesthetic fluid. This tricky procedure requires much practice so that the doctor knows when the needle has reached the epidural space and not overshot it, a mistake that could result in serious complications.

The feel a doctor has for the needle's penetration is the variable force that must be applied to advance the needle through the tissues. Figure 7-14a is a graph of the force magnitude $F$ versus displacement $x$ of the needle tip in a typical epidural procedure. (The line segments have been straightened somewhat from the original

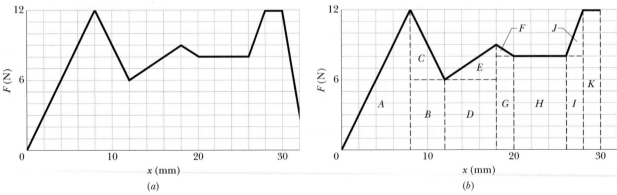

FIG. 7-14 (a) The force magnitude $F$ versus the displacement $x$ of the needle in an epidural procedure. (b) Breaking up the region between the plotted curve and the displacement axis to calculate the area.

data.) As $x$ increases from 0, the skin resists the needle, but at $x = 8.0$ mm the force is finally great enough to pierce the skin, and then the required force decreases. Similarly, the needle finally pierces the interspinous ligament at $x = 18$ mm and the relatively tough ligamentum flavum at $x = 30$ mm. The needle then enters the epidural space (where it is to deliver the anesthetic fluid), and the force drops sharply. A new doctor must learn this pattern of force versus displacement to recognize when to stop pushing on the needle. (This is the pattern to be programmed into a virtual-reality simulation of an epidural procedure.) How much work $W$ is done by the force exerted on the needle to get the needle to the epidural space at $x = 30$ mm?

---

**KEY IDEAS** (1) We can calculate the work $W$ done by a variable force $F(x)$ by integrating the force versus position $x$. Equation 7-32 tells us that

$$W = \int_{x_i}^{x_f} F(x)\, dx.$$

We want the work done by the force during the displacement from $x_i = 0$ to $x_f = 0.030$ m. (2) We can evaluate the integral by finding the area under the curve on the graph of Fig. 7-14a.

$$W = \begin{pmatrix} \text{area between force curve} \\ \text{and } x \text{ axis, from } x_i \text{ to } x_f \end{pmatrix}.$$

*Calculations:* Because our graph consists of straight-line segments, we can find the area by splitting the region below the curve into rectangular and triangular regions, as shown in Fig. 7-14b. For example, the area in triangular region $A$ is

$$\text{area}_A = \tfrac{1}{2}(0.0080 \text{ m})(12 \text{ N}) = 0.048 \text{ N} \cdot \text{m} = 0.048 \text{ J}.$$

Once we've calculated the areas for all the labeled regions in Fig. 7-14b, we find that the total work is

$W = $ (sum of the areas of regions $A$ through $K$)

$= 0.048 + 0.024 + 0.012 + 0.036 + 0.009 + 0.001$
$\quad + 0.016 + 0.048 + 0.016 + 0.004 + 0.024$

$= 0.238$ J.                  (Answer)

---

## Sample Problem  7-10

Force $\vec{F} = (3x^2 \text{ N})\hat{i} + (4 \text{ N})\hat{j}$, with $x$ in meters, acts on a particle, changing only the kinetic energy of the particle. How much work is done on the particle as it moves from coordinates (2 m, 3 m) to (3 m, 0 m)? Does the speed of the particle increase, decrease, or remain the same?

---

**KEY IDEA** The force is a variable force because its $x$ component depends on the value of $x$. Thus, we cannot use Eqs. 7-7 and 7-8 to find the work done. Instead, we must use Eq. 7-36 to integrate the force.

*Calculation:* We set up two integrals, one along each axis:

$$W = \int_2^3 3x^2\, dx + \int_3^0 4\, dy = 3\int_2^3 x^2\, dx + 4\int_3^0 dy$$

$$= 3[\tfrac{1}{3}x^3]_2^3 + 4[y]_3^0 = [3^3 - 2^3] + 4[0 - 3]$$

$$= 7.0 \text{ J}.$$                  (Answer)

The positive result means that energy is transferred to the particle by force $\vec{F}$. Thus, the kinetic energy of the particle increases and, because $K = \tfrac{1}{2}mv^2$, its speed must also increase.

---

# 7-9 | Power

The time rate at which work is done by a force is said to be the **power** due to the force. If a force does an amount of work $W$ in an amount of time $\Delta t$, the **average power** due to the force during that time interval is

$$P_{avg} = \frac{W}{\Delta t} \quad \text{(average power)}. \tag{7-42}$$

The **instantaneous power** $P$ is the instantaneous time rate of doing work, which we can write as

$$P = \frac{dW}{dt} \quad \text{(instantaneous power)}. \tag{7-43}$$

Suppose we know the work $W(t)$ done by a force as a function of time. Then to get the instantaneous power $P$ at, say, time $t = 3.0$ s during the work, we would first take the time derivative of $W(t)$ and then evaluate the result for $t = 3.0$ s.

The SI unit of power is the joule per second. This unit is used so often that it has a special name, the **watt** (W), after James Watt, who greatly improved the rate at which steam engines could do work. In the British system, the unit of power is the foot-pound per second. Often the horsepower is used. These are related by

$$1 \text{ watt} = 1 \text{ W} = 1 \text{ J/s} = 0.738 \text{ ft} \cdot \text{lb/s} \qquad (7\text{-}44)$$

and

$$1 \text{ horsepower} = 1 \text{ hp} = 550 \text{ ft} \cdot \text{lb/s} = 746 \text{ W}. \qquad (7\text{-}45)$$

Inspection of Eq. 7-42 shows that work can be expressed as power multiplied by time, as in the common unit kilowatt-hour. Thus,

$$1 \text{ kilowatt-hour} = 1 \text{ kW} \cdot \text{h} = (10^3 \text{ W})(3600 \text{ s})$$
$$= 3.60 \times 10^6 \text{ J} = 3.60 \text{ MJ}. \qquad (7\text{-}46)$$

Perhaps because they appear on our utility bills, the watt and the kilowatt-hour have become identified as electrical units. They can be used equally well as units for other examples of power and energy. Thus, if you pick up a book from the floor and put it on a tabletop, you are free to report the work that you have done as, say, $4 \times 10^{-6} \text{ kW} \cdot \text{h}$ (or more conveniently as $4 \text{ mW} \cdot \text{h}$).

We can also express the rate at which a force does work on a particle (or particle-like object) in terms of that force and the particle's velocity. For a particle that is moving along a straight line (say, an $x$ axis) and is acted on by a constant force $\vec{F}$ directed at some angle $\phi$ to that line, Eq. 7-43 becomes

$$P = \frac{dW}{dt} = \frac{F \cos \phi \, dx}{dt} = F \cos \phi \left( \frac{dx}{dt} \right),$$

or

$$P = Fv \cos \phi. \qquad (7\text{-}47)$$

Reorganizing the right side of Eq. 7-47 as the dot product $\vec{F} \cdot \vec{v}$, we may also write the equation as

$$P = \vec{F} \cdot \vec{v} \qquad \text{(instantaneous power)}. \qquad (7\text{-}48)$$

For example, the truck in Fig. 7-15 exerts a force $\vec{F}$ on the trailing load, which has velocity $\vec{v}$ at some instant. The instantaneous power due to $\vec{F}$ is the rate at which $\vec{F}$ does work on the load at that instant and is given by Eqs. 7-47 and 7-48. Saying that this power is "the power of the truck" is often acceptable, but keep in mind what is meant: Power is the rate at which the applied *force* does work.

**FIG. 7-15** The power due to the truck's applied force on the trailing load is the rate at which that force does work on the load. *(REGLAIN FREDERIC/Gamma-Presse, Inc.)*

✓**CHECKPOINT 3**     A block moves with uniform circular motion because a cord tied to the block is anchored at the center of a circle. Is the power due to the force on the block from the cord positive, negative, or zero?

---

**Sample Problem** | **7-11**

Figure 7-16 shows constant forces $\vec{F}_1$ and $\vec{F}_2$ acting on a box as the box slides rightward across a frictionless floor. Force $\vec{F}_1$ is horizontal, with magnitude 2.0 N; force $\vec{F}_2$ is angled upward by 60° to the floor and has magnitude 4.0 N. The speed $v$ of the box at a certain instant is 3.0 m/s. What is the power due to each force acting on the box at that instant, and what is the net power? Is the net power changing at that instant?

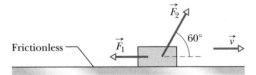

**FIG. 7-16** Two forces $\vec{F}_1$ and $\vec{F}_2$ act on a box that slides rightward across a frictionless floor. The velocity of the box is $\vec{v}$.

**KEY IDEA**     We want an instantaneous power, not an average power over a time period. Also, we know the box's velocity (rather than the work done on it).

**Calculation:** We use Eq. 7-47 for each force. For force $\vec{F}_1$, at angle $\phi_1 = 180°$ to velocity $\vec{v}$, we have

$$P_1 = F_1 v \cos \phi_1 = (2.0 \text{ N})(3.0 \text{ m/s}) \cos 180°$$
$$= -6.0 \text{ W}. \qquad \text{(Answer)}$$

This negative result tells us that force $\vec{F}_1$ is transferring energy *from* the box at the rate of 6.0 J/s.

For force $\vec{F}_2$, at angle $\phi_2 = 60°$ to velocity $\vec{v}$, we have

$$P_2 = F_2 v \cos \phi_2 = (4.0 \text{ N})(3.0 \text{ m/s}) \cos 60°$$
$$= 6.0 \text{ W}. \qquad \text{(Answer)}$$

This positive result tells us that force $\vec{F}_2$ is transferring energy *to* the box at the rate of 6.0 J/s.

The net power is the sum of the individual powers:

$$P_{net} = P_1 + P_2$$
$$= -6.0 \text{ W} + 6.0 \text{ W} = 0, \qquad \text{(Answer)}$$

which tells us that the net rate of transfer of energy to or from the box is zero. Thus, the kinetic energy $(K = \frac{1}{2}mv^2)$ of the box is not changing, and so the speed of the box will remain at 3.0 m/s. With neither the forces $\vec{F}_1$ and $\vec{F}_2$ nor the velocity $\vec{v}$ changing, we see from Eq. 7-48 that $P_1$ and $P_2$ are constant and thus so is $P_{net}$.

## Sample Problem  7-12

Provided a funny car does not lose traction, the time it takes to race from rest through a distance $D$ depends primarily on the engine's power $P$. Assuming the power is constant, derive the time in terms of $D$ and $P$.

**KEY IDEAS** (1) The power of an engine is the rate at which it can do work, as expressed by Eq. 7-43 ($P = dW/dt$). (2) We can relate the work done during the race to the kinetic energy with Eq. 7-10, the work–kinetic energy theorem ($W = K_f - K_i$).

*Power and kinetic energy:* From the work–kinetic energy theorem, a small amount of work $dW$ results in a small change $dK$ of kinetic energy: $dW = dK$. Substituting this into Eq. 7-43 and rearranging give us

$$dK = P\,dt.$$

Integrating both sides and substituting that the kinetic energy is $K = 0$ when the race starts at $t = 0$, we find

$$\int_0^K dK = \int_0^t P\,dt$$

and

$$K = Pt.$$

After substituting $\frac{1}{2}mv^2$ for $K$, we solve for $v$, the speed at the end of the race:

$$v = \left(\frac{2Pt}{m}\right)^{1/2}. \qquad (7\text{-}49)$$

*Distance and speed:* From the definition of velocity in Chapter 2, we know that $v = dx/dt$. Rearranging the definition and setting up integration on both sides, we find

$$\int_0^D dx = \int_0^t v\,dt.$$

Substituting from Eq. 7-49, we have

$$\int_0^D dx = \int_0^t \left(\frac{2Pt}{m}\right)^{1/2} dt = \left(\frac{2P}{m}\right)^{1/2} \int_0^t t^{1/2}\,dt.$$

Integrating then yields

$$D = \left(\frac{2P}{m}\right)^{1/2} \frac{2}{3} t^{3/2}.$$

Solving for $t$ tells us that a funny car's elapsed time $t$ depends on $D$ and $P$ as given by

$$t = \left(\frac{3}{2}D\right)^{2/3}\left(\frac{m}{2P}\right)^{1/3}. \qquad \text{(Answer)}$$

*Comments:* In words, the elapsed time depends on the inverse cube root of the power. If the racing crew can coax more power out of the engine, the elapsed time decreases because of the *inverse* dependence, but only modestly because of the *cube root* dependence.

## REVIEW & SUMMARY

**Kinetic Energy**  The **kinetic energy** $K$ associated with the motion of a particle of mass $m$ and speed $v$, where $v$ is well below the speed of light, is

$$K = \tfrac{1}{2}mv^2 \qquad \text{(kinetic energy)}. \qquad (7\text{-}1)$$

**Work**  Work $W$ is energy transferred to or from an object via a force acting on the object. Energy transferred to the object is positive work, and from the object, negative work.

**Work Done by a Constant Force**  The work done on a particle by a constant force $\vec{F}$ during displacement $\vec{d}$ is

$$W = Fd \cos \phi = \vec{F} \cdot \vec{d} \qquad \text{(work, constant force)}, \quad (7\text{-}7, 7\text{-}8)$$

in which $\phi$ is the constant angle between the directions of $\vec{F}$ and $\vec{d}$. Only the component of $\vec{F}$ that is along the displacement $\vec{d}$ can do work on the object. When two or more forces act on an object, their **net work** is the sum of the individual works done by the forces, which is also equal to the work that would be done on the object by the net force $\vec{F}_{net}$ of those forces.

**Work and Kinetic Energy**  For a particle, a change $\Delta K$ in the kinetic energy equals the net work $W$ done on the particle:

$$\Delta K = K_f - K_i = W \qquad \text{(work–kinetic energy theorem)}, \quad (7\text{-}10)$$

in which $K_i$ is the initial kinetic energy of the particle and $K_f$ is the kinetic energy after the work is done. Equation 7-10 rearranged gives us

$$K_f = K_i + W. \qquad (7\text{-}11)$$

**Work Done by the Gravitational Force** The work $W_g$ done by the gravitational force $\vec{F}_g$ on a particle-like object of mass $m$ as the object moves through a displacement $\vec{d}$ is given by

$$W_g = mgd\cos\phi, \qquad (7\text{-}12)$$

in which $\phi$ is the angle between $\vec{F}_g$ and $\vec{d}$.

**Work Done in Lifting and Lowering an Object** The work $W_a$ done by an applied force as a particle-like object is either lifted or lowered is related to the work $W_g$ done by the gravitational force and the change $\Delta K$ in the object's kinetic energy by

$$\Delta K = K_f - K_i = W_a + W_g. \qquad (7\text{-}15)$$

If $K_f = K_i$, then Eq. 7-15 reduces to

$$W_a = -W_g, \qquad (7\text{-}16)$$

which tells us that the applied force transfers as much energy to the object as the gravitational force transfers from it.

**Spring Force** The force $\vec{F}_s$ from a spring is

$$\vec{F}_s = -k\vec{d} \qquad \text{(Hooke's law)}, \qquad (7\text{-}20)$$

where $\vec{d}$ is the displacement of the spring's free end from its position when the spring is in its **relaxed state** (neither compressed nor extended), and $k$ is the **spring constant** (a measure of the spring's stiffness). If an $x$ axis lies along the spring, with the origin at the location of the spring's free end when the spring is in its relaxed state, Eq. 7-20 can be written as

$$F_x = -kx \qquad \text{(Hooke's law)}. \qquad (7\text{-}21)$$

A spring force is thus a variable force: It varies with the displacement of the spring's free end.

**Work Done by a Spring Force** If an object is attached to the spring's free end, the work $W_s$ done on the object by the spring force when the object is moved from an initial position $x_i$ to a final position $x_f$ is

$$W_s = \tfrac{1}{2}kx_i^2 - \tfrac{1}{2}kx_f^2. \qquad (7\text{-}25)$$

If $x_i = 0$ and $x_f = x$, then Eq. 7-25 becomes

$$W_s = -\tfrac{1}{2}kx^2. \qquad (7\text{-}26)$$

**Work Done by a Variable Force** When the force $\vec{F}$ on a particle-like object depends on the position of the object, the work done by $\vec{F}$ on the object while the object moves from an initial position $r_i$ with coordinates $(x_i, y_i, z_i)$ to a final position $r_f$ with coordinates $(x_f, y_f, z_f)$ must be found by integrating the force. If we assume that component $F_x$ may depend on $x$ but not on $y$ or $z$, component $F_y$ may depend on $y$ but not on $x$ or $z$, and component $F_z$ may depend on $z$ but not on $x$ or $y$, then the work is

$$W = \int_{x_i}^{x_f} F_x\,dx + \int_{y_i}^{y_f} F_y\,dy + \int_{z_i}^{z_f} F_z\,dz. \qquad (7\text{-}36)$$

If $\vec{F}$ has only an $x$ component, then Eq. 7-36 reduces to

$$W = \int_{x_i}^{x_f} F(x)\,dx. \qquad (7\text{-}32)$$

**Power** The **power** due to a force is the *rate* at which that force does work on an object. If the force does work $W$ during a time interval $\Delta t$, the *average power* due to the force over that time interval is

$$P_{\text{avg}} = \frac{W}{\Delta t}. \qquad (7\text{-}42)$$

Instantaneous power is the instantaneous rate of doing work:

$$P = \frac{dW}{dt}. \qquad (7\text{-}43)$$

For a force $\vec{F}$ at an angle $\phi$ to the direction of travel of the instantaneous velocity $\vec{v}$, the instantaneous power is

$$P = Fv\cos\phi = \vec{F}\cdot\vec{v}. \qquad (7\text{-}47, 7\text{-}48)$$

# QUESTIONS

**1** Is positive or negative work done by a constant force $\vec{F}$ on a particle during a straight-line displacement $\vec{d}$ if (a) the angle between $\vec{F}$ and $\vec{d}$ is 30°; (b) the angle is 100°; (c) $\vec{F} = 2\hat{i} - 3\hat{j}$ and $\vec{d} = -4\hat{i}$?

**2** In three situations, a briefly applied horizontal force changes the velocity of a hockey puck that slides over frictionless ice. The overhead views of Fig. 7-17 indicate, for each situation, the puck's initial speed $v_i$, its final speed $v_f$, and the directions of the corresponding velocity vectors. Rank the situations according to the work done on the puck by the applied force, most positive first and most negative last.

**3** Rank the following velocities according to the kinetic energy a particle will have with each velocity, greatest first: (a) $\vec{v} = 4\hat{i} + 3\hat{j}$, (b) $\vec{v} = -4\hat{i} + 3\hat{j}$, (c) $\vec{v} = -3\hat{i} + 4\hat{j}$, (d) $\vec{v} = 3\hat{i} - 4\hat{j}$, (e) $\vec{v} = 5\hat{i}$, and (f) $v = 5$ m/s at 30° to the horizontal.

**4** Figure 7-18a shows two horizontal forces that act on a block that is sliding to the right across a frictionless floor. Figure 7-18b shows three plots of the block's kinetic energy $K$

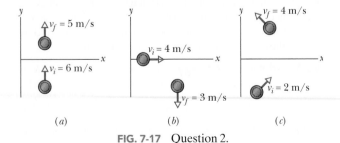

FIG. 7-17 Question 2.

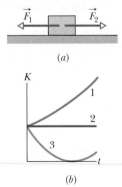

FIG. 7-18 Question 4.

versus time $t$. Which of the plots best corresponds to the following three situations: (a) $F_1 = F_2$, (b) $F_1 > F_2$, (c) $F_1 < F_2$?

**5** In Fig. 7-19, a greased pig has a choice of three frictionless slides along which to slide to the ground. Rank the slides according to how much work the gravitational force does on the pig during the descent, greatest first.

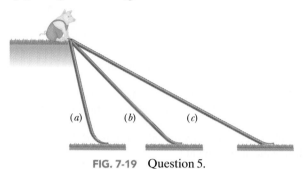

**FIG. 7-19** Question 5.

**6** Figure 7-20a shows four situations in which a horizontal force acts on the same block, which is initially at rest. The force magnitudes are $F_2 = F_4 = 2F_1 = 2F_3$. The horizontal component $v_x$ of the block's velocity is shown in Fig. 7-20b for the four situations. (a) Which plot in Fig. 7-20b best corresponds to which force in Fig. 7-20a? (b) Which plot in Fig. 7-20c (for kinetic energy $K$ versus time $t$) best corresponds to which plot in Fig. 7-20b?

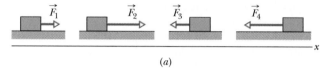

(a)

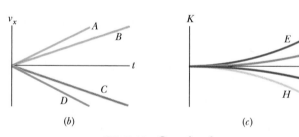

(b)                                              (c)

**FIG. 7-20** Question 6.

**7** Figure 7-21 shows four graphs (drawn to the same scale) of the $x$ component $F_x$ of a variable force (directed along an $x$ axis) versus the position $x$ of a particle on which the force acts. Rank the graphs according to the work done by the force on the particle from $x = 0$ to $x = x_1$, from most positive work first to most negative work last.

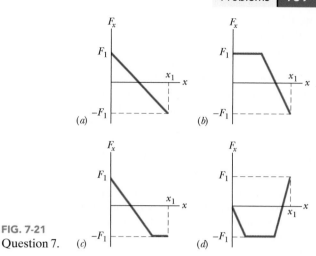

**FIG. 7-21**
Question 7.    (c)                                    (d)

**8** Figure 7-22 gives the $x$ component $F_x$ of a force that can act on a particle. If the particle begins at rest at $x = 0$, what is its coordinate when it has (a) its greatest kinetic energy, (b) its greatest speed, and (c) zero speed? (d) What is the particle's direction of travel after it reaches $x = 6$ m?

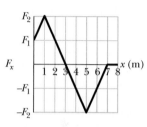

**FIG. 7-22** Question 8.

**9** Spring $A$ is stiffer than spring $B$ ($k_A > k_B$). The spring force of which spring does more work if the springs are compressed (a) the same distance and (b) by the same applied force?

**10** A glob of slime is launched or dropped from the edge of a cliff. Which of the graphs in Fig. 7-23 could possibly show how the kinetic energy of the glob changes during its flight?

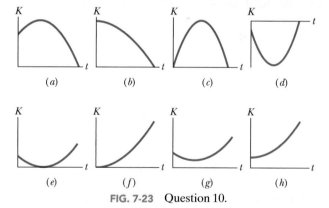

(a)          (b)          (c)          (d)

(e)          (f)          (g)          (h)

**FIG. 7-23** Question 10.

## PROBLEMS

GO   Tutoring problem available (at instructor's discretion) in *WileyPLUS* and WebAssign

SSM   Worked-out solution available in Student Solutions Manual      WWW   Worked-out solution is at ——

• – •••   Number of dots indicates level of problem difficulty            ILW   Interactive solution is at —— http://www.wiley.com/college/halliday

       Additional information available in *The Flying Circus of Physics* and at flyingcircusofphysics.com

### sec. 7-3   Kinetic Energy

•**1** On August 10, 1972, a large meteorite skipped across the atmosphere above the western United States and western Canada, much like a stone skipped across water. The accom-

panying fireball was so bright that it could be seen in the daytime sky and was brighter than the usual meteorite trail. The meteorite's mass was about $4 \times 10^6$ kg; its speed was about 15 km/s. Had it entered the atmosphere vertically, it would have hit Earth's surface with about the same speed. (a) Calculate the meteorite's loss of kinetic energy (in joules) that would have been associated with the vertical impact. (b) Express the energy as a multiple of the explosive energy of 1 megaton of TNT, which is $4.2 \times 10^{15}$ J. (c) The energy associated with the atomic bomb explosion over Hiroshima was equivalent to 13 kilotons of TNT. To how many Hiroshima bombs would the meteorite impact have been equivalent?

•2 If a Saturn V rocket with an Apollo spacecraft attached had a combined mass of $2.9 \times 10^5$ kg and reached a speed of 11.2 km/s, how much kinetic energy would it then have?

•3 A proton (mass $m = 1.67 \times 10^{-27}$ kg) is being accelerated along a straight line at $3.6 \times 10^{15}$ m/s$^2$ in a machine. If the proton has an initial speed of $2.4 \times 10^7$ m/s and travels 3.5 cm, what then is (a) its speed and (b) the increase in its kinetic energy? **SSM**

•4 A force $\vec{F}_a$ is applied to a bead as the bead is moved along a straight wire through displacement +5.0 cm. The magnitude of $\vec{F}_a$ is set at a certain value, but the angle $\phi$ between $\vec{F}_a$ and the bead's displacement can be chosen. Figure 7-24 gives the work $W$ done by $\vec{F}_a$ on the bead for a range of $\phi$ values; $W_0 =$ 25 J. How much work is done by $\vec{F}_a$ if $\phi$ is (a) 64° and (b) 147°?

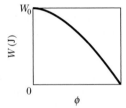

FIG. 7-24 Problem 4.

••5 A father racing his son has half the kinetic energy of the son, who has half the mass of the father. The father speeds up by 1.0 m/s and then has the same kinetic energy as the son. What are the original speeds of (a) the father and (b) the son?

••6 A bead with mass $1.8 \times 10^{-2}$ kg is moving along a wire in the positive direction of an $x$ axis. Beginning at time $t = 0$, when the bead passes through $x = 0$ with speed 12 m/s, a constant force acts on the bead. Figure 7-25 indicates the bead's position at times $t_0 = 0$, $t_1 = 1.0$ s, $t_2 = 2.0$ s, and $t_3 =$ 3.0 s. The bead momentarily stops at $t = 3.0$ s. What is the kinetic energy of the bead at $t = 10$ s?

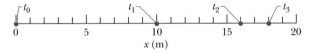

FIG. 7-25 Problem 6.

## sec. 7-5 Work and Kinetic Energy

•7 The only force acting on a 2.0 kg canister that is moving in an $xy$ plane has a magnitude of 5.0 N. The canister initially has a velocity of 4.0 m/s in the positive $x$ direction and some time later has a velocity of 6.0 m/s in the positive $y$ direction. How much work is done on the canister by the 5.0 N force during this time?

•8 A coin slides over a frictionless plane and across an $xy$ coordinate system from the origin to a point with $xy$ coordinates (3.0 m, 4.0 m) while a constant force acts on it. The force has magnitude 2.0 N and is directed at a counterclockwise angle of 100° from the positive direction of the $x$ axis. How much work is done by the force on the coin during the displacement?

•9 A 3.0 kg body is at rest on a frictionless horizontal air track when a constant horizontal force $\vec{F}$ acting in the positive

direction of an $x$ axis along the track is applied to the body. A stroboscopic graph of the position of the body as it slides to the right is shown in Fig. 7-26. The force $\vec{F}$ is applied to the body at $t = 0$, and the graph records the position of the body at 0.50 s intervals. How much work is done on the body by the applied force $\vec{F}$ between $t = 0$ and $t = 2.0$ s?

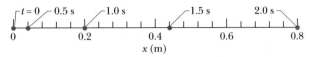

FIG. 7-26 Problem 9.

•10 A floating ice block is pushed through a displacement $\vec{d} = (15$ m$)\hat{i} - (12$ m$)\hat{j}$ along a straight embankment by rushing water, which exerts a force $\vec{F} = (210$ N$)\hat{i} - (150$ N$)\hat{j}$ on the block. How much work does the force do on the block during the displacement?

••11 A luge and its rider, with a total mass of 85 kg, emerge from a downhill track onto a horizontal straight track with an initial speed of 37 m/s. If a force slows them to a stop at a constant rate of 2.0 m/s$^2$, (a) what magnitude $F$ is required for the force, (b) what distance $d$ do they travel while slowing, and (c) what work $W$ is done on them by the force? What are (d) $F$, (e) $d$, and (f) $W$ if they, instead, slow at 4.0 m/s$^2$?

••12 An 8.0 kg object is moving in the positive direction of an $x$ axis. When it passes through $x = 0$, a constant force directed along the axis begins to act on it. Figure 7-27 gives its kinetic energy $K$ versus position $x$ as it moves from $x = 0$ to $x = 5.0$ m; $K_0 = 30.0$ J. The force continues to act. What is $v$ when the object moves back through $x = -3.0$ m?

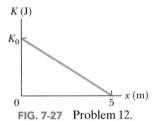

FIG. 7-27 Problem 12.

••13 Figure 7-28 shows three forces applied to a trunk that moves leftward by 3.00 m over a frictionless floor. The force magnitudes are $F_1 = 5.00$ N, $F_2 = 9.00$ N, and $F_3 = 3.00$ N, and the indicated angle is $\theta = 60.0°$. During the displacement, (a) what is the net work done on the trunk by the three forces and (b) does the kinetic energy of the trunk increase or decrease? **GO**

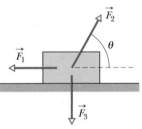

FIG. 7-28 Problem 13.

••14 A can of bolts and nuts is pushed 2.00 m along an $x$ axis by a broom along the greasy (frictionless) floor of a car repair shop in a version of shuffleboard. Figure 7-29 gives the work $W$ done on the can by the constant horizontal force from the broom, versus the can's position $x$. The scale of the figure's vertical axis is set by $W_s = 6.0$ J. (a) What is the magnitude of that force? (b) If the can had an initial kinetic energy of 3.00 J, moving in the positive direction of the $x$ axis, what is its kinetic energy at the end of the 2.00 m?

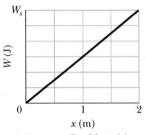

FIG. 7-29 Problem 14.

**••15** A 12.0 N force with a fixed orientation does work on a particle as the particle moves through displacement $\vec{d} = (2.00\hat{i} - 4.00\hat{j} + 3.00\hat{k})$ m. What is the angle between the force and the displacement if the change in the particle's kinetic energy is (a) +30.0 J and (b) −30.0 J?

**••16** Figure 7-30 shows an overhead view of three horizontal forces acting on a cargo canister that was initially stationary but now moves across a frictionless floor. The force magnitudes are $F_1 = 3.00$ N, $F_2 = 4.00$ N, and $F_3 = 10.0$ N, and the indicated angles are $\theta_2 = 50.0°$ and $\theta_3 = 35.0°$. What is the net work done on the canister by the three forces during the first 4.00 m of displacement?

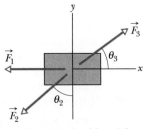

FIG. 7-30 Problem 16.

### sec. 7-6 Work Done by the Gravitational Force

**•17** A helicopter lifts a 72 kg astronaut 15 m vertically from the ocean by means of a cable. The acceleration of the astronaut is g/10. How much work is done on the astronaut by (a) the force from the helicopter and (b) the gravitational force on her? Just before she reaches the helicopter, what are her (c) kinetic energy and (d) speed? **SSM WWW**

**•18** (a) In 1975 the roof of Montreal's Velodrome, with a weight of 360 kN, was lifted by 10 cm so that it could be centered. How much work was done on the roof by the forces making the lift? (b) In 1960 a Tampa, Florida, mother reportedly raised one end of a car that had fallen onto her son when a jack failed. If her panic lift effectively raised 4000 N (about $\frac{1}{4}$ of the car's weight) by 5.0 cm, how much work did her force do on the car?

**••19** A cord is used to vertically lower an initially stationary block of mass $M$ at a constant downward acceleration of $g/4$. When the block has fallen a distance $d$, find (a) the work done by the cord's force on the block, (b) the work done by the gravitational force on the block, (c) the kinetic energy of the block, and (d) the speed of the block. **SSM**

**••20** In Fig. 7-31, a horizontal force $\vec{F}_a$ of magnitude 20.0 N is applied to a 3.00 kg psychology book as the book slides a distance $d = 0.500$ m up a frictionless ramp at angle $\theta = 30.0°$. (a) During the displacement, what is the net work done on the book by $\vec{F}_a$, the gravitational force on the book, and the normal force on the book? (b) If the book has zero kinetic energy at the start of the displacement, what is its speed at the end of the displacement? **GO**

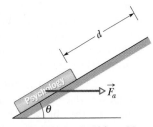

FIG. 7-31 Problem 20.

**••21** In Fig. 7-32, a constant force $\vec{F}_a$ of magnitude 82.0 N is applied to a 3.00 kg shoe box at angle $\phi = 53.0°$, causing the box to move up a frictionless ramp at constant speed. How much work is done on the box by $\vec{F}_a$ when the box has moved through vertical distance $h = 0.150$ m?

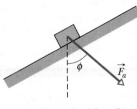

FIG. 7-32 Problem 21.

**••22** A block is sent up a frictionless ramp along which an $x$ axis extends upward. Figure 7-33 gives the kinetic energy of the block as a function of position $x$; the scale of the figure's vertical axis is set by $K_s = 40.0$ J. If the block's initial speed is 4.00 m/s, what is the normal force on the block?

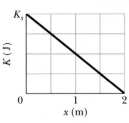

FIG. 7-33 Problem 22.

**••23** In Fig. 7-34, a block of ice slides down a frictionless ramp at angle $\theta = 50°$ while an ice worker pulls on the block (via a rope) with a force $\vec{F}_r$ that has a magnitude of 50 N and is directed up the ramp. As the block slides through distance $d = 0.50$ m along the ramp, its kinetic energy increases by 80 J. How much greater would its kinetic energy have been if the rope had not been attached to the block?

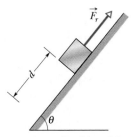

FIG. 7-34 Problem 23.

**••24** A cave rescue team lifts an injured spelunker directly upward and out of a sinkhole by means of a motor-driven cable. The lift is performed in three stages, each requiring a vertical distance of 10.0 m: (a) the initially stationary spelunker is accelerated to a speed of 5.00 m/s; (b) he is then lifted at the constant speed of 5.00 m/s; (c) finally he is decelerated to zero speed. How much work is done on the 80.0 kg rescuee by the force lifting him during each stage?

**•••25** In Fig. 7-35, a 0.250 kg block of cheese lies on the floor of a 900 kg elevator cab that is being pulled upward by a cable through distance $d_1 = 2.40$ m and then through distance $d_2 = 10.5$ m. (a) Through $d_1$, if the normal force on the block from the floor has constant magnitude $F_N = 3.00$ N, how much work is done on the cab by the force from the cable? (b) Through $d_2$, if the work done on the cab by the (constant) force from the cable is 92.61 kJ, what is the magnitude of $F_N$? **GO**

FIG. 7-35 Problem 25.

### sec. 7-7 Work Done by a Spring Force

**•26** During spring semester at MIT, residents of the parallel buildings of the East Campus dorms battle one another with large catapults that are made with surgical hose mounted on a window frame. A balloon filled with dyed water is placed in a pouch attached to the hose, which is then stretched through the width of the room. Assume that the stretching of the hose obeys Hooke's law with a spring constant of 100 N/m. If the hose is stretched by 5.00 m and then released, how much work does the force from the hose do on the balloon in the pouch by the time the hose reaches its relaxed length?

**•27** A spring and block are in the arrangement of Fig. 7-11. When the block is pulled out to $x = +4.0$ cm, we must apply a force of magnitude 360 N to hold it there. We pull the block to $x = 11$ cm and then release it. How much work does the spring do on the block as the block moves from $x_i = +5.0$ cm to (a) $x = +3.0$ cm, (b) $x = -3.0$ cm, (c) $x = -5.0$ cm, and (d) $x = -9.0$ cm?

**•28** In Fig. 7-11, we must apply a force of magnitude 80 N to hold the block stationary at $x = -2.0$ cm. From that position, we then slowly move the block so that our force does +4.0 J of work on the spring–block system; the block is then again stationary. What is the block's position? (*Hint:* There are two answers.)

••**29** The only force acting on a 2.0 kg body as it moves along a positive x axis has an x component $F_x = -6x$ N, with x in meters. The velocity at $x = 3.0$ m is 8.0 m/s. (a) What is the velocity of the body at $x = 4.0$ m? (b) At what positive value of x will the body have a velocity of 5.0 m/s? **SSM WWW**

••**30** Figure 7-36 gives spring force $F_x$ versus position x for the spring–block arrangement of Fig. 7-11. The scale is set by $F_s$ = 160.0 N. We release the block at $x = 12$ cm. How much work does the spring do on the block when the block moves from $x_i$ = +8.0 cm to (a) $x = +5.0$ cm, (b) $x = -5.0$ cm, (c) $x = -8.0$ cm, and (d) $x = -10.0$ cm?

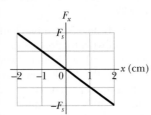

**FIG. 7-36** Problem 30.

••**31** In the arrangement of Fig. 7-11, we gradually pull the block from $x = 0$ to $x = +3.0$ cm, where it is stationary. Figure 7-37 gives the work that our force does on the block. The scale of the figure's vertical axis is set by $W_s = 1.0$ J. We then pull the block out to $x = +5.0$ cm and release it from rest. How much work does the spring do on the block when the block moves from $x_i = +5.0$ cm to (a) $x = +4.0$ cm, (b) $x = -2.0$ cm, and (c) $x = -5.0$ cm?

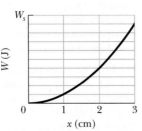

**FIG. 7-37** Problem 31.

••**32** In Fig. 7-11a, a block of mass m lies on a horizontal frictionless surface and is attached to one end of a horizontal spring (spring constant k) whose other end is fixed. The block is initially at rest at the position where the spring is

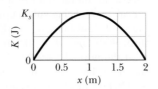

**FIG. 7-38** Problem 32.

unstretched ($x = 0$) when a constant horizontal force $\vec{F}$ in the positive direction of the x axis is applied to it. A plot of the resulting kinetic energy of the block versus its position x is shown in Fig. 7-38. The scale of the figure's vertical axis is set by $K_s = 4.0$ J. (a) What is the magnitude of $\vec{F}$? (b) What is the value of k?

•••**33** The block in Fig. 7-11a lies on a horizontal frictionless surface, and the spring constant is 50 N/m. Initially, the spring is at its relaxed length and the block is stationary at position $x = 0$. Then an applied force with a constant magnitude of 3.0 N pulls the block in the positive direction of the x axis, stretching the spring until the block stops. When that stopping point is reached, what are (a) the position of the block, (b) the work that has been done on the block by the applied force, and (c) the work that has been done on the block by the spring force? During the block's displacement, what are (d) the block's position when its kinetic energy is maximum and (e) the value of that maximum kinetic energy?

### sec. 7-8 **Work Done by a General Variable Force**

•**34** A 5.0 kg block moves in a straight line on a horizontal frictionless surface under the influence of a force that varies

with position as shown in Fig. 7-39. The scale of the figure's vertical axis is set by $F_s = 10.0$ N. How much work is done by the force as the block moves from the origin to $x = 8.0$ m?

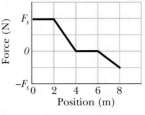

**FIG. 7-39** Problem 34.

•**35** The force on a particle is directed along an x axis and given by $F = F_0(x/x_0 - 1)$. Find the work done by the force in moving the particle from $x = 0$ to $x = 2x_0$ by (a) plotting $F(x)$ and measuring the work from the graph and (b) integrating $F(x)$. **SSM WWW**

•**36** A 10 kg brick moves along an x axis. Its acceleration as a function of its position is shown in Fig. 7-40. The scale of the figure's vertical axis is set by $a_s = 20.0$ m/s². What is the net work performed on the brick by the force causing the acceleration as the brick moves from $x = 0$ to $x = 8.0$ m? **ILW**

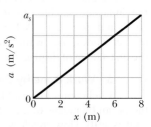

**FIG. 7-40** Problem 36.

••**37** A single force acts on a 3.0 kg particle-like object whose position is given by $x = 3.0t - 4.0t^2 + 1.0t^3$, with x in meters and t in seconds. Find the work done on the object by the force from $t = 0$ to $t = 4.0$ s.

••**38** A can of sardines is made to move along an x axis from $x = 0.25$ m to $x = 1.25$ m by a force with a magnitude given by $F = \exp(-4x^2)$, with x in meters and F in newtons. (Here exp is the exponential function.) How much work is done on the can by the force?

••**39** Figure 7-41 gives the acceleration of a 2.00 kg particle as an applied force $\vec{F}_a$ moves it from rest along an x axis from $x = 0$ to $x = 9.0$ m. The scale of the figure's vertical axis is set by $a_s = 6.0$ m/s². How much work has the force done on the particle when the particle reaches (a) $x = 4.0$ m, (b) $x = 7.0$ m, and (c) $x = 9.0$ m? What is the particle's speed and direction of travel when it reaches (d) $x = 4.0$ m, (e) $x = 7.0$ m, and (f) $x = 9.0$ m?

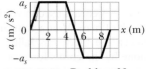

**FIG. 7-41** Problem 39.

••**40** A 1.5 kg block is initially at rest on a horizontal frictionless surface when a horizontal force along an x axis is applied to the block. The force is given by $\vec{F}(x) = (2.5 - x^2)\hat{i}$ N, where x is in meters and the initial position of the block is $x = 0$. (a) What is the kinetic energy of the block as it passes through $x = 2.0$ m? (b) What is the maximum kinetic energy of the block between $x = 0$ and $x = 2.0$ m?

••**41** A force $\vec{F} = (cx - 3.00x^2)\hat{i}$ acts on a particle as the particle moves along an x axis, with $\vec{F}$ in newtons, x in meters, and c a constant. At $x = 0$, the particle's kinetic energy is 20.0 J; at $x = 3.00$ m, it is 11.0 J. Find c. **GO**

•••**42** Figure 7-42 shows a cord attached to a cart that can slide along a frictionless horizontal rail aligned along an x axis. The left end of the cord is pulled over a pulley, of negligible mass and friction and at cord height $h = 1.20$ m, so the cart slides from $x_1 = 3.00$ m to $x_2 = 1.00$ m. During the move, the tension in the cord is a constant 25.0 N. What is the change in the kinetic energy of the cart during the move?

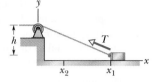

FIG. 7-42   Problem 42.

### sec. 7-9  Power

•43   A 100 kg block is pulled at a constant speed of 5.0 m/s across a horizontal floor by an applied force of 122 N directed 37° above the horizontal. What is the rate at which the force does work on the block?  **SSM ILW**

•44   The loaded cab of an elevator has a mass of $3.0 \times 10^3$ kg and moves 210 m up the shaft in 23 s at constant speed. At what average rate does the force from the cable do work on the cab?

•45   A force of 5.0 N acts on a 15 kg body initially at rest. Compute the work done by the force in (a) the first, (b) the second, and (c) the third seconds and (d) the instantaneous power due to the force at the end of the third second.  **SSM**

•46   A skier is pulled by a towrope up a frictionless ski slope that makes an angle of 12° with the horizontal. The rope moves parallel to the slope with a constant speed of 1.0 m/s. The force of the rope does 900 J of work on the skier as the skier moves a distance of 8.0 m up the incline. (a) If the rope moved with a constant speed of 2.0 m/s, how much work would the force of the rope do on the skier as the skier moved a distance of 8.0 m up the incline? At what rate is the force of the rope doing work on the skier when the rope moves with a speed of (b) 1.0 m/s and (c) 2.0 m/s?

••47   A fully loaded, slow-moving freight elevator has a cab with a total mass of 1200 kg, which is required to travel upward 54 m in 3.0 min, starting and ending at rest. The elevator's counterweight has a mass of only 950 kg, and so the elevator motor must help. What average power is required of the force the motor exerts on the cab via the cable?  **SSM**

••48   (a) At a certain instant, a particle-like object is acted on by a force $\vec{F} = (4.0\,\text{N})\hat{i} - (2.0\,\text{N})\hat{j} + (9.0\,\text{N})\hat{k}$ while the object's velocity is $\vec{v} = -(2.0\,\text{m/s})\hat{i} + (4.0\,\text{m/s})\hat{k}$. What is the instantaneous rate at which the force does work on the object? (b) At some other time, the velocity consists of only a y component. If the force is unchanged and the instantaneous power is $-12$ W, what is the velocity of the object?

••49   A machine carries a 4.0 kg package from an initial position of $\vec{d}_i = (0.50\,\text{m})\hat{i} + (0.75\,\text{m})\hat{j} + (0.20\,\text{m})\hat{k}$ at $t = 0$ to a final position of $\vec{d}_f = (7.50\,\text{m})\hat{i} + (12.0\,\text{m})\hat{j} + (7.20\,\text{m})\hat{k}$ at $t = 12$ s. The constant force applied by the machine on the package is $\vec{F} = (2.00\,\text{N})\hat{i} + (4.00\,\text{N})\hat{j} + (6.00\,\text{N})\hat{k}$. For that displacement, find (a) the work done on the package by the machine's force and (b) the average power of the machine's force on the package.

••50   A 0.30 kg ladle sliding on a horizontal frictionless surface is attached to one end of a horizontal spring ($k = 500$ N/m) whose other end is fixed. The ladle has a kinetic energy of 10 J as it passes through its equilibrium position (the point at which the spring force is zero). (a) At what rate is the spring doing work on the ladle as the ladle passes through its equilibrium position? (b) At what rate is the spring doing work on the ladle when the spring is compressed 0.10 m and the ladle is moving away from the equilibrium position?

••51   A force $\vec{F} = (3.00\,\text{N})\hat{i} + (7.00\,\text{N})\hat{j} + (7.00\,\text{N})\hat{k}$ acts on a 2.00 kg mobile object that moves from an initial position of $\vec{d}_i = (3.00\,\text{m})\hat{i} - (2.00\,\text{m})\hat{j} + (5.00\,\text{m})\hat{k}$ to a final position of $\vec{d}_f = -(5.00\,\text{m})\hat{i} + (4.00\,\text{m})\hat{j} + (7.00\,\text{m})\hat{k}$ in 4.00 s. Find (a) the work done on the object by the force in the 4.00 s interval, (b) the average power due to the force during that interval, and (c) the angle between vectors $\vec{d}_i$ and $\vec{d}_f$.

•••52   A funny car accelerates from rest through a measured track distance in time $T$ with the engine operating at a constant power $P$. If the track crew can increase the engine power by a differential amount $dP$, what is the change in the time required for the run?

### Additional Problems

53   An explosion at ground level leaves a crater with a diameter that is proportional to the energy of the explosion raised to the $\frac{1}{3}$ power; an explosion of 1 megaton of TNT leaves a crater with a 1 km diameter. Below Lake Huron in Michigan there appears to be an ancient impact crater with a 50 km diameter. What was the kinetic energy associated with that impact, in terms of (a) megatons of TNT (1 megaton yields $4.2 \times 10^{15}$ J) and (b) Hiroshima bomb equivalents (13 kilotons of TNT each)? (Ancient meteorite or comet impacts may have significantly altered Earth's climate and contributed to the extinction of the dinosaurs and other life-forms.)

54   A 250 g block is dropped onto a relaxed vertical spring that has a spring constant of $k = 2.5$ N/cm (Fig. 7-43). The block becomes attached to the spring and compresses the spring 12 cm before momentarily stopping. While the spring is being compressed, what work is done on the block by (a) the gravitational force on it and (b) the spring force? (c) What is the speed of the block just before it hits the spring? (Assume that friction is negligible.) (d) If the speed at impact is doubled, what is the maximum compression of the spring?

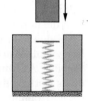

FIG. 7-43
Problem 54.

55   How much work is done by a force $\vec{F} = (2x\,\text{N})\hat{i} + (3\,\text{N})\hat{j}$, with $x$ in meters, that moves a particle from a position $\vec{r}_i = (2\,\text{m})\hat{i} + (3\,\text{m})\hat{j}$ to a position $\vec{r}_f = -(4\,\text{m})\hat{i} - (3\,\text{m})\hat{j}$?

56   To pull a 50 kg crate across a horizontal frictionless floor, a worker applies a force of 210 N, directed 20° above the horizontal. As the crate moves 3.0 m, what work is done on the crate by (a) the worker's force, (b) the gravitational force on the crate, and (c) the normal force on the crate from the floor? (d) What is the total work done on the crate?

57   In Fig. 7-44, a cord runs around two massless, frictionless pulleys. A canister with mass $m = 20$ kg hangs from one pulley, and you exert a force $\vec{F}$ on the free end of the cord. (a) What must be the magnitude of $\vec{F}$ if you are to lift the canister at a constant speed? (b) To lift the canister by 2.0 cm, how far must you

FIG. 7-44   Problem 57.

pull the free end of the cord? During that lift, what is the work done on the canister by (c) your force (via the cord) and (d) the gravitational force? (*Hint:* When a cord loops around a pulley as shown, it pulls on the pulley with a net force that is twice the tension in the cord.)

**58** A force $\vec{F} = (4.0\text{ N})\hat{i} + c\hat{j}$ acts on a particle as the particle goes through displacement $\vec{d} = (3.0\text{ m})\hat{i} - (2.0\text{ m})\hat{j}$. (Other forces also act on the particle.) What is $c$ if the work done on the particle by force $\vec{F}$ is (a) 0, (b) 17 J, and (c) −18 J?

**59** A constant force of magnitude 10 N makes an angle of 150° (measured counterclockwise) with the positive $x$ direction as it acts on a 2.0 kg object moving in an $xy$ plane. How much work is done on the object by the force as the object moves from the origin to the point having position vector $(2.0\text{ m})\hat{i} - (4.0\text{ m})\hat{j}$?

**60** An initially stationary 2.0 kg object accelerates horizontally and uniformly to a speed of 10 m/s in 3.0 s. (a) In that 3.0 s interval, how much work is done on the object by the force accelerating it? What is the instantaneous power due to that force (b) at the end of the interval and (c) at the end of the first half of the interval?

**61** If a ski lift raises 100 passengers averaging 660 N in weight to a height of 150 m in 60.0 s, at constant speed, what average power is required of the force making the lift?

**62** Boxes are transported from one location to another in a warehouse by means of a conveyor belt that moves with a constant speed of 0.50 m/s. At a certain location the conveyor belt moves for 2.0 m up an incline that makes an angle of 10° with the horizontal, then for 2.0 m horizontally, and finally for 2.0 m down an incline that makes an angle of 10° with the horizontal. Assume that a 2.0 kg box rides on the belt without slipping. At what rate is the force of the conveyor belt doing work on the box as the box moves (a) up the 10° incline, (b) horizontally, and (c) down the 10° incline?

**63** A horse pulls a cart with a force of 40 lb at an angle of 30° above the horizontal and moves along at a speed of 6.0 mi/h. (a) How much work does the force do in 10 min? (b) What is the average power (in horsepower) of the force? **SSM**

**64** An iceboat is at rest on a frictionless frozen lake when a sudden wind exerts a constant force of 200 N, toward the east, on the boat. Due to the angle of the sail, the wind causes the boat to slide in a straight line for a distance of 8.0 m in a direction 20° north of east. What is the kinetic energy of the iceboat at the end of that 8.0 m?

**65** A 230 kg crate hangs from the end of a rope of length $L = 12.0$ m. You push horizontally on the crate with a varying force $\vec{F}$ to move it distance $d = 4.00$ m to the side (Fig. 7-45). (a) What is the magnitude of $\vec{F}$ when the crate is in this final position? During the crate's displacement, what are (b) the total work done on it, (c) the work done by the gravitational force on the crate, and (d) the work done by the pull on the crate from the

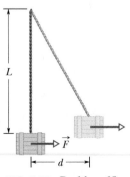

**FIG. 7-45** Problem 65.

rope? (e) Knowing that the crate is motionless before and after its displacement, use the answers to (b), (c), and (d) to find the work your force $\vec{F}$ does on the crate. (f) Why is the work of your force not equal to the product of the horizontal displacement and the answer to (a)?

**66** The only force acting on a 2.0 kg body as the body moves along an $x$ axis varies as shown in Fig. 7-46. The scale of the figure's vertical axis is set by $F_s = 4.0$ N. The velocity of the body at $x = 0$ is 4.0 m/s. (a) What is the kinetic energy of the body at $x = 3.0$ m? (b) At what value of $x$ will the body have a kinetic energy of 8.0 J? (c) What is the maximum kinetic energy of the body between $x = 0$ and $x = 5.0$ m?

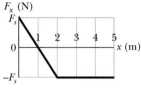

**FIG. 7-46** Problem 66.

**67** Figure 7-47 shows a cold package of hot dogs sliding rightward across a frictionless floor through a distance $d = 20.0$ cm while three forces act on the package. Two of them are horizontal and have the magnitudes $F_1 = 5.00$ N and $F_2 = 1.00$ N; the third is angled down by $\theta = 60.0°$ and has the magnitude $F_3 = 4.00$ N. (a) For the 20.0 cm displacement, what is the *net* work done on the package by the three applied forces, the gravitational force on the package, and the normal force on the package? (b) If the package has a mass of 2.0 kg and an initial kinetic energy of 0, what is its speed at the end of the displacement?

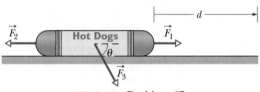

**FIG. 7-47** Problem 67.

**68** A frightened child is restrained by her mother as the child slides down a frictionless playground slide. If the force on the child from the mother is 100 N up the slide, the child's kinetic energy increases by 30 J as she moves down the slide a distance of 1.8 m. (a) How much work is done on the child by the gravitational force during the 1.8 m descent? (b) If the child is not restrained by her mother, how much will the child's kinetic energy increase as she comes down the slide that same distance of 1.8 m?

**69** To push a 25.0 kg crate up a frictionless incline, angled at 25.0° to the horizontal, a worker exerts a force of 209 N parallel to the incline. As the crate slides 1.50 m, how much work is done on the crate by (a) the worker's applied force, (b) the gravitational force on the crate, and (c) the normal force exerted by the incline on the crate? (d) What is the total work done on the crate? **SSM**

**70** If a car of mass 1200 kg is moving along a highway at 120 km/h, what is the car's kinetic energy as determined by someone standing alongside the highway?

**71** A spring with a pointer attached is hanging next to a scale marked in millimeters. Three different packages are hung from the spring, in turn, as shown in Fig. 7-48. (a) Which mark on the scale will the pointer indicate when no package is hung from the spring? (b) What is the weight $W$ of the third package? **SSM**

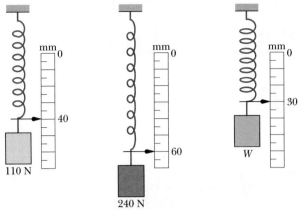

FIG. 7-48   Problem 71.

**72** A particle moves along a straight path through displacement $\vec{d} = (8\text{ m})\hat{i} + c\hat{j}$ while force $\vec{F} = (2\text{ N})\hat{i} - (4\text{ N})\hat{j}$ acts on it. (Other forces also act on the particle.) What is the value of $c$ if the work done by $\vec{F}$ on the particle is (a) zero, (b) positive, and (c) negative?

**73** An elevator cab has a mass of 4500 kg and can carry a maximum load of 1800 kg. If the cab is moving upward at full load at 3.80 m/s, what power is required of the force moving the cab to maintain that speed?   SSM

**74** A 45 kg block of ice slides down a frictionless incline 1.5 m long and 0.91 m high. A worker pushes up against the ice, parallel to the incline, so that the block slides down at constant speed. (a) Find the magnitude of the worker's force. How much work is done on the block by (b) the worker's force, (c) the gravitational force on the block, (d) the normal force on the block from the surface of the incline, and (e) the net force on the block?

**75** A force $\vec{F}$ in the positive direction of an $x$ axis acts on an object moving along the axis. If the magnitude of the force is $F = 10e^{-x/2.0}$ N, with $x$ in meters, find the work done by $\vec{F}$ as the object moves from $x = 0$ to $x = 2.0$ m by (a) plotting $F(x)$ and estimating the area under the curve and (b) integrating to find the work analytically.

**76** In Fig. 7-49a, a 2.0 N force is applied to a 4.0 kg block at a downward angle $\theta$ as the block moves rightward through 1.0 m across a frictionless floor. Find an expression for the speed $v_f$ of the block at the end of that distance if the block's initial velocity is (a) 0 and (b) 1.0 m/s to the right. (c) The situation in Fig. 7-49b is similar in that the block is initially moving at 1.0 m/s to the right, but now the 2.0 N force is directed downward to the left. Find an expression for the speed $v_f$ of the block at the end of the 1.0 m distance. (d) Graph all three expressions for $v_f$ versus downward angle $\theta$ for $\theta = 0°$ to $\theta = 90°$. Interpret the graphs.

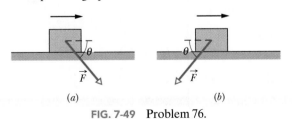

FIG. 7-49   Problem 76.

**77** A 2.0 kg lunchbox is sent sliding over a frictionless surface, in the positive direction of an $x$ axis along the surface. Beginning at time $t = 0$, a steady wind pushes on the lunchbox in the negative direction of the $x$ axis. Figure 7-50 shows the position $x$ of the lunchbox as a function of time $t$ as the wind pushes on the lunchbox. From the graph, estimate the kinetic energy of the lunchbox at (a) $t = 1.0$ s and (b) $t = 5.0$ s. (c) How much work does the force from the wind do on the lunchbox from $t = 1.0$ s to $t = 5.0$ s?   SSM

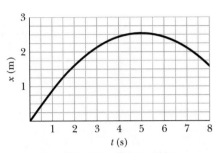

FIG. 7-50   Problem 77.

**78** *Numerical integration.* A breadbox is made to move along an $x$ axis from $x = 0.15$ m to $x = 1.20$ m by a force with a magnitude given by $F = \exp(-2x^2)$, with $x$ in meters and $F$ in newtons. (Here exp is the exponential function.) How much work is done on the breadbox by the force?

**79** As a particle moves along an $x$ axis, a force in the positive direction of the axis acts on it. Figure 7-51 shows the magnitude $F$ of the force versus position $x$ of the particle. The curve is given by $F = a/x^2$, with $a = 9.0$ N·m$^2$. Find the work done on the particle by the force as the particle moves from $x = 1.0$ m to $x = 3.0$ m by (a) estimating the work from the graph and (b) integrating the force function.

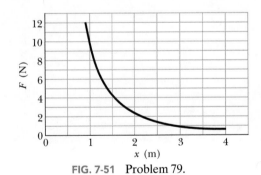

FIG. 7-51   Problem 79.

**80** A CD case slides along a floor in the positive direction of an $x$ axis while an applied force $\vec{F}_a$ acts on the case. The force is directed along the $x$ axis and has the $x$ component $F_{ax} = 9x - 3x^2$, with $x$ in meters and $F_{ax}$ in newtons. The case starts at rest at the position $x = 0$, and it moves until it is again at rest. (a) Plot the work $\vec{F}_a$ does on the case as a function of $x$. (b) At what position is the work maximum, and (c) what is that maximum value? (d) At what position has the work decreased to zero? (e) At what position is the case again at rest?

# 8 Potential Energy and Conservation of Energy

Courtesy Mark Reid, USGS

When an avalanche of rocks moves down a mountainside and into a valley, friction between the moving rocks and the land slow and eventually stop the flow. The runout (the distance the rocks can move across a valley) is typically 2/3 of the height from which the rocks descended. However, in large avalanches, where a large amount of material comes down a mountainside, the runout can be 30 times as much, which can take a seemingly safe community by fatal surprise.

## Why can large avalanches have such long runouts?

The answer is in this chapter.

## 8-1 | WHAT IS PHYSICS?

One job of physics is to identify the different types of energy in the world, especially those that are of common importance. One general type of energy is **potential energy** $U$. Technically, potential energy is energy that can be associated with the configuration (arrangement) of a system of objects that exert forces on one another.

This is a pretty formal definition of something that is actually familiar to you. An example might help better than the definition: A bungee-cord jumper plunges from a staging platform (Fig. 8-1). The system of objects consists of Earth and the jumper. The force between the objects is the gravitational force. The configuration of the system changes (the separation between the jumper and Earth decreases—that is, of course, the thrill of the jump). We can account for the jumper's motion and increase in kinetic energy by defining a **gravitational potential energy** $U$. This is the energy associated with the state of separation between two objects that attract each other by the gravitational force, here the jumper and Earth.

When the jumper begins to stretch the bungee cord near the end of the plunge, the system of objects consists of the cord and the jumper. The force between the objects is an elastic (spring-like) force. The configuration of the system changes (the cord stretches). We can account for the jumper's decrease in kinetic energy and the cord's increase in length by defining an **elastic potential energy** $U$. This is the energy associated with the state of compression or extension of an elastic object, here the bungee cord.

Physics determines how the potential energy of a system can be calculated so that energy might be stored or put to use. For example, before any particular bungee-cord jumper takes the plunge, someone (probably a mechanical engineer) must determine the correct cord to be used by calculating the gravitational and elastic potential energies that can be expected. Then the jump is only thrilling and not fatal.

FIG. 8-1    The kinetic energy of a bungee-cord jumper increases during the free fall, and then the cord begins to stretch, slowing the jumper. *(KOFUJIWARA/amana images/ Getty Images News and Sport Services)*

## 8-2 | Work and Potential Energy

In Chapter 7 we discussed the relation between work and a change in kinetic energy. Here we discuss the relation between work and a change in potential energy.

Let us throw a tomato upward (Fig. 8-2). We already know that as the tomato rises, the work $W_g$ done on the tomato by the gravitational force is negative because the force transfers energy *from* the kinetic energy of the tomato. We can now finish the story by saying that this energy is transferred by the gravitational force *to* the gravitational potential energy of the tomato–Earth system.

The tomato slows, stops, and then begins to fall back down because of the gravitational force. During the fall, the transfer is reversed: The work $W_g$ done on the tomato by the gravitational force is now positive—that force transfers energy *from* the gravitational potential energy of the tomato–Earth system *to* the kinetic energy of the tomato.

For either rise or fall, the change $\Delta U$ in gravitational potential energy is defined as being equal to the negative of the work done on the tomato by the gravitational force. Using the general symbol $W$ for work, we write this as

$$\Delta U = -W. \tag{8-1}$$

This equation also applies to a block–spring system, as in Fig. 8-3. If we abruptly shove the block to send it moving rightward, the spring force acts leftward and thus does negative work on the block, transferring energy from the kinetic energy of the block to the elastic potential energy of the spring–block

Negative work done by the gravitational force

Positive work done by the gravitational force

FIG. 8-2    A tomato is thrown upward. As it rises, the gravitational force does negative work on it, decreasing its kinetic energy. As the tomato descends, the gravitational force does positive work on it, increasing its kinetic energy.

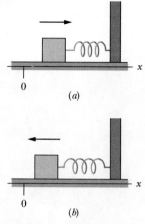

**FIG. 8-3** A block, attached to a spring and initially at rest at $x = 0$, is set in motion toward the right. (*a*) As the block moves rightward (as indicated by the arrow), the spring force does negative work on it. (*b*) Then, as the block moves back toward $x = 0$, the spring force does positive work on it.

system. The block slows and eventually stops, and then begins to move leftward because the spring force is still leftward. The transfer of energy is then reversed—it is from potential energy of the spring–block system to kinetic energy of the block.

## Conservative and Nonconservative Forces

Let us list the key elements of the two situations we just discussed:

1. The *system* consists of two or more objects.

2. A *force* acts between a particle-like object (tomato or block) in the system and the rest of the system.

3. When the system configuration changes, the force does *work* (call it $W_1$) on the particle-like object, transferring energy between the kinetic energy $K$ of the object and some other type of energy of the system.

4. When the configuration change is reversed, the force reverses the energy transfer, doing work $W_2$ in the process.

In a situation in which $W_1 = -W_2$ is always true, the other type of energy is a potential energy and the force is said to be a **conservative force.** As you might suspect, the gravitational force and the spring force are both conservative (since otherwise we could not have spoken of gravitational potential energy and elastic potential energy, as we did previously).

A force that is not conservative is called a **nonconservative force.** The kinetic frictional force and drag force are nonconservative. For an example, let us send a block sliding across a floor that is not frictionless. During the sliding, a kinetic frictional force from the floor slows the block by transferring energy from its kinetic energy to a type of energy called *thermal energy* (which has to do with the random motions of atoms and molecules). We know from experiment that this energy transfer cannot be reversed (thermal energy cannot be transferred back to kinetic energy of the block by the kinetic frictional force). Thus, although we have a system (made up of the block and the floor), a force that acts between parts of the system, and a transfer of energy by the force, the force is not conservative. Therefore, thermal energy is not a potential energy.

*When only conservative forces act on a particle-like object, we can greatly simplify otherwise difficult problems involving motion of the object.* The next section, in which we develop a test for identifying conservative forces, provides one means for simplifying such problems.

# 8-3 | Path Independence of Conservative Forces

The primary test for determining whether a force is conservative or nonconservative is this: Let the force act on a particle that moves along any *closed path,* beginning at some initial position and eventually returning to that position (so that the particle makes a *round trip* beginning and ending at the initial position). The force is conservative only if the total energy it transfers to and from the particle during the round trip along this and any other closed path is zero. In other words:

☞ The net work done by a conservative force on a particle moving around any closed path is zero.

We know from experiment that the gravitational force passes this *closed-path test.* An example is the tossed tomato of Fig. 8-2. The tomato leaves the launch point with speed $v_0$ and kinetic energy $\frac{1}{2}mv_0^2$. The gravitational force acting on the tomato slows it, stops it, and then causes it to fall back down. When the tomato returns to the launch point, it again has speed $v_0$ and kinetic energy $\frac{1}{2}mv_0^2$. Thus, the gravitational force transfers as much energy *from* the tomato during the ascent as it transfers *to* the tomato during the descent back to the

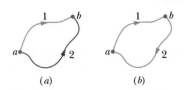

launch point. The net work done on the tomato by the gravitational force during the round trip is zero.

An important result of the closed-path test is that:

> The work done by a conservative force on a particle moving between two points does not depend on the path taken by the particle.

For example, suppose that a particle moves from point $a$ to point $b$ in Fig. 8-4$a$ along either path 1 or path 2. If only a conservative force acts on the particle, then the work done on the particle is the same along the two paths. In symbols, we can write this result as

$$W_{ab,1} = W_{ab,2}, \tag{8-2}$$

**FIG. 8-4** (*a*) As a conservative force acts on it, a particle can move from point *a* to point *b* along either path 1 or path 2. (*b*) The particle moves in a round trip, from point *a* to point *b* along path 1 and then back to point *a* along path 2.

where the subscript $ab$ indicates the initial and final points, respectively, and the subscripts 1 and 2 indicate the path.

This result is powerful because it allows us to simplify difficult problems when only a conservative force is involved. Suppose you need to calculate the work done by a conservative force along a given path between two points, and the calculation is difficult or even impossible without additional information. You can find the work by substituting some other path between those two points for which the calculation is easier and possible. Sample Problem 8-1 gives an example, but first we need to prove Eq. 8-2.

## Proof of Equation 8-2

Figure 8-4$b$ shows an arbitrary round trip for a particle that is acted upon by a single force. The particle moves from an initial point $a$ to point $b$ along path 1 and then back to point $a$ along path 2. The force does work on the particle as the particle moves along each path. Without worrying about where positive work is done and where negative work is done, let us just represent the work done from $a$ to $b$ along path 1 as $W_{ab,1}$ and the work done from $b$ back to $a$ along path 2 as $W_{ba,2}$. If the force is conservative, then the net work done during the round trip must be zero:

$$W_{ab,1} + W_{ba,2} = 0,$$

and thus

$$W_{ab,1} = -W_{ba,2}. \tag{8-3}$$

In words, the work done along the outward path must be the negative of the work done along the path back.

Let us now consider the work $W_{ab,2}$ done on the particle by the force when the particle moves from $a$ to $b$ along path 2, as indicated in Fig. 8-4$a$. If the force is conservative, that work is the negative of $W_{ba,2}$:

$$W_{ab,2} = -W_{ba,2}. \tag{8-4}$$

Substituting $W_{ab,2}$ for $-W_{ba,2}$ in Eq. 8-3, we obtain

$$W_{ab,1} = W_{ab,2},$$

which is what we set out to prove.

✓**CHECKPOINT 1**    The figure shows three paths connecting points $a$ and $b$. A single force $\vec{F}$ does the indicated work on a particle moving along each path in the indicated direction. On the basis of this information, is force $\vec{F}$ conservative?

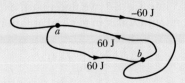

Figure 8-5a shows a 2.0 kg block of slippery cheese that slides along a frictionless track from point $a$ to point $b$. The cheese travels through a total distance of 2.0 m along the track, and a net vertical distance of 0.80 m. How much work is done on the cheese by the gravitational force during the slide?

**KEY IDEAS** (1) We *cannot* calculate the work by using Eq. 7-12 ($W_g = mgd \cos \phi$). The reason is that the angle $\phi$ between the directions of the gravitational force $\vec{F_g}$ and the displacement $\vec{d}$ varies along the track in an unknown way. (Even if we did know the shape of the track and could calculate $\phi$ along it, the calculation could be very difficult.) (2) Because $\vec{F_g}$ is a conservative force, we can find the work by choosing some other path between $a$ and $b$—one that makes the calculation easy.

**Calculations:** Let us choose the dashed path in Fig. 8-5b; it consists of two straight segments. Along the horizontal segment, the angle $\phi$ is a constant 90°. Even though we do not know the displacement along that horizontal segment, Eq. 7-12 tells us that the work $W_h$ done there is

$$W_h = mgd \cos 90° = 0.$$

Along the vertical segment, the displacement $d$ is 0.80 m and, with $\vec{F_g}$ and $\vec{d}$ both downward, the angle $\phi$ is a con-

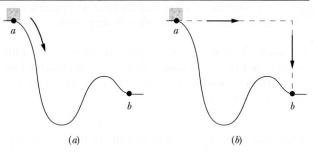

(a)           (b)

**FIG. 8-5** (a) A block of cheese slides along a frictionless track from point $a$ to point $b$. (b) Finding the work done on the cheese by the gravitational force is easier along the dashed path than along the actual path taken by the cheese; the result is the same for both paths.

stant 0°. Thus, Eq. 7-12 gives us, for the work $W_v$ done along the vertical part of the dashed path,

$$W_v = mgd \cos 0°$$
$$= (2.0 \text{ kg})(9.8 \text{ m/s}^2)(0.80 \text{ m})(1) = 15.7 \text{ J}.$$

The total work done on the cheese by $\vec{F_g}$ as the cheese moves from point $a$ to point $b$ along the dashed path is then

$$W = W_h + W_v = 0 + 15.7 \text{ J} \approx 16 \text{ J}. \quad \text{(Answer)}$$

This is also the work done as the cheese slides along the track from $a$ to $b$.

## 8-4 | Determining Potential Energy Values

Here we find equations that give the value of the two types of potential energy discussed in this chapter: gravitational potential energy and elastic potential energy. However, first we must find a general relation between a conservative force and the associated potential energy.

Consider a particle-like object that is part of a system in which a conservative force $\vec{F}$ acts. When that force does work $W$ on the object, the change $\Delta U$ in the potential energy associated with the system is the negative of the work done. We wrote this fact as Eq. 8-1 ($\Delta U = -W$). For the most general case, in which the force may vary with position, we may write the work $W$ as in Eq. 7-32:

$$W = \int_{x_i}^{x_f} F(x) \, dx. \tag{8-5}$$

This equation gives the work done by the force when the object moves from point $x_i$ to point $x_f$, changing the configuration of the system. (Because the force is conservative, the work is the same for all paths between those two points.)

Substituting Eq. 8-5 into Eq. 8-1, we find that the change in potential energy due to the change in configuration is, in general notation,

$$\Delta U = -\int_{x_i}^{x_f} F(x) \, dx. \tag{8-6}$$

## Gravitational Potential Energy

We first consider a particle with mass $m$ moving vertically along a $y$ axis (the positive direction is upward). As the particle moves from point $y_i$ to point $y_f$, the gravitational force $\vec{F}_g$ does work on it. To find the corresponding change in the gravitational potential energy of the particle–Earth system, we use Eq. 8-6 with two changes: (1) We integrate along the $y$ axis instead of the $x$ axis, because the gravitational force acts vertically. (2) We substitute $-mg$ for the force symbol $F$, because $\vec{F}_g$ has the magnitude $mg$ and is directed down the $y$ axis. We then have

$$\Delta U = -\int_{y_i}^{y_f} (-mg)\, dy = mg \int_{y_i}^{y_f} dy = mg \Big[ y \Big]_{y_i}^{y_f},$$

which yields

$$\Delta U = mg(y_f - y_i) = mg\, \Delta y. \qquad (8\text{-}7)$$

Only *changes* $\Delta U$ in gravitational potential energy (or any other type of potential energy) are physically meaningful. However, to simplify a calculation or a discussion, we sometimes would like to say that a certain gravitational potential value $U$ is associated with a certain particle–Earth system when the particle is at a certain height $y$. To do so, we rewrite Eq. 8-7 as

$$U - U_i = mg(y - y_i). \qquad (8\text{-}8)$$

Then we take $U_i$ to be the gravitational potential energy of the system when it is in a **reference configuration** in which the particle is at a **reference point** $y_i$. Usually we take $U_i = 0$ and $y_i = 0$. Doing this changes Eq. 8-8 to

$$U(y) = mgy \qquad \text{(gravitational potential energy).} \qquad (8\text{-}9)$$

This equation tells us:

> The gravitational potential energy associated with a particle–Earth system depends only on the vertical position $y$ (or height) of the particle relative to the reference position $y = 0$, not on the horizontal position.

## Elastic Potential Energy

We next consider the block–spring system shown in Fig. 8-3, with the block moving on the end of a spring of spring constant $k$. As the block moves from point $x_i$ to point $x_f$, the spring force $F_x = -kx$ does work on the block. To find the corresponding change in the elastic potential energy of the block–spring system, we substitute $-kx$ for $F(x)$ in Eq. 8-6. We then have

$$\Delta U = -\int_{x_i}^{x_f} (-kx)\, dx = k \int_{x_i}^{x_f} x\, dx = \tfrac{1}{2}k \Big[ x^2 \Big]_{x_i}^{x_f},$$

or

$$\Delta U = \tfrac{1}{2}kx_f^2 - \tfrac{1}{2}kx_i^2. \qquad (8\text{-}10)$$

To associate a potential energy value $U$ with the block at position $x$, we choose the reference configuration to be when the spring is at its relaxed length and the block is at $x_i = 0$. Then the elastic potential energy $U_i$ is 0, and Eq. 8-10 becomes

$$U - 0 = \tfrac{1}{2}kx^2 - 0,$$

which gives us

$$U(x) = \tfrac{1}{2}kx^2 \qquad \text{(elastic potential energy).} \qquad (8\text{-}11)$$

**✓CHECKPOINT 2**     A particle is to move along an $x$ axis from $x = 0$ to $x_1$ while a conservative force, directed along the $x$ axis, acts on the particle. The figure shows three situations in which the $x$ component of that force varies with $x$. The force has the same maximum magnitude $F_1$ in all three situations. Rank the situations according to the change in the associated potential energy during the particle's motion, most positive first.

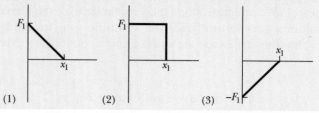

(1)          (2)          (3)

---

**PROBLEM-SOLVING TACTICS**

**Tactic 1: Using the Term "Potential Energy"** A potential energy is associated with a system as a whole. However, you might see statements that associate it with only part of the system. For example, you might read, "An apple hanging in a tree has a gravitational potential energy of 30 J." Such statements are often acceptable, but you should always keep in mind that the potential energy is actually associated with a system—here the apple–Earth system. Also keep in mind that assigning a particular potential energy value, such as 30 J here, to an object or even a system makes sense *only* if the reference potential energy value is known, as explored in Sample Problem 8-2.

---

## Sample Problem   8-2

A 2.0 kg sloth hangs 5.0 m above the ground (Fig. 8-6).

(a) What is the gravitational potential energy $U$ of the sloth–Earth system if we take the reference point $y = 0$ to be (1) at the ground, (2) at a balcony floor that is 3.0 m above the ground, (3) at the limb, and (4) 1.0 m above the limb? Take the gravitational potential energy to be zero at $y = 0$.

**KEY IDEA**    Once we have chosen the reference point for $y = 0$, we can calculate the gravitational potential energy $U$ of the system *relative to that reference point* with Eq. 8-9.

**Calculations:** For choice (1) the sloth is at $y = 5.0$ m, and

$$U = mgy = (2.0\text{ kg})(9.8\text{ m/s}^2)(5.0\text{ m})$$
$$= 98\text{ J}. \hspace{2cm} \text{(Answer)}$$

For the other choices, the values of $U$ are

(2)   $U = mgy = mg(2.0\text{ m}) = 39\text{ J}$,
(3)   $U = mgy = mg(0) = 0\text{ J}$,
(4)   $U = mgy = mg(-1.0\text{ m})$
$$= -19.6\text{ J} \approx -20\text{ J}. \hspace{1cm} \text{(Answer)}$$

(b) The sloth drops to the ground. For each choice of reference point, what is the change $\Delta U$ in the potential energy of the sloth–Earth system due to the fall?

**KEY IDEA**    The *change* in potential energy does not depend on the choice of the reference point for $y = 0$; instead, it depends on the change in height $\Delta y$.

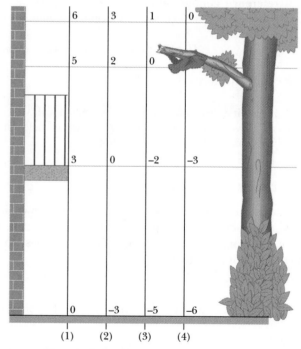

**FIG. 8-6**   Four choices of reference point $y = 0$. Each $y$ axis is marked in units of meters. The choice affects the value of the potential energy $U$ of the sloth–Earth system. However, it does not affect the change $\Delta U$ in potential energy of the system if the sloth moves by, say, falling.

**Calculation:** For all four situations, we have the same $\Delta y = -5.0$ m. Thus, for (1) to (4), Eq. 8-7 tells us that

$$\Delta U = mg\,\Delta y = (2.0\text{ kg})(9.8\text{ m/s}^2)(-5.0\text{ m})$$
$$= -98\text{ J}. \hspace{2cm} \text{(Answer)}$$

# 8-5 | Conservation of Mechanical Energy

The **mechanical energy** $E_{mec}$ of a system is the sum of its potential energy $U$ and the kinetic energy $K$ of the objects within it:

$$E_{mec} = K + U \quad \text{(mechanical energy)}. \tag{8-12}$$

In this section, we examine what happens to this mechanical energy when only conservative forces cause energy transfers within the system—that is, when frictional and drag forces do not act on the objects in the system. Also, we shall assume that the system is *isolated* from its environment; that is, no *external force* from an object outside the system causes energy changes inside the system.

When a conservative force does work $W$ on an object within the system, that force transfers energy between kinetic energy $K$ of the object and potential energy $U$ of the system. From Eq. 7-10, the change $\Delta K$ in kinetic energy is

$$\Delta K = W \tag{8-13}$$

and from Eq. 8-1, the change $\Delta U$ in potential energy is

$$\Delta U = -W. \tag{8-14}$$

Combining Eqs. 8-13 and 8-14, we find that

$$\Delta K = -\Delta U. \tag{8-15}$$

In words, one of these energies increases exactly as much as the other decreases. We can rewrite Eq. 8-15 as

$$K_2 - K_1 = -(U_2 - U_1), \tag{8-16}$$

where the subscripts refer to two different instants and thus to two different arrangements of the objects in the system. Rearranging Eq. 8-16 yields

$$K_2 + U_2 = K_1 + U_1 \quad \text{(conservation of mechanical energy)}. \tag{8-17}$$

In words, this equation says:

$$\begin{pmatrix} \text{the sum of } K \text{ and } U \text{ for} \\ \text{any state of a system} \end{pmatrix} = \begin{pmatrix} \text{the sum of } K \text{ and } U \text{ for} \\ \text{any other state of the system} \end{pmatrix},$$

when the system is isolated and only conservative forces act on the objects in the system. In other words:

> In an isolated system where only conservative forces cause energy changes, the kinetic energy and potential energy can change, but their sum, the mechanical energy $E_{mec}$ of the system, cannot change.

This result is called the **principle of conservation of mechanical energy.** (Now you can see where *conservative* forces got their name.) With the aid of Eq. 8-15, we can write this principle in one more form, as

$$\Delta E_{mec} = \Delta K + \Delta U = 0. \tag{8-18}$$

The principle of conservation of mechanical energy allows us to solve problems that would be quite difficult to solve using only Newton's laws:

> When the mechanical energy of a system is conserved, we can relate the sum of kinetic energy and potential energy at one instant to that at another instant *without considering the intermediate motion* and *without finding the work done by the forces involved.*

Figure 8-7 shows an example in which the principle of conservation of mechanical energy can be applied: As a pendulum swings, the energy of the

In olden days, a person would be tossed via a blanket to be able to see farther over the flat terrain. Nowadays, it is done just for fun. During the ascent of the person in the photograph, energy is transferred from kinetic energy to gravitational potential energy. The maximum height is reached when that transfer is complete. Then the transfer is reversed during the fall. (©AP/Wide World Photos)

$v = +v_{max}$

(a)

(b)

$v = 0$

(h)

$v = 0$

(g)

(c)

**FIG. 8-7** A pendulum, with its mass concentrated in a bob at the lower end, swings back and forth. One full cycle of the motion is shown. During the cycle the values of the potential and kinetic energies of the pendulum–Earth system vary as the bob rises and falls, but the mechanical energy $E_{mec}$ of the system remains constant. The energy $E_{mec}$ can be described as continuously shifting between the kinetic and potential forms. In stages $(a)$ and $(e)$, all the energy is kinetic energy. The bob then has its greatest speed and is at its lowest point. In stages $(c)$ and $(g)$, all the energy is potential energy. The bob then has zero speed and is at its highest point. In stages $(b)$, $(d)$, $(f)$, and $(h)$, half the energy is kinetic energy and half is potential energy. If the swinging involved a frictional force at the point where the pendulum is attached to the ceiling, or a drag force due to the air, then $E_{mec}$ would not be conserved, and eventually the pendulum would stop.

(f)

$v = -v_{max}$

(d)

(e)

pendulum–Earth system is transferred back and forth between kinetic energy $K$ and gravitational potential energy $U$, with the sum $K + U$ being constant. If we know the gravitational potential energy when the pendulum bob is at its highest point (Fig. 8-7c), Eq. 8-17 gives us the kinetic energy of the bob at the lowest point (Fig. 8-7e).

For example, let us choose the lowest point as the reference point, with the gravitational potential energy $U_2 = 0$. Suppose then that the potential energy at the highest point is $U_1 = 20$ J relative to the reference point. Because the bob momentarily stops at its highest point, the kinetic energy there is $K_1 = 0$. Putting these values into Eq. 8-17 gives us the kinetic energy $K_2$ at the lowest point:

$$K_2 + 0 = 0 + 20 \text{ J} \quad \text{or} \quad K_2 = 20 \text{ J}.$$

Note that we get this result without considering the motion between the highest and lowest points (such as in Fig. 8-7d) and without finding the work done by any forces involved in the motion.

✓**CHECKPOINT 3** The figure shows four situations—one in which an initially stationary block is dropped and three in which the block is allowed to slide down frictionless ramps. (a) Rank the situations according to the kinetic energy of the block at point $B$, greatest first. (b) Rank them according to the speed of the block at point $B$, greatest first.

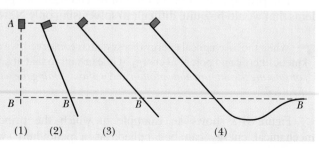

In Fig. 8-8, a child of mass $m$ is released from rest at the top of a water slide, at height $h = 8.5$ m above the bottom of the slide. Assuming that the slide is frictionless because of the water on it, find the child's speed at the bottom of the slide.

**KEY IDEAS** (1) We cannot find her speed at the bottom by using her acceleration along the slide as we might have in earlier chapters because we do not know the slope (angle) of the slide. However, because that speed is related to her kinetic energy, perhaps we can use the principle of conservation of mechanical energy to get the speed. Then we would not need to know the slope. (2) Mechanical energy is conserved in a system *if* the system is isolated and *if* only conservative forces cause energy transfers within it. Let's check.

*Forces:* Two forces act on the child. The *gravitational force,* a conservative force, does work on her. The *normal force* on her from the slide does no work because its direction at any point during the descent is always perpendicular to the direction in which the child moves.

*System:* Because the only force doing work on the child is the gravitational force, we choose the child–Earth system as our system, which we can take to be isolated.

Thus, we have only a conservative force doing work in an isolated system, so we *can* use the principle of conservation of mechanical energy.

**Calculations:** Let the mechanical energy be $E_{\text{mec},t}$ when the child is at the top of the slide and $E_{\text{mec},b}$ when she is at the bottom. Then the conservation principle tells us

$$E_{\text{mec},b} = E_{\text{mec},t}. \qquad (8\text{-}19)$$

FIG. 8-8    A child slides down a water slide as she descends a height $h$.

To show both kinds of mechanical energy, we have

$$K_b + U_b = K_t + U_t, \qquad (8\text{-}20)$$

or     $\frac{1}{2}mv_b^2 + mgy_b = \frac{1}{2}mv_t^2 + mgy_t.$

Dividing by $m$ and rearranging yield

$$v_b^2 = v_t^2 + 2g(y_t - y_b).$$

Putting $v_t = 0$ and $y_t - y_b = h$ leads to

$$v_b = \sqrt{2gh} = \sqrt{(2)(9.8 \text{ m/s}^2)(8.5 \text{ m})}$$
$$= 13 \text{ m/s}. \qquad \text{(Answer)}$$

This is the same speed that the child would reach if she fell 8.5 m vertically. On an actual slide, some frictional forces would act and the child would not be moving quite so fast.

**Comments:** Although this problem is hard to solve directly with Newton's laws, using conservation of mechanical energy makes the solution much easier. However, if we were asked to find the time taken for the child to reach the bottom of the slide, energy methods would be of no use; we would need to know the shape of the slide, and we would have a difficult problem.

**PROBLEM-SOLVING TACTICS**

*Tactic 2:* *Conservation of Mechanical Energy* Asking the following questions will help you to solve problems involving the principle of conservation of mechanical energy.

*For what system is mechanical energy conserved?* You should be able to separate your system from its environment. Imagine drawing a closed surface such that whatever is inside is your system and whatever is outside is the environment of that system.

*Is friction or drag present?* If friction or drag is present, mechanical energy is not conserved.

*Is your system isolated?* The principle of conservation of mechanical energy applies only to isolated systems. That means that no *external forces* (forces exerted by objects outside the system) should do work on the objects in the system.

*What are the initial and final states of your system?* The system changes from some initial configuration to some final configuration. You apply the principle of conservation of mechanical energy by saying that $E_{\text{mec}}$ has the same value in both these configurations. Be very clear about what these two configurations are.

# 8-6 | Reading a Potential Energy Curve

Once again we consider a particle that is part of a system in which a conservative force acts. This time suppose that the particle is constrained to move along an

$x$ axis while the conservative force does work on it. We can learn a lot about the motion of the particle from a plot of the system's potential energy $U(x)$. However, before we discuss such plots, we need one more relationship.

### Finding the Force Analytically

Equation 8-6 tells us how to find the change $\Delta U$ in potential energy between two points in a one-dimensional situation if we know the force $F(x)$. Now we want to go the other way; that is, we know the potential energy function $U(x)$ and want to find the force.

For one-dimensional motion, the work $W$ done by a force that acts on a particle as the particle moves through a distance $\Delta x$ is $F(x)\,\Delta x$. We can then write Eq. 8-1 as

$$\Delta U(x) = -W = -F(x)\,\Delta x. \tag{8-21}$$

Solving for $F(x)$ and passing to the differential limit yield

$$F(x) = -\frac{dU(x)}{dx} \qquad \text{(one-dimensional motion),} \tag{8-22}$$

which is the relation we sought.

We can check this result by putting $U(x) = \frac{1}{2}kx^2$, which is the elastic potential energy function for a spring force. Equation 8-22 then yields, as expected, $F(x) = -kx$, which is Hooke's law. Similarly, we can substitute $U(x) = mgx$, which is the gravitational potential energy function for a particle–Earth system, with a particle of mass $m$ at height $x$ above Earth's surface. Equation 8-22 then yields $F = -mg$, which is the gravitational force on the particle.

### The Potential Energy Curve

Figure 8-9a is a plot of a potential energy function $U(x)$ for a system in which a particle is in one-dimensional motion while a conservative force $F(x)$ does work

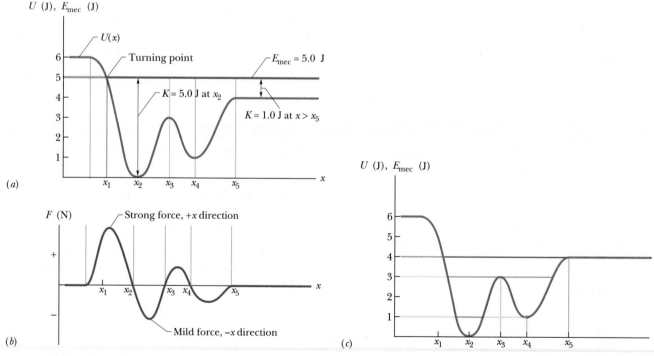

FIG. 8-9   (a) A plot of $U(x)$, the potential energy function of a system containing a particle confined to move along an $x$ axis. There is no friction, so mechanical energy is conserved. (b) A plot of the force $F(x)$ acting on the particle, derived from the potential energy plot by taking its slope at various points. (c) The $U(x)$ plot of (a) with three possible values of $E_{\text{mec}}$ shown.

on it. We can easily find $F(x)$ by (graphically) taking the slope of the $U(x)$ curve at various points. (Equation 8-22 tells us that $F(x)$ is the negative of the slope of the $U(x)$ curve.) Figure 8-9$b$ is a plot of $F(x)$ found in this way.

## Turning Points

In the absence of a nonconservative force, the mechanical energy $E$ of a system has a constant value given by

$$U(x) + K(x) = E_{mec}. \tag{8-23}$$

Here $K(x)$ is the *kinetic energy function* of a particle in the system (this $K(x)$ gives the kinetic energy as a function of the particle's location $x$). We may rewrite Eq. 8-23 as

$$K(x) = E_{mec} - U(x). \tag{8-24}$$

Suppose that $E_{mec}$ (which has a constant value, remember) happens to be 5.0 J. It would be represented in Fig. 8-9$a$ by a horizontal line that runs through the value 5.0 J on the energy axis. (It is, in fact, shown there.)

Equation 8-24 tells us how to determine the kinetic energy $K$ for any location $x$ of the particle: On the $U(x)$ curve, find $U$ for that location $x$ and then subtract $U$ from $E_{mec}$. For example, if the particle is at any point to the right of $x_5$, then $K = 1.0$ J. The value of $K$ is greatest (5.0 J) when the particle is at $x_2$ and least (0 J) when the particle is at $x_1$.

Since $K$ can never be negative (because $v^2$ is always positive), the particle can never move to the left of $x_1$, where $E_{mec} - U$ is negative. Instead, as the particle moves toward $x_1$ from $x_2$, $K$ decreases (the particle slows) until $K = 0$ at $x_1$ (the particle stops there).

Note that when the particle reaches $x_1$, the force on the particle, given by Eq. 8-22, is positive (because the slope $dU/dx$ is negative). This means that the particle does not remain at $x_1$ but instead begins to move to the right, opposite its earlier motion. Hence $x_1$ is a **turning point**, a place where $K = 0$ (because $U = E$) and the particle changes direction. There is no turning point (where $K = 0$) on the right side of the graph. When the particle heads to the right, it will continue indefinitely.

## Equilibrium Points

Figure 8-9$c$ shows three different values for $E_{mec}$ superposed on the plot of the potential energy function $U(x)$ of Fig. 8-9$a$. Let us see how they change the situation. If $E_{mec} = 4.0$ J (purple line), the turning point shifts from $x_1$ to a point between $x_1$ and $x_2$. Also, at any point to the right of $x_5$, the system's mechanical energy is equal to its potential energy; thus, the particle has no kinetic energy and (by Eq. 8-22) no force acts on it, and so it must be stationary. A particle at such a position is said to be in **neutral equilibrium.** (A marble placed on a horizontal tabletop is in that state.)

If $E_{mec} = 3.0$ J (pink line), there are two turning points: One is between $x_1$ and $x_2$, and the other is between $x_4$ and $x_5$. In addition, $x_3$ is a point at which $K = 0$. If the particle is located exactly there, the force on it is also zero, and the particle remains stationary. However, if it is displaced even slightly in either direction, a nonzero force pushes it farther in the same direction, and the particle continues to move. A particle at such a position is said to be in **unstable equilibrium.** (A marble balanced on top of a bowling ball is an example.)

Next consider the particle's behavior if $E_{mec} = 1.0$ J (green line). If we place it at $x_4$, it is stuck there. It cannot move left or right on its own because to do so would require a negative kinetic energy. If we push it slightly left or right, a restoring force appears that moves it back to $x_4$. A particle at such a position is said to be in **stable equilibrium.** (A marble placed at the bottom of a hemispherical bowl is an example.) If we place the particle in the cup-like *potential well* centered at $x_2$, it is between two turning points. It can still move somewhat, but only partway to $x_1$ or $x_3$.

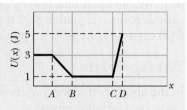

✔**CHECKPOINT 4** The figure gives the potential energy function $U(x)$ for a system in which a particle is in one-dimensional motion. (a) Rank regions $AB$, $BC$, and $CD$ according to the magnitude of the force on the particle, greatest first. (b) What is the direction of the force when the particle is in region $AB$?

## Sample Problem 8-4

A 2.00 kg particle moves along an $x$ axis in one-dimensional motion while a conservative force along that axis acts on it. The potential energy $U(x)$ associated with the force is plotted in Fig. 8-10$a$. That is, if the particle were placed at any position between $x = 0$ and $x = 7.00$ m, it would have the plotted value of $U$. At $x = 6.5$ m, the particle has velocity $v_0 = (-4.00$ m/s$)\hat{i}$.

(a) From Fig. 8-10$a$, determine the particle's speed at $x_1 = 4.5$ m.

**KEY IDEAS** (1) The particle's kinetic energy is given by Eq. 7-1 ($K = \frac{1}{2}mv^2$). (2) Because only a conservative force acts on the particle, the mechanical energy $E_{mec}(= K + U)$ is conserved as the particle moves. (3) Therefore, on a plot of $U(x)$ such as Fig. 8-10$a$, the kinetic energy is equal to the difference between $E_{mec}$ and $U$.

**Calculations:** At $x = 6.5$ m, the particle has kinetic energy

$$K_0 = \frac{1}{2}mv_0^2 = \frac{1}{2}(2.00 \text{ kg})(4.00 \text{ m/s})^2$$
$$= 16.0 \text{ J}.$$

Because the potential energy there is $U = 0$, the mechanical energy is

$$E_{mec} = K_0 + U_0 = 16.0 \text{ J} + 0 = 16.0 \text{ J}.$$

This value for $E_{mec}$ is plotted as a horizontal line in Fig. 8-10$a$. From that figure we see that at $x = 4.5$ m, the potential energy is $U_1 = 7.0$ J. The kinetic energy $K_1$ is the difference between $E_{mec}$ and $U_1$:

$$K_1 = E_{mec} - U_1 = 16.0 \text{ J} - 7.0 \text{ J} = 9.0 \text{ J}.$$

Because $K_1 = \frac{1}{2}mv_1^2$, we find

$$v_1 = 3.0 \text{ m/s}. \qquad \text{(Answer)}$$

(b) Where is the particle's turning point located?

**KEY IDEA** The turning point is where the force momentarily stops and then reverses the particle's motion. That is, it is where the particle momentarily has $v = 0$ and thus $K = 0$.

**Calculations:** Because $K$ is the difference between $E_{mec}$ and $U$, we want the point in Fig. 8-10$a$ where the plot of $U$ rises to meet the horizontal line of $E_{mec}$, as shown in Fig. 8-10$b$. Because the plot of $U$ is a straight line in Fig. 8-10$b$, we can draw nested right triangles as shown and then write the proportionality of distances

$$\frac{16 - 7.0}{d} = \frac{20 - 7.0}{4.0 - 1.0},$$

which gives us $d = 2.08$ m. Thus, the turning point is at

$$x = 4.0 \text{ m} - d = 1.9 \text{ m}. \qquad \text{(Answer)}$$

(c) Evaluate the force acting on the particle when it is in the region $1.9 \text{ m} < x < 4.0 \text{ m}$.

**KEY IDEA** The force is given by Eq. 8-22 ($F(x) = -dU(x)/dx$). The equation states that the force is equal to the negative of the slope on a graph of $U(x)$.

**Calculations:** For the graph of Fig. 8-10$b$, we see that for the range $1.0 \text{ m} < x < 4.0 \text{ m}$ the force is

$$F = -\frac{20 \text{ J} - 7.0 \text{ J}}{1.0 \text{ m} - 4.0 \text{ m}} = 4.3 \text{ N}. \qquad \text{(Answer)}$$

Thus, the force has magnitude 4.3 N and is in the positive direction of the $x$ axis. This result is consistent with the fact that the initially leftward-moving particle is stopped by the force and then sent rightward.

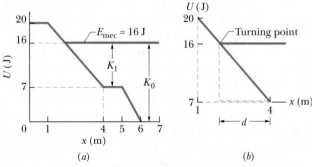

FIG. 8-10 ($a$) A plot of potential energy $U$ versus position $x$. ($b$) A section of the plot used to find where the particle turns around.

## 8-7 | Work Done on a System by an External Force

In Chapter 7, we defined work as being energy transferred to or from an object by means of a force acting on the object. We can now extend that definition to an external force acting on a system of objects.

> Work is energy transferred to or from a system by means of an external force acting on that system.

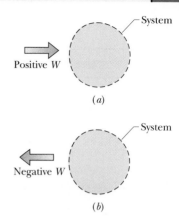

FIG. 8-11 (a) Positive work $W$ done on an arbitrary system means a transfer of energy to the system. (b) Negative work $W$ means a transfer of energy from the system.

Figure 8-11a represents positive work (a transfer of energy *to* a system), and Fig. 8-11b represents negative work (a transfer of energy *from* a system). When more than one force acts on a system, their *net work* is the energy transferred to or from the system.

These transfers are like transfers of money to and from a bank account. If a system consists of a single particle or particle-like object, as in Chapter 7, the work done on the system by a force can change only the kinetic energy of the system. The energy statement for such transfers is the work–kinetic energy theorem of Eq. 7-10 ($\Delta K = W$); that is, a single particle has only one energy account, called kinetic energy. External forces can transfer energy into or out of that account. If a system is more complicated, however, an external force can change other forms of energy (such as potential energy); that is, a more complicated system can have multiple energy accounts.

Let us find energy statements for such systems by examining two basic situations, one that does not involve friction and one that does.

### No Friction Involved

To compete in a bowling-ball-hurling contest, you first squat and cup your hands under the ball on the floor. Then you rapidly straighten up while also pulling your hands up sharply, launching the ball upward at about face level. During your upward motion, your applied force on the ball obviously does work; that is, it is an external force that transfers energy, but to what system?

To answer, we check to see which energies change. There is a change $\Delta K$ in the ball's kinetic energy and, because the ball and Earth become more separated, there is a change $\Delta U$ in the gravitational potential energy of the ball–Earth system. To include both changes, we need to consider the ball–Earth system. Then your force is an external force doing work on that system, and the work is

$$W = \Delta K + \Delta U, \tag{8-25}$$

or $\qquad W = \Delta E_{\text{mec}} \qquad$ (work done on system, no friction involved), $\qquad$ (8-26)

where $\Delta E_{\text{mec}}$ is the change in the mechanical energy of the system. These two equations, which are represented in Fig. 8-12, are equivalent energy statements for work done on a system by an external force when friction is not involved.

### Friction Involved

We next consider the example in Fig. 8-13a. A constant horizontal force $\vec{F}$ pulls a block along an $x$ axis and through a displacement of magnitude $d$, increasing the block's velocity from $\vec{v}_0$ to $\vec{v}$. During the motion, a constant kinetic frictional force $\vec{f}_k$ from the floor acts on the block. Let us first choose the block as our system and apply Newton's second law to it. We can write that law for components along the $x$ axis ($F_{\text{net}, x} = ma_x$) as

$$F - f_k = ma. \tag{8-27}$$

Because the forces are constant, the acceleration $\vec{a}$ is also constant. Thus, we can use Eq. 2-16 to write

$$v^2 = v_0^2 + 2ad.$$

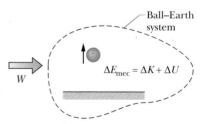

FIG. 8-12 Positive work $W$ is done on a system of a bowling ball and Earth, causing a change $\Delta E_{\text{mec}}$ in the mechanical energy of the system, a change $\Delta K$ in the ball's kinetic energy, and a change $\Delta U$ in the system's gravitational potential energy.

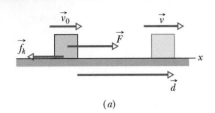

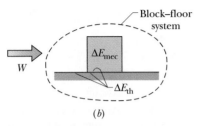

FIG. 8-13 (a) A block is pulled across a floor by force $\vec{F}$ while a kinetic frictional force $\vec{f}_k$ opposes the motion. The block has velocity $\vec{v}_0$ at the start of a displacement $\vec{d}$ and velocity $\vec{v}$ at the end of the displacement. (b) Positive work $W$ is done on the block–floor system by force $\vec{F}$, resulting in a change $\Delta E_{mec}$ in the block's mechanical energy and a change $\Delta E_{th}$ in the thermal energy of the block and floor.

Solving this equation for $a$, substituting the result into Eq. 8-27, and rearranging then give us

$$Fd = \tfrac{1}{2}mv^2 - \tfrac{1}{2}mv_0^2 + f_k d \tag{8-28}$$

or, because $\tfrac{1}{2}mv^2 - \tfrac{1}{2}mv_0^2 = \Delta K$ for the block,

$$Fd = \Delta K + f_k d. \tag{8-29}$$

In a more general situation (say, one in which the block is moving up a ramp), there can be a change in potential energy. To include such a possible change, we generalize Eq. 8-29 by writing

$$Fd = \Delta E_{mec} + f_k d. \tag{8-30}$$

By experiment we find that the block and the portion of the floor along which it slides become warmer as the block slides. As we shall discuss in Chapter 18, the temperature of an object is related to the object's thermal energy $E_{th}$ (the energy associated with the random motion of the atoms and molecules in the object). Here, the thermal energy of the block and floor increases because (1) there is friction between them and (2) there is sliding. Recall that friction is due to the cold-welding between two surfaces. As the block slides over the floor, the sliding causes repeated tearing and re-forming of the welds between block and the floor, which makes the block and floor warmer. Thus, the sliding increases their thermal energy $E_{th}$.

Through experiment, we find that the increase $\Delta E_{th}$ in thermal energy is equal to the product of the magnitudes $f_k$ and $d$:

$$\Delta E_{th} = f_k d \quad \text{(increase in thermal energy by sliding).} \tag{8-31}$$

Thus, we can rewrite Eq. 8-30 as

$$Fd = \Delta E_{mec} + \Delta E_{th}. \tag{8-32}$$

$Fd$ is the work $W$ done by the external force $\vec{F}$ (the energy transferred by the force), but on which system is the work done (where are the energy transfers made)? To answer, we check to see which energies change. The block's mechanical energy changes, and the thermal energies of the block and floor also change. Therefore, the work done by force $\vec{F}$ is done on the block–floor system. That work is

$$W = \Delta E_{mec} + \Delta E_{th} \quad \text{(work done on system, friction involved).} \tag{8-33}$$

This equation, which is represented in Fig. 8-13b, is the energy statement for the work done on a system by an external force when friction is involved.

✓ **CHECKPOINT 5** In three trials, a block is pushed by a horizontal applied force across a floor that is not frictionless, as in Fig. 8-13a. The magnitudes $F$ of the applied force and the results of the pushing on the block's speed are given in the table. In all three trials, the block is pushed through the same distance $d$. Rank the three trials according to the change in the thermal energy of the block and floor that occurs in that distance $d$, greatest first.

| Trial | $F$ | Result on Block's Speed |
|---|---|---|
| a | 5.0 N | decreases |
| b | 7.0 N | remains constant |
| c | 8.0 N | increases |

**Sample Problem** **8-5**

The prehistoric people of Easter Island carved hundreds of gigantic stone statues in a quarry and then moved them to sites all over the island (Fig. 8-14). How they managed to move the statues by as much as 10 km without the use

**FIG. 8-14** Easter Island stone statues. *(©LMR Group/Alamy Images)*

of sophisticated machines has been hotly debated. They most likely cradled each statue in a wooden sled and then pulled the sled over a "runway" consisting of almost identical logs acting as rollers. In a modern reenactment of this technique, 25 men were able to move a 9000 kg Easter Island–type statue 45 m over level ground in 2 min.

**(a)** Estimate the work the net force $\vec{F}$ from the men did during the 45 m displacement of the statue, and determine the system on which that force did the work.

**KEY IDEAS** (1) We can calculate the work done with Eq. 7-7 ($W = Fd \cos \phi$). (2) To determine the system on which the work is done we see which energies change.

**Calculations:** In Eq. 7-7, $d$ is the distance 45 m, $F$ is the magnitude of the net force on the statue from the 25 men, and $\phi = 0°$. Let us estimate that each man pulled with a force magnitude equal to twice his weight, which we take to be the same value $mg$ for all the men. Thus, the magnitude of the net force from the men was

$F = (25)(2mg) = 50mg$. Estimating a man's mass as 80 kg, we can then write Eq. 7-7 as

$$W = Fd \cos \phi = 50mgd \cos \phi$$
$$= (50)(80 \text{ kg})(9.8 \text{ m/s}^2)(45 \text{ m}) \cos 0°$$
$$= 1.8 \times 10^6 \text{ J} \approx 2 \text{ MJ}. \qquad \text{(Answer)}$$

Because the statue moved, there was certainly a change $\Delta K$ in its kinetic energy during the motion. We can easily guess that there must have been considerable kinetic friction between the sled, logs, and ground, resulting in a change $\Delta E_{th}$ in their thermal energies. Thus, the system on which the work was done consisted of the statue, sled, logs, and ground.

**(b)** What was the increase $\Delta E_{th}$ in the thermal energy of the system during the 45 m displacement?

**KEY IDEA** We can relate $\Delta E_{th}$ to the work $W$ done by $\vec{F}$ with the energy statement of Eq. 8-33 for a system that involves friction:

$$W = \Delta E_{mec} + \Delta E_{th}.$$

**Calculations:** We know the value of $W$ from (a). The change $\Delta E_{mec}$ in the statue's mechanical energy was zero because the statue was stationary at the beginning and the end of the move and its elevation did not change. Thus, we find

$$\Delta E_{th} = W = 1.8 \times 10^6 \text{ J} \approx 2 \text{ MJ}. \quad \text{(Answer)}$$

**(c)** Estimate the work that would have been done by the 25 men if they had moved the statue 10 km across level ground on Easter Island. Also estimate the total change $\Delta E_{th}$ that would have occurred in the statue–sled–logs–ground system.

**Calculation:** We calculate $W$ as in (a), but with $1 \times 10^4$ m substituted for $d$. Also, we again equate $\Delta E_{th}$ to $W$. We get

$$W = \Delta E_{th} = 3.9 \times 10^8 \text{ J} \approx 400 \text{ MJ}. \quad \text{(Answer)}$$

This would have been a staggering amount of energy for the men to have transferred during the movement of a statue. Still, the 25 men *could* have moved the statue 10 km without some mysterious energy source.

---

**Sample Problem** 8-6

A food shipper pushes a wood crate of cabbage heads (total mass $m = 14$ kg) across a concrete floor with a constant horizontal force $\vec{F}$ of magnitude 40 N. In a straight-line displacement of magnitude $d = 0.50$ m, the speed of the crate decreases from $v_0 = 0.60$ m/s to $v = 0.20$ m/s.

**(a)** How much work is done by force $\vec{F}$, and on what system does it do the work?

**KEY IDEA** Because the applied force $\vec{F}$ is constant,

we can calculate the work it does by using Eq. 7-7 ($W = Fd \cos \phi$).

**Calculation:** Substituting given data, including the fact that force $\vec{F}$ and displacement $\vec{d}$ are in the same direction, we find

$$W = Fd \cos \phi = (40 \text{ N})(0.50 \text{ m}) \cos 0°$$
$$= 20 \text{ J}. \qquad \text{(Answer)}$$

*Reasoning:* We can determine the system on which the work is done to see which energies change. Because the crate's speed changes, there is certainly a change $\Delta K$ in the crate's kinetic energy. Is there friction between the floor and the crate, and thus a change in thermal energy? Note that $\vec{F}$ and the crate's velocity have the same direction. Thus, if there is no friction, then $\vec{F}$ should be accelerating the crate to a *greater* speed. However, the crate is *slowing,* so there must be friction and a change $\Delta E_{th}$ in thermal energy of the crate and the floor. Therefore, the system on which the work is done is the crate–floor system, because both energy changes occur in that system.

(b) What is the increase $\Delta E_{th}$ in the thermal energy of the crate and floor?

**KEY IDEA** We can relate $\Delta E_{th}$ to the work $W$ done by $\vec{F}$ with the energy statement of Eq. 8-33 for a system that involves friction:

$$W = \Delta E_{mec} + \Delta E_{th}. \qquad (8\text{-}34)$$

*Calculations:* We know the value of $W$ from (a). The change $\Delta E_{mec}$ in the crate's mechanical energy is just the change in its kinetic energy because no potential energy changes occur, so we have

$$\Delta E_{mec} = \Delta K = \tfrac{1}{2}mv^2 - \tfrac{1}{2}mv_0^2.$$

Substituting this into Eq. 8-34 and solving for $\Delta E_{th}$, we find

$$\Delta E_{th} = W - (\tfrac{1}{2}mv^2 - \tfrac{1}{2}mv_0^2) = W - \tfrac{1}{2}m(v^2 - v_0^2)$$
$$= 20\ \text{J} - \tfrac{1}{2}(14\ \text{kg})[(0.20\ \text{m/s})^2 - (0.60\ \text{m/s})^2]$$
$$= 22.2\ \text{J} \approx 22\ \text{J}. \qquad \text{(Answer)}$$

## 8-8 | Conservation of Energy

We now have discussed several situations in which energy is transferred to or from objects and systems, much like money is transferred between accounts. In each situation we assume that the energy that was involved could always be accounted for; that is, energy could not magically appear or disappear. In more formal language, we assumed (correctly) that energy obeys a law called the **law of conservation of energy,** which is concerned with the **total energy** $E$ of a system. That total is the sum of the system's mechanical energy, thermal energy, and any type of *internal energy* in addition to thermal energy. (We have not yet discussed other types of internal energy.) The law states that

> The total energy $E$ of a system can change only by amounts of energy that are transferred to or from the system.

The only type of energy transfer that we have considered is work $W$ done on a system. Thus, for us at this point, this law states that

$$W = \Delta E = \Delta E_{mec} + \Delta E_{th} + \Delta E_{int}, \qquad (8\text{-}35)$$

where $\Delta E_{mec}$ is any change in the mechanical energy of the system, $\Delta E_{th}$ is any change in the thermal energy of the system, and $\Delta E_{int}$ is any change in any other type of internal energy of the system. Included in $\Delta E_{mec}$ are changes $\Delta K$ in kinetic energy and changes $\Delta U$ in potential energy (elastic, gravitational, or any other type we might find).

This law of conservation of energy is *not* something we have derived from basic physics principles. Rather, it is a law based on countless experiments. Scientists and engineers have never found an exception to it.

### Isolated System

If a system is isolated from its environment, there can be no energy transfers to or from it. For that case, the law of conservation of energy states:

> The total energy $E$ of an isolated system cannot change.

Many energy transfers may be going on *within* an isolated system—between, say, kinetic energy and a potential energy or between kinetic energy and thermal

**FIG. 8-15** To descend, the rock climber must transfer energy from the gravitational potential energy of a system consisting of him, his gear, and Earth. He has wrapped the rope around metal rings so that the rope rubs against the rings. This allows most of the transferred energy to go to the thermal energy of the rope and rings rather than to his kinetic energy. *(Tyler Stableford/The Image Bank/ Getty Images)*

energy. However, the total of all the types of energy in the system cannot change.

We can use the rock climber in Fig. 8-15 as an example, approximating him, his gear, and Earth as an isolated system. As he rappels down the rock face, changing the configuration of the system, he needs to control the transfer of energy from the gravitational potential energy of the system. (That energy cannot just disappear.) Some of it is transferred to his kinetic energy. However, he obviously does not want very much transferred to that type or he will be moving too quickly, so he has wrapped the rope around metal rings to produce friction between the rope and the rings as he moves down. The sliding of the rings on the rope then transfers the gravitational potential energy of the system to thermal energy of the rings and rope in a way that he can control. The total energy of the climber–gear–Earth system (the total of its gravitational potential energy, kinetic energy, and thermal energy) does not change during his descent.

For an isolated system, the law of conservation of energy can be written in two ways. First, by setting $W = 0$ in Eq. 8-35, we get

$$\Delta E_{mec} + \Delta E_{th} + \Delta E_{int} = 0 \qquad \text{(isolated system)}. \qquad (8\text{-}36)$$

We can also let $\Delta E_{mec} = E_{mec,2} - E_{mec,1}$, where the subscripts 1 and 2 refer to two different instants—say, before and after a certain process has occurred. Then Eq. 8-36 becomes

$$E_{mec,2} = E_{mec,1} - \Delta E_{th} - \Delta E_{int}. \qquad (8\text{-}37)$$

Equation 8-37 tells us:

> ☞ In an isolated system, we can relate the total energy at one instant to the total energy at another instant *without considering the energies at intermediate times.*

This fact can be a very powerful tool in solving problems about isolated systems when you need to relate energies of a system before and after a certain process occurs in the system.

In Section 8-5, we discussed a special situation for isolated systems—namely, the situation in which nonconservative forces (such as a kinetic frictional force) do not act within them. In that special situation, $\Delta E_{th}$ and $\Delta E_{int}$ are both zero, and so Eq. 8-37 reduces to Eq. 8-18. In other words, the mechanical energy of an isolated system is conserved when nonconservative forces do not act in it.

### External Forces and Internal Energy Transfers

An external force can change the kinetic energy or potential energy of an object without doing work on the object—that is, without transferring energy to the object. Instead, the force is responsible for transfers of energy from one type to another inside the object.

Figure 8-16 shows an example. An initially stationary ice-skater pushes away from a railing and then slides over the ice (Figs. 8-16a and b). Her kinetic energy increases because of an external force $\vec{F}$ on her from the rail. However, that force does not transfer energy from the rail to her. Thus, the force does no work on her. Rather, her kinetic energy increases as a result of internal transfers from the biochemical energy in her muscles.

Figure 8-17 shows another example. An engine increases the speed of a car with four-wheel drive (all four wheels are made to turn by the engine). During the acceleration, the engine causes the tires to push backward on the road surface. This push produces frictional forces $\vec{f}$ that act on each tire in the forward direction. The net external force $\vec{F}$ from the road, which is the sum of these frictional forces, accelerates the car, increasing its kinetic energy. However, $\vec{F}$ does not transfer energy from the road to the car and so does no work on the car.

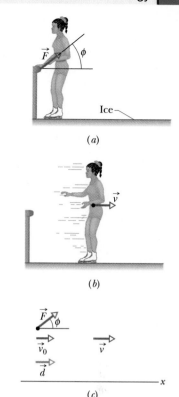

(a)

(b)

(c)

**FIG. 8-16** (a) As a skater pushes herself away from a railing, the force on her from the railing is $\vec{F}$. (b) After the skater leaves the railing, she has velocity $\vec{v}$. (c) External force $\vec{F}$ acts on the skater, at angle $\phi$ with a horizontal $x$ axis. When the skater goes through displacement $\vec{d}$, her velocity is changed from $\vec{v}_0 (= 0)$ to $\vec{v}$ by the horizontal component of $\vec{F}$.

**FIG. 8-17** A vehicle accelerates to the right using four-wheel drive. The road exerts four frictional forces (two of them shown) on the bottom surfaces of the tires. Taken together, these four forces make up the net external force $\vec{F}$ acting on the car.

Rather, the car's kinetic energy increases as a result of internal transfers from the energy stored in the fuel.

In situations like these two, we can sometimes relate the external force $\vec{F}$ on an object to the change in the object's mechanical energy if we can simplify the situation. Consider the ice-skater example. During her push through distance $d$ in Fig. 8-16c, we can simplify by assuming that the acceleration is constant, her speed changing from $v_0 = 0$ to $v$. (That is, we assume $\vec{F}$ has constant magnitude $F$ and angle $\phi$.) After the push, we can simplify the skater as being a particle and neglect the fact that the exertions of her muscles have increased the thermal energy in her muscles and changed other physiological features. Then we can apply Eq. 7-5 ($\frac{1}{2}mv^2 - \frac{1}{2}mv_0^2 = F_x d$) to write

$$K - K_0 = (F \cos \phi)d,$$

or
$$\Delta K = Fd \cos \phi. \tag{8-38}$$

If the situation also involves a change in the elevation of an object, we can include the change $\Delta U$ in gravitational potential energy by writing

$$\Delta U + \Delta K = Fd \cos \phi. \tag{8-39}$$

The force on the right side of this equation does no work on the object but is still responsible for the changes in energy shown on the left side.

### Power

Now that you have seen how energy can be transferred from one type to another, we can expand the definition of power given in Section 7-9. There power is defined as the rate at which work is done by a force. In a more general sense, power $P$ is the rate at which energy is transferred by a force from one type to another. If an amount of energy $\Delta E$ is transferred in an amount of time $\Delta t$, the **average power** due to the force is

$$P_{\text{avg}} = \frac{\Delta E}{\Delta t}. \tag{8-40}$$

Similarly, the **instantaneous power** due to the force is

$$P = \frac{dE}{dt}. \tag{8-41}$$

---

**Sample Problem** | **8-7**

In Fig. 8-18, a 2.0 kg package of tamales slides along a floor with speed $v_1 = 4.0$ m/s. It then runs into and compresses a spring, until the package momentarily stops. Its path to the initially relaxed spring is frictionless, but as it compresses the spring, a kinetic frictional force from the floor, of magnitude 15 N, acts on the package. If $k = 10\,000$ N/m, by what distance $d$ is the spring compressed when the package stops?

---

**KEY IDEAS** We need to examine all the forces and then to determine whether we have an isolated system or a system on which an external force is doing work.

*Forces:* The normal force on the package from the floor does no work on the package because the direction of this force is always perpendicular to the direction of the package's displacement. For the same rea-

son, the gravitational force on the package does no work. As the spring is compressed, however, a spring force does work on the package, transferring energy to elastic potential energy of the spring. The spring force also pushes against a rigid wall. Because there is friction between the package and the floor, the sliding of

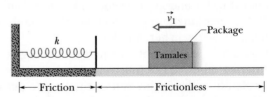

**FIG. 8-18** A package slides across a frictionless floor with velocity $\vec{v}_1$ toward a spring of spring constant $k$. When the package reaches the spring, a frictional force from the floor acts on the package.

the package across the floor increases their thermal energies.

*System:* The package–spring–floor–wall system includes all these forces and energy transfers in one isolated system. Therefore, because the system is isolated, its total energy cannot change. We can then apply the law of conservation of energy in the form of Eq. 8-37 to the system:

$$E_{mec,2} = E_{mec,1} - \Delta E_{th}. \qquad (8\text{-}42)$$

*Calculations:* In Eq. 8-42, let subscript 1 correspond to the initial state of the sliding package and subscript 2 correspond to the state in which the package is momentarily stopped and the spring is compressed by distance $d$. For both states the mechanical energy of the system is the sum of the package's kinetic energy ($K = \frac{1}{2}mv^2$) and the spring's potential energy ($U = \frac{1}{2}kx^2$). For state 1,

$U = 0$ (because the spring is not compressed), and the package's speed is $v_1$. Thus, we have

$$E_{mec,1} = K_1 + U_1 = \tfrac{1}{2}mv_1^2 + 0.$$

For state 2, $K = 0$ (because the package is stopped), and the compression distance is $d$. Therefore, we have

$$E_{mec,2} = K_2 + U_2 = 0 + \tfrac{1}{2}kd^2.$$

Finally, by Eq. 8-31, we can substitute $f_k d$ for the change $\Delta E_{th}$ in the thermal energy of the package and the floor. We can now rewrite Eq. 8-42 as

$$\tfrac{1}{2}kd^2 = \tfrac{1}{2}mv_1^2 - f_k d.$$

Rearranging and substituting known data give us

$$5000d^2 + 15d - 16 = 0.$$

Solving this quadratic equation yields

$$d = 0.055 \text{ m} = 5.5 \text{ cm}. \qquad \text{(Answer)}$$

---

## Sample Problem | 8-8

Figure 8-19a shows the mountain slope and the valley along which a rock avalanche moves. The rocks have a total mass $m$, fall from a height $y = H$, move a distance $d_1$ along a slope of angle $\theta = 45°$, and then move a distance $d_2$ along a flat valley. What is the ratio $d_2/H$ of the runout to the fall height if the coefficient of kinetic friction has the reasonable value of 0.60?

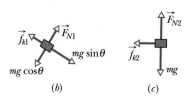

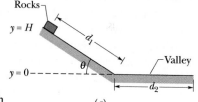

**FIG. 8-19** (*a*) The path a rock avalanche takes down a mountainside and across a valley floor. The forces on the rock material along (*b*) the mountainside and (*c*) the valley floor.

**KEY IDEAS** (1) The mechanical energy $E_{mec}$ of the rocks–Earth system is the sum of the kinetic energy ($K = \frac{1}{2}mv^2$) and the gravitational potential energy ($U = mgy$). (2) The mechanical energy is not conserved during the slide because a (nonconservative) frictional force acts on the rocks, transferring an amount of energy $\Delta E_{th}$ to the thermal energy of the rocks and ground. (3) The transferred energy $\Delta E_{th}$ is related to the magnitude of the kinetic frictional force and the distance of sliding by Eq. 8-31 ($\Delta E_{th} = f_k d$). (4) The mechanical energy $E_{mec,2}$ at any point during the slide is related to the initial mechanical energy $E_{mec,1}$ and the transferred energy $\Delta E_{th}$ by Eq. 8-37, which can be rewritten as $E_{mec,2} = E_{mec,1} - \Delta E_{th}$.

*Calculations:* The final mechanical energy $E_{mec,2}$ is equal to the initial mechanical energy $E_{mec,1}$ minus the amount $\Delta E_{th}$ lost to thermal energy:

$$E_{mec,2} = E_{mec,1} - \Delta E_{th}. \qquad (8\text{-}43)$$

Initially the rocks have potential energy $U = mgH$ and kinetic energy $K = 0$, and so the initial mechanical energy is $E_{mec,1} = mgH$. Finally (when the rocks stop) the rocks have potential energy $U = 0$ and kinetic energy $K = 0$, and so $E_{mec,2} = 0$. The amount of energy transferred to thermal energy is $\Delta E_{th,1} = f_{k1}d_1$ during the

slide down the slope and $\Delta E_{th,2} = f_{k2}d_2$ during the runout across the valley. Substituting these expressions into Eq. 8-43, we have

$$0 = mgH - f_{k1}d_1 - F_{k2}d_2. \qquad (8\text{-}44)$$

From Fig. 8-19a, we see that $d_1 = H/(\sin \theta)$. To obtain expressions for the kinetic frictional forces, we use Eq. 6-2 ($f_k = \mu_k F_N$). Recall from Chapter 6 that on an inclined plane the normal force offsets the component $mg \cos \theta$ of the gravitational force (Fig. 8-19b). Similarly, recall from Chapter 5 that on a horizontal surface the normal force offsets the full magnitude $mg$ of the gravitational force (Fig. 8-19c). Substituting these expressions into Eq. 8-44 and solving for the ratio $d_2/H$, we find

$$0 = mgH - \mu_k(mg \cos \theta)\frac{H}{\sin \theta} - \mu_k mgd_2$$

and

$$\frac{d_2}{H} = \left(\frac{1}{\mu_k} - \frac{1}{\tan \theta}\right). \qquad (8\text{-}45)$$

Substituting $\mu_k = 0.60$ and $\theta = 45°$, we find

$$\frac{d_2}{H} = 0.67. \qquad \text{(Answer)}$$

**Comments:** Our result is typical for a small avalanche. However, for a large avalanche, the ratio $d_2/H$ may be as large as 20. If you substitute this ratio into Eq. 8-45 and

solve for the coefficient of kinetic friction, you find $\mu_k = 0.05$. Researchers do not understand how a large avalanche of jagged, tumbling rocks can have a value of $\mu_k$ small enough to rival that of very slippery ice. One of the most promising ideas is that the material is continuously levitated by a thin layer of small oscillating debris and thus almost never touches the mountain slope or valley floor until the avalanche stops.

---

**Sample Problem | 8-9**

In Fig. 8-20a, a 20 kg block is about to collide with a spring at its relaxed length. As the block compresses the spring, a kinetic frictional force between the block and floor acts on the block. Figure 8-20b gives the block's kinetic energy $K(x)$ and the spring's potential energy $U(x)$ as functions of the block's position $x$ as the spring is compressed. What is the coefficient of kinetic friction $\mu_k$ between the block and floor?

**KEY IDEAS** (1) The mechanical energy $E_{mec}\,(= K + U)$ is not conserved during the compression because the nonconservative frictional force acts on the block, transferring an amount of energy $\Delta E_{th}$ to the thermal energy of the block and floor. (2) The transferred energy $\Delta E_{th}$ is related to the magnitude of the kinetic frictional force and the distance of sliding by Eq. 8-31

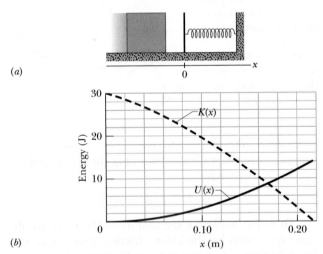

(a)

(b)

**FIG. 8-20** (a) Block on the verge of colliding with a spring. (b) The change of kinetic energy $K$ and potential energy $U$ as the spring is compressed and the block slows to a stop.

($\Delta E_{th} = f_k d$). (3) The mechanical energy $E_{mec,2}$ at any point during the compression is related to the initial mechanical energy $E_{mec,1}$ and $\Delta E_{th}$ by Eq. 8-37, which can be rewritten as $E_{mec,2} = E_{mec,1} - \Delta E_{th}$.

**Finding $\Delta E_{th}$:** From Fig. 8-20b, we see that when the block is at $x = 0$ and is on the verge of compressing the spring, its kinetic energy is $K = 30$ J and the spring's potential energy is $U = 0$. Thus, the sum of $K$ and $U$ is

$$E_{mec,1} = 30 \text{ J}.$$

The spring is at its maximum compression when the block stops—that is, when the kinetic energy drops to 0. From the figure we see that this occurs at $x \approx 0.215$ m, at which $K = 0$ and $U = 14$ J. Thus, at the stopping point, the sum of $K$ and $U$ is

$$E_{mec,2} = 14 \text{ J}.$$

To find the amount of energy transferred to thermal energy, we write $E_{mec,2} = E_{mec,1} - \Delta E_{th}$ as

$$14 \text{ J} = 30 \text{ J} - \Delta E_{th}$$

or

$$\Delta E_{th} = 16 \text{ J}.$$

**Finding $\mu_k$:** From Eq. 6-2, we know that a kinetic frictional force is given by $f_k = \mu_k F_N$, where the normal force is given by Eq. 5-14 ($F_N = mg$). Here, the frictional force $f_k$ transfers 16 J to thermal energy over distance $d = 0.215$ m, according to $\Delta E_{th} = f_k d$. Pulling these several expressions together, we write

$$\Delta E_{th} = f_k d = \mu_k F_N d = \mu_k mgd$$

and then substitute the data $\Delta E_{th} = 16$ J, $m = 20$ kg, $g = 9.8$ m/s², and $d = 0.215$ m, finding

$$\mu_k = 0.38. \qquad \text{(Answer)}$$

---

## REVIEW & SUMMARY

**Conservative Forces** A force is a **conservative force** if the net work it does on a particle moving around any closed path, from an initial point and then back to that point, is zero. Equivalently, a force is conservative if the net work it does on a particle moving between two points does not depend on the path taken by the particle. The gravitational force and the

spring force are conservative forces; the kinetic frictional force is a **nonconservative force.**

**Potential Energy** A **potential energy** is energy that is associated with the configuration of a system in which a conservative force acts. When the conservative force does work $W$

on a particle within the system, the change $\Delta U$ in the potential energy of the system is

$$\Delta U = -W. \tag{8-1}$$

If the particle moves from point $x_i$ to point $x_f$, the change in the potential energy of the system is

$$\Delta U = -\int_{x_i}^{x_f} F(x)\, dx. \tag{8-6}$$

**Gravitational Potential Energy** The potential energy associated with a system consisting of Earth and a nearby particle is **gravitational potential energy.** If the particle moves from height $y_i$ to height $y_f$, the change in the gravitational potential energy of the particle–Earth system is

$$\Delta U = mg(y_f - y_i) = mg\,\Delta y. \tag{8-7}$$

If the **reference point** of the particle is set as $y_i = 0$ and the corresponding gravitational potential energy of the system is set as $U_i = 0$, then the gravitational potential energy $U$ when the particle is at any height $y$ is

$$U(y) = mgy. \tag{8-9}$$

**Elastic Potential Energy** Elastic potential energy is the energy associated with the state of compression or extension of an elastic object. For a spring that exerts a spring force $F = -kx$ when its free end has displacement $x$, the elastic potential energy is

$$U(x) = \tfrac{1}{2}kx^2. \tag{8-11}$$

The **reference configuration** has the spring at its relaxed length, at which $x = 0$ and $U = 0$.

**Mechanical Energy** The **mechanical energy** $E_{mec}$ of a system is the sum of its kinetic energy $K$ and potential energy $U$:

$$E_{mec} = K + U. \tag{8-12}$$

An *isolated system* is one in which no *external force* causes energy changes. If only conservative forces do work within an isolated system, then the mechanical energy $E_{mec}$ of the system cannot change. This **principle of conservation of mechanical energy** is written as

$$K_2 + U_2 = K_1 + U_1, \tag{8-17}$$

in which the subscripts refer to different instants during an energy transfer process. This conservation principle can also be written as

$$\Delta E_{mec} = \Delta K + \Delta U = 0. \tag{8-18}$$

**Potential Energy Curves** If we know the potential energy function $U(x)$ for a system in which a one-dimensional force $F(x)$ acts on a particle, we can find the force as

$$F(x) = -\frac{dU(x)}{dx}. \tag{8-22}$$

If $U(x)$ is given on a graph, then at any value of $x$, the force $F(x)$ is the negative of the slope of the curve there and the kinetic energy of the particle is given by

$$K(x) = E_{mec} - U(x), \tag{8-24}$$

where $E_{mec}$ is the mechanical energy of the system. A **turning point** is a point $x$ at which the particle reverses its motion (there, $K = 0$). The particle is in **equilibrium** at points where the slope of the $U(x)$ curve is zero (there, $F(x) = 0$).

**Work Done on a System by an External Force** Work $W$ is energy transferred to or from a system by means of an external force acting on the system. When more than one force acts on a system, their *net work* is the transferred energy. When friction is not involved, the work done on the system and the change $\Delta E_{mec}$ in the mechanical energy of the system are equal:

$$W = \Delta E_{mec} = \Delta K + \Delta U. \tag{8-26, 8-25}$$

When a kinetic frictional force acts within the system, then the thermal energy $E_{th}$ of the system changes. (This energy is associated with the random motion of atoms and molecules in the system.) The work done on the system is then

$$W = \Delta E_{mec} + \Delta E_{th}. \tag{8-33}$$

The change $\Delta E_{th}$ is related to the magnitude $f_k$ of the frictional force and the magnitude $d$ of the displacement caused by the external force by

$$\Delta E_{th} = f_k d. \tag{8-31}$$

**Conservation of Energy** The **total energy** $E$ of a system (the sum of its mechanical energy and its internal energies, including thermal energy) can change only by amounts of energy that are transferred to or from the system. This experimental fact is known as the **law of conservation of energy.** If work $W$ is done on the system, then

$$W = \Delta E = \Delta E_{mec} + \Delta E_{th} + \Delta E_{int}. \tag{8-35}$$

If the system is isolated ($W = 0$), this gives

$$\Delta E_{mec} + \Delta E_{th} + \Delta E_{int} = 0 \tag{8-36}$$

and

$$E_{mec,2} = E_{mec,1} - \Delta E_{th} - \Delta E_{int}, \tag{8-37}$$

where the subscripts 1 and 2 refer to two different instants.

**Power** The **power** due to a force is the *rate* at which that force transfers energy. If an amount of energy $\Delta E$ is transferred by a force in an amount of time $\Delta t$, the **average power** of the force is

$$P_{avg} = \frac{\Delta E}{\Delta t}. \tag{8-40}$$

The **instantaneous power** due to a force is

$$P = \frac{dE}{dt}. \tag{8-41}$$

## QUESTIONS

**1** Figure 8-21 shows one direct path and four indirect paths from point $i$ to point $f$. Along the direct path and three of the indirect paths, only a conservative force $F_c$ acts on a certain object. Along the fourth indirect path, both $F_c$ and a noncon-

servative force $F_{nc}$ act on the object. The change $\Delta E_{mec}$ in the object's mechanical energy (in joules) in going from $i$ to $f$ is indicated along each straight-line segment of the indirect paths. What is $\Delta E_{mec}$ (a) from $i$ to $f$ along the direct path and

(b) due to $F_{nc}$ along the one path where it acts?

**2** In Fig. 8-22, a small, initially stationary block is released on a frictionless ramp at a height of 3.0 m. Hill heights along the ramp are as shown. The hills have identical circular tops, and the block does not fly off any hill. (a) Which hill is the first the block cannot cross? (b) What does the block do after failing to cross that hill? On which hilltop is (c) the centripetal acceleration of the block greatest and (d) the normal force on the block least?

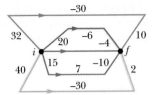

**FIG. 8-21** Question 1.

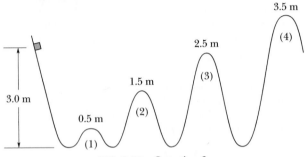

**FIG. 8-22** Question 2.

**3** In Fig. 8-23, a horizontally moving block can take three frictionless routes, differing only in elevation, to reach the dashed finish line. Rank the routes according to (a) the speed of the block at the finish line and (b) the travel time of the block to the finish line, greatest first.

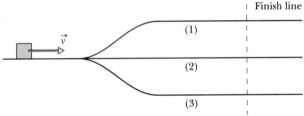

**FIG. 8-23** Question 3.

**4** Figure 8-24 gives the potential energy function of a particle. (a) Rank regions *AB, BC, CD,* and *DE* according to the magnitude of the force on the particle, greatest first. What value must the mechanical energy $E_{mec}$ of the particle not exceed if the particle is to be (b) trapped in the potential well at the left, (c) trapped in the potential well at the right, and (d) able to move between the two potential wells but not to the right of point *H*? For the situation of (d), in which of regions *BC, DE,* and *FG* will the particle have (e) the greatest kinetic energy and (f) the least speed?

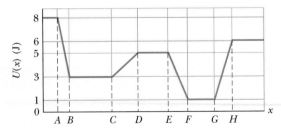

**FIG. 8-24** Question 4.

**5** Figure 8-25 shows three situations involving a plane that is not frictionless and a block sliding along the plane. The block begins with the same speed in all three situations and slides until the kinetic frictional force has stopped it. Rank the situations according to the increase in thermal energy due to the sliding, greatest first.

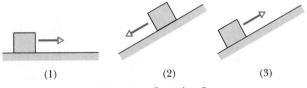

**FIG. 8-25** Question 5.

**6** In Fig. 8-26a, you pull upward on a rope that is attached to a cylinder on a vertical rod. Because the cylinder fits tightly on the rod, the cylinder slides along the rod with considerable friction. Your force does work $W = +100$ J on the cylinder–rod–Earth system (Fig. 8-26b). An "energy statement" for the system is shown in Fig. 8-26c: the kinetic energy $K$ increases by 50 J, and the gravitational potential energy $U_g$ increases by 20 J. The only other change in energy within the system is for the thermal energy $E_{th}$. What is the change $\Delta E_{th}$?

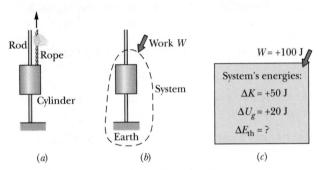

**FIG. 8-26** Question 6.

**7** The arrangement shown in Fig. 8-27 is similar to that in Question 6. Here you pull downward on the rope that is attached to the cylinder, which fits tightly on the rod. Also, as the cylinder descends, it pulls on a block via a second rope, and the block slides over a lab table. Again consider the cylinder–rod–Earth system, similar to that shown in Fig. 8-26b. Your work on the system is 200 J. The system does work of 60 J on the block. Within the system, the kinetic energy increases by 130 J and the gravitational potential energy decreases by 20 J. (a) Draw an "energy statement" for the system, as in Fig. 8-26c. (b) What is the change in the thermal energy within the system?

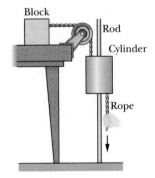

**FIG. 8-27** Question 7.

**8** In Fig. 8-28, a block slides along a track that descends through distance *h*. The track is frictionless except for the lower section. There the block slides to a stop in a certain distance *D* because of friction. (a) If we decrease *h*, will the block now slide to a stop in a distance that is greater than, less than, or equal to $D$? (b) If, instead, we increase the mass of the block, will the stopping distance now be greater than, less than, or equal to $D$?

FIG. 8-28   Question 8.

**9** In Fig. 8-29, a block slides from $A$ to $C$ along a frictionless ramp, and then it passes through horizontal region $CD$, where a frictional force acts on it. Is the block's kinetic energy

increasing, decreasing, or constant in (a) region $AB$, (b) region $BC$, and (c) region $CD$? (d) Is the block's mechanical energy increasing, decreasing, or constant in those regions?

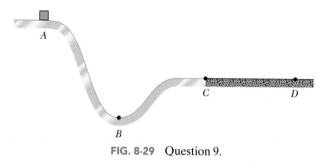

FIG. 8-29   Question 9.

---

## PROBLEMS

**GO**   Tutoring problem available (at instructor's discretion) in *WileyPLUS* and WebAssign

**SSM**   Worked-out solution available in Student Solutions Manual

**• – •••**   Number of dots indicates level of problem difficulty

Additional information available in *The Flying Circus of Physics* and at flyingcircusofphysics.com

**WWW**   Worked-out solution is at

**ILW**   Interactive solution is at — http://www.wiley.com/college/halliday

### sec. 8-4   Determining Potential Energy Values

**•1** You drop a 2.00 kg book to a friend who stands on the ground at distance $D = 10.0$ m below. If your friend's outstretched hands are at distance $d = 1.50$ m above the ground (Fig. 8-30), (a) how much work $W_g$ does the gravitational force do on the book as it drops to her hands? (b) What is the change $\Delta U$ in the gravitational potential energy of the book–Earth system during the drop? If the gravitational potential energy $U$ of that system is taken to be zero at ground level, what is $U$ (c) when the book is released and (d) when it reaches her hands? Now take $U$ to be 100 J at ground level and again find (e) $W_g$, (f) $\Delta U$, (g) $U$ at the release point, and (h) $U$ at her hands.

FIG. 8-30   Problems 1 and 10.

**•2** Figure 8-31 shows a ball with mass $m = 0.341$ kg attached to the end of a thin rod with length $L = 0.452$ m and negligible mass. The other end of the rod is pivoted so that the ball can move in a vertical circle. The rod is held horizontally as shown and then given enough of a downward push to cause the ball to swing down and around and just reach the vertically up position, with zero speed there. How much work is done on the ball by the gravitational force from the initial point to (a) the lowest point, (b) the highest point, and (c) the point on the right level with the initial point? If the gravitational potential energy of the ball–Earth system is taken to be zero at the initial point, what is it when the ball reaches (d) the lowest point, (e) the highest point, and (f) the

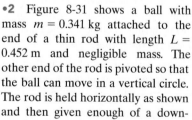

FIG. 8-31   Problems 2 and 12.

point on the right level with the initial point? (g) Suppose the rod were pushed harder so that the ball passed through the highest point with a nonzero speed. Would $\Delta U_g$ from the lowest point to the highest point then be greater than, less than, or the same as it was when the ball stopped at the highest point?

**•3** In Fig. 8-32, a 2.00 g ice flake is released from the edge of a hemispherical bowl whose radius $r$ is 22.0 cm. The flake–bowl contact is frictionless. (a) How much work is done on the flake by the gravitational force during the flake's descent to the bottom of the bowl? (b) What is the change in the potential energy of the flake–Earth system during that descent? (c) If that potential energy is taken to be zero at the bottom of the bowl, what is its value when the flake is released? (d) If, instead, the potential energy is taken to be zero at the release point, what is its value when the flake reaches the bottom of the bowl? (e) If the mass of the flake were doubled, would the magnitudes of the answers to (a) through (d) increase, decrease, or remain the same? **SSM**

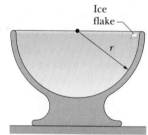

FIG. 8-32
Problems 3 and 11.

**•4** In Fig. 8-33, a frictionless roller-coaster car of mass $m = 825$ kg tops the first hill with speed $v_0 = 17.0$ m/s at height $h = 42.0$ m. How much work does the gravitational

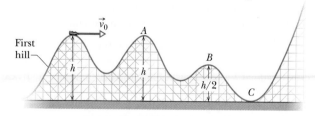

FIG. 8-33   Problems 4 and 13.

force do on the car from that point to (a) point $A$, (b) point $B$, and (c) point $C$? If the gravitational potential energy of the car–Earth system is taken to be zero at $C$, what is its value when the car is at (d) $B$ and (e) $A$? (f) If mass $m$ were doubled, would the change in the gravitational potential energy of the system between points $A$ and $B$ increase, decrease, or remain the same?

•5   What is the spring constant of a spring that stores 25 J of elastic potential energy when compressed by 7.5 cm?   SSM

•6   A 1.50 kg snowball is fired from a cliff 12.5 m high. The snowball's initial velocity is 14.0 m/s, directed 41.0° above the horizontal. (a) How much work is done on the snowball by the gravitational force during its flight to the flat ground below the cliff? (b) What is the change in the gravitational potential energy of the snowball–Earth system during the flight? (c) If that gravitational potential energy is taken to be zero at the height of the cliff, what is its value when the snowball reaches the ground?

••7   Figure 8-34 shows a thin rod, of length $L = 2.00$ m and negligible mass, that can pivot about one end to rotate in a vertical circle. A ball of mass $m = 5.00$ kg is attached to the other end. The rod is pulled aside to angle $\theta_0 = 30.0°$ and released with initial velocity $\vec{v}_0 = 0$. As the ball descends to its lowest point, (a) how much work does the gravitational force do on it and (b) what is the change in the gravitational potential energy of the ball–Earth system? (c) If the gravitational potential energy is taken to be zero at the lowest point, what is its value just as the ball is released? (d) Do the magnitudes of the answers to (a) through (c) increase, decrease, or remain the same if angle $\theta$ is increased?

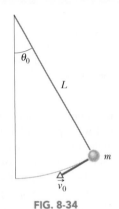

**FIG. 8-34**
Problems 7, 16, and 17.

••8   In Fig. 8-35, a small block of mass $m = 0.032$ kg can slide along the frictionless loop-the-loop, with loop radius $R = 12$ cm. The block is released from rest at point $P$, at height $h = 5.0R$ above the bottom of the loop. How much work does the gravitational force do on the block as the block travels from point $P$ to (a) point $Q$ and (b) the top of the loop? If the gravitational potential energy of the block–Earth system is taken to be zero at the bottom of the loop, what is that potential energy when the block is (c) at point $P$, (d) at point $Q$, and (e) at the top of the loop? (f) If, instead of merely being released, the block is given some initial speed downward along the track, do the answers to (a) through (e) increase, decrease, or remain the same?

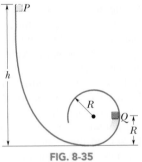

**FIG. 8-35**
Problems 8 and 19.

### sec. 8-5  Conservation of Mechanical Energy

•9   In Fig. 8-36, a runaway truck with failed brakes is moving downgrade at 130 km/h just before the driver steers the truck up a frictionless emergency escape ramp with an inclination of $\theta = 15°$. The truck's mass is $1.2 \times 10^4$ kg. (a) What minimum length $L$ must the ramp have if the truck is to stop (momen-

tarily) along it? (Assume the truck is a particle, and justify that assumption.) Does the minimum length $L$ increase, decrease, or remain the same if (b) the truck's mass is decreased and (c) its speed is decreased?   SSM

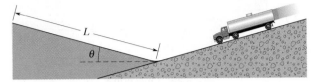

**FIG. 8-36**   Problem 9.

•10   (a) In Problem 1, what is the speed of the book when it reaches the hands? (b) If we substituted a second book with twice the mass, what would its speed be? (c) If, instead, the book were thrown down, would the answer to (a) increase, decrease, or remain the same?

•11   (a) In Problem 3, what is the speed of the flake when it reaches the bottom of the bowl? (b) If we substituted a second flake with twice the mass, what would its speed be? (c) If, instead, we gave the flake an initial downward speed along the bowl, would the answer to (a) increase, decrease, or remain the same?   SSM WWW

•12   (a) In Problem 2, what initial speed must be given the ball so that it reaches the vertically upward position with zero speed? What then is its speed at (b) the lowest point and (c) the point on the right at which the ball is level with the initial point? (d) If the ball's mass were doubled, would the answers to (a) through (c) increase, decrease, or remain the same?

•13   In Problem 4, what is the speed of the car at (a) point $A$, (b) point $B$, and (c) point $C$? (d) How high will the car go on the last hill, which is too high for it to cross? (e) If we substitute a second car with twice the mass, what then are the answers to (a) through (d)?   GO

•14   (a) In Problem 6, using energy techniques rather than the techniques of Chapter 4, find the speed of the snowball as it reaches the ground below the cliff. What is that speed (b) if the launch angle is changed to 41.0° *below* the horizontal and (c) if the mass is changed to 2.50 kg?

•15   A 5.0 g marble is fired vertically upward using a spring gun. The spring must be compressed 8.0 cm if the marble is to just reach a target 20 m above the marble's position on the compressed spring. (a) What is the change $\Delta U_g$ in the gravitational potential energy of the marble–Earth system during the 20 m ascent? (b) What is the change $\Delta U_s$ in the elastic potential energy of the spring during its launch of the marble? (c) What is the spring constant of the spring?   SSM

••16   (a) In Problem 7, what is the speed of the ball at the lowest point? (b) Does the speed increase, decrease, or remain the same if the mass is increased?

••17   Figure 8-34 shows a pendulum of length $L = 1.25$ m. Its bob (which effectively has all the mass) has speed $v_0$ when the cord makes an angle $\theta_0 = 40.0°$ with the vertical. (a) What is the speed of the bob when it is in its lowest position if $v_0 = 8.00$ m/s? What is the least value that $v_0$ can have if the pendulum is to swing down and then up (b) to a horizontal position, and (c) to a vertical position with the cord remaining straight? (d) Do the answers to (b) and (c) increase, decrease, or remain the same if $\theta_0$ is increased by a few degrees?

••18 A 700 g block is released from rest at height $h_0$ above a vertical spring with spring constant $k = 400$ N/m and negligible mass. The block sticks to the spring and momentarily stops after compressing the spring 19.0 cm. How much work is done (a) by the block on the spring and (b) by the spring on the block? (c) What is the value of $h_0$? (d) If the block were released from height $2.00h_0$ above the spring, what would be the maximum compression of the spring?

••19 In Problem 8, what are the magnitudes of (a) the horizontal component and (b) the vertical component of the *net* force acting on the block at point $Q$? (c) At what height $h$ should the block be released from rest so that it is on the verge of losing contact with the track at the top of the loop? (*On the verge of losing contact* means that the normal force on the block from the track has just then become zero.) (d) Graph the magnitude of the normal force on the block at the top of the loop versus initial height $h$, for the range $h = 0$ to $h = 6R$.

••20 A single conservative force $\vec{F} = (6.0x - 12)\hat{i}$ N, where $x$ is in meters, acts on a particle moving along an $x$ axis. The potential energy $U$ associated with this force is assigned a value of 27 J at $x = 0$. (a) Write an expression for $U$ as a function of $x$, with $U$ in joules and $x$ in meters. (b) What is the maximum positive potential energy? At what (c) negative value and (d) positive value of $x$ is the potential energy equal to zero?

••21 The string in Fig. 8-37 is $L = 120$ cm long, has a ball attached to one end, and is fixed at its other end. The distance $d$ from the fixed end to a fixed peg at point $P$ is 75.0 cm. When the initially stationary ball is released with the string horizontal as shown, it will swing along the dashed arc. What is its speed when it reaches (a) its lowest point and (b) its highest point after the string catches on the peg? ILW

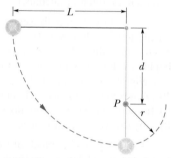

FIG. 8-37 Problems 21 and 68.

••22 A block of mass $m = 2.0$ kg is dropped from height $h = 40$ cm onto a spring of spring constant $k = 1960$ N/m (Fig. 8-38). Find the maximum distance the spring is compressed.

••23 At $t = 0$ a 1.0 kg ball is thrown from a tall tower with $\vec{v} = (18$ m/s$)\hat{i} + (24$ m/s$)\hat{j}$. What is $\Delta U$ of the ball–Earth system between $t = 0$ and $t = 6.0$ s (still free fall)?

••24 A 60 kg skier starts from rest at

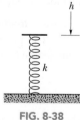

FIG. 8-38
Problem 22.

height $H = 20$ m above the end of a ski-jump ramp (Fig. 8-39) and leaves the ramp at angle $\theta = 28°$. Neglect the effects of air resistance and assume the ramp is frictionless. (a) What is the maximum height $h$ of his jump above the end of the ramp? (b) If he increased his weight by putting on a backpack, would $h$ then be greater, less, or the same?

••25 Tarzan, who weighs 688 N, swings from a cliff at the end of a vine 18 m long (Fig. 8-40). From the top of the cliff to the bottom of the swing, he descends by 3.2 m. The vine will break if the force on it exceeds 950 N. (a) Does the vine break? (b) If no, what is the greatest force on it during the swing? If yes, at what angle with the vertical does it break?

FIG. 8-40 Problem 25.

••26 A pendulum consists of a 2.0 kg stone swinging on a 4.0 m string of negligible mass. The stone has a speed of 8.0 m/s when it passes its lowest point. (a) What is the speed when the string is at 60° to the vertical? (b) What is the greatest angle with the vertical that the string will reach during the stone's motion? (c) If the potential energy of the pendulum–Earth system is taken to be zero at the stone's lowest point, what is the total mechanical energy of the system?

••27 Figure 8-41 shows an 8.00 kg stone at rest on a spring. The spring is compressed 10.0 cm by the stone. (a) What is the spring constant? (b) The stone is pushed down an additional 30.0 cm and released. What is the elastic potential energy of the compressed spring just before that release? (c) What is the change in the gravitational potential energy of the stone–Earth system when the stone moves from the release point to its maximum height? (d) What is that maximum height, measured from the release point? GO

FIG. 8-41
Problem 27.

••28 A 2.0 kg breadbox on a frictionless incline of angle $\theta = 40°$ is connected, by a cord that runs over a pulley, to a light spring of spring constant $k = 120$ N/m, as shown in Fig. 8-42. The box is released from rest when the spring is unstretched. Assume that the pulley is massless and frictionless. (a) What is the speed of the box when it has moved 10 cm down the incline? (b) How far down the incline from its point of release does the box slide before momentarily stopping, and what are the (c) magnitude and (d) direction (up or down the incline) of the box's acceleration at the instant the box momentarily stops?

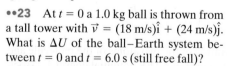

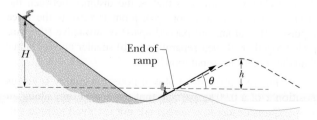

FIG. 8-39 Problem 24.

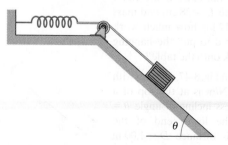

FIG. 8-42 Problem 28.

••**29** A block with mass $m = 2.00$ kg is placed against a spring on a frictionless incline with angle $\theta = 30.0°$ (Fig. 8-43). (The block is not attached to the spring.) The spring, with spring constant $k = 19.6$ N/cm, is compressed 20.0 cm and then released. (a) What is the elastic potential energy of the compressed spring? (b) What is the change in the gravitational potential energy of the block–Earth system as the block moves from the release point to its highest point on the incline? (c) How far along the incline is the highest point from the release point? **ILW**

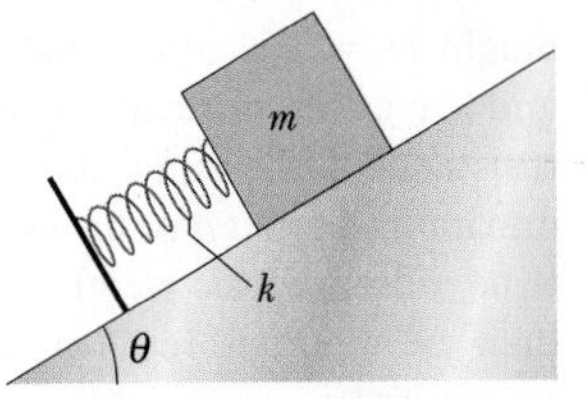

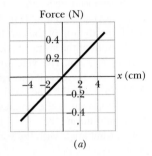

**FIG. 8-43** Problem 29.

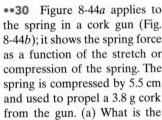

(a)

••**30** Figure 8-44a applies to the spring in a cork gun (Fig. 8-44b); it shows the spring force as a function of the stretch or compression of the spring. The spring is compressed by 5.5 cm and used to propel a 3.8 g cork from the gun. (a) What is the speed of the cork if it is released as the spring passes through its relaxed position? (b) Suppose, instead, that the cork sticks to the spring and stretches it 1.5 cm before separation occurs. What now is the speed of the cork at the time of release?

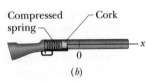

(b)

**FIG. 8-44** Problem 30.

••**31** In Fig. 8-45, a block of mass $m = 12$ kg is released from rest on a frictionless incline of angle $\theta = 30°$. Below the block is a spring that can be compressed 2.0 cm by a force of 270 N. The block momentarily stops when it compresses the spring by 5.5 cm. (a) How far does the block move down the incline from its rest position to this stopping point? (b) What is the speed of the block just as it touches the spring? **SSM WWW**

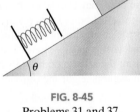

**FIG. 8-45**
Problems 31 and 37.

••**32** In Fig. 8-46, a chain is held on a frictionless table with one-fourth of its length hanging over the edge. If the chain has length $L = 28$ cm and mass $m = 0.012$ kg, how much work is required to pull the hanging part back onto the table?

**FIG. 8-46** Problem 32.

••**33** In Fig. 8-47, a spring with $k = 170$ N/m is at the top of a frictionless incline of angle $\theta = 37.0°$. The lower end of the incline is distance $D = 1.00$ m from the end of the spring,

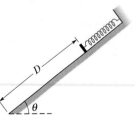

**FIG. 8-47** Problem 33.

which is at its relaxed length. A 2.00 kg canister is pushed against the spring until the spring is compressed 0.200 m and released from rest. (a) What is the speed of the canister at the instant the spring returns to its relaxed length (which is when the canister loses contact with the spring)? (b) What is the speed of the canister when it reaches the lower end of the incline? **GO**

•••**34** Two children are playing a game in which they try to hit a small box on the floor with a marble fired from a spring-loaded gun that is mounted on a table. The target box is horizontal distance $D = 2.20$ m from the edge of the table; see Fig. 8-48. Bobby compresses the spring 1.10 cm, but the center of the marble falls 27.0 cm short of the center of the box. How far should Rhoda compress the spring to score a direct hit? Assume that neither the spring nor the ball encounters friction in the gun.

**FIG. 8-48** Problem 34.

•••**35** A uniform cord of length 25 cm and mass 15 g is initially stuck to a ceiling. Later, it hangs vertically from the ceiling with only one end still stuck. What is the change in the gravitational potential energy of the cord with this change in orientation? (*Hint:* Consider a differential slice of the cord and then use integral calculus.)

•••**36** A boy is initially seated on the top of a hemispherical ice mound of radius $R = 13.8$ m. He begins to slide down the ice, with a negligible initial speed (Fig. 8-49). Approximate the ice as being frictionless. At what height does the boy lose contact with the ice?

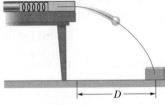

**FIG. 8-49** Problem 36.

•••**37** In Fig. 8-45, a block of mass $m = 3.20$ kg slides from rest a distance $d$ down a frictionless incline at angle $\theta = 30.0°$ where it runs into a spring of spring constant 431 N/m. When the block momentarily stops, it has compressed the spring by 21.0 cm. What are (a) distance $d$ and (b) the distance between the point of the first block–spring contact and the point where the block's speed is greatest?

### sec. 8-6 Reading a Potential Energy Curve

••**38** The potential energy of a diatomic molecule (a two-atom system like $H_2$ or $O_2$) is given by

$$U = \frac{A}{r^{12}} - \frac{B}{r^6},$$

where $r$ is the separation of the two atoms of the molecule and $A$ and $B$ are positive constants. This potential energy is associated with the force that binds the two atoms together. (a) Find the *equilibrium separation*—that is, the distance between the atoms at which the force on each atom is zero. Is the force repulsive (the atoms are pushed apart) or attractive (they are pulled together) if their separation is (b) smaller and (c) larger than the equilibrium separation?

••**39** Figure 8-50 shows a plot of potential energy $U$ versus position $x$ of a 0.90 kg particle that can travel only along an $x$ axis. (Nonconservative forces are not involved.) Three values are $U_A = 15.0$ J, $U_B = 35.0$ J, and $U_C = 45.0$ J. The particle is released at $x = 4.5$ m with an initial speed of 7.0 m/s, headed in the negative $x$ direction. (a) If the particle can reach $x = 1.0$ m, what is its speed there, and if it cannot, what is its turning point? What are the (b) magnitude and (c) direction of the force on the particle as it begins to move to the left of $x = 4.0$ m? Suppose, instead, the particle is headed in the positive $x$ direction when it is released at $x = 4.5$ m at speed 7.0 m/s. (d) If the particle can reach $x = 7.0$ m, what is its speed there, and if it cannot, what is its turning point? What are the (e) magnitude and (f) direction of the force on the particle as it begins to move to the right of $x = 5.0$ m?

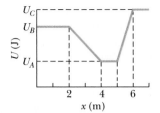

**FIG. 8-50**   Problem 39.

•• **40**   Figure 8-51 shows a plot of potential energy $U$ versus position $x$ of a 0.200 kg particle that can travel only along an $x$ axis under the influence of a conservative force. The graph has these values: $U_A = 9.00$ J, $U_C = 20.00$ J, and $U_D = 24.00$ J. The particle is released at the point where $U$ forms a "potential hill" of "height" $U_B = 12.00$ J, with kinetic energy 4.00 J. What is the speed of the particle at (a) $x = 3.5$ m and (b) $x = 6.5$ m? What is the position of the turning point on (c) the right side and (d) the left side?

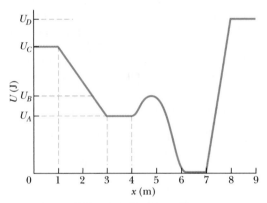

**FIG. 8-51**   Problem 40.

••• **41**   A single conservative force $F(x)$ acts on a 1.0 kg particle that moves along an $x$ axis. The potential energy $U(x)$ associated with $F(x)$ is given by

$$U(x) = -4x\, e^{-x/4}\ \text{J},$$

where $x$ is in meters. At $x = 5.0$ m the particle has a kinetic energy of 2.0 J. (a) What is the mechanical energy of the system? (b) Make a plot of $U(x)$ as a function of $x$ for $0 \le x \le 10$ m, and on the same graph draw the line that represents the mechanical energy of the system. Use part (b) to determine (c) the least value of $x$ the particle can reach and (d) the greatest value of $x$ the particle can reach. Use part (b) to determine (e) the maximum kinetic energy of the particle and (f) the value of $x$ at which it occurs. (g) Determine an expression in newtons and meters for $F(x)$ as a function of $x$. (h) For what (finite) value of $x$ does $F(x) = 0$?

### sec. 8-7   Work Done on a System by an External Force

• **42**   A worker pushed a 27 kg block 9.2 m along a level floor at constant speed with a force directed 32° below the horizontal. If the coefficient of kinetic friction between block and floor was 0.20, what were (a) the work done by the worker's force and (b) the increase in thermal energy of the block–floor system?

• **43**   A collie drags its bed box across a floor by applying a horizontal force of 8.0 N. The kinetic frictional force acting on the box has magnitude 5.0 N. As the box is dragged through 0.70 m along the way, what are (a) the work done by the collie's applied force and (b) the increase in thermal energy of the bed and floor?

•• **44**   A horizontal force of magnitude 35.0 N pushes a block of mass 4.00 kg across a floor where the coefficient of kinetic friction is 0.600. (a) How much work is done by that applied force on the block–floor system when the block slides through a displacement of 3.00 m across the floor? (b) During that displacement, the thermal energy of the block increases by 40.0 J. What is the increase in thermal energy of the floor? (c) What is the increase in the kinetic energy of the block?

•• **45**   A rope is used to pull a 3.57 kg block at constant speed 4.06 m along a horizontal floor. The force on the block from the rope is 7.68 N and directed 15.0° above the horizontal. What are (a) the work done by the rope's force, (b) the increase in thermal energy of the block–floor system, and (c) the coefficient of kinetic friction between the block and floor?   SSM

### sec. 8-8   Conservation of Energy

• **46**   A 60 kg skier leaves the end of a ski-jump ramp with a velocity of 24 m/s directed 25° above the horizontal. Suppose that as a result of air drag the skier returns to the ground with a speed of 22 m/s, landing 14 m vertically below the end of the ramp. From the launch to the return to the ground, by how much is the mechanical energy of the skier–Earth system reduced because of air drag?

• **47**   A 25 kg bear slides, from rest, 12 m down a lodgepole pine tree, moving with a speed of 5.6 m/s just before hitting the ground. (a) What change occurs in the gravitational potential energy of the bear–Earth system during the slide? (b) What is the kinetic energy of the bear just before hitting the ground? (c) What is the average frictional force that acts on the sliding bear?   SSM ILW

• **48**   An outfielder throws a baseball with an initial speed of 81.8 mi/h. Just before an infielder catches the ball at the same level, the ball's speed is 110 ft/s. In foot-pounds, by how much is the mechanical energy of the ball–Earth system reduced because of air drag? (The weight of a baseball is 9.0 oz.)

• **49**   A 75 g Frisbee is thrown from a point 1.1 m above the ground with a speed of 12 m/s. When it has reached a height of 2.1 m, its speed is 10.5 m/s. What was the reduction in $E_{\text{mec}}$ of the Frisbee–Earth system because of air drag?

• **50**   In Fig. 8-52, a block slides down an incline. As it moves from point $A$ to point $B$, which are 5.0 m apart, force $\vec{F}$ acts on the block, with magnitude 2.0 N and directed down the incline. The magnitude of the frictional force acting on the block is 10 N. If the kinetic energy of the block increases by 35 J between $A$ and $B$, how much work is done on the block by the gravitational force as the block moves from $A$ to $B$?

**FIG. 8-52**
Problems 50 and 69.

• **51**   During a rockslide, a 520 kg rock slides from rest down a hillside that is 500 m long and 300 m high. The coefficient of kinetic friction between the rock and the hill surface is 0.25. (a) If the gravitational potential energy $U$ of the rock–Earth system is zero at the bottom of the hill, what is the value of $U$ just before the slide? (b) How much energy is transferred to thermal energy during the slide? (c) What is the kinetic energy of the rock as it reaches the bottom of the hill? (d) What is its speed then?

••**52** You push a 2.0 kg block against a horizontal spring, compressing the spring by 15 cm. Then you release the block, and the spring sends it sliding across a tabletop. It stops 75 cm from where you released it. The spring constant is 200 N/m. What is the block–table coefficient of kinetic friction?

••**53** In Fig. 8-53, a block slides along a track from one level to a higher level after passing through an intermediate valley. The track is frictionless until the block reaches the higher level. There a frictional force stops the block in a distance $d$. The block's initial speed $v_0$ is 6.0 m/s, the height difference $h$ is 1.1 m, and $\mu_k$ is 0.60. Find $d$. ⊙

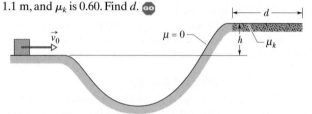

**FIG. 8-53** Problem 53.

••**54** A large fake cookie sliding on a horizontal surface is attached to one end of a horizontal spring with spring constant $k = 400$ N/m; the other end of the spring is fixed in place. The cookie has a kinetic energy of 20.0 J as it passes through the spring's equilibrium position. As the cookie slides, a frictional force of magnitude 10.0 N acts on it. (a) How far will the cookie slide from the equilibrium position before coming momentarily to rest? (b) What will be the kinetic energy of the cookie as it slides back through the equilibrium position?

••**55** In Fig. 8-54, a 3.5 kg block is accelerated from rest by a compressed spring of spring constant 640 N/m. The block leaves the spring at the spring's relaxed length and then travels over a horizontal

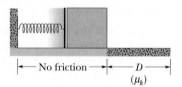

**FIG. 8-54** Problem 55.

floor with a coefficient of kinetic friction $\mu_k = 0.25$. The frictional force stops the block in distance $D = 7.8$ m. What are (a) the increase in the thermal energy of the block–floor system, (b) the maximum kinetic energy of the block, and (c) the original compression distance of the spring? ⊙

••**56** A 4.0 kg bundle starts up a 30° incline with 128 J of kinetic energy. How far will it slide up the incline if the coefficient of kinetic friction between bundle and incline is 0.30?

••**57** When a click beetle is upside down on its back, it jumps upward by suddenly arching its back, transferring energy stored in a muscle to mechanical energy. This launching mechanism produces an audible click, giving the beetle its name. Videotape of a certain click-beetle jump shows that a beetle of mass $m = 4.0 \times 10^{-6}$ kg moved directly upward by 0.77 mm during the launch and then to a maximum height of

$h = 0.30$ m. During the launch, what are the average magnitudes of (a) the external force on the beetle's back from the floor and (b) the acceleration of the beetle in terms of $g$? ✈

••**58** A child whose weight is 267 N slides down a 6.1 m playground slide that makes an angle of 20° with the horizontal. The coefficient of kinetic friction between slide and child is 0.10. (a) How much energy is transferred to thermal energy? (b) If she starts at the top with a speed of 0.457 m/s, what is her speed at the bottom?

••**59** In Fig. 8-55, a block of mass $m = 2.5$ kg slides head on into a spring of spring constant $k = 320$ N/m. When the block stops, it has compressed the spring by 7.5 cm. The coefficient of kinetic friction between block and floor is 0.25. While the block is in contact with the spring and being brought to rest, what are (a) the work done by the spring force and (b) the increase in thermal energy of the block–floor system? (c) What is the block's speed just as it reaches the spring? ILW

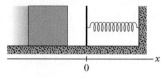

**FIG. 8-55** Problem 59.

••**60** A cookie jar is moving up a 40° incline. At a point 55 cm from the bottom of the incline (measured along the incline), the jar has a speed of 1.4 m/s. The coefficient of kinetic friction between jar and incline is 0.15. (a) How much farther up the incline will the jar move? (b) How fast will it be going when it has slid back to the bottom of the incline? (c) Do the answers to (a) and (b) increase, decrease, or remain the same if we decrease the coefficient of kinetic friction (but do not change the given speed or location)?

••**61** A stone with a weight of 5.29 N is launched vertically from ground level with an initial speed of 20.0 m/s, and the air drag on it is 0.265 N throughout the flight. What are (a) the maximum height reached by the stone and (b) its speed just before it hits the ground?

•••**62** In Fig. 8-56, a block is released from rest at height $d = 40$ cm and slides down a frictionless ramp and onto a first plateau, which has length $d$ and where the coefficient of kinetic friction is 0.50. If the block is still moving, it then slides down a second frictionless ramp through height $d/2$ and onto a lower plateau, which has length $d/2$ and where the coefficient of kinetic friction is again 0.50. If the block is still moving, it then slides up a frictionless ramp until it (momentarily) stops. Where does the block stop? If its final stop is on a plateau, state which one and give the distance $L$ from the left edge of that plateau. If the block reaches the ramp, give the height $H$ above the lower plateau where it momentarily stops.

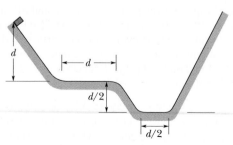

**FIG. 8-56** Problem 62.

**•••63** A particle can slide along a track with elevated ends and a flat central part, as shown in Fig. 8-57. The flat part has length $L = 40$ cm. The curved portions of the track are frictionless, but for the flat part the coefficient of kinetic friction is $\mu_k = 0.20$. The particle is released from rest at point $A$, which is at height $h = L/2$. How far from the left edge of the flat part does the particle finally stop?

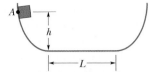

FIG. 8-57 Problem 63.

**•••64** In Fig. 8-58, a block slides along a path that is without friction until the block reaches the section of length $L = 0.75$ m, which begins at height $h = 2.0$ m on a ramp of angle $\theta = 30°$. In that section, the coefficient of kinetic friction is 0.40. The block passes through point $A$ with a speed of 8.0 m/s. If the block can reach point $B$ (where the friction ends), what is its speed there, and if it cannot, what is its greatest height above $A$?

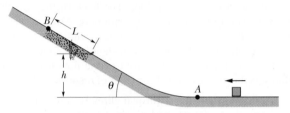

FIG. 8-58 Problem 64.

**•••65** The cable of the 1800 kg elevator cab in Fig. 8-59 snaps when the cab is at rest at the first floor, where the cab bottom is a distance $d = 3.7$ m above a spring of spring constant $k = 0.15$ MN/m. A safety device clamps the cab against guide rails so that a constant frictional force of 4.4 kN opposes the cab's motion. (a) Find the speed of the cab just before it hits the spring. (b) Find the maximum distance $x$ that the spring is compressed (the frictional force still acts during this compression). (c) Find the distance that the cab will bounce back up the shaft. (d) Using conservation of energy, find the approximate total distance that the cab will move before coming to rest. (Assume that the frictional force on the cab is negligible when the cab is stationary.)

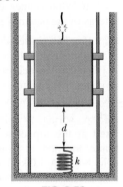

FIG. 8-59
Problem 65.

**Additional Problems**

**66** At a certain factory, 300 kg crates are dropped vertically from a packing machine onto a conveyor belt moving at 1.20 m/s (Fig. 8-60). (A motor maintains the belt's constant speed.) The coefficient of kinetic friction between the belt

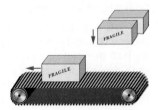

FIG. 8-60 Problem 66.

and each crate is 0.400. After a short time, slipping between the belt and the crate ceases, and the crate then moves along with the belt. For the period of time during which the crate is being brought to rest relative to the belt, calculate, for a coordinate system at rest in the factory, (a) the kinetic energy supplied to the crate, (b) the magnitude of the kinetic frictional force acting on the crate, and (c) the energy supplied by the motor. (d) Explain why answers (a) and (c) differ.

**67** A playground slide is in the form of an arc of a circle that has a radius of 12 m. The maximum height of the slide is $h = 4.0$ m, and the ground is tangent to the circle (Fig. 8-61). A 25 kg child starts from rest at the top of the slide and has a speed of 6.2 m/s at the bottom. (a) What is the length of the slide? (b) What average frictional force acts on the child over this distance? If, instead of the ground, a vertical line through the *top of the slide* is tangent to the circle, what are (c) the length of the slide and (d) the average frictional force on the child?

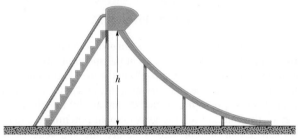

FIG. 8-61 Problem 67.

**68** In Fig. 8-37, the string is $L = 120$ cm long, has a ball attached to one end, and is fixed at its other end. A fixed peg is at point $P$. Released from rest, the ball swings down until the string catches on the peg; then the ball swings up, around the peg. If the ball is to swing completely around the peg, what value must distance $d$ exceed? (*Hint:* The ball must still be moving at the top of its swing. Do you see why?)

**69** In Fig. 8-52, a block is sent sliding down a frictionless ramp. Its speeds at points $A$ and $B$ are 2.00 m/s and 2.60 m/s, respectively. Next, it is again sent sliding down the ramp, but this time its speed at point $A$ is 4.00 m/s. What then is its speed at point $B$? **SSM**

**70** A certain spring is found *not* to conform to Hooke's law. The force (in newtons) it exerts when stretched a distance $x$ (in meters) is found to have magnitude $52.8x + 38.4x^2$ in the direction opposing the stretch. (a) Compute the work required to stretch the spring from $x = 0.500$ m to $x = 1.00$ m. (b) With one end of the spring fixed, a particle of mass 2.17 kg is attached to the other end of the spring when it is stretched by an amount $x = 1.00$ m. If the particle is then released from rest, what is its speed at the instant the stretch in the spring is $x = 0.500$ m? (c) Is the force exerted by the spring conservative or nonconservative? Explain.

**71** A factory worker accidentally releases a 180 kg crate that was being held at rest at the top of a ramp that is 3.7 m long and inclined at 39° to the horizontal. The coefficient of kinetic friction between the crate and the ramp, and between the crate and the horizontal factory floor, is 0.28. (a) How fast is the crate moving as it reaches the bottom of the ramp? (b) How far will it subsequently slide across the floor? (Assume that the crate's kinetic energy does not change as it moves from the ramp onto the floor.) (c) Do the answers to (a) and (b) increase, decrease, or remain the same if we halve the mass of the crate?

**72** In Fig. 8-62, a small block is sent through point $A$ with a speed of 7.0 m/s. Its path is without friction until it reaches the section of length $L = 12$ m, where the coefficient of kinetic friction is 0.70. The indicated heights are $h_1 = 6.0$ m and $h_2 = 2.0$ m. What are the speeds of the block at (a) point $B$ and (b) point $C$? (c) Does the block reach point $D$? If so, what is its speed there; if not, how far through the section of friction does it travel?

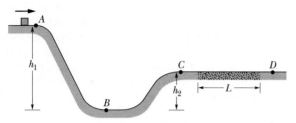

FIG. 8-62   Problem 72.

**73**   A 2.50 kg beverage can is thrown directly downward from a height of 4.00 m, with an initial speed of 3.00 m/s. The air drag on the can is negligible. What is the kinetic energy of the can (a) as it reaches the ground at the end of its fall and (b) when it is halfway to the ground? What are (c) the kinetic energy of the can and (d) the gravitational potential energy of the can–Earth system 0.200 s before the can reaches the ground? For the latter, take the reference point $y = 0$ to be at the ground.

**74**   A 1.50 kg water balloon is shot straight up with an initial speed of 3.00 m/s. (a) What is the kinetic energy of the balloon just as it is launched? (b) How much work does the gravitational force do on the balloon during the balloon's full ascent? (c) What is the change in the gravitational potential energy of the balloon–Earth system during the full ascent? (d) If the gravitational potential energy is taken to be zero at the launch point, what is its value when the balloon reaches its maximum height? (e) If, instead, the gravitational potential energy is taken to be zero at the maximum height, what is its value at the launch point? (f) What is the maximum height?

**75**   In Fig. 8-63, the pulley has negligible mass, and both it and the inclined plane are frictionless. Block $A$ has a mass of 1.0 kg, block $B$ has a mass of 2.0 kg, and angle $\theta$ is 30°. If the blocks are released from rest with the connecting cord taut, what is their total kinetic energy when block $B$ has fallen 25 cm?   SSM

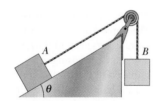

FIG. 8-63   Problem 75.

**76**   From the edge of a cliff, a 0.55 kg projectile is launched with an initial kinetic energy of 1550 J. The projectile's maximum upward displacement from the launch point is +140 m. What are the (a) horizontal and (b) vertical components of its launch velocity? (c) At the instant the vertical component of its velocity is 65 m/s, what is its vertical displacement from the launch point?

**77**   The only force acting on a particle is conservative force $\vec{F}$. If the particle is at point $A$, the potential energy of the system associated with $\vec{F}$ and the particle is 40 J. If the particle moves from point $A$ to point $B$, the work done on the particle by $\vec{F}$ is +25 J. What is the potential energy of the system with the particle at $B$?

**78**   A constant horizontal force moves a 50 kg trunk 6.0 m up a 30° incline at constant speed. The coefficient of kinetic fric-

tion between the trunk and the incline is 0.20. What are (a) the work done by the applied force and (b) the increase in the thermal energy of the trunk and incline?

**79**   Two blocks, of masses $M = 2.0$ kg and $2M$, are connected to a spring of spring constant $k = 200$ N/m that has one end fixed, as shown in Fig. 8-64. The horizontal surface and the pulley are frictionless, and the pulley has negligible mass. The blocks are released from rest with the spring relaxed. (a) What is the combined kinetic energy of the two blocks when the hanging block has fallen 0.090 m? (b) What is the kinetic energy of the hanging block when it has fallen that 0.090 m? (c) What maximum distance does the hanging block fall before momentarily stopping?

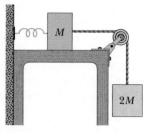

FIG. 8-64   Problem 79.

**80**   A volcanic ash flow is moving across horizontal ground when it encounters a 10° upslope. The front of the flow then travels 920 m up the slope before stopping. Assume that the gases entrapped in the flow lift the flow and thus make the frictional force from the ground negligible; assume also that the mechanical energy of the front of the flow is conserved. What was the initial speed of the front of the flow?

**81**   A 0.50 kg banana is thrown directly upward with an initial speed of 4.00 m/s and reaches a maximum height of 0.80 m. What change does air drag cause in the mechanical energy of the banana–Earth system during the ascent?

**82**   If a 70 kg baseball player steals home by sliding into the plate with an initial speed of 10 m/s just as he hits the ground, (a) what is the decrease in the player's kinetic energy and (b) what is the increase in the thermal energy of his body and the ground along which he slides?

**83**   A spring ($k = 200$ N/m) is fixed at the top of a frictionless plane inclined at angle $\theta = 40°$ (Fig. 8-65). A 1.0 kg block is projected up the plane, from an initial position that is distance $d = 0.60$ m from the end of the relaxed spring, with an initial kinetic energy of 16 J. (a) What is the kinetic energy of the block at the instant it has compressed the spring 0.20 m? (b) With what kinetic energy must the block be projected up the plane if it is to stop momentarily when it has compressed the spring by 0.40 m?   SSM

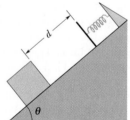

FIG. 8-65   Problem 83.

**84**   A 3.2 kg sloth hangs 3.0 m above the ground. (a) What is the gravitational potential energy of the sloth–Earth system if we take the reference point $y = 0$ to be at the ground? If the sloth drops to the ground and air drag on it is assumed to be negligible, what are the (b) kinetic energy and (c) speed of the sloth just before it reaches the ground?

**85**   A machine pulls a 40 kg trunk 2.0 m up a 40° ramp at constant velocity, with the machine's force on the trunk directed parallel to the ramp. The coefficient of kinetic friction between the trunk and the ramp is 0.40. What are (a) the work done on the trunk by the machine's force and (b) the increase in thermal energy of the trunk and the ramp?

**86**   The luxury liner *Queen Elizabeth 2* has a diesel-electric power plant with a maximum power of 92 MW at a cruising speed of 32.5 knots. What forward force is exerted on the ship at this speed? (1 knot = 1.852 km/h.)

**87** The temperature of a plastic cube is monitored while the cube is pushed 3.0 m across a floor at constant speed by a horizontal force of 15 N. The thermal energy of the cube increases by 20 J. What is the increase in the thermal energy of the floor along which the cube slides? **SSM**

**88** Two snowy peaks are at heights $H = 850$ m and $h = 750$ m above the valley between them. A ski run extends between the peaks, with a total length of 3.2 km and an average slope of $\theta = 30°$ (Fig. 8-66). (a) A skier starts from rest at the top of the higher peak. At what speed will he arrive at the top of the lower peak if he coasts without using ski poles? Ignore friction. (b) Approximately what coefficient of kinetic friction between snow and skis would make him stop just at the top of the lower peak?

FIG. 8-66   Problem 88.

**89** A swimmer moves through the water at an average speed of 0.22 m/s. The average drag force is 110 N. What average power is required of the swimmer?

**90** An automobile with passengers has weight 16 400 N and is moving at 113 km/h when the driver brakes, sliding to a stop. The frictional force on the wheels from the road has a magnitude of 8230 N. Find the stopping distance.

**91** A 0.63 kg ball thrown directly upward with an initial speed of 14 m/s reaches a maximum height of 8.1 m. What is the change in the mechanical energy of the ball–Earth system during the ascent of the ball to that maximum height?

**92** The summit of Mount Everest is 8850 m above sea level. (a) How much energy would a 90 kg climber expend against the gravitational force on him in climbing to the summit from sea level? (b) How many candy bars, at 1.25 MJ per bar, would supply an energy equivalent to this? Your answer should suggest that work done against the gravitational force is a very small part of the energy expended in climbing a mountain.

**93** A sprinter who weighs 670 N runs the first 7.0 m of a race in 1.6 s, starting from rest and accelerating uniformly. What are the sprinter's (a) speed and (b) kinetic energy at the end of the 1.6 s? (c) What average power does the sprinter generate during the 1.6 s interval?

**94** In Fig. 8-67, a 1400 kg block of granite is pulled up an incline at a constant speed of 1.34 m/s by a cable and winch. The indicated distances are $d_1 = 40$ m and $d_2 = 30$ m. The coefficient of kinetic friction between the block and the incline is 0.40. What is the power due to the force applied to the block by the cable?

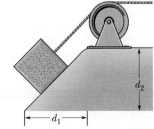

FIG. 8-67   Problem 94.

**95** A 1.50 kg snowball is shot

upward at an angle of 34.0° to the horizontal with an initial speed of 20.0 m/s. (a) What is its initial kinetic energy? (b) By how much does the gravitational potential energy of the snowball–Earth system change as the snowball moves from the launch point to the point of maximum height? (c) What is that maximum height?

**96** A 20 kg object is acted on by a conservative force given by $F = -3.0x - 5.0x^2$, with $F$ in newtons and $x$ in meters. Take the potential energy associated with the force to be zero when the object is at $x = 0$. (a) What is the potential energy of the system associated with the force when the object is at $x = 2.0$ m? (b) If the object has a velocity of 4.0 m/s in the negative direction of the $x$ axis when it is at $x = 5.0$ m, what is its speed when it passes through the origin? (c) What are the answers to (a) and (b) if the potential energy of the system is taken to be $-8.0$ J when the object is at $x = 0$?

**97** A 9.40 kg projectile is fired vertically upward. Air drag decreases the mechanical energy of the projectile–Earth system by 68.0 kJ during the projectile's ascent. How much higher would the projectile have gone were air drag negligible?

**98** A metal tool is sharpened by being held against the rim of a wheel on a grinding machine by a force of 180 N. The frictional forces between the rim and the tool grind off small pieces of the tool. The wheel has a radius of 20.0 cm and rotates at 2.50 rev/s. The coefficient of kinetic friction between the wheel and the tool is 0.320. At what rate is energy being transferred from the motor driving the wheel to the thermal energy of the wheel and tool and to the kinetic energy of the material thrown from the tool?

**99** A spring with a spring constant of 3200 N/m is initially stretched until the elastic potential energy of the spring is 1.44 J. ($U = 0$ for the relaxed spring.) What is $\Delta U$ if the initial stretch is changed to (a) a stretch of 2.0 cm, (b) a compression of 2.0 cm, and (c) a compression of 4.0 cm?

**100** The spring in the muzzle of a child's spring gun has a spring constant of 700 N/m. To shoot a ball from the gun, first the spring is compressed and then the ball is placed on it. The gun's trigger then releases the spring, which pushes the ball through the muzzle. The ball leaves the spring just as it leaves the outer end of the muzzle. When the gun is inclined upward by 30° to the horizontal, a 57 g ball is shot to a maximum height of 1.83 m above the gun's muzzle. Assume air drag on the ball is negligible. (a) At what speed does the spring launch the ball? (b) Assuming that friction on the ball within the gun can be neglected, find the spring's initial compression distance.

**101** A 60.0 kg circus performer slides 4.00 m down a pole to the circus floor, starting from rest. What is the kinetic energy of the performer as she reaches the floor if the frictional force on her from the pole (a) is negligible (she will be hurt) and (b) has a magnitude of 500 N?

**102** In 1981, Daniel Goodwin climbed 443 m up the *exterior* of the Sears Building in Chicago using suction cups and metal clips. (a) Approximate his mass and then compute how much energy he had to transfer from biomechanical (internal) energy to the gravitational potential energy of the Earth–Goodwin system to lift himself to that height. (b) How much energy would he have had to transfer if he had, instead, taken the stairs inside the building (to the same height)?

**103** A 30 g bullet moving a horizontal velocity of 500 m/s comes to a stop 12 cm within a solid wall. (a) What is the change in the bullet's mechanical energy? (b) What is the magnitude of the average force from the wall stopping it?

**104** Resistance to the motion of an automobile consists of road friction, which is almost independent of speed, and air drag, which is proportional to speed-squared. For a certain car with a weight of 12 000 N, the total resistant force $F$ is given by $F = 300 + 1.8v^2$, with $F$ in newtons and $v$ in meters per second. Calculate the power (in horsepower) required to accelerate the car at 0.92 m/s$^2$ when the speed is 80 km/h.

**105** A locomotive with a power capability of 1.5 MW can accelerate a train from a speed of 10 m/s to 25 m/s in 6.0 min. (a) Calculate the mass of the train. Find (b) the speed of the train and (c) the force accelerating the train as functions of time (in seconds) during the 6.0 min interval. (d) Find the distance moved by the train during the interval.

**106** A 5.0 kg block is projected at 5.0 m/s up a plane that is inclined at 30° with the horizontal. How far up along the plane does the block go (a) if the plane is frictionless and (b) if the coefficient of kinetic friction between the block and the plane is 0.40? (c) In the latter case, what is the increase in thermal energy of block and plane during the block's ascent? (d) If the block then slides back down against the frictional force, what is the block's speed when it reaches the original projection point?

**107** A 20 kg block on a horizontal surface is attached to a horizontal spring of spring constant $k = 4.0$ kN/m. The block is pulled to the right so that the spring is stretched 10 cm beyond its relaxed length, and the block is then released from rest. The frictional force between the sliding block and the surface has a magnitude of 80 N. (a) What is the kinetic energy of the block when it has moved 2.0 cm from its point of release? (b) What is the kinetic energy of the block when it first slides back through the point at which the spring is relaxed? (c) What is the maximum kinetic energy attained by the block as it slides from its point of release to the point at which the spring is relaxed?

**108** A 70.0 kg man jumping from a window lands in an elevated fire rescue net 11.0 m below the window. He momentarily stops when he has stretched the net by 1.50 m. Assuming that mechanical energy is conserved during this process and that the net functions like an ideal spring, find the elastic potential energy of the net when it is stretched by 1.50 m.

**109** To form a pendulum, a 0.092 kg ball is attached to one end of a rod of length 0.62 m and negligible mass, and the other end of the rod is mounted on a pivot. The rod is rotated until it is straight up, and then it is released from rest so that it swings down around the pivot. When the ball reaches its lowest point, what are (a) its speed and (b) the tension in the rod? Next, the rod is rotated until it is horizontal, and then it is again released from rest. (c) At what angle from the vertical does the tension in the rod equal the weight of the ball? (d) If the mass of the ball is increased, does the answer to (c) increase, decrease, or remain the same? **SSM**

**110** A skier weighing 600 N goes over a frictionless circular hill of radius $R = 20$ m (Fig. 8-68). Assume that the effects of air resistance on the skier are negligible. As she comes up the hill, her speed is 8.0 m/s at point $B$, at angle $\theta = 20°$. (a) What is her speed at the hilltop (point $A$) if she coasts without using

her poles? (b) What minimum speed can she have at $B$ and still coast to the hilltop? (c) Do the answers to these two questions increase, decrease, or remain the same if the skier weighs 700 N?

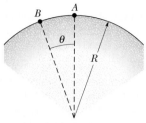

FIG. 8-68 Problem 110.

**111** A 50 g ball is thrown from a window with an initial velocity of 8.0 m/s at an angle of 30° above the horizontal. Using energy methods, determine (a) the kinetic energy of the ball at the top of its flight and (b) its speed when it is 3.0 m below the window. Does the answer to (b) depend on either (c) the mass of the ball or (d) the initial angle? **SSM**

**112** A 68 kg sky diver falls at a constant terminal speed of 59 m/s. (a) At what rate is the gravitational potential energy of the Earth–sky diver system being reduced? (b) At what rate is the system's mechanical energy being reduced?

**113** A river descends 15 m through rapids. The speed of the water is 3.2 m/s upon entering the rapids and 13 m/s upon leaving. What percentage of the gravitational potential energy of the water–Earth system is transferred to kinetic energy during the descent? (*Hint:* Consider the descent of, say, 10 kg of water.)

**114** The magnitude of the gravitational force between a particle of mass $m_1$ and one of mass $m_2$ is given by

$$F(x) = G\frac{m_1 m_2}{x^2},$$

where $G$ is a constant and $x$ is the distance between the particles. (a) What is the corresponding potential energy function $U(x)$? Assume that $U(x) \to 0$ as $x \to \infty$ and that $x$ is positive. (b) How much work is required to increase the separation of the particles from $x = x_1$ to $x = x_1 + d$?

**115** Approximately $5.5 \times 10^6$ kg of water falls 50 m over Niagara Falls each second. (a) What is the decrease in the gravitational potential energy of the water–Earth system each second? (b) If all this energy could be converted to electrical energy (it cannot be), at what rate would electrical energy be supplied? (The mass of 1 m$^3$ of water is 1000 kg.) (c) If the electrical energy were sold at 1 cent/kW·h, what would be the yearly income?

**116** A 1500 kg car starts from rest on a horizontal road and gains a speed of 72 km/h in 30 s. (a) What is its kinetic energy at the end of the 30 s? (b) What is the average power required of the car during the 30 s interval? (c) What is the instantaneous power at the end of the 30 s interval, assuming that the acceleration is constant?

**117** A particle can move along only an $x$ axis, where conservative forces act on it (Fig. 8-69 and the following table). The particle is released at $x = 5.00$ m with a kinetic energy of $K = 14.0$ J and a potential energy of $U = 0$. If its motion is in the negative direction of the $x$ axis, what are its (a) $K$ and (b) $U$ at $x = 2.00$ m and its (c) $K$ and (d) $U$ at $x = 0$? If its motion is in the positive direction of the $x$ axis, what are its (e) $K$ and (f) $U$ at $x = 11.0$ m, its (g) $K$ and (h) $U$ at $x = 12.0$ m, and its (i) $K$ and (j) $U$ at $x = 13.0$ m? (k) Plot $U(x)$ versus $x$ for the range $x = 0$ to $x = 13.0$ m.

FIG. 8-69   Problems 117 and 118.

Next, the particle is released from rest at $x = 0$. What are (l) its kinetic energy at $x = 5.0$ m and (m) the maximum positive position $x_{max}$ it reaches? (n) What does the particle do after it reaches $x_{max}$?

| Range | Force |
|---|---|
| 0 to 2.00 m | $\vec{F}_1 = +(3.00 \text{ N})\hat{i}$ |
| 2.00 m to 3.00 m | $\vec{F}_2 = +(5.00 \text{ N})\hat{i}$ |
| 3.00 m to 8.00 m | $F = 0$ |
| 8.00 m to 11.0 m | $\vec{F}_3 = -(4.00 \text{ N})\hat{i}$ |
| 11.0 m to 12.0 m | $\vec{F}_4 = -(1.00 \text{ N})\hat{i}$ |
| 12.0 m to 15.0 m | $F = 0$ |

**118**   For the arrangement of forces in Problem 117, a 2.00 kg particle is released at $x = 5.00$ m with an initial velocity of 3.45 m/s in the negative direction of the $x$ axis. (a) If the particle can reach $x = 0$ m, what is its speed there, and if it cannot, what is its turning point? Suppose, instead, the particle is headed in the positive $x$ direction when it is released at $x = 5.00$ m at speed 3.45 m/s. (b) If the particle can reach $x = 13.0$ m, what is its speed there, and if it cannot, what is its turning point?

**119**   A 0.42 kg shuffleboard disk is initially at rest when a player uses a cue to increase its speed to 4.2 m/s at constant acceleration. The acceleration takes place over a 2.0 m distance, at the end of which the cue loses contact with the disk. Then the disk slides an additional 12 m before stopping. Assume that the shuffleboard court is level and that the force of friction on the disk is constant. What is the increase in the thermal energy of the disk–court system (a) for that additional 12 m and (b) for the entire 14 m distance? (c) How much work is done on the disk by the cue?   SSM

**120**   We move a particle along an $x$ axis, first outward from $x = 1.0$ m to $x = 4.0$ m and then back to $x = 1.0$ m, while an external force acts on it. That force is directed along the $x$ axis, and its $x$ component can have different values for the outward trip and for the return trip. Here are the values (in newtons) for four situations, where $x$ is in meters:

| Outward | Inward |
|---|---|
| (a) +3.0 | −3.0 |
| (b) +5.0 | +5.0 |
| (c) +2.0x | −2.0x |
| (d) +3.0x² | +3.0x² |

Find the net work done on the particle by the external force *for the round trip* for each of the four situations. (e) For which, if any, is the external force conservative?

**121**   A conservative force $F(x)$ acts on a 2.0 kg particle that moves along an $x$ axis. The potential energy $U(x)$ associated with $F(x)$ is graphed in Fig. 8-70. When the particle is at $x = 2.0$ m, its velocity is −1.5 m/s. What are the (a) magnitude and

(b) direction of $F(x)$ at this position? Between what positions on the (c) left and (d) right does the particle move? (e) What is the particle's speed at $x = 7.0$ m?   SSM

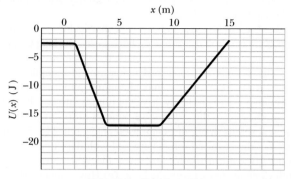

FIG. 8-70   Problem 121.

**122**   To make a pendulum, a 300 g ball is attached to one end of a string that has a length of 1.4 m and negligible mass. (The other end of the string is fixed.) The ball is pulled to one side until the string makes an angle of 30.0° with the vertical; then (with the string taut) the ball is released from rest. Find (a) the speed of the ball when the string makes an angle of 20.0° with the vertical and (b) the maximum speed of the ball. (c) What is the angle between the string and the vertical when the speed of the ball is one-third its maximum value?

**123**   A 1500 kg car begins sliding down a 5.0° inclined road with a speed of 30 km/h. The engine is turned off, and the only forces acting on the car are a net frictional force from the road and the gravitational force. After the car has traveled 50 m along the road, its speed is 40 km/h. (a) How much is the mechanical energy of the car reduced because of the net frictional force? (b) What is the magnitude of that net frictional force?   SSM

**124**   In a circus act, a 60 kg clown is shot from a cannon with an initial velocity of 16 m/s at some unknown angle above the horizontal. A short time later the clown lands in a net that is 3.9 m vertically above the clown's initial position. Disregard air drag. What is the kinetic energy of the clown as he lands in the net?

**125**   The maximum force you can exert on an object with one of your back teeth is about 750 N. Suppose that as you gradually bite on a clump of licorice, the licorice resists compression by one of your teeth by acting like a spring for which $k = 2.5 \times 10^5$ N/m. Find (a) the distance the licorice is com-

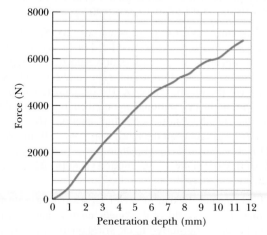

FIG. 8-71   Problem 125.

pressed by your tooth and (b) the work the tooth does on the licorice during the compression. (c) Plot the magnitude of your force versus the compression distance. (d) If there is a potential energy associated with this compression, plot it versus compression distance.

In the 1990s the pelvis of a particular *Triceratops* dinosaur was found to have deep bite marks. The shape of the marks suggested that they were made by a *Tyrannosaurus rex* dinosaur. To test the idea, researchers made a replica of a *T. rex* tooth from bronze and aluminum and then used a hydraulic press to gradually drive the replica into cow bone to the depth seen in the *Triceratops* bone. A graph of the force required versus depth of penetration is given in Fig. 8-71 for one trial; the required force increased with depth because, as the nearly conical tooth penetrated the bone, more of the tooth came in contact with the bone. (e) How much work was done by the hydraulic press—and thus presumably by the *T. rex*—in such a penetration? (f) Is there a potential energy associated with this penetration? (The large biting force and energy expenditure attributed to the *T. rex* by this research suggest that the animal was a predator and not a scavenger.)

**126** A 70 kg firefighter slides, from rest, 4.3 m down a vertical pole. (a) If the firefighter holds onto the pole lightly, so that the frictional force of the pole on her is negligible, what is her speed just before reaching the ground floor? (b) If the firefighter grasps the pole more firmly as she slides, so that the average frictional force of the pole on her is 500 N upward, what is her speed just before reaching the ground floor?

**127** A 15 kg block is accelerated at 2.0 m/s² along a horizontal frictionless surface, with the speed increasing from 10 m/s to 30 m/s. What are (a) the change in the block's mechanical energy and (b) the average rate at which energy is transferred to the block? What is the instantaneous rate of that transfer when the block's speed is (c) 10 m/s and (d) 30 m/s? **SSM**

**128** Repeat Problem 127, but now with the block accelerated up a frictionless plane inclined at 5.0° to the horizontal.

**129** The surface of the continental United States has an area of about $8 \times 10^6$ km² and an average elevation of about 500 m (above sea level). The average yearly rainfall is 75 cm. The fraction of this rainwater that returns to the atmosphere by evaporation is $\frac{2}{3}$; the rest eventually flows into the ocean. If the decrease in gravitational potential energy of the water–Earth system associated with that flow could be fully converted to electrical energy, what would be the average power? (The mass of 1 m³ of water is 1000 kg.)

**130** A spring with spring constant $k = 200$ N/m is suspended vertically with its upper end fixed to the ceiling and its lower end at position $y = 0$. A block of weight 20 N is attached to the lower end, held still for a moment, and then released. What are (a) the kinetic energy $K$, (b) the change (from the initial value) in the gravitational potential energy $\Delta U_g$, and (c) the change in the elastic potential energy $\Delta U_e$ of the spring–block system when the block is at $y = -5.0$ cm? What are (d) $K$, (e) $\Delta U_g$, and (f) $\Delta U_e$ when $y = -10$ cm, (g) $K$, (h) $\Delta U_g$, and (i) $\Delta U_e$ when $y = -15$ cm, and (j) $K$, (k) $\Delta U_g$, and (l) $\Delta U_e$ when $y = -20$ cm?

**131** Each second, 1200 m³ of water passes over a waterfall 100 m high. Three-fourths of the kinetic energy gained by the water in falling is transferred to electrical energy by a hydro-

electric generator. At what rate does the generator produce electrical energy? (The mass of 1 m³ of water is 1000 kg.) **SSM**

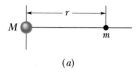

(a)

**132** Figure 8-72a shows a molecule consisting of two atoms of masses $m$ and $M$ (with $m \ll M$) and separation $r$. Figure 8-72b shows the potential energy $U(r)$ of the molecule as a function of $r$. Describe the motion of the atoms (a) if the total mechanical energy $E$ of the two-atom system is greater than zero (as is $E_1$), and (b) if $E$ is less than zero (as is $E_2$). For $E_1 = 1 \times 10^{-19}$ J and $r = 0.3$ nm, find (c) the potential energy of the system, (d) the total kinetic energy of the atoms, and (e) the force (magnitude and direction) acting on each atom. For what values of $r$ is the force (f) repulsive, (g) attractive, and (h) zero?

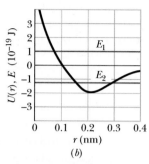

**FIG. 8-72** Problem 132.

**133** A massless rigid rod of length $L$ has a ball of mass $m$ attached to one end (Fig. 8-73). The other end is pivoted in such a way that the ball will move in a vertical circle. First, assume that there is no friction at the pivot. The system is launched downward from the horizontal position $A$ with initial speed $v_0$. The ball just barely reaches point $D$ and then stops. (a) Derive an expression for $v_0$ in terms of $L, m,$ and $g$. (b) What is the tension in the rod when the ball passes through $B$? (c) A little grit is placed on the pivot to increase the friction there. Then the ball just barely reaches $C$ when launched from $A$ with the same speed as before. What is the decrease in the mechanical energy during this motion? (d) What is the decrease in the mechanical energy by the time the ball finally comes to rest at $B$ after several oscillations? **SSM**

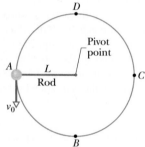

**FIG. 8-73** Problem 133.

**134** Conservative force $F(x)$ acts on a particle that moves along an $x$ axis. Figure 8-74 shows how the potential energy $U(x)$ associated with force $F(x)$ varies with the position of the particle. (a) Plot $F(x)$ for the range $0 < x < 6$ m. (b) The mechanical energy $E$ of the system is 4.0 J. Plot the kinetic energy $K(x)$ of the particle directly on Fig. 8-74.

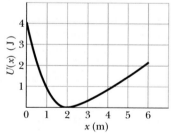

**FIG. 8-74** Problem 134.

**135** Fasten one end of a vertical spring to a ceiling, attach a cabbage to the other end, and then slowly lower the cabbage until the upward force on it from the spring balances the gravitational force on it. Show that the loss of gravitational potential energy of the cabbage–Earth system equals twice the gain in the spring's potential energy.

# Center of Mass and Linear Momentum

©franzfoto.com/Alamy Images

The males of the bighorn sheep (Ovis canadensis) fight one another to gain the attention of the females. Two males repeatedly run at each other at full speed with their heads lowered so that the horns collide, until one of them finally gives up. Such clashes can be costly because if a horn breaks, the male will likely be seriously hurt or killed in the next collision. But even without a broken horn, both sheep should seemingly fall to the ground with concussions.

*How can the head-banging sheep withstand such violent clashes?*

The answer is in this chapter.

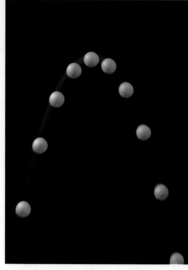

(a)

(b)

FIG. 9-1  (a) A ball tossed into the air follows a parabolic path. (b) The center of mass (black dot) of a baseball bat flipped into the air follows a parabolic path, but all other points of the bat follow more complicated curved paths. *(a: Richard Megna/Fundamental Photographs)*

## 9-1 | WHAT IS PHYSICS?

Every mechanical engineer hired as an expert witness to reconstruct a traffic accident uses physics. Every trainer who coaches a ballerina on how to leap uses physics. Indeed, analyzing complicated motion of any sort requires simplification via an understanding of physics. In this chapter we discuss how the complicated motion of a system of objects, such as a car or a ballerina, can be simplified if we determine a special point of the system—the *center of mass* of that system.

Here is a quick example. If you toss a ball into the air without much spin on the ball (Fig. 9-1a), its motion is simple—it follows a parabolic path, as we discussed in Chapter 4, and the ball can be treated as a particle. If, instead, you flip a baseball bat into the air (Fig. 9-1b), its motion is more complicated. Because every part of the bat moves differently, along paths of many different shapes, you cannot represent the bat as a particle. Instead, it is a system of particles each of which follows its own path through the air. However, the bat has one special point—the center of mass—that *does* move in a simple parabolic path. The other parts of the bat move around the center of mass. (To locate the center of mass, balance the bat on an outstretched finger; the point is above your finger, on the bat's central axis.)

You cannot make a career of flipping baseball bats into the air, but you can make a career of advising long-jumpers or dancers on how to leap properly into the air while either moving their arms and legs or rotating their torso. Your starting point would be the person's center of mass because of its simple motion.

## 9-2 | The Center of Mass

We define the **center of mass** (com) of a system of particles (such as a person) in order to predict the possible motion of the system.

> ☞ The center of mass of a system of particles is the point that moves as though (1) all of the system's mass were concentrated there and (2) all external forces were applied there.

In this section we discuss how to determine where the center of mass of a system of particles is located. We start with a system of only a few particles, and then we consider a system of a great many particles (a solid body, such as a baseball bat). Later in the chapter, we discuss how the center of mass of a system moves when external forces act on the system.

### Systems of Particles

Figure 9-2a shows two particles of masses $m_1$ and $m_2$ separated by distance $d$. We have arbitrarily chosen the origin of an $x$ axis to coincide with the particle of mass $m_1$. We *define* the position of the center of mass (com) of this two-particle system to be

$$x_{com} = \frac{m_2}{m_1 + m_2} d. \tag{9-1}$$

Suppose, as an example, that $m_2 = 0$. Then there is only one particle, of mass $m_1$, and the center of mass must lie at the position of that particle; Eq. 9-1 dutifully reduces to $x_{com} = 0$. If $m_1 = 0$, there is again only one particle (of mass $m_2$), and we have, as we expect, $x_{com} = d$. If $m_1 = m_2$, the center of mass should be halfway between the two particles; Eq. 9-1 reduces to $x_{com} = \frac{1}{2}d$, again as we expect. Finally, Eq. 9-1 tells us that if neither $m_1$ nor $m_2$ is zero, $x_{com}$ can have only values that lie between zero and $d$; that is, the center of mass must lie somewhere between the two particles.

Figure 9-2b shows a more generalized situation, in which the coordinate system has been shifted leftward. The position of the center of mass is now defined

as

$$x_{com} = \frac{m_1 x_1 + m_2 x_2}{m_1 + m_2}. \tag{9-2}$$

Note that if we put $x_1 = 0$, then $x_2$ becomes $d$ and Eq. 9-2 reduces to Eq. 9-1, as it must. Note also that in spite of the shift of the coordinate system, the center of mass is still the same distance from each particle.

We can rewrite Eq. 9-2 as

$$x_{\text{com}} = \frac{m_1 x_1 + m_2 x_2}{M}, \tag{9-3}$$

in which $M$ is the total mass of the system. (Here, $M = m_1 + m_2$.) We can extend this equation to a more general situation in which $n$ particles are strung out along the $x$ axis. Then the total mass is $M = m_1 + m_2 + \cdots + m_n$, and the location of the center of mass is

$$x_{\text{com}} = \frac{m_1 x_1 + m_2 x_2 + m_3 x_3 + \cdots + m_n x_n}{M}$$

$$= \frac{1}{M} \sum_{i=1}^{n} m_i x_i. \tag{9-4}$$

The subscript $i$ is an index that takes on all integer values from 1 to $n$.

If the particles are distributed in three dimensions, the center of mass must be identified by three coordinates. By extension of Eq. 9-4, they are

$$x_{\text{com}} = \frac{1}{M} \sum_{i=1}^{n} m_i x_i, \qquad y_{\text{com}} = \frac{1}{M} \sum_{i=1}^{n} m_i y_i, \qquad z_{\text{com}} = \frac{1}{M} \sum_{i=1}^{n} m_i z_i. \tag{9-5}$$

We can also define the center of mass with the language of vectors. First recall that the position of a particle at coordinates $x_i$, $y_i$, and $z_i$ is given by a position vector:

$$\vec{r}_i = x_i \hat{i} + y_i \hat{j} + z_i \hat{k}. \tag{9-6}$$

Here the index identifies the particle, and $\hat{i}$, $\hat{j}$, and $\hat{k}$ are unit vectors pointing, respectively, in the positive direction of the $x$, $y$, and $z$ axes. Similarly, the position of the center of mass of a system of particles is given by a position vector:

$$\vec{r}_{\text{com}} = x_{\text{com}} \hat{i} + y_{\text{com}} \hat{j} + z_{\text{com}} \hat{k}. \tag{9-7}$$

The three scalar equations of Eq. 9-5 can now be replaced by a single vector equation,

$$\vec{r}_{\text{com}} = \frac{1}{M} \sum_{i=1}^{n} m_i \vec{r}_i, \tag{9-8}$$

where again $M$ is the total mass of the system. You can check that this equation is correct by substituting Eqs. 9-6 and 9-7 into it, and then separating out the $x$, $y$, and $z$ components. The scalar relations of Eq. 9-5 result.

### Solid Bodies

An ordinary object, such as a baseball bat, contains so many particles (atoms) that we can best treat it as a continuous distribution of matter. The "particles" then become differential mass elements $dm$, the sums of Eq. 9-5 become integrals, and the coordinates of the center of mass are defined as

$$x_{\text{com}} = \frac{1}{M} \int x \, dm, \qquad y_{\text{com}} = \frac{1}{M} \int y \, dm, \qquad z_{\text{com}} = \frac{1}{M} \int z \, dm, \tag{9-9}$$

where $M$ is now the mass of the object.

Evaluating these integrals for most common objects (such as a television set or a moose) would be difficult, so here we consider only *uniform* objects. Such objects have uniform *density,* or mass per unit volume; that is, the density $\rho$

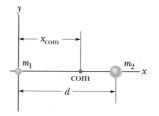

(a)

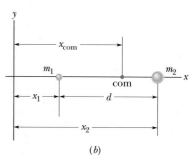

(b)

**FIG. 9-2** (*a*) Two particles of masses $m_1$ and $m_2$ are separated by distance $d$. The dot labeled com shows the position of the center of mass, calculated from Eq. 9-1. (*b*) The same as (*a*) except that the origin is located farther from the particles. The position of the center of mass is calculated from Eq. 9-2. The location of the center of mass with respect to the particles is the same in both cases.

(Greek letter rho) is the same for any given element of an object as for the whole object. From Eq. 1-8, we can write

$$\rho = \frac{dm}{dV} = \frac{M}{V}, \tag{9-10}$$

where $dV$ is the volume occupied by a mass element $dm$, and $V$ is the total volume of the object. Substituting $dm = (M/V)\, dV$ from Eq. 9-10 into Eq. 9-9 gives

$$x_{com} = \frac{1}{V}\int x\, dV, \qquad y_{com} = \frac{1}{V}\int y\, dV, \qquad z_{com} = \frac{1}{V}\int z\, dV. \tag{9-11}$$

You can bypass one or more of these integrals if an object has a point, a line, or a plane of symmetry. The center of mass of such an object then lies at that point, on that line, or in that plane. For example, the center of mass of a uniform sphere (which has a point of symmetry) is at the center of the sphere (which is the point of symmetry). The center of mass of a uniform cone (whose axis is a line of symmetry) lies on the axis of the cone. The center of mass of a banana (which has a plane of symmetry that splits it into two equal parts) lies somewhere in that plane.

The center of mass of an object need not lie within the object. There is no dough at the com of a doughnut, and no iron at the com of a horseshoe.

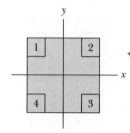

✓ **CHECKPOINT 1**    The figure shows a uniform square plate from which four identical squares at the corners will be removed. (a) Where is the center of mass of the plate originally? Where is it after the removal of (b) square 1; (c) squares 1 and 2; (d) squares 1 and 3; (e) squares 1, 2, and 3; (f) all four squares? Answer in terms of quadrants, axes, or points (without calculation, of course).

---

**Sample Problem    9-1**

Three particles of masses $m_1 = 1.2$ kg, $m_2 = 2.5$ kg, and $m_3 = 3.4$ kg form an equilateral triangle of edge length $a = 140$ cm. Where is the center of mass of this system?

**KEY IDEA**    We are dealing with particles instead of an extended solid body, so we can use Eq. 9-5 to locate their center of mass. The particles are in the plane of the equilateral triangle, so we need only the first two equations.

*Calculations:* We can simplify the calculations by choosing the $x$ and $y$ axes so that one of the particles is located at the origin and the $x$ axis coincides with one of the triangle's sides (Fig. 9-3). The three particles then have the following coordinates:

| Particle | Mass (kg) | x (cm) | y (cm) |
|----------|-----------|--------|--------|
| 1 | 1.2 | 0 | 0 |
| 2 | 2.5 | 140 | 0 |
| 3 | 3.4 | 70 | 120 |

The total mass $M$ of the system is 7.1 kg.

From Eq. 9-5, the coordinates of the center of mass are

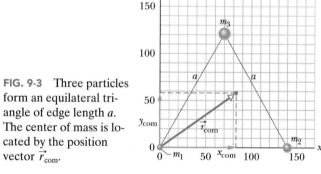

**FIG. 9-3**    Three particles form an equilateral triangle of edge length $a$. The center of mass is located by the position vector $\vec{r}_{com}$.

$$x_{com} = \frac{1}{M}\sum_{i=1}^{3} m_i x_i = \frac{m_1 x_1 + m_2 x_2 + m_3 x_3}{M}$$

$$= \frac{(1.2 \text{ kg})(0) + (2.5 \text{ kg})(140 \text{ cm}) + (3.4 \text{ kg})(70 \text{ cm})}{7.1 \text{ kg}}$$

$$= 83 \text{ cm} \qquad \text{(Answer)}$$

and $$y_{com} = \frac{1}{M}\sum_{i=1}^{3} m_i y_i = \frac{m_1 y_1 + m_2 y_2 + m_3 y_3}{M}$$

$$= \frac{(1.2 \text{ kg})(0) + (2.5 \text{ kg})(0) + (3.4 \text{ kg})(120 \text{ cm})}{7.1 \text{ kg}}$$

$$= 58 \text{ cm}. \qquad \text{(Answer)}$$

In Fig. 9-3, the center of mass is located by the position vector $\vec{r}_{com}$, which has components $x_{com}$ and $y_{com}$.

Figure 9-4a shows a uniform metal plate P of radius 2R from which a disk of radius R has been stamped out (removed) in an assembly line. Using the xy coordinate system shown, locate the center of mass com$_P$ of the plate.

**KEY IDEAS** (1) Let us roughly locate the center of plate P by using symmetry. We note that the plate is symmetric about the x axis (we get the portion below that axis by rotating the upper portion about the axis). Thus, com$_P$ must be on the x axis. The plate (with the disk removed) is not symmetric about the y axis. However, because there is somewhat more mass on the right of the y axis, com$_P$ must be somewhat to the right of that axis. Thus, the location of com$_P$ should be roughly as indicated in Fig. 9-4a.

(2) Plate P is an extended solid body, so we can use Eqs. 9-11 to find the actual coordinates of com$_P$. However, that procedure is difficult. Here is a much easier way: In working with centers of mass, we can assume that the mass of a *uniform* object is concentrated in a particle at the object's center of mass.

*Calculations:* First, put the stamped-out disk (call it disk S) back into place (Fig. 9-4b) to form the original composite plate (call it plate C). Because of its circular symmetry, the center of mass com$_S$ for disk S is at the center of S, at $x = -R$ (as shown). Similarly, the center of mass com$_C$ for composite plate C is at the center of C, at the origin (as shown). We then have the following:

| Plate | Center of Mass | Location of com | Mass |
|-------|----------------|-----------------|------|
| P | com$_P$ | $x_P = ?$ | $m_P$ |
| S | com$_S$ | $x_S = -R$ | $m_S$ |
| C | com$_C$ | $x_C = 0$ | $m_C = m_S + m_P$ |

Assume that mass $m_S$ of disk S is concentrated in a particle at $x_S = -R$, and mass $m_P$ is concentrated in a particle at $x_P$ (Fig. 9-4c). Next treat these two particles as a two-particle system, using Eq. 9-2 to find their center of mass $x_{S+P}$. We get

$$x_{S+P} = \frac{m_S x_S + m_P x_P}{m_S + m_P}. \qquad (9\text{-}12)$$

Next note that the combination of disk S and plate P is composite plate C. Thus, the position $x_{S+P}$ of com$_{S+P}$ must coincide with the position $x_C$ of com$_C$, which is at the origin; so $x_{S+P} = x_C = 0$. Substituting this into Eq. 9-12 and solving for $x_P$, we get

$$x_P = -x_S \frac{m_S}{m_P}. \qquad (9\text{-}13)$$

We can relate these masses to the face areas of S and

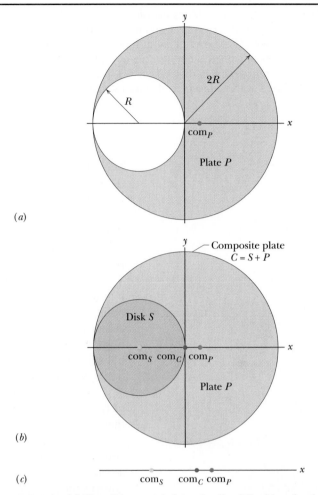

(a)

(b)

(c)

**FIG. 9-4** (a) Plate P is a metal plate of radius 2R, with a circular hole of radius R. The center of mass of P is at point com$_P$. (b) Disk S has been put back into place to form a composite plate C. The center of mass com$_S$ of disk S and the center of mass com$_C$ of plate C are shown. (c) The center of mass com$_{S+P}$ of the combination of S and P coincides with com$_C$, which is at $x = 0$.

P by noting that

$$\text{mass} = \text{density} \times \text{volume}$$
$$= \text{density} \times \text{thickness} \times \text{area}.$$

Then
$$\frac{m_S}{m_P} = \frac{\text{density}_S}{\text{density}_P} \times \frac{\text{thickness}_S}{\text{thickness}_P} \times \frac{\text{area}_S}{\text{area}_P}.$$

Because the plate is uniform, the densities and thicknesses are equal; we are left with

$$\frac{m_S}{m_P} = \frac{\text{area}_S}{\text{area}_P} = \frac{\text{area}_S}{\text{area}_C - \text{area}_S}$$

$$= \frac{\pi R^2}{\pi(2R)^2 - \pi R^2} = \frac{1}{3}.$$

Substituting this and $x_S = -R$ into Eq. 9-13, we have

$$x_P = \tfrac{1}{3}R. \qquad \text{(Answer)}$$

*Tactic 1: Center-of-Mass Problems* Sample Problems 9-1 and 9-2 provide three strategies for simplifying center-of-mass problems. (1) Make full use of the symmetry of the object, be it about a point, a line, or a plane. (2) If the object can be divided into several parts, treat each of these parts as a particle, located at its own center of mass. (3) Choose your axes wisely: If your system is a group of particles, choose one of the particles as your origin. If your system is a body with a line of symmetry, let that be your $x$ or $y$ axis. The choice of origin can be completely arbitrary because the location of the center of mass is the same regardless of the origin from which it is measured.

## 9-3 I Newton's Second Law for a System of Particles

Now that we know how to locate the center of mass of a system of particles, we discuss how external forces can move a center of mass. Let us start with a simple system of two billiard balls.

If you roll a cue ball at a second billiard ball that is at rest, you expect that the two-ball system will continue to have some forward motion after impact. You would be surprised, for example, if both balls came back toward you or if both moved to the right or to the left.

What continues to move forward, its steady motion completely unaffected by the collision, is the center of mass of the two-ball system. If you focus on this point—which is always halfway between these bodies because they have identical masses—you can easily convince yourself by trial at a billiard table that this is so. No matter whether the collision is glancing, head-on, or somewhere in between, the center of mass continues to move forward, as if the collision had never occurred. Let us look into this center-of-mass motion in more detail.

To do so, we replace the pair of billiard balls with an assemblage of $n$ particles of (possibly) different masses. We are interested not in the individual motions of these particles but *only* in the motion of the center of mass of the assemblage. Although the center of mass is just a point, it moves like a particle whose mass is equal to the total mass of the system; we can assign a position, a velocity, and an acceleration to it. We state (and shall prove next) that the vector equation that governs the motion of the center of mass of such a system of particles is

$$\vec{F}_{net} = M\vec{a}_{com} \quad \text{(system of particles).} \quad (9\text{-}14)$$

This equation is Newton's second law for the motion of the center of mass of a system of particles. Note that its form is the same as the form of the equation ($\vec{F}_{net} = m\vec{a}$) for the motion of a single particle. However, the three quantities that appear in Eq. 9-14 must be evaluated with some care:

1. $\vec{F}_{net}$ is the net force of *all external forces* that act on the system. Forces on one part of the system from another part of the system (*internal forces*) are not included in Eq. 9-14.

2. $M$ is the *total mass* of the system. We assume that no mass enters or leaves the system as it moves, so that $M$ remains constant. The system is said to be **closed.**

3. $\vec{a}_{com}$ is the acceleration of the *center of mass* of the system. Equation 9-14 gives no information about the acceleration of any other point of the system.

Equation 9-14 is equivalent to three equations involving the components of $\vec{F}_{net}$ and $\vec{a}_{com}$ along the three coordinate axes. These equations are

$$F_{net,x} = Ma_{com,x} \qquad F_{net,y} = Ma_{com,y} \qquad F_{net,z} = Ma_{com,z}. \quad (9\text{-}15)$$

Now we can go back and examine the behavior of the billiard balls. Once the cue ball has begun to roll, no net external force acts on the (two-ball) system. Thus, because $\vec{F}_{net} = 0$, Eq. 9-14 tells us that $\vec{a}_{com} = 0$ also. Because acceleration is the rate of change of velocity, we conclude that the velocity of the center of mass of the system of two balls does not change. When the two balls collide, the forces that come into play are *internal* forces, on one ball from the other. Such forces do not contribute to the net force $\vec{F}_{net}$, which remains zero. Thus, the center of mass of the system, which was moving forward before the collision, must continue to move forward after the collision, with the same speed and in the same direction.

Equation 9-14 applies not only to a system of particles but also to a solid body, such as the bat of Fig. 9-1*b*. In that case, *M* in Eq. 9-14 is the mass of the bat and $\vec{F}_{net}$ is the gravitational force on the bat. Equation 9-14 then tells us that $\vec{a}_{com} = \vec{g}$. In other words, the center of mass of the bat moves as if the bat were a single particle of mass *M*, with force $\vec{F}_g$ acting on it.

Figure 9-5 shows another interesting case. Suppose that at a fireworks display, a rocket is launched on a parabolic path. At a certain point, it explodes into fragments. If the explosion had not occurred, the rocket would have continued along the trajectory shown in the figure. The forces of the explosion are *internal* to the system (at first the system is just the rocket, and later it is its fragments); that is, they are forces on parts of the system from other parts. If we ignore air drag, the net *external* force $\vec{F}_{net}$ acting on the system is the gravitational force on the system, regardless of whether the rocket explodes. Thus, from Eq. 9-14, the acceleration $\vec{a}_{com}$ of the center of mass of the fragments (while they are in flight) remains equal to $\vec{g}$. This means that the center of mass of the fragments follows the same parabolic trajectory that the rocket would have followed had it not exploded.

When a ballet dancer leaps across the stage in a grand jeté, she raises her arms and stretches her legs out horizontally as soon as her feet leave the stage (Fig. 9-6). These actions shift her center of mass upward through her body. Although the shifting center of mass faithfully follows a parabolic path across the stage, its movement relative to the body decreases the height that is attained by her head and torso, relative to that of a normal jump. The result is that the head and torso follow a nearly horizontal path, giving an illusion that the dancer is floating.

FIG. 9-5 A fireworks rocket explodes in flight. In the absence of air drag, the center of mass of the fragments would continue to follow the original parabolic path, until fragments began to hit the ground.

### Proof of Equation 9-14

Now let us prove this important equation. From Eq. 9-8 we have, for a system of *n* particles,

$$M\vec{r}_{com} = m_1\vec{r}_1 + m_2\vec{r}_2 + m_3\vec{r}_3 + \cdots + m_n\vec{r}_n, \qquad (9\text{-}16)$$

FIG. 9-6 A grand jeté. (Adapted from *The Physics of Dance,* by Kenneth Laws, Schirmer Books, 1984.)

in which $M$ is the system's total mass and $\vec{r}_{com}$ is the vector locating the position of the system's center of mass.

Differentiating Eq. 9-16 with respect to time gives

$$M\vec{v}_{com} = m_1\vec{v}_1 + m_2\vec{v}_2 + m_3\vec{v}_3 + \cdots + m_n\vec{v}_n. \qquad (9\text{-}17)$$

Here $\vec{v}_i\ (= d\vec{r}_i/dt)$ is the velocity of the $i$th particle, and $\vec{v}_{com}\ (= d\vec{r}_{com}/dt)$ is the velocity of the center of mass.

Differentiating Eq. 9-17 with respect to time leads to

$$M\vec{a}_{com} = m_1\vec{a}_1 + m_2\vec{a}_2 + m_3\vec{a}_3 + \cdots + m_n\vec{a}_n. \qquad (9\text{-}18)$$

Here $\vec{a}_i\ (= d\vec{v}_i/dt)$ is the acceleration of the $i$th particle, and $\vec{a}_{com}\ (= d\vec{v}_{com}/dt)$ is the acceleration of the center of mass. Although the center of mass is just a geometrical point, it has a position, a velocity, and an acceleration, as if it were a particle.

From Newton's second law, $m_i\vec{a}_i$ is equal to the resultant force $\vec{F}_i$ that acts on the $i$th particle. Thus, we can rewrite Eq. 9-18 as

$$M\vec{a}_{com} = \vec{F}_1 + \vec{F}_2 + \vec{F}_3 + \cdots + \vec{F}_n. \qquad (9\text{-}19)$$

Among the forces that contribute to the right side of Eq. 9-19 will be forces that the particles of the system exert on each other (internal forces) and forces exerted on the particles from outside the system (external forces). By Newton's third law, the internal forces form third-law force pairs and cancel out in the sum that appears on the right side of Eq. 9-19. What remains is the vector sum of all the *external* forces that act on the system. Equation 9-19 then reduces to Eq. 9-14, the relation that we set out to prove.

✓ **CHECKPOINT 2**    Two skaters on frictionless ice hold opposite ends of a pole of negligible mass. An axis runs along the pole, and the origin of the axis is at the center of mass of the two-skater system. One skater, Fred, weighs twice as much as the other skater, Ethel. Where do the skaters meet if (a) Fred pulls hand over hand along the pole so as to draw himself to Ethel, (b) Ethel pulls hand over hand to draw herself to Fred, and (c) both skaters pull hand over hand?

## Sample Problem 9-3

The three particles in Fig. 9-7a are initially at rest. Each experiences an *external* force due to bodies outside the three-particle system. The directions are indicated, and the magnitudes are $F_1 = 6.0$ N, $F_2 = 12$ N, and $F_3 = 14$ N. What is the acceleration of the center of mass of the system, and in what direction does it move?

**KEY IDEAS**    The position of the center of mass, calculated by the method of Sample Problem 9-1, is marked by a dot in the figure. We can treat the center of mass as if it were a real particle, with a mass equal to the system's total mass $M = 16$ kg. We can also treat the

**FIG. 9-7**   (a) Three particles, initially at rest in the positions shown, are acted on by the external forces shown. The center of mass (com) of the system is marked. (b) The forces are now transferred to the center of mass of the system, which behaves like a particle with a mass $M$ equal to the total mass of the system. The net external force $\vec{F}_{net}$ and the acceleration $\vec{a}_{com}$ of the center of mass are shown.

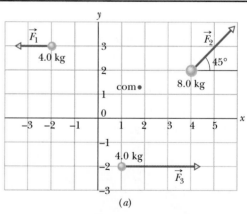

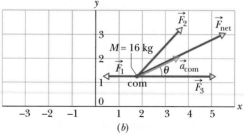

three external forces as if they act at the center of mass (Fig. 9-7b).

**Calculations:** We can now apply Newton's second law ($\vec{F}_{net} = m\vec{a}$) to the center of mass, writing

$$\vec{F}_{net} = M\vec{a}_{com} \qquad (9\text{-}20)$$

or

$$\vec{F}_1 + \vec{F}_2 + \vec{F}_3 = M\vec{a}_{com}$$

so

$$\vec{a}_{com} = \frac{\vec{F}_1 + \vec{F}_2 + \vec{F}_3}{M}. \qquad (9\text{-}21)$$

Equation 9-20 tells us that the acceleration $\vec{a}_{com}$ of the center of mass is in the same direction as the net external force $\vec{F}_{net}$ on the system (Fig. 9-7b). Because the particles are initially at rest, the center of mass must also be at rest. As the center of mass then begins to accelerate, it must move off in the common direction of $\vec{a}_{com}$ and $\vec{F}_{net}$.

We can evaluate the right side of Eq. 9-21 directly on a vector-capable calculator, or we can rewrite Eq. 9-21 in component form, find the components of $\vec{a}_{com}$, and then find $\vec{a}_{com}$. Along the $x$ axis, we have

$$a_{com,x} = \frac{F_{1x} + F_{2x} + F_{3x}}{M}$$

$$= \frac{-6.0\text{ N} + (12\text{ N})\cos 45° + 14\text{ N}}{16\text{ kg}} = 1.03\text{ m/s}^2.$$

Along the $y$ axis, we have

$$a_{com,y} = \frac{F_{1y} + F_{2y} + F_{3y}}{M}$$

$$= \frac{0 + (12\text{ N})\sin 45° + 0}{16\text{ kg}} = 0.530\text{ m/s}^2.$$

From these components, we find that $\vec{a}_{com}$ has the magnitude

$$a_{com} = \sqrt{(a_{com,x})^2 + (a_{com,y})^2}$$
$$= 1.16\text{ m/s}^2 \approx 1.2\text{ m/s}^2 \qquad \text{(Answer)}$$

and the angle (from the positive direction of the $x$ axis)

$$\theta = \tan^{-1}\frac{a_{com,y}}{a_{com,x}} = 27°. \qquad \text{(Answer)}$$

## 9-4 | Linear Momentum

In this section, we discuss only a single particle instead of a system of particles, in order to define two important quantities. Then in Section 9-5, we extend those definitions to systems of many particles.

The first definition concerns a familiar word—*momentum*—that has several meanings in everyday language but only a single precise meaning in physics and engineering. The **linear momentum** of a particle is a vector quantity $\vec{p}$ that is defined as

$$\vec{p} = m\vec{v} \qquad \text{(linear momentum of a particle)}, \qquad (9\text{-}22)$$

in which $m$ is the mass of the particle and $\vec{v}$ is its velocity. (The adjective *linear* is often dropped, but it serves to distinguish $\vec{p}$ from *angular* momentum, which is introduced in Chapter 11 and which is associated with rotation.) Since $m$ is always a positive scalar quantity, Eq. 9-22 tells us that $\vec{p}$ and $\vec{v}$ have the same direction. From Eq. 9-22, the SI unit for momentum is the kilogram-meter per second (kg·m/s).

Newton expressed his second law of motion in terms of momentum:

The time rate of change of the momentum of a particle is equal to the net force acting on the particle and is in the direction of that force.

In equation form this becomes

$$\vec{F}_{net} = \frac{d\vec{p}}{dt}. \qquad (9\text{-}23)$$

In words, Eq. 9-23 says that the net external force $\vec{F}_{net}$ on a particle changes the particle's linear momentum $\vec{p}$. Conversely, the linear momentum can be changed only by a net external force. If there is no net external force, $\vec{p}$ *cannot* change. As we shall see in Section 9-7, this last fact can be an extremely powerful tool in solving problems.

Manipulating Eq. 9-23 by substituting for $\vec{p}$ from Eq. 9-22 gives, for constant mass $m$,

$$\vec{F}_{net} = \frac{d\vec{p}}{dt} = \frac{d}{dt}(m\vec{v}) = m\frac{d\vec{v}}{dt} = m\vec{a}.$$

Thus, the relations $\vec{F}_{net} = d\vec{p}/dt$ and $\vec{F}_{net} = m\vec{a}$ are equivalent expressions of Newton's second law of motion for a particle.

✓ **CHECKPOINT 3** The figure gives the magnitude $p$ of the linear momentum versus time $t$ for a particle moving along an axis. A force directed along the axis acts on the particle. (a) Rank the four regions indicated according to the magnitude of the force, greatest first. (b) In which region is the particle slowing?

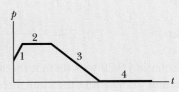

## 9-5 | The Linear Momentum of a System of Particles

Let's extend the definition of linear momentum to a system of particles. Consider a system of $n$ particles, each with its own mass, velocity, and linear momentum. The particles may interact with each other, and external forces may act on them. The system as a whole has a total linear momentum $\vec{P}$, which is defined to be the vector sum of the individual particles' linear momenta. Thus,

$$\begin{aligned}\vec{P} &= \vec{p}_1 + \vec{p}_2 + \vec{p}_3 + \cdots + \vec{p}_n \\ &= m_1\vec{v}_1 + m_2\vec{v}_2 + m_3\vec{v}_3 + \cdots + m_n\vec{v}_n.\end{aligned} \qquad (9\text{-}24)$$

If we compare this equation with Eq. 9-17, we see that

$$\vec{P} = M\vec{v}_{com} \qquad \text{(linear momentum, system of particles)}, \qquad (9\text{-}25)$$

which is another way to define the linear momentum of a system of particles:

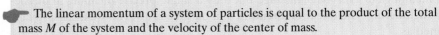 The linear momentum of a system of particles is equal to the product of the total mass $M$ of the system and the velocity of the center of mass.

If we take the time derivative of Eq. 9-25, we find

$$\frac{d\vec{P}}{dt} = M\frac{d\vec{v}_{com}}{dt} = M\vec{a}_{com}. \qquad (9\text{-}26)$$

Comparing Eqs. 9-14 and 9-26 allows us to write Newton's second law for a system of particles in the equivalent form

$$\vec{F}_{net} = \frac{d\vec{P}}{dt} \qquad \text{(system of particles)}, \qquad (9\text{-}27)$$

where $\vec{F}_{net}$ is the net external force acting on the system. This equation is the generalization of the single-particle equation $\vec{F}_{net} = d\vec{p}/dt$ to a system of many particles. In words, the equation says that the net external force $\vec{F}_{net}$ on a system of particles changes the linear momentum $\vec{P}$ of the system. Conversely, the linear momentum can be changed only by a net external force. If there is no net external force, $\vec{P}$ *cannot* change.

## 9-6 | Collision and Impulse

The momentum $\vec{p}$ of any particle-like body cannot change unless a net external force changes it. For example, we could push on the body to change its momentum. More dramatically, we could arrange for the body to collide with a

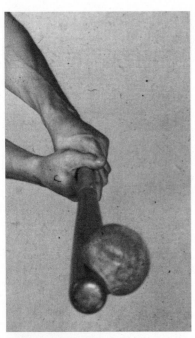

The collision of a ball with a bat collapses part of the ball. *(Photo by Harold E. Edgerton. ©The Harold and Esther Edgerton Family Trust, courtesy of Palm Press, Inc.)*

baseball bat. In such a *collision* (or *crash*), the external force on the body is brief, has large magnitude, and suddenly changes the body's momentum. Collisions occur commonly in our world, but before we get to them, we need to consider a simple collision in which a moving particle-like body (a *projectile*) collides with some other body (a *target*).

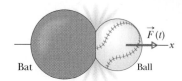

**FIG. 9-8** Force $\vec{F}(t)$ acts on a ball as the ball and a bat collide.

## Single Collision

Let the projectile be a ball and the target be a bat. The collision is brief, and the ball experiences a force that is great enough to slow, stop, or even reverse its motion. Figure 9-8 depicts the collision at one instant. The ball experiences a force $\vec{F}(t)$ that varies during the collision and changes the linear momentum $\vec{p}$ of the ball. That change is related to the force by Newton's second law written in the form $\vec{F} = d\vec{p}/dt$. Thus, in time interval $dt$, the change in the ball's momentum is

$$d\vec{p} = \vec{F}(t)\,dt. \tag{9-28}$$

We can find the net change in the ball's momentum due to the collision if we integrate both sides of Eq. 9-28 from a time $t_i$ just before the collision to a time $t_f$ just after the collision:

$$\int_{t_i}^{t_f} d\vec{p} = \int_{t_i}^{t_f} \vec{F}(t)\,dt. \tag{9-29}$$

The left side of this equation gives us the change in momentum: $\vec{p}_f - \vec{p}_i = \Delta\vec{p}$. The right side, which is a measure of both the magnitude and the duration of the collision force, is called the **impulse** $\vec{J}$ of the collision:

$$\vec{J} = \int_{t_i}^{t_f} \vec{F}(t)\,dt \qquad \text{(impulse defined).} \tag{9-30}$$

Thus, the change in an object's momentum is equal to the impulse on the object:

$$\Delta\vec{p} = \vec{J} \qquad \text{(linear momentum–impulse theorem).} \tag{9-31}$$

This expression can also be written in the vector form

$$\vec{p}_f - \vec{p}_i = \vec{J} \tag{9-32}$$

and in such component forms as

$$\Delta p_x = J_x \tag{9-33}$$

and

$$p_{fx} - p_{ix} = \int_{t_i}^{t_f} F_x\,dt. \tag{9-34}$$

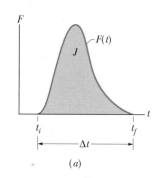

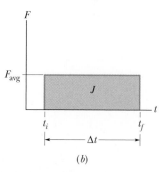

If we have a function for $\vec{F}(t)$, we can evaluate $\vec{J}$ (and thus the change in momentum) by integrating the function. If we have a plot of $\vec{F}$ versus time $t$, we can evaluate $\vec{J}$ by finding the area between the curve and the $t$ axis, such as in Fig. 9-9a. In many situations we do not know how the force varies with time but we do know the average magnitude $F_{\text{avg}}$ of the force and the duration $\Delta t$ ($= t_f - t_i$) of the collision. Then we can write the magnitude of the impulse as

$$J = F_{\text{avg}}\,\Delta t. \tag{9-35}$$

The average force is plotted versus time as in Fig. 9-9b. The area under that curve is equal to the area under the curve for the actual force $F(t)$ in Fig. 9-9a because both areas are equal to impulse magnitude $J$.

Instead of the ball, we could have focused on the bat in Fig. 9-8. At any instant, Newton's third law tells us that the force on the bat has the same magnitude but the opposite direction as the force on the ball. From Eq. 9-30, this means that the impulse on the bat has the same magnitude but the opposite direction as the impulse on the ball.

**FIG. 9-9** (a) The curve shows the magnitude of the time-varying force $F(t)$ that acts on the ball in the collision of Fig. 9-8. The area under the curve is equal to the magnitude of the impulse $\vec{J}$ on the ball in the collision. (b) The height of the rectangle represents the average force $F_{\text{avg}}$ acting on the ball over the time interval $\Delta t$. The area within the rectangle is equal to the area under the curve in (a) and thus is also equal to the magnitude of the impulse $\vec{J}$ in the collision.

✓**CHECKPOINT 4**   A paratrooper whose chute fails to open lands in snow; he is hurt slightly. Had he landed on bare ground, the stopping time would have been 10 times shorter and the collision lethal. Does the presence of the snow increase, decrease, or leave unchanged the values of (a) the paratrooper's change in momentum, (b) the impulse stopping the paratrooper, and (c) the force stopping the paratrooper?

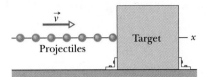

**FIG. 9-10**  A steady stream of projectiles, with identical linear momenta, collides with a target, which is fixed in place. The average force $F_{avg}$ on the target is to the right and has a magnitude that depends on the rate at which the projectiles collide with the target or, equivalently, the rate at which mass collides with the target.

### Series of Collisions

Now let's consider the force on a body when it undergoes a series of identical, repeated collisions. For example, as a prank, we might adjust one of those machines that fire tennis balls to fire them at a rapid rate directly at a wall. Each collision would produce a force on the wall, but that is not the force we are seeking. We want the average force $F_{avg}$ on the wall during the bombardment—that is, the average force during a large number of collisions.

In Fig. 9-10, a steady stream of projectile bodies, with identical mass $m$ and linear momenta $m\vec{v}$, moves along an $x$ axis and collides with a target body that is fixed in place. Let $n$ be the number of projectiles that collide in a time interval $\Delta t$. Because the motion is along only the $x$ axis, we can use the components of the momenta along that axis. Thus, each projectile has initial momentum $mv$ and undergoes a change $\Delta p$ in linear momentum because of the collision. The total change in linear momentum for $n$ projectiles during interval $\Delta t$ is $n\,\Delta p$. The resulting impulse $\vec{J}$ on the target during $\Delta t$ is along the $x$ axis and has the same magnitude of $n\,\Delta p$ but is in the opposite direction. We can write this relation in component form as

$$J = -n\,\Delta p, \tag{9-36}$$

where the minus sign indicates that $J$ and $\Delta p$ have opposite directions.

By rearranging Eq. 9-35 and substituting Eq. 9-36, we find the average force $F_{avg}$ acting on the target during the collisions:

$$F_{avg} = \frac{J}{\Delta t} = -\frac{n}{\Delta t}\Delta p = -\frac{n}{\Delta t} m\,\Delta v. \tag{9-37}$$

This equation gives us $F_{avg}$ in terms of $n/\Delta t$, the rate at which the projectiles collide with the target, and $\Delta v$, the change in the velocity of those projectiles.

If the projectiles stop upon impact, then in Eq. 9-37 we can substitute, for $\Delta v$,

$$\Delta v = v_f - v_i = 0 - v = -v, \tag{9-38}$$

where $v_i\ (=v)$ and $v_f\ (=0)$ are the velocities before and after the collision, respectively. If, instead, the projectiles bounce (rebound) directly backward from the target with no change in speed, then $v_f = -v$ and we can substitute

$$\Delta v = v_f - v_i = -v - v = -2v. \tag{9-39}$$

In time interval $\Delta t$, an amount of mass $\Delta m = nm$ collides with the target. With this result, we can rewrite Eq. 9-37 as

$$F_{avg} = -\frac{\Delta m}{\Delta t}\Delta v. \tag{9-40}$$

This equation gives the average force $F_{avg}$ in terms of $\Delta m/\Delta t$, the rate at which mass collides with the target. Here again we can substitute for $\Delta v$ from Eq. 9-38 or 9-39 depending on what the projectiles do.

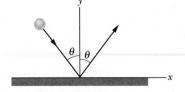

✓**CHECKPOINT 5**   The figure shows an overhead view of a ball bouncing from a vertical wall without any change in its speed. Consider the change $\Delta\vec{p}$ in the ball's linear momentum. (a) Is $\Delta p_x$ positive, negative, or zero? (b) Is $\Delta p_y$ positive, negative, or zero? (c) What is the direction of $\Delta\vec{p}$?

## Sample Problem 9-4

When a male bighorn sheep runs head-first into another male, the rate at which its speed drops to zero is dramatic. Figure 9-11 gives a typical graph of the acceleration $a$ versus time $t$ for such a collision, with the acceleration taken as negative to correspond to an initially positive velocity. The peak acceleration has magnitude 34 m/s² and the duration of the collision is 0.27 s. Assume that the sheep's mass is 90.0 kg. What are the magnitudes of the impulse and average force due to the collision?

**KEY IDEAS** (1) Impulse is defined as being the integration of force with respect to time, according to Eq. 9-30 ($\vec{J} = \int \vec{F}(t)\, dt$). (2) Average force is related to the impulse and the elapsed time by Eq. 9-35 ($J = F_{avg}\, \Delta t$).

*Calculations:* We do not have a function for the force that we can integrate. However, we do have a graph of $a$ versus $t$ that we can transform to be $F$ versus $t$ by multiplying the scale on the acceleration axis by the

mass of 90.0 kg. Then we can graphically integrate by finding the area between the plot and the time axis. Since the plot is in the shape of a triangle, we have for the impulse magnitude

$$J = \text{area} = \tfrac{1}{2}(0.27\ \text{s})(90.0\ \text{kg})(34.0\ \text{m/s}^2)$$

$$= 4.13 \times 10^2\ \text{N·s} \approx 4.1 \times 10^2\ \text{N·s}. \quad \text{(Answer)}$$

For the magnitude of the average force, we can write

$$F_{avg} = \frac{J}{\Delta t} = \frac{4.13 \times 10^2\ \text{N·s}}{0.27\ \text{s}}$$

$$= 1.5 \times 10^3\ \text{N}. \quad \text{(Answer)}$$

*Comment:* The impulse is equal to the change in the sheep's momentum during the collision. So, the size of the impulse depends on the sheep's mass and its speed right before the collision. To win the fight, a male wants a large momentum magnitude. However, if the sheep were to hit skull-to-skull or skull-to-horn, the collision duration would be 1/10 of what we just used and thus the average force would be 10 times what we just calculated. Such a large force would result in concussion or even death; neither result would win the favors of onlooking female sheep. A male avoids such results by having flexible horns that yield somewhat during the collision. Such yielding prolongs the collision and decreases the force to about 1500 N, which the skull, brain, and muscles can withstand. Thus, if a horn breaks during a collision, the next collision could be fatal.

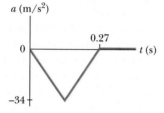

**FIG. 9-11** The acceleration versus time of a bighorn sheep during a collision with another male.

## Sample Problem 9-5

*Race-car wall collision.* Figure 9-12a is an overhead view of the path taken by a race car driver as his car collides with the racetrack wall. Just before the collision, he is traveling at speed $v_i = 70$ m/s along a straight line at 30° from the wall. Just after the collision, he is traveling at speed $v_f = 50$ m/s along a straight line at 10° from the wall. His mass $m$ is 80 kg.

**(a)** What is the impulse $\vec{J}$ on the driver due to the collision?

**KEY IDEAS** We can treat the driver as a particle-like body and thus apply the physics of this section. However, we cannot calculate $\vec{J}$ directly from Eq. 9-30 because we do not know anything about the force $\vec{F}(t)$ on the driver during the collision. That is, we do not have a function of $\vec{F}(t)$ or a plot for it and thus cannot integrate to find $\vec{J}$. However, we *can* find $\vec{J}$ from the change in the driver's linear momentum $\vec{p}$ via Eq. 9-32 ($\vec{J} = \vec{p}_f - \vec{p}_i$).

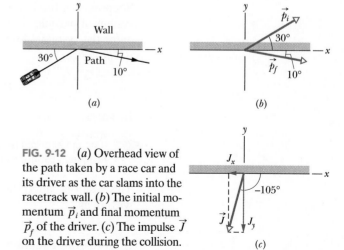

**FIG. 9-12** (*a*) Overhead view of the path taken by a race car and its driver as the car slams into the racetrack wall. (*b*) The initial momentum $\vec{p}_i$ and final momentum $\vec{p}_f$ of the driver. (*c*) The impulse $\vec{J}$ on the driver during the collision.

*Calculations:* Figure 9-12b shows the driver's momentum $\vec{p}_i$ before the collision (at angle 30° from the positive $x$ direction) and his momentum $\vec{p}_f$ after the collision (at angle −10°). From Eqs. 9-32 and 9-22 ($\vec{p} =$

$m\vec{v}$), we can write

$$\vec{J} = \vec{p}_f - \vec{p}_i = m\vec{v}_f - m\vec{v}_i = m(\vec{v}_f - \vec{v}_i). \quad (9\text{-}41)$$

We could evaluate the right side of this equation directly on a vector-capable calculator because we know $m$ is 80 kg, $\vec{v}_f$ is 50 m/s at $-10°$, and $\vec{v}_i$ is 70 m/s at 30°. Instead, here we evaluate Eq. 9-41 in component form.

**x component:** Along the $x$ axis we have

$$
\begin{aligned}
J_x &= m(v_{fx} - v_{ix}) \\
&= (80 \text{ kg})[(50 \text{ m/s}) \cos(-10°) - (70 \text{ m/s}) \cos 30°] \\
&= -910 \text{ kg} \cdot \text{m/s}.
\end{aligned}
$$

**y component:** Along the $y$ axis,

$$
\begin{aligned}
J_y &= m(v_{fy} - v_{iy}) \\
&= (80 \text{ kg})[(50 \text{ m/s}) \sin(-10°) - (70 \text{ m/s}) \sin 30°] \\
&= -3495 \text{ kg} \cdot \text{m/s} \approx -3500 \text{ kg} \cdot \text{m/s}.
\end{aligned}
$$

**Impulse:** The impulse is then

$$\vec{J} = (-910\hat{i} - 3500\hat{j}) \text{ kg} \cdot \text{m/s}, \quad \text{(Answer)}$$

which means the impulse magnitude is

$$J = \sqrt{J_x^2 + J_y^2} = 3616 \text{ kg} \cdot \text{m/s} \approx 3600 \text{ kg} \cdot \text{m/s}.$$

The angle of $\vec{J}$ is given by

$$\theta = \tan^{-1} \frac{J_y}{J_x}, \quad \text{(Answer)}$$

which a calculator evaluates as 75.4°. Recall that the physically correct result of an inverse tangent might be the displayed answer plus 180°. We can tell which is correct here by drawing the components of $\vec{J}$ (Fig. 9-12c). We find that $\theta$ is actually $75.4° + 180° = 255.4°$, which we can write as

$$\theta = -105°. \quad \text{(Answer)}$$

**(b)** The collision lasts for 14 ms. What is the magnitude of the average force on the driver during the collision?

**KEY IDEA** From Eq. 9-35 ($J = F_{avg} \Delta t$), the magnitude $F_{avg}$ of the average force is the ratio of the impulse magnitude $J$ to the duration $\Delta t$ of the collision.

**Calculations:** We have

$$
\begin{aligned}
F_{avg} &= \frac{J}{\Delta t} = \frac{3616 \text{ kg} \cdot \text{m/s}}{0.014 \text{ s}} \\
&= 2.583 \times 10^5 \text{ N} \approx 2.6 \times 10^5 \text{ N}. \quad \text{(Answer)}
\end{aligned}
$$

Using $F = ma$ with $m = 80$ kg, you can show that the magnitude of the driver's average acceleration during the collision is about $3.22 \times 10^3 \text{ m/s}^2 = 329g$. We can guess that the collision will probably be fatal.

**Surviving:** Mechanical engineers attempt to reduce the chances of a fatality by designing and building racetrack walls with more "give," so that a collision lasts longer. For example, if the collision here lasted 10 times longer and the other data remained the same, the magnitudes of the average force and average acceleration would be 10 times less and probably survivable.

## 9-7 | Conservation of Linear Momentum

Suppose that the net external force $\vec{F}_{net}$ (and thus the net impulse $\vec{J}$) acting on a system of particles is zero (the system is isolated) and that no particles leave or enter the system (the system is closed). Putting $\vec{F}_{net} = 0$ in Eq. 9-27 then yields $d\vec{P}/dt = 0$, or

$$\vec{P} = \text{constant} \qquad \text{(closed, isolated system).} \qquad (9\text{-}42)$$

In words,

☛ If no net external force acts on a system of particles, the total linear momentum $\vec{P}$ of the system cannot change.

This result is called the **law of conservation of linear momentum.** It can also be written as

$$\vec{P}_i = \vec{P}_f \qquad \text{(closed, isolated system).} \qquad (9\text{-}43)$$

In words, this equation says that, for a closed, isolated system,

$$\begin{pmatrix} \text{total linear momentum} \\ \text{at some initial time } t_i \end{pmatrix} = \begin{pmatrix} \text{total linear momentum} \\ \text{at some later time } t_f \end{pmatrix}.$$

*Caution:* Momentum should not be confused with energy. In the sample problems of this section, momentum is conserved but energy is definitely not.

Equations 9-42 and 9-43 are vector equations and, as such, each is equivalent to three equations corresponding to the conservation of linear momentum in three mutually perpendicular directions as in, say, an *xyz* coordinate system. Depending on the forces acting on a system, linear momentum might be conserved in one or two directions but not in all directions. However,

> If the component of the net *external* force on a closed system is zero along an axis, then the component of the linear momentum of the system along that axis cannot change.

As an example, suppose that you toss a grapefruit across a room. During its flight, the only external force acting on the grapefruit (which we take as the system) is the gravitational force $\vec{F}_g$, which is directed vertically downward. Thus, the vertical component of the linear momentum of the grapefruit changes, but since no horizontal external force acts on the grapefruit, the horizontal component of the linear momentum cannot change.

Note that we focus on the external forces acting on a closed system. Although internal forces can change the linear momentum of portions of the system, they cannot change the total linear momentum of the entire system.

The sample problems in this section involve explosions that are either one-dimensional (meaning that the motions before and after the explosion are along a single axis) or two-dimensional (meaning that they are in a plane containing two axes). In the following sections we consider one-dimensional and two-dimensional collisions.

✓**CHECKPOINT 6**  An initially stationary device lying on a frictionless floor explodes into two pieces, which then slide across the floor. One piece slides in the positive direction of an *x* axis. (a) What is the sum of the momenta of the two pieces after the explosion? (b) Can the second piece move at an angle to the *x* axis? (c) What is the direction of the momentum of the second piece?

## Sample Problem  9-6

*One-dimensional explosion:* A ballot box with mass $m = 6.0$ kg slides with speed $v = 4.0$ m/s across a frictionless floor in the positive direction of an *x* axis. The box explodes into two pieces. One piece, with mass $m_1 = 2.0$ kg, moves in the positive direction of the *x* axis at $v_1 = 8.0$ m/s. What is the velocity of the second piece, with mass $m_2$?

**KEY IDEAS**  (1) We could get the velocity of the second piece if we knew its momentum, because we already know its mass is $m_2 = m - m_1 = 4.0$ kg. (2) We can relate the momenta of the two pieces to the original momentum of the box if momentum is conserved.

**Calculations:** Our reference frame will be that of the floor. Our system, which consists initially of the box and then of the two pieces, is closed but is not isolated, because the box and pieces each experience a normal force from the floor and a gravitational force. However, those forces are both vertical and thus cannot change the horizontal component of the momentum of the system. Neither can the forces produced by the explosion, because those forces are internal to the system. Thus, the horizontal component of the momentum of the system is conserved, and we can apply Eq. 9-43 along the *x* axis.

The initial momentum of the system is that of the box:

$$\vec{P}_i = m\vec{v}.$$

Similarly, we can write the final momenta of the two pieces as

$$\vec{P}_{f1} = m_1\vec{v}_1 \quad \text{and} \quad \vec{P}_{f2} = m_2\vec{v}_2.$$

The final total momentum $\vec{P}_f$ of the system is the vector sum of the momenta of the two pieces:

$$\vec{P}_f = \vec{P}_{f1} + \vec{P}_{f2} = m_1\vec{v}_1 + m_2\vec{v}_2.$$

Since all the velocities and momenta in this problem are vectors along the *x* axis, we can write them in terms of their *x* components. Doing so while applying Eq. 9-43, we now obtain

$$P_i = P_f$$

or $\qquad mv = m_1v_1 + m_2v_2.$

Inserting known data, we find

$$(6.0 \text{ kg})(4.0 \text{ m/s}) = (2.0 \text{ kg})(8.0 \text{ m/s}) + (4.0 \text{ kg})v_2$$

and thus $\qquad v_2 = 2.0$ m/s.  (Answer)

Since the result is positive, the second piece moves in the positive direction of the *x* axis.

## Sample Problem    9-7

*One-dimensional explosion:* Figure 9-13a shows a space hauler and cargo module, of total mass $M$, traveling along an $x$ axis in deep space. They have an initial velocity $\vec{v}_i$ of magnitude 2100 km/h relative to the Sun. With a small explosion, the hauler ejects the cargo module, of mass $0.20M$ (Fig. 9-13b). The hauler then travels 500 km/h faster than the module along the $x$ axis; that is, the relative speed $v_{\text{rel}}$ between the hauler and the module is 500 km/h. What then is the velocity $\vec{v}_{HS}$ of the hauler relative to the Sun?

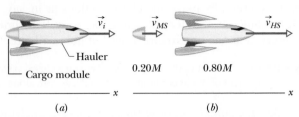

**FIG. 9-13** (a) A space hauler, with a cargo module, moving at initial velocity $\vec{v}_i$. (b) The hauler has ejected the cargo module. Now the velocities relative to the Sun are $\vec{v}_{MS}$ for the module and $\vec{v}_{HS}$ for the hauler.

**KEY IDEA**   Because the hauler–module system is closed and isolated, its total linear momentum is conserved; that is,

$$\vec{P}_i = \vec{P}_f, \tag{9-44}$$

where the subscripts $i$ and $f$ refer to values before and after the ejection, respectively.

**Calculations:** Because the motion is along a single axis, we can write momenta and velocities in terms of their $x$ components, using a sign to indicate direction. Before the ejection, we have

$$P_i = Mv_i. \tag{9-45}$$

Let $v_{MS}$ be the velocity of the ejected module relative to the Sun. The total linear momentum of the system after the ejection is then

$$P_f = (0.20M)v_{MS} + (0.80M)v_{HS}, \tag{9-46}$$

where the first term on the right is the linear momentum of the module and the second term is that of the hauler.

We do not know the velocity $v_{MS}$ of the module rel-

ative to the Sun, but we can relate it to the known velocities with

$$\begin{pmatrix} \text{velocity of} \\ \text{hauler relative} \\ \text{to Sun} \end{pmatrix} = \begin{pmatrix} \text{velocity of} \\ \text{hauler relative} \\ \text{to module} \end{pmatrix} + \begin{pmatrix} \text{velocity of} \\ \text{module relative} \\ \text{to Sun} \end{pmatrix}.$$

In symbols, this gives us

$$v_{HS} = v_{\text{rel}} + v_{MS} \tag{9-47}$$

or

$$v_{MS} = v_{HS} - v_{\text{rel}}.$$

Substituting this expression for $v_{MS}$ into Eq. 9-46, and then substituting Eqs. 9-45 and 9-46 into Eq. 9-44, we find

$$Mv_i = 0.20M(v_{HS} - v_{\text{rel}}) + 0.80Mv_{HS},$$

which gives us

$$v_{HS} = v_i + 0.20v_{\text{rel}},$$

or

$$v_{HS} = 2100 \text{ km/h} + (0.20)(500 \text{ km/h})$$

$$= 2200 \text{ km/h}. \qquad \text{(Answer)}$$

## Sample Problem    9-8

*Two-dimensional explosion:* A firecracker placed inside a coconut of mass $M$, initially at rest on a frictionless floor, blows the coconut into three pieces that slide across the floor. An overhead view is shown in Fig. 9-14a. Piece $C$, with mass $0.30M$, has final speed $v_{fC} = 5.0$ m/s.

(a) What is the speed of piece $B$, with mass $0.20M$?

**KEY IDEA**   First we need to see whether linear momentum is conserved. We note that (1) the coconut and its pieces form a closed system, (2) the explosion forces are internal to that system, and (3) no net external force acts on the system. Therefore, the linear momentum of the system is conserved.

**Calculations:** To get started, we superimpose an $xy$ coordinate system as shown in Fig. 9-14b, with the negative direction of the $x$ axis coinciding with the direction

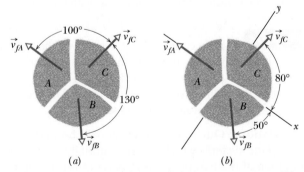

**FIG. 9-14** Three pieces of an exploded coconut move off in three directions along a frictionless floor. (a) An overhead view of the event. (b) The same with a two-dimensional axis system imposed.

of $\vec{v}_{fA}$. The $x$ axis is at 80° with the direction of $\vec{v}_{fC}$ and 50° with the direction of $\vec{v}_{fB}$.

Linear momentum is conserved separately along

each axis. Let's use the $y$ axis and write

$$P_{iy} = P_{fy}, \qquad (9\text{-}48)$$

where subscript $i$ refers to the initial value (before the explosion), and subscript $y$ refers to the $y$ component of $\vec{P}_i$ or $\vec{P}_f$.

The component $P_{iy}$ of the initial linear momentum is zero, because the coconut is initially at rest. To get an expression for $P_{fy}$, we find the $y$ component of the final linear momentum of each piece, using the $y$-component version of Eq. 9-22 ($p_y = mv_y$):

$$p_{fA,y} = 0,$$
$$p_{fB,y} = -0.20Mv_{fB,y} = -0.20Mv_{fB}\sin 50°,$$
$$p_{fC,y} = 0.30Mv_{fC,y} = 0.30Mv_{fC}\sin 80°.$$

(Note that $p_{fA,y} = 0$ because of our choice of axes.) Equation 9-48 can now be written as

$$P_{iy} = P_{fy} = p_{fA,y} + p_{fB,y} + p_{fC,y}.$$

Then, with $v_{fC} = 5.0$ m/s, we have

$$0 = 0 - 0.20Mv_{fB}\sin 50° + (0.30M)(5.0\text{ m/s})\sin 80°,$$

from which we find

$$v_{fB} = 9.64\text{ m/s} \approx 9.6\text{ m/s}. \qquad \text{(Answer)}$$

(b) What is the speed of piece $A$?

*Calculations:* Because linear momentum is also conserved along the $x$ axis, we have

$$P_{ix} = P_{fx}, \qquad (9\text{-}49)$$

where $P_{ix} = 0$ because the coconut is initially at rest. To get $P_{fx}$, we find the $x$ components of the final momenta, using the fact that piece $A$ must have a mass of $0.50M$ ($= M - 0.20M - 0.30M$):

$$p_{fA,x} = -0.50Mv_{fA},$$
$$p_{fB,x} = 0.20Mv_{fB,x} = 0.20Mv_{fB}\cos 50°,$$
$$p_{fC,x} = 0.30Mv_{fC,x} = 0.30Mv_{fC}\cos 80°.$$

Equation 9-49 can now be written as

$$P_{ix} = P_{fx} = p_{fA,x} + p_{fB,x} + p_{fC,x}.$$

Then, with $v_{fC} = 5.0$ m/s and $v_{fB} = 9.64$ m/s, we have

$$0 = -0.50Mv_{fA} + 0.20M(9.64\text{ m/s})\cos 50°$$
$$+ 0.30M(5.0\text{ m/s})\cos 80°,$$

from which we find

$$v_{fA} = 3.0\text{ m/s}. \qquad \text{(Answer)}$$

---

**PROBLEM-SOLVING TACTICS**

*Tactic 2: Conservation of Linear Momentum* For problems involving the conservation of linear momentum, first make sure that you have chosen a closed, isolated system. *Closed* means that no matter (no particles) passes through the system boundary in any direction. *Isolated* means that the net external force acting on the system is zero. If it is not isolated, then remember that each component of linear momentum is conserved separately if the corresponding component of the net external force is zero. So, you might conserve one component and not another.

Next, select two appropriate states of the system (which you may choose to call the initial state and the final state) and write expressions for the linear momentum of the system in each of these two states. In writing these expressions, make sure that you know what inertial reference frame you are using, and make sure also that you include the entire system, not missing any part of it and not including objects that do not belong to your system.

Finally, set your expressions for $\vec{P}_i$ and $\vec{P}_f$ equal to each other and solve for what is requested.

## 9-8 I Momentum and Kinetic Energy in Collisions

In Section 9-6, we considered the collision of two particle-like bodies but focused on only one of the bodies at a time. For the next several sections we switch our focus to the system itself, with the assumption that the system is closed and isolated. In Section 9-7, we discussed a rule about such a system: The total linear momentum $\vec{P}$ of the system cannot change because there is no net external force to change it. This is a very powerful rule because it can allow us to determine the results of a collision *without* knowing the details of the collision (such as how much damage is done).

We shall also be interested in the total kinetic energy of a system of two colliding bodies. If that total happens to be unchanged by the collision, then the kinetic energy of the system is *conserved* (it is the same before and after the collision). Such a collision is called an **elastic collision.** In everyday collisions of common bodies, such as two cars or a ball and a bat, some energy is always transferred from kinetic energy to other forms of energy, such as thermal energy or energy of sound. Thus, the kinetic energy of the system is *not* conserved. Such a collision is called an **inelastic collision.**

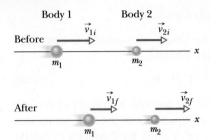

**FIG. 9-15** Bodies 1 and 2 move along an $x$ axis, before and after they have an inelastic collision.

However, in some situations, we can *approximate* a collision of common bodies as elastic. Suppose that you drop a Superball onto a hard floor. If the collision between the ball and floor (or Earth) were elastic, the ball would lose no kinetic energy because of the collision and would rebound to its original height. However, the actual rebound height is somewhat short, showing that at least some kinetic energy is lost in the collision and thus that the collision is somewhat inelastic. Still, we might choose to neglect that small loss of kinetic energy to approximate the collision as elastic.

The inelastic collision of two bodies always involves a loss in the kinetic energy of the system. The greatest loss occurs if the bodies stick together, in which case the collision is called a **completely inelastic collision.** The collision of a baseball and a bat is inelastic. However, the collision of a wet putty ball and a bat is completely inelastic because the putty sticks to the bat.

## 9-9 | Inelastic Collisions in One Dimension

### One-Dimensional Inelastic Collision

Figure 9-15 shows two bodies just before and just after they have a one-dimensional collision. The velocities before the collision (subscript $i$) and after the collision (subscript $f$) are indicated. The two bodies form our system, which is closed and isolated. We can write the law of conservation of linear momentum for this two-body system as

$$\begin{pmatrix} \text{total momentum } \vec{P}_i \\ \text{before the collision} \end{pmatrix} = \begin{pmatrix} \text{total momentum } \vec{P}_f \\ \text{after the collision} \end{pmatrix},$$

which we can symbolize as

$$\vec{p}_{1i} + \vec{p}_{2i} = \vec{p}_{1f} + \vec{p}_{2f} \qquad \text{(conservation of linear momentum)}. \qquad (9\text{-}50)$$

Because the motion is one-dimensional, we can drop the overhead arrows for vectors and use only components along the axis, indicating direction with a sign. Thus, from $p = mv$, we can rewrite Eq. 9-50 as

$$m_1 v_{1i} + m_2 v_{2i} = m_1 v_{1f} + m_2 v_{2f}. \qquad (9\text{-}51)$$

If we know values for, say, the masses, the initial velocities, and one of the final velocities, we can find the other final velocity with Eq. 9-51.

### One-Dimensional Completely Inelastic Collision

Figure 9-16 shows two bodies before and after they have a completely inelastic collision (meaning they stick together). The body with mass $m_2$ happens to be initially at rest ($v_{2i} = 0$). We can refer to that body as the *target* and to the incoming body as the *projectile*. After the collision, the stuck-together bodies move with velocity $V$. For this situation, we can rewrite Eq. 9-51 as

$$m_1 v_{1i} = (m_1 + m_2) V \qquad (9\text{-}52)$$

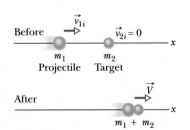

**FIG. 9-16** A completely inelastic collision between two bodies. Before the collision, the body with mass $m_2$ is at rest and the body with mass $m_1$ moves directly toward it. After the collision, the stuck-together bodies move with the same velocity $\vec{V}$.

or

$$V = \frac{m_1}{m_1 + m_2} v_{1i}. \qquad (9\text{-}53)$$

If we know values for, say, the masses and the initial velocity $v_{1i}$ of the projectile, we can find the final velocity $V$ with Eq. 9-53. Note that $V$ must be less than $v_{1i}$ because the mass ratio $m_1/(m_1 + m_2)$ must be less than unity.

### Velocity of the Center of Mass

In a closed, isolated system, the velocity $\vec{v}_{\text{com}}$ of the center of mass of the system cannot be changed by a collision because, with the system isolated, there is no net external force to change it. To get an expression for $\vec{v}_{com}$, let us return to the two-body system and one-dimensional collision of Fig. 9-15. From Eq. 9-25 ($\vec{P} = M\vec{v}_{com}$), we can relate $\vec{v}_{com}$ to the total linear momentum $\vec{P}$ of that two-body system by writing

$$\vec{P} = M\vec{v}_{com} = (m_1 + m_2)\vec{v}_{com}. \tag{9-54}$$

The total linear momentum $\vec{P}$ is conserved during the collision; so it is given by either side of Eq. 9-50. Let us use the left side to write

$$\vec{P} = \vec{p}_{1i} + \vec{p}_{2i}. \tag{9-55}$$

Substituting this expression for $\vec{P}$ in Eq. 9-54 and solving for $\vec{v}_{com}$ give us

$$\vec{v}_{com} = \frac{\vec{P}}{m_1 + m_2} = \frac{\vec{p}_{1i} + \vec{p}_{2i}}{m_1 + m_2}. \tag{9-56}$$

The right side of this equation is a constant, and $\vec{v}_{com}$ has that same constant value before and after the collision.

For example, Fig. 9-17 shows, in a series of freeze-frames, the motion of the center of mass for the completely inelastic collision of Fig. 9-16. Body 2 is the target, and its initial linear momentum in Eq. 9-56 is $\vec{p}_{2i} = m_2\vec{v}_{2i} = 0$. Body 1 is the projectile, and its initial linear momentum in Eq. 9-56 is $\vec{p}_{1i} = m_1\vec{v}_{1i}$. Note that as the series of freeze-frames progresses to and then beyond the collision, the center of mass moves at a constant velocity to the right. After the collision, the common final speed $V$ of the bodies is equal to $\vec{v}_{com}$ because then the center of mass travels with the stuck-together bodies.

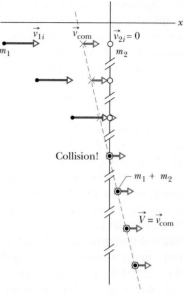

✓ **CHECKPOINT 7**    Body 1 and body 2 are in a completely inelastic one-dimensional collision. What is their final momentum if their initial momenta are, respectively, (a) 10 kg·m/s and 0; (b) 10 kg·m/s and 4 kg·m/s; (c) 10 kg·m/s and −4 kg·m/s?

**FIG. 9-17** Some freeze-frames of the two-body system in Fig. 9-16, which undergoes a completely inelastic collision. The system's center of mass is shown in each freeze-frame. The velocity $\vec{v}_{com}$ of the center of mass is unaffected by the collision. Because the bodies stick together after the collision, their common velocity $\vec{V}$ must be equal to $\vec{v}_{com}$.

---

**Sample Problem   9-9   Build your skill**

The *ballistic pendulum* was used to measure the speeds of bullets before electronic timing devices were developed. The version shown in Fig. 9-18 consists of a large block of wood of mass $M = 5.4$ kg, hanging from two long cords. A bullet of mass $m = 9.5$ g is fired into the block, coming quickly to rest. The *block + bullet* then swing upward, their center of mass rising a vertical distance $h = 6.3$ cm before the pendulum comes momentarily to rest at the end of its arc. What is the speed of the bullet just prior to the collision?

**KEY IDEAS**   We can see that the bullet's speed $v$ must determine the rise height $h$. However, we cannot use the conservation of mechanical energy to relate these two quantities because surely energy is transferred from mechanical energy to other forms (such as thermal energy and energy to break apart the wood) as the bullet penetrates the block. Nevertheless, we can split this complicated motion into two steps that we can separately analyze: (1) the bullet–block collision and (2) the bullet–block rise, during which mechanical energy *is* conserved.

**Reasoning step 1:** Because the collision within the bullet–block system is so brief, we can make two im-

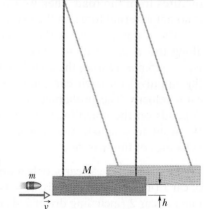

**FIG. 9-18** A ballistic pendulum, used to measure the speeds of bullets.

portant assumptions: (1) During the collision, the gravitational force on the block and the force on the block from the cords are still balanced. Thus, during the collision, the net external impulse on the bullet–block system is zero. Therefore, the system is isolated and its total linear momentum is conserved. (2) The collision is one-dimensional in the sense that the direction of the bullet and block *just after the collision* is in the bullet's original direction of motion.

Because the collision is one-dimensional, the block is initially at rest, and the bullet sticks in the block, we use Eq. 9-53 to express the conservation of linear momentum. Replacing the symbols there with the corresponding symbols here, we have

$$V = \frac{m}{m + M} v. \quad (9\text{-}57)$$

*Reasoning step 2:* As the bullet and block now swing up together, the mechanical energy of the bullet–block–Earth system is conserved. (This mechanical energy is not changed by the force of the cords on the block, because that force is always directed perpendicular to the block's direction of travel.) Let's take the block's initial level as our reference level of zero gravitational potential energy. Then conservation of mechanical energy means that the system's kinetic energy at the start of the swing must equal its gravitational potential energy at the highest point of the swing. Because

the speed of the bullet and block at the start of the swing is the speed $V$ immediately after the collision, we may write this conservation as

$$\tfrac{1}{2}(m + M)V^2 = (m + M)gh.$$

*Combining steps:* Substituting for $V$ from Eq. 9-57 leads to

$$v = \frac{m + M}{m} \sqrt{2gh}$$

$$= \left( \frac{0.0095 \text{ kg} + 5.4 \text{ kg}}{0.0095 \text{ kg}} \right) \sqrt{(2)(9.8 \text{ m/s}^2)(0.063 \text{ m})}$$

$$= 630 \text{ m/s}. \quad \text{(Answer)}$$

The ballistic pendulum is a kind of "transformer," exchanging the high speed of a light object (the bullet) for the low—and thus more easily measurable—speed of a massive object (the block).

---

**Sample Problem** | **9-10** | **Build your skill**

The most dangerous type of collision between two cars is a head-on collision. Surprisingly, data suggest that the risk of fatality to a driver is less if that driver has a passenger in the car. Let's see why.

Figure 9-19 represents two identical cars about to collide head-on in a completely inelastic, one-dimensional collision along an $x$ axis. During the collision, the two cars form a closed system. Let's make the reasonable assumption that during the collision the impulse between the cars is so great that we can neglect the relatively minor impulses due to the frictional forces on the tires from the road. Then we can assume that there is no net external force on the two-car system.

The $x$ component of the initial velocity of car 1 along the $x$ axis is $v_{1i} = +25$ m/s, and that of car 2 is $v_{2i} = -25$ m/s. During the collision, the force (and thus the impulse) on each car causes a change $\Delta v$ in the car's velocity. The probability of a driver being killed depends on the magnitude of $\Delta v$ for that driver's car. We want to calculate the changes $\Delta v_1$ and $\Delta v_2$ in the velocities of the two cars.

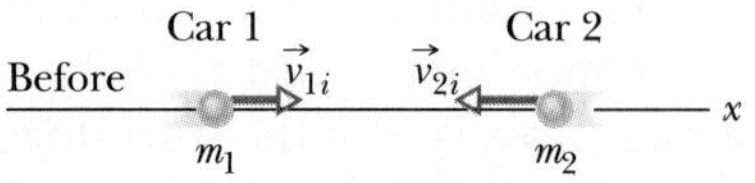

**(a)** First, let each car carry only a driver. The total mass of car 1 (including driver 1) is $m_1 = 1400$ kg, and the total mass of car 2 (including driver 2) is $m_2 = 1400$ kg. What are the changes $\Delta v_1$ and $\Delta v_2$ in the velocities of the cars?

**KEY IDEA** Because the system is closed and isolated, its total linear momentum is conserved.

*Calculations:* From Eq. 9-51, we can write this as

$$m_1 v_{1i} + m_2 v_{2i} = m_1 v_{1f} + m_2 v_{2f}. \quad (9\text{-}58)$$

Since the collision is completely inelastic, the two cars stick together and thus have the same velocity $V$ after

FIG. 9-19 Two cars about to collide head-on.

the collision. Substituting $V$ for $v_{1f}$ and $v_{2f}$ into Eq. 9-58 and solving for $V$, we have

$$V = \frac{m_1 v_{1i} + m_2 v_{2i}}{m_1 + m_2}. \quad (9\text{-}59)$$

Substitution of the given data then results in

$$V = \frac{(1400 \text{ kg})(+25 \text{ m/s}) + (1400 \text{ kg})(-25 \text{ m/s})}{1400 \text{ kg} + 1400 \text{ kg}} = 0.$$

Thus, the change in the velocity of car 1 is

$$\Delta v_1 = v_{1f} - v_{1i} = V - v_{1i}$$

$$= 0 - (+25 \text{ m/s}) = -25 \text{ m/s}, \quad \text{(Answer)}$$

and the change in the velocity of car 2 is

$$\Delta v_2 = v_{2f} - v_{2i} = V - v_{2i}$$

$$= 0 - (-25 \text{ m/s}) = +25 \text{ m/s}. \quad \text{(Answer)}$$

**(b)** Next, we reconsider the collision, but this time with an 80 kg passenger in car 1. What are $\Delta v_1$ and $\Delta v_2$ now?

*Calculations:* Repeating our steps but now substituting $m_1 = 1480$ kg, we find that

$$V = 0.694 \text{ m/s},$$

which gives

$$\Delta v_1 = -24.3 \text{ m/s}$$

and

$$\Delta v_2 = +25.7 \text{ m/s}. \quad \text{(Answer)}$$

**(c)** The magnitude of $\Delta v_1$ is less with the passenger in the car. Because the probability of a driver being killed depends on $\Delta v_1$, we can reason that the probability is less for driver 1.

The data on head-on car collisions do not include values of $\Delta v$, but they do include the car masses and whether or not a collision was fatal. Fitting a function to the collected data, researchers have found that the fatality risk $r_1$ of driver 1 is given by

$$r_1 = c\left(\frac{m_2}{m_1}\right)^{1.79}, \qquad (9\text{-}60)$$

where $c$ is a constant. Justify why the ratio $m_2/m_1$ appears in this equation, and then use the equation to compare the fatality risks for driver 1 with and without the passenger.

**Calculations:** We first rewrite Eq. 9-58 as

$$m_1(v_{1f} - v_{1i}) = -m_2(v_{2f} - v_{2i}).$$

Substituting $\Delta v_1 = v_{1f} - v_{1i}$ and $\Delta v_2 = v_{2f} - v_{2i}$ and rearranging give us

$$\frac{m_2}{m_1} = -\frac{\Delta v_1}{\Delta v_2}. \qquad (9\text{-}61)$$

A driver's fatality risk depends on the change $\Delta v$ for that driver. In Eq. 9-61, we see that the ratio of $\Delta v$ values in a collision is the inverse of the ratio of the masses, and this is the reason researchers can link fatality risk to the ratio of masses in Eq. 9-60.

From part (a) and Eq. 9-60, *without* the passenger, driver 1 has a fatality risk of

$$r_1 = c\left(\frac{1400 \text{ kg}}{1400 \text{ kg}}\right)^{1.79} = c. \qquad (9\text{-}62)$$

From part (b) and Eq. 9-60, *with* the passenger, driver 1 has a fatality risk of

$$r_1' = c\left(\frac{1400 \text{ kg}}{1400 \text{ kg} + 80 \text{ kg}}\right)^{1.79} = 0.9053c.$$

Substituting for $c$ from Eq. 9-62, we find

$$r_1' = 0.9053r_1 \approx 0.91r_1. \qquad \text{(Answer)}$$

In words, the fatality risk for driver 1 is about 9% less when a passenger is in the car.

## 9-10 | Elastic Collisions in One Dimension

As we discussed in Section 9-8, everyday collisions are inelastic but we can approximate some of them as being elastic; that is, we can approximate that the total kinetic energy of the colliding bodies is conserved and is not transferred to other forms of energy:

$$\left(\begin{array}{c}\text{total kinetic energy}\\\text{before the collision}\end{array}\right) = \left(\begin{array}{c}\text{total kinetic energy}\\\text{after the collision}\end{array}\right).$$

This does not mean that the kinetic energy of each colliding body cannot change. Rather, it means this:

In an elastic collision, the kinetic energy of each colliding body may change, but the total kinetic energy of the system does not change.

For example, the collision of a cue ball with an object ball in a game of pool can be approximated as being an elastic collision. If the collision is head-on (the cue ball heads directly toward the object ball), the kinetic energy of the cue ball can be transferred almost entirely to the object ball. (Still, the fact that the collision makes a sound means that at least a little of the kinetic energy is transferred to the energy of the sound.)

### Stationary Target

Figure 9-20 shows two bodies before and after they have a one-dimensional collision, like a head-on collision between pool balls. A projectile body of mass $m_1$ and initial velocity $v_{1i}$ moves toward a target body of mass $m_2$ that is initially at rest ($v_{2i} = 0$). Let's assume that this two-body system is closed and isolated. Then the net linear momentum of the system is conserved, and from Eq. 9-51 we can write that conservation as

$$m_1v_{1i} = m_1v_{1f} + m_2v_{2f} \qquad \text{(linear momentum).} \qquad (9\text{-}63)$$

If the collision is also elastic, then the total kinetic energy is conserved and we can write that conservation as

$$\tfrac{1}{2}m_1v_{1i}^2 = \tfrac{1}{2}m_1v_{1f}^2 + \tfrac{1}{2}m_2v_{2f}^2 \qquad \text{(kinetic energy).} \qquad (9\text{-}64)$$

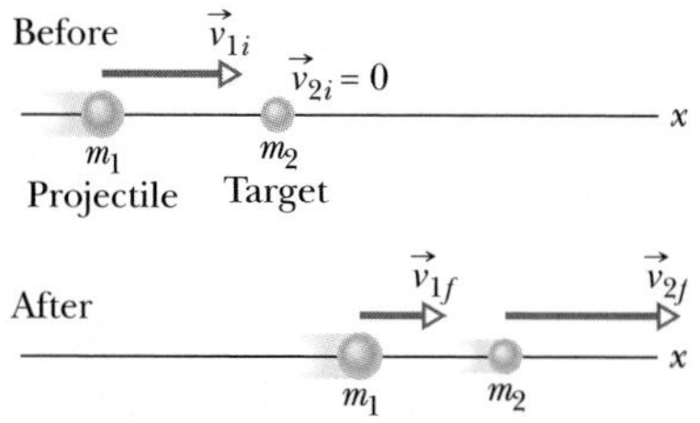

**FIG. 9-20** Body 1 moves along an $x$ axis before having an elastic collision with body 2, which is initially at rest. Both bodies move along that axis after the collision.

In each of these equations, the subscript $i$ identifies the initial velocities and the subscript $f$ the final velocities of the bodies. If we know the masses of the bodies and if we also know $v_{1i}$, the initial velocity of body 1, the only unknown quantities are $v_{1f}$ and $v_{2f}$, the final velocities of the two bodies. With two equations at our disposal, we should be able to find these two unknowns.

To do so, we rewrite Eq. 9-63 as

$$m_1(v_{1i} - v_{1f}) = m_2 v_{2f} \tag{9-65}$$

and Eq. 9-64 as*

$$m_1(v_{1i} - v_{1f})(v_{1i} + v_{1f}) = m_2 v_{2f}^2. \tag{9-66}$$

After dividing Eq. 9-66 by Eq. 9-65 and doing some more algebra, we obtain

$$v_{1f} = \frac{m_1 - m_2}{m_1 + m_2} v_{1i} \tag{9-67}$$

and

$$v_{2f} = \frac{2m_1}{m_1 + m_2} v_{1i}. \tag{9-68}$$

We note from Eq. 9-68 that $v_{2f}$ is always positive (the initially stationary target body with mass $m_2$ always moves forward). From Eq. 9-67 we see that $v_{1f}$ may be of either sign (the projectile body with mass $m_1$ moves forward if $m_1 > m_2$ but rebounds if $m_1 < m_2$).

Let us look at a few special situations.

1. **Equal masses**   If $m_1 = m_2$, Eqs. 9-67 and 9-68 reduce to

$$v_{1f} = 0 \quad \text{and} \quad v_{2f} = v_{1i},$$

which we might call a pool player's result. It predicts that after a head-on collision of bodies with equal masses, body 1 (initially moving) stops dead in its tracks and body 2 (initially at rest) takes off with the initial speed of body 1. In head-on collisions, bodies of equal mass simply exchange velocities. This is true even if body 2 is not initially at rest.

2. **A massive target**   In Fig. 9-20, a massive target means that $m_2 \gg m_1$. For example, we might fire a golf ball at a stationary cannonball. Equations 9-67 and 9-68 then reduce to

$$v_{1f} \approx -v_{1i} \quad \text{and} \quad v_{2f} \approx \left(\frac{2m_1}{m_2}\right) v_{1i}. \tag{9-69}$$

This tells us that body 1 (the golf ball) simply bounces back along its incoming path, its speed essentially unchanged. Initially stationary body 2 (the cannonball) moves forward at a low speed, because the quantity in parentheses in Eq. 9-69 is much less than unity. All this is what we should expect.

3. **A massive projectile**   This is the opposite case; that is, $m_1 \gg m_2$. This time, we fire a cannonball at a stationary golf ball. Equations 9-67 and 9-68 reduce to

$$v_{1f} \approx v_{1i} \quad \text{and} \quad v_{2f} \approx 2v_{1i}. \tag{9-70}$$

Equation 9-70 tells us that body 1 (the cannonball) simply keeps on going, scarcely slowed by the collision. Body 2 (the golf ball) charges ahead at twice the speed of the cannonball.

You may wonder: Why twice the speed? As a starting point in thinking about the matter, recall the collision described by Eq. 9-69, in which the velocity of the incident light body (the golf ball) changed from $+v$ to $-v$, a velocity *change* of $2v$. The same *change* in velocity (but now from zero to $2v$) occurs in this example also.

---

*In this step, we use the identity $a^2 - b^2 = (a - b)(a + b)$. It reduces the amount of algebra needed to solve the simultaneous equations Eqs. 9-65 and 9-66.

## Moving Target

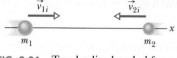

Now that we have examined the elastic collision of a projectile and a stationary target, let us examine the situation in which both bodies are moving before they undergo an elastic collision.

**FIG. 9-21** Two bodies headed for a one-dimensional elastic collision.

For the situation of Fig. 9-21, the conservation of linear momentum is written as

$$m_1 v_{1i} + m_2 v_{2i} = m_1 v_{1f} + m_2 v_{2f}, \tag{9-71}$$

and the conservation of kinetic energy is written as

$$\tfrac{1}{2} m_1 v_{1i}^2 + \tfrac{1}{2} m_2 v_{2i}^2 = \tfrac{1}{2} m_1 v_{1f}^2 + \tfrac{1}{2} m_2 v_{2f}^2. \tag{9-72}$$

To solve these simultaneous equations for $v_{1f}$ and $v_{2f}$, we first rewrite Eq. 9-71 as

$$m_1(v_{1i} - v_{1f}) = -m_2(v_{2i} - v_{2f}), \tag{9-73}$$

and Eq. 9-72 as

$$m_1(v_{1i} - v_{1f})(v_{1i} + v_{1f}) = -m_2(v_{2i} - v_{2f})(v_{2i} + v_{2f}). \tag{9-74}$$

After dividing Eq. 9-74 by Eq. 9-73 and doing some more algebra, we obtain

$$v_{1f} = \frac{m_1 - m_2}{m_1 + m_2} v_{1i} + \frac{2m_2}{m_1 + m_2} v_{2i} \tag{9-75}$$

and

$$v_{2f} = \frac{2m_1}{m_1 + m_2} v_{1i} + \frac{m_2 - m_1}{m_1 + m_2} v_{2i}. \tag{9-76}$$

Note that the assignment of subscripts 1 and 2 to the bodies is arbitrary. If we exchange those subscripts in Fig. 9-21 and in Eqs. 9-75 and 9-76, we end up with the same set of equations. Note also that if we set $v_{2i} = 0$, body 2 becomes a stationary target as in Fig. 9-20, and Eqs. 9-75 and 9-76 reduce to Eqs. 9-67 and 9-68, respectively.

**✓ CHECKPOINT 8** What is the final linear momentum of the target in Fig. 9-20 if the initial linear momentum of the projectile is 6 kg·m/s and the final linear momentum of the projectile is (a) 2 kg·m/s and (b) −2 kg·m/s? (c) What is the final kinetic energy of the target if the initial and final kinetic energies of the projectile are, respectively, 5 J and 2 J?

## Sample Problem 9-11

Two metal spheres, suspended by vertical cords, initially just touch, as shown in Fig. 9-22. Sphere 1, with mass $m_1 = 30$ g, is pulled to the left to height $h_1 = 8.0$ cm, and then released from rest. After swinging down, it undergoes an elastic collision with sphere 2, whose mass $m_2 = 75$ g. What is the velocity $v_{1f}$ of sphere 1 just after the collision?

**KEY IDEA** We can split this complicated motion into two steps that we can analyze separately: (1) the descent of sphere 1 (in which mechanical energy is conserved) and (2) the two-sphere collision (in which momentum is conserved).

**Step 1:** As sphere 1 swings down, the mechanical energy of the sphere–Earth system is conserved. (The mechanical energy is not changed by the force of the cord on sphere 1 because that force is always directed perpendicular to the sphere's direction of travel.)

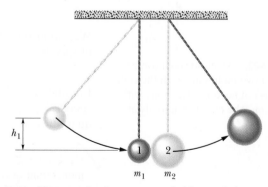

**FIG. 9-22** Two metal spheres suspended by cords just touch when they are at rest. Sphere 1, with mass $m_1$, is pulled to the left to height $h_1$ and then released.

**Calculation:** Let's take the lowest level as our reference level of zero gravitational potential energy. Then the kinetic energy of sphere 1 at the lowest level must equal the gravitational potential energy of the system when sphere 1 is at height $h_1$. Thus,

$$\tfrac{1}{2}m_1v_{1i}^2 = m_1gh_1,$$

which we solve for the speed $v_{1i}$ of sphere 1 just before the collision:

$$v_{1i} = \sqrt{2gh_1} = \sqrt{(2)(9.8 \text{ m/s}^2)(0.080 \text{ m})}$$
$$= 1.252 \text{ m/s}.$$

**Step 2:** Here we can make two assumptions in addition to the assumption that the collision is elastic. First, we can assume that the collision is one-dimensional because the motions of the spheres are approximately horizontal from just before the collision to just after it. Second, because the collision is so brief, we can assume that the

two-sphere system is closed and isolated. This means that the total linear momentum of the system is conserved.

**Calculation:** Thus, we can use Eq. 9-67 to find the velocity of sphere 1 just after the collision:

$$v_{1f} = \frac{m_1 - m_2}{m_1 + m_2} v_{1i}$$
$$= \frac{0.030 \text{ kg} - 0.075 \text{ kg}}{0.030 \text{ kg} + 0.075 \text{ kg}} (1.252 \text{ m/s})$$
$$= -0.537 \text{ m/s} \approx -0.54 \text{ m/s}. \qquad \text{(Answer)}$$

The minus sign tells us that sphere 1 moves to the left just after the collision.

## 9-11 | Collisions in Two Dimensions

When two bodies collide, the impulse between them determines the directions in which they then travel. In particular, when the collision is not head-on, the bodies do not end up traveling along their initial axis. For such two-dimensional collisions in a closed, isolated system, the total linear momentum must still be conserved:

$$\vec{P}_{1i} + \vec{P}_{2i} = \vec{P}_{1f} + \vec{P}_{2f}. \qquad (9\text{-}77)$$

If the collision is also elastic (a special case), then the total kinetic energy is also conserved:

$$K_{1i} + K_{2i} = K_{1f} + K_{2f}. \qquad (9\text{-}78)$$

Equation 9-77 is often more useful for analyzing a two-dimensional collision if we write it in terms of components on an $xy$ coordinate system. For example, Fig. 9-23 shows a *glancing collision* (it is not head-on) between a projectile body and a target body initially at rest. The impulses between the bodies have sent the bodies off at angles $\theta_1$ and $\theta_2$ to the $x$ axis, along which the projectile initially traveled. In this situation we would rewrite Eq. 9-77 for components along the $x$ axis as

$$m_1v_{1i} = m_1v_{1f} \cos \theta_1 + m_2v_{2f} \cos \theta_2, \qquad (9\text{-}79)$$

and along the $y$ axis as

$$0 = -m_1v_{1f} \sin \theta_1 + m_2v_{2f} \sin \theta_2. \qquad (9\text{-}80)$$

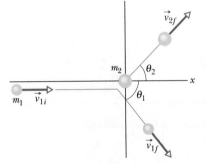

We can also write Eq. 9-78 (for the special case of an elastic collision) in terms of speeds:

$$\tfrac{1}{2}m_1v_{1i}^2 = \tfrac{1}{2}m_1v_{1f}^2 + \tfrac{1}{2}m_2v_{2f}^2 \qquad \text{(kinetic energy).} \qquad (9\text{-}81)$$

**FIG. 9-23** An elastic collision between two bodies in which the collision is not head-on. The body with mass $m_2$ (the target) is initially at rest.

Equations 9-79 to 9-81 contain seven variables: two masses, $m_1$ and $m_2$; three speeds, $v_{1i}$, $v_{1f}$, and $v_{2f}$; and two angles, $\theta_1$ and $\theta_2$. If we know any four of these quantities, we can solve the three equations for the remaining three quantities.

✓**CHECKPOINT 9** In Fig. 9-23, suppose that the projectile has an initial momentum of 6 kg·m/s, a final $x$ component of momentum of 4 kg·m/s, and a final $y$ component of momentum of $-3$ kg·m/s. For the target, what then are (a) the final $x$ component of momentum and (b) the final $y$ component of momentum?

## 9-12 | Systems with Varying Mass: A Rocket

In the systems we have dealt with so far, we have assumed that the total mass of the system remains constant. Sometimes, as in a rocket, it does not. Most of the mass of a rocket on its launching pad is fuel, all of which will eventually be burned and ejected from the nozzle of the rocket engine.

We handle the variation of the mass of the rocket as the rocket accelerates by applying Newton's second law, not to the rocket alone but to the rocket and its ejected combustion products taken together. The mass of *this* system does *not* change as the rocket accelerates.

### Finding the Acceleration

Assume that we are at rest relative to an inertial reference frame, watching a rocket accelerate through deep space with no gravitational or atmospheric drag forces acting on it. For this one-dimensional motion, let $M$ be the mass of the rocket and $v$ its velocity at an arbitrary time $t$ (see Fig. 9-24a).

Figure 9-24b shows how things stand a time interval $dt$ later. The rocket now has velocity $v + dv$ and mass $M + dM$, where the change in mass $dM$ is a *negative quantity*. The exhaust products released by the rocket during interval $dt$ have mass $-dM$ and velocity $U$ relative to our inertial reference frame.

Our system consists of the rocket and the exhaust products released during interval $dt$. The system is closed and isolated, so the linear momentum of the system must be conserved during $dt$; that is,

$$P_i = P_f, \tag{9-82}$$

where the subscripts $i$ and $f$ indicate the values at the beginning and end of time interval $dt$. We can rewrite Eq. 9-82 as

$$Mv = -dM\,U + (M + dM)(v + dv), \tag{9-83}$$

where the first term on the right is the linear momentum of the exhaust products released during interval $dt$ and the second term is the linear momentum of the rocket at the end of interval $dt$.

We can simplify Eq. 9-83 by using the relative speed $v_{\text{rel}}$ between the rocket and the exhaust products, which is related to the velocities relative to the frame with

$$\begin{pmatrix}\text{velocity of rocket} \\ \text{relative to frame}\end{pmatrix} = \begin{pmatrix}\text{velocity of rocket} \\ \text{relative to products}\end{pmatrix} + \begin{pmatrix}\text{velocity of products} \\ \text{relative to frame}\end{pmatrix}.$$

In symbols, this means

$$(v + dv) = v_{\text{rel}} + U,$$

or

$$U = v + dv - v_{\text{rel}}. \tag{9-84}$$

Substituting this result for $U$ into Eq. 9-83 yields, with a little algebra,

$$-dM\,v_{\text{rel}} = M\,dv. \tag{9-85}$$

Dividing each side by $dt$ gives us

$$-\frac{dM}{dt}\,v_{\text{rel}} = M\,\frac{dv}{dt}. \tag{9-86}$$

We replace $dM/dt$ (the rate at which the rocket loses mass) by $-R$, where $R$ is the (positive) mass rate of fuel consumption, and we recognize that $dv/dt$ is the acceleration of the rocket. With these changes, Eq. 9-86 becomes

$$Rv_{\text{rel}} = Ma \qquad \text{(first rocket equation)}. \tag{9-87}$$

Equation 9-87 holds at any instant, with the mass $M$, the fuel consumption rate $R$, and the acceleration $a$ evaluated at that instant.

Note the left side of Eq. 9-87 has the dimensions of force (kg/s · m/s = kg · m/s² = N) and depends only on design characteristics of the rocket engine — namely, the rate $R$ at which it consumes fuel mass and the speed $v_{\text{rel}}$ with which that mass is ejected relative to the rocket. We call this term $Rv_{\text{rel}}$ the **thrust** of the rocket engine and represent it with $T$. Newton's second law emerges clearly

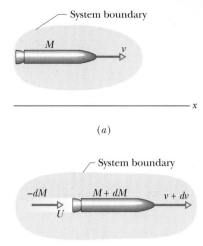

**FIG. 9-24** (*a*) An accelerating rocket of mass $M$ at time $t$, as seen from an inertial reference frame. (*b*) The same but at time $t + dt$. The exhaust products released during interval $dt$ are shown.

if we write Eq. 9-87 as $T = Ma$, in which $a$ is the acceleration of the rocket at the time that its mass is $M$.

### Finding the Velocity

How will the velocity of a rocket change as it consumes its fuel? From Eq. 9-85 we have

$$dv = -v_{rel} \frac{dM}{M}.$$

Integrating leads to

$$\int_{v_i}^{v_f} dv = -v_{rel} \int_{M_i}^{M_f} \frac{dM}{M},$$

in which $M_i$ is the initial mass of the rocket and $M_f$ its final mass. Evaluating the integrals then gives

$$v_f - v_i = v_{rel} \ln \frac{M_i}{M_f} \qquad \text{(second rocket equation)} \qquad (9\text{-}88)$$

for the increase in the speed of the rocket during the change in mass from $M_i$ to $M_f$. (The symbol "ln" in Eq. 9-88 means the *natural logarithm*.) We see here the advantage of multistage rockets, in which $M_f$ is reduced by discarding successive stages when their fuel is depleted. An ideal rocket would reach its destination with only its payload remaining.

### Sample Problem    9-12

A rocket whose initial mass $M_i$ is 850 kg consumes fuel at the rate $R = 2.3$ kg/s. The speed $v_{rel}$ of the exhaust gases relative to the rocket engine is 2800 m/s. What thrust does the rocket engine provide?

**KEY IDEA**    Thrust $T$ is equal to the product of the fuel consumption rate $R$ and the relative speed $v_{rel}$ at which exhaust gases are expelled, as given by Eq. 9-87.

**Calculation:** Here we find

$$T = Rv_{rel} = (2.3 \text{ kg/s})(2800 \text{ m/s})$$
$$= 6440 \text{ N} \approx 6400 \text{ N}. \qquad \text{(Answer)}$$

(b) What is the initial acceleration of the rocket?

**KEY IDEA**    We can relate the thrust $T$ of a rocket to the magnitude $a$ of the resulting acceleration with

$T = Ma$, where $M$ is the rocket's mass. However, $M$ decreases and $a$ increases as fuel is consumed. Because we want the initial value of $a$ here, we must use the initial value $M_i$ of the mass.

**Calculation:** We find

$$a = \frac{T}{M_i} = \frac{6440 \text{ N}}{850 \text{ kg}} = 7.6 \text{ m/s}^2. \qquad \text{(Answer)}$$

To be launched from Earth's surface, a rocket must have an initial acceleration greater than $g = 9.8$ m/s$^2$. Put another way, the thrust $T$ of the rocket engine must exceed the initial gravitational force on the rocket, which here has the magnitude $M_i g$, which gives us (850 kg)(9.8 m/s$^2$), or 8330 N. Because the acceleration or thrust requirement is not met (here $T = 6400$ N), our rocket could not be launched from Earth's surface by itself; it would require another, more powerful, rocket.

## REVIEW & SUMMARY

**Center of Mass**    The **center of mass** of a system of $n$ particles is defined to be the point whose coordinates are given by

$$x_{com} = \frac{1}{M} \sum_{i=1}^{n} m_i x_i, \quad y_{com} = \frac{1}{M} \sum_{i=1}^{n} m_i y_i, \quad z_{com} = \frac{1}{M} \sum_{i=1}^{n} m_i z_i, \qquad (9\text{-}5)$$

or

$$\vec{r}_{com} = \frac{1}{M} \sum_{i=1}^{n} m_i \vec{r}_i, \qquad (9\text{-}8)$$

where $M$ is the total mass of the system.

**Newton's Second Law for a System of Particles**    The motion of the center of mass of any system of particles is governed by **Newton's second law for a system of particles,** which is

$$\vec{F}_{net} = M\vec{a}_{com}. \qquad (9\text{-}14)$$

Here $\vec{F}_{net}$ is the net force of all the *external* forces acting on the system, $M$ is the total mass of the system, and $\vec{a}_{com}$ is the acceleration of the system's center of mass.

**Linear Momentum and Newton's Second Law** For a single particle, we define a quantity $\vec{p}$ called its **linear momentum** as

$$\vec{p} = m\vec{v}, \tag{9-22}$$

and can write Newton's second law in terms of this momentum:

$$\vec{F}_{\text{net}} = \frac{d\vec{p}}{dt}. \tag{9-23}$$

For a system of particles these relations become

$$\vec{P} = M\vec{v}_{\text{com}} \quad \text{and} \quad \vec{F}_{\text{net}} = \frac{d\vec{P}}{dt}. \tag{9-25, 9-27}$$

**Collision and Impulse** Applying Newton's second law in momentum form to a particle-like body involved in a collision leads to the **impulse–linear momentum theorem:**

$$\vec{p}_f - \vec{p}_i = \Delta\vec{p} = \vec{J}, \tag{9-31, 9-32}$$

where $\vec{p}_f - \vec{p}_i = \Delta\vec{p}$ is the change in the body's linear momentum, and $\vec{J}$ is the **impulse** due to the force $\vec{F}(t)$ exerted on the body by the other body in the collision:

$$\vec{J} = \int_{t_i}^{t_f} \vec{F}(t)\, dt. \tag{9-30}$$

If $F_{\text{avg}}$ is the average magnitude of $\vec{F}(t)$ during the collision and $\Delta t$ is the duration of the collision, then for one-dimensional motion

$$J = F_{\text{avg}}\, \Delta t. \tag{9-35}$$

When a steady stream of bodies, each with mass $m$ and speed $v$, collides with a body whose position is fixed, the average force on the fixed body is

$$F_{\text{avg}} = -\frac{n}{\Delta t}\, \Delta p = -\frac{n}{\Delta t}\, m\, \Delta v, \tag{9-37}$$

where $n/\Delta t$ is the rate at which the bodies collide with the fixed body, and $\Delta v$ is the change in velocity of each colliding body. This average force can also be written as

$$F_{\text{avg}} = -\frac{\Delta m}{\Delta t}\, \Delta v, \tag{9-40}$$

where $\Delta m/\Delta t$ is the rate at which mass collides with the fixed body. In Eqs. 9-37 and 9-40, $\Delta v = -v$ if the bodies stop upon impact and $\Delta v = -2v$ if they bounce directly backward with no change in their speed.

**Conservation of Linear Momentum** If a system is isolated so that no net *external* force acts on it, the linear momentum $\vec{P}$ of the system remains constant:

$$\vec{P} = \text{constant} \quad \text{(closed, isolated system).} \tag{9-42}$$

This can also be written as

$$\vec{P}_i = \vec{P}_f \quad \text{(closed, isolated system),} \tag{9-43}$$

where the subscripts refer to the values of $\vec{P}$ at some initial time and at a later time. Equations 9-42 and 9-43 are equivalent statements of the **law of conservation of linear momentum.**

**Inelastic Collision in One Dimension** In an *inelastic collision* of two bodies, the kinetic energy of the two-body system is not conserved. If the system is closed and isolated,

the total linear momentum of the system *must* be conserved, which we can write in vector form as

$$\vec{p}_{1i} + \vec{p}_{2i} = \vec{p}_{1f} + \vec{p}_{2f}, \tag{9-50}$$

where subscripts $i$ and $f$ refer to values just before and just after the collision, respectively.

If the motion of the bodies is along a single axis, the collision is one-dimensional and we can write Eq. 9-50 in terms of velocity components along that axis:

$$m_1 v_{1i} + m_2 v_{2i} = m_1 v_{1f} + m_2 v_{2f}. \tag{9-51}$$

If the bodies stick together, the collision is a *completely inelastic collision* and the bodies have the same final velocity $V$ (because they *are* stuck together).

**Motion of the Center of Mass** The center of mass of a closed, isolated system of two colliding bodies is not affected by a collision. In particular, the velocity $\vec{v}_{\text{com}}$ of the center of mass cannot be changed by the collision.

**Elastic Collisions in One Dimension** An *elastic collision* is a special type of collision in which the kinetic energy of a system of colliding bodies is conserved. If the system is closed and isolated, its linear momentum is also conserved. For a one-dimensional collision in which body 2 is a target and body 1 is an incoming projectile, conservation of kinetic energy and linear momentum yield the following expressions for the velocities immediately after the collision:

$$v_{1f} = \frac{m_1 - m_2}{m_1 + m_2}\, v_{1i} \tag{9-67}$$

and

$$v_{2f} = \frac{2m_1}{m_1 + m_2}\, v_{1i}. \tag{9-68}$$

**Collisions in Two Dimensions** If two bodies collide and their motion is not along a single axis (the collision is not head-on), the collision is two-dimensional. If the two-body system is closed and isolated, the law of conservation of momentum applies to the collision and can be written as

$$\vec{P}_{1i} + \vec{P}_{2i} = \vec{P}_{1f} + \vec{P}_{2f}. \tag{9-77}$$

In component form, the law gives two equations that describe the collision (one equation for each of the two dimensions). If the collision is also elastic (a special case), the conservation of kinetic energy during the collision gives a third equation:

$$K_{1i} + K_{2i} = K_{1f} + K_{2f}. \tag{9-78}$$

**Variable-Mass Systems** In the absence of external forces a rocket accelerates at an instantaneous rate given by

$$Rv_{\text{rel}} = Ma \quad \text{(first rocket equation),} \tag{9-87}$$

in which $M$ is the rocket's instantaneous mass (including unexpended fuel), $R$ is the fuel consumption rate, and $v_{\text{rel}}$ is the fuel's exhaust speed relative to the rocket. The term $Rv_{\text{rel}}$ is the **thrust** of the rocket engine. For a rocket with constant $R$ and $v_{\text{rel}}$, whose speed changes from $v_i$ to $v_f$ when its mass changes from $M_i$ to $M_f$,

$$v_f - v_i = v_{\text{rel}} \ln \frac{M_i}{M_f} \quad \text{(second rocket equation).} \tag{9-88}$$

## QUESTIONS

**1** Figure 9-25 shows an overhead view of three particles on which external forces act. The magnitudes and directions of the forces on two of the particles are indicated. What are the magnitude and direction of the force acting on the third particle if the center of mass of the three-particle system is (a) stationary, (b) moving at a constant velocity rightward, and (c) accelerating rightward?

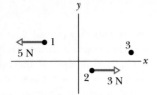

FIG. 9-25   Question 1.

**2** Figure 9-26 shows an overhead view of four particles of equal mass sliding over a frictionless surface at constant velocity. The directions of the velocities are indicated; their magnitudes are equal. Consider pairing the particles. Which pairs form a system with a center of mass that (a) is stationary, (b) is stationary and at the origin, and (c) passes through the origin?

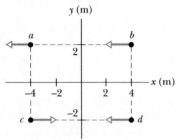

FIG. 9-26   Question 2.

**3** The free-body diagrams in Fig. 9-27 give, from overhead views, the horizontal forces acting on three boxes of chocolates as the boxes move over a frictionless confectioner's counter. For each box, is its linear momentum conserved along the $x$ axis and the $y$ axis?

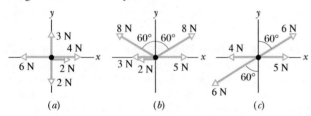

FIG. 9-27   Question 3.

**4** Figure 9-28 shows four groups of three or four identical

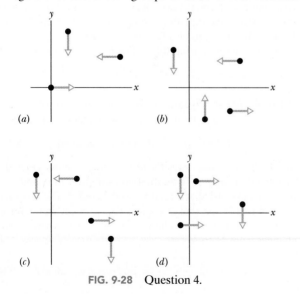

FIG. 9-28   Question 4.

particles that move parallel to either the $x$ axis or the $y$ axis, at identical speeds. Rank the groups according to center-of-mass speed, greatest first.

**5** Consider a box, like that in Sample Problem 9-6, which explodes into two pieces while moving with a constant positive velocity along an $x$ axis. If one piece, with mass $m_1$, ends up with positive velocity $\vec{v}_1$, then the second piece, with mass $m_2$, could end up with (a) a positive velocity $\vec{v}_2$ (Fig. 9-29a), (b) a negative velocity $\vec{v}_2$ (Fig. 9-29b), or (c) zero velocity (Fig. 9-29c). Rank those three possible results for the second piece according to the corresponding magnitude of $\vec{v}_1$, greatest first.

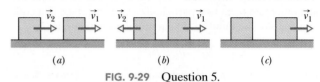

FIG. 9-29   Question 5.

**6** Figure 9-30 shows graphs of force magnitude versus time for a body involved in a collision. Rank the graphs according to the magnitude of the impulse on the body, greatest first.

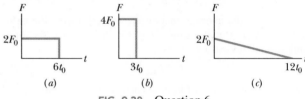

FIG. 9-30   Question 6.

**7** Two bodies have undergone an elastic one-dimensional collision along an $x$ axis. Figure 9-31 is a graph of position versus time for those bodies and for their center of mass. (a) Were both bodies initially moving, or was one initially stationary?

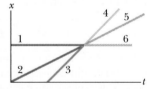

FIG. 9-31   Question 7.

Which line segment corresponds to the motion of the center of mass (b) before the collision and (c) after the collision? (d) Is the mass of the body that was moving faster before the collision greater than, less than, or equal to that of the other body?

**8** Figure 9-32: A block on a horizontal floor is initially either stationary, sliding in the positive direction of an $x$ axis, or sliding in the negative direction of that axis. Then the block ex-

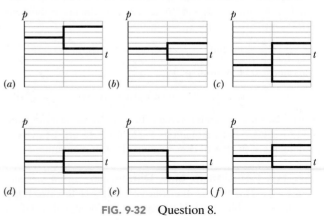

FIG. 9-32   Question 8.

plodes into two pieces that slide along the *x* axis. Assume the block and the two pieces form a closed, isolated system. Six choices for a graph of the momenta of the block and the pieces are given, all versus time *t*. Determine which choices represent physically impossible situations and explain why.

**9** Block 1 with mass $m_1$ slides along an *x* axis across a frictionless floor and then undergoes an elastic collision with a stationary block 2 with mass $m_2$. Figure 9-33 shows a plot of position *x* versus time *t* of block 1 until the collision occurs at position $x_c$ and time $t_c$. In which of the lettered regions on the graph will the plot be continued (after the collision) if (a) $m_1 < m_2$ and (b) $m_1 > m_2$? (c) Along which of the numbered dashed lines will the plot be continued if $m_1 = m_2$?

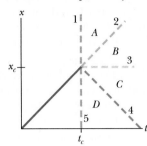

FIG. 9-33   Question 9.

**10** Figure 9-34 shows four graphs of position versus time for two bodies and their center of mass. The two bodies form a closed, isolated system and undergo a completely inelastic, one-dimensional collision on an *x* axis. In graph 1, are (a) the two bodies and (b) the center of mass moving in the positive or negative direction of the *x* axis? (c) Which graphs correspond to a physically impossible situation? Explain.

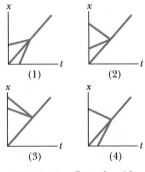

FIG. 9-34   Question 10.

**11** A block slides along a frictionless floor and into a stationary second block with the same mass. Figure 9-35 shows four choices for a graph of the kinetic energies *K* of the blocks. (a) Determine which represent physically impossible situations. Of the others, which best represents (b) an elastic collision and (c) an inelastic collision?

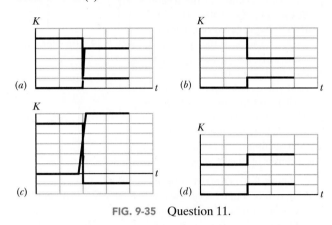

FIG. 9-35   Question 11.

**12** Figure 9-36 shows a snapshot of block 1 as it slides along an *x* axis on a frictionless floor, before it undergoes an elastic collision with stationary block 2. The figure also shows three possible positions of the center of mass (com) of the two-block system at the time of the snapshot. (Point *B* is halfway between the centers of the two blocks.) Is block 1 stationary, moving forward, or moving backward after the collision if the com is located in the snapshot at (a) *A*, (b) *B*, and (c) *C*?

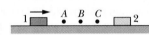

FIG. 9-36   Question 12.

## PROBLEMS

**GO**   Tutoring problem available (at instructor's discretion) in *WileyPLUS* and WebAssign

**SSM**   Worked-out solution available in Student Solutions Manual          **WWW**   Worked-out solution is at ———

• – •••   Number of dots indicates level of problem difficulty          **ILW**   Interactive solution is at ——— http://www.wiley.com/college/halliday

Additional information available in *The Flying Circus of Physics* and at flyingcircusofphysics.com

### sec. 9-2   The Center of Mass

•**1** A 2.00 kg particle has the *xy* coordinates (−1.20 m, 0.500 m), and a 4.00 kg particle has the *xy* coordinates (0.600 m, −0.750 m). Both lie on a horizontal plane. At what (a) *x* and (b) *y* coordinates must you place a 3.00 kg particle such that the center of mass of the three-particle system has the coordinates (−0.500 m, −0.700 m)?

•**2** Figure 9-37 shows a three-particle system, with masses $m_1 = 3.0$ kg, $m_2 = 4.0$ kg, and $m_3 = 8.0$ kg. The scales on the axes are set by $x_s = 2.0$ m and $y_s = 2.0$ m. What are (a) the *x* coordinate and (b) the *y* coordinate of the system's center of mass? (c) If $m_3$ is gradually increased, does the center of mass of the system shift toward or away from that particle, or does it remain stationary?

••**3** What are (a) the *x* coordinate and (b) the *y* coordinate of the center of mass for the uniform plate shown in Fig. 9-38 if *L* = 5.0 cm? **GO**

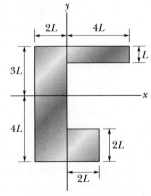

FIG. 9-38   Problem 3.

••**4** In Fig. 9-39, three uniform thin rods, each of length *L* = 22 cm, form an inverted U. The vertical rods each have a mass of 14 g; the horizontal rod has a mass of 42 g. What are (a) the *x* coordinate and (b) the

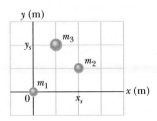

FIG. 9-37   Problem 2.

*y* coordinate of the system's center of mass?

**••5** In the ammonia ($NH_3$) molecule of Fig. 9-40, three hydrogen (H) atoms form an equilateral triangle, with the center of the triangle at distance $d = 9.40 \times 10^{-11}$ m from each hydrogen atom. The nitrogen (N) atom is at the apex of a pyramid, with the three hydrogen atoms forming the base. The nitrogen-to-hydrogen atomic mass ratio is 13.9, and the nitrogen-to-hydrogen distance is $L = 10.14 \times 10^{-11}$ m. What are the (a) *x* and (b) *y* coordinates of the molecule's center of mass? **ILW**

**••6** Figure 9-41 shows a cubical box that has been constructed from uniform metal plate of negligible thickness. The box is open at the top and has edge length $L = 40$ cm. Find (a) the *x* coordinate, (b) the *y* coordinate, and (c) the *z* coordinate of the center of mass of the box.

**••7** Figure 9-42 shows a slab with dimensions $d_1 = 11.0$ cm, $d_2 = 2.80$ cm, and $d_3 = 13.0$ cm. Half the slab consists of aluminum (density $= 2.70$ g/cm³) and half consists of iron (density $= 7.85$ g/cm³). What are (a) the *x* coordinate, (b) the *y* coordinate, and (c) the *z* coordinate of the slab's center of mass?

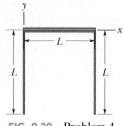

FIG. 9-39 Problem 4.

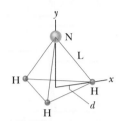

FIG. 9-40 Problem 5.

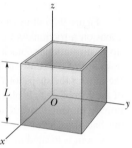

FIG. 9-41 Problem 6.

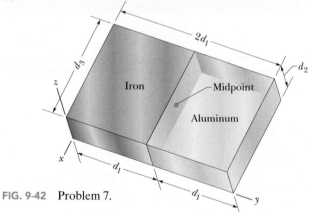

FIG. 9-42 Problem 7.

**•••8** A uniform soda can of mass 0.140 kg is 12.0 cm tall and filled with 1.31 kg of soda (Fig. 9-43). Then small holes are drilled in the top and bottom (with negligible loss of metal) to drain the soda. What is the height *h* of the com of the can and contents (a) initially and (b) after the can loses all the soda? (c) What happens to *h* as the soda drains out? (d) If *x* is the height of the remaining soda at any given instant, find *x* when the com reaches its lowest point.

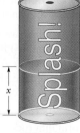

FIG. 9-43
Problem 8.

**sec. 9-3 Newton's Second Law for a System of Particles**

**•9** A big olive ($m = 0.50$ kg) lies at the origin of an *xy* coordinate system, and a big Brazil nut ($M = 1.5$ kg) lies at the point (1.0, 2.0) m. At $t = 0$, a force $\vec{F}_o = (2.0\hat{i} + 3.0\hat{j})$ N begins to act on the olive, and a force $\vec{F}_n = (-3.0\hat{i} - 2.0\hat{j})$ N begins to act on the nut. In unit-vector notation, what is the displacement of the center of mass of the olive–nut system at $t = 4.0$ s, with respect to its position at $t = 0$?

**•10** Two skaters, one with mass 65 kg and the other with mass 40 kg, stand on an ice rink holding a pole of length 10 m and negligible mass. Starting from the ends of the pole, the skaters pull themselves along the pole until they meet. How far does the 40 kg skater move?

**•11** A stone is dropped at $t = 0$. A second stone, with twice the mass of the first, is dropped from the same point at $t = 100$ ms. (a) How far below the release point is the center of mass of the two stones at $t = 300$ ms? (Neither stone has yet reached the ground.) (b) How fast is the center of mass of the two-stone system moving at that time? **ILW**

**•12** A 1000 kg automobile is at rest at a traffic signal. At the instant the light turns green, the automobile starts to move with a constant acceleration of 4.0 m/s². At the same instant a 2000 kg truck, traveling at a constant speed of 8.0 m/s, overtakes and passes the automobile. (a) How far is the com of the automobile–truck system from the traffic light at $t = 3.0$ s? (b) What is the speed of the com then?

**••13** Figure 9-44 shows an arrangement with an air track, in which a cart is connected by a cord to a hanging block. The cart has mass $m_1 = 0.600$ kg, and its center is initially at *xy* coordinates $(-0.500$ m, 0 m); the block has mass $m_2 = 0.400$ kg, and its center is initially at *xy* coordinates $(0, -0.100$ m). The mass of the cord and pulley are negligible. The cart is released from rest, and both cart and block move until the cart hits the pulley. The friction between the cart and the air track and between the pulley and its axle is negligible. (a) In unit-vector notation, what is the acceleration of the center of mass of the cart–block system? (b) What is the velocity of the com as a function of time *t*? (c) Sketch the path taken by the com. (d) If the path is curved, determine whether it bulges upward to the right or downward to the left, and if it is straight, find the angle between it and the *x* axis.

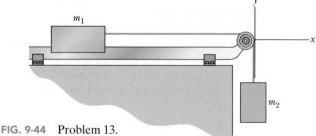

FIG. 9-44 Problem 13.

**••14** In Figure 9-45, two particles are launched from the origin of the coordinate system at time $t = 0$. Particle 1 of mass $m_1 = 5.00$ g is shot directly along the *x* axis on a frictionless floor, with constant speed 10.0 m/s. Particle 2 of mass $m_2 = 3.00$ g is shot with a velocity of magnitude 20.0 m/s, at an upward angle such that it

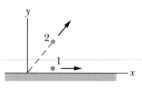

FIG. 9-45 Problem 14.

always stays directly above particle 1. (a) What is the maximum height $H_{max}$ reached by the com of the two-particle system? In unit-vector notation, what are the (b) velocity and (c) acceleration of the com when the com reaches $H_{max}$?

••**15** A shell is shot with an initial velocity $\vec{v}_0$ of 20 m/s, at an angle of $\theta_0 = 60°$ with the horizontal. At the top of the trajectory, the shell explodes into two fragments of equal mass (Fig. 9-46). One fragment, whose speed immediately after the explosion is zero, falls vertically. How far from the gun does the other fragment land, assuming that the terrain is level and that air drag is negligible? SSM

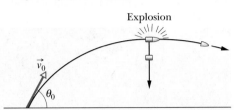

Explosion

**FIG. 9-46**
Problem 15.

•••**16** Ricardo, of mass 80 kg, and Carmelita, who is lighter, are enjoying Lake Merced at dusk in a 30 kg canoe. When the canoe is at rest in the placid water, they exchange seats, which are 3.0 m apart and symmetrically located with respect to the canoe's center. If the canoe moves 40 cm horizontally relative to a pier post, what is Carmelita's mass?

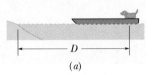

(a)

•••**17** In Fig. 9-47a, a 4.5 kg dog stands on an 18 kg flatboat at distance $D = 6.1$ m from the shore. It walks 2.4 m along the boat toward shore and then stops. Assuming no friction between the boat and the water, find how far the dog is then from the shore. (*Hint:* See Fig. 9-47b.)

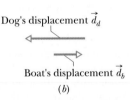

Dog's displacement $\vec{d}_d$

Boat's displacement $\vec{d}_b$

(b)

**FIG. 9-47** Problem 17.

### sec. 9-5 The Linear Momentum of a System of Particles

•**18** A 0.70 kg ball moving horizontally at 5.0 m/s strikes a vertical wall and rebounds with speed 2.0 m/s. What is the magnitude of the change in its linear momentum?

•**19** A 2100 kg truck traveling north at 41 km/h turns east and accelerates to 51 km/h. (a) What is the change in the truck's kinetic energy? What are the (b) magnitude and (c) direction of the change in its momentum? ILW

••**20** Figure 9-48 gives an overhead view of the path taken by a 0.165 kg cue ball as it bounces from a rail of a pool table. The ball's initial speed is 2.00 m/s, and the angle $\theta_1$ is 30.0°. The bounce reverses the y component of the ball's velocity but does not alter the x component. What are (a) angle $\theta_2$ and (b) the change in the ball's linear momentum in unit-vector notation? (The fact that the ball rolls is irrelevant to the problem.)

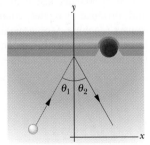

**FIG. 9-48** Problem 20.

••**21** A 0.30 kg softball has a velocity of 15 m/s at an angle of 35° below the horizontal just before making contact with the bat.

What is the magnitude of the change in momentum of the ball while in contact with the bat if the ball leaves with a velocity of (a) 20 m/s, vertically downward, and (b) 20 m/s, horizontally back toward the pitcher?

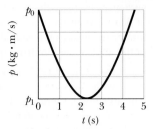

**FIG. 9-49** Problem 22.

••**22** At time $t = 0$, a ball is struck at ground level and sent over level ground. Figure 9-49 gives momentum $p$ versus $t$ during the flight ($p_0 = 6.0$ kg·m/s and $p_1 = 4.0$ kg·m/s). At what initial angle is the ball launched?

### sec. 9-6 Collision and Impulse

•**23** A force in the negative direction of an x axis is applied for 27 ms to a 0.40 kg ball initially moving at 14 m/s in the positive direction of the axis. The force varies in magnitude, and the impulse has magnitude 32.4 N·s. What are the ball's (a) speed and (b) direction of travel just after the force is applied? What are (c) the average magnitude of the force and (d) the direction of the impulse on the ball? SSM

•**24** In a common but dangerous prank, a chair is pulled away as a person is moving downward to sit on it, causing the victim to land hard on the floor. Suppose the victim falls by 0.50 m, the mass that moves downward is 70 kg, and the collision on the floor lasts 0.082 s. What are the magnitudes of the (a) impulse and (b) average force acting on the victim from the floor during the collision?

•**25** Until his seventies, Henri LaMothe (Fig. 9-50) excited

**FIG. 9-50** Problem 25. Belly-flopping into 30 cm of water. (*George Long/ Sports Illustrated/©Time, Inc.*)

audiences by belly-flopping from a height of 12 m into 30 cm of water. Assuming that he stops just as he reaches the bottom of the water and estimating his mass, find the magnitude of the impulse on him from the water.

•26 In February 1955, a paratrooper fell 370 m from an airplane without being able to open his chute but happened to land in snow, suffering only minor injuries. Assume that his speed at impact was 56 m/s (terminal speed), that his mass (including gear) was 85 kg, and that the magnitude of the force on him from the snow was at the survivable limit of $1.2 \times 10^5$ N. What are (a) the minimum depth of snow that would have stopped him safely and (b) the magnitude of the impulse on him from the snow?

•27 A 1.2 kg ball drops vertically onto a floor, hitting with a speed of 25 m/s. It rebounds with an initial speed of 10 m/s. (a) What impulse acts on the ball during the contact? (b) If the ball is in contact with the floor for 0.020 s, what is the magnitude of the average force on the floor from the ball?

•28 In tae-kwon-do, a hand is slammed down onto a target at a speed of 13 m/s and comes to a stop during the 5.0 ms collision. Assume that during the impact the hand is independent of the arm and has a mass of 0.70 kg. What are the magnitudes of the (a) impulse and (b) average force on the hand from the target?

•29 Suppose a gangster sprays Superman's chest with 3 g bullets at the rate of 100 bullets/min, and the speed of each bullet is 500 m/s. Suppose too that the bullets rebound straight back with no change in speed. What is the magnitude of the average force on Superman's chest?

••30 A 5.0 kg toy car can move along an x axis; Fig. 9-51 gives $F_x$ of the force acting on the car, which begins at rest at time $t$ = 0. The scale on the $F_x$ axis is set by $F_{xs}$ = 5.0 N. In unit-vector notation, what is $\vec{p}$ at (a) $t$ = 4.0 s and (b) $t$ = 7.0 s, and (c) what is $\vec{v}$ at $t$ = 9.0 s?

••31 Figure 9-52 shows a 0.300 kg baseball just before and just after it collides with a bat. Just before, the ball has velocity $\vec{v}_1$ of magnitude 12.0 m/s and angle $\theta_1$ = 35.0°. Just after, it is traveling directly upward with velocity $\vec{v}_2$ of magnitude 10.0 m/s. The duration of the collision is 2.00 ms. What are the (a) magnitude and (b) direction (relative to the positive direction of the x axis) of the impulse on the ball from the bat? What are the (c) magnitude and (d) direction of the average force on the ball from the bat?

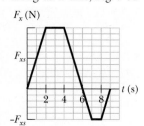

FIG. 9-51 Problem 30.

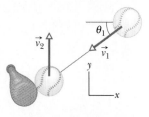

FIG. 9-52 Problem 31.

••32 Basilisk lizards can run across the top of a water surface (Fig. 9-53). With each step, a lizard first slaps its foot against the water and then pushes it down into the water rapidly enough to form an air cavity around the top of the foot. To avoid having to pull the foot back up against water drag in order to complete the step, the lizard withdraws the foot before water can flow into the air cavity. If the lizard is

not to sink, the average upward impulse on the lizard during this full action of slap, downward push, and withdrawal must match the downward impulse due to the gravitational force. Suppose the mass of a basilisk lizard is 90.0 g, the mass of each foot is 3.00 g, the speed of a foot as it slaps the water is 1.50 m/s, and the time for a single step is 0.600 s. (a) What is the magnitude of the impulse on the lizard during the slap? (Assume this impulse is directly upward.) (b) During the 0.600 s duration of a step, what is the downward impulse on the lizard due to the gravitational force? (c) Which action, the slap or the push, provides the primary support for the lizard, or are they approximately equal in their support?

FIG. 9-53 Problem 32. Lizard running across water. (Stephen Dalton/Photo Researchers)

••33 Jumping up before the elevator hits. After the cable snaps and the safety system fails, an elevator cab free-falls from a height of 36 m. During the collision at the bottom of the elevator shaft, a 90 kg passenger is stopped in 5.0 ms. (Assume that neither the passenger nor the cab rebounds.) What are the magnitudes of the (a) impulse and (b) average force on the passenger during the collision? If the passenger were to jump upward with a speed of 7.0 m/s relative to the cab floor just before the cab hits the bottom of the shaft, what are the magnitudes of the (c) impulse and (d) average force (assuming the same stopping time)?

••34 Two average forces. A steady stream of 0.250 kg snowballs is shot perpendicularly into a wall at a speed of 4.00 m/s. Each ball sticks to the wall. Figure 9-54 gives the magnitude $F$ of the force on the wall as a function of time $t$ for two of the snowball impacts. Impacts occur with a repetition time interval $\Delta t_r$ = 50.0 ms, last a duration time interval $\Delta t_d$ = 10 ms, and produce isosceles triangles on the graph, with each impact reaching a force maximum $F_{max}$ = 200 N. During each impact, what are the magnitudes of (a) the impulse and (b) the average force on the wall? (c) During a time interval

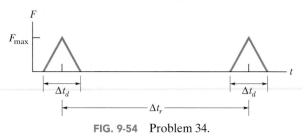

FIG. 9-54 Problem 34.

of many impacts, what is the magnitude of the average force on the wall?

**••35** A soccer player kicks a soccer ball of mass 0.45 kg that is initially at rest. The player's foot is in contact with the ball for $3.0 \times 10^{-3}$ s, and the force of the kick is given by

$$F(t) = [(6.0 \times 10^6)t - (2.0 \times 10^9)t^2] \text{ N}$$

for $0 \le t \le 3.0 \times 10^{-3}$ s, where $t$ is in seconds. Find the magnitudes of (a) the impulse on the ball due to the kick, (b) the average force on the ball from the player's foot during the period of contact, (c) the maximum force on the ball from the player's foot during the period of contact, and (d) the ball's velocity immediately after it loses contact with the player's foot. SSM

**••36** In the overhead view of Fig. 9-55, a 300 g ball with a speed $v$ of 6.0 m/s strikes a wall at an angle $\theta$ of 30° and then rebounds with the same speed and angle. It is in contact with the wall for 10 ms. In unit-vector notation, what are (a) the impulse on the ball from the wall and (b) the average force on the wall from the ball?

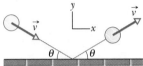

**FIG. 9-55** Problem 36.

**••37** Figure 9-56 shows an approximate plot of force magnitude $F$ versus time $t$ during the collision of a 58 g Superball with a wall. The initial velocity of the ball is 34 m/s perpendicular to the wall; the ball rebounds directly back with approximately the same speed, also perpendicular to the wall. What is $F_{max}$, the maximum magnitude of the force on the ball from the wall during the collision? GO

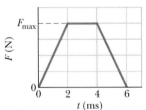

**FIG. 9-56** Problem 37.

**••38** A 0.25 kg puck is initially stationary on an ice surface with negligible friction. At time $t = 0$, a horizontal force begins to move the puck. The force is given by $\vec{F} = (12.0 - 3.00t^2)\hat{i}$, with $\vec{F}$ in newtons and $t$ in seconds, and it acts until its magnitude is zero. (a) What is the magnitude of the impulse on the puck from the force between $t = 0.500$ s and $t = 1.25$ s? (b) What is the change in momentum of the puck between $t = 0$ and the instant at which $F = 0$?

### sec. 9-7 Conservation of Linear Momentum

**•39** A 91 kg man lying on a surface of negligible friction shoves a 68 g stone away from himself, giving it a speed of 4.0 m/s. What speed does the man acquire as a result? SSM

**•40** A space vehicle is traveling at 4300 km/h relative to Earth when the exhausted rocket motor (mass $4m$) is disengaged and sent backward with a speed of 82 km/h relative to the command module (mass $m$). What is the speed of the command module relative to Earth just after the separation?

**••41** In the Olympiad of 708 B.C., some athletes competing in the standing long jump used handheld weights called *halteres* to lengthen their jumps (Fig. 9-57). The weights were swung up in front just before liftoff and then swung down and thrown backward during the flight. Suppose a modern 78 kg long jumper similarly uses two 5.50 kg halteres, throwing them horizontally to the rear at his maximum height such that

their horizontal velocity is zero relative to the ground. Let his liftoff velocity be $\vec{v} = (9.5\hat{i} + 4.0\hat{j})$ m/s with or without the halteres, and assume that he lands at the liftoff level. What distance would the use of the halteres add to his range?

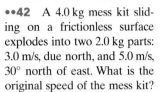

**FIG. 9-57** Problem 41. *(Réunion des Musées Nationaux/Art Resource)*

**••42** A 4.0 kg mess kit sliding on a frictionless surface explodes into two 2.0 kg parts: 3.0 m/s, due north, and 5.0 m/s, 30° north of east. What is the original speed of the mess kit?

**••43** Figure 9-58 shows a two-ended "rocket" that is initially stationary on a frictionless floor, with its center at the origin of an $x$ axis. The rocket consists of a central block $C$ (of mass $M = 6.00$ kg) and blocks $L$ and $R$ (each of mass $m = 2.00$ kg) on the left and right sides. Small explosions can shoot either of the side blocks away from block $C$ and along the $x$ axis. Here is the sequence: (1) At time $t = 0$, block $L$ is shot to the left with a speed of 3.00 m/s *relative* to the velocity that the explosion gives the rest of the rocket. (2) Next, at time $t = 0.80$ s, block $R$ is shot to the right with a speed of 3.00 m/s *relative* to the velocity that block $C$ then has. At $t = 2.80$ s, what are (a) the velocity of block $C$ and (b) the position of its center?

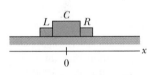

**FIG. 9-58** Problem 43.

**••44** An object, with mass $m$ and speed $v$ relative to an observer, explodes into two pieces, one three times as massive as the other; the explosion takes place in deep space. The less massive piece stops relative to the observer. How much kinetic energy is added to the system during the explosion, as measured in the observer's reference frame?

**••45** A vessel at rest at the origin of an $xy$ coordinate system explodes into three pieces. Just after the explosion, one piece, of mass $m$, moves with velocity $(-30 \text{ m/s})\hat{i}$ and a second piece, also of mass $m$, moves with velocity $(-30 \text{ m/s})\hat{j}$. The third piece has mass $3m$. Just after the explosion, what are the (a) magnitude and (b) direction of the velocity of the third piece?

**••46** In Fig. 9-59, a stationary block explodes into two pieces $L$ and $R$ that slide across a frictionless floor and then into regions with friction, where they stop. Piece $L$, with a mass of 2.0 kg, encounters a coefficient of kinetic friction $\mu_L = 0.40$ and slides to a stop in distance $d_L = 0.15$ m. Piece $R$ encounters a coefficient of kinetic friction $\mu_R = 0.50$ and slides to a stop in distance $d_R = 0.25$ m. What was the mass of the block?

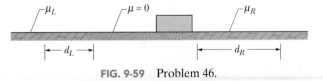

**FIG. 9-59** Problem 46.

**••47** A 20.0 kg body is moving through space in the positive direction of an $x$ axis with a speed of 200 m/s when, due to an internal explosion, it breaks into three parts. One part, with a mass of 10.0 kg, moves away from the point of explosion with a speed of 100 m/s in the positive $y$ direction. A second part, with a mass of 4.00 kg, moves in the negative $x$ direction with a speed of 500 m/s. (a) In unit-vector notation, what is the velocity of the third part? (b) How much energy is released in the explosion? Ignore effects due to the gravitational force. **SSM WWW**

•••**48** Particle $A$ and particle $B$ are held together with a compressed spring between them. When they are released, the spring pushes them apart, and they then fly off in opposite directions, free of the spring. The mass of $A$ is 2.00 times the mass of $B$, and the energy stored in the spring was 60 J. Assume that the spring has negligible mass and that all its stored energy is transferred to the particles. Once that transfer is complete, what are the kinetic energies of (a) particle $A$ and (b) particle $B$?

### sec. 9-9 Inelastic Collisions in One Dimension

•**49** A bullet of mass 10 g strikes a ballistic pendulum of mass 2.0 kg. The center of mass of the pendulum rises a vertical distance of 12 cm. Assuming that the bullet remains embedded in the pendulum, calculate the bullet's initial speed.

•**50** A 5.20 g bullet moving at 672 m/s strikes a 700 g wooden block at rest on a frictionless surface. The bullet emerges, traveling in the same direction with its speed reduced to 428 m/s. (a) What is the resulting speed of the block? (b) What is the speed of the bullet–block center of mass?

••**51** In Anchorage, collisions of a vehicle with a moose are so common that they are referred to with the abbreviation MVC. Suppose a 1000 kg car slides into a stationary 500 kg moose on a very slippery road, with the moose being thrown through the windshield (a common MVC result). (a) What percent of the original kinetic energy is lost in the collision to other forms of energy? A similar danger occurs in Saudi Arabia because of camel–vehicle collisions (CVC). (b) What percent of the original kinetic energy is lost if the car hits a 300 kg camel? (c) Generally, does the percent loss increase or decrease if the animal mass decreases?

••**52** In the "before" part of Fig. 9-60, car $A$ (mass 1100 kg) is stopped at a traffic light when it is rear-ended by car $B$ (mass 1400 kg). Both cars then slide with locked wheels until the frictional force from the slick road (with a low $\mu_k$ of 0.13) stops them, at distances $d_A = 8.2$ m and $d_B = 6.1$ m. What are the speeds of (a) car $A$ and (b) car $B$ at the start of the sliding, just after the collision? (c) Assuming that linear momentum is conserved during the collision, find the speed of car $B$

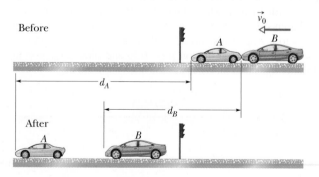

**FIG. 9-60**  Problem 52.

just before the collision. (d) Explain why this assumption may be invalid.

••**53** In Fig. 9-61$a$, a 3.50 g bullet is fired horizontally at two blocks at rest on a frictionless table. The bullet passes through block 1 (mass 1.20 kg) and embeds itself in block 2 (mass 1.80 kg). The blocks end up with speeds $v_1 = 0.630$ m/s and $v_2 = 1.40$ m/s (Fig. 9-61$b$). Neglecting the material removed from block 1 by the bullet, find the speed of the bullet as it (a) leaves and (b) enters block 1. **GO**

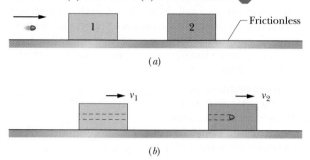

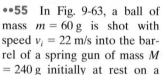

(b)

**FIG. 9-61**  Problem 53.

••**54** In Fig. 9-62, a 10 g bullet moving directly upward at 1000 m/s strikes and passes through the center of mass of a 5.0 kg block initially at rest. The bullet emerges from the block moving directly upward at 400 m/s. To what maximum height does the block then rise above its initial position?

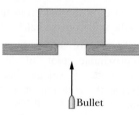

**FIG. 9-62**  Problem 54.

••**55** In Fig. 9-63, a ball of mass $m = 60$ g is shot with speed $v_i = 22$ m/s into the barrel of a spring gun of mass $M = 240$ g initially at rest on a

**FIG. 9-63**  Problem 55.

frictionless surface. The ball sticks in the barrel at the point of maximum compression of the spring. Assume that the increase in thermal energy due to friction between the ball and the barrel is negligible. (a) What is the speed of the spring gun after the ball stops in the barrel? (b) What fraction of the initial kinetic energy of the ball is stored in the spring? **GO**

••**56** A completely inelastic collision occurs between two balls of wet putty that move directly toward each other along a vertical axis. Just before the collision, one ball, of mass 3.0 kg, is moving upward at 20 m/s and the other ball, of mass 2.0 kg, is moving downward at 12 m/s. How high do the combined two balls of putty rise above the collision point? (Neglect air drag.)

••**57** A 5.0 kg block with a speed of 3.0 m/s collides with a 10 kg block that has a speed of 2.0 m/s in the same direction. After the collision, the 10 kg block travels in the original direction with a speed of 2.5 m/s. (a) What is the velocity of the 5.0 kg block immediately after the collision? (b) By how much does the total kinetic energy of the system of two blocks change because of the collision? (c) Suppose, instead, that the 10 kg block ends up with a speed of 4.0 m/s. What then is the change in the total kinetic energy? (d) Account for the result you obtained in (c). **ILW**

**•••58** In Fig. 9-64, block 2 (mass 1.0 kg) is at rest on a frictionless surface and touching the end of an unstretched spring of spring constant 200

FIG. 9-64   Problem 58.

N/m. The other end of the spring is fixed to a wall. Block 1 (mass 2.0 kg), traveling at speed $v_1 = 4.0$ m/s, collides with block 2, and the two blocks stick together. When the blocks momentarily stop, by what distance is the spring compressed?

**•••59** In Fig. 9-65, block 1 (mass 2.0 kg) is moving rightward at 10 m/s and block 2 (mass 5.0 kg) is moving rightward at 3.0 m/s. The surface is frictionless, and a spring with a

FIG. 9-65   Problems 59 and 126.

spring constant of 1120 N/m is fixed to block 2. When the blocks collide, the compression of the spring is maximum at the instant the blocks have the same velocity. Find the maximum compression.   ILW

### sec. 9-10   Elastic Collisions in One Dimension

**•60** Two titanium spheres approach each other head-on with the same speed and collide elastically. After the collision, one of the spheres, whose mass is 300 g, remains at rest. (a) What is the mass of the other sphere? (b) What is the speed of the two-sphere center of mass if the initial speed of each sphere is 2.00 m/s?

**•61** A cart with mass 340 g moving on a frictionless linear air track at an initial speed of 1.2 m/s undergoes an elastic collision with an initially stationary cart of unknown mass. After the collision, the first cart continues in its original direction at 0.66 m/s. (a) What is the mass of the second cart? (b) What is its speed after impact? (c) What is the speed of the two-cart center of mass?   SSM

**•62** In Fig. 9-66, block $A$ (mass 1.6 kg) slides into block $B$ (mass 2.4 kg), along a frictionless surface. The directions of three velocities before ($i$) and after ($f$) the collision are indicated; the corresponding speeds are $v_{Ai} = 5.5$ m/s, $v_{Bi} = 2.5$ m/s, and $v_{Bf} = 4.9$ m/s. What are the (a) speed and (b) direction (left or right) of velocity $\vec{v}_{Af}$? (c) Is the collision elastic?

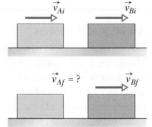

FIG. 9-66   Problem 62.

**••63** A body of mass 2.0 kg makes an elastic collision with another body at rest and continues to move in the original direction but with one-fourth of its original speed. (a) What is the mass of the other body? (b) What is the speed of the two-body center of mass if the initial speed of the 2.0 kg body was 4.0 m/s?   SSM

**••64** Block 1, with mass $m_1$ and speed 4.0 m/s, slides along an $x$ axis on a frictionless floor and then undergoes a one-dimensional elastic collision with stationary block 2, with mass $m_2 = 0.40m_1$. The two blocks then slide into a region where the coefficient of kinetic friction is 0.50; there they stop. How far into that region do (a) block 1 and (b) block 2 slide?

**••65** In Fig. 9-67, particle 1 of mass $m_1 = 0.30$ kg slides rightward along an $x$ axis on a frictionless floor with a speed

of 2.0 m/s. When it reaches $x = 0$, it undergoes a one-dimensional elastic collision with stationary particle 2 of mass $m_2 = 0.40$ kg. When particle 2 then reaches a wall at $x_w = 70$ cm, it bounces from the wall with no loss of speed. At what position on the $x$ axis does particle 2 then collide with particle 1?

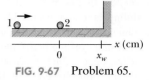

FIG. 9-67   Problem 65.

**••66** A steel ball of mass 0.500 kg is fastened to a cord that is 70.0 cm long and fixed at the far end. The ball is then released when the cord is horizontal (Fig. 9-68). At the bottom of its path, the ball strikes a

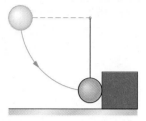

FIG. 9-68   Problem 66.

2.50 kg steel block initially at rest on a frictionless surface. The collision is elastic. Find (a) the speed of the ball and (b) the speed of the block, both just after the collision.

**••67** Block 1 of mass $m_1$ slides along a frictionless floor and into a one-dimensional elastic collision with stationary block 2 of mass $m_2 = 3m_1$. Prior to the collision, the center of mass of the two-block system had a speed of 3.00 m/s. Afterward, what are the speeds of (a) the center of mass and (b) block 2?

**••68** In Fig. 9-69, block 1 of mass $m_1$ slides from rest along a frictionless ramp from height $h = 2.50$ m and then collides with stationary block 2, which has mass $m_2 = 2.00m_1$. After the collision, block 2 slides into a region where the coefficient of kinetic friction $\mu_k$ is 0.500 and comes to a stop in distance $d$ within that region. What is the value of distance $d$ if the collision is (a) elastic and (b) completely inelastic?

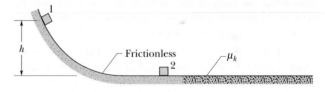

FIG. 9-69   Problem 68.

**•••69** A small ball of mass $m$ is aligned above a larger ball of mass $M = 0.63$ kg (with a slight separation, as with the baseball and basketball of Fig. 9-70$a$), and the two are dropped simultaneously from a height of $h = 1.8$ m. (Assume the radius of each ball is negligible relative to $h$.) (a) If the larger ball rebounds elastically from the floor and then the small ball rebounds elastically from the larger ball, what value of $m$ results in the larger ball stopping when it collides with the small ball? (b) What height does the small ball then reach (Fig. 9-70$b$)?

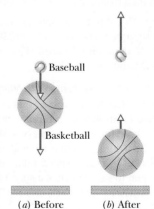

FIG. 9-70   Problem 69.

**•••70** In Fig. 9-71, puck 1 of mass $m_1 = 0.20$ kg is sent sliding across a frictionless lab bench, to undergo a one-dimensional elastic collision with stationary puck 2. Puck 2 then slides off the bench and lands a distance $d$ from the base of the bench. Puck 1 rebounds from the collision and slides off the opposite edge of the bench, landing a distance $2d$ from the base of the bench. What is the mass of puck 2? (*Hint:* Be careful with signs.)

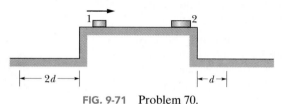

FIG. 9-71   Problem 70.

### sec. 9-11   Collisions in Two Dimensions

**••71** A projectile proton with a speed of 500 m/s collides elastically with a target proton initially at rest. The two protons then move along perpendicular paths, with the projectile path at 60° from the original direction. After the collision, what are the speeds of (a) the target proton and (b) the projectile proton?

**••72** Two 2.0 kg bodies, $A$ and $B$, collide. The velocities before the collision are $\vec{v}_A = (15\hat{i} + 30\hat{j})$ m/s and $\vec{v}_B = (-10\hat{i} + 5.0\hat{j})$ m/s. After the collision, $\vec{v}'_A = (-5.0\hat{i} + 20\hat{j})$ m/s. What are (a) the final velocity of $B$ and (b) the change in the total kinetic energy (including sign)?

**••73** In Fig. 9-23, projectile particle 1 is an alpha particle and target particle 2 is an oxygen nucleus. The alpha particle is scattered at angle $\theta_1 = 64.0°$ and the oxygen nucleus recoils with speed $1.20 \times 10^5$ m/s and at angle $\theta_2 = 51.0°$. In atomic mass units, the mass of the alpha particle is 4.00 u and the mass of the oxygen nucleus is 16.0 u. What are the (a) final and (b) initial speeds of the alpha particle?  ILW

**••74** Ball $B$, moving in the positive direction of an $x$ axis at speed $v$, collides with stationary ball $A$ at the origin. $A$ and $B$ have different masses. After the collision, $B$ moves in the negative direction of the $y$ axis at speed $v/2$. (a) In what direction does $A$ move? (b) Show that the speed of $A$ cannot be determined from the given information.

**•••75** After a completely inelastic collision, two objects of the same mass and same initial speed move away together at half their initial speed. Find the angle between the initial velocities of the objects.

### sec. 9-12   Systems with Varying Mass: A Rocket

**•76** Consider a rocket that is in deep space and at rest relative to an inertial reference frame. The rocket's engine is to be fired for a certain interval. What must be the rocket's *mass ratio* (ratio of initial to final mass) over that interval if the rocket's original speed relative to the inertial frame is to be equal to (a) the exhaust speed (speed of the exhaust products relative to the rocket) and (b) 2.0 times the exhaust speed?

**•77** A rocket that is in deep space and initially at rest relative to an inertial reference frame has a mass of $2.55 \times 10^5$ kg, of which $1.81 \times 10^5$ kg is fuel. The rocket engine is then fired for 250 s while fuel is consumed at the rate of 480 kg/s. The speed of the exhaust products relative to the

rocket is 3.27 km/s. (a) What is the rocket's thrust? After the 250 s firing, what are (b) the mass and (c) the speed of the rocket?  SSM ILW

**•78** A 6090 kg space probe moving nose-first toward Jupiter at 105 m/s relative to the Sun fires its rocket engine, ejecting 80.0 kg of exhaust at a speed of 253 m/s relative to the space probe. What is the final velocity of the probe?

**•79** In Fig. 9-72, two long barges are moving in the same direction in still water, one with a speed of 10 km/h and the other with a speed of 20 km/h. While they are passing each other, coal is shoveled from the slower to the faster one at a rate of 1000 kg/min. How much additional force must be provided by the driving engines of (a) the faster barge and (b) the slower barge if neither is to change speed? Assume that the shoveling is always perfectly sideways and that the frictional forces between the barges and the water do not depend on the mass of the barges.  SSM

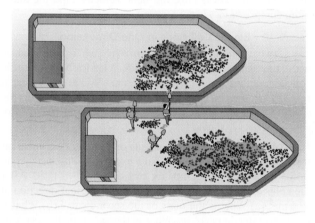

FIG. 9-72   Problem 79.

### Additional Problems

**80** *Speed amplifier.* In Fig. 9-73, block 1 of mass $m_1$ slides along an $x$ axis on a frictionless floor with a speed of $v_{1i} = 4.00$ m/s. Then it undergoes a one-dimensional elastic collision with stationary block 2 of mass $m_2 = 0.500m_1$. Next,

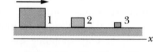

FIG. 9-73   Problem 80.

block 2 undergoes a one-dimensional elastic collision with stationary block 3 of mass $m_3 = 0.500m_2$. (a) What then is the speed of block 3? Are (b) the speed, (c) the kinetic energy, and (d) the momentum of block 3 greater than, less than, or the same as the initial values for block 1?

**81** *Speed deamplifier.* In Fig. 9-74, block 1 of mass $m_1$ slides along an $x$ axis on a frictionless floor at speed 4.00 m/s. Then it undergoes a one-dimensional elastic collision with stationary block 2 of mass $m_2 = 2.00m_1$. Next, block 2 undergoes a one-dimensional elastic collision with stationary block 3 of mass $m_3 = 2.00m_2$. (a) What then is the speed of block 3? Are (b) the speed, (c) the kinetic energy, and (d) the momentum of block 3 greater than, less than, or the same as the initial values for block 1?

FIG. 9-74   Problem 81.

**82** Figure 9-75 shows an overhead view of two particles sliding at constant velocity over a frictionless surface. The particles have the same mass and the same initial speed $v = 4.00$ m/s, and they collide where their paths intersect. An $x$ axis is arranged to bisect the angle between their incoming paths, so that $\theta = 40.0°$. The region to the right of the collision is divided into four lettered sections by the $x$ axis and four numbered dashed lines. In what region or along what line do the particles travel if the collision is (a) completely inelastic, (b) elastic, and (c) inelastic? What are their final speeds if the collision is (d) completely inelastic and (e) elastic?

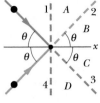

FIG. 9-75
Problem 82.

**83** *"Relative" is an important word.* In Fig. 9-76, block $L$ of mass $m_L = 1.00$ kg and block $R$ of mass $m_R = 0.500$ kg are held in place with a compressed

FIG. 9-76    Problem 83.

spring between them. When the blocks are released, the spring sends them sliding across a frictionless floor. (The spring has negligible mass and falls to the floor after the blocks leave it.) (a) If the spring gives block $L$ a release speed of 1.20 m/s *relative* to the floor, how far does block $R$ travel in the next 0.800 s? (b) If, instead, the spring gives block $L$ a release speed of 1.20 m/s *relative* to the velocity that the spring gives block $R$, how far does block $R$ travel in the next 0.800 s?

**84** *Pancake collapse of a tall building.* In the section of a tall building shown in Fig. 9-77$a$, the infrastructure of any given floor $K$ must support the weight $W$ of all higher floors. Normally the infrastructure is constructed with a safety factor $s$ so that it can withstand an even greater downward force of $sW$. If, however, the

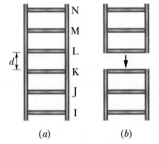

FIG. 9-77    Problem 84.

support columns between $K$ and $L$ suddenly collapse and allow the higher floors to free-fall together onto floor $K$ (Fig. 9-77$b$), the force in the collision can exceed $sW$ and, after a brief pause, cause $K$ to collapse onto floor $J$, which collapses on floor $I$, and so on until the ground is reached. Assume that the floors are separated by $d = 4.0$ m and have the same mass. Also assume that when the floors above $K$ free-fall onto $K$, the collision lasts 1.5 ms. Under these simplified conditions, what value must the safety factor $s$ exceed to prevent pancake collapse of the building?

**85** A railroad car moves under a grain elevator at a constant speed of 3.20 m/s. Grain drops into the car at the rate of 540 kg/min. What is the magnitude of the force needed to keep the car moving at constant speed if friction is negligible?

**86** *Tyrannosaurus rex* may have known from experience not to run particularly fast because of the danger of tripping, in which case its short forearms would have been no help in cushioning the fall. Suppose a *T. rex* of mass $m$ trips while walking, toppling over, with its center of mass falling freely a

distance of 1.5 m. Then its center of mass descends an additional 0.30 m due to compression of its body and the ground. (a) In multiples of the dinosaur's weight, what is the approximate magnitude of the average vertical force on the dinosaur during its collision with the ground (during the descent of 0.30 m)? Now assume that the dinosaur is running at a speed of 19 m/s (fast) when it trips, falls to the ground, and then slides to a stop with a coefficient of kinetic friction of 0.6. Assume also that the average vertical force during the collision and sliding is that in (a). What, approximately, are (b) the magnitude of the average total force on the dinosaur from the ground (again in multiples of its weight) and (c) the sliding distance? The force magnitudes of (a) and (b) strongly suggest that the collision would injure the torso of the dinosaur. The head, which would fall farther, would suffer even greater injury.

**87** A man (weighing 915 N) stands on a long railroad flatcar (weighing 2415 N) as it rolls at 18.2 m/s in the positive direction of an $x$ axis, with negligible friction. Then the man runs along the flatcar in the negative $x$ direction at 4.00 m/s relative to the flatcar. What is the resulting increase in the speed of the flatcar?

**88** Figure 9-78 shows a uniform square plate of edge length $6d = 6.0$ m from which a square piece of edge length $2d$ has been removed. What are (a) the $x$ coordinate and (b) the $y$ coordinate of the center of mass of the remaining piece?

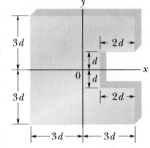

FIG. 9-78    Problem 88.

**89** The last stage of a rocket, which is traveling at a speed of 7600 m/s, consists of two parts that are clamped together: a rocket case with a mass of 290.0 kg and a payload capsule with a mass of 150.0 kg. When the clamp is released, a compressed spring causes the two parts to separate with a relative speed of 910.0 m/s. What are the speeds of (a) the rocket case and (b) the payload after they have separated? Assume that all velocities are along the same line. Find the total kinetic energy of the two parts (c) before and (d) after they separate. (e) Account for the difference.

**90** An object is tracked by a radar station and found to have a position vector given by $\vec{r} = (3500 - 160t)\hat{i} + 2700\hat{j} + 300\hat{k}$, with $\vec{r}$ in meters and $t$ in seconds. The radar station's $x$ axis points east, its $y$ axis north, and its $z$ axis vertically up. If the object is a 250 kg meteorological missile, what are (a) its linear momentum, (b) its direction of motion, and (c) the net force on it?

**91** A pellet gun fires ten 2.0 g pellets per second with a speed of 500 m/s. The pellets are stopped by a rigid wall. What are (a) the magnitude of the momentum of each pellet, (b) the kinetic energy of each pellet, and (c) the magnitude of the average force on the wall from the stream of pellets? (d) If each pellet is in contact with the wall for 0.60 ms, what is the magnitude of the average force on the wall from each pellet during contact? (e) Why is this average force so different from the average force calculated in (c)? **SSM**

**92** A body is traveling at 2.0 m/s along the positive direction of an $x$ axis; no net force acts on the body. An internal explosion separates the body into two parts, each of 4.0 kg, and increases the total kinetic energy by 16 J. The forward part continues to move in the original direction of motion. What are the speeds of (a) the rear part and (b) the forward part?

**93** A 1400 kg car moving at 5.3 m/s is initially traveling north along the positive direction of a $y$ axis. After completing a 90° right-hand turn in 4.6 s, the inattentive operator drives into a tree, which stops the car in 350 ms. In unit-vector notation, what is the impulse on the car (a) due to the turn and (b) due to the collision? What is the magnitude of the average force that acts on the car (c) during the turn and (d) during the collision? (e) What is the direction of the average force during the turn? **SSM**

**94** A spacecraft is separated into two parts by detonating the explosive bolts that hold them together. The masses of the parts are 1200 kg and 1800 kg; the magnitude of the impulse on each part from the bolts is 300 N·s. With what relative speed do the two parts separate because of the detonation?

**95** A ball having a mass of 150 g strikes a wall with a speed of 5.2 m/s and rebounds with only 50% of its initial kinetic energy. (a) What is the speed of the ball immediately after rebounding? (b) What is the magnitude of the impulse on the wall from the ball? (c) If the ball is in contact with the wall for 7.6 ms, what is the magnitude of the average force on the ball from the wall during this time interval?

**96** An old Chrysler with mass 2400 kg is moving along a straight stretch of road at 80 km/h. It is followed by a Ford with mass 1600 kg moving at 60 km/h. How fast is the center of mass of the two cars moving?

**97** A railroad freight car of mass $3.18 \times 10^4$ kg collides with a stationary caboose car. They couple together, and 27.0% of the initial kinetic energy is transferred to thermal energy, sound, vibrations, and so on. Find the mass of the caboose. **SSM**

**98** Two blocks of masses 1.0 kg and 3.0 kg are connected by a spring and rest on a frictionless surface. They are given velocities toward each other such that the 1.0 kg block travels initially at 1.7 m/s toward the center of mass, which remains at rest. What is the initial speed of the other block?

**99** A 75 kg man is riding on a 39 kg cart traveling at a velocity of 2.3 m/s. He jumps off with zero horizontal velocity relative to the ground. What is the resulting change in the cart's velocity, including sign?

**100** A certain radioactive (parent) nucleus transforms to a different (daughter) nucleus by emitting an electron and a neutrino. The parent nucleus was at rest at the origin of an $xy$ coordinate system. The electron moves away from the origin with linear momentum $(-1.2 \times 10^{-22}$ kg·m/s$)\hat{i}$; the neutrino moves away from the origin with linear momentum $(-6.4 \times 10^{-23}$ kg·m/s$)\hat{j}$. What are the (a) magnitude and (b) direction of the linear momentum of the daughter nucleus? (c) If the daughter nucleus has a mass of $5.8 \times 10^{-26}$ kg, what is its kinetic energy? **ILW**

**101** In the arrangement of Fig. 9-23, billiard ball 1 moving at a speed of 2.2 m/s undergoes a glancing collision with identical billiard ball 2 that is at rest. After the collision, ball 2

moves at speed 1.1 m/s, at an angle of $\theta_2 = 60°$. What are (a) the magnitude and (b) the direction of the velocity of ball 1 after the collision? (c) Do the given data suggest the collision is elastic or inelastic? **SSM**

**102** A rocket is moving away from the solar system at a speed of $6.0 \times 10^3$ m/s. It fires its engine, which ejects exhaust with a speed of $3.0 \times 10^3$ m/s relative to the rocket. The mass of the rocket at this time is $4.0 \times 10^4$ kg, and its acceleration is 2.0 m/s². (a) What is the thrust of the engine? (b) At what rate, in kilograms per second, is exhaust ejected during the firing?

**103** The three balls in the overhead view of Fig. 9-79 are identical. Balls 2 and 3 touch each other and are aligned perpendicular to the path of ball 1. The velocity of ball 1 has magnitude $v_0 = 10$ m/s and is directed at the contact point of balls 1 and 2. After the collision, what are the (a) speed and (b) direction of the velocity of ball 2, the (c) speed and (d) direction of the velocity of ball 3, and the (e) speed and (f) direction of the velocity of ball 1? (*Hint:* With friction absent, each impulse is directed along the line connecting the centers of the colliding balls, normal to the colliding surfaces.)

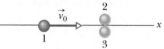

FIG. 9-79   Problem 103.

**104** In a game of pool, the cue ball strikes another ball of the same mass and initially at rest. After the collision, the cue ball moves at 3.50 m/s along a line making an angle of 22.0° with the cue ball's original direction of motion, and the second ball has a speed of 2.00 m/s. Find (a) the angle between the direction of motion of the second ball and the original direction of motion of the cue ball and (b) the original speed of the cue ball. (c) Is kinetic energy (of the centers of mass, don't consider the rotation) conserved?

**105** In Fig. 9-80, two identical containers of sugar are connected by a cord that passes over a frictionless pulley. The cord and pulley have negligible mass, each container and its sugar together have a mass of 500 g, the centers of the containers are separated by 50 mm, and the containers are held fixed at the same height. What is the horizontal distance between the center of container 1 and the center of mass of the two-container system (a) initially and (b) after 20 g of sugar is transferred from container 1 to container 2? After the transfer and after the containers are released, (c) in what direction and (d) at what acceleration magnitude does the center of mass move?

FIG. 9-80   Problem 105.

**106** A 0.15 kg ball hits a wall with a velocity of $(5.00$ m/s$)\hat{i} + (6.50$ m/s$)\hat{j} + (4.00$ m/s$)\hat{k}$. It rebounds from the wall with a velocity of $(2.00$ m/s$)\hat{i} + (3.50$ m/s$)\hat{j} + (-3.20$ m/s$)\hat{k}$. What are (a) the change in the ball's momentum, (b) the impulse on the ball, and (c) the impulse on the wall?

**107** At time $t = 0$, force $\vec{F}_1 = (-4.00\hat{i} + 5.00\hat{j})$ N acts on an initially stationary particle of mass $2.00 \times 10^{-3}$ kg and force $\vec{F}_2 = (2.00\hat{i} - 4.00\hat{j})$ N acts on an initially stationary particle of mass $4.00 \times 10^{-3}$ kg. From time $t = 0$ to $t = 2.00$ ms, what are the (a) magnitude and (b) angle (relative to the positive direction of the $x$ axis) of the displacement of the center of mass of the two-particle system? (c) What is the kinetic energy of the center of mass at $t = 2.00$ ms? **SSM**

**108**  A 0.550 kg ball falls directly down onto concrete, hitting it with a speed of 12.0 m/s and rebounding directly upward with a speed of 3.00 m/s. Extend a $y$ axis upward. In unit-vector notation, what are (a) the change in the ball's momentum, (b) the impulse on the ball, and (c) the impulse on the concrete?

**109**  A collision occurs between a 2.00 kg particle traveling with velocity $\vec{v}_1 = (-4.00 \text{ m/s})\hat{i} + (-5.00 \text{ m/s})\hat{j}$ and a 4.00 kg particle traveling with velocity $\vec{v}_2 = (6.00 \text{ m/s})\hat{i} + (-2.00 \text{ m/s})\hat{j}$. The collision connects the two particles. What then is their velocity in (a) unit-vector notation and as a (b) magnitude and (c) angle?

**110**  An atomic nucleus at rest at the origin of an $xy$ coordinate system transforms into three particles. Particle 1, mass $16.7 \times 10^{-27}$ kg, moves away from the origin at velocity $(6.00 \times 10^6 \text{ m/s})\hat{i}$; particle 2, mass $8.35 \times 10^{-27}$ kg, moves away at velocity $(-8.00 \times 10^6 \text{ m/s})\hat{j}$. (a) In unit-vector notation, what is the linear momentum of the third particle, mass $11.7 \times 10^{-27}$ kg? (b) How much kinetic energy appears in this transformation?

**111**  An electron undergoes a one-dimensional elastic collision with an initially stationary hydrogen atom. What percentage of the electron's initial kinetic energy is transferred to kinetic energy of the hydrogen atom? (The mass of the hydrogen atom is 1840 times the mass of the electron.)

**112**  The script for an action movie calls for a small race car (of mass 1500 kg and length 3.0 m) to accelerate along a flattop boat (of mass

FIG. 9-81   Problem 112.

4000 kg and length 14 m), from one end of the boat to the other, where the car will then jump the gap between the boat and a somewhat lower dock. You are the technical advisor for the movie. The boat will initially touch the dock, as in Fig. 9-81; the boat can slide through the water without significant resistance; both the car and the boat can be approximated as uniform in their mass distribution. Determine what the width of the gap will be just as the car is about to make the jump.

**113**  A rocket sled with a mass of 2900 kg moves at 250 m/s on a set of rails. At a certain point, a scoop on the sled dips into a trough of water located between the tracks and scoops water into an empty tank on the sled. By applying the principle of conservation of linear momentum, determine the speed of the sled after 920 kg of water has been scooped up. Ignore any retarding force on the scoop. **SSM**

**114**  A 140 g ball with speed 7.8 m/s strikes a wall perpendicularly and rebounds in the opposite direction with the same speed. The collision lasts 3.80 ms. What are the magnitudes of the (a) impulse and (b) average force on the wall from the ball?

**115**  (a) How far is the center of mass of the Earth–Moon system from the center of Earth? (Appendix C gives the masses of Earth and the Moon and the distance between the two.) (b) What percentage of Earth's radius is that distance? **SSM**

**116**  A 500.0 kg module is attached to a 400.0 kg shuttle craft, which moves at 1000 m/s relative to the stationary main

spaceship. Then a small explosion sends the module backward with speed 100.0 m/s relative to the new speed of the shuttle craft. As measured by someone on the main spaceship, by what fraction did the kinetic energy of the module and shuttle craft increase because of the explosion?

**117**  A 6100 kg rocket is set for vertical firing from the ground. If the exhaust speed is 1200 m/s, how much gas must be ejected each second if the thrust (a) is to equal the magnitude of the gravitational force on the rocket and (b) is to give the rocket an initial upward acceleration of 21 m/s²? **SSM**

**118**  A 2140 kg railroad flatcar, which can move with negligible friction, is motionless next to a platform. A 242 kg sumo wrestler runs at 5.3 m/s along the platform (parallel to the track) and then jumps onto the flatcar. What is the speed of the flatcar if he then (a) stands on it, (b) runs at 5.3 m/s relative to it in his original direction, and (c) turns and runs at 5.3 m/s relative to the flatcar opposite his original direction?

**119**  In Fig. 9-82, block 1 slides along an $x$ axis on a frictionless floor with a speed of 0.75 m/s. When it reaches stationary block 2, the two blocks undergo an elastic collision. The following table

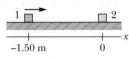

FIG. 9-82   Problem 119.

gives the mass and length of the (uniform) blocks and also the locations of their centers at time $t = 0$. Where is the center of mass of the two-block system located (a) at $t = 0$, (b) when the two blocks first touch, and (c) at $t = 4.0$ s?

| Block | Mass (kg) | Length (cm) | Center at $t = 0$ |
|-------|-----------|-------------|-------------------|
| 1 | 0.25 | 5.0 | $x = -1.50$ m |
| 2 | 0.50 | 6.0 | $x = 0$ |

**120**  In Fig. 9-83, an 80 kg man is on a ladder hanging from a balloon that has a total mass of 320 kg (including the basket passenger). The balloon is initially stationary relative to the ground. If the man on the ladder begins to climb at 2.5 m/s relative to the ladder, (a) in what direction and (b) at what speed does the balloon move? (c) If the man then stops climbing, what is the speed of the balloon?

FIG. 9-83
Problem 120.

**121**  Particle 1 of mass 200 g and speed 3.00 m/s undergoes a one-dimensional collision with stationary particle 2 of mass 400 g. What is the magnitude of the impulse on particle 1 if the collision is (a) elastic and (b) completely inelastic?

**122**  During a lunar mission, it is necessary to increase the speed of a spacecraft by 2.2 m/s when it is moving at 400 m/s relative to the Moon. The speed of the exhaust products from the rocket engine is 1000 m/s relative to the spacecraft. What fraction of the initial mass of the spacecraft must be burned and ejected to accomplish the speed increase?

**123**  In Fig. 9-84, a 3.2 kg box of running shoes slides on a horizontal frictionless table and collides with a 2.0 kg box of ballet slippers initially at rest on the edge of the table, at height $h = 0.40$ m. The speed of the 3.2 kg box is 3.0 m/s just before the collision. If the two boxes stick together because of packing tape on their sides, what is their kinetic energy just before they strike the floor?

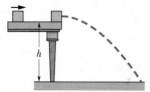

FIG. 9-84   Problem 123.

**124** In the two-sphere arrangement of Sample Problem 9-11, assume that sphere 1 has a mass of 50 g and an initial height of $h_1 = 9.0$ cm, and that sphere 2 has a mass of 85 g. After sphere 1 is released and collides elastically with sphere 2, what height is reached by (a) sphere 1 and (b) sphere 2? After the next (elastic) collision, what height is reached by (c) sphere 1 and (d) sphere 2? (*Hint:* Do not use rounded-off values.)

**125** A 3000 kg block falls vertically through 6.0 m and then collides with a 500 kg pile, driving it 3.0 cm into bedrock. Assuming that the block–pile collision is completely inelastic, find the magnitude of the average force on the pile from the bedrock during the 3.0 cm descent.

**126** In Fig. 9-65, block 1 (mass 6.0 kg) is moving rightward at 8.0 m/s and block 2 (mass 4.0 kg) is moving rightward at 2.0 m/s. The surface is frictionless, and a spring with a spring constant of 8000 N/m is fixed to block 2. Eventually block 1 overtakes block 2. At the instant block 1 is moving rightward at 6.4 m/s, what are (a) the speed of block 2 and (b) the elastic potential energy of the spring?

**127** An electron (mass $m_1 = 9.11 \times 10^{-31}$ kg) and a proton (mass $m_2 = 1.67 \times 10^{-27}$ kg) attract each other via an electrical force. Suppose that an electron and a proton are released from rest with an initial separation $d = 3.0 \times 10^{-6}$ m. When their separation has decreased to $1.0 \times 10^{-6}$ m, what is the ratio of (a) the electron's linear momentum magnitude to the proton's linear momentum magnitude, (b) the electron's speed to the proton's speed, and (c) the electron's kinetic energy to the proton's kinetic energy? (d) As the separation continues to decrease, do the answers to (a) through (c) increase, decrease, or remain the same?

**128** A railroad freight car weighing 280 kN and traveling at 1.52 m/s overtakes one weighing 210 kN and traveling at 0.914 m/s in the same direction. If the cars couple together, find (a) the speed of the cars after the collision and (b) the loss of kinetic energy during the collision. If instead, as is very unlikely, the collision is elastic, find the after-collision speed of (c) the lighter car and (d) the heavier car.

**129** A 3.0 kg object moving at 8.0 m/s in the positive direction of an $x$ axis has a one-dimensional elastic collision with an object of mass $M$, initially at rest. After the collision the object of mass $M$ has a velocity of 6.0 m/s in the positive direction of the axis. What is mass $M$?   SSM

**130** Two particles $P$ and $Q$ are released from rest 1.0 m apart. $P$ has a mass of 0.10 kg, and $Q$ a mass of 0.30 kg. $P$ and $Q$ attract each other with a constant force of $1.0 \times 10^{-2}$ N. No external forces act on the system. (a) What is the speed of the center of mass of $P$ and $Q$ when the separation is 0.50 m? (b) At what distance from $P$'s original position do the particles collide?

**131** In Fig. 9-85, block 1 of mass $m_1 = 6.6$ kg is at rest on a long frictionless table that is up against a wall. Block 2 of

mass $m_2$ is placed between block 1 and the wall and sent sliding to the left, toward block 1, with constant speed $v_{2i}$. Find the value of $m_2$ for which both blocks move with the same velocity after block 2 has collided once with block 1 and once with the wall. Assume all collisions are elastic (the collision with the wall does not change the speed of block 2).

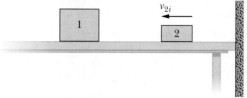

FIG. 9-85   Problem 131.

**132** A rocket of mass $M$ moves along an $x$ axis at the constant speed $v_i = 40$ m/s. A small explosion separates the rocket into a rear section (of mass $m_1$) and a front section; both sections move along the $x$ axis. The relative speed between the rear and front sections is 20 m/s. What are (a) the minimum possible value of final speed $v_f$ of the front section and (b) for what limiting value of $m_1$ does it occur? (c) What is the maximum possible value of $v_f$ and (d) for what limiting value of $m_1$ does it occur?

**133** A 2.65 kg stationary package explodes into three parts that then slide across a frictionless floor. The package had been at the origin of a coordinate system. Part 1 has mass $m_1 = 0.500$ kg and velocity $(10.0\hat{i} + 12.0\hat{j})$ m/s. Part 2 has mass $m_2 = 0.750$ kg and a speed of 14.0 m/s, and travels at an angle 110° (counterclockwise from the positive direction of the $x$ axis). (a) What is the speed of part 3? (b) In what direction does it travel?

**134** Particle 1 with mass 3.0 kg and velocity $(5.0$ m/s$)\hat{i}$ undergoes a one-dimensional elastic collision with particle 2 with mass 2.0 kg and velocity $(-6.0$ m/s$)\hat{i}$. After the collision, what are the velocities of (a) particle 1 and (b) particle 2?

**135** At a certain instant, four particles have the $xy$ coordinates and velocities given in the following table. At that instant, what are the (a) $x$ and (b) $y$ coordinates of their center of mass and (c) the velocity of their center of mass?

| Particle | Mass (kg) | Position (m) | Velocity (m/s) |
|---|---|---|---|
| 1 | 2.0 | 0, 3.0 | $-9.0\hat{j}$ |
| 2 | 4.0 | 3.0, 0 | $6.0\hat{i}$ |
| 3 | 3.0 | 0, $-2.0$ | $6.0\hat{j}$ |
| 4 | 12 | $-1.0, 0$ | $-2.0\hat{i}$ |

**136** Figure 9-86 shows two 22.7 kg ice sleds that are placed a short distance apart, one directly behind the other. A 3.63 kg cat initially standing on one sled jumps to the other one and then back to the first. Both jumps are made at a speed of 3.05 m/s relative to the ice. What are the final speeds of (a) the first sled and (b) the other sled?

FIG. 9-86   Problem 136.

# Rotation

David Wrobel/Visuals Unlimited

*The snapping shrimp stuns its prey (tiny crabs) by clamping its oversized claw shut but not on the prey itself. Instead, the prey is stunned by a powerful sound wave generated by the moving part of the claw as it nears the stationary part of the claw. The sound (a snap resembling the pop of popcorn) can be heard by a diver and, with enough shrimp, can be loud enough to hide a submarine from sonar detection. The sound wave can also produce very faint flashes of light, what is generally called sound-produced light or sonoluminescence. However, some researchers have dubbed the shrimp-produced light shrimpoluminescence.*

## How can a shrimp claw generate such a powerful sound wave?

The answer is in this chapter.

## 10-1 | WHAT IS PHYSICS?

As we have discussed, one focus of physics is motion. However, so far we have examined only the motion of **translation,** in which an object moves along a straight or curved line, as in Fig. 10-1*a*. We now turn to the motion of **rotation,** in which an object turns about an axis, as in Fig. 10-1*b*.

You see rotation in nearly every machine, you use it every time you open a beverage can with a pull tab, and you pay to experience it every time you go to an amusement park. Rotation is the key to many fun activities, such as hitting a long drive in golf (the ball needs to rotate in order for the air to keep it aloft longer) and throwing a curveball in baseball (the ball needs to rotate in order for the air to push it left or right). Rotation is also the key to more serious matters, such as metal failure in aging airplanes.

We begin our discussion of rotation by defining the variables for the motion, just as we did for translation in Chapter 2.

## 10-2 | The Rotational Variables

We wish to examine the rotation of a rigid body about a fixed axis. A **rigid body** is a body that can rotate with all its parts locked together and without any change in its shape. A **fixed axis** means that the rotation occurs about an axis that does not move. Thus, we shall not examine an object like the Sun, because the parts of the Sun (a ball of gas) are not locked together. We also shall not examine an object like a bowling ball rolling along a lane, because the ball rotates about a moving axis (the ball's motion is a mixture of rotation and translation).

Figure 10-2 shows a rigid body of arbitrary shape in rotation about a fixed axis, called the **axis of rotation** or the **rotation axis.** In pure rotation (*angular motion*), every point of the body moves in a circle whose center lies on the axis of rotation, and every point moves through the same angle during a particular time interval. In pure translation (*linear motion*), every point of the body moves in a straight line, and every point moves through the same *linear distance* during a particular time interval.

We deal now—one at a time—with the angular equivalents of the linear quantities position, displacement, velocity, and acceleration.

### Angular Position

Figure 10-2 shows a *reference line*, fixed in the body, perpendicular to the rotation axis and rotating with the body. The **angular position** of this line is the angle of the line relative to a fixed direction, which we take as the **zero angular position.** In Fig. 10-3, the angular position $\theta$ is measured relative to the positive direction of the *x* axis. From geometry, we know that $\theta$ is given by

$$\theta = \frac{s}{r} \quad \text{(radian measure).} \tag{10-1}$$

Here *s* is the length of a circular arc that extends from the *x* axis (the zero angular position) to the reference line, and *r* is the radius of the circle.

An angle defined in this way is measured in **radians** (rad) rather than in revolutions (rev) or degrees. The radian, being the ratio of two lengths, is a pure number and thus has no dimension. Because the circumference of a circle of radius *r* is $2\pi r$, there are $2\pi$ radians in a complete circle:

$$1 \text{ rev} = 360° = \frac{2\pi r}{r} = 2\pi \text{ rad}, \tag{10-2}$$

and thus

$$1 \text{ rad} = 57.3° = 0.159 \text{ rev.} \tag{10-3}$$

(*a*)

(*b*)

**FIG. 10-1** Figure skater Sasha Cohen in motion of (*a*) pure translation in a fixed direction and (*b*) pure rotation about a vertical axis. (*a: Mike Segar/Reuters/Landov LLC; b: Elsa/Getty Images, Inc.*)

We do *not* reset $\theta$ to zero with each complete rotation of the reference line about the rotation axis. If the reference line completes two revolutions from the zero angular position, then the angular position $\theta$ of the line is $\theta = 4\pi$ rad.

For pure translation along an $x$ axis, we can know all there is to know about a moving body if we know $x(t)$, its position as a function of time. Similarly, for pure rotation, we can know all there is to know about a rotating body if we know $\theta(t)$, the angular position of the body's reference line as a function of time.

### Angular Displacement

If the body of Fig. 10-3 rotates about the rotation axis as in Fig. 10-4, changing the angular position of the reference line from $\theta_1$ to $\theta_2$, the body undergoes an **angular displacement** $\Delta\theta$ given by

$$\Delta\theta = \theta_2 - \theta_1. \qquad (10\text{-}4)$$

This definition of angular displacement holds not only for the rigid body as a whole but also for *every particle within that body.*

If a body is in translational motion along an $x$ axis, its displacement $\Delta x$ is either positive or negative, depending on whether the body is moving in the positive or negative direction of the axis. Similarly, the angular displacement $\Delta\theta$ of a rotating body is either positive or negative, according to the following rule:

> An angular displacement in the counterclockwise direction is positive, and one in the clockwise direction is negative.

The phrase "*clocks are negative*" can help you remember this rule (they certainly are negative when their alarms sound off early in the morning).

✓**CHECKPOINT 1**   A disk can rotate about its central axis like a merry-go-round. Which of the following pairs of values for its initial and final angular positions, respectively, give a negative angular displacement: (a) $-3$ rad, $+5$ rad, (b) $-3$ rad, $-7$ rad, (c) 7 rad, $-3$ rad?

### Angular Velocity

Suppose that our rotating body is at angular position $\theta_1$ at time $t_1$ and at angular position $\theta_2$ at time $t_2$ as in Fig. 10-4. We define the **average angular velocity** of the body in the time interval $\Delta t$ from $t_1$ to $t_2$ to be

$$\omega_{\text{avg}} = \frac{\theta_2 - \theta_1}{t_2 - t_1} = \frac{\Delta\theta}{\Delta t}, \qquad (10\text{-}5)$$

in which $\Delta\theta$ is the angular displacement that occurs during $\Delta t$ ($\omega$ is the lowercase Greek letter omega).

The **(instantaneous) angular velocity** $\omega$, with which we shall be most concerned, is the limit of the ratio in Eq. 10-5 as $\Delta t$ approaches zero. Thus,

$$\omega = \lim_{\Delta t \to 0} \frac{\Delta\theta}{\Delta t} = \frac{d\theta}{dt}. \qquad (10\text{-}6)$$

If we know $\theta(t)$, we can find the angular velocity $\omega$ by differentiation.

Equations 10-5 and 10-6 hold not only for the rotating rigid body as a whole but also for *every particle of that body* because the particles are all locked together. The unit of angular velocity is commonly the radian per second (rad/s) or the revolution per second (rev/s). Another measure of angular velocity was used during at least the first three decades of rock: Music was produced by vinyl (phonograph) records that were played on turntables at "$33\frac{1}{3}$ rpm" or "45 rpm," meaning at $33\frac{1}{3}$ rev/min or 45 rev/min.

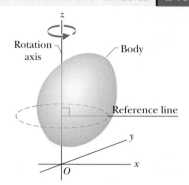

**FIG. 10-2**   A rigid body of arbitrary shape in pure rotation about the $z$ axis of a coordinate system. The position of the *reference line* with respect to the rigid body is arbitrary, but it is perpendicular to the rotation axis. It is fixed in the body and rotates with the body.

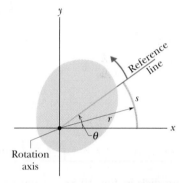

**FIG. 10-3**   The rotating rigid body of Fig. 10-2 in cross section, viewed from above. The plane of the cross section is perpendicular to the rotation axis, which now extends out of the page, toward you. In this position of the body, the reference line makes an angle $\theta$ with the $x$ axis.

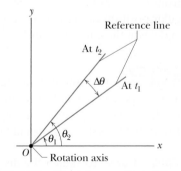

**FIG. 10-4**   The reference line of the rigid body of Figs. 10-2 and 10-3 is at angular position $\theta_1$ at time $t_1$ and at angular position $\theta_2$ at a later time $t_2$. The quantity $\Delta\theta$ ($= \theta_2 - \theta_1$) is the angular displacement that occurs during the interval $\Delta t$ ($= t_2 - t_1$). The body itself is not shown.

If a particle moves in translation along an $x$ axis, its linear velocity $v$ is either positive or negative, depending on whether the particle is moving in the positive or negative direction of the axis. Similarly, the angular velocity $\omega$ of a rotating rigid body is either positive or negative, depending on whether the body is rotating counterclockwise (positive) or clockwise (negative). ("Clocks are negative" still works.) The magnitude of an angular velocity is called the **angular speed,** which is also represented with $\omega$.

## Angular Acceleration

If the angular velocity of a rotating body is not constant, then the body has an angular acceleration. Let $\omega_2$ and $\omega_1$ be its angular velocities at times $t_2$ and $t_1$, respectively. The **average angular acceleration** of the rotating body in the interval from $t_1$ to $t_2$ is defined as

$$\alpha_{avg} = \frac{\omega_2 - \omega_1}{t_2 - t_1} = \frac{\Delta\omega}{\Delta t}, \qquad (10\text{-}7)$$

in which $\Delta\omega$ is the change in the angular velocity that occurs during the time interval $\Delta t$. The **(instantaneous) angular acceleration** $\alpha$, with which we shall be most concerned, is the limit of this quantity as $\Delta t$ approaches zero. Thus,

$$\alpha = \lim_{\Delta t \to 0} \frac{\Delta\omega}{\Delta t} = \frac{d\omega}{dt}. \qquad (10\text{-}8)$$

Equations 10-7 and 10-8 also hold for *every particle of that body.* The unit of angular acceleration is commonly the radian per second-squared ($rad/s^2$) or the revolution per second-squared ($rev/s^2$).

---

## Sample Problem | 10-1 | Build your skill

The disk in Fig. 10-5a is rotating about its central axis like a merry-go-round. The angular position $\theta(t)$ of a reference line on the disk is given by

$$\theta = -1.00 - 0.600t + 0.250t^2, \qquad (10\text{-}9)$$

with $t$ in seconds, $\theta$ in radians, and the zero angular position as indicated in the figure.

(a) Graph the angular position of the disk versus time from $t = -3.0$ s to $t = 5.4$ s. Sketch the disk and its angular position reference line at $t = -2.0$ s, 0 s, and 4.0 s, and when the curve crosses the $t$ axis.

**KEY IDEA** The angular position of the disk is the angular position $\theta(t)$ of its reference line, which is given by Eq. 10-9 as a function of time $t$. So we graph Eq. 10-9; the result is shown in Fig. 10-5b.

*Calculations:* To sketch the disk and its reference line at a particular time, we need to determine $\theta$ for that time. To do so, we substitute the time into Eq. 10-9. For $t = -2.0$ s, we get

$$\theta = -1.00 - (0.600)(-2.0) + (0.250)(-2.0)^2$$

$$= 1.2 \text{ rad} = 1.2 \text{ rad} \frac{360°}{2\pi \text{ rad}} = 69°.$$

This means that at $t = -2.0$ s the reference line on the disk is rotated counterclockwise from the zero position by 1.2 rad = 69° (counterclockwise because $\theta$ is positive). Sketch 1 in Fig. 10-5b shows this position of the reference line.

Similarly, for $t = 0$, we find $\theta = -1.00$ rad $= -57°$, which means that the reference line is rotated clockwise from the zero angular position by 1.0 rad, or 57°, as shown in sketch 3. For $t = 4.0$ s, we find $\theta = 0.60$ rad $= 34°$ (sketch 5). Drawing sketches for when the curve crosses the $t$ axis is easy, because then $\theta = 0$ and the reference line is momentarily aligned with the zero angular position (sketches 2 and 4).

(b) At what time $t_{min}$ does $\theta(t)$ reach the minimum value shown in Fig. 10-5b? What is that minimum value?

**KEY IDEA** To find the extreme value (here the minimum) of a function, we take the first derivative of the function and set the result to zero.

*Calculations:* The first derivative of $\theta(t)$ is

$$\frac{d\theta}{dt} = -0.600 + 0.500t. \qquad (10\text{-}10)$$

Setting this to zero and solving for $t$ give us the time at

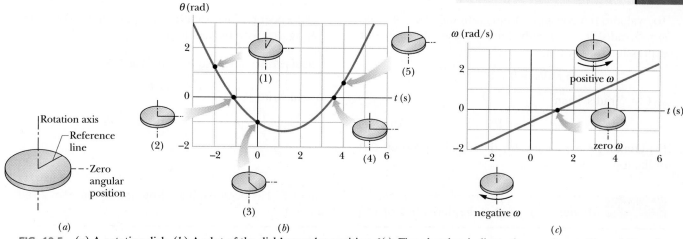

**FIG. 10-5** (*a*) A rotating disk. (*b*) A plot of the disk's angular position $\theta(t)$. Five sketches indicate the angular position of the reference line on the disk for five points on the curve. (*c*) A plot of the disk's angular velocity $\omega(t)$. Positive values of $\omega$ correspond to counterclockwise rotation, and negative values to clockwise rotation.

which $\theta(t)$ is minimum:

$$t_{min} = 1.20 \text{ s.} \qquad \text{(Answer)}$$

To get the minimum value of $\theta$, we next substitute $t_{min}$ into Eq. 10-9, finding

$$\theta = -1.36 \text{ rad} \approx -77.9°. \qquad \text{(Answer)}$$

This *minimum* of $\theta(t)$ (the bottom of the curve in Fig. 10-5*b*) corresponds to the *maximum clockwise* rotation of the disk from the zero angular position, somewhat more than is shown in sketch 3.

(c) Graph the angular velocity $\omega$ of the disk versus time from $t = -3.0$ s to $t = 6.0$ s. Sketch the disk and indicate the direction of turning and the sign of $\omega$ at $t = -2.0$ s, 4.0 s, and $t_{min}$.

**KEY IDEA** From Eq. 10-6, the angular velocity $\omega$ is equal to $d\theta/dt$ as given in Eq. 10-10. So, we have

$$\omega = -0.600 + 0.500t. \qquad (10\text{-}11)$$

The graph of this function $\omega(t)$ is shown in Fig. 10-5*c*.

**Calculations:** To sketch the disk at $t = -2.0$ s, we substitute that value into Eq. 10-11, obtaining

$$\omega = -1.6 \text{ rad/s.} \qquad \text{(Answer)}$$

The minus sign tells us that at $t = -2.0$ s, the disk is turning clockwise (the lowest sketch in Fig. 10-5*c*).

Substituting $t = 4.0$ s into Eq. 10-11 gives us

$$\omega = 1.4 \text{ rad/s.} \qquad \text{(Answer)}$$

The implied plus sign tells us that at $t = 4.0$ s, the disk is turning counterclockwise (the highest sketch in Fig. 10-5*c*).

For $t_{min}$, we already know that $d\theta/dt = 0$. So, we must also have $\omega = 0$. That is, the disk momentarily stops when the reference line reaches the minimum value of $\theta$ in Fig. 10-5*b*, as suggested by the center sketch in Fig. 10-5*c*.

(d) Use the results in parts (a) through (c) to describe the motion of the disk from $t = -3.0$ s to $t = 6.0$ s.

**Description:** When we first observe the disk at $t = -3.0$ s, it has a positive angular position and is turning clockwise but slowing. It stops at angular position $\theta = -1.36$ rad and then begins to turn counterclockwise, with its angular position eventually becoming positive again.

---

**Sample Problem** **10-2**

A child's top is spun with angular acceleration

$$\alpha = 5t^3 - 4t,$$

with $t$ in seconds and $\alpha$ in radians per second-squared. At $t = 0$, the top has angular velocity 5 rad/s, and a reference line on it is at angular position $\theta = 2$ rad.

(a) Obtain an expression for the angular velocity $\omega(t)$ of the top.

**KEY IDEA** By definition, $\alpha(t)$ is the derivative of $\omega(t)$

with respect to time. Thus, we can find $\omega(t)$ by integrating $\alpha(t)$ with respect to time.

**Calculations:** Equation 10-8 tells us

$$d\omega = \alpha \, dt,$$

so

$$\int d\omega = \int \alpha \, dt.$$

From this we find

$$\omega = \int (5t^3 - 4t) \, dt = \tfrac{5}{4}t^4 - \tfrac{4}{2}t^2 + C.$$

To evaluate the constant of integration $C$, we note that $\omega = 5$ rad/s at $t = 0$. Substituting these values in our expression for $\omega$ yields

$$5 \text{ rad/s} = 0 - 0 + C,$$

so $C = 5$ rad/s. Then

$$\omega = \tfrac{5}{4}t^4 - 2t^2 + 5. \qquad \text{(Answer)}$$

(b) Obtain an expression for the angular position $\theta(t)$ of the top.

---

**KEY IDEA** By definition, $\omega(t)$ is the derivative of $\theta(t)$ with respect to time. Therefore, we can find $\theta(t)$ by integrating $\omega(t)$ with respect to time.

**Calculations:** Since Eq. 10-6 tells us that

$$d\theta = \omega\, dt,$$

we can write

$$\theta = \int \omega\, dt = \int (\tfrac{5}{4}t^4 - 2t^2 + 5)\, dt$$

$$= \tfrac{1}{4}t^5 - \tfrac{2}{3}t^3 + 5t + C'$$

$$= \tfrac{1}{4}t^5 - \tfrac{2}{3}t^3 + 5t + 2, \qquad \text{(Answer)}$$

where $C'$ has been evaluated by noting that $\theta = 2$ rad at $t = 0$.

---

## 10-3 | Are Angular Quantities Vectors?

We can describe the position, velocity, and acceleration of a single particle by means of vectors. If the particle is confined to a straight line, however, we do not really need vector notation. Such a particle has only two directions available to it, and we can indicate these directions with plus and minus signs.

In the same way, a rigid body rotating about a fixed axis can rotate only clockwise or counterclockwise as seen along the axis, and again we can select between the two directions by means of plus and minus signs. The question arises: "Can we treat the angular displacement, velocity, and acceleration of a rotating body as vectors?" The answer is a qualified "yes" (see the caution below, in connection with angular displacements).

Consider the angular velocity. Figure 10-6a shows a vinyl record rotating on a turntable. The record has a constant angular speed $\omega\ (= 33\tfrac{1}{3}$ rev/min) in the clockwise direction. We can represent its angular velocity as a vector $\vec{\omega}$ pointing along the axis of rotation, as in Fig. 10-6b. Here's how: We choose the length of this vector according to some convenient scale, for example, with 1 cm corresponding to 10 rev/min. Then we establish a direction for the vector $\vec{\omega}$ by using a **right-hand rule,** as Fig. 10-6c shows: Curl your right hand about the rotating record, your fingers pointing *in the direction of rotation.* Your extended thumb will then point in the direction of the angular velocity vector. If the record were to rotate in the opposite sense, the right-hand rule would tell you that the angular velocity vector then points in the opposite direction.

It is not easy to get used to representing angular quantities as vectors. We instinctively expect that something should be moving *along* the direction of a vector. That is not the case here. Instead, something (the rigid body) is rotating *around* the direction of the vector. In the world of pure rotation, a vector defines

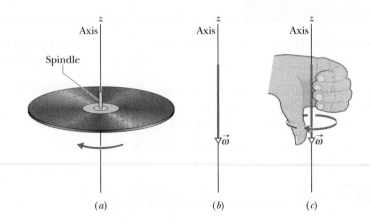

**FIG. 10-6** (a) A record rotating about a vertical axis that coincides with the axis of the spindle. (b) The angular velocity of the rotating record can be represented by the vector $\vec{\omega}$, lying along the axis and pointing down, as shown. (c) We establish the direction of the angular velocity vector as downward by using a right-hand rule. When the fingers of the right hand curl around the record and point the way it is moving, the extended thumb points in the direction of $\vec{\omega}$.

an axis of rotation, not a direction in which something moves. Nonetheless, the vector also defines the motion. Furthermore, it obeys all the rules for vector manipulation discussed in Chapter 3. The angular acceleration $\vec{\alpha}$ is another vector, and it too obeys those rules.

In this chapter we consider only rotations that are about a fixed axis. For such situations, we need not consider vectors—we can represent angular velocity with $\omega$ and angular acceleration with $\alpha$, and we can indicate direction with an implied plus sign for counterclockwise or an explicit minus sign for clockwise.

Now for the caution: Angular *displacements* (unless they are very small) *cannot* be treated as vectors. Why not? We can certainly give them both magnitude and direction, as we did for the angular velocity vector in Fig. 10-6. However, to be represented as a vector, a quantity must *also* obey the rules of vector addition, one of which says that if you add two vectors, the order in which you add them does not matter. Angular displacements fail this test.

Figure 10-7 gives an example. An initially horizontal book is given two 90° angular displacements, first in the order of Fig. 10-7a and then in the order of Fig. 10-7b. Although the two angular displacements are identical, their order is not, and the book ends up with different orientations. Here's another example. Hold your right arm downward, palm toward your thigh. Keeping your wrist rigid, (1) lift the arm forward until it is horizontal, (2) move it horizontally until it points toward the right, and (3) then bring it down to your side. Your palm faces forward. If you start over, but reverse the steps, which way does your palm end up facing? From either example, we must conclude that the addition of two angular displacements depends on their order and they cannot be vectors.

## 10-4 I Rotation with Constant Angular Acceleration

In pure translation, motion with a *constant linear acceleration* (for example, that of a falling body) is an important special case. In Table 2-1, we displayed a series of equations that hold for such motion.

In pure rotation, the case of *constant angular acceleration* is also important, and a parallel set of equations holds for this case also. We shall not derive them here, but simply write them from the corresponding linear equations, substituting equivalent angular quantities for the linear ones. This is done in Table 10-1, which lists both sets of equations (Eqs. 2-11 and 2-15 to 2-18; 10-12 to 10-16).

Recall that Eqs. 2-11 and 2-15 are basic equations for constant linear acceleration—the other equations in the Linear list can be derived from them. Similarly, Eqs. 10-12 and 10-13 are the basic equations for constant angular acceleration, and the other equations in the Angular list can be derived from them. To solve a simple problem involving constant angular acceleration, you can usually use an equation from the Angular list (*if you have the list*). Choose an equation for which the only unknown variable will be the variable requested in the problem. A better plan is to remember only Eqs. 10-12 and 10-13, and then solve them as simultaneous equations whenever needed. An example is given in Sample Problem 10-4.

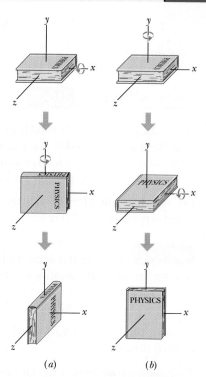

**FIG. 10-7** (*a*) From its initial position, at the top, the book is given two successive 90° rotations, first about the (horizontal) *x* axis and then about the (vertical) *y* axis. (*b*) The book is given the same rotations, but in the reverse order.

**TABLE 10-1**

**Equations of Motion for Constant Linear Acceleration and for Constant Angular Acceleration**

| Equation Number | Linear Equation | Missing Variable | | Angular Equation | Equation Number |
|---|---|---|---|---|---|
| (2-11) | $v = v_0 + at$ | $x - x_0$ | $\theta - \theta_0$ | $\omega = \omega_0 + \alpha t$ | (10-12) |
| (2-15) | $x - x_0 = v_0 t + \frac{1}{2}at^2$ | $v$ | $\omega$ | $\theta - \theta_0 = \omega_0 t + \frac{1}{2}\alpha t^2$ | (10-13) |
| (2-16) | $v^2 = v_0^2 + 2a(x - x_0)$ | $t$ | $t$ | $\omega^2 = \omega_0^2 + 2\alpha(\theta - \theta_0)$ | (10-14) |
| (2-17) | $x - x_0 = \frac{1}{2}(v_0 + v)t$ | $a$ | $\alpha$ | $\theta - \theta_0 = \frac{1}{2}(\omega_0 + \omega)t$ | (10-15) |
| (2-18) | $x - x_0 = vt - \frac{1}{2}at^2$ | $v_0$ | $\omega_0$ | $\theta - \theta_0 = \omega t - \frac{1}{2}\alpha t^2$ | (10-16) |

✓**CHECKPOINT 2** In four situations, a rotating body has angular position $\theta(t)$ given by (a) $\theta = 3t - 4$, (b) $\theta = -5t^3 + 4t^2 + 6$, (c) $\theta = 2/t^2 - 4/t$, and (d) $\theta = 5t^2 - 3$. To which situations do the angular equations of Table 10-1 apply?

## Sample Problem 10-3

A grindstone (Fig. 10-8) rotates at constant angular acceleration $\alpha = 0.35$ rad/s$^2$. At time $t = 0$, it has an angular velocity of $\omega_0 = -4.6$ rad/s and a reference line on it is horizontal, at the angular position $\theta_0 = 0$.

**(a)** At what time after $t = 0$ is the reference line at the angular position $\theta = 5.0$ rev?

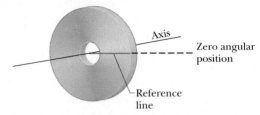

**FIG. 10-8** A grindstone. At $t = 0$ the reference line (which we imagine to be marked on the stone) is horizontal.

**KEY IDEA** The angular acceleration is constant, so we can use the rotation equations of Table 10-1. We choose Eq. 10-13,

$$\theta - \theta_0 = \omega_0 t + \tfrac{1}{2}\alpha t^2,$$

because the only unknown variable it contains is the desired time $t$.

**Calculations:** Substituting known values and setting $\theta_0 = 0$ and $\theta = 5.0$ rev $= 10\pi$ rad give us

$$10\pi \text{ rad} = (-4.6 \text{ rad/s})t + \tfrac{1}{2}(0.35 \text{ rad/s}^2)t^2.$$

(We converted 5.0 rev to $10\pi$ rad to keep the units consistent.) Solving this quadratic equation for $t$, we find

$$t = 32 \text{ s.} \qquad \text{(Answer)}$$

**(b)** Describe the grindstone's rotation between $t = 0$ and $t = 32$ s.

**Description:** The wheel is initially rotating in the negative (clockwise) direction with angular velocity $\omega_0 = -4.6$ rad/s, but its angular acceleration $\alpha$ is positive. This

initial opposition of the signs of angular velocity and angular acceleration means that the wheel slows in its rotation in the negative direction, stops, and then reverses to rotate in the positive direction. After the reference line comes back through its initial orientation of $\theta = 0$, the wheel turns an additional 5.0 rev by time $t = 32$ s.

**(c)** At what time $t$ does the grindstone momentarily stop?

**Calculation:** We again go to the table of equations for constant angular acceleration, and again we need an equation that contains only the desired unknown variable $t$. However, now the equation must also contain the variable $\omega$, so that we can set it to 0 and then solve for the corresponding time $t$. We choose Eq. 10-12, which yields

$$t = \frac{\omega - \omega_0}{\alpha} = \frac{0 - (-4.6 \text{ rad/s})}{0.35 \text{ rad/s}^2} = 13 \text{ s.} \qquad \text{(Answer}$$

## Sample Problem 10-4

While you are operating a Rotor (the rotating cylindrical ride discussed in Sample Problem 6-8), you spot a passenger in acute distress and decrease the angular velocity of the cylinder from 3.40 rad/s to 2.00 rad/s in 20.0 rev, at constant angular acceleration. (The passenger is obviously more of a "translation person" than a "rotation person.")

**(a)** What is the constant angular acceleration during this decrease in angular speed?

**KEY IDEA** Because the cylinder's angular acceleration is constant, we can relate it to the angular velocity and angular displacement via the basic equations for constant angular acceleration (Eqs. 10-12 and 10-13).

**Calculations:** The initial angular velocity is $\omega_0 = 3.40$ rad/s, the angular displacement is $\theta - \theta_0 = 20.0$ rev, and the angular velocity at the end of that displacement is $\omega = 2.00$ rad/s. But we do not know the angular acceleration $\alpha$ and time $t$, which are in both basic equations.

To eliminate the unknown $t$, we use Eq. 10-12 to write

$$t = \frac{\omega - \omega_0}{\alpha},$$

which we then substitute into Eq. 10-13 to write

$$\theta - \theta_0 = \omega_0\left(\frac{\omega - \omega_0}{\alpha}\right) + \tfrac{1}{2}\alpha\left(\frac{\omega - \omega_0}{\alpha}\right)^2.$$

Solving for $\alpha$, substituting known data, and converting 20 rev to 125.7 rad, we find

$$\alpha = \frac{\omega^2 - \omega_0^2}{2(\theta - \theta_0)} = \frac{(2.00 \text{ rad/s})^2 - (3.40 \text{ rad/s})^2}{2(125.7 \text{ rad})}$$

$$= -0.0301 \text{ rad/s}^2. \qquad \text{(Answer)}$$

(b) How much time did the speed decrease take?

**Calculation:** Now that we know $\alpha$, we can use Eq. 10-12 to solve for $t$:

$$t = \frac{\omega - \omega_0}{\alpha} = \frac{2.00 \text{ rad/s} - 3.40 \text{ rad/s}}{-0.0301 \text{ rad/s}^2}$$

$$= 46.5 \text{ s}. \qquad \text{(Answer)}$$

## 10-5 | Relating the Linear and Angular Variables

In Section 4-7, we discussed uniform circular motion, in which a particle travels at constant linear speed $v$ along a circle and around an axis of rotation. When a rigid body, such as a merry-go-round, rotates around an axis, each particle in the body moves in its own circle around that axis. Since the body is rigid, all the particles make one revolution in the same amount of time; that is, they all have the same angular speed $\omega$.

However, the farther a particle is from the axis, the greater the circumference of its circle is, and so the faster its linear speed $v$ must be. You can notice this on a merry-go-round. You turn with the same angular speed $\omega$ regardless of your distance from the center, but your linear speed $v$ increases noticeably if you move to the outside edge of the merry-go-round.

We often need to relate the linear variables $s$, $v$, and $a$ for a particular point in a rotating body to the angular variables $\theta$, $\omega$, and $\alpha$ for that body. The two sets of variables are related by $r$, the *perpendicular distance* of the point from the rotation axis. This perpendicular distance is the distance between the point and the rotation axis, measured along a perpendicular to the axis. It is also the radius $r$ of the circle traveled by the point around the axis of rotation.

### The Position

If a reference line on a rigid body rotates through an angle $\theta$, a point within the body at a position $r$ from the rotation axis moves a distance $s$ along a circular arc, where $s$ is given by Eq. 10-1:

$$s = \theta r \qquad \text{(radian measure).} \qquad (10\text{-}17)$$

This is the first of our linear–angular relations. *Caution:* The angle $\theta$ here must be measured in radians because Eq. 10-17 is itself the definition of angular measure in radians.

### The Speed

Differentiating Eq. 10-17 with respect to time—with $r$ held constant—leads to

$$\frac{ds}{dt} = \frac{d\theta}{dt} r.$$

However, $ds/dt$ is the linear speed (the magnitude of the linear velocity) of the point in question, and $d\theta/dt$ is the angular speed $\omega$ of the rotating body. So

$$v = \omega r \qquad \text{(radian measure).} \qquad (10\text{-}18)$$

*Caution:* The angular speed $\omega$ must be expressed in radian measure.

Equation 10-18 tells us that since all points within the rigid body have the same angular speed $\omega$, points with greater radius $r$ have greater linear speed $v$. Figure 10-9a reminds us that the linear velocity is always tangent to the circular path of the point in question.

If the angular speed $\omega$ of the rigid body is constant, then Eq. 10-18 tells us that the linear speed $v$ of any point within it is also constant. Thus, each point within the body undergoes uniform circular motion. The period of revolution $T$

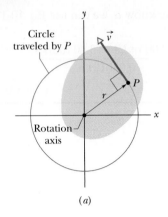

(a)

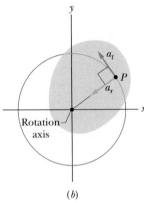

(b)

**FIG. 10-9** The rotating rigid body of Fig. 10-2, shown in cross section viewed from above. Every point of the body (such as P) moves in a circle around the rotation axis. (a) The linear velocity $\vec{v}$ of every point is tangent to the circle in which the point moves. (b) The linear acceleration $\vec{a}$ of the point has (in general) two components: tangential $a_t$ and radial $a_r$.

for the motion of each point and for the rigid body itself is given by Eq. 4-35:

$$T = \frac{2\pi r}{v}. \qquad (10\text{-}19)$$

This equation tells us that the time for one revolution is the distance $2\pi r$ traveled in one revolution divided by the speed at which that distance is traveled. Substituting for $v$ from Eq. 10-18 and canceling $r$, we find also that

$$T = \frac{2\pi}{\omega} \qquad \text{(radian measure)}. \qquad (10\text{-}20)$$

This equivalent equation says that the time for one revolution is the angular distance $2\pi$ rad traveled in one revolution divided by the angular speed (or rate) at which that angle is traveled.

### The Acceleration

Differentiating Eq. 10-18 with respect to time—again with $r$ held constant—leads to

$$\frac{dv}{dt} = \frac{d\omega}{dt} r. \qquad (10\text{-}21)$$

Here we run up against a complication. In Eq. 10-21, $dv/dt$ represents only the part of the linear acceleration that is responsible for changes in the *magnitude v* of the linear velocity $\vec{v}$. Like $\vec{v}$, that part of the linear acceleration is tangent to the path of the point in question. We call it the *tangential component $a_t$* of the linear acceleration of the point, and we write

$$a_t = \alpha r \qquad \text{(radian measure)}, \qquad (10\text{-}22)$$

where $\alpha = d\omega/dt$. *Caution:* The angular acceleration $\alpha$ in Eq. 10-22 must be expressed in radian measure.

In addition, as Eq. 4-34 tells us, a particle (or point) moving in a circular path has a *radial component* of linear acceleration, $a_r = v^2/r$ (directed radially inward), that is responsible for changes in the *direction* of the linear velocity $\vec{v}$. By substituting for $v$ from Eq. 10-18, we can write this component as

$$a_r = \frac{v^2}{r} = \omega^2 r \qquad \text{(radian measure)}. \qquad (10\text{-}23)$$

Thus, as Fig. 10-9b shows, the linear acceleration of a point on a rotating rigid body has, in general, two components. The radially inward component $a_r$ (given by Eq. 10-23) is present whenever the angular velocity of the body is not zero. The tangential component $a_t$ (given by Eq. 10-22) is present whenever the angular acceleration is not zero.

✓**CHECKPOINT 3** A cockroach rides the rim of a rotating merry-go-round. If the angular speed of this system (*merry-go-round + cockroach*) is constant, does the cockroach have (a) radial acceleration and (b) tangential acceleration? If $\omega$ is decreasing, does the cockroach have (c) radial acceleration and (d) tangential acceleration?

**Sample Problem** | 10-5

In spite of the extreme care taken in engineering a roller coaster, an unlucky few of the millions of people who ride roller coasters each year end up with a medical condition called *roller-coaster headache*. Symptoms, which might not appear for several days, include ver-

tigo and headache, both severe enough to require medical treatment.

Let's investigate the probable cause by designing the track for our own *induction roller coaster* (which can be accelerated by magnetic forces even on a hori-

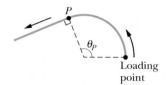

**FIG. 10-10** An overhead view of a horizontal track for a roller coaster. The track begins as a circular arc at the loading point and then, at point $P$, continues along a tangent to the arc.

zontal track). To create an initial thrill, we want each passenger to leave the loading point with acceleration $g$ along the horizontal track. To increase the thrill, we also want that first section of track to form a circular arc (Fig. 10-10), so that the passenger also experiences a centripetal acceleration. As the passenger accelerates along the arc, the magnitude of this centripetal acceleration increases alarmingly. When the magnitude $a$ of the net acceleration reaches $4g$ at some point $P$ and angle $\theta_P$ along the arc, we want the passenger then to move in a straight line, along a tangent to the arc.

**(a)** What angle $\theta_P$ should the arc subtend so that $a$ is $4g$ at point $P$?

**KEY IDEAS** (1) At any given time, the passenger's net acceleration $\vec{a}$ is the vector sum of the tangential acceleration $\vec{a}_t$ along the track and the radial acceleration $\vec{a}_r$ toward the arc's center of curvature (as in Fig. 10-9$b$). (2) The value of $a_r$ at any given time depends on the angular speed $\omega$ according to Eq. 10-23 ($a_r = \omega^2 r$, where $r$ is the radius of the circular arc). (3) An angular acceleration $\alpha$ around the arc is associated with the tangential acceleration $a_t$ along the track according to Eq. 10-22 ($a_t = \alpha r$). (4) Because $a_t$ and $r$ are constant, so is $\alpha$ and thus we can use the constant angular-acceleration equations.

**Calculations:** Because we are trying to determine a value for angular position $\theta$, let's choose Eq. 10-14 from among the constant angular-acceleration equations:

$$\omega^2 = \omega_0^2 + 2\alpha(\theta - \theta_0). \tag{10-24}$$

For the angular acceleration $\alpha$, we substitute from Eq. 10-22:

$$\alpha = \frac{a_t}{r}. \tag{10-25}$$

We also substitute $\omega_0 = 0$ and $\theta_0 = 0$, and we find

$$\omega^2 = \frac{2a_t\theta}{r}. \tag{10-26}$$

Substituting this result for $\omega^2$ into

$$a_r = \omega^2 r \tag{10-27}$$

gives a relation between the radial acceleration, the tangential acceleration, and the angular position $\theta$:

$$a_r = 2a_t\theta. \tag{10-28}$$

Because $\vec{a}_t$ and $\vec{a}_r$ are perpendicular vectors, their sum has the magnitude

$$a = \sqrt{a_t^2 + a_r^2}. \tag{10-29}$$

Substituting for $a_r$ from Eq. 10-28 and solving for $\theta$ lead to

$$\theta = \frac{1}{2}\sqrt{\frac{a^2}{a_t^2} - 1}. \tag{10-30}$$

When $a$ reaches the design value of $4g$, angle $\theta$ is the angle $\theta_P$ we want. Substituting $a = 4g$, $\theta = \theta_P$, and $a_t = g$ into Eq. 10-30, we find

$$\theta_P = \frac{1}{2}\sqrt{\frac{(4g)^2}{g^2} - 1} = 1.94 \text{ rad} = 111°. \quad \text{(Answer)}$$

**(b)** What is the magnitude $a$ of the passenger's net acceleration at point $P$ and after point $P$?

**Reasoning:** At $P$, $a$ has the design value of $4g$. Just after $P$ is reached, the passenger moves in a straight line and no longer has centripetal acceleration. Thus, the passenger has only the acceleration magnitude $g$ along the track. Hence,

$$a = 4g \text{ at } P \quad \text{and} \quad a = g \text{ after } P. \quad \text{(Answer)}$$

Roller-coaster headache can occur when a passenger's head undergoes an abrupt change in acceleration, with the acceleration magnitude large before or after the change. The reason is that the change can cause the brain to move relative to the skull, tearing the veins that bridge the brain and skull. Our design to increase the acceleration from $g$ to $4g$ along the path to $P$ might harm the passenger, but the abrupt change in acceleration as the passenger passes through point $P$ is more likely to cause roller-coaster headache.

---

**PROBLEM-SOLVING TACTICS**

**Tactic 1:** *Units for Angular Variables* In Eq. 10-1 ($\theta = s/r$), we began the use of radian measure for all angular variables whenever we are using equations that contain both angular and linear variables. Thus, we must express angular displacements in radians, angular velocities in rad/s and rad/min, and angular accelerations in rad/s² and rad/min². Equations 10-17, 10-18, 10-20, 10-22, and 10-23 are marked to emphasize this. The only exceptions to this rule are equations that involve

*only* angular variables, such as the angular equations listed in Table 10-1. Here you are free to use any unit you wish for the angular variables; that is, you may use radians, degrees, or revolutions, as long as you use them consistently.

In equations where radian measure must be used, you need not keep track of the unit "radian" (rad) algebraically, as you must do for other units. You can add or delete it at will, to suit the context.

## 10-6 | Kinetic Energy of Rotation

The rapidly rotating blade of a table saw certainly has kinetic energy due to that rotation. How can we express the energy? We cannot apply the familiar formula $K = \frac{1}{2}mv^2$ to the saw as a whole because that would give us the kinetic energy only of the saw's center of mass, which is zero.

Instead, we shall treat the table saw (and any other rotating rigid body) as a collection of particles with different speeds. We can then add up the kinetic energies of all the particles to find the kinetic energy of the body as a whole. In this way we obtain, for the kinetic energy of a rotating body,

$$K = \tfrac{1}{2}m_1v_1^2 + \tfrac{1}{2}m_2v_2^2 + \tfrac{1}{2}m_3v_3^2 + \cdots$$
$$= \sum \tfrac{1}{2}m_iv_i^2, \tag{10-31}$$

in which $m_i$ is the mass of the $i$th particle and $v_i$ is its speed. The sum is taken over all the particles in the body.

The problem with Eq. 10-31 is that $v_i$ is not the same for all particles. We solve this problem by substituting for $v$ from Eq. 10-18 ($v = \omega r$), so that we have

$$K = \sum \tfrac{1}{2}m_i(\omega r_i)^2 = \tfrac{1}{2}\left(\sum m_ir_i^2\right)\omega^2, \tag{10-32}$$

in which $\omega$ *is* the same for all particles.

The quantity in parentheses on the right side of Eq. 10-32 tells us how the mass of the rotating body is distributed about its axis of rotation. We call that quantity the **rotational inertia** (or **moment of inertia**) $I$ of the body with respect to the axis of rotation. It is a constant for a particular rigid body and a particular rotation axis. (That axis must always be specified if the value of $I$ is to be meaningful.)

We may now write

$$I = \sum m_ir_i^2 \qquad \text{(rotational inertia)} \tag{10-33}$$

and substitute into Eq. 10-32, obtaining

$$K = \tfrac{1}{2}I\omega^2 \qquad \text{(radian measure)} \tag{10-34}$$

as the expression we seek. Because we have used the relation $v = \omega r$ in deriving Eq. 10-34, $\omega$ must be expressed in radian measure. The SI unit for $I$ is the kilogram–square meter ($\text{kg} \cdot \text{m}^2$).

Equation 10-34, which gives the kinetic energy of a rigid body in pure rotation, is the angular equivalent of the formula $K = \frac{1}{2}Mv_{\text{com}}^2$, which gives the kinetic energy of a rigid body in pure translation. In both formulas there is a factor of $\frac{1}{2}$. Where mass $M$ appears in one equation, $I$ (which involves both mass and its distribution) appears in the other. Finally, each equation contains as a factor the square of a speed—translational or rotational as appropriate. The kinetic energies of translation and of rotation are not different kinds of energy. They are both kinetic energy, expressed in ways that are appropriate to the motion at hand.

We noted previously that the rotational inertia of a rotating body involves not only its mass but also how that mass is distributed. Here is an example that you can literally feel. Rotate a long, fairly heavy rod (a pole, a length of lumber, or something similar), first around its central (longitudinal) axis (Fig. 10-11a) and then around an axis perpendicular to the rod and through the center (Fig. 10-11b). Both rotations involve the very same mass, but the first rotation is much easier than the second. The reason is that the mass is distributed much closer to the rotation axis in the first rotation. As a result, the rotational inertia of the rod is much smaller in Fig. 10-11a than in Fig. 10-11b. In general, smaller rotational inertia means easier rotation.

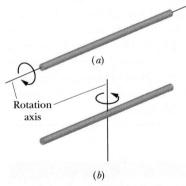

FIG. 10-11 A long rod is much easier to rotate about (a) its central (longitudinal) axis than about (b) an axis through its center and perpendicular to its length. The reason for the difference is that the mass is distributed closer to the rotation axis in (a) than in (b).

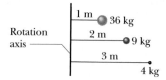

✓**CHECKPOINT 4**    The figure shows three small spheres that rotate about a vertical axis. The perpendicular distance between the axis and the center of each sphere is given. Rank the three spheres according to their rotational inertia about that axis, greatest first.

## 10-7 | Calculating the Rotational Inertia

If a rigid body consists of a few particles, we can calculate its rotational inertia about a given rotation axis with Eq. 10-33 ($I = \Sigma\, m_i r_i^2$); that is, we can find the product $mr^2$ for each particle and then sum the products. (Recall that $r$ is the perpendicular distance a particle is from the given rotation axis.)

If a rigid body consists of a great many adjacent particles (it is *continuous,* like a Frisbee), using Eq. 10-33 would require a computer. Thus, instead, we replace the sum in Eq. 10-33 with an integral and define the rotational inertia of the body as

$$I = \int r^2\, dm \qquad \text{(rotational inertia, continuous body).} \qquad (10\text{-}35)$$

Table 10-2 gives the results of such integration for nine common body shapes and the indicated axes of rotation.

### *Parallel-Axis Theorem*

Suppose we want to find the rotational inertia $I$ of a body of mass $M$ about a given axis. In principle, we can always find $I$ with the integration of Eq. 10-35. However, there is a shortcut if we happen to already know the rotational inertia $I_{com}$ of the body about a *parallel* axis that extends through the body's center of mass. Let $h$ be the perpendicular distance between the given axis and the axis

**TABLE 10-2**

**Some Rotational Inertias**

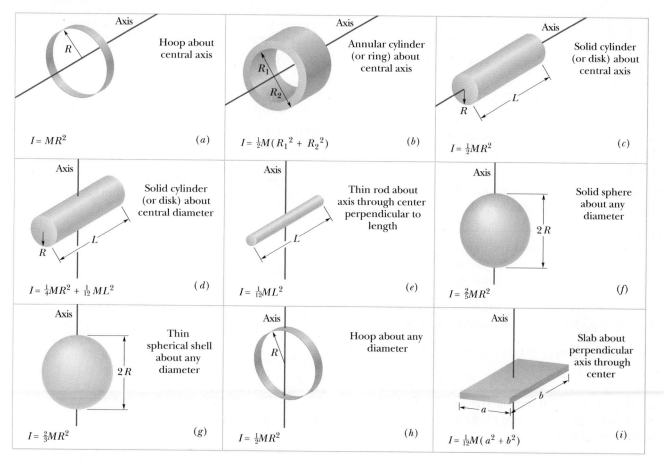

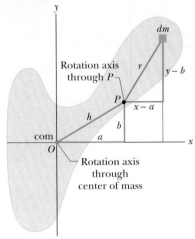

**FIG. 10-12** A rigid body in cross section, with its center of mass at $O$. The parallel-axis theorem (Eq. 10-36) relates the rotational inertia of the body about an axis through $O$ to that about a parallel axis through a point such as $P$, a distance $h$ from the body's center of mass. Both axes are perpendicular to the plane of the figure.

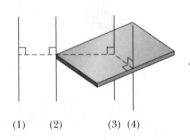

through the center of mass (remember these two axes must be parallel). Then the rotational inertia $I$ about the given axis is

$$I = I_{\text{com}} + Mh^2 \qquad \text{(parallel-axis theorem).} \qquad (10\text{-}36)$$

This equation is known as the **parallel-axis theorem.** We shall now prove it and then put it to use in Checkpoint 5 and Sample Problem 10-6.

### Proof of the Parallel-Axis Theorem

Let $O$ be the center of mass of the arbitrarily shaped body shown in cross section in Fig. 10-12. Place the origin of the coordinates at $O$. Consider an axis through $O$ perpendicular to the plane of the figure, and another axis through point $P$ parallel to the first axis. Let the $x$ and $y$ coordinates of $P$ be $a$ and $b$.

Let $dm$ be a mass element with the general coordinates $x$ and $y$. The rotational inertia of the body about the axis through $P$ is then, from Eq. 10-35,

$$I = \int r^2 \, dm = \int [(x - a)^2 + (y - b)^2] \, dm,$$

which we can rearrange as

$$I = \int (x^2 + y^2) \, dm - 2a \int x \, dm - 2b \int y \, dm + \int (a^2 + b^2) \, dm. \quad (10\text{-}37)$$

From the definition of the center of mass (Eq. 9-9), the middle two integrals of Eq. 10-37 give the coordinates of the center of mass (multiplied by a constant) and thus must each be zero. Because $x^2 + y^2$ is equal to $R^2$, where $R$ is the distance from $O$ to $dm$, the first integral is simply $I_{\text{com}}$, the rotational inertia of the body about an axis through its center of mass. Inspection of Fig. 10-12 shows that the last term in Eq. 10-37 is $Mh^2$, where $M$ is the body's total mass. Thus, Eq. 10-37 reduces to Eq. 10-36, which is the relation that we set out to prove.

✔ **CHECKPOINT 5** The figure shows a book-like object (one side is longer than the other) and four choices of rotation axes, all perpendicular to the face of the object. Rank the choices according to the rotational inertia of the object about the axis, greatest first.

(1)  (2)  (3) (4)

---

**Sample Problem** | **10-6**

Figure 10-13*a* shows a rigid body consisting of two particles of mass $m$ connected by a rod of length $L$ and negligible mass.

(a) What is the rotational inertia $I_{\text{com}}$ about an axis through the center of mass, perpendicular to the rod as shown?

**KEY IDEA** Because we have only two particles with mass, we can find the body's rotational inertia $I_{\text{com}}$ by using Eq. 10-33 rather than by integration.

**Calculations:** For the two particles, each at perpendicular distance $\frac{1}{2}L$ from the rotation axis, we have

$$I = \sum m_i r_i^2 = (m)(\tfrac{1}{2}L)^2 + (m)(\tfrac{1}{2}L)^2$$
$$= \tfrac{1}{2}mL^2. \qquad \text{(Answer)}$$

(b) What is the rotational inertia $I$ of the body about an axis through the left end of the rod and parallel to the first axis (Fig. 10-13*b*)?

**KEY IDEAS** This situation is simple enough that we can find $I$ using either of two techniques. The first is similar to the one used in part (a). The other, more powerful one is to apply the parallel-axis theorem.

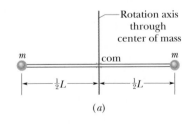

**FIG. 10-13** A rigid body consisting of two particles of mass $m$ joined by a rod of negligible mass.

**First technique:** We calculate $I$ as in part (a), except here the perpendicular distance $r_i$ is zero for the particle on the left and $L$ for the particle on the right. Now Eq. 10-33 gives us

$$I = m(0)^2 + mL^2 = mL^2. \quad \text{(Answer)}$$

**Second technique:** Because we already know $I_{com}$

about an axis through the center of mass and because the axis here is parallel to that "com axis," we can apply the parallel-axis theorem (Eq. 10-36). We find

$$I = I_{com} + Mh^2 = \tfrac{1}{2}mL^2 + (2m)(\tfrac{1}{2}L)^2$$
$$= mL^2. \quad \text{(Answer)}$$

---

## Sample Problem | 10-7

Figure 10-14 shows a thin, uniform rod of mass $M$ and length $L$, on an $x$ axis with the origin at the rod's center.

(a) What is the rotational inertia of the rod about the perpendicular rotation axis through the center?

**KEY IDEAS** (1) Because the rod is uniform, its center of mass is at its center. Therefore, we are looking for $I_{com}$. (2) Because the rod is a continuous object, we must use the integral of Eq. 10-35,

$$I = \int r^2 \, dm, \quad \text{(10-38)}$$

to find the rotational inertia.

**Calculations:** We want to integrate with respect to coordinate $x$ (not mass $m$ as indicated in the integral), so we must relate the mass $dm$ of an element of the rod to its length $dx$ along the rod. (Such an element is shown in Fig. 10-14.) Because the rod is uniform, the ratio of mass to length is the same for all the elements and for the rod as a whole. Thus, we can write

$$\frac{\text{element's mass } dm}{\text{element's length } dx} = \frac{\text{rod's mass } M}{\text{rod's length } L}$$

or

$$dm = \frac{M}{L} \, dx.$$

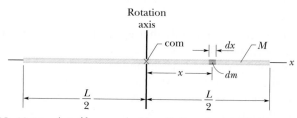

Rotation axis

FIG. 10-14    A uniform rod of length $L$ and mass $M$. An element of mass $dm$ and length $dx$ is represented.

We can now substitute this result for $dm$ and $x$ for $r$ in Eq. 10-38. Then we integrate from end to end of the rod (from $x = -L/2$ to $x = L/2$) to include all the elements. We find

$$I = \int_{x=-L/2}^{x=+L/2} x^2 \left(\frac{M}{L}\right) dx$$
$$= \frac{M}{3L}\left[x^3\right]_{-L/2}^{+L/2} = \frac{M}{3L}\left[\left(\frac{L}{2}\right)^3 - \left(-\frac{L}{2}\right)^3\right]$$
$$= \tfrac{1}{12}ML^2. \quad \text{(Answer)}$$

This agrees with the result given in Table 10-2e.

(b) What is the rod's rotational inertia $I$ about a new rotation axis that is perpendicular to the rod and through the left end?

**KEY IDEAS** We can find $I$ by shifting the origin of the $x$ axis to the left end of the rod and then integrating from $x = 0$ to $x = L$. However, here we shall use a more powerful (and easier) technique by applying the parallel-axis theorem (Eq. 10-36).

**Calculations:** If we place the axis at the rod's end so that it is parallel to the axis through the center of mass, then we can use the parallel-axis theorem (Eq. 10-36). We know from part (a) that $I_{com}$ is $\tfrac{1}{12}ML^2$. From Fig. 10-14, the perpendicular distance $h$ between the new rotation axis and the center of mass is $\tfrac{1}{2}L$. Equation 10-36 then gives us

$$I = I_{com} + Mh^2 = \tfrac{1}{12}ML^2 + (M)(\tfrac{1}{2}L)^2$$
$$= \tfrac{1}{3}ML^2. \quad \text{(Answer)}$$

Actually, this result holds for any axis through the left or right end that is perpendicular to the rod, whether it is parallel to the axis shown in Fig. 10-14 or not.

---

## Sample Problem | 10-8

Large machine components that undergo prolonged, high-speed rotation are first examined for the possibility of failure in a *spin test system*. In this system, a component is *spun up* (brought up to high speed) while inside a cylindrical arrangement of lead bricks and containment liner, all within a steel shell that is closed

by a lid clamped into place. If the rotation causes the component to shatter, the soft lead bricks are supposed to catch the pieces for later analysis.

In 1985, Test Devices, Inc. (www.testdevices.com) was spin testing a sample of a solid steel rotor (a disk) of mass $M = 272$ kg and radius $R = 38.0$ cm. When the

**FIG. 10-15**   Some of the destruction caused by the explosion of a rapidly rotating steel disk. *(Courtesy Test Devices, Inc)*

sample reached an angular speed $\omega$ of 14 000 rev/min, the test engineers heard a dull thump from the test system, which was located one floor down and one room over from them. Investigating, they found that lead bricks had been thrown out in the hallway leading to the test room, a door to the room had been hurled into the adjacent parking lot, one lead brick had shot from the test site through the wall of a neighbor's kitchen, the structural beams of the test building had been damaged, the concrete floor beneath the spin chamber had been shoved downward by about 0.5 cm, and the 900 kg lid had been blown upward through the ceiling and had then crashed back onto the test equipment (Fig. 10-15). The exploding pieces had not penetrated the room of the test engineers only by luck.

How much energy was released in the explosion of the rotor?

**KEY IDEA**   The released energy was equal to the rotational kinetic energy $K$ of the rotor just as it reached the angular speed of 14 000 rev/min.

**Calculations:** We can find $K$ with Eq. 10-34 ($K = \frac{1}{2}I\omega^2$), but first we need an expression for the rotational inertia $I$. Because the rotor was a disk that rotated like a merry-go-round, $I$ is given by the expression in Table 10-2c ($I = \frac{1}{2}MR^2$). Thus, we have

$$I = \tfrac{1}{2}MR^2 = \tfrac{1}{2}(272 \text{ kg})(0.38 \text{ m})^2 = 19.64 \text{ kg} \cdot \text{m}^2.$$

The angular speed of the rotor was

$$\omega = (14\,000 \text{ rev/min})(2\pi \text{ rad/rev})\left(\frac{1 \text{ min}}{60 \text{ s}}\right)$$

$$= 1.466 \times 10^3 \text{ rad/s}.$$

Now we can use Eq. 10-34 to write

$$K = \tfrac{1}{2}I\omega^2 = \tfrac{1}{2}(19.64 \text{ kg} \cdot \text{m}^2)(1.466 \times 10^3 \text{ rad/s})^2$$

$$= 2.1 \times 10^7 \text{ J}. \qquad \text{(Answer)}$$

Being near this explosion was quite dangerous.

## 10-8 | Torque

A doorknob is located as far as possible from the door's hinge line for a good reason. If you want to open a heavy door, you must certainly apply a force; that alone, however, is not enough. Where you apply that force and in what direction you push are also important. If you apply your force nearer to the hinge line than the knob, or at any angle other than 90° to the plane of the door, you must use a greater force to move the door than if you apply the force at the knob and perpendicular to the door's plane.

Figure 10-16*a* shows a cross section of a body that is free to rotate about an axis passing through $O$ and perpendicular to the cross section. A force $\vec{F}$ is applied at point $P$, whose position relative to $O$ is defined by a position vector $\vec{r}$. The directions of vectors $\vec{F}$ and $\vec{r}$ make an angle $\phi$ with each other. (For simplicity, we consider only forces that have no component parallel to the rotation axis; thus, $\vec{F}$ is in the plane of the page.)

To determine how $\vec{F}$ results in a rotation of the body around the rotation axis, we resolve $\vec{F}$ into two components (Fig. 10-16*b*). One component, called the *radial component* $F_r$, points along $\vec{r}$. This component does not cause rotation, because it acts along a line that extends through $O$. (If you pull on a door parallel to the plane of the door, you do not rotate the door.) The other component of $\vec{F}$, called the *tangential component* $F_t$, is perpendicular to $\vec{r}$ and has magnitude $F_t = F \sin \phi$. This component *does* cause rotation. (If you pull on a door perpendicular to its plane, you can rotate the door.)

The ability of $\vec{F}$ to rotate the body depends not only on the magnitude of its tangential component $F_t$, but also on just how far from $O$ the force is applied. To include both these factors, we define a quantity called **torque** $\tau$ as the product of the two factors and write it as

$$\tau = (r)(F \sin \phi). \qquad (10\text{-}39)$$

Two equivalent ways of computing the torque are

$$\tau = (r)(F \sin \phi) = rF_t \qquad (10\text{-}40)$$

and

$$\tau = (r \sin \phi)(F) = r_\perp F, \qquad (10\text{-}41)$$

where $r_\perp$ is the perpendicular distance between the rotation axis at $O$ and an extended line running through the vector $\vec{F}$ (Fig. 10-16c). This extended line is called the **line of action** of $\vec{F}$, and $r_\perp$ is called the **moment arm** of $\vec{F}$. Figure 10-16b shows that we can describe $r$, the magnitude of $\vec{r}$, as being the moment arm of the force component $F_t$.

Torque, which comes from the Latin word meaning "to twist," may be loosely identified as the turning or twisting action of the force $\vec{F}$. When you apply a force to an object—such as a screwdriver or torque wrench—with the purpose of turning that object, you are applying a torque. The SI unit of torque is the newton-meter (N · m). *Caution:* The newton-meter is also the unit of work. Torque and work, however, are quite different quantities and must not be confused. Work is often expressed in joules (1 J = 1 N · m), but torque never is.

In the next chapter we shall discuss torque in a general way as being a vector quantity. Here, however, because we consider only rotation around a single axis, we do not need vector notation. Instead, a torque has either a positive or negative value depending on the direction of rotation it would give a body initially at rest: If the body would rotate counterclockwise, the torque is positive. If the object would rotate clockwise, the torque is negative. (The phrase "clocks are negative" from Section 10-2 still works.)

Torques obey the superposition principle that we discussed in Chapter 5 for forces: When several torques act on a body, the **net torque** (or **resultant torque**) is the sum of the individual torques. The symbol for net torque is $\tau_{\text{net}}$.

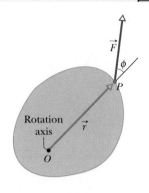
(a)

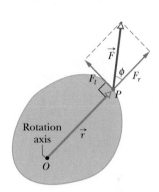
(b)

**✓ CHECKPOINT 6** The figure shows an overhead view of a meter stick that can pivot about the dot at the position marked 20 (for 20 cm). All five forces on the stick are horizontal and have the same magnitude. Rank the forces according to the magnitude of the torque they produce, greatest first.

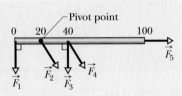

## 10-9 | Newton's Second Law for Rotation

A torque can cause rotation of a rigid body, as when you use a torque to rotate a door. Here we want to relate the net torque $\tau_{\text{net}}$ on a rigid body to the angular acceleration $\alpha$ that torque causes about a rotation axis. We do so by analogy with Newton's second law ($F_{\text{net}} = ma$) for the acceleration $a$ of a body of mass $m$ due to a net force $F_{\text{net}}$ along a coordinate axis. We replace $F_{\text{net}}$ with $\tau_{\text{net}}$, $m$ with $I$, and $a$ with $\alpha$ in radian measure, writing

$$\tau_{\text{net}} = I\alpha \qquad \text{(Newton's second law for rotation).} \qquad (10\text{-}42)$$

### Proof of Equation 10-42

We prove Eq. 10-42 by first considering the simple situation shown in Fig. 10-17. The rigid body there consists of a particle of mass $m$ on one end of a massless rod of length $r$. The rod can move only by rotating about its other end, around a rotation axis (an axle) that is perpendicular to the plane of the page. Thus, the particle can move only in a circular path that has the rotation axis at its center.

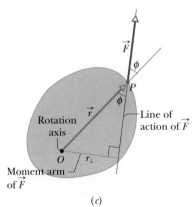

**FIG. 10-16** (a) A force $\vec{F}$ acts at point $P$ on a rigid body that is free to rotate about an axis through $O$; the axis is perpendicular to the plane of the cross section shown here. (b) The torque due to this force is $(r)(F \sin \phi)$. We can also write it as $rF_t$, where $F_t$ is the tangential component of $\vec{F}$. (c) The torque can also be written as $r_\perp F$, where $r_\perp$ is the moment arm of $\vec{F}$.

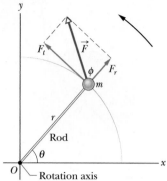

**FIG. 10-17** A simple rigid body, free to rotate about an axis through O, consists of a particle of mass $m$ fastened to the end of a rod of length $r$ and negligible mass. An applied force $\vec{F}$ causes the body to rotate.

A force $\vec{F}$ acts on the particle. However, because the particle can move only along the circular path, only the tangential component $F_t$ of the force (the component that is tangent to the circular path) can accelerate the particle along the path. We can relate $F_t$ to the particle's tangential acceleration $a_t$ along the path with Newton's second law, writing

$$F_t = ma_t.$$

The torque acting on the particle is, from Eq. 10-40,

$$\tau = F_t r = ma_t r.$$

From Eq. 10-22 ($a_t = \alpha r$) we can write this as

$$\tau = m(\alpha r)r = (mr^2)\alpha. \tag{10-43}$$

The quantity in parentheses on the right is the rotational inertia of the particle about the rotation axis (see Eq. 10-33). Thus, Eq. 10-43 reduces to

$$\tau = I\alpha \quad \text{(radian measure).} \tag{10-44}$$

For the situation in which more than one force is applied to the particle, we can generalize Eq. 10-44 as

$$\tau_{\text{net}} = I\alpha \quad \text{(radian measure),} \tag{10-45}$$

which we set out to prove. We can extend this equation to any rigid body rotating about a fixed axis, because any such body can always be analyzed as an assembly of single particles.

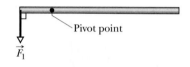

✓**CHECKPOINT 7**    The figure shows an overhead view of a meter stick that can pivot about the point indicated, which is to the left of the stick's midpoint. Two horizontal forces, $\vec{F_1}$ and $\vec{F_2}$, are applied to the stick. Only $\vec{F_1}$ is shown. Force $\vec{F_2}$ is perpendicular to the stick and is applied at the right end. If the stick is not to turn, (a) what should be the direction of $\vec{F_2}$, and (b) should $F_2$ be greater than, less than, or equal to $F_1$?

---

**Sample Problem**  **10-9**  **Build your skill**

Figure 10-18a shows a uniform disk, with mass $M = 2.5$ kg and radius $R = 20$ cm, mounted on a fixed horizontal axle. A block with mass $m = 1.2$ kg hangs from a massless cord that is wrapped around the rim of the disk. Find the acceleration of the falling block, the angular acceleration of the disk, and the tension in the cord. The cord does not slip, and there is no friction at the axle.

**KEY IDEAS**  (1) Taking the block as a system, we can relate its acceleration $a$ to the forces acting on it with Newton's second law ($\vec{F}_{\text{net}} = m\vec{a}$). (2) Taking the disk as a system, we can relate its angular acceleration $\alpha$ to the torque acting on it with Newton's second law for rotation ($\tau_{\text{net}} = I\alpha$). (3) To combine the motions of block and disk, we use the fact that the linear acceleration $a$ of the block and the (tangential) linear acceleration $a_t$ of the disk rim are equal.

**Forces on block:** The forces are shown in the block's free-body diagram in Fig. 10-18b: The force from the

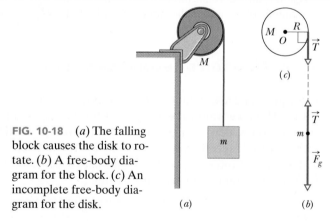

**FIG. 10-18**  (a) The falling block causes the disk to rotate. (b) A free-body diagram for the block. (c) An incomplete free-body diagram for the disk.

cord is $\vec{T}$, and the gravitational force is $\vec{F_g}$, of magnitude $mg$. We can now write Newton's second law for components along a vertical $y$ axis ($F_{\text{net},y} = ma_y$) as

$$T - mg = ma. \tag{10-46}$$

However, we cannot solve this equation for $a$ because it also contains the unknown $T$.

**Torque on disk:** Previously, when we got stuck on the $y$ axis, we switched to the $x$ axis. Here, we switch to the rotation of the disk. To calculate the torques and the rotational inertia $I$, we take the rotation axis to be perpendicular to the disk and through its center, at point $O$ in Fig. 10-18$c$.

The torques are then given by Eq. 10-40 ($\tau = rF_t$). The gravitational force on the disk and the force on the disk from the axle both act at the center of the disk and thus at distance $r = 0$, so their torques are zero. The force $\vec{T}$ on the disk due to the cord acts at distance $r = R$ and is tangent to the rim of the disk. Therefore, its torque is $-RT$, negative because the torque rotates the disk clockwise from rest. From Table 10-2$c$, the rotational inertia $I$ of the disk is $\frac{1}{2}MR^2$. Thus we can write $\tau_{net} = I\alpha$ as

$$-RT = \tfrac{1}{2}MR^2\alpha. \qquad (10\text{-}47)$$

This equation seems useless because it has two unknowns, $\alpha$ and $T$, neither of which is the desired $a$. However, mustering physics courage, we can make it useful with this fact: Because the cord does not slip, the linear acceleration $a$ of the block and the (tangential) linear acceleration $a_t$ of the rim of the disk are equal. Then, by Eq. 10-22 ($a_t = \alpha r$) we see that here $\alpha = a/R$.

Substituting this in Eq. 10-47 yields

$$T = -\tfrac{1}{2}Ma. \qquad (10\text{-}48)$$

**Combining results:** Combining Eqs. 10-46 and 10-48 leads to

$$a = -g\,\frac{2m}{M + 2m} = -(9.8 \text{ m/s}^2)\,\frac{(2)(1.2 \text{ kg})}{2.5 \text{ kg} + (2)(1.2 \text{ kg})}$$

$$= -4.8 \text{ m/s}^2. \qquad \text{(Answer)}$$

We then use Eq. 10-48 to find $T$:

$$T = -\tfrac{1}{2}Ma = -\tfrac{1}{2}(2.5 \text{ kg})(-4.8 \text{ m/s}^2)$$

$$= 6.0 \text{ N}. \qquad \text{(Answer)}$$

As we should expect, acceleration $a$ of the falling block is less than $g$, and tension $T$ in the cord ($= 6.0$ N) is less than the gravitational force on the hanging block ($= mg = 11.8$ N). We see also that $a$ and $T$ depend on the mass of the disk but not on its radius. As a check, we note that the formulas derived above predict $a = -g$ and $T = 0$ for the case of a massless disk ($M = 0$). This is what we would expect; the block simply falls as a free body. From Eq. 10-22, the angular acceleration of the disk is

$$\alpha = \frac{a}{R} = \frac{-4.8 \text{ m/s}^2}{0.20 \text{ m}} = -24 \text{ rad/s}^2. \qquad \text{(Answer)}$$

---

**Sample Problem    10-10**

To throw an 80 kg opponent with a basic judo hip throw, you intend to pull his uniform with a force $\vec{F}$ and a moment arm $d_1 = 0.30$ m from a pivot point (rotation axis) on your right hip (Fig. 10-19). You wish to rotate him about the pivot point with an angular acceleration $\alpha$ of $-6.0$ rad/s$^2$—that is, with an angular acceleration that is *clockwise* in the figure. Assume that his rotational inertia $I$ relative to the pivot point is 15 kg $\cdot$ m$^2$.

(a) What must the magnitude of $\vec{F}$ be if, before you throw him, you bend your opponent forward to bring his center of mass to your hip (Fig. 10-19$a$)?

**KEY IDEA**    We can relate your pull $\vec{F}$ on your opponent to the given angular acceleration $\alpha$ via Newton's second law for rotation ($\tau_{net} = I\alpha$).

**Calculations:** As his feet leave the floor, we can assume that only three forces act on him: your pull $\vec{F}$, a force $\vec{N}$ on him from you at the pivot point (this force is not indicated in Fig. 10-19), and the gravitational force $\vec{F}_g$. To use $\tau_{net} = I\alpha$, we need the corresponding three torques, each about the pivot point.

From Eq. 10-41 ($\tau = r_\perp F$), the torque due to your pull $\vec{F}$ is equal to $-d_1 F$, where $d_1$ is the moment arm $r_\perp$ and the sign indicates the clockwise rotation this torque

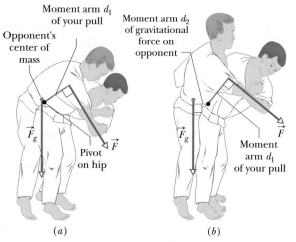

**FIG. 10-19**    A judo hip throw ($a$) correctly executed and ($b$) incorrectly executed.

tends to cause. The torque due to $\vec{N}$ is zero, because $\vec{N}$ acts at the pivot point and thus has moment arm $r_\perp = 0$.

To evaluate the torque due to $\vec{F}_g$, we can assume that $\vec{F}_g$ acts at your opponent's center of mass. With the center of mass at the pivot point, $\vec{F}_g$ has moment arm $r_\perp = 0$ and thus the torque due to $\vec{F}_g$ is zero. Thus, the only torque on your opponent is due to your pull $\vec{F}$, and we can write $\tau_{net} = I\alpha$ as

$$-d_1 F = I\alpha.$$

We then find

$$F = \frac{-I\alpha}{d_1} = \frac{-(15 \text{ kg} \cdot \text{m}^2)(-6.0 \text{ rad/s}^2)}{0.30 \text{ m}}$$

$$= 300 \text{ N}. \qquad \text{(Answer)}$$

(b) What must the magnitude of $\vec{F}$ be if your opponent remains upright before you throw him, so that $\vec{F}_g$ has a moment arm $d_2 = 0.12$ m (Fig. 10-19b)?

**KEY IDEA** Because the moment arm for $\vec{F}_g$ is no longer zero, the torque due to $\vec{F}_g$ is now equal to $d_2 mg$ and is positive because the torque attempts counterclockwise rotation.

*Calculations:* Now we write $\tau_{\text{net}} = I\alpha$ as

$$-d_1 F + d_2 mg = I\alpha,$$

which gives

$$F = -\frac{I\alpha}{d_1} + \frac{d_2 mg}{d_1}.$$

From (a), we know that the first term on the right is equal to 300 N. Substituting this and the given data, we have

$$F = 300 \text{ N} + \frac{(0.12 \text{ m})(80 \text{ kg})(9.8 \text{ m/s}^2)}{0.30 \text{ m}}$$

$$= 613.6 \text{ N} \approx 610 \text{ N}. \qquad \text{(Answer)}$$

The results indicate that you will have to pull much harder if you do not initially bend your opponent to bring his center of mass to your hip. A good judo fighter knows this lesson from physics.

## 10-10 | Work and Rotational Kinetic Energy

As we discussed in Chapter 7, when a force $F$ causes a rigid body of mass $m$ to accelerate along a coordinate axis, the force does work $W$ on the body. Thus, the body's kinetic energy ($K = \frac{1}{2}mv^2$) can change. Suppose it is the only energy of the body that changes. Then we relate the change $\Delta K$ in kinetic energy to the work $W$ with the work–kinetic energy theorem (Eq. 7-10), writing

$$\Delta K = K_f - K_i = \tfrac{1}{2}mv_f^2 - \tfrac{1}{2}mv_i^2 = W \qquad \text{(work–kinetic energy theorem).} \qquad (10\text{-}49)$$

For motion confined to an $x$ axis, we can calculate the work with Eq. 7-32,

$$W = \int_{x_i}^{x_f} F \, dx \qquad \text{(work, one-dimensional motion).} \qquad (10\text{-}50)$$

This reduces to $W = Fd$ when $F$ is constant and the body's displacement is $d$. The rate at which the work is done is the power, which we can find with Eqs. 7-43 and 7-48,

$$P = \frac{dW}{dt} = Fv \qquad \text{(power, one-dimensional motion).} \qquad (10\text{-}51)$$

Now let us consider a rotational situation that is similar. When a torque accelerates a rigid body in rotation about a fixed axis, the torque does work $W$ on the body. Therefore, the body's rotational kinetic energy ($K = \frac{1}{2}I\omega^2$) can change. Suppose that it is the only energy of the body that changes. Then we can still relate the change $\Delta K$ in kinetic energy to the work $W$ with the work–kinetic energy theorem, except now the kinetic energy is a rotational kinetic energy:

$$\Delta K = K_f - K_i = \tfrac{1}{2}I\omega_f^2 - \tfrac{1}{2}I\omega_i^2 = W \qquad \text{(work–kinetic energy theorem).} \qquad (10\text{-}52)$$

Here, $I$ is the rotational inertia of the body about the fixed axis and $\omega_i$ and $\omega_f$ are the angular speeds of the body before and after the work is done, respectively.

Also, we can calculate the work with a rotational equivalent of Eq. 10-50,

$$W = \int_{\theta_i}^{\theta_f} \tau \, d\theta \qquad \text{(work, rotation about fixed axis),} \qquad (10\text{-}53)$$

where $\tau$ is the torque doing the work $W$, and $\theta_i$ and $\theta_f$ are the body's angular positions before and after the work is done, respectively. When $\tau$ is constant, Eq. 10-53 reduces to

$$W = \tau(\theta_f - \theta_i) \qquad \text{(work, constant torque).} \qquad (10\text{-}54)$$

**TABLE 10-3**

**Some Corresponding Relations for Translational and Rotational Motion**

| Pure Translation (Fixed Direction) | | Pure Rotation (Fixed Axis) | |
| --- | --- | --- | --- |
| Position | $x$ | Angular position | $\theta$ |
| Velocity | $v = dx/dy$ | Angular velocity | $\omega = d\theta/dt$ |
| Acceleration | $a = dv/dt$ | Angular acceleration | $\alpha = d\omega/dt$ |
| Mass | $m$ | Rotational inertia | $I$ |
| Newton's second law | $F_{net} = ma$ | Newton's second law | $\tau_{net} = I\alpha$ |
| Work | $W = \int F\,dx$ | Work | $W = \int \tau\,d\theta$ |
| Kinetic energy | $K = \frac{1}{2}mv^2$ | Kinetic energy | $K = \frac{1}{2}I\omega^2$ |
| Power (constant force) | $P = Fv$ | Power (constant torque) | $P = \tau\omega$ |
| Work–kinetic energy theorem | $W = \Delta K$ | Work–kinetic energy theorem | $W = \Delta K$ |

The rate at which the work is done is the power, which we can find with the rotational equivalent of Eq. 10-51,

$$P = \frac{dW}{dt} = \tau\omega \qquad \text{(power, rotation about fixed axis).} \qquad (10\text{-}55)$$

Table 10-3 summarizes the equations that apply to the rotation of a rigid body about a fixed axis and the corresponding equations for translational motion.

### Proof of Eqs. 10-52 through 10-55

Let us again consider the situation of Fig. 10-17, in which force $\vec{F}$ rotates a rigid body consisting of a single particle of mass $m$ fastened to the end of a massless rod. During the rotation, force $\vec{F}$ does work on the body. Let us assume that the only energy of the body that is changed by $\vec{F}$ is the kinetic energy. Then we can apply the work–kinetic energy theorem of Eq. 10-49:

$$\Delta K = K_f - K_i = W. \qquad (10\text{-}56)$$

Using $K = \frac{1}{2}mv^2$ and Eq. 10-18 ($v = \omega r$), we can rewrite Eq. 10-56 as

$$\Delta K = \frac{1}{2}mr^2\omega_f^2 - \frac{1}{2}mr^2\omega_i^2 = W. \qquad (10\text{-}57)$$

From Eq. 10-33, the rotational inertia for this one-particle body is $I = mr^2$. Substituting this into Eq. 10-57 yields

$$\Delta K = \frac{1}{2}I\omega_f^2 - \frac{1}{2}I\omega_i^2 = W,$$

which is Eq. 10-52. We derived it for a rigid body with one particle, but it holds for any rigid body rotated about a fixed axis.

We next relate the work $W$ done on the body in Fig. 10-17 to the torque $\tau$ on the body due to force $\vec{F}$. When the particle moves a distance $ds$ along its circular path, only the tangential component $F_t$ of the force accelerates the particle along the path. Therefore, only $F_t$ does work on the particle. We write that work $dW$ as $F_t\,ds$. However, we can replace $ds$ with $r\,d\theta$, where $d\theta$ is the angle through which the particle moves. Thus we have

$$dW = F_t r\,d\theta. \qquad (10\text{-}58)$$

From Eq. 10-40, we see that the product $F_t r$ is equal to the torque $\tau$, so we can rewrite Eq. 10-58 as

$$dW = \tau\,d\theta. \qquad (10\text{-}59)$$

The work done during a finite angular displacement from $\theta_i$ to $\theta_f$ is then

$$W = \int_{\theta_i}^{\theta_f} \tau\,d\theta,$$

which is Eq. 10-53. It holds for any rigid body rotating about a fixed axis. Equation 10-54 comes directly from Eq. 10-53.

We can find the power $P$ for rotational motion from Eq. 10-59:

$$P = \frac{dW}{dt} = \tau \frac{d\theta}{dt} = \tau\omega,$$

which is Eq. 10-55

## Sample Problem   10-11

Let the disk in Sample Problem 10-9 and Fig. 10-18 start from rest at time $t = 0$. What is its rotational kinetic energy $K$ at $t = 2.5$ s?

**KEY IDEA**   We can find $K$ with Eq. 10-34 ($K = \frac{1}{2}I\omega^2$). We already know that $I = \frac{1}{2}MR^2$, but we do not yet know $\omega$ at $t = 2.5$ s. However, because the angular acceleration $\alpha$ has the constant value of $-24$ rad/s$^2$, we can apply the equations for constant angular acceleration in Table 10-1.

**Calculations:** Because we want $\omega$ and know $\alpha$ and $\omega_0$ ($= 0$), we use Eq. 10-12:

$$\omega = \omega_0 + \alpha t = 0 + \alpha t = \alpha t.$$

Substituting $\omega = \alpha t$ and $I = \frac{1}{2}MR^2$ into Eq. 10-34, we find

$$K = \tfrac{1}{2}I\omega^2 = \tfrac{1}{2}(\tfrac{1}{2}MR^2)(\alpha t)^2 = \tfrac{1}{4}M(R\alpha t)^2$$
$$= \tfrac{1}{4}(2.5 \text{ kg})[(0.20 \text{ m})(-24 \text{ rad/s}^2)(2.5 \text{ s})]^2$$
$$= 90 \text{ J}. \qquad \text{(Answer)}$$

**KEY IDEA**   We can also get this answer by finding the disk's kinetic energy from the work done on the disk.

**Calculations:** First, we relate the *change* in the kinetic energy of the disk to the net work $W$ done on the disk, using the work–kinetic energy theorem of Eq. 10-52 ($K_f - K_i = W$). With $K$ substituted for $K_f$ and 0 for $K_i$, we get

$$K = K_i + W = 0 + W = W. \qquad (10\text{-}60)$$

Next we want to find the work $W$. We can relate $W$ to the torques acting on the disk with Eq. 10-53 or 10-54. The only torque causing angular acceleration and doing work is the torque due to force $\vec{T}$ on the disk from the cord. From Sample Problem 10-9, this torque is equal to $-TR$. Because $\alpha$ is constant, this torque also must be constant. Thus, we can use Eq. 10-54 to write

$$W = \tau(\theta_f - \theta_i) = -TR(\theta_f - \theta_i). \qquad (10\text{-}61)$$

Because $\alpha$ is constant, we can use Eq. 10-13 to find $\theta_f - \theta_i$. With $\omega_i = 0$, we have

$$\theta_f - \theta_i = \omega_i t + \tfrac{1}{2}\alpha t^2 = 0 + \tfrac{1}{2}\alpha t^2 = \tfrac{1}{2}\alpha t^2.$$

Now we substitute this into Eq. 10-61 and then substitute the result into Eq. 10-60. With $T = 6.0$ N and $\alpha = -24$ rad/s$^2$ (from Sample Problem 10-9), we have

$$K = W = -TR(\theta_f - \theta_i) = -TR(\tfrac{1}{2}\alpha t^2) = -\tfrac{1}{2}TR\alpha t^2$$
$$= -\tfrac{1}{2}(6.0 \text{ N})(0.20 \text{ m})(-24 \text{ rad/s}^2)(2.5 \text{ s})^2$$
$$= 90 \text{ J}. \qquad \text{(Answer)}$$

## Sample Problem   10-12

A tall, cylindrical chimney will fall over when its base is ruptured. Treat the chimney as a thin rod of length $L = 55.0$ m (Fig. 10-20a). At the instant it makes an angle of $\theta = 35.0°$ with the vertical, what is its angular speed $\omega_f$?

**KEY IDEAS**   (1) During the rotation, the mechanical energy (the sum of the rotational kinetic energy $K$ and the gravitational potential energy $U$) does not change. (2) The rotational kinetic energy is given by Eq. 10-34 ($K = \frac{1}{2}I\omega^2$).

**Conservation of mechanical energy:** As the center of mass of the chimney falls, energy is transferred from gravitational potential energy $U$ to rotational kinetic energy $K$ but the total amount does not change. We can write this fact as

$$K_f + U_f = K_i + U_i. \qquad (10\text{-}62)$$

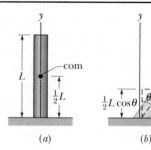

**FIG. 10-20**   (a) A cylindrical chimney. (b) The height of its center of mass is determined with the right triangle.

**Rotational kinetic energy:** The kinetic energy $K$ is initially zero but its value thereafter ($= \frac{1}{2}I\omega^2$) depends on the rotational inertia $I$. If we had a thin rod rotating around its center of mass (at its center), we know from Table 10-2 that $I_{com} = \frac{1}{12}mL^2$, where $m$ is the rod's mass and $L$ is the rod's length. However, our rod-like chimney rotates around one end, at a distance of $L/2$ from the center, and so we use the parallel-axis theorem to find

$$I = \tfrac{1}{12}mL + m\left(\frac{L}{2}\right)^2 = \tfrac{1}{3}mL^2. \qquad (10\text{-}63)$$

Substituting this into $K = \tfrac{1}{2}I\omega^2$ tells us

$$K_f = \tfrac{1}{2}(\tfrac{1}{3}mL^2)\omega^2. \qquad (10\text{-}64)$$

**Potential energy:** The potential energy $U\,(= mgy)$ depends on the height of each segment of the chimney. However, we can calculate $U$ by assuming that all of the mass is concentrated at the chimney's com, which is initially at height $\tfrac{1}{2}L$. So the initial potential energy is

$$U_i = \tfrac{1}{2}mgL. \qquad (10\text{-}65)$$

When the chimney has rotated through angle $\theta$, Fig.

10-20$b$ tells us that the center is at height $\tfrac{1}{2}L\cos\theta$. So, the potential energy is now

$$U_f = \tfrac{1}{2}mgL\cos\theta. \qquad (10\text{-}66)$$

**Angular speed:** After substituting Eqs. 10-66, 10-65, and 10-64 into Eq. 10-62 and setting $K_i = 0$, we solve for $\omega_f$ and find

$$\omega = \sqrt{\frac{3g}{L}(1 - \cos\theta)} = \sqrt{\frac{3(9.8\text{ m/s}^2)}{55.0\text{ m}}(1 - \cos 35.0°)}$$

$$= 0.311\text{ rad/s}. \qquad \text{(Answer)}$$

**Comments:** The bottom portion of a chimney tends to rotate around the base faster than the top portion, and the chimney is likely to break apart during the rotation, with the top half lagging behind the bottom half.

---

**Sample Problem  10-13**

In the oversized claw of a snapping shrimp, the dactylus (the large, mobile section of the claw) is drawn away from the propodius (the opposing, stationary part of the claw) by a muscle that is gradually put under tension (Fig. 10-21). Energy stored in the muscle increases as the tension increases. The sudden release of the dactylus allows it to rotate about a pivot point, to slam shut on the propodius in a time interval $\Delta t$ of only 290 $\mu$s. In particular, the *plunger* on the dactylus runs into a cavity on the propodius, causing water to squirt out of the cavity so quickly that the water undergoes *cavitation*. That is, the water vaporizes to form bubbles of water vapor. These bubbles rapidly grow as they enter the surrounding water and then they suddenly collapse, emitting an intense sound wave. The combination of these sound waves from many bubbles can stun the shrimp's prey.

The peak angular speed $\omega$ of the dactylus is about $2 \times 10^3$ rad/s and its rotational inertia $I$ is about $3 \times 10^{-11}$ kg·m$^2$. At what akverage rate is energy transferred from the muscle to the rotation?

**KEY IDEA** (1) Rotational kinetic energy is given by Eq. 10-34 ($K = \tfrac{1}{2}I\omega^2$). (2) Average power is given by Eq. 8-40 ($P_{\text{avg}} = \Delta E/\Delta t$).

**Calculations:** When the angular speed reaches its peak

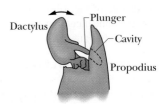

**FIG. 10-21** The oversized claw of a snapping shrimp. The dactylus is first pulled away from the opposing section of the propodius and then allowed to snap back to it, thrusting the plunger into the cavity.

value, the rotational kinetic energy is

$$K = \tfrac{1}{2}I\omega^2 = \tfrac{1}{2}(3 \times 10^{-11}\text{ kg·m}^2)(2 \times 10^3\text{ rad/s})^2$$

$$= 6 \times 10^{-5}\text{ J}.$$

The average power is then

$$P_{\text{avg}} = \frac{\Delta E}{\Delta t} = \frac{6 \times 10^{-5}\text{ J}}{290 \times 10^{-6}\text{ s}}$$

$$= 0.2\text{ W}. \qquad \text{(Answer)}$$

This power greatly exceeds what any fast-acting muscle in the shrimp can produce. However, in the claw the shrimp effectively locks the dactylus against a spring so that it can gradually increase the tension and stored energy (the power of this stage is low). Then, once the stored energy is high, the dactylus is released and the spring-like muscle slams it shut (the power is now very high). Many other animals make use of such low-power storing of energy and then a high-power release of the energy that allows them to capture lunch or to avoid becoming lunch.

---

## REVIEW & SUMMARY

**Angular Position**  To describe the rotation of a rigid body about a fixed axis, called the **rotation axis,** we assume a **reference line** is fixed in the body, perpendicular to that axis and rotating with the body. We measure the **angular position** $\theta$ of this

line relative to a fixed direction. When $\theta$ is measured in **radians,**

$$\theta = \frac{s}{r} \qquad \text{(radian measure)}, \qquad (10\text{-}1)$$

where $s$ is the arc length of a circular path of radius $r$ and angle $\theta$. Radian measure is related to angle measure in revolutions and degrees by

$$1 \text{ rev} = 360° = 2\pi \text{ rad.} \tag{10-2}$$

**Angular Displacement**   A body that rotates about a rotation axis, changing its angular position from $\theta_1$ to $\theta_2$, undergoes an **angular displacement**

$$\Delta\theta = \theta_2 - \theta_1, \tag{10-4}$$

where $\Delta\theta$ is positive for counterclockwise rotation and negative for clockwise rotation.

**Angular Velocity and Speed**   If a body rotates through an angular displacement $\Delta\theta$ in a time interval $\Delta t$, its **average angular velocity** $\omega_{avg}$ is

$$\omega_{avg} = \frac{\Delta\theta}{\Delta t}. \tag{10-5}$$

The **(instantaneous) angular velocity** $\omega$ of the body is

$$\omega = \frac{d\theta}{dt}. \tag{10-6}$$

Both $\omega_{avg}$ and $\omega$ are vectors, with directions given by the **right-hand rule** of Fig. 10-6. They are positive for counterclockwise rotation and negative for clockwise rotation. The magnitude of the body's angular velocity is the **angular speed.**

**Angular Acceleration**   If the angular velocity of a body changes from $\omega_1$ to $\omega_2$ in a time interval $\Delta t = t_2 - t_1$, the **average angular acceleration** $\alpha_{avg}$ of the body is

$$\alpha_{avg} = \frac{\omega_2 - \omega_1}{t_2 - t_1} = \frac{\Delta\omega}{\Delta t}. \tag{10-7}$$

The **(instantaneous) angular acceleration** $\alpha$ of the body is

$$\alpha = \frac{d\omega}{dt}. \tag{10-8}$$

Both $\alpha_{avg}$ and $\alpha$ are vectors.

**The Kinematic Equations for Constant Angular Acceleration**   Constant angular acceleration ($\alpha = $ constant) is an important special case of rotational motion. The appropriate kinematic equations, given in Table 10-1, are

$$\omega = \omega_0 + \alpha t, \tag{10-12}$$

$$\theta - \theta_0 = \omega_0 t + \tfrac{1}{2}\alpha t^2, \tag{10-13}$$

$$\omega^2 = \omega_0^2 + 2\alpha(\theta - \theta_0), \tag{10-14}$$

$$\theta - \theta_0 = \tfrac{1}{2}(\omega_0 + \omega)t, \tag{10-15}$$

$$\theta - \theta_0 = \omega t - \tfrac{1}{2}\alpha t^2. \tag{10-16}$$

**Linear and Angular Variables Related**   A point in a rigid rotating body, at a *perpendicular distance $r$* from the rotation axis, moves in a circle with radius $r$. If the body rotates through an angle $\theta$, the point moves along an arc with length $s$ given by

$$s = \theta r \qquad \text{(radian measure),} \tag{10-17}$$

where $\theta$ is in radians.

The linear velocity $\vec{v}$ of the point is tangent to the circle; the point's linear speed $v$ is given by

$$v = \omega r \qquad \text{(radian measure),} \tag{10-18}$$

where $\omega$ is the angular speed (in radians per second) of the body.

The linear acceleration $\vec{a}$ of the point has both *tangential* and *radial* components. The tangential component is

$$a_t = \alpha r \qquad \text{(radian measure),} \tag{10-22}$$

where $\alpha$ is the magnitude of the angular acceleration (in radians per second-squared) of the body. The radial component of $\vec{a}$ is

$$a_r = \frac{v^2}{r} = \omega^2 r \qquad \text{(radian measure).} \tag{10-23}$$

If the point moves in uniform circular motion, the period $T$ of the motion for the point and the body is

$$T = \frac{2\pi r}{v} = \frac{2\pi}{\omega} \qquad \text{(radian measure).} \tag{10-19, 10-20}$$

**Rotational Kinetic Energy and Rotational Inertia**   The kinetic energy $K$ of a rigid body rotating about a fixed axis is given by

$$K = \tfrac{1}{2}I\omega^2 \qquad \text{(radian measure),} \tag{10-34}$$

in which $I$ is the **rotational inertia** of the body, defined as

$$I = \sum m_i r_i^2 \tag{10-33}$$

for a system of discrete particles and defined as

$$I = \int r^2 \, dm \tag{10-35}$$

for a body with continuously distributed mass. The $r$ and $r_i$ in these expressions represent the perpendicular distance from the axis of rotation to each mass element in the body.

**The Parallel-Axis Theorem**   The *parallel-axis theorem* relates the rotational inertia $I$ of a body about any axis to that of the same body about a parallel axis through the center of mass:

$$I = I_{com} + Mh^2. \tag{10-36}$$

Here $h$ is the perpendicular distance between the two axes.

**Torque**   *Torque* is a turning or twisting action on a body about a rotation axis due to a force $\vec{F}$. If $\vec{F}$ is exerted at a point given by the position vector $\vec{r}$ relative to the axis, then the magnitude of the torque is

$$\tau = rF_t = r_\perp F = rF \sin\phi, \qquad \text{(10-40, 10-41, 10-39)}$$

where $F_t$ is the component of $\vec{F}$ perpendicular to $\vec{r}$ and $\phi$ is the angle between $\vec{r}$ and $\vec{F}$. The quantity $r_\perp$ is the perpendicular distance between the rotation axis and an extended line running through the $\vec{F}$ vector. This line is called the **line of action** of $\vec{F}$, and $r_\perp$ is called the **moment arm** of $\vec{F}$. Similarly, $r$ is the moment arm of $F_t$.

The SI unit of torque is the newton-meter (N·m). A torque $\tau$ is positive if it tends to rotate a body at rest counterclockwise and negative if it tends to rotate the body in the clockwise direction.

**Newton's Second Law in Angular Form**   The rotational analog of Newton's second law is

$$\tau_{net} = I\alpha, \tag{10-45}$$

where $\tau_{net}$ is the net torque acting on a particle or rigid body, $I$ is the rotational inertia of the particle or body about the rotation axis, and $\alpha$ is the resulting angular acceleration about that axis.

**Work and Rotational Kinetic Energy**   The equations used for calculating work and power in rotational motion correspond to equations used for translational motion and are

$$W = \int_{\theta_i}^{\theta_f} \tau \, d\theta \qquad (10\text{-}53)$$

and $$P = \frac{dW}{dt} = \tau\omega. \qquad (10\text{-}55)$$

When $\tau$ is constant, Eq. 10-53 reduces to

$$W = \tau(\theta_f - \theta_i). \qquad (10\text{-}54)$$

The form of the work–kinetic energy theorem used for rotating bodies is

$$\Delta K = K_f - K_i = \tfrac{1}{2}I\omega_f^2 - \tfrac{1}{2}I\omega_i^2 = W. \qquad (10\text{-}52)$$

## QUESTIONS

**1** Figure 10-22 is a graph of the angular velocity versus time for a disk rotating like a merry-go-round. For a point on the disk rim, rank the instants $a$, $b$, $c$, and $d$ according to the magnitude of the (a) tangential and (b) radial acceleration, greatest first.

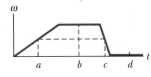

**FIG. 10-22** Question 1.

**2** Figure 10-23$b$ is a graph of the angular position of the rotating disk of Fig. 10-23$a$. Is the angular velocity of the disk positive, negative, or zero at (a) $t = 1$ s, (b) $t = 2$ s, and (c) $t = 3$ s? (d) Is the angular acceleration positive or negative?

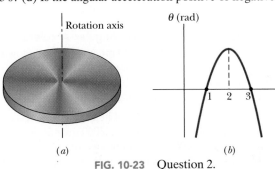

**FIG. 10-23** Question 2.

**3** Figure 10-24 shows a uniform metal plate that had been square before 25% of it was snipped off. Three lettered points are indicated. Rank them according to the rotational inertia of the plate around a perpendicular axis through them, greatest first.

**FIG. 10-24** Question 3.

**4** Figure 10-25 shows plots of angular position $\theta$ versus time $t$ for three cases in which a disk is rotated like a merry-go-round. In each case, the rotation direction changes at a certain angular position $\theta_{change}$. (a) For each case, determine whether $\theta_{change}$ is clockwise or counterclockwise from $\theta = 0$, or whether it is at $\theta = 0$. For each case, determine (b) whether $\omega$ is zero before, after, or at $t = 0$ and (c) whether $\alpha$ is positive, negative, or zero.

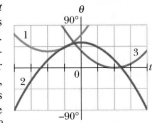

**FIG. 10-25** Question 4.

**5** Figure 10-26$a$ is an overhead view of a horizontal bar that can pivot; two horizontal forces act on the bar, but it is stationary. If the angle between the bar and $\vec{F}_2$ is now decreased from 90° and the bar is still not to turn, should $F_2$ be made larger, made smaller, or left the same?

**6** Figure 10-26$b$ shows an overhead view of a horizontal bar

that is rotated about the pivot point by two horizontal forces, $\vec{F}_1$ and $\vec{F}_2$, with $\vec{F}_2$ at angle $\phi$ to the bar. Rank the following values of $\phi$ according to the magnitude of the angular acceleration of the bar, greatest first: 90°, 70°, and 110°.

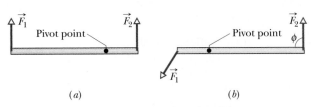

**FIG. 10-26** Questions 5 and 6.

**7** In Fig. 10-27, two forces $\vec{F}_1$ and $\vec{F}_2$ act on a disk that turns about its center like a merry-go-round. The forces maintain the indicated angles during the rotation, which is counterclockwise and at a constant rate. However, we are to decrease the angle $\theta$ of $\vec{F}_1$ without changing the magnitude of $\vec{F}_1$. (a) To keep the angular speed constant, should we increase, decrease, or maintain the magnitude of $\vec{F}_2$? Do forces (b) $\vec{F}_1$ and (c) $\vec{F}_2$ tend to rotate the disk clockwise or counterclockwise?

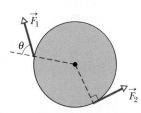

**FIG. 10-27** Question 7.

**8** In the overhead view of Fig. 10-28, five forces of the same magnitude act on a strange merry-go-round; it is a square that can rotate about point $P$, at midlength along one of the edges. Rank the forces according to the magnitude of the torque they create about point $P$, greatest first.

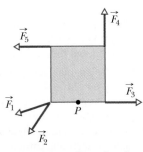

**FIG. 10-28** Question 8.

**9** A force is applied to the rim of a disk that can rotate like a merry-go-round, so as to change its angular velocity. Its initial and final angular velocities, respectively, for four situations are: (a) −2 rad/s, 5 rad/s; (b) 2 rad/s, 5 rad/s; (c) −2 rad/s, −5 rad/s; and (d) 2 rad/s, −5 rad/s. Rank the situations according to the work done by the torque due to the force, greatest first.

**10** Figure 10-29 shows three flat disks (of the same radius) that can rotate about their centers like merry-go-rounds. Each disk consists of the same two materials, one denser than the other (density is mass per unit volume). In disks 1 and 3, the denser material forms the outer half of the disk area. In disk 2, it forms the inner half of the disk area. Forces with identical magnitudes are applied tangentially to the disk, either at the outer edge or at the interface of the two materials, as shown. Rank the disks according to (a) the torque about the disk center, (b) the rotational inertia about the disk center, and (c) the angular acceleration of the disk, greatest first.

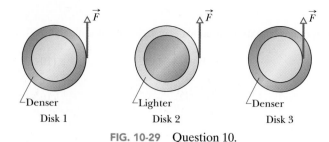

**FIG. 10-29** Question 10.

## PROBLEMS

| | |
|---|---|
| **GO** | Tutoring problem available (at instructor's discretion) in *WileyPLUS* and WebAssign |
| **SSM** | Worked-out solution available in Student Solutions Manual |
| • – ••• | Number of dots indicates level of problem difficulty |
| ✈ | Additional information available in *The Flying Circus of Physics* and at flyingcircusofphysics.com |

**WWW** Worked-out solution is at —
**ILW** Interactive solution is at —
http://www.wiley.com/college/halliday

### sec. 10-2  The Rotational Variables

•1  A good baseball pitcher can throw a baseball toward home plate at 85 mi/h with a spin of 1800 rev/min. How many revolutions does the baseball make on its way to home plate? For simplicity, assume that the 60 ft path is a straight line.

•2  What is the angular speed of (a) the second hand, (b) the minute hand, and (c) the hour hand of a smoothly running analog watch? Answer in radians per second.

••3  A diver makes 2.5 revolutions on the way from a 10-m-high platform to the water. Assuming zero initial vertical velocity, find the average angular velocity during the dive.  **ILW**

••4  The angular position of a point on the rim of a rotating wheel is given by $\theta = 4.0t - 3.0t^2 + t^3$, where $\theta$ is in radians and $t$ is in seconds. What are the angular velocities at (a) $t = 2.0$ s and (b) $t = 4.0$ s? (c) What is the average angular acceleration for the time interval that begins at $t = 2.0$ s and ends at $t = 4.0$ s? What are the instantaneous angular accelerations at (d) the beginning and (e) the end of this time interval?

••5  When a slice of buttered toast is accidentally pushed over the edge of a counter, it rotates as it falls. If the distance to the floor is 76 cm and for rotation less than 1 rev, what are the (a) smallest and (b) largest angular speeds that cause the toast to hit and then topple to be butter-side down?  ✈

•••7  The wheel in Fig. 10-30 has eight equally spaced spokes and a radius of 30 cm. It is mounted on a fixed axle and is spinning at 2.5 rev/s. You want to shoot a 20-cm-long arrow parallel to this axle and through the wheel without hitting any of the spokes. Assume that the arrow and the spokes are very thin. (a) What minimum speed must the arrow have? (b) Does it matter where between the axle and rim of the wheel you aim? If so, what is the best location?

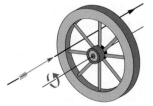

**FIG. 10-30**  Problem 7.

••6  The angular position of a point on a rotating wheel is given by $\theta = 2.0 + 4.0t^2 + 2.0t^3$, where $\theta$ is in radians and $t$ is in seconds. At $t = 0$, what are (a) the point's angular position and (b) its angular velocity? (c) What is its angular velocity at $t = 4.0$ s? (d) Calculate its angular acceleration at $t = 2.0$ s. (e) Is its angular acceleration constant?

•••8  The angular acceleration of a wheel is $\alpha = 6.0t^4 - 4.0t^2$, with $\alpha$ in radians per second-squared and $t$ in seconds. At time $t = 0$, the wheel has an angular velocity of $+2.0$ rad/s and an angular position of $+1.0$ rad. Write expressions for (a) the angular velocity (rad/s) and (b) the angular position (rad) as functions of time (s).

### sec. 10-4  Rotation with Constant Angular Acceleration

•9  A disk, initially rotating at 120 rad/s, is slowed down with a constant angular acceleration of magnitude 4.0 rad/s². (a) How much time does the disk take to stop? (b) Through what angle does the disk rotate during that time?

•10  The angular speed of an automobile engine is increased at a constant rate from 1200 rev/min to 3000 rev/min in 12 s. (a) What is its angular acceleration in revolutions per minute-squared? (b) How many revolutions does the engine make during this 12 s interval?

•11  A drum rotates around its central axis at an angular velocity of 12.60 rad/s. If the drum then slows at a constant rate of 4.20 rad/s², (a) how much time does it take and (b) through what angle does it rotate in coming to rest?

•12  Starting from rest, a disk rotates about its central axis with constant angular acceleration. In 5.0 s, it rotates 25 rad. During that time, what are the magnitudes of (a) the angular acceleration and (b) the average angular velocity? (c) What is the instantaneous angular velocity of the disk at the end of the 5.0 s? (d) With the angular acceleration unchanged, through what additional angle will the disk turn during the next 5.0 s?

••13  A wheel has a constant angular acceleration of 3.0 rad/s². During a certain 4.0 s interval, it turns through an angle of 120 rad. Assuming that the wheel started from rest, how long has it been in motion at the start of this 4.0 s interval?  **SSM**

••14  A merry-go-round rotates from rest with an angular acceleration of 1.50 rad/s². How long does it take to rotate through (a) the first 2.00 rev and (b) the next 2.00 rev?

••15  At $t = 0$, a flywheel has an angular velocity of 4.7 rad/s, a constant angular acceleration of $-0.25$ rad/s², and a reference line at $\theta_0 = 0$. (a) Through what maximum angle $\theta_{max}$ will the reference line turn in the positive direction? What are the (b) first and (c) second times the reference line will be at $\theta = \frac{1}{2}\theta_{max}$? At what (d) negative time and (e) positive time will the reference line be at $\theta = -10.5$ rad? (f) Graph $\theta$ versus $t$, and indicate the answers to (a) through (e) on the graph.

••**16** A disk rotates about its central axis starting from rest and accelerates with constant angular acceleration. At one time it is rotating at 10 rev/s; 60 revolutions later, its angular speed is 15 rev/s. Calculate (a) the angular acceleration, (b) the time required to complete the 60 revolutions, (c) the time required to reach the 10 rev/s angular speed, and (d) the number of revolutions from rest until the time the disk reaches the 10 rev/s angular speed.

••**17** A flywheel turns through 40 rev as it slows from an angular speed of 1.5 rad/s to a stop. (a) Assuming a constant angular acceleration, find the time for it to come to rest. (b) What is its angular acceleration? (c) How much time is required for it to complete the first 20 of the 40 revolutions? **ILW**

### sec. 10-5 Relating the Linear and Angular Variables

•**18** A vinyl record is played by rotating the record so that an approximately circular groove in the vinyl slides under a stylus. Bumps in the groove run into the stylus, causing it to oscillate. The equipment converts those oscillations to electrical signals and then to sound. Suppose that a record turns at the rate of $33\frac{1}{3}$ rev/min, the groove being played is at a radius of 10.0 cm, and the bumps in the groove are uniformly separated by 1.75 mm. At what rate (hits per second) do the bumps hit the stylus?

•**19** Between 1911 and 1990, the top of the leaning bell tower at Pisa, Italy, moved toward the south at an average rate of 1.2 mm/y. The tower is 55 m tall. In radians per second, what is the average angular speed of the tower's top about its base?

•**20** An astronaut is being tested in a centrifuge. The centrifuge has a radius of 10 m and, in starting, rotates according to $\theta = 0.30t^2$, where $t$ is in seconds and $\theta$ is in radians. When $t = 5.0$ s, what are the magnitudes of the astronaut's (a) angular velocity, (b) linear velocity, (c) tangential acceleration, and (d) radial acceleration?

•**21** A flywheel with a diameter of 1.20 m is rotating at an angular speed of 200 rev/min. (a) What is the angular speed of the flywheel in radians per second? (b) What is the linear speed of a point on the rim of the flywheel? (c) What constant angular acceleration (in revolutions per minute-squared) will increase the wheel's angular speed to 1000 rev/min in 60.0 s? (d) How many revolutions does the wheel make during that 60.0 s? **SSM WWW**

•**22** If an airplane propeller rotates at 2000 rev/min while the airplane flies at a speed of 480 km/h relative to the ground, what is the linear speed of a point on the tip of the propeller, at radius 1.5 m, as seen by (a) the pilot and (b) an observer on the ground? The plane's velocity is parallel to the propeller's axis of rotation.

•**23** What are the magnitudes of (a) the angular velocity, (b) the radial acceleration, and (c) the tangential acceleration of a spaceship taking a circular turn of radius 3220 km at a speed of 29 000 km/h?

•**24** An object rotates about a fixed axis, and the angular position of a reference line on the object is given by $\theta = 0.40e^{2t}$, where $\theta$ is in radians and $t$ is in seconds. Consider a point on the object that is 4.0 cm from the axis of rotation. At $t =$

0, what are the magnitudes of the point's (a) tangential component of acceleration and (b) radial component of acceleration?

••**25** A disk, with a radius of 0.25 m, is to be rotated like a merry-go-round through 800 rad, starting from rest, gaining angular speed at the constant rate $\alpha_1$ through the first 400 rad and then losing angular speed at the constant rate $-\alpha_1$ until it is again at rest. The magnitude of the centripetal acceleration of any portion of the disk is not to exceed 400 m/s². (a) What is the least time required for the rotation? (b) What is the corresponding value of $\alpha_1$?

••**26** A gyroscope flywheel of radius 2.83 cm is accelerated from rest at 14.2 rad/s² until its angular speed is 2760 rev/min. (a) What is the tangential acceleration of a point on the rim of the flywheel during this spin-up process? (b) What is the radial acceleration of this point when the flywheel is spinning at full speed? (c) Through what distance does a point on the rim move during the spin-up?

••**27** An early method of measuring the speed of light makes use of a rotating slotted wheel. A beam of light passes through one of the slots at the outside edge of the wheel, as in Fig. 10-31, travels to a distant mirror, and returns to the wheel just in time to pass through the next slot in the wheel. One such slotted wheel has a radius of 5.0 cm and 500 slots around its edge. Measurements taken when the mirror is $L = 500$ m from the wheel indicate a speed of light of $3.0 \times 10^5$ km/s. (a) What is the (constant) angular speed of the wheel? (b) What is the linear speed of a point on the edge of the wheel?

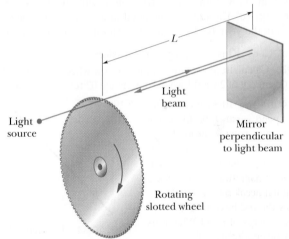

**FIG. 10-31** Problem 27.

••**28** The flywheel of a steam engine runs with a constant angular velocity of 150 rev/min. When steam is shut off, the friction of the bearings and of the air stops the wheel in 2.2 h. (a) What is the constant angular acceleration, in revolutions per minute-squared, of the wheel during the slowdown? (b) How many revolutions does the wheel make before stopping? (c) At the instant the flywheel is turning at 75 rev/min, what is the tangential component of the linear acceleration of a flywheel particle that is 50 cm from the axis of rotation? (d) What is the magnitude of the net linear acceleration of the particle in (c)?

••**29** (a) What is the angular speed $\omega$ about the polar axis of a point on Earth's surface at latitude 40° N? (Earth rotates about that axis.) (b) What is the linear speed $v$ of the point? What are (c) $\omega$ and (d) $v$ for a point at the equator? **SSM**

**••30** In Fig. 10-32, wheel $A$ of radius $r_A = 10$ cm is coupled by belt $B$ to wheel $C$ of radius $r_C = 25$ cm. The angular speed of wheel $A$ is increased from rest at a constant rate of 1.6 rad/s². Find the time needed for wheel $C$ to reach

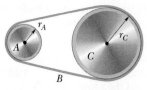

FIG. 10-32  Problem 30.

an angular speed of 100 rev/min, assuming the belt does not slip. (*Hint:* If the belt does not slip, the linear speeds at the two rims must be equal.)

**••31** A record turntable is rotating at $33\frac{1}{3}$ rev/min. A watermelon seed is on the turntable 6.0 cm from the axis of rotation. (a) Calculate the acceleration of the seed, assuming that it does not slip. (b) What is the minimum value of the coefficient of static friction between the seed and the turntable if the seed is not to slip? (c) Suppose that the turntable achieves its angular speed by starting from rest and undergoing a constant angular acceleration for 0.25 s. Calculate the minimum coefficient of static friction required for the seed not to slip during the acceleration period.

**•••32** A pulsar is a rapidly rotating neutron star that emits a radio beam the way a lighthouse emits a light beam. We receive a radio pulse for each rotation of the star. The period $T$ of rotation is found by measuring the time between pulses. The pulsar in the Crab nebula has a period of rotation of $T = 0.033$ s that is increasing at the rate of $1.26 \times 10^{-5}$ s/y. (a) What is the pulsar's angular acceleration $\alpha$? (b) If $\alpha$ is constant, how many years from now will the pulsar stop rotating? (c) The pulsar originated in a supernova explosion seen in the year 1054. Assuming constant $\alpha$, find the initial $T$.

### sec. 10-6  Kinetic Energy of Rotation

**•33** Calculate the rotational inertia of a wheel that has a kinetic energy of 24 400 J when rotating at 602 rev/min. **SSM**

**•34** Figure 10-33 gives angular speed versus time for a thin rod that rotates around one end. The scale on the $\omega$ axis is set by $\omega_s = 6.0$ rad/s. (a) What is the magnitude of the rod's angular acceleration? (b) At $t = 4.0$ s, the rod has a rotational kinetic energy of 1.60 J. What is its kinetic energy at $t = 0$?

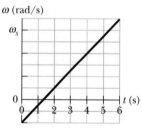

FIG. 10-33  Problem 34.

### sec. 10-7  Calculating the Rotational Inertia

**•35** Calculate the rotational inertia of a meter stick, with mass 0.56 kg, about an axis perpendicular to the stick and located at the 20 cm mark. (Treat the stick as a thin rod.) **SSM**

**•36** Figure 10-34 shows three 0.0100 kg particles that have been glued to a rod of length $L = 6.00$ cm and negligible mass. The assembly can rotate around a perpendicular axis through point $O$ at the left end. If we remove one particle (that is, 33% of the mass), by what percentage does the rotational inertia of the assembly around the rotation axis decrease when that removed particle is (a) the innermost one and (b) the outermost one?

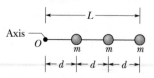

FIG. 10-34  Problems 36 and 64.

**•37** Two uniform solid cylinders, each rotating about its central (longitudinal) axis at 235 rad/s, have the same mass of 1.25 kg but differ in radius. What is the rotational kinetic energy of (a) the smaller cylinder, of radius 0.25 m, and (b) the larger cylinder, of radius 0.75 m? **SSM**

**•38** Figure 10-35a shows a disk that can rotate about an axis at a radial distance $h$ from the center of the disk. Figure 10-35b gives the rotational inertia $I$ of the disk about the axis as a function of that distance $h$, from the center out to the edge of the disk. The scale on the $I$ axis is set by $I_A = 0.050$ kg·m² and $I_B = 0.150$ kg·m². What is the mass of the disk?

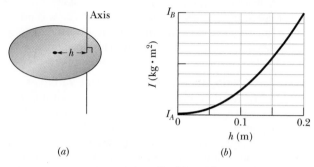

FIG. 10-35  Problem 38.

**••39** In Fig. 10-36, two particles, each with mass $m = 0.85$ kg, are fastened to each other, and to a rotation axis at $O$, by two thin rods, each with length $d = 5.6$ cm and mass $M = 1.2$ kg. The combination rotates around the rotation axis with angular speed $\omega = 0.30$ rad/s. Measured about $O$, what are the combination's (a) rotational inertia and (b) kinetic energy? **GO**

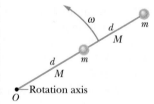

FIG. 10-36  Problem 39.

**••40** Four identical particles of mass 0.50 kg each are placed at the vertices of a 2.0 m × 2.0 m square and held there by four massless rods, which form the sides of the square. What is the rotational inertia of this rigid body about an axis that (a) passes through the midpoints of opposite sides and lies in the plane of the square, (b) passes through the midpoint of one of the sides and is perpendicular to the plane of the square, and (c) lies in the plane of the square and passes through two diagonally opposite particles?

**••41** The uniform solid block in Fig. 10-37 has mass 0.172 kg and edge lengths $a = 3.5$ cm, $b = 8.4$ cm, and $c = 1.4$ cm. Calculate its rotational inertia about an axis through one corner and perpendicular to the large faces. **SSM** **WWW**

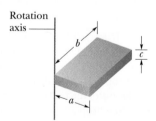

FIG. 10-37  Problem 41.

**••42** Figure 10-38 shows an arrangement of 15 identical disks that have been glued together in a rod-like shape of length $L = 1.0000$ m and (total) mass $M = 100.0$ mg. The disk arrangement can rotate about a perpendicular axis through its central disk at point $O$. (a) What is the rotational inertia of the arrangement about that axis? (b) If we approximated the arrangement as being a uniform rod of mass $M$ and length $L$, what percentage error would we make in using the formula in Table 10-2e to calculate the rotational inertia?

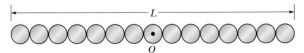

**FIG. 10-38**   Problem 42.

**••43**   Trucks can be run on energy stored in a rotating flywheel, with an electric motor getting the flywheel up to its top speed of $200\pi$ rad/s. One such flywheel is a solid, uniform cylinder with a mass of 500 kg and a radius of 1.0 m. (a) What is the kinetic energy of the flywheel after charging? (b) If the truck uses an average power of 8.0 kW, for how many minutes can it operate between chargings?

**••44**   The masses and coordinates of four particles are as follows: 50 g, $x = 2.0$ cm, $y = 2.0$ cm; 25 g, $x = 0$, $y = 4.0$ cm; 25 g, $x = -3.0$ cm, $y = -3.0$ cm; 30 g, $x = -2.0$ cm, $y = 4.0$ cm. What are the rotational inertias of this collection about the (a) $x$, (b) $y$, and (c) $z$ axes? (d) Suppose the answers to (a) and (b) are $A$ and $B$, respectively. Then what is the answer to (c) in terms of $A$ and $B$?

#### sec. 10-8   Torque

**•45**   A small ball of mass 0.75 kg is attached to one end of a 1.25-m-long massless rod, and the other end of the rod is hung from a pivot. When the resulting pendulum is 30° from the vertical, what is the magnitude of the gravitational torque calculated about the pivot?   **SSM**

**•46**   The length of a bicycle pedal arm is 0.152 m, and a downward force of 111 N is applied to the pedal by the rider. What is the magnitude of the torque about the pedal arm's pivot when the arm is at angle (a) 30°, (b) 90°, and (c) 180° with the vertical?

**•47**   The body in Fig. 10-39 is pivoted at $O$, and two forces act on it as shown. If $r_1 = 1.30$ m, $r_2 = 2.15$ m, $F_1 = 4.20$ N, $F_2 = 4.90$ N, $\theta_1 = 75.0°$, and $\theta_2 = 60.0°$, what is the net torque about the pivot?   **SSM ILW**

**FIG. 10-39**   Problem 47.

**•48**   The body in Fig. 10-40 is pivoted at $O$. Three forces act on it: $F_A = 10$ N at point $A$, 8.0 m from $O$; $F_B = 16$ N at $B$, 4.0 m from $O$; and $F_C = 19$ N at $C$, 3.0 m from $O$. What is the net torque about $O$?

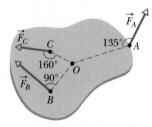

**FIG. 10-40**   Problem 48.

#### sec. 10-9   Newton's Second Law for Rotation

**•49**   During the launch from a board, a diver's angular speed about her center of mass changes from zero to 6.20 rad/s in 220 ms. Her rotational inertia about her center of mass is 12.0 kg·m². During the launch, what are the magnitudes of (a) her average angular acceleration and (b) the average external torque on her from the board?   **SSM ILW**

**•50**   If a 32.0 N·m torque on a wheel causes angular acceleration 25.0 rad/s², what is the wheel's rotational inertia?

**••51**   Figure 10-41 shows a uniform disk that can rotate around its center like a merry-go-round. The disk has a radius of 2.00 cm and a mass of 20.0 grams and is initially at rest. Starting at time $t = 0$, two forces are to be applied tangentially to the rim as indicated, so that at time $t = 1.25$ s the disk has an angular velocity of 250 rad/s counterclockwise. Force $\vec{F}_1$ has a magnitude of 0.100 N. What is magnitude $F_2$?   **GO**

**FIG. 10-41**
Problem 51.

**••52**   Figure 10-42 shows particles 1 and 2, each of mass $m$, attached to the ends of a rigid massless rod of length $L_1 + L_2$, with $L_1 = 20$ cm and $L_2 = 80$ cm. The rod is held horizontally on the fulcrum and then released. What are the magnitudes of the initial accelerations of (a) particle 1 and (b) particle 2?

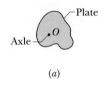

**FIG. 10-42**   Problem 52.

**••53**   In Fig. 10-43a, an irregularly shaped plastic plate with uniform thickness and density (mass per unit volume) is to be rotated around an axle that is perpendicular to the plate face and through point $O$. The rotational inertia of the plate about that axle is measured with the following method. A circular disk of mass 0.500 kg and radius 2.00 cm is glued to the plate, with its center aligned with point $O$ (Fig. 10-43b). A string is wrapped around the edge of the disk the way a string is wrapped around a top. Then the string is pulled for 5.00 s. As a result, the disk and plate are rotated by a constant force of 0.400 N that is applied by the string tangentially to the edge of the disk. The resulting angular speed is 114 rad/s. What is the rotational inertia of the plate about the axle?   **GO**

**FIG. 10-43**
Problem 53.

**••54**   In Fig. 10-44, a cylinder having a mass of 2.0 kg can rotate about its central axis through point $O$. Forces are applied as shown: $F_1 = 6.0$ N, $F_2 = 4.0$ N, $F_3 = 2.0$ N, and $F_4 = 5.0$ N. Also, $r = 5.0$ cm and $R = 12$ cm. Find the (a) magnitude and (b) direction of the angular acceleration of the cylinder. (During the rotation, the forces maintain their same angles relative to the cylinder.)

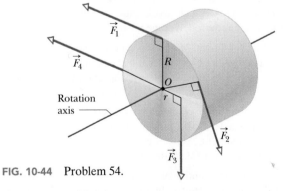

**FIG. 10-44**   Problem 54.

**••55**   In Fig. 10-45, block 1 has mass $m_1 = 460$ g, block 2 has mass $m_2 = 500$ g, and the pulley, which is mounted on a horizontal axle with negligible friction, has radius $R = 5.00$ cm. When released from rest, block 2 falls 75.0 cm in 5.00 s without the cord slipping on the pulley. (a) What is the magnitude of the acceleration of the blocks? What are (b) tension $T_2$ and (c) tension $T_1$? (d) What is the magnitude of the pulley's angular acceleration? (e) What is its rotational inertia? **GO**

**••56** In a judo foot-sweep move, you sweep your opponent's left foot out from under him while pulling on his gi (uniform) toward that side. As a result, your opponent rotates around his right foot and onto the mat. Figure 10-46 shows a simplified diagram of your opponent as you face him, with his left foot swept out. The rotational axis is through point $O$. The gravitational force $\vec{F}_g$ on him effectively acts at his center of mass, which is a horizontal distance $d = 28$ cm from point $O$. His mass is 70 kg, and his rotational inertia about point $O$ is 65 kg·m². What is the magnitude of his initial angular acceleration about point $O$ if your pull $\vec{F}_a$ on his gi is (a) negligible and (b) horizontal with a magnitude of 300 N and applied at height $h = 1.4$ m?

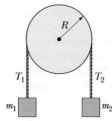

FIG. 10-45 Problems 55 and 73.

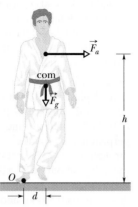

FIG. 10-46 Problem 56.

**•••57** A pulley, with a rotational inertia of $1.0 \times 10^{-3}$ kg·m² about its axle and a radius of 10 cm, is acted on by a force applied tangentially at its rim. The force magnitude varies in time as $F = 0.50t + 0.30t^2$, with $F$ in newtons and $t$ in seconds. The pulley is initially at rest. At $t = 3.0$ s what are (a) its angular acceleration and (b) its angular speed?

### sec. 10-10 Work and Rotational Kinetic Energy

**•58** A thin rod of length 0.75 m and mass 0.42 kg is suspended freely from one end. It is pulled to one side and then allowed to swing like a pendulum, passing through its lowest position with angular speed 4.0 rad/s. Neglecting friction and air resistance, find (a) the rod's kinetic energy at its lowest position and (b) how far above that position the center of mass rises.

**•59** A 32.0 kg wheel, essentially a thin hoop with radius 1.20 m, is rotating at 280 rev/min. It must be brought to a stop in 15.0 s. (a) How much work must be done to stop it? (b) What is the required average power?

**•60** (a) If $R = 12$ cm, $M = 400$ g, and $m = 50$ g in Fig. 10-18, find the speed of the block after it has descended 50 cm starting from rest. Solve the problem using energy conservation principles. (b) Repeat (a) with $R = 5.0$ cm.

**•61** An automobile crankshaft transfers energy from the engine to the axle at the rate of 100 hp (= 74.6 kW) when rotating at a speed of 1800 rev/min. What torque (in newton-meters) does the crankshaft deliver?

**••62** A uniform cylinder of radius 10 cm and mass 20 kg is mounted so as to rotate freely about a horizontal axis that is parallel to and 5.0 cm from the central longitudinal axis of the cylinder. (a) What is the rotational inertia of the cylinder about the axis of rotation? (b) If the cylinder is released from rest with its central longitudinal axis at the same height as the axis about which the cylinder rotates, what is the angular speed of the cylinder as it passes through its lowest position?

**••63** A meter stick is held vertically with one end on the floor and is then allowed to fall. Find the speed of the other end just before it hits the floor, assuming that the end on the floor does not slip. (*Hint:* Consider the stick to be a thin rod and use the conservation of energy principle.) **SSM ILW**

**••64** In Fig. 10-34, three 0.0100 kg particles have been glued to a rod of length $L = 6.00$ cm and negligible mass and can rotate around a perpendicular axis through point $O$ at one end. How much work is required to change the rotational rate (a) from 0 to 20.0 rad/s, (b) from 20.0 rad/s to 40.0 rad/s, and (c) from 40.0 rad/s to 60.0 rad/s? (d) What is the slope of a plot of the assembly's kinetic energy (in joules) versus the square of its rotation rate (in radians-squared per second-squared)?

**•••65** Figure 10-47 shows a rigid assembly of a thin hoop (of mass $m$ and radius $R = 0.150$ m) and a thin radial rod (of mass $m$ and length $L = 2.00R$). The assembly is upright, but if we give it a slight nudge, it will rotate around a horizontal axis in the plane of the rod and hoop, through the lower end of the rod. Assuming that the energy given to the assembly in such a nudge is negligible, what would be the assembly's angular speed about the rotation axis when it passes through the upside-down (inverted) orientation? **GO**

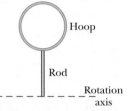

FIG. 10-47 Problem 65.

**•••66** A uniform spherical shell of mass $M = 4.5$ kg and radius $R = 8.5$ cm can rotate about a vertical axis on frictionless bearings (Fig. 10-48). A massless cord passes around the equator of the shell, over a pulley of rotational inertia $I = 3.0 \times 10^{-3}$ kg·m² and radius $r = 5.0$ cm, and is attached to a small object of mass $m = 0.60$ kg. There is no friction on the pulley's axle; the cord does not slip on the pulley. What is the speed of the object when it has fallen 82 cm after being released from rest? Use energy considerations.

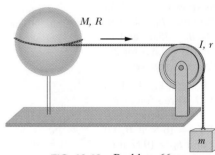

FIG. 10-48 Problem 66.

**•••67** A tall, cylindrical chimney falls over when its base is ruptured. Treat the chimney as a thin rod of length 55.0 m. At the instant it makes an angle of 35.0° with the vertical as it falls, what are (a) the radial acceleration of the top, and (b) the tangential acceleration of the top. (*Hint:* Use energy considerations, not a torque.) (c) At what angle $\theta$ is the tangential acceleration equal to $g$?

### Additional Problems

**68** George Washington Gale Ferris, Jr., a civil engineering graduate from Rensselaer Polytechnic Institute, built the original Ferris wheel for the 1893 World's Columbian Exposition in Chicago. The wheel, an astounding engineering construction at the time, carried 36 wooden cars, each holding up to 60 passengers, around a circle 76 m in diameter. The cars were loaded 6 at a time, and once all 36 cars were full, the wheel made a complete rotation at constant angular speed in about 2 min. Estimate the amount of work that was required of the machinery to rotate the passengers alone.

**69** In Fig. 10-49, two 6.20 kg blocks are connected by a massless string over a pulley of radius 2.40 cm and rotational inertia $7.40 \times 10^{-4}$ kg·m$^2$. The string does not slip on the pulley; it is not known whether there is friction between the table and the sliding block; the pulley's axis is frictionless. When this system is released from rest, the pulley turns through 1.30 rad in 91.0 ms and the acceleration of the blocks is constant. What are (a) the magnitude of the pulley's angular acceleration, (b) the magnitude of either block's acceleration, (c) string tension $T_1$, and (d) string tension $T_2$? **SSM**

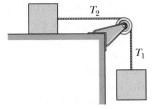

FIG. 10-49  Problem 69.

**70** Figure 10-50 shows a flat construction of two circular rings that have a common center and are held together by three rods of negligible mass. The construction, which is initially at rest, can rotate around the common center (like a merry-go-round), where another rod of negligible mass lies. The mass, inner radius, and outer radius of the rings are given in the following table. A tangential force of magnitude 12.0 N is applied to the outer edge of the outer ring for 0.300 s. What is the change in the angular speed of the construction during that time interval?

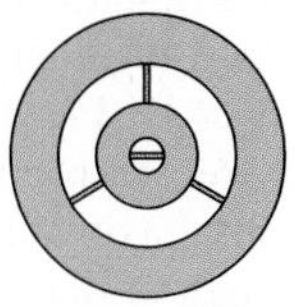

FIG. 10-50 Problem 70.

| Ring | Mass (kg) | Inner Radius (m) | Outer Radius (m) |
|---|---|---|---|
| 1 | 0.120 | 0.0160 | 0.0450 |
| 2 | 0.240 | 0.0900 | 0.1400 |

**71** In Fig. 10-51, a small disk of radius $r = 2.00$ cm has been glued to the edge of a larger disk of radius $R = 4.00$ cm so that the disks lie in the same plane. The disks can be rotated around a perpendicular axis through point $O$ at the center of the larger disk. The disks both have a uniform density (mass per unit volume) of $1.40 \times 10^3$ kg/m$^3$ and a uniform thickness of 5.00 mm. What is the rotational inertia of the two-disk assembly about the rotation axis through $O$?

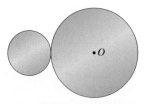

FIG. 10-51  Problem 71.

**72** At 7:14 A.M. on June 30, 1908, a huge explosion occurred above remote central Siberia, at latitude 61° N and longitude 102° E; the fireball thus created was the brightest flash seen by anyone before nuclear weapons. The *Tunguska Event*, which according to one chance witness "covered an enormous part of the sky," was probably the explosion of a *stony asteroid* about 140 m wide. (a) Considering only Earth's rotation, determine how much later the asteroid would have had to arrive to put the explosion above Helsinki at longitude 25° E. This would have obliterated the city. (b) If the asteroid had, instead, been a *metallic asteroid,* it could have reached Earth's surface. How much later would such an asteroid have had to arrive to put the impact in the Atlantic Ocean at longi-

tude 20° W? (The resulting tsunamis would have wiped out coastal civilization on both sides of the Atlantic.)

**73** In Fig. 10-45, two blocks, of mass $m_1 = 400$ g and $m_2 = 600$ g, are connected by a massless cord that is wrapped around a uniform disk of mass $M = 500$ g and radius $R = 12.0$ cm. The disk can rotate without friction about a fixed horizontal axis through its center; the cord cannot slip on the disk. The system is released from rest. Find (a) the magnitude of the acceleration of the blocks, (b) the tension $T_1$ in the cord at the left, and (c) the tension $T_2$ in the cord at the right.

**74** Attached to each end of a thin steel rod of length 1.20 m and mass 6.40 kg is a small ball of mass 1.06 kg. The rod is constrained to rotate in a horizontal plane about a vertical axis through its midpoint. At a certain instant, it is rotating at 39.0 rev/s. Because of friction, it slows to a stop in 32.0 s. Assuming a constant retarding torque due to friction, compute (a) the angular acceleration, (b) the retarding torque, (c) the total energy transferred from mechanical energy to thermal energy by friction, and (d) the number of revolutions rotated during the 32.0 s. (e) Now suppose that the retarding torque is known not to be constant. If any of the quantities (a), (b), (c), and (d) can still be computed without additional information, give its value.

**75** A uniform helicopter rotor blade is 7.80 m long, has a mass of 110 kg, and is attached to the rotor axle by a single bolt. (a) What is the magnitude of the force on the bolt from the axle when the rotor is turning at 320 rev/min? (*Hint:* For this calculation the blade can be considered to be a point mass at its center of mass. Why?) (b) Calculate the torque that must be applied to the rotor to bring it to full speed from rest in 6.70 s. Ignore air resistance. (The blade cannot be considered to be a point mass for this calculation. Why not? Assume the mass distribution of a uniform thin rod.) (c) How much work does the torque do on the blade in order for the blade to reach a speed of 320 rev/min?

**76** A wheel, starting from rest, rotates with a constant angular acceleration of 2.00 rad/s$^2$. During a certain 3.00 s interval, it turns through 90.0 rad. (a) What is the angular velocity of the wheel at the start of the 3.00 s interval? (b) How long has the wheel been turning before the start of the 3.00 s interval?

**77** A golf ball is launched at an angle of 20° to the horizontal, with a speed of 60 m/s and a rotation rate of 90 rad/s. Neglecting air drag, determine the number of revolutions the ball makes by the time it reaches maximum height.

**78** Two uniform solid spheres have the same mass of 1.65 kg, but one has a radius of 0.226 m and the other has a radius of 0.854 m. Each can rotate about an axis through its center. (a) What is the magnitude $\tau$ of the torque required to bring the smaller sphere from rest to an angular speed of 317 rad/s in 15.5 s? (b) What is the magnitude $F$ of the force that must be applied tangentially at the sphere's equator to give that torque? What are the corresponding values of (c) $\tau$ and (d) $F$ for the larger sphere?

**79** The thin uniform rod in Fig. 10-52 has length 2.0 m and can pivot about a horizontal, frictionless pin through one end. It is released from rest at angle $\theta = 40°$ above the horizontal. Use the principle of conservation of energy to determine the angular speed of the rod as it passes through the horizontal position.

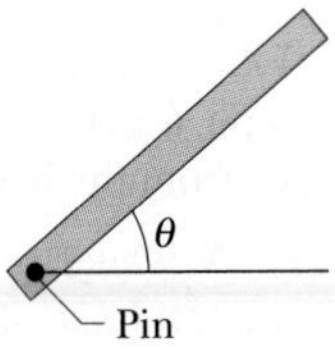

FIG. 10-52  Problem 79.

**80** The flywheel of an engine is rotating at 25.0 rad/s. When the engine is turned off, the flywheel slows at a constant rate and stops in 20.0 s. Calculate (a) the angular acceleration of the flywheel, (b) the angle through which the flywheel rotates in stopping, and (c) the number of revolutions made by the flywheel in stopping.

**81** A small ball with mass 1.30 kg is mounted on one end of a rod 0.780 m long and of negligible mass. The system rotates in a horizontal circle about the other end of the rod at 5010 rev/min. (a) Calculate the rotational inertia of the system about the axis of rotation. (b) There is an air drag of $2.30 \times 10^{-2}$ N on the ball, directed opposite its motion. What torque must be applied to the system to keep it rotating at constant speed?

**82** Starting from rest at $t = 0$, a wheel undergoes a constant angular acceleration. When $t = 2.0$ s, the angular velocity of the wheel is 5.0 rad/s. The acceleration continues until $t = 20$ s, when it abruptly ceases. Through what angle does the wheel rotate in the interval $t = 0$ to $t = 40$ s?

**83** A high-wire walker always attempts to keep his center of mass over the wire (or rope). He normally carries a long, heavy pole to help: If he leans, say, to his right (his com moves to the right) and is in danger of rotating around the wire, he moves the pole to his left (its com moves to the left) to slow the rotation and allow himself time to adjust his balance. Assume that the walker has a mass of 70.0 kg and a rotational inertia of 15.0 kg·m² about the wire. What is the magnitude of his angular acceleration about the wire if his com is 5.0 cm to the right of the wire and (a) he carries no pole and (b) the 14.0 kg pole he carries has its com 10 cm to the left of the wire?

**84** *Racing disks.* Figure 10-53 shows two disks that can rotate about their centers like a merry-go-round. At time $t = 0$, the reference lines of the two disks have the same orientation. Disk *A* is already rotating, with a constant angular

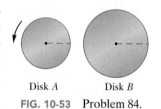

Disk *A*          Disk *B*

**FIG. 10-53** Problem 84.

velocity of 9.5 rad/s. Disk *B* has been stationary but now begins to rotate at a constant angular acceleration of 2.2 rad/s². (a) At what time $t$ will the reference lines of the two disks momentarily have the same angular displacement $\theta$? (b) Will that time $t$ be the first time since $t = 0$ that the reference lines are momentarily aligned?

**85** A bicyclist of mass 70 kg puts all his mass on each downward-moving pedal as he pedals up a steep road. Take the diameter of the circle in which the pedals rotate to be 0.40 m, and determine the magnitude of the maximum torque he exerts about the rotation axis of the pedals.

**86** A disk rotates at constant angular acceleration, from angular position $\theta_1 = 10.0$ rad to angular position $\theta_2 = 70.0$ rad in 6.00 s. Its angular velocity at $\theta_2$ is 15.0 rad/s. (a) What was its angular velocity at $\theta_1$? (b) What is the angular acceleration? (c) At what angular position was the disk initially at rest? (d) Graph $\theta$ versus time $t$ and angular speed $\omega$ versus $t$ for the disk, from the beginning of the motion (let $t = 0$ then).

**87** A wheel of radius 0.20 m is mounted on a frictionless horizontal axis. The rotational inertia of the wheel about the axis is 0.050 kg·m². A massless cord wrapped around the wheel is attached to a 2.0 kg block that slides on a hori-

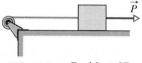

zontal frictionless surface. If a horizontal force of magnitude $P = 3.0$ N is applied to the block as shown in Fig. 10-54, what is the magnitude of the angular acceleration of the wheel? Assume the cord does not slip on the wheel. **SSM**

**FIG. 10-54** Problem 87.

**88** Our Sun is $2.3 \times 10^4$ ly (light-years) from the center of our Milky Way galaxy and is moving in a circle around that center at a speed of 250 km/s. (a) How long does it take the Sun to make one revolution about the galactic center? (b) How many revolutions has the Sun completed since it was formed about $4.5 \times 10^9$ years ago?

**89** A record turntable rotating at $33\frac{1}{3}$ rev/min slows down and stops in 30 s after the motor is turned off. (a) Find its (constant) angular acceleration in revolutions per minute-squared. (b) How many revolutions does it make in this time? **SSM**

**90** A rigid body is made of three identical thin rods, each with length $L = 0.600$ m, fastened together in the form of a letter **H** (Fig. 10-55). The body is free to rotate about a horizontal axis that runs along the

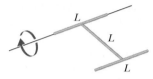

**FIG. 10-55** Problem 90.

length of one of the legs of the **H**. The body is allowed to fall from rest from a position in which the plane of the **H** is horizontal. What is the angular speed of the body when the plane of the **H** is vertical?

**91** (a) Show that the rotational inertia of a solid cylinder of mass $M$ and radius $R$ about its central axis is equal to the rotational inertia of a thin hoop of mass $M$ and radius $R/\sqrt{2}$ about its central axis. (b) Show that the rotational inertia $I$ of any given body of mass $M$ about any given axis is equal to the rotational inertia of an *equivalent hoop* about that axis, if the hoop has the same mass $M$ and a radius $k$ given by

$$k = \sqrt{\frac{I}{M}}.$$

The radius $k$ of the equivalent hoop is called the *radius of gyration* of the given body. **SSM**

**92** A thin spherical shell has a radius of 1.90 m. An applied torque of 960 N·m gives the shell an angular acceleration of 6.20 rad/s² about an axis through the center of the shell. What are (a) the rotational inertia of the shell about that axis and (b) the mass of the shell?

**93** In Fig. 10-56, a wheel of radius 0.20 m is mounted on a frictionless horizontal axle. A massless cord is wrapped around the wheel and attached to a 2.0 kg box that slides on a

**FIG. 10-56** Problem 93.

frictionless surface inclined at angle $\theta = 20°$ with the horizontal. The box accelerates down the surface at 2.0 m/s². What is the rotational inertia of the wheel about the axle?

**94** The method by which the massive lintels (top stones) were lifted to the top of the upright stones at Stonehenge has long been debated. One possible method was tested in a small Czech town. A concrete block of mass 5124 kg was pulled up along two oak beams whose top surfaces had been debarked and then lubricated with fat (Fig. 10-57). The beams were 10 m long, and each extended from the ground to the top of one of the two upright pillars onto which the block was to be raised. The pillars were 3.9 m high; the coefficient of static friction between block and beams was 0.22. The pull on the block was via ropes wrapped around it and around the top ends of two spruce logs of length 4.5 m. A platform was strung at the opposite end of each log. When enough workers sat or stood on a platform, the attached spruce log would pivot about the top of its upright pillar and pull one end of the block a short distance up a beam. For each log, the rope holding the block was approximately perpendicular to the log; the distance between the pivot point and the point where the rope wrapped around the log was 0.70 m. Assuming that each worker had a mass of 85 kg, find the smallest number of workers needed on the two platforms so that the block begins to move up along the beams. (About half this number could actually move the block by moving first one end of it and then the other.)

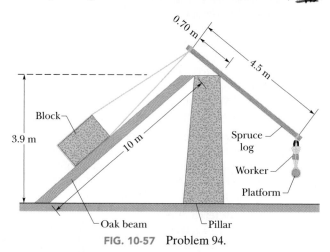

FIG. 10-57  Problem 94.

**95** Figure 10-58 shows a propeller blade that rotates at 2000 rev/min about a perpendicular axis at point B. Point A is at the outer tip of the blade, at radial distance 1.50 m. (a) What is the difference in the magnitudes *a* of the centripetal acceleration of point A and of a point at radial distance 0.150 m? (b) Find the slope of a plot of *a* versus radial distance along the blade.

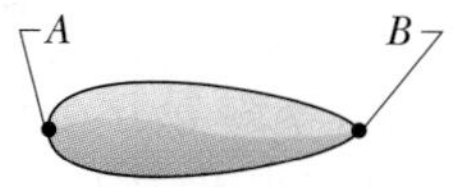

FIG. 10-58
Problem 95.

**96** A yo-yo-shaped device mounted on a horizontal frictionless axis is used to lift a 30 kg box as shown in Fig. 10-59. The outer radius R of the device is 0.50 m, and the radius r of the hub is 0.20 m. When a constant horizontal force $\vec{F}_{app}$ of magnitude 140 N is applied to a rope wrapped around the outside of the device, the box, which is suspended from a rope wrapped around the hub, has an upward acceleration of magnitude 0.80 m/s². What is the rotational inertia of the device about its axis of rotation?

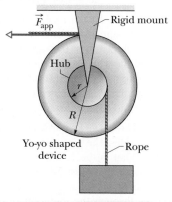

FIG. 10-59  Problem 96.

**97** The rigid body shown in Fig. 10-60 consists of three particles connected by massless rods. It is to be rotated about an axis perpendicular to its plane through point P. If M = 0.40 kg, a = 30 cm, and b = 50 cm, how much work is required to take the body from rest to an angular speed of 5.0 rad/s?

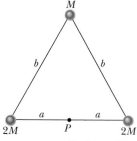

FIG. 10-60  Problem 97.

**98** *Beverage engineering.* The pull tab was a major advance in the engineering design of beverage containers. The tab pivots on a central bolt in the can's top. When you pull upward on one end of the tab, the other end presses downward on a portion of the can's top that has been scored. If you pull upward with a 10 N force, approximately what is the magnitude of the force applied to the scored section? (You will need to examine a can with a pull tab.)

**99** Cheetahs running at top speed have been reported at an astounding 114 km/h (about 71 mi/h) by observers driving alongside the animals. Imagine trying to measure a cheetah's speed by keeping your vehicle abreast of the animal while also glancing at your speedometer, which is registering 114 km/h. You keep the vehicle a constant 8.0 m from the cheetah, but the noise of the vehicle causes the cheetah to continuously veer away from you along a circular path of radius 92 m. Thus, you travel along a circular path of radius 100 m. (a) What is the angular speed of you and the cheetah around the circular paths? (b) What is the linear speed of the cheetah along its path? (If you did not account for the circular motion, you would conclude erroneously that the cheetah's speed is 114 km/h, and that type of error was apparently made in the published reports.)

**100** A point on the rim of a 0.75-m-diameter grinding wheel changes speed at a constant rate from 12 m/s to 25 m/s in 6.2 s. What is the average angular acceleration of the wheel?

**101** In Fig. 10-61, a thin uniform rod (mass 3.0 kg, length 4.0 m) rotates freely about a horizontal axis A that is perpendicular to the rod and passes through a point at distance d = 1.0 m from the end of the rod. The kinetic energy of the rod as it passes through the vertical position is 20 J. (a) What is the rotational inertia of the rod about axis A? (b) What is the (linear) speed of the end B of the rod as the rod passes through the vertical position? (c) At what angle θ will the rod momentarily stop in its upward swing?

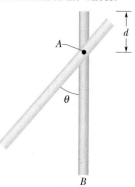

FIG. 10-61  Problem 101.

**102** A car starts from rest and moves around a circular track of radius 30.0 m. Its speed increases at the constant rate of 0.500 m/s². (a) What is the magnitude of its *net* linear acceleration 15.0 s later? (b) What angle does this net acceleration vector make with the car's velocity at this time?

**103** A pulley wheel that is 8.0 cm in diameter has a 5.6-m-long cord wrapped around its periphery. Starting from rest, the wheel is given a constant angular acceleration of 1.5 rad/s². (a) Through what angle must the wheel turn for the cord to unwind completely? (b) How long will this take?

**104** A heavy flywheel rotating on its central axis is slowing down because of friction in its bearings. At the end of the first minute of slowing, its angular speed is 0.900 of its initial angular speed of 250 rev/min. Assuming a constant angular acceleration, find its angular speed at the end of the second minute.

**105** Figure 10-62 shows a communications satellite that is a solid cylinder with mass 1210 kg, diameter 1.21 m, and length 1.75 m. Prior to launch from the shuttle cargo bay, the satellite is set spinning at 1.52 rev/s about its long axis. What are (a) its rotational inertia about the rotation axis and (b) its rotational kinetic energy?

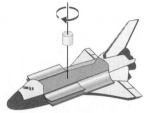
**FIG. 10-62** Problem 105.

**106** A vinyl record on a turntable rotates at $33\frac{1}{3}$ rev/min. (a) What is its angular speed in radians per second? What is the linear speed of a point on the record (b) 15 cm and (c) 7.4 cm from the turntable axis?

**107** What is the angular speed of a car traveling at 50 km/h and rounding a circular turn of radius 110 m?

**108** Calculate (a) the torque, (b) the energy, and (c) the average power required to accelerate Earth in 1 day from rest to its present angular speed about its axis.

**109** The oxygen molecule $O_2$ has a mass of $5.30 \times 10^{-26}$ kg and a rotational inertia of $1.94 \times 10^{-46}$ kg·m² about an axis through the center of the line joining the atoms and perpendicular to that line. Suppose the center of mass of an $O_2$ molecule in a gas has a translational speed of 500 m/s and the molecule has a rotational kinetic energy that is $\frac{2}{3}$ of the translational kinetic energy of its center of mass. What then is the molecule's angular speed about the center of mass?

**110** The rigid object shown in Fig. 10-63 consists of three balls and three connecting rods, with $M = 1.6$ kg, $L = 0.60$ m, and $\theta = 30°$. The balls may be treated as particles, and the connecting rods have negligible mass. Determine the rotational kinetic energy of the object if it has an angular speed of 1.2 rad/s about (a) an axis that passes through point $P$ and is perpendicular to the plane of the figure and (b) an axis that passes through point $P$, is perpendicular to the rod of length $2L$, and lies in the plane of the figure.

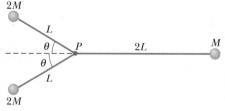
**FIG. 10-63** Problem 110.

**111** In Fig. 10-64, four pulleys are connected by two belts. Pulley $A$ (radius 15 cm) is the drive pulley, and it rotates at 10 rad/s. Pulley $B$ (radius 10 cm) is connected by belt 1 to pulley $A$. Pulley $B'$ (radius 5 cm) is concentric with pulley $B$ and is rigidly attached to it. Pulley $C$ (radius 25 cm) is connected by

belt 2 to pulley $B'$. Calculate (a) the linear speed of a point on belt 1, (b) the angular speed of pulley $B$, (c) the angular speed of pulley $B'$, (d) the linear speed of a point on belt 2, and (e) the angular speed of pulley $C$. (*Hint:* If the belt between two pulleys does not slip, the linear speeds at the rims of the two pulleys must be equal.)

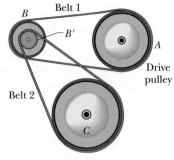
**FIG. 10-64** Problem 111.

**112** Four particles, each of mass 0.20 kg, are placed at the vertices of a square with sides of length 0.50 m. The particles are connected by rods of negligible mass. This rigid body can rotate in a vertical plane about a horizontal axis $A$ that passes through one of the particles. The body is released from rest with rod $AB$ horizontal (Fig. 10-65). (a) What is the rotational inertia of the body about axis $A$? (b) What is the angular speed of the body about axis $A$ when rod $AB$ swings through the vertical position?

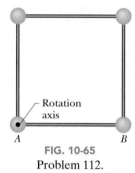
**FIG. 10-65** Problem 112.

**113** The turntable of a record player has an angular speed of 8.0 rad/s at the instant it is switched off. Three seconds later, the turntable has an angular speed of 2.6 rad/s. Through how many radians does the turntable rotate from the time it is turned off until it stops? (Assume constant $\alpha$.)

**114** Two thin rods (each of mass 0.20 kg) are joined together to form a rigid body as shown in Fig. 10-66. One of the rods has length $L_1 = 0.40$ m, and the other has length $L_2 = 0.50$ m. What is the rotational inertia of this rigid body about (a) an axis that is perpendicular to the plane of the paper and passes through the center of the shorter rod and (b) an axis that is perpendicular to the plane of the paper and passes through the center of the longer rod?

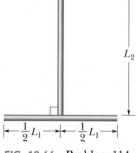
**FIG. 10-66** Problem 114.

**115** In Fig. 10-18a, a wheel of radius 0.20 m is mounted on a frictionless horizontal axis. The rotational inertia of the wheel about the axis is 0.40 kg·m². A massless cord wrapped around the wheel's circumference is attached to a 6.0 kg box. The system is released from rest. When the box has a kinetic energy of 6.0 J, what are (a) the wheel's rotational kinetic energy and (b) the distance the box has fallen? **SSM**

**116** Three 0.50 kg particles form an equilateral triangle with 0.60 m sides. The particles are connected by rods of negligible mass. What is the rotational inertia of this rigid body about (a) an axis that passes through one of the particles and is parallel to the rod connecting the other two, (b) an axis that passes through the midpoint of one of the sides and is perpendicular to the plane of the triangle, and (c) an axis that is parallel to one side of the triangle and passes through the midpoints of the other two sides?

# Rolling, Torque, and Angular Momentum

*Ballet has several types of jumps but a tour jeté is the most enchanting. After leaping straight up in that jump, a ballet performer suddenly begins to rotate as if spun by an invisible hand. After half a turn, the rotation vanishes and then the performer lands. Even if an audience knows nothing of Newton's laws, they know that rotation cannot suddenly turn on and off while the performer is in midair. Hence, what they see is magical.*

## *What accounts for the magic of a tour jeté?*

The answer is in this chapter.

©2005 Lois Greenfield

## 11-1 | WHAT IS PHYSICS?

As we discussed in Chapter 10, physics includes the study of rotation. Arguably, the most important application of that physics is in the rolling motion of wheels and wheel-like objects. This applied physics has long been used. For example, when the prehistoric people of Easter Island moved their gigantic stone statues from the quarry and across the island, they dragged them over logs acting as rollers. Much later, when settlers moved westward across America in the 1800s, they rolled their possessions first by wagon and then later by train. Today, like it or not, the world is filled with cars, trucks, motorcycles, bicycles, and other rolling vehicles.

The physics and engineering of rolling have been around for so long that you might think no fresh ideas remain to be developed. However, skateboards and in-line skates were invented and engineered fairly recently, to become huge financial successes. Street luge is now catching on, and the self-righting Segway (Fig. 11-1) may change the way people move around in large cities. Applying the physics of rolling can still lead to surprises and rewards. Our starting point in exploring that physics is to simplify rolling motion.

**FIG. 11-1** The self-righting Segway Human Transporter. *(Justin Sullivan/Getty Images News and Sport Services)*

## 11-2 | Rolling as Translation and Rotation Combined

Here we consider only objects that *roll smoothly* along a surface; that is, the objects roll without slipping or bouncing on the surface. Figure 11-2 shows how complicated smooth rolling motion can be: Although the center of the object moves in a straight line parallel to the surface, a point on the rim certainly does not. However, we can study this motion by treating it as a combination of translation of the center of mass and rotation of the rest of the object around that center.

To see how we do this, pretend you are standing on a sidewalk watching the bicycle wheel of Fig. 11-3 as it rolls along a street. As shown, you see the center of mass $O$ of the wheel move forward at constant speed $v_{\text{com}}$. The point $P$ on the street where the wheel makes contact with the street surface also moves forward at speed $v_{\text{com}}$, so that $P$ always remains directly below $O$.

During a time interval $t$, you see both $O$ and $P$ move forward by a distance $s$. The bicycle rider sees the wheel rotate through an angle $\theta$ about the center of the wheel, with the point of the wheel that was touching the street at the beginning of $t$ moving through arc length $s$. Equation 10-17 relates the arc length $s$ to the rotation angle $\theta$:

$$s = \theta R, \tag{11-1}$$

where $R$ is the radius of the wheel. The linear speed $v_{\text{com}}$ of the center of the wheel (the center of mass of this uniform wheel) is $ds/dt$. The angular speed $\omega$ of the wheel about its center is $d\theta/dt$. Thus, differentiating Eq. 11-1 with respect to time (with $R$ held constant) gives us

$$v_{\text{com}} = \omega R \quad \text{(smooth rolling motion).} \tag{11-2}$$

Figure 11-4 shows that the rolling motion of a wheel is a combination of purely translational and purely rotational motions. Figure 11-4*a* shows the purely rotational motion (as if the rotation axis through the center were stationary): Every point on the wheel rotates about the center with angular speed $\omega$. (This is the type of motion we considered in Chapter 10.) Every point on the outside edge of the wheel has linear speed $v_{\text{com}}$ given by Eq. 11-2. Figure 11-4*b* shows the purely translational motion (as if the wheel did not rotate at all): Every point on the wheel moves to the right with speed $v_{\text{com}}$.

The combination of Figs. 11-4*a* and 11-4*b* yields the actual rolling motion of the wheel, Fig. 11-4*c*. Note that in this combination of motions, the portion of the

**FIG. 11-2** A time-exposure photograph of a rolling disk. Small lights have been attached to the disk, one at its center and one at its edge. The latter traces out a curve called a *cycloid*. *(Richard Megna/Fundamental Photographs)*

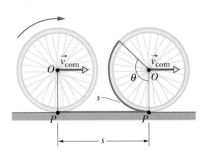

**FIG. 11-3** The center of mass $O$ of a rolling wheel moves a distance $s$ at velocity $\vec{v}_{\text{com}}$ while the wheel rotates through angle $\theta$. The point $P$ at which the wheel makes contact with the surface over which the wheel rolls also moves a distance $s$.

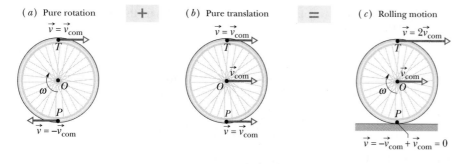

**FIG. 11-4** Rolling motion of a wheel as a combination of purely rotational motion and purely translational motion. (a) The purely rotational motion: All points on the wheel move with the same angular speed $\omega$. Points on the outside edge of the wheel all move with the same linear speed $v = v_{com}$. The linear velocities $\vec{v}$ of two such points, at top ($T$) and bottom ($P$) of the wheel, are shown. (b) The purely translational motion: All points on the wheel move to the right with the same linear velocity $\vec{v}_{com}$. (c) The rolling motion of the wheel is the combination of (a) and (b).

wheel at the bottom (at point $P$) is stationary and the portion at the top (at point $T$) is moving at speed $2v_{com}$, faster than any other portion of the wheel. These results are demonstrated in Fig. 11-5, which is a time exposure of a rolling bicycle wheel. You can tell that the wheel is moving faster near its top than near its bottom because the spokes are more blurred at the top than at the bottom.

The motion of any round body rolling smoothly over a surface can be separated into purely rotational and purely translational motions, as in Figs. 11-4a and 11-4b.

### Rolling as Pure Rotation

Figure 11-6 suggests another way to look at the rolling motion of a wheel— namely, as pure rotation about an axis that always extends through the point where the wheel contacts the street as the wheel moves. We consider the rolling motion to be pure rotation about an axis passing through point $P$ in Fig. 11-4c and perpendicular to the plane of the figure. The vectors in Fig. 11-6 then represent the instantaneous velocities of points on the rolling wheel.

**Question:** What angular speed about this new axis will a stationary observer assign to a rolling bicycle wheel?

**Answer:** The same $\omega$ that the rider assigns to the wheel as she or he observes it in pure rotation about an axis through its center of mass.

To verify this answer, let us use it to calculate the linear speed of the top of the rolling wheel from the point of view of a stationary observer. If we call the wheel's radius $R$, the top is a distance $2R$ from the axis through $P$ in Fig. 11-6, so the linear speed at the top should be (using Eq. 11-2)

$$v_{top} = (\omega)(2R) = 2(\omega R) = 2v_{com},$$

in exact agreement with Fig. 11-4c. You can similarly verify the linear speeds shown for the portions of the wheel at points $O$ and $P$ in Fig. 11-4c.

**FIG. 11-5** A photograph of a rolling bicycle wheel. The spokes near the wheel's top are more blurred than those near the bottom because the top ones are moving faster, as Fig. 11-4c shows. (*Courtesy Alice Halliday*)

✓ **CHECKPOINT 1**     The rear wheel on a clown's bicycle has twice the radius of the front wheel. (a) When the bicycle is moving, is the linear speed at the very top of the rear wheel greater than, less than, or the same as that of the very top of the front wheel? (b) Is the angular speed of the rear wheel greater than, less than, or the same as that of the front wheel?

## 11-3 I The Kinetic Energy of Rolling

Let us now calculate the kinetic energy of the rolling wheel as measured by the stationary observer. If we view the rolling as pure rotation about an axis through $P$ in Fig. 11-6, then from Eq. 10-34 we have

$$K = \tfrac{1}{2}I_P\omega^2, \tag{11-3}$$

in which $\omega$ is the angular speed of the wheel and $I_P$ is the rotational inertia of

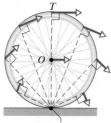

Rotation axis at $P$

**FIG. 11-6** Rolling can be viewed as pure rotation, with angular speed $\omega$, about an axis that always extends through $P$. The vectors show the instantaneous linear velocities of selected points on the rolling wheel. You can obtain the vectors by combining the translational and rotational motions as in Fig. 11-4.

the wheel about the axis through $P$. From the parallel-axis theorem of Eq. 10-36 $(I = I_{com} + Mh^2)$, we have

$$I_P = I_{com} + MR^2, \qquad (11\text{-}4)$$

in which $M$ is the mass of the wheel, $I_{com}$ is its rotational inertia about an axis through its center of mass, and $R$ (the wheel's radius) is the perpendicular distance $h$. Substituting Eq. 11-4 into Eq. 11-3, we obtain

$$K = \tfrac{1}{2}I_{com}\omega^2 + \tfrac{1}{2}MR^2\omega^2,$$

and using the relation $v_{com} = \omega R$ (Eq. 11-2) yields

$$K = \tfrac{1}{2}I_{com}\omega^2 + \tfrac{1}{2}Mv_{com}^2. \qquad (11\text{-}5)$$

We can interpret the term $\tfrac{1}{2}I_{com}\omega^2$ as the kinetic energy associated with the rotation of the wheel about an axis through its center of mass (Fig. 11-4a), and the term $\tfrac{1}{2}Mv_{com}^2$ as the kinetic energy associated with the translational motion of the wheel's center of mass (Fig. 11-4b). Thus, we have the following rule:

> ☞ A rolling object has two types of kinetic energy: a rotational kinetic energy $(\tfrac{1}{2}I_{com}\omega^2)$ due to its rotation about its center of mass and a translational kinetic energy $(\tfrac{1}{2}Mv_{com}^2)$ due to translation of its center of mass.

**Sample Problem** | **11-1**

The current land-speed record was set in the Black Rock Desert of Nevada in 1997 by the jet-powered car *Thrust SSC*. The car's speed was 1222 km/h in one direction and 1233 km/h in the opposite direction. Both speeds exceeded the speed of sound at that location (1207 km/h).

Setting the land-speed record was obviously very dangerous for many reasons. One of them had to do with the car's wheels. Approximate each wheel on the car *Thrust SSC* as a disk of uniform thickness and mass $M = 170$ kg, and assume smooth rolling. When the car's speed was 1233 km/h, what was the kinetic energy of each wheel? ✈

**KEY IDEAS** Equation 11-5 gives the kinetic energy of a rolling object, but we need three ideas to use it:

1. When we speak of the speed of a rolling object, we always mean the speed of the center of mass, so here $v_{com} = 1233$ km/h $= 342.5$ m/s.

2. Equation 11-5 requires the angular speed $\omega$ of the rolling object, which we can relate to $v_{com}$ with Eq. 11-2, writing $\omega = v_{com}/R$, where $R$ is the wheel's radius.

3. Equation 11-5 also requires the rotational inertia $I_{com}$ of the object about its center of mass. From Table 10-2c, we find that, for a uniform disk, $I_{com} = \tfrac{1}{2}MR^2$.

**Calculations:** Now Eq. 11-5 gives us

$$\begin{aligned}
K &= \tfrac{1}{2}I_{com}\omega^2 + \tfrac{1}{2}Mv_{com}^2 \\
&= (\tfrac{1}{2})(\tfrac{1}{2}MR^2)(v_{com}/R)^2 + \tfrac{1}{2}Mv_{com}^2 = \tfrac{3}{4}Mv_{com}^2 \\
&= \tfrac{3}{4}(170 \text{ kg})(342.5 \text{ m/s})^2 \\
&= 1.50 \times 10^7 \text{ J}. \qquad \text{(Answer)}
\end{aligned}$$

(Note that the wheel's radius $R$ cancels out of the calculation.)

This answer gives one measure of the danger when the land-speed record was set by *Thrust SSC*: The kinetic energy of each (cast aluminum) wheel on the car was huge, almost as much as the kinetic energy ($2.1 \times 10^7$ J) of the spinning steel disk that exploded in Sample Problem 10-8. Had a wheel hit any hard obstacle along the car's path, the wheel would have exploded the way the steel disk did, with the car and driver moving faster than sound!

## 11-4 | The Forces of Rolling

### Friction and Rolling

If a wheel rolls at constant speed, as in Fig. 11-3, it has no tendency to slide at the point of contact $P$, and thus no frictional force acts there. However, if a net force acts on the rolling wheel to speed it up or to slow it, then that net force causes acceleration $\vec{a}_{com}$ of the center of mass along the direction of travel. It also causes the wheel to rotate faster or slower, which means it causes an angu-

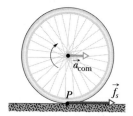

lar acceleration $\alpha$. These accelerations tend to make the wheel slide at $P$. Thus, a frictional force must act on the wheel at $P$ to oppose that tendency.

If the wheel *does not* slide, the force is a *static* frictional force $\vec{f}_s$ and the motion is smooth rolling. We can then relate the magnitudes of the linear acceleration $\vec{a}_{com}$ and the angular acceleration $\alpha$ by differentiating Eq. 11-2 with respect to time (with $R$ held constant). On the left side, $dv_{com}/dt$ is $a_{com}$, and on the right side $d\omega/dt$ is $\alpha$. So, for smooth rolling we have

$$a_{com} = \alpha R \qquad \text{(smooth rolling motion)}. \qquad (11\text{-}6)$$

If the wheel *does* slide when the net force acts on it, the frictional force that acts at $P$ in Fig. 11-3 is a *kinetic* frictional force $\vec{f}_k$. The motion then is not smooth rolling, and Eq. 11-6 does not apply to the motion. In this chapter we discuss only smooth rolling motion.

Figure 11-7 shows an example in which a wheel is being made to rotate faster while rolling to the right along a flat surface, as on a bicycle at the start of a race. The faster rotation tends to make the bottom of the wheel slide to the left at point $P$. A frictional force at $P$, directed to the right, opposes this tendency to slide. If the wheel does not slide, that frictional force is a static frictional force $\vec{f}_s$ (as shown), the motion is smooth rolling, and Eq. 11-6 applies to the motion. (Without friction, bicycle races would be stationary and very boring.)

If the wheel in Fig. 11-7 were made to rotate slower, as on a slowing bicycle, we would change the figure in two ways: The directions of the center-of-mass acceleration $\vec{a}_{com}$ and the frictional force $\vec{f}_s$ at point $P$ would now be to the left.

**FIG. 11-7** A wheel rolls horizontally without sliding while accelerating with linear acceleration $\vec{a}_{com}$. A static frictional force $\vec{f}_s$ acts on the wheel at $P$, opposing its tendency to slide.

### Rolling Down a Ramp

Figure 11-8 shows a round uniform body of mass $M$ and radius $R$ rolling smoothly down a ramp at angle $\theta$, along an $x$ axis. We want to find an expression for the body's acceleration $a_{com,x}$ down the ramp. We do this by using Newton's second law in both its linear version ($F_{net} = Ma$) and its angular version ($\tau_{net} = I\alpha$).

We start by drawing the forces on the body as shown in Fig. 11-8:

1. The gravitational force $\vec{F}_g$ on the body is directed downward. The tail of the vector is placed at the center of mass of the body. The component along the ramp is $F_g \sin\theta$, which is equal to $Mg \sin\theta$.

2. A normal force $\vec{F}_N$ is perpendicular to the ramp. It acts at the point of contact $P$, but in Fig. 11-8 the vector has been shifted along its direction until its tail is at the body's center of mass.

3. A static frictional force $\vec{f}_s$ acts at the point of contact $P$ and is directed up the ramp. (Do you see why? If the body were to slide at $P$, it would slide *down* the ramp. Thus, the frictional force opposing the sliding must be *up* the ramp.)

We can write Newton's second law for components along the $x$ axis in Fig. 11-8 ($F_{net,x} = ma_x$) as

$$f_s - Mg \sin\theta = Ma_{com,x}. \qquad (11\text{-}7)$$

This equation contains two unknowns, $f_s$ and $a_{com,x}$. (We should *not* assume that $f_s$ is at its maximum value $f_{s,max}$. All we know is that the value of $f_s$ is just right for the body to roll smoothly down the ramp, without sliding.)

We now wish to apply Newton's second law in angular form to the body's rotation about its center of mass. First, we shall use Eq. 10-41 ($\tau = r_\perp F$) to write the torques on the body about that point. The frictional force $\vec{f}_s$ has moment arm $R$ and thus produces a torque $Rf_s$, which is positive because it tends to rotate the body counterclockwise in Fig. 11-8. Forces $\vec{F}_g$ and $\vec{F}_N$ have zero moment arms about the center of mass and thus produce zero torques. So we can write the angular form of Newton's second law ($\tau_{net} = I\alpha$) about an axis through the body's center of mass as

$$Rf_s = I_{com}\alpha. \qquad (11\text{-}8)$$

This equation contains two unknowns, $f_s$ and $\alpha$.

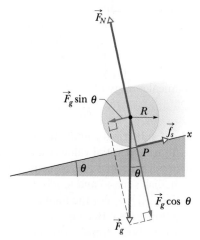

**FIG. 11-8** A round uniform body of radius $R$ rolls down a ramp. The forces that act on it are the gravitational force $\vec{F}_g$, a normal force $\vec{F}_N$, and a frictional force $\vec{f}_s$ pointing up the ramp. (For clarity, vector $\vec{F}_N$ has been shifted in the direction it points until its tail is at the center of the body.)

Because the body is rolling smoothly, we can use Eq. 11-6 ($a_{com} = \alpha R$) to relate the unknowns $a_{com,x}$ and $\alpha$. But we must be cautious because here $a_{com,x}$ is negative (in the negative direction of the $x$ axis) and $\alpha$ is positive (counter-clockwise). Thus we substitute $-a_{com,x}/R$ for $\alpha$ in Eq. 11-8. Then, solving for $f_s$, we obtain

$$f_s = -I_{com}\frac{a_{com,x}}{R^2}. \tag{11-9}$$

Substituting the right side of Eq. 11-9 for $f_s$ in Eq. 11-7, we then find

$$a_{com,x} = -\frac{g\sin\theta}{1 + I_{com}/MR^2}. \tag{11-10}$$

We can use this equation to find the linear acceleration $a_{com,x}$ of any body rolling along an incline of angle $\theta$ with the horizontal.

✓ **CHECKPOINT 2** Disks $A$ and $B$ are identical and roll across a floor with equal speeds. Then disk $A$ rolls up an incline, reaching a maximum height $h$, and disk $B$ moves up an incline that is identical except that it is frictionless. Is the maximum height reached by disk $B$ greater than, less than, or equal to $h$?

**Sample Problem** | **11-2** | **Build your skill**

A uniform ball, of mass $M = 6.00$ kg and radius $R$, rolls smoothly from rest down a ramp at angle $\theta = 30.0°$ (Fig. 11-8).

(a) The ball descends a vertical height $h = 1.20$ m to reach the bottom of the ramp. What is its speed at the bottom?

**KEY IDEAS** The mechanical energy $E$ of the ball–Earth system is conserved as the ball rolls down the ramp. The reason is that the only force doing work on the ball is the gravitational force, a conservative force. The normal force on the ball from the ramp does zero work because it is perpendicular to the ball's path. The frictional force on the ball from the ramp does not transfer any energy to thermal energy because the ball does not slide (it *rolls smoothly*).

Therefore, we can write the conservation of mechanical energy ($E_f = E_i$) as

$$K_f + U_f = K_i + U_i, \tag{11-11}$$

where subscripts $f$ and $i$ refer to the final values (at the bottom) and initial values (at rest), respectively. The gravitational potential energy is initially $U_i = Mgh$ (where $M$ is the ball's mass) and finally $U_f = 0$. The kinetic energy is initially $K_i = 0$. For the final kinetic energy $K_f$, we need an additional idea: Because the ball rolls, the kinetic energy involves both translation *and* rotation, so we include them both by using the right side of Eq. 11-5.

*Calculations:* Substituting into Eq. 11-11 gives us

$$(\tfrac{1}{2}I_{com}\omega^2 + \tfrac{1}{2}Mv^2_{com}) + 0 = 0 + Mgh, \tag{11-12}$$

where $I_{com}$ is the ball's rotational inertia about an axis through its center of mass, $v_{com}$ is the requested speed at the bottom, and $\omega$ is the angular speed there.

Because the ball rolls smoothly, we can use Eq. 11-2 to substitute $v_{com}/R$ for $\omega$ to reduce the unknowns in Eq. 11-12. Doing so, substituting $\tfrac{2}{5}MR^2$ for $I_{com}$ (from Table 10-2f), and then solving for $v_{com}$ give us

$$v_{com} = \sqrt{(\tfrac{10}{7})gh} = \sqrt{(\tfrac{10}{7})(9.8 \text{ m/s}^2)(1.20 \text{ m})}$$
$$= 4.10 \text{ m/s}. \qquad\qquad \text{(Answer)}$$

Note that the answer does not depend on $M$ or $R$.

(b) What are the magnitude and direction of the frictional force on the ball as it rolls down the ramp?

**KEY IDEA** Because the ball rolls smoothly, Eq. 11-9 gives the frictional force on the ball.

*Calculations:* Before we can use Eq. 11-9, we need the ball's acceleration $a_{com,x}$ from Eq. 11-10:

$$a_{com,x} = -\frac{g\sin\theta}{1 + I_{com}/MR^2} = -\frac{g\sin\theta}{1 + \tfrac{2}{5}MR^2/MR^2}$$
$$= -\frac{(9.8 \text{ m/s}^2)\sin 30.0°}{1 + \tfrac{2}{5}} = -3.50 \text{ m/s}^2.$$

Note that we needed neither mass $M$ nor radius $R$ to find $a_{com,x}$. Thus, any size ball with any uniform mass would have this acceleration down a 30.0° ramp, provided the ball rolls smoothly.

We can now solve Eq. 11-9 as

$$f_s = -I_{com}\frac{a_{com,x}}{R^2} = -\tfrac{2}{5}MR^2\frac{a_{com,x}}{R^2} = -\tfrac{2}{5}Ma_{com,x}$$
$$= -\tfrac{2}{5}(6.00 \text{ kg})(-3.50 \text{ m/s}^2) = 8.40 \text{ N}. \quad \text{(Answer)}$$

Note that we needed mass $M$ but not radius $R$. Thus, the frictional force on any 6.00 kg ball rolling smoothly down a 30.0° ramp would be 8.40 N regardless of the ball's radius.

## 11-5 | The Yo-Yo

A yo-yo is a physics lab that you can fit in your pocket. If a yo-yo rolls down its string for a distance $h$, it loses potential energy in amount $mgh$ but gains kinetic energy in both translational ($\frac{1}{2}Mv_{com}^2$) and rotational ($\frac{1}{2}I_{com}\omega^2$) forms. As it climbs back up, it loses kinetic energy and regains potential energy.

In a modern yo-yo, the string is not tied to the axle but is looped around it. When the yo-yo "hits" the bottom of its string, an upward force on the axle from the string stops the descent. The yo-yo then spins, axle inside loop, with only rotational kinetic energy. The yo-yo keeps spinning ("sleeping") until you "wake it" by jerking on the string, causing the string to catch on the axle and the yo-yo to climb back up. The rotational kinetic energy of the yo-yo at the bottom of its string (and thus the sleeping time) can be considerably increased by throwing the yo-yo downward so that it starts down the string with initial speeds $v_{com}$ and $\omega$ instead of rolling down from rest.

To find an expression for the linear acceleration $a_{com}$ of a yo-yo rolling down a string, we could use Newton's second law just as we did for the body rolling down a ramp in Fig. 11-8. The analysis is the same except for the following:

1. Instead of rolling down a ramp at angle $\theta$ with the horizontal, the yo-yo rolls down a string at angle $\theta = 90°$ with the horizontal.

2. Instead of rolling on its outer surface at radius $R$, the yo-yo rolls on an axle of radius $R_0$ (Fig. 11-9a).

3. Instead of being slowed by frictional force $\vec{f}_s$, the yo-yo is slowed by the force $\vec{T}$ on it from the string (Fig. 11-9b).

The analysis would again lead us to Eq. 11-10. Therefore, let us just change the notation in Eq. 11-10 and set $\theta = 90°$ to write the linear acceleration as

$$a_{com} = -\frac{g}{1 + I_{com}/MR_0^2},  \qquad (11\text{-}13)$$

where $I_{com}$ is the yo-yo's rotational inertia about its center and $M$ is its mass. A yo-yo has the same downward acceleration when it is climbing back up.

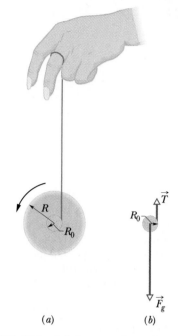

**FIG. 11-9** (a) A yo-yo, shown in cross section. The string, of assumed negligible thickness, is wound around an axle of radius $R_0$. (b) A free-body diagram for the falling yo-yo. Only the axle is shown.

## 11-6 | Torque Revisited

In Chapter 10 we defined torque $\tau$ for a rigid body that can rotate around a fixed axis, with each particle in the body forced to move in a path that is a circle centered on that axis. We now expand the definition of torque to apply it to an individual particle that moves along any path relative to a fixed *point* (rather than a fixed axis). The path need no longer be a circle, and we must write the torque as a vector $\vec{\tau}$ that may have any direction.

Figure 11-10a shows such a particle at point $A$ in an $xy$ plane. A single force $\vec{F}$ in that plane acts on the particle, and the particle's position relative to the origin $O$ is given by position vector $\vec{r}$. The torque $\vec{\tau}$ acting on the particle relative to the fixed point $O$ is a vector quantity defined as

$$\vec{\tau} = \vec{r} \times \vec{F} \qquad \text{(torque defined)}. \qquad (11\text{-}14)$$

We can evaluate the vector (or cross) product in this definition of $\vec{\tau}$ by using the rules for such products given in Section 3-8. To find the direction of $\vec{\tau}$, we slide the vector $\vec{F}$ (without changing its direction) until its tail is at the origin $O$, so that the two vectors in the vector product are tail to tail as in Fig. 11-10b. We then use the right-hand rule for vector products in Fig. 3-21a, sweeping the fingers of the right hand from $\vec{r}$ (the first vector in the product) into $\vec{F}$ (the second vector). The outstretched right thumb then gives the direction of $\vec{\tau}$. In Fig. 11-10b, the direction of $\vec{\tau}$ is in the positive direction of the $z$ axis.

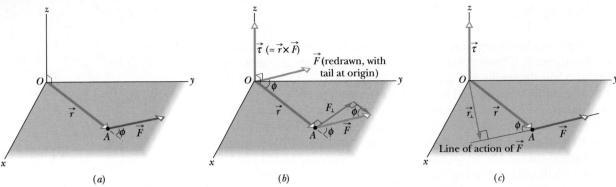

(a)        (b)        (c)

**FIG. 11-10** Defining torque. (*a*) A force $\vec{F}$, lying in an *xy* plane, acts on a particle at point *A*. (*b*) This force produces a torque $\vec{\tau}\,(=\vec{r}\times\vec{F})$ on the particle with respect to the origin *O*. By the right-hand rule for vector (cross) products, the torque vector points in the positive direction of *z*. Its magnitude is given by $rF_{\perp}$ in (*b*) and by $r_{\perp}F$ in (*c*).

To determine the magnitude of $\vec{\tau}$, we apply the general result of Eq. 3-27 ($c = ab\sin\phi$), finding

$$\tau = rF\sin\phi, \qquad (11\text{-}15)$$

where $\phi$ is the smaller angle between the directions of $\vec{r}$ and $\vec{F}$ when the vectors are tail to tail. From Fig. 11-10*b*, we see that Eq. 11-15 can be rewritten as

$$\tau = rF_{\perp}, \qquad (11\text{-}16)$$

where $F_{\perp}\,(=F\sin\phi)$ is the component of $\vec{F}$ perpendicular to $\vec{r}$. From Fig. 11-10*c*, we see that Eq. 11-15 can also be rewritten as

$$\tau = r_{\perp}F, \qquad (11\text{-}17)$$

where $r_{\perp}\,(=r\sin\phi)$ is the moment arm of $\vec{F}$ (the perpendicular distance between *O* and the line of action of $\vec{F}$).

✓ **CHECKPOINT 3** The position vector $\vec{r}$ of a particle points along the positive direction of a *z* axis. If the torque on the particle is (a) zero, (b) in the negative direction of *x*, and (c) in the negative direction of *y*, in what direction is the force causing the torque?

---

**Sample Problem** | **11-3**

In Fig. 11-11*a*, three forces, each of magnitude 2.0 N, act on a particle. The particle is in the *xz* plane at point *A* given by position vector $\vec{r}$, where $r = 3.0$ m and $\theta = 30°$. Force $\vec{F}_1$ is parallel to the *x* axis, force $\vec{F}_2$ is parallel to the *z* axis, and force $\vec{F}_3$ is parallel to the *y* axis. What is the torque, about the origin *O*, due to each force?

**KEY IDEA** Because the three force vectors do not lie in a plane, we cannot evaluate their torques as in Chapter 10. Instead, we must use vector (or cross) products, with magnitudes given by Eq. 11-15 ($\tau = rF\sin\phi$) and directions given by the right-hand rule for vector products.

**Calculations:** Because we want the torques with respect to the origin *O*, the vector $\vec{r}$ required for each cross product is the given position vector. To determine the angle $\phi$ between the direction of $\vec{r}$ and the direction of each force, we shift the force vectors of Fig. 11-11*a*, each in

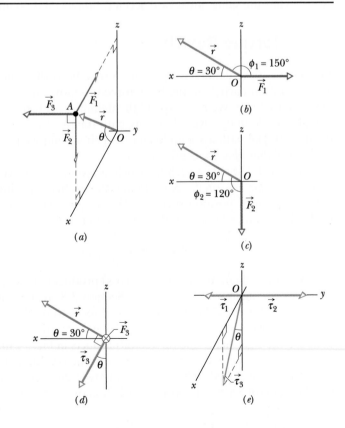

**FIG. 11-11** (*a*) A particle at point *A* is acted on by three forces, each parallel to a coordinate axis. The angle $\phi$ (used in finding torque) is shown (*b*) for $\vec{F}_1$ and (*c*) for $\vec{F}_2$. (*d*) Torque $\vec{\tau}_3$ is perpendicular to both $\vec{r}$ and $\vec{F}_3$ (force $\vec{F}_3$ is directed into the plane of the figure). (*e*) The torques (relative to the origin *O*) acting on the particle.

turn, so that their tails are at the origin. Figures 11-11*b*, *c*, and *d*, which are direct views of the *xz* plane, show the shifted force vectors $\vec{F}_1$, $\vec{F}_2$, and $\vec{F}_3$, respectively. (Note how much easier the angles are to see.) In Fig. 11-11*d*, the angle between the directions of $\vec{r}$ and $\vec{F}_3$ is 90° and the symbol $\otimes$ means $\vec{F}_3$ is directed into the page. If it were directed out of the page, it would be represented with the symbol $\odot$.

Now, applying Eq. 11-15 for each force, we find the magnitudes of the torques to be

$$\tau_1 = rF_1 \sin \phi_1 = (3.0\ \text{m})(2.0\ \text{N})(\sin 150°) = 3.0\ \text{N·m},$$

$$\tau_2 = rF_2 \sin \phi_2 = (3.0\ \text{m})(2.0\ \text{N})(\sin 120°) = 5.2\ \text{N·m},$$

and
$$\tau_3 = rF_3 \sin \phi_3 = (3.0\ \text{m})(2.0\ \text{N})(\sin 90°)$$
$$= 6.0\ \text{N·m}. \qquad \text{(Answer)}$$

To find the directions of these torques, we use the right-hand rule, placing the fingers of the right hand so as to rotate $\vec{r}$ into $\vec{F}$ through the *smaller* of the two angles between their directions. The thumb points in the direction of the torque. Thus $\vec{\tau}_1$ is directed into the page in Fig. 11-11*b*; $\vec{\tau}_2$ is directed out of the page in Fig. 11-11*c*; and $\vec{\tau}_3$ is directed as shown in Fig. 11-11*d*. All three torque vectors are shown in Fig. 11-11*e*.

---

**PROBLEM-SOLVING TACTICS**

*Tactic 1: Vector Products and Torques* Equation 11-15 for torques is our first application of the vector (or cross) product. Section 3-8, where the rules for the vector product are given, lists many common errors in finding the direction of a vector product.

Keep in mind that a torque is calculated *with respect to* (or *about*) a point, which must be known if the value of the torque is to be meaningful. Changing the point can change the torque in both magnitude and direction. For example, in Sample Problem 11-3, the torques due to the three forces are calculated about the origin *O*. You can show that the torques due to the same three forces are all zero if they are calculated about point *A* (at the position of the particle), because then $r = 0$ for each force.

---

# 11-7 | Angular Momentum

Recall that the concept of linear momentum $\vec{p}$ and the principle of conservation of linear momentum are extremely powerful tools. They allow us to predict the outcome of, say, a collision of two cars without knowing the details of the collision. Here we begin a discussion of the angular counterpart of $\vec{p}$, winding up in Section 11-11 with the angular counterpart of the conservation principle.

Figure 11-12 shows a particle of mass *m* with linear momentum $\vec{p}\,(= m\vec{v})$ as it passes through point *A* in an *xy* plane. The **angular momentum** $\vec{\ell}$ of this particle with respect to the origin *O* is a vector quantity defined as

$$\vec{\ell} = \vec{r} \times \vec{p} = m(\vec{r} \times \vec{v}) \quad \text{(angular momentum defined),} \quad (11\text{-}18)$$

where $\vec{r}$ is the position vector of the particle with respect to *O*. As the particle moves relative to *O* in the direction of its momentum $\vec{p}\,(= m\vec{v})$, position vector $\vec{r}$ rotates around *O*. Note carefully that to have angular momentum about *O*, the particle does *not* itself have to rotate around *O*. Comparison of Eqs. 11-14 and 11-18 shows that angular momentum bears the same relation to linear momentum that torque does to force. The SI unit of angular momentum is the kilogram-meter-squared per second (kg·m²/s), equivalent to the joule-second (J·s).

To find the direction of the angular momentum vector $\vec{\ell}$ in Fig. 11-12, we slide the vector $\vec{p}$ until its tail is at the origin *O*. Then we use the right-hand rule for vector products, sweeping the fingers from $\vec{r}$ into $\vec{p}$. The outstretched thumb then shows that the direction of $\vec{\ell}$ is in the positive direction of the *z* axis in Fig. 11-12. This positive direction is consistent with the counterclockwise rotation of position vector $\vec{r}$ about the *z* axis, as the particle moves. (A negative direction of $\vec{\ell}$ would be consistent with a clockwise rotation of $\vec{r}$ about the *z* axis.)

To find the magnitude of $\vec{\ell}$, we use the general result of Eq. 3-27 to write

$$\ell = rmv \sin \phi, \quad (11\text{-}19)$$

where $\phi$ is the smaller angle between $\vec{r}$ and $\vec{p}$ when these two vectors are tail

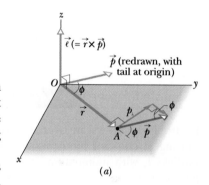

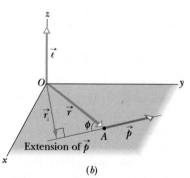

**FIG. 11-12** Defining angular momentum. A particle passing through point *A* has linear momentum $\vec{p}\,(= m\vec{v})$, with the vector $\vec{p}$ lying in an *xy* plane. The particle has angular momentum $\vec{\ell}\,(= \vec{r} \times \vec{p})$ with respect to the origin *O*. By the right-hand rule, the angular momentum vector points in the positive direction of *z*. (*a*) The magnitude of $\vec{\ell}$ is given by $\ell = rp_\perp = rmv_\perp$. (*b*) The magnitude of $\vec{\ell}$ is also given by $\ell = r_\perp p = r_\perp mv$.

to tail. From Fig. 11-12a, we see that Eq. 11-19 can be rewritten as

$$\ell = rp_{\perp} = rmv_{\perp}, \qquad (11\text{-}20)$$

where $p_{\perp}$ is the component of $\vec{p}$ perpendicular to $\vec{r}$ and $v_{\perp}$ is the component of $\vec{v}$ perpendicular to $\vec{r}$. From Fig. 11-12b, we see that Eq. 11-19 can also be rewritten as

$$\ell = r_{\perp}p = r_{\perp}mv, \qquad (11\text{-}21)$$

where $r_{\perp}$ is the perpendicular distance between $O$ and the extension of $\vec{p}$.

Just as is true for torque, angular momentum has meaning only with respect to a specified origin. Moreover, if the particle in Fig. 11-12 did not lie in the $xy$ plane, or if the linear momentum $\vec{p}$ of the particle did not also lie in that plane, the angular momentum $\vec{\ell}$ would not be parallel to the $z$ axis. The direction of the angular momentum vector is always perpendicular to the plane formed by the position and linear momentum vectors $\vec{r}$ and $\vec{p}$.

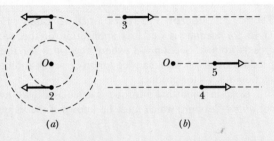

✓ **CHECKPOINT 4** In part $a$ of the figure, particles 1 and 2 move around point $O$ in opposite directions, in circles with radii 2 m and 4 m. In part $b$, particles 3 and 4 travel in the same direction, along straight lines at perpendicular distances of 4 m and 2 m from point $O$. Particle 5 moves directly away from $O$. All five particles have the same mass and the same constant speed. (a) Rank the particles according to the magnitudes of their angular momentum about point $O$, greatest first. (b) Which particles have negative angular momentum about point $O$?

---

**Sample Problem 11-4**

Figure 11-13 shows an overhead view of two particles moving at constant momentum along horizontal paths. Particle 1, with momentum magnitude $p_1 = 5.0 \text{ kg} \cdot \text{m/s}$, has position vector $\vec{r}_1$ and will pass 2.0 m from point $O$. Particle 2, with momentum magnitude $p_2 = 2.0 \text{ kg} \cdot \text{m/s}$, has position vector $\vec{r}_2$ and will pass 4.0 m from point $O$. What are the magnitude and direction of the net angular momentum $\vec{L}$ about point $O$ of the two-particle system?

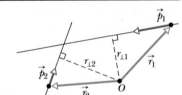

**FIG. 11-13** Two particles pass near point $O$.

**KEY IDEA** To find $\vec{L}$, we can first find the individual angular momenta $\vec{\ell}_1$ and $\vec{\ell}_2$ and then add them. To evaluate their magnitudes, we can use any one of Eqs. 11-18 through 11-21. However, Eq. 11-21 is easiest, because we are given the perpendicular distances $r_{\perp 1}\ (= 2.0 \text{ m})$ and $r_{\perp 2}\ (= 4.0 \text{ m})$ and the momentum magnitudes $p_1$ and $p_2$.

**Calculations:** For particle 1, Eq. 11-21 yields

$$\ell_1 = r_{\perp 1}p_1 = (2.0 \text{ m})(5.0 \text{ kg} \cdot \text{m/s})$$
$$= 10 \text{ kg} \cdot \text{m}^2\text{/s}.$$

To find the direction of vector $\vec{\ell}_1$, we use Eq. 11-18 and the right-hand rule for vector products. For $\vec{r}_1 \times \vec{p}_1$, the vector product is out of the page, perpendicular to the plane of Fig. 11-13. This is the positive direction, consistent with the counterclockwise rotation of the particle's

position vector $\vec{r}_1$ around $O$ as particle 1 moves. Thus, the angular momentum vector for particle 1 is

$$\ell_1 = +10 \text{ kg} \cdot \text{m}^2\text{/s}.$$

Similarly, the magnitude of $\vec{\ell}_2$ is

$$\ell_2 = r_{\perp 2}p_2 = (4.0 \text{ m})(2.0 \text{ kg} \cdot \text{m/s})$$
$$= 8.0 \text{ kg} \cdot \text{m}^2\text{/s},$$

and the vector product $\vec{r}_2 \times \vec{p}_2$ is into the page, which is the negative direction, consistent with the clockwise rotation of $\vec{r}_2$ around $O$ as particle 2 moves. Thus, the angular momentum vector for particle 2 is

$$\ell_2 = -8.0 \text{ kg} \cdot \text{m}^2\text{/s}.$$

The net angular momentum for the two-particle system is

$$L = \ell_1 + \ell_2 = +10 \text{ kg} \cdot \text{m}^2\text{/s} + (-8.0 \text{ kg} \cdot \text{m}^2\text{/s})$$
$$= +2.0 \text{ kg} \cdot \text{m}^2\text{/s}. \qquad \text{(Answer)}$$

The plus sign means that the system's net angular momentum about point $O$ is out of the page.

# 11-8 | Newton's Second Law in Angular Form

Newton's second law written in the form

$$\vec{F}_{net} = \frac{d\vec{p}}{dt} \quad \text{(single particle)} \tag{11-22}$$

expresses the close relation between force and linear momentum for a single particle. We have seen enough of the parallelism between linear and angular quantities to be pretty sure that there is also a close relation between torque and angular momentum. Guided by Eq. 11-22, we can even guess that it must be

$$\vec{\tau}_{net} = \frac{d\vec{\ell}}{dt} \quad \text{(single particle).} \tag{11-23}$$

Equation 11-23 is indeed an angular form of Newton's second law for a single particle:

> The (vector) sum of all the torques acting on a particle is equal to the time rate of change of the angular momentum of that particle.

Equation 11-23 has no meaning unless the torques $\vec{\tau}$ and the angular momentum $\vec{\ell}$ are defined with respect to the same origin.

## Proof of Equation 11-23

We start with Eq. 11-18, the definition of the angular momentum of a particle:

$$\vec{\ell} = m(\vec{r} \times \vec{v}),$$

where $\vec{r}$ is the position vector of the particle and $\vec{v}$ is the velocity of the particle. Differentiating* each side with respect to time $t$ yields

$$\frac{d\vec{\ell}}{dt} = m\left(\vec{r} \times \frac{d\vec{v}}{dt} + \frac{d\vec{r}}{dt} \times \vec{v}\right). \tag{11-24}$$

However, $d\vec{v}/dt$ is the acceleration $\vec{a}$ of the particle, and $d\vec{r}/dt$ is its velocity $\vec{v}$. Thus, we can rewrite Eq. 11-24 as

$$\frac{d\vec{\ell}}{dt} = m(\vec{r} \times \vec{a} + \vec{v} \times \vec{v}).$$

Now $\vec{v} \times \vec{v} = 0$ (the vector product of any vector with itself is zero because the angle between the two vectors is necessarily zero). This leads to

$$\frac{d\vec{\ell}}{dt} = m(\vec{r} \times \vec{a}) = \vec{r} \times m\vec{a}.$$

We now use Newton's second law ($\vec{F}_{net} = m\vec{a}$) to replace $m\vec{a}$ with its equal, the vector sum of the forces that act on the particle, obtaining

$$\frac{d\vec{\ell}}{dt} = \vec{r} \times \vec{F}_{net} = \sum(\vec{r} \times \vec{F}). \tag{11-25}$$

Here the symbol $\Sigma$ indicates that we must sum the vector products $\vec{r} \times \vec{F}$ for all the forces. However, from Eq. 11-14, we know that each one of those vector products is the torque associated with one of the forces. Therefore, Eq. 11-25 tells us that

$$\vec{\tau}_{net} = \frac{d\vec{\ell}}{dt}.$$

This is Eq. 11-23, the relation that we set out to prove.

---

*In differentiating a vector product, be sure not to change the order of the two quantities (here $\vec{r}$ and $\vec{v}$) that form that product. (See Eq. 3-28.)

✓**CHECKPOINT 5** The figure shows the position vector $\vec{r}$ of a particle at a certain instant, and four choices for the direction of a force that is to accelerate the particle. All four choices lie in the $xy$ plane. (a) Rank the choices according to the magnitude of the time rate of change $(d\vec{\ell}/dt)$ they produce in the angular momentum of the particle about point $O$, greatest first. (b) Which choice results in a negative rate of change about $O$?

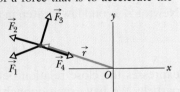

---

**Sample Problem** | **11-5** | **Build your skill**

In Fig. 11-14, a penguin of mass $m$ falls from rest at point $A$, a horizontal distance $D$ from the origin $O$ of an $xyz$ coordinate system. (The positive direction of the $z$ axis is directly outward from the plane of the figure.)

(a) What is the angular momentum $\vec{\ell}$ of the falling penguin about $O$?

**KEY IDEA** We can treat the penguin as a particle, and thus its angular momentum $\vec{\ell}$ is given by Eq. 11-18 ($\vec{\ell} = \vec{r} \times \vec{p}$), where $\vec{r}$ is the penguin's position vector (extending from $O$ to the penguin) and $\vec{p}$ is the penguin's linear momentum. (The penguin has *angular* momentum about $O$ even though it moves in a straight line, because vector $\vec{r}$ rotates about $O$ as the penguin falls.)

***Calculations:*** To find the magnitude of $\vec{\ell}$, we can use any one of the scalar equations derived from Eq. 11-18—namely, Eqs. 11-19 through 11-21. However, Eq. 11-21 ($\ell = r_\perp mv$) is easiest because the perpendicular distance $r_\perp$ between $O$ and an extension of vector $\vec{p}$ is the given distance $D$. The speed of an object that has fallen from rest for a time $t$ is $v = gt$. We can now write Eq. 11-21 in terms of given quantities as

$$\ell = r_\perp mv = Dmgt. \qquad \text{(Answer)}$$

To find the direction of $\vec{\ell}$, we use the right-hand

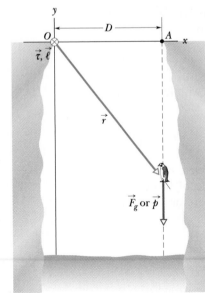

**FIG. 11-14** A penguin falls vertically from point $A$. The torque $\vec{\tau}$ and the angular momentum $\vec{\ell}$ of the falling penguin with respect to the origin $O$ are directed into the plane of the figure at $O$.

rule for the vector product $\vec{r} \times \vec{p}$ in Eq. 11-18. Mentally shift $\vec{p}$ until its tail is at the origin, and then use the fingers of your right hand to rotate $\vec{r}$ into $\vec{p}$ through the smaller angle between the two vectors. Your outstretched thumb then points into the plane of the figure, indicating that the product $\vec{r} \times \vec{p}$ and thus also $\vec{\ell}$ are directed into that plane, in the negative direction of the $z$ axis. We represent $\vec{\ell}$ with an encircled cross $\otimes$ at $O$. The vector $\vec{\ell}$ changes with time in magnitude only; its direction remains unchanged.

(b) About the origin $O$, what is the torque $\vec{\tau}$ on the penguin due to the gravitational force $\vec{F}_g$?

**KEY IDEAS** (1) The torque is given by Eq. 11-14 ($\vec{\tau} = \vec{r} \times \vec{F}$), where now the force is $\vec{F}_g$. (2) Force $\vec{F}_g$ causes a torque on the penguin, even though the penguin moves in a straight line, because $\vec{r}$ rotates about $O$ as the penguin moves.

***Calculations:*** To find the magnitude of $\vec{\tau}$, we can use any one of the scalar equations derived from Eq. 11-14—namely, Eqs. 11-15 through 11-17. However, Eq. 11-17 ($\tau = r_\perp F$) is easiest because the perpendicular distance $r_\perp$ between $O$ and the line of action of $\vec{F}_g$ is the given distance $D$. So, substituting $D$ and using $mg$ for the magnitude of $\vec{F}_g$, we can write Eq. 11-17 as

$$\tau = DF_g = Dmg. \qquad \text{(Answer)}$$

Using the right-hand rule for the vector product $\vec{r} \times \vec{F}$ in Eq. 11-14, we find that the direction of $\vec{\tau}$ is the negative direction of the $z$ axis, the same as $\vec{\ell}$.

The results we obtained in parts (a) and (b) must be consistent with Newton's second law in the angular form of Eq. 11-23 ($\vec{\tau}_{net} = d\vec{\ell}/dt$). To check the magnitudes we got, we write Eq. 11-23 in component form for the $z$ axis and then substitute our result $\ell = Dmgt$. We find

$$\tau = \frac{d\ell}{dt} = \frac{d(Dmgt)}{dt} = Dmg,$$

which is the magnitude we found for $\vec{\tau}$. To check the directions, we note that Eq. 11-23 tells us that $\vec{\tau}$ and $d\vec{\ell}/dt$ must have the same direction. So $\vec{\tau}$ and $\vec{\ell}$ must also have the same direction, which is what we found.

## 11-9 | The Angular Momentum of a System of Particles

Now we turn our attention to the angular momentum of a system of particles with respect to an origin. The total angular momentum $\vec{L}$ of the system is the (vector) sum of the angular momenta $\vec{\ell}$ of the individual particles (here with label $i$):

$$\vec{L} = \vec{\ell}_1 + \vec{\ell}_2 + \vec{\ell}_3 + \cdots + \vec{\ell}_n = \sum_{i=1}^{n} \vec{\ell}_i. \qquad (11\text{-}26)$$

With time, the angular momenta of individual particles may change because of interactions between the particles or with the outside. We can find the resulting change in $\vec{L}$ by taking the time derivative of Eq. 11-26. Thus,

$$\frac{d\vec{L}}{dt} = \sum_{i=1}^{n} \frac{d\vec{\ell}_i}{dt}. \qquad (11\text{-}27)$$

From Eq. 11-23, we see that $d\vec{\ell}_i/dt$ is equal to the net torque $\vec{\tau}_{\text{net},i}$ on the $i$th particle. We can rewrite Eq. 11-27 as

$$\frac{d\vec{L}}{dt} = \sum_{i=1}^{n} \vec{\tau}_{\text{net},i}. \qquad (11\text{-}28)$$

That is, the rate of change of the system's angular momentum $\vec{L}$ is equal to the vector sum of the torques on its individual particles. Those torques include *internal torques* (due to forces between the particles) and *external torques* (due to forces on the particles from bodies external to the system). However, the forces between the particles always come in third-law force pairs so their torques sum to zero. Thus, the only torques that can change the total angular momentum $\vec{L}$ of the system are the external torques acting on the system.

Let $\vec{\tau}_{\text{net}}$ represent the net external torque, the vector sum of all external torques on all particles in the system. Then we can write Eq. 11-28 as

$$\vec{\tau}_{\text{net}} = \frac{d\vec{L}}{dt} \qquad \text{(system of particles)}, \qquad (11\text{-}29)$$

which is Newton's second law in angular form. It says:

> ☞ The net external torque $\vec{\tau}_{\text{net}}$ acting on a system of particles is equal to the time rate of change of the system's total angular momentum $\vec{L}$.

Equation 11-29 is analogous to $\vec{F}_{\text{net}} = d\vec{P}/dt$ (Eq. 9-27) but requires extra caution: Torques and the system's angular momentum must be measured relative to the same origin. If the center of mass of the system is not accelerating relative to an inertial frame, that origin can be any point. However, if the center of mass of the system *is* accelerating, the origin can be only at that center of mass. As an example, consider a wheel as the system of particles. If the wheel is rotating about an axis that is fixed relative to the ground, then the origin for applying Eq. 11-29 can be any point that is stationary relative to the ground. However, if the wheel is rotating about an axis that is accelerating (such as when the wheel rolls down a ramp), then the origin can be only at the wheel's center of mass.

## 11-10 | The Angular Momentum of a Rigid Body Rotating About a Fixed Axis

We next evaluate the angular momentum of a system of particles that form a rigid body that rotates about a fixed axis. Figure 11-15a shows such a body. The fixed axis of rotation is a $z$ axis, and the body rotates about it with constant angular speed $\omega$. We wish to find the angular momentum of the body about that axis.

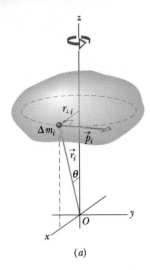

(a)

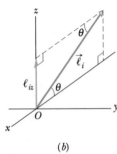

(b)

**FIG. 11-15** (a) A rigid body rotates about a z axis with angular speed $\omega$. A mass element of mass $\Delta m_i$ within the body moves about the z axis in a circle with radius $r_{\perp i}$. The mass element has linear momentum $\vec{p}_i$, and it is located relative to the origin O by position vector $\vec{r}_i$. Here the mass element is shown when $r_{\perp i}$ is parallel to the x axis. (b) The angular momentum $\vec{\ell}_i$, with respect to O, of the mass element in (a). The z component $\ell_{iz}$ is also shown.

We can find the angular momentum by summing the z components of the angular momenta of the mass elements in the body. In Fig. 11-15a, a typical mass element, of mass $\Delta m_i$, moves around the z axis in a circular path. The position of the mass element is located relative to the origin O by position vector $\vec{r}_i$. The radius of the mass element's circular path is $r_{\perp i}$, the perpendicular distance between the element and the z axis.

The magnitude of the angular momentum $\vec{\ell}_i$ of this mass element, with respect to O, is given by Eq. 11-19:

$$\ell_i = (r_i)(p_i)(\sin 90°) = (r_i)(\Delta m_i v_i),$$

where $p_i$ and $v_i$ are the linear momentum and linear speed of the mass element, and 90° is the angle between $\vec{r}_i$ and $\vec{p}_i$. The angular momentum vector $\vec{\ell}_i$ for the mass element in Fig. 11-15a is shown in Fig. 11-15b; its direction must be perpendicular to those of $\vec{r}_i$ and $\vec{p}_i$.

We are interested in the component of $\vec{\ell}_i$ that is parallel to the rotation axis, here the z axis. That z component is

$$\ell_{iz} = \ell_i \sin \theta = (r_i \sin \theta)(\Delta m_i v_i) = r_{\perp i} \Delta m_i v_i.$$

The z component of the angular momentum for the rotating rigid body as a whole is found by adding up the contributions of all the mass elements that make up the body. Thus, because $v = \omega r_\perp$, we may write

$$L_z = \sum_{i=1}^{n} \ell_{iz} = \sum_{i=1}^{n} \Delta m_i v_i r_{\perp i} = \sum_{i=1}^{n} \Delta m_i (\omega r_{\perp i}) r_{\perp i}$$

$$= \omega \left( \sum_{i=1}^{n} \Delta m_i r_{\perp i}^2 \right). \tag{11-30}$$

We can remove $\omega$ from the summation here because it has the same value for all points of the rotating rigid body.

The quantity $\sum \Delta m_i r_{\perp i}^2$ in Eq. 11-30 is the rotational inertia I of the body about the fixed axis (see Eq. 10-33). Thus Eq. 11-30 reduces to

$$L = I\omega \quad \text{(rigid body, fixed axis).} \tag{11-31}$$

We have dropped the subscript z, but you must remember that the angular momentum defined by Eq. 11-31 is the angular momentum about the rotation axis. Also, I in that equation is the rotational inertia about that same axis.

Table 11-1, which supplements Table 10-3, extends our list of corresponding linear and angular relations.

**TABLE 11-1**

**More Corresponding Variables and Relations for Translational and Rotational Motion[a]**

| Translational | | Rotational | |
|---|---|---|---|
| Force | $\vec{F}$ | Torque | $\vec{\tau} \ (= \vec{r} \times \vec{F})$ |
| Linear momentum | $\vec{p}$ | Angular momentum | $\vec{\ell} \ (= \vec{r} \times \vec{p})$ |
| Linear momentum[b] | $\vec{P} \ (= \Sigma \vec{p}_i)$ | Angular momentum[b] | $\vec{L} \ (= \Sigma \vec{\ell}_i)$ |
| Linear momentum[b] | $\vec{P} = M\vec{v}_{com}$ | Angular momentum[c] | $L = I\omega$ |
| Newton's second law[b] | $\vec{F}_{net} = \dfrac{d\vec{P}}{dt}$ | Newton's second law[b] | $\vec{\tau}_{net} = \dfrac{d\vec{L}}{dt}$ |
| Conservation law[d] | $\vec{P} = $ a constant | Conservation law[d] | $\vec{L} = $ a constant |

[a]See also Table 10-3.

[b]For systems of particles, including rigid bodies.

[c]For a rigid body about a fixed axis, with L being the component along that axis.

[d]For a closed, isolated system.

✓**CHECKPOINT 6** In the figure, a disk, a hoop, and a solid sphere are made to spin about fixed central axes (like a top) by

 Disk  Hoop  Sphere

means of strings wrapped around them, with the strings producing the same constant tangential force $\vec{F}$ on all three objects. The three objects have the same mass and radius, and they are initially stationary. Rank the objects according to (a) their angular momentum about their central axes and (b) their angular speed, greatest first, when the strings have been pulled for a certain time $t$.

## Sample Problem    11-6

George Washington Gale Ferris, Jr., a civil engineering graduate from Rensselaer Polytechnic Institute, built the original Ferris wheel (Fig. 11-16) for the 1893 World's Columbian Exposition in Chicago. The wheel, an astounding engineering construction at the time, carried 36 wooden cars, each holding as many as 60 passengers, around a circle of radius $R = 38$ m. The mass of each car was about $1.1 \times 10^4$ kg. The mass of the wheel's structure was about $6.0 \times 10^5$ kg, which was mostly in the circular grid from which the cars were suspended. The wheel made a complete rotation at an angular speed $\omega_F$ in about 2 min.

(a) Estimate the magnitude $L$ of the angular momentum of the wheel and its passengers while the wheel rotated at $\omega_F$.

**KEY IDEA** We can treat the wheel, cars, and passengers as a rigid object rotating about a fixed axis, at the wheel's axle. Then Eq. 11-31 ($L = I\omega$) gives the magnitude of the angular momentum of that object. We need to find $\omega_F$ and the rotational inertia $I$ of this object.

**Rotational inertia:** To find $I$, let us start with the loaded cars. Because we can treat them as particles, at distance $R$ from the axis of rotation, we know from Eq. 10-33 that their rotational inertia is $I_{pc} = M_{pc}R^2$, where $M_{pc}$ is their total mass. Let us assume that the 36 cars are each filled with 60 passengers, each of mass 70 kg. Then their total mass is

$$M_{pc} = 36[1.1 \times 10^4 \text{ kg} + 60(70 \text{ kg})] = 5.47 \times 10^5 \text{ kg}$$

and their rotational inertia is

$$I_{pc} = M_{pc}R^2 = (5.47 \times 10^5 \text{ kg})(38 \text{ m})^2 = 7.90 \times 10^8 \text{ kg} \cdot \text{m}^2.$$

Next we consider the structure of the wheel. Let us assume that the rotational inertia of the structure is due mainly to the circular grid suspending the cars. Further, let us assume that the grid forms a hoop of radius $R$, with a mass $M_{hoop}$ of $3.0 \times 10^5$ kg (half the wheel's mass). From Table 10-2a, the rotational inertia of the hoop is

$$I_{hoop} = M_{hoop}R^2 = (3.0 \times 10^5 \text{ kg})(38 \text{ m})^2$$
$$= 4.33 \times 10^8 \text{ kg} \cdot \text{m}^2.$$

The combined rotational inertia $I$ of the cars, passen-

**FIG. 11-16** The original Ferris wheel. *(From "Shepp's World's Fair Photographed" by James W. Shepp and Daniel P. Shepp, Globe Publishing Co., Chicago and Philadelphia, 1893)*

gers, and hoop is then

$$I = I_{pc} + I_{hoop} = 7.90 \times 10^8 \text{ kg} \cdot \text{m}^2 + 4.33 \times 10^8 \text{ kg} \cdot \text{m}^2$$
$$= 1.22 \times 10^9 \text{ kg} \cdot \text{m}^2.$$

**Angular speed:** To find the rotational speed $\omega_F$, we use Eq. 10-5 ($\omega_{avg} = \Delta\theta/\Delta t$). Here the wheel goes through an angular displacement of $\Delta\theta = 2\pi$ rad in a time period $\Delta t = 2$ min. Thus, we have

$$\omega_F = \frac{2\pi \text{ rad}}{(2 \text{ min})(60 \text{ s/min})} = 0.0524 \text{ rad/s}.$$

**Angular momentum:** Now we can find the magnitude $L$ of the angular momentum with Eq. 11-31:

$$L = I\omega_F = (1.22 \times 10^9 \text{ kg} \cdot \text{m}^2)(0.0524 \text{ rad/s})$$
$$= 6.39 \times 10^7 \text{ kg} \cdot \text{m}^2/\text{s} \approx 6.4 \times 10^7 \text{ kg} \cdot \text{m}^2/\text{s}. \quad \text{(Answer)}$$

(b) If the fully loaded wheel is rotated from rest to $\omega_F$ in a time period $\Delta t_1 = 5.0$ s, what is the magnitude $\tau_{avg}$ of the average net external torque acting on it?

**KEY IDEA** The average net external torque is related

to the change $\Delta L$ in the angular momentum of the loaded wheel by Eq. 11-29 ($\vec{\tau}_{net} = d\vec{L}/dt$).

**Calculation:** Because the wheel rotates about a fixed axis to reach angular speed $\omega_F$ in time period $\Delta t_1$, we can rewrite Eq. 11-29 as $\tau_{avg} = \Delta L/\Delta t_1$. The change $\Delta L$ is

from zero to the answer for part (a). Thus, we have

$$\tau_{avg} = \frac{\Delta L}{\Delta t_1} = \frac{6.39 \times 10^7 \text{ kg} \cdot \text{m}^2/\text{s} - 0}{5.0 \text{ s}}$$

$$\approx 1.3 \times 10^7 \text{ N} \cdot \text{m}. \qquad \text{(Answer)}$$

## 11-11 | Conservation of Angular Momentum

So far we have discussed two powerful conservation laws, the conservation of energy and the conservation of linear momentum. Now we meet a third law of this type, involving the conservation of angular momentum. We start from Eq. 11-29 ($\vec{\tau}_{net} = d\vec{L}/dt$), which is Newton's second law in angular form. If no net external torque acts on the system, this equation becomes $d\vec{L}/dt = 0$, or

$$\vec{L} = \text{a constant} \qquad \text{(isolated system).} \qquad (11\text{-}32)$$

This result, called the **law of conservation of angular momentum,** can also be written as

$$\begin{pmatrix} \text{net angular momentum} \\ \text{at some initial time } t_i \end{pmatrix} = \begin{pmatrix} \text{net angular momentum} \\ \text{at some later time } t_f \end{pmatrix},$$

or $\qquad\qquad \vec{L}_i = \vec{L}_f \qquad \text{(isolated system).} \qquad (11\text{-}33)$

Equations 11-32 and 11-33 tell us:

> If the net external torque acting on a system is zero, the angular momentum $\vec{L}$ of the system remains constant, no matter what changes take place within the system.

Equations 11-32 and 11-33 are vector equations; as such, they are equivalent to three component equations corresponding to the conservation of angular momentum in three mutually perpendicular directions. Depending on the torques acting on a system, the angular momentum of the system might be conserved in only one or two directions but not in all directions:

> If the component of the net *external* torque on a system along a certain axis is zero, then the component of the angular momentum of the system along that axis cannot change, no matter what changes take place within the system.

We can apply this law to the isolated body in Fig. 11-15, which rotates around the $z$ axis. Suppose that the initially rigid body somehow redistributes its mass relative to that rotation axis, changing its rotational inertia about that axis. Equations 11-32 and 11-33 state that the angular momentum of the body cannot change. Substituting Eq. 11-31 (for the angular momentum along the rotational axis) into Eq. 11-33, we write this conservation law as

$$I_i\omega_i = I_f\omega_f. \qquad (11\text{-}34)$$

Here the subscripts refer to the values of the rotational inertia $I$ and angular speed $\omega$ before and after the redistribution of mass.

Like the other two conservation laws that we have discussed, Eqs. 11-32 and 11-33 hold beyond the limitations of Newtonian mechanics. They hold for particles whose speeds approach that of light (where the theory of special relativity reigns), and they remain true in the world of subatomic particles (where quantum physics reigns). No exceptions to the law of conservation of angular momentum have ever been found.

We now discuss four examples involving this law.

**1. The spinning volunteer** Figure 11-17 shows a student seated on a stool that

**FIG. 11-17** (*a*) The student has a relatively large rotational inertia about the rotation axis and a relatively small angular speed. (*b*) By decreasing his rotational inertia, the student automatically increases his angular speed. The angular momentum $\vec{L}$ of the rotating system remains unchanged.

can rotate freely about a vertical axis. The student, who has been set into rotation at a modest initial angular speed $\omega_i$, holds two dumbbells in his outstretched hands. His angular momentum vector $\vec{L}$ lies along the vertical rotation axis, pointing upward.

The instructor now asks the student to pull in his arms; this action reduces his rotational inertia from its initial value $I_i$ to a smaller value $I_f$ because he moves mass closer to the rotation axis. His rate of rotation increases markedly, from $\omega_i$ to $\omega_f$. The student can then slow down by extending his arms once more, moving the dumbbells outward.

No net external torque acts on the system consisting of the student, stool, and dumbbells. Thus, the angular momentum of that system about the rotation axis must remain constant, no matter how the student maneuvers the dumbbells. In Fig. 11-17a, the student's angular speed $\omega_i$ is relatively low and his rotational inertia $I_i$ is relatively high. According to Eq. 11-34, his angular speed in Fig. 11-17b must be greater to compensate for the decreased $I_f$.

2. ***The springboard diver*** Figure 11-18 shows a diver doing a forward one-and-a-half-somersault dive. As you should expect, her center of mass follows a parabolic path. She leaves the springboard with a definite angular momentum $\vec{L}$ about an axis through her center of mass, represented by a vector pointing into the plane of Fig. 11-18, perpendicular to the page. When she is in the air, no net external torque acts on her about her center of mass, so her angular momentum about her center of mass cannot change. By pulling her arms and

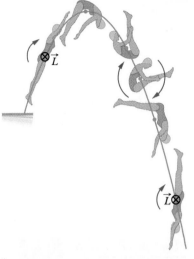

FIG. 11-18 The diver's angular momentum $\vec{L}$ is constant throughout the dive, being represented by the tail ⊗ of an arrow that is perpendicular to the plane of the figure. Note also that her center of mass (see the dots) follows a parabolic path.

FIG. 11-19 Windmill motion of the arms during a long jump helps maintain body orientation for a proper landing.

legs into the closed *tuck position,* she can considerably reduce her rotational inertia about the same axis and thus, according to Eq. 11-34, considerably increase her angular speed. Pulling out of the tuck position (into the *open layout position*) at the end of the dive increases her rotational inertia and thus slows her rotation rate so she can enter the water with little splash. Even in a more complicated dive involving both twisting and somersaulting, the angular momentum of the diver must be conserved, in both magnitude *and* direction, throughout the dive.

3. ***Long jump*** When an athlete takes off from the ground in a running long jump, the forces on the launching foot give the athlete an angular momentum with a forward rotation around a horizontal axis. Such rotation would not allow the jumper to land properly: In the landing, the legs should be together and extended forward at an angle so that the heels mark the sand at the greatest distance. Once airborne, the angular momentum cannot change (it is conserved) because no external torque acts to change it. However, the jumper can shift most of the angular momentum to the arms by rotating them in windmill fashion (Fig. 11-19). Then the body remains upright and in the proper orientation for landing.

4. ***Tour jeté*** In a tour jeté, a ballet performer leaps with a small twisting motion on the floor with one foot while holding the other leg perpendicular to the body (Fig. 11-20a). The angular speed is so small that it may not be perceptible to the audience. As the performer ascends, the outstretched leg is brought down and the other leg is brought up, with both ending up at angle $\theta$ to the

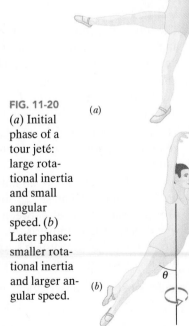

FIG. 11-20 (a) Initial phase of a tour jeté: large rotational inertia and small angular speed. (b) Later phase: smaller rotational inertia and larger angular speed.

body (Fig. 11-20b). The motion is graceful, but it also serves to increase the rotation because bringing in the initially outstretched leg decreases the performer's rotational inertia. Since no external torque acts on the airborne performer, the angular momentum cannot change. Thus, with a decrease in rotational inertia, the angular speed must increase. When the jump is well executed, the performer seems to suddenly begin to spin and rotates 180° before the initial leg orientations are reversed in preparation for the landing. Once a leg is again outstretched, the rotation seems to vanish.

✓ **CHECKPOINT 7**    A rhinoceros beetle rides the rim of a small disk that rotates like a merry-go-round. If the beetle crawls toward the center of the disk, do the following (each relative to the central axis) increase, decrease, or remain the same for the beetle–disk system: (a) rotational inertia, (b) angular momentum, and (c) angular speed?

**Sample Problem** | **11-7**

Figure 11-21a shows a student, again sitting on a stool that can rotate freely about a vertical axis. The student, initially at rest, is holding a bicycle wheel whose rim is loaded with lead and whose rotational inertia $I_{wh}$ about its central axis is 1.2 kg·m². The wheel is rotating at an angular speed $\omega_{wh}$ of 3.9 rev/s; as seen from overhead, the rotation is counterclockwise. The axis of the wheel is vertical, and the angular momentum $\vec{L}_{wh}$ of the wheel points vertically upward. The student now inverts the wheel (Fig. 11-21b) so that, as seen from overhead, it is rotating clockwise. Its angular momentum is now $-\vec{L}_{wh}$. The inversion results in the student, the stool, and the wheel's center rotating together as a composite rigid body about the stool's rotation axis, with rotational inertia $I_b = 6.8$ kg·m². (The fact that the wheel is also rotating about its center does not affect the mass distribution of this composite body; thus, $I_b$ has the same value whether or not the wheel rotates.) With what angular speed $\omega_b$ and in what direction does the composite body rotate after the inversion of the wheel?

**KEY IDEAS**

1. The angular speed $\omega_b$ we seek is related to the final angular momentum $\vec{L}_b$ of the composite body about the stool's rotation axis by Eq. 11-31 ($L = I\omega$).

2. The initial angular speed $\omega_{wh}$ of the wheel is related to the angular momentum $\vec{L}_{wh}$ of the wheel's rotation about its center by the same equation.

3. The vector addition of $\vec{L}_b$ and $\vec{L}_{wh}$ gives the total angular momentum $\vec{L}_{tot}$ of the system of student, stool, and wheel.

4. As the wheel is inverted, no net *external* torque acts on that system to change $\vec{L}_{tot}$ about any vertical axis. (Torques due to forces between the student and the wheel as the student inverts the wheel are *internal* to the system.) So, the system's total angular momentum is conserved about any vertical axis.

*Calculations:* The conservation of $\vec{L}_{tot}$ is represented

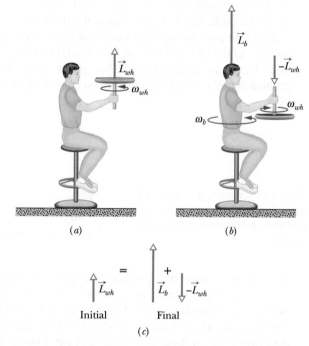

**FIG. 11-21**   (a) A student holds a bicycle wheel rotating around a vertical axis. (b) The student inverts the wheel, setting himself into rotation. (c) The net angular momentum of the system must remain the same in spite of the inversion.

with vectors in Fig. 11-21c. We can also write this conservation in terms of components along a vertical axis as

$$L_{b,f} + L_{wh,f} = L_{b,i} + L_{wh,i}, \qquad (11-35)$$

where $i$ and $f$ refer to the initial state (before inversion of the wheel) and the final state (after inversion). Because inversion of the wheel inverted the angular momentum vector of the wheel's rotation, we substitute $-L_{wh,i}$ for $L_{wh,f}$. Then, if we set $L_{b,i} = 0$ (because the student, the stool, and the wheel's center were initially at rest), Eq. 11-35 yields

$$L_{b,f} = 2L_{wh,i}.$$

Using Eq. 11-31, we next substitute $I_b\omega_b$ for $L_{b,f}$ and

$I_{wh}\omega_{wh}$ for $L_{wh,i}$ and solve for $\omega_b$, finding

$$\omega_b = \frac{2I_{wh}}{I_b}\,\omega_{wh}$$

$$= \frac{(2)(1.2\ \text{kg}\cdot\text{m}^2)(3.9\ \text{rev/s})}{6.8\ \text{kg}\cdot\text{m}^2} = 1.4\ \text{rev/s.}\quad\text{(Answer)}$$

This positive result tells us that the student rotates counterclockwise about the stool axis as seen from overhead. If the student wishes to stop rotating, he has only to invert the wheel once more.

---

**Sample Problem    11-8**

In Fig. 11-22, a cockroach with mass $m$ rides on a disk of mass $6.00m$ and radius $R$. The disk rotates like a merry-go-round around its central axis at angular speed $\omega_i = 1.50\ \text{rad/s}$. The cockroach is initially at radius $r = 0.800R$, but then it crawls out to the rim of the disk. Treat the cockroach as a particle. What then is the angular speed?

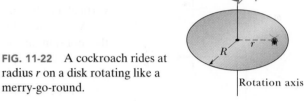

**FIG. 11-22** A cockroach rides at radius $r$ on a disk rotating like a merry-go-round.

**KEY IDEAS** (1) The cockroach's crawl changes the mass distribution (and thus the rotational inertia) of the cockroach–disk system. (2) The angular momentum of the system does not change because there is no external torque to change it. (The forces and torques due to the cockroach's crawl are internal to the system.) (3) The magnitude of the angular momentum of a rigid body or a particle is given by Eq. 11-31 ($L = I\omega$).

**Calculations:** We want to find the final angular speed. Our key is to equate the final angular momentum $L_f$ to the initial angular momentum $L_i$, because both involve angular speed. They also involve rotational inertia $I$. So, let's start by finding the rotational inertia of the system of cockroach and disk before and after the crawl.

The rotational inertia of a disk rotating about its central axis is given by Table 10-2c as $\frac{1}{2}MR^2$. Substituting $6.00m$ for the mass $M$, our disk here has rotational inertia

$$I_d = 3.00mR^2. \tag{11-36}$$

(We don't have values for $m$ and $R$, but we shall continue with physics courage.)

From Eq. 10-33, we know that the rotational inertia of the cockroach (a particle) is equal to $mr^2$. Substituting the cockroach's initial radius ($r = 0.800R$) and final

radius ($r = R$), we find that its initial rotational inertia about the rotation axis is

$$I_{ci} = 0.64mR^2 \tag{11-37}$$

and its final rotational inertia about the rotation axis is

$$I_{cf} = mR^2. \tag{11-38}$$

So, the cockroach–disk system initially has the rotational inertia

$$I_i = I_d + I_{ci} = 3.64mR^2, \tag{11-39}$$

and finally has the rotational inertia

$$I_f = I_d + I_{cf} = 4.00mR^2. \tag{11-40}$$

Next, we use Eq. 11-31 ($L = I\omega$) to write the fact that the system's final angular momentum $L_f$ is equal to the system's initial angular momentum $L_i$:

$$I_f\omega_f = I_i\omega_i$$

or     $4.00mR^2\omega_f = 3.64mR^2(1.50\ \text{rad/s}).$

After canceling the unknowns $m$ and $R$, we come to

$$\omega_f = 1.37\ \text{rad/s.}\quad\text{(Answer)}$$

Note that the angular speed decreased because part of the mass moved outward from the rotation axis.

---

# 11-12 | Precession of a Gyroscope

A simple gyroscope consists of a wheel fixed to a shaft and free to spin about the axis of the shaft. If one end of the shaft of a *nonspinning* gyroscope is placed on a support as in Fig. 11-23a and the gyroscope is released, the gyroscope falls by rotating downward about the tip of the support. Since the fall involves rotation, it is governed by Newton's second law in angular form, which is given by Eq. 11-29:

$$\vec{\tau} = \frac{d\vec{L}}{dt}. \tag{11-41}$$

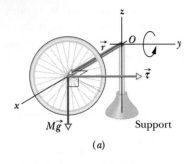

(a)

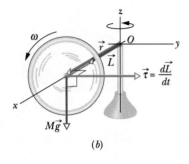

(b)

Circular path
taken by head
of $\vec{L}$ vector

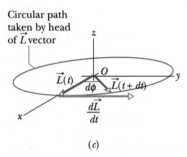

(c)

**FIG. 11-23** (a) A nonspinning gyroscope falls by rotating in an $xz$ plane because of torque $\vec{\tau}$. (b) A rapidly spinning gyroscope, with angular momentum $\vec{L}$, precesses around the $z$ axis. Its precessional motion is in the $xy$ plane. (c) The change $d\vec{L}/dt$ in angular momentum leads to a rotation of $\vec{L}$ about O.

This equation tells us that the torque causing the downward rotation (the fall) changes the angular momentum $\vec{L}$ of the gyroscope from its initial value of zero. The torque $\vec{\tau}$ is due to the gravitational force $M\vec{g}$ acting at the gyroscope's center of mass, which we take to be at the center of the wheel. The moment arm relative to the support tip, located at O in Fig. 11-23a, is $\vec{r}$. The magnitude of $\vec{\tau}$ is

$$\tau = Mgr \sin 90° = Mgr \tag{11-42}$$

(because the angle between $M\vec{g}$ and $\vec{r}$ is 90°), and its direction is as shown in Fig. 11-23a.

A rapidly spinning gyroscope behaves differently. Assume it is released with the shaft angled slightly upward. It first rotates slightly downward but then, while it is still spinning about its shaft, it begins to rotate horizontally about a vertical axis through support point O in a motion called **precession.**

Why does the spinning gyroscope stay aloft instead of falling over like the nonspinning gyroscope? The clue is that when the spinning gyroscope is released, the torque due to $M\vec{g}$ must change not an initial angular momentum of zero but rather some already existing nonzero angular momentum due to the spin.

To see how this nonzero initial angular momentum leads to precession, we first consider the angular momentum $\vec{L}$ of the gyroscope due to its spin. To simplify the situation, we assume the spin rate is so rapid that the angular momentum due to precession is negligible relative to $\vec{L}$. We also assume the shaft is horizontal when precession begins, as in Fig. 11-23b. The magnitude of $\vec{L}$ is given by Eq. 11-31:

$$L = I\omega, \tag{11-43}$$

where $I$ is the rotational moment of the gyroscope about its shaft and $\omega$ is the angular speed at which the wheel spins about the shaft. The vector $\vec{L}$ points along the shaft, as in Fig. 11-23b. Since $\vec{L}$ is parallel to $\vec{r}$, torque $\vec{\tau}$ must be perpendicular to $\vec{L}$.

According to Eq. 11-41, torque $\vec{\tau}$ causes an incremental change $d\vec{L}$ in the angular momentum of the gyroscope in an incremental time interval $dt$; that is,

$$d\vec{L} = \vec{\tau}\, dt. \tag{11-44}$$

However, for a *rapidly spinning* gyroscope, the magnitude of $\vec{L}$ is fixed by Eq. 11-43. Thus the torque can change only the direction of $\vec{L}$, not its magnitude.

From Eq. 11-44 we see that the direction of $d\vec{L}$ is in the direction of $\vec{\tau}$, perpendicular to $\vec{L}$. The only way that $\vec{L}$ can be changed in the direction of $\vec{\tau}$ without the magnitude $L$ being changed is for $\vec{L}$ to rotate around the $z$ axis as shown in Fig. 11-23c. $\vec{L}$ maintains its magnitude, the head of the $\vec{L}$ vector follows a circular path, and $\vec{\tau}$ is always tangent to that path. Since $\vec{L}$ must always point along the shaft, the shaft must rotate about the $z$ axis in the direction of $\vec{\tau}$. Thus we have precession. Because the spinning gyroscope must obey Newton's law in angular form in response to any change in its initial angular momentum, it must precess instead of merely toppling over.

We can find the **precession rate** $\Omega$ by first using Eqs. 11-44 and 11-42 to get the magnitude of $d\vec{L}$:

$$dL = \tau\, dt = Mgr\, dt. \tag{11-45}$$

As $\vec{L}$ changes by an incremental amount in an incremental time interval $dt$, the shaft and $\vec{L}$ precess around the $z$ axis through incremental angle $d\phi$. (In Fig. 11-23c, angle $d\phi$ is exaggerated for clarity.) With the aid of Eqs. 11-43 and 11-45, we find that $d\phi$ is given by

$$d\phi = \frac{dL}{L} = \frac{Mgr\, dt}{I\omega}.$$

Dividing this expression by $dt$ and setting the rate $\Omega = d\phi/dt$, we obtain

$$\Omega = \frac{Mgr}{I\omega} \qquad \text{(precession rate).} \tag{11-46}$$

This result is valid under the assumption that the spin rate $\omega$ is rapid. Note that $\Omega$ decreases as $\omega$ is increased. Note also that there would be no precession if the gravitational force $M\vec{g}$ did not act on the gyroscope, but because $I$ is a function of $M$, mass cancels from Eq. 11-46; thus $\Omega$ is independent of the mass.

Equation 11-46 also applies if the shaft of a spinning gyroscope is at an angle to the horizontal. It holds as well for a spinning top, which is essentially a spinning gyroscope at an angle to the horizontal.

## REVIEW & SUMMARY

**Rolling Bodies**   For a wheel of radius $R$ rolling smoothly,

$$v_{com} = \omega R,$$   (11-2)

where $v_{com}$ is the linear speed of the wheel's center of mass and $\omega$ is the angular speed of the wheel about its center. The wheel may also be viewed as rotating instantaneously about the point $P$ of the "road" that is in contact with the wheel. The angular speed of the wheel about this point is the same as the angular speed of the wheel about its center. The rolling wheel has kinetic energy

$$K = \tfrac{1}{2}I_{com}\omega^2 + \tfrac{1}{2}Mv_{com}^2,$$   (11-5)

where $I_{com}$ is the rotational moment of the wheel about its center of mass and $M$ is the mass of the wheel. If the wheel is being accelerated but is still rolling smoothly, the acceleration of the center of mass $\vec{a}_{com}$ is related to the angular acceleration $\alpha$ about the center with

$$a_{com} = \alpha R.$$   (11-6)

If the wheel rolls smoothly down a ramp of angle $\theta$, its acceleration along an $x$ axis extending up the ramp is

$$a_{com,x} = -\frac{g \sin \theta}{1 + I_{com}/MR^2}.$$   (11-10)

**Torque as a Vector**   In three dimensions, *torque* $\vec{\tau}$ is a vector quantity defined relative to a fixed point (usually an origin); it is

$$\vec{\tau} = \vec{r} \times \vec{F},$$   (11-14)

where $\vec{F}$ is a force applied to a particle and $\vec{r}$ is a position vector locating the particle relative to the fixed point. The magnitude of $\vec{\tau}$ is given by

$$\tau = rF \sin \phi = rF_\perp = r_\perp F,$$   (11-15, 11-16, 11-17)

where $\phi$ is the angle between $\vec{F}$ and $\vec{r}$, $F_\perp$ is the component of $\vec{F}$ perpendicular to $\vec{r}$, and $r_\perp$ is the moment arm of $\vec{F}$. The direction of $\vec{\tau}$ is given by the right-hand rule.

**Angular Momentum of a Particle**   The *angular momentum* $\vec{\ell}$ of a particle with linear momentum $\vec{p}$, mass $m$, and linear velocity $\vec{v}$ is a vector quantity defined relative to a fixed point (usually an origin) as

$$\vec{\ell} = \vec{r} \times \vec{p} = m(\vec{r} \times \vec{v}).$$   (11-18)

The magnitude of $\vec{\ell}$ is given by

$$\ell = rmv \sin \phi$$   (11-19)
$$= rp_\perp = rmv_\perp$$   (11-20)
$$= r_\perp p = r_\perp mv,$$   (11-21)

where $\phi$ is the angle between $\vec{r}$ and $\vec{p}$, $p_\perp$ and $v_\perp$ are the components of $\vec{p}$ and $\vec{v}$ perpendicular to $\vec{r}$, and $r_\perp$ is the perpendicular distance between the fixed point and the extension of $\vec{p}$. The direction of $\vec{\ell}$ is given by the right-hand rule for cross products.

**Newton's Second Law in Angular Form**   Newton's second law for a particle can be written in angular form as

$$\vec{\tau}_{net} = \frac{d\vec{\ell}}{dt},$$   (11-23)

where $\vec{\tau}_{net}$ is the net torque acting on the particle and $\vec{\ell}$ is the angular momentum of the particle.

**Angular Momentum of a System of Particles**   The angular momentum $\vec{L}$ of a system of particles is the vector sum of the angular momenta of the individual particles:

$$\vec{L} = \vec{\ell}_1 + \vec{\ell}_2 + \cdots + \vec{\ell}_n = \sum_{i=1}^{n} \vec{\ell}_i.$$   (11-26)

The time rate of change of this angular momentum is equal to the net external torque on the system (the vector sum of the torques due to interactions of the particles of the system with particles external to the system):

$$\vec{\tau}_{net} = \frac{d\vec{L}}{dt} \qquad \text{(system of particles)}.$$   (11-29)

**Angular Momentum of a Rigid Body**   For a rigid body rotating about a fixed axis, the component of its angular momentum parallel to the rotation axis is

$$L = I\omega \qquad \text{(rigid body, fixed axis)}.$$   (11-31)

**Conservation of Angular Momentum**   The angular momentum $\vec{L}$ of a system remains constant if the net external torque acting on the system is zero:

$$\vec{L} = \text{a constant} \qquad \text{(isolated system)}$$   (11-32)

or   $$\vec{L}_i = \vec{L}_f \qquad \text{(isolated system)}.$$   (11-33)

This is the **law of conservation of angular momentum.**

**Precession of a Gyroscope**   A spinning gyroscope can precess about a vertical axis through its support at the rate

$$\Omega = \frac{Mgr}{I\omega},$$   (11-46)

where $M$ is the gyroscope's mass, $r$ is the moment arm, $I$ is the rotational inertia, and $\omega$ is the spin rate.

## QUESTIONS

**1** In Fig. 11-24, three forces of the same magnitude are applied to a particle at the origin ($\vec{F}_1$ acts directly into the plane of the figure). Rank the forces according to the magnitudes of the torques they create about (a) point $P_1$, (b) point $P_2$, and (c) point $P_3$, greatest first.

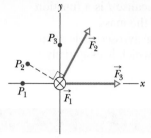

**FIG. 11-24** Question 1.

**2** The position vector $\vec{r}$ of a particle relative to a certain point has a magnitude of 3 m, and the force $\vec{F}$ on the particle has a magnitude of 4 N. What is the angle between the directions of $\vec{r}$ and $\vec{F}$ if the magnitude of the associated torque equals (a) zero and (b) 12 N·m?

**3** What happens to the initially stationary yo-yo in Fig. 11-25 if you pull it via its string with (a) force $\vec{F}_2$ (the line of action passes through the point of contact on the table, as indicated), (b) force $\vec{F}_1$ (the line of action passes above the point of contact), and (c) force $\vec{F}_3$ (the line of action passes to the right of the point of contact)?

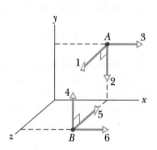

**FIG. 11-25** Question 3.

**4** Figure 11-26 shows two particles $A$ and $B$ at $xyz$ coordinates (1 m, 1 m, 0) and (1 m, 0, 1 m). Acting on each particle are three numbered forces, all of the same magnitude and each directed parallel to an axis. (a) Which of the forces produce a torque about the origin that is directed parallel to $y$? (b) Rank the forces according to the magnitudes of the torques they produce on the particles about the origin, greatest first.

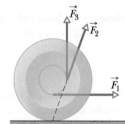

**FIG. 11-26** Question 4.

**5** Figure 11-27 shows three particles of the same mass and the same constant speed moving as indicated by the velocity vectors. Points $a$, $b$, $c$, and $d$ form a square, with point $e$ at the center. Rank the points according to the magnitude of the net angular momentum of the three-particle system when measured about the points, greatest first.

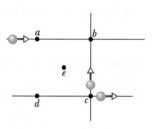

**FIG. 11-27** Question 5.

**6** Figure 11-28 shows a particle moving at constant velocity $\vec{v}$ and five points with their $xy$ coordinates. Rank the points according to the magnitude of the angular momentum of the particle measured about them, greatest first.

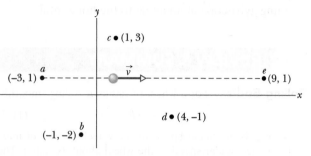

**FIG. 11-28** Question 6.

**7** Figure 11-29 gives the angular momentum magnitude $L$ of a wheel versus time $t$. Rank the four lettered time intervals according to the magnitude of the torque acting on the wheel, greatest first.

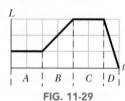

**FIG. 11-29** Question 7.

**8** The angular momenta $\ell(t)$ of a particle in four situations are (1) $\ell = 3t + 4$; (2) $\ell = -6t^2$; (3) $\ell = 2$; (4) $\ell = 4/t$. In which situation is the net torque on the particle (a) zero, (b) positive and constant, (c) negative and increasing in magnitude ($t > 0$), and (d) negative and decreasing in magnitude ($t > 0$)?

**9** A rhinoceros beetle rides the rim of a horizontal disk rotating counterclockwise like a merry-go-round. If the beetle then walks along the rim in the direction of the rotation, will the magnitudes of the following quantities (each measured about the rotation axis) increase, decrease, or remain the same (the disk is still rotating in the counterclockwise direction): (a) the angular momentum of the beetle–disk system, (b) the angular momentum and angular velocity of the beetle, and (c) the angular momentum and angular velocity of the disk? (d) What are your answers if the beetle walks in the direction opposite the rotation?

**10** Figure 11-30 shows an overhead view of a rectangular slab that can spin like a merry-go-round about its center at $O$. Also shown are seven paths along which wads of bubble gum can be thrown (all with the same speed and mass) to stick onto the stationary slab. (a) Rank the paths according to the angular speed that the slab (and gum) will have after the gum sticks, greatest first. (b) For which paths will the angular momentum of the slab (and gum) about $O$ be negative from the view of Fig. 11-30?

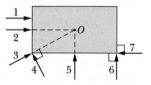

**FIG. 11-30** Question 10.

# PROBLEMS

**GO** Tutoring problem available (at instructor's discretion) in *WileyPLUS* and WebAssign

**SSM** Worked-out solution available in Student Solutions Manual   **WWW** Worked-out solution is at ⎯⎯⎯

**• – •••** Number of dots indicates level of problem difficulty   **ILW** Interactive solution is at ⎯⎯⎯ http://www.wiley.com/college/halliday

Additional information available in *The Flying Circus of Physics* and at flyingcircusofphysics.com

### sec. 11-2   Rolling as Translation and Rotation Combined

•1   A car travels at 80 km/h on a level road in the positive direction of an *x* axis. Each tire has a diameter of 66 cm. Relative to a woman riding in the car and in unit-vector notation, what are the velocity $\vec{v}$ at the (a) center, (b) top, and (c) bottom of the tire and the magnitude *a* of the acceleration at the (d) center, (e) top, and (f) bottom of each tire? Relative to a hitchhiker sitting next to the road and in unit-vector notation, what are the velocity $\vec{v}$ at the (g) center, (h) top, and (i) bottom of the tire and the magnitude *a* of the acceleration at the (j) center, (k) top, and (l) bottom of each tire?

•2   An automobile traveling at 80.0 km/h has tires of 75.0 cm diameter. (a) What is the angular speed of the tires about their axles? (b) If the car is brought to a stop uniformly in 30.0 complete turns of the tires (without skidding), what is the magnitude of the angular acceleration of the wheels? (c) How far does the car move during the braking?

### sec. 11-4   The Forces of Rolling

•3   A 1000 kg car has four 10 kg wheels. When the car is moving, what fraction of its total kinetic energy is due to rotation of the wheels about their axles? Assume that the wheels have the same rotational inertia as uniform disks of the same mass and size. Why do you not need to know the radius of the wheels?   **ILW**

•4   A uniform solid sphere rolls down an incline. (a) What must be the incline angle if the linear acceleration of the center of the sphere is to have a magnitude of 0.10*g*? (b) If a frictionless block were to slide down the incline at that angle, would its acceleration magnitude be more than, less than, or equal to 0.10*g*? Why?

•5   A 140 kg hoop rolls along a horizontal floor so that the hoop's center of mass has a speed of 0.150 m/s. How much work must be done on the hoop to stop it?   **SSM**

••6   A hollow sphere of radius 0.15 m, with rotational inertia $I = 0.040$ kg·m² about a line through its center of mass, rolls without slipping up a surface inclined at 30° to the horizontal. At a certain initial position, the sphere's total kinetic energy is 20 J. (a) How much of this initial kinetic energy is rotational? (b) What is the speed of the center of mass of the sphere at the initial position? When the sphere has moved 1.0 m up the incline from its initial position, what are (c) its total kinetic energy and (d) the speed of its center of mass?

••7   In Fig. 11-31, a constant horizontal force $\vec{F}_{app}$ of magnitude 10 N is applied to a wheel of mass 10 kg and radius 0.30 m. The wheel rolls smoothly on the horizontal surface, and the acceleration of its center of mass has magnitude

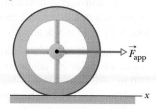

**FIG. 11-31**   Problem 7.

0.60 m/s². (a) In unit-vector notation, what is the frictional force on the wheel? (b) What is the rotational inertia of the wheel about the rotation axis through its center of mass?

••8   In Fig. 11-32, a solid brass ball of mass 0.280 g will roll smoothly along a loop-the-loop track when released from rest along the straight section. The circular loop has radius $R = 14.0$ cm, and the ball has radius $r \ll R$. (a) What is *h* if the ball is on the verge of leaving the track when it reaches the top of the loop? If the ball is released at height $h = 6.00R$, what are the (b) magnitude and (c) direction of the horizontal force component acting on the ball at point *Q*?

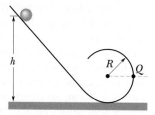

**FIG. 11-32**   Problem 8.

••9   In Fig. 11-33, a solid cylinder of radius 10 cm and mass 12 kg starts from rest and rolls without slipping a distance $L = 6.0$ m down a roof that is inclined at the angle $\theta = 30°$. (a) What is the angular speed of the cylinder about its center as it leaves the roof? (b) The roof's edge is at height $H = 5.0$ m. How far horizontally from the roof's edge does the cylinder hit the level ground?   **ILW**

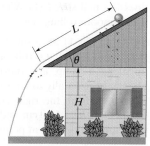

**FIG. 11-33**   Problem 9.

••10   Figure 11-34 gives the speed *v* versus time *t* for a 0.500 kg object of radius 6.00 cm that rolls smoothly down a 30° ramp. The scale on the velocity axis is set by $v_s = 4.0$ m/s. What is the rotational inertia of the object?

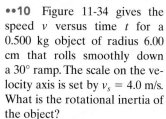

**FIG. 11-34**   Problem 10.

••11   In Fig. 11-35, a solid ball rolls smoothly from rest (starting at height $H = 6.0$ m) until it leaves the horizontal section at the end of the track, at height $h = 2.0$ m. How far horizontally from point *A* does the ball hit the floor?   **GO**

••12   Figure 11-36 shows the potential energy $U(x)$ of a solid ball that can roll along an *x* axis. The scale on the *U* axis is set by $U_s = 100$ J. The ball is uniform, rolls smoothly, and has a mass of 0.400 kg. It is released at $x = 7.0$ m headed in the negative direction of the *x* axis with a mechanical energy of 75 J.

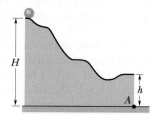

**FIG. 11-35**   Problem 11.

each $x = 0$ here, and if its turning stead, it is e direction is released at $x = 7.0$ m ...... J. (b) If the ball can reach $x = 13$ m, what is its speed there, and if it cannot, what is its turning point?

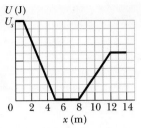

FIG. 11-36  Problem 12.

**••13** A bowler throws a bowling ball of radius $R = 11$ cm along a lane. The ball (Fig. 11-37) slides on the lane with initial speed $v_{com,0} = 8.5$ m/s and initial angular speed $\omega_0 = 0$.

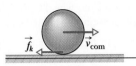

FIG. 11-37  Problem 13.

The coefficient of kinetic friction between the ball and the lane is 0.21. The kinetic frictional force $\vec{f}_k$ acting on the ball causes a linear acceleration of the ball while producing a torque that causes an angular acceleration of the ball. When speed $v_{com}$ has decreased enough and angular speed $\omega$ has increased enough, the ball stops sliding and then rolls smoothly. (a) What then is $v_{com}$ in terms of $\omega$? During the sliding, what are the ball's (b) linear acceleration and (c) angular acceleration? (d) How long does the ball slide? (e) How far does the ball slide? (f) What is the linear speed of the ball when smooth rolling begins?

**••14** In Fig. 11-38, a small, solid, uniform ball is to be shot from point $P$ so that it rolls smoothly along a horizontal path, up along a ramp, and onto a plateau. Then it leaves the plateau horizontally to land on a game board, at a horizontal distance $d$ from the right edge of the plateau. The vertical heights are $h_1 = 5.00$ cm and $h_2 = 1.60$ cm. With what speed must the ball be shot at point $P$ for it to land at $d = 6.00$ cm?

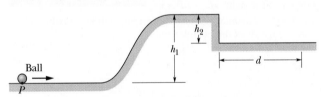

FIG. 11-38  Problem 14.

**•••15** *Nonuniform ball.* In Fig. 11-39, a ball of mass $M$ and radius $R$ rolls smoothly from rest down a ramp and onto a circular loop of radius 0.48 m. The initial height of the ball is $h = 0.36$ m. At the loop bottom, the magnitude of the normal force on the ball is $2.00Mg$. The ball consists of an outer spherical shell (of a certain uniform density) that is glued to a central sphere (of a different uniform density). The rotational inertia of the ball can be expressed in the general form $I = \beta MR^2$, but $\beta$ is not 0.4 as it is for a ball of uniform density. Determine $\beta$.

FIG. 11-39  Problem 15.

**•••16** *Nonuniform cylindrical object.* In Fig. 11-40, a cylindrical object of mass $M$ and radius $R$ rolls smoothly from rest down a ramp and onto a horizontal section. From there it rolls off the ramp and onto the floor, landing a horizontal dis-

tance $d = 0.506$ m from the end of the ramp. The initial height of the object is $H = 0.90$ m; the end of the ramp is at height $h = 0.10$ m. The object consists of an outer cylindrical shell (of a certain uniform density) that is glued to a central cylinder (of a different uniform density). The rotational inertia of the object can be expressed in the general form $I = \beta MR^2$, but $\beta$ is not 0.5 as it is for a cylinder of uniform density. Determine $\beta$.

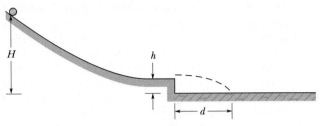

FIG. 11-40  Problem 16.

**sec. 11-5  The Yo-Yo**

**•17** A yo-yo has a rotational inertia of 950 g·cm² and a mass of 120 g. Its axle radius is 3.2 mm, and its string is 120 cm long. The yo-yo rolls from rest down to the end of the string. (a) What is the magnitude of its linear acceleration? (b) How long does it take to reach the end of the string? As it reaches the end of the string, what are its (c) linear speed, (d) translational kinetic energy, (e) rotational kinetic energy, and (f) angular speed? **SSM**

**•18** In 1980, over San Francisco Bay, a large yo-yo was released from a crane. The 116 kg yo-yo consisted of two uniform disks of radius 32 cm connected by an axle of radius 3.2 cm. What was the magnitude of the acceleration of the yo-yo during (a) its fall and (b) its rise? (c) What was the tension in the cord on which it rolled? (d) Was that tension near the cord's limit of 52 kN? Suppose you build a scaled-up version of the yo-yo (same shape and materials but larger). (e) Will the magnitude of your yo-yo's acceleration as it falls be greater than, less than, or the same as that of the San Francisco yo-yo? (f) How about the tension in the cord?

**sec. 11-6  Torque Revisited**

**•19** In unit-vector notation, what is the torque about the origin on a particle located at coordinates $(0, -4.0 \text{ m}, 3.0 \text{ m})$ if that torque is due to (a) force $\vec{F}_1$ with components $F_{1x} = 2.0$ N, $F_{1y} = F_{1z} = 0$, and (b) force $\vec{F}_2$ with components $F_{2x} = 0$, $F_{2y} = 2.0$ N, $F_{2z} = 4.0$ N?

**•20** A plum is located at coordinates $(-2.0 \text{ m}, 0, 4.0 \text{ m})$. In unit-vector notation, what is the torque about the origin on the plum if that torque is due to a force $\vec{F}$ whose only component is (a) $F_x = 6.0$ N, (b) $F_x = -6.0$ N, (c) $F_z = 6.0$ N, and (d) $F_z = -6.0$ N?

**•21** In unit-vector notation, what is the net torque about the origin on a flea located at coordinates $(0, -4.0 \text{ m}, 5.0 \text{ m})$ when forces $\vec{F}_1 = (3.0 \text{ N})\hat{k}$ and $\vec{F}_2 = (-2.0 \text{ N})\hat{j}$ act on the flea?

**••22** In unit-vector notation, what is the torque about the origin on a jar of jalapeño peppers located at coordinates $(3.0 \text{ m}, -2.0 \text{ m}, 4.0 \text{ m})$ due to (a) force $\vec{F}_1 = (3.0 \text{ N})\hat{i} - (4.0 \text{ N})\hat{j} + (5.0 \text{ N})\hat{k}$, (b) force $\vec{F}_2 = (-3.0 \text{ N})\hat{i} - (4.0 \text{ N})\hat{j} - (5.0 \text{ N})\hat{k}$, and (c) the vector sum of $\vec{F}_1$ and $\vec{F}_2$? (d) Repeat part

(c) for the torque about the point with coordinates (3.0 m, 2.0 m, 4.0 m).

**••23** Force $\vec{F} = (-8.0\text{ N})\hat{i} + (6.0\text{ N})\hat{j}$ acts on a particle with position vector $\vec{r} = (3.0\text{ m})\hat{i} + (4.0\text{ m})\hat{j}$. What are (a) the torque on the particle about the origin, in unit-vector notation, and (b) the angle between the directions of $\vec{r}$ and $\vec{F}$? **SSM**

**••24** A particle moves through an *xyz* coordinate system while a force acts on the particle. When the particle has the position vector $\vec{r} = (2.00\text{ m})\hat{i} - (3.00\text{ m})\hat{j} + (2.00\text{ m})\hat{k}$, the force is $\vec{F} = F_x\hat{i} + (7.00\text{ N})\hat{j} - (6.00\text{ N})\hat{k}$ and the corresponding torque about the origin is $\vec{\tau} = (4.00\text{ N}\cdot\text{m})\hat{i} + (2.00\text{ N}\cdot\text{m})\hat{j} - (1.00\text{ N}\cdot\text{m})\hat{k}$. Determine $F_x$.

**••25** Force $\vec{F} = (2.0\text{ N})\hat{i} - (3.0\text{ N})\hat{k}$ acts on a pebble with position vector $\vec{r} = (0.50\text{ m})\hat{j} - (2.0\text{ m})\hat{k}$ relative to the origin. In unit-vector notation, what is the resulting torque on the pebble about (a) the origin and (b) the point (2.0 m, 0, −3.0 m)?

### sec. 11-7 Angular Momentum

**•26** A 2.0 kg particle-like object moves in a plane with velocity components $v_x = 30$ m/s and $v_y = 60$ m/s as it passes through the point with (x, y) coordinates of (3.0, −4.0) m. Just then, in unit-vector notation, what is its angular momentum relative to (a) the origin and (b) the point (−2.0, −2.0) m?

**•27** In the instant of Fig. 11-41, two particles move in an *xy* plane. Particle $P_1$ has mass 6.5 kg and speed $v_1 = 2.2$ m/s, and it is at distance $d_1 = 1.5$ m from point *O*. Particle $P_2$ has mass 3.1 kg and speed $v_2 = 3.6$ m/s, and it is at distance $d_2 = 2.8$ m from point *O*. What are the (a) magnitude and (b) direction of the net angular momentum of the two particles about *O*? **ILW**

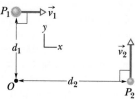

**FIG. 11-41** Problem 27.

**•28** At the instant of Fig. 11-42, a 2.0 kg particle *P* has a position vector $\vec{r}$ of magnitude 3.0 m and angle $\theta_1 = 45°$ and a velocity vector $\vec{v}$ of magnitude 4.0 m/s and angle $\theta_2 = 30°$. Force $\vec{F}$, of magnitude 2.0 N and angle $\theta_3 = 30°$, acts on *P*. All three vectors lie in the *xy* plane. About the origin, what are the (a) magnitude and (b) direction of the angular momentum of *P* and the (c) magnitude and (d) direction of the torque acting on *P*?

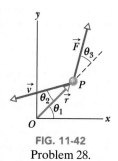

**FIG. 11-42** Problem 28.

**•29** At one instant, force $\vec{F} = 4.0\hat{j}$ N acts on a 0.25 kg object that has position vector $\vec{r} = (2.0\hat{i} - 2.0\hat{k})$ m and velocity vector $\vec{v} = (-5.0\hat{i} + 5.0\hat{k})$ m/s. About the origin and in unit-vector notation, what are (a) the object's angular momentum and (b) the torque acting on the object? **SSM**

**••30** At the instant the displacement of a 2.00 kg object relative to the origin is $\vec{d} = (2.00\text{ m})\hat{i} + (4.00\text{ m})\hat{j} - (3.00\text{ m})\hat{k}$, its velocity is $\vec{v} = -(6.00\text{ m/s})\hat{i} + (3.00\text{ m/s})\hat{j} + (3.00\text{ m/s})\hat{k}$ and it is subject to a force $\vec{F} = (6.00\text{ N})\hat{i} - (8.00\text{ N})\hat{j} + (4.00\text{ N})\hat{k}$. Find (a) the acceleration of the object, (b) the angular momentum of the object about the origin, (c) the torque about the origin acting on the object, and (d) the angle between the velocity of the object and the force acting on the object.

**••31** In Fig. 11-43, a 0.400 kg ball is shot directly upward at initial speed 40.0 m/s. What is its angular momentum about *P*, 2.00 m horizontally from the launch point, when the ball is (a) at maximum height and (b) halfway back to the ground? What is the torque on the ball about *P* due to the gravitational force when the ball is (c) at maximum height and (d) halfway back to the ground?

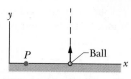

**FIG. 11-43** Problem 31.

### sec. 11-8 Newton's Second Law in Angular Form

**•32** A particle is to move in an *xy* plane, clockwise around the origin as seen from the positive side of the *z* axis. In unit-vector notation, what torque acts on the particle if the magnitude of its angular momentum about the origin is (a) 4.0 kg·m²/s, (b) $4.0t^2$ kg·m²/s, (c) $4.0\sqrt{t}$ kg·m²/s, and (d) $4.0/t^2$ kg·m²/s?

**•33** A 3.0 kg particle with velocity $\vec{v} = (5.0\text{ m/s})\hat{i} - (6.0\text{ m/s})\hat{j}$ is at x = 3.0 m, y = 8.0 m. It is pulled by a 7.0 N force in the negative *x* direction. About the origin, what are (a) the particle's angular momentum, (b) the torque acting on the particle, and (c) the rate at which the angular momentum is changing? **SSM ILW WWW**

**•34** A particle is acted on by two torques about the origin: $\vec{\tau}_1$ has a magnitude of 2.0 N·m and is directed in the positive direction of the *x* axis, and $\vec{\tau}_2$ has a magnitude of 4.0 N·m and is directed in the negative direction of the *y* axis. In unit-vector notation, find $d\vec{\ell}/dt$, where $\vec{\ell}$ is the angular momentum of the particle about the origin.

**••35** At time t, $\vec{r} = 4.0t^2\hat{i} - (2.0t + 6.0t^2)\hat{j}$ gives the position of a 3.0 kg particle relative to the origin of an *xy* coordinate system ($\vec{r}$ is in meters and *t* is in seconds). (a) Find an expression for the torque acting on the particle relative to the origin. (b) Is the magnitude of the particle's angular momentum relative to the origin increasing, decreasing, or unchanging?

### sec. 11-10 The Angular Momentum of a Rigid Body Rotating About a Fixed Axis

**•36** A sanding disk with rotational inertia $1.2 \times 10^{-3}$ kg·m² is attached to an electric drill whose motor delivers a torque of magnitude 16 N·m about the central axis of the disk. About that axis and with the torque applied for 33 ms, what is the magnitude of the (a) angular momentum and (b) angular velocity of the disk?

**•37** The angular momentum of a flywheel having a rotational inertia of 0.140 kg·m² about its central axis decreases from 3.00 to 0.800 kg·m²/s in 1.50 s. (a) What is the magnitude of the average torque acting on the flywheel about its central axis during this period? (b) Assuming a constant angular acceleration, through what angle does the flywheel turn? (c) How much work is done on the wheel? (d) What is the average power of the flywheel? **SSM**

**•38** Figure 11-44 shows three rotating, uniform disks that are coupled by belts. One belt runs around the rims of disks *A* and *C*. Another belt runs around a central hub on disk *A* and the rim of disk *B*. The belts move smoothly without slippage on the rims and hub. Disk *A* has radius *R*; its hub has radius 0.5000*R*; disk *B* has radius 0.2500*R*; and disk *C* has radius 2.000*R*. Disks *B* and *C* have the same density (mass per unit volume) and thickness. What is the ratio of the magnitude of the angular momentum of disk $C$ to that of disk $B$?

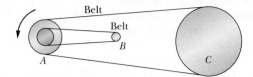

**FIG. 11-44** Problem 38.

•**39** In Fig. 11-45, three particles of mass $m = 23$ g are fastened to three rods of length $d = 12$ cm and negligible mass. The rigid assembly rotates around point $O$ at angular speed $\omega = 0.85$ rad/s. About $O$, what are (a) the rotational inertia of the assembly, (b) the magnitude of the angular momentum of the middle particle, and (c) the magnitude of the angular momentum of the assssembly?

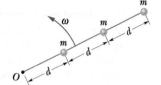

**FIG. 11-45** Problem 39.

••**40** Figure 11-46 gives the torque $\tau$ that acts on an initially stationary disk that can rotate about its center like a merry-go-round. The scale on the $\tau$ axis is set by $\tau_s = 4.0$ N·m. What is the angular momentum of the disk about the rotation axis at times (a) $t = 7.0$ s and (b) $t = 20$ s?

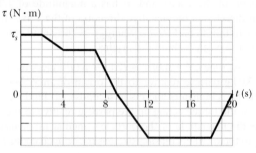

**FIG. 11-46** Problem 40.

••**41** Figure 11-47 shows a rigid structure consisting of a circular hoop of radius $R$ and mass $m$, and a square made of four thin bars, each of length $R$ and mass $m$. The rigid structure rotates at a constant speed about a vertical axis, with a period of rotation of 2.5 s. Assuming $R = 0.50$ m and $m = 2.0$ kg, calculate (a) the structure's rotational inertia about the axis of rotation and (b) its angular momentum about that axis. **GO**

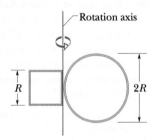

**FIG. 11-47** Problem 41.

••**42** A disk with a rotational inertia of 7.00 kg·m² rotates like a merry-go-round while undergoing a torque given by $\tau = (5.00 + 2.00t)$ N·m. At time $t = 1.00$ s, its angular momentum is 5.00 kg·m²/s. What is its angular momentum at $t = 3.00$ s?

### sec. 11-11 Conservation of Angular Momentum

•**43** A man stands on a platform that is rotating (without friction) with an angular speed of 1.2 rev/s; his arms are outstretched and he holds a brick in each hand. The rotational inertia of the system consisting of the man, bricks, and platform about the central vertical axis of the platform is 6.0 kg·m². If by moving the bricks the man decreases the rotational inertia of the system to 2.0 kg·m², what are (a) the resulting angular speed of the platform and (b) the ratio of the new kinetic energy of the system to the original kinetic energy? (c) What source provided the added kinetic energy? **SSM WWW**

•**44** The rotor of an electric motor has rotational inertia $I_m = 2.0 \times 10^{-3}$ kg·m² about its central axis. The motor is used to change the orientation of the space probe in which it is mounted. The motor axis is mounted along the central axis of the probe; the probe has rotational inertia $I_p = 12$ kg·m² about this axis. Calculate the number of revolutions of the rotor required to turn the probe through 30° about its central axis.

•**45** A wheel is rotating freely at angular speed 800 rev/min on a shaft whose rotational inertia is negligible. A second wheel, initially at rest and with twice the rotational inertia of the first, is suddenly coupled to the same shaft. (a) What is the angular speed of the resultant combination of the shaft and two wheels? (b) What fraction of the original rotational kinetic energy is lost? **SSM ILW**

•**46** A Texas cockroach first rides at the center of a circular disk that rotates freely like a merry-go-round without external torques. The cockroach then walks out to the edge of the disk, at radius $R$. Figure 11-48 gives the angular speed $\omega$ of the cockroach–disk system during the walk. The scale on the $\omega$ axis is set by $\omega_a = 5.0$ rad/s and $\omega_b = 6.0$ rad/s. When the cockroach is on the edge at radius $R$, what is the ratio of the bug's rotational inertia to that of the disk, both calculated about the rotation axis?

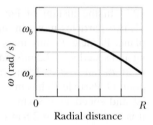

**FIG. 11-48** Problem 46.

•**47** Two disks are mounted (like a merry-go-round) on low-friction bearings on the same axle and can be brought together so that they couple and rotate as one unit. The first disk, with rotational inertia 3.30 kg·m² about its central axis, is set spinning counterclockwise at 450 rev/min. The second disk, with rotational inertia 6.60 kg·m² about its central axis, is set spinning counterclockwise at 900 rev/min. They then couple together. (a) What is their angular speed after coupling? If instead the second disk is set spinning clockwise at 900 rev/min, what are their (b) angular speed and (c) direction of rotation after they couple together?

•**48** The rotational inertia of a collapsing spinning star drops to $\frac{1}{3}$ its initial value. What is the ratio of the new rotational kinetic energy to the initial rotational kinetic energy?

•**49** A track is mounted on a large wheel that is free to turn with negligible friction about a vertical axis (Fig. 11-49). A toy train of mass $m$ is placed on the track and, with the system initially at rest, the train's electrical power is turned on. The train reaches speed 0.15 m/s with respect to the track. What is the angular speed of the wheel if its mass is $1.1m$ and its radius is 0.43 m? (Treat the wheel as a hoop, and neglect the mass of the spokes and hub.) **SSM**

FIG. 11-49 Problem 49.

•**50** A Texas cockroach of mass 0.17 kg runs counterclockwise around the rim of a lazy Susan (a circular disk mounted on a vertical axle) that has radius 15 cm, rotational inertia $5.0 \times 10^{-3} \text{ kg} \cdot \text{m}^2$, and frictionless bearings. The cockroach's speed (relative to the ground) is 2.0 m/s, and the lazy Susan turns clockwise with angular velocity $\omega_0 = 2.8$ rad/s. The cockroach finds a bread crumb on the rim and, of course, stops. (a) What is the angular speed of the lazy Susan after the cockroach stops? (b) Is mechanical energy conserved as it stops?

•**51** In Fig. 11-50, two skaters, each of mass 50 kg, approach each other along parallel paths separated by 3.0 m. They have opposite velocities of 1.4 m/s each. One skater carries one end of a long pole of negligible mass, and the other skater grabs the other end as she passes. The skaters then rotate around the center of the pole. Assume that the friction between skates and ice is negligible. What are (a) the radius of the circle, (b) the angular speed of the skaters, and (c) the kinetic energy of the two-skater system? Next, the skaters pull along the pole until they are separated by 1.0 m. What then are (d) their angular speed and (e) the kinetic energy of the system? (f) What provided the energy for the increased kinetic energy?

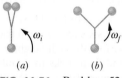

FIG. 11-50 Problem 51.

•**52** A bola consists of three massive, identical spheres connected to a common point by identical lengths of sturdy string (Fig. 11-51a). To launch this native South American weapon, you hold one of the spheres overhead and then rotate that hand about its wrist so as to rotate the other two spheres in a horizontal path about the hand. Once you manage sufficient rotation, you cast the weapon at a target. Initially the bola rotates around the previously held sphere at angular speed $\omega_i$ but then quickly changes so that the spheres rotate around the common connection point at angular speed $\omega_f$ (Fig. 11-51b). (a) What is the ratio $\omega_f/\omega_i$? (b) In the center-of-mass frame, what is the ratio $K_f/K_i$ of the corresponding rotational kinetic energies?

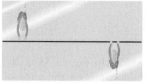

(a)          (b)

FIG. 11-51 Problem 52.

•**53** A horizontal vinyl record of mass 0.10 kg and radius 0.10 m rotates freely about a vertical axis through its center with an angular speed of 4.7 rad/s. The rotational inertia of the record about its axis of rotation is $5.0 \times 10^{-4} \text{ kg} \cdot \text{m}^2$. A wad of wet putty of mass 0.020 kg drops vertically onto the record from above and sticks to the edge of the record. What is the angular speed of the record immediately after the putty sticks to it?

••**54** In a long jump, an athlete leaves the ground with an initial angular momentum that tends to rotate her body forward, threatening to ruin her landing. To counter this tendency, she rotates her outstretched arms to "take up" the angular momentum (Fig. 11-19). In 0.700 s, one arm sweeps

through 0.500 rev and the other arm sweeps through 1.000 rev. Treat each arm as a thin rod of mass 4.0 kg and length 0.60 m, rotating around one end. In the athlete's reference frame, what is the magnitude of the total angular momentum of the arms around the common rotation axis through the shoulders?

••**55** A uniform thin rod of length 0.500 m and mass 4.00 kg can rotate in a horizontal plane about a vertical axis through its center. The rod is at rest when a 3.00 g bullet traveling in the rotation plane is fired into one end of the rod. As viewed from above, the bullet's path makes angle $\theta = 60.0°$ with the rod (Fig. 11-52). If the bullet lodges in the rod and the angular velocity of the rod is 10 rad/s immediately after the collision, what is the bullet's speed just before impact? **GO**

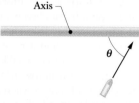

FIG. 11-52 Problem 55.

••**56** A cockroach of mass $m$ lies on the rim of a uniform disk of mass $4.00m$ that can rotate freely about its center like a merry-go-round. Initially the cockroach and disk rotate together with an angular velocity of 0.260 rad/s. Then the cockroach walks halfway to the center of the disk. (a) What then is the angular velocity of the cockroach–disk system? (b) What is the ratio $K/K_0$ of the new kinetic energy of the system to its initial kinetic energy? (c) What accounts for the change in the kinetic energy?

••**57** Figure 11-53 is an overhead view of a thin uniform rod of length 0.800 m and mass $M$ rotating horizontally at angular speed 20.0 rad/s about an axis through its center. A particle of mass $M/3.00$ initially attached to one end is ejected from the rod and travels along a path that is perpendicular to the rod at the instant of ejection. If the particle's speed $v_p$ is 6.00 m/s greater than the speed of the rod end just after ejection, what is the value of $v_p$?

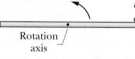

FIG. 11-53 Problem 57.

••**58** In Fig. 11-54, a 1.0 g bullet is fired into a 0.50 kg block attached to the end of a 0.60 m nonuniform rod of mass 0.50 kg. The block–rod–bullet system then rotates in the plane of the figure, about a fixed axis at $A$. The rotational inertia of the rod alone about that axis at $A$ is 0.060 kg·m². Treat the block as a particle. (a) What then is the rotational inertia of the block–rod–bullet system about point $A$? (b) If the angular speed of the system about $A$ just after impact is 4.5 rad/s, what is the bullet's speed just before impact?

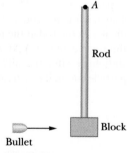

FIG. 11-54 Problem 58.

••**59** A uniform disk of mass $10m$ and radius $3.0r$ can rotate freely about its fixed center like a merry-go-round. A smaller uniform disk of mass $m$ and radius $r$ lies on top of the larger disk, concentric with it. Initially the two disks rotate together with an angular velocity of 20 rad/s. Then a slight disturbance causes the smaller disk to slide outward across the larger disk, until the outer edge of the smaller disk catches on the outer edge of the larger disk. Afterward, the two disks again rotate together (without further sliding). (a) What then is their angular velocity about the center of the larger disk? (b) What is the ratio $K/K_0$ of the new kinetic energy of the two-disk system to the system's initial kinetic energy?

••60   A horizontal platform in the shape of a circular disk rotates on a frictionless bearing about a vertical axle through the center of the disk. The platform has a mass of 150 kg, a radius of 2.0 m, and a rotational inertia of 300 kg·m² about the axis of rotation. A 60 kg student walks slowly from the rim of the platform toward the center. If the angular speed of the system is 1.5 rad/s when the student starts at the rim, what is the angular speed when she is 0.50 m from the center?

••61   The uniform rod (length 0.60 m, mass 1.0 kg) in Fig. 11-55 rotates in the plane of the figure about an axis through one end, with a rotational inertia of 0.12 kg·m². As the rod swings through its lowest position, it collides with a 0.20 kg putty wad that sticks to the end of the rod. If the rod's angular speed just before collision is 2.4 rad/s, what is the angular speed of the rod–putty system immediately after collision?

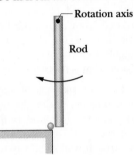

FIG. 11-55   Problem 61.

•••62   A ballerina begins a tour jeté (Fig. 11-20a) with angular speed $\omega_i$ and a rotational inertia consisting of two parts: $I_{leg} = 1.44$ kg·m² for her leg extended outward at angle $\theta = 90.0°$ to her body and $I_{trunk} = 0.660$ kg·m² for the rest of her body (primarily her trunk). Near her maximum height she holds both legs at angle $\theta = 30.0°$ to her body and has angular speed $\omega_f$ (Fig. 11-20b). Assuming that $I_{trunk}$ has not changed, what is the ratio $\omega_f/\omega_i$?

•••63   Figure 11-56 is an overhead view of a thin uniform rod of length 0.600 m and mass $M$ rotating horizontally at 80.0 rad/s counterclockwise about an axis through its center. A particle of mass $M/3.00$ and traveling horizontally at speed 40.0 m/s hits the rod and sticks. The particle's path is perpendicular to the rod at the instant of the hit, at a distance $d$ from the rod's center. (a) At what value of $d$ are rod and particle stationary after the hit? (b) In which direction do rod and particle rotate if $d$ is greater than this value?

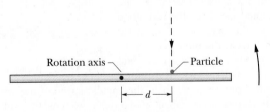

FIG. 11-56   Problem 63.

•••64   During a jump to his partner, an aerialist is to make a quadruple somersault lasting a time $t = 1.87$ s. For the first and last quarter-revolution, he is in the extended orientation shown in Fig. 11-57, with rotational inertia $I_1 = 19.9$ kg·m² around his center of mass (the dot). During the rest of the flight he is in a tight tuck, with rotational inertia $I_2 = 3.93$ kg·m². What must be his angular speed $\omega_2$ around his center of mass during the tuck?

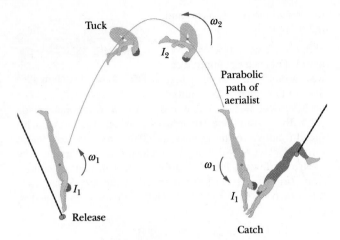

FIG. 11-57   Problem 64.

•••65   In Fig. 11-58, a 30 kg child stands on the edge of a stationary merry-go-round of mass 100 kg and radius 2.0 m. The rotational inertia of the merry-go-round about its rotation axis is 150 kg·m². The child catches a ball of mass 1.0 kg thrown by a friend. Just before the ball is caught, it has a horizontal velocity $\vec{v}$ of magnitude 12 m/s, at angle $\phi = 37°$ with a line tangent to the outer edge of the merry-go-round, as shown. What is the angular speed of the merry-go-round just after the ball is caught?

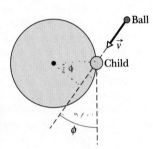

FIG. 11-58   Problem 65.

•••66   In Fig. 11-59, a small 50 g block slides down a frictionless surface through height $h = 20$ cm and then sticks to a uniform rod of mass 100 g and length 40 cm. The rod pivots about point $O$ through angle $\theta$ before momentarily stopping. Find $\theta$.

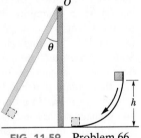

FIG. 11-59   Problem 66.

•••67   Two 2.00 kg balls are attached to the ends of a thin rod of length 50.0 cm and negligible mass. The rod is free to rotate in a vertical plane without friction about a horizontal axis through its center. With the rod initially horizontal (Fig. 11-60), a 50.0 g wad of wet putty drops onto one of the balls, hitting it

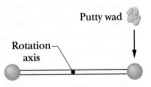

FIG. 11-60   Problem 67.

with a speed of 3.00 m/s and then sticking to it. (a) What is the angular speed of the system just after the putty wad hits? (b) What is the ratio of the kinetic energy of the system after the collision to that of the putty wad just before? (c) Through what angle will the system rotate before it momentarily stops? SSM WWW

## sec. 11-12   Precession of a Gyroscope

••68   A top spins at 30 rev/s about an axis that makes an angle of 30° with the vertical. The mass of the top is 0.50 kg, its rotational inertia about its central axis is $5.0 \times 10^{-4}$ kg·m², and its center of mass is 4.0 cm from the pivot point.

If the spin is clockwise from an overhead view, what are the (a) precession rate and (b) direction of the precession as viewed from overhead?

••69  A certain gyroscope consists of a uniform disk with a 50 cm radius mounted at the center of an axle that is 11 cm long and of negligible mass. The axle is horizontal and supported at one end. If the disk is spinning around the axle at 1000 rev/min, what is the precession rate?

### Additional Problems

70  A uniform block of granite in the shape of a book has face dimensions of 20 cm and 15 cm and a thickness of 1.2 cm. The density (mass per unit volume) of granite is 2.64 g/cm³. The block rotates around an axis that is perpendicular to its face and halfway between its center and a corner. Its angular momentum about that axis is 0.104 kg·m²/s. What is its rotational kinetic energy about that axis?

71  Figure 11-61 shows an overhead view of a ring that can rotate about its center like a merry-go-round. Its outer radius $R_2$ is 0.800 m, its inner radius $R_1$ is $R_2/2.00$, its mass $M$ is 8.00 kg, and the mass of the crossbars at its center is negligible. It initially rotates at an angular speed of 8.00 rad/s with a cat of mass $m = M/4.00$

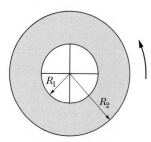

FIG. 11-61  Problem 71.

on its outer edge, at radius $R_2$. By how much does the cat increase the kinetic energy of the cat–ring system if the cat crawls to the inner edge, at radius $R_1$? GO

72  A 2.50 kg particle that is moving horizontally over a floor with velocity $(-3.00 \text{ m/s})\hat{j}$ undergoes a completely inelastic collision with a 4.00 kg particle that is moving horizontally over the floor with velocity $(4.50 \text{ m/s})\hat{i}$. The collision occurs at $xy$ coordinates $(-0.500 \text{ m}, -0.100 \text{ m})$. After the collision and in unit-vector notation, what is the angular momentum of the stuck-together particles with respect to the origin?

73  Two particles, each of mass $2.90 \times 10^{-4}$ kg and speed 5.46 m/s, travel in opposite directions along parallel lines separated by 4.20 cm. (a) What is the magnitude $L$ of the angular momentum of the two-particle system around a point midway between the two lines? (b) Does the value of $L$ change if the point about which it is calculated is not midway between the lines? If the direction of travel for one of the particles is reversed, what would be (c) the answer to part (a) and (d) the answer to part (b)? SSM

74  A uniform rod rotates in a horizontal plane about a vertical axis through one end. The rod is 6.00 m long, weighs 10.0 N, and rotates at 240 rev/min. Calculate (a) its rotational inertia about the axis of rotation and (b) the magnitude of its angular momentum about that axis.

75  Wheels $A$ and $B$ in Fig. 11-62 are connected by a belt that does not slip. The radius of $B$ is 3.00 times the radius of $A$. What would be the ratio of the rotational inertias $I_A/I_B$ if the two wheels had (a) the same

FIG. 11-62  Problem 75.

angular momentum about their central axes and (b) the same rotational kinetic energy?

76  At time $t = 0$, a 2.0 kg particle has position vector $\vec{r} = (4.0 \text{ m})\hat{i} - (2.0 \text{ m})\hat{j}$ relative to the origin. Its velocity is given by $\vec{v} = (-6.0t^2 \text{ m/s})\hat{i}$ for $t \geq 0$ in seconds. About the origin, what are (a) the particle's angular momentum $\vec{L}$ and (b) the torque $\vec{\tau}$ acting on the particle, both in unit-vector notation and for $t > 0$? About the point $(-2.0 \text{ m}, -3.0 \text{ m}, 0)$, what are (c) $\vec{L}$ and (d) $\vec{\tau}$ for $t > 0$?

77  A uniform wheel of mass 10.0 kg and radius 0.400 m is mounted rigidly on an axle through its center (Fig. 11-63). The radius of the axle is 0.200 m, and the rotational inertia of the wheel–axle combination about its central axis is 0.600 kg·m². The wheel is initially at rest at the top of a surface that is inclined at angle $\theta = 30.0°$ with the horizontal; the axle rests on the surface while the wheel extends into a groove in the surface without touching the surface. Once released, the axle rolls down along the surface smoothly and without slipping. When the wheel–axle combination has moved down the surface by 2.00 m, what are (a) its rotational kinetic energy and (b) its translational kinetic energy? SSM

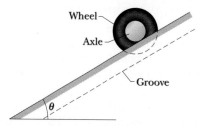

FIG. 11-63  Problem 77.

78  Suppose that the yo-yo in Problem 17, instead of rolling from rest, is thrown so that its initial speed down the string is 1.3 m/s. (a) How long does the yo-yo take to reach the end of the string? As it reaches the end of the string, what are its (b) total kinetic energy, (c) linear speed, (d) translational kinetic energy, (e) angular speed, and (f) rotational kinetic energy?

79  A small solid sphere with radius 0.25 cm and mass 0.56 g rolls without slipping on the inside of a large fixed hemisphere with radius 15 cm and a vertical axis of symmetry. The sphere starts at the top from rest. (a) What is its kinetic energy at the bottom? (b) What fraction of its kinetic energy at the bottom is associated with rotation about an axis through its com? (c) What is the magnitude of the normal force on the hemisphere from the sphere when the sphere reaches the bottom?

80  A uniform solid ball rolls smoothly along a floor, then up a ramp inclined at 15.0°. It momentarily stops when it has rolled 1.50 m along the ramp. What was its initial speed?

81  A body of radius $R$ and mass $m$ is rolling smoothly with speed $v$ on a horizontal surface. It then rolls up a hill to a maximum height $h$. (a) If $h = 3v^2/4g$, what is the body's rotational inertia about the rotational axis through its center of mass? (b) What might the body be?

82  A wheel of radius 0.250 m, which is moving initially at 43.0 m/s, rolls to a stop in 225 m. Calculate the magnitudes of (a) its linear acceleration and (b) its angular acceleration. (c) The wheel's rotational inertia is 0.155 kg·m² about its

central axis. Calculate the magnitude of the torque about the central axis due to friction on the wheel.

**83** If Earth's polar ice caps fully melted and the water returned to the oceans, the oceans would be deeper by about 30 m. What effect would this have on Earth's rotation? Make an estimate of the resulting change in the length of the day.

**84** A 1200 kg airplane is flying in a straight line at 80 m/s, 1.3 km above the ground. What is the magnitude of its angular momentum with respect to a point on the ground directly under the path of the plane?

**85** In a playground, there is a small merry-go-round of radius 1.20 m and mass 180 kg. Its radius of gyration (see Problem 91 of Chapter 10) is 91.0 cm. A child of mass 44.0 kg runs at a speed of 3.00 m/s along a path that is tangent to the rim of the initially stationary merry-go-round and then jumps on. Neglect friction between the bearings and the shaft of the merry-go-round. Calculate (a) the rotational inertia of the merry-go-round about its axis of rotation, (b) the magnitude of the angular momentum of the running child about the axis of rotation of the merry-go-round, and (c) the angular speed of the merry-go-round and child after the child has jumped onto the merry-go-round. **SSM**

**86** A wheel rotates clockwise about its central axis with an angular momentum of 600 kg·m²/s. At time $t = 0$, a torque of magnitude 50 N·m is applied to the wheel to reverse the rotation. At what time $t$ is the angular speed zero?

**87** A 3.0 kg toy car moves along an $x$ axis with a velocity given by $\vec{v} = -2.0t^3\hat{i}$ m/s, with $t$ in seconds. For $t > 0$, what are (a) the angular momentum $\vec{L}$ of the car and (b) the torque $\vec{\tau}$ on the car, both calculated about the origin? What are (c) $\vec{L}$ and (d) $\vec{\tau}$ about the point (2.0 m, 5.0 m, 0)? What are (e) $\vec{L}$ and (f) $\vec{\tau}$ about the point (2.0 m, −5.0 m, 0)? **SSM**

**88** A thin-walled pipe rolls along the floor. What is the ratio of its translational kinetic energy to its rotational kinetic energy about the central axis parallel to its length?

**89** A solid sphere of weight 36.0 N rolls up an incline at an angle of 30.0°. At the bottom of the incline the center of mass of the sphere has a translational speed of 4.90 m/s. (a) What is the kinetic energy of the sphere at the bottom of the incline? (b) How far does the sphere travel up along the incline? (c) Does the answer to (b) depend on the sphere's mass?

**90** An automobile has a total mass of 1700 kg. It accelerates from rest to 40 km/h in 10 s. Assume each wheel is a uniform 32 kg disk. Find, for the end of the 10 s interval, (a) the rotational kinetic energy of each wheel about its axle, (b) the total kinetic energy of each wheel, and (c) the total kinetic energy of the automobile.

**91** With axle and spokes of negligible mass and a thin rim, a certain bicycle wheel has a radius of 0.350 m and weighs 37.0 N; it can turn on its axle with negligible friction. A man holds the wheel above his head with the axle vertical while he stands on a turntable that is free to rotate without friction; the wheel rotates clockwise, as seen from above, with an angular speed of 57.7 rad/s, and the turntable is initially at rest. The rotational inertia of *wheel + man + turntable* about the common axis of rotation is 2.10 kg·m². The man's free hand suddenly stops the rotation of the wheel (relative to the

turntable). Determine the resulting (a) angular speed and (b) direction of rotation of the system.

**92** For an 84 kg person standing at the equator, what is the magnitude of the angular momentum about Earth's center due to Earth's rotation?

**93** A girl of mass $M$ stands on the rim of a frictionless merry-go-round of radius $R$ and rotational inertia $I$ that is not moving. She throws a rock of mass $m$ horizontally in a direction that is tangent to the outer edge of the merry-go-round. The speed of the rock, relative to the ground, is $v$. Afterward, what are (a) the angular speed of the merry-go-round and (b) the linear speed of the girl?

**94** A 4.0 kg particle moves in an $xy$ plane. At the instant when the particle's position and velocity are $\vec{r} = (2.0\hat{i} + 4.0\hat{j})$ m and $\vec{v} = -4.0\hat{j}$ m/s, the force on the particle is $\vec{F} = -3.0\hat{i}$ N. At this instant, determine (a) the particle's angular momentum about the origin, (b) the particle's angular momentum about the point $x = 0, y = 4.0$ m, (c) the torque acting on the particle about the origin, and (d) the torque acting on the particle about the point $x = 0, y = 4.0$ m.

**95** In Fig. 11-64, a constant horizontal force $\vec{F}_{app}$ of magnitude 12 N is applied to a uniform solid cylinder by fishing line wrapped around the cylinder. The mass of the cylinder is 10 kg, its radius is 0.10 m, and the cylinder rolls smoothly

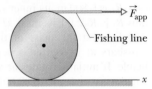

**FIG. 11-64** Problem 95.

on the horizontal surface. (a) What is the magnitude of the acceleration of the center of mass of the cylinder? (b) What is the magnitude of the angular acceleration of the cylinder about the center of mass? (c) In unit-vector notation, what is the frictional force acting on the cylinder? **SSM**

**96** (a) In Sample Problem 10-8, when the rotor exploded, how much angular momentum, calculated about the rotation axis, was released to the surroundings? (b) If we assume that most of the pieces of the rotor were stopped within 0.025 s after the explosion, what was the magnitude of the average torque acting on those pieces, calculated about the rotation axis?

**97** A particle of mass $M = 0.25$ kg is dropped from a point that is at height $h = 1.80$ m above the ground and horizontal distance $s = 0.45$ m from an observation point $O$, as shown in Fig. 11-65. What is the magnitude of the angular momentum of the particle with respect to point $O$ when the particle has fallen half the distance to the ground?

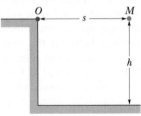

**FIG. 11-65** Problem 97.

**98** At one instant, a 0.80 kg particle is located at the position $\vec{r} = (2.0 \text{ m})\hat{i} + (3.0 \text{ m})\hat{j}$. The linear momentum of the particle lies in the $xy$ plane and has a magnitude of 2.4 kg·m/s and a direction of 115° measured counterclockwise from the positive direction of $x$. What is the angular momentum of the particle about the origin, in unit-vector notation?

# Equilibrium and Elasticity

**12**

PhotoDisc, Inc/Getty Images, Inc.

*The famous tower in Pisa, Italy, began to lean toward the south even during its construction, which spanned two centuries. The leaning increased with time but only at the snail's pace of 0.001° per year. In recent years, when the tilt reached 5.5°, the tower was closed to tourists because authorities feared that it would soon collapse. But doesn't collapse require that the tower's center of mass move out beyond the base of the tower? That would not have happened for many more years.*

## So, what was the danger to the tower?

The answer is in this chapter.

**FIG. 12-1** A balancing rock. Although its perch seems precarious, the rock is in static equilibrium. *(Symon Lobsang/Photis/Jupiter Images Corp.)*

## 12-1 | WHAT IS PHYSICS?

Human constructions are supposed to be stable in spite of the forces that act on them. A building, for example, should be stable in spite of the gravitational force and wind forces on it, and a bridge should be stable in spite of the gravitational force pulling it downward and the repeated jolting it receives from cars and trucks.

One focus of physics is on what allows an object to be stable in spite of any forces acting on it. In this chapter we examine the two main aspects of stability: the *equilibrium* of the forces and torques acting on rigid objects and the *elasticity* of nonrigid objects, a property that governs how such objects can deform. When this physics is done correctly, it is the subject of countless articles in physics and engineering journals; when it is done incorrectly, it is the subject of countless articles in newspapers and legal journals.

## 12-2 | Equilibrium

Consider these objects: (1) a book resting on a table, (2) a hockey puck sliding with constant velocity across a frictionless surface, (3) the rotating blades of a ceiling fan, and (4) the wheel of a bicycle that is traveling along a straight path at constant speed. For each of these four objects,

1. The linear momentum $\vec{P}$ of its center of mass is constant.
2. Its angular momentum $\vec{L}$ about its center of mass, or about any other point, is also constant.

We say that such objects are in **equilibrium.** The two requirements for equilibrium are then

$$\vec{P} = \text{a constant} \quad \text{and} \quad \vec{L} = \text{a constant.} \qquad (12\text{-}1)$$

Our concern in this chapter is with situations in which the constants in Eq. 12-1 are zero; that is, we are concerned largely with objects that are not moving in any way—either in translation or in rotation—in the reference frame from which we observe them. Such objects are in **static equilibrium.** Of the four objects mentioned at the beginning of this section, only one—the book resting on the table—is in static equilibrium.

The balancing rock of Fig. 12-1 is another example of an object that, for the present at least, is in static equilibrium. It shares this property with countless other structures, such as cathedrals, houses, filing cabinets, and taco stands, that remain stationary over time.

As we discussed in Section 8-6, if a body returns to a state of static equilibrium after having been displaced from that state by a force, the body is said to be in *stable* static equilibrium. A marble placed at the bottom of a hemispherical bowl is an example. However, if a small force can displace the body and end the equilibrium, the body is in *unstable* static equilibrium.

For example, suppose we balance a domino with the domino's center of mass vertically above the supporting edge, as in Fig. 12-2*a*. The torque about the supporting edge due to the gravitational force $\vec{F}_g$ on the domino is zero because the line of action of $\vec{F}_g$ is through that edge. Thus, the domino is in equilibrium. Of course, even a slight force on it due to some chance disturbance ends the equilibrium. As the line of action of $\vec{F}_g$ moves to one side of the supporting edge (as in Fig. 12-2*b*), the torque due to $\vec{F}_g$ increases the rotation of the domino. Therefore, the domino in Fig. 12-2*a* is in unstable static equilibrium.

The domino in Fig. 12-2*c* is not quite as unstable. To topple this domino, a force would have to rotate it through and then beyond the balance position of Fig. 12-2*a*, in which the center of mass is above a supporting edge. A slight force will not topple this domino, but a vigorous flick of the finger against the domino

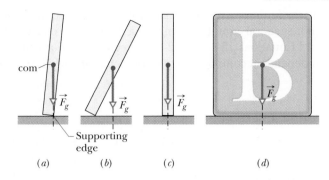

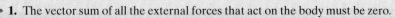

(a)          (b)          (c)          (d)

FIG. 12-2  (a) A domino balanced on one edge, with its center of mass vertically above that edge. The gravitational force $\vec{F}_g$ on the domino is directed through the supporting edge. (b) If the domino is rotated even slightly from the balanced orientation, then $\vec{F}_g$ causes a torque that increases the rotation. (c) A domino upright on a narrow side is somewhat more stable than the domino in (a). (d) A square block is even more stable.

certainly will. (If we arrange a chain of such upright dominos, a finger flick against the first can cause the whole chain to fall.)

The child's square block in Fig. 12-2d is even more stable because its center of mass would have to be moved even farther to get it to pass above a supporting edge. A flick of the finger may not topple the block. (This is why you never see a chain of toppling square blocks.) The worker in Fig. 12-3 is like both the domino and the square block: Parallel to the beam, his stance is wide and he is stable; perpendicular to the beam, his stance is narrow and he is unstable (and at the mercy of a chance gust of wind).

The analysis of static equilibrium is very important in engineering practice. The design engineer must isolate and identify all the external forces and torques that may act on a structure and, by good design and wise choice of materials, ensure that the structure will remain stable under these loads. Such analysis is necessary to ensure, for example, that bridges do not collapse under their traffic and wind loads and that the landing gear of aircraft will function after the shock of rough landings.

## 12-3 | The Requirements of Equilibrium

The translational motion of a body is governed by Newton's second law in its linear momentum form, given by Eq. 9-27 as

$$\vec{F}_{net} = \frac{d\vec{P}}{dt}. \tag{12-2}$$

If the body is in translational equilibrium—that is, if $\vec{P}$ is a constant—then $d\vec{P}/dt = 0$ and we must have

$$\vec{F}_{net} = 0 \quad \text{(balance of forces).} \tag{12-3}$$

The rotational motion of a body is governed by Newton's second law in its angular momentum form, given by Eq. 11-29 as

$$\vec{\tau}_{net} = \frac{d\vec{L}}{dt}. \tag{12-4}$$

If the body is in rotational equilibrium—that is, if $\vec{L}$ is a constant—then $d\vec{L}/dt = 0$ and we must have

$$\vec{\tau}_{net} = 0 \quad \text{(balance of torques).} \tag{12-5}$$

Thus, the two requirements for a body to be in equilibrium are as follows:

**1.** The vector sum of all the external forces that act on the body must be zero.

**2.** The vector sum of all external torques that act on the body, measured about *any* possible point, must also be zero.

FIG. 12-3   A construction worker balanced on a steel beam is in static equilibrium but is more stable parallel to the beam than perpendicular to it. *(Robert Brenner/PhotoEdit)*

These requirements obviously hold for *static* equilibrium. They also hold for the more general equilibrium in which $\vec{P}$ and $\vec{L}$ are constant but not zero.

Equations 12-3 and 12-5, as vector equations, are each equivalent to three independent component equations, one for each direction of the coordinate axes:

| Balance of forces | Balance of torques | |
| --- | --- | --- |
| $F_{\text{net},x} = 0$ | $\tau_{\text{net},x} = 0$ | |
| $F_{\text{net},y} = 0$ | $\tau_{\text{net},y} = 0$ | (12-6) |
| $F_{\text{net},z} = 0$ | $\tau_{\text{net},z} = 0$ | |

We shall simplify matters by considering only situations in which the forces that act on the body lie in the *xy* plane. This means that the only torques that can act on the body must tend to cause rotation around an axis parallel to the *z* axis. With this assumption, we eliminate one force equation and two torque equations from Eqs. 12-6, leaving

$$F_{\text{net},x} = 0 \qquad \text{(balance of forces)}, \qquad (12\text{-}7)$$

$$F_{\text{net},y} = 0 \qquad \text{(balance of forces)}, \qquad (12\text{-}8)$$

$$\tau_{\text{net},z} = 0 \qquad \text{(balance of torques)}. \qquad (12\text{-}9)$$

Here, $\tau_{\text{net},z}$ is the net torque that the external forces produce either about the *z* axis or about *any* axis parallel to it.

A hockey puck sliding at constant velocity over ice satisfies Eqs. 12-7, 12-8, and 12-9 and is thus in equilibrium *but not in static equilibrium*. For static equilibrium, the linear momentum $\vec{P}$ of the puck must be not only constant but also zero; the puck must be at rest on the ice. Thus, there is another requirement for static equilibrium:

> **3.** The linear momentum $\vec{P}$ of the body must be zero.

 **CHECKPOINT 1**    The figure gives six overhead views of a uniform rod on which two or more forces act perpendicularly to the rod. If the magnitudes of the forces are adjusted properly (but kept nonzero), in which situations can the rod be in static equilibrium?

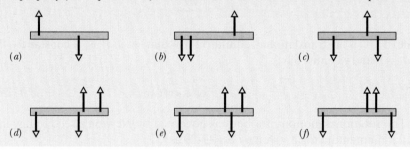

## 12-4 | The Center of Gravity

The gravitational force on an extended body is the vector sum of the gravitational forces acting on the individual elements (the atoms) of the body. Instead of considering all those individual elements, we can say that

> The gravitational force $\vec{F}_g$ on a body effectively acts at a single point, called the **center of gravity** (cog) of the body.

Here the word "effectively" means that if the forces on the individual elements were somehow turned off and force $\vec{F}_g$ at the center of gravity were turned on, the net force and the net torque (about any point) acting on the body would not change.

Until now, we have assumed that the gravitational force $\vec{F}_g$ acts at the center of mass (com) of the body. This is equivalent to assuming that the center of gravity is at the center of mass. Recall that, for a body of mass $M$, the force $\vec{F}_g$ is equal to $M\vec{g}$, where $\vec{g}$ is the acceleration that the force would produce if the body were to fall freely. In the proof that follows, we show that

☛ If $\vec{g}$ is the same for all elements of a body, then the body's center of gravity (cog) is coincident with the body's center of mass (com).

This is approximately true for everyday objects because $\vec{g}$ varies only a little along Earth's surface and decreases in magnitude only slightly with altitude. Thus, for objects like a mouse or a moose, we have been justified in assuming that the gravitational force acts at the center of mass. After the following proof, we shall resume that assumption.

### Proof

First, we consider the individual elements of the body. Figure 12-4$a$ shows an extended body, of mass $M$, and one of its elements, of mass $m_i$. A gravitational force $\vec{F}_{gi}$ acts on each such element and is equal to $m_i\vec{g}_i$. The subscript on $\vec{g}_i$ means $\vec{g}_i$ is the gravitational acceleration *at the location of the element i* (it can be different for other elements).

In Fig. 12-4$a$, each force $\vec{F}_{gi}$ produces a torque $\tau_i$ on the element about the origin $O$, with moment arm $x_i$. Using Eq. 10-41 ($\tau = r_\perp F$), we can write torque $\tau_i$ as

$$\tau_i = x_i F_{gi}. \tag{12-10}$$

The net torque on all the elements of the body is then

$$\tau_{\text{net}} = \sum \tau_i = \sum x_i F_{gi}. \tag{12-11}$$

Next, we consider the body as a whole. Figure 12-4$b$ shows the gravitational force $\vec{F}_g$ acting at the body's center of gravity. This force produces a torque $\tau$ on the body about $O$, with moment arm $x_{\text{cog}}$. Again using Eq. 10-41, we can write this torque as

$$\tau = x_{\text{cog}} F_g. \tag{12-12}$$

The gravitational force $\vec{F}_g$ on the body is equal to the sum of the gravitational forces $\vec{F}_{gi}$ on all its elements, so we can substitute $\sum F_{gi}$ for $F_g$ in Eq. 12-12 to write

$$\tau = x_{\text{cog}} \sum F_{gi}. \tag{12-13}$$

Now recall that the torque due to force $\vec{F}_g$ acting at the center of gravity is equal to the net torque due to all the forces $\vec{F}_{gi}$ acting on all the elements of the body. (That is how we defined the center of gravity.) Thus, $\tau$ in Eq. 12-13 is equal to $\tau_{\text{net}}$ in Eq. 12-11. Putting those two equations together, we can write

$$x_{\text{cog}} \sum F_{gi} = \sum x_i F_{gi}.$$

Substituting $m_i g_i$ for $F_{gi}$ gives us

$$x_{\text{cog}} \sum m_i g_i = \sum x_i m_i g_i. \tag{12-14}$$

Now here is a key idea: If the accelerations $g_i$ at all the locations of the elements are the same, we can cancel $g_i$ from this equation to write

$$x_{\text{cog}} \sum m_i = \sum x_i m_i. \tag{12-15}$$

The sum $\sum m_i$ of the masses of all the elements is the mass $M$ of the body. Therefore, we can rewrite Eq. 12-15 as

$$x_{\text{cog}} = \frac{1}{M} \sum x_i m_i. \tag{12-16}$$

The right side of this equation gives the coordinate $x_{\text{com}}$ of the body's center of

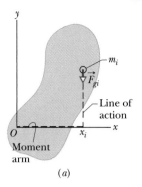

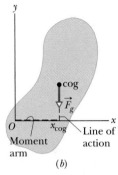

**FIG. 12-4** (*a*) An element of mass $m_i$ in an extended body. The gravitational force $\vec{F}_{gi}$ on the element has moment arm $x_i$ about the origin $O$ of the coordinate system. (*b*) The gravitational force $\vec{F}_g$ on a body is said to act at the center of gravity (cog) of the body. Here $\vec{F}_g$ has moment arm $x_{\text{cog}}$ about origin $O$.

mass (Eq. 9-4). We now have what we sought to prove:

$$x_{\text{cog}} = x_{\text{com}}. \qquad (12\text{-}17)$$

## 12-5 | Some Examples of Static Equilibrium

In this section we examine four sample problems involving static equilibrium. In each, we select a system of one or more objects to which we apply the equations of equilibrium (Eqs. 12-7, 12-8, and 12-9). The forces involved in the equilibrium are all in the $xy$ plane, which means that the torques involved are parallel to the $z$ axis. Thus, in applying Eq. 12-9, the balance of torques, we select an axis parallel to the $z$ axis about which to calculate the torques. Although Eq. 12-9 is satisfied for *any* such choice of axis, you will see that certain choices simplify the application of Eq. 12-9 by eliminating one or more unknown force terms.

✓**CHECKPOINT 2**   The figure gives an overhead view of a uniform rod in static equilibrium. (a) Can you find the magnitudes of unknown forces $\vec{F_1}$ and $\vec{F_2}$ by balancing the forces? (b) If you wish to find the magnitude of force $\vec{F_2}$ by using a balance of torques equation, where should you place a rotational axis to eliminate $\vec{F_1}$ from the equation? (c) The magnitude of $\vec{F_2}$ turns out to be 65 N. What then is the magnitude of $\vec{F_1}$?

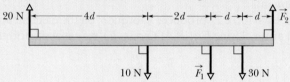

---

**Sample Problem** | **12-1** | **Build your skill**

In Fig. 12-5*a*, a uniform beam, of length $L$ and mass $m = 1.8$ kg, is at rest on two scales. A uniform block, with mass $M = 2.7$ kg, is at rest on the beam, with its center a distance $L/4$ from the beam's left end. What do the scales read?

**KEY IDEAS**   The first steps in the solution of *any* problem about static equilibrium are these: Clearly define the system to be analyzed and then draw a free-body diagram of it, indicating all the forces on the system. Here, let us choose the system as the beam and block taken together. Then the forces on the system are shown in the free-body diagram of Fig. 12-5*b*. (Choosing the system takes experience, and often there can be more than one good choice; see item 2 of Problem-Solving Tactic 1 below.) Because the system is in static equilibrium, we can apply the balance of forces equations (Eqs. 12-7 and 12-8) and the balance of torques equation (Eq. 12-9) to it.

**Calculations:** The normal forces on the beam from the scales are $\vec{F_l}$ on the left and $\vec{F_r}$ on the right. The scale readings that we want are equal to the magnitudes of those forces. The gravitational force $\vec{F}_{g,beam}$ on the beam acts at the beam's center of mass and is equal to $m\vec{g}$. Similarly, the gravitational force $\vec{F}_{g,block}$ on the block acts at the block's center of mass and is equal to $M\vec{g}$. However, to simplify Fig. 12-5*b*, the block is represented by a dot within the boundary of the beam and $\vec{F}_{g,block}$ is

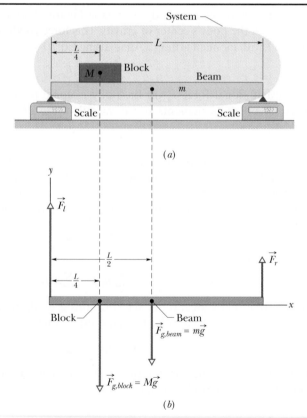

**FIG. 12-5**   (*a*) A beam of mass $m$ supports a block of mass $M$. (*b*) A free-body diagram, showing the forces that act on the system *beam* + *block*.

drawn with its tail on that dot. (This shift of vector $\vec{F}_{g,block}$ along its line of action does not alter the torque due to $\vec{F}_{g,block}$ about any axis perpendicular to the figure.)

The forces have no $x$ components, so Eq. 12-7 ($F_{net,x} = 0$) provides no information. For the $y$ components, Eq. 12-8 ($F_{net,y} = 0$) gives us

$$F_l + F_r - Mg - mg = 0. \qquad (12\text{-}18)$$

This equation contains two unknowns, the forces $F_l$ and $F_r$, so we also need to use Eq. 12-9, the balance of torques equation. We can apply it to *any* rotation axis perpendicular to the plane of Fig. 12-5. Let us choose a rotation axis through the left end of the beam. We shall also use our general rule for assigning signs to torques: If a torque would cause an initially stationary body to rotate clockwise about the rotation axis, the torque is negative. If the rotation would be counterclockwise, the torque is positive. Finally, we shall write the torques in the form $r_\perp F$, where the moment arm $r_\perp$ is 0 for $\vec{F}_l$, $L/4$ for $M\vec{g}$, $L/2$ for $m\vec{g}$, and $L$ for $\vec{F}_r$.

We now can write the balancing equation ($\tau_{net,z} = 0$) as

$$(0)(F_l) - (L/4)(Mg) - (L/2)(mg) + (L)(F_r) = 0,$$

which gives us

$$\begin{aligned}
F_r &= \tfrac{1}{4}Mg + \tfrac{1}{2}mg \\
&= \tfrac{1}{4}(2.7\ \text{kg})(9.8\ \text{m/s}^2) + \tfrac{1}{2}(1.8\ \text{kg})(9.8\ \text{m/s}^2) \\
&= 15.44\ \text{N} \approx 15\ \text{N}. \qquad \text{(Answer)}
\end{aligned}$$

Now, solving Eq. 12-18 for $F_l$ and substituting this result, we find

$$\begin{aligned}
F_l &= (M + m)g - F_r \\
&= (2.7\ \text{kg} + 1.8\ \text{kg})(9.8\ \text{m/s}^2) - 15.44\ \text{N} \\
&= 28.66\ \text{N} \approx 29\ \text{N}. \qquad \text{(Answer)}
\end{aligned}$$

*Notice the strategy in the solution:* When we wrote an equation for the balance of force components, we got stuck with two unknowns. If we had written an equation for the balance of torques around some *arbitrary* axis, we would have again gotten stuck with those two unknowns. However, because we chose the axis to pass through the point of application of one of the unknown forces, here $\vec{F}_l$, we did not get stuck. Our choice neatly eliminated that force from the torque equation, allowing us to solve for the other unknown force magnitude $F_r$. Then we returned to the equation for the balance of force components to find the remaining unknown force magnitude.

---

**Sample Problem**   **12-2**

In Fig. 12-6a, a ladder of length $L = 12$ m and mass $m = 45$ kg leans against a slick (frictionless) wall. The ladder's upper end is at height $h = 9.3$ m above the pavement on which the lower end rests (the pavement is not frictionless). The ladder's center of mass is $L/3$ from the lower end. A firefighter of mass $M = 72$ kg climbs the ladder until her center of mass is $L/2$ from the lower end. What then are the magnitudes of the forces on the ladder from the wall and the pavement?

**KEY IDEAS**   First, we choose our system as being the firefighter and ladder, together, and then we draw the free-body diagram of Fig. 12-6b. Because the system is in static equilibrium, the balancing equations (Eqs. 12-7 through 12-9) apply to it.

*Calculations:* In Fig. 12-6b, the firefighter is represented with a dot within the boundary of the ladder. The gravitational force on her is represented with its equivalent $M\vec{g}$, and that vector has been shifted along its line of action, so that its tail is on the dot. (The shift does not alter a torque due to $M\vec{g}$ about any axis perpendicular to the figure.)

The only force on the ladder from the wall is the horizontal force $\vec{F}_w$ (there cannot be a frictional force along a frictionless wall). The force $\vec{F}_p$ on the ladder from the pavement has a horizontal component $\vec{F}_{px}$ that is a static frictional force and a vertical component $\vec{F}_{py}$ that is a normal force.

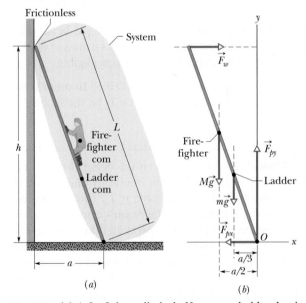

FIG. 12-6   (*a*) A firefighter climbs halfway up a ladder that is leaning against a frictionless wall. The pavement beneath the ladder is not frictionless. (*b*) A free-body diagram, showing the forces that act on the *firefighter + ladder* system. The origin $O$ of a coordinate system is placed at the point of application of the unknown force $\vec{F}_p$ (whose vector components $\vec{F}_{px}$ and $\vec{F}_{py}$ are shown).

To apply the balancing equations, let's start with Eq. 12-9 ($\tau_{net,z} = 0$). To choose an axis about which to calculate the torques, note that we have unknown forces

($\vec{F}_w$ and $\vec{F}_p$) at the two ends of the ladder. To eliminate, say, $\vec{F}_p$ from the calculation, we place the axis at point $O$, perpendicular to the figure. We also place the origin of an $xy$ coordinate system at $O$. We can find torques about $O$ with any of Eqs. 10-39 through 10-41, but Eq. 10-41 ($\tau = r_\perp F$) is easiest to use here.

To find the moment arm $r_\perp$ of $\vec{F}_w$, we draw a line of action through that vector (horizontal dashed line in Fig. 12-6b). Then $r_\perp$ is the perpendicular distance between $O$ and the line of action. In Fig. 12-6b, $r_\perp$ extends along the $y$ axis and is equal to the height $h$. We similarly draw lines of action for $M\vec{g}$ and $m\vec{g}$ and see that their moment arms extend along the $x$ axis. For the distance $a$ shown in Fig. 12-6a, the moment arms are $a/2$ (the firefighter is halfway up the ladder) and $a/3$ (the ladder's center of mass is one-third of the way up the ladder), respectively. The moment arms for $\vec{F}_{px}$ and $\vec{F}_{py}$ are zero.

Now, with torques written in the form $r_\perp F$, the balancing equation $\tau_{net,z} = 0$ becomes

$$-(h)(F_w) + (a/2)(Mg) + (a/3)(mg)$$
$$+ (0)(F_{px}) + (0)(F_{py}) = 0. \quad (12\text{-}19)$$

(Recall our rule: A positive torque corresponds to counterclockwise rotation and a negative torque corresponds to clockwise rotation.)

Using the Pythagorean theorem, we find that

$$a = \sqrt{L^2 - h^2} = 7.58 \text{ m}.$$

Then Eq. 12-19 gives us

$$F_w = \frac{ga(M/2 + m/3)}{h}$$

$$= \frac{(9.8 \text{ m/s}^2)(7.58 \text{ m})(72/2 \text{ kg} + 45/3 \text{ kg})}{9.3 \text{ m}}$$

$$= 407 \text{ N} \approx 410 \text{ N}. \quad \text{(Answer)}$$

Now we need to use the force balancing equations. The equation $F_{net,x} = 0$ gives us

$$F_w - F_{px} = 0,$$

so $\quad\quad F_{px} = F_w = 410 \text{ N}. \quad \text{(Answer)}$

The equation $F_{net,y} = 0$ gives us

$$F_{py} - Mg - mg = 0,$$

so $\quad F_{py} = (M + m)g = (72 \text{ kg} + 45 \text{ kg})(9.8 \text{ m/s}^2)$

$$= 1146.6 \text{ N} \approx 1100 \text{ N}. \quad \text{(Answer)}$$

---

**Sample Problem** | **12-3**

Figure 12-7a shows a safe, of mass $M = 430$ kg, hanging by a rope from a boom with dimensions $a = 1.9$ m and $b = 2.5$ m. The boom consists of a hinged beam and a horizontal cable. The uniform beam has a mass $m$ of 85 kg; the masses of the cable and rope are negligible.

(a) What is the tension $T_c$ in the cable? In other words, what is the magnitude of the force $\vec{T}_c$ on the beam from the cable?

**KEY IDEAS** The system here is the beam alone, and the forces on it are shown in the free-body diagram of Fig. 12-7b. The force from the cable is $\vec{T}_c$. The gravitational force on the beam acts at the beam's center of mass (at the beam's center) and is represented by its equivalent $m\vec{g}$. The vertical component of the force on the beam from the hinge is $\vec{F}_v$, and the horizontal component of the force from the hinge is $\vec{F}_h$. The force from the rope supporting the safe is $\vec{T}_r$. Because beam, rope, and safe are stationary, the magnitude of $\vec{T}_r$ is equal to the weight of the safe: $T_r = Mg$. We place the origin $O$ of an $xy$ coordinate system at the hinge. Because the system is in static equilibrium, the balancing equations apply to it.

**Calculations:** Let us start with Eq. 12-9 ($\tau_{net,z} = 0$). Note that we are asked for the magnitude of force $\vec{T}_c$ and not of forces $\vec{F}_h$ and $\vec{F}_v$ acting at the hinge, at point $O$. To eliminate $\vec{F}_h$ and $\vec{F}_v$ from the torque calculation,

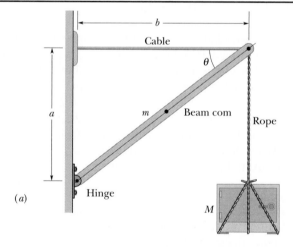

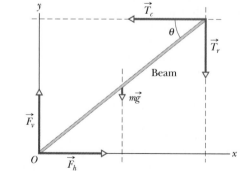

**FIG. 12-7** (a) A heavy safe is hung from a boom consisting of a horizontal steel cable and a uniform beam. (b) A free-body diagram for the beam.

we should calculate torques about an axis that is perpendicular to the figure at point $O$. Then $\vec{F}_h$ and $\vec{F}_v$ will have moment arms of zero. The lines of action for $\vec{T}_c$, $\vec{T}_r$, and $m\vec{g}$ are dashed in Fig. 12-7b. The corresponding moment arms are $a$, $b$, and $b/2$.

Writing torques in the form of $r_{\perp}F$ and using our rule about signs for torques, the balancing equation $\tau_{\text{net},z} = 0$ becomes

$$(a)(T_c) - (b)(T_r) - (\tfrac{1}{2}b)(mg) = 0.$$

Substituting $Mg$ for $T_r$ and solving for $T_c$, we find that

$$T_c = \frac{gb(M + \tfrac{1}{2}m)}{a}$$

$$= \frac{(9.8 \text{ m/s}^2)(2.5 \text{ m})(430 \text{ kg} + 85/2 \text{ kg})}{1.9 \text{ m}}$$

$$= 6093 \text{ N} \approx 6100 \text{ N}. \qquad \text{(Answer)}$$

**(b)** Find the magnitude $F$ of the net force on the beam from the hinge.

---

**KEY IDEA** Now we want $F_h$ and $F_v$ so we can combine

them to get $F$. Because we know $T_c$, we apply the force balancing equations to the beam.

**Calculations:** For the horizontal balance, we write $F_{\text{net},x} = 0$ as

$$F_h - T_c = 0,$$

and so $\qquad F_h = T_c = 6093 \text{ N}.$

For the vertical balance, we write $F_{\text{net},y} = 0$ as

$$F_v - mg - T_r = 0.$$

Substituting $Mg$ for $T_r$ and solving for $F_v$, we find that

$$F_v = (m + M)g = (85 \text{ kg} + 430 \text{ kg})(9.8 \text{ m/s}^2)$$
$$= 5047 \text{ N}.$$

From the Pythagorean theorem, we now have

$$F = \sqrt{F_h^2 + F_v^2}$$
$$= \sqrt{(6093 \text{ N})^2 + (5047 \text{ N})^2} \approx 7900 \text{ N}. \quad \text{(Answer)}$$

Note that $F$ is substantially greater than either the combined weights of the safe and the beam, 5000 N, or the tension in the horizontal wire, 6100 N.

---

**Sample Problem** 12-4

Let's assume that the Tower of Pisa is a uniform hollow cylinder of radius $R = 9.8$ m and height $h = 60$ m. The center of mass is located at height $h/2$, along the cylinder's central axis. In Fig. 12-8a, the cylinder is upright. In Fig. 12-8b, it leans rightward (toward the tower's southern wall) by $\theta = 5.5°$, which shifts the com by a distance $d$. Let's assume that the ground exerts only two forces on the tower. A normal force $\vec{F}_{NL}$ acts on the left (northern) wall, and a normal force $\vec{F}_{NR}$ acts on the right (southern) wall. By what percent does the magnitude $F_{NR}$ increase because of the leaning?

---

**KEY IDEA** Because the tower is still standing, it is in equilibrium and thus the sum of torques calculated around any point must be zero.

**Calculations:** Because we want to calculate $F_{NR}$ on the right side and do not know or want $F_{NL}$ on the left side, we use a pivot point on the left side to calculate torques. The forces on the upright tower are represented in Fig. 12-8c. The gravitational force $m\vec{g}$, taken to act at the com, has a vertical line of action and a moment arm of $R$ (the perpendicular distance from the pivot to the line of action). About the pivot, the torque associated with this force would tend to create clockwise rotation and thus is negative. The normal force $\vec{F}_{NR}$ on the southern wall also has a vertical line of action, and its moment arm is $2R$. About the pivot, the torque associated with this force would tend to create counterclockwise rotation and thus is positive. We now can

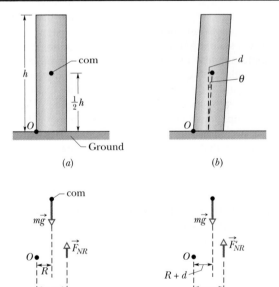

**FIG. 12-8** A cylinder modeling the Tower of Pisa: (a) upright and (b) leaning, with the center of mass shifted rightward. The forces and moment arms to find torques about a pivot at point $O$ for the cylinder (c) upright and (d) leaning.

write the torque-balancing equation ($\tau_{\text{net},z} = 0$) as

$$-(R)(mg) + (2R)(F_{NR}) = 0,$$

which yields

$$F_{NR} = \tfrac{1}{2}mg. \qquad (12\text{-}20)$$

We should have been able to guess this result: With the center of mass located on the central axis (the cylinder's line of symmetry), the right side supports half the cylinder's weight.

In Fig. 12-8b, the com is shifted rightward by distance

$$d = \tfrac{1}{2}h \tan \theta.$$

The only change in the balance of torques equation is that the moment arm for the gravitational force is now $R + d$ and the normal force at the right has a new magnitude $F'_{NR}$ (Fig. 12-8d). Thus, we write

$$-(R + d)(mg) + (2R)(F'_{NR}) = 0,$$

which gives us

$$F'_{NR} = \frac{(R + d)}{2R} mg. \qquad (12\text{-}21)$$

Dividing Eq. 12-21 by Eq. 12-20 and then substituting for $d$, we obtain

$$\frac{F'_{NR}}{F_{NR}} = \frac{R + d}{R} = 1 + \frac{d}{R} = 1 + \frac{0.5h \tan \theta}{R}.$$

Substituting the values of $h = 60$ m, $R = 9.8$ m, and $\theta = 5.5°$ leads to

$$\frac{F'_{NR}}{F_{NR}} = 1.29.$$

Thus, our simple model predicts that, although the tilt is modest, the normal force on the tower's southern wall has increased by about 30%. One danger to the tower is that the force may cause the southern wall to buckle and explode outward.

---

*Tactic 1: Static Equilibrium Problems* Here is a list of steps for solving static equilibrium problems:

1. Draw a *sketch* of the problem.

2. Select the *system* to which you will apply the laws of equilibrium, drawing a closed curve around the system on your sketch to fix it clearly in your mind. In some situations you can select a single object as the system; it is the object you wish to be in equilibrium. In other situations, you might include additional objects in the system *if* their inclusion simplifies the calculations for equilibrium. For example, suppose in Sample Problem 12-2 you select only the ladder as the system. Then in Fig. 12-6b you will have to account for additional unknown forces exerted on the ladder by the hands and feet of the firefighter. These additional unknowns complicate the equilibrium calculations. The system of Fig. 12-6 was chosen to include the firefighter so that those unknown forces are *internal* to the system and thus need not be found in order to solve Sample Problem 12-2.

3. Draw a *free-body diagram* of the system. Show all the forces that act on the system, labeling them and making sure that their points of application and lines of action are correctly shown.

4. Draw in the *x and y axes* of a coordinate system with at least one axis parallel to one or more unknown force. Resolve into components the forces that do not lie along one of the axes. In all our sample problems it made sense to choose the *x* axis horizontal and the *y* axis vertical.

5. Write the two *balance of forces equations,* using symbols throughout.

6. Choose one or more rotation axes perpendicular to the plane of the figure and write the *balance of torques equation* for each axis. If you choose an axis that passes through the line of action of an unknown force, the equation will be simplified because that force will not appear in it.

7. *Solve* your equations *algebraically* for the unknowns. Some students feel more confident in substituting numbers with units in the independent equations at this stage, especially if the algebra is particularly involved. However, experienced problem solvers prefer the algebra, which reveals the dependence of solutions on the various variables.

8. Finally, *substitute numbers* with units in your algebraic solutions, obtaining numerical values for the unknowns.

9. Look at your answer—does it make sense? Is it obviously too large or too small? Is the sign correct? Are the units appropriate?

## 12-6 | Indeterminate Structures

For the problems of this chapter, we have only three independent equations at our disposal, usually two balance of forces equations and one balance of torques equation about a given rotation axis. Thus, if a problem has more than three unknowns, we cannot solve it.

It is easy to find such problems. In Sample Problem 12-2, for example, we could have assumed that there is friction between the wall and the top of the ladder. Then there would have been a vertical frictional force acting where the ladder touches the wall, making a total of four unknown forces. With only three equations, we could not have solved this problem.

Consider also an unsymmetrically loaded car. What are the forces—all different—on the four tires? Again, we cannot find them because we have only three independent equations. Similarly, we can solve an equilibrium problem for a table with three legs but not for one with four legs. Problems like these, in which there are more unknowns than equations, are called **indeterminate.**

Yet solutions to indeterminate problems exist in the real world. If you rest the tires of the car on four platform scales, each scale will register a definite reading, the sum of the readings being the weight of the car. What is eluding us in our efforts to find the individual forces by solving equations?

The problem is that we have assumed—without making a great point of it—that the bodies to which we apply the equations of static equilibrium are perfectly rigid. By this we mean that they do not deform when forces are applied to them. Strictly, there are no such bodies. The tires of the car, for example, deform easily under load until the car settles into a position of static equilibrium.

We have all had experience with a wobbly restaurant table, which we usually level by putting folded paper under one of the legs. If a big enough elephant sat on such a table, however, you may be sure that if the table did not collapse, it would deform just like the tires of a car. Its legs would all touch the floor, the forces acting upward on the table legs would all assume definite (and different) values as in Fig. 12-9, and the table would no longer wobble. How do we find the values of those forces acting on the legs?

To solve such indeterminate equilibrium problems, we must supplement equilibrium equations with some knowledge of *elasticity,* the branch of physics and engineering that describes how real bodies deform when forces are applied to them. The next section provides an introduction to this subject.

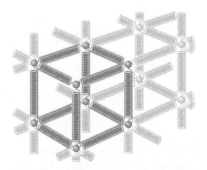

**FIG. 12-9** The table is an indeterminate structure. The four forces on the table legs differ from one another in magnitude and cannot be found from the laws of static equilibrium alone.

✓**CHECKPOINT 3**    A horizontal uniform bar of weight 10 N is to hang from a ceiling by two wires that exert upward forces $\vec{F}_1$ and $\vec{F}_2$ on the bar. The figure shows four arrangements for the wires. Which arrangements, if any, are indeterminate (so that we cannot solve for numerical values of $\vec{F}_1$ and $\vec{F}_2$)?

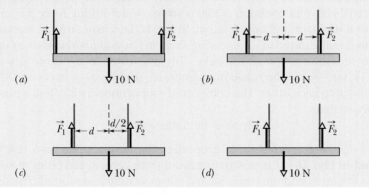

## 12-7 | Elasticity

When a large number of atoms come together to form a metallic solid, such as an iron nail, they settle into equilibrium positions in a three-dimensional *lattice,* a repetitive arrangement in which each atom is a well-defined equilibrium distance from its nearest neighbors. The atoms are held together by interatomic forces that are modeled as tiny springs in Fig. 12-10. The lattice is remarkably rigid, which is another way of saying that the "interatomic springs" are extremely stiff. It is for this reason that we perceive many ordinary objects, such as metal ladders, tables, and spoons, as perfectly rigid. Of course, some ordinary objects, such as garden hoses or rubber gloves, do not strike us as rigid at all. The atoms that make up these objects *do not* form a rigid lattice like that of Fig. 12-10 but are aligned in long, flexible molecular chains, each chain being only loosely bound to its neighbors.

**FIG. 12-10** The atoms of a metallic solid are distributed on a repetitive three-dimensional lattice. The springs represent interatomic forces.

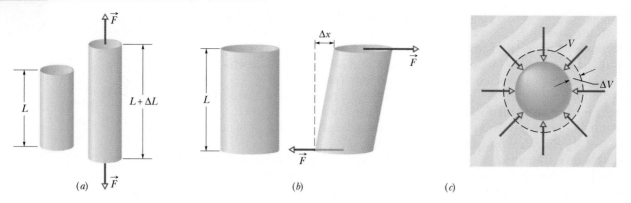

(a)        (b)        (c)

**FIG. 12-11** (*a*) A cylinder subject to *tensile stress* stretches by an amount $\Delta L$. (*b*) A cylinder subject to *shearing stress* deforms by an amount $\Delta x$, somewhat like a pack of playing cards would. (*c*) A solid sphere subject to uniform *hydraulic stress* from a fluid shrinks in volume by an amount $\Delta V$. All the deformations shown are greatly exaggerated.

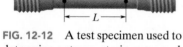

**FIG. 12-12** A test specimen used to determine a stress–strain curve such as that of Fig. 12-13. The change $\Delta L$ that occurs in a certain length $L$ is measured in a tensile stress–strain test.

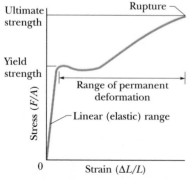

**FIG. 12-13** A stress–strain curve for a steel test specimen such as that of Fig. 12-12. The specimen deforms permanently when the stress is equal to the *yield strength* of the specimen's material. It ruptures when the stress is equal to the *ultimate strength* of the material.

All real "rigid" bodies are to some extent **elastic**, which means that we can change their dimensions slightly by pulling, pushing, twisting, or compressing them. To get a feeling for the orders of magnitude involved, consider a vertical steel rod 1 m long and 1 cm in diameter attached to a factory ceiling. If you hang a subcompact car from the free end of such a rod, the rod will stretch but only by about 0.5 mm, or 0.05%. Furthermore, the rod will return to its original length when the car is removed.

If you hang two cars from the rod, the rod will be permanently stretched and will not recover its original length when you remove the load. If you hang three cars from the rod, the rod will break. Just before rupture, the elongation of the rod will be less than 0.2%. Although deformations of this size seem small, they are important in engineering practice. (Whether a wing under load will stay on an airplane is obviously important.)

Figure 12-11 shows three ways in which a solid might change its dimensions when forces act on it. In Fig. 12-11*a*, a cylinder is stretched. In Fig. 12-11*b*, a cylinder is deformed by a force perpendicular to its long axis, much as we might deform a pack of cards or a book. In Fig. 12-11*c*, a solid object placed in a fluid under high pressure is compressed uniformly on all sides. What the three deformation types have in common is that a **stress,** or deforming force per unit area, produces a **strain,** or unit deformation. In Fig. 12-11, *tensile stress* (associated with stretching) is illustrated in (*a*), *shearing stress* in (*b*), and *hydraulic stress* in (*c*).

The stresses and the strains take different forms in the three situations of Fig. 12-11, but — over the range of engineering usefulness — stress and strain are proportional to each other. The constant of proportionality is called a **modulus of elasticity,** so that

$$\text{stress} = \text{modulus} \times \text{strain}. \qquad (12\text{-}22)$$

In a standard test of tensile properties, the tensile stress on a test cylinder (like that in Fig. 12-12) is slowly increased from zero to the point at which the cylinder fractures, and the strain is carefully measured and plotted. The result is a graph of stress versus strain like that in Fig. 12-13. For a substantial range of applied stresses, the stress–strain relation is linear, and the specimen recovers its original dimensions when the stress is removed; it is here that Eq. 12-22 applies. If the stress is increased beyond the **yield strength** $S_y$ of the specimen, the specimen becomes permanently deformed. If the stress continues to increase, the specimen eventually ruptures, at a stress called the **ultimate strength** $S_u$.

## Tension and Compression

For simple tension or compression, the stress on the object is defined as $F/A$, where $F$ is the magnitude of the force applied perpendicularly to an area $A$ on the object. The strain, or unit deformation, is then the dimensionless quantity $\Delta L/L$, the fractional (or sometimes percentage) change in a length of the specimen. If the specimen is a long rod and the stress does not exceed the yield strength, then not only the entire rod but also every section of it experiences the same strain when a given stress is applied. Because the strain is dimensionless, the modulus in Eq. 12-22 has the same dimensions as the stress—namely, force per unit area.

The modulus for tensile and compressive stresses is called the **Young's modulus** and is represented in engineering practice by the symbol $E$. Equation 12-22 becomes

$$\frac{F}{A} = E \frac{\Delta L}{L}. \tag{12-23}$$

The strain $\Delta L/L$ in a specimen can often be measured conveniently with a *strain gage* (Fig. 12-14). This simple and useful device, which can be attached directly to operating machinery with an adhesive, is based on the principle that its electrical properties are dependent on the strain it undergoes.

Although the Young's modulus for an object may be almost the same for tension and compression, the object's ultimate strength may well be different for the two types of stress. Concrete, for example, is very strong in compression but is so weak in tension that it is almost never used in that manner. Table 12-1 shows the Young's modulus and other elastic properties for some materials of engineering interest.

**FIG. 12-14** A strain gage of overall dimensions 9.8 mm by 4.6 mm. The gage is fastened with adhesive to the object whose strain is to be measured; it experiences the same strain as the object. The electrical resistance of the gage varies with the strain, permitting strains up to 3% to be measured. *(Courtesy Vishay Micro-Measurements Group, Raleigh, NC)*

### Shearing

In the case of shearing, the stress is also a force per unit area, but the force vector lies in the plane of the area rather than perpendicular to it. The strain is the dimensionless ratio $\Delta x/L$, with the quantities defined as shown in Fig. 12-11$b$. The corresponding modulus, which is given the symbol $G$ in engineering practice, is called the **shear modulus.** For shearing, Eq. 12-22 is written as

$$\frac{F}{A} = G \frac{\Delta x}{L}. \tag{12-24}$$

Shearing stresses play a critical role in the buckling of shafts that rotate under load and in bone fractures caused by bending.

### Hydraulic Stress

In Fig. 12-11$c$, the stress is the fluid pressure $p$ on the object, and, as you will see in Chapter 14, pressure is a force per unit area. The strain is $\Delta V/V$, where $V$ is the original volume of the specimen and $\Delta V$ is the absolute value of the change in volume. The corresponding modulus, with symbol $B$, is called the **bulk modulus** of the material. The object is said to be under *hydraulic compression*, and the pressure can be called the *hydraulic stress.* For this situation, we write Eq. 12-22 as

$$p = B \frac{\Delta V}{V}. \tag{12-25}$$

**TABLE 12-1**

**Some Elastic Properties of Selected Materials of Engineering Interest**

| Material | Density $\rho$ (kg/m³) | Young's Modulus $E$ ($10^9$ N/m²) | Ultimate Strength $S_u$ ($10^6$ N/m²) | Yield Strength $S_y$ ($10^6$ N/m²) |
|---|---|---|---|---|
| Steel[a] | 7860 | 200 | 400 | 250 |
| Aluminum | 2710 | 70 | 110 | 95 |
| Glass | 2190 | 65 | 50[b] | — |
| Concrete[c] | 2320 | 30 | 40[b] | — |
| Wood[d] | 525 | 13 | 50[b] | — |
| Bone | 1900 | 9[b] | 170[b] | — |
| Polystyrene | 1050 | 3 | 48 | — |

[a]Structural steel (ASTM-A36).   [c]High strength.
[b]In compression.   [d]Douglas fir.

The bulk modulus is $2.2 \times 10^9$ N/m² for water and $1.6 \times 10^{11}$ N/m² for steel. The pressure at the bottom of the Pacific Ocean, at its average depth of about 4000 m, is $4.0 \times 10^7$ N/m². The fractional compression $\Delta V/V$ of a volume of water due to this pressure is 1.8%; that for a steel object is only about 0.025%. In general, solids—with their rigid atomic lattices—are less compressible than liquids, in which the atoms or molecules are less tightly coupled to their neighbors.

## Sample Problem  12-5

One end of a steel rod of radius $R = 9.5$ mm and length $L = 81$ cm is held in a vise. A force of magnitude $F = 62$ kN is then applied perpendicularly to the end face (uniformly across the area) at the other end. What are the stress on the rod and the elongation $\Delta L$ and strain of the rod?

**KEY IDEAS** (1) The stress is the ratio of the magnitude $F$ of the perpendicular force to the area $A$. The ratio is the left side of Eq. 12-23. (2) The elongation $\Delta L$ is related to the stress and Young's modulus $E$ by Eq. 12-23 ($F/A = E\,\Delta L/L$). (3) Strain is the ratio of the elongation to the initial length $L$.

*Calculations:* To find the stress, we write

$$\text{stress} = \frac{F}{A} = \frac{F}{\pi R^2} = \frac{6.2 \times 10^4 \text{ N}}{(\pi)(9.5 \times 10^{-3} \text{ m})^2}$$

$$= 2.2 \times 10^8 \text{ N/m}^2. \qquad \text{(Answer)}$$

The yield strength for structural steel is $2.5 \times 10^8$ N/m², so this rod is dangerously close to its yield strength.

We find the value of Young's modulus for steel in Table 12-1. Then from Eq. 12-23 we find the elongation:

$$\Delta L = \frac{(F/A)L}{E} = \frac{(2.2 \times 10^8 \text{ N/m}^2)(0.81 \text{ m})}{2.0 \times 10^{11} \text{ N/m}^2}$$

$$= 8.9 \times 10^{-4} \text{ m} = 0.89 \text{ mm}. \qquad \text{(Answer)}$$

For the strain, we have

$$\frac{\Delta L}{L} = \frac{8.9 \times 10^{-4} \text{ m}}{0.81 \text{ m}}$$

$$= 1.1 \times 10^{-3} = 0.11\%. \qquad \text{(Answer)}$$

## Sample Problem  12-6

A table has three legs that are 1.00 m in length and a fourth leg that is longer by $d = 0.50$ mm, so that the table wobbles slightly. A steel cylinder with mass $M = 290$ kg is placed on the table (which has a mass much less than $M$) so that all four legs are compressed but unbuckled and the table is level but no longer wobbles. The legs are wooden cylinders with cross-sectional area $A = 1.0$ cm²; Young's modulus is $E = 1.3 \times 10^{10}$ N/m². What are the magnitudes of the forces on the legs from the floor?

**KEY IDEAS** We take the table plus steel cylinder as our system. The situation is like that in Fig. 12-9, except now we have a steel cylinder on the table. If the tabletop remains level, the legs must be compressed in the following ways: Each of the short legs must be compressed by the same amount (call it $\Delta L_3$) and thus by the same force of magnitude $F_3$. The single long leg must be compressed by a larger amount $\Delta L_4$ and thus by a force with a larger magnitude $F_4$. In other words, for a level tabletop, we must have

$$\Delta L_4 = \Delta L_3 + d. \qquad (12\text{-}26)$$

From Eq. 12-23, we can relate a change in length to the force causing the change with $\Delta L = FL/AE$, where $L$ is the original length of a leg. We can use this relation to replace $\Delta L_4$ and $\Delta L_3$ in Eq. 12-26. However, note that

we can approximate the original length $L$ as being the same for all four legs.

*Calculations:* Making those replacements and that approximation gives us

$$\frac{F_4 L}{AE} = \frac{F_3 L}{AE} + d. \qquad (12\text{-}27)$$

We cannot solve this equation because it has two unknowns, $F_4$ and $F_3$.

To get a second equation containing $F_4$ and $F_3$, we can use a vertical $y$ axis and then write the balance of vertical forces ($F_{\text{net},y} = 0$) as

$$3F_3 + F_4 - Mg = 0, \qquad (12\text{-}28)$$

where $Mg$ is equal to the magnitude of the gravitational force on the system. (*Three* legs have force $\vec{F}_3$ on them.) To solve the simultaneous equations 12-27 and 12-28 for, say, $F_3$, we first use Eq. 12-28 to find that $F_4 = Mg - 3F_3$. Substituting that into Eq. 12-27 then yields, after some algebra,

$$F_3 = \frac{Mg}{4} - \frac{dAE}{4L}$$

$$= \frac{(290 \text{ kg})(9.8 \text{ m/s}^2)}{4}$$

$$- \frac{(5.0 \times 10^{-4} \text{ m})(10^{-4} \text{ m}^2)(1.3 \times 10^{10} \text{ N/m}^2)}{(4)(1.00 \text{ m})}$$

$$= 548 \text{ N} \approx 5.5 \times 10^2 \text{ N}. \qquad \text{(Answer)}$$

From Eq. 12-28 we then find

$$F_4 = Mg - 3F_3 = (290 \text{ kg})(9.8 \text{ m/s}^2) - 3(548 \text{ N})$$
$$\approx 1.2 \text{ kN.} \qquad \text{(Answer)}$$

You can show that to reach their equilibrium configuration, the three short legs are each compressed by 0.42 mm and the single long leg by 0.92 mm.

## REVIEW & SUMMARY

**Static Equilibrium**  A rigid body at rest is said to be in **static equilibrium.** For such a body, the vector sum of the external forces acting on it is zero:

$$\vec{F}_{\text{net}} = 0 \qquad \text{(balance of forces).} \qquad (12\text{-}3)$$

If all the forces lie in the $xy$ plane, this vector equation is equivalent to two component equations:

$$F_{\text{net},x} = 0 \quad \text{and} \quad F_{\text{net},y} = 0 \qquad \text{(balance of forces).} \quad (12\text{-}7, 12\text{-}8)$$

Static equilibrium also implies that the vector sum of the external torques acting on the body about *any* point is zero, or

$$\vec{\tau}_{\text{net}} = 0 \qquad \text{(balance of torques).} \qquad (12\text{-}5)$$

If the forces lie in the $xy$ plane, all torque vectors are parallel to the $z$ axis, and Eq. 12-5 is equivalent to the single component equation

$$\tau_{\text{net},z} = 0 \qquad \text{(balance of torques).} \qquad (12\text{-}9)$$

**Center of Gravity**  The gravitational force acts individually on each element of a body. The net effect of all individual actions may be found by imagining an equivalent total gravitational force $\vec{F}_g$ acting at the **center of gravity.** If the gravitational acceleration $\vec{g}$ is the same for all the elements of the body, the center of gravity is at the center of mass.

**Elastic Moduli**  Three **elastic moduli** are used to describe the elastic behavior (deformations) of objects as they respond to forces that act on them. The **strain** (fractional change in length) is linearly related to the applied **stress** (force per unit area) by the proper modulus, according to the general relation

$$\text{stress} = \text{modulus} \times \text{strain.} \qquad (12\text{-}22)$$

**Tension and Compression**  When an object is under tension or compression, Eq. 12-22 is written as

$$\frac{F}{A} = E \frac{\Delta L}{L}, \qquad (12\text{-}23)$$

where $\Delta L/L$ is the tensile or compressive strain of the object, $F$ is the magnitude of the applied force $\vec{F}$ causing the strain, $A$ is the cross-sectional area over which $\vec{F}$ is applied (perpendicular to $A$, as in Fig. 12-11a), and $E$ is the **Young's modulus** for the object. The stress is $F/A$.

**Shearing**  When an object is under a shearing stress, Eq. 12-22 is written as

$$\frac{F}{A} = G \frac{\Delta x}{L}, \qquad (12\text{-}24)$$

where $\Delta x/L$ is the shearing strain of the object, $\Delta x$ is the displacement of one end of the object in the direction of the applied force $\vec{F}$ (as in Fig. 12-11b), and $G$ is the **shear modulus** of the object. The stress is $F/A$.

**Hydraulic Stress**  When an object undergoes *hydraulic compression* due to a stress exerted by a surrounding fluid, Eq. 12-22 is written as

$$p = B \frac{\Delta V}{V}, \qquad (12\text{-}25)$$

where $p$ is the pressure (*hydraulic stress*) on the object due to the fluid, $\Delta V/V$ (the strain) is the absolute value of the fractional change in the object's volume due to that pressure, and $B$ is the **bulk modulus** of the object.

## QUESTIONS

**1**  Figure 12-15 shows four overhead views of rotating uniform disks that are sliding across a frictionless floor. Three forces, of magnitude $F$, $2F$, or $3F$, act on each disk, either at the rim, at the center, or halfway between rim and center. The force vectors rotate along with the disks, and, in the "snapshots" of Fig. 12-15, point left or right. Which disks are in equilibrium?

**2**  Figure 12-16 shows an overhead view of a uniform stick on which four forces act. Suppose we choose a rotational axis through point $O$, calculate the torques about that axis due to the forces, and find that these torques balance. Will the torques balance if, instead, the rotational axis is chosen to be at (a) point $A$ (on the stick), (b) point $B$ (on line with the stick), or (c) point $C$ (off to one side of the stick)? (d) Suppose, instead, that we find that the torques about point $O$ do not balance. Is there another point about which the torques will balance?

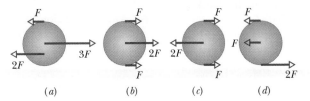

FIG. 12-15  Question 1.

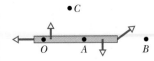

FIG. 12-16  Question 2.

**3** Figure 12-17 shows a mobile of toy penguins hanging from a ceiling. Each crossbar is horizontal, has negligible mass, and extends three times as far to the right of the wire supporting it as to the left. Penguin 1 has mass $m_1 = 48$ kg. What are the masses of (a) penguin 2, (b) penguin 3, and (c) penguin 4?

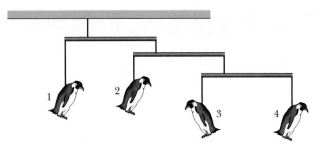

FIG. 12-17   Question 3.

**4** In Fig. 12-18, a rigid beam is attached to two posts that are fastened to a floor. A small but heavy safe is placed at the six positions indicated, in turn. Assume that the mass of the beam is negligible compared to that of the safe. (a) Rank the positions according to the force on post $A$ due to the safe, greatest compression first, greatest tension last, and indicate where, if anywhere, the force is zero. (b) Rank them according to the force on post $B$.

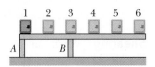

FIG. 12-18   Question 4.

**5** Figure 12-19 shows three situations in which the same horizontal rod is supported by a hinge on a wall at one end and a cord at its other end. Without written calculation, rank the situations according to the magnitudes of (a) the force on the rod from the cord, (b) the vertical force on the rod from the hinge, and (c) the horizontal force on the rod from the hinge, greatest first.

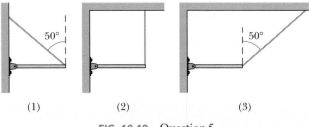

FIG. 12-19   Question 5.

**6** A ladder leans against a frictionless wall but is prevented from falling because of friction between it and the ground. Suppose you shift the base of the ladder toward the wall. Determine whether the following become larger, smaller, or stay the same (in magnitude): (a) the normal force on the ladder from the ground, (b) the force on the ladder from the wall, (c) the static frictional force on the ladder from the ground, and (d) the maximum value $f_{s,\text{max}}$ of the static frictional force.

**7** In Fig. 12-20, a vertical rod is hinged at its lower end and attached to a cable at its upper end. A horizontal force $\vec{F}_a$ is to be applied to the rod as shown. If the point at which the force is applied is moved up the rod, does the tension in the cable increase, decrease, or remain the same?

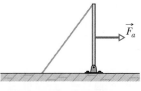

FIG. 12-20   Question 7.

**8** Three piñatas hang from the (stationary) assembly of massless pulleys and cords seen in Fig. 12-21. One long cord runs from the ceiling at the right to the lower pulley at the left, looping halfway around all the pulleys. Several shorter cords suspend pulleys from the ceiling or piñatas from the pulleys. The weights (in newtons) of two piñatas are given. (a) What is the weight of the third piñata? (*Hint:* A cord that loops halfway around a pulley pulls on the pulley with a net force that is twice the tension in the cord.) (b) What is the tension in the short cord labeled with $T$?

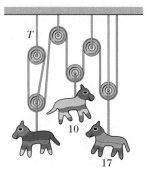

FIG. 12-21   Question 8.

**9** In Fig. 12-22, a stationary 5 kg rod $AC$ is held against a wall by a rope and friction between rod and wall. The uniform rod is 1 m long, and angle $\theta = 30°$. (a) If you are to find the magnitude of the force $\vec{T}$ on the rod from the rope with a single equation, at what labeled point should a rotational axis be placed? With that choice of axis and counterclockwise torques positive, what is the sign of (b) the torque $\tau_w$ due to the rod's weight and (c) the torque $\tau_r$ due to the pull on the rod by the rope? (d) Is the magnitude of $\tau_r$ greater than, less than, or equal to the magnitude of $\tau_w$?

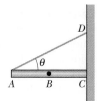

FIG. 12-22
Question 9.

**10** Figure 12-23 shows a horizontal block that is suspended by two wires, $A$ and $B$, which are identical except for their original lengths. The center of mass of the block is closer to wire $B$ than to wire $A$. (a) Measuring torques about the block's center of mass, state whether the magnitude of the torque due to wire $A$ is greater than, less than, or equal to the magnitude of the torque due to wire $B$. (b) Which wire exerts more force on the block? (c) If the wires are now equal in length, which one was originally shorter (before the block was suspended)?

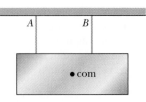

FIG. 12-23   Question 10.

## PROBLEMS

| GO | Tutoring problem available (at instructor's discretion) in *WileyPLUS* and WebAssign |
| --- | --- |
| SSM | Worked-out solution available in Student Solutions Manual |
| • – ••• | Number of dots indicates level of problem difficulty |
| | Additional information available in *The Flying Circus of Physics* and at flyingcircusofphysics.com |

WWW   Worked-out solution is at —
ILW   Interactive solution is at —
http://www.wiley.com/college/halliday

### sec. 12-4   The Center of Gravity

•1   Because *g* varies so little over the extent of most structures, any structure's center of gravity effectively coincides with its center of mass. Here is a fictitious example where *g* varies more significantly. Figure 12-24 shows an array of six particles, each with mass *m*, fixed to the edge of a rigid structure of negligible mass. The distance between adjacent particles along the edge is 2.00 m. The following table gives the value of $g$ ($m/s^2$) at each particle's location.

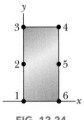

**FIG. 12-24**
Problem 1.

Using the coordinate system shown, find (a) the *x* coordinate $x_{com}$ and (b) the *y* coordinate $y_{com}$ of the center of mass of the six-particle system. Then find (c) the *x* coordinate $x_{cog}$ and (d) the *y* coordinate $y_{cog}$ of the center of gravity of the six-particle system.

| Particle | $g$ | Particle | $g$ |
| --- | --- | --- | --- |
| 1 | 8.00 | 4 | 7.40 |
| 2 | 7.80 | 5 | 7.60 |
| 3 | 7.60 | 6 | 7.80 |

### sec. 12-5   Some Examples of Static Equilibrium

•2   An archer's bow is drawn at its midpoint until the tension in the string is equal to the force exerted by the archer. What is the angle between the two halves of the string?

•3   A rope of negligible mass is stretched horizontally between two supports that are 3.44 m apart. When an object of weight 3160 N is hung at the center of the rope, the rope is observed to sag by 35.0 cm. What is the tension in the rope?   ILW

•4   A physics Brady Bunch, whose weights in newtons are indicated in Fig. 12-25, is balanced on a seesaw. What is the number of the person who causes the largest torque about the rotation axis at *fulcrum f* directed (a) out of the page and (b) into the page?

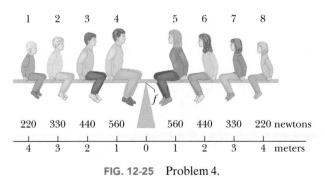

**FIG. 12-25**   Problem 4.

•5   In Fig. 12-26, a uniform sphere of mass *m* = 0.85 kg and radius *r* = 4.2 cm is held in place by a massless rope attached to a frictionless wall a distance *L* = 8.0 cm above the center of the sphere. Find (a) the tension in the rope and (b) the force on the sphere from the wall.   SSM   WWW

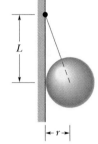

**FIG. 12-26**
Problem 5.

•6   An automobile with a mass of 1360 kg has 3.05 m between the front and rear axles. Its center of gravity is located 1.78 m behind the front axle. With the automobile on level ground, determine the magnitude of the force from the ground on (a) each front wheel (assuming equal forces on the front wheels) and (b) each rear wheel (assuming equal forces on the rear wheels).

•7   A diver of weight 580 N stands at the end of a diving board of length *L* = 4.5 m and negligible mass (Fig. 12-27). The board is fixed to two pedestals separated by distance *d* = 1.5 m. Of the forces acting on the board, what are the (a) magnitude and (b) direction (up or down) of the force from the left pedestal and the (c) magnitude and (d) direction (up or down) of the force from the right pedestal? (e) Which pedestal (left or right) is being stretched, and (f) which is being compressed?   SSM

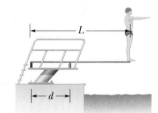

**FIG. 12-27**   Problem 7.

•8   A scaffold of mass 60 kg and length 5.0 m is supported in a horizontal position by a vertical cable at each end. A window washer of mass 80 kg stands at a point 1.5 m from one end. What is the tension in (a) the nearer cable and (b) the farther cable?

•9   A 75 kg window cleaner uses a 10 kg ladder that is 5.0 m long. He places one end on the ground 2.5 m from a wall, rests the upper end against a cracked window, and climbs the ladder. He is 3.0 m up along the ladder when the window breaks. Neglect friction between the ladder and window and assume that the base of the ladder does not slip. When the window is on the verge of breaking, what are (a) the magnitude of the force on the window from the ladder, (b) the magnitude of the force on the ladder from the ground, and (c) the angle (relative to the horizontal) of that force on the ladder?

•10   In Fig. 12-28, a man is trying to get his car out of mud on the shoulder of a road. He ties one end of a rope tightly around the front bumper and the other end tightly around a utility pole 18 m away. He then pushes sideways on the rope at its midpoint with a force of 550 N, displacing the center of the rope 0.30 m from its previous position, and the car barely moves. What is the magnitude of the force on the car from the rope? (The rope stretches somewhat.)

**FIG. 12-28** Problem 10.

•**11** A meter stick balances horizontally on a knife-edge at the 50.0 cm mark. With two 5.00 g coins stacked over the 12.0 cm mark, the stick is found to balance at the 45.5 cm mark. What is the mass of the meter stick? **SSM**

•**12** The system in Fig. 12-29 is in equilibrium, with the string in the center exactly horizontal. Block $A$ weighs 40 N, block $B$ weighs 50 N, and angle $\phi$ is 35°. Find (a) tension $T_1$, (b) tension $T_2$, (c) tension $T_3$, and (d) angle $\theta$. **GO**

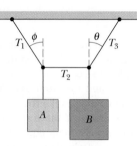

**FIG. 12-29** Problem 12.

•**13** Forces $\vec{F}_1$, $\vec{F}_2$, and $\vec{F}_3$ act on the structure of Fig. 12-30, shown in an overhead view. We wish to put the structure in equilibrium by applying a fourth force, at a point such as $P$. The fourth force has vector components $\vec{F}_h$ and $\vec{F}_v$. We are given that $a = 2.0$ m, $b = 3.0$ m, $c = 1.0$ m, $F_1 = 20$ N, $F_2 = 10$ N, and $F_3 = 5.0$ N. Find (a) $F_h$, (b) $F_v$, and (c) $d$. **ILW**

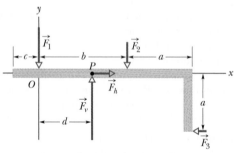

**FIG. 12-30** Problem 13.

•**14** A uniform cubical crate is 0.750 m on each side and weighs 500 N. It rests on a floor with one edge against a very small, fixed obstruction. At what least height above the floor must a horizontal force of magnitude 350 N be applied to the crate to tip it?

•**15** To crack a certain nut in a nutcracker, forces with magnitudes of at least 40 N must act on its shell from both sides. For the nutcracker of Fig. 12-31, with distances $L = 12$ cm and $d = 2.6$ cm, what are the force components $F_\perp$ (perpendicular to the handles) corresponding to that 40 N?

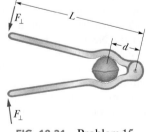

**FIG. 12-31** Problem 15.

•**16** In Fig. 12-32, a horizontal scaffold, of length 2.00 m and

**FIG. 12-32** Problem 16.

uniform mass 50.0 kg, is suspended from a building by two cables. The scaffold has dozens of paint cans stacked on it at various points. The total mass of the paint cans is 75.0 kg. The tension in the cable at the right is 722 N. How far horizontally from *that* cable is the center of mass of the system of paint cans?

•**17** Figure 12-33 shows the anatomical structures in the lower leg and foot that are involved in standing on tiptoe, with the heel raised slightly off the floor so that the foot effectively contacts the floor only at point $P$. Assume distance $a = 5.0$ cm, distance $b = 15$ cm, and the person's weight $W = 900$ N. Of the forces acting on the foot, what are the (a) magnitude and (b) direction (up or down) of the force at point $A$ from the calf muscle and the (c) magnitude and (d) direction (up or down) of the force at point $B$ from the lower leg bones?

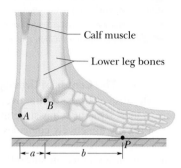

**FIG. 12-33** Problem 17.

•**18** A bowler holds a bowling ball ($M = 7.2$ kg) in the palm of his hand (Fig. 12-34). His upper arm is vertical, his lower arm (1.8 kg) is horizontal. What is the magnitude of (a) the force of the biceps muscle on the lower arm and (b) the force between the bony structures at the elbow contact point?

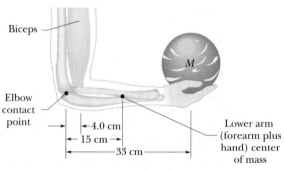

**FIG. 12-34** Problem 18.

•**19** In Fig. 12-35, a uniform beam of weight 500 N and length 3.0 m is suspended horizontally. On the left it is hinged to a wall; on the right it is supported by a cable bolted to the wall at distance $D$ above the beam. The least tension that will snap the cable is 1200 N. (a) What value of $D$ corresponds to that tension? (b) To prevent the cable from snapping, should $D$ be increased or decreased from that value?

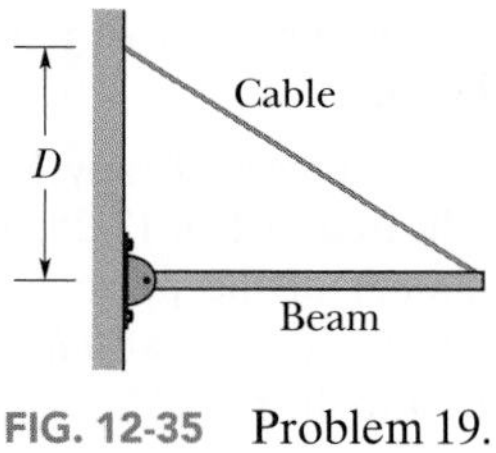

**FIG. 12-35** Problem 19.

•**20** In Fig. 12-36, horizontal scaffold 2, with uniform mass $m_2 = 30.0$ kg and length $L_2 = 2.00$ m, hangs from horizontal scaffold 1, with uniform mass $m_1 = 50.0$ kg. A 20.0 kg box of nails lies on scaffold 2, centered at distance $d = 0.500$ m from the left end. What is the tension $T$ in the cable indicated? **GO**

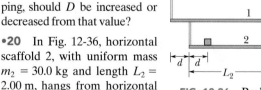

**FIG. 12-36** Problem 20.

••**21** In Fig. 12-37, what magnitude of (constant) force $\vec{F}$ applied horizontally at the axle of the wheel is necessary to raise the wheel over an obstacle of height $h = 3.00$ cm? The wheel's radius is $r = 6.00$ cm, and its mass is $m = 0.800$ kg.
**SSM WWW**

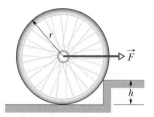

**FIG. 12-37** Problem 21.

••**22** In Fig. 12-38, a climber with a weight of 533.8 N is held by a belay rope connected to her climbing harness and belay device; the force of the rope on her has a line of action through her center of mass. The indicated angles are $\theta = 40.0°$ and $\phi = 30.0°$. If her feet are on the verge of sliding on the vertical wall, what is the coefficient of static friction between her climbing shoes and the wall? ✈

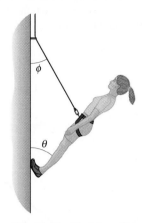

**FIG. 12-38** Problem 22.

••**23** In Fig. 12-39, a 15 kg block is held in place via a pulley system. The person's upper arm is vertical; the forearm is at angle $\theta = 30°$ with the horizontal. Forearm and hand together have a mass of 2.0 kg, with a center of mass at distance $d_1 = 15$ cm from the contact point of the forearm bone and the upper-arm bone (humerus). The triceps muscle pulls vertically upward on the forearm at distance $d_2 = 2.5$ cm behind that contact point. Distance $d_3$ is 35 cm. What are the (a) magnitude and (b) direction (up or down) of the force on the forearm from the triceps muscle and the (c) magnitude and (d) direction (up or down) of the force on the forearm from the humerus?

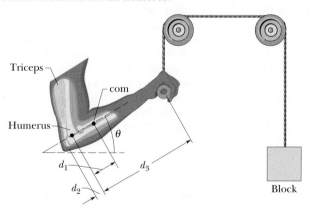

**FIG. 12-39** Problem 23.

••**24** In Fig. 12-40, a climber leans out against a vertical ice wall that has negligible friction. Distance $a$ is 0.914 m and distance $L$ is 2.10 m. His center of mass is distance $d = 0.940$ m from the feet–ground contact point. If he is on the verge of sliding, what is the coefficient of static friction between feet and ground? ✈

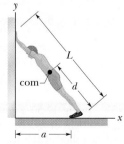

**FIG. 12-40** Problem 24.

••**25** In Fig. 12-41, one end of a uniform beam of weight 222 N is hinged to a wall; the other end is supported by a wire that makes angles $\theta = 30.0°$ with both wall and beam. Find (a) the tension in the wire and the (b) horizontal and (c) vertical components of the force of the hinge on the beam.

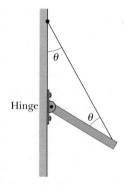

**FIG. 12-41** Problem 25.

••**26** In Fig. 12-42, a 55 kg rock climber is in a lie-back climb along a fissure, with hands pulling on one side of the fissure and feet pressed against the opposite side. The fissure has width $w = 0.20$ m, and the center of mass of the climber is a horizontal distance $d = 0.40$ m from the fissure. The coefficient of static friction between hands and rock is $\mu_1 = 0.40$, and between boots and rock it is $\mu_2 = 1.2$. (a) What is the least horizontal pull by the hands and push by the feet that will keep the climber stable? (b) For the horizontal pull of (a), what must be the vertical distance $h$ between hands and feet? If the climber encounters wet rock, so that $\mu_1$ and $\mu_2$ are reduced, what happens to (c) the answer to (a) and (d) the answer to (b)? **GO** ✈

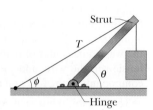

**FIG. 12-42** Problem 26.

••**27** The system in Fig. 12-43 is in equilibrium. A concrete block of mass 225 kg hangs from the end of the uniform strut of mass 45.0 kg. For angles $\phi = 30.0°$ and $\theta = 45.0°$, find (a) the tension $T$ in the cable and the (b) horizontal and (c) vertical components of the force on the strut from the hinge. **ILW**

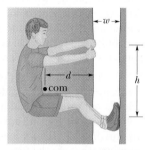

**FIG. 12-43** Problem 27.

••**28** In Fig. 12-44, a 50.0 kg uniform square sign, of edge length $L = 2.00$ m, is hung from a horizontal rod of length $d_h = 3.00$ m and negligible mass. A cable is attached to the end of the rod and to a point on the wall at distance $d_v = 4.00$ m above the point where the rod is hinged to the wall. (a) What is the tension in the cable? What are the (b) magnitude and (c) direction (left or right) of the horizontal component of the force on the rod from the wall, and the (d) magnitude and (e) direction (up or down) of the vertical component of this force?

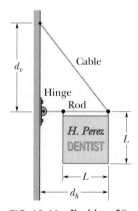

**FIG. 12-44** Problem 28.

••**29** In Fig. 12-45, a nonuniform bar is suspended at rest in a horizontal position by two

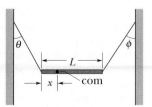

**FIG. 12-45** Problem 29.

massless cords. One cord makes the angle $\theta = 36.9°$ with the vertical; the other makes the angle $\phi = 53.1°$ with the vertical. If the length $L$ of the bar is 6.10 m, compute the distance $x$ from the left end of the bar to its center of mass.

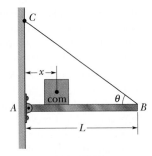

**FIG. 12-46**
Problems 30 and 32.

••**30** In Fig. 12-46, suppose the length $L$ of the uniform bar is 3.00 m and its weight is 200 N. Also, let the block's weight $W = 300$ N and the angle $\theta = 30.0°$. The wire can withstand a maximum tension of 500 N. (a) What is the maximum possible distance $x$ before the wire breaks? With the block placed at this maximum $x$, what are the (b) horizontal and (c) vertical components of the force on the bar from the hinge at $A$?

••**31** A door has a height of 2.1 m along a $y$ axis that extends vertically upward and a width of 0.91 m along an $x$ axis that extends outward from the hinged edge of the door. A hinge 0.30 m from the top and a hinge 0.30 m from the bottom each support half the door's mass, which is 27 kg. In unit-vector notation, what are the forces on the door at (a) the top hinge and (b) the bottom hinge?

••**32** In Fig. 12-46, a thin horizontal bar $AB$ of negligible weight and length $L$ is hinged to a vertical wall at $A$ and supported at $B$ by a thin wire $BC$ that makes an angle $\theta$ with the horizontal. A block of weight $W$ can be moved anywhere along the bar; its position is defined by the distance $x$ from the wall to its center of mass. As a function of $x$, find (a) the tension in the wire, and the (b) horizontal and (c) vertical components of the force on the bar from the hinge at $A$.

••**33** A cubical box is filled with sand and weighs 890 N. We wish to "roll" the box by pushing horizontally on one of the upper edges. (a) What minimum force is required? (b) What minimum coefficient of static friction between box and floor is required? (c) If there is a more efficient way to roll the box, find the smallest possible force that would have to be applied directly to the box to roll it. (*Hint:* At the onset of tipping, where is the normal force located?) SSM WWW

••**34** Figure 12-47 shows a 70 kg climber hanging by only the *crimp hold* of one hand on the edge of a shallow horizontal ledge in a rock wall. (The fingers are pressed down to gain purchase.) Her feet touch the rock wall at distance $H = 2.0$ m directly below her crimped fingers but do not provide any support. Her center of mass is distance $a = 0.20$ m from the wall. Assume that the force from the ledge supporting her fingers is equally shared by the four fingers. What are the values of the (a) horizontal component $F_h$ and (b) vertical component $F_v$ of the force on *each* fingertip?

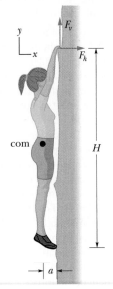

**FIG. 12-47**
Problem 34.

••**35** Figure 12-48a shows a vertical uniform beam of length $L$ that is hinged at its lower end. A horizontal force $\vec{F}_a$ is applied to the beam at distance $y$ from the lower end. The beam remains vertical because of a cable attached at the upper end, at angle $\theta$ with the horizontal. Figure 12-48b gives the tension $T$ in the cable as a function of the position of the applied force given as a fraction $y/L$ of the beam length. The scale of the $T$ axis is set by $T_s = 600$ N. Figure 12-48c gives the magnitude $F_h$ of the horizontal force on the beam from the hinge, also as a function of $y/L$. Evaluate (a) angle $\theta$ and (b) the magnitude of $\vec{F}_a$.

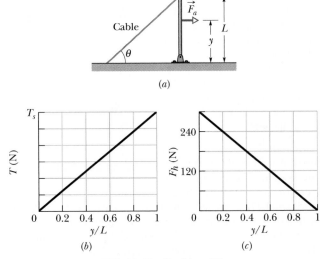

**FIG. 12-48** Problem 35.

••**36** In Fig. 12-49, the driver of a car on a horizontal road makes an emergency stop by applying the brakes so that all four wheels lock and skid along the road. The coefficient of kinetic friction between tires and road is 0.40. The separation between the front and rear axles is $L = 4.2$ m, and the center of mass of the car is located at distance $d = 1.8$ m behind the front axle and distance $h = 0.75$ m above the road. The car weighs 11 kN. Find the magnitude of (a) the braking acceleration of the car, (b) the normal force on each rear wheel, (c) the normal force on each front wheel, (d) the braking force on each rear wheel, and (e) the braking force on each front wheel. (*Hint:* Although the car is not in translational equilibrium, it *is* in rotational equilibrium.)

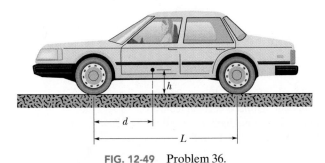

**FIG. 12-49** Problem 36.

••**37** In Fig. 12-50, a uniform plank, with a length $L$ of 6.10 m and a weight of 445 N, rests on the ground and against a frictionless roller at the top of a wall of height $h = 3.05$ m. The plank remains in equilibrium for any value of $\theta \geq 70°$ but slips if $\theta < 70°$. Find the coefficient of static friction between the plank and the ground.

•• **38** In Fig. 12-51, uniform beams $A$ and $B$ are attached to a wall with hinges and loosely bolted together (there is no torque of one on the other). Beam $A$ has length $L_A = 2.40$ m and mass 54.0 kg; beam $B$ has mass 68.0 kg. The two hinge points are separated by distance $d = 1.80$ m. In unit-vector notation, what is the force on (a) beam $A$ due to its hinge, (b) beam $A$ due to the bolt, (c) beam $B$ due to its hinge, and (d) beam $B$ due to the bolt?

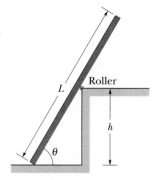

**FIG. 12-50** Problem 37.

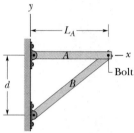

**FIG. 12-51** Problem 38.

••• **39** A crate, in the form of a cube with edge lengths of 1.2 m, contains a piece of machinery; the center of mass of the crate and its contents is located 0.30 m above the crate's geometrical center. The crate rests on a ramp that makes an angle $\theta$ with the horizontal. As $\theta$ is increased from zero, an angle will be reached at which the crate will either tip over or start to slide down the ramp. If the coefficient of static friction $\mu_s$ between ramp and crate is 0.60, (a) does the crate tip or slide and (b) at what angle $\theta$ does this occur? If $\mu_s = 0.70$, (c) does the crate tip or slide and (d) at what angle $\theta$ does this occur? (*Hint:* At the onset of tipping, where is the normal force located?)

••• **40** In Sample Problem 12-2, let the coefficient of static friction $\mu_s$ between the ladder and the pavement be 0.53. How far (in percent) up the ladder must the firefighter go to put the ladder on the verge of sliding?

••• **41** For the stepladder shown in Fig. 12-52, sides $AC$ and $CE$ are each 2.44 m long and hinged at $C$. Bar $BD$ is a tie-rod 0.762 m long, halfway up. A man weighing 854 N climbs 1.80 m along the ladder. Assuming that the floor is frictionless and neglecting the mass of the ladder, find (a) the tension in the tie-rod and the magnitudes of the forces on the ladder from the floor at (b) $A$ and (c) $E$. (*Hint:* Isolate parts of the ladder in applying the equilibrium conditions.)

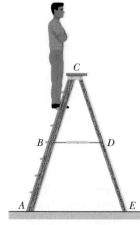

**FIG. 12-52** Problem 41.

••• **42** Figure 12-53a shows a horizontal uniform beam of mass $m_b$ and length $L$ that is supported on the left by a hinge attached to a wall and on the right by a cable at angle $\theta$ with the horizontal. A package of mass $m_p$ is positioned on the beam at a distance $x$ from the left end. The total mass is $m_b + m_p = 61.22$ kg. Figure 12-53b gives the tension $T$ in the cable as a function of the package's position given as a fraction $x/L$ of the beam length. The scale of

the $T$ axis is set by $T_a = 500$ N and $T_b = 700$ N. Evaluate (a) angle $\theta$, (b) mass $m_b$, and (c) mass $m_p$.

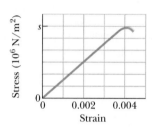

**FIG. 12-53** Problem 42.

### sec. 12-7 Elasticity

• **43** A horizontal aluminum rod 4.8 cm in diameter projects 5.3 cm from a wall. A 1200 kg object is suspended from the end of the rod. The shear modulus of aluminum is $3.0 \times 10^{10}$ N/m². Neglecting the rod's mass, find (a) the shear stress on the rod and (b) the vertical deflection of the end of the rod. **SSM ILW**

• **44** Figure 12-54 shows the stress–strain curve for a material. The scale of the stress axis is set by $s = 300$, in units of $10^6$ N/m². What are (a) the Young's modulus and (b) the approximate yield strength for this material?

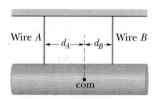

**FIG. 12-54** Problem 44.

•• **45** In Fig. 12-55, a 103 kg uniform log hangs by two steel wires, $A$ and $B$, both of radius 1.20 mm. Initially, wire $A$ was 2.50 m long and 2.00 mm shorter than wire $B$. The log is now horizontal. What are the magnitudes of the forces on it from (a) wire $A$ and (b) wire $B$? (c) What is the ratio $d_A/d_B$? **GO**

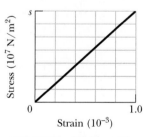

**FIG. 12-55** Problem 45.

•• **46** Figure 12-56 shows the stress versus strain plot for an aluminum wire that is stretched by a machine pulling in opposite directions at the two ends of the wire. The scale of the stress axis is set by $s = 7.0$, in units of $10^7$ N/m². The wire has an initial length of 0.800 m and an initial cross-sectional area of $2.00 \times 10^{-6}$ m². How much work does the force from the machine do on the wire to produce a strain of $1.00 \times 10^{-3}$?

•• **47** In Fig. 12-57, a lead brick rests horizontally on cylinders $A$ and $B$. The areas of the top faces of the cylinders are related by $A_A = 2A_B$; the Young's moduli of the cylinders are related by $E_A = 2E_B$. The

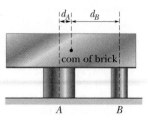

**FIG. 12-57** Problem 47.

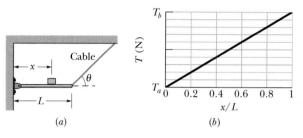

cylinders had identical lengths before the brick was placed on them. What fraction of the brick's mass is supported (a) by cylinder $A$ and (b) by cylinder $B$? The horizontal distances between the center of mass of the brick and the centerlines of the cylinders are $d_A$ for cylinder $A$ and $d_B$ for cylinder $B$. (c) What is the ratio $d_A/d_B$?

••48  Figure 12-58 shows an approximate plot of stress versus strain for a spider-web thread, out to the point of breaking at a strain of 2.00. The vertical axis scale is set by $a = 0.12$ GN/m$^2$, $b = 0.30$ GN/m$^2$, and $c = 0.80$ GN/m$^2$. Assume that the thread has an initial length of 0.80 cm, an initial cross-sectional area of $8.0 \times 10^{-12}$ m$^2$, and (during stretching) a constant volume. Assume also that when the single thread snares a flying insect, the insect's kinetic energy is transferred to the stretching of the thread. (a) How much kinetic energy would put the thread on the verge of breaking? What is the kinetic energy of (b) a fruit fly of mass 6.00 mg and speed 1.70 m/s and (c) a bumble bee of mass 0.388 g and speed 0.420 m/s? Would (d) the fruit fly and (e) the bumble bee break the thread?

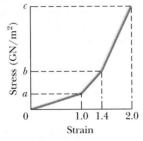

FIG. 12-58    Problem 48.

••49   A tunnel of length $L = 150$ m, height $H = 7.2$ m, and width 5.8 m (with a flat roof) is to be constructed at distance $d = 60$ m beneath the ground. (See Fig. 12-59.) The tunnel roof is to be supported entirely by square steel columns, each with a cross-sectional area of 960 cm$^2$. The mass of 1.0 cm$^3$ of the ground material is 2.8 g. (a) What is the total weight of the ground material the columns must support? (b) How many columns are needed to keep the compressive stress on each column at one-half its ultimate strength?

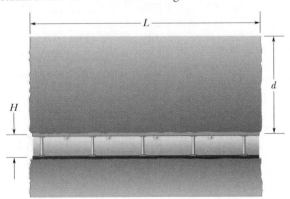

FIG. 12-59    Problem 49.

•••50  Figure 12-60 represents an insect caught at the midpoint of a spider-web thread. The thread breaks under a stress of $8.20 \times 10^8$ N/m$^2$ and a strain of 2.00. Initially, it was horizontal and had a length of 2.00 cm and a cross-sectional area of $8.00 \times 10^{-12}$ m$^2$. As the thread was stretched under the weight of the insect, its volume remained constant. If the weight of the insect puts the thread on the verge of breaking, what is the insect's mass? (A spider's web is built to

FIG. 12-60    Problem 50.

break if a potentially harmful insect, such as a bumble bee, becomes snared in the web.)

•••51  Figure 12-61 is an overhead view of a rigid rod that turns about a vertical axle until the identical rubber stoppers $A$ and $B$ are forced against rigid walls at distances $r_A = 7.0$ cm and $r_B = 4.0$ cm from the axle. Initially the stoppers touch the walls without being compressed. Then force $\vec{F}$ of magnitude 220 N is applied perpendicular to the rod at a distance $R = 5.0$ cm from the axle. Find the magnitude of the force compressing (a) stopper $A$ and (b) stopper $B$.

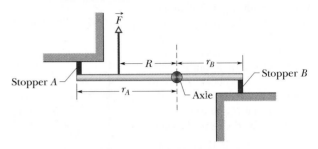

FIG. 12-61    Problem 51.

**Additional Problems**

52   Figure 12-62$a$ shows a uniform ramp between two buildings that allows for motion between the buildings due to strong winds. At its left end, it is hinged to the building wall; at its right end, it has a roller that can roll along the building wall. There is no vertical force on the roller from the building, only a horizontal force with magnitude $F_h$. The horizontal distance between the buildings is $D = 4.00$ m. The rise of the ramp is $h = 0.490$ m. A man walks across the ramp from the left. Figure 12-62$b$ gives $F_h$ as a function of the horizontal distance $x$ of the man from the building at the left. The scale of the $F_h$ axis is set by $a = 20$ kN and $b = 25$ kN. What are the masses of (a) the ramp and (b) the man?

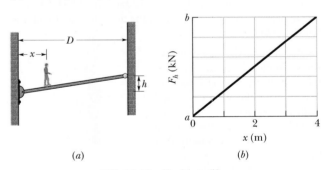

(a)                                        (b)

FIG. 12-62    Problem 52.

53   In Fig. 12-63, a 10 kg sphere is supported on a frictionless plane inclined at angle $\theta = 45°$ from the horizontal. Angle $\phi$ is 25°. Calculate the tension in the cable.

54   In Fig. 12-64$a$, a uniform 40.0 kg beam is centered over two rollers. Vertical lines across the beam mark off equal lengths. Two of the lines are centered over the rollers; a 10.0 kg package of tamales is centered over roller $B$. What

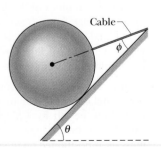

FIG. 12-63    Problem 53.

are the magnitudes of the forces on the beam from (a) roller $A$ and (b) roller $B$? The beam is then rolled to the left until the right-hand end is centered over roller $B$ (Fig. 12-64b). What now are the magnitudes of the forces on the beam from (c) roller $A$ and (d) roller $B$? Next, the beam is rolled to the right. Assume that it has a length of 0.800 m. (e) What horizontal distance between the package and roller $B$ puts the beam on the verge of losing contact with roller $A$?

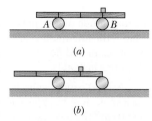

**FIG. 12-64** Problem 54.

**55** In Fig. 12-65, an 817 kg construction bucket is suspended by a cable $A$ that is attached at $O$ to two other cables $B$ and $C$, making angles $\theta_1 = 51.0°$ and $\theta_2 = 66.0°$ with the horizontal. Find the tensions in (a) cable $A$, (b) cable $B$, and (c) cable $C$. (*Hint:* To avoid solving two equations in two unknowns, position the axes as shown in the figure.) **SSM**

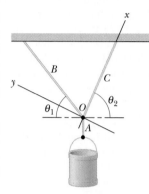

**FIG. 12-65** Problem 55.

**56** In Fig. 12-66, a package of mass $m$ hangs from a short cord that is tied to the wall via cord 1 and to the ceiling via cord 2. Cord 1 is at angle $\phi = 40°$ with the horizontal; cord 2 is at angle $\theta$. (a) For what value of $\theta$ is the tension in cord 2 minimized? (b) In terms of $mg$, what is the minimum tension in cord 2?

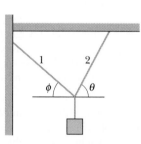

**FIG. 12-66** Problem 56.

**57** The force $\vec{F}$ in Fig. 12-67 keeps the 6.40 kg block and the pulleys in equilibrium. The pulleys have negligible mass and friction. Calculate the tension $T$ in the upper cable. (*Hint:* When a cable wraps halfway around a pulley as here, the magnitude of its net force on the pulley is twice the tension in the cable.) **ILW**

**58** In Fig. 12-68, two identical, uniform, and frictionless spheres, each of mass $m$, rest in a rigid rectangular container. A line connecting their centers is at 45° to the horizontal. Find the magnitudes of the forces on the spheres from (a) the bottom of the container, (b) the left side of the container, (c) the right side of the container, and (d) each other. (*Hint:* The force of one sphere on the other is directed along the center–center line.)

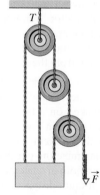

**FIG. 12-67**
Problem 57.

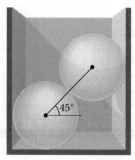

**FIG. 12-68** Problem 58.

**59** Four bricks of length $L$, identical and uniform, are stacked on top of one another (Fig. 12-69) in such a way that part of each extends beyond the one beneath. Find, in terms of $L$, the maximum values of (a) $a_1$, (b) $a_2$, (c) $a_3$, (d) $a_4$, and (e) $h$, such that the stack is in equilibrium.

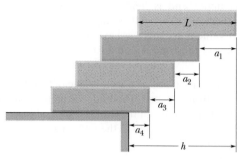

**FIG. 12-69** Problem 59.

**60** After a fall, a 95 kg rock climber finds himself dangling from the end of a rope that had been 15 m long and 9.6 mm in diameter but has stretched by 2.8 cm. For the rope, calculate (a) the strain, (b) the stress, and (c) the Young's modulus.

**61** In Fig. 12-70, a rectangular slab of slate rests on a bedrock surface inclined at angle $\theta = 26°$. The slab has length $L = 43$ m, thickness $T = 2.5$ m, and width $W = 12$ m, and 1.0 cm³ of it has a mass of 3.2 g. The coefficient of static friction between slab and bedrock is 0.39. (a) Calculate the

**FIG. 12-70** Problem 61.

component of the gravitational force on the slab parallel to the bedrock surface. (b) Calculate the magnitude of the static frictional force on the slab. By comparing (a) and (b), you can see that the slab is in danger of sliding. This is prevented only by chance protrusions of bedrock. (c) To stabilize the slab, bolts are to be driven perpendicular to the bedrock surface (two bolts are shown). If each bolt has a cross-sectional area of 6.4 cm² and will snap under a shearing stress of $3.6 \times 10^8$ N/m², what is the minimum number of bolts needed? Assume that the bolts do not affect the normal force. **SSM**

**62** A uniform ladder whose length is 5.0 m and whose weight is 400 N leans against a frictionless vertical wall. The coefficient of static friction between the level ground and the foot of the ladder is 0.46. What is the greatest distance the foot of the ladder can be placed from the base of the wall without the ladder immediately slipping?

**63** In Fig. 12-71, block $A$ (mass 10 kg) is in equilibrium, but it would slip if block $B$ (mass 5.0 kg) were any heavier. For angle $\theta = 30°$, what is the coefficient of static friction between block $A$ and the surface below it? **SSM**

**64** A mine elevator is supported by a single steel cable 2.5 cm in diameter. The total mass of the elevator cage and occupants is 670 kg. By how

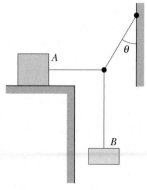

**FIG. 12-71** Problem 63.

much does the cable stretch when the elevator hangs by (a) 12 m of cable and (b) 362 m of cable? (Neglect the mass of the cable.)

**65** In Fig. 12-72, a uniform rod of mass $m$ is hinged to a building at its lower end, while its upper end is held in place by a rope attached to the wall. If angle $\theta_1 = 60°$, what value must angle $\theta_2$ have so that the tension in the rope is equal to $mg/2$? **SSM**

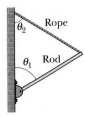

FIG. 12-72
Problem 65.

**66** A 73 kg man stands on a level bridge of length $L$. He is at distance $L/4$ from one end. The bridge is uniform and weighs 2.7 kN. What are the magnitudes of the vertical forces on the bridge from its supports at (a) the end farther from him and (b) the nearer end?

**67** A makeshift swing is constructed by making a loop in one end of a rope and tying the other end to a tree limb. A child is sitting in the loop with the rope hanging vertically when the child's father pulls on the child with a horizontal force and displaces the child to one side. Just before the child is released from rest, the rope makes an angle of 15° with the vertical and the tension in the rope is 280 N. (a) How much does the child weigh? (b) What is the magnitude of the (horizontal) force of the father on the child just before the child is released? (c) If the maximum horizontal force the father can exert on the child is 93 N, what is the maximum angle with the vertical the rope can make while the father is pulling horizontally?

**68** The system in Fig. 12-73 is in equilibrium. The angles are $\theta_1 = 60°$ and $\theta_2 = 20°$, and the ball has mass $M = 2.0$ kg. What is the tension in (a) string $ab$ and (b) string $bc$?

**69** Figure 12-74 shows a stationary arrangement of two crayon boxes and three cords. Box $A$ has a mass of 11.0 kg and is on a ramp at angle $\theta = 30.0°$; box $B$ has a mass of 7.00 kg and hangs on a cord. The cord connected to box $A$ is parallel to the ramp, which is frictionless. (a) What is the tension in the upper cord, and (b) what angle does that cord make with the horizontal?

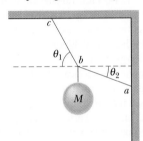

FIG. 12-73   Problem 68.

**70** A construction worker attempts to lift a uniform beam off the floor and raise it to a vertical position. The beam is 2.50 m long and weighs 500 N. At a certain instant the worker holds the beam momentarily at rest with one end at distance $d = 1.50$ m above the floor, as shown in Fig. 12-75, by exerting a force $\vec{P}$ on the beam, perpendicular to the beam. (a) What is the magnitude $P$? (b) What is

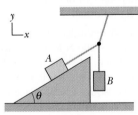

FIG. 12-74   Problem 69.

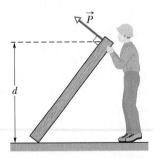

FIG. 12-75   Problem 70.

the magnitude of the (net) force of the floor on the beam? (c) What is the minimum value the coefficient of static friction between beam and floor can have in order for the beam not to slip at this instant?

**71** A solid copper cube has an edge length of 85.5 cm. How much stress must be applied to the cube to reduce the edge length to 85.0 cm? The bulk modulus of copper is $1.4 \times 10^{11}$ N/m².

**72** A uniform beam is 5.0 m long and has a mass of 53 kg. In Fig. 12-76, the beam is supported in a horizontal position by a hinge and a cable, with angle $\theta = 60°$. In unit-vector notation, what is the force on the beam from the hinge?

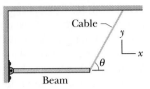

FIG. 12-76   Problem 72.

**73** In Fig. 12-77, a uniform beam with a weight of 60 N and a length of 3.2 m is hinged at its lower end, and a horizontal force $\vec{F}$ of magnitude 50 N acts at its upper end. The beam is held vertical by a cable that makes angle $\theta = 25°$ with the ground and is attached to the beam at height $h = 2.0$ m. What are (a) the tension in the cable and (b) the force on the beam from the hinge in unit-vector notation?

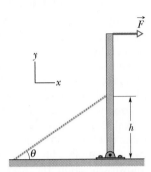

FIG. 12-77   Problem 73.

**74** In Fig. 12-78, a uniform beam of length 12.0 m is supported by a horizontal cable and a hinge at angle $\theta = 50.0°$. The tension in the cable is 400 N. In unit-vector notation, what are (a) the gravitational force on the beam and (b) the force on the beam from the hinge?

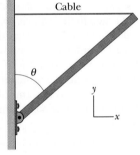

FIG. 12-78   Problem 74.

**75** Four bricks of length $L$, identical and uniform, are stacked on a table in two ways, as shown in Fig. 12-79 (compare with Problem 59). We seek to maximize the overhang distance $h$ in both arrangements. Find the optimum distances $a_1$, $a_2$, $b_1$, and $b_2$, and calculate $h$ for the two arrangements.

**76** A pan balance is made up of a rigid, massless rod with a hanging pan attached at each end. The rod is supported at and free to rotate about a point not at its center. It is balanced by unequal masses placed in the two pans. When an unknown mass $m$ is placed in the left pan, it is balanced by a mass $m_1$ placed in the right pan; when the mass $m$ is placed in the right pan, it is balanced by a mass $m_2$ in the left pan. Show that $m = \sqrt{m_1 m_2}$.

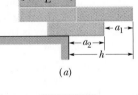

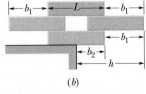

FIG. 12-79   Problem 75.

**77** The rigid square frame in Fig. 12-80 consists of the four side bars *AB*, *BC*, *CD*, and *DA* plus two diagonal bars *AC* and *BD*, which pass each other freely at *E*. By means of the turnbuckle *G*, bar *AB* is put under tension, as if its ends were subject to horizontal, outward forces $\vec{T}$ of magnitude 535 N. (a) Which of the other bars are in tension? What are the magnitudes of (b) the forces causing the tension in those bars and (c) the forces causing compression in the other bars? (*Hint:* Symmetry considerations can lead to considerable simplification in this problem.)

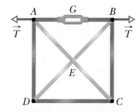

FIG. 12-80  Problem 77.

**78** A gymnast with mass 46.0 kg stands on the end of a uniform balance beam as shown in Fig. 12-81. The beam is 5.00 m long and has a mass of 250 kg (excluding the mass of the two supports). Each support is 0.540 m from its end of the beam. In unit-vector notation, what are the forces on the beam due to (a) support 1 and (b) support 2?

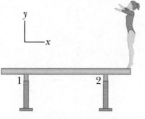

FIG. 12-81  Problem 78.

**79** Figure 12-82 shows a 300 kg cylinder that is horizontal. Three steel wires support the cylinder from a ceiling. Wires 1 and 3 are attached at the ends of the cylinder, and wire 2 is attached at the center. The wires each have a cross-sectional area of $2.00 \times 10^{-6}$ m². Initially (before the cylinder was put in place) wires 1 and 3 were 2.0000 m long and wire 2 was 6.00 mm longer than that. Now (with the cylinder in place) all three wires have been stretched. What is the tension in (a) wire 1 and (b) wire 2?

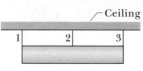

FIG. 12-82  Problem 79.

**80** Figure 12-83*a* shows details of a finger in the crimp hold of the climber in Fig. 12-47. A tendon that runs from muscles in the forearm is attached to the far bone in the finger. Along the way, the tendon runs through several guiding sheaths called *pulleys*. The A2 pulley is attached to the first finger bone; the A4 pulley is attached to the second finger bone. To pull the finger toward the palm, the forearm muscles pull the tendon through the pulleys, much like strings on a marionette

can be pulled to move parts of the marionette. Figure 12-83*b* is a simplified diagram of the second finger bone, which has length *d*. The tendon's pull $\vec{F_t}$ on the bone acts at the point where the tendon enters the A4 pulley, at distance *d*/3 along the bone. If the force components on each of the four crimped fingers in Fig. 12-47 are $F_h = 13.4$ N and $F_v = 162.4$ N, what is the magnitude of $\vec{F_t}$? The result is probably tolerable, but if the climber hangs by only one or two fingers, the A2 and A4 pulleys can be ruptured, a common ailment among rock climbers. ✈

**81** A uniform cube of side length 8.0 cm rests on a horizontal floor. The coefficient of static friction between cube and floor is $\mu$. A horizontal pull $\vec{P}$ is applied perpendicular to one of the vertical faces of the cube, at a distance 7.0 cm above the floor on the vertical midline of the cube face. The magnitude of $\vec{P}$ is gradually increased. During that increase, for what values of $\mu$ will the cube eventually (a) begin to slide and (b) begin to tip? (*Hint:* At the onset of tipping, where is the normal force located?) SSM

**82** A cylindrical aluminum rod, with an initial length of 0.8000 m and radius 1000.0 $\mu$m, is clamped in place at one end and then stretched by a machine pulling parallel to its length at its other end. Assuming that the rod's density (mass per unit volume) does not change, find the force magnitude that is required of the machine to decrease the radius to 999.9 $\mu$m. (The yield strength is not exceeded.)

**83** A beam of length *L* is carried by three men, one man at one end and the other two supporting the beam between them on a crosspiece placed so that the load of the beam is equally divided among the three men. How far from the beam's free end is the crosspiece placed? (Neglect the mass of the crosspiece.)

**84** A trap door in a ceiling is 0.91 m square, has a mass of 11 kg, and is hinged along one side, with a catch at the opposite side. If the center of gravity of the door is 10 cm toward the hinged side from the door's center, what are the magnitudes of the forces exerted by the door on (a) the catch and (b) the hinge?

**85** A uniform ladder is 10 m long and weighs 200 N. In Fig. 12-84, the ladder leans against a vertical, frictionless wall at height $h = 8.0$ m above the ground. A horizontal force $\vec{F}$ is applied to the ladder at distance $d = 2.0$ m from its base (measured along the ladder). (a) If force magnitude $F = 50$ N, what is the force of the ground on the ladder, in unit-vector notation? (b) If $F = 150$ N, what is the force of the ground on the ladder, also in unit-vector notation? (c) Suppose the coefficient of static friction between the ladder and the ground is 0.38; for what minimum value of the force magnitude $F$ will the base of the ladder just barely start to move toward the wall? SSM

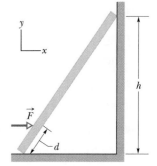

FIG. 12-84  Problem 85.

**86** If the (square) beam in Fig. 12-7*a* is of Douglas fir, what must be its thickness to keep the compressive stress on it to $\frac{1}{6}$ of its ultimate strength? (See Sample Problem 12-3.)

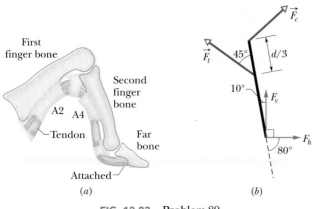

First finger bone

Second finger bone

A2  A4

Tendon

Far bone

Attached

(a)

$\vec{F_c}$

$\vec{F_t}$  45°  *d*/3

10°

$F_v$

$F_h$

80°

(b)

FIG. 12-83  Problem 80.

# 13 Gravitation

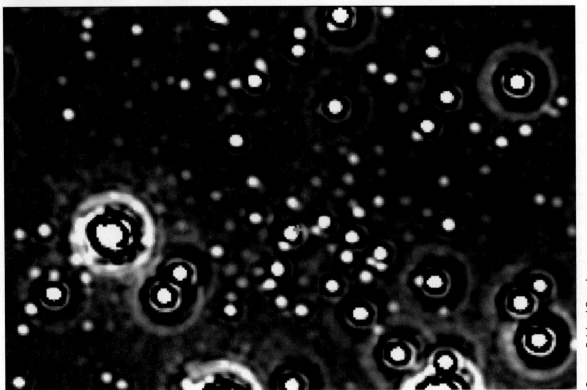

Courtesy Reinhard Genzel

This is an image of the stars near our Milky Way galaxy's center, which is marked with a small cross. Note that nothing shows up exactly at the center, but slightly off center (at the 8:00 position) there is a small circle. That circle is the image of a star known as S2. The other circles are also images of stars (the halos around them are artificially produced by the method of processing the images). Most stars in our galaxy move so slowly that we cannot actually see them move relative to one another, not even over a lifetime of observations. However, S2 is very different— we can see it move. In fact, it is moving so rapidly that it makes a complete trip around the Galaxy's center in only 15.2 years. There must be something huge at the center, yet we see nothing there.

## What monster lies at our galaxy's center?

The answer is in this chapter.

## 13-1 WHAT IS PHYSICS?

One of the long-standing goals of physics is to understand the gravitational force—the force that holds you to Earth, holds the Moon in orbit around Earth, and holds Earth in orbit around the Sun. It also reaches out through the whole of our Milky Way galaxy, holding together the billions and billions of stars in the Galaxy and the countless molecules and dust particles between stars. We are located somewhat near the edge of this disk-shaped collection of stars and other matter, $2.6 \times 10^4$ light-years ($2.5 \times 10^{20}$ m) from the galactic center, around which we slowly revolve.

The gravitational force also reaches across intergalactic space, holding together the Local Group of galaxies, which includes, in addition to the Milky Way, the Andromeda Galaxy (Fig. 13-1) at a distance of $2.3 \times 10^6$ light-years away from Earth, plus several closer dwarf galaxies, such as the Large Magellanic Cloud. The Local Group is part of the Local Supercluster of galaxies that is being drawn by the gravitational force toward an exceptionally massive region of space called the Great Attractor. This region appears to be about $3.0 \times 10^8$ light-years from Earth, on the opposite side of the Milky Way. And the gravitational force is even more far-reaching because it attempts to hold together the entire universe, which is expanding.

This force is also responsible for some of the most mysterious structures in the universe: *black holes*. When a star considerably larger than our Sun burns out, the gravitational force between all its particles can cause the star to collapse in on itself and thereby to form a black hole. The gravitational force at the surface of such a collapsed star is so strong that neither particles nor light can escape from the surface (thus the term "black hole"). Any star coming too near a black hole can be ripped apart by the strong gravitational force and pulled into the hole. Enough captures like this yields a *supermassive black hole*. Such mysterious monsters appear to be common in the universe.

Although the gravitational force is still not fully understood, the starting point in our understanding of it lies in the *law of gravitation* of Isaac Newton.

FIG. 13-1    The Andromeda Galaxy. Located $2.3 \times 10^6$ light-years from us, and faintly visible to the naked eye, it is very similar to our home galaxy, the Milky Way. *(Courtesy NASA)*

## 13-2 | Newton's Law of Gravitation

Physicists like to study seemingly unrelated phenomena to show that a relationship can be found if the phenomena are examined closely enough. This search for unification has been going on for centuries. In 1665, the 23-year-old Isaac Newton made a basic contribution to physics when he showed that the force that holds the Moon in its orbit is the same force that makes an apple fall. We take this knowledge so much for granted now that it is not easy for us to comprehend the ancient belief that the motions of earthbound bodies and heavenly bodies were different in kind and were governed by different laws.

Newton concluded not only that Earth attracts both apples and the Moon but also that every body in the universe attracts every other body; this tendency of bodies to move toward each other is called **gravitation.** Newton's conclusion takes a little getting used to, because the familiar attraction of Earth for earthbound bodies is so great that it overwhelms the attraction that earthbound bodies have for each other. For example, Earth attracts an apple with a force magnitude of about 0.8 N. You also attract a nearby apple (and it attracts you), but the force of attraction has less magnitude than the weight of a speck of dust.

Newton proposed a *force law* that we call **Newton's law of gravitation:** Every particle attracts any other particle with a **gravitational force** of magnitude

$$F = G\,\frac{m_1 m_2}{r^2} \qquad \text{(Newton's law of gravitation).} \qquad (13\text{-}1)$$

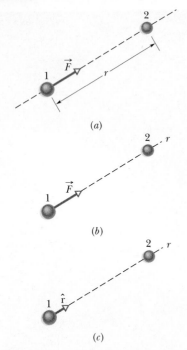

FIG. 13-2  (a) The gravitational force $\vec{F}$ on particle 1 due to particle 2 is an attractive force because particle 1 is attracted to particle 2. (b) Force $\vec{F}$ is directed along a radial coordinate axis $r$ extending from particle 1 through particle 2. (c) $\vec{F}$ is in the direction of a unit vector $\hat{r}$ along the $r$ axis.

Here $m_1$ and $m_2$ are the masses of the particles, $r$ is the distance between them, and $G$ is the **gravitational constant,** with a value that is now known to be

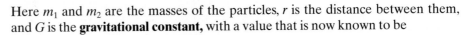

$$G = 6.67 \times 10^{-11} \, \text{N} \cdot \text{m}^2/\text{kg}^2$$
$$= 6.67 \times 10^{-11} \, \text{m}^3/\text{kg} \cdot \text{s}^2. \qquad (13\text{-}2)$$

In Fig. 13-2a, $\vec{F}$ is the gravitational force acting on particle 1 (mass $m_1$) due to particle 2 (mass $m_2$). The force is directed toward particle 2 and is said to be an *attractive force* because particle 1 is attracted toward particle 2. The magnitude of the force is given by Eq. 13-1.

We can describe $\vec{F}$ as being in the positive direction of an $r$ axis extending radially from particle 1 through particle 2 (Fig. 13-2b). We can also describe $\vec{F}$ by using a radial unit vector $\hat{r}$ (a dimensionless vector of magnitude 1) that is directed away from particle 1 along the $r$ axis (Fig. 13-2c). From Eq. 13-1, the force on particle 1 is then

$$\vec{F} = G \frac{m_1 m_2}{r^2} \hat{r}. \qquad (13\text{-}3)$$

The gravitational force on particle 2 due to particle 1 has the same magnitude as the force on particle 1 but the opposite direction. These two forces form a third-law force pair, and we can speak of the gravitational force *between* the two particles as having a magnitude given by Eq. 13-1. This force between two particles is not altered by other objects, even if they are located between the particles. Put another way, no object can shield either particle from the gravitational force due to the other particle.

The strength of the gravitational force—that is, how strongly two particles with given masses at a given separation attract each other—depends on the value of the gravitational constant $G$. If $G$—by some miracle—were suddenly multiplied by a factor of 10, you would be crushed to the floor by Earth's attraction. If $G$ were divided by this factor, Earth's attraction would be so weak that you could jump over a building.

Although Newton's law of gravitation applies strictly to particles, we can also apply it to real objects as long as the sizes of the objects are small relative to the distance between them. The Moon and Earth are far enough apart so that, to a good approximation, we can treat them both as particles—but what about an apple and Earth? From the point of view of the apple, the broad and level Earth, stretching out to the horizon beneath the apple, certainly does not look like a particle.

Newton solved the apple–Earth problem by proving an important theorem called the *shell theorem:*

> A uniform spherical shell of matter attracts a particle that is outside the shell as if all the shell's mass were concentrated at its center.

Earth can be thought of as a nest of such shells, one within another and each shell attracting a particle outside Earth's surface as if the mass of that shell were at the center of the shell. Thus, from the apple's point of view, Earth *does* behave like a particle, one that is located at the center of Earth and has a mass equal to that of Earth.

Suppose that, as in Fig. 13-3, Earth pulls down on an apple with a force of magnitude 0.80 N. The apple must then pull up on Earth with a force of magnitude 0.80 N, which we take to act at the center of Earth. Although the forces are matched in magnitude, they produce different accelerations when the apple is released. The accelerations of the apple is about 9.8 m/s², the familiar acceleration of a falling body near Earth's surface. The acceleration of Earth, however, measured in a reference frame attached to the center of mass of the apple–Earth system, is only about $1 \times 10^{-25}$ m/s².

$\nabla F = 0.80$ N
$\triangle F = 0.80$ N

FIG. 13-3  The apple pulls up on Earth just as hard as Earth pulls down on the apple.

✓**CHECKPOINT 1**    A particle is to be placed, in turn, outside four objects, each of mass $m$: (1) a large uniform solid sphere, (2) a large uniform spherical shell, (3) a small uniform solid sphere, and (4) a small uniform shell. In each situation, the distance between the particle and the center of the object is $d$. Rank the objects according to the magnitude of the gravitational force they exert on the particle, greatest first.

## 13-3 | Gravitation and the Principle of Superposition

Given a group of particles, we find the net (or resultant) gravitational force on any one of them from the others by using the **principle of superposition.** This is a general principle that says a net effect is the sum of the individual effects. Here, the principle means that we first compute the individual gravitational forces that act on our selected particle due to each of the other particles. We then find the net force by adding these forces vectorially, as usual.

For $n$ interacting particles, we can write the principle of superposition for the gravitational forces on particle 1 as

$$\vec{F}_{1,\text{net}} = \vec{F}_{12} + \vec{F}_{13} + \vec{F}_{14} + \vec{F}_{15} + \cdots + \vec{F}_{1n}. \tag{13-4}$$

Here $\vec{F}_{1,\text{net}}$ is the net force on particle 1 and, for example, $\vec{F}_{13}$ is the force on particle 1 from particle 3. We can express this equation more compactly as a vector sum:

$$\vec{F}_{1,\text{net}} = \sum_{i=2}^{n} \vec{F}_{1i}. \tag{13-5}$$

What about the gravitational force on a particle from a real (extended) object? This force is found by dividing the object into parts small enough to treat as particles and then using Eq. 13-5 to find the vector sum of the forces on the particle from all the parts. In the limiting case, we can divide the extended object into differential parts each of mass $dm$ and each producing a differential force $d\vec{F}$ on the particle. In this limit, the sum of Eq. 13-5 becomes an integral and we have

$$\vec{F}_1 = \int d\vec{F}, \tag{13-6}$$

in which the integral is taken over the entire extended object and we drop the subscript "net." If the extended object is a uniform sphere or a spherical shell, we can avoid the integration of Eq. 13-6 by assuming that the object's mass is concentrated at the object's center and using Eq. 13-1.

✓**CHECKPOINT 2**    The figure shows four arrangements of three particles of equal masses. (a) Rank the arrangements according to the magnitude of the net gravitational force on the particle labeled $m$, greatest first. (b) In arrangement 2, is the direction of the net force closer to the line of length $d$ or to the line of length $D$?

## Sample Problem    13-1

Figure 13-4a shows an arrangement of three particles, particle 1 of mass $m_1 = 6.0$ kg and particles 2 and 3 of mass $m_2 = m_3 = 4.0$ kg, and distance $a = 2.0$ cm. What is the net gravitational force $\vec{F}_{1,\text{net}}$ on particle 1 due to the other particles?

**KEY IDEAS**    (1) Because we have particles, the magnitude of the gravitational force on particle 1 due to either of the other particles is given by Eq. 13-1 ($F = Gm_1m_2/r^2$). (2) The direction of either gravitational force on particle 1 is toward the particle responsible for it. (3) Because the forces are not along a single axis, we *cannot* simply add or subtract their magnitudes or their components to get the net force. Instead, we must add them as vectors.

**Calculations:** From Eq. 13-1, the magnitude of the force $\vec{F}_{12}$ on particle 1 from particle 2 is

$$F_{12} = \frac{Gm_1m_2}{a^2}$$

$$= \frac{(6.67 \times 10^{-11}\ \text{m}^3/\text{kg}\cdot\text{s}^2)(6.0\ \text{kg})(4.0\ \text{kg})}{(0.020\ \text{m})^2}$$

$$= 4.00 \times 10^{-6}\ \text{N}.$$

Similarly, the magnitude of force $\vec{F}_{13}$ on particle 1 from particle 3 is

$$F_{13} = \frac{Gm_1m_3}{(2a)^2}$$

$$= \frac{(6.67 \times 10^{-11}\ \text{m}^3/\text{kg}\cdot\text{s}^2)(6.0\ \text{kg})(4.0\ \text{kg})}{(0.040\ \text{m})^2}$$

$$= 1.00 \times 10^{-6}\ \text{N}.$$

Force $\vec{F}_{12}$ is directed in the positive direction of the $y$ axis (Fig. 13-4b) and has only the $y$ component $F_{12}$. Similarly, $\vec{F}_{13}$ is directed in the negative direction of the $x$ axis and has only the $x$ component $-F_{13}$.

To find the net force $\vec{F}_{1,\text{net}}$ on particle 1, we must add the two forces as vectors. We can do so on a vector-capable calculator. However, here we note that $-F_{13}$ and $F_{12}$ are actually the $x$ and $y$ components of $\vec{F}_{1,\text{net}}$.

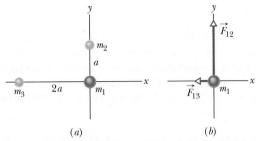

**FIG. 13-4** (a) An arrangement of three particles. (b) The forces acting on the particle of mass $m_1$ due to the other particles.

Therefore, we can use Eq. 3-6 to find first the magnitude and then the direction of $\vec{F}_{1,\text{net}}$. The magnitude is

$$F_{1,\text{net}} = \sqrt{(F_{12})^2 + (-F_{13})^2}$$

$$= \sqrt{(4.00 \times 10^{-6}\ \text{N})^2 + (-1.00 \times 10^{-6}\ \text{N})^2}$$

$$= 4.1 \times 10^{-6}\ \text{N}. \qquad \text{(Answer)}$$

Relative to the positive direction of the $x$ axis, Eq. 3-6 gives the direction of $\vec{F}_{1,\text{net}}$ as

$$\theta = \tan^{-1}\frac{F_{12}}{-F_{13}} = \tan^{-1}\frac{4.00 \times 10^{-6}\ \text{N}}{-1.00 \times 10^{-6}\ \text{N}} = -76°.$$

Is this a reasonable direction? No, because the direction of $\vec{F}_{1,\text{net}}$ must be between the directions of $\vec{F}_{12}$ and $\vec{F}_{13}$. Recall from Chapter 3 (Problem-Solving Tactic 3) that a calculator displays only one of the two possible answers to a $\tan^{-1}$ function. We find the other answer by adding 180°:

$$-76° + 180° = 104°, \qquad \text{(Answer)}$$

which *is* a reasonable direction for $\vec{F}_{1,\text{net}}$.

---

**Sample Problem** **13-2** **Build your skill**

Figure 13-5a shows an arrangement of five particles, with masses $m_1 = 8.0$ kg, $m_2 = m_3 = m_4 = m_5 = 2.0$ kg, and with $a = 2.0$ cm and $\theta = 30°$. What is the net gravitational force $\vec{F}_{1,\text{net}}$ on particle 1 due to the other particles?

**KEY IDEAS** (1) Because we have particles, the magnitude of the gravitational force on particle 1 due to either of the other particles is given by Eq. 13-1 ($F = Gm_1m_2/r^2$). (2) The direction of a gravitational force on particle 1 is toward the particle responsible for the force. (3) We can use symmetry to eliminate unneeded calculations.

**Calculations:** For the magnitudes of the forces on particle 1, first note that particles 2 and 4 have equal masses and equal distances of $r = 2a$ from particle 1. Thus, from Eq. 13-1, we find

$$F_{12} = F_{14} = \frac{Gm_1m_2}{(2a)^2}. \qquad (13\text{-}7)$$

Similarly, since particles 3 and 5 have equal masses and are both distance $r = a$ from particle 1, we find

$$F_{13} = F_{15} = \frac{Gm_1m_3}{a^2}. \qquad (13\text{-}8)$$

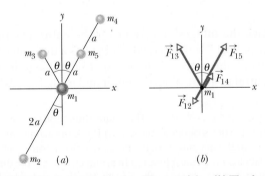

**FIG. 13-5** (a) An arrangement of five particles. (b) The forces acting on the particle of mass $m_1$ due to the other four particles.

We could now substitute known data into these two equations to evaluate the magnitudes of the forces, indicate the directions of the forces on the free-body diagram of Fig. 13-5b, and then find the net force either (1) by resolving the vectors into $x$ and $y$ components, finding the net $x$ and net $y$ components, and then vectorially combining them or (2) by adding the vectors directly on a vector-capable calculator.

Instead, however, we shall make further use of the symmetry of the problem. First, we note that $\vec{F}_{12}$ and $\vec{F}_{14}$ are equal in magnitude but opposite in direction; thus, those forces *cancel*. Inspection of Fig. 13-5*b* and Eq. 13-8 reveals that the *x* components of $\vec{F}_{13}$ and $\vec{F}_{15}$ also *cancel*, and that their *y* components are identical in magnitude and both act in the positive direction of the *y* axis. Thus, $\vec{F}_{1,\text{net}}$ acts in that same direction, and its magnitude is twice the *y* component of $\vec{F}_{13}$:

$$F_{1,\text{net}} = 2F_{13}\cos\theta = 2\,\frac{Gm_1 m_3}{a^2}\cos\theta$$

$$= \frac{2(6.67\times10^{-11}\,\text{m}^3/\text{kg}\cdot\text{s}^2)(8.0\,\text{kg})(2.0\,\text{kg})}{(0.020\,\text{m})^2}\cos 30°$$

$$= 4.6\times10^{-6}\,\text{N}. \qquad\qquad\text{(Answer)}$$

Note that the presence of particle 5 along the line between particles 1 and 4 does not alter the gravitational force on particle 1 from particle 4.

---

**PROBLEM-SOLVING TACTICS**

*Tactic 1: Drawing Gravitational Force Vectors* When you are given a diagram of particles, such as Fig. 13-4*a*, and asked to find the net gravitational force on one of them, you should usually draw a free-body diagram showing only the particle of concern and the forces on *it alone*, as in Fig. 13-4*b*. If, instead, you choose to superimpose the force vectors on the given diagram, be sure to draw the vectors with either all tails (preferably) or all heads on the particle experiencing those forces. If you draw the vectors elsewhere, you invite confusion—and confusion is guaranteed if you draw the vectors on the particles *causing* the forces.

*Tactic 2: Simplifying a Sum of Forces with Symmetry* In Sample Problem 13-2 we used the symmetry of the situation: By realizing that particles 2 and 4 are positioned symmetrically about particle 1, and thus that $\vec{F}_{12}$ and $\vec{F}_{14}$ cancel, we avoided calculating either force. By realizing that the *x* components of $\vec{F}_{13}$ and $\vec{F}_{15}$ cancel and that their *y* components are identical and add, we saved even more effort.

In problems with symmetry, you can save much effort and reduce the chance of error by identifying which calculations are not needed because of the symmetry. Such identification is a skill acquired only by doing many homework problems.

## 13-4 | Gravitation Near Earth's Surface

Let us assume that Earth is a uniform sphere of mass *M*. The magnitude of the gravitational force from Earth on a particle of mass *m*, located outside Earth a distance *r* from Earth's center, is then given by Eq. 13-1 as

$$F = G\,\frac{Mm}{r^2}. \qquad (13\text{-}9)$$

If the particle is released, it will fall toward the center of Earth, as a result of the gravitational force $\vec{F}$, with an acceleration we shall call the **gravitational acceleration** $\vec{a}_g$. Newton's second law tells us that magnitudes *F* and $a_g$ are related by

$$F = ma_g. \qquad (13\text{-}10)$$

Now, substituting *F* from Eq. 13-9 into Eq. 13-10 and solving for $a_g$, we find

$$a_g = \frac{GM}{r^2}. \qquad (13\text{-}11)$$

Table 13-1 shows values of $a_g$ computed for various altitudes above Earth's surface. Notice $a_g$ is significant even at 400 km.

Since Section 5-4, we have assumed that Earth is an inertial frame by neglecting its rotation. This simplification has allowed us to assume that the free-fall acceleration *g* of a particle is the same as the particle's gravitational acceleration (which we now call $a_g$). Furthermore, we assumed that *g* has the constant value 9.8 m/s² any place on Earth's surface. However, any *g* value measured at a given location will differ from the $a_g$ value calculated with Eq. 13-11 for that location for three reasons: (1) Earth's mass is not distributed uniformly, (2) Earth is not a perfect sphere, and (3) Earth rotates. Moreover, because *g* differs from $a_g$, the same three reasons mean that the measured weight *mg* of a particle differs from the magnitude of the gravitational force on the particle as given by Eq. 13-9. Let us now examine those reasons.

**TABLE 13-1**

**Variation of $a_g$ with Altitude**

| Altitude (km) | $a_g$ (m/s²) | Altitude Example |
|---|---|---|
| 0 | 9.83 | Mean Earth surface |
| 8.8 | 9.80 | Mt. Everest |
| 36.6 | 9.71 | Highest crewed balloon |
| 400 | 8.70 | Space shuttle orbit |
| 35 700 | 0.225 | Communications satellite |

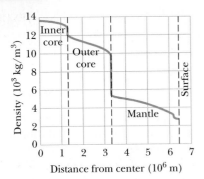

**FIG. 13-6** The density of Earth as a function of distance from the center. The limits of the solid inner core, the largely liquid outer core, and the solid mantle are shown, but the crust of Earth is too thin to show clearly on this plot.

1. **Earth's mass is not uniformly distributed.** The density (mass per unit volume) of Earth varies radially as shown in Fig. 13-6, and the density of the crust (outer section) varies from region to region over Earth's surface. Thus, g varies from region to region over the surface.

2. **Earth is not a sphere.** Earth is approximately an ellipsoid, flattened at the poles and bulging at the equator. Its equatorial radius is greater than its polar radius by 21 km. Thus, a point at the poles is closer to the dense core of Earth than is a point on the equator. This is one reason the free-fall acceleration g increases as one proceeds, at sea level, from the equator toward either pole.

3. **Earth is rotating.** The rotation axis runs through the north and south poles of Earth. An object located on Earth's surface anywhere except at those poles must rotate in a circle about the rotation axis and thus must have a centripetal acceleration directed toward the center of the circle. This centripetal acceleration requires a centripetal net force that is also directed toward that center.

To see how Earth's rotation causes g to differ from $a_g$, let us analyze a simple situation in which a crate of mass m is on a scale at the equator. Figure 13-7a shows this situation as viewed from a point in space above the north pole.

Figure 13-7b, a free-body diagram for the crate, shows the two forces on the crate, both acting along a radial r axis that extends from Earth's center. The normal force $\vec{F}_N$ on the crate from the scale is directed outward, in the positive direction of the r axis. The gravitational force, represented with its equivalent $m\vec{a}_g$, is directed inward. Because it travels in a circle about the center of Earth as Earth turns, the crate has a centripetal acceleration $\vec{a}$ directed toward Earth's center. From Eq. 10-23 ($a_r = \omega^2 r$), we know this acceleration is equal to $\omega^2 R$, where $\omega$ is Earth's angular speed and R is the circle's radius (approximately Earth's radius). Thus, we can write Newton's second law for forces along the r axis ($F_{net,r} = ma_r$) as

$$F_N - ma_g = m(-\omega^2 R). \qquad (13\text{-}12)$$

The magnitude $F_N$ of the normal force is equal to the weight mg read on the scale. With mg substituted for $F_N$, Eq. 13-12 gives us

$$mg = ma_g - m(\omega^2 R), \qquad (13\text{-}13)$$

which says

$$\begin{pmatrix}\text{measured}\\\text{weight}\end{pmatrix} = \begin{pmatrix}\text{magnitude of}\\\text{gravitational force}\end{pmatrix} - \begin{pmatrix}\text{mass times}\\\text{centripetal acceleration}\end{pmatrix}.$$

Thus, the measured weight is less than the magnitude of the gravitational force on the crate, because of Earth's rotation.

To find a corresponding expression for g and $a_g$, we cancel m from Eq. 13-13 to write

$$g = a_g - \omega^2 R, \qquad (13\text{-}14)$$

which says

$$\begin{pmatrix}\text{free-fall}\\\text{acceleration}\end{pmatrix} = \begin{pmatrix}\text{gravitational}\\\text{acceleration}\end{pmatrix} - \begin{pmatrix}\text{centripetal}\\\text{acceleration}\end{pmatrix}.$$

Thus, the measured free-fall acceleration is less than the gravitational acceleration because of Earth's rotation.

The difference between accelerations g and $a_g$ is equal to $\omega^2 R$ and is greatest on the equator (for one reason, the radius of the circle traveled by the crate is

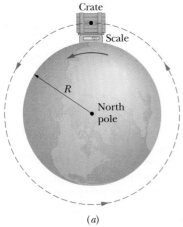

(a)

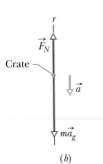

(b)

**FIG. 13-7** (a) A crate sitting on a scale at Earth's equator, as seen by an observer positioned on Earth's rotation axis at some point above the north pole. (b) A free-body diagram for the crate, with a radial r axis extending from Earth's center. The gravitational force on the crate is represented with its equivalent $m\vec{a}_g$. The normal force on the crate from the scale is $\vec{F}_N$. Because of Earth's rotation, the crate has a centripetal acceleration $\vec{a}$ that is directed toward Earth's center.

greatest there). To find the difference, we can use Eq. 10-5 ($\omega = \Delta\theta/\Delta t$) and Earth's radius $R = 6.37 \times 10^6$ m. For one rotation of Earth, $\theta$ is $2\pi$ rad and the time period $\Delta t$ is about 24 h. Using these values (and converting hours to seconds), we find that $g$ is less than $a_g$ by only about 0.034 m/s² (small compared to 9.8 m/s²). Therefore, neglecting the difference in accelerations $g$ and $a_g$ is often justified. Similarly, neglecting the difference between weight and the magnitude of the gravitational force is also often justified.

## Sample Problem 13-3

(a) An astronaut whose height $h$ is 1.70 m floats "feet down" in an orbiting space shuttle at distance $r = 6.77 \times 10^6$ m away from the center of Earth. What is the difference between the gravitational acceleration at her feet and at her head?

**KEY IDEAS** We can approximate Earth as a uniform sphere of mass $M_E$. Then, from Eq. 13-11, the gravitational acceleration at any distance $r$ from the center of Earth is

$$a_g = \frac{GM_E}{r^2}. \tag{13-15}$$

We might simply apply this equation twice, first with $r = 6.77 \times 10^6$ m for the feet and then with $r = 6.77 \times 10^6$ m + 1.70 m for the head. However, a calculator may give us the same value for $a_g$ twice, and thus a difference of zero, because $h$ is so much smaller than $r$. Here's a more promising approach: Because we have a differential change $dr$ in $r$ between the astronaut's feet and head, we should differentiate Eq. 13-15 with respect to $r$.

**Calculations:** The differentiation gives us

$$da_g = -2\frac{GM_E}{r^3}\,dr, \tag{13-16}$$

where $da_g$ is the differential change in the gravitational acceleration due to the differential change $dr$ in $r$. For the astronaut, $dr = h$ and $r = 6.77 \times 10^6$ m. Substituting data into Eq. 13-16, we find

$$da_g = -2\frac{(6.67 \times 10^{-11}\ \text{m}^3/\text{kg}\cdot\text{s}^2)(5.98 \times 10^{24}\ \text{kg})}{(6.77 \times 10^6\ \text{m})^3}(1.70\ \text{m})$$
$$= -4.37 \times 10^{-6}\ \text{m/s}^2, \quad\text{(Answer)}$$

where the $M_E$ value is taken from Appendix C. This result means that the gravitational acceleration of the astronaut's feet toward Earth is slightly greater than the gravitational acceleration of her head toward Earth. This difference in acceleration tends to stretch her body, but the difference is so small that the stretching is unnoticeable.

(b) If the astronaut is now "feet down" at the same orbital radius $r = 6.77 \times 10^6$ m about a black hole of mass $M_h = 1.99 \times 10^{31}$ kg (10 times our Sun's mass), what is the difference between the gravitational acceleration at her feet and at her head? The black hole has a mathematical surface (*event horizon*) of radius $R_h = 2.95 \times 10^4$ m. Nothing, not even light, can escape from that surface or anywhere inside it. Note that the astronaut is well outside the surface (at $r = 229R_h$).

**Calculations:** We again have a differential change $dr$ in $r$ between the astronaut's feet and head, so we can again use Eq. 13-16. However, now we substitute $M_h = 1.99 \times 10^{31}$ kg for $M_E$. We find

$$da_g = -2\frac{(6.67 \times 10^{-11}\ \text{m}^3/\text{kg}\cdot\text{s}^2)(1.99 \times 10^{31}\ \text{kg})}{(6.77 \times 10^6\ \text{m})^3}(1.70\ \text{m})$$
$$= -14.5\ \text{m/s}^2. \quad\text{(Answer)}$$

This means that the gravitational acceleration of the astronaut's feet toward the black hole is noticeably larger than that of her head. The resulting tendency to stretch her body would be bearable but quite painful. If she drifted closer to the black hole, the stretching tendency would increase drastically.

## 13-5 | Gravitation Inside Earth

Newton's shell theorem can also be applied to a situation in which a particle is located *inside* a uniform shell, to show the following:

> A uniform shell of matter exerts no *net* gravitational force on a particle located inside it.

*Caution:* This statement does *not* mean that the gravitational forces on the particle from the various elements of the shell magically disappear. Rather, it means that the *sum* of the force vectors on the particle from all the elements is zero.

If Earth's mass were uniformly distributed, the gravitational force acting on a particle would be a maximum at Earth's surface and would decrease as the particle moved outward, away from the planet. If the particle were to move inward, perhaps down a deep mine shaft, the gravitational force would change for two reasons. (1) It would tend to increase because the particle would be moving closer to the center of Earth. (2) It would tend to decrease because the thickening shell of material lying outside the particle's radial position would not exert any net force on the particle.

For a uniform Earth, the second influence would prevail and the force on the particle would steadily decrease to zero as the particle approached the center of Earth. However, for the real (nonuniform) Earth, the force on the particle actually increases as the particle begins to descend. The force reaches a maximum at a certain depth and then decreases as the particle descends farther.

## Sample Problem 13-4

In *Pole to Pole*, an early science fiction story by George Griffith, three explorers attempt to travel by capsule through a naturally formed (and, of course, fictional) tunnel directly from the south pole to the north pole (Fig. 13-8). According to the story, as the capsule approaches Earth's center, the gravitational force on the explorers becomes alarmingly large and then, exactly at the center, it suddenly but only momentarily disappears. Then the capsule travels through the second half of the tunnel, to the north pole.

Check Griffith's description by finding the gravitational force on the capsule of mass $m$ when it reaches a distance $r$ from Earth's center. Assume that Earth is a sphere of uniform density $\rho$ (mass per unit volume).

**KEY IDEAS** Newton's shell theorem gives us three ideas:

1. When the capsule is at radius $r$ from Earth's center, the portion of Earth that lies outside a sphere of radius $r$ does *not* produce a net gravitational force on the capsule.

2. The portion of Earth that lies inside that sphere *does* produce a net gravitational force on the capsule.

3. We can treat the mass $M_{ins}$ of that inside portion of Earth as being the mass of a particle located at Earth's center.

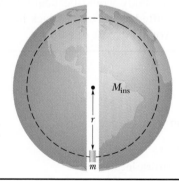

FIG. 13-8 A capsule of mass $m$ falls from rest through a tunnel that connects Earth's south and north poles. When the capsule is at distance $r$ from Earth's center, the portion of Earth's mass that is contained in a sphere of that radius is $M_{ins}$.

*Calculations:* All three ideas tell us that we can write Eq. 13-1, for the magnitude of the gravitational force on the capsule, as

$$F = \frac{GmM_{ins}}{r^2}. \qquad (13\text{-}17)$$

To write the mass $M_{ins}$ in terms of the radius $r$, we note that the volume $V_{ins}$ containing this mass is $\frac{4}{3}\pi r^3$. Also, because we're assuming an Earth of uniform density, the density $\rho_{ins} = M_{ins}/V_{ins}$ is Earth's density $\rho$. Thus, we have

$$M_{ins} = \rho V_{ins} = \rho \frac{4\pi r^3}{3}. \qquad (13\text{-}18)$$

Then, after substituting this expression into Eq. 13-17 and canceling, we have

$$F = \frac{4\pi Gm\rho}{3} r. \qquad \text{(Answer)} \quad (13\text{-}19)$$

This equation tells us that the force magnitude $F$ depends linearly on the capsule's distance $r$ from Earth's center. Thus, as $r$ decreases, $F$ also decreases (opposite of Griffith's description), until it is zero at Earth's center. At least Griffith got the zero-at-the-center detail correct.

Equation 13-19 can also be written in terms of the force vector $\vec{F}$ and the capsule's position vector $\vec{r}$ along a radial axis extending from Earth's center. Let $K$ represent the collection of constants $4\pi Gm\rho/3$. Then, Eq. 13-19 becomes

$$\vec{F} = -K\vec{r}, \qquad (13\text{-}20)$$

in which we have inserted a minus sign to indicate that $\vec{F}$ and $\vec{r}$ have opposite directions. Equation 13-20 has the form of Hooke's law (Eq. 7-20, $\vec{F} = -k\vec{d}$). Thus, under the idealized conditions of the story, the capsule would oscillate like a block on a spring, with the center of the oscillation at Earth's center. After the capsule had fallen from the south pole to Earth's center, it would travel from the center to the north pole (as Griffith said) and then back again, repeating the cycle forever.

## 13-6 | Gravitational Potential Energy

In Section 8-4, we discussed the gravitational potential energy of a particle–Earth system. We were careful to keep the particle near Earth's surface, so that we could regard the gravitational force as constant. We then chose some reference configuration of the system as having a gravitational potential energy of zero. Often, in this configuration the particle was on Earth's surface. For particles not on Earth's surface, the gravitational potential energy decreased when the separation between the particle and Earth decreased.

Here, we broaden our view and consider the gravitational potential energy $U$ of two particles, of masses $m$ and $M$, separated by a distance $r$. We again choose a reference configuration with $U$ equal to zero. However, to simplify the equations, the separation distance $r$ in the reference configuration is now large enough to be approximated as *infinite*. As before, the gravitational potential energy decreases when the separation decreases. Since $U = 0$ for $r = \infty$, the potential energy is negative for any finite separation and becomes progressively more negative as the particles move closer together.

With these facts in mind and as we shall justify next, we take the gravitational potential energy of the two-particle system to be

$$U = -\frac{GMm}{r} \qquad \text{(gravitational potential energy).} \qquad (13\text{-}21)$$

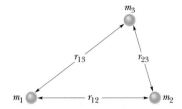

FIG. 13-9 A system consisting of three particles. The gravitational potential energy *of the system* is the sum of the gravitational potential energies of all three pairs of particles.

Note that $U(r)$ approaches zero as $r$ approaches infinity and that for any finite value of $r$, the value of $U(r)$ is negative.

The potential energy given by Eq. 13-21 is a property of the system of two particles rather than of either particle alone. There is no way to divide this energy and say that so much belongs to one particle and so much to the other. However, if $M \gg m$, as is true for Earth (mass $M$) and a baseball (mass $m$), we often speak of "the potential energy of the baseball." We can get away with this because, when a baseball moves in the vicinity of Earth, changes in the potential energy of the baseball–Earth system appear almost entirely as changes in the kinetic energy of the baseball, since changes in the kinetic energy of Earth are too small to be measured. Similarly, in Section 13-8 we shall speak of "the potential energy of an artificial satellite" orbiting Earth, because the satellite's mass is so much smaller than Earth's mass. When we speak of the potential energy of bodies of comparable mass, however, we have to be careful to treat them as a system.

If our system contains more than two particles, we consider each pair of particles in turn, calculate the gravitational potential energy of that pair with Eq. 13-21 as if the other particles were not there, and then algebraically sum the results. Applying Eq. 13-21 to each of the three pairs of Fig. 13-9, for example, gives the potential energy of the system as

$$U = -\left( \frac{Gm_1m_2}{r_{12}} + \frac{Gm_1m_3}{r_{13}} + \frac{Gm_2m_3}{r_{23}} \right). \qquad (13\text{-}22)$$

## Proof of Equation 13-21

Let us shoot a baseball directly away from Earth along the path in Fig. 13-10. We want to find an expression for the gravitational potential energy $U$ of the ball at point $P$ along its path, at radial distance $R$ from Earth's center. To do so, we first find the work $W$ done on the ball by the gravitational force as the ball travels from point $P$ to a great (infinite) distance from Earth. Because the gravitational force $\vec{F}(r)$ is a variable force (its magnitude depends on $r$), we must use the techniques of Section 7-8 to find the work. In vector notation, we can write

$$W = \int_R^\infty \vec{F}(r) \cdot d\vec{r}. \qquad (13\text{-}23)$$

The integral contains the scalar (or dot) product of the force $\vec{F}(r)$ and the differential displacement vector $d\vec{r}$ along the ball's path. We can expand that product as

$$\vec{F}(r) \cdot d\vec{r} = F(r) \, dr \cos \phi, \qquad (13\text{-}24)$$

where $\phi$ is the angle between the directions of $\vec{F}(r)$ and $d\vec{r}$. When we substitute

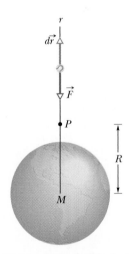

FIG. 13-10 A baseball is shot directly away from Earth, through point $P$ at radial distance $R$ from Earth's center. The gravitational force $\vec{F}$ on the ball and a differential displacement vector $d\vec{r}$ are shown, both directed along a radial $r$ axis.

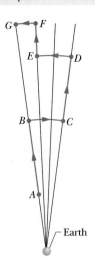

**FIG. 13-11** Near Earth, a baseball is moved from point $A$ to point $G$ along a path consisting of radial lengths and circular arcs.

180° for $\phi$ and Eq. 13-1 for $F(r)$, Eq. 13-24 becomes

$$\vec{F}(r) \cdot d\vec{r} = -\frac{GMm}{r^2}\, dr,$$

where $M$ is Earth's mass and $m$ is the mass of the ball.

Substituting this into Eq. 13-23 and integrating give us

$$W = -GMm \int_R^\infty \frac{1}{r^2}\, dr = \left[\frac{GMm}{r}\right]_R^\infty$$

$$= 0 - \frac{GMm}{R} = -\frac{GMm}{R}, \tag{13-25}$$

where $W$ is the work required to move the ball from point $P$ (at distance $R$) to infinity. Equation 8-1 ($\Delta U = -W$) tells us that we can also write that work in terms of potential energies as

$$U_\infty - U = -W.$$

Because the potential energy $U_\infty$ at infinity is zero, $U$ is the potential energy at $P$, and $W$ is given by Eq. 13-25, this equation becomes

$$U = W = -\frac{GMm}{R}.$$

Switching $R$ to $r$ gives us Eq. 13-21, which we set out to prove.

### Path Independence

In Fig. 13-11, we move a baseball from point $A$ to point $G$ along a path consisting of three radial lengths and three circular arcs (centered on Earth). We are interested in the total work $W$ done by Earth's gravitational force $\vec{F}$ on the ball as it moves from $A$ to $G$. The work done along each circular arc is zero, because the direction of $\vec{F}$ is perpendicular to the arc at every point. Thus, $W$ is the sum of only the works done by $\vec{F}$ along the three radial lengths.

Now, suppose we mentally shrink the arcs to zero. We would then be moving the ball directly from $A$ to $G$ along a single radial length. Does that change $W$? No. Because no work was done along the arcs, eliminating them does not change the work. The path taken from $A$ to $G$ now is clearly different, but the work done by $\vec{F}$ is the same.

We discussed such a result in a general way in Section 8-3. Here is the point: The gravitational force is a conservative force. Thus, the work done by the gravitational force on a particle moving from an initial point $i$ to a final point $f$ is independent of the path taken between the points. From Eq. 8-1, the change $\Delta U$ in the gravitational potential energy from point $i$ to point $f$ is given by

$$\Delta U = U_f - U_i = -W. \tag{13-26}$$

Since the work $W$ done by a conservative force is independent of the actual path taken, the change $\Delta U$ in gravitational potential energy is *also independent* of the path taken.

### Potential Energy and Force

In the proof of Eq. 13-21, we derived the potential energy function $U(r)$ from the force function $\vec{F}(r)$. We should be able to go the other way—that is, to start from the potential energy function and derive the force function. Guided by Eq. 8-22 ($F(x) = -dU(x)/dx$), we can write

$$F = -\frac{dU}{dr} = -\frac{d}{dr}\left(-\frac{GMm}{r}\right)$$

$$= -\frac{GMm}{r^2}. \tag{13-27}$$

This is Newton's law of gravitation (Eq. 13-1). The minus sign indicates that the force on mass $m$ points radially inward, toward mass $M$.

## Escape Speed

If you fire a projectile upward, usually it will slow, stop momentarily, and return to Earth. There is, however, a certain minimum initial speed that will cause it to move upward forever, theoretically coming to rest only at infinity. This minimum initial speed is called the (Earth) **escape speed.**

Consider a projectile of mass $m$, leaving the surface of a planet (or some other astronomical body or system) with escape speed $v$. The projectile has a kinetic energy $K$ given by $\frac{1}{2}mv^2$ and a potential energy $U$ given by Eq. 13-21:

$$U = -\frac{GMm}{R},$$

in which $M$ is the mass of the planet and $R$ is its radius.

When the projectile reaches infinity, it stops and thus has no kinetic energy. It also has no potential energy because an infinite separation between two bodies is our zero-potential-energy configuration. Its total energy at infinity is therefore zero. From the principle of conservation of energy, its total energy at the planet's surface must also have been zero, and so

$$K + U = \tfrac{1}{2}mv^2 + \left(-\frac{GMm}{R}\right) = 0.$$

This yields

$$v = \sqrt{\frac{2GM}{R}}. \tag{13-28}$$

Note that $v$ does not depend on the direction in which a projectile is fired from a planet. However, attaining that speed is easier if the projectile is fired in the direction the launch site is moving as the planet rotates about its axis. For example, rockets are launched eastward at Cape Canaveral to take advantage of the Cape's eastward speed of 1500 km/h due to Earth's rotation.

Equation 13-28 can be applied to find the escape speed of a projectile from any astronomical body, provided we substitute the mass of the body for $M$ and the radius of the body for $R$. Table 13-2 shows some escape speeds.

✓**CHECKPOINT 3**    You move a ball of mass $m$ away from a sphere of mass $M$. (a) Does the gravitational potential energy of the ball–sphere system increase or decrease? (b) Is positive or negative work done by the gravitational force between the ball and the sphere?

**TABLE 13-2**

**Some Escape Speeds**

| Body | Mass (kg) | Radius (m) | Escape Speed (km/s) |
|---|---|---|---|
| Ceres[a] | $1.17 \times 10^{21}$ | $3.8 \times 10^5$ | 0.64 |
| Earth's moon[a] | $7.36 \times 10^{22}$ | $1.74 \times 10^6$ | 2.38 |
| Earth | $5.98 \times 10^{24}$ | $6.37 \times 10^6$ | 11.2 |
| Jupiter | $1.90 \times 10^{27}$ | $7.15 \times 10^7$ | 59.5 |
| Sun | $1.99 \times 10^{30}$ | $6.96 \times 10^8$ | 618 |
| Sirius B[b] | $2 \times 10^{30}$ | $1 \times 10^7$ | 5200 |
| Neutron star[c] | $2 \times 10^{30}$ | $1 \times 10^4$ | $2 \times 10^5$ |

[a]The most massive of the asteroids.

[b]A *white dwarf* (a star in a final stage of evolution) that is a companion of the bright star Sirius.

[c]The collapsed core of a star that remains after that star has exploded in a *supernova* event.

An asteroid, headed directly toward Earth, has a speed of 12 km/s relative to the planet when the asteroid is 10 Earth radii from Earth's center. Neglecting the effects of Earth's atmosphere on the asteroid, find the asteroid's speed $v_f$ when it reaches Earth's surface.

**KEY IDEAS** Because we are to neglect the effects of the atmosphere on the asteroid, the mechanical energy of the asteroid–Earth system is conserved during the fall. Thus, the final mechanical energy (when the asteroid reaches Earth's surface) is equal to the initial mechanical energy. With kinetic energy $K$ and gravitational potential energy $U$, we can write this as

$$K_f + U_f = K_i + U_i. \qquad (13\text{-}29)$$

Also, if we assume the system is isolated, the system's linear momentum must be conserved during the fall. Therefore, the momentum change of the asteroid and that of Earth must be equal in magnitude and opposite in sign. However, because Earth's mass is so much greater than the asteroid's mass, the change in Earth's speed is negligible relative to the change in the asteroid's speed. So, the change in Earth's kinetic energy is also negligible. Thus, we can assume that the kinetic energies in Eq. 13-29 are those of the asteroid alone.

*Calculations:* Let $m$ represent the asteroid's mass and $M$ represent Earth's mass ($5.98 \times 10^{24}$ kg). The asteroid is initially at distance $10R_E$ and finally at distance $R_E$,

where $R_E$ is Earth's radius ($6.37 \times 10^6$ m). Substituting Eq. 13-21 for $U$ and $\frac{1}{2}mv^2$ for $K$, we rewrite Eq. 13-29 as

$$\tfrac{1}{2}mv_f^2 - \frac{GMm}{R_E} = \tfrac{1}{2}mv_i^2 - \frac{GMm}{10R_E}.$$

Rearranging and substituting known values, we find

$$
\begin{aligned}
v_f^2 &= v_i^2 + \frac{2GM}{R_E}\left(1 - \frac{1}{10}\right) \\
&= (12 \times 10^3 \text{ m/s})^2 \\
&\quad + \frac{2(6.67 \times 10^{-11} \text{ m}^3/\text{kg}\cdot\text{s}^2)(5.98 \times 10^{24} \text{ kg})}{6.37 \times 10^6 \text{ m}}\,0.9 \\
&= 2.567 \times 10^8 \text{ m}^2/\text{s}^2,
\end{aligned}
$$

and

$$v_f = 1.60 \times 10^4 \text{ m/s} = 16 \text{ km/s}. \qquad \text{(Answer)}$$

At this speed, the asteroid would not have to be particularly large to do considerable damage at impact. If it were only 5 m across, the impact could release about as much energy as the nuclear explosion at Hiroshima. Alarmingly, about 500 million asteroids of this size are near Earth's orbit, and in 1994 one of them apparently penetrated Earth's atmosphere and exploded 20 km above the South Pacific (setting off nuclear-explosion warnings on six military satellites). The impact of an asteroid 500 m across (there may be a million of them near Earth's orbit) could end modern civilization and almost eliminate humans worldwide.

## 13-7 | Planets and Satellites: Kepler's Laws

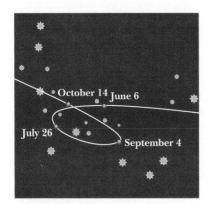

FIG. 13-12 The path seen from Earth for the planet Mars as it moved against a background of the constellation Capricorn during 1971. The planet's position on four days is marked. Both Mars and Earth are moving in orbits around the Sun so that we see the position of Mars relative to us; this relative motion sometimes results in an apparent loop in the path of Mars.

The motions of the planets, as they seemingly wander against the background of the stars, have been a puzzle since the dawn of history. The "loop-the-loop" motion of Mars, shown in Fig. 13-12, was particularly baffling. Johannes Kepler (1571–1630), after a lifetime of study, worked out the empirical laws that govern these motions. Tycho Brahe (1546–1601), the last of the great astronomers to make observations without the help of a telescope, compiled the extensive data from which Kepler was able to derive the three laws of planetary motion that now bear Kepler's name. Later, Newton (1642–1727) showed that his law of gravitation leads to Kepler's laws.

In this section we discuss each of Kepler's three laws. Although here we apply the laws to planets orbiting the Sun, they hold equally well for satellites, either natural or artificial, orbiting Earth or any other massive central body.

➤ **1. THE LAW OF ORBITS:** All planets move in elliptical orbits, with the Sun at one focus.

Figure 13-13 shows a planet of mass $m$ moving in such an orbit around the Sun, whose mass is $M$. We assume that $M \gg m$, so that the center of mass of the planet–Sun system is approximately at the center of the Sun.

The orbit in Fig. 13-13 is described by giving its **semimajor axis** $a$ and its **eccentricity** $e$, the latter defined so that $ea$ is the distance from the center of the ellipse to either focus $F$ or $F'$. *An eccentricity of zero corresponds to a circle, in* which the two foci merge to a single central point. The eccentricities of the planetary orbits are not large; so if the orbits are drawn to scale, they look circular. The eccentricity of the ellipse of Fig. 13-13, which has been exaggerated for clarity, is 0.74. The eccentricity of Earth's orbit is only 0.0167.

2. **THE LAW OF AREAS:** A line that connects a planet to the Sun sweeps out equal areas in the plane of the planet's orbit in equal time intervals; that is, the rate $dA/dt$ at which it sweeps out area $A$ is constant.

Qualitatively, this second law tells us that the planet will move most slowly when it is farthest from the Sun and most rapidly when it is nearest to the Sun. As it turns out, Kepler's second law is totally equivalent to the law of conservation of angular momentum. Let us prove it.

The area of the shaded wedge in Fig. 13-14a closely approximates the area swept out in time $\Delta t$ by a line connecting the Sun and the planet, which are separated by distance $r$. The area $\Delta A$ of the wedge is approximately the area of a triangle with base $r\Delta\theta$ and height $r$. Since the area of a triangle is one-half of the base times the height, $\Delta A \approx \frac{1}{2}r^2\,\Delta\theta$. This expression for $\Delta A$ becomes more exact as $\Delta t$ (hence $\Delta\theta$) approaches zero. The instantaneous rate at which area is being swept out is then

$$\frac{dA}{dt} = \tfrac{1}{2}r^2\frac{d\theta}{dt} = \tfrac{1}{2}r^2\omega, \tag{13-30}$$

in which $\omega$ is the angular speed of the rotating line connecting Sun and planet.

Figure 13-14b shows the linear momentum $\vec{p}$ of the planet, along with the radial and perpendicular components of $\vec{p}$. From Eq. 11-20 ($L = rp_\perp$), the magnitude of the angular momentum $\vec{L}$ of the planet about the Sun is given by the product of $r$ and $p_\perp$, the component of $\vec{p}$ perpendicular to $r$. Here, for a planet of mass $m$,

$$L = rp_\perp = (r)(mv_\perp) = (r)(m\omega r)$$
$$= mr^2\omega, \tag{13-31}$$

where we have replaced $v_\perp$ with its equivalent $\omega r$ (Eq. 10-18). Eliminating $r^2\omega$ between Eqs. 13-30 and 13-31 leads to

$$\frac{dA}{dt} = \frac{L}{2m}. \tag{13-32}$$

If $dA/dt$ is constant, as Kepler said it is, then Eq. 13-32 means that $L$ must also be constant—angular momentum is conserved. Kepler's second law is indeed equivalent to the law of conservation of angular momentum.

3. **THE LAW OF PERIODS:** The square of the period of any planet is proportional to the cube of the semimajor axis of its orbit.

To see this, consider the circular orbit of Fig. 13-15, with radius $r$ (the radius of a circle is equivalent to the semimajor axis of an ellipse). Applying Newton's

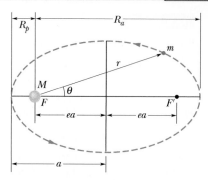

**FIG. 13-13** A planet of mass $m$ moving in an elliptical orbit around the Sun. The Sun, of mass $M$, is at one focus $F$ of the ellipse. The other focus is $F'$, which is located in empty space. Each focus is a distance $ea$ from the ellipse's center, with $e$ being the eccentricity of the ellipse. The semimajor axis $a$ of the ellipse, the perihelion (nearest the Sun) distance $R_p$, and the aphelion (farthest from the Sun) distance $R_a$ are also shown.

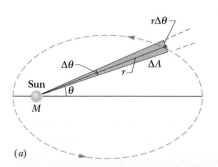

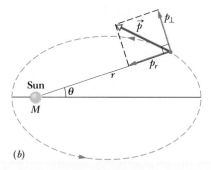

**FIG. 13-14** (a) In time $\Delta t$, the line $r$ connecting the planet to the Sun moves through an angle $\Delta\theta$, sweeping out an area $\Delta A$ (shaded). (b) The linear momentum $\vec{p}$ of the planet and the components of $\vec{p}$.

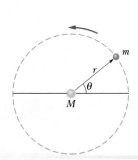

**FIG. 13-15** A planet of mass $m$ moving around the Sun in a circular orbit of radius $r$.

**TABLE 13-3**

**Kepler's Law of Periods for the Solar System**

| Planet | Semimajor Axis $a$ ($10^{10}$ m) | Period $T$ (y) | $T^2/a^3$ ($10^{-34}$ y²/m³) |
|---|---|---|---|
| Mercury | 5.79 | 0.241 | 2.99 |
| Venus | 10.8 | 0.615 | 3.00 |
| Earth | 15.0 | 1.00 | 2.96 |
| Mars | 22.8 | 1.88 | 2.98 |
| Jupiter | 77.8 | 11.9 | 3.01 |
| Saturn | 143 | 29.5 | 2.98 |
| Uranus | 287 | 84.0 | 2.98 |
| Neptune | 450 | 165 | 2.99 |
| Pluto | 590 | 248 | 2.99 |

second law ($F = ma$) to the orbiting planet in Fig. 13-15 yields

$$\frac{GMm}{r^2} = (m)(\omega^2 r). \tag{13-33}$$

Here we have substituted from Eq. 13-1 for the force magnitude $F$ and used Eq. 10-23 to substitute $\omega^2 r$ for the centripetal acceleration. If we now use Eq. 10-20 to replace $\omega$ with $2\pi/T$, where $T$ is the period of the motion, we obtain Kepler's third law:

$$T^2 = \left(\frac{4\pi^2}{GM}\right)r^3 \quad \text{(law of periods).} \tag{13-34}$$

The quantity in parentheses is a constant that depends only on the mass $M$ of the central body about which the planet orbits.

Equation 13-34 holds also for elliptical orbits, provided we replace $r$ with $a$, the semimajor axis of the ellipse. This law predicts that the ratio $T^2/a^3$ has essentially the same value for every planetary orbit around a given massive body. Table 13-3 shows how well it holds for the orbits of the planets of the solar system.

✓**CHECKPOINT 4**  Satellite 1 is in a certain circular orbit around a planet, while satellite 2 is in a larger circular orbit. Which satellite has (a) the longer period and (b) the greater speed?

**Sample Problem  13-6**

Comet Halley orbits the Sun with a period of 76 years and, in 1986, had a distance of closest approach to the Sun, its *perihelion distance* $R_p$, of $8.9 \times 10^{10}$ m. Table 13-3 shows that this is between the orbits of Mercury and Venus.

(a) What is the comet's farthest distance from the Sun, which is called its *aphelion distance* $R_a$?

**KEY IDEAS**  From Fig. 13-13, we see that $R_a + R_p = 2a$, where $a$ is the semimajor axis of the orbit. Thus, we can find $R_a$ if we first find $a$. We can relate $a$ to the given period via the law of periods (Eq. 13-34) if we simply substitute the semimajor axis $a$ for $r$.

**Calculations:** Making that substitution and then solving for $a$, we have

$$a = \left(\frac{GMT^2}{4\pi^2}\right)^{1/3}. \tag{13-35}$$

If we substitute the mass $M$ of the Sun, $1.99 \times 10^{30}$ kg, and the period $T$ of the comet, 76 years or $2.4 \times 10^9$ s, into Eq. 13-35, we find that $a = 2.7 \times 10^{12}$ m. Now we

have

$$R_a = 2a - R_p$$
$$= (2)(2.7 \times 10^{12} \text{ m}) - 8.9 \times 10^{10} \text{ m}$$
$$= 5.3 \times 10^{12} \text{ m.} \quad \text{(Answer)}$$

Table 13-3 shows that this is a little less than the semimajor axis of the orbit of Pluto. Thus, the comet does not get farther from the Sun than Pluto.

(b) What is the eccentricity $e$ of the orbit of comet Halley?

**KEY IDEA**  We can relate $e$, $a$, and $R_p$ via Fig. 13-13, in which we see that $ea = a - R_p$.

**Calculation:** We have

$$e = \frac{a - R_p}{a} = 1 - \frac{R_p}{a}$$
$$= 1 - \frac{8.9 \times 10^{10} \text{ m}}{2.7 \times 10^{12} \text{ m}} = 0.97. \quad \text{(Answer)}$$

This tells us that, with an eccentricity approaching unity, this orbit must be a long thin ellipse.

**Sample Problem  13-7**

Let's return to the story that opens this chapter. Figure 13-16 shows the observed orbit of the star S2 as the star moves around a mysterious and unobserved object called Sagittarius A* (pronounced "A star"), which is at the center of the Milky Way galaxy. S2 orbits Sagittarius A* with a period of $T = 15.2$ y and with a semimajor axis of $a = 5.50$ light-days ($= 1.42 \times 10^{14}$ m). What is the mass $M$ of Sagittarius A*? What is Sagittarius A*?

**KEY IDEA** The period $T$ and the semimajor axis $a$ of the orbit are related to the mass $M$ of Sagittarius A* according to Kepler's law of periods. From Eq. 13-34, with $a$ replacing the radius $r$ of a circular orbit, we have

$$T^2 = \left(\frac{4\pi^2}{GM}\right)a^3. \qquad (13\text{-}36)$$

**Calculations:** Solving Eq. 13-36 for $M$ and substituting the given data lead us to

$$M = \frac{4\pi^2 a^3}{GT^2}$$

$$= \frac{4\pi^2(1.42 \times 10^{14}\ \text{m})^3}{(6.67 \times 10^{-11}\ \text{N} \cdot \text{m}^2/\text{kg}^2)[(15.2\ \text{y})(3.16 \times 10^7\ \text{s/y})]^2}$$

$$= 7.35 \times 10^{36}\ \text{kg}. \qquad \text{(Answer)}$$

To figure out what Sagittarius A* might be, let's divide this mass by the mass of our Sun ($M_{\text{Sun}} = 1.99 \times 10^{30}$ kg) to find that

$$M = (3.7 \times 10^6)M_{\text{Sun}}.$$

Sagittarius A* has a mass of 3.7 million Suns! However, it cannot be seen. Thus, it is an extremely compact ob-

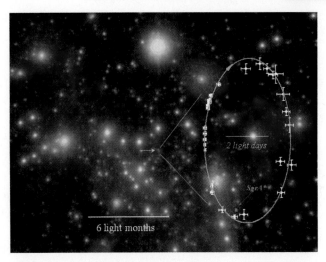

**FIG. 13-16** The orbit of star S2 about Sagittarius A* (Sgr A*). The elliptical orbit appears skewed because we do not see it from directly above the orbital plane. Uncertainties in the location of S2 are indicated by the crossbars. *(Courtesy Reinhard Genzel)*

ject. Such a huge mass in such a small object leads to the reasonable conclusion that this object is a *supermassive black hole*. In fact, evidence is mounting that a supermassive black hole lurks at the center of most galaxies. (Movies of the stars orbiting Sagittarius A* are available on the Web; search under "black hole galactic center.")

## 13-8 | Satellites: Orbits and Energy

As a satellite orbits Earth in an elliptical path, both its speed, which fixes its kinetic energy $K$, and its distance from the center of Earth, which fixes its gravitational potential energy $U$, fluctuate with fixed periods. However, the mechanical energy $E$ of the satellite remains constant. (Since the satellite's mass is so much smaller than Earth's mass, we assign $U$ and $E$ for the Earth–satellite system to the satellite alone.)

The potential energy of the system is given by Eq. 13-21:

$$U = -\frac{GMm}{r}$$

(with $U = 0$ for infinite separation). Here $r$ is the radius of the satellite's orbit, assumed for the time being to be circular, and $M$ and $m$ are the masses of Earth and the satellite, respectively.

To find the kinetic energy of a satellite in a circular orbit, we write Newton's second law ($F = ma$) as

$$\frac{GMm}{r^2} = m\frac{v^2}{r}, \qquad (13\text{-}37)$$

where $v^2/r$ is the centripetal acceleration of the satellite. Then, from Eq. 13-37, the kinetic energy is

$$K = \tfrac{1}{2}mv^2 = \frac{GMm}{2r}, \qquad (13\text{-}38)$$

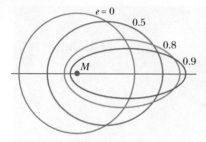

FIG. 13-17 Four orbits with different eccentricities $e$ about an object of mass $M$. All four orbits have the same semimajor axis $a$ and thus correspond to the same total mechanical energy $E$.

which shows us that for a satellite in a circular orbit,

$$K = -\frac{U}{2} \qquad \text{(circular orbit).} \tag{13-39}$$

The total mechanical energy of the orbiting satellite is

$$E = K + U = \frac{GMm}{2r} - \frac{GMm}{r}$$

or

$$E = -\frac{GMm}{2r} \qquad \text{(circular orbit).} \tag{13-40}$$

This tells us that for a satellite in a circular orbit, the total energy $E$ is the negative of the kinetic energy $K$:

$$E = -K \qquad \text{(circular orbit).} \tag{13-41}$$

For a satellite in an elliptical orbit of semimajor axis $a$, we can substitute $a$ for $r$ in Eq. 13-40 to find the mechanical energy:

$$E = -\frac{GMm}{2a} \qquad \text{(elliptical orbit).} \tag{13-42}$$

Equation 13-42 tells us that the total energy of an orbiting satellite depends only on the semimajor axis of its orbit and not on its eccentricity $e$. For example, four orbits with the same semimajor axis are shown in Fig. 13-17; the same satellite would have the same total mechanical energy $E$ in all four orbits. Figure 13-18 shows the variation of $K$, $U$, and $E$ with $r$ for a satellite moving in a circular orbit about a massive central body.

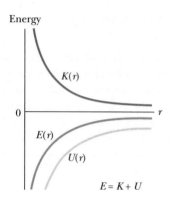

FIG. 13-18 The variation of kinetic energy $K$, potential energy $U$, and total energy $E$ with radius $r$ for a satellite in a circular orbit. For any value of $r$, the values of $U$ and $E$ are negative, the value of $K$ is positive, and $E = -K$. As $r \rightarrow \infty$, all three energy curves approach a value of zero.

✓CHECKPOINT 5    In the figure here, a space shuttle is initially in a circular orbit of radius $r$ about Earth. At point $P$, the pilot briefly fires a forward-pointing thruster to decrease the shuttle's kinetic energy $K$ and mechanical energy $E$. (a) Which of the dashed elliptical orbits shown in the figure will the shuttle then take? (b) Is the orbital period $T$ of the shuttle (the time to return to $P$) then greater than, less than, or the same as in the circular orbit?

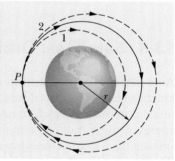

---

**Sample Problem** | **13-8**

---

A playful astronaut releases a bowling ball, of mass $m = 7.20$ kg, into circular orbit about Earth at an altitude $h$ of 350 km.

(a) What is the mechanical energy $E$ of the ball in its orbit?

---

**KEY IDEA**    We can get $E$ from the orbital energy, given by Eq. 13-40 ($E = -GMm/2r$), if we first find the orbital radius $r$.

**Calculations:** The orbital radius must be

$$r = R + h = 6370 \text{ km} + 350 \text{ km} = 6.72 \times 10^6 \text{ m},$$

in which $R$ is the radius of Earth. Then, from Eq. 13-40,

the mechanical energy is

$$E = -\frac{GMm}{2r}$$

$$= -\frac{(6.67 \times 10^{-11} \text{ N} \cdot \text{m}^2/\text{kg}^2)(5.98 \times 10^{24} \text{ kg})(7.20 \text{ kg})}{(2)(6.72 \times 10^6 \text{ m})}$$

$$= -2.14 \times 10^8 \text{ J} = -214 \text{ MJ}. \qquad \text{(Answer)}$$

(b) What is the mechanical energy $E_0$ of the ball on the launchpad at Cape Canaveral? From there to the orbit, what is the change $\Delta E$ in the ball's mechanical energy?

---

**KEY IDEA**    On the launchpad, the ball is *not* in orbit and thus Eq. 13-40 does *not* apply. Instead, we must find

$E_0 = K_0 + U_0$, where $K_0$ is the ball's kinetic energy and $U_0$ is the gravitational potential energy of the ball–Earth system.

**Calculations:** To find $U_0$, we use Eq. 13-21 to write

$$U_0 = -\frac{GMm}{R}$$

$$= -\frac{(6.67 \times 10^{-11} \text{ N} \cdot \text{m}^2/\text{kg}^2)(5.98 \times 10^{24} \text{ kg})(7.20 \text{ kg})}{6.37 \times 10^6 \text{ m}}$$

$$= -4.51 \times 10^8 \text{ J} = -451 \text{ MJ}.$$

The kinetic energy $K_0$ of the ball is due to the ball's motion with Earth's rotation. You can show that $K_0$ is less than 1 MJ,

which is negligible relative to $U_0$. Thus, the mechanical energy of the ball on the launchpad is

$$E_0 = K_0 + U_0 \approx 0 - 451 \text{ MJ} = -451 \text{ MJ}. \quad \text{(Answer)}$$

The *increase* in the mechanical energy of the ball from launchpad to orbit is

$$\Delta E = E - E_0 = (-214 \text{ MJ}) - (-451 \text{ MJ})$$

$$= 237 \text{ MJ}. \quad \text{(Answer)}$$

This is worth a few dollars at your utility company. Obviously the high cost of placing objects into orbit is not due to their required mechanical energy.

## 13-9 | Einstein and Gravitation

### Principle of Equivalence

Albert Einstein once said: "I was . . . in the patent office at Bern when all of a sudden a thought occurred to me: 'If a person falls freely, he will not feel his own weight.' I was startled. This simple thought made a deep impression on me. It impelled me toward a theory of gravitation."

Thus Einstein tells us how he began to form his **general theory of relativity.** The fundamental postulate of this theory about gravitation (the gravitating of objects toward each other) is called the **principle of equivalence,** which says that gravitation and acceleration are equivalent. If a physicist were locked up in a small box as in Fig. 13-19, he would not be able to tell whether the box was at rest on Earth (and subject only to Earth's gravitational force), as in Fig. 13-19a, or accelerating through interstellar space at 9.8 m/s² (and subject only to the force producing that acceleration), as in Fig. 13-19b. In both situations he would feel the same and would read the same value for his weight on a scale. Moreover, if he watched an object fall past him, the object would have the same acceleration relative to him in both situations.

### Curvature of Space

We have thus far explained gravitation as due to a force between masses. Einstein showed that, instead, gravitation is due to a curvature of space that is caused by the masses. (As is discussed later in this book, space and time are entangled, so the curvature of which Einstein spoke is really a curvature of *spacetime,* the combined four dimensions of our universe.)

Picturing how space (such as vacuum) can have curvature is difficult. An analogy might help: Suppose that from orbit we watch a race in which two boats begin on Earth's equator with a separation of 20 km and head due south (Fig. 13-20a). To the sailors, the boats travel along flat, parallel paths. However, with time the boats draw together until, nearer the south pole, they touch. The sailors in the boats can interpret this drawing together in terms of a force acting on the boats. Looking on from space, however, we can see that the boats draw together simply because of the curvature of Earth's surface. We can see this because we are viewing the race from "outside" that surface.

Figure 13-20b shows a similar race: Two horizontally separated apples are dropped from the same height above Earth. Although the apples may appear to travel along parallel paths, they actually move toward each other because they both fall toward Earth's center. We can interpret the motion of the apples in terms of the gravitational force on the apples from Earth. We can also interpret the motion in terms of a curvature of the space near Earth, a curvature due to the

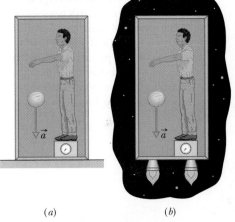

(a)                    (b)

**FIG. 13-19** (a) A physicist in a box resting on Earth sees a cantaloupe falling with acceleration $a = 9.8$ m/s². (b) If he and the box accelerate in deep space at 9.8 m/s², the cantaloupe has the same acceleration relative to him. It is not possible, by doing experiments within the box, for the physicist to tell which situation he is in. For example, the platform scale on which he stands reads the same weight in both situations.

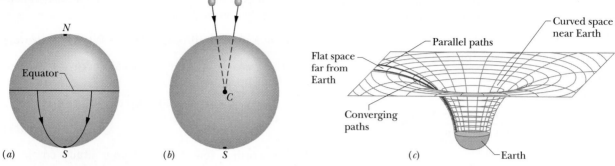

**FIG. 13-20** (*a*) Two objects moving along lines of longitude toward the south pole converge because of the curvature of Earth's surface. (*b*) Two objects falling freely near Earth move along lines that converge toward the center of Earth because of the curvature of space near Earth. (*c*) Far from Earth (and other masses), space is flat and parallel paths remain parallel. Close to Earth, the parallel paths begin to converge because space is curved by Earth's mass.

presence of Earth's mass. This time we cannot see the curvature because we cannot get "outside" the curved space, as we got "outside" the curved Earth in the boat example. However, we can depict the curvature with a drawing like Fig. 13-20*c*; there the apples would move along a surface that curves toward Earth because of Earth's mass.

When light passes near Earth, the path of the light bends slightly because of the curvature of space there, an effect called *gravitational lensing*. When light passes a more massive structure, like a galaxy or a black hole having large mass, its path can be bent more. If such a massive structure is between us and a quasar (an extremely bright, extremely distant source of light), the light from the quasar can bend around the massive structure and toward us (Fig. 13-21*a*). Then, because the light seems to be coming to us from a number of slightly different directions in the sky, we see the same quasar in all those different directions. In some situations, the quasars we see blend together to form a giant luminous arc, which is called an *Einstein ring* (Fig. 13-21*b*).

Should we attribute gravitation to the curvature of spacetime due to the presence of masses or to a force between masses? Or should we attribute it to the actions of a type of fundamental particle called a *graviton*, as conjectured in some modern physics theories? We just don't know.

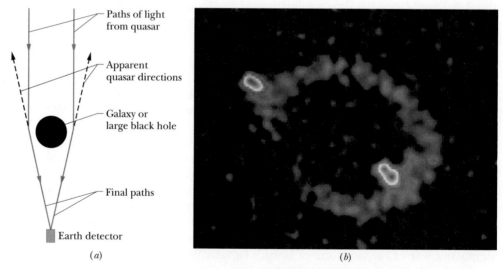

**FIG. 13-21** (*a*) Light from a distant quasar follows curved paths around a galaxy or a large black hole because the mass of the galaxy or black hole has curved the adjacent space. If the light is detected, it appears to have originated along the backward extensions of the final paths (dashed lines). (*b*) The Einstein ring known as MG1131+0456 on the computer screen of a telescope. The source of the light (actually, radio waves, which are a form of invisible light) is far behind the large, unseen galaxy that produces the ring; a portion of the source appears as the two bright spots seen along the ring. (*Courtesy National Radio Astronomy Observatory*)

**The Law of Gravitation** Any particle in the universe attracts any other particle with a **gravitational force** whose magnitude is

$$F = G \frac{m_1 m_2}{r^2} \quad \text{(Newton's law of gravitation)}, \quad (13\text{-}1)$$

where $m_1$ and $m_2$ are the masses of the particles, $r$ is their separation, and $G$ ($= 6.67 \times 10^{-11} \text{ N} \cdot \text{m}^2/\text{kg}^2$) is the *gravitational constant*.

**Gravitational Behavior of Uniform Spherical Shells** Equation 13-1 holds only for particles. The gravitational force between extended bodies must generally be found by adding (integrating) the individual forces on individual particles within the bodies. However, if either of the bodies is a uniform spherical shell or a spherically symmetric solid, the net gravitational force it exerts on an *external* object may be computed as if all the mass of the shell or body were located at its center.

**Superposition** Gravitational forces obey the **principle of superposition;** that is, if $n$ particles interact, the net force $\vec{F}_{1,\text{net}}$ on a particle labeled particle 1 is the sum of the forces on it from all the other particles taken one at a time:

$$\vec{F}_{1,\text{net}} = \sum_{i=2}^{n} \vec{F}_{1i}, \quad (13\text{-}5)$$

in which the sum is a vector sum of the forces $\vec{F}_{1i}$ on particle 1 from particles 2, 3, . . . , $n$. The gravitational force $\vec{F}_1$ on a particle from an extended body is found by dividing the body into units of differential mass $dm$, each of which produces a differential force $d\vec{F}$ on the particle, and then integrating to find the sum of those forces:

$$\vec{F}_1 = \int d\vec{F}. \quad (13\text{-}6)$$

**Gravitational Acceleration** The *gravitational acceleration* $a_g$ of a particle (of mass $m$) is due solely to the gravitational force acting on it. When the particle is at distance $r$ from the center of a uniform, spherical body of mass $M$, the magnitude $F$ of the gravitational force on the particle is given by Eq. 13-1. Thus, by Newton's second law,

$$F = ma_g, \quad (13\text{-}10)$$

which gives

$$a_g = \frac{GM}{r^2}. \quad (13\text{-}11)$$

**Free-Fall Acceleration and Weight** Because Earth's mass is not distributed uniformly, because the planet is not perfectly spherical, and because it rotates, the actual free-fall acceleration $\vec{g}$ of a particle near Earth differs slightly from the gravitational acceleration $\vec{a}_g$, and the particle's weight (equal to $mg$) differs from the magnitude of the gravitational force acting on the particle as computed with Eq. 13-1.

**Gravitation Within a Spherical Shell** A uniform shell of matter exerts no net gravitational force on a particle located inside it. This means that if a particle is located inside a uniform solid sphere at distance $r$ from its center, the gravita-

tional force exerted on the particle is due only to the mass $M_{\text{ins}}$ that lies inside a sphere of radius $r$. This mass is given by

$$M_{\text{ins}} = \rho \frac{4\pi r^3}{3}, \quad (13\text{-}18)$$

where $\rho$ is the density of the sphere.

**Gravitational Potential Energy** The gravitational potential energy $U(r)$ of a system of two particles, with masses $M$ and $m$ and separated by a distance $r$, is the negative of the work that would be done by the gravitational force of either particle acting on the other if the separation between the particles were changed from infinite (very large) to $r$. This energy is

$$U = -\frac{GMm}{r} \quad \text{(gravitational potential energy)}. \quad (13\text{-}21)$$

**Potential Energy of a System** If a system contains more than two particles, its total gravitational potential energy $U$ is the sum of terms representing the potential energies of all the pairs. As an example, for three particles, of masses $m_1, m_2$, and $m_3$,

$$U = -\left( \frac{Gm_1 m_2}{r_{12}} + \frac{Gm_1 m_3}{r_{13}} + \frac{Gm_2 m_3}{r_{23}} \right). \quad (13\text{-}22)$$

**Escape Speed** An object will escape the gravitational pull of an astronomical body of mass $M$ and radius $R$ (that is, it will reach an infinite distance) if the object's speed near the body's surface is at least equal to the **escape speed,** given by

$$v = \sqrt{\frac{2GM}{R}}. \quad (13\text{-}28)$$

**Kepler's Laws** Gravitational attraction holds the solar system together and makes possible orbiting Earth satellites, both natural and artificial. Such motions are governed by Kepler's three laws of planetary motion, all of which are direct consequences of Newton's laws of motion and gravitation:

1. *The law of orbits.* All planets move in elliptical orbits with the Sun at one focus.

2. *The law of areas.* A line joining any planet to the Sun sweeps out equal areas in equal time intervals. (This statement is equivalent to conservation of angular momentum.)

3. *The law of periods.* The square of the period $T$ of any planet is proportional to the cube of the semimajor axis $a$ of its orbit. For circular orbits with radius $r$,

$$T^2 = \left( \frac{4\pi^2}{GM} \right) r^3 \quad \text{(law of periods)}, \quad (13\text{-}34)$$

where $M$ is the mass of the attracting body—the Sun in the case of the solar system. For elliptical planetary orbits, the semimajor axis $a$ is substituted for $r$.

**Energy in Planetary Motion** When a planet or satellite with mass $m$ moves in a circular orbit with radius $r$, its potential energy $U$ and kinetic energy $K$ are given by

$$U = -\frac{GMm}{r} \quad \text{and} \quad K = \frac{GMm}{2r}. \quad \text{(13-21, 13-38)}$$

The mechanical energy $E = K + U$ is then

$$E = -\frac{GMm}{2r}. \quad \text{(13-40)}$$

For an elliptical orbit of semimajor axis $a$,

$$E = -\frac{GMm}{2a}. \quad \text{(13-42)}$$

**Einstein's View of Gravitation** Einstein pointed out that gravitation and acceleration are equivalent. This **principle of equivalence** led him to a theory of gravitation (the **general theory of relativity**) that explains gravitational effects in terms of a curvature of space.

# QUESTIONS

**1** In Fig. 13-22, a central particle is surrounded by two circular rings of particles, at radii $r$ and $R$, with $R > r$. All the particles have mass $m$. What are the magnitude and direction of the net gravitational force on the central particle due to the particles in the rings?

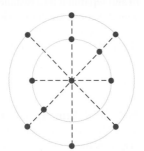

FIG. 13-22 Question 1.

**2** In Fig. 13-23, two particles, of masses $m$ and $2m$, are fixed in place on an axis. (a) Where on the axis can a third particle of mass $3m$ be placed (other than at infinity) so that the net gravitational force on it from the first two particles is zero: to the left of the first two particles, to their right, between them but closer to the more massive particle, or between them but closer to the less massive particle? (b) Does the answer change if the third particle has, instead, a mass of $16m$? (c) Is there a point off the axis (other than infinity) at which the net force on the third particle would be zero?

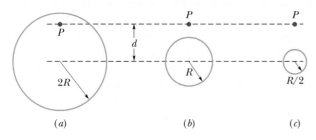

FIG. 13-23 Question 2.

**3** Figure 13-24 shows three situations involving a point particle $P$ with mass $m$ and a spherical shell with a uniformly distributed mass $M$. The radii of the shells are given. Rank the situations according to the magnitude of the gravitational force on particle $P$ due to the shell, greatest first.

FIG. 13-24 Question 3.

**4** Figure 13-25 shows three arrangements of the same identical particles, with three of them placed on a circle of radius 0.20 m and the fourth one placed at the center of the circle. (a) Rank the arrangements

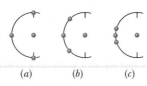

FIG. 13-25 Question 4.

according to the magnitude of the net gravitational force on the central particle due to the other three particles, greatest first. (b) Rank them according to the gravitational potential energy of the four-particle system, least negative first.

**5** In Fig. 13-26, a central particle of mass $M$ is surrounded by a square array of other particles, separated by either distance $d$ or distance $d/2$ along the perimeter of the square. What are the magnitude and direction of the net gravitational force on the central particle due to the other particles?

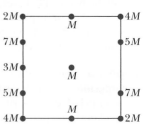

FIG. 13-26 Question 5.

**6** Figure 13-27 gives the gravitational acceleration $a_g$ for four planets as a function of the radial distance $r$ from the center of the planet, starting at the surface of the planet (at radius $R_1$, $R_2$, $R_3$, or $R_4$). Plots 1 and 2 coincide for $r \geq R_2$; plots 3 and 4 coincide for $r \geq R_4$. Rank the four planets according to (a) mass and (b) mass per unit volume, greatest first.

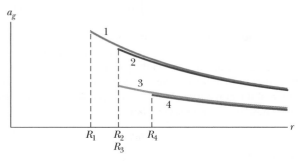

FIG. 13-27 Question 6.

**7** Figure 13-28 shows three particles initially fixed in place, with $B$ and $C$ identical and positioned symmetrically about the $y$ axis, at distance $d$ from $A$. (a) In what direction is the net gravitational force $\vec{F}_{net}$ on $A$? (b) If we move $C$ directly away from the origin, does $\vec{F}_{net}$ change in direction? If so, how and what is the limit of the change?

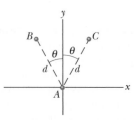

FIG. 13-28 Question 7.

**8** In Fig. 13-29, three particles are fixed in place. The mass of $B$ is greater than the mass of $C$. Can a fourth particle (particle $D$) be placed somewhere so that the net gravitational force on particle *A* from particles *B, C,* and *D* is zero? If so, in which quadrant should it be placed and which axis should it be near?

**9** Rank the four systems of equal-mass particles shown in Checkpoint 2 according to the absolute value of the gravitational potential energy of the system, greatest first.

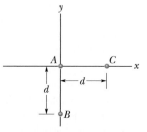

**FIG. 13-29** Question 8.

**10** In Fig. 13-30, a particle of mass *m* (not shown) is to be moved from an infinite distance to one of the three possible locations *a, b,* and *c.* Two other particles, of masses *m* and 2*m,* are fixed in place. Rank the three possible locations according to the work done by the net gravitational force on the moving particle due to the fixed particles, greatest first.

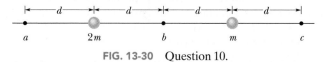

**FIG. 13-30** Question 10.

**11** Figure 13-31 shows three uniform spherical planets that are identical in size and mass. The periods of rotation *T* for the planets are given, and six lettered points are indicated—three points are on the equators of the planets and three points are on the north poles. Rank the points according to the value of the free-fall acceleration *g* at them, greatest first.

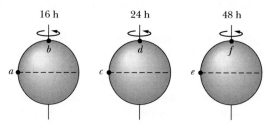

**FIG. 13-31** Question 11.

**12** Figure 13-32 shows six paths by which a rocket orbiting a moon might move from point *a* to point *b.* Rank the paths according to (a) the corresponding change in the gravitational potential energy of the rocket–moon system and (b) the net work done on the rocket by the gravitational force from the moon, greatest first.

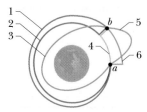

**FIG. 13-32** Question 12.

---

## PROBLEMS

**GO**   Tutoring problem available (at instructor's discretion) in *WileyPLUS* and WebAssign

**SSM**   Worked-out solution available in Student Solutions Manual    **WWW**   Worked-out solution is at

• – •••   Number of dots indicates level of problem difficulty    **ILW**   Interactive solution is at   http://www.wiley.com/college/halliday

   Additional information available in *The Flying Circus of Physics* and at flyingcircusofphysics.com

### sec. 13-2 Newton's Law of Gravitation

•**1** What must the separation be between a 5.2 kg particle and a 2.4 kg particle for their gravitational attraction to have a magnitude of $2.3 \times 10^{-12}$ N? **SSM**

•**2** The Sun and Earth each exert a gravitational force on the Moon. What is the ratio $F_{Sun}/F_{Earth}$ of these two forces? (The average Sun–Moon distance is equal to the Sun–Earth distance.)

•**3** A mass *M* is split into two parts, *m* and *M* − *m,* which are then separated by a certain distance. What ratio *m*/*M* maximizes the magnitude of the gravitational force between the parts? **ILW**

•**4** *Moon effect.* Some people believe that the Moon controls their activities. If the Moon moves from being directly on the opposite side of Earth from you to being directly overhead, by what percent does (a) the Moon's gravitational pull on you increase and (b) your weight (as measured on a scale) decrease? Assume that the Earth–Moon (center-to-center) distance is $3.82 \times 10^8$ m and Earth's radius is $6.37 \times 10^6$ m.

### sec. 13-3 Gravitation and the Principle of Superposition

•**5** *One dimension.* In Fig. 13-33, two point particles are fixed on an *x* axis separated by distance *d.* Particle *A* has mass $m_A$ and particle *B* has mass $3.00m_A$. A

**FIG. 13-33** Problem 5.

third particle *C,* of mass $75.0m_A$, is to be placed on the *x* axis and near particles *A* and *B.* In terms of distance *d,* at what *x* coordinate should *C* be placed so that the net gravitational force on particle *A* from particles *B* and *C* is zero?

•**6** In Fig. 13-34, three 5.00 kg spheres are located at distances $d_1 = 0.300$ m and $d_2 = 0.400$ m. What are the (a) magnitude and (b) direction (relative to the positive direction of the *x* axis) of the net gravitational force on sphere *B* due to spheres *A* and *C*?

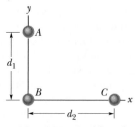

**FIG. 13-34** Problem 6.

•**7** How far from Earth must a space probe be along a line toward the Sun so that the Sun's gravitational pull on the probe balances Earth's pull? **SSM WWW**

•**8** In Fig. 13-35, a square of edge length 20.0 cm is formed by four spheres of masses $m_1 = 5.00$ g, $m_2 = 3.00$ g, $m_3 = 1.00$ g, and $m_4 = 5.00$ g. In unit-vector notation, what is the net gravitational force from them on a central sphere with mass $m_5 = 2.50$ g? **GO**

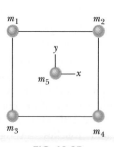

**FIG. 13-35** Problem 8.

•9 *Miniature black holes.* Left over from the big-bang beginning of the universe, tiny black holes might still wander through the universe. If one with a mass of $1 \times 10^{11}$ kg (and a radius of only $1 \times 10^{-16}$ m) reached Earth, at what distance from your head would its gravitational pull on you match that of Earth's?

••10 In Fig. 13-36a, particle $A$ is fixed in place at $x = -0.20$ m on the $x$ axis and particle $B$, with a mass of 1.0 kg, is fixed in place at the origin. Particle $C$ (not shown) can be moved along the $x$ axis, between particle $B$ and $x = \infty$. Figure 13-36b shows the $x$ component $F_{net,x}$ of the net gravitational force on particle $B$ due to particles $A$ and $C$, as a function of position $x$ of particle $C$. The plot actually extends to the right, approaching an asymptote of $-4.17 \times 10^{-10}$ N as $x \to \infty$. What are the masses of (a) particle $A$ and (b) particle $C$?

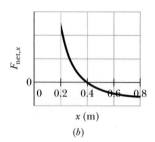

(a)                              (b)

**FIG. 13-36** Problem 10.

••11 As seen in Fig. 13-37, two spheres of mass $m$ and a third sphere of mass $M$ form an equilateral triangle, and a fourth sphere of mass $m_4$ is at the center of the triangle. The net gravitational force on that central sphere from the three other spheres is zero. (a) What is $M$ in terms of $m$? (b) If we double the value of $m_4$, what then is the magnitude of the net gravitational force on the central sphere?

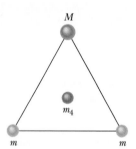

**FIG. 13-37** Problem 11.

••12 Three point particles are fixed in position in an $xy$ plane. Two of them, particle $A$ of mass 6.00 g and particle $B$ of mass 12.0 g, are shown in Fig. 13-38, with a separation of $d_{AB} = 0.500$ m at angle $\theta = 30°$. Particle $C$, with mass 8.00 g, is not shown. The net gravitational force acting on particle $A$ due to particles $B$ and $C$ is 2.77 × $10^{-14}$ N at an angle of $-163.8°$ from the positive direction of the $x$ axis. What are (a) the $x$ coordinate and (b) the $y$ coordinate of particle $C$?

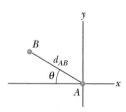

**FIG. 13-38** Problem 12.

••13 Figure 13-39 shows a spherical hollow inside a lead sphere of radius $R = 4.00$ cm; the surface of the hollow passes through the center of the sphere and "touches" the right side of the sphere. The mass of the sphere before hollowing was $M = 2.95$ kg. With what gravita-

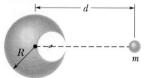

**FIG. 13-39** Problem 13.

tional force does the hollowed-out lead sphere attract a small sphere of mass $m = 0.431$ kg that lies at a distance $d = 9.00$ cm from the center of the lead sphere, on the straight line connecting the centers of the spheres and of the hollow?

••14 *Two dimensions.* In Fig. 13-40, three point particles are fixed in place in an $xy$ plane. Particle $A$ has mass $m_A$, particle $B$ has mass $2.00m_A$, and particle $C$ has mass $3.00m_A$. A fourth particle $D$, with mass $4.00m_A$, is to be placed near the other three particles. In terms of distance $d$, at what (a) $x$ coordinate and (b) $y$ coordinate should particle $D$ be placed so that the net gravitational force on particle $A$ from particles $B$, $C$, and $D$ is zero? GO

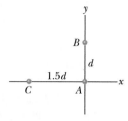

**FIG. 13-40** Problem 14.

•••15 *Three dimensions.* Three point particles are fixed in place in an $xyz$ coordinate system. Particle $A$, at the origin, has mass $m_A$. Particle $B$, at $xyz$ coordinates $(2.00d, 1.00d, 2.00d)$, has mass $2.00m_A$, and particle $C$, at coordinates $(-1.00d, 2.00d, -3.00d)$, has mass $3.00m_A$. A fourth particle $D$, with mass $4.00m_A$, is to be placed near the other particles. In terms of distance $d$, at what (a) $x$, (b) $y$, and (c) $z$ coordinate should $D$ be placed so that the net gravitational force on $A$ from $B$, $C$, and $D$ is zero?

•••16 In Fig. 13-41, a particle of mass $m_1 = 0.67$ kg is a distance $d = 23$ cm from one end of a uniform rod with length $L = 3.0$ m and mass $M = 5.0$ kg. What is the magnitude of the gravitational force $\vec{F}$ on the particle from the rod?

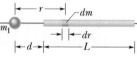

**FIG. 13-41** Problem 16.

### sec. 13-4 Gravitation Near Earth's Surface

•17 At what altitude above Earth's surface would the gravitational acceleration be 4.9 m/s²? SSM

•18 *Mile-high building.* In 1956, Frank Lloyd Wright proposed the construction of a mile-high building in Chicago. Suppose the building had been constructed. Ignoring Earth's rotation, find the change in your weight if you were to ride an elevator from the street level, where you weigh 600 N, to the top of the building.

•19 (a) What will an object weigh on the Moon's surface if it weighs 100 N on Earth's surface? (b) How many Earth radii must this same object be from the center of Earth if it is to weigh the same as it does on the Moon?

•20 *Mountain pull.* A large mountain can slightly affect the direction of "down" as determined by a plumb line. Assume that we can model a mountain as a sphere of radius $R = 2.00$ km and density (mass per unit volume) $2.6 \times 10^3$ kg/m³. Assume also that we hang a 0.50 m plumb line at a distance of $3R$ from the sphere's center and such that the sphere pulls horizontally on the lower end. How far would the lower end move toward the sphere?

••21 One model for a certain planet has a core of radius $R$ and mass $M$ surrounded by an outer shell of inner radius $R$, outer radius $2R$, and mass $4M$. If $M = 4.1 \times 10^{24}$ kg and $R = 6.0 \times 10^6$ m, what is the gravitational acceleration of a particle at points (a) $R$ and (b) $3R$ from the center of the planet?

**•22** The radius $R_h$ and mass $M_h$ of a black hole are related by $R_h = 2GM_h/c^2$, where $c$ is the speed of light. Assume that the gravitational acceleration $a_g$ of an object at a distance $r_o = 1.001R_h$ from the center of a black hole is given by Eq. 13-11 (it is, for large black holes). (a) In terms of $M_h$, find $a_g$ at $r_o$. (b) Does $a_g$ at $r_o$ increase or decrease as $M_h$ increases? (c) What is $a_g$ at $r_o$ for a very large black hole whose mass is $1.55 \times 10^{12}$ times the solar mass of $1.99 \times 10^{30}$ kg? (d) If the astronaut of Sample Problem 13-3 is at $r_o$ with her feet toward this black hole, what is the difference in gravitational acceleration between her head and her feet? (e) Is the tendency to stretch the astronaut severe?

**•23** Certain neutron stars (extremely dense stars) are believed to be rotating at about 1 rev/s. If such a star has a radius of 20 km, what must be its minimum mass so that material on its surface remains in place during the rapid rotation? **ILW**

### sec. 13-5 Gravitation Inside Earth

**•24** Two concentric spherical shells with uniformly distributed masses $M_1$ and $M_2$ are situated as shown in Fig. 13-42. Find the magnitude of the net gravitational force on a particle of mass $m$, due to the shells, when the particle is located at radial distance (a) $a$, (b) $b$, and (c) $c$.

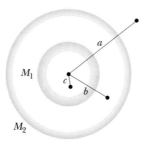

**FIG. 13-42** Problem 24.

**••25** Figure 13-43 shows, not to scale, a cross section through the interior of Earth. Rather than being uniform throughout, Earth is divided into three zones: an outer *crust*, a *mantle*, and an inner *core*. The dimensions of these zones and the masses contained within them are shown on the figure. Earth has a total mass of $5.98 \times 10^{24}$ kg and a radius of 6370 km. Ignore rotation and assume that Earth is spherical. (a) Calculate $a_g$ at the surface. (b) Suppose that a bore hole (the *Mohole*) is driven to the crust–mantle interface at a depth of 25.0 km; what would be the value of $a_g$ at the bottom of the hole? (c) Suppose that Earth were a uniform sphere with the same total mass and size. What would be the value of $a_g$ at a depth of 25.0 km? (Precise measurements of $a_g$ are sensitive probes of the interior structure of Earth, although results can be clouded by local variations in mass distribution.)

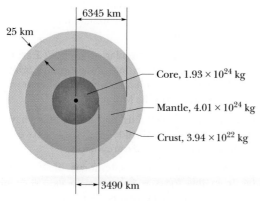

**FIG. 13-43** Problem 25.

**••26** Assume a planet is a uniform sphere of radius $R$ that (somehow) has a narrow radial tunnel through its center (Fig. 13-8). Also assume we can position an apple anywhere along the tunnel or outside the sphere. Let $F_R$ be the magnitude of the gravitational force on the apple when it is located at the planet's surface. How far from the surface is there a point where the magnitude of the gravitational force on the apple is $\frac{1}{2}F_R$ if we move the apple (a) away from the planet and (b) into the tunnel?

**••27** A solid uniform sphere has a mass of $1.0 \times 10^4$ kg and a radius of 1.0 m. What is the magnitude of the gravitational force due to the sphere on a particle of mass $m$ located at a distance of (a) 1.5 m and (b) 0.50 m from the center of the sphere? (c) Write a general expression for the magnitude of the gravitational force on the particle at a distance $r \le 1.0$ m from the center of the sphere.

**••28** Consider a pulsar, a collapsed star of extremely high density, with a mass $M$ equal to that of the Sun ($1.98 \times 10^{30}$ kg), a radius $R$ of only 12 km, and a rotational period $T$ of 0.041 s. By what percentage does the free-fall acceleration $g$ differ from the gravitational acceleration $a_g$ at the equator of this spherical star?

### sec. 13-6 Gravitational Potential Energy

**•29** The mean diameters of Mars and Earth are $6.9 \times 10^3$ km and $1.3 \times 10^4$ km, respectively. The mass of Mars is 0.11 times Earth's mass. (a) What is the ratio of the mean density (mass per unit volume) of Mars to that of Earth? (b) What is the value of the gravitational acceleration on Mars? (c) What is the escape speed on Mars? **SSM**

**•30** (a) What is the gravitational potential energy of the two-particle system in Problem 1? If you triple the separation between the particles, how much work is done (b) by the gravitational force between the particles and (c) by you?

**•31** What multiple of the energy needed to escape from Earth gives the energy needed to escape from (a) the Moon and (b) Jupiter?

**•32** Figure 13-44 gives the potential energy function $U(r)$ of a projectile, plotted outward from the surface of a planet of radius $R_s$. If the projectile is launched radially outward from the surface with a mechanical energy of $-2.0 \times 10^9$ J, what are (a) its kinetic energy at radius $r = 1.25R_s$ and (b) its *turning point* (see Section 8-6) in terms of $R_s$?

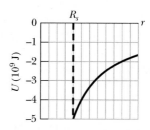

**FIG. 13-44** Problems 32 and 33.

**•33** Figure 13-44 gives the potential energy function $U(r)$ of a projectile, plotted outward from the surface of a planet of radius $R_s$. What least kinetic energy is required of a projectile launched at the surface if the projectile is to "escape" the planet?

**•34** In Problem 3, what ratio $m/M$ gives the least gravitational potential energy for the system?

**••35** The three spheres in Fig. 13-45, with masses $m_A = 80$ g, $m_B = 10$ g, and $m_C = 20$ g, have their centers on a common line, with $L = 12$ cm and $d = 4.0$ cm. You move sphere

*B* along the line until its center-to-center separation from *C* is *d* = 4.0 cm. How much work is done on sphere *B* (a) by you and (b) by the net gravitational force on *B* due to spheres *A* and *C*? **GO**

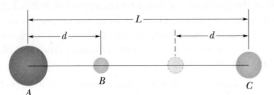

**FIG. 13-45** Problem 35.

••**36** A projectile is shot directly away from Earth's surface. Neglect the rotation of Earth. What multiple of Earth's radius $R_E$ gives the radial distance a projectile reaches if (a) its initial speed is 0.500 of the escape speed from Earth and (b) its initial kinetic energy is 0.500 of the kinetic energy required to escape Earth? (c) What is the least initial mechanical energy required at launch if the projectile is to escape Earth?

••**37** (a) What is the escape speed on a spherical asteroid whose radius is 500 km and whose gravitational acceleration at the surface is 3.0 m/s²? (b) How far from the surface will a particle go if it leaves the asteroid's surface with a radial speed of 1000 m/s? (c) With what speed will an object hit the asteroid if it is dropped from 1000 km above the surface? **SSM**

••**38** Zero, a hypothetical planet, has a mass of $5.0 \times 10^{23}$ kg, a radius of $3.0 \times 10^6$ m, and no atmosphere. A 10 kg space probe is to be launched vertically from its surface. (a) If the probe is launched with an initial energy of $5.0 \times 10^7$ J, what will be its kinetic energy when it is $4.0 \times 10^6$ m from the center of Zero? (b) If the probe is to achieve a maximum distance of $8.0 \times 10^6$ m from the center of Zero, with what initial kinetic energy must it be launched from the surface of Zero?

••**39** Two neutron stars are separated by a distance of $1.0 \times 10^{10}$ m. They each have a mass of $1.0 \times 10^{30}$ kg and a radius of $1.0 \times 10^5$ m. They are initially at rest with respect to each other. As measured from that rest frame, how fast are they moving when (a) their separation has decreased to one-half its initial value and (b) they are about to collide? **SSM**

••**40** In deep space, sphere *A* of mass 20 kg is located at the origin of an *x* axis and sphere *B* of mass 10 kg is located on the axis at *x* = 0.80 m. Sphere *B* is released from rest while sphere *A* is held at the origin. (a) What is the gravitational potential energy of the two-sphere system just as *B* is released? (b) What is the kinetic energy of *B* when it has moved 0.20 m toward *A*?

••**41** Figure 13-46 shows four particles, each of mass 20.0 g, that form a square with an edge length of *d* = 0.600 m. If *d* is reduced to 0.200 m, what is the change in the gravitational potential energy of the four-particle system? **GO**

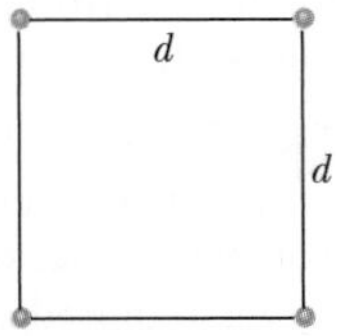

**FIG. 13-46** Problem 41.

••**42** Figure 13-47*a* shows a particle *A* that can be moved along a *y* axis from an infinite distance to the origin. That origin lies at the midpoint between particles *B* and *C*, which have identical masses, and the *y* axis is a perpendicular bisector between them. Distance *D* is 0.3057 m. Figure 13-47*b* shows the potential energy *U* of the three-particle system as a function of the position of particle *A* along the *y* axis. The curve actually extends rightward and approaches an asymptote of $-2.7 \times 10^{-11}$ J as *y* → ∞. What are the masses of (a) particles *B* and *C* and (b) particle *A*?

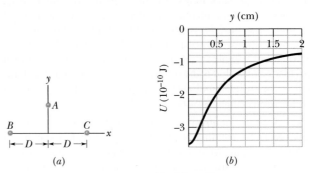

**FIG. 13-47** Problem 42.

### sec. 13-7 Planets and Satellites: Kepler's Laws

•**43** The Martian satellite Phobos travels in an approximately circular orbit of radius $9.4 \times 10^6$ m with a period of 7 h 39 min. Calculate the mass of Mars from this information.

•**44** The first known collision between space debris and a functioning satellite occurred in 1996: At an altitude of 700 km, a year-old French spy satellite was hit by a piece of an Ariane rocket that had been in orbit for 10 years. A stabilizing boom on the satellite was demolished, and the satellite was sent spinning out of control. Just before the collision and in kilometers per hour, what was the speed of the rocket piece relative to the satellite if both were in circular orbits and the collision was (a) head-on and (b) along perpendicular paths?

•**45** The Sun, which is $2.2 \times 10^{20}$ m from the center of the Milky Way galaxy, revolves around that center once every $2.5 \times 10^8$ years. Assuming each star in the Galaxy has a mass equal to the Sun's mass of $2.0 \times 10^{30}$ kg, the stars are distributed uniformly in a sphere about the galactic center, and the Sun is at the edge of that sphere, estimate the number of stars in the Galaxy. **SSM WWW**

•**46** The mean distance of Mars from the Sun is 1.52 times that of Earth from the Sun. From Kepler's law of periods, calculate the number of years required for Mars to make one revolution around the Sun; compare your answer with the value given in Appendix C.

•**47** A satellite, moving in an elliptical orbit, is 360 km above Earth's surface at its farthest point and 180 km above at its closest point. Calculate (a) the semimajor axis and (b) the eccentricity of the orbit. **SSM**

•**48** A satellite is put in a circular orbit about Earth with a radius equal to one-half the radius of the Moon's orbit. What is its period of revolution in lunar months? (A lunar month is the period of revolution of the Moon.)

•**49** (a) What linear speed must an Earth satellite have to be in a circular orbit at an altitude of 160 km above Earth's surface? (b) What is the period of revolution?

•**50** The Sun's center is at one focus of Earth's orbit. How far from this focus is the other focus, (a) in meters and (b) in terms of the solar radius, $6.96 \times 10^8$ m? The eccentricity is 0.0167, and the semimajor axis is $1.50 \times 10^{11}$ m.

**•51** A comet that was seen in April 574 by Chinese astronomers on a day known by them as the Woo Woo day was spotted again in May 1994. Assume the time between observations is the period of the Woo Woo day comet and take its eccentricity as 0.11. What are (a) the semimajor axis of the comet's orbit and (b) its greatest distance from the Sun in terms of the mean orbital radius $R_P$ of Pluto?

**•52** An orbiting satellite stays over a certain spot on the equator of (rotating) Earth. What is the altitude of the orbit (called a *geosynchronous orbit*)?

**••53** In 1610, Galileo used his telescope to discover four prominent moons around Jupiter. Their mean orbital radii $a$ and periods $T$ are as follows:
(a) Plot log $a$ ($y$ axis) against log $T$ ($x$ axis) and show that you get a straight line. (b) Measure the slope of the line and compare it with the value that you expect from Kepler's third law. (c) Find the mass of Jupiter from the intercept of this line with the $y$ axis.

| Name | $a$ ($10^8$ m) | $T$ (days) |
|---|---|---|
| Io | 4.22 | 1.77 |
| Europa | 6.71 | 3.55 |
| Ganymede | 10.7 | 7.16 |
| Callisto | 18.8 | 16.7 |

**••54** In 1993 the spacecraft *Galileo* sent home an image (Fig. 13-48) of asteroid 243 Ida and a tiny orbiting moon (now known as Dactyl), the first confirmed example of an asteroid–moon system. In the image, the moon, which is 1.5 km wide, is 100 km from the center of the asteroid, which is 55 km long. The shape of the moon's orbit is not well known; assume it is circular with a period of 27 h. (a) What is the mass of the asteroid? (b) The volume of the asteroid, measured from the *Galileo* images, is 14 100 km³. What is the density (mass per unit volume) of the asteroid?

**FIG. 13-48** Problem 54. A tiny moon (at right) orbits asteroid 243 Ida. *(Courtesy NASA)*

**••55** In a certain binary-star system, each star has the same mass as our Sun, and they revolve about their center of mass. The distance between them is the same as the distance between Earth and the Sun. What is their period of revolution in years? **ILW**

**••56** *Hunting a black hole.* Observations of the light from a certain star indicate that it is part of a binary (two-star) sys-

tem. This visible star has orbital speed $v = 270$ km/s, orbital period $T = 1.70$ days, and approximate mass $m_1 = 6M_s$, where $M_s$ is the Sun's mass, $1.99 \times 10^{30}$ kg. Assume that the visible star and its companion star, which is dark and unseen, are both in circular orbits (Fig. 13-49). What multiple of $M_s$ gives the approximate mass $m_2$ of the dark star?

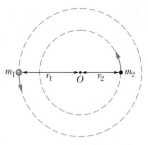

**FIG. 13-49** Problem 56.

**••57** A 20 kg satellite has a circular orbit with a period of 2.4 h and a radius of $8.0 \times 10^6$ m around a planet of unknown mass. If the magnitude of the gravitational acceleration on the surface of the planet is 8.0 m/s², what is the radius of the planet?

**•••58** The presence of an unseen planet orbiting a distant star can sometimes be inferred from the motion of the star as we see it. As the star and planet orbit the center of mass of the star–planet system, the star moves toward and away from us with what is called the *line of sight velocity*, a motion that can be detected. Figure 13-50 shows a graph of the line of sight velocity versus time for the star 14 Herculis. The star's mass is believed to be 0.90 of the mass of our Sun. Assume that only one planet orbits the star and that our view is along the plane of the orbit. Then approximate (a) the planet's mass in terms of Jupiter's mass $m_J$ and (b) the planet's orbital radius in terms of Earth's orbital radius $r_E$.

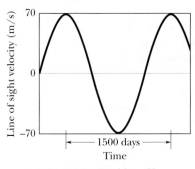

**FIG. 13-50** Problem 58.

**•••59** Three identical stars of mass $M$ form an equilateral triangle that rotates around the triangle's center as the stars move in a common circle about that center. The triangle has edge length $L$. What is the speed of the stars?

### sec. 13-8 Satellites: Orbits and Energy
**•60** Two Earth satellites, $A$ and $B$, each of mass $m$, are to be launched into circular orbits about Earth's center. Satellite $A$ is to orbit at an altitude of 6370 km. Satellite $B$ is to orbit at an altitude of 19 110 km. The radius of Earth $R_E$ is 6370 km. (a) What is the ratio of the potential energy of satellite $B$ to that of satellite $A$, in orbit? (b) What is the ratio of the kinetic energy of satellite $B$ to that of satellite $A$, in orbit? (c) Which satellite has the greater total energy if each has a mass of 14.6 kg? (d) By how much?

**•61** An asteroid, whose mass is $2.0 \times 10^{-4}$ times the mass of Earth, revolves in a circular orbit around the Sun at a distance that is twice Earth's distance from the Sun. (a) Calculate the period of revolution of the asteroid in years. (b) What is the ratio of the kinetic energy of the asteroid to the kinetic energy of Earth? **SSM** **WWW**

•62 A satellite orbits a planet of unknown mass in a circle of radius $2.0 \times 10^7$ m. The magnitude of the gravitational force on the satellite from the planet is $F = 80$ N. (a) What is the kinetic energy of the satellite in this orbit? (b) What would $F$ be if the orbit radius were increased to $3.0 \times 10^7$ m?

•63 (a) At what height above Earth's surface is the energy required to lift a satellite to that height equal to the kinetic energy required for the satellite to be in orbit at that height? (b) For greater heights, which is greater, the energy for lifting or the kinetic energy for orbiting?

•64 In Fig. 13-51, two satellites, $A$ and $B$, both of mass $m = 125$ kg, move in the same circular orbit of radius $r = 7.87 \times 10^6$ m around Earth but in opposite senses of rotation and therefore on a collision course. (a) Find the total mechanical energy $E_A + E_B$ of the *two satellites + Earth* system before the collision. (b) If the collision is completely inelastic so that the wreckage remains as one piece of tangled material (mass = $2m$), find the total mechanical energy immediately after the collision. (c) Just after the collision, is the wreckage falling directly toward Earth's center or orbiting around Earth?

**FIG. 13-51** Problem 64.

••65 A satellite is in a circular Earth orbit of radius $r$. The area $A$ enclosed by the orbit depends on $r^2$ because $A = \pi r^2$. Determine how the following properties of the satellite depend on $r$: (a) period, (b) kinetic energy, (c) angular momentum, and (d) speed.

••66 One way to attack a satellite in Earth orbit is to launch a swarm of pellets in the same orbit as the satellite but in the opposite direction. Suppose a satellite in a circular orbit 500 km above Earth's surface collides with a pellet having mass 4.0 g. (a) What is the kinetic energy of the pellet in the reference frame of the satellite just before the collision? (b) What is the ratio of this kinetic energy to the kinetic energy of a 4.0 g bullet from a modern army rifle with a muzzle speed of 950 m/s?

•••67 What are (a) the speed and (b) the period of a 220 kg satellite in an approximately circular orbit 640 km above the surface of Earth? Suppose the satellite loses mechanical energy at the average rate of $1.4 \times 10^5$ J per orbital revolution. Adopting the reasonable approximation that the satellite's orbit becomes a "circle of slowly diminishing radius," determine the satellite's (c) altitude, (d) speed, and (e) period at the end of its 1500th revolution. (f) What is the magnitude of the average retarding force on the satellite? Is angular momentum around Earth's center conserved for (g) the satellite and (h) the satellite–Earth system (assuming that system is isolated)?

•••68 Two small spaceships, each with mass $m = 2000$ kg, are in the circular Earth orbit of Fig. 13-52, at an altitude $h$ of 400 km. Igor, the commander of one of the ships, arrives at any fixed point in the orbit 90 s ahead of Picard, the commander of the other ship. What are the (a) period $T_0$ and (b) speed $v_0$ of

the ships? At point $P$ in Fig. 13-52, Picard fires an instantaneous burst in the forward direction, *reducing* his ship's speed by 1.00%. After this burst, he follows the elliptical orbit shown dashed in the figure. What are the (c) kinetic energy and (d) potential energy of his ship immediately after the burst? In Picard's new elliptical orbit, what are

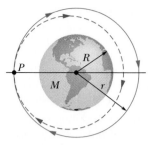

**FIG. 13-52** Problem 68.

(e) the total energy $E$, (f) the semimajor axis $a$, and (g) the orbital period $T$? (h) How much earlier than Igor will Picard return to $P$? **GO**

### sec. 13-9 Einstein and Gravitation

•69 In Fig. 13-19b, the scale on which the 60 kg physicist stands reads 220 N. How long will the cantaloupe take to reach the floor if the physicist drops it (from rest relative to himself) at a height of 2.1 m above the floor?

### Additional Problems

70 The mysterious visitor that appears in the enchanting story *The Little Prince* was said to come from a planet that "was scarcely any larger than a house!" Assume that the mass per unit volume of the planet is about that of Earth and that the planet does not appreciably spin. Approximate (a) the free-fall acceleration on the planet's surface and (b) the escape speed from the planet.

71 Figure 13-53 is a graph of the kinetic energy $K$ of an asteroid versus its distance $r$ from Earth's center, as the asteroid falls directly in toward that center. (a) What is the (approximate) mass of the asteroid? (b) What is its speed at $r = 1.945 \times 10^7$ m?

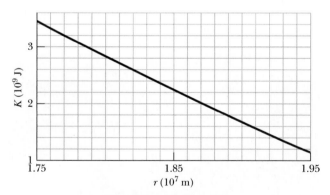

**FIG. 13-53** Problem 71.

72 The radius $R_h$ of a black hole is the radius of a mathematical sphere, called the event horizon, that is centered on the black hole. Information from events inside the event horizon cannot reach the outside world. According to Einstein's general theory of relativity, $R_h = 2GM/c^2$, where $M$ is the mass of the black hole and $c$ is the speed of light.

Suppose that you wish to study a black hole near it, at a radial distance of $50R_h$. However, you do not want the difference in gravitational acceleration between your feet and your head to exceed 10 m/s² when you are feet down (or head down) toward the black hole. (a) As a multiple of our Sun's mass $M_S$, approximately what is the limit to the mass of the black hole you can tolerate at the given radial distance? (You need to estimate your height.) (b) Is the limit an upper limit (you can tolerate smaller masses) or a lower limit (you can tolerate larger masses)?

**73** Sphere $A$ with mass 80 kg is located at the origin of an $xy$ coordinate system; sphere $B$ with mass 60 kg is located at coordinates (0.25 m, 0); sphere $C$ with mass 0.20 kg is located in the first quadrant 0.20 m from $A$ and 0.15 m from $B$. In unit-vector notation, what is the gravitational force on $C$ due to $A$ and $B$?

**74** A satellite is in elliptical orbit with a period of $8.00 \times 10^4$ s about a planet of mass $7.00 \times 10^{24}$ kg. At aphelion, at radius $4.5 \times 10^7$ m, the satellite's angular speed is $7.158 \times 10^{-5}$ rad/s. What is its angular speed at perihelion?

**75** In a shuttle craft of mass $m = 3000$ kg, Captain Janeway orbits a planet of mass $M = 9.50 \times 10^{25}$ kg, in a circular orbit of radius $r = 4.20 \times 10^7$ m. What are (a) the period of the orbit and (b) the speed of the shuttle craft? Janeway briefly fires a forward-pointing thruster, reducing her speed by 2.00%. Just then, what are (c) the speed, (d) the kinetic energy, (e) the gravitational potential energy, and (f) the mechanical energy of the shuttle craft? (g) What is the semimajor axis of the elliptical orbit now taken by the craft? (h) What is the difference between the period of the original circular orbit and that of the new elliptical orbit? (i) Which orbit has the smaller period? **SSM**

**76** A typical neutron star may have a mass equal to that of the Sun but a radius of only 10 km. (a) What is the gravitational acceleration at the surface of such a star? (b) How fast would an object be moving if it fell from rest through a distance of 1.0 m on such a star? (Assume the star does not rotate.)

**77** Four uniform spheres, with masses $m_A = 40$ kg, $m_B = 35$ kg, $m_C = 200$ kg, and $m_D = 50$ kg, have $(x, y)$ coordinates of (0, 50 cm), (0, 0), (−80 cm, 0), and (40 cm, 0), respectively. In unit-vector notation, what is the net gravitational force on sphere $B$ due to the other spheres? **GO**

**78** (a) In Problem 77, remove sphere $A$ and calculate the gravitational potential energy of the remaining three-particle system. (b) If $A$ is then put back in place, is the potential energy of the four-particle system more or less than that of the system in (a)? (c) In (a), is the work done by you to remove $A$ positive or negative? (d) In (b), is the work done by you to replace $A$ positive or negative?

**79** A very early, simple satellite consisted of an inflated spherical aluminum balloon 30 m in diameter and of mass 20 kg. Suppose a meteor having a mass of 7.0 kg passes within 3.0 m of the surface of the satellite. What is the magnitude of the gravitational force on the meteor from the satellite at the closest approach? **SSM**

**80** A uniform solid sphere of radius $R$ produces a gravitational acceleration of $a_g$ on its surface. At what distance from the sphere's center are there points (a) inside and (b) outside the sphere where the gravitational acceleration is $a_g/3$?

**81** A projectile is fired vertically from Earth's surface with an initial speed of 10 km/s. Neglecting air drag, how far above the surface of Earth will it go? **ILW**

**82** A 50 kg satellite circles planet Cruton every 6.0 h. The magnitude of the gravitational force exerted on the satellite

by Cruton is 80 N. (a) What is the radius of the orbit? (b) What is the kinetic energy of the satellite? (c) What is the mass of planet Cruton?

**83** In a double-star system, two stars of mass $3.0 \times 10^{30}$ kg each rotate about the system's center of mass at radius $1.0 \times 10^{11}$ m. (a) What is their common angular speed? (b) If a meteoroid passes through the system's center of mass perpendicular to their orbital plane, what minimum speed must it have at the center of mass if it is to escape to "infinity" from the two-star system? **SSM**

**84** An object lying on Earth's equator is accelerated (a) toward the center of Earth because Earth rotates, (b) toward the Sun because Earth revolves around the Sun in an almost circular orbit, and (c) toward the center of our galaxy because the Sun moves around the galactic center. For the latter, the period is $2.5 \times 10^8$ y and the radius is $2.2 \times 10^{20}$ m. Calculate these three accelerations as multiples of $g = 9.8$ m/s$^2$.

**85** The masses and coordinates of three spheres are as follows: 20 kg, $x = 0.50$ m, $y = 1.0$ m; 40 kg, $x = −1.0$ m, $y = −1.0$ m; 60 kg, $x = 0$ m, $y = −0.50$ m. What is the magnitude of the gravitational force on a 20 kg sphere located at the origin due to these three spheres? **ILW**

**86** With what speed would mail pass through the center of Earth if falling in the tunnel of Sample Problem 13-4?

**87** The orbit of Earth around the Sun is *almost* circular: The closest and farthest distances are $1.47 \times 10^8$ km and $1.52 \times 10^8$ km respectively. Determine the corresponding variations in (a) total energy, (b) gravitational potential energy, (c) kinetic energy, and (d) orbital speed. (*Hint:* Use conservation of energy and conservation of angular momentum.) **SSM**

**88** A spaceship is on a straight-line path between Earth and the Moon. At what distance from Earth is the net gravitational force on the spaceship zero?

**89** An object of mass $m$ is initially held in place at radial distance $r = 3R_E$ from the center of Earth, where $R_E$ is the radius of Earth. Let $M_E$ be the mass of Earth. A force is applied to the object to move it to a radial distance $r = 4R_E$, where it again is held in place. Calculate the work done by the applied force during the move by integrating the force magnitude.

**90** The fastest possible rate of rotation of a planet is that for which the gravitational force on material at the equator just barely provides the centripetal force needed for the rotation. (Why?) (a) Show that the corresponding shortest period of rotation is

$$T = \sqrt{\frac{3\pi}{G\rho}},$$

where $\rho$ is the uniform density (mass per unit volume) of the spherical planet. (b) Calculate the rotation period assuming a density of 3.0 g/cm$^3$, typical of many planets, satellites, and asteroids. No astronomical object has ever been found to be spinning with a period shorter than that determined by this analysis.

**91** (a) If the legendary apple of Newton could be released from rest at a height of 2 m from the surface of a neutron star with a mass 1.5 times that of our Sun and a radius of 20 km, what would be the apple's speed when it reached the surface of the star? (b) If the apple could rest on the surface of the star, what would be the approximate difference between the gravitational acceleration at the top and at the bottom of the apple? (Choose a reasonable size for an apple; the answer indicates that an apple would never survive near a neutron star.)

**92** Some people believe that the positions of the planets at the time of birth influence the newborn. Others deride this belief and claim that the gravitational force exerted on a baby by the obstetrician is greater than the force exerted by the planets. To check this claim, calculate the magnitude of the gravitational force exerted on a 3 kg baby (a) by a 70 kg obstetrician who is 1 m away and roughly approximated as a point mass, (b) by the massive planet Jupiter ($m = 2 \times 10^{27}$ kg) at its closest approach to Earth ($= 6 \times 10^{11}$ m), and (c) by Jupiter at its greatest distance from Earth ($= 9 \times 10^{11}$ m). (d) Is the claim correct?

**93** A certain triple-star system consists of two stars, each of mass $m$, revolving in the same circular orbit of radius $r$ around a central star of mass $M$ (Fig. 13-54). The two orbiting stars are always at opposite ends of a diameter of the orbit. Derive an expression for the period of revolution of the stars. **SSM**

FIG. 13-54
Problem 93.

**94** A 150.0 kg rocket moving radially outward from Earth has a speed of 3.70 km/s when its engine shuts off 200 km above Earth's surface. (a) Assuming negligible air drag, find the rocket's kinetic energy when the rocket is 1000 km above Earth's surface. (b) What maximum height above the surface is reached by the rocket?

**95** Planet Roton, with a mass of $7.0 \times 10^{24}$ kg and a radius of 1600 km, gravitationally attracts a meteorite that is initially at rest relative to the planet, at a distance great enough to take as infinite. The meteorite falls toward the planet. Assuming the planet is airless, find the speed of the meteorite when it reaches the planet's surface.

**96** Two 20 kg spheres are fixed in place on a $y$ axis, one at $y = 0.40$ m and the other at $y = -0.40$ m. A 10 kg ball is then released from rest at a point on the $x$ axis that is at a great distance (effectively infinite) from the spheres. If the only forces acting on the ball are the gravitational forces from the spheres, then when the ball reaches the $(x, y)$ point $(0.30$ m, $0)$, what are (a) its kinetic energy and (b) the net force on it from the spheres, in unit-vector notation?

**97** A satellite of mass 125 kg is in a circular orbit of radius $7.00 \times 10^6$ m around a planet. If the period is 8050 s, what is the mechanical energy of the satellite?

**98** In his 1865 science fiction novel *From the Earth to the Moon*, Jules Verne described how three astronauts are shot to the Moon by means of a huge gun. According to Verne, the aluminum capsule containing the astronauts is accelerated by ignition of nitrocellulose to a speed of 11 km/s along the gun barrel's length of 220 m. (a) In $g$ units, what is the average acceleration of the capsule and astronauts in the gun barrel? (b) Is that acceleration tolerable or deadly to the astronauts?

A modern version of such gun-launched spacecraft (although without passengers) has been proposed. In this modern version, called the SHARP (Super High Altitude Research Project) gun, ignition of methane and air shoves a piston along the gun's tube, compressing hydrogen gas that then launches a rocket. During this launch, the rocket moves 3.5 km and reaches a speed of 7.0 km/s. Once launched, the rocket can be fired to gain additional speed. (c) In $g$ units, what would be the average acceleration of the rocket within the launcher? (d) How much additional speed is needed (via the rocket engine) if the rocket is to orbit Earth at an altitude of 700 km?

**99** Several planets (Jupiter, Saturn, Uranus) are encircled by rings, perhaps composed of material that failed to form a satellite. In addition, many galaxies contain ring-like structures. Consider a homogeneous thin ring of mass $M$ and outer radius $R$ (Fig. 13-55). (a) What gravitational attraction does it exert on a particle of mass $m$ located on the ring's central axis a distance $x$ from the ring center? (b) Suppose the particle falls from rest as a result of the attraction of the ring of matter. What is the speed with which it passes through the center of the ring?

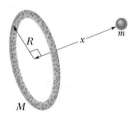

FIG. 13-55   Problem 99.

**100** Four identical 1.5 kg particles are placed at the corners of a square with sides equal to 20 cm. What is the magnitude of the net gravitational force on any one of the particles due to the others?

**101** We watch two identical astronomical bodies $A$ and $B$, each of mass $m$, fall toward each other from rest because of the gravitational force on each from the other. Their initial center-to-center separation is $R_i$. Assume that we are in an inertial reference frame that is stationary with respect to the center of mass of this two-body system. Use the principle of conservation of mechanical energy ($K_f + U_f = K_i + U_i$) to find the following when the center-to-center separation is $0.5R_i$: (a) the total kinetic energy of the system, (b) the kinetic energy of each body, (c) the speed of each body relative to us, and (d) the speed of body $B$ relative to body $A$.

Next assume that we are in a reference frame attached to body $A$ (we ride on the body). Now we see body $B$ fall from rest toward us. From this reference frame, again use $K_f + U_f = K_i + U_i$ to find the following when the center-to-center separation is $0.5R_i$: (e) the kinetic energy of body $B$ and (f) the speed of body $B$ relative to body $A$. (g) Why are the answers to (d) and (f) different? Which answer is correct?

**102** What is the percentage change in the acceleration of Earth toward the Sun when the alignment of Earth, Sun, and Moon changes from an eclipse of the Sun (with the Moon between Earth and Sun) to an eclipse of the Moon (Earth between Moon and Sun)?

**103** A planet requires 300 (Earth) days to complete its circular orbit around its sun, which has a mass of $6.0 \times 10^{30}$ kg. What are the planet's (a) orbital radius and (b) orbital speed?

**104** A particle of comet dust with mass $m$ is a distance $R$ from Earth's center and a distance $r$ from the Moon's center. If Earth's mass is $M_E$ and the Moon's mass is $M_m$, what is the sum of the gravitational potential energy of the particle–Earth system and the gravitational potential energy of the particle–Moon system?

# Fluids

Warren Bolster/Stone/Getty Images

*A surfer patiently kneels on his surfboard to catch the next big wave. When he spots a wave building in height as it heads his way, he rapidly paddles toward shore so that he is moving almost as fast as the wave as the wave begins to sweep under him. Then he stands up on the board, continuously adjusting his balance. How does he manage to continue riding the wave? How, instead, can he move up or down the front of the wave?*

*In short, how does a surfer surf?*

The answer is in this chapter.

## 14-1 | WHAT IS PHYSICS?

The physics of fluids is the basis of hydraulic engineering, a branch of engineering that is applied in a great many fields. A nuclear engineer might study the fluid flow in the hydraulic system of an aging nuclear reactor, while a medical engineer might study the blood flow in the arteries of an aging patient. An environmental engineer might be concerned about the drainage from waste sites or the efficient irrigation of farmlands. A naval engineer might be concerned with the dangers faced by a deep-sea diver or with the possibility of a crew escaping from a downed submarine. An aeronautical engineer might design the hydraulic systems controlling the wing flaps that allow a jet airplane to land. Hydraulic engineering is also applied in many Broadway and Las Vegas shows, where huge sets are quickly put up and brought down by hydraulic systems.

Before we can study any such application of the physics of fluids, we must first answer the question "What is a fluid?"

## 14-2 | What Is a Fluid?

A **fluid,** in contrast to a solid, is a substance that can flow. Fluids conform to the boundaries of any container in which we put them. They do so because a fluid cannot sustain a force that is tangential to its surface. (In the more formal language of Section 12-7, a fluid is a substance that flows because it cannot withstand a shearing stress. It can, however, exert a force in the direction perpendicular to its surface.) Some materials, such as pitch, take a long time to conform to the boundaries of a container, but they do so eventually; thus, we classify even those materials as fluids.

You may wonder why we lump liquids and gases together and call them fluids. After all (you may say), liquid water is as different from steam as it is from ice. Actually, it is not. Ice, like other crystalline solids, has its constituent atoms organized in a fairly rigid three-dimensional array called a crystalline lattice. In neither steam nor liquid water, however, is there any such orderly long-range arrangement.

## 14-3 | Density and Pressure

When we discuss rigid bodies, we are concerned with particular lumps of matter, such as wooden blocks, baseballs, or metal rods. Physical quantities that we find useful, and in whose terms we express Newton's laws, are *mass* and *force*. We might speak, for example, of a 3.6 kg block acted on by a 25 N force.

With fluids, we are more interested in the extended substance and in properties that can vary from point to point in that substance. It is more useful to speak of **density** and **pressure** than of mass and force.

### Density

To find the density $\rho$ of a fluid at any point, we isolate a small volume element $\Delta V$ around that point and measure the mass $\Delta m$ of the fluid contained within that element. The **density** is then

$$\rho = \frac{\Delta m}{\Delta V}. \qquad (14\text{-}1)$$

In theory, the density at any point in a fluid is the limit of this ratio as the volume element $\Delta V$ at that point is made smaller and smaller. In practice, we assume that a fluid sample is large relative to atomic dimensions and thus is "smooth" (with uni-

**TABLE 14-1**

**Some Densities**

| Material or Object | Density (kg/m³) | Material or Object | Density (kg/m³) |
|---|---|---|---|
| Interstellar space | $10^{-20}$ | Iron | $7.9 \times 10^3$ |
| Best laboratory vacuum | $10^{-17}$ | Mercury (the metal, not the planet) | $13.6 \times 10^3$ |
| Air: 20°C and 1 atm pressure | 1.21 | Earth: average | $5.5 \times 10^3$ |
| 20°C and 50 atm | 60.5 | core | $9.5 \times 10^3$ |
| Styrofoam | $1 \times 10^2$ | crust | $2.8 \times 10^3$ |
| Ice | $0.917 \times 10^3$ | Sun: average | $1.4 \times 10^3$ |
| Water: 20°C and 1 atm | $0.998 \times 10^3$ | core | $1.6 \times 10^5$ |
| 20°C and 50 atm | $1.000 \times 10^3$ | White dwarf star (core) | $10^{10}$ |
| Seawater: 20°C and 1 atm | $1.024 \times 10^3$ | Uranium nucleus | $3 \times 10^{17}$ |
| Whole blood | $1.060 \times 10^3$ | Neutron star (core) | $10^{18}$ |

form density), rather than "lumpy" with atoms. This assumption allows us to write Eq. 14-1 as

$$\rho = \frac{m}{V} \quad \text{(uniform density)}, \quad (14\text{-}2)$$

where $m$ and $V$ are the mass and volume of the sample.

Density is a scalar property; its SI unit is the kilogram per cubic meter. Table 14-1 shows the densities of some substances and the average densities of some objects. Note that the density of a gas (see Air in the table) varies considerably with pressure, but the density of a liquid (see Water) does not; that is, gases are readily *compressible* but liquids are not.

### Pressure

Let a small pressure-sensing device be suspended inside a fluid-filled vessel, as in Fig. 14-1$a$. The sensor (Fig. 14-1$b$) consists of a piston of surface area $\Delta A$ riding in a close-fitting cylinder and resting against a spring. A readout arrangement allows us to record the amount by which the (calibrated) spring is compressed by the surrounding fluid, thus indicating the magnitude $\Delta F$ of the force that acts normal to the piston. We define the **pressure** on the piston from the fluid as

$$p = \frac{\Delta F}{\Delta A}. \quad (14\text{-}3)$$

In theory, the pressure at any point in the fluid is the limit of this ratio as the surface area $\Delta A$ of the piston, centered on that point, is made smaller and smaller. However, if the force is uniform over a flat area $A$, we can write Eq. 14-3 as

$$p = \frac{F}{A} \quad \text{(pressure of uniform force on flat area)}, \quad (14\text{-}4)$$

where $F$ is the magnitude of the normal force on area $A$. (When we say a force is uniform over an area, we mean that the force is evenly distributed over every point of the area.)

We find by experiment that at a given point in a fluid at rest, the pressure $p$ defined by Eq. 14-4 has the same value no matter how the pressure sensor is oriented. Pressure is a scalar, having no directional properties. It is true that the force acting on the piston of our pressure sensor is a vector quantity, but Eq. 14-4 involves only the *magnitude* of that force, a scalar quantity.

The SI unit of pressure is the newton per square meter, which is given a special name, the **pascal** (Pa). In metric countries, tire pressure gauges are calibrated in kilopascals. The pascal is related to some other common (non-SI)

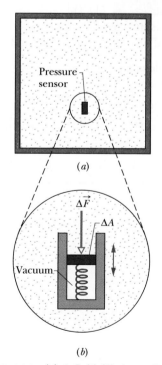

(a)

(b)

**FIG. 14-1** (*a*) A fluid-filled vessel containing a small pressure sensor, shown in (*b*). The pressure is measured by the relative position of the movable piston in the sensor.

**Some Pressures**

| | Pressure (Pa) | | Pressure (Pa) |
|---|---|---|---|
| Center of the Sun | $2 \times 10^{16}$ | Automobile tire[a] | $2 \times 10^5$ |
| Center of Earth | $4 \times 10^{11}$ | Atmosphere at sea level | $1.0 \times 10^5$ |
| Highest sustained laboratory pressure | $1.5 \times 10^{10}$ | Normal blood systolic pressure[a,b] | $1.6 \times 10^4$ |
| Deepest ocean trench (bottom) | $1.1 \times 10^8$ | Best laboratory vacuum | $10^{-12}$ |
| Spike heels on a dance floor | $10^6$ | | |

[a]Pressure in excess of atmospheric pressure.
[b]Equivalent to 120 torr on the physician's pressure gauge.

pressure units as follows:

$$1 \text{ atm} = 1.01 \times 10^5 \text{ Pa} = 760 \text{ torr} = 14.7 \text{ lb/in.}^2.$$

The *atmosphere* (atm) is, as the name suggests, the approximate average pressure of the atmosphere at sea level. The *torr* (named for Evangelista Torricelli, who invented the mercury barometer in 1674) was formerly called the *millimeter of mercury* (mm Hg). The pound per square inch is often abbreviated psi. Table 14-2 shows some pressures.

**Sample Problem    14-1**

A living room has floor dimensions of 3.5 m and 4.2 m and a height of 2.4 m.

**(a)** What does the air in the room weigh when the air pressure is 1.0 atm?

**KEY IDEAS**  (1) The air's weight is equal to $mg$, where $m$ is its mass. (2) Mass $m$ is related to the air density $\rho$ and the air volume $V$ by Eq. 14-2 ($\rho = m/V$).

*Calculation:* Putting the two ideas together and taking the density of air at 1.0 atm from Table 14-1, we find

$mg = (\rho V)g$
$\quad = (1.21 \text{ kg/m}^3)(3.5 \text{ m} \times 4.2 \text{ m} \times 2.4 \text{ m})(9.8 \text{ m/s}^2)$
$\quad = 418 \text{ N} \approx 420 \text{ N}.$         (Answer)

This is the weight of about 110 cans of Pepsi.

**(b)** What is the magnitude of the atmosphere's downward force on the top of your head, which we take to have an area of 0.040 m²?

**KEY IDEA**  When the fluid pressure $p$ on a surface of area $A$ is uniform, the fluid force on the surface can be obtained from Eq. 14-4 ($p = F/A$).

*Calculation:* Although air pressure varies daily, we can approximate that $p = 1.0$ atm. Then Eq. 14-4 gives

$$F = pA = (1.0 \text{ atm})\left(\frac{1.01 \times 10^5 \text{ N/m}^2}{1.0 \text{ atm}}\right)(0.040 \text{ m}^2)$$

$$= 4.0 \times 10^3 \text{ N}. \qquad \text{(Answer)}$$

This large force is equal to the weight of the column of air that covers the top of your head and extends all the way to the top of the atmosphere.

## 14-4 | Fluids at Rest

Figure 14-2a shows a tank of water—or other liquid—open to the atmosphere. As every diver knows, the pressure *increases* with depth below the air–water interface. The diver's depth gauge, in fact, is a pressure sensor much like that of Fig. 14-1b. As every mountaineer knows, the pressure *decreases* with altitude as one ascends into the atmosphere. The pressures encountered by the diver and the mountaineer are usually called *hydrostatic pressures,* because they are due to fluids that are static (at rest). Here we want to find an expression for hydrostatic pressure as a function of depth or altitude.

Let us look first at the increase in pressure with depth below the water's surface. We set up a vertical $y$ axis in the tank, with its origin at the air–water interface and the positive direction upward. We next consider a water sample contained in an imaginary right circular cylinder of horizontal base (or face) area $A$, such that $y_1$ and $y_2$ (both of which are *negative* numbers) are the depths below the surface of the upper and lower cylinder faces, respectively.

Figure 14-2*b* shows a free-body diagram for the water in the cylinder. The water is in *static equilibrium;* that is, it is stationary and the forces on it balance. Three forces act on it vertically: Force $\vec{F}_1$ acts at the top surface of the cylinder and is due to the water above the cylinder. Similarly, force $\vec{F}_2$ acts at the bottom surface of the cylinder and is due to the water below the cylinder. The gravitational force on the water in the cylinder is represented by $m\vec{g}$, where $m$ is the mass of the water in the cylinder. The balance of these forces is written as

$$F_2 = F_1 + mg. \tag{14-5}$$

We want to transform Eq. 14-5 into an equation involving pressures. From Eq. 14-4, we know that

$$F_1 = p_1 A \quad \text{and} \quad F_2 = p_2 A. \tag{14-6}$$

The mass $m$ of the water in the cylinder is, from Eq. 14-2, $m = \rho V$, where the cylinder's volume $V$ is the product of its face area $A$ and its height $y_1 - y_2$. Thus, $m$ is equal to $\rho A(y_1 - y_2)$. Substituting this and Eq. 14-6 into Eq. 14-5, we find

$$p_2 A = p_1 A + \rho A g(y_1 - y_2)$$

or

$$p_2 = p_1 + \rho g(y_1 - y_2). \tag{14-7}$$

This equation can be used to find pressure both in a liquid (as a function of depth) and in the atmosphere (as a function of altitude or height). For the former, suppose we seek the pressure $p$ at a depth $h$ below the liquid surface. Then we choose level 1 to be the surface, level 2 to be a distance $h$ below it (as in Fig. 14-3), and $p_0$ to represent the atmospheric pressure on the surface. We then substitute

$$y_1 = 0, \quad p_1 = p_0 \quad \text{and} \quad y_2 = -h, \quad p_2 = p$$

into Eq. 14-7, which becomes

$$p = p_0 + \rho g h \quad \text{(pressure at depth } h\text{)}. \tag{14-8}$$

Note that the pressure at a given depth in the liquid depends on that depth but not on any horizontal dimension.

> 👉 The pressure at a point in a fluid in static equilibrium depends on the depth of that point but not on any horizontal dimension of the fluid or its container.

Thus, Eq. 14-8 holds no matter what the shape of the container. If the bottom surface of the container is at depth $h$, then Eq. 14-8 gives the pressure $p$ there.

In Eq. 14-8, $p$ is said to be the total pressure, or **absolute pressure,** at level 2. To see why, note in Fig. 14-3 that the pressure $p$ at level 2 consists of two contributions: (1) $p_0$, the pressure due to the atmosphere, which bears down on the liquid, and (2) $\rho g h$, the pressure due to the liquid above level 2, which bears down on level 2. In general, the difference between an absolute pressure and an atmospheric pressure is called the **gauge pressure.** (The name comes from the use of a gauge to measure this difference in pressures.) For the situation of Fig. 14-3, the gauge pressure is $\rho g h$.

Equation 14-7 also holds above the liquid surface: It gives the atmospheric pressure at a given distance above level 1 in terms of the atmospheric pressure $p_1$ at level 1 (*assuming* that the atmospheric density is uniform over that distance). For example, to find the atmospheric pressure at a distance $d$ above level 1 in Fig. 14-3, we substitute

$$y_1 = 0, \quad p_1 = p_0 \quad \text{and} \quad y_2 = d, \quad p_2 = p.$$

Then with $\rho = \rho_{\text{air}}$, we obtain

$$p = p_0 - \rho_{\text{air}} g d.$$

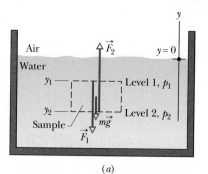

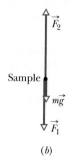

(*b*)

**FIG. 14-2** (*a*) A tank of water in which a sample of water is contained in an imaginary cylinder of horizontal base area $A$. Force $\vec{F}_1$ acts at the top surface of the cylinder; force $\vec{F}_2$ acts at the bottom surface of the cylinder; the gravitational force on the water in the cylinder is represented by $m\vec{g}$. (*b*) A free-body diagram of the water sample.

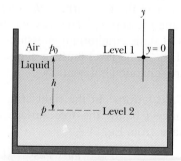

**FIG. 14-3** The pressure $p$ increases with depth $h$ below the liquid surface according to Eq. 14-8.

✔**CHECKPOINT 1**
The figure shows four containers of olive oil. Rank them according to the pressure at depth $h$, greatest first.

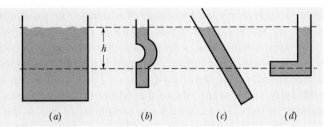

$(a)$  $(b)$  $(c)$  $(d)$

---

## Sample Problem 14-2

A novice scuba diver practicing in a swimming pool takes enough air from his tank to fully expand his lungs before abandoning the tank at depth $L$ and swimming to the surface. He ignores instructions and fails to exhale during his ascent. When he reaches the surface, the difference between the external pressure on him and the air pressure in his lungs is 9.3 kPa. From what depth does he start? What potentially lethal danger does he face?

**KEY IDEA** The pressure at depth $h$ in a liquid of density $\rho$ is given by Eq. 14-8 ($p = p_0 + \rho gh$), where the gauge pressure $\rho gh$ is added to the atmospheric pressure $p_0$.

**Calculations:** Here, when the diver fills his lungs at depth $L$, the external pressure on him (and thus the air pressure within his lungs) is greater than normal and given by Eq. 14-8 as

$$p = p_0 + \rho gL,$$

where $p_0$ is atmospheric pressure and $\rho$ is the water's density (998 kg/m³, from Table 14-1). As he ascends, the external pressure on him decreases, until it is atmospheric pressure $p_0$ at the surface. His blood pressure also decreases, until it is normal. However, because he does not exhale, the air pressure in his lungs remains at the value it had at depth $L$. At the surface, the pressure difference between the higher pressure in his lungs and the lower pressure on his chest is

$$\Delta p = p - p_0 = \rho gL,$$

from which we find

$$L = \frac{\Delta p}{\rho g} = \frac{9300 \text{ Pa}}{(998 \text{ kg/m}^3)(9.8 \text{ m/s}^2)}$$
$$= 0.95 \text{ m.} \qquad \text{(Answer)}$$

This is not deep! Yet, the pressure difference of 9.3 kPa (about 9% of atmospheric pressure) is sufficient to rupture the diver's lungs and force air from them into the depressurized blood, which then carries the air to the heart, killing the diver. If the diver follows instructions and gradually exhales as he ascends, he allows the pressure in his lungs to equalize with the external pressure, and then there is no danger.

---

## Sample Problem 14-3

The U-tube in Fig. 14-4 contains two liquids in static equilibrium: Water of density $\rho_w$ (= 998 kg/m³) is in the right arm, and oil of unknown density $\rho_x$ is in the left. Measurement gives $l = 135$ mm and $d = 12.3$ mm. What is the density of the oil?

**KEY IDEAS** (1) The pressure $p_{int}$ at the level of the oil–water interface in the left arm depends on the density $\rho_x$ and height of the oil above the interface. (2) The water in the right arm *at the same level* must be at the same pressure $p_{int}$. The reason is that, because the water is in static equilibrium, pressures at points in the water at the same level must be the same even if the points are separated horizontally.

**Calculations:** In the right arm, the interface is a distance $l$ below the free surface of the *water,* and we have, from Eq. 14-8,

$$p_{int} = p_0 + \rho_w gl \qquad \text{(right arm).}$$

In the left arm, the interface is a distance $l + d$ below the free surface of the *oil,* and we have, again from Eq. 14-8,

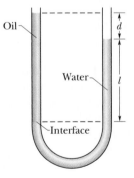

**FIG. 14-4** The oil in the left arm stands higher than the water in the right arm because the oil is less dense than the water. Both fluid columns produce the same pressure $p_{int}$ at the level of the interface.

$$p_{int} = p_0 + \rho_x g(l + d) \qquad \text{(left arm).}$$

Equating these two expressions and solving for the unknown density yield

$$\rho_x = \rho_w \frac{l}{l + d} = (998 \text{ kg/m}^3) \frac{135 \text{ mm}}{135 \text{ mm} + 12.3 \text{ mm}}$$
$$= 915 \text{ kg/m}^3. \qquad \text{(Answer)}$$

Note that the answer does not depend on the atmospheric pressure $p_0$ or the free-fall acceleration $g$.

# 14-5 | Measuring Pressure

## The Mercury Barometer

Figure 14-5a shows a very basic *mercury barometer*, a device used to measure the pressure of the atmosphere. The long glass tube is filled with mercury and inverted with its open end in a dish of mercury, as the figure shows. The space above the mercury column contains only mercury vapor, whose pressure is so small at ordinary temperatures that it can be neglected.

We can use Eq. 14-7 to find the atmospheric pressure $p_0$ in terms of the height $h$ of the mercury column. We choose level 1 of Fig. 14-2 to be that of the air–mercury interface and level 2 to be that of the top of the mercury column, as labeled in Fig. 14-5a. We then substitute

$$y_1 = 0, \quad p_1 = p_0 \quad \text{and} \quad y_2 = h, \quad p_2 = 0$$

into Eq. 14-7, finding that

$$p_0 = \rho g h, \tag{14-9}$$

where $\rho$ is the density of the mercury.

For a given pressure, the height $h$ of the mercury column does not depend on the cross-sectional area of the vertical tube. The fanciful mercury barometer of Fig. 14-5b gives the same reading as that of Fig. 14-5a; all that counts is the vertical distance $h$ between the mercury levels.

Equation 14-9 shows that, for a given pressure, the height of the column of mercury depends on the value of $g$ at the location of the barometer and on the density of mercury, which varies with temperature. The height of the column (in millimeters) is numerically equal to the pressure (in torr) *only* if the barometer is at a place where $g$ has its accepted standard value of 9.80665 m/s² *and* the temperature of the mercury is 0°C. If these conditions do not prevail (and they rarely do), small corrections must be made before the height of the mercury column can be transformed into a pressure.

## The Open-Tube Manometer

An *open-tube manometer* (Fig. 14-6) measures the gauge pressure $p_g$ of a gas. It consists of a U-tube containing a liquid, with one end of the tube connected to the vessel whose gauge pressure we wish to measure and the other end open to the atmosphere. We can use Eq. 14-7 to find the gauge pressure in terms of the height $h$ shown in Fig. 14-6. Let us choose levels 1 and 2 as shown in Fig. 14-6. We then substitute

$$y_1 = 0, \quad p_1 = p_0 \quad \text{and} \quad y_2 = -h, \quad p_2 = p$$

into Eq. 14-7, finding that

$$p_g = p - p_0 = \rho g h, \tag{14-10}$$

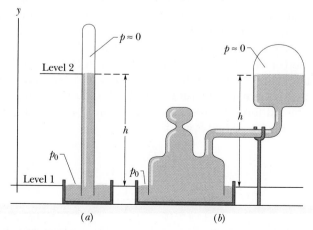

(a)      (b)

FIG. 14-5   (a) A mercury barometer. (b) Another mercury barometer. The distance $h$ is the same in both cases.

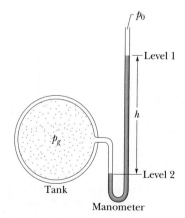

FIG. 14-6   An open-tube manometer, connected to measure the gauge pressure of the gas in the tank on the left. The right arm of the U-tube is open to the atmosphere.

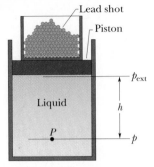

FIG. 14-7 Lead shot (small balls of lead) loaded onto the piston create a pressure $p_{ext}$ at the top of the enclosed (incompressible) liquid. If $p_{ext}$ is increased, by adding more lead shot, the pressure increases by the same amount at all points within the liquid.

where $\rho$ is the density of the liquid in the tube. The gauge pressure $p_g$ is directly proportional to $h$.

The gauge pressure can be positive or negative, depending on whether $p > p_0$ or $p < p_0$. In inflated tires or the human circulatory system, the (absolute) pressure is greater than atmospheric pressure, so the gauge pressure is a positive quantity, sometimes called the *overpressure*. If you suck on a straw to pull fluid up the straw, the (absolute) pressure in your lungs is actually less than atmospheric pressure. The gauge pressure in your lungs is then a negative quantity.

## 14-6 | Pascal's Principle

When you squeeze one end of a tube to get toothpaste out the other end, you are watching **Pascal's principle** in action. This principle is also the basis for the Heimlich maneuver, in which a sharp pressure increase properly applied to the abdomen is transmitted to the throat, forcefully ejecting food lodged there. The principle was first stated clearly in 1652 by Blaise Pascal (for whom the unit of pressure is named):

> A change in the pressure applied to an enclosed incompressible fluid is transmitted undiminished to every portion of the fluid and to the walls of its container.

### Demonstrating Pascal's Principle

Consider the case in which the incompressible fluid is a liquid contained in a tall cylinder, as in Fig. 14-7. The cylinder is fitted with a piston on which a container of lead shot rests. The atmosphere, container, and shot exert pressure $p_{ext}$ on the piston and thus on the liquid. The pressure $p$ at any point $P$ in the liquid is then

$$p = p_{ext} + \rho g h. \tag{14-11}$$

Let us add a little more lead shot to the container to increase $p_{ext}$ by an amount $\Delta p_{ext}$. The quantities $\rho$, $g$, and $h$ in Eq. 14-11 are unchanged, so the pressure change at $P$ is

$$\Delta p = \Delta p_{ext}. \tag{14-12}$$

This pressure change is independent of $h$, so it must hold for all points within the liquid, as Pascal's principle states.

### Pascal's Principle and the Hydraulic Lever

Figure 14-8 shows how Pascal's principle can be made the basis of a hydraulic lever. In operation, let an external force of magnitude $F_i$ be directed downward on the left-hand (or input) piston, whose surface area is $A_i$. An incompressible liquid in the device then produces an upward force of magnitude $F_o$ on the right-hand (or output) piston, whose surface area is $A_o$. To keep the system in equilibrium, there must be a downward force of magnitude $F_o$ on the output piston from an external load (not shown). The force $\vec{F}_i$ applied on the left and the downward force $\vec{F}_o$ from the load on the right produce a change $\Delta p$ in the pressure of the liquid that is given by

$$\Delta p = \frac{F_i}{A_i} = \frac{F_o}{A_o},$$

so

$$F_o = F_i \frac{A_o}{A_i}. \tag{14-13}$$

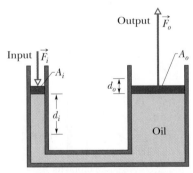

FIG. 14-8 A hydraulic arrangement that can be used to magnify a force $\vec{F}_i$. The work done is, however, not magnified and is the same for both the input and output forces.

Equation 14-13 shows that the output force $F_o$ on the load must be greater than the input force $F_i$ if $A_o > A_i$, as is the case in Fig. 14-8.

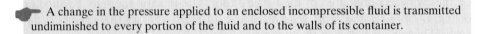

If we move the input piston downward a distance $d_i$, the output piston moves upward a distance $d_o$, such that the same volume $V$ of the incompressible liquid is displaced at both pistons. Then

$$V = A_i d_i = A_o d_o,$$

which we can write as

$$d_o = d_i \frac{A_i}{A_o}. \qquad (14\text{-}14)$$

This shows that, if $A_o > A_i$ (as in Fig. 14-8), the output piston moves a smaller distance than the input piston moves.

From Eqs. 14-13 and 14-14 we can write the output work as

$$W = F_o d_o = \left( F_i \frac{A_o}{A_i} \right)\left( d_i \frac{A_i}{A_o} \right) = F_i d_i, \qquad (14\text{-}15)$$

which shows that the work $W$ done *on* the input piston by the applied force is equal to the work $W$ done *by* the output piston in lifting the load placed on it.

The advantage of a hydraulic lever is this:

> ☞ With a hydraulic lever, a given force applied over a given distance can be transformed to a greater force applied over a smaller distance.

The product of force and distance remains unchanged so that the same work is done. However, there is often tremendous advantage in being able to exert the larger force. Most of us, for example, cannot lift an automobile directly but can with a hydraulic jack, even though we have to pump the handle farther than the automobile rises and in a series of small strokes.

## 14-7 | Archimedes' Principle

Figure 14-9 shows a student in a swimming pool, manipulating a very thin plastic sack (of negligible mass) that is filled with water. She finds that the sack and its contained water are in static equilibrium, tending neither to rise nor to sink. The downward gravitational force $\vec{F}_g$ on the contained water must be balanced by a net upward force from the water surrounding the sack.

This net upward force is a **buoyant force** $\vec{F}_b$. It exists because the pressure in the surrounding water increases with depth below the surface. Thus, the pressure near the bottom of the sack is greater than the pressure near the top, which means the forces on the sack due to this pressure are greater in magnitude near the bottom of the sack than near the top. Some of the forces are represented in Fig. 14-10a, where the space occupied by the sack has been left empty. Note that the force vectors drawn near the bottom of that space (with upward components) have longer lengths than those drawn near the top of the sack (with downward components). If we vectorially add all the forces on the sack from the water, the horizontal components cancel and the vertical components add to yield the upward buoyant force $\vec{F}_b$ on the sack. (Force $\vec{F}_b$ is shown to the right of the pool in Fig. 14-10a.)

Because the sack of water is in static equilibrium, the magnitude of $\vec{F}_b$ is equal to the magnitude $m_f g$ of the gravitational force $\vec{F}_g$ on the sack of water: $F_b = m_f g$. (Subscript $f$ refers to *fluid*, here the water.) In words, the magnitude of the buoyant force is equal to the weight of the water in the sack.

In Fig. 14-10b, we have replaced the sack of water with a stone that exactly fills the hole in Fig. 14-10a. The stone is said to *displace* the water, meaning that the stone occupies space that would otherwise be occupied by water. We have changed nothing about the shape of the hole, so the forces at the hole's surface must be the same as when the water-filled sack was in place. Thus, the same upward buoyant force that acted on the water-filled sack now acts on the stone;

**FIG. 14-9** A thin-walled plastic sack of water is in static equilibrium in the pool. The gravitational force on the sack must be balanced by a net upward force on it from the surrounding water.

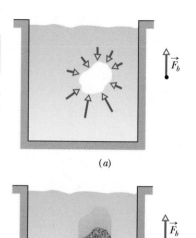

(a)

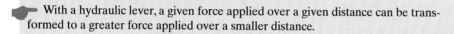

(b)

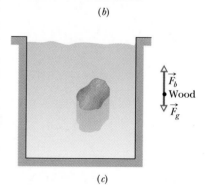

(c)

**FIG. 14-10** (a) The water surrounding the hole in the water produces a net upward buoyant force on whatever fills the hole. (b) For a stone of the same volume as the hole, the gravitational force exceeds the buoyant force in magnitude. (c) For a lump of wood of the same volume, the gravitational force is less than the buoyant force in magnitude.

that is, the magnitude $F_b$ of the buoyant force is equal to $m_f g$, the weight of the water displaced by the stone.

Unlike the water-filled sack, the stone is not in static equilibrium. The downward gravitational force $\vec{F}_g$ on the stone is greater in magnitude than the upward buoyant force, as is shown in the free-body diagram in Fig. 14-10b. The stone thus accelerates downward, sinking to the bottom of the pool.

Let us next exactly fill the hole in Fig. 14-10a with a block of lightweight wood, as in Fig. 14-10c. Again, nothing has changed about the forces at the hole's surface, so the magnitude $F_b$ of the buoyant force is still equal to $m_f g$, the weight of the displaced water. Like the stone, the block is not in static equilibrium. However, this time the gravitational force $\vec{F}_g$ is lesser in magnitude than the buoyant force (as shown to the right of the pool), and so the block accelerates upward, rising to the top surface of the water.

Our results with the sack, stone, and block apply to all fluids and are summarized in **Archimedes' principle:**

> When a body is fully or partially submerged in a fluid, a buoyant force $\vec{F}_b$ from the surrounding fluid acts on the body. The force is directed upward and has a magnitude equal to the weight $m_f g$ of the fluid that has been displaced by the body.

The buoyant force on a body in a fluid has the magnitude

$$F_b = m_f g \quad \text{(buoyant force)}, \qquad (14\text{-}16)$$

where $m_f$ is the mass of the fluid that is displaced by the body.

## Floating

When we release a block of lightweight wood just above the water in a pool, the block moves into the water because the gravitational force on it pulls it downward. As the block displaces more and more water, the magnitude $F_b$ of the upward buoyant force acting on it increases. Eventually, $F_b$ is large enough to equal the magnitude $F_g$ of the downward gravitational force on the block, and the block comes to rest. The block is then in static equilibrium and is said to be *floating* in the water. In general,

> When a body floats in a fluid, the magnitude $F_b$ of the buoyant force on the body is equal to the magnitude $F_g$ of the gravitational force on the body.

We can write this statement as

$$F_b = F_g \quad \text{(floating)}. \qquad (14\text{-}17)$$

From Eq. 14-16, we know that $F_b = m_f g$. Thus,

> When a body floats in a fluid, the magnitude $F_g$ of the gravitational force on the body is equal to the weight $m_f g$ of the fluid that has been displaced by the body.

We can write this statement as

$$F_g = m_f g \quad \text{(floating)}. \qquad (14\text{-}18)$$

In other words, a floating body displaces its own weight of fluid.

## Apparent Weight in a Fluid

If we place a stone on a scale that is calibrated to measure weight, then the reading on the scale is the stone's weight. However, if we do this underwater, the upward buoyant force on the stone from the water decreases the reading. That reading is then an apparent weight. In general, an **apparent weight** is related to the actual weight of a body and the buoyant force on the body by

$$\begin{pmatrix} \text{apparent} \\ \text{weight} \end{pmatrix} = \begin{pmatrix} \text{actual} \\ \text{weight} \end{pmatrix} - \begin{pmatrix} \text{magnitude of} \\ \text{buoyant force} \end{pmatrix},$$

which we can write as

$$\text{weight}_{app} = \text{weight} - F_b \qquad \text{(apparent weight).} \qquad (14\text{-}19)$$

If, in some test of strength, you had to lift a heavy stone, you could do it more easily with the stone underwater. Then your applied force would need to exceed only the stone's apparent weight, not its larger actual weight, because the upward buoyant force would help you lift the stone.

The magnitude of the buoyant force on a floating body is equal to the body's weight. Equation 14-19 thus tells us that a floating body has an apparent weight of zero—the body would produce a reading of zero on a scale. (When astronauts prepare to perform a complex task in space, they practice the task floating underwater, where their apparent weight is zero as it is in space.)

✓ **CHECKPOINT 2**    A penguin floats first in a fluid of density $\rho_0$, then in a fluid of density $0.95\rho_0$, and then in a fluid of density $1.1\rho_0$. (a) Rank the densities according to the magnitude of the buoyant force on the penguin, greatest first. (b) Rank the densities according to the amount of fluid displaced by the penguin, greatest first.

---

**Sample Problem    14-4**

In Fig. 14-11a, a surfer rides on the front side of a wave, at a point where a tangent to the wave has a slope of $\theta = 30.0°$. The combined mass of surfer and surfboard is $m = 83.0$ kg, and the board has a submerged volume of $V = 2.50 \times 10^{-2}$ m³. The surfer maintains his position on the wave as the wave moves at constant speed toward shore. What are the magnitude and direction (relative to the positive direction of the x axis in Fig. 14-11b) of the drag force on the surfboard from the water?

**KEY IDEAS**    (1) The buoyancy force on the surfer has a magnitude $F_b$ equal to the weight of the seawater displaced by the submerged volume of the surfboard. The direction of the force is perpendicular to the surface at the surfer's location. (2) By Newton's second law, because the surfer moves at constant speed toward the shore, the (vector) sum of the buoyancy force $\vec{F}_b$, the gravitational force $\vec{F}_g$, and the drag force $\vec{F}_d$ must be 0.

**Calculations:** The forces and their components are shown in the free-body diagram of Fig. 14-11b. The gravi-

tational force $\vec{F}_g$ is downward and (as we saw in Chapter 5) has a component of $mg \sin \theta$ down the slope and a component of $mg \cos \theta$ perpendicular to the slope. A drag force $\vec{F}_d$ from the water acts on the surfboard because water is continuously forced up into the wave as the wave continues to move toward the shore. This push on the surfboard is upward and to the rear, at angle $\phi$ to the x axis. The buoyancy force $\vec{F}_b$ is perpendicular to the water surface; its magnitude depends on the mass $m_f$ of the water displaced by the surfboard, as given by Eq. 14-16 ($F_b = m_f g$). From Eq. 14-2 ($\rho = m/V$), we can write the mass in terms of the seawater density $\rho_w$ and the submerged volume V of the surfboard: $m_f = \rho_w V$. From Table 14-1, seawater density $\rho_w$ is $1.024 \times 10^3$ kg/m³. Thus, the magnitude of the buoyant force is

$$\begin{aligned} F_b &= m_f g = \rho_w V g \\ &= (1.024 \times 10^3 \text{ kg/m}^3)(2.50 \times 10^{-2} \text{ m}^3)(9.8 \text{ m/s}^2) \\ &= 2.509 \times 10^2 \text{ N.} \end{aligned}$$

So, Newton's second law for the y axis,

$$F_{dy} + F_b - mg \cos \theta = m(0),$$

becomes

$$F_{dy} + 2.509 \times 10^2 \text{ N} - (83 \text{ kg})(9.8 \text{ m/s}^2) \cos 30.0° = 0,$$

yielding

$$F_{dy} = 453.5 \text{ N.}$$

Similarly, Newton's second law $\vec{F}_{net} = m\vec{a}$ for the x axis,

$$F_{dx} - mg \sin \theta = m(0),$$

yields

$$F_{dx} = 406.7 \text{ N.}$$

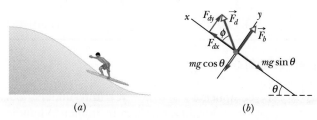

(a)

(b)

**FIG. 14-11**    (a) Surfer. (b) Free-body diagram showing the forces on the surfer–surfboard system.

Combining the two components of the drag force tells us that the force has magnitude

$$F_d = \sqrt{(406.7 \text{ N})^2 + (453.5 \text{ N})^2}$$
$$= 609 \text{ N} \qquad \text{(Answer)}$$

and angle

$$\phi = \tan^{-1}\left(\frac{453.5 \text{ N}}{406.7 \text{ N}}\right) = 48.1°. \qquad \text{(Answer)}$$

**Wipeout avoided:** If the surfer tilts the board slightly forward, the magnitude of the drag force decreases and angle $\phi$ changes. The result is that the net force is no

longer zero and the surfer moves down the face of the wave. The descent is somewhat self-adjusting because, as the surfer descends, the tilt angle $\theta$ of the wave surface decreases and thus so does the component of the gravitational force $mg \sin \theta$ pulling the surfer down the slope. So, the surfer can adjust the board to re-establish equilibrium, now lower on the wave. Similarly, by tilting the board slightly backward, the surfer increases the drag and moves up the face of the wave. If the surfer is still on the lower part of the wave, then both $\theta$ and $mg \sin \theta$ increase and again the surfer can control the forces and re-establish equilibrium.

---

**Sample Problem  14-5**

In Fig. 14-12, a block of density $\rho = 800 \text{ kg/m}^3$ floats face down in a fluid of density $\rho_f = 1200 \text{ kg/m}^3$. The block has height $H = 6.0$ cm.

(a) By what depth $h$ is the block submerged?

**KEY IDEA** (1) Floating requires that the upward buoyant force on the block match the downward gravitational force on the block. (2) The buoyant force is equal to the weight $m_f g$ of the fluid displaced by the submerged portion of the block.

**Calculations:** From Eq. 14-16, we know that the buoyant force has the magnitude $F_b = m_f g$, where $m_f$ is the mass of the fluid displaced by the block's submerged volume $V_f$. From Eq. 14-2 ($\rho = m/V$), we know that the mass of the displaced fluid is $m_f = \rho_f V_f$. We don't know $V_f$ but if we symbolize the block's face length as $L$ and its width as $W$, then from Fig. 14-12 we see that the submerged volume must be $V_f = LWh$. If we now combine our three expressions, we find that the upward buoyant force has magnitude

$$F_b = m_f g = \rho_f V_f g = \rho_f LWhg. \qquad (14\text{-}20)$$

Similarly, we can write the magnitude $F_g$ of the gravitational force on the block, first in terms of the block's mass $m$, then in terms of the block's density $\rho$ and (full) volume $V$, and then in terms of the block's dimensions $L$, $W$, and $H$ (the full height):

$$F_g = mg = \rho Vg = \rho_f LWHg. \qquad (14\text{-}21)$$

The floating block is stationary. Thus, writing Newton's second law for components along a vertical $y$ axis ($F_{net,y} = ma_y$), we have

$$F_b - F_g = m(0),$$

or from Eqs. 14-20 and 14-21,

$$\rho_f LWhg - \rho LWHg = 0,$$

which gives us

$$h = \frac{\rho}{\rho_f} H = \frac{800 \text{ kg/m}^3}{1200 \text{ kg/m}^3}(6.0 \text{ cm})$$
$$= 4.0 \text{ cm}. \qquad \text{(Answer)}$$

(b) If the block is held fully submerged and then released, what is the magnitude of its acceleration?

**Calculations:** The gravitational force on the block is the same but now, with the block fully submerged, the volume of the displaced water is $V = LWH$. (The full height of the block is used.) This means that the value of $F_b$ is now larger, and the block will no longer be stationary but will accelerate upward. Now Newton's second law yields

$$F_b - F_g = ma,$$

or

$$\rho_f LWHg - \rho LWHg = \rho LWHa,$$

where we inserted $\rho LWH$ for the mass $m$ of the block. Solving for $a$ leads to

$$a = \left(\frac{\rho_f}{\rho} - 1\right)g = \left(\frac{1200 \text{ kg/m}^3}{800 \text{ kg/m}^3} - 1\right)(9.8 \text{ m/s}^2)$$
$$= 4.9 \text{ m/s}^2. \qquad \text{(Answer)}$$

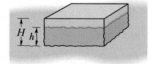

**FIG. 14-12** Block of height $H$ floats in a fluid, to a depth of $h$.

---

# 14-8 | Ideal Fluids in Motion

The motion of *real fluids* is very complicated and not yet fully understood. Instead, we shall discuss the motion of an **ideal fluid,** which is simpler to handle mathematically and yet provides useful results. Here are four assumptions that we make about our ideal fluid; they all are concerned with *flow:*

1. ***Steady flow***   In *steady* (or *laminar*) *flow*, the velocity of the moving fluid at any fixed point does not change with time, either in magnitude or in direction. The gentle flow of water near the center of a quiet stream is steady; the flow in a chain of rapids is not. Figure 14-13 shows a transition from steady flow to *nonsteady* (or *nonlaminar* or *turbulent*) *flow* for a rising stream of smoke. The speed of the smoke particles increases as they rise and, at a certain critical speed, the flow changes from steady to nonsteady.

2. ***Incompressible flow***   We assume, as for fluids at rest, that our ideal fluid is incompressible; that is, its density has a constant, uniform value.

3. ***Nonviscous flow***   Roughly speaking, the viscosity of a fluid is a measure of how resistive the fluid is to flow. For example, thick honey is more resistive to flow than water, and so honey is said to be more viscous than water. Viscosity is the fluid analog of friction between solids; both are mechanisms by which the kinetic energy of moving objects can be transferred to thermal energy. In the absence of friction, a block could glide at constant speed along a horizontal surface. In the same way, an object moving through a nonviscous fluid would experience no *viscous drag force*—that is, no resistive force due to viscosity; it could move at constant speed through the fluid. The British scientist Lord Rayleigh noted that in an ideal fluid a ship's propeller would not work, but, on the other hand, in an ideal fluid a ship (once set into motion) would not need a propeller!

4. ***Irrotational flow***   Although it need not concern us further, we also assume that the flow is *irrotational*. To test for this property, let a tiny grain of dust move with the fluid. Although this test body may (or may not) move in a circular path, in irrotational flow the test body will not rotate about an axis through its own center of mass. For a loose analogy, the motion of a Ferris wheel is rotational; that of its passengers is irrotational.

We can make the flow of a fluid visible by adding a *tracer*. This might be a dye injected into many points across a liquid stream (Fig. 14-14) or smoke particles added to a gas flow (Fig. 14-13). Each bit of a tracer follows a *streamline*, which is the path that a tiny element of the fluid would take as the fluid flows. Recall from Chapter 4 that the velocity of a particle is always tangent to the path taken by the particle. Here the particle is the fluid element, and its velocity $\vec{v}$ is always tangent to a streamline (Fig. 14-15). For this reason, two streamlines can never intersect; if they did, then an element arriving at their intersection would have two different velocities simultaneously—an impossibility.

**FIG. 14-13**   At a certain point, the rising flow of smoke and heated gas changes from steady to turbulent. *(Will McIntyre/Photo Researchers)*

## 14-9 | The Equation of Continuity

You may have noticed that you can increase the speed of the water emerging from a garden hose by partially closing the hose opening with your thumb. Apparently the speed *v* of the water depends on the cross-sectional area *A* through which the water flows.

**FIG. 14-14**   The steady flow of a fluid around a cylinder, as revealed by a dye tracer that was injected into the fluid upstream of the cylinder. *(Courtesy D.H. Peregrine, University of Bristol)*

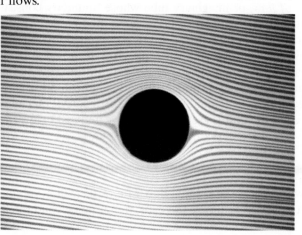

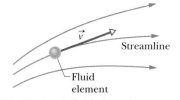

**FIG. 14-15**   A fluid element traces out a streamline as it moves. The velocity vector of the element is tangent to the streamline at every point.

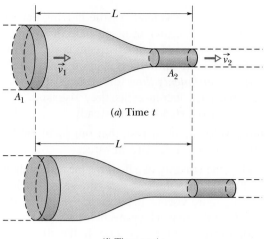

*(a)* Time *t*

*(b)* Time *t* + Δ*t*

**FIG. 14-16** Fluid flows from left to right at a steady rate through a tube segment of length *L*. The fluid's speed is $v_1$ at the left side and $v_2$ at the right side. The tube's cross-sectional area is $A_1$ at the left side and $A_2$ at the right side. From time *t* in (*a*) to time *t* + Δ*t* in (*b*), the amount of fluid shown in purple enters at the left side and the equal amount of fluid shown in green emerges at the right side.

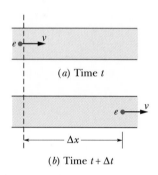

*(a)* Time *t*

*(b)* Time *t* + Δ*t*

**FIG. 14-17** Fluid flows at a constant speed *v* through a tube. (*a*) At time *t*, fluid element *e* is about to pass the dashed line. (*b*) At time *t* + Δ*t*, element *e* is a distance Δ*x* = *v* Δ*t* from the dashed line.

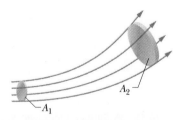

**FIG. 14-18** A tube of flow is defined by the streamlines that form the boundary of the tube. The volume flow rate must be the same for all cross sections of the tube of flow.

Here we wish to derive an expression that relates *v* and *A* for the steady flow of an ideal fluid through a tube with varying cross section, like that in Fig. 14-16. The flow there is toward the right, and the tube segment shown (part of a longer tube) has length *L*. The fluid has speeds $v_1$ at the left end of the segment and $v_2$ at the right end. The tube has cross-sectional areas $A_1$ at the left end and $A_2$ at the right end. Suppose that in a time interval Δ*t* a volume Δ*V* of fluid enters the tube segment at its left end (that volume is colored purple in Fig. 14-16). Then, because the fluid is incompressible, an identical volume Δ*V* must emerge from the right end of the segment (it is colored green in Fig. 14-16).

We can use this common volume Δ*V* to relate the speeds and areas. To do so, we first consider Fig. 14-17, which shows a side view of a tube of *uniform* cross-sectional area *A*. In Fig. 14-17*a*, a fluid element *e* is about to pass through the dashed line drawn across the tube width. The element's speed is *v*, so during a time interval Δ*t*, the element moves along the tube a distance Δ*x* = *v* Δ*t*. The volume Δ*V* of fluid that has passed through the dashed line in that time interval Δ*t* is

$$\Delta V = A\, \Delta x = Av\, \Delta t. \tag{14-22}$$

Applying Eq. 14-22 to both the left and right ends of the tube segment in Fig. 14-16, we have

$$\Delta V = A_1 v_1\, \Delta t = A_2 v_2\, \Delta t$$

or

$$A_1 v_1 = A_2 v_2 \quad \text{(equation of continuity)}. \tag{14-23}$$

This relation between speed and cross-sectional area is called the **equation of continuity** for the flow of an ideal fluid. It tells us that the flow speed increases when we decrease the cross-sectional area through which the fluid flows (as when we partially close off a garden hose with a thumb).

Equation 14-23 applies not only to an actual tube but also to any so-called *tube of flow,* or imaginary tube whose boundary consists of streamlines. Such a tube acts like a real tube because no fluid element can cross a streamline; thus, all the fluid within a tube of flow must remain within its boundary. Figure 14-18 shows a tube of flow in which the cross-sectional area increases from area $A_1$ to area $A_2$ along the flow direction. From Eq. 14-23 we know that, with the increase in area, the speed must decrease, as is indicated by the greater spacing between streamlines at the right in Fig. 14-18. Similarly, you can see that in Fig. 14-14 the speed of the flow is greatest just above and just below the cylinder.

We can rewrite Eq. 14-23 as

$$R_V = Av = \text{a constant} \quad \text{(volume flow rate, equation of continuity)}, \tag{14-24}$$

in which $R_V$ is the **volume flow rate** of the fluid (volume past a given point per unit time). Its SI unit is the cubic meter per second (m³/s). If the density $\rho$ of the fluid is uniform, we can multiply Eq. 14-24 by that density to get the **mass flow rate** $R_m$ (mass per unit time):

$$R_m = \rho R_V = \rho A v = \text{a constant} \quad \text{(mass flow rate).} \quad (14\text{-}25)$$

The SI unit of mass flow rate is the kilogram per second (kg/s). Equation 14-25 says that the mass that flows into the tube segment of Fig. 14-16 each second must be equal to the mass that flows out of that segment each second.

✓**CHECKPOINT 3**
The figure shows a pipe and gives the volume flow rate (in cm³/s) and the direction of flow for all but one section. What are the volume flow rate and the direction of flow for that section?

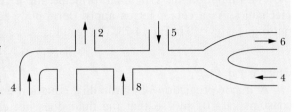

## Sample Problem | 14-6

Figure 14-19 shows how the stream of water emerging from a faucet "necks down" as it falls. The indicated cross-sectional areas are $A_0 = 1.2$ cm² and $A = 0.35$ cm². The two levels are separated by a vertical distance $h = 45$ mm. What is the volume flow rate from the tap?

**KEY IDEA** The volume flow rate through the higher cross section must be the same as that through the lower cross section.

**FIG. 14-19** As water falls from a tap, its speed increases. Because the volume flow rate must be the same at all horizontal cross sections of the stream, the stream must "neck down" (narrow).

**Calculations:** From Eq. 14-24, we have

$$A_0 v_0 = A v, \quad (14\text{-}26)$$

where $v_0$ and $v$ are the water speeds at the levels corresponding to $A_0$ and $A$. From Eq. 2-16 we can also write, because the water is falling freely with acceleration $g$,

$$v^2 = v_0^2 + 2gh. \quad (14\text{-}27)$$

Eliminating $v$ between Eqs. 14-26 and 14-27 and solving for $v_0$, we obtain

$$v_0 = \sqrt{\frac{2ghA^2}{A_0^2 - A^2}}$$

$$= \sqrt{\frac{(2)(9.8 \text{ m/s}^2)(0.045 \text{ m})(0.35 \text{ cm}^2)^2}{(1.2 \text{ cm}^2)^2 - (0.35 \text{ cm}^2)^2}}$$

$$= 0.286 \text{ m/s} = 28.6 \text{ cm/s}.$$

From Eq. 14-24, the volume flow rate $R_V$ is then

$$R_V = A_0 v_0 = (1.2 \text{ cm}^2)(28.6 \text{ cm/s})$$
$$= 34 \text{ cm}^3/\text{s}. \quad \text{(Answer)}$$

# 14-10 | Bernoulli's Equation

Figure 14-20 represents a tube through which an ideal fluid is flowing at a steady rate. In a time interval $\Delta t$, suppose that a volume of fluid $\Delta V$, colored purple in Fig. 14-20, enters the tube at the left (or input) end and an identical volume, colored green in Fig. 14-20, emerges at the right (or output) end. The emerging volume must be the same as the entering volume because the fluid is incompressible, with an assumed constant density $\rho$.

Let $y_1$, $v_1$, and $p_1$ be the elevation, speed, and pressure of the fluid entering at the left, and $y_2$, $v_2$, and $p_2$ be the corresponding quantities for the fluid emerging at the right. By applying the principle of conservation of energy to the fluid, we shall show that these quantities are related by

$$p_1 + \tfrac{1}{2}\rho v_1^2 + \rho g y_1 = p_2 + \tfrac{1}{2}\rho v_2^2 + \rho g y_2. \quad (14\text{-}28)$$

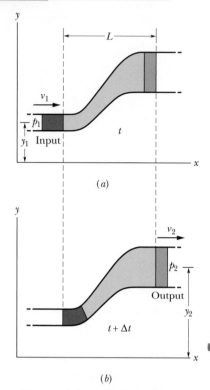

(a)

(b)

**FIG. 14-20** Fluid flows at a steady rate through a length $L$ of a tube, from the input end at the left to the output end at the right. From time $t$ in (a) to time $t + \Delta t$ in (b), the amount of fluid shown in purple enters the input end and the equal amount shown in green emerges from the output end.

In general, the term $\frac{1}{2}\rho v^2$ is called the fluid's **kinetic energy density** (kinetic energy per unit volume). We can also write Eq. 14-28 as

$$p + \tfrac{1}{2}\rho v^2 + \rho g y = \text{a constant} \qquad \text{(Bernoulli's equation).} \qquad (14\text{-}29)$$

Equations 14-28 and 14-29 are equivalent forms of **Bernoulli's equation,** after Daniel Bernoulli, who studied fluid flow in the 1700s.* Like the equation of continuity (Eq. 14-24), Bernoulli's equation is not a new principle but simply the reformulation of a familiar principle in a form more suitable to fluid mechanics. As a check, let us apply Bernoulli's equation to fluids at rest, by putting $v_1 = v_2 = 0$ in Eq. 14-28. The result is

$$p_2 = p_1 + \rho g(y_1 - y_2),$$

which is Eq. 14-7.

A major prediction of Bernoulli's equation emerges if we take $y$ to be a constant ($y = 0$, say) so that the fluid does not change elevation as it flows. Equation 14-28 then becomes

$$p_1 + \tfrac{1}{2}\rho v_1^2 = p_2 + \tfrac{1}{2}\rho v_2^2, \qquad (14\text{-}30)$$

which tells us that:

> If the speed of a fluid element increases as the element travels along a horizontal streamline, the pressure of the fluid must decrease, and conversely.

Put another way, where the streamlines are relatively close together (where the velocity is relatively great), the pressure is relatively low, and conversely.

The link between a change in speed and a change in pressure makes sense if you consider a fluid element. When the element nears a narrow region, the higher pressure behind it accelerates it so that it then has a greater speed in the narrow region. When it nears a wide region, the higher pressure ahead of it decelerates it so that it then has a lesser speed in the wide region.

Bernoulli's equation is strictly valid only to the extent that the fluid is ideal. If viscous forces are present, thermal energy will be involved. We take no account of this in the derivation that follows.

### Proof of Bernoulli's Equation

Let us take as our system the entire volume of the (ideal) fluid shown in Fig. 14-20. We shall apply the principle of conservation of energy to this system as it moves from its initial state (Fig. 14-20a) to its final state (Fig. 14-20b). The fluid lying between the two vertical planes separated by a distance $L$ in Fig. 14-20 does not change its properties during this process; we need be concerned only with changes that take place at the input and output ends.

First, we apply energy conservation in the form of the work–kinetic energy theorem,

$$W = \Delta K, \qquad (14\text{-}31)$$

which tells us that the change in the kinetic energy of our system must equal the net work done on the system. The change in kinetic energy results from the change in speed between the ends of the tube and is

$$\Delta K = \tfrac{1}{2}\Delta m\, v_2^2 - \tfrac{1}{2}\Delta m\, v_1^2$$
$$= \tfrac{1}{2}\rho\, \Delta V(v_2^2 - v_1^2), \qquad (14\text{-}32)$$

in which $\Delta m$ ($= \rho\, \Delta V$) is the mass of the fluid that enters at the input end and leaves at the output end during a small time interval $\Delta t$.

*For irrotational flow (which we assume), the constant in Eq. 14-29 has the same value for all points within the tube of flow; the points do not have to lie along the same streamline. Similarly, the points 1 and 2 in Eq. 14-28 can lie anywhere within the tube of flow.

The work done on the system arises from two sources. The work $W_g$ done by the gravitational force ($\Delta m \, \vec{g}$) on the fluid of mass $\Delta m$ during the vertical lift of the mass from the input level to the output level is

$$W_g = -\Delta m \, g(y_2 - y_1)$$
$$= -\rho g \, \Delta V(y_2 - y_1). \tag{14-33}$$

This work is negative because the upward displacement and the downward gravitational force have opposite directions.

Work must also be done *on* the system (at the input end) to push the entering fluid into the tube and *by* the system (at the output end) to push forward the fluid that is located ahead of the emerging fluid. In general, the work done by a force of magnitude $F$, acting on a fluid sample contained in a tube of area $A$ to move the fluid through a distance $\Delta x$, is

$$F \, \Delta x = (pA)(\Delta x) = p(A \, \Delta x) = p \, \Delta V.$$

The work done on the system is then $p_1 \, \Delta V$, and the work done by the system is $-p_2 \, \Delta V$. Their sum $W_p$ is

$$W_p = -p_2 \, \Delta V + p_1 \, \Delta V$$
$$= -(p_2 - p_1) \, \Delta V. \tag{14-34}$$

The work–kinetic energy theorem of Eq. 14-31 now becomes

$$W = W_g + W_p = \Delta K.$$

Substituting from Eqs. 14-32, 14-33, and 14-34 yields

$$-\rho g \, \Delta V(y_2 - y_1) - \Delta V(p_2 - p_1) = \tfrac{1}{2}\rho \, \Delta V(v_2^2 - v_1^2).$$

This, after a slight rearrangement, matches Eq. 14-28, which we set out to prove.

✓**CHECKPOINT 4**    Water flows smoothly through the pipe shown in the figure, descending in the process. Rank the four numbered sections of pipe according to (a) the volume flow rate $R_V$ through them, (b) the flow speed $v$ through them, and (c) the water pressure $p$ within them, greatest first.

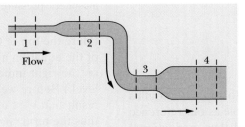

---

## Sample Problem 14-7

Ethanol of density $\rho = 791 \text{ kg/m}^3$ flows smoothly through a horizontal pipe that tapers (as in Fig. 14-16) in cross-sectional area from $A_1 = 1.20 \times 10^{-3} \text{ m}^2$ to $A_2 = A_1/2$. The pressure difference between the wide and narrow sections of pipe is 4120 Pa. What is the volume flow rate $R_V$ of the ethanol?

**KEY IDEAS**  (1) Because the fluid flowing through the wide section of pipe must entirely pass through the narrow section, the volume flow rate $R_V$ must be the same in the two sections. Thus, from Eq. 14-24,

$$R_V = v_1 A_1 = v_2 A_2. \tag{14-35}$$

However, with two unknown speeds, we cannot evaluate this equation for $R_V$. (2) Because the flow is smooth, we can apply Bernoulli's equation. From Eq. 14-28, we can write

$$p_1 + \tfrac{1}{2}\rho v_1^2 + \rho g y = p_2 + \tfrac{1}{2}\rho v_2^2 + \rho g y, \tag{14-36}$$

where subscripts 1 and 2 refer to the wide and narrow sections of pipe, respectively, and $y$ is their common elevation. This equation hardly seems to help because it does not contain the desired $R_V$ and it contains the unknown speeds $v_1$ and $v_2$.

*Calculations:* There is a neat way to make Eq. 14-36 work for us: First, we can use Eq. 14-35 and the fact that $A_2 = A_1/2$ to write

$$v_1 = \frac{R_V}{A_1} \quad \text{and} \quad v_2 = \frac{R_V}{A_2} = \frac{2R_V}{A_1}. \tag{14-37}$$

Then we can substitute these expressions into Eq. 14-36 to eliminate the unknown speeds and introduce the desired volume flow rate. Doing this and solving for $R_V$ yield

$$R_V = A_1 \sqrt{\frac{2(p_1 - p_2)}{3\rho}}. \tag{14-38}$$

We still have a decision to make: We know that the pressure difference between the two sections is 4120 Pa, but does that mean that $p_1 - p_2$ is 4120 Pa or $-4120$ Pa? We could guess the former is true, or otherwise the square root in Eq. 14-38 would give us an imaginary number. Instead of guessing, however, let's try some reasoning. From Eq. 14-35 we see that speed $v_2$ in the narrow section (small $A_2$) must be greater than speed $v_1$ in the wider section (larger $A_1$). Recall that if the

speed of a fluid increases as the fluid travels along a horizontal path (as here), the pressure of the fluid must decrease. Thus, $p_1$ is greater than $p_2$, and $p_1 - p_2 = 4120$ Pa. Inserting this and known data into Eq. 14-38 gives

$$R_V = 1.20 \times 10^{-3}\,\text{m}^2 \sqrt{\frac{(2)(4120\,\text{Pa})}{(3)(791\,\text{kg/m}^3)}}$$

$$= 2.24 \times 10^{-3}\,\text{m}^3/\text{s}. \qquad \text{(Answer)}$$

---

## Sample Problem | 14-8

In the old West, a desperado fires a bullet into an open water tank (Fig. 14-21), creating a hole a distance $h$ below the water surface. What is the speed $v$ of the water exiting the tank?

**KEY IDEAS** (1) This situation is essentially that of water moving (downward) with speed $v_0$ through a wide pipe (the tank) of cross-sectional area $A$ and then moving (horizontally) with speed $v$ through a narrow pipe (the hole) of cross-sectional area $a$. (2) Because the water flowing through the wide pipe must entirely pass through the narrow pipe, the volume flow rate $R_V$ must be the same in the two "pipes." (3) We can also relate $v$ to $v_0$ (and to $h$) through Bernoulli's equation (Eq. 14-28).

**Calculations:** From Eq. 14-24,

$$R_V = av = Av_0$$

and thus

$$v_0 = \frac{a}{A}\,v.$$

Because $a \ll A$, we see that $v_0 \ll v$. To apply Bernoulli's equation, we take the level of the hole as our reference level for measuring elevations (and thus gravitational potential energy). Noting that the pressure at the top of the tank and at the bullet hole is the atmospheric pressure $p_0$ (because both places are exposed to the atmosphere), we write Eq. 14-28 as

$$p_0 + \tfrac{1}{2}\rho v_0^2 + \rho gh = p_0 + \tfrac{1}{2}\rho v^2 + \rho g(0). \quad \text{(14-39)}$$

**FIG. 14-21** Water pours through a hole in a water tank, at a distance $h$ below the water surface. The pressure at the water surface and at the hole is atmospheric pressure $p_0$.

(Here the top of the tank is represented by the left side of the equation and the hole by the right side. The zero on the right indicates that the hole is at our reference level.) Before we solve Eq. 14-39 for $v$, we can use our result that $v_0 \ll v$ to simplify it: We assume that $v_0^2$, and thus the term $\tfrac{1}{2}\rho v_0^2$ in Eq. 14-39, is negligible relative to the other terms, and we drop it. Solving the remaining equation for $v$ then yields

$$v = \sqrt{2gh}. \qquad \text{(Answer)}$$

This is the same speed that an object would have when falling a height $h$ from rest.

---

## Sample Problem | 14-9 | Build your skill

Many types of race cars depend on *negative lift* (or *downforce*) to push them down against the track surface so they can take turns quickly without sliding out into the track wall. Part of the negative lift is the *ground force*, which is a force due to the airflow beneath the car. As the race car in Fig. 14-22a moves forward at 27.25 m/s, air is forced to flow over and under the car (Fig. 14-22a). The air forced to flow under the car enters through a vertical cross-sectional area $A_0 = 0.0330$ m$^2$ at the front of the car (Fig. 14-22b) and then flows beneath the car where the vertical cross-sectional area is $A_1 = 0.0310$ m$^2$. Treat this flow as steady flow through a

stationary horizontal pipe that decreases in cross-sectional area from $A_0$ to $A_1$ (Fig. 14-22c).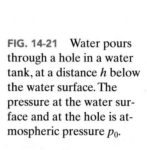

(a) At the moment it passes through $A_0$, the air is at atmospheric pressure $p_0$. At what pressure $p_1$ is the air as it moves through $A_1$?

**KEY IDEAS** (1) Because the flow is steady, we can apply Bernoulli's equation (Eq. 14-28) to the flow. To be consistent with the given subscripts, we write the equation as

$$p_0 + \tfrac{1}{2}\rho v_0^2 + \rho gy = p_1 + \tfrac{1}{2}\rho v_1^2 + \rho gy, \quad \text{(14-40)}$$

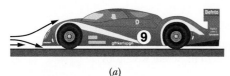

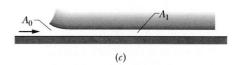

(a)                    (b)                    (c)

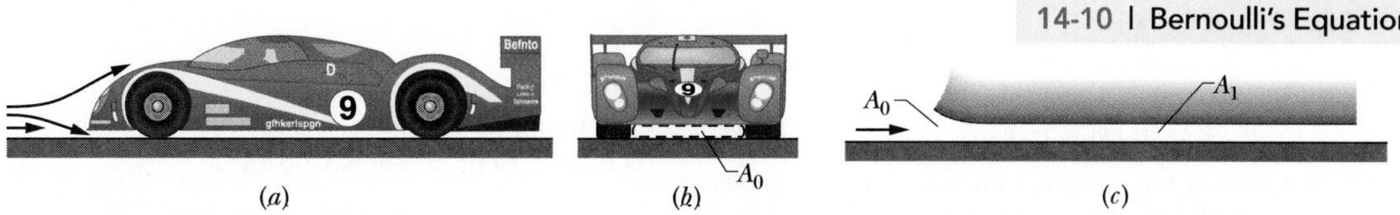

**FIG. 14-22** (a) Air flows above and below a race car. (b) The flow beneath the car enters through vertical cross-sectional area $A_0$. (c) The flow is then constrained as in a pipe that narrows to vertical cross-sectional area $A_1$.

where $\rho$ is the air density and $y$ is the distance above the ground of the flowing air. (2) Because all the air entering through cross-sectional area $A_0$ flows through cross-sectional area $A_1$, the volume flow rate $R_V$ through the two areas must be the same.

*Calculations:* From Eq. 14-24, we can write

$$A_0 v_0 = A_1 v_1,$$

or

$$v_1 = v_0 \frac{A_0}{A_1}. \qquad (14\text{-}41)$$

Substituting Eq. 14-41 into Eq. 14-40 and rearranging give us

$$p_1 = p_0 - \tfrac{1}{2}\rho v_0^2 \left( \frac{A_0^2}{A_1^2} - 1 \right). \qquad (14\text{-}42)$$

The speed of the air as it enters $A_0$ at the front of the car is equal to 27.25 m/s, the speed of the car as it moves forward through the air. Substituting this speed, the air density $\rho = 1.21$ kg/m$^3$, and the values for $A_0$ and $A_1$ into Eq. 14-42, we find

$$p_1 = p_0 - \tfrac{1}{2}(1.21 \text{ kg/m}^3)(27.25 \text{ m/s})^2 \left( \frac{(0.0330 \text{ m}^2)^2}{(0.0310 \text{ m}^2)^2} - 1 \right)$$

$$= p_0 - 59.838 \text{ Pa} \approx p_0 - 59.8 \text{ Pa}. \qquad \text{(Answer)}$$

Thus, the air pressure beneath the car is 59.8 Pa less than atmospheric pressure.

(b) If the horizontal cross-sectional area of the car is $A_h = 4.86$ m$^2$, what is the magnitude of the net vertical force $F_{net,y}$ on the car due to the air pressures above and below the car?

**KEY IDEA** The pressure on a surface is the force per unit area, as given by Eq. 14-4 ($p = F/A$).

*Calculations:* Here we are concerned with the top and bottom surfaces of the car, where we take both surfaces to have area $A_h$. Above the car the air is at atmospheric pressure $p_0$ and presses down on the car with a vertical component

$$F_{y,\text{above}} = -p_0 A_h.$$

Below the car, the air is at pressure $p_1 = p_0 - 59.838$ Pa and presses up on the car with a vertical component

$$F_{y,\text{below}} = (p_0 - 59.838 \text{ Pa})A_h.$$

The net vertical force is then

$$\begin{aligned} F_{\text{net},y} &= F_{y,\text{below}} + F_{y,\text{above}} \\ &= (p_0 - 59.838 \text{ Pa})A_h - p_0 A_h \\ &= -(59.838 \text{ Pa})(4.86 \text{ m}^2) = -291 \text{ N.} \quad \text{(Answer)} \end{aligned}$$

*Danger of drafting:* This net downward force, which is due to the reduced air pressure beneath the car, is the ground effect acting on the car. It is about 30% of the total negative lift that helps hold the car on the track. Without negative lift, a car must greatly slow for turns or else it will slide outward into the track wall. In a race, a driver can reduce the air drag on his car by closely following another car, a procedure known as *drafting*. However, the leading car disrupts the steady flow of air under the trailing car, eliminating the ground effect on the trailing car. If the trailing driver does not anticipate that elimination and slow accordingly, sliding into the track wall may be unavoidable.

## REVIEW & SUMMARY

**Density** The **density** $\rho$ of any material is defined as the material's mass per unit volume:

$$\rho = \frac{\Delta m}{\Delta V}. \qquad (14\text{-}1)$$

Usually, where a material sample is much larger than atomic dimensions, we can write Eq. 14-1 as

$$\rho = \frac{m}{V}. \qquad (14\text{-}2)$$

**Fluid Pressure** A **fluid** is a substance that can flow; it conforms to the boundaries of its container because it cannot withstand shearing stress. It can, however, exert a force perpendicular to its surface. That force is described in terms of **pressure** $p$:

$$p = \frac{\Delta F}{\Delta A}, \qquad (14\text{-}3)$$

in which $\Delta F$ is the force acting on a surface element of area $\Delta A$. If the force is uniform over a flat area, Eq. 14-3 can be written as

$$p = \frac{F}{A}. \qquad (14\text{-}4)$$

The force resulting from fluid pressure at a particular point in a fluid has the same magnitude in all directions. **Gauge pressure** is the difference between the actual pressure (or *absolute pressure*) at a point and the atmospheric pressure.

**Pressure Variation with Height and Depth** Pressure in a fluid at rest varies with vertical position $y$. For $y$ measured positive upward,

$$p_2 = p_1 + \rho g(y_1 - y_2). \qquad (14\text{-}7)$$

The pressure in a fluid is the same for all points at the same level. If $h$ is the *depth* of a fluid sample below some reference level at which the pressure is $p_0$, Eq. 14-7 becomes

$$p = p_0 + \rho gh, \qquad (14\text{-}8)$$

where $p$ is the pressure in the sample.

**Pascal's Principle** A change in the pressure applied to an enclosed fluid is transmitted undiminished to every portion of the fluid and to the walls of the containing vessel.

**Archimedes' Principle** When a body is fully or partially submerged in a fluid, a buoyant force $\vec{F}_b$ from the surrounding fluid acts on the body. The force is directed upward and has a magnitude given by

$$F_b = m_f g, \qquad (14\text{-}16)$$

where $m_f$ is the mass of the fluid that has been displaced by the body (that is, the fluid that has been pushed out of the way by the body).

When a body floats in a fluid, the magnitude $F_b$ of the (upward) buoyant force on the body is equal to the magni-

tude $F_g$ of the (downward) gravitational force on the body. The **apparent weight** of a body on which a buoyant force acts is related to its actual weight by

$$\text{weight}_{\text{app}} = \text{weight} - F_b. \qquad (14\text{-}19)$$

**Flow of Ideal Fluids** An **ideal fluid** is incompressible and lacks viscosity, and its flow is steady and irrotational. A *streamline* is the path followed by an individual fluid particle. A *tube of flow* is a bundle of streamlines. The flow within any tube of flow obeys the **equation of continuity:**

$$R_V = Av = \text{a constant}, \qquad (14\text{-}24)$$

in which $R_V$ is the **volume flow rate**, $A$ is the cross-sectional area of the tube of flow at any point, and $v$ is the speed of the fluid at that point. The **mass flow rate** $R_m$ is

$$R_m = \rho R_V = \rho Av = \text{a constant}. \qquad (14\text{-}25)$$

**Bernoulli's Equation** Applying the principle of conservation of mechanical energy to the flow of an ideal fluid leads to **Bernoulli's equation:**

$$p + \tfrac{1}{2}\rho v^2 + \rho gy = \text{a constant} \qquad (14\text{-}29)$$

along any tube of flow.

## QUESTIONS

**1** *The teapot effect.* Water poured slowly from a teapot spout can double back under the spout for a considerable distance before detaching and falling. (The water layer is held against the underside of the spout by atmospheric pressure.) In Fig. 14-23, in the water layer inside the spout, point $a$ is at the top of the layer and point $b$ is at the bottom of the layer; in the water layer outside the spout, point $c$ is at the top of the layer and point $d$ is at the bottom of the layer. Rank those four points according to the gauge pressure in the water there, most positive first.

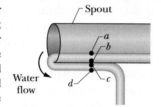

FIG. 14-23 Question 1.

**2** Figure 14-24 shows a tank filled with water. Five horizontal floors and ceilings are indicated; all have the same area and are located at distances $L$, $2L$, or $3L$ below the top of the tank. Rank them according to the force on them due to the water, greatest first.

**3** We fully submerge an irregular 3 kg lump of material in a certain fluid. The fluid that would have been in the space now occupied by the lump has a mass of 2 kg. (a) When we release the lump, does it move upward, move downward, or remain in place? (b) If we next fully submerge the lump in a less dense fluid and again release it, what does it do?

**4** Figure 14-25 shows four situations in which a red liquid

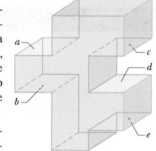

FIG. 14-24 Question 2.

and a gray liquid are in a U-tube. In one situation the liquids cannot be in static equilibrium. (a) Which situation is that? (b) For the other three situations, assume static equilibrium. For each, is the density of the red liquid greater than, less than, or equal to the density of the gray liquid?

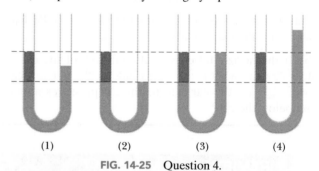

FIG. 14-25 Question 4.

**5** A boat with an anchor on board floats in a swimming pool that is somewhat wider than the boat. Does the pool water level move up, move down, or remain the same if the anchor is (a) dropped into the water or (b) thrown onto the surrounding ground? (c) Does the water level in the pool move upward, move downward, or remain the same if, instead, a cork is dropped from the boat into the water, where it floats?

**6** Figure 14-26 shows three identical open-top containers filled to the brim with water; toy ducks float in two of them. Rank the

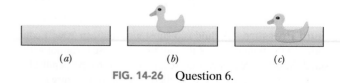

FIG. 14-26 Question 6.

containers and contents according to their weight, greatest first.

**7** Water flows smoothly in a horizontal pipe. Figure 14-27 shows the kinetic energy $K$ of a water element as it moves along an $x$ axis that runs along the pipe. Rank the three lettered sections of the pipe according to the pipe radius, greatest first.

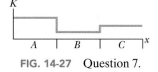

**FIG. 14-27** Question 7.

**8** The gauge pressure $p_g$ versus depth $h$ is plotted in Fig. 14-28 for three liquids. For a rigid plastic bead fully submerged in each liquid, rank the plots according to the magnitude of the buoyant force on the bead, greatest first.

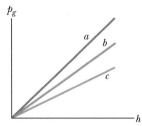

**FIG. 14-28** Question 8.

**9** Figure 14-29 shows four arrangements of pipes through which water flows smoothly toward the right. The radii of the pipe sections are indicated. In which arrangements is the net work done on a unit volume of water moving from the leftmost section to the rightmost section (a) zero, (b) positive, and (c) negative?

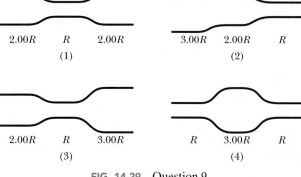

(1)    2.00R   R   2.00R      (2)   3.00R   2.00R   R

(3)    2.00R   R   3.00R      (4)   R   3.00R   R

**FIG. 14-29** Question 9.

**10** A rectangular block is pushed face-down into three liquids, in turn. The apparent weight $W_{app}$ of the block versus depth $h$ in the three liquids is plotted in Fig. 14-30. Rank the liquids according to their weight per unit volume, greatest first.

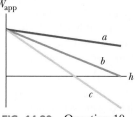

**FIG. 14-30** Question 10.

# PROBLEMS

**GO**   Tutoring problem available (at instructor's discretion) in *WileyPLUS* and WebAssign

**SSM**   Worked-out solution available in Student Solutions Manual    **WWW**   Worked-out solution is at

•–•••   Number of dots indicates level of problem difficulty    **ILW**   Interactive solution is at    http://www.wiley.com/college/halliday

  Additional information available in *The Flying Circus of Physics* and at flyingcircusofphysics.com

## sec. 14-3 Density and Pressure

**•1** Find the pressure increase in the fluid in a syringe when a nurse applies a force of 42 N to the syringe's circular piston, which has a radius of 1.1 cm. **SSM**

**•2** Three liquids that will not mix are poured into a cylindrical container. The volumes and densities of the liquids are 0.50 L, 2.6 g/cm³; 0.25 L, 1.0 g/cm³; and 0.40 L, 0.80 g/cm³. What is the force on the bottom of the container due to these liquids? One liter = 1 L = 1000 cm³. (Ignore the contribution due to the atmosphere.)

**•3** An office window has dimensions 3.4 m by 2.1 m. As a result of the passage of a storm, the outside air pressure drops to 0.96 atm, but inside the pressure is held at 1.0 atm. What net force pushes out on the window? **SSM**

**•4** You inflate the front tires on your car to 28 psi. Later, you measure your blood pressure, obtaining a reading of 120/80, the readings being in mm Hg. In metric countries (which is to say, most of the world), these pressures are customarily reported in kilopascals (kPa). In kilopascals, what are (a) your tire pressure and (b) your blood pressure?

**•5** A fish maintains its depth in fresh water by adjusting the air content of porous bone or air sacs to make its average density the same as that of the water. Suppose that with its air sacs collapsed, a fish has a density of 1.08 g/cm³. To what fraction of its expanded body volume must the fish inflate the air sacs to reduce its density to that of water? **ILW**

**•6** A partially evacuated airtight container has a tight-fitting lid of surface area 77 m² and negligible mass. If the force required to remove the lid is 480 N and the atmospheric pressure is $1.0 \times 10^5$ Pa, what is the internal air pressure?

**••7** In 1654 Otto von Guericke, inventor of the air pump, gave a demonstration before the noblemen of the Holy Roman Empire in which two teams of eight horses could not pull apart two evacuated brass hemispheres. (a) Assuming the

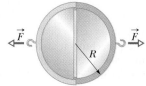

**FIG. 14-31** Problem 7.

hemispheres have (strong) thin walls, so that $R$ in Fig. 14-31 may be considered both the inside and outside radius, show that the force $\vec{F}$ required to pull apart the hemispheres has magnitude $F = \pi R^2 \, \Delta p$, where $\Delta p$ is the difference between the pressures outside and inside the sphere. (b) Taking $R$ as 30 cm, the inside pressure as 0.10 atm, and the outside pressure as 1.00 atm, find the force magnitude the teams of horses would have had to exert to pull apart the hemispheres. (c) Explain why one team of horses could have proved the point just as well if the hemispheres were attached to a sturdy wall.

## sec. 14-4 Fluids at Rest

**•8** Calculate the hydrostatic difference in blood pressure between the brain and the foot in a person of height 1.83 m. The density of blood is $1.06 \times 10^3$ kg/m³.

•9 At a depth of 10.9 km, the Challenger Deep in the Marianas Trench of the Pacific Ocean is the deepest site in any ocean. Yet, in 1960, Donald Walsh and Jacques Piccard reached the Challenger Deep in the bathyscaph *Trieste*. Assuming that seawater has a uniform density of 1024 kg/m³, approximate the hydrostatic pressure (in atmospheres) that the *Trieste* had to withstand.

•10 The maximum depth $d_{max}$ that a diver can snorkel is set by the density of the water and the fact that human lungs can function against a maximum pressure difference (between inside and outside the chest cavity) of 0.050 atm. What is the difference in $d_{max}$ for fresh water and the water of the Dead Sea (the saltiest natural water in the world, with a density of $1.5 \times 10^3$ kg/m³)?

•11 Crew members attempt to escape from a damaged submarine 100 m below the surface. What force must be applied to a pop-out hatch, which is 1.2 m by 0.60 m, to push it out at that depth? Assume that the density of the ocean water is 1024 kg/m³ and the internal air pressure is at 1.00 atm. **SSM**

•12 The plastic tube in Fig. 14-32 has a cross-sectional area of 5.00 cm². The tube is filled with water until the short arm (of length $d = 0.800$ m) is full. Then the short arm is sealed and more water is gradually poured into the long arm. If the seal will pop off when the force on it exceeds 9.80 N, what total height of water in the long arm will put the seal on the verge of popping?

**FIG. 14-32**
Problems 12 and 75.

•13 What gauge pressure must a machine produce in order to suck mud of density 1800 kg/m³ up a tube by a height of 1.5 m?

•14 *The bends during flight.* Anyone who scuba dives is advised not to fly within the next 24 h because the air mixture for diving can introduce nitrogen to the bloodstream. Without allowing the nitrogen to come out of solution slowly, any sudden air-pressure reduction (such as during airplane ascent) can result in the nitrogen forming bubbles in the blood, creating the *bends*, which can be painful and even fatal. Military special operation forces are especially at risk. What is the change in pressure on such a special-op soldier who must scuba dive at a depth of 20 m in seawater one day and parachute at an altitude of 7.6 km the next day? Assume that the average air density within the altitude range is 0.87 kg/m³.

•15 *Giraffe bending to drink.* In a giraffe with its head 2.0 m above its heart, and its heart 2.0 m above its feet, the (hydrostatic) gauge pressure in the blood at its heart is 250 torr. Assume that the giraffe stands upright and the blood density is $1.06 \times 10^3$ kg/m³. In torr (or mm Hg), find the (gauge) blood pressure (a) at the brain (the pressure is enough to perfuse the brain with blood, to keep the giraffe from fainting) and (b) at the feet (the pressure must be countered by tight-fitting skin acting like a pressure stocking). (c) If the giraffe were to lower its head to drink from a pond without splaying its legs and moving slowly, what would be the increase in the blood pressure in the brain? (Such action would probably be lethal.)

•16 In Fig. 14-33, an open tube of length $L = 1.8$ m and cross-sectional area $A = 4.6$ cm² is fixed to the top of a cylindrical barrel of diameter $D = 1.2$ m and height $H = 1.8$ m.

The barrel and tube are filled with water (to the top of the tube). Calculate the ratio of the hydrostatic force on the bottom of the barrel to the gravitational force on the water contained in the barrel. Why is that ratio not equal to 1.0? (You need not consider the atmospheric pressure.)

**FIG. 14-33**
Problem 16.

•17 *Blood pressure in Argentinosaurus.* (a) If this long-necked, gigantic sauropod had a head height of 21 m and a heart height of 9.0 m, what (hydrostatic) gauge pressure in its blood was required at the heart such that the blood pressure at the brain was 80 torr (just enough to perfuse the brain with blood)? Assume the blood had a density of $1.06 \times 10^3$ kg/m³. (b) What was the blood pressure (in torr or mm Hg) at the feet?

•18 *Snorkeling by humans and elephants.* When a person snorkels, the lungs are connected directly to the atmosphere through the snorkel tube and thus are at atmospheric pressure. In atmospheres, what is the difference $\Delta p$ between this internal air pressure and

**FIG. 14-34** Problem 18.

the water pressure against the body if the length of the snorkel tube is (a) 20 cm (standard situation) and (b) 4.0 m (probably lethal situation)? In the latter, the pressure difference causes blood vessels on the walls of the lungs to rupture, releasing blood into the lungs. As depicted in Fig. 14-34, an elephant can safely snorkel through its trunk while swimming with its lungs 4.0 m below the water surface because the membrane around its lungs contains connective tissue that holds and protects the blood vessels, preventing rupturing.

••19 Two identical cylindrical vessels with their bases at the same level each contain a liquid of density $1.30 \times 10^3$ kg/m³. The area of each base is 4.00 cm², but in one vessel the liquid height is 0.854 m and in the other it is 1.560 m. Find the work done by the gravitational force in equalizing the levels when the two vessels are connected. **SSM**

••20 *g-LOC in dogfights.* When a pilot takes a tight turn at high speed in a modern fighter airplane, the blood pressure at the brain level decreases, blood no longer perfuses the brain, and the blood in the brain drains. If the heart maintains the (hydrostatic) gauge pressure in the aorta at 120 torr (or mm Hg) when the pilot undergoes a horizontal centripetal acceleration of 4g, what is the blood pressure (in torr) at the brain, 30 cm radially inward from the heart? The perfusion in the brain is small enough that the vision switches to black and white and narrows to "tunnel vision" and the pilot can undergo g-LOC ("g-induced loss of consciousness"). Blood density is $1.06 \times 10^3$ kg/m³.

••21 In analyzing certain geological features, it is often appropriate to assume that the pressure at some horizontal *level of compensation,* deep inside Earth, is the same over a large region and is equal to the pressure due to the gravitational force on the overlying material. Thus, the pressure on the level of compensation is given by the fluid pressure formula. This model requires, for one thing, that mountains have *roots* of continental rock extending into the denser mantle (Fig. 14-35). Consider a mountain of height $H$ = 6.0 km on a continent of thickness $T$ = 32 km. The continental rock has a density of 2.9 g/cm³, and beneath this rock the mantle has a density of 3.3 g/cm³. Calculate the depth $D$ of the root. (*Hint:* Set the pressure at points $a$ and $b$ equal; the depth $y$ of the level of compensation will cancel out.)

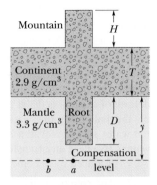

**FIG. 14-35** Problem 21.

**••22** The L-shaped tank shown in Fig. 14-36 is filled with water and is open at the top. If $d$ = 5.0 m, what is the force due to the water (a) on face $A$ and (b) on face $B$?

**••23** A large aquarium of height 5.00 m is filled with fresh water to a depth of 2.00 m. One wall of the aquarium consists of thick plastic 8.00 m wide. By how much does the total force on that wall increase if the aquarium is next filled to a depth of 4.00 m? GO

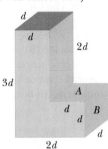

**FIG. 14-36** Problem 22.

**•••24** In Fig. 14-37, water stands at depth $D$ = 35.0 m behind the vertical upstream face of a dam of width $W$ = 314 m. Find (a) the net horizontal force on the dam from the gauge pressure of the water and (b) the net torque due to that force about a line through $O$ parallel to the width of the dam. (c) Find the moment arm of this torque. GO

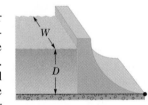

**FIG. 14-37** Problem 24.

### sec. 14-5 Measuring Pressure

**•25** In one observation, the column in a mercury barometer (as is shown in Fig. 14-5a) has a measured height $h$ of 740.35 mm. The temperature is −5.0°C, at which temperature the density of mercury $\rho$ is 1.3608 × 10⁴ kg/m³. The free-fall acceleration $g$ at the site of the barometer is 9.7835 m/s². What is the atmospheric pressure at that site in pascals and in torr (which is the common unit for barometer readings)?

**•26** To suck lemonade of density 1000 kg/m³ up a straw to a maximum height of 4.0 cm, what minimum gauge pressure (in atmospheres) must you produce in your lungs?

**••27** What would be the height of the atmosphere if the air density (a) were uniform and (b) decreased linearly to zero with height? Assume that at sea level the air pressure is 1.0 atm and the air density is 1.3 kg/m³. SSM

### sec. 14-6 Pascal's Principle

**•28** A piston of cross-sectional area $a$ is used in a hydraulic press to exert a small force of magnitude $f$ on the enclosed liquid. A connecting pipe leads to a larger piston of cross-sectional area $A$ (Fig. 14-38). (a) What force magnitude $F$ will

the larger piston sustain without moving? (b) If the piston diameters are 3.80 cm and 53.0 cm, what force magnitude on the small piston will balance a 20.0 kN force on the large piston?

**••29** In Fig. 14-39, a spring of spring constant 3.00 × 10⁴ N/m is between a rigid beam and the output piston of a hydraulic lever. An empty container with negligible mass sits on the input piston. The input piston has area $A_i$, and the output piston has area 18.0$A_i$. Initially the spring is at its rest length. How many kilograms of sand must be (slowly) poured into the container to compress the spring by 5.00 cm?

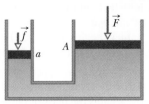

**FIG. 14-38** Problem 28.

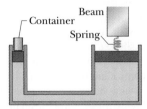

**FIG. 14-39** Problem 29.

### sec. 14-7 Archimedes' Principle

**•30** In Fig. 14-40, a cube of edge length $L$ = 0.600 m and mass 450 kg is suspended by a rope in an open tank of liquid of density 1030 kg/m³. Find (a) the magnitude of the total downward force on the top of the cube from the liquid and the atmosphere, assuming atmospheric pressure is 1.00 atm, (b) the magnitude of the total upward force on the bottom of the cube, and (c) the tension in the rope. (d) Calculate the magnitude of the buoyant force on the cube using Archimedes' principle. What relation exists among all these quantities?

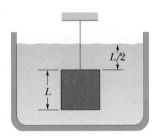

**FIG. 14-40** Problem 30.

**•31** An iron anchor of density 7870 kg/m³ appears 200 N lighter in water than in air. (a) What is the volume of the anchor? (b) How much does it weigh in air? SSM

**•32** A boat floating in fresh water displaces water weighing 35.6 kN. (a) What is the weight of the water this boat displaces when floating in salt water of density 1.10 × 10³ kg/m³? (b) What is the difference between the volume of fresh water displaced and the volume of salt water displaced?

**•33** Three children, each of weight 356 N, make a log raft by lashing together logs of diameter 0.30 m and length 1.80 m. How many logs will be needed to keep them afloat in fresh water? Take the density of the logs to be 800 kg/m³.

**•34** A 5.00 kg object is released from rest while fully submerged in a liquid. The liquid displaced by the submerged object has a mass of 3.00 kg. How far and in what direction does the object move in 0.200 s, assuming that it moves freely and that the drag force on it from the liquid is negligible?

**•35** A block of wood floats in fresh water with two-thirds of its volume $V$ submerged and in oil with 0.90$V$ submerged. Find the density of (a) the wood and (b) the oil. SSM

**••36** A flotation device is in the shape of a right cylinder, with a height of 0.500 m and a face area of 4.00 m² on top and bottom, and its density is 0.400 times that of fresh water. It is initially held fully submerged in fresh water, with its top face at the water surface. Then it is allowed to ascend gradually until it begins to float. How much work does the buoyant force do on the device during the ascent?

**••37** A hollow sphere of inner radius 8.0 cm and outer radius 9.0 cm floats half-submerged in a liquid of density 800 kg/m³. (a) What is the mass of the sphere? (b) Calculate the density of the material of which the sphere is made. **SSM WWW**

**••38** *Lurking alligators.* An alligator waits for prey by floating with only the top of its head exposed, so that the prey cannot easily see it. One way it can adjust the extent of sinking is by

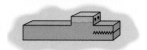
FIG. 14-41 Problem 38.

controlling the size of its lungs. Another way may be by swallowing stones (*gastrolithes*) that then reside in the stomach. Figure 14-41 shows a highly simplified model (a "rhombohedron gater") of mass 130 kg that roams with its head partially exposed. The top head surface has area 0.20 m². If the alligator were to swallow stones with a total mass of 1.0% of its body mass (a typical amount), how far would it sink?

**••39** What fraction of the volume of an iceberg (density 917 kg/m³) would be visible if the iceberg floats (a) in the ocean (salt water, density 1024 kg/m³) and (b) in a river (fresh water, density 1000 kg/m³)? (When salt water freezes to form ice, the salt is excluded. So, an iceberg could provide fresh water to a community.)

**••40** A small solid ball is released from rest while fully submerged in a liquid and then its kinetic energy is measured when it has moved 4.0 cm in the liquid. Figure 14-42 gives the results after many liquids are used: The kinetic energy $K$ is plotted versus the liquid density $\rho_{\text{liq}}$, and $K_s = 1.60$ J sets the scale on the vertical axis. What are (a) the density and (b) the volume of the ball? **GO**

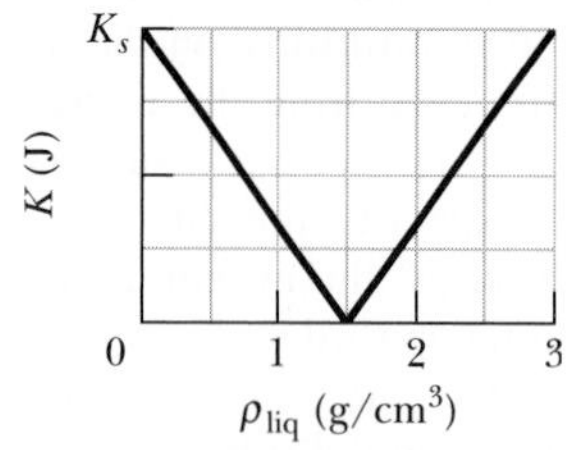

FIG. 14-42 Problem 40.

**••41** A hollow spherical iron shell floats almost completely submerged in water. The outer diameter is 60.0 cm, and the density of iron is 7.87 g/cm³. Find the inner diameter. **ILW**

**••42** In Fig. 14-43a, a rectangular block is gradually pushed face-down into a liquid. The block has height $d$; on the bottom and top the face area is $A = 5.67$ cm². Figure 14-43b gives the apparent weight $W_{\text{app}}$ of the block as a function of the depth $h$ of its lower face. The scale on the vertical axis is set by $W_s = 0.20$ N. What is the density of the liquid?

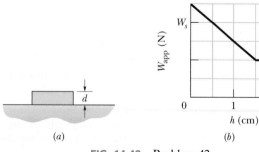

FIG. 14-43 Problem 42.

**••43** An iron casting containing a number of cavities weighs 6000 N in air and 4000 N in water. What is the total volume of all the cavities in the casting? The density of iron (that is, a sample with no cavities) is 7.87 g/cm³.

**••44** Suppose that you release a small ball from rest at a depth of 0.600 m below the surface in a pool of water. If the density of the ball is 0.300 that of water and if the drag force on the ball from the water is negligible, how high above the water surface will the ball shoot as it emerges from the water? (Neglect any transfer of energy to the splashing and waves produced by the emerging ball.)

**••45** The volume of air space in the passenger compartment of an 1800 kg car is 5.00 m³. The volume of the motor and front wheels is 0.750 m³, and the volume of the rear wheels, gas tank, and trunk is 0.800 m³; water cannot enter these two regions. The car rolls into a lake. (a) At first, no water enters the passenger compartment. How much of the car, in cubic meters, is below the water surface with the car floating (Fig. 14-44)? (b) As water slowly enters, the car sinks. How many cubic meters of water are in the car as it disappears below the water surface? (The car, with a heavy load in the trunk, remains horizontal.)

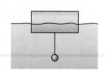

FIG. 14-44 Problem 45.

**••46** A block of wood has a mass of 3.67 kg and a density of 600 kg/m³. It is to be loaded with lead ($1.14 \times 10^4$ kg/m³) so that it will float in water with 0.900 of its volume submerged. What mass of lead is needed if the lead is attached to (a) the top of the wood and (b) the bottom of the wood?

**••47** When researchers find a reasonably complete fossil of a dinosaur, they can determine the mass and weight of the living dinosaur with a scale model sculpted from plastic and based on the dimensions of the fossil bones. The scale of the model is 1/20; that is, lengths are 1/20 actual length, areas are (1/20)² ac-

FIG. 14-45 Problem 47.

tual areas, and volumes are (1/20)³ actual volumes. First, the model is suspended from one arm of a balance and weights are added to the other arm until equilibrium is reached. Then the model is fully submerged in water and enough weights are removed from the second arm to reestablish equilibrium (Fig. 14-45). For a model of a particular *T. rex* fossil, 637.76 g had to be removed to reestablish equilibrium. What was the volume of (a) the model and (b) the actual *T. rex*? (c) If the density of *T. rex* was approximately the density of water, what was its mass?

**•••48** Figure 14-46 shows an iron ball suspended by thread of negligible mass from an upright cylinder that floats partially submerged in water. The cylinder

FIG. 14-46
Problem 48.

has a height of 6.00 cm, a face area of 12.0 cm² on the top and bottom, and a density of 0.30 g/cm³, and 2.00 cm of its height is above the water surface. What is the radius of the iron ball? GO

### sec. 14-9 The Equation of Continuity

•49 A garden hose with an internal diameter of 1.9 cm is connected to a (stationary) lawn sprinkler that consists merely of a container with 24 holes, each 0.13 cm in diameter. If the water in the hose has a speed of 0.91 m/s, at what speed does it leave the sprinkler holes? SSM

•50 Two streams merge to form a river. One stream has a width of 8.2 m, depth of 3.4 m, and current speed of 2.3 m/s. The other stream is 6.8 m wide and 3.2 m deep, and flows at 2.6 m/s. If the river has width 10.5 m and speed 2.9 m/s, what is its depth?

•51 *Canal effect.* Figure 14-47 shows an anchored barge that extends across a canal by distance $d = 30$ m and into the water by distance $b = 12$ m. The canal has a width $D = 55$ m, a water depth $H = 14$ m, and a uniform water-flow speed $v_i = 1.5$ m/s. Assume that the flow around the barge is uniform. As the water passes the bow, the water level undergoes a dramatic dip known as the canal effect. If the dip has depth $h = 0.80$ m, what is the water speed alongside the boat through the vertical cross sections at (a) point $a$ and (b) point $b$? The erosion due to the speed increase is a common concern to hydraulic engineers.

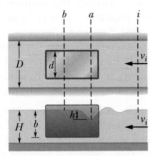

FIG. 14-47 Problem 51.

•52 Figure 14-48 shows two sections of an old pipe system that runs through a hill, with distances $d_A = d_B = 30$ m and $D = 110$ m. On each

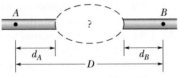

FIG. 14-48 Problem 52.

side of the hill, the pipe radius is 2.00 cm. However, the radius of the pipe inside the hill is no longer known. To determine it, hydraulic engineers first establish that water flows through the left and right sections at 2.50 m/s. Then they release a dye in the water at point $A$ and find that it takes 88.8 s to reach point $B$. What is the average radius of the pipe within the hill?

••53 Water is pumped steadily out of a flooded basement at a speed of 5.0 m/s through a uniform hose of radius 1.0 cm. The hose passes out through a window 3.0 m above the waterline. What is the power of the pump? SSM

••54 The water flowing through a 1.9 cm (inside diameter) pipe flows out through three 1.3 cm pipes. (a) If the flow rates in the three smaller pipes are 26, 19, and 11 L/min, what is the flow rate in the 1.9 cm pipe? (b) What is the ratio of the speed in the 1.9 cm pipe to that in the pipe carrying 26 L/min?

### sec. 14-10 Bernoulli's Equation

•55 Water is moving with a speed of 5.0 m/s through a pipe with a cross-sectional area of 4.0 cm². The water gradually descends 10 m as the pipe cross-sectional area increases to 8.0 cm². (a) What is the speed at the lower level? (b) If the pressure at the upper level is $1.5 \times 10^5$ Pa, what is the pressure at the lower level? SSM

•56 The intake in Fig. 14-49 has cross-sectional area of 0.74 m² and water flow at 0.40 m/s. At the outlet, distance $D = 180$ m below the intake, the cross-sectional area is smaller than at the intake and the water flows out at 9.5 m/s. What is the pressure difference between inlet and outlet?

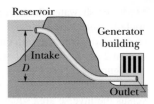

FIG. 14-49 Problem 56.

•57 A water pipe having a 2.5 cm inside diameter carries water into the basement of a house at a speed of 0.90 m/s and a pressure of 170 kPa. If the pipe tapers to 1.2 cm and rises to the second floor 7.6 m above the input point, what are the (a) speed and (b) water pressure at the second floor? ILW

•58 Models of torpedoes are sometimes tested in a horizontal pipe of flowing water, much as a wind tunnel is used to test model airplanes. Consider a circular pipe of internal diameter 25.0 cm and a torpedo model aligned along the long axis of the pipe. The model has a 5.00 cm diameter and is to be tested with water flowing past it at 2.50 m/s. (a) With what speed must the water flow in the part of the pipe that is unconstricted by the model? (b) What will the pressure difference be between the constricted and unconstricted parts of the pipe?

•59 A cylindrical tank with a large diameter is filled with water to a depth $D = 0.30$ m. A hole of cross-sectional area $A = 6.5$ cm² in the bottom of the tank allows water to drain out. (a) What is the rate at which water flows out, in cubic meters per second? (b) At what distance below the bottom of the tank is the cross-sectional area of the stream equal to one-half the area of the hole? SSM

•60 Suppose that two tanks, 1 and 2, each with a large opening at the top, contain different liquids. A small hole is made in the side of each tank at the same depth $h$ below the liquid surface, but the hole in tank 1 has half the cross-sectional area of the hole in tank 2. (a) What is the ratio $\rho_1/\rho_2$ of the densities of the liquids if the mass flow rate is the same for the two holes? (b) What is the ratio $R_{V1}/R_{V2}$ of the volume flow rates from the two tanks? (c) At one instant, the liquid in tank 1 is 12.0 cm above the hole. If the tanks are to have *equal* volume flow rates, what height above the hole must the liquid in tank 2 be just then?

•61 How much work is done by pressure in forcing 1.4 m³ of water through a pipe having an internal diameter of 13 mm if the difference in pressure at the two ends of the pipe is 1.0 atm?

••62 In Fig. 14-50, water flows through a horizontal pipe and then out into the atmosphere at a speed $v_1 = 15$ m/s. The diameters of the left and right sections of the pipe are 5.0 cm and 3.0 cm. (a) What volume of water flows into the atmosphere during a 10 min period? In the left section of the pipe, what are (b) the speed $v_2$ and (c) the gauge pressure? GO

FIG. 14-50 Problem 62.

••63 In Fig. 14-51, the fresh water behind a reservoir dam has depth $D = 15$ m. A horizontal pipe 4.0 cm in diameter passes through the dam at depth $d = 6.0$ m. A plug secures the pipe opening. (a) Find the magnitude of the frictional force between plug and pipe wall. (b) The plug is removed. What water volume exits the pipe in 3.0 h? ILW

••**64** Fresh water flows horizontally from pipe section 1 of cross-sectional area $A_1$ into pipe section 2 of cross-sectional area $A_2$. Figure 14-52 gives a plot of the pressure difference $p_2 - p_1$ versus the inverse area squared $A_1^{-2}$ that would be expected for a volume flow rate of a certain value if the water flow were laminar under all circumstances. The scale on the vertical axis is set by $\Delta p_s = 300$ kN/m². For the conditions of the figure, what are the values of (a) $A_2$ and (b) the volume flow rate?

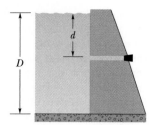

**FIG. 14-51** Problem 63.

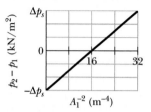

**FIG. 14-52** Problem 64.

••**65** Figure 14-53 shows a stream of water flowing through a hole at depth $h = 10$ cm in a tank holding water to height $H = 40$ cm. (a) At what distance $x$ does the stream strike the floor? (b) At what depth should a second hole be made to give the same value of $x$? (c) At what depth should a hole be made to maximize $x$?

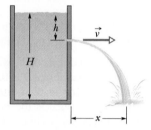

**FIG. 14-53** Problem 65.

••**66** In Fig. 14-54, water flows steadily from the left pipe section (radius $r_1 = 2.00R$), through the middle section (radius $R$), and into the right section (radius $r_3 = 3.00R$). The speed of the water in the middle section is 0.500 m/s. What is the net work done on 0.400 m³ of the water as it moves from the left section to the right section? GO

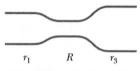

**FIG. 14-54** Problem 66.

••**67** A *venturi meter* is used to measure the flow speed of a fluid in a pipe. The meter is connected between two sections of the pipe (Fig. 14-55); the cross-sectional area $A$ of the entrance and exit of the meter matches the pipe's cross-sectional area. Between the entrance and exit, the fluid flows from the pipe with speed $V$ and then through a narrow "throat" of cross-sectional area $a$ with speed $v$. A manometer connects the wider portion of the meter to the narrower portion. The change in the fluid's speed is accompanied by a change $\Delta p$ in the fluid's pressure, which causes a height dif-

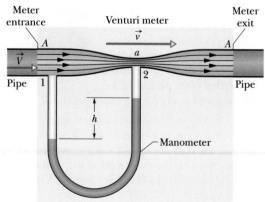

**FIG. 14-55** Problems 67 and 68.

ference $h$ of the liquid in the two arms of the manometer. (Here $\Delta p$ means pressure in the throat minus pressure in the pipe.) (a) By applying Bernoulli's equation and the equation of continuity to points 1 and 2 in Fig. 14-55, show that

$$V = \sqrt{\frac{2a^2 \, \Delta p}{\rho(a^2 - A^2)}},$$

where $\rho$ is the density of the fluid. (b) Suppose that the fluid is fresh water, that the cross-sectional areas are 64 cm² in the pipe and 32 cm² in the throat, and that the pressure is 55 kPa in the pipe and 41 kPa in the throat. What is the rate of water flow in cubic meters per second? SSM WWW

••**68** Consider the venturi tube of Problem 67 and Fig. 14-55 without the manometer. Let $A$ equal $5a$. Suppose the pressure $p_1$ at $A$ is 2.0 atm. Compute the values of (a) the speed $V$ at $A$ and (b) the speed $v$ at $a$ that make the pressure $p_2$ at $a$ equal to zero. (c) Compute the corresponding volume flow rate if the diameter at $A$ is 5.0 cm. The phenomenon that occurs at $a$ when $p_2$ falls to nearly zero is known as cavitation. The water vaporizes into small bubbles.

••**69** A liquid of density 900 kg/m³ flows through a horizontal pipe that has a cross-sectional area of $1.90 \times 10^{-2}$ m² in region $A$ and a cross-sectional area of $9.50 \times 10^{-2}$ m² in region $B$. The pressure difference between the two regions is $7.20 \times 10^3$ Pa. What are (a) the volume flow rate and (b) the mass flow rate?

••**70** A pitot tube (Fig. 14-56) is used to determine the airspeed of an airplane. It consists of an outer tube with a number of small holes $B$ (four are shown) that allow air into the tube; that tube is connected to one arm of a U-tube. The other arm of the U-tube is connected to hole $A$ at the front end of the device, which points in the direction the plane is headed. At $A$ the air becomes stagnant so that $v_A = 0$. At $B$, however, the speed of the air presumably equals the airspeed $v$ of the plane. (a) Use Bernoulli's equation to show that

$$v = \sqrt{\frac{2\rho g h}{\rho_{air}}},$$

where $\rho$ is the density of the liquid in the U-tube and $h$ is the difference in the liquid levels in that tube. (b) Suppose that the tube contains alcohol and the level difference $h$ is 26.0 cm. What is the plane's speed relative to the air? The density of the air is 1.03 kg/m³ and that of alcohol is 810 kg/m³.

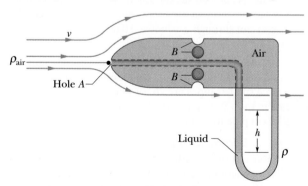

**FIG. 14-56** Problems 70 and 71.

••**71** A pitot tube (see Problem 70) on a high-altitude aircraft measures a differential pressure of 180 Pa. What is the aircraft's airspeed if the density of the air is 0.031 kg/m³?

**•••72** A very simplified schematic of the rain drainage system for a home is shown in Fig. 14-57. Rain falling on the slanted roof runs off into gutters around the roof edge; it then drains through downspouts (only one is shown) into a main drainage pipe $M$ below the basement,

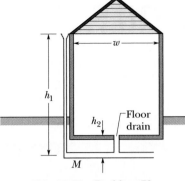

**FIG. 14-57** Problem 72.

which carries the water to an even larger pipe below the street. In Fig. 14-57, a floor drain in the basement is also connected to drainage pipe $M$. Suppose the following apply:

1. the downspouts have height $h_1 = 11$ m,
2. the floor drain has height $h_2 = 1.2$ m,
3. pipe $M$ has radius 3.0 cm,
4. the house has side width $w = 30$ m and front length $L = 60$ m,
5. all the water striking the roof goes through pipe $M$,
6. the initial speed of the water in a downspout is negligible,
7. the wind speed is negligible (the rain falls vertically).

At what rainfall rate, in centimeters per hour, will water from pipe $M$ reach the height of the floor drain and threaten to flood the basement?

**Additional Problems**

**73** A glass ball of radius 2.00 cm sits at the bottom of a container of milk that has a density of 1.03 g/cm³. The normal force on the ball from the container's lower surface has magnitude $9.48 \times 10^{-2}$ N. What is the mass of the ball?

**74** When you cough, you expel air at high speed through the trachea and upper bronchi so that the air will remove excess mucus lining the pathway. You produce the high speed by this procedure: You breathe in a large amount of air, trap it by closing the glottis (the narrow opening in the larynx), increase the air pressure by contracting the lungs, partially collapse the trachea and upper bronchi to narrow the pathway, and then expel the air through the pathway by suddenly reopening the glottis. Assume that during the expulsion the volume flow rate is $7.0 \times 10^{-3}$ m³/s. What multiple of the speed of sound $v_s$ ($= 343$ m/s) is the airspeed through the trachea if the trachea diameter (a) remains its normal value of 14 mm and (b) contracts to 5.2 mm?

**75** Figure 14-32 shows a modified U-tube: the right arm is shorter than the left arm. The open end of the right arm is height $d = 10.0$ cm above the laboratory bench. The radius throughout the tube is 1.50 cm. Water is gradually poured into the open end of the left arm until the water begins to flow out the open end of the right arm. Then a liquid of density 0.80 g/cm³ is gradually added to the left arm until its height in that arm is 8.0 cm (it does not mix with the water). How much water flows out of the right arm? **SSM**

**76** Caught in an avalanche, a skier is fully submerged in flowing snow of density 96 kg/m³. Assume that the average density of the skier, clothing, and skiing equipment is 1020 kg/m³. What percentage of the gravitational force on the skier is offset by the buoyant force from the snow?

**77** Figure 14-58 shows a *siphon*, which is a device for removing liquid from a container. Tube $ABC$ must initially be filled, but once this has been done, liquid will flow through the tube until the liquid surface in the container is level with the tube opening at $A$. The liquid has density 1000 kg/m³ and negligible viscosity. The distances shown are $h_1 = 25$ cm, $d = 12$ cm, and $h_2 = 40$ cm. (a) With what speed does the liquid emerge from the tube at $C$? (b) If the atmospheric pressure is $1.0 \times 10^5$ Pa, what is the pressure in the liquid at the topmost point $B$? (c) Theoretically, what is the greatest possible height $h_1$ that a siphon can lift water?

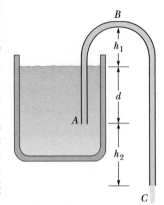

**FIG. 14-58** Problem 77.

**78** Suppose that your body has a uniform density of 0.95 times that of water. (a) If you float in a swimming pool, what fraction of your body's volume is above the water surface?

Quicksand is a fluid produced when water is forced up into sand, moving the sand grains away from one another so they are no longer locked together by friction. Pools of quicksand can form when water drains underground from hills into valleys where there are sand pockets. (b) If you float in a deep pool of quicksand that has a density 1.6 times that of water, what fraction of your body's volume is above the quicksand surface? (c) In particular, are you submerged enough to be unable to breathe?

**79** If a bubble in sparkling water accelerates upward at the rate of 0.225 m/s² and has a radius of 0.500 mm, what is its mass? Assume that the drag force on the bubble is negligible.

**80** What is the acceleration of a rising hot-air balloon if the ratio of the air density outside the balloon to that inside is 1.39? Neglect the mass of the balloon fabric and the basket.

**81** A tin can has a total volume of 1200 cm³ and a mass of 130 g. How many grams of lead shot of density 11.4 g/cm³ could it carry without sinking in water?

**82** A simple open U-tube contains mercury. When 11.2 cm of water is poured into the right arm of the tube, how high above its initial level does the mercury rise in the left arm?

**83** An object hangs from a spring balance. The balance registers 30 N in air, 20 N when this object is immersed in water, and 24 N when the object is immersed in another liquid of unknown density. What is the density of that other liquid?

**84** In an experiment, a rectangular block with height $h$ is allowed to float in four separate liquids. In the first liquid, which is water, it floats fully submerged. In liquids $A$, $B$, and $C$, it floats with heights $h/2$, $2h/3$, and $h/4$ above the liquid surface, respectively. What are the *relative densities* (the densities relative to that of water) of (a) $A$, (b) $B$, and (c) $C$?

**85** About one-third of the body of a person floating in the Dead Sea will be above the waterline. Assuming that the human body density is 0.98 g/cm³, find the density of the water in the Dead Sea. (Why is it so much greater than 1.0 g/cm³?)

# Oscillations

If a tall building sways slowly in a wind, the occupants may not even notice the motion, but if the swaying repeats more than 10 times per second, it becomes annoying and may even cause motion sickness. One reason is that when a person is standing, the head tends to sway even more than the feet, setting off motion sensors in the balancing region of the inner ear. Various mechanisms are employed to decrease a building's sway. For example, the large ball ($5.4 \times 10^5$ kg) seen in this photograph hangs on the 92nd floor of one of the world's tallest buildings.

## How can the ball counter the building's sway?

The answer is in this chapter.

REUTERS/Richard Chung/Landov LLC

## 15-1 | WHAT IS PHYSICS?

Our world is filled with oscillations in which objects move back and forth repeatedly. Many oscillations are merely amusing or annoying, but many others are financially important or dangerous. Here are a few examples: When a bat hits a baseball, the bat may oscillate enough to sting the batter's hands or even to break apart. When wind blows past a power line, the line may oscillate ("gallop" in electrical engineering terms) so severely that it rips apart, shutting off the power supply to a community. When an airplane is in flight, the turbulence of the air flowing past the wings makes them oscillate, eventually leading to metal fatigue and even failure. When a train travels around a curve, its wheels oscillate horizontally ("hunt" in mechanical engineering terms) as they are forced to turn in new directions (you can hear the oscillations).

When an earthquake occurs near a city, buildings may be set oscillating so severely that they are shaken apart. When an arrow is shot from a bow, the feathers at the end of the arrow manage to snake around the bow staff without hitting it because the arrow oscillates. When a coin drops into a metal collection plate, the coin oscillates with such a familiar ring that the coin's denomination can be determined from the sound. When a rodeo cowboy rides a bull, the cowboy oscillates wildly as the bull jumps and turns (at least the cowboy hopes to be oscillating).

The study and control of oscillations are two of the primary goals of both physics and engineering. In this chapter we discuss a basic type of oscillation called *simple harmonic motion.*

## 15-2 | Simple Harmonic Motion

Figure 15-1a shows a sequence of "snapshots" of a simple oscillating system, a particle moving repeatedly back and forth about the origin of an $x$ axis. In this section we simply describe the motion. Later, we shall discuss how to attain such motion.

One important property of oscillatory motion is its **frequency,** or number of oscillations that are completed each second. The symbol for frequency is $f$, and its SI unit is the **hertz** (abbreviated Hz), where

$$1 \text{ hertz} = 1 \text{ Hz} = 1 \text{ oscillation per second} = 1 \text{ s}^{-1}. \qquad (15\text{-}1)$$

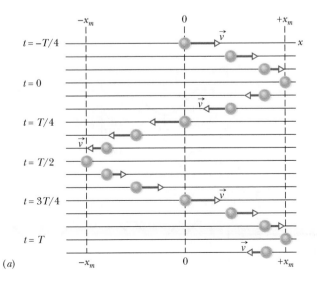

FIG. 15-1 (a) A sequence of "snapshots" (taken at equal time intervals) showing the position of a particle as it oscillates back and forth about the origin of an $x$ axis, between the limits $+x_m$ and $-x_m$. The vector arrows are scaled to indicate the speed of the particle. The speed is maximum when the particle is at the origin and zero when it is at $\pm x_m$. If the time $t$ is chosen to be zero when the particle is at $+x_m$, then the particle returns to $+x_m$ at $t = T$, where $T$ is the period of the motion. The motion is then repeated. (b) A graph of $x$ as a function of time for the motion of (a).

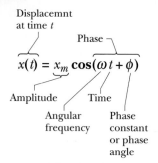

$$\underbrace{x(t)}_{} = \underbrace{x_m}_{} \underbrace{\cos(}_{}\overbrace{\omega}^{} \overbrace{t}^{} + \overbrace{\phi}^{})$$

FIG. 15-2 A handy reference to the quantities in Eq. 15-3 for simple harmonic motion.

Related to the frequency is the **period** $T$ of the motion, which is the time for one complete oscillation (or **cycle**); that is,

$$T = \frac{1}{f}. \qquad (15\text{-}2)$$

Any motion that repeats itself at regular intervals is called **periodic motion** or **harmonic motion.** We are interested here in motion that repeats itself in a particular way—namely, like that in Fig. 15-1a. For such motion the displacement $x$ of the particle from the origin is given as a function of time by

$$x(t) = x_m \cos(\omega t + \phi) \quad \text{(displacement)}, \qquad (15\text{-}3)$$

in which $x_m$, $\omega$, and $\phi$ are constants. This motion is called **simple harmonic motion** (SHM), a term that means the periodic motion is a sinusoidal function of time. Equation 15-3, in which the sinusoidal function is a cosine function, is graphed in Fig. 15-1b. (You can get that graph by rotating Fig. 15-1a counterclockwise by 90° and then connecting the successive locations of the particle with a curve.) The quantities that determine the shape of the graph are displayed in Fig. 15-2 with their names. We now shall define those quantities.

The quantity $x_m$, called the **amplitude** of the motion, is a positive constant whose value depends on how the motion was started. The subscript $m$ stands for *maximum* because the amplitude is the magnitude of the maximum displacement of the particle in either direction. The cosine function in Eq. 15-3 varies between the limits $\pm1$; so the displacement $x(t)$ varies between the limits $\pm x_m$.

The time-varying quantity $(\omega t + \phi)$ in Eq. 15-3 is called the **phase** of the motion, and the constant $\phi$ is called the **phase constant** (or **phase angle**). The value of $\phi$ depends on the displacement and velocity of the particle at time $t = 0$. For the $x(t)$ plots of Fig. 15-3a, the phase constant $\phi$ is zero.

To interpret the constant $\omega$, called the **angular frequency** of the motion, we first note that the displacement $x(t)$ must return to its initial value after one period $T$ of the motion; that is, $x(t)$ must equal $x(t + T)$ for all $t$. To simplify this analysis, let us put $\phi = 0$ in Eq. 15-3. From that equation we then can write

$$x_m \cos \omega t = x_m \cos \omega(t + T). \qquad (15\text{-}4)$$

The cosine function first repeats itself when its argument (the phase) has increased by $2\pi$ rad; so Eq. 15-4 gives us

$$\omega(t + T) = \omega t + 2\pi$$

or

$$\omega T = 2\pi.$$

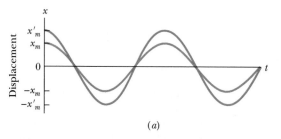

(a)

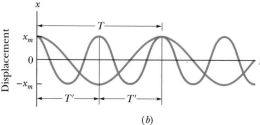

(b)

FIG. 15-3 In all three cases, the blue curve is obtained from Eq. 15-3 with $\phi = 0$. (a) The red curve differs from the blue curve *only* in that the red-curve amplitude $x'_m$ is greater (the red-curve extremes of displacement are higher and lower). (b) The red curve differs from the blue curve *only* in that the red-curve period is $T' = T/2$ (the red curve is compressed horizontally). (c) The red curve differs from the blue curve *only* in that for the red curve $\phi = -\pi/4$ rad rather than zero (the negative value of $\phi$ shifts the red curve to the right).

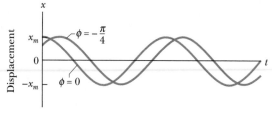

(c)

Thus, from Eq. 15-2 the angular frequency is

$$\omega = \frac{2\pi}{T} = 2\pi f. \quad (15\text{-}5)$$

The SI unit of angular frequency is the radian per second. (To be consistent, then, $\phi$ must be in radians.) Figure 15-3 compares $x(t)$ for two simple harmonic motions that differ either in amplitude, in period (and thus in frequency and angular frequency), or in phase constant.

✔ **CHECKPOINT 1**    A particle undergoing simple harmonic oscillation of period $T$ (like that in Fig. 15-1) is at $-x_m$ at time $t = 0$. Is it at $-x_m$, at $+x_m$, at 0, between $-x_m$ and 0, or between 0 and $+x_m$ when (a) $t = 2.00T$, (b) $t = 3.50T$, and (c) $t = 5.25T$?

## The Velocity of SHM

By differentiating Eq. 15-3, we can find an expression for the velocity of a particle moving with simple harmonic motion; that is,

$$v(t) = \frac{dx(t)}{dt} = \frac{d}{dt}[x_m \cos(\omega t + \phi)]$$

or        $$v(t) = -\omega x_m \sin(\omega t + \phi) \quad \text{(velocity)}. \quad (15\text{-}6)$$

Figure 15-4a is a plot of Eq. 15-3 with $\phi = 0$. Figure 15-4b shows Eq. 15-6, also with $\phi = 0$. Analogous to the amplitude $x_m$ in Eq. 15-3, the positive quantity $\omega x_m$ in Eq. 15-6 is called the **velocity amplitude** $v_m$. As you can see in Fig. 15-4b, the velocity of the oscillating particle varies between the limits $\pm v_m = \pm \omega x_m$. Note also in that figure that the curve of $v(t)$ is *shifted* (to the left) from the curve of $x(t)$ by one-quarter period; when the magnitude of the displacement is greatest (that is, $x(t) = x_m$), the magnitude of the velocity is least (that is, $v(t) = 0$). When the magnitude of the displacement is least (that is, zero), the magnitude of the velocity is greatest (that is, $v_m = \omega x_m$).

## The Acceleration of SHM

Knowing the velocity $v(t)$ for simple harmonic motion, we can find an expression for the acceleration of the oscillating particle by differentiating once more. Thus, we have, from Eq. 15-6,

$$a(t) = \frac{dv(t)}{dt} = \frac{d}{dt}[-\omega x_m \sin(\omega t + \phi)]$$

or        $$a(t) = -\omega^2 x_m \cos(\omega t + \phi) \quad \text{(acceleration)}. \quad (15\text{-}7)$$

Figure 15-4c is a plot of Eq. 15-7 for the case $\phi = 0$. The positive quantity $\omega^2 x_m$ in Eq. 15-7 is called the **acceleration amplitude** $a_m$; that is, the acceleration of the particle varies between the limits $\pm a_m = \pm \omega^2 x_m$, as Fig. 15-4c shows. Note also that the acceleration curve $a(t)$ is shifted (to the left) by $\frac{1}{4}T$ relative to the velocity curve $v(t)$.

We can combine Eqs. 15-3 and 15-7 to yield

$$a(t) = -\omega^2 x(t), \quad (15\text{-}8)$$

which is the hallmark of simple harmonic motion:

☞ In SHM, the acceleration is proportional to the displacement but opposite in sign, and the two quantities are related by the square of the angular frequency.

Thus, as Fig. 15-4 shows, when the displacement has its greatest positive value, the acceleration has its greatest negative value, and conversely. When the displacement is zero, the acceleration is also zero.

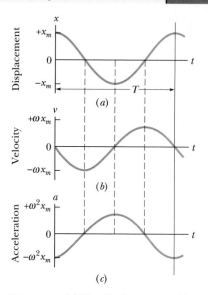

**FIG. 15-4** (a) The displacement $x(t)$ of a particle oscillating in SHM with phase angle $\phi$ equal to zero. The period $T$ marks one complete oscillation. (b) The velocity $v(t)$ of the particle. (c) The acceleration $a(t)$ of the particle.

*Tactic 1:* *Phase Angles* Note the effect of the phase angle $\phi$ on a plot of $x(t)$. When $\phi = 0$, $x(t)$ has a graph like that in Fig. 15-4a, a typical cosine curve. Increasing $\phi$ shifts the curve leftward along the $t$ axis. (You might remember this with the symbol $\leftarrow\uparrow\phi$, where the up arrow indicates an increase in $\phi$ and the left arrow indicates the resulting shift in the curve.) Decreasing $\phi$ shifts the curve rightward, as in Fig. 15-3c for $\phi = -\pi/4$.

Two plots of SHM with different phase angles are said to have a *phase difference;* each is said to be *phase-shifted* from the other, or *out of phase* with the other. The curves in Fig. 15-3c, for example, have a phase difference of $\pi/4$ rad; that is, one curve is phase-shifted from the other by $\pi/4$ rad.

Because SHM repeats after each period $T$ and the cosine function repeats after each $2\pi$ rad, one period $T$ represents a phase difference of $2\pi$ rad. In Fig. 15-4, $x(t)$ is phase-shifted to the right from $v(t)$ by one-quarter period, or $-\pi/2$ rad; it is shifted to the right from $a(t)$ by one-half period, or $-\pi$ rad. A phase shift of $2\pi$ rad causes a curve of SHM to coincide with itself; that is, it looks unchanged.

## 15-3 | The Force Law for Simple Harmonic Motion

Once we know how the acceleration of a particle varies with time, we can use Newton's second law to learn what force must act on the particle to give it that acceleration. If we combine Newton's second law and Eq. 15-8, we find, for simple harmonic motion,

$$F = ma = -(m\omega^2)x. \qquad (15\text{-}9)$$

This result—a restoring force that is proportional to the displacement but opposite in sign—is familiar. It is Hooke's law,

$$F = -kx, \qquad (15\text{-}10)$$

for a spring, the spring constant here being

$$k = m\omega^2. \qquad (15\text{-}11)$$

We can in fact take Eq. 15-10 as an alternative definition of simple harmonic motion. It says:

> ☞ Simple harmonic motion is the motion executed by a particle subject to a force that is proportional to the displacement of the particle but opposite in sign.

The block–spring system of Fig. 15-5 forms a **linear simple harmonic oscillator** (linear oscillator, for short), where "linear" indicates that $F$ is proportional to $x$ rather than to some other power of $x$. The angular frequency $\omega$ of the simple harmonic motion of the block is related to the spring constant $k$ and the mass $m$ of the block by Eq. 15-11, which yields

$$\omega = \sqrt{\frac{k}{m}} \qquad \text{(angular frequency).} \qquad (15\text{-}12)$$

By combining Eqs. 15-5 and 15-12, we can write, for the **period** of the linear oscillator of Fig. 15-5,

$$T = 2\pi\sqrt{\frac{m}{k}} \qquad \text{(period).} \qquad (15\text{-}13)$$

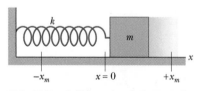

**FIG. 15-5** A linear simple harmonic oscillator. The surface is frictionless. Like the particle of Fig. 15-1, the block moves in simple harmonic motion once it has been either pulled or pushed away from the $x = 0$ position and released. Its displacement is then given by Eq. 15-3.

Equations 15-12 and 15-13 tell us that a large angular frequency (and thus a small period) goes with a stiff spring (large $k$) and a light block (small $m$).

Every oscillating system, be it a diving board or a violin string, has some element of "springiness" and some element of "inertia" or mass, and thus resembles a linear oscillator. In the linear oscillator of Fig. 15-5, these elements are located in separate parts of the system: The springiness is entirely in the spring, which we assume to be massless, and the inertia is entirely in the block, which we assume to be rigid. In a violin string, however, the two elements are both within the string, as you will see in Chapter 16.

✓ **CHECKPOINT 2**   Which of the following relationships between the force $F$ on a particle and the particle's position $x$ implies simple harmonic oscillation: (a) $F = -5x$, (b) $F = -400x^2$, (c) $F = 10x$, (d) $F = 3x^2$?

## Sample Problem   15-1

A block whose mass $m$ is 680 g is fastened to a spring whose spring constant $k$ is 65 N/m. The block is pulled a distance $x = 11$ cm from its equilibrium position at $x = 0$ on a frictionless surface and released from rest at $t = 0$.

(a) What are the angular frequency, the frequency, and the period of the resulting motion?

**KEY IDEA**   The block–spring system forms a linear simple harmonic oscillator, with the block undergoing SHM.

**Calculations:** The angular frequency is given by Eq. 15-12:

$$\omega = \sqrt{\frac{k}{m}} = \sqrt{\frac{65 \text{ N/m}}{0.68 \text{ kg}}} = 9.78 \text{ rad/s}$$

$$\approx 9.8 \text{ rad/s}. \qquad \text{(Answer)}$$

The frequency follows from Eq. 15-5, which yields

$$f = \frac{\omega}{2\pi} = \frac{9.78 \text{ rad/s}}{2\pi \text{ rad}} = 1.56 \text{ Hz} \approx 1.6 \text{ Hz}. \quad \text{(Answer)}$$

The period follows from Eq. 15-2, which yields

$$T = \frac{1}{f} = \frac{1}{1.56 \text{ Hz}} = 0.64 \text{ s} = 640 \text{ ms}. \quad \text{(Answer)}$$

(b) What is the amplitude of the oscillation?

**KEY IDEA**   With no friction involved, the mechanical energy of the spring–block system is conserved.

**Reasoning:** The block is released from rest 11 cm from its equilibrium position, with zero kinetic energy and the elastic potential energy of the system at a maximum. Thus, the block will have zero kinetic energy whenever it is again 11 cm from its equilibrium position, which means it will never be farther than 11 cm from that position. Its maximum displacement is 11 cm:

$$x_m = 11 \text{ cm}. \qquad \text{(Answer)}$$

(c) What is the maximum speed $v_m$ of the oscillating block, and where is the block when it has this speed?

**KEY IDEA**   The maximum speed $v_m$ is the velocity amplitude $\omega x_m$ in Eq. 15-6.

**Calculation:** Thus, we have

$$v_m = \omega x_m = (9.78 \text{ rad/s})(0.11 \text{ m})$$

$$= 1.1 \text{ m/s}. \qquad \text{(Answer)}$$

This maximum speed occurs when the oscillating block is rushing through the origin; compare Figs. 15-4a and 15-4b, where you can see that the speed is a maximum whenever $x = 0$.

(d) What is the magnitude $a_m$ of the maximum acceleration of the block?

**KEY IDEA**   The magnitude $a_m$ of the maximum acceleration is the acceleration amplitude $\omega^2 x_m$ in Eq. 15-7.

**Calculation:** So, we have

$$a_m = \omega^2 x_m = (9.78 \text{ rad/s})^2(0.11 \text{ m})$$

$$= 11 \text{ m/s}^2. \qquad \text{(Answer)}$$

This maximum acceleration occurs when the block is at the ends of its path. At those points, the force acting on the block has its maximum magnitude; compare Figs. 15-4a and 15-4c, where you can see that the magnitudes of the displacement and acceleration are maximum at the same times.

(e) What is the phase constant $\phi$ for the motion?

**Calculations:** Equation 15-3 gives the displacement of the block as a function of time. We know that at time $t = 0$, the block is located at $x = x_m$. Substituting these *initial conditions,* as they are called, into Eq. 15-3 and canceling $x_m$ give us

$$1 = \cos \phi. \qquad (15\text{-}14)$$

Taking the inverse cosine then yields

$$\phi = 0 \text{ rad}. \qquad \text{(Answer)}$$

(Any angle that is an integer multiple of $2\pi$ rad also satisfies Eq. 15-14; we chose the smallest angle.)

(f) What is the displacement function $x(t)$ for the spring–block system?

**Calculation:** The function $x(t)$ is given in general form by Eq. 15-3. Substituting known quantities into that equation gives us

$$x(t) = x_m \cos(\omega t + \phi)$$

$$= (0.11 \text{ m}) \cos[(9.8 \text{ rad/s})t + 0]$$

$$= 0.11 \cos(9.8t), \qquad \text{(Answer)}$$

where $x$ is in meters and $t$ is in seconds.

At $t = 0$, the displacement $x(0)$ of the block in a linear oscillator like that of Fig. 15-5 is $-8.50$ cm. (Read $x(0)$ as "$x$ at time zero.") The block's velocity $v(0)$ then is $-0.920$ m/s, and its acceleration $a(0)$ is $+47.0$ m/s$^2$.

(a) What is the angular frequency $\omega$ of this system?

**KEY IDEA** With the block in SHM, Eqs. 15-3, 15-6, and 15-7 give its displacement, velocity, and acceleration, respectively, and each contains $\omega$.

**Calculations:** Let's substitute $t = 0$ into each to see whether we can solve any one of them for $\omega$. We find

$$x(0) = x_m \cos \phi, \qquad (15\text{-}15)$$
$$v(0) = -\omega x_m \sin \phi, \qquad (15\text{-}16)$$
and $$a(0) = -\omega^2 x_m \cos \phi. \qquad (15\text{-}17)$$

In Eq. 15-15, $\omega$ has disappeared. In Eqs. 15-16 and 15-17, we know values for the left sides, but we do not know $x_m$ and $\phi$. However, if we divide Eq. 15-17 by Eq. 15-15, we neatly eliminate both $x_m$ and $\phi$ and can then solve for $\omega$ as

$$\omega = \sqrt{-\frac{a(0)}{x(0)}} = \sqrt{-\frac{47.0 \text{ m/s}^2}{-0.0850 \text{ m}}}$$
$$= 23.5 \text{ rad/s.} \qquad \text{(Answer)}$$

(b) What are the phase constant $\phi$ and amplitude $x_m$?

**Calculations:** We know $\omega$ and want $\phi$ and $x_m$. If we divide Eq. 15-16 by Eq. 15-15, we find

$$\frac{v(0)}{x(0)} = \frac{-\omega x_m \sin \phi}{x_m \cos \phi} = -\omega \tan \phi.$$

Solving for $\tan \phi$, we find

$$\tan \phi = -\frac{v(0)}{\omega x(0)} = -\frac{-0.920 \text{ m/s}}{(23.5 \text{ rad/s})(-0.0850 \text{ m})}$$
$$= -0.461.$$

This equation has two solutions:

$$\phi = -25° \quad \text{and} \quad \phi = 180° + (-25°) = 155°.$$

(Normally only the first solution here is displayed by a calculator.) To choose the proper solution, we test them both by using them to compute values for the amplitude $x_m$. From Eq. 15-15, we find that if $\phi = -25°$, then

$$x_m = \frac{x(0)}{\cos \phi} = \frac{-0.0850 \text{ m}}{\cos(-25°)} = -0.094 \text{ m.}$$

We find similarly that if $\phi = 155°$, then $x_m = 0.094$ m. Because the amplitude of SHM must be a positive constant, the correct phase constant and amplitude here are

$$\phi = 155° \quad \text{and} \quad x_m = 0.094 \text{ m} = 9.4 \text{ cm.} \quad \text{(Answer)}$$

---

**PROBLEM-SOLVING TACTICS**

**Tactic 2: Identifying SHM** In linear SHM the acceleration $a$ and displacement $x$ of the system are related by an equation of the form

$$a = -(\text{a positive constant})x,$$

which says that the acceleration is proportional to the displacement from the equilibrium position but is in the opposite direction. Once you find such an expression for an oscillating system, you can immediately compare it with Eq. 15-8, identify the positive constant as being equal to $\omega^2$, and so quickly get an expression for the angular frequency of the motion. With Eq. 15-5 you then can find the period $T$ and the frequency $f$.

In some problems you might derive an expression for the force $F$ as a function of displacement $x$. If the motion is linear SHM, the force and displacement are related by

$$F = -(\text{a positive constant})x,$$

which says that the force is proportional to the displacement but is in the opposite direction. Once you have found such an expression, you can immediately compare it with Eq. 15-10 and identify the positive constant as being $k$. If you know the mass that is involved, you can then use Eqs. 15-12, 15-13, and 15-5 to find the angular frequency $\omega$, period $T$, and frequency $f$.

## 15-4 | Energy in Simple Harmonic Motion

In Chapter 8 we saw that the energy of a linear oscillator transfers back and forth between kinetic energy and potential energy, while the sum of the two—the mechanical energy $E$ of the oscillator—remains constant. We now consider this situation quantitatively.

The potential energy of a linear oscillator like that of Fig. 15-5 is associated entirely with the spring. Its value depends on how much the spring is stretched or compressed—that is, on $x(t)$. We can use Eqs. 8-11 and 15-3 to find

$$U(t) = \tfrac{1}{2}kx^2 = \tfrac{1}{2}kx_m^2 \cos^2(\omega t + \phi). \qquad (15\text{-}18)$$

*Caution:* A function written in the form $\cos^2 A$ (as here) means $(\cos A)^2$ and is *not* the same as one written $\cos A^2$, which means $\cos(A^2)$.

The kinetic energy of the system of Fig. 15-5 is associated entirely with the block. Its value depends on how fast the block is moving—that is, on $v(t)$. We can use Eq. 15-6 to find

$$K(t) = \tfrac{1}{2}mv^2 = \tfrac{1}{2}m\omega^2 x_m^2 \sin^2(\omega t + \phi). \qquad (15\text{-}19)$$

If we use Eq. 15-12 to substitute $k/m$ for $\omega^2$, we can write Eq. 15-19 as

$$K(t) = \tfrac{1}{2}mv^2 = \tfrac{1}{2}kx_m^2 \sin^2(\omega t + \phi). \qquad (15\text{-}20)$$

The mechanical energy follows from Eqs. 15-18 and 15-20 and is

$$\begin{aligned}
E &= U + K \\
&= \tfrac{1}{2}kx_m^2 \cos^2(\omega t + \phi) + \tfrac{1}{2}kx_m^2 \sin^2(\omega t + \phi) \\
&= \tfrac{1}{2}kx_m^2 [\cos^2(\omega t + \phi) + \sin^2(\omega t + \phi)].
\end{aligned}$$

For any angle $\alpha$,

$$\cos^2 \alpha + \sin^2 \alpha = 1.$$

Thus, the quantity in the square brackets above is unity and we have

$$E = U + K = \tfrac{1}{2}kx_m^2. \qquad (15\text{-}21)$$

The mechanical energy of a linear oscillator is indeed constant and independent of time. The potential energy and kinetic energy of a linear oscillator are shown as functions of time $t$ in Fig. 15-6a, and they are shown as functions of displacement $x$ in Fig. 15-6b.

You might now understand why an oscillating system normally contains an element of springiness and an element of inertia: The former stores its potential energy and the latter stores its kinetic energy.

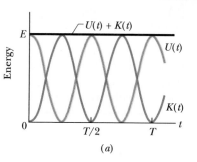

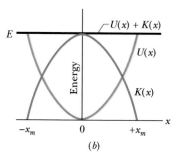

**FIG. 15-6** (a) Potential energy $U(t)$, kinetic energy $K(t)$, and mechanical energy $E$ as functions of time $t$ for a linear harmonic oscillator. Note that all energies are positive and that the potential energy and the kinetic energy peak twice during every period. (b) Potential energy $U(x)$, kinetic energy $K(x)$, and mechanical energy $E$ as functions of position $x$ for a linear harmonic oscillator with amplitude $x_m$. For $x = 0$ the energy is all kinetic, and for $x = \pm x_m$ it is all potential.

✓**CHECKPOINT 3**  In Fig. 15-5, the block has a kinetic energy of 3 J and the spring has an elastic potential energy of 2 J when the block is at $x = +2.0$ cm. (a) What is the kinetic energy when the block is at $x = 0$? What is the elastic potential energy when the block is at (b) $x = -2.0$ cm and (c) $x = -x_m$?

---

**Sample Problem | 15-3**

The huge ball that appears in this chapter's opening photograph hangs from four cables and swings like a pendulum when the building sways in the wind. When the building sways—say, eastward—the massive pendulum does also but delayed enough so that as it finally swings eastward, the building is swaying westward. Thus, the pendulum's motion is out of step with the building's motion, tending to counter it.

Many other buildings have other types of *mass dampers*, as these anti-sway devices are called. Some, like the John Hancock building in Boston, have a large block oscillating at the end of a spring and on a lubricated track. The principle is the same as with the pendulum: The motion of the oscillator is out of step with the motion of the building.

Suppose the block has mass $m = 2.72 \times 10^5$ kg and is designed to oscillate at frequency $f = 10.0$ Hz and with amplitude $x_m = 20.0$ cm.

(a) What is the total mechanical energy $E$ of the spring–block system?

**KEY IDEA**  The mechanical energy $E$ (the sum of the kinetic energy $K = \tfrac{1}{2}mv^2$ of the block and the potential energy $U = \tfrac{1}{2}kx^2$ of the spring) is constant throughout the motion of the oscillator. Thus, we can evaluate $E$ at any point during the motion.

*Calculations:* Because we are given amplitude $x_m$ of the oscillations, let's evaluate $E$ when the block is at position $x = x_m$, where it has velocity $v = 0$. However, to evaluate $U$ at that point, we first need to find the spring constant $k$. From Eq. 15-12 ($\omega = \sqrt{k/m}$) and Eq. 15-5 ($\omega = 2\pi f$), we find

$$\begin{aligned}
k &= m\omega^2 = m(2\pi f)^2 \\
&= (2.72 \times 10^5 \text{ kg})(2\pi)^2(10.0 \text{ Hz})^2 \\
&= 1.073 \times 10^9 \text{ N/m.}
\end{aligned}$$

We can now evaluate $E$ as

$$\begin{aligned}
E &= K + U = \tfrac{1}{2}mv^2 + \tfrac{1}{2}kx^2 \\
&= 0 + \tfrac{1}{2}(1.073 \times 10^9 \text{ N/m})(0.20 \text{ m})^2 \\
&= 2.147 \times 10^7 \text{ J} \approx 2.1 \times 10^7 \text{ J.} \qquad \text{(Answer)}
\end{aligned}$$

(b) What is the block's speed as it passes through the equilibrium point?

**Calculations:** We want the speed at $x = 0$, where the potential energy is $U = \frac{1}{2}kx^2 = 0$ and the mechanical energy is entirely kinetic energy. So, we can write

$$E = K + U = \tfrac{1}{2}mv^2 + \tfrac{1}{2}kx^2$$

$$2.147 \times 10^7 \text{ J} = \tfrac{1}{2}(2.72 \times 10^5 \text{ kg})v^2 + 0,$$

or $\qquad\qquad v = 12.6 \text{ m/s}.$ (Answer)

Because $E$ is entirely kinetic energy, this is the maximum speed $v_m$.

## 15-5 | An Angular Simple Harmonic Oscillator

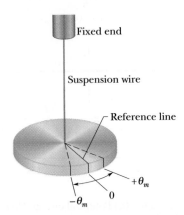

**FIG. 15-7** A torsion pendulum is an angular version of a linear simple harmonic oscillator. The disk oscillates in a horizontal plane; the reference line oscillates with angular amplitude $\theta_m$. The twist in the suspension wire stores potential energy as a spring does and provides the restoring torque.

Figure 15-7 shows an angular version of a simple harmonic oscillator; the element of springiness or elasticity is associated with the twisting of a suspension wire rather than the extension and compression of a spring as we previously had. The device is called a **torsion pendulum,** with *torsion* referring to the twisting.

If we rotate the disk in Fig. 15-7 by some angular displacement $\theta$ from its rest position (where the reference line is at $\theta = 0$) and release it, it will oscillate about that position in **angular simple harmonic motion.** Rotating the disk through an angle $\theta$ in either direction introduces a restoring torque given by

$$\tau = -\kappa\theta. \qquad (15\text{-}22)$$

Here $\kappa$ (Greek *kappa*) is a constant, called the **torsion constant,** that depends on the length, diameter, and material of the suspension wire.

Comparison of Eq. 15-22 with Eq. 15-10 leads us to suspect that Eq. 15-22 is the angular form of Hooke's law, and that we can transform Eq. 15-13, which gives the period of linear SHM, into an equation for the period of angular SHM: We replace the spring constant $k$ in Eq. 15-13 with its equivalent, the constant $\kappa$ of Eq. 15-22, and we replace the mass $m$ in Eq. 15-13 with *its* equivalent, the rotational inertia $I$ of the oscillating disk. These replacements lead to

$$T = 2\pi \sqrt{\frac{I}{\kappa}} \quad \text{(torsion pendulum),} \qquad (15\text{-}23)$$

which is the correct equation for the period of an angular simple harmonic oscillator, or torsion pendulum.

---

**PROBLEM-SOLVING TACTICS**

**Tactic 3: Identifying Angular SHM** When a system undergoes angular simple harmonic motion, its angular acceleration $\alpha$ and angular displacement $\theta$ are related by an equation of the form

$$\alpha = -(\text{a positive constant})\theta.$$

This equation is the angular equivalent of Eq. 15-8 ($a = -\omega^2 x$). It says that the angular acceleration $\alpha$ is proportional to the angular displacement $\theta$ from the equilibrium position but is in the direction opposite the displacement. If you have an expression of this form, you can identify the positive constant as being $\omega^2$, and then you can determine $\omega, f,$ and $T$.

You can also identify angular SHM if you have an expression for the torque $\tau$ in terms of the angular displacement, because that expression must be in the form of Eq. 15-22 ($\tau = -\kappa\theta$) or

$$\tau = -(\text{a positive constant})\theta.$$

This equation is the angular equivalent of Eq. 15-10 ($F = -kx$). It says that the torque $\tau$ is proportional to the angular displacement $\theta$ from the equilibrium position but tends to rotate the system in the opposite direction. If you have an expression of this form, then you can identify the positive constant as being the system's torsion constant $\kappa$. If you know the rotational inertia $I$ of the system, you can then determine $T$.

---

**Sample Problem** 15-4

Figure 15-8*a* shows a thin rod whose length $L$ is 12.4 cm and whose mass $m$ is 135 g, suspended at its midpoint from a long wire. Its period $T_a$ of angular SHM is measured to be 2.53 s. An irregularly shaped object, which we call object $X$, is then hung from the same wire, as in Fig.

15-8*b*, and its period $T_b$ is found to be 4.76 s. What is the rotational inertia of object $X$ about its suspension axis?

**KEY IDEA** The rotational inertia of either the rod or object $X$ is related to the measured period by Eq. 15-23.

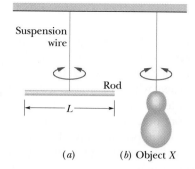

FIG. 15-8 Two torsion pendulums, consisting of (a) a wire and a rod and (b) the same wire and an irregularly shaped object.

(a)    (b) Object X

**Calculations:** In Table 10-2e, the rotational inertia of a thin rod about a perpendicular axis through its midpoint is given as $\frac{1}{12}mL^2$. Thus, we have, for the rod in Fig. 15-8a,

$$I_a = \tfrac{1}{12}mL^2 = (\tfrac{1}{12})(0.135 \text{ kg})(0.124 \text{ m})^2$$
$$= 1.73 \times 10^{-4} \text{ kg} \cdot \text{m}^2.$$

Now let us write Eq. 15-23 twice, once for the rod and once for object X:

$$T_a = 2\pi\sqrt{\frac{I_a}{\kappa}} \quad \text{and} \quad T_b = 2\pi\sqrt{\frac{I_b}{\kappa}}.$$

The constant $\kappa$, which is a property of the wire, is the same for both figures; only the periods and the rotational inertias differ.

Let us square each of these equations, divide the second by the first, and solve the resulting equation for $I_b$. The result is

$$I_b = I_a \frac{T_b^2}{T_a^2} = (1.73 \times 10^{-4} \text{ kg} \cdot \text{m}^2) \frac{(4.76 \text{ s})^2}{(2.53 \text{ s})^2}$$
$$= 6.12 \times 10^{-4} \text{ kg} \cdot \text{m}^2. \qquad \text{(Answer)}$$

---

## 15-6 I Pendulums

We turn now to a class of simple harmonic oscillators in which the springiness is associated with the gravitational force rather than with the elastic properties of a twisted wire or a compressed or stretched spring.

### The Simple Pendulum

If an apple swings on a long thread, does it have simple harmonic motion? If so, what is the period $T$? To answer, we consider a **simple pendulum,** which consists of a particle of mass $m$ (called the *bob* of the pendulum) suspended from one end of an unstretchable, massless string of length $L$ that is fixed at the other end, as in Fig. 15-9a. The bob is free to swing back and forth in the plane of the page, to the left and right of a vertical line through the pendulum's pivot point.

The forces acting on the bob are the force $\vec{T}$ from the string and the gravitational force $\vec{F}_g$, as shown in Fig. 15-9b, where the string makes an angle $\theta$ with the vertical. We resolve $\vec{F}_g$ into a radial component $F_g \cos\theta$ and a component $F_g \sin\theta$ that is tangent to the path taken by the bob. This tangential component produces a restoring torque about the pendulum's pivot point because the component always acts opposite the displacement of the bob so as to bring the bob back toward its central location. That location is called the *equilibrium position* ($\theta = 0$) because the pendulum would be at rest there were it not swinging.

From Eq. 10-41 ($\tau = r_\perp F$), we can write this restoring torque as

$$\tau = -L(F_g \sin\theta), \qquad (15\text{-}24)$$

where the minus sign indicates that the torque acts to reduce $\theta$ and $L$ is the moment arm of the force component $F_g \sin\theta$ about the pivot point. Substituting Eq. 15-24 into Eq. 10-44 ($\tau = I\alpha$) and then substituting $mg$ as the magnitude of $F_g$, we obtain

$$-L(mg \sin\theta) = I\alpha, \qquad (15\text{-}25)$$

where $I$ is the pendulum's rotational inertia about the pivot point and $\alpha$ is its angular acceleration about that point.

We can simplify Eq. 15-25 if we assume the angle $\theta$ is small, for then we can approximate $\sin\theta$ with $\theta$ (expressed in radian measure). (As an example, if $\theta = 5.00° = 0.0873$ rad, then $\sin\theta = 0.0872$, a difference of only about 0.1%.) With that approximation and some rearranging, we then have

$$\alpha = -\frac{mgL}{I}\,\theta. \qquad (15\text{-}26)$$

This equation is the angular equivalent of Eq. 15-8, the hallmark of SHM. It tells us

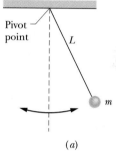

(a)

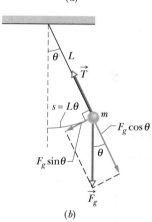

(b)

FIG. 15-9 (a) A simple pendulum. (b) The forces acting on the bob are the gravitational force $\vec{F}_g$ and the force $\vec{T}$ from the string. The tangential component $F_g \sin\theta$ of the gravitational force is a restoring force that tends to bring the pendulum back to its central position.

that the angular acceleration $\alpha$ of the pendulum is proportional to the angular displacement $\theta$ but opposite in sign. Thus, as the pendulum bob moves to the right, as in Fig. 15-9a, its acceleration *to the left* increases until the bob stops and begins moving to the left. Then, when it is to the left of the equilibrium position, its acceleration to the right tends to return it to the right, and so on, as it swings back and forth in SHM. More precisely, the motion of a *simple pendulum swinging through only small angles* is approximately SHM. We can state this restriction to small angles another way: The **angular amplitude** $\theta_m$ of the motion (the maximum angle of swing) must be small.

Comparing Eqs. 15-26 and 15-8, we see that the angular frequency of the pendulum is $\omega = \sqrt{mgL/I}$. Next, if we substitute this expression for $\omega$ into Eq. 15-5 ($\omega = 2\pi/T$), we see that the period of the pendulum may be written as

$$T = 2\pi \sqrt{\frac{I}{mgL}}. \qquad (15\text{-}27)$$

All the mass of a simple pendulum is concentrated in the mass $m$ of the particle-like bob, which is at radius $L$ from the pivot point. Thus, we can use Eq. 10-33 ($I = mr^2$) to write $I = mL^2$ for the rotational inertia of the pendulum. Substituting this into Eq. 15-27 and simplifying then yield

$$T = 2\pi \sqrt{\frac{L}{g}} \qquad \text{(simple pendulum, small amplitude).} \qquad (15\text{-}28)$$

We assume small-angle swinging in this chapter.

### The Physical Pendulum

A real pendulum, usually called a **physical pendulum,** can have a complicated distribution of mass, much different from that of a simple pendulum. Does a physical pendulum also undergo SHM? If so, what is its period?

Figure 15-10 shows an arbitrary physical pendulum displaced to one side by angle $\theta$. The gravitational force $\vec{F}_g$ acts at its center of mass $C$, at a distance $h$ from the pivot point $O$. Comparison of Figs. 15-10 and 15-9b reveals only one important difference between an arbitrary physical pendulum and a simple pendulum. For a physical pendulum the restoring component $F_g \sin \theta$ of the gravitational force has a moment arm of distance $h$ about the pivot point, rather than of string length $L$. In all other respects, an analysis of the physical pendulum would duplicate our analysis of the simple pendulum up through Eq. 15-27. Again (for small $\theta_m$), we would find that the motion is approximately SHM.

If we replace $L$ with $h$ in Eq. 15-27, we can write the period as

$$T = 2\pi \sqrt{\frac{I}{mgh}} \qquad \text{(physical pendulum, small amplitude).} \qquad (15\text{-}29)$$

As with the simple pendulum, $I$ is the rotational inertia of the pendulum about $O$. However, now $I$ is not simply $mL^2$ (it depends on the shape of the physical pendulum), but it is still proportional to $m$.

A physical pendulum will not swing if it pivots at its center of mass. Formally, this corresponds to putting $h = 0$ in Eq. 15-29. That equation then predicts $T \to \infty$, which implies that such a pendulum will never complete one swing.

Corresponding to any physical pendulum that oscillates about a given pivot point $O$ with period $T$ is a simple pendulum of length $L_0$ with the same period $T$. We can find $L_0$ with Eq. 15-28. The point along the physical pendulum at distance $L_0$ from point $O$ is called the *center of oscillation* of the physical pendulum for the given suspension point.

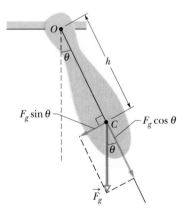

**FIG. 15-10** A physical pendulum. The restoring torque is $hF_g \sin \theta$. When $\theta = 0$, center of mass $C$ hangs directly below pivot point $O$.

## Measuring g

We can use a physical pendulum to measure the free-fall acceleration $g$ at a particular location on Earth's surface. (Countless thousands of such measurements have been made during geophysical prospecting.)

To analyze a simple case, take the pendulum to be a uniform rod of length $L$, suspended from one end. For such a pendulum, $h$ in Eq. 15-29, the distance between the pivot point and the center of mass, is $\frac{1}{2}L$. Table 10-2e tells us that the rotational inertia of this pendulum about a perpendicular axis through its center of mass is $\frac{1}{12}mL^2$. From the parallel-axis theorem of Eq. 10-36 ($I = I_{com} + Mh^2$), we then find that the rotational inertia about a perpendicular axis through one end of the rod is

$$I = I_{com} + mh^2 = \tfrac{1}{12}mL^2 + m(\tfrac{1}{2}L)^2 = \tfrac{1}{3}mL^2. \tag{15-30}$$

If we put $h = \frac{1}{2}L$ and $I = \frac{1}{3}mL^2$ in Eq. 15-29 and solve for $g$, we find

$$g = \frac{8\pi^2 L}{3T^2}. \tag{15-31}$$

Thus, by measuring $L$ and the period $T$, we can find the value of $g$ at the pendulum's location. (If precise measurements are to be made, a number of refinements are needed, such as swinging the pendulum in an evacuated chamber.)

✓**CHECKPOINT 4**    Three physical pendulums, of masses $m_0$, $2m_0$, and $3m_0$, have the same shape and size and are suspended at the same point. Rank the masses according to the periods of the pendulums, greatest first.

## Sample Problem   15-5

In Fig. 15-11a, a meter stick swings about a pivot point at one end, at distance $h$ from the stick's center of mass.

(a) What is the period of oscillation $T$?

**KEY IDEA**   The stick is not a simple pendulum because its mass is not concentrated in a bob at the end opposite the pivot point—so the stick is a physical pendulum.

*Calculations:* The period for a physical pendulum is given by Eq. 15-29, for which we need the rotational inertia $I$ of the stick about the pivot point. We can treat the stick as a uniform rod of length $L$ and mass $m$. Then Eq. 15-30 tells us that $I = \frac{1}{3}mL^2$, and the distance $h$ in Eq. 15-29 is $\frac{1}{2}L$. Substituting these quantities into Eq. 15-29, we find

$$T = 2\pi\sqrt{\frac{I}{mgh}} = 2\pi\sqrt{\frac{\frac{1}{3}mL^2}{mg(\frac{1}{2}L)}} = 2\pi\sqrt{\frac{2L}{3g}} \tag{15-32}$$

$$= 2\pi\sqrt{\frac{(2)(1.00\text{ m})}{(3)(9.8\text{ m/s}^2)}} = 1.64\text{ s}. \qquad \text{(Answer)}$$

Note the result is independent of the pendulum's mass $m$.

(b) What is the distance $L_0$ between the pivot point $O$ of the stick and the center of oscillation of the stick?

*Calculations:* We want the length $L_0$ of the simple pendulum (drawn in Fig. 15-11b) that has the same period as the physical pendulum (the stick) of Fig. 15-11a.

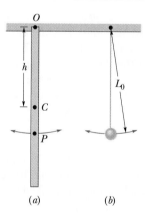

*(a)*                 *(b)*

**FIG. 15-11**   (a) A meter stick suspended from one end as a physical pendulum. (b) A simple pendulum whose length $L_0$ is chosen so that the periods of the two pendulums are equal. Point $P$ on the pendulum of (a) marks the center of oscillation.

Setting Eqs. 15-28 and 15-32 equal yields

$$T = 2\pi\sqrt{\frac{L_0}{g}} = 2\pi\sqrt{\frac{2L}{3g}}.$$

You can see by inspection that

$$L_0 = \tfrac{2}{3}L = (\tfrac{2}{3})(100\text{ cm}) = 66.7\text{ cm}. \quad \text{(Answer)}$$

In Fig. 15-11a, point $P$ marks this distance from suspension point $O$. Thus, point $P$ is the stick's center of oscillation for the given suspension point.

## Sample Problem | 15-6 | Build your skill

A competition diving board sits on a fulcrum about one-third of the way out from the fixed end of the board (Fig. 15-12a). In a running dive, a diver takes three quick steps along the board, out past the fulcrum so as to rotate the board's free end downward. As the board rebounds back through the horizontal, the diver leaps upward and toward the board's free end (Fig. 15-12b). A skilled diver trains to land on the free end just as the board has completed 2.5 oscillations during the leap. With such timing, the diver lands as the free end is moving downward with greatest speed (Fig. 15-12c). The landing then drives the free end down substantially, and the rebound catapults the diver high into the air.

Figure 15-12d shows a simple but realistic model of a competition board. The board section beyond the fulcrum is treated as a stiff rod of length $L$ that can rotate about a hinge at the fulcrum, compressing an (imaginary) spring under the board's free end. If the rod's mass is $m = 20.0$ kg and the diver's leap lasts $t_{fl} = 0.620$ s, what spring constant $k$ is required of the spring for a proper landing?

**KEY IDEA** If the rod is in SHM, then the acceleration and displacement of the oscillating end of the rod must be related by an expression in the form of Eq. 15-8 ($a = -\omega^2 x$). If so, we shall be able to find $\omega$ and then the desired $k$ from the expression.

**Torque and force:** Because the rod rotates about the hinge as the free end oscillates, we are concerned with a torque $\vec{\tau}$ on the rod about the hinge. That torque is due to the force $\vec{F}$ on the rod from the spring. Because $\vec{F}$ varies with time, $\vec{\tau}$ must also. However, at any given instant we can relate the magnitudes of $\vec{\tau}$ and $\vec{F}$ with Eq. 10-39 ($\tau = rF\sin\phi$). Here we have

$$\tau = LF\sin 90°, \qquad (15\text{-}33)$$

where $L$ is the moment arm of force $\vec{F}$ and $90°$ is the angle between the moment arm and the force's line of action. Combining Eq. 15-33 with Eq. 10-44 ($\tau = I\alpha$) gives us

$$I\alpha = LF, \qquad (15\text{-}34)$$

where $I$ is the rod's rotational inertia about the hinge and $\alpha$ is its angular acceleration about that point. From Eq. 15-30, the rod's rotational inertia $I$ is $\frac{1}{3}mL^2$.

Now let us mentally erect a vertical $x$ axis through the oscillating right end of the rod, with the positive direction upward. Then the force on the right end of the rod from the spring is $F = -kx$, where $x$ is the vertical displacement of the right end.

Substituting these expressions for $I$ and $F$ into Eq. 15-34 gives us

$$\frac{mL^2\alpha}{3} = -Lkx. \qquad (15\text{-}35)$$

**Mixture:** We now have a mixture of linear displacement $x$ (vertically) and rotational acceleration $\alpha$ (about the

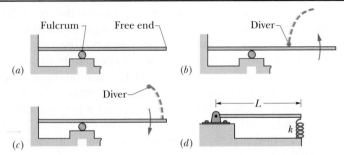

FIG. 15-12 (a) A diving board. (b) The diver leaps upward and forward as the board moves through the horizontal. (c) The diver lands 2.5 oscillations later. (d) A spring-oscillator model of the oscillating board.

hinge). We can replace $\alpha$ in Eq. 15-35 with the (linear) acceleration $a$ along the $x$ axis by substituting according to Eq. 10-22 ($a_t = \alpha r$) for tangential acceleration. Here the tangential acceleration is $a$ and the radius of rotation $r$ is $L$, so $\alpha = a/L$. With that substitution, Eq. 15-35 becomes

$$\frac{mL^2 a}{3L} = -Lkx,$$

which yields

$$a = -\frac{3k}{m}x. \qquad (15\text{-}36)$$

Equation 15-36 is, in fact, of the same form as Eq. 15-8 ($a = -\omega^2 x$). Therefore, the rod does indeed undergo SHM, and comparison of Eqs. 15-36 and 15-8 shows that

$$\omega^2 = \frac{3k}{m}.$$

Solving for $k$ and substituting for $\omega$ from Eq. 15-5 ($\omega = 2\pi/T$) give us

$$k = \frac{m}{3}\left(\frac{2\pi}{T}\right)^2, \qquad (15\text{-}37)$$

where $T$ is the period of the board's oscillation. We want the time of flight $t_{fl}$ to last for 2.5 oscillations of the board and thus also for 2.5 oscillations of our rod. Thus we want $t_{fl} = 2.5T$. Substituting this and given data into Eq. 15-37 leads to

$$k = \frac{m}{3}\left(\frac{2\pi}{t_{fl}}2.5\right)^2 \qquad (15\text{-}38)$$

$$= \frac{(20.0\text{ kg})}{3}\left(\frac{2\pi}{0.620\text{ s}}2.5\right)^2$$

$$= 4.28 \times 10^3 \text{ N/m}. \qquad \text{(Answer)}$$

This is the effective spring constant $k$ of the diving board.

**Diving skills:** From Eq. 15-38, we see that a diver with a longer flight time $t_{fl}$ requires a smaller spring constant $k$ in order to land at the proper instant at the end of the board. The value of $k$ can be increased by moving the fulcrum toward the free end or decreased by moving it in the opposite direction. A skilled diver trains to leap with a certain flight time $t_{fl}$ and knows how to set the fulcrum position accordingly.

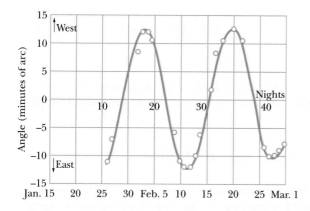

**FIG. 15-13** The angle between Jupiter and its moon Callisto as seen from Earth. The circles are based on Galileo's 1610 measurements. The curve is a best fit, strongly suggesting simple harmonic motion. At Jupiter's mean distance from Earth, 10 minutes of arc corresponds to about $2 \times 10^6$ km. (Adapted from A. P. French, *Newtonian Mechanics*, W. W. Norton & Company, New York, 1971, p. 288.)

## 15-7 | Simple Harmonic Motion and Uniform Circular Motion

In 1610, Galileo, using his newly constructed telescope, discovered the four principal moons of Jupiter. Over weeks of observation, each moon seemed to him to be moving back and forth relative to the planet in what today we would call simple harmonic motion; the disk of the planet was the midpoint of the motion. The record of Galileo's observations, written in his own hand, is still available. A. P. French of MIT used Galileo's data to work out the position of the moon Callisto relative to Jupiter. In the results shown in Fig. 15-13, the circles are based on Galileo's observations and the curve is a best fit to the data. The curve strongly suggests Eq. 15-3, the displacement function for SHM. A period of about 16.8 days can be measured from the plot.

*Actually,* Callisto moves with essentially constant speed in an essentially circular orbit around Jupiter. Its true motion—far from being simple harmonic—is uniform circular motion. What Galileo saw—and what you can see with a good pair of binoculars and a little patience—is the projection of this uniform circular motion on a line in the plane of the motion. We are led by Galileo's remarkable observations to the conclusion that simple harmonic motion is uniform circular motion viewed edge-on. In more formal language:

> ☞ Simple harmonic motion is the projection of uniform circular motion on a diameter of the circle in which the circular motion occurs.

Figure 15-14a gives an example. It shows a *reference particle P'* moving in uniform circular motion with (constant) angular speed $\omega$ in a *reference circle*. The radius $x_m$ of the circle is the magnitude of the particle's position vector. At any time $t$, the angular position of the particle is $\omega t + \phi$, where $\phi$ is its angular position at $t = 0$.

The projection of particle $P'$ onto the $x$ axis is a point $P$, which we take to be a second particle. The projection of the position vector of particle $P'$ onto the $x$ axis gives the location $x(t)$ of $P$. Thus, we find

$$x(t) = x_m \cos(\omega t + \phi),$$

which is precisely Eq. 15-3. Our conclusion is correct. If reference particle $P'$ moves in uniform circular motion, its projection particle $P$ moves in simple harmonic motion along a diameter of the circle.

Figure 15-14b shows the velocity $\vec{v}$ of the reference particle. From Eq. 10-18 ($v = \omega r$), the magnitude of the velocity vector is $\omega x_m$; its projection on the $x$ axis is

$$v(t) = -\omega x_m \sin(\omega t + \phi),$$

which is exactly Eq. 15-6. The minus sign appears because the velocity component of $P$ in Fig. 15-14b is directed to the left, in the negative direction of $x$.

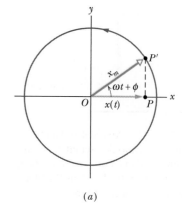

(a)

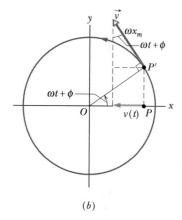

(b)

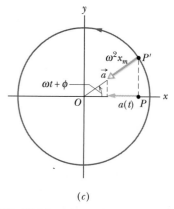

(c)

**FIG. 15-14** (a) A reference particle $P'$ moving with uniform circular motion in a reference circle of radius $x_m$. Its projection $P$ on the $x$ axis executes simple harmonic motion. (b) The projection of the velocity $\vec{v}$ of the reference particle is the velocity of SHM. (c) The projection of the radial acceleration $\vec{a}$ of the reference particle is the acceleration of SHM.

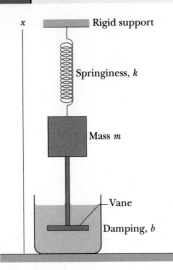

x

Rigid support

Springiness, *k*

Mass *m*

Vane

Damping, *b*

**FIG. 15-15** An idealized damped simple harmonic oscillator. A vane immersed in a liquid exerts a damping force on the block as the block oscillates parallel to the *x* axis.

Figure 15-14*c* shows the radial acceleration $\vec{a}$ of the reference particle. From Eq. 10-23 ($a_r = \omega^2 r$), the magnitude of the radial acceleration vector is $\omega^2 x_m$; its projection on the *x* axis is

$$a(t) = -\omega^2 x_m \cos(\omega t + \phi),$$

which is exactly Eq. 15-7. Thus, whether we look at the displacement, the velocity, or the acceleration, the projection of uniform circular motion is indeed simple harmonic motion.

## 15-8 | Damped Simple Harmonic Motion

A pendulum will swing only briefly underwater, because the water exerts on the pendulum a drag force that quickly eliminates the motion. A pendulum swinging in air does better, but still the motion dies out eventually, because the air exerts a drag force on the pendulum (and friction acts at its support point), transferring energy from the pendulum's motion.

When the motion of an oscillator is reduced by an external force, the oscillator and its motion are said to be **damped.** An idealized example of a damped oscillator is shown in Fig. 15-15, where a block with mass *m* oscillates vertically on a spring with spring constant *k*. From the block, a rod extends to a vane (both assumed massless) that is submerged in a liquid. As the vane moves up and down, the liquid exerts an inhibiting drag force on it and thus on the entire oscillating system. With time, the mechanical energy of the block–spring system decreases, as energy is transferred to thermal energy of the liquid and vane.

Let us assume the liquid exerts a **damping force** $\vec{F}_d$ that is proportional to the velocity $\vec{v}$ of the vane and block (an assumption that is accurate if the vane moves slowly). Then, for components along the *x* axis in Fig. 15-15, we have

$$F_d = -bv, \tag{15-39}$$

where *b* is a **damping constant** that depends on the characteristics of both the vane and the liquid and has the SI unit of kilogram per second. The minus sign indicates that $\vec{F}_d$ opposes the motion.

The force on the block from the spring is $F_s = -kx$. Let us assume that the gravitational force on the block is negligible relative to $F_d$ and $F_s$. Then we can write Newton's second law for components along the *x* axis ($F_{\text{net},x} = ma_x$) as

$$-bv - kx = ma. \tag{15-40}$$

Substituting $dx/dt$ for *v* and $d^2x/dt^2$ for *a* and rearranging give us the differential equation

$$m\frac{d^2x}{dt^2} + b\frac{dx}{dt} + kx = 0. \tag{15-41}$$

The solution of this equation is

$$x(t) = x_m e^{-bt/2m} \cos(\omega' t + \phi), \tag{15-42}$$

where $x_m$ is the amplitude and $\omega'$ is the angular frequency of the damped oscillator. This angular frequency is given by

$$\omega' = \sqrt{\frac{k}{m} - \frac{b^2}{4m^2}}. \tag{15-43}$$

If $b = 0$ (there is no damping), then Eq. 15-43 reduces to Eq. 15-12 ($\omega = \sqrt{k/m}$) for the angular frequency of an undamped oscillator, and Eq. 15-42 reduces to Eq. 15-3 for the displacement of an undamped oscillator. If the damping constant is small but not zero (so that $b \ll \sqrt{km}$), then $\omega' \approx \omega$.

We can regard Eq. 15-42 as a cosine function whose amplitude, which is $x_m e^{-bt/2m}$, gradually decreases with time, as Fig. 15-16 suggests. For an undamped

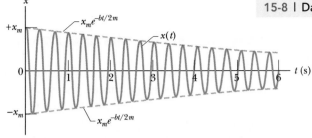

**FIG. 15-16** The displacement function $x(t)$ for the damped oscillator of Fig. 15-15, with values given in Sample Problem 15-7. The amplitude, which is $x_m e^{-bt/2m}$, decreases exponentially with time.

oscillator, the mechanical energy is constant and is given by Eq. 15-21 ($E = \frac{1}{2}kx_m^2$). If the oscillator is damped, the mechanical energy is not constant but decreases with time. If the damping is small, we can find $E(t)$ by replacing $x_m$ in Eq. 15-21 with $x_m e^{-bt/2m}$, the amplitude of the damped oscillations. By doing so, we find that

$$E(t) \approx \tfrac{1}{2}kx_m^2 e^{-bt/m}, \tag{15-44}$$

which tells us that, like the amplitude, the mechanical energy decreases exponentially with time.

✓**CHECKPOINT 5**    Here are three sets of values for the spring constant, damping constant, and mass for the damped oscillator of Fig. 15-15. Rank the sets according to the time required for the mechanical energy to decrease to one-fourth of its initial value, greatest first.

| | | | |
|---|---|---|---|
| Set 1 | $2k_0$ | $b_0$ | $m_0$ |
| Set 2 | $k_0$ | $6b_0$ | $4m_0$ |
| Set 3 | $3k_0$ | $3b_0$ | $m_0$ |

## Sample Problem  15-7

For the damped oscillator of Fig. 15-15, $m = 250$ g, $k = 85$ N/m, and $b = 70$ g/s.

(a) What is the period of the motion?

**KEY IDEA**    Because $b \ll \sqrt{km} = 4.6$ kg/s, the period is approximately that of the undamped oscillator.

*Calculation:* From Eq. 15-13, we then have

$$T = 2\pi\sqrt{\frac{m}{k}} = 2\pi\sqrt{\frac{0.25 \text{ kg}}{85 \text{ N/m}}} = 0.34 \text{ s.} \quad \text{(Answer)}$$

(b) How long does it take for the amplitude of the damped oscillations to drop to half its initial value?

**KEY IDEA**    The amplitude at time $t$ is displayed in Eq. 15-42 as $x_m e^{-bt/2m}$.

*Calculations:* The amplitude has the value $x_m$ at $t = 0$. Thus, we must find the value of $t$ for which

$$x_m e^{-bt/2m} = \tfrac{1}{2}x_m.$$

Canceling $x_m$ and taking the natural logarithm of the equation that remains, we have $\ln\frac{1}{2}$ on the right side and

$$\ln(e^{-bt/2m}) = -bt/2m$$

on the left side. Thus,

$$t = \frac{-2m \ln\frac{1}{2}}{b} = \frac{-(2)(0.25 \text{ kg})(\ln\frac{1}{2})}{0.070 \text{ kg/s}}$$

$$= 5.0 \text{ s.} \quad \text{(Answer)}$$

Because $T = 0.34$ s, this is about 15 periods of oscillation.

(c) How long does it take for the mechanical energy to drop to one-half its initial value?

**KEY IDEA**    From Eq. 15-44, the mechanical energy at time $t$ is $\frac{1}{2}kx_m^2 e^{-bt/m}$.

*Calculations:* The mechanical energy has the value $\frac{1}{2}kx_m^2$ at $t = 0$. Thus, we must find the value of $t$ for which

$$\tfrac{1}{2}kx_m^2 e^{-bt/m} = \tfrac{1}{2}(\tfrac{1}{2}kx_m^2).$$

If we divide both sides of this equation by $\frac{1}{2}kx_m^2$ and solve for $t$ as we did above, we find

$$t = \frac{-m \ln\frac{1}{2}}{b} = \frac{-(0.25 \text{ kg})(\ln\frac{1}{2})}{0.070 \text{ kg/s}} = 2.5 \text{ s.} \quad \text{(Answer)}$$

This is exactly half the time we calculated in (b), or about 7.5 periods of oscillation. Figure 15-16 was drawn to illustrate this sample problem.

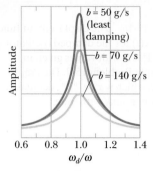

**FIG. 15-17** The displacement amplitude $x_m$ of a forced oscillator varies as the angular frequency $\omega_d$ of the driving force is varied. The curves here correspond to three values of the damping constant $b$.

# 15-9 | Forced Oscillations and Resonance

A person swinging in a swing without anyone pushing it is an example of *free oscillation.* However, if someone pushes the swing periodically, the swing has *forced,* or *driven, oscillations. Two* angular frequencies are associated with a system undergoing driven oscillations: (1) the *natural* angular frequency $\omega$ of the system, which is the angular frequency at which it would oscillate if it were suddenly disturbed and then left to oscillate freely, and (2) the angular frequency $\omega_d$ of the external driving force causing the driven oscillations.

We can use Fig. 15-15 to represent an idealized forced simple harmonic oscillator if we allow the structure marked "rigid support" to move up and down at a variable angular frequency $\omega_d$. Such a forced oscillator oscillates at the angular frequency $\omega_d$ of the driving force, and its displacement $x(t)$ is given by

$$x(t) = x_m \cos(\omega_d t + \phi), \qquad (15\text{-}45)$$

where $x_m$ is the amplitude of the oscillations.

How large the displacement amplitude $x_m$ is depends on a complicated function of $\omega_d$ and $\omega$. The velocity amplitude $v_m$ of the oscillations is easier to describe: it is greatest when

$$\omega_d = \omega \qquad \text{(resonance),} \qquad (15\text{-}46)$$

a condition called **resonance.** Equation 15-46 is also *approximately* the condition at which the displacement amplitude $x_m$ of the oscillations is greatest. Thus, if you push a swing at its natural angular frequency, the displacement and velocity amplitudes will increase to large values, a fact that children learn quickly by trial and error. If you push at other angular frequencies, either higher or lower, the displacement and velocity amplitudes will be smaller.

Figure 15-17 shows how the displacement amplitude of an oscillator depends on the angular frequency $\omega_d$ of the driving force, for three values of the damping coefficient $b$. Note that for all three the amplitude is approximately greatest when $\omega_d/\omega = 1$ (the resonance condition of Eq. 15-46). The curves of Fig. 15-17 show that less damping gives a taller and narrower *resonance peak.*

All mechanical structures have one or more natural angular frequencies, and if a structure is subjected to a strong external driving force that matches one of these angular frequencies, the resulting oscillations of the structure may rupture it. Thus, for example, aircraft designers must make sure that none of the natural angular frequencies at which a wing can oscillate matches the angular frequency of the engines in flight. A wing that flaps violently at certain engine speeds would obviously be dangerous.

Resonance appears to be one reason buildings in Mexico City collapsed in September 1985 when a major earthquake (8.1 on the Richter scale) occurred on the western coast of Mexico. The seismic waves from the earthquake should have been too weak to cause extensive damage when they reached Mexico City about 400 km away. However, Mexico City is largely built on an ancient lake bed, where the soil is still soft with water. Although the amplitude of the seismic waves was small in the firmer ground en route to Mexico City, their amplitude substantially increased in the loose soil of the city. Acceleration amplitudes of the waves were as much as $0.20g$, and the angular frequency was (surprisingly) concentrated around 3 rad/s. Not only was the ground severely oscillated, but many intermediate-height buildings had resonant angular frequencies of about 3 rad/s. Most of those buildings collapsed during the shaking (Fig. 15-18), while shorter buildings (with higher resonant angular frequencies) and taller buildings (with lower resonant angular frequencies) remained standing.

**FIG. 15-18** In 1985, buildings of intermediate height collapsed in Mexico City as a result of an earthquake far from the city. Taller and shorter buildings remained standing. *(John T. Barr/Getty Images News and Sport Services)*

## REVIEW & SUMMARY

**Frequency**  The *frequency f* of periodic, or oscillatory, motion is the number of oscillations per second. In the SI system, it is measured in hertz:

$$1 \text{ hertz} = 1 \text{ Hz} = 1 \text{ oscillation per second} = 1 \text{ s}^{-1}. \quad (15\text{-}1)$$

**Period**  The *period T* is the time required for one complete oscillation, or **cycle.** It is related to the frequency by

$$T = \frac{1}{f}. \quad (15\text{-}2)$$

**Simple Harmonic Motion**  In *simple harmonic motion* (SHM), the displacement $x(t)$ of a particle from its equilibrium position is described by the equation

$$x = x_m \cos(\omega t + \phi) \quad \text{(displacement)}, \quad (15\text{-}3)$$

in which $x_m$ is the **amplitude** of the displacement, the quantity $(\omega t + \phi)$ is the **phase** of the motion, and $\phi$ is the **phase constant.** The **angular frequency** $\omega$ is related to the period and frequency of the motion by

$$\omega = \frac{2\pi}{T} = 2\pi f \quad \text{(angular frequency)}. \quad (15\text{-}5)$$

Differentiating Eq. 15-3 leads to equations for the particle's SHM velocity and acceleration as functions of time:

$$v = -\omega x_m \sin(\omega t + \phi) \quad \text{(velocity)} \quad (15\text{-}6)$$

and

$$a = -\omega^2 x_m \cos(\omega t + \phi) \quad \text{(acceleration)}. \quad (15\text{-}7)$$

In Eq. 15-6, the positive quantity $\omega x_m$ is the **velocity amplitude** $v_m$ of the motion. In Eq. 15-7, the positive quantity $\omega^2 x_m$ is the **acceleration amplitude** $a_m$ of the motion.

**The Linear Oscillator**  A particle with mass $m$ that moves under the influence of a Hooke's law restoring force given by $F = -kx$ exhibits simple harmonic motion with

$$\omega = \sqrt{\frac{k}{m}} \quad \text{(angular frequency)} \quad (15\text{-}12)$$

and

$$T = 2\pi \sqrt{\frac{m}{k}} \quad \text{(period)}. \quad (15\text{-}13)$$

Such a system is called a **linear simple harmonic oscillator.**

**Energy**  A particle in simple harmonic motion has, at any time, kinetic energy $K = \frac{1}{2}mv^2$ and potential energy $U = \frac{1}{2}kx^2$. If no friction is present, the mechanical energy $E = K + U$ remains constant even though $K$ and $U$ change.

**Pendulums**  Examples of devices that undergo simple harmonic motion are the **torsion pendulum** of Fig. 15-7, the **simple pendulum** of Fig. 15-9, and the **physical pendulum** of Fig. 15-10. Their periods of oscillation for small oscillations are, respectively,

$$T = 2\pi \sqrt{I/\kappa} \quad \text{(torsion pendulum)}, \quad (15\text{-}23)$$

$$T = 2\pi \sqrt{L/g} \quad \text{(simple pendulum)}, \quad (15\text{-}28)$$

$$T = 2\pi \sqrt{I/mgh} \quad \text{(physical pendulum)}. \quad (15\text{-}29)$$

**Simple Harmonic Motion and Uniform Circular Motion**  Simple harmonic motion is the projection of uniform circular motion onto the diameter of the circle in which the circular motion occurs. Figure 15-14 shows that all parameters of circular motion (position, velocity, and acceleration) project to the corresponding values for simple harmonic motion.

**Damped Harmonic Motion**  The mechanical energy $E$ in a real oscillating system decreases during the oscillations because external forces, such as a drag force, inhibit the oscillations and transfer mechanical energy to thermal energy. The real oscillator and its motion are then said to be **damped.** If the **damping force** is given by $\vec{F}_d = -b\vec{v}$, where $\vec{v}$ is the velocity of the oscillator and $b$ is a **damping constant,** then the displacement of the oscillator is given by

$$x(t) = x_m e^{-bt/2m} \cos(\omega' t + \phi), \quad (15\text{-}42)$$

where $\omega'$, the angular frequency of the damped oscillator, is given by

$$\omega' = \sqrt{\frac{k}{m} - \frac{b^2}{4m^2}}. \quad (15\text{-}43)$$

If the damping constant is small ($b \ll \sqrt{km}$), then $\omega' \approx \omega$, where $\omega$ is the angular frequency of the undamped oscillator. For small $b$, the mechanical energy $E$ of the oscillator is given by

$$E(t) \approx \tfrac{1}{2}kx_m^2 e^{-bt/m}. \quad (15\text{-}44)$$

**Forced Oscillations and Resonance**  If an external driving force with angular frequency $\omega_d$ acts on an oscillating system with *natural* angular frequency $\omega$, the system oscillates with angular frequency $\omega_d$. The velocity amplitude $v_m$ of the system is greatest when

$$\omega_d = \omega, \quad (15\text{-}46)$$

a condition called **resonance.** The amplitude $x_m$ of the system is (approximately) greatest under the same condition.

## QUESTIONS

**1**  The acceleration $a(t)$ of a particle undergoing SHM is graphed in Fig. 15-19. (a) Which of the labeled points corresponds to the particle at $-x_m$? (b) At point 4, is the velocity of the particle positive, negative, or zero? (c) At point 5, is the

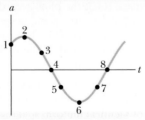

**FIG. 15-19**   Question 1.

particle at $-x_m$, at $+x_m$, at 0, between $-x_m$ and 0, or between 0 and $+x_m$?

**2**  Which of the following relationships between the acceleration $a$ and the displacement $x$ of a particle involve SHM: (a) $a = 0.5x$, (b) $a = 400x^2$, (c) $a = -20x$, (d) $a = -3x^2$?

**3**  Which of the following describe $\phi$ for the SHM of Fig. 15-20a:

(a) $-\pi < \phi < -\pi/2$,

(b) $\pi < \phi < 3\pi/2$,

(c) $-3\pi/2 < \phi < -\pi$?

**4** The velocity $v(t)$ of a particle undergoing SHM is graphed in Fig. 15-20b. Is the particle momentarily stationary, headed toward $-x_m$, or headed toward $+x_m$ at (a) point $A$ on the graph and (b) point $B$? Is the particle at $-x_m$, at $+x_m$, at 0, between $-x_m$ and 0, or between 0 and $+x_m$ when its velocity is represented by (c) point $A$ and (d) point $B$? Is the speed of the particle increasing or decreasing at (e) point $A$ and (f) point $B$?

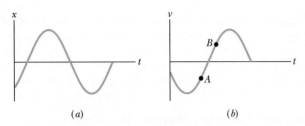

**FIG. 15-20** Questions 3 and 4.

**5** Figure 15-21 shows the $x(t)$ curves for three experiments involving a particular spring–box system oscillating in SHM. Rank the curves according to (a) the system's angular frequency, (b) the spring's potential energy at time $t = 0$, (c) the box's kinetic energy at $t = 0$, (d) the box's speed at $t = 0$, and (e) the box's maximum kinetic energy, greatest first.

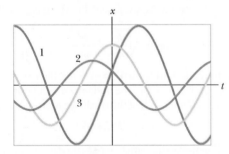

**FIG. 15-21** Question 5.

**6** Figure 15-22 gives, for three situations, the displacements $x(t)$ of a pair of simple harmonic oscillators ($A$ and $B$) that are identical except for phase. For each pair, what phase shift (in radians and in degrees) is needed to shift the curve for $A$ to coincide with the curve for $B$? Of the many possible answers, choose the shift with the smallest absolute magnitude.

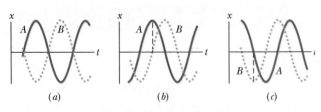

**FIG. 15-22** Question 6.

**7** You are to complete Fig. 15-23a so that it is a plot of velocity $v$ versus time $t$ for the spring–block oscillator that is shown in Fig. 15-23b for $t = 0$. (a) In Fig. 15-23a, at which lettered point or in what region between the points should the (vertical) $v$ axis intersect the $t$ axis? (For example, should it intersect at point $A$, or maybe in the region between points $A$ and $B$?) (b) If the block's velocity is given by $v = -v_m \sin(\omega t + \phi)$, what is the value of $\phi$? Make it positive, and if you cannot specify the value (such as $+\pi/2$ rad), then give a range of values (such as between 0 and $\pi/2$).

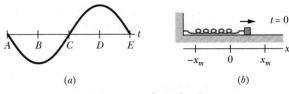

**FIG. 15-23** Question 7.

**8** You are to complete Fig. 15-24a so that it is a plot of acceleration $a$ versus time $t$ for the spring–block oscillator that is shown in Fig. 15-24b for $t = 0$. (a) In Fig. 15-24a, at which lettered point or in what region between the points should the (vertical) $a$ axis intersect the $t$ axis? (For example, should it intersect at point $A$, or maybe in the region between points $A$ and $B$?) (b) If the block's acceleration is given by $a = -a_m \cos(\omega t + \phi)$, what is the value of $\phi$? Make it positive, and if you cannot specify the value (such as $+\pi/2$ rad), then give a range of values (such as between 0 and $\pi/2$).

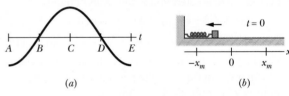

**FIG. 15-24** Question 8.

**9** In Fig. 15-25, a spring–block system is put into SHM in two experiments. In the first, the block is pulled from the equilibrium position through a displacement $d_1$ and then released. In the second, it is pulled from the equilibrium position through a greater displacement $d_2$ and then released. Are the (a) amplitude, (b) period, (c) frequency, (d) maximum kinetic energy, and (e) maximum potential energy in the second experiment greater than, less than, or the same as those in the first experiment?

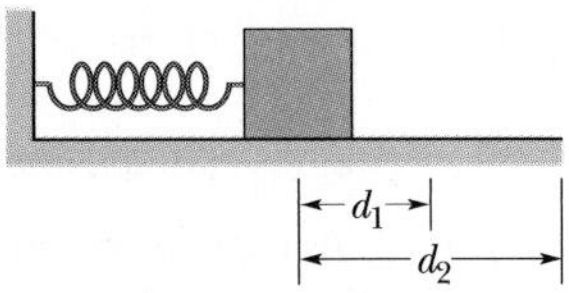

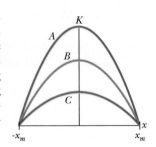

**FIG. 15-25** Question 9.

**10** Figure 15-26 shows plots of the kinetic energy $K$ versus position $x$ for three harmonic oscillators that have the same mass. Rank the plots according to (a) the corresponding spring constant and (b) the corresponding period of the oscillator, greatest first.

**11** Figure 15-27 shows three physical pendulums consisting of identical uniform spheres of the same mass that are rigidly connected by identical rods of negligible mass. Each pendulum is vertical and can pivot about suspension point $O$. Rank

**FIG. 15-26** Question 10.

the pendulums according to period of oscillation, greatest first.

**12** You are to build the oscillation transfer device shown in Fig. 15-28. It consists of two spring–block systems hanging from a flexible rod. When the spring of system 1 is stretched and then released, the resulting SHM of system 1 at frequency $f_1$ oscillates the rod. The rod then exerts a driving force on system 2, at the same frequency $f_1$. You can choose from four springs with spring constants $k$ of 1600, 1500, 1400, and 1200 N/m, and four blocks with masses $m$

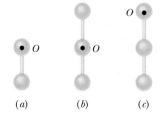

(a)      (b)      (c)

**FIG. 15-27** Question 11.

of 800, 500, 400, and 200 kg. Mentally determine which spring should go with which block in each of the two systems to maximize the amplitude of oscillations in system 2.

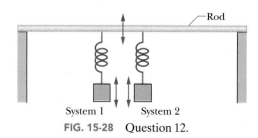

**FIG. 15-28** Question 12.

# PROBLEMS

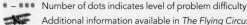

GO    Tutoring problem available (at instructor's discretion) in *WileyPLUS* and WebAssign
SSM   Worked-out solution available in Student Solutions Manual    WWW   Worked-out solution is at
•–•••  Number of dots indicates level of problem difficulty          ILW    Interactive solution is at ── http://www.wiley.com/college/halliday
      Additional information available in *The Flying Circus of Physics* and at flyingcircusofphysics.com

## sec. 15-3 The Force Law for Simple Harmonic Motion

**•1** What is the maximum acceleration of a platform that oscillates at amplitude 2.20 cm and frequency 6.60 Hz?

**•2** A particle with a mass of $1.00 \times 10^{-20}$ kg is oscillating with simple harmonic motion with a period of $1.00 \times 10^{-5}$ s and a maximum speed of $1.00 \times 10^3$ m/s. Calculate (a) the angular frequency and (b) the maximum displacement of the particle.

**•3** In an electric shaver, the blade moves back and forth over a distance of 2.0 mm in simple harmonic motion, with frequency 120 Hz. Find (a) the amplitude, (b) the maximum blade speed, and (c) the magnitude of the maximum blade acceleration. **SSM**

**•4** A 0.12 kg body undergoes simple harmonic motion of amplitude 8.5 cm and period 0.20 s. (a) What is the magnitude of the maximum force acting on it? (b) If the oscillations are produced by a spring, what is the spring constant?

**•5** An object undergoing simple harmonic motion takes 0.25 s to travel from one point of zero velocity to the next such point. The distance between those points is 36 cm. Calculate the (a) period, (b) frequency, and (c) amplitude of the motion.

**•6** An automobile can be considered to be mounted on four identical springs as far as vertical oscillations are concerned. The springs of a certain car are adjusted so that the oscillations have a frequency of 3.00 Hz. (a) What is the spring constant of each spring if the mass of the car is 1450 kg and the mass is evenly distributed over the springs? (b) What will be the oscillation frequency if five passengers, averaging 73.0 kg each, ride in the car with an even distribution of mass?

**•7** An oscillator consists of a block of mass 0.500 kg connected to a spring. When set into oscillation with amplitude 35.0 cm, the oscillator repeats its motion every 0.500 s. Find the (a) period, (b) frequency, (c) angular frequency, (d) spring constant, (e) maximum speed, and (f) magnitude of the maximum force on the block from the spring. **SSM**

**•8** An oscillating block–spring system takes 0.75 s to begin repeating its motion. Find (a) the period, (b) the frequency in hertz, and (c) the angular frequency in radians per second.

**•9** A loudspeaker produces a musical sound by means of the oscillation of a diaphragm whose amplitude is limited to 1.00 μm. (a) At what frequency is the magnitude $a$ of the diaphragm's acceleration equal to $g$? (b) For greater frequencies, is $a$ greater than or less than $g$? **SSM**

**•10** What is the phase constant for the harmonic oscillator with the position function $x(t)$ given in Fig. 15-29 if the position function has the form $x = x_m \cos(\omega t + \phi)$? The vertical axis scale is set by $x_s = 6.0$ cm.

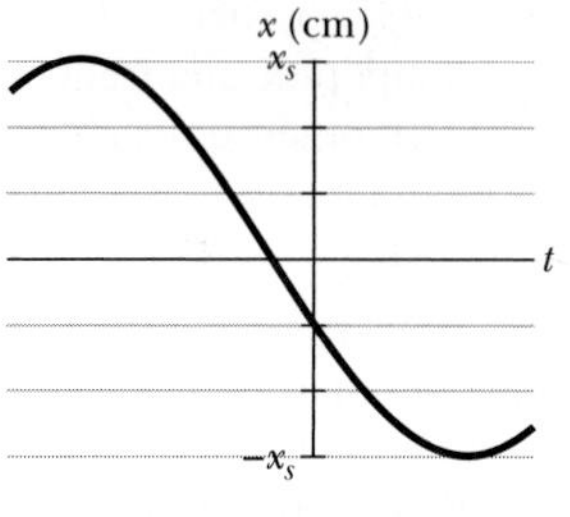

**•11** The function $x = (6.0 \text{ m})$ $\cos[(3\pi \text{ rad/s})t + \pi/3 \text{ rad}]$ gives the simple harmonic motion of a body. At $t = 2.0$ s, what are the (a) displacement, (b) velocity, (c) acceleration, and (d) phase of the motion? Also, what are the (e) frequency and (f) period of the motion?

**FIG. 15-29** Problem 10.

**•12** What is the phase constant for the harmonic oscillator with the velocity function $v(t)$ given in Fig. 15-30 if the position function $x(t)$ has the form $x = x_m \cos(\omega t + \phi)$? The vertical axis scale is set by $v_s = 4.0$ cm/s.

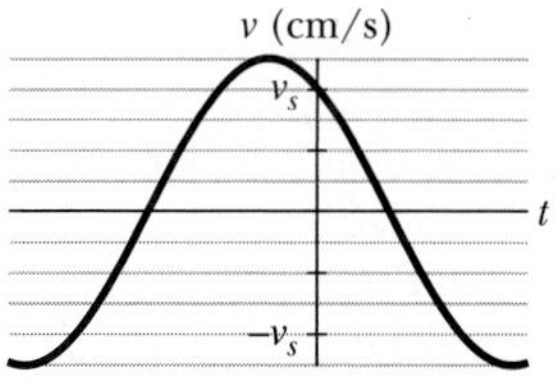

**FIG. 15-30** Problem 12.

**•13** In Fig. 15-31, two identical springs of spring constant 7580 N/m are attached to a block of mass 0.245 kg. What is the frequency of oscillation on the frictionless floor?

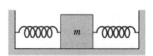

**FIG. 15-31**
Problems 13 and 23.

**••14** Figure 15-32 shows block 1 of mass 0.200 kg sliding to the right over a frictionless elevated surface at a speed of 8.00 m/s. The block undergoes an elastic collision with stationary block 2, which is attached to a spring of spring constant 1208.5 N/m. (Assume that the spring does not affect the collision.) After the collision, block 2 oscillates in SHM with a period of 0.140 s, and block 1 slides off the opposite end of the elevated surface, landing a distance $d$ from the base of that surface after falling height $h = 4.90$ m. What is the value of $d$?

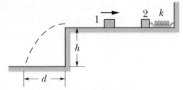

FIG. 15-32   Problem 14.

**••15** An oscillator consists of a block attached to a spring ($k = 400$ N/m). At some time $t$, the position (measured from the system's equilibrium location), velocity, and acceleration of the block are $x = 0.100$ m, $v = -13.6$ m/s, and $a = -123$ m/s$^2$. Calculate (a) the frequency of oscillation, (b) the mass of the block, and (c) the amplitude of the motion.   **ILW**

**••16** At a certain harbor, the tides cause the ocean surface to rise and fall a distance $d$ (from highest level to lowest level) in simple harmonic motion, with a period of 12.5 h. How long does it take for the water to fall a distance $0.250d$ from its highest level?

**••17** A block is on a horizontal surface (a shake table) that is moving back and forth horizontally with simple harmonic motion of frequency 2.0 Hz. The coefficient of static friction between block and surface is 0.50. How great can the amplitude of the SHM be if the block is not to slip along the surface?   **SSM   WWW**

**••18** Two particles execute simple harmonic motion of the same amplitude and frequency along close parallel lines. They pass each other moving in opposite directions each time their displacement is half their amplitude. What is their phase difference?

**••19** Two particles oscillate in simple harmonic motion along a common straight-line segment of length $A$. Each particle has a period of 1.5 s, but they differ in phase by $\pi/6$ rad. (a) How far apart are they (in terms of $A$) 0.50 s after the lagging particle leaves one end of the path? (b) Are they then moving in the same direction, toward each other, or away from each other?   **SSM**

**••20** Figure 15-33$a$ is a partial graph of the position function $x(t)$ for a simple harmonic oscillator with an angular frequency of 1.20 rad/s; Fig. 15-33$b$ is a partial graph of the corresponding velocity function $v(t)$. The vertical axis scales are set by $x_s = 5.0$ cm and $v_s = 5.0$ cm/s. What is the phase constant of the SHM if the position function $x(t)$ is given the form $x = x_m \cos(\omega t + \phi)$?

**••21** A block rides on a piston that is moving vertically with simple harmonic motion.

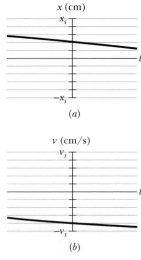

FIG. 15-33   Problem 20.

(a) If the SHM has period 1.0 s, at what amplitude of motion will the block and piston separate? (b) If the piston has an amplitude of 5.0 cm, what is the maximum frequency for which the block and piston will be in contact continuously?

**••22** A simple harmonic oscillator consists of a block of mass 2.00 kg attached to a spring of spring constant 100 N/m. When $t = 1.00$ s, the position and velocity of the block are $x = 0.129$ m and $v = 3.415$ m/s. (a) What is the amplitude of the oscillations? What were the (b) position and (c) velocity of the block at $t = 0$ s?

**••23** In Fig. 15-31, two springs are attached to a block that can oscillate over a frictionless floor. If the left spring is removed, the block oscillates at a frequency of 30 Hz. If, instead, the spring on the right is removed, the block oscillates at a frequency of 45 Hz. At what frequency does the block oscillate with both springs attached?   **ILW**

**•••24** In Fig. 15-34, two blocks ($m = 1.8$ kg and $M = 10$ kg) and a spring ($k = 200$ N/m) are arranged on a horizontal, frictionless surface. The coefficient of static friction between the two blocks is 0.40. What amplitude of simple harmonic motion of the spring–blocks system puts the smaller block on the verge of slipping over the larger block?   **GO**

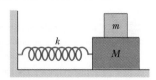

FIG. 15-34   Problem 24.

**•••25** In Fig. 15-35, a block weighing 14.0 N, which can slide without friction on an incline at angle $\theta = 40.0°$, is connected to the top of the incline by a massless spring of unstretched length 0.450 m and spring constant 120 N/m. (a) How far from the top of the incline is the block's equilibrium point? (b) If the block is pulled slightly down the incline and released, what is the period of the resulting oscillations?   **GO**

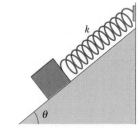

FIG. 15-35   Problem 25.

**•••26** In Fig. 15-36, two springs are joined and connected to a block of mass 0.245 kg that is set oscillating over a frictionless floor. The springs each have spring constant $k = 6430$ N/m. What is the frequency of the oscillations?

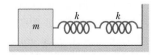

FIG. 15-36   Problem 26.

**sec. 15-4   Energy in Simple Harmonic Motion**

**•27** Find the mechanical energy of a block–spring system having a spring constant of 1.3 N/cm and an oscillation amplitude of 2.4 cm.   **SSM**

**•28** An oscillating block–spring system has a mechanical energy of 1.00 J, an amplitude of 10.0 cm, and a maximum speed of 1.20 m/s. Find (a) the spring constant, (b) the mass of the block, and (c) the frequency of oscillation.

**•29** When the displacement in SHM is one-half the amplitude $x_m$, what fraction of the total energy is (a) kinetic energy and (b) potential energy? (c) At what displacement, in terms of the amplitude, is the energy of the system half kinetic energy and half potential energy?   **SSM**

**•30** Figure 15-37 gives the one-dimensional potential energy well for a 2.0 kg particle (the function $U(x)$ has the form $bx^2$ and the vertical axis scale is set by $U_s = 2.0$ J). (a) If the particle passes through the equilibrium position with a velocity of 85 cm/s, will it be turned back before it reaches $x = 15$ cm? (b) If yes, at what position, and if no, what is the speed of the particle at $x = 15$ cm?

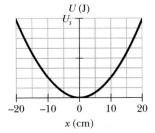

**FIG. 15-37**  Problem 30.

**•31** A 5.00 kg object on a horizontal frictionless surface is attached to a spring with $k = 1000$ N/m. The object is displaced from equilibrium 50.0 cm horizontally and given an initial velocity of 10.0 m/s back toward the equilibrium position. What are (a) the motion's frequency, (b) the initial potential energy of the block–spring system, (c) the initial kinetic energy, and (d) the motion's amplitude? **ILW**

**•32** Figure 15-38 shows the kinetic energy $K$ of a simple harmonic oscillator versus its position $x$. The vertical axis scale is set by $K_s = 4.0$ J. What is the spring constant?

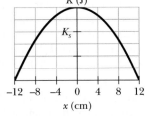

**FIG. 15-38**  Problem 32.

**••33** A 10 g particle undergoes SHM with an amplitude of 2.0 mm, a maximum acceleration of magnitude $8.0 \times 10^3$ m/s², and an unknown phase constant $\phi$. What are (a) the period of the motion, (b) the maximum speed of the particle, and (c) the total mechanical energy of the oscillator? What is the magnitude of the force on the particle when the particle is at (d) its maximum displacement and (e) half its maximum displacement?

**••34** If the phase angle for a block–spring system in SHM is $\pi/6$ rad and the block's position is given by $x = x_m \cos(\omega t + \phi)$, what is the ratio of the kinetic energy to the potential energy at time $t = 0$?

**••35** A block of mass $M = 5.4$ kg, at rest on a horizontal frictionless table, is attached to a rigid support by a spring of constant $k = 6000$ N/m. A bullet of mass $m = 9.5$ g and velocity $\vec{v}$ of magnitude 630 m/s strikes and is embedded in the block (Fig. 15-39). Assuming the compression of the spring is negligible until the bullet is embedded, determine (a) the speed of the block immediately after the collision and (b) the amplitude of the resulting simple harmonic motion. **GO**

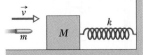

**FIG. 15-39**  Problem 35.

**••36** In Fig. 15-40, block 2 of mass 2.0 kg oscillates on the end of a spring in SHM with a period of 20 ms. The position of the block is given by $x = (1.0 \text{ cm}) \cos(\omega t + \pi/2)$. Block 1 of mass 4.0 kg slides toward block 2 with a velocity of magnitude 6.0 m/s, directed along the spring's length. The two blocks undergo a completely inelastic collision at time $t = 5.0$ ms. (The duration of the collision is much less than the period

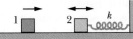

**FIG. 15-40**  Problem 36.

of motion.) What is the amplitude of the SHM after the collision? **GO**

**•••37** A massless spring hangs from the ceiling with a small object attached to its lower end. The object is initially held at rest in a position $y_i$ such that the spring is at its rest length. The object is then released from $y_i$ and oscillates up and down, with its lowest position being 10 cm below $y_i$. (a) What is the frequency of the oscillation? (b) What is the speed of the object when it is 8.0 cm below the initial position? (c) An object of mass 300 g is attached to the first object, after which the system oscillates with half the original frequency. What is the mass of the first object? (d) How far below $y_i$ is the new equilibrium (rest) position with both objects attached to the spring?

### sec. 15-5  An Angular Simple Harmonic Oscillator

**•38** A 95 kg solid sphere with a 15 cm radius is suspended by a vertical wire. A torque of 0.20 N · m is required to rotate the sphere through an angle of 0.85 rad and then maintain that orientation. What is the period of the oscillations that result when the sphere is then released?

**••39** The balance wheel of an old-fashioned watch oscillates with angular amplitude $\pi$ rad and period 0.500 s. Find (a) the maximum angular speed of the wheel, (b) the angular speed at displacement $\pi/2$ rad, and (c) the magnitude of the angular acceleration at displacement $\pi/4$ rad. **SSM WWW**

### sec. 15-6  Pendulums

**•40** Suppose that a simple pendulum consists of a small 60.0 g bob at the end of a cord of negligible mass. If the angle $\theta$ between the cord and the vertical is given by

$$\theta = (0.0800 \text{ rad}) \cos[(4.43 \text{ rad/s})t + \phi],$$

what are (a) the pendulum's length and (b) its maximum kinetic energy?

**•41** (a) If the physical pendulum of Sample Problem 15-5 is inverted and suspended at point $P$, what is its period of oscillation? (b) Is the period now greater than, less than, or equal to its previous value?

**•42** In Sample Problem 15-5, we saw that a physical pendulum has a center of oscillation at distance $2L/3$ from its point of suspension. Show that the distance between the point of suspension and the center of oscillation for a physical pendulum of any form is $I/mh$, where $I$ and $h$ have the meanings assigned to them in Eq. 15-29 and $m$ is the mass of the pendulum.

**•43** In Fig. 15-41, the pendulum consists of a uniform disk with radius $r = 10.0$ cm and mass 500 g attached to a uniform rod with length $L = 500$ mm and mass 270 g. (a) Calculate the rotational inertia of the pendulum about the pivot point. (b) What is the distance between the pivot point and the center of mass of the pendulum? (c) Calculate the period of oscillation. **SSM**

**•44** A physical pendulum consists of a meter stick that is pivoted at a small hole drilled through the

**FIG. 15-41**  Problem 43.

stick a distance $d$ from the 50 cm mark. The period of oscillation is 2.5 s. Find $d$. **ILW**

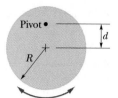

**FIG. 15-42**
Problem 45.

•45 In Fig. 15-42, a physical pendulum consists of a uniform solid disk (of radius $R = 2.35$ cm) supported in a vertical plane by a pivot located a distance $d = 1.75$ cm from the center of the disk. The disk is displaced by a small angle and released. What is the period of the resulting simple harmonic motion?

•46 A physical pendulum consists of two meter-long sticks joined together as shown in Fig. 15-43. What is the pendulum's period of oscillation about a pin inserted through point $A$ at the center of the horizontal stick?

**FIG. 15-43**
Problem 46.

•47 A performer seated on a trapeze is swinging back and forth with a period of 8.85 s. If she stands up, thus raising the center of mass of the *trapeze + performer* system by 35.0 cm, what will be the new period of the system? Treat *trapeze + performer* as a simple pendulum.

••48 A thin uniform rod (mass = 0.50 kg) swings about an axis that passes through one end of the rod and is perpendicular to the plane of the swing. The rod swings with a period of 1.5 s and an angular amplitude of 10°. (a) What is the length of the rod? (b) What is the maximum kinetic energy of the rod as it swings?

••49 In Fig. 15-44, a stick of length $L = 1.85$ m oscillates as a physical pendulum. (a) What value of distance $x$ between the stick's center of mass and its pivot point $O$ gives the least period? (b) What is that least period?

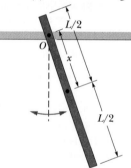

**FIG. 15-44** Problem 49.

••50 The 3.00 kg cube in Fig. 15-45 has edge lengths $d = 6.00$ cm and is mounted on an axle through its center. A spring ($k = 1200$ N/m) connects the cube's upper corner to a rigid wall. Initially the spring is at its rest length. If the cube is rotated 3° and released, what is the period of the resulting SHM? **GO**

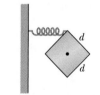

**FIG. 15-45** Problem 50.

••51 In the overhead view of Fig. 15-46, a long uniform rod of mass 0.600 kg is free to rotate in a horizontal plane about a vertical axis through its center. A spring with force constant $k = 1850$ N/m is connected horizontally between one end of the rod and a fixed wall. When the rod is in equilibrium, it is parallel to the wall. What is the period of the small oscillations that result when the rod is rotated slightly and released? **SSM ILW**

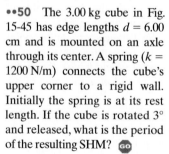

**FIG. 15-46** Problem 51.

••52 A rectangular block, with face lengths $a = 35$ cm and $b = 45$ cm, is to be suspended on a thin horizontal rod running through a narrow hole in the block. The block is then to be set swinging about the rod like a pendulum, through small angles so that it is in SHM. Figure 15-47 shows one possible position of the hole, at distance $r$ from the block's center, along a line connecting the center with a corner. (a) Plot the period of the pendulum versus distance $r$ along that line such that the minimum in the curve is apparent. (b) For what value of $r$ does that minimum occur? There is actually a line of points around the block's center for which the period of swinging has the same minimum value. (c) What shape does that line make?

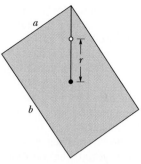

**FIG. 15-47** Problem 52.

••53 The angle of the pendulum of Fig. 15-9b is given by $\theta = \theta_m \cos[(4.44 \text{ rad/s})t + \phi]$. If at $t = 0$, $\theta = 0.040$ rad and $d\theta/dt = -0.200$ rad/s, what are (a) the phase constant $\phi$ and (b) the maximum angle $\theta_m$? (*Hint:* Don't confuse the rate $d\theta/dt$ at which $\theta$ changes with the $\omega$ of the SHM.)

••54 In Fig. 15-48a, a metal plate is mounted on an axle through its center of mass. A spring with $k = 2000$ N/m connects a wall with a point on the rim a distance $r = 2.5$ cm from the center of mass. Initially the spring is at its rest length. If the plate is rotated by 7° and released, it rotates about the axle in SHM, with its angular position given by Fig. 15-48b. The horizontal axis scale is set by $t_s = 20$ ms. What is the rotational inertia of the plate about its center of mass? **GO**

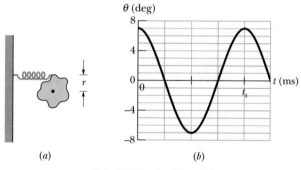

**FIG. 15-48** Problem 54.

•••55 A pendulum is formed by pivoting a long thin rod about a point on the rod. In a series of experiments, the period is measured as a function of the distance $x$ between the pivot point and the rod's center. (a) If the rod's length is $L = 2.20$ m and its mass is $m = 22.1$ g, what is the minimum period? (b) If $x$ is chosen to minimize the period and then $L$ is increased, does the period increase, decrease, or remain the same? (c) If, instead, $m$ is increased without $L$ increasing, does the period increase, decrease, or remain the same?

•••56 In Fig. 15-49, a 2.50 kg disk of diameter $D = 42.0$ cm is supported by a rod of length $L = 76.0$ cm and negligible mass that is pivoted at its end. (a) With the massless torsion spring unconnected, what is the period of oscillation? (b) With the torsion spring connected, the rod is vertical at equilibrium.

What is the torsion constant of the spring if the period of oscillation has been decreased by 0.500 s?

### sec. 15-8 Damped Simple Harmonic Motion

•57   In Fig. 15-15, the block has a mass of 1.50 kg and the spring constant is 8.00 N/m. The damping force is given by $-b(dx/dt)$, where $b = 230$ g/s. The block is pulled down 12.0 cm and released. (a) Calculate the time required for the amplitude of the resulting oscillations to fall to one-third of its initial value. (b) How many oscillations are made by the block in this time?   SSM WWW

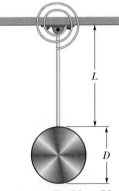

FIG. 15-49   Problem 56.

•58   In Sample Problem 15-7, what is the ratio of the amplitude of the damped oscillations to the initial amplitude at the end of 20 cycles?

•59   The amplitude of a lightly damped oscillator decreases by 3.0% during each cycle. What percentage of the mechanical energy of the oscillator is lost in each cycle?

••60   The suspension system of a 2000 kg automobile "sags" 10 cm when the chassis is placed on it. Also, the oscillation amplitude decreases by 50% each cycle. Estimate the values of (a) the spring constant $k$ and (b) the damping constant $b$ for the spring and shock absorber system of one wheel, assuming each wheel supports 500 kg.

### sec. 15-9   Forced Oscillations and Resonance

•61   For Eq. 15-45, suppose the amplitude $x_m$ is given by

$$x_m = \frac{F_m}{[m^2(\omega_d^2 - \omega^2)^2 + b^2\omega_d^2]^{1/2}},$$

where $F_m$ is the (constant) amplitude of the external oscillating force exerted on the spring by the rigid support in Fig. 15-15. At resonance, what are the (a) amplitude and (b) velocity amplitude of the oscillating object?

•62   Hanging from a horizontal beam are nine simple pendulums of the following lengths: (a) 0.10, (b) 0.30, (c) 0.40, (d) 0.80, (e) 1.2, (f) 2.8, (g) 3.5, (h) 5.0, and (i) 6.2 m. Suppose the beam undergoes horizontal oscillations with angular frequencies in the range from 2.00 rad/s to 4.00 rad/s. Which of the pendulums will be (strongly) set in motion?

••63   A 1000 kg car carrying four 82 kg people travels over a "washboard" dirt road with corrugations 4.0 m apart. The car bounces with maximum amplitude when its speed is 16 km/h. When the car stops, and the people get out, by how much does the car body rise on its suspension?

### Additional Problems

64   A block is in SHM on the end of a spring, with position given by $x = x_m \cos(\omega t + \phi)$. If $\phi = \pi/5$ rad, then at $t = 0$ what percentage of the total mechanical energy is potential energy?

65   Figure 15-50 gives the position of a 20 g block oscillating in SHM on the end of a spring. The horizontal axis scale is set by $t_s = 40.0$ ms. What are (a) the maximum kinetic energy of the block and (b) the number of times per second

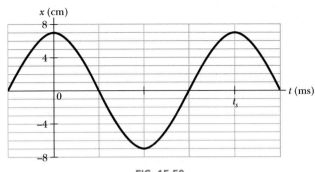

FIG. 15-50
Problems 65 and 66.

that maximum is reached? (*Hint:* Measuring a slope will probably not be very accurate. Find another approach.)

66   Figure 15-50 gives the position $x(t)$ of a block oscillating in SHM on the end of a spring ($t_s = 40.0$ ms). What are (a) the speed and (b) the magnitude of the radial acceleration of a particle in the corresponding uniform circular motion?

67   Figure 15-51 shows the kinetic energy $K$ of a simple pendulum versus its angle $\theta$ from the vertical. The vertical axis scale is set by $K_s = 10.0$ mJ. The pendulum bob has mass 0.200 kg. What is the length of the pendulum?

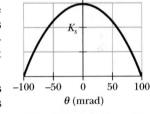

FIG. 15-51   Problem 67.

68   Although California is known for earthquakes, it has large regions dotted with precariously balanced rocks that would be easily toppled by even a mild earthquake. The rocks have stood this way for thousands of years, suggesting that major earthquakes have not occurred in those regions during that time. If an earthquake were to put such a rock into sinusoidal oscillation (parallel to the ground) with a frequency of 2.2 Hz, an oscillation amplitude of 1.0 cm would cause the rock to topple. What would be the magnitude of the maximum acceleration of the oscillation, in terms of $g$?

69   A 4.00 kg block is suspended from a spring with $k = 500$ N/m. A 50.0 g bullet is fired into the block from directly below with a speed of 150 m/s and becomes embedded in the block. (a) Find the amplitude of the resulting SHM. (b) What percentage of the original kinetic energy of the bullet is transferred to mechanical energy of the oscillator?

70   A 55.0 g block oscillates in SHM on the end of a spring with $k = 1500$ N/m according to $x = x_m \cos(\omega t + \phi)$. How long does the block take to move from position $+0.800x_m$ to (a) position $+0.600x_m$ and (b) position $-0.800x_m$?

71   A loudspeaker diaphragm is oscillating in simple harmonic motion with a frequency of 440 Hz and a maximum displacement of 0.75 mm. What are the (a) angular frequency, (b) maximum speed, and (c) magnitude of the maximum acceleration?

72   The tip of one prong of a tuning fork undergoes SHM of frequency 1000 Hz and amplitude 0.40 mm. For this tip, what is the magnitude of the (a) maximum acceleration, (b) maximum velocity, (c) acceleration at tip displacement 0.20 mm, and (d) velocity at tip displacement 0.20 mm?

**73** A flat uniform circular disk has a mass of 3.00 kg and a radius of 70.0 cm. It is suspended in a horizontal plane by a vertical wire attached to its center. If the disk is rotated 2.50 rad about the wire, a torque of 0.0600 N·m is required to maintain that orientation. Calculate (a) the rotational inertia of the disk about the wire, (b) the torsion constant, and (c) the angular frequency of this torsion pendulum when it is set oscillating.

**74** A uniform circular disk whose radius $R$ is 12.6 cm is suspended as a physical pendulum from a point on its rim. (a) What is its period? (b) At what radial distance $r < R$ is there a pivot point that gives the same period?

**75** What is the frequency of a simple pendulum 2.0 m long (a) in a room, (b) in an elevator accelerating upward at a rate of 2.0 m/s², and (c) in free fall? **SSM**

**76** A particle executes linear SHM with frequency 0.25 Hz about the point $x = 0$. At $t = 0$, it has displacement $x = 0.37$ cm and zero velocity. For the motion, determine the (a) period, (b) angular frequency, (c) amplitude, (d) displacement $x(t)$, (e) velocity $v(t)$, (f) maximum speed, (g) magnitude of the maximum acceleration, (h) displacement at $t = 3.0$ s, and (i) speed at $t = 3.0$ s.

**77** A 50.0 g stone is attached to the bottom of a vertical spring and set vibrating. If the maximum speed of the stone is 15.0 cm/s and the period is 0.500 s, find the (a) spring constant of the spring, (b) amplitude of the motion, and (c) frequency of oscillation.

**78** A 2.00 kg block hangs from a spring. A 300 g body hung below the block stretches the spring 2.00 cm farther. (a) What is the spring constant? (b) If the 300 g body is removed and the block is set into oscillation, find the period of the motion.

**79** The end point of a spring oscillates with a period of 2.0 s when a block with mass $m$ is attached to it. When this mass is increased by 2.0 kg, the period is found to be 3.0 s. Find $m$.

**80** A 0.10 kg block oscillates back and forth along a straight line on a frictionless horizontal surface. Its displacement from the origin is given by

$$x = (10 \text{ cm}) \cos[(10 \text{ rad/s})t + \pi/2 \text{ rad}].$$

(a) What is the oscillation frequency? (b) What is the maximum speed acquired by the block? (c) At what value of $x$ does this occur? (d) What is the magnitude of the maximum acceleration of the block? (e) At what value of $x$ does this occur? (f) What force, applied to the block by the spring, results in the given oscillation?

**81** A 3.0 kg particle is in simple harmonic motion in one dimension and moves according to the equation

$$x = (5.0 \text{ m}) \cos[(\pi/3 \text{ rad/s})t - \pi/4 \text{ rad}],$$

with $t$ in seconds. (a) At what value of $x$ is the potential energy of the particle equal to half the total energy? (b) How long does the particle take to move to this position $x$ from the equilibrium position?

**82** A massless spring with spring constant 19 N/m hangs vertically. A body of mass 0.20 kg is attached to its free end and then released. Assume that the spring was unstretched before the body was released. Find (a) how far below the initial position the body descends, and the (b) frequency and (c) amplitude of the resulting SHM.

**83** The piston in the cylinder head of a locomotive has a stroke (twice the amplitude) of 0.76 m. If the piston moves with simple harmonic motion with an angular frequency of 180 rev/min, what is its maximum speed? **SSM**

**84** A wheel is free to rotate about its fixed axle. A spring is attached to one of its spokes a distance $r$ from the axle, as shown in Fig. 15-52. (a) Assuming that the wheel is a hoop of mass $m$ and radius $R$, what is the angular frequency $\omega$ of small oscillations of this system in terms of $m$, $R$, $r$, and the spring constant $k$? What is $\omega$ if (b) $r = R$ and (c) $r = 0$?

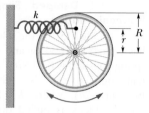

**FIG. 15-52** Problem 84.

**85** The scale of a spring balance that reads from 0 to 15.0 kg is 12.0 cm long. A package suspended from the balance is found to oscillate vertically with a frequency of 2.00 Hz. (a) What is the spring constant? (b) How much does the package weigh?

**86** A uniform spring with $k = 8600$ N/m is cut into pieces 1 and 2 of unstretched lengths $L_1 = 7.0$ cm and $L_2 = 10$ cm. What are (a) $k_1$ and (b) $k_2$? A block attached to the original spring as in Fig. 15-5 oscillates at 200 Hz. What is the oscillation frequency of the block attached to (c) piece 1 and (d) piece 2?

**87** In Fig. 15-53, three 10 000 kg ore cars are held at rest on a mine railway using a cable that is parallel to the rails, which are inclined at angle $\theta = 30°$. The cable stretches 15 cm just before the coupling between the two lower cars breaks, detaching the lowest car. Assuming that the cable obeys Hooke's law, find the (a) frequency and (b) amplitude of the resulting oscillations of the remaining two cars.

**FIG. 15-53** Problem 87.

**88** A simple pendulum of length 20 cm and mass 5.0 g is suspended in a race car traveling with constant speed 70 m/s around a circle of radius 50 m. If the pendulum undergoes small oscillations in a radial direction about its equilibrium position, what is the frequency of oscillation?

**89** A vertical spring stretches 9.6 cm when a 1.3 kg block is hung from its end. (a) Calculate the spring constant. This block is then displaced an additional 5.0 cm downward and released from rest. Find the (b) period, (c) frequency, (d) amplitude, and (e) maximum speed of the resulting SHM. **SSM**

**90** A block weighing 20 N oscillates at one end of a vertical spring for which $k = 100$ N/m; the other end of the spring is attached to a ceiling. At a certain instant the spring is stretched 0.30 m beyond its relaxed length (the length when no object is attached) and the block has zero velocity. (a) What is the net force on the block at this instant? What are the (b) amplitude and (c) period of the resulting simple harmonic motion? (d) What is the maximum kinetic energy of the block as it oscillates?

**91** A 1.2 kg block sliding on a horizontal frictionless surface is attached to a horizontal spring with $k = 480$ N/m. Let $x$ be the displacement of the block from the position at which the spring is unstretched. At $t = 0$ the block passes through $x = 0$ with a speed of 5.2 m/s in the positive $x$ direction. What are the (a) frequency and (b) amplitude of the block's motion? (c) Write an expression for $x$ as a function of time. **SSM**

**92** A simple harmonic oscillator consists of an 0.80 kg block attached to a spring ($k = 200$ N/m). The block slides on a horizontal frictionless surface about the equilibrium point $x = 0$ with a total mechanical energy of 4.0 J. (a) What is the amplitude of the oscillation? (b) How many oscillations does the block complete in 10 s? (c) What is the maximum kinetic energy attained by the block? (d) What is the speed of the block at $x = 0.15$ m?

**93** An engineer has an odd-shaped 10 kg object and needs to find its rotational inertia about an axis through its center of mass. The object is supported on a wire stretched along the desired axis. The wire has a torsion constant $\kappa = 0.50$ N·m. If this torsion pendulum oscillates through 20 cycles in 50 s, what is the rotational inertia of the object?

**94** A grandfather clock has a pendulum that consists of a thin brass disk of radius $r = 15.00$ cm and mass 1.000 kg that is attached to a long thin rod of negligible mass. The pendulum swings freely about an axis perpendicular to the rod and through the end of the rod opposite the disk, as shown in Fig. 15-54. If the pendulum is to have a period of 2.000 s for small oscillations at a place where $g = 9.800$ m/s², what must be the rod length $L$ to the nearest tenth of a millimeter?

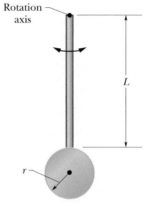

FIG. 15-54    Problem 94.

**95** A block sliding on a horizontal frictionless surface is attached to a horizontal spring with a spring constant of 600 N/m. The block executes SHM about its equilibrium position with a period of 0.40 s and an amplitude of 0.20 m. As the block slides through its equilibrium position, a 0.50 kg putty wad is dropped vertically onto the block. If the putty wad sticks to the block, determine (a) the new period of the motion and (b) the new amplitude of the motion.

**96** When a 20 N can is hung from the bottom of a vertical spring, it causes the spring to stretch 20 cm. (a) What is the spring constant? (b) This spring is now placed horizontally on a frictionless table. One end of it is held fixed, and the other end is attached to a 5.0 N can. The can is then moved (stretching the spring) and released from rest. What is the period of the resulting oscillation?

**97** A 4.00 kg block hangs from a spring, extending it 16.0 cm from its unstretched position. (a) What is the spring constant? (b) The block is removed, and a 0.500 kg body is hung from the same spring. If the spring is then stretched and released, what is its period of oscillation?

**98** A damped harmonic oscillator consists of a block ($m = 2.00$ kg), a spring ($k = 10.0$ N/m), and a damping force ($F = -bv$). Initially, it oscillates with an amplitude of 25.0 cm; because of the damping, the amplitude falls to three-fourths of this initial value at the completion of four oscillations. (a) What is the value of $b$? (b) How much energy has been "lost" during these four oscillations?

**99** A common device for entertaining a toddler is a *jump seat* that hangs from the horizontal portion of a doorframe via elastic cords (Fig. 15-55). Assume that only one cord is on each side in spite of the more realistic arrangement shown. When a child is placed in the seat, they both descend by a distance $d_s$ as the cords stretch (treat them as springs). Then the seat is pulled down an extra distance $d_m$ and released, so that the child oscillates vertically, like a block on the end of a spring. Suppose you are the safety engineer for the manufacturer of the seat. You do not want the magnitude of the child's acceleration to exceed $0.20g$ for fear of hurting the child's neck. If $d_m = 10$ cm, what value of $d_s$ corresponds to that acceleration magnitude?

FIG. 15-55    Problem 99.

**100** What is the phase constant for SMH with $a(t)$ given in Fig. 15-56 if the position function $x(t)$ has the form $x = x_m \cos(\omega t + \phi)$ and $a_s = 4.0$ m/s²?

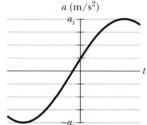

FIG. 15-56    Problem 100.

**101** A torsion pendulum consists of a metal disk with a wire running through its center and soldered in place. The wire is mounted vertically on clamps and pulled taut. Figure 15-57a gives the magnitude $\tau$ of the torque needed to rotate the disk about its center (and thus twist the wire) versus the rotation angle $\theta$. The vertical axis scale is set by $\tau_s = 4.0 \times 10^{-3}$ N·m. The disk is rotated to $\theta = 0.200$ rad and then released. Figure 15-57b shows the resulting oscillation in terms of angular position $\theta$ versus time $t$. The horizontal axis scale is set by $t_s = 0.40$ s. (a) What is the rotational inertia of the disk about its center? (b) What is the maximum angular speed $d\theta/dt$ of the disk? (*Caution:* Do not confuse the (constant) angular frequency of the SHM with the (varying) angular speed of the rotating disk, even though they usually have the same symbol $\omega$. *Hint:* The potential energy $U$ of a torsion pendulum is equal to $\frac{1}{2}\kappa\theta^2$, analogous to $U = \frac{1}{2}kx^2$ for a spring.)

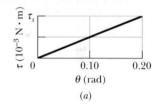

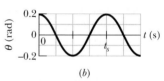

FIG. 15-57    Problem 101.

**102** A spider can tell when its web has captured, say, a fly because the fly's thrashing causes the web threads to oscillate. A spider can even determine the size of the fly by the frequency of the oscillations. Assume that a fly oscillates on the *capture thread* on which it is caught like a block on a spring. What is the ratio of oscillation frequency for a fly with mass $m$ to a fly with mass $2.5m$?

**103** For a simple pendulum, find the angular amplitude $\theta_m$ at which the restoring torque required for simple harmonic motion deviates from the actual restoring torque by 1.0%. (See "Trigonometric Expansions" in Appendix E.)

**104** A simple harmonic oscillator consists of a block attached to a spring with $k = 200$ N/m. The block slides on a frictionless surface, with equilibrium point $x = 0$ and amplitude 0.20 m. A graph of the block's velocity $v$ as a function of time $t$ is shown in Fig. 15-58. The horizontal scale is set by $t_s = 0.20$ s. What are (a) the period of the SHM, (b) the block's mass, (c) its displacement at $t = 0$, (d) its acceleration at $t = 0.10$ s, and (e) its maximum kinetic energy?

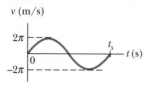

FIG. 15-58 Problem 104.

**105** A simple harmonic oscillator consists of a 0.50 kg block attached to a spring. The block slides back and forth along a straight line on a frictionless surface with equilibrium point $x = 0$. At $t = 0$ the block is at $x = 0$ and moving in the positive $x$ direction. A graph of the magnitude of the net force $\vec{F}$ on the block as a function of its position is shown in Fig. 15-59. The vertical scale is set by $F_s = 75.0$ N. What are (a) the amplitude and (b) the period of the motion, (c) the magnitude of the maximum acceleration, and (d) the maximum kinetic energy?

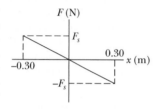

FIG. 15-59 Problem 105.

**106** In Fig. 15-60, a solid cylinder attached to a horizontal spring ($k = 3.00$ N/m) rolls without slipping along a horizontal surface. If the system is released from rest when the spring is stretched by 0.250 m, find (a) the translational kinetic energy and (b) the rotational kinetic energy of the cylinder as it passes through the equilibrium position. (c) Show that under these conditions the cylinder's center of mass executes simple harmonic motion with period

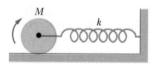

FIG. 15-60 Problem 106.

$$T = 2\pi \sqrt{\frac{3M}{2k}},$$

where $M$ is the cylinder mass. (*Hint:* Find the time derivative of the total mechanical energy.)

**107** A block weighing 10.0 N is attached to the lower end of a vertical spring ($k = 200.0$ N/m), the other end of which is attached to a ceiling. The block oscillates vertically and has a kinetic energy of 2.00 J as it passes through the point at which the spring is unstretched. (a) What is the period of the oscillation? (b) Use the law of conservation of energy to determine the maximum distance the block moves both above and below the point at which the spring is unstretched. (These are not necessarily the same.) (c) What is the amplitude of the oscillation? (d) What is the maximum kinetic energy of the block as it oscillates?

**108** A 2.0 kg block executes SHM while attached to a horizontal spring of spring constant 200 N/m. The maximum speed of the block as it slides on a horizontal frictionless surface is 3.0 m/s. What are (a) the amplitude of the block's motion, (b) the magnitude of its maximum acceleration, and (c) the magnitude of its minimum acceleration? (d) How long does the block take to complete 7.0 cycles of its motion?

**109** The vibration frequencies of atoms in solids at normal temperatures are of the order of $10^{13}$ Hz. Imagine the atoms to be connected to one another by springs. Suppose that a single silver atom in a solid vibrates with this frequency and that all the other atoms are at rest. Compute the effective spring constant. One mole of silver ($6.02 \times 10^{23}$ atoms) has a mass of 108 g.

**110** In Fig. 15-61, a 2500 kg demolition ball swings from the end of a crane. The length of the swinging segment of cable is 17 m. (a) Find the period of the swinging, assuming that the system can be treated as a simple pendulum. (b) Does the period depend on the ball's mass?

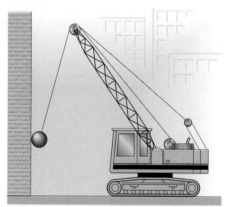

FIG. 15-61 Problem 110.

**111** The center of oscillation of a physical pendulum has this interesting property: If an impulse (assumed horizontal and in the plane of oscillation) acts at the center of oscillation, no oscillations are felt at the point of support. Baseball players (and players of many other sports) know that unless the ball hits the bat at this point (called the "sweet spot" by athletes), the oscillations due to the impact will sting their hands. To prove this property, let the stick in Fig. 15-11a simulate a baseball bat. Suppose that a horizontal force $\vec{F}$ (due to impact with the ball) acts toward the right at $P$, the center of oscillation. The batter is assumed to hold the bat at $O$, the pivot point of the stick. (a) What acceleration does the point $O$ undergo as a result of $\vec{F}$? (b) What angular acceleration is produced by $\vec{F}$ about the center of mass of the stick? (c) As a result of the angular acceleration in (b), what linear acceleration does point $O$ undergo? (d) Considering the magnitudes and directions of the accelerations in (a) and (c), convince yourself that $P$ is indeed the "sweet spot."

**112** A 2.0 kg block is attached to the end of a spring with a spring constant of 350 N/m and forced to oscillate by an applied force $F = (15$ N$) \sin(\omega_d t)$, where $\omega_d = 35$ rad/s. The damping constant is $b = 15$ kg/s. At $t = 0$, the block is at rest with the spring at its rest length. (a) Use numerical integration to plot the displacement of the block for the first 1.0 s. Use the motion near the end of the 1.0 s interval to estimate the amplitude, period, and angular frequency. Repeat the calculation for (b) $\omega_d = \sqrt{k/m}$ and (c) $\omega_d = 20$ rad/s.

# Waves—I

Sion Touhig/Getty Images News and Sport Services

Bruno Vincent/Reportage/Getty Images, Inc.

*In 2001, a sleek footbridge across the Thames River was opened in London to connect the Tate Modern art gallery with the vicinity of St. Paul's Cathedral and to mark the new millennium. However, soon after the first surge of pedestrians began to walk over it, the Millennium Bridge, as it is called, began to oscillate so much that many of the pedestrians had difficulty keeping their balance. Similar oscillations can occur in a mosh pit on a standard wood floor, especially if the participants are pogoing or body slamming.*

## What causes such oscillations, which can be the nightmare of structural engineers?

The answer is in this chapter.

413

## 16-1 | WHAT IS PHYSICS?

One of the primary subjects of physics is waves. To see how important waves are in the modern world, just consider the music industry. Every piece of music you hear, from some retro-punk band playing in a campus dive to the most eloquent concerto playing on the Web, depends on performers producing waves and your detecting those waves. In between production and detection, the information carried by the waves might need to be transmitted (as in a live performance on the Web) or recorded and then reproduced (as with CDs, DVDs, or the other devices currently being developed in engineering labs worldwide). The financial importance of controlling music waves is staggering, and the rewards to engineers who develop new control techniques can be rich.

This chapter focuses on waves traveling along a stretched string, such as on a guitar. The next chapter focuses on sound waves, such as those produced by a guitar string being played. Before we do all this, though, our first job is to classify the countless waves of the everyday world into basic types.

## 16-2 | Types of Waves

Waves are of three main types:

1. **Mechanical waves.** These waves are most familiar because we encounter them almost constantly; common examples include water waves, sound waves, and seismic waves. All these waves have two central features: They are governed by Newton's laws, and they can exist only within a material medium, such as water, air, and rock.

2. **Electromagnetic waves.** These waves are less familiar, but you use them constantly; common examples include visible and ultraviolet light, radio and television waves, microwaves, x rays, and radar waves. These waves require no material medium to exist. Light waves from stars, for example, travel through the vacuum of space to reach us. All electromagnetic waves travel through a vacuum at the same speed $c = 299\,792\,458$ m/s.

3. **Matter waves.** Although these waves are commonly used in modern technology, they are probably very unfamiliar to you. These waves are associated with electrons, protons, and other fundamental particles, and even atoms and molecules. Because we commonly think of these particles as constituting matter, such waves are called matter waves.

Much of what we discuss in this chapter applies to waves of all kinds. However, for specific examples we shall refer to mechanical waves.

## 16-3 | Transverse and Longitudinal Waves

A wave sent along a stretched, taut string is the simplest mechanical wave. If you give one end of a stretched string a single up-and-down jerk, a wave in the form of a single *pulse* travels along the string, as in Fig. 16-1a. This pulse and its motion can occur because the string is under tension. When you pull your end of the string upward, it begins to pull upward on the adjacent section of the string via tension between the two sections. As the adjacent section moves upward, it begins to pull the next section upward, and so on. Meanwhile, you have pulled down on your end of the string. As each section moves upward in turn, it begins to be pulled back downward by neighboring sections that are already on the way down. The net result is that a distortion in the string's shape (the pulse) moves along the string at some velocity $\vec{v}$.

If you move your hand up and down in continuous simple harmonic motion, a continuous wave travels along the string at velocity $\vec{v}$. Because the motion of your hand is a sinusoidal function of time, the wave has a sinusoidal shape at any given instant, as in Fig. 16-1b; that is, the wave has the shape of a sine curve or a cosine curve.

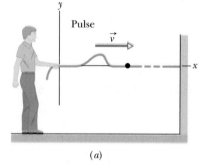

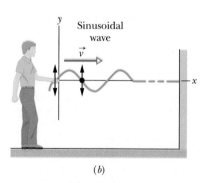

**FIG. 16-1** (a) A single pulse is sent along a stretched string. A typical string element (marked with a dot) moves up once and then down as the pulse passes. The element's motion is perpendicular to the wave's direction of travel, so the pulse is a *transverse wave*. (b) A sinusoidal wave is sent along the string. A typical string element moves up and down continuously as the wave passes. This too is a transverse wave.

We consider here only an "ideal" string, in which no friction-like forces within the string cause the wave to die out as it travels along the string. In addition, we assume that the string is so long that we need not consider a wave rebounding from the far end.

One way to study the waves of Fig. 16-1 is to monitor the **wave forms** (shapes of the waves) as they move to the right. Alternatively, we could monitor the motion of an element of the string as the element oscillates up and down while a wave passes through it. We would find that the displacement of every such oscillating string element is *perpendicular* to the direction of travel of the wave, as indicated in Fig. 16-1*b*. This motion is said to be **transverse,** and the wave is said to be a **transverse wave.**

Figure 16-2 shows how a sound wave can be produced by a piston in a long, air-filled pipe. If you suddenly move the piston rightward and then leftward, you can send a pulse of sound along the pipe. The rightward motion of the piston moves the elements of air next to it rightward, changing the air pressure there. The increased air pressure then pushes rightward on the elements of air somewhat farther along the pipe. Moving the piston leftward reduces the air pressure next to it. As a result, first the elements nearest the piston and then farther elements move leftward. Thus, the motion of the air and the change in air pressure travel rightward along the pipe as a pulse.

If you push and pull on the piston in simple harmonic motion, as is being done in Fig. 16-2, a sinusoidal wave travels along the pipe. Because the motion of the elements of air is parallel to the direction of the wave's travel, the motion is said to be **longitudinal,** and the wave is said to be a **longitudinal wave.** In this chapter we focus on transverse waves, and string waves in particular; in Chapter 17 we focus on longitudinal waves, and sound waves in particular.

Both a transverse wave and a longitudinal wave are said to be **traveling waves** because they both travel from one point to another, as from one end of the string to the other end in Fig. 16-1 and from one end of the pipe to the other end in Fig. 16-2. Note that it is the wave that moves from end to end, not the material (string or air) through which the wave moves.

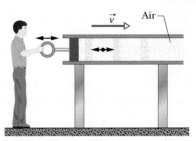

**FIG. 16-2** A sound wave is set up in an air-filled pipe by moving a piston back and forth. Because the oscillations of an element of the air (represented by the dot) are parallel to the direction in which the wave travels, the wave is a *longitudinal wave.*

## Sample Problem 16-1

Seismic waves are waves that travel either through Earth's interior or along the ground. Seismology stations are set up mainly to record seismic waves generated by earthquakes, but they also record seismic waves generated by any large release of energy near Earth's surface, such as an explosion. As the seismic waves travel past a station, they oscillate a recording pen and the pen traces out a graph. Figure 16-3*a* is one of the graphs created by seismic waves from the mysterious sinking of the Russian submarine *Kursk* in August 2000. The first oscillations of the pen are marked with an arrow and were of small amplitude. Much stronger oscillations began about 134 s later.

From this, analysts concluded that the first seismic waves were generated by an onboard explosion, possibly a torpedo that failed to launch when fired. The explosion presumably breached the hull, started a fire, and sank the submarine. The later, much stronger seismic waves were generated after the submarine was sunk and were possibly generated when the fire caused several of the powerful missiles on board to explode simultaneously.

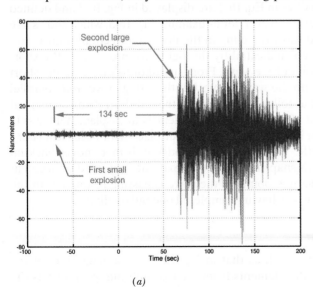

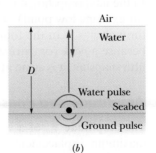

**FIG. 16-3** (*a*) Graph made by a recording device as seismic waves from the *Kursk* passed the device. Amplitude is plotted vertically; time increases rightward. *(Courtesy of Jay Pulli/BBN Technologies)* (*b*) With the submarine sitting at depth *D*, the large explosion sent pulses both into the ground and up through the water.

(*a*)

(*b*)

These stronger waves arrived at seismology stations as pulses separated by a time interval $\Delta t$ of about 0.11 s. To what depth $D$ did the submarine sink?

**KEY IDEA** The speed of a sound is equal to the distance traveled divided by the travel time.

**Calculations:** Let us assume that the stronger explosion occurred after the *Kursk* was sitting on the ocean floor (the seabed). That explosion sent a pulse into the seabed and a pulse upward through the water (Fig. 16-3b). The pulse traveling through the water "bounced" several times between the water surface and the seabed. Each time it hit the seabed, it sent another pulse into the ground, and seismology stations detected those ground pulses as they arrived one after another. Thus, the time $\Delta t$ between any two detected ground pulses was equal to the round-trip time for the water pulse to travel up to the water surface and back to the seabed. From Eq. 2-2 ($v_{avg} = \Delta x/\Delta t$), we can relate the speed $v$ of the water pulse to the round-trip distance $2D$ and the round-trip time $\Delta t$ as

$$v = \frac{2D}{\Delta t},$$

or

$$D = \frac{v\,\Delta t}{2}. \tag{16-1}$$

Waves travel through seawater at about 1500 m/s. Substituting this value and $\Delta t = 0.11$ s in Eq. 16-1, we find that the seismic data predicted the sunken *Kursk* was at a depth of approximately

$$D = \frac{(1500 \text{ m/s})(0.11 \text{ s})}{2}$$

$$= 82.5 \text{ m} \approx 83 \text{ m}. \qquad \text{(Answer)}$$

In fact, the wreck was discovered at a depth of about 115 m.

## 16-4 | Wavelength and Frequency

To completely describe a wave on a string (and the motion of any element along its length), we need a function that gives the shape of the wave. This means that we need a relation in the form $y = h(x, t)$, in which $y$ is the transverse displacement of any string element as a function $h$ of the time $t$ and the position $x$ of the element along the string. In general, a sinusoidal shape like the wave in Fig. 16-1b can be described with $h$ being either a sine or cosine function; both give the same general shape for the wave. In this chapter we use the sine function.

Imagine a sinusoidal wave like that of Fig. 16-1b traveling in the positive direction of an $x$ axis. As the wave sweeps through succeeding elements (that is, very short sections) of the string, the elements oscillate parallel to the $y$ axis. At time $t$, the displacement $y$ of the element located at position $x$ is given by

$$y(x, t) = y_m \sin(kx - \omega t). \tag{16-2}$$

Because this equation is written in terms of position $x$, it can be used to find the displacements of all the elements of the string as a function of time. Thus, it can tell us the shape of the wave at any given time and how that shape changes as the wave moves along the string.

The names of the quantities in Eq. 16-2 are displayed in Fig. 16-4 and defined next. Before we discuss them, however, let us examine Fig. 16-5, which shows five "snapshots" of a sinusoidal wave traveling in the positive direction of an $x$ axis. The movement of the wave is indicated by the rightward progress of the short arrow pointing to a high point of the wave. From snapshot to snapshot, the short arrow moves to the right with the wave shape, but the string moves *only* parallel to the $y$ axis. To see that, let us follow the motion of the red-dyed string element at $x = 0$. In the first snapshot (Fig. 16-5a), this element is at displacement $y = 0$. In the next snapshot, it is at its extreme downward displacement because a *valley* (or extreme low point) of the wave is passing through it. It then moves back up through $y = 0$. In the fourth snapshot, it is at its extreme upward displacement because a *peak* (or extreme high point) of the wave is passing through it. In the fifth snapshot, it is again at $y = 0$, having completed one full oscillation.

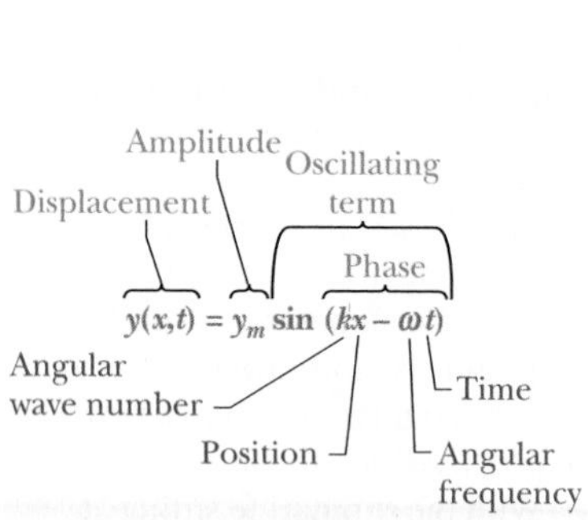

**FIG. 16-4** The names of the quantities in Eq. 16-2, for a transverse sinusoidal wave.

### Amplitude and Phase

The **amplitude** $y_m$ of a wave, such as that in Fig. 16-5, is the magnitude of the maximum displacement of the elements from their equilibrium positions as the wave passes through them. (The subscript $m$ stands for maximum.) Because $y_m$ is a magnitude, it is always a positive quantity, even if it is measured downward instead of upward as drawn in Fig. 16-5a.

The **phase** of the wave is the *argument* $kx - \omega t$ of the sine in Eq. 16-2. As the wave sweeps through a string element at a particular position $x$, the phase changes linearly with time $t$. This means that the sine also changes, oscillating between $+1$ and $-1$. Its extreme positive value ($+1$) corresponds to a peak of the wave moving through the element; at that instant the value of $y$ at position $x$ is $y_m$. Its extreme negative value ($-1$) corresponds to a valley of the wave moving through the element; at that instant the value of $y$ at position $x$ is $-y_m$. Thus, the sine function and the time-dependent phase of a wave correspond to the oscillation of a string element, and the amplitude of the wave determines the extremes of the element's displacement.

## Wavelength and Angular Wave Number

The **wavelength** $\lambda$ of a wave is the distance (parallel to the direction of the wave's travel) between repetitions of the shape of the wave (or *wave shape*). A typical wavelength is marked in Fig. 16-5a, which is a snapshot of the wave at time $t = 0$. At that time, Eq. 16-2 gives, for the description of the wave shape,

$$y(x,0) = y_m \sin kx. \qquad (16\text{-}3)$$

By definition, the displacement $y$ is the same at both ends of this wavelength—that is, at $x = x_1$ and $x = x_1 + \lambda$. Thus, by Eq. 16-3,

$$y_m \sin kx_1 = y_m \sin k(x_1 + \lambda)$$
$$= y_m \sin(kx_1 + k\lambda). \qquad (16\text{-}4)$$

A sine function begins to repeat itself when its angle (or argument) is increased by $2\pi$ rad, so in Eq. 16-4 we must have $k\lambda = 2\pi$, or

$$k = \frac{2\pi}{\lambda} \qquad \text{(angular wave number)}. \qquad (16\text{-}5)$$

We call $k$ the **angular wave number** of the wave; its SI unit is the radian per meter, or the inverse meter. (Note that the symbol $k$ here does *not* represent a spring constant as previously.)

Notice that the wave in Fig. 16-5 moves to the right by $\frac{1}{4}\lambda$ from one snapshot to the next. Thus, by the fifth snapshot, it has moved to the right by $1\lambda$.

## Period, Angular Frequency, and Frequency

Figure 16-6 shows a graph of the displacement $y$ of Eq. 16-2 versus time $t$ at a certain position along the string, taken to be $x = 0$. If you were to monitor the string, you would see that the single element of the string at that position moves up and down in simple harmonic motion given by Eq. 16-2 with $x = 0$:

$$y(0,t) = y_m \sin(-\omega t)$$
$$= -y_m \sin \omega t \qquad (x = 0). \qquad (16\text{-}6)$$

Here we have made use of the fact that $\sin(-\alpha) = -\sin \alpha$, where $\alpha$ is any angle. Figure 16-6 is a graph of this equation; it *does not* show the shape of the wave.

We define the **period** of oscillation $T$ of a wave to be the time any string element takes to move through one full oscillation. A typical period is marked on the graph of Fig. 16-6. Applying Eq. 16-6 to both ends of this time interval and equating the results yield

$$-y_m \sin \omega t_1 = -y_m \sin \omega(t_1 + T)$$
$$= -y_m \sin(\omega t_1 + \omega T). \qquad (16\text{-}7)$$

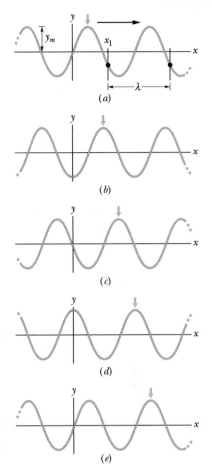

**FIG. 16-5** Five "snapshots" of a string wave traveling in the positive direction of an $x$ axis. The amplitude $y_m$ is indicated. A typical wavelength $\lambda$, measured from an arbitrary position $x_1$, is also indicated.

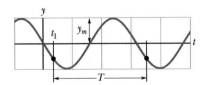

**FIG. 16-6** A graph of the displacement of the string element at $x = 0$ as a function of time, as the sinusoidal wave of Fig. 16-5 passes through the element. The amplitude $y_m$ is indicated. A typical period $T$, measured from an arbitrary time $t_1$, is also indicated.

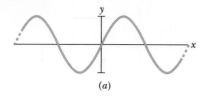

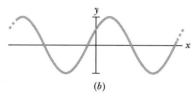

**FIG. 16-7** A sinusoidal traveling wave at $t = 0$ with a phase constant $\phi$ of (a) 0 and (b) $\pi/5$ rad.

This can be true only if $\omega T = 2\pi$, or if

$$\omega = \frac{2\pi}{T} \qquad \text{(angular frequency).} \qquad (16\text{-}8)$$

We call $\omega$ the **angular frequency** of the wave; its SI unit is the radian per second.

Look back at the five snapshots of a traveling wave in Fig. 16-5. The time between snapshots is $\frac{1}{4}T$. Thus, by the fifth snapshot, every string element has made one full oscillation.

The **frequency** $f$ of a wave is defined as $1/T$ and is related to the angular frequency $\omega$ by

$$f = \frac{1}{T} = \frac{\omega}{2\pi} \qquad \text{(frequency).} \qquad (16\text{-}9)$$

Like the frequency of simple harmonic motion in Chapter 15, this frequency $f$ is a number of oscillations per unit time—here, the number made by a string element as the wave moves through it. As in Chapter 15, $f$ is usually measured in hertz or its multiples, such as kilohertz.

✓**CHECKPOINT 1**    The figure is a composite of three snapshots, each of a wave traveling along a particular string. The phases for the waves are given by (a) $2x - 4t$, (b) $4x - 8t$, and (c) $8x - 16t$. Which phase corresponds to which wave in the figure?

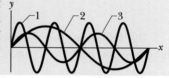

### Phase Constant

When a sinusoidal traveling wave is given by the wave function of Eq. 16-2, the wave near $x = 0$ looks like Fig. 16-7a when $t = 0$. Note that at $x = 0$, the displacement is $y = 0$ and the slope is at its maximum positive value. We can generalize Eq. 16-2 by inserting a **phase constant** $\phi$ in the wave function:

$$y = y_m \sin(kx - \omega t + \phi). \qquad (16\text{-}10)$$

The value of $\phi$ can be chosen so that the function gives some other displacement and slope at $x = 0$ when $t = 0$. For example, a choice of $\phi = +\pi/5$ rad gives the displacement and slope shown in Fig. 16-7b when $t = 0$. The wave is still sinusoidal with the same values of $y_m$, $k$, and $\omega$, but it is now shifted from what you see in Fig. 16-7a (where $\phi = 0$).

## 16-5 | The Speed of a Traveling Wave

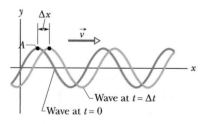

**FIG. 16-8** Two snapshots of the wave of Fig. 16-5, at time $t = 0$ and then at time $t = \Delta t$. As the wave moves to the right at velocity $\vec{v}$, the entire curve shifts a distance $\Delta x$ during $\Delta t$. Point $A$ "rides" with the wave form, but the string elements move only up and down.

Figure 16-8 shows two snapshots of the wave of Eq. 16-2, taken a small time interval $\Delta t$ apart. The wave is traveling in the positive direction of $x$ (to the right in Fig. 16-8), the entire wave pattern moving a distance $\Delta x$ in that direction during the interval $\Delta t$. The ratio $\Delta x/\Delta t$ (or, in the differential limit, $dx/dt$) is the **wave speed** $v$. How can we find its value?

As the wave in Fig. 16-8 moves, each point of the moving wave form, such as point $A$ marked on a peak, retains its displacement $y$. (Points on the string do not retain their displacement, but points on the wave *form* do.) If point $A$ retains its displacement as it moves, the phase in Eq. 16-2 giving it that displacement must remain a constant:

$$kx - \omega t = \text{a constant.} \qquad (16\text{-}11)$$

Note that although this argument is constant, both $x$ and $t$ are changing. In fact, as $t$ increases, $x$ must also, to keep the argument constant. This confirms that the wave pattern is moving in the positive direction of $x$.

To find the wave speed $v$, we take the derivative of Eq. 16-11, getting

$$k\frac{dx}{dt} - \omega = 0$$

or

$$\frac{dx}{dt} = v = \frac{\omega}{k}. \tag{16-12}$$

Using Eq. 16-5 ($k = 2\pi/\lambda$) and Eq. 16-8 ($\omega = 2\pi/T$), we can rewrite the wave speed as

$$v = \frac{\omega}{k} = \frac{\lambda}{T} = \lambda f \quad \text{(wave speed).} \tag{16-13}$$

The equation $v = \lambda/T$ tells us that the wave speed is one wavelength per period; the wave moves a distance of one wavelength in one period of oscillation.

Equation 16-2 describes a wave moving in the positive direction of $x$. We can find the equation of a wave traveling in the opposite direction by replacing $t$ in Eq. 16-2 with $-t$. This corresponds to the condition

$$kx + \omega t = \text{a constant,} \tag{16-14}$$

which (compare Eq. 16-11) requires that $x$ *decrease* with time. Thus, a wave traveling in the negative direction of $x$ is described by the equation

$$y(x, t) = y_m \sin(kx + \omega t). \tag{16-15}$$

If you analyze the wave of Eq. 16-15 as we have just done for the wave of Eq. 16-2, you will find for its velocity

$$\frac{dx}{dt} = -\frac{\omega}{k}. \tag{16-16}$$

The minus sign (compare Eq. 16-12) verifies that the wave is indeed moving in the negative direction of $x$ and justifies our switching the sign of the time variable.

Consider now a wave of arbitrary shape, given by

$$y(x, t) = h(kx \pm \omega t), \tag{16-17}$$

where $h$ represents *any* function, the sine function being one possibility. Our previous analysis shows that all waves in which the variables $x$ and $t$ enter into the combination $kx \pm \omega t$ are traveling waves. Furthermore, all traveling waves *must* be of the form of Eq. 16-17. Thus, $y(x, t) = \sqrt{ax + bt}$ represents a possible (though perhaps physically a little bizarre) traveling wave. The function $y(x, t) = \sin(ax^2 - bt)$, on the other hand, does *not* represent a traveling wave.

✓ **CHECKPOINT 2**    Here are the equations of three waves:
(1) $y(x, t) = 2 \sin(4x - 2t)$,   (2) $y(x, t) = \sin(3x - 4t)$,   (3) $y(x, t) = 2 \sin(3x - 3t)$.
Rank the waves according to their (a) wave speed and (b) maximum speed perpendicular to the wave's direction of travel (the transverse speed), greatest first.

---

**Sample Problem | 16-2**

A wave traveling along a string is described by

$$y(x, t) = 0.00327 \sin(72.1x - 2.72t), \tag{16-18}$$

in which the numerical constants are in SI units (0.00327 m, 72.1 rad/m, and 2.72 rad/s).

(a) What is the amplitude of this wave?

**KEY IDEA**    Equation 16-18 is of the same form as Eq. 16-2,

$$y = y_m \sin(kx - \omega t), \tag{16-19}$$

so we have a sinusoidal wave. By comparing the two equations, we can find the amplitude.

*Calculation:* We see that

$$y_m = 0.00327 \text{ m} = 3.27 \text{ mm.} \quad \text{(Answer)}$$

(b) What are the wavelength, period, and frequency of this wave?

*Calculations:* By comparing Eqs. 16-18 and 16-19, we see that the angular wave number and angular frequency are

$$k = 72.1 \text{ rad/m} \quad \text{and} \quad \omega = 2.72 \text{ rad/s}.$$

We then relate wavelength $\lambda$ to $k$ via Eq. 16-5:

$$\lambda = \frac{2\pi}{k} = \frac{2\pi \text{ rad}}{72.1 \text{ rad/m}}$$

$$= 0.0871 \text{ m} = 8.71 \text{ cm.} \quad \text{(Answer)}$$

Next, we relate $T$ to $\omega$ with Eq. 16-8:

$$T = \frac{2\pi}{\omega} = \frac{2\pi \text{ rad}}{2.72 \text{ rad/s}} = 2.31 \text{ s,} \quad \text{(Answer)}$$

and from Eq. 16-9 we have

$$f = \frac{1}{T} = \frac{1}{2.31 \text{ s}} = 0.433 \text{ Hz.} \quad \text{(Answer)}$$

(c) What is the velocity of this wave?

**Calculation:** The speed of the wave is given by Eq. 16-13:

$$v = \frac{\omega}{k} = \frac{2.72 \text{ rad/s}}{72.1 \text{ rad/m}} = 0.0377 \text{ m/s}$$

$$= 3.77 \text{ cm/s.} \quad \text{(Answer)}$$

Because the phase in Eq. 16-18 contains the position variable $x$, the wave is moving along the $x$ axis. Also, because the wave equation is written in the form of Eq. 16-2, the *minus* sign in front of the $\omega t$ term indicates that the wave is moving in the *positive* direction of the $x$ axis. (Note that the quantities calculated in (b) and (c) are independent of the amplitude of the wave.)

(d) What is the displacement $y$ at $x = 22.5$ cm and $t = 18.9$ s?

**Calculation:** Equation 16-18 gives the displacement as a function of position $x$ and time $t$. Substituting the given values into the equation yields

$$y = 0.00327 \sin(72.1 \times 0.225 - 2.72 \times 18.9)$$

$$= (0.00327 \text{ m}) \sin(-35.1855 \text{ rad})$$

$$= (0.00327 \text{ m})(0.588)$$

$$= 0.00192 \text{ m} = 1.92 \text{ mm.} \quad \text{(Answer)}$$

Thus, the displacement is positive. (Be sure to change your calculator mode to radians before evaluating the sine.)

---

**Sample Problem** 16-3

In Sample Problem 16-2d, we showed that at $t = 18.9$ s the transverse displacement $y$ of the element of the string at $x = 0.255$ m due to the wave of Eq. 16-18 is 1.92 mm.

(a) What is $u$, the transverse velocity of the same element of the string, at that time? (This velocity, which is associated with the transverse oscillation of an element of the string, is in the $y$ direction. Do not confuse it with $v$, the constant velocity at which the *wave form* travels along the $x$ axis.)

**KEY IDEAS** The transverse velocity $u$ is the rate at which the displacement $y$ of the element is changing. In general, that displacement is given by

$$y(x, t) = y_m \sin(kx - \omega t). \quad (16\text{-}20)$$

For an element at a certain location $x$, we find the rate of change of $y$ by taking the derivative of Eq. 16-20 with respect to $t$ while treating $x$ as a constant. A derivative taken while one (or more) of the variables is treated as a constant is called a *partial derivative* and is represented by the symbol $\partial/\partial x$ rather than $d/dx$.

**Calculations:** Here we have

$$u = \frac{\partial y}{\partial t} = -\omega y_m \cos(kx - \omega t). \quad (16\text{-}21)$$

Next, substituting numerical values from Sample Problem 16-2, we obtain

$$u = (-2.72 \text{ rad/s})(3.27 \text{ mm}) \cos(-35.1855 \text{ rad})$$

$$= 7.20 \text{ mm/s.} \quad \text{(Answer)}$$

Thus, at $t = 18.9$ s, the element of the string at $x = 22.5$ cm is moving in the positive direction of $y$ with a speed of 7.20 mm/s.

(b) What is the transverse acceleration $a_y$ of the same element at that time?

**KEY IDEA** The transverse acceleration $a_y$ is the rate at which the transverse velocity of the element is changing.

**Calculations:** From Eq. 16-21, again treating $x$ as a constant but allowing $t$ to vary, we find

$$a_y = \frac{\partial u}{\partial t} = -\omega^2 y_m \sin(kx - \omega t).$$

Comparison with Eq. 16-20 shows that we can write this as

$$a_y = -\omega^2 y.$$

We see that the transverse acceleration of an oscillating string element is proportional to its transverse displacement but opposite in sign. This is completely consistent with the action of the element itself—namely, that it is moving transversely in simple harmonic motion. Substituting numerical values yields

$$a_y = -(2.72 \text{ rad/s})^2(1.92 \text{ mm})$$

$$= -14.2 \text{ mm/s}^2. \quad \text{(Answer)}$$

Thus, at $t = 18.9$ s, the element of string at $x = 22.5$ cm is displaced from its equilibrium position by 1.92 mm in the positive $y$ direction and has an acceleration of magnitude 14.2 mm/s$^2$ in the negative $y$ direction.

*Tactic 1: Evaluating Large Phases* Sometimes, as in Sample Problems 16-2d and 16-3, an angle much greater than $2\pi$ rad (or 360°) crops up and you are asked to find its sine or cosine. Adding or subtracting an integral multiple of $2\pi$ rad to such an angle does not change the value of any of its trigonometric functions. In Sample Problem 16-2d, for example, we used the angle $-35.1855$ rad in a sine function. Adding $(6)(2\pi$ rad) to this angle yields

$$-35.1855 \text{ rad} + (6)(2\pi \text{ rad}) = 2.51361 \text{ rad},$$

an angle of less than $2\pi$ rad that has the same trigonometric functions as $-35.1855$ rad (Fig. 16-9). As an example, the sine of both 2.51361 rad and $-35.1855$ rad is 0.588.

Your calculator will reduce such large angles for you automatically. *Caution:* Do not round off large angles if you intend to take their sines or cosines. In taking the sine of a very large angle, you are throwing away most of the angle and taking the sine of what is left over. If, for example, you were to round

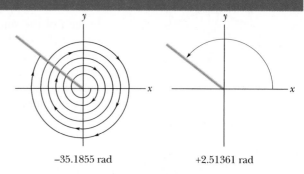

**FIG. 16-9** These two angles are different, but all their trigonometric functions are identical.

$-35.1855$ rad to $-35$ rad (a change of 0.5% and normally a reasonable step), you would be changing the sine of the angle by 27%. Also, if you change a large angle from degrees to radians, be sure to use an exact conversion factor (such as $180° = \pi$ rad) rather than an approximate one (such as $57.3° \approx 1$ rad).

# 16-6 | Wave Speed on a Stretched String

The speed of a wave is related to the wave's wavelength and frequency by Eq. 16-13, but *it is set by the properties of the medium.* If a wave is to travel through a medium such as water, air, steel, or a stretched string, it must cause the particles of that medium to oscillate as it passes. For that to happen, the medium must possess both mass (so that there can be kinetic energy) and elasticity (so that there can be potential energy). Thus, the medium's mass and elasticity properties determine how fast the wave can travel in the medium. Conversely, it should be possible to calculate the speed of the wave through the medium in terms of these properties. We do so now for a stretched string, in two ways.

## Dimensional Analysis

In dimensional analysis we carefully examine the dimensions of all the physical quantities that enter into a given situation to determine the quantities they produce. In this case, we examine mass and elasticity to find a speed $v$, which has the dimension of length divided by time, or $LT^{-1}$.

For the mass, we use the mass of a string element, which is the mass $m$ of the string divided by the length $l$ of the string. We call this ratio the *linear density* $\mu$ of the string. Thus, $\mu = m/l$, its dimension being mass divided by length, $ML^{-1}$.

You cannot send a wave along a string unless the string is under tension, which means that it has been stretched and pulled taut by forces at its two ends. The tension $\tau$ in the string is equal to the common magnitude of those two forces. As a wave travels along the string, it displaces elements of the string by causing additional stretching, with adjacent sections of string pulling on each other because of the tension. Thus, we can associate the tension in the string with the stretching (elasticity) of the string. The tension and the stretching forces it produces have the dimension of a force—namely, $MLT^{-2}$ (from $F = ma$).

We need to combine $\mu$ (dimension $ML^{-1}$) and $\tau$ (dimension $MLT^{-2}$) to get $v$ (dimension $LT^{-1}$). A little juggling of various combinations suggests

$$v = C\sqrt{\frac{\tau}{\mu}}, \tag{16-22}$$

in which $C$ is a dimensionless constant that cannot be determined with dimensional analysis. In our second approach to determining wave speed, you will see that Eq. 16-22 is indeed correct and that $C = 1$.

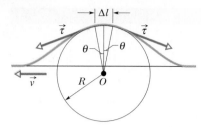

FIG. 16-10 A symmetrical pulse, viewed from a reference frame in which the pulse is stationary and the string appears to move right to left with speed $v$. We find speed $v$ by applying Newton's second law to a string element of length $\Delta l$, located at the top of the pulse.

## Derivation from Newton's Second Law

Instead of the sinusoidal wave of Fig. 16-1b, let us consider a single symmetrical pulse such as that of Fig. 16-10, moving from left to right along a string with speed $v$. For convenience, we choose a reference frame in which the pulse remains stationary; that is, we run along with the pulse, keeping it constantly in view. In this frame, the string appears to move past us, from right to left in Fig. 16-10, with speed $v$.

Consider a small string element of length $\Delta l$ within the pulse, an element that forms an arc of a circle of radius $R$ and subtending an angle $2\theta$ at the center of that circle. A force $\vec{\tau}$ with a magnitude equal to the tension in the string pulls tangentially on this element at each end. The horizontal components of these forces cancel, but the vertical components add to form a radial restoring force $\vec{F}$. In magnitude,

$$F = 2(\tau \sin \theta) \approx \tau(2\theta) = \tau \frac{\Delta l}{R} \qquad \text{(force)}, \qquad (16\text{-}23)$$

where we have approximated $\sin \theta$ as $\theta$ for the small angles $\theta$ in Fig. 16-10. From that figure, we have also used $2\theta = \Delta l/R$. The mass of the element is given by

$$\Delta m = \mu \, \Delta l \qquad \text{(mass)}, \qquad (16\text{-}24)$$

where $\mu$ is the string's linear density.

At the moment shown in Fig. 16-10, the string element $\Delta l$ is moving in an arc of a circle. Thus, it has a centripetal acceleration toward the center of that circle, given by

$$a = \frac{v^2}{R} \qquad \text{(acceleration)}. \qquad (16\text{-}25)$$

Equations 16-23, 16-24, and 16-25 contain the elements of Newton's second law. Combining them in the form

$$\text{force} = \text{mass} \times \text{acceleration}$$

gives

$$\frac{\tau \, \Delta l}{R} = (\mu \, \Delta l) \frac{v^2}{R}.$$

Solving this equation for the speed $v$ yields

$$v = \sqrt{\frac{\tau}{\mu}} \qquad \text{(speed)}, \qquad (16\text{-}26)$$

in exact agreement with Eq. 16-22 if the constant $C$ in that equation is given the value unity. Equation 16-26 gives the speed of the pulse in Fig. 16-10 and the speed of *any* other wave on the same string under the same tension.

Equation 16-26 tells us:

> The speed of a wave along a stretched ideal string depends only on the tension and linear density of the string and not on the frequency of the wave.

The *frequency* of the wave is fixed entirely by whatever generates the wave (for example, the person in Fig. 16-1b). The *wavelength* of the wave is then fixed by Eq. 16-13 in the form $\lambda = v/f$.

✓ **CHECKPOINT 3**   You send a traveling wave along a particular string by oscillating one end. If you increase the frequency of the oscillations, do (a) the speed of the wave and (b) the wavelength of the wave increase, decrease, or remain the same? If, instead, you increase the tension in the string, do (c) the speed of the wave and (d) the wavelength of the wave increase, decrease, or remain the same?

## Sample Problem | 16-4

In Fig. 16-11, two strings have been tied together with a knot and then stretched between two rigid supports. The strings have linear densities $\mu_1 = 1.4 \times 10^{-4}$ kg/m and $\mu_2 = 2.8 \times 10^{-4}$ kg/m. Their lengths are $L_1 = 3.0$ m and $L_2 = 2.0$ m, and string 1 is under a tension of 400 N. Simultaneously, on each string a pulse is sent from the rigid support end, toward the knot. Which pulse reaches the knot first?

KEY IDEAS

**1.** The time $t$ taken by a pulse to travel a length $L$ is $t = L/v$, where $v$ is the constant speed of the pulse.

**2.** The speed of a pulse on a stretched string depends on the string's tension $\tau$ and linear density $\mu$, and is given by Eq. 16-26 ($v = \sqrt{\tau/\mu}$).

**3.** Because the two strings are stretched together, they must both be under the same tension $\tau$ ($= 400$ N).

*Calculations:* Putting these three ideas together gives us, as the time for the pulse on string 1 to reach the knot,

$$t_1 = \frac{L_1}{v_1} = L_1\sqrt{\frac{\mu_1}{\tau}} = (3.0 \text{ m})\sqrt{\frac{1.4 \times 10^{-4} \text{ kg/m}}{400 \text{ N}}}$$

$$= 1.77 \times 10^{-3} \text{ s}.$$

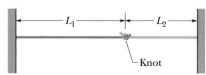

FIG. 16-11 Two strings, of lengths $L_1$ and $L_2$, tied together with a knot and stretched between two rigid supports.

Similarly, the data for the pulse on string 2 give us

$$t_2 = L_2\sqrt{\frac{\mu_2}{\tau}} = 1.67 \times 10^{-3} \text{ s}.$$

Thus, the pulse on string 2 reaches the knot first.

Now look back at the second key idea. The linear density of string 2 is greater than that of string 1, so the pulse on string 2 must be slower than that on string 1. Could we have guessed the answer from that fact alone? No, because from the first key idea we see that the distance traveled by a pulse also matters.

## 16-7 | Energy and Power of a Wave Traveling Along a String

When we set up a wave on a stretched string, we provide energy for the motion of the string. As the wave moves away from us, it transports that energy as both kinetic energy and elastic potential energy. Let us consider each form in turn.

### Kinetic Energy

A string element of mass $dm$, oscillating transversely in simple harmonic motion as the wave passes through it, has kinetic energy associated with its transverse velocity $\vec{u}$. When the element is rushing through its $y = 0$ position (element $b$ in Fig. 16-12), its transverse velocity—and thus its kinetic energy—is a maximum. When the element is at its extreme position $y = y_m$ (as is element $a$), its transverse velocity—and thus its kinetic energy—is zero.

### Elastic Potential Energy

To send a sinusoidal wave along a previously straight string, the wave must necessarily stretch the string. As a string element of length $dx$ oscillates transversely, its length must increase and decrease in a periodic way if the string element is to fit the sinusoidal wave form. Elastic potential energy is associated with these length changes, just as for a spring.

When the string element is at its $y = y_m$ position (element $a$ in Fig. 16-12), its length has its normal undisturbed value $dx$, so its elastic potential energy is zero. However, when the element is rushing through its $y = 0$ position, it has maximum stretch and thus maximum elastic potential energy.

### Energy Transport

The oscillating string element thus has both its maximum kinetic energy and its maximum elastic potential energy at $y = 0$. In the snapshot of Fig. 16-12, the regions of the string at maximum displacement have no energy, and the regions at zero displacement have maximum energy. As the wave travels along the string, forces due to the tension in the string continuously do work to transfer energy from regions with energy to regions with no energy.

Suppose we set up a wave on a string stretched along a horizontal $x$ axis so that Eq. 16-2 describes the string's displacement. We might send a wave along the

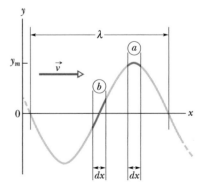

FIG. 16-12 A snapshot of a traveling wave on a string at time $t = 0$. String element $a$ is at displacement $y = y_m$, and string element $b$ is at displacement $y = 0$. The kinetic energy of the string element at each position depends on the transverse velocity of the element. The potential energy depends on the amount by which the string element is stretched as the wave passes through it.

string by continuously oscillating one end of the string, as in Fig. 16-1*b*. In doing so, we continuously provide energy for the motion and stretching of the string — as the string sections oscillate perpendicularly to the *x* axis, they have kinetic energy and elastic potential energy. As the wave moves into sections that were previously at rest, energy is transferred into those new sections. Thus, we say that the wave *transports* the energy along the string.

### The Rate of Energy Transmission

The kinetic energy *dK* associated with a string element of mass *dm* is given by

$$dK = \tfrac{1}{2} dm \, u^2, \tag{16-27}$$

where *u* is the transverse speed of the oscillating string element. To find *u*, we differentiate Eq. 16-2 with respect to time while holding *x* constant:

$$u = \frac{\partial y}{\partial t} = -\omega y_m \cos(kx - \omega t). \tag{16-28}$$

Using this relation and putting $dm = \mu \, dx$, we rewrite Eq. 16-27 as

$$dK = \tfrac{1}{2}(\mu \, dx)(-\omega y_m)^2 \cos^2(kx - \omega t). \tag{16-29}$$

Dividing Eq. 16-29 by *dt* gives the rate at which kinetic energy passes through a string element, and thus the rate at which kinetic energy is carried along by the wave. The ratio *dx/dt* that then appears on the right of Eq. 16-29 is the wave speed *v*, so we obtain

$$\frac{dK}{dt} = \tfrac{1}{2}\mu v \omega^2 y_m^2 \cos^2(kx - \omega t). \tag{16-30}$$

The *average* rate at which kinetic energy is transported is

$$\left(\frac{dK}{dt}\right)_{\text{avg}} = \tfrac{1}{2}\mu v \omega^2 y_m^2 \left[\cos^2(kx - \omega t)\right]_{\text{avg}}$$

$$= \tfrac{1}{4}\mu v \omega^2 y_m^2. \tag{16-31}$$

Here we have taken the average over an integer number of wavelengths and have used the fact that the average value of the square of a cosine function over an integer number of periods is $\tfrac{1}{2}$.

Elastic potential energy is also carried along with the wave, and at the same average rate given by Eq. 16-31. Although we shall not examine the proof, you should recall that, in an oscillating system such as a pendulum or a spring–block system, the average kinetic energy and the average potential energy are equal.

The **average power,** which is the average rate at which energy of both kinds is transmitted by the wave, is then

$$P_{\text{avg}} = 2\left(\frac{dK}{dt}\right)_{\text{avg}} \tag{16-32}$$

or, from Eq. 16-31,

$$P_{\text{avg}} = \tfrac{1}{2}\mu v \omega^2 y_m^2 \qquad \text{(average power).} \tag{16-33}$$

The factors $\mu$ and *v* in this equation depend on the material and tension of the string. The factors $\omega$ and $y_m$ depend on the process that generates the wave. The dependence of the average power of a wave on the square of its amplitude and also on the square of its angular frequency is a general result, true for waves of all types.

## Sample Problem    16-5

A string has linear density $\mu = 525$ g/m and tension $\tau = 45$ N. We send a sinusoidal wave with frequency $f = 120$ Hz and amplitude $y_m = 8.5$ mm along the string. At what average rate does the wave transport energy?

**KEY IDEA**  The average rate of energy transport is the average power $P_{avg}$ as given by Eq. 16-33.

**Calculations:** To use Eq. 16-33, we first must calculate angular frequency $\omega$ and wave speed $v$. From Eq. 16-9,

$$\omega = 2\pi f = (2\pi)(120 \text{ Hz}) = 754 \text{ rad/s}.$$

From Eq. 16-26 we have

$$v = \sqrt{\frac{\tau}{\mu}} = \sqrt{\frac{45 \text{ N}}{0.525 \text{ kg/m}}} = 9.26 \text{ m/s}.$$

Equation 16-33 then yields

$$P_{avg} = \tfrac{1}{2}\mu v \omega^2 y_m^2$$
$$= (\tfrac{1}{2})(0.525 \text{ kg/m})(9.26 \text{ m/s})(754 \text{ rad/s})^2(0.0085 \text{ m})^2$$
$$\approx 100 \text{ W.} \qquad\qquad \text{(Answer)}$$

## 16-8 | The Wave Equation

As a wave passes through any element on a stretched string, the element moves perpendicularly to the wave's direction of travel. By applying Newton's second law to the element's motion, we can derive a general differential equation, called the *wave equation*, that governs the travel of waves of any type.

Figure 16-13a shows a snapshot of a string element of mass $dm$ and length $\ell$ as a wave travels along a string of linear density $\mu$ that is stretched along a horizontal $x$ axis. Let us assume that the wave amplitude is small so that the element can be tilted only slightly from the $x$ axis as the wave passes. The force $\vec{F}_2$ on the right end of the element has a magnitude equal to tension $\tau$ in the string and is directed slightly upward. The force $\vec{F}_1$ on the left end of the element also has a magnitude equal to the tension $\tau$ but is directed slightly downward. Because of the slight curvature of the element, these two forces produce a net force that causes the element to have an upward acceleration $a_y$. Newton's second law written for $y$ components ($F_{net,y} = ma_y$) gives us

$$F_{2y} - F_{1y} = dm\, a_y. \qquad (16\text{-}34)$$

Let's analyze this equation in parts.

*Mass.* The element's mass $dm$ can be written in terms of the string's linear density $\mu$ and the element's length $\ell$ as $dm = \mu\ell$. Because the element can have only a slight tilt, $\ell \approx dx$ (Fig. 16-13a) and we have the approximation

$$dm = \mu\, dx. \qquad (16\text{-}35)$$

*Acceleration.* The acceleration $a_y$ in Eq. 16-34 is the second derivative of the displacement $y$ with respect to time:

$$a_y = \frac{d^2y}{dt^2}. \qquad (16\text{-}36)$$

*Forces.* Figure 16-13b shows that $\vec{F}_2$ is tangent to the string at the right end of the string element. Thus we can relate the components of the force to the string slope $S_2$ at the right end as

$$\frac{F_{2y}}{F_{2x}} = S_2. \qquad (16\text{-}37)$$

We can also relate the components to the magnitude $F_2$ ($= \tau$) with

$$F_2 = \sqrt{F_{2x}^2 + F_{2y}^2}$$

or

$$\tau = \sqrt{F_{2x}^2 + F_{2y}^2}. \qquad (16\text{-}38)$$

However, because we assume that the element is only slightly tilted, $F_{2y} \ll F_{2x}$ and therefore we can rewrite Eq. 16-38 as

$$\tau = F_{2x}. \qquad (16\text{-}39)$$

Substituting this into Eq. 16-37 and solving for $F_{2y}$ yield

$$F_{2y} = \tau S_2. \qquad (16\text{-}40)$$

Similar analysis at the left end of the string element gives us

$$F_{1y} = \tau S_1. \qquad (16\text{-}41)$$

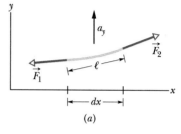

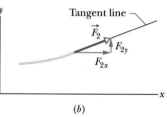

**FIG. 16-13** (a) A string element as a sinusoidal transverse wave travels on a stretched string. Forces $\vec{F}_1$ and $\vec{F}_2$ act at the left and right ends, producing acceleration $\vec{a}$ having a vertical component $a_y$. (b) The force at the element's right end is directed along a tangent to the element's right side.

We can now substitute Eqs. 16-35, 16-36, 16-40, and 16-41 into Eq. 16-34 to write

$$\tau S_2 - \tau S_1 = (\mu\, dx)\frac{d^2y}{dt^2},$$

or

$$\frac{S_2 - S_1}{dx} = \frac{\mu}{\tau}\frac{d^2y}{dt^2}. \tag{16-42}$$

Because the string element is short, slopes $S_2$ and $S_1$ differ by only a differential amount $dS$, where $S$ is the slope at any point:

$$S = \frac{dy}{dx}. \tag{16-43}$$

First replacing $S_2 - S_1$ in Eq. 16-42 with $dS$ and then using Eq. 16-43 to substitute $dy/dx$ for $S$, we find

$$\frac{dS}{dx} = \frac{\mu}{\tau}\frac{d^2y}{dt^2},$$

$$\frac{d(dy/dx)}{dx} = \frac{\mu}{\tau}\frac{d^2y}{dt^2},$$

and

$$\frac{\partial^2 y}{\partial x^2} = \frac{\mu}{\tau}\frac{\partial^2 y}{\partial t^2}. \tag{16-44}$$

In the last step, we switched to the notation of partial derivatives because on the left we differentiate only with respect to $x$ and on the right we differentiate only with respect to $t$. Finally, substituting from Eq. 16-26 ($v = \sqrt{\tau/\mu}$), we find

$$\frac{\partial^2 y}{\partial x^2} = \frac{1}{v^2}\frac{\partial^2 y}{\partial t^2} \qquad \text{(wave equation)}. \tag{16-45}$$

This is the general differential equation that governs the travel of waves of all types.

## 16-9 I The Principle of Superposition for Waves

It often happens that two or more waves pass simultaneously through the same region. When we listen to a concert, for example, sound waves from many instruments fall simultaneously on our eardrums. The electrons in the antennas of our radio and television receivers are set in motion by the net effect of many electromagnetic waves from many different broadcasting centers. The water of a lake or harbor may be churned up by waves in the wakes of many boats.

Suppose that two waves travel simultaneously along the same stretched string. Let $y_1(x, t)$ and $y_2(x, t)$ be the displacements that the string would experience if each wave traveled alone. The displacement of the string when the waves overlap is then the algebraic sum

$$y'(x, t) = y_1(x, t) + y_2(x, t). \tag{16-46}$$

This summation of displacements along the string means that

Overlapping waves algebraically add to produce a **resultant wave** (or **net wave**).

This is another example of the **principle of superposition,** which says that when several effects occur simultaneously, their net effect is the sum of the individual effects.

Figure 16-14 shows a sequence of snapshots of two pulses traveling in opposite directions on the same stretched string. When the pulses overlap, the resultant pulse is their sum. Moreover,

Overlapping waves do not in any way alter the travel of each other.

# 16-10 | Interference of Waves

Suppose we send two sinusoidal waves of the same wavelength and amplitude in the same direction along a stretched string. The superposition principle applies. What resultant wave does it predict for the string?

The resultant wave depends on the extent to which the waves are *in phase* (in step) with respect to each other—that is, how much one wave form is shifted from the other wave form. If the waves are exactly in phase (so that the peaks and valleys of one are exactly aligned with those of the other), they combine to double the displacement of either wave acting alone. If they are exactly out of phase (the peaks of one are exactly aligned with the valleys of the other), they combine to cancel everywhere, and the string remains straight. We call this phenomenon of combining waves **interference,** and the waves are said to **interfere.** (These terms refer only to the wave displacements; the travel of the waves is unaffected.)

Let one wave traveling along a stretched string be given by

$$y_1(x, t) = y_m \sin(kx - \omega t) \tag{16-47}$$

and another, shifted from the first, by

$$y_2(x, t) = y_m \sin(kx - \omega t + \phi). \tag{16-48}$$

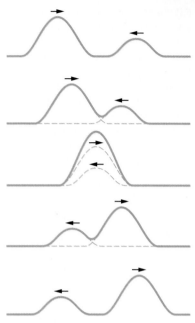

**FIG. 16-14** A series of snapshots that show two pulses traveling in opposite directions along a stretched string. The superposition principle applies as the pulses move through each other.

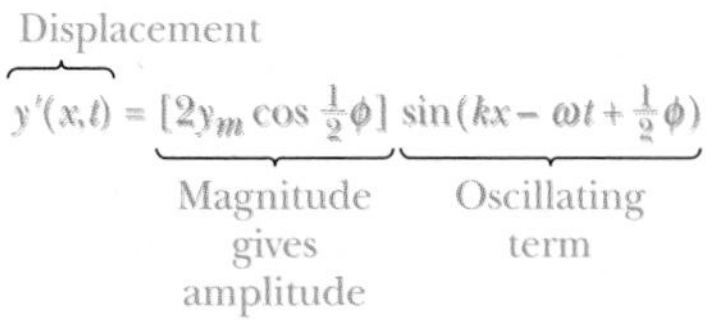

These waves have the same angular frequency $\omega$ (and thus the same frequency $f$), the same angular wave number $k$ (and thus the same wavelength $\lambda$), and the same amplitude $y_m$. They both travel in the positive direction of the $x$ axis, with the same speed, given by Eq. 16-26. They differ only by a constant angle $\phi$, the phase constant. These waves are said to be *out of phase* by $\phi$ or to have a *phase difference* of $\phi$, or one wave is said to be *phase-shifted* from the other by $\phi$.

From the principle of superposition (Eq. 16-46), the resultant wave is the algebraic sum of the two interfering waves and has displacement

$$y'(x, t) = y_1(x, t) + y_2(x, t)$$
$$= y_m \sin(kx - \omega t) + y_m \sin(kx - \omega t + \phi). \tag{16-49}$$

In Appendix E we see that we can write the sum of the sines of two angles $\alpha$ and $\beta$ as

$$\sin \alpha + \sin \beta = 2 \sin \tfrac{1}{2}(\alpha + \beta) \cos \tfrac{1}{2}(\alpha - \beta). \tag{16-50}$$

Applying this relation to Eq. 16-49 leads to

$$y'(x, t) = [2y_m \cos \tfrac{1}{2}\phi] \sin(kx - \omega t + \tfrac{1}{2}\phi). \tag{16-51}$$

As Fig. 16-15 shows, the resultant wave is also a sinusoidal wave traveling in the direction of increasing $x$. It is the only wave you would actually see on the string (you would *not* see the two interfering waves of Eqs. 16-47 and 16-48).

> If two sinusoidal waves of the same amplitude and wavelength travel in the *same* direction along a stretched string, they interfere to produce a resultant sinusoidal wave traveling in that direction.

The resultant wave differs from the interfering waves in two respects: (1) its phase constant is $\tfrac{1}{2}\phi$, and (2) its amplitude $y'_m$ is the magnitude of the quantity in the brackets in Eq. 16-51:

$$y'_m = |2y_m \cos \tfrac{1}{2}\phi| \qquad \text{(amplitude)}. \tag{16-52}$$

If $\phi = 0$ rad (or 0°), the two interfering waves are exactly in phase, as in Fig. 16-16a. Then Eq. 16-51 reduces to

$$y'(x, t) = 2y_m \sin(kx - \omega t) \qquad (\phi = 0). \tag{16-53}$$

This resultant wave is plotted in Fig. 16-16d. Note from both that figure and Eq. 16-53 that the amplitude of the resultant wave is twice the amplitude of either

Displacement

$$\overbrace{y'(x,t)}^{} = \underbrace{[2y_m \cos \tfrac{1}{2}\phi]}_{\substack{\text{Magnitude} \\ \text{gives} \\ \text{amplitude}}} \underbrace{\sin(kx - \omega t + \tfrac{1}{2}\phi)}_{\substack{\text{Oscillating} \\ \text{term}}}$$

**FIG. 16-15** The resultant wave of Eq. 16-51, due to the interference of two sinusoidal transverse waves, is also a sinusoidal transverse wave, with an amplitude and an oscillating term.

**FIG. 16-16** Two identical sinusoidal waves, $y_1(x, t)$ and $y_2(x, t)$, travel along a string in the positive direction of an $x$ axis. They interfere to give a resultant wave $y'(x, t)$. The resultant wave is what is actually seen on the string. The phase difference $\phi$ between the two interfering waves is (a) 0 rad or 0°, (b) $\pi$ rad or 180°, and (c) $\frac{2}{3}\pi$ rad or 120°. The corresponding resultant waves are shown in (d), (e), and (f).

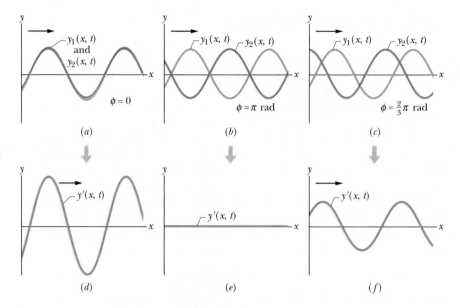

interfering wave. That is the greatest amplitude the resultant wave can have, because the cosine term in Eqs. 16-51 and 16-52 has its greatest value (unity) when $\phi = 0$. Interference that produces the greatest possible amplitude is called *fully constructive interference.*

If $\phi = \pi$ rad (or 180°), the interfering waves are exactly out of phase as in Fig. 16-16b. Then $\cos\frac{1}{2}\phi$ becomes $\cos \pi/2 = 0$, and the amplitude of the resultant wave as given by Eq. 16-52 is zero. We then have, for all values of $x$ and $t$,

$$y'(x, t) = 0 \qquad (\phi = \pi\,\text{rad}). \qquad (16\text{-}54)$$

The resultant wave is plotted in Fig. 16-16e. Although we sent two waves along the string, we see no motion of the string. This type of interference is called *fully destructive interference.*

Because a sinusoidal wave repeats its shape every $2\pi$ rad, a phase difference $\phi = 2\pi$ rad (or 360°) corresponds to a shift of one wave relative to the other wave by a distance equivalent to one wavelength. Thus, phase differences can be described in terms of wavelengths as well as angles. For example, in Fig. 16-16b the waves may be said to be 0.50 wavelength out of phase. Table 16-1 shows some other examples of phase differences and the interference they produce. Note that when interference is neither fully constructive nor fully destructive, it is called *intermediate interference.* The amplitude of the resultant wave is then intermediate between 0 and $2y_m$. For example, from Table 16-1, if the interfering waves

**TABLE 16-1**

**Phase Difference and Resulting Interference Types**[a]

| Phase Difference, in | | | Amplitude of Resultant Wave | Type of Interference |
|---|---|---|---|---|
| Degrees | Radians | Wavelengths | | |
| 0 | 0 | 0 | $2y_m$ | Fully constructive |
| 120 | $\frac{2}{3}\pi$ | 0.33 | $y_m$ | Intermediate |
| 180 | $\pi$ | 0.50 | 0 | Fully destructive |
| 240 | $\frac{4}{3}\pi$ | 0.67 | $y_m$ | Intermediate |
| 360 | $2\pi$ | 1.00 | $2y_m$ | Fully constructive |
| 865 | 15.1 | 2.40 | $0.60y_m$ | Intermediate |

[a]The phase difference is between two otherwise identical waves, with amplitude $y_m$, moving in the same direction.

have a phase difference of 120° ($\phi = \frac{2}{3}\pi$ rad = 0.33 wavelength), then the resultant wave has an amplitude of $y_m$, the same as that of the interfering waves (see Figs. 16-16c and f).

Two waves with the same wavelength are in phase if their phase difference is zero or any integer number of wavelengths. Thus, the integer part of any phase difference *expressed in wavelengths* may be discarded. For example, a phase difference of 0.40 wavelength is equivalent in every way to one of 2.40 wavelengths, and so the simpler of the two numbers can be used in computations.

√ **CHECKPOINT 4**    Here are four possible phase differences between two identical waves, expressed in wavelengths: 0.20, 0.45, 0.60, and 0.80. Rank them according to the amplitude of the resultant wave, greatest first.

---

**Sample Problem**    16-6

Two identical sinusoidal waves, moving in the same direction along a stretched string, interfere with each other. The amplitude $y_m$ of each wave is 9.8 mm, and the phase difference $\phi$ between them is 100°.

(a) What is the amplitude $y'_m$ of the resultant wave due to the interference, and what is the type of this interference?

**KEY IDEA**    These are identical sinusoidal waves traveling in the *same direction* along a string, so they interfere to produce a sinusoidal traveling wave.

*Calculations:* Because they are identical, the waves have the *same amplitude*. Thus, the amplitude $y'_m$ of the resultant wave is given by Eq. 16-52:

$$y'_m = |2y_m \cos \tfrac{1}{2}\phi| = |(2)(9.8 \text{ mm}) \cos(100°/2)|$$
$$= 13 \text{ mm}. \qquad \text{(Answer)}$$

We can tell that the interference is *intermediate* in two ways. The phase difference is between 0 and 180°, and, correspondingly, the amplitude $y'_m$ is between 0 and $2y_m$ (= 19.6 mm).

(b) What phase difference, in radians and wavelengths, will give the resultant wave an amplitude of 4.9 mm?

*Calculations:* Now we are given $y'_m$ and seek $\phi$. From Eq. 16-52,

$$y'_m = |2y_m \cos \tfrac{1}{2}\phi|,$$

we now have

$$4.9 \text{ mm} = (2)(9.8 \text{ mm}) \cos \tfrac{1}{2}\phi,$$

which gives us (with a calculator in the radian mode)

$$\phi = 2 \cos^{-1} \frac{4.9 \text{ mm}}{(2)(9.8 \text{ mm})}$$
$$= \pm 2.636 \text{ rad} \approx \pm 2.6 \text{ rad}. \qquad \text{(Answer)}$$

There are two solutions because we can obtain the same resultant wave by letting the first wave *lead* (travel ahead of) or *lag* (travel behind) the second wave by 2.6 rad. In wavelengths, the phase difference is

$$\frac{\phi}{2\pi \text{ rad/wavelength}} = \frac{\pm 2.636 \text{ rad}}{2\pi \text{ rad/wavelength}}$$
$$= \pm 0.42 \text{ wavelength.} \quad \text{(Answer)}$$

---

## 16-11 | Phasors

We can represent a string wave (or any other type of wave) vectorially with a **phasor.** In essence, a phasor is a vector that has a magnitude equal to the amplitude of the wave and that rotates around an origin; the angular speed of the phasor is equal to the angular frequency $\omega$ of the wave. For example, the wave

$$y_1(x, t) = y_{m1} \sin(kx - \omega t) \qquad (16\text{-}55)$$

is represented by the phasor shown in Fig. 16-17a. The magnitude of the phasor is the amplitude $y_{m1}$ of the wave. As the phasor rotates around the origin at angular speed $\omega$, its projection $y_1$ on the vertical axis varies sinusoidally, from a maximum of $y_{m1}$ through zero to a minimum of $-y_{m1}$ and then back to $y_{m1}$. This variation corresponds to the sinusoidal variation in the displacement $y_1$ of any point along the string as the wave passes through that point.

When two waves travel along the same string in the same direction, we can represent them and their resultant wave in a *phasor diagram*. The phasors in Fig.

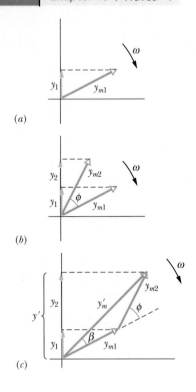

(a)

(b)

(c)

FIG. 16-17 (a) A phasor of magnitude $y_{m1}$ rotating about an origin at angular speed $\omega$ represents a sinusoidal wave. The phasor's projection $y_1$ on the vertical axis represents the displacement of a point through which the wave passes. (b) A second phasor, also of angular speed $\omega$ but of magnitude $y_{m2}$ and rotating at a constant angle $\phi$ from the first phasor, represents a second wave, with a phase constant $\phi$. (c) The resultant wave is represented by the vector sum $y'_m$ of the two phasors.

16-17b represent the wave of Eq. 16-55 and a second wave given by

$$y_2(x, t) = y_{m2} \sin(kx - \omega t + \phi). \qquad (16\text{-}56)$$

This second wave is phase-shifted from the first wave by phase constant $\phi$. Because the phasors rotate at the same angular speed $\omega$, the angle between the two phasors is always $\phi$. If $\phi$ is a *positive* quantity, then the phasor for wave 2 *lags* the phasor for wave 1 as they rotate, as drawn in Fig. 16-17b. If $\phi$ is a negative quantity, then the phasor for wave 2 *leads* the phasor for wave 1.

Because waves $y_1$ and $y_2$ have the same angular wave number $k$ and angular frequency $\omega$, we know from Eqs. 16-51 and 16-52 that their resultant is of the form

$$y'(x, t) = y'_m \sin(kx - \omega t + \beta), \qquad (16\text{-}57)$$

where $y'_m$ is the amplitude of the resultant wave and $\beta$ is its phase constant. To find the values of $y'_m$ and $\beta$, we would have to sum the two combining waves, as we did to obtain Eq. 16-51. To do this on a phasor diagram, we vectorially add the two phasors at any instant during their rotation, as in Fig. 16-17c where phasor $y_{m2}$ has been shifted to the head of phasor $y_{m1}$. The magnitude of the vector sum equals the amplitude $y'_m$ in Eq. 16-57. The angle between the vector sum and the phasor for $y_1$ equals the phase constant $\beta$ in Eq. 16-57.

Note that, in contrast to the method of Section 16-10:

☞  We can use phasors to combine waves *even if their amplitudes are different.*

Two sinusoidal waves $y_1(x, t)$ and $y_2(x, t)$ have the same wavelength and travel together in the same direction along a string. Their amplitudes are $y_{m1} = 4.0$ mm and $y_{m2} = 3.0$ mm, and their phase constants are 0 and $\pi/3$ rad, respectively. What are the amplitude $y'_m$ and phase constant $\beta$ of the resultant wave? Write the resultant wave in the form of Eq. 16-57.

**KEY IDEAS**  (1) The two waves have a number of properties in common: Because they travel along the same string, they must have the same speed $v$, as set by the tension and linear density of the string according to Eq. 16-26. With the same wavelength $\lambda$, they must have the same angular wave number $k$ ($= 2\pi/\lambda$). Also, with the same wave number $k$ and speed $v$, they must have the same angular frequency $\omega$ ($= kv$).

(2) The waves (call them waves 1 and 2) can be represented by phasors rotating at the same angular speed $\omega$ about an origin. Because the phase constant for wave 2 is *greater* than that for wave 1 by $\pi/3$, phasor 2 must *lag* phasor 1 by $\pi/3$ rad in their clockwise rotation, as shown in Fig. 16-18a. The resultant wave due to the interference of waves 1 and 2 can then be represented by a phasor that is the vector sum of phasors 1 and 2.

*Calculations:* To simplify the vector summation, we drew phasors 1 and 2 in Fig. 16-18a at the instant when phasor 1 lies along the horizontal axis. We then drew lagging phasor 2 at positive angle $\pi/3$ rad. In Fig. 16-18b we shifted phasor 2 so its tail is at the head of phasor 1. Then we can draw the phasor $y'_m$ of the resultant wave from the tail of phasor 1 to the head of phasor 2. The phase constant $\beta$ is the angle phasor $y'_m$ makes with phasor 1.

To find values for $y'_m$ and $\beta$, we can sum phasors 1 and 2 directly on a vector-capable calculator (by adding a vector of magnitude 4.0 and angle 0 rad to a vector of

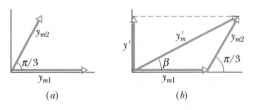

(a)                      (b)

FIG. 16-18  (a) Two phasors of magnitudes $y_{m1}$ and $y_{m2}$ and with phase difference $\pi/3$. (b) Vector addition of these phasors at any instant during their rotation gives the magnitude $y'_m$ of the phasor for the resultant wave.

magnitude 3.0 and angle $\pi/3$ rad) or we can add the vectors by components. For the horizontal components we have

$$y'_{mh} = y_{m1}\cos 0 + y_{m2}\cos \pi/3$$
$$= 4.0 \text{ mm} + (3.0 \text{ mm})\cos \pi/3 = 5.50 \text{ mm}.$$

For the vertical components we have

$$y'_{mv} = y_{m1}\sin 0 + y_{m2}\sin \pi/3$$
$$= 0 + (3.0 \text{ mm})\sin \pi/3 = 2.60 \text{ mm}.$$

Thus, the resultant wave has an amplitude of

$$y'_m = \sqrt{(5.50 \text{ mm})^2 + (2.60 \text{ mm})^2}$$
$$= 6.1 \text{ mm} \qquad \text{(Answer)}$$

and a phase constant of

$$\beta = \tan^{-1}\frac{2.60 \text{ mm}}{5.50 \text{ mm}} = 0.44 \text{ rad.} \quad \text{(Answer)}$$

From Fig. 16-18b, phase constant $\beta$ is a *positive* angle relative to phasor 1. Thus, the resultant wave *lags* wave 1 in their travel by phase constant $\beta = +0.44$ rad. From Eq. 16-57, we can write the resultant wave as

$$y'(x, t) = (6.1 \text{ mm})\sin(kx - \omega t + 0.44 \text{ rad}). \quad \text{(Answer)}$$

## 16-12 I Standing Waves

In Section 16-10, we discussed two sinusoidal waves of the same wavelength and amplitude traveling *in the same direction* along a stretched string. What if they travel in opposite directions? We can again find the resultant wave by applying the superposition principle.

Figure 16-19 suggests the situation graphically. It shows the two combining waves, one traveling to the left in Fig. 16-19a, the other to the right in Fig. 16-19b. Figure 16-19c shows their sum, obtained by applying the superposition principle graphically. The outstanding feature of the resultant wave is that there are places along the string, called **nodes,** where the string never moves. Four such nodes are marked by dots in Fig. 16-19c. Halfway between adjacent nodes are **antinodes,** where the amplitude of the resultant wave is a maximum. Wave patterns such as that of Fig. 16-19c are called **standing waves** because the wave patterns do not move left or right; the locations of the maxima and minima do not change.

☞ If two sinusoidal waves of the same amplitude and wavelength travel in *opposite* directions along a stretched string, their interference with each other produces a standing wave.

To analyze a standing wave, we represent the two combining waves with the equations

$$y_1(x, t) = y_m \sin(kx - \omega t) \tag{16-58}$$

and

$$y_2(x, t) = y_m \sin(kx + \omega t). \tag{16-59}$$

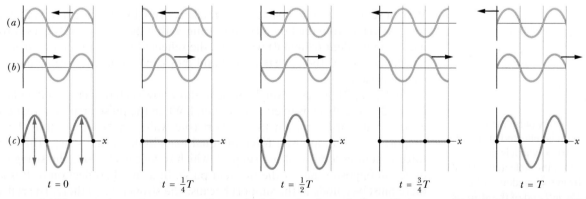

(a)

(b)

(c)

$t = 0$      $t = \frac{1}{4}T$      $t = \frac{1}{2}T$      $t = \frac{3}{4}T$      $t = T$

**FIG. 16-19** (a) Five snapshots of a wave traveling to the left, at the times $t$ indicated below part (c) ($T$ is the period of oscillation). (b) Five snapshots of a wave identical to that in (a) but traveling to the right, at the same times $t$. (c) Corresponding snapshots for the superposition of the two waves on the same string. At $t = 0, \frac{1}{2}T$, and $T$, fully constructive interference occurs because of the alignment of peaks with peaks and valleys with valleys. At $t = \frac{1}{4}T$ and $\frac{3}{4}T$, fully destructive interference occurs because of the alignment of peaks with valleys. Some points (the nodes, marked with dots) never oscillate; some points (the antinodes) oscillate the most.

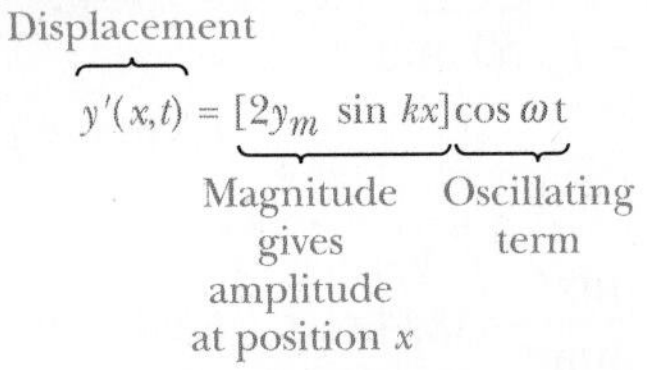

Displacement

$$y'(x,t) = \underbrace{[2y_m \sin kx]}_{\substack{\text{Magnitude} \\ \text{gives} \\ \text{amplitude} \\ \text{at position } x}} \underbrace{\cos \omega t}_{\substack{\text{Oscillating} \\ \text{term}}}$$

**FIG. 16-20** The resultant wave of Eq. 16-60 is a standing wave and is due to the interference of two sinusoidal waves of the same amplitude and wavelength that travel in opposite directions.

*(a)*        *(b)*

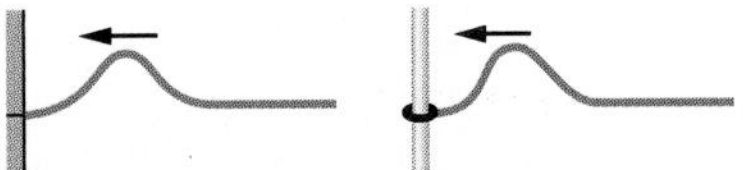

**FIG. 16-21** *(a)* A pulse incident from the right is reflected at the left end of the string, which is tied to a wall. Note that the reflected pulse is inverted from the incident pulse. *(b)* Here the left end of the string is tied to a ring that can slide without friction up and down the rod. Now the pulse is not inverted by the reflection.

The principle of superposition gives, for the combined wave,

$$y'(x,t) = y_1(x,t) + y_2(x,t) = y_m \sin(kx - \omega t) + y_m \sin(kx + \omega t).$$

Applying the trigonometric relation of Eq. 16-50 leads to

$$y'(x,t) = [2y_m \sin kx] \cos \omega t, \qquad (16\text{-}60)$$

which is displayed in Fig. 16-20. This equation does not describe a traveling wave because it is not of the form of Eq. 16-17. Instead, it describes a standing wave.

The quantity $2y_m \sin kx$ in the brackets of Eq. 16-60 can be viewed as the amplitude of oscillation of the string element that is located at position $x$. However, since an amplitude is always positive and $\sin kx$ can be negative, we take the absolute value of the quantity $2y_m \sin kx$ to be the amplitude at $x$.

In a traveling sinusoidal wave, the amplitude of the wave is the same for all string elements. That is not true for a standing wave, in which the amplitude *varies with position*. In the standing wave of Eq. 16-60, for example, the amplitude is zero for values of $kx$ that give $\sin kx = 0$. Those values are

$$kx = n\pi, \qquad \text{for } n = 0, 1, 2, \ldots . \qquad (16\text{-}61)$$

Substituting $k = 2\pi/\lambda$ in this equation and rearranging, we get

$$x = n\frac{\lambda}{2}, \qquad \text{for } n = 0, 1, 2, \ldots \qquad \text{(nodes)}, \qquad (16\text{-}62)$$

as the positions of zero amplitude—the nodes—for the standing wave of Eq. 16-60. Note that adjacent nodes are separated by $\lambda/2$, half a wavelength.

The amplitude of the standing wave of Eq. 16-60 has a maximum value of $2y_m$, which occurs for values of $kx$ that give $|\sin kx| = 1$. Those values are

$$kx = \tfrac{1}{2}\pi, \tfrac{3}{2}\pi, \tfrac{5}{2}\pi, \ldots$$
$$= (n + \tfrac{1}{2})\pi, \qquad \text{for } n = 0, 1, 2, \ldots . \qquad (16\text{-}63)$$

Substituting $k = 2\pi/\lambda$ in Eq. 16-63 and rearranging, we get

$$x = \left(n + \frac{1}{2}\right)\frac{\lambda}{2}, \qquad \text{for } n = 0, 1, 2, \ldots \qquad \text{(antinodes)}, \qquad (16\text{-}64)$$

as the positions of maximum amplitude—the antinodes—of the standing wave of Eq. 16-60. The antinodes are separated by $\lambda/2$ and are located halfway between pairs of nodes.

### Reflections at a Boundary

We can set up a standing wave in a stretched string by allowing a traveling wave to be reflected from the far end of the string so that the wave travels back through itself. The incident (original) wave and the reflected wave can then be described by Eqs. 16-58 and 16-59, respectively, and they can combine to form a pattern of standing waves.

In Fig. 16-21, we use a single pulse to show how such reflections take place. In Fig. 16-21a, the string is fixed at its left end. When the pulse arrives at that end, it exerts an upward force on the support (the wall). By Newton's third law, the support exerts an opposite force of equal magnitude on the string. This second force generates a pulse at the support, which travels back along the string in the direction opposite that of the incident pulse. In a "hard" reflection of this kind, there must be a node at the support because the string is fixed there. The reflected and incident pulses must have opposite signs, so as to cancel each other at that point.

In Fig. 16-21b, the left end of the string is fastened to a light ring that is free to slide without friction along a rod. When the incident pulse arrives, the ring moves up the rod. As the ring moves, it pulls on the string, stretching the string and producing a reflected pulse with the same sign and amplitude as the incident pulse. Thus, in such a "soft" reflection, the incident and reflected pulses reinforce each other, creating an antinode at the end of the string; the maximum displacement of the ring is twice the amplitude of either of these pulses.

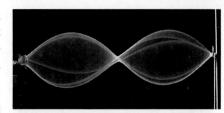

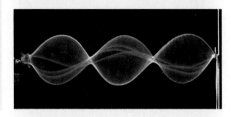

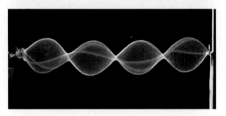

> **✓CHECKPOINT 5**    Two waves with the same amplitude and wavelength interfere in three different situations to produce resultant waves with the following equations:
>
> (1) $y'(x,t) = 4\sin(5x - 4t)$
> (2) $y'(x,t) = 4\sin(5x)\cos(4t)$
> (3) $y'(x,t) = 4\sin(5x + 4t)$
>
> In which situation are the two combining waves traveling (a) toward positive $x$, (b) toward negative $x$, and (c) in opposite directions?

## 16-13 | Standing Waves and Resonance

Consider a string, such as a guitar string, that is stretched between two clamps. Suppose we send a continuous sinusoidal wave of a certain frequency along the string, say, toward the right. When the wave reaches the right end, it reflects and begins to travel back to the left. That left-going wave then overlaps the wave that is still traveling to the right. When the left-going wave reaches the left end, it reflects again and the newly reflected wave begins to travel to the right, overlapping the left-going and right-going waves. In short, we very soon have many overlapping traveling waves, which interfere with one another.

For certain frequencies, the interference produces a standing wave pattern (or **oscillation mode**) with nodes and large antinodes like those in Fig. 16-22. Such a standing wave is said to be produced at **resonance,** and the string is said to *resonate* at these certain frequencies, called **resonant frequencies.** If the string is oscillated at some frequency other than a resonant frequency, a standing wave is not set up. Then the interference of the right-going and left-going traveling waves results in only small (perhaps imperceptible) oscillations of the string.

Let a string be stretched between two clamps separated by a fixed distance $L$. To find expressions for the resonant frequencies of the string, we note that a node must exist at each of its ends, because each end is fixed and cannot oscillate. The simplest pattern that meets this key requirement is that in Fig. 16-23a, which shows the string at both its extreme displacements (one solid and one dashed, together forming a single "loop"). There is only one antinode, which is at the center of the string. Note that half a wavelength spans the length $L$, which we take to be the string's length. Thus, for this pattern, $\lambda/2 = L$. This condition tells us that if the left-going and right-going traveling waves are to set up this pattern by their interference, they must have the wavelength $\lambda = 2L$.

A second simple pattern meeting the requirement of nodes at the fixed ends is shown in Fig. 16-23b. This pattern has three nodes and two antinodes and is said to be a two-loop pattern. For the left-going and right-going waves to set it up, they must have a wavelength $\lambda = L$. A third pattern is shown in Fig. 16-23c. It has four nodes, three antinodes, and three loops, and the wavelength is $\lambda = \frac{2}{3}L$. We could continue this progression by drawing increasingly more complicated patterns. In each step of the progression, the pattern would have one more node and one more antinode than the preceding step, and an additional $\lambda/2$ would be fitted into the distance $L$.

Thus, a standing wave can be set up on a string of length $L$ by a wave with a wavelength equal to one of the values

$$\lambda = \frac{2L}{n}, \qquad \text{for } n = 1, 2, 3, \dots . \tag{16-65}$$

The resonant frequencies that correspond to these wavelengths follow from

**FIG. 16-22** Stroboscopic photographs reveal (imperfect) standing wave patterns on a string being made to oscillate by an oscillator at the left end. The patterns occur at certain frequencies of oscillation. *(Richard Megna/Fundamental Photographs)*

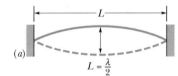

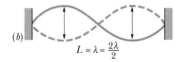

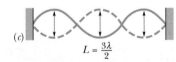

**FIG. 16-23** A string, stretched between two clamps, is made to oscillate in standing wave patterns. (*a*) The simplest possible pattern consists of one *loop,* which refers to the composite shape formed by the string in its extreme displacements (the solid and dashed lines). (*b*) The next simplest pattern has two loops. (*c*) The next has three loops.

**FIG. 16-24** One of many possible standing wave patterns for a kettle-drum head, made visible by dark powder sprinkled on the drumhead. As the head is set into oscillation at a single frequency by a mechanical oscillator at the upper left of the photograph, the powder collects at the nodes, which are circles and straight lines in this two-dimensional example. *(Courtesy Thomas D. Rossing, Northern Illinois University)*

Eq. 16-13:

$$f = \frac{v}{\lambda} = n\frac{v}{2L}, \quad \text{for } n = 1, 2, 3, \ldots. \tag{16-66}$$

Here $v$ is the speed of traveling waves on the string.

Equation 16-66 tells us that the resonant frequencies are integer multiples of the lowest resonant frequency, $f = v/2L$, which corresponds to $n = 1$. The oscillation mode with that lowest frequency is called the *fundamental mode* or the *first harmonic*. The *second harmonic* is the oscillation mode with $n = 2$, the *third harmonic* is that with $n = 3$, and so on. The frequencies associated with these modes are often labeled $f_1, f_2, f_3$, and so on. The collection of all possible oscillation modes is called the **harmonic series,** and $n$ is called the **harmonic number** of the $n$th harmonic.

For a given string under a given tension, each resonant frequency corresponds to a particular oscillation pattern. Thus, if the frequency is in the audible range, you can hear the shape of the string. Resonance can also occur in two dimensions (such as on the surface of the kettledrum in Fig. 16-24) and in three dimensions (such as in the wind-induced swaying and twisting of a tall building).

## Footbridges and Mosh Pits

When the Millennium Bridge over the Thames River was first opened to pedestrians, the bridge did not initially exhibit oscillations. The footsteps of the pedestrians produced vertical and horizontal forces on the bridge that tended to set up the second harmonic on the bridge (which is much like the second harmonic on a string), but the pedestrians were few and their walking was uncoordinated. However, once their number exceeded a critical value, the second harmonic suddenly became noticeable and walking became difficult. To keep their balance, many pedestrians began to time their steps to the bridge's swaying, which then became even worse and led to the bridge being closed until damping devices (see Sample Problem 15-3) could be added.

Similar structural oscillations occur when an audience on a football stadium deck or in a concert hall begins to sway or stomp in a coordinated fashion. Perhaps the worse situation can occur in a mosh pit on a lightweight suspended floor. When the crowded dancers begin to pogo, in which their coordinated jumps are in time with the music's beat, they can set up resonance in the floor, which typically has a resonant frequency of 2 Hz. Resonance can then quickly build as more people are forced to coordinate their motion with the oscillations, which might become large enough to break the floor and cause it to collapse. To avoid that possibility, modern building codes commonly require that suspended dance floors be built with resonant frequencies no lower than 5 Hz.

✓ **CHECKPOINT 6**  In the following series of resonant frequencies, one frequency (lower than 400 Hz) is missing: 150, 225, 300, 375 Hz. (a) What is the missing frequency? (b) What is the frequency of the seventh harmonic?

---

**Sample Problem** | **16-8** | **Build your skill**

Figure 16-25 shows a pattern of resonant oscillation of a string of mass $m = 2.500$ g and length $L = 0.800$ m and that is under tension $\tau = 325.0$ N. What is the wavelength $\lambda$ of the transverse waves producing the standing-wave pattern, and what is the harmonic number $n$? What is the frequency $f$ of the transverse waves and of the oscillations of the moving string elements? What is the maximum magnitude of the transverse velocity $u_m$ of the element oscillating at coordinate $x = 0.180$ m (note the $x$

axis in the figure)? At what point during the element's oscillation is the transverse velocity maximum?

**KEY IDEAS** (1) The traverse waves that produce a standing-wave pattern must have a wavelength such that an integer number $n$ of half-wavelengths fit into the length $L$ of the string. (2) The frequency of those waves and of the oscillations of the string elements is given by Eq. 16-66 ($f = nv/2L$). (3) The displacement of a string el-

**FIG. 16-25** Resonant oscillation of a string under tension.

ement as a function of position $x$ and time $t$ is given by Eq. 16-60:

$$y'(x, t) = [2y_m \sin kx] \cos \omega t. \qquad (16\text{-}67)$$

**Wavelength and harmonic number:** In Fig. 16-25, the solid line, which is effectively a snapshot (or freeze frame) of the oscillations, reveals that 2 full wavelengths fit into the length $L = 0.800$ m of the string. Thus, we have

$$2\lambda = L,$$

or $$\lambda = \frac{L}{2}. \qquad (16\text{-}68)$$

$$= \frac{0.800 \text{ m}}{2} = 0.400 \text{ m}. \qquad \text{(Answer)}$$

By counting the number of loops (or half-wavelengths) in Fig. 16-25, we see that the harmonic number is

$$n = 4. \qquad \text{(Answer)}$$

We reach the same conclusion by comparing Eqs. 16-68 and 16-65 ($\lambda = 2L/n$). Thus, the string is oscillating in its fourth harmonic.

**Frequency:** We can get the frequency $f$ of the transverse waves from Eq. 16-13 ($v = \lambda f$) if we first find the speed $v$ of the waves. That speed is given by Eq. 16-26, but we must substitute $m/L$ for the unknown linear density $\mu$. We obtain

$$v = \sqrt{\frac{\tau}{\mu}} = \sqrt{\frac{\tau}{m/L}} = \sqrt{\frac{\tau L}{m}}$$

$$= \sqrt{\frac{(325 \text{ N})(0.800 \text{ m})}{2.50 \times 10^{-3} \text{ kg}}} = 322.49 \text{ m/s}.$$

After rearranging Eq. 16-13, we write

$$f = \frac{v}{\lambda} = \frac{322.49 \text{ m/s}}{0.400 \text{ m}}$$

$$= 806.2 \text{ Hz} \approx 806 \text{ Hz}. \qquad \text{(Answer)}$$

Note that we get the same answer by substituting into Eq. 16-66:

$$f = n\frac{v}{2L} = 4\frac{322.49 \text{ m/s}}{2(0.800 \text{ m})}$$

$$= 806 \text{ Hz}. \qquad \text{(Answer)}$$

Now note that this 806 Hz is not only the frequency of the waves producing the fourth harmonic but also the frequency of the string elements that oscillate vertically in Fig. 16-25. It is also the frequency of the sound you would hear from the string.

**Transverse velocity:** The displacement $y'$ of the string element located at coordinate $x$ is given by Eq. 16-67 as a function of time $t$. The term $\cos \omega t$ contains the dependence on time and thus provides the "motion" of the standing wave. The term $2y_m \sin kx$ sets the extent of the motion—that is, the amplitude. The greatest amplitude occurs at an antinode, where $\sin kx$ is +1 or −1 and thus the greatest amplitude is $2y_m$. From Fig. 16-25, we see that $2y_m = 4.00$ mm, which tells us that $y_m = 2.00$ mm.

We want the transverse velocity—the velocity of a string element parallel to the $y$ axis. To find it, we take the time derivative of Eq. 16-67:

$$u(x, t) = \frac{\partial y'}{\partial t} = \frac{\partial}{\partial t}[(2y_m \sin kx) \cos \omega t]$$

$$= [-2y_m \omega \sin kx] \sin \omega t. \qquad (16\text{-}69)$$

Here the term $\sin \omega t$ provides the variation with time and the term $-2y_m \omega \sin kx$ provides the extent of that variation. We want the absolute magnitude of that extent:

$$u_m = |-2y_m \omega \sin kx|.$$

To evaluate this for the element at $x = 0.180$ m, we first note that $y_m = 2.00$ mm, $k = 2\pi/\lambda = 2\pi/(0.400 \text{ m})$, and $\omega = 2\pi f = 2\pi(806.2 \text{ Hz})$. Then the maximum speed of the element at $x = 0.180$ m is

$$u_m = \left| -2(2.00 \times 10^{-3} \text{ m})(2\pi)(806.2 \text{ Hz}) \right.$$

$$\left. \times \sin\left(\frac{2\pi}{0.400 \text{ m}}(0.180 \text{ m})\right) \right|$$

$$= 6.26 \text{ m/s}. \qquad \text{(Answer)}$$

To determine when the string element has this maximum speed, we could investigate Eq. 16-69. However, a little thought can save a lot of work. The element is undergoing simple harmonic motion and must come to a momentary stop at its extreme upward position and extreme downward position. It has the greatest speed as it zips through the midpoint of its oscillation, just as a block does in a block–spring oscillator.

---

**PROBLEM-SOLVING TACTICS**

*Tactic 2:* **Harmonics on a String** When you need to obtain information about a certain harmonic on a stretched string of given length $L$, first draw that harmonic (as in Fig. 16-23). If you are asked about, say, the fifth harmonic, you need to draw five loops between the fixed support points. That would mean that five loops, each of length $\lambda/2$, occupy the length $L$ of

the string. Thus, $5(\lambda/2) = L$, and $\lambda = 2L/5$. You can then use Eq. 16-13 ($f = v/\lambda$) to find the frequency of the harmonic.

Keep in mind that the wavelength of a harmonic is set only by the length $L$ of the string, but the frequency depends also on the wave speed $v$, which is set by the tension and the linear density of the string via Eq. 16-26.

## REVIEW & SUMMARY

**Transverse and Longitudinal Waves** Mechanical waves can exist only in material media and are governed by Newton's laws. **Transverse** mechanical waves, like those on a stretched string, are waves in which the particles of the medium oscillate perpendicular to the wave's direction of travel. Waves in which the particles of the medium oscillate parallel to the wave's direction of travel are **longitudinal** waves.

**Sinusoidal Waves** A sinusoidal wave moving in the positive direction of an $x$ axis has the mathematical form

$$y(x, t) = y_m \sin(kx - \omega t), \qquad (16\text{-}2)$$

where $y_m$ is the **amplitude** of the wave, $k$ is the **angular wave number**, $\omega$ is the **angular frequency**, and $kx - \omega t$ is the **phase**. The **wavelength** $\lambda$ is related to $k$ by

$$k = \frac{2\pi}{\lambda}. \qquad (16\text{-}5)$$

The **period** $T$ and **frequency** $f$ of the wave are related to $\omega$ by

$$\frac{\omega}{2\pi} = f = \frac{1}{T}. \qquad (16\text{-}9)$$

Finally, the **wave speed** $v$ is related to these other parameters by

$$v = \frac{\omega}{k} = \frac{\lambda}{T} = \lambda f. \qquad (16\text{-}13)$$

**Equation of a Traveling Wave** Any function of the form

$$y(x, t) = h(kx \pm \omega t) \qquad (16\text{-}17)$$

can represent a **traveling wave** with a wave speed given by Eq. 16-13 and a wave shape given by the mathematical form of $h$. The plus sign denotes a wave traveling in the negative direction of the $x$ axis, and the minus sign a wave traveling in the positive direction.

**Wave Speed on Stretched String** The speed of a wave on a stretched string is set by properties of the string. The speed on a string with tension $\tau$ and linear density $\mu$ is

$$v = \sqrt{\frac{\tau}{\mu}}. \qquad (16\text{-}26)$$

**Power** The **average power** of, or average rate at which energy is transmitted by, a sinusoidal wave on a stretched string is given by

$$P_{\text{avg}} = \tfrac{1}{2}\mu v \omega^2 y_m^2. \qquad (16\text{-}33)$$

**Superposition of Waves** When two or more waves traverse the same medium, the displacement of any particle of the medium is the sum of the displacements that the individual waves would give it.

**Interference of Waves** Two sinusoidal waves on the same string exhibit **interference**, adding or canceling according to the principle of superposition. If the two are traveling in the same direction and have the same amplitude $y_m$ and frequency (hence the same wavelength) but differ in phase by a **phase constant** $\phi$, the result is a single wave with this same frequency:

$$y'(x, t) = [2y_m \cos \tfrac{1}{2}\phi] \sin(kx - \omega t + \tfrac{1}{2}\phi). \qquad (16\text{-}51)$$

If $\phi = 0$, the waves are exactly in phase and their interference is fully constructive; if $\phi = \pi$ rad, they are exactly out of phase and their interference is fully destructive.

**Phasors** A wave $y(x, t)$ can be represented with a **phasor**. This is a vector that has a magnitude equal to the amplitude $y_m$ of the wave and that rotates about an origin with an angular speed equal to the angular frequency $\omega$ of the wave. The projection of the rotating phasor on a vertical axis gives the displacement $y$ of a point along the wave's travel.

**Standing Waves** The interference of two identical sinusoidal waves moving in opposite directions produces **standing waves**. For a string with fixed ends, the standing wave is given by

$$y'(x, t) = [2y_m \sin kx] \cos \omega t. \qquad (16\text{-}60)$$

Standing waves are characterized by fixed locations of zero displacement called **nodes** and fixed locations of maximum displacement called **antinodes**.

**Resonance** Standing waves on a string can be set up by reflection of traveling waves from the ends of the string. If an end is fixed, it must be the position of a node. This limits the frequencies at which standing waves will occur on a given string. Each possible frequency is a **resonant frequency**, and the corresponding standing wave pattern is an **oscillation mode**. For a stretched string of length $L$ with fixed ends, the resonant frequencies are

$$f = \frac{v}{\lambda} = n \frac{v}{2L}, \qquad \text{for } n = 1, 2, 3, \ldots . \qquad (16\text{-}66)$$

The oscillation mode corresponding to $n = 1$ is called the *fundamental mode* or the *first harmonic*; the mode corresponding to $n = 2$ is the *second harmonic*; and so on.

## QUESTIONS

**1** Figure 16-26a gives a snapshot of a wave traveling in the direction of positive $x$ along a string under tension. Four string elements are indicated by the lettered points. For each of those elements, determine whether, at the instant of the snapshot, the element is moving upward or downward or is momentarily at rest. (*Hint:* Imagine the wave as it moves through the four string elements, as if you were watching a video of the wave as it traveled rightward.)

Figure 16-26b gives the displacement of a string element located at, say, $x = 0$ as a function of time. At the lettered times, is the element moving upward or downward or is it momentarily at rest?

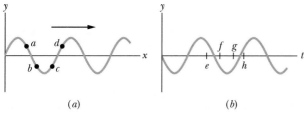

**FIG. 16-26**   Question 1.

**2**   Figure 16-27 shows three waves that are *separately* sent along a string that is stretched under a certain tension along an $x$ axis. Rank the waves according to their (a) wavelengths, (b) speeds, and (c) angular frequencies, greatest first.

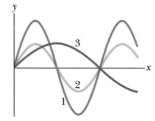

**FIG. 16-27**   Question 2.

**3**   The following four waves are sent along strings with the same linear densities ($x$ is in meters and $t$ is in seconds). Rank the waves according to (a) their wave speed and (b) the tension in the strings along which they travel, greatest first:

(1)  $y_1 = (3 \text{ mm}) \sin(x - 3t)$,   (3)  $y_3 = (1 \text{ mm}) \sin(4x - t)$,

(2)  $y_2 = (6 \text{ mm}) \sin(2x - t)$,   (4)  $y_4 = (2 \text{ mm}) \sin(x - 2t)$.

**4**   In Fig. 16-28, wave 1 consists of a rectangular peak of height 4 units and width $d$, and a rectangular valley of depth 2 units and width $d$. The wave travels rightward along an $x$ axis. Choices 2, 3, and 4 are similar waves, with the same heights, depths, and widths, that will travel leftward along that axis and through wave 1. Right-going wave 1 and one of the left-going waves will interfere as they pass through each other. With which left-going wave will the interference give, for an instant, (a) the deepest valley, (b) a flat line, and (c) a flat peak $2d$ wide?

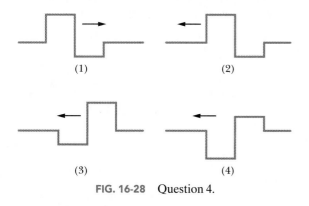

**FIG. 16-28**   Question 4.

**5**   A sinusoidal wave is sent along a cord under tension, transporting energy at the average rate of $P_{\text{avg},1}$. Two waves, identical to that first one, are then to be sent along the cord with a phase difference $\phi$ of either 0, 0.2 wavelength, or 0.5 wavelength. (a) With only mental calculation, rank those choices of $\phi$ according to the average rate at which the waves will transport energy, greatest first. (b) For the first choice of $\phi$, what is the average rate in terms of $P_{\text{avg},1}$?

**6**   The amplitudes and phase differences for four pairs of waves of equal wavelengths are (a) 2 mm, 6 mm, and $\pi$ rad; (b) 3 mm, 5 mm, and $\pi$ rad; (c) 7 mm, 9 mm, and $\pi$ rad; (d) 2 mm, 2 mm, and 0 rad. Each pair travels in the same direction along the same string. Without written calculation, rank the four pairs according to the amplitude of their resultant wave, greatest first. (*Hint:* Construct phasor diagrams.)

**7**   If you start with two sinusoidal waves of the same amplitude traveling in phase on a string and then somehow phase-shift one of them by 5.4 wavelengths, what type of interference will occur on the string?

**8**   If you set up the seventh harmonic on a string, (a) how many nodes are present, and (b) is there a node, antinode, or some intermediate state at the midpoint? If you next set up the sixth harmonic, (c) is its resonant wavelength longer or shorter than that for the seventh harmonic, and (d) is the resonant frequency higher or lower?

**9**   Figure 16-29 shows phasor diagrams for three situations in which two waves travel along the same string. All six waves have the same amplitude. Rank the situations according to the amplitude of the net wave on the string, greatest first.

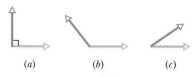

**FIG. 16-29**   Question 9.

**10**   (a) If a standing wave on a string is given by

$$y'(t) = (3 \text{ mm}) \sin(5x) \cos(4t),$$

is there a node or an antinode of the oscillations of the string at $x = 0$? (b) If the standing wave is given by

$$y'(t) = (3 \text{ mm}) \sin(5x + \pi/2) \cos(4t),$$

is there a node or an antinode at $x = 0$?

**11**   Strings $A$ and $B$ have identical lengths and linear densities, but string $B$ is under greater tension than string $A$. Figure 16-30 shows four situations, ($a$) through ($d$), in which standing wave patterns exist on the two strings. In which situations is there the possibility that strings $A$ and $B$ are oscillating at the same resonant frequency?

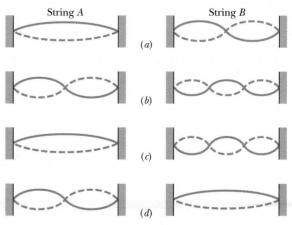

**FIG. 16-30**   Question 11.

## PROBLEMS

**GO** Tutoring problem available (at instructor's discretion) in *WileyPLUS* and WebAssign

**SSM** Worked-out solution available in Student Solutions Manual

• – ••• Number of dots indicates level of problem difficulty

**WWW** Worked-out solution is at —

**ILW** Interactive solution is at — http://www.wiley.com/college/halliday

✈ Additional information available in *The Flying Circus of Physics* and at flyingcircusofphysics.com

### sec. 16-5 The Speed of a Traveling Wave

•1 A wave has an angular frequency of 110 rad/s and a wavelength of 1.80 m. Calculate (a) the angular wave number and (b) the speed of the wave.

•2 A sand scorpion can detect the motion of a nearby beetle (its prey) by the waves the motion sends along the sand surface (Fig. 16-31). The waves are of two types: transverse waves traveling at $v_t = 50$ m/s and longitudinal waves traveling at $v_l = 150$ m/s. If a sudden motion sends out such waves, a scorpion can tell the distance of the beetle from the difference $\Delta t$ in the arrival times of the waves at its leg nearest the beetle. If $\Delta t = 4.0$ ms, what is the beetle's distance? ✈

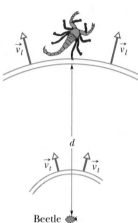

**FIG. 16-31** Problem 2.

•3 A sinusoidal wave travels along a string. The time for a particular point to move from maximum displacement to zero is 0.170 s. What are the (a) period and (b) frequency? (c) The wavelength is 1.40 m; what is the wave speed?

•4 *A human wave.* During sporting events within large, densely packed stadiums, spectators will send a wave (or pulse) around the stadium (Fig. 16-32). As the wave reaches a group of spectators, they stand with a cheer and then sit. At

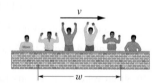

**FIG. 16-32** Problem 4.

any instant, the width $w$ of the wave is the distance from the leading edge (people are just about to stand) to the trailing edge (people have just sat down). Suppose a human wave travels a distance of 853 seats around a stadium in 39 s, with spectators requiring about 1.8 s to respond to the wave's passage by standing and then sitting. What are (a) the wave speed $v$ (in seats per second) and (b) width $w$ (in number of seats)? ✈

•5 If $y(x, t) = (6.0$ mm$) \sin(kx + (600$ rad/s$)t + \phi)$ describes a wave traveling along a string, how much time does any given point on the string take to move between displacements $y = +2.0$ mm and $y = -2.0$ mm?

••6 Figure 16-33 shows the transverse velocity $u$ versus time $t$ of the point on a string at $x = 0$, as a wave passes through it. The scale on the vertical axis is set by $u_s = 4.0$ m/s. The wave has form $y(x, t) = y_m \sin(kx - \omega t + \phi)$.

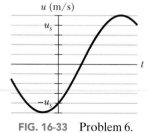

**FIG. 16-33** Problem 6.

What is $\phi$? (*Caution:* A calculator does not always give the proper inverse trig function, so check your answer by substituting it and an assumed value of $\omega$ into $y(x, t)$ and then plotting the function.)

••7 A sinusoidal wave of frequency 500 Hz has a speed of 350 m/s. (a) How far apart are two points that differ in phase by $\pi/3$ rad? (b) What is the phase difference between two displacements at a certain point at times 1.00 ms apart? **ILW**

••8 The equation of a transverse wave traveling along a very long string is $y = 6.0 \sin(0.020\pi x + 4.0\pi t)$, where $x$ and $y$ are expressed in centimeters and $t$ is in seconds. Determine (a) the amplitude, (b) the wavelength, (c) the frequency, (d) the speed, (e) the direction of propagation of the wave, and (f) the maximum transverse speed of a particle in the string. (g) What is the transverse displacement at $x = 3.5$ cm when $t = 0.26$ s?

••9 A transverse sinusoidal wave is moving along a string in the positive direction of an $x$ axis with a speed of 80 m/s. At $t = 0$, the string particle at $x = 0$ has a transverse displacement of 4.0 cm from its equilibrium position and is not moving. The maximum transverse speed of the string particle at $x = 0$ is 16 m/s. (a) What is the frequency of the wave? (b) What is the wavelength of the wave? If the wave equation is of the form $y(x, t) = y_m \sin(kx \pm \omega t + \phi)$, what are (c) $y_m$, (d) $k$, (e) $\omega$, (f) $\phi$, and (g) the correct choice of sign in front of $\omega$?

••10 The function $y(x, t) = (15.0$ cm$) \cos(\pi x - 15\pi t)$, with $x$ in meters and $t$ in seconds, describes a wave on a taut string. What is the transverse speed for a point on the string at an instant when that point has the displacement $y = +12.0$ cm?

••11 A sinusoidal wave moving along a string is shown twice in Fig. 16-34, as crest $A$ travels in the positive direction of an $x$ axis by distance $d = 6.0$ cm in 4.0 ms. The tick marks along the axis are separated by 10 cm; height $H = 6.00$ mm. If the wave equation is of the form $y(x, t) = y_m \sin(kx \pm \omega t)$, what are (a) $y_m$, (b) $k$, (c) $\omega$, and (d) the correct choice of sign in front of $\omega$?

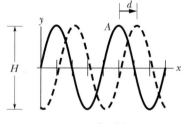

**FIG. 16-34** Problem 11.

••12 A sinusoidal wave travels along a string under tension. Figure 16-35 gives the slopes along the string at time $t = 0$. The scale of the $x$ axis is set by $x_s = 0.80$ m. What is the amplitude of the wave? **GO**

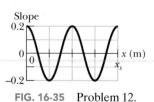

**FIG. 16-35** Problem 12.

**••13** A sinusoidal transverse wave of wavelength 20 cm travels along a string in the positive direction of an x axis. The displacement y of the string particle at $x = 0$ is given in Fig. 16-36 as a function of time t. The scale of the vertical

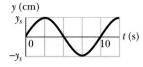

**FIG. 16-36** Problem 13.

axis is set by $y_s = 4.0$ cm. The wave equation is to be in the form $y(x, t) = y_m \sin(kx \pm \omega t + \phi)$. (a) At $t = 0$, is a plot of y versus x in the shape of a positive sine function or a negative sine function? What are (b) $y_m$, (c) k, (d) $\omega$, (e) $\phi$, (f) the sign in front of $\omega$, and (g) the speed of the wave? (h) What is the transverse velocity of the particle at $x = 0$ when $t = 5.0$ s? **GO**

### sec. 16-6 Wave Speed on a Stretched String

**•14** The tension in a wire clamped at both ends is doubled without appreciably changing the wire's length between the clamps. What is the ratio of the new to the old wave speed for transverse waves traveling along this wire?

**•15** What is the speed of a transverse wave in a rope of length 2.00 m and mass 60.0 g under a tension of 500 N? **SSM**

**•16** The heaviest and lightest strings on a certain violin have linear densities of 3.0 and 0.29 g/m. What is the ratio of the diameter of the heaviest string to that of the lightest string, assuming that the strings are of the same material?

**•17** A stretched string has a mass per unit length of 5.00 g/cm and a tension of 10.0 N. A sinusoidal wave on this string has an amplitude of 0.12 mm and a frequency of 100 Hz and is traveling in the negative direction of an x axis. If the wave equation is of the form $y(x, t) = y_m \sin(kx \pm \omega t)$, what are (a) $y_m$, (b) k, (c) $\omega$, and (d) the correct choice of sign in front of $\omega$? **SSM WWW**

**•18** The speed of a transverse wave on a string is 170 m/s when the string tension is 120 N. To what value must the tension be changed to raise the wave speed to 180 m/s?

**•19** The linear density of a string is $1.6 \times 10^{-4}$ kg/m. A transverse wave on the string is described by the equation

$$y = (0.021 \text{ m}) \sin[(2.0 \text{ m}^{-1})x + (30 \text{ s}^{-1})t].$$

What are (a) the wave speed and (b) the tension in the string?

**•20** The equation of a transverse wave on a string is

$$y = (2.0 \text{ mm}) \sin[(20 \text{ m}^{-1})x - (600 \text{ s}^{-1})t].$$

The tension in the string is 15 N. (a) What is the wave speed? (b) Find the linear density of this string in grams per meter.

**••21** A sinusoidal transverse wave is traveling along a string in the negative direction of an x axis. Figure 16-37 shows a plot of the displacement as a function of position at time $t = 0$; the scale of the y axis is set by $y_s = 4.0$ cm. The string tension is 3.6 N, and its linear density is 25 g/m. Find the (a) amplitude, (b) wavelength, (c) wave speed, and (d) period of the wave. (e) Find the maximum transverse speed of a particle in the string. If the wave is of the form $y(x, t) = y_m \sin(kx \pm \omega t + \phi)$, what are (f) k, (g) $\omega$, (h) $\phi$, and (i) the correct choice of sign in front of $\omega$? **SSM ILW**

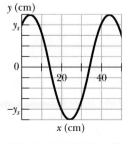

**FIG. 16-37** Problem 21.

**••22** A sinusoidal wave is traveling on a string with speed 40 cm/s. The displacement of the particles of the string at $x = 10$ cm is found to vary with time according to the equation $y = (5.0 \text{ cm}) \sin[1.0 - (4.0 \text{ s}^{-1})t]$. The linear density of the string is 4.0 g/cm. What are (a) the frequency and (b) the wavelength of the wave? If the wave equation is of the form $y(x, t) = y_m \sin(kx \pm \omega t)$, what are (c) $y_m$, (d) k, (e) $\omega$, and (f) the correct choice of sign in front of $\omega$? (g) What is the tension in the string?

**••23** A 100 g wire is held under a tension of 250 N with one end at $x = 0$ and the other at $x = 10.0$ m. At time $t = 0$, pulse 1 is sent along the wire from the end at $x = 10.0$ m. At time $t = 30.0$ ms, pulse 2 is sent along the wire from the end at $x = 0$. At what position x do the pulses begin to meet? **ILW**

**•••24** In Fig. 16-38a, string 1 has a linear density of 3.00 g/m, and string 2 has a linear density of 5.00 g/m. They are under tension due to the hanging block of mass $M = 500$ g. Calculate the wave speed on (a) string 1 and (b) string 2. (*Hint:* When a string loops halfway around a pulley, it pulls on the pulley with a net force that is twice the tension in the string.) Next the block is divided into two blocks (with $M_1 + M_2 = M$) and the apparatus is rearranged as shown in Fig. 16-38b. Find (c) $M_1$ and (d) $M_2$ such that the wave speeds in the two strings are equal.

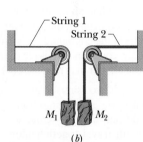

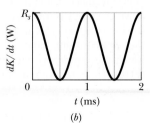

**FIG. 16-38** Problem 24.

**•••25** A uniform rope of mass m and length L hangs from a ceiling. (a) Show that the speed of a transverse wave on the rope is a function of y, the distance from the lower end, and is given by $v = \sqrt{gy}$. (b) Show that the time a transverse wave takes to travel the length of the rope is given by $t = 2\sqrt{L/g}$.

### sec. 16-7 Energy and Power of a Wave Traveling Along a String

**•26** A string along which waves can travel is 2.70 m long and has a mass of 260 g. The tension in the string is 36.0 N. What must be the frequency of traveling waves of amplitude 7.70 mm for the average power to be 85.0 W?

**••27** A sinusoidal wave is sent along a string with a linear density of 2.0 g/m. As it travels, the kinetic energies of the mass elements along the string vary. Figure 16-39a gives

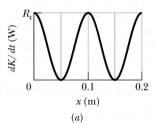

**FIG. 16-39** Problem 27.

the rate $dK/dt$ at which kinetic energy passes through the string elements at a particular instant, plotted as a function of distance $x$ along the string. Figure 16-39b is similar except that it gives the rate at which kinetic energy passes through a particular mass element (at a particular location), plotted as a function of time $t$. For both figures, the scale on the vertical (rate) axis is set by $R_s = 10$ W. What is the amplitude of the wave? **GO**

### sec. 16-8    The Wave Equation

•28   Use the wave equation to find the speed of a wave given by

$$y(x, t) = (3.00 \text{ mm}) \sin[(4.00 \text{ m}^{-1})x - (7.00 \text{ s}^{-1})t].$$

••29   Use the wave equation to find the speed of a wave given by

$$y(x, t) = (2.00 \text{ mm})[(20 \text{ m}^{-1})x - (4.0 \text{ s}^{-1})t]^{0.5}.$$

•••30   Use the wave equation to find the speed of a wave given in terms of the general function $h(x, t)$:

$$y(x, t) = (4.00 \text{ mm}) h[(30 \text{ m}^{-1})x + (6.0 \text{ s}^{-1})t].$$

### sec. 16-10    Interference of Waves

•31   Two identical traveling waves, moving in the same direction, are out of phase by $\pi/2$ rad. What is the amplitude of the resultant wave in terms of the common amplitude $y_m$ of the two combining waves? **SSM**

•32   What phase difference between two identical traveling waves, moving in the same direction along a stretched string, results in the combined wave having an amplitude 1.50 times that of the common amplitude of the two combining waves? Express your answer in (a) degrees, (b) radians, and (c) wavelengths.

••33   Two sinusoidal waves with the same amplitude of 9.00 mm and the same wavelength travel together along a string that is stretched along an $x$ axis. Their resultant wave is shown twice in Fig. 16-40, as valley $A$ travels in the negative direction of the $x$ axis by distance $d = 56.0$ cm in 8.0 ms. The tick marks along the axis are separated by 10 cm, and height $H$ is 8.0 mm. Let the equation for one wave be of the form $y(x, t) = y_m \sin(kx \pm \omega t + \phi_1)$, where $\phi_1 = 0$ and you must choose the correct sign in front of $\omega$. For the equation for the other wave, what are (a) $y_m$, (b) $k$, (c) $\omega$, (d) $\phi_2$, and (e) the sign in front of $\omega$?

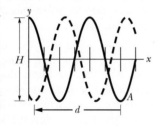

**FIG. 16-40**    Problem 33.

•••34   A sinusoidal wave of angular frequency 1200 rad/s and amplitude 3.00 mm is sent along a cord with linear density 2.00 g/m and tension 1200 N. (a) What is the average rate at which energy is transported by the wave to the opposite end of the cord? (b) If, simultaneously, an identical wave travels along an adjacent, identical cord, what is the total average rate at which energy is transported to the opposite ends of the two cords by the waves? If, instead, those two waves are sent along the *same* cord simultaneously, what is the total average rate at which they transport energy when their phase difference is (c) 0, (d) $0.4\pi$ rad, and (e) $\pi$ rad?

### sec. 16-11    Phasors

•35   Two sinusoidal waves of the same frequency travel in the same direction along a string. If $y_{m1} = 3.0$ cm, $y_{m2} =$

4.0 cm, $\phi_1 = 0$, and $\phi_2 = \pi/2$ rad, what is the amplitude of the resultant wave? **SSM**

••36   Two sinusoidal waves of the same frequency are to be sent in the same direction along a taut string. One wave has an amplitude of 5.0 mm, the other 8.0 mm. (a) What phase difference $\phi_1$ between the two waves results in the smallest amplitude of the resultant wave? (b) What is that smallest amplitude? (c) What phase difference $\phi_2$ results in the largest amplitude of the resultant wave? (d) What is that largest amplitude? (e) What is the resultant amplitude if the phase angle is $(\phi_1 - \phi_2)/2$?

••37   Two sinusoidal waves of the same period, with amplitudes of 5.0 and 7.0 mm, travel in the same direction along a stretched string; they produce a resultant wave with an amplitude of 9.0 mm. The phase constant of the 5.0 mm wave is 0. What is the phase constant of the 7.0 mm wave?

••38   Four waves are to be sent along the same string, in the same direction:

$$y_1(x, t) = (4.00 \text{ mm}) \sin(2\pi x - 400\pi t)$$
$$y_2(x, t) = (4.00 \text{ mm}) \sin(2\pi x - 400\pi t + 0.7\pi)$$
$$y_3(x, t) = (4.00 \text{ mm}) \sin(2\pi x - 400\pi t + \pi)$$
$$y_4(x, t) = (4.00 \text{ mm}) \sin(2\pi x - 400\pi t + 1.7\pi).$$

What is the amplitude of the resultant wave?

••39   These two waves travel along the same string:

$$y_1(x, t) = (4.60 \text{ mm}) \sin(2\pi x - 400\pi t)$$
$$y_2(x, t) = (5.60 \text{ mm}) \sin(2\pi x - 400\pi t + 0.80\pi \text{ rad}).$$

What are (a) the amplitude and (b) the phase angle (relative to wave 1) of the resultant wave? (c) If a third wave of amplitude 5.00 mm is also to be sent along the string in the same direction as the first two waves, what should be its phase angle in order to maximize the amplitude of the new resultant wave? **GO**

### sec. 16-13    Standing Waves and Resonance

•40   A 125 cm length of string has mass 2.00 g and tension 7.00 N. (a) What is the wave speed for this string? (b) What is the lowest resonant frequency of this string?

•41   What are (a) the lowest frequency, (b) the second lowest frequency, and (c) the third lowest frequency for standing waves on a wire that is 10.0 m long, has a mass of 100 g, and is stretched under a tension of 250 N? **SSM WWW**

•42   String $A$ is stretched between two clamps separated by distance $L$. String $B$, with the same linear density and under the same tension as string $A$, is stretched between two clamps separated by distance $4L$. Consider the first eight harmonics of string $B$. For which of these eight harmonics of $B$ (if any) does the frequency match the frequency of (a) $A$'s first harmonic, (b) $A$'s second harmonic, and (c) $A$'s third harmonic?

•43   A string fixed at both ends is 8.40 m long and has a mass of 0.120 kg. It is subjected to a tension of 96.0 N and set oscillating. (a) What is the speed of the waves on the string? (b) What is the longest possible wavelength for a standing wave? (c) Give the frequency of that wave. **SSM**

•44   Two sinusoidal waves with identical wavelengths and amplitudes travel in opposite directions along a string with a speed of 10 cm/s. If the time interval between instants when the string is flat is 0.50 s, what is the wavelength of the waves?

•45  A nylon guitar string has a linear density of 7.20 g/m and is under a tension of 150 N. The fixed supports are distance $D =$ 90.0 cm apart. The string is oscil-

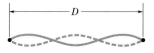

FIG. 16-41  Problem 45.

lating in the standing wave pattern shown in Fig. 16-41. Calculate the (a) speed, (b) wavelength, and (c) frequency of the traveling waves whose superposition gives this standing wave.  ILW

•46  A string under tension $\tau_i$ oscillates in the third harmonic at frequency $f_3$, and the waves on the string have wavelength $\lambda_3$. If the tension is increased to $\tau_f = 4\tau_i$ and the string is again made to oscillate in the third harmonic, what then are (a) the frequency of oscillation in terms of $f_3$ and (b) the wavelength of the waves in terms of $\lambda_3$?

•47  A string that is stretched between fixed supports separated by 75.0 cm has resonant frequencies of 420 and 315 Hz, with no intermediate resonant frequencies. What are (a) the lowest resonant frequency and (b) the wave speed?  SSM ILW

•48  If a transmission line in a cold climate collects ice, the increased diameter tends to cause vortex formation in a passing wind. The air pressure variations in the vortexes tend to cause the line to oscillate (*gallop*), especially if the frequency of the variations matches a resonant frequency of the line. In long lines, the resonant frequencies are so close that almost any wind speed can set up a resonant mode vigorous enough to pull down support towers or cause the line to *short out* with an adjacent line. If a transmission line has a length of 347 m, a linear density of 3.35 kg/m, and a tension of 65.2 MN, what are (a) the frequency of the fundamental mode and (b) the frequency difference between successive modes?

•49  One of the harmonic frequencies for a particular string under tension is 325 Hz. The next higher harmonic frequency is 390 Hz. What harmonic frequency is next higher after the harmonic frequency 195 Hz?

••50  A rope, under a tension of 200 N and fixed at both ends, oscillates in a second-harmonic standing wave pattern. The displacement of the rope is given by

$$y = (0.10 \text{ m})(\sin \pi x/2) \sin 12\pi t,$$

where $x = 0$ at one end of the rope, $x$ is in meters, and $t$ is in seconds. What are (a) the length of the rope, (b) the speed of the waves on the rope, and (c) the mass of the rope? (d) If the rope oscillates in a third-harmonic standing wave pattern, what will be the period of oscillation?

••51  A string oscillates according to the equation

$$y' = (0.50 \text{ cm}) \sin\left[\left(\frac{\pi}{3} \text{ cm}^{-1}\right)x\right] \cos[(40\pi \text{ s}^{-1})t].$$

What are the (a) amplitude and (b) speed of the two waves (identical except for direction of travel) whose superposition gives this oscillation? (c) What is the distance between nodes? (d) What is the transverse speed of a particle of the string at the position $x = 1.5$ cm when $t = \frac{9}{8}$ s?

••52  A standing wave pattern on a string is described by

$$y(x, t) = 0.040 (\sin 5\pi x)(\cos 40\pi t),$$

where $x$ and $y$ are in meters and $t$ is in seconds. For $x \geq 0$, what is the location of the node with the (a) smallest, (b) second

smallest, and (c) third smallest value of $x$? (d) What is the period of the oscillatory motion of any (nonnode) point? What are the (e) speed and (f) amplitude of the two traveling waves that interfere to produce this wave? For $t \geq 0$, what are the (g) first, (h) second, and (i) third time that all points on the string have zero transverse velocity?

••53  Two waves are generated on a string of length 3.0 m to produce a three-loop standing wave with an amplitude of 1.0 cm. The wave speed is 100 m/s. Let the equation for one of the waves be of the form $y(x, t) = y_m \sin(kx + \omega t)$. In the equation for the other wave, what are (a) $y_m$, (b) $k$, (c) $\omega$, and (d) the sign in front of $\omega$?  SSM WWW

••54  For a certain transverse standing wave on a long string, an antinode is at $x = 0$ and an adjacent node is at $x = 0.10$ m. The displacement $y(t)$ of the string particle at $x = 0$ is shown in Fig. 16-42, where the scale of the $y$ axis is set by $y_s = 4.0$ cm. When $t = 0.50$ s, what is the dis-

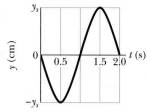

FIG. 16-42  Problem 54.

placement of the string particle at (a) $x = 0.20$ m and (b) $x = 0.30$ m? What is the transverse velocity of the string particle at $x = 0.20$ m at (c) $t = 0.50$ s and (d) $t = 1.0$ s? (e) Sketch the standing wave at $t = 0.50$ s for the range $x = 0$ to $x = 0.40$ m.

••55  A generator at one end of a very long string creates a wave given by

$$y = (6.0 \text{ cm}) \cos \frac{\pi}{2} [(2.00 \text{ m}^{-1})x + (8.00 \text{ s}^{-1})t],$$

and a generator at the other end creates the wave

$$y = (6.0 \text{ cm}) \cos \frac{\pi}{2} [(2.00 \text{ m}^{-1})x - (8.00 \text{ s}^{-1})t].$$

Calculate the (a) frequency, (b) wavelength, and (c) speed of each wave. For $x \geq 0$, what is the location of the node having the (d) smallest, (e) second smallest, and (f) third smallest value of $x$? For $x \geq 0$, what is the location of the antinode having the (g) smallest, (h) second smallest, and (i) third smallest value of $x$?

••56  Two sinusoidal waves with the same amplitude and wavelength travel through each other along a string that is stretched along an $x$ axis. Their resultant wave is shown twice in Fig. 16-43, as the antinode $A$ travels from an extreme upward displacement to an extreme downward displacement in 6.0 ms. The tick marks along the axis are separated by 10 cm; height $H$ is 1.80 cm. Let the equation for one of the two waves be of the form $y(x, t) = y_m \sin(kx + \omega t)$. In the equation for the other wave, what are (a) $y_m$, (b) $k$, (c) $\omega$, and (d) the sign in front of $\omega$?  GO

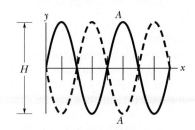

FIG. 16-43  Problem 56.

**••57** The following two waves are sent in opposite directions on a horizontal string so as to create a standing wave in a vertical plane:

$$y_1(x, t) = (6.00 \text{ mm}) \sin(4.00\pi x - 400\pi t)$$

$$y_2(x, t) = (6.00 \text{ mm}) \sin(4.00\pi x + 400\pi t),$$

with $x$ in meters and $t$ in seconds. An antinode is located at point $A$. In the time interval that point takes to move from maximum upward displacement to maximum downward displacement, how far does each wave move along the string?

**••58** In Fig. 16-44, a string, tied to a sinusoidal oscillator at $P$ and running over a support at $Q$, is stretched by a block of mass $m$. Separation $L = 1.20$ m, linear density $\mu = 1.6$ g/m, and the oscillator frequency $f = 120$ Hz. The amplitude of the motion at $P$ is small enough for that point to be considered a node. A node also exists at $Q$. (a) What mass $m$ allows the oscillator to set up the fourth harmonic on the string? (b) What standing wave mode, if any, can be set up if $m = 1.00$ kg?

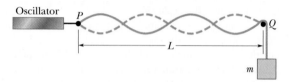

FIG. 16-44    Problems 58 and 60.

**•••59** In Fig. 16-45, an aluminum wire, of length $L_1 = 60.0$ cm, cross-sectional area $1.00 \times 10^{-2}$ cm², and density 2.60 g/cm³, is joined to a steel wire, of density 7.80 g/cm³ and the same cross-sectional area. The compound wire, loaded with a block of mass $m = 10.0$ kg, is arranged so that the distance $L_2$ from the joint to the supporting pulley is 86.6 cm. Transverse waves are set up on the wire by an external source of variable frequency; a node is located at the pulley. (a) Find the lowest frequency that generates a standing wave having the joint as one of the nodes. (b) How many nodes are observed at this frequency?

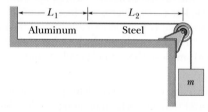

FIG. 16-45
Problem 59.

**•••60** In Fig. 16-44, a string, tied to a sinusoidal oscillator at $P$ and running over a support at $Q$, is stretched by a block of mass $m$. The separation $L$ between $P$ and $Q$ is 1.20 m, and the frequency $f$ of the oscillator is fixed at 120 Hz. The amplitude of the motion at $P$ is small enough for that point to be considered a node. A node also exists at $Q$. A standing wave appears when the mass of the hanging block is 286.1 g or 447.0 g, but not for any intermediate mass. What is the linear density of the string? **GO**

**Additional Problems**

**61** Three sinusoidal waves of the same frequency travel along a string in the positive direction of an $x$ axis. Their amplitudes are $y_1$, $y_1/2$, and $y_1/3$, and their phase constants are 0, $\pi/2$, and $\pi$, respectively. What are the (a) amplitude and (b) phase constant of the resultant wave? (c) Plot the wave form of the resultant wave at $t = 0$, and discuss its behavior as $t$ increases. **SSM**

**62** Figure 16-46 shows the displacement $y$ versus time $t$ of the point on a string at $x = 0$, as a wave passes through that point. The scale of the $y$ axis is set by $y_s = 6.0$ mm. The wave has form $y(x, t) = y_m \sin(kx - \omega t + \phi)$. What is $\phi$? (*Caution:* A calculator does not always give the proper inverse trig function, so check your answer by substituting it and an assumed value of $\omega$ into $y(x, t)$ and then plotting the function.)

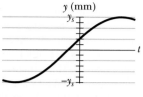

FIG. 16-46    Problem 62.

**63** Two sinusoidal waves, identical except for phase, travel in the same direction along a string, producing a net wave $y'(x, t) = (3.0 \text{ mm}) \sin(20x - 4.0t + 0.820 \text{ rad})$, with $x$ in meters and $t$ in seconds. What are (a) the wavelength $\lambda$ of the two waves, (b) the phase difference between them, and (c) their amplitude $y_m$?

**64** Figure 16-47 shows transverse acceleration $a_y$ versus time $t$ of the point on a string at $x = 0$, as a wave written in the form $y(x, t) = y_m \sin(kx - \omega t + \phi)$ passes through that point. The scale of the vertical axis is set by $a_s = 400$ m/s². What is $\phi$? (*Caution:* A calculator does not always give the proper inverse trig function, so check your answer by substituting it and an assumed value of $\omega$ into $y(x, t)$ and then plotting the function.)

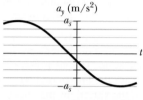

FIG. 16-47    Problem 64.

**65** At time $t = 0$ and at position $x = 0$ m along a string, a traveling sinusoidal wave with an angular frequency of 440 rad/s has displacement $y = +4.5$ mm and transverse velocity $u = -0.75$ m/s. If the wave has general form $y(x, t) = y_m \sin(kx - \omega t + \phi)$, what is phase constant $\phi$?

**66** A single pulse, given by $h(x - 5.0t)$, is shown in Fig. 16-48 for $t = 0$. The scale of the vertical axis is set by $h_s = 2$. Here $x$ is in centimeters and $t$ is in seconds. What are the (a) speed and (b) direction of travel of the pulse? (c) Plot $h(x - 5t)$ as a function of $x$ for $t = 2$ s. (d) Plot $h(x - 5t)$ as a function of $t$ for $x = 10$ cm.

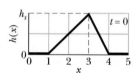

FIG. 16-48    Problem 66.

**67** A transverse sinusoidal wave is generated at one end of a long, horizontal string by a bar that moves up and down through a distance of 1.00 cm. The motion is continuous and is repeated regularly 120 times per second. The string has linear density 120 g/m and is kept under a tension of 90.0 N. Find the maximum value of (a) the transverse speed $u$ and (b) the transverse component of the tension $\tau$.

(c) Show that the two maximum values calculated above occur at the same phase values for the wave. What is the transverse displacement $y$ of the string at these phases? (d) What is the maximum rate of energy transfer along the string? (e) What is the transverse displacement $y$ when this maximum transfer occurs? (f) What is the minimum rate of energy transfer along the string? (g) What is the transverse displacement $y$ when this minimum transfer occurs?

**68** Two sinusoidal 120 Hz waves, of the same frequency and amplitude, are to be sent in the positive direction of an $x$ axis that is directed along a cord under tension. The waves can be sent in phase, or they can be phase-shifted. Figure 16-49 shows the amplitude $y'$ of the resulting wave versus the distance of the shift (how far one wave is shifted from the other wave). The scale of the vertical

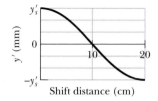

FIG. 16-49   Problem 68.

axis is set by $y'_s = 6.0$ mm. If the equations for the two waves are of the form $y(x, t) = y_m \sin(kx \pm \omega t)$, what are (a) $y_m$, (b) $k$, (c) $\omega$, and (d) the correct choice of sign in front of $\omega$?

**69**   A sinusoidal transverse wave of amplitude $y_m$ and wavelength $\lambda$ travels on a stretched cord. (a) Find the ratio of the maximum particle speed (the speed with which a single particle in the cord moves transverse to the wave) to the wave speed. (b) Does this ratio depend on the material of which the cord is made?   SSM

**70**   A sinusoidal transverse wave traveling in the positive direction of an $x$ axis has an amplitude of 2.0 cm, a wavelength of 10 cm, and a frequency of 400 Hz. If the wave equation is of the form $y(x,t) = y_m \sin(kx \pm \omega t)$, what are (a) $y_m$, (b) $k$, (c) $\omega$, and (d) the correct choice of sign in front of $\omega$? What are (e) the maximum transverse speed of a point on the cord and (f) the speed of the wave?

**71**   A sinusoidal transverse wave traveling in the negative direction of an $x$ axis has an amplitude of 1.00 cm, a frequency of 550 Hz, and a speed of 330 m/s. If the wave equation is of the form $y(x, t) = y_m \sin(kx \pm \omega t)$, what are (a) $y_m$, (b) $\omega$, (c) $k$, and (d) the correct choice of sign in front of $\omega$?

**72**   Two sinusoidal waves of the same wavelength travel in the same direction along a stretched string. For wave 1, $y_m = 3.0$ mm and $\phi = 0$; for wave 2, $y_m = 5.0$ mm and $\phi = 70°$. What are the (a) amplitude and (b) phase constant of the resultant wave?

**73**   A wave has a speed of 240 m/s and a wavelength of 3.2 m. What are the (a) frequency and (b) period of the wave?

**74**   When played in a certain manner, the lowest resonant frequency of a certain violin string is concert A (440 Hz). What is the frequency of the (a) second and (b) third harmonic of the string?

**75**   A 120 cm length of string is stretched between fixed supports. What are the (a) longest, (b) second longest, and (c) third longest wavelength for waves traveling on the string if standing waves are to be set up? (d) Sketch those standing waves.

**76**   The equation of a transverse wave traveling along a string is

$$y = 0.15 \sin(0.79x - 13t),$$

in which $x$ and $y$ are in meters and $t$ is in seconds. (a) What is the displacement $y$ at $x = 2.3$ m, $t = 0.16$ s? A second wave is to be added to the first wave to produce standing waves on the string. If the wave equation for the second wave is of the form $y(x, t) = y_m \sin(kx \pm \omega t)$, what are (b) $y_m$, (c) $k$, (d) $\omega$, and (e) the correct choice of sign in front of $\omega$ for this second wave? (f) What is the displacement of the resultant standing wave at $x = 2.3$ m, $t = 0.16$ s?

**77**   A 1.50 m wire has a mass of 8.70 g and is under a tension of 120 N. The wire is held rigidly at both ends and set into oscillation. (a) What is the speed of waves on the wire? What is the wave-

length of the waves that produce (b) one-loop and (c) two-loop standing waves? What is the frequency of the waves that produce (d) one-loop and (e) two-loop standing waves?   SSM

**78**   Energy is transmitted at rate $P_1$ by a wave of frequency $f_1$ on a string under tension $\tau_1$. What is the new energy transmission rate $P_2$ in terms of $P_1$ (a) if the tension is increased to $\tau_2 = 4\tau_1$ and (b) if, instead, the frequency is decreased to $f_2 = f_1/2$?

**79**   The equation of a transverse wave traveling along a string is

$$y = (2.0 \text{ mm}) \sin[(20 \text{ m}^{-1})x - (600 \text{ s}^{-1})t].$$

Find the (a) amplitude, (b) frequency, (c) velocity (including sign), and (d) wavelength of the wave. (e) Find the maximum transverse speed of a particle in the string.

**80**   Oscillation of a 600 Hz tuning fork sets up standing waves in a string clamped at both ends. The wave speed for the string is 400 m/s. The standing wave has four loops and an amplitude of 2.0 mm. (a) What is the length of the string? (b) Write an equation for the displacement of the string as a function of position and time.

**81**   In an experiment on standing waves, a string 90 cm long is attached to the prong of an electrically driven tuning fork that oscillates perpendicular to the length of the string at a frequency of 60 Hz. The mass of the string is 0.044 kg. What tension must the string be under (weights are attached to the other end) if it is to oscillate in four loops?

**82**   *Body armor.* When a high-speed projectile such as a bullet or bomb fragment strikes modern body armor, the fabric of the armor stops the projectile and prevents penetration by quickly spreading the projectile's energy over a large area. This spreading is done by longitudinal and transverse pulses that move *radially* from the impact point, where the projectile pushes a cone-shaped dent into the fabric. The longitudinal pulse, racing along the fibers of the fabric at speed $v_l$ ahead of the denting, causes the fibers to thin and stretch, with

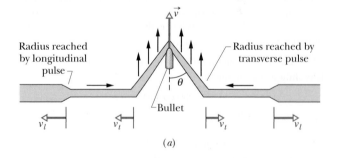

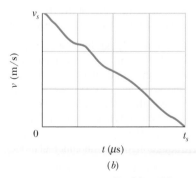

FIG. 16-50   Problem 82.

material flowing radially inward into the dent. One such radial fiber is shown in Fig. 16-50a. Part of the projectile's energy goes into this motion and stretching. The transverse pulse, moving at a slower speed $v_t$, is due to the denting. As the projectile increases the dent's depth, the dent increases in radius, causing the material in the fibers to move in the same direction as the projectile (perpendicular to the transverse pulse's direction of travel). The rest of the projectile's energy goes into this motion. All the energy that does not eventually go into permanently deforming the fibers ends up as thermal energy.

Figure 16-50b is a graph of speed $v$ versus time $t$ for a bullet of mass 10.2 g fired from a .38 Special revolver directly into body armor. The scales of the vertical and horizontal axes are set by $v_s = 300$ m/s and $t_s = 40.0$ $\mu$s. Take $v_l = 2000$ m/s, and assume that the half-angle $\theta$ of the conical dent is 60°. At the end of the collision, what are the radii of (a) the thinned region and (b) the dent (assuming that the person wearing the armor remains stationary)?

**83** (a) What is the fastest transverse wave that can be sent along a steel wire? For safety reasons, the maximum tensile stress to which steel wires should be subjected is $7.00 \times 10^8$ N/m². The density of steel is 7800 kg/m³. (b) Does your answer depend on the diameter of the wire?

**84** (a) Write an equation describing a sinusoidal transverse wave traveling on a cord in the positive direction of a $y$ axis with an angular wave number of 60 cm⁻¹, a period of 0.20 s, and an amplitude of 3.0 mm. Take the transverse direction to be the $z$ direction. (b) What is the maximum transverse speed of a point on the cord?

**85** A wave on a string is described by

$$y(x, t) = 15.0 \sin(\pi x/8 - 4\pi t),$$

where $x$ and $y$ are in centimeters and $t$ is in seconds. (a) What is the transverse speed for a point on the string at $x = 6.00$ cm when $t = 0.250$ s? (b) What is the maximum transverse speed of any point on the string? (c) What is the magnitude of the transverse acceleration for a point on the string at $x = 6.00$ cm when $t = 0.250$ s? (d) What is the magnitude of the maximum transverse acceleration for any point on the string?

**86** A standing wave results from the sum of two transverse traveling waves given by

$$y_1 = 0.050 \cos(\pi x - 4\pi t)$$

and

$$y_2 = 0.050 \cos(\pi x + 4\pi t),$$

where $x$, $y_1$, and $y_2$ are in meters and $t$ is in seconds. (a) What is the smallest positive value of $x$ that corresponds to a node? Beginning at $t = 0$, what is the value of the (b) first, (c) second, and (d) third time the particle at $x = 0$ has zero velocity?

**87** In a demonstration, a 1.2 kg horizontal rope is fixed in place at its two ends ($x = 0$ and $x = 2.0$ m) and made to oscillate up and down in the fundamental mode, at frequency 5.0 Hz. At $t = 0$, the point at $x = 1.0$ m has zero displacement and is moving upward in the positive direction of a $y$ axis with a transverse velocity of 5.0 m/s. What are (a) the amplitude of the motion of that point and (b) the tension in the rope? (c) Write the standing wave equation for the fundamental mode. **SSM**

**88** A certain transverse sinusoidal wave of wavelength 20 cm is moving in the positive direction of an $x$ axis. The transverse velocity of the particle at $x = 0$ as a function of time is shown in Fig.

16-51, where the scale of the vertical axis is set by $u_s = 5.0$ cm/s. What are the (a) wave speed, (b) amplitude, and (c) frequency? (d) Sketch the wave between $x = 0$ and $x = 20$ cm at $t = 2.0$ s.

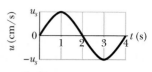

**FIG. 16-51** Problem 88.

**89** The type of rubber band used inside some baseballs and golf balls obeys Hooke's law over a wide range of elongation of the band. A segment of this material has an unstretched length $\ell$ and a mass $m$. When a force $F$ is applied, the band stretches an additional length $\Delta\ell$. (a) What is the speed (in terms of $m$, $\Delta\ell$, and the spring constant $k$) of transverse waves on this stretched rubber band? (b) Using your answer to (a), show that the time required for a transverse pulse to travel the length of the rubber band is proportional to $1/\sqrt{\Delta\ell}$ if $\Delta\ell \ll \ell$ and is constant if $\Delta\ell \gg \ell$. **SSM**

**90** Two waves,

$$y_1 = (2.50 \text{ mm}) \sin[(25.1 \text{ rad/m})x - (440 \text{ rad/s})t]$$

and $y_2 = (1.50 \text{ mm}) \sin[(25.1 \text{ rad/m})x + (440 \text{ rad/s})t],$

travel along a stretched string. (a) Plot the resultant wave as a function of $t$ for $x = 0$, $\lambda/8$, $\lambda/4$, $3\lambda/8$, and $\lambda/2$, where $\lambda$ is the wavelength. The graphs should extend from $t = 0$ to a little over one period. (b) The resultant wave is the superposition of a standing wave and a traveling wave. In which direction does the traveling wave move? (c) How can you change the original waves so the resultant wave is the superposition of standing and traveling waves with the same amplitudes as before but with the traveling wave moving in the opposite direction? Next, use your graphs to find the place at which the oscillation amplitude is (d) maximum and (e) minimum. (f) How is the maximum amplitude related to the amplitudes of the original two waves? (g) How is the minimum amplitude related to the amplitudes of the original two waves?

**91** Two waves are described by

$$y_1 = 0.30 \sin[\pi(5x - 200)t]$$

and $y_2 = 0.30 \sin[\pi(5x - 200t) + \pi/3],$

where $y_1$, $y_2$, and $x$ are in meters and $t$ is in seconds. When these two waves are combined, a traveling wave is produced. What are the (a) amplitude, (b) wave speed, and (c) wavelength of that traveling wave?

**92** The speed of electromagnetic waves (which include visible light, radio, and x rays) in vacuum is $3.0 \times 10^8$ m/s. (a) Wavelengths of visible light waves range from about 400 nm in the violet to about 700 nm in the red. What is the range of frequencies of these waves? (b) The range of frequencies for shortwave radio (for example, FM radio and VHF television) is 1.5 to 300 MHz. What is the corresponding wavelength range? (c) X ray wavelengths range from about 5.0 nm to about $1.0 \times 10^{-2}$ nm. What is the frequency range for x rays?

**93** A traveling wave on a string is described by

$$y = 2.0 \sin\left[2\pi\left(\frac{t}{0.40} + \frac{x}{80}\right)\right],$$

where $x$ and $y$ are in centimeters and $t$ is in seconds. (a) For $t = 0$, plot $y$ as a function of $x$ for $0 \le x \le 160$ cm. (b) Repeat (a) for $t = 0.05$ s and $t = 0.10$ s. From your graphs, determine (c) the wave speed and (d) the direction in which the wave is traveling.

# Waves—II

Andoni Canela/Age Fotostock America, Inc.

*Echoes can be enchanting in certain outdoor settings and annoying in rooms where they make speech unintelligible, but they always mimic the source of the sound. For example, the sound of a handclap returns as the sound of a handclap. However, an echo in front of these steps up the side of a pyramid in the Mayan ruins at Chichen Itza, Mexico, is remarkably different because the handclap returns as a musical note descending in frequency.*

*What causes this musical echo, said to be a chirped echo?*

The answer is in this chapter.

445

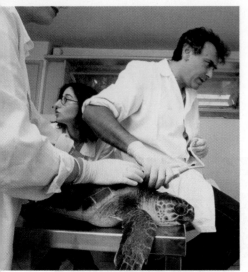

**FIG. 17-1** A loggerhead turtle is being checked with ultrasound (which has a frequency above your hearing range); an image of its interior is being produced on a monitor off to the right. *(Mauro Fermariello/SPL/Photo Researchers)*

# 17-1 | WHAT IS PHYSICS?

The physics of sound waves is the basis of countless studies in the research journals of many fields. Here are just a few examples. Some physiologists are concerned with how speech is produced, how speech impairment might be corrected, how hearing loss can be alleviated, and even how snoring is produced. Some acoustic engineers are concerned with improving the acoustics of cathedrals and concert halls, with reducing noise near freeways and road construction, and with reproducing music by speaker systems. Some aviation engineers are concerned with the shock waves produced by supersonic aircraft and the aircraft noise produced in communities near an airport. Some medical researchers are concerned with how noises produced by the heart and lungs can signal a medical problem in a patient. Some paleontologists are concerned with how a dinosaur's fossil might reveal the dinosaur's vocalizations. Some military engineers are concerned with how the sounds of sniper fire might allow a soldier to pinpoint the sniper's location, and, on the gentler side, some biologists are concerned with how a cat purrs.

To begin our discussion of the physics of sound, we must first answer the question "What *are* sound waves?"

## 17-2 | Sound Waves

As we saw in Chapter 16, mechanical waves are waves that require a material medium to exist. There are two types of mechanical waves: *Transverse waves* involve oscillations perpendicular to the direction in which the wave travels; *longitudinal waves* involve oscillations parallel to the direction of wave travel.

In this book, a **sound wave** is defined roughly as any longitudinal wave. Seismic prospecting teams use such waves to probe Earth's crust for oil. Ships carry sound-ranging gear (sonar) to detect underwater obstacles. Submarines use sound waves to stalk other submarines, largely by listening for the characteristic noises produced by the propulsion system. Figure 17-1 suggests how sound waves can be used to explore the soft tissues of an animal or human body. In this chapter we shall focus on sound waves that travel through the air and that are audible to people.

Figure 17-2 illustrates several ideas that we shall use in our discussions. Point *S* represents a tiny sound source, called a *point source,* that emits sound waves in all directions. The *wavefronts* and *rays* indicate the direction of travel and the spread of the sound waves. **Wavefronts** are surfaces over which the oscillations due to the sound wave have the same value; such surfaces are represented by whole or partial circles in a two-dimensional drawing for a point source. **Rays** are directed lines perpendicular to the wavefronts that indicate the direction of travel of the wavefronts. The short double arrows superimposed on the rays of Fig. 17-2 indicate that the longitudinal oscillations of the air are parallel to the rays.

Near a point source like that of Fig. 17-2, the wavefronts are spherical and spread out in three dimensions, and there the waves are said to be *spherical.* As the wavefronts move outward and their radii become larger, their curvature decreases. Far from the source, we approximate the wavefronts as planes (or lines on two-dimensional drawings), and the waves are said to be *planar.*

## 17-3 | The Speed of Sound

The speed of any mechanical wave, transverse or longitudinal, depends on both an inertial property of the medium (to store kinetic energy) and an elastic property of

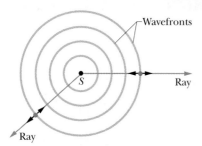

**FIG. 17-2** A sound wave travels from a point source *S* through a three-dimensional medium. The wavefronts form spheres centered on *S*; the rays are radial to *S*. The short, double-headed arrows indicate that elements of the medium oscillate parallel to the rays.

the medium (to store potential energy). Thus, we can generalize Eq. 16-26, which gives the speed of a transverse wave along a stretched string, by writing

$$v = \sqrt{\frac{\tau}{\mu}} = \sqrt{\frac{\text{elastic property}}{\text{inertial property}}}, \qquad (17\text{-}1)$$

where (for transverse waves) $\tau$ is the tension in the string and $\mu$ is the string's linear density. If the medium is air and the wave is longitudinal, we can guess that the inertial property, corresponding to $\mu$, is the volume density $\rho$ of air. What shall we put for the elastic property?

In a stretched string, potential energy is associated with the periodic stretching of the string elements as the wave passes through them. As a sound wave passes through air, potential energy is associated with periodic compressions and expansions of small volume elements of the air. The property that determines the extent to which an element of a medium changes in volume when the pressure (force per unit area) on it changes is the **bulk modulus** $B$, defined (from Eq. 12-25) as

$$B = -\frac{\Delta p}{\Delta V/V} \qquad \text{(definition of bulk modulus).} \qquad (17\text{-}2)$$

Here $\Delta V/V$ is the fractional change in volume produced by a change in pressure $\Delta p$. As explained in Section 14-3, the SI unit for pressure is the newton per square meter, which is given a special name, the *pascal* (Pa). From Eq. 17-2 we see that the unit for $B$ is also the pascal. The signs of $\Delta p$ and $\Delta V$ are always opposite: When we increase the pressure on an element ($\Delta p$ is positive), its volume decreases ($\Delta V$ is negative). We include a minus sign in Eq. 17-2 so that $B$ is always a positive quantity. Now substituting $B$ for $\tau$ and $\rho$ for $\mu$ in Eq. 17-1 yields

$$v = \sqrt{\frac{B}{\rho}} \qquad \text{(speed of sound)} \qquad (17\text{-}3)$$

as the speed of sound in a medium with bulk modulus $B$ and density $\rho$. Table 17-1 lists the speed of sound in various media.

The density of water is almost 1000 times greater than the density of air. If this were the only relevant factor, we would expect from Eq. 17-3 that the speed of sound in water would be considerably less than the speed of sound in air. However, Table 17-1 shows us that the reverse is true. We conclude (again from Eq. 17-3) that the bulk modulus of water must be more than 1000 times greater than that of air. This is indeed the case. Water is much more incompressible than air, which (see Eq. 17-2) is another way of saying that its bulk modulus is much greater.

## Formal Derivation of Eq. 17-3

We now derive Eq. 17-3 by direct application of Newton's laws. Let a single pulse in which air is compressed travel (from right to left) with speed $v$ through the air in a long tube, like that in Fig. 16-2. Let us run along with the pulse at that speed, so that the pulse appears to stand still in our reference frame. Figure 17-3a shows the situation as it is viewed from that frame. The pulse is standing still, and air is moving at speed $v$ through it from left to right.

Let the pressure of the undisturbed air be $p$ and the pressure inside the pulse be $p + \Delta p$, where $\Delta p$ is positive due to the compression. Consider an element of air of thickness $\Delta x$ and face area $A$, moving toward the pulse at speed $v$. As this element enters the pulse, the leading face of the element encounters a region of higher pressure, which slows the element to speed $v + \Delta v$, in which $\Delta v$ is negative. This slowing is complete when the rear face of the element reaches the pulse, which requires time interval

$$\Delta t = \frac{\Delta x}{v}. \qquad (17\text{-}4)$$

**TABLE 17-1**

**The Speed of Sound**[a]

| Medium | Speed (m/s) |
|---|---|
| *Gases* | |
| Air (0°C) | 331 |
| Air (20°C) | 343 |
| Helium | 965 |
| Hydrogen | 1284 |
| *Liquids* | |
| Water (0°C) | 1402 |
| Water (20°C) | 1482 |
| Seawater[b] | 1522 |
| *Solids* | |
| Aluminum | 6420 |
| Steel | 5941 |
| Granite | 6000 |

[a]At 0°C and 1 atm pressure, except where noted.

[b]At 20°C and 3.5% salinity.

**FIG. 17-3** A compression pulse is sent from right to left down a long air-filled tube. The reference frame of the figure is chosen so that the pulse is at rest and the air moves from left to right. (*a*) An element of air of width $\Delta x$ moves toward the pulse with speed $v$. (*b*) The leading face of the element enters the pulse. The forces acting on the leading and trailing faces (due to air pressure) are shown.

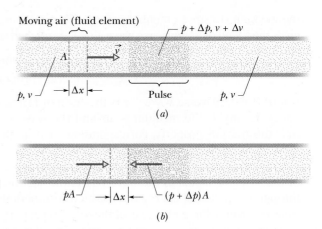

Let us apply Newton's second law to the element. During $\Delta t$, the average force on the element's trailing face is $pA$ toward the right, and the average force on the leading face is $(p + \Delta p)A$ toward the left (Fig. 17-3*b*). Therefore, the average net force on the element during $\Delta t$ is

$$F = pA - (p + \Delta p)A$$
$$= -\Delta p\, A \quad \text{(net force)}. \tag{17-5}$$

The minus sign indicates that the net force on the air element is directed to the left in Fig. 17-3*b*. The volume of the element is $A\,\Delta x$, so with the aid of Eq. 17-4, we can write its mass as

$$\Delta m = \rho\, \Delta V = \rho A\, \Delta x = \rho A v\, \Delta t \quad \text{(mass)}. \tag{17-6}$$

The average acceleration of the element during $\Delta t$ is

$$a = \frac{\Delta v}{\Delta t} \quad \text{(acceleration)}. \tag{17-7}$$

Thus, from Newton's second law ($F = ma$), we have, from Eqs. 17-5, 17-6, and 17-7,

$$-\Delta p\, A = (\rho A v\, \Delta t)\frac{\Delta v}{\Delta t},$$

which we can write as

$$\rho v^2 = -\frac{\Delta p}{\Delta v/v}. \tag{17-8}$$

The air that occupies a volume $V\,(= Av\,\Delta t)$ outside the pulse is compressed by an amount $\Delta V\,(= A\,\Delta v\,\Delta t)$ as it enters the pulse. Thus,

$$\frac{\Delta V}{V} = \frac{A\,\Delta v\,\Delta t}{Av\,\Delta t} = \frac{\Delta v}{v}. \tag{17-9}$$

Substituting Eq. 17-9 and then Eq. 17-2 into Eq. 17-8 leads to

$$\rho v^2 = -\frac{\Delta p}{\Delta v/v} = -\frac{\Delta p}{\Delta V/V} = B.$$

Solving for $v$ yields Eq. 17-3 for the speed of the air toward the right in Fig. 17-3, and thus for the actual speed of the pulse toward the left.

## Sample Problem | 17-1

When a sound pulse, as from a handclap, is produced at the foot of the stairs at the Mayan pyramid shown in the chapter's opening photograph, the sound waves reflect from the steps in succession, the closest (lowest) one first (Fig. 17-4*a*) and the farthest (highest) one last (Fig. 17-4*b*). The depth and height of the steps are $d = 0.263$ m, and the speed of sound is 343 m/s. The paths taken by the sound waves to and from the steps near the bottom of the stairs are approximately horizontal. The slanted paths taken by the sound waves to and from the

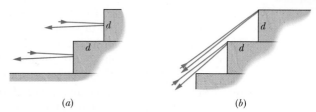

steps near the top are approximately 45° to the horizontal. At what frequency $f_{bot}$ do the echo pulses arrive at the listener from the bottom steps? At what frequency $f_{top}$ do they arrive from the top steps a short time later?

**KEY IDEAS** (1) The frequency $f$ at which the pulses return to the listener is the inverse of the time $\Delta t$ between successive pulses. (2) The time interval $\Delta t$ required by sound to travel a given distance $L$ is related to the speed of sound $v$ by $v = L/\Delta t$.

*Calculations:* In Fig. 17-4a at the bottom of the stairs, the sound wave that reflects from the higher step travels a distance $L = 2d$ more than the sound wave that reflects from the lower step. (The higher wave must travel twice across the step's depth.) So, the arrivals of the echo pulses at the listener are separated by the time interval

$$\Delta t_{bot} = \frac{L}{v} = \frac{2d}{v} \tag{17-10}$$

$$= \frac{2(0.263 \text{ m})}{343 \text{ m/s}} = 1.533 \times 10^{-3}\text{s}.$$

The frequency $f_{bot}$ at which the pulses arrive at the listener is

$$f_{bot} = \frac{1}{\Delta t_{bot}} \tag{17-11}$$

$$= \frac{1}{1.533 \times 10^{-3}\text{ s}} = 652 \text{ Hz.} \quad \text{(Answer)}$$

The time interval $\Delta t_{bot}$ is too short for a listener to distinguish the individual pulses. Instead, the frequency $f_{bot}$ is brought to consciousness—the listener hears a musical note of frequency 652 Hz.

In Fig. 17-4b at the top of the stairs, the slanted approach and return of the sound waves means that the wave reflected from the higher step travels a distance $L = 2\sqrt{2}d$ more than the wave that reflects from the lower step. (The travel is twice along the hypotenuse of a right triangle with equal legs of length $d$.) So, now the arrivals of the echo pulses at the listener are separated by the time interval

$$\Delta t_{top} = \frac{L}{v} = \frac{2\sqrt{2}d}{v} \tag{17-12}$$

$$= \frac{2\sqrt{2}(0.263 \text{ m})}{343 \text{ m/s}} = 2.168 \times 10^{-3}\text{ s},$$

and the frequency that is brought to consciousness is

$$f_{top} = \frac{1}{\Delta t_{top}}$$

$$= \frac{1}{2.168 \times 10^{-3}\text{ s}} = 461 \text{ Hz.} \quad \text{(Answer)}$$

Thus, a handclap in front of the stairs produces an echo that begins with a frequency of 652 Hz and ends with a frequency of 461 Hz. You might be able to hear such a musical echo from other stairs or even from a picket fence if you stand alongside it.

**FIG. 17-4** Sound waves reflect from (a) the bottom steps and (b) the top steps of a tall flight of stairs.

## 17-4 | Traveling Sound Waves

Here we examine the displacements and pressure variations associated with a sinusoidal sound wave traveling through air. Figure 17-5a displays such a wave traveling rightward through a long air-filled tube. Recall from Chapter 16 that

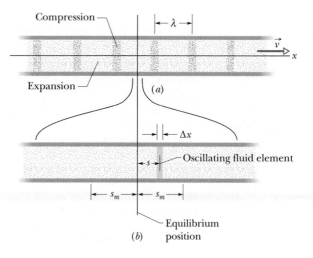

**FIG. 17-5** (a) A sound wave, traveling through a long air-filled tube with speed $v$, consists of a moving, periodic pattern of expansions and compressions of the air. The wave is shown at an arbitrary instant. (b) A horizontally expanded view of a short piece of the tube. As the wave passes, an air element of thickness $\Delta x$ oscillates left and right in simple harmonic motion about its equilibrium position. At the instant shown in (b), the element happens to be displaced a distance $s$ to the right of its equilibrium position. Its maximum displacement, either right or left, is $s_m$.

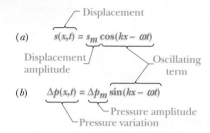

FIG. 17-6 (a) The displacement function and (b) the pressure-variation function of a traveling sound wave consist of an amplitude and an oscillating term.

we can produce such a wave by sinusoidally moving a piston at the left end of the tube (as in Fig. 16-2). The piston's rightward motion moves the element of air next to the piston face and compresses that air; the piston's leftward motion allows the element of air to move back to the left and the pressure to decrease. As each element of air pushes on the next element in turn, the right–left motion of the air and the change in its pressure travel along the tube as a sound wave.

Consider the thin element of air of thickness $\Delta x$ shown in Fig. 17-5b. As the wave travels through this portion of the tube, the element of air oscillates left and right in simple harmonic motion about its equilibrium position. Thus, the oscillations of each air element due to the traveling sound wave are like those of a string element due to a transverse wave, except that the air element oscillates *longitudinally* rather than *transversely*. Because string elements oscillate parallel to the $y$ axis, we write their displacements in the form $y(x, t)$. Similarly, because air elements oscillate parallel to the $x$ axis, we could write their displacements in the confusing form $x(x, t)$, but we shall use $s(x, t)$ instead.

To show that the displacements $s(x, t)$ are sinusoidal functions of $x$ and $t$, we can use either a sine function or a cosine function. In this chapter we use a cosine function, writing

$$s(x, t) = s_m \cos(kx - \omega t). \qquad (17\text{-}13)$$

Figure 17-6a labels the various parts of this equation. In it, $s_m$ is the **displacement amplitude**—that is, the maximum displacement of the air element to either side of its equilibrium position (see Fig. 17-5b). The angular wave number $k$, angular frequency $\omega$, frequency $f$, wavelength $\lambda$, speed $v$, and period $T$ for a sound (longitudinal) wave are defined and interrelated exactly as for a transverse wave, except that $\lambda$ is now the distance (again along the direction of travel) in which the pattern of compression and expansion due to the wave begins to repeat itself (see Fig. 17-5a). (We assume $s_m$ is much less than $\lambda$.)

As the wave moves, the air pressure at any position $x$ in Fig. 17-5a varies sinusoidally, as we prove next. To describe this variation we write

$$\Delta p(x, t) = \Delta p_m \sin(kx - \omega t). \qquad (17\text{-}14)$$

Figure 17-6b labels the various parts of this equation. A negative value of $\Delta p$ in Eq. 17-14 corresponds to an expansion of the air, and a positive value to a compression. Here $\Delta p_m$ is the **pressure amplitude,** which is the maximum increase or decrease in pressure due to the wave; $\Delta p_m$ is normally very much less than the pressure $p$ present when there is no wave. As we shall prove, the pressure amplitude $\Delta p_m$ is related to the displacement amplitude $s_m$ in Eq. 17-13 by

$$\Delta p_m = (v \rho \omega) s_m. \qquad (17\text{-}15)$$

Figure 17-7 shows plots of Eqs. 17-13 and 17-14 at $t = 0$; with time, the two curves would move rightward along the horizontal axes. Note that the displacement and pressure variation are $\pi/2$ rad (or 90°) out of phase. Thus, for example, the pressure variation $\Delta p$ at any point along the wave is zero when the displacement there is a maximum.

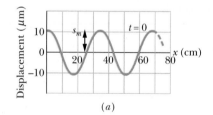

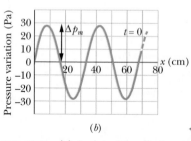

FIG. 17-7 (a) A plot of the displacement function (Eq. 17-13) for $t = 0$. (b) A similar plot of the pressure-variation function (Eq. 17-14). Both plots are for a 1000 Hz sound wave whose pressure amplitude is at the threshold of pain; see Sample Problem 17-2.

✓ **CHECKPOINT 1** When the oscillating air element in Fig. 17-5b is moving rightward through the point of zero displacement, is the pressure in the element at its equilibrium value, just beginning to increase, or just beginning to decrease?

## Derivation of Eqs. 17-14 and 17-15

Figure 17-5b shows an oscillating element of air of cross-sectional area $A$ and thickness $\Delta x$, with its center displaced from its equilibrium position by distance $s$. From Eq. 17-2 we can write, for the pressure variation in the displaced element,

$$\Delta p = -B \frac{\Delta V}{V}. \qquad (17\text{-}16)$$

The quantity $V$ in Eq. 17-16 is the volume of the element, given by

$$V = A\,\Delta x. \qquad (17\text{-}17)$$

The quantity $\Delta V$ in Eq. 17-16 is the change in volume that occurs when the element is displaced. This volume change comes about because the displacements of the two faces of the element are not quite the same, differing by some amount $\Delta s$. Thus, we can write the change in volume as

$$\Delta V = A\,\Delta s. \qquad (17\text{-}18)$$

Substituting Eqs. 17-17 and 17-18 into Eq. 17-16 and passing to the differential limit yield

$$\Delta p = -B \frac{\Delta s}{\Delta x} = -B \frac{\partial s}{\partial x}. \qquad (17\text{-}19)$$

The symbols $\partial$ indicate that the derivative in Eq. 17-19 is a *partial derivative*, which tells us how $s$ changes with $x$ when the time $t$ is fixed. From Eq. 17-13 we then have, treating $t$ as a constant,

$$\frac{\partial s}{\partial x} = \frac{\partial}{\partial x}\left[s_m \cos(kx - \omega t)\right] = -k s_m \sin(kx - \omega t).$$

Substituting this quantity for the partial derivative in Eq. 17-19 yields

$$\Delta p = B k s_m \sin(kx - \omega t).$$

Setting $\Delta p_m = B k s_m$, this yields Eq. 17-14, which we set out to prove.
   Using Eq. 17-3, we can now write

$$\Delta p_m = (Bk)s_m = (v^2 \rho k)s_m.$$

Equation 17-15, which we also wanted to prove, follows at once if we substitute $\omega/v$ for $k$ from Eq. 16-13.

## Sample Problem    17-2

The maximum pressure amplitude $\Delta p_m$ that the human ear can tolerate in loud sounds is about 28 Pa (which is very much less than the normal air pressure of about $10^5$ Pa). What is the displacement amplitude $s_m$ for such a sound in air of density $\rho = 1.21$ kg/m³, at a frequency of 1000 Hz and a speed of 343 m/s?

**KEY IDEA**   The displacement amplitude $s_m$ of a sound wave is related to the pressure amplitude $\Delta p_m$ of the wave according to Eq. 17-15.

**Calculations:** Solving that equation for $s_m$ yields

$$s_m = \frac{\Delta p_m}{v \rho \omega} = \frac{\Delta p_m}{v \rho (2\pi f)}.$$

Substituting known data then gives us

$$s_m = \frac{28 \text{ Pa}}{(343 \text{ m/s})(1.21 \text{ kg/m}^3)(2\pi)(1000 \text{ Hz})}$$
$$= 1.1 \times 10^{-5} \text{ m} = 11 \ \mu\text{m}. \qquad \text{(Answer)}$$

That is only about one-seventh the thickness of this page. Obviously, the displacement amplitude of even the loudest sound that the ear can tolerate is very small.

The pressure amplitude $\Delta p_m$ for the *faintest* detectable sound at 1000 Hz is $2.8 \times 10^{-5}$ Pa. Proceeding as above leads to $s_m = 1.1 \times 10^{-11}$ m or 11 pm, which is about one-tenth the radius of a typical atom. The ear is indeed a sensitive detector of sound waves.

# 17-5 | Interference

Like transverse waves, sound waves can undergo interference. Let us consider, in particular, the interference between two identical sound waves traveling in

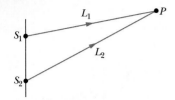

**FIG. 17-8** Two point sources $S_1$ and $S_2$ emit spherical sound waves in phase. The rays indicate that the waves pass through a common point $P$.

the same direction. Figure 17-8 shows how we can set up such a situation: Two point sources $S_1$ and $S_2$ emit sound waves that are in phase and of identical wavelength $\lambda$. Thus, the sources themselves are said to be in phase; that is, as the waves emerge from the sources, their displacements are always identical. We are interested in the waves that then travel through point $P$ in Fig. 17-8. We assume that the distance to $P$ is much greater than the distance between the sources so that we can approximate the waves as traveling in the same direction at $P$.

If the waves traveled along paths with identical lengths to reach point $P$, they would be in phase there. As with transverse waves, this means that they would undergo fully constructive interference there. However, in Fig. 17-8, path $L_2$ traveled by the wave from $S_2$ is longer than path $L_1$ traveled by the wave from $S_1$. The difference in path lengths means that the waves may not be in phase at point $P$. In other words, their phase difference $\phi$ at $P$ depends on their **path length difference $\Delta L = |L_2 - L_1|$**.

To relate phase difference $\phi$ to path length difference $\Delta L$, we recall (from Section 16-4) that a phase difference of $2\pi$ rad corresponds to one wavelength. Thus, we can write the proportion

$$\frac{\phi}{2\pi} = \frac{\Delta L}{\lambda}, \tag{17-20}$$

from which

$$\phi = \frac{\Delta L}{\lambda} 2\pi. \tag{17-21}$$

Fully constructive interference occurs when $\phi$ is zero, $2\pi$, or any integer multiple of $2\pi$. We can write this condition as

$$\phi = m(2\pi), \quad \text{for } m = 0, 1, 2, \ldots \quad \text{(fully constructive interference).} \tag{17-22}$$

From Eq. 17-21, this occurs when the ratio $\Delta L/\lambda$ is

$$\frac{\Delta L}{\lambda} = 0, 1, 2, \ldots \quad \text{(fully constructive interference).} \tag{17-23}$$

For example, if the path length difference $\Delta L = |L_2 - L_1|$ in Fig. 17-8 is equal to $2\lambda$, then $\Delta L/\lambda = 2$ and the waves undergo fully constructive interference at point $P$. The interference is fully constructive because the wave from $S_2$ is phase-shifted relative to the wave from $S_1$ by $2\lambda$, putting the two waves *exactly in phase* at $P$.

Fully destructive interference occurs when $\phi$ is an odd multiple of $\pi$, a condition we can write as

$$\phi = (2m + 1)\pi, \quad \text{for } m = 0, 1, 2, \ldots \quad \text{(fully destructive interference).} \tag{17-24}$$

From Eq. 17-21, this occurs when the ratio $\Delta L/\lambda$ is

$$\frac{\Delta L}{\lambda} = 0.5, 1.5, 2.5, \ldots \quad \text{(fully destructive interference).} \tag{17-25}$$

For example, if the path length difference $\Delta L = |L_2 - L_1|$ in Fig. 17-8 is equal to $2.5\lambda$, then $\Delta L/\lambda = 2.5$ and the waves undergo fully destructive interference at point $P$. The interference is fully destructive because the wave from $S_2$ is phase-shifted relative to the wave from $S_1$ by 2.5 wavelengths, which puts the two waves *exactly out of phase* at $P$.

Of course, two waves could produce intermediate interference as, say, when $\Delta L/\lambda = 1.2$. This would be closer to fully constructive interference ($\Delta L/\lambda = 1.0$) than to fully destructive interference ($\Delta L/\lambda = 1.5$).

**Sample Problem** **17-3** **Build your skill**

In Fig. 17-9a, two point sources $S_1$ and $S_2$, which are in phase and separated by distance $D = 1.5\lambda$, emit identical sound waves of wavelength $\lambda$.

(a) What is the path length difference of the waves from $S_1$ and $S_2$ at point $P_1$, which lies on the perpendicular bisector of distance $D$, at a distance greater than $D$ from the sources? What type of interference occurs at $P_1$?

**Reasoning:** Because the waves travel identical distances to reach $P_1$, their path length difference is

$$\Delta L = 0. \qquad \text{(Answer)}$$

From Eq. 17-23, this means that the waves undergo fully constructive interference at $P_1$.

(b) What are the path length difference and type of interference at point $P_2$ in Fig. 17-9a?

**Reasoning:** The wave from $S_1$ travels the extra distance $D (= 1.5\lambda)$ to reach $P_2$. Thus, the path length difference is

$$\Delta L = 1.5\lambda. \qquad \text{(Answer)}$$

From Eq. 17-25, this means that the waves are exactly out of phase at $P_2$ and undergo fully destructive interference there.

(c) Figure 17-9b shows a circle with a radius much greater than $D$, centered on the midpoint between sources $S_1$ and $S_2$. What is the number of points $N$ around this circle at which the interference is fully constructive?

**Reasoning:** Imagine that, starting at point $a$, we move clockwise along the circle to point $d$. As we move to point $d$, the path length difference $\Delta L$ increases and so the type of interference changes. From (a), we know that the path length difference is $\Delta L = 0\lambda$ at point $a$. From (b), we know that $\Delta L = 1.5\lambda$ at point $d$. Thus, there must be one point along the circle between $a$ and $d$ at which $\Delta L = \lambda$, as indicated in Fig. 17-9b. From Eq. 17-23, fully constructive interference occurs at that point. Also, there can be no other point along the way

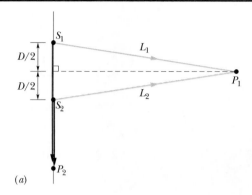

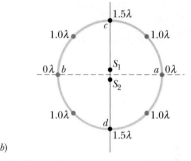

**FIG. 17-9** (a) Two point sources $S_1$ and $S_2$, separated by distance $D$, emit spherical sound waves in phase. The waves travel equal distances to reach point $P_1$. Point $P_2$ is on the line extending through $S_1$ and $S_2$. (b) The path length difference (in terms of wavelength) between the waves from $S_1$ and $S_2$, at eight points on a large circle around the sources.

from point $a$ to point $d$ at which fully constructive interference occurs, because there is no other integer than 1 between 0 and 1.5.

We can now use symmetry to locate the other points of fully constructive interference along the rest of the circle. Symmetry about line $cd$ gives us point $b$, at which $\Delta L = 0\lambda$. Also, there are three more points at which $\Delta L = \lambda$. In all we have

$$N = 6. \qquad \text{(Answer)}$$

## 17-6 I Intensity and Sound Level

If you have ever tried to sleep while someone played loud music nearby, you are well aware that there is more to sound than frequency, wavelength, and speed. There is also intensity. The **intensity** $I$ of a sound wave at a surface is the average rate per unit area at which energy is transferred by the wave through or onto the surface. We can write this as

$$I = \frac{P}{A}, \qquad (17\text{-}26)$$

where $P$ is the time rate of energy transfer (the power) of the sound wave and $A$ is the area of the surface intercepting the sound. As we shall derive shortly, the intensity $I$ is related to the displacement amplitude $s_m$ of the sound wave by

$$I = \tfrac{1}{2}\rho v \omega^2 s_m^2. \qquad (17\text{-}27)$$

Sound can cause the wall of a drinking glass to oscillate. If the sound produces a standing wave of oscillations and if the intensity of the sound is large enough, the glass will shatter. *(Ben Rose/The Image Bank/Getty Images)*

## Variation of Intensity with Distance

How intensity varies with distance from a real sound source is often complex. Some real sources (like loudspeakers) may transmit sound only in particular directions, and the environment usually produces echoes (reflected sound waves) that overlap the direct sound waves. In some situations, however, we can ignore echoes and assume that the sound source is a point source that emits the sound *isotropically*—that is, with equal intensity in all directions. The wavefronts spreading from such an isotropic point source $S$ at a particular instant are shown in Fig. 17-10.

Let us assume that the mechanical energy of the sound waves is conserved as they spread from this source. Let us also center an imaginary sphere of radius $r$ on the source, as shown in Fig. 17-10. All the energy emitted by the source must pass through the surface of the sphere. Thus, the time rate at which energy is transferred through the surface by the sound waves must equal the time rate at which energy is emitted by the source (that is, the power $P_s$ of the source). From Eq. 17-26, the intensity $I$ at the sphere must then be

$$I = \frac{P_s}{4\pi r^2}, \tag{17-28}$$

where $4\pi r^2$ is the area of the sphere. Equation 17-28 tells us that the intensity of sound from an isotropic point source decreases with the square of the distance $r$ from the source.

✓**CHECKPOINT 2**　The figure indicates three small patches 1, 2, and 3 that lie on the surfaces of two imaginary spheres; the spheres are centered on an isotropic point source $S$ of sound. The rates at which energy is transmitted through the three patches by the sound waves are equal. Rank the patches according to (a) the intensity of the sound on them and (b) their area, greatest first.

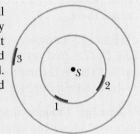

## The Decibel Scale

You saw in Sample Problem 17-2 that the displacement amplitude at the human ear ranges from about $10^{-5}$ m for the loudest tolerable sound to about $10^{-11}$ m for the faintest detectable sound, a ratio of $10^6$. From Eq. 17-27 we see that the intensity of a sound varies as the *square* of its amplitude, so the ratio of intensities at these two limits of the human auditory system is $10^{12}$. Humans can hear over an enormous range of intensities.

We deal with such an enormous range of values by using logarithms. Consider the relation

$$y = \log x,$$

in which $x$ and $y$ are variables. It is a property of this equation that if we *multiply* $x$ by 10, then $y$ increases by 1. To see this, we write

$$y' = \log(10x) = \log 10 + \log x = 1 + y.$$

Similarly, if we multiply $x$ by $10^{12}$, $y$ increases by only 12.

Thus, instead of speaking of the intensity $I$ of a sound wave, it is much more convenient to speak of its **sound level** $\beta$, defined as

$$\beta = (10 \text{ dB}) \log \frac{I}{I_0}. \tag{17-29}$$

Here dB is the abbreviation for **decibel**, the unit of sound level, a name that was chosen to recognize the work of Alexander Graham Bell. $I_0$ in Eq. 17-29 is a

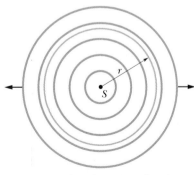

FIG. 17-10　A point source $S$ emits sound waves uniformly in all directions. The waves pass through an imaginary sphere of radius $r$ that is centered on $S$.

standard reference intensity ($= 10^{-12}$ W/m$^2$), chosen because it is near the lower limit of the human range of hearing. For $I = I_0$, Eq. 17-29 gives $\beta = 10 \log 1 = 0$, so our standard reference level corresponds to zero decibels. Then $\beta$ increases by 10 dB every time the sound intensity increases by an order of magnitude (a factor of 10). Thus, $\beta = 40$ corresponds to an intensity that is $10^4$ times the standard reference level. Table 17-2 lists the sound levels for a variety of environments.

**TABLE 17-2**

**Some Sound Levels (dB)**

| | |
|---|---|
| Hearing threshold | 0 |
| Rustle of leaves | 10 |
| Conversation | 60 |
| Rock concert | 110 |
| Pain threshold | 120 |
| Jet engine | 130 |

## Derivation of Eq. 17-27

Consider, in Fig. 17-5a, a thin slice of air of thickness $dx$, area $A$, and mass $dm$, oscillating back and forth as the sound wave of Eq. 17-13 passes through it. The kinetic energy $dK$ of the slice of air is

$$dK = \tfrac{1}{2} dm \, v_s^2. \tag{17-30}$$

Here $v_s$ is not the speed of the wave but the speed of the oscillating element of air, obtained from Eq. 17-13 as

$$v_s = \frac{\partial s}{\partial t} = -\omega s_m \sin(kx - \omega t).$$

Using this relation and putting $dm = \rho A \, dx$ allow us to rewrite Eq. 17-30 as

$$dK = \tfrac{1}{2}(\rho A \, dx)(-\omega s_m)^2 \sin^2(kx - \omega t). \tag{17-31}$$

Dividing Eq. 17-31 by $dt$ gives the rate at which kinetic energy moves along with the wave. As we saw in Chapter 16 for transverse waves, $dx/dt$ is the wave speed $v$, so we have

$$\frac{dK}{dt} = \tfrac{1}{2}\rho A v \omega^2 s_m^2 \sin^2(kx - \omega t). \tag{17-32}$$

The *average* rate at which kinetic energy is transported is

$$\left(\frac{dK}{dt}\right)_{\text{avg}} = \tfrac{1}{2}\rho A v \omega^2 s_m^2 [\sin^2(kx - \omega t)]_{\text{avg}}$$

$$= \tfrac{1}{4}\rho A v \omega^2 s_m^2. \tag{17-33}$$

To obtain this equation, we have used the fact that the average value of the square of a sine (or a cosine) function over one full oscillation is $\tfrac{1}{2}$.

We assume that *potential* energy is carried along with the wave at this same average rate. The wave intensity $I$, which is the average rate per unit area at which energy of both kinds is transmitted by the wave, is then, from Eq. 17-33,

$$I = \frac{2(dK/dt)_{\text{avg}}}{A} = \tfrac{1}{2}\rho v \omega^2 s_m^2,$$

which is Eq. 17-27, the equation we set out to derive.

**Sample Problem** 17-4

An electric spark jumps along a straight line of length $L = 10$ m, emitting a pulse of sound that travels radially outward from the spark. (The spark is said to be a *line source* of sound.) The power of the emission is $P_s = 1.6 \times 10^4$ W.

(a) What is the intensity $I$ of the sound when it reaches a distance $r = 12$ m from the spark?

**FIG. 17-11** A spark along a straight line of length $L$ emits sound waves radially outward. The waves pass through an imaginary cylinder of radius $r$ and length $L$ that is centered on the spark.

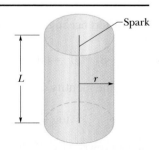

**KEY IDEAS** (1) Let us center an imaginary cylinder of radius $r = 12$ m and length $L = 10$ m (open at both ends) on the spark, as shown in Fig. 17-11. Then the intensity $I$ at the cylindrical surface is the ratio $P/A$,

where $P$ is the time rate at which sound energy passes through the surface and $A$ is the surface area. (2) We assume that the principle of conservation of energy applies to the sound energy. This means that the rate $P$ at which energy is transferred through the cylinder must equal the rate $P_s$ at which energy is emitted by the source.

*Calculations:* Putting these ideas together and noting that the area of the cylindrical surface is $A = 2\pi rL$, we have

$$I = \frac{P}{A} = \frac{P_s}{2\pi rL}. \qquad (17\text{-}34)$$

This tells us that the intensity of the sound from a line source decreases with distance $r$ (and not with the square of distance $r$ as for a point source). Substituting the given data, we find

$$I = \frac{1.6 \times 10^4 \text{ W}}{2\pi(12 \text{ m})(10 \text{ m})}$$

$$= 21.2 \text{ W/m}^2 \approx 21 \text{ W/m}^2. \qquad \text{(Answer)}$$

(b) At what time rate $P_d$ is sound energy intercepted by an acoustic detector of area $A_d = 2.0 \text{ cm}^2$, aimed at the spark and located a distance $r = 12$ m from the spark?

*Calculations:* We know that the intensity of sound at the detector is the ratio of the energy transfer rate $P_d$ there to the detector's area $A_d$:

$$I = \frac{P_d}{A_d}. \qquad (17\text{-}35)$$

We can imagine that the detector lies on the cylindrical surface of (a). Then the sound intensity at the detector is the intensity $I$ $(= 21.2 \text{ W/m}^2)$ at the cylindrical surface. Solving Eq. 17-35 for $P_d$ gives us

$$P_d = (21.2 \text{ W/m}^2)(2.0 \times 10^{-4} \text{ m}^2) = 4.2 \text{ mW}. \quad \text{(Answer)}$$

---

**Sample Problem** | **17-5**

Many veteran rockers suffer from acute hearing damage because of the high sound levels they endured for years while playing music near loudspeakers or listening to music on headphones. Some, like Ted Nugent, can no longer hear in a damaged ear. Others, like Peter Townshend of the Who, have a continuous ringing sensation (tinnitus). Recently, many rockers, such as Lars Ulrich of Metallica (Fig. 17-12), began wearing special earplugs to protect their hearing during performances. If an earplug decreases the sound level of the sound waves by 20 dB, what is the ratio of the final intensity $I_f$ of the waves to their initial intensity $I_i$?

**KEY IDEA** For both the final and initial waves, the sound level $\beta$ is related to the intensity by the definition of sound level in Eq. 17-29.

*Calculations:* For the final waves we have

$$\beta_f = (10 \text{ dB}) \log \frac{I_f}{I_0},$$

and for the initial waves we have

$$\beta_i = (10 \text{ dB}) \log \frac{I_i}{I_0}.$$

The difference in the sound levels is

$$\beta_f - \beta_i = (10 \text{ dB})\left(\log\frac{I_f}{I_0} - \log\frac{I_i}{I_0}\right). \qquad (17\text{-}36)$$

Using the identity

$$\log\frac{a}{b} - \log\frac{c}{d} = \log\frac{ad}{bc},$$

we can rewrite Eq. 17-36 as

$$\beta_f - \beta_i = (10 \text{ dB}) \log \frac{I_f}{I_i}. \qquad (17\text{-}37)$$

**FIG. 17-12** Lars Ulrich of Metallica is an advocate for the organization HEAR (Hearing Education and Awareness for Rockers), which warns about the damage high sound levels can have on hearing. *(Tim Mosenfelder/ Getty Images News and Sport Services)*

Rearranging and then substituting the given decrease in sound level as $\beta_f - \beta_i = -20$ dB, we find

$$\log\frac{I_f}{I_i} = \frac{\beta_f - \beta_i}{10 \text{ dB}} = \frac{-20 \text{ dB}}{10 \text{ dB}} = -2.0.$$

We next take the antilog of the far left and far right sides of this equation. (Although the antilog $10^{-2.0}$ can be evaluated mentally, you could use a calculator by keying in 10^-2.0 or using the $10^x$ key.) We find

$$\frac{I_f}{I_i} = \log^{-1}(-2.0) = 0.010. \qquad \text{(Answer)}$$

Thus, the earplug reduces the intensity of the sound waves to 0.010 of their initial intensity, which is a decrease of two orders of magnitude.

# 17-7 | Sources of Musical Sound

Musical sounds can be set up by oscillating strings (guitar, piano, violin), membranes (kettledrum, snare drum), air columns (flute, oboe, pipe organ, and the digeridoo of Fig. 17-13), wooden blocks or steel bars (marimba, xylophone), and many other oscillating bodies. Most common instruments involve more than a single oscillating part.

Recall from Chapter 16 that standing waves can be set up on a stretched string that is fixed at both ends. They arise because waves traveling along the string are reflected back onto the string at each end. If the wavelength of the waves is suitably matched to the length of the string, the superposition of waves traveling in opposite directions produces a standing wave pattern (or oscillation mode). The wavelength required of the waves for such a match is one that corresponds to a *resonant frequency* of the string. The advantage of setting up standing waves is that the string then oscillates with a large, sustained amplitude, pushing back and forth against the surrounding air and thus generating a noticeable sound wave with the same frequency as the oscillations of the string. This production of sound is of obvious importance to, say, a guitarist.

We can set up standing waves of sound in an air-filled pipe in a similar way. As sound waves travel through the air in the pipe, they are reflected at each end and travel back through the pipe. (The reflection occurs even if an end is open, but the reflection is not as complete as when the end is closed.) If the wavelength of the sound waves is suitably matched to the length of the pipe, the superposition of waves traveling in opposite directions through the pipe sets up a standing wave pattern. The wavelength required of the sound waves for such a match is one that corresponds to a resonant frequency of the pipe. The advantage of such a standing wave is that the air in the pipe oscillates with a large, sustained amplitude, emitting at any open end a sound wave that has the same frequency as the oscillations in the pipe. This emission of sound is of obvious importance to, say, an organist.

Many other aspects of standing sound wave patterns are similar to those of string waves: The closed end of a pipe is like the fixed end of a string in that there must be a node (zero displacement) there, and the open end of a pipe is like the end of a string attached to a freely moving ring, as in Fig. 16-21b, in that there must be an antinode there. (Actually, the antinode for the open end of a pipe is located slightly beyond the end, but we shall not dwell on that detail.)

The simplest standing wave pattern that can be set up in a pipe with two open ends is shown in Fig. 17-14a. There is an antinode across each open end, as required. There is also a node across the middle of the pipe. An easier way of representing this standing longitudinal sound wave is shown in Fig. 17-14b—by drawing it as a standing transverse string wave.

The standing wave pattern of Fig. 17-14a is called the *fundamental mode* or *first harmonic*. For it to be set up, the sound waves in a pipe of length L must have a wavelength given by $L = \lambda/2$, so that $\lambda = 2L$. Several more standing sound wave patterns for a pipe with two open ends are shown in Fig. 17-15a using string wave representations. The *second harmonic* requires sound waves of wavelength $\lambda = L$, the *third harmonic* requires wavelength $\lambda = 2L/3$, and so on.

More generally, the resonant frequencies for a pipe of length L with two open ends correspond to the wavelengths

$$\lambda = \frac{2L}{n}, \quad \text{for } n = 1, 2, 3, \ldots, \quad (17\text{-}38)$$

where n is called the *harmonic number*. Letting v be the speed of sound, we write the resonant frequencies for a pipe with two open ends as

$$f = \frac{v}{\lambda} = \frac{nv}{2L}, \quad \text{for } n = 1, 2, 3, \ldots \quad \text{(pipe, two open ends).} \quad (17\text{-}39)$$

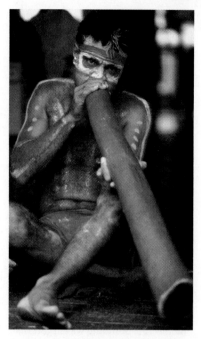

**FIG. 17-13** The air column within a digeridoo ("a pipe") oscillates when the instrument is played. *(Alamy Images)*

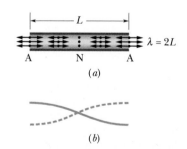

**FIG. 17-14** (a) The simplest standing wave pattern of displacement for (longitudinal) sound waves in a pipe with both ends open has an antinode (A) across each end and a node (N) across the middle. (The longitudinal displacements represented by the double arrows are greatly exaggerated.) (b) The corresponding standing wave pattern for (transverse) string waves.

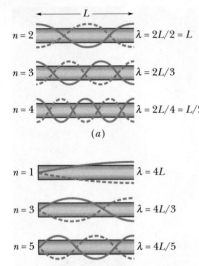

$n = 2$    $\lambda = 2L/2 = L$

$n = 3$    $\lambda = 2L/3$

$n = 4$    $\lambda = 2L/4 = L/2$

(a)

$n = 1$    $\lambda = 4L$

$n = 3$    $\lambda = 4L/3$

$n = 5$    $\lambda = 4L/5$

$n = 7$    $\lambda = 4L/7$

(b)

**FIG. 17-15** Standing wave patterns for string waves superimposed on pipes to represent standing sound wave patterns in the pipes. (a) With *both* ends of the pipe open, any harmonic can be set up in the pipe. (b) With only *one* end open, only odd harmonics can be set up.

Figure 17-15*b* shows (using string wave representations) some of the standing sound wave patterns that can be set up in a pipe with only one open end. As required, across the open end there is an antinode and across the closed end there is a node. The simplest pattern requires sound waves having a wavelength given by $L = \lambda/4$, so that $\lambda = 4L$. The next simplest pattern requires a wavelength given by $L = 3\lambda/4$, so that $\lambda = 4L/3$, and so on.

More generally, the resonant frequencies for a pipe of length $L$ with only one open end correspond to the wavelengths

$$\lambda = \frac{4L}{n}, \qquad \text{for } n = 1, 3, 5, \ldots, \tag{17-40}$$

in which the harmonic number $n$ *must be an odd number*. The resonant frequencies are then given by

$$f = \frac{v}{\lambda} = \frac{nv}{4L}, \qquad \text{for } n = 1, 3, 5, \ldots \quad \text{(pipe, one open end)}. \tag{17-41}$$

Note again that only odd harmonics can exist in a pipe with one open end. For example, the second harmonic, with $n = 2$, cannot be set up in such a pipe. Note also that for such a pipe the adjective in a phrase such as "the third harmonic" still refers to the harmonic number $n$ (and not to, say, the third possible harmonic).

The length of a musical instrument reflects the range of frequencies over which the instrument is designed to function, and smaller length implies higher frequencies. Figure 17-16, for example, shows the saxophone and violin families, with their frequency ranges suggested by the piano keyboard. Note that, for every instrument, there is overlap with its higher- and lower-frequency neighbors.

In any oscillating system that gives rise to a musical sound, whether it is a violin string or the air in an organ pipe, the fundamental and one or more of the higher harmonics are usually generated simultaneously. Thus, you hear them together—that is, superimposed as a net wave. When different instruments are played at the same note, they produce the same fundamental frequency but different intensities for the higher harmonics. For example, the fourth harmonic of middle C might be relatively loud on one instrument and relatively quiet or even missing on another. Thus, because different instruments produce different net waves, they sound different to you even when they are played at the same note. That would be the case for the two net waves shown in Fig. 17-17, which were produced at the same note by different instruments.

✓**CHECKPOINT 3**    Pipe *A*, with length *L*, and pipe *B*, with length 2*L*, both have two open ends. Which harmonic of pipe *B* has the same frequency as the fundamental of pipe *A*?

**FIG. 17-16** The saxophone and violin families, showing the relations between instrument length and frequency range. The frequency range of each instrument is indicated by a horizontal bar along a frequency scale suggested by the keyboard at the bottom; the frequency increases toward the right.

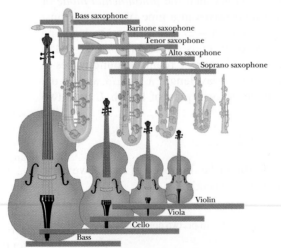

Bass saxophone
Baritone saxophone
Tenor saxophone
Alto saxophone
Soprano saxophone

Violin
Viola
Cello
Bass

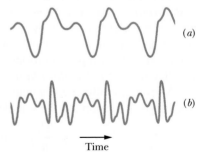

(a)

(b)

Time

**FIG. 17-17** The wave forms produced by (a) a flute and (b) an oboe when played at the same note, with the same first harmonic frequency.

## Sample Problem   17-6

Weak background noises from a room set up the fundamental standing wave in a cardboard tube of length $L = 67.0$ cm with two open ends. Assume that the speed of sound in the air within the tube is 343 m/s.

**(a)** What frequency do you hear from the tube?

**KEY IDEA**   With both pipe ends open, we have a symmetric situation in which the standing wave has an antinode at each end of the tube. The standing wave pattern (in string wave style) is that of Fig. 17-14b.

**Calculation:** The frequency is given by Eq. 17-39 with $n = 1$ for the fundamental mode:

$$f = \frac{nv}{2L} = \frac{(1)(343 \text{ m/s})}{2(0.670 \text{ m})} = 256 \text{ Hz}. \quad \text{(Answer)}$$

If the background noises set up any higher harmonics, such as the second harmonic, you also hear frequencies that are *integer* multiples of 256 Hz.

**(b)** If you jam your ear against one end of the tube, what fundamental frequency do you hear from the tube?

**KEY IDEA**   With your ear effectively closing one end of the tube, we have an asymmetric situation—an antinode still exists at the open end, but a node is now at the other (closed) end. The standing wave pattern is the top one in Fig. 17-15b.

**Calculation:** The frequency is given by Eq. 17-41 with $n = 1$ for the fundamental mode:

$$f = \frac{nv}{4L} = \frac{(1)(343 \text{ m/s})}{4(0.670 \text{ m})} = 128 \text{ Hz}. \quad \text{(Answer)}$$

If the background noises set up any higher harmonics, they will be *odd* multiples of 128 Hz. That means that the frequency of 256 Hz (which is an even multiple) cannot now occur.

## 17-8 I Beats

If we listen, a few minutes apart, to two sounds whose frequencies are, say, 552 and 564 Hz, most of us cannot tell one from the other. However, if the sounds reach our ears simultaneously, what we hear is a sound whose frequency is 558 Hz, the *average* of the two combining frequencies. We also hear a striking variation in the intensity of this sound—it increases and decreases in slow, wavering **beats** that repeat at a frequency of 12 Hz, the *difference* between the two combining frequencies. Figure 17-18 shows this beat phenomenon.

Let the time-dependent variations of the displacements due to two sound waves of equal amplitude $s_m$ be

$$s_1 = s_m \cos \omega_1 t \quad \text{and} \quad s_2 = s_m \cos \omega_2 t, \quad (17\text{-}42)$$

where $\omega_1 > \omega_2$. From the superposition principle, the resultant displacement is

$$s = s_1 + s_2 = s_m(\cos \omega_1 t + \cos \omega_2 t).$$

Using the trigonometric identity (see Appendix E)

$$\cos \alpha + \cos \beta = 2 \cos[\tfrac{1}{2}(\alpha - \beta)] \cos[\tfrac{1}{2}(\alpha + \beta)]$$

allows us to write the resultant displacement as

$$s = 2 s_m \cos[\tfrac{1}{2}(\omega_1 - \omega_2)t] \cos[\tfrac{1}{2}(\omega_1 + \omega_2)t]. \quad (17\text{-}43)$$

If we write

$$\omega' = \tfrac{1}{2}(\omega_1 - \omega_2) \quad \text{and} \quad \omega = \tfrac{1}{2}(\omega_1 + \omega_2), \quad (17\text{-}44)$$

we can then write Eq. 17-43 as

$$s(t) = [2 s_m \cos \omega' t] \cos \omega t. \quad (17\text{-}45)$$

We now assume that the angular frequencies $\omega_1$ and $\omega_2$ of the combining waves are almost equal, which means that $\omega \gg \omega'$ in Eq. 17-44. We can then regard Eq. 17-45 as a cosine function whose angular frequency is $\omega$ and whose amplitude (which is not constant but varies with angular frequency $\omega'$) is the absolute value of the quantity in the brackets.

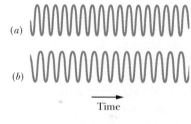

(a)

(b)

Time

(c)

**FIG. 17-18**   (a, b) The pressure variations $\Delta p$ of two sound waves as they would be detected separately. The frequencies of the waves are nearly equal. (c) The resultant pressure variation if the two waves are detected simultaneously.

A maximum amplitude will occur whenever cos $\omega't$ in Eq. 17-45 has the value $+1$ or $-1$, which happens twice in each repetition of the cosine function. Because cos $\omega't$ has angular frequency $\omega'$, the angular frequency $\omega_{beat}$ at which beats occur is $\omega_{beat} = 2\omega'$. Then, with the aid of Eq. 17-44, we can write

$$\omega_{beat} = 2\omega' = (2)(\tfrac{1}{2})(\omega_1 - \omega_2) = \omega_1 - \omega_2.$$

Because $\omega = 2\pi f$, we can recast this as

$$f_{beat} = f_1 - f_2 \quad \text{(beat frequency).} \tag{17-46}$$

Musicians use the beat phenomenon in tuning instruments. If an instrument is sounded against a standard frequency (for example, the note called "concert A" played on an orchestra's first oboe) and tuned until the beat disappears, the instrument is in tune with that standard. In musical Vienna, concert A (440 Hz) is available as a telephone service for the city's many musicians.

## Sample Problem 17-7

When an emperor penguin returns from a search for food, how can it find its mate among the thousands of penguins huddled together for warmth in the harsh Antarctic weather? It is not by sight, because penguins all look alike, even to a penguin.

The answer lies in the way penguins vocalize. Most birds vocalize by using only one side of their two-sided vocal organ, called the *syrinx*. Emperor penguins, however, vocalize by using both sides simultaneously. Each side sets up acoustic standing waves in the bird's throat and mouth, much like in a pipe with two open ends. Suppose that the frequency of the first harmonic produced by side $A$ is $f_{A1} = 432$ Hz and the frequency of the first harmonic produced by side $B$ is $f_{B1} = 371$ Hz. What is the beat frequency between those two first-harmonic frequencies and between the two second-harmonic frequencies?

**KEY IDEA** The beat frequency between two frequencies is their difference, as given by Eq. 17-46 ($f_{beat} = f_1 - f_2$).

**Calculations:** For the two first-harmonic frequencies $f_{A1}$ and $f_{B1}$, the beat frequency is

$$f_{beat,1} = f_{A1} - f_{B1} = 432 \text{ Hz} - 371 \text{ Hz}$$
$$= 61 \text{ Hz.} \quad \text{(Answer)}$$

Because the standing waves in the penguin are effectively in a pipe with two open ends, the resonant frequencies are given by Eq. 17-39 ($f = nv/2L$), in which $L$ is the (unknown) length of the effective pipe. The first-harmonic frequency is $f_1 = v/2L$, and the second-harmonic frequency is $f_2 = 2v/2L$. Comparing these two frequencies, we see that, in general,

$$f_2 = 2f_1.$$

For the penguin, the second harmonic of side $A$ has frequency $f_{A2} = 2f_{A1}$ and the second harmonic of side $B$ has frequency $f_{B2} = 2f_{B1}$. Using Eq. 17-46 with frequencies $f_{A2}$ and $f_{B2}$, we find that the corresponding beat frequency is

$$f_{beat,2} = f_{A2} - f_{B2} = 2f_{A1} - 2f_{B1}$$
$$= 2(432 \text{ Hz}) - 2(371 \text{ Hz})$$
$$= 122 \text{ Hz.} \quad \text{(Answer)}$$

Experiments indicate that penguins can perceive such large beat frequencies (humans cannot). Thus, a penguin's cry can be rich with different harmonics and different beat frequencies, allowing the voice to be recognized even among the voices of thousands of other, closely huddled penguins.

## 17-9 | The Doppler Effect

A police car is parked by the side of the highway, sounding its 1000 Hz siren. If you are also parked by the highway, you will hear that same frequency. However, if there is relative motion between you and the police car, either toward or away from each other, you will hear a different frequency. For example, if you are driving *toward* the police car at 120 km/h (about 75 mi/h), you will hear a *higher* frequency (1096 Hz, an *increase* of 96 Hz). If you are driving *away from* the police car at that same speed, you will hear a *lower* frequency (904 Hz, a *decrease* of 96 Hz).

These motion-related frequency changes are examples of the **Doppler effect.** The effect was proposed (although not fully worked out) in 1842 by Austrian physicist Johann Christian Doppler. It was tested experimentally in

1845 by Buys Ballot in Holland, "using a locomotive drawing an open car with several trumpeters."

The Doppler effect holds not only for sound waves but also for electromagnetic waves, including microwaves, radio waves, and visible light. Here, however, we shall consider only sound waves, and we shall take as a reference frame the body of air through which these waves travel. This means that we shall measure the speeds of a source $S$ of sound waves and a detector $D$ of those waves *relative to that body of air.* (Unless otherwise stated, the body of air is stationary relative to the ground, so the speeds can also be measured relative to the ground.) We shall assume that $S$ and $D$ move either directly toward or directly away from each other, at speeds less than the speed of sound.

If either the detector or the source is moving, or both are moving, the emitted frequency $f$ and the detected frequency $f'$ are related by

$$f' = f\frac{v \pm v_D}{v \pm v_S} \qquad \text{(general Doppler effect)}, \qquad (17\text{-}47)$$

where $v$ is the speed of sound through the air, $v_D$ is the detector's speed relative to the air, and $v_S$ is the source's speed relative to the air. The choice of plus or minus signs is set by this rule:

> When the motion of detector or source is toward the other, the sign on its speed must give an upward shift in frequency. When the motion of detector or source is away from the other, the sign on its speed must give a downward shift in frequency.

In short, *toward* means *shift up,* and *away* means *shift down.*

Here are some examples of the rule. If the detector moves toward the source, use the plus sign in the numerator of Eq. 17-47 to get a shift up in the frequency. If it moves away, use the minus sign in the numerator to get a shift down. If it is stationary, substitute 0 for $v_D$. If the source moves toward the detector, use the minus sign in the denominator of Eq. 17-47 to get a shift up in the frequency. If it moves away, use the plus sign in the denominator to get a shift down. If the source is stationary, substitute 0 for $v_S$.

Next, we derive equations for the Doppler effect for the following two specific situations and then derive Eq. 17-47 for the general situation.

1. When the detector moves relative to the air and the source is stationary relative to the air, the motion changes the frequency at which the detector intercepts wavefronts and thus changes the detected frequency of the sound wave.

2. When the source moves relative to the air and the detector is stationary relative to the air, the motion changes the wavelength of the sound wave and thus changes the detected frequency (recall that frequency is related to wavelength).

## Detector Moving, Source Stationary

In Fig. 17-19, a detector $D$ (represented by an ear) is moving at speed $v_D$ toward a stationary source $S$ that emits spherical wavefronts, of wavelength $\lambda$ and fre-

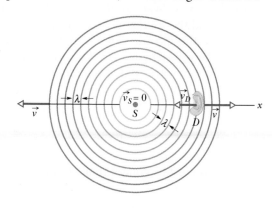

FIG. 17-19   A stationary source of sound $S$ emits spherical wavefronts, shown one wavelength apart, that expand outward at speed $v$. A sound detector $D$, represented by an ear, moves with velocity $\vec{v}_D$ toward the source. The detector senses a higher frequency because of its motion.

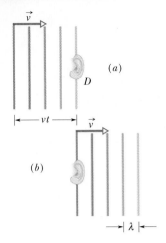

**FIG. 17-20** The wavefronts of Fig. 17-19, assumed planar, (a) reach and (b) pass a stationary detector $D$; they move a distance $vt$ to the right in time $t$.

quency $f$, moving at the speed $v$ of sound in air. The wavefronts are drawn one wavelength apart. The frequency detected by detector $D$ is the rate at which $D$ intercepts wavefronts (or individual wavelengths). If $D$ were stationary, that rate would be $f$, but since $D$ is moving into the wavefronts, the rate of interception is greater, and thus the detected frequency $f'$ is greater than $f$.

Let us for the moment consider the situation in which $D$ is stationary (Fig. 17-20). In time $t$, the wavefronts move to the right a distance $vt$. The number of wavelengths in that distance $vt$ is the number of wavelengths intercepted by $D$ in time $t$, and that number is $vt/\lambda$. The rate at which $D$ intercepts wavelengths, which is the frequency $f$ detected by $D$, is

$$f = \frac{vt/\lambda}{t} = \frac{v}{\lambda}. \tag{17-48}$$

In this situation, with $D$ stationary, there is no Doppler effect—the frequency detected by $D$ is the frequency emitted by $S$.

Now let us again consider the situation in which $D$ moves in the direction opposite the wavefront velocity (Fig. 17-21). In time $t$, the wavefronts move to the right a distance $vt$ as previously, but now $D$ moves to the left a distance $v_D t$. Thus, in this time $t$, the distance moved by the wavefronts relative to $D$ is $vt + v_D t$. The number of wavelengths in this relative distance $vt + v_D t$ is the number of wavelengths intercepted by $D$ in time $t$ and is $(vt + v_D t)/\lambda$. The *rate* at which $D$ intercepts wavelengths in this situation is the frequency $f'$, given by

$$f' = \frac{(vt + v_D t)/\lambda}{t} = \frac{v + v_D}{\lambda}. \tag{17-49}$$

From Eq. 17-48, we have $\lambda = v/f$. Then Eq. 17-49 becomes

$$f' = \frac{v + v_D}{v/f} = f \frac{v + v_D}{v}. \tag{17-50}$$

Note that in Eq. 17-50, $f' > f$ unless $v_D = 0$ (the detector is stationary).

Similarly, we can find the frequency detected by $D$ if $D$ moves away from the source. In this situation, the wavefronts move a distance $vt - v_D t$ relative to $D$ in time $t$, and $f'$ is given by

$$f' = f \frac{v - v_D}{v}. \tag{17-51}$$

In Eq. 17-51, $f' < f$ unless $v_D = 0$. We can summarize Eqs. 17-50 and 17-51 with

$$f' = f \frac{v \pm v_D}{v} \qquad \text{(detector moving, source stationary).} \tag{17-52}$$

### Source Moving, Detector Stationary

Let detector $D$ be stationary with respect to the body of air, and let source $S$ move toward $D$ at speed $v_S$ (Fig. 17-22). The motion of $S$ changes the wavelength of the sound waves it emits and thus the frequency detected by $D$.

To see this change, let $T\ (= 1/f)$ be the time between the emission of any pair of successive wavefronts $W_1$ and $W_2$. During $T$, wavefront $W_1$ moves a distance $vT$ and the source moves a distance $v_S T$. At the end of $T$, wavefront $W_2$ is emitted. In the direction in which $S$ moves, the distance between $W_1$ and $W_2$, which is the wavelength $\lambda'$ of the waves moving in that direction, is $vT - v_S T$. If $D$ detects those waves, it detects frequency $f'$ given by

$$f' = \frac{v}{\lambda'} = \frac{v}{vT - v_S T} = \frac{v}{v/f - v_S/f}$$

$$= f \frac{v}{v - v_S}. \tag{17-53}$$

Note that $f'$ must be greater than $f$ unless $v_S = 0$.

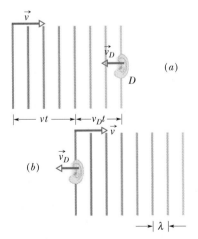

**FIG. 17-21** Wavefronts traveling to the right (a) reach and (b) pass detector $D$, which moves in the opposite direction. In time $t$, the wavefronts move a distance $vt$ to the right and $D$ moves a distance $v_D t$ to the left.

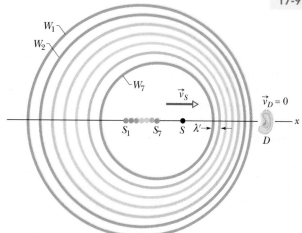

**FIG. 17-22** A detector $D$ is stationary, and a source $S$ is moving toward it at speed $v_S$. Wavefront $W_1$ was emitted when the source was at $S_1$, wavefront $W_7$ when it was at $S_7$. At the moment depicted, the source is at $S$. The detector senses a higher frequency because the moving source, chasing its own wavefronts, emits a reduced wavelength $\lambda'$ in the direction of its motion.

In the direction opposite that taken by $S$, the wavelength $\lambda'$ of the waves is $vT + v_S T$. If $D$ detects those waves, it detects frequency $f'$ given by

$$f' = f\frac{v}{v + v_S}. \qquad (17\text{-}54)$$

Now $f'$ must be less than $f$ unless $v_S = 0$.

We can summarize Eqs. 17-53 and 17-54 with

$$f' = f\frac{v}{v \pm v_S} \qquad \text{(source moving, detector stationary).} \qquad (17\text{-}55)$$

## General Doppler Effect Equation

We can now derive the general Doppler effect equation by replacing $f$ in Eq. 17-55 (the source frequency) with $f'$ of Eq. 17-52 (the frequency associated with motion of the detector). The result is Eq. 17-47 for the general Doppler effect.

That general equation holds not only when both detector and source are moving but also in the two specific situations we just discussed. For the situation in which the detector is moving and the source is stationary, substitution of $v_S = 0$ into Eq. 17-47 gives us Eq. 17-52, which we previously found. For the situation in which the source is moving and the detector is stationary, substitution of $v_D = 0$ into Eq. 17-47 gives us Eq. 17-55, which we previously found. Thus, Eq. 17-47 is the equation to remember.

✓**CHECKPOINT 4** The figure indicates the directions of motion of a sound source and a detector for six situations in stationary air. For each situation, is the detected frequency greater than or less than the emitted frequency, or can't we tell without more information about the actual speeds?

| | Source | Detector | | Source | Detector |
|---|---|---|---|---|---|
| (a) | $\longrightarrow$ | • 0 speed | (d) | $\longleftarrow$ | $\longleftarrow$ |
| (b) | $\longleftarrow$ | • 0 speed | (e) | $\longrightarrow$ | $\longleftarrow$ |
| (c) | $\longrightarrow$ | $\longrightarrow$ | (f) | $\longleftarrow$ | $\longrightarrow$ |

**Sample Problem** | **17-8** | **Build your skill**

Bats navigate and search out prey by emitting, and then detecting reflections of, ultrasonic waves, which are sound waves with frequencies greater than can be heard by a human. Suppose a bat emits ultrasound at frequency $f_{be} = 82.52$ kHz while flying with velocity $\vec{v}_b = (9.00 \text{ m/s})\hat{i}$ as it chases a moth that flies with velocity $\vec{v}_m = (8.00 \text{ m/s})\hat{i}$. What frequency $f_{md}$ does the moth detect? What frequency $f_{bd}$ does the bat detect in the returning echo from the moth?

**KEY IDEAS** The frequency is shifted by the relative motion of the bat and moth. Because they move along a single axis, the shifted frequency is given by Eq. 17-47 for the general Doppler effect. Motion *toward* tends to shift the frequency *up*, and motion *away* tends to shift the frequency *down*.

**Detection by moth:** The general Doppler equation is

$$f' = f\frac{v \pm v_D}{v \pm v_S}. \qquad (17\text{-}56)$$

Here, the detected frequency $f'$ that we want is the frequency $f_{md}$ detected by the moth. On the right side of the equation, the emitted frequency $f$ is the bat's emission frequency $f_{be} = 82.52$ kHz, the speed of sound is $v = 343$ m/s, the speed $v_D$ of the detector is the moth's speed $v_m = 8.00$ m/s, and the speed $v_S$ of the source is the bat's speed $v_b = 9.00$ m/s.

These substitutions into Eq. 17-56 are easy to make. However, the decisions about the plus and minus signs can be tricky. Think in terms of *toward* and *away*. We have the speed of the moth (the detector) in the numerator of Eq. 17-56. The moth moves *away* from the bat, which tends to lower the detected frequency. Because the speed is in the numerator, we choose the minus sign to meet that tendency (the numerator becomes smaller). These reasoning steps are shown in Table 17-3.

We have the speed of the bat in the denominator of Eq. 17-56. The bat moves *toward* the moth, which tends to increase the detected frequency. Because the speed is in the denominator, we choose the minus sign to meet that tendency (the denominator becomes smaller).

With these substitutions and decisions, we have

$$f_{md} = f_{be}\frac{v - v_m}{v - v_b}$$

$$= (82.52 \text{ kHz})\frac{343 \text{ m/s} - 8.00 \text{ m/s}}{343 \text{ m/s} - 9.00 \text{ m/s}}$$

$$= 82.767 \text{ kHz} \approx 82.8 \text{ kHz}. \qquad \text{(Answer)}$$

**Detection of echo by bat:** In the echo back to the bat, the moth acts as a source of sound, emitting at the frequency $f_{md}$ we just calculated. So now the moth is the source (moving *away*) and the bat is the detector (moving *toward*). The reasoning steps are shown in Table 17-3. To find the frequency $f_{bd}$ detected by the bat, we write Eq. 17-56 as

$$f_{bd} = f_{md}\frac{v + v_b}{v + v_m}$$

$$= (82.767 \text{ kHz})\frac{343 \text{ m/s} + 9.00 \text{ m/s}}{343 \text{ m/s} + 8.00 \text{ m/s}}$$

$$= 83.00 \text{ kHz} \approx 83.0 \text{ kHz}. \qquad \text{(Answer)}$$

Some moths evade bats by "jamming" the detection system with ultrasonic clicks.

**TABLE 17-3**

| Bat to Moth | | | Echo Back to Bat | |
| --- | --- | --- | --- | --- |
| Detector | Source | | Detector | Source |
| moth | bat | | bat | moth |
| speed $v_D = v_m$ | speed $v_S = v_b$ | | speed $v_D = v_b$ | speed $v_S = v_m$ |
| away | toward | | toward | away |
| shift down | shift up | | shift up | shift down |
| numerator | denominator | | numerator | denominator |
| minus | minus | | plus | plus |

## 17-10 | Supersonic Speeds, Shock Waves

If a source is moving toward a stationary detector at a speed equal to the speed of sound—that is, if $v_S = v$—Eqs. 17-47 and 17-55 predict that the detected frequency $f'$ will be infinitely great. This means that the source is moving so fast that it keeps pace with its own spherical wavefronts, as Fig. 17-23a suggests. What happens when the speed of the source *exceeds* the speed of sound?

For such *supersonic* speeds, Eqs. 17-47 and 17-55 no longer apply. Figure 17-23b depicts the spherical wavefronts that originated at various positions of the source. The radius of any wavefront in this figure is $vt$, where $v$ is the speed of sound and $t$ is the time that has elapsed since the source emitted that wavefront. Note that all the wavefronts bunch along a V-shaped envelope in the two-dimensional drawing of Fig. 17-23b. The wavefronts actually extend in three dimensions, and the bunching actually forms a cone called the *Mach cone*. A *shock wave* is said to exist along the surface of this cone, because the bunching of wavefronts causes an abrupt rise and fall of air pressure as the surface

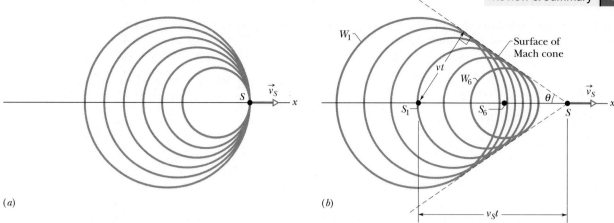

**FIG. 17-23** (*a*) A source of sound *S* moves at speed $v_S$ equal to the speed of sound and thus as fast as the wavefronts it generates. (*b*) A source *S* moves at speed $v_S$ faster than the speed of sound and thus faster than the wavefronts. When the source was at position $S_1$ it generated wavefront $W_1$, and at position $S_6$ it generated $W_6$. All the spherical wavefronts expand at the speed of sound *v* and bunch along the surface of a cone called the Mach cone, forming a shock wave. The surface of the cone has half-angle $\theta$ and is tangent to all the wavefronts.

passes through any point. From Fig. 17-23*b*, we see that the half-angle $\theta$ of the cone, called the *Mach cone angle*, is given by

$$\sin \theta = \frac{vt}{v_S t} = \frac{v}{v_S} \quad \text{(Mach cone angle)}. \tag{17-57}$$

The ratio $v_S/v$ is called the *Mach number*. When you hear that a particular plane has flown at Mach 2.3, it means that its speed was 2.3 times the speed of sound in the air through which the plane was flying. The shock wave generated by a supersonic aircraft (Fig. 17-24) or projectile produces a burst of sound, called a *sonic boom,* in which the air pressure first suddenly increases and then suddenly decreases below normal before returning to normal. Part of the sound that is heard when a rifle is fired is the sonic boom produced by the bullet. A sonic boom can also be heard from a long bullwhip when it is snapped quickly: Near the end of the whip's motion, its tip is moving faster than sound and produces a small sonic boom—the *crack* of the whip.

**FIG. 17-24** Shock waves produced by the wings of a Navy FA 18 jet. The shock waves are visible because the sudden decrease in air pressure in them caused water molecules in the air to condense, forming a fog. (*U.S. Navy photo by Ensign John Gay*)

## REVIEW & SUMMARY

**Sound Waves** Sound waves are longitudinal mechanical waves that can travel through solids, liquids, or gases. The speed *v* of a sound wave in a medium having **bulk modulus** *B* and density $\rho$ is

$$v = \sqrt{\frac{B}{\rho}} \quad \text{(speed of sound)}. \tag{17-3}$$

In air at 20°C, the speed of sound is 343 m/s.

A sound wave causes a longitudinal displacement *s* of a mass element in a medium as given by

$$s = s_m \cos(kx - \omega t), \tag{17-13}$$

where $s_m$ is the **displacement amplitude** (maximum displacement) from equilibrium, $k = 2\pi/\lambda$, and $\omega = 2\pi f$, $\lambda$ and *f* being the wavelength and frequency, respectively, of the sound wave.

The sound wave also causes a pressure change $\Delta p$ of the medium from the equilibrium pressure:

$$\Delta p = \Delta p_m \sin(kx - \omega t), \tag{17-14}$$

where the **pressure amplitude** is

$$\Delta p_m = (v\rho\omega)s_m. \tag{17-15}$$

**Interference** The interference of two sound waves with identical wavelengths passing through a common point depends on their phase difference $\phi$ there. If the sound waves were emitted in phase and are traveling in approximately the same direction, $\phi$ is given by

$$\phi = \frac{\Delta L}{\lambda} 2\pi, \tag{17-21}$$

where $\Delta L$ is their **path length difference** (the difference in the distances traveled by the waves to reach the common point). Fully constructive interference occurs when $\phi$ is an integer multiple of $2\pi$,

$$\phi = m(2\pi), \qquad \text{for } m = 0, 1, 2, \ldots, \qquad (17\text{-}22)$$

and, equivalently, when $\Delta L$ is related to wavelength $\lambda$ by

$$\frac{\Delta L}{\lambda} = 0, 1, 2, \ldots \qquad (17\text{-}23)$$

Fully destructive interference occurs when $\phi$ is an odd multiple of $\pi$,

$$\phi = (2m + 1)\pi, \qquad \text{for } m = 0, 1, 2, \ldots, \qquad (17\text{-}24)$$

and, equivalently, when $\Delta L$ is related to $\lambda$ by

$$\frac{\Delta L}{\lambda} = 0.5, 1.5, 2.5, \ldots \qquad (17\text{-}25)$$

**Sound Intensity** The **intensity** $I$ of a sound wave at a surface is the average rate per unit area at which energy is transferred by the wave through or onto the surface:

$$I = \frac{P}{A}, \qquad (17\text{-}26)$$

where $P$ is the time rate of energy transfer (power) of the sound wave and $A$ is the area of the surface intercepting the sound. The intensity $I$ is related to the displacement amplitude $s_m$ of the sound wave by

$$I = \tfrac{1}{2}\rho v \omega^2 s_m^2. \qquad (17\text{-}27)$$

The intensity at a distance $r$ from a point source that emits sound waves of power $P_s$ is

$$I = \frac{P_s}{4\pi r^2}. \qquad (17\text{-}28)$$

**Sound Level in Decibels** The *sound level* $\beta$ in *decibels* (dB) is defined as

$$\beta = (10 \text{ dB}) \log \frac{I}{I_0}, \qquad (17\text{-}29)$$

where $I_0$ ($= 10^{-12}$ W/m$^2$) is a reference intensity level to which all intensities are compared. For every factor-of-10 increase in intensity, 10 dB is added to the sound level.

**Standing Wave Patterns in Pipes** Standing sound wave patterns can be set up in pipes. A pipe open at both ends will resonate at frequencies

$$f = \frac{v}{\lambda} = \frac{nv}{2L}, \qquad n = 1, 2, 3, \ldots, \qquad (17\text{-}39)$$

where $v$ is the speed of sound in the air in the pipe. For a pipe closed at one end and open at the other, the resonant frequencies are

$$f = \frac{v}{\lambda} = \frac{nv}{4L}, \qquad n = 1, 3, 5, \ldots. \qquad (17\text{-}41)$$

**Beats** *Beats* arise when two waves having slightly different frequencies, $f_1$ and $f_2$, are detected together. The beat frequency is

$$f_{\text{beat}} = f_1 - f_2. \qquad (17\text{-}46)$$

**The Doppler Effect** The *Doppler effect* is a change in the observed frequency of a wave when the source or the detector moves relative to the transmitting medium (such as air). For sound the observed frequency $f'$ is given in terms of the source frequency $f$ by

$$f' = f \frac{v \pm v_D}{v \pm v_S} \qquad \text{(general Doppler effect)}, \qquad (17\text{-}47)$$

where $v_D$ is the speed of the detector relative to the medium, $v_S$ is that of the source, and $v$ is the speed of sound in the medium. The signs are chosen such that $f'$ tends to be *greater* for motion toward and *less* for motion away.

**Shock Wave** If the speed of a source relative to the medium exceeds the speed of sound in the medium, the Doppler equation no longer applies. In such a case, shock waves result. The half-angle $\theta$ of the Mach cone is given by

$$\sin \theta = \frac{v}{v_S} \qquad \text{(Mach cone angle)}. \qquad (17\text{-}57)$$

# QUESTIONS

**1** In Fig. 17-25, three long tubes ($A$, $B$, and $C$) are filled with different gases under different pressures. The ratio of the bulk modulus to the density is indicated for each gas in terms of a basic value $B_0/\rho_0$. Each tube has a piston at its left end that can send a sound pulse through the tube (as in Fig. 16-2). The three pulses are sent simultaneously. Rank the tubes according to the time of arrival of the pulses at the open right ends of the tubes, earliest first.

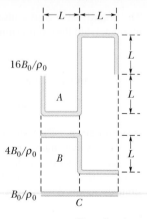

**FIG. 17-25** Question 1.

**2** In Fig. 17-26, two point sources $S_1$ and $S_2$, which are in phase, emit identical sound waves of wavelength 2.0 m. In terms of wavelengths, what is the phase difference between the waves arriving at point $P$ if (a) $L_1 = 38$ m and $L_2 = 34$ m, and (b) $L_1 = 39$ m and $L_2 = 36$ m? (c) Assuming that the source separation is much smaller than $L_1$ and $L_2$, what type of interference occurs at $P$ in situations (a) and (b)?

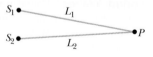

**FIG. 17-26** Question 2.

**3** In a first experiment, a sinusoidal sound wave is sent through a long tube of air, transporting energy at the average rate of $P_{\text{avg,1}}$. In a second experiment, two other sound waves, identical to the first one, are to be sent simultaneously through the tube with a phase difference $\phi$ of either 0,

0.2 wavelength, or 0.5 wavelength between the waves. (a) With only mental calculation, rank those choices of $\phi$ according to the average rate at which the waves will transport energy, greatest first. (b) For the first choice of $\phi$, what is the average rate in terms of $P_{\text{avg},1}$?

**4** Pipe $A$ has length $L$ and one open end. Pipe $B$ has length $2L$ and two open ends. Which harmonics of pipe $B$ have a frequency that matches a resonant frequency of pipe $A$?

**5** For a particular tube, here are four of the six harmonic frequencies below 1000 Hz: 300, 600, 750, and 900 Hz. What two frequencies are missing from the list?

**6** The sixth harmonic is set up in a pipe. (a) How many open ends does the pipe have (it has at least one)? (b) Is there a node, antinode, or some intermediate state at the midpoint?

**7** In Fig. 17-27, pipe $A$ is made to oscillate in its third harmonic by a small internal sound source. Sound emitted at the right end happens to resonate four nearby pipes, each with only one open end (they are *not* drawn to scale). Pipe $B$ oscillates in its lowest harmonic, pipe $C$ in its second lowest harmonic, pipe $D$ in its third lowest harmonic, and pipe $E$ in its fourth lowest harmonic. Without computation, rank all five pipes according to their length, greatest first. (*Hint:* Draw the standing waves to scale and then draw the pipes to scale.)

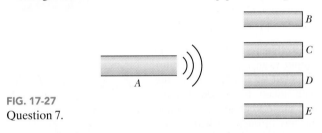

**FIG. 17-27**
Question 7.

**8** Figure 17-28 shows a stretched string of length $L$ and pipes $a$, $b$, $c$, and $d$ of lengths $L$, $2L$, $L/2$, and $L/2$, respectively.

The string's tension is adjusted until the speed of waves on the string equals the speed of sound waves in the air. The fundamental mode of oscillation is then set up on the string. In which pipe will the sound produced by the string cause resonance, and what oscillation mode will that sound set up?

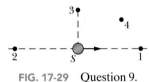

**FIG. 17-28** Question 8.

**9** Figure 17-29 shows a moving sound source $S$ that emits at a certain frequency, and four stationary sound detectors. Rank the detectors according to the frequency of the sound they detect from the source, greatest first.

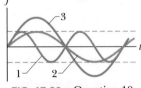

**FIG. 17-29** Question 9.

**10** A friend rides, in turn, the rims of three fast merry-go-rounds while holding a sound source that emits isotropically at a certain frequency. You stand far from each merry-go-round. The frequency you hear for each of your friend's three rides varies as the merry-go-round rotates. The variations in frequency for the three rides are given by the three curves in Fig. 17-30. Rank the curves according to (a) the linear speed $v$ of the sound source, (b) the angular speeds $\omega$ of the merry-go-rounds, and (c) the radii $r$ of the merry-go-rounds, greatest first.

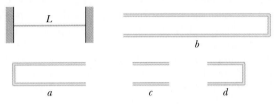

**FIG. 17-30** Question 10.

## PROBLEMS

| | |
|---|---|
| **GO** | Tutoring problem available (at instructor's discretion) in *WileyPLUS* and WebAssign |
| **SSM** | Worked-out solution available in Student Solutions Manual |
| • – ••• | Number of dots indicates level of problem difficulty |
| | Additional information available in *The Flying Circus of Physics* and at flyingcircusofphysics.com |

**WWW** Worked-out solution is at ⸺
**ILW** Interactive solution is at ⸺ http://www.wiley.com/college/halliday

*Where needed in the problems, use*

$$\text{speed of sound in air} = 343 \text{ m/s}$$

*and* $\quad$ density of air $= 1.21 \text{ kg/m}^3$

*unless otherwise specified.*

### sec. 17-3 The Speed of Sound

•**1** When the door of the Chapel of the Mausoleum in Hamilton, Scotland, is slammed shut, the last echo heard by someone standing just inside the door reportedly comes 15 s later. (a) If that echo were due to a single reflection off a wall opposite the door, how far from the door would that wall be? (b) If, instead, the wall is 25.7 m away, how many reflections (back and forth) correspond to the last echo?

•**2** A column of soldiers, marching at 120 paces per minute, keep in step with the beat of a drummer at the head of the column. The soldiers in the rear end of the column are striding forward with the left foot when the drummer is advancing with the right foot. What is the approximate length of the column?

•**3** Two spectators at a soccer game in Montjuic Stadium see, and a moment later hear, the ball being kicked on the playing field. The time delay for spectator $A$ is 0.23 s, and for spectator $B$ it is 0.12 s. Sight lines from the two spectators to the player kicking the ball meet at an angle of 90°. How far are (a) spectator $A$ and (b) spectator $B$ from the player? (c) How far are the spectators from each other?

•**4** What is the bulk modulus of oxygen if 32.0 g of oxygen occupies 22.4 L and the speed of sound in the oxygen is 317 m/s?

**••5** A stone is dropped into a well. The splash is heard 3.00 s later. What is the depth of the well? SSM WWW

**••6** *Hot chocolate effect.* Tap a metal spoon inside a mug of water and note the frequency $f_i$ you hear. Then add a spoonful of powder (say, chocolate mix or instant coffee) and tap again as you stir the powder. The frequency you hear has a lower value $f_s$ because the tiny air bubbles released by the powder change the water's bulk modulus. As the bubbles reach the water surface and disappear, the frequency gradually shifts back to its initial value. During the effect, the bubbles don't appreciably change the water's density or volume or the sound's wavelength. Rather, they change the value of $dV/dp$—that is, the differential change in volume due to the differential change in the pressure caused by the sound wave in the water. If $f_s/f_i = 0.333$, what is the ratio $(dV/dp)_s/(dV/dp)_i$?

**••7** Earthquakes generate sound waves inside Earth. Unlike a gas, Earth can experience both transverse (S) and longitudinal (P) sound waves. Typically, the speed of S waves is about 4.5 km/s, and that of P waves 8.0 km/s. A seismograph records P and S waves from an earthquake. The first P waves arrive 3.0 min before the first S waves. If the waves travel in a straight line, how far away does the earthquake occur? SSM ILW

**••8** A man strikes one end of a thin rod with a hammer. The speed of sound in the rod is 15 times the speed of sound in air. A woman, at the other end with her ear close to the rod, hears the sound of the blow twice with a 0.12 s interval between; one sound comes through the rod and the other comes through the air alongside the rod. If the speed of sound in air is 343 m/s, what is the length of the rod?

## sec. 17-4 Traveling Sound Waves

**•9** Diagnostic ultrasound of frequency 4.50 MHz is used to examine tumors in soft tissue. (a) What is the wavelength in air of such a sound wave? (b) If the speed of sound in tissue is 1500 m/s, what is the wavelength of this wave in tissue? SSM

**•10** The pressure in a traveling sound wave is given by the equation

$$\Delta p = (1.50 \text{ Pa}) \sin \pi[(0.900 \text{ m}^{-1})x - (315 \text{ s}^{-1})t].$$

Find the (a) pressure amplitude, (b) frequency, (c) wavelength, and (d) speed of the wave.

**•11** If the form of a sound wave traveling through air is

$$s(x, t) = (6.0 \text{ nm}) \cos(kx + (3000 \text{ rad/s})t + \phi),$$

how much time does any given air molecule along the path take to move between displacements $s = +2.0$ nm and $s = -2.0$ nm?

**•12** *Underwater illusion.* One clue used by your brain to determine the direction of a source of sound is the time delay $\Delta t$ between the arrival of the sound at the ear closer to the source and the arrival at the farther ear. Assume that the source is distant so that a wavefront from it is approximately planar when it reaches you, and let $D$ represent the separation between your ears. (a) If the source is located at angle $\theta$ in front of you (Fig. 17-31), what is $\Delta t$ in terms of $D$ and the speed of sound $v$ in air? (b) If you are submerged in water and the sound source is directly to your right, what is $\Delta t$ in terms of $D$ and the speed of sound $v_w$ in water? (c) Based on the time-delay clue, your brain interprets the submerged sound to arrive

at an angle $\theta$ from the forward direction. Evaluate $\theta$ for fresh water at 20°C.

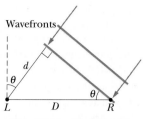

FIG. 17-31   Problem 12.

**••13** A handclap on stage in an amphitheater sends out sound waves that scatter from terraces of width $w = 0.75$ m (Fig. 17-32). The sound returns to the stage as a periodic series of pulses, one from each terrace; the parade of pulses sounds like a played note. (a) Assuming that all the rays in Fig. 17-32 are horizontal, find the frequency at which the pulses return (that is, the frequency of the perceived note). (b) If the width $w$ of the terraces were smaller, would the frequency be higher or lower?

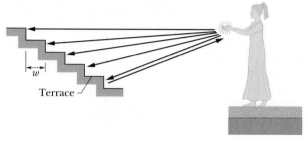

**FIG. 17-32**   Problem 13.

**••14** Figure 17-33 shows the output from a pressure monitor mounted at a point along the path taken by a sound wave of a single frequency traveling at 343 m/s through air with a uniform density of 1.21 kg/m³. The vertical axis scale is set by $\Delta p_s = 4.0$ mPa. If the *displacement* function of the wave is written as $s(x, t) = s_m \cos(kx - \omega t)$, what are (a) $s_m$, (b) $k$, and (c) $\omega$? The air is then cooled so that its density is 1.35 kg/m³ and the speed of a sound wave through it is 320 m/s. The sound source again emits the sound wave at the same frequency and same pressure amplitude. What now are (d) $s_m$, (e) $k$, and (f) $\omega$?

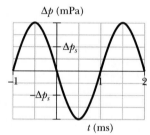

**FIG. 17-33**   Problem 14.

**••15** A sound wave of the form $s = s_m \cos(kx - \omega t + \phi)$ travels at 343 m/s through air in a long horizontal tube. At one instant, air molecule $A$ at $x = 2.000$ m is at its maximum positive displacement of 6.00 nm and air molecule $B$ at $x = 2.070$ m is at a positive displacement of 2.00 nm. All the molecules between $A$ and $B$ are at intermediate displacements. What is the frequency of the wave?

## sec. 17-5 Interference

**•16** Two sound waves, from two different sources with the same frequency, 540 Hz, travel in the same direction at 330 m/s. The sources are in phase. What is the phase difference of the waves at a point that is 4.40 m from one source and 4.00 m from the other?

**••17** Figure 17-34 shows two isotropic point sources of sound, $S_1$ and $S_2$. The sources

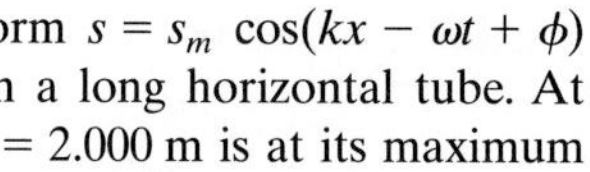

FIG. 17-34
Problems 17 and 107.

emit waves in phase at wavelength 0.50 m; they are separated by $D = 1.75$ m. If we move a sound detector along a large circle centered at the midpoint between the sources, at how many points do waves arrive at the detector (a) exactly in phase and (b) exactly out of phase?

••**18** In Fig. 17-35, sound with a 40.0 cm wavelength travels rightward from a source and through a tube that consists of a straight portion and a half-circle. Part of the sound wave travels through the half-circle and then rejoins the rest of the wave, which goes directly through the straight portion. This rejoining results in interference. What is the smallest radius $r$ that results in an intensity minimum at the detector?

**FIG. 17-35** Problem 18.

••**19** In Fig. 17-36, two speakers separated by distance $d_1 = 2.00$ m are in phase. Assume the amplitudes of the sound waves from the speakers are approximately the same at the listener's ear at distance $d_2 = 3.75$ m directly in front of one speaker. Consider the full audible range for normal hearing, 20 Hz to 20 kHz. (a) What is the lowest frequency $f_{min,1}$ that gives minimum signal (destructive interference) at the listener's ear? By what number must $f_{min,1}$ be multiplied to get (b) the second lowest frequency $f_{min,2}$ that gives minimum signal and (c) the third lowest frequency $f_{min,3}$ that gives minimum signal? (d) What is the lowest frequency $f_{max,1}$ that gives maximum signal (constructive interference) at the listener's ear? By what number must $f_{max,1}$ be multiplied to get (e) the second lowest frequency $f_{max,2}$ that gives maximum signal and (f) the third lowest frequency $f_{max,3}$ that gives maximum signal? SSM

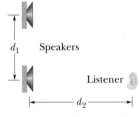

**FIG. 17-36** Problem 19.

••**20** In Fig. 17-37, sound waves $A$ and $B$, both of wavelength $\lambda$, are initially in phase and traveling rightward, as indicated by the two rays. Wave $A$ is reflected from four surfaces but ends up traveling in its original direction. Wave $B$ ends in that direction after reflecting from two surfaces. Let distance $L$ in the figure be expressed as a multiple $q$ of $\lambda$: $L = q\lambda$. What are the (a) smallest and (b) second smallest values of $q$ that put $A$ and $B$ exactly out of phase with each other after the reflections?

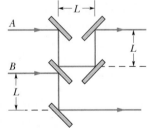

**FIG. 17-37** Problem 20.

••**21** Two loudspeakers are located 3.35 m apart on an outdoor stage. A listener is 18.3 m from one and 19.5 m from the other. During the sound check, a signal generator drives the two speakers in phase with the same amplitude and frequency. The transmitted frequency is swept through the audible range (20 Hz to 20 kHz). (a) What is the lowest frequency $f_{min,1}$ that gives minimum signal (destructive interference) at the listener's location? By what number must $f_{min,1}$ be multiplied to get (b) the second lowest frequency $f_{min,2}$ that gives minimum signal and (c) the third lowest frequency $f_{min,3}$ that gives minimum signal? (d) What is the lowest frequency $f_{max,1}$ that gives

maximum signal (constructive interference) at the listener's location? By what number must $f_{max,1}$ be multiplied to get (e) the second lowest frequency $f_{max,2}$ that gives maximum signal and (f) the third lowest frequency $f_{max,3}$ that gives maximum signal? ILW

••**22** Figure 17-38 shows four isotropic point sources of sound that are uniformly spaced on an $x$ axis. The sources emit sound at the same wavelength $\lambda$ and same amplitude $s_m$, and they emit in phase. A point $P$ is shown on the $x$ axis. Assume that as the sound waves travel to $P$, the decrease in their amplitude is negligible. What multiple of $s_m$ is the amplitude of the net wave at $P$ if distance $d$ in the figure is (a) $\lambda/4$, (b) $\lambda/2$, and (c) $\lambda$?

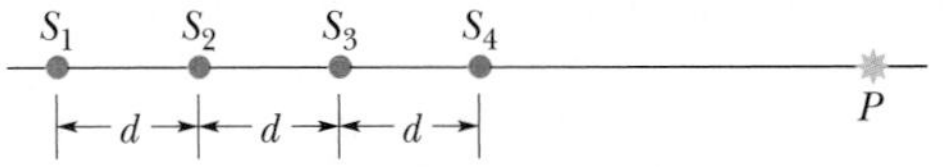

**FIG. 17-38** Problem 22.

•••**23** Figure 17-39 shows two point sources $S_1$ and $S_2$ that emit sound of wavelength $\lambda = 2.00$ m. The emissions are isotropic and in phase, and the separation between the sources is $d = 16.0$ m. At any point $P$ on the $x$ axis, the wave from $S_1$ and the wave from $S_2$ interfere. When $P$ is very far away ($x \approx \infty$), what are (a) the phase difference between the arriving waves from $S_1$ and $S_2$ and (b) the type of interference they produce? Now move point $P$ along the $x$ axis toward $S_1$. (c) Does the phase difference between the waves increase or decrease? At what distance $x$ do the waves have a phase difference of (d) $0.50\lambda$, (e) $1.00\lambda$, and (f) $1.50\lambda$? GO

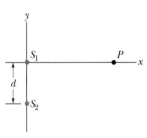

**FIG. 17-39** Problem 23.

### sec. 17-6 Intensity and Sound Level

•**24** A 1.0 W point source emits sound waves isotropically. Assuming that the energy of the waves is conserved, find the intensity (a) 1.0 m from the source and (b) 2.5 m from the source.

•**25** A source emits sound waves isotropically. The intensity of the waves 2.50 m from the source is $1.91 \times 10^{-4}$ W/m². Assuming that the energy of the waves is conserved, find the power of the source. SSM

•**26** Two sounds differ in sound level by 1.00 dB. What is the ratio of the greater intensity to the smaller intensity?

•**27** A sound wave of frequency 300 Hz has an intensity of 1.00 $\mu$W/m². What is the amplitude of the air oscillations caused by this wave?

•**28** The source of a sound wave has a power of 1.00 $\mu$W. If it is a point source, (a) what is the intensity 3.00 m away and (b) what is the sound level in decibels at that distance?

•**29** A certain sound source is increased in sound level by 30.0 dB. By what multiple is (a) its intensity increased and (b) its pressure amplitude increased? SSM WWW

•**30** Suppose that the sound level of a conversation is initially at an angry 70 dB and then drops to a soothing 50 dB. Assuming that the frequency of the sound is 500 Hz, determine the (a) initial and (b) final sound intensities and the (c) initial and (d) final sound wave amplitudes.

•31   Male *Rana catesbeiana* bullfrogs are known for their loud mating call. The call is emitted not by the frog's mouth but by its eardrums, which lie on the surface of the head. And, surprisingly, the sound has nothing to do with the frog's inflated throat. If the emitted sound has a frequency of 260 Hz and a sound level of 85 dB (near the eardrum), what is the amplitude of the eardrum's oscillation? The air density is 1.21 kg/m³.

•32   Approximately a third of people with normal hearing have ears that continuously emit a low-intensity sound outward through the ear canal. A person with such *spontaneous otoacoustic emission* is rarely aware of the sound, except perhaps in a noise-free environment, but occasionally the emission is loud enough to be heard by someone else nearby. In one observation, the sound wave had a frequency of 1665 Hz and a pressure amplitude of $1.13 \times 10^{-3}$ Pa. What were (a) the displacement amplitude and (b) the intensity of the wave emitted by the ear?

•33   When you "crack" a knuckle, you suddenly widen the knuckle cavity, allowing more volume for the synovial fluid inside it and causing a gas bubble suddenly to appear in the fluid. The sudden production of the bubble, called "cavitation," produces a sound pulse—the cracking sound. Assume that the sound is transmitted uniformly in all directions and that it fully passes from the knuckle interior to the outside. If the pulse has a sound level of 62 dB at your ear, estimate the rate at which energy is produced by the cavitation.

••34   *Party hearing.* As the number of people at a party increases, you must raise your voice for a listener to hear you against the *background noise* of the other partygoers. However, once you reach the level of yelling, the only way you can be heard is if you move closer to your listener, into the listener's "personal space." Model the situation by replacing you with an isotropic point source of fixed power P and replacing your listener with a point that absorbs part of your sound waves. These points are initially separated by $r_i = 1.20$ m. If the background noise increases by $\Delta\beta = 5$ dB, the sound level at your listener must also increase. What separation $r_f$ is then required?

••35   A point source emits 30.0 W of sound isotropically. A small microphone intercepts the sound in an area of 0.750 cm², 200 m from the source. Calculate (a) the sound intensity there and (b) the power intercepted by the microphone.

••36   Two atmospheric sound sources A and B emit isotropically at constant power. The sound levels β of their emissions are plotted in Fig. 17-40 versus the radial distance r from the sources. The vertical axis scale is set by $\beta_1 = 85.0$ dB and $\beta_2 = 65.0$ dB. What are (a) the ratio of the larger power to the smaller power and (b) the sound level difference at r = 10 m?

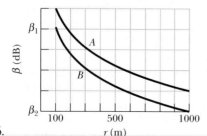

FIG. 17-40
Problem 36.

•••37   A sound source sends a sinusoidal sound wave of angular frequency 3000 rad/s and amplitude 12.0 nm through

a tube of air. The internal radius of the tube is 2.00 cm. (a) What is the average rate at which energy (the sum of the kinetic and potential energies) is transported to the opposite end of the tube? (b) If, simultaneously, an identical wave travels along an adjacent, identical tube, what is the total average rate at which energy is transported to the opposite ends of the two tubes by the waves? If, instead, those two waves are sent along the *same* tube simultaneously, what is the total average rate at which they transport energy when their phase difference is (c) 0, (d) $0.40\pi$ rad, and (e) $\pi$ rad?

### sec. 17-7   Sources of Musical Sound

•38   The crest of a *Parasaurolophus* dinosaur skull contains a nasal passage in the shape of a long, bent tube open at both ends. The dinosaur may have used the passage to produce sound by setting up the fundamental mode in it. (a) If the nasal passage in a certain *Parasaurolophus* fossil is 2.0 m long, what frequency would have been produced? (b) If that dinosaur could be recreated (as in *Jurassic Park*), would a person with a hearing range of 60 Hz to 20 kHz be able to hear that fundamental mode and, if so, would the sound be high or low frequency? Fossil skulls that contain shorter nasal passages are thought to be those of the female *Parasaurolophus*. (c) Would that make the female's fundamental frequency higher or lower than the male's?

•39   A violin string 15.0 cm long and fixed at both ends oscillates in its n = 1 mode. The speed of waves on the string is 250 m/s, and the speed of sound in air in 348 m/s. What are the (a) frequency and (b) wavelength of the emitted sound wave?

•40   A sound wave in a fluid medium is reflected at a barrier so that a standing wave is formed. The distance between nodes is 3.8 cm, and the speed of propagation is 1500 m/s. Find the frequency of the sound wave.

•41   In pipe A, the ratio of a particular harmonic frequency to the next lower harmonic frequency is 1.2. In pipe B, the ratio of a particular harmonic frequency to the next lower harmonic frequency is 1.4. How many open ends are in (a) pipe A and (b) pipe B?

•42   Organ pipe A, with both ends open, has a fundamental frequency of 300 Hz. The third harmonic of organ pipe B, with one end open, has the same frequency as the second harmonic of pipe A. How long are (a) pipe A and (b) pipe B?

•43   (a) Find the speed of waves on a violin string of mass 800 mg and length 22.0 cm if the fundamental frequency is 920 Hz. (b) What is the tension in the string? For the fundamental, what is the wavelength of (c) the waves on the string and (d) the sound waves emitted by the string?   **SSM ILW**

•44   The water level in a vertical glass tube 1.00 m long can be adjusted to any position in the tube. A tuning fork vibrating at 686 Hz is held just over the open top end of the tube, to set up a standing wave of sound in the air-filled top portion of the tube. (That air-filled top portion acts as a tube with one end closed and the other end open.) (a) For how many different positions of the water level will sound from the fork set up resonance in the tube's air-filled portion, which acts as a pipe with one end closed (by the water) and the other end open? What are the (b) least and (c) second least water heights in the tube for resonance to occur?

•45   In Fig. 17-41, S is a small loudspeaker driven by an audio oscillator with a frequency that is varied from 1000 Hz to 2000 Hz, and *D* is a cylindrical pipe with two open ends and a length of 45.7 cm. The speed of sound in the air-filled pipe is 344 m/s. (a) At how many frequencies does the sound from the loudspeaker set up resonance in the pipe? What are the (b) lowest and (c) second lowest frequencies at which resonance occurs? SSM

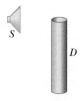

FIG. 17-41
Problem 45.

••46 One of the harmonic frequencies of tube *A* with two open ends is 325 Hz. The next-highest harmonic frequency is 390 Hz. (a) What harmonic frequency is next highest after the harmonic frequency 195 Hz? (b) What is the number of this next-highest harmonic?

One of the harmonic frequencies of tube *B* with only one open end is 1080 Hz. The next-highest harmonic frequency is 1320 Hz. (c) What harmonic frequency is next highest after the harmonic frequency 600 Hz? (d) What is the number of this next-highest harmonic?

••47 A violin string 30.0 cm long with linear density 0.650 g/m is placed near a loudspeaker that is fed by an audio oscillator of variable frequency. It is found that the string is set into oscillation only at the frequencies 880 and 1320 Hz as the frequency of the oscillator is varied over the range 500–1500 Hz. What is the tension in the string? SSM

••48 A tube 1.20 m long is closed at one end. A stretched wire is placed near the open end. The wire is 0.330 m long and has a mass of 9.60 g. It is fixed at both ends and oscillates in its fundamental mode. By resonance, it sets the air column in the tube into oscillation at that column's fundamental frequency. Find (a) that frequency and (b) the tension in the wire. GO

••49 A well with vertical sides and water at the bottom resonates at 7.00 Hz and at no lower frequency. (The air-filled portion of the well acts as a tube with one closed end and one open end.) The air in the well has a density of 1.10 kg/m³ and a bulk modulus of $1.33 \times 10^5$ Pa. How far down in the well is the water surface?

••50 Pipe *A*, which is 1.20 m long and open at both ends, oscillates at its third lowest harmonic frequency. It is filled with air for which the speed of sound is 343 m/s. Pipe *B*, which is closed at one end, oscillates at its second lowest harmonic frequency. This frequency of *B* happens to match the frequency of *A*. An *x* axis extends along the interior of *B*, with *x* = 0 at the closed end. (a) How many nodes are along that axis? What are the (b) smallest and (c) second smallest value of *x* locating those nodes? (d) What is the fundamental frequency of *B*? GO

### sec. 17-8 Beats

•51 The A string of a violin is a little too tightly stretched. Beats at 4.00 per second are heard when the string is sounded together with a tuning fork that is oscillating accurately at concert A (440 Hz). What is the period of the violin string oscillation?

•52 A tuning fork of unknown frequency makes 3.00 beats per second with a standard fork of frequency 384 Hz. The beat frequency decreases when a small piece of wax is put on a prong of the first fork. What is the frequency of this fork?

••53 Two identical piano wires have a fundamental frequency of 600 Hz when kept under the same tension. What

fractional increase in the tension of one wire will lead to the occurrence of 6.0 beats/s when both wires oscillate simultaneously? SSM

••54 You have five tuning forks that oscillate at close but different frequencies. What are the (a) maximum and (b) minimum number of different beat frequencies you can produce by sounding the forks two at a time, depending on how the frequencies differ?

### sec. 17-9 The Doppler Effect

•55 A state trooper chases a speeder along a straight road; both vehicles move at 160 km/h. The siren on the trooper's vehicle produces sound at a frequency of 500 Hz. What is the Doppler shift in the frequency heard by the speeder?

•56 An ambulance with a siren emitting a whine at 1600 Hz overtakes and passes a cyclist pedaling a bike at 2.44 m/s. After being passed, the cyclist hears a frequency of 1590 Hz. How fast is the ambulance moving?

•57 A whistle of frequency 540 Hz moves in a circle of radius 60.0 cm at an angular speed of 15.0 rad/s. What are the (a) lowest and (b) highest frequencies heard by a listener a long distance away, at rest with respect to the center of the circle? ILW

••58 A stationary motion detector sends sound waves of frequency 0.150 MHz toward a truck approaching at a speed of 45.0 m/s. What is the frequency of the waves reflected back to the detector?

••59 An acoustic burglar alarm consists of a source emitting waves of frequency 28.0 kHz. What is the beat frequency between the source waves and the waves reflected from an intruder walking at an average speed of 0.950 m/s directly away from the alarm? ILW

••60 A sound source *A* and a reflecting surface *B* move directly toward each other. Relative to the air, the speed of source *A* is 29.9 m/s, the speed of surface *B* is 65.8 m/s, and the speed of sound is 329 m/s. The source emits waves at frequency 1200 Hz as measured in the source frame. In the reflector frame, what are the (a) frequency and (b) wavelength of the arriving sound waves? In the source frame, what are the (c) frequency and (d) wavelength of the sound waves reflected back to the source?

••61 In Fig. 17-42, a French submarine and a U.S. submarine move toward each other during maneuvers in motionless water in the North Atlantic. The French sub moves at speed $v_F$ = 50.00 km/h, and the U.S. sub at $v_{US}$ = 70.00 km/h. The French sub sends out a sonar signal (sound wave in water) at $1.000 \times 10^3$ Hz. Sonar waves travel at 5470 km/h. (a) What is the signal's frequency as detected by the U.S. sub? (b) What frequency is detected by the French sub in the signal reflected back to it by the U.S. sub? GO

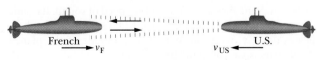

French $v_F$    $v_{US}$ U.S.

FIG. 17-42 Problem 61.

••62 A stationary detector measures the frequency of a sound source that first moves at constant velocity directly toward the detector and then (after passing the detector) directly away from it. The emitted frequency is $f$. During the approach the detected frequency is $f'_{app}$ and during the recession it is $f'_{rec}$. If $(f'_{app} - f'_{rec})/f = 0.500$, what is the ratio $v_s/v$ of the speed of the source to the speed of sound?

••**63** A bat is flitting about in a cave, navigating via ultrasonic bleeps. Assume that the sound emission frequency of the bat is 39 000 Hz. During one fast swoop directly toward a flat wall surface, the bat is moving at 0.025 times the speed of sound in air. What frequency does the bat hear reflected off the wall?

••**64** Figure 17-43 shows four tubes with lengths 1.0 m or 2.0 m, with one or two open ends as drawn. The third harmonic is set up in each tube, and some of the sound that escapes from them is detected by detector $D$, which moves directly away from the tubes. In terms of the speed of sound $v$, what speed must the detector have such that the detected frequency of the sound from (a) tube 1, (b) tube 2, (c) tube 3, and (d) tube 4 is equal to the tube's fundamental frequency?

**FIG. 17-43** Problem 64.

•••**65** A girl is sitting near the open window of a train that is moving at a velocity of 10.00 m/s to the east. The girl's uncle stands near the tracks and watches the train move away. The locomotive whistle emits sound at frequency 500.0 Hz. The air is still. (a) What frequency does the uncle hear? (b) What frequency does the girl hear? A wind begins to blow from the east at 10.00 m/s. (c) What frequency does the uncle now hear? (d) What frequency does the girl now hear? SSM WWW

•••**66** Two trains are traveling toward each other at 30.5 m/s relative to the ground. One train is blowing a whistle at 500 Hz. (a) What frequency is heard on the other train in still air? (b) What frequency is heard on the other train if the wind is blowing at 30.5 m/s toward the whistle and away from the listener? (c) What frequency is heard if the wind direction is reversed?

•••**67** A 2000 Hz siren and a civil defense official are both at rest with respect to the ground. What frequency does the official hear if the wind is blowing at 12 m/s (a) from source to official and (b) from official to source? GO

**sec. 17-10 Supersonic Speeds, Shock Waves**

•**68** The shock wave off the cockpit of the FA 18 in Fig. 17-24 has an angle of about 60°. The airplane was traveling at about 1350 km/h when the photograph was taken. Approximately what was the speed of sound at the airplane's altitude?

••**69** A jet plane passes over you at a height of 5000 m and a speed of Mach 1.5. (a) Find the Mach cone angle (the sound speed is 331 m/s). (b) How long after the jet passes directly overhead does the shock wave reach you? SSM

••**70** A plane flies at 1.25 times the speed of sound. Its sonic boom reaches a man on the ground 1.00 min after the plane passes directly overhead. What is the altitude of the plane? Assume the speed of sound to be 330 m/s.

**Additional Problems**

**71** In Fig. 17-44, sound of wavelength 0.850 m is emitted isotropically by point source $S$. Sound ray 1 extends directly to detector $D$, at distance $L = 10.0$ m. Sound ray 2 extends to $D$

via a reflection (effectively, a "bouncing") of the sound at a flat surface. That reflection occurs on a perpendicular bisector to the $SD$ line, at distance $d$ from the line. Assume that the reflection shifts the sound wave by $0.500\lambda$. For what least value of $d$ (other than zero) do the direct sound and the reflected sound arrive at $D$ (a) exactly out of phase and (b) exactly in phase?

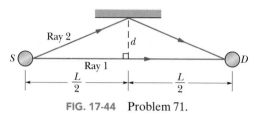

**FIG. 17-44** Problem 71.

**72** A detector initially moves at constant velocity directly toward a stationary sound source and then (after passing it) directly from it. The emitted frequency is $f$. During the approach the detected frequency is $f'_{app}$ and during the recession it is $f'_{rec}$. If the frequencies are related by $(f'_{app} - f'_{rec})/f = 0.500$, what is the ratio $v_D/v$ of the speed of the detector to the speed of sound?

**73** Two sound waves with an amplitude of 12 nm and a wavelength of 35 cm travel in the same direction through a long tube, with a phase difference of $\pi/3$ rad. What are the (a) amplitude and (b) wavelength of the net sound wave produced by their interference? If, instead, the sound waves travel through the tube in opposite directions, what are the (c) amplitude and (d) wavelength of the net wave?

**74** A sinusoidal sound wave moves at 343 m/s through air in the positive direction of an $x$ axis. At one instant, air molecule $A$ is at its maximum displacement in the negative direction of the axis while air molecule $B$ is at its equilibrium position. The separation between those molecules is 15.0 cm, and the molecules between $A$ and $B$ have intermediate displacements in the negative direction of the axis. (a) What is the frequency of the sound wave?

In a similar arrangement, for a different sinusoidal sound wave, air molecule $C$ is at its maximum displacement in the positive direction while molecule $D$ is at its maximum displacement in the negative direction. The separation between the molecules is again 15.0 cm, and the molecules between $C$ and $D$ have intermediate displacements. (b) What is the frequency of the sound wave?

**75** In Fig. 17-45, sound waves $A$ and $B$, both of wavelength $\lambda$, are initially in phase and traveling rightward, as indicated by the two rays. Wave $A$ is reflected from four surfaces but ends up traveling in its original direction. What multiple of wavelength $\lambda$ is the smallest value of distance $L$ in the figure that puts $A$ and $B$ exactly out of phase with each other after the reflections?

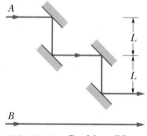

**FIG. 17-45** Problem 75.

**76** A trumpet player on a moving railroad flatcar moves toward a second trumpet player standing alongside the track while both play a 440 Hz note. The sound waves heard by a stationary observer between the two players have a beat frequency of 4.0 beats/s. What is the flatcar's speed?

**77** A siren emitting a sound of frequency 1000 Hz moves away from you toward the face of a cliff at a speed of 10 m/s. Take the speed of sound in air as 330 m/s. (a) What is the frequency of the sound you hear coming directly from the siren? (b) What is the frequency of the sound you hear reflected off the cliff? (c) What is the beat frequency between the two sounds? Is it perceptible (less than 20 Hz)? **SSM**

**78** A sound source moves along an *x* axis, between detectors *A* and *B*. The wavelength of the sound detected at *A* is 0.500 that of the sound detected at *B*. What is the ratio $v_s/v$ of the speed of the source to the speed of sound?

**79** A certain loudspeaker system emits sound isotropically with a frequency of 2000 Hz and an intensity of 0.960 mW/m² at a distance of 6.10 m. Assume that there are no reflections. (a) What is the intensity at 30.0 m? At 6.10 m, what are (b) the displacement amplitude and (c) the pressure amplitude?

**80** At a certain point, two waves produce pressure variations given by $\Delta p_1 = \Delta p_m \sin \omega t$ and $\Delta p_2 = \Delta p_m \sin(\omega t - \phi)$. At this point, what is the ratio $\Delta p_r/\Delta p_m$, where $\Delta p_r$ is the pressure amplitude of the resultant wave, if $\phi$ is (a) 0, (b) $\pi/2$, (c) $\pi/3$, and (d) $\pi/4$?

**81** The sound intensity is 0.0080 W/m² at a distance of 10 m from an isotropic point source of sound. (a) What is the power of the source? (b) What is the sound intensity 5.0 m from the source? (c) What is the sound level 10 m from the source? **SSM**

**82** The average density of Earth's crust 10 km beneath the continents is 2.7 g/cm³. The speed of longitudinal seismic waves at that depth, found by timing their arrival from distant earthquakes, is 5.4 km/s. Use this information to find the bulk modulus of Earth's crust at that depth. For comparison, the bulk modulus of steel is about $16 \times 10^{10}$ Pa.

**83** Two identical tuning forks can oscillate at 440 Hz. A person is located somewhere on the line between them. Calculate the beat frequency as measured by this individual if (a) she is standing still and the tuning forks move in the same direction along the line at 3.00 m/s, and (b) the tuning forks are stationary and the listener moves along the line at 3.00 m/s.

**84** You can estimate your distance from a lightning stroke by counting the seconds between the flash you see and the thunder you later hear. By what integer should you divide the number of seconds to get the distance in kilometers?

**85** (a) If two sound waves, one in air and one in (fresh) water, are equal in intensity and angular frequency, what is the ratio of the pressure amplitude of the wave in water to that of the wave in air? Assume the water and the air are at 20°C. (See Table 14-1.) (b) If the pressure amplitudes are equal instead, what is the ratio of the intensities of the waves? **SSM**

**86** Find the ratios (greater to smaller) of the (a) intensities, (b) pressure amplitudes, and (c) particle displacement amplitudes for two sounds whose sound levels differ by 37 dB.

**87** Figure 17-46 shows an air-filled, acoustic interferometer, used to demonstrate the interference of sound waves. Sound source *S* is an oscillating diaphragm; *D* is a sound detector, such as the ear or a microphone. Path *SBD* can be varied in length, but path *SAD* is fixed. At *D*, the sound wave coming along path *SBD* interferes with that coming along path *SAD*. In one demonstration, the sound intensity at *D* has a minimum value of 100 units at one position of the movable arm and con-

tinuously climbs to a maximum value of 900 units when that arm is shifted by 1.65 cm. Find (a) the frequency of the sound emitted by the source and (b) the ratio of the amplitude at *D* of the *SAD* wave to that of the *SBD* wave. (c) How can it happen that these waves have different amplitudes, considering that they originate at the same source? **SSM**

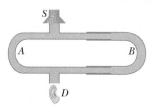
**FIG. 17-46** Problem 87.

**88** A bullet is fired with a speed of 685 m/s. Find the angle made by the shock cone with the line of motion of the bullet.

**89** A sperm whale (Fig. 17-47a) vocalizes by producing a series of clicks. Actually, the whale makes only a single sound near the front of its head to start the series. Part of that sound then emerges from the head into the water to become the first click of the series. The rest of the sound travels backward through the spermaceti sac (a body of fat), reflects from the frontal sac (an air layer), and then travels forward through the spermaceti sac. When it reaches the distal sac (another air layer) at the front of the head, some of the sound escapes into the water to form the second click, and the rest is sent back through the spermaceti sac (and ends up forming later clicks).

Figure 17-47b shows a strip-chart recording of a series of clicks. A unit time interval of 1.0 ms is indicated on the chart. Assuming that the speed of sound in the spermaceti sac is 1372 m/s, find the length of the spermaceti sac. From such a calculation, marine scientists estimate the length of a whale from its click series.

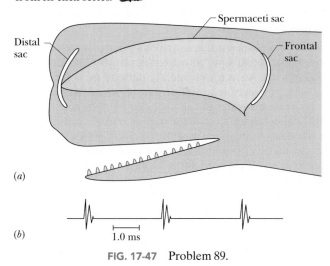
**FIG. 17-47** Problem 89.

**90** A continuous sinusoidal longitudinal wave is sent along a very long coiled spring from an attached oscillating source. The wave travels in the negative direction of an *x* axis; the source frequency is 25 Hz; at any instant the distance between successive points of maximum expansion in the spring is 24 cm; the maximum longitudinal displacement of a spring particle is 0.30 cm; and the particle at *x* = 0 has zero displacement at time *t* = 0. If the wave is written in the form $s(x, t) = s_m \cos(kx \pm \omega t)$, what are (a) $s_m$, (b) $k$, (c) $\omega$, (d) the wave speed, and (e) the correct choice of sign in front of $\omega$?

**91** At a distance of 10 km, a 100 Hz horn, assumed to be an isotropic point source, is barely audible. At what distance would it begin to cause pain?

**92** The speed of sound in a certain metal is $v_m$. One end of a long pipe of that metal of length $L$ is struck a hard blow. A listener at the other end hears two sounds, one from the wave that travels along the pipe's metal wall and the other from the wave that travels through the air inside the pipe. (a) If $v$ is the speed of sound in air, what is the time interval $\Delta t$ between the arrivals of the two sounds at the listener's ear? (b) If $\Delta t = 1.00$ s and the metal is steel, what is the length $L$?

**93** A pipe 0.60 m long and closed at one end is filled with an unknown gas. The third lowest harmonic frequency for the pipe is 750 Hz. (a) What is the speed of sound in the unknown gas? (b) What is the fundamental frequency for this pipe when it is filled with the unknown gas?

**94** Four sound waves are to be sent through the same tube of air, in the same direction:

$$s_1(x, t) = (9.00 \text{ nm}) \cos(2\pi x - 700\pi t)$$
$$s_2(x, t) = (9.00 \text{ nm}) \cos(2\pi x - 700\pi t + 0.7\pi)$$
$$s_3(x, t) = (9.00 \text{ nm}) \cos(2\pi x - 700\pi t + \pi)$$
$$s_4(x, t) = (9.00 \text{ nm}) \cos(2\pi x - 700\pi t + 1.7\pi).$$

What is the amplitude of the resultant wave? (*Hint:* Use a phasor diagram to simplify the problem.)

**95** Straight line $AB$ connects two point sources that are 5.00 m apart, emit 300 Hz sound waves of the same amplitude, and emit exactly out of phase. (a) What is the shortest distance between the midpoint of $AB$ and a point on $AB$ where the interfering waves cause maximum oscillation of the air molecules? What are the (b) second and (c) third shortest distances?

**96** A point source that is stationary on an $x$ axis emits a sinusoidal sound wave at a frequency of 686 Hz and speed 343 m/s. The wave travels radially outward from the source, causing air molecules to oscillate radially inward and outward. Let us define a wavefront as a line that connects points where the air molecules have the maximum, radially outward displacement. At any given instant, the wavefronts are concentric circles that are centered on the source. (a) Along $x$, what is the adjacent wavefront separation? Next, the source moves along $x$ at a speed of 110 m/s. Along $x$, what are the wavefront separations (b) in front of and (c) behind the source?

**97** You are standing at a distance $D$ from an isotropic point source of sound. You walk 50.0 m toward the source and observe that the intensity of the sound has doubled. Calculate the distance $D$.

**98** On July 10, 1996, a granite block broke away from a wall in Yosemite Valley and, as it began to slide down the wall, was launched into projectile motion. Seismic waves produced by its impact with the ground triggered seismographs as far away as 200 km. Later measurements indicated that the block had a mass between $7.3 \times 10^7$ kg and $1.7 \times 10^8$ kg and that it landed 500 m vertically below the launch point and 30 m horizontally from it. (The launch angle is not known.) (a) Estimate the block's kinetic energy just before it landed.

Consider two types of seismic waves that spread from the impact point—a hemispherical *body wave* traveled through the ground in an expanding hemisphere and a cylindrical *surface wave* traveled along the ground in an expanding shallow vertical cylinder (Fig. 17-48). Assume that the impact lasted 0.50 s, the vertical cylinder had a depth $d$ of 5.0 m, and each wave type received 20% of the energy the block had just before impact. Neglecting any mechanical energy loss the waves experienced as they traveled, determine the intensities of (b) the body wave and (c) the surface wave when they reached a seismograph 200 km away. (d) On the basis of these results, which wave is more easily detected on a distant seismograph?

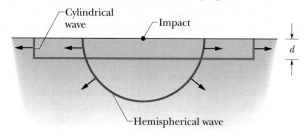

**FIG. 17-48** Problem 98.

**99** An avalanche of sand along some rare desert sand dunes can produce a booming that is loud enough to be heard 10 km away. The booming apparently results from a periodic oscillation of the sliding layer of sand—the layer's thickness expands and contracts. If the emitted frequency is 90 Hz, what are (a) the period of the thickness oscillation and (b) the wavelength of the sound?

**100** Passengers in an auto traveling at 16.0 m/s toward the east hear a siren frequency of 950 Hz from an emergency vehicle approaching them from behind at a speed (relative to the air and ground) of 40.0 m/s. The speed of sound in air is 340 m/s. (a) What siren frequency does a passenger riding in the emergency vehicle hear? (b) What frequency do the passengers in the auto hear after the emergency vehicle passes them?

**101** Ultrasound, which consists of sound waves with frequencies above the human audible range, can be used to produce an image of the interior of a human body. Moreover, ultrasound can be used to measure the speed of the blood in the body; it does so by comparing the frequency of the ultrasound sent into the body with the frequency of the ultrasound reflected back to the body's surface by the blood. As the blood pulses, this detected frequency varies.

Suppose that an ultrasound image of the arm of a patient shows an artery that is angled at $\theta = 20°$ to the ultrasound's line of travel (Fig. 17-49). Suppose also that the frequency of the ultrasound reflected by the blood in the artery is increased by a maximum of 5495 Hz from the original ultrasound frequency of 5.000 000 MHz. (a) In Fig. 17-49, is the direction of the blood flow rightward or leftward? (b) The speed of sound in the human arm is 1540 m/s. What is the maximum speed of the blood? (*Hint:* The Doppler effect is caused by the component of the blood's velocity along the ultrasound's direction of travel.) (c) If angle $\theta$ were greater, would the reflected frequency be greater or less? SSM

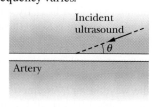

**FIG. 17-49** Problem 101.

**102** Pipe $A$ has only one open end; pipe $B$ is four times as long and has two open ends. Of the lowest 10 harmonic numbers $n_B$ of pipe $B$, what are the (a) smallest, (b) second smallest, and (c)

third smallest values at which a harmonic frequency of $B$ matches one of the harmonic frequencies of $A$?

**103** *Waterfall acoustics.* The turbulent impact of the water in a waterfall causes the surrounding ground to oscillate in a wide range of low frequencies. If the water falls freely (instead of hitting rock on the way down), the oscillations are greatest in amplitude at a particular frequency $f_m$. This fact suggests that acoustic resonance is involved and $f_m$ is the fundamental frequency. The following table gives, for nine U.S. and Canadian waterfalls, measured values for $f_m$ and for the length $L$ of the water's free fall. Determine how to plot the data to get the speed of sound in the water of a waterfall. From the plot, find the speed of sound if waterfall resonance is effectively like that in a tube with (a) two open ends and (b) only one open end. The speed of sound in turbulent water filled with air bubbles can be about 25% less than the speed of 1400 m/s in still water. (c) From the answers to (a) and (b), determine how many open ends are effectively involved in waterfall resonance.

| WATERFALL | 1 | 2 | 3 | 4 | 5 | 6 | 7 | 8 | 9 |
|---|---|---|---|---|---|---|---|---|---|
| $f_m$ (Hz) | 5.6 | 3.8 | 8.0 | 6.1 | 8.8 | 6.0 | 19 | 21 | 40 |
| $L$ (m) | 97 | 71 | 53 | 49 | 35 | 24 | 13 | 11 | 8 |

**104** A person on a railroad car blows a trumpet note at 440 Hz. The car is moving toward a wall at 20.0 m/s. Calculate the frequency of (a) the sound as received at the wall and (b) the reflected sound arriving back at the trumpeter.

**105** A police car is chasing a speeding Porsche 911. Assume that the Porsche's maximum speed is 80.0 m/s and the police car's is 54.0 m/s. At the moment both cars reach their maximum speed, what frequency will the Porsche driver hear if the frequency of the police car's siren is 440 Hz? Take the speed of sound in air to be 340 m/s.

**106** A sound wave travels out uniformly in all directions from a point source. (a) Justify the following expression for the displacement $s$ of the transmitting medium at any distance $r$ from the source:

$$s = \frac{b}{r} \sin k(r - vt),$$

where $b$ is a constant. Consider the speed, direction of propagation, periodicity, and intensity of the wave. (b) What is the dimension of the constant $b$?

**107** In Fig. 17-34, $S_1$ and $S_2$ are two isotropic point sources of sound. They emit waves in phase at wavelength 0.50 m; they are separated by $D = 1.60$ m. If we move a sound detector along a large circle centered at the midpoint between the sources, at how many points do waves arrive at the detector (a) exactly in phase and (b) exactly out of phase?

**108** Suppose a spherical loudspeaker emits sound isotropically at 10 W into a room with completely absorbent walls, floor, and ceiling (an *anechoic chamber*). (a) What is the intensity of the sound at distance $d = 3.0$ m from the center of the source? (b) What is the ratio of the wave amplitude at $d = 4.0$ m to that at $d = 3.0$ m?

**109** To search for a fossilized dinosaur embedded in rock, paleontologists can use sound waves to produce a computer image of the dinosaur. The image then guides the paleontolo-

gists as they dig the dinosaur out of the rock. (The technique is shown in the opening scenes of the movie *Jurassic Park*.) The basic idea of the detection technique is that a strong pulse of sound is emitted by a source (a seismic gun) at ground level and then detected by hydrophones that lie at evenly spaced depths in a bore hole drilled into the ground. The source and one hydrophone are shown in Fig. 17-50.

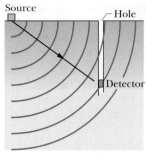

**FIG. 17-50** Problem 109.

If the sound wave travels from the source to the hydrophone through only rock as in Fig. 17-50, it travels at a known speed $V$ and takes a certain time $T$. If, instead, it travels through a fossilized bone along the way, it takes slightly more time because it travels more slowly in the bone than in the rock. By measuring the difference $\Delta t$ between the expected and measured travel times, the distance $d$ traveled in the bone can be determined. After this procedure is repeated for many locations of the source and hydrophones, a computer can transform the many computed distances $d$ into an image of the fossil.

(a) Let the speed of sound through fossilized bone be $V - \Delta V$, where $\Delta V$ is small relative to $V$. Show that the distance $d$ is given by

$$d \approx \frac{V^2 \, \Delta t}{\Delta V}.$$

(b) For $V = 5000$ m/s and $\Delta V = 200$ m/s, what typical value of $\Delta t$ can be expected if the sound passes along the diameter of a leg bone of an adult *T. rex*? (Estimate the bone's diameter.)

**110** The period of a pulsating variable star may be estimated by considering the star to be executing *radial* longitudinal pulsations in the fundamental standing wave mode; that is, the star's radius varies periodically with time, with a displacement antinode at the star's surface. (a) Would you expect the center of the star to be a displacement node or antinode? (b) By analogy with a pipe with one open end, show that the period of pulsation $T$ is given by $T = 4R/v$, where $R$ is the equilibrium radius of the star and $v$ is the average sound speed in the material of the star. (c) Typical white dwarf stars are composed of material with a bulk modulus of $1.33 \times 10^{22}$ Pa and a density of $10^{10}$ kg/m³. They have radii equal to $9.0 \times 10^{-3}$ solar radius. What is the approximate pulsation period of a white dwarf?

**111** A listener at rest (with respect to the air and the ground) hears a signal of frequency $f_1$ from a source moving toward him with a velocity of 15 m/s, due east. If the listener then moves toward the approaching source with a velocity of 25 m/s, due west, he hears a frequency $f_2$ that differs from $f_1$ by 37 Hz. What is the frequency of the source? (Take the speed of sound in air to be 340 m/s.)

**112** A guitar player tunes the fundamental frequency of a guitar string to 440 Hz. (a) What will be the fundamental frequency if she then increases the tension in the string by 20%? (b) What will it be if, instead, she decreases the length along which the string oscillates by sliding her finger from the tuning key one-third of the way down the string toward the bridge at the lower end?

# 18 Temperature, Heat, and the First Law of Thermodynamics

*The fairly small Melanophila beetles are known for a bizarre behavior: They fly toward forest fires and mate near them, and then the females fly into the still smoldering ruins to lay their eggs under burnt bark. This is the ideal environment for the larvae that hatch from the eggs, because the tree can no longer protect itself from the larvae by rosin or chemical means. If a beetle were at the periphery of a fire, detecting the fire would be easy, of course. However, these beetles can detect a fairly large fire from as far away as 12 km. They do this without seeing or smelling the fire.*

## How can the beetles detect a distant fire?

The answer is in this chapter.

Courtesy Nathan Schiff, Ph. D., USDA Forest Service, Center for Bottomland Hardwoods Research, Stoneville, MS

## 18-1 | WHAT IS PHYSICS?

One of the principal branches of physics and engineering is **thermodynamics,** which is the study and application of the *thermal energy* (often called the *internal energy*) of systems. One of the central concepts of thermodynamics is *temperature,* which we begin to explore in the next section. Since childhood, you have been developing a working knowledge of thermal energy and temperature. For example, you know to be cautious with hot foods and hot stoves and to store perishable foods in cool or cold compartments. You also know how to control the temperature inside home and car, and how to protect yourself from wind chill and heat stroke.

Examples of how thermodynamics figures into everyday engineering and science are countless. Automobile engineers are concerned with the heating of a car engine, such as during a NASCAR race. Food engineers are concerned both with the proper heating of foods, such as pizzas being microwaved, and with the proper cooling of foods, such as TV dinners being quickly frozen at a processing plant. Geologists are concerned with the transfer of thermal energy in an El Niño event and in the gradual warming of ice expanses in the Arctic and Antarctic. Agricultural engineers are concerned with the weather conditions that determine whether the agriculture of a country thrives or vanishes. Medical engineers are concerned with how a patient's temperature might distinguish between a benign viral infection and a cancerous growth.

The starting point in our discussion of thermodynamics is the concept of temperature and how it is measured.

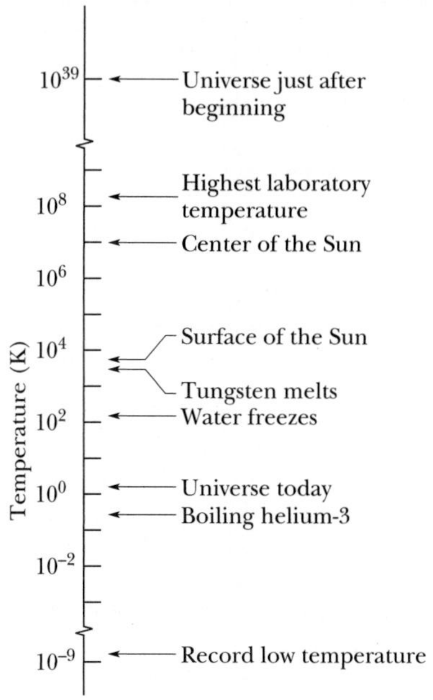

**FIG. 18-1** Some temperatures on the Kelvin scale. Temperature $T = 0$ corresponds to $10^{-\infty}$ and cannot be plotted on this logarithmic scale.

## 18-2 | Temperature

Temperature is one of the seven SI base quantities. Physicists measure temperature on the **Kelvin scale,** which is marked in units called *kelvins.* Although the temperature of a body apparently has no upper limit, it does have a lower limit; this limiting low temperature is taken as the zero of the Kelvin temperature scale. Room temperature is about 290 kelvins, or 290 K as we write it, above this *absolute zero.* Figure 18-1 shows a wide range of temperatures.

When the universe began 13.7 billion years ago, its temperature was about $10^{39}$ K. As the universe expanded it cooled, and it has now reached an average temperature of about 3 K. We on Earth are a little warmer than that because we happen to live near a star. Without our Sun, we too would be at 3 K (or, rather, we could not exist).

## 18-3 | The Zeroth Law of Thermodynamics

The properties of many bodies change as we alter their temperature, perhaps by moving them from a refrigerator to a warm oven. To give a few examples: As their temperature increases, the volume of a liquid increases, a metal rod grows a little longer, and the electrical resistance of a wire increases, as does the pressure exerted by a confined gas. We can use any one of these properties as the basis of an instrument that will help us pin down the concept of temperature.

Figure 18-2 shows such an instrument. Any resourceful engineer could design and construct it, using any one of the properties listed above. The instrument is fitted with a digital readout display and has the following properties: If you heat it (say, with a Bunsen burner), the displayed number starts to increase; if you then put it into a refrigerator, the displayed number starts to decrease. The instrument is not calibrated in any way, and the numbers have (as yet) no physical meaning. The device is a *thermoscope* but not (as yet) a *thermometer.*

Suppose that, as in Fig. 18-3a, we put the thermoscope (which we shall call body $T$) into intimate contact with another body (body $A$). The entire system is confined within a thick-walled insulating box. The numbers displayed by the

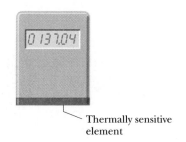

**FIG. 18-2** A thermoscope. The numbers increase when the device is heated and decrease when it is cooled. The thermally sensitive element could be—among many possibilities—a coil of wire whose electrical resistance is measured and displayed.

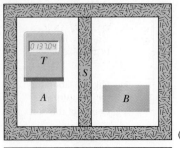

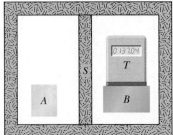

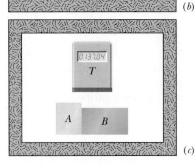

**FIG. 18-3** (*a*) Body *T* (a thermoscope) and body *A* are in thermal equilibrium. (Body *S* is a thermally insulating screen.) (*b*) Body *T* and body *B* are also in thermal equilibrium, at the same reading of the thermoscope. (*c*) If (*a*) and (*b*) are true, the zeroth law of thermodynamics states that body *A* and body *B* are also in thermal equilibrium.

thermoscope roll by until, eventually, they come to rest (let us say the reading is "137.04") and no further change takes place. In fact, we suppose that every measurable property of body *T* and of body *A* has assumed a stable, unchanging value. Then we say that the two bodies are in *thermal equilibrium* with each other. Even though the displayed readings for body *T* have not been calibrated, we conclude that bodies *T* and *A* must be at the same (unknown) temperature.

Suppose that we next put body *T* into intimate contact with body *B* (Fig. 18-3*b*) and find that the two bodies come to thermal equilibrium *at the same reading of the thermoscope*. Then bodies *T* and *B* must be at the same (still unknown) temperature. If we now put bodies *A* and *B* into intimate contact (Fig. 18-3*c*), are they immediately in thermal equilibrium with each other? Experimentally, we find that they are.

The experimental fact shown in Fig. 18-3 is summed up in the **zeroth law of thermodynamics:**

> If bodies *A* and *B* are each in thermal equilibrium with a third body *T*, then *A* and *B* are in thermal equilibrium with each other.

In less formal language, the message of the zeroth law is: "Every body has a property called **temperature.** When two bodies are in thermal equilibrium, their temperatures are equal. And vice versa." We can now make our thermoscope (the third body *T*) into a thermometer, confident that its readings will have physical meaning. All we have to do is calibrate it.

We use the zeroth law constantly in the laboratory. If we want to know whether the liquids in two beakers are at the same temperature, we measure the temperature of each with a thermometer. We do not need to bring the two liquids into intimate contact and observe whether they are or are not in thermal equilibrium.

The zeroth law, which has been called a logical afterthought, came to light only in the 1930s, long after the first and second laws of thermodynamics had been discovered and numbered. Because the concept of temperature is fundamental to those two laws, the law that establishes temperature as a valid concept should have the lowest number—hence the zero.

## 18-4 | Measuring Temperature

Here we first define and measure temperatures on the Kelvin scale. Then we calibrate a thermoscope so as to make it a thermometer.

### The Triple Point of Water

To set up a temperature scale, we pick some reproducible thermal phenomenon and, quite arbitrarily, assign a certain Kelvin temperature to its environment; that is, we select a *standard fixed point* and give it a standard fixed-point *temperature.* We could, for example, select the freezing point or the boiling point of water but, for technical reasons, we select instead the **triple point of water.**

Liquid water, solid ice, and water vapor (gaseous water) can coexist, in thermal equilibrium, at only one set of values of pressure and temperature. Figure 18-4 shows a triple-point cell, in which this so-called triple point of water can be achieved in the laboratory. By international agreement, the triple point of water has been assigned a value of 273.16 K as the standard fixed-point temperature for the calibration of thermometers; that is,

$$T_3 = 273.16 \text{ K} \qquad \text{(triple-point temperature)}, \qquad (18\text{-}1)$$

in which the subscript 3 means "triple point." This agreement also sets the size of the kelvin as 1/273.16 of the difference between absolute zero and the triple-point temperature of water.

Note that we do not use a degree mark in reporting Kelvin temperatures. It is 300 K (not 300°K), and it is read "300 kelvins" (not "300 degrees Kelvin"). The usual SI prefixes apply. Thus, 0.0035 K is 3.5 mK. No distinction in nomenclature is made between Kelvin temperatures and temperature differences, so we can write, "the boiling point of sulfur is 717.8 K" and "the temperature of this water bath was raised by 8.5 K."

## The Constant-Volume Gas Thermometer

The standard thermometer, against which all other thermometers are calibrated, is based on the pressure of a gas in a fixed volume. Figure 18-5 shows such a **constant-volume gas thermometer;** it consists of a gas-filled bulb connected by a tube to a mercury manometer. By raising and lowering reservoir $R$, the mercury level in the left arm of the U-tube can always be brought to the zero of the scale to keep the gas volume constant (variations in the gas volume can affect temperature measurements).

The temperature of any body in thermal contact with the bulb (such as the liquid surrounding the bulb in Fig. 18-5) is then defined to be

$$T = Cp, \tag{18-2}$$

in which $p$ is the pressure exerted by the gas and $C$ is a constant. From Eq. 14-10, the pressure $p$ is

$$p = p_0 - \rho g h, \tag{18-3}$$

in which $p_0$ is the atmospheric pressure, $\rho$ is the density of the mercury in the manometer, and $h$ is the measured difference between the mercury levels in the two arms of the tube.* (The minus sign is used in Eq. 18-3 because pressure $p$ is measured *above* the level at which the pressure is $p_0$.)

If we next put the bulb in a triple-point cell (Fig. 18-4), the temperature now being measured is

$$T_3 = Cp_3, \tag{18-4}$$

in which $p_3$ is the gas pressure now. Eliminating $C$ between Eqs. 18-2 and 18-4 gives us the temperature as

$$T = T_3 \left( \frac{p}{p_3} \right) = (273.16 \text{ K}) \left( \frac{p}{p_3} \right) \quad \text{(provisional).} \tag{18-5}$$

We still have a problem with this thermometer. If we use it to measure, say, the boiling point of water, we find that different gases in the bulb give slightly different results. However, as we use smaller and smaller amounts of gas to fill the bulb, the readings converge nicely to a single temperature, no matter what gas we use. Figure 18-6 shows this convergence for three gases.

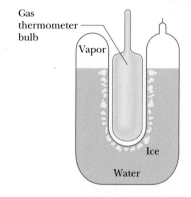

**FIG. 18-4** A triple-point cell, in which solid ice, liquid water, and water vapor coexist in thermal equilibrium. By international agreement, the temperature of this mixture has been defined to be 273.16 K. The bulb of a constant-volume gas thermometer is shown inserted into the well of the cell.

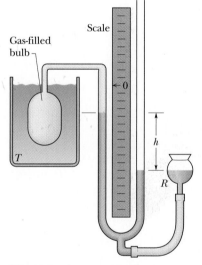

**FIG. 18-5** A constant-volume gas thermometer, its bulb immersed in a liquid whose temperature $T$ is to be measured.

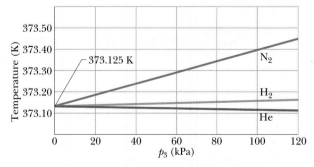

**FIG. 18-6** Temperatures measured by a constant-volume gas thermometer, with its bulb immersed in boiling water. For temperature calculations using Eq. 18-5, pressure $p_3$ was measured at the triple point of water. Three different gases in the thermometer bulb gave generally different results at different gas pressures, but as the amount of gas was decreased (decreasing $p_3$), all three curves converged to 373.125 K.

---

*For pressure units, we shall use units introduced in Section 14-3. The SI unit for pressure is the newton per square meter, which is called the pascal (Pa). The pascal is related to other common pressure units by

$$1 \text{ atm} = 1.01 \times 10^5 \text{ Pa} = 760 \text{ torr} = 14.7 \text{ lb/in.}^2.$$

Thus the recipe for measuring a temperature with a gas thermometer is

$$T = (273.16 \text{ K}) \left( \lim_{\text{gas} \to 0} \frac{p}{p_3} \right). \tag{18-6}$$

The recipe instructs us to measure an unknown temperature $T$ as follows: Fill the thermometer bulb with an arbitrary amount of *any* gas (for example, nitrogen) and measure $p_3$ (using a triple-point cell) and $p$, the gas pressure at the temperature being measured. (Keep the gas volume the same.) Calculate the ratio $p/p_3$. Then repeat both measurements with a smaller amount of gas in the bulb, and again calculate this ratio. Continue this way, using smaller and smaller amounts of gas, until you can extrapolate to the ratio $p/p_3$ that you would find if there were approximately no gas in the bulb. Calculate the temperature $T$ by substituting that extrapolated ratio into Eq. 18-6. (The temperature is called the *ideal gas temperature*.)

## 18-5 | The Celsius and Fahrenheit Scales

So far, we have discussed only the Kelvin scale, used in basic scientific work. In nearly all countries of the world, the Celsius scale (formerly called the centigrade scale) is the scale of choice for popular and commercial use and much scientific use. Celsius temperatures are measured in degrees, and the Celsius degree has the same size as the kelvin. However, the zero of the Celsius scale is shifted to a more convenient value than absolute zero. If $T_C$ represents a Celsius temperature and $T$ a Kelvin temperature, then

$$T_C = T - 273.15°. \tag{18-7}$$

In expressing temperatures on the Celsius scale, the degree symbol is commonly used. Thus, we write 20.00°C for a Celsius reading but 293.15 K for a Kelvin reading.

The Fahrenheit scale, used in the United States, employs a smaller degree than the Celsius scale and a different zero of temperature. You can easily verify both these differences by examining an ordinary room thermometer on which both scales are marked. The relation between the Celsius and Fahrenheit scales is

$$T_F = \tfrac{9}{5} T_C + 32°, \tag{18-8}$$

where $T_F$ is Fahrenheit temperature. Converting between these two scales can be done easily by remembering a few corresponding points, such as the freezing and boiling points of water (Table 18-1). Figure 18-7 compares the Kelvin, Celsius, and Fahrenheit scales.

We use the letters C and F to distinguish measurements and degrees on the two scales. Thus,

$$0°C = 32°F$$

**TABLE 18-1**

**Some Corresponding Temperatures**

| Temperature | °C | °F |
|---|---|---|
| Boiling point of water[a] | 100 | 212 |
| Normal body temperature | 37.0 | 98.6 |
| Accepted comfort level | 20 | 68 |
| Freezing point of water[a] | 0 | 32 |
| Zero of Fahrenheit scale | $\approx -18$ | 0 |
| Scales coincide | -40 | -40 |

[a]Strictly, the boiling point of water on the Celsius scale is 99.975°C, and the freezing point is 0.00°C. Thus, there is slightly less than 100 C° between those two points.

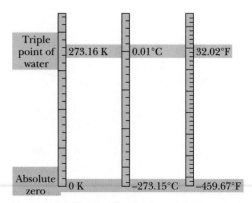

**FIG. 18-7**   The Kelvin, Celsius, and Fahrenheit temperature scales compared.

means that 0° on the Celsius scale measures the same temperature as 32° on the Fahrenheit scale, whereas

$$5 \, C° = 9 \, F°$$

means that a temperature difference of 5 Celsius degrees (note the degree symbol appears *after* C) is equivalent to a temperature difference of 9 Fahrenheit degrees.

✓ **CHECKPOINT 1**    The figure here shows three linear temperature scales with the freezing and boiling points of water indicated. (a) Rank the degrees on these scales by size, greatest first. (b) Rank the following temperatures, highest first: 50°X, 50°W, and 50°Y.

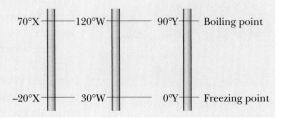

## Sample Problem  18-1

Suppose you come across old scientific notes that describe a temperature scale called Z on which the boiling point of water is 65.0°Z and the freezing point is −14.0°Z. To what temperature on the Fahrenheit scale would a temperature of $T = -98.0°Z$ correspond? Assume that the Z scale is linear; that is, the size of a Z degree is the same everywhere on the Z scale.

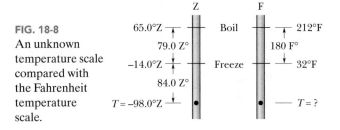

**FIG. 18-8** An unknown temperature scale compared with the Fahrenheit temperature scale.

**KEY IDEA**    A conversion factor between two (linear) temperature scales can be calculated by using two known (benchmark) temperatures, such as the boiling and freezing points of water. The number of degrees between the known temperatures on one scale is equivalent to the number of degrees between them on the other scale.

*Calculations:* We begin by relating the given temperature $T$ to *either* known temperature on the Z scale. Since $T = -98.0°Z$ is closer to the freezing point (−14.0°Z) than to the boiling point (65.0°Z), we use the freezing point. Then we note that $T$ is *below this point* by $-14.0°Z - (-98.0°Z) = 84.0 \, Z°$ (Fig. 18-8). (Read this difference as "84.0 Z degrees.")

Next, we set up a conversion factor between the Z and Fahrenheit scales to convert this difference. To do

so, we use *both* known temperatures on the Z scale and the corresponding temperatures on the Fahrenheit scale. On the Z scale, the difference between the boiling and freezing points is $65.0°Z - (-14.0°Z) = 79.0 \, Z°$. On the Fahrenheit scale, it is $212°F - 32.0°F = 180 \, F°$. Thus, a temperature difference of 79.0 Z° is equivalent to a temperature difference of 180 F° (Fig. 18-8), and we can use the ratio $(180 \, F°)/(79.0 \, Z°)$ as our conversion factor.

Now, since $T$ is below the freezing point by 84.0 Z°, it must also be below the freezing point by

$$(84.0 \, Z°) \, \frac{180 \, F°}{79.0 \, Z°} = 191 \, F°.$$

Because the freezing point is at 32.0°F, this means that

$$T = 32.0°F - 191 \, F° = -159°F. \quad \text{(Answer)}$$

## PROBLEM-SOLVING TACTICS

*Tactic 1:  Temperature Changes*    Between the boiling and freezing points of water, there are (approximately) 100 kelvins and 100 Celsius degrees. Thus, a kelvin is the same size as a Celsius degree. From this or from Eq. 18-7, we then know that any temperature change is the same number whether expressed in kelvins or Celsius degrees. For example, a temperature change of 10 K is exactly equivalent to a temperature change of 10 C°.

Between the boiling and freezing points of water, there are 180 Fahrenheit degrees. Thus, 180 F° = 100 K, and a Fahrenheit degree must be 100/180, or 5/9, the size of a kelvin or Celsius degree. From this or from Eq. 18-8, we then know that any temperature change expressed in Fahrenheit degrees

must be $\frac{9}{5}$ times that same temperature change expressed in either kelvins or Celsius degrees. For example, in Fahrenheit degrees, a temperature change of 10 K is (9/5)(10 K), or 18 F°.

Take care not to confuse a *temperature* with a temperature *change* or *difference*. A temperature of 10 K is certainly not the same as one of 10°C or 18°F but, as noted above, a temperature *change* of 10 K is the same as one of 10 C° or 18 F°. This distinction is very important in an equation containing a temperature $T$ instead of a temperature change or difference such as $T_2 - T_1$: A temperature $T$ by itself should generally be in kelvins and not degrees Celsius or Fahrenheit. In short, beware the "bare $T$."

**FIG. 18-9** When a Concorde flew faster than the speed of sound, thermal expansion due to the rubbing by passing air increased the aircraft's length by about 12.5 cm. (The temperature increased to about 128°C at the aircraft nose and about 90°C at the tail, and cabin windows were noticeably warm to the touch.) *(Hugh Thomas/BWP Media/Getty Images News and Sport Services)*

## 18-6 | Thermal Expansion

You can often loosen a tight metal jar lid by holding it under a stream of hot water. Both the metal of the lid and the glass of the jar expand as the hot water adds energy to their atoms. (With the added energy, the atoms can move a bit farther from one another than usual, against the spring-like interatomic forces that hold every solid together.) However, because the atoms in the metal move farther apart than those in the glass, the lid expands more than the jar and thus is loosened.

Such **thermal expansion** of materials with an increase in temperature must be anticipated in many common situations. When a bridge is subject to large seasonal changes in temperature, for example, sections of the bridge are separated by *expansion slots* so that the sections have room to expand on hot days without the bridge buckling. When a dental cavity is filled, the filling material must have the same thermal expansion properties as the surrounding tooth; otherwise, consuming cold ice cream and then hot coffee would be very painful. When the Concorde aircraft (Fig. 18-9) was built, the design had to allow for the thermal expansion of the fuselage during supersonic flight because of frictional heating by the passing air.

The thermal expansion properties of some materials can be put to common use. Thermometers and thermostats may be based on the differences in expansion between the components of a *bimetal strip* (Fig. 18-10). Also, the familiar liquid-in-glass thermometers are based on the fact that liquids such as mercury and alcohol expand to a different (greater) extent than their glass containers.

### Linear Expansion

If the temperature of a metal rod of length $L$ is raised by an amount $\Delta T$, its length is found to increase by an amount

$$\Delta L = L\alpha\,\Delta T, \tag{18-9}$$

in which $\alpha$ is a constant called the **coefficient of linear expansion.** The coefficient $\alpha$ has the unit "per degree" or "per kelvin" and depends on the material. Although $\alpha$ varies somewhat with temperature, for most practical purposes it can be taken as constant for a particular material. Table 18-2 shows some coefficients of linear expansion. Note that the unit C° there could be replaced with the unit K.

The thermal expansion of a solid is like photographic enlargement except it is in three dimensions. Figure 18-11*b* shows the (exaggerated) thermal expansion of a steel ruler. Equation 18-9 applies to every linear dimension of the ruler, including its edge, thickness, diagonals, and the diameters of the circle etched on it and the circular hole cut in it. If the disk cut from that hole originally fits snugly in the hole, it will continue to fit snugly if it undergoes the same temperature increase as the ruler.

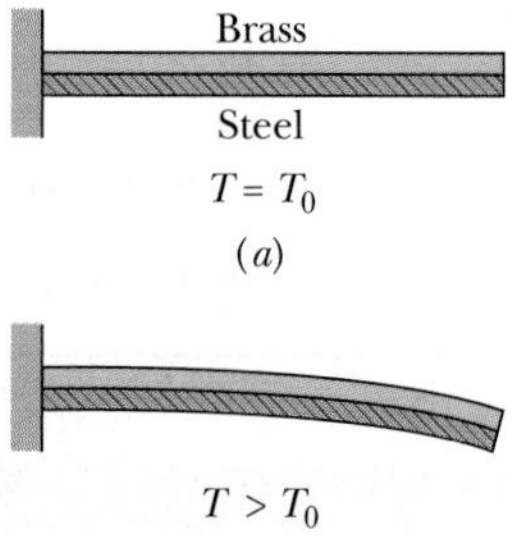

Brass

Steel

$T = T_0$

(a)

$T > T_0$

(b)

**FIG. 18-10** (*a*) A bimetal strip, consisting of a strip of brass and a strip of steel welded together, at temperature $T_0$. (*b*) The strip bends as shown at temperatures above this reference temperature. Below the reference temperature the strip bends the other way. Many thermostats operate on this principle, making and breaking an electrical contact as the temperature rises and falls.

### Volume Expansion

If all dimensions of a solid expand with temperature, the volume of that solid must also expand. For liquids, volume expansion is the only meaningful expansion parameter. If the temperature of a solid or liquid whose volume is $V$ is increased by an amount $\Delta T$, the increase in volume is found to be

$$\Delta V = V\beta\,\Delta T, \tag{18-10}$$

where $\beta$ is the **coefficient of volume expansion** of the solid or liquid. The coefficients of volume expansion and linear expansion for a solid are related by

$$\beta = 3\alpha. \tag{18-11}$$

**TABLE 18-2**

**Some Coefficients of Linear Expansion**[a]

| Substance | $\alpha$ $(10^{-6}/\text{C}°)$ | Substance | $\alpha$ $(10^{-6}/\text{C}°)$ |
|---|---|---|---|
| Ice (at 0°C) | 51 | Steel | 11 |
| Lead | 29 | Glass (ordinary) | 9 |
| Aluminum | 23 | Glass (Pyrex) | 3.2 |
| Brass | 19 | Diamond | 1.2 |
| Copper | 17 | Invar[b] | 0.7 |
| Concrete | 12 | Fused quartz | 0.5 |

[a]Room temperature values except for the listing for ice.

[b]This alloy was designed to have a low coefficient of expansion. The word is a shortened form of "invariable."

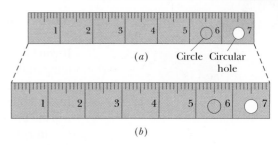

**FIG. 18-11** The same steel ruler at two different temperatures. When it expands, the scale, the numbers, the thickness, and the diameters of the circle and circular hole are all increased by the same factor. (The expansion has been exaggerated for clarity.)

The most common liquid, water, does not behave like other liquids. Above about 4°C, water expands as the temperature rises, as we would expect. Between 0 and about 4°C, however, water *contracts* with increasing temperature. Thus, at about 4°C, the density of water passes through a maximum. At all other temperatures, the density of water is less than this maximum value.

This behavior of water is the reason lakes freeze from the top down rather than from the bottom up. As water on the surface is cooled from, say, 10°C toward the freezing point, it becomes denser ("heavier") than lower water and sinks to the bottom. Below 4°C, however, further cooling makes the water then on the surface *less* dense ("lighter") than the lower water, so it stays on the surface until it freezes. Thus the surface freezes while the lower water is still liquid. If lakes froze from the bottom up, the ice so formed would tend not to melt completely during the summer, because it would be insulated by the water above. After a few years, many bodies of open water in the temperate zones of Earth would be frozen solid all year round—and aquatic life could not exist.

✓**CHECKPOINT 2**  The figure here shows four rectangular metal plates, with sides of $L$, $2L$, or $3L$. They are all made of the same material, and their temperature is to be increased by the same amount. Rank the plates according to the expected increase in (a) their vertical heights and (b) their areas, greatest first.

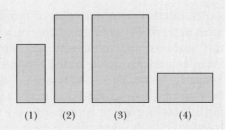

(1)   (2)   (3)   (4)

**Sample Problem** 18-2

On a hot day in Las Vegas, an oil trucker loaded 37 000 L of diesel fuel. He encountered cold weather on the way to Payson, Utah, where the temperature was 23.0 K lower than in Las Vegas, and where he delivered his entire load. How many liters did he deliver? The coefficient of volume expansion for diesel fuel is $9.50 \times 10^{-4}/\text{C}°$, and the coefficient of linear expansion for his steel truck tank is $11 \times 10^{-6}/\text{C}°$.

**KEY IDEA**  The volume of the diesel fuel depends directly on the temperature. Thus, because the tempera-

ture decreased, the volume of the fuel did also, as given by Eq. 18-10 ($\Delta V = V\beta\,\Delta T$).

**Calculations:** We find

$$\Delta V = (37\,000\text{ L})(9.50 \times 10^{-4}/\text{C}°)(-23.0\text{ K}) = -808\text{ L}.$$

Thus, the amount delivered was

$$V_{\text{del}} = V + \Delta V = 37\,000\text{ L} - 808\text{ L}$$
$$= 36\,190\text{ L}. \quad\quad\quad \text{(Answer)}$$

Note that the thermal expansion of the steel tank has nothing to do with the problem. Question: Who paid for the "missing" diesel fuel?

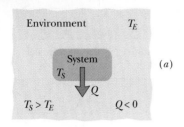

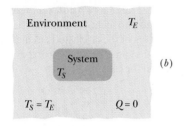

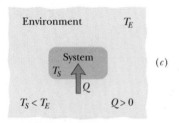

**FIG. 18-12** If the temperature of a system exceeds that of its environment as in (a), heat $Q$ is lost by the system to the environment until thermal equilibrium (b) is established. (c) If the temperature of the system is below that of the environment, heat is absorbed by the system until thermal equilibrium is established.

## 18-7 | Temperature and Heat

If you take a can of cola from the refrigerator and leave it on the kitchen table, its temperature will rise—rapidly at first but then more slowly—until the temperature of the cola equals that of the room (the two are then in thermal equilibrium). In the same way, the temperature of a cup of hot coffee, left sitting on the table, will fall until it also reaches room temperature.

In generalizing this situation, we describe the cola or the coffee as a *system* (with temperature $T_S$) and the relevant parts of the kitchen as the *environment* (with temperature $T_E$) of that system. Our observation is that if $T_S$ is not equal to $T_E$, then $T_S$ will change ($T_E$ can also change some) until the two temperatures are equal and thus thermal equilibrium is reached.

Such a change in temperature is due to a change in the thermal energy of the system because of a transfer of energy between the system and the system's environment. (Recall that *thermal energy* is an internal energy that consists of the kinetic and potential energies associated with the random motions of the atoms, molecules, and other microscopic bodies within an object.) The transferred energy is called **heat** and is symbolized $Q$. Heat is *positive* when energy is transferred to a system's thermal energy from its environment (we say that heat is absorbed by the system). Heat is *negative* when energy is transferred from a system's thermal energy to its environment (we say that heat is released or lost by the system).

This transfer of energy is shown in Fig. 18-12. In the situation of Fig. 18-12a, in which $T_S > T_E$, energy is transferred from the system to the environment, so $Q$ is negative. In Fig. 18-12b, in which $T_S = T_E$, there is no such transfer, $Q$ is zero, and heat is neither released nor absorbed. In Fig. 18-12c, in which $T_S < T_E$, the transfer is to the system from the environment; so $Q$ is positive.

We are led then to this definition of heat:

> Heat is the energy transferred between a system and its environment because of a temperature difference that exists between them.

Recall that energy can also be transferred between a system and its environment as *work W* via a force acting on a system. Heat and work, unlike temperature, pressure, and volume, are not intrinsic properties of a system. They have meaning only as they describe the transfer of energy into or out of a system. Similarly, the phrase "a $600 transfer" has meaning if it describes the transfer to or from an account, not what is in the account, because the account holds money, not a transfer. Here, it is proper to say: "During the last 3 min, 15 J of heat was transferred to the system from its environment" or "During the last minute, 12 J of work was done on the system by its environment." It is meaningless to say: "This system contains 450 J of heat" or "This system contains 385 J of work."

Before scientists realized that heat is transferred energy, heat was measured in terms of its ability to raise the temperature of water. Thus, the **calorie** (cal) was defined as the amount of heat that would raise the temperature of 1 g of water from 14.5°C to 15.5°C. In the British system, the corresponding unit of heat was the **British thermal unit** (Btu), defined as the amount of heat that would raise the temperature of 1 lb of water from 63°F to 64°F.

In 1948, the scientific community decided that since heat (like work) is transferred energy, the SI unit for heat should be the one we use for energy—namely, the **joule.** The calorie is now defined to be 4.1868 J (exactly), with no reference to the heating of water. (The "calorie" used in nutrition, sometimes called the Calorie (Cal), is really a kilocalorie.) The relations among the various heat units are

$$1 \text{ cal} = 3.968 \times 10^{-3} \text{ Btu} = 4.1868 \text{ J.} \quad (18\text{-}12)$$

# 18-8 | The Absorption of Heat by Solids and Liquids

## Heat Capacity

The **heat capacity** $C$ of an object is the proportionality constant between the heat $Q$ that the object absorbs or loses and the resulting temperature change $\Delta T$ of the object; that is,

$$Q = C \Delta T = C(T_f - T_i), \qquad (18\text{-}13)$$

in which $T_i$ and $T_f$ are the initial and final temperatures of the object. Heat capacity $C$ has the unit of energy per degree or energy per kelvin. The heat capacity $C$ of, say, a marble slab used in a bun warmer might be 179 cal/C°, which we can also write as 179 cal/K or as 749 J/K.

The word "capacity" in this context is really misleading in that it suggests analogy with the capacity of a bucket to hold water. *That analogy is false,* and you should not think of the object as "containing" heat or being limited in its ability to absorb heat. Heat transfer can proceed without limit as long as the necessary temperature difference is maintained. The object may, of course, melt or vaporize during the process.

## Specific Heat

Two objects made of the same material—say, marble—will have heat capacities proportional to their masses. It is therefore convenient to define a "heat capacity per unit mass" or **specific heat** $c$ that refers not to an object but to a unit mass of the material of which the object is made. Equation 18-13 then becomes

$$Q = cm\,\Delta T = cm(T_f - T_i). \qquad (18\text{-}14)$$

Through experiment we would find that although the heat capacity of a particular marble slab might be 179 cal/C° (or 749 J/K), the specific heat of marble itself (in that slab or in any other marble object) is 0.21 cal/g · C° (or 880 J/kg · K).

From the way the calorie and the British thermal unit were initially defined, the specific heat of water is

$$c = 1\ \text{cal/g} \cdot \text{C}° = 1\ \text{Btu/lb} \cdot \text{F}° = 4190\ \text{J/kg} \cdot \text{K}. \qquad (18\text{-}15)$$

Table 18-3 shows the specific heats of some substances at room temperature. Note that the value for water is relatively high. The specific heat of any substance actually depends somewhat on temperature, but the values in Table 18-3 apply reasonably well in a range of temperatures near room temperature.

✓**CHECKPOINT 3** A certain amount of heat $Q$ will warm 1 g of material $A$ by 3 C° and 1 g of material $B$ by 4 C°. Which material has the greater specific heat?

## Molar Specific Heat

In many instances the most convenient unit for specifying the amount of a substance is the mole (mol), where

$$1\ \text{mol} = 6.02 \times 10^{23}\ \text{elementary units}$$

of *any* substance. Thus 1 mol of aluminum means $6.02 \times 10^{23}$ atoms (the atom is the elementary unit), and 1 mol of aluminum oxide means $6.02 \times 10^{23}$ molecules (the molecule is the elementary unit of the compound).

When quantities are expressed in moles, specific heats must also involve moles (rather than a mass unit); they are then called **molar specific heats.** Table 18-3 shows the values for some elemental solids (each consisting of a single element) at room temperature.

**TABLE 18-3**

**Some Specific Heats and Molar Specific Heats at Room Temperature**

| Substance | Specific Heat cal g·K | Specific Heat J kg·K | Molar Specific Heat J mol·K |
|---|---|---|---|
| *Elemental Solids* | | | |
| Lead | 0.0305 | 128 | 26.5 |
| Tungsten | 0.0321 | 134 | 24.8 |
| Silver | 0.0564 | 236 | 25.5 |
| Copper | 0.0923 | 386 | 24.5 |
| Aluminum | 0.215 | 900 | 24.4 |
| *Other Solids* | | | |
| Brass | 0.092 | 380 | |
| Granite | 0.19 | 790 | |
| Glass | 0.20 | 840 | |
| Ice ($-10$°C) | 0.530 | 2220 | |
| *Liquids* | | | |
| Mercury | 0.033 | 140 | |
| Ethyl alcohol | 0.58 | 2430 | |
| Seawater | 0.93 | 3900 | |
| Water | 1.00 | 4180 | |

### An Important Point

In determining and then using the specific heat of any substance, we need to know the conditions under which energy is transferred as heat. For solids and liquids, we usually assume that the sample is under constant pressure (usually atmospheric) during the transfer. It is also conceivable that the sample is held at constant volume while the heat is absorbed. This means that thermal expansion of the sample is prevented by applying external pressure. For solids and liquids, this is very hard to arrange experimentally, but the effect can be calculated, and it turns out that the specific heats under constant pressure and constant volume for any solid or liquid differ usually by no more than a few percent. Gases, as you will see, have quite different values for their specific heats under constant-pressure conditions and under constant-volume conditions.

### Heats of Transformation

When energy is absorbed as heat by a solid or liquid, the temperature of the sample does not necessarily rise. Instead, the sample may change from one *phase,* or *state,* to another. Matter can exist in three common states: In the *solid state,* the molecules of a sample are locked into a fairly rigid structure by their mutual attraction. In the *liquid state,* the molecules have more energy and move about more. They may form brief clusters, but the sample does not have a rigid structure and can flow or settle into a container. In the *gas,* or *vapor, state,* the molecules have even more energy, are free of one another, and can fill up the full volume of a container.

To *melt* a solid means to change it from the solid state to the liquid state. The process requires energy because the molecules of the solid must be freed from their rigid structure. Melting an ice cube to form liquid water is a common example. To *freeze* a liquid to form a solid is the reverse of melting and requires that energy be removed from the liquid, so that the molecules can settle into a rigid structure.

To *vaporize* a liquid means to change it from the liquid state to the vapor (gas) state. This process, like melting, requires energy because the molecules must be freed from their clusters. Boiling liquid water to transfer it to water vapor (or steam—a gas of individual water molecules) is a common example. *Condensing* a gas to form a liquid is the reverse of vaporizing; it requires that energy be removed from the gas, so that the molecules can cluster instead of flying away from one another.

The amount of energy per unit mass that must be transferred as heat when a sample completely undergoes a phase change is called the **heat of transformation** $L$. Thus, when a sample of mass $m$ completely undergoes a phase change, the total energy transferred is

$$Q = Lm. \qquad (18\text{-}16)$$

When the phase change is from liquid to gas (then the sample must absorb heat) or from gas to liquid (then the sample must release heat), the heat of transformation is called the **heat of vaporization** $L_V$. For water at its normal boiling or condensation temperature,

$$L_V = 539 \text{ cal/g} = 40.7 \text{ kJ/mol} = 2256 \text{ kJ/kg}. \qquad (18\text{-}17)$$

When the phase change is from solid to liquid (then the sample must absorb heat) or from liquid to solid (then the sample must release heat), the heat of transformation is called the **heat of fusion** $L_F$. For water at its normal freezing or melting temperature,

$$L_F = 79.5 \text{ cal/g} = 6.01 \text{ kJ/mol} = 333 \text{ kJ/kg}. \qquad (18\text{-}18)$$

Table 18-4 shows the heats of transformation for some substances.

TABLE 18-4

**TABLE 18-4**

**Some Heats of Transformation**

| | Melting | | Boiling | |
|---|---|---|---|---|
| Substance | Melting Point (K) | Heat of Fusion $L_F$ (kJ/kg) | Boiling Point (K) | Heat of Vaporization $L_V$ (kJ/kg) |
| Hydrogen | 14.0 | 58.0 | 20.3 | 455 |
| Oxygen | 54.8 | 13.9 | 90.2 | 213 |
| Mercury | 234 | 11.4 | 630 | 296 |
| Water | 273 | 333 | 373 | 2256 |
| Lead | 601 | 23.2 | 2017 | 858 |
| Silver | 1235 | 105 | 2323 | 2336 |
| Copper | 1356 | 207 | 2868 | 4730 |

**Sample Problem    18-3**

(a) How much heat must be absorbed by ice of mass $m = 720$ g at $-10°C$ to take it to liquid state at $15°C$?

**KEY IDEAS** The heating process is accomplished in three steps: (1) The ice cannot melt at a temperature below the freezing point—so initially, any energy transferred to the ice as heat can only increase the temperature of the ice, until $0°C$ is reached. (2) The temperature then cannot increase until all the ice melts—so any energy transferred to the ice as heat now can only change ice to liquid water, until all the ice melts. (3) Now the energy transferred to the liquid water as heat can only increase the temperature of the liquid water.

**Warming the ice:** The heat $Q_1$ needed to increase the temperature of the ice from the initial value $T_i = -10°C$ to a final value $T_f = 0°C$ (so that the ice can then melt) is given by Eq. 18-14 ($Q = cm \Delta T$). Using the specific heat of ice $c_{ice}$ in Table 18-3 gives us

$$Q_1 = c_{ice}m(T_f - T_i)$$
$$= (2220 \text{ J/kg} \cdot \text{K})(0.720 \text{ kg})[0°C - (-10°C)]$$
$$= 15\,984 \text{ J} \approx 15.98 \text{ kJ}.$$

**Melting the ice:** The heat $Q_2$ needed to melt all the ice is given by Eq. 18-16 ($Q = Lm$). Here $L$ is the heat of fusion $L_F$, with the value given in Eq. 18-18 and Table 18-4. We find

$$Q_2 = L_F m = (333 \text{ kJ/kg})(0.720 \text{ kg}) \approx 239.8 \text{ kJ}.$$

**Warming the liquid:** The heat $Q_3$ needed to increase the temperature of the water from the initial value $T_i = 0°C$ to the final value $T_f = 15°C$ is given by Eq. 18-14 (with the specific heat of liquid water $c_{liq}$):

$$Q_3 = c_{liq}m(T_f - T_i)$$
$$= (4190 \text{ J/kg} \cdot \text{K})(0.720 \text{ kg})(15°C - 0°C)$$
$$= 45\,252 \text{ J} \approx 45.25 \text{ kJ}.$$

**Total:** The total required heat $Q_{tot}$ is the sum of the amounts required in the three steps:

$$Q_{tot} = Q_1 + Q_2 + Q_3$$
$$= 15.98 \text{ kJ} + 239.8 \text{ kJ} + 45.25 \text{ kJ}$$
$$\approx 300 \text{ kJ}. \qquad \text{(Answer)}$$

Note that the heat required to melt the ice is much greater than the heat required to raise the temperature of either the ice or the liquid water.

(b) If we supply the ice with a total energy of only 210 kJ (as heat), what then are the final state and temperature of the water?

**KEY IDEA** From step 1, we know that 15.98 kJ is needed to raise the temperature of the ice to the melting point. The remaining heat $Q_{rem}$ is then 210 kJ − 15.98 kJ, or about 194 kJ. From step 2, we can see that this amount of heat is insufficient to melt all the ice. Because the melting of the ice is incomplete, we must end up with a mixture of ice and liquid; the temperature of the mixture must be the freezing point, $0°C$.

**Calculations:** We can find the mass $m$ of ice that is melted by the available energy $Q_{rem}$ by using Eq. 18-16 with $L_F$:

$$m = \frac{Q_{rem}}{L_F} = \frac{194 \text{ kJ}}{333 \text{ kJ/kg}} = 0.583 \text{ kg} \approx 580 \text{ g}.$$

Thus, the mass of the ice that remains is 720 g − 580 g, or 140 g, and we have

580 g water    and    140 g ice,    at $0°C$.    (Answer)

A copper slug whose mass $m_c$ is 75 g is heated in a laboratory oven to a temperature $T$ of 312°C. The slug is then dropped into a glass beaker containing a mass $m_w = 220$ g of water. The heat capacity $C_b$ of the beaker is 45 cal/K. The initial temperature $T_i$ of the water and the beaker is 12°C. Assuming that the slug, beaker, and water are an isolated system and the water does not vaporize, find the final temperature $T_f$ of the system at thermal equilibrium.

**KEY IDEAS** (1) Because the system is isolated, the system's total energy cannot change and only internal transfers of thermal energy can occur. (2) Because nothing in the system undergoes a phase change, the thermal energy transfers can only change the temperatures.

*Calculations:* To relate the transfers to the temperature changes, we can use Eqs. 18-13 and 18-14 to write

$$\text{for the water:} \quad Q_w = c_w m_w (T_f - T_i); \quad (18\text{-}19)$$

$$\text{for the beaker:} \quad Q_b = C_b (T_f - T_i); \quad (18\text{-}20)$$

$$\text{for the copper:} \quad Q_c = c_c m_c (T_f - T). \quad (18\text{-}21)$$

Because the total energy of the system cannot change, the sum of these three energy transfers is zero:

$$Q_w + Q_b + Q_c = 0. \quad (18\text{-}22)$$

Substituting Eqs. 18-19 through 18-21 into Eq. 18-22 yields

$$c_w m_w (T_f - T_i) + C_b (T_f - T_i) + c_c m_c (T_f - T) = 0. \quad (18\text{-}23)$$

Temperatures are contained in Eq. 18-23 only as differences. Thus, because the differences on the Celsius and Kelvin scales are identical, we can use either of those scales in this equation. Solving it for $T_f$, we obtain

$$T_f = \frac{c_c m_c T + C_b T_i + c_w m_w T_i}{c_w m_w + C_b + c_c m_c}.$$

Using Celsius temperatures and taking values for $c_c$ and $c_w$ from Table 18-3, we find the numerator to be

$$(0.0923 \text{ cal/g}\cdot\text{K})(75 \text{ g})(312°\text{C}) + (45 \text{ cal/K})(12°\text{C})$$
$$+ (1.00 \text{ cal/g}\cdot\text{K})(220 \text{ g})(12°\text{C}) = 5339.8 \text{ cal},$$

and the denominator to be

$$(1.00 \text{ cal/g}\cdot\text{K})(220 \text{ g}) + 45 \text{ cal/K}$$
$$+ (0.0923 \text{ cal/g}\cdot\text{K})(75 \text{ g}) = 271.9 \text{ cal/C°}.$$

We then have

$$T_f = \frac{5339.8 \text{ cal}}{271.9 \text{ cal/C°}} = 19.6°\text{C} \approx 20°\text{C}. \quad \text{(Answer)}$$

From the given data you can show that

$$Q_w \approx 1670 \text{ cal}, \quad Q_b \approx 342 \text{ cal}, \quad Q_c \approx -2020 \text{ cal}.$$

Apart from rounding errors, the algebraic sum of these three heat transfers is indeed zero, as Eq. 18-22 requires.

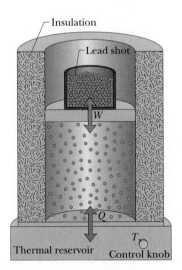

Insulation
Lead shot
$W$
$Q$
$T$
Thermal reservoir    Control knob

**FIG. 18-13** A gas is confined to a cylinder with a movable piston. Heat $Q$ can be added to or withdrawn from the gas by regulating the temperature $T$ of the adjustable thermal reservoir. Work $W$ can be done by the gas by raising or lowering the piston.

## 18-9 | A Closer Look at Heat and Work

Here we look in some detail at how energy can be transferred as heat and work between a system and its environment. Let us take as our system a gas confined to a cylinder with a movable piston, as in Fig. 18-13. The upward force on the piston due to the pressure of the confined gas is equal to the weight of lead shot loaded onto the top of the piston. The walls of the cylinder are made of insulating material that does not allow any transfer of energy as heat. The bottom of the cylinder rests on a reservoir for thermal energy, a *thermal reservoir* (perhaps a hot plate) whose temperature $T$ you can control by turning a knob.

The system (the gas) starts from an *initial state i*, described by a pressure $p_i$, a volume $V_i$, and a temperature $T_i$. You want to change the system to a *final state f*, described by a pressure $p_f$, a volume $V_f$, and a temperature $T_f$. The procedure by which you change the system from its initial state to its final state is called a *thermodynamic process*. During such a process, energy may be transferred into the system from the thermal reservoir (positive heat) or vice versa (negative heat). Also, work can be done by the system to raise the loaded piston (positive work) or lower it (negative work). We assume that all such changes occur slowly, with the result that the system is always in (approximate) thermal equilibrium (that is, every part of the system is always in thermal equilibrium with every other part).

Suppose that you remove a few lead shot from the piston of Fig. 18-13, allowing the gas to push the piston and remaining shot upward through a differential displacement $d\vec{s}$ with an upward force $\vec{F}$. Since the displacement is tiny, we can assume that $\vec{F}$ is constant during the displacement. Then $\vec{F}$ has a magnitude that is equal to $pA$, where $p$ is the pressure of the gas and $A$ is the face area of the piston. The differential work $dW$ done by the gas during the displacement is

$$dW = \vec{F} \cdot d\vec{s} = (pA)(ds) = p(A\,ds)$$
$$= p\,dV, \tag{18-24}$$

in which $dV$ is the differential change in the volume of the gas due to the movement of the piston. When you have removed enough shot to allow the gas to change its volume from $V_i$ to $V_f$, the total work done by the gas is

$$W = \int dW = \int_{V_i}^{V_f} p\,dV. \tag{18-25}$$

During the change in volume, the pressure and temperature of the gas may also change. To evaluate the integral in Eq. 18-25 directly, we would need to know how pressure varies with volume for the actual process by which the system changes from state $i$ to state $f$.

There are actually many ways to take the gas from state $i$ to state $f$. One way is shown in Fig. 18-14$a$, which is a plot of the pressure of the gas versus its volume and which is called a $p$-$V$ diagram. In Fig. 18-14$a$, the curve indicates that the pressure decreases as the volume increases. The integral in Eq. 18-25 (and thus the work $W$ done by the gas) is represented by the shaded area under the curve between points $i$ and $f$. Regardless of what exactly we do to take the gas along the curve, that work is positive, due to the fact that the gas increases its volume by forcing the piston upward.

Another way to get from state $i$ to state $f$ is shown in Fig. 18-14$b$. There the change takes place in two steps—the first from state $i$ to state $a$, and the second from state $a$ to state $f$.

Step $ia$ of this process is carried out at constant pressure, which means that you leave undisturbed the lead shot that ride on top of the piston in Fig. 18-13. You cause the volume to increase (from $V_i$ to $V_f$) by slowly turning up the temperature control knob, raising the temperature of the gas to some higher value $T_a$. (Increasing the temperature increases the force from the gas on the piston, moving it upward.) During this step, positive work is done by the expanding gas (to lift the loaded piston) and heat is absorbed by the system from the thermal reservoir (in response to the arbitrarily small temperature differences that you create as you turn up the temperature). This heat is positive because it is added to the system.

Step $af$ of the process of Fig. 18-14$b$ is carried out at constant volume, so you must wedge the piston, preventing it from moving. Then as you use the control knob to decrease the temperature, you find that the pressure drops from $p_a$ to its final value $p_f$. During this step, heat is lost by the system to the thermal reservoir.

For the overall process $iaf$, the work $W$, which is positive and is carried out only during step $ia$, is represented by the shaded area under the curve. Energy is transferred as heat during both steps $ia$ and $af$, with a net energy transfer $Q$.

Figure 18-14$c$ shows a process in which the previous two steps are carried out in reverse order. The work $W$ in this case is smaller than for Fig. 18-14$b$, as is the

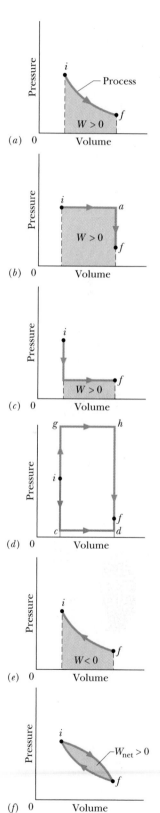

FIG. 18-14 (a) The shaded area represents the work $W$ done by a system as it goes from an initial state $i$ to a final state $f$. Work $W$ is positive because the system's volume increases. (b) $W$ is still positive, but now greater. (c) $W$ is still positive, but now smaller. (d) $W$ can be even smaller (path $icdf$) or larger (path $ighf$). (e) Here the system goes from state $f$ to state $i$ as the gas is compressed to less volume by an external force. The work $W$ done by the system is now negative. (f) The net work $W_{\text{net}}$ done by the system during a complete cycle is represented by the shaded area.

net heat absorbed. Figure 18-14*d* suggests that you can make the work done by the gas as small as you want (by following a path like *icdf*) or as large as you want (by following a path like *ighf*).

To sum up: A system can be taken from a given initial state to a given final state by an infinite number of processes. Heat may or may not be involved, and in general, the work $W$ and the heat $Q$ will have different values for different processes. We say that heat and work are *path-dependent* quantities.

Figure 18-14*e* shows an example in which negative work is done by a system as some external force compresses the system, reducing its volume. The absolute value of the work done is still equal to the area beneath the curve, but because the gas is *compressed,* the work done by the gas is negative.

Figure 18-14*f* shows a *thermodynamic cycle* in which the system is taken from some initial state *i* to some other state *f* and then back to *i*. The net work done by the system during the cycle is the sum of the *positive* work done during the expansion and the *negative* work done during the compression. In Fig. 18-14*f*, the net work is positive because the area under the expansion curve (*i* to *f*) is greater than the area under the compression curve (*f* to *i*).

✓ **CHECKPOINT 4**   The *p-V* diagram here shows six curved paths (connected by vertical paths) that can be followed by a gas. Which two of the curved paths should be part of a closed cycle (those curved paths plus connecting vertical paths) if the net work done by the gas during the cycle is to be at its maximum positive value?

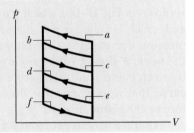

## 18-10 | The First Law of Thermodynamics

You have just seen that when a system changes from a given initial state to a given final state, both the work $W$ and the heat $Q$ depend on the nature of the process. Experimentally, however, we find a surprising thing. *The quantity $Q - W$ is the same for all processes.* It depends only on the initial and final states and does not depend at all on how the system gets from one to the other. All other combinations of $Q$ and $W$, including $Q$ alone, $W$ alone, $Q + W$, and $Q - 2W$, are *path dependent;* only the quantity $Q - W$ is not.

The quantity $Q - W$ must represent a change in some intrinsic property of the system. We call this property the *internal energy* $E_{int}$ and we write

$$\Delta E_{int} = E_{int,f} - E_{int,i} = Q - W \quad \text{(first law).} \tag{18-26}$$

Equation 18-26 is the **first law of thermodynamics.** If the thermodynamic system undergoes only a differential change, we can write the first law as*

$$dE_{int} = dQ - dW \quad \text{(first law).} \tag{18-27}$$

☞ The internal energy $E_{int}$ of a system tends to increase if energy is added as heat $Q$ and tends to decrease if energy is lost as work $W$ done by the system.

In Chapter 8, we discussed the principle of energy conservation as it applies to isolated systems—that is, to systems in which no energy enters or leaves the system. The first law of thermodynamics is an extension of that principle to

---

*Here $dQ$ and $dW$, unlike $dE_{int}$, are not true differentials; that is, there are no such functions as $Q(p, V)$ and $W(p, V)$ that depend only on the state of the system. The quantities $dQ$ and $dW$ are called *inexact differentials* and are usually represented by the symbols $đQ$ and $đW$. For our purposes, we can treat them simply as infinitesimally small energy transfers.

systems that are *not* isolated. In such cases, energy may be transferred into or out of the system as either work $W$ or heat $Q$. In our statement of the first law of thermodynamics above, we assume that there are no changes in the kinetic energy or the potential energy of the system as a whole; that is, $\Delta K = \Delta U = 0$.

Before this chapter, the term *work* and the symbol $W$ always meant the work done *on* a system. However, starting with Eq. 18-24 and continuing through the next two chapters about thermodynamics, we focus on the work done *by* a system, such as the gas in Fig. 18-13.

The work done *on* a system is always the negative of the work done *by* the system, so if we rewrite Eq. 18-26 in terms of the work $W_{on}$ done *on* the system, we have $\Delta E_{int} = Q + W_{on}$. This tells us the following: The internal energy of a system tends to increase if heat is absorbed by the system or if positive work is done *on* the system. Conversely, the internal energy tends to decrease if heat is lost by the system or if negative work is done *on* the system.

> ✓ **CHECKPOINT 5**    The figure here shows four paths on a *p-V* diagram along which a gas can be taken from state $i$ to state $f$. Rank the paths according to (a) the change $\Delta E_{int}$ in the internal energy of the gas, (b) the work $W$ done by the gas, and (c) the magnitude of the energy transferred as heat $Q$ between the gas and its environment, greatest first.

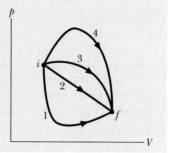

## 18-11 | Some Special Cases of the First Law of Thermodynamics

Here we look at four different thermodynamic processes to see what consequences follow when we apply the first law of thermodynamics. The results are summarized in Table 18-5.

1. ***Adiabatic processes.*** An adiabatic process is one that occurs so rapidly or occurs in a system that is so well insulated that *no transfer of energy as heat* occurs between the system and its environment. Putting $Q = 0$ in the first law (Eq. 18-26) yields

$$\Delta E_{int} = -W \qquad \text{(adiabatic process).} \qquad (18\text{-}28)$$

This tells us that if work is done *by* the system (that is, if $W$ is positive), the internal energy of the system decreases by the amount of work. Conversely, if work is done *on* the system (that is, if $W$ is negative), the internal energy of the system increases by that amount.

Figure 18-15 shows an idealized adiabatic process. Heat cannot enter or leave the system because of the insulation. Thus, the only way energy can be

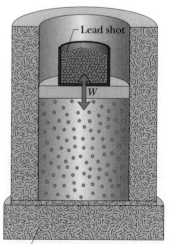

**FIG. 18-15** An adiabatic expansion can be carried out by slowly removing lead shot from the top of the piston. Adding lead shot reverses the process at any stage.

**TABLE 18-5**

**The First Law of Thermodynamics: Four Special Cases**

| | | |
|---|---|---|
| *The Law:* $\Delta E_{int} = Q - W$ (Eq. 18-26) | | |
| Process | Restriction | Consequence |
| Adiabatic | $Q = 0$ | $\Delta E_{int} = -W$ |
| Constant volume | $W = 0$ | $\Delta E_{int} = Q$ |
| Closed cycle | $\Delta E_{int} = 0$ | $Q = W$ |
| Free expansion | $Q = W = 0$ | $\Delta E_{int} = 0$ |

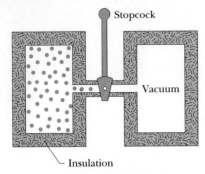

FIG. 18-16 The initial stage of a free-expansion process. After the stopcock is opened, the gas fills both chambers and eventually reaches an equilibrium state.

transferred between the system and its environment is by work. If we remove shot from the piston and allow the gas to expand, the work done by the system (the gas) is positive and the internal energy of the gas decreases. If, instead, we add shot and compress the gas, the work done by the system is negative and the internal energy of the gas increases.

2. **Constant-volume processes.** If the volume of a system (such as a gas) is held constant, that system can do no work. Putting $W = 0$ in the first law (Eq. 18-26) yields

$$\Delta E_{int} = Q \qquad \text{(constant-volume process)}. \qquad (18\text{-}29)$$

Thus, if heat is absorbed by a system (that is, if $Q$ is positive), the internal energy of the system increases. Conversely, if heat is lost during the process (that is, if $Q$ is negative), the internal energy of the system must decrease.

3. **Cyclical processes.** There are processes in which, after certain interchanges of heat and work, the system is restored to its initial state. In that case, no intrinsic property of the system—including its internal energy—can possibly change. Putting $\Delta E_{int} = 0$ in the first law (Eq. 18-26) yields

$$Q = W \qquad \text{(cyclical process)}. \qquad (18\text{-}30)$$

Thus, the net work done during the process must exactly equal the net amount of energy transferred as heat; the store of internal energy of the system remains unchanged. Cyclical processes form a closed loop on a $p$-$V$ plot, as shown in Fig. 18-14f. We discuss such processes in detail in Chapter 20.

4. **Free expansions.** These are adiabatic processes in which no transfer of heat occurs between the system and its environment and no work is done on or by the system. Thus, $Q = W = 0$, and the first law requires that

$$\Delta E_{int} = 0 \qquad \text{(free expansion)}. \qquad (18\text{-}31)$$

Figure 18-16 shows how such an expansion can be carried out. A gas, which is in thermal equilibrium within itself, is initially confined by a closed stopcock to one half of an insulated double chamber; the other half is evacuated. The stopcock is opened, and the gas expands freely to fill both halves of the chamber. No heat is transferred to or from the gas because of the insulation. No work is done by the gas because it rushes into a vacuum and thus does not meet any pressure.

A free expansion differs from all other processes we have considered because it cannot be done slowly and in a controlled way. As a result, at any given instant during the sudden expansion, the gas is not in thermal equilibrium and its pressure is not uniform. Thus, although we can plot the initial and final states on a $p$-$V$ diagram, we cannot plot the expansion itself.

✓ **CHECKPOINT 6** For one complete cycle as shown in the $p$-$V$ diagram here, are (a) $\Delta E_{int}$ for the gas and (b) the net energy transferred as heat $Q$ positive, negative, or zero?

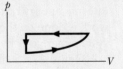

---

**Sample Problem | 18-5**

Let 1.00 kg of liquid water at 100°C be converted to steam at 100°C by boiling at standard atmospheric pressure (which is 1.00 atm or $1.01 \times 10^5$ Pa) in the arrangement of Fig. 18-17. The volume of that water changes from an initial value of $1.00 \times 10^{-3}$ m$^3$ as a liquid to 1.671 m$^3$ as steam.

(a) How much work is done by the system during this process?

**KEY IDEAS** (1) The system must do positive work because the volume increases. (2) We calculate the work $W$ done by integrating the pressure with respect to the volume (Eq. 18-25).

**Calculation:** Because here the pressure is constant at $1.01 \times 10^5$ Pa, we can take $p$ outside the integral. We then have

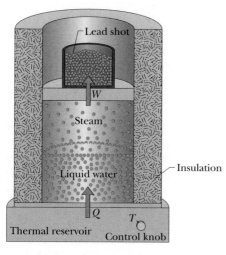

**FIG. 18-17**
Water boiling at constant pressure. Energy is transferred from the thermal reservoir as heat until the liquid water has changed completely into steam. Work is done by the expanding gas as it lifts the loaded piston.

$$W = \int_{V_i}^{V_f} p \, dV = p \int_{V_i}^{V_f} dV = p(V_f - V_i)$$
$$= (1.01 \times 10^5 \text{ Pa})(1.671 \text{ m}^3 - 1.00 \times 10^{-3} \text{ m}^3)$$
$$= 1.69 \times 10^5 \text{ J} = 169 \text{ kJ.} \qquad \text{(Answer)}$$

(b) How much energy is transferred as heat during the process?

**KEY IDEA** Because the heat causes only a phase change and not a change in temperature, it is given fully by Eq. 18-16 ($Q = Lm$).

*Calculation:* Because the change is from liquid to gaseous phase, $L$ is the heat of vaporization $L_V$, with the value given in Eq. 18-17 and Table 18-4. We find

$$Q = L_V m = (2256 \text{ kJ/kg})(1.00 \text{ kg})$$
$$= 2256 \text{ kJ} \approx 2260 \text{ kJ.} \qquad \text{(Answer)}$$

(c) What is the change in the system's internal energy during the process?

**KEY IDEA** The change in the system's internal energy is related to the heat (here, this is energy transferred into the system) and the work (here, this is energy transferred out of the system) by the first law of thermodynamics (Eq. 18-26).

*Calculation:* We write the first law as

$$\Delta E_{\text{int}} = Q - W = 2256 \text{ kJ} - 169 \text{ kJ}$$
$$\approx 2090 \text{ kJ} = 2.09 \text{ MJ.} \qquad \text{(Answer)}$$

This quantity is positive, indicating that the internal energy of the system has increased during the boiling process. This energy goes into separating the $H_2O$ molecules, which strongly attract one another in the liquid state. We see that, when water is boiled, about 7.5% ($= 169 \text{ kJ}/2260 \text{ kJ}$) of the heat goes into the work of pushing back the atmosphere. The rest of the heat goes into the system's internal energy.

## 18-12 I Heat Transfer Mechanisms

We have discussed the transfer of energy as heat between a system and its environment, but we have not yet described how that transfer takes place. There are three transfer mechanisms: conduction, convection, and radiation.

### Conduction

If you leave the end of a metal poker in a fire for enough time, its handle will get hot. Energy is transferred from the fire to the handle by (thermal) **conduction** along the length of the poker. The vibration amplitudes of the atoms and electrons of the metal at the fire end of the poker become relatively large because of the high temperature of their environment. These increased vibrational amplitudes, and thus the associated energy, are passed along the poker, from atom to atom, during collisions between adjacent atoms. In this way, a region of rising temperature extends itself along the poker to the handle.

Consider a slab of face area $A$ and thickness $L$, whose faces are maintained at temperatures $T_H$ and $T_C$ by a hot reservoir and a cold reservoir, as in Fig. 18-18. Let $Q$ be the energy that is transferred as heat through the slab, from its hot face to its cold face, in time $t$. Experiment shows that the *conduction rate* $P_{\text{cond}}$ (the amount of energy transferred per unit time) is

$$P_{\text{cond}} = \frac{Q}{t} = kA \frac{T_H - T_C}{L}, \qquad (18\text{-}32)$$

in which $k$, called the *thermal conductivity*, is a constant that depends on the

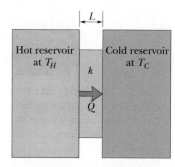

**FIG. 18-18** Thermal conduction. Energy is transferred as heat from a reservoir at temperature $T_H$ to a cooler reservoir at temperature $T_C$ through a conducting slab of thickness $L$ and thermal conductivity $k$.

**TABLE 18-6**

**Some Thermal Conductivities**

| Substance | $k$ (W/m·K) |
| --- | --- |
| *Metals* | |
| Stainless steel | 14 |
| Lead | 35 |
| Iron | 67 |
| Brass | 109 |
| Aluminum | 235 |
| Copper | 401 |
| Silver | 428 |
| *Gases* | |
| Air (dry) | 0.026 |
| Helium | 0.15 |
| Hydrogen | 0.18 |
| *Building Materials* | |
| Polyurethane foam | 0.024 |
| Rock wool | 0.043 |
| Fiberglass | 0.048 |
| White pine | 0.11 |
| Window glass | 1.0 |

material of which the slab is made. A material that readily transfers energy by conduction is a *good thermal conductor* and has a high value of $k$. Table 18-6 gives the thermal conductivities of some common metals, gases, and building materials.

## Thermal Resistance to Conduction (R-Value)

If you are interested in insulating your house or in keeping cola cans cold on a picnic, you are more concerned with poor heat conductors than with good ones. For this reason, the concept of *thermal resistance R* has been introduced into engineering practice. The *R*-value of a slab of thickness $L$ is defined as

$$R = \frac{L}{k}. \tag{18-33}$$

The lower the thermal conductivity of the material of which a slab is made, the higher the *R*-value of the slab; so something that has a high *R*-value is a *poor thermal conductor* and thus a *good thermal insulator*.

Note that $R$ is a property attributed to a slab of a specified thickness, not to a material. The commonly used unit for $R$ (which, in the United States at least, is almost never stated) is the square foot–Fahrenheit degree–hour per British thermal unit (ft$^2$·F°·h/Btu). (Now you know why the unit is rarely stated.)

## Conduction Through a Composite Slab

Figure 18-19 shows a composite slab, consisting of two materials having different thicknesses $L_1$ and $L_2$ and different thermal conductivities $k_1$ and $k_2$. The temperatures of the outer surfaces of the slab are $T_H$ and $T_C$. Each face of the slab has area $A$. Let us derive an expression for the conduction rate through the slab under the assumption that the transfer is a *steady-state* process; that is, the temperatures everywhere in the slab and the rate of energy transfer do not change with time.

In the steady state, the conduction rates through the two materials must be equal. This is the same as saying that the energy transferred through one material in a certain time must be equal to that transferred through the other material in the same time. If this were not true, temperatures in the slab would be changing and we would not have a steady-state situation. Letting $T_X$ be the temperature of the interface between the two materials, we can now use Eq. 18-32 to write

$$P_{\text{cond}} = \frac{k_2 A(T_H - T_X)}{L_2} = \frac{k_1 A(T_X - T_C)}{L_1}. \tag{18-34}$$

Solving Eq. 18-34 for $T_X$ yields, after a little algebra,

$$T_X = \frac{k_1 L_2 T_C + k_2 L_1 T_H}{k_1 L_2 + k_2 L_1}. \tag{18-35}$$

Substituting this expression for $T_X$ into either equality of Eq. 18-34 yields

$$P_{\text{cond}} = \frac{A(T_H - T_C)}{L_1/k_1 + L_2/k_2}. \tag{18-36}$$

We can extend Eq. 18-36 to apply to any number $n$ of materials making up a slab:

$$P_{\text{cond}} = \frac{A(T_H - T_C)}{\Sigma (L/k)}. \tag{18-37}$$

The summation sign in the denominator tells us to add the values of $L/k$ for all the materials.

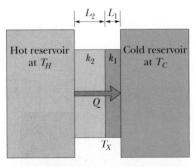

**FIG. 18-19** Heat is transferred at a steady rate through a composite slab made up of two different materials with different thicknesses and different thermal conductivities.
The steady-state temperature at the interface of the two materials is $T_X$.

**✓CHECKPOINT 7**  The figure shows the face and interface temperatures of a composite slab consisting of four materials, of identical thicknesses, through which the heat transfer is steady. Rank the materials according to their thermal conductivities, greatest first.

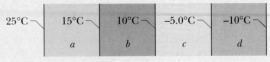

## Convection

When you look at the flame of a candle or a match, you are watching thermal energy being transported upward by **convection.** Such energy transfer occurs when a fluid, such as air or water, comes in contact with an object whose temperature is higher than that of the fluid. The temperature of the part of the fluid that is in contact with the hot object increases, and (in most cases) that fluid expands and thus becomes less dense. Because this expanded fluid is now lighter than the surrounding cooler fluid, buoyant forces cause it to rise. Some of the surrounding cooler fluid then flows so as to take the place of the rising warmer fluid, and the process can then continue.

Convection is part of many natural processes. Atmospheric convection plays a fundamental role in determining global climate patterns and daily weather variations. Glider pilots and birds alike seek rising thermals (convection currents of warm air) that keep them aloft. Huge energy transfers take place within the oceans by the same process. Finally, energy is transported to the surface of the Sun from the nuclear furnace at its core by enormous cells of convection, in which hot gas rises to the surface along the cell core and cooler gas around the core descends below the surface.

## Radiation

The third method by which an object and its environment can exchange energy as heat is via electromagnetic waves (visible light is one kind of electromagnetic wave). Energy transferred in this way is often called **thermal radiation** to distinguish it from electromagnetic *signals* (as in, say, television broadcasts) and from nuclear radiation (energy and particles emitted by nuclei). (To "radiate" generally means to emit.) When you stand in front of a big fire, you are warmed by absorbing thermal radiation from the fire; that is, your thermal energy increases as the fire's thermal energy decreases. No medium is required for heat transfer via radiation—the radiation can travel through vacuum from, say, the Sun to you.

The rate $P_{rad}$ at which an object emits energy via electromagnetic radiation depends on the object's surface area $A$ and the temperature $T$ of that area in kelvins and is given by

$$P_{rad} = \sigma \varepsilon A T^4. \tag{18-38}$$

Here $\sigma = 5.6704 \times 10^{-8} \ \text{W/m}^2 \cdot \text{K}^4$ is called the *Stefan–Boltzmann constant* after Josef Stefan (who discovered Eq. 18-38 experimentally in 1879) and Ludwig Boltzmann (who derived it theoretically soon after). The symbol $\varepsilon$ represents the *emissivity* of the object's surface, which has a value between 0 and 1, depending on the composition of the surface. A surface with the maximum emissivity of 1.0 is said to be a *blackbody radiator,* but such a surface is an ideal limit and does not occur in nature. Note again that the temperature in Eq. 18-38 must be in kelvins so that a temperature of absolute zero corresponds to no radiation. Note also that every object whose temperature is above 0 K—including you—emits thermal radiation. (See Fig. 18-20.)

The rate $P_{abs}$ at which an object absorbs energy via thermal radiation from its environment, which we take to be at uniform temperature $T_{env}$ (in kelvins), is

$$P_{abs} = \sigma \varepsilon A T_{env}^4. \tag{18-39}$$

The emissivity $\varepsilon$ in Eq. 18-39 is the same as that in Eq. 18-38. An idealized blackbody radiator, with $\varepsilon = 1$, will absorb all the radiated energy it intercepts (rather than sending a portion back away from itself through reflection or scattering).

Because an object will radiate energy to the environment while it absorbs energy from the environment, the object's net rate $P_{net}$ of energy exchange due to thermal radiation is

$$P_{net} = P_{abs} - P_{rad} = \sigma \varepsilon A (T_{env}^4 - T^4). \tag{18-40}$$

**FIG. 18-20** A false-color thermogram reveals the rate at which energy is radiated by a cat. The rate is color-coded, with white and red indicating the greatest radiation rate. The nose is cool. (*Edward Kinsman/Photo Researchers*)

**FIG. 18-21** A rattlesnake's face has thermal radiation detectors, allowing the snake to strike at an animal even in complete darkness. *(David A. Northcott/Corbis Images)*

$P_{net}$ is positive if net energy is being absorbed via radiation and negative if it is being lost via radiation.

Let's now return to the story about the ability of a *Melanophila* beetle to detect a fairly large fire from a distance of 12 km without seeing or smelling it. A pair of organs along each side of the beetle's body can detect even low-level thermal radiation. Each organ contains about 70 small knob-like sensors that expand very slightly when they absorb thermal radiation from the fire; the expansion causes them to press down on sensory cells. Thus, the detector is a mechanism that transfers energy from the thermal radiation to the energy of a mechanical device. The beetle can locate the fire by orienting itself so that all four infrared-detecting organs are affected, and then it flies toward the fire so that the response of the organs increases.

Thermal radiation is also involved in the numerous medical cases of a *dead* rattlesnake striking a hand reaching toward it. Pits between each eye and nostril of a rattlesnake (Fig. 18-21) serve as sensors of thermal radiation. When, say, a mouse moves close to a rattlesnake's head, the thermal radiation from the mouse triggers these sensors, causing a reflex action in which the snake strikes the mouse with its fangs and injects its venom. The thermal radiation from a reaching hand can cause the same reflex action even if the snake has been dead for as long as 30 min because the snake's nervous system continues to function. As one snake expert advised, if you must remove a recently killed rattlesnake, use a long stick rather than your hand.

## Sample Problem  18-6

Figure 18-22 shows the cross section of a wall made of white pine of thickness $L_a$ and brick of thickness $L_d$ $(= 2.0L_a)$, sandwiching two layers of unknown material with identical thicknesses and thermal conductivities. The thermal conductivity of the pine is $k_a$ and that of the brick is $k_d$ $(= 5.0k_a)$. The face area $A$ of the wall is unknown. Thermal conduction through the wall has reached the steady state; the only known interface temperatures are $T_1 = 25°C$, $T_2 = 20°C$, and $T_5 = -10°C$. What is interface temperature $T_4$?

**KEY IDEAS** (1) Temperature $T_4$ helps determine the rate $P_d$ at which energy is conducted through the brick, as given by Eq. 18-32. However, we lack enough data to solve Eq. 18-32 for $T_4$. (2) Because the conduction is steady, the conduction rate $P_d$ through the brick must equal the conduction rate $P_a$ through the pine. That gets us going.

**Calculations:** From Eq. 18-32 and Fig. 18-22, we can write

$$P_a = k_a A \frac{T_1 - T_2}{L_a} \quad \text{and} \quad P_d = k_d A \frac{T_4 - T_5}{L_d}.$$

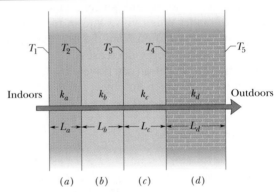

**FIG. 18-22** A wall of four layers through which there is steady-state heat transfer.

Setting $P_a = P_d$ and solving for $T_4$ yield

$$T_4 = \frac{k_a L_d}{k_d L_a}(T_1 - T_2) + T_5.$$

Letting $L_d = 2.0L_a$ and $k_d = 5.0k_a$, and inserting the known temperatures, we find

$$T_4 = \frac{k_a(2.0L_a)}{(5.0k_a)L_a}(25°C - 20°C) + (-10°C)$$
$$= -8.0°C. \qquad \text{(Answer)}$$

## Sample Problem  18-7  **Build your skill**

During an extended wilderness hike, you have a terrific craving for ice. Unfortunately, the air temperature drops to only 6.0°C each night—too high to freeze water. However, because a clear, moonless night sky acts like a blackbody radiator at a temperature of $T_s = -23°C$, perhaps you can make ice by letting a shallow layer of water radiate energy to such a sky. To start, you thermally insulate a container from the ground by placing a poorly conducting layer of, say, foam rubber or straw beneath it. Then you pour water into the container, forming a thin, uniform layer with mass $m = 4.5$ g, top surface area $A = 9.0$ cm$^2$, depth $d = 5.0$ mm, emissivity $\varepsilon = 0.90$, and initial temperature 6.0°C. Find the time required for the water to freeze via radiation. Can the freezing be accomplished during one night?

**KEY IDEAS** (1) The water cannot freeze at a temperature above the freezing point. Therefore, the radiation must first remove an amount of energy $Q_1$ to reduce the water temperature from 6.0°C to the freezing point of 0°C. (2) The radiation then must remove an additional amount of energy $Q_2$ to freeze all the water. (3) Throughout this process, the water is also absorbing energy radiated to it from the sky. We want a net loss of energy.

**Cooling the water:** Using Eq. 18-14 and Table 18-3, we find that cooling the water to 0°C requires an energy loss of

$$Q_1 = cm(T_f - T_i)$$
$$= (4190 \text{ J/kg} \cdot \text{K})(4.5 \times 10^{-3} \text{ kg})(0°C - 6.0°C)$$
$$= -113 \text{ J}.$$

Thus, 113 J must be radiated away by the water to drop its temperature to the freezing point.

**Freezing the water:** Using Eq. 18-16 ($Q = mL$) with the value of $L$ being $L_F$ from Eq. 18-18 or Table 18-4, and inserting a minus sign to indicate an energy loss, we find

$$Q_2 = -mL_F = -(4.5 \times 10^{-3} \text{ kg})(3.33 \times 10^5 \text{ J/kg})$$
$$= -1499 \text{ J}.$$

The total required energy loss is thus

$$Q_{tot} = Q_1 + Q_2 = -113 \text{ J} - 1499 \text{ J} = -1612 \text{ J}.$$

**Radiation:** While the water loses energy by radiating to the sky, it also absorbs energy radiated to it from the sky. In a total time $t$, we want the net energy of this exchange to be the energy loss $Q_{tot}$; so we want the power of this exchange to be

$$\text{power} = \frac{\text{net energy}}{\text{time}} = \frac{Q_{tot}}{t}. \qquad (18\text{-}41)$$

The power of such an energy exchange is also the net rate $P_{net}$ of thermal radiation, as given by Eq. 18-40; so the time $t$ required for the energy loss to be $Q_{tot}$ is

$$t = \frac{Q}{P_{net}} = \frac{Q}{\sigma \varepsilon A(T_s^4 - T^4)}. \qquad (18\text{-}42)$$

Although the temperature $T$ of the water decreases slightly while the water is cooling, we can approximate $T$ as being the freezing point, 273 K. With $T_s = 250$ K, the denominator of Eq. 18-42 is

$$(5.67 \times 10^{-8} \text{ W/m}^2 \cdot \text{K}^4)(0.90)(9.0 \times 10^{-4} \text{ m}^2)$$
$$\times [(250 \text{ K})^4 - (273 \text{ K})^4] = -7.57 \times 10^{-2} \text{ J/s},$$

and Eq. 18-42 gives us

$$t = \frac{-1612 \text{ J}}{-7.57 \times 10^{-2} \text{ J/s}}$$
$$= 2.13 \times 10^4 \text{ s} = 5.9 \text{ h}. \qquad (\text{Answer})$$

Because $t$ is less than a night, freezing water by having it radiate to the dark sky is feasible. In fact, in some parts of the world people used this technique long before the introduction of electric freezers.

## REVIEW & SUMMARY

**Temperature; Thermometers** Temperature is an SI base quantity related to our sense of hot and cold. It is measured with a thermometer, which contains a working substance with a measurable property, such as length or pressure, that changes in a regular way as the substance becomes hotter or colder.

**Zeroth Law of Thermodynamics** When a thermometer and some other object are placed in contact with each other, they eventually reach thermal equilibrium. The reading of the thermometer is then taken to be the temperature of the other object. The process provides consistent and useful temperature measurements because of the **zeroth law of thermodynamics:** If bodies $A$ and $B$ are each in thermal equilibrium with a third body $C$ (the thermometer), then $A$ and $B$ are in thermal equilibrium with each other.

**The Kelvin Temperature Scale** In the SI system, temperature is measured on the **Kelvin scale,** which is based on the *triple point* of water (273.16 K). Other temperatures are then defined by use of a *constant-volume gas thermometer,* in

which a sample of gas is maintained at constant volume so its pressure is proportional to its temperature. We define the *temperature $T$* as measured with a gas thermometer to be

$$T = (273.16 \text{ K}) \left( \lim_{\text{gas} \to 0} \frac{p}{p_3} \right). \qquad (18\text{-}6)$$

Here $T$ is in kelvins, and $p_3$ and $p$ are the pressures of the gas at 273.16 K and the measured temperature, respectively.

**Celsius and Fahrenheit Scales** The Celsius temperature scale is defined by

$$T_C = T - 273.15°, \qquad (18\text{-}7)$$

with $T$ in kelvins. The Fahrenheit temperature scale is defined by

$$T_F = \tfrac{9}{5}T_C + 32°. \qquad (18\text{-}8)$$

**Thermal Expansion** All objects change size with changes in temperature. For a temperature change $\Delta T$, a change $\Delta L$ in any linear dimension $L$ is given by

$$\Delta L = L\alpha \, \Delta T, \qquad (18\text{-}9)$$

in which $\alpha$ is the **coefficient of linear expansion.** The change $\Delta V$ in the volume $V$ of a solid or liquid is

$$\Delta V = V\beta \Delta T. \qquad (18\text{-}10)$$

Here $\beta = 3\alpha$ is the material's **coefficient of volume expansion.**

**Heat** Heat $Q$ is energy that is transferred between a system and its environment because of a temperature difference between them. It can be measured in **joules** (J), **calories** (cal), **kilocalories** (Cal or kcal), or **British thermal units** (Btu), with

$$1\text{ cal} = 3.968 \times 10^{-3}\text{ Btu} = 4.1868\text{ J}. \qquad (18\text{-}12)$$

**Heat Capacity and Specific Heat** If heat $Q$ is absorbed by an object, the object's temperature change $T_f - T_i$ is related to $Q$ by

$$Q = C(T_f - T_i), \qquad (18\text{-}13)$$

in which $C$ is the **heat capacity** of the object. If the object has mass $m$, then

$$Q = cm(T_f - T_i), \qquad (18\text{-}14)$$

where $c$ is the **specific heat** of the material making up the object. The **molar specific heat** of a material is the heat capacity per mole, which means per $6.02 \times 10^{23}$ elementary units of the material.

**Heat of Transformation** Heat absorbed by a material may change the material's physical state—for example, from solid to liquid or from liquid to gas. The amount of energy required per unit mass to change the state (but not the temperature) of a particular material is its **heat of transformation** $L$. Thus,

$$Q = Lm. \qquad (18\text{-}16)$$

The **heat of vaporization** $L_V$ is the amount of energy per unit mass that must be added to vaporize a liquid or that must be removed to condense a gas. The **heat of fusion** $L_F$ is the amount of energy per unit mass that must be added to melt a solid or that must be removed to freeze a liquid.

**Work Associated with Volume Change** A gas may exchange energy with its surroundings through work. The amount of work $W$ done *by* a gas as it expands or contracts from an initial volume $V_i$ to a final volume $V_f$ is given by

$$W = \int dW = \int_{V_i}^{V_f} p\, dV. \qquad (18\text{-}25)$$

The integration is necessary because the pressure $p$ may vary during the volume change.

**First Law of Thermodynamics** The principle of conservation of energy for a thermodynamic process is expressed in the **first law of thermodynamics,** which may assume either of the forms

$$\Delta E_{\text{int}} = E_{\text{int},f} - E_{\text{int},i} = Q - W \quad \text{(first law)} \quad (18\text{-}26)$$

or

$$dE_{\text{int}} = dQ - dW \quad \text{(first law)}. \qquad (18\text{-}27)$$

$E_{\text{int}}$ represents the internal energy of the material, which depends only on the material's state (temperature, pressure, and volume). $Q$ represents the energy exchanged as heat between the system and its surroundings; $Q$ is positive if the system absorbs heat and negative if the system loses heat. $W$ is the work done *by* the system; $W$ is positive if the system expands against an external force from the surroundings and negative if the system contracts because of an external force. $Q$ and $W$ are path dependent; $\Delta E_{\text{int}}$ is path independent.

**Applications of the First Law** The first law of thermodynamics finds application in several special cases:

$$\begin{aligned}
\textit{adiabatic processes:} &\quad Q = 0, \quad \Delta E_{\text{int}} = -W \\
\textit{constant-volume processes:} &\quad W = 0, \quad \Delta E_{\text{int}} = Q \\
\textit{cyclical processes:} &\quad \Delta E_{\text{int}} = 0, \quad Q = W \\
\textit{free expansions:} &\quad Q = W = \Delta E_{\text{int}} = 0
\end{aligned}$$

**Conduction, Convection, and Radiation** The rate $P_{\text{cond}}$ at which energy is *conducted* through a slab whose faces are maintained at temperatures $T_H$ and $T_C$ is

$$P_{\text{cond}} = \frac{Q}{t} = kA\frac{T_H - T_C}{L}, \qquad (18\text{-}32)$$

in which $A$ and $L$ are the face area and length of the slab, and $k$ is the thermal conductivity of the material.

*Convection* occurs when temperature differences cause an energy transfer by motion within a fluid. *Radiation* is an energy transfer via the emission of electromagnetic energy. The rate $P_{\text{rad}}$ at which an object emits energy via thermal radiation is

$$P_{\text{rad}} = \sigma\varepsilon AT^4, \qquad (18\text{-}38)$$

where $\sigma\ (= 5.6704 \times 10^{-8}\text{ W/m}^2\cdot\text{K}^4)$ is the Stefan–Boltzmann constant, $\varepsilon$ is the emissivity of the object's surface, $A$ is its surface area, and $T$ is its surface temperature (in kelvins). The rate $P_{\text{abs}}$ at which an object absorbs energy via thermal radiation from its environment, which is at the uniform temperature $T_{\text{env}}$ (in kelvins), is

$$P_{\text{abs}} = \sigma\varepsilon AT_{\text{env}}^4. \qquad (18\text{-}39)$$

## QUESTIONS

**1** Materials $A$, $B$, and $C$ are solids that are at their melting temperatures. Material $A$ requires 200 J to melt 4 kg, material $B$ requires 300 J to melt 5 kg, and material $C$ requires 300 J to melt 6 kg. Rank the materials according to their heats of fusion, greatest first.

**2** Figure 18-23 shows three linear temperature scales, with the freezing and boiling points of water indicated. Rank the

three scales according to the size of one degree on them, greatest first.

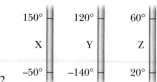

FIG. 18-23 Question 2.

**3** The initial length $L$, change in temperature $\Delta T$, and change in length $\Delta L$ of four rods are given in the following table. Rank the rods according to their coefficients of thermal expansion, greatest first.

| Rod | $L$ (m) | $\Delta T$ (C°) | $\Delta L$ (m) |
|-----|---------|-----------------|----------------|
| $a$ | 2 | 10 | $4 \times 10^{-4}$ |
| $b$ | 1 | 20 | $4 \times 10^{-4}$ |
| $c$ | 2 | 10 | $8 \times 10^{-4}$ |
| $d$ | 4 | 5 | $4 \times 10^{-4}$ |

**4** Figure 18-24 shows three different arrangements of materials 1, 2, and 3 to form a wall. The thermal conductivities are $k_1 > k_2 > k_3$. The left side of the wall is 20 C° higher than the right side. Rank the arrangements according to (a) the (steady state) rate of energy conduction through the wall and (b) the temperature difference across material 1, greatest first.

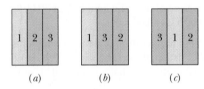

FIG. 18-24   Question 4.

**5** Figure 18-25 shows two closed cycles on $p$-$V$ diagrams for a gas. The three parts of cycle 1 are of the same length and shape as those of cycle 2. For each cycle, should the cycle be traversed clockwise or counterclockwise if (a) the net work $W$ done by the gas is to be positive and (b) the net energy transferred by the gas as heat $Q$ is to be positive?

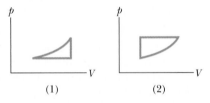

FIG. 18-25   Questions 5 and 6.

**6** For which cycle in Fig. 18-25, traversed clockwise, is (a) $W$ greater and (b) $Q$ greater?

**7** A hot object is dropped into a thermally insulated container of water, and the object and water are then allowed to come to thermal equilibrium. The experiment is repeated twice, with different hot objects. All three objects have the same mass and initial temperature, and the mass and initial temperature of the water are the same in the three experiments. For each of the experiments, Fig. 18-26 gives graphs of the temperatures $T$ of

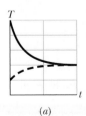
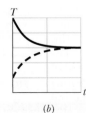
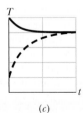

FIG. 18-26   Question 7.

the object and the water versus time $t$. Rank the graphs according to the specific heats of the objects, greatest first.

**8** A sample $A$ of liquid water and a sample $B$ of ice, of identical mass, are placed in a thermally insulated container and allowed to come to thermal equilibrium. Figure 18-27a is a sketch of the temperature $T$ of the samples versus time $t$. (a) Is the equilibrium temperature above, below, or at the freezing point of water? (b) In reaching equilibrium, does the liquid partly freeze, fully freeze, or undergo no freezing? (c) Does the ice partly melt, fully melt, or undergo no melting?

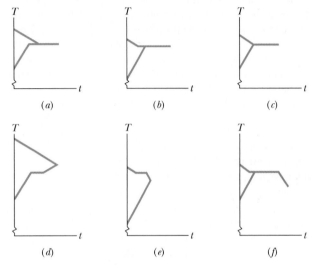

FIG. 18-27   Questions 8 and 9.

**9** Question 8 continued: Graphs $b$ through $f$ of Fig. 18-27 are additional sketches of $T$ versus $t$, of which one or more are impossible to produce. (a) Which is impossible and why? (b) In the possible ones, is the equilibrium temperature above, below, or at the freezing point of water? (c) As the possible situations reach equilibrium, does the liquid partly freeze, fully freeze, or undergo no freezing? Does the ice partly melt, fully melt, or undergo no melting?

**10** A solid cube of edge length $r$, a solid sphere of radius $r$, and a solid hemisphere of radius $r$, all made of the same material, are maintained at temperature 300 K in an environment at temperature 350 K. Rank the objects according to the net rate at which thermal radiation is exchanged with the environment, greatest first.

**11** Three different materials of identical mass are placed one at a time in a special freezer that can extract energy from a material at a certain constant rate. During the cooling process, each material begins in the liquid state and ends in the solid state; Fig. 18-28 shows the temperature $T$ versus time $t$. (a) For material 1, is the specific heat for the liquid state greater than or less than that for the solid state? Rank the materials according to (b) freezing-point temperature, (c) specific heat in the liquid state, (d) specific heat in the solid state, and (e) heat of fusion, all greatest first.

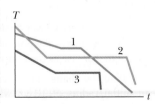

FIG. 18-28   Question 11.

## PROBLEMS

**GO**    Tutoring problem available (at instructor's discretion) in *WileyPLUS* and WebAssign
**SSM**    Worked-out solution available in Student Solutions Manual        **WWW**    Worked-out solution is at ──
• – •••    Number of dots indicates level of problem difficulty        **ILW**    Interactive solution is at ──        http://www.wiley.com/college/halliday
✈    Additional information available in *The Flying Circus of Physics* and at flyingcircusofphysics.com

### sec. 18-4    Measuring Temperature

•1    A gas thermometer is constructed of two gas-containing bulbs, each in a water bath, as shown in Fig. 18-29. The pressure difference between the two bulbs is measured by a mercury manometer as shown. Appropriate reservoirs, not shown in the diagram, maintain constant gas volume in the two bulbs. There is no difference in pressure when both baths are at the triple point of water. The pressure difference is 120 torr when one bath is at the triple point and the other is at the boiling point of water. It is 90.0 torr when one bath is at the triple point and the other is at an unknown temperature to be measured. What is the unknown temperature?

FIG. 18-29    Problem 1.

•2    Two constant-volume gas thermometers are assembled, one with nitrogen and the other with hydrogen. Both contain enough gas so that $p_3 = 80$ kPa. (a) What is the difference between the pressures in the two thermometers if both bulbs are in boiling water? (*Hint:* See Fig. 18-6.) (b) Which gas is at higher pressure?

•3    Suppose the temperature of a gas is 373.15 K when it is at the boiling point of water. What then is the limiting value of the ratio of the pressure of the gas at that boiling point to its pressure at the triple point of water? (Assume the volume of the gas is the same at both temperatures.)

### sec. 18-5    The Celsius and Fahrenheit Scales

•4    (a) In 1964, the temperature in the Siberian village of Oymyakon reached −71°C. What temperature is this on the Fahrenheit scale? (b) The highest officially recorded temperature in the continental United States was 134°F in Death Valley, California. What is this temperature on the Celsius scale?

•5    At what temperature is the Fahrenheit scale reading equal to (a) twice that of the Celsius scale and (b) half that of the Celsius scale?

••6    On a linear X temperature scale, water freezes at −125.0°X and boils at 375.0°X. On a linear Y temperature scale, water freezes at −70.00°Y and boils at −30.00°Y. A temperature of 50.00°Y corresponds to what temperature on the X scale?

••7    Suppose that on a linear temperature scale X, water boils at −53.5°X and freezes at −170°X. What is a temperature of 340 K on the X scale? (Approximate water's boiling point as 373 K.)    **ILW**

### sec. 18-6    Thermal Expansion

•8    An aluminum flagpole is 33 m high. By how much does its length increase as the temperature increases by 15 C°?

•9    Find the change in volume of an aluminum sphere with an initial radius of 10 cm when the sphere is heated from 0.0°C to 100°C.    **SSM**

•10    An aluminum-alloy rod has a length of 10.000 cm at 20.000°C and a length of 10.015 cm at the boiling point of water. (a) What is the length of the rod at the freezing point of water? (b) What is the temperature if the length of the rod is 10.009 cm?

•11    A circular hole in an aluminum plate is 2.725 cm in diameter at 0.000°C. What is its diameter when the temperature of the plate is raised to 100.0°C?    **ILW**

•12    At 20°C, a brass cube has an edge length of 30 cm. What is the increase in the cube's surface area when it is heated from 20°C to 75°C?

•13    What is the volume of a lead ball at 30.00°C if the ball's volume at 60.00°C is 50.00 cm³?

••14    When the temperature of a metal cylinder is raised from 0.0°C to 100°C, its length increases by 0.23%. (a) Find the percent change in density. (b) What is the metal? Use Table 18-2.

••15    An aluminum cup of 100 cm³ capacity is completely filled with glycerin at 22°C. How much glycerin, if any, will spill out of the cup if the temperature of both the cup and the glycerin is increased to 28°C? (The coefficient of volume expansion of glycerin is $5.1 \times 10^{-4}/C°$.)    **SSM WWW**

••16    At 20°C, a rod is exactly 20.05 cm long on a steel ruler. Both the rod and the ruler are placed in an oven at 270°C, where the rod now measures 20.11 cm on the same ruler. What is the coefficient of linear expansion for the material of which the rod is made?

••17    A steel rod is 3.000 cm in diameter at 25.00°C. A brass ring has an interior diameter of 2.992 cm at 25.00°C. At what common temperature will the ring just slide onto the rod?    **ILW**

••18    When the temperature of a copper coin is raised by 100 C°, its diameter increases by 0.18%. To two significant figures, give the percent increase in (a) the area of a face, (b) the thickness, (c) the volume, and (d) the mass of the coin. (e) Calculate the coefficient of linear expansion of the coin.

••19    A vertical glass tube of length $L = 1.280\,000$ m is half filled with a liquid at 20.000 000°C. How much will the height of the liquid column change when the tube is heated to 30.000 000°C? Take $\alpha_{glass} = 1.000\,000 \times 10^{-5}$/K and $\beta_{liquid} = 4.000\,000 \times 10^{-5}$/K.    **GO**

••20    In a certain experiment, a small radioactive source must move at selected, extremely slow speeds. This motion is accomplished by fastening the source to one end of an aluminum rod and heating the central section of the rod in a controlled way. If the effective heated section of the rod in Fig. 18-30 has length $d = 2.00$ cm, at what constant rate must the temperature of the rod be changed if the source is to

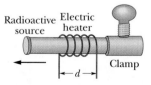

FIG. 18-30    Problem 20.

move at a constant speed of 100 nm/s?

•••**21** As a result of a temperature rise of 32 C°, a bar with a crack at its center buckles upward (Fig. 18-31). If the fixed distance $L_0$ is 3.77 m and the coefficient of linear expansion of the bar is $25 \times 10^{-6}/C°$, find the rise $x$ of the center. **SSM ILW**

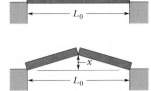

**FIG. 18-31** Problem 21.

### sec. 18-8 The Absorption of Heat by Solids and Liquids

•**22** A certain substance has a mass per mole of 50.0 g/mol. When 314 J is added as heat to a 30.0 g sample, the sample's temperature rises from 25.0°C to 45.0°C. What are the (a) specific heat and (b) molar specific heat of this substance? (c) How many moles are present?

•**23** A certain diet doctor encourages people to diet by drinking ice water. His theory is that the body must burn off enough fat to raise the temperature of the water from 0.00°C to the body temperature of 37.0°C. How many liters of ice water would have to be consumed to burn off 454 g (about 1 lb) of fat, assuming that burning this much fat requires 3500 Cal be transferred to the ice water? Why is it not advisable to follow this diet? (One liter $= 10^3$ cm³. The density of water is 1.00 g/cm³.)

•**24** How much water remains unfrozen after 50.2 kJ is transferred as heat from 260 g of liquid water initially at its freezing point?

•**25** Calculate the minimum amount of energy, in joules, required to completely melt 130 g of silver initially at 15.0°C. **SSM**

•**26** One way to keep the contents of a garage from becoming too cold on a night when a severe subfreezing temperature is forecast is to put a tub of water in the garage. If the mass of the water is 125 kg and its initial temperature is 20°C, (a) how much energy must the water transfer to its surroundings in order to freeze completely and (b) what is the lowest possible temperature of the water and its surroundings until that happens?

•**27** A small electric immersion heater is used to heat 100 g of water for a cup of instant coffee. The heater is labeled "200 watts" (it converts electrical energy to thermal energy at this rate). Calculate the time required to bring all this water from 23.0°C to 100°C, ignoring any heat losses. **SSM**

•**28** What mass of butter, which has a usable energy content of 6.0 Cal/g ($= 6000$ cal/g), would be equivalent to the change in gravitational potential energy of a 73.0 kg man who ascends from sea level to the top of Mt. Everest, at elevation 8.84 km? Assume that the average $g$ for the ascent is 9.80 m/s².

••**29** What mass of steam at 100°C must be mixed with 150 g of ice at its melting point, in a thermally insulated container, to produce liquid water at 50°C? **ILW**

••**30** A 150 g copper bowl contains 220 g of water, both at 20.0°C. A very hot 300 g copper cylinder is dropped into the water, causing the water to boil, with 5.00 g being converted to steam. The final temperature of the system is 100°C. Neglect energy transfers with the environment. (a) How much energy (in calories) is transferred to the water as heat? (b) How much to the bowl? (c) What is the original temperature of the cylinder?

••**31** *Nonmetric version:* (a) How long does a $2.0 \times 10^5$

Btu/h water heater take to raise the temperature of 40 gal of water from 70°F to 100°F? *Metric version:* (b) How long does a 59 kW water heater take to raise the temperature of 150 L of water from 21°C to 38°C?

••**32** Samples $A$ and $B$ are at different initial temperatures when they are placed in a thermally insulated container and allowed to come to thermal equilibrium. Figure 18-32a gives their temperatures $T$ versus time $t$. Sample $A$ has a mass of 5.0 kg; sample $B$ has a mass of 1.5 kg. Figure 18-32b is a general plot for the material of sample $B$. It shows the temperature change $\Delta T$ that the material undergoes when energy is transferred to it as heat $Q$. The change $\Delta T$ is plotted versus the energy $Q$ per unit mass of the material, and the scale of the vertical axis is set by $\Delta T_s = 4.0$ C°. What is the specific heat of sample $A$? **GO**

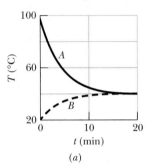
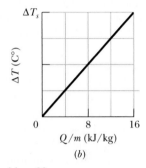

**FIG. 18-32** Problem 32.

••**33** In a solar water heater, energy from the Sun is gathered by water that circulates through tubes in a rooftop collector. The solar radiation enters the collector through a transparent cover and warms the water in the tubes; this water is pumped into a holding tank. Assume that the efficiency of the overall system is 20% (that is, 80% of the incident solar energy is lost from the system). What collector area is necessary to raise the temperature of 200 L of water in the tank from 20°C to 40°C in 1.0 h when the intensity of incident sunlight is 700 W/m²?

••**34** A 0.400 kg sample is placed in a cooling apparatus that removes energy as heat at a constant rate. Figure 18-33 gives the temperature $T$ of the sample versus time $t$; the horizontal scale is set by $t_s = 80.0$ min. The sample freezes during the energy removal. The specific heat of the sample in its initial liquid phase is 3000 J/kg·K. What are (a) the sample's heat of fusion and (b) its specific heat in the frozen phase?

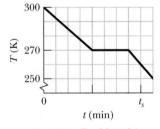

**FIG. 18-33** Problem 34.

••**35** An insulated Thermos contains 130 cm³ of hot coffee at 80.0°C. You put in a 12.0 g ice cube at its melting point to cool the coffee. By how many degrees has your coffee cooled once the ice has melted and equilibrium is reached? Treat the coffee as though it were pure water and neglect energy exchanges with the environment.

••**36** A 0.530 kg sample of liquid water and a sample of ice are placed in a thermally insulated container. The container also contains a device that transfers energy as heat from the liquid water to the ice at a constant rate $P$, until thermal equilibrium is reached. The temperatures $T$ of the liquid water and the ice are given in Fig. 18-34 as functions of time $t$; the horizontal scale is set by $t_s = 80.0$ min. (a) What is rate $P$? (b) What is the initial mass of the ice in the container? (c) When thermal equilibrium is reached, what is the mass of the ice produced in this process?

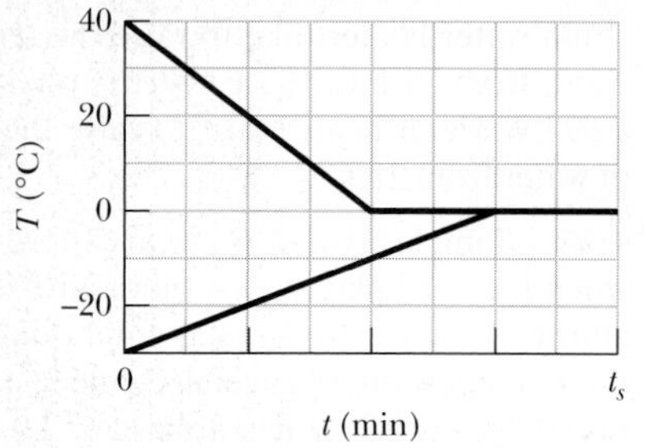

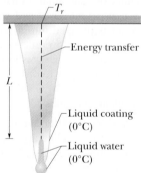

FIG. 18-34    Problem 36.

••**37**  Ethyl alcohol has a boiling point of 78.0°C, a freezing point of −114°C, a heat of vaporization of 879 kJ/kg, a heat of fusion of 109 kJ/kg, and a specific heat of 2.43 kJ/kg·K. How much energy must be removed from 0.510 kg of ethyl alcohol that is initially a gas at 78.0°C so that it becomes a solid at −114°C? **GO**

••**38**  The specific heat of a substance varies with temperature according to $c = 0.20 + 0.14T + 0.023T^2$, with $T$ in °C and $c$ in cal/g·K. Find the energy required to raise the temperature of 2.0 g of this substance from 5.0°C to 15°C.

••**39**  A person makes a quantity of iced tea by mixing 500 g of hot tea (essentially water) with an equal mass of ice at its melting point. Assume the mixture has negligible energy exchanges with its environment. If the tea's initial temperature is $T_i = 90$°C, when thermal equilibrium is reached what are (a) the mixture's temperature $T_f$ and (b) the remaining mass $m_f$ of ice? If $T_i = 70$°C, when thermal equilibrium is reached what are (c) $T_f$ and (d) $m_f$?

•••**40**  *Icicles.* Liquid water coats an active (growing) icicle and extends up a short, narrow tube along the central axis (Fig. 18-35). Because the water–ice interface must have a temperature of 0°C, the water in the tube cannot lose energy through the sides of the icicle or down through the tip because there is no temperature change in those directions. It can lose energy and freeze only by sending energy up (through distance $L$) to the top of the icicle, where the temperature $T_r$ can be below 0°C. Take $L = 0.12$ m and $T_r = −5$°C. Assume that the central tube and the upward conduction path both have cross-sectional area $A$. In terms of $A$, what rate is (a) energy conducted upward and (b) mass converted from liquid to ice at the top of the central tube? (c) At what rate does the top of the tube move downward because of water freezing there? The thermal conductivity of ice is 0.400 W/m·K, and the density of liquid water is 1000 kg/m³.

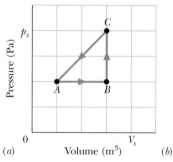

FIG. 18-35    Problem 40.

•••**41**  (a) Two 50 g ice cubes are dropped into 200 g of water in a thermally insulated container. If the water is initially at 25°C, and the ice comes directly from a freezer at −15°C, what is the final temperature at thermal equilibrium? (b) What is the final temperature if only one ice cube is used? **SSM WWW**

•••**42**  A 20.0 g copper ring at 0.000°C has an inner diameter of $D = 2.54000$ cm. An aluminum sphere at 100.0°C has a diameter of $d = 2.545\ 08$ cm. The sphere is placed on top of the ring (Fig. 18-36), and the two are allowed to come to thermal equilibrium, with no heat lost to the surroundings. The sphere just passes through the ring at the equilibrium temperature. What is the mass of the sphere?

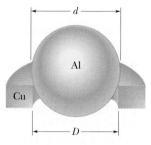

FIG. 18-36    Problem 42.

### sec. 18-11   Some Special Cases of the First Law of Thermodynamics

•**43**  A gas within a closed chamber undergoes the cycle shown in the p-V diagram of Fig. 18-37. The horizontal scale is set by $V_s = 4.0$ m³. Calculate the net energy added to the system as heat during one complete cycle. **SSM ILW**

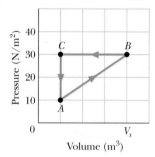

FIG. 18-37    Problem 43.

•**44**  Suppose 200 J of work is done on a system and 70.0 cal is extracted from the system as heat. In the sense of the first law of thermodynamics, what are the values (including algebraic signs) of (a) $W$, (b) $Q$, and (c) $\Delta E_{int}$?

•**45**  In Fig. 18-38, a gas sample expands from $V_0$ to $4.0V_0$ while its pressure decreases from $p_0$ to $p_0/4.0$. If $V_0 = 1.0$ m³ and $p_0 = 40$ Pa, how much work is done by the gas if its pressure changes with volume via (a) path $A$, (b) path $B$, and (c) path $C$?

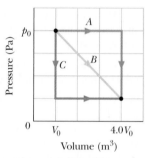

FIG. 18-38    Problem 45.

•**46**  A thermodynamic system is taken from state $A$ to state $B$ to state $C$, and then back to $A$, as shown in the p-V diagram of Fig. 18-39a. The vertical scale is set by $p_s = 40$ Pa, and the horizontal scale is set by $V_s = 4.0$ m³. (a)–(g) Complete the table in Fig. 18-39b by inserting a plus sign, a minus sign, or a zero in each indicated cell. (h) What is the net work done by the system as it moves once through the cycle $ABCA$?

|   | Q | W | $\Delta E_{int}$ |
|---|---|---|---|
| $A \longrightarrow B$ | (a) | (b) | + |
| $B \longrightarrow C$ | + | (c) | (d) |
| $C \longrightarrow A$ | (e) | (f) | (g) |

(a)    Volume (m³)    (b)

FIG. 18-39    Problem 46.

••**47**  Figure 18-40 displays a closed cycle for a gas (the figure is not drawn to scale). The change in the internal energy of the gas as it moves from *a* to *c* along the path *abc* is −200 J. As it moves from *c* to *d*, 180 J must be transferred to it as heat. An additional transfer of 80 J to it as heat is needed as it moves from *d* to *a*. How much work is done on the gas as it moves from *c* to *d*? **GO**

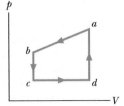

**FIG. 18-40**   Problem 47.

**••48**  A sample of gas is taken through cycle *abca* shown in the *p-V* diagram of Fig. 18-41. The net work done is +1.2 J. Along path *ab*, the change in the internal energy is +3.0 J and the magnitude of the work done is 5.0 J. Along path *ca*, the energy transferred to the gas as heat is +2.5 J. How much energy is transferred as heat along (a) path *ab* and (b) path *bc*?

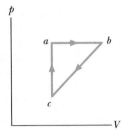

**FIG. 18-41**   Problem 48.

**••49**  When a system is taken from state *i* to state *f* along path *iaf* in Fig. 18-42, $Q = 50$ cal and $W = 20$ cal. Along path *ibf*, $Q = 36$ cal. (a) What is $W$ along path *ibf*? (b) If $W = -13$ cal for the return path *fi*, what is $Q$ for this path? (c) If $E_{int,i} = 10$ cal, what is $E_{int,f}$? If $E_{int,b} = 22$ cal, what is $Q$ for (d) path *ib* and (e) path *bf*? **SSM WWW**

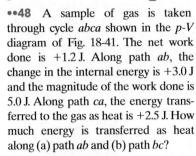

**FIG. 18-42**   Problem 49.

**••50**  Gas within a chamber passes through the cycle shown in Fig. 18-43. Determine the energy transferred by the system as heat during process *CA* if the energy added as heat $Q_{AB}$ during process *AB* is 20.0 J, no energy is transferred as heat during process *BC*, and the net work done during the cycle is 15.0 J.

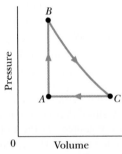

**FIG. 18-43**   Problem 50.

### sec. 18-12   Heat Transfer Mechanisms

**•51**  Consider the slab shown in Fig. 18-18. Suppose that $L = 25.0$ cm, $A = 90.0$ cm², and the material is copper. If $T_H = 125°C$, $T_C = 10.0°C$, and a steady state is reached, find the conduction rate through the slab.   **SSM**

**•52**  If you were to walk briefly in space without a spacesuit while far from the Sun (as an astronaut does in the movie *2001, A Space Odyssey*), you would feel the cold of space—while you radiated energy, you would absorb almost none from your environment. (a) At what rate would you lose energy? (b) How much energy would you lose in 30 s? Assume that your emissivity is 0.90, and estimate other data needed in the calculations.

**•53**  A cylindrical copper rod of length 1.2 m and cross-sectional area 4.8 cm² is insulated to prevent heat loss through its surface. The ends are maintained at a temperature difference of 100 C° by having one end in a water–ice mixture and the other in a mixture of boiling water and steam. (a) At what rate is energy conducted along the rod? (b) At what rate does ice melt at the cold end?   **ILW**

**•54**  The ceiling of a single-family dwelling in a cold climate should have an *R*-value of 30. To give such insulation, how thick would a layer of (a) polyurethane foam and (b) silver have to be?

**•55**  A sphere of radius 0.500 m, temperature 27.0°C, and emissivity 0.850 is located in an environment of temperature 77.0°C. At what rate does the sphere (a) emit and (b) absorb thermal radiation? (c) What is the sphere's net rate of energy exchange?

**••56**  A solid cylinder of radius $r_1 = 2.5$ cm, length $h_1 = 5.0$ cm, emissivity 0.85, and temperature 30°C is suspended in an environment of temperature 50°C. (a) What is the cylinder's net thermal radiation transfer rate $P_1$? (b) If the cylinder is stretched until its radius is $r_2 = 0.50$ cm, its net thermal radiation transfer rate becomes $P_2$. What is the ratio $P_2/P_1$?

**••57**  In Fig. 18-44*a*, two identical rectangular rods of metal are welded end to end, with a temperature of $T_1 = 0°C$ on the left side and a temperature of $T_2 = 100°C$ on the right side. In 2.0 min, 10 J is conducted at a constant rate from the right side to the left side. How much time would be required to conduct 10 J if the rods were welded side to side as in Fig. 18-44*b*?

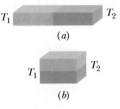

**FIG. 18-44**   Problem 57.

**••58**  Figure 18-45 shows the cross section of a wall made of three layers. The thicknesses of the layers are $L_1$, $L_2 = 0.700L_1$, and $L_3 = 0.350L_1$. The thermal conductivities are $k_1$, $k_2 = 0.900k_1$, and $k_3 = 0.800k_1$. The

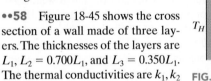

**FIG. 18-45**   Problem 58.

temperatures at the left and right sides of the wall are $T_H = 30.0°C$ and $T_C = -15.0°C$, respectively. Thermal conduction is steady. (a) What is the temperature difference $\Delta T_2$ across layer 2 (between the left and right sides of the layer)? If $k_2$ were, instead, equal to $1.1k_1$, (b) would the rate at which energy is conducted through the wall be greater than, less than, or the same as previously, and (c) what would be the value of $\Delta T_2$?

**••59**  (a) What is the rate of energy loss in watts per square meter through a glass window 3.0 mm thick if the outside temperature is −20°F and the inside temperature is +72°F? (b) A storm window having the same thickness of glass is installed parallel to the first window, with an air gap of 7.5 cm between the two windows. What now is the rate of energy loss if conduction is the only important energy-loss mechanism?

**••60**  The giant hornet *Vespa mandarinia japonica* preys on Japanese bees. However, if one of the hornets attempts to invade a beehive, several hundred of the bees quickly form a compact ball around the hornet to stop it. They don't sting, bite, crush, or suffocate it. Rather they overheat it by quickly raising their body temperatures from the normal 35°C to 47°C or 48°C, which is lethal to the hornet but not to the bees (Fig. 18-46). Assume the following: 500 bees form a ball of radius $R = 2.0$

**FIG. 18-46**
Problem 60.
*(© Dr. Masato Ono, Tamagawa University)*

cm for a time $t = 20$ min, the primary loss of energy by the ball is by thermal radiation, the ball's surface has emissivity $\varepsilon = 0.80$, and the ball has a uniform temperature. On average, how much additional energy must each bee produce during the 20 min to maintain 47°C?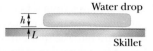

••61 Figure 18-47 shows (in cross section) a wall consisting of four layers, with thermal conductivities $k_1 = 0.060$ W/m·K, $k_3 = 0.040$ W/m·K, and $k_4 = 0.12$ W/m·K ($k_2$ is not known). The layer thicknesses are $L_1 = 1.5$ cm, $L_3 = 2.8$ cm, and $L_4 = 3.5$ cm ($L_2$ is not known). The known temperatures are $T_1 = 30°C$, $T_{12} = 25°C$, and $T_4 = -10°C$. Energy transfer through the wall is steady. What is interface temperature $T_{34}$? **GO**

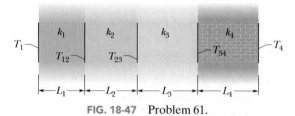

FIG. 18-47   Problem 61.

••62   *Penguin huddling.* To withstand the harsh weather of the Antarctic, emperor penguins huddle in groups (Fig. 18-48). Assume that a penguin is a circular cylinder with a top surface area $a = 0.34$ m² and height $h = 1.1$ m. Let $P_r$ be the rate at which an individual penguin radiates energy to the environment (through the top and the sides); thus $NP_r$ is the rate at which $N$ identical, well-separated penguins radiate. If the penguins huddle closely to form a *huddled cylinder* with top surface area $Na$ and height $h$, the cylinder radiates at the rate $P_h$. If $N = 1000$, (a) what is the value of the fraction $P_h/NP_r$ and (b) by what percentage does huddling reduce the total radiation loss?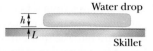

FIG. 18-48   Problem 62.
*(Alain Torterotot/Peter Arnold, Inc.)*

••63   Ice has formed on a shallow pond, and a steady state has been reached, with the air above the ice at $-5.0°C$ and the bottom of the pond at $4.0°C$. If the total depth of *ice + water* is 1.4 m, how thick is the ice? (Assume that the thermal conductivities of ice and water are 0.40 and 0.12 cal/m·C°·s, respectively.)

••64   *Leidenfrost effect.* A water drop that is slung onto a skillet with a temperature between 100°C and about 200°C will last about 1 s. However, if the skillet is much hotter, the drop can last several minutes, an effect named after an early

investigator. The longer lifetime is due to the support of a thin layer of air and water vapor that separates the drop from the metal (by distance $L$ in Fig. 18-49). Let $L = 0.100$ mm, and assume that the drop is flat with height $h = 1.50$ mm and bottom face area $A = 4.00 \times 10^{-6}$ m². Also assume that the skillet has a constant temperature $T_s = 300°C$ and the drop has a temperature of 100°C. Water has density $\rho = 1000$ kg/m³, and the supporting layer has thermal conductivity $k = 0.026$ W/m·K. (a) At what rate is energy conducted from the skillet to the drop through the drop's bottom surface? (b) If conduction is the primary way energy moves from the skillet to the drop, how long will the drop last?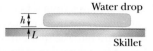

FIG. 18-49   Problem 64.

••65   A tank of water has been outdoors in cold weather, and a slab of ice 5.0 cm thick has formed on its surface (Fig. 18-50). The air above the ice is at $-10°C$. Calculate the rate of ice formation (in centimeters per hour) on the ice slab. Take the thermal conductivity of ice to be 0.0040 cal/s·cm·C° and its density to be 0.92 g/cm³. Assume no energy transfer through the tank walls or bottom. **SSM**

•••66   *Evaporative cooling of beverages.* A cold beverage can be kept cold even on a warm day if it is slipped into a porous ceramic container that has been soaked in water. Assume that energy lost to evaporation matches the net energy gained via the radiation exchange through the top and side surfaces. The container and beverage have temperature $T = 15°C$, the environment has temperature $T_{env} = 32°C$, and the container is a cylinder with radius $r = 2.2$ cm and height 10 cm. Approximate the emissivity as $\varepsilon = 1$, and neglect other energy exchanges. At what rate $dm/dt$ is the container losing water mass? **GO**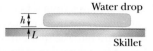

FIG. 18-50   Problem 65.

**Additional Problems**

67   In the extrusion of cold chocolate from a tube, work is done on the chocolate by the pressure applied by a ram forcing the chocolate through the tube. The work per unit mass of extruded chocolate is equal to $p/\rho$, where $p$ is the difference between the applied pressure and the pressure where the chocolate emerges from the tube, and $\rho$ is the density of the chocolate. Rather than increasing the temperature of the chocolate, this work melts cocoa fats in the chocolate. These fats have a heat of fusion of 150 kJ/kg. Assume that all of the work goes into that melting and that these fats make up 30% of the chocolate's mass. What percentage of the fats melt during the extrusion if $p = 5.5$ MPa and $\rho = 1200$ kg/m³?

68   In a series of experiments, block $B$ is to be placed in a thermally insulated container with block $A$, which has the same mass as block $B$. In each experiment, block $B$ is initially at a certain temperature $T_B$, but temperature $T_A$ of block $A$ is changed from experiment to experiment. Let $T_f$ represent the final temperature of the two blocks when they reach ther-

mal equilibrium in any of the experiments. Figure 18-51 gives temperature $T_f$ versus the initial temperature $T_A$ for a range of possible values of $T_A$, from $T_{A1}$ = 0 K to $T_{A2}$ = 500 K. The vertical axis scale is set by $T_{fs}$ = 400 K. What are (a) temperature $T_B$ and (b) the ratio $c_B/c_A$ of the specific heats of the blocks?

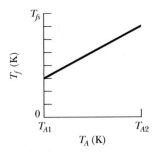

FIG. 18-51   Problem 68.

**69** A 0.300 kg sample is placed in a cooling apparatus that removes energy as heat at a constant rate of 2.81 W. Figure 18-52 gives the temperature $T$ of the sample versus time $t$. The temperature scale is set by $T_s$ = 30°C and the time scale is set by $t_s$ = 20 min. What is the specific heat of the sample?

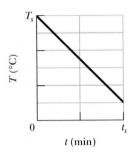

FIG. 18-52   Problem 69.

**70** Calculate the specific heat of a metal from the following data. A container made of the metal has a mass of 3.6 kg and contains 14 kg of water. A 1.8 kg piece of the metal initially at a temperature of 180°C is dropped into the water. The container and water initially have a temperature of 16.0°C, and the final temperature of the entire system is 18.0°C.

**71** What is the volume increase of an aluminum cube 5.00 cm on an edge when heated from 10.0°C to 60.0°C?

**72** A copper rod, an aluminum rod, and a brass rod, each of 6.00 m length and 1.00 cm diameter, are placed end to end with the aluminum rod between the other two. The free end of the copper rod is maintained at water's boiling point, and the free end of the brass rod is maintained at water's freezing point. What is the steady-state temperature of (a) the copper–aluminum junction and (b) the aluminum–brass junction?

**73** A sample of gas undergoes a transition from an initial state $a$ to a final state $b$ by three different paths (processes), as shown in the $p$-$V$ diagram in Fig. 18-53, where $V_b$ = 5.00$V_i$. The energy transferred to the gas as heat in process 1 is 10$p_iV_i$. In terms of $p_iV_i$, what are (a) the energy transferred to the gas as heat in process 2 and (b) the change in internal energy that the gas undergoes in process 3? SSM

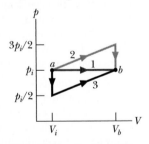

FIG. 18-53   Problem 73.

**74** The average rate at which energy is conducted outward through the ground surface in North America is 54.0 mW/m², and the average thermal conductivity of the near-surface rocks is 2.50 W/m·K. Assuming a surface temperature of 10.0°C, find the temperature at a depth of 35.0 km (near the base of the crust). Ignore the heat generated by the presence of radioactive elements.

**75** The temperature of a Pyrex disk is changed from 10.0°C to 60.0°C. Its initial radius is 8.00 cm; its initial thickness is 0.500 cm. Take these data as being exact. What is the change in the volume of the disk? (See Table 18-2.) SSM

**76** In a certain solar house, energy from the Sun is stored in barrels filled with water. In a particular winter stretch of five cloudy days, $1.00 \times 10^6$ kcal is needed to maintain the inside of the house at 22.0°C. Assuming that the water in the barrels is at 50.0°C and that the water has a density of $1.00 \times 10^3$ kg/m³, what volume of water is required?

**77** A sample of gas expands from an initial pressure and volume of 10 Pa and 1.0 m³ to a final volume of 2.0 m³. During the expansion, the pressure and volume are related by the equation $p = aV^2$, where $a$ = 10 N/m⁸. Determine the work done by the gas during this expansion. SSM

**78** (a) Calculate the rate at which body heat is conducted through the clothing of a skier in a steady-state process, given the following data: the body surface area is 1.8 m², and the clothing is 1.0 cm thick; the skin surface temperature is 33°C and the outer surface of the clothing is at 1.0°C; the thermal conductivity of the clothing is 0.040 W/m·K. (b) If, after a fall, the skier's clothes became soaked with water of thermal conductivity 0.60 W/m·K, by how much is the rate of conduction multiplied?

**79** Figure 18-54 displays a closed cycle for a gas. From $c$ to $b$, 40 J is transferred from the gas as heat. From $b$ to $a$, 130 J is transferred from the gas as heat, and the magnitude of the work done by the gas is 80 J. From $a$ to $c$, 400 J is transferred to the gas as heat. What is the work done by the gas from $a$ to $c$? (Hint: You need to supply the plus and minus signs for the given data.)

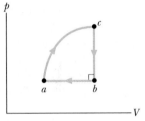

FIG. 18-54   Problem 79.

**80** A glass window pane is exactly 20 cm by 30 cm at 10°C. By how much has its area increased when its temperature is 40°C, assuming that it can expand freely?

**81** A 2.50 kg lump of aluminum is heated to 92.0°C and then dropped into 8.00 kg of water at 5.00°C. Assuming that the lump–water system is thermally isolated, what is the system's equilibrium temperature? SSM

**82** Figure 18-55a shows a cylinder containing gas and closed by a movable piston. The cylinder is kept submerged in an ice–water mixture. The piston is *quickly* pushed down from position 1 to position 2 and then held at position 2 until the gas is again at the temperature of the ice–water mixture; it then is *slowly* raised back to position 1. Figure 18-55b is a p-V diagram for the process. If 100 g of ice is melted during the cycle, how much work has been done *on* the gas?

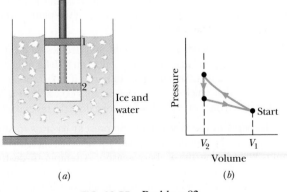

FIG. 18-55   Problem 82.

**83** The temperature of a 0.700 kg cube of ice is decreased to −150°C. Then energy is gradually transferred to the cube as heat while it is otherwise thermally isolated from its environment. The total transfer is 0.6993 MJ. Assume the value of $c_{ice}$ given in Table 18-3 is valid for temperatures from −150°C to 0°C. What is the final temperature of the water? SSM

**84** A steel rod at 25.0°C is bolted at both ends and then cooled. At what temperature will it rupture? Use Table 12-1.

**85** Suppose that you intercept $5.0 \times 10^{-3}$ of the energy radiated by a hot sphere that has a radius of 0.020 m, an emissivity of 0.80, and a surface temperature of 500 K. How much energy do you intercept in 2.0 min?

**86** Three equal-length straight rods, of aluminum, Invar, and steel, all at 20.0°C, form an equilateral triangle with hinge pins at the vertices. At what temperature will the angle opposite the Invar rod be 59.95°? See Appendix E for needed trigonometric formulas and Table 18-2 for needed data.

**87** It is possible to melt ice by rubbing one block of it against another. How much work, in joules, would you have to do to get 1.00 g of ice to melt?

**88** A thermometer of mass 0.0550 kg and of specific heat 0.837 kJ/kg·K reads 15.0°C. It is then completely immersed in 0.300 kg of water, and it comes to the same final temperature as the water. If the thermometer then reads 44.4°C, what was the temperature of the water before insertion of the thermometer?

**89** A recruit can join the semi-secret "300 F" club at the Amundsen–Scott South Pole Station only when the outside temperature is below −70°C. On such a day, the recruit first basks in a hot sauna and then runs outside wearing only shoes. (This is, of course, extremely dangerous, but the rite is effectively a protest against the constant danger of the cold.)

Assume that upon stepping out of the sauna, the recruit's skin temperature is 102°F and the walls, ceiling, and floor of the sauna room have a temperature of 30°C. Estimate the recruit's surface area, and take the skin emissivity to be 0.80. (a) What is the approximate net rate $P_{net}$ at which the recruit loses energy via thermal radiation exchanges with the room? Next, assume that when outdoors, half the recruit's surface area exchanges thermal radiation with the sky at a temperature of −25°C and the other half exchanges thermal radiation with the snow and ground at a temperature of −80°C. What is the approximate net rate at which the recruit loses energy via thermal radiation exchanges with (b) the sky and (c) the snow and ground?

**90** A rectangular plate of glass initially has the dimensions 0.200 m by 0.300 m. The coefficient of linear expansion for the glass is $9.00 \times 10^{-6}$/K. What is the change in the plate's area if its temperature is increased by 20.0 K?

**91** An athlete needs to lose weight and decides to do it by "pumping iron." (a) How many times must an 80.0 kg weight be lifted a distance of 1.00 m in order to burn off 1.00 lb of fat, assuming that that much fat is equivalent to 3500 Cal? (b) If the weight is lifted once every 2.00 s, how long does the task take?

**92** Icebergs in the North Atlantic present hazards to shipping, causing the lengths of shipping routes to be increased by about 30% during the iceberg season. Attempts to destroy icebergs include planting explosives, bombing, torpedoing, shelling, ramming, and coating with black soot. Suppose that

direct melting of the iceberg, by placing heat sources in the ice, is tried. How much energy as heat is required to melt 10% of an iceberg that has a mass of 200 000 metric tons? (Use 1 metric ton = 1000 kg.)

**93** A sample of gas expands from $V_1 = 1.0$ m³ and $p_1 =$ 40 Pa to $V_2 = 4.0$ m³ and $p_2 = 10$ Pa along path $B$ in the $p$-$V$ diagram in Fig. 18-56. It is then compressed back to $V_1$ along either path $A$ or path $C$. Compute the net work done by the gas for the complete cycle along (a) path $BA$ and (b) path $BC$.

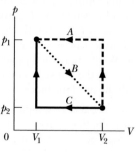

FIG. 18-56    Problem 93.

**94** Soon after Earth was formed, heat released by the decay of radioactive elements raised the average internal temperature from 300 to 3000 K, at about which value it remains today. Assuming an average coefficient of volume expansion of $3.0 \times 10^{-5}$ K⁻¹, by how much has the radius of Earth increased since the planet was formed?

**95** Figure 18-57 displays a closed cycle for a gas. The change in internal energy along path $ca$ is −160 J. The energy transferred to the gas as heat is 200 J along path $ab$, and 40 J along path $bc$. How much work is done by the gas along (a) path $abc$ and (b) path $ab$?

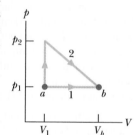

FIG. 18-57
Problem 95.

**96** The $p$-$V$ diagram in Fig. 18-58 shows two paths along which a sample of gas can be taken from state $a$ to state $b$, where $V_b = 3.0V_1$. Path 1 requires that energy equal to $5.0p_1V_1$ be transferred to the gas as heat. Path 2 requires that energy equal to $5.5p_1V_1$ be transferred to the gas as heat. What is the ratio $p_2/p_1$?

**97** A cube of edge length $6.0 \times 10^{-6}$ m, emissivity 0.75, and temperature −100°C floats in an environment at −150°C. What is the cube's net thermal radiation transfer rate?

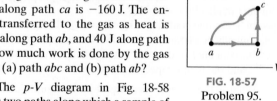

FIG. 18-58    Problem 96.

**98** A *flow calorimeter* is a device used to measure the specific heat of a liquid. Energy is added as heat at a known rate to a stream of the liquid as it passes through the calorimeter at a known rate. Measurement of the resulting temperature difference between the inflow and the outflow points of the liquid stream enables us to compute the specific heat of the liquid. Suppose a liquid of density 0.85 g/cm³ flows through a calorimeter at the rate of 8.0 cm³/s. When energy is added at the rate of 250 W by means of an electric heating coil, a temperature difference of 15 C° is established in steady-state conditions between the inflow and the outflow points. What is the specific heat of the liquid?

**99** An object of mass 6.00 kg falls through a height of 50.0 m and, by means of a mechanical linkage, rotates a paddle wheel that stirs 0.600 kg of water. Assume that the initial gravitational potential energy of the object is fully transferred to thermal energy of the water, which is initially at 15.0°C. What is the temperature rise of the water?

# The Kinetic Theory of Gases

Tom Branch/Photo Researchers

When a container of cold champagne, soda pop, or any other carbonated drink is opened, a slight fog forms around the opening and some of the liquid sprays outward. In the photograph, the fog is the white cloud that surrounds the stopper, and the spray has formed streaks within the cloud.

## What causes the fog?

The answer is in this chapter.

## 19-1 | WHAT IS PHYSICS?

One of the main subjects in thermodynamics is the physics of gases. A gas consists of atoms (either individually or bound together as molecules) that fill their container's volume and exert pressure on the container's walls. We can usually assign a temperature to such a contained gas. These three variables associated with a gas—volume, pressure, and temperature—are all a consequence of the motion of the atoms. The volume is a result of the freedom the atoms have to spread throughout the container, the pressure is a result of the collisions of the atoms with the container's walls, and the temperature has to do with the kinetic energy of the atoms. The **kinetic theory of gases,** the focus of this chapter, relates the motion of the atoms to the volume, pressure, and temperature of the gas.

Applications of the kinetic theory of gases are countless. Automobile engineers are concerned with the combustion of vaporized fuel (a gas) in the automobile engines. Food engineers are concerned with the production rate of the fermentation gas that causes bread to rise as it bakes. Beverage engineers are concerned with how gas can produce the head in a glass of beer or shoot a cork from a champagne bottle. Medical engineers and physiologists are concerned with calculating how long a scuba diver must pause during ascent to eliminate nitrogen gas from the bloodstream (to avoid the *bends*). Environmental scientists are concerned with how heat exchanges between the oceans and the atmosphere can affect weather conditions.

The first step in our discussion of the kinetic theory of gases deals with measuring the amount of a gas present in a sample, for which we use Avogadro's number.

## 19-2 | Avogadro's Number

When our thinking is slanted toward atoms and molecules, it makes sense to measure the sizes of our samples in moles. If we do so, we can be certain that we are comparing samples that contain the same number of atoms or molecules. The *mole* is one of the seven SI base units and is defined as follows:

➤ One mole is the number of atoms in a 12 g sample of carbon-12.

The obvious question now is: "How many atoms or molecules are there in a mole?" The answer is determined experimentally and, as you saw in Chapter 18, is

$$N_A = 6.02 \times 10^{23} \text{ mol}^{-1} \quad \text{(Avogadro's number)}, \tag{19-1}$$

where mol$^{-1}$ represents the inverse mole or "per mole," and mol is the abbreviation for mole. The number $N_A$ is called **Avogadro's number** after Italian scientist Amedeo Avogadro (1776–1856), who suggested that all gases occupying the same volume under the same conditions of temperature and pressure contain the same number of atoms or molecules.

The number of moles $n$ contained in a sample of any substance is equal to the ratio of the number of molecules $N$ in the sample to the number of molecules $N_A$ in 1 mol:

$$n = \frac{N}{N_A}. \tag{19-2}$$

(*Caution:* The three symbols in this equation can easily be confused with one another, so you should sort them with their meanings now, before you end in "N-confusion.") We can find the number of moles $n$ in a sample from the mass $M_{sam}$ of the sample and either the *molar mass M* (the mass of 1 mol) or the molecular mass $m$ (the mass of one molecule):

$$n = \frac{M_{sam}}{M} = \frac{M_{sam}}{mN_A}. \tag{19-3}$$

In Eq. 19-3, we used the fact that the mass $M$ of 1 mol is the product of the mass $m$ of one molecule and the number of molecules $N_A$ in 1 mol:

$$M = mN_A. \tag{19-4}$$

**PROBLEM-SOLVING TACTICS**

*Tactic 1: Avogadro's Number of What?* In Eq. 19-1, Avogadro's number is expressed in terms of $mol^{-1}$, which is the inverse mole, or 1/mol. We could instead explicitly state the elementary unit involved in a given situation. For example, if the elementary unit is an atom, we might write $N_A = 6.02 \times 10^{23}$ atoms/mol. If, instead, the elementary unit is a molecule, then we might write $N_A = 6.02 \times 10^{23}$ molecules/mol. Being explicit about the unit is usually the better style of writing.

## 19-3 | Ideal Gases

Our goal in this chapter is to explain the macroscopic properties of a gas—such as its pressure and its temperature—in terms of the behavior of the molecules that make it up. However, there is an immediate problem: which gas? Should it be hydrogen, oxygen, or methane, or perhaps uranium hexafluoride? They are all different. Experimenters have found, though, that if we confine 1 mol samples of various gases in boxes of identical volume and hold the gases at the same temperature, then their measured pressures are nearly—though not exactly—the same. If we repeat the measurements at lower gas densities, then these small differences in the measured pressures tend to disappear. Further experiments show that, at low enough densities, all real gases tend to obey the relation

$$pV = nRT \quad \text{(ideal gas law)}, \tag{19-5}$$

in which $p$ is the absolute (not gauge) pressure, $n$ is the number of moles of gas present, and $T$ is the temperature in kelvins. The symbol $R$ is a constant called the **gas constant** that has the same value for all gases—namely,

$$R = 8.31 \text{ J/mol} \cdot \text{K}. \tag{19-6}$$

Equation 19-5 is called the **ideal gas law.** Provided the gas density is low, this law holds for any single gas or for any mixture of different gases. (For a mixture, $n$ is the total number of moles in the mixture.)

We can rewrite Eq. 19-5 in an alternative form, in terms of a constant called the **Boltzmann constant** $k$, which is defined as

$$k = \frac{R}{N_A} = \frac{8.31 \text{ J/mol} \cdot \text{K}}{6.02 \times 10^{23} \text{ mol}^{-1}} = 1.38 \times 10^{-23} \text{ J/K}. \tag{19-7}$$

This allows us to write $R = kN_A$. Then, with Eq. 19-2 ($n = N/N_A$), we see that

$$nR = Nk. \tag{19-8}$$

Substituting this into Eq. 19-5 gives a second expression for the ideal gas law:

$$pV = NkT \quad \text{(ideal gas law)}. \tag{19-9}$$

(*Caution*: Note the difference between the two expressions for the ideal gas law—Eq. 19-5 involves the number of moles $n$, and Eq. 19-9 involves the number of molecules $N$.)

You may well ask, "What is an *ideal gas,* and what is so 'ideal' about it?" The answer lies in the simplicity of the law (Eqs. 19-5 and 19-9) that governs its macroscopic properties. Using this law—as you will see—we can deduce many

**FIG. 19-1** A railroad tank car crushed overnight. *(Photo courtesy www.Houston.RailFan.net)*

properties of the ideal gas in a simple way. Although there is no such thing in nature as a truly ideal gas, *all real* gases approach the ideal state at low enough densities—that is, under conditions in which their molecules are far enough apart that they do not interact with one another. Thus, the ideal gas concept allows us to gain useful insights into the limiting behavior of real gases.

The interior of the railroad tank car in Fig. 19-1 was being cleaned with steam by a crew late one afternoon. Because the job was unfinished at the end of their work shift, they sealed the car and left for the night. When they returned the next morning, they discovered that something had crushed the car in spite of its extremely strong steel walls, as if some giant creature from a grade B science fiction movie had stepped on it during a rampage that night.

With Eq. 19-9, we can explain what actually crushed the railroad tank car. When the car was being cleaned, its interior was filled with very hot steam, which is a gas of water molecules. The cleaning crew left the steam inside the car when they closed all the valves on the car at the end of their work shift. At that point the pressure of the gas in the car was equal to atmospheric pressure because the valves had been opened to the atmosphere during the cleaning. As the car cooled during the night, the steam cooled and much of it condensed, which means that the number $N$ of gas molecules and the temperature $T$ of the gas both decreased. Thus, the right side of Eq. 19-9 decreased, and because volume $V$ was constant, the gas pressure $p$ on the left side also decreased. At some point during the night, the gas pressure inside the car reached such a low value that the external atmospheric pressure was able to crush the car's steel walls. The cleaning crew could have prevented this accident by leaving the valves open, so that air could enter the car to keep the internal pressure equal to the external atmospheric pressure.

## Work Done by an Ideal Gas at Constant Temperature

Suppose we put an ideal gas in a piston–cylinder arrangement like those in Chapter 18. Suppose also that we allow the gas to expand from an initial volume $V_i$ to a final volume $V_f$ while we keep the temperature $T$ of the gas constant. Such a process, at *constant temperature,* is called an **isothermal expansion** (and the reverse is called an **isothermal compression**).

On a *p-V* diagram, an *isotherm* is a curve that connects points that have the same temperature. Thus, it is a graph of pressure versus volume for a gas whose temperature $T$ is held constant. For $n$ moles of an ideal gas, it is a graph of the equation

$$p = nRT\frac{1}{V} = (\text{a constant})\frac{1}{V}. \tag{19-10}$$

Figure 19-2 shows three isotherms, each corresponding to a different (constant) value of $T$. (Note that the values of $T$ for the isotherms increase upward to the right.) Superimposed on the middle isotherm is the path followed by a gas during an isothermal expansion from state $i$ to state $f$ at a constant temperature of 310 K.

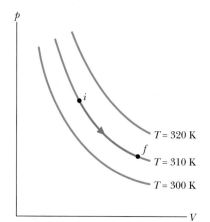

**FIG. 19-2** Three isotherms on a *p-V* diagram. The path shown along the middle isotherm represents an isothermal expansion of a gas from an initial state $i$ to a final state $f$. The path from $f$ to $i$ along the isotherm would represent the reverse process—that is, an isothermal compression.

To find the work done by an ideal gas during an isothermal expansion, we start with Eq. 18-25,

$$W = \int_{V_i}^{V_f} p\, dV. \tag{19-11}$$

This is a general expression for the work done during any change in volume of any gas. For an ideal gas, we can use Eq. 19-5 ($pV = nRT$) to substitute for $p$, obtaining

$$W = \int_{V_i}^{V_f} \frac{nRT}{V}\, dV. \tag{19-12}$$

Because we are considering an isothermal expansion, $T$ is constant, so we can move it in front of the integral sign to write

$$W = nRT \int_{V_i}^{V_f} \frac{dV}{V} = nRT \left[ \ln V \right]_{V_i}^{V_f}. \tag{19-13}$$

By evaluating the expression in brackets at the limits and then using the relationship $\ln a - \ln b = \ln(a/b)$, we find that

$$W = nRT \ln \frac{V_f}{V_i} \qquad \text{(ideal gas, isothermal process).} \tag{19-14}$$

Recall that the symbol ln specifies a *natural* logarithm, which has base $e$.

For an expansion, $V_f$ is greater than $V_i$, so the ratio $V_f/V_i$ in Eq. 19-14 is greater than unity. The natural logarithm of a quantity greater than unity is positive, and so the work $W$ done by an ideal gas during an isothermal expansion is positive, as we expect. For a compression, $V_f$ is less than $V_i$, so the ratio of volumes in Eq. 19-14 is less than unity. The natural logarithm in that equation—hence the work $W$—is negative, again as we expect.

## Work Done at Constant Volume and at Constant Pressure

Equation 19-14 does not give the work $W$ done by an ideal gas during *every* thermodynamic process. Instead, it gives the work only for a process in which the temperature is held constant. If the temperature varies, then the symbol $T$ in Eq. 19-12 cannot be moved in front of the integral symbol as in Eq. 19-13, and thus we do not end up with Eq. 19-14.

However, we can always go back to Eq. 19-11 to find the work $W$ done by an ideal gas (or any other gas) during any process, such as a constant-volume process and a constant-pressure process. If the volume of the gas is constant, then Eq. 19-11 yields

$$W = 0 \qquad \text{(constant-volume process).} \tag{19-15}$$

If, instead, the volume changes while the pressure $p$ of the gas is held constant, then Eq. 19-11 becomes

$$W = p(V_f - V_i) = p \, \Delta V \qquad \text{(constant-pressure process).} \tag{19-16}$$

✓**CHECKPOINT 1** An ideal gas has an initial pressure of 3 pressure units and an initial volume of 4 volume units. The table gives the final pressure and volume of the gas (in those same units) in five processes. Which processes start and end on the same isotherm?

| | a | b | c | d | e |
|---|---|---|---|---|---|
| $p$ | 12 | 6 | 5 | 4 | 1 |
| $V$ | 1 | 2 | 7 | 3 | 12 |

## Sample Problem 19-1

A cylinder contains 12 L of oxygen at 20°C and 15 atm. The temperature is raised to 35°C, and the volume is reduced to 8.5 L. What is the final pressure of the gas in atmospheres? Assume that the gas is ideal.

**KEY IDEA** Because the gas is ideal, its pressure, volume, temperature, and number of moles are related by the ideal gas law, both in the initial state $i$ and in the final state $f$ (after the changes).

**Calculations:** From Eq. 19-5 we can write

$$p_i V_i = nRT_i \quad \text{and} \quad p_f V_f = nRT_f.$$

Dividing the second equation by the first equation and solving for $p_f$ yields

$$p_f = \frac{p_i T_f V_i}{T_i V_f}. \tag{19-17}$$

Note here that if we converted the given initial and final volumes from liters to the proper units of cubic meters, the multiplying conversion factors would cancel out of Eq. 19-17. The same would be true for conversion factors that convert the pressures from atmospheres to the proper pascals. However, to convert the given temperatures to kelvins requires the addition of an amount that would not cancel and thus must be included. Hence, we must write and

$$T_i = (273 + 20)\ K = 293\ K$$
$$T_f = (273 + 35)\ K = 308\ K.$$

Inserting the given data into Eq. 19-17 then yields

$$p_f = \frac{(15\ atm)(308\ K)(12\ L)}{(293\ K)(8.5\ L)} = 22\ atm.\quad\text{(Answer)}$$

One mole of oxygen (assume it to be an ideal gas) expands at a constant temperature $T$ of 310 K from an initial volume $V_i$ of 12 L to a final volume $V_f$ of 19 L. How much work is done by the gas during the expansion?

**KEY IDEA** Generally we find the work by integrating the gas pressure with respect to the gas volume, using Eq. 19-11. However, because the gas here is ideal and the expansion is isothermal, that integration leads to Eq. 19-14.

*Calculation:* Therefore, we can write

$$W = nRT \ln \frac{V_f}{V_i}$$

$$= (1\ mol)(8.31\ J/mol \cdot K)(310\ K) \ln \frac{19\ L}{12\ L}$$

$$= 1180\ J.\qquad\text{(Answer)}$$

The expansion is graphed in the p-V diagram of Fig. 19-3. The work done by the gas during the expansion is represented by the area beneath the curve *if*.

You can show that if the expansion is now reversed, with the gas undergoing an isothermal compression

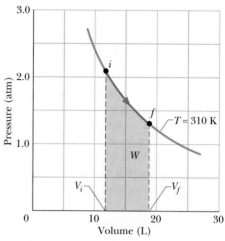

**FIG. 19-3** The shaded area represents the work done by 1 mol of oxygen in expanding from $V_i$ to $V_f$ at a constant temperature $T$ of 310 K.

from 19 L to 12 L, the work done by the gas will be −1180 J. Thus, an external force would have to do 1180 J of work on the gas to compress it.

## 19-4 | Pressure, Temperature, and RMS Speed

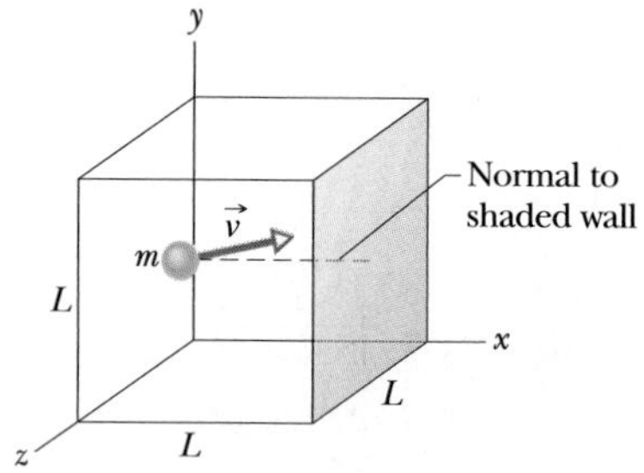

Here is our first kinetic theory problem. Let $n$ moles of an ideal gas be confined in a cubical box of volume $V$, as in Fig. 19-4. The walls of the box are held at temperature $T$. What is the connection between the pressure $p$ exerted by the gas on the walls and the speeds of the molecules?

The molecules of gas in the box are moving in all directions and with various speeds, bumping into one another and bouncing from the walls of the box like balls in a racquetball court. We ignore (for the time being) collisions of the molecules with one another and consider only elastic collisions with the walls.

Figure 19-4 shows a typical gas molecule, of mass $m$ and velocity $\vec{v}$, that is about to collide with the shaded wall. Because we assume that any collision of a molecule with a wall is elastic, when this molecule collides with the shaded wall, the only component of its velocity that is changed is the $x$ component, and that component is reversed. This means that the only change in the particle's momentum is along the $x$ axis, and that change is

$$\Delta p_x = (-mv_x) - (mv_x) = -2mv_x.$$

Hence, the momentum $\Delta p_x$ delivered to the wall by the molecule during the collision is $+2mv_x$. (Because in this book the symbol $p$ represents both momentum and pressure, we must be careful to note that here $p$ represents momentum and is a vector quantity.)

The molecule of Fig. 19-4 will hit the shaded wall repeatedly. The time $\Delta t$ between collisions is the time the molecule takes to travel to the opposite wall and back again (a distance $2L$) at speed $v_x$. Thus, $\Delta t$ is equal to $2L/v_x$. (Note that this re-

**FIG. 19-4** A cubical box of edge length $L$, containing $n$ moles of an ideal gas. A molecule of mass $m$ and velocity $\vec{v}$ is about to collide with the shaded wall of area $L^2$. A normal to that wall is shown.

sult holds even if the molecule bounces off any of the other walls along the way, because those walls are parallel to $x$ and so cannot change $v_x$.) Therefore, the average rate at which momentum is delivered to the shaded wall by this single molecule is

$$\frac{\Delta p_x}{\Delta t} = \frac{2mv_x}{2L/v_x} = \frac{mv_x^2}{L}.$$

From Newton's second law ($\vec{F} = d\vec{p}/dt$), the rate at which momentum is delivered to the wall is the force acting on that wall. To find the total force, we must add up the contributions of all the molecules that strike the wall, allowing for the possibility that they all have different speeds. Dividing the magnitude of the total force $F_x$ by the area of the wall ($= L^2$) then gives the pressure $p$ on that wall, where now and in the rest of this discussion, $p$ represents pressure. Thus, using the expression for $\Delta p_x/\Delta t$, we can write this pressure as

$$p = \frac{F_x}{L^2} = \frac{mv_{x1}^2/L + mv_{x2}^2/L + \cdots + mv_{xN}^2/L}{L^2}$$

$$= \left(\frac{m}{L^3}\right)(v_{x1}^2 + v_{x2}^2 + \cdots + v_{xN}^2), \qquad (19\text{-}18)$$

where $N$ is the number of molecules in the box.

Since $N = nN_A$, there are $nN_A$ terms in the second set of parentheses of Eq. 19-18. We can replace that quantity by $nN_A(v_x^2)_{avg}$, where $(v_x^2)_{avg}$ is the average value of the square of the $x$ components of all the molecular speeds. Equation 19-18 then becomes

$$p = \frac{nmN_A}{L^3}(v_x^2)_{avg}.$$

However, $mN_A$ is the molar mass $M$ of the gas (that is, the mass of 1 mol of the gas). Also, $L^3$ is the volume of the box, so

$$p = \frac{nM(v_x^2)_{avg}}{V}. \qquad (19\text{-}19)$$

For any molecule, $v^2 = v_x^2 + v_y^2 + v_z^2$. Because there are many molecules and because they are all moving in random directions, the average values of the squares of their velocity components are equal, so that $v_x^2 = \frac{1}{3}v^2$. Thus, Eq. 19-19 becomes

$$p = \frac{nM(v^2)_{avg}}{3V}. \qquad (19\text{-}20)$$

The square root of $(v^2)_{avg}$ is a kind of average speed, called the **root-mean-square speed** of the molecules and symbolized by $v_{rms}$. Its name describes it rather well: You *square* each speed, you find the *mean* (that is, the average) of all these squared speeds, and then you take the square *root* of that mean. With $\sqrt{(v^2)_{avg}} = v_{rms}$, we can then write Eq. 19-20 as

$$p = \frac{nMv_{rms}^2}{3V}. \qquad (19\text{-}21)$$

Equation 19-21 is very much in the spirit of kinetic theory. It tells us how the pressure of the gas (a purely macroscopic quantity) depends on the speed of the molecules (a purely microscopic quantity).

We can turn Eq. 19-21 around and use it to calculate $v_{rms}$. Combining Eq. 19-21 with the ideal gas law ($pV = nRT$) leads to

$$v_{rms} = \sqrt{\frac{3RT}{M}}. \qquad (19\text{-}22)$$

Table 19-1 shows some rms speeds calculated from Eq. 19-22. The speeds are sur-

**TABLE 19-1**

**Some RMS Speeds at Room Temperature ($T = 300$ K)**[a]

| Gas | Molar Mass ($10^{-3}$ kg/mol) | $v_{rms}$ (m/s) |
|---|---|---|
| Hydrogen ($H_2$) | 2.02 | 1920 |
| Helium (He) | 4.0 | 1370 |
| Water vapor ($H_2O$) | 18.0 | 645 |
| Nitrogen ($N_2$) | 28.0 | 517 |
| Oxygen ($O_2$) | 32.0 | 483 |
| Carbon dioxide ($CO_2$) | 44.0 | 412 |
| Sulfur dioxide ($SO_2$) | 64.1 | 342 |

[a]For convenience, we often set room temperature equal to 300 K even though (at 27°C or 81°F) that represents a fairly warm room.

prisingly high. For hydrogen molecules at room temperature (300 K), the rms speed is 1920 m/s, or 4300 mi/h—faster than a speeding bullet! On the surface of the Sun, where the temperature is $2 \times 10^6$ K, the rms speed of hydrogen molecules would be 82 times greater than at room temperature were it not for the fact that at such high speeds, the molecules cannot survive collisions among themselves. Remember too that the rms speed is only a kind of average speed; many molecules move much faster than this, and some much slower.

The speed of sound in a gas is closely related to the rms speed of the molecules of that gas. In a sound wave, the disturbance is passed on from molecule to molecule by means of collisions. The wave cannot move any faster than the "average" speed of the molecules. In fact, the speed of sound must be somewhat less than this "average" molecular speed because not all molecules are moving in exactly the same direction as the wave. As examples, at room temperature, the rms speeds of hydrogen and nitrogen molecules are 1920 m/s and 517 m/s, respectively. The speeds of sound in these two gases at this temperature are 1350 m/s and 350 m/s, respectively.

A question often arises: If molecules move so fast, why does it take as long as a minute or so before you can smell perfume when someone opens a bottle across a room? The answer is that, as we shall discuss in Section 19-6, each perfume molecule may have a high speed but it moves away from the bottle only very slowly because its repeated collisions with other molecules prevent it from moving directly across the room to you.

**Sample Problem   19-3**

Here are five numbers: 5, 11, 32, 67, and 89.

(a) What is the average value $n_{avg}$ of these numbers?

**Calculation:** We find this from

$$n_{avg} = \frac{5 + 11 + 32 + 67 + 89}{5} = 40.8. \quad \text{(Answer)}$$

(b) What is the rms value $n_{rms}$ of these numbers?

**Calculation:** We find this from

$$n_{rms} = \sqrt{\frac{5^2 + 11^2 + 32^2 + 67^2 + 89^2}{5}}$$

$$= 52.1. \quad \text{(Answer)}$$

The rms value is greater than the average value because the larger numbers—being squared—are relatively more important in forming the rms value. To test this, let us replace 89 in our set of five numbers by 300. The average value of the new set of five numbers (as you should show) is 2.0 times the previous average value. The rms value, however, is 2.7 times the previous rms value.

## 19-5 | Translational Kinetic Energy

We again consider a single molecule of an ideal gas as it moves around in the box of Fig. 19-4, but we now assume that its speed changes when it collides with other molecules. Its translational kinetic energy at any instant is $\frac{1}{2}mv^2$. Its *average* translational kinetic energy over the time that we watch it is

$$K_{avg} = \left(\tfrac{1}{2}mv^2\right)_{avg} = \tfrac{1}{2}m(v^2)_{avg} = \tfrac{1}{2}mv_{rms}^2, \quad (19\text{-}23)$$

in which we make the assumption that the average speed of the molecule during our observation is the same as the average speed of all the molecules at any given time. (Provided the total energy of the gas is not changing and provided we observe our molecule for long enough, this assumption is appropriate.) Substituting for $v_{rms}$ from Eq. 19-22 leads to

$$K_{avg} = \left(\tfrac{1}{2}m\right)\frac{3RT}{M}.$$

However, $M/m$, the molar mass divided by the mass of a molecule, is simply Avogadro's number. Thus,

$$K_{avg} = \frac{3RT}{2N_A}.$$

Using Eq. 19-7 ($k = R/N_A$), we can then write

$$K_{avg} = \tfrac{3}{2}kT. \qquad (19\text{-}24)$$

This equation tells us something unexpected:

➤ At a given temperature $T$, all ideal gas molecules—no matter what their mass—have the same average translational kinetic energy—namely, $\tfrac{3}{2}kT$. When we measure the temperature of a gas, we are also measuring the average translational kinetic energy of its molecules.

**✓ CHECKPOINT 2** A gas mixture consists of molecules of types 1, 2, and 3, with molecular masses $m_1 > m_2 > m_3$. Rank the three types according to (a) average kinetic energy and (b) rms speed, greatest first.

## 19-6 | Mean Free Path

We continue to examine the motion of molecules in an ideal gas. Figure 19-5 shows the path of a typical molecule as it moves through the gas, changing both speed and direction abruptly as it collides elastically with other molecules. Between collisions, the molecule moves in a straight line at constant speed. Although the figure shows the other molecules as stationary, they are also moving.

One useful parameter to describe this random motion is the **mean free path** $\lambda$ of the molecules. As its name implies, $\lambda$ is the average distance traversed by a molecule between collisions. We expect $\lambda$ to vary inversely with $N/V$, the number of molecules per unit volume (or density of molecules). The larger $N/V$ is, the more collisions there should be and the smaller the mean free path. We also expect $\lambda$ to vary inversely with the size of the molecules—with their diameter $d$, say. (If the molecules were points, as we have assumed them to be, they would never collide and the mean free path would be infinite.) Thus, the larger the molecules are, the smaller the mean free path. We can even predict that $\lambda$ should vary (inversely) as the *square* of the molecular diameter because the cross section of a molecule—not its diameter—determines its effective target area.

The expression for the mean free path does, in fact, turn out to be

$$\lambda = \frac{1}{\sqrt{2}\,\pi d^2\, N/V} \qquad \text{(mean free path).} \qquad (19\text{-}25)$$

To justify Eq. 19-25, we focus attention on a single molecule and assume—as Fig. 19-5 suggests—that our molecule is traveling with a constant speed $v$ and that all the other molecules are at rest. Later, we shall relax this assumption.

We assume further that the molecules are spheres of diameter $d$. A collision will then take place if the centers of two molecules come within a distance $d$ of each other, as in Fig. 19-6a. Another, more helpful way to look at the situation is to consider our single molecule to have a *radius* of $d$ and all the other molecules to be *points,* as in Fig. 19-6b. This does not change our criterion for a collision.

As our single molecule zigzags through the gas, it sweeps out a short cylinder of cross-sectional area $\pi d^2$ between successive collisions. If we watch this molecule for a time interval $\Delta t$, it moves a distance $v\,\Delta t$, where $v$ is its assumed speed. Thus, if we align all the short cylinders swept out in interval $\Delta t$, we form a composite cylinder (Fig. 19-7) of length $v\,\Delta t$ and volume $(\pi d^2)(v\,\Delta t)$. The number of collisions that occur in time $\Delta t$ is then equal to the number of (point) molecules that lie within this cylinder.

Since $N/V$ is the number of molecules per unit volume, the number of molecules in the cylinder is $N/V$ times the volume of the cylinder, or $(N/V)(\pi d^2 v\,\Delta t)$. This is also the number of collisions in time $\Delta t$. The mean free path is the length of the path (and

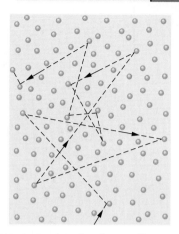

**FIG. 19-5** A molecule traveling through a gas, colliding with other gas molecules in its path. Although the other molecules are shown as stationary, they are also moving in a similar fashion.

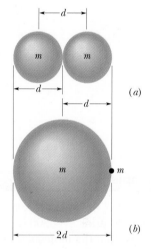

**FIG. 19-6** (a) A collision occurs when the centers of two molecules come within a distance $d$ of each other, $d$ being the molecular diameter. (b) An equivalent but more convenient representation is to think of the moving molecule as having a *radius $d$* and all other molecules as being points. The condition for a collision is unchanged.

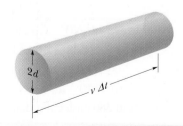

**FIG. 19-7** In time $\Delta t$ the moving molecule effectively sweeps out a cylinder of length $v\,\Delta t$ and radius $d$.

of the cylinder) divided by this number:

$$\lambda = \frac{\text{length of path during } \Delta t}{\text{number of collisions in } \Delta t} \approx \frac{v\,\Delta t}{\pi d^2 v\,\Delta t\,N/V}$$

$$= \frac{1}{\pi d^2\,N/V}. \tag{19-26}$$

This equation is only approximate because it is based on the assumption that all the molecules except one are at rest. In fact, *all* the molecules are moving; when this is taken properly into account, Eq. 19-25 results. Note that it differs from the (approximate) Eq. 19-26 only by a factor of $1/\sqrt{2}$.

The approximation in Eq. 19-26 involves the two $v$ symbols we canceled. The $v$ in the numerator is $v_{avg}$, the mean speed of the molecules *relative to the container*. The $v$ in the denominator is $v_{rel}$, the mean speed of our single molecule *relative to the other molecules,* which are moving. It is this latter average speed that determines the number of collisions. A detailed calculation, taking into account the actual speed distribution of the molecules, gives $v_{rel} = \sqrt{2}\,v_{avg}$ and thus the factor $\sqrt{2}$.

The mean free path of air molecules at sea level is about 0.1 $\mu$m. At an altitude of 100 km, the density of air has dropped to such an extent that the mean free path rises to about 16 cm. At 300 km, the mean free path is about 20 km. A problem faced by those who would study the physics and chemistry of the upper atmosphere in the laboratory is the unavailability of containers large enough to hold gas samples that simulate upper atmospheric conditions. Yet studies of the concentrations of Freon, carbon dioxide, and ozone in the upper atmosphere are of vital public concern.

✓ **CHECKPOINT 3**   One mole of gas $A$, with molecular diameter $2d_0$ and average molecular speed $v_0$, is placed inside a certain container. One mole of gas $B$, with molecular diameter $d_0$ and average molecular speed $2v_0$ (the molecules of $B$ are smaller but faster), is placed in an identical container. Which gas has the greater average collision rate within its container?

## Sample Problem    19-4

(a) What is the mean free path $\lambda$ for oxygen molecules at temperature $T = 300$ K and pressure $p = 1.0$ atm? Assume that the molecular diameter is $d = 290$ pm and the gas is ideal.

**KEY IDEA**   Each oxygen molecule moves among other *moving* oxygen molecules in a zigzag path due to the resulting collisions. Thus, we use Eq. 19-25 for the mean free path.

*Calculation:* We first need the number of molecules per unit volume, $N/V$. Because we assume the gas is ideal, we can use the ideal gas law of Eq. 19-9 ($pV = NkT$) to write $N/V = p/kT$. Substituting this into Eq. 19-25, we find

$$\lambda = \frac{1}{\sqrt{2}\pi d^2\,N/V} = \frac{kT}{\sqrt{2}\pi d^2 p}$$

$$= \frac{(1.38 \times 10^{-23}\,\text{J/K})(300\,\text{K})}{\sqrt{2}\pi(2.9 \times 10^{-10}\,\text{m})^2(1.01 \times 10^5\,\text{Pa})}$$

$$= 1.1 \times 10^{-7}\,\text{m.} \qquad \text{(Answer)}$$

This is about 380 molecular diameters.

(b) Assume the average speed of the oxygen molecules is $v = 450$ m/s. What is the average time $t$ between successive collisions for any given molecule? At what rate does the molecule collide; that is, what is the frequency $f$ of its collisions?

**KEY IDEAS**   (1) Between collisions, the molecule travels, on average, the mean free path $\lambda$ at speed $v$. (2) The average rate or frequency at which the collisions occur is the inverse of the time $t$ between collisions.

*Calculations:* From the first key idea, the average time between collisions is

$$t = \frac{\text{distance}}{\text{speed}} = \frac{\lambda}{v} = \frac{1.1 \times 10^{-7}\,\text{m}}{450\,\text{m/s}}$$

$$= 2.44 \times 10^{-10}\,\text{s} \approx 0.24\,\text{ns.} \qquad \text{(Answer)}$$

This tells us that, on average, any given oxygen molecule has less than a nanosecond between collisions.

From the second key idea, the collision frequency is

$$f = \frac{1}{t} = \frac{1}{2.44 \times 10^{-10}\,\text{s}} = 4.1 \times 10^9\,\text{s}^{-1}. \qquad \text{(Answer)}$$

This tells us that, on average, any given oxygen molecule makes about 4 billion collisions per second.

# 19-7 | The Distribution of Molecular Speeds

The root-mean-square speed $v_{rms}$ gives us a general idea of molecular speeds in a gas at a given temperature. We often want to know more. For example, what fraction of the molecules have speeds greater than the rms value? What fraction have speeds greater than twice the rms value? To answer such questions, we need to know how the possible values of speed are distributed among the molecules. Figure 19-8a shows this distribution for oxygen molecules at room temperature ($T = 300$ K); Fig. 19-8b compares it with the distribution at $T = 80$ K.

In 1852, Scottish physicist James Clerk Maxwell first solved the problem of finding the speed distribution of gas molecules. His result, known as **Maxwell's speed distribution law,** is

$$P(v) = 4\pi \left( \frac{M}{2\pi RT} \right)^{3/2} v^2 e^{-Mv^2/2RT}. \qquad (19\text{-}27)$$

Here $M$ is the molar mass of the gas, $R$ is the gas constant, $T$ is the gas temperature, and $v$ is the molecular speed. It is this equation that is plotted in Fig. 19-8a, b. The quantity $P(v)$ in Eq. 19-27 and Fig. 19-8 is a *probability distribution function:* For any speed $v$, the product $P(v)\, dv$ (a dimensionless quantity) is the fraction of molecules with speeds in the interval $dv$ centered on speed $v$.

As Fig. 19-8a shows, this fraction is equal to the area of a strip with height $P(v)$ and width $dv$. The total area under the distribution curve corresponds to the fraction of the molecules whose speeds lie between zero and infinity. All molecules fall into this category, so the value of this total area is unity; that is,

$$\int_0^\infty P(v)\, dv = 1. \qquad (19\text{-}28)$$

The fraction (frac) of molecules with speeds in an interval of, say, $v_1$ to $v_2$ is then

$$\text{frac} = \int_{v_1}^{v_2} P(v)\, dv. \qquad (19\text{-}29)$$

## Average, RMS, and Most Probable Speeds

In principle, we can find the **average speed** $v_{avg}$ of the molecules in a gas with the following procedure: We *weight* each value of $v$ in the distribution; that is, we mul-

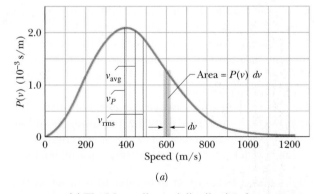

(a)

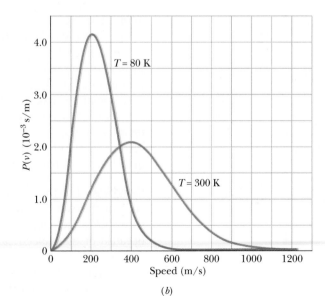

(b)

**FIG. 19-8**  (a) The Maxwell speed distribution for oxygen molecules at $T = 300$ K. The three characteristic speeds are marked. (b) The curves for 300 K and 80 K. Note that the molecules move more slowly at the lower temperature. Because these are probability distributions, the area under each curve has a numerical value of unity.

tiply it by the fraction $P(v)\, dv$ of molecules with speeds in a differential interval $dv$ centered on $v$. Then we add up all these values of $v\, P(v)\, dv$. The result is $v_{avg}$. In practice, we do all this by evaluating

$$v_{avg} = \int_0^\infty v\, P(v)\, dv. \tag{19-30}$$

Substituting for $P(v)$ from Eq. 19-27 and using generic integral 20 from the list of integrals in Appendix E, we find

$$v_{avg} = \sqrt{\frac{8RT}{\pi M}} \qquad \text{(average speed).} \tag{19-31}$$

Similarly, we can find the average of the square of the speeds $(v^2)_{avg}$ with

$$(v^2)_{avg} = \int_0^\infty v^2\, P(v)\, dv. \tag{19-32}$$

Substituting for $P(v)$ from Eq. 19-27 and using generic integral 16 from the list of integrals in Appendix E, we find

$$(v^2)_{avg} = \frac{3RT}{M}. \tag{19-33}$$

The square root of $(v^2)_{avg}$ is the root-mean-square speed $v_{rms}$. Thus,

$$v_{rms} = \sqrt{\frac{3RT}{M}} \qquad \text{(rms speed),} \tag{19-34}$$

which agrees with Eq. 19-22.

The **most probable speed** $v_P$ is the speed at which $P(v)$ is maximum (see Fig. 19-8$a$). To calculate $v_P$, we set $dP/dv = 0$ (the slope of the curve in Fig. 19-8$a$ is zero at the maximum of the curve) and then solve for $v$. Doing so, we find

$$v_P = \sqrt{\frac{2RT}{M}} \qquad \text{(most probable speed).} \tag{19-35}$$

A molecule is more likely to have speed $v_P$ than any other speed, but some molecules will have speeds that are many times $v_P$. These molecules lie in the *high-speed tail* of a distribution curve like that in Fig. 19-8$a$. We should be thankful for these few, higher speed molecules because they make possible both rain and sunshine (without which we could not exist). We next see why.

***Rain*** The speed distribution of water molecules in, say, a pond at summertime temperatures can be represented by a curve similar to that of Fig. 19-8$a$. Most of the molecules do not have nearly enough kinetic energy to escape from the water through its surface. However, small numbers of very fast molecules with speeds far out in the high-speed tail of the curve can do so. It is these water molecules that evaporate, making clouds and rain a possibility.

As the fast water molecules leave the surface, carrying energy with them, the temperature of the remaining water is maintained by heat transfer from the surroundings. Other fast molecules—produced in particularly favorable collisions—quickly take the place of those that have left, and the speed distribution is maintained.

***Sunshine*** Let the distribution curve of Fig. 19-8$a$ now refer to protons in the core of the Sun. The Sun's energy is supplied by a nuclear fusion process that starts with the merging of two protons. However, protons repel each other because of their electrical charges, and protons of average speed do not have enough kinetic energy to overcome the repulsion and get close enough to merge. Very fast protons with speeds in the high-speed tail of the distribution curve can do so, however, and for that reason the Sun can shine.

## Sample Problem    19-5

A container is filled with oxygen gas maintained at room temperature (300 K). What fraction of the molecules have speeds in the interval 599 to 601 m/s? The molar mass $M$ of oxygen is 0.0320 kg/mol.

### KEY IDEAS

1. The speeds of the molecules are distributed over a wide range of values, with the distribution $P(v)$ of Eq. 19-27.
2. The fraction of molecules with speeds in a differential interval $dv$ is $P(v)\,dv$.
3. For a larger interval, the fraction is found by integrating $P(v)$ over the interval.
4. However, the interval $\Delta v = 2$ m/s here is small compared to the speed $v = 600$ m/s on which it is centered.

**Calculations:** Because $\Delta v$ is small, we can avoid the integration by approximating the fraction as

$$\text{frac} = P(v)\,\Delta v = 4\pi\left(\frac{M}{2\pi RT}\right)^{3/2} v^2 e^{-Mv^2/2RT}\,\Delta v.$$

The function $P(v)$ is plotted in Fig. 19-8a. The total area between the curve and the horizontal axis represents the total fraction of molecules (unity). The area of the thin gold strip represents the fraction we seek.

To evaluate frac in parts, we can write

$$\text{frac} = (4\pi)(A)(v^2)(e^B)(\Delta v), \qquad (19\text{-}36)$$

where

$$A = \left(\frac{M}{2\pi RT}\right)^{3/2} = \left(\frac{0.0320 \text{ kg/mol}}{(2\pi)(8.31 \text{ J/mol·K})(300 \text{ K})}\right)^{3/2}$$
$$= 2.92 \times 10^{-9} \text{ s}^3/\text{m}^3$$

and $B = -\dfrac{Mv^2}{2RT} = -\dfrac{(0.0320 \text{ kg/mol})(600 \text{ m/s})^2}{(2)(8.31 \text{ J/mol·K})(300 \text{ K})}$

$$= -2.31.$$

Substituting $A$ and $B$ into Eq. 19-36 yields

$$\text{frac} = (4\pi)(A)(v^2)(e^B)(\Delta v)$$
$$= (4\pi)(2.92 \times 10^{-9} \text{ s}^3/\text{m}^3)(600 \text{ m/s})^2(e^{-2.31})(2 \text{ m/s})$$
$$= 2.62 \times 10^{-3}. \qquad \text{(Answer)}$$

Thus, at room temperature, 0.262% of the oxygen molecules will have speeds that lie in the narrow range between 599 and 601 m/s. If the gold strip of Fig. 19-8a were drawn to the scale of this problem, it would be a very thin strip indeed.

## Sample Problem    19-6

The molar mass $M$ of oxygen is 0.0320 kg/mol.

(a) What is the average speed $v_{avg}$ of oxygen gas molecules at $T = 300$ K?

### KEY IDEA
To find the average speed, we must weight speed $v$ with the distribution function $P(v)$ of Eq. 19-27 and then integrate the resulting expression over the range of possible speeds (0 to $\infty$).

**Calculation:** We end up with Eq. 19-31, which gives us

$$v_{avg} = \sqrt{\frac{8RT}{\pi M}}$$
$$= \sqrt{\frac{8(8.31 \text{ J/mol·K})(300 \text{ K})}{\pi(0.0320 \text{ kg/mol})}}$$
$$= 445 \text{ m/s.} \qquad \text{(Answer)}$$

This result is plotted in Fig. 19-8a.

(b) What is the root-mean-square speed $v_{rms}$ at 300 K?

### KEY IDEA
To find $v_{rms}$, we must first find $(v^2)_{avg}$ by weighting $v^2$ with the distribution function $P(v)$ of Eq. 19-27 and then integrating the expression over the range of possible speeds. Then we must take the square root of the result.

**Calculation:** We end up with Eq. 19-34, which gives us

$$v_{rms} = \sqrt{\frac{3RT}{M}}$$
$$= \sqrt{\frac{3(8.31 \text{ J/mol·K})(300 \text{ K})}{0.0320 \text{ kg/mol}}}$$
$$= 483 \text{ m/s.} \qquad \text{(Answer)}$$

This result, plotted in Fig. 19-8a, is greater than $v_{avg}$ because the greater speed values influence the calculation more when we integrate the $v^2$ values than when we integrate the $v$ values.

(c) What is the most probable speed $v_P$ at 300 K?

### KEY IDEA
Speed $v_P$ corresponds to the maximum of the distribution function $P(v)$, which we obtain by setting the derivative $dP/dv = 0$ and solving the result for $v$.

**Calculation:** We end up with Eq. 19-35, which gives us

$$v_P = \sqrt{\frac{2RT}{M}}$$
$$= \sqrt{\frac{2(8.31 \text{ J/mol·K})(300 \text{ K})}{0.0320 \text{ kg/mol}}}$$
$$= 395 \text{ m/s.} \qquad \text{(Answer)}$$

This result is also plotted in Fig. 19-8a.

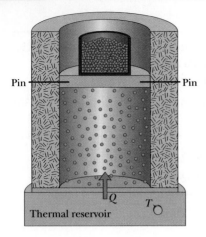

Pin — — — Pin

Q

T

Thermal reservoir

(a)

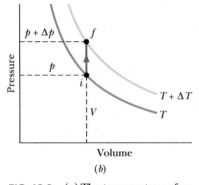

Pressure

$p + \Delta p$    f

$p$    i

$V$    T

$T + \Delta T$

Volume

(b)

**FIG. 19-9** (a) The temperature of an ideal gas is raised from $T$ to $T + \Delta T$ in a constant-volume process. Heat is added, but no work is done. (b) The process on a $p$-$V$ diagram.

## 19-8 I The Molar Specific Heats of an Ideal Gas

In this section, we want to derive from molecular considerations an expression for the internal energy $E_{int}$ of an ideal gas. In other words, we want an expression for the energy associated with the random motions of the atoms or molecules in the gas. We shall then use that expression to derive the molar specific heats of an ideal gas.

### Internal Energy $E_{int}$

Let us first assume that our ideal gas is a *monatomic gas* (which has individual atoms rather than molecules), such as helium, neon, or argon. Let us also assume that the internal energy $E_{int}$ of our ideal gas is simply the sum of the translational kinetic energies of its atoms. (As explained by quantum theory, individual atoms do not have rotational kinetic energy.)

The average translational kinetic energy of a single atom depends only on the gas temperature and is given by Eq. 19-24 as $K_{avg} = \frac{3}{2}kT$. A sample of $n$ moles of such a gas contains $nN_A$ atoms. The internal energy $E_{int}$ of the sample is then

$$E_{int} = (nN_A)K_{avg} = (nN_A)(\tfrac{3}{2}kT). \qquad (19\text{-}37)$$

Using Eq. 19-7 ($k = R/N_A$), we can rewrite this as

$$E_{int} = \tfrac{3}{2}nRT \qquad \text{(monatomic ideal gas).} \qquad (19\text{-}38)$$

☞ The internal energy $E_{int}$ of an ideal gas is a function of the gas temperature *only*; it does not depend on any other variable.

With Eq. 19-38 in hand, we are now able to derive an expression for the molar specific heat of an ideal gas. Actually, we shall derive two expressions. One is for the case in which the volume of the gas remains constant as energy is transferred to or from it as heat. The other is for the case in which the pressure of the gas remains constant as energy is transferred to or from it as heat. The symbols for these two molar specific heats are $C_V$ and $C_p$, respectively. (By convention, the capital letter $C$ is used in both cases, even though $C_V$ and $C_p$ represent types of specific heat and not heat capacities.)

### Molar Specific Heat at Constant Volume

Figure 19-9a shows $n$ moles of an ideal gas at pressure $p$ and temperature $T$, confined to a cylinder of fixed volume $V$. This *initial state i* of the gas is marked on the $p$-$V$ diagram of Fig. 19-9b. Suppose now that you add a small amount of energy to the gas as heat $Q$ by slowly turning up the temperature of the thermal reservoir. The gas temperature rises a small amount to $T + \Delta T$, and its pressure rises to $p + \Delta p$, bringing the gas to *final state f*. In such experiments, we would find that the heat $Q$ is related to the temperature change $\Delta T$ by

$$Q = nC_V \Delta T \qquad \text{(constant volume),} \qquad (19\text{-}39)$$

where $C_V$ is a constant called the **molar specific heat at constant volume**. Substituting this expression for $Q$ into the first law of thermodynamics as given by Eq. 18-26 ($\Delta E_{int} = Q - W$) yields

$$\Delta E_{int} = nC_V \Delta T - W. \qquad (19\text{-}40)$$

With the volume held constant, the gas cannot expand and thus cannot do any work. Therefore, $W = 0$, and Eq. 19-40 gives us

$$C_V = \frac{\Delta E_{int}}{n \Delta T}. \qquad (19\text{-}41)$$

From Eq. 19-38, the change in internal energy must be

$$\Delta E_{int} = \tfrac{3}{2}nR \, \Delta T. \qquad (19\text{-}42)$$

Substituting this result into Eq. 19-41 yields

$$C_V = \tfrac{3}{2}R = 12.5 \text{ J/mol} \cdot \text{K} \qquad \text{(monatomic gas)}. \qquad (19\text{-}43)$$

As Table 19-2 shows, this prediction of the kinetic theory (for ideal gases) agrees very well with experiment for real monatomic gases, the case that we have assumed. The (predicted and) experimental values of $C_V$ for *diatomic gases* (which have molecules with two atoms) and *polyatomic gases* (which have molecules with more than two atoms) are greater than those for monatomic gases for reasons that will be suggested in Section 19-9.

We can now generalize Eq. 19-38 for the internal energy of any ideal gas by substituting $C_V$ for $\tfrac{3}{2}R$; we get

$$E_{\text{int}} = nC_V T \qquad \text{(any ideal gas)}. \qquad (19\text{-}44)$$

This equation applies not only to an ideal monatomic gas but also to diatomic and polyatomic ideal gases, provided the appropriate value of $C_V$ is used. Just as with Eq. 19-38, we see that the internal energy of a gas depends on the temperature of the gas but not on its pressure or density.

When an ideal gas that is confined to a container undergoes a temperature change $\Delta T$, then from either Eq. 19-41 or Eq. 19-44 we can write the resulting change in its internal energy as

$$\Delta E_{\text{int}} = nC_V\,\Delta T \qquad \text{(ideal gas, any process)}. \qquad (19\text{-}45)$$

This equation tells us:

> A change in the internal energy $E_{\text{int}}$ of a confined ideal gas depends on the change in the gas temperature only; it does *not* depend on what type of process produces the change in the temperature.

As examples, consider the three paths between the two isotherms in the *p-V* diagram of Fig. 19-10. Path 1 represents a constant-volume process. Path 2 represents a constant-pressure process (that we are about to examine). Path 3 represents a process in which no heat is exchanged with the system's environment (we discuss this in Section 19-11). Although the values of heat $Q$ and work $W$ associated with these three paths differ, as do $p_f$ and $V_f$, the values of $\Delta E_{\text{int}}$ associated with the three paths are identical and are all given by Eq. 19-45, because they all involve the same temperature change $\Delta T$. Therefore, no matter what path is actually taken between $T$ and $T + \Delta T$, we can *always* use path 1 and Eq. 19-45 to compute $\Delta E_{\text{int}}$ easily.

## Molar Specific Heat at Constant Pressure

We now assume that the temperature of our ideal gas is increased by the same small amount $\Delta T$ as previously but now the necessary energy (heat $Q$) is added with the gas under constant pressure. An experiment for doing this is shown in Fig. 19-11a; the *p-V* diagram for the process is plotted in Fig. 19-11b. From such experiments we find that the heat $Q$ is related to the temperature change $\Delta T$ by

$$Q = nC_p\,\Delta T \qquad \text{(constant pressure)}, \qquad (19\text{-}46)$$

where $C_p$ is a constant called the **molar specific heat at constant pressure**. This $C_p$ is *greater* than the molar specific heat at constant volume $C_V$, because energy must now be supplied not only to raise the temperature of the gas but also for the gas to do work—that is, to lift the weighted piston of Fig. 19-11a.

To relate molar specific heats $C_p$ and $C_V$, we start with the first law of thermodynamics (Eq. 18-26):

$$\Delta E_{\text{int}} = Q - W. \qquad (19\text{-}47)$$

We next replace each term in Eq. 19-47. For $\Delta E_{\text{int}}$, we substitute from Eq. 19-45. For $Q$, we substitute from Eq. 19-46. To replace $W$, we first note that since

**TABLE 19-2**

**Molar Specific Heats at Constant Volume**

| Molecule | Example | | $C_V$ (J/mol·K) |
|---|---|---|---|
| Monatomic | Ideal | | $\tfrac{3}{2}R = 12.5$ |
| | Real | He | 12.5 |
| | | Ar | 12.6 |
| Diatomic | Ideal | | $\tfrac{5}{2}R = 20.8$ |
| | Real | $N_2$ | 20.7 |
| | | $O_2$ | 20.8 |
| Polyatomic | Ideal | | $3R = 24.9$ |
| | Real | $NH_4$ | 29.0 |
| | | $CO_2$ | 29.7 |

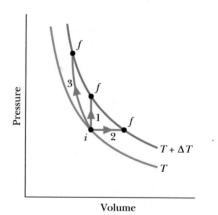

**FIG. 19-10** Three paths representing three different processes that take an ideal gas from an initial state *i* at temperature $T$ to some final state *f* at temperature $T + \Delta T$. The change $\Delta E_{\text{int}}$ in the internal energy of the gas is the same for these three processes and for any others that result in the same change of temperature.

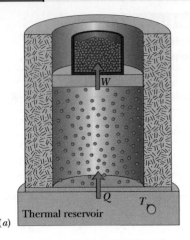

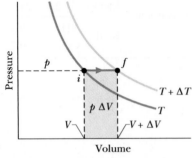

(a) Thermal reservoir $T$

(b) Pressure / Volume

**FIG. 19-11** (a) The temperature of an ideal gas is raised from $T$ to $T + \Delta T$ in a constant-pressure process. Heat is added and work is done in lifting the loaded piston. (b) The process on a $p$-$V$ diagram. The work $p \Delta V$ is given by the shaded area.

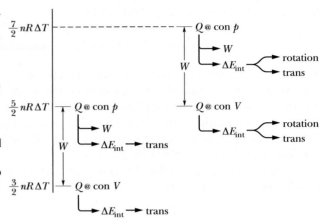

**FIG. 19-12** The relative values of $Q$ for a monatomic gas (left side) and a diatomic gas undergoing a constant-volume process (labeled "con $V$") and a constant-pressure process (labeled "con $p$"). The transfer of the energy into work $W$ and internal energy ($\Delta E_{int}$) is noted.

the pressure remains constant, Eq. 19-16 tells us that $W = p \Delta V$. Then we note that, using the ideal gas equation ($pV = nRT$), we can write

$$W = p \Delta V = nR \Delta T. \tag{19-48}$$

Making these substitutions in Eq. 19-47 and then dividing through by $n \Delta T$, we find

$$C_V = C_p - R$$

and then

$$C_p = C_V + R. \tag{19-49}$$

This prediction of kinetic theory agrees well with experiment, not only for monatomic gases but also for gases in general, as long as their density is low enough so that we may treat them as ideal.

The left side of Fig. 19-12 shows the relative values of $Q$ for a monatomic gas undergoing either a constant-volume process ($Q = \frac{3}{2}nR \Delta T$) or a constant-pressure process ($Q = \frac{5}{2}nR \Delta T$). Note that for the latter, the value of $Q$ is higher by the amount $W$, the work done by the gas in the expansion. Note also that for the constant-volume process, the energy added as $Q$ goes entirely into the change in internal energy $\Delta E_{int}$ and for the constant-pressure process, the energy added as $Q$ goes into both $\Delta E_{int}$ and the work $W$.

✓ **CHECKPOINT 4** The figure here shows five paths traversed by a gas on a $p$-$V$ diagram. Rank the paths according to the change in internal energy of the gas, greatest first.

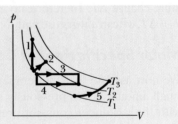

---

**Sample Problem** | **19-7** | **Build your skill**

A bubble of 5.00 mol of helium is submerged at a certain depth in liquid water when the water (and thus the helium) undergoes a temperature increase $\Delta T$ of 20.0 C° at constant pressure. As a result, the bubble expands. The helium is monatomic and ideal.

(a) How much energy is added to the helium as heat during the increase and expansion?

**KEY IDEA** Heat $Q$ is related to the temperature change $\Delta T$ by a molar specific heat of the gas.

**Calculations:** Because the pressure $p$ is held constant during the addition of energy, we use the molar specific heat at constant pressure $C_p$ and Eq. 19-46,

$$Q = nC_p \Delta T, \tag{19-50}$$

to find $Q$. To evaluate $C_p$ we go to Eq. 19-49, which tells us that for any ideal gas, $C_p = C_V + R$. Then from Eq. 19-43, we know that for any *monatomic* gas (like the helium here), $C_V = \frac{3}{2}R$. Thus, Eq. 19-50 gives us

$$Q = n(C_V + R)\,\Delta T = n(\tfrac{3}{2}R + R)\,\Delta T = n(\tfrac{5}{2}R)\,\Delta T$$
$$= (5.00\text{ mol})(2.5)(8.31\text{ J/mol} \cdot \text{K})(20.0\text{ C}°)$$
$$= 2077.5\text{ J} \approx 2080\text{ J}. \qquad \text{(Answer)}$$

**(b)** What is the change $\Delta E_{int}$ in the internal energy of the helium during the temperature increase?

**KEY IDEA** Because the bubble expands, this is not a constant-volume process. However, the helium is nonetheless confined (to the bubble). Thus, the change $\Delta E_{int}$ is the same as *would occur* in a constant-volume process with the same temperature change $\Delta T$.

**Calculation:** We can now easily find the constant-volume change $\Delta E_{int}$ with Eq. 19-45:

$$\Delta E_{int} = nC_V\,\Delta T = n(\tfrac{3}{2}R)\,\Delta T$$
$$= (5.00\text{ mol})(1.5)(8.31\text{ J/mol} \cdot \text{K})(20.0\text{ C}°)$$
$$= 1246.5\text{ J} \approx 1250\text{ J}. \qquad \text{(Answer)}$$

**(c)** How much work $W$ is done by the helium as it expands against the pressure of the surrounding water during the temperature increase?

**KEY IDEAS** The work done by *any* gas expanding against the pressure from its environment is given by Eq. 19-11, which tells us to integrate $p\,dV$. When the pressure is constant (as here), we can simplify that to $W = p\,\Delta V$. When the gas is *ideal* (as here), we can use the ideal gas law (Eq. 19-5) to write $p\,\Delta V = nR\,\Delta T$.

**Calculation:** We end up with

$$W = nR\,\Delta T$$
$$= (5.00\text{ mol})(8.31\text{ J/mol} \cdot \text{K})(20.0\text{ C}°)$$
$$= 831\text{ J}. \qquad \text{(Answer)}$$

**Another way:** Because we happen to know $Q$ and $\Delta E_{int}$, we can work this problem another way: We can account for the energy changes of the gas with the first law of thermodynamics, writing

$$W = Q - \Delta E_{int} = 2077.5\text{ J} - 1246.5\text{ J}$$
$$= 831\text{ J}. \qquad \text{(Answer)}$$

**The transfers:** Let's follow the energy. Of the 2077.5 J transferred to the helium as heat $Q$, 831 J goes into the work $W$ required for the expansion and 1246.5 J goes into the internal energy $E_{int}$, which, for a monatomic gas, is entirely the kinetic energy of the atoms in their translational motion. These several results are suggested on the left side of Fig. 19-12.

## 19-9 | Degrees of Freedom and Molar Specific Heats

As Table 19-2 shows, the prediction that $C_V = \frac{3}{2}R$ agrees with experiment for monatomic gases but fails for diatomic and polyatomic gases. Let us try to explain the discrepancy by considering the possibility that molecules with more than one atom can store internal energy in forms other than translational kinetic energy.

Figure 19-13 shows common models of helium (a *monatomic* molecule, containing a single atom), oxygen (a *diatomic* molecule, containing two atoms), and methane (a *polyatomic* molecule). From such models, we would assume that all three types of molecules can have translational motions (say, moving left–right and up–down) and rotational motions (spinning about an axis like a top). In addition, we would assume that the diatomic and polyatomic molecules can have oscillatory motions, with the atoms oscillating slightly toward and away from one another, as if attached to opposite ends of a spring.

To keep account of the various ways in which energy can be stored in a gas, James Clerk Maxwell introduced the theorem of the **equipartition of energy:**

> Every kind of molecule has a certain number $f$ of *degrees of freedom*, which are independent ways in which the molecule can store energy. Each such degree of freedom has associated with it—on average—an energy of $\frac{1}{2}kT$ per molecule (or $\frac{1}{2}RT$ per mole).

Let us apply the theorem to the translational and rotational motions of the molecules in Fig. 19-13. (We discuss oscillatory motion in the next section.) For the translational motion, superimpose an $xyz$ coordinate system on any gas. The molecules will, in general, have velocity components along all three axes. Thus, gas molecules of all types have three degrees of translational freedom (three ways to move in translation) and, on average, an associated energy of $3(\frac{1}{2}kT)$ per molecule.

For the rotational motion, imagine the origin of our $xyz$ coordinate system at the center of each molecule in Fig. 19-13. In a gas, each molecule should be able

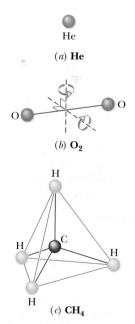

**FIG. 19-13** Models of molecules as used in kinetic theory: (*a*) helium, a typical monatomic molecule; (*b*) oxygen, a typical diatomic molecule; and (*c*) methane, a typical polyatomic molecule. The spheres represent atoms, and the lines between them represent bonds. Two rotation axes are shown for the oxygen molecule.

**TABLE 19-3**

**Degrees of Freedom for Various Molecules**

| Molecule | Example | Degrees of Freedom | | | Predicted Molar Specific Heats | |
| | | Translational | Rotational | Total ($f$) | $C_V$ (Eq. 19-51) | $C_p = C_V + R$ |
| --- | --- | --- | --- | --- | --- | --- |
| Monatomic | He | 3 | 0 | 3 | $\frac{3}{2}R$ | $\frac{5}{2}R$ |
| Diatomic | $O_2$ | 3 | 2 | 5 | $\frac{5}{2}R$ | $\frac{7}{2}R$ |
| Polyatomic | $CH_4$ | 3 | 3 | 6 | $3R$ | $4R$ |

to rotate with an angular velocity component along each of the three axes, so each gas should have three degrees of rotational freedom and, on average, an additional energy of $3(\frac{1}{2}kT)$ per molecule. *However,* experiment shows this is true only for the polyatomic molecules. According to *quantum theory,* the physics dealing with the allowed motions and energies of molecules and atoms, a monatomic gas molecule does not rotate and so has no rotational energy (a single atom cannot rotate like a top). A diatomic molecule can rotate like a top only about axes perpendicular to the line connecting the atoms (the axes are shown in Fig. 19-13b) and not about that line itself. Therefore, a diatomic molecule can have only two degrees of rotational freedom and a rotational energy of only $2(\frac{1}{2}kT)$ per molecule.

To extend our analysis of molar specific heats ($C_p$ and $C_V$, in Section 19-8) to ideal diatomic and polyatomic gases, it is necessary to retrace the derivations of that analysis in detail. First, we replace Eq. 19-38 ($E_{int} = \frac{3}{2}nRT$) with $E_{int} = (f/2)nRT$, where $f$ is the number of degrees of freedom listed in Table 19-3. Doing so leads to the prediction

$$C_V = \left(\frac{f}{2}\right)R = 4.16f \text{ J/mol} \cdot \text{K,} \qquad (19\text{-}51)$$

which agrees—as it must—with Eq. 19-43 for monatomic gases ($f = 3$). As Table 19-2 shows, this prediction also agrees with experiment for diatomic gases ($f = 5$), but it is too low for polyatomic gases ($f = 6$ for molecules comparable to $CH_4$).

**Sample Problem**    **19-8**    **Build your skill**

We transfer 1000 J to a diatomic gas, allowing it to expand with the pressure held constant. The gas molecules rotate but do not oscillate. How much of the 1000 J goes into the increase of the gas's internal energy? Of that amount, how much goes into $\Delta K_{tran}$ (the kinetic energy of the translational motion of the molecules) and $\Delta K_{rot}$ (the kinetic energy of their rotational motion)?

**KEY IDEAS**

1. The transfer of energy as heat $Q$ to a gas under constant pressure is related to the resulting temperature increase $\Delta T$ via Eq. 19-46 ($Q = nC_p \Delta T$).

2. Because the gas is diatomic with molecules undergoing rotation but not oscillation, the molar specific heat is, from Fig. 19-12 and Table 19-3, $C_p = \frac{7}{2}R$.

3. The increase $\Delta E_{int}$ in the internal energy is the same as would occur with a constant-volume process resulting in the same $\Delta T$. Thus, from Eq. 19-45, $\Delta E_{int} = nC_V \Delta T$. From Fig. 19-12 and Table 19-3, we see that $C_V = \frac{5}{2}R$.

4. For the same $n$ and $\Delta T$, $\Delta E_{int}$ is greater for a diatomic gas than a monatomic gas because additional energy is required for rotation.

***Increase in $E_{int}$:*** Let's first get the temperature change $\Delta T$ due to the transfer of energy as heat. From Eq. 19-46, substituting $\frac{7}{2}R$ for $C_p$, we have

$$\Delta T = \frac{Q}{\frac{7}{2}nR}. \qquad (19\text{-}52)$$

We next find $\Delta E_{int}$ from Eq. 19-45, substituting the molar specific heat $C_V (= \frac{5}{2}R)$ for a constant-volume process and using the same $\Delta T$. Because we are dealing with a diatomic gas, let's call this change $\Delta E_{int,dia}$. Equation 19-45 gives us

$$\Delta E_{int,dia} = nC_V \Delta T = n\frac{5}{2}R\left(\frac{Q}{\frac{7}{2}nR}\right) = \frac{5}{7}Q$$

$$= 0.71428Q = 714.3 \text{ J.} \qquad \text{(Answer)}$$

In words, about 71% of the energy transferred to the gas goes into the internal energy. The rest goes into the work required to increase the volume of the gas.

***Increases in K:*** If we were to increase the temperature of a *monatomic* gas (with the same value of $n$) by the amount given in Eq. 19-52, the internal energy would change by a smaller amount, call it $\Delta E_{int,mon}$, because rotational motion is not involved. To calculate that smaller amount, we still use Eq. 19-45 but now we substitute the value of $C_V$ for a monatomic gas—namely, $C_V = \frac{3}{2}R$. So,

$$\Delta E_{int,mon} = n\frac{3}{2}R\,\Delta T.$$

Substituting for $\Delta T$ from Eq. 19-52 leads us to

$$\Delta E_{int,mon} = n\frac{3}{2}R\left(\frac{Q}{n\frac{7}{2}R}\right) = \frac{3}{7}Q$$

$$= 0.42857Q = 428.6\text{ J}.$$

For the monatomic gas, all this energy would go into the kinetic energy of the translational motion of the atoms. The important point here is that for a diatomic gas with the same values of $n$ and $\Delta T$, the same amount of energy goes into the kinetic energy of the translational motion of the molecules. The rest of $\Delta E_{int,dia}$ (that is, the additional 285.7 J) goes into the rotational motion of the molecules. Thus, for the diatomic gas,

$$\Delta K_{trans} = 428.6\text{ J} \quad \text{and} \quad \Delta K_{rot} = 285.7\text{ J}. \qquad \text{(Answer)}$$

---

A cabin of volume $V$ is filled with air (which we consider to be an ideal diatomic gas) at an initial low temperature $T_1$. After you light a wood stove, the air temperature increases to $T_2$. What is the resulting change $\Delta E_{int}$ in the internal energy of the air in the cabin?

**KEY IDEAS**  As the air temperature increases, the air pressure $p$ cannot change but must always be equal to the air pressure outside the room. The reason is that, because the room is not airtight, the air is not confined. As the temperature increases, air molecules leave through various openings and thus the number of moles $n$ of air in the room decreases. Thus, we *cannot* use Eq. 19-45 ($\Delta E_{int} = nC_V\,\Delta T$) to find $\Delta E_{int}$, because it requires constant $n$. However, we *can* relate the internal energy $E_{int}$ at any instant to $n$ and the temperature $T$ with Eq. 19-44 ($E_{int} = nC_VT$).

*Calculations:* From Eq. 19-44 we can then write

$$\Delta E_{int} = \Delta(nC_VT) = C_V\,\Delta(nT).$$

Next, using Eq. 19-5 ($pV = nRT$), we can replace $nT$ with $pV/R$, obtaining

$$\Delta E_{int} = C_V\,\Delta\left(\frac{pV}{R}\right).$$

Because $p$, $V$, and $R$ are all constants, this yields

$$\Delta E_{int} = 0, \qquad \text{(Answer)}$$

even though the temperature changes.

Why does the cabin feel more comfortable at the higher temperature? There are at least two factors involved: (1) You exchange electromagnetic radiation (thermal radiation) with surfaces inside the room, and (2) you exchange energy with air molecules that collide with you. When the room temperature is increased, (1) the amount of thermal radiation emitted by the surfaces and absorbed by you is increased, and (2) the amount of energy you gain through the collisions of air molecules with you is increased.

---

## 19-10 | A Hint of Quantum Theory

We can improve the agreement of kinetic theory with experiment by including the oscillations of the atoms in a gas of diatomic or polyatomic molecules. For example, the two atoms in the $O_2$ molecule of Fig. 19-13b can oscillate toward and away from each other, with the interconnecting bond acting like a spring. However, experiment shows that such oscillations occur only at relatively high temperatures of the gas—the motion is "turned on" only when the gas molecules have relatively large energies. Rotational motion is also subject to such "turning on," but at a lower temperature.

Figure 19-14 is of help in seeing this turning on of rotational motion and oscillatory motion. The ratio $C_V/R$ for diatomic hydrogen gas ($H_2$) is plotted there against temperature, with the temperature scale logarithmic to cover several orders of magnitude. Below about 80 K, we find that $C_V/R = 1.5$. This result implies that only the three translational degrees of freedom of hydrogen are involved in the specific heat.

As the temperature increases, the value of $C_V/R$ gradually increases to 2.5, implying that two additional degrees of freedom have become involved. Quantum theory shows that these two degrees of freedom are associated with the

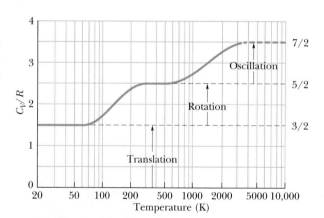

**FIG. 19-14**  $C_V/R$ versus temperature for (diatomic) hydrogen gas. Because rotational and oscillatory motions begin at certain energies, only translation is possible at very low temperatures. As the temperature increases, rotational motion can begin. At still higher temperatures, oscillatory motion can begin.

rotational motion of the hydrogen molecules and that this motion requires a certain minimum amount of energy. At very low temperatures (below 80 K), the molecules do not have enough energy to rotate. As the temperature increases from 80 K, first a few molecules and then more and more obtain enough energy to rotate, and $C_V/R$ increases, until all of them are rotating and $C_V/R = 2.5$.

Similarly, quantum theory shows that oscillatory motion of the molecules requires a certain (higher) minimum amount of energy. This minimum amount is not met until the molecules reach a temperature of about 1000 K, as shown in Fig. 19-14. As the temperature increases beyond 1000 K, more molecules have enough energy to oscillate and $C_V/R$ increases, until all of them are oscillating and $C_V/R = 3.5$. (In Fig. 19-14, the plotted curve stops at 3200 K because there the atoms of a hydrogen molecule oscillate so much that they overwhelm their bond, and the molecule then *dissociates* into two separate atoms.)

## 19-11 | The Adiabatic Expansion of an Ideal Gas

We saw in Section 17-4 that sound waves are propagated through air and other gases as a series of compressions and expansions; these variations in the transmission medium take place so rapidly that there is no time for energy to be transferred from one part of the medium to another as heat. As we saw in Section 18-11, a process for which $Q = 0$ is an *adiabatic process*. We can ensure that $Q = 0$ either by carrying out the process very quickly (as in sound waves) or by doing it (at any rate) in a well-insulated container. Let us see what the kinetic theory has to say about adiabatic processes.

Figure 19-15a shows our usual insulated cylinder, now containing an ideal gas and resting on an insulating stand. By removing mass from the piston, we can allow the gas to expand adiabatically. As the volume increases, both the pressure and the temperature drop. We shall prove next that the relation between the pressure and the volume during such an adiabatic process is

$$pV^\gamma = \text{a constant} \qquad \text{(adiabatic process)}, \qquad (19\text{-}53)$$

in which $\gamma = C_p/C_V$, the ratio of the molar specific heats for the gas. On a $p$-$V$ diagram such as that in Fig. 19-15b, the process occurs along a line (called an *adiabat*) that has the equation $p = (\text{a constant})/V^\gamma$. Since the gas goes from an initial state $i$ to a final state $f$, we can rewrite Eq. 19-53 as

$$p_iV_i^\gamma = p_fV_f^\gamma \qquad \text{(adiabatic process)}. \qquad (19\text{-}54)$$

To write an equation for an adiabatic process in terms of $T$ and $V$, we use the ideal gas equation ($pV = nRT$) to eliminate $p$ from Eq. 19-53, finding

$$\left(\frac{nRT}{V}\right)V^\gamma = \text{a constant}.$$

FIG. 19-15 (a) The volume of an ideal gas is increased by removing mass from the piston. The process is adiabatic ($Q = 0$). (b) The process proceeds from $i$ to $f$ along an adiabat on a $p$-$V$ diagram.

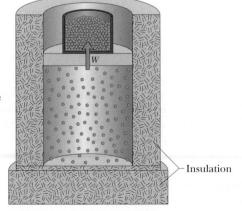

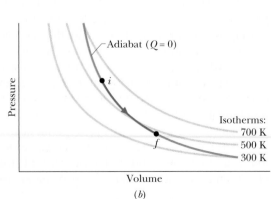

Because $n$ and $R$ are constants, we can rewrite this in the alternative form

$$TV^{\gamma-1} = \text{a constant} \quad \text{(adiabatic process),} \quad (19\text{-}55)$$

in which the constant is different from that in Eq. 19-53. When the gas goes from an initial state $i$ to a final state $f$, we can rewrite Eq. 19-55 as

$$T_i V_i^{\gamma-1} = T_f V_f^{\gamma-1} \quad \text{(adiabatic process).} \quad (19\text{-}56)$$

Understanding adiabatic processes allows you to understand why popping the cork on a cold bottle of champagne (as in this chapter's opening photograph) or the tab on a cold can of soda causes a slight fog to form at the opening of the container. At the top of any unopened carbonated drink sits a gas of carbon dioxide and water vapor. Because the gas pressure is greater than atmospheric pressure, the gas expands out into the atmosphere when the container is opened. Thus, the gas volume increases, but that means the gas must do work pushing against the atmosphere. Because the expansion is rapid, it is adiabatic, and the only source of energy for the work is the internal energy of the gas. Because the internal energy decreases, the temperature of the gas also decreases, which causes the water vapor in the gas to condense into tiny drops of fog. (Note that Eq. 19-56 also tells us that the temperature must decrease during an adiabatic expansion: $V_f$ is greater than $V_i$, and so $T_f$ must be less than $T_i$.)

## Proof of Eq. 19-53

Suppose that you remove some shot from the piston of Fig. 19-15a, allowing the ideal gas to push the piston and the remaining shot upward and thus to increase the volume by a differential amount $dV$. Since the volume change is tiny, we may assume that the pressure $p$ of the gas on the piston is constant during the change. This assumption allows us to say that the work $dW$ done by the gas during the volume increase is equal to $p \, dV$. From Eq. 18-27, the first law of thermodynamics can then be written as

$$dE_{\text{int}} = Q - p \, dV. \quad (19\text{-}57)$$

Since the gas is thermally insulated (and thus the expansion is adiabatic), we substitute 0 for $Q$. Then we use Eq. 19-45 to substitute $nC_V \, dT$ for $dE_{\text{int}}$. With these substitutions, and after some rearranging, we have

$$n \, dT = -\left(\frac{p}{C_V}\right) dV. \quad (19\text{-}58)$$

Now from the ideal gas law ($pV = nRT$) we have

$$p \, dV + V \, dp = nR \, dT. \quad (19\text{-}59)$$

Replacing $R$ with its equal, $C_p - C_V$, in Eq. 19-59 yields

$$n \, dT = \frac{p \, dV + V \, dp}{C_p - C_V}. \quad (19\text{-}60)$$

Equating Eqs. 19-58 and 19-60 and rearranging then give

$$\frac{dp}{p} + \left(\frac{C_p}{C_V}\right)\frac{dV}{V} = 0.$$

Replacing the ratio of the molar specific heats with $\gamma$ and integrating (see integral 5 in Appendix E) yield

$$\ln p + \gamma \ln V = \text{a constant.}$$

Rewriting the left side as $\ln pV^{\gamma}$ and then taking the antilog of both sides, we find

$$pV^{\gamma} = \text{a constant.} \quad (19\text{-}61)$$

## Free Expansions

Recall from Section 18-11 that a free expansion of a gas is an adiabatic process that involves no work done on or by the gas, and no change in the internal energy of the gas. A free expansion is thus quite different from the type of adiabatic process described by Eqs. 19-53 through 19-61, in which work is done and the internal energy changes. Those equations then do *not* apply to a free expansion, even though such an expansion is adiabatic.

Also recall that in a free expansion, a gas is in equilibrium only at its initial and final points; thus, we can plot only those points, but not the expansion itself, on a $p$-$V$ diagram. In addition, because $\Delta E_{int} = 0$, the temperature of the final state must be that of the initial state. Thus, the initial and final points on a $p$-$V$ diagram must be on the same isotherm, and instead of Eq. 19-56 we have

$$T_i = T_f \quad \text{(free expansion).} \tag{19-62}$$

If we next assume that the gas is ideal (so that $pV = nRT$), then because there is no change in temperature, there can be no change in the product $pV$. Thus, instead of Eq. 19-53 a free expansion involves the relation

$$p_iV_i = p_fV_f \quad \text{(free expansion).} \tag{19-63}$$

## Sample Problem 19-10

In Sample Problem 19-2, 1 mol of oxygen (assumed to be an ideal gas) expands isothermally (at 310 K) from an initial volume of 12 L to a final volume of 19 L.

**(a)** What would be the final temperature if the gas had expanded adiabatically to this same final volume? Oxygen ($O_2$) is diatomic and here has rotation but not oscillation.

**KEY IDEAS**

1. When a gas expands against the pressure of its environment, it must do work.

2. When the process is adiabatic (no energy is transferred as heat), then the energy required for the work can come only from the internal energy of the gas.

3. Because the internal energy decreases, the temperature $T$ must also decrease.

*Calculations:* We can relate the initial and final temperatures and volumes with Eq. 19-56:

$$T_iV_i^{\gamma-1} = T_fV_f^{\gamma-1}. \tag{19-64}$$

Because the molecules are diatomic and have rotation but not oscillation, we can take the molar specific heats from Table 19-3. Thus,

$$\gamma = \frac{C_p}{C_V} = \frac{\frac{7}{2}R}{\frac{5}{2}R} = 1.40.$$

Solving Eq. 19-64 for $T_f$ and inserting known data then yield

$$T_f = \frac{T_iV_i^{\gamma-1}}{V_f^{\gamma-1}} = \frac{(310\text{ K})(12\text{ L})^{1.40-1}}{(19\text{ L})^{1.40-1}}$$

$$= (310\text{ K})(\tfrac{12}{19})^{0.40} = 258\text{ K.} \quad \text{(Answer)}$$

**(b)** What would be the final temperature and pressure if, instead, the gas had expanded freely to the new volume, from an initial pressure of 2.0 Pa?

**KEY IDEA** The temperature does not change in a free expansion.

*Calculation:* Thus, the temperature is

$$T_f = T_i = 310\text{ K.} \quad \text{(Answer)}$$

We find the new pressure using Eq. 19-63, which gives us

$$p_f = p_i \frac{V_i}{V_f} = (2.0\text{ Pa})\frac{12\text{ L}}{19\text{ L}} = 1.3\text{ Pa.} \quad \text{(Answer)}$$

## PROBLEM-SOLVING TACTICS

*Tactic 2: A Graphical Summary of Four Gas Processes*
In this chapter we have discussed four special processes that an ideal gas can undergo. An example of each (for a monatomic ideal gas) is shown in Fig. 19-16, and some associated characteristics are given in Table 19-4, including two process names (isobaric and isochoric) that we have not used but that you might see in other courses.

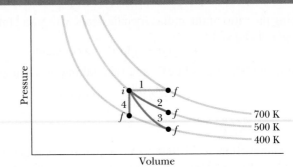

**FIG. 19-16** A $p$-$V$ diagram representing four special processes for an ideal monatomic gas.

**TABLE 19-4**

**Four Special Processes**

| Path in Fig. 19-16 | Constant Quantity | Process Type | Some Special Results ($\Delta E_{int} = Q - W$ and $\Delta E_{int} = nC_V\Delta T$ for all paths) |
|---|---|---|---|
| 1 | $p$ | Isobaric | $Q = nC_p\Delta T; W = p\,\Delta V$ |
| 2 | $T$ | Isothermal | $Q = W = nRT\ln(V_f/V_i); \Delta E_{int} = 0$ |
| 3 | $pV^\gamma, TV^{\gamma-1}$ | Adiabatic | $Q = 0; \quad W = -\Delta E_{int}$ |
| 4 | $V$ | Isochoric | $Q = \Delta E_{int} = nC_V\Delta T; \quad W = 0$ |

✓**CHECKPOINT 5**     Rank paths 1, 2, and 3 in Fig. 19-16 according to the energy transfer to the gas as heat, greatest first.

## REVIEW & SUMMARY

**Kinetic Theory of Gases**   The *kinetic theory of gases* relates the *macroscopic* properties of gases (for example, pressure and temperature) to the *microscopic* properties of gas molecules (for example, speed and kinetic energy).

**Avogadro's Number**   One mole of a substance contains $N_A$ (*Avogadro's number*) elementary units (usually atoms or molecules), where $N_A$ is found experimentally to be

$$N_A = 6.02 \times 10^{23} \text{ mol}^{-1} \quad \text{(Avogadro's number).} \quad (19\text{-}1)$$

One molar mass $M$ of any substance is the mass of one mole of the substance. It is related to the mass $m$ of the individual molecules of the substance by

$$M = mN_A. \quad (19\text{-}4)$$

The number of moles $n$ contained in a sample of mass $M_{sam}$, consisting of $N$ molecules, is given by

$$n = \frac{N}{N_A} = \frac{M_{sam}}{M} = \frac{M_{sam}}{mN_A}. \quad (19\text{-}2, 19\text{-}3)$$

**Ideal Gas**   An *ideal gas* is one for which the pressure $p$, volume $V$, and temperature $T$ are related by

$$pV = nRT \quad \text{(ideal gas law).} \quad (19\text{-}5)$$

Here $n$ is the number of moles of the gas present and $R$ is a constant (8.31 J/mol·K) called the **gas constant.** The ideal gas law can also be written as

$$pV = NkT, \quad (19\text{-}9)$$

where the **Boltzmann constant** $k$ is

$$k = \frac{R}{N_A} = 1.38 \times 10^{-23} \text{ J/K.} \quad (19\text{-}7)$$

**Work in an Isothermal Volume Change**   The work done *by* an ideal gas during an **isothermal** (constant-temperature) change from volume $V_i$ to volume $V_f$ is

$$W = nRT\ln\frac{V_f}{V_i} \quad \text{(ideal gas, isothermal process).} \quad (19\text{-}14)$$

**Pressure, Temperature, and Molecular Speed**   The pressure exerted by $n$ moles of an ideal gas, in terms of the speed of its molecules, is

$$p = \frac{nMv_{rms}^2}{3V}, \quad (19\text{-}21)$$

where $v_{rms} = \sqrt{(v^2)_{avg}}$ is the **root-mean-square speed** of the molecules of the gas. With Eq. 19-5 this gives

$$v_{rms} = \sqrt{\frac{3RT}{M}}. \quad (19\text{-}22)$$

**Temperature and Kinetic Energy**   The average translational kinetic energy $K_{avg}$ per molecule of an ideal gas is

$$K_{avg} = \tfrac{3}{2}kT. \quad (19\text{-}24)$$

**Mean Free Path**   The *mean free path* $\lambda$ of a gas molecule is its average path length between collisions and is given by

$$\lambda = \frac{1}{\sqrt{2}\pi d^2\, N/V}, \quad (19\text{-}25)$$

where $N/V$ is the number of molecules per unit volume and $d$ is the molecular diameter.

**Maxwell Speed Distribution**   The *Maxwell speed distribution* $P(v)$ is a function such that $P(v)\,dv$ gives the *fraction* of molecules with speeds in the interval $dv$ at speed $v$:

$$P(v) = 4\pi\left(\frac{M}{2\pi RT}\right)^{3/2} v^2 e^{-Mv^2/2RT}. \quad (19\text{-}27)$$

Three measures of the distribution of speeds among the molecules of a gas are

$$v_{avg} = \sqrt{\frac{8RT}{\pi M}} \quad \text{(average speed),} \quad (19\text{-}31)$$

$$v_P = \sqrt{\frac{2RT}{M}} \quad \text{(most probable speed),} \quad (19\text{-}35)$$

and the rms speed defined above in Eq. 19-22.

**Molar Specific Heats**   The molar specific heat $C_V$ of a gas at constant volume is defined as

$$C_V = \frac{Q}{n\,\Delta T} = \frac{\Delta E_{int}}{n\,\Delta T}, \quad (19\text{-}39, 19\text{-}41)$$

in which $Q$ is the energy transferred as heat to or from a sample of $n$ moles of the gas, $\Delta T$ is the resulting temperature change of the gas, and $\Delta E_{int}$ is the resulting change in the internal energy of the gas. For an ideal monatomic gas,

$$C_V = \tfrac{3}{2}R = 12.5 \text{ J/mol·K.} \quad (19\text{-}43)$$

The molar specific heat $C_p$ of a gas at constant pressure is defined to be

$$C_p = \frac{Q}{n \, \Delta T}, \tag{19-46}$$

in which $Q$, $n$, and $\Delta T$ are defined as above. $C_p$ is also given by

$$C_p = C_V + R. \tag{19-49}$$

For $n$ moles of an ideal gas,

$$E_{\text{int}} = nC_VT \quad \text{(ideal gas).} \tag{19-44}$$

If $n$ moles of a confined ideal gas undergo a temperature change $\Delta T$ due to *any* process, the change in the internal energy of the gas is

$$\Delta E_{\text{int}} = nC_V \, \Delta T \quad \text{(ideal gas, any process),} \tag{19-45}$$

in which the appropriate value of $C_V$ must be substituted, according to the type of ideal gas.

**Degrees of Freedom and $C_V$** We find $C_V$ by using the *equipartition of energy* theorem, which states that every *degree of freedom* of a molecule (that is, every independent way it can store energy) has associated with it—on average—an energy $\frac{1}{2}kT$ per molecule ($= \frac{1}{2}RT$ per mole). If $f$ is the number of degrees of freedom, then $E_{\text{int}} = (f/2)nRT$ and

$$C_V = \left(\frac{f}{2}\right)R = 4.16f \text{ J/mol·K.} \tag{19-51}$$

For monatomic gases $f = 3$ (three translational degrees); for diatomic gases $f = 5$ (three translational and two rotational degrees).

**Adiabatic Process** When an ideal gas undergoes a slow adiabatic volume change (a change for which $Q = 0$), its pressure and volume are related by

$$pV^\gamma = \text{a constant} \quad \text{(adiabatic process),} \tag{19-53}$$

in which $\gamma\, (= C_p/C_V)$ is the ratio of molar specific heats for the gas. For a free expansion, however, $pV = $ a constant.

## QUESTIONS

**1** For a temperature increase of $\Delta T_1$, a certain amount of an ideal gas requires 30 J when heated at constant volume and 50 J when heated at constant pressure. How much work is done by the gas in the second situation?

**2** The dot in Fig. 19-17a represents the initial state of a gas, and the vertical line through the dot divides the p-V diagram into regions 1 and 2. For the following processes, determine whether the work W done by the gas is positive, negative, or zero: (a) the gas moves up along the vertical line, (b) it moves down along the vertical line, (c) it moves to anywhere in region 1, and (d) it moves to anywhere in region 2.

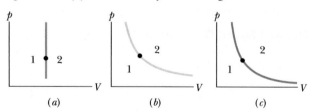

FIG. 19-17   Questions 2, 4, and 6.

**3** For four situations for an ideal gas, the table gives the energy transferred to or from the gas as heat $Q$ and either the work $W$ done by the gas or the work $W_{\text{on}}$ done on the gas, all in joules. Rank the four situations in terms of the temperature change of the gas, most positive first.

|          | a   | b   | c   | d   |
|----------|-----|-----|-----|-----|
| $Q$      | −50 | +35 | −15 | +20 |
| $W$      | −50 | +35 |     |     |
| $W_{\text{on}}$ |     |     | −40 | +40 |

**4** The dot in Fig. 19-17b represents the initial state of a gas, and the isotherm through the dot divides the p-V diagram into regions 1 and 2. For the following processes, determine whether the change $\Delta E_{\text{int}}$ in the internal energy of the gas is positive, negative, or zero: (a) the gas moves up along the isotherm, (b) it moves down along the isotherm, (c) it moves to anywhere in region 1, and (d) it moves to anywhere in region 2.

**5** A certain amount of energy is to be transferred as heat to 1 mol of a monatomic gas (a) at constant pressure and (b) at constant volume, and to 1 mol of a diatomic gas (c) at constant pressure and (d) at constant volume. Figure 19-18 shows four paths from an initial point to four final points on a p-V diagram. Which path goes with which process? (e) Are the molecules of the diatomic gas rotating?

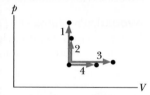

FIG. 19-18   Question 5.

**6** The dot in Fig. 19-17c represents the initial state of a gas, and the adiabat through the dot divides the p-V diagram into regions 1 and 2. For the following processes, determine whether the corresponding heat $Q$ is positive, negative, or zero: (a) the gas moves up along the adiabat, (b) it moves down along the adiabat, (c) it moves to anywhere in region 1, and (d) it moves to anywhere in region 2.

**7** An ideal diatomic gas, with molecular rotation but not oscillation, loses energy as heat $Q$. Is the resulting decrease in the internal energy of the gas greater if the loss occurs in a constant-volume process or in a constant-pressure process?

**8** In the p-V diagram of Fig. 19-19, the gas does 5 J of work when taken along isotherm ab and 4 J when taken along adiabat bc. What is the change in the internal energy of the gas when it is taken along the straight path from a to c?

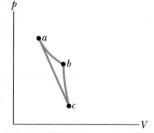

FIG. 19-19   Question 8.

**9** (a) Rank the four paths of Fig. 19-16 according to the work done by the gas, greatest first. (b) Rank paths 1, 2, and 3 according to the change in the internal energy of the gas, most positive first and most negative last.

**10** Does the temperature of an ideal gas increase, decrease, or stay the same during (a) an isothermal expansion, (b) an expansion at constant pressure, (c) an adiabatic expansion, and (d) an increase in pressure at constant volume?

# PROGRAMS

**PROBLEMS**

| GO | Tutoring problem available (at instructor's discretion) in *WileyPLUS* and WebAssign | | |
|---|---|---|---|
| SSM | Worked-out solution available in Student Solutions Manual | WWW | Worked-out solution is at |
| • – ••• | Number of dots indicates level of problem difficulty | ILW | Interactive solution is at |
|  | Additional information available in *The Flying Circus of Physics* and at flyingcircusofphysics.com | | |

http://www.wiley.com/college/halliday

### sec. 19-2 Avogadro's Number

•1 Gold has a molar mass of 197 g/mol. (a) How many moles of gold are in a 2.50 g sample of pure gold? (b) How many atoms are in the sample?

•2 Find the mass in kilograms of $7.50 \times 10^{24}$ atoms of arsenic, which has a molar mass of 74.9 g/mol.

### sec. 19-3 Ideal Gases

•3 The best laboratory vacuum has a pressure of about $1.00 \times 10^{-18}$ atm, or $1.01 \times 10^{-13}$ Pa. How many gas molecules are there per cubic centimeter in such a vacuum at 293 K?

•4 Compute (a) the number of moles and (b) the number of molecules in 1.00 cm³ of an ideal gas at a pressure of 100 Pa and a temperature of 220 K.

•5 An automobile tire has a volume of $1.64 \times 10^{-2}$ m³ and contains air at a gauge pressure (pressure above atmospheric pressure) of 165 kPa when the temperature is 0.00°C. What is the gauge pressure of the air in the tires when its temperature rises to 27.0°C and its volume increases to $1.67 \times 10^{-2}$ m³? Assume atmospheric pressure is $1.01 \times 10^5$ Pa.

•6 A quantity of ideal gas at 10.0°C and 100 kPa occupies a volume of 2.50 m³. (a) How many moles of the gas are present? (b) If the pressure is now raised to 300 kPa and the temperature is raised to 30.0°C, how much volume does the gas occupy? Assume no leaks.

•7 Oxygen gas having a volume of 1000 cm³ at 40.0°C and $1.01 \times 10^5$ Pa expands until its volume is 1500 cm³ and its pressure is $1.06 \times 10^5$ Pa. Find (a) the number of moles of oxygen present and (b) the final temperature of the sample. **SSM**

•8 A container encloses 2 mol of an ideal gas that has molar mass $M_1$ and 0.5 mol of a second ideal gas that has molar mass $M_2 = 3M_1$. What fraction of the total pressure on the container wall is attributable to the second gas? (The kinetic theory explanation of pressure leads to the experimentally discovered law of partial pressures for a mixture of gases that do not react chemically: *The total pressure exerted by the mixture is equal to the sum of the pressures that the several gases would exert separately if each were to occupy the vessel alone.*)

•9 Suppose 1.80 mol of an ideal gas is taken from a volume of 3.00 m³ to a volume of 1.50 m³ via an isothermal compression at 30°C. (a) How much energy is transferred as heat during the compression, and (b) is the transfer *to* or *from* the gas?

•10 *Water bottle in a hot car.* In the American Southwest, the temperature in a closed car parked in sunlight during the summer can be high enough to burn flesh. Suppose a bottle of water at a refrigerator temperature of 5.00°C is opened, then closed, and then left in a closed car with an internal temperature of 75.0°C. Neglecting the thermal expansion of the water and the bottle, find the pressure in the air pocket trapped in the bottle. (The pressure can be enough to push the bottle cap past the threads that are intended to keep the bottle closed.)

••11 Suppose 0.825 mol of an ideal gas undergoes an isothermal expansion as energy is added to it as heat Q. If Fig. 19-20 shows the final volume $V_f$ versus Q, what is the gas temperature? The scale of the vertical axis is set by $V_{fs} = 0.30$ m³, and the scale of the horizontal axis is set by $Q_s = 1200$ J.

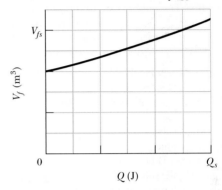

**FIG. 19-20** Problem 11.

••12 In the temperature range 310 K to 330 K, the pressure p of a certain nonideal gas is related to volume V and temperature T by

$$p = (24.9 \text{ J/K}) \frac{T}{V} - (0.00662 \text{ J/K}^2) \frac{T^2}{V}.$$

How much work is done by the gas if its temperature is raised from 315 K to 325 K while the pressure is held constant?

••13 Air that initially occupies 0.140 m³ at a gauge pressure of 103.0 kPa is expanded isothermally to a pressure of 101.3 kPa and then cooled at constant pressure until it reaches its initial volume. Compute the work done by the air. (Gauge pressure is the difference between the actual pressure and atmospheric pressure.) **SSM ILW WWW**

••14 *Submarine rescue.* When the U. S. submarine *Squalus* became disabled at a depth of 80 m, a cylindrical chamber was lowered from a ship to rescue the crew. The chamber had a radius of 1.00 m and a height of 4.00 m, was open at the bottom, and held two rescuers. It slid along a guide cable that a diver had attached to a hatch on the submarine. Once the chamber reached the hatch and clamped to the hull, the crew could escape into the chamber. During the descent, air was released from tanks to prevent water from flooding the chamber. Assume that the interior air pressure matched the water pressure at depth h as given by $p_0 + \rho gh$, where $p_0 = 1.000$ atm is the surface pressure and $\rho = 1024$ kg/m³ is the density of seawater. Assume a surface temperature of 20.0°C and a submerged water temperature of −30.0°C. (a) What is the air volume in the chamber at the surface? (b) If air had not been released from the tanks, what would have been the air volume in the chamber at depth h = 80.0 m? (c) How many moles of air were needed to be released to maintain the original air volume in the chamber? **GO**

••15 A sample of an ideal gas is taken through the cyclic

process *abca* shown in Fig. 19-21. The scale of the vertical axis is set by $p_b = 7.5$ kPa and $p_{ac} = 2.5$ kPa. At point *a*, $T = 200$ K. (a) How many moles of gas are in the sample? What are (b) the temperature of the gas at point *b*, (c) the temperature of the gas at point *c*, and (d) the net energy added to the gas as heat during the cycle? **GO**

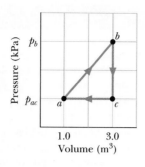

**FIG. 19-21** Problem 15.

•••**16** An air bubble of volume 20 cm³ is at the bottom of a lake 40 m deep, where the temperature is 4.0°C. The bubble rises to the surface, which is at a temperature of 20°C. Take the temperature of the bubble's air to be the same as that of the surrounding water. Just as the bubble reaches the surface, what is its volume?

•••**17** Container A in Fig. 19-22 holds an ideal gas at a pressure of $5.0 \times 10^5$ Pa and a temperature of 300 K. It is connected by a thin tube (and a closed valve) to container B, with four times the volume of A. Container B holds the same ideal gas at a pressure of $1.0 \times 10^5$ Pa and a temperature of 400 K. The valve is opened to allow the pressures to equalize, but the temperature of each container is maintained. What then is the pressure in the two containers?

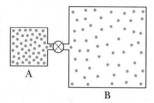

**FIG. 19-22** Problem 17.

#### sec. 19-4  Pressure, Temperature, and RMS Speed

•**18** Calculate the rms speed of helium atoms at 1000 K. See Appendix F for the molar mass of helium atoms.

•**19** The lowest possible temperature in outer space is 2.7 K. What is the rms speed of hydrogen molecules at this temperature? (The molar mass is given in Table 19-1.) **SSM**

•**20** Find the rms speed of argon atoms at 313 K. See Appendix F for the molar mass of argon atoms.

•**21** (a) Compute the rms speed of a nitrogen molecule at 20.0°C. The molar mass of nitrogen molecules ($N_2$) is given in Table 19-1. At what temperatures will the rms speed be (b) half that value and (c) twice that value?

•**22** The temperature and pressure in the Sun's atmosphere are $2.00 \times 10^6$ K and 0.0300 Pa. Calculate the rms speed of free electrons (mass $9.11 \times 10^{-31}$ kg) there, assuming they are an ideal gas.

••**23** A beam of hydrogen molecules ($H_2$) is directed toward a wall, at an angle of 55° with the normal to the wall. Each molecule in the beam has a speed of 1.0 km/s and a mass of $3.3 \times 10^{-24}$ g. The beam strikes the wall over an area of 2.0 cm², at the rate of $10^{23}$ molecules per second. What is the beam's pressure on the wall?

••**24** At 273 K and $1.00 \times 10^{-2}$ atm, the density of a gas is $1.24 \times 10^{-5}$ g/cm³. (a) Find $v_{rms}$ for the gas molecules. (b) Find the molar mass of the gas and (c) identify the gas. (*Hint:* The gas is listed in Table 19-1.)

#### sec. 19-5  Translational Kinetic Energy

•**25** Determine the average value of the translational kinetic energy of the molecules of an ideal gas at (a) 0.00°C and (b) 100°C. What is the translational kinetic energy per mole of an ideal gas at (c) 0.00°C and (d) 100°C?

•**26** What is the average translational kinetic energy of nitrogen molecules at 1600 K?

••**27** Water standing in the open at 32.0°C evaporates because of the escape of some of the surface molecules. The heat of vaporization (539 cal/g) is approximately equal to $\varepsilon n$, where $\varepsilon$ is the average energy of the escaping molecules and *n* is the number of molecules per gram. (a) Find $\varepsilon$. (b) What is the ratio of $\varepsilon$ to the average kinetic energy of $H_2O$ molecules, assuming the latter is related to temperature in the same way as it is for gases?

#### sec. 19-6  Mean Free Path

•**28** The mean free path of nitrogen molecules at 0.0°C and 1.0 atm is $0.80 \times 10^{-5}$ cm. At this temperature and pressure there are $2.7 \times 10^{19}$ molecules/cm³. What is the molecular diameter?

•**29** The atmospheric density at an altitude of 2500 km is about 1 molecule/cm³. (a) Assuming the molecular diameter of $2.0 \times 10^{-8}$ cm, find the mean free path predicted by Eq. 19-25. (b) Explain whether the predicted value is meaningful. **SSM**

•**30** At what frequency would the wavelength of sound in air be equal to the mean free path of oxygen molecules at 1.0 atm pressure and 0.00°C? Take the diameter of an oxygen molecule to be $3.0 \times 10^{-8}$ cm.

••**31** In a certain particle accelerator, protons travel around a circular path of diameter 23.0 m in an evacuated chamber, whose residual gas is at 295 K and $1.00 \times 10^{-6}$ torr pressure. (a) Calculate the number of gas molecules per cubic centimeter at this pressure. (b) What is the mean free path of the gas molecules if the molecular diameter is $2.00 \times 10^{-8}$ cm?

••**32** At 20°C and 750 torr pressure, the mean free paths for argon gas (Ar) and nitrogen gas ($N_2$) are $\lambda_{Ar} = 9.9 \times 10^{-6}$ cm and $\lambda_{N_2} = 27.5 \times 10^{-6}$ cm. (a) Find the ratio of the diameter of an Ar atom to that of an $N_2$ molecule. What is the mean free path of argon at (b) 20°C and 150 torr, and (c) −40°C and 750 torr?

#### sec. 19-7  The Distribution of Molecular Speeds

•**33** Ten particles are moving with the following speeds: four at 200 m/s, two at 500 m/s, and four at 600 m/s. Calculate their (a) average and (b) rms speeds. (c) Is $v_{rms} > v_{avg}$?

•**34** The speeds of 22 particles are as follows ($N_i$ represents the number of particles that have speed $v_i$):

| $N_i$ | 2 | 4 | 6 | 8 | 2 |
|---|---|---|---|---|---|
| $v_i$ (cm/s) | 1.0 | 2.0 | 3.0 | 4.0 | 5.0 |

What are (a) $v_{avg}$, (b) $v_{rms}$, and (c) $v_P$?

•**35** The speeds of 10 molecules are 2.0, 3.0, 4.0, . . . , 11 km/s. What are their (a) average speed and (b) rms speed? **SSM**

••**36** Figure 19-23 gives the probability distribution for nitrogen gas. The scale of the horizontal axis is set by $v_s = 1200$ m/s. What are the (a) gas temperature and (b) rms speed of the molecules?

••**37** At what temperature does the rms speed of (a) $H_2$ (molecular hydrogen) and (b)

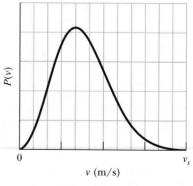

**FIG. 19-23** Problem 36.

$O_2$ (molecular oxygen) equal the escape speed from Earth (Table 13-2)? At what temperature does the rms speed of (c) $H_2$ and (d) $O_2$ equal the escape speed from the Moon (where the gravitational acceleration at the surface has magnitude $0.16g$)? Considering the answers to parts (a) and (b), should there be much (e) hydrogen and (f) oxygen high in Earth's upper atmosphere, where the temperature is about 1000 K?

**••38** Two containers are at the same temperature. The first contains gas with pressure $p_1$, molecular mass $m_1$, and rms speed $v_{rms1}$. The second contains gas with pressure $2.0p_1$, molecular mass $m_2$, and average speed $v_{avg2} = 2.0v_{rms1}$. Find the mass ratio $m_1/m_2$.

**••39** A hydrogen molecule (diameter $1.0 \times 10^{-8}$ cm), traveling at the rms speed, escapes from a 4000 K furnace into a chamber containing *cold* argon atoms (diameter $3.0 \times 10^{-8}$ cm) at a density of $4.0 \times 10^{19}$ atoms/cm³. (a) What is the speed of the hydrogen molecule? (b) If it collides with an argon atom, what is the closest their centers can be, considering each as spherical? (c) What is the initial number of collisions per second experienced by the hydrogen molecule? (*Hint*: Assume that the argon atoms are stationary. Then the mean free path of the hydrogen molecule is given by Eq. 19-26 and not Eq. 19-25.)

**••40** It is found that the most probable speed of molecules in a gas when it has (uniform) temperature $T_2$ is the same as the rms speed of the molecules in this gas when it has (uniform) temperature $T_1$. Calculate $T_2/T_1$.

**••41** Figure 19-24 shows a hypothetical speed distribution for a sample of $N$ gas particles (note that $P(v) = 0$ for speed $v > 2v_0$). What are the values of (a) $av_0$, (b) $v_{avg}/v_0$, and (c) $v_{rms}/v_0$? (d) What fraction of the particles has a speed between $1.5v_0$ and $2.0v_0$? **SSM WWW**

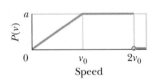

**FIG. 19-24** Problem 41.

### sec. 19-8 The Molar Specific Heats of an Ideal Gas

**•42** What is the internal energy of 1.0 mol of an ideal monatomic gas at 273 K?

**••43** The temperature of 2.00 mol of an ideal monatomic gas is raised 15.0 K at constant volume. What are (a) the work $W$ done by the gas, (b) the energy transferred as heat $Q$, (c) the change $\Delta E_{int}$ in the internal energy of the gas, and (d) the change $\Delta K$ in the average kinetic energy per atom?

**••44** Under constant pressure, the temperature of 2.00 mol of an ideal monatomic gas is raised 15.0 K. What are (a) the work $W$ done by the gas, (b) the energy transferred as heat $Q$, (c) the change $\Delta E_{int}$ in the internal energy of the gas, and (d) the change $\Delta K$ in the average kinetic energy per atom?

**••45** A container holds a mixture of three nonreacting gases: 2.40 mol of gas 1 with $C_{V1} = 12.0$ J/mol·K, 1.50 mol of gas 2 with $C_{V2} = 12.8$ J/mol·K, and 3.20 mol of gas 3 with $C_{V3} = 20.0$ J/mol·K. What is $C_V$ of the mixture? **SSM**

**••46** One mole of an ideal diatomic gas goes from $a$ to $c$ along the diagonal path in Fig. 19-25. The scale of the vertical axis is set by $p_{ab} = 5.0$ kPa and $p_c = 2.0$

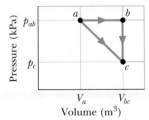

**FIG. 19-25** Problem 46.

kPa, and the scale of the horizontal axis is set by $V_{bc} = 4.0$ m³ and $V_a = 2.0$ m³. During the transition, (a) what is the change in internal energy of the gas, and (b) how much energy is added to the gas as heat? (c) How much heat is required if the gas goes from $a$ to $c$ along the indirect path $abc$? **GO**

**••47** The mass of a gas molecule can be computed from its specific heat at constant volume $c_V$. (Note that this is not $C_V$.) Take $c_V = 0.075$ cal/g·C° for argon and calculate (a) the mass of an argon atom and (b) the molar mass of argon. **ILW**

**••48** When 20.9 J was added as heat to a particular ideal gas, the volume of the gas changed from 50.0 cm³ to 100 cm³ while the pressure remained at 1.00 atm. (a) By how much did the internal energy of the gas change? If the quantity of gas present was $2.00 \times 10^{-3}$ mol, find (b) $C_p$ and (c) $C_V$.

**••49** The temperature of 3.00 mol of an ideal diatomic gas is increased by 40.0 C° without the pressure of the gas changing. The molecules in the gas rotate but do not oscillate. (a) How much energy is transferred to the gas as heat? (b) What is the change in the internal energy of the gas? (c) How much work is done by the gas? (d) By how much does the rotational kinetic energy of the gas increase? **GO**

### sec. 19-9 Degrees of Freedom and Molar Specific Heats

**•50** We give 70 J as heat to a diatomic gas, which then expands at constant pressure. The gas molecules rotate but do not oscillate. By how much does the internal energy of the gas increase?

**•51** When 1.0 mol of oxygen ($O_2$) gas is heated at constant pressure starting at 0°C, how much energy must be added to the gas as heat to double its volume? (The molecules rotate but do not oscillate.) **ILW**

**••52** Suppose 12.0 g of oxygen ($O_2$) gas is heated at constant atmospheric pressure from 25.0°C to 125°C. (a) How many moles of oxygen are present? (See Table 19-1 for the molar mass.) (b) How much energy is transferred to the oxygen as heat? (The molecules rotate but do not oscillate.) (c) What fraction of the heat is used to raise the internal energy of the oxygen?

**••53** Suppose 4.00 mol of an ideal diatomic gas, with molecular rotation but not oscillation, experienced a temperature increase of 60.0 K under constant-pressure conditions. What are (a) the energy transferred as heat $Q$, (b) the change $\Delta E_{int}$ in internal energy of the gas, (c) the work $W$ done by the gas, and (d) the change $\Delta K$ in the total translational kinetic energy of the gas? **SSM WWW**

### sec. 19-11 The Adiabatic Expansion of an Ideal Gas

**•54** Suppose 1.00 L of a gas with $\gamma = 1.30$, initially at 273 K and 1.00 atm, is suddenly compressed adiabatically to half its initial volume. Find its final (a) pressure and (b) temperature. (c) If the gas is then cooled to 273 K at constant pressure, what is its final volume?

**•55** A certain gas occupies a volume of 4.3 L at a pressure of 1.2 atm and a temperature of 310 K. It is compressed adiabatically to a volume of 0.76 L. Determine (a) the final pressure and (b) the final temperature, assuming the gas to be an ideal gas for which $\gamma = 1.4$.

**•56** We know that for an adiabatic process $pV^\gamma = $ a constant. Evaluate "a constant" for an adiabatic process involving exactly 2.0 mol of an ideal gas passing through the state having exactly $p = 1.0$ atm and $T = 300$ K. Assume a diatomic gas whose molecules rotate but do not oscillate.

**••57** Figure 19-26 shows two paths that may be taken by a gas from an initial point $i$ to a final point $f$. Path 1 consists of an isothermal expansion (work is 50 J in magnitude), an adiabatic expansion (work is 40 J in magnitude), an isothermal compression (work is 30 J in magnitude), and then an adiabatic compression (work is 25 J in magnitude). What is the change in the internal energy of the gas if the gas goes from point $i$ to point $f$ along path 2? **GO**

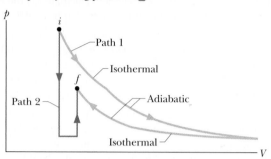

**FIG. 19-26** Problem 57.

**••58** *Adiabatic wind.* The normal airflow over the Rocky Mountains is west to east. The air loses much of its moisture content and is chilled as it climbs the western side of the mountains. When it descends on the eastern side, the increase in pressure toward lower altitudes causes the temperature to increase. The flow, then called a chinook wind, can rapidly raise the air temperature at the base of the mountains. Assume that the air pressure $p$ depends on altitude $y$ according to $p = p_0 \exp(-ay)$, where $p_0 = 1.00$ atm and $a = 1.16 \times 10^{-4}$ m$^{-1}$. Also assume that the ratio of the molar specific heats is $\gamma = \frac{4}{3}$. A parcel of air with an initial temperature of $-5.00°C$ descends adiabatically from $y_1 = 4267$ m to $y = 1567$ m. What is its temperature at the end of the descent?

**••59** A gas is to be expanded from initial state $i$ to final state $f$ along either path 1 or path 2 on a $p$-$V$ diagram. Path 1 consists of three steps: an isothermal expansion (work is 40 J in magnitude), an adiabatic expansion (work is 20 J in magnitude), and another isothermal expansion (work is 30 J in magnitude). Path 2 consists of two steps: a pressure reduction at constant volume and an expansion at constant pressure. What is the change in the internal energy of the gas along path 2?

**••60** *Opening champagne.* In a bottle of champagne, the pocket of gas (primarily carbon dioxide) between the liquid and the cork is at pressure of $p_i = 5.00$ atm. When the cork is pulled from the bottle, the gas undergoes an adiabatic expansion until its pressure matches the ambient air pressure of 1.00 atm. Assume that the ratio of the molar specific heats is $\gamma = \frac{4}{3}$. If the gas has initial temperature $T_i = 5.00°C$, what is its temperature at the end of the adiabatic expansion?

**••61** The volume of an ideal gas is adiabatically reduced from 200 L to 74.3 L. The initial pressure and temperature are 1.00 atm and 300 K. The final pressure is 4.00 atm. (a) Is the gas monatomic, diatomic, or polyatomic? (b) What is the final temperature? (c) How many moles are in the gas?

**•••62** An ideal diatomic gas, with rotation but no oscillation, undergoes an adiabatic compression. Its initial pressure and volume are 1.20 atm and 0.200 m$^3$. Its final pressure is 2.40 atm. How much work is done by the gas? **GO**

**•••63** Figure 19-27 shows a cycle undergone by 1.00 mol of an ideal monatomic gas. The temperatures are $T_1 = 300$ K,

$T_2 = 600$ K, and $T_3 = 455$ K. For $1 \rightarrow 2$, what are (a) heat $Q$, (b) the change in internal energy $\Delta E_{int}$, and (c) the work done $W$? For $2 \rightarrow 3$, what are (d) $Q$, (e) $\Delta E_{int}$, and (f) $W$? For $3 \rightarrow 1$, what are (g) $Q$, (h) $\Delta E_{int}$, and (i) $W$? For the full cycle, what are (j) $Q$, (k) $\Delta E_{int}$, and (l) $W$? The initial pressure at point 1 is 1.00 atm (= $1.013 \times 10^5$ Pa). What are the (m) volume and (n) pressure at point 2 and the (o) volume and (p) pressure at point 3?

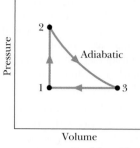

**FIG. 19-27** Problem 63.

### Additional Problems

**64** In an interstellar gas cloud at 50.0 K, the pressure is $1.00 \times 10^{-8}$ Pa. Assuming that the molecular diameters of the gases in the cloud are all 20.0 nm, what is their mean free path?

**65** The temperature of 3.00 mol of a gas with $C_V = 6.00$ cal/mol·K is to be raised 50.0 K. If the process is at *constant volume*, what are (a) the energy transferred as heat $Q$, (b) the work $W$ done by the gas, (c) the change $\Delta E_{int}$ in internal energy of the gas, and (d) the change $\Delta K$ in the total translational kinetic energy? If the process is at *constant pressure*, what are (e) $Q$, (f) $W$, (g) $\Delta E_{int}$, and (h) $\Delta K$? If the process is *adiabatic*, what are (i) $Q$, (j) $W$, (k) $\Delta E_{int}$, and (l) $\Delta K$?

**66** Oxygen (O$_2$) gas at 273 K and 1.0 atm is confined to a cubical container 10 cm on a side. Calculate $\Delta U_g/K_{avg}$, where $\Delta U_g$ is the change in the gravitational potential energy of an oxygen molecule falling the height of the box and $K_{avg}$ is the molecule's average translational kinetic energy.

**67** The envelope and basket of a hot-air balloon have a combined weight of 2.45 kN, and the envelope has a capacity (volume) of $2.18 \times 10^3$ m$^3$. When it is fully inflated, what should be the temperature of the enclosed air to give the balloon a *lifting capacity* (force) of 2.67 kN (in addition to the balloon's weight)? Assume that the surrounding air, at 20.0°C, has a weight per unit volume of 11.9 N/m$^3$ and a molecular mass of 0.028 kg/mol, and is at a pressure of 1.0 atm. **SSM**

**68** (a) An ideal gas initially at pressure $p_0$ undergoes a free expansion until its volume is 3.00 times its initial volume. What then is the ratio of its pressure to $p_0$? (b) The gas is next slowly and adiabatically compressed back to its original volume. The pressure after compression is $(3.00)^{1/3}p_0$. Is the gas monatomic, diatomic, or polyatomic? (c) What is the ratio of the average kinetic energy per molecule in this final state to that in the initial state?

**69** The temperature of 2.00 mol of an ideal monatomic gas is raised 15.0 K in an adiabatic process. What are (a) the work $W$ done by the gas, (b) the energy transferred as heat $Q$, (c) the change $\Delta E_{int}$ in internal energy of the gas, and (d) the change $\Delta K$ in the average kinetic energy per atom? **SSM**

**70** During a compression at a constant pressure of 250 Pa, the volume of an ideal gas decreases from 0.80 m$^3$ to 0.20 m$^3$. The initial temperature is 360 K, and the gas loses 210 J as heat. What are (a) the change in the internal energy of the gas and (b) the final temperature of the gas?

**71** At what frequency do molecules (diameter 290 pm) collide in (an ideal) oxygen gas (O$_2$) at temperature 400 K and pressure 2.00 atm? **SSM**

**72** An ideal gas consists of 1.50 mol of diatomic molecules that rotate but do not oscillate. The molecular diameter is 250 pm. The gas is expanded at a constant pressure of $1.50 \times 10^5$ Pa, with a transfer of 200 J as heat. What is the change in the mean free path of the molecules?

**73** An ideal monatomic gas initially has a temperature of 330 K and a pressure of 6.00 atm. It is to expand from volume 500 cm³ to volume 1500 cm³. If the expansion is isothermal, what are (a) the final pressure and (b) the work done by the gas? If, instead, the expansion is adiabatic, what are (c) the final pressure and (d) the work done by the gas?

**74** An ideal gas with 3.00 mol is initially in state 1 with pressure $p_1 = 20.0$ atm and volume $V_1 = 1500$ cm³. First it is taken to state 2 with pressure $p_2 = 1.50p_1$ and volume $V_2 = 2.00V_1$. Then it is taken to state 3 with pressure $p_3 = 2.00p_1$ and volume $V_3 = 0.500V_1$. What is the temperature of the gas in (a) state 1 and (b) state 2? (c) What is the net change in internal energy from state 1 to state 3?

**75** An ideal gas undergoes an adiabatic compression from $p = 1.0$ atm, $V = 1.0 \times 10^6$ L, $T = 0.0°C$ to $p = 1.0 \times 10^5$ atm, $V = 1.0 \times 10^3$ L. (a) Is the gas monatomic, diatomic, or polyatomic? (b) What is its final temperature? (c) How many moles of gas are present? What is the total translational kinetic energy per mole (d) before and (e) after the compression? (f) What is the ratio of the squares of the rms speeds before and after the compression?

**76** An ideal gas, at initial temperature $T_1$ and initial volume 2.0 m³, is expanded adiabatically to a volume of 4.0 m³, then expanded isothermally to a volume of 10 m³, and then compressed adiabatically back to $T_1$. What is its final volume?

**77** A sample of ideal gas expands from an initial pressure and volume of 32 atm and 1.0 L to a final volume of 4.0 L. The initial temperature is 300 K. If the gas is monatomic and the expansion isothermal, what are the (a) final pressure $p_f$, (b) final temperature $T_f$, and (c) work $W$ done by the gas? If the gas is monatomic and the expansion adiabatic, what are (d) $p_f$, (e) $T_f$, and (f) $W$? If the gas is diatomic and the expansion adiabatic, what are (g) $p_f$, (h) $T_f$, and (i) $W$? **SSM**

**78** Calculate the work done by an external agent during an isothermal compression of 1.00 mol of oxygen from a volume of 22.4 L at 0°C and 1.00 atm to a volume of 16.8 L.

**79** A steel tank contains 300 g of ammonia gas ($NH_3$) at a pressure of $1.35 \times 10^6$ Pa and a temperature of 77°C. (a) What is the volume of the tank in liters? (b) Later the temperature is 22°C and the pressure is $8.7 \times 10^5$ Pa. How many grams of gas have leaked out of the tank?

**80** At what temperature do atoms of helium gas have the same rms speed as molecules of hydrogen gas at 20.0°C? (The molar masses are given in Table 19-1.)

**81** Figure 19-28 shows a hypothetical speed distribution for particles of a certain gas: $P(v) = Cv^2$ for $0 < v \le v_0$ and $P(v) = 0$ for $v > v_0$. Find (a) an expression for $C$ in terms of $v_0$, (b) the average speed of the particles, and (c) their rms speed. **SSM**

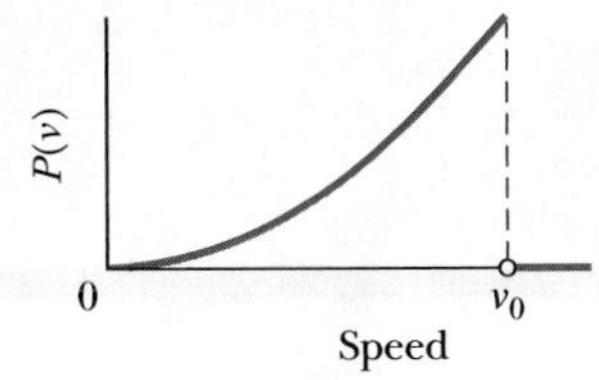

**82** In an industrial process the volume of 25.0 mol of

**FIG. 19-28** Problem 81.

a monatomic ideal gas is reduced at a uniform rate from 0.616 m³ to 0.308 m³ in 2.00 h while its temperature is increased at a uniform rate from 27.0°C to 450°C. Throughout the process, the gas passes through thermodynamic equilibrium states. What are (a) the cumulative work done on the gas, (b) the cumulative energy absorbed by the gas as heat, and (c) the molar specific heat for the process? (*Hint:* To evaluate the integral for the work, you might use

$$\int \frac{a + bx}{A + Bx} \, dx = \frac{bx}{B} + \frac{aB - bA}{B^2} \ln(A + Bx),$$

an indefinite integral.) Suppose the process is replaced with a two-step process that reaches the same final state. In step 1, the gas volume is reduced at constant temperature, and in step 2 the temperature is increased at constant volume. For this process, what are (d) the cumulative work done on the gas, (e) the cumulative energy absorbed by the gas as heat, and (f) the molar specific heat for the process?

**83** An ideal gas undergoes isothermal compression from an initial volume of 4.00 m³ to a final volume of 3.00 m³. There is 3.50 mol of the gas, and its temperature is 10.0°C. (a) How much work is done by the gas? (b) How much energy is transferred as heat between the gas and its environment? **SSM**

**84** (a) What is the number of molecules per cubic meter in air at 20°C and at a pressure of 1.0 atm ($= 1.01 \times 10^5$ Pa)? (b) What is the mass of 1.0 m³ of this air? Assume that 75% of the molecules are nitrogen ($N_2$) and 25% are oxygen ($O_2$).

**85** Figure 19-29 shows a cycle consisting of five paths: $AB$ is isothermal at 300 K, $BC$ is adiabatic with work = 5.0 J, $CD$ is at a constant pressure of 5 atm, $DE$ is isothermal, and $EA$ is adiabatic with a change in internal energy of 8.0 J. What is the change in internal energy of the gas along path $CD$?

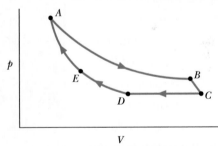

**FIG. 19-29** Problem 85.

**86** An ideal gas initially at 300 K is compressed at a constant pressure of 25 N/m² from a volume of 3.0 m³ to a volume of 1.8 m³. In the process, 75 J is lost by the gas as heat. What are (a) the change in internal energy of the gas and (b) the final temperature of the gas?

**87** An ideal gas is taken through a complete cycle in three steps: adiabatic expansion with work equal to 125 J, isothermal contraction at 325 K, and increase in pressure at constant volume. (a) Draw a $p$-$V$ diagram for the three steps. (b) How much energy is transferred as heat in step 3, and (c) is it transferred *to* or *from* the gas?

**88** (a) What is the volume occupied by 1.00 mol of an ideal gas at standard conditions—that is, 1.00 atm ($= 1.01 \times 10^5$ Pa) and 273 K? (b) Show that the number of molecules per cubic centimeter (the *Loschmidt number*) at standard conditions is $2.69 \times 10^9$.

# 20 Entropy and the Second Law of Thermodynamics

Inflating a balloon with your breath and stretching a rubber band with your hands require effort because the rubber (or rubber-like material) resists being stretched. In most materials, the resistance to stretching is due to the forces that bind the atoms and molecules together. Because any stretching tends to separate the atoms and molecules, the binding forces resist the stretching. However, rubber is very different because it is elastic and stretching does not tend to increase the separation of the atoms and molecules. Thus, its resistance is not due to binding forces. Instead, it is due to a quantity that gives direction to the flow of time.

## What causes a rubber band or balloon to resist stretching?

The answer is in this chapter.

Photo provided courtesy of Ronald P. Fowler, Jr., Flower Entertainment, www.flowerclown.com.

## 20-1 | WHAT IS PHYSICS?

Time has direction, the direction in which we age. We are accustomed to one-way processes—that is, processes that can occur only in a certain sequence (the right way) and never in the reverse sequence (the wrong way). An egg is dropped onto a floor, a pizza is baked, a car is driven into a lamppost, large waves erode a sandy beach—these one-way processes are **irreversible**, meaning that they cannot be reversed by means of only small changes in their environment.

One goal of physics is to understand why time has direction and why one-way processes are irreversible. Although this physics might seem disconnected from the practical issues of everyday life, it is in fact at the heart of any engine, such as a car engine, because it determines how well an engine can run.

The key to understanding why one-way processes cannot be reversed involves a quantity known as *entropy*.

## 20-2 | Irreversible Processes and Entropy

The one-way character of irreversible processes is so pervasive that we take it for granted. If these processes were to occur *spontaneously* (on their own) in the wrong way, we would be astonished. Yet *none* of these wrong-way events would violate the law of conservation of energy.

For example, if you were to wrap your hands around a cup of hot coffee, you would be astonished if your hands got cooler and the cup got warmer. That is obviously the wrong way for the energy transfer, but the total energy of the closed system (*hands + cup of coffee*) would be the same as the total energy if the process had run in the right way. For another example, if you popped a helium balloon, you would be astonished if, later, all the helium molecules were to gather together in the original shape of the balloon. That is obviously the wrong way for molecules to spread, but the total energy of the closed system (*molecules + room*) would be the same as for the right way.

Thus, changes in energy within a closed system do not set the direction of irreversible processes. Rather, that direction is set by another property that we shall discuss in this chapter—the *change in entropy* $\Delta S$ of the system. The change in entropy of a system is defined in the next section, but we can here state its central property, often called the *entropy postulate:*

> If an irreversible process occurs in a *closed* system, the entropy $S$ of the system always increases; it never decreases.

Entropy differs from energy in that entropy does *not* obey a conservation law. The *energy* of a closed system is conserved; it always remains constant. For irreversible processes, the *entropy* of a closed system always increases. Because of this property, the change in entropy is sometimes called "the arrow of time." For example, we associate the explosion of a popcorn kernel with the forward direction of time and with an increase in entropy. The backward direction of time (a videotape run backwards) would correspond to the exploded popcorn re-forming the original kernel. Because this backward process would result in an entropy decrease, it never happens.

There are two equivalent ways to define the change in entropy of a system: (1) in terms of the system's temperature and the energy the system gains or loses as heat, and (2) by counting the ways in which the atoms or molecules that make up the system can be arranged. We use the first approach in the next section and the second in Section 20-8.

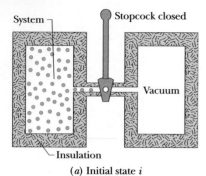

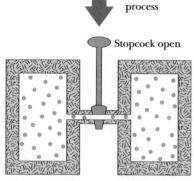

FIG. 20-1 The free expansion of an ideal gas. (*a*) The gas is confined to the left half of an insulated container by a closed stopcock. (*b*) When the stopcock is opened, the gas rushes to fill the entire container. This process is irreversible; that is, it does not occur in reverse, with the gas spontaneously collecting itself in the left half of the container.

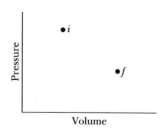

FIG. 20-2 A *p*-*V* diagram showing the initial state *i* and the final state *f* of the free expansion of Fig. 20-1. The intermediate states of the gas cannot be shown because they are not equilibrium states.

## 20-3 I Change in Entropy

Let's approach this definition of *change in entropy* by looking again at a process that we described in Sections 18-11 and 19-11: the free expansion of an ideal gas. Figure 20-1a shows the gas in its initial equilibrium state *i*, confined by a closed stopcock to the left half of a thermally insulated container. If we open the stopcock, the gas rushes to fill the entire container, eventually reaching the final equilibrium state *f* shown in Fig. 20-1b. This is an irreversible process; all the molecules of the gas will never return to the left half of the container.

The *p*-*V* plot of the process, in Fig. 20-2, shows the pressure and volume of the gas in its initial state *i* and final state *f*. Pressure and volume are *state properties*, properties that depend only on the state of the gas and not on how it reached that state. Other state properties are temperature and energy. We now assume that the gas has still another state property—its entropy. Furthermore, we define the **change in entropy** $S_f - S_i$ of a system during a process that takes the system from an initial state *i* to a final state *f* as

$$\Delta S = S_f - S_i = \int_i^f \frac{dQ}{T} \quad \text{(change in entropy defined).} \quad (20\text{-}1)$$

Here $Q$ is the energy transferred as heat to or from the system during the process, and $T$ is the temperature of the system in kelvins. Thus, an entropy change depends not only on the energy transferred as heat but also on the temperature at which the transfer takes place. Because $T$ is always positive, the sign of $\Delta S$ is the same as that of $Q$. We see from Eq. 20-1 that the SI unit for entropy and entropy change is the joule per kelvin.

There is a problem, however, in applying Eq. 20-1 to the free expansion of Fig. 20-1. As the gas rushes to fill the entire container, the pressure, temperature, and volume of the gas fluctuate unpredictably. In other words, they do not have a sequence of well-defined equilibrium values during the intermediate stages of the change from initial equilibrium state *i* to final equilibrium state *f*. Thus, we cannot trace a pressure–volume path for the free expansion on the *p*-*V* plot of Fig. 20-2 and, more important, we cannot find a relation between $Q$ and $T$ that allows us to integrate as Eq. 20-1 requires.

However, if entropy is truly a state property, the difference in entropy between states *i* and *f* must depend *only on those states* and not at all on the way the system went from one state to the other. Suppose, then, that we replace the irreversible free expansion of Fig. 20-1 with a *reversible* process that connects states *i* and *f*. With a reversible process we can trace a pressure–volume path on a *p*-*V* plot, and we can find a relation between $Q$ and $T$ that allows us to use Eq. 20-1 to obtain the entropy change.

We saw in Section 19-11 that the temperature of an ideal gas does not change during a free expansion: $T_i = T_f = T$. Thus, points *i* and *f* in Fig. 20-2 must be on the same isotherm. A convenient replacement process is then a reversible isothermal expansion from state *i* to state *f*, which actually proceeds *along* that isotherm. Furthermore, because $T$ is constant throughout a reversible isothermal expansion, the integral of Eq. 20-1 is greatly simplified.

Figure 20-3 shows how to produce such a reversible isothermal expansion. We confine the gas to an insulated cylinder that rests on a thermal reservoir maintained at the temperature $T$. We begin by placing just enough lead shot on the movable piston so that the pressure and volume of the gas are those of the initial state *i* of Fig. 20-1a. We then remove shot slowly (piece by piece) until the pressure and volume of the gas are those of the final state *f* of Fig. 20-1b. The temperature of the gas does not change because the gas remains in thermal contact with the reservoir throughout the process.

The reversible isothermal expansion of Fig. 20-3 is physically quite different from the irreversible free expansion of Fig. 20-1. However, *both processes have the same initial state and the same final state and thus must have the same change in entropy*. Because we removed the lead shot slowly, the intermediate states of the gas are equilibrium states, so we can plot them on a *p-V* diagram (Fig. 20-4).

To apply Eq. 20-1 to the isothermal expansion, we take the constant temperature *T* outside the integral, obtaining

$$\Delta S = S_f - S_i = \frac{1}{T}\int_i^f dQ.$$

Because $\int dQ = Q$, where $Q$ is the total energy transferred as heat during the process, we have

$$\Delta S = S_f - S_i = \frac{Q}{T} \qquad \text{(change in entropy, isothermal process).} \qquad (20\text{-}2)$$

To keep the temperature $T$ of the gas constant during the isothermal expansion of Fig. 20-3, heat $Q$ must have been energy transferred *from* the reservoir *to* the gas. Thus, $Q$ is positive and the entropy of the gas *increases* during the isothermal process and during the free expansion of Fig. 20-1.

To summarize:

> To find the entropy change for an irreversible process occurring in a *closed* system, replace that process with any reversible process that connects the same initial and final states. Calculate the entropy change for this reversible process with Eq. 20-1.

When the temperature change $\Delta T$ of a system is small relative to the temperature (in kelvins) before and after the process, the entropy change can be approximated as

$$\Delta S = S_f - S_i \approx \frac{Q}{T_{avg}}, \qquad (20\text{-}3)$$

where $T_{avg}$ is the average temperature of the system in kelvins during the process.

✓**CHECKPOINT 1**    Water is heated on a stove. Rank the entropy changes of the water as its temperature rises (a) from 20°C to 30°C, (b) from 30°C to 35°C, and (c) from 80°C to 85°C, greatest first.

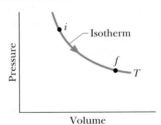

**FIG. 20-4**  A *p-V* diagram for the reversible isothermal expansion of Fig. 20-3. The intermediate states, which are now equilibrium states, are shown.

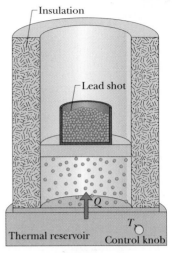

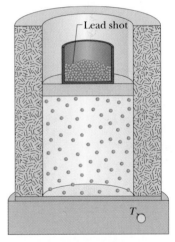

(*b*) Final state *f*

**FIG. 20-3**    The isothermal expansion of an ideal gas, done in a reversible way. The gas has the same initial state *i* and same final state *f* as in the irreversible process of Figs. 20-1 and 20-2.

### Entropy as a State Function

We have assumed that entropy, like pressure, energy, and temperature, is a property of the state of a system and is independent of how that state is reached. That entropy is indeed a *state function* (as state properties are usually called) can be deduced only by experiment. However, we can prove it is a state function for the special and important case in which an ideal gas is taken through a reversible process.

To make the process reversible, it is done slowly in a series of small steps, with the gas in an equilibrium state at the end of each step. For each small step, the energy transferred as heat to or from the gas is $dQ$, the work done by the gas is $dW$, and the change in internal energy is $dE_{int}$. These are related by the first law of thermodynamics in differential form (Eq. 18-27):

$$dE_{int} = dQ - dW.$$

Because the steps are reversible, with the gas in equilibrium states, we can use Eq. 18-24 to replace $dW$ with $p\,dV$ and Eq. 19-45 to replace $dE_{int}$ with $nC_V\,dT$. Solving for $dQ$ then leads to

$$dQ = p\,dV + nC_V\,dT.$$

Using the ideal gas law, we replace $p$ in this equation with $nRT/V$. Then we divide each term in the resulting equation by $T$, obtaining

$$\frac{dQ}{T} = nR\frac{dV}{V} + nC_V\frac{dT}{T}.$$

Now let us integrate each term of this equation between an arbitrary initial state $i$ and an arbitrary final state $f$ to get

$$\int_i^f \frac{dQ}{T} = \int_i^f nR\frac{dV}{V} + \int_i^f nC_V\frac{dT}{T}.$$

The quantity on the left is the entropy change $\Delta S\ (= S_f - S_i)$ defined by Eq. 20-1. Substituting this and integrating the quantities on the right yield

$$\Delta S = S_f - S_i = nR\ln\frac{V_f}{V_i} + nC_V\ln\frac{T_f}{T_i}. \qquad (20\text{-}4)$$

Note that we did not have to specify a particular reversible process when we integrated. Therefore, the integration must hold for all reversible processes that take the gas from state $i$ to state $f$. Thus, the change in entropy $\Delta S$ between the initial and final states of an ideal gas depends only on properties of the initial state ($V_i$ and $T_i$) and properties of the final state ($V_f$ and $T_f$); $\Delta S$ does not depend on how the gas changes between the two states.

**✓CHECKPOINT 2** An ideal gas has temperature $T_1$ at the initial state $i$ shown in the p-V diagram here. The gas has a higher temperature $T_2$ at final states $a$ and $b$, which it can reach along the paths shown. Is the entropy change along the path to state $a$ larger than, smaller than, or the same as that along the path to state $b$?

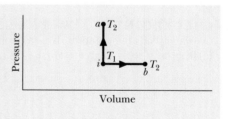

---

**Sample Problem** | **20-1**

Suppose 1.0 mol of nitrogen gas is confined to the left side of the container of Fig. 20-1a. You open the stopcock, and the volume of the gas doubles. What is the entropy change of the gas for this irreversible process? Treat the gas as ideal.

**KEY IDEAS** (1) We can determine the entropy change for the irreversible process by calculating it for a reversible process that provides the same change in volume. (2) The temperature of the gas does not change in the free expansion. Thus, the reversible process should be an isothermal expansion—namely, the one of Figs. 20-3 and 20-4.

**Calculations:** From Table 19-4, the energy $Q$ added as heat to the gas as it expands isothermally at temperature $T$ from an initial volume $V_i$ to a final volume $V_f$ is

$$Q = nRT\ln\frac{V_f}{V_i},$$

in which $n$ is the number of moles of gas present. From Eq. 20-2 the entropy change for this reversible process is

$$\Delta S_{rev} = \frac{Q}{T} = \frac{nRT\ln(V_f/V_i)}{T} = nR\ln\frac{V_f}{V_i}.$$

Substituting $n = 1.00$ mol and $V_f/V_i = 2$, we find

$$\Delta S_{rev} = nR \ln \frac{V_f}{V_i} = (1.00 \text{ mol})(8.31 \text{ J/mol} \cdot \text{K})(\ln 2)$$

$$= +5.76 \text{ J/K}.$$

Thus, the entropy change for the free expansion (and

for all other processes that connect the initial and final states shown in Fig. 20-2) is

$$\Delta S_{irrev} = \Delta S_{rev} = +5.76 \text{ J/K}. \qquad \text{(Answer)}$$

Because $\Delta S$ is positive, the entropy increases, in accordance with the entropy postulate of Section 20-2.

---

**Sample Problem** | **20-2** | **Build your skill**

Figure 20-5a shows two identical copper blocks of mass $m = 1.5$ kg: block $L$ at temperature $T_{iL} = 60°C$ and block $R$ at temperature $T_{iR} = 20°C$. The blocks are in a thermally insulated box and are separated by an insulating shutter. When we lift the shutter, the blocks eventually come to the equilibrium temperature $T_f = 40°C$ (Fig. 20-5b). What is the net entropy change of the two-block system during this irreversible process? The specific heat of copper is 386 J/kg·K.

**KEY IDEA** To calculate the entropy change, we must find a reversible process that takes the system from the initial state of Fig. 20-5a to the final state of Fig. 20-5b. We can calculate the net entropy change $\Delta S_{rev}$ of the reversible process using Eq. 20-1, and then the entropy change for the irreversible process is equal to $\Delta S_{rev}$.

*Calculations:* For the reversible process, we need a thermal reservoir whose temperature can be changed slowly (say, by turning a knob). We then take the blocks through the following two steps, illustrated in Fig. 20-6.

**Step 1** With the reservoir's temperature set at 60°C, put block $L$ on the reservoir. (Since block and reservoir are at the same temperature, they are already in thermal equilibrium.) Then slowly lower the temperature of the reservoir and the block to 40°C. As the block's temperature changes by each increment $dT$ during this process, energy $dQ$ is transferred as heat *from* the block to the reservoir. Using Eq. 18-14, we can write this transferred en-

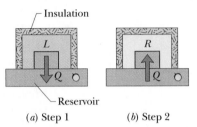

(a) Step 1      (b) Step 2

FIG. 20-6 The blocks of Fig. 20-5 can proceed from their initial state to their final state in a reversible way if we use a reservoir with a controllable temperature (a) to extract heat reversibly from block $L$ and (b) to add heat reversibly to block $R$.

ergy as $dQ = mc\, dT$, where $c$ is the specific heat of copper. According to Eq. 20-1, the entropy change $\Delta S_L$ of block $L$ during the full temperature change from initial temperature $T_{iL}$ ($= 60°C = 333$ K) to final temperature $T_f$ ($= 40°C = 313$ K) is

$$\Delta S_L = \int_i^f \frac{dQ}{T} = \int_{T_{iL}}^{T_f} \frac{mc\, dT}{T} = mc \int_{T_{iL}}^{T_f} \frac{dT}{T}$$

$$= mc \ln \frac{T_f}{T_{iL}}.$$

Inserting the given data yields

$$\Delta S_L = (1.5 \text{ kg})(386 \text{ J/kg} \cdot \text{K}) \ln \frac{313 \text{ K}}{333 \text{ K}}$$

$$= -35.86 \text{ J/K}.$$

**Step 2** With the reservoir's temperature now set at 20°C, put block $R$ on the reservoir. Then slowly raise the temperature of the reservoir and the block to 40°C. With the same reasoning used to find $\Delta S_L$, you can show that the entropy change $\Delta S_R$ of block $R$ during this process is

$$\Delta S_R = (1.5 \text{ kg})(386 \text{ J/kg} \cdot \text{K}) \ln \frac{313 \text{ K}}{293 \text{ K}}$$

$$= +38.23 \text{ J/K}.$$

The net entropy change $\Delta S_{rev}$ of the two-block system undergoing this two-step reversible process is then

$$\Delta S_{rev} = \Delta S_L + \Delta S_R$$

$$= -35.86 \text{ J/K} + 38.23 \text{ J/K} = 2.4 \text{ J/K}.$$

Thus, the net entropy change $\Delta S_{irrev}$ for the two-block system undergoing the actual irreversible process is

$$\Delta S_{irrev} = \Delta S_{rev} = 2.4 \text{ J/K}. \qquad \text{(Answer)}$$

This result is positive, in accordance with the entropy postulate of Section 20-2.

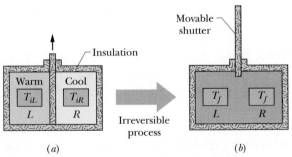

(a)      (b)

FIG. 20-5 (a) In the initial state, two copper blocks $L$ and $R$, identical except for their temperatures, are in an insulating box and are separated by an insulating shutter. (b) When the shutter is removed, the blocks exchange energy as heat and come to a final state, both with the same temperature $T_f$.

## 20-4 | The Second Law of Thermodynamics

Here is a puzzle. We saw in Sample Problem 20-1 that if we cause the reversible process of Fig. 20-3 to proceed from (*a*) to (*b*) in that figure, the change in entropy of the gas—which we take as our system—is positive. However, because the process is reversible, we can just as easily make it proceed from (*b*) to (*a*), simply by slowly adding lead shot to the piston of Fig. 20-3*b* until the original volume of the gas is restored. In this reverse process, energy must be extracted as heat *from the gas* to keep its temperature from rising. Hence *Q* is negative and so, from Eq. 20-2, the entropy of the gas must decrease.

Doesn't this decrease in the entropy of the gas violate the entropy postulate of Section 20-2, which states that entropy always increases? No, because that postulate holds only for *irreversible* processes occurring in closed systems. The procedure suggested here does not meet these requirements. The process is *not* irreversible, and (because energy is transferred as heat from the gas to the reservoir) the system—which is the gas alone—is *not* closed.

However, if we include the reservoir, along with the gas, as part of the system, then we do have a closed system. Let's check the change in entropy of the enlarged system *gas + reservoir* for the process that takes it from (*b*) to (*a*) in Fig. 20-3. During this reversible process, energy is transferred as heat from the gas to the reservoir—that is, from one part of the enlarged system to another. Let |*Q*| represent the absolute value (or magnitude) of this heat. With Eq. 20-2, we can then calculate separately the entropy changes for the gas (which loses |*Q*|) and the reservoir (which gains |*Q*|). We get

$$\Delta S_{\text{gas}} = -\frac{|Q|}{T}$$

and

$$\Delta S_{\text{gas}} = +\frac{|Q|}{T}.$$

The entropy change of the closed system is the sum of these two quantities: 0.

With this result, we can modify the entropy postulate of Section 20-2 to include both reversible and irreversible processes:

> ☞ If a process occurs in a *closed* system, the entropy of the system increases for irreversible processes and remains constant for reversible processes. It never decreases.

Although entropy may decrease in part of a closed system, there will always be an equal or larger entropy increase in another part of the system, so that the entropy of the system as a whole never decreases. This fact is one form of the **second law of thermodynamics** and can be written as

$$\Delta S \geq 0 \qquad \text{(second law of thermodynamics)}, \qquad (20\text{-}5)$$

where the greater-than sign applies to irreversible processes and the equals sign to reversible processes. Equation 20-5 applies only to closed systems.

In the real world almost all processes are irreversible to some extent because of friction, turbulence, and other factors, so the entropy of real closed systems undergoing real processes always increases. Processes in which the system's entropy remains constant are always idealizations.

### Force Due to Entropy

To understand why rubber resists being stretched, let's write the first law of thermodynamics

$$dE = dQ - dW$$

for a rubber band undergoing a small increase in length *dx* as we stretch it between our hands. The force from the rubber band has magnitude *F*, is directed inward, and does work $dW = -F \, dx$ during length increase *dx*. From Eq. 20-2 ($\Delta S = Q/T$), small changes in $Q$ and $S$ at constant temperature are related by $dS = dQ/T$, or $dQ = T \, dS$. So, now we can rewrite the first law as

$$dE = T \, dS + F \, dx. \qquad (20\text{-}6)$$

To good approximation, the change $dE$ in the internal energy of rubber is 0 if the total stretch of the rubber band is not very much. Substituting 0 for $dE$ in Eq. 20-6 leads us to an expression for the force from the rubber band:

$$F = -T \frac{dS}{dx}. \qquad (20\text{-}7)$$

This tells us that $F$ is proportional to the rate $dS/dx$ at which the rubber band's entropy changes during a small change $dx$ in the rubber band's length. Thus, you can *feel* the effect of entropy on your hands as you stretch a rubber band.

To make sense of the relation between force and entropy, let's consider a simple model of the rubber material. Rubber consists of cross-linked polymer chains (long molecules with cross links) that resemble three-dimensional zig-zags (Fig. 20-7). When the rubber band is at its rest length, the polymers are coiled up in a spaghetti-like arrangement. Because of the large disorder of the molecules, this rest state has a high value of entropy. When we stretch a rubber band, we uncoil many of those polymers, aligning them in the direction of stretch. Because the alignment decreases the disorder, the entropy of the stretched rubber band is less. That is, the change $dS/dx$ in Eq. 20-7 is a negative quantity because the entropy decreases with stretching. Thus, the force on our hands from the rubber band is due to the tendency of the polymers to return to their former disordered state and higher value of entropy.

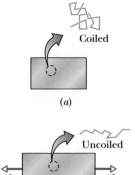

**FIG. 20-7** A section of a rubber band (*a*) unstretched and (*b*) stretched, and a polymer within it (*a*) coiled and (*b*) uncoiled.

---

**Sample Problem** | **20-3**

The force of a stretched rubber band is given approximately by Hooke's Law of Eq. 7-21 ($F_x = -kx$), in which $k$ is the *spring constant*. Suppose a rubber band with $k = 50.0$ N/m and at temperature $T = 27°C$ is stretched by $x = 1.2$ cm. For a small additional stretching, at what rate $dS/dx$ does the entropy of the rubber band decrease?

**KEY IDEA** The force of a stretched rubber is due to the change in the entropy of the polymers according to Eq. 20-7 ($F = -T \, dS/dx$).

**Calculation:** From Eq. 20-7, we know that the magnitude of the force is equal to $T|dS/dx|$. From Eq. 7-21, we know that the magnitude is also equal to $k|x|$. Thus,

$$T \left| \frac{dS}{dx} \right| = k|x|,$$

which yields

$$\left| \frac{dS}{dx} \right| = \frac{k|x|}{T} = \frac{(50.0 \text{ N/m})(0.012 \text{ m})}{(273 \text{ K} + 27 \text{ K})}$$

$$= 2.0 \times 10^{-3} \text{ J/K} \cdot \text{m}. \qquad \text{(Answer)}$$

---

# 20-5 | Entropy in the Real World: Engines

A **heat engine,** or more simply, an **engine,** is a device that extracts energy from its environment in the form of heat and does useful work. At the heart of every engine is a *working substance.* In a steam engine, the working substance is water, in both its vapor and its liquid form. In an automobile engine the working substance is a gasoline–air mixture. If an engine is to do work on a sustained basis, the working substance must operate in a *cycle;* that is, the working substance must pass through a closed series of thermodynamic processes, called *strokes,* returning again and again to each state in its cycle. Let us see what the laws of thermodynamics can tell us about the operation of engines.

## A Carnot Engine

We have seen that we can learn much about real gases by analyzing an ideal gas, which obeys the simple law $pV = nRT$. Although an ideal gas does not exist, any real gas approaches ideal behavior if its density is low enough. Similarly, we can study real engines by analyzing the behavior of an **ideal engine.**

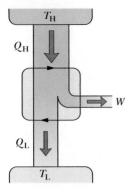

**FIG. 20-8** The elements of a Carnot engine. The two black arrowheads on the central loop suggest the working substance operating in a cycle, as if on a *p-V* plot. Energy $|Q_H|$ is transferred as heat from the high-temperature reservoir at temperature $T_H$ to the working substance. Energy $|Q_L|$ is transferred as heat from the working substance to the low-temperature reservoir at temperature $T_L$. Work $W$ is done by the engine (actually by the working substance) on something in the environment.

In an ideal engine, all processes are reversible and no wasteful energy transfers occur due to, say, friction and turbulence.

We shall focus on a particular ideal engine called a **Carnot engine** after the French scientist and engineer N. L. Sadi Carnot (pronounced "car-no"), who first proposed the engine's concept in 1824. This ideal engine turns out to be the best (in principle) at using energy as heat to do useful work. Surprisingly, Carnot was able to analyze the performance of this engine before the first law of thermodynamics and the concept of entropy had been discovered.

Figure 20-8 shows schematically the operation of a Carnot engine. During each cycle of the engine, the working substance absorbs energy $|Q_H|$ as heat from a thermal reservoir at constant temperature $T_H$ and discharges energy $|Q_L|$ as heat to a second thermal reservoir at a constant lower temperature $T_L$.

Figure 20-9 shows a *p-V* plot of the *Carnot cycle*—the cycle followed by the working substance. As indicated by the arrows, the cycle is traversed in the clockwise direction. Imagine the working substance to be a gas, confined to an insulating cylinder with a weighted, movable piston. The cylinder may be placed at will on either of the two thermal reservoirs, as in Fig. 20-6, or on an insulating slab. Figure 20-9 shows that, if we place the cylinder in contact with the high-temperature reservoir at temperature $T_H$, heat $|Q_H|$ is transferred *to* the working substance *from* this reservoir as the gas undergoes an isothermal *expansion* from volume $V_a$ to volume $V_b$. Similarly, with the working substance in contact with the low-temperature reservoir at temperature $T_L$, heat $|Q_L|$ is transferred *from* the working substance *to* the low-temperature reservoir as the gas undergoes an isothermal *compression* from volume $V_c$ to volume $V_d$.

In the engine of Fig. 20-8, we assume that heat transfers to or from the working substance can take place *only* during the isothermal processes *ab* and *cd* of Fig. 20-9. Therefore, processes *bc* and *da* in that figure, which connect the two isotherms at temperatures $T_H$ and $T_L$, must be (reversible) adiabatic processes; that is, they must be processes in which no energy is transferred as heat. To ensure this, during processes *bc* and *da* the cylinder is placed on an insulating slab as the volume of the working substance is changed.

During the consecutive processes *ab* and *bc* of Fig. 20-9, the working substance is expanding and thus doing positive work as it raises the weighted piston. This work is represented in Fig. 20-9 by the area under curve *abc*. During the consecutive processes *cd* and *da*, the working substance is being compressed, which means that it is doing negative work on its environment or, equivalently, that its environment is doing work on it as the loaded piston descends. This work is represented by the area under curve *cda*. The *net work per cycle,* which is represented by $W$ in both Figs. 20-8 and 20-9, is the difference between these two areas and is a positive quantity equal to the area enclosed by cycle *abcda* in Fig. 20-9. This work $W$ is performed on some outside object, such as a load to be lifted.

Equation 20-1 ($\Delta S = \int dQ/T$) tells us that any energy transfer as heat must involve a change in entropy. To illustrate the entropy changes for a Carnot engine, we can plot the Carnot cycle on a temperature–entropy (*T-S*) diagram as shown in Fig. 20-10. The lettered points *a*, *b*, *c*, and *d* in Fig. 20-10 correspond to the lettered points in the *p-V* diagram in Fig. 20-9. The two horizontal lines in Fig. 20-10 correspond to the two isothermal processes of the Carnot cycle (because the temperature is constant). Process *ab* is the isothermal expansion of the cycle. As the working substance (reversibly) absorbs energy $|Q_H|$ as heat at constant temperature $T_H$ during the expansion, its entropy increases. Similarly, during the isothermal compression *cd*, the working substance (reversibly) loses energy $|Q_L|$ as heat at constant temperature $T_L$, and its entropy decreases.

The two vertical lines in Fig. 20-10 correspond to the two adiabatic processes of the Carnot cycle. Because no energy is transferred as heat during the two processes, the entropy of the working substance is constant during them.

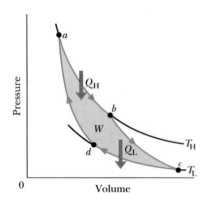

**FIG. 20-9** A pressure–volume plot of the cycle followed by the working substance of the Carnot engine in Fig. 20-8. The cycle consists of two isothermal (*ab* and *cd*) and two adiabatic processes (*bc* and *da*). The shaded area enclosed by the cycle is equal to the work $W$ per cycle done by the Carnot engine.

**The Work**   To calculate the net work done by a Carnot engine during a cycle, let us apply Eq. 18-26, the first law of thermodynamics ($\Delta E_{int} = Q - W$), to the working substance. That substance must return again and again to any arbitrarily selected state in the cycle. Thus, if $X$ represents any state property of the working substance, such as pressure, temperature, volume, internal energy, or entropy, we must have $\Delta X = 0$ for every cycle. It follows that $\Delta E_{int} = 0$ for a complete cycle of the working substance. Recalling that $Q$ in Eq. 18-26 is the *net* heat transfer per cycle and $W$ is the *net* work, we can write the first law of thermodynamics for the Carnot cycle as

$$W = |Q_H| - |Q_L|. \qquad (20\text{-}8)$$

**Entropy Changes**   In a Carnot engine, there are *two* (and only two) reversible energy transfers as heat, and thus two changes in the entropy of the working substance—one at temperature $T_H$ and one at $T_L$. The net entropy change per cycle is then

$$\Delta S = \Delta S_H + \Delta S_L = \frac{|Q_H|}{T_H} - \frac{|Q_L|}{T_L}. \qquad (20\text{-}9)$$

Here $\Delta S_H$ is positive because energy $|Q_H|$ is *added to* the working substance as heat (an increase in entropy) and $\Delta S_L$ is negative because energy $|Q_L|$ is *removed from* the working substance as heat (a decrease in entropy). Because entropy is a state function, we must have $\Delta S = 0$ for a complete cycle. Putting $\Delta S = 0$ in Eq. 20-9 requires that

$$\frac{|Q_H|}{T_H} = \frac{|Q_L|}{T_L}. \qquad (20\text{-}10)$$

Note that, because $T_H > T_L$, we must have $|Q_H| > |Q_L|$; that is, more energy is extracted as heat from the high-temperature reservoir than is delivered to the low-temperature reservoir.

We shall now use Eqs. 20-8 and 20-10 to derive an expression for the efficiency of a Carnot engine.

## Efficiency of a Carnot Engine

The purpose of any engine is to transform as much of the extracted energy $Q_H$ into work as possible. We measure its success in doing so by its **thermal efficiency** $\varepsilon$, defined as the work the engine does per cycle ("energy we get") divided by the energy it absorbs as heat per cycle ("energy we pay for"):

$$\varepsilon = \frac{\text{energy we get}}{\text{energy we pay for}} = \frac{|W|}{|Q_H|} \qquad \text{(efficiency, any engine).} \qquad (20\text{-}11)$$

For a Carnot engine we can substitute for $W$ from Eq. 20-8 to write Eq. 20-11 as

$$\varepsilon_C = \frac{|Q_H| - |Q_L|}{Q_H} = 1 - \frac{|Q_L|}{|Q_H|}. \qquad (20\text{-}12)$$

Using Eq. 20-10 we can write this as

$$\varepsilon_C = 1 - \frac{T_L}{T_H} \qquad \text{(efficiency, Carnot engine),} \qquad (20\text{-}13)$$

where the temperatures $T_L$ and $T_H$ are in kelvins. Because $T_L < T_H$, the Carnot engine necessarily has a thermal efficiency less than unity—that is, less than 100%. This is indicated in Fig. 20-8, which shows that only part of the energy extracted as heat from the high-temperature reservoir is available to do work, and the rest is delivered to the low-temperature reservoir. We shall show in Section 20-7 that no real engine can have a thermal efficiency greater than that calculated from Eq. 20-13.

Inventors continually try to improve engine efficiency by reducing the energy $|Q_L|$ that is "thrown away" during each cycle. The inventor's dream is to produce

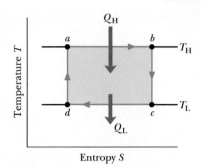

**FIG. 20-10**   The Carnot cycle of Fig. 20-9 plotted on a temperature–entropy diagram. During processes *ab* and *cd* the temperature remains constant. During processes *bc* and *da* the entropy remains constant.

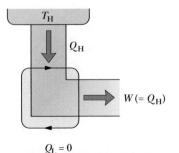

**FIG. 20-11**   The elements of a perfect engine—that is, one that converts heat $Q_H$ from a high-temperature reservoir directly to work $W$ with 100% efficiency.

FIG. 20-12 The North Anna nuclear power plant near Charlottesville, Virginia, which generates electric energy at the rate of 900 MW. At the same time, by design, it discards energy into the nearby river at the rate of 2100 MW. This plant and all others like it throw away more energy than they deliver in useful form. They are real counterparts of the ideal engine of Fig. 20-8. (© Robert Ustinich)

the *perfect engine,* diagrammed in Fig. 20-11, in which $|Q_L|$ is reduced to zero and $|Q_H|$ is converted completely into work. Such an engine on an ocean liner, for example, could extract energy as heat from the water and use it to drive the propellers, with no fuel cost. An automobile fitted with such an engine could extract energy as heat from the surrounding air and use it to drive the car, again with no fuel cost. Alas, a perfect engine is only a dream: Inspection of Eq. 20-13 shows that we can achieve 100% engine efficiency (that is, $\varepsilon = 1$) only if $T_L = 0$ or $T_H \rightarrow \infty$, impossible requirements. Instead, experience gives the following alternative version of the second law of thermodynamics, which says in short, *there are no perfect engines:*

> No series of processes is possible whose sole result is the transfer of energy as heat from a thermal reservoir and the complete conversion of this energy to work.

To summarize: The thermal efficiency given by Eq. 20-13 applies only to Carnot engines. Real engines, in which the processes that form the engine cycle are not reversible, have lower efficiencies. If your car were powered by a Carnot engine, it would have an efficiency of about 55% according to Eq. 20-13; its actual efficiency is probably about 25%. A nuclear power plant (Fig. 20-12), taken in its entirety, is an engine. It extracts energy as heat from a reactor core, does work by means of a turbine, and discharges energy as heat to a nearby river. If the power plant operated as a Carnot engine, its efficiency would be about 40%; its actual efficiency is about 30%. In designing engines of any type, there is simply no way to beat the efficiency limitation imposed by Eq. 20-13.

### Stirling Engine

Equation 20-13 applies not to all ideal engines but only to those that can be represented as in Fig. 20-9—that is, to Carnot engines. For example, Fig. 20-13 shows the operating cycle of an ideal **Stirling engine.** Comparison with the Carnot cycle of Fig. 20-9 shows that each engine has isothermal heat transfers at temperatures $T_H$ and $T_L$. However, the two isotherms of the Stirling engine cycle are connected, not by adiabatic processes as for the Carnot engine but by constant-volume processes. To increase the temperature of a gas at constant volume reversibly from $T_L$ to $T_H$ (process *da* of Fig. 20-13) requires a transfer of energy as heat to the working substance from a thermal reservoir whose temperature can be varied smoothly between those limits. Also, a reverse transfer is required in process *bc*. Thus, reversible heat transfers (and corresponding entropy changes) occur in all four of the processes that form the cycle of a Stirling engine, not just two processes as in a Carnot engine. Thus, the derivation that led to Eq. 20-13 does not apply to an ideal Stirling engine. More important, the efficiency of an ideal Stirling engine is lower than that of a Carnot engine operating between the same two temperatures. Real Stirling engines have even lower efficiencies.

The Stirling engine was developed in 1816 by Robert Stirling. This engine, long neglected, is now being developed for use in automobiles and spacecraft. A Stirling engine delivering 5000 hp (3.7 MW) has been built. Because they are quiet, Stirling engines are used on some military submarines.

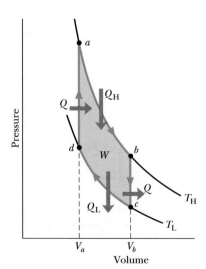

FIG. 20-13 A *p-V* plot for the working substance of an ideal Stirling engine, with the working substance assumed for convenience to be an ideal gas.

✓**CHECKPOINT 3** Three Carnot engines operate between reservoir temperatures of (a) 400 and 500 K, (b) 600 and 800 K, and (c) 400 and 600 K. Rank the engines according to their thermal efficiencies, greatest first.

## Sample Problem 20-4

Imagine a Carnot engine that operates between the temperatures $T_H = 850$ K and $T_L = 300$ K. The engine performs 1200 J of work each cycle, which takes 0.25 s.

(a) What is the efficiency of this engine?

**KEY IDEA** The efficiency $\varepsilon$ of a Carnot engine depends only on the ratio $T_L/T_H$ of the temperatures (in kelvins) of the thermal reservoirs to which it is connected.

**Calculation:** Thus, from Eq. 20-13, we have

$$\varepsilon = 1 - \frac{T_L}{T_H} = 1 - \frac{300 \text{ K}}{850 \text{ K}} = 0.647 \approx 65\%. \quad \text{(Answer)}$$

(b) What is the average power of this engine?

**KEY IDEA** The average power $P$ of an engine is the ratio of the work $W$ it does per cycle to the time $t$ that each cycle takes.

**Calculation:** For this Carnot engine, we find

$$P = \frac{W}{t} = \frac{1200 \text{ J}}{0.25 \text{ s}} = 4800 \text{ W} = 4.8 \text{ kW}. \quad \text{(Answer)}$$

(c) How much energy $|Q_H|$ is extracted as heat from the high-temperature reservoir every cycle?

**KEY IDEA** For any engine, including a Carnot engine, the efficiency $\varepsilon$ is the ratio of the work $W$ that is done per cycle to the energy $|Q_H|$ that is extracted as heat from the high-temperature reservoir per cycle ($\varepsilon = W/|Q_H|$).

**Calculation:** Here we have

$$|Q_H| = \frac{W}{\varepsilon} = \frac{1200 \text{ J}}{0.647} = 1855 \text{ J}. \quad \text{(Answer)}$$

(d) How much energy $|Q_L|$ is delivered as heat to the low-temperature reservoir every cycle?

**KEY IDEA** For a Carnot engine, the work $W$ done per cycle is equal to the difference in the energy transfers as heat: $|Q_H| - |Q_L|$, as in Eq. 20-8.

**Calculation:** Thus, we have

$$|Q_L| = |Q_H| - W$$
$$= 1855 \text{ J} - 1200 \text{ J} = 655 \text{ J}. \quad \text{(Answer)}$$

(e) By how much does the entropy of the working substance change as a result of the energy transferred to it from the high-temperature reservoir? From it to the low-temperature reservoir?

**KEY IDEA** The entropy change $\Delta S$ during a transfer of energy as heat $Q$ at constant temperature $T$ is given by Eq. 20-2 ($\Delta S = Q/T$).

**Calculations:** Thus, for the *positive* transfer of energy $Q_H$ from the high-temperature reservoir at $T_H$, the change in the entropy of the working substance is

$$\Delta S_H = \frac{Q_H}{T_H} = \frac{1855 \text{ J}}{850 \text{ K}} = +2.18 \text{ J/K}. \quad \text{(Answer)}$$

Similarly, for the *negative* transfer of energy $Q_L$ to the low-temperature reservoir at $T_L$, we have

$$\Delta S_L = \frac{Q_L}{T_L} = \frac{-655 \text{ J}}{300 \text{ K}} = -2.18 \text{ J/K}. \quad \text{(Answer)}$$

Note that the net entropy change of the working substance for one cycle is zero, as we discussed in deriving Eq. 20-10.

## Sample Problem 20-5

An inventor claims to have constructed an engine that has an efficiency of 75% when operated between the boiling and freezing points of water. Is this possible?

**KEY IDEA** The efficiency of a real engine (with its irreversible processes and wasteful energy transfers) must be less than the efficiency of a Carnot engine operating between the same two temperatures.

**Calculation:** From Eq. 20-13, we find that the efficiency of a Carnot engine operating between the boiling and freezing points of water is

$$\varepsilon = 1 - \frac{T_L}{T_H} = 1 - \frac{(0 + 273) \text{ K}}{(100 + 273) \text{ K}} = 0.268 \approx 27\%.$$

Thus, the claimed efficiency of 75% for a real engine operating between the given temperatures is impossible.

### PROBLEM-SOLVING TACTICS

*Tactic 1:* **The Language of Thermodynamics** A rich but sometimes misleading language is used in scientific and engineering studies of thermodynamics. You may see statements that say heat is added, absorbed, subtracted, extracted, rejected, discharged, discarded, withdrawn, delivered, gained, lost, transferred, or expelled, or that it flows from one body to another (as if it were a liquid). You may also see statements that describe a body as *having* heat (as if heat can be held or possessed) or that its heat is increased or decreased. You should

always keep in mind what is meant by the term *heat* in science and engineering:

> Heat is energy that is transferred from one body to another body due to a difference in the temperatures of the bodies.

When we identify one of the bodies as being our system, any such transfer of energy into it is positive heat $Q$ and out of it is negative heat $Q$.

The term *work* also requires close attention. You may see statements that say work is produced or generated, or combined with heat or changed from heat. Here is what is meant by the term *work*:

> Work is energy that is transferred from one body to another body due to a force that acts between them.

When we identify one of the bodies as being our system of interest, any such transfer of energy out of the system is either positive work $W$ done *by* the system or negative work $W$ done *on* the system. Any such transfer of energy into the system is negative work done *by* the system or positive work done *on* the system. (The preposition that is used is important.) Obviously, this can be confusing—whenever you see the term *work,* read carefully.

## 20-6 | Entropy in the Real World: Refrigerators

A **refrigerator** is a device that uses work to transfer energy from a low-temperature reservoir to a high-temperature reservoir as the device continuously repeats a set series of thermodynamic processes. In a household refrigerator, for example, work is done by an electrical compressor to transfer energy from the food storage compartment (a low-temperature reservoir) to the room (a high-temperature reservoir).

Air conditioners and heat pumps are also refrigerators. The differences are only in the nature of the high- and low-temperature reservoirs. For an air conditioner, the low-temperature reservoir is the room that is to be cooled and the high-temperature reservoir is the (presumably warmer) outdoors. A heat pump is an air conditioner that can be operated in reverse to heat a room; the room is the high-temperature reservoir, and heat is transferred to it from the (presumably cooler) outdoors.

Let us consider an *ideal refrigerator:*

> In an ideal refrigerator, all processes are reversible and no wasteful energy transfers occur as a result of, say, friction and turbulence.

Figure 20-14 shows the basic elements of an ideal refrigerator. Note that its operation is the reverse of how the Carnot engine of Fig. 20-8 operates. In other words, all the energy transfers, as either heat or work, are reversed from those of a Carnot engine. We can call such an ideal refrigerator a **Carnot refrigerator.**

The designer of a refrigerator would like to extract as much energy $|Q_L|$ as possible from the low-temperature reservoir (what we want) for the least amount of work $|W|$ (what we pay for). A measure of the efficiency of a refrigerator, then, is

$$K = \frac{\text{what we want}}{\text{what we pay for}} = \frac{|Q_L|}{|W|} \qquad \text{(coefficient of performance,} \atop \text{any refrigerator),} \qquad (20\text{-}14)$$

where $K$ is called the *coefficient of performance.* For a Carnot refrigerator, the first law of thermodynamics gives $|W| = |Q_H| - |Q_L|$, where $|Q_H|$ is the magnitude of the energy transferred as heat to the high-temperature reservoir. Equation 20-14 then becomes

$$K_C = \frac{|Q_L|}{|Q_H| - |Q_L|}. \qquad (20\text{-}15)$$

Because a Carnot refrigerator is a Carnot engine operating in reverse, we can combine Eq. 20-10 with Eq. 20-15; after some algebra we find

$$K_C = \frac{T_L}{T_H - T_L} \qquad \text{(coefficient of performance,} \atop \text{Carnot refrigerator).} \qquad (20\text{-}16)$$

For typical room air conditioners, $K \approx 2.5$. For household refrigerators, $K \approx 5$. Perversely, the value of $K$ is higher the closer the temperatures of the two reservoirs are to each other. That is why heat pumps are more effective in temperate climates than in climates where the outside temperature is much lower than the desired inside temperature.

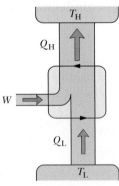

**FIG. 20-14** The elements of a refrigerator. The two black arrowheads on the central loop suggest the working substance operating in a cycle, as if on a *p-V* plot. Energy is transferred as heat $Q_L$ to the working substance from the low-temperature reservoir. Energy is transferred as heat $Q_H$ to the high-temperature reservoir from the working substance. Work $W$ is done on the refrigerator (on the working substance) by something in the environment.

It would be nice to own a refrigerator that did not require some input of work—that is, one that would run without being plugged in. Figure 20-15 represents another "inventor's dream," a *perfect refrigerator* that transfers energy as heat $Q$ from a cold reservoir to a warm reservoir without the need for work. Because the unit operates in cycles, the entropy of the working substance does not change during a complete cycle. The entropies of the two reservoirs, however, do change: The entropy change for the cold reservoir is $-|Q|/T_L$, and that for the warm reservoir is $+|Q|/T_H$. Thus, the net entropy change for the entire system is

$$\Delta S = -\frac{|Q|}{T_L} + \frac{|Q|}{T_H}.$$

Because $T_H > T_L$, the right side of this equation is negative and thus the net change in entropy per cycle for the closed system *refrigerator + reservoirs* is also negative. Because such a decrease in entropy violates the second law of thermodynamics (Eq. 20-5), a perfect refrigerator does not exist. (If you want your refrigerator to operate, you must plug it in.)

This result leads us to another (equivalent) formulation of the second law of thermodynamics:

> No series of processes is possible whose sole result is the transfer of energy as heat from a reservoir at a given temperature to a reservoir at a higher temperature.

In short, *there are no perfect refrigerators.*

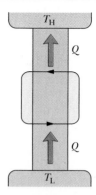

**FIG. 20-15** The elements of a perfect refrigerator—that is, one that transfers energy from a low-temperature reservoir to a high-temperature reservoir without any input of work.

✓**CHECKPOINT 4**    You wish to increase the coefficient of performance of an ideal refrigerator. You can do so by (a) running the cold chamber at a slightly higher temperature, (b) running the cold chamber at a slightly lower temperature, (c) moving the unit to a slightly warmer room, or (d) moving it to a slightly cooler room. The magnitudes of the temperature changes are to be the same in all four cases. List the changes according to the resulting coefficients of performance, greatest first.

## 20-7 | The Efficiencies of Real Engines

Let $\varepsilon_C$ be the efficiency of a Carnot engine operating between two given temperatures. In this section we prove that no real engine operating between those temperatures can have an efficiency greater than $\varepsilon_C$. If it could, the engine would violate the second law of thermodynamics.

Let us assume that an inventor, working in her garage, has constructed an engine $X$, which she claims has an efficiency $\varepsilon_X$ that is greater than $\varepsilon_C$:

$$\varepsilon_X > \varepsilon_C \quad \text{(a claim)}. \qquad (20\text{-}17)$$

Let us couple engine $X$ to a Carnot refrigerator, as in Fig. 20-16a. We adjust the

**FIG. 20-16**  (a) Engine $X$ drives a Carnot refrigerator. (b) If, as claimed, engine $X$ is more efficient than a Carnot engine, then the combination shown in (a) is equivalent to the perfect refrigerator shown here. This violates the second law of thermodynamics, so we conclude that engine $X$ cannot be more efficient than a Carnot engine.

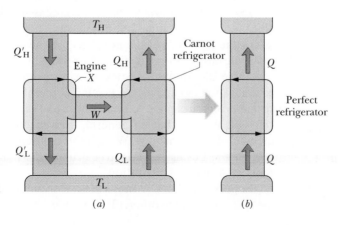

strokes of the Carnot refrigerator so that the work it requires per cycle is just equal to that provided by engine $X$. Thus, no (external) work is performed on or by the combination *engine + refrigerator* of Fig. 20-16a, which we take as our system.

If Eq. 20-17 is true, from the definition of efficiency (Eq. 20-11), we must have

$$\frac{|W|}{|Q'_H|} > \frac{|W|}{|Q_H|},$$

where the prime refers to engine $X$ and the right side of the inequality is the efficiency of the Carnot refrigerator when it operates as an engine. This inequality requires that

$$|Q_H| > |Q'_H|. \qquad (20\text{-}18)$$

Because the work done by engine $X$ is equal to the work done on the Carnot refrigerator, we have, from the first law of thermodynamics as given by Eq. 20-8,

$$|Q_H| - |Q_L| = |Q'_H| - |Q'_L|,$$

which we can write as

$$|Q_H| - |Q'_H| = |Q_L| - |Q'_L| = Q. \qquad (20\text{-}19)$$

Because of Eq. 20-18, the quantity $Q$ in Eq. 20-19 must be positive.

Comparison of Eq. 20-19 with Fig. 20-16 shows that the net effect of engine $X$ and the Carnot refrigerator working in combination is to transfer energy $Q$ as heat from a low-temperature reservoir to a high-temperature reservoir without the requirement of work. Thus, the combination acts like the perfect refrigerator of Fig. 20-15, whose existence is a violation of the second law of thermodynamics.

Something must be wrong with one or more of our assumptions, and it can only be Eq. 20-17. We conclude that *no real engine can have an efficiency greater than that of a Carnot engine when both engines work between the same two temperatures.* At most, the real engine can have an efficiency equal to that of a Carnot engine. In that case, the real engine *is* a Carnot engine.

## 20-8 | A Statistical View of Entropy

In Chapter 19 we saw that the macroscopic properties of gases can be explained in terms of their microscopic, or molecular, behavior. For one example, recall that we were able to account for the pressure exerted by a gas on the walls of its container in terms of the momentum transferred to those walls by rebounding gas molecules. Such explanations are part of a study called **statistical mechanics.**

Here we shall focus our attention on a single problem, one involving the distribution of gas molecules between the two halves of an insulated box. This problem is reasonably simple to analyze, and it allows us to use statistical mechanics to calculate the entropy change for the free expansion of an ideal gas. You will see in Sample Problem 20-7 that statistical mechanics leads to the same entropy change we obtained in Sample Problem 20-1 using thermodynamics.

Figure 20-17 shows a box that contains six identical (and thus indistinguishable) molecules of a gas. At any instant, a given molecule will be in either the left or the right half of the box; because the two halves have equal volumes, the molecule has the same likelihood, or probability, of being in either half.

Table 20-1 shows the seven possible *configurations* of the six molecules, each configuration labeled with a Roman numeral. For example, in configuration I, all six molecules are in the left half of the box ($n_1 = 6$) and none are in the right half ($n_2 = 0$). We see that, in general, a given configuration can be achieved in a number of different ways. We call these different arrangements of the molecules *microstates.* Let us see how to calculate the number of microstates that correspond to a given configuration.

Suppose we have $N$ molecules, distributed with $n_1$ molecules in one half of the box and $n_2$ in the other. (Thus $n_1 + n_2 = N$.) Let us imagine that we distribute

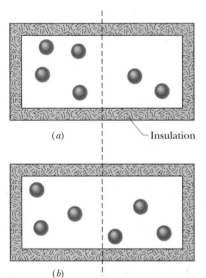

**FIG. 20-17** An insulated box contains six gas molecules. Each molecule has the same probability of being in the left half of the box as in the right half. The arrangement in (a) corresponds to configuration III in Table 20-1, and that in (b) corresponds to configuration IV.

(a)

— Insulation

(b)

**TABLE 20-1**

**Six Molecules in a Box**

| Configuration Label | $n_1$ | $n_2$ | Multiplicity $W$ (number of microstates) | Calculation of $W$ (Eq. 20-20) | Entropy $10^{-23}$ J/K (Eq. 20-21) |
|---|---|---|---|---|---|
| I | 6 | 0 | 1 | $6!/(6!\ 0!) = 1$ | 0 |
| II | 5 | 1 | 6 | $6!/(5!\ 1!) = 6$ | 2.47 |
| III | 4 | 2 | 15 | $6!/(4!\ 2!) = 15$ | 3.74 |
| IV | 3 | 3 | 20 | $6!/(3!\ 3!) = 20$ | 4.13 |
| V | 2 | 4 | 15 | $6!/(2!\ 4!) = 15$ | 3.74 |
| VI | 1 | 5 | 6 | $6!/(1!\ 5!) = 6$ | 2.47 |
| VII | 0 | 6 | 1 | $6!/(0!\ 6!) = 1$ | 0 |
| | | | Total = 64 | | |

the molecules "by hand," one at a time. If $N = 6$, we can select the first molecule in six independent ways; that is, we can pick any one of the six molecules. We can pick the second molecule in five ways, by picking any one of the remaining five molecules; and so on. The total number of ways in which we can select all six molecules is the product of these independent ways, or $6 \times 5 \times 4 \times 3 \times 2 \times 1 = 720$. In mathematical shorthand we write this product as $6! = 720$, where $6!$ is pronounced "six factorial." Your hand calculator can probably calculate factorials. For later use you will need to know that $0! = 1$. (Check this on your calculator.)

However, because the molecules are indistinguishable, these 720 arrangements are not all different. In the case that $n_1 = 4$ and $n_2 = 2$ (which is configuration III in Table 20-1), for example, the order in which you put four molecules in one half of the box does not matter, because after you have put all four in, there is no way that you can tell the order in which you did so. The number of ways in which you can order the four molecules is $4! = 24$. Similarly, the number of ways in which you can order two molecules for the other half of the box is simply $2! = 2$. To get the number of *different* arrangements that lead to the $(4, 2)$ split of configuration III, we must divide 720 by 24 and also by 2. We call the resulting quantity, which is the number of microstates that correspond to a given configuration, the *multiplicity W* of that configuration. Thus, for configuration III,

$$W_{III} = \frac{6!}{4!\ 2!} = \frac{720}{24 \times 2} = 15.$$

Thus, Table 20-1 tells us there are 15 independent microstates that correspond to configuration III. Note that, as the table also tells us, the total number of microstates for six molecules distributed over the seven configurations is 64.

Extrapolating from six molecules to the general case of $N$ molecules, we have

$$W = \frac{N!}{n_1!\ n_2!} \qquad \text{(multiplicity of configuration).} \qquad (20\text{-}20)$$

You should verify that Eq. 20-20 gives the multiplicities for all the configurations listed in Table 20-1.

The basic assumption of statistical mechanics is

All microstates are equally probable.

In other words, if we were to take a great many snapshots of the six molecules as they jostle around in the box of Fig. 20-17 and then count the number of times each microstate occurred, we would find that all 64 microstates would occur equally often. Thus the system will spend, on average, the same amount of time in each of the 64 microstates.

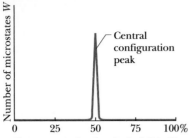

FIG. 20-18 For a *large* number of molecules in a box, a plot of the number of microstates that require various percentages of the molecules to be in the left half of the box. Nearly all the microstates correspond to an approximately equal sharing of the molecules between the two halves of the box; those microstates form the *central configuration peak* on the plot. For $N \approx 10^{22}$, the central configuration peak is much too narrow to be drawn on this plot.

Because all microstates are equally probable but different configurations have different numbers of microstates, the configurations are *not* all equally probable. In Table 20-1 configuration IV, with 20 microstates, is the *most probable configuration*, with a probability of 20/64 = 0.313. This result means that the system is in configuration IV 31.3% of the time. Configurations I and VII, in which all the molecules are in one half of the box, are the least probable, each with a probability of 1/64 = 0.016 or 1.6%. It is not surprising that the most probable configuration is the one in which the molecules are evenly divided between the two halves of the box, because that is what we expect at thermal equilibrium. However, it *is* surprising that there is *any* probability, however small, of finding all six molecules clustered in half of the box, with the other half empty.

For large values of $N$ there are extremely large numbers of microstates, but nearly all the microstates belong to the configuration in which the molecules are divided equally between the two halves of the box, as Fig. 20-18 indicates. Even though the measured temperature and pressure of the gas remain constant, the gas is churning away endlessly as its molecules "visit" all probable microstates with equal probability. However, because so few microstates lie outside the very narrow central configuration peak of Fig. 20-18, we might as well assume that the gas molecules are always divided equally between the two halves of the box. As we shall see, this is the configuration with the greatest entropy.

## Sample Problem   20-6

Suppose that there are 100 indistinguishable molecules in the box of Fig. 20-17. How many microstates are associated with the configuration $n_1 = 50$ and $n_2 = 50$, and with the configuration $n_1 = 100$ and $n_2 = 0$? Interpret the results in terms of the relative probabilities of the two configurations.

**KEY IDEA**    The multiplicity $W$ of a configuration of indistinguishable molecules in a closed box is the number of independent microstates with that configuration, as given by Eq. 20-20.

**Calculations:** For the $(n_1, n_2)$ configuration $(50, 50)$, that equation yields

$$W = \frac{N!}{n_1! \, n_2!} = \frac{100!}{50! \, 50!}$$
$$= \frac{9.33 \times 10^{157}}{(3.04 \times 10^{64})(3.04 \times 10^{64})}$$
$$= 1.01 \times 10^{29}. \qquad \text{(Answer)}$$

Similarly, for the configuration $(100, 0)$, we have

$$W = \frac{N!}{n_1! \, n_2!} = \frac{100!}{100! \, 0!} = \frac{1}{0!} = \frac{1}{1} = 1. \quad \text{(Answer)}$$

**The meaning:** Thus, a 50–50 distribution is more likely than a 100–0 distribution by the enormous factor of about $1 \times 10^{29}$. If you could count, at one per nanosecond, the number of microstates that correspond to the 50–50 distribution, it would take you about $3 \times 10^{12}$ years, which is about 200 times longer than the age of the universe. Keep in mind that the 100 molecules used in this sample problem is a very small number. Imagine what these calculated probabilities would be like for a mole of molecules, say about $N = 10^{24}$. Thus, you need never worry about suddenly finding all the air molecules clustering in one corner of your room, with you gasping for air in another corner.

## Probability and Entropy

In 1877, Austrian physicist Ludwig Boltzmann (the Boltzmann of Boltzmann's constant $k$) derived a relationship between the entropy $S$ of a configuration of a gas and the multiplicity $W$ of that configuration. That relationship is

$$S = k \ln W \qquad \text{(Boltzmann's entropy equation).} \qquad (20\text{-}21)$$

This famous formula is engraved on Boltzmann's tombstone.

It is natural that $S$ and $W$ should be related by a logarithmic function. The total entropy of two systems is the *sum* of their separate entropies. The probability of occurrence of two independent systems is the *product* of their separate probabilities. Because $\ln ab = \ln a + \ln b$, the logarithm seems the logical way to connect these quantities.

Table 20-1 displays the entropies of the configurations of the six-molecule system of Fig. 20-17, computed using Eq. 20-21. Configuration IV, which has the greatest multiplicity, also has the greatest entropy.

When you use Eq. 20-20 to calculate $W$, your calculator may signal "OVER-FLOW" if you try to find the factorial of a number greater than a few hundred. Fortunately, there is a very good approximation, known as **Stirling's approximation,** not for $N!$ but for $\ln N!$, which is exactly what is needed in Eq. 20-21. Stirling's approximation is

$$\ln N! \approx N(\ln N) - N \quad \text{(Stirling's approximation)}. \quad (20\text{-}22)$$

The Stirling of this approximation was an English mathematician and not the Robert Stirling of engine fame.

✓**CHECKPOINT 5**    A box contains 1 mol of a gas. Consider two configurations: (a) each half of the box contains half the molecules and (b) each third of the box contains one-third of the molecules. Which configuration has more microstates?

## Sample Problem  20-7

In Sample Problem 20-1 we showed that when $n$ moles of an ideal gas doubles its volume in a free expansion, the entropy increase from the initial state $i$ to the final state $f$ is $S_f - S_i = nR \ln 2$. Derive this result with statistical mechanics.

**KEY IDEA**    We can relate the entropy $S$ of any given configuration of the molecules in the gas to the multiplicity $W$ of microstates for that configuration, using Eq. 20-21 ($S = k \ln W$).

*Calculations:* We are interested in two configurations: the final configuration $f$ (with the molecules occupying the full volume of their container in Fig. 20-1$b$) and the initial configuration $i$ (with the molecules occupying the left half of the container). Because the molecules are in a closed container, we can calculate the multiplicity $W$ of their microstates with Eq. 20-20. Here we have $N$ molecules in the $n$ moles of the gas. Initially, with the molecules all in the left half of the container, their $(n_1, n_2)$ configuration is $(N, 0)$. Then, Eq. 20-20 gives their multiplicity as

$$W_i = \frac{N!}{N!\, 0!} = 1.$$

Finally, with the molecules spread through the full volume, their $(n_1, n_2)$ configuration is $(N/2, N/2)$. Then, Eq. 20-20 gives their multiplicity as

$$W_f = \frac{N!}{(N/2)!\,(N/2)!}.$$

From Eq. 20-21, the initial and final entropies are

$$S_i = k \ln W_i = k \ln 1 = 0$$

and

$$S_f = k \ln W_f = k \ln(N!) - 2k \ln[(N/2)!]. \quad (20\text{-}23)$$

In writing Eq. 20-23, we have used the relation

$$\ln \frac{a}{b^2} = \ln a - 2 \ln b.$$

Now, applying Eq. 20-22 to evaluate Eq. 20-23, we find that

$$
\begin{aligned}
S_f &= k \ln(N!) - 2k \ln[(N/2)!] \\
&= k[N(\ln N) - N] - 2k[(N/2)\ln(N/2) - (N/2)] \\
&= k[N(\ln N) - N - N\ln(N/2) + N] \\
&= k[N(\ln N) - N(\ln N - \ln 2)] = Nk \ln 2. \quad (20\text{-}24)
\end{aligned}
$$

From Eq. 19-8 we can substitute $nR$ for $Nk$, where $R$ is the universal gas constant. Equation 20-24 then becomes

$$S_f = nR \ln 2.$$

The change in entropy from the initial state to the final is thus

$$
\begin{aligned}
S_f - S_i &= nR \ln 2 - 0 \\
&= nR \ln 2, \quad \text{(Answer)}
\end{aligned}
$$

which is what we set out to show. In Sample Problem 20-1 we calculated this entropy increase for a free expansion with thermodynamics by finding an equivalent reversible process and calculating the entropy change for *that* process in terms of temperature and heat transfer. In this sample problem, we calculate the same increase in entropy with statistical mechanics using the fact that the system consists of molecules. In short, the two, very different approaches give the same answer.

## REVIEW & SUMMARY

**One-Way Processes** An **irreversible process** is one that cannot be reversed by means of small changes in the environment. The direction in which an irreversible process proceeds is set by the *change in entropy* $\Delta S$ of the system undergoing the process. Entropy $S$ is a *state property* (or *state function*) of the system; that is, it depends only on the state of the system and not on the way in which the system reached that state. The *entropy postulate* states (in part): *If an irreversible process occurs in a closed system, the entropy of the system always increases.*

**Calculating Entropy Change** The **entropy change** $\Delta S$ for an irreversible process that takes a system from an initial state $i$ to a final state $f$ is exactly equal to the entropy change $\Delta S$ for *any reversible process* that takes the system between those same two states. We can compute the latter (but not the former) with

$$\Delta S = S_f - S_i = \int_i^f \frac{dQ}{T}. \qquad (20\text{-}1)$$

Here $Q$ is the energy transferred as heat to or from the system during the process, and $T$ is the temperature of the system in kelvins during the process.

For a reversible isothermal process, Eq. 20-1 reduces to

$$\Delta S = S_f - S_i = \frac{Q}{T}. \qquad (20\text{-}2)$$

When the temperature change $\Delta T$ of a system is small relative to the temperature (in kelvins) before and after the process, the entropy change can be approximated as

$$\Delta S = S_f - S_i \approx \frac{Q}{T_{\text{avg}}}, \qquad (20\text{-}3)$$

where $T_{\text{avg}}$ is the system's average temperature during the process.

When an ideal gas changes reversibly from an initial state with temperature $T_i$ and volume $V_i$ to a final state with temperature $T_f$ and volume $V_f$, the change $\Delta S$ in the entropy of the gas is

$$\Delta S = S_f - S_i = nR \ln \frac{V_f}{V_i} + nC_V \ln \frac{T_f}{T_i}. \qquad (20\text{-}4)$$

**The Second Law of Thermodynamics** This law, which is an extension of the entropy postulate, states: *If a process occurs in a closed system, the entropy of the system increases for irreversible processes and remains constant for reversible processes. It never decreases.* In equation form,

$$\Delta S \geq 0. \qquad (20\text{-}5)$$

**Engines** An **engine** is a device that, operating in a cycle, extracts energy as heat $|Q_H|$ from a high-temperature reservoir and does a certain amount of work $|W|$. The *efficiency* $\varepsilon$ of any engine is defined as

$$\varepsilon = \frac{\text{energy we get}}{\text{energy we pay for}} = \frac{|W|}{|Q_H|}. \qquad (20\text{-}11)$$

In an **ideal engine,** all processes are reversible and no wasteful energy transfers occur due to, say, friction and turbulence. A **Carnot engine** is an ideal engine that follows the cycle of Fig. 20-9. Its efficiency is

$$\varepsilon_C = 1 - \frac{|Q_L|}{|Q_H|} = 1 - \frac{T_L}{T_H}, \qquad (20\text{-}12, 20\text{-}13)$$

in which $T_H$ and $T_L$ are the temperatures of the high- and low-temperature reservoirs, respectively. Real engines always have an efficiency lower than that given by Eq. 20-13. Ideal engines that are not Carnot engines also have lower efficiencies.

A *perfect engine* is an imaginary engine in which energy extracted as heat from the high-temperature reservoir is converted completely to work. Such an engine would violate the second law of thermodynamics, which can be restated as follows: No series of processes is possible whose sole result is the absorption of energy as heat from a thermal reservoir and the complete conversion of this energy to work.

**Refrigerators** A **refrigerator** is a device that, operating in a cycle, has work $W$ done on it as it extracts energy $|Q_L|$ as heat from a low-temperature reservoir. The coefficient of performance $K$ of a refrigerator is defined as

$$K = \frac{\text{what we want}}{\text{what we pay for}} = \frac{|Q_L|}{|W|}. \qquad (20\text{-}14)$$

A **Carnot refrigerator** is a Carnot engine operating in reverse. For a Carnot refrigerator, Eq. 20-14 becomes

$$K_C = \frac{|Q_L|}{|Q_H| - |Q_L|} = \frac{T_L}{T_H - T_L}. \qquad (20\text{-}15, 20\text{-}16)$$

A *perfect refrigerator* is an imaginary refrigerator in which energy extracted as heat from the low-temperature reservoir is converted completely to heat discharged to the high-temperature reservoir, without any need for work. Such a refrigerator would violate the second law of thermodynamics, which can be restated as follows: No series of processes is possible whose sole result is the transfer of energy as heat from a reservoir at a given temperature to a reservoir at a higher temperature.

**Entropy from a Statistical View** The entropy of a system can be defined in terms of the possible distributions of its molecules. For identical molecules, each possible distribution of molecules is called a **microstate** of the system. All equivalent microstates are grouped into a **configuration** of the system. The number of microstates in a configuration is the **multiplicity** $W$ of the configuration.

For a system of $N$ molecules that may be distributed between the two halves of a box, the multiplicity is given by

$$W = \frac{N!}{n_1! \, n_2!}, \qquad (20\text{-}20)$$

in which $n_1$ is the number of molecules in one half of the box and $n_2$ is the number in the other half. A basic assumption of **statistical mechanics** is that all the microstates are equally probable. Thus, configurations with a large multiplicity occur most often. When $N$ is very large (say, $N = 10^{22}$ molecules or more), the molecules are nearly always in the configuration in which $n_1 = n_2$.

The multiplicity $W$ of a configuration of a system and the entropy $S$ of the system in that configuration are related by Boltzmann's entropy equation:

$$S = k \ln W, \qquad (20\text{-}21)$$

where $k = 1.38 \times 10^{-23}$ J/K is the Boltzmann constant.

When $N$ is very large (the usual case), we can approximate $\ln N!$ with *Stirling's approximation:*

$$\ln N! \approx N(\ln N) - N. \qquad (20\text{-}22)$$

## QUESTIONS

**1** In four experiments, 2.5 mol of hydrogen gas undergoes reversible isothermal expansions, starting from the same volume but at different temperatures. The corresponding $p$-$V$ plots are shown in Fig. 20-19. Rank the situations according to the change in the entropy of the gas, greatest first. (*Hint:* See Sample Problem 20-1.)

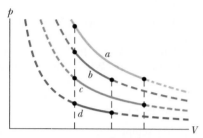

**FIG. 20-19** Question 1.

**2** In four experiments, blocks $A$ and $B$, starting at different initial temperatures, were brought together in an insulating box (as in Sample Problem 20-2) and allowed to reach a common final temperature. The entropy changes for the blocks in the four experiments had the following values (in joules per kelvin), but not necessarily in the order given. Determine which values for $A$ go with which values for $B$.

| Block | Values | | | |
|---|---|---|---|---|
| $A$ | 8 | 5 | 3 | 9 |
| $B$ | −3 | −8 | −5 | −2 |

**3** Point $i$ in Fig. 20-20 represents the initial state of an ideal gas at temperature $T$. Taking algebraic signs into account, rank the entropy changes that the gas undergoes as it moves, successively and reversibly, from point $i$ to points $a, b, c,$ and $d$, greatest first.

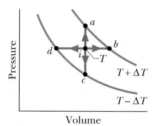

**FIG. 20-20** Question 3.

**4** An ideal monatomic gas at initial temperature $T_0$ (in kelvins) expands from initial volume $V_0$ to volume $2V_0$ by each of the five processes indicated in the $T$-$V$ diagram of Fig. 20-21. In which process is the expansion (a) isothermal, (b) isobaric (constant pressure), and (c) adiabatic? Explain your answers. (d) In which processes does the entropy of the gas decrease?

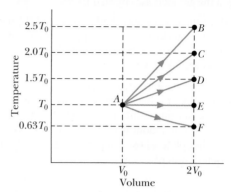

**FIG. 20-21** Question 4.

**5** A gas, confined to an insulated cylinder, is compressed adiabatically to half its volume. Does the entropy of the gas increase, decrease, or remain unchanged during this process?

**6** Three Carnot engines operate between temperature limits of (a) 400 and 500 K, (b) 500 and 600 K, and (c) 400 and 600 K. Each engine extracts the same amount of energy per cycle from the high-temperature reservoir. Rank the magnitudes of the work done by the engines per cycle, greatest first.

**7** An inventor claims to have invented four engines, each of which operates between constant-temperature reservoirs at 400 and 300 K. Data on each engine, per cycle of operation, are: engine A, $Q_H = 200$ J, $Q_L = -175$ J, and $W = 40$ J; engine B, $Q_H = 500$ J, $Q_L = -200$ J, and $W = 400$ J; engine C, $Q_H = 600$ J, $Q_L = -200$ J, and $W = 400$ J; engine D, $Q_H = 100$ J, $Q_L = -90$ J, and $W = 10$ J. Of the first and second laws of thermodynamics, which (if either) does each engine violate?

**8** Does the entropy per cycle increase, decrease, or remain the same for (a) a Carnot refrigerator, (b) a real refrigerator, and (c) a perfect refrigerator (which is, of course, impossible to build)?

**9** Does the entropy per cycle increase, decrease, or remain the same for (a) a Carnot engine, (b) a real engine, and (c) a perfect engine (which is, of course, impossible to build)?

**10** A box contains 100 atoms in a configuration that has 50 atoms in each half of the box. Suppose that you could count the different microstates associated with this configuration at the rate of 100 billion states per second, using a super-computer. Without written calculation, guess how much computing time you would need: a day, a year, or much more than a year.

## PROBLEMS

Tutoring problem available (at instructor's discretion) in *WileyPLUS* and WebAssign

**SSM** Worked-out solution available in Student Solutions Manual

• – ••• Number of dots indicates level of problem difficulty

Additional information available in *The Flying Circus of Physics* and at flyingcircusofphysics.com

**WWW** Worked-out solution is at

**ILW** Interactive solution is at

http://www.wiley.com/college/halliday

### sec. 20-3 Change in Entropy

•**1** A 2.50 mol sample of an ideal gas expands reversibly and isothermally at 360 K until its volume is doubled. What is the increase in entropy of the gas? **ILW**

•2 How much energy must be transferred as heat for a reversible isothermal expansion of an ideal gas at 132°C if the entropy of the gas increases by 46.0 J/K?

•3 Find (a) the energy absorbed as heat and (b) the change in entropy of a 2.00 kg block of copper whose temperature is increased reversibly from 25.0°C to 100°C. The specific heat of copper is 386 J/kg·K. ILW

•4 (a) What is the entropy change of a 12.0 g ice cube that melts completely in a bucket of water whose temperature is just above the freezing point of water? (b) What is the entropy change of a 5.00 g spoonful of water that evaporates completely on a hot plate whose temperature is slightly above the boiling point of water?

•5 Suppose 4.00 mol of an ideal gas undergoes a reversible isothermal expansion from volume $V_1$ to volume $V_2 = 2.00V_1$ at temperature $T = 400$ K. Find (a) the work done by the gas and (b) the entropy change of the gas. (c) If the expansion is reversible and adiabatic instead of isothermal, what is the entropy change of the gas? SSM

•6 An ideal gas undergoes a reversible isothermal expansion at 77.0°C, increasing its volume from 1.30 L to 3.40 L. The entropy change of the gas is 22.0 J/K. How many moles of gas are present?

••7 In an experiment, 200 g of aluminum (with a specific heat of 900 J/kg·K) at 100°C is mixed with 50.0 g of water at 20.0°C, with the mixture thermally isolated. (a) What is the equilibrium temperature? What are the entropy changes of (b) the aluminum, (c) the water, and (d) the aluminum–water system? SSM WWW

••8 A 364 g block is put in contact with a thermal reservoir. The block is initially at a lower temperature than the reservoir. Assume that the consequent transfer of energy as heat from the reservoir to the block is reversible. Figure 20-22 gives the change in entropy $\Delta S$ of the block until thermal equilibrium is reached. The scale of the horizontal axis is set by $T_a = 280$ K and $T_b = 380$ K. What is the specific heat of the block?

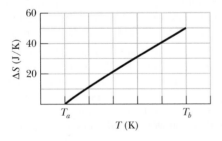

FIG. 20-22   Problem 8.

••9 In the irreversible process of Fig. 20-5, let the initial temperatures of identical blocks L and R be 305.5 and 294.5 K, respectively, and let 215 J be the energy that must be transferred between the blocks in order to reach equilibrium. For the reversible processes of Fig. 20-6, what is $\Delta S$ for (a) block L, (b) its reservoir, (c) block R, (d) its reservoir, (e) the two-block system, and (f) the system of the two blocks and the two reservoirs?

••10 A gas sample undergoes a reversible isothermal ex-

pansion. Figure 20-23 gives the change $\Delta S$ in entropy of the gas versus the final volume $V_f$ of the gas. The scale of the vertical axis is set by $\Delta S_s = 64$ J/K. How many moles are in the sample?

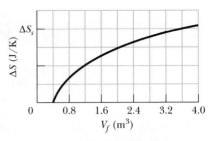

FIG. 20-23   Problem 10.

••11 A 50.0 g block of copper whose temperature is 400 K is placed in an insulating box with a 100 g block of lead whose temperature is 200 K. (a) What is the equilibrium temperature of the two-block system? (b) What is the change in the internal energy of the system between the initial state and the equilibrium state? (c) What is the change in the entropy of the system? (See Table 18-3.) ILW

••12 At very low temperatures, the molar specific heat $C_V$ of many solids is approximately $C_V = AT^3$, where A depends on the particular substance. For aluminum, $A = 3.15 \times 10^{-5}$ J/mol·K⁴. Find the entropy change for 4.00 mol of aluminum when its temperature is raised from 5.00 K to 10.0 K.

••13 In Fig. 20-24, where $V_{23} = 3.00V_1$, n moles of a diatomic ideal gas are taken through the cycle with the molecules rotating but not oscillating. What are (a) $p_2/p_1$, (b) $p_3/p_1$, and (c) $T_3/T_1$? For path $1 \rightarrow 2$, what are (d) $W/nRT_1$, (e) $Q/nRT_1$, (f) $\Delta E_{int}/nRT_1$, and (g) $\Delta S/nR$? For path $2 \rightarrow 3$, what are (h) $W/nRT_1$, (i) $Q/nRT_1$, (j) $\Delta E_{int}/nRT_1$, (k) $\Delta S/nR$? For path $3 \rightarrow 1$, what are (l) $W/nRT_1$, (m) $Q/nRT_1$, (n) $\Delta E_{int}/nRT_1$, (o) $\Delta S/nR$?

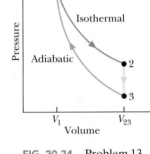

FIG. 20-24   Problem 13.

••14 A 2.0 mol sample of an ideal monatomic gas undergoes the reversible process shown in Fig. 20-25. The scale of the vertical axis is set by $T_s = 400.0$ K and the scale of the horizontal axis is set by $S_s = 20.0$ J/K. (a) How much energy is absorbed as heat by the gas? (b) What is the change in the internal energy of the gas? (c) How much work is done by the gas? GO

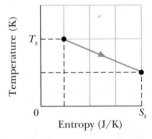

FIG. 20-25   Problem 14.

••15 A 10 g ice cube at −10°C is placed in a lake whose temperature is 15°C. Calculate the change in entropy of the cube–lake system as the ice cube comes to thermal equilibrium with the lake. The specific heat of ice is 2220 J/kg·K. (Hint: Will the ice cube affect the lake temperature?)

••16 (a) For 1.0 mol of a monatomic ideal gas taken through the cycle in Fig. 20-26, where $V_1 = 4.00V_0$, what is $W/p_0V_0$ as the gas goes from state a to state c along path abc? What is $\Delta E_{int}/p_0V_0$ in going (b) from b to c and (c) through one full cycle? What is $\Delta S$ in going (d) from $b$ to $c$ and (e) through one full cycle?

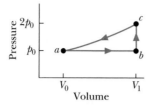

**••17** A mixture of 1773 g of water and 227 g of ice is in an initial equilibrium state at 0.000°C. The mixture is then, in a reversible process, brought to a second equilibrium state

**FIG. 20-26** Problem 16.

where the water–ice ratio, by mass, is 1.00:1.00 at 0.000°C. (a) Calculate the entropy change of the system during this process. (The heat of fusion for water is 333 kJ/kg.) (b) The system is then returned to the initial equilibrium state in an irreversible process (say, by using a Bunsen burner). Calculate the entropy change of the system during this process. (c) Are your answers consistent with the second law of thermodynamics?

**••18** An 8.0 g ice cube at $-10$°C is put into a Thermos flask containing 100 cm³ of water at 20°C. By how much has the entropy of the cube–water system changed when equilibrium is reached? The specific heat of ice is 2220 J/kg·K.

**•••19** Energy can be removed from water as heat at and even below the normal freezing point (0.0°C at atmospheric pressure) without causing the water to freeze; the water is then said to be *supercooled*. Suppose a 1.00 g water drop is supercooled until its temperature is that of the surrounding air, which is at $-5.00$°C. The drop then suddenly and irreversibly freezes, transferring energy to the air as heat. What is the entropy change for the drop? (*Hint:* Use a three-step reversible process as if the water were taken through the normal freezing point.) The specific heat of ice is 2220 J/kg·K. **GO**

**•••20** An insulated Thermos contains 130 g of water at 80.0°C. You put in a 12.0 g ice cube at 0°C to form a system of *ice + original water*. (a) What is the equilibrium temperature of the system? What are the entropy changes of the water that was originally the ice cube (b) as it melts and (c) as it warms to the equilibrium temperature? (d) What is the entropy change of the original water as it cools to the equilibrium temperature? (e) What is the net entropy change of the *ice + original water* system as it reaches the equilibrium temperature? **GO**

**•••21** Suppose 1.00 mol of a monatomic ideal gas is taken from initial pressure $p_1$ and volume $V_1$ through two steps: (1) an isothermal expansion to volume $2.00V_1$ and (2) a pressure increase to $2.00p_1$ at constant volume. What is $Q/p_1V_1$ for (a) step 1 and (b) step 2? What is $W/p_1V_1$ for (c) step 1 and (d) step 2? For the full process, what are (e) $\Delta E_{int}/p_1V_1$ and (f) $\Delta S$? The gas is returned to its initial state and again taken to the same final state but now through these two steps: (1) an isothermal compression to pressure $2.00p_1$ and (2) a volume increase to $2.00V_1$ at constant pressure. What is $Q/p_1V_1$ for (g) step 1 and (h) step 2? What is $W/p_1V_1$ for (i) step 1 and (j) step 2? For the full process, what are (k) $\Delta E_{int}/p_1V_1$ and (l) $\Delta S$?

**•••22** Expand 1.00 mol of an monatomic gas initially at 5.00 kPa and 600 K from initial volume $V_i = 1.00$ m³ to final volume $V_f = 2.00$ m³. At any instant during the expansion, the pressure $p$ and volume $V$ of the gas are related by $p = 5.00 \exp[(V_i - V)/a]$, with $p$ in kilopascals, $V_i$ and $V$ in cubic

meters, and $a = 1.00$ m³. What are the final (a) pressure and (b) temperature of the gas? (c) How much work is done by the gas during the expansion? (d) What is $\Delta S$ for the expansion? (*Hint:* Use two simple reversible processes to find $\Delta S$.)

## sec. 20-5 Entropy in the Real World: Engines

**•23** A Carnot engine has an efficiency of 22.0%. It operates between constant-temperature reservoirs differing in temperature by 75.0 C°. What is the temperature of the (a) lower-temperature and (b) higher-temperature reservoir?

**•24** In a hypothetical nuclear fusion reactor, the fuel is deuterium gas at a temperature of $7 \times 10^8$ K. If this gas could be used to operate a Carnot engine with $T_L = 100$°C, what would be the engine's efficiency? Take both temperatures to be exact and report your answer to seven significant figures.

**•25** A Carnot engine operates between 235°C and 115°C, absorbing $6.30 \times 10^4$ J per cycle at the higher temperature. (a) What is the efficiency of the engine? (b) How much work per cycle is this engine capable of performing? **SSM WWW**

**•26** A Carnot engine absorbs 52 kJ as heat and exhausts 36 kJ as heat in each cycle. Calculate (a) the engine's efficiency and (b) the work done per cycle in kilojoules.

**•27** A Carnot engine whose low-temperature reservoir is at 17°C has an efficiency of 40%. By how much should the temperature of the high-temperature reservoir be increased to increase the efficiency to 50%?

**••28** A 500 W Carnot engine operates between constant-temperature reservoirs at 100°C and 60.0°C. What is the rate at which energy is (a) taken in by the engine as heat and (b) exhausted by the engine as heat?

**••29** Figure 20-27 shows a reversible cycle through which 1.00 mol of a monatomic ideal gas is taken. Volume $V_c = 8.00V_b$. Process $bc$ is an adiabatic expansion, with $p_b = 10.0$ atm and $V_b = 1.00 \times 10^{-3}$ m³. For the cycle, find (a) the energy added to the gas as heat, (b) the energy leaving the gas as heat, (c) the net work done by the gas, and (d) the efficiency of the cycle. **SSM ILW**

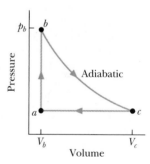

**FIG. 20-27** Problem 29.

**••30** A Carnot engine is set up to produce a certain work $W$ per cycle. In each cycle, energy in the form of heat $Q_H$ is transferred to the working substance of the engine from the higher-temperature thermal reservoir, which is at an adjustable temperature $T_H$. The lower-temperature thermal reservoir is maintained at temperature $T_L = 250$ K. Figure 20-28 gives $Q_H$ for a range of $T_H$. The scale of the vertical axis is set by $Q_{Hs} = 6.0$ kJ. If $T_H$ is set at 550 K, what is $Q_H$?

**••31** Figure 20-29 shows a reversible cycle

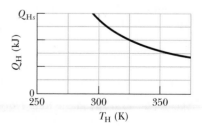

**FIG. 20-28** Problem 30.

through which 1.00 mol of a monatomic ideal gas is taken. Assume that $p = 2p_0$, $V = 2V_0$, $p_0 = 1.01 \times 10^5$ Pa, and $V_0 = 0.0225$ m$^3$. Calculate (a) the work done during the cycle, (b) the energy added as heat during stroke *abc*, and (c) the efficiency of the cycle. (d) What is the efficiency of a Carnot engine operating between the highest and lowest temperatures that occur in the cycle? (e) Is this greater than or less than the efficiency calculated in (c)? **GO**

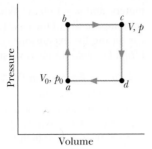

**FIG. 20-29** Problem 31.

**••32** An ideal gas (1.0 mol) is the working substance in an engine that operates on the cycle shown in Fig. 20-30. Processes *BC* and *DA* are reversible and adiabatic. (a) Is the gas monatomic, diatomic, or polyatomic? (b) What is the engine efficiency?

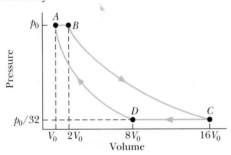

**FIG. 20-30** Problem 32.

**••33** The efficiency of a particular car engine is 25% when the engine does 8.2 kJ of work per cycle. Assume the process is reversible. What are (a) the energy the engine gains per cycle as heat $Q_{gain}$ from the fuel combustion and (b) the energy the engine loses per cycle as heat $Q_{lost}$. If a tune-up increases the efficiency to 31%, what are (c) $Q_{gain}$ and (d) $Q_{lost}$ at the same work value?

**••34** In the first stage of a two-stage Carnot engine, energy is absorbed as heat $Q_1$ at temperature $T_1$, work $W_1$ is done, and energy is expelled as heat $Q_2$ at a lower temperature $T_2$. The second stage absorbs that energy as heat $Q_2$, does work $W_2$, and expels energy as heat $Q_3$ at a still lower temperature $T_3$. Prove that the efficiency of the engine is $(T_1 - T_3)/T_1$.

**•••35** The cycle in Fig. 20-31 represents the operation of a gasoline internal combustion engine. Volume $V_3 = 4.00V_1$. Assume the gasoline–air intake mixture is an ideal gas with $\gamma = 1.30$. What are the ratios (a) $T_2/T_1$, (b) $T_3/T_1$, (c) $T_4/T_1$, (d) $p_3/p_1$, and (e) $p_4/p_1$? (f) What is the engine efficiency?

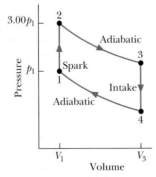

**FIG. 20-31** Problem 35.

### sec. 20-6  Entropy in the Real World: Refrigerators

**•36** The electric motor of a heat pump transfers energy as heat from the outdoors, which is at $-5.0°$C, to a room that is at $17°$C. If the heat pump were a Carnot heat pump (a Carnot en-

gine working in reverse), how much energy would be transferred as heat to the room for each joule of electric energy consumed?

**•37** A Carnot air conditioner takes energy from the thermal energy of a room at 70°F and transfers it as heat to the outdoors, which is at 96°F. For each joule of electric energy required to operate the air conditioner, how many joules are removed from the room? **SSM**

**•38** To make ice, a freezer that is a reverse Carnot engine extracts 42 kJ as heat at $-15°$C during each cycle, with coefficient of performance 5.7. The room temperature is 30.3°C. How much (a) energy per cycle is delivered as heat to the room and (b) work per cycle is required to run the freezer?

**•39** A heat pump is used to heat a building. The outside temperature is $-5.0°$C, and the temperature inside the building is to be maintained at 22°C. The pump's coefficient of performance is 3.8, and the heat pump delivers 7.54 MJ as heat to the building each hour. If the heat pump is a Carnot engine working in reverse, at what rate must work be done to run it? **SSM**

**•40** How much work must be done by a Carnot refrigerator to transfer 1.0 J as heat (a) from a reservoir at 7.0°C to one at 27°C, (b) from a reservoir at $-73°$C to one at 27°C, (c) from a reservoir at $-173°$C to one at 27°C, and (d) from a reservoir at $-223°$C to one at 27°C?

**••41** Figure 20-32 represents a Carnot engine that works between temperatures $T_1 = 400$ K and $T_2 = 150$ K and drives a Carnot refrigerator that works between temperatures $T_3 = 325$ K and $T_4 = 225$ K. What is the ratio $Q_3/Q_1$? **GO**

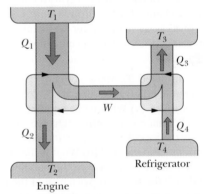

**FIG. 20-32** Problem 41.

**••42** (a) During each cycle, a Carnot engine absorbs 750 J as heat from a high-temperature reservoir at 360 K, with the low-temperature reservoir at 280 K. How much work is done per cycle? (b) The engine is then made to work in reverse to function as a Carnot refrigerator between those same two reservoirs. During each cycle, how much work is required to remove 1200 J as heat from the low-temperature reservoir?

**••43** An air conditioner operating between 93°F and 70°F is rated at 4000 Btu/h cooling capacity. Its coefficient of performance is 27% of that of a Carnot refrigerator operating between the same two temperatures. What horsepower is required of the air conditioner motor? **ILW**

**••44** The motor in a refrigerator has a power of 200 W. If the freezing compartment is at 270 K and the outside air is at 300 K, and assuming the efficiency of a Carnot refrigerator, what is the maximum amount of energy that can be extracted as heat from the freezing compartment in 10.0 min?

## sec. 20-8 A Statistical View of Entropy

•**45** Construct a table like Table 20-1 for eight molecules.

••**46** A box contains $N$ identical gas molecules equally divided between its two halves. For $N = 50$, what are (a) the multiplicity $W$ of the central configuration, (b) the total number of microstates, and (c) the percentage of the time the system spends in the central configuration? For $N = 100$, what are (d) $W$ of the central configuration, (e) the total number of microstates, and (f) the percentage of the time the system spends in the central configuration? For $N = 200$, what are (g) $W$ of the central configuration, (h) the total number of microstates, and (i) the percentage of the time the system spends in the central configuration? (j) Does the time spent in the central configuration increase or decrease with an increase in $N$?

•••**47** A box contains $N$ gas molecules. Consider the box to be divided into three equal parts. (a) By extension of Eq. 20-20, write a formula for the multiplicity of any given configuration. (b) Consider two configurations: configuration $A$ with equal numbers of molecules in all three thirds of the box, and configuration $B$ with equal numbers of molecules in each half of the box divided into two equal parts rather than three. What is the ratio $W_A/W_B$ of the multiplicity of configuration $A$ to that of configuration $B$? (c) Evaluate $W_A/W_B$ for $N = 100$. (Because 100 is not evenly divisible by 3, put 34 molecules into one of the three box parts of configuration $A$ and 33 in each of the other two parts.) SSM WWW

## Additional Problems

**48** Figure 20-33 gives the force magnitude $F$ versus stretch distance $x$ for a rubber band, with the scale of the $F$ axis set by $F_s = 1.50$ N and the scale of the $x$ axis set by $x_s = 3.50$ cm. The temperature is 2.00°C. When the rubber band is stretched by $x = 1.70$ cm, at what rate does the entropy of the rubber band change during a small additional stretch?

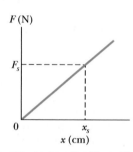

**FIG. 20-33** Problem 48.

**49** As a sample of nitrogen gas ($N_2$) undergoes a temperature increase at constant volume, the distribution of molecular speeds increases. That is, the probability distribution function $P(v)$ for the molecules spreads to higher speed values, as suggested in Fig. 19-8b. One way to report the spread in $P(v)$ is to measure the difference $\Delta v$ between the most probable speed $v_P$ and the rms speed $v_{rms}$. When $P(v)$ spreads to higher speeds, $\Delta v$ increases. Assume that the gas is ideal and the $N_2$ molecules rotate but do not oscillate. For 1.5 mol, an initial temperature of 250 K, and a final temperature of 500 K, what are (a) the initial difference $\Delta v_i$, (b) the final difference $\Delta v_f$, and (c) the entropy change $\Delta S$ for the gas? SSM

**50** A three-step cycle is undergone by 3.4 mol of an ideal diatomic gas: (1) the temperature of the gas is increased from 200 K to 500 K at constant volume; (2) the gas is then isothermally expanded to its original pressure; (3) the gas is then contracted at constant pressure back to its original volume. Throughout the cycle, the molecules rotate but do not oscillate. What is the efficiency of the cycle?

**51** Suppose that a deep shaft were drilled in Earth's crust near one of the poles, where the surface temperature is −40°C, to a depth where the temperature is 800°C. (a) What is the theoretical limit to the efficiency of an engine operating between these temperatures? (b) If all the energy released as heat into the low-temperature reservoir were used to melt ice that was initially at −40°C, at what rate could liquid water at 0°C be produced by a 100 MW power plant (treat it as an engine)? The specific heat of ice is 2220 J/kg·K; water's heat of fusion is 333 kJ/kg. (Note that the engine can operate only between 0°C and 800°C in this case. Energy exhausted at −40°C cannot warm anything above −40°C.)

**52** (a) A Carnot engine operates between a hot reservoir at 320 K and a cold one at 260 K. If the engine absorbs 500 J as heat per cycle at the hot reservoir, how much work per cycle does it deliver? (b) If the engine working in reverse functions as a refrigerator between the same two reservoirs, how much work per cycle must be supplied to remove 1000 J as heat from the cold reservoir?

**53** A 600 g lump of copper at 80.0°C is placed in 70.0 g of water at 10.0°C in an insulated container. (See Table 18-3 for specific heats.) (a) What is the equilibrium temperature of the copper–water system? What entropy changes do (b) the copper, (c) the water, and (d) the copper–water system undergo in reaching the equilibrium temperature?

**54** Suppose 0.550 mol of an ideal gas is isothermally and reversibly expanded in the four situations given below. What is the change in the entropy of the gas for each situation?

| Situation | (a) | (b) | (c) | (d) |
|---|---|---|---|---|
| Temperature (K) | 250 | 350 | 400 | 450 |
| Initial volume (cm³) | 0.200 | 0.200 | 0.300 | 0.300 |
| Final volume (cm³) | 0.800 | 0.800 | 1.20 | 1.20 |

**55** A 0.600 kg sample of water is initially ice at temperature −20°C. What is the sample's entropy change if its temperature is increased to 40°C? SSM

**56** What is the entropy change for 3.20 mol of an ideal monatomic gas undergoing a reversible increase in temperature from 380 K to 425 K at constant volume?

**57** A three-step cycle is undergone reversibly by 4.00 mol of an ideal gas: (1) an adiabatic expansion that gives the gas 2.00 times its initial volume, (2) a constant-volume process, (3) an isothermal compression back to the initial state of the gas. We do not know whether the gas is monatomic or diatomic; if it is diatomic, we do not know whether the molecules are rotating or oscillating. What are the entropy changes for (a) the cycle, (b) process 1, (c) process 3, and (d) process 2?

**58** Suppose 1.0 mol of a monatomic ideal gas initially at 10 L and 300 K is heated at constant volume to 600 K, allowed to expand isothermally to its initial pressure, and finally compressed at constant pressure to its original volume, pressure, and temperature. During the cycle, what are (a) the net energy entering the system (the gas) as heat and (b) the net work done by the gas? (c) What is the efficiency of the cycle?

**59** A 2.00 mol diatomic gas initially at 300 K undergoes this cycle: It is (1) heated at constant volume to 800 K, (2) then allowed to expand isothermally to its initial pressure, (3) then compressed at constant pressure to its initial state. Assuming the gas molecules neither rotate nor oscillate, find (a) the net energy transferred as heat to the gas, (b) the net work done by the gas, and (c) the efficiency of the cycle.

**60** A 45.0 g block of tungsten at 30.0°C and a 25.0 g block of silver at −120°C are placed together in an insulated container. (See Table 18-3 for specific heats.) (a) What is the equilibrium temperature? What entropy changes do (b) the tungsten, (c) the silver, and (d) the tungsten–silver system undergo in reaching the equilibrium temperature?

**61** A cylindrical copper rod of length 1.50 m and radius 2.00 cm is insulated to prevent heat loss through its curved surface. One end is attached to a thermal reservoir fixed at 300°C; the other is attached to a thermal reservoir fixed at 30.0°C. What is the rate at which entropy increases for the rod–reservoirs system?

**62** An ideal refrigerator does 150 J of work to remove 560 J as heat from its cold compartment. (a) What is the refrigerator's coefficient of performance? (b) How much heat per cycle is exhausted to the kitchen?

**63** A Carnot refrigerator extracts 35.0 kJ as heat during each cycle, operating with a coefficient of performance of 4.60. What are (a) the energy per cycle transferred as heat to the room and (b) the work done per cycle? **SSM**

**64** Four particles are in the insulated box of Fig. 20-17. What are (a) the least multiplicity, (b) the greatest multiplicity, (c) the least entropy, and (d) the greatest entropy of the four-particle system?

**65** A brass rod is in thermal contact with a constant-temperature reservoir at 130°C at one end and a constant-temperature reservoir at 24.0°C at the other end. (a) Compute the total change in entropy of the rod–reservoirs system when 5030 J of energy is conducted through the rod, from one reservoir to the other. (b) Does the entropy of the rod change? **GO**

**66** An apparatus that liquefies helium is in a room maintained at 300 K. If the helium in the apparatus is at 4.0 K, what is the minimum ratio $Q_{to}/Q_{from}$, where $Q_{to}$ is the energy delivered as heat to the room and $Q_{from}$ is the energy removed as heat from the helium?

**67** System A of three particles and system B of five particles are in insulated boxes like that in Fig. 20-17. What is the least multiplicity W of (a) system A and (b) system B? What is the greatest multiplicity W of (c) A and (d) B? What is the greatest entropy of (e) A and (f) B? **SSM**

**68** Calculate the efficiency of a fossil-fuel power plant that consumes 380 metric tons of coal each hour to produce useful work at the rate of 750 MW. The heat of combustion of coal (the heat due to burning it) is 28 MJ/kg.

**69** The temperature of 1.00 mol of a monatomic ideal gas is raised reversibly from 300 K to 400 K, with its volume kept constant. What is the entropy change of the gas?

**70** Repeat Problem 69, with the pressure now kept constant.

**71** Suppose that 260 J is conducted from a constant-temperature reservoir at 400 K to one at (a) 100 K, (b) 200 K, (c) 300 K, and (d) 360 K. What is the net change in entropy $\Delta S_{net}$ of the reservoirs in each case? (e) As the temperature difference of the two reservoirs decreases, does $\Delta S_{net}$ increase, decrease, or remain the same?

**72** A Carnot engine whose high-temperature reservoir is at

400 K has an efficiency of 30.0%. By how much should the temperature of the low-temperature reservoir be changed to increase the efficiency to 40.0%?

**73** A box contains N molecules. Consider two configurations: configuration A with an equal division of the molecules between the two halves of the box, and configuration B with 60.0% of the molecules in the left half of the box and 40.0% in the right half. For N = 50, what are (a) the multiplicity $W_A$ of configuration A, (b) the multiplicity $W_B$ of configuration B, and (c) the ratio $f_{B/A}$ of the time the system spends in configuration B to the time it spends in configuration A? For N = 100, what are (d) $W_A$, (e) $W_B$, and (f) $f_{B/A}$? For N = 200, what are (g) $W_A$, (h) $W_B$, and (i) $f_{B/A}$? (j) With increasing N, does f increase, decrease, or remain the same?

**74** Suppose 2.00 mol of a diatomic gas is taken reversibly around the cycle shown in the T-S diagram of Fig. 20-34, where $S_1 = 6.00$ J/K and $S_2 = 8.00$ J/K. The molecules do not rotate or oscillate. What is the energy transferred as heat Q for (a) path 1 → 2, (b) path 2 → 3, and (c) the full cycle? (d) What is the work W for the isothermal process? The volume $V_1$ in state 1 is 0.200 m³.

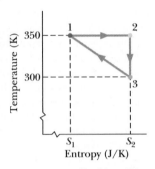

FIG. 20-34 Problem 74.

What is the volume in (e) state 2 and (f) state 3?
What is the change $\Delta E_{int}$ for (g) path 1 → 2, (h) path 2 → 3, and (i) the full cycle? (*Hint:* (h) can be done with one or two lines of calculation using Section 19-8 or with a page of calculation using Section 19-11.) (j) What is the work W for the adiabatic process?

**75** An inventor has built an engine X and claims that its efficiency $\varepsilon_X$ is greater than the efficiency $\varepsilon$ of an ideal engine operating between the same two temperatures. Suppose you couple engine X to an ideal refrigerator (Fig. 20-35a) and adjust the cycle of engine X so that the work per cycle it provides equals the work per cycle required by the ideal refrigerator. Treat this combination as a single unit and show that if the inventor's claim were true (if $\varepsilon_X > \varepsilon$), the combined unit would act as a perfect refrigerator (Fig. 20-35b), transferring energy as heat from the low-temperature reservoir to the high-temperature reservoir without the need for work.

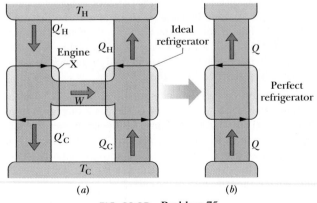

FIG. 20-35 Problem 75.

# Electric Charge

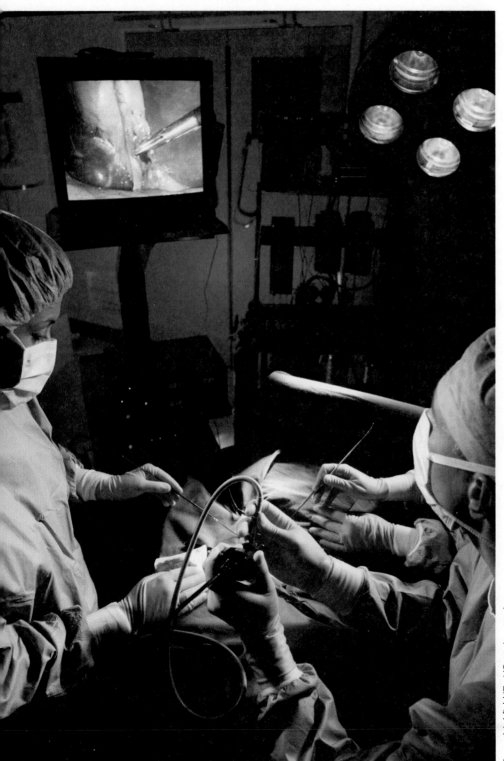

Hospital personnel go to extraordinary lengths to avoid bacterial infection of a patient. Surfaces are scrubbed, masks are donned, hands are meticulously cleaned and then gloved, and instruments are sanitized at high temperature and in alcohol baths. But there are still subtle sources of bacteria, such as possibly in this photograph.

## Can you find the bacterial source?

The answer is in this chapter.

Samuel Ashfield/Taxi/Getty Images, Inc.

## 21-1 | WHAT IS PHYSICS?

You are surrounded by devices that depend on the physics of electromagnetism, which is the combination of electric and magnetic phenomena. This physics is at the root of computers, television, radio, telecommunications, household lighting, and even the ability of food wrap to cling to a container. This physics is also the basis of the natural world. Not only does it hold together all the atoms and molecules in the world, it also produces lightning, auroras, and rainbows.

The physics of electromagnetism was first studied by the early Greek philosophers, who discovered that if a piece of amber is rubbed and then brought near bits of straw, the straw will jump to the amber. We now know that the attraction between amber and straw is due to an electric force. The Greek philosophers also discovered that if a certain type of stone (a naturally occurring magnet) is brought near bits of iron, the iron will jump to the stone. We now know that the attraction between magnet and iron is due to a magnetic force.

From these modest origins with the Greek philosophers, the sciences of electricity and magnetism developed separately for centuries—until 1820, in fact, when Hans Christian Oersted found a connection between them: an electric current in a wire can deflect a magnetic compass needle. Interestingly enough, Oersted made this discovery, a big surprise, while preparing a lecture demonstration for his physics students.

The new science of electromagnetism was developed further by workers in many countries. One of the best was Michael Faraday, a truly gifted experimenter with a talent for physical intuition and visualization. That talent is attested to by the fact that his collected laboratory notebooks do not contain a single equation. In the mid-nineteenth century, James Clerk Maxwell put Faraday's ideas into mathematical form, introduced many new ideas of his own, and put electromagnetism on a sound theoretical basis.

Our discussion of electromagnetism is spread through the next 16 chapters. We begin with electrical phenomena, and our first step is to discuss the nature of electric charge and electric force.

## 21-2 | Electric Charge

In dry weather, you can produce a spark by walking across certain types of carpet and then bringing one of your fingers near a metal doorknob, metal faucet, or even a friend. You can also produce multiple sparks when you pull, say, a sweater from your body or clothes from a dryer. Sparks and the "static cling" of clothing (similar to what is seen in Fig. 21-1) are usually just annoying. However, if you happen to pull off a sweater and then spark to a computer, the results are more than just annoying.

These examples reveal that we have electric charge in our bodies, sweaters, carpets, doorknobs, faucets, and computers. In fact, every object contains a vast amount of electric charge. **Electric charge** is an intrinsic characteristic of the fundamental particles making up those objects; that is, it is a property that comes automatically with those particles wherever they exist.

The vast amount of charge in an everyday object is usually hidden because the object contains equal amounts of the two kinds of charge: *positive charge* and *negative charge*. With such an equality—or *balance*—of charge, the object is said to be *electrically neutral*; that is, it contains no *net* charge. If the two types of charge are not in balance, then there *is* a net charge. We say that an object is *charged* to indicate that it has a charge imbalance, or net charge. The imbalance is always much smaller than the total amounts of positive charge and negative charge contained in the object.

Charged objects interact by exerting forces on one another. To show this, we first charge a glass rod by rubbing one end with silk. At points of contact between

**FIG. 21-1** Static cling, an electrical phenomenon that accompanies dry weather, causes these pieces of paper to stick to one another and to the plastic comb, and your clothing to stick to your body.
(*Fundamental Photographs*)

the rod and the silk, tiny amounts of charge are transferred from one to the other, slightly upsetting the electrical neutrality of each. (We *rub* the silk over the rod to increase the number of contact points and thus the amount, still tiny, of transferred charge.)

Suppose we now suspend the charged rod from a thread to *electrically isolate* it from its surroundings so that its charge cannot change. If we bring a second, similarly charged, glass rod nearby (Fig. 21-2*a*), the two rods *repel* each other; that is, each rod experiences a force directed away from the other rod. However, if we rub a *plastic* rod with fur and then bring the rod near the suspended glass rod (Fig. 21-2*b*), the two rods *attract* each other; that is, each rod experiences a force directed toward the other rod.

We can understand these two demonstrations in terms of positive and negative charges. When a glass rod is rubbed with silk, the glass loses some of its negative charge and then has a small unbalanced positive charge (represented by the plus signs in Fig. 21-2*a*). When the plastic rod is rubbed with fur, the plastic gains a small unbalanced negative charge (represented by the minus signs in Fig. 21-2*b*). Our two demonstrations reveal the following:

> Charges with the same electrical sign repel each other, and charges with opposite electrical signs attract each other.

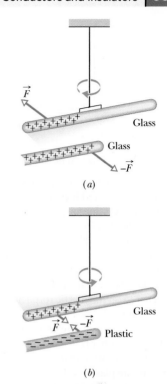

**FIG. 21-2** (*a*) Two charged rods of the same sign repel each other. (*b*) Two charged rods of opposite signs attract each other. Plus signs indicate a positive net charge, and minus signs indicate a negative net charge.

In Section 21-4, we shall put this rule into quantitative form as Coulomb's law of *electrostatic force* (or *electric force*) between charges. The term *electrostatic* is used to emphasize that, relative to each other, the charges are either stationary or moving only very slowly.

The "positive" and "negative" labels and signs for electric charge were chosen arbitrarily by Benjamin Franklin. He could easily have interchanged the labels or used some other pair of opposites to distinguish the two kinds of charge. (Franklin was a scientist of international reputation. It has even been said that Franklin's triumphs in diplomacy in France during the American War of Independence were facilitated, and perhaps even made possible, because he was so highly regarded as a scientist.)

The attraction and repulsion between charged bodies have many industrial applications, including electrostatic paint spraying and powder coating, fly-ash collection in chimneys, nonimpact ink-jet printing, and photocopying. Figure 21-3 shows a tiny carrier bead in a photocopying machine, covered with particles of black powder called *toner*, which stick to it by means of electrostatic forces. The negatively charged toner particles are eventually attracted from the carrier bead to a rotating drum, where a positively charged image of the document being copied has formed. A charged sheet of paper then attracts the toner particles from the drum to itself, after which they are heat-fused permanently in place to produce the copy.

## 21-3 I Conductors and Insulators

We can classify materials generally according to the ability of charge to move through them. **Conductors** are materials through which charge can move rather freely; examples include metals (such as copper in common lamp wire), the human body, and tap water. **Nonconductors**—also called **insulators**—are materials through which charge cannot move freely; examples include rubber (such as the insulation on common lamp wire), plastic, glass, and chemically pure water. **Semiconductors** are materials that are intermediate between conductors and insulators; examples include silicon and germanium in computer chips. **Superconductors** are materials that are *perfect* conductors, allowing charge to move without *any* hindrance. In these chapters we discuss only conductors and insulators.

Here is an example of how conduction can eliminate excess charge on an object. If you rub a copper rod with wool, charge is transferred from the wool to

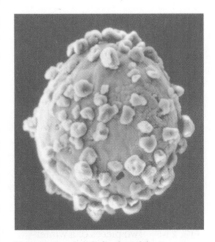

**FIG. 21-3** A carrier bead from a photocopying machine; the bead is covered with toner particles that cling to it by electrostatic attraction. The diameter of the bead is about 0.3 mm. (*Courtesy Xerox*)

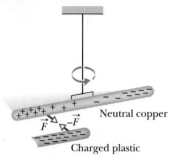

**FIG. 21-4** A neutral copper rod is electrically isolated from its surroundings by being suspended on a nonconducting thread. Either end of the copper rod will be attracted by a charged rod. Here, conduction electrons in the copper rod are repelled to the far end of that rod by the negative charge on the plastic rod. Then that negative charge attracts the remaining positive charge on the near end of the copper rod, rotating the copper rod to bring that near end closer to the plastic rod.

the rod. However, if you are holding the rod while also touching a faucet, you cannot charge the rod in spite of the transfer. The reason is that you, the rod, and the faucet are all conductors connected, via the plumbing, to Earth's surface, which is a huge conductor. Because the excess charges put on the rod by the wool repel one another, they move away from one another by moving first through the rod, then through you, and then through the faucet and plumbing to reach Earth's surface, where they can spread out. The process leaves the rod electrically neutral.

In thus setting up a pathway of conductors between an object and Earth's surface, we are said to *ground* the object, and in neutralizing the object (by eliminating an unbalanced positive or negative charge), we are said to *discharge* the object. If instead of holding the copper rod in your hand, you hold it by an insulating handle, you eliminate the conducting path to Earth, and the rod can then be charged by rubbing (the charge remains on the rod), as long as you do not touch it directly with your hand.

The properties of conductors and insulators are due to the structure and electrical nature of atoms. Atoms consist of positively charged *protons*, negatively charged *electrons*, and electrically neutral *neutrons*. The protons and neutrons are packed tightly together in a central *nucleus*.

The charge of a single electron and that of a single proton have the same magnitude but are opposite in sign. Hence, an electrically neutral atom contains equal numbers of electrons and protons. Electrons are held near the nucleus because they have the electrical sign opposite that of the protons in the nucleus and thus are attracted to the nucleus.

When atoms of a conductor like copper come together to form the solid, some of their outermost (and so most loosely held) electrons become free to wander about within the solid, leaving behind positively charged atoms (*positive ions*). We call the mobile electrons *conduction electrons*. There are few (if any) free electrons in a nonconductor.

The experiment of Fig. 21-4 demonstrates the mobility of charge in a conductor. A negatively charged plastic rod will attract either end of an isolated neutral copper rod. What happens is that many of the conduction electrons in the closer end of the copper rod are repelled by the negative charge on the plastic rod. Some of the conduction electrons move to the far end of the copper rod, leaving the near end depleted in electrons and thus with an unbalanced positive charge. This positive charge is attracted to the negative charge in the plastic rod. Although the copper rod is still neutral, it is said to have an *induced charge*, which means that some of its positive and negative charges have been separated due to the presence of a nearby charge.

Similarly, if a positively charged glass rod is brought near one end of a neutral copper rod, conduction electrons in the copper rod are attracted to that end. That end becomes negatively charged and the other end positively charged, so again an induced charge is set up in the copper rod. Although the copper rod is still neutral, it and the glass rod attract each other.

Note that only conduction electrons, with their negative charges, can move; positive ions are fixed in place. Thus, an object becomes positively charged only through the *removal of negative charges*.

### Blue Flashes from a Wintergreen LifeSaver

Indirect evidence for the attraction of charges with opposite signs can be seen with a wintergreen LifeSaver (the candy shaped in the form of a marine lifesaver). If you adapt your eyes to darkness for about 15 minutes and then have a friend chomp on a piece of the candy in the darkness, you will see a faint blue flash from your friend's mouth with each chomp. Whenever a chomp breaks a sugar crystal into pieces, each piece will probably end up with a different number of electrons. Suppose a crystal breaks into pieces $A$ and $B$, with $A$ ending up with more electrons on its surface than $B$ (Fig. 21-5). This means that $B$ has positive ions (atoms that lost electrons to $A$) on its surface. Because the electrons on $A$

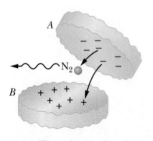

**FIG. 21-5** Two pieces of a wintergreen LifeSaver candy as they fall away from each other. Electrons jumping from the negative surface of piece $A$ to the positive surface of piece $B$ collide with nitrogen ($N_2$) molecules in the air.

are strongly attracted to the positive ions on $B$, some of those electrons jump across the gap between the pieces.

As $A$ and $B$ fall away from each other, air (primarily nitrogen, $N_2$) flows into the gap, and many of the jumping electrons collide with nitrogen molecules in the air, causing the molecules to emit ultraviolet light. You cannot see this type of light. However, the wintergreen molecules on the surfaces of the candy pieces absorb the ultraviolet light and then emit blue light, which you *can* see—it is the blue light coming from your friend's mouth. ✈

### Bacterial Contamination and Electrostatics

Electrostatic forces can play a subtle role in bacterial contamination in a hospital, such as during endoscopic surgery. In such a procedure, a surgeon sees the interior of a patient's body on the viewing screen of a video monitor. In a traditional (tube) monitor (not a "flat-screen monitor"), the screen image is produced by electrons directed toward a positively charged screen. The charged screen also attracts airborne particles floating around in the operating room, such as lint, dust, and skin cells. If an airborne particle is negatively charged, it is pulled onto the screen's exterior surface. If, instead, it is electrically neutral, some of its conduction electrons can be pulled to the side of the particle nearest the screen, giving the particle an induced charge (Fig. 21-6$a$). Such a particle is then pulled to the screen's exterior surface just as the copper rod is pulled to the charged plastic rod in Fig. 21-4.

Because many of the particles collected on the screen's exterior surface carry bacteria, the screen becomes contaminated with bacteria. Suppose a surgeon's gloved fingers come within a few centimeters of the screen, pointing to a particular part of the image, say, in explaining a surgical concern to other medical staff. The positively charged screen pulls electrons from inside the fingers to the fingertips (Fig. 21-6$b$). The negatively charged fingertips then cause particles (airborne or on the screen) to collect on the gloves at the tips. When the surgeon next touches the patient with the contaminated gloves, the bacteria end up on or (worse) inside the patient's body. To avoid this risk, surgeons are now warned not to bring fingers near a video monitor.

Similar contamination can occur with the plastic aprons commonly worn by hospital staff to protect them from a patient's blood. Those aprons can become highly charged when peeled from a dispenser or when repeatedly rubbed against underlying clothing or skin, especially in a dry environment. Once an apron becomes charged, it can pull bacteria and contaminated dust out of the surrounding air. Because a staff member occasionally touches the apron, the bacteria can easily be transferred to a patient during an examination or surgery. ✈

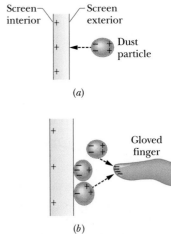

**FIG. 21-6** (*a*) A cross section of the viewing screen on a traditional (tube) video monitor. The positively charged screen produces induced charge on a nearby neutral dust particle. (*b*) A gloved finger (not to scale) brought near the screen has an induced charge and can attract dust particles from the air and from the screen.

✔ **CHECKPOINT 1**     The figure shows five pairs of plates: $A$, $B$, and $D$ are charged plastic plates and $C$ is an electrically neutral copper plate. The electrostatic forces between the pairs of plates are shown for three of the pairs. For the remaining two pairs, do the plates repel or attract each other?

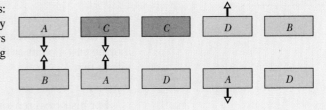

## 21-4 | Coulomb's Law

If two charged particles are brought near each other, they each exert a force on the other. If the particles have the same sign of charge, they repel each other (Figs. 21-7$a$ and $b$). That is, the force on each particle is directed away from the other particle, and if the particles can move, they move away from each other. If, instead, the particles have opposite signs of charge, they attract each other (Fig. 21-7$c$) and, if free to move, they move closer to each other.

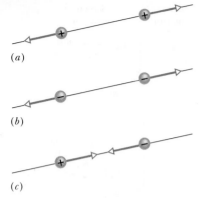

(a)

(b)

(c)

**FIG. 21-7** Two charged particles repel each other if they have the same sign of charge, either (a) both positive or (b) both negative. (c) They attract each other if they have opposite signs of charge.

This force of repulsion or attraction due to the charge properties of objects is called an **electrostatic force.** The equation giving the force for charged *particles* is called **Coulomb's law** after Charles-Augustin de Coulomb, whose experiments in 1785 led him to it. In terms of the particles in Fig. 21-8, where particle 1 has charge $q_1$ and particle 2 has charge $q_2$, the force on particle 1 is

$$\vec{F} = k\frac{q_1 q_2}{r^2}\hat{r} \qquad \text{(Coulomb's law)}, \qquad (21\text{-}1)$$

in which $\hat{r}$ is a unit vector along an axis extending through the two particles, $r$ is the distance between them, and $k$ is a constant. (As with other unit vectors, $\hat{r}$ has a magnitude of exactly 1 and no dimension or unit; its purpose is to point.) If the particles have the same signs of charge, the force on particle 1 is in the direction of $\hat{r}$; if they have opposite signs, the force is opposite $\hat{r}$.

Curiously, the form of Eq. 21-1 is the same as that of Newton's equation (Eq. 13-3) for the gravitational force between two particles with masses $m_1$ and $m_2$ that are separated by a distance $r$:

$$\vec{F} = G\frac{m_1 m_2}{r^2}\hat{r} \qquad \text{(Newton's law)}, \qquad (21\text{-}2)$$

in which $G$ is the gravitational constant.

The constant $k$ in Eq. 21-1, by analogy with the gravitational constant $G$ in Eq. 21-2, may be called the *electrostatic constant.* Both equations describe inverse square laws that involve a property of the interacting particles—the mass in one case and the charge in the other. The laws differ in that gravitational forces are always attractive but electrostatic forces may be either attractive or repulsive, depending on the signs of the two charges. This difference arises from the fact that, although there is only one kind of mass, there are two kinds of charge.

Coulomb's law has survived every experimental test; no exceptions to it have ever been found. It holds even within the atom, correctly describing the force between the positively charged nucleus and each of the negatively charged electrons, even though classical Newtonian mechanics fails in that realm and is replaced there by quantum physics. This simple law also correctly accounts for the forces that bind atoms together to form molecules, and for the forces that bind atoms and molecules together to form solids and liquids.

The SI unit of charge is the **coulomb.** For practical reasons having to do with the accuracy of measurements, the coulomb unit is derived from the SI unit *ampere* for *electric current i.* Current is the rate $dq/dt$ at which charge moves past a point or through a region. In Chapter 26 we shall discuss current in detail. Until then we shall use the relation

$$i = \frac{dq}{dt} \qquad \text{(electric current)}, \qquad (21\text{-}3)$$

in which $i$ is the current (in amperes) and $dq$ (in coulombs) is the amount of charge moving past a point or through a region in time $dt$ (in seconds). Rearranging Eq. 21-3 tells us that

$$1\,\text{C} = (1\,\text{A})(1\,\text{s}).$$

For historical reasons (and because doing so simplifies many other formulas), the electrostatic constant $k$ of Eq. 21-1 is usually written $1/4\pi\varepsilon_0$. Then the magnitude of the force in Coulomb's law becomes

$$F = \frac{1}{4\pi\varepsilon_0}\frac{|q_1||q_2|}{r^2} \qquad \text{(Coulomb's law)}. \qquad (21\text{-}4)$$

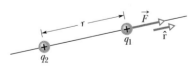

$q_2$    $q_1$    $r$    $\vec{F}$    $\hat{r}$

**FIG. 21-8** The electrostatic force on particle 1 can be described in terms of a unit vector $\hat{r}$ along an axis through the two particles.

The constants in Eqs. 21-1 and 21-4 have the value

$$k = \frac{1}{4\pi\varepsilon_0} = 8.99 \times 10^9\,\text{N}\cdot\text{m}^2/\text{C}^2. \qquad (21\text{-}5)$$

The quantity $\varepsilon_0$, called the **permittivity constant,** sometimes appears separately in equations and is

$$\varepsilon_0 = 8.85 \times 10^{-12} \text{ C}^2/\text{N} \cdot \text{m}^2. \qquad (21\text{-}6)$$

Still another parallel between the gravitational force and the electrostatic force is that both obey the principle of superposition. If we have $n$ charged particles, they interact independently in pairs, and the force on any one of them, let us say particle 1, is given by the vector sum

$$\vec{F}_{1,\text{net}} = \vec{F}_{12} + \vec{F}_{13} + \vec{F}_{14} + \vec{F}_{15} + \cdots + \vec{F}_{1n}, \qquad (21\text{-}7)$$

in which, for example, $\vec{F}_{14}$ is the force acting on particle 1 due to the presence of particle 4. An identical formula holds for the gravitational force.

Finally, the shell theorem that we found so useful in our study of gravitation has analogs in electrostatics:

> A shell of uniform charge attracts or repels a charged particle that is outside the shell as if all the shell's charge were concentrated at its center.

> If a charged particle is located inside a shell of uniform charge, there is no net electrostatic force on the particle from the shell.

(In the first theorem, we assume that the charge on the shell is much greater than that of the particle. Then any redistribution of the charge on the shell due to the presence of the particle's charge can be neglected.)

### Spherical Conductors

If excess charge is placed on a spherical shell that is made of conducting material, the excess charge spreads uniformly over the (external) surface. For example, if we place excess electrons on a spherical metal shell, those electrons repel one another and tend to move apart, spreading over the available surface until they are uniformly distributed. That arrangement maximizes the distances between all pairs of the excess electrons. According to the first shell theorem, the shell then will attract or repel an external charge as if all the excess charge on the shell were concentrated at its center.

If we remove negative charge from a spherical metal shell, the resulting positive charge of the shell is also spread uniformly over the surface of the shell. For example, if we remove $n$ electrons, there are then $n$ sites of positive charge (sites missing an electron) that are spread uniformly over the shell. According to the first shell theorem, the shell will again attract or repel an external charge as if all the shell's excess charge were concentrated at its center.

✓ **CHECKPOINT 2** The figure shows two protons (symbol p) and one electron (symbol e) on an axis. What is the direction of (a) the electrostatic force on the central proton due to the electron, (b) the electrostatic force on the central proton due to the other proton, and (c) the net electrostatic force on the central proton?

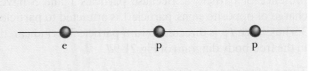

**Sample Problem** 21-1 **Build your skill**

(a) Figure 21-9a shows two positively charged particles fixed in place on an $x$ axis. The charges are $q_1 = 1.60 \times 10^{-19}$ C and $q_2 = 3.20 \times 10^{-19}$ C, and the particle separation is $R = 0.0200$ m. What are the magnitude and direction of the electrostatic force $\vec{F}_{12}$ on particle 1 from particle 2?

**FIG. 21-9 a–b** (a) Two charged particles of charges $q_1$ and $q_2$ are fixed in place on an $x$ axis. (b) The free-body diagram for particle 1, showing the electrostatic force on it from particle 2.

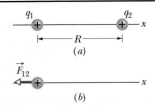

**KEY IDEAS** Because both particles are positively charged, particle 1 is repelled by particle 2, with a force magnitude given by Eq. 21-4. Thus, the direction of force

$\vec{F}_{12}$ on particle 1 is *away from* particle 2, in the negative direction of the $x$ axis, as indicated in the free-body diagram of Fig. 21-9b.

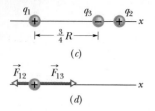

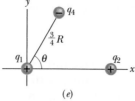

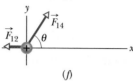

FIG. 21-9 continued (c) Particle 3 included. (d) Free-body diagram for particle 1. (e) Particle 4 included. (f) Free-body diagram for particle 1.

**Two particles:** Using Eq. 21-4 with separation $R$ substituted for $r$, we can write the magnitude $F_{12}$ of this force as

$$F_{12} = \frac{1}{4\pi\varepsilon_0} \frac{|q_1||q_2|}{R^2}$$

$$= (8.99 \times 10^9 \, \text{N} \cdot \text{m}^2/\text{C}^2)$$

$$\times \frac{(1.60 \times 10^{-19} \, \text{C})(3.20 \times 10^{-19} \, \text{C})}{(0.0200 \, \text{m})^2}$$

$$= 1.15 \times 10^{-24} \, \text{N}.$$

Thus, force $\vec{F}_{12}$ has the following magnitude and direction (relative to the positive direction of the $x$ axis):

$$1.15 \times 10^{-24} \, \text{N} \quad \text{and} \quad 180°. \quad \text{(Answer)}$$

We can also write $\vec{F}_{12}$ in unit-vector notation as

$$\vec{F}_{12} = -(1.15 \times 10^{-24} \, \text{N})\hat{\text{i}}. \quad \text{(Answer)}$$

(b) Figure 21-9c is identical to Fig. 21-9a except that particle 3 now lies on the $x$ axis between particles 1 and 2. Particle 3 has charge $q_3 = -3.20 \times 10^{-19}$ C and is at a distance $\frac{3}{4}R$ from particle 1. What is the net electrostatic force $\vec{F}_{1,\text{net}}$ on particle 1 due to particles 2 and 3?

**KEY IDEA** The presence of particle 3 does not alter the electrostatic force on particle 1 from particle 2. Thus, force $\vec{F}_{12}$ still acts on particle 1. Similarly, the force $\vec{F}_{13}$ that acts on particle 1 due to particle 3 is not affected by the presence of particle 2. Because particles 1 and 3 have charge of opposite signs, particle 1 is attracted to particle 3. Thus, force $\vec{F}_{13}$ is directed *toward* particle 3, as indicated in the free-body diagram of Fig. 21-9d.

**Three particles:** To find the magnitude of $\vec{F}_{13}$, we can rewrite Eq. 21-4 as

$$F_{13} = \frac{1}{4\pi\varepsilon_0} \frac{|q_1||q_3|}{(\frac{3}{4}R)^2}$$

$$= (8.99 \times 10^9 \, \text{N} \cdot \text{m}^2/\text{C}^2)$$

$$\times \frac{(1.60 \times 10^{-19} \, \text{C})(3.20 \times 10^{-19} \, \text{C})}{(\frac{3}{4})^2(0.0200 \, \text{m})^2}$$

$$= 2.05 \times 10^{-24} \, \text{N}.$$

We can also write $\vec{F}_{13}$ in unit-vector notation:

$$\vec{F}_{13} = (2.05 \times 10^{-24} \, \text{N})\hat{\text{i}}.$$

The net force $\vec{F}_{1,\text{net}}$ on particle 1 is the vector sum of $\vec{F}_{12}$ and $\vec{F}_{13}$; that is, from Eq. 21-7, we can write the net

force $\vec{F}_{1,\text{net}}$ on particle 1 in unit-vector notation as

$$\vec{F}_{1,\text{net}} = \vec{F}_{12} + \vec{F}_{13}$$

$$= -(1.15 \times 10^{-24} \, \text{N})\hat{\text{i}} + (2.05 \times 10^{-24} \, \text{N})\hat{\text{i}}$$

$$= (9.00 \times 10^{-25} \, \text{N})\hat{\text{i}}. \quad \text{(Answer)}$$

Thus, $\vec{F}_{1,\text{net}}$ has the following magnitude and direction (relative to the positive direction of the $x$ axis):

$$9.00 \times 10^{-25} \, \text{N} \quad \text{and} \quad 0°. \quad \text{(Answer)}$$

(c) Figure 21-9e is identical to Fig. 21-9a except that particle 4 is now included. It has charge $q_4 = -3.20 \times 10^{-19}$ C, is at a distance $\frac{3}{4}R$ from particle 1, and lies on a line that makes an angle $\theta = 60°$ with the $x$ axis. What is the net electrostatic force $\vec{F}_{1,\text{net}}$ on particle 1 due to particles 2 and 4?

**KEY IDEA** The net force $\vec{F}_{1,\text{net}}$ is the vector sum of $\vec{F}_{12}$ and a new force $\vec{F}_{14}$ acting on particle 1 due to particle 4. Because particles 1 and 4 have charge of opposite signs, particle 1 is attracted to particle 4. Thus, force $\vec{F}_{14}$ on particle 1 is directed *toward* particle 4, at angle $\theta = 60°$, as indicated in the free-body diagram of Fig. 21-9f.

**Four particles:** We can rewrite Eq. 21-4 as

$$F_{14} = \frac{1}{4\pi\varepsilon_0} \frac{|q_1||q_4|}{(\frac{3}{4}R)^2}$$

$$= (8.99 \times 10^9 \, \text{N} \cdot \text{m}^2/\text{C}^2)$$

$$\times \frac{(1.60 \times 10^{-19} \, \text{C})(3.20 \times 10^{-19} \, \text{C})}{(\frac{3}{4})^2(0.0200 \, \text{m})^2}$$

$$= 2.05 \times 10^{-24} \, \text{N}.$$

Then from Eq. 21-7, we can write the net force $\vec{F}_{1,\text{net}}$ on particle 1 as

$$\vec{F}_{1,\text{net}} = \vec{F}_{12} + \vec{F}_{14}.$$

Because the forces $\vec{F}_{12}$ and $\vec{F}_{14}$ are not directed along the same axis, we *cannot* sum simply by combining their magnitudes. Instead, we must add them as vectors, using one of the following methods.

**Method 1.** *Summing directly on a vector-capable calculator.* For $\vec{F}_{12}$, we enter the magnitude $1.15 \times 10^{-24}$ and the angle 180°. For $\vec{F}_{14}$, we enter the magnitude $2.05 \times 10^{-24}$ and the angle 60°. Then we add the vectors.

**Method 2.** *Summing in unit-vector notation.* First we rewrite $\vec{F}_{14}$ as

$$\vec{F}_{14} = (F_{14} \cos \theta)\hat{\text{i}} + (F_{14} \sin \theta)\hat{\text{j}}.$$

Substituting $2.05 \times 10^{-24}$ N for $F_{14}$ and 60° for $\theta$, this becomes

$$\vec{F}_{14} = (1.025 \times 10^{-24} \, \text{N})\hat{\text{i}} + (1.775 \times 10^{-24} \, \text{N})\hat{\text{j}}.$$

Then we sum:

$$\vec{F}_{1,\text{net}} = \vec{F}_{12} + \vec{F}_{14}$$

$$= -(1.15 \times 10^{-24} \, \text{N})\hat{\text{i}}$$

$$+ (1.025 \times 10^{-24} \, \text{N})\hat{\text{i}} + (1.775 \times 10^{-24} \, \text{N})\hat{\text{j}}$$

$$\approx (-1.25 \times 10^{-25} \, \text{N})\hat{\text{i}} + (1.78 \times 10^{-24} \, \text{N})\hat{\text{j}}.$$

$$\text{(Answer)}$$

**Method 3.** *Summing components axis by axis.* The sum of the $x$ components gives us

$$F_{1,\text{net},x} = F_{12,x} + F_{14,x} = F_{12} + F_{14}\cos 60°$$
$$= -1.15 \times 10^{-24}\,\text{N} + (2.05 \times 10^{-24}\,\text{N})(\cos 60°)$$
$$= -1.25 \times 10^{-25}\,\text{N}.$$

The sum of the $y$ components gives us

$$F_{1,\text{net},y} = F_{12,y} + F_{14,y} = 0 + F_{14}\sin 60°$$
$$= (2.05 \times 10^{-24}\,\text{N})(\sin 60°)$$
$$= 1.78 \times 10^{-24}\,\text{N}.$$

The net force $\vec{F}_{1,\text{net}}$ has the magnitude

$$F_{1,\text{net}} = \sqrt{F_{1,\text{net},x}^2 + F_{1,\text{net},y}^2} = 1.78 \times 10^{-24}\,\text{N}. \quad \text{(Answer)}$$

To find the direction of $\vec{F}_{1,\text{net}}$, we take

$$\theta = \tan^{-1}\frac{F_{1,\text{net},y}}{F_{1,\text{net},x}} = -86.0°.$$

However, this is an unreasonable result because $\vec{F}_{1,\text{net}}$ must have a direction between the directions of $\vec{F}_{12}$ and $\vec{F}_{14}$. To correct $\theta$, we add 180°, obtaining

$$-86.0° + 180° = 94.0°. \quad \text{(Answer)}$$

✓ **CHECKPOINT 3**   The figure here shows three arrangements of an electron e and two protons p. (a) Rank the arrangements according to the magnitude of the net electrostatic force on the electron due to the protons, largest first. (b) In situation $c$, is the angle between the net force on the electron and the line labeled $d$ less than or more than 45°?

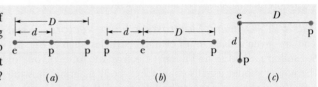

*(a)*     *(b)*     *(c)*

---

**PROBLEM-SOLVING TACTICS**

*Tactic 1:* **Symbols Representing Charge**   Here is a general guide to the symbols representing charge. If the symbol $q$, with or without a subscript, is used in a sentence when no electrical sign has been specified, the charge can be either positive or negative. Sometimes the sign is explicitly shown, as in the notation $+q$ or $-q$.

When more than one charged object is being considered, their charges might be given as multiples of a charge magnitude. As examples, the notation $+2q$ means a positive charge with magnitude twice that of some reference charge magnitude $q$, and $-3q$ means a negative charge with magnitude three times that of the reference charge magnitude $q$.

---

**Sample Problem   21-2**

Figure 21-10$a$ shows two particles fixed in place: a particle of charge $q_1 = +8q$ at the origin and a particle of charge $q_2 = -2q$ at $x = L$. At what point (other than infinitely far away) can a proton be placed so that it is in *equilibrium* (the net force on it is zero)? Is that equilibrium *stable* or *unstable*?

**KEY IDEA**   If $\vec{F}_1$ is the force on the proton due to charge $q_1$ and $\vec{F}_2$ is the force on the proton due to charge $q_2$, then the point we seek is where $\vec{F}_1 + \vec{F}_2 = 0$. Thus,

$$\vec{F}_1 = -\vec{F}_2. \quad (21\text{-}8)$$

This tells us that at the point we seek, the forces acting on the proton due to the other two particles must be of equal magnitudes,

$$F_1 = F_2, \quad (21\text{-}9)$$

and that the forces must have opposite directions.

*Reasoning:* Because a proton has a positive charge, the proton and the particle of charge $q_1$ are of the same sign, and force $\vec{F}_1$ on the proton must point away from $q_1$. Also, the proton and the particle of charge $q_2$ are of opposite signs, so force $\vec{F}_2$ on the proton must point toward $q_2$. "Away from $q_1$" and "toward $q_2$" can be in opposite directions only if the proton is located on the $x$ axis.

If the proton is on the $x$ axis at any point between $q_1$ and $q_2$, such as point $P$ in Fig. 21-10$b$, then $\vec{F}_1$ and $\vec{F}_2$ are in the same direction and not in opposite directions as required. If the proton is at any point on the $x$ axis to

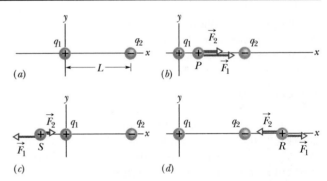

**FIG. 21-10**   (*a*) Two particles of charges $q_1$ and $q_2$ are fixed in place on an $x$ axis, with separation $L$. (*b*)–(*d*) Three possible locations $P$, $S$, and $R$ for a proton. At each location, $\vec{F}_1$ is the force on the proton from particle 1 and $\vec{F}_2$ is the force on the proton from particle 2.

the left of $q_1$, such as point $S$ in Fig. 21-10$c$, then $\vec{F}_1$ and $\vec{F}_2$ are in opposite directions. However, Eq. 21-4 tells us that $\vec{F}_1$ and $\vec{F}_2$ cannot have equal magnitudes there: $F_1$ must be greater than $F_2$, because $F_1$ is produced by a closer charge (with lesser $r$) of greater magnitude ($8q$ versus $2q$).

Finally, if the proton is at any point on the $x$ axis to the right of $q_2$, such as point $R$ in Fig. 21-10$d$, then $\vec{F}_1$ and $\vec{F}_2$ are again in opposite directions. However, because now the charge of greater magnitude ($q_1$) is *farther* away from the proton than the charge of lesser magnitude, there is a point at which $F_1$ is equal to $F_2$. Let $x$ be the coordinate of this point, and let $q_p$ be the charge of the proton.

*Calculations:* With the aid of Eq. 21-4, we can now rewrite Eq. 21-9 (which says that the forces have equal magnitudes):

$$\frac{1}{4\pi\varepsilon_0}\frac{8qq_\mathrm{p}}{x^2} = \frac{1}{4\pi\varepsilon_0}\frac{2qq_\mathrm{p}}{(x-L)^2}. \qquad (21\text{-}10)$$

(Note that only the charge magnitudes appear in Eq. 21-10.) Rearranging Eq. 21-10 gives us

$$\left(\frac{x-L}{x}\right)^2 = \frac{1}{4}.$$

After taking the square roots of both sides, we have

$$\frac{x-L}{x} = \frac{1}{2},$$

which gives us

$$x = 2L. \qquad \text{(Answer)}$$

The equilibrium at $x = 2L$ is unstable; that is, if the proton is displaced leftward from point $R$, then $F_1$ and $F_2$ both increase but $F_2$ increases more (because $q_2$ is closer than $q_1$), and a net force will drive the proton farther leftward. If the proton is displaced rightward, both $F_1$ and $F_2$ decrease but $F_2$ decreases more, and a net force will then drive the proton farther rightward. In a stable equilibrium, if the proton is displaced slightly, it returns to the equilibrium position.

---

**PROBLEM-SOLVING TACTICS**

***Tactic 2:*** *Drawing Electrostatic Force Vectors* When you are given a diagram of charged particles, such as Fig. 21-9a, and are asked to find the net electrostatic force on one of them, you should usually draw a free-body diagram showing only the particle of concern and the forces that particle experiences, as in Fig. 21-9b. If, instead, you choose to superimpose those forces on the given diagram showing all the particles, be sure to draw the force vectors with either their tails (preferably) or their heads on the particle of concern. If you draw the vectors elsewhere in the diagram, you invite confusion—and confusion is guaranteed if you draw the vectors on the particles *causing* the forces on the particle of concern.

---

**Sample Problem  21-3**

In Fig. 21-11a, two identical, electrically isolated conducting spheres $A$ and $B$ are separated by a (center-to-center) distance $a$ that is large compared to the spheres. Sphere $A$ has a positive charge of $+Q$, and sphere $B$ is electrically neutral. Initially, there is no electrostatic force between the spheres. (Assume that there is no induced charge on the spheres because of their large separation.)

(a) Suppose the spheres are connected for a moment by a conducting wire. The wire is thin enough so that any net charge on it is negligible. What is the electrostatic force between the spheres after the wire is removed?

**KEY IDEAS** (1) Because the spheres are identical, connecting them means that they end up with identical charges (same sign and same amount). (2) The initial sum of the charges (including the signs of the charges) must equal the final sum of the charges.

*Reasoning:* When the spheres are wired together, the (negative) conduction electrons on $B$, which repel one another, have a way to move away from one another (along the wire to positively charged $A$, which attracts them—Fig. 21-11b.) As $B$ loses negative charge, it becomes positively charged, and as $A$ gains negative charge, it becomes *less* positively charged. The transfer of charge stops when the charge on $B$ has increased to $+Q/2$ and the charge on $A$ has decreased to $+Q/2$, which occurs when $-Q/2$ has shifted from $B$ to $A$.

After the wire has been removed (Fig. 21-11c), we can assume that the charge on either sphere does not disturb the uniformity of the charge distribution on the other sphere, because the spheres are small relative to their separation. Thus, we can apply the first shell theorem to each sphere. By Eq. 21-4 with $q_1 = q_2 = Q/2$ and $r = a$,

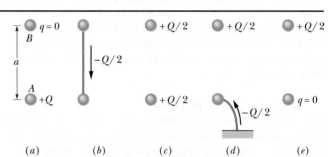

**FIG. 21-11** Two small conducting spheres $A$ and $B$. (a) To start, sphere $A$ is charged positively. (b) Negative charge is transferred from $B$ to $A$ through a connecting wire. (c) Both spheres are then charged positively. (d) Negative charge is transferred through a grounding wire to sphere $A$. (e) Sphere $A$ is then neutral.

$$F = \frac{1}{4\pi\varepsilon_0}\frac{(Q/2)(Q/2)}{a^2} = \frac{1}{16\pi\varepsilon_0}\left(\frac{Q}{a}\right)^2. \quad \text{(Answer)}$$

The spheres, now positively charged, repel each other.

(b) Next, suppose sphere $A$ is grounded momentarily, and then the ground connection is removed. What now is the electrostatic force between the spheres?

*Reasoning:* When we provide a conducting path between a charged object and the ground (which is a huge conductor), we neutralize the object. Were sphere $A$ negatively charged, the mutual repulsion between the excess electrons would cause them to move from the sphere to the ground. However, because sphere $A$ is positively charged, electrons with a total charge of $-Q/2$ move *from* the ground up onto the sphere (Fig. 21-11d), leaving the sphere with a charge of 0 (Fig. 21-11e). Thus, there is (again) no electrostatic force between the two spheres.

# 21-5 I Charge Is Quantized

TABLE 21-1

**The Charges of Three Particles**

| Particle | Symbol | Charge |
|----------|--------|--------|
| Electron | e or e⁻ | $-e$ |
| Proton | p | $+e$ |
| Neutron | n | 0 |

In Benjamin Franklin's day, electric charge was thought to be a continuous fluid—an idea that was useful for many purposes. However, we now know that fluids themselves, such as air and water, are not continuous but are made up of atoms and molecules; matter is discrete. Experiment shows that "electrical fluid" is also not continuous but is made up of multiples of a certain elementary charge. Any positive or negative charge $q$ that can be detected can be written as

$$q = ne, \quad n = \pm 1, \pm 2, \pm 3, \ldots, \tag{21-11}$$

in which $e$, the **elementary charge,** has the approximate value

$$e = 1.602 \times 10^{-19} \text{ C}. \tag{21-12}$$

The elementary charge $e$ is one of the important constants of nature. The electron and proton both have a charge of magnitude $e$ (Table 21-1). (Quarks, the constituent particles of protons and neutrons, have charges of $\pm e/3$ or $\pm 2e/3$, but they apparently cannot be detected individually. For this and for historical reasons, we do not take their charges to be the elementary charge.)

You often see phrases—such as "the charge on a sphere," "the amount of charge transferred," and "the charge carried by the electron"—that suggest that charge is a substance. (Indeed, such statements have already appeared in this chapter.) You should, however, keep in mind what is intended: *Particles* are the substance and charge happens to be one of their properties, just as mass is.

When a physical quantity such as charge can have only discrete values rather than any value, we say that the quantity is **quantized.** It is possible, for example, to find a particle that has no charge at all or a charge of $+10e$ or $-6e$, but not a particle with a charge of, say, $3.57e$.

The quantum of charge is small. In an ordinary 100 W lightbulb, for example, about $10^{19}$ elementary charges enter the bulb every second and just as many leave. However, the graininess of electricity does not show up in such large-scale phenomena (the bulb does not flicker with each electron), just as you cannot feel the individual molecules of water with your hand.

✓ **CHECKPOINT 4**     Initially, sphere $A$ has a charge of $-50e$ and sphere $B$ has a charge of $+20e$. The spheres are made of conducting material and are identical in size. If the spheres then touch, what is the resulting charge on sphere $A$?

---

**Sample Problem    21-4**

The nucleus in an iron atom has a radius of about $4.0 \times 10^{-15}$ m and contains 26 protons.

**(a)** What is the magnitude of the repulsive electrostatic force between two of the protons that are separated by $4.0 \times 10^{-15}$ m?

**KEY IDEA**    The protons can be treated as charged particles, so the magnitude of the electrostatic force on one from the other is given by Coulomb's law.

*Calculation:* Table 21-1 tells us that the charge of a proton is $+e$. Thus, Eq. 21-4 gives us

$$F = \frac{1}{4\pi\varepsilon_0} \frac{e^2}{r^2}$$

$$= \frac{(8.99 \times 10^9 \text{ N} \cdot \text{m}^2/\text{C}^2)(1.602 \times 10^{-19} \text{ C})^2}{(4.0 \times 10^{-15} \text{ m})^2}$$

$$= 14 \text{ N.} \qquad \text{(Answer)}$$

*No explosion:* This is a small force to be acting on a macroscopic object like a cantaloupe, but an enormous force to be acting on a proton. Such forces should explode the nucleus of any element but hydrogen (which has only one proton in its nucleus). However, they don't, not even in nuclei with a great many protons. Therefore, there must be some enormous attractive force to counter this enormous repulsive electrostatic force.

**(b)** What is the magnitude of the gravitational force between those same two protons?

**KEY IDEA**    Because the protons are particles, the magnitude of the gravitational force on one from the other is given by Newton's equation for the gravitational force (Eq. 21-2).

*Calculation:* With $m_p$ ($= 1.67 \times 10^{-27}$ kg) representing the mass of a proton, Eq. 21-2 gives us

$$F = G \frac{m_p^2}{r^2}$$

$$= \frac{(6.67 \times 10^{-11}\,\text{N} \cdot \text{m}^2/\text{kg}^2)(1.67 \times 10^{-27}\,\text{kg})^2}{(4.0 \times 10^{-15}\,\text{m})^2}$$

$$= 1.2 \times 10^{-35}\,\text{N}. \qquad \text{(Answer)}$$

**Weak versus strong:** This result tells us that the (attractive) gravitational force is far too weak to counter the repulsive electrostatic forces between protons in a nucleus. Instead, the protons are bound together by an enormous force called (aptly) the *strong nuclear*

*force*—a force that acts between protons (and neutrons) when they are close together, as in a nucleus.

Although the gravitational force is many times weaker than the electrostatic force, it is more important in large-scale situations because it is always attractive. This means that it can collect many small bodies into huge bodies with huge masses, such as planets and stars, that then exert large gravitational forces. The electrostatic force, on the other hand, is repulsive for charges of the same sign, so it is unable to collect either positive charge or negative charge into large concentrations that would then exert large electrostatic forces.

## 21-6 | Charge Is Conserved

If you rub a glass rod with silk, a positive charge appears on the rod. Measurement shows that a negative charge of equal magnitude appears on the silk. This suggests that rubbing does not create charge but only transfers it from one body to another, upsetting the electrical neutrality of each body during the process. This hypothesis of **conservation of charge,** first put forward by Benjamin Franklin, has stood up under close examination, both for large-scale charged bodies and for atoms, nuclei, and elementary particles. No exceptions have ever been found. Thus, we add electric charge to our list of quantities—including energy and both linear and angular momentum—that obey a conservation law.

Important examples of the conservation of charge occur in the *radioactive decay* of nuclei, in which a nucleus transforms into (becomes) a different type of nucleus. For example, a uranium-238 nucleus ($^{238}$U) transforms into a thorium-234 nucleus ($^{234}$Th) by emitting an *alpha particle*. Because that particle has the same makeup as a helium-4 nucleus, it has the symbol $^4$He. The number used in the name of a nucleus and as a superscript in the symbol for the nucleus is called the *mass number* and is the total number of the protons and neutrons in the nucleus. For example, the total number in $^{238}$U is 238. The number of protons in a nucleus is the *atomic number Z*, which is listed for all the elements in Appendix F. From that list we find that in the decay

$$^{238}\text{U} \rightarrow {}^{234}\text{Th} + {}^4\text{He}, \qquad (21\text{-}13)$$

the *parent* nucleus $^{238}$U contains 92 protons (a charge of $+92e$), the *daughter* nucleus $^{234}$Th contains 90 protons (a charge of $+90e$), and the emitted alpha particle $^4$He contains 2 protons (a charge of $+2e$). We see that the total charge is $+92e$ before and after the decay; thus, charge is conserved. (The total number of protons and neutrons is also conserved: 238 before the decay and $234 + 4 = 238$ after the decay.)

Another example of charge conservation occurs when an electron $e^-$ (whose charge is $-e$) and its antiparticle, the *positron* $e^+$ (whose charge is $+e$), undergo an *annihilation process* in which they transform into two *gamma rays* (high-energy light):

$$e^- + e^+ \rightarrow \gamma + \gamma \qquad \text{(annihilation)}. \qquad (21\text{-}14)$$

In applying the conservation-of-charge principle, we must add the charges algebraically, with due regard for their signs. In the annihilation process of Eq. 21-14 then, the net charge of the system is zero both before and after the event. Charge is conserved.

In *pair production*, the converse of annihilation, charge is also conserved. In this process a gamma ray transforms into an electron and a positron:

$$\gamma \rightarrow e^- + e^+ \qquad \text{(pair production)}. \qquad (21\text{-}15)$$

Figure 21-12 shows such a pair-production event that occurred in a bubble cham-

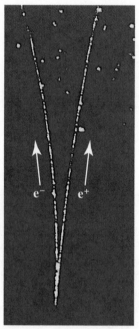

**FIG. 21-12** A photograph of trails of bubbles left in a bubble chamber by an electron and a positron. The pair of particles was produced by a gamma ray that entered the chamber directly from the bottom. Being electrically neutral, the gamma ray did not generate a telltale trail of bubbles along its path, as the electron and positron did. *(Courtesy Lawrence Berkeley Laboratory)*

ber. A gamma ray entered the chamber from the bottom and at one point transformed into an electron and a positron. Because those new particles were charged and moving, each left a trail of tiny bubbles. (The trails were curved because a magnetic field had been set up in the chamber.) The gamma ray, being electrically neutral, left no trail. Still, you can tell exactly where it underwent pair production—at the tip of the curved V, which is where the trails of the electron and positron begin.

## REVIEW & SUMMARY

**Electric Charge** The strength of a particle's electrical interaction with objects around it depends on its **electric charge,** which can be either positive or negative. Charges with the same sign repel each other, and charges with opposite signs attract each other. An object with equal amounts of the two kinds of charge is electrically neutral, whereas one with an imbalance is electrically charged.

**Conductors** are materials in which a significant number of charged particles (electrons in metals) are free to move. The charged particles in **nonconductors,** or **insulators,** are not free to move.

**The Coulomb and Ampere** The SI unit of charge is the **coulomb** (C). It is defined in terms of the unit of current, the ampere (A), as the charge passing a particular point in 1 second when there is a current of 1 ampere at that point:

$$1\text{ C} = (1\text{ A})(1\text{ s}).$$

This is based on the relation between current $i$ and the rate $dq/dt$ at which charge passes a point:

$$i = \frac{dq}{dt} \qquad \text{(electric current).} \qquad (21\text{-}3)$$

**Coulomb's Law** *Coulomb's law* describes the **electrostatic force** between small (point) electric charges $q_1$ and $q_2$ at rest

(or nearly at rest) and separated by a distance $r$:

$$F = \frac{1}{4\pi\varepsilon_0}\frac{|q_1||q_2|}{r^2} \qquad \text{(Coulomb's law).} \qquad (21\text{-}4)$$

Here $\varepsilon_0 = 8.85 \times 10^{-12}\text{ C}^2/\text{N}\cdot\text{m}^2$ is the **permittivity constant,** and $1/4\pi\varepsilon_0 = k = 8.99 \times 10^9\text{ N}\cdot\text{m}^2/\text{C}^2$.

The force of attraction or repulsion between point charges at rest acts along the line joining the two charges. If more than two charges are present, Eq. 21-4 holds for each pair of charges. The net force on each charge is then found, using the superposition principle, as the vector sum of the forces exerted on the charge by all the others.

The two shell theorems for electrostatics are

*A shell of uniform charge attracts or repels a charged particle that is outside the shell as if all the shell's charge were concentrated at its center.*

*If a charged particle is located inside a shell of uniform charge, there is no net electrostatic force on the particle from the shell.*

**The Elementary Charge** Electric charge is **quantized:** any charge can be written as $ne$, where $n$ is a positive or negative integer and $e$ is a constant of nature called the **elementary charge** ($\approx 1.602 \times 10^{-19}$ C). Electric charge is **conserved:** the net charge of any isolated system cannot change.

## QUESTIONS

**1** Figure 21-13 shows four situations in which charged particles are fixed in place on an axis. In which situations is there a point to the left of the particles where an electron will be in equilibrium?

| | |
|---|---|
| +q      −3q | −q      +3q |
| (a) | (b) |
| +3q      −q | −3q      +q |
| (c) | (d) |

**FIG. 21-13** Question 1.

**2** Figure 21-14 shows two charged particles on an axis. The charges are free to move. However, a third charged particle can be placed at a certain point such that all three particles are then in equilibrium. (a) Is that point to the left of the first two particles, to their right, or between them? (b) Should the third particle be positively or negatively charged? (c) Is the equilibrium stable or unstable?

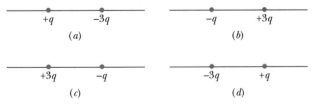

−3q        −q

**FIG. 21-14** Question 2.

**3** Figure 21-15 shows four situations in which five charged particles are evenly spaced along an axis. The charge values are indicated except for the central particle, which has the same charge in all four situations. Rank the situations according to the magnitude of the net electrostatic force on the central particle, greatest first.

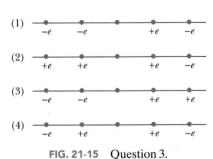

(1)  −e    −e         +e    −e

(2)  +e    +e         +e    −e

(3)  −e    −e         +e    +e

(4)  −e    +e         +e    −e

**FIG. 21-15** Question 3.

**4** Figure 21-16 shows three pairs of identical spheres that are to be touched together and then separated. The initial charges on them are indicated. Rank the pairs according to (a) the magnitude of the charge transferred during touching and (b) the charge left on the positively charged sphere, greatest first.

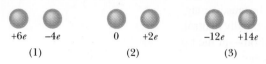

$$+6e \quad -4e \qquad 0 \quad +2e \qquad -12e \quad +14e$$
$$(1) \qquad\qquad (2) \qquad\qquad (3)$$

**FIG. 21-16** Question 4.

**5** Figure 21-17 shows three situations involving a charged particle and a uniformly charged spherical shell. The charges are given, and the radii of the shells are indicated. Rank the situations according to the magnitude of the force on the particle due to the presence of the shell, greatest first.

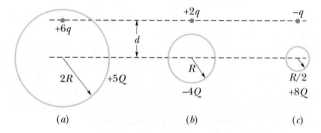

$$(a) \qquad\qquad (b) \qquad\qquad (c)$$

**FIG. 21-17** Question 5.

**6** In Fig. 21-18, a central particle of charge $-2q$ is surrounded by a square array of charged particles, separated by either distance $d$ or $d/2$ along the perimeter of the square. What are the magnitude and direction of the net electrostatic force on the central particle due to the other particles? (*Hint:* Consideration of symmetry can greatly reduce the amount of work required here.)

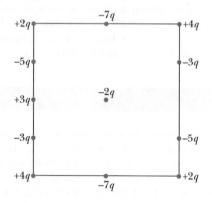

**FIG. 21-18** Question 6.

**7** In Fig. 21-19, a central particle of charge $-q$ is surrounded by two circular rings of charged particles. What are the magnitude and direction of the net electrostatic force on the central particle due to the other particles? (*Hint:* Consideration of symmetry can greatly reduce the amount of work required here.)

**8** A positively charged ball is brought close to an electrically neutral isolated conductor. The

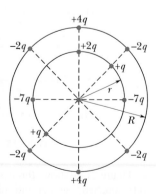

**FIG. 21-19** Question 7.

conductor is then grounded while the ball is kept close. Is the conductor charged positively, charged negatively, or neutral if (a) the ball is first taken away and then the ground connection is removed and (b) the ground connection is first removed and then the ball is taken away?

**9** Figure 21-20 shows four situations in which particles of charge $+q$ or $-q$ are fixed in place. In each situation, the particles on the $x$ axis are equidistant from the $y$ axis. First, consider the middle particle in situation 1; the middle particle experiences an electrostatic force from each of the other two particles. (a) Are the magnitudes $F$ of those forces the same or different? (b) Is the magnitude of the net force on the middle particle equal to, greater than, or less than $2F$? (c) Do the $x$ components of the two forces add or cancel? (d) Do their $y$ components add or cancel? (e) Is the direction of the net force on the middle particle that of the canceling components or the adding components? (f) What is the direction of that net force? Now consider the remaining situations: What is the direction of the net force on the middle particle in (g) situation 2, (h) situation 3, and (i) situation 4? (In each situation, consider the symmetry of the charge distribution and determine the canceling components and the adding components.)

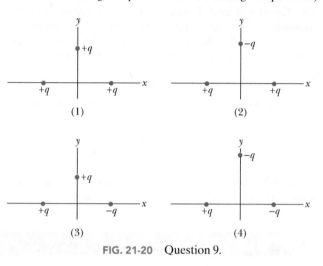

**FIG. 21-20** Question 9.

**10** Figure 21-21 shows four arrangements of charged particles. Rank the arrangements according to the magnitude of the net electrostatic force on the particle with charge $+Q$, greatest first.

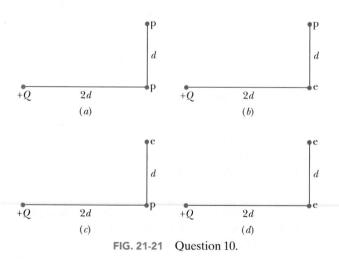

**FIG. 21-21** Question 10.

# PROBLEMS

 **GO**    Tutoring problem available (at instructor's discretion) in *WileyPLUS* and WebAssign

**SSM**    Worked-out solution available in Student Solutions Manual     **WWW**   Worked-out solution is at —

• – •••    Number of dots indicates level of problem difficulty       **ILW**    Interactive solution is at —

      Additional information available in *The Flying Circus of Physics* and at flyingcircusofphysics.com

`http://www.wiley.com/college/halliday`

### sec. 21-4 Coulomb's Law

•1 What must be the distance between point charge $q_1 = 26.0\ \mu C$ and point charge $q_2 = -47.0\ \mu C$ for the electrostatic force between them to have a magnitude of 5.70 N? **SSM**

•2 Two equally charged particles are held $3.2 \times 10^{-3}$ m apart and then released from rest. The initial acceleration of the first particle is observed to be 7.0 m/s² and that of the second to be 9.0 m/s². If the mass of the first particle is $6.3 \times 10^{-7}$ kg, what are (a) the mass of the second particle and (b) the magnitude of the charge of each particle? **ILW**

•3 A particle of charge $+3.00 \times 10^{-6}$ C is 12.0 cm distant from a second particle of charge $-1.50 \times 10^{-6}$ C. Calculate the magnitude of the electrostatic force between the particles.

•4 Identical isolated conducting spheres 1 and 2 have equal charges and are separated by a distance that is large compared with their diameters (Fig. 21-22a). The electrostatic force acting on sphere 2 due to sphere 1 is $\vec{F}$. Suppose now that a third identical sphere 3, having an insulating handle and initially neutral, is touched first to sphere 1 (Fig. 21-22b), then to sphere 2 (Fig. 21-22c), and finally removed (Fig. 21-22d). The electrostatic force that now acts on sphere 2 has magnitude $F'$. What is the ratio $F'/F$?

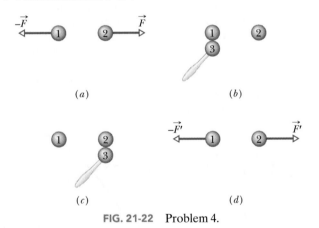

FIG. 21-22   Problem 4.

•5 Of the charge $Q$ initially on a tiny sphere, a portion $q$ is to be transferred to a second, nearby sphere. Both spheres can be treated as particles. For what value of $q/Q$ will the electrostatic force between the two spheres be maximized? **SSM ILW**

•6 In the return stroke of a typical lightning bolt, a current of $2.5 \times 10^4$ A exists for 20 μs. How much charge is transferred in this event?

••7 Two identical conducting spheres, fixed in place, attract each other with an electrostatic force of 0.108 N when their center-to-center separation is 50.0 cm. The spheres are then connected by a thin conducting wire. When the wire is removed, the spheres repel each other with an electrostatic force of 0.0360 N. Of the initial charges on the spheres, with

a positive net charge, what was (a) the negative charge on one of them and (b) the positive charge on the other? **SSM WWW**

••8 In Fig. 21-23, four particles form a square. The charges are $q_1 = q_4 = Q$ and $q_2 = q_3 = q$. (a) What is $Q/q$ if the net electrostatic force on particles 1 and 4 is zero? (b) Is there any value of $q$ that makes the net electrostatic force on each of the four particles zero? Explain. **GO**

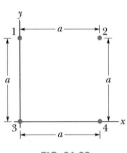

FIG. 21-23
Problems 8, 9, and 62.

••9 In Fig. 21-23, the particles have charges $q_1 = -q_2 = 100$ nC and $q_3 = -q_4 = 200$ nC, and distance $a = 5.0$ cm. What are the (a) $x$ and (b) $y$ components of the net electrostatic force on particle 3? **ILW**

••10 Three particles are fixed on an $x$ axis. Particle 1 of charge $q_1$ is at $x = -a$, and particle 2 of charge $q_2$ is at $x = +a$. If their net electrostatic force on particle 3 of charge $+Q$ is to be zero, what must be the ratio $q_1/q_2$ when particle 3 is at (a) $x = +0.500a$ and (b) $x = +1.50a$?

••11 In Fig. 21-24, three charged particles lie on an $x$ axis. Particles 1 and 2 are fixed in place. Particle 3 is free to move, but the net electrostatic force on it from particles 1 and 2 happens to be zero. If $L_{23} = L_{12}$, what is the ratio $q_1/q_2$?

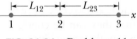

FIG. 21-24   Problems 11 and 56.

••12 Figure 21-25 shows four identical conducting spheres that are actually well separated from one another. Sphere $W$ (with an initial charge of zero) is touched to sphere $A$ and then they are separated. Next, sphere $W$ is touched to sphere $B$ (with an initial charge of $-32e$) and then they are separated. Finally, sphere $W$ is touched to sphere $C$ (with an initial charge of $+48e$), and then they are separated. The final charge on sphere $W$ is $+18e$. What was the initial charge on sphere $A$? **GO**

FIG. 21-25   Problem 12.

••13 In Fig. 21-26a, particles 1 and 2 have charge 20.0 μC each and are held at separation distance $d = 1.50$ m. (a) What is the magnitude of the electrostatic force on particle 1 due to particle 2? In Fig. 21-26b, particle 3 of charge 20.0 μC is positioned so as to complete an equilateral triangle. (b) What is the magnitude of the net electrostatic force on particle 1 due to particles 2 and 3?

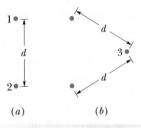

FIG. 21-26   Problem 13.

••**14** In Fig. 21-27a, particle 1 (of charge $q_1$) and particle 2 (of charge $q_2$) are fixed in place on an $x$ axis, 8.00 cm apart. Particle 3 (of charge $q_3 = +8.00 \times 10^{-19}$ C) is to be placed on the line between particles 1 and 2 so that they produce a net electrostatic force $\vec{F}_{3,net}$ on it. Figure 21-27b gives the $x$ component of that force versus the coordinate $x$ at which particle 3 is placed. The scale of the $x$ axis is set by $x_s = 8.0$ cm. What are (a) the sign of charge $q_1$ and (b) the ratio $q_2/q_1$?

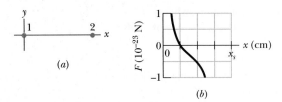

*(a)*

*(b)*

**FIG. 21-27**   Problem 14.

••**15** In Fig. 21-28, particle 1 of charge $+1.0 \,\mu$C and particle 2 of charge $-3.0 \,\mu$C are held at separation $L = 10.0$ cm on an $x$ axis. If particle 3 of unknown charge $q_3$ is to be located such that the net electrostatic force on it from particles 1 and 2 is zero, what must be the (a) $x$ and (b) $y$ coordinates of particle 3?

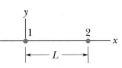

**FIG. 21-28**   Problems 15, 19, 32, 64, and 69.

••**16** In Fig. 21-29a, three positively charged particles are fixed on an $x$ axis. Particles $B$ and $C$ are so close to each other that they can be considered to be at the same distance from particle $A$. The net force on particle $A$ due to particles $B$ and $C$ is $2.014 \times 10^{-23}$ N in the negative direction of the $x$ axis. In Fig. 21-29b, particle $B$ has been moved to the opposite side of $A$ but is still at the same distance from it. The net force on $A$ is now $2.877 \times 10^{-24}$ N in the negative direction of the $x$ axis. What is the ratio $q_C/q_B$?

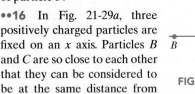

*(a)*

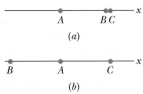

*(b)*

**FIG. 21-29**   Problem 16.

••**17** The charges and coordinates of two charged particles held fixed in an $xy$ plane are $q_1 = +3.0 \,\mu$C, $x_1 = 3.5$ cm, $y_1 = 0.50$ cm, and $q_2 = -4.0 \,\mu$C, $x_2 = -2.0$ cm, $y_2 = 1.5$ cm. Find the (a) magnitude and (b) direction of the electrostatic force on particle 2 due to particle 1. At what (c) $x$ and (d) $y$ coordinates should a third particle of charge $q_3 = +4.0 \,\mu$C be placed such that the net electrostatic force on particle 2 due to particles 1 and 3 is zero? **GO**

••**18** Two particles are fixed on an $x$ axis. Particle 1 of charge $40 \,\mu$C is located at $x = -2.0$ cm; particle 2 of charge $Q$ is located at $x = 3.0$ cm. Particle 3 of charge magnitude $20 \,\mu$C is released from rest on the $y$ axis at $y = 2.0$ cm. What is the value of $Q$ if the initial acceleration of particle 3 is in the positive direction of (a) the $x$ axis and (b) the $y$ axis?

••**19** In Fig. 21-28, particle 1 of charge $+q$ and particle 2 of charge $+4.00q$ are held at separation $L = 9.00$ cm on an $x$ axis. If particle 3 of charge $q_3$ is to be located such that the three particles remain in place when released, what must be the (a) $x$ and (b) $y$ coordinates of particle 3, and (c) the ratio $q_3/q$? **SSM WWW**

•••**20** Figure 21-30 shows an arrangement of four charged particles, with angle $\theta = 30.0°$ and distance $d = 2.00$ cm. Particle 2 has charge $q_2 = +8.00 \times 10^{-19}$ C; particles 3 and 4 have charges $q_3 = q_4 = -1.60 \times 10^{-19}$ C. (a) What is distance $D$ between the origin and particle 2 if the net electrostatic force on particle 1 due to the other particles is zero? (b) If particles 3 and 4 were moved closer to the $x$ axis but maintained their symmetry about that axis, would the required value of $D$ be greater than, less than, or the same as in part (a)? **GO**

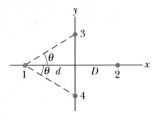

**FIG. 21-30**   Problem 20.

•••**21** In Fig. 21-31, particles 1 and 2 of charge $q_1 = q_2 = +3.20 \times 10^{-19}$ C are on a $y$ axis at distance $d = 17.0$ cm from the origin. Particle 3 of charge $q_3 = +6.40 \times 10^{-19}$ C is moved gradually along the $x$ axis from $x = 0$ to $x = +5.0$ m. At what values of $x$ will the magnitude of the electrostatic force on the third particle from the other two particles be (a) minimum and (b) maximum? What are the (c) minimum and (d) maximum magnitudes?

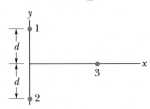

**FIG. 21-31**   Problem 21.

•••**22** Figure 21-32a shows an arrangement of three charged particles separated by distance $d$. Particles $A$ and $C$ are fixed on the $x$ axis, but particle $B$ can be moved along a circle centered on particle $A$. During the movement, a radial line between $A$ and $B$ makes an angle $\theta$ relative to the positive direction of the $x$ axis (Fig. 21-32b). The curves in Fig. 21-32c give, for two situations, the magnitude $F_{net}$ of the net electrostatic force on particle $A$ due to the other particles. That net force is given as a function of angle $\theta$ and as a multiple of a basic amount $F_0$. For example on curve 1, at $\theta = 180°$, we see that $F_{net} = 2F_0$. (a) For the situation corresponding to curve 1, what is the ratio of the charge of particle $C$ to that of particle $B$ (including sign)? (b) For the situation corresponding to curve 2, what is that ratio?

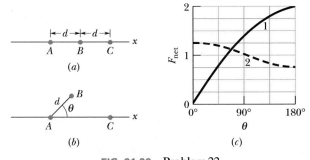

**FIG. 21-32**   Problem 22.

•••**23** A nonconducting spherical shell, with an inner radius of 4.0 cm and an outer radius of 6.0 cm, has charge spread nonuniformly through its volume between its inner and outer surfaces. The *volume charge density* $\rho$ is the charge per unit volume, with the unit coulomb per cubic meter. For this shell $\rho = b/r$, where $r$ is the distance in meters from the center of the shell and $b = 3.0 \,\mu$C/m². What is the net charge in the shell?

### sec. 21-5   Charge Is Quantized

•**24** What is the magnitude of the electrostatic force between a singly charged sodium ion (Na$^+$, of charge $+e$) and an adjacent singly charged chlorine ion (Cl$^-$, of charge $-e$) in a salt crystal if their separation is $2.82 \times 10^{-10}$ m?

•25  The magnitude of the electrostatic force between two identical ions that are separated by a distance of $5.0 \times 10^{-10}$ m is $3.7 \times 10^{-9}$ N. (a) What is the charge of each ion? (b) How many electrons are "missing" from each ion (thus giving the ion its charge imbalance? **SSM**

•26  A current of 0.300 A through your chest can send your heart into fibrillation, ruining the normal rhythm of heartbeat and disrupting the flow of blood (and thus oxygen) to your brain. If that current persists for 2.00 min, how many conduction electrons pass through your chest?

•27  How many electrons would have to be removed from a coin to leave it with a charge of $+1.0 \times 10^{-7}$ C? **ILW**

•28  Two tiny, spherical water drops, with identical charges of $-1.00 \times 10^{-16}$ C, have a center-to-center separation of 1.00 cm. (a) What is the magnitude of the electrostatic force acting between them? (b) How many excess electrons are on each drop, giving it its charge imbalance?

••29  Earth's atmosphere is constantly bombarded by *cosmic ray protons* that originate somewhere in space. If the protons all passed through the atmosphere, each square meter of Earth's surface would intercept protons at the average rate of 1500 protons per second. What would be the electric current intercepted by the total surface area of the planet? **ILW**

••30  Figure 21-33a shows charged particles 1 and 2 that are fixed in place on an x axis. Particle 1 has a charge with a magnitude of $|q_1| = 8.00e$. Particle 3 of charge $q_3 = +8.00e$ is initially on the x axis near particle 2. Then particle 3 is gradually moved in the positive direction of the x axis. As a result, the magnitude of the net electrostatic force $\vec{F}_{2,net}$ on particle 2 due to particles 1 and 3 changes. Figure 21-33b gives the x component of that net force as a function of the position x of particle 3. The scale of the x axis is set by $x_s = 0.80$ m. The plot has an asymptote of $F_{2,net} = 1.5 \times 10^{-25}$ N as $x \to \infty$. As a multiple of e and including the sign, what is the charge $q_2$ of particle 2?

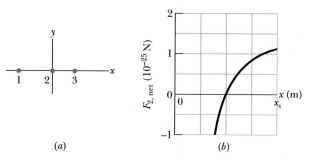

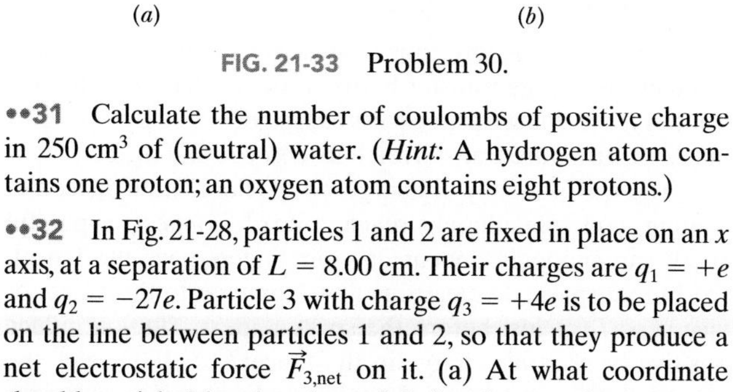

(a)                    (b)

FIG. 21-33  Problem 30.

••31  Calculate the number of coulombs of positive charge in 250 cm$^3$ of (neutral) water. (*Hint:* A hydrogen atom contains one proton; an oxygen atom contains eight protons.)

••32  In Fig. 21-28, particles 1 and 2 are fixed in place on an x axis, at a separation of $L = 8.00$ cm. Their charges are $q_1 = +e$ and $q_2 = -27e$. Particle 3 with charge $q_3 = +4e$ is to be placed on the line between particles 1 and 2, so that they produce a net electrostatic force $\vec{F}_{3,net}$ on it. (a) At what coordinate should particle 3 be placed to minimize the magnitude of that force? (b) What is that minimum magnitude?

••33  In Fig. 21-34, particles 2 and 4, of charge $-e$, are fixed in place on a y axis, at $y_2 = -10.0$ cm and $y_4 = 5.00$ cm. Particles 1 and 3, of charge $-e$, can be moved along the x axis. Particle 5, of charge $+e$, is fixed at the origin. Initially particle 1 is at $x_1 = -10.0$ cm and particle 3 is at $x_3 = 10.0$ cm. (a) To what x value must particle 1 be moved to rotate the direction of the net electric force $\vec{F}_{net}$ on particle 5 by 30° counterclockwise? (b) With particle 1 fixed at its new position, to what x value must you move particle 3 to rotate $\vec{F}_{net}$ back to its original direction?

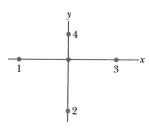

FIG. 21-34  Problem 33.

•••34  Figure 21-35 shows electrons 1 and 2 on an x axis and charged ions 3 and 4 of identical charge $-q$ and at identical angles $\theta$. Electron 2 is free to move; the other three particles are fixed in place at horizontal distances $R$ from electron 2 and are intended to hold electron 2 in place. For physically possible values of $q \le 5e$, what are the (a) smallest, (b) second smallest, and (c) third smallest values of $\theta$ for which electron 2 is held in place?

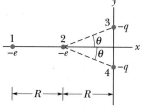

FIG. 21-35  Problem 34.

•••35  In crystals of the salt cesium chloride, cesium ions Cs$^+$ form the eight corners of a cube and a chlorine ion Cl$^-$ is at the cube's center (Fig. 21-36). The edge length of the cube is 0.40 nm. The Cs$^+$ ions are each deficient by one electron (and thus each has a charge of $+e$), and the Cl$^-$ ion has one excess electron (and thus has a charge of $-e$). (a) What is the magnitude of the net electrostatic force exerted on the Cl$^-$ ion by the eight Cs$^+$ ions at the corners of the cube? (b) If one of the Cs$^+$ ions is missing, the crystal is said to have a *defect*; what is the magnitude of the net electrostatic force exerted on the Cl$^-$ ion by the seven remaining Cs$^+$ ions? **SSM**

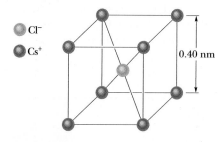

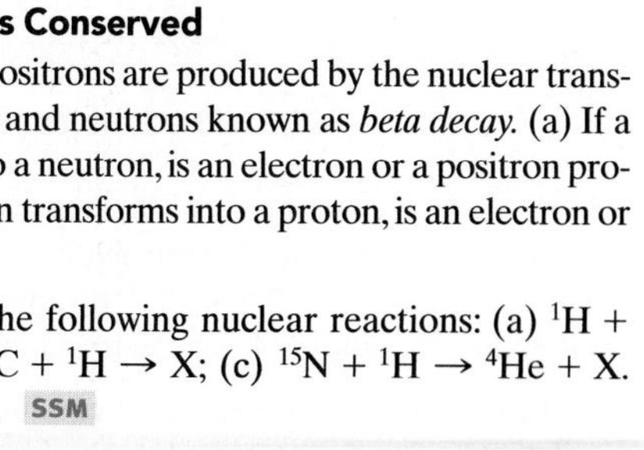

FIG. 21-36
Problem 35.

### sec. 21-6  Charge Is Conserved

•36  Electrons and positrons are produced by the nuclear transformations of protons and neutrons known as *beta decay*. (a) If a proton transforms into a neutron, is an electron or a positron produced? (b) If a neutron transforms into a proton, is an electron or a positron produced?

•37  Identify X in the following nuclear reactions: (a) $^1$H + $^9$Be → X + n; (b) $^{12}$C + $^1$H → X; (c) $^{15}$N + $^1$H → $^4$He + X. Appendix F will help. **SSM**

### Additional Problems

38  In Fig. 21-37, four particles are fixed along an x axis, sepa-

FIG. 21-37  Problem 38.

rated by distances $d = 2.00$ cm. The charges are $q_1 = +2e$, $q_2 = -e$, $q_3 = +e$, and $q_4 = +4e$, with $e = 1.60 \times 10^{-19}$ C. In unit-vector notation, what is the net electrostatic force on (a) particle 1 and (b) particle 2 due to the other particles?

**39** In Fig. 21-38, particle 1 of charge $+4e$ is above a floor by distance $d_1 = 2.00$ mm and particle 2 of charge $+6e$ is on the floor, at distance $d_2 = 6.00$ mm horizontally from particle 1. What is the $x$ component of the electrostatic force on particle 2 due to particle 1? **SSM**

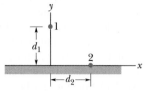

**FIG. 21-38** Problem 39.

**40** A particle of charge $Q$ is fixed at the origin of an $xy$ coordinate system. At $t = 0$ a particle ($m = 0.800$ g, $q = 4.00$ $\mu$C) is located on the $x$ axis at $x = 20.0$ cm, moving with a speed of 50.0 m/s in the positive $y$ direction. For what value of $Q$ will the moving particle execute circular motion? (Neglect the gravitational force on the particle.)

**41** A charged nonconducting rod, with a length of 2.00 m and a cross-sectional area of 4.00 cm², lies along the positive side of an $x$ axis with one end at the origin. The *volume charge density* $\rho$ is charge per unit volume in coulombs per cubic meter. How many excess electrons are on the rod if $\rho$ is (a) uniform, with a value of $-4.00$ $\mu$C/m³, and (b) nonuniform, with a value given by $\rho = bx^2$, where $b = -2.00$ $\mu$C/m⁵?

**42** A charge of 6.0 $\mu$C is to be split into two parts that are then separated by 3.0 mm. What is the maximum possible magnitude of the electrostatic force between those two parts?

**43** How many megacoulombs of positive charge are in 1.00 mol of neutral molecular-hydrogen gas ($H_2$)?

**44** Figure 21-39 shows a long, nonconducting, massless rod of length $L$, pivoted at its center and balanced with a block of weight $W$ at a distance $x$ from the left end. At the left and right ends of the rod are attached small conducting spheres with positive charges $q$ and $2q$, respectively. A distance $h$ directly beneath each of these spheres is a fixed sphere with positive charge $Q$. (a) Find the distance $x$ when the rod is horizontal and balanced. (b) What value should $h$ have so that the rod exerts no vertical force on the bearing when the rod is horizontal and balanced?

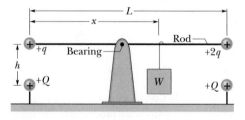

**FIG. 21-39** Problem 44.

**45** A neutron consists of one "up" quark of charge $+2e/3$ and two "down" quarks each having charge $-e/3$. If we assume that the down quarks are $2.6 \times 10^{-15}$ m apart inside the neutron, what is the magnitude of the electrostatic force between them?

**46** In Fig. 21-40, three identical conducting spheres form an equilateral triangle of side length $d = 20.0$ cm. The sphere radii are much smaller than $d$, and the sphere charges are $q_A = -2.00$ nC, $q_B = -4.00$ nC, and $q_C = +8.00$ nC. (a) What is the

magnitude of the electrostatic force between spheres $A$ and $C$? The following steps are then taken: $A$ and $B$ are connected by a thin wire and then disconnected; $B$ is grounded by the wire, and the wire is then removed; $B$ and $C$ are connected by the wire and then disconnected. What now are the magnitudes of the electrostatic force (b) between spheres $A$ and $C$ and (c) between spheres $B$ and $C$?

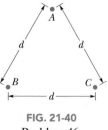

**FIG. 21-40** Problem 46.

**47** What would be the magnitude of the electrostatic force between two 1.00 C point charges separated by a distance of (a) 1.00 m and (b) 1.00 km if such point charges existed (they do not) and this configuration could be set up?

**48** In Fig. 21-41, three identical conducting spheres initially have the following charges: sphere $A$, $4Q$; sphere $B$, $-6Q$; and sphere $C$, 0. Spheres $A$ and $B$ are fixed in place, with a center-to-center separation that is much larger than the spheres. Two experiments are conducted. In experiment 1, sphere $C$ is touched to sphere $A$ and then (separately) to sphere $B$, and then it is removed. In experiment 2, starting with the same initial states, the procedure is reversed: Sphere $C$ is touched to sphere $B$ and then (separately) to sphere $A$, and then it is removed. What is the ratio of the electrostatic force between $A$ and $B$ at the end of experiment 2 to that at the end of experiment 1?

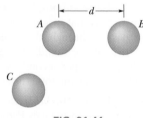

**FIG. 21-41** Problems 48 and 67.

**49** We know that the negative charge on the electron and the positive charge on the proton are equal. Suppose, however, that these magnitudes differ from each other by 0.00010%. With what force would two copper coins, placed 1.0 m apart, repel each other? Assume that each coin contains $3 \times 10^{22}$ copper atoms. (*Hint:* A neutral copper atom contains 29 protons and 29 electrons.) What do you conclude?

**50** How far apart must two protons be if the magnitude of the electrostatic force acting on either one due to the other is equal to the magnitude of the gravitational force on a proton at Earth's surface? **SSM**

**51** Of the charge $Q$ on a tiny sphere, a fraction $\alpha$ is to be transferred to a second, nearby sphere. The spheres can be treated as particles. (a) What value of $\alpha$ maximizes the magnitude $F$ of the electrostatic force between the two spheres? What are the (b) smaller and (c) larger values of $\alpha$ that put $F$ at half the maximum magnitude?

**52** If a cat repeatedly rubs against your cotton slacks on a dry day, the charge transfer between the cat hair and the cotton can leave you with an excess charge of $-2.00$ $\mu$C. (a) How many electrons are transferred between you and the cat?

You will gradually discharge via the floor, but if instead of waiting, you immediately reach toward a faucet, a painful spark can suddenly appear as your fingers near the faucet. (b) In that spark, do electrons flow from you to the faucet or vice versa? (c) Just before the spark appears, do you induce positive or negative charge in the faucet? (d) If, instead, the cat reaches a paw toward the faucet, which way do electrons flow in the resulting spark? (e) If you stroke a cat with a bare hand on a dry day, you should take care not to bring your fingers near the cat's nose or you will hurt it with a spark. Considering that cat hair is an insulator, explain how the spark can appear.

**53** (a) What equal positive charges would have to be placed on Earth and on the Moon to neutralize their gravitational attraction? (b) Why don't you need to know the lunar distance to solve this problem? (c) How many kilograms of hydrogen ions (that is, protons) would be needed to provide the positive charge calculated in (a)?

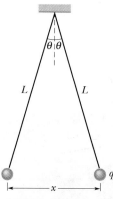

**54** In Fig. 21-42, two tiny conducting balls of identical mass $m$ and identical charge $q$ hang from non-conducting threads of length $L$. Assume that $\theta$ is so small that tan $\theta$ can be replaced by its approximate equal, sin $\theta$. (a) Show that

**FIG. 21-42** Problems 54 and 55.

$$x = \left(\frac{q^2 L}{2\pi\varepsilon_0 mg}\right)^{1/3}$$

gives the equilibrium separation $x$ of the balls. (b) If $L = 120$ cm, $m = 10$ g, and $x = 5.0$ cm, what is $|q|$?

**55** (a) Explain what happens to the balls of Problem 54 if one of them is discharged (loses its charge $q$ to, say, the ground). (b) Find the new equilibrium separation $x$, using the given values of $L$ and $m$ and the computed value of $|q|$.

**56** In Fig. 21-24, particles 1 and 2 are fixed in place, but particle 3 is free to move. If the net electrostatic force on particle 3 due to particles 1 and 2 is zero and $L_{23} = 2.00L_{12}$, what is the ratio $q_1/q_2$?

**57** What is the total charge in coulombs of 75.0 kg of electrons?

**58** In Fig. 21-43, six charged particles surround particle 7 at radial distances of either $d = 1.0$ cm or $2d$, as drawn. The charges are $q_1 = +2e$, $q_2 = +4e$, $q_3 = +e$, $q_4 = +4e$, $q_5 = +2e$, $q_6 = +8e$, $q_7 = +6e$, with $e = 1.60 \times 10^{-19}$ C. What is the magnitude of the net electrostatic force on particle 7?

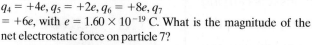

**FIG. 21-43** Problem 58.

**59** Three charged particles form a triangle: particle 1 with charge $Q_1 = 80.0$ nC is at $xy$ coordinates (0, 3.00 mm), particle 2 with charge $Q_2$ is at (0, −3.00 mm), and particle 3 with charge $q = 18.0$ nC is at (4.00 mm, 0). In unit-vector notation, what is the electrostatic force on particle 3 due to the other two particles if $Q_2$ is equal to (a) 80.0 nC and (b) −80.0 nC?

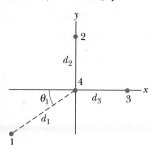

**60** In Fig. 21-44, what are the (a) magnitude and (b) direction of the net electrostatic force on

**FIG. 21-44** Problem 60.

particle 4 due to the other three particles? All four particles are fixed in the $xy$ plane, and $q_1 = -3.20 \times 10^{-19}$ C, $q_2 = +3.20 \times 10^{-19}$ C, $q_3 = +6.40 \times 10^{-19}$ C, $q_4 = +3.20 \times 10^{-19}$ C, $\theta_1 = 35.0°$, $d_1 = 3.00$ cm, and $d_2 = d_3 = 2.00$ cm. **SSM**

**61** Two point charges of 30 nC and −40 nC are held fixed on an $x$ axis, at the origin and at $x = 72$ cm, respectively. A particle with a charge of 42 $\mu$C is released from rest at $x = 28$ cm. If the initial acceleration of the particle has a magnitude of 100 km/s², what is the particle's mass?

**62** In Fig. 21-23, four particles form a square. The charges are $q_1 = +Q$, $q_2 = q_3 = q$, and $q_4 = -2.00Q$. What is $q/Q$ if the net electrostatic force on particle 1 is zero?

**63** Point charges of +6.0 $\mu$C and −4.0 $\mu$C are placed on an $x$ axis, at $x = 8.0$ m and $x = 16$ m, respectively. What charge must be placed at $x = 24$ m so that any charge placed at the origin would experience no electrostatic force? **GO**

**64** In Fig. 21-28, particle 1 of charge −80.0 $\mu$C and particle 2 of charge +40.0 $\mu$C are held at separation $L = 20.0$ cm on an $x$ axis. In unit-vector notation, what is the net electrostatic force on particle 3, of charge $q_3 = 20.0$ $\mu$C, if particle 3 is placed at (a) $x = 40.0$ cm and (b) $x = 80.0$ cm? What should be the (c) $x$ and (d) $y$ coordinates of particle 3 if the net electrostatic force on it due to particles 1 and 2 is zero?

**65** In the radioactive decay of Eq. 21-13, a $^{238}$U nucleus transforms to $^{234}$Th and an ejected $^4$He. (These are nuclei, not atoms, and thus electrons are not involved.) When the separation between $^{234}$Th and $^4$He is $9.0 \times 10^{-15}$ m, what are the magnitudes of (a) the electrostatic force between them and (b) the acceleration of the $^4$He particle?

**66** Two small, positively charged spheres have a combined charge of $5.0 \times 10^{-5}$ C. If each sphere is repelled from the other by an electrostatic force of 1.0 N when the spheres are 2.0 m apart, what is the charge on the sphere with the smaller charge?

**67** The initial charges on the three identical metal spheres in Fig. 21-41 are the following: sphere $A$, $Q$; sphere $B$, $-Q/4$; and sphere $C$, $Q/2$, where $Q = 2.00 \times 10^{-14}$ C. Spheres $A$ and $B$ are fixed in place, with a center-to-center separation of $d = 1.20$ m, which is much larger than the spheres. Sphere $C$ is touched first to sphere $A$ and then to sphere $B$ and is then removed. What then is the magnitude of the electrostatic force between spheres $A$ and $B$?

**68** An electron is in a vacuum near Earth's surface and located at $y = 0$ on a vertical $y$ axis. At what value of $y$ should a second electron be placed such that its electrostatic force on the first electron balances the gravitational force on the first electron?

**69** In Fig. 21-28, particle 1 of charge −5.00$q$ and particle 2 of charge +2.00$q$ are held at separation $L$ on an $x$ axis. If particle 3 of unknown charge $q_3$ is to be located such that the net electrostatic force on it from particles 1 and 2 is zero, what must be the (a) $x$ and (b) $y$ coordinates of particle 3? **SSM**

**70** Two engineering students, John with a mass of 90 kg and Mary with a mass of 45 kg, are 30 m apart. Suppose each has a 0.01% imbalance in the amount of positive and negative charge, one student being positive and the other negative. Find the order of magnitude of the electrostatic force of attraction between them by replacing each student with a sphere of water having the same mass as the student.

Steve Hopkin/Taxi/Getty Images, Inc.

*Reproduction in many flower species depends on insects carrying pollen from one flower to another. Honeybees commonly provide this service as they visit flowers to collect nectar. However, they do not merely brush against the flower, collecting the pollen as you would collect chalk from a dusty chalkboard by brushing against it. Instead, the pollen jumps from a flower to the bee, clings to the bee during the flight to a second flower, and then jumps to the second flower.*

*What causes the pollen to jump, first to the bee and then away from it?*

The answer is in this chapter.

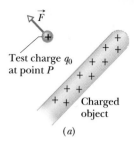

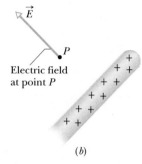

FIG. 22-1 (a) A positive test charge $q_0$ placed at point $P$ near a charged object. An electrostatic force $\vec{F}$ acts on the test charge. (b) The electric field $\vec{E}$ at point $P$ produced by the charged object.

## 22-1 | WHAT IS PHYSICS?

The physics of the preceding chapter tells us how to find the electric force on a particle 1 of charge $+q_1$ when the particle is placed near a particle 2 of charge $+q_2$. A nagging question remains: How does particle 1 "know" of the presence of particle 2? That is, since the particles do not touch, how can particle 2 push on particle 1—how can there be such an *action at a distance* with no visible connection between the particles?

One purpose of physics is to record observations about our world, such as the magnitude and direction of the push on particle 1. Another purpose is to provide a deeper explanation of what is recorded. One purpose of this chapter is to provide such a deeper explanation to our nagging questions about electric force at a distance. We can answer those questions by saying that particle 2 sets up an **electric field** in the space surrounding itself. If we place particle 1 at any given point in that space, the particle "knows" of the presence of particle 2 because it is affected by the electric field that particle 2 has already set up at that point. Thus, particle 2 pushes on particle 1 not by touching it but by means of the electric field produced by particle 2.

Our goal in this chapter is to define electric field and discuss how to calculate it for various arrangements of charged particles.

## 22-2 | The Electric Field

The temperature at every point in a room has a definite value. You can measure the temperature at any given point or combination of points by putting a thermometer there. We call the resulting distribution of temperatures a *temperature field*. In much the same way, you can imagine a *pressure field* in the atmosphere; it consists of the distribution of air pressure values, one for each point in the atmosphere. These two examples are of *scalar fields* because temperature and air pressure are scalar quantities.

The electric field is a *vector field;* it consists of a distribution of *vectors,* one for each point in the region around a charged object, such as a charged rod. In principle, we can define the electric field at some point near the charged object, such as point $P$ in Fig. 22-1a, as follows: We first place a *positive* charge $q_0$, called a *test charge,* at the point. We then measure the electrostatic force $\vec{F}$ that acts on the test charge. Finally, we define the electric field $\vec{E}$ at point $P$ due to the charged object as

$$\vec{E} = \frac{\vec{F}}{q_0} \quad \text{(electric field)}. \qquad (22\text{-}1)$$

Thus, the magnitude of the electric field $\vec{E}$ at point $P$ is $E = F/q_0$, and the direction of $\vec{E}$ is that of the force $\vec{F}$ that acts on the *positive* test charge. As shown in Fig. 22-1b, we represent the electric field at $P$ with a vector whose tail is at $P$. To define the electric field within some region, we must similarly define it at all points in the region.

The SI unit for the electric field is the newton per coulomb (N/C). Table 22-1 shows the electric fields that occur in a few physical situations.

Although we use a positive test charge to define the electric field of a charged object, that field exists independently of the test charge. The field at point $P$ in Figure 22-1b existed both before and after the test charge of Fig. 22-1a was put there. (We assume that in our defining procedure, the presence of the test charge does not affect the charge distribution on the charged object, and thus does not alter the electric field we are defining.)

To examine the role of an electric field in the interaction between charged objects, we have two tasks: (1) calculating the electric field produced by a given

**TABLE 22-1**

**Some Electric Fields**

| Field Location or Situation | Value (N/C) |
|---|---|
| At the surface of a uranium nucleus | $3 \times 10^{21}$ |
| Within a hydrogen atom, at a radius of $5.29 \times 10^{-11}$ m | $5 \times 10^{11}$ |
| Electric breakdown occurs in air | $3 \times 10^{6}$ |
| Near the charged drum of a photocopier | $10^{5}$ |
| Near a charged comb | $10^{3}$ |
| In the lower atmosphere | $10^{2}$ |
| Inside the copper wire of household circuits | $10^{-2}$ |

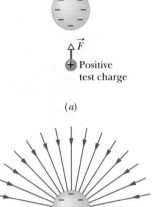

(a)

Electric
field lines

(b)

FIG. 22-2 (a) The electrostatic force $\vec{F}$ acting on a positive test charge near a sphere of uniform negative charge. (b) The electric field vector $\vec{E}$ at the location of the test charge, and the electric field lines in the space near the sphere. The field lines extend *toward* the negatively charged sphere. (They originate on distant positive charges.)

distribution of charge and (2) calculating the force that a given field exerts on a charge placed in it. We perform the first task in Sections 22-4 through 22-7 for several charge distributions. We perform the second task in Sections 22-8 and 22-9 by considering a point charge and a pair of point charges in an electric field. First, however, we discuss a way to visualize electric fields.

## 22-3 | Electric Field Lines

Michael Faraday, who introduced the idea of electric fields in the 19th century, thought of the space around a charged body as filled with *lines of force*. Although we no longer attach much reality to these lines, now usually called **electric field lines,** they still provide a nice way to visualize patterns in electric fields.

The relation between the field lines and electric field vectors is this: (1) At any point, the direction of a straight field line or the direction of the tangent to a curved field line gives the direction of $\vec{E}$ at that point, and (2) the field lines are drawn so that the number of lines per unit area, measured in a plane that is perpendicular to the lines, is proportional to the *magnitude* of $\vec{E}$. Thus, $E$ is large where field lines are close together and small where they are far apart.

Figure 22-2a shows a sphere of uniform negative charge. If we place a *positive* test charge anywhere near the sphere, an electrostatic force pointing *toward* the center of the sphere will act on the test charge as shown. In other words, the electric field vectors at all points near the sphere are directed radially toward the sphere. This pattern of vectors is neatly displayed by the field lines in Fig. 22-2b, which point in the same directions as the force and field vectors. Moreover, the spreading of the field lines with distance from the sphere tells us that the magnitude of the electric field decreases with distance from the sphere.

If the sphere of Fig. 22-2 were of uniform *positive* charge, the electric field vectors at all points near the sphere would be directed radially *away from* the sphere. Thus, the electric field lines would also extend radially away from the sphere. We then have the following rule:

> Electric field lines extend away from positive charge (where they originate) and toward negative charge (where they terminate).

Figure 22-3a shows part of an infinitely large, nonconducting *sheet* (or plane) with a uniform distribution of positive charge on one side. If we were to place a positive test charge at any point near the sheet of Fig. 22-3a, the net electrostatic force acting on the test charge would be perpendicular to the sheet, because forces acting in all other directions would cancel one another as a result of the symmetry. Moreover, the net force on the test charge would point away from the sheet as shown. Thus, the electric field vector at any point in the space on either side of the sheet is also perpendicular to the sheet and directed away from

FIG. 22-3 (a) The electrostatic force $\vec{F}$ on a positive test charge near a very large, nonconducting sheet with uniformly distributed positive charge on one side. (b) The electric field vector $\vec{E}$ at the location of the test charge, and the electric field lines in the space near the sheet. The field lines extend *away from* the positively charged sheet. (c) Side view of (b).

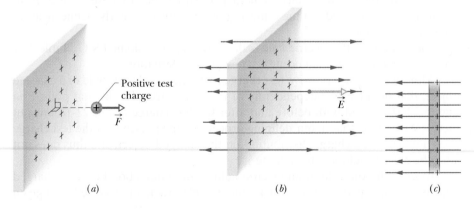

(a)                    (b)                    (c)

it (Figs. 22-3b and c). Because the charge is uniformly distributed along the sheet, all the field vectors have the same magnitude. Such an electric field, with the same magnitude and direction at every point, is a *uniform electric field*.

Of course, no real nonconducting sheet (such as a flat expanse of plastic) is infinitely large, but if we consider a region that is near the middle of a real sheet and not near its edges, the field lines through that region are arranged as in Figs. 22-3b and c.

Figure 22-4 shows the field lines for two equal positive charges. Figure 22-5 shows the pattern for two charges that are equal in magnitude but of opposite sign, a configuration that we call an **electric dipole.** Although we do not often use field lines quantitatively, they are very useful to visualize what is going on. Can you not almost "see" the charges being pushed apart in Fig. 22-4 and pulled together in Fig. 22-5?

## 22-4 | The Electric Field Due to a Point Charge

To find the electric field due to a point charge $q$ (or charged particle) at any point a distance $r$ from the point charge, we put a positive test charge $q_0$ at that point. From Coulomb's law (Eq. 21-1), the electrostatic force acting on $q_0$ is

$$\vec{F} = \frac{1}{4\pi\varepsilon_0} \frac{qq_0}{r^2} \hat{\mathbf{r}}. \tag{22-2}$$

The direction of $\vec{F}$ is directly away from the point charge if $q$ is positive, and directly toward the point charge if $q$ is negative. The electric field vector is, from Eq. 22-1,

$$\vec{E} = \frac{\vec{F}}{q_0} = \frac{1}{4\pi\varepsilon_0} \frac{q}{r^2} \hat{\mathbf{r}} \qquad \text{(point charge).} \tag{22-3}$$

The direction of $\vec{E}$ is the same as that of the force on the positive test charge: directly away from the point charge if $q$ is positive, and toward it if $q$ is negative.

Because there is nothing special about the point we chose for $q_0$, Eq. 22-3 gives the field at every point around the point charge $q$. The field for a positive point charge is shown in Fig. 22-6 in vector form (not as field lines).

We can quickly find the net, or resultant, electric field due to more than one point charge. If we place a positive test charge $q_0$ near $n$ point charges $q_1, q_2, \ldots, q_n$, then, from Eq. 21-7, the net force $\vec{F}_0$ from the $n$ point charges acting on the test charge is

$$\vec{F}_0 = \vec{F}_{01} + \vec{F}_{02} + \cdots + \vec{F}_{0n}.$$

Therefore, from Eq. 22-1, the net electric field at the position of the test charge is

$$\begin{aligned} \vec{E} &= \frac{\vec{F}_0}{q_0} = \frac{\vec{F}_{01}}{q_0} + \frac{\vec{F}_{02}}{q_0} + \cdots + \frac{\vec{F}_{0n}}{q_0} \\ &= \vec{E}_1 + \vec{E}_2 + \cdots + \vec{E}_n. \end{aligned} \tag{22-4}$$

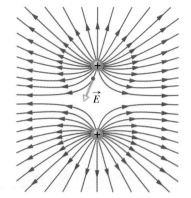

**FIG. 22-4** Field lines for two equal positive point charges. The charges repel each other. (The lines terminate on distant negative charges.) To "see" the actual three-dimensional pattern of field lines, mentally rotate the pattern shown here about an axis passing through both charges in the plane of the page. The three-dimensional pattern and the electric field it represents are said to have *rotational symmetry* about that axis. The electric field vector at one point is shown; note that it is tangent to the field line through that point.

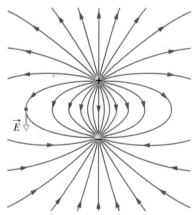

**FIG. 22-5** Field lines for a positive point charge and a nearby negative point charge that are equal in magnitude. The charges attract each other. The pattern of field lines and the electric field it represents have rotational symmetry about an axis passing through both charges in the plane of the page. The electric field vector at one point is shown; the vector is tangent to the field line through the point.

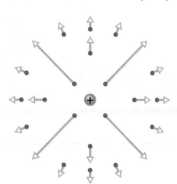

**FIG. 22-6** The electric field vectors at various points around a positive point charge.

Here $\vec{E}_i$ is the electric field that would be set up by point charge $i$ acting alone. Equation 22-4 shows us that the principle of superposition applies to electric fields as well as to electrostatic forces.

✓ **CHECKPOINT 1**    The figure here shows a proton p and an electron e on an $x$ axis. What is the direction of the electric field due to the electron at (a) point $S$ and (b) point $R$? What is the direction of the net electric field at (c) point $R$ and (d) point $S$?

---

**Sample Problem    22-1**

Figure 22-7a shows three particles with charges $q_1 = +2Q$, $q_2 = -2Q$, and $q_3 = -4Q$, each a distance $d$ from the origin. What net electric field $\vec{E}$ is produced at the origin?

**KEY IDEA**    Charges $q_1$, $q_2$, and $q_3$ produce electric field vectors $\vec{E}_1$, $\vec{E}_2$, and $\vec{E}_3$, respectively, at the origin, and the net electric field is the vector sum $\vec{E} = \vec{E}_1 + \vec{E}_2 + \vec{E}_3$. To find this sum, we first must find the magnitudes and orientations of the three field vectors.

**Magnitudes and directions:** To find the magnitude of $\vec{E}_1$, which is due to $q_1$, we use Eq. 22-3, substituting $d$ for $r$ and $2Q$ for $q$ and obtaining

$$E_1 = \frac{1}{4\pi\varepsilon_0}\frac{2Q}{d^2}.$$

Similarly, we find the magnitudes of the fields $\vec{E}_2$ and $\vec{E}_3$

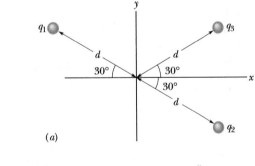

(a)

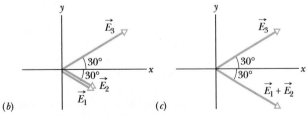

(b)                                    (c)

**FIG. 22-7**    (a) Three particles with charges $q_1$, $q_2$, and $q_3$ are at the same distance $d$ from the origin. (b) The electric field vectors $\vec{E}_1$, $\vec{E}_2$, and $\vec{E}_3$, at the origin due to the three particles. (c) The electric field vector $\vec{E}_3$ and the vector sum $\vec{E}_1 + \vec{E}_2$ at the origin.

to be

$$E_2 = \frac{1}{4\pi\varepsilon_0}\frac{2Q}{d^2} \quad \text{and} \quad E_3 = \frac{1}{4\pi\varepsilon_0}\frac{4Q}{d^2}.$$

We next must find the orientations of the three electric field vectors at the origin. Because $q_1$ is a positive charge, the field vector it produces points directly *away* from it, and because $q_2$ and $q_3$ are both negative, the field vectors they produce point directly *toward* each of them. Thus, the three electric fields produced at the origin by the three charged particles are oriented as in Fig. 22-7b. (*Caution:* Note that we have placed the tails of the vectors at the point where the fields are to be evaluated; doing so decreases the chance of error.)

**Adding the fields:** We can now add the fields vectorially as outlined for forces in Sample Problem 21-1c. However, here we can use symmetry to simplify the procedure. From Fig. 22-7b, we see that $\vec{E}_1$ and $\vec{E}_2$ have the same direction. Hence, their vector sum has that direction and has the magnitude

$$E_1 + E_2 = \frac{1}{4\pi\varepsilon_0}\frac{2Q}{d^2} + \frac{1}{4\pi\varepsilon_0}\frac{2Q}{d^2}$$

$$= \frac{1}{4\pi\varepsilon_0}\frac{4Q}{d^2},$$

which happens to equal the magnitude of field $\vec{E}_3$.

We must now combine two vectors, $\vec{E}_3$ and the vector sum $\vec{E}_1 + \vec{E}_2$, that have the same magnitude and that are oriented symmetrically about the $x$ axis, as shown in Fig. 22-7c. From the symmetry of Fig. 22-7c, we realize that the equal $y$ components of our two vectors cancel and the equal $x$ components add. Thus, the net electric field $\vec{E}$ at the origin is in the positive direction of the $x$ axis and has the magnitude

$$E = 2E_{3x} = 2E_3 \cos 30°$$

$$= (2)\frac{1}{4\pi\varepsilon_0}\frac{4Q}{d^2}(0.866) = \frac{6.93Q}{4\pi\varepsilon_0 d^2}. \quad \text{(Answer)}$$

## 22-5 | The Electric Field Due to an Electric Dipole

Figure 22-8a shows two charged particles of magnitude $q$ but of opposite sign, separated by a distance $d$. As was noted in connection with Fig. 22-5, we call this configuration an *electric dipole*. Let us find the electric field due to the dipole of Fig. 22-8a at a point $P$, a distance $z$ from the midpoint of the dipole and on the axis through the particles, which is called the *dipole axis*.

From symmetry, the electric field $\vec{E}$ at point $P$—and also the fields $\vec{E}_{(+)}$ and $\vec{E}_{(-)}$ due to the separate charges that make up the dipole—must lie along the dipole axis, which we have taken to be a $z$ axis. Applying the superposition principle for electric fields, we find that the magnitude $E$ of the electric field at $P$ is

$$
\begin{aligned}
E &= E_{(+)} - E_{(-)} \\
&= \frac{1}{4\pi\varepsilon_0} \frac{q}{r_{(+)}^2} - \frac{1}{4\pi\varepsilon_0} \frac{q}{r_{(-)}^2} \\
&= \frac{q}{4\pi\varepsilon_0(z - \frac{1}{2}d)^2} - \frac{q}{4\pi\varepsilon_0(z + \frac{1}{2}d)^2}.
\end{aligned}
\tag{22-5}
$$

After a little algebra, we can rewrite this equation as

$$
E = \frac{q}{4\pi\varepsilon_0 z^2}\left( \frac{1}{\left(1 - \dfrac{d}{2z}\right)^2} - \frac{1}{\left(1 + \dfrac{d}{2z}\right)^2} \right).
\tag{22-6}
$$

After forming a common denominator and multiplying its terms, we come to

$$
E = \frac{q}{4\pi\varepsilon_0 z^2} \frac{2d/z}{\left(1 - \left(\dfrac{d}{2z}\right)^2\right)^2} = \frac{q}{2\pi\varepsilon_0 z^3} \frac{d}{\left(1 - \left(\dfrac{d}{2z}\right)^2\right)^2}.
\tag{22-7}
$$

We are usually interested in the electrical effect of a dipole only at distances that are large compared with the dimensions of the dipole—that is, at distances such that $z \gg d$. At such large distances, we have $d/2z \ll 1$ in Eq. 22-7. Thus, in our approximation, we can neglect the $d/2z$ term in the denominator, which leaves us with

$$
E = \frac{1}{2\pi\varepsilon_0} \frac{qd}{z^3}.
\tag{22-8}
$$

The product $qd$, which involves the two intrinsic properties $q$ and $d$ of the dipole, is the magnitude $p$ of a vector quantity known as the **electric dipole moment** $\vec{p}$ of the dipole. (The unit of $\vec{p}$ is the coulomb-meter.) Thus, we can write Eq. 22-8 as

$$
E = \frac{1}{2\pi\varepsilon_0} \frac{p}{z^3} \qquad \text{(electric dipole)}.
\tag{22-9}
$$

The direction of $\vec{p}$ is taken to be from the negative to the positive end of the dipole, as indicated in Fig. 22-8b. We can use the direction of $\vec{p}$ to specify the orientation of a dipole.

Equation 22-9 shows that, if we measure the electric field of a dipole only at distant points, we can never find $q$ and $d$ separately; instead, we can find only their product. The field at distant points would be unchanged if, for example, $q$ were doubled and $d$ simultaneously halved.

Although Eq. 22-9 holds only for distant points along the dipole axis, it turns out that $E$ for a dipole varies as $1/r^3$ for *all* distant points, regardless of whether they lie on the dipole axis; here $r$ is the distance between the point in question and the dipole center.

Inspection of Fig. 22-8 and of the field lines in Fig. 22-5 shows that the direction of $\vec{E}$ for distant points on the dipole axis is always the direction of the dipole

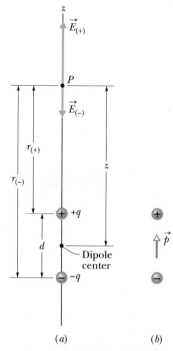

**FIG. 22-8**  (a) An electric dipole. The electric field vectors $\vec{E}_{(+)}$ and $\vec{E}_{(-)}$ at point $P$ on the dipole axis result from the dipole's two charges. Point $P$ is at distances $r_{(+)}$ and $r_{(-)}$ from the individual charges that make up the dipole. (b) The dipole moment $\vec{p}$ of the dipole points from the negative charge to the positive charge.

moment vector $\vec{p}$. This is true whether point $P$ in Fig. 22-8*a* is on the upper or the lower part of the dipole axis.

Inspection of Eq. 22-9 shows that if you double the distance of a point from a dipole, the electric field at the point drops by a factor of 8. If you double the distance from a single point charge, however (see Eq. 22-3), the electric field drops only by a factor of 4. Thus the electric field of a dipole decreases more rapidly with distance than does the electric field of a single charge. The physical reason for this rapid decrease in electric field for a dipole is that from distant points a dipole looks like two equal but opposite charges that almost—but not quite—coincide. Thus, their electric fields at distant points almost—but not quite—cancel each other.

---

**Sample Problem** | **22-2**

Sprites (Fig. 22-9*a*) are huge flashes that occur far above a large thunderstorm. They were seen for decades by pilots flying at night, but they were so brief and dim that most pilots figured they were just illusions. Then in the 1990s sprites were captured on video. They are still not well understood but are believed to be produced when especially powerful lightning occurs between the ground and storm clouds, particularly when the lightning transfers a huge amount of negative charge $-q$ from the ground to the base of the clouds (Fig. 22-9*b*).

Just after such a transfer, the ground has a complicated distribution of positive charge. However, we can model the electric field due to the charges in the clouds and the ground by assuming a vertical electric dipole that has charge $-q$ at cloud height $h$ and charge $+q$ at below-ground depth $h$ (Fig. 22-9*c*). If $q = 200$ C and $h = 6.0$ km, what is the magnitude of the dipole's electric field at altitude $z_1 = 30$ km somewhat above the clouds and altitude $z_2 = 60$ km somewhat above the stratosphere?

**KEY IDEA** We can approximate the magnitude $E$ of an electric dipole's electric field on the dipole axis with Eq. 22-8.

**Calculations:** We write that equation as

$$E = \frac{1}{2\pi\varepsilon_0}\frac{q(2h)}{z^3},$$

where $2h$ is the separation between $-q$ and $+q$ in Fig. 22-9*c*. For the electric field at altitude $z_1 = 30$ km, we find

$$E = \frac{1}{2\pi\varepsilon_0}\frac{(200\text{ C})(2)(6.0 \times 10^3\text{ m})}{(30 \times 10^3\text{ m})^3}$$

$$= 1.6 \times 10^3\text{ N/C.} \qquad \text{(Answer)}$$

Similarly, for altitude $z_2 = 60$ km, we find

$$E = 2.0 \times 10^2\text{ N/C.} \qquad \text{(Answer)}$$

As we discuss in Section 22-8, when the magnitude

of an electric field exceeds a certain critical value $E_c$, the field can pull electrons out of atoms (ionize the atoms), and then the freed electrons can run into other atoms, causing those atoms to emit light. The value of $E_c$ depends on the density of the air in which the electric field exists. At altitude $z_2 = 60$ km the density of the air is so low that $E = 2.0 \times 10^2$ N/C exceeds $E_c$, and thus light is emitted by the atoms in the air. That light forms sprites. Lower down, just above the clouds at $z_1 = 30$ km, the density of the air is much higher, $E = 1.6 \times 10^3$ N/C does not exceed $E_c$, and no light is emitted. Hence, sprites occur only far above storm clouds.

(*a*)

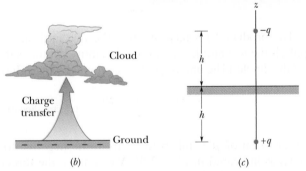

(*b*)            (*c*)

**FIG. 22-9** (*a*) Photograph of a sprite. (*Courtesy NASA*) (*b*) Lightning in which a large amount of negative charge is transferred from ground to cloud base. (*c*) The cloud–ground system modeled as a vertical electric dipole.

---

## 22-6 | The Electric Field Due to a Line of Charge

So far we have considered the electric field that is produced by one or, at most, a few point charges. We now consider charge distributions that consist of a great many closely spaced point charges (perhaps billions) that are spread along a line, over a surface, or within a volume. Such distributions are said to be **continuous** rather than discrete. Since these distributions can include an enormous number of point charges, we find the electric fields that they produce by means of calculus rather than by considering the point charges one by one. In this section we discuss the electric field caused by a line of charge. We consider a charged surface in the next section. In the next chapter, we shall find the field inside a uniformly charged sphere.

When we deal with continuous charge distributions, it is most convenient to express the charge on an object as a *charge density* rather than as a total charge. For a line of charge, for example, we would report the *linear charge density* (or charge per unit length) $\lambda$, whose SI unit is the coulomb per meter. Table 22-2 shows the other charge densities we shall be using.

Figure 22-10 shows a thin ring of radius $R$ with a uniform positive linear charge density $\lambda$ around its circumference. We may imagine the ring to be made of plastic or some other insulator, so that the charges can be regarded as fixed in place. What is the electric field $\vec{E}$ at point $P$, a distance $z$ from the plane of the ring along its central axis?

To answer, we cannot just apply Eq. 22-3, which gives the electric field set up by a point charge, because the ring is obviously not a point charge. However, we can mentally divide the ring into differential elements of charge that are so small that they are like point charges, and then we can apply Eq. 22-3 to each of them. Next, we can add the electric fields set up at $P$ by all the differential elements. The vector sum of the fields gives us the field set up at $P$ by the ring.

Let $ds$ be the (arc) length of any differential element of the ring. Since $\lambda$ is the charge per unit (arc) length, the element has a charge of magnitude

$$dq = \lambda \, ds. \tag{22-10}$$

This differential charge sets up a differential electric field $d\vec{E}$ at point $P$, which is a distance $r$ from the element. Treating the element as a point charge and using Eq. 22-10, we can rewrite Eq. 22-3 to express the magnitude of $d\vec{E}$ as

$$dE = \frac{1}{4\pi\varepsilon_0} \frac{dq}{r^2} = \frac{1}{4\pi\varepsilon_0} \frac{\lambda \, ds}{r^2}. \tag{22-11}$$

From Fig. 22-10, we can rewrite Eq. 22-11 as

$$dE = \frac{1}{4\pi\varepsilon_0} \frac{\lambda \, ds}{(z^2 + R^2)}. \tag{22-12}$$

Figure 22-10 shows that $d\vec{E}$ is at angle $\theta$ to the central axis (which we have taken to be a $z$ axis) and has components perpendicular to and parallel to that axis.

Every charge element in the ring sets up a differential field $d\vec{E}$ at $P$, with magnitude given by Eq. 22-12. All the $d\vec{E}$ vectors have identical components parallel to the central axis, in both magnitude and direction. All these $d\vec{E}$ vectors have components perpendicular to the central axis as well; these perpendicular components are identical in magnitude but point in different directions. In fact, for any perpendicular component that points in a given direction, there is another one that points in the opposite direction. The sum of this pair of components, like the sum of all other pairs of oppositely directed components, is zero.

Thus, the perpendicular components cancel and we need not consider them further. This leaves the parallel components; they all have the same direction, so the net electric field at $P$ is their sum.

The parallel component of $d\vec{E}$ shown in Fig. 22-10 has magnitude $dE \cos \theta$. The figure also shows us that

$$\cos \theta = \frac{z}{r} = \frac{z}{(z^2 + R^2)^{1/2}}. \tag{22-13}$$

Then multiplying Eq. 22-12 by Eq. 22-13 gives us, for the parallel component of $d\vec{E}$,

$$dE \cos \theta = \frac{z\lambda}{4\pi\varepsilon_0 (z^2 + R^2)^{3/2}} \, ds. \tag{22-14}$$

| TABLE 22-2 | | |
|---|---|---|
| **Some Measures of Electric Charge** | | |
| Name | Symbol | SI Unit |
| Charge | $q$ | C |
| Linear charge density | $\lambda$ | C/m |
| Surface charge density | $\sigma$ | C/m$^2$ |
| Volume charge density | $\rho$ | C/m$^3$ |

**FIG. 22-10** A ring of uniform positive charge. A differential element of charge occupies a length $ds$ (greatly exaggerated for clarity). This element sets up an electric field $d\vec{E}$ at point $P$. The component of $d\vec{E}$ along the central axis of the ring is $dE \cos \theta$.

To add the parallel components $dE \cos \theta$ produced by all the elements, we integrate Eq. 22-14 around the circumference of the ring, from $s = 0$ to $s = 2\pi R$. Since the only quantity in Eq. 22-14 that varies during the integration is $s$, the other quantities can be moved outside the integral sign. The integration then gives us

$$E = \int dE \cos \theta = \frac{z\lambda}{4\pi\varepsilon_0(z^2 + R^2)^{3/2}} \int_0^{2\pi R} ds$$

$$= \frac{z\lambda(2\pi R)}{4\pi\varepsilon_0(z^2 + R^2)^{3/2}}. \qquad (22\text{-}15)$$

Since $\lambda$ is the charge per length of the ring, the term $\lambda(2\pi R)$ in Eq. 22-15 is $q$, the total charge on the ring. We then can rewrite Eq. 22-15 as

$$E = \frac{qz}{4\pi\varepsilon_0(z^2 + R^2)^{3/2}} \qquad \text{(charged ring)}. \qquad (22\text{-}16)$$

If the charge on the ring is negative, instead of positive as we have assumed, the magnitude of the field at $P$ is still given by Eq. 22-16. However, the electric field vector then points toward the ring instead of away from it.

Let us check Eq. 22-16 for a point on the central axis that is so far away that $z \gg R$. For such a point, the expression $z^2 + R^2$ in Eq. 22-16 can be approximated as $z^2$, and Eq. 22-16 becomes

$$E = \frac{1}{4\pi\varepsilon_0} \frac{q}{z^2} \qquad \text{(charged ring at large distance)}. \qquad (22\text{-}17)$$

This is a reasonable result because from a large distance, the ring "looks like" a point charge. If we replace $z$ with $r$ in Eq. 22-17, we indeed do have Eq. 22-3, the magnitude of the electric field due to a point charge.

Let us next check Eq. 22-16 for a point at the center of the ring—that is, for $z = 0$. At that point, Eq. 22-16 tells us that $E = 0$. This is a reasonable result because if we were to place a test charge at the center of the ring, there would be no net electrostatic force acting on it; the force due to any element of the ring would be canceled by the force due to the element on the opposite side of the ring. By Eq. 22-1, if the force at the center of the ring were zero, the electric field there would also have to be zero.

---

**Sample Problem** | **22-3** | **Build your skill**

Figure 22-11a shows a plastic rod having a uniformly distributed charge $-Q$. The rod has been bent in a 120° circular arc of radius $r$. We place coordinate axes such that the axis of symmetry of the rod lies along the $x$ axis and the origin is at the center of curvature $P$ of the rod. In terms of $Q$ and $r$, what is the electric field $\vec{E}$ due to the rod at point $P$?

**KEY IDEA** Because the rod has a continuous charge distribution, we must find an expression for the electric fields due to differential elements of the rod and then sum those fields via calculus.

**An element:** Consider a differential element having arc length $ds$ and located at an angle $\theta$ above the $x$ axis

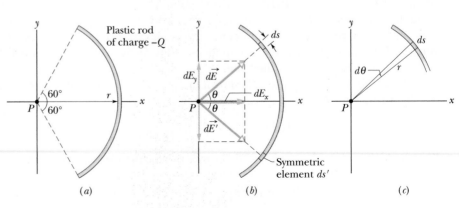

**FIG. 22-11** (a) A plastic rod of charge $-Q$ is a circular section of radius $r$ and central angle 120°; point $P$ is the center of curvature of the rod. (b) A differential element in the top half of the rod, at an angle $\theta$ to the $x$ axis and of arc length $ds$, sets up a differential electric field $d\vec{E}$ at $P$. An element $ds'$, symmetric to $ds$ about the $x$ axis, sets up a field $d\vec{E}'$ at $P$ with the same magnitude. (c) Arc length $ds$ makes an angle $d\theta$ about point $P$.

(Fig. 22-11b). If we let $\lambda$ represent the linear charge density of the rod, our element $ds$ has a differential charge of magnitude

$$dq = \lambda \, ds. \tag{22-18}$$

**The element's field:** Our element produces a differential electric field $d\vec{E}$ at point $P$, which is a distance $r$ from the element. Treating the element as a point charge, we can rewrite Eq. 22-3 to express the magnitude of $d\vec{E}$ as

$$dE = \frac{1}{4\pi\varepsilon_0} \frac{dq}{r^2} = \frac{1}{4\pi\varepsilon_0} \frac{\lambda \, ds}{r^2}. \tag{22-19}$$

The direction of $d\vec{E}$ is toward $ds$ because charge $dq$ is negative.

**Symmetric partner:** Our element has a symmetrically located (mirror image) element $ds'$ in the bottom half of the rod. The electric field $d\vec{E}'$ set up at $P$ by $ds'$ also has the magnitude given by Eq. 22-19, but the field vector points toward $ds'$ as shown in Fig. 22-11b. If we resolve the electric field vectors of $ds$ and $ds'$ into $x$ and $y$ components as shown in Fig. 22-11b, we see that their $y$ components cancel (because they have equal magnitudes and are in opposite directions). We also see that their $x$ components have equal magnitudes and are in the same direction.

**Summing:** Thus, to find the electric field set up by the rod, we need sum (via integration) only the $x$ components of the differential electric fields set up by all the differential elements of the rod. From Fig. 22-11b and Eq. 22-19, we can write the component $dE_x$ set up by $ds$ as

$$dE_x = dE \cos\theta = \frac{1}{4\pi\varepsilon_0} \frac{\lambda}{r^2} \cos\theta \, ds. \tag{22-20}$$

Equation 22-20 has two variables, $\theta$ and $s$. Before we can integrate it, we must eliminate one variable. We do so by replacing $ds$, using the relation

$$ds = r \, d\theta,$$

in which $d\theta$ is the angle at $P$ that includes arc length $ds$ (Fig. 22-11c). With this replacement, we can integrate Eq. 22-20 over the angle made by the rod at $P$, from $\theta = -60°$ to $\theta = 60°$; that will give us the magnitude of the electric field at $P$ due to the rod:

$$
\begin{aligned}
E &= \int dE_x = \int_{-60°}^{60°} \frac{1}{4\pi\varepsilon_0} \frac{\lambda}{r^2} \cos\theta \, r \, d\theta \\
&= \frac{\lambda}{4\pi\varepsilon_0 r} \int_{-60°}^{60°} \cos\theta \, d\theta = \frac{\lambda}{4\pi\varepsilon_0 r} \Big[ \sin\theta \Big]_{-60°}^{60°} \\
&= \frac{\lambda}{4\pi\varepsilon_0 r} [\sin 60° - \sin(-60°)] \\
&= \frac{1.73\lambda}{4\pi\varepsilon_0 r}. \tag{22-21}
\end{aligned}
$$

(If we had reversed the limits on the integration, we would have gotten the same result but with a minus sign. Since the integration gives only the magnitude of $\vec{E}$, we would then have discarded the minus sign.)

**Charge density:** To evaluate $\lambda$, we note that the rod subtends an angle of $120°$ and so is one-third of a full circle. Its arc length is then $2\pi r/3$, and its linear charge density must be

$$\lambda = \frac{\text{charge}}{\text{length}} = \frac{Q}{2\pi r/3} = \frac{0.477Q}{r}.$$

Substituting this into Eq. 22-21 and simplifying give us

$$
\begin{aligned}
E &= \frac{(1.73)(0.477Q)}{4\pi\varepsilon_0 r^2} \\
&= \frac{0.83Q}{4\pi\varepsilon_0 r^2}. \tag{Answer}
\end{aligned}
$$

The direction of $\vec{E}$ is toward the rod, along the axis of symmetry of the charge distribution. We can write $\vec{E}$ in unit-vector notation as

$$\vec{E} = \frac{0.83Q}{4\pi\varepsilon_0 r^2} \hat{i}.$$

---

**PROBLEM-SOLVING TACTICS**

**Tactic 1:** *A Field Guide for Lines of Charge*   Here is a generic guide for finding the electric field $\vec{E}$ produced at a point $P$ by a line of uniform charge, either circular or straight. The general strategy is to pick out an element $dq$ of the charge, find $d\vec{E}$ due to that element, and integrate $d\vec{E}$ over the entire line of charge.

**Step 1.** If the line of charge is circular, let $ds$ be the arc length of an element of the distribution. If the line is straight, run an $x$ axis along it and let $dx$ be the length of an element. Mark the element on a sketch.

**Step 2.** Relate the charge $dq$ of the element to the length of the element with either $dq = \lambda \, ds$ or $dq = \lambda \, dx$. Consider $dq$ and $\lambda$ to be positive, even if the charge is actually negative. (The sign of the charge is used in the next step.)

**Step 3.** Express the field $d\vec{E}$ produced at $P$ by $dq$ with Eq. 22-3, replacing $q$ in that equation with either $\lambda \, ds$ or $\lambda \, dx$. If the charge on the line is positive, then at $P$ draw a vector $d\vec{E}$ that points directly away from $dq$. If the charge is negative, draw the vector pointing directly toward $dq$.

**Step 4.** Always look for any symmetry in the situation. If $P$ is on an axis of symmetry of the charge distribution, resolve the field $d\vec{E}$ produced by $dq$ into components that are perpendicular and parallel to the axis of symmetry. Then consider a second element $dq'$ that is located symmetrically to $dq$ about the line of symmetry. At $P$ draw the vector $d\vec{E}'$ that this symmetrical element produces and resolve it into components. One of the components produced by $dq$ is a *canceling component*; it is canceled by the corresponding component produced by $dq'$ and

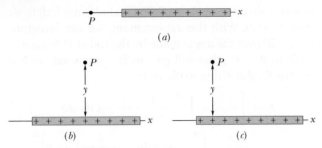

(a)

(b)                                    (c)

**FIG. 22-12** (a) Point P is on an extension of the line of charge. (b) P is on a line of symmetry of the line of charge, at perpendicular distance y from that line. (c) Same as (b) except that P is not on a line of symmetry.

needs no further attention. The other component produced by dq is an *adding component;* it adds to the corresponding component produced by dq'. Add the adding components of all the elements via integration.

**Step 5.** Here are four general types of uniform charge distributions, with strategies for simplifying the integral of step 4.

*Ring,* with point P on (central) axis of symmetry, as in Fig. 22-10. In the expression for dE, replace $r^2$ with $z^2 + R^2$, as in Eq. 22-12. Express the adding component of $d\vec{E}$ in terms of θ. That introduces cos θ, but θ is identical for all elements and thus is not a variable. Replace cos θ as in Eq. 22-13. Integrate over s, around the circumference of the ring.

*Circular arc,* with point P at the center of curvature, as in Fig. 22-11. Express the adding component of $d\vec{E}$ in terms of θ. That introduces either sin θ or cos θ. Reduce the resulting two variables s and θ to one, θ, by replacing ds with r dθ. Integrate over θ, as in Sample Problem 22-3, from one end of the arc to the other end.

*Straight line,* with point P on an extension of the line, as in Fig. 22-12a. In the expression for dE, replace r with x. Integrate over x, from end to end of the line of charge.

*Straight line,* with point P at perpendicular distance y from the line of charge, as in Fig. 22-12b. In the expression for dE, replace r with an expression involving x and y. If P is on the perpendicular bisector of the line of charge, find an expression for the adding component of $d\vec{E}$. That will introduce either sin θ or cos θ. Reduce the resulting two variables x and θ to one, x, by replacing the trigonometric function with an expression (its definition) involving x and y. Integrate over x from end to end of the line of charge. If P is not on a line of symmetry, as in Fig. 22-12c, set up an integral to sum the components $dE_x$, and integrate over x to find $E_x$. Also set up an integral to sum the components $dE_y$, and integrate over x again to find $E_y$. Use the components $E_x$ and $E_y$ in the usual way to find the magnitude E and the orientation of $\vec{E}$.

**Step 6.** One arrangement of the integration limits gives a positive result. The reverse arrangement gives the same result with a minus sign; discard the minus sign. If the result is to be stated in terms of the total charge Q of the distribution, replace λ with Q/L, in which L is the length of the distribution. For a ring, L is the ring's circumference.

✓**CHECKPOINT 2** The figure here shows three non-conducting rods, one circular and two straight. Each has a uniform charge of magnitude Q along its top half and another along its bottom half. For each rod, what is the direction of the net electric field at point P?

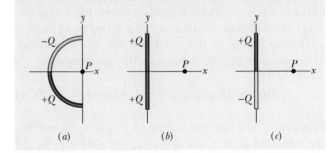

(a)          (b)          (c)

## 22-7 | The Electric Field Due to a Charged Disk

Figure 22-13 shows a circular plastic disk of radius R that has a positive surface charge of uniform density σ on its upper surface (see Table 22-2). What is the electric field at point P, a distance z from the disk along its central axis?

Our plan is to divide the disk into concentric flat rings and then to calculate the electric field at point P by adding up (that is, by integrating) the contributions of all the rings. Figure 22-13 shows one such ring, with radius r and radial width dr. Since σ is the charge per unit area, the charge on the ring is

$$dq = \sigma \, dA = \sigma \, (2\pi r \, dr),\qquad(22\text{-}22)$$

where dA is the differential area of the ring.

We have already solved the problem of the electric field due to a ring of charge. Substituting dq from Eq. 22-22 for q in Eq. 22-16, and replacing R in Eq. 22-16 with r, we obtain an expression for the electric field dE at P due to the arbitrarily chosen flat ring of charge shown in Fig. 22-13:

$$dE = \frac{z\sigma 2\pi r \, dr}{4\pi\varepsilon_0(z^2 + r^2)^{3/2}},$$

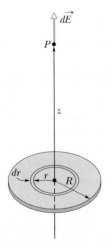

**FIG. 22-13** A disk of radius R and uniform positive charge. The ring shown has radius r and radial width dr. It sets up a differential electric field $d\vec{E}$ at point P on its central axis.

which we may write as

$$dE = \frac{\sigma z}{4\varepsilon_0} \frac{2r\, dr}{(z^2 + r^2)^{3/2}}. \tag{22-23}$$

We can now find $E$ by integrating Eq. 22-23 over the surface of the disk — that is, by integrating with respect to the variable $r$ from $r = 0$ to $r = R$. Note that $z$ remains constant during this process. We get

$$E = \int dE = \frac{\sigma z}{4\varepsilon_0} \int_0^R (z^2 + r^2)^{-3/2}(2r)\, dr. \tag{22-24}$$

To solve this integral, we cast it in the form $\int X^m\, dX$ by setting $X = (z^2 + r^2)$, $m = -\frac{3}{2}$, and $dX = (2r)\, dr$. For the recast integral we have

$$\int X^m\, dX = \frac{X^{m+1}}{m + 1},$$

and so Eq. 22-24 becomes

$$E = \frac{\sigma z}{4\varepsilon_0} \left[ \frac{(z^2 + r^2)^{-1/2}}{-\frac{1}{2}} \right]_0^R. \tag{22-25}$$

Taking the limits in Eq. 22-25 and rearranging, we find

$$E = \frac{\sigma}{2\varepsilon_0} \left( 1 - \frac{z}{\sqrt{z^2 + R^2}} \right) \qquad \text{(charged disk)} \tag{22-26}$$

as the magnitude of the electric field produced by a flat, circular, charged disk at points on its central axis. (In carrying out the integration, we assumed that $z \geq 0$.)

If we let $R \to \infty$ while keeping $z$ finite, the second term in the parentheses in Eq. 22-26 approaches zero, and this equation reduces to

$$E = \frac{\sigma}{2\varepsilon_0} \qquad \text{(infinite sheet).} \tag{22-27}$$

This is the electric field produced by an infinite sheet of uniform charge located on one side of a nonconductor such as plastic. The electric field lines for such a situation are shown in Fig. 22-3.

We also get Eq. 22-27 if we let $z \to 0$ in Eq. 22-26 while keeping $R$ finite. This shows that at points very close to the disk, the electric field set up by the disk is the same as if the disk were infinite in extent.

## 22-8 | A Point Charge in an Electric Field

In the preceding four sections we worked at the first of our two tasks: given a charge distribution, to find the electric field it produces in the surrounding space. Here we begin the second task: to determine what happens to a charged particle when it is in an electric field set up by other stationary or slowly moving charges.

What happens is that an electrostatic force acts on the particle, as given by

$$\vec{F} = q\vec{E}, \tag{22-28}$$

in which $q$ is the charge of the particle (including its sign) and $\vec{E}$ is the electric field that other charges have produced at the location of the particle. (The field is *not* the field set up by the particle itself; to distinguish the two fields, the field acting on the particle in Eq. 22-28 is often called the *external field*. A charged particle or object is not affected by its own electric field.) Equation 22-28 tells us

The electrostatic force $\vec{F}$ acting on a charged particle located in an external electric field $\vec{E}$ has the direction of $\vec{E}$ if the charge $q$ of the particle is positive and has the opposite direction if $q$ is negative.

**✓CHECKPOINT 3** (a) In the figure, what is the direction of the electrostatic force on the electron due to the external electric field shown? (b) In which direction will the electron accelerate if it is moving parallel to the *y* axis before it encounters the external field? (c) If, instead, the electron is initially moving rightward, will its speed increase, decrease, or remain constant?

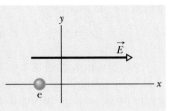

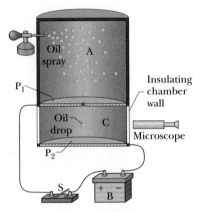

**FIG. 22-14** The Millikan oil-drop apparatus for measuring the elementary charge *e*. When a charged oil drop drifted into chamber C through the hole in plate $P_1$, its motion could be controlled by closing and opening switch S and thereby setting up or eliminating an electric field in chamber C. The microscope was used to view the drop, to permit timing of its motion.

### Measuring the Elementary Charge

Equation 22-28 played a role in the measurement of the elementary charge *e* by American physicist Robert A. Millikan in 1910–1913. Figure 22-14 is a representation of his apparatus. When tiny oil drops are sprayed into chamber A, some of them become charged, either positively or negatively, in the process. Consider a drop that drifts downward through the small hole in plate $P_1$ and into chamber C. Let us assume that this drop has a negative charge *q*.

If switch S in Fig. 22-14 is open as shown, battery B has no electrical effect on chamber C. If the switch is closed (the connection between chamber C and the positive terminal of the battery is then complete), the battery causes an excess positive charge on conducting plate $P_1$ and an excess negative charge on conducting plate $P_2$. The charged plates set up a downward-directed electric field $\vec{E}$ in chamber C. According to Eq. 22-28, this field exerts an electrostatic force on any charged drop that happens to be in the chamber and affects its motion. In particular, our negatively charged drop will tend to drift upward.

By timing the motion of oil drops with the switch opened and with it closed and thus determining the effect of the charge *q*, Millikan discovered that the values of *q* were always given by

$$q = ne, \quad \text{for } n = 0, \pm 1, \pm 2, \pm 3, \ldots, \quad (22\text{-}29)$$

in which *e* turned out to be the fundamental constant we call the *elementary charge*, $1.60 \times 10^{-19}$ C. Millikan's experiment is convincing proof that charge is quantized, and he earned the 1923 Nobel Prize in physics in part for this work. Modern measurements of the elementary charge rely on a variety of interlocking experiments, all more precise than the pioneering experiment of Millikan.

### Ink-Jet Printing

The need for high-quality, high-speed printing has caused a search for an alternative to impact printing, such as occurs in a standard typewriter. Building up letters by squirting tiny drops of ink at the paper is one such alternative.

Figure 22-15 shows a negatively charged drop moving between two conducting deflecting plates, between which a uniform, downward-directed electric field $\vec{E}$ has been set up. The drop is deflected upward according to Eq. 22-28 and then strikes the paper at a position that is determined by the magnitudes of $\vec{E}$ and the charge *q* of the drop.

In practice, *E* is held constant and the position of the drop is determined by the charge *q* delivered to the drop in the charging unit, through which the drop must pass before entering the deflecting system. The charging unit, in turn, is activated by electronic signals that encode the material to be printed.

### Electrical Breakdown and Sparking

If the magnitude of an electric field in air exceeds a certain critical value $E_c$, the air undergoes *electrical breakdown,* a process whereby the field removes electrons from the atoms in the air. The air then begins to conduct electric current because the freed electrons are propelled into motion by the field. As they move, they collide with any atoms in their path, causing those atoms to emit light. We can see the paths, commonly called sparks, taken by the freed electrons because of that

**FIG. 22-15** The essential features of an ink-jet printer. Drops are shot out from generator *G* and receive a charge in charging unit *C*. An input signal from a computer controls the charge given to each drop and thus the effect of field $\vec{E}$ on the drop and the position on the paper at which the drop lands. About 100 tiny drops are needed to form a single character.

emitted light. Figure 22-16 shows sparks above charged metal wires where the electric fields due to the wires cause electrical breakdown of the air.

## Pollination and Electrostatics

The ability of a bee to transport pollen from flower to flower depends on two features. (1) Bees become charged while flying through the air. (2) The anther of a flower (Fig. 22-17a) is electrically isolated from the ground but the stigma is electrically connected to the ground. When a bee hovers near an anther, the electric field due to the charge on the bee induces charge on a neutral pollen grain, making the closer side slightly more negative than the farther side (Fig. 22-17b). The charges on the two sides are equal but their distances from the bee are not, and the attractive force on the closer side is slightly larger than the repulsive force on the farther side. As a result, the pollen grain is pulled to the bee, where it clings to hairs during the bee's flight to the next flower.

When the bee happens to come close to a stigma in the next flower, the charge on the bee and induced charge on the grain bring conduction electrons up to the tip of the stigma (Fig. 22-17c) because the stigma is electrically connected to the ground. Those electrons attract the closer side of the grain and repel the farther side. If the grain is close enough to the stigma, the net force causes the grain to jump to the stigma, starting the fertilization of the flower. Agricultural engineers now mimic this process by spraying plants with charged pollen grains, so that the grains collect on the stigma rather than fall uselessly to the ground.

**FIG. 22-16** The metal wires are so charged that the electric fields they produce in the surrounding space cause the air there to undergo electrical breakdown. (*Adam Hart-Davis/Photo Researchers*)

**FIG. 22-17** (*a*) The anther and stigma portions of a flower. (*b*) A pollen grain near an anther has induced charge due to a bee. (*c*) Electrons collect at a stigma tip, attracting the grain.

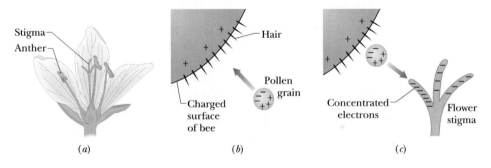

(*a*)    (*b*)    (*c*)

---

**Sample Problem** 22-4

Figure 22-18 shows the deflecting plates of an ink-jet printer, with superimposed coordinate axes. An ink drop with a mass $m$ of $1.3 \times 10^{-10}$ kg and a negative charge of magnitude $Q = 1.5 \times 10^{-13}$ C enters the region between the plates, initially moving along the $x$ axis with speed $v_x = 18$ m/s. The length $L$ of each plate is 1.6 cm. The plates are charged and thus produce an electric field at all points between them. Assume that field $\vec{E}$ is downward directed, is uniform,

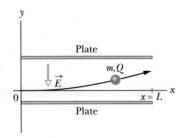

**FIG. 22-18** An ink drop of mass $m$ and charge magnitude $Q$ is deflected in the electric field of an ink-jet printer.

and has a magnitude of $1.4 \times 10^6$ N/C. What is the vertical deflection of the drop at the far edge of the plates? (The gravitational force on the drop is small relative to the electrostatic force acting on the drop and can be neglected.)

**KEY IDEA** The drop is negatively charged and the electric field is directed *downward*. From Eq. 22-28, a constant electrostatic force of magnitude $QE$ acts *upward* on the charged drop. Thus, as the drop travels parallel to the $x$ axis at constant speed $v_x$, it accelerates upward with some constant acceleration $a_y$.

**Calculations:** Applying Newton's second law ($F = ma$) for components along the $y$ axis, we find that

$$a_y = \frac{F}{m} = \frac{QE}{m}. \qquad (22\text{-}30)$$

Let $t$ represent the time required for the drop to pass through the region between the plates. During $t$ the vertical and horizontal displacements of the drop are

$$y = \tfrac{1}{2}a_y t^2 \quad \text{and} \quad L = v_x t, \qquad (22\text{-}31)$$

respectively. Eliminating $t$ between these two equations and substituting Eq. 22-30 for $a_y$, we find

$$y = \frac{QEL^2}{2mv_x^2}$$

$$= \frac{(1.5 \times 10^{-13}\ \text{C})(1.4 \times 10^6\ \text{N/C})(1.6 \times 10^{-2}\ \text{m})^2}{(2)(1.3 \times 10^{-10}\ \text{kg})(18\ \text{m/s})^2}$$

$$= 6.4 \times 10^{-4}\ \text{m}$$

$$= 0.64\ \text{mm.} \qquad \text{(Answer)}$$

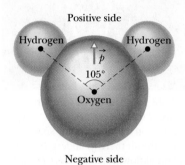

Positive side

Hydrogen

Hydrogen

$\vec{p}$

105°

Oxygen

Negative side

**FIG. 22-19**   A molecule of H₂O, showing the three nuclei (represented by dots) and the regions in which the electrons can be located. The electric dipole moment $\vec{p}$ points from the (negative) oxygen side to the (positive) hydrogen side of the molecule.

## 22-9 | A Dipole in an Electric Field

We have defined the electric dipole moment $\vec{p}$ of an electric dipole to be a vector that points from the negative to the positive end of the dipole. As you will see, the behavior of a dipole in a uniform external electric field $\vec{E}$ can be described completely in terms of the two vectors $\vec{E}$ and $\vec{p}$, with no need of any details about the dipole's structure.

A molecule of water (H₂O) is an electric dipole; Fig. 22-19 shows why. There the black dots represent the oxygen nucleus (having eight protons) and the two hydrogen nuclei (having one proton each). The colored enclosed areas represent the regions in which electrons can be located around the nuclei.

In a water molecule, the two hydrogen atoms and the oxygen atom do not lie on a straight line but form an angle of about 105°, as shown in Fig. 22-19. As a result, the molecule has a definite "oxygen side" and "hydrogen side." Moreover, the 10 electrons of the molecule tend to remain closer to the oxygen nucleus than to the hydrogen nuclei. This makes the oxygen side of the molecule slightly more negative than the hydrogen side and creates an electric dipole moment $\vec{p}$ that points along the symmetry axis of the molecule as shown. If the water molecule is placed in an external electric field, it behaves as would be expected of the more abstract electric dipole of Fig. 22-8.

To examine this behavior, we now consider such an abstract dipole in a uniform external electric field $\vec{E}$, as shown in Fig. 22-20a. We assume that the dipole is a rigid structure that consists of two centers of opposite charge, each of magnitude $q$, separated by a distance $d$. The dipole moment $\vec{p}$ makes an angle $\theta$ with field $\vec{E}$.

Electrostatic forces act on the charged ends of the dipole. Because the electric field is uniform, those forces act in opposite directions (as shown in Fig. 22-20a) and with the same magnitude $F = qE$. Thus, *because the field is uniform,* the net force on the dipole from the field is zero and the center of mass of the dipole does not move. However, the forces on the charged ends do produce a net torque $\vec{\tau}$ on the dipole about its center of mass. The center of mass lies on the line connecting the charged ends, at some distance $x$ from one end and thus a distance $d - x$ from the other end. From Eq. 10-39 ($\tau = rF \sin \phi$), we can write the magnitude of the net torque $\vec{\tau}$ as

$$\tau = Fx \sin \theta + F(d - x) \sin \theta = Fd \sin \theta. \qquad (22\text{-}32)$$

We can also write the magnitude of $\vec{\tau}$ in terms of the magnitudes of the electric field $E$ and the dipole moment $p = qd$. To do so, we substitute $qE$ for $F$ and $p/q$ for $d$ in Eq. 22-32, finding that the magnitude of $\vec{\tau}$ is

$$\tau = pE \sin \theta. \qquad (22\text{-}33)$$

We can generalize this equation to vector form as

$$\vec{\tau} = \vec{p} \times \vec{E} \quad \text{(torque on a dipole).} \qquad (22\text{-}34)$$

Vectors $\vec{p}$ and $\vec{E}$ are shown in Fig. 22-20b. The torque acting on a dipole tends to rotate $\vec{p}$ (hence the dipole) into the direction of field $\vec{E}$, thereby reducing $\theta$. In Fig. 22-20, such rotation is clockwise. As we discussed in Chapter 10, we can represent a torque that gives rise to a clockwise rotation by including a minus sign

$d$

$+q$

$\vec{F}$

$\vec{p}$

$\theta$

com

$\vec{E}$

$-q$

$-\vec{F}$

(a)

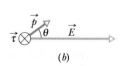

$\vec{p}$

$\theta$

$\vec{E}$

$\vec{\tau} \otimes$

(b)

**FIG. 22-20**   (a) An electric dipole in a uniform external electric field $\vec{E}$. Two centers of equal but opposite charge are separated by distance $d$. The line between them represents their rigid connection. (b) Field $\vec{E}$ causes a torque $\vec{\tau}$ on the dipole. The direction of $\vec{\tau}$ is into the page, as represented by the symbol $\otimes$.

with the magnitude of the torque. With that notation, the torque of Fig. 22-20 is

$$\tau = -pE \sin \theta. \tag{22-35}$$

## Potential Energy of an Electric Dipole

Potential energy can be associated with the orientation of an electric dipole in an electric field. The dipole has its least potential energy when it is in its equilibrium orientation, which is when its moment $\vec{p}$ is lined up with the field $\vec{E}$ (then $\vec{\tau} = \vec{p} \times \vec{E} = 0$). It has greater potential energy in all other orientations. Thus the dipole is like a pendulum, which has *its* least gravitational potential energy in *its* equilibrium orientation—at its lowest point. To rotate the dipole or the pendulum to any other orientation requires work by some external agent.

In any situation involving potential energy, we are free to define the zero-potential-energy configuration in a perfectly arbitrary way because only differences in potential energy have physical meaning. It turns out that the expression for the potential energy of an electric dipole in an external electric field is simplest if we choose the potential energy to be zero when the angle $\theta$ in Fig. 22-20 is 90°. We then can find the potential energy $U$ of the dipole at any other value of $\theta$ with Eq. 8-1 ($\Delta U = -W$) by calculating the work $W$ done by the field on the dipole when the dipole is rotated to that value of $\theta$ from 90°. With the aid of Eq. 10-53 ($W = \int \tau \, d\theta$) and Eq. 22-35, we find that the potential energy $U$ at any angle $\theta$ is

$$U = -W = -\int_{90°}^{\theta} \tau \, d\theta = \int_{90°}^{\theta} pE \sin \theta \, d\theta. \tag{22-36}$$

Evaluating the integral leads to

$$U = -pE \cos \theta. \tag{22-37}$$

We can generalize this equation to vector form as

$$U = -\vec{p} \cdot \vec{E} \quad \text{(potential energy of a dipole).} \tag{22-38}$$

Equations 22-37 and 22-38 show us that the potential energy of the dipole is least ($U = -pE$) when $\theta = 0$ ($\vec{p}$ and $\vec{E}$ are in the same direction); the potential energy is greatest ($U = pE$) when $\theta = 180°$ ($\vec{p}$ and $\vec{E}$ are in opposite directions).

When a dipole rotates from an initial orientation $\theta_i$ to another orientation $\theta_f$, the work $W$ done on the dipole by the electric field is

$$W = -\Delta U = -(U_f - U_i), \tag{22-39}$$

where $U_f$ and $U_i$ are calculated with Eq. 22-38. If the change in orientation is caused by an applied torque (commonly said to be due to an external agent), then the work $W_a$ done on the dipole by the applied torque is the negative of the work done on the dipole by the field; that is,

$$W_a = -W = (U_f - U_i). \tag{22-40}$$

**✓CHECKPOINT 4**    The figure shows four orientations of an electric dipole in an external electric field. Rank the orientations according to (a) the magnitude of the torque on the dipole and (b) the potential energy of the dipole, greatest first.

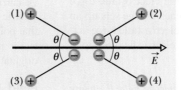

## Microwave Cooking

Food can be warmed and cooked in a microwave oven if the food contains water because water molecules are electric dipoles. When you turn on the oven, the microwave source sets up a rapidly oscillating electric field $\vec{E}$ within the oven and thus also within the food. From Eq. 22-34, we see that any electric field $\vec{E}$ produces a torque on an electric dipole moment $\vec{p}$ to align $\vec{p}$ with $\vec{E}$. Because the oven's $\vec{E}$ oscillates, the water molecules continuously flip-flop in a frustrated attempt to align with $\vec{E}$.

Energy is transferred from the electric field to the thermal energy of the water (and thus of the food) where three water molecules happened to have bonded together to form a group. The flip-flop breaks some of the bonds. When the molecules reform the bonds, energy is transferred to the random motion of the group and then to the surrounding molecules. Soon, the thermal energy of the water is enough to cook the food. Sometimes the heating is surprising. If you heat a jelly donut, for example, the jelly (which holds a lot of water) heats far more than the donut material (which holds much less water). Although the exterior of the donut may not be hot, biting into the jelly can burn you. If water molecules were not electric dipoles, we would not have microwave ovens.

---

## Sample Problem 22-5

A neutral water molecule ($H_2O$) in its vapor state has an electric dipole moment of magnitude $6.2 \times 10^{-30}$ C·m.

(a) How far apart are the molecule's centers of positive and negative charge?

**KEY IDEA** A molecule's dipole moment depends on the magnitude $q$ of the molecule's positive or negative charge and the charge separation $d$.

**Calculations:** There are 10 electrons and 10 protons in a neutral water molecule; so the magnitude of its dipole moment is

$$p = qd = (10e)(d),$$

in which $d$ is the separation we are seeking and $e$ is the elementary charge. Thus,

$$d = \frac{p}{10e} = \frac{6.2 \times 10^{-30}\text{ C·m}}{(10)(1.60 \times 10^{-19}\text{ C})}$$
$$= 3.9 \times 10^{-12}\text{ m} = 3.9\text{ pm.} \qquad \text{(Answer)}$$

This distance is not only small, but it is also actually smaller than the radius of a hydrogen atom.

(b) If the molecule is placed in an electric field of $1.5 \times 10^4$ N/C, what maximum torque can the field exert on it? (Such a field can easily be set up in the laboratory.)

**KEY IDEA** The torque on a dipole is maximum when the angle $\theta$ between $\vec{p}$ and $\vec{E}$ is 90°.

**Calculation:** Substituting $\theta = 90°$ in Eq. 22-33 yields

$$\tau = pE \sin\theta$$
$$= (6.2 \times 10^{-30}\text{ C·m})(1.5 \times 10^4\text{ N/C})(\sin 90°)$$
$$= 9.3 \times 10^{-26}\text{ N·m.} \qquad \text{(Answer)}$$

(c) How much work must an *external agent* do to rotate this molecule by 180° in this field, starting from its fully aligned position, for which $\theta = 0$?

**KEY IDEA** The work done by an external agent (by means of a torque applied to the molecule) is equal to the change in the molecule's potential energy due to the change in orientation.

**Calculation:** From Eq. 22-40, we find

$$W_a = U_{180°} - U_0$$
$$= (-pE \cos 180°) - (-pE \cos 0)$$
$$= 2pE = (2)(6.2 \times 10^{-30}\text{ C·m})(1.5 \times 10^4\text{ N/C})$$
$$= 1.9 \times 10^{-25}\text{ J.} \qquad \text{(Answer)}$$

---

## REVIEW & SUMMARY

**Electric Field** One way to explain the electrostatic force between two charges is to assume that each charge sets up an electric field in the space around it. The electrostatic force acting on any one charge is then due to the electric field set up at its location by the other charge.

**Definition of Electric Field** The *electric field* $\vec{E}$ at any point is defined in terms of the electrostatic force $\vec{F}$ that would be exerted on a positive test charge $q_0$ placed there:

$$\vec{E} = \frac{\vec{F}}{q_0}. \qquad (22\text{-}1)$$

**Electric Field Lines** *Electric field lines* provide a means for

visualizing the direction and magnitude of electric fields. The electric field vector at any point is tangent to a field line through that point. The density of field lines in any region is proportional to the magnitude of the electric field in that region. Field lines originate on positive charges and terminate on negative charges.

**Field Due to a Point Charge** The magnitude of the electric field $\vec{E}$ set up by a point charge $q$ at a distance $r$ from the charge is

$$\vec{E} = \frac{1}{4\pi\varepsilon_0}\frac{q}{r^2}\hat{r} \qquad (22\text{-}3)$$

The direction of $\vec{E}$ is away from the point charge if the charge is positive and toward it if the charge is negative.

**Field Due to an Electric Dipole** An *electric dipole* consists of two particles with charges of equal magnitude $q$ but opposite sign, separated by a small distance $d$. Their **electric dipole moment** $\vec{p}$ has magnitude $qd$ and points from the negative charge to the positive charge. The magnitude of the electric field set up by the dipole at a distant point on the dipole axis (which runs through both charges) is

$$E = \frac{1}{2\pi\varepsilon_0}\frac{p}{z^3}, \qquad (22\text{-}9)$$

where $z$ is the distance between the point and the center of the dipole.

**Field Due to a Continuous Charge Distribution** The electric field due to a *continuous charge distribution* is found by treating charge elements as point charges and then summing, via integration, the electric field vectors produced by all the charge elements.

**Force on a Point Charge in an Electric Field** When a point charge $q$ is placed in an external electric field $\vec{E}$, the electrostatic force $\vec{F}$ that acts on the point charge is

$$\vec{F} = q\vec{E}. \qquad (22\text{-}28)$$

Force $\vec{F}$ has the same direction as $\vec{E}$ if $q$ is positive and the opposite direction if $q$ is negative.

**Dipole in an Electric Field** When an electric dipole of dipole moment $\vec{p}$ is placed in an electric field $\vec{E}$, the field exerts a torque $\vec{\tau}$ on the dipole:

$$\vec{\tau} = \vec{p} \times \vec{E}. \qquad (22\text{-}34)$$

The dipole has a potential energy $U$ associated with its orientation in the field:

$$U = -\vec{p} \cdot \vec{E}. \qquad (22\text{-}38)$$

This potential energy is defined to be zero when $\vec{p}$ is perpendicular to $\vec{E}$; it is least ($U = -pE$) when $\vec{p}$ is aligned with $\vec{E}$ and greatest ($U = pE$) when $\vec{p}$ is directed opposite $\vec{E}$.

## QUESTIONS

**1** Figure 22-21 shows three arrangements of electric field lines. In each arrangement, a proton is released from rest at point $A$ and is then accelerated through point $B$ by the electric field. Points $A$ and $B$ have equal separations in the three arrangements. Rank the arrangements according to the linear momentum of the proton at point $B$, greatest first.

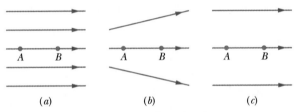

**FIG. 22-21** Question 1.

**2** Figure 22-22 shows four situations in which four charged particles are evenly spaced to the left and right of a central point. The charge values are indicated. Rank the situations according to the magnitude of the net electric field at the central point, greatest first.

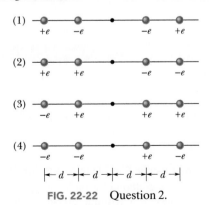

**FIG. 22-22** Question 2.

**3** Figure 22-23 shows two charged particles fixed in place on an axis. (a) Where on the

$$+q \qquad\qquad -3q$$

**FIG. 22-23** Question 3.

axis (other than at an infinite distance) is there a point at which their net electric field is zero: between the charges, to their left, or to their right? (b) Is there a point of zero net electric field anywhere *off* the axis (other than at an infinite distance)?

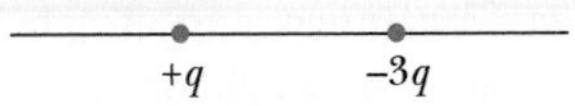

**4** Figure 22-24 shows two square arrays of charged particles. The squares, which are centered on point $P$, are misaligned. The particles are separated by either $d$ or $d/2$ along the perimeters of the squares. What are the magnitude and direction of the net electric field at $P$?

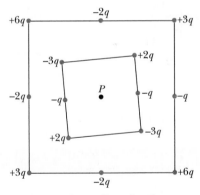

**FIG. 22-24** Question 4.

**5** In Fig. 22-25, two particles of charge $-q$ are arranged symmetrically about the $y$ axis; each produces an electric field at point $P$ on that axis. (a) Are the magnitudes of the fields at $P$ equal? (b) Is each electric field directed toward or away from the charge producing it? (c) Is the magnitude of the net electric field at $P$ equal to the sum of the magnitudes $E$ of the two field vectors (is it equal to $2E$)? (d) Do the $x$ components of those two field vectors add or cancel? (e) Do their $y$ components add or cancel? (f) Is the direction of the net field at $P$ that of the canceling components or

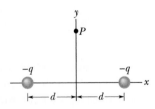

**FIG. 22-25** Question 5.

the adding components? (g) What is the direction of the net field?

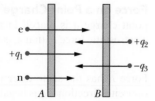

**6** In Fig. 22-26, an electron e travels through a small hole in plate $A$ and then toward plate $B$. A uniform electric field in the region between the plates

**FIG. 22-26** Question 6.

then slows the electron without deflecting it. (a) What is the direction of the field? (b) Four other particles similarly travel through small holes in either plate $A$ or plate $B$ and then into the region between the plates. Three have charges $+q_1$, $+q_2$, and $-q_3$. The fourth (labeled n) is a neutron, which is electrically neutral. Does the speed of each of those four other particles increase, decrease, or remain the same in the region between the plates?

**7** In Fig. 22-27a, a circular plastic rod with uniform charge $+Q$ produces an electric field of magnitude $E$ at the center of curvature (at the origin). In Figs. 22-27b, c, and d, more circular rods, each with identical uniform charges $+Q$, are added until the circle is complete. A fifth arrangement (which would be labeled e) is like that in d except the rod in the

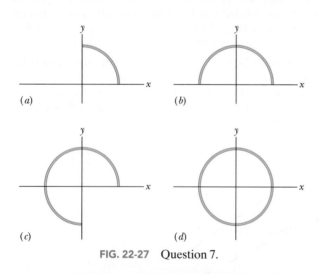

**FIG. 22-27** Question 7.

fourth quadrant has charge $-Q$. Rank the five arrangements according to the magnitude of the electric field at the center of curvature, greatest first.

**8** In Fig. 22-28, two identical circular nonconducting rings are centered on the same line. For three situations, the uniform charges on rings $A$ and $B$ are, respectively, (1) $q_0$ and $q_0$, (2) $-q_0$ and $-q_0$, and (3) $-q_0$ and $q_0$. Rank the situations according to the magnitude of the net electric field at (a) point $P_1$ midway between the rings, (b) point $P_2$ at the center of ring $B$, and (c) point $P_3$ to the right of ring $B$, greatest first.

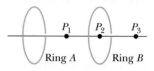

**FIG. 22-28** Question 8.

**9** The potential energies associated with four orientations of an electric dipole in an electric field are (1) $-5U_0$, (2) $-7U_0$, (3) $3U_0$, and (4) $5U_0$, where $U_0$ is positive. Rank the orientations according to (a) the angle between the electric dipole moment $\vec{p}$ and the electric field $\vec{E}$ and (b) the magnitude of the torque on the electric dipole, greatest first.

**10** (a) In Checkpoint 4, if the dipole rotates from orientation 1 to orientation 2, is the work done on the dipole by the field positive, negative, or zero? (b) If, instead, the dipole rotates from orientation 1 to orientation 4, is the work done by the field more than, less than, or the same as in (a)?

**11** Figure 22-29 shows two disks and a flat ring, each with the same uniform charge $Q$. Rank the objects according to the magnitude of the electric field they create at points $P$ (which are at the same vertical heights), greatest first.

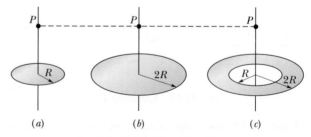

**FIG. 22-29** Question 11.

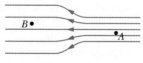

## PROBLEMS

 **GO**    Tutoring problem available (at instructor's discretion) in *WileyPLUS* and WebAssign

**SSM**    Worked-out solution available in Student Solutions Manual    **WWW**   Worked-out solution is at ____

• – •••    Number of dots indicates level of problem difficulty    **ILW**   Interactive solution is at ____ `http://www.wiley.com/college/halliday`

   Additional information available in *The Flying Circus of Physics* and at flyingcircusofphysics.com

### sec. 22-3   Electric Field Lines

•**1** In Fig. 22-30 the electric field lines on the left have twice the separation of those on the right. (a) If the magnitude of the field at $A$ is 40 N/C, what is the magnitude of the force on a

**FIG. 22-30** Problem 1.

proton at $A$? (b) What is the magnitude of the field at $B$?

•**2** Sketch qualitatively the electric field lines both between and outside two concentric conducting spherical shells when a

uniform positive charge $q_1$ is on the inner shell and a uniform negative charge $-q_2$ is on the outer. Consider the cases $q_1 > q_2$, $q_1 = q_2$, and $q_1 < q_2$.

### sec. 22-4   The Electric Field Due to a Point Charge

•**3** What is the magnitude of a point charge whose electric field 50 cm away has the magnitude 2.0 N/C? **SSM**

•**4** What is the magnitude of a point charge that would create an electric field of 1.00 N/C at points 1.00 m away?

•**5** The nucleus of a plutonium-239 atom contains 94 protons. Assume that the nucleus is a sphere with radius 6.64 fm and with the charge of the protons uniformly spread through the sphere. At the nucleus surface, what are the (a) magnitude and (b) direction (radially inward or outward) of the electric field produced by the protons? **SSM**

•6  Two particles are fixed to an $x$ axis: particle 1 of charge $-2.00 \times 10^{-7}$ C at $x = 6.00$ cm and particle 2 of charge $+2.00 \times 10^{-7}$ C at $x = 21.0$ cm. Midway between the particles, what is their net electric field in unit-vector notation?

••7  Two particles are fixed to an $x$ axis: particle 1 of charge $q_1 = 2.1 \times 10^{-8}$ C at $x = 20$ cm and particle 2 of charge $q_2 = -4.00q_1$ at $x = 70$ cm. At what coordinate on the axis is the net electric field produced by the particles equal to zero? **SSM**

••8  In Fig. 22-31, particle 1 of charge $q_1 = -5.00q$ and particle 2 of charge $q_2 = +2.00q$ are fixed to an $x$ axis. (a) As a multiple of distance $L$, at what coordinate on the axis is the net electric field of the particles zero? (b) Sketch the net electric field lines.

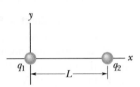

**FIG. 22-31**  Problem 8.

••9  In Fig. 22-32, the four particles form a square of edge length $a = 5.00$ cm and have charges $q_1 = +10.0$ nC,  $q_2 = -20.0$ nC,  $q_3 = +20.0$ nC, and $q_4 = -10.0$ nC. In unit-vector notation, what net electric field do the particles produce at the square's center? **SSM ILW WWW**

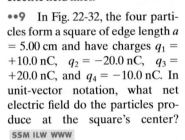

**FIG. 22-32**  Problem 9.

••10  In Fig. 22-33, the four particles are fixed in place and have charges $q_1 = q_2 = +5e$, $q_3 = +3e$, and $q_4 = -12e$. Distance $d = 5.0$ $\mu$m. What is the magnitude of the net electric field at point $P$ due to the particles? **GO**

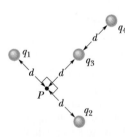

**FIG. 22-33**  Problem 10.

••11  Figure 22-34 shows two charged particles on an $x$ axis: $-q = -3.20 \times 10^{-19}$ C at $x = -3.00$ m and $q = 3.20 \times 10^{-19}$ C at $x = +3.00$ m. What are the (a) magnitude and (b) direction (relative to the positive direction of the $x$ axis) of the net electric field produced at point $P$ at $y = 4.00$ m? **GO**

••12  Figure 22-35$a$ shows two charged particles fixed in place on an $x$ axis with separation $L$. The ratio $q_1/q_2$ of their charge magnitudes is 4.00. Figure 22-35$b$ shows the $x$ component $E_{net,x}$ of their net electric field along the $x$ axis just to the right of particle 2. The $x$ axis scale is set by $x_s = 30.0$ cm. (a) At what value of $x > 0$ is $E_{net,x}$ maximum? (b) If particle 2 has charge $-q_2 = -3e$, what is the value of that maximum?

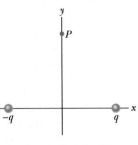

**FIG. 22-34**  Problem 11.

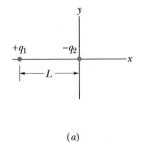

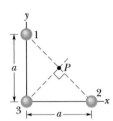

*(a)*                            *(b)*

**FIG. 22-35**  Problem 12.

••13  In Fig. 22-36, the three particles are fixed in place and have charges $q_1 = q_2 = +e$ and $q_3 = +2e$. Distance $a = 6.00$ $\mu$m. What are the (a) magnitude and (b) direction of the net electric field at point $P$ due to the particles?

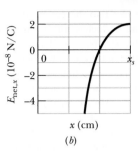

**FIG. 22-36**
Problem 13.

••14  Figure 22-37 shows an uneven arrangement of electrons (e) and protons (p) on a circular arc of radius $r = 2.00$ cm, with angles $\theta_1 = 30.0°$, $\theta_2 = 50.0°$, $\theta_3 = 30.0°$, and $\theta_4 = 20.0°$. What are the (a) magnitude and (b) direction (relative to the positive direction of the $x$ axis) of the net electric field produced at the center of the arc?

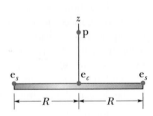

**FIG. 22-37**  Problem 14.

••15  Figure 22-38 shows a proton (p) on the central axis through a disk with a uniform charge density due to excess electrons. Three of those electrons are shown: electron $e_c$ at the disk center and electrons $e_s$ at opposite sides of the disk, at radius $R$ from the center. The proton is initially at distance $z = R = 2.00$ cm from the disk. At that location, what are the magnitudes of (a) the electric field $\vec{E}_c$ due to electron $e_c$ and (b) the net electric field $\vec{E}_{s,net}$ due to electrons $e_s$? The proton is then moved to $z = R/10.0$. What then are the magnitudes of (c) $\vec{E}_c$ and (d) $\vec{E}_{s,net}$ at the proton's location? (e) From (a) and (c) we see that as the proton gets nearer to the disk, the magnitude of $\vec{E}_c$ increases. Why does the magnitude of $\vec{E}_{s,net}$ decrease, as we see from (b) and (d)?

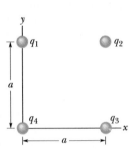

**FIG. 22-38**  Problem 15.

•••16  Figure 22-39 shows a plastic ring of radius $R = 50.0$ cm. Two small charged beads are on the ring: Bead 1 of charge $+2.00$ $\mu$C is fixed in place at the left side; bead 2 of charge $+6.00$ $\mu$C can be moved along the ring. The two beads produce a net electric field of magnitude $E$ at the center of the ring. At what (a) positive and (b) negative value of angle $\theta$ should bead 2 be positioned such that $E = 2.00 \times 10^5$ N/C?

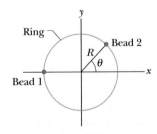

**FIG. 22-39**  Problem 16.

**•••17** Two charged beads are on the plastic ring in Fig. 22-40a. Bead 2, which is not shown, is fixed in place on the ring, which has radius $R = 60.0$ cm. Bead 1 is initially on the $x$ axis at angle $\theta = 0°$. It is then moved to the opposite side, at angle $\theta = 180°$, through the first and second quadrants of the $xy$ coordinate system. Figure 22-40b gives the $x$ component of the net electric field produced at the origin by the two beads as a function of $\theta$, and Fig. 22-40c gives the $y$ component. The vertical axis scales are set by $E_{xs} = 5.0 \times 10^4$ N/C and $E_{ys} = -9.0 \times 10^4$ N/C. (a) At what angle $\theta$ is bead 2 located? What are the charges of (b) bead 1 and (c) bead 2?

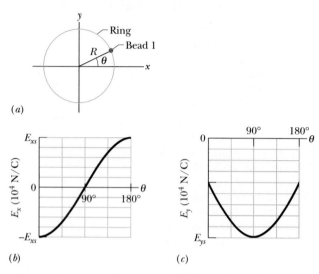

(a)

(b)    (c)

FIG. 22-40    Problem 17.

### sec. 22-5    The Electric Field Due to an Electric Dipole

**••18** Equations 22-8 and 22-9 are approximations of the magnitude of the electric field of an electric dipole, at points along the dipole axis. Consider a point $P$ on that axis at distance $z = 5.00d$ from the dipole center ($d$ is the separation distance between the particles of the dipole). Let $E_{appr}$ be the magnitude of the field at point $P$ as approximated by Eqs. 22-8 and 22-9. Let $E_{act}$ be the actual magnitude. What is the ratio $E_{appr}/E_{act}$?

**••19** Figure 22-41 shows an electric dipole. What are the (a) magnitude and (b) direction (relative to the positive direction of the $x$ axis) of the dipole's electric field at point $P$, located at distance $r \gg d$?

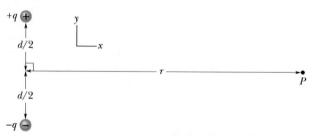

FIG. 22-41    Problem 19.

**••20** The electric field of an electric dipole along the dipole axis is approximated by Eqs. 22-8 and 22-9. If a binomial expansion is made of Eq. 22-7, what is the next term in the expression for the dipole's electric field along the dipole axis? That is, what is $E_{next}$ in the expression

$$E = \frac{1}{2\pi\varepsilon_0} \frac{qd}{z^3} + E_{next}?$$

**•••21** *Electric quadrupole.* Figure 22-42 shows an electric quadrupole. It consists of two dipoles with dipole moments that are equal in magnitude but opposite in direction. Show that the value of $E$ on the axis of the quadrupole for a point $P$ a distance $z$ from its center (assume $z \gg d$) is given by

$$E = \frac{3Q}{4\pi\varepsilon_0 z^4},$$

in which $Q$ ($= 2qd^2$) is known as the *quadrupole moment* of the charge distribution. **SSM**

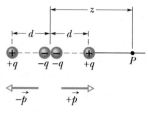

FIG. 22-42    Problem 21.

### sec. 22-6    The Electric Field Due to a Line of Charge

**•22** *Density, density, density.* (a) A charge $-300e$ is uniformly distributed along a circular arc of radius 4.00 cm, which subtends an angle of 40°. What is the linear charge density along the arc? (b) A charge $-300e$ is uniformly distributed over one face of a circular disk of radius 2.00 cm. What is the surface charge density over that face? (c) A charge $-300e$ is uniformly distributed over the surface of a sphere of radius 2.00 cm. What is the surface charge density over that surface? (d) A charge $-300e$ is uniformly spread through the volume of a sphere of radius 2.00 cm. What is the volume charge density in that sphere?

**•23** Figure 22-43 shows two parallel nonconducting rings with their central axes along a common line. Ring 1 has uniform charge $q_1$ and radius $R$; ring 2 has uniform charge $q_2$ and the same radius $R$. The rings are separated by distance $d = 3.00R$. The net electric field at point $P$ on the common line, at distance $R$ from ring 1, is zero. What is the ratio $q_1/q_2$?

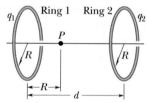

FIG. 22-43    Problem 23.

**••24** In Fig. 22-44, a thin glass rod forms a semicircle of radius $r = 5.00$ cm. Charge is uniformly distributed along the rod, with $+q = 4.50$ pC in the upper half and $-q = -4.50$ pC in the lower half. What are the (a) magnitude and (b) direction (relative to the positive direction of the $x$ axis) of the electric field $\vec{E}$ at $P$, the center of the semicircle? **ILW**

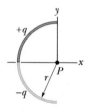

FIG. 22-44
Problem 24.

**••25** In Fig. 22-45, two curved plastic rods, one of charge $+q$ and the other of charge $-q$, form a circle of radius $R = 8.50$ cm in an $xy$ plane. The $x$ axis passes through both of the connecting points, and the charge is distributed uniformly on both rods. If $q = 15.0$ pC, what are the (a) magnitude and (b) direction (relative to the positive direction of the electric field $\vec{E}$ produced at $P$, the center of the circle?

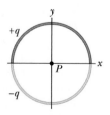

FIG. 22-45
Problem 25.

**••26** Charge is uniformly distributed around a ring of radius $R = 2.40$ cm, and the resulting electric field magnitude $E$ is measured along the ring's central axis (perpendicular to the plane of the ring). At what distance from the ring's center is $E$ maximum?

**••27** In Fig. 22-46, a nonconducting rod of length $L = 8.15$ cm has charge $-q = -4.23$

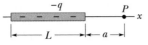

**FIG. 22-46**  Problem 27.

fC uniformly distributed along its length. (a) What is the linear charge density of the rod? What are the (b) magnitude and (c) direction (relative to the positive direction of the $x$ axis) of the electric field produced at point $P$, at distance $a = 12.0$ cm from the rod? What is the electric field magnitude produced at distance $a = 50$ m by (d) the rod and (e) a particle of charge $-q = -4.23$ fC that replaces the rod? **SSM ILW WWW**

**••28** Figure 22-47 shows two concentric rings, of radii $R$ and $R' = 3.00R$, that lie on the same plane. Point $P$ lies on the central $z$ axis, at distance $D = 2.00R$ from the center of the rings. The smaller ring has uniformly distributed charge $+Q$. In terms of $Q$, what is the uniformly distributed charge on the larger ring if the net electric field at $P$ is zero?

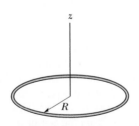

**FIG. 22-47**  Problem 28.

**••29** Figure 22-48 shows three circular arcs centered on the origin of a coordinate system. On each arc, the uniformly distributed charge is given in terms of $Q = 2.00 \mu C$. The radii are given in terms of $R = 10.0$ cm. What are the (a) magnitude and (b) direction (relative to the positive $x$ direction) of the net electric field at the origin due to the arcs?

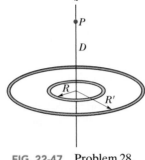

**FIG. 22-48**  Problem 29.

**••30** A thin nonconducting rod with a uniform distribution of positive charge $Q$ is bent into a circle of radius $R$ (Fig. 22-49). The central perpendicular axis through the ring is a $z$ axis, with the origin at the center of the ring. What is the magnitude of the electric field due to the rod at (a) $z = 0$ and (b) $z = \infty$?

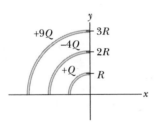

**FIG. 22-49**  Problem 30.

(c) In terms of $R$, at what positive value of $z$ is that magnitude maximum? (d) If $R = 2.00$ cm and $Q = 4.00 \mu C$, what is the maximum magnitude?

**••31** Figure 22-50a shows a nonconducting rod with a uniformly distributed charge $+Q$. The rod forms a half-circle with

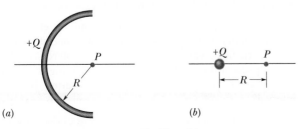

**FIG. 22-50**  Problem 31.

radius $R$ and produces an electric field of magnitude $E_{arc}$ at its center of curvature $P$. If the arc is collapsed to a point at distance $R$ from $P$ (Fig. 22-50b), by what factor is the magnitude of the electric field at $P$ multiplied?

**•••32** In Fig. 22-51, positive charge $q = 7.81$ pC is spread uniformly along a thin nonconducting rod of length $L = 14.5$ cm. What are the (a) magnitude and (b) direction (relative to the positive direction of the $x$ axis) of the electric field produced at point $P$, at distance $R = 6.00$ cm from the rod along its perpendicular bisector? **GO**

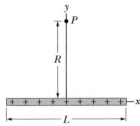

**FIG. 22-51**  Problem 32.

**•••33** In Fig. 22-52, a "semi-infinite" nonconducting rod (that is, infinite in one direction only) has uniform linear charge density $\lambda$. Show that the electric field $\vec{E}_p$ at point $P$ makes an angle of $45°$ with the rod and that this result is independent

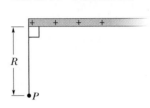

**FIG. 22-52**  Problem 33.

of the distance $R$. (*Hint*: Separately find the component of $\vec{E}_p$ parallel to the rod and the component perpendicular to the rod.)

### sec. 22-7  The Electric Field Due to a Charged Disk

**•34** A disk of radius 2.5 cm has a surface charge density of $5.3 \mu C/m^2$ on its upper face. What is the magnitude of the electric field produced by the disk at a point on its central axis at distance $z = 12$ cm from the disk?

**•35** At what distance along the central perpendicular axis of a uniformly charged plastic disk of radius 0.600 m is the magnitude of the electric field equal to one-half the magnitude of the field at the center of the surface of the disk? **SSM WWW**

**••36** Figure 22-53a shows a circular disk that is uniformly charged. The central $z$ axis is perpendicular to the disk face, with the origin at the disk. Figure 22-53b gives the magnitude of the electric field along that axis in terms of the maximum magnitude $E_m$ at the disk surface. The $z$ axis scale is set by $z_s = 8.0$ cm. What is the radius of the disk?

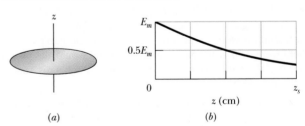

**FIG. 22-53**  Problem 36.

**••37** Suppose you design an apparatus in which a uniformly charged disk of radius $R$ is to produce an electric field. The field magnitude is most important along the central perpendicular axis of the disk, at a point $P$ at distance $2.00R$ from the disk (Fig. 22-54a). Cost analysis suggests that you

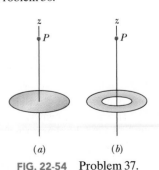

**FIG. 22-54**  Problem 37.

switch to a ring of the same outer radius $R$ but with inner radius $R/2.00$ (Fig. 22-54$b$). Assume that the ring will have the same surface charge density as the original disk. If you switch to the ring, by what percentage will you decrease the electric field magnitude at $P$?

**••38** A circular plastic disk with radius $R = 2.00$ cm has a uniformly distributed charge $Q = +(2.00 \times 10^6)e$ on one face. A circular ring of width 30 $\mu$m is centered on that face, with the center of that width at radius $r = 0.50$ cm. In coulombs, what charge is contained within the width of the ring?

### sec. 22-8 A Point Charge in an Electric Field

**•39** An electron is released from rest in a uniform electric field of magnitude $2.00 \times 10^4$ N/C. Calculate the acceleration of the electron. (Ignore gravitation.) **SSM**

**•40** An electron is accelerated eastward at $1.80 \times 10^9$ m/s$^2$ by an electric field. Determine the (a) magnitude and (b) direction of the electric field.

**•41** An electron on the axis of an electric dipole is 25 nm from the center of the dipole. What is the magnitude of the electrostatic force on the electron if the dipole moment is $3.6 \times 10^{-29}$ C·m? Assume that 25 nm is much larger than the dipole charge separation. **ILW**

**•42** An alpha particle (the nucleus of a helium atom) has a mass of $6.64 \times 10^{-27}$ kg and a charge of $+2e$. What are the (a) magnitude and (b) direction of the electric field that will balance the gravitational force on the particle?

**•43** A charged cloud system produces an electric field in the air near Earth's surface. A particle of charge $-2.0 \times 10^{-9}$ C is acted on by a downward electrostatic force of $3.0 \times 10^{-6}$ N when placed in this field. (a) What is the magnitude of the electric field? What are the (b) magnitude and (c) direction of the electrostatic force $\vec{F}_{el}$ on the proton placed in this field? (d) What is the magnitude of the gravitational force $\vec{F}_g$ on the proton? (e) What is the ratio $F_{el}/F_g$ in this case? **SSM**

**•44** Humid air breaks down (its molecules become ionized) in an electric field of $3.0 \times 10^6$ N/C. In that field, what is the magnitude of the electrostatic force on (a) an electron and (b) an ion with a single electron missing?

**•45** Beams of high-speed protons can be produced in "guns" using electric fields to accelerate the protons. (a) What acceleration would a proton experience if the gun's electric field were $2.00 \times 10^4$ N/C? (b) What speed would the proton attain if the field accelerated the proton through a distance of 1.00 cm? **SSM**

**•46** An electron with a speed of $5.00 \times 10^8$ cm/s enters an electric field of magnitude $1.00 \times 10^3$ N/C, traveling along a field line in the direction that retards its motion. (a) How far will the electron travel in the field before stopping momentarily, and (b) how much time will have elapsed? (c) If the region containing the electric field is 8.00 mm long (too short for the electron to stop within it), what fraction of the electron's initial kinetic energy will be lost in that region? **GO**

**•47** In Millikan's experiment, an oil drop of radius 1.64 $\mu$m and density 0.851 g/cm$^3$ is suspended in chamber C (Fig. 22-14) when a downward electric field of $1.92 \times 10^5$ N/C is applied. Find the charge on the drop, in terms of $e$.

**••48** At some instant the velocity components of an electron moving between two charged parallel plates are $v_x =$

$1.5 \times 10^5$ m/s and $v_y = 3.0 \times 10^3$ m/s. Suppose the electric field between the plates is given by $\vec{E} = (120$ N/C)$\hat{j}$. In unit-vector notation, what are (a) the electron's acceleration in that field and (b) the electron's velocity when its $x$ coordinate has changed by 2.0 cm?

**••49** A uniform electric field exists in a region between two oppositely charged plates. An electron is released from rest at the surface of the negatively charged plate and strikes the surface of the opposite plate, 2.0 cm away, in a time $1.5 \times 10^{-8}$ s. (a) What is the speed of the electron as it strikes the second plate? (b) What is the magnitude of the electric field $\vec{E}$? **ILW**

**••50** In Fig. 22-55, an electron is shot at an initial speed of $v_0 = 2.00 \times 10^6$ m/s, at angle $\theta_0 = 40.0°$ from an $x$ axis. It moves through a uniform electric field $\vec{E} = (5.00$ N/C)$\hat{j}$. A screen for detecting electrons is positioned parallel to the $y$ axis, at distance $x = 3.00$ m. In unit-vector notation, what is the velocity of the electron when it hits the screen? **GO**

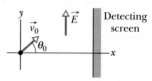

**FIG. 22-55** Problem 50.

**••51** Two large parallel copper plates are 5.0 cm apart and have a uniform electric field between them as depicted in Fig. 22-56. An electron is released from the negative plate at the same time that a proton is released from the positive plate. Neglect the force of the particles on each other and find their distance from the positive plate when they pass each other. (Does it surprise you that you need not know the electric field to solve this problem?) **GO**

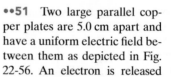

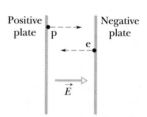

**FIG. 22-56** Problem 51.

**••52** In Fig. 22-57, an electron (e) is to be released from rest on the central axis of a uniformly charged disk of radius $R$. The surface charge density on the disk is $+4.00$ $\mu$C/m$^2$. What is the magnitude of the electron's initial acceleration if it is released at a distance (a) $R$, (b) $R/100$, and (c) $R/1000$ from the center of the disk? (d) Why does the acceleration magnitude increase only slightly as the release point is moved closer to the disk?

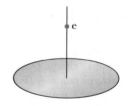

**FIG. 22-57** Problem 52.

**••53** A 10.0 g block with a charge of $+8.00 \times 10^{-5}$ C is placed in an electric field $\vec{E} = (3000\hat{i} - 600\hat{j})$ N/C. What are the (a) magnitude and (b) direction (relative to the positive direction of the $x$ axis) of the electrostatic force on the block? If the block is released from rest at the origin at time $t = 0$, what are its (c) $x$ and (d) $y$ coordinates at $t = 3.00$ s?

**••54** An electron enters a region of uniform electric field with an initial velocity of 40 km/s in the same direction as the electric field, which has magnitude $E = 50$ N/C. (a) What is the speed of the electron 1.5 ns after entering this region? (b) How far does the electron travel during the 1.5 ns interval?

**••55** Assume that a honeybee is a sphere of diameter 1.000 cm with a charge of $+45.0$ pC uniformly spread over its surface.

Assume also that a spherical pollen grain of diameter 40.0 $\mu$m is electrically held on the surface of the sphere because the bee's charge induces a charge of $-1.00$ pC on the near side of the sphere and a charge of $+1.00$ pC on the far side. (a) What is the magnitude of the net electrostatic force on the grain due to the bee? Next, assume that the bee brings the grain to a distance of 1.000 mm from the tip of a flower's stigma and that the tip is a particle of charge $-45.0$ pC. (b) What is the magnitude of the net electrostatic force on the grain due to the stigma? (c) Does the grain remain on the bee or does it move to the stigma?

### sec. 22-9 A Dipole in an Electric Field

•56 An electric dipole consists of charges $+2e$ and $-2e$ separated by 0.78 nm. It is in an electric field of strength $3.4 \times 10^6$ N/C. Calculate the magnitude of the torque on the dipole when the dipole moment is (a) parallel to, (b) perpendicular to, and (c) antiparallel to the electric field.

•57 An electric dipole consisting of charges of magnitude 1.50 nC separated by 6.20 $\mu$m is in an electric field of strength 1100 N/C. What are (a) the magnitude of the electric dipole moment and (b) the difference between the potential energies for dipole orientations parallel and antiparallel to $\vec{E}$? SSM

••58 A certain electric dipole is placed in a uniform electric field $\vec{E}$ of magnitude 40 N/C. Figure 22-58 gives the magnitude $\tau$ of the torque on the dipole versus the angle $\theta$ between field $\vec{E}$ and the dipole moment $\vec{p}$. The vertical axis scale is set by $\tau_s = 100 \times 10^{-28}$ N·m. What is the magnitude of $\vec{p}$?

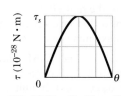

FIG. 22-58 Problem 58.

••59 Find an expression for the oscillation frequency of an electric dipole of dipole moment $\vec{p}$ and rotational inertia $I$ for small amplitudes of oscillation about its equilibrium position in a uniform electric field of magnitude $E$.

••60 A certain electric dipole is placed in a uniform electric field $\vec{E}$ of magnitude 20 N/C. Figure 22-59 gives the potential energy $U$ of the dipole versus the angle $\theta$ between $\vec{E}$ and the dipole moment $\vec{p}$. The vertical axis scale is set by $U_s = 100 \times 10^{-28}$ J. What is the magnitude of $\vec{p}$?

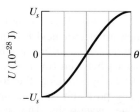

FIG. 22-59 Problem 60.

••61 How much work is required to turn an electric dipole 180° in a uniform electric field of magnitude $E = 46.0$ N/C if $p = 3.02 \times 10^{-25}$ C·m and the initial angle is 64°?

### Additional Problems

62 In one of his experiments, Millikan observed that the following measured charges, among others, appeared at different times on a single drop:

| | | |
|---|---|---|
| $6.563 \times 10^{-19}$ C | $13.13 \times 10^{-19}$ C | $19.71 \times 10^{-19}$ C |
| $8.204 \times 10^{-19}$ C | $16.48 \times 10^{-19}$ C | $22.89 \times 10^{-19}$ C |
| $11.50 \times 10^{-19}$ C | $18.08 \times 10^{-19}$ C | $26.13 \times 10^{-19}$ C |

What value for the elementary charge $e$ can be deduced from these data?

63 In Fig. 22-60a, a particle of charge $+Q$ produces an electric field of magnitude $E_{part}$ at point $P$, at distance $R$ from the particle. In Fig. 22-60b, that same amount of charge is spread uniformly along a circular arc that has radius $R$ and subtends an angle $\theta$. The charge on the arc produces an electric field of magnitude $E_{arc}$ at its center of curvature $P$. For what value of $\theta$ does $E_{arc} = 0.500E_{part}$? (Hint: You will probably resort to a graphical solution.)

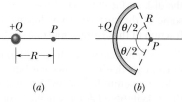

FIG. 22-60 Problem 63.

64 In Fig. 22-61, eight particles form a square in which distance $d = 2.0$ cm. The charges are $q_1 = +3e, q_2 = +e$, $q_3 = -5e$, $q_4 = -2e$, $q_5 = +3e$, $q_6 = +e$, $q_7 = -5e$, and $q_8 = +e$. In unit-vector notation, what net electric field do the particles produce at the square's center?

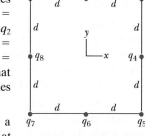

FIG. 22-61 Problem 64.

65 Two particles, each with a charge of magnitude 12 nC, are at two of the vertices of an equilateral triangle with edge length 2.0 m. What is the magnitude of the electric field at the third vertex if (a) both charges are positive and (b) one charge is positive and the other is negative?

66 Three particles, each with positive charge $Q$, form an equilateral triangle, with each side of length $d$. What is the magnitude of the electric field produced by the particles at the midpoint of any side?

67 A particle of charge $-q_1$ is at the origin of an $x$ axis. (a) At what location on the axis should a particle of charge $-4q_1$ be placed so that the net electric field is zero at $x = 2.0$ mm on the axis? (b) If, instead, a particle of charge $+4q_1$ is placed at that location, what is the direction (relative to the positive direction of the $x$ axis) of the net electric field at $x = 2.0$ mm?

68 A proton and an electron form two corners of an equilateral triangle of side length $2.0 \times 10^{-6}$ m. What is the magnitude of the net electric field these two particles produce at the third corner?

69 In Fig. 22-62, particle 1 (of charge $+1.00$ $\mu$C), particle 2 (of charge $+1.00$ $\mu$C), and particle 3 (of charge $Q$) form an equilateral triangle of edge length $a$. For what value of $Q$ (both sign and magnitude) does the net electric field produced by the particles at the center of the triangle vanish?

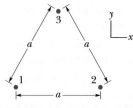

FIG. 22-62 Problems 69 and 82.

70 (a) What total (excess) charge $q$ must the disk in Fig. 22-13 have for the electric field on the surface of the disk at its center to have magnitude $3.0 \times 10^6$ N/C, the $E$ value at which air breaks down electrically, producing sparks? Take the disk radius as 2.5 cm, and use the listing for air in Table 22-1. (b) Suppose each surface atom has an effective cross-sectional area of 0.015 nm². How many atoms are needed to make up

the disk surface? (c) The charge calculated in (a) results from some of the surface atoms having one excess electron. What fraction of these atoms must be so charged?

**71** A spherical water drop 1.20 μm in diameter is suspended in calm air due to a downward-directed atmospheric electric field of magnitude $E = 462$ N/C. (a) What is the magnitude of the gravitational force on the drop? (b) How many excess electrons does it have?

**72** In Fig. 22-63, an electric dipole swings from an initial orientation $i$ ($\theta_i = 20.0°$) to a final orientation $f$ ($\theta_f = 20.0°$) in a uniform external electric field $\vec{E}$. The electric dipole moment is $1.60 \times 10^{-27}$ C·m; the field magnitude is $3.00 \times 10^6$ N/C. What is the change in the dipole's potential energy?

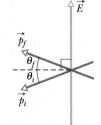

FIG. 22-63
Problem 72.

**73** A charge of 20 nC is uniformly distributed along a straight rod of length 4.0 m that is bent into a circular arc with a radius of 2.0 m. What is the magnitude of the electric field at the center of curvature of the arc?

**74** (a) What is the magnitude of an electron's acceleration in a uniform electric field of magnitude $1.40 \times 10^6$ N/C? (b) How long would the electron take, starting from rest, to attain one-tenth the speed of light? (c) How far would it travel in that time?

**75** A clock face has negative point charges $-q$, $-2q$, $-3q$, $\ldots$, $-12q$ fixed at the positions of the corresponding numerals. The clock hands do not perturb the net field due to the point charges. At what time does the hour hand point in the same direction as the electric field vector at the center of the dial? (*Hint:* Use symmetry.)

**76** An electron is constrained to the central axis of the ring of charge of radius $R$ in Fig. 22-10, with $z \ll R$. Show that the electrostatic force on the electron can cause it to oscillate through the ring center with an angular frequency

$$\omega = \sqrt{\frac{eq}{4\pi\varepsilon_0 mR^3}},$$

where $q$ is the ring's charge and $m$ is the electron's mass.

**77** An electric field $\vec{E}$ with an average magnitude of about 150 N/C points downward in the atmosphere near Earth's surface. We wish to "float" a sulfur sphere weighing 4.4 N in this field by charging the sphere. (a) What charge (both sign and magnitude) must be used? (b) Why is the experiment impractical?

**78** Calculate the electric dipole moment of an electron and a proton 4.30 nm apart.

**79** The electric field in an $xy$ plane produced by a positively charged particle is $7.2(4.0\hat{i} + 3.0\hat{j})$ N/C at the point (3.0, 3.0) cm and $100\hat{i}$ N/C at the point (2.0, 0) cm. What are the (a) $x$ and (b) $y$ coordinates of the particle? (c) What is the charge of the particle? SSM

**80** A circular rod has a radius of curvature $R = 9.00$ cm and a uniformly distributed positive charge $Q = 6.25$ pC and subtends an angle $\theta = 2.40$ rad. What is the magnitude of the electric field that $Q$ produces at the center of curvature?

**81** An electric dipole with dipole moment

$$\vec{p} = (3.00\hat{i} + 4.00\hat{j})(1.24 \times 10^{-30} \text{ C·m})$$

is in an electric field $\vec{E} = (4000$ N/C)$\hat{i}$. (a) What is the potential energy of the electric dipole? (b) What is the torque acting on it? (c) If an external agent turns the dipole until its electric dipole moment is

$$\vec{p} = (-4.00\hat{i} + 3.00\hat{j})(1.24 \times 10^{-30} \text{ C·m}),$$

how much work is done by the agent? SSM

**82** In Fig. 22-62, particle 1 (of charge +2.00 pC), particle 2 (of charge −2.00 pC), and particle 3 (of charge +5.00 pC) form an equilateral triangle of edge length $a = 9.50$ cm. (a) Relative to the positive direction of the $x$ axis, determine the direction of the force $\vec{F}_3$ on particle 3 due to the other particles by sketching electric field lines of the other particles. (b) Calculate the magnitude of force $\vec{F}_3$.

**83** A charge (uniform linear density = 9.0 nC/m) lies on a string that is stretched along an $x$ axis from $x = 0$ to $x = 3.0$ m. Determine the magnitude of the electric field at $x = 4.0$ m on the $x$ axis.

**84** Two particles, each of positive charge $q$, are fixed in place on a $y$ axis, one at $y = d$ and the other at $y = -d$. (a) Write an expression that gives the magnitude $E$ of the net electric field at points on the $x$ axis given by $x = \alpha d$. (b) Graph $E$ versus $\alpha$ for the range $0 < \alpha < 4$. From the graph, determine the values of $\alpha$ that give (c) the maximum value of $E$ and (d) half the maximum value of $E$.

**85** In Fig. 22-64, particle 1 of charge $q_1 = 1.00$ pC and particle 2 of charge $q_2 = -2.00$ pC are fixed at a distance $d = 5.00$ cm apart. In unit-vector notation, what is the net electric field at points (a) $A$, (b) $B$, and (c) $C$? (d) Sketch the electric field lines.

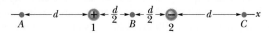

FIG. 22-64   Problem 85.

**86** In Fig. 22-65, a uniform, upward electric field $\vec{E}$ of magnitude $2.00 \times 10^3$ N/C has been set up between two horizontal plates by charging the lower plate positively and the upper plate negatively. The plates have length $L = 10.0$ cm and separation $d = 2.00$ cm. An electron is then shot between the plates from the left edge of the lower plate. The initial velocity $\vec{v}_0$ of the electron makes an angle $\theta = 45.0°$ with the lower plate and has a magnitude of $6.00 \times 10^6$ m/s. (a) Will the electron strike one of the plates? (b) If so, which plate and how far horizontally from the left edge will the electron strike?

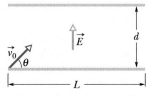

FIG. 22-65   Problem 86.

**87** For the data of Problem 62, assume that the charge $q$ on the drop is given by $q = ne$, where $n$ is an integer and $e$ is the elementary charge. (a) Find $n$ for each given value of $q$. (b) Do a linear regression fit of the values of $q$ versus the values of $n$ and then use that fit to find $e$.

**88** In Fig. 22-8, let both charges be positive. Assuming $z \gg d$, show that $E$ at point $P$ in that figure is then given by

$$E = \frac{1}{4\pi\varepsilon_0} \frac{2q}{z^2}.$$

# Gauss' Law

*Lightning storms are dangerous for several reasons: (1) If lightning strikes you or something you are touching, it will produce a fatal charge flow in your body. (2) If it strikes an object near you, a portion of the charge flow can jump through the air to you (an effect known as* side flash*). (3) If lightning strikes the ground near you, part of the charge flow it produces along the ground can be diverted through your body. Recently a fourth danger of a lightning storm was recognized. The photograph here, in which lightning strikes a tree, contains a hint.*

## What is this additional danger of a lightning storm?

The answer is in this chapter.

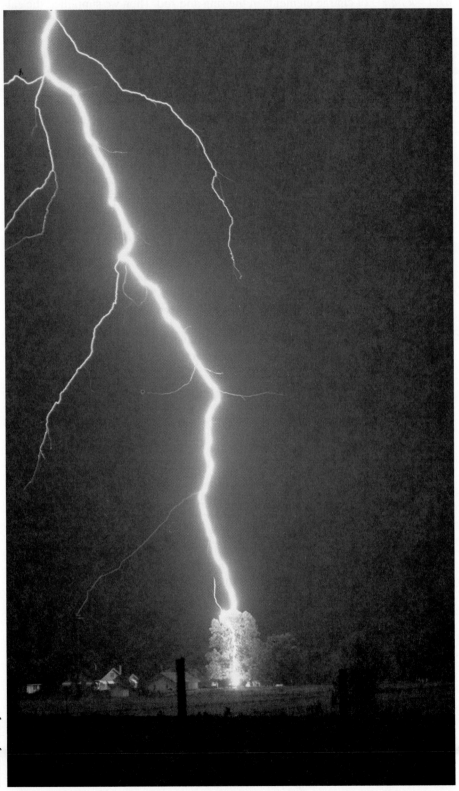

C. Johnny Autery

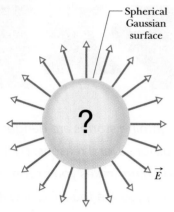

FIG. 23-1   A spherical Gaussian surface. If the electric field vectors are of uniform magnitude and point radially outward at all surface points, you can conclude that a net positive distribution of charge must lie within the surface and have spherical symmetry.

## 23-1 | WHAT IS PHYSICS?

One of the primary goals of physics is to find simple ways of solving seemingly complex problems. One of the main tools of physics in attaining this goal is the use of symmetry. For example, in finding the electric field $\vec{E}$ of the charged ring of Fig. 22-10 and the charged rod of Fig. 22-11, we considered the fields $d\vec{E}$ $(= k\,dq/r^2)$ of charge elements in the ring and rod. Then we simplified the calculation of $\vec{E}$ by using symmetry to discard the perpendicular components of the $d\vec{E}$ vectors. That saved us some work.

For certain charge distributions involving symmetry, we can save far more work by using a law called Gauss' law, developed by German mathematician and physicist Carl Friedrich Gauss (1777–1855). Instead of considering the fields $d\vec{E}$ of charge elements in a given charge distribution, Gauss' law considers a hypothetical (imaginary) closed surface enclosing the charge distribution. This **Gaussian surface**, as it is called, can have any shape, but the shape that minimizes our calculations of the electric field is one that mimics the symmetry of the charge distribution. For example, if the charge is spread uniformly over a sphere, we enclose the sphere with a spherical Gaussian surface, such as the one in Fig. 23-1, and then, as we discuss in this chapter, find the electric field on the surface by using the fact that

> Gauss' law relates the electric fields at points on a (closed) Gaussian surface to the net charge enclosed by that surface.

We can also use Gauss' law in reverse: If we know the electric field on a Gaussian surface, we can find the net charge enclosed by the surface. As a limited example, suppose that the electric field vectors in Fig. 23-1 all point radially outward from the center of the sphere and have equal magnitude. Gauss' law immediately tells us that the spherical surface must enclose a net positive charge that is either a particle or distributed spherically. However, to calculate how *much* charge is enclosed, we need a way of calculating how much electric field is intercepted by the Gaussian surface in Fig. 23-1. This measure of intercepted field is called *flux*, which we discuss next.

## 23-2 | Flux

Suppose that, as in Fig. 23-2a, you aim a wide airstream of uniform velocity $\vec{v}$ at a small square loop of area $A$. Let $\Phi$ represent the *volume flow rate* (volume per unit time) at which air flows through the loop. This rate depends on the angle between $\vec{v}$ and the plane of the loop. If $\vec{v}$ is perpendicular to the plane, the rate $\Phi$ is equal to $vA$.

If $\vec{v}$ is parallel to the plane of the loop, no air moves through the loop, so $\Phi$ is zero. For an intermediate angle $\theta$, the rate $\Phi$ depends on the component of

FIG. 23-2   (a) A uniform airstream of velocity $\vec{v}$ is perpendicular to the plane of a square loop of area $A$. (b) The component of $\vec{v}$ perpendicular to the plane of the loop is $v \cos \theta$, where $\theta$ is the angle between $\vec{v}$ and a normal to the plane. (c) The area vector $\vec{A}$ is perpendicular to the plane of the loop and makes an angle $\theta$ with $\vec{v}$. (d) The velocity field intercepted by the area of the loop.

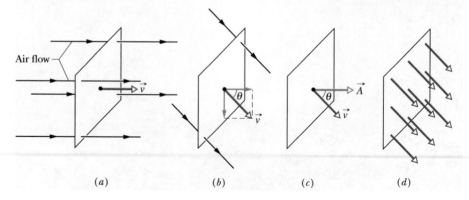

(a)           (b)           (c)           (d)

$\vec{v}$ normal to the plane (Fig. 23-2b). Since that component is $v \cos \theta$, the rate of volume flow through the loop is

$$\Phi = (v \cos \theta)A. \qquad (23\text{-}1)$$

This rate of flow through an area is an example of a **flux**—a *volume flux* in this situation.

Before we discuss a flux involved in electrostatics, we need to rewrite Eq. 23-1 in terms of vectors. To do this, we first define an *area vector* $\vec{A}$ as being a vector whose magnitude is equal to an area (here the area of the loop) and whose direction is normal to the plane of the area (Fig. 23-2c). We then rewrite Eq. 23-1 as the scalar (or dot) product of the velocity vector $\vec{v}$ of the airstream and the area vector $\vec{A}$ of the loop:

$$\Phi = vA \cos \theta = \vec{v} \cdot \vec{A}, \qquad (23\text{-}2)$$

where $\theta$ is the angle between $\vec{v}$ and $\vec{A}$.

The word "flux" comes from the Latin word meaning "to flow." That meaning makes sense if we talk about the flow of air volume through the loop. However, Eq. 23-2 can be regarded in a more abstract way. To see this different way, note that we can assign a velocity vector to each point in the airstream passing through the loop (Fig. 23-2d). Because the composite of all those vectors is a *velocity field*, we can interpret Eq. 23-2 as giving the *flux of the velocity field through the loop*. With this interpretation, flux no longer means the actual flow of something through an area—rather it means the product of an area and the field across that area.

## 23-3 | Flux of an Electric Field

To define the flux of an electric field, consider Fig. 23-3, which shows an arbitrary (asymmetric) Gaussian surface immersed in a nonuniform electric field. Let us divide the surface into small squares of area $\Delta A$, each square being small enough to permit us to neglect any curvature and to consider the individual square to be flat. We represent each such element of area with an area vector $\Delta \vec{A}$, whose magnitude is the area $\Delta A$. Each vector $\Delta \vec{A}$ is perpendicular to the Gaussian surface and directed away from the interior of the surface.

Because the squares have been taken to be arbitrarily small, the electric field $\vec{E}$ may be taken as constant over any given square. The vectors $\Delta \vec{A}$ and $\vec{E}$ for each square then make some angle $\theta$ with each other. Figure 23-3 shows an enlarged view of three squares on the Gaussian surface and the angle $\theta$ for each.

A provisional definition for the flux of the electric field for the Gaussian surface of Fig. 23-3 is

$$\Phi = \sum \vec{E} \cdot \Delta \vec{A}. \qquad (23\text{-}3)$$

This equation instructs us to visit each square on the Gaussian surface, evaluate the scalar product $\vec{E} \cdot \Delta \vec{A}$ for the two vectors $\vec{E}$ and $\Delta \vec{A}$ we find there, and sum the results algebraically (that is, with signs included) for all the squares that make up the surface. The value of each scalar product (positive, negative, or zero) determines whether the flux through its square is positive, negative, or zero. Squares like square 1 in Fig. 23-3, in which $\vec{E}$ points inward, make a negative contribution to the sum of Eq. 23-3. Squares like 2, in which $\vec{E}$ lies in the surface, make zero contribution. Squares like 3, in which $\vec{E}$ points outward, make a positive contribution.

The exact definition of the flux of the electric field through a closed surface is found by allowing the area of the squares shown in Fig. 23-3 to become smaller and smaller, approaching a differential limit $dA$. The area vectors then approach a differential limit $d\vec{A}$. The sum of Eq. 23-3 then becomes an integral and we

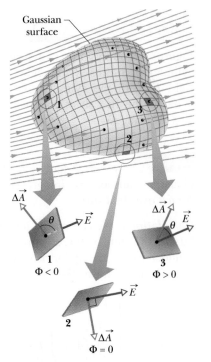

**FIG. 23-3** A Gaussian surface of arbitrary shape immersed in an electric field. The surface is divided into small squares of area $\Delta A$. The electric field vectors $\vec{E}$ and the area vectors $\Delta \vec{A}$ for three representative squares, marked 1, 2, and 3, are shown.

have, for the definition of electric flux,

$$\Phi = \oint \vec{E} \cdot d\vec{A} \qquad \text{(electric flux through a Gaussian surface).}\qquad (23\text{-}4)$$

The loop on the integral sign indicates that the integration is to be taken over the entire (closed) surface. The flux of the electric field is a scalar, and its SI unit is the newton–square-meter per coulomb ($\text{N} \cdot \text{m}^2/\text{C}$).

We can interpret Eq. 23-4 in the following way: First recall that we can use the density of electric field lines passing through an area as a proportional measure of the magnitude of the electric field $\vec{E}$ there. Specifically, the magnitude $E$ is proportional to the number of electric field lines per unit area. Thus, the scalar product $\vec{E} \cdot d\vec{A}$ in Eq. 23-4 is proportional to the number of electric field lines passing through area $d\vec{A}$. Then, because the integration in Eq. 23-4 is carried out over a Gaussian surface, which is closed, we see that

> The electric flux $\Phi$ through a Gaussian surface is proportional to the net number of electric field lines passing through that surface.

✓**CHECKPOINT 1**     The figure here shows a Gaussian cube of face area $A$ immersed in a uniform electric field $\vec{E}$ that has the positive direction of the $z$ axis. In terms of $E$ and $A$, what is the flux through (a) the front face (which is in the $xy$ plane), (b) the rear face, (c) the top face, and (d) the whole cube?

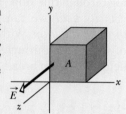

**Sample Problem | 23-1**

Figure 23-4 shows a Gaussian surface in the form of a cylinder of radius $R$ immersed in a uniform electric field $\vec{E}$, with the cylinder axis parallel to the field. What is the flux $\Phi$ of the electric field through this closed surface?

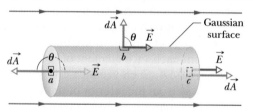

**FIG. 23-4** A cylindrical Gaussian surface, closed by end caps, is immersed in a uniform electric field. The cylinder axis is parallel to the field direction.

**KEY IDEA** We can find the flux $\Phi$ through the Gaussian surface by integrating the scalar product $\vec{E} \cdot d\vec{A}$ over that surface.

**Calculations:** We can do the integration by writing the flux as the sum of three terms: integrals over the left cylinder cap $a$, the cylindrical surface $b$, and the right cap $c$. Thus, from Eq. 23-4,

$$\Phi = \oint \vec{E} \cdot d\vec{A}$$

$$= \int_a \vec{E} \cdot d\vec{A} + \int_b \vec{E} \cdot d\vec{A} + \int_c \vec{E} \cdot d\vec{A}. \qquad (23\text{-}5)$$

For all points on the left cap, the angle $\theta$ between $\vec{E}$ and $d\vec{A}$ is 180° and the magnitude $E$ of the field is uniform. Thus,

$$\int_a \vec{E} \cdot d\vec{A} = \int E(\cos 180°)\, dA = -E \int dA = -EA,$$

where $\int dA$ gives the cap's area $A$ ($= \pi R^2$). Similarly,

for the right cap, where $\theta = 0$ for all points,

$$\int_c \vec{E} \cdot d\vec{A} = \int E(\cos 0)\, dA = EA.$$

Finally, for the cylindrical surface, where the angle $\theta$ is 90° at all points,

$$\int_b \vec{E} \cdot d\vec{A} = \int E(\cos 90°)\, dA = 0.$$

Substituting these results into Eq. 23-5 leads us to

$$\Phi = -EA + 0 + EA = 0. \qquad \text{(Answer)}$$

This result is perhaps not surprising because the field lines that represent the electric field all pass entirely through the Gaussian surface, entering through the left end cap, leaving through the right end cap, and giving a net flux of zero.

**Sample Problem** 23-2 **Build your skill**

A *nonuniform* electric field given by $\vec{E} = 3.0x\hat{i} + 4.0\hat{j}$ pierces the Gaussian cube shown in Fig. 23-5. ($E$ is in newtons per coulomb and $x$ is in meters.) What is the electric flux through the right face, the left face, and the top face? (We consider the other faces in Sample Problem 23-4.)

**KEY IDEA** We can find the flux $\Phi$ through the surface by integrating the scalar product $\vec{E} \cdot d\vec{A}$ over each face.

*Right face:* An area vector $\vec{A}$ is always perpendicular to its surface and always points away from the interior of a Gaussian surface. Thus, the vector $d\vec{A}$ for the right face of the cube must point in the positive direction of the $x$ axis. In unit-vector notation,

$$d\vec{A} = dA\hat{i}.$$

From Eq. 23-4, the flux $\Phi_r$ through the right face is then

$$\Phi_r = \int \vec{E} \cdot d\vec{A} = \int (3.0x\hat{i} + 4.0\hat{j}) \cdot (dA\hat{i})$$

$$= \int [(3.0x)(dA)\hat{i} \cdot \hat{i} + (4.0)(dA)\hat{j} \cdot \hat{i}]$$

$$= \int (3.0x\, dA + 0) = 3.0 \int x\, dA.$$

We are about to integrate over the right face, but we note that $x$ has the same value everywhere on that face—namely, $x = 3.0$ m. This means we can substitute that constant value for $x$. Then

$$\Phi_r = 3.0 \int (3.0)\, dA = 9.0 \int dA.$$

The integral $\int dA$ merely gives us the area $A = 4.0$ m$^2$ of the right face; so

$$\Phi_r = (9.0\text{ N/C})(4.0\text{ m}^2) = 36\text{ N} \cdot \text{m}^2/\text{C}. \quad \text{(Answer)}$$

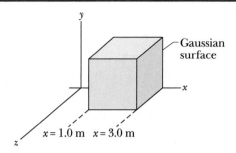

FIG. 23-5 A Gaussian cube with one edge on the $x$ axis lies within a nonuniform electric field.

*Left face:* The procedure for finding the flux through the left face is the same as that for the right face. However, two factors change. (1) The differential area vector $d\vec{A}$ points in the negative direction of the $x$ axis, and thus $d\vec{A} = -dA\hat{i}$. (2) The term $x$ again appears in our integration, and it is again constant over the face being considered. However, on the left face, $x = 1.0$ m. With these two changes, we find that the flux $\Phi_l$ through the left face is

$$\Phi_l = -12\text{ N} \cdot \text{m}^2/\text{C}. \quad \text{(Answer)}$$

*Top face:* The differential area vector $d\vec{A}$ points in the positive direction of the $y$ axis, and thus $d\vec{A} = dA\hat{j}$. The flux $\Phi_t$ through the top face is then

$$\Phi_t = \int (3.0x\hat{i} + 4.0\hat{j}) \cdot (dA\hat{j})$$

$$= \int [(3.0x)(dA)\hat{i} \cdot \hat{j} + (4.0)(dA)\hat{j} \cdot \hat{j}]$$

$$= \int (0 + 4.0\, dA) = 4.0 \int dA$$

$$= 16\text{ N} \cdot \text{m}^2/\text{C}. \quad \text{(Answer)}$$

## 23-4 | Gauss' Law

Gauss' law relates the net flux $\Phi$ of an electric field through a closed surface (a Gaussian surface) to the *net* charge $q_{enc}$ that is *enclosed* by that surface. It tells us that

$$\varepsilon_0 \Phi = q_{enc} \quad \text{(Gauss' law).} \quad (23\text{-}6)$$

By substituting Eq. 23-4, the definition of flux, we can also write Gauss' law as

$$\varepsilon_0 \oint \vec{E} \cdot d\vec{A} = q_{enc} \quad \text{(Gauss' law).} \quad (23\text{-}7)$$

Equations 23-6 and 23-7 hold only when the net charge is located in a vacuum or (what is the same for most practical purposes) in air. In Chapter 25, we modify Gauss' law to include situations in which a material such as mica, oil, or glass is present.

In Eqs. 23-6 and 23-7, the net charge $q_{enc}$ is the algebraic sum of all the *enclosed* positive and negative charges, and it can be positive, negative, or zero. We

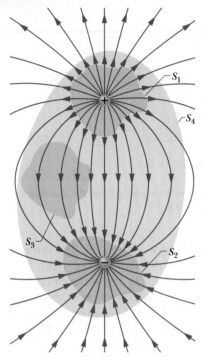

**FIG. 23-6** Two point charges, equal in magnitude but opposite in sign, and the field lines that represent their net electric field. Four Gaussian surfaces are shown in cross section. Surface $S_1$ encloses the positive charge. Surface $S_2$ encloses the negative charge. Surface $S_3$ encloses no charge. Surface $S_4$ encloses both charges and thus no net charge.

include the sign, rather than just use the magnitude of the enclosed charge, because the sign tells us something about the net flux through the Gaussian surface: If $q_{enc}$ is positive, the net flux is *outward*; if $q_{enc}$ is negative, the net flux is *inward*.

Charge outside the surface, no matter how large or how close it may be, is not included in the term $q_{enc}$ in Gauss' law. The exact form and location of the charges inside the Gaussian surface are also of no concern; the only things that matter on the right side of Eqs. 23-6 and 23-7 are the magnitude and sign of the net enclosed charge. The quantity $\vec{E}$ on the left side of Eq. 23-7, however, is the electric field resulting from *all* charges, both those inside and those outside the Gaussian surface. This may seem to be inconsistent, but keep in mind what we saw in Sample Problem 23-1: The electric field due to a charge outside the Gaussian surface contributes zero net flux *through* the surface, because as many field lines due to that charge enter the surface as leave it.

Let us apply these ideas to Fig. 23-6, which shows two point charges, equal in magnitude but opposite in sign, and the field lines describing the electric fields the charges set up in the surrounding space. Four Gaussian surfaces are also shown, in cross section. Let us consider each in turn.

**Surface $S_1$.** The electric field is outward for all points on this surface. Thus, the flux of the electric field through this surface is positive, and so is the net charge within the surface, as Gauss' law requires. (That is, in Eq. 23-6, if $\Phi$ is positive, $q_{enc}$ must be also.)

**Surface $S_2$.** The electric field is inward for all points on this surface. Thus, the flux of the electric field is negative and so is the enclosed charge, as Gauss' law requires.

**Surface $S_3$.** This surface encloses no charge, and thus $q_{enc} = 0$. Gauss' law (Eq. 23-6) requires that the net flux of the electric field through this surface be zero. That is reasonable because all the field lines pass entirely through the surface, entering it at the top and leaving at the bottom.

**Surface $S_4$.** This surface encloses no *net* charge, because the enclosed positive and negative charges have equal magnitudes. Gauss' law requires that the net flux of the electric field through this surface be zero. That is reasonable because there are as many field lines leaving surface $S_4$ as entering it.

What would happen if we were to bring an enormous charge $Q$ up close to surface $S_4$ in Fig. 23-6? The pattern of the field lines would certainly change, but the net flux for each of the four Gaussian surfaces would not change. We can understand this because the field lines associated with the added $Q$ would pass entirely through each of the four Gaussian surfaces, making no contribution to the net flux through any of them. The value of $Q$ would not enter Gauss' law in any way, because $Q$ lies outside all four of the Gaussian surfaces that we are considering.

✓**CHECKPOINT 2** The figure shows three situations in which a Gaussian cube sits in an electric field. The arrows and the values indicate the directions of the field lines and the magnitudes (in N·m²/C) of the flux through the six sides of each cube. (The lighter arrows are for the hidden faces.) In which situation does the cube enclose (a) a positive net charge, (b) a negative net charge, and (c) zero net charge?

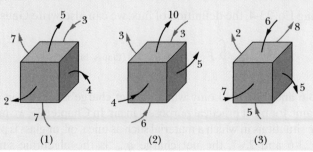

## Sample Problem   23-3

Figure 23-7 shows five charged lumps of plastic and an electrically neutral coin. The cross section of a Gaussian surface $S$ is indicated. What is the net electric flux through the surface if $q_1 = q_4 = +3.1$ nC, $q_2 = q_5 = -5.9$ nC, and $q_3 = -3.1$ nC?

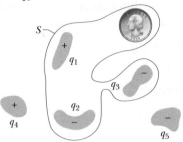

**FIG. 23-7**   Five plastic objects, each with an electric charge, and a coin, which has no net charge. A Gaussian surface, shown in cross section, encloses three of the plastic objects and the coin.

**KEY IDEA**   The *net* flux $\Phi$ through the surface depends on the *net* charge $q_{enc}$ enclosed by surface $S$.

*Calculation:* The coin does not contribute to $\Phi$ because it is neutral and thus contains equal amounts of positive and negative charge. Charges $q_4$ and $q_5$ do not contribute because they are outside surface $S$. Thus, $q_{enc}$ is $q_1 + q_2 + q_3$ and Eq. 23-6 gives us

$$\Phi = \frac{q_{enc}}{\varepsilon_0} = \frac{q_1 + q_2 + q_3}{\varepsilon_0}$$

$$= \frac{+3.1 \times 10^{-9}\,\text{C} - 5.9 \times 10^{-9}\,\text{C} - 3.1 \times 10^{-9}\,\text{C}}{8.85 \times 10^{-12}\,\text{C}^2/\text{N}\cdot\text{m}^2}$$

$$= -670\,\text{N}\cdot\text{m}^2/\text{C}. \qquad \text{(Answer)}$$

The minus sign shows that the net flux through the surface is inward and thus that the net charge within the surface is negative.

## Sample Problem   23-4   Build your skill

What is the net charge enclosed by the Gaussian cube of Sample Problem 23-2 and Fig. 23-5?

**KEY IDEA**   The net charge enclosed by a (real or mathematical) closed surface is related to the total electric flux through the surface by Gauss' law as given by Eq. 23-6 ($\varepsilon_0 \Phi = q_{enc}$).

*Flux:* To use Eq. 23-6, we need to know the flux through all six faces of the cube. We already know the flux through the right face ($\Phi_r = 36$ N·m²/C), the left face ($\Phi_l = -12$ N·m²/C), and the top face ($\Phi_t = 16$ N·m²/C).

For the bottom face, our calculation is just like that for the top face *except* that the differential area vector $d\vec{A}$ is now directed downward along the $y$ axis (recall, it must be *outward* from the Gaussian enclosure). Thus, we have $d\vec{A} = -dA\hat{j}$, and we find

$$\Phi_b = -16\,\text{N}\cdot\text{m}^2/\text{C}.$$

For the front face we have $d\vec{A} = dA\hat{k}$, and for the back face, $d\vec{A} = -dA\hat{k}$. When we take the dot product of the given electric field $\vec{E} = 3.0x\hat{i} + 4.0\hat{j}$ with either of these expressions for $d\vec{A}$, we get 0 and thus there is no flux through those faces. We can now find the total flux through the six sides of the cube:

$$\Phi = (36 - 12 + 16 - 16 + 0 + 0)\,\text{N}\cdot\text{m}^2/\text{C}$$
$$= 24\,\text{N}\cdot\text{m}^2/\text{C}.$$

*Enclosed charge:* Next, we use Gauss' law to find the charge $q_{enc}$ enclosed by the cube:

$$q_{enc} = \varepsilon_0 \Phi = (8.85 \times 10^{-12}\,\text{C}^2/\text{N}\cdot\text{m}^2)(24\,\text{N}\cdot\text{m}^2/\text{C})$$
$$= 2.1 \times 10^{-10}\,\text{C}. \qquad \text{(Answer)}$$

Thus, the cube encloses a *net* positive charge.

## 23-5 | Gauss' Law and Coulomb's Law

Because Gauss' law and Coulomb's law are different ways of describing the relation between electric charge and electric field in static situations, we should be able to derive each from the other. Here we derive Coulomb's law from Gauss' law and some symmetry considerations.

Figure 23-8 shows a positive point charge $q$, around which we have drawn a concentric spherical Gaussian surface of radius $r$. Let us divide this surface into differential areas $dA$. By definition, the area vector $d\vec{A}$ at any point is perpendicular to the surface and directed outward from the interior. From the symmetry of the situation, we know that at any point the electric field $\vec{E}$ is also perpendicular to the surface and directed outward from the interior. Thus, since the angle $\theta$ between $\vec{E}$ and $d\vec{A}$ is zero, we can rewrite Eq. 23-7 for Gauss' law as

$$\varepsilon_0 \oint \vec{E} \cdot d\vec{A} = \varepsilon_0 \oint E\,dA = q_{enc}. \qquad (23\text{-}8)$$

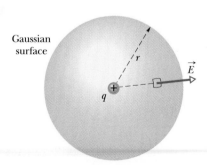

**FIG. 23-8**   A spherical Gaussian surface centered on a point charge $q$.

Here $q_{enc} = q$. Although $E$ varies radially with distance from $q$, it has the same value everywhere on the spherical surface. Since the integral in Eq. 23-8 is taken over that surface, $E$ is a constant in the integration and can be brought out in front of the integral sign. That gives us

$$\varepsilon_0 E \oint dA = q. \tag{23-9}$$

The integral is now merely the sum of all the differential areas $dA$ on the sphere and thus is just the surface area, $4\pi r^2$. Substituting this, we have

$$\varepsilon_0 E(4\pi r^2) = q$$

or

$$E = \frac{1}{4\pi\varepsilon_0} \frac{q}{r^2}. \tag{23-10}$$

This is exactly Eq. 22-3, which we found using Coulomb's law.

✓**CHECKPOINT 3** There is a certain net flux $\Phi_i$ through a Gaussian sphere of radius $r$ enclosing an isolated charged particle. Suppose the enclosing Gaussian surface is changed to (a) a larger Gaussian sphere, (b) a Gaussian cube with edge length equal to $r$, and (c) a Gaussian cube with edge length equal to $2r$. In each case, is the net flux through the new Gaussian surface greater than, less than, or equal to $\Phi_i$?

**PROBLEM-SOLVING TACTICS**

*Tactic 1:* *Choosing a Gaussian Surface* Because the derivation of Eq. 23-10 using Gauss' law is a warm-up for derivations of electric fields produced by other charge configurations, let us go back over the steps involved. We started with a given positive point charge $q$; we know that electric field lines extend radially outward from $q$ in a spherically symmetric pattern.

To use Gauss' law (Eq. 23-7) to find the magnitude $E$ of the electric field at a distance $r$, we had to place a hypothetical closed Gaussian surface around $q$, through a point that is a distance $r$ from $q$. Then we had to sum via integration the values of $\vec{E} \cdot d\vec{A}$ over the full Gaussian surface. To make this integration as simple as possible, we chose a spherical Gaussian surface (to mimic the spherical symmetry of the electric field). That choice produced three simplifying features. (1) The dot product $\vec{E} \cdot d\vec{A}$ became simple, because at

all points on the Gaussian surface the angle between $\vec{E}$ and $d\vec{A}$ is zero, and so at all points we have $\vec{E} \cdot d\vec{A} = E \, dA$. (2) The electric field magnitude $E$ is the same at all points on the spherical Gaussian surface; so $E$ was a constant in the integration and could be brought out in front of the integral sign. (3) The result was a very simple integration—a summation of the differential areas of the sphere, which we could immediately write as $4\pi r^2$.

Note that Gauss' law holds regardless of the shape of the Gaussian surface we choose to place around charge $q_{enc}$. However, if we had chosen, say, a cubical Gaussian surface, our three simplifying features would have disappeared and the integration of $\vec{E} \cdot d\vec{A}$ over the cubical surface would have been very difficult. The moral here is to choose the Gaussian surface that most simplifies the integration in Gauss' law.

## 23-6 | A Charged Isolated Conductor

Gauss' law permits us to prove an important theorem about conductors:

☞ If an excess charge is placed on an isolated conductor, that amount of charge will move entirely to the surface of the conductor. None of the excess charge will be found within the body of the conductor.

This might seem reasonable, considering that charges with the same sign repel one another. You might imagine that, by moving to the surface, the added charges are getting as far away from one another as they can. We turn to Gauss' law for verification of this speculation.

Figure 23-9a shows, in cross section, an isolated lump of copper hanging from an insulating thread and having an excess charge $q$. We place a Gaussian surface just inside the actual surface of the conductor.

The electric field inside this conductor must be zero. If this were not so, the field would exert forces on the conduction (free) electrons, which are always present in a conductor, and thus current would always exist within a conductor. (That is, charge would flow from place to place within the conductor.) Of course, there is no such perpetual current in an isolated conductor, and so the internal electric field is zero.

(An internal electric field *does* appear as a conductor is being charged. However, the added charge quickly distributes itself in such a way that the net internal electric field—the vector sum of the electric fields due to all the charges, both inside and outside—is zero. The movement of charge then ceases, because the net force on each charge is zero; the charges are then in *electrostatic equilibrium*.)

If $\vec{E}$ is zero everywhere inside our copper conductor, it must be zero for all points on the Gaussian surface because that surface, though close to the surface of the conductor, is definitely inside the conductor. This means that the flux through the Gaussian surface must be zero. Gauss' law then tells us that the net charge inside the Gaussian surface must also be zero. Then because the excess charge is not inside the Gaussian surface, it must be outside that surface, which means it must lie on the actual surface of the conductor.

### An Isolated Conductor with a Cavity

Figure 23-9*b* shows the same hanging conductor, but now with a cavity that is totally within the conductor. It is perhaps reasonable to suppose that when we scoop out the electrically neutral material to form the cavity, we do not change the distribution of charge or the pattern of the electric field that exists in Fig. 23-9*a*. Again, we must turn to Gauss' law for a quantitative proof.

We draw a Gaussian surface surrounding the cavity, close to its surface but inside the conducting body. Because $\vec{E} = 0$ inside the conductor, there can be no flux through this new Gaussian surface. Therefore, from Gauss' law, that surface can enclose no net charge. We conclude that there is no net charge on the cavity walls; all the excess charge remains on the outer surface of the conductor, as in Fig. 23-9*a*.

### The Conductor Removed

Suppose that, by some magic, the excess charges could be "frozen" into position on the conductor's surface, perhaps by embedding them in a thin plastic coating, and suppose that then the conductor could be removed completely. This is equivalent to enlarging the cavity of Fig. 23-9*b* until it consumes the entire conductor, leaving only the charges. The electric field would not change at all; it would remain zero inside the thin shell of charge and would remain unchanged for all external points. This shows us that the electric field is set up by the charges and not by the conductor. The conductor simply provides an initial pathway for the charges to take up their positions.

### The External Electric Field

You have seen that the excess charge on an isolated conductor moves entirely to the conductor's surface. However, unless the conductor is spherical, the charge does not distribute itself uniformly. Put another way, the surface charge density $\sigma$ (charge per unit area) varies over the surface of any nonspherical conductor. Generally, this variation makes the determination of the electric field set up by the surface charges very difficult.

However, the electric field just outside the surface of a conductor is easy to determine using Gauss' law. To do this, we consider a section of the surface that is small enough to permit us to neglect any curvature and thus to take the section to be flat. We then imagine a tiny cylindrical Gaussian surface to be embedded in the section as in Fig. 23-10: One end cap is fully inside the conductor, the other is fully outside, and the cylinder is perpendicular to the conductor's surface.

The electric field $\vec{E}$ at and just outside the conductor's surface must also be perpendicular to that surface. If it were not, then it would have a component along the conductor's surface that would exert forces on the surface charges,

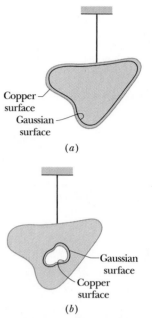

(a)

(b)

**FIG. 23-9** (*a*) A lump of copper with a charge *q* hangs from an insulating thread. A Gaussian surface is placed within the metal, just inside the actual surface. (*b*) The lump of copper now has a cavity within it. A Gaussian surface lies within the metal, close to the cavity surface.

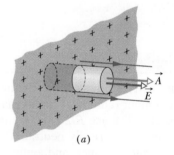

(a)

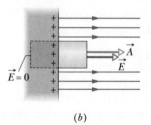

(b)

**FIG. 23-10** (*a*) Perspective view and (*b*) side view of a tiny portion of a large, isolated conductor with excess positive charge on its surface. A (closed) cylindrical Gaussian surface, embedded perpendicularly in the conductor, encloses some of the charge. Electric field lines pierce the external end cap of the cylinder, but not the internal end cap. The external end cap has area *A* and area vector $\vec{A}$.

causing them to move. However, such motion would violate our implicit assumption that we are dealing with electrostatic equilibrium. Therefore, $\vec{E}$ is perpendicular to the conductor's surface.

We now sum the flux through the Gaussian surface. There is no flux through the internal end cap, because the electric field within the conductor is zero. There is no flux through the curved surface of the cylinder, because internally (in the conductor) there is no electric field and externally the electric field is parallel to the curved portion of the Gaussian surface. The only flux through the Gaussian surface is that through the external end cap, where $\vec{E}$ is perpendicular to the plane of the cap. We assume that the cap area $A$ is small enough that the field magnitude $E$ is constant over the cap. Then the flux through the cap is $EA$, and that is the net flux $\Phi$ through the Gaussian surface.

The charge $q_{enc}$ enclosed by the Gaussian surface lies on the conductor's surface in an area $A$. If $\sigma$ is the charge per unit area, then $q_{enc}$ is equal to $\sigma A$. When we substitute $\sigma A$ for $q_{enc}$ and $EA$ for $\Phi$, Gauss' law (Eq. 23-6) becomes

$$\varepsilon_0 EA = \sigma A,$$

from which we find

$$E = \frac{\sigma}{\varepsilon_0} \quad \text{(conducting surface).} \tag{23-11}$$

Thus, the magnitude of the electric field just outside a conductor is proportional to the surface charge density on the conductor. If the charge on the conductor is positive, the electric field is directed away from the conductor as in Fig. 23-10. It is directed toward the conductor if the charge is negative.

The field lines in Fig. 23-10 must terminate on negative charges somewhere in the environment. If we bring those charges near the conductor, the charge density at any given location on the conductor's surface changes, and so does the magnitude of the electric field. However, the relation between $\sigma$ and $E$ is still given by Eq. 23-11.

## Sample Problem   23-5

Figure 23-11a shows a cross section of a spherical metal shell of inner radius $R$. A point charge of $-5.0\ \mu C$ is located at a distance $R/2$ from the center of the shell. If the shell is electrically neutral, what are the (induced) charges on its inner and outer surfaces? Are those charges uniformly distributed? What is the field pattern inside and outside the shell?

**KEY IDEAS**   Figure 23-11b shows a cross section of a spherical Gaussian surface within the metal, just outside the inner wall of the shell. The electric field must be zero inside the metal (and thus on the Gaussian surface inside the metal). This means that the electric flux through the Gaussian surface must also be zero. Gauss' law then tells us that the *net* charge enclosed by the Gaussian surface must be zero.

**Reasoning:** With a point charge of $-5.0\ \mu C$ within the shell, a charge of $+5.0\ \mu C$ must lie on the inner wall of the shell in order that the net enclosed charge be zero. If the point charge were centered, this positive charge would be uniformly distributed along the inner wall. However, since

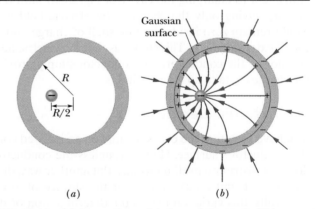

**FIG. 23-11**   (a) A negative point charge is located within a spherical metal shell that is electrically neutral. (b) As a result, positive charge is nonuniformly distributed on the inner wall of the shell, and an equal amount of negative charge is uniformly distributed on the outer wall.

the point charge is off-center, the distribution of positive charge is skewed, as suggested by Fig. 23-11b, because the positive charge tends to collect on the section of the inner wall nearest the (negative) point charge.

Because the shell is electrically neutral, its inner wall can have a charge of $+5.0\ \mu C$ only if electrons, with a total charge of $-5.0\ \mu C$, leave the inner wall and move to the outer wall. There they spread out uniformly, as is also suggested by Fig. 23-11b. This distribution of negative charge is uniform because the shell is spherical and because the skewed distribution of positive charge on the inner wall cannot produce an electric field in the shell to affect the distribution of charge on the outer wall.

The field lines inside and outside the shell are shown approximately in Fig. 23-11b. All the field lines intersect the shell and the point charge perpendicularly. Inside the shell the pattern of field lines is skewed because of the skew of the positive charge distribution. Outside the shell the pattern is the same as if the point charge were centered and the shell were missing. In fact, this would be true no matter where inside the shell the point charge happened to be located.

## 23-7 I Applying Gauss' Law: Cylindrical Symmetry

Figure 23-12 shows a section of an infinitely long cylindrical plastic rod with a uniform positive linear charge density $\lambda$. Let us find an expression for the magnitude of the electric field $\vec{E}$ at a distance $r$ from the axis of the rod.

Our Gaussian surface should match the symmetry of the problem, which is cylindrical. We choose a circular cylinder of radius $r$ and length $h$, coaxial with the rod. Because the Gaussian surface must be closed, we include two end caps as part of the surface.

Imagine now that, while you are not watching, someone rotates the plastic rod about its longitudinal axis or turns it end for end. When you look again at the rod, you will not be able to detect any change. We conclude from this symmetry that the only uniquely specified direction in this problem is along a radial line. Thus, at every point on the cylindrical part of the Gaussian surface, $\vec{E}$ must have the same magnitude $E$ and (for a positively charged rod) must be directed radially outward.

Since $2\pi r$ is the cylinder's circumference and $h$ is its height, the area $A$ of the cylindrical surface is $2\pi rh$. The flux of $\vec{E}$ through this cylindrical surface is then

$$\Phi = EA\cos\theta = E(2\pi rh)\cos 0 = E(2\pi rh).$$

There is no flux through the end caps because $\vec{E}$, being radially directed, is parallel to the end caps at every point.

The charge enclosed by the surface is $\lambda h$, which means Gauss' law,

$$\varepsilon_0 \Phi = q_{enc},$$

reduces to

$$\varepsilon_0 E(2\pi rh) = \lambda h,$$

yielding

$$E = \frac{\lambda}{2\pi\varepsilon_0 r} \qquad \text{(line of charge).} \qquad (23\text{-}12)$$

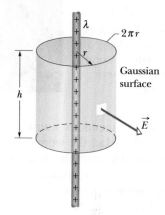

FIG. 23-12  A Gaussian surface in the form of a closed cylinder surrounds a section of a very long, uniformly charged, cylindrical plastic rod.

This is the electric field due to an infinitely long, straight line of charge, at a point that is a radial distance $r$ from the line. The direction of $\vec{E}$ is radially outward from the line of charge if the charge is positive, and radially inward if it is negative. Equation 23-12 also approximates the field of a *finite* line of charge at points that are not too near the ends (compared with the distance from the line).

### Sample Problem    23-6

*Upward streamer in a lightning storm.* The woman in Fig. 23-13 was standing on a lookout platform in the Sequoia National Park when a large storm cloud moved overhead. Some of the conduction electrons in her body were driven into the ground by the cloud's negatively charged base (Fig. 23-14a), leaving her positively charged. You can tell she was highly charged because her hair strands repelled one another and extended away from her along the electric field lines produced by the charge on her.

Lightning did not strike the woman, but she was in extreme danger because that electric field was on the verge of causing electrical breakdown in the surrounding air. Such a breakdown would have occurred along a path extending away from her in what is called an *upward streamer*. You can see a bright upward streamer near the top of the tree in the opening photograph for this chapter. An upward streamer is dangerous because the resulting ionization of molecules in the air suddenly

**FIG. 23-13** This woman has become positively charged by an overhead storm cloud. *(Courtesy NOAA)*

because we assume that the charge is uniformly distributed along this line, we can approximate the magnitude of the electric field along the side of her body with Eq. 23-12 ($E = \lambda/2\pi\varepsilon_0 r$).

**Calculations:** Substituting the critical value $E_c$ for $E$, the cylinder radius $R$ for radial distance $r$, and the ratio $Q/L$ for linear charge density $\lambda$, we have

$$E_c = \frac{Q/L}{2\pi\varepsilon_0 R},$$

or

$$Q = 2\pi\varepsilon_0 RLE_c.$$

Substituting given data then gives us

$$Q = (2\pi)(8.85 \times 10^{-12}\ \mathrm{C^2/N \cdot m^2})(0.10\ \mathrm{m})$$
$$\times (1.8\ \mathrm{m})(2.4 \times 10^6\ \mathrm{N/C})$$
$$= 2.402 \times 10^{-5}\ \mathrm{C} \approx 24\ \mu\mathrm{C}. \qquad \text{(Answer)}$$

frees a tremendous number of electrons from those molecules. Had the woman in Fig. 23-13 developed an upward streamer, the free electrons in the air would have moved to neutralize her (Fig. 23-14b), producing a large, perhaps fatal, charge flow through her body.

Let's model her body as a narrow vertical cylinder of height $L = 1.8$ m and radius $R = 0.10$ m (Fig. 23-14c). Assume that charge $Q$ was uniformly distributed along the cylinder and that electrical breakdown would have occurred if the electric field magnitude along her body had exceeded the critical value $E_c = 2.4$ MN/C. What value of $Q$ would have put the air along her body on the verge of breakdown?

**KEY IDEA** Because $R \ll L$, we can approximate the charge distribution as a long line of charge. Further,

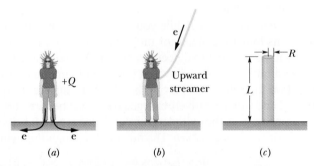

(a)  (b)  (c)

**FIG. 23-14** (a) Some of the conduction electrons in the woman's body are driven into the ground, leaving her positively charged. (b) An upward streamer develops if the air undergoes electrical breakdown, which provides a path for electrons freed from molecules in the air to move to the woman. (c) A cylinder represents the woman.

## 23-8 | Applying Gauss' Law: Planar Symmetry

### Nonconducting Sheet

Figure 23-15 shows a portion of a thin, infinite, nonconducting sheet with a uniform (positive) surface charge density $\sigma$. A sheet of thin plastic wrap, uniformly charged on one side, can serve as a simple model. Let us find the electric field $\vec{E}$ a distance $r$ in front of the sheet.

A useful Gaussian surface is a closed cylinder with end caps of area $A$, arranged to pierce the sheet perpendicularly as shown. From symmetry, $\vec{E}$ must be perpendicular to the sheet and hence to the end caps. Furthermore, since the charge is positive, $\vec{E}$ is directed *away* from the sheet, and thus the electric field lines pierce the two Gaussian end caps in an outward direction. Because the field lines do not pierce the curved surface, there is no flux through this portion of the Gaussian surface. Thus $\vec{E} \cdot d\vec{A}$ is simply $E\, dA$; then Gauss' law,

$$\varepsilon_0 \oint \vec{E} \cdot d\vec{A} = q_{\mathrm{enc}},$$

becomes

$$\varepsilon_0(EA + EA) = \sigma A,$$

where $\sigma A$ is the charge enclosed by the Gaussian surface. This gives

$$E = \frac{\sigma}{2\varepsilon_0} \quad \text{(sheet of charge).} \tag{23-13}$$

Since we are considering an infinite sheet with uniform charge density, this result holds for any point at a finite distance from the sheet. Equation 23-13 agrees with Eq. 22-27, which we found by integration of the electric field components produced by individual charges. (Look back to that time-consuming and challenging integration, and note how much more easily we obtain the result with Gauss' law. That is one reason for devoting a whole chapter to that law: for certain symmetric arrangements of charge, it is very much easier to use than integration of field components.)

## Two Conducting Plates

Figure 23-16a shows a cross section of a thin, infinite conducting plate with excess positive charge. From Section 23-6 we know that this excess charge lies on the surface of the plate. Since the plate is thin and very large, we can assume that essentially all the excess charge is on the two large faces of the plate.

If there is no external electric field to force the positive charge into some particular distribution, it will spread out on the two faces with a uniform surface charge density of magnitude $\sigma_1$. From Eq. 23-11 we know that just outside the plate this charge sets up an electric field of magnitude $E = \sigma_1/\varepsilon_0$. Because the excess charge is positive, the field is directed away from the plate.

Figure 23-16b shows an identical plate with excess negative charge having the same magnitude of surface charge density $\sigma_1$. The only difference is that now the electric field is directed toward the plate.

Suppose we arrange for the plates of Figs. 23-16a and b to be close to each other and parallel (Fig. 23-16c). Since the plates are conductors, when we bring them into this arrangement, the excess charge on one plate attracts the excess charge on the other plate, and all the excess charge moves onto the inner faces of the plates as in Fig. 23-16c. With twice as much charge now on each inner face, the new surface charge density (call it $\sigma$) on each inner face is twice $\sigma_1$. Thus, the electric field at any point between the plates has the magnitude

$$E = \frac{2\sigma_1}{\varepsilon_0} = \frac{\sigma}{\varepsilon_0}. \tag{23-14}$$

This field is directed away from the positively charged plate and toward the negatively charged plate. Since no excess charge is left on the outer faces, the electric field to the left and right of the plates is zero.

Because the charges on the plates moved when we brought the plates close to each other, Fig. 23-16c is *not* the superposition of Figs. 23-16a and b; that is, the charge distribution of the two-plate system is not merely the sum of the charge distributions of the individual plates.

You may wonder why we discuss such seemingly unrealistic situations as the field set up by an infinite line of charge, an infinite sheet of charge, or a pair of

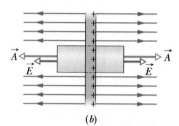

(a)

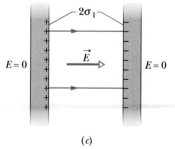

(b)

**FIG. 23-15** (a) Perspective view and (b) side view of a portion of a very large, thin plastic sheet, uniformly charged on one side to surface charge density $\sigma$. A closed cylindrical Gaussian surface passes through the sheet and is perpendicular to it.

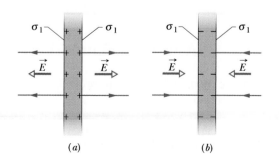

(a)          (b)          (c)

**FIG. 23-16** (a) A thin, very large conducting plate with excess positive charge. (b) An identical plate with excess negative charge. (c) The two plates arranged so they are parallel and close.

infinite plates of charge. One reason is that analyzing such situations with Gauss' law is easy. More important is that analyses for "infinite" situations yield good approximations to many real-world problems. Thus, Eq. 23-13 holds well for a finite nonconducting sheet as long as we are dealing with points close to the sheet and not too near its edges. Equation 23-14 holds well for a pair of finite conducting plates as long as we consider points that are not too close to their edges.

The trouble with the edges of a sheet or a plate, and the reason we take care not to deal with them, is that near an edge we can no longer use planar symmetry to find expressions for the fields. In fact, the field lines there are curved (said to be an *edge effect* or *fringing*), and the fields can be very difficult to express algebraically.

## Sample Problem 23-7

Figure 23-17a shows portions of two large, parallel, non-conducting sheets, each with a fixed uniform charge on one side. The magnitudes of the surface charge densities are $\sigma_{(+)} = 6.8\ \mu C/m^2$ for the positively charged sheet and $\sigma_{(-)} = 4.3\ \mu C/m^2$ for the negatively charged sheet.

Find the electric field $\vec{E}$ (a) to the left of the sheets, (b) between the sheets, and (c) to the right of the sheets.

**KEY IDEA** With the charges fixed in place (they are on nonconductors), we can find the electric field of the sheets in Fig. 23-17a by (1) finding the field of each sheet as if that sheet were isolated and (2) algebraically adding the fields of the isolated sheets via the superposition principle. (We can add the fields algebraically because they are parallel to each other.)

*Calculations:* At any point, the electric field $\vec{E}_{(+)}$ due to the positive sheet is directed *away* from the sheet and, from Eq. 23-13, has the magnitude

$$E_{(+)} = \frac{\sigma_{(+)}}{2\varepsilon_0} = \frac{6.8 \times 10^{-6}\ C/m^2}{(2)(8.85 \times 10^{-12}\ C^2/N \cdot m^2)}$$
$$= 3.84 \times 10^5\ N/C.$$

Similarly, at any point, the electric field $\vec{E}_{(-)}$ due to the

negative sheet is directed *toward* that sheet and has the magnitude

$$E_{(-)} = \frac{\sigma_{(-)}}{2\varepsilon_0} = \frac{4.3 \times 10^{-6}\ C/m^2}{(2)(8.85 \times 10^{-12}\ C^2/N \cdot m^2)}$$
$$= 2.43 \times 10^5\ N/C.$$

Figure 23-17b shows the fields set up by the sheets to the left of the sheets (L), between them (B), and to their right (R).

The resultant fields in these three regions follow from the superposition principle. To the left, the field magnitude is

$$E_L = E_{(+)} - E_{(-)}$$
$$= 3.84 \times 10^5\ N/C - 2.43 \times 10^5\ N/C$$
$$= 1.4 \times 10^5\ N/C. \qquad \text{(Answer)}$$

Because $E_{(+)}$ is larger than $E_{(-)}$, the net electric field $\vec{E}_L$ in this region is directed to the left, as Fig. 23-17c shows. To the right of the sheets, the electric field $\vec{E}_R$ has the same magnitude but is directed to the right, as Fig. 23-17c shows.

Between the sheets, the two fields add and we have

$$E_B = E_{(+)} + E_{(-)}$$
$$= 3.84 \times 10^5\ N/C + 2.43 \times 10^5\ N/C$$
$$= 6.3 \times 10^5\ N/C. \qquad \text{(Answer)}$$

The electric field $\vec{E}_B$ is directed to the right.

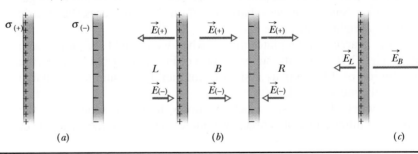

**FIG. 23-17** (a) Two large, parallel sheets, uniformly charged on one side. (b) The individual electric fields resulting from the two charged sheets. (c) The net field due to both charged sheets, found by superposition.

# 23-9 | Applying Gauss' Law: Spherical Symmetry

Here we use Gauss' law to prove the two shell theorems presented without proof in Section 21-4:

> A shell of uniform charge attracts or repels a charged particle that is outside the shell as if all the shell's charge were concentrated at the center of the shell.

👉 If a charged particle is located inside a shell of uniform charge, there is no electrostatic force on the particle from the shell.

Figure 23-18 shows a charged spherical shell of total charge $q$ and radius $R$ and two concentric spherical Gaussian surfaces, $S_1$ and $S_2$. If we followed the procedure of Section 23-5 as we applied Gauss' law to surface $S_2$, for which $r \geq R$, we would find that

$$E = \frac{1}{4\pi\varepsilon_0} \frac{q}{r^2} \qquad \text{(spherical shell, field at } r \geq R\text{).} \qquad (23\text{-}15)$$

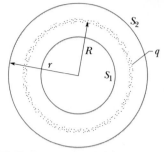

**FIG. 23-18** A thin, uniformly charged, spherical shell with total charge $q$, in cross section. Two Gaussian surfaces $S_1$ and $S_2$ are also shown in cross section. Surface $S_2$ encloses the shell, and $S_1$ encloses only the empty interior of the shell.

This field is the same as one set up by a point charge $q$ at the center of the shell of charge. Thus, the force produced by a shell of charge $q$ on a charged particle placed outside the shell is the same as the force produced by a point charge $q$ located at the center of the shell. This proves the first shell theorem.

Applying Gauss' law to surface $S_1$, for which $r < R$, leads directly to

$$E = 0 \qquad \text{(spherical shell, field at } r < R\text{),} \qquad (23\text{-}16)$$

because this Gaussian surface encloses no charge. Thus, if a charged particle were enclosed by the shell, the shell would exert no net electrostatic force on the particle. This proves the second shell theorem.

Any spherically symmetric charge distribution, such as that of Fig. 23-19, can be constructed with a nest of concentric spherical shells. For purposes of applying the two shell theorems, the volume charge density $\rho$ should have a single value for each shell but need not be the same from shell to shell. Thus, for the charge distribution as a whole, $\rho$ can vary, but only with $r$, the radial distance from the center. We can then examine the effect of the charge distribution "shell by shell."

In Fig. 23-19a, the entire charge lies within a Gaussian surface with $r > R$. The charge produces an electric field on the Gaussian surface as if the charge were a point charge located at the center, and Eq. 23-15 holds.

Figure 23-19b shows a Gaussian surface with $r < R$. To find the electric field at points on this Gaussian surface, we consider two sets of charged shells—one set inside the Gaussian surface and one set outside. Equation 23-16 says that the charge lying *outside* the Gaussian surface does not set up a net electric field on the Gaussian surface. Equation 23-15 says that the charge *enclosed* by the surface sets up an electric field as if that enclosed charge were concentrated at the center. Letting $q'$ represent that enclosed charge, we can then rewrite Eq. 23-15 as

$$E = \frac{1}{4\pi\varepsilon_0} \frac{q'}{r^2} \qquad \text{(spherical distribution, field at } r \leq R\text{).} \qquad (23\text{-}17)$$

*If the full charge $q$ enclosed within radius $R$ is uniform,* then $q'$ enclosed within radius $r$ in Fig. 23-19b is proportional to $q$:

$$\frac{\begin{pmatrix} \text{charge enclosed by} \\ \text{sphere of radius } r \end{pmatrix}}{\begin{pmatrix} \text{volume enclosed by} \\ \text{sphere of radius } r \end{pmatrix}} = \frac{\text{full charge}}{\text{full volume}}$$

or

$$\frac{q'}{\frac{4}{3}\pi r^3} = \frac{q}{\frac{4}{3}\pi R^3}. \qquad (23\text{-}18)$$

This gives us

$$q' = q\,\frac{r^3}{R^3}. \qquad (23\text{-}19)$$

Substituting this into Eq. 23-17 yields

$$E = \left(\frac{q}{4\pi\varepsilon_0 R^3}\right) r \qquad \text{(uniform charge, field at } r \leq R\text{).} \qquad (23\text{-}20)$$

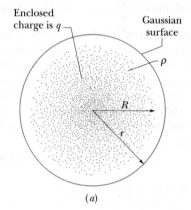

(a)

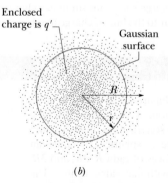

(b)

**FIG. 23-19** The dots represent a spherically symmetric distribution of charge of radius $R$, whose volume charge density $\rho$ is a function only of distance from the center. The charged object is not a conductor, and therefore the charge is assumed to be fixed in position. A concentric spherical Gaussian surface with $r > R$ is shown in (a). A similar Gaussian surface with $r < R$ is shown in (b).

**CHECKPOINT 4** The figure shows two large, parallel, nonconducting sheets with identical (positive) uniform surface charge densities, and a sphere with a uniform (positive) volume charge density. Rank the four numbered points according to the magnitude of the net electric field there, greatest first.

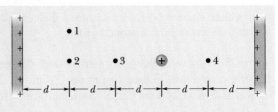

# REVIEW & SUMMARY

**Gauss' Law**  *Gauss' law* and Coulomb's law are different ways of describing the relation between charge and electric field in static situations. Gauss' law is

$$\varepsilon_0\Phi = q_{\text{enc}} \quad \text{(Gauss' law),} \tag{23-6}$$

in which $q_{\text{enc}}$ is the net charge inside an imaginary closed surface (a *Gaussian surface*) and $\Phi$ is the net *flux* of the electric field through the surface:

$$\Phi = \oint \vec{E} \cdot d\vec{A} \quad \begin{array}{l}\text{(electric flux through a}\\\text{Gaussian surface).}\end{array} \tag{23-4}$$

Coulomb's law can be derived from Gauss' law.

**Applications of Gauss' Law**  Using Gauss' law and, in some cases, symmetry arguments, we can derive several important results in electrostatic situations. Among these are:

1. An excess charge on an isolated *conductor* is located entirely on the outer surface of the conductor.
2. The external electric field near the *surface of a charged conductor* is perpendicular to the surface and has magnitude

$$E = \frac{\sigma}{\varepsilon_0} \quad \text{(conducting surface).} \tag{23-11}$$

   Within the conductor, $E = 0$.
3. The electric field at any point due to an infinite *line of charge* with uniform linear charge density $\lambda$ is perpendicular to the line of charge and has magnitude

$$E = \frac{\lambda}{2\pi\varepsilon_0 r} \quad \text{(line of charge),} \tag{23-12}$$

where $r$ is the perpendicular distance from the line of charge to the point.

4. The electric field due to an *infinite nonconducting sheet* with uniform surface charge density $\sigma$ is perpendicular to the plane of the sheet and has magnitude

$$E = \frac{\sigma}{2\varepsilon_0} \quad \text{(sheet of charge).} \tag{23-13}$$

5. The electric field *outside a spherical shell of charge* with radius $R$ and total charge $q$ is directed radially and has magnitude

$$E = \frac{1}{4\pi\varepsilon_0}\frac{q}{r^2} \quad \text{(spherical shell, for } r \geq R\text{).} \tag{23-15}$$

Here $r$ is the distance from the center of the shell to the point at which $E$ is measured. (The charge behaves, for external points, as if it were all located at the center of the sphere.) The field *inside* a uniform spherical shell of charge is exactly zero:

$$E = 0 \quad \text{(spherical shell, for } r < R\text{).} \tag{23-16}$$

6. The electric field *inside a uniform sphere of charge* is directed radially and has magnitude

$$E = \left(\frac{q}{4\pi\varepsilon_0 R^3}\right) r. \tag{23-20}$$

# QUESTIONS

**1**  Figure 23-20 shows, in cross section, a central metal ball, two spherical metal shells, and three spherical Gaussian surfaces of radii $R$, $2R$, and $3R$, all with the same center. The uniform charges on the three objects are: ball, $Q$; smaller shell, $3Q$; larger shell, $5Q$. Rank the Gaussian surfaces according to the magnitude of the electric field at any point on the surface, greatest first.

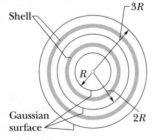

**FIG. 23-20**  Question 1.

**2**  Figure 23-21 shows, in cross section, two Gaussian spheres and two Gaussian cubes that are centered on a positively charged particle. (a) Rank the net flux through the four Gaussian surfaces, greatest first. (b) Rank the magnitudes of the electric fields on the surfaces, greatest first, and indicate whether the magnitudes are uniform or variable along each surface.

**3**  A surface has the area vector $\vec{A} = (2\hat{i} + 3\hat{j})$ m$^2$. What is the flux of a uniform electric field through it if the field is (a) $\vec{E} = 4\hat{i}$ N/C and (b) $\vec{E} = 4\hat{k}$ N/C?

**4**  Figure 23-22 shows, in cross section, three solid cylinders, each of length $L$ and uniform charge $Q$. Concentric with each cylinder is a cylindrical Gaussian surface, with all three surfaces having the same radius. Rank the Gaussian surfaces according to the electric field at any point on the surface, greatest first.

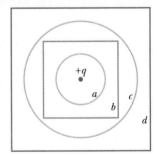

**FIG. 23-21**  Question 2.

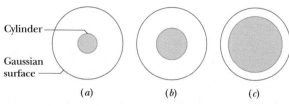

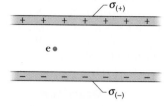

Cylinder

Gaussian surface

(a)    (b)    (c)

**FIG. 23-22** Question 4.

**5** Figure 23-23 shows four situations in which four very long rods extend into and out of the page (we see only their cross sections). The value below each cross section gives that particular rod's uniform charge density in microcoulombs per meter. The rods are separated by either $d$ or $2d$ as drawn, and a central point is shown midway between the inner rods. Rank the situations according to the magnitude of the net electric field at that central point, greatest first.

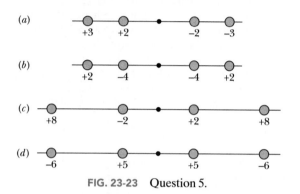

**FIG. 23-23** Question 5.

**6** A small charged ball lies within the hollow of a metallic spherical shell of radius $R$. For three situations, the net charges on the ball and shell, respectively, are (1) $+4q$, 0; (2) $-6q$, $+10q$; (3) $+16q$, $-12q$. Rank the situations according to the charge on (a) the inner surface of the shell and (b) the outer surface, most positive first.

**7** Rank the situations of Question 6 according to the magnitude of the electric field (a) halfway through the shell and (b) at a point $2R$ from the center of the shell, greatest first.

**8** Three infinite nonconducting sheets, with uniform positive surface charge densities $\sigma$, $2\sigma$, and $3\sigma$, are arranged to be parallel like the two sheets in Fig. 23-17a. What is their order, from left to right, if the electric field $\vec{E}$ produced by the arrangement has magnitude $E = 0$ in one region and $E = 2\sigma/\varepsilon_0$ in another region?

**9** In Fig. 23-24, an electron is released between two infinite nonconducting sheets that are horizontal and have uniform surface charge densities $\sigma_{(+)}$ and $\sigma_{(-)}$, as indicated. The electron is subjected to the following three situations involving surface charge densities and sheet separations. Rank the magnitudes of the electron's acceleration, greatest first.

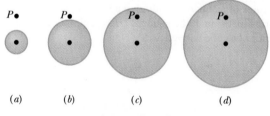

**FIG. 23-24** Question 9.

| Situation | $\sigma_{(+)}$ | $\sigma_{(-)}$ | Separation |
|-----------|----------------|----------------|------------|
| 1 | $+4\sigma$ | $-4\sigma$ | $d$ |
| 2 | $+7\sigma$ | $-\sigma$ | $4d$ |
| 3 | $+3\sigma$ | $-5\sigma$ | $9d$ |

**10** Figure 23-25 shows four solid spheres, each with charge $Q$ uniformly distributed through its volume. (a) Rank the spheres according to their volume charge density, greatest first. The figure also shows a point $P$ for each sphere, all at the same distance from the center of the sphere. (b) Rank the spheres according to the magnitude of the electric field they produce at point $P$, greatest first.

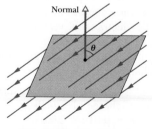

**FIG. 23-25** Question 10.

---

## PROBLEMS

| | |
|---|---|
| **GO** | Tutoring problem available (at instructor's discretion) in *WileyPLUS* and WebAssign |
| **SSM** | Worked-out solution available in Student Solutions Manual |
| • – ••• | Number of dots indicates level of problem difficulty |
| ✈ | Additional information available in *The Flying Circus of Physics* and at flyingcircusofphysics.com |

**WWW** Worked-out solution is at _____
**ILW** Interactive solution is at _____ http://www.wiley.com/college/halliday

### sec. 23-3 Flux of an Electric Field

**•1** The square surface shown in Fig. 23-26 measures 3.2 mm on each side. It is immersed in a uniform electric field with magnitude $E = 1800$ N/C and with field lines at an angle of $\theta = 35°$ with a normal to the surface, as shown. Take that

Normal

$\theta$

**FIG. 23-26** Problem 1.

normal to be directed "outward," as though the surface were one face of a box. Calculate the electric flux through the surface. **SSM**

**••2** An electric field given by $\vec{E} = 4.0\hat{i} - 3.0(y^2 + 2.0)\hat{j}$ pierces a Gaussian cube of edge length 2.0 m and positioned as shown in Fig. 23-5. (The magnitude $E$ is in newtons per coulomb and the position $x$ is in meters.) What is the electric flux through the (a) top face, (b) bottom face, (c) left face, and (d) back face? (e) What is the net electric flux through the cube?

**••3** The cube in Fig. 23-27 has edge length 1.40 m and is oriented as shown in a region of uniform electric field. Find the electric flux through the right face if the electric field, in newtons per coulomb, is given by (a) $6.00\hat{i}$, (b) $-2.00\hat{j}$, and (c) $-3.00\hat{i} + 4.00\hat{k}$. (d) What is the total flux through the cube for each field?

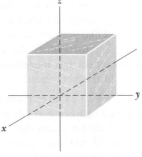

**FIG. 23-27** Problems 3, 4, and 11.

**sec. 23-4 Gauss' Law**

**•4** At each point on the surface of the cube shown in Fig. 23-27, the electric field is parallel to the z axis. The length of each edge of the cube is 3.0 m. On the top face of the cube $\vec{E} = -34\hat{k}$ N/C, and on the bottom face $\vec{E} = +20\hat{k}$ N/C. Determine the net charge contained within the cube.

**•5** A point charge of 1.8 $\mu$C is at the center of a cubical Gaussian surface 55 cm on edge. What is the net electric flux through the surface?

**•6** In Fig. 23-28, a butterfly net is in a uniform electric field of magnitude $E = 3.0$ mN/C. The rim, a circle of radius $a = 11$ cm, is aligned perpendicular to the field. The net contains no net charge. Find the electric flux through the netting.

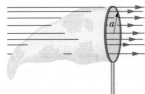

**FIG. 23-28** Problem 6.

**•7** In Fig. 23-29, a proton is a distance $d/2$ directly above the center of a square of side $d$. What is the magnitude of the electric flux through the square? (*Hint:* Think of the square as one face of a cube with edge $d$.)

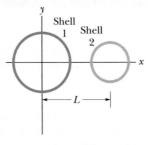

**FIG. 23-29** Problem 7.

**••8** Figure 23-30 shows two nonconducting spherical shells fixed in place. Shell 1 has uniform surface charge density $+6.0$ $\mu$C/m² on its outer surface and radius 3.0 cm; shell 2 has uniform surface charge density $+4.0$ $\mu$C/m² on its outer surface and radius 2.0 cm; the shell centers are separated by $L = 10$ cm. In unit-vector notation, what is the net electric field at $x = 2.0$ cm?

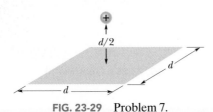

**FIG. 23-30** Problem 8.

**••9** It is found experimentally that the electric field in a certain region of Earth's atmosphere is directed vertically down. At an altitude of 300 m the field has magnitude 60.0 N/C; at an altitude of 200 m, the magnitude is 100 N/C. Find the net amount of charge contained in a cube 100 m on edge, with horizontal faces at altitudes of 200 and 300 m. **SSM**

**••10** When a shower is turned on in a closed bathroom, the splashing of the water on the bare tub can fill the room's air with negatively charged ions and produce an electric field in the air as great as 1000 N/C. Consider a bathroom with dimensions 2.5 m × 3.0 m × 2.0 m. Along the ceiling, floor, and four walls, approximate the electric field in the air as being directed perpendicular to the surface and as having a uniform magnitude of 600 N/C. Also, treat those surfaces as forming a closed Gaussian surface around the room's air. What are (a) the volume charge density $\rho$ and (b) the number of excess elementary charges $e$ per cubic meter in the room's air?

**••11** Fig. 23-27 shows a Gaussian surface in the shape of a cube with edge length 1.40 m. What are (a) the net flux $\Phi$ through the surface and (b) the net charge $q_{enc}$ enclosed by the surface if $\vec{E} = (3.00y\hat{j})$ N/C, with y in meters? What are (c) $\Phi$ and (d) $q_{enc}$ if $\vec{E} = [-4.00\hat{i} + (6.00 + 3.00y)\hat{j}]$ N/C? **ILW**

**••12** *Flux and nonconducting shells.* A charged particle is suspended at the center of two concentric spherical shells that are very thin and made of nonconducting material. Figure 23-31a shows a cross section. Figure 23-31b gives the net flux $\Phi$ through a Gaussian sphere centered on the particle, as a function of the radius r of the sphere. The scale of the vertical axis is set by $\Phi_s = 5.0 \times 10^5$ N·m²/C. (a) What is the charge of the central particle? What are the net charges of (b) shell A and (c) shell B?

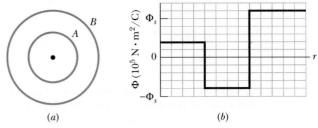

**FIG. 23-31** Problem 12.

**••13** A particle of charge $+q$ is placed at one corner of a Gaussian cube. What multiple of $q/\varepsilon_0$ gives the flux through (a) each cube face forming that corner and (b) each of the other cube faces?

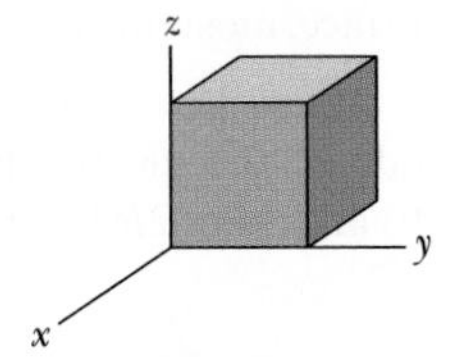

**••14** Figure 23-32 shows a closed Gaussian surface in the shape of a cube of edge length 2.00 m. It lies in a region where the electric field is given by $\vec{E} = (3.00x + 4.00)\hat{i} + 6.00\hat{j} + 7.00\hat{k}$ N/C, with x in meters. What is the net charge contained by the cube?

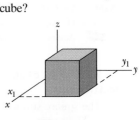

**FIG. 23-32** Problem 14.

**••15** Figure 23-33 shows a closed Gaussian surface in the shape of a cube of edge length 2.00 m, with one corner at $x_1 = 5.00$ m, $y_1 = 4.00$ m. The cube lies in a region where the electric field vector is given by $\vec{E} = -3.00\hat{i} - 4.00y^2\hat{j} + 3.00\hat{k}$ N/C, with y in meters. What is the net charge contained by the cube? **GO**

**FIG. 23-33** Problem 15.

**•••16** The box-like Gaussian surface of Fig. 23-34 encloses a net charge of $+24.0\varepsilon_0$ C and lies in an electric field given by $\vec{E} = [(10.0 + 2.00x)\hat{i} - 3.00\hat{j} + bz\hat{k}]$ N/C, with x and z in meters and b a constant. The bottom face is in the xz plane; the top face is in the horizontal plane passing through $y_2 = 1.00$

m. For $x_1 = 1.00$ m, $x_2 = 4.00$ m, $z_1 = 1.00$ m, and $z_2 = 3.00$ m, what is $b$?

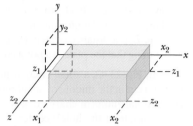

FIG. 23-34 Problem 16.

### sec. 23-6 A Charged Isolated Conductor

•17 Space vehicles traveling through Earth's radiation belts can intercept a significant number of electrons. The resulting charge buildup can damage electronic components and disrupt operations. Suppose a spherical metal satellite 1.3 m in diameter accumulates 2.4 $\mu$C of charge in one orbital revolution. (a) Find the resulting surface charge density. (b) Calculate the magnitude of the electric field just outside the surface of the satellite, due to the surface charge.

•18 *Flux and conducting shells.* A charged particle is held at the center of two concentric conducting spherical shells. Figure 23-35$a$ shows a cross section. Figure 23-35$b$ gives the net flux $\Phi$ through a Gaussian sphere centered on the particle, as a function of the radius $r$ of the sphere. The scale of the vertical axis is set by $\Phi_s = 5.0 \times 10^5$ N·m²/C. What are (a) the charge of the central particle and the net charges of (b) shell $A$ and (c) shell $B$? GO

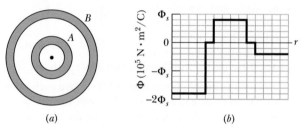

FIG. 23-35 Problem 18.

•19 A uniformly charged conducting sphere of 1.2 m diameter has a surface charge density of 8.1 $\mu$C/m². (a) Find the net charge on the sphere. (b) What is the total electric flux leaving the surface of the sphere? SSM

•20 The electric field just above the surface of the charged drum of a photocopying machine has a magnitude $E$ of $2.3 \times 10^5$ N/C. What is the surface charge density on the drum, assuming the drum is a conductor?

••21 An isolated conductor of arbitrary shape has a net charge of $+10 \times 10^{-6}$ C. Inside the conductor is a cavity within which is a point charge $q = +3.0 \times 10^{-6}$ C. What is the charge (a) on the cavity wall and (b) on the outer surface of the conductor?

### sec. 23-7 Applying Gauss' Law: Cylindrical Symmetry

•22 Figure 23-36 shows a section of a long, thin-walled metal tube of radius $R = 3.00$ cm, with a charge per unit length $\lambda = 2.00 \times 10^{-8}$ C/m. What is the magni-

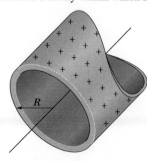

FIG. 23-36 Problem 22.

tude $E$ of the electric field at radial distance (a) $r = R/2.00$ and (b) $r = 2.00R$? (c) Graph $E$ versus $r$ for the range $r = 0$ to $2.00R$.

•23 An infinite line of charge produces a field of magnitude $4.5 \times 10^4$ N/C at a distance of 2.0 m. Calculate the linear charge density. SSM

•24 An electron is released from rest at a perpendicular distance of 9.0 cm from a line of charge on a very long nonconducting rod. That charge is uniformly distributed, with 6.0 $\mu$C per meter. What is the magnitude of the electron's initial acceleration?

•25 (a) The drum of a photocopying machine has a length of 42 cm and a diameter of 12 cm. The electric field just above the drum's surface is $2.3 \times 10^5$ N/C. What is the total charge on the drum? (b) The manufacturer wishes to produce a desktop version of the machine. This requires reducing the drum length to 28 cm and the diameter to 8.0 cm. The electric field at the drum surface must not change. What must be the charge on this new drum?

••26 In Fig. 23-37, short sections of two very long parallel lines of charge are shown, fixed in place, separated by $L = 8.0$ cm. The uniform linear charge densities are $+6.0$ $\mu$C/m for line 1 and $-2.0$ $\mu$C/m for line 2. Where along the $x$ axis shown is the net electric field from the two lines zero?

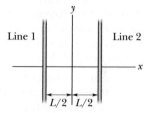

FIG. 23-37 Problem 26.

••27 Figure 23-38 is a section of a conducting rod of radius $R_1 = 1.30$ mm and length $L = 11.00$ m inside a thin-walled coaxial conducting cylindrical shell of radius $R_2 = 10.0R_1$ and the (same) length $L$. The net charge on the rod is $Q_1 = +3.40 \times 10^{-12}$ C; that on the shell is $Q_2 = -2.00Q_1$. What are the (a) magnitude $E$ and (b) direction (radially inward or outward) of the electric field at radial distance $r = 2.00R_2$? What are (c) $E$ and (d) the direction at $r = 5.00R_1$? What is the charge on the (e) interior and (f) exterior surface of the shell? SSM WWW

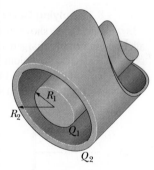

FIG. 23-38 Problem 27.

••28 Figure 23-39$a$ shows a narrow charged solid cylinder that is coaxial with a larger charged cylindrical shell. Both are nonconducting and thin and have uniform surface charge densities on their outer surfaces. Figure 23-39$b$ gives the

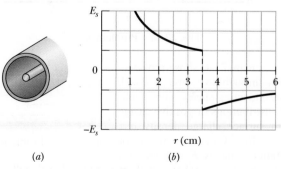

FIG. 23-39 Problem 28.

radial component $E$ of the electric field versus radial distance $r$ from the common axis. The vertical axis scale is set by $E_s = 3.0 \times 10^3$ N/C. What is the linear charge density of the shell?

••29 Two long, charged, thin-walled, concentric cylindrical shells have radii of 3.0 and 6.0 cm. The charge per unit length is $5.0 \times 10^{-6}$ C/m on the inner shell and $-7.0 \times 10^{-6}$ C/m on the outer shell. What are the (a) magnitude $E$ and (b) direction (radially inward or outward) of the electric field at radial distance $r = 4.0$ cm? What are (c) $E$ and (d) the direction at $r = 8.0$ cm? **ILW**

••30 A charge of uniform linear density 2.0 nC/m is distributed along a long, thin, nonconducting rod. The rod is coaxial with a long conducting cylindrical shell (inner radius = 5.0 cm, outer radius = 10 cm). The net charge on the shell is zero. (a) What is the magnitude of the electric field 15 cm from the axis of the shell? What is the surface charge density on the (b) inner and (c) outer surface of the shell? **GO**

••31 A long, straight wire has fixed negative charge with a linear charge density of magnitude 3.6 nC/m. The wire is to be enclosed by a coaxial, thin-walled nonconducting cylindrical shell of radius 1.5 cm. The shell is to have positive charge on its outside surface with a surface charge density $\sigma$ that makes the net external electric field zero. Calculate $\sigma$.

•••32 A long, nonconducting, solid cylinder of radius 4.0 cm has a nonuniform volume charge density $\rho$ that is a function of radial distance $r$ from the cylinder axis: $\rho = Ar^2$. For $A = 2.5$ μC/m$^5$, what is the magnitude of the electric field at (a) $r = 3.0$ cm and (b) $r = 5.0$ cm?

### sec. 23-8 Applying Gauss' Law: Planar Symmetry

•33 Figure 23-40a shows three plastic sheets that are large, parallel, and uniformly charged. Figure 23-40b gives the component of the net electric field along an $x$ axis through the sheets. The scale of the vertical axis is set by $E_s = 6.0 \times 10^5$ N/C. What is the ratio of the charge density on sheet 3 to that on sheet 2? **GO**

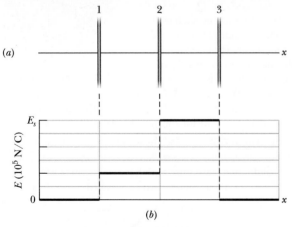

FIG. 23-40 Problem 33.

•34 Figure 23-41 shows cross sections through two large, parallel, nonconducting sheets with identical distributions of positive charge with surface charge density $\sigma = 1.77 \times 10^{-22}$ C/m$^2$. In unit-vector notation, what is $\vec{E}$ at points (a) above the sheets, (b) between them, and (c) below them?

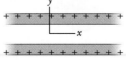

FIG. 23-41 Problem 34.

•35 A square metal plate of edge length 8.0 cm and negligible thickness has a total charge of $6.0 \times 10^{-6}$ C. (a) Estimate the magnitude $E$ of the electric field just off the center of the plate (at, say, a distance of 0.50 mm from the center) by assuming that the charge is spread uniformly over the two faces of the plate. (b) Estimate $E$ at a distance of 30 m (large relative to the plate size) by assuming that the plate is a point charge. **SSM WWW**

•36 In Fig. 23-42, a small circular hole of radius $R = 1.80$ cm has been cut in the middle of an infinite, flat, nonconducting surface that has uniform charge density $\sigma = 4.50$ pC/m$^2$. A $z$ axis, with its origin at the hole's center, is perpendicular to the surface. In unit-vector notation, what is the electric field at point $P$ at $z = 2.56$ cm? (*Hint:* See Eq. 22-26 and use superposition.)

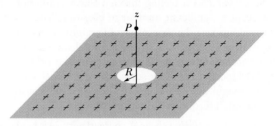

FIG. 23-42 Problem 36.

•37 In Fig. 23-43, two large, thin metal plates are parallel and close to each other. On their inner faces, the plates have excess surface charge densities of opposite signs and magnitude $7.00 \times 10^{-22}$ C/m$^2$. In unit-vector notation, what is the electric field at points (a) to the left of the plates, (b) to the right of them, and (c) between them?

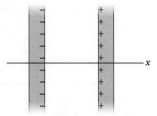

FIG. 23-43 Problem 37.

••38 Two large metal plates of area 1.0 m$^2$ face each other. They are 5.0 cm apart and have equal but opposite charges on their inner surfaces. If the magnitude $E$ of the electric field between the plates is 55 N/C, what is the magnitude of the charge on each plate? Neglect edge effects.

••39 An electron is shot directly toward the center of a large metal plate that has surface charge density $-2.0 \times 10^{-6}$ C/m$^2$. If the initial kinetic energy of the electron is $1.60 \times 10^{-17}$ J and if the electron is to stop (due to electrostatic repulsion from the plate) just as it reaches the plate, how far from the plate must the launch point be? **GO**

••40 In Fig. 23-44a, an electron is shot directly away from a uniformly charged plastic sheet, at speed $v_s = 2.0 \times 10^5$ m/s. The sheet is nonconducting, flat, and very large. Figure 23-44b gives the electron's vertical velocity component $v$ versus time $t$

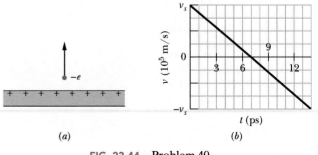

FIG. 23-44 Problem 40.

until the return to the launch point. What is the sheet's surface charge density?

**••41** In Fig. 23-45, a small, nonconducting ball of mass $m = 1.0$ mg and charge $q = 2.0 \times 10^{-8}$ C (distributed uniformly through its volume) hangs from an insulating thread that makes an angle $\theta = 30°$ with a vertical, uniformly charged nonconducting sheet (shown in cross section). Considering the gravitational force on the ball and assuming the sheet extends far vertically and into and out of the page, calculate the surface charge density $\sigma$ of the sheet. **SSM**

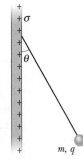

**FIG. 23-45**
**Problem 41.**

**••42** Figure 23-46 shows a very large nonconducting sheet that has a uniform surface charge density of $\sigma = -2.00$ $\mu C/m^2$; it also shows a particle of charge $Q = 6.00$ $\mu C$, at distance $d$ from the sheet. Both are fixed in place. If $d = 0.200$ m, at what (a) positive and (b) negative coordinate on the $x$ axis (other than infinity) is the net electric field $\vec{E}_{net}$ of the sheet and particle zero? (c) If $d = 0.800$ m, at what coordinate on the $x$ axis is $\vec{E}_{net} = 0$?

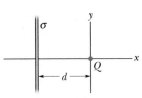

**FIG. 23-46**  **Problem 42.**

**•••43** Figure 23-47 shows a cross section through a very large nonconducting slab of thickness $d = 9.40$ mm and uniform volume charge density $\rho = 5.80$ fC/m³. The origin of an $x$ axis is at the slab's center. What is the magnitude of the slab's electric field at an $x$ coordinate of (a) 0, (b) 2.00 mm, (c) 4.70 mm, and (d) 26.0 mm?

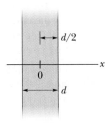

**FIG. 23-47**
**Problem 43.**

### sec. 23-9  Applying Gauss' Law: Spherical Symmetry

**•44** A point charge causes an electric flux of $-750$ N·m²/C to pass through a spherical Gaussian surface of 10.0 cm radius centered on the charge. (a) If the radius of the Gaussian surface were doubled, how much flux would pass through the surface? (b) What is the value of the point charge?

**•45** An unknown charge sits on a conducting solid sphere of radius 10 cm. If the electric field 15 cm from the center of the sphere has the magnitude $3.0 \times 10^3$ N/C and is directed radially inward, what is the net charge on the sphere? **SSM**

**•46** Figure 23-48 gives the magnitude of the electric field inside and outside a sphere with a positive charge distributed uniformly throughout its volume. The scale of the vertical axis is set by $E_s = 5.0 \times 10^7$ N/C. What is the charge on the sphere?

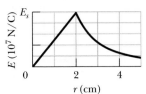

**FIG. 23-48**  **Problem 46.**

**•47** Two charged concentric spherical shells have radii 10.0 cm and 15.0 cm. The charge on the inner shell is $4.00 \times 10^{-8}$ C, and that on the outer shell is $2.00 \times 10^{-8}$ C. Find the electric field (a) at $r = 12.0$ cm and (b) at $r = 20.0$ cm.

**••48** Figure 23-49 shows two nonconducting spherical shells fixed in place on an $x$ axis. Shell 1 has uniform surface charge density $+4.0$ $\mu C/m^2$ on its outer surface and radius 0.50 cm,

and shell 2 has uniform surface charge density $-2.0$ $\mu C/m^2$ on its outer surface and radius 2.0 cm; the centers are separated by $L = 6.0$ cm. Other than at $x = \infty$, where on the $x$ axis is the net electric field equal to zero? **GO**

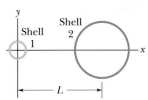

**FIG. 23-49**  **Problem 48.**

**••49** In Fig. 23-50, a nonconducting spherical shell of inner radius $a = 2.00$ cm and outer radius $b = 2.40$ cm has (within its thickness) a positive volume charge density $\rho = A/r$, where $A$ is a constant and $r$ is the distance from the center of the shell. In addition, a small ball of charge $q = 45.0$ fC is located at that center. What value should $A$ have if the electric field in the shell ($a \le r \le b$) is to be uniform? **SSM WWW**

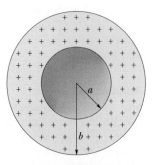

**FIG. 23-50**  **Problem 49.**

**••50** Figure 23-51 shows a spherical shell with uniform volume charge density $\rho = 1.84$ nC/m³, inner radius $a = 10.0$ cm, and outer radius $b = 2.00a$. What is the magnitude of the electric field at radial distances (a) $r = 0$; (b) $r = a/2.00$, (c) $r = a$, (d) $r = 1.50a$, (e) $r = b$, and (f) $r = 3.00b$?

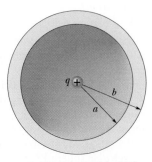

**FIG. 23-51**  **Problem 50.**

**••51** In Fig. 23-52, a solid sphere of radius $a = 2.00$ cm is concentric with a spherical conducting shell of inner radius $b = 2.00a$ and outer radius $c = 2.40a$. The sphere has a net uniform charge $q_1 = +5.00$ fC; the shell has a net charge $q_2 = -q_1$. What is the magnitude of the electric field at radial distances (a) $r = 0$, (b) $r = a/2.00$, (c) $r = a$, (d) $r = 1.50a$, (e) $r = 2.30a$, and (f) $r = 3.50a$? What is the net charge on the (g) inner and (h) outer surface of the shell?

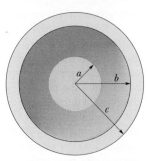

**FIG. 23-52**  **Problem 51.**

**••52** A charged particle is held at the center of a spherical shell. Figure 23-53 gives the magnitude $E$ of the electric field

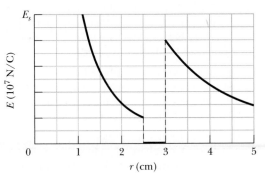

**FIG. 23-53**  **Problem 52.**

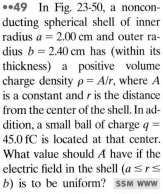
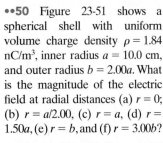

versus radial distance $r$. The scale of the vertical axis is set by $E_s = 10.0 \times 10^7$ N/C. Approximately, what is the net charge on the shell?

•••**53** A charge distribution that is spherically symmetric but not uniform radially produces an electric field of magnitude $E = Kr^4$, directed radially outward from the center of the sphere. Here $r$ is the radial distance from that center, and $K$ is a constant. What is the volume density $\rho$ of the charge distribution?

•••**54** Figure 23-54 shows, in cross section, two solid spheres with uniformly distributed charge throughout their volumes. Each has radius $R$. Point $P$ lies on a line connecting the centers of the spheres, at radial distance $R/2.00$ from the center of sphere 1. If the net electric field at point $P$ is zero, what is the ratio $q_2/q_1$ of the total charge $q_2$ in sphere 2 to the total charge $q_1$ in sphere 1?

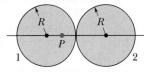

**FIG. 23-54** Problem 54.

•••**55** A solid nonconducting sphere of radius $R = 5.60$ cm has a nonuniform charge distribution of volume charge density $\rho = (14.1 \text{ pC/m}^3)r/R$, where $r$ is radial distance from the sphere's center. (a) What is the sphere's total charge? What is the magnitude $E$ of the electric field at (b) $r = 0$, (c) $r = R/2.00$, and (d) $r = R$? (e) Sketch a graph of $E$ versus $r$. ILW

### Additional Problems

**56** *The chocolate crumb mystery.* Explosions ignited by electrostatic discharges (sparks) constitute a serious danger in facilities handling grain or powder. Such an explosion occurred in chocolate crumb powder at a biscuit factory in the 1970s. Workers usually emptied newly delivered sacks of the powder into a loading bin, from which it was blown through electrically grounded plastic pipes to a silo for storage. Somewhere along this route, two conditions for an explosion were met: (1) The magnitude of an electric field became $3.0 \times 10^6$ N/C or greater, so that electrical breakdown and thus sparking could occur. (2) The energy of a spark was 150 mJ or greater so that it could ignite the powder explosively. Let us check for the first condition in the powder flow through the plastic pipes.

Suppose a stream of *negatively* charged powder was blown through a cylindrical pipe of radius $R = 5.0$ cm. Assume that the powder and its charge were spread uniformly through the pipe with a volume charge density $\rho$. (a) Using Gauss' law, find an expression for the magnitude of the electric field $\vec{E}$ in the pipe as a function of radial distance $r$ from the pipe center. (b) Does $E$ increase or decrease with increasing $r$? (c) Is $\vec{E}$ directed radially inward or outward? (d) For $\rho = 1.1 \times 10^{-3}$ C/m$^3$ (a typical value at the factory), find the maximum $E$ and determine where that maximum field occurs. (e) Could sparking occur, and if so, where? (The story continues with Problem 68 in Chapter 24.) ✈

**57** Charge $Q$ is uniformly distributed in a sphere of radius $R$. (a) What fraction of the charge is contained within radius $r = R/2.00$? (b) What is the ratio of the electric field magnitude at $r = R/2.00$ to that on the surface of the sphere?

**58** Charge of uniform volume density $\rho = 3.2\ \mu$C/m$^3$ fills a nonconducting solid sphere of radius 5.0 cm. What is the magnitude of the electric field (a) 3.5 cm and (b) 8.0 cm from the sphere's center?

**59** The electric field at point $P$ just outside the outer surface of a hollow spherical conductor of inner radius 10 cm and outer radius 20 cm has magnitude 450 N/C and is directed outward. When an unknown point charge $Q$ is introduced into the center of the sphere, the electric field at $P$ is still directed outward but is now 180 N/C. (a) What was the net charge enclosed by the outer surface before $Q$ was introduced? (b) What is charge $Q$? After $Q$ is introduced, what is the charge on the (c) inner and (d) outer surface of the conductor? SSM

**60** Assume that a ball of charged particles has a uniformly distributed negative charge density except for a narrow radial tunnel through its center, from the surface on one side to the surface on the opposite side. Also assume that we can position a proton anywhere along the tunnel or outside the ball. Let $F_R$ be the magnitude of the electrostatic force on the proton when it is located at the ball's surface, at radius $R$. As a multiple of $R$, how far from the surface is there a point where the force magnitude is $0.50F_R$ if we move the proton (a) away from the ball and (b) into the tunnel?

**61** Charge of uniform volume density $\rho = 1.2$ nC/m$^3$ fills an infinite slab between $x = -5.0$ cm and $x = +5.0$ cm. What is the magnitude of the electric field at any point with the coordinate (a) $x = 4.0$ cm and (b) $x = 6.0$ cm?

**62** A uniform surface charge of density 8.0 nC/m$^2$ is distributed over the entire $xy$ plane. What is the electric flux through a spherical Gaussian surface centered on the origin and having a radius of 5.0 cm?

**63** A thin-walled metal spherical shell has radius 25.0 cm and charge $2.00 \times 10^{-7}$ C. Find $E$ for a point (a) inside the shell, (b) just outside it, and (c) 3.00 m from the center.

**64** The electric field in a particular space is $\vec{E} = (x + 2)\hat{i}$ N/C, with $x$ in meters. Consider a cylindrical Gaussian surface of radius 20 cm that is coaxial with the $x$ axis. One end of the cylinder is at $x = 0$. (a) What is the magnitude of the electric flux through the other end of the cylinder at $x = 2.0$ m? (b) What net charge is enclosed within the cylinder?

**65** Figure 23-55 shows, in cross section, three infinitely large nonconducting sheets on which charge is uniformly spread. The surface charge densities are $\sigma_1 = +2.00\ \mu$C/m$^2$, $\sigma_2 = +4.00\ \mu$C/m$^2$, and $\sigma_3 = -5.00\ \mu$C/m$^2$, and distance $L = 1.50$ cm. In unit-vector notation, what is the net electric field at point $P$?

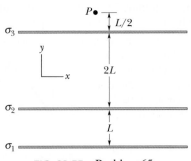

**FIG. 23-55** Problem 65.

**66** The net electric flux through each face of a die (singular of dice) has a magnitude in units of $10^3$ N·m$^2$/C that is exactly equal to the number of spots $N$ on the face (1 through 6). The flux is inward for $N$ odd and outward for $N$ even. What is the net charge inside the die?

**67** A Gaussian surface in the form of a hemisphere of radius $R = 5.68$ cm lies in a uniform electric field of magnitude $E = 2.50$ N/C. The surface encloses no net charge. At the (flat) base of the surface, the field is perpendicular to the surface and directed into the surface. What is the flux through (a) the base and (b) the curved portion of the surface?

**68** A point charge $q = 1.0 \times 10^{-7}$ C is at the center of a spherical cavity of radius 3.0 cm in a chunk of metal. Use Gauss' law to find the electric field (a) 1.5 cm from the cavity center and (b) anyplace in the metal.

**69** A thin-walled metal spherical shell of radius $a$ has a charge $q_a$. Concentric with it is a thin-walled metal spherical shell of radius $b > a$ and charge $q_b$. Find the electric field at points a distance $r$ from the common center, where (a) $r < a$, (b) $a < r < b$, and (c) $r > b$. (d) Discuss the criterion you would use to determine how the charges are distributed on the inner and outer surfaces of the shells. **SSM**

**70** What net charge is enclosed by the Gaussian cube of Problem 2?

**71** A proton with speed $v = 3.00 \times 10^5$ m/s orbits just outside a charged sphere of radius $r = 1.00$ cm. What is the charge on the sphere?

**72** Equation 23-11 ($E = \sigma/\varepsilon_0$) gives the electric field at points near a charged conducting surface. Apply this equation to a conducting sphere of radius $r$ and charge $q$, and show that the electric field outside the sphere is the same as the field of a point charge located at the center of the sphere.

**73** Figure 23-56 shows a Geiger counter, a device used to detect ionizing radiation, which causes ionization of atoms. A thin, positively charged central wire is surrounded by a concentric, circular, conducting cylindrical shell with an equal negative charge, creating a strong radial electric field. The shell contains a low-pressure inert gas. A particle of radiation entering the device through the shell wall ionizes a few of the gas atoms. The resulting free electrons (e) are drawn to the positive wire. However, the electric field is so intense that, between collisions with gas atoms, the free electrons gain energy sufficient to ionize these atoms also. More free electrons are thereby created, and the process is repeated until the electrons reach the wire. The resulting "avalanche" of electrons is collected by the wire, generating a signal that is used to record the passage of the original particle of radiation. Suppose that the radius of the central wire is 25 $\mu$m, the inner radius of the shell 1.4 cm, and the length of the shell 16 cm. If the electric field at the shell's inner wall is $2.9 \times 10^4$ N/C, what is the total positive charge on the central wire?

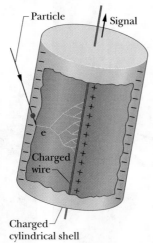

**FIG. 23-56** Problem 73.

**74** Charge is distributed uniformly throughout the volume of an infinitely long solid cylinder of radius $R$. (a) Show that, at a distance $r < R$ from the cylinder axis,

$$E = \frac{\rho r}{2\varepsilon_0},$$

where $\rho$ is the volume charge density. (b) Write an expression for $E$ when $r > R$.

**75** Water in an irrigation ditch of width $w = 3.22$ m and depth $d = 1.04$ m flows with a speed of 0.207 m/s. The *mass flux* of the flowing water through an imaginary surface is the product of the water's density (1000 kg/m³) and its volume flux through that surface. Find the mass flux through the following imaginary surfaces: (a) a surface of area $wd$, entirely in the water, perpendicular to the flow; (b) a surface with area $3wd/2$, of which $wd$ is in the water, perpendicular to the flow; (c) a surface of area $wd/2$, entirely in the water, perpendicular to the flow; (d) a surface of area $wd$, half in the water and half out, perpendicular to the flow; (e) a surface of area $wd$, entirely in the water, with its normal 34.0° from the direction of flow.

**76** A free electron is placed between two large, parallel, nonconducting plates that are horizontal and 2.3 cm apart. One plate has a uniform positive charge; the other has an equal amount of uniform negative charge. The force on the electron due to the electric field $\vec{E}$ between the plates balances the gravitational force on the electron. What are (a) the magnitude of the surface charge density on the plates and (b) the direction (up or down) of $\vec{E}$? **SSM**

**77** A nonconducting solid sphere has a uniform volume charge density $\rho$. Let $\vec{r}$ be the vector from the center of the sphere to a general point $P$ within the sphere. (a) Show that the electric field at $P$ is given by $\vec{E} = \rho\vec{r}/3\varepsilon_0$. (Note that the result is independent of the radius of the sphere.) (b) A spherical cavity is hollowed out of the sphere, as shown in Fig. 23-57. Using superposition concepts, show that the electric field at all points within the cavity is uniform and equal to $\vec{E} = \rho\vec{a}/3\varepsilon_0$, where $\vec{a}$ is the position vector from the center of the sphere to the center of the cavity. (Note that this result is independent of the radius of the sphere and the radius of the cavity.)

**FIG. 23-57** Problem 77.

**78** A uniform charge density of 500 nC/m³ is distributed throughout a spherical volume of radius 6.00 cm. Consider a cubical Gaussian surface with its center at the center of the sphere. What is the electric flux through this cubical surface if its edge length is (a) 4.00 cm and (b) 14.0 cm?

**79** A spherical conducting shell has a charge of $-14$ $\mu$C on its outer surface and a charged particle in its hollow. If the net charge on the shell is $-10$ $\mu$C, what is the charge (a) on the inner surface of the shell and (b) of the particle? **SSM**

**80** A charge of 6.00 pC is spread uniformly throughout the volume of a sphere of radius $r = 4.00$ cm. What is the magnitude of the electric field at a radial distance of (a) 6.00 cm and (b) 3.00 cm?

**81** A spherical ball of charged particles has a uniform charge density. In terms of the ball's radius $R$, at what radial distances (a) inside and (b) outside the ball is the magnitude of the ball's electric field equal to $\frac{1}{4}$ of the maximum magnitude of that field?

**82** Charge of uniform surface density 8.00 nC/m² is distributed over an entire $xy$ plane; charge of uniform surface density 3.00 nC/m² is distributed over the parallel plane defined by $z = 2.00$ m. Determine the magnitude of the electric field at any point having a $z$ coordinate of (a) 1.00 m and (b) 3.00 m.

*If a person peels off a jacket or sweater while working at a computer, the computer can be destroyed. If a child slides down a plastic slide and then reaches for another person, the child can be in for a painful surprise. If an anesthesiologist does not wear the right type of shoes during surgery, the patient can be fatally injured. If a customer at a self-serve gas station slides back into the car while the tank is being filled, a fire can erupt at the pump nozzle when the customer returns to pull the nozzle from the car.*

## What is the danger in these situations?

The answer is in this chapter.

Jim Doberman/The Image Bank/Getty Images

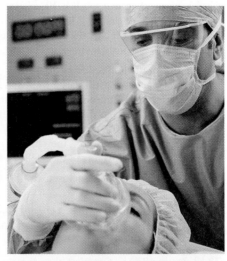

Andrew Olney/Digital Vision/AFP/Getty Images

Vladimir Pcholkin/Taxi/AFP/Getty Images

OnRequest Images/Index Stock

## 24-1 | WHAT IS PHYSICS?

One goal of physics is to identify basic forces in our world, such as the electric force we discussed in Chapter 21. A related goal is to determine whether a force is conservative—that is, whether a potential energy can be associated with it. The motivation for associating a potential energy with a force is that we can then apply the principle of the conservation of mechanical energy to closed systems involving the force. This extremely powerful principle allows us to calculate the results of experiments for which force calculations alone would be very difficult. Experimentally, physicists and engineers discovered that the electric force is conservative and thus has an associated electric potential energy. In this chapter we first define this type of potential energy and then put it to use.

## 24-2 | Electric Potential Energy

When an electrostatic force acts between two or more charged particles within a system of particles, we can assign an **electric potential energy** $U$ to the system. If the system changes its configuration from an initial state $i$ to a different final state $f$, the electrostatic force does work $W$ on the particles. From Eq. 8-1, we then know that the resulting change $\Delta U$ in the potential energy of the system is

$$\Delta U = U_f - U_i = -W. \qquad (24\text{-}1)$$

As with other conservative forces, the work done by the electrostatic force is *path independent*. Suppose a charged particle within the system moves from point $i$ to point $f$ while an electrostatic force between it and the rest of the system acts on it. Provided the rest of the system does not change, the work $W$ done by the force on the particle is the same for *all* paths between points $i$ and $f$.

For convenience, we usually take the *reference configuration* of a system of charged particles to be that in which the particles are all infinitely separated from one another. Also, we usually set the corresponding *reference potential energy* to be zero. Suppose that several charged particles come together from initially infinite separations (state $i$) to form a system of neighboring particles (state $f$). Let the initial potential energy $U_i$ be zero, and let $W_\infty$ represent the work done by the electrostatic forces between the particles during the move in from infinity. Then from Eq. 24-1, the final potential energy $U$ of the system is

$$U = -W_\infty. \qquad (24\text{-}2)$$

**✓CHECKPOINT 1**   In the figure, a proton moves from point $i$ to point $f$ in a uniform electric field directed as shown. (a) Does the electric field do positive or negative work on the proton? (b) Does the electric potential energy of the proton increase or decrease?

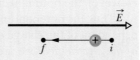

---

**PROBLEM-SOLVING TACTICS**

*Tactic 1:   Electric Potential Energy; Work Done by a Field*   An electric potential energy is associated with a system of particles as a whole. However, you will see statements (starting with Sample Problem 24-1) that associate it with only one particle within a system. For example, you might read, "An electron in an electric field has a potential energy of $10^{-7}$ J." Such statements are often acceptable, but you should always keep in mind that the potential energy is actually associated *with a system*—here, the electron plus the charged particles

that set up the electric field. Also keep in mind that it makes sense to assign a particular potential energy value, such as $10^{-7}$ J here, to a particle or even a system *only* if the reference potential energy value is known.

If the potential energy is associated with only one particle within a system, you may read that work is done on the particle *by the electric field*. This means that work is done on the particle by the force due to the charges that set up the field.

## Sample Problem 24-1

Electrons are continually being knocked out of air molecules in the atmosphere by cosmic-ray particles coming in from space. Once released, each electron experiences an electrostatic force $\vec{F}$ due to the electric field $\vec{E}$ that is produced in the atmosphere by charged particles already on Earth. Near Earth's surface the electric field has the magnitude $E = 150$ N/C and is directed downward. What is the change $\Delta U$ in the electric potential energy of a released electron when the electrostatic force causes it to move vertically upward through a distance $d = 520$ m (Fig. 24-1)?

**KEY IDEAS** (1) The change $\Delta U$ in the electric potential energy of the electron is related to the work $W$ done on the electron by the electric field. Equation 24-1 ($\Delta U = -W$) gives the relation. (2) The work done by a constant force $\vec{F}$ on a particle undergoing a displacement $\vec{d}$ is

$$W = \vec{F} \cdot \vec{d}. \qquad (24\text{-}3)$$

(3) The electrostatic force and the electric field are related by $\vec{F} = q\vec{E}$, where here $q$ is the charge of an electron ($= -1.6 \times 10^{-19}$ C).

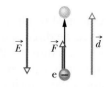

**FIG. 24-1** An electron in the atmosphere is moved upward through displacement $\vec{d}$ by an electrostatic force $\vec{F}$ due to an electric field $\vec{E}$.

**Calculations:** Substituting for $\vec{F}$ in Eq. 24-3 and taking the dot product yield

$$W = q\vec{E} \cdot \vec{d} = qEd \cos \theta, \qquad (24\text{-}4)$$

where $\theta$ is the angle between the directions of $\vec{E}$ and $\vec{d}$. The field $\vec{E}$ is directed downward and the displacement $\vec{d}$ is directed upward; so $\theta = 180°$. Substituting this and other data into Eq. 24-4, we find

$$W = (-1.6 \times 10^{-19} \text{ C})(150 \text{ N/C})(520 \text{ m}) \cos 180°$$
$$= 1.2 \times 10^{-14} \text{ J}.$$

Equation 24-1 then yields

$$\Delta U = -W = -1.2 \times 10^{-14} \text{ J}. \qquad \text{(Answer)}$$

This result tells us that during the 520 m ascent, the electric potential energy of the electron *decreases* by $1.2 \times 10^{-14}$ J.

## 24-3 | Electric Potential

As you can infer from Sample Problem 24-1, the potential energy of a charged particle in an electric field depends on the charge magnitude. However, the potential energy *per unit charge* has a unique value at any point in an electric field.

For an example of this, suppose we place a test particle of positive charge $1.60 \times 10^{-19}$ C at a point in an electric field where the particle has an electric potential energy of $2.40 \times 10^{-17}$ J. Then the potential energy per unit charge is

$$\frac{2.40 \times 10^{-17} \text{ J}}{1.60 \times 10^{-19} \text{ C}} = 150 \text{ J/C}.$$

Next, suppose we replace that test particle with one having twice as much positive charge, $3.20 \times 10^{-19}$ C. We would find that the second particle has an electric potential energy of $4.80 \times 10^{-17}$ J, twice that of the first particle. However, the potential energy per unit charge would be the same, still 150 J/C.

Thus, the potential energy per unit charge, which can be symbolized as $U/q$, is independent of the charge $q$ of the particle we happen to use and is *characteristic only of the electric field* we are investigating. The potential energy per unit charge at a point in an electric field is called the **electric potential** $V$ (or simply the **potential**) at that point. Thus,

$$V = \frac{U}{q}. \qquad (24\text{-}5)$$

*Note that electric potential is a scalar, not a vector.*

The *electric potential difference* $\Delta V$ between any two points $i$ and $f$ in an electric field is equal to the difference in potential energy per unit charge between the two points:

$$\Delta V = V_f - V_i = \frac{U_f}{q} - \frac{U_i}{q} = \frac{\Delta U}{q}. \qquad (24\text{-}6)$$

Using Eq. 24-1 to substitute $-W$ for $\Delta U$ in Eq. 24-6, we can define the potential difference between points $i$ and $f$ as

$$\Delta V = V_f - V_i = -\frac{W}{q} \quad \text{(potential difference defined)}. \quad (24\text{-}7)$$

The potential difference between two points is thus the negative of the work done by the electrostatic force to move a unit charge from one point to the other. A potential difference can be positive, negative, or zero, depending on the signs and magnitudes of $q$ and $W$.

If we set $U_i = 0$ at infinity as our reference potential energy, then by Eq. 24-5, the electric potential $V$ must also be zero there. Then from Eq. 24-7, we can define the electric potential at any point in an electric field to be

$$V = -\frac{W_\infty}{q} \quad \text{(potential defined)}, \quad (24\text{-}8)$$

where $W_\infty$ is the work done by the electric field on a charged particle as that particle moves in from infinity to point $f$. A potential $V$ can be positive, negative, or zero, depending on the signs and magnitudes of $q$ and $W_\infty$.

The SI unit for potential that follows from Eq. 24-8 is the joule per coulomb. This combination occurs so often that a special unit, the *volt* (abbreviated V), is used to represent it. Thus,

$$1 \text{ volt} = 1 \text{ joule per coulomb}. \quad (24\text{-}9)$$

This new unit allows us to adopt a more conventional unit for the electric field $\vec{E}$, which we have measured up to now in newtons per coulomb. With two unit conversions, we obtain

$$1 \text{ N/C} = \left(1\frac{\text{N}}{\text{C}}\right)\left(\frac{1 \text{ V} \cdot \text{C}}{1 \text{ J}}\right)\left(\frac{1 \text{ J}}{1 \text{ N} \cdot \text{m}}\right)$$

$$= 1 \text{ V/m}. \quad (24\text{-}10)$$

The conversion factor in the second set of parentheses comes from Eq. 24-9; that in the third set of parentheses is derived from the definition of the joule. From now on, we shall express values of the electric field in volts per meter rather than in newtons per coulomb.

Finally, we can now define an energy unit that is a convenient one for energy measurements in the atomic and subatomic domain: One *electron-volt* (eV) is the energy equal to the work required to move a single elementary charge $e$, such as that of the electron or the proton, through a potential difference of exactly one volt. Equation 24-7 tells us that the magnitude of this work is $q \, \Delta V$; so

$$1 \text{ eV} = e(1 \text{ V})$$

$$= (1.60 \times 10^{-19} \text{ C})(1 \text{ J/C}) = 1.60 \times 10^{-19} \text{ J}.$$

**PROBLEM-SOLVING TACTICS**

*Tactic 2:* **Electric Potential and Electric Potential Energy** Electric potential $V$ and electric potential energy $U$ are quite different quantities and should not be confused.

☞ *Electric potential* is a scalar property associated with an electric field, regardless of whether a charged object has been placed in that field; it is measured in joules per coulomb, or volts.

☞ *Electric potential energy* is an energy of a charged object in an external electric field (or more precisely, an energy of the system consisting of the object and the external electric field); it is measured in joules.

## Work Done by an Applied Force

Suppose we move a particle of charge $q$ from point $i$ to point $f$ in an electric field by applying a force to it. During the move, our applied force does work $W_{\text{app}}$ on the charge while the electric field does work $W$ on it. By the work–kinetic energy theorem of Eq. 7-10, the change $\Delta K$ in the kinetic energy of the particle is

$$\Delta K = K_f - K_i = W_{app} + W. \qquad (24\text{-}11)$$

Now suppose the particle is stationary before and after the move. Then $K_f$ and $K_i$ are both zero, and Eq. 24-11 reduces to

$$W_{app} = -W. \qquad (24\text{-}12)$$

In words, the work $W_{app}$ done by our applied force during the move is equal to the negative of the work $W$ done by the electric field—provided there is no change in kinetic energy.

By using Eq. 24-12 to substitute $W_{app}$ into Eq. 24-1, we can relate the work done by our applied force to the change in the potential energy of the particle during the move. We find

$$\Delta U = U_f - U_i = W_{app}. \qquad (24\text{-}13)$$

By similarly using Eq. 24-12 to substitute $W_{app}$ into Eq. 24-7, we can relate our work $W_{app}$ to the electric potential difference $\Delta V$ between the initial and final locations of the particle. We find

$$W_{app} = q\,\Delta V. \qquad (24\text{-}14)$$

$W_{app}$ can be positive, negative, or zero depending on the signs and magnitudes of $q$ and $\Delta V$.

✓**CHECKPOINT 2**     In the figure of Checkpoint 1, we move the proton from point $i$ to point $f$ in a uniform electric field directed as shown. (a) Does our force do positive or negative work? (b) Does the proton move to a point of higher or lower potential?

## 24-4 | Equipotential Surfaces

Adjacent points that have the same electric potential form an **equipotential surface,** which can be either an imaginary surface or a real, physical surface. No net work $W$ is done on a charged particle by an electric field when the particle moves between two points $i$ and $f$ on the same equipotential surface. This follows from Eq. 24-7, which tells us that $W$ must be zero if $V_f = V_i$. Because of the path independence of work (and thus of potential energy and potential), $W = 0$ for *any* path connecting points $i$ and $f$ on a given equipotential surface regardless of whether that path lies entirely on that surface.

Figure 24-2 shows a *family* of equipotential surfaces associated with the electric field due to some distribution of charges. The work done by the electric field on a charged particle as the particle moves from one end to the other of paths I and II is zero because each of these paths begins and ends on the same equipotential surface. The work done as the charged particle moves from one end to the other of paths III and IV is not zero but has the same value for both these paths because the initial and final potentials are identical for the two paths; that is, paths III and IV connect the same pair of equipotential surfaces.

From symmetry, the equipotential surfaces produced by a point charge or a spherically symmetrical charge distribution are a family of concentric spheres. For a uniform electric field, the surfaces are a family of planes perpendicular to the field lines. In fact, equipotential surfaces are always perpendicular to electric field lines and thus to $\vec{E}$, which is always tangent to these lines. If $\vec{E}$ were *not* perpendicular to an equipotential surface, it would have a component lying along that surface. This component would then do work on a charged particle as it moved along the surface. However, by Eq. 24-7 work cannot be done if the surface is truly an equipotential surface; the only possible conclusion is that $\vec{E}$ must be everywhere perpendicular to the surface. Figure 24-3 shows electric field lines and cross sections of the equipotential surfaces for a uniform electric field and for the field associated with a point charge and with an electric dipole.

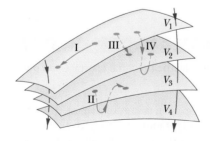

**FIG. 24-2** Portions of four equipotential surfaces at electric potentials $V_1 = 100$ V, $V_2 = 80$ V, $V_3 = 60$ V, and $V_4 = 40$ V. Four paths along which a test charge may move are shown. Two electric field lines are also indicated.

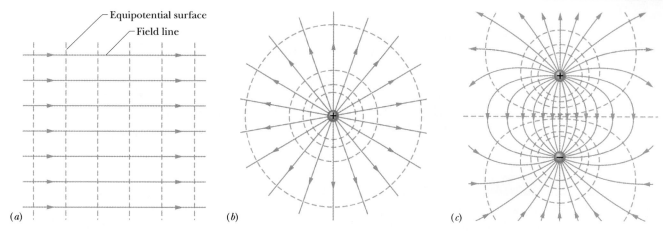

**FIG. 24-3** Electric field lines (purple) and cross sections of equipotential surfaces (gold) for (*a*) a uniform electric field, (*b*) the field due to a point charge, and (*c*) the field due to an electric dipole.

## 24-5 | Calculating the Potential from the Field

We can calculate the potential difference between any two points *i* and *f* in an electric field if we know the electric field vector $\vec{E}$ all along any path connecting those points. To make the calculation, we find the work done on a positive test charge by the field as the charge moves from *i* to *f*, and then use Eq. 24-7.

Consider an arbitrary electric field, represented by the field lines in Fig. 24-4, and a positive test charge $q_0$ that moves along the path shown from point *i* to point *f*. At any point on the path, an electrostatic force $q_0\vec{E}$ acts on the charge as it moves through a differential displacement $d\vec{s}$. From Chapter 7, we know that the differential work $dW$ done on a particle by a force $\vec{F}$ during a displacement $d\vec{s}$ is

$$dW = \vec{F} \cdot d\vec{s}. \qquad (24\text{-}15)$$

For the situation of Fig. 24-4, $\vec{F} = q_0\vec{E}$ and Eq. 24-15 becomes

$$dW = q_0\vec{E} \cdot d\vec{s}. \qquad (24\text{-}16)$$

To find the total work $W$ done on the particle by the field as the particle moves from point *i* to point *f*, we sum—via integration—the differential works done on the charge as it moves through all the displacements $d\vec{s}$ along the path:

$$W = q_0 \int_i^f \vec{E} \cdot d\vec{s}. \qquad (24\text{-}17)$$

If we substitute the total work $W$ from Eq. 24-17 into Eq. 24-7, we find

$$V_f - V_i = -\int_i^f \vec{E} \cdot d\vec{s}. \qquad (24\text{-}18)$$

Thus, the potential difference $V_f - V_i$ between any two points *i* and *f* in an electric field is equal to the negative of the *line integral* (meaning the integral along a particular path) of $\vec{E} \cdot d\vec{s}$ from *i* to *f*. However, because the electrostatic force is conservative, all paths (whether easy or difficult to use) yield the same result.

Equation 24-18 allows us to calculate the difference in potential between any two points in the field. If we set potential $V_i = 0$, then Eq. 24-18 becomes

$$V = -\int_i^f \vec{E} \cdot d\vec{s}, \qquad (24\text{-}19)$$

in which we have dropped the subscript *f* on $V_f$. Equation 24-19 gives us the potential *V* at any point *f* in the electric field *relative to the zero potential* at point *i*. If we let point *i* be at infinity, then Eq. 24-19 gives us the potential *V* at any point *f* relative to the zero potential at infinity.

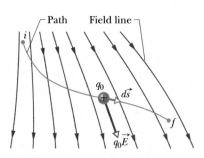

**FIG. 24-4** A test charge $q_0$ moves from point *i* to point *f* along the path shown in a nonuniform electric field. During a displacement $d\vec{s}$, an electrostatic force $q_0\vec{E}$ acts on the test charge. This force points in the direction of the field line at the location of the test charge.

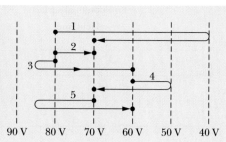

**✓ CHECKPOINT 3** The figure here shows a family of parallel equipotential surfaces (in cross section) and five paths along which we shall move an electron from one surface to another. (a) What is the direction of the electric field associated with the surfaces? (b) For each path, is the work we do positive, negative, or zero? (c) Rank the paths according to the work we do, greatest first.

## Sample Problem  24-2

(a) Figure 24-5a shows two points $i$ and $f$ in a uniform electric field $\vec{E}$. The points lie on the same electric field line (not shown) and are separated by a distance $d$. Find the potential difference $V_f - V_i$ by moving a positive test charge $q_0$ from $i$ to $f$ along the path shown, which is parallel to the field direction.

**KEY IDEA** We can find the potential difference between any two points in an electric field by integrating $\vec{E} \cdot d\vec{s}$ along a path connecting those two points according to Eq. 24-18.

**Calculations:** We begin by mentally moving a test charge $q_0$ along that path, from initial point $i$ to final point $f$. As we move such a test charge along the path in Fig. 24-5a, its differential displacement $d\vec{s}$ always has the same direction as $\vec{E}$. Thus, the angle $\theta$ between $\vec{E}$ and $d\vec{s}$ is zero and the dot product in Eq. 24-18 is

$$\vec{E} \cdot d\vec{s} = E\, ds \cos \theta = E\, ds. \qquad (24\text{-}20)$$

Equations 24-18 and 24-20 then give us

$$V_f - V_i = -\int_i^f \vec{E} \cdot d\vec{s} = -\int_i^f E\, ds. \qquad (24\text{-}21)$$

Since the field is uniform, $E$ is constant over the path and can be moved outside the integral, giving us

$$V_f - V_i = -E \int_i^f ds = -Ed, \qquad \text{(Answer)}$$

in which the integral is simply the length $d$ of the path. The minus sign in the result shows that the potential at point $f$ in Fig. 24-5a is lower than the potential at point $i$. This is a general result: The potential always decreases along a path that extends in the direction of the electric field lines.

(b) Now find the potential difference $V_f - V_i$ by moving the positive test charge $q_0$ from $i$ to $f$ along the path $icf$ shown in Fig. 24-5b.

**Calculations:** The Key Idea of (a) applies here too, except now we move the test charge along a path that consists of two lines: $ic$ and $cf$. At all points along line $ic$, the

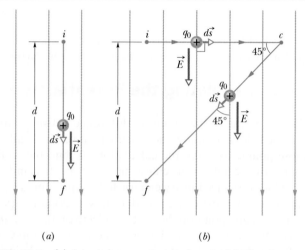

*(a)*            *(b)*

**FIG. 24-5**  (a) A test charge $q_0$ moves in a straight line from point $i$ to point $f$, along the direction of a uniform external electric field. (b) Charge $q_0$ moves along path $icf$ in the same electric field.

displacement $d\vec{s}$ of the test charge is perpendicular to $\vec{E}$. Thus, the angle $\theta$ between $\vec{E}$ and $d\vec{s}$ is 90°, and the dot product $\vec{E} \cdot d\vec{s}$ is 0. Equation 24-18 then tells us that points $i$ and $c$ are at the same potential: $V_c - V_i = 0$.

For line $cf$ we have $\theta = 45°$ and, from Eq. 24-18,

$$V_f - V_i = -\int_c^f \vec{E} \cdot d\vec{s} = -\int_c^f E(\cos 45°)\, ds$$

$$= -E(\cos 45°) \int_c^f ds.$$

The integral in this equation is just the length of line $cf$; from Fig. 24-5b, that length is $d/\sin 45°$. Thus,

$$V_f - V_i = -E(\cos 45°)\, \frac{d}{\sin 45°} = -Ed. \qquad \text{(Answer)}$$

This is the same result we obtained in (a), as it must be; the potential difference between two points does not depend on the path connecting them. Moral: When you want to find the potential difference between two points by moving a test charge between them, you can save time and work by choosing a path that simplifies the use of Eq. 24-18.

# 24-6 | Potential Due to a Point Charge

We now use Eq. 24-18 to derive, for the space around a charged particle, an expression for the electric potential $V$ relative to the zero potential at infinity. Consider a point $P$ at distance $R$ from a fixed particle of positive charge $q$ (Fig. 24-6). To use Eq. 24-18, we imagine that we move a positive test charge $q_0$ from point $P$ to infinity. Because the path we take does not matter, let us choose the simplest one—a line that extends radially from the fixed particle through $P$ to infinity.

To use Eq. 24-18, we must evaluate the dot product

$$\vec{E} \cdot d\vec{s} = E \cos \theta \, ds. \qquad (24\text{-}22)$$

The electric field $\vec{E}$ in Fig. 24-6 is directed radially outward from the fixed particle. Thus, the differential displacement $d\vec{s}$ of the test particle along its path has the same direction as $\vec{E}$. That means that in Eq. 24-22, angle $\theta = 0$ and $\cos \theta = 1$. Because the path is radial, let us write $ds$ as $dr$. Then, substituting the limits $R$ and $\infty$, we can write Eq. 24-18 as

$$V_f - V_i = -\int_R^\infty E \, dr. \qquad (24\text{-}23)$$

Next, we set $V_f = 0$ (at $\infty$) and $V_i = V$ (at $R$). Then, for the magnitude of the electric field at the site of the test charge, we substitute from Eq. 22-3:

$$E = \frac{1}{4\pi\varepsilon_0} \frac{q}{r^2}. \qquad (24\text{-}24)$$

With these changes, Eq. 24-23 then gives us

$$0 - V = -\frac{q}{4\pi\varepsilon_0} \int_R^\infty \frac{1}{r^2} \, dr = \frac{q}{4\pi\varepsilon_0} \left[\frac{1}{r}\right]_R^\infty$$

$$= -\frac{1}{4\pi\varepsilon_0} \frac{q}{R}. \qquad (24\text{-}25)$$

Solving for $V$ and switching $R$ to $r$, we then have

$$V = \frac{1}{4\pi\varepsilon_0} \frac{q}{r} \qquad (24\text{-}26)$$

as the electric potential $V$ due to a particle of charge $q$ at any radial distance $r$ from the particle.

Although we have derived Eq. 24-26 for a positively charged particle, the derivation holds also for a negatively charged particle, in which case, $q$ is a negative quantity. Note that the sign of $V$ is the same as the sign of $q$:

> 👉 A positively charged particle produces a positive electric potential. A negatively charged particle produces a negative electric potential.

Figure 24-7 shows a computer-generated plot of Eq. 24-26 for a positively charged particle; the magnitude of $V$ is plotted vertically. Note that the magnitude increases as $r \to 0$. In fact, according to Eq. 24-26, $V$ is infinite at $r = 0$, although Fig. 24-7 shows a finite, smoothed-off value there.

Equation 24-26 also gives the electric potential either *outside or on the external surface of* a spherically symmetric charge distribution. We can prove this by using one of the shell theorems of Sections 21-4 and 23-9 to replace the actual spherical charge distribution with an equal charge concentrated at its center. Then the derivation leading to Eq. 24-26 follows, provided we do not consider a point within the actual distribution.

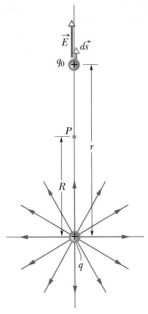

**FIG. 24-6** The positive point charge $q$ produces an electric field $\vec{E}$ and an electric potential $V$ at point $P$. We find the potential by moving a test charge $q_0$ from $P$ to infinity. The test charge is shown at distance $r$ from the point charge, during differential displacement $d\vec{s}$.

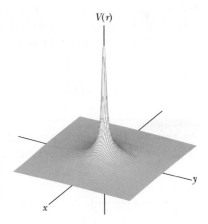

**FIG. 24-7** A computer-generated plot of the electric potential $V(r)$ due to a positive point charge located at the origin of an $xy$ plane. The potentials at points in the $xy$ plane are plotted vertically. (Curved lines have been added to help you visualize the plot.) The infinite value of $V$ predicted by Eq. 24-26 for $r = 0$ is not plotted.

*Tactic 3:* **Finding a Potential Difference** To find the potential difference $\Delta V$ between any two points in the field of an isolated point charge, evaluate Eq. 24-26 at each point and then subtract one result from the other. The value of $\Delta V$ will be the same for any choice of reference potential energy because that choice is eliminated by the subtraction.

## 24-7 | Potential Due to a Group of Point Charges

We can find the net potential at a point due to a group of point charges with the help of the superposition principle. Using Eq. 24-26 with the sign of the charge included, we calculate separately the potential resulting from each charge at the given point. Then we sum the potentials. For $n$ charges, the net potential is

$$V = \sum_{i=1}^{n} V_i = \frac{1}{4\pi\varepsilon_0} \sum_{i=1}^{n} \frac{q_i}{r_i} \qquad (n \text{ point charges}). \qquad (24\text{-}27)$$

Here $q_i$ is the value of the $i$th charge and $r_i$ is the radial distance of the given point from the $i$th charge. The sum in Eq. 24-27 is an *algebraic sum,* not a vector sum like the sum that would be used to calculate the electric field resulting from a group of point charges. Herein lies an important computational advantage of potential over electric field: It is a lot easier to sum several scalar quantities than to sum several vector quantities whose directions and components must be considered.

✓**CHECKPOINT 4** The figure here shows three arrangements of two protons. Rank the arrangements according to the net electric potential produced at point $P$ by the protons, greatest first.

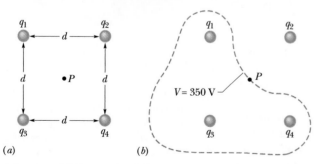

(a)    (b)    (c)

---

**Sample Problem** | **24-3**

What is the electric potential at point $P$, located at the center of the square of point charges shown in Fig. 24-8a? The distance $d$ is 1.3 m, and the charges are

$$q_1 = +12 \text{ nC}, \qquad q_3 = +31 \text{ nC},$$
$$q_2 = -24 \text{ nC}, \qquad q_4 = +17 \text{ nC}.$$

**KEY IDEA** The electric potential $V$ at $P$ is the algebraic sum of the electric potentials contributed by the four point charges. (Because electric potential is a scalar, the orientations of the point charges do not matter.)

**Calculations:** From Eq. 24-27, we have

$$V = \sum_{i=1}^{4} V_i = \frac{1}{4\pi\varepsilon_0} \left( \frac{q_1}{r} + \frac{q_2}{r} + \frac{q_3}{r} + \frac{q_4}{r} \right).$$

The distance $r$ is $d/\sqrt{2}$, which is 0.919 m, and the sum of the charges is

$$q_1 + q_2 + q_3 + q_4 = (12 - 24 + 31 + 17) \times 10^{-9} \text{ C}$$
$$= 36 \times 10^{-9} \text{ C}.$$

Thus, 
$$V = \frac{(8.99 \times 10^9 \text{ N} \cdot \text{m}^2/\text{C}^2)(36 \times 10^{-9} \text{ C})}{0.919 \text{ m}}$$
$$\approx 350 \text{ V}. \qquad \text{(Answer)}$$

Close to any of the three positive charges in Fig. 24-8a, the potential has very large positive values. Close to the single negative charge, the potential has very large negative values. Therefore, there must be points within the square that have the same intermediate potential as that at point $P$. The curve in Fig. 24-8b shows the intersection of the plane of the figure with the equipotential surface that contains point $P$. Any point along that curve has the same potential as point $P$.

**FIG. 24-8** (a) Four point charges are held fixed at the corners of a square. (b) The closed curve is a cross section, in the plane of the figure, of the equipotential surface that contains point $P$. (The curve is drawn only roughly.)

**Sample Problem** **24-4**

(a) In Fig. 24-9a, 12 electrons (of charge $-e$) are equally spaced and fixed around a circle of radius $R$. Relative to $V = 0$ at infinity, what are the electric potential and electric field at the center $C$ of the circle due to these electrons?

**KEY IDEAS** (1) The electric potential $V$ at $C$ is the algebraic sum of the electric potentials contributed by all the electrons. (Because electric potential is a scalar, the orientations of the electrons do not matter.) (2) The electric field at $C$ is a vector quantity and thus the orientation of the electrons *is* important.

**Calculations:** Because the electrons all have the same negative charge $-e$ and are all the same distance $R$ from $C$, Eq. 24-27 gives us

$$V = -12 \frac{1}{4\pi\varepsilon_0} \frac{e}{R}. \quad \text{(Answer)} \quad (24\text{-}28)$$

Because of the symmetry of the arrangement in Fig. 24-9a, the electric field vector at $C$ due to any given electron is canceled by the field vector due to the electron that is diametrically opposite it. Thus, at $C$,

$$\vec{E} = 0. \quad \text{(Answer)}$$

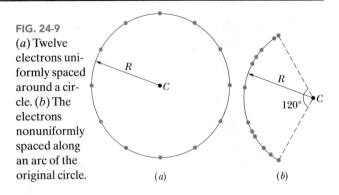

**FIG. 24-9** (a) Twelve electrons uniformly spaced around a circle. (b) The electrons nonuniformly spaced along an arc of the original circle.

(a)          (b)

(b) If the electrons are moved along the circle until they are nonuniformly spaced over a 120° arc (Fig. 24-9b), what then is the potential at $C$? How does the electric field at $C$ change (if at all)?

**Reasoning:** The potential is still given by Eq. 24-28, because the distance between $C$ and each electron is unchanged and orientation is irrelevant. The electric field is no longer zero, however, because the arrangement is no longer symmetric. A net field is now directed toward the charge distribution.

## 24-8 | Potential Due to an Electric Dipole

Now let us apply Eq. 24-27 to an electric dipole to find the potential at an arbitrary point $P$ in Fig. 24-10a. At $P$, the positive point charge (at distance $r_{(+)}$) sets up potential $V_{(+)}$ and the negative point charge (at distance $r_{(-)}$) sets up potential $V_{(-)}$. Then the net potential at $P$ is given by Eq. 24-27 as

$$V = \sum_{i=1}^{2} V_i = V_{(+)} + V_{(-)} = \frac{1}{4\pi\varepsilon_0} \left( \frac{q}{r_{(+)}} + \frac{-q}{r_{(-)}} \right)$$

$$= \frac{q}{4\pi\varepsilon_0} \frac{r_{(-)} - r_{(+)}}{r_{(-)}r_{(+)}}. \quad (24\text{-}29)$$

Naturally occurring dipoles—such as those possessed by many molecules—are quite small; so we are usually interested only in points that are relatively far from the dipole, such that $r \gg d$, where $d$ is the distance between the charges. Under those conditions, the approximations that follow from Fig. 24-10b are

$$r_{(-)} - r_{(+)} \approx d \cos \theta \quad \text{and} \quad r_{(-)}r_{(+)} \approx r^2.$$

If we substitute these quantities into Eq. 24-29, we can approximate $V$ to be

$$V = \frac{q}{4\pi\varepsilon_0} \frac{d \cos \theta}{r^2},$$

where $\theta$ is measured from the dipole axis as shown in Fig. 24-10a. We can now

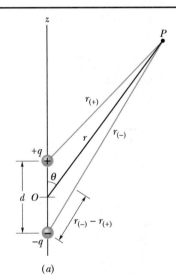

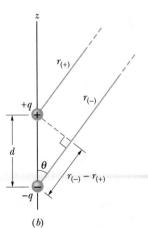

**FIG. 24-10** (a) Point $P$ is a distance $r$ from the midpoint $O$ of a dipole. The line $OP$ makes an angle $\theta$ with the dipole axis. (b) If $P$ is far from the dipole, the lines of lengths $r_{(+)}$ and $r_{(-)}$ are approximately parallel to the line of length $r$, and the dashed black line is approximately perpendicular to the line of length $r_{(-)}$.

(b)

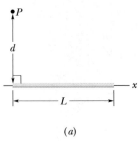

write $V$ as

$$V = \frac{1}{4\pi\varepsilon_0}\frac{p\cos\theta}{r^2} \qquad \text{(electric dipole)}, \qquad (24\text{-}30)$$

in which $p (= qd)$ is the magnitude of the electric dipole moment $\vec{p}$ defined in Section 22-5. The vector $\vec{p}$ is directed along the dipole axis, from the negative to the positive charge. (Thus, $\theta$ is measured from the direction of $\vec{p}$.)

✓**CHECKPOINT 5**  Suppose that three points are set at equal (large) distances $r$ from the center of the dipole in Fig. 24-10: Point $a$ is on the dipole axis above the positive charge, point $b$ is on the axis below the negative charge, and point $c$ is on a perpendicular bisector through the line connecting the two charges. Rank the points according to the electric potential of the dipole there, greatest (most positive) first.

### Induced Dipole Moment

Many molecules, such as water, have *permanent* electric dipole moments. In other molecules (called *nonpolar molecules*) and in every isolated atom, the centers of the positive and negative charges coincide (Fig. 24-11a) and thus no dipole moment is set up. However, if we place an atom or a nonpolar molecule in an external electric field, the field distorts the electron orbits and separates the centers of positive and negative charge (Fig. 24-11b). Because the electrons are negatively charged, they tend to be shifted in a direction opposite the field. This shift sets up a dipole moment $\vec{p}$ that points in the direction of the field. This dipole moment is said to be *induced* by the field, and the atom or molecule is then said to be *polarized* by the field (that is, it has a positive side and a negative side). When the field is removed, the induced dipole moment and the polarization disappear.

### 24-9 | Potential Due to a Continuous Charge Distribution

When a charge distribution $q$ is continuous (as on a uniformly charged thin rod or disk), we cannot use the summation of Eq. 24-27 to find the potential $V$ at a point $P$. Instead, we must choose a differential element of charge $dq$, determine the potential $dV$ at $P$ due to $dq$, and then integrate over the entire charge distribution.

Let us again take the zero of potential to be at infinity. If we treat the element of charge $dq$ as a point charge, then we can use Eq. 24-26 to express the potential $dV$ at point $P$ due to $dq$:

$$dV = \frac{1}{4\pi\varepsilon_0}\frac{dq}{r} \qquad \text{(positive or negative } dq\text{)}. \qquad (24\text{-}31)$$

Here $r$ is the distance between $P$ and $dq$. To find the total potential $V$ at $P$, we integrate to sum the potentials due to all the charge elements:

$$V = \int dV = \frac{1}{4\pi\varepsilon_0}\int \frac{dq}{r}. \qquad (24\text{-}32)$$

The integral must be taken over the entire charge distribution. Note that because the electric potential is a scalar, there are *no vector components* to consider in Eq. 24-32. We now examine two continuous charge distributions, a line and a disk.

### Line of Charge

In Fig. 24-12a, a thin nonconducting rod of length $L$ has a positive charge of uniform linear density $\lambda$. Let us determine the electric potential $V$ due to the rod at point $P$, a perpendicular distance $d$ from the left end of the rod.

**FIG. 24-11**  (a) An atom, showing the positively charged nucleus (green) and the negatively charged electrons (gold shading). The centers of positive and negative charge coincide. (b) If the atom is placed in an external electric field $\vec{E}$, the electron orbits are distorted so that the centers of positive and negative charge no longer coincide. An induced dipole moment $\vec{p}$ appears. The distortion is greatly exaggerated here.

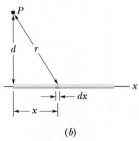

**FIG. 24-12**  (a) A thin, uniformly charged rod produces an electric potential $V$ at point $P$. (b) An element of charge produces a differential potential $dV$ at $P$.

We consider a differential element $dx$ of the rod as shown in Fig. 24-12$b$. This (or any other) element of the rod has a differential charge of

$$dq = \lambda \, dx. \tag{24-33}$$

This element produces an electric potential $dV$ at point $P$, which is a distance $r = (x^2 + d^2)^{1/2}$ from the element. Treating the element as a point charge, we can use Eq. 24-31 to write the potential $dV$ as

$$dV = \frac{1}{4\pi\varepsilon_0} \frac{dq}{r} = \frac{1}{4\pi\varepsilon_0} \frac{\lambda \, dx}{(x^2 + d^2)^{1/2}}. \tag{24-34}$$

Since the charge on the rod is positive and we have taken $V = 0$ at infinity, we know from Section 24-6 that $dV$ in Eq. 24-34 must be positive.

We now find the total potential $V$ produced by the rod at point $P$ by integrating Eq. 24-34 along the length of the rod, from $x = 0$ to $x = L$, using integral 17 in Appendix E. We find

$$V = \int dV = \int_0^L \frac{1}{4\pi\varepsilon_0} \frac{\lambda}{(x^2 + d^2)^{1/2}} \, dx$$

$$= \frac{\lambda}{4\pi\varepsilon_0} \int_0^L \frac{dx}{(x^2 + d^2)^{1/2}}$$

$$= \frac{\lambda}{4\pi\varepsilon_0} \left[ \ln\!\left( x + (x^2 + d^2)^{1/2} \right) \right]_0^L$$

$$= \frac{\lambda}{4\pi\varepsilon_0} \left[ \ln\!\left( L + (L^2 + d^2)^{1/2} \right) - \ln d \right].$$

We can simplify this result by using the general relation $\ln A - \ln B = \ln(A/B)$. We then find

$$V = \frac{\lambda}{4\pi\varepsilon_0} \ln\left[ \frac{L + (L^2 + d^2)^{1/2}}{d} \right]. \tag{24-35}$$

Because $V$ is the sum of positive values of $dV$, it should be positive—but does Eq. 24-35 give a positive $V$? Since the argument of the logarithm is greater than 1, the logarithm is a positive number and $V$ is indeed positive.

### Charged Disk

In Section 22-7, we calculated the magnitude of the electric field at points on the central axis of a plastic disk of radius $R$ that has a uniform charge density $\sigma$ on one surface. Here we derive an expression for $V(z)$, the electric potential at any point on the central axis.

In Fig. 24-13, consider a differential element consisting of a flat ring of radius $R'$ and radial width $dR'$. Its charge has magnitude

$$dq = \sigma(2\pi R')(dR'),$$

in which $(2\pi R')(dR')$ is the upper surface area of the ring. All parts of this charged element are the same distance $r$ from point $P$ on the disk's axis. With the aid of Fig. 24-13, we can use Eq. 24-31 to write the contribution of this ring to the electric potential at $P$ as

$$dV = \frac{1}{4\pi\varepsilon_0} \frac{dq}{r} = \frac{1}{4\pi\varepsilon_0} \frac{\sigma(2\pi R')(dR')}{\sqrt{z^2 + R'^2}}. \tag{24-36}$$

We find the net potential at $P$ by adding (via integration) the contributions of all the rings from $R' = 0$ to $R' = R$:

$$V = \int dV = \frac{\sigma}{2\varepsilon_0} \int_0^R \frac{R' \, dR'}{\sqrt{z^2 + R'^2}} = \frac{\sigma}{2\varepsilon_0} \left( \sqrt{z^2 + R^2} - z \right). \tag{24-37}$$

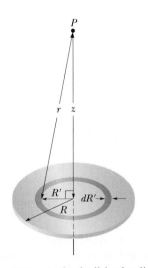

**FIG. 24-13** A plastic disk of radius $R$, charged on its top surface to a uniform surface charge density $\sigma$. We wish to find the potential $V$ at point $P$ on the central axis of the disk.

Note that the variable in the second integral of Eq. 24-37 is $R'$ and not $z$, which remains constant while the integration over the surface of the disk is carried out. (Note also that, in evaluating the integral, we have assumed that $z \geq 0$.)

**PROBLEM-SOLVING TACTICS**

*Tactic 4: Signs of Trouble with Electric Potential*
When you calculate the potential $V$ at some point $P$ due to a line of charge or any other continuous charge configuration, the signs can cause you trouble. Here is a generic guide to sort out the signs.

If the charge is negative, should the symbols $dq$ and $\lambda$ represent negative quantities, or should you explicitly show the signs, using $-dq$ and $-\lambda$? You can do either as long as you remember what your notation means, so that when you get to the final step, you can correctly interpret the sign of $V$.

Another approach, which can be used when the entire charge distribution is of a single sign, is to let the symbols $dq$ and $\lambda$ represent magnitudes only. The result of the calculation

will give you the magnitude of $V$ at $P$. Then add a sign to $V$ based on the sign of the charge. (If the zero potential is at infinity, positive charge gives a positive potential and negative charge gives a negative potential.)

If you happen to reverse the limits on the integral used to calculate a potential, you will obtain a negative value for $V$. The magnitude will be correct, but discard the minus sign. Then determine the proper sign for $V$ from the sign of the charge. As an example, we would have obtained a minus sign in Eq. 24-35 if we had reversed the limits in the integral above that equation. We would then have discarded that minus sign and noted that the potential is positive because the charge producing it is positive.

## 24-10 | Calculating the Field from the Potential

In Section 24-5, you saw how to find the potential at a point $f$ if you know the electric field along a path from a reference point to point $f$. In this section, we propose to go the other way—that is, to find the electric field when we know the potential. As Fig. 24-3 shows, solving this problem graphically is easy: If we know the potential $V$ at all points near an assembly of charges, we can draw in a family of equipotential surfaces. The electric field lines, sketched perpendicular to those surfaces, reveal the variation of $\vec{E}$. What we are seeking here is the mathematical equivalent of this graphical procedure.

Figure 24-14 shows cross sections of a family of closely spaced equipotential surfaces, the potential difference between each pair of adjacent surfaces being $dV$. As the figure suggests, the field $\vec{E}$ at any point $P$ is perpendicular to the equipotential surface through $P$.

Suppose that a positive test charge $q_0$ moves through a displacement $d\vec{s}$ from one equipotential surface to the adjacent surface. From Eq. 24-7, we see that the work the electric field does on the test charge during the move is $-q_0\,dV$. From Eq. 24-16 and Fig. 24-14, we see that the work done by the electric field may also be written as the scalar product $(q_0\vec{E}) \cdot d\vec{s}$, or $q_0 E(\cos\theta)\,ds$. Equating these two expressions for the work yields

$$-q_0\,dV = q_0 E(\cos\theta)\,ds, \qquad (24\text{-}38)$$

or

$$E\cos\theta = -\frac{dV}{ds}. \qquad (24\text{-}39)$$

Since $E\cos\theta$ is the component of $\vec{E}$ in the direction of $d\vec{s}$, Eq. 24-39 becomes

$$E_s = -\frac{\partial V}{\partial s}. \qquad (24\text{-}40)$$

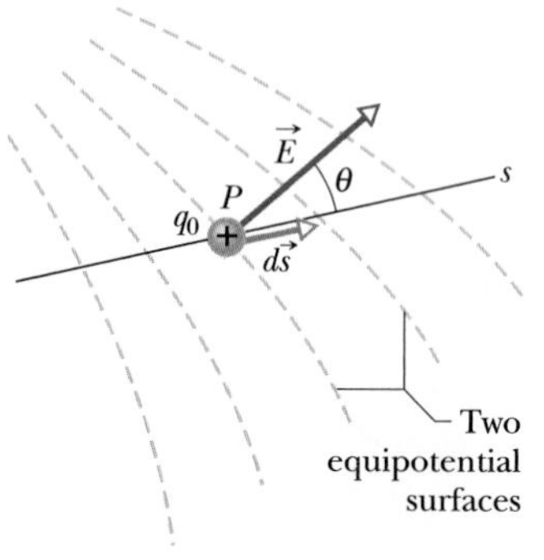

We have added a subscript to $E$ and switched to the partial derivative symbols to emphasize that Eq. 24-40 involves only the variation of $V$ along a specified axis (here called the $s$ axis) and only the component of $\vec{E}$ along that axis. In words, Eq. 24-40 (which is essentially the reverse operation of Eq. 24-18) states:

☞ The component of $\vec{E}$ in any direction is the negative of the rate at which the electric potential changes with distance in that direction.

FIG. 24-14 A test charge $q_0$ moves a distance $d\vec{s}$ from one equipotential surface to another. (The separation between the surfaces has been exaggerated for clarity.) The displacement $d\vec{s}$ makes an angle $\theta$ with the direction of the electric field $\vec{E}$.

If we take the $s$ axis to be, in turn, the $x$, $y$, and $z$ axes, we find that the $x$, $y$, and $z$ components of $\vec{E}$ at any point are

$$E_x = -\frac{\partial V}{\partial x}; \quad E_y = -\frac{\partial V}{\partial y}; \quad E_z = -\frac{\partial V}{\partial z}. \quad (24\text{-}41)$$

Thus, if we know $V$ for all points in the region around a charge distribution—that is, if we know the function $V(x, y, z)$—we can find the components of $\vec{E}$, and thus $\vec{E}$ itself, at any point by taking partial derivatives.

For the simple situation in which the electric field $\vec{E}$ is uniform, Eq. 24-40 becomes

$$E = -\frac{\Delta V}{\Delta s}, \quad (24\text{-}42)$$

where $s$ is perpendicular to the equipotential surfaces. The component of the electric field is zero in any direction parallel to the equipotential surfaces.

✓**CHECKPOINT 6**    The figure shows three pairs of parallel plates with the same separation, and the electric potential of each plate. The electric field between the plates is uniform and perpendicular to the plates. (a) Rank the pairs according to the magnitude of the electric field between the plates, greatest first. (b) For which pair is the electric field pointing rightward? (c) If an electron is released midway between the third pair of plates, does it remain there, move rightward at constant speed, move leftward at constant speed, accelerate rightward, or accelerate leftward?

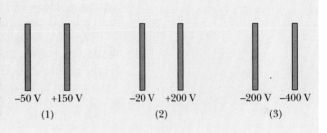

−50 V  +150 V       −20 V  +200 V       −200 V  −400 V
(1)                 (2)                 (3)

---

**Sample Problem** | **24-5**

The electric potential at any point on the central axis of a uniformly charged disk is given by Eq. 24-37,

$$V = \frac{\sigma}{2\varepsilon_0}(\sqrt{z^2 + R^2} - z).$$

Starting with this expression, derive an expression for the electric field at any point on the axis of the disk.

---

**KEY IDEAS**  We want the electric field $\vec{E}$ as a function of distance $z$ along the axis of the disk. For any value of $z$, the direction of $\vec{E}$ must be along that axis because the disk has circular symmetry about that axis. Thus, we

want the component $E_z$ of $\vec{E}$ in the direction of $z$. This component is the negative of the rate at which the electric potential changes with distance $z$.

**Calculation:** Thus, from the last of Eqs. 24-41, we can write

$$E_z = -\frac{\partial V}{\partial z} = -\frac{\sigma}{2\varepsilon_0}\frac{d}{dz}(\sqrt{z^2 + R^2} - z)$$

$$= \frac{\sigma}{2\varepsilon_0}\left(1 - \frac{z}{\sqrt{z^2 + R^2}}\right). \quad (\text{Answer})$$

This is the same expression that we derived in Section 22-7 by integration, using Coulomb's law.

---

# 24-11 | Electric Potential Energy of a System of Point Charges

In Section 24-2, we discussed the electric potential energy of a charged particle as an electrostatic force does work on it. In that section, we assumed that the charges that produced the force were fixed in place, so that neither the force nor the corresponding electric field could be influenced by the presence of the test charge. In this section we can take a broader view, to find the electric potential energy of a *system* of charges due to the electric field produced *by* those same charges.

For a simple example, suppose you push together two bodies that have charges of the same electrical sign. The work that you must do is stored as electric

FIG. 24-15 Two charges held a fixed distance $r$ apart.

potential energy in the two-body system (provided the kinetic energy of the bodies does not change). If you later release the charges, you can recover this stored energy, in whole or in part, as kinetic energy of the charged bodies as they rush away from each other.

We define the electric potential energy *of a system of point charges,* held in fixed positions by forces not specified, as follows:

> The electric potential energy of a system of fixed point charges is equal to the work that must be done by an external agent to assemble the system, bringing each charge in from an infinite distance.

We assume that the charges are stationary both in their initial infinitely distant positions and in their final assembled configuration.

Figure 24-15 shows two point charges $q_1$ and $q_2$, separated by a distance $r$. To find the electric potential energy of this two-charge system, we must mentally build the system, starting with both charges infinitely far away and at rest. When we bring $q_1$ in from infinity and put it in place, we do no work because no electrostatic force acts on $q_1$. However, when we next bring $q_2$ in from infinity and put it in place, we must do work because $q_1$ exerts an electrostatic force on $q_2$ during the move.

We can calculate that work with Eq. 24-8 by dropping the minus sign (so that the equation gives the work *we* do rather than the field's work) and substituting $q_2$ for the general charge $q$. Our work is then equal to $q_2V$, where $V$ is the potential that has been set up by $q_1$ at the point where we put $q_2$. From Eq. 24-26, that potential is

$$V = \frac{1}{4\pi\varepsilon_0}\frac{q_1}{r}.$$

Thus, from our definition, the electric potential energy of the pair of point charges of Fig. 24-15 is

$$U = W = q_2V = \frac{1}{4\pi\varepsilon_0}\frac{q_1q_2}{r}. \tag{24-43}$$

If the charges have the same sign, we have to do positive work to push them together against their mutual repulsion. Hence, as Eq. 24-43 shows, the potential energy of the system is then positive. If the charges have opposite signs, we have to do negative work against their mutual attraction to bring them together if they are to be stationary. The potential energy of the system is then negative. Sample Problem 24-6 shows how to extend this process to more than two charges.

## Sample Problem  24-6

Figure 24-16 shows three point charges held in fixed positions by forces that are not shown. What is the electric potential energy $U$ of this system of charges? Assume that $d = 12$ cm and that

$$q_1 = +q, \quad q_2 = -4q, \quad \text{and} \quad q_3 = +2q,$$

in which $q = 150$ nC.

**KEY IDEA**  The potential energy $U$ of the system is equal to the work we must do to assemble the system, bringing in each charge from an infinite distance.

**Calculations:** Let's mentally build the system of Fig. 24-16, starting with one of the point charges, say $q_1$, in

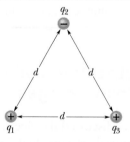

FIG. 24-16  Three charges are fixed at the vertices of an equilateral triangle. What is the electric potential energy of the system?

place and the others at infinity. Then we bring another one, say $q_2$, in from infinity and put it in place. From Eq. 24-43 with $d$ substituted for $r$, the potential energy $U_{12}$

associated with the pair of point charges $q_1$ and $q_2$ is

$$U_{12} = \frac{1}{4\pi\varepsilon_0}\frac{q_1 q_2}{d}.$$

We then bring the last point charge $q_3$ in from infinity and put it in place. The work that we must do in this last step is equal to the sum of the work we must do to bring $q_3$ near $q_1$ and the work we must do to bring it near $q_2$. From Eq. 24-43, with $d$ substituted for $r$, that sum is

$$W_{13} + W_{23} = U_{13} + U_{23} = \frac{1}{4\pi\varepsilon_0}\frac{q_1 q_3}{d} + \frac{1}{4\pi\varepsilon_0}\frac{q_2 q_3}{d}.$$

The total potential energy $U$ of the three-charge system is the sum of the potential energies associated with the three pairs of charges. This sum (which is actually independent of the order in which the charges are brought together) is

$$U = U_{12} + U_{13} + U_{23}$$

$$= \frac{1}{4\pi\varepsilon_0}\left(\frac{(+q)(-4q)}{d} + \frac{(+q)(+2q)}{d} + \frac{(-4q)(+2q)}{d}\right)$$

$$= -\frac{10q^2}{4\pi\varepsilon_0 d}$$

$$= -\frac{(8.99 \times 10^9\,\text{N}\cdot\text{m}^2/\text{C}^2)(10)(150 \times 10^{-9}\,\text{C})^2}{0.12\,\text{m}}$$

$$= -1.7 \times 10^{-2}\,\text{J} = -17\,\text{mJ}. \qquad \text{(Answer)}$$

The negative potential energy means that negative work would have to be done to assemble this structure, starting with the three charges infinitely separated and at rest. Put another way, an external agent would have to do 17 mJ of work to disassemble the structure completely, ending with the three charges infinitely far apart.

---

**Sample Problem** | **24-7** | **Build your skill**

An alpha particle (two protons, two neutrons) moves into a stationary gold atom (79 protons, 118 neutrons), passing through the electron region that surrounds the gold nucleus like a shell and headed directly toward the nucleus (Fig. 24-17). The alpha particle slows until it momentarily stops when its center is at radial distance $r = 9.23$ fm from the nuclear center. Then it moves back along its incoming path. (Because the gold nucleus is much more massive than the alpha particle, we can assume the gold nucleus does not move.) What was the kinetic energy $K_i$ of the alpha particle when it was initially far away (hence external to the gold atom)? Assume that the only force acting between the alpha particle and the gold nucleus is the (electrostatic) Coulomb force.

**KEY IDEA** During the entire process, the mechanical energy of the *alpha particle + gold atom* system is conserved.

*Reasoning:* When the alpha particle is outside the atom, the system's initial electric potential energy $U_i$ is zero because the atom has an equal number of electrons and protons, which produce a *net* electric field of zero. However, once the alpha particle passes through the electron region surrounding the nucleus on its way to the nucleus, the electric field due to the electrons goes to zero. The reason is that the electrons act like a closed spherical shell of uniform negative charge and, as discussed in Section 23-9, such a shell produces zero electric field in the space it encloses. The alpha particle still experiences the electric field of the protons in the nucleus, which produces a repulsive force on the protons within the alpha particle.

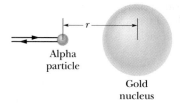

**FIG. 24-17** An alpha particle, traveling head-on toward the center of a gold nucleus, comes to a momentary stop (at which time all its kinetic energy has been transferred to electric potential energy) and then reverses its path.

As the incoming alpha particle is slowed by this repulsive force, its kinetic energy is transferred to electric potential energy of the system. The transfer is complete when the alpha particle momentarily stops and the kinetic energy is $K_f = 0$.

*Calculations:* The principle of conservation of mechanical energy tells us that

$$K_i + U_i = K_f + U_f. \qquad (24\text{-}44)$$

We know two values: $U_i = 0$ and $K_f = 0$. We also know that the potential energy $U_f$ at the stopping point is given by the right side of Eq. 24-43, with $q_1 = 2e$, $q_2 = 79e$ (in which $e$ is the elementary charge, $1.60 \times 10^{-19}$ C), and $r = 9.23$ fm. Thus, we can rewrite Eq. 24-44 as

$$K_i = \frac{1}{4\pi\varepsilon_0}\frac{(2e)(79e)}{9.23\,\text{fm}}$$

$$= \frac{(8.99 \times 10^9\,\text{N}\cdot\text{m}^2/\text{C}^2)(158)(1.60 \times 10^{-19}\,\text{C})^2}{9.23 \times 10^{-15}\,\text{m}}$$

$$= 3.94 \times 10^{-12}\,\text{J} = 24.6\,\text{MeV}. \qquad \text{(Answer)}$$

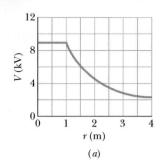

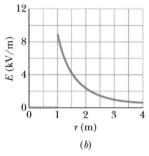

**FIG. 24-18** (a) A plot of $V(r)$ both inside and outside a charged spherical shell of radius 1.0 m. (b) A plot of $E(r)$ for the same shell.

## 24-12 | Potential of a Charged Isolated Conductor

In Section 23-6, we concluded that $\vec{E} = 0$ for all points inside an isolated conductor. We then used Gauss' law to prove that an excess charge placed on an isolated conductor lies entirely on its surface. (This is true even if the conductor has an empty internal cavity.) Here we use the first of these facts to prove an extension of the second:

An excess charge placed on an isolated conductor will distribute itself on the surface of that conductor so that all points of the conductor—whether on the surface or inside—come to the same potential. This is true even if the conductor has an internal cavity and even if that cavity contains a net charge.

Our proof follows directly from Eq. 24-18, which is

$$V_f - V_i = -\int_i^f \vec{E} \cdot d\vec{s}.$$

Since $\vec{E} = 0$ for all points within a conductor, it follows directly that $V_f = V_i$ for all possible pairs of points $i$ and $f$ in the conductor.

Figure 24-18a is a plot of potential against radial distance $r$ from the center for an isolated spherical conducting shell of 1.0 m radius, having a charge of 1.0 μC. For points outside the shell, we can calculate $V(r)$ from Eq. 24-26 because the charge $q$ behaves for such external points as if it were concentrated at the center of the shell. That equation holds right up to the surface of the shell. Now let us push a small test charge through the shell—assuming a small hole exists—to its center. No extra work is needed to do this because no net electric force acts on the test charge once it is inside the shell. Thus, the potential at all points inside the shell has the same value as that on the surface, as Fig. 24-18a shows.

Figure 24-18b shows the variation of electric field with radial distance for the same shell. Note that $E = 0$ everywhere inside the shell. The curves of Fig. 24-18b can be derived from the curve of Fig. 24-18a by differentiating with respect to $r$, using Eq. 24-40 (recall that the derivative of any constant is zero). The curve of Fig. 24-18a can be derived from the curves of Fig. 24-18b by integrating with respect to $r$, using Eq. 24-19.

### Spark Discharge from a Charged Conductor

**FIG. 24-19** A large spark jumps to a car's body and then exits by moving across the insulating left front tire (note the flash there), leaving the person inside unharmed. *(Courtesy Westinghouse Electric Corporation)*

On nonspherical conductors, a surface charge does not distribute itself uniformly over the surface of the conductor. At sharp points or sharp edges, the surface charge density—and thus the external electric field, which is proportional to it—may reach very high values. The air around such sharp points or edges may become ionized, producing the corona discharge that golfers and mountaineers see on the tips of bushes, golf clubs, and rock hammers when thunderstorms threaten. Such corona discharges, like hair that stands on end, are often the precursors of lightning strikes. In such circumstances, it is wise to enclose yourself in a cavity inside a conducting shell, where the electric field is guaranteed to be zero. A car (unless it is a convertible or made with a plastic body) is almost ideal (Fig. 24-19).

Your body is a fairly good electrical conductor and can be easily charged if you move around or change clothing. Such action produces a great many contact points between your clothing and your skin. For many types of fabrics, such contact allows some of the conduction electrons on one surface to move to the other surface. For example, you might gain conduction electrons when you peel off a sweater. If the air humidity is high, these electrons are quickly drained from you by airborne water drops. If the humidity is low, however, you may have so much excess charge that the potential difference between your body and your surroundings is 5 kV or more. If you touch a computer keyboard while charged like this, the excess charge on your body can flow through the computer's circuit chips, overloading and ruining them.

There are countless examples in which contact between a person and some other type of material leaves the person so highly charged that the person might discharge with a spark. Children sliding down a plastic slide on a dry day have been measured to have a potential of about 60 kV. If such a child reaches for any conducting object (such as another person), the child probably will discharge to the object with a very painful spark.

Such a spark discharge would be disastrous in a hospital operating room where a flammable gas (such as an anesthetic gas) is present. To drain the charge they collect as they move around, a surgical team wears conducting shoes and stands on a conducting floor. Spark discharges have also caused a number of fires at self-serve gasoline stations when a customer has slid back into a car seat to wait for the car's tank to be filled. Contact with the car seat can leave the customer with so much charge that a spark might jump between fingers and pump nozzle when the customer goes back to remove the nozzle from the car. The spark can ignite the gasoline vapor that surrounds the nozzle.

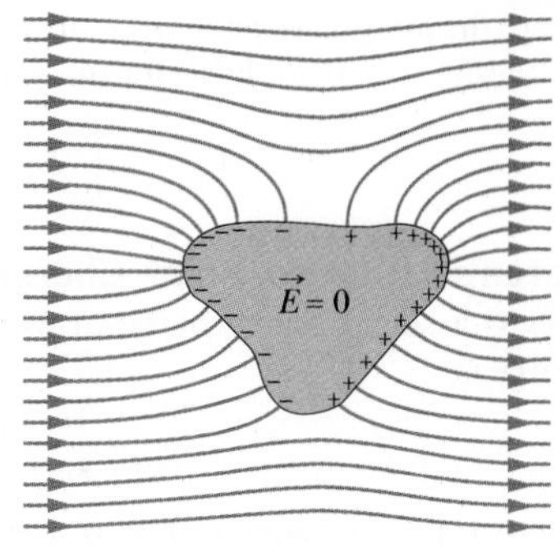

FIG. 24-20  An uncharged conductor is suspended in an external electric field. The free electrons in the conductor distribute themselves on the surface as shown, so as to reduce the net electric field inside the conductor to zero and make the net field at the surface perpendicular to the surface.

## Isolated Conductor in an External Electric Field

If an isolated conductor is placed in an *external electric field,* as in Fig. 24-20, all points of the conductor still come to a single potential regardless of whether the conductor has an excess charge. The free conduction electrons distribute themselves on the surface in such a way that the electric field they produce at interior points cancels the external electric field that would otherwise be there. Furthermore, the electron distribution causes the net electric field at all points on the surface to be perpendicular to the surface. If the conductor in Fig. 24-20 could be somehow removed, leaving the surface charges frozen in place, the pattern of the electric field would remain absolutely unchanged, for both exterior and interior points.

## REVIEW & SUMMARY

**Electric Potential Energy**  The change $\Delta U$ in the electric potential energy $U$ of a point charge as the charge moves from an initial point $i$ to a final point $f$ in an electric field is

$$\Delta U = U_f - U_i = -W, \qquad (24\text{-}1)$$

where $W$ is the work done by the electrostatic force (due to the external electric field) on the point charge during the move from $i$ to $f$. If the potential energy is defined to be zero at infinity, the **electric potential energy** $U$ of the point charge at a particular point is

$$U = -W_\infty. \qquad (24\text{-}2)$$

Here $W_\infty$ is the work done by the electrostatic force on the point charge as the charge moves from infinity to the particular point.

**Electric Potential Difference and Electric Potential**  We define the **potential difference** $\Delta V$ between two points $i$ and $f$ in an electric field as

$$\Delta V = V_f - V_i = -\frac{W}{q}, \qquad (24\text{-}7)$$

where $q$ is the charge of a particle on which work is done by the field. The **potential** at a point is

$$V = -\frac{W_\infty}{q}. \qquad (24\text{-}8)$$

The SI unit of potential is the *volt:* 1 volt = 1 joule per coulomb.

Potential and potential difference can also be written in terms of the electric potential energy $U$ of a particle of charge $q$ in an electric field:

$$V = \frac{U}{q}, \qquad (24\text{-}5)$$

$$\Delta V = V_f - V_i = \frac{U_f}{q} - \frac{U_i}{q} = \frac{\Delta U}{q}. \qquad (24\text{-}6)$$

**Equipotential Surfaces**  The points on an **equipotential surface** all have the same electric potential. The work done on a test charge in moving it from one such surface to another is independent of the locations of the initial and final points on these surfaces and of the path that joins the points. The electric field $\vec{E}$ is always directed perpendicularly to corresponding equipotential surfaces.

**Finding $V$ from $\vec{E}$**  The electric potential difference between two points $i$ and $f$ is

$$V_f - V_i = -\int_i^f \vec{E} \cdot d\vec{s}, \qquad (24\text{-}18)$$

where the integral is taken over any path connecting the points. If we choose $V_i = 0$, we have, for the potential at a particular point,

$$V = -\int_i^f \vec{E} \cdot d\vec{s}. \qquad (24\text{-}19)$$

**Potential Due to Point Charges** The electric potential due to a single point charge at a distance $r$ from that point charge is

$$V = \frac{1}{4\pi\varepsilon_0} \frac{q}{r}, \qquad (24\text{-}26)$$

where $V$ has the same sign as $q$. The potential due to a collection of point charges is

$$V = \sum_{i=1}^n V_i = \frac{1}{4\pi\varepsilon_0} \sum_{i=1}^n \frac{q_i}{r_i}. \qquad (24\text{-}27)$$

**Potential Due to an Electric Dipole** At a distance $r$ from an electric dipole with dipole moment magnitude $p = qd$, the electric potential of the dipole is

$$V = \frac{1}{4\pi\varepsilon_0} \frac{p \cos\theta}{r^2} \qquad (24\text{-}30)$$

for $r \gg d$; the angle $\theta$ is defined in Fig. 24-10.

**Potential Due to a Continuous Charge Distribution** For a continuous distribution of charge, Eq. 24-27 becomes

$$V = \frac{1}{4\pi\varepsilon_0} \int \frac{dq}{r}, \qquad (24\text{-}32)$$

in which the integral is taken over the entire distribution.

**Calculating $\vec{E}$ from $V$** The component of $\vec{E}$ in any direction is the negative of the rate at which the potential changes with distance in that direction:

$$E_s = -\frac{\partial V}{\partial s}. \qquad (24\text{-}40)$$

The $x$, $y$, and $z$ components of $\vec{E}$ may be found from

$$E_x = -\frac{\partial V}{\partial x}; \qquad E_y = -\frac{\partial V}{\partial y}; \qquad E_z = -\frac{\partial V}{\partial z}. \qquad (24\text{-}41)$$

When $\vec{E}$ is uniform, Eq. 24-40 reduces to

$$E = -\frac{\Delta V}{\Delta s}, \qquad (24\text{-}42)$$

where $s$ is perpendicular to the equipotential surfaces. The electric field is zero parallel to an equipotential surface.

**Electric Potential Energy of a System of Point Charges** The electric potential energy of a system of point charges is equal to the work needed to assemble the system with the charges initially at rest and infinitely distant from each other. For two charges at separation $r$,

$$U = W = \frac{1}{4\pi\varepsilon_0} \frac{q_1 q_2}{r}. \qquad (24\text{-}43)$$

**Potential of a Charged Conductor** An excess charge placed on a conductor will, in the equilibrium state, be located entirely on the outer surface of the conductor. The charge will distribute itself so that the entire conductor, including interior points, is at a uniform potential.

## QUESTIONS

**1** Figure 24-21 shows four pairs of charged particles. For each pair, let $V = 0$ at infinity and consider $V_{net}$ at points on the $x$ axis. For which pairs is there a point at which $V_{net} = 0$ (a) between the particles and (b) to the right of the particles? (c) At such a point is $\vec{E}_{net}$ due to the particles equal to zero? (d) For each pair, are there off-axis points (other than at infinity) where $V_{net} = 0$?

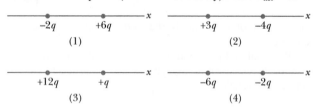

FIG. 24-21 Questions 1 and 7.

**2** Figure 24-22 shows four arrangements of charged particles, all the same distance from the origin. Rank the situations according to the net electric potential at the origin, most positive first. Take the potential to be zero at infinity.

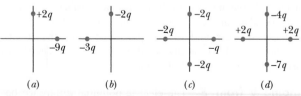

FIG. 24-22 Question 2.

**3** In Fig. 24-23, eight particles form a square, with distance $d$ between adjacent particles. What is the electric potential at point $P$ at the center of the square if the electric potential is zero at infinity?

**4** Figure 24-24 shows three sets of cross sections of equipotential surfaces; all three cover the same size region of space. (a) Rank the arrangements according to the magnitude of the electric field present in the region, greatest first. (b) In which is the electric field directed down the page?

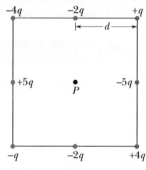

FIG. 24-23 Question 3.

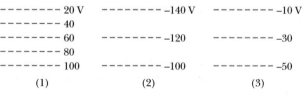

FIG. 24-24 Question 4.

**5** Figure 24-25 shows three paths along which we can move the positively charged sphere $A$ closer to positively charged sphere $B$, which is held fixed in place. (a) Would sphere $A$ be moved to a higher or lower electric potential? Is the

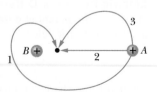

FIG. 24-25 Question 5.

work done (b) by our force and (c) by the electric field due to *B* positive, negative, or zero? (d) Rank the paths according to the work our force does, greatest first.

**6** Figure 24-26 gives the electric potential *V* as a function of *x*. (a) Rank the five regions according to the magnitude of the *x* component of the electric field within them, greatest first. What is the direction of the field along the *x* axis in (b) region 2 and (c) region 4?

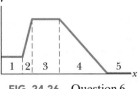
FIG. 24-26    Question 6.

**7** Figure 24-21 shows four pairs of charged particles with identical separations. (a) Rank the pairs according to their electric potential energy, greatest (most positive) first. (b) For each pair, if the separation between the particles is increased, does the potential energy of the pair increase or decrease?

**8** (a) In Fig. 24-27*a*, what is the potential at point *P* due to

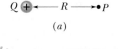
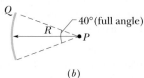
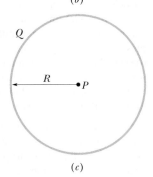
FIG. 24-27    Question 8.

charge *Q* at distance *R* from *P*? Set *V* = 0 at infinity. (b) In Fig. 24-27*b*, the same charge *Q* has been spread uniformly over a circular arc of radius *R* and central angle 40°. What is the potential at point *P*, the center of curvature of the arc? (c) In Fig. 24-27*c*, the same charge *Q* has been spread uniformly over a circle of radius *R*. What is the potential at point *P*, the center of the circle? (d) Rank the three situations according to the magnitude of the electric field that is set up at *P*, greatest first.

**9** Figure 24-28 shows a system of three charged particles. If you move the particle of charge +*q* from point *A* to point *D*, are the following positive, negative, or zero: (a) the change in the electric potential energy of the three-particle system, (b) the work done by the net electrostatic force on the particle you moved, and (c) the work done by your force? (d) What are the answers to (a) through (c) if, instead, the particle is moved from *B* to *C*?

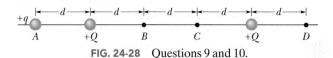
FIG. 24-28    Questions 9 and 10.

**10** In the situation of Question 9, is the work done by your force positive, negative, or zero if the particle is moved (a) from *A* to *B*, (b) from *A* to *C*, and (c) from *B* to *D*? (d) Rank those moves according to the magnitude of the work done by your force, greatest first.

## PROBLEMS

**GO**  Tutoring problem available (at instructor's discretion) in *WileyPLUS* and WebAssign
**SSM**  Worked-out solution available in Student Solutions Manual    **WWW**  Worked-out solution is at
• – •••  Number of dots indicates level of problem difficulty    **ILW**  Interactive solution is at — http://www.wiley.com/college/halliday
  Additional information available in *The Flying Circus of Physics* and at flyingcircusofphysics.com

### sec. 24-3    Electric Potential

•1  Much of the material making up Saturn's rings is in the form of tiny dust grains having radii on the order of $10^{-6}$ m. These grains are located in a region containing a dilute ionized gas, and they pick up excess electrons. As an approximation, suppose each grain is spherical, with radius $R = 1.0 \times 10^{-6}$ m. How many electrons would one grain have to pick up to have a potential of $-400$ V on its surface (taking $V = 0$ at infinity)?

•2  The electric potential difference between the ground and a cloud in a particular thunderstorm is $1.2 \times 10^9$ V. In the unit electron-volts, what is the magnitude of the change in the electric potential energy of an electron that moves between the ground and the cloud?

•3  A particular 12 V car battery can send a total charge of 84 A·h (ampere-hours) through a circuit, from one terminal to the other. (a) How many coulombs of charge does this represent? (*Hint:* See Eq. 21-3.) (b) If this entire charge undergoes a change in electric potential of 12 V, how much energy is involved?   **SSM**

### sec. 24-5    Calculating the Potential from the Field

•4  When an electron moves from *A* to *B* along an electric field line in Fig. 24-29, the electric field does $3.94 \times 10^{-19}$ J of work on

it. What are the electric potential differences (a) $V_B - V_A$, (b) $V_C - V_A$, and (c) $V_C - V_B$?

•5  An infinite nonconducting sheet has a surface charge density $\sigma = 0.10$ $\mu$C/m² on one side. How far apart are equipotential surfaces whose potentials differ by 50 V?   **SSM**

FIG. 24-29    Problem 4.

•6  Two large, parallel, conducting plates are 12 cm apart and have charges of equal magnitude and opposite sign on their facing surfaces. An electrostatic force of $3.9 \times 10^{-15}$ N acts on an electron placed anywhere between the two plates. (Neglect fringing.) (a) Find the electric field at the position of the electron. (b) What is the potential difference between the plates?

••7  An infinite nonconducting sheet has a surface charge density $\sigma = +5.80$ pC/m². (a) How much work is done by the electric field due to the sheet if a particle of charge $q = +1.60 \times 10^{-19}$ C is moved from the sheet to a point *P* at distance $d = 3.56$ cm from the sheet? (b) If the electric potential *V* is defined to be zero on the sheet, what is *V* at *P*?

**••8** A graph of the $x$ component of the electric field as a function of $x$ in a region of space is shown in Fig. 24-30. The scale of the vertical axis is set by $E_{xs} = 20.0$ N/C. The $y$ and $z$ components of the electric field are zero in this region. If the electric

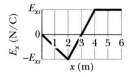

**FIG. 24-30** Problem 8.

potential at the origin is 10 V, (a) what is the electric potential at $x = 2.0$ m, (b) what is the greatest positive value of the electric potential for points on the $x$ axis for which $0 \le x \le 6.0$ m, and (c) for what value of $x$ is the electric potential zero?

**••9** The electric field in a region of space has the components $E_y = E_z = 0$ and $E_x = (4.00 \text{ N/C})x$. Point $A$ is on the $y$ axis at $y = 3.00$ m, and point $B$ is on the $x$ axis at $x = 4.00$ m. What is the potential difference $V_B - V_A$?

**•••10** Two uniformly charged, infinite, nonconducting planes are parallel to a $yz$ plane and positioned at $x = -50$ cm and $x = +50$ cm. The charge densities on the planes are $-50$ nC/m$^2$ and $+25$ nC/m$^2$, respectively. What is the magnitude of the potential difference between the origin and the point on the $x$ axis at $x = +80$ cm? (*Hint:* Use Gauss' law.)

**•••11** A nonconducting sphere has radius $R = 2.31$ cm and uniformly distributed charge $q = +3.50$ fC. Take the electric potential at the sphere's center to be $V_0 = 0$. What is $V$ at radial distance (a) $r = 1.45$ cm and (b) $r = R$. (*Hint:* See Section 23-9.)

### sec. 24-7 Potential Due to a Group of Point Charges

**•12** Consider a point charge $q = 1.0$ μC, point $A$ at distance $d_1 = 2.0$ m from $q$, and point $B$ at distance $d_2 = 1.0$ m. (a) If $A$ and $B$ are diametrically opposite each other, as in Fig. 24-31a, what is the electric potential difference $V_A - V_B$? (b) What is that electric potential difference if $A$ and $B$ are located as in Fig. 24-31b?

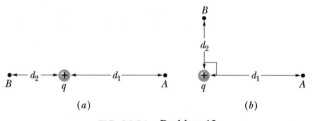

**FIG. 24-31** Problem 12.

**•13** What are (a) the charge and (b) the charge density on the surface of a conducting sphere of radius 0.15 m whose potential is 200 V (with $V = 0$ at infinity)?

**•14** As a space shuttle moves through the dilute ionized gas of Earth's ionosphere, the shuttle's potential is typically changed by $-1.0$ V during one revolution. Assuming the shuttle is a sphere of radius 10 m, estimate the amount of charge it collects.

**••15** In Fig. 24-32, what is the net electric potential at point $P$ due to the four particles if $V = 0$ at infinity, $q = 5.00$ fC, and $d = 4.00$ cm? **GO**

**••16** Two particles, of charges $q_1$ and $q_2$, are separated by distance $d$ in Fig. 24-33. The net electric field due to the particles is zero at $x = d/4$. With $V = 0$ at

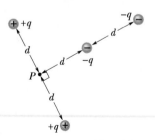

**FIG. 24-32** Problem 15.

infinity, locate (in terms of $d$) any point on the $x$ axis (other than at infinity) at which the electric potential due to the two particles is zero.

**FIG. 24-33** Problems 16, 17, and 91.

**••17** In Fig. 24-33, particles of the charges $q_1 = +5e$ and $q_2 = -15e$ are fixed in place with a separation of $d = 24.0$ cm. With $V = 0$ at infinity, what are the finite (a) positive and (b) negative values of $x$ at which the net electric potential on the $x$ axis is zero?

**••18** Figure 24-34 shows a rectangular array of charged particles fixed in place, with distance $a = 39.0$ cm and the charges shown as integer multiples of $q_1 = 3.40$ pC and $q_2 = 6.00$ pC. With $V = 0$ at infinity, what is the net electric potential at the rectangle's center? (*Hint:*

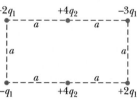

**FIG. 24-34** Problem 18.

Thoughtful examination can reduce the calculation.) **GO**

**••19** A spherical drop of water carrying a charge of 30 pC has a potential of 500 V at its surface (with $V = 0$ at infinity). (a) What is the radius of the drop? (b) If two such drops of the same charge and radius combine to form a single spherical drop, what is the potential at the surface of the new drop? **SSM ILW**

**••20** Two charged particles are shown in Fig. 24-35a. Particle 1, with charge $q_1$, is fixed in place at distance $d$. Particle 2, with charge $q_2$, can be moved along the $x$ axis. Figure 24-35b gives the net electric potential $V$ at the origin due to the two particles as a function of the $x$ coordinate of particle 2. The scale of the $x$ axis is set by $x_s = 16.0$ cm. The plot has an asymptote of $V = 5.76 \times 10^{-7}$ V as $x \to \infty$. What is $q_2$ in terms of $e$?

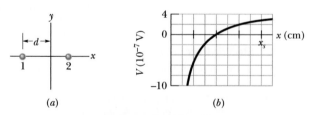

**FIG. 24-35** Problem 20.

### sec. 24-8 Potential Due to an Electric Dipole

**•21** The ammonia molecule NH$_3$ has a permanent electric dipole moment equal to 1.47 D, where 1 D = 1 debye unit = $3.34 \times 10^{-30}$ C·m. Calculate the electric potential due to an ammonia molecule at a point 52.0 nm away along the axis of the dipole. (Set $V = 0$ at infinity.) **ILW**

**••22** In Fig. 24-36a, a particle of charge $+e$ is initially at coordinate $z = 20$ nm on the dipole axis through an electric dipole, on the positive side of the dipole. (The origin of $z$ is at the dipole center.) The particle is then moved along a circular path around the dipole center until it is at coordinate $z = -20$ nm. Figure 24-36b gives the work $W_a$ done by the force moving the particle versus the angle $\theta$ that locates the particle. The scale of the vertical axis is set by $W_{as} = 4.0 \times 10^{-30}$ J. What is the magnitude of the dipole moment?

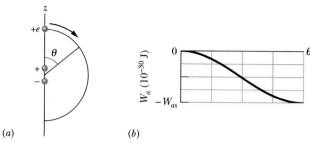

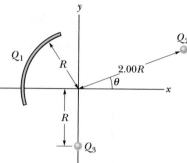

(a)

(b)

FIG. 24-36   Problem 22.

### sec. 24-9   Potential Due to a Continuous Charge Distribution

•23   A plastic rod has been bent into a circle of radius $R = 8.20$ cm. It has a charge $Q_1 = +4.20$ pC uniformly distributed along one-quarter of its circumference and a charge $Q_2 = -6Q_1$ uniformly distributed along the rest of the circumference (Fig. 24-37). With $V = 0$ at infinity, what is the electric potential (a) at the center $C$ of the circle and (b) at point $P$, which is on the central axis of the circle at distance $D = 6.71$ cm from the center?

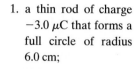

FIG. 24-37   Problem 23.

•24   In Fig. 24-38, a plastic rod having a uniformly distributed charge $Q = -25.6$ pC has been bent into a circular arc of radius $R = 3.71$ cm and central angle $\phi = 120°$. With $V = 0$ at infinity, what is the electric potential at $P$, the center of curvature of the rod?

•25   (a) Figure 24-39a shows a nonconducting rod of length $L = 6.00$ cm and uniform linear charge density $\lambda = +3.68$ pC/m. Take $V = 0$ at infinity. What is $V$ at point $P$ at distance $d = 8.00$ cm along the rod's perpendicular bisector? (b) Figure 24-39b shows an identical rod except that one half is now negatively charged. Both halves have a linear charge density of magnitude 3.68 pC/m. With $V = 0$ at infinity, what is $V$ at $P$?

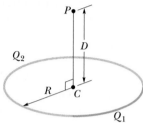

FIG. 24-38
Problem 24.

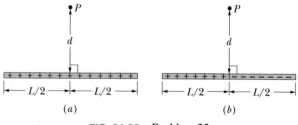

(a)

(b)

FIG. 24-39   Problem 25.

•26   A Gaussian sphere of radius 4.00 cm is centered on a ball that has a radius of 1.00 cm and a uniform charge distribution. The total (net) electric flux through the surface of the Gaussian sphere is $+5.60 \times 10^4$ N·m²/C. What is the electric potential 12.0 cm from the center of the ball?

••27   In Fig. 24-40, what is the net electric potential at the origin due to the circular arc of charge $Q_1 = +7.21$ pC and the two particles of charges $Q_2 = 4.00Q_1$ and $Q_3 = -2.00Q_1$? The arc's center of curvature is at the origin and its radius is

$R = 2.00$ m; the angle indicated is $\theta = 20.0°$.

••28   The smiling face of Fig. 24-41 consists of three items: **GO**

1. a thin rod of charge $-3.0$ μC that forms a full circle of radius 6.0 cm;

2. a second thin rod of charge 2.0 μC that forms a circular arc of radius 4.0 cm, subtending an angle of 90° about the center of the full circle;

3. an electric dipole with a dipole moment that is perpendicular to a radial line and has magnitude $1.28 \times 10^{-21}$ C·m.

What is the net electric potential at the center?

••29   A plastic disk of radius $R = 64.0$ cm is charged on one side with a uniform surface charge density $\sigma = 7.73$ fC/m², and then three quadrants of the disk are removed. The remaining quadrant is shown in Fig. 24-42. With $V = 0$ at infinity, what is the potential due to the remaining quadrant at point $P$, which is on the central axis of the original disk at distance $D = 25.9$ cm from the original center? **SSM WWW**

••30   Figure 24-43 shows a thin plastic rod of length $L = 12.0$ cm and uniform positive charge $Q = 56.1$ fC lying on an $x$ axis. With $V = 0$ at infinity, find the electric potential at point $P_1$ on the axis, at distance $d = 2.50$ cm from one end of the rod. **GO**

••31   In Fig. 24-44, three thin plastic rods form quarter-circles with a common center of curvature at the origin. The uniform charges on the rods are $Q_1 = +30$ nC, $Q_2 = +3.0Q_1$, and $Q_3 = -8.0Q_1$. What is the net electric potential at the origin due to the rods?

•••32   A nonuniform linear charge distribution given by $\lambda = bx$, where $b$ is a constant, is located along an $x$ axis from $x = 0$ to $x = 0.20$ m. If $b = 20$ nC/m² and $V = 0$ at infinity, what is the electric potential at (a) the origin and (b) the point $y = 0.15$ m on the $y$ axis?

•••33   The thin plastic rod shown in Fig. 24-43 has length $L = 12.0$ cm and a nonuniform linear charge density $\lambda = cx$, where

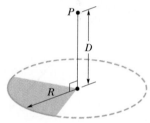

FIG. 24-40   Problem 27.

FIG. 24-41   Problem 28.

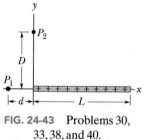

FIG. 24-42   Problem 29.

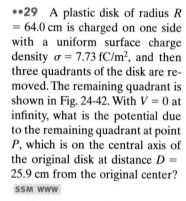

FIG. 24-43   Problems 30, 33, 38, and 40.

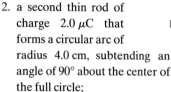

FIG. 24-44   Problem 31.

$c = 28.9$ pC/m$^2$. With $V = 0$ at infinity, find the electric potential at point $P_1$ on the axis, at distance $d = 3.00$ cm from one end.

### sec. 24-10 Calculating the Field from the Potential

•**34** The electric potential $V$ in the space between two flat parallel plates 1 and 2 is given (in volts) by $V = 1500x^2$, where $x$ (in meters) is the perpendicular distance from plate 1. At $x = 1.3$ cm, (a) what is the magnitude of the electric field and (b) is the field directed toward or away from plate 1?

•**35** The electric potential at points in an $xy$ plane is given by $V = (2.0$ V/m$^2)x^2 - (3.0$ V/m$^2)y^2$. In unit-vector notation, what is the electric field at the point (3.0 m, 2.0 m)?

•**36** Two large parallel metal plates are 1.5 cm apart and have charges of equal magnitudes but opposite signs on their facing surfaces. Take the potential of the negative plate to be zero. If the potential halfway between the plates is then $+5.0$ V, what is the electric field in the region between the plates?

••**37** An electron is placed in an $xy$ plane where the electric potential depends on $x$ and $y$ as shown in Fig. 24-45 (the potential does not depend on $z$). The scale of the vertical axis is set by $V_s = 500$ V. In unit-vector notation, what is the electric force on the electron?

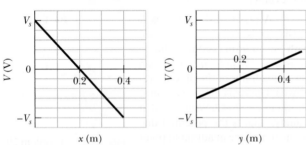

**FIG. 24-45** Problem 37.

••**38** Figure 24-43 shows a thin plastic rod of length $L = 13.5$ cm and uniform charge 43.6 fC. (a) In terms of distance $d$, find an expression for the electric potential at point $P_1$. (b) Next, substitute variable $x$ for $d$ and find an expression for the magnitude of the component $E_x$ of the electric field at $P_1$. (c) What is the direction of $E_x$ relative to the positive direction of the $x$ axis? (d) What is the value of $E_x$ at $P_1$ for $x = d = 6.20$ cm? (e) From the symmetry in Fig. 24-43, determine $E_y$ at $P_1$.

••**39** What is the magnitude of the electric field at the point $(3.00\hat{i} - 2.00\hat{j} + 4.00\hat{k})$ m if the electric potential is given by $V = 2.00xyz^2$, where $V$ is in volts and $x$, $y$, and $z$ are in meters? **SSM**

•••**40** The thin plastic rod of length $L = 10.0$ cm in Fig. 24-43 has a nonuniform linear charge density $\lambda = cx$, where $c = 49.9$ pC/m$^2$. (a) With $V = 0$ at infinity, find the electric potential at point $P_2$ on the $y$ axis at $y = D = 3.56$ cm. (b) Find the electric field component $E_y$ at $P_2$. (c) Why cannot the field component $E_x$ at $P_2$ be found using the result of (a)?

### sec. 24-11 Electric Potential Energy of a System of Point Charges

•**41** How much work is required to set up the arrangement of Fig. 24-46 if $q =$

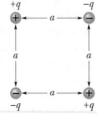

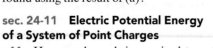

**FIG. 24-46** Problem 41.

2.30 pC, $a = 64.0$ cm, and the particles are initially infinitely far apart and at rest? **SSM ILW WWW**

•**42** In Fig. 24-47, seven charged particles are fixed in place to form a square with an edge length of 4.0 cm. How much work must we do to bring a particle of charge $+6e$ initially at rest from an infinite distance to the center of the square?

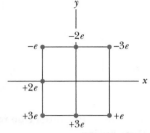

**FIG. 24-47** Problem 42.

•**43** A particle of charge $+7.5$ $\mu$C is released from rest at the point $x = 60$ cm on an $x$ axis. The particle begins to move due to the presence of a charge $Q$ that remains fixed at the origin. What is the kinetic energy of the particle at the instant it has moved 40 cm if (a) $Q = +20$ $\mu$C and (b) $Q = -20$ $\mu$C?

••**44** (a) What is the electric potential energy of two electrons separated by 2.00 nm? (b) If the separation increases, does the potential energy increase or decrease?

••**45** In the rectangle of Fig. 24-48, the sides have lengths 5.0 cm and 15 cm, $q_1 = -5.0$ $\mu$C, and $q_2 = +2.0$ $\mu$C. With $V = 0$ at infinity, what is the electric potential at (a) corner $A$ and (b) corner $B$? (c) How much work is required to move a charge $q_3 = +3.0$ $\mu$C from $B$ to $A$ along a diagonal of the rectangle? (d) Does this work increase or decrease the electric potential energy of the three-charge system? Is more, less, or the same work required if $q_3$ is moved along a path that is (e) inside the rectangle but not on a diagonal and (f) outside the rectangle? **GO**

**FIG. 24-48** Problem 45.

••**46** In Fig. 24-49, how much work must we do to bring a particle, of charge $Q = +16e$ and initially at rest, along the dashed line from infinity to the indicated point near two fixed particles of charges $q_1 = +4e$ and $q_2 = -q_1/2$? Distance $d = 1.40$ cm, $\theta_1 = 43°$, and $\theta_2 = 60°$.

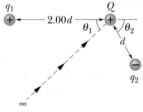

**FIG. 24-49** Problem 46.

••**47** A particle of charge $q$ is fixed at point $P$, and a second particle of mass $m$ and the same charge $q$ is initially held a distance $r_1$ from $P$. The second particle is then released. Determine its speed when it is a distance $r_2$ from $P$. Let $q = 3.1$ $\mu$C, $m = 20$ mg, $r_1 = 0.90$ mm, and $r_2 = 2.5$ mm. **ILW**

••**48** A charge of $-9.0$ nC is uniformly distributed around a thin plastic ring lying in a $yz$ plane with the ring center at the origin. A $-6.0$ pC point charge is located on the $x$ axis at $x = 3.0$ m. For a ring radius of 1.5 m, how much work must an external force do on the point charge to move it to the origin?

••**49** What is the *escape speed* for an electron initially at rest on the surface of a sphere with a radius of 1.0 cm and a uniformly distributed charge of $1.6 \times 10^{-15}$ C? That is, what initial speed must the electron have in order to reach an infinite distance from the sphere and have zero kinetic energy when it gets there?

**••50** A thin, spherical, conducting shell of radius $R$ is mounted on an isolating support and charged to a potential of $-125$ V. An electron is then fired directly toward the center of the shell, from point $P$ at distance $r$ from the center of the shell ($r \gg R$). What initial speed $v_0$ is needed for the electron to just reach the shell before reversing direction?

**••51** Two tiny metal spheres $A$ and $B$ of mass $m_A = 5.00$ g and $m_B = 10.0$ g have equal positive charge $q = 5.00$ μC. The spheres are connected by a massless nonconducting string of length $d = 1.00$ m, which is much greater than the radii of the spheres. (a) What is the electric potential energy of the system? (b) Suppose you cut the string. At that instant, what is the acceleration of each sphere? (c) A long time after you cut the string, what is the speed of each sphere?

**••52** Figure 24-50a shows three particles on an $x$ axis. Particle 1 (with a charge of $+5.0$ μC) and particle 2 (with a charge of $+3.0$ μC) are fixed in place with separation $d = 4.0$ cm. Particle 3 can be moved along the $x$ axis to the right of particle 2. Figure 24-50b gives the electric potential energy $U$ of the three-particle system as a function of the $x$ coordinate of particle 3. The scale of the vertical axis is set by $U_s = 5.0$ J. What is the charge of particle 3?

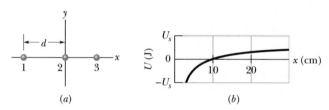

(a)                                    (b)

FIG. 24-50   Problem 52.

**••53** Two electrons are fixed 2.0 cm apart. Another electron is shot from infinity and stops midway between the two. What is its initial speed? **GO**

**••54** *Proton in a well.* Figure 24-51 shows electric potential $V$ along an $x$ axis. The scale of the vertical axis is set by $V_s = 10.0$ V. A proton is to be released at $x = 3.5$ cm with initial kinetic energy 4.00 eV. (a) If it is initially moving in the negative direction of the axis, does it reach a turning point (if so, what is the $x$ coordinate of that point) or does it escape from the plotted region (if so, what is its speed at $x = 0$)? (b) If it is initially moving in the positive direction of the axis, does it reach a turning point (if so, what is the $x$ coordinate of that point) or does it escape from the plotted region (if so, what is its speed at $x = 6.0$ cm)? What are the (c) magnitude $F$ and (d) direction (positive or negative direction of the $x$ axis) of the electric force on the proton if the proton moves just to the left of $x = 3.0$ cm? What are (e) $F$ and (f) the direction if the proton moves just to the right of $x = 5.0$ cm?

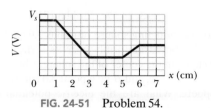

FIG. 24-51   Problem 54.

**••55** In Fig. 24-52, a charged particle (either an electron or a proton) is moving rightward between two parallel charged plates separated by distance $d = 2.00$ mm. The plate potentials are $V_1 = -70.0$ V and $V_2 = -50.0$ V. The particle is slowing from an initial speed of 90.0 km/s at the left plate. (a) Is the particle an electron or a proton? (b) What is its speed just as it reaches plate 2?

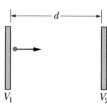

FIG. 24-52
Problem 55.

**••56** Figure 24-53a shows an electron moving along an electric dipole axis toward the negative side of the dipole. The dipole is fixed in place. The electron was initially very far from the dipole, with kinetic energy 100 eV. Figure 24-53b gives the kinetic energy $K$ of the electron versus its distance $r$ from the dipole center. The scale of the horizontal axis is set by $r_s = 0.10$ m. What is the magnitude of the dipole moment?

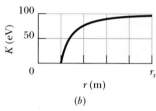

(a)

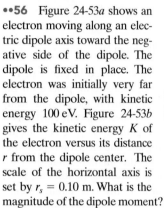

(b)

FIG. 24-53   Problem 56.

**••57** An electron is projected with an initial speed of $3.2 \times 10^5$ m/s directly toward a proton that is fixed in place. If the electron is initially a great distance from the proton, at what distance from the proton is the speed of the electron instantaneously equal to twice the initial value?

**••58** A positron (charge $+e$, mass equal to the electron mass) is moving at $1.0 \times 10^7$ m/s in the positive direction of an $x$ axis when, at $x = 0$, it encounters an electric field directed along the $x$ axis. The electric potential $V$ associated with the field is given in Fig. 24-54. The scale of the vertical axis is set by $V_s = 500.0$ V. (a) Does the positron emerge from the field at $x = 0$ (which means its motion is reversed) or at $x = 0.50$ m (which means its motion is not reversed)? (b) What is its speed when it emerges?

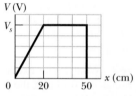

FIG. 24-54   Problem 58.

**••59** Identical 50 μC charges are fixed on an $x$ axis at $x = \pm 3.0$ m. A particle of charge $q = -15$ μC is then released from rest at a point on the positive part of the $y$ axis. Due to the symmetry of the situation, the particle moves along the $y$ axis and has kinetic energy 1.2 J as it passes through the point $x = 0$, $y = 4.0$ m. (a) What is the kinetic energy of the particle as it passes through the origin? (b) At what negative value of $y$ will the particle momentarily stop? **SSM**

**••60** In Fig. 24-55a, we move an electron from an infinite distance to a point at distance $R = 8.00$ cm from a tiny charged ball. The move requires work $W = 2.16 \times 10^{-13}$ J by us. (a) What is the charge $Q$ on the ball? In Fig. 24-55b, the ball has been sliced up and the slices spread out so that an equal amount of charge is at the hour positions on a circular clock face of radius $R = 8.00$ cm. Now the electron is brought from an infinite distance to the center of the circle. (b) With that addition of the electron to the system of 12 charged particles, what is the change in the electric potential energy of the system?

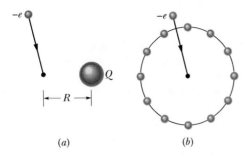

FIG. 24-55  Problem 60.

•••**61**  Suppose $N$ electrons can be placed in either of two configurations. In configuration 1, they are all placed on the circumference of a narrow ring of radius $R$ and are uniformly distributed so that the distance between adjacent electrons is the same everywhere. In configuration 2, $N - 1$ electrons are uniformly distributed on the ring and one electron is placed in the center of the ring. (a) What is the smallest value of $N$ for which the second configuration is less energetic than the first? (b) For that value of $N$, consider any one circumference electron—call it $e_0$. How many other circumference electrons are closer to $e_0$ than the central electron is?

**sec. 24-12   Potential of a Charged Isolated Conductor**

•**62**  A hollow metal sphere has a potential of +400 V with respect to ground (defined to be at $V = 0$) and a charge of $5.0 \times 10^{-9}$ C. Find the electric potential at the center of the sphere.

•**63**  What is the excess charge on a conducting sphere of radius $r = 0.15$ m if the potential of the sphere is 1500 V and $V = 0$ at infinity?  **SSM**

•**64**  Sphere 1 with radius $R_1$ has positive charge $q$. Sphere 2 with radius $2.00R_1$ is far from sphere 1 and initially uncharged. After the separated spheres are connected with a wire thin enough to retain only negligible charge, (a) is potential $V_1$ of sphere 1 greater than, less than, or equal to potential $V_2$ of sphere 2? What fraction of $q$ ends up on (b) sphere 1 and (c) sphere 2? (d) What is the ratio $\sigma_1/\sigma_2$ of the surface charge densities of the spheres?

•**65**  Two metal spheres, each of radius 3.0 cm, have a center-to-center separation of 2.0 m. Sphere 1 has charge $+1.0 \times 10^{-8}$ C; sphere 2 has charge $-3.0 \times 10^{-8}$ C. Assume that the separation is large enough for us to assume that the charge on each sphere is uniformly distributed (the spheres do not affect each other). With $V = 0$ at infinity, calculate (a) the potential at the point halfway between the centers and the potential on the surface of (b) sphere 1 and (c) sphere 2.  **SSM  WWW**

••**66**  Two isolated, concentric, conducting spherical shells have radii $R_1 = 0.500$ m and $R_2 = 1.00$ m, uniform charges $q_1 = +2.00$ $\mu$C and $q_2 = +1.00$ $\mu$C, and negligible thicknesses. What is the magnitude of the electric field $E$ at radial distance (a) $r = 4.00$ m, (b) $r = 0.700$ m, and (c) $r = 0.200$ m? With $V = 0$ at infinity, what is $V$ at (d) $r = 4.00$ m, (e) $r = 1.00$ m, (f) $r = 0.700$ m, (g) $r = 0.500$ m, (h) $r = 0.200$ m, and (i) $r = 0$? (j) Sketch $E(r)$ and $V(r)$.

••**67**  A metal sphere of radius 15 cm has a net charge of $3.0 \times 10^{-8}$ C. (a) What is the electric field at the sphere's surface? (b) If $V = 0$ at infinity, what is the electric potential at the sphere's surface? (c) At what distance from the sphere's surface has the electric potential decreased by 500 V?

**Additional Problems**

**68**  *The chocolate crumb mystery.* This story begins with Problem 56 in Chapter 23. (a) From the answer to part (a) of that problem, find an expression for the electric potential as a function of the radial distance $r$ from the center of the pipe. (The electric potential is zero on the grounded pipe wall.) (b) For the typical volume charge density $\rho = -1.1 \times 10^{-3}$ C/m$^3$, what is the difference in the electric potential between the pipe's center and its inside wall? (The story continues with Problem 56 in Chapter 25.)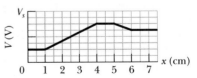

**69**  An electron is released from rest on the axis of an electric dipole that has charge $e$ and charge separation $d = 20$ pm and that is fixed in place. The release point is on the positive side of the dipole, at distance $7.0d$ from the dipole center. What is the electron's speed when it reaches a point $5.0d$ from the dipole center?

**70**  Figure 24-56 shows a ring of outer radius $R = 13.0$ cm, inner radius $r = 0.200R$, and uniform surface charge density $\sigma = 6.20$ pC/m$^2$. With $V = 0$ at infinity, find the electric potential at point $P$ on the central axis of the ring, at distance $z = 2.00R$ from the center of the ring.

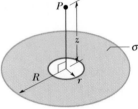

FIG. 24-56  Problem 70.

**71**  *Electron in a well.* Figure 24-57 shows electric potential $V$ along an $x$ axis. The scale of the vertical axis is set by $V_s = 8.0$ V. An electron is to be released at $x = 4.5$ cm with initial kinetic energy 3.00 eV. (a) If it is initially moving in the negative direction of the axis, does it reach a turning point (if so, what is the $x$ coordinate of that point) or does it escape from the plotted region (if so, what is its speed at $x = 0$)? (b) If it is initially moving in the positive direction of the axis, does it reach a turning point (if so, what is the $x$ coordinate of that point) or does it escape from the plotted region (if so, what is its speed at $x = 7.0$ cm)? What are the (c) magnitude $F$ and (d) direction (positive or negative direction of the $x$ axis) of the electric force on the electron if the electron moves just to the left of $x = 4.0$ cm? What are (e) $F$ and (f) the direction if it moves just to the right of $x = 5.0$ cm?

FIG. 24-57  Problem 71.

**72**  A solid conducting sphere of radius 3.0 cm has a charge of 30 nC distributed uniformly over its surface. Let $A$ be a point 1.0 cm from the center of the sphere, $S$ be a point on the surface of the sphere, and $B$ be a point 5.0 cm from the center of the sphere. What are the electric potential differences (a) $V_S - V_B$ and (b) $V_A - V_B$?

**73** In Fig. 24-58, point $P$ is at distance $d_1 = 4.00$ m from particle 1 ($q_1 = -2e$) and distance $d_2 = 2.00$ m from particle 2 ($q_2 = +2e$), with both particles fixed in place. (a) With $V = 0$ at infinity, what is $V$ at $P$? If we bring a particle of charge $q_3 = +2e$ from infinity to $P$, (b) how much work do we do and (c) what is the potential energy of the three-particle system?

FIG. 24-58  Problem 73.

**74** Figure 24-59 shows a thin rod with a uniform charge density of 2.00 $\mu$C/m. Evaluate the electric potential at point $P$ if $d = D = L/4.00$. GO

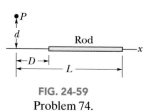

FIG. 24-59
Problem 74.

**75** Three $+0.12$ C charges form an equilateral triangle 1.7 m on a side. Using energy supplied at the rate of 0.83 kW, how many days would be required to move one of the charges to the midpoint of the line joining the other two charges? SSM

**76** The magnitude $E$ of an electric field depends on the radial distance $r$ according to $E = A/r^4$, where $A$ is a constant with the unit volt–cubic meter. As a multiple of $A$, what is the magnitude of the electric potential difference between $r = 2.00$ m and $r = 3.00$ m?

**77** A long, solid, conducting cylinder has a radius of 2.0 cm. The electric field at the surface of the cylinder is 160 N/C, directed radially outward. Let $A$, $B$, and $C$ be points that are 1.0 cm, 2.0 cm, and 5.0 cm, respectively, from the central axis of the cylinder. What are (a) the magnitude of the electric field at $C$ and the electric potential differences (b) $V_B - V_C$ and (c) $V_A - V_B$? SSM

**78** (a) If Earth had a net surface charge density of 1.0 electron/m$^2$ (a very artificial assumption), what would its potential be? (Set $V = 0$ at infinity.) What would be the (b) magnitude and (c) direction (radially inward or outward) of the electric field due to Earth just outside its surface?

**79** In Fig. 24-60, we move a particle of charge $+2e$ in from infinity to the $x$ axis. How much work do we do? Distance $D$ is 4.00 m.

**80** Figure 24-61 shows a hemisphere with a charge of 4.00 $\mu$C distributed uniformly through its volume. The hemisphere lies on an $xy$ plane the way half a grapefruit might lie face down on a kitchen table. Point $P$ is located on the plane, along a radial line from the hemisphere's center of curvature, at radial distance 15 cm. What is the electric potential at point $P$ due to the hemisphere?

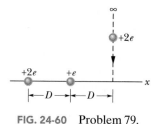

FIG. 24-61  Problem 80.

**81** Initially two electrons are fixed in place with a separation of 2.00 $\mu$m. How much work must we do to bring a third electron in from infinity to complete an equilateral triangle?

**82** Three particles, charge $q_1 = +10$ $\mu$C, $q_2 = -20$ $\mu$C, and

$q_3 = +30 \mu$C, are positioned at the vertices of an isosceles triangle as shown in Fig. 24-62. If $a = 10$ cm and $b = 6.0$ cm, how much work must an external agent do to exchange the positions of (a) $q_1$ and $q_3$ and, instead, (b) $q_1$ and $q_2$?

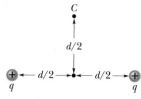

FIG. 24-62
Problem 82.

**83** (a) If an isolated conducting sphere 10 cm in radius has a net charge of 4.0 $\mu$C and if $V = 0$ at infinity, what is the potential on the surface of the sphere? (b) Can this situation actually occur, given that the air around the sphere undergoes electrical breakdown when the field exceeds 3.0 MV/m?

**84** Two charges $q = +2.0$ $\mu$C are fixed a distance $d = 2.0$ cm apart (Fig. 24-63). (a) With $V = 0$ at infinity, what is the electric potential at point $C$? (b) You bring a third charge $q = +2.0$ $\mu$C from infinity to $C$. How much work must you do? (c) What is the potential energy $U$ of the three-charge configuration when the third charge is in place?

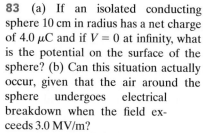

FIG. 24-63  Problem 84.

**85** A uniform charge of $+16.0$ $\mu$C is on a thin circular ring lying in an $xy$ plane and centered on the origin. The ring's radius is 3.00 cm. If point $A$ is at the origin and point $B$ is on the $z$ axis at $z = 4.00$ cm, what is $V_B - V_A$? SSM

**86** The charges and coordinates of two point charges located in an $xy$ plane are $q_1 = +3.00 \times 10^{-6}$ C, $x = +3.50$ cm, $y = +0.500$ cm and $q_2 = -4.00 \times 10^{-6}$ C, $x = -2.00$ cm, $y = +1.50$ cm. How much work must be done to locate these charges at their given positions, starting from infinite separation?

**87** Two charged, parallel, flat conducting surfaces are spaced $d = 1.00$ cm apart and produce a potential difference $\Delta V = 625$ V between them. An electron is projected from one surface directly toward the second. What is the initial speed of the electron if it stops just at the second surface?

**88** A particle of positive charge $Q$ is fixed at point $P$. A second particle of mass $m$ and negative charge $-q$ moves at constant speed in a circle of radius $r_1$, centered at $P$. Derive an expression for the work $W$ that must be done by an external agent on the second particle to increase the radius of the circle of motion to $r_2$.

**89** An electric field of approximately 100 V/m is often observed near the surface of Earth. If this were the field over the entire surface, what would be the electric potential of a point on the surface? (Set $V = 0$ at infinity.)

**90** In Fig. 24-64, point $P$ is at the center of the rectangle. With $V = 0$ at infinity, $q_1 = 5.00$ fC, $q_2 = 2.00$ fC, $q_3 = 3.00$ fC, and $d = 2.54$ cm, what is the net electric potential at $P$ due to the six charged particles?

**91** Figure 24-33 shows two

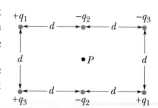

FIG. 24-64  Problem 90.

charged particles on an axis. Sketch the electric field lines and the equipotential surfaces in the plane of the page for (a) $q_1 = +q, q_2 = +2q$ and (b) $q_1 = +q, q_2 = -3q$.

**92** A charge $q$ is distributed uniformly throughout a spherical volume of radius $R$. Let $V = 0$ at infinity. What are (a) $V$ at radial distance $r < R$ and (b) the potential difference between points at $r = R$ and the point at $r = 0$?

**93** A thick-walled spherical shell of charge $Q$ and uniform volume charge density $\rho$ is bounded by radii $r_1$ and $r_2 > r_1$. With $V = 0$ at infinity, find the electric potential $V$ as a function of distance $r$ from the center of the distribution, considering regions (a) $r > r_2$, (b) $r_2 > r > r_1$, and (c) $r < r_1$. (d) Do these solutions agree with each other at $r = r_2$ and $r = r_1$? (*Hint:* See Section 23-9.) **SSM**

**94** An alpha particle (which has two protons) is sent directly toward a target nucleus containing 92 protons. The alpha particle has an initial kinetic energy of 0.48 pJ. What is the least center-to-center distance the alpha particle will be from the target nucleus, assuming the nucleus does not move?

**95** Starting from Eq. 24-30, derive an expression for the electric field due to a dipole at a point on the dipole axis. **SSM**

**96** A charge of $1.50 \times 10^{-8}$ C lies on an isolated metal sphere of radius 16.0 cm. With $V = 0$ at infinity, what is the electric potential at points on the sphere's surface?

**97** In a Millikan oil-drop experiment (Section 22-8), a uniform electric field of $1.92 \times 10^5$ N/C is maintained in the region between two plates separated by 1.50 cm. Find the potential difference between the plates.

**98** Consider a point charge $q = 1.50 \times 10^{-8}$ C, and take $V = 0$ at infinity. (a) What are the shape and dimensions of an equipotential surface having a potential of 30.0 V due to $q$ alone? (b) Are surfaces whose potentials differ by a constant amount (1.0 V, say) evenly spaced?

**99** In the quark model of fundamental particles, a proton is composed of three quarks: two "up" quarks, each having charge $+2e/3$, and one "down" quark, having charge $-e/3$. Suppose that the three quarks are equidistant from one another. Take that separation distance to be $1.32 \times 10^{-15}$ m and calculate the electric potential energy of the system of (a) only the two up quarks and (b) all three quarks.

**100** (a) A proton of kinetic energy 4.80 MeV travels head-on toward a lead nucleus. Assuming that the proton does not penetrate the nucleus and that the only force between proton and nucleus is the Coulomb force, calculate the smallest center-to-center separation $d_p$ between proton and nucleus when the proton momentarily stops. If the proton were replaced with an alpha particle (which contains two protons) of the same initial kinetic energy, the alpha particle would stop at center-to-center separation $d_\alpha$. (b) What is $d_\alpha/d_p$?

**101** (a) Using Eq. 24-32, show that the electric potential at a point on the central axis of a thin ring (of charge $q$ and radius $R$) and at distance $z$ from the ring is

$$V = \frac{1}{4\pi\varepsilon_0} \frac{q}{\sqrt{z^2 + R^2}}.$$

(b) From this result, derive an expression for $E$ at points on the ring's axis; compare your result with the calculation of $E$ in Section 22-6.

**102** What is the electric potential energy of the charge configuration of Fig. 24-8a? Use the numerical values provided in Sample Problem 24-3.

**103** A solid copper sphere whose radius is 1.0 cm has a very thin surface coating of nickel. Some of the nickel atoms are radioactive, each atom emitting an electron as it decays. Half of these electrons enter the copper sphere, each depositing 100 keV of energy there. The other half of the electrons escape, each carrying away a charge $-e$. The nickel coating has an activity of $3.70 \times 10^8$ radioactive decays per second. The sphere is hung from a long, nonconducting string and isolated from its surroundings. (a) How long will it take for the potential of the sphere to increase by 1000 V? (b) How long will it take for the temperature of the sphere to increase by 5.0 K due to the energy deposited by the electrons? The heat capacity of the sphere is 14 J/K. **SSM**

**104** Figure 24-65 shows three circular, nonconducting arcs of radius $R = 8.50$ cm. The charges on the arcs are $q_1 = 4.52$ pC, $q_2 = -2.00q_1$, $q_3 = +3.00q_1$. With $V = 0$ at infinity, what is the net electric potential of the arcs at the common center of curvature?

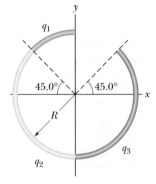

**FIG. 24-65** Problem 104.

**105** In Fig. 24-66, two particles of charges $q_1$ and $q_2$ are fixed to an $x$ axis. If a third particle, of charge $+6.0\,\mu$C, is brought from an infinite distance to point $P$, the three-particle system has the same electric potential energy as the original two-particle system. What is the charge ratio $q_1/q_2$?

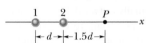

**FIG. 24-66** Problem 105.

**106** In Fig. 24-67, let the separation $d$ between the particles be 1.0 m; let their charges be $q_1 = +q$ and $q_2 = +2.0q$; and let $V = 0$ at infinity. At what finite coordinate on the $x$ axis is (a) the net electric potential due to the two particles zero and (b) the net electric field due to them zero?

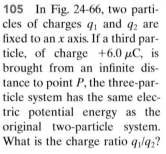

**FIG. 24-67** Problem 106.

**107** In Fig. 24-68, a particle of charge $q_2 = +5e$ is brought along the dashed line from infinity to the indicated point near two fixed particles of charges $q_1 = +2e$ and $q_3 = -q_1$. What is the ratio of the potential energy of this three-particle system to that of the original two-particle system?

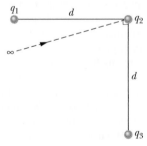

**FIG. 24-68** Problem 107.

**108** In a certain situation, the electric potential varies along an $x$ axis as shown in Fig. 24-69. The scale of the vertical axis is set by $V_s = 12.0$ V. For the intervals (a) $ab$, (b) $bc$, (c) $cd$, (d) $de$, (e) $ef$, (f) $fg$, and (g) $gh$, determine the $x$ component of the electric field, and then plot $E_x$ versus $x$. (Ignore behavior at the interval end points.)

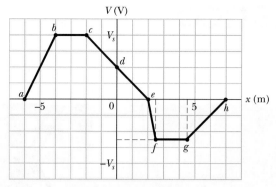

FIG. 24-69 Problem 108.

**109** A disk has radius $R = 2.20$ cm. Its surface charge density is $1.50 \times 10^{-6}$ C/m² from $r = 0$ to $r = R/2$ and $8.00 \times 10^{-7}$ C/m² from $r = R/2$ to $r = R$. (a) What is the total charge on the disk? (b) With $V = 0$ at infinity, what is the electric potential at a point on the perpendicular central axis of the disk, at distance $z = R/2$ from the center of the disk?

**110** In Fig. 24-70, particle 1 of charge $q_1 = +e$ and particle 2 of charge $q_2 = -5e$ are fixed on an $x$ axis. Distance $d = 5.60$ $\mu$m. What is the electric potential difference $V_A - V_B$?

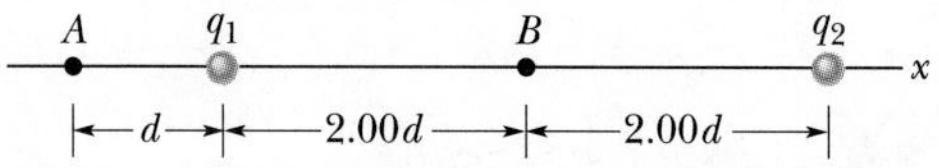

FIG. 24-70 Problem 110.

**111** Point charges of equal magnitudes (25 nC) and opposite signs are placed on diagonally opposite corners of a 60 cm × 80 cm rectangle. Point $A$ is the unoccupied corner nearest the positive charge, and point $B$ is the other unoccupied corner. Determine the potential difference $V_B - V_A$.

**112** A decade before Einstein published his theory of relativity, J. J. Thomson proposed that the electron might consist of small parts and attributed the electron's mass $m$ to the electric potential energy of the interaction of the parts. Furthermore, he suggested that the energy equals $mc^2$, where $c$ is the speed of light. Make a rough estimate of the electron mass in the following way: Assume the electron is composed of three identical parts that are brought in from infinity and placed at the vertices of an equilateral triangle having sides equal to the *classical radius* of the electron, $2.82 \times 10^{-15}$ m. (a) Find the total electric potential energy of this arrangement. (b) Divide by $c^2$ and compare your result with the accepted electron mass. (The result improves if more parts are assumed.)

**113** Figure 24-71 shows three charged particles located on a horizontal axis. For points (such as $P$) on the axis with $r \gg d$, show that the electric potential $V(r)$ is given by

$$V = \frac{1}{4\pi\varepsilon_0} \frac{q}{r} \left(1 + \frac{2d}{r}\right).$$

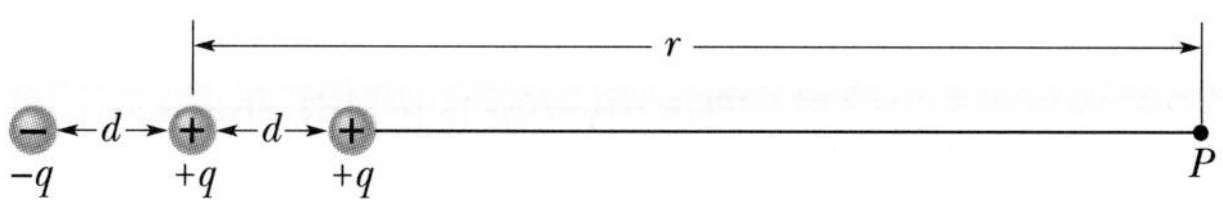

FIG. 24-71 Problem 113.

(*Hint:* The charge configuration can be viewed as the sum of an isolated charge and a dipole.)

**114** A point charge $q_1 = +6.0e$ is fixed at the origin of a rectangular coordinate system, and a point charge $q_2 = -10e$ is fixed at $x = 8.6$ nm, $y = 0$. The locus of all points in the $xy$ plane for which $V = 0$ (other than infinity) is a circle centered on the $x$ axis, as shown in Fig. 24-72. Find (a) the location $x_c$ of the center of the circle and (b) the radius $R$ of the circle. (c) Is the $xy$ cross section of the 5 V equipotential surface also a circle?

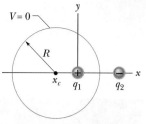

FIG. 24-72 Problem 114.

**115** Charge $q_1 = -1.2 \times 10^{-9}$ C is at the origin, and charge $q_2 = 2.5 \times 10^{-9}$ C is on the $y$ axis at $y = 0.50$ m. Take the electric potential to be zero far from both charges. (a) Plot the intersection of the $V = 5.0$ V equipotential surface with the $xy$ plane. It encloses one of the charges. (b) There are two equipotential surfaces corresponding to $V = 3.0$ V. One encloses one of the charges and the other encloses both charges. Plot their intersections with the $xy$ plane. (c) Find the value of the potential for which the pattern of the electric potential switches from one equipotential surface to two.

**116** In Fig. 24-73, three long parallel lines of charge, with the linear charge densities shown, extend perpendicular to the page in both directions. Sketch some electric field lines; also sketch the cross sections in the plane of the figure of some equipotential surfaces.

$-2\lambda$

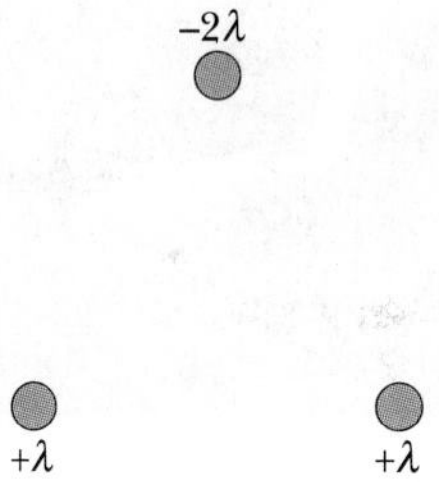

FIG. 24-73 Problem 116. $+\lambda$ $+\lambda$

**117** Two infinite lines of charge are parallel to and in the same plane as a $z$ axis. One, of charge per unit length $+\lambda$, is a distance $a$ to the right of this axis. The other, of charge per unit length $-\lambda$, is a distance $a$ to the left of this axis. Sketch some of the equipotential surfaces due to this arrangement.

**118** In a 1911, Ernest Rutherford modelled an atom as being a point of positive charge $Ze$ surrounded by a negative charge $-Ze$ uniformly distributed in a sphere of radius $R$ centered on the point. At distance $r$ within the sphere, the electric field is

$$E = \frac{Ze}{4\pi\varepsilon_0}\left(\frac{1}{r^2} - \frac{r}{R^3}\right).$$

He also gave the electric potential as

$$V = \frac{Ze}{4\pi\varepsilon_0}\left(\frac{1}{r} - \frac{3}{2R} + \frac{r^2}{2R^3}\right).$$

(a) Show how the expression for the electric field follows from the expression for $V$. (b) Why does this expression for $V$ not go to zero as $r \to \infty$?

# 25 Capacitance

Tienda Brothers/Gamma-Presse, Inc.

*Explosions of airborne dust in grain storage bins (as above), coal mines, flour mills, and many powder industries are a common occurrence, often with loss of life and much property damage. Usually the explosions are due to sparking between charged objects or between a charged object and a grounded connection. Engineers cannot eliminate the possibility of sparking, but they can take measures to reduce the chance that a spark will set off an explosion.*

What determines whether sparking will cause an explosion of airborne dust?

The answer is in this chapter.

## 25-1 | WHAT IS PHYSICS?

One goal of physics is to provide the basic science for practical devices designed by engineers. The focus of this chapter is on one extremely common example—the capacitor, a device in which electrical energy can be stored. For example, the batteries in a camera store energy in the photoflash unit by charging a capacitor. The batteries can supply energy at only a modest rate, too slowly for the photoflash unit to emit a flash of light. However, once the capacitor is charged, it can supply energy at a much greater rate when the photoflash unit is triggered—enough energy to allow the unit to emit a burst of bright light.

The physics of capacitors can be generalized to other devices and to any situation involving electric fields. For example, Earth's atmospheric electric field is modeled by meteorologists as being produced by a huge spherical capacitor that partially discharges via lightning. The charge that skis collect as they slide along snow can be modeled as being stored in a capacitor that frequently discharges as sparks (which can be seen by nighttime skiers on dry snow).

The first step in our discussion of capacitors is to determine how much charge can be stored. This "how much" is called capacitance.

**FIG. 25-1** An assortment of capacitors.

## 25-2 | Capacitance

Figure 25-1 shows some of the many sizes and shapes of capacitors. Figure 25-2 shows the basic elements of *any* capacitor—two isolated conductors of any shape. No matter what their geometry, flat or not, we call these conductors *plates*.

Figure 25-3*a* shows a less general but more conventional arrangement, called a *parallel-plate capacitor,* consisting of two parallel conducting plates of area $A$ separated by a distance $d$. The symbol we use to represent a capacitor (⊣⊢) is based on the structure of a parallel-plate capacitor but is used for capacitors of all geometries. We assume for the time being that no material medium (such as glass or plastic) is present in the region between the plates. In Section 25-6, we shall remove this restriction.

When a capacitor is *charged,* its plates have charges of equal magnitudes but opposite signs: $+q$ and $-q$. However, we refer to the *charge of a capacitor* as being $q$, the absolute value of these charges on the plates. (Note that $q$ is not the net charge on the capacitor, which is zero.)

Because the plates are conductors, they are equipotential surfaces; all points on a plate are at the same electric potential. Moreover, there is a potential difference between the two plates. For historical reasons, we represent the absolute value of this potential difference with $V$ rather than with the $\Delta V$ we used in previous notation.

The charge $q$ and the potential difference $V$ for a capacitor are proportional to each other; that is,

$$q = CV. \qquad (25\text{-}1)$$

The proportionality constant $C$ is called the **capacitance** of the capacitor. Its value depends only on the geometry of the plates and *not* on their charge or

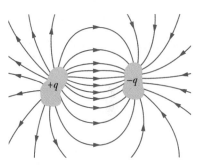

**FIG. 25-2** Two conductors, isolated electrically from each other and from their surroundings, form a *capacitor.* When the capacitor is charged, the charges on the conductors, or *plates* as they are called, have the same magnitude $q$ but opposite signs. *(Paul Silvermann/Fundamental Photographs)*

---

**FIG. 25-3** (*a*) A parallel-plate capacitor, made up of two plates of area $A$ separated by a distance $d$. The charges on the facing plate surfaces have the same magnitude $q$ but opposite signs. (*b*) As the field lines show, the electric field due to the charged plates is uniform in the central region between the plates. The field is not uniform at the edges of the plates, as indicated by the "fringing" of the field lines there.

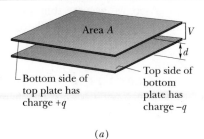

Area $A$

Bottom side of top plate has charge $+q$

Top side of bottom plate has charge $-q$

(*a*)

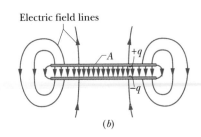

Electric field lines

(*b*)

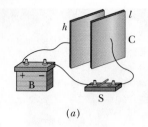

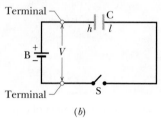

**FIG. 25-4** (a) Battery B, switch S, and plates h and l of capacitor C, connected in a circuit. (b) A schematic diagram with the *circuit elements* represented by their symbols.

potential difference. The capacitance is a measure of how much charge must be put on the plates to produce a certain potential difference between them: The *greater the capacitance, the more charge is required.*

The SI unit of capacitance that follows from Eq. 25-1 is the coulomb per volt. This unit occurs so often that it is given a special name, the *farad* (F):

$$1 \text{ farad} = 1 \text{ F} = 1 \text{ coulomb per volt} = 1 \text{ C/V}. \quad (25\text{-}2)$$

As you will see, the farad is a very large unit. Submultiples of the farad, such as the microfarad ($1 \, \mu\text{F} = 10^{-6} \text{ F}$) and the picofarad ($1 \text{ pF} = 10^{-12} \text{ F}$), are more convenient units in practice.

## Charging a Capacitor

One way to charge a capacitor is to place it in an electric circuit with a battery. An *electric circuit* is a path through which charge can flow. A *battery* is a device that maintains a certain potential difference between its *terminals* (points at which charge can enter or leave the battery) by means of internal electrochemical reactions in which electric forces can move internal charge.

In Fig. 25-4a, a battery B, a switch S, an uncharged capacitor C, and interconnecting wires form a circuit. The same circuit is shown in the *schematic diagram* of Fig. 25-4b, in which the symbols for a battery, a switch, and a capacitor represent those devices. The battery maintains potential difference V between its terminals. The terminal of higher potential is labeled + and is often called the *positive* terminal; the terminal of lower potential is labeled − and is often called the *negative* terminal.

The circuit shown in Figs. 25-4a and b is said to be *incomplete* because switch S is *open;* that is, the switch does not electrically connect the wires attached to it. When the switch is *closed*, electrically connecting those wires, the circuit is complete and charge can then flow through the switch and the wires. As we discussed in Chapter 21, the charge that can flow through a conductor, such as a wire, is that of electrons. When the circuit of Fig. 25-4 is completed, electrons are driven through the wires by an electric field that the battery sets up in the wires. The field drives electrons from capacitor plate h to the positive terminal of the battery; thus, plate h, losing electrons, becomes positively charged. The field drives just as many electrons from the negative terminal of the battery to capacitor plate l; thus, plate l, gaining electrons, becomes negatively charged *just as much* as plate h, losing electrons, becomes positively charged.

Initially, when the plates are uncharged, the potential difference between them is zero. As the plates become oppositely charged, that potential difference increases until it equals the potential difference V between the terminals of the battery. Then plate h and the positive terminal of the battery are at the same potential, and there is no longer an electric field in the wire between them. Similarly, plate l and the negative terminal reach the same potential, and there is then no electric field in the wire between them. Thus, with the field zero, there is no further drive of electrons. The capacitor is then said to be *fully charged*, with a potential difference V and charge q that are related by Eq. 25-1.

In this book we assume that during the charging of a capacitor and afterward, charge cannot pass from one plate to the other across the gap separating them. Also, we assume that a capacitor can retain (or *store*) charge indefinitely, until it is put into a circuit where it can be *discharged.*

✓**CHECKPOINT 1** Does the capacitance C of a capacitor increase, decrease, or remain the same (a) when the charge q on it is doubled and (b) when the potential difference V across it is tripled?

*Tactic 1: The Symbol V and Potential Difference* In previous chapters, the symbol $V$ represents an electric potential at a point or along an equipotential surface. However, in matters concerning electrical devices, $V$ often represents a *potential difference* between two points or two equipotential surfaces. Equation 25-1 is an example of this second use of the symbol. In Section 25-3, you will see a mixture of the two meanings of $V$. There and in later chapters, you need to be alert as to the intent of this symbol.

You will also be seeing, in this book and elsewhere, a variety of phrases regarding potential difference. A potential difference or a "potential" or a "voltage" may be *applied* to a device, or it may be *across* a device. A capacitor can be charged to a potential difference, as in "a capacitor is charged to 12 V." Also, a battery can be characterized by the potential difference across it, as in "a 12 V battery." Always keep in mind what is meant by such phrases: There is a potential difference between two points, such as two points in a circuit or at the terminals of a device such as a battery.

## 25-3 | Calculating the Capacitance

Our goal here is to calculate the capacitance of a capacitor once we know its geometry. Because we shall consider a number of different geometries, it seems wise to develop a general plan to simplify the work. In brief our plan is as follows: (1) Assume a charge $q$ on the plates; (2) calculate the electric field $\vec{E}$ between the plates in terms of this charge, using Gauss' law; (3) knowing $\vec{E}$, calculate the potential difference $V$ between the plates from Eq. 24-18; (4) calculate $C$ from Eq. 25-1.

Before we start, we can simplify the calculation of both the electric field and the potential difference by making certain assumptions. We discuss each in turn.

### Calculating the Electric Field

To relate the electric field $\vec{E}$ between the plates of a capacitor to the charge $q$ on either plate, we shall use Gauss' law:

$$\varepsilon_0 \oint \vec{E} \cdot d\vec{A} = q. \qquad (25\text{-}3)$$

Here $q$ is the charge enclosed by a Gaussian surface and $\oint \vec{E} \cdot d\vec{A}$ is the net electric flux through that surface. In all cases that we shall consider, the Gaussian surface will be such that whenever there is an electric flux through it, $\vec{E}$ will have a uniform magnitude $E$ and the vectors $\vec{E}$ and $d\vec{A}$ will be parallel. Equation 25-3 then reduces to

$$q = \varepsilon_0 E A \qquad \text{(special case of Eq. 25-3),} \qquad (25\text{-}4)$$

in which $A$ is the area of that part of the Gaussian surface through which there is a flux. For convenience, we shall always draw the Gaussian surface in such a way that it completely encloses the charge on the positive plate; see Fig. 25-5 for an example.

### Calculating the Potential Difference

In the notation of Chapter 24 (Eq. 24-18), the potential difference between the plates of a capacitor is related to the field $\vec{E}$ by

$$V_f - V_i = -\int_i^f \vec{E} \cdot d\vec{s}, \qquad (25\text{-}5)$$

in which the integral is to be evaluated along any path that starts on one plate and ends on the other. We shall always choose a path that follows an electric field line, from the negative plate to the positive plate. For this path, the vectors $\vec{E}$ and $d\vec{s}$ will have opposite directions; so the dot product $\vec{E} \cdot d\vec{s}$ will be equal to $-E\,ds$. Thus, the right side of Eq. 25-5 will then be positive. Letting $V$ represent

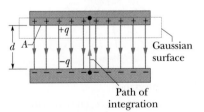

**FIG. 25-5** A charged parallel-plate capacitor. A Gaussian surface encloses the charge on the positive plate. The integration of Eq. 25-6 is taken along a path extending directly from the negative plate to the positive plate.

the difference $V_f - V_i$, we can then recast Eq. 25-5 as

$$V = \int_-^+ E \, ds \qquad \text{(special case of Eq. 25-5),} \qquad (25\text{-}6)$$

in which the $-$ and $+$ remind us that our path of integration starts on the negative plate and ends on the positive plate.

We are now ready to apply Eqs. 25-4 and 25-6 to some particular cases.

### A Parallel-Plate Capacitor

We assume, as Fig. 25-5 suggests, that the plates of our parallel-plate capacitor are so large and so close together that we can neglect the fringing of the electric field at the edges of the plates, taking $\vec{E}$ to be constant throughout the region between the plates.

We draw a Gaussian surface that encloses just the charge $q$ on the positive plate, as in Fig. 25-5. From Eq. 25-4 we can then write

$$q = \varepsilon_0 EA, \qquad (25\text{-}7)$$

where $A$ is the area of the plate.

Equation 25-6 yields

$$V = \int_-^+ E \, ds = E \int_0^d ds = Ed. \qquad (25\text{-}8)$$

In Eq. 25-8, $E$ can be placed outside the integral because it is a constant; the second integral then is simply the plate separation $d$.

If we now substitute $q$ from Eq. 25-7 and $V$ from Eq. 25-8 into the relation $q = CV$ (Eq. 25-1), we find

$$C = \frac{\varepsilon_0 A}{d} \qquad \text{(parallel-plate capacitor).} \qquad (25\text{-}9)$$

Thus, the capacitance does indeed depend only on geometrical factors—namely, the plate area $A$ and the plate separation $d$. Note that $C$ increases as we increase area $A$ or decrease separation $d$.

As an aside, we point out that Eq. 25-9 suggests one of our reasons for writing the electrostatic constant in Coulomb's law in the form $1/4\pi\varepsilon_0$. If we had not done so, Eq. 25-9—which is used more often in engineering practice than Coulomb's law—would have been less simple in form. We note further that Eq. 25-9 permits us to express the permittivity constant $\varepsilon_0$ in a unit more appropriate for use in problems involving capacitors; namely,

$$\varepsilon_0 = 8.85 \times 10^{-12} \text{ F/m} = 8.85 \text{ pF/m}. \qquad (25\text{-}10)$$

We have previously expressed this constant as

$$\varepsilon_0 = 8.85 \times 10^{-12} \text{ C}^2/\text{N} \cdot \text{m}^2. \qquad (25\text{-}11)$$

### A Cylindrical Capacitor

Figure 25-6 shows, in cross section, a cylindrical capacitor of length $L$ formed by two coaxial cylinders of radii $a$ and $b$. We assume that $L \gg b$ so that we can neglect the fringing of the electric field that occurs at the ends of the cylinders. Each plate contains a charge of magnitude $q$.

As a Gaussian surface, we choose a cylinder of length $L$ and radius $r$, closed by end caps and placed as is shown in Fig. 25-6. Equation 25-4 then yields

$$q = \varepsilon_0 EA = \varepsilon_0 E(2\pi r L),$$

in which $2\pi r L$ is the area of the curved part of the Gaussian surface. There is

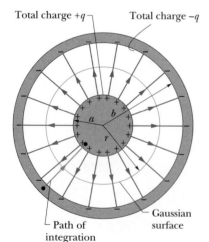

Total charge $+q$    Total charge $-q$

Path of integration

Gaussian surface

**FIG. 25-6** A cross section of a long cylindrical capacitor, showing a cylindrical Gaussian surface of radius $r$ (that encloses the positive plate) and the radial path of integration along which Eq. 25-6 is to be applied. This figure also serves to illustrate a spherical capacitor in a cross section through its center.

no flux through the end caps. Solving for $E$ yields

$$E = \frac{q}{2\pi\varepsilon_0 L r}. \qquad (25\text{-}12)$$

Substitution of this result into Eq. 25-6 yields

$$V = \int_-^+ E\, ds = -\frac{q}{2\pi\varepsilon_0 L} \int_b^a \frac{dr}{r} = \frac{q}{2\pi\varepsilon_0 L} \ln\left(\frac{b}{a}\right), \qquad (25\text{-}13)$$

where we have used the fact that here $ds = -dr$ (we integrated radially inward). From the relation $C = q/V$, we then have

$$C = 2\pi\varepsilon_0 \frac{L}{\ln(b/a)} \qquad \text{(cylindrical capacitor).} \qquad (25\text{-}14)$$

We see that the capacitance of a cylindrical capacitor, like that of a parallel-plate capacitor, depends only on geometrical factors, in this case $L$, $b$, and $a$.

## A Spherical Capacitor

Figure 25-6 can also serve as a central cross section of a capacitor that consists of two concentric spherical shells, of radii $a$ and $b$. As a Gaussian surface we draw a sphere of radius $r$ concentric with the two shells; then Eq. 25-4 yields

$$q = \varepsilon_0 E A = \varepsilon_0 E (4\pi r^2),$$

in which $4\pi r^2$ is the area of the spherical Gaussian surface. We solve this equation for $E$, obtaining

$$E = \frac{1}{4\pi\varepsilon_0} \frac{q}{r^2}, \qquad (25\text{-}15)$$

which we recognize as the expression for the electric field due to a uniform spherical charge distribution (Eq. 23-15).

If we substitute this expression into Eq. 25-6, we find

$$V = \int_-^+ E\, ds = -\frac{q}{4\pi\varepsilon_0} \int_b^a \frac{dr}{r^2} = \frac{q}{4\pi\varepsilon_0} \left(\frac{1}{a} - \frac{1}{b}\right) = \frac{q}{4\pi\varepsilon_0} \frac{b-a}{ab}, \qquad (25\text{-}16)$$

where again we have substituted $-dr$ for $ds$. If we now substitute Eq. 25-16 into Eq. 25-1 and solve for $C$, we find

$$C = 4\pi\varepsilon_0 \frac{ab}{b-a} \qquad \text{(spherical capacitor).} \qquad (25\text{-}17)$$

## An Isolated Sphere

We can assign a capacitance to a *single* isolated spherical conductor of radius $R$ by assuming that the "missing plate" is a conducting sphere of infinite radius. After all, the field lines that leave the surface of a positively charged isolated conductor must end somewhere; the walls of the room in which the conductor is housed can serve effectively as our sphere of infinite radius.

To find the capacitance of the conductor, we first rewrite Eq. 25-17 as

$$C = 4\pi\varepsilon_0 \frac{a}{1 - a/b}.$$

If we then let $b \rightarrow \infty$ and substitute $R$ for $a$, we find

$$C = 4\pi\varepsilon_0 R \qquad \text{(isolated sphere).} \qquad (25\text{-}18)$$

Note that this formula and the others we have derived for capacitance (Eqs. 25-9, 25-14, and 25-17) involve the constant $\varepsilon_0$ multiplied by a quantity that has the dimensions of a length.

**CHECKPOINT 2** For capacitors charged by the same battery, does the charge stored by the capacitor increase, decrease, or remain the same in each of the following situations? (a) The plate separation of a parallel-plate capacitor is increased. (b) The radius of the inner cylinder of a cylindrical capacitor is increased. (c) The radius of the outer spherical shell of a spherical capacitor is increased.

---

**Sample Problem** | **25-1**

In Fig. 25-7a, switch S is closed to connect the uncharged capacitor of capacitance $C = 0.25\ \mu F$ to the battery of potential difference $V = 12$ V. The lower capacitor plate has thickness $L = 0.50$ cm and face area $A = 2.0 \times 10^{-4}\ m^2$, and it consists of copper, in which the density of conduction electrons is $n = 8.49 \times 10^{28}$ electrons/$m^3$. From what depth $d$ within the plate (Fig. 25-7b) must electrons move to the plate face as the capacitor becomes charged?

**KEY IDEA** The charge collected on the plate is related to the capacitance and the potential difference across the capacitor by Eq. 25-1 ($q = CV$).

**Calculations:** Because the lower plate is connected to the negative terminal of the battery, conduction electrons move up to the face of the plate. From Eq. 25-1,

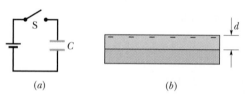

*(a)*                *(b)*

**FIG. 25-7** (*a*) A battery and capacitor circuit. (*b*) The lower capacitor plate.

the total charge magnitude that collects there is

$$q = CV = (0.25 \times 10^{-6}\ F)(12\ V)$$
$$= 3.0 \times 10^{-6}\ C.$$

Dividing this result by $e$ gives us the number $N$ of conduction electrons that come up to the face:

$$N = \frac{q}{e} = \frac{3.0 \times 10^{-6}\ C}{1.602 \times 10^{-19}\ C}$$
$$= 1.873 \times 10^{13}\ \text{electrons}.$$

These electrons come from a volume that is the product of the face area $A$ and the depth $d$ we seek. Thus, from the density of conduction electrons (number per volume), we can write

$$n = \frac{N}{Ad},$$

or

$$d = \frac{N}{An} = \frac{1.873 \times 10^{13}\ \text{electrons}}{(2.0 \times 10^{-4}\ m^2)\,(8.49 \times 10^{28}\ \text{electrons/}m^3)}$$
$$= 1.1 \times 10^{-12}\ m = 1.1\ \text{pm}. \qquad \text{(Answer)}$$

In common speech, we would say that the battery charges the capacitor by supplying the charged particles. But what the battery really does is set up an electric field in the wires and plate such that electrons very close to the plate face move up to the negative face.

---

## 25-4 | Capacitors in Parallel and in Series

When there is a combination of capacitors in a circuit, we can sometimes replace that combination with an **equivalent capacitor**—that is, a single capacitor that has the same capacitance as the actual combination of capacitors. With such a replacement, we can simplify the circuit, affording easier solutions for unknown quantities of the circuit. Here we discuss two basic combinations of capacitors that allow such a replacement.

### Capacitors in Parallel

Figure 25-8a shows an electric circuit in which three capacitors are connected *in parallel* to battery B. This description has little to do with how the capacitor plates are drawn. Rather, "in parallel" means that the capacitors are directly wired together at one plate and directly wired together at the other plate, and that the same potential difference $V$ is applied across the two groups of wired-together plates. Thus, each capacitor has the same potential difference $V$, which produces charge on the capacitor. (In Fig. 25-8a, the applied potential $V$ is maintained by the battery.) In general,

⬤► When a potential difference $V$ is applied across several capacitors connected in parallel, that potential difference $V$ is applied across each capacitor. The total charge $q$ stored on the capacitors is the sum of the charges stored on all the capacitors.

When we analyze a circuit of capacitors in parallel, we can simplify it with this mental replacement:

⬤► Capacitors connected in parallel can be replaced with an equivalent capacitor that has the same *total* charge $q$ and the same potential difference $V$ as the actual capacitors.

(You might remember this result with the nonsense word "par-V," which is close to "party," to mean "capacitors in parallel have the same $V$.") Figure 25-8b shows the equivalent capacitor (with equivalent capacitance $C_{eq}$) that has replaced the three capacitors (with actual capacitances $C_1$, $C_2$, and $C_3$) of Fig. 25-8a.

To derive an expression for $C_{eq}$ in Fig. 25-8b, we first use Eq. 25-1 to find the charge on each actual capacitor:

$$q_1 = C_1V, \quad q_2 = C_2V, \quad \text{and} \quad q_3 = C_3V.$$

The total charge on the parallel combination of Fig. 25-8a is then

$$q = q_1 + q_2 + q_3 = (C_1 + C_2 + C_3)V.$$

The equivalent capacitance, with the same total charge $q$ and applied potential difference $V$ as the combination, is then

$$C_{eq} = \frac{q}{V} = C_1 + C_2 + C_3,$$

a result that we can easily extend to any number $n$ of capacitors, as

$$C_{eq} = \sum_{j=1}^{n} C_j \qquad (n \text{ capacitors in parallel}). \tag{25-19}$$

Thus, to find the equivalent capacitance of a parallel combination, we simply add the individual capacitances.

## Capacitors in Series

Figure 25-9a shows three capacitors connected *in series* to battery B. This description has little to do with how the capacitors are drawn. Rather, "in series" means that the capacitors are wired serially, one after the other, and that a potential difference $V$ is applied across the two ends of the series. (In Fig. 25-9a, this potential difference $V$ is maintained by battery B.) The potential differences that then exist across the capacitors in the series produce identical charges $q$ on them.

⬤► When a potential difference $V$ is applied across several capacitors connected in series, the capacitors have identical charge $q$. The sum of the potential differences across all the capacitors is equal to the applied potential difference $V$.

We can explain how the capacitors end up with identical charge by following a *chain reaction* of events, in which the charging of each capacitor causes the charging of the next capacitor. We start with capacitor 3 and work upward to capacitor 1. When the battery is first connected to the series of capacitors, it produces charge $-q$ on the bottom plate of capacitor 3. That charge then repels negative charge from the top plate of capacitor 3 (leaving it with charge $+q$). The repelled negative charge moves to the bottom plate of capacitor 2 (giving it charge $-q$). That charge on the bottom plate of capacitor 2 then repels negative charge from the top plate of capacitor 2 (leaving it with charge $+q$) to the bottom plate of capacitor 1 (giving it charge $-q$). Finally, the charge on the bottom plate

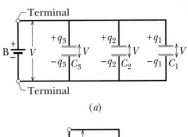

**FIG. 25-8** (*a*) Three capacitors connected in parallel to battery B. The battery maintains potential difference $V$ across its terminals and thus across *each* capacitor. (*b*) The equivalent capacitor, with capacitance $C_{eq}$, replaces the parallel combination.

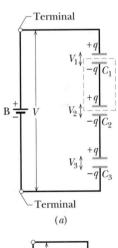

**FIG. 25-9** (*a*) Three capacitors connected in series to battery B. The battery maintains potential difference $V$ between the top and bottom plates of the series combination. (*b*) The equivalent capacitor, with capacitance $C_{eq}$, replaces the series combination.

of capacitor 1 helps move negative charge from the top plate of capacitor 1 to the battery, leaving that top plate with charge $+q$.

Here are two important points about capacitors in series:

1. When charge is shifted from one capacitor to another in a series of capacitors, it can move along only one route, such as from capacitor 3 to capacitor 2 in Fig. 25-9a. If there are additional routes, the capacitors are not in series. An example is given in Sample Problem 25-2.

2. The battery directly produces charges on only the two plates to which it is connected (the bottom plate of capacitor 3 and the top plate of capacitor 1 in Fig. 25-9a). Charges that are produced on the other plates are due merely to the shifting of charge already there. For example, in Fig. 25-9a, the part of the circuit enclosed by dashed lines is electrically isolated from the rest of the circuit. Thus, the net charge of that part cannot be changed by the battery — its charge can only be redistributed.

When we analyze a circuit of capacitors in series, we can simplify it with this mental replacement:

> Capacitors that are connected in series can be replaced with an equivalent capacitor that has the same charge $q$ and the same *total* potential difference $V$ as the actual series capacitors.

(You might remember this with the nonsense word "seri-q" to mean "capacitors in series have the same $q$.") Figure 25-9b shows the equivalent capacitor (with equivalent capacitance $C_{eq}$) that has replaced the three actual capacitors (with actual capacitances $C_1$, $C_2$, and $C_3$) of Fig. 25-9a.

To derive an expression for $C_{eq}$ in Fig. 25-9b, we first use Eq. 25-1 to find the potential difference of each actual capacitor:

$$V_1 = \frac{q}{C_1}, \quad V_2 = \frac{q}{C_2}, \quad \text{and} \quad V_3 = \frac{q}{C_3}.$$

The total potential difference $V$ due to the battery is the sum of these three potential differences. Thus,

$$V = V_1 + V_2 + V_3 = q\left(\frac{1}{C_1} + \frac{1}{C_2} + \frac{1}{C_3}\right).$$

The equivalent capacitance is then

$$C_{eq} = \frac{q}{V} = \frac{1}{1/C_1 + 1/C_2 + 1/C_3},$$

or

$$\frac{1}{C_{eq}} = \frac{1}{C_1} + \frac{1}{C_2} + \frac{1}{C_3}.$$

We can easily extend this to any number $n$ of capacitors as

$$\frac{1}{C_{eq}} = \sum_{j=1}^{n} \frac{1}{C_j} \qquad (n \text{ capacitors in series}). \qquad (25\text{-}20)$$

Using Eq. 25-20 you can show that the equivalent capacitance of a series of capacitances is always *less* than the least capacitance in the series.

**✓ CHECKPOINT 3** A battery of potential $V$ stores charge $q$ on a combination of two identical capacitors. What are the potential difference across and the charge on either capacitor if the capacitors are (a) in parallel and (b) in series?

**Sample Problem**  **25-2**

(a) Find the equivalent capacitance for the combination of capacitances shown in Fig. 25-10a, across which potential difference $V$ is applied. Assume

$$C_1 = 12.0 \ \mu F, \quad C_2 = 5.30 \ \mu F, \quad \text{and} \quad C_3 = 4.50 \ \mu F.$$

**KEY IDEA**  Any capacitors connected in series can be replaced with their equivalent capacitor, and any capacitors connected in parallel can be replaced with their equivalent capacitor. Therefore, we should first check whether any of the capacitors in Fig. 25-10a are in parallel or series.

*Finding equivalent capacitance:* Capacitors 1 and 3 are connected one after the other, but are they in series? No. The potential V that is applied to the capacitors produces charge on the bottom plate of capacitor 3. That charge causes charge to shift from the top plate of capacitor 3. However, note that the shifting charge can move to the bottom plates of both capacitor 1 and capacitor 2. Because there is more than one route for the shifting charge, capacitor 3 is not in series with capacitor 1 (or capacitor 2).

Are capacitor 1 and capacitor 2 in parallel? Yes. Their top plates are directly wired together and their bottom plates are directly wired together, and electric potential is applied between the top-plate pair and the bottom-plate pair. Thus, capacitor 1 and capacitor 2 are in parallel, and Eq. 25-19 tells us that their equivalent capacitance $C_{12}$ is

$$C_{12} = C_1 + C_2 = 12.0 \ \mu F + 5.30 \ \mu F = 17.3 \ \mu F.$$

In Fig. 25-10b, we have replaced capacitors 1 and 2 with their equivalent capacitor, called capacitor 12 (say "one two" and not "twelve"). (The connections at points $A$ and $B$ are exactly the same in Figs. 25-10a and b.)

Is capacitor 12 in series with capacitor 3? Again applying the test for series capacitances, we see that the charge that shifts from the top plate of capacitor 3 must entirely go to the bottom plate of capacitor 12. Thus, capacitor 12 and capacitor 3 are in series, and we can replace them with their equivalent $C_{123}$ ("one two three"), as shown in Fig. 25-10c. From Eq. 25-20, we have

$$\frac{1}{C_{123}} = \frac{1}{C_{12}} + \frac{1}{C_3}$$

$$= \frac{1}{17.3 \ \mu F} + \frac{1}{4.50 \ \mu F} = 0.280 \ \mu F^{-1},$$

from which

$$C_{123} = \frac{1}{0.280 \ \mu F^{-1}} = 3.57 \ \mu F. \quad \text{(Answer)}$$

(b) The potential difference applied to the input terminals in Fig. 25-10a is $V = 12.5$ V. What is the charge on $C_1$?

**KEY IDEAS**  We now need to work backwards from the equivalent capacitance to get the charge on a particular capacitor. We have two techniques for such "backwards work": (1) Seri-q: Series capacitors have the same charge as their equivalent capacitor. (2) Par-V: Parallel capacitors have the same potential difference as their equivalent capacitor.

*Working backwards:* To get the charge $q_1$ on capacitor 1, we work backwards to that capacitor, starting with the equivalent capacitor 123. Because the given potential difference $V$ ($= 12.5$ V) is applied across the actual combination of three capacitors in Fig. 25-10a, it is also applied across $C_{123}$ in Fig. 25-10c. Thus, Eq. 25-1 ($q = CV$) gives us

$$q_{123} = C_{123}V = (3.57 \ \mu F)(12.5 \ V) = 44.6 \ \mu C.$$

The series capacitors 12 and 3 in Fig. 25-10b each have the same charge as their equivalent capacitor 123. Thus, capacitor 12 has charge $q_{12} = q_{123} = 44.6 \ \mu C$. From Eq. 25-1, the potential difference across capacitor 12 must be

$$V_{12} = \frac{q_{12}}{C_{12}} = \frac{44.6 \ \mu C}{17.3 \ \mu F} = 2.58 \ V.$$

The parallel capacitors 1 and 2 each have the same potential difference as their equivalent capacitor 12. Thus, capacitor 1 has potential difference $V_1 = V_{12} = 2.58$ V, and, from Eq. 25-1, the charge on capacitor 1 must be

$$q_1 = C_1V_1 = (12.0 \ \mu F)(2.58 \ V)$$
$$= 31.0 \ \mu C. \quad \text{(Answer)}$$

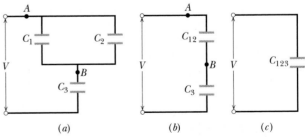

(a)          (b)          (c)

**FIG. 25-10**  (a) Three capacitors. (b) $C_1$ and $C_2$, a parallel combination, are replaced by the equivalent capacitance $C_{12}$. (c) $C_{12}$ and $C_3$, a series combination, are replaced by the equivalent capacitance $C_{123}$.

**Sample Problem** | **25-3** | **Build your skill**

Capacitor 1, with $C_1 = 3.55\ \mu F$, is charged to a potential difference $V_0 = 6.30$ V, using a 6.30 V battery. The battery is then removed, and the capacitor is connected as in Fig. 25-11 to an uncharged capacitor 2, with $C_2 = 8.95\ \mu F$. When switch S is closed, charge flows between the capacitors. Find the charge on each capacitor when equilibrium is reached.

**KEY IDEAS** The situation here differs from the previous example because here an applied electric potential is *not* maintained across a combination of capacitors by a battery or some other source. Here, just after switch S is closed, the only applied electric potential is that of capacitor 1 on capacitor 2, and that potential is decreasing. Thus, the capacitors in Fig. 25-11 are not connected *in series;* and although they are drawn parallel, in this situation they are not *in parallel.*

As the electric potential across capacitor 1 decreases, that across capacitor 2 increases. Equilibrium is reached when the two potentials are equal because, with no potential difference between connected plates of the capacitors, there is no electric field within the connecting wires to move conduction electrons. The initial charge on capacitor 1 is then shared between the two capacitors.

**Calculations:** Initially, when capacitor 1 is connected to the battery, the charge it acquires is, from Eq. 25-1,

$$q_0 = C_1 V_0 = (3.55 \times 10^{-6}\ F)\,(6.30\ V)$$
$$= 22.365 \times 10^{-6}\ C.$$

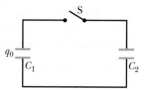

**FIG. 25-11** A potential difference $V_0$ is applied to capacitor 1 and the charging battery is removed. Switch S is then closed so that the charge on capacitor 1 is shared with capacitor 2.

When switch S in Fig. 25-11 is closed and capacitor 1 begins to charge capacitor 2, the electric potential and charge on capacitor 1 decrease and those on capacitor 2 increase until

$$V_1 = V_2 \qquad \text{(equilibrium)}.$$

From Eq. 25-1, we can rewrite this as

$$\frac{q_1}{C_1} = \frac{q_2}{C_2} \qquad \text{(equilibrium)}.$$

Because the total charge cannot magically change, the total after the transfer must be

$$q_1 + q_2 = q_0 \qquad \text{(charge conservation)};$$

thus

$$q_2 = q_0 - q_1.$$

We can now rewrite the second equilibrium equation as

$$\frac{q_1}{C_1} = \frac{q_0 - q_1}{C_2}.$$

Solving this for $q_1$ and substituting given data, we find

$$q_1 = 6.35\ \mu C. \qquad \text{(Answer)}$$

The rest of the initial charge ($q_0 = 22.365\ \mu C$) must be on capacitor 2:

$$q_2 = 16.0\ \mu C. \qquad \text{(Answer)}$$

---

**PROBLEM-SOLVING TACTICS**

*Tactic 2: Multiple-Capacitor Circuits* Let us review the procedure used in the solution of Sample Problem 25-2, in which several capacitors are connected to a battery. To find a single equivalent capacitance, we simplify the given arrangement of capacitances by replacing them, in steps, with equivalent capacitances, using Eq. 25-19 when we find capacitances in parallel and Eq. 25-20 when we find capacitances in series. Then, to find the charge stored by that single equivalent capacitance, we use Eq. 25-1 and the potential difference $V$ applied by the battery.

That result tells us the net charge stored on the actual arrangement of capacitors. However, to find the charge on, or the potential difference across, any particular capacitor in the actual arrangement, we need to reverse our steps of simplification. With each reversed step, we use these two rules: When capacitances are in parallel, they have the same potential difference as their equivalent capacitance, and we use Eq. 25-1 to find the charge on each capacitance; when they are in series, they have the same charge as their equivalent capacitance,

and we use Eq. 25-1 to find the potential difference across each capacitance.

*Tactic 3: Batteries and Capacitors* A battery maintains a certain potential difference across its terminals. Thus, when capacitor 1 of Sample Problem 25-3 is connected to the 6.30 V battery, charge flows between the capacitor and the battery until the capacitor has the same potential difference across it as the battery.

A capacitor differs from a battery in that a capacitor lacks the internal electrochemical reactions needed to release charged particles (electrons) from internal atoms and molecules. Thus, when the charged capacitor 1 of Sample Problem 25-3 is disconnected from the battery and then connected to the uncharged capacitor 2 with switch S closed, the potential difference across capacitor 1 is not maintained. The quantity that *is* maintained is the charge $q_0$ of the two-capacitor system; that is, charge obeys a conservation law, *not* electric potential.

## 25-5 | Energy Stored in an Electric Field

Work must be done by an external agent to charge a capacitor. Starting with an uncharged capacitor, for example, imagine that—using "magic tweezers"—you remove electrons from one plate and transfer them one at a time to the other plate. The electric field that builds up in the space between the plates has a direction that tends to oppose further transfer. Thus, as charge accumulates on the capacitor plates, you have to do increasingly larger amounts of work to transfer additional electrons. In practice, this work is done not by "magic tweezers" but by a battery, at the expense of its store of chemical energy.

We visualize the work required to charge a capacitor as being stored in the form of electric potential energy $U$ in the electric field between the plates. You can recover this energy at will, by discharging the capacitor in a circuit, just as you can recover the potential energy stored in a stretched bow by releasing the bowstring to transfer the energy to the kinetic energy of an arrow.

Suppose that, at a given instant, a charge $q'$ has been transferred from one plate of a capacitor to the other. The potential difference $V'$ between the plates at that instant will be $q'/C$. If an extra increment of charge $dq'$ is then transferred, the increment of work required will be, from Eq. 24-7,

$$dW = V' \, dq' = \frac{q'}{C} \, dq'.$$

The work required to bring the total capacitor charge up to a final value $q$ is

$$W = \int dW = \frac{1}{C} \int_0^q q' \, dq' = \frac{q^2}{2C}.$$

This work is stored as potential energy $U$ in the capacitor, so that

$$U = \frac{q^2}{2C} \qquad \text{(potential energy).} \qquad (25\text{-}21)$$

From Eq. 25-1, we can also write this as

$$U = \tfrac{1}{2}CV^2 \qquad \text{(potential energy).} \qquad (25\text{-}22)$$

Equations 25-21 and 25-22 hold no matter what the geometry of the capacitor is.

To gain some physical insight into energy storage, consider two parallel-plate capacitors that are identical except that capacitor 1 has twice the plate separation of capacitor 2. Then capacitor 1 has twice the volume between its plates and also, from Eq. 25-9, half the capacitance of capacitor 2. Equation 25-4 tells us that if both capacitors have the same charge $q$, the electric fields between their plates are identical. And Eq. 25-21 tells us that capacitor 1 has twice the stored potential energy of capacitor 2. Thus, of two otherwise identical capacitors with the same charge and same electric field, the one with twice the volume between its plates has twice the stored potential energy. Arguments like this tend to verify our earlier assumption:

The potential energy of a charged capacitor may be viewed as being stored in the electric field between its plates.

## Explosions in Airborne Dust

As we discussed in Section 24-12, making contact with certain materials, such as clothing, carpets, and even playground slides, can leave you with a significant electrical potential. You might become painfully aware of that potential if a spark leaps between you and a grounded object, such as a faucet. In many industries involving the production and transport of powder, such as in the cosmetic and food industries, such a spark can be disastrous. Although the powder in bulk may not burn at all, when individual powder grains are airborne and thus surrounded by oxygen, they can burn so fiercely that a cloud of the grains burns as an explosion. Safety engineers cannot eliminate all possible sources of sparks in the powder industries. Instead, they attempt to keep the amount of energy available in the sparks below the threshold value $U_t \, (\approx 150 \, \text{mJ})$ typically required to ignite airborne grains.

Suppose a person becomes charged by contact with various surfaces as he walks through an airborne powder. We can roughly model the person as a spherical capacitor of radius $R = 1.8$ m. From Eq. 25-18 ($C = 4\pi\varepsilon_0 R$) and Eq. 25-22 ($U = \frac{1}{2}CV^2$), we see that the energy of the capacitor is

$$U = \frac{1}{2}(4\pi\varepsilon_0 R)V^2.$$

From this we see that the threshold energy corresponds to a potential of

$$V = \sqrt{\frac{2U_t}{4\pi\varepsilon_0 R}} = \sqrt{\frac{2(150 \times 10^{-3} \text{ J})}{4\pi(8.85 \times 10^{-12} \text{ C}^2/\text{N} \cdot \text{m}^2)(1.8 \text{ m})}}$$

$$= 3.9 \times 10^4 \text{ V}.$$

Safety engineers attempt to keep the potential of the personnel below this level by "bleeding" off the charge through, say, a conducting floor.

### Energy Density

In a parallel-plate capacitor, neglecting fringing, the electric field has the same value at all points between the plates. Thus, the **energy density** $u$—that is, the potential energy per unit volume between the plates—should also be uniform. We can find $u$ by dividing the total potential energy by the volume $Ad$ of the space between the plates. Using Eq. 25-22, we obtain

$$u = \frac{U}{Ad} = \frac{CV^2}{2Ad}.$$

With Eq. 25-9 ($C = \varepsilon_0 A/d$), this result becomes

$$u = \frac{1}{2}\varepsilon_0 \left(\frac{V}{d}\right)^2.$$

However, from Eq. 24-42 ($E = -\Delta V/\Delta s$), $V/d$ equals the electric field magnitude $E$; so

$$u = \frac{1}{2}\varepsilon_0 E^2 \quad \text{(energy density)}. \tag{25-25}$$

Although we derived this result for the special case of a parallel-plate capacitor, it holds generally, whatever may be the source of the electric field. If an electric field $\vec{E}$ exists at any point in space, we can think of that point as a site of electric potential energy whose amount per unit volume is given by Eq. 25-25.

### Sample Problem 25-4

Often, a burn victim is treated while lying on a gurney in an enclosed chamber filled with oxygen-enriched air (a hyperbaric chamber). Once a treatment session is over, a hospital worker pulls the gurney and patient from the chamber onto a trolley, to be rolled away. On at least two occasions, the gurney caught fire at the end that was last to leave the chamber. Obviously, a burning gurney holding a patient already suffering from burns is a dangerous situation, and obviously fires burn easily in air rich in oxygen, but the question remains: What caused the gurneys to catch fire?

Investigators realized that charge separation occurred between the patient's skin, the hospital gown on the patient, and the sheet on the gurney. They also found that the gurney and the part of the chamber's metal framework below the gurney formed a parallel-plate capacitor (Fig. 25-12) of capacitance $C_i = 250$ pF. If the gurney discharged its excess charge and the associated energy by sparking, could the spark ignite the gurney? Measurements revealed that a spark could oc-

cur only if the potential difference $V$ on the gurney–framework capacitor exceeded 2000 V and that a fire could start only if the capacitor's potential energy $U$ exceeded 0.20 mJ. However, the potential difference on the gurney–framework capacitor was only $V_i = 600$ V, not enough to produce a spark.

(a) As the gurney was withdrawn from the chamber, the area of the gurney–framework overlap decreased. Thus, the plate area of the capacitor decreased from its initial value $A_i$. What was the potential difference $V_f$ when the overlap plate area was $A_f = 0.10A_i$?

**KEY IDEAS** (1) The potential difference $V$ across a capacitor is related to the charge $q$ and capacitance $C$ according to Eq. 25-1 ($q = CV$). (2) As the gurney was withdrawn from the chamber, the charge $q$ did not change. (3) The capacitance of a parallel-plate capacitor is related to the plate area according to Eq. 25-9 ($C = \varepsilon_0 A/d$).

*Calculations:* From Eq. 25-1, the charge $q$ was

$$q = C_f V_f = C_i V_i,$$

or

$$V_f = \frac{C_i}{C_f} V_i. \tag{25-23}$$

From Eq. 25-9, we can write

$$C_f = \frac{\varepsilon_0 A_f}{d} = \frac{\varepsilon_0 (0.10 A_i)}{d}$$

$$= 0.10 \frac{\varepsilon_0 A_i}{d} = 0.10 C_i. \tag{25-24}$$

Substituting this result into Eq. 25-23 brings us to

$$V_f = \frac{C_i}{0.10 C_i} V_i = 10 V_i = (10)(600 \text{ V})$$

$$= 6000 \text{ V.} \qquad \text{(Answer)}$$

As the gurney was withdrawn, the potential difference increased because the charge on the capacitor was crowded into a smaller plate area. The potential difference $V_f = 6000$ V was more than enough to produce a spark.

(b) What was the energy $U_f$ of the gurney–framework capacitor when the plate area was $0.10 A_i$?

**KEY IDEA** The potential energy $U$ stored in a capacitor is related to the capacitance $C$ and potential difference $V$ according to Eq. 25-22 ($U = \frac{1}{2}CV^2$).

*Calculation:* Using Eq. 25-24, we write

$$U_f = \frac{1}{2} C_f V_f^2 = \frac{1}{2}(0.10 C_i) V_f^2$$

$$= \frac{1}{2}(0.10)(250 \times 10^{-12} \text{ F})(6000 \text{ V})^2$$

$$= 4.5 \times 10^{-4} \text{ J} = 0.45 \text{ mJ.} \qquad \text{(Answer)}$$

This was more than enough energy to ignite the gurney.

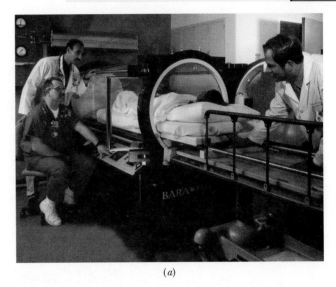

(a)

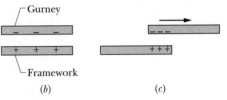

(b)          (c)

**FIG. 25-12** (*a*) A hyperbaric chamber. (*b*) A gurney and the chamber's metal framework form a capacitor that is charged by stray electrostatic charge. (*c*) As the gurney is pulled from the chamber, the charge crowds onto a smaller area. (*BARA-MED® Monoplace chamber designed and manufactured by BioMedical Systems Group, Environmental Tectonics Corp.*)

The investigators concluded that the gurney fire was due to a spark produced by the gurney–framework capacitor as the charge became crowded into a smaller area while the gurney was being withdrawn from the chamber.

---

**Sample Problem** | **25-5**

An isolated conducting sphere whose radius $R$ is 6.85 cm has a charge $q = 1.25$ nC.

(a) How much potential energy is stored in the electric field of this charged conductor?

**KEY IDEAS** (1) An isolated sphere has capacitance given by Eq. 25-18 ($C = 4\pi\varepsilon_0 R$). (2) The energy $U$ stored in a capacitor depends on the capacitor's charge $q$ and capacitance $C$ according to Eq. 25-21 ($U = q^2/2C$).

*Calculation:* Substituting $C = 4\pi\varepsilon_0 R$ into Eq. 25-21 gives us

$$U = \frac{q^2}{2C} = \frac{q^2}{8\pi\varepsilon_0 R}$$

$$= \frac{(1.25 \times 10^{-9} \text{ C})^2}{(8\pi)(8.85 \times 10^{-12} \text{ F/m})(0.0685 \text{ m})}$$

$$= 1.03 \times 10^{-7} \text{ J} = 103 \text{ nJ.} \qquad \text{(Answer)}$$

(b) What is the energy density at the surface of the sphere?

**KEY IDEA** The density $u$ of the energy stored in an electric field depends on the magnitude $E$ of the field, according to Eq. 25-25 ($u = \frac{1}{2}\varepsilon_0 E^2$).

*Calculations:* Here we must first find $E$ at the surface of the sphere, as given by Eq. 23-15:

$$E = \frac{1}{4\pi\varepsilon_0} \frac{q}{R^2}.$$

The energy density is then

$$u = \frac{1}{2}\varepsilon_0 E^2 = \frac{q^2}{32\pi^2 \varepsilon_0 R^4}$$

$$= \frac{(1.25 \times 10^{-9} \text{ C})^2}{(32\pi^2)(8.85 \times 10^{-12} \text{ C}^2/\text{N}\cdot\text{m}^2)(0.0685 \text{ m})^4}$$

$$= 2.54 \times 10^{-5} \text{ J/m}^3 = 25.4 \text{ } \mu\text{J/m}^3. \qquad \text{(Answer)}$$

**FIG. 25-13** The simple electrostatic apparatus used by Faraday. An assembled apparatus (second from left) forms a spherical capacitor consisting of a central brass ball and a concentric brass shell. Faraday placed dielectric materials in the space between the ball and the shell. *(The Royal Institute, England/Bridgeman Art Library/NY)*

## 25-6 | Capacitor with a Dielectric

If you fill the space between the plates of a capacitor with a *dielectric,* which is an insulating material such as mineral oil or plastic, what happens to the capacitance? Michael Faraday—to whom the whole concept of capacitance is largely due and for whom the SI unit of capacitance is named—first looked into this matter in 1837. Using simple equipment much like that shown in Fig. 25-13, he found that the capacitance *increased* by a numerical factor $\kappa$, which he called the **dielectric constant** of the insulating material. Table 25-1 shows some dielectric materials and their dielectric constants. The dielectric constant of a vacuum is unity by definition. Because air is mostly empty space, its measured dielectric constant is only slightly greater than unity.

Another effect of the introduction of a dielectric is to limit the potential difference that can be applied between the plates to a certain value $V_{max}$, called the *breakdown potential.* If this value is substantially exceeded, the dielectric material will break down and form a conducting path between the plates. Every dielectric material has a characteristic *dielectric strength,* which is the maximum value of the electric field that it can tolerate without breakdown. A few such values are listed in Table 25-1.

As we discussed just after Eq. 25-18, the capacitance of any capacitor can be written in the form

$$C = \varepsilon_0 \mathscr{L}, \tag{25-26}$$

in which $\mathscr{L}$ has the dimension of length. For example, $\mathscr{L} = A/d$ for a parallel-plate capacitor. Faraday's discovery was that, with a dielectric *completely* filling the space between the plates, Eq. 25-26 becomes

$$C = \kappa \varepsilon_0 \mathscr{L} = \kappa C_{air}, \tag{25-27}$$

where $C_{air}$ is the value of the capacitance with only air between the plates.

Figure 25-14 provides some insight into Faraday's experiments. In Fig. 25-14a the battery ensures that the potential difference $V$ between the plates will remain constant. When a dielectric slab is inserted between the plates, the charge $q$ on the plates increases by a factor of $\kappa$; the additional charge is delivered to the capacitor plates by the battery. In Fig. 25-14b there is no battery, and therefore the charge $q$ must remain constant when the dielectric slab is inserted; then the potential difference $V$ between the plates decreases by a factor of $\kappa$. Both these observations are consistent (through the relation $q = CV$) with the increase in capacitance caused by the dielectric.

Comparison of Eqs. 25-26 and 25-27 suggests that the effect of a dielectric can be summed up in more general terms:

**TABLE 25-1**

**Some Properties of Dielectrics**[a]

| Material | Dielectric Constant $\kappa$ | Dielectric Strength (kV/mm) |
|---|---|---|
| Air (1 atm) | 1.00054 | 3 |
| Polystyrene | 2.6 | 24 |
| Paper | 3.5 | 16 |
| Transformer oil | 4.5 | |
| Pyrex | 4.7 | 14 |
| Ruby mica | 5.4 | |
| Porcelain | 6.5 | |
| Silicon | 12 | |
| Germanium | 16 | |
| Ethanol | 25 | |
| Water (20°C) | 80.4 | |
| Water (25°C) | 78.5 | |
| Titania ceramic | 130 | |
| Strontium titanate | 310 | 8 |

For a vacuum, $\kappa$ = unity.

[a]Measured at room temperature, except for the water.

In a region completely filled by a dielectric material of dielectric constant $\kappa$, all electrostatic equations containing the permittivity constant $\varepsilon_0$ are to be modified by replacing $\varepsilon_0$ with $\kappa\varepsilon_0$.

Thus, the magnitude of the electric field produced by a point charge inside a dielectric is given by this modified form of Eq. 23-15:

$$E = \frac{1}{4\pi\kappa\varepsilon_0} \frac{q}{r^2}. \tag{25-28}$$

Also, the expression for the electric field just outside an isolated conductor immersed in a dielectric (see Eq. 23-11) becomes

$$E = \frac{\sigma}{\kappa\varepsilon_0}. \tag{25-29}$$

Because $\kappa$ is always greater than unity, both these equations show that *for a fixed distribution of charges, the effect of a dielectric is to weaken the electric field* that would otherwise be present.

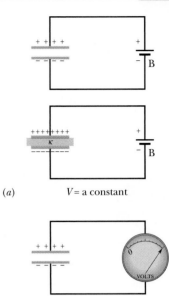

(a)     $V$ = a constant

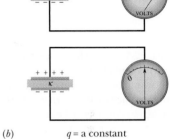

(b)     $q$ = a constant

**FIG. 25-14** (*a*) If the potential difference between the plates of a capacitor is maintained, as by battery B, the effect of a dielectric is to increase the charge on the plates. (*b*) If the charge on the capacitor plates is maintained, as in this case, the effect of a dielectric is to reduce the potential difference between the plates. The scale shown is that of a *potentiometer*, a device used to measure potential difference (here, between the plates). A capacitor cannot discharge through a potentiometer.

---

**Sample Problem | 25-6**

A parallel-plate capacitor whose capacitance $C$ is 13.5 pF is charged by a battery to a potential difference $V = 12.5$ V between its plates. The charging battery is now disconnected, and a porcelain slab ($\kappa = 6.50$) is slipped between the plates.

**(a)** What is the potential energy of the capacitor before the slab is inserted?

**KEY IDEA** We can relate the potential energy $U_i$ of the capacitor to the capacitance $C$ and either the potential $V$ (with Eq. 25-22) or the charge $q$ (with Eq. 25-21):

$$U_i = \tfrac{1}{2}CV^2 = \frac{q^2}{2C}.$$

**Calculation:** Because we are given the initial potential $V$ (= 12.5 V), we use Eq. 25-22 to find the initial stored energy:

$$U_i = \tfrac{1}{2}CV^2 = \tfrac{1}{2}(13.5 \times 10^{-12} \text{ F})(12.5 \text{ V})^2$$
$$= 1.055 \times 10^{-9} \text{ J} = 1055 \text{ pJ} \approx 1100 \text{ pJ}. \quad \text{(Answer)}$$

**(b)** What is the potential energy of the capacitor–slab device after the slab is inserted?

**KEY IDEA** Because the battery has been discon-

nected, the charge on the capacitor cannot change when the dielectric is inserted. However, the potential *does* change.

**Calculations:** Thus, we must now use Eq. 25-21 to write the final potential energy $U_f$, but now that the slab is within the capacitor, the capacitance is $\kappa C$. We then have

$$U_f = \frac{q^2}{2\kappa C} = \frac{U_i}{\kappa} = \frac{1055 \text{ pJ}}{6.50}$$
$$= 162 \text{ pJ} \approx 160 \text{ pJ}. \quad \text{(Answer)}$$

When the slab is introduced, the potential energy decreases by a factor of $\kappa$.

The "missing" energy, in principle, would be apparent to the person who introduced the slab. The capacitor would exert a tiny tug on the slab and would do work on it, in amount

$$W = U_i - U_f = (1055 - 162) \text{ pJ} = 893 \text{ pJ}.$$

If the slab were allowed to slide between the plates with no restraint and if there were no friction, the slab would oscillate back and forth between the plates with a (constant) mechanical energy of 893 pJ, and this system energy would transfer back and forth between kinetic energy of the moving slab and potential energy stored in the electric field.

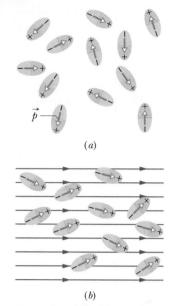

FIG. 25-15 (a) Molecules with a permanent electric dipole moment, showing their random orientation in the absence of an external electric field. (b) An electric field is applied, producing partial alignment of the dipoles. Thermal agitation prevents complete alignment.

FIG. 25-16 (a) A nonpolar dielectric slab. The circles represent the electrically neutral atoms within the slab. (b) An electric field is applied via charged capacitor plates; the field slightly stretches the atoms, separating the centers of positive and negative charge. (c) The separation produces surface charges on the slab faces. These charges set up a field $\vec{E}'$, which opposes the applied field $\vec{E}_0$. The resultant field $\vec{E}$ inside the dielectric (the vector sum of $\vec{E}_0$ and $\vec{E}'$) has the same direction as $\vec{E}_0$ but a smaller magnitude.

## 25-7 | Dielectrics: An Atomic View

What happens, in atomic and molecular terms, when we put a dielectric in an electric field? There are two possibilities, depending on the type of molecule:

1. *Polar dielectrics.* The molecules of some dielectrics, like water, have permanent electric dipole moments. In such materials (called *polar dielectrics*), the electric dipoles tend to line up with an external electric field as in Fig. 25-15. Because the molecules are continuously jostling each other as a result of their random thermal motion, this alignment is not complete, but it becomes more complete as the magnitude of the applied field is increased (or as the temperature, and thus the jostling, are decreased). The alignment of the electric dipoles produces an electric field that is directed opposite the applied field and is smaller in magnitude.

2. *Nonpolar dielectrics.* Regardless of whether they have permanent electric dipole moments, molecules acquire dipole moments by induction when placed in an external electric field. In Section 24-8 (see Fig. 24-11), we saw that this occurs because the external field tends to "stretch" the molecules, slightly separating the centers of negative and positive charge.

Figure 25-16a shows a nonpolar dielectric slab with no external electric field applied. In Fig. 25-16b, an electric field $\vec{E}_0$ is applied via a capacitor, whose plates are charged as shown. The result is a slight separation of the centers of the positive and negative charge distributions within the slab, producing positive charge on one face of the slab (due to the positive ends of dipoles there) and negative charge on the opposite face (due to the negative ends of dipoles there). The slab as a whole remains electrically neutral and—within the slab—there is no excess charge in any volume element.

Figure 25-16c shows that the induced surface charges on the faces produce an electric field $\vec{E}'$ in the direction opposite that of the applied electric field $\vec{E}_0$. The resultant field $\vec{E}$ inside the dielectric (the vector sum of fields $\vec{E}_0$ and $\vec{E}'$) has the direction of $\vec{E}_0$ but is smaller in magnitude.

Both the field $\vec{E}'$ produced by the surface charges in Fig. 25-16c and the electric field produced by the permanent electric dipoles in Fig. 25-15 act in the same way—they oppose the applied field $\vec{E}$. Thus, the effect of both polar and nonpolar dielectrics is to weaken any applied field within them, as between the plates of a capacitor.

We can now see why the dielectric porcelain slab in Sample Problem 25-6 is pulled into the capacitor: As it enters the space between the plates, the surface charge that appears on each slab face has the sign opposite that of the charge on the nearby capacitor plate. Thus, slab and plates attract each other.

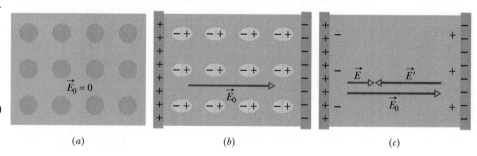

## 25-8 | Dielectrics and Gauss' Law

In our discussion of Gauss' law in Chapter 23, we assumed that the charges existed in a vacuum. Here we shall see how to modify and generalize that law if dielectric materials, such as those listed in Table 25-1, are present. Figure 25-17 shows a parallel-plate capacitor of plate area $A$, both with and without a dielectric. We assume that the charge $q$ on the plates is the same in both situations. Note that the field between the plates induces charges on the faces of the dielectric by one of the methods described in Section 25-7.

For the situation of Fig. 25-17a, without a dielectric, we can find the electric field $\vec{E}_0$ between the plates as we did in Fig. 25-5: We enclose the charge $+q$ on the top plate with a Gaussian surface and then apply Gauss' law. Letting $E_0$ represent the magnitude of the field, we find

$$\varepsilon_0 \oint \vec{E} \cdot d\vec{A} = \varepsilon_0 E_0 A = q, \tag{25-30}$$

or

$$E_0 = \frac{q}{\varepsilon_0 A}. \tag{25-31}$$

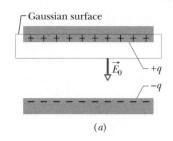

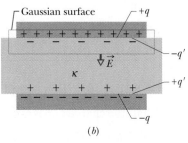

FIG. 25-17 A parallel-plate capacitor (a) without and (b) with a dielectric slab inserted. The charge $q$ on the plates is assumed to be the same in both cases.

In Fig. 25-17b, with the dielectric in place, we can find the electric field between the plates (and within the dielectric) by using the same Gaussian surface. However, now the surface encloses two types of charge: It still encloses charge $+q$ on the top plate, but it now also encloses the induced charge $-q'$ on the top face of the dielectric. The charge on the conducting plate is said to be *free charge* because it can move if we change the electric potential of the plate; the induced charge on the surface of the dielectric is not free charge because it cannot move from that surface.

The net charge enclosed by the Gaussian surface in Fig. 25-17b is $q - q'$, so Gauss' law now gives

$$\varepsilon_0 \oint \vec{E} \cdot d\vec{A} = \varepsilon_0 EA = q - q', \tag{25-32}$$

or

$$E = \frac{q - q'}{\varepsilon_0 A}. \tag{25-33}$$

The effect of the dielectric is to weaken the original field $E_0$ by a factor of $\kappa$; so we may write

$$E = \frac{E_0}{\kappa} = \frac{q}{\kappa \varepsilon_0 A}. \tag{25-34}$$

Comparison of Eqs. 25-33 and 25-34 shows that

$$q - q' = \frac{q}{\kappa}. \tag{25-35}$$

Equation 25-35 shows correctly that the magnitude $q'$ of the induced surface charge is less than that of the free charge $q$ and is zero if no dielectric is present (because then $\kappa = 1$ in Eq. 25-35).

By substituting for $q - q'$ from Eq. 25-35 in Eq. 25-32, we can write Gauss' law in the form

$$\varepsilon_0 \oint \kappa \vec{E} \cdot d\vec{A} = q \quad \text{(Gauss' law with dielectric)}. \tag{25-36}$$

This equation, although derived for a parallel-plate capacitor, is true generally and is the most general form in which Gauss' law can be written. Note:

1. The flux integral now involves $\kappa \vec{E}$, not just $\vec{E}$. (The vector $\varepsilon_0 \kappa \vec{E}$ is sometimes called the *electric displacement* $\vec{D}$, so that Eq. 25-36 can be written in the form $\oint \vec{D} \cdot d\vec{A} = q$.)

2. The charge $q$ enclosed by the Gaussian surface is now taken to be the *free charge only*. The induced surface charge is deliberately ignored on the right side of Eq. 25-36, having been taken fully into account by introducing the dielectric constant $\kappa$ on the left side.

3. Equation 25-36 differs from Eq. 23-7, our original statement of Gauss' law, only in that $\varepsilon_0$ in the latter equation has been replaced by $\kappa\varepsilon_0$. We keep $\kappa$ inside the integral of Eq. 25-36 to allow for cases in which $\kappa$ is not constant over the entire Gaussian surface.

**Sample Problem** 25-7

Figure 25-18 shows a parallel-plate capacitor of plate area $A$ and plate separation $d$. A potential difference $V_0$ is applied between the plates. The battery is then disconnected, and a dielectric slab of thickness $b$ and dielectric constant $\kappa$ is placed between the plates as shown. Assume $A = 115$ cm², $d = 1.24$ cm, $V_0 = 85.5$ V, $b = 0.780$ cm, and $\kappa = 2.61$.

(a) What is the capacitance $C_0$ before the dielectric slab is inserted?

**Calculation:** From Eq. 25-9 we have

$$C_0 = \frac{\varepsilon_0 A}{d} = \frac{(8.85 \times 10^{-12}\ \text{F/m})(115 \times 10^{-4}\ \text{m}^2)}{1.24 \times 10^{-2}\ \text{m}}$$

$$= 8.21 \times 10^{-12}\ \text{F} = 8.21\ \text{pF.} \qquad \text{(Answer)}$$

(b) What free charge appears on the plates?

**Calculation:** From Eq. 25-1,

$$q = C_0 V_0 = (8.21 \times 10^{-12}\ \text{F})(85.5\ \text{V})$$

$$= 7.02 \times 10^{-10}\ \text{C} = 702\ \text{pC.} \qquad \text{(Answer)}$$

Because the battery was disconnected before the slab was inserted, the free charge is unchanged.

(c) What is the electric field $E_0$ in the gaps between the plates and the dielectric slab?

**KEY IDEA** We need to apply Gauss' law, in the form of Eq. 25-36, to Gaussian surface I in Fig. 25-18.

**Calculations:** That surface passes through the gap, and so it encloses *only* the free charge on the upper capacitor plate. Because the area vector $d\vec{A}$ and the field vector $\vec{E}_0$ are both directed downward, the dot product in Eq. 25-36 becomes

$$\vec{E}_0 \cdot d\vec{A} = E_0\ dA \cos 0° = E_0\ dA.$$

Equation 25-36 then becomes

$$\varepsilon_0 \kappa E_0 \oint dA = q.$$

The integration now simply gives the surface area $A$ of the plate. Thus, we obtain

$$\varepsilon_0 \kappa E_0 A = q,$$

or

$$E_0 = \frac{q}{\varepsilon_0 \kappa A}.$$

We must put $\kappa = 1$ here because Gaussian surface I does not pass through the dielectric. Thus, we have

$$E_0 = \frac{q}{\varepsilon_0 \kappa A} = \frac{7.02 \times 10^{-10}\ \text{C}}{(8.85 \times 10^{-12}\ \text{F/m})(1)(115 \times 10^{-4}\ \text{m}^2)}$$

$$= 6900\ \text{V/m} = 6.90\ \text{kV/m.} \qquad \text{(Answer)}$$

Note that the value of $E_0$ does not change when the slab is introduced because the amount of charge enclosed by Gaussian surface I in Fig. 25-18 does not change.

(d) What is the electric field $E_1$ in the dielectric slab?

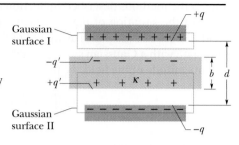

**FIG. 25-18** A parallel-plate capacitor containing a dielectric slab that only partially fills the space between the plates.

**KEY IDEA** Now we apply Gauss' law in the form of Eq. 25-36 to Gaussian surface II in Fig. 25-18.

**Calculations:** That surface encloses free charge $-q$ and induced charge $+q'$, but we ignore the latter when we use Eq. 25-36. We find

$$\varepsilon_0 \oint \kappa \vec{E}_1 \cdot d\vec{A} = -\varepsilon_0 \kappa E_1 A = -q. \qquad (25\text{-}37)$$

(The first minus sign in this equation comes from the dot product $\vec{E}_1 \cdot d\vec{A}$, because now the field vector $\vec{E}_1$ is directed downward and the area vector $d\vec{A}$ is directed upward.) Now $\kappa = 2.61$. Thus, Eq. 25-37 gives us

$$E_1 = \frac{q}{\varepsilon_0 \kappa A} = \frac{E_0}{\kappa} = \frac{6.90\ \text{kV/m}}{2.61}$$

$$= 2.64\ \text{kV/m.} \qquad \text{(Answer)}$$

(e) What is the potential difference $V$ between the plates after the slab has been introduced?

**KEY IDEA** We find $V$ by integrating along a straight line directly from the bottom plate to the top plate.

**Calculation:** Within the dielectric, the path length is $b$ and the electric field is $E_1$. Within the two gaps above and below the dielectric, the total path length is $d - b$ and the electric field is $E_0$. Equation 25-6 then yields

$$V = \int_-^+ E\ ds = E_0(d - b) + E_1 b$$

$$= (6900\ \text{V/m})(0.0124\ \text{m} - 0.00780\ \text{m})$$

$$+ (2640\ \text{V/m})(0.00780\ \text{m})$$

$$= 52.3\ \text{V.} \qquad \text{(Answer)}$$

This is less than the original potential difference of 85.5 V.

(f) What is the capacitance with the slab in place between the plates of the capacitor?

**KEY IDEA** The capacitance $C$ is related to the free charge $q$ and the potential difference $V$ via Eq. 25-1.

**Calculation:** Taking $q$ from (b) and $V$ from (e), we have

$$C = \frac{q}{V} = \frac{7.02 \times 10^{-10}\ \text{C}}{52.3\ \text{V}}$$

$$= 1.34 \times 10^{-11}\ \text{F} = 13.4\ \text{pF.} \qquad \text{(Answer)}$$

This is greater than the original capacitance of 8.21 pF.

## REVIEW & SUMMARY

**Capacitor; Capacitance** A **capacitor** consists of two isolated conductors (the *plates*) with charges $+q$ and $-q$. Its **capacitance** $C$ is defined from

$$q = CV, \qquad (25\text{-}1)$$

where $V$ is the potential difference between the plates. The SI unit of capacitance is the farad (1 farad = 1 F = 1 coulomb per volt).

**Determining Capacitance** We generally determine the capacitance of a particular capacitor configuration by (1) assuming a charge $q$ to have been placed on the plates, (2) finding the electric field $\vec{E}$ due to this charge, (3) evaluating the potential difference $V$, and (4) calculating $C$ from Eq. 25-1. Some specific results are the following:

A *parallel-plate capacitor* with flat parallel plates of area $A$ and spacing $d$ has capacitance

$$C = \frac{\varepsilon_0 A}{d}. \qquad (25\text{-}9)$$

A *cylindrical capacitor* (two long coaxial cylinders) of length $L$ and radii $a$ and $b$ has capacitance

$$C = 2\pi\varepsilon_0 \frac{L}{\ln(b/a)}. \qquad (25\text{-}14)$$

A *spherical capacitor* with concentric spherical plates of radii $a$ and $b$ has capacitance

$$C = 4\pi\varepsilon_0 \frac{ab}{b - a}. \qquad (25\text{-}17)$$

If we let $b \to \infty$ and $a = R$ in Eq. 25-17, we obtain the capacitance of an *isolated sphere* of radius $R$:

$$C = 4\pi\varepsilon_0 R. \qquad (25\text{-}18)$$

**Capacitors in Parallel and in Series** The **equivalent capacitances** $C_{\text{eq}}$ of combinations of individual capacitors connected in **parallel** and in **series** can be found from

$$C_{\text{eq}} = \sum_{j=1}^{n} C_j \qquad (n \text{ capacitors in parallel}) \qquad (25\text{-}19)$$

and

$$\frac{1}{C_{\text{eq}}} = \sum_{j=1}^{n} \frac{1}{C_j} \qquad (n \text{ capacitors in series}). \qquad (25\text{-}20)$$

Equivalent capacitances can be used to calculate the capacitances of more complicated series–parallel combinations.

**Potential Energy and Energy Density** The **electric potential energy** $U$ of a charged capacitor,

$$U = \frac{q^2}{2C} = \tfrac{1}{2}CV^2, \qquad (25\text{-}21, 25\text{-}22)$$

is equal to the work required to charge the capacitor. This energy can be associated with the capacitor's electric field $\vec{E}$. By extension we can associate stored energy with any electric field. In vacuum, the **energy density** $u$, or potential energy per unit volume, within an electric field of magnitude $E$ is given by

$$u = \tfrac{1}{2}\varepsilon_0 E^2. \qquad (25\text{-}25)$$

**Capacitance with a Dielectric** If the space between the plates of a capacitor is completely filled with a dielectric material, the capacitance $C$ is increased by a factor $\kappa$, called the **dielectric constant,** which is characteristic of the material. In a region that is completely filled by a dielectric, all electrostatic equations containing $\varepsilon_0$ must be modified by replacing $\varepsilon_0$ with $\kappa\varepsilon_0$.

The effects of adding a dielectric can be understood physically in terms of the action of an electric field on the permanent or induced electric dipoles in the dielectric slab. The result is the formation of induced charges on the surfaces of the dielectric, which results in a weakening of the field within the dielectric for a given amount of free charge on the plates.

**Gauss' Law with a Dielectric** When a dielectric is present, Gauss' law may be generalized to

$$\varepsilon_0 \oint \kappa \vec{E} \cdot d\vec{A} = q. \qquad (25\text{-}36)$$

Here $q$ is the free charge; any induced surface charge is accounted for by including the dielectric constant $\kappa$ inside the integral.

## QUESTIONS

**1** Figure 25-19 shows plots of charge versus potential difference for three parallel-plate capacitors that have the plate areas and separations given in the table. Which plot goes with which capacitor?

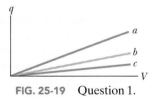

**FIG. 25-19** Question 1.

| Capacitor | Area | Separation |
|-----------|------|------------|
| 1 | $A$ | $d$ |
| 2 | $2A$ | $d$ |
| 3 | $A$ | $2d$ |

**2** Figure 25-20 shows an open switch, a battery of potential difference $V$, a current-measuring meter A, and three identical uncharged capacitors of capacitance $C$. When the switch is closed and the circuit reaches equilibrium, what are (a) the potential difference across each capacitor and (b) the charge on the left plate of each capacitor? (c) During charging, what net charge passes through the meter?

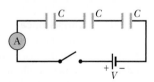

**FIG. 25-20** Question 2.

**3** For each circuit in Fig. 25-21, are the capacitors connected in series, in parallel, or in neither mode?

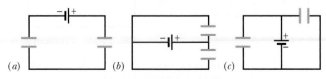

**FIG. 25-21** Question 3.

**4** What is $C_{eq}$ of three capacitors, each of capacitance $C$, if they are connected to a battery (a) in series with one another and (b) in parallel? (c) In which arrangement is there more charge on the equivalent capacitance?

**5** (a) In Fig. 25-22a, are capacitors 1 and 3 in series? (b) In the same figure, are capacitors 1 and 2 in parallel? (c) Rank the equivalent capacitances of the four circuits shown in Fig. 25-22, greatest first.

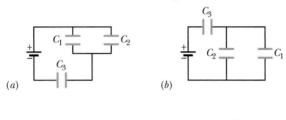
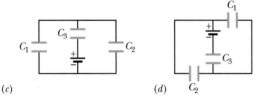
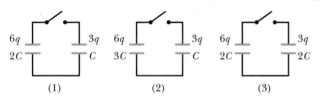

**FIG. 25-22** Question 5.

**6** Figure 25-23 shows three circuits, each consisting of a switch and two capacitors, initially charged as indicated (top plate positive). After the switches have been closed, in which circuit (if any)

will the charge on the left-hand capacitor (a) increase, (b) decrease, and (c) remain the same?

**7** Initially, a single capacitance $C_1$ is wired to a battery. Then capacitance $C_2$ is added in parallel. Are (a) the potential difference across $C_1$ and (b) the charge $q_1$ on $C_1$ now more than, less than, or the same as previously? (c) Is the equivalent capacitance $C_{12}$ of $C_1$ and $C_2$ more than, less than, or equal to $C_1$? (d) Is the charge stored on $C_1$ and $C_2$ together more than, less than, or equal to the charge stored previously on $C_1$?

**8** Repeat Question 7 for $C_2$ added in series rather than in parallel.

**9** You are to connect capacitances $C_1$ and $C_2$, with $C_1 > C_2$, to a battery, first individually, then in series, and then in parallel. Rank those arrangements according to the amount of charge stored, greatest first.

**10** When a dielectric slab is inserted between the plates of one of the two identical capacitors in Fig. 25-24, do the following properties of that capacitor increase, decrease, or remain the same: (a) capacitance, (b) charge, (c) potential difference, and (d) potential energy? (e) How about the same properties of the other capacitor?

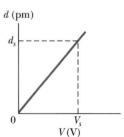

**FIG. 25-24** Question 10.

**11** A parallel-plate capacitor is connected to a battery of electric potential difference $V$. If the plate separation is decreased, do the following quantities increase, decrease, or remain the same: (a) the capacitor's capacitance, (b) the potential difference across the capacitor, (c) the charge on the capacitor, (d) the energy stored by the capacitor, (e) the magnitude of the electric field between the plates, and (f) the energy density of that electric field?

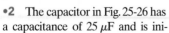

**FIG. 25-23** Question 6.

# PROBLEMS

 **GO**    Tutoring problem available (at instructor's discretion) in *WileyPLUS* and WebAssign
**SSM**    Worked-out solution available in Student Solutions Manual     **WWW** Worked-out solution is at
• – •••    Number of dots indicates level of problem difficulty     **ILW**   Interactive solution is at   http://www.wiley.com/college/halliday
   Additional information available in *The Flying Circus of Physics* and at flyingcircusofphysics.com

### sec. 25-2 Capacitance

**•1** The two metal objects in Fig. 25-25 have net charges of $+70$ pC and $-70$ pC, which result in a 20 V potential difference between them. (a) What is the capacitance of the system? (b) If the charges are changed to $+200$ pC and $-200$ pC, what does the capacitance become? (c) What does the potential difference become?

**FIG. 25-25** Problem 1.

**•2** The capacitor in Fig. 25-26 has a capacitance of $25\ \mu F$ and is initially uncharged. The battery provides a potential difference of 120 V. After switch S is closed, how much charge will pass through it?

**FIG. 25-26** Problem 2.

### sec. 25-3 Calculating the Capacitance

**•3** If an uncharged parallel-plate capacitor (capacitance $C$) is connected to a battery, one plate becomes negatively charged as electrons move to the plate face (area $A$). In Fig. 25-27, the depth $d$ from which the electrons come in the plate in a particular capacitor is plotted against a range of values for the potential difference $V$ of the battery. The vertical scale is set by $d_s = 1.00$ pm, and the horizontal scale is set by $V_s = 20.0$ V. What is the ratio $C/A$?

**FIG. 25-27** Problem 3.

**•4** You have two flat metal plates, each of area 1.00 m², with which to construct a parallel-plate capacitor. (a) If the capacitance of the device is to be 1.00 F, what must be the separation between the plates? (b) Could this capacitor actually be constructed?

•5   A parallel-plate capacitor has circular plates of 8.20 cm radius and 1.30 mm separation. (a) Calculate the capacitance. (b) What charge will appear on the plates if a potential difference of 120 V is applied?   SSM

•6   The plates of a spherical capacitor have radii 38.0 mm and 40.0 mm. (a) Calculate the capacitance. (b) What must be the plate area of a parallel-plate capacitor with the same plate separation and capacitance?

•7   What is the capacitance of a drop that results when two mercury spheres, each of radius $R = 2.00$ mm, merge?

### sec. 25-4   Capacitors in Parallel and in Series
•8   In Fig. 25-28, find the equivalent capacitance of the combination. Assume that $C_1$ is 10.0 $\mu$F, $C_2$ is 5.00 $\mu$F, and $C_3$ is 4.00 $\mu$F.

•9   In Fig. 25-29, find the equivalent capacitance of the combination. Assume that $C_1$ = 10.0 $\mu$F, $C_2$ = 5.00 $\mu$F, and $C_3$ = 4.00 $\mu$F.   ILW

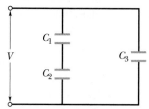

FIG. 25-28   Problems 8 and 36.

•10   How many 1.00 $\mu$F capacitors must be connected in parallel to store a charge of 1.00 C with a potential of 110 V across the capacitors?

•11   Each of the uncharged capacitors in Fig. 25-30 has a capacitance of 25.0 $\mu$F. A potential difference of $V = 4200$ V is established when the switch is closed. How many coulombs of charge then pass through meter A?

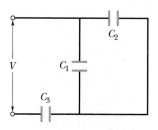

FIG. 25-29   Problems 9, 13, and 34.

••12   In Fig. 25-31, the battery has a potential difference of $V = 10.0$ V and the five capacitors each have a capacitance of 10.0 $\mu$F. What is the charge on (a) capacitor 1 and (b) capacitor 2?

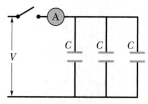

FIG. 25-30   Problem 11.

••13   In Fig. 25-29, a potential difference of $V = 100.0$ V is applied across a capacitor arrangement with capacitances $C_1$ = 10.0 $\mu$F, $C_2$ = 5.00 $\mu$F, and $C_3$ = 4.00 $\mu$F. If capacitor 3 undergoes electrical breakdown so that it becomes equivalent to conducting wire, what is the increase in (a) the charge on capacitor 1 and (b) the potential difference across capacitor 1?   GO

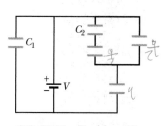

FIG. 25-31   Problem 12.

••14   Two parallel-plate capacitors, 6.0 $\mu$F each, are connected in parallel to a 10 V battery. One of the capacitors is then squeezed so that its plate separation is 50.0% of its initial value. Because of the squeezing, (a) how much additional charge is transferred to the capacitors by the battery and (b) what is the increase in the total charge stored on the capacitors?

••15   A 100 pF capacitor is charged to a potential difference of 50 V, and the charging battery is disconnected. The capacitor is then connected in parallel with a second (initially uncharged) capacitor. If the potential difference across the first capacitor drops to 35 V, what is the capacitance of this second capacitor?   SSM ILW

••16   Figure 25-32 shows a circuit section of four air-filled capacitors that is connected to a larger circuit. The graph below the section shows the electric potential $V(x)$ as a function of position $x$ along the lower part of the section, through capacitor 4. Similarly, the graph above the section shows the electric potential $V(x)$ as a function of position $x$ along the upper part of the section, through capacitors 1, 2, and 3. Capacitor 3 has a capacitance of 0.80 $\mu$F. What are the capacitances of (a) capacitor 1 and (b) capacitor 2?

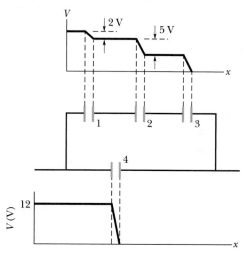

FIG. 25-32   Problem 16.

••17   In Fig. 25-33, a 20.0 V battery is connected across capacitors of capacitances $C_1 = C_6 = 3.00$ $\mu$F and $C_3 = C_5 = 2.00C_2 = 2.00C_4 = 4.00$ $\mu$F. What are (a) the equivalent capacitance $C_{eq}$ of the capacitors and (b) the charge stored by $C_{eq}$? What are (c) $V_1$ and (d) $q_1$ of capacitor 1, (e) $V_2$ and (f) $q_2$ of capacitor 2, and (g) $V_3$ and (h) $q_3$ of capacitor 3?   GO

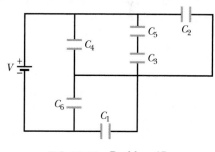

FIG. 25-33   Problem 17.

••18   Plot 1 in Fig. 25-34a gives the charge $q$ that can be

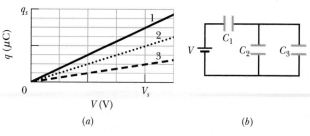

FIG. 25-34   Problem 18.

stored on capacitor 1 versus the electric potential $V$ set up across it. The vertical scale is set by $q_s = 16.0\ \mu C$, and the horizontal scale is set by $V_s = 2.0$ V. Plots 2 and 3 are similar plots for capacitors 2 and 3, respectively. Figure 25-34b shows a circuit with those three capacitors and a 6.0 V battery. What is the charge stored on capacitor 2 in that circuit?

••**19** In Fig. 25-35, the capacitances are $C_1 = 1.0\ \mu F$ and $C_2 = 3.0$ $\mu F$, and both capacitors are charged to a potential difference of $V = 100$ V but with opposite polarity as shown. Switches $S_1$ and $S_2$ are now closed. (a) What is now the potential difference between points $a$ and $b$? What now is the charge on capacitor (b) 1 and (c) 2?

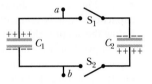

FIG. 25-35    Problem 19.

SSM WWW

••**20** In Fig. 25-36, $V = 10$ V, $C_1 = 10\ \mu F$, and $C_2 = C_3 = 20$ $\mu F$. Switch S is first thrown to the left side until capacitor 1 reaches equilibrium. Then the switch is thrown to the right. When equilibrium is again reached, how much charge is on capacitor 1?

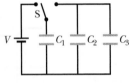

FIG. 25-36    Problem 20.

••**21** In Fig. 25-37, two parallel-plate capacitors (with air between the plates) are connected to a battery. Capacitor 1 has a plate area of 1.5 cm² and an electric field (between its plates) of magnitude 2000 V/m. Capacitor 2 has a plate area of 0.70 cm² and an electric field of magnitude 1500 V/m. What is the total charge on the two capacitors?

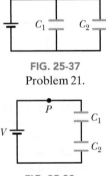

FIG. 25-37
Problem 21.

••**22** Figure 25-38 represents two air-filled cylindrical capacitors connected in series across a battery with potential $V = 10$ V. Capacitor 1 has an inner plate radius of 5.0 mm, an outer plate radius of 1.5 cm, and a length of 5.0 cm. Capacitor 2 has an inner plate radius of 2.5 mm, an outer plate radius of 1.0 cm, and a length of 9.0 cm. The outer plate of capacitor 2 is a conducting organic membrane that can be stretched, and the capacitor can be inflated to increase the plate separation. If the outer plate radius is increased to 2.5 cm by inflation, (a) how many electrons move through point $P$ and (b) do they move toward or away from the battery?

••**23** In Fig. 25-39, the battery has potential difference $V = 9.0$ V, $C_2 = 3.0\ \mu F$, $C_4 = 4.0\ \mu F$, and all the capacitors are initially uncharged. When switch S is closed, a total charge of 12 $\mu C$ passes through point $a$ and a total charge of 8.0 $\mu C$ passes through point $b$. What are (a) $C_1$ and (b) $C_3$?

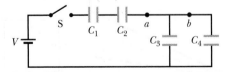

FIG. 25-39    Problem 23.

••**24** Figure 25-40 shows a variable "air gap" capacitor for manual tuning. Alternate plates are connected together; one group of plates is fixed in position, and the other group is ca-

pable of rotation. Consider a capacitor of $n = 8$ plates of alternating polarity, each plate having area $A = 1.25$ cm² and separated from adjacent plates by distance $d = 3.40$ mm. What is the maximum capacitance of the device?

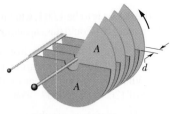

FIG. 25-40    Problem 24.

••**25** The capacitors in Fig. 25-41 are initially uncharged. The capacitances are $C_1 = 4.0\ \mu F$, $C_2 = 8.0\ \mu F$, and $C_3 = 12\ \mu F$, and the battery's potential difference is $V = 12$ V. When switch S is closed, how many electrons travel through (a) point $a$, (b) point $b$, (c) point $c$, and (d) point $d$? In the figure, do the electrons travel up or down through (e) point $b$ and (f) point $c$?

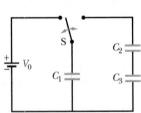

FIG. 25-41    Problem 25.

•••**26** Figure 25-42 displays a 12.0 V battery and 3 uncharged capacitors of capacitances $C_1 = 4.00\ \mu F$, $C_2 = 6.00\ \mu F$, and $C_3 = 3.00\ \mu F$. The switch is thrown to the left side until capacitor 1 is fully charged. Then the switch is thrown to the right. What is the final charge on (a) capacitor 1, (b) capacitor 2, and (c) capacitor 3?

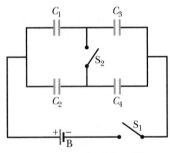

FIG. 25-42    Problem 26.

•••**27** Figure 25-43 shows a 12.0 V battery and four uncharged capacitors of capacitances $C_1 = 1.00\ \mu F$, $C_2 = 2.00\ \mu F$, $C_3 = 3.00\ \mu F$, and $C_4 = 4.00\ \mu F$. If only switch $S_1$ is closed, what is the charge on (a) capacitor 1, (b) capacitor 2, (c) capacitor 3, and (d) capacitor 4? If both switches are closed, what is the charge on (e) capacitor 1, (f) capacitor 2, (g) capacitor 3, and (h) capacitor 4? **GO**

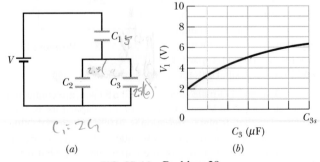

FIG. 25-43    Problem 27.

•••**28** Capacitor 3 in Fig. 25-44a is a *variable capacitor* (its capacitance $C_3$ can be varied). Figure 25-44b gives the electric potential $V_1$ across capacitor 1 versus $C_3$. The horizontal scale is set by $C_{3s} = 12.0\ \mu F$. Electric potential $V_1$ approaches an asymptote of 10 V as $C_3 \to \infty$. What are (a) the electric potential $V$ across the battery, (b) $C_1$, and (c) $C_2$?

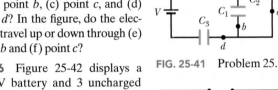

(a)

(b)

FIG. 25-44    Problem 28.

## sec. 25-5 Energy Stored in an Electric Field

•29 A 2.0 $\mu$F capacitor and a 4.0 $\mu$F capacitor are connected in parallel across a 300 V potential difference. Calculate the total energy stored in the capacitors. **SSM**

•30 A parallel-plate air-filled capacitor having area 40 cm$^2$ and plate spacing 1.0 mm is charged to a potential difference of 600 V. Find (a) the capacitance, (b) the magnitude of the charge on each plate, (c) the stored energy, (d) the electric field between the plates, and (e) the energy density between the plates.

•31 What capacitance is required to store an energy of 10 kW · h at a potential difference of 1000 V?

•32 How much energy is stored in 1.00 m$^3$ of air due to the "fair weather" electric field of magnitude 150 V/m?

••33 Assume that a stationary electron is a point of charge. What is the energy density $u$ of its electric field at radial distances (a) $r = 1.00$ mm, (b) $r = 1.00$ $\mu$m, (c) $r = 1.00$ nm, and (d) $r = 1.00$ pm? (e) What is $u$ in the limit as $r \rightarrow 0$?

••34 In Fig. 25-29, a potential difference $V = 100$ V is applied across a capacitor arrangement with capacitances $C_1 = 10.0$ $\mu$F, $C_2 = 5.00$ $\mu$F, and $C_3 = 15.0$ $\mu$F. What are (a) charge $q_3$, (b) potential difference $V_3$, and (c) stored energy $U_3$ for capacitor 3, (d) $q_1$, (e) $V_1$, and (f) $U_1$ for capacitor 1, and (g) $q_2$, (h) $V_2$, and (i) $U_2$ for capacitor 2?

••35 The parallel plates in a capacitor, with a plate area of 8.50 cm$^2$ and an air-filled separation of 3.00 mm, are charged by a 6.00 V battery. They are then disconnected from the battery and pulled apart (without discharge) to a separation of 8.00 mm. Neglecting fringing, find (a) the potential difference between the plates, (b) the initial stored energy, (c) the final stored energy, and (d) the work required to separate the plates. **SSM ILW WWW**

••36 In Fig. 25-28, a potential difference $V = 100$ V is applied across a capacitor arrangement with capacitances $C_1 = 10.0$ $\mu$F, $C_2 = 5.00$ $\mu$F, and $C_3 = 4.00$ $\mu$F. What are (a) charge $q_3$, (b) potential difference $V_3$, and (c) stored energy $U_3$ for capacitor 3, (d) $q_1$, (e) $V_1$, and (f) $U_1$ for capacitor 1, and (g) $q_2$, (h) $V_2$, and (i) $U_2$ for capacitor 2?

••37 In Fig. 25-45, $C_1 = 10.0$ $\mu$F, $C_2 = 20.0$ $\mu$F, and $C_3 = 25.0$ $\mu$F. If no capacitor can withstand a potential difference of more than 100 V without failure, what are (a) the magnitude of the maximum potential difference that can exist between points $A$ and $B$ and (b) the maximum energy that can be stored in the three-capacitor arrangement? **GO**

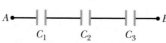

FIG. 25-45 Problem 37.

••38 As a safety engineer, you must evaluate the practice of storing flammable conducting liquids in nonconducting containers. The company supplying a certain liquid has been using a squat, cylindrical plastic container of radius $r = 0.20$ m and filling it to height $h = 10$ cm, which is not the container's full interior height (Fig. 25-46). Your investigation reveals that during handling at the company, the exterior surface of the container commonly acquires

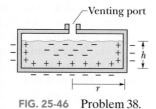

FIG. 25-46 Problem 38.

a negative charge density of magnitude 2.0 $\mu$C/m$^2$ (approximately uniform). Because the liquid is a conducting material, the charge on the container induces charge separation within the liquid. (a) How much negative charge is induced in the center of the liquid's bulk? (b) Assume the capacitance of the central portion of the liquid relative to ground is 35 pF. What is the potential energy associated with the negative charge in that effective capacitor? (c) If a spark occurs between the ground and the central portion of the liquid (through the venting port), the potential energy can be fed into the spark. The minimum spark energy needed to ignite the liquid is 10 mJ. In this situation, can a spark ignite the liquid?

••39 A charged isolated metal sphere of diameter 10 cm has a potential of 8000 V relative to $V = 0$ at infinity. Calculate the energy density in the electric field near the surface of the sphere.

## sec. 25-6 Capacitor with a Dielectric

•40 A parallel-plate air-filled capacitor has a capacitance of 50 pF. (a) If each of its plates has an area of 0.35 m$^2$, what is the separation? (b) If the region between the plates is now filled with material having $\kappa = 5.6$, what is the capacitance?

•41 Given a 7.4 pF air-filled capacitor, you are asked to convert it to a capacitor that can store up to 7.4 $\mu$J with a maximum potential difference of 652 V. Which dielectric in Table 25-1 should you use to fill the gap in the capacitor if you do not allow for a margin of error?

•42 An air-filled parallel-plate capacitor has a capacitance of 1.3 pF. The separation of the plates is doubled, and wax is inserted between them. The new capacitance is 2.6 pF. Find the dielectric constant of the wax.

•43 A coaxial cable used in a transmission line has an inner radius of 0.10 mm and an outer radius of 0.60 mm. Calculate the capacitance per meter for the cable. Assume that the space between the conductors is filled with polystyrene. **SSM**

••44 In Fig. 25-47, how much charge is stored on the parallel-plate capacitors by the 12.0 V battery? One is filled with air, and the other is filled with a dielectric for which $\kappa = 3.00$; both capacitors have a plate area of $5.00 \times 10^{-3}$ m$^2$ and a plate separation of 2.00 mm.

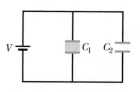

FIG. 25-47 Problem 44.

••45 A certain substance has a dielectric constant of 2.8 and a dielectric strength of 18 MV/m. If it is used as the dielectric material in a parallel-plate capacitor, what minimum area should the plates of the capacitor have to obtain a capacitance of $7.0 \times 10^{-2}$ $\mu$F and to ensure that the capacitor will be able to withstand a potential difference of 4.0 kV? **SSM ILW**

••46 You are asked to construct a capacitor having a capacitance near 1 nF and a breakdown potential in excess of 10 000 V. You think of using the sides of a tall Pyrex drinking glass as a dielectric, lining the inside and outside curved surfaces with aluminum foil to act as the plates. The glass is 15 cm tall with an inner radius of 3.6 cm and an outer radius of 3.8 cm. What are the (a) capacitance and (b) breakdown potential of this capacitor?

••47 A certain parallel-plate capacitor is filled with a dielectric for which $\kappa = 5.5$. The area of each plate is 0.034 m$^2$, and the plates are separated by 2.0 mm. The capacitor will fail (short out and burn up) if the electric field between the plates exceeds 200 kN/C. What is the maximum energy that can be stored in the capacitor?

••**48** Figure 25-48 shows a parallel-plate capacitor with a plate area $A = 5.56$ cm$^2$ and separation $d = 5.56$ mm. The left half of the gap is filled with material of dielectric constant $\kappa_1 = 7.00$; the right half is filled with material of dielectric constant $\kappa_2 = 12.0$. What is the capacitance?

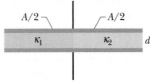

FIG. 25-48   Problem 48.

••**49** Figure 25-49 shows a parallel-plate capacitor with a plate area $A = 7.89$ cm$^2$ and plate separation $d = 4.62$ mm. The top half of the gap is filled with material of dielectric constant $\kappa_1 = 11.0$; the bottom half is filled with material of dielectric constant $\kappa_2 = 12.0$. What is the capacitance?

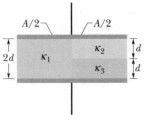

FIG. 25-49
Problem 49.

••**50** Figure 25-50 shows a parallel-plate capacitor of plate area $A = 10.5$ cm$^2$ and plate separation $2d = 7.12$ mm. The left half of the gap is filled with material of dielectric constant $\kappa_1 = 21.0$; the top of the right half is filled with material of dielectric constant $\kappa_2 = 42.0$; the bottom of the right half is filled with material of dielectric constant $\kappa_3 = 58.0$. What is the capacitance? GO

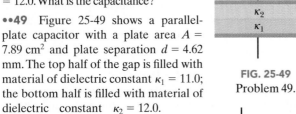

FIG. 25-50   Problem 50.

### sec. 25-8   Dielectrics and Gauss' Law

•**51** A parallel-plate capacitor has a capacitance of 100 pF, a plate area of 100 cm$^2$, and a mica dielectric ($\kappa = 5.4$) completely filling the space between the plates. At 50 V potential difference, calculate (a) the electric field magnitude $E$ in the mica, (b) the magnitude of the free charge on the plates, and (c) the magnitude of the induced surface charge on the mica. SSM WWW

•**52** In Sample Problem 25-7, suppose that the battery remains connected while the dielectric slab is being introduced. Calculate (a) the capacitance, (b) the charge on the capacitor plates, (c) the electric field in the gap, and (d) the electric field in the slab, after the slab is in place.

••**53** The space between two concentric conducting spherical shells of radii $b = 1.70$ cm and $a = 1.20$ cm is filled with a substance of dielectric constant $\kappa = 23.5$. A potential difference $V = 73.0$ V is applied across the inner and outer shells. Determine (a) the capacitance of the device, (b) the free charge $q$ on the inner shell, and (c) the charge $q'$ induced along the surface of the inner shell.

••**54** Two parallel plates of area 100 cm$^2$ are given charges of equal magnitudes $8.9 \times 10^{-7}$ C but opposite signs. The electric field within the dielectric material filling the space between the plates is $1.4 \times 10^6$ V/m. (a) Calculate the dielectric constant of the material. (b) Determine the magnitude of the charge induced on each dielectric surface.

••**55** A parallel-plate capacitor has plates of area 0.12 m$^2$

and a separation of 1.2 cm. A battery charges the plates to a potential difference of 120 V and is then disconnected. A dielectric slab of thickness 4.0 mm and dielectric constant 4.8 is then placed symmetrically between the plates. (a) What is the capacitance before the slab is inserted? (b) What is the capacitance with the slab in place? What is the free charge $q$ (c) before and (d) after the slab is inserted? What is the magnitude of the electric field (e) in the space between the plates and dielectric and (f) in the dielectric itself? (g) With the slab in place, what is the potential difference across the plates? (h) How much external work is involved in inserting the slab?

### Additional Problems

**56** *The chocolate crumb mystery.* This story begins with Problem 56 in Chapter 23. As part of the investigation of the biscuit factory explosion, the electric potentials of the workers were measured as they emptied sacks of chocolate crumb powder into the loading bin, stirring up a cloud of the powder around themselves. Each worker had an electric potential of about 7.0 kV relative to the ground, which was taken as zero potential. (a) Assuming that each worker was effectively a capacitor with a typical capacitance of 200 pF, find the energy stored in that effective capacitor. If a single spark between the worker and any conducting object connected to the ground neutralized the worker, that energy would be transferred to the spark. According to measurements, a spark that could ignite a cloud of chocolate crumb powder, and thus set off an explosion, had to have an energy of at least 150 mJ. (b) Could a spark from a worker have set off an explosion in the cloud of powder in the loading bin? (The story continues with Problem 56 in Chapter 26.)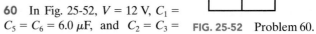

**57** Figure 25-51 shows capacitor 1 ($C_1 = 8.00$ $\mu$F), capacitor 2 ($C_2 = 6.00$ $\mu$F), and capacitor 3 ($C_3 = 8.00$ $\mu$F) connected to a 12.0 V battery. When switch S is closed so as to connect

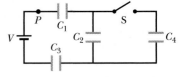

FIG. 25-51   Problem 57.

uncharged capacitor 4 ($C_4 = 6.00$ $\mu$F), (a) how much charge passes through point $P$ from the battery and (b) how much charge shows up on capacitor 4? (c) Explain the discrepancy in those two results.

**58** Two air-filled, parallel-plate capacitors are to be connected to a 10 V battery, first individually, then in series, and then in parallel. In those arrangements, the energy stored in the capacitors turns out to be, listed least to greatest: 75 $\mu$J, 100 $\mu$J, 300 $\mu$J, and 400 $\mu$J. Of the two capacitors, what is the (a) smaller and (b) greater capacitance?

**59** Two parallel-plate capacitors, 6.0 $\mu$F each, are connected in series to a 10 V battery. One of the capacitors is then squeezed so that its plate separation is halved. Because of the squeezing, (a) how much additional charge is transferred to the capacitors by the battery and (b) what is the increase in the *total* charge stored on the capacitors (the charge on the positive plate of one capacitor plus the charge on the positive plate of the other capacitor)?

**60** In Fig. 25-52, $V = 12$ V, $C_1 = C_5 = C_6 = 6.0$ $\mu$F, and $C_2 = C_3 =$

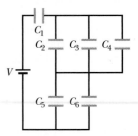

FIG. 25-52   Problem 60.

$C_4 = 4.0 \ \mu F$. What are (a) the net charge stored on the capacitors and (b) the charge on capacitor 4?

**61** In Fig. 25-53, $V = 9.0$ V, $C_1 = C_2 = 30 \ \mu F$, and $C_3 = C_4 = 15 \ \mu F$. What is the charge on capacitor 4? **SSM**

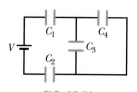

**FIG. 25-53**
Problem 61.

**62** In Fig. 25-54, the battery potential difference $V$ is 10.0 V and each of the seven capacitors has capacitance 10.0 $\mu F$. What is the charge on (a) capacitor 1 and (b) capacitor 2?

**63** In Fig. 25-55, $V = 12$ V, $C_1 = C_4 = 2.0$ $\mu F$, $C_2 = 4.0 \ \mu F$, and $C_3 = 1.0 \ \mu F$. What is the charge on capacitor 4?

**64** The capacitances of the four capacitors shown in Fig. 25-56 are given in terms of a certain quantity $C$. (a) If $C = 50 \ \mu F$, what is the equivalent capacitance between points $A$ and $B$? (*Hint:* First imagine that a battery is connected between those two points; then reduce the circuit to an equivalent capacitance.) (b) Repeat for points $A$ and $D$.

**65** A capacitor of capacitance $C_1 = 6.00 \ \mu F$ is connected in series with a capacitor of capacitance $C_2 = 4.00 \ \mu F$, and a potential difference of 200 V is applied across the pair. (a) Calculate the equivalent capacitance. What are (b) charge $q_1$ and (c) potential difference $V_1$ on capacitor 1 and (d) $q_2$ and (e) $V_2$ on capacitor 2?

**66** Repeat Problem 65 for the same two capacitors but with them now connected in parallel.

**67** A capacitor of unknown capacitance $C$ is charged to 100 V and connected across an initially uncharged 60 $\mu F$ capacitor. If the final potential difference across the 60 $\mu F$ capacitor is 40 V, what is $C$?

**68** A cylindrical capacitor has radii $a$ and $b$ as in Fig. 25-6. Show that half the stored electric potential energy lies within a cylinder whose radius is $r = \sqrt{ab}$.

**69** In Fig. 25-57, two parallel-plate capacitors $A$ and $B$ are connected in parallel across a 600 V battery. Each plate has area 80.0 cm²; the plate separations are 3.00 mm. Capacitor $A$ is filled with air; capacitor $B$ is filled with a dielectric of dielectric constant $\kappa = 2.60$. Find the magnitude of the electric field within (a) the dielectric of capacitor $B$ and (b) the air of capacitor $A$. What are the free charge densities $\sigma$ on the higher-potential plate of (c) capacitor $A$ and (d) capacitor $B$? (e) What is the induced charge density $\sigma'$ on the top surface of the dielectric? **SSM**

**70** A potential difference of 300 V is applied to a series connection of two capacitors of capacitances $C_1 = 2.00 \ \mu F$

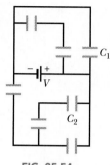

**FIG. 25-54**
Problem 62.

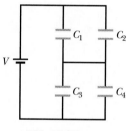

**FIG. 25-55**
Problem 63.

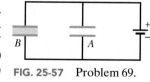

**FIG. 25-56** Problem 64.

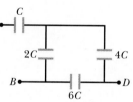

**FIG. 25-57** Problem 69.

and $C_2 = 8.00 \ \mu F$. What are (a) charge $q_1$ and (b) potential difference $V_1$ on capacitor 1 and (c) $q_2$ and (d) $V_2$ on capacitor 2? The *charged* capacitors are then disconnected from each other and from the battery. Then the capacitors are reconnected with plates of the *same* signs wired together (the battery is not used). What now are (e) $q_1$, (f) $V_1$, (g) $q_2$, and (h) $V_2$? Suppose, *instead*, the capacitors charged in part (a) are reconnected with plates of *opposite* signs wired together. What now are (i) $q_1$, (j) $V_1$, (k) $q_2$, and (l) $V_2$?

**71** A certain capacitor is charged to a potential difference $V$. If you wish to increase its stored energy by 10%, by what percentage should you increase $V$?

**72** You have two plates of copper, a sheet of mica (thickness = 0.10 mm, $\kappa = 5.4$), a sheet of glass (thickness = 2.0 mm, $\kappa = 7.0$), and a slab of paraffin (thickness = 1.0 cm, $\kappa = 2.0$). To make a parallel-plate capacitor with the largest $C$, which sheet should you place between the copper plates?

**73** In Fig. 25-58, the parallel-plate capacitor of plate area $2.00 \times 10^{-2}$ m² is filled with two dielectric slabs, each with thickness 2.00 mm. One slab has dielectric constant 3.00, and the other, 4.00. How much charge does the 7.00 V battery store on the capacitor? **SSM**

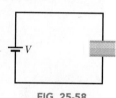

**FIG. 25-58**
Problem 73.

**74** A slab of copper of thickness $b = 2.00$ mm is thrust into a parallel-plate capacitor of plate area $A = 2.40$ cm² and plate separation $d = 5.00$ mm, as shown in Fig. 25-59; the slab is exactly halfway between the plates. (a) What is the capacitance after the slab is introduced? (b) If a charge $q = 3.40 \ \mu C$ is maintained on the plates, what is the ratio of the stored energy before to that after the slab is inserted? (c) How much work is done on the slab as it is inserted? (d) Is the slab sucked in or must it be pushed in?

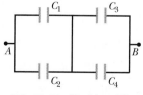

**FIG. 25-59** Problems 74 and 75.

**75** Repeat Problem 74, assuming that a potential difference $V = 85.0$ V, rather than the charge, is held constant.

**76** You have many 2.0 $\mu F$ capacitors, each capable of withstanding 200 V without undergoing electrical breakdown (in which they conduct charge instead of storing it). How would you assemble a combination having an equivalent capacitance of (a) 0.40 $\mu F$ and (b) 1.2 $\mu F$, each combination capable of withstanding 1000 V?

**77** Figure 25-60 shows a four-capacitor arrangement that is connected to a larger circuit at points $A$ and $B$. The capacitances are $C_1 = 10 \ \mu F$ and $C_2 = C_3 = C_4 = 20 \ \mu F$. The charge on capacitor 1 is 30 $\mu C$. What is the magnitude of the potential difference $V_A - V_B$?

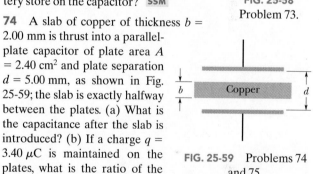

**FIG. 25-60** Problem 77.

**78** A 10 V battery is connected to a series of $n$ capacitors, each of capacitance 2.0 $\mu F$. If the total energy stored in the capacitors is 25 $\mu J$, what is $n$?

# 26 | Current and Resistance

Courtesy of The Roanoke Times/©Eric Brady

*Lightning strikes the ground less than 1 km from a Virginia Tech football game. The chance of lightning striking a person directly is slim. The much greater danger lies in the ground current—the current that spreads out from the strike point. Everyone on the field or in the stands could have been knocked down, paralyzed, or killed by the ground current. If you are caught in the open during a lightning storm like this, there is a simple procedure for reducing your risk from ground current.*

*How can you reduce your risk from ground current?*

The answer is in this chapter.

## 26-1 | WHAT IS PHYSICS?

In the last five chapters we discussed electrostatics—the physics of stationary charges. In this and the next chapter, we discuss the physics of **electric currents**—that is, charges in motion.

Examples of electric currents abound and involve many professions. Meteorologists are concerned with lightning and with the less dramatic slow flow of charge through the atmosphere. Biologists, physiologists, and engineers working in medical technology are concerned with the nerve currents that control muscles and especially with how those currents can be reestablished after spinal cord injuries. Electrical engineers are concerned with countless electrical systems, such as power systems, lightning protection systems, information storage systems, and music systems. Space engineers monitor and study the flow of charged particles from our Sun because that flow can wipe out telecommunication systems in orbit and even power transmission systems on the ground.

In this chapter we discuss the basic physics of electric currents and why they can be established in some materials but not in others. We begin with the meaning of electric current.

## 26-2 | Electric Current

Although an electric current is a stream of moving charges, not all moving charges constitute an electric current. If there is to be an electric current through a given surface, there must be a net flow of charge through that surface. Two examples clarify our meaning.

1. The free electrons (conduction electrons) in an isolated length of copper wire are in random motion at speeds of the order of $10^6$ m/s. If you pass a hypothetical plane through such a wire, conduction electrons pass through it *in both directions* at the rate of many billions per second—but there is *no net transport* of charge and thus *no current* through the wire. However, if you connect the ends of the wire to a battery, you slightly bias the flow in one direction, with the result that there now is a net transport of charge and thus an electric current through the wire.

2. The flow of water through a garden hose represents the directed flow of positive charge (the protons in the water molecules) at a rate of perhaps several million coulombs per second. There is no net transport of charge, however, because there is a parallel flow of negative charge (the electrons in the water molecules) of exactly the same amount moving in exactly the same direction.

In this chapter we restrict ourselves largely to the study—within the framework of classical physics—of *steady* currents of *conduction electrons* moving through *metallic conductors* such as copper wires.

As Fig. 26-1a reminds us, any isolated conducting loop—regardless of whether it has an excess charge—is all at the same potential. No electric field can exist within it or along its surface. Although conduction electrons are available, no net electric force acts on them and thus there is no current.

If, as in Fig. 26-1b, we insert a battery in the loop, the conducting loop is no longer at a single potential. Electric fields act inside the material making up the loop, exerting forces on the conduction electrons, causing them to move and thus establishing a current. After a very short time, the electron flow reaches a constant value and the current is in its *steady state* (it does not vary with time).

Figure 26-2 shows a section of a conductor, part of a conducting loop in which current has been established. If charge $dq$ passes through a hypothetical plane (such as $aa'$) in time $dt$, then the current $i$ through that plane is defined as

$$i = \frac{dq}{dt} \quad \text{(definition of current).} \tag{26-1}$$

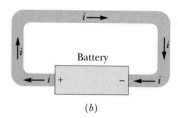

**FIG. 26-1** (*a*) A loop of copper in electrostatic equilibrium. The entire loop is at a single potential, and the electric field is zero at all points inside the copper. (*b*) Adding a battery imposes an electric potential difference between the ends of the loop that are connected to the terminals of the battery. The battery thus produces an electric field within the loop, from terminal to terminal, and the field causes charges to move around the loop. This movement of charges is a current $i$.

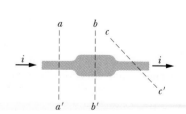

**FIG. 26-2** The current $i$ through the conductor has the same value at planes $aa'$, $bb'$, and $cc'$.

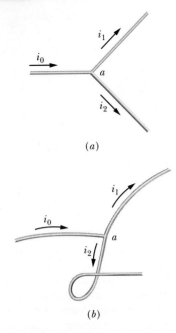

FIG. 26-3   The relation $i_0 = i_1 + i_2$ is true at junction $a$ no matter what the orientation in space of the three wires. Currents are scalars, not vectors.

We can find the charge that passes through the plane in a time interval extending from 0 to $t$ by integration:

$$q = \int dq = \int_0^t i \, dt, \tag{26-2}$$

in which the current $i$ may vary with time.

Under steady-state conditions, the current is the same for planes $aa'$, $bb'$, and $cc'$ and indeed for all planes that pass completely through the conductor, no matter what their location or orientation. This follows from the fact that charge is conserved. Under the steady-state conditions assumed here, an electron must pass through plane $aa'$ for every electron that passes through plane $cc'$. In the same way, if we have a steady flow of water through a garden hose, a drop of water must leave the nozzle for every drop that enters the hose at the other end. The amount of water in the hose is a conserved quantity.

The SI unit for current is the coulomb per second, or the ampere (A), which is an SI base unit:

$$1 \text{ ampere} = 1 \text{ A} = 1 \text{ coulomb per second} = 1 \text{ C/s}.$$

The formal definition of the ampere is discussed in Chapter 29.

Current, as defined by Eq. 26-1, is a scalar because both charge and time in that equation are scalars. Yet, as in Fig. 26-1$b$, we often represent a current with an arrow to indicate that charge is moving. Such arrows are not vectors, however, and they do not require vector addition. Figure 26-3$a$ shows a conductor with current $i_0$ splitting at a junction into two branches. Because charge is conserved, the magnitudes of the currents in the branches must add to yield the magnitude of the current in the original conductor, so that

$$i_0 = i_1 + i_2. \tag{26-3}$$

As Fig. 26-3$b$ suggests, bending or reorienting the wires in space does not change the validity of Eq. 26-3. Current arrows show only a direction (or sense) of flow along a conductor, not a direction in space.

### The Directions of Currents

In Fig. 26-1$b$ we drew the current arrows in the direction in which positively charged particles would be forced to move through the loop by the electric field. Such positive *charge carriers*, as they are often called, would move away from the positive battery terminal and toward the negative terminal. Actually, the charge carriers in the copper loop of Fig. 26-1$b$ are electrons and thus are negatively charged. The electric field forces them to move in the direction opposite the current arrows, from the negative terminal to the positive terminal. For historical reasons, however, we use the following convention:

> ☛ A current arrow is drawn in the direction in which positive charge carriers would move, even if the actual charge carriers are negative and move in the opposite direction.

We can use this convention because in *most* situations, the assumed motion of positive charge carriers in one direction has the same effect as the actual motion of negative charge carriers in the opposite direction. (When the effect is not the same, we shall drop the convention and describe the actual motion.)

✓ **CHECKPOINT 1**   The figure here shows a portion of a circuit. What are the magnitude and direction of the current $i$ in the lower right-hand wire?

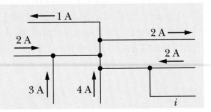

Water flows through a garden hose at a volume flow rate $dV/dt$ of 450 cm³/s. What is the current of negative charge?

**KEY IDEAS** The current $i$ of negative charge is due to the electrons in the water molecules moving through the hose. The current is the rate at which that negative charge passes through any plane that cuts completely across the hose.

*Calculations:* We can write the current in terms of the number of molecules that pass through such a plane per second as

$$i = \left(\begin{array}{c}\text{charge} \\ \text{per} \\ \text{electron}\end{array}\right)\left(\begin{array}{c}\text{electrons} \\ \text{per} \\ \text{molecule}\end{array}\right)\left(\begin{array}{c}\text{molecules} \\ \text{per} \\ \text{second}\end{array}\right)$$

or

$$i = (e)(10)\frac{dN}{dt}.$$

We substitute 10 electrons per molecule because a water ($H_2O$) molecule contains 8 electrons in the single oxygen atom and 1 electron in each of the two hydrogen atoms.

We can express the rate $dN/dt$ in terms of the given volume flow rate $dV/dt$ by first writing

$$\left(\begin{array}{c}\text{molecules} \\ \text{per} \\ \text{second}\end{array}\right) = \left(\begin{array}{c}\text{molecules} \\ \text{per} \\ \text{mole}\end{array}\right)\left(\begin{array}{c}\text{moles} \\ \text{per unit} \\ \text{mass}\end{array}\right)$$
$$\times \left(\begin{array}{c}\text{mass} \\ \text{per unit} \\ \text{volume}\end{array}\right)\left(\begin{array}{c}\text{volume} \\ \text{per} \\ \text{second}\end{array}\right).$$

"Molecules per mole" is Avogadro's number $N_A$. "Moles per unit mass" is the inverse of the mass per mole, which is the molar mass $M$ of water. "Mass per unit volume" is the (mass) density $\rho_{mass}$ of water. The volume per second is the volume flow rate $dV/dt$. Thus, we have

$$\frac{dN}{dt} = N_A\left(\frac{1}{M}\right)\rho_{mass}\left(\frac{dV}{dt}\right) = \frac{N_A\rho_{mass}}{M}\frac{dV}{dt}.$$

Substituting this into the equation for $i$, we find

$$i = 10eN_AM^{-1}\rho_{mass}\frac{dV}{dt}.$$

$N_A$ is $6.02 \times 10^{23}$ molecules/mol, or $6.02 \times 10^{23}$ mol⁻¹, and $\rho_{mass}$ is 1000 kg/m³. We can get the molar mass of water from the molar masses listed in Appendix F: We add the molar mass of oxygen (16 g/mol) to twice the molar mass of hydrogen (1 g/mol), obtaining 18 g/mol = 0.018 kg/mol. Then

$$i = (10)(1.6 \times 10^{-19}\text{ C})(6.02 \times 10^{23}\text{ mol}^{-1})$$
$$\times (0.018\text{ kg/mol})^{-1}(1000\text{ kg/m}^3)(450 \times 10^{-6}\text{ m}^3\text{/s})$$
$$= 2.41 \times 10^7\text{ C/s} = 2.41 \times 10^7\text{ A}$$
$$= 24.1\text{ MA.} \qquad\qquad\qquad \text{(Answer)}$$

This current of negative charge is exactly compensated by a current of positive charge associated with the nuclei of the three atoms that make up the water molecule. Thus, there is no net flow of charge through the hose.

## 26-3 | Current Density

Sometimes we are interested in the current $i$ in a particular conductor. At other times we take a localized view and study the flow of charge through a cross section of the conductor at a particular point. To describe this flow, we can use the **current density** $\vec{J}$, which has the same direction as the velocity of the moving charges if they are positive and the opposite direction if they are negative. For each element of the cross section, the magnitude $J$ is equal to the current per unit area through that element. We can write the amount of current through the element as $\vec{J} \cdot d\vec{A}$, where $d\vec{A}$ is the area vector of the element, perpendicular to the element. The total current through the surface is then

$$i = \int \vec{J} \cdot d\vec{A}. \qquad (26\text{-}4)$$

If the current is uniform across the surface and parallel to $d\vec{A}$, then $\vec{J}$ is also uniform and parallel to $d\vec{A}$. Then Eq. 26-4 becomes

$$i = \int J\,dA = J\int dA = JA,$$

so

$$J = \frac{i}{A}, \qquad (26\text{-}5)$$

where $A$ is the total area of the surface. From Eq. 26-4 or 26-5 we see that the SI unit for current density is the ampere per square meter (A/m²).

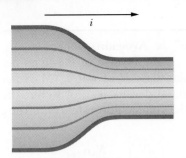

**FIG. 26-4** Streamlines representing current density in the flow of charge through a constricted conductor.

In Chapter 22 we saw that we can represent an electric field with electric field lines. Figure 26-4 shows how current density can be represented with a similar set of lines, which we can call *streamlines*. The current, which is toward the right in Fig. 26-4, makes a transition from the wider conductor at the left to the narrower conductor at the right. Because charge is conserved during the transition, the amount of charge and thus the amount of current cannot change. However, the current density does change—it is greater in the narrower conductor. The spacing of the streamlines suggests this increase in current density; streamlines that are closer together imply greater current density.

## Drift Speed

When a conductor does not have a current through it, its conduction electrons move randomly, with no net motion in any direction. When the conductor does have a current through it, these electrons actually still move randomly, but now they tend to *drift* with a **drift speed** $v_d$ in the direction opposite that of the applied electric field that causes the current. The drift speed is tiny compared with the speeds in the random motion. For example, in the copper conductors of household wiring, electron drift speeds are perhaps $10^{-5}$ or $10^{-4}$ m/s, whereas the random-motion speeds are around $10^6$ m/s.

We can use Fig. 26-5 to relate the drift speed $v_d$ of the conduction electrons in a current through a wire to the magnitude $J$ of the current density in the wire. For convenience, Fig. 26-5 shows the equivalent drift of *positive* charge carriers in the direction of the applied electric field $\vec{E}$. Let us assume that these charge carriers all move with the same drift speed $v_d$ and that the current density $J$ is uniform across the wire's cross-sectional area $A$. The number of charge carriers in a length $L$ of the wire is $nAL$, where $n$ is the number of carriers per unit volume. The total charge of the carriers in the length $L$, each with charge $e$, is then

$$q = (nAL)e.$$

Because the carriers all move along the wire with speed $v_d$, this total charge moves through any cross section of the wire in the time interval

$$t = \frac{L}{v_d}.$$

Equation 26-1 tells us that the current $i$ is the time rate of transfer of charge across a cross section, so here we have

$$i = \frac{q}{t} = \frac{nALe}{L/v_d} = nAev_d. \qquad (26\text{-}6)$$

Solving for $v_d$ and recalling Eq. 26-5 ($J = i/A$), we obtain

$$v_d = \frac{i}{nAe} = \frac{J}{ne}$$

or, extended to vector form,

$$\vec{J} = (ne)\vec{v}_d. \qquad (26\text{-}7)$$

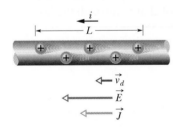

**FIG. 26-5** Positive charge carriers drift at speed $v_d$ in the direction of the applied electric field $\vec{E}$. By convention, the direction of the current density $\vec{J}$ and the sense of the current arrow are drawn in that same direction.

Here the product $ne$, whose SI unit is the coulomb per cubic meter (C/m³), is the *carrier charge density*. For positive carriers, $ne$ is positive and Eq. 26-7 predicts that $\vec{J}$ and $\vec{v}_d$ have the same direction. For negative carriers, $ne$ is negative and $\vec{J}$ and $\vec{v}_d$ have opposite directions.

✓**CHECKPOINT 2** The figure shows conduction electrons moving leftward in a wire. Are the following leftward or rightward: (a) the current $i$, (b) the current density $\vec{J}$, (c) the electric field $\vec{E}$ in the wire?

## Sample Problem 26-2

(a) The current density in a cylindrical wire of radius $R = 2.0$ mm is uniform across a cross section of the wire and is $J = 2.0 \times 10^5$ A/m². What is the current through the outer portion of the wire between radial distances $R/2$ and $R$ (Fig. 26-6a)?

**KEY IDEA**  Because the current density is uniform across the cross section, the current density $J$, the current $i$, and the cross-sectional area $A$ are related by Eq. 26-5 ($J = i/A$).

*Calculations:* We want only the current through a reduced cross-sectional area $A'$ of the wire (rather than the entire area), where

$$A' = \pi R^2 - \pi \left(\frac{R}{2}\right)^2 = \pi \left(\frac{3R^2}{4}\right)$$

$$= \frac{3\pi}{4}(0.0020 \text{ m})^2 = 9.424 \times 10^{-6} \text{ m}^2.$$

So, we rewrite Eq. 26-5 as

$$i = JA'$$

and then substitute the data to find

$$i = (2.0 \times 10^5 \text{ A/m}^2)(9.424 \times 10^{-6} \text{ m}^2)$$

$$= 1.9 \text{ A.} \qquad \text{(Answer)}$$

(b) Suppose, instead, that the current density through a cross section varies with radial distance $r$ as $J = ar^2$, in which $a = 3.0 \times 10^{11}$ A/m⁴ and $r$ is in meters. What now is the current through the same outer portion of the wire?

**KEY IDEA**  Because the current density is not uniform across a cross section of the wire, we must resort to Eq. 26-4 ($i = \int \vec{J} \cdot d\vec{A}$) and integrate the current density over the portion of the wire from $r = R/2$ to $r = R$.

*Calculations:* The current density vector $\vec{J}$ (along the wire's length) and the differential area vector $d\vec{A}$ (perpendicular to a cross section of the wire) have the same direction. Thus,

$$\vec{J} \cdot d\vec{A} = J \, dA \cos 0 = J \, dA.$$

We need to replace the differential area $dA$ with something we can actually integrate between the limits $r = R/2$ and $r = R$. The simplest replacement (because $J$ is given as a function of $r$) is the area $2\pi r \, dr$ of a thin ring of circumference $2\pi r$ and width $dr$ (Fig. 26-6b). We can then integrate with $r$ as the variable of integration. Equation 26-4 then gives us

$$i = \int \vec{J} \cdot d\vec{A} = \int J \, dA$$

$$= \int_{R/2}^{R} ar^2 \, 2\pi r \, dr = 2\pi a \int_{R/2}^{R} r^3 \, dr$$

$$= 2\pi a \left[\frac{r^4}{4}\right]_{R/2}^{R} = \frac{\pi a}{2}\left[R^4 - \frac{R^4}{16}\right] = \frac{15}{32}\pi a R^4$$

$$= \frac{15}{32}\pi(3.0 \times 10^{11} \text{ A/m}^4)(0.0020 \text{ m})^4 = 7.1 \text{ A.}$$

(Answer)

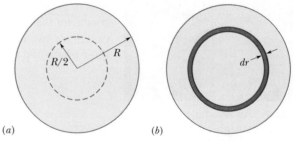

(a)                                    (b)

**FIG. 26-6** (a) Cross section of a wire of radius $R$. (b) A thin ring has width $dr$ and circumference $2\pi r$, and thus a differential area $dA = 2\pi r \, dr$.

## Sample Problem 26-3

What is the drift speed of the conduction electrons in a copper wire with radius $r = 900$ μm when it has a uniform current $i = 17$ mA? Assume that each copper atom contributes one conduction electron to the current and that the current density is uniform across the wire's cross section.

**KEY IDEAS**

1. The drift speed $v_d$ is related to the current density $\vec{J}$ and the number $n$ of conduction electrons per unit volume according to Eq. 26-7, which we can write as $J = nev_d$.

2. Because the current density is uniform, its magnitude $J$ is related to the given current $i$ and wire size by Eq.

26-5 ($J = i/A$, where $A$ is the cross-sectional area of the wire).

3. Because we assume one conduction electron per atom, the number $n$ of conduction electrons per unit volume is the same as the number of atoms per unit volume.

*Calculations:* Let us start with the third idea by writing

$$n = \begin{pmatrix} \text{atoms} \\ \text{per unit} \\ \text{volume} \end{pmatrix} = \begin{pmatrix} \text{atoms} \\ \text{per} \\ \text{mole} \end{pmatrix}\begin{pmatrix} \text{moles} \\ \text{per unit} \\ \text{mass} \end{pmatrix}\begin{pmatrix} \text{mass} \\ \text{per unit} \\ \text{volume} \end{pmatrix}.$$

The number of atoms per mole is just Avogadro's number $N_A$ (= $6.02 \times 10^{23}$ mol⁻¹). Moles per unit mass is the inverse of the mass per mole, which here is the molar mass $M$ of copper. The mass per unit volume is the (mass) density $\rho_{mass}$ of copper. Thus,

$$n = N_A\left(\frac{1}{M}\right)\rho_{mass} = \frac{N_A\rho_{mass}}{M}.$$

Taking copper's molar mass $M$ and density $\rho_{mass}$ from Appendix F, we then have (with some conversions of units)

$$n = \frac{(6.02 \times 10^{23} \text{ mol}^{-1})(8.96 \times 10^3 \text{ kg/m}^3)}{63.54 \times 10^{-3} \text{ kg/mol}}$$

$$= 8.49 \times 10^{28} \text{ electrons/m}^3$$

or

$$n = 8.49 \times 10^{28} \text{ m}^{-3}.$$

Next let us combine the first two key ideas by writing

$$\frac{i}{A} = nev_d.$$

Substituting for $A$ with $\pi r^2$ ($= 2.54 \times 10^{-6} \text{ m}^2$) and solving for $v_d$, we then find

$$v_d = \frac{i}{ne(\pi r^2)}$$

$$= \frac{17 \times 10^{-3} \text{ A}}{(8.49 \times 10^{28} \text{ m}^{-3})(1.6 \times 10^{-19} \text{ C})(2.54 \times 10^{-6} \text{ m}^2)}$$

$$= 4.9 \times 10^{-7} \text{ m/s}, \qquad \text{(Answer)}$$

which is only 1.8 mm/h, slower than a sluggish snail.

**Lights are fast:** You may well ask: "If the electrons drift so slowly, why do the room lights turn on so quickly when I throw the switch?" Confusion on this point results from not distinguishing between the drift speed of the electrons and the speed at which *changes* in the electric field configuration travel along wires. This latter speed is nearly that of light; electrons everywhere in the wire begin drifting almost at once, including into the lightbulbs. Similarly, when you open the valve on your garden hose with the hose full of water, a pressure wave travels along the hose at the speed of sound in water. The speed at which the water itself moves through the hose— measured perhaps with a dye marker—is much slower.

## 26-4 | Resistance and Resistivity

If we apply the same potential difference between the ends of geometrically similar rods of copper and of glass, very different currents result. The characteristic of the conductor that enters here is its electrical **resistance.** We determine the resistance between any two points of a conductor by applying a potential difference $V$ between those points and measuring the current $i$ that results. The resistance $R$ is then

$$R = \frac{V}{i} \qquad \text{(definition of } R\text{).} \qquad (26\text{-}8)$$

The SI unit for resistance that follows from Eq. 26-8 is the volt per ampere. This combination occurs so often that we give it a special name, the **ohm** (symbol $\Omega$); that is,

$$1 \text{ ohm} = 1 \text{ }\Omega = 1 \text{ volt per ampere}$$

$$= 1 \text{ V/A.} \qquad (26\text{-}9)$$

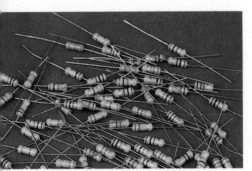

**FIG. 26-7** An assortment of resistors. The circular bands are color-coding marks that identify the value of the resistance. *(The Image Works)*

A conductor whose function in a circuit is to provide a specified resistance is called a **resistor** (see Fig. 26-7). In a circuit diagram, we represent a resistor and a resistance with the symbol ‒\/\/\‒. If we write Eq. 26-8 as

$$i = \frac{V}{R},$$

we see that, for a given $V$, the greater the resistance, the smaller the current.

The resistance of a conductor depends on the manner in which the potential difference is applied to it. Figure 26-8, for example, shows a given potential difference applied in two different ways to the same conductor. As the current density streamlines suggest, the currents in the two cases—hence the measured resistances—will be different. Unless otherwise stated, we shall assume that any given potential difference is applied as in Fig. 26-8b.

As we have done several times in other connections, we often wish to take a general view and deal not with particular objects but with materials. Here we do so by focusing not on the potential difference $V$ across a particular resistor but on the electric field $\vec{E}$ at a point in a resistive material. Instead of dealing with the current $i$

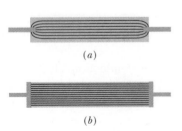

*(a)*

*(b)*

**FIG. 26-8** Two ways of applying a potential difference to a conducting rod. The gray connectors are assumed to have negligible resistance. When they are arranged as in (a) in a small region at each rod end, the measured resistance is larger than when they are arranged as in (b) to cover the entire rod end.

through the resistor, we deal with the current density $\vec{J}$ at the point in question. Instead of the resistance $R$ of an object, we deal with the **resistivity** $\rho$ of the *material:*

$$\rho = \frac{E}{J} \qquad \text{(definition of } \rho). \qquad (26\text{-}10)$$

(Compare this equation with Eq. 26-8.)

If we combine the SI units of $E$ and $J$ according to Eq. 26-10, we get, for the unit of $\rho$, the ohm-meter ($\Omega \cdot$ m):

$$\frac{\text{unit }(E)}{\text{unit }(J)} = \frac{\text{V/m}}{\text{A/m}^2} = \frac{\text{V}}{\text{A}}\,\text{m} = \Omega \cdot \text{m}.$$

(Do not confuse the *ohm-meter,* the unit of resistivity, with the *ohmmeter,* which is an instrument that measures resistance.) Table 26-1 lists the resistivities of some materials.

We can write Eq. 26-10 in vector form as

$$\vec{E} = \rho\vec{J}. \qquad (26\text{-}11)$$

Equations 26-10 and 26-11 hold only for *isotropic* materials—materials whose electrical properties are the same in all directions.

We often speak of the **conductivity** $\sigma$ of a material. This is simply the reciprocal of its resistivity, so

$$\sigma = \frac{1}{\rho} \qquad \text{(definition of } \sigma). \qquad (26\text{-}12)$$

The SI unit of conductivity is the reciprocal ohm-meter, $(\Omega \cdot \text{m})^{-1}$. The unit name mhos per meter is sometimes used (mho is ohm backwards). The definition of $\sigma$ allows us to write Eq. 26-11 in the alternative form

$$\vec{J} = \sigma\vec{E}. \qquad (26\text{-}13)$$

**TABLE 26-1**

**Resistivities of Some Materials at Room Temperature (20°C)**

| Material | Resistivity, $\rho$ ($\Omega \cdot$ m) | Temperature Coefficient of Resistivity, $\alpha$ (K$^{-1}$) |
|---|---|---|
| *Typical Metals* | | |
| Silver | $1.62 \times 10^{-8}$ | $4.1 \times 10^{-3}$ |
| Copper | $1.69 \times 10^{-8}$ | $4.3 \times 10^{-3}$ |
| Gold | $2.35 \times 10^{-8}$ | $4.0 \times 10^{-3}$ |
| Aluminum | $2.75 \times 10^{-8}$ | $4.4 \times 10^{-3}$ |
| Manganin[a] | $4.82 \times 10^{-8}$ | $0.002 \times 10^{-3}$ |
| Tungsten | $5.25 \times 10^{-8}$ | $4.5 \times 10^{-3}$ |
| Iron | $9.68 \times 10^{-8}$ | $6.5 \times 10^{-3}$ |
| Platinum | $10.6 \times 10^{-8}$ | $3.9 \times 10^{-3}$ |
| *Typical Semiconductors* | | |
| Silicon, pure | $2.5 \times 10^{3}$ | $-70 \times 10^{-3}$ |
| Silicon, $n$-type[b] | $8.7 \times 10^{-4}$ | |
| Silicon, $p$-type[c] | $2.8 \times 10^{-3}$ | |
| *Typical Insulators* | | |
| Glass | $10^{10} - 10^{14}$ | |
| Fused quartz | $\sim 10^{16}$ | |

[a]An alloy specifically designed to have a small value of $\alpha$.
[b]Pure silicon doped with phosphorus impurities to a charge carrier density of $10^{23}$ m$^{-3}$.
[c]Pure silicon doped with aluminum impurities to a charge carrier density of $10^{23}$ m$^{-3}$.

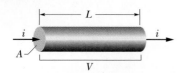

**FIG. 26-9** A potential difference $V$ is applied between the ends of a wire of length $L$ and cross section $A$, establishing a current $i$.

## Calculating Resistance from Resistivity

We have just made an important distinction:

> Resistance is a property of an object. Resistivity is a property of a material.

If we know the resistivity of a substance such as copper, we can calculate the resistance of a length of wire made of that substance. Let $A$ be the cross-sectional area of the wire, let $L$ be its length, and let a potential difference $V$ exist between its ends (Fig. 26-9). If the streamlines representing the current density are uniform throughout the wire, the electric field and the current density will be constant for all points within the wire and, from Eqs. 24-42 and 26-5, will have the values

$$E = V/L \quad \text{and} \quad J = i/A. \tag{26-14}$$

We can then combine Eqs. 26-10 and 26-14 to write

$$\rho = \frac{E}{J} = \frac{V/L}{i/A}. \tag{26-15}$$

However, $V/i$ is the resistance $R$, which allows us to recast Eq. 26-15 as

$$R = \rho \frac{L}{A}. \tag{26-16}$$

Equation 26-16 can be applied only to a homogeneous isotropic conductor of uniform cross section, with the potential difference applied as in Fig. 26-8$b$.

The macroscopic quantities $V$, $i$, and $R$ are of greatest interest when we are making electrical measurements on specific conductors. They are the quantities that we read directly on meters. We turn to the microscopic quantities $E$, $J$, and $\rho$ when we are interested in the fundamental electrical properties of materials.

✓**CHECKPOINT 3** The figure here shows three cylindrical copper conductors along with their face areas and lengths. Rank them according to the current through them, greatest first, when the same potential difference $V$ is placed across their lengths.

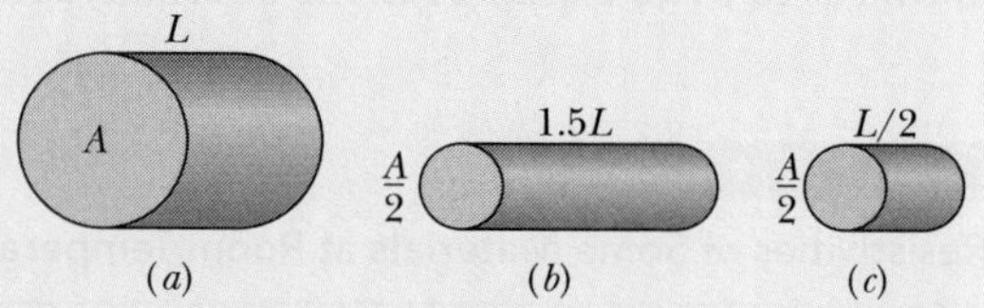

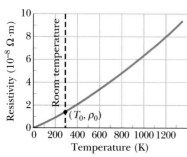

## Variation with Temperature

The values of most physical properties vary with temperature, and resistivity is no exception. Figure 26-10, for example, shows the variation of this property for copper over a wide temperature range. The relation between temperature and resistivity for copper—and for metals in general—is fairly linear over a rather broad temperature range. For such linear relations we can write an empirical approximation that is good enough for most engineering purposes:

$$\rho - \rho_0 = \rho_0 \alpha (T - T_0). \tag{26-17}$$

Here $T_0$ is a selected reference temperature and $\rho_0$ is the resistivity at that temperature. Usually $T_0 = 293$ K (room temperature), for which $\rho_0 = 1.69 \times 10^{-8}$ $\Omega \cdot$m for copper.

Because temperature enters Eq. 26-17 only as a difference, it does not matter whether you use the Celsius or Kelvin scale in that equation because the sizes of degrees on these scales are identical. The quantity $\alpha$ in Eq. 26-17, called the *temperature coefficient of resistivity,* is chosen so that the equation gives good agreement with experiment for temperatures in the chosen range. Some values of $\alpha$ for metals are listed in Table 26-1.

**FIG. 26-10** The resistivity of copper as a function of temperature. The dot on the curve marks a convenient reference point at temperature $T_0 = 293$ K and resistivity $\rho_0 = 1.69 \times 10^{-8}$ $\Omega \cdot$m.

### Sample Problem 26-4

A rectangular block of iron has dimensions 1.2 cm × 1.2 cm × 15 cm. A potential difference is to be applied to the block between parallel sides and in such a way that those sides are equipotential surfaces (as in Fig. 26-8b). What is the resistance of the block if the two parallel sides are (1) the square ends (with dimensions 1.2 cm × 1.2 cm) and (2) two rectangular sides (with dimensions 1.2 cm × 15 cm)?

**KEY IDEA** The resistance $R$ of an object depends on how the electric potential is applied to the object. In particular, it depends on the ratio $L/A$, according to Eq. 26-16 ($R = \rho L/A$), where $A$ is the area of the surfaces to which the potential difference is applied and $L$ is the distance between those surfaces.

*Calculations:* For arrangement 1, we have $L = 15$ cm = 0.15 m and

$$A = (1.2 \text{ cm})^2 = 1.44 \times 10^{-4} \text{ m}^2.$$

Substituting into Eq. 26-16 with the resistivity $\rho$ from Table 26-1, we then find that for arrangement 1,

$$R = \frac{\rho L}{A} = \frac{(9.68 \times 10^{-8} \, \Omega \cdot \text{m})(0.15 \text{ m})}{1.44 \times 10^{-4} \text{ m}^2}$$

$$= 1.0 \times 10^{-4} \, \Omega = 100 \, \mu\Omega. \qquad \text{(Answer)}$$

Similarly, for arrangement 2, with distance $L = 1.2$ cm and area $A = (1.2 \text{ cm})(15 \text{ cm})$, we obtain

$$R = \frac{\rho L}{A} = \frac{(9.68 \times 10^{-8} \, \Omega \cdot \text{m})(1.2 \times 10^{-2} \text{ m})}{1.80 \times 10^{-3} \text{ m}^2}$$

$$= 6.5 \times 10^{-7} \, \Omega = 0.65 \, \mu\Omega. \qquad \text{(Answer)}$$

### Sample Problem 26-5 | Build your skill

Figure 26-11 shows a person and a cow, each a radial distance $D = 60.0$ m from the point where lightning of current $I = 100$ kA strikes the ground. The current spreads through the ground uniformly over a hemisphere centered on the strike point. The person's feet are separated by radial distance $\Delta r_{\text{per}} = 0.50$ m; the cow's front and rear hooves are separated by radial distance $\Delta r_{\text{cow}} = 1.50$ m. The resistivity of the ground is $\rho_{\text{gr}} = 100 \, \Omega \cdot \text{m}$. The resistance both across the person, between left and right feet, and across the cow, between front and rear hooves, is $R = 4.00$ kΩ.

(a) What is the current $i_p$ through the person?

**KEY IDEAS**

1. The lightning strike sets up an electric field and an electric potential in the surrounding ground.

2. Because one foot is closer to the strike point than the other foot, a potential difference $\Delta V$ is set up across the person.

3. That $\Delta V$ drives a current $i_p$ through the person.

*Potential difference:* Because the lightning's current $I$ spreads uniformly over a hemisphere in the ground, the current density at any given radius $r$ from the strike point is, from Eq. 26-5 ($J = i/A$),

$$J = \frac{I}{2\pi r^2}, \qquad (26\text{-}18)$$

where $2\pi r^2$ is the area of the curved surface of a hemisphere. From Eq. 26-10 ($\rho = E/J$), the magnitude of the electric field is then

$$E = \rho_{\text{gr}} J = \frac{\rho_{\text{gr}} I}{2\pi r^2}. \qquad (26\text{-}19)$$

From Eq. 24-18 ($\Delta V = -\int \vec{E} \cdot d\vec{s}$), the potential differ-

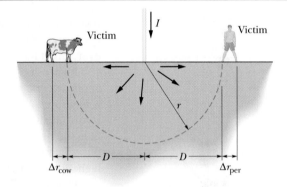

**FIG. 26-11** Current from a lightning strike spreads hemispherically through the ground and reaches a cow and a person, each located a distance $D$ from the strike point. The danger to them depends on the separation $\Delta r$.

ence $\Delta V$ between a point at radial distance $D$ and a point at radial distance $D + \Delta r$ is

$$\Delta V = -\int_D^{D+\Delta r} E \, dr. \qquad (26\text{-}20)$$

Substituting Eq. 26-19 into this integral and then integrating give us the potential difference:

$$\Delta V = -\int_D^{D+\Delta r} \frac{\rho_{\text{gr}} I}{2\pi r^2} \, dr = -\frac{\rho_{\text{gr}} I}{2\pi} \left[ -\frac{1}{r} \right]_D^{D+\Delta r}$$

$$= \frac{\rho_{\text{gr}} I}{2\pi} \left( \frac{1}{D+\Delta r} - \frac{1}{D} \right)$$

$$= -\frac{\rho_{\text{gr}} I}{2\pi} \frac{\Delta r}{D(D+\Delta r)}. \qquad (26\text{-}21)$$

*Current:* If one of the person's feet is at radial distance $D$ from the strike point and the other foot is at radial distance $D + \Delta r$, the potential difference $\Delta V$ between the feet is given by Eq. 26-21. This $\Delta V$ drives a current $i_p$

through the person. To find that current, we use Eq. 26-8 ($R = V/i$), in which $V$ represents the magnitude of the potential difference. Substituting the magnitude of $\Delta V$ from Eq. 26-21 for $V$ in Eq. 26-8, we find the current to be

$$i = \frac{V}{R} = \frac{\rho_{gr}I}{2\pi} \frac{\Delta r}{D(D + \Delta r)} \frac{1}{R}. \quad (26\text{-}22)$$

Substituting known values, including the foot-to-foot separation $\Delta r_{per} = 0.50$ m, gives the current through the person:

$$i_p = \frac{(100\ \Omega \cdot m)(100\ kA)}{2\pi}$$

$$\times \frac{0.50\ m}{(60\ m)(60.0\ m + 0.50\ m)} \frac{1}{4.00\ k\Omega}$$

$$= 0.0548\ A = 54.8\ mA. \quad \text{(Answer)}$$

This amount of current causes involuntary muscle contraction; the person will collapse but probably soon recover. Note that the person could reduce the current an order of magnitude by standing with feet together so that $\Delta r$ is only a few centimeters.

(b) What is the current $i_c$ through the cow?

**Calculation:** We again use Eq. 26-22, but now $\Delta r$ is $\Delta r_{cow} = 1.50$ m. We now find that the current through the cow is

$$i_c = 0.162\ A = 162\ mA, \quad \text{(Answer)}$$

which is fatal. The cow is in more danger from the ground current because of its greater value of $\Delta r$. The cow is, of course, unable to reduce its danger by standing with its hooves together (which would be a bizarre sight).

## 26-5 | Ohm's Law

As we just discussed in Section 26-4, a resistor is a conductor with a specified resistance. It has that same resistance no matter what the magnitude and direction (*polarity*) of the applied potential difference are. Other conducting devices, however, might have resistances that change with the applied potential difference.

Figure 26-12a shows how to distinguish such devices. A potential difference $V$ is applied across the device being tested, and the resulting current $i$ through the device is measured as $V$ is varied in both magnitude and polarity. The polarity of $V$ is arbitrarily taken to be positive when the left terminal of the device is at a higher potential than the right terminal. The direction of the resulting current (from left to right) is arbitrarily assigned a plus sign. The reverse polarity of $V$ (with the right terminal at a higher potential) is then negative; the current it causes is assigned a minus sign.

Figure 26-12b is a plot of $i$ versus $V$ for one device. This plot is a straight line passing through the origin, so the ratio $i/V$ (which is the slope of the straight line) is the same for all values of $V$. This means that the resistance $R = V/i$ of the device is independent of the magnitude and polarity of the applied potential difference $V$.

Figure 26-12c is a plot for another conducting device. Current can exist in this device only when the polarity of $V$ is positive and the applied potential difference is more than about 1.5 V. When current does exist, the relation between $i$ and $V$ is not linear; it depends on the value of the applied potential difference $V$.

We distinguish between the two types of device by saying that one obeys Ohm's law and the other does not.

> ☛ **Ohm's law** is an assertion that the current through a device is *always* directly proportional to the potential difference applied to the device.

(This assertion is correct only in certain situations; still, for historical reasons, the term "law" is used.) The device of Fig. 26-12b—which turns out to be a 1000 $\Omega$ resistor—obeys Ohm's law. The device of Fig. 26-12c—which is called a *pn* junction diode—does not.

> ☛ A conducting device obeys Ohm's law when the resistance of the device is independent of the magnitude and polarity of the applied potential difference.

Modern microelectronics depends almost totally on devices that do *not* obey Ohm's law. Your calculator, for example, is full of them.

It is often contended that $V = iR$ is a statement of Ohm's law. That is not true! This equation is the defining equation for resistance, and it applies to all

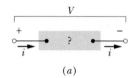

(a)

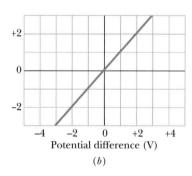

Potential difference (V)

(b)

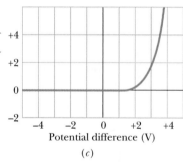

Potential difference (V)

(c)

**FIG. 26-12** (a) A potential difference $V$ is applied to the terminals of a device, establishing a current $i$. (b) A plot of current $i$ versus applied potential difference $V$ when the device is a 1000 $\Omega$ resistor. (c) A plot when the device is a semiconducting *pn* junction diode.

conducting devices, whether they obey Ohm's law or not. If we measure the potential difference $V$ across, and the current $i$ through, any device, even a *pn* junction diode, we can find its resistance *at that value of V* as $R = V/i$. The essence of Ohm's law, however, is that a plot of $i$ versus $V$ is linear; that is, $R$ is independent of $V$.

We can express Ohm's law in a more general way if we focus on conducting *materials* rather than on conducting *devices*. The relevant relation is then Eq. 26-11 ($\vec{E} = \rho\vec{J}$), which corresponds to $V = iR$.

> ☞ A conducting material obeys Ohm's law when the resistivity of the material is independent of the magnitude and direction of the applied electric field.

All homogeneous materials, whether they are conductors like copper or semiconductors like pure silicon or silicon containing special impurities, obey Ohm's law within some range of values of the electric field. If the field is too strong, however, there are departures from Ohm's law in all cases.

✔ **CHECKPOINT 4**    The following table gives the current $i$ (in amperes) through two devices for several values of potential difference $V$ (in volts). From these data, determine which device does not obey Ohm's law.

| Device 1 | | Device 2 | |
|---|---|---|---|
| $V$ | $i$ | $V$ | $i$ |
| 2.00 | 4.50 | 2.00 | 1.50 |
| 3.00 | 6.75 | 3.00 | 2.20 |
| 4.00 | 9.00 | 4.00 | 2.80 |

## 26-6 | A Microscopic View of Ohm's Law

To find out *why* particular materials obey Ohm's law, we must look into the details of the conduction process at the atomic level. Here we consider only conduction in metals, such as copper. We base our analysis on the *free-electron model,* in which we assume that the conduction electrons in the metal are free to move throughout the volume of a sample, like the molecules of a gas in a closed container. We also assume that the electrons collide not with one another but only with atoms of the metal.

According to classical physics, the electrons should have a Maxwellian speed distribution somewhat like that of the molecules in a gas (Section 19-7), and thus the average electron speed should depend on the temperature. The motions of electrons are, however, governed not by the laws of classical physics but by those of quantum physics. As it turns out, an assumption that is much closer to the quantum reality is that conduction electrons in a metal move with a single effective speed $v_{\text{eff}}$, and this speed is essentially independent of the temperature. For copper, $v_{\text{eff}} \approx 1.6 \times 10^6$ m/s.

When we apply an electric field to a metal sample, the electrons modify their random motions slightly and drift very slowly—in a direction opposite that of the field—with an average drift speed $v_d$. As we saw in Sample Problem 26-3, the drift speed in a typical metallic conductor is about $5 \times 10^{-7}$ m/s, less than the effective speed ($1.6 \times 10^6$ m/s) by many orders of magnitude. Figure 26-13 suggests the relation between these two speeds. The gray lines show a possible random path for an electron in the absence of an applied field; the electron proceeds from $A$ to $B$, making six collisions along the way. The green lines show how the same events *might* occur when an electric field $\vec{E}$ is applied. We see that the electron drifts steadily to the right, ending at $B'$ rather than at $B$. Figure 26-13 was drawn with the assumption that $v_d \approx 0.02v_{\text{eff}}$. However, because the actual value is more like $v_d \approx (10^{-13})v_{\text{eff}}$, the drift displayed in the figure is greatly exaggerated.

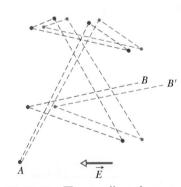

**FIG. 26-13** The gray lines show an electron moving from $A$ to $B$, making six collisions en route. The green lines show what the electron's path might be in the presence of an applied electric field $\vec{E}$. Note the steady drift in the direction of $-\vec{E}$. (Actually, the green lines should be slightly curved, to represent the parabolic paths followed by the electrons between collisions, under the influence of an electric field.)

The motion of conduction electrons in an electric field $\vec{E}$ is thus a combination of the motion due to random collisions and that due to $\vec{E}$. When we consider all the free electrons, their random motions average to zero and make no contribution to the drift speed. Thus, the drift speed is due only to the effect of the electric field on the electrons.

If an electron of mass $m$ is placed in an electric field of magnitude $E$, the electron will experience an acceleration given by Newton's second law:

$$a = \frac{F}{m} = \frac{eE}{m}. \tag{26-23}$$

The nature of the collisions experienced by conduction electrons is such that, after a typical collision, each electron will—so to speak—completely lose its memory of its previous drift velocity. Each electron will then start off fresh after every encounter, moving off in a random direction. In the average time $\tau$ between collisions, the average electron will acquire a drift speed of $v_d = a\tau$. Moreover, if we measure the drift speeds of all the electrons at any instant, we will find that their average drift speed is also $a\tau$. Thus, at any instant, on average, the electrons will have drift speed $v_d = a\tau$. Then Eq. 26-23 gives us

$$v_d = a\tau = \frac{eE\tau}{m}. \tag{26-24}$$

Combining this result with Eq. 26-7 ($\vec{J} = ne\vec{v}_d$), in magnitude form, yields

$$v_d = \frac{J}{ne} = \frac{eE\tau}{m},$$

which we can write as

$$E = \left(\frac{m}{e^2 n\tau}\right) J.$$

Comparing this with Eq. 26-11 ($\vec{E} = \rho\vec{J}$), in magnitude form, leads to

$$\rho = \frac{m}{e^2 n\tau}. \tag{26-25}$$

Equation 26-25 may be taken as a statement that metals obey Ohm's law if we can show that, for metals, their resistivity $\rho$ is a constant, independent of the strength of the applied electric field $\vec{E}$. Because $n$, $m$, and $e$ are constant, this reduces to convincing ourselves that $\tau$, the average time (or *mean free time*) between collisions, is a constant, independent of the strength of the applied electric field. Indeed, $\tau$ can be considered to be a constant because the drift speed $v_d$ caused by the field is so much smaller than the effective speed $v_{\text{eff}}$ that the electron speed—and thus $\tau$—is hardly affected by the field.

---

**Sample Problem** | **26-6**

(a) What is the mean free time $\tau$ between collisions for the conduction electrons in copper?

---

**KEY IDEAS** The mean free time $\tau$ of copper is approximately constant, and in particular does not depend on any electric field that might be applied to a sample of the copper. Thus, we need not consider any particular value of applied electric field. However, because the resistivity $\rho$ displayed by copper under an electric field depends on $\tau$, we can find the mean free time $\tau$ from Eq. 26-25 ($\rho = m/e^2 n\tau$).

*Calculations:* That equation gives us

$$\tau = \frac{m}{ne^2\rho}.$$

We take the value of $n$, the number of conduction electrons per unit volume in copper, from Sample Problem 26-3. We take the value of $\rho$ from Table 26-1. The denominator then becomes

$$(8.49 \times 10^{28} \text{ m}^{-3})(1.6 \times 10^{-19} \text{ C})^2(1.69 \times 10^{-8}\,\Omega\cdot\text{m})$$
$$= 3.67 \times 10^{-17} \text{ C}^2\cdot\Omega/\text{m}^2 = 3.67 \times 10^{-17} \text{ kg/s,}$$

where we converted units as

$$\frac{C^2 \cdot \Omega}{m^2} = \frac{C^2 \cdot V}{m^2 \cdot A} = \frac{C^2 \cdot J/C}{m^2 \cdot C/s} = \frac{kg \cdot m^2/s^2}{m^2/s} = \frac{kg}{s}.$$

Using these results and substituting for the electron mass $m$, we then have

$$\tau = \frac{9.1 \times 10^{-31}\ kg}{3.67 \times 10^{-17}\ kg/s} = 2.5 \times 10^{-14}\ s.\quad \text{(Answer)}$$

(b) The mean free path $\lambda$ of the conduction electrons in a conductor is the average distance traveled by an electron between collisions. (This definition parallels that in Section 19-6 for the mean free path of molecules in a gas.) What is $\lambda$ for the conduction electrons in copper, assuming that their effective speed $v_{eff}$ is $1.6 \times 10^6$ m/s?

**KEY IDEA** The distance $d$ any particle travels in a certain time $t$ at a constant speed $v$ is $d = vt$.

*Calculation:* For the electrons in copper, this gives us

$$\lambda = v_{eff}\tau = (1.6 \times 10^6\ m/s)(2.5 \times 10^{-14}\ s)$$
$$= 4.0 \times 10^{-8}\ m = 40\ nm.\quad \text{(Answer)}$$

This is about 150 times the distance between nearest-neighbor atoms in a copper lattice. Thus, on the average, each conduction electron passes many copper atoms before finally hitting one.

## 26-7 I Power in Electric Circuits

Figure 26-14 shows a circuit consisting of a battery B that is connected by wires, which we assume have negligible resistance, to an unspecified conducting device. The device might be a resistor, a storage battery (a rechargeable battery), a motor, or some other electrical device. The battery maintains a potential difference of magnitude $V$ across its own terminals and thus (because of the wires) across the terminals of the unspecified device, with a greater potential at terminal $a$ of the device than at terminal $b$.

Because there is an external conducting path between the two terminals of the battery, and because the potential differences set up by the battery are maintained, a steady current $i$ is produced in the circuit, directed from terminal $a$ to terminal $b$. The amount of charge $dq$ that moves between those terminals in time interval $dt$ is equal to $i\ dt$. This charge $dq$ moves through a decrease in potential of magnitude $V$, and thus its electric potential energy decreases in magnitude by the amount

$$dU = dq\ V = i\ dt\ V.$$

The principle of conservation of energy tells us that the decrease in electric potential energy from $a$ to $b$ is accompanied by a transfer of energy to some other form. The power $P$ associated with that transfer is the rate of transfer $dU/dt$, which is

$$P = iV \qquad \text{(rate of electrical energy transfer).} \qquad (26\text{-}26)$$

Moreover, this power $P$ is also the rate at which energy is transferred from the battery to the unspecified device. If that device is a motor connected to a mechanical load, the energy is transferred as work done on the load. If the device is a storage battery that is being charged, the energy is transferred to stored chemical energy in the storage battery. If the device is a resistor, the energy is transferred to internal thermal energy, tending to increase the resistor's temperature.

The unit of power that follows from Eq. 26-26 is the volt-ampere (V·A). We can write it as

$$1\ V \cdot A = \left(1\ \frac{J}{C}\right)\left(1\ \frac{C}{s}\right) = 1\ \frac{J}{s} = 1\ W.$$

As an electron moves through a resistor at constant drift speed, its average kinetic energy remains constant and its lost electric potential energy appears as thermal energy in the resistor and the surroundings. On a microscopic scale this energy transfer is due to collisions between the electron and the molecules of the resistor, which leads to an increase in the temperature of the resistor lattice. The mechanical energy thus transferred to thermal energy is *dissipated* (lost) because the transfer cannot be reversed.

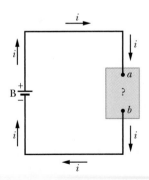

**FIG. 26-14** A battery B sets up a current $i$ in a circuit containing an unspecified conducting device.

For a resistor or some other device with resistance $R$, we can combine Eqs. 26-8 ($R = V/i$) and 26-26 to obtain, for the rate of electrical energy dissipation due to a resistance, either

$$P = i^2R \quad \text{(resistive dissipation)} \tag{26-27}$$

or

$$P = \frac{V^2}{R} \quad \text{(resistive dissipation).} \tag{26-28}$$

*Caution:* We must be careful to distinguish these two equations from Eq. 26-26: $P = iV$ applies to electrical energy transfers of all kinds; $P = i^2R$ and $P = V^2/R$ apply only to the transfer of electric potential energy to thermal energy in a device with resistance.

✓**CHECKPOINT 5**    A potential difference $V$ is connected across a device with resistance $R$, causing current $i$ through the device. Rank the following variations according to the change in the rate at which electrical energy is converted to thermal energy due to the resistance, greatest change first: (a) $V$ is doubled with $R$ unchanged, (b) $i$ is doubled with $R$ unchanged, (c) $R$ is doubled with $V$ unchanged, (d) $R$ is doubled with $i$ unchanged.

## Sample Problem | 26-7

You are given a length of uniform heating wire made of a nickel–chromium–iron alloy called Nichrome; it has a resistance $R$ of 72 $\Omega$. At what rate is energy dissipated in each of the following situations? (1) A potential difference of 120 V is applied across the full length of the wire. (2) The wire is cut in half, and a potential difference of 120 V is applied across the length of each half.

**KEY IDEA**    Current in a resistive material produces a transfer of mechanical energy to thermal energy; the rate of transfer (dissipation) is given by Eqs. 26-26 to 26-28.

*Calculations:* Because we know the potential $V$ and resistance $R$, we use Eq. 26-28, which yields, for situation 1,

$$P = \frac{V^2}{R} = \frac{(120 \text{ V})^2}{72 \ \Omega} = 200 \text{ W}. \quad \text{(Answer)}$$

In situation 2, the resistance of each half of the wire is (72 $\Omega$)/2, or 36 $\Omega$. Thus, the dissipation rate for each half is

$$P' = \frac{(120 \text{ V})^2}{36 \ \Omega} = 400 \text{ W},$$

and that for the two halves is

$$P = 2P' = 800 \text{ W}. \quad \text{(Answer)}$$

This is four times the dissipation rate of the full length of wire. Thus, you might conclude that you could buy a heating coil, cut it in half, and reconnect it to obtain four times the heat output. Why is this unwise? (What would happen to the amount of current in the coil?)

## 26-8 | Semiconductors

Semiconducting devices are at the heart of the microelectronic revolution that ushered in the information age. Table 26-2 compares the properties of silicon—a typical semiconductor—and copper—a typical metallic conductor. We see that silicon has many fewer charge carriers, a much higher resistivity, and a temperature coefficient of resistivity that is both large and negative. Thus, although the resistivity of copper increases with increasing temperature, that of pure silicon decreases.

**TABLE 26-2**

**Some Electrical Properties of Copper and Silicon**

| Property | Copper | Silicon |
|---|---|---|
| Type of material | Metal | Semiconductor |
| Charge carrier density, m$^{-3}$ | $8.49 \times 10^{28}$ | $1 \times 10^{16}$ |
| Resistivity, $\Omega \cdot$ m | $1.69 \times 10^{-8}$ | $2.5 \times 10^3$ |
| Temperature coefficient of resistivity, K$^{-1}$ | $+4.3 \times 10^{-3}$ | $-70 \times 10^{-3}$ |

Pure silicon has such a high resistivity that it is effectively an insulator and thus not of much direct use in microelectronic circuits. However, its resistivity can be greatly reduced in a controlled way by adding minute amounts of specific "impurity" atoms in a process called *doping*. Table 26-1 gives typical values of resistivity for silicon before and after doping with two different impurities.

We can roughly explain the differences in resistivity (and thus in conductivity) between semiconductors, insulators, and metallic conductors in terms of the energies of their electrons. (We need quantum physics to explain in more detail.) In a metallic conductor such as copper wire, most of the electrons are firmly locked in place within the atoms; much energy would be required to free them so they could move and participate in an electric current. However, there are also some electrons that, roughly speaking, are only loosely held in place and that require only little energy to become free. Thermal energy can supply that energy, as can an electric field applied across the conductor. The field would not only free these loosely held electrons but would also propel them along the wire; thus, the field would drive a current through the conductor.

In an insulator, significantly greater energy is required to free electrons so they can move through the material. Thermal energy cannot supply enough energy, and neither can any reasonable electric field applied to the insulator. Thus, no electrons are available to move through the insulator, and hence no current occurs even with an applied electric field.

A semiconductor is like an insulator *except* that the energy required to free some electrons is not quite so great. More important, doping can supply electrons or positive charge carriers that are very loosely held within the material and thus are easy to get moving. Moreover, by controlling the doping of a semiconductor, we can control the density of charge carriers that can participate in a current and thereby can control some of its electrical properties. Most semiconducting devices, such as transistors and junction diodes, are fabricated by the selective doping of different regions of the silicon with impurity atoms of different kinds.

Let us now look again at Eq. 26-25 for the resistivity of a conductor:

$$\rho = \frac{m}{e^2 n \tau}, \tag{26-29}$$

where $n$ is the number of charge carriers per unit volume and $\tau$ is the mean time between collisions of the charge carriers. (We derived this equation for conductors, but it also applies to semiconductors.) Let us consider how the variables $n$ and $\tau$ change as the temperature is increased.

In a conductor, $n$ is large but very nearly constant with any change in temperature. The increase of resistivity with temperature for metals (Fig. 26-10) is due to an increase in the collision rate of the charge carriers, which shows up in Eq. 26-29 as a decrease in $\tau$, the mean time between collisions.

In a semiconductor, $n$ is small but increases very rapidly with temperature as the increased thermal agitation makes more charge carriers available. This causes a *decrease* of resistivity with increasing temperature, as indicated by the negative temperature coefficient of resistivity for silicon in Table 26-2. The same increase in collision rate that we noted for metals also occurs for semiconductors, but its effect is swamped by the rapid increase in the number of charge carriers.

## 26-9 | Superconductors

In 1911, Dutch physicist Kamerlingh Onnes discovered that the resistivity of mercury absolutely disappears at temperatures below about 4 K (Fig. 26-15). This phenomenon of **superconductivity** is of vast potential importance in technology because it means that charge can flow through a superconducting conductor without losing its energy to thermal energy. Currents created in a superconducting ring, for example, have persisted for several years without loss; the electrons making up the current require a force and a source of energy at start-up time but not thereafter.

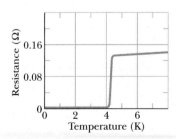

**FIG. 26-15** The resistance of mercury drops to zero at a temperature of about 4 K.

A disk-shaped magnet is levitated above a superconducting material that has been cooled by liquid nitrogen. The goldfish is along for the ride. *(Courtesy Shoji Tonaka/International Superconductivity Technology Center, Tokyo, Japan)*

Prior to 1986, the technological development of superconductivity was throttled by the cost of producing the extremely low temperatures required to achieve the effect. In 1986, however, new ceramic materials were discovered that become superconducting at considerably higher (and thus cheaper to produce) temperatures. Practical application of superconducting devices at room temperature may eventually become commonplace.

Superconductivity is a phenomenon much different from conductivity. In fact, the best of the normal conductors, such as silver and copper, cannot become superconducting at any temperature, and the new ceramic superconductors are actually good insulators when they are not at low enough temperatures to be in a superconducting state.

One explanation for superconductivity is that the electrons that make up the current move in coordinated pairs. One of the electrons in a pair may electrically distort the molecular structure of the superconducting material as it moves through, creating nearby a short-lived concentration of positive charge. The other electron in the pair may then be attracted toward this positive charge. According to the theory, such coordination between electrons would prevent them from colliding with the molecules of the material and thus would eliminate electrical resistance. The theory worked well to explain the pre-1986, lower temperature superconductors, but new theories appear to be needed for the newer, higher temperature superconductors.

## REVIEW & SUMMARY

**Current** An **electric current** $i$ in a conductor is defined by

$$i = \frac{dq}{dt}.$$ (26-1)

Here $dq$ is the amount of (positive) charge that passes in time $dt$ through a hypothetical surface that cuts across the conductor. By convention, the direction of electric current is taken as the direction in which positive charge carriers would move. The SI unit of electric current is the **ampere** (A): 1 A = 1 C/s.

**Current Density** Current (a scalar) is related to **current density** $\vec{J}$ (a vector) by

$$i = \int \vec{J} \cdot d\vec{A},$$ (26-4)

where $d\vec{A}$ is a vector perpendicular to a surface element of area $dA$ and the integral is taken over any surface cutting across the conductor. $\vec{J}$ has the same direction as the velocity of the moving charges if they are positive and the opposite direction if they are negative.

**Drift Speed of the Charge Carriers** When an electric field $\vec{E}$ is established in a conductor, the charge carriers (assumed positive) acquire a **drift speed** $v_d$ in the direction of $\vec{E}$; the velocity $\vec{v}_d$ is related to the current density by

$$\vec{J} = (ne)\vec{v}_d,$$ (26-7)

where $ne$ is the *carrier charge density*.

**Resistance of a Conductor** The **resistance** $R$ of a conductor is defined as

$$R = \frac{V}{i} \quad \text{(definition of } R\text{)},$$ (26-8)

where $V$ is the potential difference across the conductor and $i$ is the current. The SI unit of resistance is the **ohm** ($\Omega$): 1 $\Omega$ = 1 V/A. Similar equations define the **resistivity** $\rho$ and **conductivity** $\sigma$ of a material:

$$\rho = \frac{1}{\sigma} = \frac{E}{J} \quad \text{(definitions of } \rho \text{ and } \sigma\text{)}, \quad (26\text{-}12, 26\text{-}10)$$

where $E$ is the magnitude of the applied electric field. The SI unit of resistivity is the ohm-meter ($\Omega \cdot$m). Equation 26-10 corresponds to the vector equation

$$\vec{E} = \rho\vec{J}.$$ (26-11)

The resistance $R$ of a conducting wire of length $L$ and uniform cross section is

$$R = \rho\frac{L}{A},$$ (26-16)

where $A$ is the cross-sectional area.

**Change of $\rho$ with Temperature** The resistivity $\rho$ for most materials changes with temperature. For many materials, including metals, the relation between $\rho$ and temperature $T$ is approximated by the equation

$$\rho - \rho_0 = \rho_0\alpha(T - T_0).$$ (26-17)

Here $T_0$ is a reference temperature, $\rho_0$ is the resistivity at $T_0$, and $\alpha$ is the temperature coefficient of resistivity for the material.

**Ohm's Law** A given device (conductor, resistor, or any other electrical device) obeys *Ohm's law* if its resistance $R$, defined by Eq. 26-8 as $V/i$, is independent of the applied potential difference $V$. A given *material* obeys Ohm's law if its resistivity, defined by Eq. 26-10, is independent of the magnitude and direction of the applied electric field $\vec{E}$.

**Resistivity of a Metal** By assuming that the conduction electrons in a metal are free to move like the molecules of a gas, it is possible to derive an expression for the resistivity of a metal:

$$\rho = \frac{m}{e^2 n\tau}.$$ (26-25)

Here $n$ is the number of free electrons per unit volume and $\tau$ is the mean time between the collisions of an electron with the atoms of the metal. We can explain why metals obey Ohm's law by pointing out that $\tau$ is essentially independent of the magnitude $E$ of any electric field applied to a metal.

**Power** The power $P$, or rate of energy transfer, in an electrical device across which a potential difference $V$ is maintained is

$$P = iV \quad \text{(rate of electrical energy transfer).} \quad (26\text{-}26)$$

**Resistive Dissipation** If the device is a resistor, we can write Eq. 26-26 as

$$P = i^2 R = \frac{V^2}{R} \quad \text{(resistive dissipation).} \quad (26\text{-}27, 26\text{-}28)$$

In a resistor, electric potential energy is converted to internal thermal energy via collisions between charge carriers and atoms.

**Semiconductors** *Semiconductors* are materials that have few conduction electrons but can become conductors when they are *doped* with other atoms that contribute free electrons.

**Superconductors** *Superconductors* are materials that lose all electrical resistance at low temperatures. Recent research has discovered materials that are superconducting at surprisingly high temperatures.

## QUESTIONS

**1** Figure 26-16 shows four situations in which positive and negative charges move horizontally and gives the rate at which each charge moves. Rank the situations according to the effective current through the regions, greatest first.

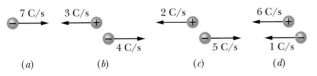

FIG. 26-16 Question 1.

**2** Figure 26-17 shows plots of the current $i$ through a certain cross section of a wire over four different time periods. Rank the periods according to the net charge that passes through the cross section during the period, greatest first.

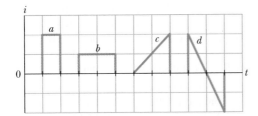

FIG. 26-17
Question 2.

**3** Figure 26-18 shows cross sections through three long conductors of the same length and material, with square cross sections of edge lengths as shown. Conductor $B$ fits snugly within conductor $A$, and conductor $C$ fits snugly within conductor $B$. Rank the following according to their end-to-end resistances, greatest first: the individual conductors and the combinations of $A + B$ ($B$ inside $A$), $B + C$ ($C$ inside $B$), and $A + B + C$ ($B$ inside $A$ inside $C$).

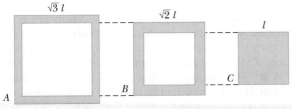

FIG. 26-18 Question 3.

**4** Figure 26-19 shows cross sections through three wires of identical length and material; the sides are given in millimeters. Rank the wires according to their resistance (measured end to end along each wire's length), greatest first.

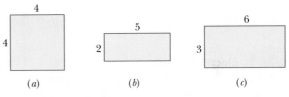

FIG. 26-19 Question 4.

**5** Figure 26-20 shows a rectangular solid conductor of edge lengths $L$, $2L$, and $3L$. A potential difference $V$ is to be applied uniformly between pairs of opposite faces of the conductor as in Fig. 26-8b. First $V$ is applied between the left–right faces, then between the top–bottom faces, and then between the front–back faces. Rank those pairs, greatest first, according to the following (within the conductor): (a) the magnitude of the electric field, (b) the current density, (c) the current, and (d) the drift speed of the electrons.

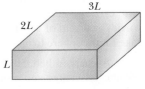

FIG. 26-20 Question 5.

**6** The following table gives the lengths of three copper rods, their diameters, and the potential differences between their ends. Rank the rods according to (a) the magnitude of the electric field within them, (b) the current density within them, and (c) the drift speed of electrons through them, greatest first.

| Rod | Length | Diameter | Potential Difference |
|-----|--------|----------|----------------------|
| 1 | $L$ | $3d$ | $V$ |
| 2 | $2L$ | $d$ | $2V$ |
| 3 | $3L$ | $2d$ | $2V$ |

**7** Figure 26-21 gives the drift speed $v_d$ of conduction electrons in a copper wire versus position $x$ along the wire. The wire consists of three sections that differ in radius. Rank the three sections according to the following

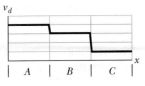

FIG. 26-21 Question 7.

quantities, greatest first: (a) radius, (b) number of conduction electrons per cubic meter, (c) magnitude of electric field, (d) conductivity.

**8** Three wires, of the same diameter, are connected in turn between two points maintained at a constant potential difference. Their resistivities and lengths are $\rho$ and $L$ (wire $A$), $1.2\rho$ and $1.2L$ (wire $B$), and $0.9\rho$ and $L$ (wire $C$). Rank the wires according to the rate at which energy is transferred to thermal energy, greatest first.

**9** Figure 26-22 gives the electric potential $V(x)$ versus position $x$ along a copper wire carrying current. The wire consists of three sections that differ in radius. Rank the three sections according to the magnitude of the (a) electric field and (b) current density, greatest first.

**10** In Fig. 26-23, a wire that carries a current consists of three sections with different radii. Rank the sections according to the following quantities, greatest first: (a) current, (b) magnitude of current density, and (c) magnitude of electric field.

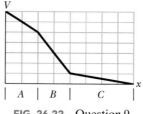

**FIG. 26-22** Question 9.

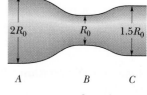

**FIG. 26-23** Question 10.

## PROBLEMS

| | |
|---|---|
| **GO** | Tutoring problem available (at instructor's discretion) in *WileyPLUS* and WebAssign |
| **SSM** | Worked-out solution available in Student Solutions Manual |
| • – ••• | Number of dots indicates level of problem difficulty |
| ✈ | Additional information available in *The Flying Circus of Physics* and at flyingcircusofphysics.com |

**WWW** Worked-out solution is at ——
**ILW** Interactive solution is at —— http://www.wiley.com/college/halliday

### sec. 26-2 Electric Current

**•1** During the 4.0 min a 5.0 A current is set up in a wire, how many (a) coulombs and (b) electrons pass through any cross section across the wire's width?

**••2** An isolated conducting sphere has a 10 cm radius. One wire carries a current of 1.000 002 0 A into it. Another wire carries a current of 1.000 000 0 A out of it. How long would it take for the sphere to increase in potential by 1000 V?

**••3** A charged belt, 50 cm wide, travels at 30 m/s between a source of charge and a sphere. The belt carries charge into the sphere at a rate corresponding to 100 $\mu$A. Compute the surface charge density on the belt.

### sec. 26-3 Current Density

**•4** A small but measurable current of $1.2 \times 10^{-10}$ A exists in a copper wire whose diameter is 2.5 mm. The number of charge carriers per unit volume is $8.49 \times 10^{28}$ m$^{-3}$. Assuming the current is uniform, calculate the (a) current density and (b) electron drift speed.

**•5** A fuse in an electric circuit is a wire that is designed to melt, and thereby open the circuit, if the current exceeds a predetermined value. Suppose that the material to be used in a fuse melts when the current density rises to 440 A/cm$^2$. What diameter of cylindrical wire should be used to make a fuse that will limit the current to 0.50 A?

**•6** The (United States) National Electric Code, which sets maximum safe currents for insulated copper wires of various diameters, is given (in part) in the table. Plot the safe current density as a function of diameter. Which wire gauge has the maximum safe current density? ("Gauge" is a way of identifying wire diameters, and 1 mil = $10^{-3}$ in.)

| Gauge | 4 | 6 | 8 | 10 | 12 | 14 | 16 | 18 |
|---|---|---|---|---|---|---|---|---|
| Diameter, mils | 204 | 162 | 129 | 102 | 81 | 64 | 51 | 40 |
| Safe current, A | 70 | 50 | 35 | 25 | 20 | 15 | 6 | 3 |

**•7** A beam contains $2.0 \times 10^8$ doubly charged positive ions per cubic centimeter, all of which are moving north with a speed of $1.0 \times 10^5$ m/s. What are the (a) magnitude and (b) direction of the current density $\vec{J}$? (c) What additional quantity do you need to calculate the total current $i$ in this ion beam? **SSM WWW**

**•8** A certain cylindrical wire carries current. We draw a circle of radius $r$ around its central axis in Fig. 26-24$a$ to determine the current $i$ within the circle. Figure 26-24$b$ shows current $i$ as a function of $r^2$. The vertical scale is set by $i_s = 4.0$ mA, and the horizontal scale is set by $r_s^2 = 4.0$ mm$^2$. (a) Is the current density uniform? (b) If so, what is its magnitude?

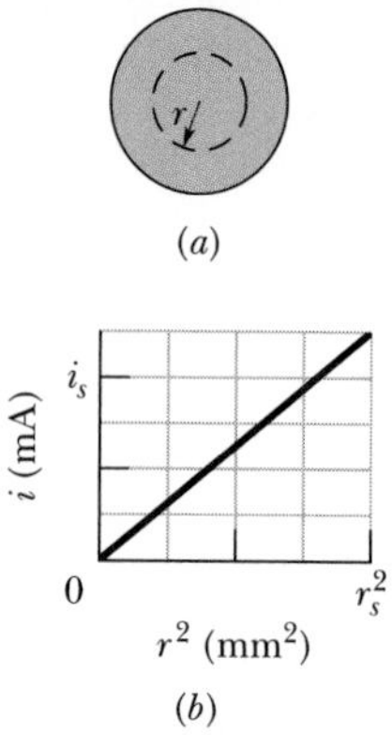

(a)

**••9** How long does it take electrons to get from a car battery to the starting motor? Assume the current is 300 A and the electrons travel through a copper wire with cross-sectional area 0.21 cm$^2$ and length 0.85 m. The number of charge carriers per unit volume is $8.49 \times 10^{28}$ m$^{-3}$. **ILW**

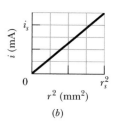

(b)

**FIG. 26-24**
Problem 8.

**••10** Near Earth, the density of protons in the solar wind (a stream of particles from the Sun) is 8.70 cm$^{-3}$, and their speed is 470 km/s. (a) Find the current density of these protons. (b) If Earth's magnetic field did not deflect the protons, what total current would Earth receive?

**••11** The magnitude $J(r)$ of the current density in a certain cylindrical wire is given as a function of radial distance from the center of the wire's cross section as $J(r) = Br$, where $r$ is in meters, $J$ is in amperes per square meter, and $B = 2.00 \times 10^5$ A/m$^3$. This function applies out to the wire's radius of 2.00 mm. How much current is contained within the width of a thin ring concentric with the wire if the ring has a radial width of 10.0 $\mu$m and is at a radial distance of 1.20 mm?

**••12** The magnitude $J$ of the current density in a certain wire with a circular cross section of radius $R = 2.00$ mm is given by $J = (3.00 \times 10^8)r^2$, with $J$ in amperes per square meter and radial distance $r$ in meters. What is the current through the outer section bounded by $r = 0.900R$ and $r = R$?

•**13** What is the current in a wire of radius $R = 3.40$ mm if the magnitude of the current density is given by (a) $J_a = J_0r/R$ and (b) $J_b = J_0(1 - r/R)$, in which $r$ is the radial distance and $J_0 = 5.50 \times 10^4$ A/m$^2$? (c) Which function maximizes the current density near the wire's surface?

### sec. 26-4  Resistance and Resistivity

•**14** Copper and aluminum are being considered for a high-voltage transmission line that must carry a current of 60.0 A. The resistance per unit length is to be 0.150 $\Omega$/km. The densities of copper and aluminum are 8960 and 2600 kg/m$^3$, respectively. Compute (a) the magnitude $J$ of the current density and (b) the mass per unit length $\lambda$ for a copper cable and (c) $J$ and (d) $\lambda$ for an aluminum cable.

•**15** A wire of Nichrome (a nickel–chromium–iron alloy commonly used in heating elements) is 1.0 m long and 1.0 mm$^2$ in cross-sectional area. It carries a current of 4.0 A when a 2.0 V potential difference is applied between its ends. Calculate the conductivity $\sigma$ of Nichrome.

•**16** A wire 4.00 m long and 6.00 mm in diameter has a resistance of 15.0 m$\Omega$. A potential difference of 23.0 V is applied between the ends. (a) What is the current in the wire? (b) What is the magnitude of the current density? (c) Calculate the resistivity of the wire material. (d) Using Table 26-1, identify the material.

•**17** What is the resistivity of a wire of 1.0 mm diameter, 2.0 m length, and 50 m$\Omega$ resistance?  **SSM**

•**18** A certain wire has a resistance $R$. What is the resistance of a second wire, made of the same material, that is half as long and has half the diameter?

•**19** A coil is formed by winding 250 turns of insulated 16-gauge copper wire (diameter = 1.3 mm) in a single layer on a cylindrical form of radius 12 cm. What is the resistance of the coil? Neglect the thickness of the insulation. (Use Table 26-1.)  **SSM**

•**20** A human being can be electrocuted if a current as small as 50 mA passes near the heart. An electrician working with sweaty hands makes good contact with the two conductors he is holding, one in each hand. If his resistance is 2000 $\Omega$, what might the fatal voltage be?

••**21** A wire with a resistance of 6.0 $\Omega$ is drawn out through a die so that its new length is three times its original length. Find the resistance of the longer wire, assuming that the resistivity and density of the material are unchanged.  **SSM ILW**

••**22** Figure 26-25a gives the magnitude $E(x)$ of the electric

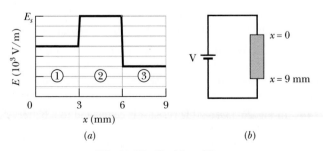

FIG. 26-25  Problem 22.

fields that have been set up by a battery along a resistive rod of length 9.00 mm (Fig. 26-25b). The vertical scale is set by $E_s = 4.00 \times 10^3$ V/m. The rod consists of three sections of the same material but with different radii. (The schematic diagram of Fig. 26-25b does not indicate the different radii.) The radius of section 3 is 2.00 mm. What is the radius of (a) section 1 and (b) section 2?

••**23** Two conductors are made of the same material and have the same length. Conductor $A$ is a solid wire of diameter 1.0 mm. Conductor $B$ is a hollow tube of outside diameter 2.0 mm and inside diameter 1.0 mm. What is the resistance ratio $R_A/R_B$, measured between their ends?  **SSM WWW**

••**24** Figure 26-26 gives the electric potential $V(x)$ along a copper wire carrying uniform current, from a point of higher potential $V_s = 12.0 \ \mu$V at $x = 0$ to a point of zero potential at $x_s = 3.00$ m. The wire has a radius of 2.00 mm. What is the current in the wire?  **GO**

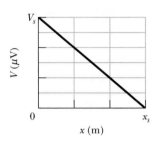

FIG. 26-26  Problem 24.

••**25** A common flashlight bulb is rated at 0.30 A and 2.9 V (the values of the current and voltage under operating conditions). If the resistance of the tungsten bulb filament at room temperature (20°C) is 1.1 $\Omega$, what is the temperature of the filament when the bulb is on?  **ILW**

••**26** *Kiting during a storm.* The legend that Benjamin Franklin flew a kite as a storm approached is only a legend—he was neither stupid nor suicidal. Suppose a kite string of radius 2.00 mm extends directly upward by 0.800 km and is coated with a 0.500 mm layer of water having resistivity 150$\Omega \cdot$m. If the potential difference between the two ends of the string is 160 MV, what is the current through the water layer? The danger is not this current but the chance that the string draws a lightning strike, which can have a current as large as 500 000 A (way beyond just being lethal).

••**27** A potential difference of 3.00 nV is set up across a 2.00 cm length of copper wire that has a radius of 2.00 mm. How much charge drifts through a cross section in 3.00 ms?

••**28** In Fig. 26-27a, a 9.00 V battery is connected to a resistive strip that consists of three sections with the same cross-sectional areas but different conductivities. Figure 26-27b gives the electric potential $V(x)$ versus position $x$ along the strip. The horizontal scale is set by $x_s = 8.00$ mm. Section 3 has conductivity $3.00 \times 10^7$ $(\Omega \cdot$m$)^{-1}$. What is the conductivity of section (a) 1 and (b) 2?

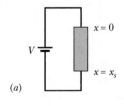

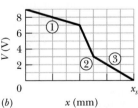

FIG. 26-27  Problem 28.

••**29** When 115 V is applied across a wire that is 10 m long and has a 0.30 mm radius, the magnitude of the current density is $1.4 \times 10^4$ A/m$^2$. Find the resistivity of the wire.

••**30** Earth's lower atmosphere contains negative and positive ions that are produced by radioactive elements in the soil and cosmic rays from space. In a certain region, the atmospheric electric field strength is 120 V/m and the field is directed vertically down. This field causes singly charged positive ions, at a density of 620 cm$^{-3}$, to drift downward and singly charged negative ions, at a density of 550 cm$^{-3}$, to drift upward (Fig. 26-28). The measured conductivity of the air in that region is $2.70 \times 10^{-14}$ $(\Omega \cdot \mathrm{m})^{-1}$. Calculate (a) the magnitude of the current density and (b) the ion drift speed, assumed to be the same for positive and negative ions.

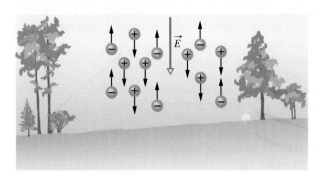

**FIG. 26-28**    Problem 30.

**••31**    A block in the shape of a rectangular solid has a cross-sectional area of 3.50 cm$^2$ across its width, a front-to-rear length of 15.8 cm, and a resistance of 935 $\Omega$. The block's material contains $5.33 \times 10^{22}$ conduction electrons/m$^3$. A potential difference of 35.8 V is maintained between its front and rear faces. (a) What is the current in the block? (b) If the current density is uniform, what is its magnitude? What are (c) the drift velocity of the conduction electrons and (d) the magnitude of the electric field in the block?

**••32**    If the gauge number of a wire is increased by 6, the diameter is halved; if a gauge number is increased by 1, the diameter decreases by the factor $2^{1/6}$ (see the table in Problem 6). Knowing this, and knowing that 1000 ft of 10-gauge copper wire has a resistance of approximately 1.00 $\Omega$, estimate the resistance of 25 ft of 22-gauge copper wire.

**••33**    An electrical cable consists of 125 strands of fine wire, each having 2.65 $\mu\Omega$ resistance. The same potential difference is applied between the ends of all the strands and results in a total current of 0.750 A. (a) What is the current in each strand? (b) What is the applied potential difference? (c) What is the resistance of the cable?

**••34**    *Swimming during a storm.* Figure 26-29 shows a swimmer at distance $D = 35.0$ m from a lightning strike to the water, with current $I = 78$ kA. The water has resistivity 30 $\Omega \cdot \mathrm{m}$, the width of the swimmer along a radial line from the strike is 0.70 m, and his resistance across that width is 4.00 k$\Omega$. Assume that the current spreads through the water over a hemisphere centered on the strike point. What is the current through the swimmer?

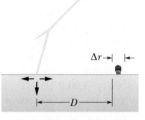

**FIG. 26-29**    Problem 34.

**•••35**    In Fig. 26-30, current is set up through a truncated right circular cone of resistivity 731 $\Omega \cdot \mathrm{m}$, left radius $a = 2.00$ mm, right radius $b = 2.30$ mm, and length $L = 1.94$ cm. Assume

that the current density is uniform across any cross section taken perpendicular to the length. What is the resistance of the cone?

**FIG. 26-30**    Problem 35.

**•••36**    Figure 26-31 shows wire section 1 of diameter $D_1 = 4.00R$ and wire section 2 of diameter $D_2 = 2.00R$, connected by a tapered section. The wire is copper and carries a current. Assume that the current is uniformly distributed across any cross-sectional area through the wire's width. The electric potential change $V$ along the length $L = 2.00$ m shown in section 2 is 10.0 $\mu$V. The number of charge carriers per unit volume is $8.49 \times 10^{28}$ m$^{-3}$. What is the drift speed of the conduction electrons in section 1?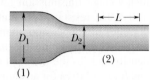

**FIG. 26-31**    Problem 36.

### sec. 26-6    A Microscopic View of Ohm's Law
**••37**    Show that, according to the free-electron model of electrical conduction in metals and classical physics, the resistivity of metals should be proportional to $\sqrt{T}$, where $T$ is the temperature in kelvins. (See Eq. 19-31.)

### sec. 26-7    Power in Electric Circuits
**•38**    A student kept his 9.0 V, 7.0 W radio turned on at full volume from 9:00 P.M. until 2:00 A.M. How much charge went through it?

**•39**    A 120 V potential difference is applied to a space heater whose resistance is 14 $\Omega$ when hot. (a) At what rate is electrical energy transferred to thermal energy? (b) What is the cost for 5.0 h at US$0.05/kW $\cdot$ h?    **SSM**

**•40**    In Fig. 26-32, a battery of potential difference $V = 12$ V is connected to a resistive strip of resistance $R = 6.0$ $\Omega$. When an electron moves through the strip from one end to the other, (a) in which direction in the figure does the electron move, (b) how much work is done on the electron by the electric field in the strip, and (c) how much energy is transferred to the thermal energy of the strip by the electron?

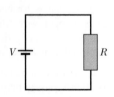

**FIG. 26-32**
Problem 40.

**•41**    An unknown resistor is connected between the terminals of a 3.00 V battery. Energy is dissipated in the resistor at the rate of 0.540 W. The same resistor is then connected between the terminals of a 1.50 V battery. At what rate is energy now dissipated?    **ILW**

**•42**    Thermal energy is produced in a resistor at a rate of 100 W when the current is 3.00 A. What is the resistance?

**•43**    A 1250 W radiant heater is constructed to operate at 115 V. (a) What is the current in the heater when the unit is operating? (b) What is the resistance of the heating coil? (c) How much thermal energy is produced in 1.0 h?    **SSM ILW**

**•44**    In Fig. 26-33a, a 20 $\Omega$ resistor is connected to a battery. Figure 26-33b shows the increase of thermal energy $E_{\mathrm{th}}$ in the resistor as a function of time $t$. The vertical scale is set by $E_{\mathrm{th},s} = 2.50$ mJ, and the horizontal scale is set by $t_s = 4.0$ s. What is the electric potential across the battery?

**•45** A certain brand of hot-dog cooker works by applying a potential difference of 120 V across opposite ends of a hot dog and allowing it to cook by means of the thermal energy produced. The current is 10.0 A, and the energy required to cook one hot dog is 60.0 kJ. If the rate at which energy is supplied is unchanged, how long will it take to cook three hot dogs simultaneously?

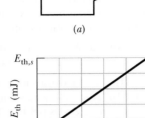

(a)

(b)

**FIG. 26-33** Problem 44.

**••46** *Exploding shoes.* The rain-soaked shoes of a person may explode if ground current from nearby lightning vaporizes the water. The sudden conversion of water to water vapor causes a dramatic expansion that can rip apart shoes. Water has density 1000 kg/m³ and requires 333 kJ/kg to be vaporized. If horizontal current lasts 2.00 ms and encounters water with resistivity 150 $\Omega \cdot$m, length 12.0 cm, and vertical cross-sectional area $15 \times 10^{-5}$ m², what average current is required to vaporize the water?

**••47** A 120 V potential difference is applied to a space heater that dissipates 500 W during operation. (a) What is its resistance during operation? (b) At what rate do electrons flow through any cross section of the heater element?

**••48** The current through the battery and resistors 1 and 2 in Fig. 26-34a is 2.00 A. Energy is transferred from the current to thermal energy $E_{th}$ in both resistors. Curves 1 and 2 in Fig. 26-34b give that thermal energy $E_{th}$ for resistors 1 and 2, respectively, as a function of time $t$. The vertical scale is set by $E_{th,s} = 40.0$ mJ, and the horizontal scale is set by $t_s = 5.00$ s. What is the power of the battery?

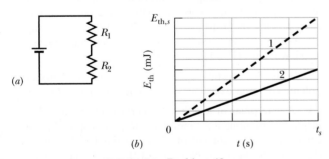

(a)

(b)          $t$ (s)

**FIG. 26-34** Problem 48.

**••49** A heating element is made by maintaining a potential difference of 75.0 V across the length of a Nichrome wire that has a $2.60 \times 10^{-6}$ m² cross section. Nichrome has a resistivity of $5.00 \times 10^{-7}$ $\Omega \cdot$m. (a) If the element dissipates 5000 W, what is its length? (b) If 100 V is used to obtain the same dissipation rate, what should the length be?

**••50** The current-density magnitude in a certain circular wire is $J = (2.75 \times 10^{10}$ A/m⁴$)r^2$, where $r$ is the radial distance out to the wire's radius of 3.00 mm. The potential applied to the wire (end to end) is 60.0 V. How much energy is converted to thermal energy in 1.00 h? **GO**

**••51** A 100 W lightbulb is plugged into a standard 120 V outlet. (a) How much does it cost per 31-day month to leave the light turned on continuously? Assume electrical energy costs US$0.06/kW $\cdot$ h. (b) What is the resistance of the bulb? (c) What is the current in the bulb?

**••52** A copper wire of cross-sectional area $2.00 \times 10^{-6}$ m² and length 4.00 m has a current of 2.00 A uniformly distributed across that area. (a) What is the magnitude of the electric field along the wire? (b) How much electrical energy is transferred to thermal energy in 30 min? **GO**

**••53** Wire $C$ and wire $D$ are made from different materials and have length $L_C = L_D = 1.0$ m. The resistivity and diameter of wire $C$ are $2.0 \times 10^{-6}$ $\Omega \cdot$m and 1.00 mm, and those of wire $D$ are $1.0 \times 10^{-6}$ $\Omega \cdot$m and 0.50 mm. The wires are joined as shown in Fig. 26-35, and a current of 2.0 A is set up in them. What is the electric potential difference between (a) points 1 and 2 and (b) points 2 and 3? What is the rate at which energy is dissipated between (c) points 1 and 2 and (d) points 2 and 3? **SSM WWW**

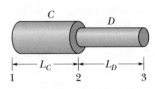

**FIG. 26-35** Problem 53.

**•••54** Figure 26-36a shows a rod of resistive material. The resistance per unit length of the rod increases in the positive direction of the $x$ axis. At any position $x$ along the rod, the resistance $dR$ of a narrow (differential) section of width $dx$ is given by $dR = 5.00x \, dx$, where $dR$ is in ohms and $x$ is in meters. Figure 26-36b shows such a narrow section. You are to slice off a length of the rod between $x = 0$ and some position $x = L$ and then connect that length to a battery with potential difference $V = 5.0$ V (Fig. 26-36c). You want the current in the length to transfer energy to thermal energy at the rate of 200 W. At what position $x = L$ should you cut the rod?

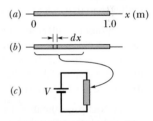

**FIG. 26-36** Problem 54.

**Additional Problems**

**55** A steady beam of alpha particles ($q = +2e$) traveling with constant kinetic energy 20 MeV carries a current of 0.25 $\mu$A. (a) If the beam is directed perpendicular to a flat surface, how many alpha particles strike the surface in 3.0 s? (b) At any instant, how many alpha particles are there in a given 20 cm length of the beam? (c) Through what potential difference is it necessary to accelerate each alpha particle from rest to bring it to an energy of 20 MeV? **SSM**

**56** *The chocolate crumb mystery.* This story begins with Problem 56 in Chapter 23 and continues through Chapters 24 and 25. The chocolate crumb powder moved to the silo through a pipe of radius $R$ with uniform speed $v$ and uniform charge density $\rho$. (a) Find an expression for the current $i$ (the rate at which charge on the powder moved) through a perpendicular cross section of the pipe. (b) Evaluate $i$ for the conditions at the factory: pipe radius $R = 5.0$ cm, speed $v = 2.0$ m/s, and charge density $\rho = 1.1 \times 10^{-3}$ C/m³.

If the powder were to flow through a change $V$ in electric potential, its energy could be transferred to a spark at the rate $P = iV$. (c) Could there be such a transfer within the pipe due to the radial potential difference discussed in Problem 68 of Chapter 24?

As the powder flowed from the pipe into the silo, the electric potential of the powder changed. The magnitude of that change was at least equal to the radial potential difference within the pipe (as evaluated in Problem 68 of Chapter 24). (d) Assuming that value for the potential difference and using the current found in (b) above, find the rate at which energy could have been transferred from the powder to a spark as the powder exited the pipe. (e) If a spark did occur at the exit and lasted for 0.20 s (a reasonable expectation), how much energy would have been transferred to the spark?

Recall from Problem 56 in Chapter 23 that a minimum energy transfer of 150 mJ is needed to cause an explosion. (f) Where did the powder explosion most likely occur: in the powder cloud at the unloading bin (Problem 56 of Chapter 25), within the pipe, or at the exit of the pipe into the silo? ✈

**57** A 500 W heating unit is designed to operate with an applied potential difference of 115 V. (a) By what percentage will its heat output drop if the applied potential difference drops to 110 V? Assume no change in resistance. (b) If you took the variation of resistance with temperature into account, would the actual drop in heat output be larger or smaller than that calculated in (a)?

**58** A potential difference of 1.20 V will be applied to a 33.0 m length of 18-gauge copper wire (diameter = 0.0400 in.). Calculate (a) the current, (b) the magnitude of the current density, (c) the magnitude of the electric field within the wire, and (d) the rate at which thermal energy will appear in the wire.

**59** A Nichrome heater dissipates 500 W when the applied potential difference is 110 V and the wire temperature is 800°C. What would be the dissipation rate if the wire temperature were held at 200°C by immersing the wire in a bath of cooling oil? The applied potential difference remains the same, and $\alpha$ for Nichrome at 800°C is $4.0 \times 10^{-4}\,\mathrm{K}^{-1}$. **SSM**

**60** An aluminum rod with a square cross section is 1.3 m long and 5.2 mm on edge. (a) What is the resistance between its ends? (b) What must be the diameter of a cylindrical copper rod of length 1.3 m if its resistance is to be the same as that of the aluminum rod?

**61** A cylindrical metal rod is 1.60 m long and 5.50 mm in diameter. The resistance between its two ends (at 20°C) is $1.09 \times 10^{-3}\,\Omega$. (a) What is the material? (b) A round disk, 2.00 cm in diameter and 1.00 mm thick, is formed of the same material. What is the resistance between the round faces, assuming that each face is an equipotential surface?

**62** A cylindrical resistor of radius 5.0 mm and length 2.0 cm is made of material that has a resistivity of $3.5 \times 10^{-5}\,\Omega \cdot \mathrm{m}$. What are (a) the magnitude of the current density and (b) the potential difference when the energy dissipation rate in the resistor is 1.0 W?

**63** A potential difference $V$ is applied to a wire of cross-sectional area $A$, length $L$, and resistivity $\rho$. You want to change the applied potential difference and stretch the wire so that the energy dissipation rate is multiplied by 30.0 and the current is multiplied by 4.00. Assuming the wire's density does not change, what are (a) the ratio of the new length to $L$ and (b) the ratio of the new cross-sectional area to $A$?

**64** The headlights of a moving car require about 10 A from the 12 V alternator, which is driven by the engine. Assume the alternator is 80% efficient (its output electrical power is 80% of its input mechanical power), and calculate the horsepower the engine must supply to run the lights.

**65** An 18.0 W device has 9.00 V across it. How much charge goes through the device in 4.00 h?

**66** The current density in a wire is uniform and has magnitude $2.0 \times 10^6\,\mathrm{A/m}^2$, the wire's length is 5.0 m, and the density of conduction electrons is $8.49 \times 10^{28}\,\mathrm{m}^{-3}$. How long does an electron take (on the average) to travel the length of the wire? **GO**

**67** How much energy is consumed in 2.00 h by an electrical resistance of 400 $\Omega$ when the potential applied across it is 90.0 V?

**68** A resistor with a potential difference of 200 V across it transfers electrical energy to thermal energy at the rate of 3000 W. What is the resistance of the resistor?

**69** A coil of current-carrying Nichrome wire is immersed in a liquid. (Nichrome is a nickel–chromium–iron alloy commonly used in heating elements.) When the potential difference across the coil is 12 V and the current through the coil is 5.2 A, the liquid evaporates at the steady rate of 21 mg/s. Calculate the heat of vaporization of the liquid (see Section 18-8).

**70** A current is established in a gas discharge tube when a sufficiently high potential difference is applied across the two electrodes in the tube. The gas ionizes; electrons move toward the positive terminal and singly charged positive ions toward the negative terminal. (a) What is the current in a hydrogen discharge tube in which $3.1 \times 10^{18}$ electrons and $1.1 \times 10^{18}$ protons move past a cross-sectional area of the tube each second? (b) Is the direction of the current density $\vec{J}$ toward or away from the negative terminal?

**71** A 2.0 kW heater element from a dryer has a length of 80 cm. If a 10 cm section is removed, what power is used by the now shortened element at 120 V?

**72** The copper windings of a motor have a resistance of 50 $\Omega$ at 20°C when the motor is idle. After the motor has run for several hours, the resistance rises to 58 $\Omega$. What is the temperature of the windings now? Ignore changes in the dimensions of the windings. (Use Table 26-1.)

**73** A certain x-ray tube operates at a current of 7.00 mA and a potential difference of 80.0 kV. What is its power in watts?

**74** A caterpillar of length 4.0 cm crawls in the direction of electron drift along a 5.2-mm-diameter bare copper wire that carries a uniform current of 12 A. (a) What is the potential difference between the two ends of the caterpillar? (b) Is its tail positive or negative relative to its head? (c) How much time does the caterpillar take to crawl 1.0 cm if it crawls at the drift speed of the electrons in the wire? (The number of charge carriers per unit volume is $8.49 \times 10^{28}\,\mathrm{m}^{-3}$.)

**75** (a) At what temperature would the resistance of a copper conductor be double its resistance at 20.0°C? (Use 20.0°C as the reference point in Eq. 26-17; compare your answer with Fig. 26-10.) (b) Does this same "doubling temperature" hold for all copper conductors, regardless of shape or size? **SSM**

**76** A steel trolley-car rail has a cross-sectional area of 56.0 cm². What is the resistance of 10.0 km of rail? The resistivity of the steel is $3.00 \times 10^{-7}\,\Omega \cdot \mathrm{m}$.

# Circuits

© AP/Wide World Photos

When a race car pulls in for a pit stop, the pit crew scurries to make any needed car adjustments and to refuel the car. The crew members pride themselves on their speed because even a fraction of a second wasted during the pit stop can cost the driver the race. However, rapid application of the fuel dispenser seems dangerous because the car may be charged to a potential of −30 kV when it arrives in the pit. If the car sparks through any of the fuel vapor, a fire can break out.

## What precaution must be taken to prevent such a fire?

The answer is in this chapter.

## 27-1 | WHAT IS PHYSICS?

You are surrounded by electric circuits. You might take pride in the number of electrical devices you own and might even carry a mental list of the devices you wish you owned. Every one of those devices, as well as the electrical grid that powers your home, depends on modern electrical engineering. We cannot easily estimate the current financial worth of electrical engineering and its products, but we can be certain that the financial worth continues to grow yearly as more and more tasks are handled electrically. Radios are now tuned electronically instead of manually. Messages are now sent by email instead of through the postal system. Research journals are now read on a computer instead of in a library building, and research papers are now copied and filed electronically instead of photocopied and tucked into a filing cabinet.

The basic science of electrical engineering is physics. In this chapter we cover the physics of electric circuits that are combinations of resistors and batteries (and, in Section 27-9, capacitors). We restrict our discussion to circuits through which charge flows in one direction, which are called either *direct-current circuits* or *DC circuits*. We begin with the question: How can you get charges to flow?

## 27-2 | "Pumping" Charges

If you want to make charge carriers flow through a resistor, you must establish a potential difference between the ends of the device. One way to do this is to connect each end of the resistor to one plate of a charged capacitor. The trouble with this scheme is that the flow of charge acts to discharge the capacitor, quickly bringing the plates to the same potential. When that happens, there is no longer an electric field in the resistor, and thus the flow of charge stops.

To produce a steady flow of charge, you need a "charge pump," a device that—by doing work on the charge carriers—maintains a potential difference between a pair of terminals. We call such a device an **emf device,** and the device is said to provide an **emf** $\mathscr{E}$, which means that it does work on charge carriers. An emf device is sometimes called a *seat of emf*. The term *emf* comes from the outdated phrase *electromotive force*, which was adopted before scientists clearly understood the function of an emf device.

In Chapter 26, we discussed the motion of charge carriers through a circuit in terms of the electric field set up in the circuit—the field produces forces that move the charge carriers. In this chapter we take a different approach: We discuss the motion of the charge carriers in terms of the required energy—an emf device supplies the energy for the motion via the work it does.

A common emf device is the *battery,* used to power a wide variety of machines from wristwatches to submarines. The emf device that most influences our daily lives, however, is the *electric generator,* which, by means of electrical connections (wires) from a generating plant, creates a potential difference in our homes and workplaces. The emf devices known as *solar cells,* long familiar as the wing-like panels on spacecraft, also dot the countryside for domestic applications. Less familiar emf devices are the *fuel cells* that power the space shuttles and the *thermopiles* that provide onboard electrical power for some spacecraft and for remote stations in Antarctica and elsewhere. An emf device does not have to be an instrument—living systems, ranging from electric eels and human beings to plants, have physiological emf devices.

Although the devices we have listed differ widely in their modes of operation, they all perform the same basic function—they do work on charge carriers and thus maintain a potential difference between their terminals.

The world's largest battery energy storage plant (dismantled in 1996) connected over 8000 large lead-acid batteries in 8 strings at 1000 V each with a capability of 10 MW of power for 4 hours. Charged up at night, the batteries were then put to use during peak power demands on the electrical system. (*Courtesy Southern California Edison Company*)

## 27-3 | Work, Energy, and Emf

Figure 27-1 shows an emf device (consider it to be a battery) that is part of a simple circuit containing a single resistance $R$ (the symbol for resistance and a resistor is -\/\/\-). The emf device keeps one of its terminals (called the positive terminal and often labeled +) at a higher electric potential than the other terminal (called the negative terminal and labeled −). We can represent the emf of the device with an arrow that points from the negative terminal toward the positive terminal as in Fig. 27-1. A small circle on the tail of the emf arrow distinguishes it from the arrows that indicate current direction.

When an emf device is not connected to a circuit, the internal chemistry of the device does not cause any net flow of charge carriers within it. However, when it is connected to a circuit as in Fig. 27-1, its internal chemistry causes a net flow of positive charge carriers from the negative terminal to the positive terminal, in the direction of the emf arrow. This flow is part of the current that is set up around the circuit in that same direction (clockwise in Fig. 27-1).

Within the emf device, positive charge carriers move from a region of low electric potential and thus low electric potential energy (at the negative terminal) to a region of higher electric potential and higher electric potential energy (at the positive terminal). This motion is just the opposite of what the electric field between the terminals (which is directed from the positive terminal toward the negative terminal) would cause the charge carriers to do.

Thus, there must be some source of energy within the device, enabling it to do work on the charges by forcing them to move as they do. The energy source may be chemical, as in a battery or a fuel cell. It may involve mechanical forces, as in an electric generator. Temperature differences may supply the energy, as in a thermopile; or the Sun may supply it, as in a solar cell.

Let us now analyze the circuit of Fig. 27-1 from the point of view of work and energy transfers. In any time interval $dt$, a charge $dq$ passes through any cross section of this circuit, such as $aa'$. This same amount of charge must enter the emf device at its low-potential end and leave at its high-potential end. The device must do an amount of work $dW$ on the charge $dq$ to force it to move in this way. We define the emf of the emf device in terms of this work:

$$\mathscr{E} = \frac{dW}{dq} \quad \text{(definition of } \mathscr{E}\text{)}. \quad (27\text{-}1)$$

In words, the emf of an emf device is the work per unit charge that the device does in moving charge from its low-potential terminal to its high-potential terminal. The SI unit for emf is the joule per coulomb; in Chapter 24 we defined that unit as the *volt*.

An **ideal emf device** is one that lacks any internal resistance to the internal movement of charge from terminal to terminal. The potential difference between the terminals of an ideal emf device is equal to the emf of the device. For example, an ideal battery with an emf of 12.0 V always has a potential difference of 12.0 V between its terminals.

A **real emf device,** such as any real battery, has internal resistance to the internal movement of charge. When a real emf device is not connected to a circuit, and thus does not have current through it, the potential difference between its terminals is equal to its emf. However, when that device has current through it, the potential difference between its terminals differs from its emf. We shall discuss such real batteries in Section 27-5.

When an emf device is connected to a circuit, the device transfers energy to the charge carriers passing through it. This energy can then be transferred from the charge carriers to other devices in the circuit, for example, to light a bulb.

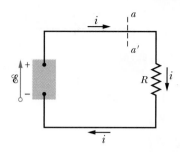

**FIG. 27-1** A simple electric circuit, in which a device of emf $\mathscr{E}$ does work on the charge carriers and maintains a steady current $i$ in a resistor of resistance $R$.

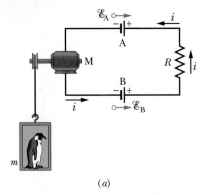

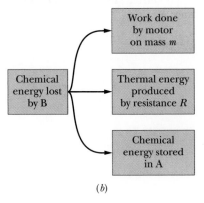

**FIG. 27-2** (*a*) In the circuit, $\mathscr{E}_B > \mathscr{E}_A$; so battery B determines the direction of the current. (*b*) The energy transfers in the circuit.

Figure 27-2*a* shows a circuit containing two ideal rechargeable (*storage*) batteries A and B, a resistance *R*, and an electric motor M that can lift an object by using energy it obtains from charge carriers in the circuit. Note that the batteries are connected so that they tend to send charges around the circuit in opposite directions. The actual direction of the current in the circuit is determined by the battery with the larger emf, which happens to be battery B, so the chemical energy within battery B is decreasing as energy is transferred to the charge carriers passing through it. However, the chemical energy within battery A is increasing because the current in it is directed from the positive terminal to the negative terminal. Thus, battery B is charging battery A. Battery B is also providing energy to motor M and energy that is being dissipated by resistance *R*. Figure 27-2*b* shows all three energy transfers from battery B; each decreases that battery's chemical energy.

## 27-4 | Calculating the Current in a Single-Loop Circuit

We discuss here two equivalent ways to calculate the current in the simple *single-loop* circuit of Fig. 27-3; one method is based on energy conservation considerations, and the other on the concept of potential. The circuit consists of an ideal battery B with emf $\mathscr{E}$, a resistor of resistance *R*, and two connecting wires. (Unless otherwise indicated, we assume that wires in circuits have negligible resistance. Their function, then, is merely to provide pathways along which charge carriers can move.)

### Energy Method

Equation 26-27 ($P = i^2R$) tells us that in a time interval *dt* an amount of energy given by $i^2R \, dt$ will appear in the resistor of Fig. 27-3 as thermal energy. As noted in Section 26-7, this energy is said to be *dissipated*. (Because we assume the wires to have negligible resistance, no thermal energy will appear in them.) During the same interval, a charge $dq = i \, dt$ will have moved through battery B, and the work that the battery will have done on this charge, according to Eq. 27-1, is

$$dW = \mathscr{E} \, dq = \mathscr{E}i \, dt.$$

From the principle of conservation of energy, the work done by the (ideal) battery must equal the thermal energy that appears in the resistor:

$$\mathscr{E}i \, dt = i^2R \, dt.$$

This gives us

$$\mathscr{E} = iR.$$

The emf $\mathscr{E}$ is the energy per unit charge transferred to the moving charges by the battery. The quantity *iR* is the energy per unit charge transferred *from* the moving charges to thermal energy within the resistor. Therefore, this equation means that the energy per unit charge transferred to the moving charges is equal to the energy per unit charge transferred from them. Solving for *i*, we find

$$i = \frac{\mathscr{E}}{R}. \tag{27-2}$$

### Potential Method

Suppose we start at any point in the circuit of Fig. 27-3 and mentally proceed around the circuit in either direction, adding algebraically the potential differences that we encounter. Then when we return to our starting point, we must also have returned to our starting potential. Before actually doing so, we shall formalize this idea in a statement that holds not only for single-loop circuits such as that of Fig. 27-3 but also for any complete loop in a *multiloop* circuit, as we

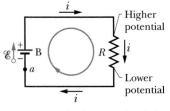

**FIG. 27-3** A single-loop circuit in which a resistance *R* is connected across an ideal battery B with emf $\mathscr{E}$. The resulting current *i* is the same throughout the circuit.

shall discuss in Section 27-7:

> **LOOP RULE:** The algebraic sum of the changes in potential encountered in a complete traversal of any loop of a circuit must be zero.

This is often referred to as *Kirchhoff's loop rule* (or *Kirchhoff's voltage law*), after German physicist Gustav Robert Kirchhoff. This rule is equivalent to saying that each point on a mountain has only one elevation above sea level. If you start from any point and return to it after walking around the mountain, the algebraic sum of the changes in elevation that you encounter must be zero.

In Fig. 27-3, let us start at point *a*, whose potential is $V_a$, and mentally walk clockwise around the circuit until we are back at *a*, keeping track of potential changes as we move. Our starting point is at the low-potential terminal of the battery. Because the battery is ideal, the potential difference between its terminals is equal to $\mathscr{E}$. When we pass through the battery to the high-potential terminal, the change in potential is $+\mathscr{E}$.

As we walk along the top wire to the top end of the resistor, there is no potential change because the wire has negligible resistance; it is at the same potential as the high-potential terminal of the battery. So too is the top end of the resistor. When we pass through the resistor, however, the potential changes according to Eq. 26-8 (which we can rewrite as $V = iR$). Moreover, the potential must decrease because we are moving from the higher potential side of the resistor. Thus, the change in potential is $-iR$.

We return to point *a* by moving along the bottom wire. Because this wire also has negligible resistance, we again find no potential change. Back at point *a*, the potential is again $V_a$. Because we traversed a complete loop, our initial potential, as modified for potential changes along the way, must be equal to our final potential; that is,

$$V_a + \mathscr{E} - iR = V_a.$$

The value of $V_a$ cancels from this equation, which becomes

$$\mathscr{E} - iR = 0.$$

Solving this equation for *i* gives us the same result, $i = \mathscr{E}/R$, as the energy method (Eq. 27-2).

If we apply the loop rule to a complete *counterclockwise* walk around the circuit, the rule gives us

$$-\mathscr{E} + iR = 0$$

and we again find that $i = \mathscr{E}/R$. Thus, you may mentally circle a loop in either direction to apply the loop rule.

To prepare for circuits more complex than that of Fig. 27-3, let us set down two rules for finding potential differences as we move around a loop:

> **RESISTANCE RULE:** For a move through a resistance in the direction of the current, the change in potential is $-iR$; in the opposite direction it is $+iR$.

> **EMF RULE:** For a move through an ideal emf device in the direction of the emf arrow, the change in potential is $+\mathscr{E}$; in the opposite direction it is $-\mathscr{E}$.

✓**CHECKPOINT 1**    The figure shows the current *i* in a single-loop circuit with a battery B and a resistance *R* (and wires of negligible resistance). (a) Should the emf arrow at B be drawn pointing leftward or rightward? At points *a*, *b*, and *c*, rank (b) the magnitude of the current, (c) the electric potential, and (d) the electric potential energy of the charge carriers, greatest first.

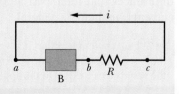

**FIG. 27-4** (a) A single-loop circuit containing a real battery having internal resistance $r$ and emf $\mathscr{E}$. (b) The same circuit, now spread out in a line. The potentials encountered in traversing the circuit clockwise from $a$ are also shown. The potential $V_a$ is arbitrarily assigned a value of zero, and other potentials in the circuit are graphed relative to $V_a$.

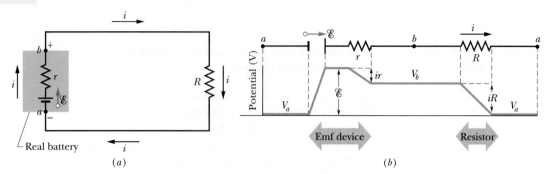

(a)

(b)

## 27-5 | Other Single-Loop Circuits

In this section we extend the simple circuit of Fig. 27-3 in two ways.

### Internal Resistance

Figure 27-4a shows a real battery, with internal resistance $r$, wired to an external resistor of resistance $R$. The internal resistance of the battery is the electrical resistance of the conducting materials of the battery and thus is an unremovable feature of the battery. In Fig. 27-4a, however, the battery is drawn as if it could be separated into an ideal battery with emf $\mathscr{E}$ and a resistor of resistance $r$. The order in which the symbols for these separated parts are drawn does not matter.

If we apply the loop rule clockwise beginning at point $a$, the *changes* in potential give us

$$\mathscr{E} - ir - iR = 0. \tag{27-3}$$

Solving for the current, we find

$$i = \frac{\mathscr{E}}{R + r}. \tag{27-4}$$

Note that this equation reduces to Eq. 27-2 if the battery is ideal—that is, if $r = 0$.

Figure 27-4b shows graphically the changes in electric potential around the circuit. (To better link Fig. 27-4b with the *closed circuit* in Fig. 27-4a, imagine curling the graph into a cylinder with point $a$ at the left overlapping point $a$ at the right.) Note how traversing the circuit is like walking around a (potential) mountain back to your starting point—you return to the starting elevation.

In this book, when a battery is not described as real or if no internal resistance is indicated, you can generally assume that it is ideal—but, of course, in the real world batteries are always real and have internal resistance.

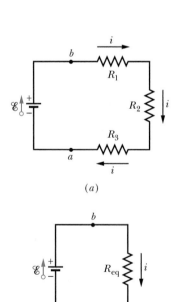

(a)

(b)

**FIG. 27-5** (a) Three resistors are connected in series between points $a$ and $b$. (b) An equivalent circuit, with the three resistors replaced with their equivalent resistance $R_{\text{eq}}$.

### Resistances in Series

Figure 27-5a shows three resistances connected **in series** to an ideal battery with emf $\mathscr{E}$. This description has little to do with how the resistances are drawn. Rather, "in series" means that the resistances are wired one after another and that a potential difference $V$ is applied across the two ends of the series. In Fig. 27-5a, the resistances are connected one after another between $a$ and $b$, and a potential difference is maintained across $a$ and $b$ by the battery. The potential differences that then exist across the resistances in the series produce identical currents $i$ in them. In general,

> When a potential difference $V$ is applied across resistances connected in series, the resistances have identical currents $i$. The sum of the potential differences across the resistances is equal to the applied potential difference $V$.

Note that charge moving through the series resistances can move along only a single route. If there are additional routes, so that the currents in different resistances are different, the resistances are not connected in series.

☞ Resistances connected in series can be replaced with an equivalent resistance $R_{eq}$ that has the same current $i$ and the same *total* potential difference $V$ as the actual resistances.

You might remember that $R_{eq}$ and all the actual series resistances have the same current $i$ with the nonsense word "ser-i." Figure 27-5b shows the equivalent resistance $R_{eq}$ that can replace the three resistances of Fig. 27-5a.

To derive an expression for $R_{eq}$ in Fig. 27-5b, we apply the loop rule to both circuits. For Fig. 27-5a, starting at $a$ and going clockwise around the circuit, we find

$$\mathscr{E} - iR_1 - iR_2 - iR_3 = 0,$$

or
$$i = \frac{\mathscr{E}}{R_1 + R_2 + R_3}. \tag{27-5}$$

For Fig. 27-5b, with the three resistances replaced with a single equivalent resistance $R_{eq}$, we find

$$\mathscr{E} - iR_{eq} = 0,$$

or
$$i = \frac{\mathscr{E}}{R_{eq}}. \tag{27-6}$$

Comparison of Eqs. 27-5 and 27-6 shows that

$$R_{eq} = R_1 + R_2 + R_3.$$

The extension to $n$ resistances is straightforward and is

$$R_{eq} = \sum_{j=1}^{n} R_j \qquad (n \text{ resistances in series}). \tag{27-7}$$

Note that when resistances are in series, their equivalent resistance is greater than any of the individual resistances.

✓ **CHECKPOINT 2**   In Fig. 27-5a, if $R_1 > R_2 > R_3$, rank the three resistances according to (a) the current through them and (b) the potential difference across them, greatest first.

## 27-6 | Potential Difference Between Two Points

We often want to find the potential difference between two points in a circuit. For example, in Fig. 27-6, what is the potential difference $V_b - V_a$ between points $a$ and $b$? To find out, let's start at point $a$ (at potential $V_a$) and move through the battery to point $b$ (at potential $V_b$) while keeping track of the potential changes we encounter. When we pass through the battery's emf, the potential increases by $\mathscr{E}$. When we pass through the battery's internal resistance $r$, we move in the direction of the current and thus the potential decreases by $ir$. We are then at the potential of point $b$ and we have

$$V_a + \mathscr{E} - ir = V_b,$$

or
$$V_b - V_a = \mathscr{E} - ir. \tag{27-8}$$

To evaluate this expression, we need the current $i$. Note that the circuit is the same as in Fig. 27-4a, for which Eq. 27-4 gives the current as

$$i = \frac{\mathscr{E}}{R + r}. \tag{27-9}$$

Substituting this equation into Eq. 27-8 gives us

$$V_b - V_a = \mathscr{E} - \frac{\mathscr{E}}{R + r} r$$

$$= \frac{\mathscr{E}}{R + r} R. \tag{27-10}$$

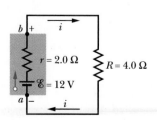

**FIG. 27-6**   Points $a$ and $b$, which are at the terminals of a real battery, differ in potential.

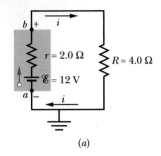

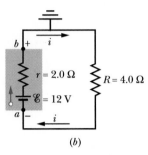

FIG. 27-7    (a) Point a is directly connected to ground. (b) Point b is directly connected to ground.

Now substituting the data given in Fig. 27-6, we have

$$V_b - V_a = \frac{12 \text{ V}}{4.0\ \Omega + 2.0\ \Omega}\ 4.0\ \Omega = 8.0 \text{ V}. \qquad (27\text{-}11)$$

Suppose, instead, we move from a to b counterclockwise, passing through resistor R rather than through the battery. Because we move opposite the current, the potential increases by iR. Thus,

$$V_a + iR = V_b$$

or

$$V_b - V_a = iR. \qquad (27\text{-}12)$$

Substituting for i from Eq. 27-9, we again find Eq. 27-10. Hence, substitution of the data in Fig. 27-6 yields the same result, $V_b - V_a = 8.0$ V. In general,

☞ To find the potential between any two points in a circuit, start at one point and traverse the circuit to the other point, following any path, and add algebraically the changes in potential you encounter.

## Potential Difference Across a Real Battery

In Fig. 27-6, points a and b are located at the terminals of the battery. Thus, the potential difference $V_b - V_a$ is the terminal-to-terminal potential difference V across the battery. From Eq. 27-8, we see that

$$V = \mathscr{E} - ir. \qquad (27\text{-}13)$$

If the internal resistance r of the battery in Fig. 27-6 were zero, Eq. 27-13 tells us that V would be equal to the emf $\mathscr{E}$ of the battery—namely, 12 V. However, because $r = 2.0\ \Omega$, Eq. 27-13 tells us that V is less than $\mathscr{E}$. From Eq. 27-11, we know that V is only 8.0 V. Note that the result depends on the value of the current through the battery. If the same battery were in a different circuit and had a different current through it, V would have some other value.

## Grounding a Circuit

Figure 27-7a shows the same circuit as Fig. 27-6 except that here point a is directly connected to *ground*, as indicated by the common symbol ⏚. *Grounding a circuit* usually means connecting the circuit to a conducting path to Earth's surface (actually to the electrically conducting moist dirt and rock below ground). Here, such a connection means only that the potential is defined to be zero at the grounding point in the circuit. Thus in Fig. 27-7a, the potential at a is defined to be $V_a = 0$. Equation 27-11 then tells us that the potential at b is $V_b = 8.0$ V.

Figure 27-7b is the same circuit except that point b is now directly connected to ground. Thus, the potential there is defined to be $V_b = 0$. Equation 27-11 now tells us that the potential at a is $V_a = -8.0$ V.

## Power, Potential, and Emf

When a battery or some other type of emf device does work on the charge carriers to establish a current i, the device transfers energy from its source of energy (such as the chemical source in a battery) to the charge carriers. Because a real emf device has an internal resistance r, it also transfers energy to internal thermal energy via resistive dissipation (Section 26-7). Let us relate these transfers.

The net rate P of energy transfer from the emf device to the charge carriers is given by Eq. 26-26:

$$P = iV, \qquad (27\text{-}14)$$

where V is the potential across the terminals of the emf device. From Eq. 27-13, we can substitute $V = \mathscr{E} - ir$ into Eq. 27-14 to find

$$P = i(\mathscr{E} - ir) = i\mathscr{E} - i^2r. \qquad (27\text{-}15)$$

From Eq. 26-27, we recognize the term $i^2r$ in Eq. 27-15 as the rate $P_r$ of energy transfer to thermal energy within the emf device:

$$P_r = i^2r \qquad \text{(internal dissipation rate)}. \qquad (27\text{-}16)$$

Then the term $i\mathscr{E}$ in Eq. 27-15 must be the rate $P_{\text{emf}}$ at which the emf device transfers energy *both* to the charge carriers and to internal thermal energy. Thus,

$$P_{\text{emf}} = i\mathscr{E} \qquad \text{(power of emf device)}. \qquad (27\text{-}17)$$

If a battery is being *recharged,* with a "wrong way" current through it, the energy transfer is then *from* the charge carriers *to* the battery—both to the battery's chemical energy and to the energy dissipated in the internal resistance $r$. The rate of change of the chemical energy is given by Eq. 27-17, the rate of dissipation is given by Eq. 27-16, and the rate at which the carriers supply energy is given by Eq. 27-14.

✓**CHECKPOINT 3**    A battery has an emf of 12 V and an internal resistance of 2 Ω. Is the terminal-to-terminal potential difference greater than, less than, or equal to 12 V if the current in the battery is (a) from the negative to the positive terminal, (b) from the positive to the negative terminal, and (c) zero?

---

**Sample Problem** | **27-1**

---

The emfs and resistances in the circuit of Fig. 27-8a have the following values:

$$\mathscr{E}_1 = 4.4 \text{ V}, \quad \mathscr{E}_2 = 2.1 \text{ V},$$
$$r_1 = 2.3 \text{ }\Omega, \quad r_2 = 1.8 \text{ }\Omega, \quad R = 5.5 \text{ }\Omega.$$

**(a)** What is the current $i$ in the circuit?

**KEY IDEA**    We can get an expression involving the current $i$ in this single-loop circuit by applying the loop rule.

**Calculations:** Although knowing the direction of $i$ is not necessary, we can easily determine it from the emfs of the two batteries. Because $\mathscr{E}_1$ is greater than $\mathscr{E}_2$, battery 1 controls the direction of $i$, so the direction is clockwise. Let us then apply the loop rule by going counterclockwise—against the current—and starting at point $a$. We find

$$-\mathscr{E}_1 + ir_1 + iR + ir_2 + \mathscr{E}_2 = 0.$$

Check that this equation also results if we apply the loop rule clockwise or start at some point other than $a$. Also, take the time to compare this equation term by term with Fig. 27-8b, which shows the potential changes graphically (with the potential at point $a$ arbitrarily taken to be zero).

Solving the above loop equation for the current $i$, we obtain

$$i = \frac{\mathscr{E}_1 - \mathscr{E}_2}{R + r_1 + r_2} = \frac{4.4 \text{ V} - 2.1 \text{ V}}{5.5 \text{ }\Omega + 2.3 \text{ }\Omega + 1.8 \text{ }\Omega}$$
$$= 0.2396 \text{ A} \approx 240 \text{ mA}. \qquad \text{(Answer)}$$

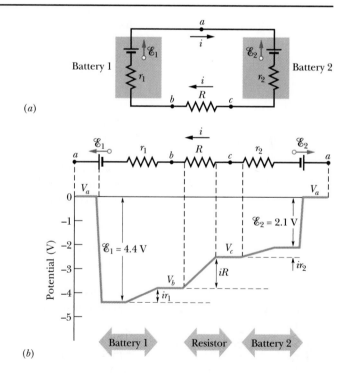

(a)

(b)

**FIG. 27-8**    (a) A single-loop circuit containing two real batteries and a resistor. The batteries oppose each other; that is, they tend to send current in opposite directions through the resistor. (b) A graph of the potentials, counterclockwise from point $a$, with the potential at $a$ arbitrarily taken to be zero. (To better link the circuit with the graph, mentally cut the circuit at $a$ and then unfold the left side of the circuit toward the left and the right side of the circuit toward the right.)

(b) What is the potential difference between the terminals of battery 1 in Fig. 27-8a?

**KEY IDEA** We need to sum the potential differences between points $a$ and $b$.

**Calculations:** Let us start at point $b$ (effectively the negative terminal of battery 1) and travel clockwise through battery 1 to point $a$ (effectively the positive terminal), keeping track of potential changes. We find that

$$V_b - ir_1 + \mathscr{E}_1 = V_a,$$

which gives us

$$V_a - V_b = -ir_1 + \mathscr{E}_1$$
$$= -(0.2396 \text{ A})(2.3 \text{ }\Omega) + 4.4 \text{ V}$$
$$= +3.84 \text{ V} \approx 3.8 \text{ V}, \qquad \text{(Answer)}$$

which is less than the emf of the battery. You can verify this result by starting at point $b$ in Fig. 27-8a and traversing the circuit counterclockwise to point $a$.

---

**PROBLEM-SOLVING TACTICS**

**Tactic 1: Assuming the Direction of a Current Can Be an Arbitrary Decision** In solving circuit problems, you do not need to know the direction of a current in advance. Instead, you can just assume its direction, although that may take some physics courage. To show this, assume the current in Fig. 27-8a is counterclockwise; that is, reverse the direction of the current arrows shown. Applying the loop rule counterclockwise from point $a$ now yields

$$-\mathscr{E}_1 - ir_1 - iR - ir_2 + \mathscr{E}_2 = 0,$$

or

$$i = -\frac{\mathscr{E}_1 - \mathscr{E}_2}{R + r_1 + r_2}.$$

Substituting the numerical values of Sample Problem 27-1 yields $i = -240$ mA for the current. The minus sign is a signal that the current is opposite the direction we initially assumed.

---

## 27-7 | Multiloop Circuits

Figure 27-9 shows a circuit containing more than one loop. For simplicity, we assume the batteries are ideal. There are two *junctions* in this circuit, at $b$ and $d$, and there are three *branches* connecting these junctions. The branches are the left branch (*bad*), the right branch (*bcd*), and the central branch (*bd*). What are the currents in the three branches?

We arbitrarily label the currents, using a different subscript for each branch. Current $i_1$ has the same value everywhere in branch *bad*, $i_2$ has the same value everywhere in branch *bcd*, and $i_3$ is the current through branch *bd*. The directions of the currents are assumed arbitrarily.

Consider junction $d$ for a moment: Charge comes into that junction via incoming currents $i_1$ and $i_3$, and it leaves via outgoing current $i_2$. Because there is no variation in the charge at the junction, the total incoming current must equal the total outgoing current:

$$i_1 + i_3 = i_2. \qquad (27\text{-}18)$$

You can easily check that applying this condition to junction $b$ leads to exactly the same equation. Equation 27-18 thus suggests a general principle:

> **JUNCTION RULE:** The sum of the currents entering any junction must be equal to the sum of the currents leaving that junction.

This rule is often called *Kirchhoff's junction rule* (or *Kirchhoff's current law*). It is simply a statement of the conservation of charge for a steady flow of charge — there is neither a buildup nor a depletion of charge at a junction. Thus, our basic tools for solving complex circuits are the *loop rule* (based on the conservation of energy) and the *junction rule* (based on the conservation of charge).

Equation 27-18 is a single equation involving three unknowns. To solve the circuit completely (that is, to find all three currents), we need two more equations involving those same unknowns. We obtain them by applying the loop rule twice. In the circuit of Fig. 27-9, we have three loops from which to choose: the left-hand loop (*badb*), the right-hand loop (*bcdb*), and the big loop (*badcb*). Which two loops we choose does not matter — let's choose the left-hand loop and the right-hand loop.

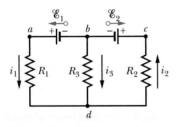

**FIG. 27-9** A multiloop circuit consisting of three branches: left-hand branch *bad*, right-hand branch *bcd*, and central branch *bd*. The circuit also consists of three loops: left-hand loop *badb*, right-hand loop *bcdb*, and big loop *badcb*.

If we traverse the left-hand loop in a counterclockwise direction from point $b$, the loop rule gives us

$$\mathscr{E}_1 - i_1 R_1 + i_3 R_3 = 0. \tag{27-19}$$

If we traverse the right-hand loop in a counterclockwise direction from point $b$, the loop rule gives us

$$-i_3 R_3 - i_2 R_2 - \mathscr{E}_2 = 0. \tag{27-20}$$

We now have three equations (Eqs. 27-18, 27-19, and 27-20) in the three unknown currents, and they can be solved by a variety of techniques.

If we had applied the loop rule to the big loop, we would have obtained (moving counterclockwise from $b$) the equation

$$\mathscr{E}_1 - i_1 R_1 - i_2 R_2 - \mathscr{E}_2 = 0.$$

This equation may look like fresh information, but in fact it is only the sum of Eqs. 27-19 and 27-20. (It would, however, yield the proper results when used with Eq. 27-18 and either Eq. 27-19 or Eq. 27-20.)

### Resistances in Parallel

Figure 27-10a shows three resistances connected *in parallel* to an ideal battery of emf $\mathscr{E}$. The term "in parallel" means that the resistances are directly wired together on one side and directly wired together on the other side, and that a potential difference $V$ is applied across the pair of connected sides. Thus, all three resistances have the same potential difference $V$ across them, producing a current through each. In general,

> ☞ When a potential difference $V$ is applied across resistances connected in parallel, the resistances all have that same potential difference $V$.

In Fig. 27-10a, the applied potential difference $V$ is maintained by the battery. In Fig. 27-10b, the three parallel resistances have been replaced with an equivalent resistance $R_{eq}$.

> ☞ Resistances connected in parallel can be replaced with an equivalent resistance $R_{eq}$ that has the same potential difference $V$ and the same *total* current $i$ as the actual resistances.

You might remember that $R_{eq}$ and all the actual parallel resistances have the same potential difference $V$ with the nonsense word "par-V."

To derive an expression for $R_{eq}$ in Fig. 27-10b, we first write the current in each actual resistance in Fig. 27-10a as

$$i_1 = \frac{V}{R_1}, \quad i_2 = \frac{V}{R_2}, \quad \text{and} \quad i_3 = \frac{V}{R_3},$$

where $V$ is the potential difference between $a$ and $b$. If we apply the junction rule at point $a$ in Fig. 27-10a and then substitute these values, we find

$$i = i_1 + i_2 + i_3 = V\left(\frac{1}{R_1} + \frac{1}{R_2} + \frac{1}{R_3}\right). \tag{27-21}$$

If we replaced the parallel combination with the equivalent resistance $R_{eq}$ (Fig. 27-10b), we would have

$$i = \frac{V}{R_{eq}}. \tag{27-22}$$

Comparing Eqs. 27-21 and 27-22 leads to

$$\frac{1}{R_{eq}} = \frac{1}{R_1} + \frac{1}{R_2} + \frac{1}{R_3}. \tag{27-23}$$

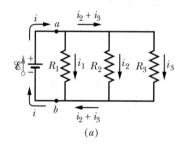

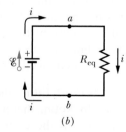

**FIG. 27-10** (a) Three resistors connected in parallel across points $a$ and $b$. (b) An equivalent circuit, with the three resistors replaced with their equivalent resistance $R_{eq}$.

**TABLE 27-1**

**Series and Parallel Resistors and Capacitors**

| Series | Parallel | Series | Parallel |
|---|---|---|---|
| Resistors | | Capacitors | |
| $R_{eq} = \sum\limits_{j=1}^{n} R_j$  Eq. 27-7 | $\dfrac{1}{R_{eq}} = \sum\limits_{j=1}^{n} \dfrac{1}{R_j}$  Eq. 27-24 | $\dfrac{1}{C_{eq}} = \sum\limits_{j=1}^{n} \dfrac{1}{C_j}$  Eq. 25-20 | $C_{eq} = \sum\limits_{j=1}^{n} C_j$  Eq. 25-19 |
| Same current through all resistors | Same potential difference across all resistors | Same charge on all capacitors | Same potential difference across all capacitors |

Extending this result to the case of $n$ resistances, we have

$$\frac{1}{R_{eq}} = \sum_{j=1}^{n} \frac{1}{R_j} \qquad (n \text{ resistances in parallel}). \qquad (27\text{-}24)$$

For the case of two resistances, the equivalent resistance is their product divided by their sum; that is,

$$R_{eq} = \frac{R_1 R_2}{R_1 + R_2}. \qquad (27\text{-}25)$$

If you accidentally took the equivalent resistance to be the sum divided by the product, you would notice that this result would be dimensionally incorrect.

Note that when two or more resistances are connected in parallel, the equivalent resistance is smaller than any of the combining resistances. Table 27-1 summarizes the equivalence relations for resistors and capacitors in series and in parallel.

✓ **CHECKPOINT 4**    A battery, with potential $V$ across it, is connected to a combination of two identical resistors and then has current $i$ through it. What are the potential difference across and the current through either resistor if the resistors are (a) in series and (b) in parallel?

---

**Sample Problem**  **27-2**  **Build your skill**

Figure 27-11$a$ shows a multiloop circuit containing one ideal battery and four resistances with the following values:

$$R_1 = 20\ \Omega, \quad R_2 = 20\ \Omega, \quad \mathcal{E} = 12\ \text{V},$$

$$R_3 = 30\ \Omega, \quad R_4 = 8.0\ \Omega.$$

(a)  What is the current through the battery?

**KEY IDEA**    Noting that the current through the battery must also be the current through $R_1$, we see that we might find the current by applying the loop rule to a loop that includes $R_1$ because the current would be included in the potential difference across $R_1$.

**Incorrect method:** Either the left-hand loop or the big loop should do. Noting that the emf arrow of the battery points upward, so the current the battery supplies is clockwise, we might apply the loop rule to the left-hand loop, clockwise from point $a$. With $i$ being the current through the battery, we would get

$$+\mathcal{E} - iR_1 - iR_2 - iR_4 = 0 \qquad \text{(incorrect)}.$$

However, this equation is incorrect because it assumes that $R_1$, $R_2$, and $R_4$ all have the same current $i$. Resistances $R_1$ and $R_4$ do have the same current, because the current passing through $R_4$ must pass through the battery and then through $R_1$ with no change in value. However, that current splits at junction point $b$ — only part passes through $R_2$, the rest through $R_3$.

**Dead-end method:** To distinguish the several currents in the circuit, we must label them individually as in Fig. 27-11$b$. Then, circling clockwise from $a$, we can write the loop rule for the left-hand loop as

$$+\mathcal{E} - i_1R_1 - i_2R_2 - i_1R_4 = 0.$$

Unfortunately, this equation contains two unknowns, $i_1$ and $i_2$; we would need at least one more equation to find them.

**Successful method:** A much easier option is to simplify the circuit of Fig. 27-11$b$ by finding equivalent resistances. Note carefully that $R_1$ and $R_2$ are *not* in series and thus cannot be replaced with an equivalent resistance. However, $R_2$ and $R_3$ are in parallel, so we can

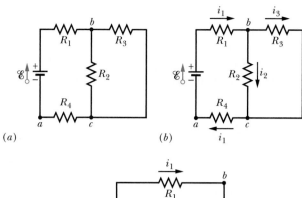

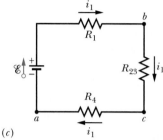

**FIG. 27-11** (*a*) A multiloop circuit with an ideal battery of emf $\mathscr{E}$ and four resistances. (*b*) Assumed currents through the resistances. (*c*) A simplification of the circuit, with resistances $R_2$ and $R_3$ replaced with their equivalent resistance $R_{23}$. The current through $R_{23}$ is equal to that through $R_1$ and $R_4$.

use either Eq. 27-24 or Eq. 27-25 to find their equivalent resistance $R_{23}$. From the latter,

$$R_{23} = \frac{R_2 R_3}{R_2 + R_3} = \frac{(20\ \Omega)(30\ \Omega)}{50\ \Omega} = 12\ \Omega.$$

We can now redraw the circuit as in Fig. 27-11*c*; note that the current through $R_{23}$ must be $i_1$ because charge that moves through $R_1$ and $R_4$ must also move through $R_{23}$. For this simple one-loop circuit, the loop rule (applied clockwise from point *a*) yields

$$+\mathscr{E} - i_1 R_1 - i_1 R_{23} - i_1 R_4 = 0.$$

Substituting the given data, we find

$$12\ \text{V} - i_1(20\ \Omega) - i_1(12\ \Omega) - i_1(8.0\ \Omega) = 0,$$

which gives us

$$i_1 = \frac{12\ \text{V}}{40\ \Omega} = 0.30\ \text{A}. \qquad \text{(Answer)}$$

**(b)** What is the current $i_2$ through $R_2$?

**KEY IDEA** (1) We must now work backward from the equivalent circuit of Fig. 27-11*c*, where $R_{23}$ has replaced $R_2$ and $R_3$. (2) Because $R_2$ and $R_3$ are in parallel, they both have the same potential difference across them as $R_{23}$.

**Working backward:** We know the current through $R_{23}$ is $i_1 = 0.30$ A. Thus, we can use Eq. 26-8 ($R = V/i$) to find the potential difference $V_{23}$ across $R_{23}$:

$$V_{23} = i_1 R_{23} = (0.30\ \text{A})(12\ \Omega) = 3.6\ \text{V}.$$

The potential difference across $R_2$ is thus 3.6 V, so the current $i_2$ in $R_2$ must be, by Eq. 26-8,

$$i_2 = \frac{V_2}{R_2} = \frac{3.6\ \text{V}}{20\ \Omega} = 0.18\ \text{A}. \qquad \text{(Answer)}$$

**(c)** What is the current $i_3$ through $R_3$?

**KEY IDEA** We can answer by using the same technique as in (b), or we can use this idea: The junction rule tells us that at point *b* in Fig. 27-11*b*, the incoming current $i_1$ and the outgoing currents $i_2$ and $i_3$ are related by

$$i_1 = i_2 + i_3.$$

**Calculation:** Rearranging this junction-rule result yields

$$i_3 = i_1 - i_2 = 0.30\ \text{A} - 0.18\ \text{A}$$
$$= 0.12\ \text{A}. \qquad \text{(Answer)}$$

---

**Sample Problem** | **27-3**

Figure 27-12 shows a circuit whose elements have the following values:

$$\mathscr{E}_1 = 3.0\ \text{V}, \quad \mathscr{E}_2 = 6.0\ \text{V},$$
$$R_1 = 2.0\ \Omega, \quad R_2 = 4.0\ \Omega.$$

The three batteries are ideal batteries. Find the magnitude and direction of the current in each of the three branches.

**KEY IDEAS** It is not worthwhile to try to simplify this circuit, because no two resistors are in parallel, and the resistors that are in series (those in the right branch or those in the left branch) present no problem. So, our plan is to apply the junction and loop rules.

**Junction rule:** Using arbitrarily chosen directions for the currents as shown in Fig. 27-12, we apply the junction rule at point *a* by writing

**FIG. 27-12** A multiloop circuit with three ideal batteries and five resistances.

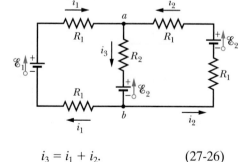

$$i_3 = i_1 + i_2. \qquad (27\text{-}26)$$

An application of the junction rule at junction *b* gives only the same equation, so we next apply the loop rule to any two of the three loops of the circuit.

**Left-hand loop:** We first arbitrarily choose the left-hand loop, arbitrarily start at point *b*, and arbitrarily traverse the loop in the clockwise direction, obtaining

$$-i_1R_1 + \mathscr{E}_1 - i_1R_1 - (i_1 + i_2)R_2 - \mathscr{E}_2 = 0,$$

where we have used $(i_1 + i_2)$ instead of $i_3$ in the middle branch. Substituting the given data and simplifying yield

$$i_1(8.0\,\Omega) + i_2(4.0\,\Omega) = -3.0\,\text{V}. \quad (27\text{-}27)$$

**Right-hand loop:** For our second application of the loop rule, we arbitrarily choose to traverse the right-hand loop counterclockwise from point $b$, finding

$$-i_2R_1 + \mathscr{E}_2 - i_2R_1 - (i_1 + i_2)R_2 - \mathscr{E}_2 = 0.$$

Substituting the given data and simplifying yield

$$i_1(4.0\,\Omega) + i_2(8.0\,\Omega) = 0. \quad (27\text{-}28)$$

**Combining equations:** We now have a system of two equations (Eqs. 27-27 and 27-28) in two unknowns ($i_1$ and $i_2$) to solve either "by hand" (which is easy enough here) or with a "math package." (One solution technique is Cramer's rule, given in Appendix E.) We find

$$i_1 = -0.50\,\text{A}. \quad (27\text{-}29)$$

(The minus sign signals that our arbitrary choice of direction for $i_1$ in Fig. 27-12 is wrong, but we must wait to correct it.) Substituting $i_1 = -0.50$ A into Eq. 27-28 and solving for $i_2$ then give us

$$i_2 = 0.25\,\text{A}. \quad \text{(Answer)}$$

With Eq. 27-26 we then find that

$$i_3 = i_1 + i_2 = -0.50\,\text{A} + 0.25\,\text{A}$$
$$= -0.25\,\text{A}.$$

The positive answer we obtained for $i_2$ signals that our choice of direction for that current is correct. However, the negative answers for $i_1$ and $i_3$ indicate that our choices for those currents are wrong. Thus, as a *last step* here, we correct the answers by reversing the arrows for $i_1$ and $i_3$ in Fig. 27-12 and then writing

$$i_1 = 0.50\,\text{A} \quad \text{and} \quad i_3 = 0.25\,\text{A}. \quad \text{(Answer)}$$

*Caution:* Always make any such correction as the last step and not before calculating *all* the currents.

---

**Sample Problem    27-4**

Electric fish are able to generate current with biological cells called *electroplaques,* which are physiological emf devices. The electroplaques in the type of electric fish known as a South American eel are arranged in 140

rows, each row stretching horizontally along the body and each containing 5000 electroplaques. The arrangement is suggested in Fig. 27-13a; each electroplaque has an emf $\mathscr{E}$ of 0.15 V and an internal resistance $r$ of 0.25 $\Omega$.

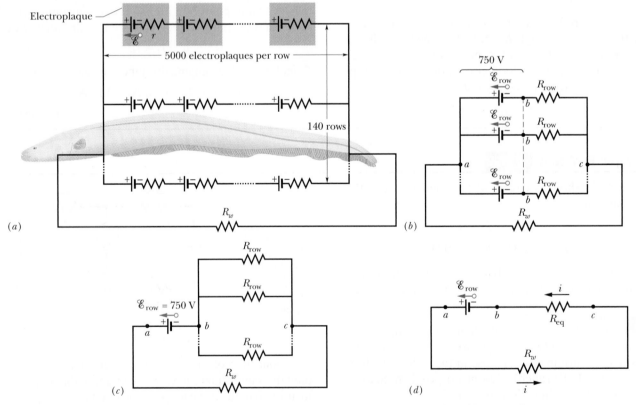

FIG. 27-13    (*a*) A model of the electric circuit of an eel in water. Each electroplaque of the eel has an emf $\mathscr{E}$ and internal resistance $r$. Along each of 140 rows extending from the head to the tail of the eel, there are 5000 electroplaques. The surrounding water has resistance $R_w$. (*b*) The emf $\mathscr{E}_{\text{row}}$ and resistance $R_{\text{row}}$ of each row. (*c*) The emf between points $a$ and $b$ is $\mathscr{E}_{\text{row}}$. Between points $b$ and $c$ are 140 parallel resistances $R_{\text{row}}$. (*d*) The simplified circuit, with $R_{\text{eq}}$ replacing the parallel combination.

The water surrounding the eel completes a circuit between the two ends of the electroplaque array, one end at the animal's head and the other near its tail.

(a) If the water surrounding the eel has resistance $R_w = 800\,\Omega$, how much current can the eel produce in the water?

**KEY IDEA**  We can simplify the circuit of Fig. 27-13a by replacing combinations of emfs and internal resistances with equivalent emfs and resistances.

*Calculations:* We first consider a single row. The total emf $\mathscr{E}_{row}$ along a row of 5000 electroplaques is the sum of the emfs:

$$\mathscr{E}_{row} = 5000\mathscr{E} = (5000)(0.15\,\text{V}) = 750\,\text{V}.$$

The total resistance $R_{row}$ along a row is the sum of the internal resistances of the 5000 electroplaques:

$$R_{row} = 5000r = (5000)(0.25\,\Omega) = 1250\,\Omega.$$

We can now represent each of the 140 identical rows as having a single emf $\mathscr{E}_{row}$ and a single resistance $R_{row}$ (Fig. 27-13b).

In Fig. 27-13b, the emf between point $a$ and point $b$ on any row is $\mathscr{E}_{row} = 750\,\text{V}$. Because the rows are identical and because they are all connected together at the left in Fig. 27-13b, all points $b$ in that figure are at the same electric potential. Thus, we can consider them to be connected so that there is only a single point $b$. The emf between point $a$ and this single point $b$ is $\mathscr{E}_{row} = 750\,\text{V}$, so we can draw the circuit as shown in Fig. 27-13c.

Between points $b$ and $c$ in Fig. 27-13c are 140 resistances $R_{row} = 1250\,\Omega$, all in parallel. The equivalent resistance $R_{eq}$ of this combination is given by Eq. 27-24 as

$$\frac{1}{R_{eq}} = \sum_{j=1}^{140} \frac{1}{R_j} = 140\frac{1}{R_{row}},$$

or

$$R_{eq} = \frac{R_{row}}{140} = \frac{1250\,\Omega}{140} = 8.93\,\Omega.$$

Replacing the parallel combination with $R_{eq}$, we obtain the simplified circuit of Fig. 27-13d. Applying the loop rule to this circuit counterclockwise from point $b$, we have

$$\mathscr{E}_{row} - iR_w - iR_{eq} = 0.$$

Solving for $i$ and substituting the known data, we find

$$i = \frac{\mathscr{E}_{row}}{R_w + R_{eq}} = \frac{750\,\text{V}}{800\,\Omega + 8.93\,\Omega}$$
$$= 0.927\,\text{A} \approx 0.93\,\text{A}. \qquad \text{(Answer)}$$

If the head or tail of the eel is near a fish, some of this current could pass along a narrow path through the fish, stunning or killing it.

(b) How much current $i_{row}$ travels through each row of Fig. 27-13a?

**KEY IDEA**  Because the rows are identical, the current into and out of the eel is evenly divided among them.

*Calculation:* Thus, we write

$$i_{row} = \frac{i}{140} = \frac{0.927\,\text{A}}{140} = 6.6 \times 10^{-3}\,\text{A}. \quad \text{(Answer)}$$

Thus, the current through each row is small, about two orders of magnitude smaller than the current through the water. This tends to spread the current through the eel's body, so that the eel need not stun or kill itself when it stuns or kills a fish.

---

**PROBLEM-SOLVING TACTICS**

**Tactic 2:  Solving Circuits of Batteries and Resistors**  Here are two general techniques for solving circuits for unknown currents or potential differences.

1. If you can simplify a circuit by replacing resistors in series or in parallel with their equivalents, do so. If you can reduce the circuit to a single loop, then you can find the current through the battery with that loop, as in Sample Problem 27-2a. You may then have to work backward, undoing the resistor simplification process, to find the current or potential difference for any particular resistor, as in Sample Problem 27-2b.

2. If a circuit cannot be simplified to a single loop, use the junction rule and the loop rule to write a set of simultaneous equations, as in Sample Problem 27-3. You need have only as many independent equations as there are unknowns in those equations. If you have to find the current or potential difference for a particular resistor, you can ensure that its current or potential difference appears in the equations by having at least one of the loops pass through it.

**Tactic 3:  Arbitrary Choices in Solving Circuit Problems**  In Sample Problem 27-3, we made several arbitrary choices before we wrote even a single equation. (1) We assumed directions for the currents in Fig. 27-12 arbitrarily. (2) We chose which of the three possible loops to write equations for arbitrarily. (3) We chose the direction in which to traverse each loop arbitrarily. (4) We chose the starting and ending point for each traversal arbitrarily.

Such arbitrariness often worries a beginning circuit solver, but an experienced circuit solver knows that it does not matter. Just keep two rules firmly in mind. First, make sure you traverse each chosen loop completely. Second, once you have chosen a direction for a current, stick with that direction until you get numerical values for all the currents. If you were wrong about a direction, the algebra will signal you with a minus sign. Then you can make a correction by simply erasing the minus sign and reversing the arrow representing that current in the circuit diagram. However, *you should not make this correction until you have completed all the required calculations for the circuit,* as we did in the very last step of Sample Problem 27-3.

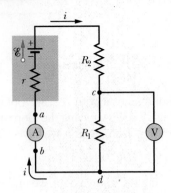

**FIG. 27-14** A single-loop circuit, showing how to connect an ammeter (A) and a voltmeter (V).

## 27-8 | The Ammeter and the Voltmeter

An instrument used to measure currents is called an *ammeter*. To measure the current in a wire, you usually have to break or cut the wire and insert the ammeter so that the current to be measured passes through the meter. (In Fig. 27-14, ammeter A is set up to measure current $i$.)

It is essential that the resistance $R_A$ of the ammeter be very much smaller than other resistances in the circuit. Otherwise, the very presence of the meter will change the current to be measured.

A meter used to measure potential differences is called a *voltmeter*. To find the potential difference between any two points in the circuit, the voltmeter terminals are connected between those points without breaking or cutting the wire. (In Fig. 27-14, voltmeter V is set up to measure the voltage across $R_1$.)

It is essential that the resistance $R_V$ of a voltmeter be very much larger than the resistance of any circuit element across which the voltmeter is connected. Otherwise, the meter itself becomes an important circuit element and alters the potential difference that is to be measured.

Often a single meter is packaged so that, by means of a switch, it can be made to serve as either an ammeter or a voltmeter—and usually also as an *ohmmeter*, designed to measure the resistance of any element connected between its terminals. Such a versatile unit is called a *multimeter*.

## 27-9 | RC Circuits

In preceding sections we dealt only with circuits in which the currents did not vary with time. Here we begin a discussion of time-varying currents.

### Charging a Capacitor

The capacitor of capacitance $C$ in Fig. 27-15 is initially uncharged. To charge it, we close switch S on point $a$. This completes an *RC series circuit* consisting of the capacitor, an ideal battery of emf $\mathscr{E}$, and a resistance $R$.

From Section 25-2, we already know that as soon as the circuit is complete, charge begins to flow (current exists) between a capacitor plate and a battery terminal on each side of the capacitor. This current increases the charge $q$ on the plates and the potential difference $V_C$ ($= q/C$) across the capacitor. When that potential difference equals the potential difference across the battery (which here is equal to the emf $\mathscr{E}$), the current is zero. From Eq. 25-1 ($q = CV$), the equilibrium (final) *charge* on the then fully charged capacitor is equal to $C\mathscr{E}$.

Here we want to examine the charging process. In particular we want to know how the charge $q(t)$ on the capacitor plates, the potential difference $V_C(t)$ across the capacitor, and the current $i(t)$ in the circuit vary with time during the charging process. We begin by applying the loop rule to the circuit, traversing it clockwise from the negative terminal of the battery. We find

$$\mathscr{E} - iR - \frac{q}{C} = 0. \qquad (27\text{-}30)$$

The last term on the left side represents the potential difference across the capacitor. The term is negative because the capacitor's top plate, which is connected to the battery's positive terminal, is at a higher potential than the lower plate. Thus, there is a drop in potential as we move down through the capacitor.

We cannot immediately solve Eq. 27-30 because it contains two variables, $i$ and $q$. However, those variables are not independent but are related by

$$i = \frac{dq}{dt}. \qquad (27\text{-}31)$$

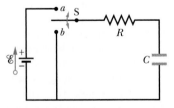

**FIG. 27-15** When switch S is closed on $a$, the capacitor is *charged* through the resistor. When the switch is afterward closed on $b$, the capacitor *discharges* through the resistor.

Substituting this for $i$ in Eq. 27-30 and rearranging, we find

$$R\frac{dq}{dt} + \frac{q}{C} = \mathscr{E} \quad \text{(charging equation).} \quad (27\text{-}32)$$

This differential equation describes the time variation of the charge $q$ on the capacitor in Fig. 27-15. To solve it, we need to find the function $q(t)$ that satisfies this equation and also satisfies the condition that the capacitor be initially uncharged; that is, $q = 0$ at $t = 0$.

We shall soon show that the solution to Eq. 27-32 is

$$q = C\mathscr{E}(1 - e^{-t/RC}) \quad \text{(charging a capacitor).} \quad (27\text{-}33)$$

(Here $e$ is the exponential base, $2.718\ldots$, and not the elementary charge.) Note that Eq. 27-33 does indeed satisfy our required initial condition, because at $t = 0$ the term $e^{-t/RC}$ is unity; so the equation gives $q = 0$. Note also that as $t$ goes to infinity (that is, a long time later), the term $e^{-t/RC}$ goes to zero; so the equation gives the proper value for the full (equilibrium) charge on the capacitor—namely, $q = C\mathscr{E}$. A plot of $q(t)$ for the charging process is given in Fig. 27-16a.

The derivative of $q(t)$ is the current $i(t)$ charging the capacitor:

$$i = \frac{dq}{dt} = \left(\frac{\mathscr{E}}{R}\right)e^{-t/RC} \quad \text{(charging a capacitor).} \quad (27\text{-}34)$$

A plot of $i(t)$ for the charging process is given in Fig. 27-16b. Note that the current has the initial value $\mathscr{E}/R$ and that it decreases to zero as the capacitor becomes fully charged.

➤ A capacitor that is being charged initially acts like ordinary connecting wire relative to the charging current. A long time later, it acts like a broken wire.

By combining Eq. 25-1 ($q = CV$) and Eq. 27-33, we find that the potential difference $V_C(t)$ across the capacitor during the charging process is

$$V_C = \frac{q}{C} = \mathscr{E}(1 - e^{-t/RC}) \quad \text{(charging a capacitor).} \quad (27\text{-}35)$$

This tells us that $V_C = 0$ at $t = 0$ and that $V_C = \mathscr{E}$ when the capacitor becomes fully charged as $t \to \infty$.

## The Time Constant

The product $RC$ that appears in Eqs. 27-33, 27-34, and 27-35 has the dimensions of time (both because the argument of an exponential must be dimensionless and because, in fact, $1.0\ \Omega \times 1.0\ \text{F} = 1.0\ \text{s}$). The product $RC$ is called the **capacitive time constant** of the circuit and is represented with the symbol $\tau$:

$$\tau = RC \quad \text{(time constant).} \quad (27\text{-}36)$$

From Eq. 27-33, we can now see that at time $t = \tau\,(= RC)$, the charge on the initially uncharged capacitor of Fig. 27-15 has increased from zero to

$$q = C\mathscr{E}(1 - e^{-1}) = 0.63C\mathscr{E}. \quad (27\text{-}37)$$

In words, during the first time constant $\tau$ the charge has increased from zero to 63% of its final value $C\mathscr{E}$. In Fig. 27-16, the small triangles along the time axes mark successive intervals of one time constant during the charging of the capacitor. The charging times for RC circuits are often stated in terms of $\tau$; the greater $\tau$ is, the greater is the charging time.

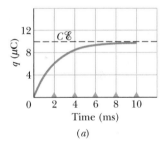

(a)

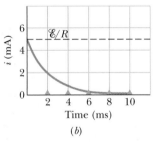

(b)

**FIG. 27-16** (a) A plot of Eq. 27-33, which shows the buildup of charge on the capacitor of Fig. 27-15. (b) A plot of Eq. 27-34, which shows the decline of the charging current in the circuit of Fig. 27-15. The curves are plotted for $R = 2000\ \Omega$, $C = 1\ \mu\text{F}$, and $\mathscr{E} = 10$ V; the small triangles represent successive intervals of one time constant $\tau$.

### Discharging a Capacitor

Assume now that the capacitor of Fig. 27-15 is fully charged to a potential $V_0$ equal to the emf $\mathscr{E}$ of the battery. At a new time $t = 0$, switch S is thrown from $a$ to $b$ so that the capacitor can *discharge* through resistance $R$. How do the charge $q(t)$ on the capacitor and the current $i(t)$ through the discharge loop of capacitor and resistance now vary with time?

The differential equation describing $q(t)$ is like Eq. 27-32 except that now, with no battery in the discharge loop, $\mathscr{E} = 0$. Thus,

$$R\frac{dq}{dt} + \frac{q}{C} = 0 \qquad \text{(discharging equation)}. \tag{27-38}$$

The solution to this differential equation is

$$q = q_0 e^{-t/RC} \qquad \text{(discharging a capacitor)}, \tag{27-39}$$

where $q_0 \, (= CV_0)$ is the initial charge on the capacitor. You can verify by substitution that Eq. 27-39 is indeed a solution of Eq. 27-38.

Equation 27-39 tells us that $q$ decreases exponentially with time, at a rate that is set by the capacitive time constant $\tau = RC$. At time $t = \tau$, the capacitor's charge has been reduced to $q_0 e^{-1}$, or about 37% of the initial value. Note that a greater $\tau$ means a greater discharge time.

Differentiating Eq. 27-39 gives us the current $i(t)$:

$$i = \frac{dq}{dt} = -\left(\frac{q_0}{RC}\right)e^{-t/RC} \qquad \text{(discharging a capacitor)}. \tag{27-40}$$

This tells us that the current also decreases exponentially with time, at a rate set by $\tau$. The initial current $i_0$ is equal to $q_0/RC$. Note that you can find $i_0$ by simply applying the loop rule to the circuit at $t = 0$; just then the capacitor's initial potential $V_0$ is connected across the resistance $R$, so the current must be $i_0 = V_0/R = (q_0/C)/R = q_0/RC$. The minus sign in Eq. 27-40 can be ignored; it merely means that the capacitor's charge $q$ is decreasing.

### Derivation of Eq. 27-33

To solve Eq. 27-32, we first rewrite it as

$$\frac{dq}{dt} + \frac{q}{RC} = \frac{\mathscr{E}}{R}. \tag{27-41}$$

The general solution to this differential equation is of the form

$$q = q_p + Ke^{-at}, \tag{27-42}$$

where $q_p$ is a *particular solution* of the differential equation, $K$ is a constant to be evaluated from the initial conditions, and $a = 1/RC$ is the coefficient of $q$ in Eq. 27-41. To find $q_p$, we set $dq/dt = 0$ in Eq. 27-41 (corresponding to the final condition of no further charging), let $q = q_p$, and solve, obtaining

$$q_p = C\mathscr{E}. \tag{27-43}$$

To evaluate $K$, we first substitute this into Eq. 27-42 to get

$$q = C\mathscr{E} + Ke^{-at}.$$

Then substituting the initial conditions $q = 0$ and $t = 0$ yields

$$0 = C\mathscr{E} + K,$$

or $K = -C\mathscr{E}$. Finally, with the values of $q_p$, $a$, and $K$ inserted, Eq. 27-42 becomes

$$q = C\mathscr{E} - C\mathscr{E}e^{-t/RC},$$

which, with a slight modification, is Eq. 27-33.

✓**CHECKPOINT 5**    The table gives four sets of values for the circuit elements in Fig. 27-15. Rank the sets according to (a) the initial current (as the switch is closed on *a*) and (b) the time required for the current to decrease to half its initial value, greatest first.

| | 1 | 2 | 3 | 4 |
|---|---|---|---|---|
| $\mathscr{E}$ (V) | 12 | 12 | 10 | 10 |
| $R$ (Ω) | 2 | 3 | 10 | 5 |
| $C$ (μF) | 3 | 2 | 0.5 | 2 |

## Sample Problem    27-5

As a car rolls along pavement, electrons move from the pavement first onto the tires and then onto the car body. The car stores this excess charge and the associated electric potential energy as if the car body were one plate of a capacitor and the pavement were the other plate (Fig. 27-17*a*). When the car stops, it discharges its excess charge and energy through the tires, just as a capacitor can discharge through a resistor. If a conducting object comes within a few centimeters of the car before the car is discharged, the remaining energy can be suddenly transferred to a spark between the car and the object. Suppose the conducting object is a fuel dispenser. The spark will not ignite the fuel and cause a fire if the spark energy is less than the critical value $U_{fire} = 50$ mJ.

When the car of Fig. 27-17*a* stops at time $t = 0$, the car–ground potential difference is $V_0 = 30$ kV. The

car–ground capacitance is $C = 500$ pF, and the resistance of *each* tire is $R_{tire} = 100$ GΩ. How much time does the car take to discharge through the tires to drop below the critical value $U_{fire}$?

**KEY IDEAS**   (1) At any time $t$, a capacitor's stored electric potential energy $U$ is related to its stored charge $q$ according to Eq. 25-21 ($U = q^2/2C$). (2) While a capacitor is discharging, the charge decreases with time according to Eq. 27-39 ($q = q_0 e^{-t/RC}$).

*Calculations:* We can treat the tires as resistors that are connected to one another at their tops via the car body and at their bottoms via the pavement. Figure 27-17*b* shows how the four resistors are connected in parallel across the car's capacitance, and Fig. 27-17*c* shows their equivalent resistance $R$. From Eq. 27-24, $R$ is given by

$$\frac{1}{R} = \frac{1}{R_{tire}} + \frac{1}{R_{tire}} + \frac{1}{R_{tire}} + \frac{1}{R_{tire}},$$

or    $R = \dfrac{R_{tire}}{4} = \dfrac{100 \times 10^9 \, \Omega}{4} = 25 \times 10^9 \, \Omega.$    (27-44)

When the car stops, it discharges its excess charge and energy through $R$.

We now use our two Key Ideas to analyze the discharge. Substituting Eq. 27-39 into Eq. 25-21 gives

$$U = \frac{q^2}{2C} = \frac{(q_0 e^{-t/RC})^2}{2C}$$

$$= \frac{q_0^2}{2C} e^{-2t/RC}. \tag{27-45}$$

From Eq. 25-1 ($q = CV$), we can relate the initial charge $q_0$ on the car to the given initial potential difference $V_0$: $q_0 = CV_0$. Substituting this equation into Eq. 27-45 brings us to

$$U = \frac{(CV_0)^2}{2C} e^{-2t/RC} = \frac{CV_0^2}{2} e^{-2t/RC},$$

or         $e^{-2t/RC} = \dfrac{2U}{CV_0^2}.$    (27-46)

Taking the natural logarithms of both sides, we obtain

$$-\frac{2t}{RC} = \ln\left(\frac{2U}{CV_0^2}\right),$$

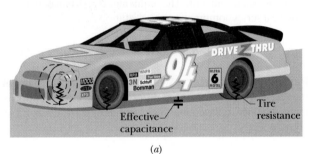

Effective capacitance      Tire resistance

(a)

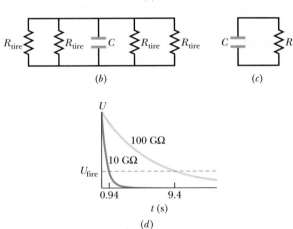

(b)             (c)

(d)

**FIG. 27-17**   (*a*) A charged car and the pavement acts like a capacitor that can discharge through the tires. (*b*) The effective circuit of the car–pavement capacitor, with four tire resistances $R_{tire}$ connected in parallel. (*c*) The equivalent resistance $R$ of the tires. (*d*) The electric potential energy $U$ in the car–pavement capacitor decreases during discharge.

or
$$t = -\frac{RC}{2} \ln\left(\frac{2U}{CV_0^2}\right). \qquad (27\text{-}47)$$

Substituting the given data, we find that the time the car takes to discharge to the energy level $U_{fire} = 50$ mJ is

$$t = -\frac{(25 \times 10^9\,\Omega)(500 \times 10^{-12}\,\text{F})}{2}$$

$$\times \ln\left(\frac{2(50 \times 10^{-3}\,\text{J})}{(500 \times 10^{-12}\,\text{F})(30 \times 10^3\,\text{V})^2}\right)$$

$$= 9.4\,\text{s}. \qquad \text{(Answer)}$$

**Fire or no fire:** This car requires at least 9.4 s before fuel or a fuel dispenser can be brought safely near it. During a race, a pit crew cannot wait that long. Instead, tires for race cars include some type of conducting material (such as carbon black) to lower the tire resistance and thus increase the car's discharge rate. Figure 27-17d shows the stored energy $U$ versus time $t$ for tire resistances of $R = 100$ GΩ (the value we used in our calculations here) and $R = 10$ GΩ. Note how much more rapidly a car discharges to level $U_{fire}$ with the lower $R$ value.

## REVIEW & SUMMARY

**Emf** An **emf device** does work on charges to maintain a potential difference between its output terminals. If $dW$ is the work the device does to force positive charge $dq$ from the negative to the positive terminal, then the **emf** (work per unit charge) of the device is

$$\mathcal{E} = \frac{dW}{dq} \qquad \text{(definition of } \mathcal{E}\text{).} \qquad (27\text{-}1)$$

The volt is the SI unit of emf as well as of potential difference. An **ideal emf device** is one that lacks any internal resistance. The potential difference between its terminals is equal to the emf. A **real emf device** has internal resistance. The potential difference between its terminals is equal to the emf only if there is no current through the device.

**Analyzing Circuits** The change in potential in traversing a resistance $R$ in the direction of the current is $-iR$; in the opposite direction it is $+iR$ (resistance rule). The change in potential in traversing an ideal emf device in the direction of the emf arrow is $+\mathcal{E}$; in the opposite direction it is $-\mathcal{E}$ (emf rule). Conservation of energy leads to the loop rule:

*Loop Rule.   The algebraic sum of the changes in potential encountered in a complete traversal of any loop of a circuit must be zero.*

Conservation of charge gives us the junction rule:

*Junction Rule.   The sum of the currents entering any junction must be equal to the sum of the currents leaving that junction.*

**Single-Loop Circuits** The current in a single-loop circuit containing a single resistance $R$ and an emf device with emf $\mathcal{E}$ and internal resistance $r$ is

$$i = \frac{\mathcal{E}}{R + r}, \qquad (27\text{-}4)$$

which reduces to $i = \mathcal{E}/R$ for an ideal emf device with $r = 0$.

**Power** When a real battery of emf $\mathcal{E}$ and internal resistance $r$ does work on the charge carriers in a current $i$ through the battery, the rate $P$ of energy transfer to the charge carriers is

$$P = iV, \qquad (27\text{-}14)$$

where $V$ is the potential across the terminals of the battery. The rate $P_r$ at which energy is dissipated as thermal energy in the battery is

$$P_r = i^2 r. \qquad (27\text{-}16)$$

The rate $P_{emf}$ at which the chemical energy in the battery changes is

$$P_{emf} = i\mathcal{E}. \qquad (27\text{-}17)$$

**Series Resistances** When resistances are in **series**, they have the same current. The equivalent resistance that can replace a series combination of resistances is

$$R_{eq} = \sum_{j=1}^{n} R_j \qquad (n \text{ resistances in series).} \qquad (27\text{-}7)$$

**Parallel Resistances** When resistances are in **parallel**, they have the same potential difference. The equivalent resistance that can replace a parallel combination of resistances is given by

$$\frac{1}{R_{eq}} = \sum_{j=1}^{n} \frac{1}{R_j} \qquad (n \text{ resistances in parallel).} \qquad (27\text{-}24)$$

**RC Circuits** When an emf $\mathcal{E}$ is applied to a resistance $R$ and capacitance $C$ in series, as in Fig. 27-15 with the switch at $a$, the charge on the capacitor increases according to

$$q = C\mathcal{E}(1 - e^{-t/RC}) \qquad \text{(charging a capacitor),} \qquad (27\text{-}33)$$

in which $C\mathcal{E} = q_0$ is the equilibrium (final) charge and $RC = \tau$ is the **capacitive time constant** of the circuit. During the charging, the current is

$$i = \frac{dq}{dt} = \left(\frac{\mathcal{E}}{R}\right)e^{-t/RC} \qquad \text{(charging a capacitor).} \qquad (27\text{-}34)$$

When a capacitor discharges through a resistance $R$, the charge on the capacitor decays according to

$$q = q_0 e^{-t/RC} \qquad \text{(discharging a capacitor).} \qquad (27\text{-}39)$$

During the discharging, the current is

$$i = \frac{dq}{dt} = -\left(\frac{q_0}{RC}\right)e^{-t/RC} \qquad \text{(discharging a capacitor).} \qquad (27\text{-}40)$$

## QUESTIONS

**1** For each circuit in Fig. 27-18, are the resistors connected in series, in parallel, or neither?

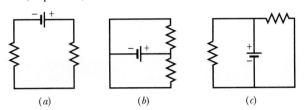

**FIG. 27-18** Question 1.

**2** In Fig. 27-19, a circuit consists of a battery and two uniform resistors, and the section lying along an $x$ axis is divided into five segments of equal lengths. (a) Assume that $R_1 = R_2$ and rank the segments according to the magnitude of the average electric field in them, greatest first. (b) Now assume that $R_1 > R_2$ and then again rank the segments. (c) What is the direction of the electric field along the $x$ axis?

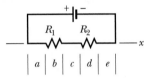

**FIG. 27-19** Question 2.

**3** (a) In Fig. 27-20$a$, with $R_1 > R_2$, is the potential difference across $R_2$ more than, less than, or equal to that across $R_1$? (b) Is the current through resistor $R_2$ more than, less than, or equal to that through resistor $R_1$?

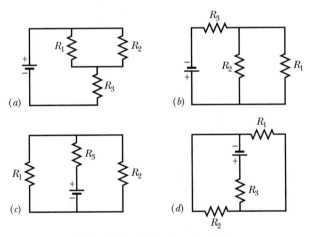

**FIG. 27-20** Questions 3 and 4.

**4** (a) In Fig. 27-20$a$, are resistors $R_1$ and $R_3$ in series? (b) Are resistors $R_1$ and $R_2$ in parallel? (c) Rank the equivalent resistances of the four circuits shown in Fig. 27-20, greatest first.

**5** You are to connect resistors $R_1$ and $R_2$, with $R_1 > R_2$, to a battery, first individually, then in series, and then in parallel. Rank those arrangements according to the amount of current through the battery, greatest first.

**6** *Cap-monster maze.* In Fig. 27-21, all the capacitors have a capacitance of 6.0 $\mu$F, and all the batteries have an emf of 10 V. What is the charge on capacitor $C$? (If you can find the proper loop through this maze, you can answer the question with a few seconds of mental calculation.)

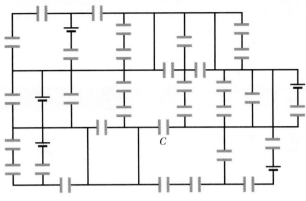

**FIG. 27-21** Question 6.

**7** Initially, a single resistor $R_1$ is wired to a battery. Then resistor $R_2$ is added in parallel. Are (a) the potential difference across $R_1$ and (b) the current $i_1$ through $R_1$ now more than, less than, or the same as previously? (c) Is the equivalent resistance $R_{12}$ of $R_1$ and $R_2$ more than, less than, or equal to $R_1$? (d) Is the total current through $R_1$ and $R_2$ together more than, less than, or equal to the current through $R_1$ previously?

**8** *Res-monster maze.* In Fig. 27-22, all the resistors have a resistance of 4.0 $\Omega$ and all the (ideal) batteries have an emf of 4.0 V. What is the current through resistor $R$? (If you can find the proper loop through this maze, you can answer the question with a few seconds of mental calculation.)

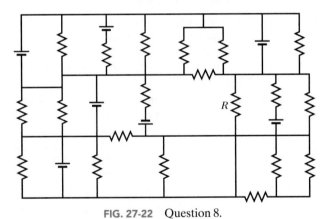

**FIG. 27-22** Question 8.

**9** A resistor $R_1$ is wired to a battery, then resistor $R_2$ is added in series. Are (a) the potential difference across $R_1$ and (b) the current $i_1$ through $R_1$ now more than, less than, or the same as previously? (c) Is the equivalent resistance $R_{12}$ of $R_1$ and $R_2$ more than, less than, or equal to $R_1$?

**10** After the switch in Fig. 27-15 is closed on point $a$, there is current $i$ through resistance $R$. Figure 27-23 gives that current for four sets of values of $R$ and capacitance $C$: (1) $R_0$ and $C_0$, (2) $2R_0$ and $C_0$, (3) $R_0$ and $2C_0$, (4) $2R_0$ and $2C_0$. Which set goes with which curve?

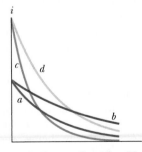

**FIG. 27-23** Question 10.

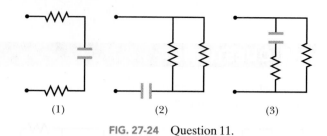

**11** Figure 27-24 shows three sections of circuit that are to be connected in turn to the same battery via a switch as in Fig. 27-15. The resistors are all identical, as are the capacitors. Rank the sections according to (a) the final (equilibrium) charge on the capacitor and (b) the time required for the capacitor to reach 50% of its final charge, greatest first.

**FIG. 27-24** Question 11.

## PROBLEMS

**GO**     Tutoring problem available (at instructor's discretion) in *WileyPLUS* and WebAssign
**SSM**    Worked-out solution available in Student Solutions Manual          **WWW**  Worked-out solution is at ⎯
• – •••    Number of dots indicates level of problem difficulty                **ILW**   Interactive solution is at ⎯ http://www.wiley.com/college/halliday
 Additional information available in *The Flying Circus of Physics* and at flyingcircusofphysics.com

### sec. 27-6  Potential Difference Between Two Points

•1  A wire of resistance 5.0 Ω is connected to a battery whose emf $\mathscr{E}$ is 2.0 V and whose internal resistance is 1.0 Ω. In 2.0 min, how much energy is (a) transferred from chemical form in the battery, (b) dissipated as thermal energy in the wire, and (c) dissipated as thermal energy in the battery?

•2  A certain car battery with a 12.0 V emf has an initial charge of 120 A · h. Assuming that the potential across the terminals stays constant until the battery is completely discharged, for how many hours can it deliver energy at the rate of 100 W?

•3  A 5.0 A current is set up in a circuit for 6.0 min by a rechargeable battery with a 6.0 V emf. By how much is the chemical energy of the battery reduced?

•4  A standard flashlight battery can deliver about 2.0 W · h of energy before it runs down. (a) If a battery costs US$0.80, what is the cost of operating a 100 W lamp for 8.0 h using batteries? (b) What is the cost if energy is provided at the rate of US$0.06 per kilowatt-hour?

•5  A car battery with a 12 V emf and an internal resistance of 0.040 Ω is being charged with a current of 50 A. What are (a) the potential difference $V$ across the terminals, (b) the rate $P_r$ of energy dissipation inside the battery, and (c) the rate $P_{emf}$ of energy conversion to chemical form? When the battery is used to supply 50 A to the starter motor, what are (d) $V$ and (e) $P_r$?  **ILW**

•6  In Fig. 27-25, the ideal batteries have emfs $\mathscr{E}_1 = 150$ V and $\mathscr{E}_2 = 50$ V and the resistances are $R_1 = 3.0$ Ω and $R_2 = 2.0$ Ω. If the potential at $P$ is defined to be 100 V, what is the potential at $Q$?

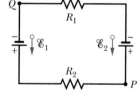

**FIG. 27-25** Problem 6.

•7  In Fig. 27-26, the ideal batteries have emfs $\mathscr{E}_1 = 12$ V and $\mathscr{E}_2 = 6.0$ V and the resistors have resistances $R_1 = 4.0$ Ω and $R_2 = 8.0$ Ω. What are (a) the current, the dissipation rate in (b) resistor 1 and (c) resistor 2, and the energy transfer rate in (d) battery 1 and (e) battery 2? Is energy being supplied or absorbed by (f) battery 1 and (g) battery 2?  **SSM WWW**

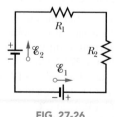

**FIG. 27-26** Problem 7.

•8  Figure 27-27 shows a circuit of four resistors that are connected to a larger circuit. The graph below the circuit shows the electric potential $V(x)$ as a function of position $x$ along the lower branch of the circuit, through resistor 4; the potential $V_A$ is 12.0 V. The graph above the circuit shows the electric potential $V(x)$ versus position $x$ along the upper branch of the circuit, through resistors 1, 2, and 3; the potential differences are $\Delta V_B = 2.00$ V and $\Delta V_C = 5.00$ V. Resistor 3 has a resistance of 200 Ω. What is the resistance of (a) resistor 1 and (b) resistor 2?  **GO**

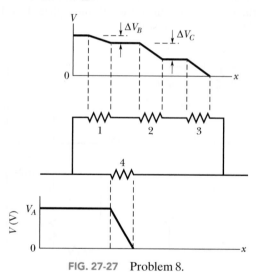

**FIG. 27-27** Problem 8.

•9  A total resistance of 3.00 Ω is to be produced by connecting an unknown resistance to a 12.0 Ω resistance. (a) What must be the value of the unknown resistance, and (b) should it be connected in series or in parallel?

•10  (a) In electron-volts, how much work does an ideal battery with a 12.0 V emf do on an electron that passes through the battery from the positive to the negative terminal? (b) If $3.40 \times 10^{18}$ electrons pass through each second, what is the power of the battery in watts?

••11  *Side flash.* Figure 27-28 indicates one reason no one should stand under a tree during a lightning storm. If lightning comes down the side of the tree, a portion can jump over to the person, especially if the current on the tree reaches a dry region on the bark and thereafter must travel through air to reach the ground. In the figure, part of the lightning jumps through distance $d$ in air and then travels through the person (who has negligible resistance relative to that of air). The rest of the current travels through air alongside the tree, for a distance $h$. If $d/h = 0.400$ and the total current is $I = 5000$ A, what is the current through the person?

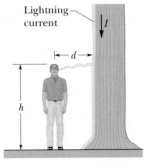

**FIG. 27-28** Problem 11.

**••12** (a) In Fig. 27-29, what value must $R$ have if the current in the circuit is to be 1.0 mA? Take $\mathscr{E}_1 = 2.0$ V, $\mathscr{E}_2 = 3.0$ V, and $r_1 = r_2 = 3.0 \ \Omega$. (b) What is the rate at which thermal energy appears in $R$?

**••13** In Fig. 27-30, circuit section $AB$ absorbs energy at a rate of 50 W when current $i = 1.0$ A through it is in the indicated direction. Resistance $R = 2.0 \ \Omega$. (a) What is the potential difference between $A$ and $B$? Emf device $X$ lacks internal resistance. (b) What is its emf? (c) Is point $B$ connected to the positive terminal of $X$ or to the negative terminal? **SSM**

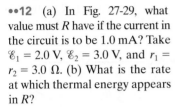

**FIG. 27-29** Problem 12.

**FIG. 27-30** Problem 13.

**••14** Figure 27-31 shows a battery connected across a uniform resistor $R_0$. A sliding contact can move across the resistor from $x = 0$ at the left to $x = 10$ cm at the right. Moving the contact changes how much resistance is to the left of the contact and how much is to the right. Find the rate at which energy is dissipated in resistor $R$ as a function of $x$. Plot the function for $\mathscr{E} = 50$ V, $R = 2000 \ \Omega$, and $R_0 = 100 \ \Omega$.

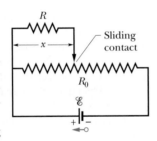

**FIG. 27-31** Problem 14.

**••15** A 10-km-long underground cable extends east to west and consists of two parallel wires, each of which has resistance 13 $\Omega$/km. An electrical short develops at distance $x$ from the west end when a conducting path of resistance $R$ connects the wires (Fig. 27-32). The resistance of the wires and the short is then 100 $\Omega$ when measured from the east end and 200 $\Omega$ when measured from the west end. What are (a) $x$ and (b) $R$?

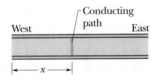

**FIG. 27-32** Problem 15.

**••16** The ideal battery in Fig. 27-33a has emf $\mathscr{E} = 6.0$ V. Plot 1 in Fig. 27-33b gives the electric potential difference $V$ that can appear across resistor 1 of the circuit versus the current $i$ in that resistor. The scale of the $V$ axis is set by $V_s = 18.0$ V, and the scale of the $i$ axis is set by $i_s = 3.00$ mA. Plots 2 and 3 are similar plots for resistors 2 and 3, respectively. What is the current in resistor 2?

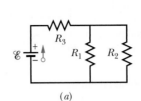

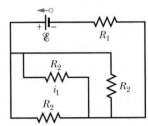

**FIG. 27-33** Problem 16.

**••17** In Fig. 27-34, $R_1 = 6.00 \ \Omega$, $R_2 = 18.0 \ \Omega$, and the ideal battery has emf $\mathscr{E} = 12.0$ V. What are the (a) size and (b) direction (left or right) of current $i_1$? (c) How much energy is dissipated by all four resistors in 1.00 min?

**••18** Figure 27-35 shows a resistor of resistance $R = 6.00 \ \Omega$ connected to an ideal battery of emf $\mathscr{E} = 12.0$ V by means of two copper wires. Each wire has length 20.0 cm and radius 1.00 mm. In dealing with such circuits in this chapter, we generally neglect the potential differences along the wires and the transfer of energy to thermal energy in them. Check the validity of this neglect for the circuit of Fig. 27-35: What is the potential difference across (a) the resistor and (b) each of the two sections of wire? At what rate is energy lost to thermal energy in (c) the resistor and (d) each section of wire?

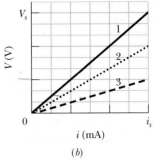

**FIG. 27-34** Problem 17.

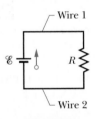

**FIG. 27-35**
Problem 18.

**••19** The current in a single-loop circuit with one resistance $R$ is 5.0 A. When an additional resistance of 2.0 $\Omega$ is inserted in series with $R$, the current drops to 4.0 A. What is $R$? **ILW**

**••20** In Fig. 27-36a, both batteries have emf $\mathscr{E} = 1.20$ V and the external resistance $R$ is a variable resistor. Figure 27-36b gives the electric potentials $V$ between the terminals of each battery as functions of $R$: Curve 1 corresponds to battery 1, and curve 2 corresponds to battery 2. The horizontal scale is set by $R_s = 0.20 \ \Omega$. What is the internal resistance of (a) battery 1 and (b) battery 2?

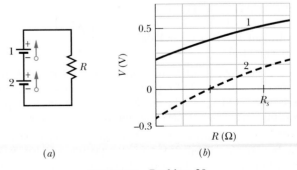

**FIG. 27-36** Problem 20.

•••**21** In Fig. 27-37, battery 1 has emf $\mathcal{E}_1 = 12.0$ V and internal resistance $r_1 = 0.016\ \Omega$ and battery 2 has emf $\mathcal{E}_2 = 12.0$ V and internal resistance $r_2 = 0.012\ \Omega$. The batteries are connected in series with an external resistance $R$. (a) What $R$ value makes the terminal-to-terminal potential difference of one of the batteries zero? (b) Which battery is that? **SSM**

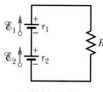

**FIG. 27-37**
Problem 21.

•••**22** A solar cell generates a potential difference of 0.10 V when a 500 $\Omega$ resistor is connected across it, and a potential difference of 0.15 V when a 1000 $\Omega$ resistor is substituted. What are the (a) internal resistance and (b) emf of the solar cell? (c) The area of the cell is 5.0 cm², and the rate per unit area at which it receives energy from light is 2.0 mW/cm². What is the efficiency of the cell for converting light energy to thermal energy in the 1000 $\Omega$ external resistor?

### sec. 27-7 Multiloop Circuits

•**23** In Fig. 27-38, $R_1 = R_2 = 4.00\ \Omega$ and $R_3 = 2.50\ \Omega$. Find the equivalent resistance between points $D$ and $E$. (*Hint:* Imagine that a battery is connected across those points.)

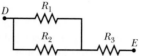

**FIG. 27-38** Problem 23.

•**24** When resistors 1 and 2 are connected in series, the equivalent resistance is 16.0 $\Omega$. When they are connected in parallel, the equivalent resistance is 3.0 $\Omega$. What are (a) the smaller resistance and (b) the larger resistance of these two resistors?

•**25** Four 18.0 $\Omega$ resistors are connected in parallel across a 25.0 V ideal battery. What is the current through the battery?

•**26** Figure 27-39 shows five 5.00 $\Omega$ resistors. Find the equivalent resistance between points (a) $F$ and $H$ and (b) $F$ and $G$. (*Hint:* For each pair of points, imagine that a battery is connected across the pair.)

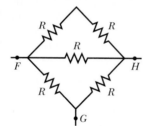

**FIG. 27-39** Problem 26.

•**27** In Fig. 27-40, $R_1 = 100\ \Omega$, $R_2 = 50\ \Omega$, and the ideal batteries have emfs $\mathcal{E}_1 = 6.0$ V, $\mathcal{E}_2 = 5.0$ V, and $\mathcal{E}_3 = 4.0$ V. Find (a) the current in resistor 1, (b) the current in resistor 2, and (c) the potential difference between points $a$ and $b$.

•**28** In Fig. 27-9, what is the potential difference $V_d - V_c$ between points $d$ and $c$ if $\mathcal{E}_1 = 4.0$ V, $\mathcal{E}_2 = 1.0$ V, $R_1 = R_2 = 10\ \Omega$, and $R_3 = 5.0\ \Omega$, and the battery is ideal?

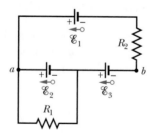

**FIG. 27-40** Problem 27.

•**29** Nine copper wires of length $l$ and diameter $d$ are connected in parallel to form a single composite conductor of resistance $R$. What must be the diameter $D$ of a single copper wire of length $l$ if it is to have the same resistance? **SSM**

••**30** The resistances in Figs. 27-41$a$ and $b$ are all 6.0 $\Omega$, and the batteries are ideal 12 V batteries. (a) When switch S in Fig. 27-41$a$ is closed, what is the change in the electric potential $V_1$ across resistor 1, or does $V_1$ remain the same? (b) When switch S in Fig. 27-41$b$ is closed, what is the change in $V_1$ across resistor 1, or does $V_1$ remain the same?

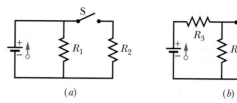

(a)   (b)

**FIG. 27-41** Problem 30.

••**31** In Fig. 27-42, the current in resistance 6 is $i_6 = 1.40$ A and the resistances are $R_1 = R_2 = R_3 = 2.00\ \Omega$, $R_4 = 16.0\ \Omega$, $R_5 = 8.00\ \Omega$, and $R_6 = 4.00\ \Omega$. What is the emf of the ideal battery? **GO**

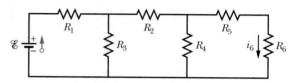

**FIG. 27-42** Problem 31.

••**32** In Fig. 27-43, the ideal batteries have emfs $\mathcal{E}_1 = 10.0$ V and $\mathcal{E}_2 = 0.500\mathcal{E}_1$, and the resistances are each 4.00 $\Omega$. What is the current in (a) resistance 2 and (b) resistance 3?

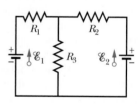

**FIG. 27-43** Problems 32, 43, and 98.

••**33** In Fig. 27-44, the ideal batteries have emfs $\mathcal{E}_1 = 5.0$ V and $\mathcal{E}_2 = 12$ V, the resistances are each 2.0 $\Omega$, and the potential is defined to be zero at the grounded point of the circuit. What are potentials (a) $V_1$ and (b) $V_2$ at the indicated points? **SSM**

••**34** Figure 27-45 shows a section of a circuit. The resistances are $R_1 = 2.0\ \Omega$, $R_2 = 4.0\ \Omega$, and $R_3 = 6.0\ \Omega$, and the indicated current is $i = 6.0$ A. The electric potential difference between points $A$ and $B$ that connect the section to the rest of the circuit is $V_A - V_B = 78$ V. (a) Is the device represented by "Box" absorbing or providing energy to the circuit, and (b) at what rate?

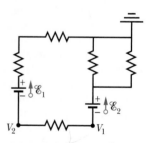

**FIG. 27-44** Problem 33.

••**35** In Fig. 27-46, $R_1 = 2.00\ \Omega$, $R_2 = 5.00\ \Omega$, and the battery is ideal. What value of $R_3$ maximizes the dissipation rate in resistance 3?

••**36** Both batteries in Fig. 27-47$a$ are ideal. Emf $\mathcal{E}_1$ of battery 1 has a fixed value, but emf $\mathcal{E}_2$ of battery 2 can be varied

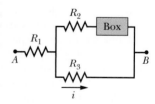

**FIG. 27-45** Problem 34.

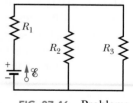

**FIG. 27-46** Problems 35, 99, 106, and 107.

between 1.0 V and 10 V. The plots in Fig. 27-47b give the currents through the two batteries as a function of $\mathscr{E}_2$. The vertical scale is set by $i_s = 0.20$ A. You must decide which plot corresponds to which battery, but for both plots, a negative current occurs when the direction of the current through the battery is opposite the direction of that battery's emf. What are (a) emf $\mathscr{E}_1$, (b) resistance $R_1$, and (c) resistance $R_2$?

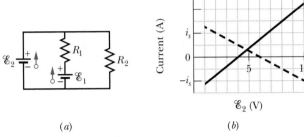
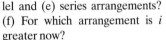

(a)    (b)

**FIG. 27-47**   Problem 36.

**••37** In Fig. 27-48, the resistances are $R_1 = 1.0$ Ω and $R_2 = 2.0$ Ω, and the ideal batteries have emfs $\mathscr{E}_1 = 2.0$ V and $\mathscr{E}_2 = \mathscr{E}_3 = 4.0$ V. What are the (a) size and (b) direction (up or down) of the current in battery 1, the (c) size and (d) direction of the current in battery 2, and the (e) size and (f) direction of the current in battery 3? (g) What is the potential difference $V_a - V_b$? **ILW**

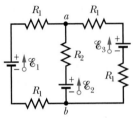

**FIG. 27-48**   Problem 37.

**••38** In Fig. 27-49, $\mathscr{E}_1 = 6.00$ V, $\mathscr{E}_2 = 12.0$ V, $R_1 = 100$ Ω, $R_2 = 200$ Ω, and $R_3 = 300$ Ω. One point of the circuit is grounded $(V = 0)$. What are the (a) size and (b) direction (up or down) of the current through resistance 1, the (c) size and (d) direction (left or right) of the current through resistance 2, and the (e) size and (f) direction of the current through resistance 3? (g) What is the electric potential at point $A$? **GO**

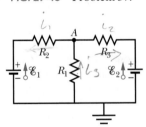

**FIG. 27-49**   Problem 38.

**••39** In Fig. 27-50, $\mathscr{E} = 12.0$ V, $R_1 = 2000$ Ω, $R_2 = 3000$ Ω, and $R_3 = 4000$ Ω. What are the potential differences (a) $V_A - V_B$, (b) $V_B - V_C$, (c) $V_C - V_D$, and (d) $V_A - V_C$?

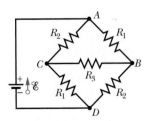

**FIG. 27-50**   Problem 39.

**••40** In Fig. 27-51, $R_1 = 100$ Ω, $R_2 = R_3 = 50.0$ Ω, $R_4 = 75.0$ Ω, and the ideal battery has emf $\mathscr{E} = 6.00$ V. (a) What is the equivalent resistance? What is $i$ in (b) resistance 1, (c) resistance 2, (d) resistance 3, and (e) resistance 4?

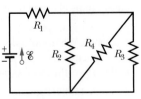

**FIG. 27-51**
Problems 40 and 48.

**••41** In Fig. 27-52, two batteries of emf $\mathscr{E} = 12.0$ V and in-

ternal resistance $r = 0.300$ Ω are connected in parallel across a resistance $R$. (a) For what value of $R$ is the dissipation rate in the resistor a maximum? (b) What is that maximum?

**••42** Two identical batteries of emf $\mathscr{E} = 12.0$ V and internal resistance $r = 0.200$ Ω are to be connected to an external resistance $R$, either in parallel (Fig. 27-52) or in series (Fig. 27-53). If $R = 2.00r$, what is the current $i$ in the external resistance in the (a) parallel and (b) series arrangements? (c) For which arrangement is $i$ greater? If $R = r/2.00$, what is $i$ in the external resistance in the (d) parallel and (e) series arrangements? (f) For which arrangement is $i$ greater now?

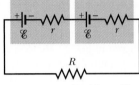

**FIG. 27-52**
Problems 41 and 42.

**••43** In Fig. 27-43, $\mathscr{E}_1 = 3.00$ V, $\mathscr{E}_2 = 1.00$ V, $R_1 = 4.00$ Ω, $R_2 = 2.00$ Ω, $R_3 = 5.00$ Ω, and both batteries are ideal. What is the rate at which energy is dissipated in (a) $R_1$, (b) $R_2$, and (c) $R_3$? What is the power of (d) battery 1 and (e) battery 2?

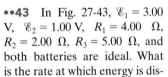

**FIG. 27-53**   Problem 42.

**••44** In Fig. 27-54a, resistor 3 is a variable resistor and the ideal battery has emf $\mathscr{E} = 12$ V. Figure 27-54b gives the current $i$ through the battery as a function of $R_3$. The horizontal scale is set by $R_{3s} = 20$ Ω. The curve has an asymptote of 2.0 mA as $R_3 \to \infty$. What are (a) resistance $R_1$ and (b) resistance $R_2$?

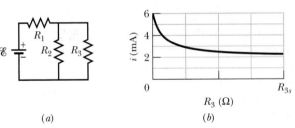

(a)    (b)

**FIG. 27-54**   Problem 44.

**••45** You are given a number of 10 Ω resistors, each capable of dissipating only 1.0 W without being destroyed. What is the minimum number of such resistors that you need to combine in series or in parallel to make a 10 Ω resistance that is capable of dissipating at least 5.0 W?

**••46** In Fig. 27-55, an array of $n$ parallel resistors is connected in series to a resistor and an ideal battery. All the resistors have the same resistance. If an identical resistor were added in parallel to the parallel array, the current through the battery would change by 1.25%. What is the value of $n$?

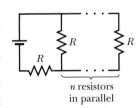

**FIG. 27-55**   Problem 46.

**•••47** A copper wire of radius $a = 0.250$ mm has an aluminum jacket of outer radius $b = 0.380$ mm. There is a current $i = 2.00$ A in the composite wire. Using Table 26-1, calculate the current in (a) the copper and (b) the aluminum. (c) If a potential difference $V = 12.0$ V between the ends maintains the current, what is the length of the composite wire? **SSM**

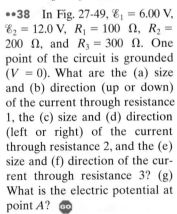

•••**48** In Fig. 27-51, $R_1 = 7.00\ \Omega$, $R_2 = 12.0\ \Omega$, $R_3 = 4.00\ \Omega$, and the ideal battery's emf is $\mathscr{E} = 24.0$ V. For what value of $R_4$ will the rate at which the battery transfers energy to the resistors equal (a) 60.0 W, (b) the maximum possible rate $P_{max}$, and (c) the minimum possible rate $P_{min}$? What are (d) $P_{max}$ and (e) $P_{min}$?

### sec. 27-8  The Ammeter and the Voltmeter

••**49** In Fig. 27-14, assume that $\mathscr{E} = 3.0$ V, $r = 100\ \Omega$, $R_1 = 250\ \Omega$, and $R_2 = 300\ \Omega$. If the voltmeter resistance $R_V$ is 5.0 k$\Omega$, what percent error does it introduce into the measurement of the potential difference across $R_1$? Ignore the presence of the ammeter.

••**50** A simple ohmmeter is made by connecting a 1.50 V flashlight battery in series with a resistance $R$ and an ammeter that reads from 0 to 1.00 mA, as shown in Fig. 27-56. Resistance $R$ is adjusted so that when the clip leads are shorted together, the meter deflects to its full-scale value of 1.00 mA. What external resistance across the leads results in a deflection of (a) 10.0%, (b) 50.0%, and (c) 90.0% of full scale? (d) If the ammeter has a resistance of 20.0 $\Omega$ and the internal resistance of the battery is negligible, what is the value of $R$?

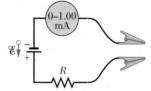

FIG. 27-56  Problem 50.

••**51** (a) In Fig. 27-57, what does the ammeter read if $\mathscr{E} = 5.0$ V (ideal battery), $R_1 = 2.0\ \Omega$, $R_2 = 4.0\ \Omega$, and $R_3 = 6.0\ \Omega$? (b) The ammeter and battery are now interchanged. Show that the ammeter reading is unchanged. **ILW**

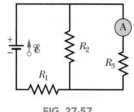

FIG. 27-57
Problem 51.

••**52** When the lights of a car are switched on, an ammeter in series with them reads 10.0 A and a voltmeter connected across them reads 12.0 V (Fig. 27-58). When the electric starting motor is turned on, the ammeter reading drops to 8.00 A and the lights dim somewhat. If the internal resistance of the battery is 0.0500 $\Omega$ and that of the ammeter is negligible, what are (a) the emf of the battery and (b) the current through the starting motor when the lights are on?

••**53** In Fig. 27-59, a voltmeter of resistance $R_V = 300\ \Omega$ and an ammeter of resistance $R_A = 3.00\ \Omega$ are being used to measure a resistance $R$ in a circuit that also contains a resistance $R_0 = 100\ \Omega$ and an ideal battery of emf $\mathscr{E} = 12.0$ V. Resistance $R$ is given by $R = V/i$, where $V$ is the potential across $R$ and $i$ is the ammeter reading. The voltmeter reading is $V'$, which is $V$ plus the potential

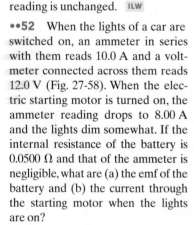

FIG. 27-58
Problem 52.

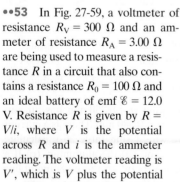

FIG. 27-59  Problem 53.

difference across the ammeter. Thus, the ratio of the two meter readings is not $R$ but only an *apparent* resistance $R' = V'/i$. If $R = 85.0\ \Omega$, what are (a) the ammeter reading, (b) the voltmeter reading, and (c) $R'$? (d) If $R_A$ is decreased, does the difference between $R'$ and $R$ increase, decrease, or remain the same?

••**54** In Fig. 27-60, $R_1 = 2.00R$, the ammeter resistance is zero, and the battery is ideal. What multiple of $\mathscr{E}/R$ gives the current in the ammeter?

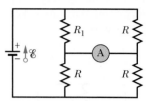

FIG. 27-60  Problem 54.

••**55** In Fig. 27-61, $R_s$ is to be adjusted in value by moving the sliding contact across it until points $a$ and $b$ are brought to the same potential. (One tests for this condition by momentarily connecting a sensitive ammeter between $a$ and $b$; if these points are at the same potential, the ammeter will not deflect.) Show that when this adjustment is made, the following relation holds: $R_x = R_s R_2/R_1$. An unknown resistance $(R_x)$ can be measured in terms of a standard $(R_s)$ using this device, which is called a Wheatstone bridge.

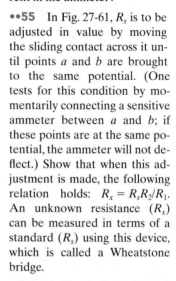

FIG. 27-61
Problems 55 and 111.

••**56** In Fig. 27-62, a voltmeter of resistance $R_V = 300\ \Omega$ and an ammeter of resistance $R_A = 3.00\ \Omega$ are being used to measure a resistance $R$ in a circuit that also contains a resistance $R_0 = 100\ \Omega$ and an ideal battery of emf $\mathscr{E} = 12.0$ V. Resistance $R$ is given by $R = V/i$, where $V$ is the voltmeter reading and $i$ is the current in resistance $R$. However, the ammeter reading is not $i$ but rather $i'$, which is $i$ plus the current through the voltmeter. Thus, the ratio of the two meter readings is not $R$ but only an *apparent* resistance $R' = V/i'$. If $R = 85.0\ \Omega$, what are (a) the ammeter reading, (b) the voltmeter reading, and (c) $R'$? (d) If $R_V$ is increased, does the difference between $R'$ and $R$ increase, decrease, or remain the same?

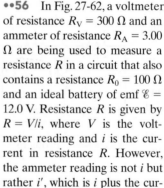

FIG. 27-62
Problem 56.

### sec. 27-9  RC Circuits

•**57** What multiple of the time constant $\tau$ gives the time taken by an initially uncharged capacitor in an *RC* series circuit to be charged to 99.0% of its final charge? **SSM**

•**58** A capacitor with initial charge $q_0$ is discharged through a resistor. What multiple of the time constant $\tau$ gives the time the capacitor takes to lose (a) the first one-third of its charge and (b) two-thirds of its charge?

•**59** A 15.0 k$\Omega$ resistor and a capacitor are connected in series, and then a 12.0 V potential difference is suddenly applied across them. The potential difference across the capacitor rises to 5.00 V in 1.30 $\mu$s. (a) Calculate the time constant of the circuit. (b) Find the capacitance of the capacitor. **ILW**

•60   In an $RC$ series circuit, $\mathscr{E} = 12.0$ V, $R = 1.40$ M$\Omega$, and $C = 1.80$ $\mu$F. (a) Calculate the time constant. (b) Find the maximum charge that will appear on the capacitor during charging. (c) How long does it take for the charge to build up to 16.0 $\mu$C?

•61   Switch S in Fig. 27-63 is closed at time $t = 0$, to begin charging an initially uncharged capacitor of capacitance $C = 15.0$ $\mu$F through a resistor of resistance $R = 20.0$ $\Omega$. At what time is the potential across the capacitor equal to that across the resistor?

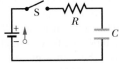

**FIG. 27-63**
Problems 61 and 76.

••62   A capacitor with an initial potential difference of 100 V is discharged through a resistor when a switch between them is closed at $t = 0$. At $t = 10.0$ s, the potential difference across the capacitor is 1.00 V. (a) What is the time constant of the circuit? (b) What is the potential difference across the capacitor at $t = 17.0$ s?

••63   The potential difference between the plates of a leaky (meaning that charge leaks from one plate to the other) 2.0 $\mu$F capacitor drops to one-fourth its initial value in 2.0 s. What is the equivalent resistance between the capacitor plates?

••64   A 1.0 $\mu$F capacitor with an initial stored energy of 0.50 J is discharged through a 1.0 M$\Omega$ resistor. (a) What is the initial charge on the capacitor? (b) What is the current through the resistor when the discharge starts? Find an expression that gives, as a function of time $t$, (c) the potential difference $V_C$ across the capacitor, (d) the potential difference $V_R$ across the resistor, and (e) the rate at which thermal energy is produced in the resistor.

••65   In the circuit of Fig. 27-64, $\mathscr{E} = 1.2$ kV, $C = 6.5$ $\mu$F, $R_1 = R_2 = R_3 = 0.73$ M$\Omega$. With $C$ completely uncharged, switch S is suddenly closed (at $t = 0$). At $t = 0$, what are (a) current $i_1$ in resistor 1, (b) current $i_2$ in resistor 2, and (c) current $i_3$ in resistor 3? At $t = \infty$ (that is, after many time constants), what are (d) $i_1$, (e) $i_2$, and (f) $i_3$? What is the potential difference $V_2$ across resistor 2 at (g) $t = 0$ and (h) $t = \infty$? (i) Sketch $V_2$ versus $t$ between these two extreme times. **SSM WWW**

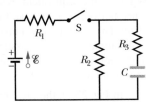

**FIG. 27-64**   Problem 65.

••66   Figure 27-65 shows the circuit of a flashing lamp, like those attached to barrels at highway construction sites. The fluorescent lamp L (of negligible capacitance) is connected in parallel across the capacitor $C$ of an $RC$ circuit. There is a current through the lamp only when the potential difference across it reaches the breakdown voltage $V_L$; then the capacitor discharges completely through the lamp and the lamp flashes briefly. For a lamp with breakdown voltage $V_L = 72.0$ V, wired to a 95.0 V ideal battery and a 0.150 $\mu$F capacitor, what resistance $R$ is needed for two flashes per second?

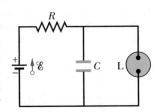

**FIG. 27-65**   Problem 66.

••67   In Fig. 27-66, $R_1 = 10.0$ k$\Omega$, $R_2 = 15.0$ k$\Omega$, $C = 0.400$ $\mu$F,

and the ideal battery has emf $\mathscr{E} = 20.0$ V. First, the switch is closed a long time so that the steady state is reached. Then the switch is opened at time $t = 0$. What is the current in resistor 2 at $t = 4.00$ ms? **GO**

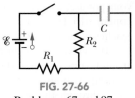

**FIG. 27-66**
Problems 67 and 97.

•••68   Figure 27-67 displays two circuits with a charged capacitor that is to be discharged through a resistor when a switch is closed. In Fig. 27-67a, $R_1 = 20.0$ $\Omega$ and $C_1 = 5.00$ $\mu$F. In Fig. 27-67b, $R_2 = 10.0$ $\Omega$ and $C_2 = 8.00$ $\mu$F. The ratio of the initial charges on the two capacitors is $q_{02}/q_{01} = 1.50$. At time $t = 0$, both switches are closed. At what time $t$ do the two capacitors have the same charge?

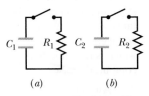

**FIG. 27-67**   Problem 68.

•••69   A 3.00 M$\Omega$ resistor and a 1.00 $\mu$F capacitor are connected in series with an ideal battery of emf $\mathscr{E} = 4.00$ V. At 1.00 s after the connection is made, what is the rate at which (a) the charge of the capacitor is increasing, (b) energy is being stored in the capacitor, (c) thermal energy is appearing in the resistor, and (d) energy is being delivered by the battery?

**Additional Problems**

70   What are the (a) size and (b) direction (up or down) of current $i$ in Fig. 27-68, where all resistances are 4.0 $\Omega$ and all batteries are ideal and have an emf of 10 V? (*Hint:* This can be answered using only mental calculation.)

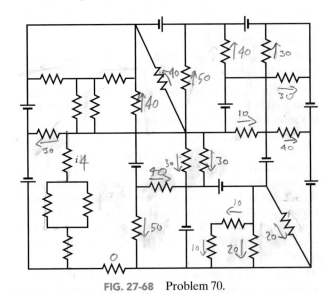

**FIG. 27-68**   Problem 70.

71   Suppose that, while you are sitting in a chair, charge separation between your clothing and the chair puts you at a potential of 200 V, with the capacitance between you and the chair at 150 pF. When you stand up, the increased separation between your body and the chair decreases the capacitance to 10 pF. (a) What then is the potential of your body? That potential is reduced over time, as the charge on you drains through your body and shoes (you are a capacitor discharging through a resistance). Assume that the resistance along that route is 300 G$\Omega$. If you touch an electrical component while

your potential is greater than 100 V, you could ruin the component. (b) How long must you wait until your potential reaches the safe level of 100 V?

If you wear a conducting wrist strap that is connected to ground, your potential does not increase as much when you stand up; you also discharge more rapidly because the resistance through the grounding connection is much less than through your body and shoes. (c) Suppose that when you stand up, your potential is 1400 V and the chair-to-you capacitance is 10 pF. What resistance in that wrist-strap grounding connection will allow you to discharge to 100 V in 0.30 s, which is less time than you would need to reach for, say, your computer?

**72** An automobile gasoline gauge is shown schematically in Fig. 27-69. The indicator (on the dashboard) has a resistance of 10 Ω. The tank unit is a float connected to a variable resistor whose resistance varies linearly with the volume of gasoline. The resistance is 140 Ω when the tank is empty and 20 Ω when the tank is full. Find the current in the circuit when the tank is (a) empty, (b) half-full, and (c) full. Treat the battery as ideal.

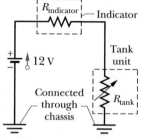

**FIG. 27-69** Problem 72.

**73** A controller on an electronic arcade game consists of a variable resistor connected across the plates of a 0.220 μF capacitor. The capacitor is charged to 5.00 V, then discharged through the resistor. The time for the potential difference across the plates to decrease to 0.800 V is measured by a clock inside the game. If the range of discharge times that can be handled effectively is from 10.0 μs to 6.00 ms, what should be the (a) lower value and (b) higher value of the resistance range of the resistor? **SSM**

**74** (a) In Fig. 27-4a, show that the rate at which energy is dissipated in $R$ as thermal energy is a maximum when $R = r$. (b) Show that this maximum power is $P = \mathscr{E}^2/4r$.

**75** Wires $A$ and $B$, having equal lengths of 40.0 m and equal diameters of 2.60 mm, are connected in series. A potential difference of 60.0 V is applied between the ends of the composite wire. The resistances are $R_A = 0.127$ Ω and $R_B = 0.729$ Ω. For wire $A$, what are (a) magnitude $J$ of the current density and (b) potential difference $V$? (c) Of what type material is wire $A$ made (see Table 26-1)? For wire $B$, what are (d) $J$ and (e) $V$? (f) Of what type material is $B$ made? **SSM**

**76** Figure 27-63 shows an ideal battery of emf $\mathscr{E} = 12$ V, a resistor of resistance $R = 4.0$ Ω, and an uncharged capacitor of capacitance $C = 4.0$ μF. After switch S is closed, what is the current through the resistor when the charge on the capacitor is 8.0 μC?

**77** The starting motor of a car is turning too slowly, and the mechanic has to decide whether to replace the motor, the cable, or the battery. The car's manual says that the 12 V battery should have no more than 0.020 Ω internal resistance, the motor no more than 0.200 Ω resistance, and the cable no more than 0.040 Ω resistance. The mechanic turns on the motor and measures 11.4 V across the battery, 3.0 V

across the cable, and a current of 50 A. Which part is defective? **SSM**

**78** Figure 27-70 shows a portion of a circuit through which there is a current $I = 6.00$ A. The resistances are $R_1 = R_2 = 2.00R_3 = 2.00R_4 = 4.00$ Ω. What is the current $i_1$ through resistor 1?

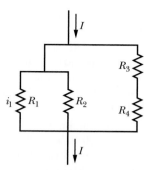

**FIG. 27-70** Problem 78.

**79** In Fig. 27-71, $R_1 = 20.0$ Ω, $R_2 = 10.0$ Ω, and the ideal battery has emf $\mathscr{E} = 120$ V. What is the current at point $a$ if we close (a) only switch $S_1$, (b) only switches $S_1$ and $S_2$, and (c) all three switches?

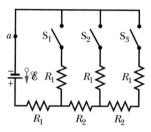

**FIG. 27-71** Problem 79.

**80** In Fig. 27-72, the ideal batteries have emfs $\mathscr{E}_1 = 20.0$ V, $\mathscr{E}_2 = 10.0$ V, $\mathscr{E}_3 = 5.00$ V, and $\mathscr{E}_4 = 5.00$ V, and the resistances are each 2.00 Ω. What are the (a) size and (b) direction (left or right) of current $i_1$ and the (c) size and (d) direction of current $i_2$? (This can be answered with only mental calculation.) (e) At what rate is energy being transferred in battery 4, and (f) is the energy being supplied or absorbed by the battery?

**81** In Fig. 27-73, $R = 10$ Ω. What is the equivalent resistance between points $A$ and $B$? (*Hint:* This circuit section might look simpler if you first assume that points $A$ and $B$ are connected to a battery.)

**82** In Fig. 27-74, the ideal battery has emf $\mathscr{E} = 30.0$ V, and the resistances are $R_1 = R_2 = 14$ Ω, $R_3 = R_4 = R_5 = 6.0$ Ω, $R_6 = 2.0$ Ω, and $R_7 = 1.5$ Ω. What are currents (a) $i_2$, (b) $i_4$, (c) $i_1$, (d) $i_3$, and (e) $i_5$?

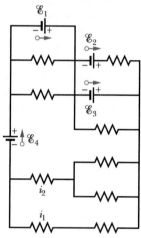

**FIG. 27-72** Problem 80.

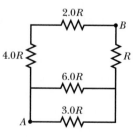

**FIG. 27-73** Problem 81.

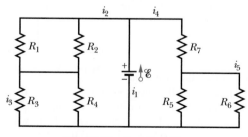

**FIG. 27-74** Problem 82.

**83** In Fig. 27-75, the ideal batteries have emfs $\mathscr{E}_1 = 12.0$ V and $\mathscr{E}_2 = 4.00$ V, and the resistances are each 4.00 Ω. What are the (a) size and (b) direction (up or down) of $i_1$ and the (c) size and (d) direction of $i_2$? (e) Does battery 1 supply or absorb energy, and (f) what is its energy transfer rate? (g) Does battery 2 supply or absorb energy, and (h) what is its energy transfer rate?

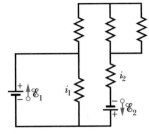

FIG. 27-75   Problem 83.

**84**   In Fig. 27-76, the ideal batteries have emfs $\mathscr{E}_1 = 20.0$ V, $\mathscr{E}_2 = 10.0$ V, and $\mathscr{E}_3 = 5.00$ V, and the resistances are each 2.00 Ω. What are the (a) size and (b) direction (left or right) of current $i_1$? (c) Does battery 1 supply or absorb energy, and (d) what is its power? (e) Does battery 2 supply or absorb energy, and (f) what is its power? (g) Does battery 3 supply or absorb energy, and (h) what is its power? **GO**

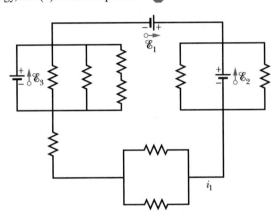

FIG. 27-76   Problem 84.

**85**   A temperature-stable resistor is made by connecting a resistor made of silicon in series with one made of iron. If the required total resistance is 1000 Ω in a wide temperature range around 20°C, what should be the resistance of the (a) silicon resistor and (b) iron resistor? (See Table 26-1.)   **SSM**

**86**   In Fig. 27-77, an ideal battery of emf $\mathscr{E} = 12.0$ V is connected to a network of resistances $R_1 = 6.00$ Ω, $R_2 = 12.0$ Ω, $R_3 = 4.00$ Ω, $R_4 = 3.00$ Ω, and $R_5 = 5.00$ Ω. What is the potential difference across resistance 5?

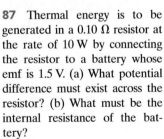

**87**   Thermal energy is to be generated in a 0.10 Ω resistor at the rate of 10 W by connecting the resistor to a battery whose emf is 1.5 V. (a) What potential difference must exist across the resistor? (b) What must be the internal resistance of the battery?

FIG. 27-77   Problem 86.

**88**   Figure 27-78 shows three 20.0 Ω resistors. Find the equivalent resistance between points (a) A and B, (b) A and C, and (c) B and C. (*Hint:* Imagine that a battery is connected between a given pair of points.)

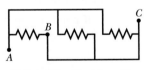

FIG. 27-78   Problem 88.

**89**   The circuit of Fig. 27-79 shows a capacitor, two ideal batteries, two resistors, and a switch S. Initially S has been open for a long time. If it is then closed for a long time, what is the change in the charge on the capacitor? Assume $C = 10$ μF, $\mathscr{E}_1 = 1.0$ V, $\mathscr{E}_2 = 3.0$ V, $R_1 = 0.20$ Ω, and $R_2 = 0.40$ Ω.

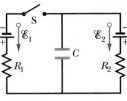

FIG. 27-79   Problem 89.

**90**   In Fig. 27-8a, calculate the potential difference between *a* and *c* by considering a path that contains $R$, $r_1$, and $\mathscr{E}_1$.

**91**   In Fig. 27-5a, find the potential difference across $R_2$ if $\mathscr{E} = 12$ V, $R_1 = 3.0$ Ω, $R_2 = 4.0$ Ω, and $R_3 = 5.0$ Ω.

**92**   In Fig. 27-14, assume that $\mathscr{E} = 5.0$ V, $r = 2.0$ Ω, $R_1 = 5.0$ Ω, and $R_2 = 4.0$ Ω. If the ammeter resistance $R_A$ is 0.10 Ω, what percent error does it introduce into the measurement of the current? Assume that the voltmeter is not present.

**93**   A 120 V power line is protected by a 15 A fuse. What is the maximum number of 500 W lamps that can be simultaneously operated in parallel on this line without "blowing" the fuse because of an excess of current?

**94**   In Fig. 27-80, $R_1 = 5.00$ Ω, $R_2 = 10.0$ Ω, $R_3 = 15.0$ Ω, $C_1 = 5.00$ μF, $C_2 = 10.0$ μF, and the ideal battery has emf $\mathscr{E} = 20.0$ V. Assuming that the circuit is in the steady state, what is the total energy stored in the two capacitors?

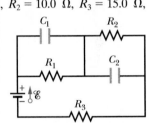

FIG. 27-80   Problem 94.

**95**   An initially uncharged capacitor C is fully charged by a device of constant emf $\mathscr{E}$ connected in series with a resistor R. (a) Show that the final energy stored in the capacitor is half the energy supplied by the emf device. (b) By direct integration of $i^2R$ over the charging time, show that the thermal energy dissipated by the resistor is also half the energy supplied by the emf device.   **SSM**

**96**   Two resistors $R_1$ and $R_2$ may be connected either in series or in parallel across an ideal battery with emf $\mathscr{E}$. We desire the rate of energy dissipation of the parallel combination to be five times that of the series combination. If $R_1 = 100$ Ω, what are the (a) smaller and (b) larger of the two values of $R_2$ that result in that dissipation rate?

**97**   In Fig. 27-66, the ideal battery has emf $\mathscr{E} = 30$ V, the resistances are $R_1 = 20$ kΩ and $R_2 = 10$ kΩ, and the capacitor is uncharged. When the switch is closed at time $t = 0$, what is the current in (a) resistance 1 and (b) resistance 2? (c) A long time later, what is the current in resistance 2?   **SSM**

**98**   In Fig. 27-43, $R_1 = 10.0$ Ω, $R_2 = 20.0$ Ω, and the ideal batteries have emfs $\mathscr{E}_1 = 20.0$ V and $\mathscr{E}_2 = 50.0$ V. What value of $R_3$ results in no current through battery 1?

**99**   In Fig. 27-46, $R_1 = R_2 = 10.0$ Ω, and the ideal battery has emf $\mathscr{E} = 12.0$ V. (a) What value of $R_3$ maximizes the rate at which the battery supplies energy and (b) what is that maximum rate?   **SSM**

**100**   Each of the six real batteries in Fig. 27-81 has an emf of 20 V and a resistance of 4.0 Ω. (a) What is the current

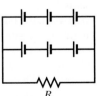

FIG. 27-81
Problem 100.

through the (external) resistance $R = 4.0 \, \Omega$? (b) What is the potential difference across each battery? (c) What is the power of each battery? (d) At what rate does each battery transfer energy to internal thermal energy? **GO**

**101** A group of $N$ identical batteries of emf $\mathscr{E}$ and internal resistance $r$ may be connected all in series (Fig. 27-82$a$) or all in parallel (Fig. 27-82$b$) and then across a resistor $R$. Show that both arrangements give the same current in $R$ if $R = r$. **SSM**

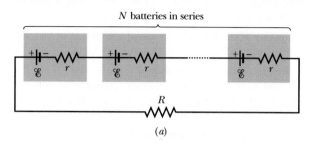

*N batteries in series*

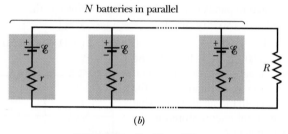

(a)

*N batteries in parallel*

(b)

**FIG. 27-82** Problem 101.

**102** The following table gives the electric potential difference $V_T$ across the terminals of a battery as a function of current $i$ being drawn from the battery. (a) Write an equation that represents the relationship between $V_T$ and $i$. Enter the data into your graphing calculator and perform a linear regression fit of $V_T$ versus $i$. From the parameters of the fit, find (b) the battery's emf and (c) its internal resistance.

| $i$ (A): | 50.0 | 75.0 | 100 | 125 | 150 | 175 | 200 |
|---|---|---|---|---|---|---|---|
| $V_T$ (V): | 10.7 | 9.00 | 7.70 | 6.00 | 4.80 | 3.00 | 1.70 |

**103** In Fig. 27-83, $\mathscr{E}_1 = 6.00$ V, $\mathscr{E}_2 = 12.0$ V, $R_1 = 200 \, \Omega$, and $R_2 = 100 \, \Omega$. What are the (a) size and (b) direction (up or down) of the current through resistance 1, the (c) size and (d) direction of the current through resistance 2, and the (e) size and (f) direction of the current through battery 2?

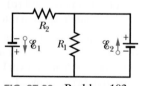

**FIG. 27-83** Problem 103.

**104** A three-way 120 V lamp bulb that contains two filaments is rated for 100-200-300 W. One filament burns out. Afterward, the bulb operates at the same intensity (dissipates energy at the same rate) on its lowest as on its highest switch positions but does not operate at all on the middle position. (a) How are the two filaments wired to the three switch positions? What are the (b) smaller and (c) larger values of the filament resistances?

**105** In Fig. 27-84, $R_1 = R_2 = 2.0 \, \Omega$, $R_3 = 4.0 \, \Omega$, $R_4 = 3.0 \, \Omega$, $R_5 = 1.0 \, \Omega$, and $R_6 = R_7 = R_8 = 8.0 \, \Omega$, and the ideal batteries

have emfs $\mathscr{E}_1 = 16$ V and $\mathscr{E}_2 = 8.0$ V. What are the (a) size and (b) direction (up or down) of current $i_1$ and the (c) size and (d) direction of current $i_2$? What is the energy transfer rate in (e) battery 1 and (f) battery 2? Is energy being supplied or absorbed in (g) battery 1 and (h) battery 2?

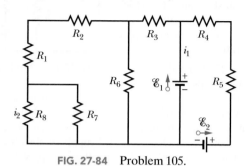

**FIG. 27-84** Problem 105.

**106** In Fig. 27-46, $\mathscr{E} = 6.00$ V, $R_1 = 100 \, \Omega$, $R_2 = 300 \, \Omega$, and $R_3 = 600 \, \Omega$. What are (a) the equivalent resistance of the three resistors, (b) the electric potential across resistance 1, and (c) the current through resistance 3?

**107** In Fig. 27-46, the ideal battery has emf $\mathscr{E} = 12.0$ V and the resistances are each $4.00 \, \Omega$. What are (a) the current in resistance 3 and (b) the rate at which energy is supplied?

**108** A capacitor with capacitance $C_0$, after having been connected to a battery with emf $\mathscr{E}_0$ for a long time, is discharged through a $200\,000 \, \Omega$ resistor at time $t = 0$. The potential difference across the capacitor is then measured as a function of time for a brief time interval; the results are recorded below. (a) Write an equation that describes the potential difference across the capacitor as a function of time. Enter the data into your calculator and have the calculator perform a linear regression fit of $\ln V_C$ versus $t$. From the parameters of the fit, determine (b) the emf $\mathscr{E}_0$ of the battery and (c) the time constant $\tau$ for the circuit. (d) Finally, determine the value of capacitance $C_0$.

| $V_C$ (V): | 9.9 | 7.2 | 5.7 | 4.4 | 3.4 | 2.7 | 2.0 |
|---|---|---|---|---|---|---|---|
| $t$ (s): | 0.20 | 0.40 | 0.60 | 0.80 | 1.0 | 1.2 | 1.4 |

**109** Power is supplied by a device of emf $\mathscr{E}$ to a transmission line with resistance $R$. Find the ratio of the power dissipated in the line for $\mathscr{E} = 110\,000$ V to that dissipated for $\mathscr{E} = 110$ V, assuming the power supplied is the same for the two cases.

**110** A battery of emf $\mathscr{E} = 2.00$ V and internal resistance $r = 0.500 \, \Omega$ is driving a motor. The motor is lifting a 2.00 N mass at constant speed $v = 0.500$ m/s. Two combinations of current $i$ in the battery–motor circuit and potential difference $V$ across the motor terminals allow this. Of the two possible pairs of answers, what are the (a) larger $i$ and (b) corresponding $V$ and the (c) smaller $i$ and (d) corresponding $V$?

**111** (a) If points $a$ and $b$ in Fig. 27-61 are connected by a wire of resistance $r$, show that the current in the wire is

$$i = \frac{\mathscr{E}(R_s - R_x)}{(R + 2r)(R_s + R_x) + 2R_s R_x},$$

where $\mathscr{E}$ is the emf of the ideal battery and $R = R_1 = R_2$. Assume that $R_0$ equals zero. (b) Is this formula consistent with the result of Problem 55?

# Magnetic Fields

Tom Walker/Getty Images, Inc.

*If you are outside on a dark night in the middle to high latitudes, you might be able to see an aurora, a ghostly "curtain" of light that hangs down from the sky. The curtain is not only local; it may be 200 km high and 4000 km long, stretching around Earth in an arc. However, it is only about 100 m thick.*

## What produces this huge display, and what makes it so thin?

The answers are in this chapter.

**FIG. 28-1** Using an electromagnet to collect and transport scrap metal at a steel mill. *(Digital Vision/Getty Images)*

## 28-1 | WHAT IS PHYSICS?

As we have discussed, one major goal of physics is the study of how an *electric field* can produce an *electric force* on a charged object. A closely related goal is the study of how a *magnetic field* can produce a *magnetic force* on a (moving) charged particle or on a magnetic object such as a magnet. You may already have a hint of what a magnetic field is if you have ever attached a note to a refrigerator door with a small magnet or accidentally erased a credit card by moving it near a magnet. The magnet acts on the door or credit card via its magnetic field.

The applications of magnetic fields and magnetic forces are countless and changing rapidly every year. Here are just a few examples. For decades, the entertainment industry depended on the magnetic recording of music and images on audiotape and videotape. Although digital technology has largely replaced magnetic recording, the industry still depends on the magnets that control CD and DVD players and computer hard drives; magnets also drive the speaker cones in headphones, TVs, computers, and telephones. A modern car comes equipped with dozens of magnets because they are required in the motors for engine ignition, automatic window control, sunroof control, and windshield wiper control. Most security alarm systems, doorbells, and automatic door latches employ magnets. In short, you are surrounded by magnets.

The science of magnetic fields is physics; the application of magnetic fields is engineering. Both the science and the application begin with the question "What produces a magnetic field?"

## 28-2 | What Produces a Magnetic Field?

Because an electric field $\vec{E}$ is produced by an electric charge, we might reasonably expect that a magnetic field $\vec{B}$ is produced by a magnetic charge. Although individual magnetic charges (called *magnetic monopoles*) are predicted by certain theories, their existence has not been confirmed. How then are magnetic fields produced? There are two ways.

One way is to use moving electrically charged particles, such as a current in a wire, to make an **electromagnet.** The current produces a magnetic field that can be used, for example, to control a computer hard drive or to sort scrap metal (Fig. 28-1). In Chapter 29, we discuss the magnetic field due to a current.

The other way to produce a magnetic field is by means of elementary particles such as electrons because these particles have an *intrinsic* magnetic field around them. That is, the magnetic field is a basic characteristic of each particle just as mass and electric charge (or lack of charge) are basic characteristics. As we discuss in Chapter 32, the magnetic fields of the electrons in certain materials add together to give a net magnetic field around the material. Such addition is the reason why a **permanent magnet**, the type used to hang refrigerator notes, has a permanent magnetic field. In other materials, the magnetic fields of the electrons cancel out, giving no net magnetic field surrounding the material. Such cancellation is the reason you do not have a permanent field around your body, which is good because otherwise you might be slammed up against a refrigerator door every time you passed one.

Our first job in this chapter is to define the magnetic field $\vec{B}$. We do so by using the experimental fact that when a charged particle moves through a magnetic field, a magnetic force $\vec{F}_B$ acts on the particle.

## 28-3 | The Definition of $\vec{B}$

We determined the electric field $\vec{E}$ at a point by putting a test particle of charge $q$ at rest at that point and measuring the electric force $\vec{F}_E$ acting on the particle.

We then defined $\vec{E}$ as

$$\vec{E} = \frac{\vec{F}_E}{q}. \qquad (28\text{-}1)$$

If a magnetic monopole were available, we could define $\vec{B}$ in a similar way. Because such particles have not been found, we must define $\vec{B}$ in another way, in terms of the magnetic force $\vec{F}_B$ exerted on a moving electrically charged test particle.

In principle, we do this by firing a charged particle through the point at which $\vec{B}$ is to be defined, using various directions and speeds for the particle and determining the force $\vec{F}_B$ that acts on the particle at that point. After many such trials we would find that when the particle's velocity $\vec{v}$ is along a particular axis through the point, force $\vec{F}_B$ is zero. For all other directions of $\vec{v}$, the magnitude of $\vec{F}_B$ is always proportional to $v \sin \phi$, where $\phi$ is the angle between the zero-force axis and the direction of $\vec{v}$. Furthermore, the direction of $\vec{F}_B$ is always perpendicular to the direction of $\vec{v}$. (These results suggest that a cross product is involved.)

We can then define a **magnetic field** $\vec{B}$ to be a vector quantity that is directed along the zero-force axis. We can next measure the magnitude of $\vec{F}_B$ when $\vec{v}$ is directed perpendicular to that axis and then define the magnitude of $\vec{B}$ in terms of that force magnitude:

$$B = \frac{F_B}{|q|v},$$

where $q$ is the charge of the particle.

We can summarize all these results with the following vector equation:

$$\vec{F}_B = q\vec{v} \times \vec{B}; \qquad (28\text{-}2)$$

that is, the force $\vec{F}_B$ on the particle is equal to the charge $q$ times the cross product of its velocity $\vec{v}$ and the field $\vec{B}$ (all measured in the same reference frame). Using Eq. 3-27 for the cross product, we can write the magnitude of $\vec{F}_B$ as

$$F_B = |q|vB \sin \phi, \qquad (28\text{-}3)$$

where $\phi$ is the angle between the directions of velocity $\vec{v}$ and magnetic field $\vec{B}$.

### Finding the Magnetic Force on a Particle

Equation 28-3 tells us that the magnitude of the force $\vec{F}_B$ acting on a particle in a magnetic field is proportional to the charge $q$ and speed $v$ of the particle. Thus, the force is equal to zero if the charge is zero or if the particle is stationary. Equation 28-3 also tells us that the magnitude of the force is zero if $\vec{v}$ and $\vec{B}$ are either parallel ($\phi = 0°$) or antiparallel ($\phi = 180°$), and the force is at its maximum when $\vec{v}$ and $\vec{B}$ are perpendicular to each other.

Equation 28-2 tells us all this plus the direction of $\vec{F}_B$. From Section 3-8, we know that the cross product $\vec{v} \times \vec{B}$ in Eq. 28-2 is a vector that is perpendicular to the two vectors $\vec{v}$ and $\vec{B}$. The right-hand rule (Fig. 28-2a) tells us that the thumb of the right hand points in the direction of $\vec{v} \times \vec{B}$ when the fingers sweep $\vec{v}$ into $\vec{B}$. If $q$ is positive, then (by Eq. 28-2) the force $\vec{F}_B$ has the same sign as $\vec{v} \times \vec{B}$ and thus must be in the same direction; that is, for positive $q$, $\vec{F}_B$ is directed along the thumb (Fig. 28-2b). If $q$ is negative, then the force $\vec{F}_B$ and cross product $\vec{v} \times \vec{B}$ have opposite signs and thus must be in opposite directions. For negative $q$, $\vec{F}_B$ is directed opposite the thumb (Fig. 28-2c).

Regardless of the sign of the charge, however,

> The force $\vec{F}_B$ acting on a charged particle moving with velocity $\vec{v}$ through a magnetic field $\vec{B}$ is *always* perpendicular to $\vec{v}$ and $\vec{B}$.

Thus, $\vec{F}_B$ *never* has a component parallel to $\vec{v}$. This means that $\vec{F}_B$ cannot change

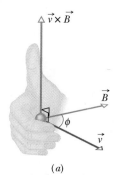

(a)

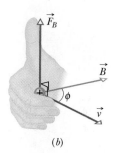

(b)

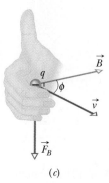

(c)

FIG. 28-2   (a) The right-hand rule (in which $\vec{v}$ is swept into $\vec{B}$ through the smaller angle $\phi$ between them) gives the direction of $\vec{v} \times \vec{B}$ as the direction of the thumb. (b) If $q$ is positive, then the direction of $\vec{F}_B = q\vec{v} \times \vec{B}$ is in the direction of $\vec{v} \times \vec{B}$. (c) If $q$ is negative, then the direction of $\vec{F}_B$ is opposite that of $\vec{v} \times \vec{B}$.

**FIG. 28-3** The tracks of two electrons (e⁻) and a positron (e⁺) in a bubble chamber that is immersed in a uniform magnetic field that is directed out of the plane of the page. *(Lawrence Berkeley Laboratory/Photo Researchers)*

**TABLE 28-1**

**Some Approximate Magnetic Fields**

| | |
|---|---|
| At surface of neutron star | $10^8$ T |
| Near big electromagnet | 1.5 T |
| Near small bar magnet | $10^{-2}$ T |
| At Earth's surface | $10^{-4}$ T |
| In interstellar space | $10^{-10}$ T |
| Smallest value in magnetically shielded room | $10^{-14}$ T |

the particle's speed $v$ (and thus it cannot change the particle's kinetic energy). The force can change only the direction of $\vec{v}$ (and thus the direction of travel); only in this sense can $\vec{F}_B$ accelerate the particle.

To develop a feeling for Eq. 28-2, consider Fig. 28-3, which shows some tracks left by charged particles moving rapidly through a *bubble chamber*. The chamber, which is filled with liquid hydrogen, is immersed in a strong uniform magnetic field that is directed out of the plane of the figure. An incoming gamma ray particle—which leaves no track because it is uncharged—transforms into an electron (spiral track marked e⁻) and a positron (track marked e⁺) while it knocks an electron out of a hydrogen atom (long track marked e⁻). Check with Eq. 28-2 and Fig. 28-2 that the three tracks made by these two negative particles and one positive particle curve in the proper directions.

The SI unit for $\vec{B}$ that follows from Eqs. 28-2 and 28-3 is the newton per coulomb-meter per second. For convenience, this is called the **tesla** (T):

$$1 \text{ tesla} = 1 \text{ T} = 1 \frac{\text{newton}}{(\text{coulomb})(\text{meter/second})}.$$

Recalling that a coulomb per second is an ampere, we have

$$1 \text{ T} = 1 \frac{\text{newton}}{(\text{coulomb/second})(\text{meter})} = 1 \frac{\text{N}}{\text{A} \cdot \text{m}}. \quad (28\text{-}4)$$

An earlier (non-SI) unit for $\vec{B}$, still in common use, is the *gauss* (G), and

$$1 \text{ tesla} = 10^4 \text{ gauss}. \quad (28\text{-}5)$$

Table 28-1 lists the magnetic fields that occur in a few situations. Note that Earth's magnetic field near the planet's surface is about $10^{-4}$ T ($= 100 \mu$T or 1 G).

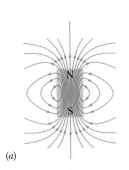

(a)

(b)

**FIG. 28-4** (*a*) The magnetic field lines for a bar magnet. (*b*) A "cow magnet"—a bar magnet that is intended to be slipped down into the rumen of a cow to prevent accidentally ingested bits of scrap iron from reaching the cow's intestines. The iron filings at its ends reveal the magnetic field lines. *(Courtesy Dr. Richard Cannon, Southeast Missouri State University, Cape Girardeau)*

✓ **CHECKPOINT 1**
The figure shows three situations in which a charged particle with velocity $\vec{v}$ travels through a uniform magnetic field $\vec{B}$. In each situation, what is the direction of the magnetic force $\vec{F}_B$ on the particle?

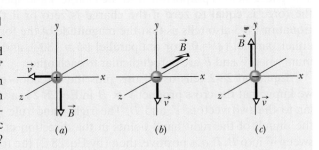

## Magnetic Field Lines

We can represent magnetic fields with field lines, as we did for electric fields. Similar rules apply: (1) the direction of the tangent to a magnetic field line at any point gives the direction of $\vec{B}$ at that point, and (2) the spacing of the lines represents the magnitude of $\vec{B}$—the magnetic field is stronger where the lines are closer together, and conversely.

Figure 28-4a shows how the magnetic field near a *bar magnet* (a permanent magnet in the shape of a bar) can be represented by magnetic field lines. The lines all pass through the magnet, and they all form closed loops (even those that are not shown closed in the figure). The external magnetic effects of a bar magnet are strongest near its ends, where the field lines are most closely spaced. Thus, the magnetic field of the bar magnet in Fig. 28-4b collects the iron filings mainly near the two ends of the magnet.

The (closed) field lines enter one end of a magnet and exit the other end. The end of a magnet from which the field lines emerge is called the *north pole* of the magnet; the other end, where field lines enter the magnet, is called the *south pole*. Because a magnet has two poles, it is said to be a **magnetic dipole.** The magnets we use to fix notes on refrigerators are short bar magnets. Figure 28-5 shows two other common shapes for magnets: a *horseshoe magnet* and a magnet that has been bent around into the shape of a **C** so that the *pole faces* are facing each other. (The magnetic field between the pole faces can then be approximately uniform.) Regardless of the shape of the magnets, if we place two of them near each other we find:

☞ Opposite magnetic poles attract each other, and like magnetic poles repel each other.

Earth has a magnetic field that is produced in its core by still unknown mechanisms. On Earth's surface, we can detect this magnetic field with a compass, which is essentially a slender bar magnet on a low-friction pivot. This bar magnet, or this needle, turns because its north-pole end is attracted toward the Arctic region of Earth. Thus, the *south* pole of Earth's magnetic field must be located toward the Arctic. Logically, we then should call the pole there a south pole. However, because we call that direction north, we are trapped into the statement that Earth has a *geomagnetic north pole* in that direction.

With more careful measurement we would find that in the Northern Hemisphere, the magnetic field lines of Earth generally point down into Earth and toward the Arctic. In the Southern Hemisphere, they generally point up out of Earth and away from the Antarctic—that is, away from Earth's *geomagnetic south pole*.

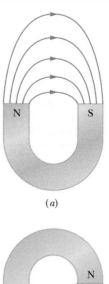

(a)

(b)

**FIG. 28-5**  (a) A horseshoe magnet and (b) a **C**-shaped magnet. (Only some of the external field lines are shown.)

---

**Sample Problem**   **28-1**

A uniform magnetic field $\vec{B}$, with magnitude 1.2 mT, is directed vertically upward throughout the volume of a laboratory chamber. A proton with kinetic energy 5.3 MeV enters the chamber, moving horizontally from south to north. What magnetic deflecting force acts on the proton as it enters the chamber? The proton mass is $1.67 \times 10^{-27}$ kg. (Neglect Earth's magnetic field.)

**KEY IDEAS**  Because the proton is charged and moving through a magnetic field, a magnetic force $\vec{F}_B$ can act on it. Because the initial direction of the proton's velocity is not along a magnetic field line, $\vec{F}_B$ is not simply zero.

**Magnitude:** To find the magnitude of $\vec{F}_B$, we can use Eq. 28-3 ($F_B = |q|vB \sin \phi$) provided we first find the proton's speed $v$. We can find $v$ from the given kinetic energy because $K = \frac{1}{2}mv^2$. Solving for $v$, we obtain

$$v = \sqrt{\frac{2K}{m}} = \sqrt{\frac{(2)(5.3 \text{ MeV})(1.60 \times 10^{-13} \text{ J/MeV})}{1.67 \times 10^{-27} \text{ kg}}}$$

$$= 3.2 \times 10^7 \text{ m/s}.$$

Equation 28-3 then yields

$$F_B = |q|vB \sin \phi$$
$$= (1.60 \times 10^{-19} \text{ C})(3.2 \times 10^7 \text{ m/s})$$
$$\times (1.2 \times 10^{-3} \text{ T})(\sin 90°)$$
$$= 6.1 \times 10^{-15} \text{ N}. \qquad \text{(Answer)}$$

This may seem like a small force, but it acts on a particle of small mass, producing a large acceleration; namely,

$$a = \frac{F_B}{m} = \frac{6.1 \times 10^{-15} \text{ N}}{1.67 \times 10^{-27} \text{ kg}} = 3.7 \times 10^{12} \text{ m/s}^2.$$

**Direction:** To find the direction of $\vec{F}_B$, we use the fact that $\vec{F}_B$ has the direction of the cross product $q\vec{v} \times \vec{B}$. Because the charge $q$ is positive, $\vec{F}_B$ must have the same direction as $\vec{v} \times \vec{B}$, which can be determined with the right-hand rule for cross products (as in Fig. 28-2b). We know that $\vec{v}$ is directed horizontally from south to north and $\vec{B}$ is directed vertically up. The right-hand rule shows us that the deflecting force $\vec{F}_B$ must be directed horizontally from west to east, as Fig. 28-6 shows. (The array of dots in the figure represents a magnetic

**FIG. 28-6** An overhead view of a proton moving from south to north with velocity $\vec{v}$ in a chamber. A magnetic field is directed vertically upward in the chamber, as represented by the array of dots (which resemble the tips of arrows). The proton is deflected toward the east.

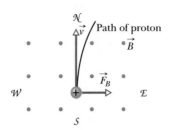

field directed out of the plane of the figure. An array of **X**s would have represented a magnetic field directed into that plane.)

If the charge of the particle were negative, the magnetic deflecting force would be directed in the opposite direction—that is, horizontally from east to west. This is predicted automatically by Eq. 28-2 if we substitute a negative value for $q$.

## 28-4 I Crossed Fields: Discovery of the Electron

Both an electric field $\vec{E}$ and a magnetic field $\vec{B}$ can produce a force on a charged particle. When the two fields are perpendicular to each other, they are said to be *crossed fields*. Here we shall examine what happens to charged particles—namely, electrons—as they move through crossed fields. We use as our example the experiment that led to the discovery of the electron in 1897 by J. J. Thomson at Cambridge University.

Figure 28-7 shows a modern, simplified version of Thomson's experimental apparatus—a *cathode ray tube* (which is like the picture tube in a standard television set). Charged particles (which we now know as electrons) are emitted by a hot filament at the rear of the evacuated tube and are accelerated by an applied potential difference $V$. After they pass through a slit in screen C, they form a narrow beam. They then pass through a region of crossed $\vec{E}$ and $\vec{B}$ fields, headed toward a fluorescent screen S, where they produce a spot of light (on a television screen the spot is part of the picture). The forces on the charged particles in the crossed-fields region can deflect them from the center of the screen. By controlling the magnitudes and directions of the fields, Thomson could thus control where the spot of light appeared on the screen. Recall that the force on a negatively charged particle due to an electric field is directed opposite the field. Thus, for the arrangement of Fig. 28-7, electrons are forced up the page by electric field $\vec{E}$ and down the page by magnetic field $\vec{B}$; that is, the forces are *in opposition*. Thomson's procedure was equivalent to the following series of steps.

1. Set $E = 0$ and $B = 0$ and note the position of the spot on screen S due to the undeflected beam.

2. Turn on $\vec{E}$ and measure the resulting beam deflection.

3. Maintaining $\vec{E}$, now turn on $\vec{B}$ and adjust its value until the beam returns to the undeflected position. (With the forces in opposition, they can be made to cancel.)

We discussed the deflection of a charged particle moving through an electric field $\vec{E}$ between two plates (step 2 here) in Sample Problem 22-4. We found that

**FIG. 28-7** A modern version of J. J. Thomson's apparatus for measuring the ratio of mass to charge for the electron. An electric field $\vec{E}$ is established by connecting a battery across the deflecting-plate terminals. The magnetic field $\vec{B}$ is set up by means of a current in a system of coils (not shown). The magnetic field shown is into the plane of the figure, as represented by the array of **X**s (which resemble the feathered ends of arrows).

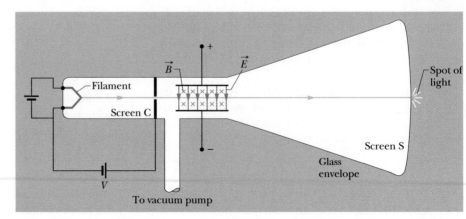

the deflection of the particle at the far end of the plates is

$$y = \frac{|q|EL^2}{2mv^2},$$   (28-6)

where $v$ is the particle's speed, $m$ its mass, and $q$ its charge, and $L$ is the length of the plates. We can apply this same equation to the beam of electrons in Fig. 28-7; if need be, we can calculate the deflection by measuring the deflection of the beam on screen S and then working back to calculate the deflection $y$ at the end of the plates. (Because the direction of the deflection is set by the sign of the particle's charge, Thomson was able to show that the particles that were lighting up his screen were negatively charged.)

When the two fields in Fig. 28-7 are adjusted so that the two deflecting forces cancel (step 3), we have from Eqs. 28-1 and 28-3

$$|q|E = |q|vB \sin(90°) = |q|vB$$

or   $$v = \frac{E}{B}.$$   (28-7)

Thus, the crossed fields allow us to measure the speed of the charged particles passing through them. Substituting Eq. 28-7 for $v$ in Eq. 28-6 and rearranging yield

$$\frac{m}{|q|} = \frac{B^2L^2}{2yE},$$   (28-8)

in which all quantities on the right can be measured. Thus, the crossed fields allow us to measure the ratio $m/|q|$ of the particles moving through Thomson's apparatus.

Thomson claimed that these particles are found in all matter. He also claimed that they are lighter than the lightest known atom (hydrogen) by a factor of more than 1000. (The exact ratio proved later to be 1836.15.) His $m/|q|$ measurement, coupled with the boldness of his two claims, is considered to be the "discovery of the electron."

✓**CHECKPOINT 2**     The figure shows four directions for the velocity vector $\vec{v}$ of a positively charged particle moving through a uniform electric field $\vec{E}$ (directed out of the page and represented with an encircled dot) and a uniform magnetic field $\vec{B}$. (a) Rank directions 1, 2, and 3 according to the magnitude of the net force on the particle, greatest first. (b) Of all four directions, which might result in a net force of zero?

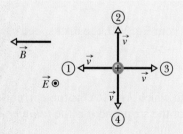

## 28-5 | Crossed Fields: The Hall Effect

As we just discussed, a beam of electrons in a vacuum can be deflected by a magnetic field. Can the drifting conduction electrons in a copper wire also be deflected by a magnetic field? In 1879, Edwin H. Hall, then a 24-year-old graduate student at the Johns Hopkins University, showed that they can. This **Hall effect** allows us to find out whether the charge carriers in a conductor are positively or negatively charged. Beyond that, we can measure the number of such carriers per unit volume of the conductor.

Figure 28-8a shows a copper strip of width $d$, carrying a current $i$ whose conventional direction is from the top of the figure to the bottom. The charge carriers are electrons and, as we know, they drift (with drift speed $v_d$) in the

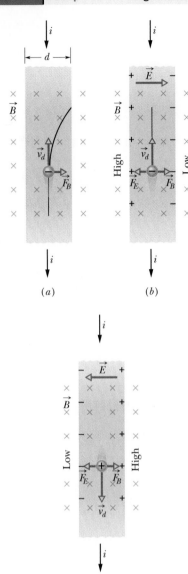

(a)    (b)

(c)

**FIG. 28-8** A strip of copper carrying a current $i$ is immersed in a magnetic field $\vec{B}$. (a) The situation immediately after the magnetic field is turned on. The curved path that will then be taken by an electron is shown. (b) The situation at equilibrium, which quickly follows. Note that negative charges pile up on the right side of the strip, leaving uncompensated positive charges on the left. Thus, the left side is at a higher potential than the right side. (c) For the same current direction, if the charge carriers were positively charged, *they* would pile up on the right side, and the right side would be at the higher potential.

opposite direction, from bottom to top. At the instant shown in Fig. 28-8a, an external magnetic field $\vec{B}$, pointing into the plane of the figure, has just been turned on. From Eq. 28-2 we see that a magnetic deflecting force $\vec{F}_B$ will act on each drifting electron, pushing it toward the right edge of the strip.

As time goes on, electrons move to the right, mostly piling up on the right edge of the strip, leaving uncompensated positive charges in fixed positions at the left edge. The separation of positive and negative charges produces an electric field $\vec{E}$ within the strip, pointing from left to right in Fig. 28-8b. This field exerts an electric force $\vec{F}_E$ on each electron, tending to push it to the left.

An equilibrium quickly develops in which the electric force on each electron builds up until it just cancels the magnetic force. When this happens, as Fig. 28-8b shows, the force due to $\vec{B}$ and the force due to $\vec{E}$ are in balance. The drifting electrons then move along the strip toward the top of the page at velocity $\vec{v}_d$ with no further collection of electrons on the right edge of the strip and thus no further increase in the electric field $\vec{E}$.

A *Hall potential difference V* is associated with the electric field across strip width $d$. From Eq. 24-42, the magnitude of that potential difference is

$$V = Ed. \tag{28-9}$$

By connecting a voltmeter across the width, we can measure the potential difference between the two edges of the strip. Moreover, the voltmeter can tell us which edge is at higher potential. For the situation of Fig. 28-8b, we would find that the left edge is at higher potential, which is consistent with our assumption that the charge carriers are negatively charged.

For a moment, let us make the opposite assumption, that the charge carriers in current $i$ are positively charged (Fig. 28-8c). Convince yourself that as these charge carriers move from top to bottom in the strip, they are pushed to the right edge by $\vec{F}_B$ and thus that the *right* edge is at higher potential. Because that last statement is contradicted by our voltmeter reading, the charge carriers must be negatively charged.

Now for the quantitative part. When the electric and magnetic forces are in balance (Fig. 28-8b), Eqs. 28-1 and 28-3 give us

$$eE = ev_dB. \tag{28-10}$$

From Eq. 26-7, the drift speed $v_d$ is

$$v_d = \frac{J}{ne} = \frac{i}{neA}, \tag{28-11}$$

in which $J \, (= i/A)$ is the current density in the strip, $A$ is the cross-sectional area of the strip, and $n$ is the *number density* of charge carriers (their number per unit volume).

In Eq. 28-10, substituting for $E$ with Eq. 28-9 and substituting for $v_d$ with Eq. 28-11, we obtain

$$n = \frac{Bi}{Vle}, \tag{28-12}$$

in which $l \, (= A/d)$ is the thickness of the strip. With this equation we can find $n$ from measurable quantities.

It is also possible to use the Hall effect to measure directly the drift speed $v_d$ of the charge carriers, which you may recall is of the order of centimeters per hour. In this clever experiment, the metal strip is moved mechanically through the magnetic field in a direction opposite that of the drift velocity of the charge carriers. The speed of the moving strip is then adjusted until the Hall potential difference vanishes. At this condition, with no Hall effect, the velocity of the charge carriers *with respect to the laboratory frame* must be zero, so the velocity of the strip must be equal in magnitude but opposite the direction of the velocity of the negative charge carriers.

Figure 28-9 shows a solid metal cube, of edge length $d =$ 1.5 cm, moving in the positive $y$ direction at a constant velocity $\vec{v}$ of magnitude 4.0 m/s. The cube moves through a uniform magnetic field $\vec{B}$ of magnitude 0.050 T in the positive $z$ direction.

(a) Which cube face is at a lower electric potential and which is at a higher electric potential because of the motion through the field?

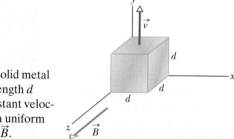

**FIG. 28-9**  A solid metal cube of edge length $d$ moving at constant velocity $\vec{v}$ through a uniform magnetic field $\vec{B}$.

**KEY IDEAS**  Because the cube is moving through a magnetic field $\vec{B}$, a magnetic force $\vec{F}_B$ acts on its charged particles, including its conduction electrons.

*Reasoning:* When the cube first begins to move through the magnetic field, its electrons do also. Because each electron has charge $q$ and is moving through a magnetic field with velocity $\vec{v}$, the magnetic force $\vec{F}_B$ acting on the electron is given by Eq. 28-2. Because $q$ is negative, the direction of $\vec{F}_B$ is opposite the cross product $\vec{v} \times \vec{B}$, which is in the positive direction of the $x$ axis in Fig. 28-9. Thus, $\vec{F}_B$ acts in the negative direction of the $x$ axis, toward the left face of the cube (which is hidden from view in Fig. 28-9).

Most of the electrons are fixed in place in the atoms of the cube. However, because the cube is a metal, it contains conduction electrons that are free to move. Some of those conduction electrons are deflected by $\vec{F}_B$ to the left cube face, making that face negatively charged and leaving the right face positively charged. This charge separation produces an electric field $\vec{E}$ directed from the positively charged right face to the negatively charged left face. Thus, the left face is at a lower electric potential, and the right face is at a higher electric potential.

(b) What is the potential difference between the faces of higher and lower electric potential?

**KEY IDEAS**

1. The electric field $\vec{E}$ created by the charge separation produces an electric force $\vec{F}_E = q\vec{E}$ on each electron. Because $q$ is negative, this force is directed opposite the field $\vec{E}$—that is, rightward. Thus on each electron, $\vec{F}_E$ acts toward the right and $\vec{F}_B$ acts toward the left.

2. When the cube had just begun to move through the magnetic field and the charge separation had just begun, the magnitude of $\vec{E}$ began to increase from zero. Thus, the magnitude of $\vec{F}_E$ also began to increase from zero and was initially smaller than the magnitude $\vec{F}_B$. During this early stage, the net force on any electron was dominated by $\vec{F}_B$, which continuously moved additional electrons to the left cube face, increasing the charge separation.

3. However, as the charge separation increased, eventually magnitude $F_E$ became equal to magnitude $F_B$. The net force on any electron was then zero, and no additional electrons were moved to the left cube face. Thus, the magnitude of $\vec{F}_E$ could not increase further, and the electrons were then in equilibrium.

*Calculations:* We seek the potential difference $V$ between the left and right cube faces after equilibrium was reached (which occurred quickly). We can obtain $V$ with Eq. 28-9 ($V = Ed$) provided we first find the magnitude $E$ of the electric field at equilibrium. We can do so with the equation for the balance of forces ($F_E = F_B$).

For $F_E$, we substitute $|q|E$, and then for $F_B$, we substitute $|q|vB \sin \phi$ from Eq. 28-3. From Fig. 28-9, we see that the angle $\phi$ between vectors $\vec{v}$ and $\vec{B}$ is 90°; thus $\sin \phi = 1$ and $F_E = F_B$ yields

$$|q|E = |q|vB \sin 90° = |q|vB.$$

This gives us $E = vB$; so $V = Ed$ becomes

$$V = vBd. \qquad (28\text{-}13)$$

Substituting known values gives us

$$V = (4.0 \text{ m/s})(0.050 \text{ T})(0.015 \text{ m})$$
$$= 0.0030 \text{ V} = 3.0 \text{ mV}. \qquad \text{(Answer)}$$

# 28-6 | A Circulating Charged Particle

If a particle moves in a circle at constant speed, we can be sure that the net force acting on the particle is constant in magnitude and points toward the center of the circle, always perpendicular to the particle's velocity. Think of a stone tied to a string and whirled in a circle on a smooth horizontal surface, or of a satellite moving in a circular orbit around Earth. In the first case, the

**FIG. 28-10** Electrons circulating in a chamber containing gas at low pressure (their path is the glowing circle). A uniform magnetic field $\vec{B}$, pointing directly out of the plane of the page, fills the chamber. Note the radially directed magnetic force $\vec{F}_B$; for circular motion to occur, $\vec{F}_B$ *must* point toward the center of the circle. Use the right-hand rule for cross products to confirm that $\vec{F}_B = q\vec{v} \times \vec{B}$ gives $\vec{F}_B$ the proper direction. (Don't forget the sign of $q$.) *(Courtesy John Le P. Webb, Sussex University, England)*

tension in the string provides the necessary force and centripetal acceleration. In the second case, Earth's gravitational attraction provides the force and acceleration.

Figure 28-10 shows another example: A beam of electrons is projected into a chamber by an *electron gun* G. The electrons enter in the plane of the page with speed $v$ and then move in a region of uniform magnetic field $\vec{B}$ directed out of that plane. As a result, a magnetic force $\vec{F}_B = q\vec{v} \times \vec{B}$ continuously deflects the electrons, and because $\vec{v}$ and $\vec{B}$ are always perpendicular to each other, this deflection causes the electrons to follow a circular path. The path is visible in the photo because atoms of gas in the chamber emit light when some of the circulating electrons collide with them.

We would like to determine the parameters that characterize the circular motion of these electrons, or of any particle of charge magnitude $|q|$ and mass $m$ moving perpendicular to a uniform magnetic field $\vec{B}$ at speed $v$. From Eq. 28-3, the force acting on the particle has a magnitude of $|q|vB$. From Newton's second law ($\vec{F} = m\vec{a}$) applied to uniform circular motion (Eq. 6-18),

$$F = m\frac{v^2}{r}, \qquad (28\text{-}14)$$

we have

$$|q|vB = \frac{mv^2}{r}. \qquad (28\text{-}15)$$

Solving for $r$, we find the radius of the circular path as

$$r = \frac{mv}{|q|B} \qquad \text{(radius)}. \qquad (28\text{-}16)$$

The period $T$ (the time for one full revolution) is equal to the circumference divided by the speed:

$$T = \frac{2\pi r}{v} = \frac{2\pi}{v}\frac{mv}{|q|B} = \frac{2\pi m}{|q|B} \qquad \text{(period)}. \qquad (28\text{-}17)$$

The frequency $f$ (the number of revolutions per unit time) is

$$f = \frac{1}{T} = \frac{|q|B}{2\pi m} \qquad \text{(frequency)}. \qquad (28\text{-}18)$$

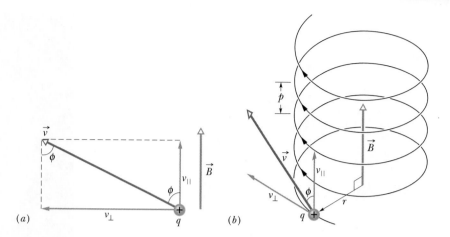

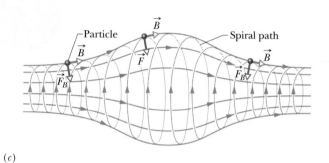

FIG. 28-11 (a) A charged particle moves in a uniform magnetic field $\vec{B}$, the particle's velocity $\vec{v}$ making an angle $\phi$ with the field direction. (b) The particle follows a helical path of radius $r$ and pitch $p$. (c) A charged particle spiraling in a nonuniform magnetic field. (The particle can become trapped, spiraling back and forth between the strong field regions at either end.) Note that the magnetic force vectors at the left and right sides have a component pointing toward the center of the figure.

The angular frequency $\omega$ of the motion is then

$$\omega = 2\pi f = \frac{|q|B}{m} \qquad \text{(angular frequency).} \qquad (28\text{-}19)$$

The quantities $T$, $f$, and $\omega$ do not depend on the speed of the particle (provided the speed is much less than the speed of light). Fast particles move in large circles and slow ones in small circles, but all particles with the same charge-to-mass ratio $|q|/m$ take the same time $T$ (the period) to complete one round trip. Using Eq. 28-2, you can show that if you are looking in the direction of $\vec{B}$, the direction of rotation for a positive particle is always counterclockwise, and the direction for a negative particle is always clockwise.

## Helical Paths

If the velocity of a charged particle has a component parallel to the (uniform) magnetic field, the particle will move in a helical path about the direction of the field vector. Figure 28-11a, for example, shows the velocity vector $\vec{v}$ of such a particle resolved into two components, one parallel to $\vec{B}$ and one perpendicular to it:

$$v_{\parallel} = v \cos \phi \quad \text{and} \quad v_{\perp} = v \sin \phi. \qquad (28\text{-}20)$$

The parallel component determines the *pitch p* of the helix—that is, the distance between adjacent turns (Fig. 28-11b). The perpendicular component determines the radius of the helix and is the quantity to be substituted for $v$ in Eq. 28-16.

Figure 28-11c shows a charged particle spiraling in a nonuniform magnetic field. The more closely spaced field lines at the left and right sides indicate that the magnetic field is stronger there. When the field at an end is strong enough, the particle "reflects" from that end. If the particle reflects from both ends, it is said to be trapped in a *magnetic bottle*.

Electrons and protons are trapped in this way by the terrestrial magnetic field; the trapped particles form the *Van Allen radiation belts*, which loop well

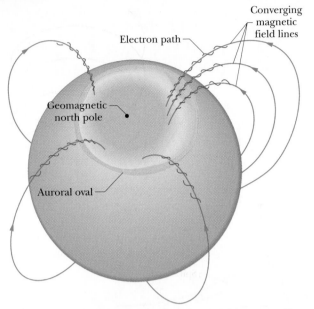

FIG. 28-12    The auroral oval surrounding Earth's geomagnetic north pole (which is currently located in northwestern Greenland). Magnetic field lines converge toward that pole. Electrons moving toward Earth are "caught by" and spiral around these field lines, entering the terrestrial atmosphere at high latitudes and producing auroras within the oval.

FIG. 28-13    A false-color image of auroras inside the north auroral oval, recorded by the satellite *Dynamic Explorer,* using ultraviolet light emitted by oxygen atoms excited in the aurora. The sunlit portion of Earth is the crescent at the left. *(Courtesy Dr. L.A. Frank, University of Iowa)*

above Earth's atmosphere between Earth's north and south geomagnetic poles. These particles bounce back and forth, from one end of this magnetic bottle to the other, within a few seconds.

When a large solar flare shoots additional energetic electrons and protons into the radiation belts, an electric field is produced in the region where electrons normally reflect. This field eliminates the reflection and instead drives electrons down into the atmosphere, where they collide with atoms and molecules of air, causing that air to emit light. This light forms the aurora—a curtain of light that hangs down to an altitude of about 100 km. Green light is emitted by oxygen atoms, and pink light is emitted by nitrogen molecules, but often the light is so dim that we perceive only white light.

Auroras extend in arcs above Earth and can occur in a region called the *auroral oval* that is shown in Figs. 28-12 and 28-13 as seen from space. Although an aurora is long, it is only about 100 m thick (north to south) because the paths of the electrons producing it converge as the electrons spiral down the converging magnetic field lines (Fig. 28-12).

✓ **CHECKPOINT 3**    The figure here shows the circular paths of two particles that travel at the same speed in a uniform magnetic field $\vec{B}$, which is directed into the page. One particle is a proton; the other is an electron (which is less massive). (a) Which particle follows the smaller circle, and (b) does that particle travel clockwise or counterclockwise?

**Sample Problem    28-3**

Figure 28-14 shows the essentials of a *mass spectrometer,* which can be used to measure the mass of an ion; an ion

of mass $m$ (to be measured) and charge $q$ is produced in source $S$. The initially stationary ion is accelerated by the

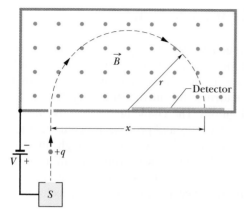

**FIG. 28-14** Essentials of a mass spectrometer. A positive ion, after being accelerated from its source $S$ by a potential difference $V$, enters a chamber of uniform magnetic field $\vec{B}$. There it travels through a semicircle of radius $r$ and strikes a detector at a distance $x$ from where it entered the chamber.

electric field due to a potential difference $V$. The ion leaves $S$ and enters a separator chamber in which a uniform magnetic field $\vec{B}$ is perpendicular to the path of the ion. A wide detector lines the bottom wall of the chamber, and the $\vec{B}$ causes the ion to move in a semicircle and thus strike the detector. Suppose that $B = 80.000$ mT, $V = 1000.0$ V, and ions of charge $q = +1.6022 \times 10^{-19}$ C strike the detector at a point that lies at $x = 1.6254$ m. What is the mass $m$ of the individual ions, in atomic mass units (Eq. 1-7: 1 u $= 1.6605 \times 10^{-27}$ kg)?

**KEY IDEAS** (1) Because the (uniform) magnetic field causes the (charged) ion to follow a circular path, we can relate the ion's mass $m$ to the path's radius $r$ with Eq. 28-16 ($r = mv/|q|B$). From Fig. 28-14 we see that $r = x/2$, and we are given the magnitude $B$ of the magnetic field. However, we lack the ion's speed $v$ in the magnetic

field after the ion has been accelerated due to the potential difference $V$. (2) To relate $v$ and $V$, we use the fact that mechanical energy ($E_{\text{mec}} = K + U$) is conserved during the acceleration.

**Finding speed:** When the ion emerges from the source, its kinetic energy is approximately zero. At the end of the acceleration, its kinetic energy is $\frac{1}{2}mv^2$. Also, during the acceleration, the positive ion moves through a change in potential of $-V$. Thus, because the ion has positive charge $q$, its potential energy changes by $-qV$. If we now write the conservation of mechanical energy as

$$\Delta K + \Delta U = 0,$$

we get

$$\tfrac{1}{2}mv^2 - qV = 0$$

or

$$v = \sqrt{\frac{2qV}{m}}. \qquad (28\text{-}21)$$

**Finding mass:** Substituting this value for $v$ into Eq. 28-16 gives us

$$r = \frac{mv}{qB} = \frac{m}{qB}\sqrt{\frac{2qV}{m}} = \frac{1}{B}\sqrt{\frac{2mV}{q}}.$$

Thus,

$$x = 2r = \frac{2}{B}\sqrt{\frac{2mV}{q}}.$$

Solving this for $m$ and substituting the given data yield

$$m = \frac{B^2qx^2}{8V}$$

$$= \frac{(0.080000 \text{ T})^2(1.6022 \times 10^{-19} \text{ C})(1.6254 \text{ m})^2}{8(1000.0 \text{ V})}$$

$$= 3.3863 \times 10^{-25} \text{ kg} = 203.93 \text{ u}. \qquad \text{(Answer)}$$

---

**Sample Problem** 28-4

An electron with a kinetic energy of 22.5 eV moves into a region of uniform magnetic field $\vec{B}$ of magnitude $4.55 \times 10^{-4}$ T. The angle between the directions of $\vec{B}$ and the electron's velocity $\vec{v}$ is 65.5°. What is the pitch of the helical path taken by the electron?

**KEY IDEAS** (1) The pitch $p$ is the distance the electron travels parallel to the magnetic field $\vec{B}$ during one period $T$ of circulation. (2) The period $T$ is given by Eq. 28-17 regardless of the angle between the directions of $\vec{v}$ and $\vec{B}$ (provided the angle is not zero, for which there is no circulation of the electron).

**Calculations:** Using Eqs. 28-20 and 28-17, we find

$$p = v_{\parallel}T = (v \cos \phi)\frac{2\pi m}{|q|B}. \qquad (28\text{-}22)$$

We can calculate the electron's speed $v$ from its kinetic energy as we did for the proton in Sample Problem 28-1. We find that $v = 2.81 \times 10^6$ m/s. Substituting this and known data in Eq. 28-22 gives us

$$p = (2.81 \times 10^6 \text{ m/s})(\cos 65.5°)$$

$$\times \frac{2\pi(9.11 \times 10^{-31} \text{ kg})}{(1.60 \times 10^{-19} \text{ C})(4.55 \times 10^{-4} \text{ T})}$$

$$= 9.16 \text{ cm}. \qquad \text{(Answer)}$$

## 28-7 | Cyclotrons and Synchrotrons

Beams of high-energy particles, such as high-energy electrons and protons, have been enormously useful in probing atoms and nuclei to reveal the fundamental structure of matter. Such beams were instrumental in the discovery that atomic nuclei consist of protons and neutrons and in the discovery that protons and neutrons consist of quarks and gluons. The challenge of such beams is how to make and control them. Because electrons and protons are charged, they can be accelerated to the required high energy if they move through large potential differences. Because electrons have low mass, accelerating them in this way can be done in a reasonable distance. However, because protons (and other charged particles) have greater mass, the distance required for the acceleration is too long.

A clever solution to this problem is first to let protons and other massive particles move through a modest potential difference (so that they gain a modest amount of energy) and then use a magnetic field to cause them to circle back and move through a modest potential difference again. If this procedure is repeated thousands of times, the particles end up with a very large energy.

Here we discuss two *accelerators* that employ a magnetic field to repeatedly bring particles back to an accelerating region, where they gain more and more energy until they finally emerge as a high-energy beam.

### The Cyclotron

Figure 28-15 is a top view of the region of a *cyclotron* in which the particles (protons, say) circulate. The two hollow **D**-shaped objects (each open on its straight edge) are made of sheet copper. These *dees*, as they are called, are part of an electrical oscillator that alternates the electric potential difference across the gap between the dees. The electrical signs of the dees are alternated so that the electric field in the gap alternates in direction, first toward one dee and then toward the other dee, back and forth. The dees are immersed in a large magnetic field directed out of the plane of the page. The magnitude $B$ of this field is set via a control on the electromagnet producing the field.

Suppose that a proton, injected by source $S$ at the center of the cyclotron in Fig. 28-15, initially moves toward a negatively charged dee. It will accelerate toward this dee and enter it. Once inside, it is shielded from electric fields by the copper walls of the dee; that is, the electric field does not enter the dee. The magnetic field, however, is not screened by the (nonmagnetic) copper dee, so the proton moves in a circular path whose radius, which depends on its speed, is given by Eq. 28-16 ($r = mv/|q|B$).

Let us assume that at the instant the proton emerges into the center gap from the first dee, the potential difference between the dees is reversed. Thus, the proton *again* faces a negatively charged dee and is *again* accelerated. This process continues, the circulating proton always being in step with the oscillations of the dee potential, until the proton has spiraled out to the edge of the dee system. There a deflector plate sends it out through a portal.

The key to the operation of the cyclotron is that the frequency $f$ at which the proton circulates in the magnetic field (and that does *not* depend on its speed) must be equal to the fixed frequency $f_{osc}$ of the electrical oscillator, or

$$f = f_{osc} \quad \text{(resonance condition)}. \quad (28\text{-}23)$$

This *resonance condition* says that, if the energy of the circulating proton is to increase, energy must be fed to it at a frequency $f_{osc}$ that is equal to the natural frequency $f$ at which the proton circulates in the magnetic field.

Combining Eqs. 28-18 ($f = |q|B/2\pi m$) and 28-23 allows us to write the

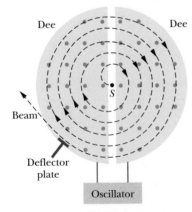

**FIG. 28-15** The elements of a cyclotron, showing the particle source $S$ and the dees. A uniform magnetic field is directed up from the plane of the page. Circulating protons spiral outward within the hollow dees, gaining energy every time they cross the gap between the dees.

resonance condition as

$$|q|B = 2\pi m f_{osc}. \qquad (28\text{-}24)$$

For the proton, $q$ and $m$ are fixed. The oscillator (we assume) is designed to work at a single fixed frequency $f_{osc}$. We then "tune" the cyclotron by varying $B$ until Eq. 28-24 is satisfied, and then many protons circulate through the magnetic field, to emerge as a beam.

Other types of charged particles can be accelerated by a cyclotron to emerge as a high-energy beam. For example, *deuterons* (bundles of one proton and one neutron) can be accelerated because they each have the same charge as a proton. However, electrically neutral particles, such as neutrons, cannot be accelerated by a cyclotron. Still, a beam of high-energy neutrons can be produced by having a deuteron beam smash into a beryllium target. This is the arrangement used in hospitals providing fast-neutron cancer therapy. The deuteron beam emerging from a cyclotron at the hospital is sent into a beryllium target in front of the patient's cancerous region. As deuterons collide with beryllium nuclei, neutrons are knocked out of the beryllium nuclei and into the cancerous region, where they break bonds in the DNA of the cancer cells, "killing" the cells (Fig. 28-16). All this can be done within a hospital because the deuterons can be accelerated to high energy while circling in a cyclotron instead of along an extremely long straight path.

**FIG. 28-16** Because it is invisible, a beam of neutrons from the portal at the right is aligned via laser crosshairs, seen superimposed on the patient. *(Fermilab/Science Photo Library/Photo Researchers)*

### The Proton Synchrotron

At proton energies above 50 MeV, the conventional cyclotron begins to fail because one of the assumptions of its design—that the frequency of revolution of a charged particle circulating in a magnetic field is independent of the particle's speed—is true only for speeds that are much less than the speed of light. At greater proton speeds (above about 10% of the speed of light), we must treat the problem relativistically. According to relativity theory, as the speed of a circulating proton approaches that of light, the proton's frequency of revolution decreases steadily. Thus, the proton gets out of step with the cyclotron's oscillator—whose frequency remains fixed at $f_{osc}$—and eventually the energy of the still circulating proton stops increasing.

There is another problem. For a 500 GeV proton in a magnetic field of 1.5 T, the path radius is 1.1 km. The corresponding magnet for a conventional cyclotron of the proper size would be impossibly expensive, the area of its pole faces being about $4 \times 10^6 \text{ m}^2$.

The *proton synchrotron* is designed to meet these two difficulties. The magnetic field $B$ and the oscillator frequency $f_{osc}$, instead of having fixed values as in the conventional cyclotron, are made to vary with time during the accelerating cycle. When this is done properly, (1) the frequency of the circulating protons remains in step with the oscillator at all times, and (2) the protons follow a circular—not a spiral—path. Thus, the magnet need extend only along that circular path, not over some $4 \times 10^6 \text{ m}^2$. The circular path, however, still must be large if high energies are to be achieved. The proton synchrotron at the Fermi National Accelerator Laboratory (Fermilab) in Illinois has a circumference of 6.3 km and can produce protons with energies of about 1 TeV $(= 10^{12} \text{ eV})$.

---

**Sample Problem** | **28-5**

Suppose a cyclotron is operated at an oscillator frequency of 12 MHz and has a dee radius $R = 53$ cm.

(a) What is the magnitude of the magnetic field needed for deuterons to be accelerated in the cyclotron? The deuteron mass is $m = 3.34 \times 10^{-27}$ kg.

**KEY IDEA** For a given oscillator frequency $f_{osc}$, the magnetic field magnitude $B$ required to accelerate any particle in a cyclotron depends on the ratio $m/|q|$ of mass to charge for the particle, according to Eq. 28-24 $(|q|B = 2\pi m f_{osc})$.

**Calculation:** For deuterons and the oscillator frequency $f_{osc} = 12$ MHz, we find

$$B = \frac{2\pi m f_{osc}}{|q|} = \frac{(2\pi)(3.34 \times 10^{-27} \text{ kg})(12 \times 10^6 \text{ s}^{-1})}{1.60 \times 10^{-19} \text{ C}}$$

$$= 1.57 \text{ T} \approx 1.6 \text{ T}. \hspace{2cm} \text{(Answer)}$$

Note that, to accelerate protons, $B$ would have to be reduced by a factor of 2, provided the oscillator frequency remained fixed at 12 MHz.

**(b)** What is the resulting kinetic energy of the deuterons?

**KEY IDEAS** (1) The kinetic energy ($\frac{1}{2}mv^2$) of a deuteron exiting the cyclotron is equal to the kinetic energy it had just before exiting, when it was traveling in a circular path with a radius approximately equal to the radius $R$ of the

cyclotron dees. (2) We can find the speed $v$ of the deuteron in that circular path with Eq. 28-16 ($r = mv/|q|B$).

**Calculations:** Solving that equation for $v$, substituting $R$ for $r$, and then substituting known data, we find

$$v = \frac{R|q|B}{m} = \frac{(0.53 \text{ m})(1.60 \times 10^{-19} \text{ C})(1.57 \text{ T})}{3.34 \times 10^{-27} \text{ kg}}$$

$$= 3.99 \times 10^7 \text{ m/s}.$$

This speed corresponds to a kinetic energy of

$$K = \frac{1}{2}mv^2$$
$$= \frac{1}{2}(3.34 \times 10^{-27} \text{ kg})(3.99 \times 10^7 \text{ m/s})^2$$
$$= 2.7 \times 10^{-12} \text{ J}, \hspace{1cm} \text{(Answer)}$$

or about 17 MeV.

---

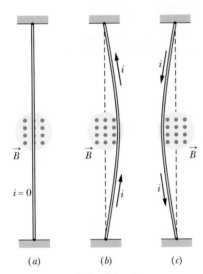

**FIG. 28-17** A flexible wire passes between the pole faces of a magnet (only the farther pole face is shown). (*a*) Without current in the wire, the wire is straight. (*b*) With upward current, the wire is deflected rightward. (*c*) With downward current, the deflection is leftward. The connections for getting the current into the wire at one end and out of it at the other end are not shown.

## 28-8 | Magnetic Force on a Current-Carrying Wire

We have already seen (in connection with the Hall effect) that a magnetic field exerts a sideways force on electrons moving in a wire. This force must then be transmitted to the wire itself, because the conduction electrons cannot escape sideways out of the wire.

In Fig. 28-17*a*, a vertical wire, carrying no current and fixed in place at both ends, extends through the gap between the vertical pole faces of a magnet. The magnetic field between the faces is directed outward from the page. In Fig. 28-17*b*, a current is sent upward through the wire; the wire deflects to the right. In Fig. 28-17*c*, we reverse the direction of the current and the wire deflects to the left.

Figure 28-18 shows what happens inside the wire of Fig. 28-17*b*. We see one of the conduction electrons, drifting downward with an assumed drift speed $v_d$. Equation 28-3, in which we must put $\phi = 90°$, tells us that a force $\vec{F}_B$ of magnitude $ev_dB$ must act on each such electron. From Eq. 28-2 we see that this force must be directed to the right. We expect then that the wire as a whole will experience a force to the right, in agreement with Fig. 28-17*b*.

If, in Fig. 28-18, we were to reverse *either* the direction of the magnetic field *or* the direction of the current, the force on the wire would reverse, being directed now to the left. Note too that it does not matter whether we consider negative charges drifting downward in the wire (the actual case) or positive charges drifting upward. The direction of the deflecting force on the wire is the same. We are safe then in dealing with a current of positive charge.

Consider a length $L$ of the wire in Fig. 28-18. All the conduction electrons in this section of wire will drift past plane $xx$ in Fig. 28-18 in a time $t = L/v_d$. Thus, in that time a charge given by

$$q = it = i\frac{L}{v_d}$$

will pass through that plane. Substituting this into Eq. 28-3 yields

$$F_B = qv_dB \sin\phi = \frac{iL}{v_d}v_dB \sin 90°$$

or

$$F_B = iLB. \hspace{2cm} (28\text{-}25)$$

This equation gives the magnetic force that acts on a length $L$ of straight wire carrying a current $i$ and immersed in a magnetic field $\vec{B}$ that is perpendicular to the wire.

If the magnetic field is *not* perpendicular to the wire, as in Fig. 28-19, the magnetic force is given by a generalization of Eq. 28-25:

$$\vec{F}_B = i\vec{L} \times \vec{B} \qquad \text{(force on a current).} \qquad (28\text{-}26)$$

Here $\vec{L}$ is a *length vector* that has magnitude $L$ and is directed along the wire segment in the direction of the (conventional) current. The force magnitude $F_B$ is

$$F_B = iLB \sin \phi, \qquad (28\text{-}27)$$

where $\phi$ is the angle between the directions of $\vec{L}$ and $\vec{B}$. The direction of $\vec{F}_B$ is that of the cross product $\vec{L} \times \vec{B}$ because we take current $i$ to be a positive quantity. Equation 28-26 tells us that $\vec{F}_B$ is always perpendicular to the plane defined by vectors $\vec{L}$ and $\vec{B}$, as indicated in Fig. 28-19.

Equation 28-26 is equivalent to Eq. 28-2 in that either can be taken as the defining equation for $\vec{B}$. In practice, we define $\vec{B}$ from Eq. 28-26 because it is much easier to measure the magnetic force acting on a wire than that on a single moving charge.

If a wire is not straight or the field is not uniform, we can imagine the wire broken up into small straight segments and apply Eq. 28-26 to each segment. The force on the wire as a whole is then the vector sum of all the forces on the segments that make it up. In the differential limit, we can write

$$d\vec{F}_B = i \, d\vec{L} \times \vec{B}, \qquad (28\text{-}28)$$

and we can find the resultant force on any given arrangement of currents by integrating Eq. 28-28 over that arrangement.

In using Eq. 28-28, bear in mind that there is no such thing as an isolated current-carrying wire segment of length $dL$. There must always be a way to introduce the current into the segment at one end and take it out at the other end.

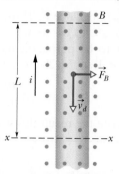

FIG. 28-18 A close-up view of a section of the wire of Fig. 28-17b. The current direction is upward, which means that electrons drift downward. A magnetic field that emerges from the plane of the page causes the electrons and the wire to be deflected to the right.

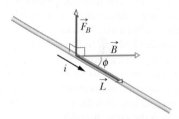

FIG. 28-19 A wire carrying current $i$ makes an angle $\phi$ with magnetic field $\vec{B}$. The wire has length $L$ in the field and length vector $\vec{L}$ (in the direction of the current). A magnetic force $\vec{F}_B = i\vec{L} \times \vec{B}$ acts on the wire.

✔ **CHECKPOINT 4** The figure shows a current $i$ through a wire in a uniform magnetic field $\vec{B}$, as well as the magnetic force $\vec{F}_B$ acting on the wire. The field is oriented so that the force is maximum. In what direction is the field?

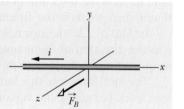

---

**Sample Problem | 28-6**

A straight, horizontal length of copper wire has a current $i = 28$ A through it. What are the magnitude and direction of the minimum magnetic field $\vec{B}$ needed to suspend the wire—that is, to balance the gravitational force on it? The linear density (mass per unit length) of the wire is 46.6 g/m.

**KEY IDEAS** (1) Because the wire carries a current, a magnetic force $\vec{F}_B$ can act on the wire if we place it in a magnetic field $\vec{B}$. To balance the downward gravitational force $\vec{F}_g$ on the wire, we want $\vec{F}_B$ to be directed upward (Fig. 28-20). (2) The direction of $\vec{F}_B$ is related to the directions of $\vec{B}$ and the wire's length vector $\vec{L}$ by Eq. 28-26 ($\vec{F}_B = i\vec{L} \times \vec{B}$).

**Calculations:** Because $\vec{L}$ is directed horizontally (and the current is taken to be positive), Eq. 28-26 and the

right-hand rule for cross products tell us that $\vec{B}$ must be horizontal and rightward (in Fig. 28-20) to give the required upward $\vec{F}_B$.

The magnitude of $\vec{F}_B$ is $F_B = iLB \sin \phi$ (Eq. 28-27). Because we want $\vec{F}_B$ to balance $\vec{F}_g$, we want

$$iLB \sin \phi = mg, \qquad (28\text{-}29)$$

where $mg$ is the magnitude of $\vec{F}_g$ and $m$ is the mass of the wire. We also want the minimal field magnitude $B$

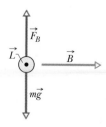

FIG. 28-20 A wire (shown in cross section) carrying current out of the page.

for $\vec{F}_B$ to balance $\vec{F}_g$. Thus, we need to maximize $\sin\phi$ in Eq. 28-29. To do so, we set $\phi = 90°$, thereby arranging for $\vec{B}$ to be perpendicular to the wire. We then have $\sin\phi = 1$, so Eq. 28-29 yields

$$B = \frac{mg}{iL\sin\phi} = \frac{(m/L)g}{i}. \qquad (28\text{-}30)$$

We write the result this way because we know $m/L$, the linear density of the wire. Substituting known data then gives us

$$B = \frac{(46.6 \times 10^{-3}\,\text{kg/m})(9.8\,\text{m/s}^2)}{28\,\text{A}}$$

$$= 1.6 \times 10^{-2}\,\text{T}. \qquad (\text{Answer})$$

This is about 160 times the strength of Earth's magnetic field.

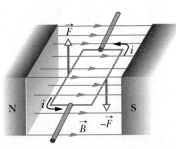

FIG. 28-21   The elements of an electric motor. A rectangular loop of wire, carrying a current and free to rotate about a fixed axis, is placed in a magnetic field. Magnetic forces on the wire produce a torque that rotates it. A commutator (not shown) reverses the direction of the current every half-revolution so that the torque always acts in the same direction.

## 28-9 | Torque on a Current Loop

Much of the world's work is done by electric motors. The forces behind this work are the magnetic forces that we studied in the preceding section—that is, the forces that a magnetic field exerts on a wire that carries a current.

Figure 28-21 shows a simple motor, consisting of a single current-carrying loop immersed in a magnetic field $\vec{B}$. The two magnetic forces $\vec{F}$ and $-\vec{F}$ produce a torque on the loop, tending to rotate it about its central axis. Although many essential details have been omitted, the figure does suggest how the action of a magnetic field on a current loop produces rotary motion. Let us analyze that action.

Figure 28-22a shows a rectangular loop of sides $a$ and $b$, carrying current $i$ through uniform magnetic field $\vec{B}$. We place the loop in the field so that its long sides, labeled 1 and 3, are perpendicular to the field direction (which is into the page), but its short sides, labeled 2 and 4, are not. Wires to lead the current into and out of the loop are needed but, for simplicity, are not shown.

To define the orientation of the loop in the magnetic field, we use a normal vector $\vec{n}$ that is perpendicular to the plane of the loop. Figure 28-22b shows a right-hand rule for finding the direction of $\vec{n}$. Point or curl the fingers of your right hand in the direction of the current at any point on the loop. Your extended thumb then points in the direction of the normal vector $\vec{n}$.

In Fig. 28-22c, the normal vector of the loop is shown at an arbitrary angle $\theta$ to the direction of the magnetic field $\vec{B}$. We wish to find the net force and net torque acting on the loop in this orientation.

The net force on the loop is the vector sum of the forces acting on its four sides. For side 2 the vector $\vec{L}$ in Eq. 28-26 points in the direction of the current and has magnitude $b$. The angle between $\vec{L}$ and $\vec{B}$ for side 2 (see Fig. 28-22c) is $90° - \theta$. Thus, the magnitude of the force acting on this side is

$$F_2 = ibB\sin(90° - \theta) = ibB\cos\theta. \qquad (28\text{-}31)$$

You can show that the force $\vec{F}_4$ acting on side 4 has the same magnitude as $\vec{F}_2$ but the opposite direction. Thus, $\vec{F}_2$ and $\vec{F}_4$ cancel out exactly. Their net force is zero and, because their common line of action is through the center of the loop, their net torque is also zero.

The situation is different for sides 1 and 3. For them, $\vec{L}$ is perpendicular to $\vec{B}$, so the forces $\vec{F}_1$ and $\vec{F}_3$ have the common magnitude $iaB$. Because these two forces have opposite directions, they do not tend to move the loop up or down. However, as Fig. 28-22c shows, these two forces do *not* share the same line of action; so they *do* produce a net torque. The torque tends to rotate the loop so as to align its normal vector $\vec{n}$ with the direction of the magnetic field $\vec{B}$. That torque has moment arm $(b/2)\sin\theta$ about the central axis of the loop. The magnitude $\tau'$ of the torque due to forces $\vec{F}_1$ and $\vec{F}_3$ is then (see Fig. 28-22c)

$$\tau' = \left(iaB\frac{b}{2}\sin\theta\right) + \left(iaB\frac{b}{2}\sin\theta\right) = iabB\sin\theta. \qquad (28\text{-}32)$$

Suppose we replace the single loop of current with a *coil* of $N$ loops, or *turns*. Further, suppose that the turns are wound tightly enough that they can be

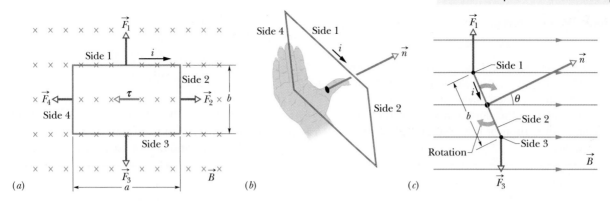

**FIG. 28-22** A rectangular loop, of length $a$ and width $b$ and carrying a current $i$, is located in a uniform magnetic field. A torque $\tau$ acts to align the normal vector $\vec{n}$ with the direction of the field. (a) The loop as seen by looking in the direction of the magnetic field. (b) A perspective of the loop showing how the right-hand rule gives the direction of $\vec{n}$, which is perpendicular to the plane of the loop. (c) A side view of the loop, from side 2. The loop rotates as indicated.

approximated as all having the same dimensions and lying in a plane. Then the turns form a *flat coil,* and a torque $\tau'$ with the magnitude given in Eq. 28-32 acts on each of them. The total torque on the coil then has magnitude

$$\tau = N\tau' = NiabB \sin \theta = (NiA)B \sin \theta, \qquad (28\text{-}33)$$

in which $A \, (= ab)$ is the area enclosed by the coil. The quantities in parentheses $(NiA)$ are grouped together because they are all properties of the coil: its number of turns, its area, and the current it carries. Equation 28-33 holds for all flat coils, no matter what their shape, provided the magnetic field is uniform.

Instead of focusing on the motion of the coil, it is simpler to keep track of the vector $\vec{n}$, which is normal to the plane of the coil. Equation 28-33 tells us that a current-carrying flat coil placed in a magnetic field will tend to rotate so that $\vec{n}$ has the same direction as the field. In a motor, the current in the coil is reversed as $\vec{n}$ begins to line up with the field direction, so that a torque continues to rotate the coil. This automatic reversal of the current is done via a commutator that electrically connects the rotating coil with the stationary contacts on the wires that supply the current from some source.

## Sample Problem 28-7

Analog voltmeters and ammeters work by measuring the torque exerted by a magnetic field on a current-carrying coil. The reading is displayed by means of the deflection of a pointer over a scale. Figure 28-23 shows the essentials of a *galvanometer,* a device on which both analog ammeters and analog voltmeters are based. Assume the coil is 2.1 cm high and 1.2 cm wide, has 250 turns, and is mounted so that it can rotate about an axis (into the page) in a uniform *radial* magnetic field with $B = 0.23$ T. For any orientation of the coil, the net magnetic field through the coil is perpendicular to the normal vector of the coil (and thus parallel to the plane of the coil). A spring Sp provides a countertorque that balances the magnetic torque, so that a given steady current $i$ in the coil results in a steady angular deflection $\phi$. The greater the current is, the greater the deflection is, and thus the greater the torque required of the spring is. If a current of 100 $\mu$A produces an angular deflection of 28°, what must be the torsional constant $\kappa$ of the spring, as used in Eq. 15-22 ($\tau = -\kappa\phi$)?

**KEY IDEA** With a constant current through the device, the resulting magnetic torque (Eq. 28-33) is balanced by the spring torque. Thus, the magnitudes of those torques are equal.

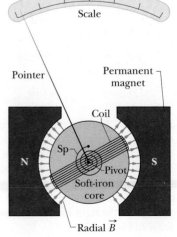

**FIG. 28-23** The elements of a galvanometer. Depending on the external circuit, this device can be wired up as either a voltmeter or an ammeter.

**Calculations:** We write this balance as

$$NiAB \sin \theta = \kappa \phi. \qquad (28\text{-}34)$$

Here $\phi$ is the angular deflection of the coil and pointer, and $A$ (= $2.52 \times 10^{-4}$ m$^2$) is the area encircled by the coil. Because the net magnetic field through the coil is always perpendicular to the normal vector of the coil, $\theta = 90°$ for any orientation of the pointer. Solving Eq. 28-34 for $\kappa$, we find

$$\begin{aligned}
\kappa &= \frac{NiAB \sin \theta}{\phi} \\
&= (250)(100 \times 10^{-6}\,\text{A})(2.52 \times 10^{-4}\,\text{m}^2) \\
&\quad \times \frac{(0.23\,\text{T})(\sin 90°)}{28°} \\
&= 5.2 \times 10^{-8}\,\text{N·m/degree.} \qquad \text{(Answer)}
\end{aligned}$$

Many modern ammeters and voltmeters are of the digital, direct-reading type and do not use a moving coil.

## 28-10 | The Magnetic Dipole Moment

As we have just discussed, a torque acts to rotate a current-carrying coil placed in a magnetic field. In that sense, the coil behaves like a bar magnet placed in the magnetic field. Thus, like a bar magnet, a current-carrying coil is said to be a *magnetic dipole*. Moreover, to account for the torque on the coil due to the magnetic field, we assign a **magnetic dipole moment** $\vec{\mu}$ to the coil. The direction of $\vec{\mu}$ is that of the normal vector $\vec{n}$ to the plane of the coil and thus is given by the same right-hand rule shown in Fig. 28-22. That is, grasp the coil with the fingers of your right hand in the direction of current $i$; the outstretched thumb of that hand gives the direction of $\vec{\mu}$. The magnitude of $\vec{\mu}$ is given by

$$\mu = NiA \quad \text{(magnetic moment),} \qquad (28\text{-}35)$$

in which $N$ is the number of turns in the coil, $i$ is the current through the coil, and $A$ is the area enclosed by each turn of the coil. From this equation, with $i$ in amperes and $A$ in square meters, we see that the unit of $\vec{\mu}$ is the ampere–square meter (A · m$^2$).

Using $\vec{\mu}$, we can rewrite Eq. 28-33 for the torque on the coil due to a magnetic field as

$$\tau = \mu B \sin \theta, \qquad (28\text{-}36)$$

in which $\theta$ is the angle between the vectors $\vec{\mu}$ and $\vec{B}$.

We can generalize this to the vector relation

$$\vec{\tau} = \vec{\mu} \times \vec{B}, \qquad (28\text{-}37)$$

which reminds us very much of the corresponding equation for the torque exerted by an *electric* field on an *electric* dipole—namely, Eq. 22-34:

$$\vec{\tau} = \vec{p} \times \vec{E}.$$

In each case the torque due to the field—either magnetic or electric—is equal to the vector product of the corresponding dipole moment and the field vector.

A magnetic dipole in an external magnetic field has a **magnetic potential energy** that depends on the dipole's orientation in the field. For electric dipoles we have shown (Eq. 22-38) that

$$U(\theta) = -\vec{p} \cdot \vec{E}.$$

In strict analogy, we can write for the magnetic case

$$U(\theta) = -\vec{\mu} \cdot \vec{B}. \qquad (28\text{-}38)$$

A magnetic dipole has its lowest energy (= $-\mu B \cos 0 = -\mu B$) when its dipole moment $\vec{\mu}$ is lined up with the magnetic field (Fig. 28-24). It has its highest energy (= $-\mu B \cos 180° = +\mu B$) when $\vec{\mu}$ is directed opposite the field. From Eq. 28-38, with $U$ in joules and $\vec{B}$ in teslas, we see that the unit of $\vec{\mu}$ can be the joule per tesla (J/T) instead of the ampere–square meter as suggested by Eq. 28-35.

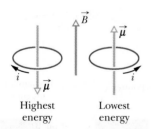

Highest energy        Lowest energy

**FIG. 28-24** The orientations of highest and lowest energy of a magnetic dipole (here a coil carrying current) in an external magnetic field $\vec{B}$. The direction of the current $i$ gives the direction of the magnetic dipole moment $\vec{\mu}$ via the right-hand rule shown for $\vec{n}$ in Fig. 28-22b.

If an applied torque (due to "an external agent") rotates a magnetic dipole from an initial orientation $\theta_i$ to another orientation $\theta_f$, then work $W_a$ is done on the dipole by the applied torque. If the dipole is stationary before and after the change in its orientation, then work $W_a$ is

$$W_a = U_f - U_i, \tag{28-39}$$

where $U_f$ and $U_i$ are calculated with Eq. 28-38.

So far, we have identified only a current-carrying coil as a magnetic dipole. However, a simple bar magnet is also a magnetic dipole, as is a rotating sphere of charge. Earth itself is (approximately) a magnetic dipole. Finally, most subatomic particles, including the electron, the proton, and the neutron, have magnetic dipole moments. As you will see in Chapter 32, all these quantities can be viewed as current loops. For comparison, some approximate magnetic dipole moments are shown in Table 28-2.

**TABLE 28-2**

**Some Magnetic Dipole Moments**

| | |
|---|---|
| Small bar magnet | 5 J/T |
| Earth | $8.0 \times 10^{22}$ J/T |
| Proton | $1.4 \times 10^{-26}$ J/T |
| Electron | $9.3 \times 10^{-24}$ J/T |

✓ **CHECKPOINT 5**     The figure shows four orientations, at angle $\theta$, of a magnetic dipole moment $\vec{\mu}$ in a magnetic field. Rank the orientations according to (a) the magnitude of the torque on the dipole and (b) the potential energy of the dipole, greatest first.

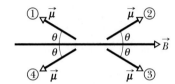

---

**Sample Problem    28-8**

Figure 28-25 shows a circular coil with 250 turns, an area $A$ of $2.52 \times 10^{-4}$ m², and a current of 100 $\mu$A. The coil is at rest in a uniform magnetic field of magnitude $B = 0.85$ T, with its magnetic dipole moment $\vec{\mu}$ initially aligned with $\vec{B}$.

**(a)** In Fig. 28-25, what is the direction of the current in the coil?

**Right-hand rule:** Imagine cupping the coil with your right hand so that your right thumb is outstretched in the direction of $\vec{\mu}$. The direction in which your fingers curl around the coil is the direction of the current in the coil. Thus, in the wires on the near side of the coil—those we see in Fig. 28-25—the current is from top to bottom.

**(b)** How much work would the torque applied by an external agent have to do on the coil to rotate it 90° from its initial orientation, so that $\vec{\mu}$ is perpendicular to $\vec{B}$ and the coil is again at rest?

**FIG. 28-25**   A side view of a circular coil carrying a current and oriented so that its magnetic dipole moment $\vec{\mu}$ is aligned with magnetic field $\vec{B}$.

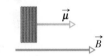

**KEY IDEA**    The work $W_a$ done by the applied torque would be equal to the change in the coil's potential energy due to its change in orientation.

**Calculations:** From Eq. 28-39 ($W_a = U_f - U_i$), we find

$$W_a = U(90°) - U(0°)$$
$$= -\mu B \cos 90° - (-\mu B \cos 0°) = 0 + \mu B$$
$$= \mu B.$$

Substituting for $\mu$ from Eq. 28-35 ($\mu = NiA$), we find that

$$W_a = (NiA)B$$
$$= (250)(100 \times 10^{-6} \text{ A})(2.52 \times 10^{-4} \text{ m}^2)(0.85 \text{ T})$$
$$= 5.355 \times 10^{-6} \text{ J} \approx 5.4 \ \mu\text{J}. \qquad \text{(Answer)}$$

---

## REVIEW & SUMMARY

**Magnetic Field $\vec{B}$**    A **magnetic field** $\vec{B}$ is defined in terms of the force $\vec{F}_B$ acting on a test particle with charge $q$ moving through the field with velocity $\vec{v}$:

$$\vec{F}_B = q\vec{v} \times \vec{B}. \tag{28-2}$$

The SI unit for $\vec{B}$ is the **tesla** (T): 1 T = 1 N/(A·m) = $10^4$ gauss.

**The Hall Effect**    When a conducting strip of thickness $l$ carrying a current $i$ is placed in a uniform magnetic field $\vec{B}$, some charge carriers (with charge $e$) build up on one side of the conductor, creating a potential difference $V$ across the strip. The po-

larities of the sides indicate the sign of the charge carriers; the number density $n$ of charge carriers can be calculated with

$$n = \frac{Bi}{Vle}. \tag{28-12}$$

**A Charged Particle Circulating in a Magnetic Field**    A charged particle with mass $m$ and charge magnitude $|q|$ moving with velocity $\vec{v}$ perpendicular to a uniform magnetic field $\vec{B}$ will travel in a circle. Applying Newton's second law to the circular motion yields

$$|q|vB = \frac{mv^2}{r}, \tag{28-15}$$

from which we find the radius $r$ of the circle to be

$$r = \frac{mv}{|q|B}. \qquad (28\text{-}16)$$

The frequency of revolution $f$, the angular frequency $\omega$, and the period of the motion $T$ are given by

$$f = \frac{\omega}{2\pi} = \frac{1}{T} = \frac{|q|B}{2\pi m}. \qquad (28\text{-}19, 28\text{-}18, 28\text{-}17)$$

**Magnetic Force on a Current-Carrying Wire** A straight wire carrying a current $i$ in a uniform magnetic field experiences a sideways force

$$\vec{F}_B = i\vec{L} \times \vec{B}. \qquad (28\text{-}26)$$

The force acting on a current element $i\, d\vec{L}$ in a magnetic field is

$$d\vec{F}_B = i\, d\vec{L} \times \vec{B}. \qquad (28\text{-}28)$$

The direction of the length vector $\vec{L}$ or $d\vec{L}$ is that of the current $i$.

**Torque on a Current-Carrying Coil** A coil (of area $A$ and $N$ turns, carrying current $i$) in a uniform magnetic field $\vec{B}$ will experience a torque $\vec{\tau}$ given by

$$\vec{\tau} = \vec{\mu} \times \vec{B}. \qquad (28\text{-}37)$$

Here $\vec{\mu}$ is the **magnetic dipole moment** of the coil, with magnitude $\mu = NiA$ and direction given by the right-hand rule.

**Orientation Energy of a Magnetic Dipole** The **magnetic potential energy** of a magnetic dipole in a magnetic field is

$$U(\theta) = -\vec{\mu} \cdot \vec{B}. \qquad (28\text{-}38)$$

If an external agent rotates a magnetic dipole from an initial orientation $\theta_i$ to some other orientation $\theta_f$ and the dipole is stationary both initially and finally, the work $W_a$ done on the dipole by the agent is

$$W_a = \Delta U = U_f - U_i. \qquad (28\text{-}39)$$

# QUESTIONS

**1** In Section 28-4, we discussed a charged particle moving through crossed fields with the forces $\vec{F}_E$ and $\vec{F}_B$ in opposition. We found that the particle moves in a straight line (that is, neither force dominates the motion) if its speed is given by Eq. 28-7 ($v = E/B$). Which of the two forces dominates if the speed of the particle is (a) $v < E/B$ and (b) $v > E/B$?

**2** Figure 28-26 shows a wire that carries current to the right through a uniform magnetic field. It also shows four choices for the direction of that field. (a) Rank the choices according to the magnitude of the electric potential difference that would

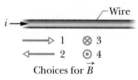

**FIG. 28-26** Question 2.

be set up across the width of the wire, greatest first. (b) For which choice is the top side of the wire at higher potential than the bottom side of the wire?

**3** Figure 28-27 shows three situations in which a positively charged particle moves at velocity $\vec{v}$ through a uniform magnetic field $\vec{B}$ and experiences a magnetic force $\vec{F}_B$. In each situation, determine whether the orientations of the vectors are physically reasonable.

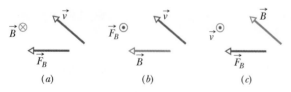

**FIG. 28-27** Question 3.

**4** Figure 28-28 shows crossed uniform electric and magnetic fields $\vec{E}$ and $\vec{B}$ and, at a certain instant, the velocity vectors of the 10 charged particles listed in Table 28-3. (The vectors are not drawn to scale.) The speeds given in the table are either less than or greater than $E/B$ (see Question 1). Which particles will move out of the page toward you after the instant shown in Fig. 28-28?

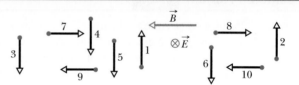

**FIG. 28-28** Question 4.

**TABLE 28-3**

**Question 4**

| Particle | Charge | Speed | Particle | Charge | Speed |
|----------|--------|---------|----------|--------|---------|
| 1 | + | Less | 6 | − | Greater |
| 2 | + | Greater | 7 | + | Less |
| 3 | + | Less | 8 | + | Greater |
| 4 | + | Greater | 9 | − | Less |
| 5 | − | Less | 10 | − | Greater |

**5** Figure 28-29 shows a metallic, rectangular solid that is to move at a certain speed $v$ through the uniform magnetic field $\vec{B}$. The dimensions of the solid are multiples of $d$, as shown. You have six choices for the direction of the velocity: parallel to $x$, $y$, or $z$ in either the positive or negative direction. (a) Rank the six choices according to the

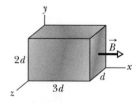

**FIG. 28-29** Question 5.

potential difference set up across the solid, greatest first. (b) For which choice is the front face at lower potential?

**6** Figure 28-30 shows the path of a particle through six regions of uniform magnetic field, where the path is either a half-circle or a quarter-circle. Upon leaving the last region, the particle travels between two charged, parallel plates and is deflected toward the plate of higher potential. What is the direction of the magnetic field in each of the six regions?

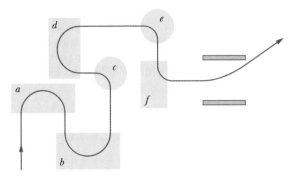

FIG. 28-30   Question 6.

**7** In Fig. 28-31, a charged particle enters a uniform magnetic field $\vec{B}$ with speed $v_0$, moves through a half-circle in time $T_0$, and then leaves the field. (a) Is the charge positive or negative? (b) Is the final speed of the particle greater than, less than, or equal to $v_0$? (c) If the initial speed had been $0.5v_0$, would the time spent in field $\vec{B}$ have been greater than, less than, or equal to $T_0$? (d) Would the path have been a half-circle, more than a half-circle, or less than a half-circle?

FIG. 28-31   Question 7.

**8** *Particle roundabout.* Figure 28-32 shows 11 paths through a region of uniform magnetic field. One path is a straight line; the rest are half-circles. Table 28-4 gives the masses, charges, and speeds of 11 particles that take these paths through the field in the directions shown. Which path in the figure corresponds to which particle in the table?

| TABLE 28-4 | | | |
|---|---|---|---|
| **Question 8** | | | |
| Particle | Mass | Charge | Speed |
| 1 | $2m$ | $q$ | $v$ |
| 2 | $m$ | $2q$ | $v$ |
| 3 | $m/2$ | $q$ | $2v$ |
| 4 | $3m$ | $3q$ | $3v$ |
| 5 | $2m$ | $q$ | $2v$ |
| 6 | $m$ | $-q$ | $2v$ |
| 7 | $m$ | $-4q$ | $v$ |
| 8 | $m$ | $-q$ | $v$ |
| 9 | $2m$ | $-2q$ | $3v$ |
| 10 | $m$ | $-2q$ | $8v$ |
| 11 | $3m$ | $0$ | $3v$ |

**9** Figure 28-33 shows the path of an electron that passes through two regions containing uniform magnetic fields of magnitudes $B_1$ and $B_2$. Its path in each region is a half-circle. (a) Which field is stronger? (b) What is the direction of each field? (c) Is the time spent by the electron in the $\vec{B}_1$ region greater than, less than, or the same as the time spent in the $\vec{B}_2$ region?

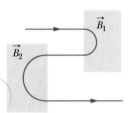

FIG. 28-33
Question 9.

**10** Figure 28-34 shows the path of an electron in a region of uniform magnetic field. The path consists of two straight sections, each between a pair of uniformly charged plates, and two half-circles. Which plate is at the higher electric potential in (a) the top pair of plates and (b) the bottom pair? (c) What is the direction of the magnetic field?

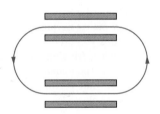

FIG. 28-34   Question 10.

**11** (a) In Checkpoint 5, if the dipole moment $\vec{\mu}$ is rotated from orientation 2 to orientation 1 by an external agent, is the work done on the dipole by the agent positive, negative, or zero? (b) Rank the work done on the dipole by the agent for these three rotations, greatest first: $2 \rightarrow 1, 2 \rightarrow 4, 2 \rightarrow 3$.

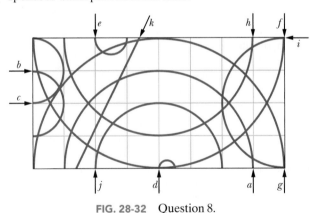

FIG. 28-32   Question 8.

---

## PROBLEMS

**GO** Tutoring problem available (at instructor's discretion) in *WileyPLUS* and WebAssign
**SSM** Worked-out solution available in Student Solutions Manual
• – ••• Number of dots indicates level of problem difficulty
**WWW** Worked-out solution is at _____
**ILW** Interactive solution is at _____ http://www.wiley.com/college/halliday
Additional information available in *The Flying Circus of Physics* and at flyingcircusofphysics.com

### sec. 28-3   The Definition of $\vec{B}$

•**1** An electron that has velocity

$$\vec{v} = (2.0 \times 10^6 \text{ m/s})\hat{i} + (3.0 \times 10^6 \text{ m/s})\hat{j}$$

moves through the uniform magnetic field $\vec{B} = (0.030 \text{ T})\hat{i} - (0.15 \text{ T})\hat{j}$. (a) Find the force on the electron due to the magnetic field. (b) Repeat your calculation for a proton having the same velocity.

•2 An alpha particle travels at a velocity $\vec{v}$ of magnitude 550 m/s through a uniform magnetic field $\vec{B}$ of magnitude 0.045 T. (An alpha particle has a charge of $+3.2 \times 10^{-19}$ C and a mass of $6.6 \times 10^{-27}$ kg.) The angle between $\vec{v}$ and $\vec{B}$ is 52°. What is the magnitude of (a) the force $\vec{F}_B$ acting on the particle due to the field and (b) the acceleration of the particle due to $\vec{F}_B$? (c) Does the speed of the particle increase, decrease, or remain the same?

•3 A proton traveling at 23.0° with respect to the direction of a magnetic field of strength 2.60 mT experiences a magnetic force of $6.50 \times 10^{-17}$ N. Calculate (a) the proton's speed and (b) its kinetic energy in electron-volts. **SSM ILW**

•4 A particle of mass 10 g and charge 80 $\mu$C moves through a uniform magnetic field, in a region where the free-fall acceleration is $-9.8\hat{j}$ m/s$^2$. The velocity of the particle is a constant $20\hat{i}$ km/s, which is perpendicular to the magnetic field. What, then, is the magnetic field?

••5 An electron moves through a uniform magnetic field given by $\vec{B} = B_x\hat{i} + (3.0B_x)\hat{j}$. At a particular instant, the electron has velocity $\vec{v} = (2.0\hat{i} + 4.0\hat{j})$ m/s and the magnetic force acting on it is $(6.4 \times 10^{-19}$ N$)\hat{k}$. Find $B_x$.

••6 A proton moves through a uniform magnetic field given by $\vec{B} = (10\hat{i} - 20\hat{j} + 30\hat{k})$ mT. At time $t_1$, the proton has a velocity given by $\vec{v} = v_x\hat{i} + v_y\hat{j} + (2.0$ km/s$)\hat{k}$ and the magnetic force on the proton is $\vec{F}_B = (4.0 \times 10^{-17}$ N$)\hat{i} + (2.0 \times 10^{-17}$ N$)\hat{j}$. At that instant, what are (a) $v_x$ and (b) $v_y$?

### sec. 28-4 Crossed Fields: Discovery of the Electron

•7 In Fig. 28-35, an electron accelerated from rest through potential difference $V_1 = 1.00$ kV enters the gap between two parallel plates having separation $d = 20.0$ mm and potential difference $V_2 = 100$ V. The lower plate is at the lower potential. Neglect fringing and assume that the electron's velocity vector is perpendicular to the electric field vector between the plates. In unit-vector notation, what uniform magnetic field allows the electron to travel in a straight line in the gap? **ILW**

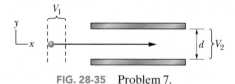

**FIG. 28-35** Problem 7.

•8 An electric field of 1.50 kV/m and a perpendicular magnetic field of 0.400 T act on a moving electron to produce no net force. What is the electron's speed?

•9 An electron has an initial velocity of $(12.0\hat{j} + 15.0\hat{k})$ km/s and a constant acceleration of $(2.00 \times 10^{12}$ m/s$^2)\hat{i}$ in a region in which uniform electric and magnetic fields are present. If $\vec{B} = (400 \, \mu$T$)\hat{i}$, find the electric field $\vec{E}$.

••10 A proton travels through uniform magnetic and electric fields. The magnetic field is $\vec{B} = -2.50\hat{i}$ mT. At one instant the velocity of the proton is $\vec{v} = 2000\hat{j}$ m/s. At that instant and in unit-vector notation, what is the net force acting on the proton if the electric field is (a) $4.00\hat{k}$ V/m, (b) $-4.00\hat{k}$ V/m, and (c) $4.00\hat{i}$ V/m?

••11 An ion source is producing $^6$Li ions, which have charge $+e$ and mass $9.99 \times 10^{-27}$ kg. The ions are accelerated by a potential difference of 10 kV and pass horizontally into a region in which there is a uniform vertical magnetic field of magnitude $B = 1.2$ T. Calculate the strength of the smallest electric field, to be set up over the same region, that will allow the $^6$Li ions to pass through undeflected.

•••12 At time $t_1$, an electron is sent along the positive direction of an $x$ axis, through both an electric field $\vec{E}$ and a magnetic field $\vec{B}$, with $\vec{E}$ directed parallel to the $y$ axis. Figure 28-36 gives the $y$ component $F_{net,y}$ of the net force on the electron due to the two fields, as a function of the electron's speed $v$ at time $t_1$. The scale of the velocity axis is set by $v_s = 100.0$ m/s. The $x$ and $z$ components of the net force are zero at $t_1$. Assuming $B_x = 0$, find (a) the magnitude $E$ and (b) $\vec{B}$ in unit-vector notation. **GO**

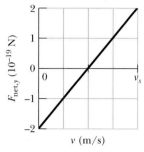

**FIG. 28-36** Problem 12.

### sec. 28-5 Crossed Fields: The Hall Effect

•13 A strip of copper 150 $\mu$m thick and 4.5 mm wide is placed in a uniform magnetic field $\vec{B}$ of magnitude 0.65 T, with $\vec{B}$ perpendicular to the strip. A current $i = 23$ A is then sent through the strip such that a Hall potential difference $V$ appears across the width of the strip. Calculate $V$. (The number of charge carriers per unit volume for copper is $8.47 \times 10^{28}$ electrons/m$^3$.)

•14 A metal strip 6.50 cm long, 0.850 cm wide, and 0.760 mm thick moves with constant velocity $\vec{v}$ through a uniform magnetic field $B = 1.20$ mT directed perpendicular to the strip, as shown in Fig. 28-37. A potential difference of 3.90 $\mu$V is measured between points $x$ and $y$ across the strip. Calculate the speed $v$.

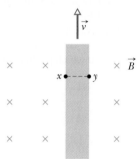

**FIG. 28-37** Problem 14.

••15 In Fig. 28-38, a conducting rectangular solid of dimensions $d_x = 5.00$ m, $d_y = 3.00$ m, and $d_z = 2.00$ m moves at constant velocity $\vec{v} = (20.0$ m/s$)\hat{i}$ through a uniform magnetic field $\vec{B} = (30.0$ mT$)\hat{j}$. What are the resulting (a) electric field within the solid, in unit-vector notation, and (b) potential difference across the solid?

•••16 Figure 28-38 shows a metallic block, with its faces parallel to coordinate axes. The block is in a uniform magnetic field of magnitude 0.020 T. One edge length of the block is 25 cm; the block is *not* drawn to scale. The block is moved at 3.0 m/s parallel to each axis, in turn, and the resulting potential difference $V$ that appears across the block is measured. With the motion parallel to the $y$ axis, $V = 12$ mV; with the motion parallel to the $z$ axis, $V = 18$ mV; with the motion parallel to the $x$ axis, $V = 0$. What are the block lengths (a) $d_x$, (b) $d_y$, and (c) $d_z$? **GO**

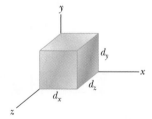

**FIG. 28-38** Problems 15 and 16.

### sec. 28-6    A Circulating Charged Particle

•17    An electron of kinetic energy 1.20 keV circles in a plane perpendicular to a uniform magnetic field. The orbit radius is 25.0 cm. Find (a) the electron's speed, (b) the magnetic field magnitude, (c) the circling frequency, and (d) the period of the motion.  **SSM**

•18    An electron is accelerated from rest by a potential difference of 350 V. It then enters a uniform magnetic field of magnitude 200 mT with its velocity perpendicular to the field. Calculate (a) the speed of the electron and (b) the radius of its path in the magnetic field.

•19    What uniform magnetic field, applied perpendicular to a beam of electrons moving at $1.30 \times 10^6$ m/s, is required to make the electrons travel in a circular arc of radius 0.350 m?

•20    In a nuclear experiment a proton with kinetic energy 1.0 MeV moves in a circular path in a uniform magnetic field. What energy must (a) an alpha particle ($q = +2e, m = 4.0$ u) and (b) a deuteron ($q = +e, m = 2.0$ u) have if they are to circulate in the same circular path?

•21    (a) Find the frequency of revolution of an electron with an energy of 100 eV in a uniform magnetic field of magnitude 35.0 $\mu$T. (b) Calculate the radius of the path of this electron if its velocity is perpendicular to the magnetic field.

•22    An electron is accelerated from rest through potential difference $V$ and then enters a region of uniform magnetic field, where it undergoes uniform circular motion. Figure 28-39 gives the radius $r$ of that motion versus $V^{1/2}$. The vertical axis scale is set by $r_s = 3.0$ mm, and the horizontal axis scale is set by $V_s^{1/2} = 40.0$ $V^{1/2}$. What is the magnitude of the magnetic field?

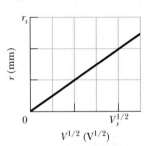

FIG. 28-39    Problem 22.

•23    A certain particle is sent into a uniform magnetic field, with the particle's velocity vector perpendicular to the direction of the field. Figure 28-40 gives the period $T$ of the particle's motion versus the *inverse* of the field magnitude $B$. The vertical axis scale is set by $T_s = 40.0$ ns, and the horizontal axis scale is set by $B_s^{-1} = 5.0$ T$^{-1}$. What is the ratio $m/q$ of the particle's mass to the magnitude of its charge?

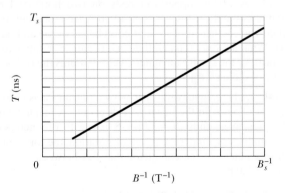

FIG. 28-40    Problem 23.

•24    In Fig. 28-41, a particle moves along a circle in a region of uniform magnetic field of magnitude $B = 4.00$ mT. The par-

ticle is either a proton or an electron (you must decide which). It experiences a magnetic force of magnitude $3.20 \times 10^{-15}$ N. What are (a) the particle's speed, (b) the radius of the circle, and (c) the period of the motion?  **GO**

FIG. 28-41
Problem 24.

•25    An alpha particle ($q = +2e, m = 4.00$ u) travels in a circular path of radius 4.50 cm in a uniform magnetic field with $B = 1.20$ T. Calculate (a) its speed, (b) its period of revolution, (c) its kinetic energy, and (d) the potential difference through which it would have to be accelerated to achieve this energy.

••26    A particle undergoes uniform circular motion of radius 26.1 $\mu$m in a uniform magnetic field. The magnetic force on the particle has a magnitude of $1.60 \times 10^{-17}$ N. What is the kinetic energy of the particle?

••27    An electron follows a helical path in a uniform magnetic field of magnitude 0.300 T. The pitch of the path is 6.00 $\mu$m, and the magnitude of the magnetic force on the electron is $2.00 \times 10^{-15}$ N. What is the electron's speed?

••28    In Fig. 28-42, a charged particle moves into a region of uniform magnetic field $\vec{B}$, goes through half a circle, and then exits that region. The particle is either a proton or an electron (you must decide which). It spends 130 ns in the region.

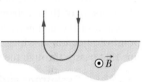

FIG. 28-42    Problem 28.

(a) What is the magnitude of $\vec{B}$? (b) If the particle is sent back through the magnetic field (along the same initial path) but with 2.00 times its previous kinetic energy, how much time does it spend in the field during this trip?

••29    A positron with kinetic energy 2.00 keV is projected into a uniform magnetic field $\vec{B}$ of magnitude 0.100 T, with its velocity vector making an angle of 89.0° with $\vec{B}$. Find (a) the period, (b) the pitch $p$, and (c) the radius $r$ of its helical path.  **SSM  WWW**

••30    An electron follows a helical path in a uniform magnetic field given by $\vec{B} = (20\hat{i} - 50\hat{j} - 30\hat{k})$ mT. At time $t = 0$, the electron's velocity is given by $\vec{v} = (20\hat{i} - 30\hat{j} + 50\hat{k})$ m/s. (a) What is the angle $\phi$ between $\vec{v}$ and $\vec{B}$? The electron's velocity changes with time. Do (b) its speed and (c) the angle $\phi$ change with time? (d) What is the radius of the helical path?

••31    A certain commercial mass spectrometer (see Sample Problem 28-3) is used to separate uranium ions of mass $3.92 \times 10^{-25}$ kg and charge $3.20 \times 10^{-19}$ C from related species. The ions are accelerated through a potential difference of 100 kV and then pass into a uniform magnetic field, where they are bent in a path of radius 1.00 m. After traveling through 180° and passing through a slit of width 1.00 mm and height 1.00 cm, they are collected in a cup. (a) What is the magnitude of the (perpendicular) magnetic field in the separator? If the machine is used to separate out 100 mg of material per hour, calculate (b) the current of the desired ions in the machine and (c) the thermal energy produced in the cup in 1.00 h.

••32    In Fig. 28-43, an electron with an initial kinetic energy of 4.0 keV enters region 1 at time $t = 0$. That region contains a uniform magnetic field directed into the page, with magnitude

0.010 T. The electron goes through a half-circle and then exits region 1, headed toward region 2 across a gap of 25.0 cm. There is an electric potential difference $\Delta V = 2000$ V across the gap, with a polarity such that the electron's speed increases uniformly as it traverses the gap. Region 2 contains a uniform magnetic field directed out of the page, with magnitude 0.020 T. The electron goes through a half-circle and then leaves region 2. At what time $t$ does it leave? **GO**

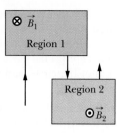

FIG. 28-43
Problem 32.

**••33** A particular type of fundamental particle decays by transforming into an electron $e^-$ and a positron $e^+$. Suppose the decaying particle is at rest in a uniform magnetic field $\vec{B}$ of magnitude 3.53 mT and the $e^-$ and $e^+$ move away from the decay point in paths lying in a plane perpendicular to $\vec{B}$. How long after the decay do the $e^-$ and $e^+$ collide?

**••34** A source injects an electron of speed $v = 1.5 \times 10^7$ m/s into a uniform magnetic field of magnitude $B = 1.0 \times 10^{-3}$ T. The velocity of the electron makes an angle $\theta = 10°$ with the direction of the magnetic field. Find the distance $d$ from the point of injection at which the electron next crosses the field line that passes through the injection point.

### sec. 28-7 Cyclotrons and Synchrotrons

**••35** Estimate the total path length traveled by a deuteron in the cyclotron of Sample Problem 28-5 during the (entire) acceleration process. Assume that the accelerating potential between the dees is 80 kV.

**••36** In a certain cyclotron a proton moves in a circle of radius 0.500 m. The magnitude of the magnetic field is 1.20 T. (a) What is the oscillator frequency? (b) What is the kinetic energy of the proton, in electron-volts?

**••37** A proton circulates in a cyclotron, beginning approximately at rest at the center. Whenever it passes through the gap between dees, the electric potential difference between the dees is 200 V. (a) By how much does its kinetic energy increase with each passage through the gap? (b) What is its kinetic energy as it completes 100 passes through the gap? Let $r_{100}$ be the radius of the proton's circular path as it completes those 100 passes and enters a dee, and let $r_{101}$ be its next radius, as it enters a dee the next time. (c) By what percentage does the radius increase when it changes from $r_{100}$ to $r_{101}$? That is, what is

$$\text{percentage increase} = \frac{r_{101} - r_{100}}{r_{100}} 100\%?$$

**••38** A cyclotron with dee radius 53.0 cm is operated at an oscillator frequency of 12.0 MHz to accelerate protons. (a) What magnitude $B$ of magnetic field is required to achieve resonance? (b) At that field magnitude, what is the kinetic energy of a proton emerging from the cyclotron? Suppose, instead, that $B = 1.57$ T. (c) What oscillator frequency is required to achieve resonance now? (d) At that frequency, what is the kinetic energy of an emerging proton?

### sec. 28-8 Magnetic Force on a Current-Carrying Wire

**•39** A 13.0 g wire of length $L = 62.0$ cm is suspended by a pair of flexible leads in a uniform magnetic field of magnitude

0.440 T (Fig. 28-44). What are the (a) magnitude and (b) direction (left or right) of the current required to remove the tension in the supporting leads? **ILW**

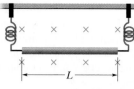

FIG. 28-44   Problem 39.

**•40** The bent wire shown in Fig. 28-45 lies in a uniform magnetic field. Each straight section is 2.0 m long and makes an angle of $\theta = 60°$ with the $x$ axis, and the wire carries a current of 2.0 A. What is the net magnetic force on the wire in unit-vector notation if the magnetic field is given by (a) $4.0\hat{k}$ T and (b) $4.0\hat{i}$ T?

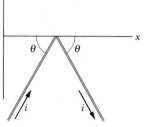

FIG. 28-45   Problem 40.

**•41** A horizontal power line carries a current of 5000 A from south to north. Earth's magnetic field ($60.0\ \mu$T) is directed toward the north and inclined downward at 70.0° to the horizontal. Find the (a) magnitude and (b) direction of the magnetic force on 100 m of the line due to Earth's field. **SSM**

**•42** A wire 1.80 m long carries a current of 13.0 A and makes an angle of 35.0° with a uniform magnetic field of magnitude $B = 1.50$ T. Calculate the magnetic force on the wire.

**••43** A wire 50.0 cm long carries a 0.500 A current in the positive direction of an $x$ axis through a magnetic field $\vec{B} = (3.00\ \text{mT})\hat{j} + (10.0\ \text{mT})\hat{k}$. In unit-vector notation, what is the magnetic force on the wire?

**••44** In Fig. 28-46, a metal wire of mass $m = 24.1$ mg can slide with negligible friction on two horizontal parallel rails separated by distance $d = 2.56$ cm. The track lies in a vertical uniform magnetic field of magnitude 56.3 mT. At time $t = 0$, device $G$ is connected to the rails, producing a constant current $i = 9.13$ mA in the wire and rails (even as the wire moves). At $t = 61.1$ ms, what are the wire's (a) speed and (b) direction of motion (left or right)?

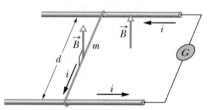

FIG. 28-46   Problem 44.

**•••45** A 1.0 kg copper rod rests on two horizontal rails 1.0 m apart and carries a current of 50 A from one rail to the other. The coefficient of static friction between rod and rails is 0.60. What are the (a) magnitude and (b) angle (relative to the vertical) of the smallest magnetic field that puts the rod on the verge of sliding? **GO**

**•••46** A long, rigid conductor, lying along an $x$ axis, carries a current of 5.0 A in the negative $x$ direction. A magnetic field $\vec{B}$ is present, given by $\vec{B} = 3.0\hat{i} + 8.0x^2\hat{j}$, with $x$ in meters and $\vec{B}$ in milliteslas. Find, in unit-vector notation, the force on the 2.0 m segment of the conductor that lies between $x = 1.0$ m and $x = 3.0$ m.

### sec. 28-9 Torque on a Current Loop

**•47** Figure 28-47 shows a rectangular 20-turn coil of wire, of dimensions 10 cm by 5.0 cm. It carries a current of 0.10 A and is hinged along one long side. It is mounted in the $xy$ plane, at angle $\theta = 30°$ to the direction of a uniform magnetic field of magnitude 0.50 T. In unit-vector notation, what is the torque acting on the coil about the hinge line? **SSM**

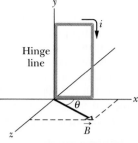
FIG. 28-47   Problem 47.

•**48**  A single-turn current loop, carrying a current of 4.00 A, is in the shape of a right triangle with sides 50.0, 120, and 130 cm. The loop is in a uniform magnetic field of magnitude 75.0 mT whose direction is parallel to the current in the 130 cm side of the loop. What is the magnitude of the magnetic force on (a) the 130 cm side, (b) the 50.0 cm side, and (c) the 120 cm side? (d) What is the magnitude of the net force on the loop?

••**49**  Figure 28-48 shows a wire ring of radius $a = 1.8$ cm that is perpendicular to the general direction of a radially symmetric, diverging magnetic field. The magnetic field at the ring is everywhere of the same magnitude $B = 3.4$ mT, and its direction at the ring everywhere makes an angle $\theta = 20°$ with a normal to the plane of the ring. The twisted lead wires have no effect on the problem. Find the magnitude of the force the field exerts on the ring if the ring carries a current $i = 4.6$ mA.

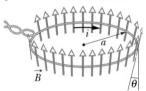
FIG. 28-48   Problem 49.

••**50**  In Fig. 28-49, a rectangular loop carrying current lies in the plane of a uniform magnetic field of magnitude 0.040 T. The loop consists of a single turn of flexible conducting wire that is wrapped around a flexible mount such that the dimensions of the rectangle can be changed. (The total length of the wire is not changed.) As edge length $x$ is varied from approximately zero to its maximum value of approximately 4.0 cm, the magnitude $\tau$ of the torque on the loop changes. The maximum value of $\tau$ is $4.80 \times 10^{-8}$ N·m. What is the current in the loop?

FIG. 28-49
Problem 50.

••**51**  The coil of a certain galvanometer (see Sample Problem 28-7) has a resistance of 75.3 Ω; its needle shows a full-scale deflection when a current of 1.62 mA passes through the coil. (a) Determine the value of the auxiliary resistance required to convert the galvanometer to a voltmeter that reads 1.00 V at full-scale deflection. (b) Should this resistance be connected in series or in parallel with the galvanometer? (c) Determine the value of the auxiliary resistance required to convert the galvanometer to an ammeter that reads 50.0 mA at full-scale deflection. (d) Should this resistance be connected in series or in parallel?

••**52**  An electron moves in a circle of radius $r = 5.29 \times 10^{-11}$ m with speed $2.19 \times 10^6$ m/s. Treat the circular path as a current loop with a constant current equal to the ratio of the electron's charge magnitude to the period of the motion. If the circle lies in a uniform magnetic field of magnitude $B = 7.10$ mT, what is the maximum possible magnitude of the torque produced on the loop by the field?

••**53**  Figure 28-50 shows a wood cylinder of mass $m = 0.250$ kg and length $L = 0.100$ m, with $N = 10.0$ turns of wire wrapped around it longitudinally, so that the plane of the wire coil contains the long central axis of the cylinder. The cylinder is released on a plane inclined at an angle $\theta$ to the horizontal, with the plane of the coil parallel to the incline plane. If there is a vertical uniform magnetic field of magnitude 0.500 T, what is the least current $i$ through the coil that keeps the cylinder from rolling down the plane?

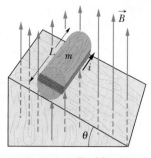
FIG. 28-50   Problem 53.

### sec. 28-10   The Magnetic Dipole Moment

•**54**  A circular wire loop of radius 15.0 cm carries a current of 2.60 A. It is placed so that the normal to its plane makes an angle of 41.0° with a uniform magnetic field of magnitude 12.0 T. (a) Calculate the magnitude of the magnetic dipole moment of the loop. (b) What is the magnitude of the torque acting on the loop?

•**55**  A circular coil of 160 turns has a radius of 1.90 cm. (a) Calculate the current that results in a magnetic dipole moment of magnitude 2.30 A·m². (b) Find the maximum magnitude of the torque that the coil, carrying this current, can experience in a uniform 35.0 mT magnetic field.  **SSM**

•**56**  The magnetic dipole moment of Earth has magnitude $8.00 \times 10^{22}$ J/T. Assume that this is produced by charges flowing in Earth's molten outer core. If the radius of their circular path is 3500 km, calculate the current they produce.

•**57**  A current loop, carrying a current of 5.0 A, is in the shape of a right triangle with sides 30, 40, and 50 cm. The loop is in a uniform magnetic field of magnitude 80 mT whose direction is parallel to the current in the 50 cm side of the loop. Find the magnitude of (a) the magnetic dipole moment of the loop and (b) the torque on the loop.

•**58**  A magnetic dipole with a dipole moment of magnitude 0.020 J/T is released from rest in a uniform magnetic field of magnitude 52 mT. The rotation of the dipole due to the magnetic force on it is unimpeded. When the dipole rotates through the orientation where its dipole moment is aligned with the magnetic field, its kinetic energy is 0.80 mJ. (a) What is the initial angle between the dipole moment and the magnetic field? (b) What is the angle when the dipole is next (momentarily) at rest?

•**59**  Two concentric, circular wire loops, of radii $r_1 = 20.0$ cm and $r_2 = 30.0$ cm, are located in an $xy$ plane; each carries a clockwise current of 7.00 A (Fig. 28-51). (a) Find the magnitude of the net magnetic dipole moment of the system. (b) Repeat for reversed current in the inner loop.  **SSM**

••**60**  Figure 28-52 gives the potential energy $U$ of a mag-

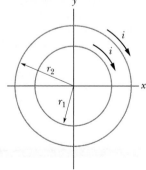
FIG. 28-51   Problem 59.

netic dipole in an external magnetic field $\vec{B}$, as a function of angle $\phi$ between the directions of $\vec{B}$ and the dipole moment. The vertical axis scale is set by $U_s = 2.0 \times 10^{-4}$ J. The dipole can be rotated about an axle with negligible friction so as to change $\phi$. Counterclockwise rotation from $\phi = 0$ yields positive values of $\phi$, and clockwise

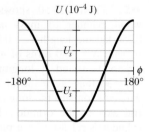

**FIG. 28-52** Problem 60.

rotations yield negative values. The dipole is to be released at angle $\phi = 0$ with a rotational kinetic energy of $6.7 \times 10^{-4}$ J, so that it rotates counterclockwise. To what maximum value of $\phi$ will it rotate? (In the language of Section 8-6, what value $\phi$ is the turning point in the potential well of Fig. 28-52?) **GO**

**••61** A circular loop of wire having a radius of 8.0 cm carries a current of 0.20 A. A vector of unit length and parallel to the dipole moment $\vec{\mu}$ of the loop is given by $0.60\hat{i} - 0.80\hat{j}$. If the loop is located in a uniform magnetic field given by $\vec{B} = (0.25$ T$)\hat{i} + (0.30$ T$)\hat{k}$, find (a) the torque on the loop (in unit-vector notation) and (b) the magnetic potential energy of the loop.

**••62** Figure 28-53 shows a current loop ABCDEFA carrying a current $i = 5.00$ A. The sides of the loop are parallel to the coordinate axes shown, with $AB = 20.0$ cm, $BC = 30.0$ cm, and $FA = 10.0$ cm. In unit-vector notation, what is the magnetic dipole moment of this loop? (*Hint:* Imagine equal and opposite currents $i$ in the line segment $AD$; then treat the two rectangular loops $ABCDA$ and $ADEFA$.)

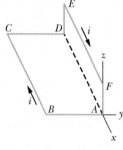

**FIG. 28-53**
Problem 62.

**••63** A wire of length 25.0 cm carrying a current of 4.51 mA is to be formed into a circular coil and placed in a uniform magnetic field $\vec{B}$ of magnitude 5.71 mT. If the torque on the coil from the field is maximized, what are (a) the angle between $\vec{B}$ and the coil's magnetic dipole moment and (b) the number of turns in the coil? (c) What is the magnitude of that maximum torque? **SSM ILW**

**••64** In Fig. 28-54a, two concentric coils, lying in the same plane, carry currents in opposite directions. The current in the larger coil 1 is fixed. Current $i_2$ in coil 2 can be varied. Figure 28-54b gives the net magnetic moment of the two-coil system as a function of $i_2$. The vertical axis scale is set by $\mu_{net,s} = 2.0 \times 10^{-5}$ A·m$^2$, and the horizontal axis scale is set by $i_{2s} = 10.0$ mA. If the current in coil 2 is then reversed, what is the magnitude of the net magnetic moment of the two-coil system when $i_2 = 7.0$ mA?

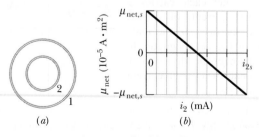

**FIG. 28-54** Problem 64.

**••65** The coil in Fig. 28-55 carries current $i = 2.00$ A in the direction indicated, is parallel to an $xz$ plane, has 3.00 turns and an area of $4.00 \times 10^{-3}$ m$^2$, and lies in a uniform magnetic field

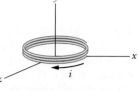

**FIG. 28-55** Problem 65.

$\vec{B} = (2.00\hat{i} - 3.00\hat{j} - 4.00\hat{k})$ mT. What are (a) the magnetic potential energy of the coil–magnetic field system and (b) the magnetic torque (in unit-vector notation) on the coil? **SSM**

**Additional Problems**

**66** A wire lying along a $y$ axis from $y = 0$ to $y = 0.250$ m carries a current of 2.00 mA in the negative direction of the axis. The wire fully lies in a nonuniform magnetic field given by $\vec{B} = (0.300$ T/m$)y\hat{i} + (0.400$ T/m$)y\hat{j}$. In unit-vector notation, what is the magnetic force on the wire?

**67** Physicist S. A. Goudsmit devised a method for measuring the mass of heavy ions by timing their period of revolution in a known magnetic field. A singly charged ion of iodine makes 7.00 rev in a 45.0 mT field in 1.29 ms. Calculate its mass in atomic mass units.

**68** An electron in an old-fashioned TV camera tube is moving at $7.20 \times 10^6$ m/s in a magnetic field of strength 83.0 mT. What is the (a) maximum and (b) minimum magnitude of the force acting on the electron due to the field? (c) At one point the electron has an acceleration of magnitude $4.90 \times 10^{14}$ m/s$^2$. What is the angle between the electron's velocity and the magnetic field?

**69** A stationary circular wall clock has a face with a radius of 15 cm. Six turns of wire are wound around its perimeter; the wire carries a current of 2.0 A in the clockwise direction. The clock is located where there is a constant, uniform external magnetic field of magnitude 70 mT (but the clock still keeps perfect time). At exactly 1:00 P.M., the hour hand of the clock points in the direction of the external magnetic field. (a) After how many minutes will the minute hand point in the direction of the torque on the winding due to the magnetic field? (b) Find the torque magnitude.

**70** In a Hall-effect experiment, a current of 3.0 A sent lengthwise through a conductor 1.0 cm wide, 4.0 cm long, and 10 $\mu$m thick produces a transverse (across the width) Hall potential difference of 10 $\mu$V when a magnetic field of 1.5 T is passed perpendicularly through the thickness of the conductor. From these data, find (a) the drift velocity of the charge carriers and (b) the number density of charge carriers. (c) Show on a diagram the polarity of the Hall potential difference with assumed current and magnetic field directions, assuming also that the charge carriers are electrons.

**71** Atom 1 of mass 35 u and atom 2 of mass 37 u are both singly ionized with a charge of $+e$. After being introduced into a mass spectrometer (Fig. 28-14) and accelerated from rest through a potential difference $V = 7.3$ kV, each ion follows a circular path in a uniform magnetic field of magnitude $B = 0.50$ T. What is the distance $\Delta x$ between the points where the ions strike the detector?

**72** An electron with kinetic energy 2.5 keV moving along the positive direction of an $x$ axis enters a region in which a uniform electric field of magnitude 10 kV/m is in the negative direction of the $y$ axis. A uniform magnetic field $\vec{B}$ is to be set up to keep the electron moving along the $x$ axis, and the direction of $\vec{B}$ is to be chosen to minimize the required magnitude of $\vec{B}$. In unit-vector notation, what $\vec{B}$ should be set up?

**73** In Fig. 28-56, an electron moves at speed $v = 100$ m/s along an $x$ axis through uniform electric and magnetic fields. The magnetic field $\vec{B}$ is directed into the page and has magnitude 5.00 T. In unit-vector notation, what is the electric field? SSM

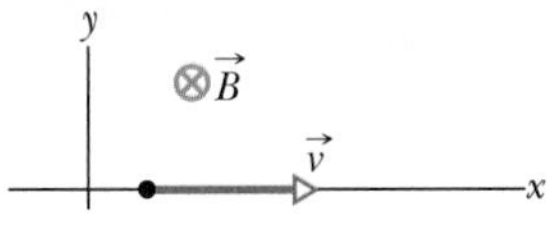

$\otimes \vec{B}$

$\vec{v}$

FIG. 28-56 Problem 73.

**74** A beam of electrons whose kinetic energy is $K$ emerges from a thin-foil "window" at the end of an accelerator tube. A metal plate at distance $d$ from this window is perpendicular to the direction of the emerging beam (Fig. 28-57). (a) Show that we can prevent the beam from hitting the plate if we apply a uniform magnetic field $\vec{B}$ such that

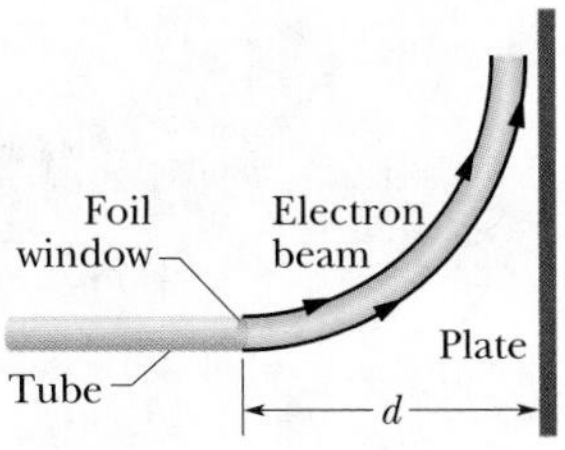

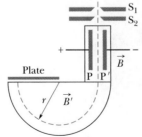

FIG. 28-57 Problem 74.

$$B \geq \sqrt{\frac{2mK}{e^2 d^2}},$$

in which $m$ and $e$ are the electron mass and charge. (b) How should $\vec{B}$ be oriented?

**75** A proton, a deuteron ($q = +e$, $m = 2.0$ u), and an alpha particle ($q = +2e$, $m = 4.0$ u) are accelerated through the same potential difference and then enter the same region of uniform magnetic field $\vec{B}$, moving perpendicular to $\vec{B}$. What is the ratio of (a) the proton's kinetic energy $K_p$ to the alpha particle's kinetic energy $K_\alpha$ and (b) the deuteron's kinetic energy $K_d$ to $K_\alpha$? If the radius of the proton's circular path is 10 cm, what is the radius of (c) the deuteron's path and (d) the alpha particle's path? SSM

**76** A proton of charge $+e$ and mass $m$ enters a uniform magnetic field $\vec{B} = B\hat{i}$ with an initial velocity $\vec{v} = v_{0x}\hat{i} + v_{0y}\hat{j}$. Find an expression in unit-vector notation for its velocity $\vec{v}$ at any later time $t$.

**77** A particle of mass 6.0 g moves at 4.0 km/s in an $xy$ plane, in a region with a uniform magnetic field given by $5.0\hat{i}$ mT. At one instant, when the particle's velocity is directed 37° counterclockwise from the positive direction of the $x$ axis, the magnetic force on the particle is $0.48\hat{k}$ N. What is the particle's charge? SSM

**78** Bainbridge's mass spectrometer, shown in Fig. 28-58, separates ions having the same velocity. The ions, after entering through slits, $S_1$ and $S_2$, pass through a velocity selector composed of an electric field produced by the charged plates P and P', and a magnetic field $\vec{B}$ perpendicular to the electric field and the ion path. The ions that then pass undeviated through the crossed $\vec{E}$ and $\vec{B}$ fields enter into a region where a second magnetic field $\vec{B}'$ exists, where they are made to follow circular paths. A photographic plate (or a modern detector) registers their arrival. Show

that, for the ions, $q/m = E/rBB'$, where $r$ is the radius of the circular orbit.

**79** At one instant, $\vec{v} = (-2.00\hat{i} + 4.00\hat{j} - 6.00\hat{k})$ m/s is the velocity of a proton in a uniform magnetic field $\vec{B} = (2.00\hat{i} - 4.00\hat{j} + 8.00\hat{k})$ mT. At that instant, what are (a) the magnetic force $\vec{F}$ on the proton, in unit-vector notation, (b) the angle between $\vec{v}$ and $\vec{F}$, and (c) the angle between $\vec{v}$ and $\vec{B}$?

**80** (a) In Fig. 28-8, show that the ratio of the Hall electric field magnitude $E$ to the magnitude $E_C$ of the electric field responsible for moving charge (the current) along the length of the strip is

$$\frac{E}{E_C} = \frac{B}{ne\rho},$$

where $\rho$ is the resistivity of the material and $n$ is the number density of the charge carriers. (b) Compute this ratio numerically for Problem 13. (See Table 26-1.)

**81** At time $t = 0$, an electron with kinetic energy 12 keV moves through $x = 0$ in the positive direction of an $x$ axis that is parallel to the horizontal component of Earth's magnetic field $\vec{B}$. The field's vertical component is downward and has magnitude 55.0 $\mu$T. (a) What is the magnitude of the electron's acceleration due to $\vec{B}$? (b) What is the electron's distance from the $x$ axis when the electron reaches coordinate $x = 20$ cm? SSM

**82** An electron has velocity $\vec{v} = (32\hat{i} + 40\hat{j})$ km/s as it enters a uniform magnetic field $\vec{B} = 60\hat{i}$ $\mu$T. What are (a) the radius of the helical path taken by the electron and (b) the pitch of that path? (c) To an observer looking into the magnetic field region from the entrance point of the electron, does the electron spiral clockwise or counterclockwise as it moves?

**83** A proton, a deuteron ($q = +e$, $m = 2.0$ u), and an alpha particle ($q = +2e$, $m = 4.0$ u) all having the same kinetic energy enter a region of uniform magnetic field $\vec{B}$, moving perpendicular to $\vec{B}$. What is the ratio of (a) the radius $r_d$ of the deuteron path to the radius $r_p$ of the proton path and (b) the radius $r_\alpha$ of the alpha particle path to $r_p$?

**84** A particle with charge 2.0 C moves through a uniform magnetic field. At one instant the velocity of the particle is $(2.0\hat{i} + 4.0\hat{j} + 6.0\hat{k})$ m/s and the magnetic force on the particle is $(4.0\hat{i} - 20\hat{j} + 12\hat{k})$ N. The $x$ and $y$ components of the magnetic field are equal. What is $\vec{B}$?

**85** A 5.0 $\mu$C particle moves through a region containing the magnetic field $-20\hat{i}$ mT and the electric field $300\hat{j}$ V/m. At one instant the velocity of the particle is $(17\hat{i} - 11\hat{j} + 7.0\hat{k})$ km/s. At that instant and in unit-vector notation, what is the net electromagnetic force (the sum of the electric and magnetic forces) on the particle?

**86** A wire lying along an $x$ axis from $x = 0$ to $x = 1.00$ m carries a current of 3.00 A in the positive $x$ direction. The wire is immersed in a nonuniform magnetic field given by $\vec{B} = (4.00$ T/m$^2)x^2\hat{i} - (0.600$ T/m$^2)x^2\hat{j}$. In unit-vector notation, what is the magnetic force on the wire?

**87** Prove that the relation $\tau = NiAB \sin \theta$ holds not only for the rectangular loop of Fig. 28-22 but also for a closed loop of any shape. (*Hint:* Replace the loop of arbitrary shape with an assembly of adjacent long, thin, approximately rectangular loops that are nearly equivalent to the loop of arbitrary shape as far as the distribution of current is concerned.)

Foil window

Electron beam

Plate

Tube

$d$

Plate

$S_1$

$S_2$

P P'

$\vec{B}$

$r$ $\vec{B}'$

FIG. 28-58 Problem 78.

When you read this sentence, a certain region of your brain is activated. When you smell a rose or feel a pencil in your grip, other regions are activated. One of the best ways to determine which regions are activated is to detect the magnetic field produced by the activation. The apparatus in the photograph can detect the magnetic field set up by a person's brain so that a map of brain activity can be correlated with what the person does. However, there are no magnetic materials in the brain.

## So, how can brain activation produce a magnetic field?

The answer is in this chapter.

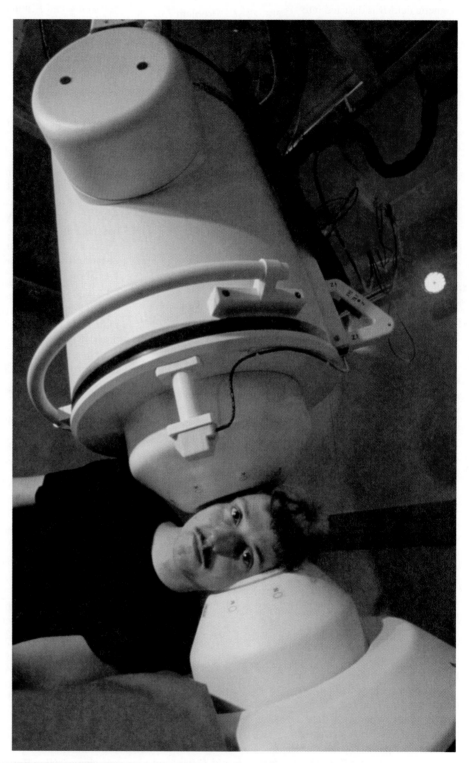

Jurgen Scriba/Photo Researchers

One basic observation of physics is that a moving charged particle produces a magnetic field around itself. Thus a current of moving charged particles produces a magnetic field around the current. This feature of *electromagnetism,* which is the combined study of electric and magnetic effects, came as a surprise to the people who discovered it. Surprise or not, this feature has become enormously important in everyday life because it is the basis of countless electromagnetic devices. For example, a magnetic field produced by an electric current can be found in machines that record or read magnetically encoded tapes and disks such as old-fashioned computer floppies, cassette audio tapes, and videotapes. Such a field can also be found in magnetically lifted trains (maglev trains) and other devices used to lift heavy loads.

Our first step in this chapter is to determine the magnetic field due to the current in a very small section of current-carrying wire. Then we shall calculate the magnetic field due to the entire wire for several different arrangements of the wire.

## 29-2 | Calculating the Magnetic Field Due to a Current

Figure 29-1 shows a wire of arbitrary shape carrying a current $i$. We want to find the magnetic field $\vec{B}$ at a nearby point $P$. We first mentally divide the wire into differential elements $ds$ and then define for each element a length vector $d\vec{s}$ that has length $ds$ and whose direction is the direction of the current in $ds$. We can then define a differential *current-length element* to be $i\,d\vec{s}$; we wish to calculate the field $d\vec{B}$ produced at $P$ by a typical current-length element. From experiment we find that magnetic fields, like electric fields, can be superimposed to find a net field. Thus, we can calculate the net field $\vec{B}$ at $P$ by summing, via integration, the contributions $d\vec{B}$ from all the current-length elements. However, this summation is more challenging than the process associated with electric fields because of a complexity; whereas a charge element $dq$ producing an electric field is a scalar, a current-length element $i\,d\vec{s}$ producing a magnetic field is a vector, being the product of a scalar and a vector.

The magnitude of the field $d\vec{B}$ produced at point $P$ at distance $r$ by a current-length element $i\,d\vec{s}$ turns out to be

$$dB = \frac{\mu_0}{4\pi}\frac{i\,ds\sin\theta}{r^2}, \tag{29-1}$$

where $\theta$ is the angle between the directions of $d\vec{s}$ and $\hat{r}$, a unit vector that points from $ds$ toward $P$. Symbol $\mu_0$ is a constant, called the *permeability constant,* whose value is defined to be exactly

$$\mu_0 = 4\pi \times 10^{-7}\,\text{T}\cdot\text{m/A} \approx 1.26 \times 10^{-6}\,\text{T}\cdot\text{m/A}. \tag{29-2}$$

The direction of $d\vec{B}$, shown as being into the page in Fig. 29-1, is that of the cross product $d\vec{s} \times \hat{r}$. We can therefore write Eq. 29-1 in vector form as

$$d\vec{B} = \frac{\mu_0}{4\pi}\frac{i\,d\vec{s} \times \hat{r}}{r^2} \qquad \text{(Biot–Savart law)}. \tag{29-3}$$

This vector equation and its scalar form, Eq. 29-1, are known as the **law of Biot and Savart** (rhymes with "Leo and bazaar"). The law, which is experimentally deduced, is an inverse-square law. We shall use this law to calculate the net magnetic field $\vec{B}$ produced at a point by various distributions of current.

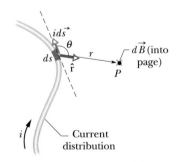

FIG. 29-1   A current-length element $i\,d\vec{s}$ produces a differential magnetic field $d\vec{B}$ at point $P$. The green $\times$ (the tail of an arrow) at the dot for point $P$ indicates that $d\vec{B}$ is directed *into* the page there.

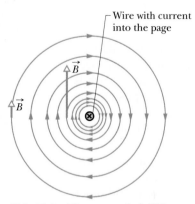

Wire with current into the page

$\vec{B}$

$\vec{B}$

**FIG. 29-2**   The magnetic field lines produced by a current in a long straight wire form concentric circles around the wire. Here the current is into the page, as indicated by the ×.

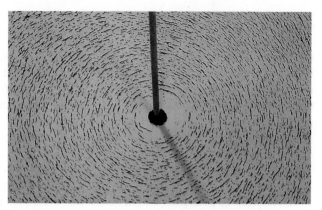

**FIG. 29-3**   Iron filings that have been sprinkled onto cardboard collect in concentric circles when current is sent through the central wire. The alignment, which is along magnetic field lines, is caused by the magnetic field produced by the current. *(Courtesy Education Development Center)*

(a)

$\vec{B}$

$i$

$i$

$\vec{B}$

(b)

**FIG. 29-4**   A right-hand rule gives the direction of the magnetic field due to a current in a wire. (a) The situation of Fig. 29-2, seen from the side. The magnetic field $\vec{B}$ at any point to the left of the wire is perpendicular to the dashed radial line and directed into the page, in the direction of the fingertips, as indicated by the ×. (b) If the current is reversed, $\vec{B}$ at any point to the left is still perpendicular to the dashed radial line but now is directed out of the page, as indicated by the dot.

## Magnetic Field Due to a Current in a Long Straight Wire

Shortly we shall use the law of Biot and Savart to prove that the magnitude of the magnetic field at a perpendicular distance $R$ from a long (infinite) straight wire carrying a current $i$ is given by

$$B = \frac{\mu_0 i}{2\pi R} \qquad \text{(long straight wire).} \qquad (29\text{-}4)$$

The field magnitude $B$ in Eq. 29-4 depends only on the current and the perpendicular distance $R$ of the point from the wire. We shall show in our derivation that the field lines of $\vec{B}$ form concentric circles around the wire, as Fig. 29-2 shows and as the iron filings in Fig. 29-3 suggest. The increase in the spacing of the lines in Fig. 29-2 with increasing distance from the wire represents the $1/R$ decrease in the magnitude of $\vec{B}$ predicted by Eq. 29-4. The lengths of the two vectors $\vec{B}$ in the figure also show the $1/R$ decrease.

Here is a simple right-hand rule for finding the direction of the magnetic field set up by a current-length element, such as a section of a long wire:

> *Right-hand rule:* Grasp the element in your right hand with your extended thumb pointing in the direction of the current. Your fingers will then naturally curl around in the direction of the magnetic field lines due to that element.

The result of applying this right-hand rule to the current in the straight wire of Fig. 29-2 is shown in a side view in Fig. 29-4a. To determine the direction of the magnetic field $\vec{B}$ set up at any particular point by this current, mentally wrap your right hand around the wire with your thumb in the direction of the current. Let your fingertips pass through the point; their direction is then the direction of the magnetic field at that point. In the view of Fig. 29-2, $\vec{B}$ at any point is *tangent to a magnetic field line;* in the view of Fig. 29-4, it is *perpendicular to a dashed radial line connecting the point and the current.*

## Proof of Equation 29-4

Figure 29-5, which is just like Fig. 29-1 except that now the wire is straight and of infinite length, illustrates the task at hand. We seek the field $\vec{B}$ at point $P$, a perpendicular distance $R$ from the wire. The magnitude of the differential magnetic field produced at $P$ by the current-length element $i\,d\vec{s}$ located a distance $r$ from $P$ is given by Eq. 29-1:

$$dB = \frac{\mu_0}{4\pi}\frac{i\,ds\,\sin\theta}{r^2}.$$

The direction of $d\vec{B}$ in Fig. 29-5 is that of the vector $d\vec{s} \times \hat{r}$—namely, directly into the page.

Note that $d\vec{B}$ at point $P$ has this same direction for all the current-length elements into which the wire can be divided. Thus, we can find the magnitude of the magnetic field produced at $P$ by the current-length elements in the upper half of the infinitely long wire by integrating $dB$ in Eq. 29-1 from 0 to $\infty$.

Now consider a current-length element in the lower half of the wire, one that is as far below $P$ as $d\vec{s}$ is above $P$. By Eq. 29-3, the magnetic field produced at $P$ by this current-length element has the same magnitude and direction as that from element $i\,d\vec{s}$ in Fig. 29-5. Further, the magnetic field produced by the lower half of the wire is exactly the same as that produced by the upper half. To find the magnitude of the *total* magnetic field $\vec{B}$ at $P$, we need only multiply the result of our integration by 2. We get

$$B = 2\int_0^\infty dB = \frac{\mu_0 i}{2\pi}\int_0^\infty \frac{\sin\theta\,ds}{r^2}. \tag{29-5}$$

The variables $\theta, s,$ and $r$ in this equation are not independent; Fig. 29-5 shows that they are related by

$$r = \sqrt{s^2 + R^2}$$

and

$$\sin\theta = \sin(\pi - \theta) = \frac{R}{\sqrt{s^2 + R^2}}.$$

With these substitutions and integral 19 in Appendix E, Eq. 29-5 becomes

$$B = \frac{\mu_0 i}{2\pi}\int_0^\infty \frac{R\,ds}{(s^2 + R^2)^{3/2}}$$

$$= \frac{\mu_0 i}{2\pi R}\left[\frac{s}{(s^2 + R^2)^{1/2}}\right]_0^\infty = \frac{\mu_0 i}{2\pi R}, \tag{29-6}$$

which is the relation we set out to prove. Note that the magnetic field at $P$ due to either the lower half or the upper half of the infinite wire in Fig. 29-5 is half this value; that is,

$$B = \frac{\mu_0 i}{4\pi R} \quad \text{(semi-infinite straight wire).} \tag{29-7}$$

### Magnetic Field Due to a Current in a Circular Arc of Wire

To find the magnetic field produced at a point by a current in a curved wire, we would again use Eq. 29-1 to write the magnitude of the field produced by a single current-length element, and we would again integrate to find the net field produced by all the current-length elements. That integration can be difficult, depending on the shape of the wire; it is fairly straightforward, however, when the wire is a circular arc and the point is the center of curvature.

Figure 29-6a shows such an arc-shaped wire with central angle $\phi$, radius $R$, and center $C$, carrying current $i$. At $C$, each current-length element $i\,d\vec{s}$ of the wire produces a magnetic field of magnitude $dB$ given by Eq. 29-1. Moreover, as Fig. 29-6b shows, no matter where the element is located on the wire, the angle $\theta$ between the vectors $d\vec{s}$ and $\hat{r}$ is 90°; also, $r = R$. Thus, by substituting $R$ for $r$ and

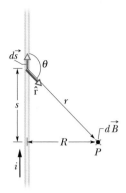

FIG. 29-5  Calculating the magnetic field produced by a current $i$ in a long straight wire. The field $d\vec{B}$ at $P$ associated with the current-length element $i\,d\vec{s}$ is directed into the page, as shown.

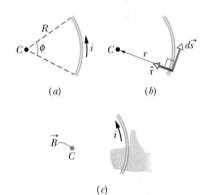

FIG. 29-6  (a) A wire in the shape of a circular arc with center $C$ carries current $i$. (b) For any element of wire along the arc, the angle between the directions of $d\vec{s}$ and $\hat{r}$ is 90°. (c) Determining the direction of the magnetic field at the center $C$ due to the current in the wire; the field is out of the page, in the direction of the fingertips, as indicated by the colored dot at $C$.

90° for $\theta$ in Eq. 29-1, we obtain

$$dB = \frac{\mu_0}{4\pi} \frac{i\, ds \sin 90°}{R^2} = \frac{\mu_0}{4\pi} \frac{i\, ds}{R^2}. \qquad (29\text{-}8)$$

The field at $C$ due to each current-length element in the arc has this magnitude.

An application of the right-hand rule anywhere along the wire (as in Fig. 29-6c) will show that all the differential fields $d\vec{B}$ have the same direction at $C$— directly out of the page. Thus, the total field at $C$ is simply the sum (via integration) of all the differential fields $d\vec{B}$. We use the identity $ds = R\, d\phi$ to change the variable of integration from $ds$ to $d\phi$ and obtain, from Eq. 29-8,

$$B = \int dB = \int_0^\phi \frac{\mu_0}{4\pi} \frac{iR\, d\phi}{R^2} = \frac{\mu_0 i}{4\pi R} \int_0^\phi d\phi.$$

Integrating, we find that

$$B = \frac{\mu_0 i \phi}{4\pi R} \qquad \text{(at center of circular arc).} \qquad (29\text{-}9)$$

Note that this equation gives us the magnetic field *only* at the center of curvature of a circular arc of current. When you insert data into the equation, you must be careful to express $\phi$ in radians rather than degrees. For example, to find the magnitude of the magnetic field at the center of a full circle of current, you would substitute $2\pi$ rad for $\phi$ in Eq. 29-9, finding

$$B = \frac{\mu_0 i (2\pi)}{4\pi R} = \frac{\mu_0 i}{2R} \qquad \text{(at center of full circle).} \qquad (29\text{-}10)$$

## Magnetic Field Due to Brain Activity

Scientists, medical researchers, and physiologists would like to understand how the human brain works. One of their new tools is *magnetoencephalography* (MEG), a procedure in which magnetic fields of a person's brain are monitored as the person performs a task, such as reading a word. The task activates part of the brain, such as the part that processes reading, causing weak electrical pulses to be sent along conducting paths between brain cells. As with any other current, each pulse produces a magnetic field.

The magnetic fields detected in MEG are probably produced by pulses along the walls of the fissures (crevices) on the brain surface (Fig. 29-7). Let's use Eq. 29-1 to estimate the magnitude of such a field at a point $P$ located a distance $r = 2$ cm from the pulse. Let the path of the pulse be tangent to the brain surface so that the angle $\theta$ in Eq. 29-1 is 90°. In a typical pulse, the current is $i = 10$ $\mu$A and the conducting path length is about 1 mm. Let's assume that this length is short enough to allow us to substitute 1 mm for the element length $ds$ in Eq. 29-1. Then Eq. 29-1 gives us

$$B = \frac{(4\pi \times 10^{-7}\ \text{T·m/A})}{4\pi} \frac{(10 \times 10^{-6}\ \text{A})(1 \times 10^{-3}\ \text{m})}{(2 \times 10^{-2}\ \text{m})^2} \sin 90°$$

$$= 2.5 \times 10^{-12}\ \text{T} \approx 3\ \text{pT}.$$

This is a *very* small magnetic field, more than a million times weaker than Earth's magnetic field at your location. Thus, if you want to detect the magnetic field of a brain, you cannot simply hold a compass near the brain and hope that brain activity will turn the compass needle. Instead, you need extremely sensitive instruments called SQUIDs (superconducting quantum interference devices) that can measure fields of magnitude less than 1 pT, and even then you must take care to eliminate other sources of varying magnetic fields in the detection area.

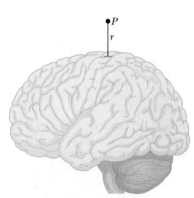

**FIG. 29-7** A pulse along a fissure wall on the brain surface produces a magnetic field at point $P$ at distance $r$.

## Sample Problem  29-1  Build your skill

The wire in Fig. 29-8a carries a current $i$ and consists of a circular arc of radius $R$ and central angle $\pi/2$ rad, and two straight sections whose extensions intersect the center $C$ of the arc. What magnetic field $\vec{B}$ does the current produce at $C$?

**KEY IDEAS**   We can find the magnetic field $\vec{B}$ at point $C$ by applying the Biot–Savart law of Eq. 29-3 to the wire. The application of Eq. 29-3 can be simplified by evaluating $\vec{B}$ separately for the three distinguishable sections of the wire—namely, (1) the straight section at the left, (2) the straight section at the right, and (3) the circular arc.

*Straight sections:* For any current-length element in section 1, the angle $\theta$ between $d\vec{s}$ and $\hat{r}$ is zero (Fig. 29-8b); so Eq. 29-1 gives us

$$dB_1 = \frac{\mu_0}{4\pi}\frac{i\,ds\,\sin\theta}{r^2} = \frac{\mu_0}{4\pi}\frac{i\,ds\,\sin 0}{r^2} = 0.$$

Thus, the current along the entire length of straight section 1 contributes no magnetic field at $C$:

$$B_1 = 0.$$

The same situation prevails in straight section 2, where the angle $\theta$ between $d\vec{s}$ and $\hat{r}$ for any current-length element is 180°. Thus,

$$B_2 = 0.$$

*Circular arc:* Application of the Biot–Savart law to evaluate the magnetic field at the center of a circular arc leads to Eq. 29-9 ($B = \mu_0 i\phi/4\pi R$). Here the central angle $\phi$ of the arc is $\pi/2$ rad. Thus from Eq. 29-9, the magnitude of the magnetic field $\vec{B}_3$ at the arc's center $C$ is

$$B_3 = \frac{\mu_0 i(\pi/2)}{4\pi R} = \frac{\mu_0 i}{8R}.$$

To find the direction of $\vec{B}_3$, we apply the right-hand rule displayed in Fig. 29-4. Mentally grasp the circular arc with your right hand as in Fig. 29-8c, with your

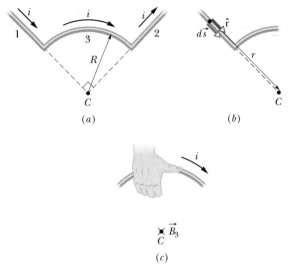

**FIG. 29-8**   (a) A wire consists of two straight sections (1 and 2) and a circular arc (3), and carries current $i$. (b) For a current-length element in section 1, the angle between $d\vec{s}$ and $\hat{r}$ is zero. (c) Determining the direction of magnetic field $\vec{B}_3$ at $C$ due to the current in the circular arc; the field is into the page there.

thumb in the direction of the current. The direction in which your fingers curl around the wire indicates the direction of the magnetic field lines around the wire. In the region of point $C$ (inside the circular arc), your fingertips point *into the plane* of the page. Thus, $\vec{B}_3$ is directed into that plane.

*Net field:* Generally, when we must combine two or more magnetic fields to find the net magnetic field, we must combine the fields as vectors and not simply add their magnitudes. Here, however, only the circular arc produces a magnetic field at point $C$. Thus, we can write the magnitude of the net field $\vec{B}$ as

$$B = B_1 + B_2 + B_3 = 0 + 0 + \frac{\mu_0 i}{8R} = \frac{\mu_0 i}{8R}. \quad \text{(Answer)}$$

The direction of $\vec{B}$ is the direction of $\vec{B}_3$—namely, into the plane of Fig. 29-8.

## Sample Problem  29-2  Build your skill

Figure 29-9a shows two long parallel wires carrying currents $i_1$ and $i_2$ in opposite directions. What are the magnitude and direction of the net magnetic field at point $P$? Assume the following values: $i_1 = 15$ A, $i_2 = 32$ A, and $d = 5.3$ cm.

**KEY IDEAS**   (1) The net magnetic field $\vec{B}$ at point $P$ is the vector sum of the magnetic fields due to the currents in the two wires. (2) We can find the magnetic field due

to any current by applying the Biot–Savart law to the current. For points near the current in a long straight wire, that law leads to Eq. 29-4.

*Finding the vectors:* In Fig. 29-9a, point $P$ is distance $R$ from both currents $i_1$ and $i_2$. Thus, Eq. 29-4 tells us that at point $P$ those currents produce magnetic fields $\vec{B}_1$ and $\vec{B}_2$ with magnitudes

$$B_1 = \frac{\mu_0 i_1}{2\pi R} \quad \text{and} \quad B_2 = \frac{\mu_0 i_2}{2\pi R}.$$

In the right triangle of Fig. 29-9a, note that the base angles (between sides $R$ and $d$) are both 45°. This allows us to write $\cos 45° = R/d$ and replace $R$ with $d \cos 45°$. Then the field magnitudes $B_1$ and $B_2$ become

$$B_1 = \frac{\mu_0 i_1}{2\pi d \cos 45°} \quad \text{and} \quad B_2 = \frac{\mu_0 i_2}{2\pi d \cos 45°}.$$

We want to combine $\vec{B}_1$ and $\vec{B}_2$ to find their vector sum, which is the net field $\vec{B}$ at $P$. To find the directions of $\vec{B}_1$ and $\vec{B}_2$, we apply the right-hand rule of Fig. 29-4 to each current in Fig. 29-9a. For wire 1, with current out of the page, we mentally grasp the wire with the right hand, with the thumb pointing out of the page. Then the curled fingers indicate that the field lines run counterclockwise. In particular, in the region of point $P$, they are directed upward to the left. Recall that the magnetic field at a point near a long, straight current-carrying wire must be directed perpendicular to a radial line between the point and the current. Thus, $\vec{B}_1$ must be directed upward to the left, as drawn in Fig. 29-9b. (Note carefully the perpendicular symbol between vector $\vec{B}_1$ and the line connecting point $P$ and wire 1.)

Repeating this analysis for the current in wire 2, we find that $\vec{B}_2$ is directed upward to the right, as drawn in Fig. 29-9b. (Note the perpendicular symbol between vector $\vec{B}_2$ and the line connecting point $P$ and wire 2.)

**Adding the vectors:** We can now vectorially add $\vec{B}_1$ and $\vec{B}_2$ to find the net magnetic field $\vec{B}$ at point $P$, either by using a vector-capable calculator or by resolving the vectors into components and then combining the components of $\vec{B}$. However, in Fig. 29-9b, there is a third method: Because $\vec{B}_1$ and $\vec{B}_2$ are perpendicular to each other, they form the legs of a right triangle, with $\vec{B}$ as the hypotenuse. The Pythagorean theorem then gives us

$$B = \sqrt{B_1^2 + B_2^2} = \frac{\mu_0}{2\pi d(\cos 45°)} \sqrt{i_1^2 + i_2^2}$$

$$= \frac{(4\pi \times 10^{-7}\,\text{T·m/A})\sqrt{(15\,\text{A})^2 + (32\,\text{A})^2}}{(2\pi)(5.3 \times 10^{-2}\,\text{m})(\cos 45°)}$$

$$= 1.89 \times 10^{-4}\,\text{T} \approx 190\,\mu\text{T}. \quad \text{(Answer)}$$

The angle $\phi$ between the directions of $\vec{B}$ and $\vec{B}_2$ in Fig. 29-9b follows from

$$\phi = \tan^{-1} \frac{B_1}{B_2},$$

which, with $B_1$ and $B_2$ as given above, yields

$$\phi = \tan^{-1} \frac{i_1}{i_2} = \tan^{-1} \frac{15\,\text{A}}{32\,\text{A}} = 25°.$$

The angle between the direction of $\vec{B}$ and the $x$ axis shown in Fig. 29-9b is then

$$\phi + 45° = 25° + 45° = 70°. \quad \text{(Answer)}$$

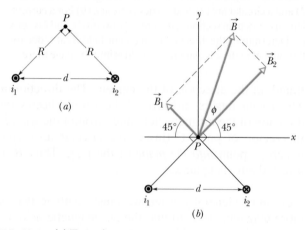

(a)

(b)

**FIG. 29-9** (a) Two wires carry currents $i_1$ and $i_2$ in opposite directions (out of and into the page). Note the right angle at $P$. (b) The separate fields $\vec{B}_1$ and $\vec{B}_2$ are combined vectorially to yield the net field $\vec{B}$.

---

**PROBLEM-SOLVING TACTICS**

**Tactic 1: Right-Hand Rules** To help you sort out the right-hand rules you have now seen (and the ones coming up), here is a review.

**Right-Hand Rule for Cross Products.** Introduced in Section 3-8, this is a way to determine the direction of the vector that results from a cross product. You point the fingers of your right hand so as to sweep the first vector expressed in the product into the second vector, through the smaller angle between the two vectors. Your outstretched thumb gives you the direction of the vector resulting from the cross product. In Chapter 11, we used this right-hand rule to find the directions of torque and angular momentum vectors; in Chapter 28, we used it to find the direction of the force on a current-carrying wire in a magnetic field.

**Curled–Straight Right-Hand Rules for Magnetism.** In many situations involving magnetism, you need to relate a "curled" element and a "straight" element. You can do so with the (curled) fingers and the (straight) thumb on your right hand. You have already seen an example in Section 28-9, where we related the current around a loop (curled element) to the normal vector $\vec{n}$ (straight element) of the loop: You curl the fingers of your right hand around in the direction of the current along the loop; your outstretched thumb then gives the direction of $\vec{n}$. This is also the direction of the magnetic dipole moment $\vec{\mu}$ of the loop.

In this section you were introduced to a second curled–straight right-hand rule for magnetism. To determine the direction of the magnetic field lines around a current-length element, you point the outstretched thumb of your right hand in the direction of the current. The fingers then naturally curl around the current-length element in the direction of the field lines around the element.

# 29-3 | Force Between Two Parallel Currents

Two long parallel wires carrying currents exert forces on each other. Figure 29-10 shows two such wires, separated by a distance $d$ and carrying currents $i_a$ and $i_b$. Let us analyze the forces on these wires due to each other.

We seek first the force on wire $b$ in Fig. 29-10 due to the current in wire $a$. That current produces a magnetic field $\vec{B}_a$, and it is this magnetic field that actually causes the force we seek. To find the force, then, we need the magnitude and direction of the field $\vec{B}_a$ *at the site of wire b*. The magnitude of $\vec{B}_a$ at every point of wire $b$ is, from Eq. 29-4,

$$B_a = \frac{\mu_0 i_a}{2\pi d}. \tag{29-11}$$

The curled–straight right-hand rule tells us that the direction of $\vec{B}_a$ at wire $b$ is down, as Fig. 29-10 shows.

Now that we have the field, we can find the force it produces on wire $b$. Equation 28-26 tells us that the force $\vec{F}_{ba}$ on a length $L$ of wire $b$ due to the external magnetic field $\vec{B}_a$ is

$$\vec{F}_{ba} = i_b \vec{L} \times \vec{B}_a, \tag{29-12}$$

where $\vec{L}$ is the length vector of the wire. In Fig. 29-10, vectors $\vec{L}$ and $\vec{B}_a$ are perpendicular to each other, and so with Eq. 29-11, we can write

$$F_{ba} = i_b L B_a \sin 90° = \frac{\mu_0 L i_a i_b}{2\pi d}. \tag{29-13}$$

The direction of $\vec{F}_{ba}$ is the direction of the cross product $\vec{L} \times \vec{B}_a$. Applying the right-hand rule for cross products to $\vec{L}$ and $\vec{B}_a$ in Fig. 29-10, we see that $\vec{F}_{ba}$ is directly toward wire $a$, as shown.

The general procedure for finding the force on a current-carrying wire is this:

> ☛ To find the force on a current-carrying wire due to a second current-carrying wire, first find the field due to the second wire at the site of the first wire. Then find the force on the first wire due to that field.

We could now use this procedure to compute the force on wire $a$ due to the current in wire $b$. We would find that the force is directly toward wire $b$; hence, the two wires with parallel currents attract each other. Similarly, if the two currents were antiparallel, we could show that the two wires repel each other. Thus,

> ☛ Parallel currents attract each other, and antiparallel currents repel each other.

The force acting between currents in parallel wires is the basis for the definition of the ampere, which is one of the seven SI base units. The definition, adopted in 1946, is this: The ampere is that constant current which, if maintained in two straight, parallel conductors of infinite length, of negligible circular cross section, and placed 1 m apart in vacuum, would produce on each of these conductors a force of magnitude $2 \times 10^{-7}$ newton per meter of wire length.

## Rail Gun

One application of the physics of Eq. 29-13 is a rail gun. In this device, a magnetic force accelerates a projectile to a high speed in a short time. The basics of a rail gun are shown in Fig. 29-11a. A large current is sent out along one of two parallel conducting rails, across a conducting "fuse" (such as a narrow piece of copper) between the rails, and then back to the current source along the second rail. The

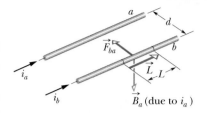

**FIG. 29-10** Two parallel wires carrying currents in the same direction attract each other. $\vec{B}_a$ is the magnetic field at wire $b$ produced by the current in wire $a$. $\vec{F}_{ba}$ is the resulting force acting on wire $b$ because it carries current in $\vec{B}_a$.

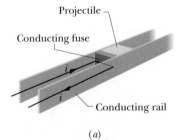

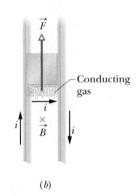

**FIG. 29-11** (a) A rail gun, as a current $i$ is set up in it. The current rapidly causes the conducting fuse to vaporize. (b) The current produces a magnetic field $\vec{B}$ between the rails, and the field causes a force $\vec{F}$ to act on the conducting gas, which is part of the current path. The gas propels the projectile along the rails, launching it.

projectile to be fired lies on the far side of the fuse and fits loosely between the rails. Immediately after the current begins, the fuse element melts and vaporizes, creating a conducting gas between the rails where the fuse had been.

The curled–straight right-hand rule of Fig. 29-4 reveals that the currents in the rails of Fig. 29-11*a* produce magnetic fields that are directed downward between the rails. The net magnetic field $\vec{B}$ exerts a force $\vec{F}$ on the gas due to the current *i* through the gas (Fig. 29-11*b*). With Eq. 29-12 and the right-hand rule for cross products, we find that $\vec{F}$ points outward along the rails. As the gas is forced outward along the rails, it pushes the projectile, accelerating it by as much as $5 \times 10^6 g$, and then launches it with a speed of 10 km/s, all within 1 ms. Some-day rail guns may be used to launch materials into space from mining operations on the Moon or an asteroid.

✓ **CHECKPOINT 1** The figure here shows three long, straight, parallel, equally spaced wires with identical currents either into or out of the page. Rank the wires according to the magnitude of the force on each due to the currents in the other two wires, greatest first.

## 29-4 | Ampere's Law

We can find the net electric field due to *any* distribution of charges by first writing the differential electric field $d\vec{E}$ due to a charge element and then summing the contributions of $d\vec{E}$ from all the elements. However, if the distribution is compli-cated, we may have to use a computer. Recall, however, that if the distribution has planar, cylindrical, or spherical symmetry, we can apply Gauss' law to find the net electric field with considerably less effort.

Similarly, we can find the net magnetic field due to *any* distribution of cur-rents by first writing the differential magnetic field $d\vec{B}$ (Eq. 29-3) due to a current-length element and then summing the contributions of $d\vec{B}$ from all the ele-ments. Again we may have to use a computer for a complicated distribution. However, if the distribution has some symmetry, we may be able to apply **Am-pere's law** to find the magnetic field with considerably less effort. This law, which can be derived from the Biot–Savart law, has traditionally been credited to An-dré-Marie Ampère (1775–1836), for whom the SI unit of current is named. How-ever, the law actually was advanced by English physicist James Clerk Maxwell.

Ampere's law is

$$\oint \vec{B} \cdot d\vec{s} = \mu_0 i_{enc} \quad \text{(Ampere's law).} \tag{29-14}$$

The loop on the integral sign means that the scalar (dot) product $\vec{B} \cdot d\vec{s}$ is to be integrated around a *closed* loop, called an *Amperian loop*. The current $i_{enc}$ is the *net* current encircled by that closed loop.

To see the meaning of the scalar product $\vec{B} \cdot d\vec{s}$ and its integral, let us first apply Ampere's law to the general situation of Fig. 29-12. The figure shows cross sections of three long straight wires that carry currents $i_1$, $i_2$, and $i_3$ either directly into or directly out of the page. An arbitrary Amperian loop lying in the plane of the page encircles two of the currents but not the third. The counterclockwise direction marked on the loop indicates the arbitrarily chosen direction of integra-tion for Eq. 29-14.

To apply Ampere's law, we mentally divide the loop into differential vector elements $d\vec{s}$ that are everywhere directed along the tangent to the loop in the direction of integration. Assume that at the location of the element $d\vec{s}$ shown in

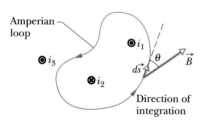

**FIG. 29-12** Ampere's law applied to an arbitrary Amperian loop that encircles two long straight wires but excludes a third wire. Note the direc-tions of the currents.

Fig. 29-12, the net magnetic field due to the three currents is $\vec{B}$. Because the wires are perpendicular to the page, we know that the magnetic field at $d\vec{s}$ due to each current is in the plane of Fig. 29-12; thus, their net magnetic field $\vec{B}$ at $d\vec{s}$ must also be in that plane. However, we do not know the orientation of $\vec{B}$ within the plane. In Fig. 29-12, $\vec{B}$ is arbitrarily drawn at an angle $\theta$ to the direction of $d\vec{s}$.

The scalar product $\vec{B} \cdot d\vec{s}$ on the left side of Eq. 29-14 is equal to $B \cos \theta \, ds$. Thus, Ampere's law can be written as

$$\oint \vec{B} \cdot d\vec{s} = \oint B \cos \theta \, ds = \mu_0 i_{enc}. \qquad (29\text{-}15)$$

We can now interpret the scalar product $\vec{B} \cdot d\vec{s}$ as being the product of a length $ds$ of the Amperian loop and the field component $B \cos \theta$ tangent to the loop. Then we can interpret the integration as being the summation of all such products around the entire loop.

When we can actually perform this integration, we do not need to know the direction of $\vec{B}$ before integrating. Instead, we arbitrarily assume $\vec{B}$ to be generally in the direction of integration (as in Fig. 29-12). Then we use the following curled–straight right-hand rule to assign a plus sign or a minus sign to each of the currents that make up the net encircled current $i_{enc}$:

> Curl your right hand around the Amperian loop, with the fingers pointing in the direction of integration. A current through the loop in the general direction of your outstretched thumb is assigned a plus sign, and a current generally in the opposite direction is assigned a minus sign.

Finally, we solve Eq. 29-15 for the magnitude of $\vec{B}$. If $B$ turns out positive, then the direction we assumed for $\vec{B}$ is correct. If it turns out negative, we neglect the minus sign and redraw $\vec{B}$ in the opposite direction.

In Fig. 29-13 we apply the curled–straight right-hand rule for Ampere's law to the situation of Fig. 29-12. With the indicated counterclockwise direction of integration, the net current encircled by the loop is

$$i_{enc} = i_1 - i_2.$$

(Current $i_3$ is not encircled by the loop.) We can then rewrite Eq. 29-15 as

$$\oint B \cos \theta \, ds = \mu_0(i_1 - i_2). \qquad (29\text{-}16)$$

You might wonder why, since current $i_3$ contributes to the magnetic-field magnitude $B$ on the left side of Eq. 29-16, it is not needed on the right side. The answer is that the contributions of current $i_3$ to the magnetic field cancel out because the integration in Eq. 29-16 is made around the full loop. In contrast, the contributions of an encircled current to the magnetic field do not cancel out.

We cannot solve Eq. 29-16 for the magnitude $B$ of the magnetic field because for the situation of Fig. 29-12 we do not have enough information to simplify and solve the integral. However, we do know the outcome of the integration; it must be equal to $\mu_0(i_1 - i_2)$, the value of which is set by the net current passing through the loop.

We shall now apply Ampere's law to two situations in which symmetry does allow us to simplify and solve the integral, hence to find the magnetic field.

### Magnetic Field Outside a Long Straight Wire with Current

Figure 29-14 shows a long straight wire that carries current $i$ directly out of the page. Equation 29-4 tells us that the magnetic field $\vec{B}$ produced by the current has the same magnitude at all points that are the same distance $r$ from the wire; that is, the field $\vec{B}$ has cylindrical symmetry about the wire. We can take advan-

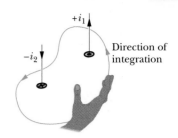

FIG. 29-13   A right-hand rule for Ampere's law, to determine the signs for currents encircled by an Amperian loop. The situation is that of Fig. 29-12.

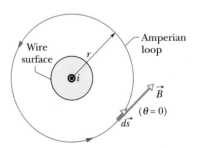

FIG. 29-14   Using Ampere's law to find the magnetic field that a current $i$ produces outside a long straight wire of circular cross section. The Amperian loop is a concentric circle that lies outside the wire.

tage of that symmetry to simplify the integral in Ampere's law (Eqs. 29-14 and 29-15) if we encircle the wire with a concentric circular Amperian loop of radius $r$, as in Fig. 29-14. The magnetic field $\vec{B}$ then has the same magnitude $B$ at every point on the loop. We shall integrate counterclockwise, so that $d\vec{s}$ has the direction shown in Fig. 29-14.

We can further simplify the quantity $B \cos \theta$ in Eq. 29-15 by noting that $\vec{B}$ is tangent to the loop at every point along the loop, as is $d\vec{s}$. Thus, $\vec{B}$ and $d\vec{s}$ are either parallel or antiparallel at each point of the loop, and we shall arbitrarily assume the former. Then at every point the angle $\theta$ between $d\vec{s}$ and $\vec{B}$ is $0°$, so $\cos \theta = \cos 0° = 1$. The integral in Eq. 29-15 then becomes

$$\oint \vec{B} \cdot d\vec{s} = \oint B \cos \theta \, ds = B \oint ds = B(2\pi r).$$

Note that $\oint ds$ is the summation of all the line segment lengths $ds$ around the circular loop; that is, it simply gives the circumference $2\pi r$ of the loop.

Our right-hand rule gives us a plus sign for the current of Fig. 29-14. The right side of Ampere's law becomes $+\mu_0 i$, and we then have

$$B(2\pi r) = \mu_0 i$$

or
$$B = \frac{\mu_0 i}{2\pi r} \qquad \text{(outside straight wire).} \qquad (29\text{-}17)$$

With a slight change in notation, this is Eq. 29-4, which we derived earlier—with considerably more effort—using the law of Biot and Savart. In addition, because the magnitude $B$ turned out positive, we know that the correct direction of $\vec{B}$ must be the one shown in Fig. 29-14.

### Magnetic Field Inside a Long Straight Wire with Current

Figure 29-15 shows the cross section of a long straight wire of radius $R$ that carries a uniformly distributed current $i$ directly out of the page. Because the current is uniformly distributed over a cross section of the wire, the magnetic field $\vec{B}$ produced by the current must be cylindrically symmetrical. Thus, to find the magnetic field at points inside the wire, we can again use an Amperian loop of radius $r$, as shown in Fig. 29-15, where now $r < R$. Symmetry again suggests that $\vec{B}$ is tangent to the loop, as shown; so the left side of Ampere's law again yields

$$\oint \vec{B} \cdot d\vec{s} = B \oint ds = B(2\pi r). \qquad (29\text{-}18)$$

To find the right side of Ampere's law, we note that because the current is uniformly distributed, the current $i_{\text{enc}}$ encircled by the loop is proportional to the area encircled by the loop; that is,

$$i_{\text{enc}} = i \frac{\pi r^2}{\pi R^2}. \qquad (29\text{-}19)$$

Our right-hand rule tells us that $i_{\text{enc}}$ gets a plus sign. Then Ampere's law gives us

$$B(2\pi r) = \mu_0 i \frac{\pi r^2}{\pi R^2}$$

or
$$B = \left(\frac{\mu_0 i}{2\pi R^2}\right) r \qquad \text{(inside straight wire).} \qquad (29\text{-}20)$$

Thus, inside the wire, the magnitude $B$ of the magnetic field is proportional to $r$; that magnitude is zero at the center and a maximum at the surface, where $r = R$. Note that Eqs. 29-17 and 29-20 give the same value for $B$ at $r = R$; that is, the expressions for the magnetic field outside the wire and inside the wire yield the same result at the surface of the wire.

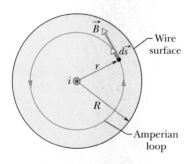

**FIG. 29-15** Using Ampere's law to find the magnetic field that a current $i$ produces inside a long straight wire of circular cross section. The current is uniformly distributed over the cross section of the wire and emerges from the page. An Amperian loop is drawn inside the wire.

✓**CHECKPOINT 2**    The figure here shows three equal currents $i$ (two parallel and one antiparallel) and four Amperian loops. Rank the loops according to the magnitude of $\oint \vec{B} \cdot d\vec{s}$ along each, greatest first.

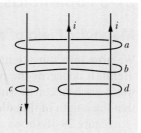

## Sample Problem  29-3

Figure 29-16$a$ shows the cross section of a long conducting cylinder with inner radius $a = 2.0$ cm and outer radius $b = 4.0$ cm. The cylinder carries a current out of the page, and the magnitude of the current density in the cross section is given by $J = cr^2$, with $c = 3.0 \times 10^6$ A/m$^4$ and $r$ in meters. What is the magnetic field $\vec{B}$ at a point that is 3.0 cm from the central axis of the cylinder?

**KEY IDEAS**    The point at which we want to evaluate $\vec{B}$ is inside the material of the conducting cylinder, between its inner and outer radii. We note that the current distribution has cylindrical symmetry (it is the same all around the cross section for any given radius). Thus, the symmetry allows us to use Ampere's law to find $\vec{B}$ at the point. We first draw the Amperian loop shown in Fig. 29-16$b$. The loop is concentric with the cylinder and has radius $r = 3.0$ cm because we want to evaluate $\vec{B}$ at that distance from the cylinder's central axis.

Next, we must compute the current $i_{enc}$ that is encircled by the Amperian loop. However, we *cannot* set up a proportionality as in Eq. 29-19, because here the current is not uniformly distributed. Instead, following the procedure of Sample Problem 26-2b, we must integrate the current density magnitude from the cylinder's inner radius $a$ to the loop radius $r$.

*Calculations:* We write the integral as

$$i_{enc} = \int J \, dA = \int_a^r cr^2 (2\pi r \, dr)$$

$$= 2\pi c \int_a^r r^3 \, dr = 2\pi c \left[ \frac{r^4}{4} \right]_a^r$$

$$= \frac{\pi c (r^4 - a^4)}{2}.$$

The direction of integration indicated in Fig. 29-16$b$ is (arbitrarily) clockwise. Applying the right-hand rule for Ampere's law to that loop, we find that we should take $i_{enc}$ as negative because the current is directed out of the page but our thumb is directed into the page.

We next evaluate the left side of Ampere's law exactly as we did in Fig. 29-15, and we again obtain

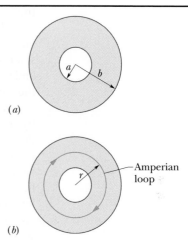

(a)

(b)

**FIG. 29-16**    (a) Cross section of a conducting cylinder of inner radius $a$ and outer radius $b$. (b) An Amperian loop of radius $r$ is added to compute the magnetic field at points that are a distance $r$ from the central axis.

Eq. 29-18. Then Ampere's law,

$$\oint \vec{B} \cdot d\vec{s} = \mu_0 i_{enc},$$

gives us

$$B(2\pi r) = -\frac{\mu_0 \pi c}{2} (r^4 - a^4).$$

Solving for $B$ and substituting known data yield

$$B = -\frac{\mu_0 c}{4r} (r^4 - a^4)$$

$$= -\frac{(4\pi \times 10^{-7} \text{ T·m/A})(3.0 \times 10^6 \text{ A/m}^4)}{4(0.030 \text{ m})}$$

$$\times [(0.030 \text{ m})^4 - (0.020 \text{ m})^4]$$

$$= -2.0 \times 10^{-5} \text{ T}.$$

Thus, the magnetic field $\vec{B}$ at a point 3.0 cm from the central axis has magnitude

$$B = 2.0 \times 10^{-5} \text{ T} \qquad \text{(Answer)}$$

and forms magnetic field lines that are directed opposite our direction of integration, hence counterclockwise in Fig. 29-16$b$.

FIG. 29-17 A solenoid carrying current $i$.

## 29-5 | Solenoids and Toroids

### Magnetic Field of a Solenoid

We now turn our attention to another situation in which Ampere's law proves useful. It concerns the magnetic field produced by the current in a long, tightly wound helical coil of wire. Such a coil is called a **solenoid** (Fig. 29-17). We assume that the length of the solenoid is much greater than the diameter.

Figure 29-18 shows a section through a portion of a "stretched-out" solenoid. The solenoid's magnetic field is the vector sum of the fields produced by the individual turns (*windings*) that make up the solenoid. For points very close to a turn, the wire behaves magnetically almost like a long straight wire, and the lines of $\vec{B}$ there are almost concentric circles. Figure 29-18 suggests that the field tends to cancel between adjacent turns. It also suggests that, at points inside the solenoid and reasonably far from the wire, $\vec{B}$ is approximately parallel to the (central) solenoid axis. In the limiting case of an *ideal solenoid,* which is infinitely long and consists of tightly packed (*close-packed*) turns of square wire, the field inside the coil is uniform and parallel to the solenoid axis.

At points above the solenoid, such as $P$ in Fig. 29-18, the magnetic field set up by the upper parts of the solenoid turns (these upper turns are marked $\odot$) is directed to the left (as drawn near $P$) and tends to cancel the field set up at $P$ by the lower parts of the turns (these lower turns are marked $\otimes$), which is directed to the right (not drawn). In the limiting case of an ideal solenoid, the magnetic field outside the solenoid is zero. Taking the external field to be zero is an excellent assumption for a real solenoid if its length is much greater than its diameter and if we consider external points such as point $P$ that are not at either end of the solenoid. The direction of the magnetic field along the solenoid axis is given by a curled–straight right-hand rule: Grasp the solenoid with your right hand so that your fingers follow the direction of the current in the windings; your extended right thumb then points in the direction of the axial magnetic field.

Figure 29-19 shows the lines of $\vec{B}$ for a real solenoid. The spacing of these lines in the central region shows that the field inside the coil is fairly strong and uniform over the cross section of the coil. The external field, however, is relatively weak.

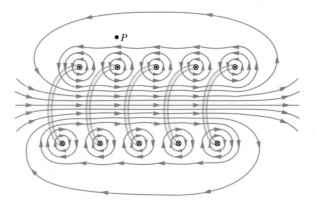

FIG. 29-18 A vertical cross section through the central axis of a "stretched-out" solenoid. The back portions of five turns are shown, as are the magnetic field lines due to a current through the solenoid. Each turn produces circular magnetic field lines near itself. Near the solenoid's axis, the field lines combine into a net magnetic field that is directed along the axis. The closely spaced field lines there indicate a strong magnetic field. Outside the solenoid the field lines are widely spaced; the field there is very weak.

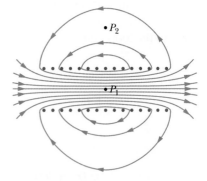

FIG. 29-19 Magnetic field lines for a real solenoid of finite length. The field is strong and uniform at interior points such as $P_1$ but relatively weak at external points such as $P_2$.

Let us now apply Ampere's law,

$$\oint \vec{B} \cdot d\vec{s} = \mu_0 i_{enc}, \tag{29-21}$$

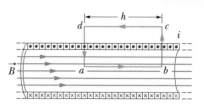

**FIG. 29-20** Application of Ampere's law to a section of a long ideal solenoid carrying a current $i$. The Amperian loop is the rectangle *abcda*.

to the ideal solenoid of Fig. 29-20, where $\vec{B}$ is uniform within the solenoid and zero outside it, using the rectangular Amperian loop *abcda*. We write $\oint \vec{B} \cdot d\vec{s}$ as the sum of four integrals, one for each loop segment:

$$\oint \vec{B} \cdot d\vec{s} = \int_a^b \vec{B} \cdot d\vec{s} + \int_b^c \vec{B} \cdot d\vec{s} + \int_c^d \vec{B} \cdot d\vec{s} + \int_d^a \vec{B} \cdot d\vec{s}. \tag{29-22}$$

The first integral on the right of Eq. 29-22 is $Bh$, where $B$ is the magnitude of the uniform field $\vec{B}$ inside the solenoid and $h$ is the (arbitrary) length of the segment from $a$ to $b$. The second and fourth integrals are zero because for every element $ds$ of these segments, $\vec{B}$ either is perpendicular to $ds$ or is zero, and thus the product $\vec{B} \cdot d\vec{s}$ is zero. The third integral, which is taken along a segment that lies outside the solenoid, is zero because $B = 0$ at all external points. Thus, $\oint \vec{B} \cdot d\vec{s}$ for the entire rectangular loop has the value $Bh$.

The net current $i_{enc}$ encircled by the rectangular Amperian loop in Fig. 29-20 is not the same as the current $i$ in the solenoid windings because the windings pass more than once through this loop. Let $n$ be the number of turns per unit length of the solenoid; then the loop encloses $nh$ turns and

$$i_{enc} = i(nh).$$

Ampere's law then gives us

$$Bh = \mu_0 inh$$

or

$$B = \mu_0 in \qquad \text{(ideal solenoid).} \tag{29-23}$$

Although we derived Eq. 29-23 for an infinitely long ideal solenoid, it holds quite well for actual solenoids if we apply it only at interior points and well away from the solenoid ends. Equation 29-23 is consistent with the experimental fact that the magnetic field magnitude $B$ within a solenoid does not depend on the diameter or the length of the solenoid and that $B$ is uniform over the solenoidal cross section. A solenoid thus provides a practical way to set up a known uniform magnetic field for experimentation, just as a parallel-plate capacitor provides a practical way to set up a known uniform electric field.

### Magnetic Field of a Toroid

Figure 29-21*a* shows a **toroid,** which we may describe as a (hollow) solenoid that has been curved until its two ends meet, forming a sort of hollow bracelet. What magnetic field $\vec{B}$ is set up inside the toroid (inside the hollow of the bracelet)? We can find out from Ampere's law and the symmetry of the bracelet.

From the symmetry, we see that the lines of $\vec{B}$ form concentric circles inside the toroid, directed as shown in Fig. 29-21*b*. Let us choose a concentric circle of radius $r$ as an Amperian loop and traverse it in the clockwise direction. Ampere's law (Eq. 29-14) yields

$$(B)(2\pi r) = \mu_0 iN,$$

where $i$ is the current in the toroid windings (and is positive for those windings enclosed by the Amperian loop) and $N$ is the total number of turns. This gives

$$B = \frac{\mu_0 iN}{2\pi} \frac{1}{r} \qquad \text{(toroid).} \tag{29-24}$$

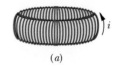

(a)

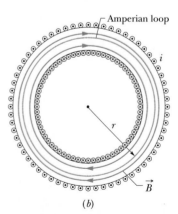

(b)

**FIG. 29-21**   (*a*) A toroid carrying a current $i$. (*b*) A horizontal cross section of the toroid. The interior magnetic field (inside the bracelet-shaped tube) can be found by applying Ampere's law with the Amperian loop shown.

In contrast to the situation for a solenoid, $B$ is not constant over the cross section of a toroid.

It is easy to show, with Ampere's law, that $B = 0$ for points outside an ideal toroid (as if the toroid were made from an ideal solenoid). The direction of the magnetic field within a toroid follows from our curled–straight right-hand rule: Grasp the toroid with the fingers of your right hand curled in the direction of the current in the windings; your extended right thumb points in the direction of the magnetic field.

<hr>

**Sample Problem** | **29-4**

A solenoid has length $L = 1.23$ m and inner diameter $d = 3.55$ cm, and it carries a current $i = 5.57$ A. It consists of five close-packed layers, each with 850 turns along length $L$. What is $B$ at its center?

**KEY IDEA** The magnitude $B$ of the magnetic field along the solenoid's central axis is related to the solenoid's current $i$ and number of turns per unit length $n$ by Eq. 29-23 ($B = \mu_0 in$).

**Calculation:** Because $B$ does not depend on the diameter of the windings, the value of $n$ for five identical layers is simply five times the value for each layer. Equation 29-23 then tells us

$$B = \mu_0 in = (4\pi \times 10^{-7}\,\text{T·m/A})(5.57\,\text{A})\frac{5 \times 850 \text{ turns}}{1.23 \text{ m}}$$

$$= 2.42 \times 10^{-2}\,\text{T} = 24.2\,\text{mT}. \qquad \text{(Answer)}$$

<hr>

## 29-6 | A Current-Carrying Coil as a Magnetic Dipole

So far we have examined the magnetic fields produced by current in a long straight wire, a solenoid, and a toroid. We turn our attention here to the field produced by a coil carrying a current. You saw in Section 28-10 that such a coil behaves as a magnetic dipole in that, if we place it in an external magnetic field $\vec{B}$, a torque $\vec{\tau}$ given by

$$\vec{\tau} = \vec{\mu} \times \vec{B} \qquad (29\text{-}25)$$

acts on it. Here $\vec{\mu}$ is the magnetic dipole moment of the coil and has the magnitude $NiA$, where $N$ is the number of turns, $i$ is the current in each turn, and $A$ is the area enclosed by each turn. (*Caution:* Don't confuse the magnetic dipole moment $\vec{\mu}$ with the permeability constant $\mu_0$.)

Recall that the direction of $\vec{\mu}$ is given by a curled–straight right-hand rule: Grasp the coil so that the fingers of your right hand curl around it in the direction of the current; your extended thumb then points in the direction of the dipole moment $\vec{\mu}$.

### Magnetic Field of a Coil

We turn now to the other aspect of a current-carrying coil as a magnetic dipole. What magnetic field does *it* produce at a point in the surrounding space? The problem does not have enough symmetry to make Ampere's law useful; so we must turn to the law of Biot and Savart. For simplicity, we first consider only a coil with a single circular loop and only points on its perpendicular central axis, which we take to be a $z$ axis. We shall show that the magnitude of the magnetic field at such points is

$$B(z) = \frac{\mu_0 i R^2}{2(R^2 + z^2)^{3/2}}, \qquad (29\text{-}26)$$

in which $R$ is the radius of the circular loop and $z$ is the distance of the point in question from the center of the loop. Furthermore, the direction of the magnetic field $\vec{B}$ is the same as the direction of the magnetic dipole moment $\vec{\mu}$ of the loop.

For axial points far from the loop, we have $z \gg R$ in Eq. 29-26. With that approximation, the equation reduces to

$$B(z) \approx \frac{\mu_0 iR^2}{2z^3}.$$

Recalling that $\pi R^2$ is the area $A$ of the loop and extending our result to include a coil of $N$ turns, we can write this equation as

$$B(z) = \frac{\mu_0}{2\pi} \frac{NiA}{z^3}.$$

Further, because $\vec{B}$ and $\vec{\mu}$ have the same direction, we can write the equation in vector form, substituting from the identity $\mu = NiA$:

$$\vec{B}(z) = \frac{\mu_0}{2\pi} \frac{\vec{\mu}}{z^3} \qquad \text{(current-carrying coil).} \qquad (29\text{-}27)$$

Thus, we have two ways in which we can regard a current-carrying coil as a magnetic dipole: (1) it experiences a torque when we place it in an external magnetic field; (2) it generates its own intrinsic magnetic field, given, for distant points along its axis, by Eq. 29-27. Figure 29-22 shows the magnetic field of a current loop; one side of the loop acts as a north pole (in the direction of $\vec{\mu}$) and the other side as a south pole, as suggested by the lightly drawn magnet in the figure.

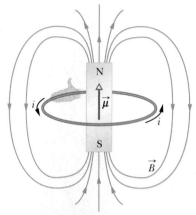

FIG. 29-22   A current loop produces a magnetic field like that of a bar magnet and thus has associated north and south poles. The magnetic dipole moment $\vec{\mu}$ of the loop, its direction given by a curled–straight right-hand rule, points from the south pole to the north pole, in the direction of the field $\vec{B}$ within the loop.

✓ **CHECKPOINT 3**   The figure here shows four arrangements of circular loops of radius $r$ or $2r$, centered on vertical axes (perpendicular to the loops) and carrying identical currents in the directions indicated. Rank the arrangements according to the magnitude of the net magnetic field at the dot, midway between the loops on the central axis, greatest first.

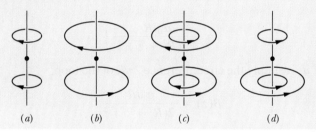

(a)        (b)        (c)        (d)

## Proof of Equation 29-26

Figure 29-23 shows the back half of a circular loop of radius $R$ carrying a current $i$. Consider a point $P$ on the central axis of the loop, a distance $z$ from its plane. Let us apply the law of Biot and Savart to a differential element $ds$ of the loop, located at the left side of the loop. The length vector $d\vec{s}$ for this element points perpendicularly out of the page. The angle $\theta$ between $d\vec{s}$ and $\hat{r}$ in Fig. 29-23 is $90°$; the plane formed by these two vectors is perpendicular to the plane of the page and contains both $\hat{r}$ and $d\vec{s}$. From the law of Biot and Savart (and the right-hand rule), the differential field $d\vec{B}$ produced at point $P$ by the current in this element is perpendicular to this plane and thus is directed in the plane of the figure, perpendicular to $\hat{r}$, as indicated in Fig. 29-23.

Let us resolve $d\vec{B}$ into two components: $dB_{\parallel}$ along the axis of the loop and $dB_{\perp}$ perpendicular to this axis. From the symmetry, the vector sum of all the perpendicular components $dB_{\perp}$ due to all the loop elements $ds$ is zero. This leaves only the axial components $dB_{\parallel}$ and we have

$$B = \int dB_{\parallel}.$$

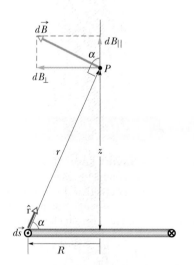

FIG. 29-23   Cross section through a current loop of radius $R$. The plane of the loop is perpendicular to the page, and only the back half of the loop is shown. We use the law of Biot and Savart to find the magnetic field at point $P$ on the central perpendicular axis of the loop.

For the element $d\vec{s}$ in Fig. 29-23, the law of Biot and Savart (Eq. 29-1) tells us that the magnetic field at distance $r$ is

$$dB = \frac{\mu_0}{4\pi} \frac{i \, ds \sin 90°}{r^2}.$$

We also have

$$dB_\parallel = dB \cos \alpha.$$

Combining these two relations, we obtain

$$dB_\parallel = \frac{\mu_0 i \cos \alpha \, ds}{4\pi r^2}. \qquad (29\text{-}28)$$

Figure 29-23 shows that $r$ and $\alpha$ are related to each other. Let us express each in terms of the variable $z$, the distance between point $P$ and the center of the loop. The relations are

$$r = \sqrt{R^2 + z^2} \qquad (29\text{-}29)$$

and

$$\cos \alpha = \frac{R}{r} = \frac{R}{\sqrt{R^2 + z^2}}. \qquad (29\text{-}30)$$

Substituting Eqs. 29-29 and 29-30 into Eq. 29-28, we find

$$dB_\parallel = \frac{\mu_0 i R}{4\pi (R^2 + z^2)^{3/2}} \, ds.$$

Note that $i$, $R$, and $z$ have the same values for all elements $ds$ around the loop; so when we integrate this equation, we find that

$$B = \int dB_\parallel$$

$$= \frac{\mu_0 i R}{4\pi (R^2 + z^2)^{3/2}} \int ds$$

or, because $\int ds$ is simply the circumference $2\pi R$ of the loop,

$$B(z) = \frac{\mu_0 i R^2}{2(R^2 + z^2)^{3/2}}.$$

This is Eq. 29-26, the relation we sought to prove.

## REVIEW & SUMMARY

**The Biot–Savart Law**   The magnetic field set up by a current-carrying conductor can be found from the *Biot–Savart law*. This law asserts that the contribution $d\vec{B}$ to the field produced by a current-length element $i \, d\vec{s}$ at a point $P$ located a distance $r$ from the current element is

$$d\vec{B} = \frac{\mu_0}{4\pi} \frac{i \, d\vec{s} \times \hat{r}}{r^2} \qquad \text{(Biot–Savart law).} \qquad (29\text{-}3)$$

Here $\hat{r}$ is a unit vector that points from the element toward $P$. The quantity $\mu_0$, called the permeability constant, has the value $4\pi \times 10^{-7}$ T·m/A $\approx 1.26 \times 10^{-6}$ T·m/A.

**Magnetic Field of a Long Straight Wire**   For a long straight wire carrying a current $i$, the Biot–Savart law gives, for the magnitude of the magnetic field at a perpendicular distance $R$ from the wire,

$$B = \frac{\mu_0 i}{2\pi R} \qquad \text{(long straight wire).} \qquad (29\text{-}4)$$

**Magnetic Field of a Circular Arc**   The magnitude of the magnetic field at the center of a circular arc, of radius $R$ and central angle $\phi$ (in radians), carrying current $i$, is

$$B = \frac{\mu_0 i \phi}{4\pi R} \qquad \text{(at center of circular arc).} \qquad (29\text{-}9)$$

**Force Between Parallel Currents**   Parallel wires carrying currents in the same direction attract each other, whereas parallel wires carrying currents in opposite directions repel each other. The magnitude of the force on a length $L$ of either

wire is

$$F_{ba} = i_b L B_a \sin 90° = \frac{\mu_0 L i_a i_b}{2\pi d}, \qquad (29\text{-}13)$$

where $d$ is the wire separation, and $i_a$ and $i_b$ are the currents in the wires.

### Ampere's Law   Ampere's law states that

$$\oint \vec{B} \cdot d\vec{s} = \mu_0 i_{enc} \qquad \text{(Ampere's law)}. \qquad (29\text{-}14)$$

The line integral in this equation is evaluated around a closed loop called an *Amperian loop*. The current $i$ is the *net* current encircled by the loop. For some current distributions, Eq. 29-14 is easier to use than Eq. 29-3 to calculate the magnetic field due to the currents.

### Fields of a Solenoid and a Toroid   Inside a *long sole-noid* carrying current $i$, at points not near its ends, the magnitude $B$ of the magnetic field is

$$B = \mu_0 i n \qquad \text{(ideal solenoid)}, \qquad (29\text{-}23)$$

where $n$ is the number of turns per unit length. At a point inside a *toroid*, the magnitude $B$ of the magnetic field is

$$B = \frac{\mu_0 i N}{2\pi} \frac{1}{r} \qquad \text{(toroid)}, \qquad (29\text{-}24)$$

where $r$ is the distance from the center of the toroid to the point.

### Field of a Magnetic Dipole   The magnetic field produced by a current-carrying coil, which is a *magnetic dipole*, at a point $P$ located a distance $z$ along the coil's perpendicular central axis is parallel to the axis and is given by

$$\vec{B}(z) = \frac{\mu_0}{2\pi} \frac{\vec{\mu}}{z^3}, \qquad (29\text{-}27)$$

where $\vec{\mu}$ is the dipole moment of the coil. This equation applies only when $z$ is much greater than the dimensions of the coil.

## QUESTIONS

**1** Figure 29-24 shows four arrangements in which long parallel wires carry equal currents directly into or out of the page at the corners of identical squares. Rank the arrangements according to the magnitude of the net magnetic field at the center of the square, greatest first.

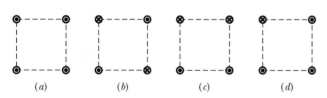

**FIG. 29-24** Question 1.

**2** Figure 29-25 shows cross sections of two long straight wires; the left-hand wire carries current $i_1$ directly out of the page. If the net magnetic field due to the two currents is to be zero at point $P$, (a) should the direction of current $i_2$ in the right-hand wire be directly into or out of the page and (b) should $i_2$ be greater than, less than, or equal to $i_1$?

**FIG. 29-25** Question 2.

**3** Figure 29-26 shows three circuits consisting of straight radial lengths and concentric circular arcs (either half- or quarter-circles of radii $r$, $2r$, and $3r$). The circuits carry the same current. Rank them according to the magnitude of the magnetic field produced at the center of curvature (the dot), greatest first.

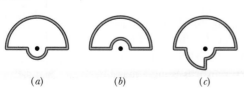

**FIG. 29-26** Question 3.

**4** Figure 29-27 represents a snapshot of the velocity vectors of four electrons near a wire carrying current $i$. The four velocities have the same magnitude; velocity $\vec{v}_2$ is directed into the page. Electrons 1 and 2 are at the same distance from the wire, as are electrons 3 and 4. Rank the electrons according to the magnitudes of the magnetic forces on them due to current $i$, greatest first.

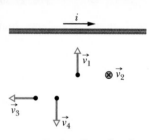

**FIG. 29-27** Question 4.

**5** Figure 29-28 shows three circuits, each consisting of two radial lengths and two concentric circular arcs, one of radius $r$ and the other of radius $R$ > $r$. The circuits have the same current through them and the same angle between the two radial lengths. Rank the circuits according to the magnitude of the net magnetic field at the center, greatest first.

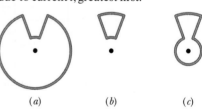

**FIG. 29-28** Question 5.

**6** Figure 29-29 shows four arrangements in which long,

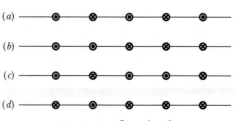

**FIG. 29-29** Question 6.

parallel, equally spaced wires carry equal currents directly into or out of the page. Rank the arrangements according to the magnitude of the net force on the central wire due to the currents in the other wires, greatest first.

**7** Figure 29-30 shows three arrangements of three long straight wires carrying equal currents directly into or out of the page. (a) Rank the arrangements according to the magnitude of the net force on wire *A* due to the currents in the other wires, greatest first. (b) In arrangement 3, is the angle between the net force on wire *A* and the dashed line equal to, less than, or more than 45°?

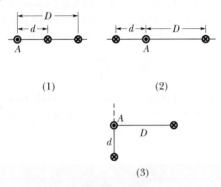

**FIG. 29-30** Question 7.

**8** Figure 29-31 shows four identical currents *i* and five Amperian paths (*a* through *e*) encircling them. Rank the paths according to the value of $\oint \vec{B} \cdot d\vec{s}$ taken in the directions shown, most positive first.

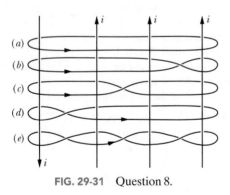

**FIG. 29-31** Question 8.

**9** Figure 29-32 shows four circular Amperian loops (*a, b, c, d*) concentric with a wire whose current is directed out of the page. The current is uniform across the wire's circular cross section (the shaded region). Rank the loops according to the magnitude of $\oint \vec{B} \cdot d\vec{s}$ around each, greatest first.

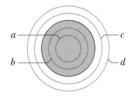

**FIG. 29-32** Question 9.

**10** Figure 29-33 gives, as a function of radial distance *r*, the magnitude *B* of the magnetic field inside and outside four wires (*a, b, c,* and *d*), each of which carries a current that is uniformly distributed across the wire's cross section. Overlapping portions of the plots are indicated by double labels. Rank the wires according to (a) radius, (b) the magnitude of the magnetic field on the surface, and (c) the value of the current, greatest first. (d) Is the magnitude of the current density in wire *a* greater than, less than, or equal to that in wire *c*?

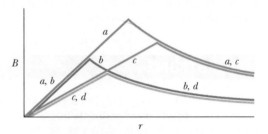

**FIG. 29-33** Question 10.

**11** Figure 29-34 shows four circular Amperian loops (*a, b, c, d*) and, in cross section, four long circular conductors (the shaded regions), all of which are concentric. Three of the conductors are hollow cylinders; the central conductor is a solid cylinder. The currents in the conductors are, from smallest radius to largest radius, 4 A out of the page, 9 A into the page, 5 A out of the page, and 3 A into the page. Rank the Amperian loops according to the magnitude of $\oint \vec{B} \cdot d\vec{s}$ around each, greatest first.

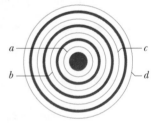

**FIG. 29-34** Question 11.

---

## PROBLEMS

**GO** Tutoring problem available (at instructor's discretion) in *WileyPLUS* and WebAssign

**SSM** Worked-out solution available in Student Solutions Manual

**● – ●●●** Number of dots indicates level of problem difficulty

**WWW** Worked-out solution is at

**ILW** Interactive solution is at

http://www.wiley.com/college/halliday

Additional information available in *The Flying Circus of Physics* and at flyingcircusofphysics.com

### sec. 29-2 Calculating the Magnetic Field Due to a Current

•**1** At a certain location in the Philippines, Earth's magnetic field of 39 $\mu$T is horizontal and directed due north. Suppose the net field is zero exactly 8.0 cm above a long, straight, horizontal wire that carries a constant current. What are the (a) magnitude and (b) direction of the current? **SSM**

•**2** A straight conductor carrying current *i* = 5.0 A splits into identical semicircular arcs as shown in Fig. 29-35. What is the magnetic field at the center *C* of the resulting circular loop?

•**3** A surveyor is using a magnetic compass 6.1 m below a power line in which there is a

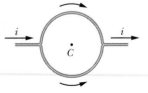

**FIG. 29-35** Problem 2.

steady current of 100 A. (a) What is the magnetic field at the site of the compass due to the power line? (b) Will this field interfere seriously with the compass reading? The horizontal component of Earth's magnetic field at the site is 20 $\mu$T.

•4 Figure 29-36a shows an element of length $ds = 1.00$ $\mu$m in a very long straight wire carrying current. The current in that element sets up a differential magnetic field $d\vec{B}$ at points in the surrounding space. Figure 29-36b gives the magnitude $dB$ of the field for points 2.5 cm from the element, as a function of angle $\theta$ between the wire and a straight line to the point. The vertical scale is set by $dB_s = 60.0$ pT. What is the magnitude of the magnetic field set up by the entire wire at perpendicular distance 2.5 cm from the wire?

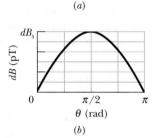

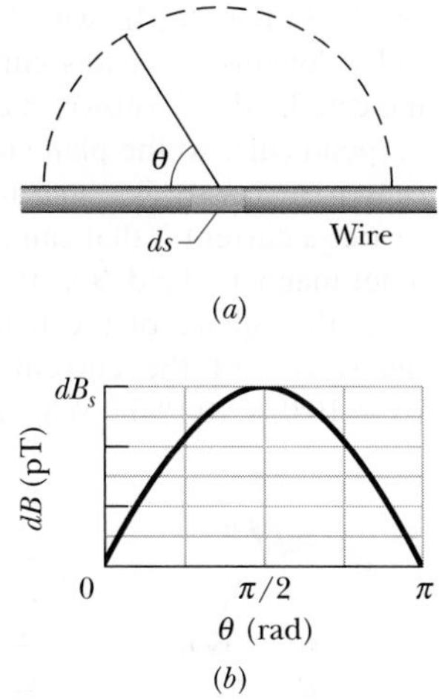

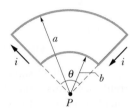

FIG. 29-36 Problem 4.

•5 In Fig. 29-37, two circular arcs have radii $a = 13.5$ cm and $b = 10.7$ cm, subtend angle $\theta = 74.0°$, carry current $i = 0.411$ A, and share the same center of curvature $P$. What are the (a) magnitude and (b) direction (into or out of the page) of the net magnetic field at $P$? GO

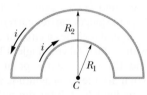

FIG. 29-37 Problem 5.

•6 In Fig. 29-38, two semicircular arcs have radii $R_2 = 7.80$ cm and $R_1 = 3.15$ cm, carry current $i = 0.281$ A, and share the same center of curvature $C$. What are the (a) magnitude and (b) direction (into or out of the page) of the net magnetic field at $C$?

FIG. 29-38 Problem 6.

•7 Two long straight wires are parallel and 8.0 cm apart. They are to carry equal currents such that the magnetic field at a point halfway between them has magnitude 300 $\mu$T. (a) Should the currents be in the same or opposite directions? (b) How much current is needed? SSM

•8 In Fig. 29-39, a wire forms a semicircle of radius $R = 9.26$ cm and two (radial) straight segments each of length $L = 13.1$ cm. The wire carries current $i = 34.8$ mA. What are the (a) magnitude and (b) direction (into or out of the page) of the net magnetic field at the semicircle's center of curvature $C$?

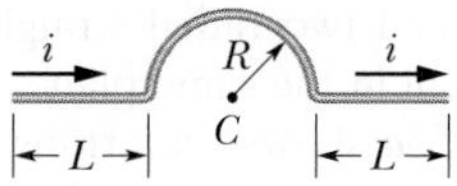

FIG. 29-39 Problem 8.

•9 In Fig. 29-40, two long straight wires are perpendicular to the page and separated by distance $d_1 = 0.75$ cm. Wire 1 carries 6.5 A into the page. What are the (a) magnitude and (b) direction (into or out of the page) of the current in wire 2 if the net magnetic field due to the two currents is zero at point $P$ located at distance $d_2 = 1.50$ cm from wire 2?

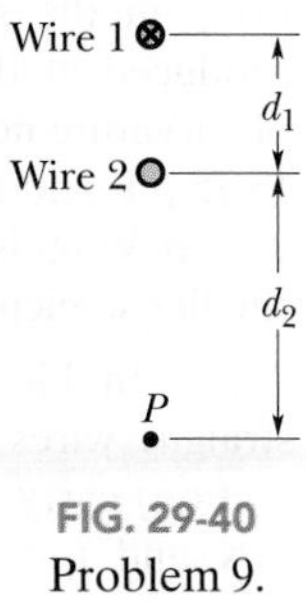

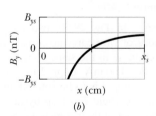

FIG. 29-40 Problem 9.

•10 In Fig. 29-41, two long straight wires at separation $d = 16.0$ cm carry currents $i_1 = 3.61$ mA and $i_2 = 3.00i_1$ out of the page. (a) At what point on the x axis shown is the net magnetic field due to the currents equal to zero? (b) If the two currents are doubled, is the point of zero magnetic field shifted toward wire 1, shifted toward wire 2, or unchanged?

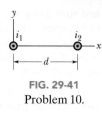

FIG. 29-41 Problem 10.

•11 In Fig. 29-42, a current $i = 10$ A is set up in a long hairpin conductor formed by bending a wire into a semicircle of radius $R = 5.0$ mm. Point b is midway between the straight sections and so distant from the semicircle that each straight section can be approximated as being an infinite wire. What are the (a) magnitude and (b) direction (into or out of the page) of $\vec{B}$ at a and the (c) magnitude and (d) direction of $\vec{B}$ at b?

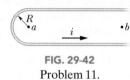

FIG. 29-42 Problem 11.

•12 In Fig. 29-43, point $P$ is at perpendicular distance $R = 2.00$ cm from a very long straight wire carrying a current. The magnetic field $\vec{B}$ set up at point $P$ is due to contributions from all the identical current-length elements $i$ $d\vec{s}$ along the wire. What is the distance $s$ to the element making (a) the greatest contribution to field $\vec{B}$ and (b) 10.0% of the greatest contribution?

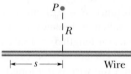

FIG. 29-43 Problem 12.

••13 Figure 29-44 shows a proton moving at velocity $\vec{v} = (-200$ m/s$)\hat{j}$ toward a long straight wire with current $i = 350$ mA. At the instant shown, the proton's distance from the wire is $d = 2.89$ cm. In unit-vector notation, what is the magnetic force on the proton due to the current? ILW

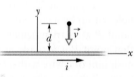

FIG. 29-44 Problem 13.

••14 Figure 29-45a shows, in cross section, two long, parallel wires carrying current and separated by distance $L$. The ratio $i_1/i_2$ of their currents is 4.00; the directions of the currents are not indicated. Figure 29-45b shows the y component $B_y$ of their net magnetic field along the x axis to the right of wire 2. The vertical scale is set by $B_{ys} = 4.0$ nT, and the horizontal scale is set by $x_s = 20.0$ cm. (a) At what value of $x > 0$ is $B_y$ maximum? (b) If $i_2 = 3$ mA, what is the value of that maximum? What is the direction (into or out of the page) of (c) $i_1$ and (d) $i_2$?

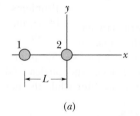

FIG. 29-45 Problem 14.

••15 A wire with current $i = 3.00$ A is shown in Fig. 29-46. Two semi-infinite straight sections, both tangent to the same circle, are connected by a circular arc that has a central angle $\theta$

and runs along the circumference of the circle. The arc and the two straight sections all lie in the same plane. If $B = 0$ at the circle's center, what is $\theta$? **SSM**

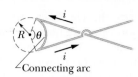

**FIG. 29-46**
Problem 15.

**••16** Figure 29-47 shows, in cross section, four thin wires that are parallel, straight, and very long. They carry identical currents in the directions indicated. Initially all four wires are at distance $d = 15.0$ cm from the origin of the coordinate system, where they create a net magnetic field $\vec{B}$. (a) To what value of $x$ must you move wire 1 along the $x$ axis in order to rotate $\vec{B}$ counterclockwise by 30°? (b) With wire 1 in that new position, to what value of $x$ must you move wire 3 along the $x$ axis to rotate $\vec{B}$ by 30° back to its initial orientation? **GO**

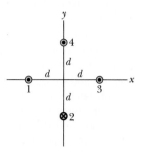

**FIG. 29-47** Problem 16.

**••17** In Fig. 29-48, point $P_1$ is at distance $R = 13.1$ cm on the perpendicular bisector of a straight wire of length $L = 18.0$ cm carrying current $i = 58.2$ mA. (Note that the wire is *not* long.) What is the magnitude of the magnetic field at $P_1$ due to $i$?

**••18** Equation 29-4 gives the magnitude $B$ of the magnetic field set up by a current in an *infinitely long* straight wire, at a point $P$ at perpendicular distance $R$ from the wire. Suppose that point $P$ is actually at perpendicular distance $R$ from the midpoint of a wire with a *finite* length $L$. Using Eq. 29-4 to calculate $B$ then results in a certain percentage error. What value must the ratio $L/R$ exceed if the percentage error is to be less than 1.00%? That is, what $L/R$ gives

$$\frac{(B \text{ from Eq. 29-4}) - (B \text{ actual})}{(B \text{ actual})} (100\%) = 1.00\%?$$

**••19** In Fig. 29-49, four long straight wires are perpendicular to the page, and their cross sections form a square of edge length $a = 20$ cm. The currents are out of the page in wires 1 and 4 and into the page in wires 2 and 3, and each wire carries 20 A. In unit-vector notation, what is the net magnetic field at the square's center? **SSM**

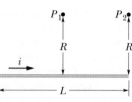

**FIG. 29-48** Problems 17 and 21.

**FIG. 29-49**
Problems 19, 36, and 39.

**••20** In Fig. 29-50, two concentric circular loops of wire carrying current in the same direction lie in the same plane. Loop 1 has radius 1.50 cm and carries 4.00 mA. Loop 2 has radius 2.50 cm and carries 6.00 mA. Loop 2 is to be rotated about a diameter while the net magnetic field $\vec{B}$ set up by the two loops at their common center is measured. Through what angle must loop 2 be rotated so that the magnitude of that net field is 100 nT? **GO**

**FIG. 29-50** Problem 20.

**••21** In Fig. 29-48, point $P_2$ is at perpendicular distance $R = 25.1$ cm from one end of straight wire of length $L = 13.6$ cm carrying current $i = 0.693$ A. (Note that the wire is *not* long.) What is the magnitude of the magnetic field at $P_2$? **SSM**

**••22** In Fig. 29-51a, wire 1 consists of a circular arc and two radial lengths; it carries current $i_1 = 0.50$ A in the direction indicated. Wire 2, shown in cross section, is long, straight, and perpendicular to the plane of the figure. Its distance from the center of the arc is equal to the radius $R$ of the arc, and it carries a current $i_2$ that can be varied. The two currents set up a net magnetic field $\vec{B}$ at the center of the arc. Figure 29-51b gives the square of the field's magnitude $B^2$ plotted versus the square of the current $i_2^2$. The vertical scale is set by $B_s^2 = 10.0 \times 10^{-10}\,\text{T}^2$. What angle is subtended by the arc?

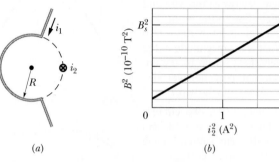

(a)                                    (b)

**FIG. 29-51** Problem 22.

**••23** Figure 29-52 shows two current segments. The lower segment carries current $i_1 = 0.40$ A and includes a circular arc with radius 5.0 cm, angle 180°, and center point $P$. The upper segment carries current $i_2 = 2i_1$ and includes a circular arc with radius 4.0 cm, angle 120°, and the same center point $P$. What are the (a) magnitude and (b) direction of the net magnetic field $\vec{B}$ at $P$ for the indicated current directions? What are the (c) magnitude and (d) direction of $\vec{B}$ if $i_1$ is reversed?

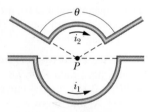

**FIG. 29-52** Problem 23.

**••24** A current is set up in a wire loop consisting of a semicircle of radius 4.00 cm, a smaller concentric semicircle, and two radial straight lengths, all in the same plane. Figure 29-53a shows the arrangement but is not drawn to scale. The magnitude of the magnetic field produced at the center of curvature is 47.25 $\mu$T. The smaller semicircle is then flipped over (rotated) until the loop is again entirely in the same plane (Fig. 29-53b). The magnetic field produced at the (same) center of curvature now has magnitude 15.75 $\mu$T, and its direction is reversed. What is the radius of the smaller semicircle?

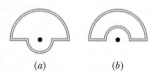

(a)            (b)

**FIG. 29-53** Problem 24.

**••25** In Fig. 29-54, two long straight wires (shown in cross section) carry currents $i_1 = 30.0$ mA and $i_2 = 40.0$ mA directly

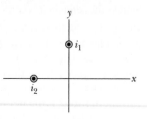

**FIG. 29-54** Problem 25.

out of the page. They are equal distances from the origin, where they set up a magnetic field $\vec{B}$. To what value must current $i_1$ be changed in order to rotate $\vec{B}$ 20.0° clockwise?

**••26** Figure 29-55a shows two wires, each carrying a current. Wire 1 consists of a circular arc of radius $R$ and two radial lengths; it carries current $i_1 = 2.0$ A in the direction indicated. Wire 2 is long and straight; it carries a current $i_2$ that can be varied; and it is at distance $R/2$ from the center of the arc. The net magnetic field $\vec{B}$ due to the two currents is measured at the center of curvature of the arc. Figure 29-55b is a plot of the component of $\vec{B}$ in the direction perpendicular to the figure as a function of current $i_2$. The horizontal scale is set by $i_{2s} = 1.00$ A. What is the angle subtended by the arc? **GO**

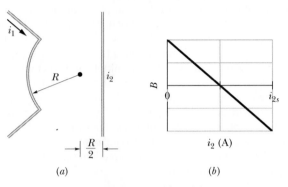

*(a)*

*(b)*

**FIG. 29-55** Problem 26.

**••27** One long wire lies along an $x$ axis and carries a current of 30 A in the positive $x$ direction. A second long wire is perpendicular to the $xy$ plane, passes through the point $(0, 4.0$ m, $0)$, and carries a current of 40 A in the positive $z$ direction. What is the magnitude of the resulting magnetic field at the point $(0, 2.0$ m, $0)$?

**••28** In Fig. 29-56, part of a long insulated wire carrying current $i = 5.78$ mA is bent into a circular section of radius $R = 1.89$ cm. In unit-vector notation, what is the magnetic field at the center of curvature $C$ if the cir-

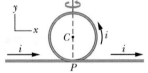

**FIG. 29-56** Problem 28.

cular section (a) lies in the plane of the page as shown and (b) is perpendicular to the plane of the page after being rotated 90° counterclockwise as indicated?

**••29** Figure 29-57 shows two very long straight wires (in cross section) that each carry a current of 4.00 A directly out of the page. Distance $d_1 = 6.00$ m and distance $d_2 = 4.00$ m. What is the magnitude of the net magnetic field at point $P$, which lies on a perpendicular bisector to the wires?

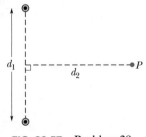

**FIG. 29-57** Problem 29.

**•••30** The current-carrying wire loop in Fig. 29-58a lies all in one plane and consists of a semicircle of radius 10.0 cm, a smaller semicircle with the same center, and two radial lengths. The smaller semicircle is rotated out of that plane by angle $\theta$, until it is perpendicular to the plane (Fig. 29-58b). Figure 29-58c gives the magnitude of the net magnetic field at

the center of curvature versus angle $\theta$. The vertical scale is set by $B_a = 10.0$ $\mu$T and $B_b = 12.0$ $\mu$T. What is the radius of the smaller semicircle?

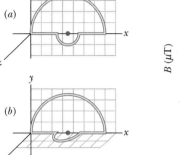

*(a)*

*(b)*

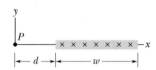

*(c)*

**FIG. 29-58** Problem 30.

**•••31** Figure 29-59 shows a cross section of a long thin ribbon of width $w = 4.91$ cm that is carrying a uniformly distributed total current $i = 4.61$ $\mu$A into the page. In unit-vector notation, what is the magnetic field $\vec{B}$ at a point $P$ in the

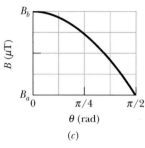

**FIG. 29-59** Problem 31.

plane of the ribbon at a distance $d = 2.16$ cm from its edge? (*Hint:* Imagine the ribbon as being constructed from many long, thin, parallel wires.) **SSM ILW**

**•••32** Figure 29-60 shows, in cross section, two long straight wires held against a plastic cylinder of radius 20.0 cm. Wire 1 carries current $i_1 = 60.0$ mA out of the page and is fixed in place at the left side of the cylinder. Wire 2 carries current $i_2 = 40.0$ mA out of the page and can be moved around the cylinder. At what (positive) angle $\theta_2$ should wire 2 be positioned such that, at the origin, the net magnetic field due to the two currents has magnitude 80.0 nT?

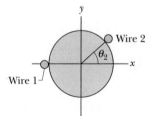

**FIG. 29-60** Problem 32.

**•••33** In Fig. 29-61, length $a$ is 4.7 cm (short) and current $i$ is 13 A. What are the (a) magnitude and (b) direction (into or out of the page) of the magnetic field at point $P$?

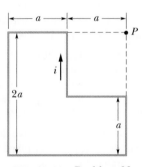

**FIG. 29-61** Problem 33.

**•••34** Two long straight thin wires with current lie against an equally long plastic cylinder, at radius $R = 20.0$ cm from the cylinder's central axis. Figure 29-62a shows, in cross section, the cylinder and wire 1 but not wire 2. With wire 2 fixed in place, wire 1 is moved around the cylinder, from angle $\theta_1 = 0°$ to angle $\theta_1 = 180°$, through the first and second quadrants of the $xy$ coordinate system. The net magnetic field $\vec{B}$ at the center of the cylinder is measured as a function of $\theta_1$. Figure

29-62b gives the x component $B_x$ of that field as a function of $\theta_1$ (the vertical scale is set by $B_{xs} = 6.0\ \mu T$), and Fig. 29-62c gives the y component $B_y$ (the vertical scale is set by $B_{ys} = 4.0\ \mu T$). (a) At what angle $\theta_2$ is wire 2 located? What are the (b) size and (c) direction (into or out of the page) of the current in wire 1 and the (d) size and (e) direction of the current in wire 2?

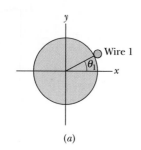

(a)

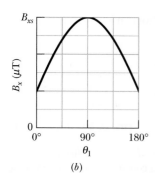 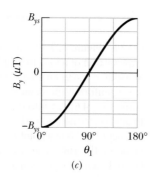

(b)   (c)

**FIG. 29-62** Problem 34.

## sec. 29-3   Force Between Two Parallel Currents

•**35**   Figure 29-63 shows wire 1 in cross section; the wire is long and straight, carries a current of 4.00 mA out of the page, and is at distance $d_1 = 2.40$ cm from a surface. Wire 2, which is parallel to wire 1 and also long, is at horizontal distance $d_2 = 5.00$ cm from wire 1 and carries a current of 6.80 mA into the page. What is the x component of the magnetic force *per unit length* on wire 2 due to wire 1? **SSM**

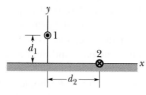

**FIG. 29-63**   Problem 35.

••**36**   In Fig. 29-49, four long straight wires are perpendicular to the page, and their cross sections form a square of edge length $a = 8.50$ cm. Each wire carries 15.0 A, and all the currents are out of the page. In unit-vector notation, what is the net magnetic force *per meter of wire length* on wire 1?

••**37**   In Fig. 29-64, five long parallel wires in an xy plane are separated by distance $d = 50.0$ cm. The currents into the page are $i_1 = 2.00$ A, $i_3 = 0.250$ A, $i_4 = 4.00$ A, and $i_5 = 2.00$ A; the current out of the page is $i_2 = 4.00$ A. What is the magnitude of the net force *per unit length* acting on wire 3 due to the currents in the other wires? **GO**

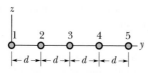

**FIG. 29-64**   Problems 37 and 38.

••**38**   In Fig. 29-64, five long parallel wires in an xy plane are

separated by distance $d = 8.00$ cm, have lengths of 10.0 m, and carry identical currents of 3.00 A out of the page. Each wire experiences a magnetic force due to the other wires. In unit-vector notation, what is the net magnetic force on (a) wire 1, (b) wire 2, (c) wire 3, (d) wire 4, and (e) wire 5?

••**39**   In Fig. 29-49, four long straight wires are perpendicular to the page, and their cross sections form a square of edge length $a = 13.5$ cm. Each wire carries 7.50 A, and the currents are out of the page in wires 1 and 4 and into the page in wires 2 and 3. In unit-vector notation, what is the net magnetic force *per meter of wire length* on wire 4? **GO**

••**40**   Figure 29-65a shows, in cross section, three current-carrying wires that are long, straight, and parallel to one another. Wires 1 and 2 are fixed in place on an x axis, with separation d. Wire 1 has a current of 0.750 A, but the direction of the current is not given. Wire 3, with a current of 0.250 A out of the page, can be moved along the x axis to the right of wire 2. As wire 3 is moved, the magnitude of the net magnetic force $\vec{F}_2$ on wire 2 due to the currents in wires 1 and 3 changes. The y component of that force is $F_{2y}$ and the value per unit length of wire 2 is $F_{2y}/L_2$. Figure 29-65b gives $F_{2y}/L_2$ versus the position x of wire 3. The plot has an asymptote $F_{2y}/L_2 = -0.627\ \mu N/m$ as $x \to \infty$. The horizontal scale is set by $x_s = 12.0$ cm. What are the (a) size and (b) direction (into or out of the page) of the current in wire 2?

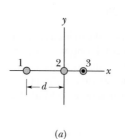 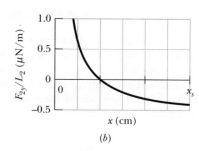

(a)   (b)

**FIG. 29-65**   Problem 40.

•••**41**   In Fig. 29-66, a long straight wire carries a current $i_1 = 30.0$ A and a rectangular loop carries current $i_2 = 20.0$ A. Take $a = 1.00$ cm, $b = 8.00$ cm, and $L = 30.0$ cm. In unit-vector notation, what is the net force on the loop due to $i_1$? **ILW**

## sec. 29-4   Ampere's Law

•**42**   Figure 29-67 shows two closed paths wrapped around two conducting loops carrying currents $i_1 = 5.0$ A and $i_2 = 3.0$ A. What is the value of the integral $\oint \vec{B} \cdot d\vec{s}$ for (a) path 1 and (b) path 2?

•**43**   Each of the eight conductors in Fig. 29-68 carries 2.0 A of current into or out of the page. Two paths are indicated

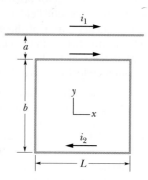

**FIG. 29-66**   Problem 41.

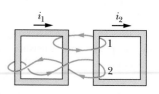

**FIG. 29-67**   Problem 42.

for the line integral $\oint \vec{B} \cdot d\vec{s}$. What is the value of the integral for (a) path 1 and (b) path 2? **SSM**

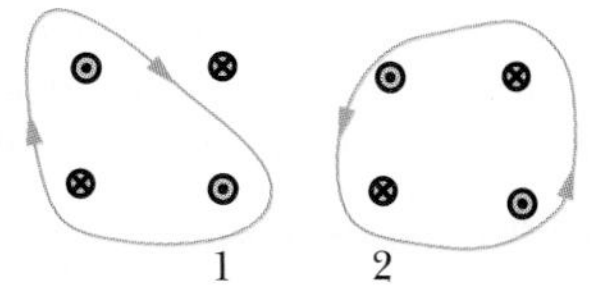

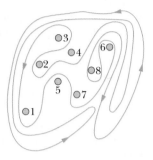

**FIG. 29-68** Problem 43.

**•44** Eight wires cut the page perpendicularly at the points shown in Fig. 29-69. A wire labeled with the integer $k$ ($k = 1, 2, \ldots, 8$) carries the current $ki$, where $i = 4.50$ mA. For those wires with odd $k$, the current is out of the page; for those with even $k$, it is into the page. Evaluate $\oint \vec{B} \cdot d\vec{s}$ along the closed path in the direction shown.

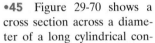

**FIG. 29-69** Problem 44.

**•45** Figure 29-70 shows a cross section across a diameter of a long cylindrical conductor of radius $a = 2.00$ cm carrying uniform current 170 A. What is the magnitude of the current's magnetic field at radial distance (a) 0, (b) 1.00 cm, (c) 2.00 cm (wire's surface), and (d) 4.00 cm?

**•46** In a particular region there is a uniform current density of 15 A/m$^2$ in the positive $z$ direction. What is the value of $\oint \vec{B} \cdot d\vec{s}$ when that line integral is calculated along the three straight-line segments from $(x, y, z)$ coordinates $(4d, 0, 0)$ to $(4d, 3d, 0)$ to $(0, 0, 0)$ to $(4d, 0, 0)$, where $d = 20$ cm?

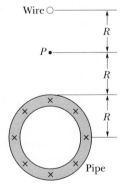

**FIG. 29-70**
Problem 45.

**••47** The current density $\vec{J}$ inside a long, solid, cylindrical wire of radius $a = 3.1$ mm is in the direction of the central axis, and its magnitude varies linearly with radial distance $r$ from the axis according to $J = J_0 r/a$, where $J_0 = 310$ A/m$^2$. Find the magnitude of the magnetic field at (a) $r = 0$, (b) $r = a/2$, and (c) $r = a$.  **ILW**

**••48** In Fig. 29-71, a long circular pipe with outside radius $R = 2.6$ cm carries a (uniformly distributed) current $i = 8.00$ mA into the page. A wire runs parallel to the pipe at a distance of $3.00R$ from center to center. Find the (a) magnitude and (b) direction (into or out of the page) of the current in the wire such that the net magnetic field at point $P$ has the same magnitude as the net magnetic field at the center of the pipe but is in the opposite direction.

### sec. 29-5 Solenoids and Toroids

**•49** A 200-turn solenoid having a length of 25 cm and a diameter of 10 cm carries a current of 0.29 A. Calculate the magnitude of the magnetic field $\vec{B}$ inside the solenoid.

**•50** A solenoid 1.30 m long and 2.60 cm in diameter carries a current of 18.0 A. The magnetic field inside the solenoid is 23.0 mT. Find the length of the wire forming the solenoid.

**FIG. 29-71**
Problem 48.

**•51** A toroid having a square cross section, 5.00 cm on a side, and an inner radius of 15.0 cm has 500 turns and carries a current of 0.800 A. (It is made up of a square solenoid—instead of a round one as in Fig. 29-17—bent into a doughnut shape.) What is the magnetic field inside the toroid at (a) the inner radius and (b) the outer radius?

**•52** A solenoid that is 95.0 cm long has a radius of 2.00 cm and a winding of 1200 turns; it carries a current of 3.60 A. Calculate the magnitude of the magnetic field inside the solenoid.

**••53** A long solenoid with 10.0 turns/cm and a radius of 7.00 cm carries a current of 20.0 mA. A current of 6.00 A exists in a straight conductor located along the central axis of the solenoid. (a) At what radial distance from the axis will the direction of the resulting magnetic field be at 45.0° to the axial direction? (b) What is the magnitude of the magnetic field there? **SSM ILW WWW**

**••54** An electron is shot into one end of a solenoid. As it enters the uniform magnetic field within the solenoid, its speed is 800 m/s and its velocity vector makes an angle of 30° with the central axis of the solenoid. The solenoid carries 4.0 A and has 8000 turns along its length. How many revolutions does the electron make along its helical path within the solenoid by the time it emerges from the solenoid's opposite end? (In a real solenoid, where the field is not uniform at the two ends, the number of revolutions would be slightly less than the answer here.)

**••55** A long solenoid has 100 turns/cm and carries current $i$. An electron moves within the solenoid in a circle of radius 2.30 cm perpendicular to the solenoid axis. The speed of the electron is $0.0460c$ ($c$ = speed of light). Find the current $i$ in the solenoid.

### sec. 29-6 A Current-Carrying Coil as a Magnetic Dipole

**•56** Figure 29-72a shows a length of wire carrying a current $i$ and bent into a circular coil of one turn. In Fig. 29-72b the same length of wire has been bent to give a coil of two turns, each of half the original radius. (a) If $B_a$ and $B_b$ are the magnitudes of the magnetic fields at the centers of the two coils, what is the ratio $B_b/B_a$? (b) What is the ratio $\mu_b/\mu_a$ of the dipole moment magnitudes of the coils?

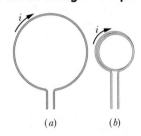

**FIG. 29-72** Problem 56.

**•57** What is the magnitude of the magnetic dipole moment $\vec{\mu}$ of the solenoid described in Problem 49? **SSM**

**•58** Figure 29-73 shows an arrangement known as a Helmholtz coil. It consists of two circular coaxial coils, each of 200 turns and radius $R = 25.0$ cm, separated by a distance $s = R$. The two coils carry equal currents $i = 12.2$ mA in the same direction. Find the magnitude of the net magnetic field at $P$, midway between the coils.

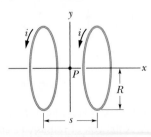

**FIG. 29-73** Problems 58 and 86.

**•59** A student makes a short electromagnet by winding 300 turns of wire around a wooden cylinder of diameter $d = 5.0$ cm. The coil is connected to a battery producing a current of 4.0 A in the wire. (a) What is the magnitude of the magnetic dipole moment of this device? (b) At what axial distance $z \gg d$ will the magnetic field have the magnitude 5.0 $\mu$T (approximately one-tenth that of Earth's magnetic field)? **SSM**

**••60** In Fig. 29-74, current $i = 56.2$ mA is set up in a loop having two radial lengths and two semicircles of radii $a = 5.72$ cm and $b = 9.36$ cm with a common center $P$. What are the (a) magnitude and (b) direction (into or out of the page) of the magnetic field at $P$ and the (c) magnitude and (d) direction of the loop's magnetic dipole moment?

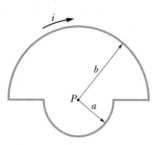

FIG. 29-74   Problem 60.

**••61** In Fig. 29-75, a conductor carries 6.0 A along the closed path $abcdefgha$ running along 8 of the 12 edges of a cube of edge length 10 cm. (a) Taking the path to be a combination of three square current loops ($bcfgb$, $abgha$, and $cdefc$), find the net magnetic moment of the path in unit-vector notation. (b) What is the magnitude of the net magnetic field at the $xyz$ coordinates of $(0, 5.0 \text{ m}, 0)$?

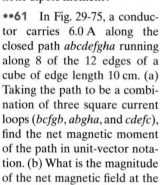

FIG. 29-75   Problem 61.

**••62** In Fig. 29-76a, two circular loops, with different currents but the same radius of 4.0 cm, are centered on a $y$ axis. They are initially separated by distance $L = 3.0$ cm, with loop 2 positioned at the origin of the axis. The currents in the two loops produce a net magnetic field at the origin, with $y$ component $B_y$. That component is to be measured as loop 2 is gradually moved in the positive direction of the $y$ axis. Figure 29-76b gives $B_y$ as a function of the position $y$ of loop 2. The curve approaches an asymptote of $B_y = 7.20$ $\mu$T as $y \to \infty$. The horizontal scale is set by $y_s = 10.0$ cm. What are (a) current $i_1$ in loop 1 and (b) current $i_2$ in loop 2?

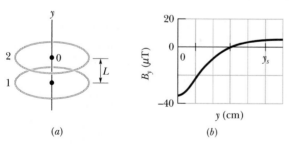

FIG. 29-76   Problem 62.

**••63** A circular loop of radius 12 cm carries a current of 15 A. A flat coil of radius 0.82 cm, having 50 turns and a current of 1.3 A, is concentric with the loop. The plane of the loop is perpendicular to the plane of the coil. Assume the loop's mag-

netic field is uniform across the coil. What is the magnitude of (a) the magnetic field produced by the loop at its center and (b) the torque on the coil due to the loop?

**Additional Problems**

**64** Figure 29-77 shows a closed loop with current $i = 2.00$ A. The loop consists of a half-circle of radius 4.00 m, two quarter-circles each of radius 2.00 m, and three radial straight wires. What is the magnitude of the net magnetic field at the common center of the circular sections?

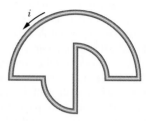

FIG. 29-77   Problem 64.

**65** Figure 29-78 shows a cross section of a long cylindrical conductor of radius $a = 4.00$ cm containing a long cylindrical hole of radius $b = 1.50$ cm. The central axes of the cylinder and hole are parallel and are distance $d = 2.00$ cm apart; current $i = 5.25$ A is uniformly distributed over the tinted area. (a) What is the magnitude of the magnetic field at the center of the hole? (b) Discuss the two special cases $b = 0$ and $d = 0$.

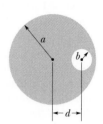

FIG. 29-78   Problem 65.

**66** The magnitude of the magnetic field 88.0 cm from the axis of a long straight wire is 7.30 $\mu$T. What is the current in the wire?

**67** Three long wires are parallel to a $z$ axis, and each carries a current of 10 A in the positive $z$ direction. Their points of intersection with the $xy$ plane form an equilateral triangle with sides of 50 cm, as shown in Fig. 29-79. A fourth wire (wire $b$) passes through the midpoint of the base of the triangle and is parallel to the other three wires. If the net magnetic force on wire $a$ is zero, what are the (a) size and (b) direction ($+z$ or $-z$) of the current in wire $b$?

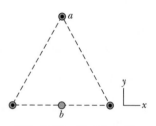

FIG. 29-79   Problem 67.

**68** Figure 29-80 shows, in cross section, two long parallel wires spaced by distance $d = 10.0$ cm; each carries 100 A, out of the page in wire 1. Point $P$ is on a perpendicular bisector of the line connecting the wires. In unit-vector notation, what is the net magnetic field at $P$ if the current in wire 2 is (a) out of the page and (b) into the page?

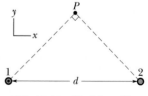

FIG. 29-80   Problem 68.

**69** A 10-gauge bare copper wire (2.6 mm in diameter) can carry a current of 50 A without overheating. For this current, what is the magnitude of the magnetic field at the surface of the wire?

**70** A long vertical wire carries an unknown current. Coaxial with the wire is a long, thin, cylindrical conducting surface that carries a current of 30 mA upward. The cylindrical surface has a radius of 3.0 mm. If the magnitude of the magnetic field at a point 5.0 mm from the wire is 1.0 $\mu$T, what are the (a) size and (b) direction of the current in the wire?

**71** Figure 29-81 shows a wire segment of length $\Delta s = 3.0$ cm, centered at the origin, carrying current $i = 2.0$ A in the positive $y$ direction (as part of some complete circuit). To calculate the magnitude of the magnetic field $\vec{B}$ produced by the segment at a point several meters from the

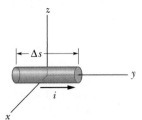

FIG. 29-81 Problem 71.

origin, we can use $B = (\mu_0/4\pi)i\,\Delta s\,(\sin\theta)/r^2$ as the Biot–Savart law. This is because $r$ and $\theta$ are essentially constant over the segment. Calculate $\vec{B}$ (in unit-vector notation) at the $(x, y, z)$ coordinates (a) $(0, 0, 5.0$ m$)$, (b) $(0, 6.0$ m$, 0)$, (c) $(7.0$ m$, 7.0$ m$, 0)$, and (d) $(-3.0$ m$, -4.0$ m$, 0)$. SSM

**72** In Fig. 29-82, a closed loop carries current $i = 200$ mA. The loop consists of two radial straight wires and two concentric circular arcs of radii 2.00 m and 4.00 m. The angle $\theta$ is $\pi/4$ rad. What are the (a) magnitude and (b) direction (into or out of the page) of the net magnetic field at the center of curvature $P$?

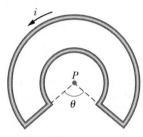

FIG. 29-82 Problem 72.

**73** A cylindrical cable of radius 8.00 mm carries a current of 25.0 A, uniformly spread over its cross-sectional area. At what distance from the center of the wire is there a point within the wire where the magnetic field magnitude is 0.100 mT?

**74** A long wire carrying 100 A is perpendicular to the magnetic field lines of a uniform magnetic field of magnitude 5.0 mT. At what distance from the wire is the net magnetic field equal to zero?

**75** Two wires, both of length $L$, are formed into a circle and a square, and each carries current $i$. Show that the square produces a greater magnetic field at its center than the circle produces at its center.

**76** A long straight wire carries a current of 50 A. An electron, traveling at $1.0 \times 10^7$ m/s, is 5.0 cm from the wire. What is the magnitude of the magnetic force on the electron if the electron velocity is directed (a) toward the wire, (b) parallel to the wire in the direction of the current, and (c) perpendicular to the two directions defined by (a) and (b)?

**77** In unit-vector notation, what is the magnetic field at point $P$ in Fig. 29-83 if $i = 10$ A and $a = 8.0$ cm? (Note that the wires are *not* long.) SSM

**78** Figure 29-84 shows, in cross section, two long parallel wires spaced by distance $d = 18.6$ cm. Each carries 4.23 A,

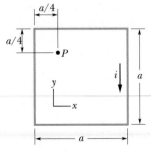

FIG. 29-83 Problem 77.

out of the page in wire 1 and into the page in wire 2. In unit-vector notation, what is the magnetic field at point $P$ at distance $R = 34.2$ cm?

**79** Figure 29-85 shows a cross section of an infinite conducting sheet carrying a current per unit $x$-length of $\lambda$; the current emerges perpendicularly out of the page. (a) Use the Biot–Savart law and

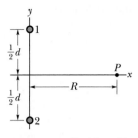

FIG. 29-84 Problem 78.

symmetry to show that for all points $P$ above the sheet and all points $P'$ below it, the magnetic field $\vec{B}$ is parallel to the sheet and directed as shown. (b) Use Ampere's law to prove that $B = \frac{1}{2}\mu_0\lambda$ at all points $P$ and $P'$. SSM

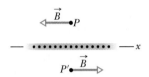

FIG. 29-85 Problem 79.

**80** Two long wires lie in an $xy$ plane, and each carries a current in the positive direction of the $x$ axis. Wire 1 is at $y = 10.0$ cm and carries 6.00 A; wire 2 is at $y = 5.00$ cm and carries 10.0 A. (a) In unit-vector notation, what is the net magnetic field $\vec{B}$ at the origin? (b) At what value of $y$ does $\vec{B} = 0$? (c) If the current in wire 1 is reversed, at what value of $y$ does $\vec{B} = 0$?

**81** Figure 29-86 shows a cross section of a hollow cylindrical conductor of radii $a$ and $b$, carrying a uniformly distributed current $i$. (a) Show that the magnetic field magnitude $B(r)$ for the radial distance $r$ in the range $b < r < a$ is given by

$$B = \frac{\mu_0 i}{2\pi(a^2 - b^2)} \frac{r^2 - b^2}{r}.$$

(b) Show that when $r = a$, this equation gives the magnetic field magnitude $B$ at the surface of a long straight wire carrying current $i$; when $r = b$, it gives zero magnetic field; and when $b = 0$, it gives the magnetic field inside a solid conductor of radius $a$ carrying current $i$. (c) Assume that $a = 2.0$ cm, $b = 1.8$ cm, and $i = 100$ A, and then plot $B(r)$ for the range $0 < r < 6$ cm. SSM

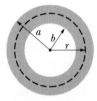

FIG. 29-86 Problem 81.

**82** Three long wires all lie in an $xy$ plane parallel to the $x$ axis. They are spaced equally, 10 cm apart. The two outer wires each carry a current of 5.0 A in the positive $x$ direction. What is the magnitude of the force on a 3.0 m section of either of the outer wires if the current in the center wire is 3.2 A (a) in the positive $x$ direction and (b) in the negative $x$ direction?

**83** In Fig. 29-87, two infinitely long wires carry equal currents $i$. Each follows a 90° arc on the circumference of the same circle of radius $R$.

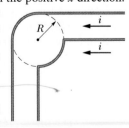

FIG. 29-87 Problem 83.

Show that the magnetic field $\vec{B}$ at the center of the circle is the same as the field $\vec{B}$ a distance $R$ below an infinite straight wire carrying a current $i$ to the left.

**84** A long wire is known to have a radius greater than 4.0 mm and to carry a current that is uniformly distributed over its cross section. The magnitude of the magnetic field due to that current is 0.28 mT at a point 4.0 mm from the axis of the wire, and 0.20 mT at a point 10 mm from the axis of the wire. What is the radius of the wire?

**85** A long, hollow, cylindrical conductor (inner radius 2.0 mm, outer radius 4.0 mm) carries a current of 24 A distributed uniformly across its cross section. A long thin wire that is co-axial with the cylinder carries a current of 24 A in the opposite direction. What is the magnitude of the magnetic field (a) 1.0 mm, (b) 3.0 mm, and (c) 5.0 mm from the central axis of the wire and cylinder?

**86** In Fig. 29-73, an arrangement known as Helmholtz coils consists of two circular coaxial coils, each of $N$ turns and radius $R$, separated by distance $s$. The two coils carry equal currents $i$ in the same direction. (a) Show that the first derivative of the magnitude of the net magnetic field of the coils ($dB/dx$) vanishes at the midpoint $P$ regardless of the value of $s$. Why would you expect this to be true from symmetry? (b) Show that the second derivative ($d^2B/dx^2$) also vanishes at $P$, provided $s = R$. This accounts for the uniformity of $B$ near $P$ for this particular coil separation.

**87** A square loop of wire of edge length $a$ carries current $i$. Show that, at the center of the loop, the magnitude of the magnetic field produced by the current is

$$B = \frac{2\sqrt{2}\mu_0 i}{\pi a}.$$

**88** Show that the magnitude of the magnetic field produced at the center of a rectangular loop of wire of length $L$ and width $W$, carrying a current $i$, is

$$B = \frac{2\mu_0 i}{\pi} \frac{(L^2 + W^2)^{1/2}}{LW}.$$

**89** A square loop of wire of edge length $a$ carries current $i$. Show that the magnitude of the magnetic field produced at a point on the central perpendicular axis of the loop and a distance $x$ from its center is

$$B(x) = \frac{4\mu_0 i a^2}{\pi(4x^2 + a^2)(4x^2 + 2a^2)^{1/2}}.$$

Prove that this result is consistent with the result shown in Problem 87. **SSM**

**90** Figure 29-88 is an idealized schematic drawing of a rail gun. Projectile $P$ sits between two wide rails of circular cross section; a source of current sends current through the rails and through the (conducting) projectile (a fuse is not used). (a) Let $w$ be the distance between the rails, $R$ the radius of each rail, and $i$ the current. Show that the force on the projectile is directed to the right along the rails and is given approximately by

$$F = \frac{i^2\mu_0}{2\pi} \ln \frac{w + R}{R}.$$

(b) If the projectile starts from the left end of the rails at rest, find the speed $v$ at which it is expelled at the right. Assume that $i = 450$ kA, $w = 12$ mm, $R = 6.7$ cm, $L = 4.0$ m, and the projectile mass is 10 g.

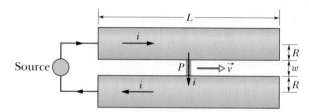

**FIG. 29-88** Problem 90.

**91** Show that a uniform magnetic field $\vec{B}$ cannot drop abruptly to zero (as is suggested by the lack of field lines to the right of point $a$ in Fig. 29-89) as one moves perpendicular to $\vec{B}$, say along the horizontal arrow in the figure. (*Hint:* Apply Ampere's law to the rectangular path shown by the dashed lines.) In actual magnets, "fringing" of the magnetic field lines always occurs, which means that $\vec{B}$ approaches zero in a gradual manner. Modify the field lines in the figure to indicate a more realistic situation. **SSM**

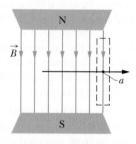

**FIG. 29-89** Problem 91.

**92** Show that if the thickness of a toroid is much smaller than its radius of curvature (a very skinny toroid), then Eq. 29-24 for the field inside a toroid reduces to Eq. 29-23 for the field inside a solenoid. Explain why this result is to be expected.

**93** Figure 29-90 shows a cross section of a long conducting coaxial cable and gives its radii ($a$, $b$, $c$). Equal but opposite currents $i$ are uniformly distributed in the two conductors. Derive expressions for $B(r)$ with radial distance $r$ in the ranges (a) $r < c$, (b) $c < r < b$, (c) $b < r < a$, and (d) $r > a$. (e) Test these expressions for all the special cases that occur to you. (f) Assume that $a = 2.0$ cm, $b = 1.8$ cm, $c = 0.40$ cm, and $i = 120$ A and plot the function $B(r)$ over the range $0 < r < 3$ cm.

**FIG. 29-90** Problem 93.

# Induction and Inductance

Mark Joseph/Getty Images, Inc.

Traditionally, foundries used flame-heated furnaces to melt metals. However, many modern foundries avoid the resulting air pollution by using an induction furnace in which the metal is heated by the current in insulated wires wrapped around the crucible that holds the metal. However, the wires themselves do not get hot enough to melt the metal (or they would also melt). Indeed, they are kept cool by a water bath.

## How then does the metal become hot enough to melt?

The answer is in this chapter.

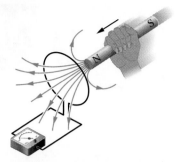

FIG. 30-1 An ammeter registers a current in the wire loop when the magnet is moving with respect to the loop.

## 30-1 | WHAT IS PHYSICS?

In Chapter 29 we discussed the fact that a current produces a magnetic field. That fact came as a surprise to the scientists who discovered the effect. Perhaps even more surprising was the discovery of the reverse effect: A magnetic field can produce an electric field that can drive a current. This link between a magnetic field and the electric field it produces (*induces*) is now called *Faraday's law of induction.*

The observations by Michael Faraday and other scientists that led to this law were at first just basic science, interesting in that they revealed another facet of how our universe works. Today, however, applications of that basic science are almost everywhere. For example, induction is the basis of the electric guitars that revolutionized early rock and still drive heavy metal and punk today. It is also the basis of the electric generators that power cities and transportation lines. Although induction stoves for the kitchen have yet to catch on with either professional chefs or amateur cooks, huge induction furnaces are commonplace in foundries where large amounts of metal must be melted rapidly.

Before we get to applications like the electric guitar, we must examine two simple experiments about Faraday's law of induction.

## 30-2 | Two Experiments

Let us examine two simple experiments to prepare for our discussion of Faraday's law of induction.

**First Experiment.** Figure 30-1 shows a conducting loop connected to a sensitive ammeter. Because there is no battery or other source of emf included, there is no current in the circuit. However, if we move a bar magnet toward the loop, a current suddenly appears in the circuit. The current disappears when the magnet stops. If we then move the magnet away, a current again suddenly appears, but now in the opposite direction. If we experimented for a while, we would discover the following:

1. A current appears only if there is relative motion between the loop and the magnet (one must move relative to the other); the current disappears when the relative motion between them ceases.

2. Faster motion produces a greater current.

3. If moving the magnet's north pole toward the loop causes, say, clockwise current, then moving the north pole away causes counterclockwise current. Moving the south pole toward or away from the loop also causes currents, but in the reversed directions.

The current produced in the loop is called an **induced current;** the work done per unit charge to produce that current (to move the conduction electrons that constitute the current) is called an **induced emf;** and the process of producing the current and emf is called **induction.**

**Second Experiment.** For this experiment we use the apparatus of Fig. 30-2, with the two conducting loops close to each other but not touching. If we close switch S, to turn on a current in the right-hand loop, the meter suddenly and briefly registers a current—an induced current—in the left-hand loop. If we then open the switch, another sudden and brief induced current appears in the left-hand loop, but in the opposite direction. We get an induced current (and thus an induced emf) only when the current in the right-hand loop is changing (either turning on or turning off) and not when it is constant (even if it is large).

The induced emf and induced current in these experiments are apparently caused when something changes—but what is that "something"? Faraday knew.

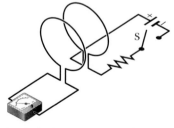

FIG. 30-2 An ammeter registers a current in the left-hand wire loop just as switch S is closed (to turn on the current in the right-hand wire loop) or opened (to turn off the current in the right-hand loop). No motion of the coils is involved.

# 30-3 I Faraday's Law of Induction

Faraday realized that an emf and a current can be induced in a loop, as in our two experiments, by changing the *amount of magnetic field* passing through the loop. He further realized that the "amount of magnetic field" can be visualized in terms of the magnetic field lines passing through the loop. **Faraday's law of induction,** stated in terms of our experiments, is this:

> An emf is induced in the loop at the left in Figs. 30-1 and 30-2 when the number of magnetic field lines that pass through the loop is changing.

The actual number of field lines passing through the loop does not matter; the values of the induced emf and induced current are determined by the *rate* at which that number changes.

In our first experiment (Fig. 30-1), the magnetic field lines spread out from the north pole of the magnet. Thus, as we move the north pole closer to the loop, the number of field lines passing through the loop increases. That increase apparently causes conduction electrons in the loop to move (the induced current) and provides energy (the induced emf) for their motion. When the magnet stops moving, the number of field lines through the loop no longer changes and the induced current and induced emf disappear.

In our second experiment (Fig. 30-2), when the switch is open (no current), there are no field lines. However, when we turn on the current in the right-hand loop, the increasing current builds up a magnetic field around that loop and at the left-hand loop. While the field builds, the number of magnetic field lines through the left-hand loop increases. As in the first experiment, the increase in field lines through that loop apparently induces a current and an emf there. When the current in the right-hand loop reaches a final, steady value, the number of field lines through the left-hand loop no longer changes, and the induced current and induced emf disappear.

Faraday's law does not explain *why* a current and an emf are induced in either experiment; it is just a statement that helps us visualize the induction.

## A Quantitative Treatment

To put Faraday's law to work, we need a way to calculate the *amount of magnetic field* that passes through a loop. In Chapter 23, in a similar situation, we needed to calculate the amount of electric field that passes through a surface. There we defined an electric flux $\Phi_E = \int \vec{E} \cdot d\vec{A}$. Here we define a *magnetic flux:* Suppose a loop enclosing an area $A$ is placed in a magnetic field $\vec{B}$. Then the **magnetic flux** through the loop is

$$\Phi_B = \int \vec{B} \cdot d\vec{A} \quad \text{(magnetic flux through area } A). \tag{30-1}$$

As in Chapter 23, $d\vec{A}$ is a vector of magnitude $dA$ that is perpendicular to a differential area $dA$.

As a special case of Eq. 30-1, suppose that the loop lies in a plane and that the magnetic field is perpendicular to the plane of the loop. Then we can write the dot product in Eq. 30-1 as $B \, dA \cos 0° = B \, dA$. If the magnetic field is also uniform, then $B$ can be brought out in front of the integral sign. The remaining $\int dA$ then gives just the area $A$ of the loop. Thus, Eq. 30-1 reduces to

$$\Phi_B = BA \quad (\vec{B} \perp \text{area } A, \vec{B} \text{ uniform}). \tag{30-2}$$

From Eqs. 30-1 and 30-2, we see that the SI unit for magnetic flux is the tesla–square meter, which is called the *weber* (abbreviated Wb):

$$1 \text{ weber} = 1 \text{ Wb} = 1 \text{ T} \cdot \text{m}^2. \tag{30-3}$$

With the notion of magnetic flux, we can state Faraday's law in a more quantitative and useful way:

> The magnitude of the emf $\mathscr{E}$ induced in a conducting loop is equal to the rate at which the magnetic flux $\Phi_B$ through that loop changes with time.

As you will see in the next section, the induced emf $\mathscr{E}$ tends to oppose the flux change, so Faraday's law is formally written as

$$\mathscr{E} = -\frac{d\Phi_B}{dt} \quad \text{(Faraday's law)}, \tag{30-4}$$

with the minus sign indicating that opposition. We often neglect the minus sign in Eq. 30-4, seeking only the magnitude of the induced emf.

If we change the magnetic flux through a coil of $N$ turns, an induced emf appears in every turn and the total emf induced in the coil is the sum of these individual induced emfs. If the coil is tightly wound (*closely packed*), so that the same magnetic flux $\Phi_B$ passes through all the turns, the total emf induced in the coil is

$$\mathscr{E} = -N\frac{d\Phi_B}{dt} \quad \text{(coil of $N$ turns)}. \tag{30-5}$$

Here are the general means by which we can change the magnetic flux through a coil:

1. Change the magnitude $B$ of the magnetic field within the coil.

2. Change either the total area of the coil or the portion of that area that lies within the magnetic field (for example, by expanding the coil or sliding it into or out of the field).

3. Change the angle between the direction of the magnetic field $\vec{B}$ and the plane of the coil (for example, by rotating the coil so that field $\vec{B}$ is first perpendicular to the plane of the coil and then is along that plane).

✓ **CHECKPOINT 1** The graph gives the magnitude $B(t)$ of a uniform magnetic field that exists throughout a conducting loop, with the direction of the field perpendicular to the plane of the loop. Rank the five regions of the graph according to the magnitude of the emf induced in the loop, greatest first.

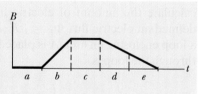

---

**Sample Problem** 30-1 **Build your skill**

The long solenoid S shown (in cross section) in Fig. 30-3 has 220 turns/cm and carries a current $i = 1.5$ A; its diameter $D$ is 3.2 cm. At its center we place a 130-turn closely packed coil C of diameter $d = 2.1$ cm. The current in the solenoid is reduced to zero at a steady rate in 25 ms. What is the magnitude of the emf that is induced in coil C while the current in the solenoid is changing?

**KEY IDEAS**

1. Because it is located in the interior of the solenoid, coil C lies within the magnetic field produced by current $i$ in the solenoid; thus, there is a magnetic flux $\Phi_B$ through coil C.

2. Because current $i$ decreases, flux $\Phi_B$ also decreases.

3. As $\Phi_B$ decreases, emf $\mathscr{E}$ is induced in coil C, according to Faraday's law.

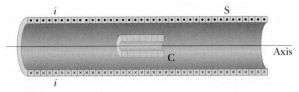

**FIG. 30-3** A coil C is located inside a solenoid S, which carries current $i$.

4. The flux through each turn of coil C depends on the area $A$ and orientation of that turn in the solenoid's magnetic field $\vec{B}$. Because $\vec{B}$ is uniform and directed perpendicular to area $A$, the flux is given by Eq. 30-2 ($\Phi_B = BA$).

5. The magnitude $B$ of the magnetic field in the interior of a solenoid depends on the solenoid's current $i$ and its number $n$ of turns per unit length, according to Eq. 29-23 ($B = \mu_0 in$).

*Calculations:* Because coil C consists of more than one turn, we apply Faraday's law in the form of Eq. 30-5 ($\mathscr{E} = -N\, d\Phi_B/dt$), where the number of turns $N$ is 130 and $d\Phi_B/dt$ is the rate at which the flux changes.

Because the current in the solenoid decreases at a steady rate, flux $\Phi_B$ also decreases at a steady rate, and so we can write $d\Phi_B/dt$ as $\Delta\Phi_B/\Delta t$. Then, to evaluate $\Delta\Phi_B$, we need the final and initial flux values. The final flux $\Phi_{B,f}$ is zero because the final current in the solenoid is zero. To find the initial flux $\Phi_{B,i}$, we first note that the area $A$ is $\frac{1}{4}\pi d^2$ ($= 3.464 \times 10^{-4}$ m$^2$) and the number $n$ is 220 turns/cm, or 22 000 turns/m. Substituting Eq. 29-23 into Eq. 30-2 then leads to

$$\Phi_{B,i} = BA = (\mu_0 in)A$$
$$= (4\pi \times 10^{-7}\,\text{T·m/A})(1.5\,\text{A})(22\,000\,\text{turns/m})$$
$$\times (3.464 \times 10^{-4}\,\text{m}^2)$$
$$= 1.44 \times 10^{-5}\,\text{Wb}.$$

Now we can write

$$\frac{d\Phi_B}{dt} = \frac{\Delta\Phi_B}{\Delta t} = \frac{\Phi_{B,f} - \Phi_{B,i}}{\Delta t}$$
$$= \frac{(0 - 1.44 \times 10^{-5}\,\text{Wb})}{25 \times 10^{-3}\,\text{s}}$$
$$= -5.76 \times 10^{-4}\,\text{Wb/s} = -5.76 \times 10^{-4}\,\text{V}.$$

We are interested only in magnitudes; so we ignore the minus signs here and in Eq. 30-5, writing

$$\mathscr{E} = N\frac{d\Phi_B}{dt} = (130\,\text{turns})(5.76 \times 10^{-4}\,\text{V})$$
$$= 7.5 \times 10^{-2}\,\text{V} = 75\,\text{mV}. \qquad \text{(Answer)}$$

## 30-4 | Lenz's Law

Soon after Faraday proposed his law of induction, Heinrich Friedrich Lenz devised a rule—now known as **Lenz's law**—for determining the direction of an induced current in a loop:

> ☞  An induced current has a direction such that the magnetic field due to *the current* opposes the change in the magnetic flux that induces the current.

Furthermore, the direction of an induced emf is that of the induced current. To get a feel for Lenz's law, let us apply it in two different but equivalent ways to Fig. 30-4, where the north pole of a magnet is being moved toward a conducting loop.

1. **Opposition to Pole Movement.** The approach of the magnet's north pole in Fig. 30-4 increases the magnetic flux through the loop and thereby induces a current in the loop. From Fig. 29-22, we know that the loop then acts as a magnetic dipole with a south pole and a north pole, and that its magnetic dipole moment $\vec{\mu}$ is directed from south to north. To *oppose* the magnetic flux increase being caused by the approaching magnet, the loop's north pole (and thus $\vec{\mu}$) must face *toward* the approaching north pole so as to repel it (Fig. 30-4). Then the curled–straight right-hand rule for $\vec{\mu}$ (Fig. 29-22) tells us that the current induced in the loop must be counterclockwise in Fig. 30-4.

   If we next pull the magnet away from the loop, a current will again be induced in the loop. Now, however, the loop will have a south pole facing the retreating north pole of the magnet, so as to oppose the retreat. Thus, the induced current will be clockwise.

2. **Opposition to Flux Change.** In Fig. 30-4, with the magnet initially distant, no magnetic flux passes through the loop. As the north pole of the magnet then nears the loop with its magnetic field $\vec{B}$ directed *downward*, the flux through

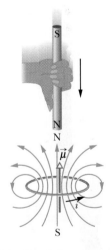

**FIG. 30-4** Lenz's law at work. As the magnet is moved toward the loop, a current is induced in the loop. The current produces its own magnetic field, with magnetic dipole moment $\vec{\mu}$ oriented so as to oppose the motion of the magnet. Thus, the induced current must be counterclockwise as shown.

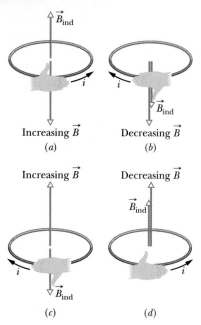

**FIG. 30-5** The direction of the current $i$ induced in a loop is such that the current's magnetic field $\vec{B}_{ind}$ opposes the *change* in the magnetic field $\vec{B}$ inducing $i$. The field $\vec{B}_{ind}$ is always directed opposite an increasing field $\vec{B}$ ($a$, $c$) and in the same direction as a decreasing field $\vec{B}$ ($b$, $d$). The curled–straight right-hand rule gives the direction of the induced current based on the direction of the induced field.

the loop increases. To oppose this increase in flux, the induced current $i$ must set up its own field $\vec{B}_{ind}$ directed *upward* inside the loop, as shown in Fig. 30-5$a$; then the upward flux of field $\vec{B}_{ind}$ opposes the increasing downward flux of field $\vec{B}$. The curled–straight right-hand rule of Fig. 29-22 then tells us that $i$ must be counterclockwise in Fig. 30-5$a$.

Note carefully that the flux of $\vec{B}_{ind}$ always opposes the *change* in the flux of $\vec{B}$, but that does not always mean that $\vec{B}_{ind}$ points opposite $\vec{B}$. For example, if we next pull the magnet away from the loop in Fig. 30-4, the flux $\Phi_B$ from the magnet is still directed downward through the loop, but it is now decreasing. The flux of $\vec{B}_{ind}$ must now be downward inside the loop, to oppose the *decrease* in $\Phi_B$, as shown in Fig. 30-5$b$. Thus, $\vec{B}_{ind}$ and $\vec{B}$ are now in the same direction.

Figures 30-5$c$ and $d$ show the situations in which the south pole of the magnet approaches and retreats from the loop, respectively.

## Electric Guitars

Figure 30-6 shows a Fender® Stratocaster®, one type of electric guitar. Whereas an acoustic guitar depends for its sound on the acoustic resonance produced in the hollow body of the instrument by the oscillations of the strings, an electric guitar is a solid instrument, so there is no body resonance. Instead, the oscillations of the metal strings are sensed by electric "pickups" that send signals to an amplifier and a set of speakers.

The basic construction of a pickup is shown in Fig. 30-7. Wire connecting the instrument to the amplifier is coiled around a small magnet. The magnetic field of the magnet produces a north and south pole in the section of the metal string just above the magnet. That section of string then has its own magnetic field. When the string is plucked and thus made to oscillate, its motion relative to the coil changes the flux of its magnetic field through the coil, inducing a current in the coil. As the string oscillates toward and away from the coil, the induced current changes direction at the same frequency as the string's oscillations, thus relaying the frequency of oscillation to the amplifier and speaker.

On a Stratocaster, there are three groups of pickups, placed at the near end of the strings (on the wide part of the body). The group closest to the near end better detects the high-frequency oscillations of the strings; the group farthest from the near end better detects the low-frequency oscillations. By throwing a toggle switch on the guitar, the musician can select which group or which pair of groups will send signals to the amplifier and speakers.

To gain further control over his music, the legendary Jimi Hendrix sometimes rewrapped the wire in the pickup coils of his guitar to change the number of turns. In this way, he altered the amount of emf induced in the coils and thus their relative sensitivity to string oscillations. Even without this additional measure, you can see that the electric guitar offers far more control over the sound that is produced than can be obtained with an acoustic guitar.

**FIG. 30-6** A Fender Stratocaster guitar. *(Courtesy Fender Musical Instruments Corporation)*

✓**CHECKPOINT 2** The figure shows three situations in which identical circular conducting loops are in uniform magnetic fields that are either increasing (Inc) or decreasing (Dec) in magnitude at identical rates. In each, the dashed line coincides with a diameter. Rank the situations according to the magnitude of the current induced in the loops, greatest first.

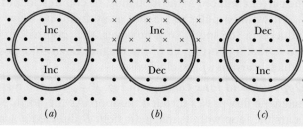

**FIG. 30-7** A side view of an electric guitar pickup. When the metal string (which acts like a magnet) is made to oscillate, it causes a variation in magnetic flux that induces a current in the coil.

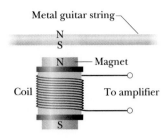

Metal guitar string

N
S

N — Magnet

Coil

To amplifier

S

## Sample Problem 30-2

Figure 30-8 shows a conducting loop consisting of a half-circle of radius $r = 0.20$ m and three straight sections. The half-circle lies in a uniform magnetic field $\vec{B}$ that is directed out of the page; the field magnitude is given by $B = 4.0t^2 + 2.0t + 3.0$, with $B$ in teslas and $t$ in seconds. An ideal battery with emf $\mathscr{E}_{bat} = 2.0$ V is connected to the loop. The resistance of the loop is 2.0 Ω.

(a) What are the magnitude and direction of the emf $\mathscr{E}_{ind}$ induced around the loop by field $\vec{B}$ at $t = 10$ s?

### KEY IDEAS

1. According to Faraday's law, the magnitude of $\mathscr{E}_{ind}$ is equal to the rate $d\Phi_B/dt$ at which the magnetic flux through the loop changes.

2. The flux through the loop depends on how much of the loop's area lies within the flux and how the area is oriented in the magnetic field $\vec{B}$.

3. Because $\vec{B}$ is uniform and is perpendicular to the plane of the loop, the flux is given by Eq. 30-2 ($\Phi_B = BA$). (We don't need to integrate $B$ over the area to get the flux.)

4. The induced field $B_{ind}$ (due to the induced current) must always oppose the *change* in the magnetic flux.

**Magnitude:** Using Eq. 30-2 and realizing that only the field magnitude $B$ changes in time (not the area $A$), we rewrite Faraday's law, Eq. 30-4, as

$$\mathscr{E}_{ind} = \frac{d\Phi_B}{dt} = \frac{d(BA)}{dt} = A\frac{dB}{dt}.$$

Because the flux penetrates the loop only within the half-circle, the area $A$ in this equation is $\frac{1}{2}\pi r^2$. Substituting this and the given expression for $B$ yields

$$\mathscr{E}_{ind} = A\frac{dB}{dt} = \frac{\pi r^2}{2}\frac{d}{dt}(4.0t^2 + 2.0t + 3.0)$$

$$= \frac{\pi r^2}{2}(8.0t + 2.0).$$

At $t = 10$ s, then,

$$\mathscr{E}_{ind} = \frac{\pi (0.20 \text{ m})^2}{2}[8.0(10) + 2.0]$$

$$= 5.152 \text{ V} \approx 5.2 \text{ V}. \qquad \text{(Answer)}$$

**Direction:** To find the direction of $\mathscr{E}_{ind}$, we first note that in Fig. 30-8 the flux through the loop is out of the page and increasing. Because the induced field $B_{ind}$ (due to the induced current) must oppose that increase, it must be *into* the page. Using the curled–straight right-hand rule (Fig. 30-5$c$), we find that the induced current is clockwise around the loop, and thus so is the induced emf $\mathscr{E}_{ind}$.

(b) What is the current in the loop at $t = 10$ s?

### KEY IDEA  The point here is that *two* emfs tend to move charges around the loop.

**Calculation:** The induced emf $\mathscr{E}_{ind}$ tends to drive a current clockwise around the loop; the battery's emf $\mathscr{E}_{bat}$ tends to drive a current counterclockwise. Because $\mathscr{E}_{ind}$ is greater than $\mathscr{E}_{bat}$, the net emf $\mathscr{E}_{net}$ is clockwise, and thus so is the current. To find the current at $t = 10$ s, we use Eq. 27-2 ($i = \mathscr{E}/R$):

$$i = \frac{\mathscr{E}_{net}}{R} = \frac{\mathscr{E}_{ind} - \mathscr{E}_{bat}}{R}$$

$$= \frac{5.152 \text{ V} - 2.0 \text{ V}}{2.0 \text{ Ω}} = 1.58 \text{ A} \approx 1.6 \text{ A}. \quad \text{(Answer)}$$

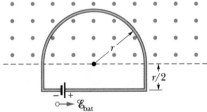

**FIG. 30-8** A battery is connected to a conducting loop that includes a half-circle of radius $r$ lying in a uniform magnetic field. The field is directed out of the page; its magnitude is changing.

## Sample Problem 30-3

Figure 30-9 shows a rectangular loop of wire immersed in a nonuniform and varying magnetic field $\vec{B}$ that is perpendicular to and directed into the page. The field's magnitude is given by $B = 4t^2x^2$, with $B$ in

teslas, $t$ in seconds, and $x$ in meters. The loop has width $W = 3.0$ m and height $H = 2.0$ m. What are the magnitude and direction of the induced emf $\mathscr{E}$ around the loop at $t = 0.10$ s?

**KEY IDEAS**

1. Because the magnitude of the magnetic field $\vec{B}$ is changing with time, the magnetic flux $\Phi_B$ through the loop is also changing.

2. The changing flux induces an emf $\mathcal{E}$ in the loop according to Faraday's law, which we can write as $\mathcal{E} = d\Phi_B/dt$.

3. To use that law, we need an expression for the flux $\Phi_B$ at any time $t$. However, because $B$ is *not* uniform over the area enclosed by the loop, we *cannot* use Eq. 30-2 ($\Phi_B = BA$) to find that expression; instead we must use Eq. 30-1 ($\Phi_B = \int \vec{B} \cdot d\vec{A}$).

*Calculations:* In Fig. 30-9, $\vec{B}$ is perpendicular to the plane of the loop (and hence parallel to the differential area vector $d\vec{A}$); so the dot product in Eq. 30-1 gives $B \, dA$. Because the magnetic field varies with the coordinate $x$ but not with the coordinate $y$, we can take the differential area $dA$ to be the area of a vertical strip of height $H$ and width $dx$ (as shown in Fig. 30-9). Then $dA = H \, dx$, and the flux through the loop is

$$\Phi_B = \int \vec{B} \cdot d\vec{A} = \int B \, dA = \int BH \, dx = \int 4t^2 x^2 H \, dx.$$

Treating $t$ as a constant for this integration and inserting the integration limits $x = 0$ and $x = 3.0$ m, we obtain

$$\Phi_B = 4t^2 H \int_0^{3.0} x^2 \, dx = 4t^2 H \left[ \frac{x^3}{3} \right]_0^{3.0} = 72t^2,$$

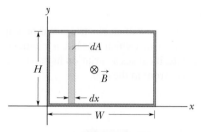

FIG. 30-9 A closed conducting loop, of width $W$ and height $H$, lies in a nonuniform, varying magnetic field that points directly into the page. To apply Faraday's law, we use the vertical strip of height $H$, width $dx$, and area $dA$.

where we have substituted $H = 2.0$ m and $\Phi_B$ is in webers. Now we can use Faraday's law to find the magnitude of $\mathcal{E}$ at any time $t$:

$$\mathcal{E} = \frac{d\Phi_B}{dt} = \frac{d(72t^2)}{dt} = 144t,$$

in which $\mathcal{E}$ is in volts. At $t = 0.10$ s,

$$\mathcal{E} = (144 \text{ V/s})(0.10 \text{ s}) \approx 14 \text{ V}. \quad \text{(Answer)}$$

The flux of $\vec{B}$ through the loop is into the page in Fig. 30-9 and is increasing in magnitude because $B$ is increasing in magnitude with time. By Lenz's law, the field $B_i$ of the induced current opposes this increase and so is directed out of the page. The curled–straight right-hand rule in Fig. 30-5 then tells us that the induced current is counterclockwise around the loop, and thus so is the induced emf $\mathcal{E}$.

## 30-5 | Induction and Energy Transfers

By Lenz's law, whether you move the magnet toward or away from the loop in Fig. 30-1, a magnetic force resists the motion, requiring your applied force to do positive work. At the same time, thermal energy is produced in the material of the loop because of the material's electrical resistance to the current that is induced by the motion. The energy you transfer to the closed *loop + magnet* system via your applied force ends up in this thermal energy. (For now, we neglect energy that is radiated away from the loop as electromagnetic waves during the induction.) The faster you move the magnet, the more rapidly your applied force does work and the greater the rate at which your energy is transferred to thermal energy in the loop; that is, the power of the transfer is greater.

Regardless of how current is induced in a loop, energy is always transferred to thermal energy during the process because of the electrical resistance of the loop (unless the loop is superconducting). For example, in Fig. 30-2, when switch S is closed and a current is briefly induced in the left-hand loop, energy is transferred from the battery to thermal energy in that loop.

Figure 30-10 shows another situation involving induced current. A rectangular loop of wire of width $L$ has one end in a uniform external magnetic field that is directed perpendicularly into the plane of the loop. This field may be produced, for example, by a large electromagnet. The dashed lines in Fig. 30-10 show the assumed limits of the magnetic field; the fringing of the field at its edges is neglected. You are to pull this loop to the right at a constant velocity $\vec{v}$.

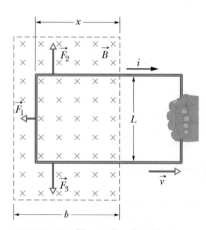

FIG. 30-10 You pull a closed conducting loop out of a magnetic field at constant velocity $\vec{v}$. While the loop is moving, a clockwise current $i$ is induced in the loop, and the loop segments still within the magnetic field experience forces $\vec{F}_1$, $\vec{F}_2$, and $\vec{F}_3$.

The situation of Fig. 30-10 does not differ in any essential way from that of Fig. 30-1. In each case a magnetic field and a conducting loop are in relative motion; in each case the flux of the field through the loop is changing with time. It is true that in Fig. 30-1 the flux is changing because $\vec{B}$ is changing and in Fig. 30-10 the flux is changing because the area of the loop still in the magnetic field is changing, but that difference is not important. The important difference between the two arrangements is that the arrangement of Fig. 30-10 makes calculations easier. Let us now calculate the rate at which you do mechanical work as you pull steadily on the loop in Fig. 30-10.

As you will see, to pull the loop at a constant velocity $\vec{v}$, you must apply a constant force $\vec{F}$ to the loop because a magnetic force of equal magnitude but opposite direction acts on the loop to oppose you. From Eq. 7-48, the rate at which you do work—that is, the power—is then

$$P = Fv, \tag{30-6}$$

where $F$ is the magnitude of your force. We wish to find an expression for $P$ in terms of the magnitude $B$ of the magnetic field and the characteristics of the loop—namely, its resistance $R$ to current and its dimension $L$.

As you move the loop to the right in Fig. 30-10, the portion of its area within the magnetic field decreases. Thus, the flux through the loop also decreases and, according to Faraday's law, a current is produced in the loop. It is the presence of this current that causes the force that opposes your pull.

To find the current, we first apply Faraday's law. When $x$ is the length of the loop still in the magnetic field, the area of the loop still in the field is $Lx$. Then from Eq. 30-2, the magnitude of the flux through the loop is

$$\Phi_B = BA = BLx. \tag{30-7}$$

As $x$ decreases, the flux decreases. Faraday's law tells us that with this flux decrease, an emf is induced in the loop. Dropping the minus sign in Eq. 30-4 and using Eq. 30-7, we can write the magnitude of this emf as

$$\mathcal{E} = \frac{d\Phi_B}{dt} = \frac{d}{dt} BLx = BL \frac{dx}{dt} = BLv, \tag{30-8}$$

in which we have replaced $dx/dt$ with $v$, the speed at which the loop moves.

Figure 30-11 shows the loop as a circuit: induced emf $\mathcal{E}$ is represented on the left, and the collective resistance $R$ of the loop is represented on the right. The direction of the induced current $i$ is obtained with a right-hand rule as in Fig. 30-5$b$ for decreasing flux; applying the rule tells us that the current must be clockwise, and $\mathcal{E}$ must have the same direction.

To find the magnitude of the induced current, we cannot apply the loop rule for potential differences in a circuit because, as you will see in Section 30-6, we cannot define a potential difference for an induced emf. However, we can apply the equation $i = \mathcal{E}/R$, as we did in Sample Problem 30-2. With Eq. 30-8, this becomes

$$i = \frac{BLv}{R}. \tag{30-9}$$

Because three segments of the loop in Fig. 30-10 carry this current through the magnetic field, sideways deflecting forces act on those segments. From Eq. 28-26 we know that such a deflecting force is, in general notation,

$$\vec{F}_d = i\vec{L} \times \vec{B}. \tag{30-10}$$

In Fig. 30-10, the deflecting forces acting on the three segments of the loop are marked $\vec{F}_1$, $\vec{F}_2$, and $\vec{F}_3$. Note, however, that from the symmetry, forces $\vec{F}_2$ and $\vec{F}_3$ are equal in magnitude and cancel. This leaves only force $\vec{F}_1$, which is directed opposite your force $\vec{F}$ on the loop and thus is the force opposing you. So, $\vec{F} = -\vec{F}_1$.

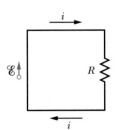

**FIG. 30-11** A circuit diagram for the loop of Fig. 30-10 while the loop is moving.

**FIG. 30-12**    A patient in an MRI apparatus. *(Michael Rosenfeld/Getty Images, Inc.)*

Using Eq. 30-10 to obtain the magnitude of $\vec{F}_1$ and noting that the angle between $\vec{B}$ and the length vector $\vec{L}$ for the left segment is 90°, we write

$$F = F_1 = iLB \sin 90° = iLB. \tag{30-11}$$

Substituting Eq. 30-9 for $i$ in Eq. 30-11 then gives us

$$F = \frac{B^2 L^2 v}{R}. \tag{30-12}$$

Because $B$, $L$, and $R$ are constants, the speed $v$ at which you move the loop is constant if the magnitude $F$ of the force you apply to the loop is also constant.

By substituting Eq. 30-12 into Eq. 30-6, we find the rate at which you do work on the loop as you pull it from the magnetic field:

$$P = Fv = \frac{B^2 L^2 v^2}{R} \qquad \text{(rate of doing work).} \tag{30-13}$$

To complete our analysis, let us find the rate at which thermal energy appears in the loop as you pull it along at constant speed. We calculate it from Eq. 26-27,

$$P = i^2 R. \tag{30-14}$$

Substituting for $i$ from Eq. 30-9, we find

$$P = \left(\frac{BLv}{R}\right)^2 R = \frac{B^2 L^2 v^2}{R} \qquad \text{(thermal energy rate),} \tag{30-15}$$

which is exactly equal to the rate at which you are doing work on the loop (Eq. 30-13). Thus, the work that you do in pulling the loop through the magnetic field appears as thermal energy in the loop.

### Burns During MRI Scans

A patient undergoing an MRI scan (Fig. 30-12) lies in an apparatus containing two magnetic fields: a large constant field $\vec{B}_{con}$ and a small sinusoidally varying field $\vec{B}(t)$. Normally the scan requires the patient to lie motionless for a long time. Any patient unable to lie motionless, such as a child, say, is sedated. Because sedation, especially a general anesthetic, can be dangerous, a sedated patient must be carefully monitored, usually with a *pulse oximeter,* a device that measures the oxygen level in the patient's blood. This device includes a probe attached to one of the patient's fingers and a cable running from the probe to a monitor located outside the MRI apparatus.

MRI scans should be perfectly harmless to a patient. In a few cases, however, disregard of Faraday's law of induction led to a sedated patient receiving severe burns. In those cases, the oximeter cable was allowed to touch the patient's arm (Fig. 30-13). The cable and the lower part of the arm then formed a closed loop through which the varying magnetic field $\vec{B}(t)$ produced a varying flux. This flux variation induced an emf around the loop. Although the cable insulation and the skin both had high electrical resistance, the induced emf was large enough to drive a significant current around the loop. As with any other circuit in which there is resistance, the current transferred energy to thermal energy at the points of resistance. In this way, the finger and the skin where the cable touched the lower arm were burned. MRI staff are now trained to keep any monitor cable from touching a patient at more than one point.

**FIG. 30-13**    A probe attached to a finger of a patient undergoing an MRI scan, in which a vertical magnetic field $\vec{B}(t)$ varies sinusoidally. The probe cable touches the patient's skin along the arm, and the cable and the lower part of the arm form a closed loop.

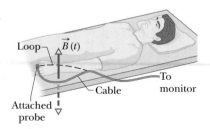

## Eddy Currents

Suppose we replace the conducting loop of Fig. 30-10 with a solid conducting plate. If we then move the plate out of the magnetic field as we did the loop (Fig. 30-14a), the relative motion of the field and the conductor again induces a current in the conductor. Thus, we again encounter an opposing force and must do work because of the induced current. With the plate, however, the conduction electrons making up the induced current do not follow one path as they do with the loop. Instead, the electrons swirl about within the plate as if they were caught in an eddy (whirlpool) of water. Such a current is called an *eddy current* and can be represented, as it is in Fig. 30-14a, *as if* it followed a single path.

As with the conducting loop of Fig. 30-10, the current induced in the plate results in mechanical energy being dissipated as thermal energy. The dissipation is more apparent in the arrangement of Fig. 30-14b; a conducting plate, free to rotate about a pivot, is allowed to swing down through a magnetic field like a pendulum. Each time the plate enters and leaves the field, a portion of its mechanical energy is transferred to its thermal energy. After several swings, no mechanical energy remains and the warmed-up plate just hangs from its pivot.

Similar dissipation of energy is the basis of an induction furnace. Figure 30-15 shows the basic design: Metal is held within a crucible around which insulated wires are wrapped. The (AC) current in the wires alternates in direction and magnitude. Thus, the magnetic field due to the current continuously varies in direction and magnitude. This changing field $\vec{B}(t)$ creates eddy currents within the metal, and electrical energy is dissipated as thermal energy at the rate given by Eq. 30-14 ($P = i^2R$). The dissipation increases the temperature of the metal to the melting point, and then the molten metal can be poured.

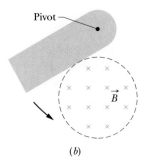

**FIG. 30-14** (a) As you pull a solid conducting plate out of a magnetic field, *eddy currents* are induced in the plate. A typical loop of eddy current is shown. (b) A conducting plate is allowed to swing like a pendulum about a pivot and into a region of magnetic field. As it enters and leaves the field, eddy currents are induced in the plate.

✓**CHECKPOINT 3**    The figure shows four wire loops, with edge lengths of either $L$ or $2L$. All four loops will move through a region of uniform magnetic field $\vec{B}$ (directed out of the page) at the same constant velocity. Rank the four loops according to the maximum magnitude of the emf induced as they move through the field, greatest first.

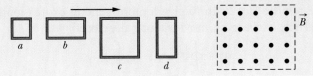

## 30-6 | Induced Electric Fields

Let us place a copper ring of radius $r$ in a uniform external magnetic field, as in Fig. 30-16a. The field—neglecting fringing—fills a cylindrical volume of radius $R$. Suppose that we increase the strength of this field at a steady rate, perhaps by increasing—in an appropriate way—the current in the windings of the electromagnet that produces the field. The magnetic flux through the ring will then change at a steady rate and—by Faraday's law—an induced emf and thus an induced current will appear in the ring. From Lenz's law we can deduce that the direction of the induced current is counterclockwise in Fig. 30-16a.

If there is a current in the copper ring, an electric field must be present along the ring because an electric field is needed to do the work of moving the conduction electrons. Moreover, the electric field must have been produced by the changing magnetic flux. This **induced electric field** $\vec{E}$ is just as real as an electric field produced by static charges; either field will exert a force $q_0\vec{E}$ on a particle of charge $q_0$.

By this line of reasoning, we are led to a useful and informative restatement of Faraday's law of induction:

☞ A changing magnetic field produces an electric field.

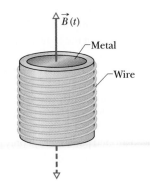

**FIG. 30-15** An induction furnace.

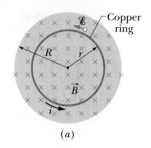

(a)

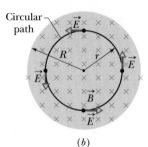

(b)

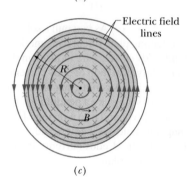

(c)

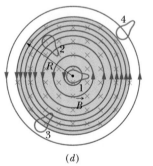

(d)

FIG. 30-16 (a) If the magnetic field increases at a steady rate, a constant induced current appears, as shown, in the copper ring of radius r.(b) An induced electric field exists even when the ring is removed; the electric field is shown at four points. (c) The complete picture of the induced electric field, displayed as field lines. (d) Four similar closed paths that enclose identical areas. Equal emfs are induced around paths 1 and 2, which lie entirely within the region of changing magnetic field. A smaller emf is induced around path 3, which only partially lies in that region. No net emf is induced around path 4, which lies entirely outside the magnetic field.

The striking feature of this statement is that the electric field is induced even if there is no copper ring.

To fix these ideas, consider Fig. 30-16b, which is just like Fig. 30-16a except the copper ring has been replaced by a hypothetical circular path of radius r. We assume, as previously, that the magnetic field $\vec{B}$ is increasing in magnitude at a constant rate $dB/dt$. The electric field induced at various points around the circular path must—from the symmetry—be tangent to the circle, as Fig. 30-16b shows.* Hence, the circular path is an electric field line. There is nothing special about the circle of radius r, so the electric field lines produced by the changing magnetic field must be a set of concentric circles, as in Fig. 30-16c.

As long as the magnetic field is *increasing* with time, the electric field represented by the circular field lines in Fig. 30-16c will be present. If the magnetic field remains *constant* with time, there will be no induced electric field and thus no electric field lines. If the magnetic field is *decreasing* with time (at a constant rate), the electric field lines will still be concentric circles as in Fig. 30-16c, but they will now have the opposite direction. All this is what we have in mind when we say "A changing magnetic field produces an electric field."

## A Reformulation of Faraday's Law

Consider a particle of charge $q_0$ moving around the circular path of Fig. 30-16b. The work W done on it in one revolution by the induced electric field is $W = \mathscr{E}q_0$, where $\mathscr{E}$ is the induced emf—that is, the work done per unit charge in moving the test charge around the path. From another point of view, the work is

$$W = \int \vec{F} \cdot d\vec{s} = (q_0 E)(2\pi r),\qquad(30\text{-}16)$$

where $q_0 E$ is the magnitude of the force acting on the test charge and $2\pi r$ is the distance over which that force acts. Setting these two expressions for W equal to each other and canceling $q_0$, we find that

$$\mathscr{E} = 2\pi r E.\qquad(30\text{-}17)$$

Next we rewrite Eq. 30-16 to give a more general expression for the work done on a particle of charge $q_0$ moving along any closed path:

$$W = \oint \vec{F} \cdot d\vec{s} = q_0 \oint \vec{E} \cdot d\vec{s}.\qquad(30\text{-}18)$$

(The loop on each integral sign indicates that the integral is to be taken around the closed path.) Substituting $\mathscr{E}q_0$ for W, we find that

$$\mathscr{E} = \oint \vec{E} \cdot d\vec{s}.\qquad(30\text{-}19)$$

This integral reduces at once to Eq. 30-17 if we evaluate it for the special case of Fig. 30-16b.

With Eq. 30-19, we can expand the meaning of induced emf. Up to this point, induced emf has meant the work per unit charge done in maintaining current due to a changing magnetic flux, or it has meant the work done per unit charge on a charged particle that moves around a closed path in a changing magnetic flux. However, with Fig. 30-16b and Eq. 30-19, an induced emf can exist without the need of a current or particle: An induced emf is the sum—via integration—of quantities $\vec{E} \cdot d\vec{s}$ around a closed path, where $\vec{E}$ is the electric field induced by a changing magnetic flux and $d\vec{s}$ is a differential length vector along the path.

If we combine Eq. 30-19 with Faraday's law in Eq. 30-4 ($\mathscr{E} = -d\Phi_B/dt$), we

*Arguments of symmetry would also permit the lines of $\vec{E}$ around the circular path to be *radial*, rather than tangential. However, such radial lines would imply that there are free charges, distributed symmetrically about the axis of symmetry, on which the electric field lines could begin or end; there are no such charges.

can rewrite Faraday's law as

$$\oint \vec{E} \cdot d\vec{s} = -\frac{d\Phi_B}{dt} \qquad \text{(Faraday's law).} \qquad (30\text{-}20)$$

This equation says simply that a changing magnetic field induces an electric field. The changing magnetic field appears on the right side of this equation, the electric field on the left.

Faraday's law in the form of Eq. 30-20 can be applied to *any* closed path that can be drawn in a changing magnetic field. Figure 30-16d, for example, shows four such paths, all having the same shape and area but located in different positions in the changing field. The induced emfs $\mathcal{E}$ ($= \oint \vec{E} \cdot d\vec{s}$) for paths 1 and 2 are equal because these paths lie entirely in the magnetic field and thus have the same value of $d\Phi_B/dt$. This is true even though the electric field vectors at points along these paths are different, as indicated by the patterns of electric field lines in the figure. For path 3 the induced emf is smaller because the enclosed flux $\Phi_B$ (hence $d\Phi_B/dt$) is smaller, and for path 4 the induced emf is zero even though the electric field is not zero at any point on the path.

## A New Look at Electric Potential

Induced electric fields are produced not by static charges but by a changing magnetic flux. Although electric fields produced in either way exert forces on charged particles, there is an important difference between them. The simplest evidence of this difference is that the field lines of induced electric fields form closed loops, as in Fig. 30-16c. Field lines produced by static charges never do so but must start on positive charges and end on negative charges.

In a more formal sense, we can state the difference between electric fields produced by induction and those produced by static charges in these words:

> Electric potential has meaning only for electric fields that are produced by static charges; it has no meaning for electric fields that are produced by induction.

You can understand this statement qualitatively by considering what happens to a charged particle that makes a single journey around the circular path in Fig. 30-16b. It starts at a certain point and, on its return to that same point, has experienced an emf $\mathcal{E}$ of, let us say, 5 V; that is, work of 5 J/C has been done on the particle, and thus the particle should then be at a point that is 5 V greater in potential. However, that is impossible because the particle is back at the same point, which cannot have two different values of potential. Thus, potential has no meaning for electric fields that are set up by changing magnetic fields.

We can take a more formal look by recalling Eq. 24-18, which defines the potential difference between two points $i$ and $f$ in an electric field $\vec{E}$:

$$V_f - V_i = -\int_i^f \vec{E} \cdot d\vec{s}. \qquad (30\text{-}21)$$

In Chapter 24 we had not yet encountered Faraday's law of induction; so the electric fields involved in the derivation of Eq. 24-18 were those due to static charges. If $i$ and $f$ in Eq. 30-21 are the same point, the path connecting them is a closed loop, $V_i$ and $V_f$ are identical, and Eq. 30-21 reduces to

$$\oint \vec{E} \cdot d\vec{s} = 0. \qquad (30\text{-}22)$$

However, when a changing magnetic flux is present, this integral is *not* zero but is $-d\Phi_B/dt$, as Eq. 30-20 asserts. Thus, assigning electric potential to an induced electric field leads us to a contradiction. We must conclude that electric potential has no meaning for electric fields associated with induction.

✓ **CHECKPOINT 4**    The figure shows five lettered regions in which a uniform magnetic field extends either directly out of the page or into the page, with the direction indicated only for region $a$. The field is increasing in magnitude at the same steady rate in all five regions; the regions are identical in area. Also shown are four numbered paths along which $\oint \vec{E} \cdot d\vec{s}$ has the magnitudes given below in terms of a quantity "mag." Determine whether the magnetic field is directed into or out of the page for regions $b$ through $e$.

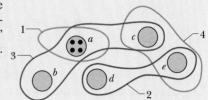

| Path | 1 | 2 | 3 | 4 |
|------|---|---|---|---|
| $\oint \vec{E} \cdot d\vec{s}$ | mag | 2(mag) | 3(mag) | 0 |

---

**Sample Problem   30-4**

---

In Fig. 30-16b, take $R = 8.5$ cm and $dB/dt = 0.13$ T/s.

**(a)** Find an expression for the magnitude $E$ of the induced electric field at points within the magnetic field, at radius $r$ from the center of the magnetic field. Evaluate the expression for $r = 5.2$ cm.

**KEY IDEA**    An electric field is induced by the changing magnetic field, according to Faraday's law.

**Calculations:** To calculate the field magnitude $E$, we apply Faraday's law in the form of Eq. 30-20. We use a circular path of integration with radius $r \le R$ because we want $E$ for points within the magnetic field. We assume from the symmetry that $\vec{E}$ in Fig. 30-16b is tangent to the circular path at all points. The path vector $d\vec{s}$ is also always tangent to the circular path; so the dot product $\vec{E} \cdot d\vec{s}$ in Eq. 30-20 must have the magnitude $E\,ds$ at all points on the path. We can also assume from the symmetry that $E$ has the same value at all points along the circular path. Then the left side of Eq. 30-20 becomes

$$\oint \vec{E} \cdot d\vec{s} = \oint E\,ds = E \oint ds = E(2\pi r). \quad (30\text{-}23)$$

(The integral $\oint ds$ is the circumference $2\pi r$ of the circular path.)

Next, we need to evaluate the right side of Eq. 30-20. Because $\vec{B}$ is uniform over the area $A$ encircled by the path of integration and is directed perpendicular to that area, the magnetic flux is given by Eq. 30-2:

$$\Phi_B = BA = B(\pi r^2). \quad (30\text{-}24)$$

Substituting this and Eq. 30-23 into Eq. 30-20 and dropping the minus sign, we find that

$$E(2\pi r) = (\pi r^2)\frac{dB}{dt}$$

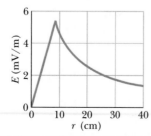

**FIG. 30-17**  A plot of the induced electric field $E(r)$ for the conditions of Sample Problem 30-4.

or $$E = \frac{r}{2}\frac{dB}{dt}. \quad \text{(Answer)} \quad (30\text{-}25)$$

Equation 30-25 gives the magnitude of the electric field at any point for which $r \le R$ (that is, within the magnetic field). Substituting given values yields, for the magnitude of $\vec{E}$ at $r = 5.2$ cm,

$$E = \frac{(5.2 \times 10^{-2}\text{ m})}{2}(0.13\text{ T/s})$$

$$= 0.0034\text{ V/m} = 3.4\text{ mV/m}. \quad \text{(Answer)}$$

**(b)** Find an expression for the magnitude $E$ of the induced electric field at points that are outside the magnetic field, at radius $r$. Evaluate the expression for $r = 12.5$ cm.

**KEY IDEAS**    The idea of part (a) applies here also, except that we use a circular path of integration with radius $r \ge R$ because we want to evaluate $E$ for points outside the magnetic field. Proceeding as in (a), we again obtain Eq. 30-23. However, we do not then obtain Eq. 30-24 because the new path of integration is now outside the magnetic field, and so the magnetic flux encircled by the new path is only that in the area $\pi R^2$ of the magnetic field region.

**Calculations:** We can now write

$$\Phi_B = BA = B(\pi R^2). \quad (30\text{-}26)$$

Substituting this and Eq. 30-23 into Eq. 30-20 (without the minus sign) and solving for $E$ yield

$$E = \frac{R^2}{2r}\frac{dB}{dt}. \quad \text{(Answer)} \quad (30\text{-}27)$$

Because $E$ is not zero here, we know that an electric field is induced even at points that are outside the changing magnetic field, an important result that (as you will see in Section 31-11) makes transformers possible.

With the given data, Eq. 30-27 yields the magnitude of $\vec{E}$ at $r = 12.5$ cm:

$$E = \frac{(8.5 \times 10^{-2}\text{ m})^2}{(2)(12.5 \times 10^{-2}\text{ m})}(0.13\text{ T/s})$$

$$= 3.8 \times 10^{-3}\text{ V/m} = 3.8\text{ mV/m}. \quad \text{(Answer)}$$

Equations 30-25 and 30-27 give the same result for $r = R$. Figure 30-17 shows a plot of $E(r)$.

# 30-7 | Inductors and Inductance

We found in Chapter 25 that a capacitor can be used to produce a desired electric field. We considered the parallel-plate arrangement as a basic type of capacitor. Similarly, an **inductor** (symbol ⅏) can be used to produce a desired magnetic field. We shall consider a long solenoid (more specifically, a short length near the middle of a long solenoid) as our basic type of inductor.

If we establish a current $i$ in the windings (turns) of the solenoid we are taking as our inductor, the current produces a magnetic flux $\Phi_B$ through the central region of the inductor. The **inductance** of the inductor is then

$$L = \frac{N\Phi_B}{i} \qquad \text{(inductance defined)}, \qquad (30\text{-}28)$$

in which $N$ is the number of turns. The windings of the inductor are said to be *linked* by the shared flux, and the product $N\Phi_B$ is called the *magnetic flux linkage*. The inductance $L$ is thus a measure of the flux linkage produced by the inductor per unit of current.

Because the SI unit of magnetic flux is the tesla–square meter, the SI unit of inductance is the tesla–square meter per ampere ($T\cdot m^2/A$). We call this the **henry** (H), after American physicist Joseph Henry, the codiscoverer of the law of induction and a contemporary of Faraday. Thus,

$$1 \text{ henry} = 1 \text{ H} = 1 \text{ T}\cdot m^2/A. \qquad (30\text{-}29)$$

The crude inductors with which Michael Faraday discovered the law of induction. In those days amenities such as insulated wire were not commercially available. It is said that Faraday insulated his wires by wrapping them with strips cut from one of his wife's petticoats. *(The Royal Institution/Bridgeman Art Library/NY)*

Through the rest of this chapter we assume that all inductors, no matter what their geometric arrangement, have no magnetic materials such as iron in their vicinity. Such materials would distort the magnetic field of an inductor.

## Inductance of a Solenoid

Consider a long solenoid of cross-sectional area $A$. What is the inductance per unit length near its middle?

To use the defining equation for inductance (Eq. 30-28), we must calculate the flux linkage set up by a given current in the solenoid windings. Consider a length $l$ near the middle of this solenoid. The flux linkage for this section of the solenoid is

$$N\Phi_B = (nl)(BA),$$

in which $n$ is the number of turns per unit length of the solenoid and $B$ is the magnitude of the magnetic field within the solenoid.

The magnitude $B$ is given by Eq. 29-23,

$$B = \mu_0 in,$$

and so from Eq. 30-28,

$$L = \frac{N\Phi_B}{i} = \frac{(nl)(BA)}{i} = \frac{(nl)(\mu_0 in)(A)}{i}$$

$$= \mu_0 n^2 lA. \qquad (30\text{-}30)$$

Thus, the inductance per unit length near the center of a long solenoid is

$$\frac{L}{l} = \mu_0 n^2 A \qquad \text{(solenoid)}. \qquad (30\text{-}31)$$

Inductance—like capacitance—depends only on the geometry of the device. The dependence on the square of the number of turns per unit length is to be expected. If you, say, triple $n$, you not only triple the number of turns ($N$) but you also triple the flux ($\Phi_B = BA = \mu_0 inA$) through each turn, multiplying the flux linkage $N\Phi_B$ and thus the inductance $L$ by a factor of 9.

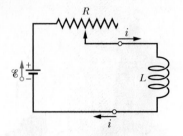

**FIG. 30-18** If the current in a coil is changed by varying the contact position on a variable resistor, a self-induced emf $\mathscr{E}_L$ will appear in the coil *while the current is changing.*

If the solenoid is very much longer than its radius, then Eq. 30-30 gives its inductance to a good approximation. This approximation neglects the spreading of the magnetic field lines near the ends of the solenoid, just as the parallel-plate capacitor formula ($C = \varepsilon_0 A/d$) neglects the fringing of the electric field lines near the edges of the capacitor plates.

From Eq. 30-30, and recalling that $n$ is a number per unit length, we can see that an inductance can be written as a product of the permeability constant $\mu_0$ and a quantity with the dimensions of a length. This means that $\mu_0$ can be expressed in the unit henry per meter:

$$\mu_0 = 4\pi \times 10^{-7}\,\text{T}\cdot\text{m/A}$$
$$= 4\pi \times 10^{-7}\,\text{H/m}. \tag{30-32}$$

## 30-8 | Self-Induction

If two coils—which we can now call inductors—are near each other, a current $i$ in one coil produces a magnetic flux $\Phi_B$ through the second coil. We have seen that if we change this flux by changing the current, an induced emf appears in the second coil according to Faraday's law. An induced emf appears in the first coil as well.

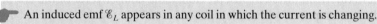

An induced emf $\mathscr{E}_L$ appears in any coil in which the current is changing.

This process (see Fig. 30-18) is called **self-induction,** and the emf that appears is called a **self-induced emf.** It obeys Faraday's law of induction just as other induced emfs do.

For any inductor, Eq. 30-28 tells us that

$$N\Phi_B = Li. \tag{30-33}$$

Faraday's law tells us that

$$\mathscr{E}_L = -\frac{d(N\Phi_B)}{dt}. \tag{30-34}$$

By combining Eqs. 30-33 and 30-34 we can write

$$\mathscr{E}_L = -L\frac{di}{dt} \quad \text{(self-induced emf)}. \tag{30-35}$$

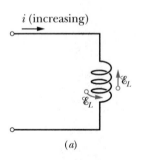

(a)

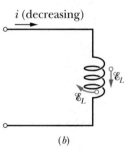

(b)

**FIG. 30-19** (a) The current $i$ is increasing, and the self-induced emf $\mathscr{E}_L$ appears along the coil in a direction such that it opposes the increase. The arrow representing $\mathscr{E}_L$ can be drawn along a turn of the coil or alongside the coil. Both are shown. (b) The current $i$ is decreasing, and the self-induced emf appears in a direction such that it opposes the decrease.

Thus, in any inductor (such as a coil, a solenoid, or a toroid) a self-induced emf appears whenever the current changes with time. The magnitude of the current has no influence on the magnitude of the induced emf; only the rate of change of the current counts.

You can find the *direction* of a self-induced emf from Lenz's law. The minus sign in Eq. 30-35 indicates that—as the law states—the self-induced emf $\mathscr{E}_L$ has the orientation such that it opposes the change in current $i$. We can drop the minus sign when we want only the magnitude of $\mathscr{E}_L$.

Suppose that, as in Fig. 30-19a, you set up a current $i$ in a coil and arrange to have the current increase with time at a rate $di/dt$. In the language of Lenz's law, this increase in the current is the "change" that the self-induction must oppose. For such opposition to occur, a self-induced emf must appear in the coil, pointing—as the figure shows—so as to oppose the increase in the current. If you cause the current to decrease with time, as in Fig. 30-19b, the self-induced emf must point in a direction that tends to oppose the decrease in the current, as the figure shows.

In Section 30-6 we saw that we cannot define an electric potential for an electric field (and thus for an emf) that is induced by a changing magnetic flux. This means that when a self-induced emf is produced in the inductor of Fig. 30-18, we cannot define an electric potential within the inductor itself, where the flux is changing. However, potentials can still be defined at points of the circuit that are not within the inductor—points where the electric fields are due to charge distributions and their associated electric potentials.

Moreover, we can define a self-induced potential difference $V_L$ *across an inductor* (between its terminals, which we assume to be outside the region of changing flux). For an *ideal inductor* (its wire has negligible resistance), the magnitude of $V_L$ is equal to the magnitude of the self-induced emf $\mathscr{E}_L$.

If, instead, the wire in the inductor has resistance $r$, we mentally separate the inductor into a resistance $r$ (which we take to be outside the region of changing flux) and an ideal inductor of self-induced emf $\mathscr{E}_L$. As with a real battery of emf $\mathscr{E}$ and internal resistance $r$, the potential difference across the terminals of a real inductor then differs from the emf. Unless otherwise indicated, we assume here that inductors are ideal.

✓ **CHECKPOINT 5**   The figure shows an emf $\mathscr{E}_L$ induced in a coil. Which of the following can describe the current through the coil: (a) constant and rightward, (b) constant and leftward, (c) increasing and rightward, (d) decreasing and rightward, (e) increasing and leftward, (f) decreasing and leftward?

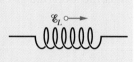

## 30-9 | *RL* Circuits

In Section 27-9 we saw that if we suddenly introduce an emf $\mathscr{E}$ into a single-loop circuit containing a resistor $R$ and a capacitor $C$, the charge on the capacitor does not build up immediately to its final equilibrium value $C\mathscr{E}$ but approaches it in an exponential fashion:

$$q = C\mathscr{E}(1 - e^{-t/\tau_C}). \tag{30-36}$$

The rate at which the charge builds up is determined by the capacitive time constant $\tau_C$, defined in Eq. 27-36 as

$$\tau_C = RC. \tag{30-37}$$

If we suddenly remove the emf from this same circuit, the charge does not immediately fall to zero but approaches zero in an exponential fashion:

$$q = q_0 e^{-t/\tau_C}. \tag{30-38}$$

The time constant $\tau_C$ describes the fall of the charge as well as its rise.

An analogous slowing of the rise (or fall) of the current occurs if we introduce an emf $\mathscr{E}$ into (or remove it from) a single-loop circuit containing a resistor $R$ and an inductor $L$. When the switch S in Fig. 30-20 is closed on $a$, for example, the current in the resistor starts to rise. If the inductor were not present, the current would rise rapidly to a steady value $\mathscr{E}/R$. Because of the inductor, however, a self-induced emf $\mathscr{E}_L$ appears in the circuit; from Lenz's law, this emf opposes the rise of the current, which means that it opposes the battery emf $\mathscr{E}$ in polarity. Thus, the current in the resistor responds to the difference between two emfs, a constant $\mathscr{E}$ due to the battery and a variable $\mathscr{E}_L$ $(= -L\, di/dt)$ due to self-induction. As long as $\mathscr{E}_L$ is present, the current will be less than $\mathscr{E}/R$.

As time goes on, the rate at which the current increases becomes less rapid and the magnitude of the self-induced emf, which is proportional to $di/dt$, becomes smaller. Thus, the current in the circuit approaches $\mathscr{E}/R$ asymptotically.

We can generalize these results as follows:

☛ Initially, an inductor acts to oppose changes in the current through it. A long time later, it acts like ordinary connecting wire.

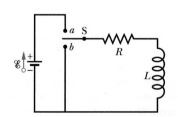

**FIG. 30-20**   An *RL* circuit. When switch S is closed on $a$, the current rises and approaches a limiting value $\mathscr{E}/R$.

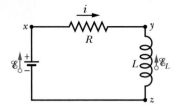

**FIG. 30-21** The circuit of Fig. 30-20 with the switch closed on *a*. We apply the loop rule for circuits clockwise, starting at *x*.

Now let us analyze the situation quantitatively. With the switch S in Fig. 30-20 thrown to *a*, the circuit is equivalent to that of Fig. 30-21. Let us apply the loop rule, starting at point *x* in this figure and moving clockwise around the loop along with current *i*.

1. *Resistor.* Because we move through the resistor in the direction of current *i*, the electric potential decreases by *iR*. Thus, as we move from point *x* to point *y*, we encounter a potential change of $-iR$.

2. *Inductor.* Because current *i* is changing, there is a self-induced emf $\mathscr{E}_L$ in the inductor. The magnitude of $\mathscr{E}_L$ is given by Eq. 30-35 as $L\,di/dt$. The direction of $\mathscr{E}_L$ is upward in Fig. 30-21 because current *i* is downward through the inductor *and* increasing. Thus, as we move from point *y* to point *z*, opposite the direction of $\mathscr{E}_L$, we encounter a potential change of $-L\,di/dt$.

3. *Battery.* As we move from point *z* back to starting point *x*, we encounter a potential change of $+\mathscr{E}$ due to the battery's emf.

Thus, the loop rule gives us

$$-iR - L\frac{di}{dt} + \mathscr{E} = 0$$

or

$$L\frac{di}{dt} + Ri = \mathscr{E} \qquad \text{(RL circuit).} \tag{30-39}$$

Equation 30-39 is a differential equation involving the variable *i* and its first derivative $di/dt$. To solve it, we seek the function $i(t)$ such that when $i(t)$ and its first derivative are substituted in Eq. 30-39, the equation is satisfied and the initial condition $i(0) = 0$ is satisfied.

Equation 30-39 and its initial condition are of exactly the form of Eq. 27-32 for an *RC* circuit, with *i* replacing *q*, *L* replacing *R*, and *R* replacing $1/C$. The solution of Eq. 30-39 must then be of exactly the form of Eq. 27-33 with the same replacements. That solution is

$$i = \frac{\mathscr{E}}{R}\left(1 - e^{-Rt/L}\right), \tag{30-40}$$

which we can rewrite as

$$i = \frac{\mathscr{E}}{R}\left(1 - e^{-t/\tau_L}\right) \qquad \text{(rise of current).} \tag{30-41}$$

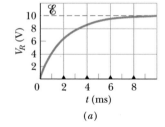

(a)

Here $\tau_L$, the **inductive time constant,** is given by

$$\tau_L = \frac{L}{R} \qquad \text{(time constant).} \tag{30-42}$$

Let's examine Eq. 30-41 for when the switch is closed (at time $t = 0$) and for a time long after the switch is closed ($t \to \infty$). If we substitute $t = 0$ into Eq. 30-41, the exponential becomes $e^{-0} = 1$. Thus, Eq. 30-41 tells us that the current is initially $i = 0$, as we expected. Next, if we let *t* go to ∞, then the exponential goes to $e^{-\infty} = 0$. Thus, Eq. 30-41 tells us that the current goes to its equilibrium value of $\mathscr{E}/R$.

We can also examine the potential differences in the circuit. For example, Fig. 30-22 shows how the potential differences $V_R \,(= iR)$ across the resistor and $V_L \,(= L\,di/dt)$ across the inductor vary with time for particular values of $\mathscr{E}$, *L*, and *R*. Compare this figure carefully with the corresponding figure for an *RC* circuit (Fig. 27-18).

To show that the quantity $\tau_L \,(= L/R)$ has the dimension of time, we convert from henries per ohm as follows:

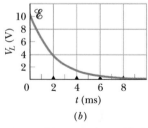

(b)

**FIG. 30-22** The variation with time of (a) $V_R$, the potential difference across the resistor in the circuit of Fig. 30-21, and (b) $V_L$, the potential difference across the inductor in that circuit. The small triangles represent successive intervals of one inductive time constant $\tau_L = L/R$. The figure is plotted for $R = 2000\ \Omega$, $L = 4.0$ H, and $\mathscr{E} = 10$ V.

$$1\,\frac{\text{H}}{\Omega} = 1\,\frac{\text{H}}{\Omega}\left(\frac{1\,\text{V}\cdot\text{s}}{1\,\text{H}\cdot\text{A}}\right)\left(\frac{1\,\Omega\cdot\text{A}}{1\,\text{V}}\right) = 1\text{ s.}$$

The first quantity in parentheses is a conversion factor based on Eq. 30-35, and the second one is a conversion factor based on the relation $V = iR$.

The physical significance of the time constant follows from Eq. 30-41. If we put $t = \tau_L = L/R$ in this equation, it reduces to

$$i = \frac{\mathcal{E}}{R}(1 - e^{-1}) = 0.63\,\frac{\mathcal{E}}{R}. \tag{30-43}$$

Thus, the time constant $\tau_L$ is the time it takes the current in the circuit to reach about 63% of its final equilibrium value $\mathcal{E}/R$. Since the potential difference $V_R$ across the resistor is proportional to the current $i$, a graph of the increasing current versus time has the same shape as that of $V_R$ in Fig. 30-22a.

If the switch S in Fig. 30-20 is closed on $a$ long enough for the equilibrium current $\mathcal{E}/R$ to be established and then is thrown to $b$, the effect will be to remove the battery from the circuit. (The connection to $b$ must actually be made an instant before the connection to $a$ is broken. A switch that does this is called a *make-before-break* switch.) With the battery gone, the current through the resistor will decrease. However, it cannot drop immediately to zero but must decay to zero over time. The differential equation that governs the decay can be found by putting $\mathcal{E} = 0$ in Eq. 30-39:

$$L\frac{di}{dt} + iR = 0. \tag{30-44}$$

By analogy with Eqs. 27-38 and 27-39, the solution of this differential equation that satisfies the initial condition $i(0) = i_0 = \mathcal{E}/R$ is

$$i = \frac{\mathcal{E}}{R}e^{-t/\tau_L} = i_0 e^{-t/\tau_L} \quad \text{(decay of current).} \tag{30-45}$$

We see that both current rise (Eq. 30-41) and current decay (Eq. 30-45) in an $RL$ circuit are governed by the same inductive time constant, $\tau_L$.

We have used $i_0$ in Eq. 30-45 to represent the current at time $t = 0$. In our case that happened to be $\mathcal{E}/R$, but it could be any other initial value.

✓ **CHECKPOINT 6** The figure shows three circuits with identical batteries, inductors, and resistors. Rank the circuits according to the current through the battery (a) just after the switch is closed and (b) a long time later, greatest first. (If you have trouble here, work through the next sample problem and then try again.)

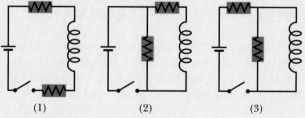

(1)　　　　　　(2)　　　　　　(3)

## Sample Problem 30-5

Figure 30-23a shows a circuit that contains three identical resistors with resistance $R = 9.0\ \Omega$, two identical inductors with inductance $L = 2.0$ mH, and an ideal battery with emf $\mathcal{E} = 18$ V.

(a) What is the current $i$ through the battery just after the switch is closed?

**KEY IDEA** Just after the switch is closed, the inductor acts to oppose a change in the current through it.

**Calculations:** Because the current through each inductor is zero before the switch is closed, it will also be zero just afterward. Thus, immediately after the switch is closed, the inductors act as broken wires, as indicated in Fig. 30-23b. We then have a single-loop circuit for which the loop rule gives us

$$\mathcal{E} - iR = 0.$$

Substituting given data, we find that

$$i = \frac{\mathcal{E}}{R} = \frac{18 \text{ V}}{9.0 \text{ } \Omega} = 2.0 \text{ A}. \qquad \text{(Answer)}$$

(b) What is the current $i$ through the battery long after the switch has been closed?

**KEY IDEA** Long after the switch has been closed, the currents in the circuit have reached their equilibrium values, and the inductors act as simple connecting wires, as indicated in Fig. 30-23c.

**Calculations:** We now have a circuit with three identical resistors in parallel; from Eq. 27-23, their equivalent resistance is $R_{eq} = R/3 = (9.0 \text{ }\Omega)/3 = 3.0 \text{ }\Omega$. The equivalent circuit shown in Fig. 30-23d then yields the loop

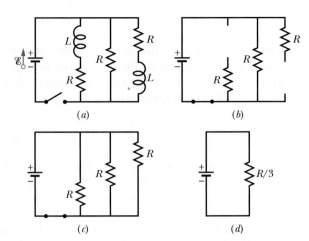

**FIG. 30-23** (a) A multiloop $RL$ circuit with an open switch. (b) The equivalent circuit just after the switch has been closed. (c) The equivalent circuit a long time later. (d) The single-loop circuit that is equivalent to circuit (c).

equation $\mathcal{E} - iR_{eq} = 0$, or

$$i = \frac{\mathcal{E}}{R_{eq}} = \frac{18 \text{ V}}{3.0 \text{ }\Omega} = 6.0 \text{ A}. \qquad \text{(Answer)}$$

---

**Sample Problem** | **30-6**

A solenoid has an inductance of 53 mH and a resistance of 0.37 $\Omega$. If the solenoid is connected to a battery, how long will the current take to reach half its final equilibrium value?

**KEY IDEA** We can mentally separate the solenoid into a resistance and an inductance that are wired in series with a battery, as in Fig. 30-21. Then application of the loop rule leads to Eq. 30-39, which has the solution of Eq. 30-41 for the current $i$ in the circuit.

**Calculations:** According to that solution, current $i$ increases exponentially from zero to its final equilibrium

value of $\mathcal{E}/R$. Let $t_0$ be the time that current $i$ takes to reach half its equilibrium value. Then Eq. 30-41 gives us

$$\frac{1}{2} \frac{\mathcal{E}}{R} = \frac{\mathcal{E}}{R} (1 - e^{-t_0/\tau_L}).$$

We solve for $t_0$ by canceling $\mathcal{E}/R$, isolating the exponential, and taking the natural logarithm of each side. We find

$$t_0 = \tau_L \ln 2 = \frac{L}{R} \ln 2 = \frac{53 \times 10^{-3} \text{ H}}{0.37 \text{ }\Omega} \ln 2$$

$$= 0.10 \text{ s}. \qquad \text{(Answer)}$$

---

## 30-10 | Energy Stored in a Magnetic Field

When we pull two charged particles of opposite signs away from each other, we say that the resulting electric potential energy is stored in the electric field of the particles. We get it back from the field by letting the particles move closer together again. In the same way we say energy is stored in a magnetic field.

To derive a quantitative expression for that stored energy, consider again Fig. 30-21, which shows a source of emf $\mathcal{E}$ connected to a resistor $R$ and an inductor $L$. Equation 30-39, restated here for convenience,

$$\mathcal{E} = L \frac{di}{dt} + iR, \tag{30-46}$$

is the differential equation that describes the growth of current in this circuit. Recall that this equation follows immediately from the loop rule and that the loop rule in turn is an expression of the principle of conservation of energy for single-loop circuits. If we multiply each side of Eq. 30-46 by $i$, we obtain

$$\mathcal{E}i = Li \frac{di}{dt} + i^2 R, \tag{30-47}$$

which has the following physical interpretation in terms of work and energy:

1. If a differential amount of charge $dq$ passes through the battery of emf $\mathcal{E}$ in Fig. 30-21 in time $dt$, the battery does work on it in the amount $\mathcal{E}\,dq$. The rate at which the battery does work is $(\mathcal{E}\,dq)/dt$, or $\mathcal{E}i$. Thus, the left side of Eq. 30-47 represents the rate at which the emf device delivers energy to the rest of the circuit.

2. The rightmost term in Eq. 30-47 represents the rate at which energy appears as thermal energy in the resistor.

3. Energy that is delivered to the circuit but does not appear as thermal energy must, by the conservation-of-energy hypothesis, be stored in the magnetic field of the inductor. Because Eq. 30-47 represents the principle of conservation of energy for *RL* circuits, the middle term must represent the rate $dU_B/dt$ at which magnetic potential energy $U_B$ is stored in the magnetic field.

Thus

$$\frac{dU_B}{dt} = Li\,\frac{di}{dt}. \tag{30-48}$$

We can write this as

$$dU_B = Li\,di.$$

Integrating yields

$$\int_0^{U_B} dU_B = \int_0^i Li\,di$$

or
$$U_B = \tfrac{1}{2}Li^2 \quad \text{(magnetic energy)}, \tag{30-49}$$

which represents the total energy stored by an inductor $L$ carrying a current $i$. Note the similarity in form between this expression and the expression for the energy stored by a capacitor with capacitance $C$ and charge $q$; namely,

$$U_E = \frac{q^2}{2C}. \tag{30-50}$$

(The variable $i^2$ corresponds to $q^2$, and the constant $L$ corresponds to $1/C$.)

---

**Sample Problem    30-7**

---

A coil has an inductance of 53 mH and a resistance of 0.35 $\Omega$.

(a) If a 12 V emf is applied across the coil, how much energy is stored in the magnetic field after the current has built up to its equilibrium value?

---

**KEY IDEA**    The energy stored in the magnetic field of a coil at any time depends on the current through the coil at that time, according to Eq. 30-49 ($U_B = \tfrac{1}{2}Li^2$).

**Calculations:** Thus, to find the energy $U_{B\infty}$ stored at equilibrium, we must first find the equilibrium current. From Eq. 30-41, the equilibrium current is

$$i_\infty = \frac{\mathcal{E}}{R} = \frac{12\ \text{V}}{0.35\ \Omega} = 34.3\ \text{A}. \tag{30-51}$$

Then substitution yields

$$U_{B\infty} = \tfrac{1}{2}Li_\infty^2 = (\tfrac{1}{2})(53 \times 10^{-3}\ \text{H})(34.3\ \text{A})^2$$

$$= 31\ \text{J}. \qquad\qquad \text{(Answer)}$$

(b) After how many time constants will half this equilibrium energy be stored in the magnetic field?

**Calculations:** Now we are being asked: At what time $t$ will the relation

$$U_B = \tfrac{1}{2}U_{B\infty}$$

be satisfied? Using Eq. 30-49 twice allows us to rewrite this energy condition as

$$\tfrac{1}{2}Li^2 = (\tfrac{1}{2})\tfrac{1}{2}Li_\infty^2$$

or
$$i = \left(\frac{1}{\sqrt{2}}\right)i_\infty. \tag{30-52}$$

However, $i$ is given by Eq. 30-41 and $i_\infty$ (see Eq. 30-51) is $\mathcal{E}/R$; so Eq. 30-52 becomes

$$\frac{\mathcal{E}}{R}(1 - e^{-t/\tau_L}) = \frac{\mathcal{E}}{\sqrt{2}R}.$$

By canceling $\mathcal{E}/R$ and rearranging, we can write this as

$$e^{-t/\tau_L} = 1 - \frac{1}{\sqrt{2}} = 0.293,$$

which yields

$$\frac{t}{\tau_L} = -\ln 0.293 = 1.23$$

or

$$t \approx 1.2\tau_L. \qquad \text{(Answer)}$$

Thus, the energy stored in the magnetic field of the coil by the current will reach half its equilibrium value 1.2 time constants after the emf is applied.

## 30-11 | Energy Density of a Magnetic Field

Consider a length $l$ near the middle of a long solenoid of cross-sectional area $A$ carrying current $i$; the volume associated with this length is $Al$. The energy $U_B$ stored by the length $l$ of the solenoid must lie entirely within this volume because the magnetic field outside such a solenoid is approximately zero. Moreover, the stored energy must be uniformly distributed within the solenoid because the magnetic field is (approximately) uniform everywhere inside.

Thus, the energy stored per unit volume of the field is

$$u_B = \frac{U_B}{Al}$$

or, since

$$U_B = \tfrac{1}{2}Li^2,$$

we have

$$u_B = \frac{Li^2}{2Al} = \frac{L}{l}\frac{i^2}{2A}.$$

Here $L$ is the inductance of length $l$ of the solenoid.

Substituting for $L/l$ from Eq. 30-31, we find

$$u_B = \tfrac{1}{2}\mu_0 n^2 i^2, \qquad (30\text{-}53)$$

where $n$ is the number of turns per unit length. From Eq. 29-23 ($B = \mu_0 in$) we can write this *energy density* as

$$u_B = \frac{B^2}{2\mu_0} \qquad \text{(magnetic energy density).} \qquad (30\text{-}54)$$

This equation gives the density of stored energy at any point where the magnitude of the magnetic field is $B$. Even though we derived it by considering the special case of a solenoid, Eq. 30-54 holds for all magnetic fields, no matter how they are generated. The equation is comparable to Eq. 25-25,

$$u_E = \tfrac{1}{2}\varepsilon_0 E^2, \qquad (30\text{-}55)$$

which gives the energy density (in a vacuum) at any point in an electric field. Note that both $u_B$ and $u_E$ are proportional to the square of the appropriate field magnitude, $B$ or $E$.

✓**CHECKPOINT 7**  The table lists the number of turns per unit length, current, and cross-sectional area for three solenoids. Rank the solenoids according to the magnetic energy density within them, greatest first.

| Solenoid | Turns per Unit Length | Current | Area |
|---|---|---|---|
| $a$ | $2n_1$ | $i_1$ | $2A_1$ |
| $b$ | $n_1$ | $2i_1$ | $A_1$ |
| $c$ | $n_1$ | $i_1$ | $6A_1$ |

## Sample Problem | 30-8

A long coaxial cable (Fig. 30-24) consists of two thin-walled concentric conducting cylinders with radii $a$ and $b$. The inner cylinder carries a steady current $i$, and the outer cylinder provides the return path for that current. The current sets up a magnetic field between the two cylinders.

**(a)** Calculate the energy stored in the magnetic field for a length $\ell$ of the cable.

### KEY IDEAS

1. We can calculate the (total) energy $U_B$ stored in the magnetic field from the energy density $u_B$ of the field.

2. That energy density depends on the magnitude $B$ of the magnetic field according to Eq. 30-54 ($u_B = B^2/2\mu_0$).

3. Because of the circular symmetry of the cable, we can find $B$ by using Ampere's law with the given current $i$.

*Finding B:* To apply these ideas, we begin with Ampere's law, using a circular path of integration with radius $r$ such that $a < r < b$ (between the two cylinders, as indicated by the dashed line in Fig. 30-24). The only current enclosed by this path is current $i$ on the inner cylinder. Thus, we can write Ampere's law as

$$\oint \vec{B} \cdot d\vec{s} = \mu_0 i. \qquad (30\text{-}56)$$

Next, we simplify the integral: Because of the circular symmetry, at all points along the circular path, $\vec{B}$ is tangent to the path and has the same magnitude $B$. Let us take the direction of integration along the path as the direction of the magnetic field around the path. Then we can replace $\vec{B} \cdot d\vec{s}$ with $B\,ds\cos 0 = B\,ds$ and then move magnitude $B$ in front of the integration symbol. The integral that remains is $\oint ds$, which just gives the circumference $2\pi r$ of the path. Thus, Eq. 30-56 simplifies to

$$B(2\pi r) = \mu_0 i$$

or

$$B = \frac{\mu_0 i}{2\pi r}. \qquad (30\text{-}57)$$

*Finding $u_B$:* Next, to obtain the energy density, we substitute Eq. 30-57 into Eq. 30-54:

$$u_B = \frac{B^2}{2\mu_0} = \frac{\mu_0 i^2}{8\pi^2 r^2}. \qquad (30\text{-}58)$$

*Finding $U_B$:* Note that $u_B$ is not uniform in the volume between the two cylinders, but instead depends on the radial distance $r$. Thus, to find the total energy $U_B$ stored between the cylinders, we must integrate $u_B$ over that volume.

Because the volume between the cylinders has circular symmetry about the cable's central axis, we consider the volume $dV$ of a cylindrical shell located

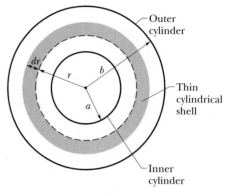

**FIG. 30-24** A cross section of a long coaxial cable consisting of two thin-walled conducting cylinders, the inner cylinder of radius $a$ and the outer cylinder of radius $b$.

between the cylinders; the shell has inner radius $r$, outer radius $r + dr$ (Fig. 30-24), and length $\ell$. The shell's cross-sectional area is the product of its circumference $2\pi r$ and thickness $dr$. Thus, the shell's volume $dV$ is $(2\pi r)(dr)(\ell)$; that is, $dV = 2\pi r\ell\,dr$.

Because points within this shell are all at approximately the same radial distance $r$, they all have approximately the same energy density $u_B$. Thus, the total energy $dU_B$ contained in the shell of volume $dV$ is given by

$$\text{energy} = \left( \begin{array}{c} \text{energy per} \\ \text{unit volume} \end{array} \right) (\text{volume})$$

or

$$dU_B = u_B\,dV.$$

Substituting Eq. 30-58 for $u_B$ and $2\pi r\ell\,dr$ for $dV$, we obtain

$$dU_B = \frac{\mu_0 i^2}{8\pi^2 r^2}(2\pi r\ell)\,dr = \frac{\mu_0 i^2 \ell}{4\pi}\frac{dr}{r}.$$

To find the total energy contained between the two cylinders, we integrate this equation over the volume between the two cylinders:

$$U_B = \int dU_B = \frac{\mu_0 i^2 \ell}{4\pi} \int_a^b \frac{dr}{r}$$

$$= \frac{\mu_0 i^2 \ell}{4\pi} \ln\frac{b}{a}. \qquad (\text{Answer}) \ (30\text{-}59)$$

No energy is stored outside the outer cylinder or inside the inner cylinder because the magnetic field is zero in both locations, as you can show with Ampere's law.

**(b)** What is the stored energy per unit length of the cable if $a = 1.2$ mm, $b = 3.5$ mm, and $i = 2.7$ A?

*Calculation:* From Eq. 30-59 we have

$$\frac{U_B}{\ell} = \frac{\mu_0 i^2}{4\pi} \ln\frac{b}{a}$$

$$= \frac{(4\pi \times 10^{-7}\ \text{H/m})(2.7\ \text{A})^2}{4\pi} \ln\frac{3.5\ \text{mm}}{1.2\ \text{mm}}$$

$$= 7.8 \times 10^{-7}\ \text{J/m} = 780\ \text{nJ/m}. \qquad (\text{Answer})$$

## 30-12 | Mutual Induction

In this section we return to the case of two interacting coils, which we first discussed in Section 30-2, and we treat it in a somewhat more formal manner. We saw earlier that if two coils are close together as in Fig. 30-2, a steady current $i$ in one coil will set up a magnetic flux $\Phi$ through the other coil (*linking* the other coil). If we change $i$ with time, an emf $\mathcal{E}$ given by Faraday's law appears in the second coil; we called this process *induction*. We could better have called it **mutual induction,** to suggest the mutual interaction of the two coils and to distinguish it from *self-induction,* in which only one coil is involved.

Let us look a little more quantitatively at mutual induction. Figure 30-25$a$ shows two circular close-packed coils near each other and sharing a common central axis. With the variable resistor set at a particular resistance $R$, the battery produces a steady current $i_1$ in coil 1. This current creates a magnetic field represented by the lines of $\vec{B}_1$ in the figure. Coil 2 is connected to a sensitive meter but contains no battery; a magnetic flux $\Phi_{21}$ (the flux through coil 2 associated with the current in coil 1) links the $N_2$ turns of coil 2.

We define the mutual inductance $M_{21}$ of coil 2 with respect to coil 1 as

$$M_{21} = \frac{N_2 \Phi_{21}}{i_1}, \qquad (30\text{-}60)$$

which has the same form as Eq. 30-28 ($L = N\Phi/i$), the definition of inductance. We can recast Eq. 30-60 as

$$M_{21} i_1 = N_2 \Phi_{21}.$$

If we cause $i_1$ to vary with time by varying $R$, we have

$$M_{21} \frac{di_1}{dt} = N_2 \frac{d\Phi_{21}}{dt}.$$

The right side of this equation is, according to Faraday's law, just the magnitude of the emf $\mathcal{E}_2$ appearing in coil 2 due to the changing current in coil 1. Thus, with a minus sign to indicate direction,

$$\mathcal{E}_2 = -M_{21} \frac{di_1}{dt}, \qquad (30\text{-}61)$$

which you should compare with Eq. 30-35 for self-induction ($\mathcal{E} = -L\,di/dt$).

Let us now interchange the roles of coils 1 and 2, as in Fig. 30-25$b$; that is, we set up a current $i_2$ in coil 2 by means of a battery, and this produces a magnetic flux $\Phi_{12}$ that links coil 1. If we change $i_2$ with time by varying $R$, we then have, by the argument given above,

$$\mathcal{E}_1 = -M_{12} \frac{di_2}{dt}. \qquad (30\text{-}62)$$

Thus, we see that the emf induced in either coil is proportional to the rate of change of current in the other coil. The proportionality constants $M_{21}$ and $M_{12}$ seem to be different. We assert, without proof, that they are in fact the same so that no subscripts are needed. (This conclusion is true but is in no way obvious.) Thus, we have

$$M_{21} = M_{12} = M, \qquad (30\text{-}63)$$

and we can rewrite Eqs. 30-61 and 30-62 as

$$\mathcal{E}_2 = -M \frac{di_1}{dt} \qquad (30\text{-}64)$$

and

$$\mathcal{E}_1 = -M \frac{di_2}{dt}. \qquad (30\text{-}65)$$

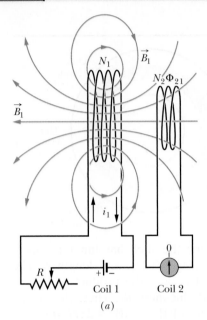

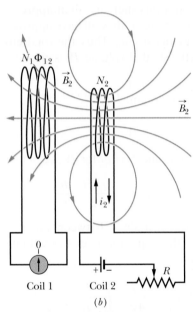

**FIG. 30-25** Mutual induction. ($a$) The magnetic field $\vec{B}_1$ produced by current $i_1$ in coil 1 extends through coil 2. If $i_1$ is varied (by varying resistance $R$), an emf is induced in coil 2 and current registers on the meter connected to coil 2. ($b$) The roles of the coils interchanged.

Figure 30-26 shows two circular close-packed coils, the smaller (radius $R_2$, with $N_2$ turns) being coaxial with the larger (radius $R_1$, with $N_1$ turns) and in the same plane.

(a) Derive an expression for the mutual inductance $M$ for this arrangement of these two coils, assuming that $R_1 \gg R_2$.

**KEY IDEA** The mutual inductance $M$ for these coils is the ratio of the flux linkage ($N\Phi$) through one coil to the current $i$ in the other coil, which produces that flux linkage. Thus, we need to assume that currents exist in the coils; then we need to calculate the flux linkage in one of the coils.

*Calculations:* The magnetic field through the larger coil due to the smaller coil is nonuniform in both magnitude and direction; so the flux through the larger coil due to the smaller coil is nonuniform and difficult to calculate. However, the smaller coil is small enough for us to assume that the magnetic field through it due to the larger coil is approximately uniform. Thus, the flux through it due to the larger coil is also approximately uniform. Hence, to find $M$ we shall assume a current $i_1$ in the larger coil and calculate the flux linkage $N_2\Phi_{21}$ in the smaller coil:

$$M = \frac{N_2\Phi_{21}}{i_1}. \qquad (30\text{-}66)$$

The flux $\Phi_{21}$ through each turn of the smaller coil is, from Eq. 30-2,

$$\Phi_{21} = B_1 A_2,$$

where $B_1$ is the magnitude of the magnetic field at points within the small coil due to the larger coil and $A_2 \, (= \pi R_2^2)$ is the area enclosed by the turn. Thus, the flux linkage in the smaller coil (with its $N_2$ turns) is

$$N_2\Phi_{21} = N_2 B_1 A_2. \qquad (30\text{-}67)$$

To find $B_1$ at points within the smaller coil, we can use Eq. 29-26,

$$B(z) = \frac{\mu_0 i R^2}{2(R^2 + z^2)^{3/2}},$$

with $z$ set to 0 because the smaller coil is in the plane of the larger coil. That equation tells us that each turn of the larger coil produces a magnetic field of magnitude $\mu_0 i_1/2R_1$ at points within the smaller coil. Thus, the larger coil (with its $N_1$ turns) produces a total magnetic field of magnitude

$$B_1 = N_1 \frac{\mu_0 i_1}{2R_1} \qquad (30\text{-}68)$$

at points within the smaller coil.

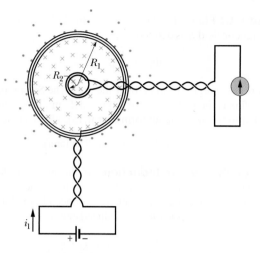

**FIG. 30-26** A small coil is located at the center of a large coil. The mutual inductance of the coils can be determined by sending current $i_1$ through the large coil.

Substituting Eq. 30-68 for $B_1$ and $\pi R_2^2$ for $A_2$ in Eq. 30-67 yields

$$N_2\Phi_{21} = \frac{\pi\mu_0 N_1 N_2 R_2^2 i_1}{2R_1}.$$

Substituting this result into Eq. 30-66, we find

$$M = \frac{N_2\Phi_{21}}{i_1} = \frac{\pi\mu_0 N_1 N_2 R_2^2}{2R_1}. \quad \text{(Answer)} \quad (30\text{-}69)$$

(b) What is the value of $M$ for $N_1 = N_2 = 1200$ turns, $R_2 = 1.1$ cm, and $R_1 = 15$ cm?

*Calculations:* Equation 30-69 yields

$$M = \frac{(\pi)(4\pi \times 10^{-7}\,\text{H/m})(1200)(1200)(0.011\,\text{m})^2}{(2)(0.15\,\text{m})}$$

$$= 2.29 \times 10^{-3}\,\text{H} \approx 2.3\,\text{mH}. \quad \text{(Answer)}$$

Consider the situation if we reverse the roles of the two coils—that is, if we produce a current $i_2$ in the smaller coil and try to calculate $M$ from Eq. 30-60 in the form

$$M = \frac{N_1\Phi_{12}}{i_2}.$$

The calculation of $\Phi_{12}$ (the nonuniform flux of the smaller coil's magnetic field encompassed by the larger coil) is not simple. If we were to do the calculation numerically using a computer, we would find $M$ to be 2.3 mH, as above! This emphasizes that Eq. 30-63 ($M_{21} = M_{12} = M$) is not obvious.

## REVIEW & SUMMARY

**Magnetic Flux** The *magnetic flux* $\Phi_B$ through an area $A$ in a magnetic field $\vec{B}$ is defined as

$$\Phi_B = \int \vec{B} \cdot d\vec{A}, \tag{30-1}$$

where the integral is taken over the area. The SI unit of magnetic flux is the weber, where $1\text{ Wb} = 1\text{ T} \cdot \text{m}^2$. If $\vec{B}$ is perpendicular to the area and uniform over it, Eq. 30-1 becomes

$$\Phi_B = BA \qquad (\vec{B} \perp A, \vec{B} \text{ uniform}). \tag{30-2}$$

**Faraday's Law of Induction** If the magnetic flux $\Phi_B$ through an area bounded by a closed conducting loop changes with time, a current and an emf are produced in the loop; this process is called *induction*. The induced emf is

$$\mathscr{E} = -\frac{d\Phi_B}{dt} \qquad \text{(Faraday's law)}. \tag{30-4}$$

If the loop is replaced by a closely packed coil of $N$ turns, the induced emf is

$$\mathscr{E} = -N\frac{d\Phi_B}{dt}. \tag{30-5}$$

**Lenz's Law** An induced current has a direction such that the magnetic field *due to the current* opposes the change in the magnetic flux that induces the current. The induced emf has the same direction as the induced current.

**Emf and the Induced Electric Field** An emf is induced by a changing magnetic flux even if the loop through which the flux is changing is not a physical conductor but an imaginary line. The changing magnetic field induces an electric field $\vec{E}$ at every point of such a loop; the induced emf is related to $\vec{E}$ by

$$\mathscr{E} = \oint \vec{E} \cdot d\vec{s}, \tag{30-19}$$

where the integration is taken around the loop. From Eq. 30-19 we can write Faraday's law in its most general form,

$$\oint \vec{E} \cdot d\vec{s} = -\frac{d\Phi_B}{dt} \qquad \text{(Faraday's law)}. \tag{30-20}$$

*A changing magnetic field induces an electric field $\vec{E}$.*

**Inductors** An **inductor** sis a device that can be used to produce a known magnetic field in a specified region. If a current $i$ is established through each of the $N$ windings of an inductor, a magnetic flux $\Phi_B$ links those windings. The **inductance** $L$ of the inductor is

$$L = \frac{N\Phi_B}{i} \qquad \text{(inductance defined)}. \tag{30-28}$$

The SI unit of inductance is the **henry** (H), where $1$ henry $= 1\text{ H} = 1\text{ T} \cdot \text{m}^2/\text{A}$. The inductance per unit length near the middle of a long solenoid of cross-sectional area $A$ and $n$ turns per unit length is

$$\frac{L}{l} = \mu_0 n^2 A \qquad \text{(solenoid)}. \tag{30-31}$$

**Self-Induction** If a current $i$ in a coil changes with time, an emf is induced in the coil. This self-induced emf is

$$\mathscr{E}_L = -L\frac{di}{dt}. \tag{30-35}$$

The direction of $\mathscr{E}_L$ is found from Lenz's law: The self-induced emf acts to oppose the change that produces it.

**Series RL Circuits** If a constant emf $\mathscr{E}$ is introduced into a single-loop circuit containing a resistance $R$ and an inductance $L$, the current rises to an equilibrium value of $\mathscr{E}/R$ according to

$$i = \frac{\mathscr{E}}{R}(1 - e^{-t/\tau_L}) \qquad \text{(rise of current)}. \tag{30-41}$$

Here $\tau_L\ (= L/R)$ governs the rate of rise of the current and is called the **inductive time constant** of the circuit. When the source of constant emf is removed, the current decays from a value $i_0$ according to

$$i = i_0 e^{-t/\tau_L} \qquad \text{(decay of current)}. \tag{30-45}$$

**Magnetic Energy** If an inductor $L$ carries a current $i$, the inductor's magnetic field stores an energy given by

$$U_B = \tfrac{1}{2}Li^2 \qquad \text{(magnetic energy)}. \tag{30-49}$$

If $B$ is the magnitude of a magnetic field at any point (in an inductor or anywhere else), the density of stored magnetic energy at that point is

$$u_B = \frac{B^2}{2\mu_0} \qquad \text{(magnetic energy density)}. \tag{30-54}$$

**Mutual Induction** If coils 1 and 2 are near each other, a changing current in either coil can induce an emf in the other. This mutual induction is described by

$$\mathscr{E}_2 = -M\frac{di_1}{dt} \tag{30-64}$$

and

$$\mathscr{E}_1 = -M\frac{di_2}{dt}, \tag{30-65}$$

where $M$ (measured in henries) is the mutual inductance.

## QUESTIONS

**1** In Fig. 30-27, a long straight wire with current $i$ passes (without touching) three rectangular wire loops with edge lengths $L$, $1.5L$, and $2L$. The loops are widely spaced (so as not to affect one another). Loops 1 and 3 are symmetric about the long wire. Rank the loops according to the size of the current induced in them if current $i$ is (a) constant and (b) increasing, greatest first.

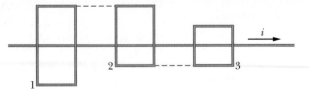

**FIG. 30-27** Question 1.

**2** Figure 30-28 shows two circuits in which a conducting bar is slid at the same speed $v$ through the same uniform magnetic field and along a U-shaped wire. The parallel lengths of the wire are separated by $2L$ in circuit 1 and by $L$ in circuit 2. The current induced in circuit 1 is counterclockwise. (a) Is the magnetic field into or out of the page? (b) Is the current induced in circuit 2 clockwise or counterclockwise? (c) Is the emf induced in circuit 1 larger than, smaller than, or the same as that in circuit 2?

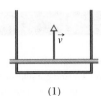

(1)

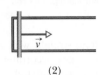

(2)

FIG. 30-28 Question 2.

**3** If the circular conductor in Fig. 30-29 undergoes thermal expansion while it is in a uniform magnetic field, a current is induced clockwise around it. Is the magnetic field directed into or out of the page?

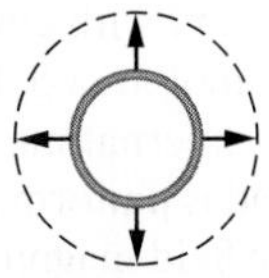

FIG. 30-29 Question 3.

**4** The wire loop in Fig. 30-30a is subjected, in turn, to six uniform magnetic fields, each directed parallel to the $z$ axis. Figure 30-30b gives the $z$ components $B_z$ of the fields versus time $t$. (Plots 1 and 3 are parallel; so are plots 4 and 6.) Rank the six plots according to the emf induced in the loop, greatest clockwise emf first, greatest counterclockwise emf last.

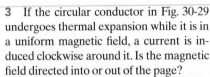

(a)                    (b)

FIG. 30-30 Question 4.

**5** Figure 30-31 shows a circular region in which a decreasing uniform magnetic field is directed out of the page, as well as four concentric circular paths. Rank the paths according to the magnitude of $\oint \vec{E} \cdot d\vec{s}$ evaluated along them, greatest first.

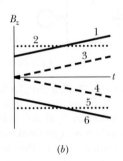

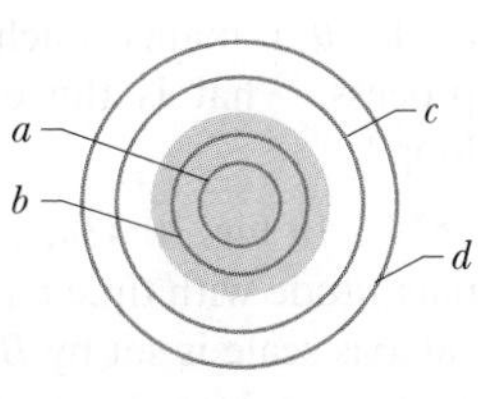

FIG. 30-31 Question 5.

**6** In Fig. 30-32, a wire loop has been bent so that it has three segments: segment $bc$ (a quarter-circle), $ac$ (a square corner), and $ab$ (straight). Here are three choices for a magnetic field through the loop:

(1) $\vec{B}_1 = 3\hat{i} + 7\hat{j} - 5t\hat{k}$,
(2) $\vec{B}_2 = 5t\hat{i} - 4\hat{j} - 15\hat{k}$,
(3) $\vec{B}_3 = 2\hat{i} - 5t\hat{j} - 12\hat{k}$,

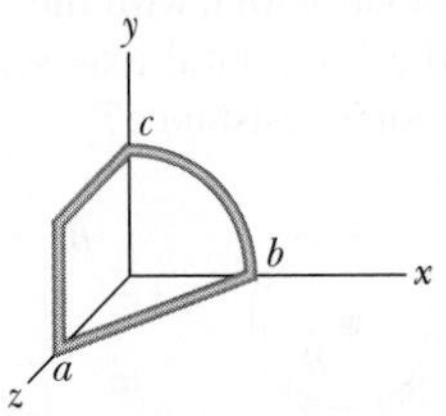

where $\vec{B}$ is in milliteslas and $t$ is in seconds. Without written calculation, rank the choices accord-

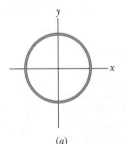

FIG. 30-32 Question 6.

ing to (a) the work done per unit charge in setting up the induced current and (b) that induced current, greatest first. (c) For each choice, what is the direction of the induced current in the figure?

**7** Figure 30-33 shows three circuits with identical batteries, inductors, and resistors. Rank the circuits, greatest first, according to the current through the resistor labeled $R$ (a) long after the switch is closed, (b) just after the switch is reopened a long time later, and (c) long after it is reopened.

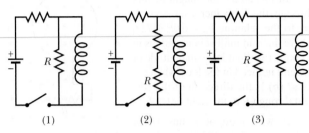

(1)                (2)                (3)

FIG. 30-33 Question 7.

**8** Figure 30-34 gives the variation with time of the potential difference $V_R$ across a resistor in three circuits wired as shown in Fig. 30-21. The circuits contain the same resistance $R$ and emf $\mathscr{E}$ but differ in the inductance $L$. Rank the circuits according to the value of $L$, greatest first.

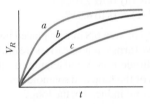

FIG. 30-34 Question 8.

**9** Figure 30-35 shows a circuit with two identical resistors and an ideal inductor. Is the current through the central resistor more than, less than, or the same as that through the other resistor (a) just after the closing of switch S, (b) a long time after that, (c) just after S is reopened a long time later, and (d) a long time after that?

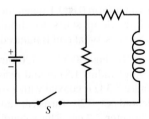

FIG. 30-35 Question 9.

**10** The switch in the circuit of Fig. 30-20 has been closed on $a$ for a very long time when it is then thrown to $b$. The resulting current through the inductor is indicated in Fig. 30-36 for four sets of values for the resistance $R$ and inductance $L$: (1) $R_0$ and $L_0$, (2) $2R_0$ and $L_0$, (3) $R_0$ and $2L_0$, (4) $2R_0$ and $2L_0$. Which set goes with which curve?

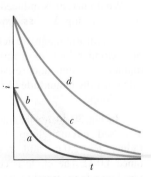

FIG. 30-36 Question 10.

## PROBLEMS

GO   Tutoring problem available (at instructor's discretion) in *WileyPLUS* and WebAssign
SSM   Worked-out solution available in Student Solutions Manual      WWW   Worked-out solution is at ____
• – •••   Number of dots indicates level of problem difficulty      ILW   Interactive solution is at ____   http://www.wiley.com/college/halliday
✈   Additional information available in *The Flying Circus of Physics* and at flyingcircusofphysics.com

### sec. 30-4   Lenz's Law

•1   In Fig. 30-37, the magnetic flux through the loop increases according to the relation $\Phi_B = 6.0t^2 + 7.0t$, where $\Phi_B$ is in milliwebers and $t$ is in seconds. (a) What is the magnitude of the emf induced in the loop when $t = 2.0$ s? (b) Is the direction of the current through $R$ to the right or left?

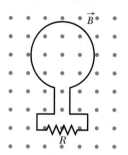

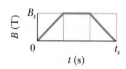

•2   A wire loop of radius 12 cm and resistance 8.5 Ω is located in a uniform magnetic field $\vec{B}$ that changes in magnitude as given in Fig. 30-38. The vertical axis scale is set by $B_s = 0.50$ T, and the horizontal axis scale is set by $t_s = 6.00$ s. The loop's plane is perpendicular to $\vec{B}$. What emf is induced in the loop during time intervals (a) 0 to 2.0 s, (b) 2.0 s to 4.0 s, and (c) 4.0 s to 6.0 s?

**FIG. 30-37**   Problem 1.

**FIG. 30-38**   Problem 2.

•3   A small loop of area 6.8 mm² is placed inside a long solenoid that has 854 turns/cm and carries a sinusoidally varying current $i$ of amplitude 1.28 A and angular frequency 212 rad/s. The central axes of the loop and solenoid coincide. What is the amplitude of the emf induced in the loop?

•4   An elastic conducting material is stretched into a circular loop of 12.0 cm radius. It is placed with its plane perpendicular to a uniform 0.800 T magnetic field. When released, the radius of the loop starts to shrink at an instantaneous rate of 75.0 cm/s. What emf is induced in the loop at that instant?

•5   In Fig. 30-39, a 120-turn coil of radius 1.8 cm and resistance 5.3 Ω is coaxial with a solenoid of 220 turns/cm and diameter 3.2 cm. The solenoid current drops from 1.5 A to zero in time interval $\Delta t = 25$ ms. What current is induced in the coil during $\Delta t$?   SSM WWW

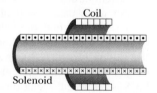

**FIG. 30-39**   Problem 5.

•6   A uniform magnetic field $\vec{B}$ is perpendicular to the plane of a circular loop of diameter 10 cm formed from wire of diameter 2.5 mm and resistivity $1.69 \times 10^{-8}$ Ω·m. At what rate must the magnitude of $\vec{B}$ change to induce a 10 A current in the loop?

•7   In Fig. 30-40, a wire forms a closed circular loop, with radius $R = 2.0$ m and resistance 4.0 Ω. The circle is centered on a long straight wire; at time $t = 0$, the current in the

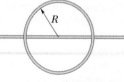

**FIG. 30-40**   Problem 7.

long straight wire is 5.0 A rightward. Thereafter, the current changes according to $i = 5.0$ A $- (2.0$ A/s²$)t^2$. (The straight wire is insulated; so there is no electrical contact between it and the wire of the loop.) What is the magnitude of the current induced in the loop at times $t > 0$?

•8   Figure 30-41a shows a circuit consisting of an ideal battery with emf $\mathcal{E} = 6.00$ µV, a resistance $R$, and a small wire loop of area 5.0 cm². For the time interval $t = 10$ s to $t = 20$ s, an external magnetic field is set up throughout the loop. The field is uniform, its direction is into the page in Fig. 30-41a, and the field magnitude is given by $B = at$, where $B$ is in teslas, $a$ is a constant, and $t$ is in seconds. Figure 30-41b gives the current $i$ in the circuit before, during, and after the external field is set up. The vertical axis scale is set by $i_s = 2.0$ mA. Find the constant $a$ in the equation for the field magnitude.

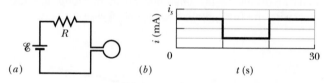

**FIG. 30-41**   Problem 8.

•9   In Fig. 30-42, a circular loop of wire 10 cm in diameter (seen edge-on) is placed with its normal $\vec{N}$ at an angle $\theta = 30°$ with the direction of a uniform magnetic field $\vec{B}$ of magnitude 0.50 T. The loop is then rotated such that $\vec{N}$ rotates in a cone about the field direction at the rate 100 rev/min; angle $\theta$ remains unchanged during the process. What is the emf induced in the loop?

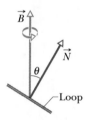

**FIG. 30-42**
Problem 9.

••10   In Fig. 30-43a, a uniform magnetic field $\vec{B}$ increases in magnitude with time $t$ as given by Fig. 30-43b, where the vertical axis scale is set by $B_s = 9.0$ mT and the horizontal scale is set by $t_s = 3.0$ s. A circular conducting loop of area $8.0 \times 10^{-4}$ m² lies in the field, in the plane of the page. The amount of charge $q$ passing point $A$ on the loop is given in Fig. 30-43c as a function of $t$, with the vertical axis scale set by $q_s = 6.0$ mC and the horizontal axis scale again set by $t_s = 3.0$ s. What is the loop's resistance?

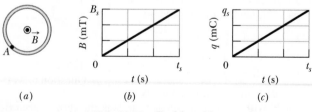

**FIG. 30-43**   Problem 10.

**••11** A square wire loop with 2.00 m sides is perpendicular to a uniform magnetic field, with half the area of the loop in the field as shown in Fig. 30-44. The loop contains an ideal battery with emf $\mathscr{E} = 20.0$ V. If the magnitude of the field varies with time according to $B = 0.0420 - 0.870t$, with $B$ in teslas and $t$ in seconds, what are (a) the net emf in the circuit and (b) the direction of the (net) current around the loop? **GO**

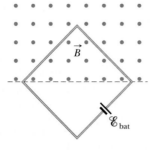

FIG. 30-44  Problem 11.

**••12** Figure 30-45$a$ shows a wire that forms a rectangle ($W = 20$ cm, $H = 30$ cm) and has a resistance of 5.0 m$\Omega$. Its interior is split into three equal areas, with magnetic fields $\vec{B}_1$, $\vec{B}_2$, and $\vec{B}_3$. The fields are uniform within each region and directly out of or into the page as indicated. Figure 30-45$b$ gives the change in the $z$ components $B_z$ of the three fields with time $t$; the vertical axis scale is set by $B_s = 4.0$ $\mu$T and $B_b = -2.5B_s$, and the horizontal axis scale is set by $t_s = 2.0$ s. What are the (a) magnitude and (b) direction of the current induced in the wire? **GO**

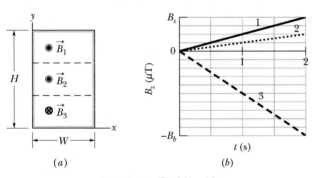

FIG. 30-45  Problem 12.

**••13** A rectangular coil of $N$ turns and of length $a$ and width $b$ is rotated at frequency $f$ in a uniform magnetic field $\vec{B}$, as indicated in Fig. 30-46. The coil is connected to co-rotating cylinders, against which metal brushes slide to make contact. (a) Show that the emf induced in the coil is given (as a function of time $t$) by

$$\mathscr{E} = 2\pi f NabB \sin(2\pi ft) = \mathscr{E}_0 \sin(2\pi ft).$$

This is the principle of the commercial alternating-current generator. (b) What value of $Nab$ gives an emf with $\mathscr{E}_0 = 150$ V when the loop is rotated at 60.0 rev/s in a uniform magnetic field of 0.500 T?

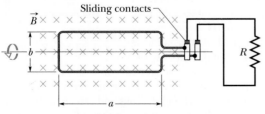

FIG. 30-46  Problem 13.

**••14** Figure 30-47 shows a closed loop of wire that consists of a pair of equal semicircles, of radius 3.7 cm, lying in mutu-

ally perpendicular planes. The loop was formed by folding a flat circular loop along a diameter until the two halves became perpendicular to each other. A uniform magnetic field $\vec{B}$ of magnitude 76 mT is directed perpendicular to the fold diameter and makes equal angles (of 45°) with the planes of the semicircles. The magnetic field is reduced to zero at a uniform rate during a time interval of 4.5 ms. During this interval, what are the (a) magnitude and (b) direction (clockwise or counterclockwise when viewed along the direction of $\vec{B}$) of the emf induced in the loop?

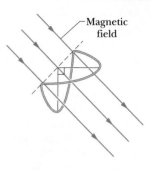

FIG. 30-47  Problem 14.

**••15** In Fig. 30-48, a stiff wire bent into a semicircle of radius $a$ = 2.0 cm is rotated at constant angular speed 40 rev/s in a uniform 20 mT magnetic field. What are the (a) frequency and (b) amplitude of the emf induced in the loop?

**••16** At a certain place, Earth's magnetic field has magnitude $B = 0.590$ gauss and is inclined downward at an angle of 70.0° to the horizontal. A flat horizontal circular coil of wire with a radius of 10.0 cm has 1000 turns and a total resistance of 85.0 $\Omega$. It is connected in series to a meter with 140 $\Omega$ resistance. The coil is flipped through a half-revolution about a diameter, so that it is again horizontal. How much charge flows through the meter during the flip?

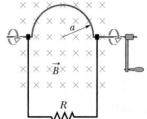

FIG. 30-48  Problem 15.

**••17** An electric generator contains a coil of 100 turns of wire, each forming a rectangular loop 50.0 cm by 30.0 cm. The coil is placed entirely in a uniform magnetic field with magnitude $B = 3.50$ T and with $\vec{B}$ initially perpendicular to the coil's plane. What is the maximum value of the emf produced when the coil is spun at 1000 rev/min about an axis perpendicular to $\vec{B}$? **ILW**

**••18** In Fig. 30-49, a wire loop of dimensions $L = 40.0$ cm and $W = 25.0$ cm lies in a magnetic field $\vec{B}$. What are the (a) magnitude $\mathscr{E}$ and (b) direction (clockwise or counterclockwise—or "none" if $\mathscr{E} = 0$) of the emf induced in the loop if $\vec{B} = (4.00 \times 10^{-2}$ T/m$)y\hat{k}$? What are (c) $\mathscr{E}$ and (d) the direction if $\vec{B} = (6.00 \times 10^{-2}$ T/s$)t\hat{k}$? What are (e) $\mathscr{E}$ and (f) the direction if $\vec{B} = (8.00 \times 10^{-2}$ T/m·s$)yt\hat{k}$? What are (g) $\mathscr{E}$ and (h) the direction if $\vec{B} = (3.00 \times 10^{-2}$ T/m·s$)xt\hat{j}$? What are (i) $\mathscr{E}$ and (j) the direction if $\vec{B} = (5.00 \times 10^{-2}$ T/m·s$)yt\hat{i}$?

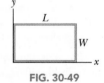

FIG. 30-49
Problem 18.

**••19** One hundred turns of (insulated) copper wire are wrapped around a wooden cylindrical core of cross-sectional area $1.20 \times 10^{-3}$ m². The two ends of the wire are connected to a resistor. The total resistance in the circuit is 13.0 $\Omega$. If an externally applied uniform longitudinal magnetic field in the core changes from 1.60 T in one direction to 1.60 T in the opposite direction, how much charge flows through a point in the circuit during the change? **ILW**

**••20** A rectangular loop (area $= 0.15$ m$^2$) turns in a uniform magnetic field, $B = 0.20$ T. When the angle between the field and the normal to the plane of the loop is $\pi/2$ rad and increasing at 0.60 rad/s, what emf is induced in the loop?

**••21** Figure 30-50 shows two parallel loops of wire having a common axis. The smaller loop (radius $r$) is above the larger loop (radius $R$) by a distance $x \gg R$. Consequently, the magnetic field due to the counterclockwise current $i$ in the larger loop is nearly uniform throughout the smaller loop. Suppose that $x$ is increasing at the constant rate $dx/dt = v$. (a) Find an expression for the magnetic flux through the area of the smaller loop as a function of $x$. (*Hint:* See Eq. 29-27.) In the smaller loop, find (b) an expression for the induced emf and (c) the direction of the induced current. **SSM**

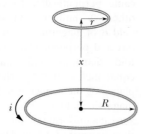

**FIG. 30-50** Problem 21.

**••22** A wire is bent into three circular segments, each of radius $r = 10$ cm, as shown in Fig. 30-51. Each segment is a quadrant of a circle, $ab$ lying in the $xy$ plane, $bc$ lying in the $yz$ plane, and $ca$ lying in the $zx$ plane. (a) If a uniform magnetic field $\vec{B}$ points in the positive $x$ direction, what is the magnitude of the emf developed in the wire when $B$ increases at the rate of 3.0 mT/s? (b) What is the direction of the current in segment $bc$?

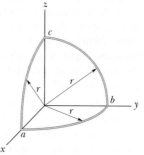

**FIG. 30-51** Problem 22.

**••23** A small circular loop of area 2.00 cm$^2$ is placed in the plane of, and concentric with, a large circular loop of radius 1.00 m. The current in the large loop is changed at a constant rate from 200 A to $-200$ A (a change in direction) in a time of 1.00 s, beginning at $t = 0$. What is the magnitude of the magnetic field $\vec{B}$ at the center of the small loop due to the current in the large loop at (a) $t = 0$, (b) $t = 0.500$ s, and (c) $t = 1.00$ s? (d) From $t = 0$ to $t = 1.00$ s, is $\vec{B}$ reversed? Because the inner loop is small, assume $\vec{B}$ is uniform over its area. (e) What emf is induced in the small loop at $t = 0.500$ s?

**•••24** For the wire arrangement in Fig. 30-52, $a = 12.0$ cm and $b = 16.0$ cm. The current in the long straight wire is given by $i = 4.50t^2 - 10.0t$, where $i$ is in amperes and $t$ is in seconds. (a) Find the emf in the square loop at $t = 3.00$ s. (b) What is the direction of the induced current in the loop?

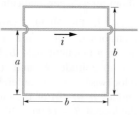

**FIG. 30-52** Problem 24.

**•••25** As seen in Fig. 30-53, a square loop of wire has sides of length 2.0 cm. A magnetic field is directed out of the page; its magnitude is given by $B = 4.0t^2y$, where $B$ is in teslas, $t$ is in seconds, and $y$ is in meters. At $t$

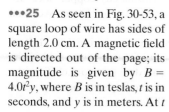

**FIG. 30-53** Problem 25.

$= 2.5$ s, what are the (a) magnitude and (b) direction of the emf induced in the loop? **ILW**

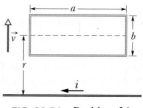

**•••26** In Fig. 30-54, a rectangular loop of wire with length $a = 2.2$ cm, width $b = 0.80$ cm, and resistance $R = 0.40$ m$\Omega$ is placed near an infinitely long wire carrying current $i = 4.7$ A. The loop is then moved away from the wire at constant speed $v = 3.2$ mm/s. When the center of the loop is at distance $r = 1.5b$, what are (a) the magnitude of the magnetic flux through the loop and (b) the current induced in the loop?

**FIG. 30-54** Problem 26.

**•••27** Two long, parallel copper wires of diameter 2.5 mm carry currents of 10 A in opposite directions. (a) Assuming that their central axes are 20 mm apart, calculate the magnetic flux per meter of wire that exists in the space between those axes. (b) What percentage of this flux lies inside the wires? (c) Repeat part (a) for parallel currents. **GO**

### sec. 30-5 Induction and Energy Transfers

**•28** In Fig. 30-55$a$, a circular loop of wire is concentric with a solenoid and lies in a plane perpendicular to the solenoid's central axis. The loop has radius 6.00 cm. The solenoid has radius 2.00 cm, consists of 8000 turns/m, and has a current $i_{sol}$ varying with time $t$ as given in Fig. 30-55$b$, where the vertical axis scale is set by $i_s = 1.00$ A and the horizontal axis scale is set by $t_s = 2.0$ s. Figure 30-55$c$ shows, as a function of time, the energy $E_{th}$ that is transferred to thermal energy of the loop; the vertical axis scale is set by $E_s = 100.0$ nJ. What is the loop's resistance?

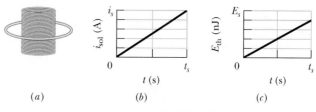

(a)   (b)   (c)

**FIG. 30-55** Problem 28.

**•29** If 50.0 cm of copper wire (diameter $= 1.00$ mm) is formed into a circular loop and placed perpendicular to a uniform magnetic field that is increasing at the constant rate of 10.0 mT/s, at what rate is thermal energy generated in the loop? **SSM ILW**

**•30** A loop antenna of area 2.00 cm$^2$ and resistance 5.21 $\mu\Omega$ is perpendicular to a uniform magnetic field of magnitude 17.0 $\mu$T. The field magnitude drops to zero in 2.96 ms. How much thermal energy is produced in the loop by the change in field?

**•31** In Fig. 30-56, a metal rod is forced to move with constant velocity $\vec{v}$ along two parallel metal rails, connected with a strip of metal at one end. A magnetic field of magnitude $B = 0.350$ T points out of the

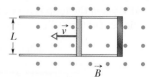

**FIG. 30-56** Problem 31.

page. (a) If the rails are separated by $L = 25.0$ cm and the speed of the rod is 55.0 cm/s, what emf is generated? (b) If the rod has a resistance of 18.0 $\Omega$ and the rails and connector have negligible resistance, what is the current in the rod? (c) At what rate is energy being transferred to thermal energy?

**••32** In Fig. 30-57, two straight conducting rails form a right angle. A conducting bar in contact with the rails starts at the vertex at time $t = 0$ and moves with a constant velocity of 5.20 m/s along them. A magnetic field with $B = 0.350$ T is directed out of the page. Calculate (a) the flux through the triangle formed by the rails and bar at $t = 3.00$ s and (b) the emf around the triangle at that time. (c) If the emf is $\mathcal{E} = at^n$, where $a$ and $n$ are constants, what is the value of $n$?

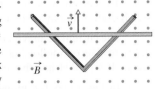

FIG. 30-57   Problem 32.

**••33**   The conducting rod shown in Fig. 30-56 has length $L$ and is being pulled along horizontal, frictionless conducting rails at a constant velocity $\vec{v}$. The rails are connected at one end with a metal strip. A uniform magnetic field $\vec{B}$, directed out of the page, fills the region in which the rod moves. Assume that $L = 10$ cm, $v = 5.0$ m/s, and $B = 1.2$ T. What are the (a) magnitude and (b) direction (up or down the page) of the emf induced in the rod? What are the (c) size and (d) direction of the current in the conducting loop? Assume that the resistance of the rod is 0.40 Ω and that the resistance of the rails and metal strip is negligibly small. (e) At what rate is thermal energy being generated in the rod? (f) What external force on the rod is needed to maintain $\vec{v}$? (g) At what rate does this force do work on the rod?

**••34**   In Fig. 30-58, a long rectangular conducting loop, of width $L$, resistance $R$, and mass $m$, is hung in a horizontal, uniform magnetic field $\vec{B}$ that is directed into the page and that exists only above line $aa$. The loop is then dropped; during its fall, it accelerates until it reaches a certain terminal speed $v_t$. Ignoring air drag, find an expression for $v_t$.

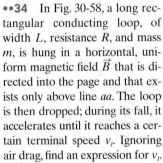

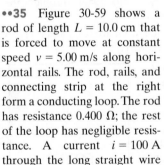

FIG. 30-58   Problem 34.

**••35**   Figure 30-59 shows a rod of length $L = 10.0$ cm that is forced to move at constant speed $v = 5.00$ m/s along horizontal rails. The rod, rails, and connecting strip at the right form a conducting loop. The rod has resistance 0.400 Ω; the rest of the loop has negligible resistance. A current $i = 100$ A through the long straight wire at distance $a = 10.0$ mm from the loop sets up a (nonuniform) magnetic field through the loop. Find the (a) emf and (b) current induced in the loop. (c) At what rate is thermal energy generated in the rod? (d) What is the magnitude of the force that must be applied to the rod to make it move at constant speed? (e) At what rate does this force do work on the rod?

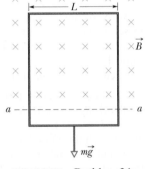

FIG. 30-59   Problem 35.

### sec. 30-6   Induced Electric Fields

**•36**   Figure 30-60 shows two circular regions $R_1$ and $R_2$ with

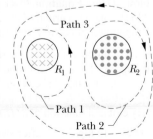

FIG. 30-60   Problem 36.

radii $r_1 = 20.0$ cm and $r_2 = 30.0$ cm. In $R_1$ there is a uniform magnetic field of magnitude $B_1 = 50.0$ mT directed into the page, and in $R_2$ there is a uniform magnetic field of magnitude $B_2 = 75.0$ mT directed out of the page (ignore fringing). Both fields are decreasing at the rate of 8.50 mT/s. Calculate $\oint \vec{E} \cdot d\vec{s}$ for (a) path 1, (b) path 2, and (c) path 3.

**•37**   A long solenoid has a diameter of 12.0 cm. When a current $i$ exists in its windings, a uniform magnetic field of magnitude $B = 30.0$ mT is produced in its interior. By decreasing $i$, the field is caused to decrease at the rate of 6.50 mT/s. Calculate the magnitude of the induced electric field (a) 2.20 cm and (b) 8.20 cm from the axis of the solenoid.   SSM ILW

**••38**   A circular region in an $xy$ plane is penetrated by a uniform magnetic field in the positive direction of the $z$ axis. The field's magnitude $B$ (in teslas) increases with time $t$ (in seconds) according to $B = at$, where $a$ is a constant. The magnitude $E$ of the electric field set up by that increase in the magnetic field is given by Fig. 30-61 versus radial distance $r$; the vertical axis scale is set by $E_s = 300$ $\mu$N/C, and the horizontal axis scale is set by $r_s = 4.00$ cm. Find $a$.

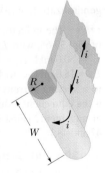

FIG. 30-61   Problem 38.

**••39**   The magnetic field of a cylindrical magnet that has a pole-face diameter of 3.3 cm can be varied sinusoidally between 29.6 T and 30.0 T at a frequency of 15 Hz. At a radial distance of 1.6 cm, what is the amplitude of the electric field induced by the variation?

### sec. 30-7   Inductors and Inductance

**•40**   The inductance of a closely packed coil of 400 turns is 8.0 mH. Calculate the magnetic flux through the coil when the current is 5.0 mA.

**•41**   A circular coil has a 10.0 cm radius and consists of 30.0 closely wound turns of wire. An externally produced magnetic field of magnitude 2.60 mT is perpendicular to the coil. (a) If no current is in the coil, what magnetic flux links its turns? (b) When the current in the coil is 3.80 A in a certain direction, the net flux through the coil is found to vanish. What is the inductance of the coil?

**••42**   Figure 30-62 shows a copper strip of width $W = 16.0$ cm that has been bent to form a shape that consists of a tube of radius $R = 1.8$ cm plus two parallel flat extensions. Current $i = 35$ mA is distributed uniformly across the width so that the tube is effectively a one-turn solenoid. Assume that the magnetic field outside the tube is negligible and the field inside the tube is uniform. What are (a) the magnetic field magnitude inside the tube and (b) the inductance of the tube (excluding the flat extensions)?

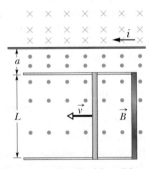

FIG. 30-62   Problem 42.

**••43**   Two identical long wires of radius $a = 1.53$ mm are parallel and carry identical currents in opposite directions. Their center-to-center separation is $d = 14.2$ cm. Neglect the flux within the wires but consider the flux in the region between the wires. What is the inductance per unit length of the wires?   GO

## sec. 30-8 Self-Induction

**•44** A 12 H inductor carries a current of 2.0 A. At what rate must the current be changed to produce a 60 V emf in the inductor?

**•45** At a given instant the current and self-induced emf in an inductor are directed as indicated in Fig. 30-63. (a) Is the current increasing or decreasing? (b) The induced emf is 17 V, and the rate of change of the current is 25 kA/s; find the inductance.

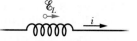

FIG. 30-63 Problem 45.

**••46** The current $i$ through a 4.6 H inductor varies with time $t$ as shown by the graph of Fig. 30-64, where the vertical axis scale is set by $i_s = 8.0$ A and the horizontal axis scale is set by $t_s = 6.0$ ms. The inductor has a resistance of 12 Ω. Find the magnitude of the induced emf $\mathscr{E}$ during time intervals (a) 0 to 2 ms, (b) 2 ms to 5 ms, and (c) 5 ms to 6 ms. (Ignore the behavior at the ends of the intervals.)

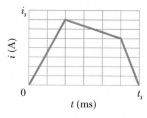

FIG. 30-64 Problem 46.

**••47** *Inductors in series.* Two inductors $L_1$ and $L_2$ are connected in series and are separated by a large distance so that the magnetic field of one cannot affect the other. (a) Show that the equivalent inductance is given by

$$L_{eq} = L_1 + L_2.$$

(*Hint:* Review the derivations for resistors in series and capacitors in series. Which is similar here?) (b) What is the generalization of (a) for $N$ inductors in series?

**••48** *Inductors in parallel.* Two inductors $L_1$ and $L_2$ are connected in parallel and separated by a large distance so that the magnetic field of one cannot affect the other. (a) Show that the equivalent inductance is given by

$$\frac{1}{L_{eq}} = \frac{1}{L_1} + \frac{1}{L_2}.$$

(*Hint:* Review the derivations for resistors in parallel and capacitors in parallel. Which is similar here?) (b) What is the generalization of (a) for $N$ inductors in parallel?

**••49** The inductor arrangement of Fig. 30-65, with $L_1 = 30.0$ mH, $L_2 = 50.0$ mH, $L_3 = 20.0$ mH, and $L_4 = 15.0$ mH, is to be connected to a varying current source. What is the equivalent inductance of the arrangement? (First see Problems 47 and 48.)

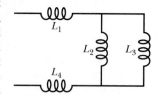

FIG. 30-65 Problem 49.

## sec. 30-9 RL Circuits

**•50** The switch in Fig. 30-20 is closed on $a$ at time $t = 0$. What is the ratio $\mathscr{E}_L/\mathscr{E}$ of the inductor's self-induced emf to the battery's emf (a) just after $t = 0$ and (b) at $t = 2.00\tau_L$? (c) At what multiple of $\tau_L$ will $\mathscr{E}_L/\mathscr{E} = 0.500$?

**•51** A battery is connected to a series RL circuit at time $t = 0$. At what multiple of $\tau_L$ will the current be 0.100% less than its equilibrium value? **SSM**

**•52** The current in an RL circuit builds up to one-third of its steady-state value in 5.00 s. Find the inductive time constant.

**•53** The current in an RL circuit drops from 1.0 A to 10 mA in the first second following removal of the battery from the circuit. If $L$ is 10 H, find the resistance $R$ in the circuit. **ILW**

**•54** In Fig. 30-66, the inductor has 25 turns and the ideal battery has an emf of 16 V. Figure 30-67 gives the magnetic flux Φ through each turn versus the current $i$ through the inductor. The vertical axis scale is set by $\Phi_s = 4.0 \times 10^{-4}$ T·m², and the horizontal axis scale is set by $i_s = 2.00$ A. If switch S is closed at time $t = 0$, at what rate $di/dt$ will the current be changing at $t = 1.5\tau_L$?

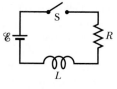
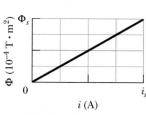

FIG. 30-66
Problems 54, 83, 87, 94, and 99.

FIG. 30-67 Problem 54.

**•55** A solenoid having an inductance of 6.30 μH is connected in series with a 1.20 kΩ resistor. (a) If a 14.0 V battery is connected across the pair, how long will it take for the current through the resistor to reach 80.0% of its final value? (b) What is the current through the resistor at time $t = 1.0\tau_L$? **SSM**

**•56** In Fig. 30-68, $\mathscr{E} = 100$ V, $R_1 = 10.0$ Ω, $R_2 = 20.0$ Ω, $R_3 = 30.0$ Ω, and $L = 2.00$ H. Immediately after switch S is closed, what are (a) $i_1$ and (b) $i_2$? (Let currents in the indicated directions have positive values and currents in the opposite directions have negative values.) A long time later, what are (c) $i_1$ and (d) $i_2$? The switch is then reopened. Just then, what are (e) $i_1$ and (f) $i_2$? A long time later, what are (g) $i_1$ and (h) $i_2$?

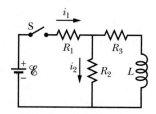

FIG. 30-68 Problem 56.

**••57** In Fig. 30-69, $R = 15$ Ω, $L = 5.0$ H, the ideal battery has $\mathscr{E} = 10$ V, and the fuse in the upper branch is an ideal 3.0 A fuse. It has zero resistance as long as the current through it remains less than 3.0 A. If the current reaches 3.0 A, the fuse "blows" and thereafter has infinite resistance. Switch S is closed at time $t = 0$. (a) When does the fuse blow? (*Hint:* Equation 30-41 does not apply. Rethink Eq. 30-39.) (b) Sketch a graph of the current $i$ through the inductor as a function of time. Mark the time at which the fuse blows.

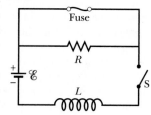

FIG. 30-69 Problem 57.

**••58** Suppose the emf of the battery in the circuit shown in Fig. 30-21 varies with time $t$ so that the current is given by $i(t) = 3.0 + 5.0t$, where $i$ is in amperes and $t$ is in seconds. Take $R = 4.0$ Ω and $L = 6.0$ H, and find an expression for the battery emf as a function of $t$. (*Hint:* Apply the loop rule.) **GO**

**•••59** In Fig. 30-70, after switch S is closed at time $t = 0$, the emf of the source is automatically adjusted to maintain a constant current $i$ through S. (a) Find the current through the inductor as a function of time. (b) At what time is the current through the resistor equal to the current through the inductor? **SSM WWW**

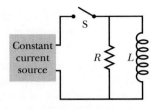

FIG. 30-70    Problem 59.

•••**60**    A wooden toroidal core with a square cross section has an inner radius of 10 cm and an outer radius of 12 cm. It is wound with one layer of wire (of diameter 1.0 mm and resistance per meter 0.020 Ω/m). What are (a) the inductance and (b) the inductive time constant of the resulting toroid? Ignore the thickness of the insulation on the wire.

### sec. 30-10    Energy Stored in a Magnetic Field

•**61**    At $t = 0$, a battery is connected to a series arrangement of a resistor and an inductor. If the inductive time constant is 37.0 ms, at what time is the rate at which energy is dissipated in the resistor equal to the rate at which energy is stored in the inductor's magnetic field?  **ILW**

•**62**    At $t = 0$, a battery is connected to a series arrangement of a resistor and an inductor. At what multiple of the inductive time constant will the energy stored in the inductor's magnetic field be 0.500 its steady-state value?

•**63**    A coil is connected in series with a 10.0 kΩ resistor. An ideal 50.0 V battery is applied across the two devices, and the current reaches a value of 2.00 mA after 5.00 ms. (a) Find the inductance of the coil. (b) How much energy is stored in the coil at this same moment?  **SSM**

•**64**    A coil with an inductance of 2.0 H and a resistance of 10 Ω is suddenly connected to an ideal battery with $\mathscr{E} = 100$ V. At 0.10 s after the connection is made, what is the rate at which (a) energy is being stored in the magnetic field, (b) thermal energy is appearing in the resistance, and (c) energy is being delivered by the battery?

••**65**    For the circuit of Fig. 30-21, assume that $\mathscr{E} = 10.0$ V, $R = 6.70$ Ω, and $L = 5.50$ H. The ideal battery is connected at time $t = 0$. (a) How much energy is delivered by the battery during the first 2.00 s? (b) How much of this energy is stored in the magnetic field of the inductor? (c) How much of this energy is dissipated in the resistor?  **GO**

••**66**    Figure 30-71a shows, in cross section, two wires that are straight, parallel, and very long. The ratio $i_1/i_2$ of the current carried by wire 1 to that carried by wire 2 is 1/3. Wire 1 is fixed in place. Wire 2 can be moved along the positive side of the $x$ axis so as to change the magnetic energy density $u_B$ set up by the two currents at the origin. Figure 30-71b gives $u_B$ as a function of the position $x$ of wire 2. The curve has an asymptote of $u_B = 1.96$ nJ/m³ as $x \to \infty$, and the horizontal axis scale is set by $x_s = 60.0$ cm. What is the value of (a) $i_1$ and (b) $i_2$?

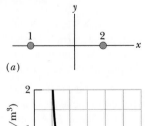

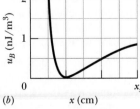

FIG. 30-71    Problem 66.

### sec. 30-11    Energy Density of a Magnetic Field

•**67**    What must be the magnitude of a uniform electric field if it is to have the same energy density as that possessed by a 0.50 T magnetic field?  **ILW**

•**68**    A toroidal inductor with an inductance of 90.0 mH encloses a volume of 0.0200 m³. If the average energy density in the toroid is 70.0 J/m³, what is the current through the inductor?

•**69**    A solenoid that is 85.0 cm long has a cross-sectional area of 17.0 cm². There are 950 turns of wire carrying a current of 6.60 A. (a) Calculate the energy density of the magnetic field inside the solenoid. (b) Find the total energy stored in the magnetic field there (neglect end effects).  **SSM**

•**70**    A circular loop of wire 50 mm in radius carries a current of 100 A. Find the (a) magnetic field strength and (b) energy density at the center of the loop.

••**71**    A length of copper wire carries a current of 10 A uniformly distributed through its cross section. Calculate the energy density of (a) the magnetic field and (b) the electric field at the surface of the wire. The wire diameter is 2.5 mm, and its resistance per unit length is 3.3 Ω/km.

### sec. 30-12    Mutual Induction

•**72**    Two solenoids are part of the spark coil of an automobile. When the current in one solenoid falls from 6.0 A to zero in 2.5 ms, an emf of 30 kV is induced in the other solenoid. What is the mutual inductance $M$ of the solenoids?

•**73**    Two coils are at fixed locations. When coil 1 has no current and the current in coil 2 increases at the rate 15.0 A/s, the emf in coil 1 is 25.0 mV. (a) What is their mutual inductance? (b) When coil 2 has no current and coil 1 has a current of 3.60 A, what is the flux linkage in coil 2?  **SSM**

•**74**    Coil 1 has $L_1 = 25$ mH and $N_1 = 100$ turns. Coil 2 has $L_2 = 40$ mH and $N_2 = 200$ turns. The coils are fixed in place; their mutual inductance $M$ is 3.0 mH. A 6.0 mA current in coil 1 is changing at the rate of 4.0 A/s. (a) What magnetic flux $\Phi_{12}$ links coil 1, and (b) what self-induced emf appears in that coil? (c) What magnetic flux $\Phi_{21}$ links coil 2, and (d) what mutually induced emf appears in that coil?

••**75**    Two coils connected as shown in Fig. 30-72 separately have inductances $L_1$ and $L_2$. Their mutual inductance is $M$. (a) Show that this combination can be replaced by a single coil of equivalent inductance given by

$$L_{eq} = L_1 + L_2 + 2M.$$

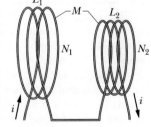

FIG. 30-72    Problem 75.

(b) How could the coils in Fig. 30-72 be reconnected to yield an equivalent inductance of

$$L_{eq} = L_1 + L_2 - 2M?$$

(This problem is an extension of Problem 47, but the requirement that the coils be far apart has been removed.)  **SSM**

••**76**    A coil C of $N$ turns is placed around a long solenoid S of radius $R$ and $n$ turns per unit length, as in Fig. 30-73. (a) Show that the mutual inductance for the coil–solenoid combination is given by $M = \mu_0 \pi R^2 nN$. (b) Explain why $M$ does not depend on the shape, size, or possible lack of close packing of the coil.

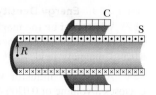

**FIG. 30-73** Problem 76.

**••77** A rectangular loop of $N$ closely packed turns is positioned near a long straight wire as shown in Fig. 30-74. What is the mutual inductance $M$ for the loop–wire combination if $N = 100$, $a = 1.0$ cm, $b = 8.0$ cm, and $l = 30$ cm? **ILW**

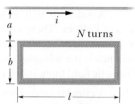

**FIG. 30-74** Problem 77.

### Additional Problems

**78** Figure 30-75a shows two concentric circular regions in which uniform magnetic fields can change. Region 1, with radius $r_1 = 1.0$ cm, has an outward magnetic field $\vec{B}_1$ that is increasing in magnitude. Region 2, with radius $r_2 = 2.0$ cm, has an outward magnetic field $\vec{B}_2$ that may also be changing. Imagine that a conducting ring of radius $R$ is centered on the two regions and then the emf $\mathscr{E}$ around the ring is determined. Figure 30-75b gives emf $\mathscr{E}$ as a function of the square $R^2$ of the ring's radius, to the outer edge of region 2. The vertical axis scale is set by $\mathscr{E}_s = 20.0$ nV. What are the rates (a) $dB_1/dt$ and (b) $dB_2/dt$? (c) Is the magnitude of $\vec{B}_2$ increasing, decreasing, or remaining constant?

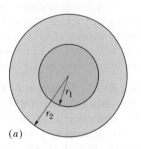

(a)

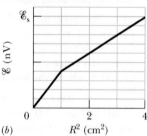

(b)

**FIG. 30-75** Problem 78.

**79** Figure 30-76 shows a uniform magnetic field $\vec{B}$ confined to a cylindrical volume of radius $R$. The magnitude of $\vec{B}$ is decreasing at a constant rate of 10 mT/s. In unit-vector notation, what is the initial acceleration of an electron released at (a) point $a$ (radial distance $r = 5.0$ cm), (b) point $b$ ($r = 0$), and (c) point $c$ ($r = 5.0$ cm)? **SSM**

**80** The inductance of a closely wound coil is such that an emf of 3.00 mV is induced when the current changes at the rate of 5.00 A/s. A steady current of 8.00 A produces a magnetic flux of 40.0 $\mu$Wb through each turn. (a) Calculate the inductance of the coil. (b) How many turns does the coil have?

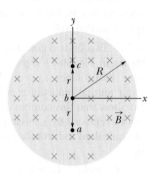

**FIG. 30-76** Problem 79.

**81** A square wire loop 20 cm on a side, with resistance 20 m$\Omega$, has its plane normal to a uniform magnetic field of magnitude $B = 2.0$ T. If you pull two opposite sides of the loop away from each other, the other two sides automatically draw toward each other, reducing the area enclosed by the

loop. If the area is reduced to zero in time $\Delta t = 0.20$ s, what are (a) the average emf and (b) the average current induced in the loop during $\Delta t$? **SSM**

**82** A square loop of wire is held in a uniform 0.24 T magnetic field directed perpendicular to the plane of the loop. The length of each side of the square is decreasing at a constant rate of 5.0 cm/s. What emf is induced in the loop when the length is 12 cm?

**83** In Fig. 30-66, a 12.0 V ideal battery, a 20.0 $\Omega$ resistor, and an inductor are connected by a switch at time $t = 0$. At what rate is the battery transferring energy to the inductor's field at $t = 1.61\tau_L$?

**84** How long would it take, following the removal of the battery, for the potential difference across the resistor in an $RL$ circuit (with $L = 2.00$ H, $R = 3.00$ $\Omega$) to decay to 10.0% of its initial value?

**85** In Fig. 30-77, the battery is ideal, $\mathscr{E} = 10$ V, $R_1 = 5.0$ $\Omega$, $R_2 = 10$ $\Omega$, and $L = 5.0$ H. Switch S is closed at time $t = 0$. Just afterwards, what are (a) $i_1$, (b) $i_2$, (c) the current $i_S$ through the switch, (d) the potential difference $V_2$ across resistor 2, (e) the potential difference $V_L$ across the inductor, and (f) the rate of change $di_2/dt$? A long time later, what are (g) $i_1$, (h) $i_2$, (i) $i_S$, (j) $V_2$, (k) $V_L$, and (l) $di_2/dt$? **SSM**

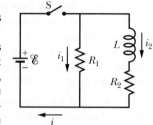

**FIG. 30-77** Problem 85.

**86** In Fig. 30-78a, switch S has been closed on $A$ long enough to establish a steady current in the inductor of inductance $L_1 = 5.00$ mH and the resistor of resistance $R_1 = 25.0$ $\Omega$. Similarly, in Fig. 30-78b, switch S has been closed on $A$ long enough to establish a steady current in the inductor of inductance $L_2 = 3.00$ mH and the resistor of resistance $R_2 = 30.0$ $\Omega$. The ratio $\Phi_{02}/\Phi_{01}$ of the magnetic flux through a turn in inductor 2 to that in inductor 1 is 1.50. At time $t = 0$, the two switches are closed on $B$. At what time $t$ is the flux through a turn in the two inductors equal?

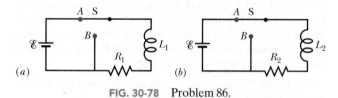

**FIG. 30-78** Problem 86.

**87** Switch S in Fig. 30-66 is closed at time $t = 0$, initiating the buildup of current in the 15.0 mH inductor and the 20.0 $\Omega$ resistor. At what time is the emf across the inductor equal to the potential difference across the resistor?

**88** At time $t = 0$, a 12.0 V potential difference is suddenly applied to the leads of a coil of inductance 23.0 mH and a certain resistance $R$. At time $t = 0.150$ ms, the current through the inductor is changing at the rate of 280 A/s. Evaluate $R$.

**89** At time $t = 0$, a 45 V potential difference is suddenly applied to the leads of a coil with inductance $L = 50$ mH and resistance $R = 180$ $\Omega$. At what rate is the current through the coil increasing at $t = 1.2$ ms?

**90** A coil with 150 turns has a magnetic flux of 50.0 nT·m² through each turn when the current is 2.00 mA. (a) What is the inductance of the coil? What are the (b) inductance and (c) flux through each turn when the current is increased to 4.00 mA? (d) What is the maximum emf $\mathscr{E}$ across the coil when the current through it is given by $i = (3.00 \text{ mA}) \cos(377t)$, with $t$ in seconds?

**91** A coil with an inductance of 2.0 H and a resistance of 10 $\Omega$ is suddenly connected to an ideal battery with $\mathscr{E} = 100$ V. (a) What is the equilibrium current? (b) How much energy is stored in the magnetic field when this current exists in the coil?

**92** A long cylindrical solenoid with 100 turns/cm has a radius of 1.6 cm. Assume that the magnetic field it produces is parallel to its axis and is uniform in its interior. (a) What is its inductance per meter of length? (b) If the current changes at the rate of 13 A/s, what emf is induced per meter?

**93** In Fig. 30-79, $R_1 = 8.0$ $\Omega$, $R_2$ = 10 $\Omega$, $L_1 = 0.30$ H, $L_2 = 0.20$ H, and the ideal battery has $\mathscr{E} =$ 6.0 V. (a) Just after switch S is closed, at what rate is the current in inductor 1 changing? (b) When the circuit is in the steady state, what is the current in inductor 1?

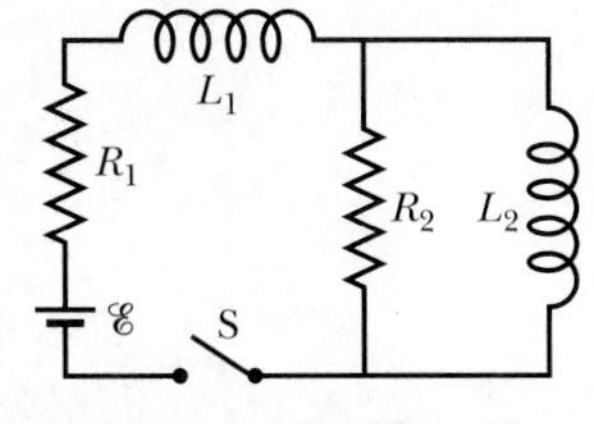

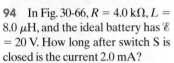

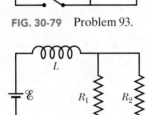

**FIG. 30-79** Problem 93.

**94** In Fig. 30-66, $R = 4.0$ k$\Omega$, $L =$ 8.0 $\mu$H, and the ideal battery has $\mathscr{E}$ = 20 V. How long after switch S is closed is the current 2.0 mA?

**95** In the circuit of Fig. 30-80, $R_1 = 20$ k$\Omega$, $R_2 = 20$ $\Omega$, $L =$ 50 mH, and the ideal battery has $\mathscr{E} = 40$ V. Switch S has been open for a long time when it is closed at

**FIG. 30-80** Problem 95.

time $t = 0$. Just after the switch is closed, what are (a) the current $i_{\text{bat}}$ through the battery and (b) the rate $di_{\text{bat}}/dt$? At $t = 3.0$ $\mu$s, what are (c) $i_{\text{bat}}$ and (d) $di_{\text{bat}}/dt$? A long time later, what are (e) $i_{\text{bat}}$ and (f) $di_{\text{bat}}/dt$? **SSM**

**96** The flux linkage through a certain coil of 0.75 $\Omega$ resistance would be 26 mWb if there were a current of 5.5 A in it. (a) Calculate the inductance of the coil. (b) If a 6.0 V ideal battery were suddenly connected across the coil, how long would it take for the current to rise from 0 to 2.5 A?

**97** Figure 30-81*a* shows a rectangular conducting loop of resistance $R = 0.020$ $\Omega$, height $H$ = 1.5 cm, and length $D = 2.5$ cm being pulled at constant speed $v$ = 40 cm/s through two regions of uniform magnetic field. Figure 30-81*b* gives the current $i$ induced in the loop as a function of the position $x$ of the right side of the loop. The vertical axis scale is set by $i_s = 3.0$ $\mu$A. For example, a current equal to $i_s$ is induced clockwise as the loop enters region 1. What are the (a) magnitude and (b) direction (into or out of the page) of the magnetic field in region 1? What are the (c) magnitude and (d) direction of the magnetic field in region 2? **SSM**

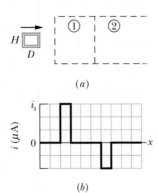

**FIG. 30-81** Problem 97.

**98** The magnetic potential energy stored in a certain inductor is 25.0 mJ when the current in the inductor is 60.0 mA. (a) Calculate the inductance. (b) What current is required for the stored energy to be 100 mJ?

**99** Once the switch S is closed in Fig. 30-66, the time required for the current to reach any obtainable value depends, in part, on the value of resistance $R$. Suppose the emf $\mathscr{E}$ of the ideal battery is 12 V and the inductance of the ideal (resistanceless) inductor is 18 mH. How much time is needed for the current to reach 2.00 A if $R$ is (a) 1.00 $\Omega$, (b) 5.00 $\Omega$, and (c) 6.00 $\Omega$? (d) For what value of $R$ is the time required for the current to reach 2.00 A least? (e) What is that least time? (*Hint:* Rethink Eq. 30-39.)

**100** A uniform magnetic field $\vec{B}$ is perpendicular to the plane of a circular wire loop of radius $r$. The magnitude of the field varies with time according to $B = B_0 e^{-t/\tau}$, where $B_0$ and $\tau$ are constants. Find an expression for the emf in the loop as a function of time.

**101** In the circuit shown in Fig. 30-82, the battery is ideal, $\mathscr{E} = 12.0$ V, $L = 10.0$ mH, $R_1 =$ 10.0 $\Omega$, and $R_2 = 20.0$ $\Omega$. The switch has been open for a long time before it is closed at $t = 0$. At what rate is the current in the inductor changing (a) immediately after the switch is closed and (b) when the current in the battery is 0.50 A? (c) What is the current in the battery when the circuit reaches its steady-state condition?

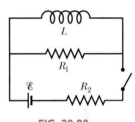

**FIG. 30-82**
Problem 101.

**102** Figure 30-83 shows a coil of $N_2$ turns wound as shown around part of a toroid of $N_1$ turns. The toroid's inner radius is $a$, its outer radius is $b$, and its height is $h$. Show that the mutual inductance $M$ for the toroid–coil combination is given by

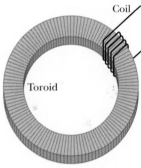

**FIG. 30-83** Problem 102.

$$M = \frac{\mu_0 N_1 N_2 h}{2\pi} \ln \frac{b}{a}.$$

**103** A circular loop (radius = 14 cm) of wire is placed in a magnetic field that makes an angle of 30° with the normal to the plane of the loop. The magnitude of this field increases at a constant rate from 30 mT to 60 mT in 15 ms. If the loop has a resistance of 5.0 $\Omega$, what is the magnitude of the current induced in the loop when the field magnitude is 50 mT?

**104** A 50-turn circular coil (radius = 15 cm) with a total resistance of 4.0 $\Omega$ is placed in a uniform magnetic field directed perpendicular to the plane of the coil. The magnitude of this field varies with time according to $B = A \sin(\omega t)$, where the amplitude is $A = 80$ $\mu$T and the angular frequency is $\omega = 50\pi$ rad/s. What is the magnitude of the current induced in the coil at $t = 20$ ms?

# 31 Electromagnetic Oscillations and Alternating Current

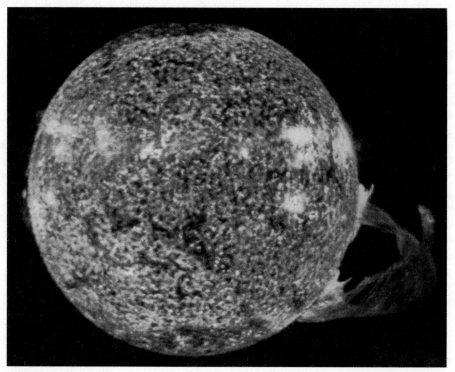

Courtesy NASA/JSC

At 2:45 A.M. on March 13, 1989, the entire power-grid system for the Canadian province of Quebec failed, leaving millions of people without power on that cold night. In fact, many power-grid systems in the Northern Hemisphere malfunctioned that night, creating a nightmare situation for the engineers who maintained the systems. The cause was not a sudden overtaxing demand for power or a failure of aging equipment. Rather, the cause was an explosion that had occurred on the Sun's surface three days earlier.

*How can a solar explosion shut down a power-grid system?*

The answer is in this chapter.

## 31-1 | WHAT IS PHYSICS?

We have explored the basic physics of electric and magnetic fields and how energy can be stored in capacitors and inductors. We next turn to the associated applied physics, in which the energy stored in one location can be transferred to another location so that it can be put to use. For example, energy produced at a power plant can show up at your home to run a computer. The total worth of this applied physics is now so high that its estimation is almost impossible. Indeed, modern civilization would be impossible without this applied physics.

In most parts of the world, electrical energy is transferred not as a direct current but as a sinusoidally oscillating current (alternating current, or ac). The challenge to both physicists and engineers is to design ac systems that transfer energy efficiently and to build appliances that make use of that energy.

In our discussion of electrically oscillating systems in this chapter, our first step is to examine oscillations in a simple circuit consisting of inductance $L$ and capacitance $C$.

## 31-2 | *LC Oscillations, Qualitatively*

Of the three circuit elements resistance $R$, capacitance $C$, and inductance $L$, we have so far discussed the series combinations $RC$ (in Section 27-9) and $RL$ (in Section 30-9). In these two kinds of circuit we found that the charge, current, and potential difference grow and decay exponentially. The time scale of the growth or decay is given by a *time constant* $\tau$, which is either capacitive or inductive.

We now examine the remaining two-element circuit combination $LC$. You will see that in this case the charge, current, and potential difference do not decay exponentially with time but vary sinusoidally (with period $T$ and angular frequency $\omega$). The resulting oscillations of the capacitor's electric field and the inductor's magnetic field are said to be **electromagnetic oscillations.** Such a circuit is said to oscillate.

Parts *a* through *h* of Fig. 31-1 show succeeding stages of the oscillations in a simple $LC$ circuit. From Eq. 25-21, the energy stored in the electric field of the capacitor at any time is

$$U_E = \frac{q^2}{2C}, \tag{31-1}$$

where $q$ is the charge on the capacitor at that time. From Eq. 30-49, the energy stored in the magnetic field of the inductor at any time is

$$U_B = \frac{Li^2}{2}, \tag{31-2}$$

where $i$ is the current through the inductor at that time.

We now adopt the convention of representing *instantaneous values* of the electrical quantities of a sinusoidally oscillating circuit with small letters, such as $q$, and the *amplitudes* of those quantities with capital letters, such as $Q$. With this convention in mind, let us assume that initially the charge $q$ on the capacitor in Fig. 31-1 is at its maximum value $Q$ and that the current $i$ through the inductor is zero. This initial state of the circuit is shown in Fig. 31-1*a*. The bar graphs for energy included there indicate that at this instant, with zero current through the inductor and maximum charge on the capacitor, the energy $U_B$ of the magnetic field is zero and the energy $U_E$ of the electric field is a maximum.

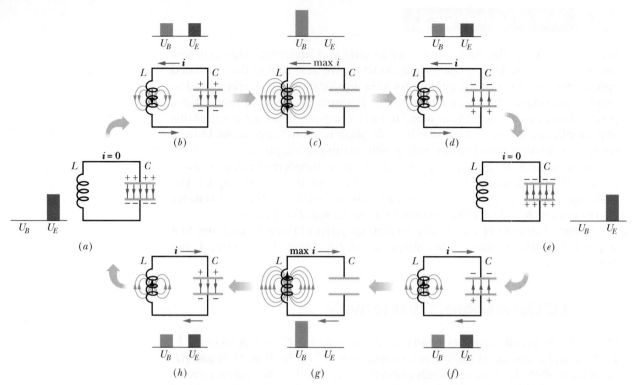

**FIG. 31-1** Eight stages in a single cycle of oscillation of a resistanceless *LC* circuit. The bar graphs by each figure show the stored magnetic and electrical energies. The magnetic field lines of the inductor and the electric field lines of the capacitor are shown. (*a*) Capacitor with maximum charge, no current. (*b*) Capacitor discharging, current increasing. (*c*) Capacitor fully discharged, current maximum. (*d*) Capacitor charging but with polarity opposite that in (*a*), current decreasing. (*e*) Capacitor with maximum charge having polarity opposite that in (*a*), no current. (*f*) Capacitor discharging, current increasing with direction opposite that in (*b*). (*g*) Capacitor fully discharged, current maximum. (*h*) Capacitor charging, current decreasing.

The capacitor now starts to discharge through the inductor, positive charge carriers moving counterclockwise, as shown in Fig. 31-1*b*. This means that a current *i*, given by *dq/dt* and pointing down in the inductor, is established. As the capacitor's charge decreases, the energy stored in the electric field within the capacitor also decreases. This energy is transferred to the magnetic field that appears around the inductor because of the current *i* that is building up there. Thus, the electric field decreases and the magnetic field builds up as energy is transferred from the electric field to the magnetic field.

The capacitor eventually loses all its charge (Fig. 31-1*c*) and thus also loses its electric field and the energy stored in that field. The energy has then been fully transferred to the magnetic field of the inductor. The magnetic field is then at its maximum magnitude, and the current through the inductor is then at its maximum value *I*.

Although the charge on the capacitor is now zero, the counterclockwise current must continue because the inductor does not allow it to change suddenly to zero. The current continues to transfer positive charge from the top plate to the bottom plate through the circuit (Fig. 31-1*d*). Energy now flows from the inductor back to the capacitor as the electric field within the capacitor builds up again. The current gradually decreases during this energy transfer. When, eventually, the energy has been transferred completely back to the capacitor (Fig. 31-1*e*), the current has decreased to zero (momentarily). The situation of Fig. 31-1*e* is like the initial situation, except that the capacitor is now charged oppositely.

The capacitor then starts to discharge again but now with a clockwise current (Fig. 31-1$f$). Reasoning as before, we see that the clockwise current builds to a maximum (Fig. 31-1$g$) and then decreases (Fig. 31-1$h$), until the circuit eventually returns to its initial situation (Fig. 31-1$a$). The process then repeats at some frequency $f$ and thus at an angular frequency $\omega = 2\pi f$. In the ideal $LC$ circuit with no resistance, there are no energy transfers other than that between the electric field of the capacitor and the magnetic field of the inductor. Because of the conservation of energy, the oscillations continue indefinitely. The oscillations need not begin with the energy all in the electric field; the initial situation could be any other stage of the oscillation.

To determine the charge $q$ on the capacitor as a function of time, we can put in a voltmeter to measure the time-varying potential difference (or *voltage*) $v_C$ that exists across the capacitor $C$. From Eq. 25-1 we can write

$$v_C = \left(\frac{1}{C}\right) q,$$

which allows us to find $q$. To measure the current, we can connect a small resistance $R$ in series with the capacitor and inductor and measure the time-varying potential difference $v_R$ across it; $v_R$ is proportional to $i$ through the relation

$$v_R = iR.$$

We assume here that $R$ is so small that its effect on the behavior of the circuit is negligible. The variations in time of $v_C$ and $v_R$, and thus of $q$ and $i$, are shown in Fig. 31-2. All four quantities vary sinusoidally.

In an actual $LC$ circuit, the oscillations will not continue indefinitely because there is always some resistance present that will drain energy from the electric and magnetic fields and dissipate it as thermal energy (the circuit may become warmer). The oscillations, once started, will die away as Fig. 31-3 suggests. Compare this figure with Fig. 15-16, which shows the decay of mechanical oscillations caused by frictional damping in a block–spring system.

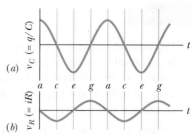

**FIG. 31-2** (a) The potential difference across the capacitor of the circuit of Fig. 31-1 as a function of time. This quantity is proportional to the charge on the capacitor. (b) A potential proportional to the current in the circuit of Fig. 31-1. The letters refer to the correspondingly labeled oscillation stages in Fig. 31-1.

✓**CHECKPOINT 1**    A charged capacitor and an inductor are connected in series at time $t = 0$. In terms of the period $T$ of the resulting oscillations, determine how much later the following reach their maximum value: (a) the charge on the capacitor; (b) the voltage across the capacitor, with its original polarity; (c) the energy stored in the electric field; and (d) the current.

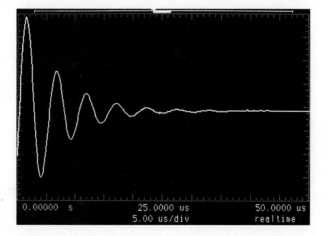

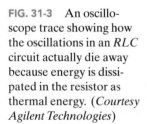

**FIG. 31-3** An oscilloscope trace showing how the oscillations in an $RLC$ circuit actually die away because energy is dissipated in the resistor as thermal energy. (*Courtesy Agilent Technologies*)

**Sample Problem**  **31-1**

A 1.5 $\mu$F capacitor is charged to 57 V. The charging battery is then disconnected, and a 12 mH coil is connected in series with the capacitor so that $LC$ oscillations occur. What is the maximum current in the coil? Assume that the circuit contains no resistance, which would cause the energy to be dissipated.

1. Because the circuit contains no resistance, the electromagnetic energy of the circuit is conserved as the energy is transferred back and forth between the electric field of the capacitor and the magnetic field of the coil (inductor).

2. At any time $t$, the energy $U_B(t)$ of the magnetic field is related to the current $i(t)$ through the coil by Eq. 31-2 ($U_B = Li^2/2$). When all the energy is stored as magnetic energy, the current is at its maximum value $I$ and that energy is $U_{B,\max} = LI^2/2$.

3. At any time $t$, the energy $U_E(t)$ of the electric field is related to the charge $q(t)$ on the capacitor by Eq. 31-1 ($U_E = q^2/2C$). When all the energy is stored as electrical energy, the charge is at its maximum value $Q$ and that energy is $U_{E,\max} = Q^2/2C$.

**Calculations:** With these ideas, we can now write the conservation of energy as

$$U_{B,\max} = U_{E,\max}$$

or

$$\frac{LI^2}{2} = \frac{Q^2}{2C}.$$

Solving for $I$ gives us

$$I = \sqrt{\frac{Q^2}{LC}}.$$

We know $L$ and $C$, but not $Q$. However, with Eq. 25-1 ($q = CV$) we can relate $Q$ to the maximum potential difference $V$ across the capacitor, which is the initial potential difference of 57 V. Thus, substituting $Q = CV$ leads to

$$I = V\sqrt{\frac{C}{L}} = (57\text{ V})\sqrt{\frac{1.5 \times 10^{-6}\text{ F}}{12 \times 10^{-3}\text{ H}}}$$

$$= 0.637\text{ A} \approx 640\text{ mA}. \qquad \text{(Answer)}$$

## 31-3 | The Electrical–Mechanical Analogy

Let us look a little closer at the analogy between the oscillating $LC$ system of Fig. 31-1 and an oscillating block–spring system. Two kinds of energy are involved in the block–spring system. One is potential energy of the compressed or extended spring; the other is kinetic energy of the moving block. These two energies are given by the formulas in the left energy column in Table 31-1.

The table also shows, in the right energy column, the two kinds of energy involved in $LC$ oscillations. By looking across the table, we can see an analogy between the forms of the two pairs of energies—the mechanical energies of the block–spring system and the electromagnetic energies of the $LC$ oscillator. The equations for $v$ and $i$ at the bottom of the table help us see the details of the analogy. They tell us that $q$ corresponds to $x$ and $i$ corresponds to $v$ (in both equations, the former is differentiated to obtain the latter). These correspondences then suggest that, in the energy expressions, $1/C$ corresponds to $k$ and $L$ corresponds to $m$. Thus,

$$q \text{ corresponds to } x, \qquad 1/C \text{ corresponds to } k,$$

$$i \text{ corresponds to } v, \quad \text{and} \quad L \text{ corresponds to } m.$$

These correspondences suggest that in an $LC$ oscillator, the capacitor is mathematically like the spring in a block–spring system and the inductor is like the block.

**TABLE 31-1**

**Comparison of the Energy in Two Oscillating Systems**

| Block–Spring System | | LC Oscillator | |
|---|---|---|---|
| Element | Energy | Element | Energy |
| Spring | Potential, $\frac{1}{2}kx^2$ | Capacitor | Electrical, $\frac{1}{2}(1/C)q^2$ |
| Block | Kinetic, $\frac{1}{2}mv^2$ | Inductor | Magnetic, $\frac{1}{2}Li^2$ |
| | $v = dx/dt$ | | $i = dq/dt$ |

In Section 15-3 we saw that the angular frequency of oscillation of a (frictionless) block–spring system is

$$\omega = \sqrt{\frac{k}{m}} \quad \text{(block–spring system).} \tag{31-3}$$

The correspondences listed above suggest that to find the angular frequency of oscillation for an ideal (resistanceless) $LC$ circuit, $k$ should be replaced by $1/C$ and $m$ by $L$, yielding

$$\omega = \frac{1}{\sqrt{LC}} \quad \text{($LC$ circuit).} \tag{31-4}$$

We derive this result in the next section.

## 31-4 | LC Oscillations, Quantitatively

Here we want to show explicitly that Eq. 31-4 for the angular frequency of $LC$ oscillations is correct. At the same time, we want to examine even more closely the analogy between $LC$ oscillations and block–spring oscillations. We start by extending somewhat our earlier treatment of the mechanical block–spring oscillator.

### The Block–Spring Oscillator

We analyzed block–spring oscillations in Chapter 15 in terms of energy transfers and did not—at that early stage—derive the fundamental differential equation that governs those oscillations. We do so now.

We can write, for the total energy $U$ of a block–spring oscillator at any instant,

$$U = U_b + U_s = \tfrac{1}{2}mv^2 + \tfrac{1}{2}kx^2, \tag{31-5}$$

where $U_b$ and $U_s$ are, respectively, the kinetic energy of the moving block and the potential energy of the stretched or compressed spring. If there is no friction—which we assume—the total energy $U$ remains constant with time, even though $v$ and $x$ vary. In more formal language, $dU/dt = 0$. This leads to

$$\frac{dU}{dt} = \frac{d}{dt}\left(\tfrac{1}{2}mv^2 + \tfrac{1}{2}kx^2\right) = mv\frac{dv}{dt} + kx\frac{dx}{dt} = 0. \tag{31-6}$$

However, $v = dx/dt$ and $dv/dt = d^2x/dt^2$. With these substitutions, Eq. 31-6 becomes

$$m\frac{d^2x}{dt^2} + kx = 0 \quad \text{(block–spring oscillations).} \tag{31-7}$$

Equation 31-7 is the fundamental *differential equation* that governs the frictionless block–spring oscillations.

The general solution to Eq. 31-7—that is, the function $x(t)$ that describes the block–spring oscillations—is (as we saw in Eq. 15-3)

$$x = X\cos(\omega t + \phi) \quad \text{(displacement),} \tag{31-8}$$

in which $X$ is the amplitude of the mechanical oscillations ($x_m$ in Chapter 15), $\omega$ is the angular frequency of the oscillations, and $\phi$ is a phase constant.

### The LC Oscillator

Now let us analyze the oscillations of a resistanceless $LC$ circuit, proceeding exactly as we just did for the block–spring oscillator. The total energy $U$ present at any instant in an oscillating $LC$ circuit is given by

$$U = U_B + U_E = \frac{Li^2}{2} + \frac{q^2}{2C}, \tag{31-9}$$

in which $U_B$ is the energy stored in the magnetic field of the inductor and $U_E$ is the energy stored in the electric field of the capacitor. Since we have assumed the circuit resistance to be zero, no energy is transferred to thermal energy and $U$ remains constant with time. In more formal language, $dU/dt$ must be zero. This leads to

$$\frac{dU}{dt} = \frac{d}{dt}\left( \frac{Li^2}{2} + \frac{q^2}{2C} \right) = Li\frac{di}{dt} + \frac{q}{C}\frac{dq}{dt} = 0. \tag{31-10}$$

However, $i = dq/dt$ and $di/dt = d^2q/dt^2$. With these substitutions, Eq. 31-10 becomes

$$L\frac{d^2q}{dt^2} + \frac{1}{C}q = 0 \qquad (LC \text{ oscillations}). \tag{31-11}$$

This is the *differential equation* that describes the oscillations of a resistanceless $LC$ circuit. Equations 31-11 and 31-7 are exactly of the same mathematical form.

## Charge and Current Oscillations

Since the differential equations are mathematically identical, their solutions must also be mathematically identical. Because $q$ corresponds to $x$, we can write the general solution of Eq. 31-11, by analogy to Eq. 31-8, as

$$q = Q \cos(\omega t + \phi) \qquad (\text{charge}), \tag{31-12}$$

where $Q$ is the amplitude of the charge variations, $\omega$ is the angular frequency of the electromagnetic oscillations, and $\phi$ is the phase constant.

Taking the first derivative of Eq. 31-12 with respect to time gives us the current of the $LC$ oscillator:

$$i = \frac{dq}{dt} = -\omega Q \sin(\omega t + \phi) \qquad (\text{current}). \tag{31-13}$$

The amplitude $I$ of this sinusoidally varying current is

$$I = \omega Q, \tag{31-14}$$

and so we can rewrite Eq. 31-13 as

$$i = -I \sin(\omega t + \phi). \tag{31-15}$$

## Angular Frequencies

We can test whether Eq. 31-12 is a solution of Eq. 31-11 by substituting Eq. 31-12 and its second derivative with respect to time into Eq. 31-11. The first derivative of Eq. 31-12 is Eq. 31-13. The second derivative is then

$$\frac{d^2q}{dt^2} = -\omega^2 Q \cos(\omega t + \phi).$$

Substituting for $q$ and $d^2q/dt^2$ in Eq. 31-11, we obtain

$$-L\omega^2 Q \cos(\omega t + \phi) + \frac{1}{C}Q \cos(\omega t + \phi) = 0.$$

Canceling $Q \cos(\omega t + \phi)$ and rearranging lead to

$$\omega = \frac{1}{\sqrt{LC}}.$$

Thus, Eq. 31-12 is indeed a solution of Eq. 31-11 if $\omega$ has the constant value $1/\sqrt{LC}$. Note that this expression for $\omega$ is exactly that given by Eq. 31-4, which we arrived at by examining correspondences.

The phase constant $\phi$ in Eq. 31-12 is determined by the conditions that exist at any certain time—say, $t = 0$. If the conditions yield $\phi = 0$ at $t = 0$, Eq. 31-12 requires that $q = Q$ and Eq. 31-13 requires that $i = 0$; these are the initial conditions represented by Fig. 31-1a.

## Electrical and Magnetic Energy Oscillations

The electrical energy stored in the $LC$ circuit at time $t$ is, from Eqs. 31-1 and 31-12,

$$U_E = \frac{q^2}{2C} = \frac{Q^2}{2C} \cos^2(\omega t + \phi). \qquad (31\text{-}16)$$

The magnetic energy is, from Eqs. 31-2 and 31-13,

$$U_B = \tfrac{1}{2}Li^2 = \tfrac{1}{2}L\omega^2 Q^2 \sin^2(\omega t + \phi).$$

Substituting for $\omega$ from Eq. 31-4 then gives us

$$U_B = \frac{Q^2}{2C} \sin^2(\omega t + \phi). \qquad (31\text{-}17)$$

Figure 31-4 shows plots of $U_E(t)$ and $U_B(t)$ for the case of $\phi = 0$. Note that

1. The maximum values of $U_E$ and $U_B$ are both $Q^2/2C$.
2. At any instant the sum of $U_E$ and $U_B$ is equal to $Q^2/2C$, a constant.
3. When $U_E$ is maximum, $U_B$ is zero, and conversely.

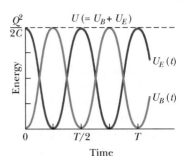

**FIG. 31-4** The stored magnetic energy and electrical energy in the circuit of Fig. 31-1 as a function of time. Note that their sum remains constant. $T$ is the period of oscillation.

✓**CHECKPOINT 2** A capacitor in an $LC$ oscillator has a maximum potential difference of 17 V and a maximum energy of 160 $\mu$J. When the capacitor has a potential difference of 5 V and an energy of 10 $\mu$J, what are (a) the emf across the inductor and (b) the energy stored in the magnetic field?

---

**Sample Problem** **31-2**

---

For the situation described in Sample Problem 31-1, let the coil (inductor) be connected to the charged capacitor at time $t = 0$. The result is an $LC$ circuit like that in Fig. 31-1.

(a) What is the potential difference $v_L(t)$ across the inductor as a function of time?

**KEY IDEAS** (1) The current and potential differences of the circuit undergo sinusoidal oscillations. (2) We can still apply the loop rule to this oscillating circuit—just as we did for the nonoscillating circuits of Chapter 27.

**Calculations:** At any time $t$ during the oscillations, the loop rule and Fig. 31-1 give us

$$v_L(t) = v_C(t); \qquad (31\text{-}18)$$

that is, the potential difference $v_L$ across the inductor must always be equal to the potential difference $v_C$ across the capacitor, so that the net potential difference around the

circuit is zero. Thus, we will find $v_L(t)$ if we can find $v_C(t)$, and we can find $v_C(t)$ from $q(t)$ with Eq. 25-1 ($q = CV$).

Because the potential difference $v_C(t)$ is maximum when the oscillations begin at time $t = 0$, the charge $q$ on the capacitor must also be maximum then. Thus, phase constant $\phi$ must be zero; so Eq. 31-12 gives us

$$q = Q \cos \omega t. \qquad (31\text{-}19)$$

(Note that this cosine function does indeed yield maximum $q$ ($= Q$) when $t = 0$.) To get the potential difference $v_C(t)$, we divide both sides of Eq. 31-19 by $C$ to write

$$\frac{q}{C} = \frac{Q}{C} \cos \omega t,$$

and then use Eq. 25-1 to write

$$v_C = V_C \cos \omega t. \qquad (31\text{-}20)$$

Here, $V_C$ is the amplitude of the oscillations in the potential difference $v_C$ across the capacitor.

Next, substituting $v_C = v_L$ from Eq. 31-18, we find

$$v_L = V_C \cos \omega t. \qquad (31\text{-}21)$$

We can evaluate the right side of this equation by first noting that the amplitude $V_C$ is equal to the initial (maximum) potential difference of 57 V across the capacitor. Then, using the values of $L$ and $C$ from Sample Problem 31-1, we find $\omega$ with Eq. 31-4:

$$\omega = \frac{1}{\sqrt{LC}} = \frac{1}{[(0.012 \text{ H})(1.5 \times 10^{-6} \text{ F})]^{0.5}}$$

$$= 7454 \text{ rad/s} \approx 7500 \text{ rad/s}.$$

Thus, Eq. 31-21 becomes

$$v_L = (57 \text{ V}) \cos(7500 \text{ rad/s})t. \qquad \text{(Answer)}$$

**(b)** What is the maximum rate $(di/dt)_{max}$ at which the current $i$ changes in the circuit?

**KEY IDEA** With the charge on the capacitor oscillating as in Eq. 31-12, the current is in the form of Eq. 31-13. Because $\phi = 0$, that equation gives us

$$i = -\omega Q \sin \omega t.$$

**Calculations:** Taking the derivative, we have

$$\frac{di}{dt} = \frac{d}{dt}(-\omega Q \sin \omega t) = -\omega^2 Q \cos \omega t.$$

We can simplify this equation by substituting $CV_C$ for $Q$ (because we know $C$ and $V_C$ but not $Q$) and $1/\sqrt{LC}$ for $\omega$ according to Eq. 31-4. We get

$$\frac{di}{dt} = -\frac{1}{LC}CV_C \cos \omega t = -\frac{V_C}{L} \cos \omega t.$$

This tells us that the current changes at a varying (sinusoidal) rate, with its maximum rate of change being

$$\frac{V_C}{L} = \frac{57 \text{ V}}{0.012 \text{ H}} = 4750 \text{ A/s} \approx 4800 \text{ A/s}. \qquad \text{(Answer)}$$

**FIG. 31-5** A series *RLC* circuit. As the charge contained in the circuit oscillates back and forth through the resistance, electromagnetic energy is dissipated as thermal energy, damping (decreasing the amplitude of) the oscillations.

## 31-5 | Damped Oscillations in an *RLC* Circuit

A circuit containing resistance, inductance, and capacitance is called an *RLC circuit*. We shall here discuss only *series RLC circuits* like that shown in Fig. 31-5. With a resistance $R$ present, the total *electromagnetic energy* $U$ of the circuit (the sum of the electrical energy and magnetic energy) is no longer constant; instead, it decreases with time as energy is transferred to thermal energy in the resistance. Because of this loss of energy, the oscillations of charge, current, and potential difference continuously decrease in amplitude, and the oscillations are said to be *damped*. As you will see, they are damped in exactly the same way as those of the damped block–spring oscillator of Section 15-8.

To analyze the oscillations of this circuit, we write an equation for the total electromagnetic energy $U$ in the circuit at any instant. Because the resistance does not store electromagnetic energy, we can use Eq. 31-9:

$$U = U_B + U_E = \frac{Li^2}{2} + \frac{q^2}{2C}. \qquad (31\text{-}22)$$

Now, however, this total energy decreases as energy is transferred to thermal energy. The rate of that transfer is, from Eq. 26-27,

$$\frac{dU}{dt} = -i^2 R, \qquad (31\text{-}23)$$

where the minus sign indicates that $U$ decreases. By differentiating Eq. 31-22 with respect to time and then substituting the result in Eq. 31-23, we obtain

$$\frac{dU}{dt} = Li\frac{di}{dt} + \frac{q}{C}\frac{dq}{dt} = -i^2 R.$$

Substituting $dq/dt$ for $i$ and $d^2q/dt^2$ for $di/dt$, we obtain

$$L\frac{d^2q}{dt^2} + R\frac{dq}{dt} + \frac{1}{C}q = 0 \qquad (RLC \text{ circuit}), \qquad (31\text{-}24)$$

which is the differential equation for damped oscillations in an *RLC* circuit. The solution to Eq. 31-24 is

$$q = Qe^{-Rt/2L}\cos(\omega't + \phi), \qquad (31\text{-}25)$$

in which

$$\omega' = \sqrt{\omega^2 - (R/2L)^2},\qquad(31\text{-}26)$$

where $\omega = 1/\sqrt{LC}$, as with an undamped oscillator. Equation 31-25 tells us how the charge on the capacitor oscillates in a damped $RLC$ circuit; that equation is the electromagnetic counterpart of Eq. 15-42, which gives the displacement of a damped block–spring oscillator.

Equation 31-25 describes a sinusoidal oscillation (the cosine function) with an *exponentially decaying amplitude* $Qe^{-Rt/2L}$ (the factor that multiplies the cosine). The angular frequency $\omega'$ of the damped oscillations is always less than the angular frequency $\omega$ of the undamped oscillations; however, we shall here consider only situations in which $R$ is small enough for us to replace $\omega'$ with $\omega$.

Let us next find an expression for the total electromagnetic energy $U$ of the circuit as a function of time. One way to do so is to monitor the energy of the electric field in the capacitor, which is given by Eq. 31-1 ($U_E = q^2/2C$). By substituting Eq. 31-25 into Eq. 31-1, we obtain

$$U_E = \frac{q^2}{2C} = \frac{[Qe^{-Rt/2L}\cos(\omega' t + \phi)]^2}{2C} = \frac{Q^2}{2C}e^{-Rt/L}\cos^2(\omega' t + \phi).\quad(31\text{-}27)$$

Thus, the energy of the electric field oscillates according to a cosine-squared term, and the amplitude of that oscillation decreases exponentially with time.

## Sample Problem  31-3

A series $RLC$ circuit has inductance $L = 12$ mH, capacitance $C = 1.6\ \mu$F, and resistance $R = 1.5\ \Omega$.

**(a)** At what time $t$ will the amplitude of the charge oscillations in the circuit be 50% of its initial value?

**KEY IDEA** The amplitude of the charge oscillations decreases exponentially with time $t$: According to Eq. 31-25, the charge amplitude at any time $t$ is $Qe^{-Rt/2L}$, in which $Q$ is the amplitude at time $t = 0$.

***Calculations:*** We want the time when the charge amplitude has decreased to $0.50Q$, that is, when

$$Qe^{-Rt/2L} = 0.50Q.$$

Canceling $Q$ and taking the natural logarithms of both sides, we have

$$-\frac{Rt}{2L} = \ln 0.50.$$

Solving for $t$ and then substituting given data yield

$$t = -\frac{2L}{R}\ln 0.50 = -\frac{(2)(12 \times 10^{-3}\,\text{H})(\ln 0.50)}{1.5\ \Omega}$$

$$= 0.0111\ \text{s} \approx 11\ \text{ms}.\qquad\text{(Answer)}$$

**(b)** How many oscillations are completed within this time?

**KEY IDEA** The time for one complete oscillation is the period $T = 2\pi/\omega$, where the angular frequency for $LC$ oscillations is given by Eq. 31-4 ($\omega = 1/\sqrt{LC}$).

***Calculation:*** In the time interval $\Delta t = 0.0111$ s, the number of complete oscillations is

$$\frac{\Delta t}{T} = \frac{\Delta t}{2\pi\sqrt{LC}}$$

$$= \frac{0.0111\ \text{s}}{2\pi[(12 \times 10^{-3}\,\text{H})(1.6 \times 10^{-6}\,\text{F})]^{1/2}} \approx 13.$$

$$\text{(Answer)}$$

Thus, the amplitude decays by 50% in about 13 complete oscillations. This damping is less severe than that shown in Fig. 31-3, where the amplitude decays by a little more than 50% in one oscillation.

# 31-6 | Alternating Current

The oscillations in an $RLC$ circuit will not damp out if an external emf device supplies enough energy to make up for the energy dissipated as thermal energy in the resistance $R$. Circuits in homes, offices, and factories, including countless $RLC$ circuits, receive such energy from local power companies. In most countries

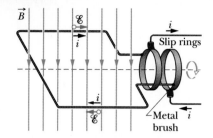

**FIG. 31-6** The basic mechanism of an alternating-current generator is a conducting loop rotated in an external magnetic field. In practice, the alternating emf induced in a coil of many turns of wire is made accessible by means of slip rings attached to the rotating loop. Each ring is connected to one end of the loop wire and is electrically connected to the rest of the generator circuit by a conducting brush against which the ring slips as the loop (and it) rotates.

the energy is supplied via oscillating emfs and currents—the current is said to be an **alternating current,** or **ac** for short. (The nonoscillating current from a battery is said to be a **direct current,** or **dc.**) These oscillating emfs and currents vary sinusoidally with time, reversing direction (in North America) 120 times per second and thus having frequency $f = 60$ Hz.

At first sight this may seem to be a strange arrangement. We have seen that the drift speed of the conduction electrons in household wiring may typically be $4 \times 10^{-5}$ m/s. If we now reverse their direction every $\frac{1}{120}$ s, such electrons can move only about $3 \times 10^{-7}$ m in a half-cycle. At this rate, a typical electron can drift past no more than about 10 atoms in the wiring before it is required to reverse its direction. How, you may wonder, can the electron ever get anywhere?

Although this question may be worrisome, it is a needless concern. The conduction electrons do not have to "get anywhere." When we say that the current in a wire is one ampere, we mean that charge passes through any plane cutting across that wire at the rate of one coulomb per second. The speed at which the charge carriers cross that plane does not matter directly; one ampere may correspond to many charge carriers moving very slowly or to a few moving very rapidly. Furthermore, the signal to the electrons to reverse directions—which originates in the alternating emf provided by the power company's generator—is propagated along the conductor at a speed close to that of light. All electrons, no matter where they are located, get their reversal instructions at about the same instant. Finally, we note that for many devices, such as lightbulbs and toasters, the direction of motion is unimportant as long as the electrons do move so as to transfer energy to the device via collisions with atoms in the device.

The basic advantage of alternating current is this: *As the current alternates, so does the magnetic field that surrounds the conductor.* This makes possible the use of Faraday's law of induction, which, among other things, means that we can step up (increase) or step down (decrease) the magnitude of an alternating potential difference at will, using a device called a transformer, as we shall discuss later. Moreover, alternating current is more readily adaptable to rotating machinery such as generators and motors than is (nonalternating) direct current.

Figure 31-6 shows a simple model of an ac generator. As the conducting loop is forced to rotate through the external magnetic field $\vec{B}$, a sinusoidally oscillating emf $\mathscr{E}$ is induced in the loop:

$$\mathscr{E} = \mathscr{E}_m \sin \omega_d t. \tag{31-28}$$

The *angular frequency* $\omega_d$ of the emf is equal to the angular speed with which the loop rotates in the magnetic field, the *phase* of the emf is $\omega_d t$, and the *amplitude* of the emf is $\mathscr{E}_m$ (where the subscript stands for maximum). When the rotating loop is part of a closed conducting path, this emf produces (*drives*) a sinusoidal (alternating) current along the path with the same angular frequency $\omega_d$, which then is called the **driving angular frequency.** We can write the current as

$$i = I \sin(\omega_d t - \phi), \tag{31-29}$$

in which $I$ is the amplitude of the driven current. (The phase $\omega_d t - \phi$ of the current is traditionally written with a minus sign instead of as $\omega_d t + \phi$.) We include a phase constant $\phi$ in Eq. 31-29 because the current $i$ may not be in phase with the emf $\mathscr{E}$. (As you will see, the phase constant depends on the circuit to which the generator is connected.) We can also write the current $i$ in terms of the **driving frequency** $f_d$ of the emf, by substituting $2\pi f_d$ for $\omega_d$ in Eq. 31-29.

## 31-7 | Forced Oscillations

We have seen that once started, the charge, potential difference, and current in both undamped *LC* circuits and damped *RLC* circuits (with small enough *R*) oscillate at angular frequency $\omega = 1/\sqrt{LC}$. Such oscillations are said to be *free oscillations* (free of any external emf), and the angular frequency $\omega$ is said to be the circuit's **natural angular frequency.**

When the external alternating emf of Eq. 31-28 is connected to an *RLC* circuit, the oscillations of charge, potential difference, and current are said to be *driven oscillations* or *forced oscillations.* These oscillations always occur at the driving angular frequency $\omega_d$:

> ☞ Whatever the natural angular frequency $\omega$ of a circuit may be, forced oscillations of charge, current, and potential difference in the circuit always occur at the driving angular frequency $\omega_d$.

However, as you will see in Section 31-9, the amplitudes of the oscillations very much depend on how close $\omega_d$ is to $\omega$. When the two angular frequencies match— a condition known as **resonance**—the amplitude $I$ of the current in the circuit is maximum.

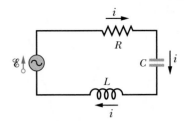

**FIG. 31-7** A single-loop circuit containing a resistor, a capacitor, and an inductor. A generator, represented by a sine wave in a circle, produces an alternating emf that establishes an alternating current; the directions of the emf and current are indicated here at only one instant.

## 31-8 | Three Simple Circuits

Later in this chapter, we shall connect an external alternating emf device to a series *RLC* circuit as in Fig. 31-7. We shall then find expressions for the amplitude $I$ and phase constant $\phi$ of the sinusoidally oscillating current in terms of the amplitude $\mathscr{E}_m$ and angular frequency $\omega_d$ of the external emf. First, let's consider three simpler circuits, each having an external emf and only one other circuit element: $R$, $C$, or $L$. We start with a resistive element (a purely *resistive load*).

### A Resistive Load

Figure 31-8 shows a circuit containing a resistance element of value $R$ and an ac generator with the alternating emf of Eq. 31-28. By the loop rule, we have

$$\mathscr{E} - v_R = 0.$$

With Eq. 31-28, this gives us

$$v_R = \mathscr{E}_m \sin \omega_d t.$$

Because the amplitude $V_R$ of the alternating potential difference (or voltage) across the resistance is equal to the amplitude $\mathscr{E}_m$ of the alternating emf, we can write this as

$$v_R = V_R \sin \omega_d t. \tag{31-30}$$

From the definition of resistance ($R = V/i$), we can now write the current $i_R$ in the resistance as

$$i_R = \frac{v_R}{R} = \frac{V_R}{R} \sin \omega_d t. \tag{31-31}$$

From Eq. 31-29, we can also write this current as

$$i_R = I_R \sin(\omega_d t - \phi), \tag{31-32}$$

where $I_R$ is the amplitude of the current $i_R$ in the resistance. Comparing Eqs. 31-31 and 31-32, we see that for a purely resistive load the phase constant $\phi = 0°$. We also see that the voltage amplitude and current amplitude are related by

$$V_R = I_R R \quad \text{(resistor).} \tag{31-33}$$

Although we found this relation for the circuit of Fig. 31-8, it applies to any resistance in any ac circuit.

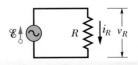

**FIG. 31-8** A resistor is connected across an alternating-current generator.

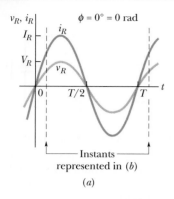

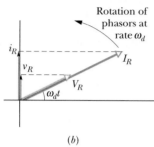

**FIG. 31-9** (a) The current $i_R$ and the potential difference $v_R$ across the resistor are plotted on the same graph, both versus time $t$. They are in phase and complete one cycle in one period $T$. (b) A phasor diagram shows the same thing as (a).

By comparing Eqs. 31-30 and 31-31, we see that the time-varying quantities $v_R$ and $i_R$ are both functions of sin $\omega_d t$ with $\phi = 0°$. Thus, these two quantities are *in phase*, which means that their corresponding maxima (and minima) occur at the same times. Figure 31-9a, which is a plot of $v_R(t)$ and $i_R(t)$, illustrates this fact. Note that $v_R$ and $i_R$ do not decay here because the generator supplies energy to the circuit to make up for the energy dissipated in $R$.

The time-varying quantities $v_R$ and $i_R$ can also be represented geometrically by *phasors*. Recall from Section 16-11 that phasors are vectors that rotate around an origin. Those that represent the voltage across and current in the resistor of Fig. 31-8 are shown in Fig. 31-9b at an arbitrary time $t$. Such phasors have the following properties:

**Angular speed:** Both phasors rotate counterclockwise about the origin with an angular speed equal to the angular frequency $\omega_d$ of $v_R$ and $i_R$.

**Length:** The length of each phasor represents the amplitude of the alternating quantity: $V_R$ for the voltage and $I_R$ for the current.

**Projection:** The projection of each phasor on the *vertical* axis represents the value of the alternating quantity at time $t$: $v_R$ for the voltage and $i_R$ for the current.

**Rotation angle:** The rotation angle of each phasor is equal to the phase of the alternating quantity at time $t$. In Fig. 31-9b, the voltage and current are in phase; so their phasors always have the same phase $\omega_d t$ and the same rotation angle, and thus they rotate together.

Mentally follow the rotation. Can you see that when the phasors have rotated so that $\omega_d t = 90°$ (they point vertically upward), they indicate that just then $v_R = V_R$ and $i_R = I_R$? Equations 31-30 and 31-32 give the same results.

✓**CHECKPOINT 3** If we increase the driving frequency in a circuit with a purely resistive load, do (a) amplitude $V_R$ and (b) amplitude $I_R$ increase, decrease, or remain the same?

---

**Sample Problem** | **31-4**

*Purely resistive load.* In Fig. 31-8, resistance $R$ is 200 $\Omega$ and the sinusoidal alternating emf device operates at amplitude $\mathcal{E}_m = 36.0$ V and frequency $f_d = 60.0$ Hz.

(a) What is the potential difference $v_R(t)$ across the resistance as a function of time $t$, and what is the amplitude $V_R$ of $v_R(t)$?

**KEY IDEA** In a circuit with a purely resistive load, the potential difference $v_R(t)$ across the resistance is always equal to the potential difference $\mathcal{E}(t)$ across the emf device.

**Calculations:** Here we have $v_R(t) = \mathcal{E}(t)$ and $V_R = \mathcal{E}_m$. Since $\mathcal{E}_m$ is given, we can write

$$V_R = \mathcal{E}_m = 36.0 \text{ V.} \quad \text{(Answer)}$$

To find $v_R(t)$, we use Eq. 31-28 to write

$$v_R(t) = \mathcal{E}(t) = \mathcal{E}_m \sin \omega_d t \quad (31\text{-}34)$$

and then substitute $\mathcal{E}_m = 36.0$ V and

$$\omega_d = 2\pi f_d = 2\pi(60 \text{ Hz}) = 120\pi$$

to obtain

$$v_R = (36.0 \text{ V}) \sin(120\pi t). \quad \text{(Answer)}$$

We can leave the argument of the sine in this form for convenience, or we can write it as $(377 \text{ rad/s})t$ or as $(377 \text{ s}^{-1})t$.

(b) What are the current $i_R(t)$ in the resistance and the amplitude $I_R$ of $i_R(t)$?

**KEY IDEA** In an ac circuit with a purely resistive load, the alternating current $i_R(t)$ in the resistance is *in phase* with the alternating potential difference $v_R(t)$ across the resistance; that is, the phase constant $\phi$ for the current is zero.

**Calculations:** Here we can write Eq. 31-29 as

$$i_R = I_R \sin(\omega_d t - \phi) = I_R \sin \omega_d t. \quad (31\text{-}35)$$

From Eq. 31-33, the amplitude $I_R$ is

$$I_R = \frac{V_R}{R} = \frac{36.0 \text{ V}}{200 \, \Omega} = 0.180 \text{ A.} \quad \text{(Answer)}$$

Substituting this and $\omega_d = 2\pi f_d = 120\pi$ into Eq. 31-35, we have

$$i_R = (0.180 \text{ A}) \sin(120\pi t). \quad \text{(Answer)}$$

## A Capacitive Load

Figure 31-10 shows a circuit containing a capacitance and a generator with the alternating emf of Eq. 31-28. Using the loop rule and proceeding as we did when we obtained Eq. 31-30, we find that the potential difference across the capacitor is

$$v_C = V_C \sin \omega_d t, \tag{31-36}$$

where $V_C$ is the amplitude of the alternating voltage across the capacitor. From the definition of capacitance we can also write

$$q_C = C v_C = C V_C \sin \omega_d t. \tag{31-37}$$

Our concern, however, is with the current rather than the charge. Thus, we differentiate Eq. 31-37 to find

$$i_C = \frac{dq_C}{dt} = \omega_d C V_C \cos \omega_d t. \tag{31-38}$$

We now modify Eq. 31-38 in two ways. First, for reasons of symmetry of notation, we introduce the quantity $X_C$, called the **capacitive reactance** of a capacitor, defined as

$$X_C = \frac{1}{\omega_d C} \qquad \text{(capacitive reactance).} \tag{31-39}$$

Its value depends not only on the capacitance but also on the driving angular frequency $\omega_d$. We know from the definition of the capacitive time constant ($\tau = RC$) that the SI unit for $C$ can be expressed as seconds per ohm. Applying this to Eq. 31-39 shows that the SI unit of $X_C$ is the *ohm*, just as for resistance $R$.

Second, we replace $\cos \omega_d t$ in Eq. 31-38 with a phase-shifted sine:

$$\cos \omega_d t = \sin(\omega_d t + 90°).$$

You can verify this identity by shifting a sine curve 90° in the negative direction.

With these two modifications, Eq. 31-38 becomes

$$i_C = \left( \frac{V_C}{X_C} \right) \sin(\omega_d t + 90°). \tag{31-40}$$

From Eq. 31-29, we can also write the current $i_C$ in the capacitor of Fig. 31-10 as

$$i_C = I_C \sin(\omega_d t - \phi), \tag{31-41}$$

where $I_C$ is the amplitude of $i_C$. Comparing Eqs. 31-40 and 31-41, we see that for a purely capacitive load the phase constant $\phi$ for the current is $-90°$. We also see that the voltage amplitude and current amplitude are related by

$$V_C = I_C X_C \qquad \text{(capacitor).} \tag{31-42}$$

Although we found this relation for the circuit of Fig. 31-10, it applies to any capacitance in any ac circuit.

Comparison of Eqs. 31-36 and 31-40, or inspection of Fig. 31-11a, shows that the quantities $v_C$ and $i_C$ are 90°, $\pi/2$ rad, or one-quarter cycle, out of phase. Furthermore, we see that $i_C$ *leads* $v_C$, which means that, if you monitored the current $i_C$ and the potential difference $v_C$ in the circuit of Fig. 31-10, you would find that $i_C$ reaches its maximum *before* $v_C$ does, by one-quarter cycle.

This relation between $i_C$ and $v_C$ is illustrated by the phasor diagram of Fig. 31-11b. As the phasors representing these two quantities rotate counterclockwise together, the phasor labeled $I_C$ does indeed lead that labeled $V_C$, and by an angle of 90°; that is, the phasor $I_C$ coincides with the vertical axis one-quarter cycle before the phasor $V_C$ does. Be sure to convince yourself that the phasor diagram of Fig. 31-11b is consistent with Eqs. 31-36 and 31-40.

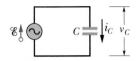

**FIG. 31-10** A capacitor is connected across an alternating-current generator.

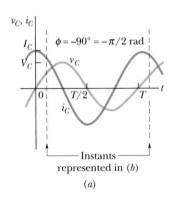

(a)

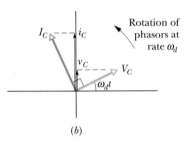

(b)

**FIG. 31-11** (a) The current in the capacitor leads the voltage by 90° (= $\pi/2$ rad). (b) A phasor diagram shows the same thing.

✓**CHECKPOINT 4** The figure shows, in (*a*), a sine curve $S(t) = \sin(\omega_d t)$ and three other sinusoidal curves $A(t)$, $B(t)$, and $C(t)$, each of the form $\sin(\omega_d t - \phi)$. (a) Rank the three other curves according to the value of $\phi$, most positive first and most negative last. (b) Which curve corresponds to which phasor in (*b*) of the figure? (c) Which curve leads the others?

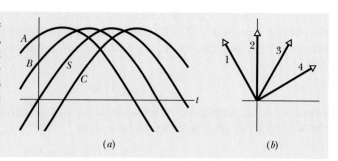

(*a*)  (*b*)

---

## Sample Problem 31-5

*Purely capacitive load.* In Fig. 31-10, capacitance $C$ is 15.0 $\mu$F and the sinusoidal alternating emf device operates at amplitude $\mathcal{E}_m = 36.0$ V and frequency $f_d = 60.0$ Hz.

**(a)** What are the potential difference $v_C(t)$ across the capacitance and the amplitude $V_C$ of $v_C(t)$?

**KEY IDEA** In a circuit with a purely capacitive load, the potential difference $v_C(t)$ across the capacitance is always equal to the potential difference $\mathcal{E}(t)$ across the emf device.

*Calculations:* Here we have $v_C(t) = \mathcal{E}(t)$ and $V_C = \mathcal{E}_m$. Since $\mathcal{E}_m$ is given, we have

$$V_C = \mathcal{E}_m = 36.0 \text{ V}. \quad \text{(Answer)}$$

To find $v_C(t)$, we use Eq. 31-28 to write

$$v_C(t) = \mathcal{E}(t) = \mathcal{E}_m \sin \omega_d t. \quad (31\text{-}43)$$

Then, substituting $\mathcal{E}_m = 36.0$ V and $\omega_d = 2\pi f_d = 120\pi$ into Eq. 31-43, we have

$$v_C = (36.0 \text{ V}) \sin(120\pi t). \quad \text{(Answer)}$$

**(b)** What are the current $i_C(t)$ in the circuit as a function of time and the amplitude $I_C$ of $i_C(t)$?

**KEY IDEA** In an ac circuit with a purely capacitive load, the alternating current $i_C(t)$ in the capacitance leads the alternating potential difference $v_C(t)$ by 90°; that is, the phase constant $\phi$ for the current is $-90°$, or $-\pi/2$ rad.

*Calculations:* Thus, we can write Eq. 31-29 as

$$i_C = I_C \sin(\omega_d t - \phi) = I_C \sin(\omega_d t + \pi/2). \quad (31\text{-}44)$$

We can find the amplitude $I_C$ from Eq. 31-42 ($V_C = I_C X_C$) if we first find the capacitive reactance $X_C$. From Eq. 31-39 ($X_C = 1/\omega_d C$), with $\omega_d = 2\pi f_d$, we can write

$$X_C = \frac{1}{2\pi f_d C} = \frac{1}{(2\pi)(60.0 \text{ Hz})(15.0 \times 10^{-6} \text{ F})}$$
$$= 177 \ \Omega.$$

Then Eq. 31-42 tells us that the current amplitude is

$$I_C = \frac{V_C}{X_C} = \frac{36.0 \text{ V}}{177 \ \Omega} = 0.203 \text{ A}. \quad \text{(Answer)}$$

Substituting this and $\omega_d = 2\pi f_d = 120\pi$ into Eq. 31-44, we have

$$i_C = (0.203 \text{ A}) \sin(120\pi t + \pi/2). \quad \text{(Answer)}$$

---

### An Inductive Load

Figure 31-12 shows a circuit containing an inductance and a generator with the alternating emf of Eq. 31-28. Using the loop rule and proceeding as we did to obtain Eq. 31-30, we find that the potential difference across the inductance is

$$v_L = V_L \sin \omega_d t, \quad (31\text{-}45)$$

where $V_L$ is the amplitude of $v_L$. From Eq. 30-35 ($\mathcal{E}_L = -L \, di/dt$), we can write the potential difference across an inductance $L$ in which the current is changing at the rate $di_L/dt$ as

$$v_L = L \frac{di_L}{dt}. \quad (31\text{-}46)$$

If we combine Eqs. 31-45 and 31-46, we have

$$\frac{di_L}{dt} = \frac{V_L}{L} \sin \omega_d t. \quad (31\text{-}47)$$

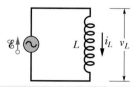

**FIG. 31-12** An inductor is connected across an alternating-current generator.

Our concern, however, is with the current rather than with its time derivative. We find the former by integrating Eq. 31-47, obtaining

$$i_L = \int di_L = \frac{V_L}{L} \int \sin \omega_d t \, dt = -\left(\frac{V_L}{\omega_d L}\right) \cos \omega_d t. \qquad (31\text{-}48)$$

We now modify this equation in two ways. First, for reasons of symmetry of notation, we introduce the quantity $X_L$, called the **inductive reactance** of an inductor, which is defined as

$$X_L = \omega_d L \qquad \text{(inductive reactance)}. \qquad (31\text{-}49)$$

The value of $X_L$ depends on the driving angular frequency $\omega_d$. The unit of the inductive time constant $\tau_L$ indicates that the SI unit of $X_L$ is the *ohm*, just as it is for $X_C$ and for $R$.

Second, we replace $-\cos \omega_d t$ in Eq. 31-48 with a phase-shifted sine:

$$-\cos \omega_d t = \sin(\omega_d t - 90°).$$

You can verify this identity by shifting a sine curve $90°$ in the positive direction. With these two changes, Eq. 31-48 becomes

$$i_L = \left(\frac{V_L}{X_L}\right) \sin(\omega_d t - 90°). \qquad (31\text{-}50)$$

From Eq. 31-29, we can also write this current in the inductance as

$$i_L = I_L \sin(\omega_d t - \phi), \qquad (31\text{-}51)$$

where $I_L$ is the amplitude of the current $i_L$. Comparing Eqs. 31-50 and 31-51, we see that for a purely inductive load the phase constant $\phi$ for the current is $+90°$. We also see that the voltage amplitude and current amplitude are related by

$$V_L = I_L X_L \qquad \text{(inductor)}. \qquad (31\text{-}52)$$

Although we found this relation for the circuit of Fig. 31-12, it applies to any inductance in any ac circuit.

Comparison of Eqs. 31-45 and 31-50, or inspection of Fig. 31-13*a*, shows that the quantities $i_L$ and $v_L$ are $90°$ out of phase. In this case, however, $i_L$ *lags* $v_L$; that is, monitoring the current $i_L$ and the potential difference $v_L$ in the circuit of Fig. 31-12 shows that $i_L$ reaches its maximum value *after* $v_L$ does, by one-quarter cycle.

The phasor diagram of Fig. 31-13*b* also contains this information. As the phasors rotate counterclockwise in the figure, the phasor labeled $I_L$ does indeed lag that labeled $V_L$, and by an angle of $90°$. Be sure to convince yourself that Fig. 31-13*b* represents Eqs. 31-45 and 31-50.

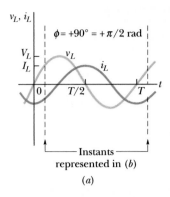

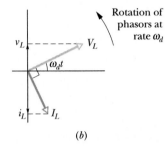

FIG. 31-13 (*a*) The current in the inductor lags the voltage by $90°$ ($= \pi/2$ rad). (*b*) A phasor diagram shows the same thing.

✓**CHECKPOINT 5**    If we increase the driving frequency in a circuit with a purely capacitive load, do (a) amplitude $V_C$ and (b) amplitude $I_C$ increase, decrease, or remain the same? If, instead, the circuit has a purely inductive load, do (c) amplitude $V_L$ and (d) amplitude $I_L$ increase, decrease, or remain the same?

**PROBLEM-SOLVING TACTICS**

*Tactic 1:  Leading and Lagging in AC Circuits* Table 31-2 summarizes the relations between the current $i$ and the voltage $v$ for each of the three kinds of circuit elements we have considered. When an applied alternating voltage produces an alternating current in these elements, the current is always in phase with the voltage across a resistor, always leads the voltage across a capacitor, and always lags the voltage across an inductor.

Many students remember these results with the mnemonic "*ELI* the *ICE* man." *ELI* contains the letter $L$

(for inductor), and in it the letter $I$ (for current) comes *after* the letter $E$ (for emf or voltage). Thus, for an inductor, the current *lags* (comes after) the voltage. Similarly *ICE* (which contains a $C$ for capacitor) means that the current *leads* (comes before) the voltage. You might also use the modified mnemonic "*ELI positively* is the *ICE* man" to remember that the phase constant $\phi$ is positive for an inductor.

If you have difficulty in remembering whether $X_C$ is equal to $\omega_d C$ (wrong) or $1/\omega_d C$ (right), try remembering that $C$ is in the "cellar"—that is, in the denominator.

**TABLE 31-2**

**Phase and Amplitude Relations for Alternating Currents and Voltages**

| Circuit Element | Symbol | Resistance or Reactance | Phase of the Current | Phase Constant (or Angle) $\phi$ | Amplitude Relation |
|---|---|---|---|---|---|
| Resistor | $R$ | $R$ | In phase with $v_R$ | $0°$ ($= 0$ rad) | $V_R = I_R R$ |
| Capacitor | $C$ | $X_C = 1/\omega_d C$ | Leads $v_C$ by $90°$ ($= \pi/2$ rad) | $-90°$ ($= -\pi/2$ rad) | $V_C = I_C X_C$ |
| Inductor | $L$ | $X_L = \omega_d L$ | Lags $v_L$ by $90°$ ($= \pi/2$ rad) | $+90°$ ($= +\pi/2$ rad) | $V_L = I_L X_L$ |

---

**Sample Problem** | **31-6**

*Purely inductive load.* In Fig. 31-12, inductance $L$ is 230 mH and the sinusoidal alternating emf device operates at amplitude $\mathcal{E}_m = 36.0$ V and frequency $f_d = 60.0$ Hz.

(a) What are the potential difference $v_L(t)$ across the inductance and the amplitude $V_L$ of $v_L(t)$?

**KEY IDEA** In a circuit with a purely inductive load, the potential difference $v_L(t)$ across the inductance is always equal to the potential difference $\mathcal{E}(t)$ across the emf device.

**Calculations:** Here we have $v_L(t) = \mathcal{E}(t)$ and $V_L = \mathcal{E}_m$. Since $\mathcal{E}_m$ is given, we know that

$$V_L = \mathcal{E}_m = 36.0 \text{ V.} \qquad \text{(Answer)}$$

To find $v_L(t)$, we use Eq. 31-28 to write

$$v_L(t) = \mathcal{E}(t) = \mathcal{E}_m \sin \omega_d t. \qquad (31\text{-}53)$$

Then, substituting $\mathcal{E}_m = 36.0$ V and $\omega_d = 2\pi f_d = 120\pi$ into Eq. 31-53, we have

$$v_L = (36.0 \text{ V}) \sin(120\pi t). \qquad \text{(Answer)}$$

(b) What are the current $i_L(t)$ in the circuit as a function of time and the amplitude $I_L$ of $i_L(t)$?

**KEY IDEA** In an ac circuit with a purely inductive load, the alternating current $i_L(t)$ in the inductance lags the alternating potential difference $v_L(t)$ by $90°$. (In the mnemonic of Problem-Solving Tactic 1, this circuit is "positively an *ELI* circuit," which tells us that the emf $E$ leads the current $I$ and that $\phi$ is *positive*.)

**Calculations:** Because the phase constant $\phi$ for the current is $+90°$, or $+\pi/2$ rad, we can write Eq. 31-29 as

$$i_L = I_L \sin(\omega_d t - \phi) = I_L \sin(\omega_d t - \pi/2). \quad (31\text{-}54)$$

We can find the amplitude $I_L$ from Eq. 31-52 ($V_L = I_L X_L$) if we first find the inductive reactance $X_L$. From Eq. 31-49 ($X_L = \omega_d L$), with $\omega_d = 2\pi f_d$, we can write

$$X_L = 2\pi f_d L = (2\pi)(60.0 \text{ Hz})(230 \times 10^{-3} \text{ H})$$
$$= 86.7 \; \Omega.$$

Then Eq. 31-52 tells us that the current amplitude is

$$I_L = \frac{V_L}{X_L} = \frac{36.0 \text{ V}}{86.7 \; \Omega} = 0.415 \text{ A.} \quad \text{(Answer)}$$

Substituting this and $\omega_d = 2\pi f_d = 120\pi$ into Eq. 31-54, we have

$$i_L = (0.415 \text{ A}) \sin(120\pi t - \pi/2). \quad \text{(Answer)}$$

---

## 31-9 | The Series *RLC* Circuit

We are now ready to apply the alternating emf of Eq. 31-28,

$$\mathcal{E} = \mathcal{E}_m \sin \omega_d t \qquad \text{(applied emf),} \qquad (31\text{-}55)$$

to the full *RLC* circuit of Fig. 31-7. Because $R$, $L$, and $C$ are in series, the same current

$$i = I \sin(\omega_d t - \phi) \qquad (31\text{-}56)$$

is driven in all three of them. We wish to find the current amplitude $I$ and the phase constant $\phi$. The solution is simplified by the use of phasor diagrams.

### The Current Amplitude

We start with Fig. 31-14a, which shows the phasor representing the current of Eq. 31-56 at an arbitrary time $t$. The length of the phasor is the current amplitude $I$, the projection of the phasor on the vertical axis is the current $i$ at time $t$, and the angle of rotation of the phasor is the phase $\omega_d t - \phi$ of the current at time $t$.

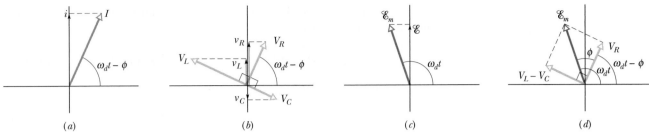

**FIG. 31-14** (*a*) A phasor representing the alternating current in the driven *RLC* circuit of Fig. 31-7 at time *t*. The amplitude *I*, the instantaneous value *i*, and the phase ($\omega_d t - \phi$) are shown. (*b*) Phasors representing the voltages across the inductor, resistor, and capacitor, oriented with respect to the current phasor in (*a*). (*c*) A phasor representing the alternating emf that drives the current of (*a*). (*d*) The emf phasor is equal to the vector sum of the three voltage phasors of (*b*). Here, voltage phasors $V_L$ and $V_C$ have been added vectorially to yield their net phasor ($V_L - V_C$).

Figure 31-14*b* shows the phasors representing the voltages across *R*, *L*, and *C* at the same time *t*. Each phasor is oriented relative to the angle of rotation of current phasor *I* in Fig. 31-14*a*, based on the information in Table 31-2:

***Resistor:*** Here current and voltage are in phase; so the angle of rotation of voltage phasor $V_R$ is the same as that of phasor *I*.

***Capacitor:*** Here current leads voltage by 90°; so the angle of rotation of voltage phasor $V_C$ is 90° less than that of phasor *I*.

***Inductor:*** Here current lags voltage by 90°; so the angle of rotation of voltage phasor $v_L$ is 90° greater than that of phasor *I*.

Figure 31-14*b* also shows the instantaneous voltages $v_R$, $v_C$, and $v_L$ across *R*, *C*, and *L* at time *t*; those voltages are the projections of the corresponding phasors on the vertical axis of the figure.

Figure 31-14*c* shows the phasor representing the applied emf of Eq. 31-55. The length of the phasor is the emf amplitude $\mathscr{E}_m$, the projection of the phasor on the vertical axis is the emf $\mathscr{E}$ at time *t*, and the angle of rotation of the phasor is the phase $\omega_d t$ of the emf at time *t*.

From the loop rule we know that at any instant the sum of the voltages $v_R$, $v_C$, and $v_L$ is equal to the applied emf $\mathscr{E}$:

$$\mathscr{E} = v_R + v_C + v_L. \qquad (31\text{-}57)$$

Thus, at time *t* the projection $\mathscr{E}$ in Fig. 31-14*c* is equal to the algebraic sum of the projections $v_R$, $v_C$, and $v_L$ in Fig. 31-14*b*. In fact, as the phasors rotate together, this equality always holds. This means that phasor $\mathscr{E}_m$ in Fig. 31-14*c* must be equal to the vector sum of the three voltage phasors $V_R$, $V_C$, and $V_L$ in Fig. 31-14*b*.

That requirement is indicated in Fig. 31-14*d*, where phasor $\mathscr{E}_m$ is drawn as the sum of phasors $V_R$, $V_L$, and $V_C$. Because phasors $V_L$ and $V_C$ have opposite directions in the figure, we simplify the vector sum by first combining $V_L$ and $V_C$ to form the single phasor $V_L - V_C$. Then we combine that single phasor with $V_R$ to find the net phasor. Again, the net phasor must coincide with phasor $\mathscr{E}_m$, as shown.

Both triangles in Fig. 31-14*d* are right triangles. Applying the Pythagorean theorem to either one yields

$$\mathscr{E}_m^2 = V_R^2 + (V_L - V_C)^2. \qquad (31\text{-}58)$$

From the amplitude information displayed in Table 31-2 we can rewrite this as

$$\mathscr{E}_m^2 = (IR)^2 + (IX_L - IX_C)^2, \qquad (31\text{-}59)$$

and then rearrange it to the form

$$I = \frac{\mathscr{E}_m}{\sqrt{R^2 + (X_L - X_C)^2}}. \qquad (31\text{-}60)$$

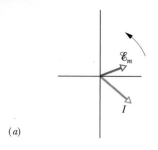

(a)

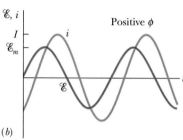

(b)

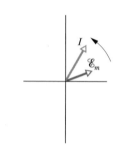

(c)

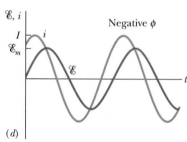

(d)

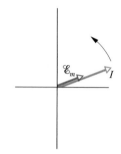

(e)

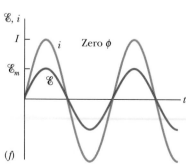

(f)

The denominator in Eq. 31-60 is called the **impedance** $Z$ of the circuit for the driving angular frequency $\omega_d$:

$$Z = \sqrt{R^2 + (X_L - X_C)^2} \qquad \text{(impedance defined).} \qquad (31\text{-}61)$$

We can then write Eq. 31-60 as

$$I = \frac{\mathscr{E}_m}{Z}. \qquad (31\text{-}62)$$

If we substitute for $X_C$ and $X_L$ from Eqs. 31-39 and 31-49, we can write Eq. 31-60 more explicitly as

$$I = \frac{\mathscr{E}_m}{\sqrt{R^2 + (\omega_d L - 1/\omega_d C)^2}} \qquad \text{(current amplitude).} \qquad (31\text{-}63)$$

We have now accomplished half our goal: We have obtained an expression for the current amplitude $I$ in terms of the sinusoidal driving emf and the circuit elements in a series $RLC$ circuit.

The value of $I$ depends on the difference between $\omega_d L$ and $1/\omega_d C$ in Eq. 31-63 or, equivalently, the difference between $X_L$ and $X_C$ in Eq. 31-60. In either equation, it does not matter which of the two quantities is greater because the difference is always squared.

The current that we have been describing in this section is the *steady-state current* that occurs after the alternating emf has been applied for some time. When the emf is first applied to a circuit, a brief *transient current* occurs. Its duration (before settling down into the steady-state current) is determined by the time constants $\tau_L = L/R$ and $\tau_C = RC$ as the inductive and capacitive elements "turn on." This transient current can, for example, destroy a motor on start-up if it is not properly taken into account in the motor's circuit design.

### The Phase Constant

From the right-hand phasor triangle in Fig. 31-14$d$ and from Table 31-2 we can write

$$\tan \phi = \frac{V_L - V_C}{V_R} = \frac{IX_L - IX_C}{IR}, \qquad (31\text{-}64)$$

which gives us

$$\tan \phi = \frac{X_L - X_C}{R} \qquad \text{(phase constant).} \qquad (31\text{-}65)$$

This is the other half of our goal: an equation for the phase constant $\phi$ in a sinusoidally driven series $RLC$ circuit. In essence, it gives us three different results for the phase constant, depending on the relative values of $X_L$ and $X_C$:

$X_L > X_C$: The circuit is said to be *more inductive than capacitive.* Equation 31-65 tells us that $\phi$ is positive for such a circuit, which means that phasor $I$ rotates behind phasor $\mathscr{E}_m$ (Fig. 31-15$a$). A plot of $\mathscr{E}$ and $i$ versus time is like that in Fig. 31-15$b$. (Figures 31-14$c$ and $d$ were drawn assuming $X_L > X_C$.)

$X_C > X_L$: The circuit is said to be *more capacitive than inductive.* Equation 31-65 tells

**FIG. 31-15** Phasor diagrams and graphs of the alternating emf $\mathscr{E}$ and current $i$ for the driven $RLC$ circuit of Fig. 31-7. In the phasor diagram of ($a$) and the graph of ($b$), the current $i$ lags the driving emf $\mathscr{E}$ and the current's phase constant $\phi$ is positive. In ($c$) and ($d$), the current $i$ leads the driving emf $\mathscr{E}$ and its phase constant $\phi$ is negative. In ($e$) and ($f$), the current $i$ is in phase with the driving emf $\mathscr{E}$ and its phase constant $\phi$ is zero.

us that $\phi$ is negative for such a circuit, which means that phasor $I$ rotates ahead of phasor $\mathscr{E}_m$ (Fig. 31-15c). A plot of $\mathscr{E}$ and $i$ versus time is like that in Fig. 31-15d.

$X_C = X_L$: The circuit is said to be in *resonance*, a state that is discussed next. Equation 31-65 tells us that $\phi = 0°$ for such a circuit, which means that phasors $\mathscr{E}_m$ and $I$ rotate together (Fig. 31-15e). A plot of $\mathscr{E}$ and $i$ versus time is like that in Fig. 31-15f.

As illustration, let us reconsider two extreme circuits: In the *purely inductive circuit* of Fig. 31-12, where $X_L$ is nonzero and $X_C = R = 0$, Eq. 31-65 tells us that $\phi = +90°$ (the greatest value of $\phi$), consistent with Fig. 31-13b. In the *purely capacitive circuit* of Fig. 31-10, where $X_C$ is nonzero and $X_L = R = 0$, Eq. 31-65 tells us that $\phi = -90°$ (the least value of $\phi$), consistent with Fig. 31-11b.

## Resonance

Equation 31-63 gives the current amplitude $I$ in an *RLC* circuit as a function of the driving angular frequency $\omega_d$ of the external alternating emf. For a given resistance $R$, that amplitude is a maximum when the quantity $\omega_d L - 1/\omega_d C$ in the denominator is zero—that is, when

$$\omega_d L = \frac{1}{\omega_d C}$$

or
$$\omega_d = \frac{1}{\sqrt{LC}} \qquad \text{(maximum } I\text{).} \qquad (31\text{-}66)$$

Because the natural angular frequency $\omega$ of the *RLC* circuit is also equal to $1/\sqrt{LC}$, the maximum value of $I$ occurs when the driving angular frequency matches the natural angular frequency—that is, at resonance. Thus, in an *RLC* circuit, resonance and maximum current amplitude $I$ occur when

$$\omega_d = \omega = \frac{1}{\sqrt{LC}} \qquad \text{(resonance).} \qquad (31\text{-}67)$$

Figure 31-16 shows three *resonance curves* for sinusoidally driven oscillations in three series *RLC* circuits differing only in $R$. Each curve peaks at its maximum current amplitude $I$ when the ratio $\omega_d/\omega$ is 1.00, but the maximum value of $I$ decreases with increasing $R$. (The maximum $I$ is always $\mathscr{E}_m/R$; to see why, com-

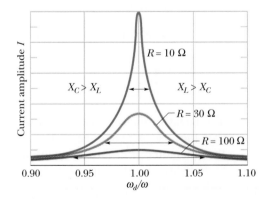

**SFIG. 31-16**   *Resonance curves* for the driven *RLC* circuit of Fig. 31-7 with $L = 100$ $\mu$H, $C = 100$ pF, and three values of $R$. The current amplitude $I$ of the alternating current depends on how close the driving angular frequency $\omega_d$ is to the natural angular frequency $\omega$. The horizontal arrow on each curve measures the curve's *half-width*, which is the width at the half-maximum level and is a measure of the sharpness of the resonance. To the left of $\omega_d/\omega = 1.00$, the circuit is mainly capacitive, with $X_C > X_L$; to the right, it is mainly inductive, with $X_L > X_C$.

bine Eqs. 31-61 and 31-62.) In addition, the curves increase in width (measured in Fig. 31-16 at half the maximum value of $I$) with increasing $R$.

To make physical sense of Fig. 31-16, consider how the reactances $X_L$ and $X_C$ change as we increase the driving angular frequency $\omega_d$, starting with a value much less than the natural frequency $\omega$. For small $\omega_d$, reactance $X_L\,(=\omega_d L)$ is small and reactance $X_C\,(=1/\omega_d C)$ is large. Thus, the circuit is mainly capacitive and the impedance is dominated by the large $X_C$, which keeps the current low.

As we increase $\omega_d$, reactance $X_C$ remains dominant but decreases while reactance $X_L$ increases. The decrease in $X_C$ decreases the impedance, allowing the current to increase, as we see on the left side of any resonance curve in Fig. 31-16. When the increasing $X_L$ and the decreasing $X_C$ reach equal values, the current is greatest and the circuit is in resonance, with $\omega_d = \omega$.

As we continue to increase $\omega_d$, the increasing reactance $X_L$ becomes progressively more dominant over the decreasing reactance $X_C$. The impedance increases because of $X_L$ and the current decreases, as on the right side of any resonance curve in Fig. 31-16. In summary, then: The low-angular-frequency side of a resonance curve is dominated by the capacitor's reactance, the high-angular-frequency side is dominated by the inductor's reactance, and resonance occurs in the middle.

**CHECKPOINT 6**     Here are the capacitive reactance and inductive reactance, respectively, for three sinusoidally driven series *RLC* circuits: (1) 50 Ω, 100 Ω; (2) 100 Ω, 50 Ω; (3) 50 Ω, 50 Ω. (a) For each, does the current lead or lag the applied emf, or are the two in phase? (b) Which circuit is in resonance?

---

**Sample Problem    31-7**

In Fig. 31-7, let $R = 200\ \Omega$, $C = 15.0\ \mu F$, $L = 230\ mH$, $f_d = 60.0\ Hz$, and $\mathscr{E}_m = 36.0\ V$. (These parameters are those used in Sample Problems 31-4, 31-5, and 31-6.)

(a) What is the current amplitude $I$?

**KEY IDEA**    The current amplitude $I$ depends on the amplitude $\mathscr{E}_m$ of the driving emf and on the impedance $Z$ of the circuit, according to Eq. 31-62 ($I = \mathscr{E}_m/Z$).

**Calculations:** So, we need to find $Z$, which depends on resistance $R$, capacitive reactance $X_C$, and inductive reactance $X_L$. The circuit's only resistance is the given resistance $R$. Its only capacitive reactance is due to the given capacitance, and, from Sample Problem 31-5, $X_C = 177\ \Omega$. Its only inductive reactance is due to the given inductance, and, from Sample Problem 31-6, $X_L = 86.7\ \Omega$. Thus, the circuit's impedance is

$$Z = \sqrt{R^2 + (X_L - X_C)^2}$$
$$= \sqrt{(200\ \Omega)^2 + (86.7\ \Omega - 177\ \Omega)^2}$$
$$= 219\ \Omega.$$

We then find

$$I = \frac{\mathscr{E}_m}{Z} = \frac{36.0\ V}{219\ \Omega} = 0.164\ A.\quad\text{(Answer)}$$

(b) What is the phase constant $\phi$ of the current in the circuit relative to the driving emf?

**KEY IDEA**    The phase constant depends on the inductive reactance, the capacitive reactance, and the resistance of the circuit, according to Eq. 31-65.

**Calculation:** Solving Eq. 31-65 for $\phi$ leads to

$$\phi = \tan^{-1}\frac{X_L - X_C}{R} = \tan^{-1}\frac{86.7\ \Omega - 177\ \Omega}{200\ \Omega}$$
$$= -24.3° = -0.424\ rad.\quad\text{(Answer)}$$

The negative phase constant is consistent with the fact that the load is mainly capacitive; that is, $X_C > X_L$. In the mnemonic of Problem-Solving Tactic 1, this circuit is an *ICE* circuit—the current *leads* the driving emf.

---

## 31-10 | Power in Alternating-Current Circuits

In the *RLC* circuit of Fig. 31-7, the source of energy is the alternating-current generator. Some of the energy that it provides is stored in the electric field in the capacitor, some is stored in the magnetic field in the inductor, and some is dissipated as thermal energy in the resistor. In steady-state operation—which we assume—the average energy stored in the capacitor and inductor together remains constant. The net transfer of energy is thus from the generator to the resistor, where electromagnetic energy is dissipated as thermal energy.

The instantaneous rate at which energy is dissipated in the resistor can be written, with the help of Eqs. 26-27 and 31-29, as

$$P = i^2 R = [I \sin(\omega_d t - \phi)]^2 R = I^2 R \sin^2(\omega_d t - \phi). \qquad (31\text{-}68)$$

The *average* rate at which energy is dissipated in the resistor, however, is the average of Eq. 31-68 over time. Over one complete cycle, the average value of $\sin \theta$, where $\theta$ is any variable, is zero (Fig. 31-17a) but the average value of $\sin^2 \theta$ is $\frac{1}{2}$ (Fig. 31-17b). (Note in Fig. 31-17b how the shaded areas under the curve but above the horizontal line marked $+\frac{1}{2}$ exactly fill in the unshaded spaces below that line.) Thus, we can write, from Eq. 31-68,

$$P_{\text{avg}} = \frac{I^2 R}{2} = \left(\frac{I}{\sqrt{2}}\right)^2 R. \qquad (31\text{-}69)$$

The quantity $I/\sqrt{2}$ is called the **root-mean-square,** or **rms,** value of the current $i$:

$$I_{\text{rms}} = \frac{I}{\sqrt{2}} \qquad \text{(rms current).} \qquad (31\text{-}70)$$

We can now rewrite Eq. 31-69 as

$$P_{\text{avg}} = I_{\text{rms}}^2 R \qquad \text{(average power).} \qquad (31\text{-}71)$$

Equation 31-71 looks much like Eq. 26-27 ($P = i^2 R$); the message is that if we switch to the rms current, we can compute the average rate of energy dissipation for alternating-current circuits just as for direct-current circuits.

We can also define rms values of voltages and emfs for alternating-current circuits:

$$V_{\text{rms}} = \frac{V}{\sqrt{2}} \quad \text{and} \quad \mathscr{E}_{\text{rms}} = \frac{\mathscr{E}_m}{\sqrt{2}} \qquad \text{(rms voltage; rms emf).} \qquad (31\text{-}72)$$

Alternating-current instruments, such as ammeters and voltmeters, are usually calibrated to read $I_{\text{rms}}$, $V_{\text{rms}}$, and $\mathscr{E}_{\text{rms}}$. Thus, if you plug an alternating-current voltmeter into a household electrical outlet and it reads 120 V, that is an rms voltage. The *maximum* value of the potential difference at the outlet is $\sqrt{2} \times (120\text{ V})$, or 170 V.

Because the proportionality factor $1/\sqrt{2}$ in Eqs. 31-70 and 31-72 is the same for all three variables, we can write Eqs. 31-62 and 31-60 as

$$I_{\text{rms}} = \frac{\mathscr{E}_{\text{rms}}}{Z} = \frac{\mathscr{E}_{\text{rms}}}{\sqrt{R^2 + (X_L - X_C)^2}}, \qquad (31\text{-}73)$$

and, indeed, this is the form that we almost always use.

We can use the relationship $I_{\text{rms}} = \mathscr{E}_{\text{rms}}/Z$ to recast Eq. 31-71 in a useful equivalent way. We write

$$P_{\text{avg}} = \frac{\mathscr{E}_{\text{rms}}}{Z} I_{\text{rms}} R = \mathscr{E}_{\text{rms}} I_{\text{rms}} \frac{R}{Z}. \qquad (31\text{-}74)$$

From Fig. 31-14d, Table 31-2, and Eq. 31-62, however, we see that $R/Z$ is just the cosine of the phase constant $\phi$:

$$\cos \phi = \frac{V_R}{\mathscr{E}_m} = \frac{IR}{IZ} = \frac{R}{Z}. \qquad (31\text{-}75)$$

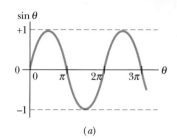

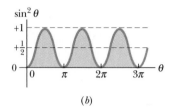

(a)

(b)

**FIG. 31-17** (a) A plot of $\sin \theta$ versus $\theta$. The average value over one cycle is zero. (b) A plot of $\sin^2 \theta$ versus $\theta$. The average value over one cycle is $\frac{1}{2}$.

Equation 31-74 then becomes

$$P_{avg} = \mathscr{E}_{rms} I_{rms} \cos \phi \quad \text{(average power)}, \quad (31\text{-}76)$$

in which the term $\cos \phi$ is called the **power factor.** Because $\cos \phi = \cos(-\phi)$, Eq. 31-76 is independent of the sign of the phase constant $\phi$.

To maximize the rate at which energy is supplied to a resistive load in an $RLC$ circuit, we should keep the power factor $\cos \phi$ as close to unity as possible. This is equivalent to keeping the phase constant $\phi$ in Eq. 31-29 as close to zero as possible. If, for example, the circuit is highly inductive, it can be made less so by putting more capacitance in the circuit, connected in series. (Recall that putting an additional capacitance into a series of capacitances decreases the equivalent capacitance $C_{eq}$ of the series.) Thus, the resulting decrease in $C_{eq}$ in the circuit reduces the phase constant and increases the power factor in Eq. 31-76. Power companies place series-connected capacitors throughout their transmission systems to get these results.

> ✓**CHECKPOINT 7** (a) If the current in a sinusoidally driven series $RLC$ circuit leads the emf, would we increase or decrease the capacitance to increase the rate at which energy is supplied to the resistance? (b) Would this change bring the resonant angular frequency of the circuit closer to the angular frequency of the emf or put it farther away?

## Sample Problem 31-8

A series $RLC$ circuit, driven with $\mathscr{E}_{rms} = 120$ V at frequency $f_d = 60.0$ Hz, contains a resistance $R = 200 \, \Omega$, an inductance with $X_L = 80.0 \, \Omega$, and a capacitance with $X_C = 150 \, \Omega$.

**(a)** What are the power factor $\cos \phi$ and phase constant $\phi$ of the circuit?

**KEY IDEA** The power factor $\cos \phi$ can be found from the resistance $R$ and impedance $Z$ via Eq. 31-75 ($\cos \phi = R/Z$).

**Calculations:** To calculate $Z$, we use Eq. 31-61:

$$Z = \sqrt{R^2 + (X_L - X_C)^2}$$
$$= \sqrt{(200 \, \Omega)^2 + (80.0 \, \Omega - 150 \, \Omega)^2} = 211.90 \, \Omega.$$

Equation 31-75 then gives us

$$\cos \phi = \frac{R}{Z} = \frac{200 \, \Omega}{211.90 \, \Omega} = 0.9438 \approx 0.944. \quad \text{(Answer)}$$

Taking the inverse cosine then yields

$$\phi = \cos^{-1} 0.944 = \pm 19.3°.$$

Both $+19.3°$ and $-19.3°$ have a cosine of 0.944. To determine which sign is correct, we must consider whether the current leads or lags the driving emf. Because $X_C > X_L$, this circuit is mainly capacitive, with the current leading the emf. Thus, $\phi$ must be negative:

$$\phi = -19.3°. \quad \text{(Answer)}$$

We could, instead, have found $\phi$ with Eq. 31-65. A calculator would then have given us the answer with the minus sign.

**(b)** What is the average rate $P_{avg}$ at which energy is dissipated in the resistance?

**KEY IDEAS** There are two ways and two ideas to use: (1) Because the circuit is assumed to be in steady-state operation, the rate at which energy is dissipated in the resistance is equal to the rate at which energy is supplied to the circuit, as given by Eq. 31-76 ($P_{avg} = \mathscr{E}_{rms} I_{rms} \cos \phi$). (2) The rate at which energy is dissipated in a resistance $R$ depends on the square of the rms current $I_{rms}$ through it, according to Eq. 31-71 ($P_{avg} = I_{rms}^2 R$).

**First way:** We are given the rms driving emf $\mathscr{E}_{rms}$ and we already know $\cos \phi$ from part (a). The rms current $I_{rms}$ is determined by the rms value of the driving emf and the circuit's impedance $Z$ (which we know), according to Eq. 31-73:

$$I_{rms} = \frac{\mathscr{E}_{rms}}{Z}.$$

Substituting this into Eq. 31-76 then leads to

$$P_{avg} = \mathscr{E}_{rms} I_{rms} \cos \phi = \frac{\mathscr{E}_{rms}^2}{Z} \cos \phi$$
$$= \frac{(120 \, \text{V})^2}{211.90 \, \Omega} (0.9438) = 64.1 \, \text{W}. \quad \text{(Answer)}$$

**Second way:** We then find

$$P_{avg} = I_{rms}^2 R = \frac{\mathscr{E}_{rms}^2}{Z^2} R$$
$$= \frac{(120 \, \text{V})^2}{(211.90 \, \Omega)^2} (200 \, \Omega) = 64.1 \, \text{W}. \quad \text{(Answer)}$$

**(c)** What new capacitance $C_{new}$ is needed to maximize $P_{avg}$ if the other parameters of the circuit are not changed?

**KEY IDEAS** (1) The average rate $P_{avg}$ at which energy is supplied and dissipated is maximized if the circuit is brought into resonance with the driving emf. (2) Resonance occurs when $X_C = X_L$.

**Calculations:** From the given data, we have $X_C > X_L$. Thus, we must decrease $X_C$ to reach resonance. From Eq. 31-39 ($X_C = 1/\omega_d C$), we see that this means we must increase $C$ to the new value $C_{new}$.

Using Eq. 31-39, we can write the resonance condition $X_C = X_L$ as

$$\frac{1}{\omega_d C_{new}} = X_L.$$

Substituting $2\pi f_d$ for $\omega_d$ (because we are given $f_d$ and not $\omega_d$) and then solving for $C_{new}$, we find

$$C_{new} = \frac{1}{2\pi f_d X_L} = \frac{1}{(2\pi)(60\ \text{Hz})(80.0\ \Omega)}$$

$$= 3.32 \times 10^{-5}\ \text{F} = 33.2\ \mu\text{F}. \qquad \text{(Answer)}$$

Following the procedure of part (b), you can show that with $C_{new}$, $P_{avg}$ would then be at its maximum value of 72.0 W.

## 31-11 | Transformers

### Energy Transmission Requirements

When an ac circuit has only a resistive load, the power factor in Eq. 31-76 is $\cos 0° = 1$ and the applied rms emf $\mathscr{E}_{rms}$ is equal to the rms voltage $V_{rms}$ across the load. Thus, with an rms current $I_{rms}$ in the load, energy is supplied and dissipated at the average rate of

$$P_{avg} = \mathscr{E}I = IV. \qquad (31\text{-}77)$$

(In Eq. 31-77 and the rest of this section, we follow conventional practice and drop the subscripts identifying rms quantities. Engineers and scientists assume that all time-varying currents and voltages are reported as rms values; that is what the meters read.) Equation 31-77 tells us that, to satisfy a given power requirement, we have a range of choices, from a relatively large current $I$ and a relatively small voltage $V$ to just the reverse, provided only that the product $IV$ is as required.

In electrical power distribution systems it is desirable for reasons of safety and for efficient equipment design to deal with relatively low voltages at both the generating end (the electrical power plant) and the receiving end (the home or factory). Nobody wants an electric toaster or a child's electric train to operate at, say, 10 kV. On the other hand, in the transmission of electrical energy from the generating plant to the consumer, we want the lowest practical current (hence the largest practical voltage) to minimize $I^2 R$ losses (often called *ohmic losses*) in the transmission line.

As an example, consider the 735 kV line used to transmit electrical energy from the La Grande 2 hydroelectric plant in Quebec to Montreal, 1000 km away. Suppose that the current is 500 A and the power factor is close to unity. Then from Eq. 31-77, energy is supplied at the average rate

$$P_{avg} = \mathscr{E}I = (7.35 \times 10^5\ \text{V})(500\ \text{A}) = 368\ \text{MW}.$$

The resistance of the transmission line is about 0.220 $\Omega$/km; thus, there is a total resistance of about 220 $\Omega$ for the 1000 km stretch. Energy is dissipated due to that resistance at a rate of about

$$P_{avg} = I^2 R = (500\ \text{A})^2(220\ \Omega) = 55.0\ \text{MW},$$

which is nearly 15% of the supply rate.

Imagine what would happen if we doubled the current and halved the voltage. Energy would be supplied by the plant at the same average rate of 368 MW as previously, but now energy would be dissipated at the rate of about

$$P_{avg} = I^2 R = (1000\ \text{A})^2(220\ \Omega) = 220\ \text{MW},$$

which is *almost 60% of the supply rate*. Hence the general energy transmission rule: Transmit at the highest possible voltage and the lowest possible current.

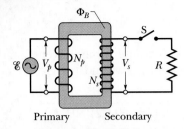

**FIG. 31-18** An ideal transformer (two coils wound on an iron core) in a basic transformer circuit. An ac generator produces current in the coil at the left (the *primary*). The coil at the right (the *secondary*) is connected to the resistive load *R* when switch S is closed.

## The Ideal Transformer

The transmission rule leads to a fundamental mismatch between the requirement for efficient high-voltage transmission and the need for safe low-voltage generation and consumption. We need a device with which we can raise (for transmission) and lower (for use) the ac voltage in a circuit, keeping the product current × voltage essentially constant. The **transformer** is such a device. It has no moving parts, operates by Faraday's law of induction, and has no simple direct-current counterpart.

The *ideal transformer* in Fig. 31-18 consists of two coils, with different numbers of turns, wound around an iron core. (The coils are insulated from the core.) In use, the primary winding, of $N_p$ turns, is connected to an alternating-current generator whose emf $\mathscr{E}$ at any time $t$ is given by

$$\mathscr{E} = \mathscr{E}_m \sin \omega t. \qquad (31\text{-}78)$$

The secondary winding, of $N_s$ turns, is connected to load resistance $R$, but its circuit is an open circuit as long as switch S is open (which we assume for the present). Thus, there can be no current through the secondary coil. We assume further for this ideal transformer that the resistances of the primary and secondary windings are negligible. Well-designed, high-capacity transformers can have energy losses as low as 1%; so our assumptions are reasonable.

For the assumed conditions, the primary winding (or *primary*) is a pure inductance and the primary circuit is like that in Fig. 31-12. Thus, the (very small) primary current, also called the *magnetizing current* $I_{\text{mag}}$, lags the primary voltage $V_p$ by 90°; the primary's power factor (= cos $\phi$ in Eq. 31-76) is zero; so no power is delivered from the generator to the transformer.

However, the small sinusoidally changing primary current $I_{\text{mag}}$ produces a sinusoidally changing magnetic flux $\Phi_B$ in the iron core. The core acts to strengthen the flux and to bring it through the secondary winding (or *secondary*). Because $\Phi_B$ varies, it induces an emf $\mathscr{E}_{\text{turn}}$ (= $d\Phi_B/dt$) in each turn of the secondary. In fact, this emf per turn $\mathscr{E}_{\text{turn}}$ is the same in the primary and the secondary. Across the primary, the voltage $V_p$ is the product of $\mathscr{E}_{\text{turn}}$ and the number of turns $N_p$; that is, $V_p = \mathscr{E}_{\text{turn}} N_p$. Similarly, across the secondary the voltage is $V_s = \mathscr{E}_{\text{turn}} N_s$. Thus, we can write

$$\mathscr{E}_{\text{turn}} = \frac{V_p}{N_p} = \frac{V_s}{N_s},$$

or $$V_s = V_p \frac{N_s}{N_p} \qquad \text{(transformation of voltage).} \qquad (31\text{-}79)$$

If $N_s > N_p$, the transformer is called a *step-up transformer* because it steps the primary's voltage $V_p$ up to a higher voltage $V_s$. Similarly, if $N_s < N_p$, the device is a *step-down transformer*.

So far, with switch S open, no energy is transferred from the generator to the rest of the circuit. Now let us close S to connect the secondary to the resistive load $R$. (In general, the load would also contain inductive and capacitive elements, but here we consider just resistance $R$.) We find that now energy *is* transferred from the generator. Let us see why.

Several things happen when we close switch S.

1. An alternating current $I_s$ appears in the secondary circuit, with corresponding energy dissipation rate $I_s^2 R$ (= $V_s^2/R$) in the resistive load.

2. This current produces its own alternating magnetic flux in the iron core, and this flux induces (from Faraday's law and Lenz's law) an opposing emf in the primary windings.

3. The voltage $V_p$ of the primary, however, cannot change in response to this opposing emf because it must always be equal to the emf $\mathscr{E}$ that is provided by the generator; closing switch S cannot change this fact.

**4.** To maintain $V_p$, the generator now produces (in addition to $I_{mag}$) an alternating current $I_p$ in the primary circuit; the magnitude and phase constant of $I_p$ are just those required for the emf induced by $I_p$ in the primary to exactly cancel the emf induced there by $I_s$. Because the phase constant of $I_p$ is not 90° like that of $I_{mag}$, this current $I_p$ can transfer energy to the primary.

We want to relate $I_s$ to $I_p$. However, rather than analyze the foregoing complex process in detail, let us just apply the principle of conservation of energy. The rate at which the generator transfers energy to the primary is equal to $I_p V_p$. The rate at which the primary then transfers energy to the secondary (via the alternating magnetic field linking the two coils) is $I_s V_s$. Because we assume that no energy is lost along the way, conservation of energy requires that

$$I_p V_p = I_s V_s.$$

Substituting for $V_s$ from Eq. 31-79, we find that

$$I_s = I_p \frac{N_p}{N_s} \qquad \text{(transformation of currents).} \qquad (31\text{-}80)$$

This equation tells us that the current $I_s$ in the secondary can differ from the current $I_p$ in the primary, depending on the *turns ratio* $N_p/N_s$.

Current $I_p$ appears in the primary circuit because of the resistive load $R$ in the secondary circuit. To find $I_p$, we substitute $I_s = V_s/R$ into Eq. 31-80 and then we substitute for $V_s$ from Eq. 31-79. We find

$$I_p = \frac{1}{R}\left(\frac{N_s}{N_p}\right)^2 V_p. \qquad (31\text{-}81)$$

This equation has the form $I_p = V_p/R_{eq}$, where equivalent resistance $R_{eq}$ is

$$R_{eq} = \left(\frac{N_p}{N_s}\right)^2 R. \qquad (31\text{-}82)$$

This $R_{eq}$ is the value of the load resistance as "seen" by the generator; the generator produces the current $I_p$ and voltage $V_p$ as if the generator were connected to a resistance $R_{eq}$.

## Impedance Matching

Equation 31-82 suggests still another function for the transformer. For maximum transfer of energy from an emf device to a resistive load, the resistance of the emf device must equal the resistance of the load. The same relation holds for ac circuits except that the *impedance* (rather than just the resistance) of the generator must equal that of the load. Often this condition is not met. For example, in a music-playing system, the amplifier has high impedance and the speaker set has low impedance. We can match the impedances of the two devices by coupling them through a transformer that has a suitable turns ratio $N_p/N_s$.

## Solar Activity and Power-Grid Systems

In a *solar flare,* a huge loop of electrons and protons extends outward from the surface of the Sun, as shown in the photograph that opens this chapter. Some solar flares explode, shooting those charged particles into space. On March 10, 1989, a gigantic solar flare exploded toward Earth. When the particles arrived three days later, they produced a $10^6$ A current, called an *electrojet,* in the high-altitude atmosphere above the Northern Hemisphere.

Because it is a current, an electrojet sets up a magnetic field $\vec{B}$ around itself, including along Earth's surface. Using the right-hand rule of Fig. 29-4, we see that the electrojet in Fig. 31-19 sets up a magnetic field component $B_x$ along Earth's

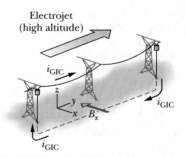

**FIG. 31-19** An electrojet (current) in the ionosphere produces a magnetic field $B_x$ through a vertical loop formed by a transmission line, the ground, and the wires grounding the transformers (located inside the cylinders at the ends of the transmission lines). Variations in $B_x$ induce current $i_{GIC}$ around the loop.

surface, directed perpendicular to the long power transmission line shown there. Grounded step-up or step-down transformers are attached at each end of the transmission line. Note that the transmission line, the ground, and the wires grounding the transformers form a conducting loop. A magnetic flux $\Phi$ due to $B_x$ penetrates that loop.

An electrojet varies in both size and location, and the resulting variations in $\Phi$ induce emf and current in the loop. The current $i_{GIC}$, called the geomagnetically induced current (GIC), is directed along the transmission line and (more important) through the transformers.

Transmission of power by a power-grid system depends on the proper sinusoidal variations in current and voltage throughout the system. The presence of $i_{GIC}$ through a transformer ruins the ability of the transformer's core to transfer the sinusoidal variations in the primary to the secondary. The reason is that the added flux in the core due to the $i_{GIC}$ *saturates* the core, making it unable to respond properly to sinusoidal variations in the primary. The result is that the current and voltage in the secondary are highly distorted and no longer sinusoidal, and this distortion disrupts the power transmission.

On March 13, 1989, this type of disruption caused the power-grid system of Quebec province to shut down. Today, whenever a solar flare explodes toward Earth, astronomers immediately warn power-grid engineers so that the engineers can brace for grid disruptions.

---

**CHECKPOINT 8** An alternating-current emf device in a certain circuit has a smaller resistance than that of the resistive load in the circuit; to increase the transfer of energy from the device to the load, a transformer will be connected between the two. (a) Should $N_s$ be greater than or less than $N_p$? (b) Will that make it a step-up or step-down transformer?

---

**Sample Problem** | **31-9**

A transformer on a utility pole operates at $V_p = 8.5$ kV on the primary side and supplies electrical energy to a number of nearby houses at $V_s = 120$ V, both quantities being rms values. Assume an ideal step-down transformer, a purely resistive load, and a power factor of unity.

**(a)** What is the turns ratio $N_p/N_s$ of the transformer?

**KEY IDEA** The turns ratio $N_p/N_s$ is related to the (given) rms primary and secondary voltages via Eq. 31-79 ($V_s = V_p N_s/N_p$).

**Calculation:** We can write Eq. 31-79 as

$$\frac{V_s}{V_p} = \frac{N_s}{N_p}. \tag{31-83}$$

(Note that the right side of this equation is the *inverse* of the turns ratio.) Inverting both sides of Eq. 31-83 gives us

$$\frac{N_p}{N_s} = \frac{V_p}{V_s} = \frac{8.5 \times 10^3 \text{ V}}{120 \text{ V}} = 70.83 \approx 71. \quad \text{(Answer)}$$

**(b)** The average rate of energy consumption (or dissipation) in the houses served by the transformer is 78 kW. What are the rms currents in the primary and secondary of the transformer?

**KEY IDEA** For a purely resistive load, the power factor $\cos \phi$ is unity; thus, the average rate at which energy is supplied and dissipated is given by Eq. 31-77 ($P_{avg} = \mathscr{E}I = IV$).

**Calculations:** In the primary circuit, with $V_p = 8.5$ kV, Eq. 31-77 yields

$$I_p = \frac{P_{avg}}{V_p} = \frac{78 \times 10^3 \text{ W}}{8.5 \times 10^3 \text{ V}} = 9.176 \text{ A} \approx 9.2 \text{ A}.$$
$$\text{(Answer)}$$

Similarly, in the secondary circuit,

$$I_s = \frac{P_{avg}}{V_s} = \frac{78 \times 10^3 \text{ W}}{120 \text{ V}} = 650 \text{ A}. \quad \text{(Answer)}$$

You can check that $I_s = I_p(N_p/N_s)$ as required by Eq. 31-80.

**(c)** What is the resistive load $R_s$ in the secondary circuit? What is the corresponding resistive load $R_p$ in the primary circuit?

**One way:** We can use $V = IR$ to relate the resistive load to the rms voltage and current. For the secondary circuit, we find

$$R_s = \frac{V_s}{I_s} = \frac{120 \text{ V}}{650 \text{ A}} = 0.1846 \ \Omega \approx 0.18 \ \Omega. \quad \text{(Answer)}$$

Similarly, for the primary circuit we find

$$R_p = \frac{V_p}{I_p} = \frac{8.5 \times 10^3 \text{ V}}{9.176 \text{ A}} = 926 \ \Omega \approx 930 \ \Omega. \quad \text{(Answer)}$$

**Second way:** We use the fact that $R_p$ equals the equivalent resistive load "seen" from the primary side of the

transformer, as given by Eq. 31-82 ($R_{eq} = (N_p/N_s)^2 R$). If we substitute $R_p$ for $R_{eq}$ and $R_s$ for $R$, that equation yields

$$R_p = \left(\frac{N_p}{N_s}\right)^2 R_s = (70.83)^2 (0.1846 \ \Omega)$$

$$= 926 \ \Omega \approx 930 \ \Omega. \quad \text{(Answer)}$$

## REVIEW & SUMMARY

**LC Energy Transfers** In an oscillating $LC$ circuit, energy is shuttled periodically between the electric field of the capacitor and the magnetic field of the inductor; instantaneous values of the two forms of energy are

$$U_E = \frac{q^2}{2C} \quad \text{and} \quad U_B = \frac{Li^2}{2}, \quad (31\text{-}1, 31\text{-}2)$$

where $q$ is the instantaneous charge on the capacitor and $i$ is the instantaneous current through the inductor. The total energy $U (= U_E + U_B)$ remains constant.

**LC Charge and Current Oscillations** The principle of conservation of energy leads to

$$L \frac{d^2q}{dt^2} + \frac{1}{C} q = 0 \quad (LC \text{ oscillations}) \quad (31\text{-}11)$$

as the differential equation of $LC$ oscillations (with no resistance). The solution of Eq. 31-11 is

$$q = Q \cos(\omega t + \phi) \quad \text{(charge)}, \quad (31\text{-}12)$$

in which $Q$ is the *charge amplitude* (maximum charge on the capacitor) and the angular frequency $\omega$ of the oscillations is

$$\omega = \frac{1}{\sqrt{LC}}. \quad (31\text{-}4)$$

The phase constant $\phi$ in Eq. 31-12 is determined by the initial conditions (at $t = 0$) of the system.

The current $i$ in the system at any time $t$ is

$$i = -\omega Q \sin(\omega t + \phi) \quad \text{(current)}, \quad (31\text{-}13)$$

in which $\omega Q$ is the *current amplitude I*.

**Damped Oscillations** Oscillations in an $LC$ circuit are damped when a dissipative element $R$ is also present in the circuit. Then

$$L \frac{d^2q}{dt^2} + R \frac{dq}{dt} + \frac{1}{C} q = 0 \quad (RLC \text{ circuit}). \quad (31\text{-}24)$$

The solution of this differential equation is

$$q = Q e^{-Rt/2L} \cos(\omega' t + \phi), \quad (31\text{-}25)$$

where

$$\omega' = \sqrt{\omega^2 - (R/2L)^2}. \quad (31\text{-}26)$$

We consider only situations with small $R$ and thus small damping; then $\omega' \approx \omega$.

**Alternating Currents; Forced Oscillations** A series $RLC$ circuit may be set into *forced oscillation* at a *driving angular frequency* $\omega_d$ by an external alternating emf

$$\mathcal{E} = \mathcal{E}_m \sin \omega_d t. \quad (31\text{-}28)$$

The current driven in the circuit by the emf is

$$i = I \sin(\omega_d t - \phi), \quad (31\text{-}29)$$

where $\phi$ is the phase constant of the current.

**Resonance** The current amplitude $I$ in a series $RLC$ circuit driven by a sinusoidal external emf is a maximum ($I = \mathcal{E}_m/R$) when the driving angular frequency $\omega_d$ equals the natural angular frequency $\omega$ of the circuit (that is, at *resonance*). Then $X_C = X_L, \phi = 0$, and the current is in phase with the emf.

**Single Circuit Elements** The alternating potential difference across a resistor has amplitude $V_R = IR$; the current is in phase with the potential difference.

For a *capacitor*, $V_C = IX_C$, in which $X_C = 1/\omega_d C$ is the **capacitive reactance;** the current here leads the potential difference by 90° ($\phi = -90° = -\pi/2$ rad).

For an *inductor*, $V_L = IX_L$, in which $X_L = \omega_d L$ is the **inductive reactance;** the current here lags the potential difference by 90° ($\phi = +90° = +\pi/2$ rad).

**Series RLC Circuits** For a series $RLC$ circuit with external emf given by Eq. 31-28 and current given by Eq. 31-29,

$$I = \frac{\mathcal{E}_m}{\sqrt{R^2 + (X_L - X_C)^2}}$$

$$= \frac{\mathcal{E}_m}{\sqrt{R^2 + (\omega_d L - 1/\omega_d C)^2}} \quad \text{(current amplitude)} \quad (31\text{-}60, 31\text{-}63)$$

and

$$\tan \phi = \frac{X_L - X_C}{R} \quad \text{(phase constant).} \quad (31\text{-}65)$$

Defining the impedance $Z$ of the circuit as

$$Z = \sqrt{R^2 + (X_L - X_C)^2} \quad \text{(impedance)} \quad (31\text{-}61)$$

allows us to write Eq. 31-60 as $I = \mathcal{E}_m/Z$.

**Power** In a series $RLC$ circuit, the **average power** $P_{avg}$ of the generator is equal to the production rate of thermal energy in the resistor:

$$P_{avg} = I_{rms}^2 R = \mathcal{E}_{rms} I_{rms} \cos \phi. \quad (31\text{-}71, 31\text{-}76)$$

Here rms stands for **root-mean-square;** the rms quantities are related to the maximum quantities by $I_{rms} = I/\sqrt{2}$, $V_{rms} = V/\sqrt{2}$, and $\mathcal{E}_{rms} = \mathcal{E}_m/\sqrt{2}$. The term $\cos \phi$ is called the **power factor** of the circuit.

**Transformers** A *transformer* (assumed to be ideal) is an iron core on which are wound a primary coil of $N_p$ turns and a secondary coil of $N_s$ turns. If the primary coil is connected across an alternating-current generator, the primary and secondary voltages are related by

$$V_s = V_p \frac{N_s}{N_p} \quad \text{(transformation of voltage).} \quad (31\text{-}79)$$

The currents through the coils are related by

$$I_s = I_p \frac{N_p}{N_s} \quad \text{(transformation of currents)}, \quad (31\text{-}80)$$

and the equivalent resistance of the secondary circuit, as seen by the generator, is

$$R_{eq} = \left(\frac{N_p}{N_s}\right)^2 R, \quad (31\text{-}82)$$

where $R$ is the resistive load in the secondary circuit. The ratio $N_p/N_s$ is called the transformer's *turns ratio*.

## QUESTIONS

**1** A charged capacitor and an inductor are connected at time $t = 0$. In terms of the period $T$ of the resulting oscillations, what is the first later time at which the following reach a maximum: (a) $U_B$, (b) the magnetic flux through the inductor, (c) $di/dt$, and (d) the emf of the inductor?

**2** What values of phase constant $\phi$ in Eq. 31-12 allow situations $(a), (c), (e)$, and $(g)$ of Fig. 31-1 to occur at $t = 0$?

**3** Figure 31-20 shows three oscillating $LC$ circuits with identical inductors and capacitors. Rank the circuits according to the time taken to fully discharge the capacitors during the oscillations, greatest first.

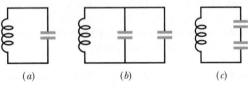

(a)        (b)        (c)

**FIG. 31-20** Question 3.

**4** Figure 31-21 shows graphs of capacitor voltage $v_C$ for $LC$ circuits 1 and 2, which contain identical capacitances and have the same maximum charge $Q$. Are (a) the inductance $L$ and (b) the maximum current $I$ in circuit 1 greater than, less than, or the same as those in circuit 2?

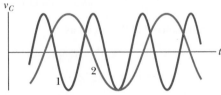

**FIG. 31-21** Question 4.

**5** Curve $a$ in Fig. 31-22 gives the impedance $Z$ of a driven $RC$ circuit versus the driving angular frequency $\omega_d$. The other two curves are similar but for different values of resistance $R$ and capacitance $C$. Rank the three curves according to the corresponding value of $R$, greatest first.

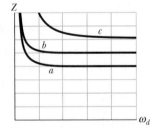

**FIG. 31-22** Question 5.

**6** Charges on the capacitors in three oscillating $LC$ circuits vary as: (1) $q = 2 \cos 4t$, (2) $q = 4 \cos t$, (3) $q = 3 \cos 4t$ (with $q$ in coulombs and $t$ in seconds). Rank the circuits according to (a) the current amplitude and (b) the period, greatest first.

**7** An alternating emf source with a certain emf amplitude is connected, in turn, to a resistor, a capacitor, and then an inductor. Once connected to one of the devices, the driving frequency $f_d$ is varied and the amplitude $I$ of the resulting current through the device is measured and plotted. Which of the three plots in Fig. 31-23 corresponds to which of the three devices?

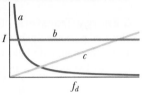

**FIG. 31-23** Question 7.

**8** Figure 31-24 shows the current $i$ and driving emf $\mathscr{E}$ for a series $RLC$ circuit. (a) Does the current lead or lag the emf? (b) Is the circuit's load mainly capacitive or mainly inductive? (c) Is the angular frequency $\omega_d$ of the emf greater than or less than the natural angular frequency $\omega$?

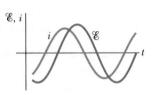

**FIG. 31-24** Questions 8 and 9.

**9** Figure 31-24 shows the current $i$ and driving emf $\mathscr{E}$ for a series $RLC$ circuit. Relative to the emf curve, does the current curve shift leftward or rightward and does the amplitude of that curve increase or decrease if we slightly increase (a) $L$, (b) $C$, and (c) $\omega_d$?

**10** The values of the phase constant $\phi$ for four sinusoidally driven series $RLC$ circuits are (1) $-15°$, (2) $+35°$, (3) $\pi/3$ rad, and (4) $-\pi/6$ rad. (a) In which is the load primarily capacitive? (b) In which does the current lag the alternating emf?

**11** Figure 31-25 shows the current $i$ and driving emf $\mathscr{E}$ for a series $RLC$ circuit. (a) Is the phase constant positive or negative? (b) To increase the rate at which energy is transferred to the resistive load, should $L$ be increased or decreased? (c) Should, instead, $C$ be increased or decreased?

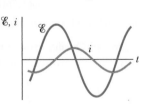

**FIG. 31-25** Question 11.

**12** Figure 31-26 shows three situations like those of Fig. 31-15. Is the driving angular frequency greater than, less than, or equal to the resonant angular frequency of the circuit in (a) situation 1, (b) situation 2, and (c) situation 3?

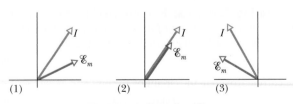

(1)        (2)        (3)

**FIG. 31-26** Question 12.

# PROBLEMS

**GO** Tutoring problem available (at instructor's discretion) in *WileyPLUS* and WebAssign

**SSM** Worked-out solution available in Student Solutions Manual     **WWW** Worked-out solution is at —

• – ••• Number of dots indicates level of problem difficulty     **ILW** Interactive solution is at —     http://www.wiley.com/college/halliday

Additional information available in *The Flying Circus of Physics* and at flyingcircusofphysics.com

## sec. 31-2  LC Oscillations, Qualitatively

•1  In a certain oscillating $LC$ circuit, the total energy is converted from electrical energy in the capacitor to magnetic energy in the inductor in 1.50 $\mu s$. What are (a) the period of oscillation and (b) the frequency of oscillation? (c) How long after the magnetic energy is a maximum will it be a maximum again?

•2  What is the capacitance of an oscillating $LC$ circuit if the maximum charge on the capacitor is 1.60 $\mu C$ and the total energy is 140 $\mu J$?

•3  In an oscillating $LC$ circuit, $L = 1.10$ mH and $C = 4.00$ $\mu F$. The maximum charge on the capacitor is 3.00 $\mu C$. Find the maximum current.

•4  The frequency of oscillation of a certain $LC$ circuit is 200 kHz. At time $t = 0$, plate $A$ of the capacitor has maximum positive charge. At what earliest time $t > 0$ will (a) plate $A$ again have maximum positive charge, (b) the other plate of the capacitor have maximum positive charge, and (c) the inductor have maximum magnetic field?

•5  An oscillating $LC$ circuit consists of a 75.0 mH inductor and a 3.60 $\mu F$ capacitor. If the maximum charge on the capacitor is 2.90 $\mu C$, what are (a) the total energy in the circuit and (b) the maximum current?

## sec. 31-3  The Electrical–Mechanical Analogy

•6  A 0.50 kg body oscillates in SHM on a spring that, when extended 2.0 mm from its equilibrium position, has an 8.0 N restoring force. What are (a) the angular frequency of oscillation, (b) the period of oscillation, and (c) the capacitance of an $LC$ circuit with the same period if $L$ is 5.0 H?

••7  The energy in an oscillating $LC$ circuit containing a 1.25 H inductor is 5.70 $\mu J$. The maximum charge on the capacitor is 175 $\mu C$. For a mechanical system with the same period, find the (a) mass, (b) spring constant, (c) maximum displacement, and (d) maximum speed. **SSM**

## sec. 31-4  LC Oscillations, Quantitatively

•8  $LC$ oscillators have been used in circuits connected to loudspeakers to create some of the sounds of electronic music. What inductance must be used with a 6.7 $\mu F$ capacitor to produce a frequency of 10 kHz, which is near the middle of the audible range of frequencies?

•9  In an oscillating $LC$ circuit with $L = 50$ mH and $C = 4.0$ $\mu F$, the current is initially a maximum. How long will it take before the capacitor is fully charged for the first time? **ILW**

•10  A single loop consists of inductors ($L_1, L_2, \ldots$), capacitors ($C_1, C_2, \ldots$), and resistors ($R_1, R_2, \ldots$) connected in series as shown, for example, in Fig. 31-27$a$. Show that regardless of the sequence of these circuit elements in the loop, the behav-

ior of this circuit is identical to that of the simple $LC$ circuit shown in Fig. 31-27$b$. (*Hint:* Consider the loop rule and see Problem 47 in Chapter 30.)

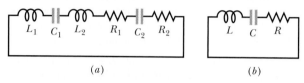

(a)                    (b)

**FIG. 31-27**  Problem 10.

••11  In Fig. 31-28, $R = 14.0$ $\Omega$, $C = 6.20$ $\mu F$, and $L = 54.0$ mH, and the ideal battery has emf $\mathscr{E} = 34.0$ V. The switch is kept in position $a$ for a long time and then thrown to position $b$. What are the (a) frequency and (b) current amplitude of the resulting oscillations? **ILW**

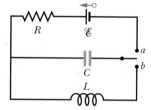

**FIG. 31-28**  Problem 11.

••12  To construct an oscillating $LC$ system, you can choose from a 10 mH inductor, a 5.0 $\mu F$ capacitor, and a 2.0 $\mu F$ capacitor. What are the (a) smallest, (b) second smallest, (c) second largest, and (d) largest oscillation frequency that can be set up by these elements in various combinations?

••13  An oscillating $LC$ circuit consisting of a 1.0 nF capacitor and a 3.0 mH coil has a maximum voltage of 3.0 V. What are (a) the maximum charge on the capacitor, (b) the maximum current through the circuit, and (c) the maximum energy stored in the magnetic field of the coil? **ILW**

••14  An inductor is connected across a capacitor whose capacitance can be varied by turning a knob. We wish to make the frequency of oscillation of this $LC$ circuit vary linearly with the angle of rotation of the knob, going from $2 \times 10^5$ to $4 \times 10^5$ Hz as the knob turns through 180°. If $L = 1.0$ mH, plot the required capacitance $C$ as a function of the angle of rotation of the knob.

••15  A variable capacitor with a range from 10 to 365 pF is used with a coil to form a variable-frequency $LC$ circuit to tune the input to a radio. (a) What is the ratio of maximum frequency to minimum frequency that can be obtained with such a capacitor? If this circuit is to obtain frequencies from 0.54 MHz to 1.60 MHz, the ratio computed in (a) is too large. By adding a capacitor in parallel to the variable capacitor, this range can be adjusted. To obtain the desired frequency range, (b) what capacitance should be added and (c) what inductance should the coil have? **SSM WWW**

••16  A series circuit containing inductance $L_1$ and capacitance $C_1$ oscillates at angular frequency $\omega$. A second series circuit, containing inductance $L_2$ and capacitance $C_2$, oscillates at the same angular frequency. In terms of $\omega$, what is the angular frequency of oscillation of a series circuit containing all four of these elements? Neglect resistance. (*Hint:* Use the formulas for equivalent capacitance and equivalent inductance; see Section 25-4 and Problem 47 in Chapter 30.)

**••17** In an oscillating $LC$ circuit with $C = 64.0 \mu F$, the current is given by $i = (1.60) \sin(2500t + 0.680)$, where $t$ is in seconds, $i$ in amperes, and the phase constant in radians. (a) How soon after $t = 0$ will the current reach its maximum value? What are (b) the inductance $L$ and (c) the total energy? **ILW**

**••18** In an oscillating $LC$ circuit, when 75.0% of the total energy is stored in the inductor's magnetic field, (a) what multiple of the maximum charge is on the capacitor and (b) what multiple of the maximum current is in the inductor?

**••19** In an oscillating $LC$ circuit, $L = 25.0$ mH and $C = 7.80 \mu F$. At time $t = 0$ the current is 9.20 mA, the charge on the capacitor is 3.80 $\mu C$, and the capacitor is charging. What are (a) the total energy in the circuit, (b) the maximum charge on the capacitor, and (c) the maximum current? (d) If the charge on the capacitor is given by $q = Q \cos(\omega t + \phi)$, what is the phase angle $\phi$? (e) Suppose the data are the same, except that the capacitor is discharging at $t = 0$. What then is $\phi$?

**••20** An oscillating $LC$ circuit has a current amplitude of 7.50 mA, a potential amplitude of 250 mV, and a capacitance of 220 nF. What are (a) the period of oscillation, (b) the maximum energy stored in the capacitor, (c) the maximum energy stored in the inductor, (d) the maximum rate at which the current changes, and (e) the maximum rate at which the inductor gains energy?

**••21** In an oscillating $LC$ circuit, $L = 3.00$ mH and $C = 2.70$ $\mu F$. At $t = 0$ the charge on the capacitor is zero and the current is 2.00 A. (a) What is the maximum charge that will appear on the capacitor? (b) At what earliest time $t > 0$ is the rate at which energy is stored in the capacitor greatest, and (c) what is that greatest rate?

**••22** In an oscillating $LC$ circuit in which $C = 4.00 \mu F$, the maximum potential difference across the capacitor during the oscillations is 1.50 V and the maximum current through the inductor is 50.0 mA. What are (a) the inductance $L$ and (b) the frequency of the oscillations? (c) How much time is required for the charge on the capacitor to rise from zero to its maximum value? **GO**

**••23** Using the loop rule, derive the differential equation for an $LC$ circuit (Eq. 31-11).

### sec. 31-5 Damped Oscillations in an *RLC* Circuit

**••24** In an oscillating series $RLC$ circuit, find the time required for the maximum energy present in the capacitor during an oscillation to fall to half its initial value. Assume $q = Q$ at $t = 0$.

**••25** What resistance $R$ should be connected in series with an inductance $L = 220$ mH and capacitance $C = 12.0 \mu F$ for the maximum charge on the capacitor to decay to 99.0% of its initial value in 50.0 cycles? (Assume $\omega' \approx \omega$.) **ILW**

**••26** A single-loop circuit consists of a 7.20 $\Omega$ resistor, a 12.0 H inductor, and a 3.20 $\mu F$ capacitor. Initially the capacitor has a charge of 6.20 $\mu C$ and the current is zero. Calculate

the charge on the capacitor $N$ complete cycles later for (a) $N = 5$, (b) $N = 10$, and (c) $N = 100$. **GO**

**•••27** In an oscillating series $RLC$ circuit, show that $\Delta U/U$, the fraction of the energy lost per cycle of oscillation, is given to a close approximation by $2\pi R/\omega L$. The quantity $\omega L/R$ is often called the $Q$ of the circuit (for *quality*). A high-$Q$ circuit has low resistance and a low fractional energy loss ($= 2\pi/Q$) per cycle. **SSM**

### sec. 31-8 Three Simple Circuits

**•28** A 50.0 $\Omega$ resistor is connected as in Fig. 31-8 to an ac generator with $\mathcal{E}_m = 30.0$ V. What is the amplitude of the resulting alternating current if the frequency of the emf is (a) 1.00 kHz and (b) 8.00 kHz?

**•29** (a) At what frequency would a 6.0 mH inductor and a 10 $\mu F$ capacitor have the same reactance? (b) What would the reactance be? (c) Show that this frequency would be the natural frequency of an oscillating circuit with the same $L$ and $C$.

**•30** A 1.50 $\mu F$ capacitor is connected as in Fig. 31-10 to an ac generator with $\mathcal{E}_m = 30.0$ V. What is the amplitude of the resulting alternating current if the frequency of the emf is (a) 1.00 kHz and (b) 8.00 kHz?

**•31** A 50.0 mH inductor is connected as in Fig. 31-12 to an ac generator with $\mathcal{E}_m = 30.0$ V. What is the amplitude of the resulting alternating current if the frequency of the emf is (a) 1.00 kHz and (b) 8.00 kHz? **ILW**

**••32** An ac generator with emf $\mathcal{E} = \mathcal{E}_m \sin \omega_d t$, where $\mathcal{E}_m = 25.0$ V and $\omega_d = 377$ rad/s, is connected to a 4.15 $\mu F$ capacitor. (a) What is the maximum value of the current? (b) When the current is a maximum, what is the emf of the generator? (c) When the emf of the generator is $-12.5$ V and increasing in magnitude, what is the current? **GO**

**••33** An ac generator has emf $\mathcal{E} = \mathcal{E}_m \sin(\omega_d t - \pi/4)$, where $\mathcal{E}_m = 30.0$ V and $\omega_d = 350$ rad/s. The current produced in a connected circuit is $i(t) = I \sin(\omega_d t - 3\pi/4)$, where $I = 620$ mA. At what time after $t = 0$ does (a) the generator emf first reach a maximum and (b) the current first reach a maximum? (c) The circuit contains a single element other than the generator. Is it a capacitor, an inductor, or a resistor? Justify your answer. (d) What is the value of the capacitance, inductance, or resistance, as the case may be? **SSM**

**••34** An ac generator has emf $\mathcal{E} = \mathcal{E}_m \sin \omega_d t$, with $\mathcal{E}_m = 25.0$ V and $\omega_d = 377$ rad/s. It is connected to a 12.7 H inductor. (a) What is the maximum value of the current? (b) When the current is a maximum, what is the emf of the generator? (c) When the emf of the generator is $-12.5$ V and increasing in magnitude, what is the current? **GO**

### sec. 31-9 The Series *RLC* Circuit

**•35** Remove the inductor from the circuit in Fig. 31-7 and set $R = 200 \Omega$, $C = 15.0 \mu F$, $f_d = 60.0$ Hz, and $\mathcal{E}_m = 36.0$ V. What are (a) $Z$, (b) $\phi$, and (c) $I$? (d) Draw a phasor diagram.

**•36** The current amplitude $I$ versus driving angular frequency $\omega_d$ for a driven $RLC$ circuit is given in Fig. 31-29, where the vertical axis scale is

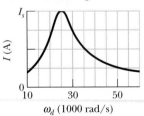

**FIG. 31-29** Problem 36.

set by $I_s = 4.00$ A. The inductance is 200 $\mu$H, and the emf amplitude is 8.0 V. What are (a) $C$ and (b) $R$?

**•37** Remove the capacitor from the circuit in Fig. 31-7 and set $R = 200$ $\Omega$, $L = 230$ mH, $f_d = 60.0$ Hz, and $\mathscr{E}_m = 36.0$ V. What are (a) $Z$, (b) $\phi$, and (c) $I$? (d) Draw a phasor diagram.

**•38** An alternating source with a variable frequency, an inductor with inductance $L$, and a resistor with resistance $R$ are connected in series. Figure 31-30 gives the impedance $Z$ of the circuit versus the driving angular frequency $\omega_d$, with the horizontal axis scale set by $\omega_{ds} = 1600$ rad/s. The figure also gives the reactance $X_L$ for the inductor versus $\omega_d$. What are (a) $R$ and (b) $L$?

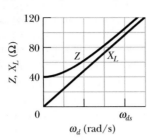

FIG. 31-30   Problem 38.

**•39** In Fig. 31-7, set $R = 200$ $\Omega$, $C = 70.0$ $\mu$F, $L = 230$ mH, $f_d = 60.0$ Hz, and $\mathscr{E}_m = 36.0$ V. What are (a) $Z$, (b) $\phi$, and (c) $I$? (d) Draw a phasor diagram.   **SSM**

**•40** An alternating source with a variable frequency, a capacitor with capacitance $C$, and a resistor with resistance $R$ are connected in series. Figure 31-31 gives the impedance $Z$ of the circuit versus the driving angular frequency $\omega_d$; the curve reaches an asymptote of 500 $\Omega$, and the horizontal scale is set by $\omega_{ds} = 300$ rad/s. The figure also gives the reactance $X_C$ for the capacitor versus $\omega_d$. What are (a) $R$ and (b) $C$?

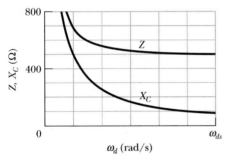

FIG. 31-31   Problem 40.

**•41** An electric motor has an effective resistance of 32.0 $\Omega$ and an inductive reactance of 45.0 $\Omega$ when working under load. The rms voltage across the alternating source is 420 V. Calculate the rms current.

**•42** An alternating source drives a series $RLC$ circuit with an emf amplitude of 6.00 V, at a phase angle of $+30.0°$. When the potential difference across the capacitor reaches its maximum positive value of $+5.00$ V, what is the potential difference across the inductor (sign included)?

**•43** A coil of inductance 88 mH and unknown resistance and a 0.94 $\mu$F capacitor are connected in series with an alternating emf of frequency 930 Hz. If the phase constant between the applied voltage and the current is 75°, what is the resistance of the coil?   **ILW**

**••44** An alternating emf source with a variable frequency $f_d$ is connected in series with a 50.0 $\Omega$ resistor and a 20.0 $\mu$F capacitor. The emf amplitude is 12.0 V. (a) Draw a phasor diagram for phasor $V_R$ (the potential across the resistor) and phasor $V_C$

(the potential across the capacitor). (b) At what driving frequency $f_d$ do the two phasors have the same length? At that driving frequency, what are (c) the phase angle in degrees, (d) the angular speed at which the phasors rotate, and (e) the current amplitude?

**••45** An $RLC$ circuit such as that of Fig. 31-7 has $R = 5.00$ $\Omega$, $C = 20.0$ $\mu$F, $L = 1.00$ H, and $\mathscr{E}_m = 30.0$ V. (a) At what angular frequency $\omega_d$ will the current amplitude have its maximum value, as in the resonance curves of Fig. 31-16? (b) What is this maximum value? At what (c) lower angular frequency $\omega_{d1}$ and (d) higher angular frequency $\omega_{d2}$ will the current amplitude be half this maximum value? (e) What is the fractional half-width $(\omega_{d1} - \omega_{d2})/\omega$ of the resonance curve for this circuit?   **SSM WWW**

**••46** Figure 31-32 shows a driven $RLC$ circuit that contains two identical capacitors and two switches. The emf amplitude is set at 12.0 V, and the driving frequency is set at 60.0 Hz. With both switches open, the current leads the emf by 30.9°. With switch $S_1$ closed and switch $S_2$ still open, the emf leads the current by 15.0°. With both switches closed, the current amplitude is 447 mA. What are (a) $R$, (b) $C$, and (c) $L$?   **GO**

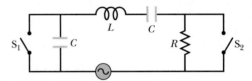

FIG. 31-32   Problem 46.

**••47** (a) In an $RLC$ circuit, can the amplitude of the voltage across an inductor be greater than the amplitude of the generator emf? (b) Consider an $RLC$ circuit with $\mathscr{E}_m = 10$ V, $R = 10$ $\Omega$, $L = 1.0$ H, and $C = 1.0$ $\mu$F. Find the amplitude of the voltage across the inductor at resonance.   **ILW**

**••48** An alternating emf source with a variable frequency $f_d$ is connected in series with an 80.0 $\Omega$ resistor and a 40.0 mH inductor. The emf amplitude is 6.00 V. (a) Draw a phasor diagram for phasor $V_R$ (the potential across the resistor) and phasor $V_L$ (the potential across the inductor). (b) At what driving frequency $f_d$ do the two phasors have the same length? At that driving frequency, what are (c) the phase angle in degrees, (d) the angular speed at which the phasors rotate, and (e) the current amplitude?

**••49** The fractional half-width $\Delta\omega_d$ of a resonance curve, such as the ones in Fig. 31-16, is the width of the curve at half the maximum value of $I$. Show that $\Delta\omega_d/\omega = R(3C/L)^{1/2}$, where $\omega$ is the angular frequency at resonance. Note that the ratio $\Delta\omega_d/\omega$ increases with $R$, as Fig. 31-16 shows.   **SSM**

**••50** An ac generator with $\mathscr{E}_m = 220$ V and operating at 400 Hz causes oscillations in a series $RLC$ circuit having $R = 220$ $\Omega$, $L = 150$ mH, and $C = 24.0$ $\mu$F. Find (a) the capacitive reactance $X_C$, (b) the impedance $Z$, and (c) the current amplitude $I$. A second capacitor of the same capacitance is then connected in series with the other components. Determine whether the values of (d) $X_C$, (e) $Z$, and (f) $I$ increase, decrease, or remain the same.   **GO**

**••51** In Fig. 31-33, a generator with an adjustable frequency of oscillation is connected to resistance $R = 100$ $\Omega$, inductances $L_1 = 1.70$ mH and $L_2 = 2.30$ mH, and capacitances $C_1 = 4.00\ \mu\text{F}$, $C_2 = 2.50\ \mu\text{F}$, and $C_3 = 3.50\ \mu\text{F}$. (a) What is the resonant frequency of the circuit? (*Hint:* See Problem 47 in Chapter 30.) What happens to the resonant frequency if (b) $R$ is increased, (c) $L_1$ is increased, and (d) $C_3$ is removed from the circuit?

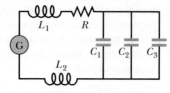

**FIG. 31-33** Problem 51.

### sec. 31-10 Power in Alternating-Current Circuits

•**52** What is the maximum value of an ac voltage whose rms value is 100 V?

•**53** What direct current will produce the same amount of thermal energy, in a particular resistor, as an alternating current that has a maximum value of 2.60 A?

•**54** An ac voltmeter with large impedance is connected in turn across the inductor, the capacitor, and the resistor in a series circuit having an alternating emf of 100 V (rms); the meter gives the same reading in volts in each case. What is this reading?

•**55** An air conditioner connected to a 120 V rms ac line is equivalent to a 12.0 Ω resistance and a 1.30 Ω inductive reactance in series. Calculate (a) the impedance of the air conditioner and (b) the average rate at which energy is supplied to the appliance. **SSM**

••**56** For Fig. 31-34, show that the average rate at which energy is dissipated in resistance $R$ is a maximum when $R$ is equal to the internal resistance $r$ of the ac generator. (In the text discussion we tacitly assumed that $r = 0$.)

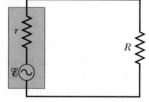

**FIG. 31-34** Problems 56 and 88.

••**57** Figure 31-35 shows an ac generator connected to a "black box" through a pair of terminals. The box contains an $RLC$ circuit, possibly even a multiloop circuit, whose elements and connections we do not know. Measurements outside the box reveal that

$$\mathscr{E}(t) = (75.0\ \text{V}) \sin \omega_d t$$

and

$$i(t) = (1.20\ \text{A}) \sin(\omega_d t + 42.0°).$$

(a) What is the power factor? (b) Does the current lead or lag the emf? (c) Is the circuit in the box largely inductive or largely capacitive? (d) Is the circuit in the box in resonance? (e) Must there be a capacitor in the box? (f) An inductor? (g) A resistor? (h) At what average rate is energy delivered to the box by the generator? (i) Why don't you need to know $\omega_d$ to answer all these questions? **SSM WWW**

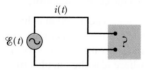

**FIG. 31-35** Problem 57.

••**58** In a series oscillating $RLC$ circuit, $R = 16.0$ Ω, $C = 31.2\ \mu\text{F}$, $L = 9.20$ mH, and $\mathscr{E} = \mathscr{E}_m \sin \omega_d t$ with $\mathscr{E}_m = 45.0$ V and $\omega_d = 3000$ rad/s. For time $t = 0.442$ ms find (a) the rate $P_g$ at which energy is being supplied by the generator, (b) the rate $P_C$ at which the energy in the capacitor is changing, (c) the rate $P_L$ at which the energy in the inductor is changing, and (d) the rate $P_R$ at which energy is being dissipated in the

resistor. (e) Is the sum of $P_C$, $P_L$, and $P_R$ greater than, less than, or equal to $P_g$? **GO**

••**59** In an $RLC$ circuit such as that of Fig. 31-7 assume that $R = 5.00$ Ω, $L = 60.0$ mH, $f_d = 60.0$ Hz, and $\mathscr{E}_m = 30.0$ V. For what values of the capacitance would the average rate at which energy is dissipated in the resistance be (a) a maximum and (b) a minimum? What are (c) the maximum dissipation rate and the corresponding (d) phase angle and (e) power factor? What are (f) the minimum dissipation rate and the corresponding (g) phase angle and (h) power factor?

••**60** A typical light dimmer used to dim the stage lights in a theater consists of a variable inductor $L$ (whose inductance is adjustable between zero and $L_{max}$) connected in series with a lightbulb B, as shown in Fig. 31-36. The electrical supply is 120 V (rms) at 60.0 Hz; the lightbulb is rated at 120 V, 1000 W. (a) What $L_{max}$ is required if the rate of energy dissipation in the lightbulb is to be varied by a factor of 5 from its upper limit of 1000 W? Assume that the resistance of the lightbulb is independent of its temperature. (b) Could one use a variable resistor (adjustable between zero and $R_{max}$) instead of an inductor? (c) If so, what $R_{max}$ is required? (d) Why isn't this done?

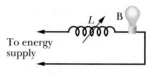

**FIG. 31-36** Problem 60.

••**61** In Fig. 31-7, $R = 15.0$ Ω, $C = 4.70\ \mu\text{F}$, and $L = 25.0$ mH. The generator provides an emf with rms voltage 75.0 V and frequency 550 Hz. (a) What is the rms current? What is the rms voltage across (b) $R$, (c) $C$, (d) $L$, (e) $C$ and $L$ together, and (f) $R$, $C$, and $L$ together? At what average rate is energy dissipated by (g) $R$, (h) $C$, and (i) $L$?

### sec. 31-11 Transformers

•**62** Figure 31-37 shows an "auto-transformer." It consists of a single coil (with an iron core). Three taps $T_i$ are provided. Between taps $T_1$ and $T_2$ there are 200 turns, and between taps $T_2$ and $T_3$ there are 800 turns. Any two taps can be chosen as the primary terminals, and any two taps can be chosen as the secondary terminals. For choices producing a step-up transformer, what are the (a) smallest, (b) second smallest, and (c) largest values of the ratio $V_s/V_p$? For a step-down transformer, what are the (d) smallest, (e) second smallest, and (f) largest values of $V_s/V_p$?

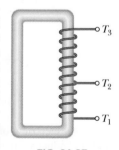

**FIG. 31-37** Problem 62.

•**63** A transformer has 500 primary turns and 10 secondary turns. (a) If $V_p$ is 120 V (rms), what is $V_s$ with an open circuit? If the secondary now has a resistive load of 15 Ω, what is the current in the (b) primary and (c) secondary? **SSM ILW**

•**64** A generator supplies 100 V to a transformer's primary coil, which has 50 turns. If the secondary coil has 500 turns, what is the secondary voltage?

••**65** An ac generator provides emf to a resistive load in a remote factory over a two-cable transmission line. At the factory a step-down transformer reduces the voltage from its (rms) transmission value $V_t$ to a much lower value that is safe and convenient for use in the factory. The transmission line resistance is 0.30 $\Omega$/cable, and the power of the generator is 250 kW. If $V_t$ = 80 kV, what are (a) the voltage decrease $\Delta V$ along the transmission line and (b) the rate $P_d$ at which energy is dissipated in the line as thermal energy? If $V_t$ = 8.0 kV, what are (c) $\Delta V$ and (d) $P_d$? If $V_t$ = 0.80 kV, what are (e) $\Delta V$ and (f) $P_d$?

### Additional Problems

**66** An electric motor connected to a 120 V, 60.0 Hz ac outlet does mechanical work at the rate of 0.100 hp (1 hp = 746 W). (a) If the motor draws an rms current of 0.650 A, what is its effective resistance, relative to power transfer? (b) Is this the same as the resistance of the motor's coils, as measured with an ohmmeter with the motor disconnected from the outlet?

**67** In Fig. 31-38, a three-phase generator G produces electrical power that is transmitted by means of three wires. The electric potentials (each relative to a common reference level) are $V_1 = A \sin \omega_d t$ for wire 1, $V_2 = A \sin(\omega_d t - 120°)$ for wire 2, and $V_3 = A \sin(\omega_d t - 240°)$ for wire 3. Some types of industrial equipment (for example, motors) have three terminals and are designed to be connected directly to these three wires. To use a more conventional two-terminal device (for example, a lightbulb), one connects it to any two of the three wires. Show that the potential difference between *any two* of the wires (a) oscillates sinusoidally with angular frequency $\omega_d$ and (b) has an amplitude of $A\sqrt{3}$.  **SSM**

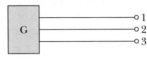

Three-wire transmission line

**FIG. 31-38** Problem 67.

**68** A 1.50 $\mu$F capacitor has a capacitive reactance of 12.0 $\Omega$. (a) What must be its operating frequency? (b) What will be the capacitive reactance if the frequency is doubled?

**69** For a certain driven series *RLC* circuit, the maximum generator emf is 125 V and the maximum current is 3.20 A. If the current leads the generator emf by 0.982 rad, what are the (a) impedance and (b) resistance of the circuit? (c) Is the circuit predominantly capacitive or inductive?

**70** An oscillating *LC* circuit has an inductance of 3.00 mH and a capacitance of 10.0 $\mu$F. Calculate the (a) angular frequency and (b) period of the oscillation. (c) At time $t = 0$, the capacitor is charged to 200 $\mu$C and the current is zero. Roughly sketch the charge on the capacitor as a function of time.

**71** In a certain series *RLC* circuit being driven at a frequency of 60.0 Hz, the maximum voltage across the inductor is 2.00 times the maximum voltage across the resistor and 2.00 times the maximum voltage across the capacitor. (a) By what angle does the current lag the generator emf? (b) If the maximum generator emf is 30.0 V, what should be the resistance of the circuit to obtain a maximum current of 300 mA?  **SSM**

**72** What capacitance would you connect across a 1.30 mH inductor to make the resulting oscillator resonate at 3.50 kHz?

**73** An *LC* circuit oscillates at a frequency of 10.4 kHz. (a) If the capacitance is 340 $\mu$F, what is the inductance? (b) If the maximum current is 7.20 mA, what is the total energy in the circuit? (c) What is the maximum charge on the capacitor?  **SSM**

**74** A series *RLC* circuit has a resonant frequency of 6.00 kHz. When it is driven at 8.00 kHz, it has an impedance of 1.00 k$\Omega$ and a phase constant of 45°. What are (a) $R$, (b) $L$, and (c) $C$ for this circuit?

**75** A capacitor of capacitance 158 $\mu$F and an inductor form an *LC* circuit that oscillates at 8.15 kHz, with a current amplitude of 4.21 mA. What are (a) the inductance, (b) the total energy in the circuit, and (c) the maximum charge on the capacitor?

**76** A series *RLC* circuit is driven in such a way that the maximum voltage across the inductor is 1.50 times the maximum voltage across the capacitor and 2.00 times the maximum voltage across the resistor. (a) What is $\phi$ for the circuit? (b) Is the circuit inductive, capacitive, or in resonance? The resistance is 49.9 $\Omega$, and the current amplitude is 200 mA. (c) What is the amplitude of the driving emf?

**77** An *RLC* circuit is driven by a generator with an emf amplitude of 80.0 V and a current amplitude of 1.25 A. The current leads the emf by 0.650 rad. What are the (a) impedance and (b) resistance of the circuit? (c) Is the circuit inductive, capacitive, or in resonance?

**78** A 45.0 mH inductor has a reactance of 1.30 k$\Omega$. (a) What is its operating frequency? (b) What is the capacitance of a capacitor with the same reactance at that frequency? If the frequency is doubled, what is the new reactance of (c) the inductor and (d) the capacitor?

**79** A generator of frequency 3000 Hz drives a series *RLC* circuit with an emf amplitude of 120 V. The resistance is 40.0 $\Omega$, the capacitance is 1.60 $\mu$F, and the inductance is 850 $\mu$H. What are (a) the phase constant in radians and (b) the current amplitude? (c) Is the circuit capacitive, inductive, or in resonance?

**80** A 1.50 mH inductor in an oscillating *LC* circuit stores a maximum energy of 10.0 $\mu$J. What is the maximum current?

**81** A generator with an adjustable frequency of oscillation is wired in series to an inductor of $L = 2.50$ mH and a capacitor of $C = 3.00$ $\mu$F. At what frequency does the generator produce the largest possible current amplitude in the circuit?

**82** A series *RLC* circuit is driven by an alternating source at a frequency of 400 Hz and an emf amplitude of 90.0 V. The resistance is 20.0 $\Omega$, the capacitance is 12.1 $\mu$F, and the inductance is 24.2 mH. What is the rms potential difference across (a) the resistor, (b) the capacitor, and (c) the inductor? (d) What is the average rate at which energy is dissipated?

**83** (a) In an oscillating *LC* circuit, in terms of the maximum charge $Q$ on the capacitor, what is the charge there when the energy in the electric field is 50.0% of that in the magnetic field? (b) What fraction of a period must elapse following the time the capacitor is fully charged for this condition to occur?  **SSM**

**84** When under load and operating at an rms voltage of 220 V, a certain electric motor draws an rms current of 3.00 A. It has a resistance of 24.0 $\Omega$ and no capacitive reactance. What is its inductive reactance?

**85** For a sinusoidally driven series *RLC* circuit, show that over one complete cycle with period $T$ (a) the energy stored in the capacitor does not change; (b) the energy stored in the inductor does not change; (c) the driving emf device supplies energy $(\frac{1}{2}T)\mathscr{E}_m I \cos \phi$; and (d) the resistor dissipates energy $(\frac{1}{2}T)RI^2$. (e) Show that the quantities found in (c) and (d) are equal.  **SSM**

**86** When the generator emf in Sample Problem 31-7 is a maximum, what is the voltage across (a) the generator, (b) the resistance, (c) the capacitance, and (d) the inductance? (e) By summing these with appropriate signs, verify that the loop rule is satisfied.

**87** An ac generator produces emf $\mathscr{E} = \mathscr{E}_m \sin(\omega_d t - \pi/4)$, where $\mathscr{E}_m = 30.0$ V and $\omega_d = 350$ rad/s. The current in the circuit attached to the generator is $i(t) = I \sin(\omega_d t + \pi/4)$, where $I = 620$ mA. (a) At what time after $t = 0$ does the generator emf first reach a maximum? (b) At what time after $t = 0$ does the current first reach a maximum? (c) The circuit contains a single element other than the generator. Is it a capacitor, an inductor, or a resistor? Justify your answer. (d) What is the value of the capacitance, inductance, or resistance, as the case may be?

**88** In Fig. 31-34, let the rectangular box on the left represent the (high-impedance) output of an audio amplifier, with $r = 1000 \ \Omega$. Let $R = 10 \ \Omega$ represent the (low-impedance) coil of a loudspeaker. For maximum transfer of energy to the load $R$ we must have $R = r$, and that is not true in this case. However, a transformer can be used to "transform" resistances, making them behave electrically as if they were larger or smaller than they actually are. (a) Sketch the primary and secondary coils of a transformer that can be introduced between the amplifier and the speaker in Fig. 31-34 to match the impedances. (b) What must be the turns ratio?

**89** A series circuit with resistor–inductor–capacitor combination $R_1, L_1, C_1$ has the same resonant frequency as a second circuit with a different combination $R_2, L_2, C_2$. You now connect the two combinations in series. Show that this new circuit has the same resonant frequency as the separate circuits.

**90** Consider the circuit shown in Fig. 31-39. With switch $S_1$ closed and the other two switches open, the circuit has a time constant $\tau_C$. With switch $S_2$ closed and the other two switches open, the circuit has a time constant $\tau_L$. With switch $S_3$ closed and the other two switches open, the circuit oscillates with a period $T$. Show that $T = 2\pi\sqrt{\tau_C \tau_L}$.

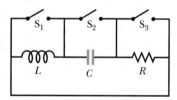

FIG. 31-39 Problem 90.

**91** The ac generator in Fig. 31-40 supplies 120 V at 60.0 Hz. With the switch open as in the diagram, the current leads the generator emf by 20.0°. With the switch in position 1, the current lags the generator emf by 10.0°. When the switch is in position 2, the current amplitude is 2.00 A. What are (a) $R$, (b) $L$, and (c) $C$?

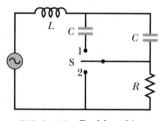

FIG. 31-40 Problem 91.

**92** In an oscillating $LC$ circuit, $L = 8.00$ mH and $C = 1.40 \ \mu$F. At time $t = 0$, the current is maximum at 12.0 mA. (a) What is the maximum charge on the capacitor during the oscillations? (b) At what earliest time $t > 0$ is the rate of change of energy in the capacitor maximum? (c) What is that maximum rate of change?

**93** A series $RLC$ circuit is driven by a generator at frequency 1050 Hz. The inductance is 90.0 mH; the capacitance is 0.500 $\mu$F; and the phase constant has a magnitude of 60.0° (you should supply the appropriate sign for the angle). (a) What is the resistance? To increase the current amplitude in the circuit, should we increase or decrease (b) the driving frequency, (c) the inductance, and (d) the capacitance?

**94** A series $RLC$ circuit is driven by a generator at a frequency of 2000 Hz and an emf amplitude of 170 V. The inductance is 60.0 mH, the capacitance is 0.400 $\mu$F, and the resistance is 200 $\Omega$. (a) What is the phase constant in radians? (b) What is the current amplitude?

**95** A 7.00 $\mu$F capacitor has an initial potential of 12.0 V when it is connected across an inductor. The combination then oscillates at a frequency of 715 Hz. What is the inductance of the inductor?

**96** An $LC$ oscillator consists of a 2.00 nF capacitor and a 2.00 mH inductor. The maximum voltage is 4.00 V. What are (a) the frequency of the oscillations, (b) the maximum current, (c) the maximum energy stored in the inductor, and (d) the maximum rate $di/dt$ at which the current changes?

**97** Figure 31-41 shows an $RLC$ circuit that is driven by an emf source of fixed amplitude $\mathscr{E}_m$. Initially the circuit consists of one resistor of resistance $R$, one inductor of inductance $L$, and one capacitor of capacitance $C$, and the driving frequency matches the natural frequency. Then switches $S_1, S_2, S_3$, and $S_4$ are closed, in that order. The closings bring in capacitors identical to the first one or resistors identical to the first one.

Let $\mathscr{E}_m = 12.0$ V, $C = 2.00 \ \mu$F, $L = 2.00$ mH, and $R = 12.0 \ \Omega$. (a) Fill in the first blank column of the following table with the values of the equivalent capacitance $C_{eq}$ of the circuit after each switch closing. Similarly, fill in the other columns with values for (b) the resonance frequency $f$, (c) the equivalent resistance $R_{eq}$, (d) the impedance $Z$, and (e) the current amplitude $I$. (Avoid rounding off the numbers until all calculations are finished.)

| Closing | $C_{eq}$ | $f$ | $R_{eq}$ | $Z$ | $I$ |
|---|---|---|---|---|---|
| $S_1$ | | | | | |
| $S_2$ | | | | | |
| $S_3$ | | | | | |
| $S_4$ | | | | | |

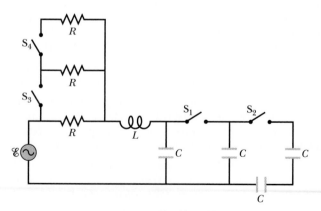

FIG. 31-41 Problem 97.

# Maxwell's Equations; Magnetism of Matter

Scala/Art Resource

*Because Earth's magnetic field gradually but continuously changes, the direction of north indicated by a compass also changes. For many reasons, researchers want to know the direction of north at specific times in the past, but finding historic records of compass readings is rare. However, certain paintings can help. For example, these murals in a hall of the Vatican (Bibliotheca Apostolica Vaticana) faithfully recorded the direction of north when they were painted in 1740.*

## How can a painting record the direction of Earth's magnetic field?

The answer is in this chapter.

FIG. 32-1 Some of the inks used for tattoos contain magnetic particles. *(Oliver Strewe/Getty Images, Inc.)*

## 32-1 | WHAT IS PHYSICS?

This chapter reveals some of the breadth of physics because it ranges from the basic science of electric and magnetic fields to the applied science and engineering of magnetic materials. First, we conclude our basic discussion of electric and magnetic fields, finding that most of the physics principles in the last 11 chapters can be summarized in only *four* equations, known as Maxwell's equations.

Second, we examine the science and engineering of magnetic materials. The careers of many scientists and engineers are focused on understanding why some materials are magnetic and others are not and on how existing magnetic materials can be improved. These researchers wonder why Earth has a magnetic field but you do not. They find countless applications for inexpensive magnetic materials in cars, kitchens, offices, and hospitals, and magnetic materials often show up in unexpected ways. For example, if you have a tattoo (Fig. 32-1) and undergo an MRI (magnetic resonance imaging) scan, the large magnetic field used in the scan may noticeably tug on your tattooed skin because some tattoo inks contain magnetic particles. In another example, some breakfast cereals are advertised as being "iron fortified" because they contain small bits of iron for you to ingest. Because these iron bits are magnetic, you can collect them by passing a magnet over a slurry of water and cereal.

Our first step here is to revisit Gauss' law, but this time for magnetic fields.

## 32-2 | Gauss' Law for Magnetic Fields

Figure 32-2 shows iron powder that has been sprinkled onto a transparent sheet placed above a bar magnet. The powder grains, trying to align themselves with the magnet's magnetic field, have fallen into a pattern that reveals the field. One end of the magnet is a *source* of the field (the field lines diverge from it) and the other end is a *sink* of the field (the field lines converge toward it). By convention, we call the source the *north pole* of the magnet and the sink the *south pole*, and we say that the magnet, with its two poles, is an example of a **magnetic dipole.**

Suppose we break a bar magnet into pieces the way we can break a piece of chalk (Fig. 32-3). We should, it seems, be able to isolate a single magnetic pole, called a *magnetic monopole.* However, we cannot—not even if we break the magnet down to its individual atoms and then to its electrons and nuclei. Each

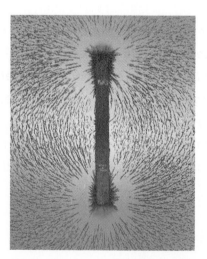

FIG. 32-2 A bar magnet is a magnetic dipole. The iron filings suggest the magnetic field lines. (Colored light fills the background.) *(Runk/Schoenberger/Grant Heilman Photography)*

FIG. 32-3 If you break a magnet, each fragment becomes a separate magnet, with its own north and south poles.

fragment has a north pole and a south pole. Thus:

> The simplest magnetic structure that can exist is a magnetic dipole. Magnetic monopoles do not exist (as far as we know).

Gauss' law for magnetic fields is a formal way of saying that magnetic monopoles do not exist. The law asserts that the net magnetic flux $\Phi_B$ through any closed Gaussian surface is zero:

$$\Phi_B = \oint \vec{B} \cdot d\vec{A} = 0 \qquad \text{(Gauss' law for magnetic fields).} \qquad (32\text{-}1)$$

Contrast this with Gauss' law for electric fields,

$$\Phi_E = \oint \vec{E} \cdot d\vec{A} = \frac{q_{enc}}{\varepsilon_0} \qquad \text{(Gauss' law for electric fields).}$$

In both equations, the integral is taken over a *closed* Gaussian surface. Gauss' law for electric fields says that this integral (the net electric flux through the surface) is proportional to the net electric charge $q_{enc}$ enclosed by the surface. Gauss' law for magnetic fields says that there can be no net magnetic flux through the surface because there can be no net "magnetic charge" (individual magnetic poles) enclosed by the surface. The simplest magnetic structure that can exist and thus be enclosed by a Gaussian surface is a dipole, which consists of both a source and a sink for the field lines. Thus, there must always be as much magnetic flux into the surface as out of it, and the net magnetic flux must always be zero.

Gauss' law for magnetic fields holds for structures more complicated than a magnetic dipole, and it holds even if the Gaussian surface does not enclose the entire structure. Gaussian surface II near the bar magnet of Fig. 32-4 encloses no poles, and we can easily conclude that the net magnetic flux through it is zero. Gaussian surface I is more difficult. It may seem to enclose only the north pole of the magnet because it encloses the label N and not the label S. However, a south pole must be associated with the lower boundary of the surface because magnetic field lines enter the surface there. (The enclosed section is like one piece of the broken bar magnet in Fig. 32-3.) Thus, Gaussian surface I encloses a magnetic dipole, and the net flux through the surface is zero.

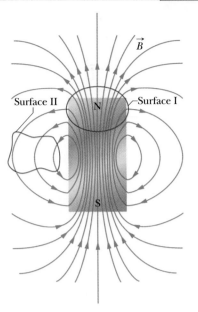

**FIG. 32-4** The field lines for the magnetic field $\vec{B}$ of a short bar magnet. The red curves represent cross sections of closed, three-dimensional Gaussian surfaces.

✓ **CHECKPOINT 1**    The figure here shows four closed surfaces with flat top and bottom faces and curved sides. The table gives the areas $A$ of the faces and the magnitudes $B$ of the uniform and perpendicular magnetic fields through those faces; the units of $A$ and $B$ are arbitrary but consistent. Rank the surfaces according to the magnitudes of the magnetic flux through their curved sides, greatest first.

| Surface | $A_{top}$ | $B_{top}$ | $A_{bot}$ | $B_{bot}$ |
|---------|-----------|-----------|-----------|-----------|
| $a$ | 2 | 6, outward | 4 | 3, inward |
| $b$ | 2 | 1, inward | 4 | 2, inward |
| $c$ | 2 | 6, inward | 2 | 8, outward |
| $d$ | 2 | 3, outward | 3 | 2, outward |

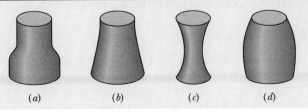

(a)      (b)      (c)      (d)

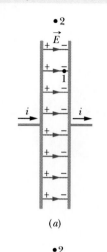

(a)

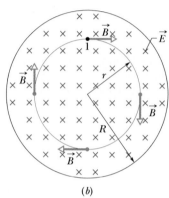

(b)

FIG. 32-5 (a) A circular parallel-plate capacitor, shown in side view, is being charged by a constant current $i$. (b) A view from within the capacitor, looking toward the plate at the right in (a). The electric field $\vec{E}$ is uniform, is directed into the page (toward the plate), and grows in magnitude as the charge on the capacitor increases. The magnetic field $\vec{B}$ induced by this changing electric field is shown at four points on a circle with a radius $r$ less than the plate radius $R$.

## 32-3 | Induced Magnetic Fields

In Chapter 30 you saw that a changing magnetic flux induces an electric field, and we ended up with Faraday's law of induction in the form

$$\oint \vec{E} \cdot d\vec{s} = -\frac{d\Phi_B}{dt} \qquad \text{(Faraday's law of induction).} \qquad (32\text{-}2)$$

Here $\vec{E}$ is the electric field induced along a closed loop by the changing magnetic flux $\Phi_B$ encircled by that loop. Because symmetry is often so powerful in physics, we should be tempted to ask whether induction can occur in the opposite sense; that is, can a changing electric flux induce a magnetic field?

The answer is that it can; furthermore, the equation governing the induction of a magnetic field is almost symmetric with Eq. 32-2. We often call it Maxwell's law of induction after James Clerk Maxwell, and we write it as

$$\oint \vec{B} \cdot d\vec{s} = \mu_0 \varepsilon_0 \frac{d\Phi_E}{dt} \qquad \text{(Maxwell's law of induction).} \qquad (32\text{-}3)$$

Here $\vec{B}$ is the magnetic field induced along a closed loop by the changing electric flux $\Phi_E$ in the region encircled by that loop.

As an example of this sort of induction, we consider the charging of a parallel-plate capacitor with circular plates (Fig. 32-5a). (Although we shall focus on this arrangement, a changing electric flux will always induce a magnetic field whenever it occurs.) We assume that the charge on the capacitor is being increased at a steady rate by a constant current $i$ in the connecting wires. Then the electric field magnitude between the plates must also be increasing at a steady rate.

Figure 32-5b is a view of the right-hand plate of Fig. 32-5a from between the plates. The electric field is directed into the page. Let us consider a circular loop through point 1 in Figs. 32-5a and b, a loop that is concentric with the capacitor plates and has a radius smaller than that of the plates. Because the electric field through the loop is changing, the electric flux through the loop must also be changing. According to Eq. 32-3, this changing electric flux induces a magnetic field around the loop.

Experiment proves that a magnetic field $\vec{B}$ is indeed induced around such a loop, directed as shown. This magnetic field has the same magnitude at every point around the loop and thus has circular symmetry about the *central axis* of the capacitor plates (the axis extending from one plate center to the other).

If we now consider a larger loop—say, through point 2 outside the plates in Figs. 32-5a and b—we find that a magnetic field is induced around that loop as well. Thus, while the electric field is changing, magnetic fields are induced between the plates, both inside and outside the gap. When the electric field stops changing, these induced magnetic fields disappear.

Although Eq. 32-3 is similar to Eq. 32-2, the equations differ in two ways. First, Eq. 32-3 has the two extra symbols $\mu_0$ and $\varepsilon_0$, but they appear only because we employ SI units. Second, Eq. 32-3 lacks the minus sign of Eq. 32-2, meaning that the induced electric field $\vec{E}$ and the induced magnetic field $\vec{B}$ have opposite directions when they are produced in otherwise similar situations. To see this opposition, examine Fig. 32-6, in which an increasing magnetic field $\vec{B}$, directed into the page, induces an electric field $\vec{E}$. The induced field $\vec{E}$ is counterclockwise, opposite the induced magnetic field $\vec{B}$ in Fig. 32-5b.

### Ampere–Maxwell Law

Now recall that the left side of Eq. 32-3, the integral of the dot product $\vec{B} \cdot d\vec{s}$ around a closed loop, appears in another equation—namely, Ampere's law:

$$\oint \vec{B} \cdot d\vec{s} = \mu_0 i_{\text{enc}} \qquad \text{(Ampere's law),} \qquad (32\text{-}4)$$

where $i_{enc}$ is the current encircled by the closed loop. Thus, our two equations that specify the magnetic field $\vec{B}$ produced by means other than a magnetic material (that is, by a current and by a changing electric field) give the field in exactly the same form. We can combine the two equations into the single equation

$$\oint \vec{B} \cdot d\vec{s} = \mu_0 \varepsilon_0 \frac{d\Phi_E}{dt} + \mu_0 i_{enc} \qquad \text{(Ampere–Maxwell law)}. \qquad (32\text{-}5)$$

When there is a current but no change in electric flux (such as with a wire carrying a constant current), the first term on the right side of Eq. 32-5 is zero, and so Eq. 32-5 reduces to Eq. 32-4, Ampere's law. When there is a change in electric flux but no current (such as inside or outside the gap of a charging capacitor), the second term on the right side of Eq. 32-5 is zero, and so Eq. 32-5 reduces to Eq. 32-3, Maxwell's law of induction.

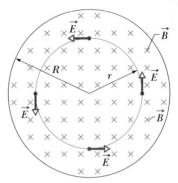

**FIG. 32-6** A uniform magnetic field $\vec{B}$ in a circular region. The field, directed into the page, is increasing in magnitude. The electric field $\vec{E}$ induced by the changing magnetic field is shown at four points on a circle concentric with the circular region. Compare this situation with that of Fig. 32-5b.

✓ **CHECKPOINT 2** The figure shows graphs of the electric field magnitude $E$ versus time $t$ for four uniform electric fields, all contained within identical circular regions as in Fig. 32-5b. Rank the fields according to the magnitudes of the magnetic fields they induce at the edge of the region, greatest first.

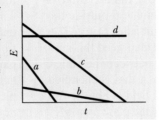

---

**Sample Problem** 32-1

A parallel-plate capacitor with circular plates of radius $R$ is being charged as in Fig. 32-5a.

**(a)** Derive an expression for the magnetic field at radius $r$ for the case $r \leq R$.

**KEY IDEAS** A magnetic field can be set up by a current and by induction due to a changing electric flux; both effects are included in Eq. 32-5. There is no current between the capacitor plates of Fig. 32-5, but the electric flux there is changing. Thus, Eq. 32-5 reduces to

$$\oint \vec{B} \cdot d\vec{s} = \mu_0 \varepsilon_0 \frac{d\Phi_E}{dt}. \qquad (32\text{-}6)$$

We shall separately evaluate the left and right sides of this equation.

**Left side of Eq. 32-6:** We choose a circular Amperian loop with a radius $r \leq R$ as shown in Fig. 32-5b because we want to evaluate the magnetic field for $r \leq R$—that is, inside the capacitor. The magnetic field $\vec{B}$ at all points along the loop is tangent to the loop, as is the path element $d\vec{s}$. Thus, $\vec{B}$ and $d\vec{s}$ are either parallel or antiparallel at each point of the loop. For simplicity, assume they are parallel (the choice does not alter our outcome here). Then

$$\oint \vec{B} \cdot d\vec{s} = \oint B \, ds \cos 0° = \oint B \, ds.$$

Due to the circular symmetry of the plates, we can also assume that $\vec{B}$ has the same magnitude at every point

around the loop. Thus, $B$ can be taken outside the integral on the right side of the above equation. The integral that remains is $\oint ds$, which simply gives the circumference $2\pi r$ of the loop. The left side of Eq. 32-6 is then $(B)(2\pi r)$.

**Right side of Eq. 32-6:** We assume that the electric field $\vec{E}$ is uniform between the capacitor plates and directed perpendicular to the plates. Then the electric flux $\Phi_E$ through the Amperian loop is $EA$, where $A$ is the area encircled by the loop within the electric field. Thus, the right side of Eq. 32-6 is $\mu_0 \varepsilon_0 \, d(EA)/dt$.

**Combining results:** Substituting our results for the left and right sides into Eq. 32-6, we get

$$(B)(2\pi r) = \mu_0 \varepsilon_0 \frac{d(EA)}{dt}.$$

Because $A$ is a constant, we write $d(EA)$ as $A \, dE$; so we have

$$(B)(2\pi r) = \mu_0 \varepsilon_0 A \frac{dE}{dt}. \qquad (32\text{-}7)$$

The area $A$ that is encircled by the Amperian loop within the electric field is the *full* area $\pi r^2$ of the loop because the loop's radius $r$ is less than (or equal to) the plate radius $R$. Substituting $\pi r^2$ for $A$ in Eq. 32-7 and solving the result for $B$ give us, for $r \leq R$,

$$B = \frac{\mu_0 \varepsilon_0 r}{2} \frac{dE}{dt}. \qquad \text{(Answer)} \quad (32\text{-}8)$$

This equation tells us that, inside the capacitor, $B$ increases linearly with increased radial distance $r$, from 0 at the central axis to a maximum value at plate radius $R$.

(b) Evaluate the field magnitude $B$ for $r = R/5 = 11.0$ mm and $dE/dt = 1.50 \times 10^{12}$ V/m·s.

**Calculation:** From the answer to (a), we have

$$B = \frac{1}{2} \mu_0 \varepsilon_0 r \frac{dE}{dt}$$

$$= \tfrac{1}{2}(4\pi \times 10^{-7}\,\text{T} \cdot \text{m/A})(8.85 \times 10^{-12}\,\text{C}^2/\text{N} \cdot \text{m}^2)$$

$$\times (11.0 \times 10^{-3}\,\text{m})(1.50 \times 10^{12}\,\text{V/m} \cdot \text{s})$$

$$= 9.18 \times 10^{-8}\,\text{T}. \qquad \text{(Answer)}$$

(c) Derive an expression for the induced magnetic field for the case $r \geq R$.

**Calculation:** Our procedure is the same as in (a) except we now use an Amperian loop with a radius $r$ that is greater than the plate radius $R$, to evaluate $B$ outside the capacitor. Evaluating the left and right sides of Eq. 32-6 again leads to Eq. 32-7. However, we then need this subtle point: The electric field exists only between the plates, not outside the plates. Thus, the area $A$ that is encircled by the Amperian loop in the electric field is *not* the full area $\pi r^2$ of the loop. Rather, $A$ is only the plate area $\pi R^2$.

Substituting $\pi R^2$ for $A$ in Eq. 32-7 and solving the result for $B$ give us, for $r \geq R$,

$$B = \frac{\mu_0 \varepsilon_0 R^2}{2r} \frac{dE}{dt}. \qquad \text{(Answer)} \quad (32\text{-}9)$$

This equation tells us that, outside the capacitor, $B$ decreases with increased radial distance $r$, from a maximum value at the plate edges (where $r = R$). By substituting $r = R$ into Eqs. 32-8 and 32-9, you can show that these equations are consistent; that is, they give the same maximum value of $B$ at the plate radius.

The magnitude of the induced magnetic field calculated in (b) is so small that it can scarcely be measured with simple apparatus. This is in sharp contrast to the magnitudes of induced electric fields (Faraday's law), which can be measured easily. This experimental difference exists partly because induced emfs can easily be multiplied by using a coil of many turns. No technique of comparable simplicity exists for multiplying induced magnetic fields. In any case, the experiment suggested by this sample problem has been done, and the presence of the induced magnetic fields has been verified quantitatively.

## 32-4 | Displacement Current

If you compare the two terms on the right side of Eq. 32-5, you will see that the product $\varepsilon_0(d\Phi_E/dt)$ must have the dimension of a current. In fact, that product has been treated as being a fictitious current called the **displacement current** $i_d$:

$$i_d = \varepsilon_0 \frac{d\Phi_E}{dt} \quad \text{(displacement current).} \quad (32\text{-}10)$$

"Displacement" is poorly chosen in that nothing is being displaced, but we are stuck with the word. Nevertheless, we can now rewrite Eq. 32-5 as

$$\oint \vec{B} \cdot d\vec{s} = \mu_0 i_{d,\text{enc}} + \mu_0 i_{\text{enc}} \quad \text{(Ampere–Maxwell law),} \quad (32\text{-}11)$$

in which $i_{d,\text{enc}}$ is the displacement current that is encircled by the integration loop.

Let us again focus on a charging capacitor with circular plates, as in Fig. 32-7a. The real current $i$ that is charging the plates changes the electric field $\vec{E}$ between the plates. The fictitious displacement current $i_d$ between the plates is associated with that changing field $\vec{E}$. Let us relate these two currents.

The charge $q$ on the plates at any time is related to the magnitude $E$ of the field between the plates at that time by Eq. 25-4:

$$q = \varepsilon_0 A E, \quad (32\text{-}12)$$

in which $A$ is the plate area. To get the real current $i$, we differentiate Eq. 32-12 with respect to time, finding

$$\frac{dq}{dt} = i = \varepsilon_0 A \frac{dE}{dt}. \quad (32\text{-}13)$$

To get the displacement current $i_d$, we can use Eq. 32-10. Assuming that the electric field $\vec{E}$ between the two plates is uniform (we neglect any fringing), we can

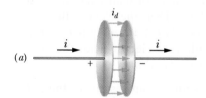

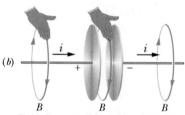

$B$      $B$      $B$
Field due   Field due   Field due
to current $i$   to current $i_d$   to current $i$

**FIG. 32-7** (*a*) The displacement current $i_d$ between the plates of a capacitor that is being charged by a current $i$. (*b*) The right-hand rule for finding the direction of the magnetic field around a wire with a real current (as at the left) also gives the direction of the magnetic field around a displacement current (as in the center).

replace the electric flux $\Phi_E$ in that equation with $EA$. Then Eq. 32-10 becomes

$$i_d = \varepsilon_0 \frac{d\Phi_E}{dt} = \varepsilon_0 \frac{d(EA)}{dt} = \varepsilon_0 A \frac{dE}{dt}. \qquad (32\text{-}14)$$

Comparing Eqs. 32-13 and 32-14, we see that the real current $i$ charging the capacitor and the fictitious displacement current $i_d$ between the plates have the same magnitude:

$$i_d = i \qquad \text{(displacement current in a capacitor)}. \qquad (32\text{-}15)$$

Thus, we can consider the fictitious displacement current $i_d$ to be simply a continuation of the real current $i$ from one plate, across the capacitor gap, to the other plate. Because the electric field is uniformly spread over the plates, the same is true of this fictitious displacement current $i_d$, as suggested by the spread of current arrows in Fig. 32-7a. Although no charge actually moves across the gap between the plates, the idea of the fictitious current $i_d$ can help us to quickly find the direction and magnitude of an induced magnetic field, as follows.

## Finding the Induced Magnetic Field

In Chapter 29 we found the direction of the magnetic field produced by a real current $i$ by using the right-hand rule of Fig. 29-4. We can apply the same rule to find the direction of an induced magnetic field produced by a fictitious displacement current $i_d$, as is shown in the center of Fig. 32-7b for a capacitor.

We can also use $i_d$ to find the magnitude of the magnetic field induced by a charging capacitor with parallel circular plates of radius $R$. We simply consider the space between the plates to be an imaginary circular wire of radius $R$ carrying the imaginary current $i_d$. Then, from Eq. 29-20, the magnitude of the magnetic field at a point inside the capacitor at radius $r$ from the center is

$$B = \left(\frac{\mu_0 i_d}{2\pi R^2}\right) r \qquad \text{(inside a circular capacitor)}. \qquad (32\text{-}16)$$

Similarly, from Eq. 29-17, the magnitude of the magnetic field at a point outside the capacitor at radius $r$ is

$$B = \frac{\mu_0 i_d}{2\pi r} \qquad \text{(outside a circular capacitor)}. \qquad (32\text{-}17)$$

✓ **CHECKPOINT 3**   The figure is a view of one plate of a parallel-plate capacitor from within the capacitor. The dashed lines show four integration paths (path $b$ follows the edge of the plate). Rank the paths according to the magnitude of $\oint \vec{B} \cdot d\vec{s}$ along the paths during the discharging of the capacitor, greatest first.

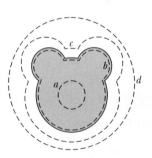

---

**Sample Problem**   **32-2**

The circular parallel-plate capacitor in Sample Problem 32-1 is being charged with a current $i$.

**(a)** Between the plates, what is the magnitude of $\oint \vec{B} \cdot d\vec{s}$, in terms of $\mu_0$ and $i$, at a radius $r = R/5$ from their center?

**KEY IDEA**   A magnetic field can be set up by a current and by induction due to a changing electric flux; both effects are included in Eq. 32-5. There is no current be-

tween the capacitor plates of Fig. 32-5, but the electric flux there is changing. However, now we can replace the product $\varepsilon_0\, d\Phi_E/dt$ in Eq. 32-5 with a fictitious displacement current $i_d$. Then integral $\oint \vec{B} \cdot d\vec{s}$ is given by Eq. 32-11, but because there is no real current $i$ between the capacitor plates, the equation reduces to

$$\oint \vec{B} \cdot d\vec{s} = \mu_0 i_{d,\text{enc}}. \qquad (32\text{-}18)$$

**Calculations:** Because we want to evaluate $\oint \vec{B} \cdot d\vec{s}$ at radius $r = R/5$ (within the capacitor), the integration loop encircles only a portion $i_{d,enc}$ of the total displacement current $i_d$. Let's assume that $i_d$ is uniformly spread over the full plate area. Then the portion of the displacement current encircled by the loop is proportional to the area encircled by the loop:

$$\frac{\left(\begin{array}{c}\text{encircled displacement}\\ \text{current } i_{d,enc}\end{array}\right)}{\left(\begin{array}{c}\text{total displacement}\\ \text{current } i_d\end{array}\right)} = \frac{\text{encircled area } \pi r^2}{\text{full plate area } \pi R^2}.$$

This gives us

$$i_{d,enc} = i_d \frac{\pi r^2}{\pi R^2}.$$

Substituting this into Eq. 32-18, we obtain

$$\oint \vec{B} \cdot d\vec{s} = \mu_0 i_d \frac{\pi r^2}{\pi R^2}. \qquad (32\text{-}19)$$

Now substituting $i_d = i$ (from Eq. 32-15) and $r = R/5$ into Eq. 32-19 leads to

$$\oint \vec{B} \cdot d\vec{s} = \mu_0 i \frac{(R/5)^2}{R^2} = \frac{\mu_0 i}{25}. \qquad \text{(Answer)}$$

**(b)** In terms of the maximum induced magnetic field, what is the magnitude of the magnetic field induced at $r = R/5$, inside the capacitor?

**KEY IDEA** Because the capacitor has parallel circular plates, we can treat the space between the plates as an imaginary wire of radius $R$ carrying the imaginary current $i_d$. Then we can use Eq. 32-16 to find the induced magnetic field magnitude $B$ at any point inside the capacitor.

**Calculations:** At $r = R/5$, Eq. 32-16 yields

$$B = \left(\frac{\mu_0 i_d}{2\pi R^2}\right) r = \frac{\mu_0 i_d (R/5)}{2\pi R^2} = \frac{\mu_0 i_d}{10\pi R}. \qquad (32\text{-}20)$$

From Eq. 32-16, the maximum field magnitude $B_{max}$ within the capacitor occurs at $r = R$. It is

$$B_{max} = \left(\frac{\mu_0 i_d}{2\pi R^2}\right) R = \frac{\mu_0 i_d}{2\pi R}. \qquad (32\text{-}21)$$

Dividing Eq. 32-20 by Eq. 32-21 and rearranging the result, we find

$$B = \frac{B_{max}}{5}. \qquad \text{(Answer)}$$

We should be able to obtain this result with a little reasoning and less work. Equation 32-16 tells us that inside the capacitor, $B$ increases linearly with $r$. Therefore, a point $\frac{1}{5}$ the distance out to the full radius $R$ of the plates, where $B_{max}$ occurs, should have a field $B$ that is $\frac{1}{5}B_{max}$.

# 32-5 | Maxwell's Equations

Equation 32-5 is the last of the four fundamental equations of electromagnetism, called *Maxwell's equations* and displayed in Table 32-1. These four equations explain a diverse range of phenomena, from why a compass needle points north to why a car starts when you turn the ignition key. They are the basis for the functioning of such electromagnetic devices as electric motors, television transmitters and receivers, telephones, fax machines, radar, and microwave ovens.

Maxwell's equations are the basis from which many of the equations you have seen since Chapter 21 can be derived. They are also the basis of many of the equations you will see in Chapters 33 through 36 concerning optics.

**TABLE 32-1**

**Maxwell's Equations**[a]

| Name | Equation | |
|---|---|---|
| Gauss' law for electricity | $\oint \vec{E} \cdot d\vec{A} = q_{enc}/\varepsilon_0$ | Relates net electric flux to net enclosed electric charge |
| Gauss' law for magnetism | $\oint \vec{B} \cdot d\vec{A} = 0$ | Relates net magnetic flux to net enclosed magnetic charge |
| Faraday's law | $\oint \vec{E} \cdot d\vec{s} = -\dfrac{d\Phi_B}{dt}$ | Relates induced electric field to changing magnetic flux |
| Ampere–Maxwell law | $\oint \vec{B} \cdot d\vec{s} = \mu_0 \varepsilon_0 \dfrac{d\Phi_E}{dt} + \mu_0 i_{enc}$ | Relates induced magnetic field to changing electric flux and to current |

[a]Written on the assumption that no dielectric or magnetic materials are present.

# 32-6 | Magnets

The first known magnets were *lodestones,* which are stones that have been *magnetized* (made magnetic) naturally. When the ancient Greeks and ancient Chinese discovered these rare stones, they were amused by the stones' ability to attract metal over a short distance, as if by magic. Only much later did they learn to use lodestones (and artificially magnetized pieces of iron) in compasses to determine direction.

Today, magnets and magnetic materials are ubiquitous. Their magnetic properties can be traced to their atoms and electrons. In fact, the inexpensive magnet you might use to hold a note on the refrigerator door is a direct result of the quantum physics taking place in the atomic and subatomic material within the magnet. Before we explore some of this physics, let's briefly discuss the largest magnet we commonly use—namely, Earth itself.

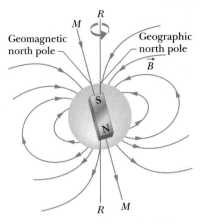

**FIG. 32-8** Earth's magnetic field represented as a dipole field. The dipole axis *MM* makes an angle of 11.5° with Earth's rotational axis *RR*. The south pole of the dipole is in Earth's Northern Hemisphere.

## The Magnetism of Earth

Earth is a huge magnet; for points near Earth's surface, its magnetic field can be approximated as the field of a huge bar magnet—a magnetic dipole—that straddles the center of the planet. Figure 32-8 is an idealized symmetric depiction of the dipole field, without the distortion caused by passing charged particles from the Sun.

Because Earth's magnetic field is that of a magnetic dipole, a magnetic dipole moment $\vec{\mu}$ is associated with the field. For the idealized field of Fig. 32-8, the magnitude of $\vec{\mu}$ is $8.0 \times 10^{22}$ J/T and the direction of $\vec{\mu}$ makes an angle of 11.5° with the rotation axis (*RR*) of Earth. The *dipole axis* (*MM* in Fig. 32-8) lies along $\vec{\mu}$ and intersects Earth's surface at the *geomagnetic north pole* in northwest Greenland and the *geomagnetic south pole* in Antarctica. The lines of the magnetic field $\vec{B}$ generally emerge in the Southern Hemisphere and reenter Earth in the Northern Hemisphere. Thus, the magnetic pole that is in Earth's Northern Hemisphere and known as a "north magnetic pole" *is really the south pole of Earth's magnetic dipole.*

The direction of the magnetic field at any location on Earth's surface is commonly specified in terms of two angles. The **field declination** is the angle (left or right) between geographic north (which is toward 90° latitude) and the horizontal component of the field. The **field inclination** is the angle (up or down) between a horizontal plane and the field's direction.

*Magnetometers* measure these angles and determine the field with much precision. However, you can do reasonably well with just a *compass* and a *dip meter.* A compass is simply a needle-shaped magnet that is mounted so it can rotate freely about a vertical axis. When it is held in a horizontal plane, the north-pole end of the needle points, generally, toward the geomagnetic north pole (really a south magnetic pole, remember). The angle between the needle and geographic north is the field declination. A dip meter is a similar magnet that can rotate freely about a horizontal axis. When its vertical plane of rotation is aligned with the direction of the compass, the angle between the meter's needle and the horizontal is the field inclination.

At any point on Earth's surface, the measured magnetic field may differ appreciably, in both magnitude and direction, from the idealized dipole field of Fig. 32-8. In fact, the point where the field is actually perpendicular to Earth's surface and inward is not located at the geomagnetic north pole in Greenland as we would expect; instead, this so-called *dip north pole* is located in the Queen Elizabeth Islands in northern Canada, far from Greenland.

In addition, the field observed at any location on the surface of Earth varies with time, by measurable amounts over a period of a few years and by substantial amounts over, say, 100 years. For example, between 1580 and 1820 the direction indicated by compass needles in London changed by 35°.

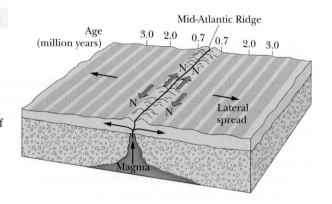

**FIG. 32-9** A magnetic profile of the seafloor on either side of the Mid-Atlantic Ridge. The seafloor, extruded through the ridge and spreading out as part of the tectonic drift system, displays a record of the past magnetic history of Earth's core. The direction of the magnetic field produced by the core reverses about every million years.

In spite of these local variations, the average dipole field changes only slowly over such relatively short time periods. Variations over longer periods can be studied by measuring the weak magnetism of the ocean floor on either side of the Mid-Atlantic Ridge (Fig. 32-9). This floor has been formed by molten magma that oozed up through the ridge from Earth's interior, solidified, and was pulled away from the ridge (by the drift of tectonic plates) at the rate of a few centimeters per year. As the magma solidified, it became weakly magnetized with its magnetic field in the direction of Earth's magnetic field at the time of solidification. Study of this solidified magma across the ocean floor reveals that Earth's field has reversed its *polarity* (directions of the north pole and south pole) about every million years. The reason for the reversals is not known. In fact, the mechanism that produces Earth's magnetic field is only vaguely understood.

## 32-7 I Magnetism and Electrons

Magnetic materials, from lodestones to videotapes, are magnetic because of the electrons within them. We have already seen one way in which electrons can generate a magnetic field: Send them through a wire as an electric current, and their motion produces a magnetic field around the wire. There are two more ways, each involving a magnetic dipole moment that produces a magnetic field in the surrounding space. However, their explanation requires quantum physics that is beyond the physics presented in this book, and so here we shall only outline the results.

### Spin Magnetic Dipole Moment

An electron has an intrinsic angular momentum called its **spin angular momentum** (or just **spin**) $\vec{S}$; associated with this spin is an intrinsic **spin magnetic dipole moment** $\vec{\mu}_s$. (By *intrinsic*, we mean that $\vec{S}$ and $\vec{\mu}_s$ are basic characteristics of an electron, like its mass and electric charge.) Vectors $\vec{S}$ and $\vec{\mu}_s$ are related by

$$\vec{\mu}_s = -\frac{e}{m}\vec{S}, \tag{32-22}$$

in which $e$ is the elementary charge ($1.60 \times 10^{-19}$ C) and $m$ is the mass of an electron ($9.11 \times 10^{-31}$ kg). The minus sign means that $\vec{\mu}_s$ and $\vec{S}$ are oppositely directed.

Spin $\vec{S}$ is different from the angular momenta of Chapter 11 in two respects:

1. Spin $\vec{S}$ itself cannot be measured. However, its component along any axis can be measured.

2. A measured component of $\vec{S}$ is *quantized,* which is a general term that means it is restricted to certain values. A measured component of $\vec{S}$ can have only two values, which differ only in sign.

Let us assume that the component of spin $\vec{S}$ is measured along the $z$ axis of a coordinate system. Then the measured component $S_z$ can have only the two values given by

$$S_z = m_s \frac{h}{2\pi}, \qquad \text{for } m_s = \pm\tfrac{1}{2}, \qquad (32\text{-}23)$$

where $m_s$ is called the *spin magnetic quantum number* and $h$ (= $6.63 \times 10^{-34}$ J·s) is the Planck constant, the ubiquitous constant of quantum physics. The signs given in Eq. 32-23 have to do with the direction of $S_z$ along the $z$ axis. When $S_z$ is parallel to the $z$ axis, $m_s$ is $+\tfrac{1}{2}$ and the electron is said to be *spin up*. When $S_z$ is antiparallel to the $z$ axis, $m_s$ is $-\tfrac{1}{2}$ and the electron is said to be *spin down*.

The spin magnetic dipole moment $\vec{\mu}_s$ of an electron also cannot be measured; only its component along any axis can be measured, and that component too is quantized, with two possible values of the same magnitude but different signs. We can relate the component $\mu_{s,z}$ measured on the $z$ axis to $S_z$ by rewriting Eq. 32-22 in component form for the $z$ axis as

$$\mu_{s,z} = -\frac{e}{m} S_z.$$

Substituting for $S_z$ from Eq. 32-23 then gives us

$$\mu_{s,z} = \pm \frac{eh}{4\pi m}, \qquad (32\text{-}24)$$

where the plus and minus signs correspond to $\mu_{s,z}$ being parallel and antiparallel to the $z$ axis, respectively.

The quantity on the right side of Eq. 32-24 is called the *Bohr magneton* $\mu_B$:

$$\mu_B = \frac{eh}{4\pi m} = 9.27 \times 10^{-24} \text{ J/T} \qquad \text{(Bohr magneton)}. \qquad (32\text{-}25)$$

Spin magnetic dipole moments of electrons and other elementary particles can be expressed in terms of $\mu_B$. For an electron, the magnitude of the measured $z$ component of $\vec{\mu}_s$ is

$$|\mu_{s,z}| = 1\mu_B. \qquad (32\text{-}26)$$

(The quantum physics of the electron, called *quantum electrodynamics*, or QED, reveals that $\mu_{s,z}$ is actually slightly greater than $1\mu_B$, but we shall neglect that fact.)

When an electron is placed in an external magnetic field $\vec{B}_{ext}$, a potential energy $U$ can be associated with the orientation of the electron's spin magnetic dipole moment $\vec{\mu}_s$ just as a potential energy can be associated with the orientation of the magnetic dipole moment $\vec{\mu}$ of a current loop placed in $\vec{B}_{ext}$. From Eq. 28-38, the potential energy for the electron is

$$U = -\vec{\mu}_s \cdot \vec{B}_{ext} = -\mu_{s,z} B_{ext}, \qquad (32\text{-}27)$$

where the $z$ axis is taken to be in the direction of $\vec{B}_{ext}$.

If we imagine an electron to be a microscopic sphere (which it is not), we can represent the spin $\vec{S}$, the spin magnetic dipole moment $\vec{\mu}_s$, and the associated magnetic dipole field as in Fig. 32-10. Although we use the word "spin" here, electrons do not spin like tops. How, then, can something have angular momentum without actually rotating? Again, we would need quantum physics to provide the answer.

Protons and neutrons also have an intrinsic angular momentum called spin and an associated intrinsic spin magnetic dipole moment. For a proton those two vectors have the same direction, and for a neutron they have opposite directions. We shall not examine the contributions of these dipole moments to the magnetic fields of atoms because they are about a thousand times smaller than that due to an electron.

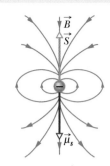

FIG. 32-10 The spin $\vec{S}$, spin magnetic dipole moment $\vec{\mu}_s$, and magnetic dipole field $\vec{B}$ of an electron represented as a microscopic sphere.

✓ **CHECKPOINT 4**    The figure here shows the spin orientations of two particles in an external magnetic field $\vec{B}_{ext}$. (a) If the particles are electrons, which spin orientation is at lower potential energy? (b) If, instead, the particles are protons, which spin orientation is at lower potential energy?

(1)          (2)

## Orbital Magnetic Dipole Moment

When it is in an atom, an electron has an additional angular momentum called its **orbital angular momentum** $\vec{L}_{orb}$. Associated with $\vec{L}_{orb}$ is an **orbital magnetic dipole moment** $\vec{\mu}_{orb}$; the two are related by

$$\vec{\mu}_{orb} = -\frac{e}{2m}\vec{L}_{orb}. \qquad (32\text{-}28)$$

The minus sign means that $\vec{\mu}_{orb}$ and $\vec{L}_{orb}$ have opposite directions.

Orbital angular momentum $\vec{L}_{orb}$ cannot be measured; only its component along any axis can be measured, and that component is quantized. The component along, say, a z axis can have only the values given by

$$L_{orb,z} = m_\ell \frac{h}{2\pi}, \qquad \text{for } m_\ell = 0, \pm 1, \pm 2, \ldots, \pm(\text{limit}), \qquad (32\text{-}29)$$

in which $m_\ell$ is called the *orbital magnetic quantum number* and "limit" refers to some largest allowed integer value for $m_\ell$. The signs in Eq. 32-29 have to do with the direction of $L_{orb,z}$ along the z axis.

The orbital magnetic dipole moment $\vec{\mu}_{orb}$ of an electron also cannot itself be measured; only its component along an axis can be measured, and that component is quantized. By writing Eq. 32-28 for a component along the same z axis as above and then substituting for $L_{orb,z}$ from Eq. 32-29, we can write the z component $\mu_{orb,z}$ of the orbital magnetic dipole moment as

$$\mu_{orb,z} = -m_\ell \frac{eh}{4\pi m} \qquad (32\text{-}30)$$

and, in terms of the Bohr magneton, as

$$\mu_{orb,z} = -m_\ell \mu_B. \qquad (32\text{-}31)$$

When an atom is placed in an external magnetic field $\vec{B}_{ext}$, a potential energy $U$ can be associated with the orientation of the orbital magnetic dipole moment of each electron in the atom. Its value is

$$U = -\vec{\mu}_{orb} \cdot \vec{B}_{ext} = -\mu_{orb,z} B_{ext}, \qquad (32\text{-}32)$$

where the z axis is taken in the direction of $\vec{B}_{ext}$.

Although we have used the words "orbit" and "orbital" here, electrons do not orbit the nucleus of an atom like planets orbiting the Sun. How can an electron have an orbital angular momentum without orbiting in the common meaning of the term? Once again, this can be explained only with quantum physics.

## Loop Model for Electron Orbits

We can obtain Eq. 32-28 with the nonquantum derivation that follows, in which we assume that an electron moves along a circular path with a radius that is much larger than an atomic radius (hence the name "loop model"). However, the derivation does not apply to an electron within an atom (for which we need quantum physics).

We imagine an electron moving at constant speed $v$ in a circular path of radius $r$, counterclockwise as shown in Fig. 32-11. The motion of the negative charge of the electron is equivalent to a conventional current $i$ (of positive charge) that is clockwise, as also shown in Fig. 32-11. The magnitude of the orbital magnetic dipole moment of such a *current loop* is obtained from Eq. 28-35 with $N = 1$:

$$\mu_{orb} = iA,  \tag{32-33}$$

where $A$ is the area enclosed by the loop. The direction of this magnetic dipole moment is, from the right-hand rule of Fig. 29-22, downward in Fig. 32-11.

To evaluate Eq. 32-33, we need the current $i$. Current is, generally, the rate at which charge passes some point in a circuit. Here, the charge of magnitude $e$ takes a time $T = 2\pi r/v$ to circle from any point back through that point, so

$$i = \frac{\text{charge}}{\text{time}} = \frac{e}{2\pi r/v}.  \tag{32-34}$$

Substituting this and the area $A = \pi r^2$ of the loop into Eq. 32-33 gives us

$$\mu_{orb} = \frac{e}{2\pi r/v}\,\pi r^2 = \frac{evr}{2}.  \tag{32-35}$$

To find the electron's orbital angular momentum $\vec{L}_{orb}$, we use Eq. 11-18, $\vec{\ell} = m(\vec{r} \times \vec{v})$. Because $\vec{r}$ and $\vec{v}$ are perpendicular, $\vec{L}_{orb}$ has the magnitude

$$L_{orb} = mrv\sin 90° = mrv.  \tag{32-36}$$

The vector $\vec{L}_{orb}$ is directed upward in Fig. 32-11 (see Fig. 11-12). Combining Eqs. 32-35 and 32-36, generalizing to a vector formulation, and indicating the opposite directions of the vectors with a minus sign yield

$$\vec{\mu}_{orb} = -\frac{e}{2m}\,\vec{L}_{orb},$$

which is Eq. 32-28. Thus, by "classical" (nonquantum) analysis we have obtained the same result, in both magnitude and direction, given by quantum physics. You might wonder, seeing as this derivation gives the correct result for an electron within an atom, why the derivation is invalid for that situation. The answer is that this line of reasoning yields other results that are contradicted by experiments.

### Loop Model in a Nonuniform Field

We continue to consider an electron orbit as a current loop, as we did in Fig. 32-11. Now, however, we draw the loop in a nonuniform magnetic field $\vec{B}_{ext}$ as shown in Fig. 32-12a. (This field could be the diverging field near the north pole of the magnet in Fig. 32-4.) We make this change to prepare for the next several sections, in which we shall discuss the forces that act on magnetic materials when the materials are placed in a nonuniform magnetic field. We shall discuss these forces by assuming that the electron orbits in the materials are tiny current loops like that in Fig. 32-12a.

Here we assume that the magnetic field vectors all around the electron's circular path have the same magnitude and form the same angle with the vertical, as shown in Figs. 32-12b and d. We also assume that all the electrons in an atom move either counterclockwise (Fig. 32-12b) or clockwise (Fig. 32-12d). The associated conventional current $i$ around the current loop and the orbital magnetic dipole moment $\vec{\mu}_{orb}$ produced by $i$ are shown for each direction of motion.

Figures 32-12c and e show diametrically opposite views of a length element $d\vec{L}$ of the loop that has the same direction as $i$, as seen from the plane of the orbit. Also shown are the field $\vec{B}_{ext}$ and the resulting magnetic force $d\vec{F}$ on $d\vec{L}$. Recall

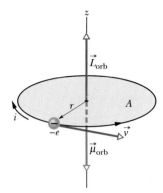

**FIG. 32-11** An electron moving at constant speed $v$ in a circular path of radius $r$ that encloses an area $A$. The electron has an orbital angular momentum $\vec{L}_{orb}$ and an associated orbital magnetic dipole moment $\vec{\mu}_{orb}$. A clockwise current $i$ (of positive charge) is equivalent to the counterclockwise circulation of the negatively charged electron.

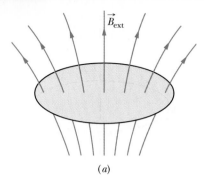

(a)

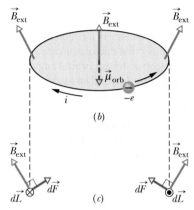

(b)

(c)

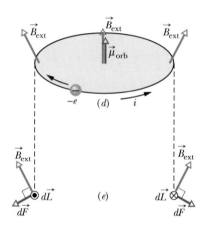

(d)

(e)

**FIG. 32-12** (a) A loop model for an electron orbiting in an atom while in a nonuniform magnetic field $\vec{B}_{ext}$. (b) Charge $-e$ moves counterclockwise; the associated conventional current $i$ is clockwise. (c) The magnetic forces $d\vec{F}$ on the left and right sides of the loop, as seen from the plane of the loop. The net force on the loop is upward. (d) Charge $-e$ now moves clockwise. (e) The net force on the loop is now downward.

that a current along an element $d\vec{L}$ in a magnetic field $\vec{B}_{ext}$ experiences a magnetic force $d\vec{F}$ as given by Eq. 28-28:

$$d\vec{F} = i \, d\vec{L} \times \vec{B}_{ext}. \qquad (32\text{-}37)$$

On the left side of Fig. 32-12c, Eq. 32-37 tells us that the force $d\vec{F}$ is directed upward and rightward. On the right side, the force $d\vec{F}$ is just as large and is directed upward and leftward. Because their angles are the same, the horizontal components of these two forces cancel and the vertical components add. The same is true at any other two symmetric points on the loop. Thus, the net force on the current loop of Fig. 32-12b must be upward. The same reasoning leads to a downward net force on the loop in Fig. 32-12d. We shall use these two results shortly when we examine the behavior of magnetic materials in nonuniform magnetic fields.

## 32-8 | Magnetic Materials

Each electron in an atom has an orbital magnetic dipole moment and a spin magnetic dipole moment that combine vectorially. The resultant of these two vector quantities combines vectorially with similar resultants for all other electrons in the atom, and the resultant for each atom combines with those for all the other atoms in a sample of a material. If the combination of all these magnetic dipole moments produces a magnetic field, then the material is magnetic. There are three general types of magnetism: diamagnetism, paramagnetism, and ferromagnetism.

1. **Diamagnetism** is exhibited by all common materials but is so feeble that it is masked if the material also exhibits magnetism of either of the other two types. In diamagnetism, weak magnetic dipole moments are produced in the atoms of the material when the material is placed in an external magnetic field $\vec{B}_{ext}$; the combination of all those induced dipole moments gives the material as a whole only a feeble net magnetic field. The dipole moments and thus their net field disappear when $\vec{B}_{ext}$ is removed. The term *diamagnetic material* usually refers to materials that exhibit only diamagnetism.

2. **Paramagnetism** is exhibited by materials containing transition elements, rare earth elements, and actinide elements (see Appendix G). Each atom of such a material has a permanent resultant magnetic dipole moment, but the moments are randomly oriented in the material and the material as a whole lacks a net magnetic field. However, an external magnetic field $\vec{B}_{ext}$ can partially align the atomic magnetic dipole moments to give the material a net magnetic field. The alignment and thus its field disappear when $\vec{B}_{ext}$ is removed. The term *paramagnetic material* usually refers to materials that exhibit primarily paramagnetism.

3. **Ferromagnetism** is a property of iron, nickel, and certain other elements (and of compounds and alloys of these elements). Some of the electrons in these materials have their resultant magnetic dipole moments aligned, which produces regions with strong magnetic dipole moments. An external field $\vec{B}_{ext}$ can then align the magnetic moments of such regions, producing a strong magnetic field for a sample of the material; the field partially persists when $\vec{B}_{ext}$ is removed. We usually use the terms *ferromagnetic material* and *magnetic material* to refer to materials that exhibit primarily ferromagnetism.

The next three sections explore these three types of magnetism.

## 32-9 | Diamagnetism

We cannot yet discuss the quantum physical explanation of diamagnetism, but we can provide a classical explanation with the loop model of Figs. 32-11 and 32-12. To begin, we assume that in an atom of a diamagnetic material each electron can orbit only clockwise as in Fig. 32-12d or counterclockwise as in Fig. 32-12b. To account for the lack of magnetism in the absence of an external magnetic field $\vec{B}_{ext}$, we assume the atom lacks a net magnetic dipole moment. This implies that before $\vec{B}_{ext}$ is applied, the number of electrons orbiting in one direction is the same as that orbiting in the opposite direction, with the result that the net upward magnetic dipole moment of the atom equals the net downward magnetic dipole moment.

Now let's turn on the nonuniform field $\vec{B}_{ext}$ of Fig. 32-12a, in which $\vec{B}_{ext}$ is directed upward but is diverging (the magnetic field lines are diverging). We could do this by increasing the current through an electromagnet or by moving the north pole of a bar magnet closer to, and below, the orbits. As the magnitude of $\vec{B}_{ext}$ increases from zero to its final maximum, steady-state value, a clockwise electric field is induced around each electron's orbital loop according to Faraday's law and Lenz's law. Let us see how this induced electric field affects the orbiting electrons in Figs. 32-12b and d.

In Fig. 32-12b, the counterclockwise electron is accelerated by the clockwise electric field. Thus, as the magnetic field $\vec{B}_{ext}$ increases to its maximum value, the electron speed increases to a maximum value. This means that the associated conventional current $i$ and the downward magnetic dipole moment $\vec{\mu}$ due to $i$ also *increase.*

In Fig. 32-12d, the clockwise electron is decelerated by the clockwise electric field. Thus, here, the electron speed, the associated current $i$, and the upward magnetic dipole moment $\vec{\mu}$ due to $i$ all *decrease.* By turning on field $\vec{B}_{ext}$, we have given the atom a *net* magnetic dipole moment that is downward. This would also be so if the magnetic field were uniform.

The nonuniformity of field $\vec{B}_{ext}$ also affects the atom. Because the current $i$ in Fig. 32-12b increases, the upward magnetic forces $d\vec{F}$ in Fig. 32-12c also increase, as does the net upward force on the current loop. Because current $i$ in Fig. 32-12d decreases, the downward magnetic forces $d\vec{F}$ in Fig. 32-12e also decrease, as does the net downward force on the current loop. Thus, by turning on the *nonuniform* field $\vec{B}_{ext}$, we have produced a net force on the atom; moreover, that force is directed *away* from the region of greater magnetic field.

We have argued with fictitious electron orbits (current loops), but we have ended up with exactly what happens to a diamagnetic material: If we apply the magnetic field of Fig. 32-12, the material develops a downward magnetic dipole moment and experiences an upward force. When the field is removed, both the dipole moment and the force disappear. The external field need not be positioned as shown in Fig. 32-12; similar arguments can be made for other orientations of $\vec{B}_{ext}$. In general,

> A diamagnetic material placed in an external magnetic field $\vec{B}_{ext}$ develops a magnetic dipole moment directed opposite $\vec{B}_{ext}$. If the field is nonuniform, the diamagnetic material is repelled *from* a region of greater magnetic field *toward* a region of lesser field.

The frog in Fig. 32-13 is diamagnetic (as is any other animal). When the frog was placed in the diverging magnetic field near the top end of a vertical current-carrying solenoid, every atom in the frog was repelled upward, away from the region of stronger magnetic field at that end of the solenoid. The frog moved upward into weaker and weaker magnetic field until the upward magnetic force balanced the gravitational force on it, and there it hung in midair. If we built a solenoid that was large enough, we could similarly levitate a person in midair due to the person's diamagnetism.

**FIG. 32-13** An overhead view of a frog that is being levitated in a magnetic field produced by current in a vertical solenoid below the frog. (The frog is not in discomfort; the sensation is like floating in water, which frogs like very much.) *(Courtesy A. K. Gein, High Field Magnet Laboratory, University of Nijmegen, The Netherlands)*

✓ **CHECKPOINT 5** The figure shows two diamagnetic spheres located near the south pole of a bar magnet. Are (a) the magnetic forces on the spheres and (b) the magnetic dipole moments of the spheres directed toward or away from the bar magnet? (c) Is the magnetic force on sphere 1 greater than, less than, or equal to that on sphere 2?

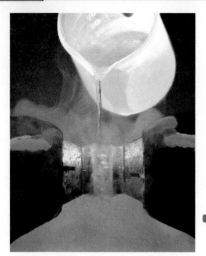

Liquid oxygen is suspended between the two pole faces of a magnet because the liquid is paramagnetic and is magnetically attracted to the magnet. *(Richard Megna/Fundamental Photographs)*

## 32-10 | Paramagnetism

In paramagnetic materials, the spin and orbital magnetic dipole moments of the electrons in each atom do not cancel but add vectorially to give the atom a net (and permanent) magnetic dipole moment $\vec{\mu}$. In the absence of an external magnetic field, these atomic dipole moments are randomly oriented, and the net magnetic dipole moment of the material is zero. However, if a sample of the material is placed in an external magnetic field $\vec{B}_{ext}$, the magnetic dipole moments tend to line up with the field, which gives the sample a net magnetic dipole moment. This alignment with the external field is the opposite of what we saw with diamagnetic materials.

A paramagnetic material placed in an external magnetic field $\vec{B}_{ext}$ develops a magnetic dipole moment in the direction of $\vec{B}_{ext}$. If the field is nonuniform, the paramagnetic material is attracted *toward* a region of greater magnetic field *from* a region of lesser field.

A paramagnetic sample with $N$ atoms would have a magnetic dipole moment of magnitude $N\mu$ if alignment of its atomic dipoles were complete. However, random collisions of atoms due to their thermal agitation transfer energy among the atoms, disrupting their alignment and thus reducing the sample's magnetic dipole moment.

The importance of thermal agitation may be measured by comparing two energies. One, given by Eq. 19-24, is the mean translational kinetic energy $K\left(=\frac{3}{2}kT\right)$ of an atom at temperature $T$, where $k$ is the Boltzmann constant ($1.38 \times 10^{-23}$ J/K) and $T$ is in kelvins (not Celsius degrees). The other, derived from Eq. 28-38, is the difference in energy $\Delta U_B (= 2\mu B_{ext})$ between parallel alignment and antiparallel alignment of the magnetic dipole moment of an atom and the external field. As we shall show below, $K \gg \Delta U_B$, even for ordinary temperatures and field magnitudes. Thus, energy transfers during collisions among atoms can significantly disrupt the alignment of the atomic dipole moments, keeping the magnetic dipole moment of a sample much less than $N\mu$.

We can express the extent to which a given paramagnetic sample is magnetized by finding the ratio of its magnetic dipole moment to its volume $V$. This vector quantity, the magnetic dipole moment per unit volume, is the **magnetization** $\vec{M}$ of the sample, and its magnitude is

$$M = \frac{\text{measured magnetic moment}}{V}. \qquad (32\text{-}38)$$

The unit of $\vec{M}$ is the ampere–square meter per cubic meter, or ampere per meter (A/m). Complete alignment of the atomic dipole moments, called *saturation* of the sample, corresponds to the maximum value $M_{max} = N\mu/V$.

In 1895 Pierre Curie discovered experimentally that the magnetization of a paramagnetic sample is directly proportional to the magnitude of the external

**FIG. 32-14** A *magnetization curve* for potassium chromium sulfate, a paramagnetic salt. The ratio of magnetization $M$ of the salt to the maximum possible magnetization $M_{max}$ is plotted versus the ratio of the applied magnetic field magnitude $B_{ext}$ to the temperature $T$. Curie's law fits the data at the left; quantum theory fits all the data. After W. E. Henry.

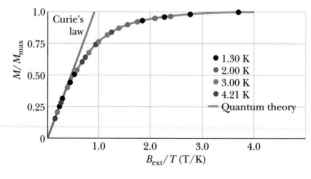

magnetic field $\vec{B}_{\text{ext}}$ and inversely proportional to the temperature $T$ in kelvins:

$$M = C\frac{B_{\text{ext}}}{T}. \tag{32-39}$$

Equation 32-39 is known as *Curie's law,* and $C$ is called the *Curie constant.* Curie's law is reasonable in that increasing $B_{\text{ext}}$ tends to align the atomic dipole moments in a sample and thus to increase $M$, whereas increasing $T$ tends to disrupt the alignment via thermal agitation and thus to decrease $M$. However, the law is actually an approximation that is valid only when the ratio $B_{\text{ext}}/T$ is not too large.

Figure 32-14 shows the ratio $M/M_{\text{max}}$ as a function of $B_{\text{ext}}/T$ for a sample of the salt potassium chromium sulfate, in which chromium ions are the paramagnetic substance. The plot is called a *magnetization curve.* The straight line for Curie's law fits the experimental data at the left, for $B_{\text{ext}}/T$ below about 0.5 T/K. The curve that fits all the data points is based on quantum physics. The data on the right side, near saturation, are very difficult to obtain because they require very strong magnetic fields (about 100 000 times Earth's field), even at very low temperatures.

**✓ CHECKPOINT 6**    The figure here shows two paramagnetic spheres located near the south pole of a bar magnet. Are (a) the magnetic forces on the spheres and (b) the magnetic dipole moments of the spheres directed toward or away from the bar magnet? (c) Is the magnetic force on sphere 1 greater than, less than, or equal to that on sphere 2?

**Sample Problem  32-3**

A paramagnetic gas at room temperature ($T = 300$ K) is placed in an external uniform magnetic field of magnitude $B = 1.5$ T; the atoms of the gas have magnetic dipole moment $\mu = 1.0\mu_B$. Calculate the mean translational kinetic energy $K$ of an atom of the gas and the energy difference $\Delta U_B$ between parallel alignment and antiparallel alignment of the atom's magnetic dipole moment with the external field.

**KEY IDEAS**   (1) The mean translational kinetic energy $K$ of an atom in a gas depends on the temperature of the gas. (2) The potential energy $U_B$ of a magnetic dipole $\vec{\mu}$ in an external magnetic field $\vec{B}$ depends on the angle $\theta$ between the directions of $\vec{\mu}$ and $\vec{B}$.

*Calculations:* From Eq. 19-24, we have

$$K = \tfrac{3}{2}kT = \tfrac{3}{2}(1.38 \times 10^{-23} \text{ J/K})(300 \text{ K})$$
$$= 6.2 \times 10^{-21} \text{ J} = 0.039 \text{ eV}. \quad \text{(Answer)}$$

From Eq. 28-38 ($U_B = -\vec{\mu} \cdot \vec{B}$), we can write the difference $\Delta U_B$ between parallel alignment ($\theta = 0°$) and antiparallel alignment ($\theta = 180°$) as

$$\Delta U_B = -\mu B \cos 180° - (-\mu B \cos 0°) = 2\mu B$$
$$= 2\mu_B B = 2(9.27 \times 10^{-24} \text{ J/T})(1.5 \text{ T})$$
$$= 2.8 \times 10^{-23} \text{ J} = 0.000 \, 17 \text{ eV}. \quad \text{(Answer)}$$

Here $K$ is about 230 times $\Delta U_B$; so energy exchanges among the atoms during their collisions with one another can easily reorient any magnetic dipole moments that might be aligned with the external magnetic field. The magnetic dipole moment exhibited by the paramagnetic gas must then be due to fleeting partial alignments of the atomic dipole moments.

## 32-11 I Ferromagnetism

When we speak of magnetism in everyday conversation, we almost always have a mental picture of a bar magnet or a disk magnet (probably clinging to a refrigerator door). That is, we picture a ferromagnetic material having strong, permanent magnetism, and not a diamagnetic or paramagnetic material having weak, temporary magnetism.

Iron, cobalt, nickel, gadolinium, dysprosium, and alloys containing these elements exhibit ferromagnetism because of a quantum physical effect called *exchange coupling* in which the electron spins of one atom interact with those

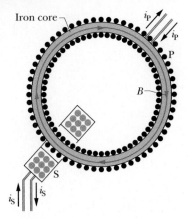

**FIG. 32-15** A Rowland ring. A primary coil P has a core made of the ferromagnetic material to be studied (here iron). The core is magnetized by a current $i_P$ sent through coil P. (The turns of the coil are represented by dots.) The extent to which the core is magnetized determines the total magnetic field $\vec{B}$ within coil P. Field $\vec{B}$ can be measured by means of a secondary coil S.

of neighboring atoms. The result is alignment of the magnetic dipole moments of the atoms, in spite of the randomizing tendency of atomic collisions due to thermal agitation. This persistent alignment is what gives ferromagnetic materials their permanent magnetism.

If the temperature of a ferromagnetic material is raised above a certain critical value, called the *Curie temperature,* the exchange coupling ceases to be effective. Most such materials then become simply paramagnetic; that is, the dipoles still tend to align with an external field but much more weakly, and thermal agitation can now more easily disrupt the alignment. The Curie temperature for iron is 1043 K ($= 770°C$).

The magnetization of a ferromagnetic material such as iron can be studied with an arrangement called a *Rowland ring* (Fig. 32-15). The material is formed into a thin toroidal core of circular cross section. A primary coil P having $n$ turns per unit length is wrapped around the core and carries current $i_P$. (The coil is essentially a long solenoid bent into a circle.) If the iron core were not present, the magnitude of the magnetic field inside the coil would be, from Eq. 29-23,

$$B_0 = \mu_0 i_P n. \qquad (32\text{-}40)$$

However, with the iron core present, the magnetic field $\vec{B}$ inside the coil is greater than $\vec{B}_0$, usually by a large amount. We can write the magnitude of this field as

$$B = B_0 + B_M, \qquad (32\text{-}41)$$

where $B_M$ is the magnitude of the magnetic field contributed by the iron core. This contribution results from the alignment of the atomic dipole moments within the iron, due to exchange coupling and to the applied magnetic field $B_0$, and is proportional to the magnetization $M$ of the iron. That is, the contribution $B_M$ is proportional to the magnetic dipole moment per unit volume of the iron. To determine $B_M$ we use a secondary coil S to measure $B$, compute $B_0$ with Eq. 32-40, and subtract as suggested by Eq. 32-41.

Figure 32-16 shows a magnetization curve for a ferromagnetic material in a Rowland ring: The ratio $B_M/B_{M,\text{max}}$, where $B_{M,\text{max}}$ is the maximum possible value of $B_M$, corresponding to saturation, is plotted versus $B_0$. The curve is like Fig. 32-14, the magnetization curve for a paramagnetic substance: Both curves show the extent to which an applied magnetic field can align the atomic dipole moments of a material.

For the ferromagnetic core yielding Fig. 32-16, the alignment of the dipole moments is about 70% complete for $B_0 \approx 1 \times 10^{-3}$ T. If $B_0$ were increased to 1 T, the alignment would be almost complete (but $B_0 = 1$ T, and thus almost complete saturation, is quite difficult to obtain).

### Magnetic Domains

Exchange coupling produces strong alignment of adjacent atomic dipoles in a ferromagnetic material at a temperature below the Curie temperature. Why, then, isn't the material naturally at saturation even when there is no applied magnetic field $B_0$? That is, why isn't every piece of iron, such as an iron nail, a naturally strong magnet?

To understand this, consider a specimen of a ferromagnetic material such as iron that is in the form of a single crystal; that is, the arrangement of the atoms that make it up—its crystal lattice—extends with unbroken regularity throughout the volume of the specimen. Such a crystal will, in its normal state, be made up of a number of *magnetic domains.* These are regions of the crystal throughout which the alignment of the atomic dipoles is essentially perfect. The domains, however, are not all aligned. For the crystal as a whole, the domains are so oriented that they largely cancel with one another as far as their external magnetic effects are concerned.

Figure 32-17 is a magnified photograph of such an assembly of domains in a single crystal of nickel. It was made by sprinkling a colloidal suspension of finely

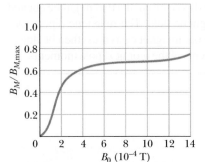

**FIG. 32-16** A magnetization curve for a ferromagnetic core material in the Rowland ring of Fig. 32-15. On the vertical axis, 1.0 corresponds to complete alignment (saturation) of the atomic dipoles within the material.

powdered iron oxide on the surface of the crystal. The domain boundaries, which are thin regions in which the alignment of the elementary dipoles changes from a certain orientation in one of the domains forming the boundary to a different orientation in the other domain, are the sites of intense, but highly localized and nonuniform, magnetic fields. The suspended colloidal particles are attracted to these boundaries and show up as the white lines (not all the domain boundaries are apparent in Fig. 32-17). Although the atomic dipoles in each domain are completely aligned as shown by the arrows, the crystal as a whole may have only a very small resultant magnetic moment.

Actually, a piece of iron as we ordinarily find it is not a single crystal but an assembly of many tiny crystals, randomly arranged; we call it a *polycrystalline solid.* Each tiny crystal, however, has its array of variously oriented domains, just as in Fig. 32-17. If we magnetize such a specimen by placing it in an external magnetic field of gradually increasing strength, we produce two effects; together they produce a magnetization curve of the shape shown in Fig. 32-16. One effect is a growth in size of the domains that are oriented along the external field at the expense of those that are not. The second effect is a shift of the orientation of the dipoles within a domain, as a unit, to become closer to the field direction.

Exchange coupling and domain shifting give us the following result:

> A ferromagnetic material placed in an external magnetic field $\vec{B}_{ext}$ develops a strong magnetic dipole moment in the direction of $\vec{B}_{ext}$. If the field is nonuniform, the ferromagnetic material is attracted *toward* a region of greater magnetic field *from* a region of lesser field.

## Mural Paintings Record Earth's Magnetic Field

The red pigments used in many mural paintings, such as the ones shown in this chapter's opening photograph, contain grains of the iron oxide hematite. Each grain consists of a single domain having a particular magnetic dipole moment. Artists' pigments are a suspension of various solids in a liquid carrier. When a pigment is applied to a wall as a mural is being created, each grain rotates in the liquid until its dipole moment aligns with Earth's magnetic field. When the paint dries, the moments are locked into place and thus record the direction of Earth's magnetic field at the time of the painting. Figure 32-18 suggests the alignment of the moments in a mural painted in 1740, when geomagnetic north was in the direction indicated by $N_{1740}$.

A researcher can determine Earth's field direction at the time a mural was painted by determining the orientation of the magnetic moments in the paint. A short section of sticky tape is applied to a portion of the mural, and the orientation of the tape is carefully measured relative to the horizontal and to today's geomagnetic north ($N_{today}$). When the tape is peeled off the wall, it carries a thin layer of the paint. In a laboratory, the tape section is mounted in an apparatus to determine the orientation of the dipole moments in that layer of paint. Evidence from mural paintings and many other sources reveal that the direction of geomagnetic north has varied gradually but continuously over recorded history.

## Hysteresis

Magnetization curves for ferromagnetic materials are not retraced as we increase and then decrease the external magnetic field $B_0$. Figure 32-19 is a plot of $B_M$ versus $B_0$ during the following operations with a Rowland ring: (1) Starting with the iron unmagnetized (point *a*), increase the current in the toroid until $B_0$ ($= \mu_0 in$) has the value corresponding to point *b*; (2) reduce the current in the toroid winding (and thus $B_0$) back to zero (point *c*); (3) reverse the toroid current and increase it in magnitude until $B_0$ has the value corresponding to point *d*; (4) reduce the current to zero again (point *e*); (5) reverse the current once more until point *b* is reached again.

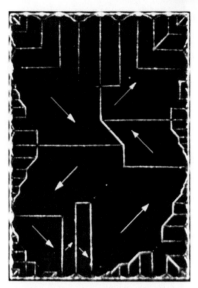

FIG. 32-17  A photograph of domain patterns within a single crystal of nickel; white lines reveal the boundaries of the domains. The white arrows superimposed on the photograph show the orientations of the magnetic dipoles within the domains and thus the orientations of the net magnetic dipoles of the domains. The crystal as a whole is unmagnetized if the net magnetic field (the vector sum over all the domains) is zero. *(Courtesy Ralph W. DeBlois)*

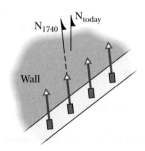

FIG. 32-18  Overhead view of a cross section of a thin layer of paint in a Vatican mural painting. The magnetic moments of hematite grains in the red pigments are aligned in the direction Earth's magnetic field had when the mural was painted in 1740. Geomagnetic north (as indicated by a horizontal compass) is shown for today and for 1740.

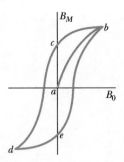

FIG. 32-19 A magnetization curve (ab) for a ferromagnetic specimen and an associated hysteresis loop (bcdeb).

The lack of retraceability shown in Fig. 32-19 is called **hysteresis,** and the curve bcdeb is called a *hysteresis loop.* Note that at points c and e the iron core is magnetized, even though there is no current in the toroid windings; this is the familiar phenomenon of permanent magnetism.

Hysteresis can be understood through the concept of magnetic domains. Evidently the motions of the domain boundaries and the reorientations of the domain directions are not totally reversible. When the applied magnetic field $B_0$ is increased and then decreased back to its initial value, the domains do not return completely to their original configuration but retain some "memory" of their alignment after the initial increase. This memory of magnetic materials is essential for the magnetic storage of information, as on magnetic tapes and disks.

This memory of the alignment of domains can also occur naturally. When lightning sends currents along multiple tortuous paths through the ground, the currents produce intense magnetic fields that can suddenly magnetize any ferromagnetic material in nearby rock. Because of hysteresis, such rock material retains some of that magnetization after the lightning strike (after the currents disappear). Pieces of the rock—later exposed, broken, and loosened by weathering—are then lodestones.

## Sample Problem | 32-4 | Build your skill

A compass needle made of pure iron (density 7900 kg/m³) has a length L of 3.0 cm, a width of 1.0 mm, and a thickness of 0.50 mm. The magnitude of the magnetic dipole moment of an iron atom is $\mu_{Fe} = 2.1 \times 10^{-23}$ J/T.

(a) If the magnetization of the needle is equivalent to the alignment of 10% of the atoms in the needle, what is the magnitude of the needle's magnetic dipole moment $\vec{\mu}$?

**KEY IDEAS** (1) Alignment of all N atoms in the needle would give a magnitude of $N\mu_{Fe}$ for the needle's magnetic dipole moment $\vec{\mu}$. However, the needle has only 10% alignment (the random orientation of the rest does not give any net contribution to $\vec{\mu}$). Thus,

$$\mu = 0.10N\mu_{Fe}. \qquad (32\text{-}42)$$

(2) We can find the number of atoms N in the needle from the needle's mass:

$$N = \frac{\text{needle's mass}}{\text{iron's atomic mass}}. \qquad (32\text{-}43)$$

**Finding N:** Iron's atomic mass is not listed in Appendix F, but its molar mass M is. Thus, we write

$$\text{iron's atomic mass} = \frac{\text{iron's molar mass } M}{\text{Avogadro's number } N_A}. \qquad (32\text{-}44)$$

Equation 32-43 then becomes

$$N = \frac{mN_A}{M}. \qquad (32\text{-}45)$$

The needle's mass m is the product of its density and its volume. The volume works out to be $1.5 \times 10^{-8}$ m³; so

needle's mass m = (needle's density)(needle's volume)
$$= (7900 \text{ kg/m}^3)(1.5 \times 10^{-8} \text{ m}^3)$$
$$= 1.185 \times 10^{-4} \text{ kg}.$$

Substituting into Eq. 32-45 with this value for m, and also 55.847 g/mol (= 0.055 847 kg/mol) for M and $6.02 \times 10^{23}$ for $N_A$, we find

$$N = \frac{(1.185 \times 10^{-4} \text{ kg})(6.02 \times 10^{23})}{0.055 \ 847 \text{ kg/mol}}$$
$$= 1.2774 \times 10^{21}.$$

**Finding $\mu$:** Substituting our value of N and the value of $\mu_{Fe}$ into Eq. 32-42 then yields

$$\mu = (0.10)(1.2774 \times 10^{21})(2.1 \times 10^{-23} \text{ J/T})$$
$$= 2.682 \times 10^{-3} \text{ J/T} \approx 2.7 \times 10^{-3} \text{ J/T}. \quad (\text{Answer})$$

(b) If the compass needle is jarred slightly from its (horizontal) north–south equilibrium position, it oscillates about that position. If the period of oscillation is 2.2 s, what is the horizontal component of the local magnetic field?

**KEY IDEAS** (1) The compass needle is a magnet with a magnetic dipole moment $\vec{\mu}$, which is directed along the needle's length, from its south pole to its north pole. (2) When the horizontal needle is jarred from its equilibrium position, Earth's magnetic field $\vec{B}$ produces a torque on the needle, about the needle's pivot axis.

**Calculations:** Because the needle is free to rotate only horizontally, only the horizontal component $B_h$ of Earth's field produces a torque to rotate the needle back toward its equilibrium position. From Eq. 28-36 ($\tau = \mu B \sin \theta$), we can write this torque as

$$\tau = -\mu B_h \sin \theta, \qquad (32\text{-}46)$$

in which the minus sign indicates that $\tau$ opposes the angular displacement $\theta$. Because the rotation angle is small, we may write $\sin \theta \approx \theta$, so that

$$\tau = -\mu B_h \theta. \qquad (32\text{-}47)$$

Because $\mu$ and $B_h$ are both constant, Eq. 32-47 tells us that the restoring torque is proportional to the negative of the angular displacement. This kind of relation is the hallmark of angular simple harmonic motion, as we saw in Section 15-5. From Eqs. 15-22 ($\tau = -\kappa\theta$) and 15-23 ($T = 2\pi(I/\kappa)^{1/2}$), the period of oscillation may then be written as

$$T = 2\pi\sqrt{\frac{I}{\mu B_h}},$$

which yields

$$B_h = \frac{I}{\mu}\left(\frac{2\pi}{T}\right)^2, \qquad (32\text{-}48)$$

where $I$ is the rotational inertia of the needle.

Approximating the needle as being a uniform thin rod, we use Table 10-2e to find

$$I = \frac{mL^2}{12} = \frac{(1.185 \times 10^{-4}\,\text{kg})(0.030\,\text{m})^2}{12}$$
$$= 8.888 \times 10^{-9}\,\text{kg} \cdot \text{m}^2.$$

Substituting this value, the value we obtained for $\mu$, and the given value for $T$ into Eq. 32-48, we find

$$B_h = \frac{8.888 \times 10^{-9}\,\text{kg} \cdot \text{m}^2}{2.682 \times 10^{-3}\,\text{J/T}}\left(\frac{2\pi}{2.2\,\text{s}}\right)^2$$
$$= 2.7 \times 10^{-5}\,\text{T}. \qquad \text{(Answer)}$$

Thus, even with an inexpensive compass, we can measure a local magnetic field by jarring the needle and timing its oscillations.

## REVIEW & SUMMARY

**Gauss' Law for Magnetic Fields** The simplest magnetic structures are magnetic dipoles. Magnetic monopoles do not exist (as far as we know). **Gauss' law** for magnetic fields,

$$\Phi_B = \oint \vec{B} \cdot d\vec{A} = 0, \qquad (32\text{-}1)$$

states that the net magnetic flux through any (closed) Gaussian surface is zero. It implies that magnetic monopoles do not exist.

**Maxwell's Extension of Ampere's Law** A changing electric flux induces a magnetic field $\vec{B}$. Maxwell's law,

$$\oint \vec{B} \cdot d\vec{s} = \mu_0\varepsilon_0\frac{d\Phi_E}{dt} \quad \text{(Maxwell's law of induction)}, \qquad (32\text{-}3)$$

relates the magnetic field induced along a closed loop to the changing electric flux $\Phi_E$ through the loop. Ampere's law, $\oint \vec{B} \cdot d\vec{s} = \mu_0 i_{enc}$ (Eq. 32-4), gives the magnetic field generated by a current $i_{enc}$ encircled by a closed loop. Maxwell's law and Ampere's law can be written as the single equation

$$\oint \vec{B} \cdot d\vec{s} = \mu_0\varepsilon_0\frac{d\Phi_E}{dt} + \mu_0 i_{enc}$$
$$\text{(Ampere–Maxwell law)}. \qquad (32\text{-}5)$$

**Displacement Current** We define the fictitious *displacement current* due to a changing electric field as

$$i_d = \varepsilon_0 \frac{d\Phi_E}{dt}. \qquad (32\text{-}10)$$

Equation 32-5 then becomes

$$\oint \vec{B} \cdot d\vec{s} = \mu_0 i_{d,enc} + \mu_0 i_{enc} \quad \text{(Ampere–Maxwell law)}, \quad (32\text{-}11)$$

where $i_{d,enc}$ is the displacement current encircled by the integration loop. The idea of a displacement current allows us to retain the notion of continuity of current through a capacitor. However, displacement current is *not* a transfer of charge.

**Maxwell's Equations** Maxwell's equations, displayed in Table 32-1, summarize electromagnetism and form its foundation.

**Earth's Magnetic Field** Earth's magnetic field can be approximated as being that of a magnetic dipole whose dipole moment makes an angle of 11.5° with Earth's rotation axis, and with the south pole of the dipole in the Northern Hemisphere. The direction of the local magnetic field at any point on Earth's surface is given by the *field declination* (the angle left or right from geographic north) and the *field inclination* (the angle up or down from the horizontal).

**Spin Magnetic Dipole Moment** An electron has an intrinsic angular momentum called *spin angular momentum* (or *spin*) $\vec{S}$, with which an intrinsic *spin magnetic dipole moment* $\vec{\mu}_s$ is associated:

$$\vec{\mu}_s = -\frac{e}{m}\vec{S}. \qquad (32\text{-}22)$$

Spin $\vec{S}$ cannot itself be measured, but any component can be measured. Assuming that the measurement is along the $z$ axis of a coordinate system, the component $S_z$ can have only the values given by

$$S_z = m_s\frac{h}{2\pi}, \quad \text{for } m_s = \pm\tfrac{1}{2}, \qquad (32\text{-}23)$$

where $h$ ($= 6.63 \times 10^{-34}\,\text{J} \cdot \text{s}$) is the Planck constant. Similarly, the electron's spin magnetic dipole moment $\vec{\mu}_s$ cannot itself be measured but its component can be measured. Along a $z$ axis, the component is

$$\mu_{s,z} = \pm\frac{eh}{4\pi m} = \pm\mu_B, \qquad (32\text{-}24, 32\text{-}26)$$

where $\mu_B$ is the *Bohr magneton:*

$$\mu_B = \frac{eh}{4\pi m} = 9.27 \times 10^{-24}\,\text{J/T}. \qquad (32\text{-}25)$$

The potential energy $U$ associated with the orientation of the spin magnetic dipole moment in an external magnetic field $\vec{B}_{ext}$ is

$$U = -\vec{\mu}_s \cdot \vec{B}_{ext} = -\mu_{s,z}B_{ext}. \qquad (32\text{-}27)$$

**Orbital Magnetic Dipole Moment** An electron in an atom has an additional angular momentum called its *orbital angular momentum* $\vec{L}_{orb}$, with which an *orbital magnetic dipole moment* $\vec{\mu}_{orb}$ is associated:

$$\vec{\mu}_{orb} = -\frac{e}{2m}\vec{L}_{orb}. \qquad (32\text{-}28)$$

Orbital angular momentum is quantized and can have only values given by

$$L_{orb,z} = m_\ell \frac{h}{2\pi},$$

$$\text{for } m_\ell = 0, \pm 1, \pm 2, \dots, \pm (\text{limit}). \qquad (32\text{-}29)$$

Thus, the magnitude of the orbital angular momentum is

$$\mu_{orb,z} = -m_\ell \frac{eh}{4\pi m} = -m_\ell \mu_B. \qquad (32\text{-}30, 32\text{-}31)$$

The potential energy $U$ associated with the orientation of the orbital magnetic dipole moment in an external magnetic field $\vec{B}_{ext}$ is

$$U = -\vec{\mu}_{orb} \cdot \vec{B}_{ext} = -\mu_{orb,z}B_{ext}. \qquad (32\text{-}32)$$

**Diamagnetism** *Diamagnetic materials* do not exhibit magnetism until they are placed in an external magnetic field $\vec{B}_{ext}$. They then develop a magnetic dipole moment directed opposite $\vec{B}_{ext}$. If the field is nonuniform, the diamagnetic material is repelled from regions of greater magnetic field. This property is called *diamagnetism*.

**Paramagnetism** In a *paramagnetic material*, each atom has a permanent magnetic dipole moment $\vec{\mu}$, but the dipole moments are randomly oriented and the material as a whole lacks a magnetic field. However, an external magnetic field $\vec{B}_{ext}$ can partially align the atomic dipole moments to give the material a net magnetic dipole moment in the direction of $\vec{B}_{ext}$. If $\vec{B}_{ext}$ is nonuniform, the material is attracted to regions of greater magnetic field. These properties are called *paramagnetism*.

The alignment of the atomic dipole moments increases with an increase in $\vec{B}_{ext}$ and decreases with an increase in temperature $T$. The extent to which a sample of volume $V$ is magnetized is given by its *magnetization* $\vec{M}$, whose magnitude is

$$M = \frac{\text{measured magnetic moment}}{V}. \qquad (32\text{-}38)$$

Complete alignment of all $N$ atomic magnetic dipoles in a sample, called *saturation* of the sample, corresponds to the maximum magnetization value $M_{max} = N\mu/V$. For low values of the ratio $B_{ext}/T$, we have the approximation

$$M = C\frac{B_{ext}}{T} \qquad (\text{Curie's law}), \qquad (32\text{-}39)$$

where $C$ is called the *Curie constant*.

**Ferromagnetism** In the absence of an external magnetic field, some of the electrons in a ferromagnetic material have their magnetic dipole moments aligned by means of a quantum physical interaction called *exchange coupling*, producing regions (domains) within the material with strong magnetic dipole moments. An external field $\vec{B}_{ext}$ can align the magnetic dipole moments of those regions, producing a strong net magnetic dipole moment for the material as a whole, in the direction of $\vec{B}_{ext}$. This net magnetic dipole moment can partially persist when field $\vec{B}_{ext}$ is removed. If $\vec{B}_{ext}$ is nonuniform, the ferromagnetic material is attracted to regions of greater magnetic field. These properties are called *ferromagnetism*. Exchange coupling disappears when a sample's temperature exceeds its *Curie temperature*.

## QUESTIONS

**1** Figure 32-20 shows, in two situations, an electric field vector $\vec{E}$ and an induced magnetic field line. In each, is the magnitude of $\vec{E}$ increasing or decreasing?

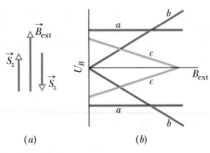

(a)                (b)

**FIG. 32-20** Question 1.

**2** Figure 32-21*a* shows a pair of opposite spin orientations for an electron in an external magnetic field $\vec{B}_{ext}$. Figure 32-21*b* gives three choices for the graph of the potential energies associated with those orientations

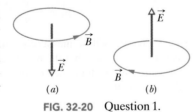

(a)                (b)

**FIG. 32-21** Question 2.

as a function of the magnitude of $\vec{B}_{ext}$. Choices *b* and *c* consist of intersecting lines, choice *a* of parallel lines. Which is the correct choice?

**3** Figure 32-22*a* shows a capacitor, with circular plates, that is being charged. Point *a* (near one of the connecting wires) and point *b* (inside the capacitor gap) are equidistant from the central axis, as are point *c* (not so near the wire) and point *d* (between the plates but outside the gap). In Fig. 32-22*b*, one curve gives the variation with distance *r* of the magnitude of the magnetic field inside and outside the wire. The other curve gives the variation with distance *r* of the magnitude of the magnetic field inside and outside the gap. The two curves

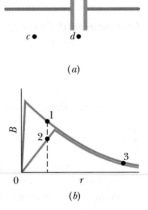

**FIG. 32-22** Question 3.

partially overlap. Which of the three points on the curves correspond to which of the four points of Fig. 32-22a?

**4** Figure 32-23 shows a parallel-plate capacitor and the current in the connecting wires that is discharging the capacitor. Are the directions of (a) electric field $\vec{E}$ and (b) displacement current $i_d$ leftward or rightward between the plates? (c) Is the magnetic field at point $P$ into or out of the page?

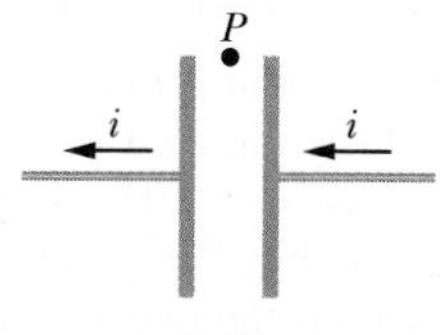

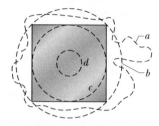

**FIG. 32-23**
Question 4.

**5** Figure 32-24 shows a face-on view of one of the two square plates of a parallel-plate capacitor, as well as four loops that are located between the plates. The capacitor is being discharged. (a) Neglecting fringing of the magnetic field, rank the loops according to the magnitude of $\oint \vec{B} \cdot d\vec{s}$ along them, greatest first. (b) Along which loop, if any, is the angle between the directions of $\vec{B}$ and $d\vec{s}$ constant (so that their dot product can easily be evaluated)? (c) Along which loop, if any, is $B$ constant (so that $B$ can be brought in front of the integral sign in Eq. 32-3)?

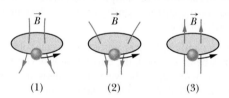

**FIG. 32-24**   Question 5.

**6** Figure 32-25 shows three loop models of an electron orbiting counterclockwise within a magnetic field. The fields are nonuniform for models 1 and 2 and uniform for model 3. For each model, are (a) the magnetic dipole moment of the loop and (b) the magnetic force on the loop directed up, directed down, or zero?

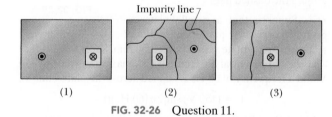

**FIG. 32-25**   Questions 6, 7, and 8.

**7** Replace the current loops of Question 6 and Fig. 32-25 with diamagnetic spheres. For each field, are (a) the magnetic

dipole moment of the sphere and (b) the magnetic force on the sphere directed up, directed down, or zero?

**8** Replace the current loops of Question 6 and Fig. 32-25 with paramagnetic spheres. For each field, are (a) the magnetic dipole moment of the sphere and (b) the magnetic force on the sphere directed up, directed down, or zero?

**9** An electron in an external magnetic field $\vec{B}_{ext}$ has its spin angular momentum $S_z$ antiparallel to $\vec{B}_{ext}$. If the electron undergoes a *spin-flip* so that $S_z$ is then parallel with $\vec{B}_{ext}$, must energy be supplied to or lost by the electron?

**10** Does the magnitude of the net force on the current loop of Figs. 32-12a and b increase, decrease, or remain the same if we increase (a) the magnitude of $\vec{B}_{ext}$ and (b) the divergence of $\vec{B}_{ext}$?

**11** Figure 32-26 represents three rectangular samples of a ferromagnetic material in which the magnetic dipoles of the domains have been directed out of the page (encircled dot) by a very strong applied field $B_0$. In each sample, an island domain still has its magnetic field directed into the page (encircled ×). Sample 1 is one (pure) crystal. The other samples contain impurities collected along lines; domains cannot easily spread across such lines.

The applied field is now to be reversed and its magnitude kept moderate. The change causes the island domain to grow. (a) Rank the three samples according to the success of that growth, greatest growth first. Ferromagnetic materials in which the magnetic dipoles are easily changed are said to be *magnetically soft;* when the changes are difficult, requiring strong applied fields, the materials are said to be *magnetically hard.* (b) Of the three samples, which is the most magnetically hard?

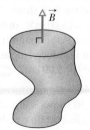

Impurity line

(1)          (2)          (3)

**FIG. 32-26**   Question 11.

## PROBLEMS

**GO**   Tutoring problem available (at instructor's discretion) in *WileyPLUS* and WebAssign
**SSM**   Worked-out solution available in Student Solutions Manual          **WWW**   Worked-out solution is at
•–•••   Number of dots indicates level of problem difficulty          **ILW**   Interactive solution is at——  http://www.wiley.com/college/halliday
Additional information available in *The Flying Circus of Physics* and at flyingcircusofphysics.com

### sec. 32-2   Gauss' Law for Magnetic Fields

•**1** The magnetic flux through each of five faces of a die (singular of "dice") is given by $\Phi_B = \pm N$ Wb, where $N$ (= 1 to 5) is the number of spots on the face. The flux is positive (outward) for $N$ even and negative (inward) for $N$ odd. What is the flux through the sixth face of the die?

•**2** Figure 32-27 shows a closed surface. Along the flat top face, which has a radius of 2.0 cm, a perpendicular magnetic field $\vec{B}$ of magnitude 0.30 T is directed outward. Along the flat

bottom face, a magnetic flux of 0.70 mWb is directed outward. What are the (a) magnitude and (b) direction (inward or outward) of the magnetic flux through the curved part of the surface?

••**3** A Gaussian surface in the shape of a right circular cylinder with end caps has a radius of 12.0 cm and

**FIG. 32-27** Problem 2.

a length of 80.0 cm. Through one end there is an inward magnetic flux of 25.0 $\mu$Wb. At the other end there is a uniform magnetic field of 1.60 mT, normal to the surface and directed outward. What are the (a) magnitude and (b) direction (inward or outward) of the net magnetic flux through the curved surface? **SSM ILW**

•••**4** Two wires, parallel to a $z$ axis and a distance $4r$ apart, carry equal currents $i$ in opposite directions, as shown in Fig. 32-28. A circular cylinder of radius $r$ and length $L$ has its axis on the $z$ axis, midway between the wires. Use Gauss' law

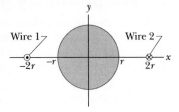

FIG. 32-28    Problem 4.

for magnetism to derive an expression for the net outward magnetic flux through the half of the cylindrical surface above the $x$ axis. (*Hint:* Find the flux through the portion of the $xz$ plane that lies within the cylinder.)

### sec. 32-3    Induced Magnetic Fields

•**5** The induced magnetic field at radial distance 6.0 mm from the central axis of a circular parallel-plate capacitor is $2.0 \times 10^{-7}$ T. The plates have radius 3.0 mm. At what rate $d\vec{E}/dt$ is the electric field between the plates changing? **SSM**

•**6** A capacitor with square plates of edge length $L$ is being discharged by a current of 0.75 A. Figure 32-29 is a head-on view of one of the plates from inside the capacitor. A dashed rectangular path is shown. If $L = 12$ cm, $W = 4.0$ cm, and $H = 2.0$ cm, what is the value of $\oint \vec{B} \cdot d\vec{s}$ around the dashed path?

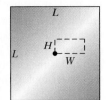

FIG. 32-29
Problem 6.

•••**7** Suppose that a parallel-plate capacitor has circular plates with radius $R = 30$ mm and a plate separation of 5.0 mm. Suppose also that a sinusoidal potential difference with a maximum value of 150 V and a frequency of 60 Hz is applied across the plates; that is,

$$V = (150 \text{ V}) \sin[2\pi(60 \text{ Hz})t].$$

(a) Find $B_{max}(R)$, the maximum value of the induced magnetic field that occurs at $r = R$. (b) Plot $B_{max}(r)$ for $0 < r < 10$ cm.

•••**8** A parallel-plate capacitor with circular plates of radius 40 mm is being discharged by a current of 6.0 A. At what radius (a) inside and (b) outside the capacitor gap is the magnitude of the induced magnetic field equal to 75% of its maximum value? (c) What is that maximum value?

•••**9** *Uniform electric flux.* Figure 32-30 shows a circular region of radius $R = 3.00$ cm in which a uniform electric flux is directed out of the plane of the page. The total electric flux through the region is given by $\Phi_E = (3.00 \text{ mV} \cdot \text{m/s})t$, where $t$ is in seconds. What is the magnitude of the magnetic field that is induced at radial distances (a) 2.00 cm and (b) 5.00 cm? **GO**

FIG. 32-30
Problems 9–12
and 25–28.

•••**10** *Nonuniform electric flux.* Figure 32-30 shows a circular region of radius $R = 3.00$ cm in which an electric flux is directed out of the plane of the page. The flux encircled by a concentric circle of radius $r$ is given by

$\Phi_{E,enc} = (0.600 \text{ V} \cdot \text{m/s})(r/R)t$, where $r \leq R$ and $t$ is in seconds. What is the magnitude of the induced magnetic field at radial distances (a) 2.00 cm and (b) 5.00 cm? **GO**

•••**11** *Uniform electric field.* In Fig. 32-30, a uniform electric field is directed out of the page within a circular region of radius $R = 3.00$ cm. The magnitude of the electric field is given by $E = (4.50 \times 10^{-3} \text{ V/m} \cdot \text{s})t$, where $t$ is in seconds. What is the magnitude of the induced magnetic field at radial distances (a) 2.00 cm and (b) 5.00 cm? **GO**

•••**12** *Nonuniform electric field.* In Fig. 32-30, an electric field is directed out of the page within a circular region of radius $R = 3.00$ cm. The magnitude of the electric field is given by $E = (0.500 \text{ V/m} \cdot \text{s})(1 - r/R)t$, where $t$ is in seconds and $r$ is the radial distance $(r \leq R)$. What is the magnitude of the induced magnetic field at radial distances (a) 2.00 cm and (b) 5.00 cm? **GO**

### sec. 32-4    Displacement Current

•**13** Prove that the displacement current in a parallel-plate capacitor of capacitance $C$ can be written as $i_d = C(dV/dt)$, where $V$ is the potential difference between the plates. **SSM**

•**14** A parallel-plate capacitor with circular plates of radius 0.10 m is being discharged. A circular loop of radius 0.20 m is concentric with the capacitor and halfway between the plates. The displacement current through the loop is 2.0 A. At what rate is the electric field between the plates changing?

•**15** At what rate must the potential difference between the plates of a parallel-plate capacitor with a 2.0 $\mu$F capacitance be changed to produce a displacement current of 1.5 A?

•**16** For the situation of Sample Problem 32-1, show that the magnitude of the current density of the displacement current is $J_d = \varepsilon_0(dE/dt)$ for $r \leq R$.

•••**17** As a parallel-plate capacitor with circular plates 20 cm in diameter is being charged, the current density of the displacement current in the region between the plates is uniform and has a magnitude of 20 A/m². (a) Calculate the magnitude $B$ of the magnetic field at a distance $r = 50$ mm from the axis of symmetry of this region. (b) Calculate $dE/dt$ in this region. **ILW**

•••**18** The magnitude of the electric field between the two circular parallel plates in Fig. 32-31 is $E = (4.0 \times 10^5) - (6.0 \times 10^4 t)$, with $E$ in volts per meter and $t$ in seconds. At $t = 0$, $\vec{E}$ is upward. The plate area is $4.0 \times 10^{-2}$ m². For $t \geq 0$, what are the (a) magnitude and (b) direction (up or down) of the displacement current between the plates and (c) is the direction of the induced magnetic field clockwise or counterclockwise in the figure?

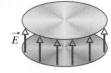

FIG. 32-31
Problem 18.

•••**19** In Fig. 32-32, a uniform electric field $\vec{E}$ collapses. The vertical axis scale is set by $E_s = 6.0 \times 10^5$ N/C, and the horizontal axis scale is set by $t_s = 12.0 \ \mu$s. Calculate the magnitude of the displacement current through a 1.6 m² area perpendicular to the field during each of the time intervals $a$, $b$, and $c$ shown on the graph. (Ignore the behavior at the ends of the intervals.) **ILW**

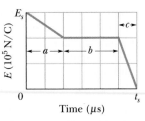

FIG. 32-32    Problem 19.

**••20** A capacitor with parallel circular plates of radius $R = 1.20$ cm is discharging via a current of 12.0 A. Consider a loop of radius $R/3$ that is centered on the central axis between the plates. (a) How much displacement current is encircled by the loop? The maximum induced magnetic field has a magnitude of 12.0 mT. At what radius (b) inside and (c) outside the capacitor gap is the magnitude of the induced magnetic field 3.00 mT?

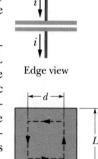

Edge view

FIG. 32-33
Problem 21.

**••21** In Fig. 32-33, a parallel-plate capacitor has square plates of edge length $L = 1.0$ m. A current of 2.0 A charges the capacitor, producing a uniform electric field $\vec{E}$ between the plates, with $\vec{E}$ perpendicular to the plates. (a) What is the displacement current $i_d$ through the region between the plates? (b) What is $dE/dt$ in this region? (c) What is the displacement current encircled by the square dashed path of edge length $d = 0.50$ m? (d) What is $\oint \vec{B} \cdot d\vec{s}$ around this square dashed path? SSM ILW

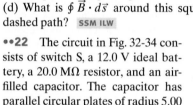

**••22** The circuit in Fig. 32-34 consists of switch S, a 12.0 V ideal battery, a 20.0 MΩ resistor, and an air-filled capacitor. The capacitor has parallel circular plates of radius 5.00 cm, separated by 3.00 mm. At time $t = 0$, switch S is closed to begin charging the capacitor. The electric field between the plates is uniform. At $t = 250$ μs, what is the magnitude of the magnetic field within the capacitor, at radial distance 3.00 cm?

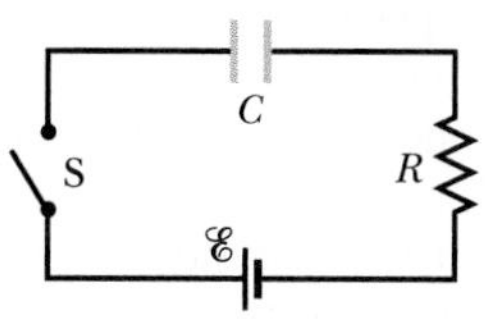

FIG. 32-34   Problem 22.

**••23** A silver wire has resistivity $\rho = 1.62 \times 10^{-8}$ Ω·m and a cross-sectional area of 5.00 mm². The current in the wire is uniform and changing at the rate of 2000 A/s when the current is 100 A. (a) What is the magnitude of the (uniform) electric field in the wire when the current in the wire is 100 A? (b) What is the displacement current in the wire at that time? (c) What is the ratio of the magnitude of the magnetic field due to the displacement current to that due to the current at a distance $r$ from the wire?

**••24** Figure 32-35*a* shows the current $i$ that is produced in a wire of resistivity $1.62 \times 10^{-8}$ Ω·m. The magnitude of the current versus time $t$ is shown in Fig. 32-35*b*. The vertical axis scale is set by $i_s = 10.0$ A, and the horizontal axis scale is set by $t_s = 50.0$ ms. Point $P$ is at radial distance 9.00 mm from the wire's center. Determine the magnitude of the magnetic field $\vec{B}_i$ at point $P$ due to the actual current $i$ in the wire at (a) $t = 20$ ms, (b) $t = 40$ ms, and (c) $t = 60$ ms. Next, assume that the electric field driving the current is confined to the wire. Then determine the magnitude of the magnetic field $\vec{B}_{id}$ at point $P$ due to the

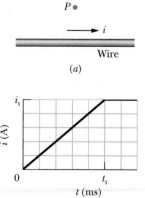

FIG. 32-35   Problem 24.

displacement current $i_d$ in the wire at (d) $t = 20$ ms, (e) $t = 40$ ms, and (f) $t = 60$ ms. At point $P$ at $t = 20$ s, what is the direction (into or out of the page) of (g) $\vec{B}_i$ and (h) $\vec{B}_{id}$? GO

**••25** *Uniform displacement-current density.* Figure 32-30 shows a circular region of radius $R = 3.00$ cm in which a displacement current is directed out of the page. The displacement current has a uniform density of magnitude $J_d = 6.00$ A/m². What is the magnitude of the magnetic field due to the displacement current at radial distances (a) 2.00 cm and (b) 5.00 cm?

**••26** *Uniform displacement current.* Figure 32-30 shows a circular region of radius $R = 3.00$ cm in which a uniform displacement current $i_d = 0.500$ A is out of the page. What is the magnitude of the magnetic field due to the displacement current at radial distances (a) 2.00 cm and (b) 5.00 cm?

**••27** *Nonuniform displacement-current density.* Figure 32-30 shows a circular region of radius $R = 3.00$ cm in which a displacement current is directed out of the page. The magnitude of the density of this displacement current is given by $J_d = (4.00 \text{ A/m}^2)(1 - r/R)$, where $r$ is the radial distance $(r \leq R)$. What is the magnitude of the magnetic field due to the displacement current at (a) $r = 2.00$ cm and (b) $r = 5.00$ cm?

**••28** *Nonuniform displacement current.* Figure 32-30 shows a circular region of radius $R = 3.00$ cm in which a displacement current $i_d$ is directed out of the page. The magnitude of the displacement current is given by $i_d = (3.00 \text{ A})(r/R)$, where $r$ is the radial distance $(r \leq R)$. What is the magnitude of the magnetic field due to $i_d$ at radial distances (a) 2.00 cm and (b) 5.00 cm?

**•••29** In Fig. 32-36, a capacitor with circular plates of radius $R = 18.0$ cm is connected to a source of emf $\mathcal{E} = \mathcal{E}_m \sin \omega t$, where $\mathcal{E}_m = 220$ V and $\omega = 130$ rad/s. The maximum value of the displacement current is $i_d = 7.60$ μA. Neglect fringing of the electric field at the edges of the plates. (a) What is the maximum value of the current $i$ in the circuit? (b) What is the maximum value of $d\Phi_E/dt$, where $\Phi_E$ is the electric flux through the region between the plates? (c) What is the separation $d$ between the plates? (d) Find the maximum value of the magnitude of $\vec{B}$ between the plates at a distance $r = 11.0$ cm from the center.

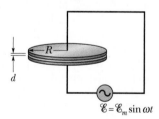

FIG. 32-36   Problem 29.

**sec. 32-6   Magnets**
**•30** Assume the average value of the vertical component of Earth's magnetic field is 43 μT (downward) for all of Arizona, which has an area of $2.95 \times 10^5$ km². What then are the (a) magnitude and (b) direction (inward or outward) of the net magnetic flux through the rest of Earth's surface (the entire surface excluding Arizona)?

**•31** In New Hampshire the average horizontal component of Earth's magnetic field in 1912 was 16 μT, and the average inclination or "dip" was 73°. What was the corresponding magnitude of Earth's magnetic field?

**sec. 32-7   Magnetism and Electrons**
**•32** What is the energy difference between parallel and antiparallel alignment of the $z$ component of an electron's

spin magnetic dipole moment with an external magnetic field of magnitude 0.25 T, directed parallel to the z axis?

•33 What is the measured component of the orbital magnetic dipole moment of an electron with (a) $m_\ell = 1$ and (b) $m_\ell = -2$?

•34 An electron is placed in a magnetic field $\vec{B}$ that is directed along a z axis. The energy difference between parallel and antiparallel alignments of the z component of the electron's spin magnetic moment with $\vec{B}$ is $6.00 \times 10^{-25}$ J. What is the magnitude of $\vec{B}$?

•35 If an electron in an atom has an orbital angular momentum with $m_\ell = 0$, what are the components (a) $L_{orb,z}$ and (b) $\mu_{orb,z}$? If the atom is in an external magnetic field $\vec{B}$ that has magnitude 35 mT and is directed along the z axis, what are (c) the potential energy $U_{orb}$ associated with $\vec{\mu}_{orb}$ and (d) the potential energy $U_{spin}$ associated with $\vec{\mu}_s$? If, instead, the electron has $m_\ell = -3$, what are (e) $L_{orb,z}$, (f) $\mu_{orb,z}$, (g) $U_{orb}$, and (h) $U_{spin}$? SSM WWW

•36 Figure 32-37a is a one-axis graph along which two of the allowed energy values (levels) of an atom are plotted. When the atom is placed in a magnetic field of 0.500 T, the graph changes to that of Fig. 32-37b because of the energy associated with $\vec{\mu}_{orb} \cdot \vec{B}$. (We neglect $\vec{\mu}_s$.) Level $E_1$ is unchanged, but level $E_2$ splits into a (closely spaced) triplet of levels. What are the allowed values of $m_\ell$ associated with (a) energy level $E_1$ and (b) energy level $E_2$? (c) In joules, what amount of energy is represented by the spacing between the triplet levels?

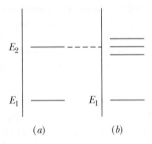

(a)          (b)

**FIG. 32-37** Problem 36.

### sec. 32-9 Diamagnetism

•37 Figure 32-38 shows a loop model (loop $L$) for a diamagnetic material. (a) Sketch the magnetic field lines within and about the material due to the bar magnet. What is the direction of (b) the loop's net magnetic dipole moment $\vec{\mu}$, (c) the conventional current $i$ in the loop (clockwise or counterclockwise in the figure), and (d) the magnetic force on the loop?

**FIG. 32-38** Problems 37 and 73.

•••38 Assume that an electron of mass $m$ and charge magnitude $e$ moves in a circular orbit of radius $r$ about a nucleus. A uniform magnetic field $\vec{B}$ is then established perpendicular to the plane of the orbit. Assuming also that the radius of the orbit does not change and that the change in the speed of the electron due to field $\vec{B}$ is small, find an expression for the change in the orbital magnetic dipole moment of the electron due to the field.

### sec. 32-10 Paramagnetism

•39 A magnet in the form of a cylindrical rod has a length of 5.00 cm and a diameter of 1.00 cm. It has a uniform magnetization of $5.30 \times 10^3$ A/m. What is its magnetic dipole moment? SSM ILW

•40 A 0.50 T magnetic field is applied to a paramagnetic gas whose atoms have an intrinsic magnetic dipole moment of

$1.0 \times 10^{-23}$ J/T. At what temperature will the mean kinetic energy of translation of the atoms equal the energy required to reverse such a dipole end for end in this magnetic field?

•41 A sample of the paramagnetic salt to which the magnetization curve of Fig. 32-14 applies is to be tested to see whether it obeys Curie's law. The sample is placed in a uniform 0.50 T magnetic field that remains constant throughout the experiment. The magnetization $M$ is then measured at temperatures ranging from 10 to 300 K. Will it be found that Curie's law is valid under these conditions?

•42 A sample of the paramagnetic salt to which the magnetization curve of Fig. 32-14 applies is held at room temperature (300 K). At what applied magnetic field will the degree of magnetic saturation of the sample be (a) 50% and (b) 90%? (c) Are these fields attainable in the laboratory?

••43 An electron with kinetic energy $K_e$ travels in a circular path that is perpendicular to a uniform magnetic field, which is in the positive direction of a z axis. The electron's motion is subject only to the force due to the field. (a) Show that the magnetic dipole moment of the electron due to its orbital motion has magnitude $\mu = K_e/B$ and that it is in the direction opposite that of $\vec{B}$. What are the (b) magnitude and (c) direction of the magnetic dipole moment of a positive ion with kinetic energy $K_i$ under the same circumstances? (d) An ionized gas consists of $5.3 \times 10^{21}$ electrons/m³ and the same number density of ions. Take the average electron kinetic energy to be $6.2 \times 10^{-20}$ J and the average ion kinetic energy to be $7.6 \times 10^{-21}$ J. Calculate the magnetization of the gas when it is in a magnetic field of 1.2 T.

••44 Figure 32-39 gives the magnetization curve for a paramagnetic material. The vertical axis scale is set by $a = 0.15$, and the horizontal axis scale is set by $b = 0.2$ T/K. Let $\mu_{sam}$ be the measured net magnetic moment of a sample of the material and $\mu_{max}$ be the maximum possible net magnetic moment of that sample. According to Curie's law, what would be the ratio $\mu_{sam}/\mu_{max}$ were the sample placed in a uniform magnetic field of magnitude 0.800 T, at a temperature of 2.00 K?

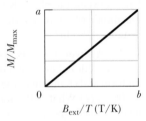

**FIG. 32-39** Problem 44.

•••45 Consider a solid containing $N$ atoms per unit volume, each atom having a magnetic dipole moment $\vec{\mu}$. Suppose the direction of $\vec{\mu}$ can be only parallel or antiparallel to an externally applied magnetic field $\vec{B}$ (this will be the case if $\vec{\mu}$ is due to the spin of a single electron). According to statistical mechanics, the probability of an atom being in a state with energy $U$ is proportional to $e^{-U/kT}$, where $T$ is the temperature and $k$ is Boltzmann's constant. Thus, because energy $U$ is $-\vec{\mu} \cdot \vec{B}$, the fraction of atoms whose dipole moment is parallel to $\vec{B}$ is proportional to $e^{\mu B/kT}$ and the fraction of atoms whose dipole moment is antiparallel to $\vec{B}$ is proportional to $e^{-\mu B/kT}$. (a) Show that the magnitude of the magnetization of this solid is $M = N\mu \tanh(\mu B/kT)$. Here tanh is the hyperbolic tangent function: $\tanh(x) = (e^x - e^{-x})/(e^x + e^{-x})$. (b) Show that the result given in (a) reduces to $M = N\mu^2 B/kT$ for $\mu B \ll kT$. (c) Show that the result of (a) reduces to $M = N\mu$ for $\mu B \gg kT$. (d) Show that both (b) and (c) agree qualitatively with Fig. 32-14. SSM

## sec. 32-11 **Ferromagnetism**

•46 The magnitude of the dipole moment associated with an atom of iron in an iron bar is $2.1 \times 10^{-23}$ J/T. Assume that all the atoms in the bar, which is 5.0 cm long and has a cross-sectional area of 1.0 cm$^2$, have their dipole moments aligned. (a) What is the dipole moment of the bar? (b) What torque must be exerted to hold this magnet perpendicular to an external field of magnitude 1.5 T? (The density of iron is 7.9 g/cm$^3$.)

•47 The exchange coupling mentioned in Section 32-11 as being responsible for ferromagnetism is *not* the mutual magnetic interaction between two elementary magnetic dipoles. To show this, calculate (a) the magnitude of the magnetic field a distance of 10 nm away, along the dipole axis, from an atom with magnetic dipole moment $1.5 \times 10^{-23}$ J/T (cobalt), and (b) the minimum energy required to turn a second identical dipole end for end in this field. (c) By comparing the latter with the results of Sample Problem 32-3, what can you conclude? SSM

•48 Measurements in mines and boreholes indicate that Earth's interior temperature increases with depth at the average rate of 30 C°/km. Assuming a surface temperature of 10°C, at what depth does iron cease to be ferromagnetic? (The Curie temperature of iron varies very little with pressure.)

•49 The saturation magnetization $M_{max}$ of the ferromagnetic metal nickel is $4.70 \times 10^5$ A/m. Calculate the magnetic dipole moment of a single nickel atom. (The density of nickel is 8.90 g/cm$^3$, and its molar mass is 58.71 g/mol.)

••50 You place a magnetic compass on a horizontal surface, allow the needle to settle into its equilibrium position, and then give the compass a gentle wiggle to cause the needle to oscillate about that equilibrium position. The frequency of oscillation is 0.312 Hz. Earth's magnetic field at the location of the compass has a horizontal component of 18.0 $\mu$T. The needle has a magnetic moment of 0.680 mJ/T. What is the needle's rotational inertia about its (vertical) axis of rotation? GO

••51 A Rowland ring is formed of ferromagnetic material. It is circular in cross section, with an inner radius of 5.0 cm and an outer radius of 6.0 cm, and is wound with 400 turns of wire. (a) What current must be set up in the windings to attain a toroidal field of magnitude $B_0 = 0.20$ mT? (b) A secondary coil wound around the toroid has 50 turns and resistance 8.0 $\Omega$. If, for this value of $B_0$, we have $B_M = 800B_0$, how much charge moves through the secondary coil when the current in the toroid windings is turned on?

••52 A magnetic rod with length 6.00 cm, radius 3.00 mm, and (uniform) magnetization $2.70 \times 10^3$ A/m can turn about its center like a compass needle. It is placed in a uniform magnetic field $\vec{B}$ of magnitude 35.0 mT, such that the directions of its dipole moment and $\vec{B}$ make an angle of 68.0°. (a) What is the magnitude of the torque on the rod due to $\vec{B}$? (b) What is the change in the magnetic potential energy of the rod if the angle changes to 34.0°?

••53 The magnitude of the magnetic dipole moment of Earth is $8.0 \times 10^{22}$ J/T. (a) If the origin of this magnetism were a magnetized iron sphere at the center of Earth, what would be its radius? (b) What fraction of the volume of Earth would such a sphere occupy? Assume complete alignment of the dipoles. The density of Earth's inner core is 14 g/cm$^3$. The mag-

netic dipole moment of an iron atom is $2.1 \times 10^{-23}$ J/T. (*Note:* Earth's inner core is in fact thought to be in both liquid and solid forms and partly iron, but a permanent magnet as the source of Earth's magnetism has been ruled out by several considerations. For one, the temperature is certainly above the Curie point.) SSM ILW WWW

## Additional Problems

54 Two plates (as in Fig. 32-7) are being discharged by a constant current. Each plate has a radius of 4.00 cm. During the discharging, at a point between the plates at radial distance 2.00 cm from the central axis, the magnetic field has a magnitude of 12.5 nT. (a) What is the magnitude of the magnetic field at radial distance 6.00 cm? (b) What is the current in the wires attached to the plates?

55 The magnetic field of Earth can be approximated as the magnetic field of a dipole. The horizontal and vertical components of this field at any distance $r$ from Earth's center are given by

$$B_h = \frac{\mu_0 \mu}{4\pi r^3} \cos \lambda_m, \qquad B_v = \frac{\mu_0 \mu}{2\pi r^3} \sin \lambda_m,$$

where $\lambda_m$ is the *magnetic latitude* (this type of latitude is measured from the geomagnetic equator toward the north or south geomagnetic pole). Assume that Earth's magnetic dipole moment has magnitude $\mu = 8.00 \times 10^{22}$ A·m$^2$. (a) Show that the magnitude of Earth's field at latitude $\lambda_m$ is given by

$$B = \frac{\mu_0 \mu}{4\pi r^3} \sqrt{1 + 3 \sin^2 \lambda_m}.$$

(b) Show that the inclination $\phi_i$ of the magnetic field is related to the magnetic latitude $\lambda_m$ by $\tan \phi_i = 2 \tan \lambda_m$. SSM

56 Use the results displayed in Problem 55 to predict the (a) magnitude and (b) inclination of Earth's magnetic field at the geomagnetic equator, the (c) magnitude and (d) inclination at geomagnetic latitude 60.0°, and the (e) magnitude and (f) inclination at the north geomagnetic pole.

57 Earth has a magnetic dipole moment of $8.0 \times 10^{22}$ J/T. (a) What current would have to be produced in a single turn of wire extending around Earth at its geomagnetic equator if we wished to set up such a dipole? Could such an arrangement be used to cancel out Earth's magnetism (b) at points in space well above Earth's surface or (c) on Earth's surface?

58 Using the approximations given in Problem 55, find (a) the altitude above Earth's surface where the magnitude of its magnetic field is 50.0% of the surface value at the same latitude; (b) the maximum magnitude of the magnetic field at the core–mantle boundary, 2900 km below Earth's surface; and the (c) magnitude and (d) inclination of Earth's magnetic field at the north geographic pole. (e) Suggest why the values you calculated for (c) and (d) differ from measured values.

59 A parallel-plate capacitor with circular plates of radius 55.0 mm is being charged. At what radius (a) inside and (b) outside the capacitor gap is the magnitude of the induced magnetic field equal to 50.0% of its maximum value?

60 A charge $q$ is distributed uniformly around a thin ring of radius $r$. The ring is rotating about an axis through its center and perpendicular to its plane, at an angular speed $\omega$. (a) Show

that the magnetic moment due to the rotating charge has magnitude $\mu = \frac{1}{2}q\omega r^2$. (b) What is the direction of this magnetic moment if the charge is positive?

**61** If an electron in an atom has orbital angular momentum with $m_\ell$ values limited by $\pm 3$, how many values of (a) $L_{\text{orb},z}$ and (b) $\mu_{\text{orb},z}$ can the electron have? In terms of $h$, $m$, and $e$, what is the greatest allowed magnitude for (c) $L_{\text{orb},z}$ and (d) $\mu_{\text{orb},z}$? (e) What is the greatest allowed magnitude for the $z$ component of the electron's *net* angular momentum (orbital plus spin)? (f) How many values (signs included) are allowed for the $z$ component of its net angular momentum? **SSM**

**62** The capacitor in Fig. 32-7 is being charged with a 2.50 A current. The wire radius is 1.50 mm, and the plate radius is 2.00 cm. Assume that the current $i$ in the wire and the displacement current $i_d$ in the capacitor gap are both uniformly distributed. What is the magnitude of the magnetic field due to $i$ at the following radial distances from the wire's center: (a) 1.00 mm (inside the wire), (b) 3.00 mm (outside the wire), and (c) 2.20 cm (outside the wire)? What is the magnitude of the magnetic field due to $i_d$ at the following radial distances from the central axis between the plates: (d) 1.00 mm (inside the gap), (e) 3.00 mm (inside the gap), and (f) 2.20 cm (outside the gap)? (g) Explain why the fields at the two smaller radii are so different for the wire and the gap but the fields at the largest radius are not.

**63** Suppose that $\pm 4$ are the limits to the values of $m_\ell$ for an electron in an atom. (a) How many different values of the $z$ component $\mu_{\text{orb},z}$ of the electron's orbital magnetic dipole moment are possible? (b) What is the greatest magnitude of those possible values? Next, suppose that the atom is in a magnetic field of magnitude 0.250 T, in the positive direction of the $z$ axis. What are (c) the maximum potential energy and (d) the minimum potential energy associated with those possible values of $\mu_{\text{orb},z}$?

**64** What is the measured component of the orbital magnetic dipole moment of an electron with the values (a) $m_\ell = 3$ and (b) $m_\ell = -4$?

**65** A magnetic compass has its needle, of mass 0.050 kg and length 4.0 cm, aligned with the horizontal component of Earth's magnetic field at a place where that component has the value $B_h = 16\ \mu$T. After the compass is given a momentary gentle shake, the needle oscillates with angular frequency $\omega = 45$ rad/s. Assuming that the needle is a uniform thin rod mounted at its center, find the magnitude of its magnetic dipole moment.

**66** A parallel-plate capacitor with circular plates is being charged. Consider a circular loop centered on the central axis and located between the plates. If the loop radius of 3.00 cm is greater than the plate radius, what is the displacement current between the plates when the magnetic field along the loop has magnitude 2.00 $\mu$T?

**67** In Fig. 32-40, a parallel-plate capacitor is being discharged by a current $i = 5.0$ A. The plates are square with edge length $L = 8.0$ mm. (a) What is the rate at which the electric field

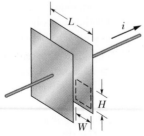

**FIG. 32-40** Problem 67.

between the plates is changing? (b) What is the value of $\oint \vec{B} \cdot d\vec{s}$ around the dashed path, where $H = 2.0$ mm and $W = 3.0$ mm?

**68** A magnetic flux of 7.0 mWb is directed outward through the flat bottom face of the closed surface shown in Fig. 32-41. Along the flat top face (which has a radius of 4.2 cm) there is a 0.40 T magnetic field $\vec{B}$ directed perpendicular to the face. What are the (a) magnitude and (b) direction (inward or outward) of the magnetic flux through the curved part of the surface?

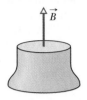

**FIG. 32-41** Problem 68.

**69** A parallel-plate capacitor with circular plates of radius $R = 16$ mm and gap width $d = 5.0$ mm has a uniform electric field between the plates. Starting at time $t = 0$, the potential difference between the two plates is $V = (100$ V$)e^{-t/\tau}$, where the time constant $\tau = 12$ ms. At radial distance $r = 0.80R$ from the central axis, what is the magnetic field magnitude (a) as a function of time for $t \geq 0$ and (b) at time $t = 3\tau$?

**70** A sample of the paramagnetic salt to which the magnetization curve of Fig. 32-14 applies is immersed in a uniform magnetic field of 2.0 T. At what temperature will the degree of magnetic saturation of the sample be (a) 50% and (b) 90%?

**71** A parallel-plate capacitor with circular plates of radius $R$ is being discharged. The displacement current through a central circular area, parallel to the plates and with radius $R/2$, is 2.0 A. What is the discharging current?

**72** Figure 32-42 gives the variation of an electric field that is perpendicular to a circular area of 2.0 m². During the time period shown, what is the greatest displacement current through the area?

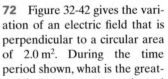

**FIG. 32-42** Problem 72.

**73** Figure 32-38 shows a loop model (loop $L$) for a paramagnetic material. (a) Sketch the field lines through and about the material due to the magnet. What is the direction of (b) the loop's net magnetic dipole moment $\vec{\mu}$, (c) the conventional current $i$ in the loop (clockwise or counterclockwise in the figure), and (d) the magnetic force acting on the loop?

**74** In the lowest energy state of the hydrogen atom, the most probable distance of the single electron from the central proton (the nucleus) is $5.2 \times 10^{-11}$ m. (a) Compute the magnitude of the proton's electric field at that distance. The component $\mu_{s,z}$ of the proton's spin magnetic dipole moment measured on a $z$ axis is $1.4 \times 10^{-26}$ J/T. (b) Compute the magnitude of the proton's magnetic field at the distance $5.2 \times 10^{-11}$ m on the $z$ axis. (*Hint:* Use Eq. 29-27.) (c) What is the ratio of the spin magnetic dipole moment of the electron to that of the proton?

**75** In Fig. 32-43, a bar magnet lies near a paper cylinder. (a) Sketch the magnetic field lines that pass through the surface of the cylinder. (b) What is the sign of $\vec{B} \cdot d\vec{A}$ for every area $d\vec{A}$ on the surface? (c) Does this contradict Gauss' law for magnetism? Explain.

**FIG. 32-43** Problem 75.

# Electromagnetic Waves

Norbert Rosing/National Geographic/Getty Images, Inc.

*On bright days the sky can display a variety of bright spots and arcs (halos). Probably the most common display is a bright spot called a sun dog that appears to the left or right of the Sun and sometimes on both sides. Because sun dogs often show colors, they can be mistaken for rainbows. However, sun dogs are not produced by water drops as rainbows are.*

## What produces sun dogs?

The answer is in this chapter.

## 33-1 | WHAT IS PHYSICS?

The information age in which we live is based almost entirely on the physics of electromagnetic waves. Like it or not, we are now globally connected by television, telephones, and the Web. And like it or not, we are constantly immersed in those signals because of television, radio, and telephone transmitters.

Much of this global interconnection of information processors was not imagined by even the most visionary engineers of 20 years ago. The challenge for today's engineers is trying to envision what the global interconnection will be like 20 years from now. The starting point in meeting that challenge is understanding the basic physics of electromagnetic waves, which come in so many different types that they are poetically said to form *Maxwell's rainbow*.

## 33-2 | Maxwell's Rainbow

The crowning achievement of James Clerk Maxwell (see Chapter 32) was to show that a beam of light is a traveling wave of electric and magnetic fields—an **electromagnetic wave**—and thus that optics, the study of visible light, is a branch of electromagnetism. In this chapter we move from one to the other: we conclude our discussion of strictly electrical and magnetic phenomena, and we build a foundation for optics.

In Maxwell's time (the mid 1800s), the visible, infrared, and ultraviolet forms of light were the only electromagnetic waves known. Spurred on by Maxwell's work, however, Heinrich Hertz discovered what we now call radio waves and verified that they move through the laboratory at the same speed as visible light.

As Fig. 33-1 shows, we now know a wide *spectrum* (or range) of electromagnetic waves: Maxwell's rainbow. Consider the extent to which are immersed in electromagnetic waves throughout this spectrum. The Sun, whose radiations define the environment in which we as a species have evolved and adapted, is the dominant source. We are also crisscrossed by radio and television signals. Microwaves from radar systems and from telephone relay systems may reach us. There are electromagnetic waves from lightbulbs, from the heated engine blocks of automobiles, from x-ray machines, from lightning flashes, and from buried

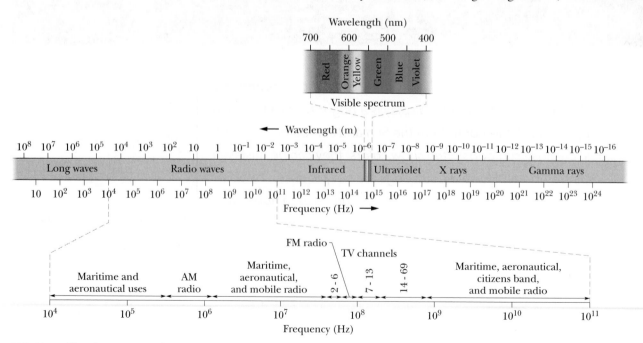

**FIG. 33-1**  The electromagnetic spectrum.

radioactive materials. Beyond this, radiation reaches us from stars and other objects in our galaxy and from other galaxies. Electromagnetic waves also travel in the other direction. Television signals, transmitted from Earth since about 1950, have now taken news about us (along with episodes of *I Love Lucy,* albeit *very* faintly) to whatever technically sophisticated inhabitants there may be on whatever planets may encircle the nearest 400 or so stars.

In the wavelength scale in Fig. 33-1 (and similarly the corresponding frequency scale), each scale marker represents a change in wavelength (and correspondingly in frequency) by a factor of 10. The scale is open-ended; the wavelengths of electromagnetic waves have no inherent upper or lower bound.

Certain regions of the electromagnetic spectrum in Fig. 33-1 are identified by familiar labels, such as *x rays* and *radio waves.* These labels denote roughly defined wavelength ranges within which certain kinds of sources and detectors of electromagnetic waves are in common use. Other regions of Fig. 33-1, such as those labeled TV channels and AM radio, represent specific wavelength bands assigned by law for certain commercial or other purposes. There are no gaps in the electromagnetic spectrum—and all electromagnetic waves, no matter where they lie in the spectrum, travel through *free space* (vacuum) with the same speed c.

The visible region of the spectrum is of course of particular interest to us. Figure 33-2 shows the relative sensitivity of the human eye to light of various wavelengths. The center of the visible region is about 555 nm, which produces the sensation that we call yellow-green.

The limits of this visible spectrum are not well defined because the eye-sensitivity curve approaches the zero-sensitivity line asymptotically at both long and short wavelengths. If we take the limits, arbitrarily, as the wavelengths at which eye sensitivity has dropped to 1% of its maximum value, these limits are about 430 and 690 nm; however, the eye can detect electromagnetic waves somewhat beyond these limits if they are intense enough.

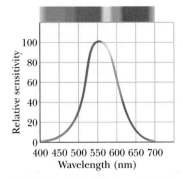

**FIG. 33-2** The relative sensitivity of the average human eye to electromagnetic waves at different wavelengths. This portion of the electromagnetic spectrum to which the eye is sensitive is called *visible light.*

## 33-3 | The Traveling Electromagnetic Wave, Qualitatively

Some electromagnetic waves, including x rays, gamma rays, and visible light, are *radiated* (emitted) from sources that are of atomic or nuclear size, where quantum physics rules. Here we discuss how other electromagnetic waves are generated. To simplify matters, we restrict ourselves to that region of the spectrum (wavelength $\lambda \approx 1$ m) in which the source of the *radiation* (the emitted waves) is both macroscopic and of manageable dimensions.

Figure 33-3 shows, in broad outline, the generation of such waves. At its heart is an *LC oscillator,* which establishes an angular frequency $\omega$ ($= 1/\sqrt{LC}$). Charges and currents in this circuit vary sinusoidally at this frequency, as depicted in Fig. 31-1. An external source—possibly an ac generator—must be included to supply energy to compensate both for thermal losses in the circuit and for energy carried away by the radiated electromagnetic wave.

The *LC* oscillator of Fig. 33-3 is coupled by a transformer and a transmission line to an *antenna,* which consists essentially of two thin, solid, conducting rods.

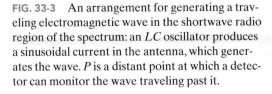

**FIG. 33-3** An arrangement for generating a traveling electromagnetic wave in the shortwave radio region of the spectrum: an *LC* oscillator produces a sinusoidal current in the antenna, which generates the wave. *P* is a distant point at which a detector can monitor the wave traveling past it.

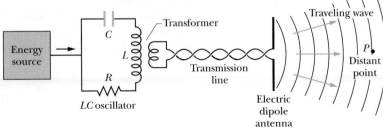

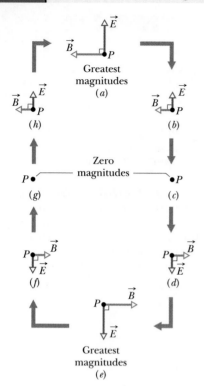

**FIG. 33-4** (a)–(h) The variation in the electric field $\vec{E}$ and the magnetic field $\vec{B}$ at the distant point $P$ of Fig. 33-3 as one wavelength of the electromagnetic wave travels past it. In this perspective, the wave is traveling directly out of the page. The two fields vary sinusoidally in magnitude and direction. Note that they are always perpendicular to each other and to the wave's direction of travel.

Through this coupling, the sinusoidally varying current in the oscillator causes charge to oscillate sinusoidally along the rods of the antenna at the angular frequency $\omega$ of the $LC$ oscillator. The current in the rods associated with this movement of charge also varies sinusoidally, in magnitude and direction, at angular frequency $\omega$. The antenna has the effect of an electric dipole whose electric dipole moment varies sinusoidally in magnitude and direction along the antenna.

Because the dipole moment varies in magnitude and direction, the electric field produced by the dipole varies in magnitude and direction. Also, because the current varies, the magnetic field produced by the current varies in magnitude and direction. However, the changes in the electric and magnetic fields do not happen everywhere instantaneously; rather, the changes travel outward from the antenna at the speed of light $c$. Together the changing fields form an electromagnetic wave that travels away from the antenna at speed $c$. The angular frequency of this wave is $\omega$, the same as that of the $LC$ oscillator.

Figure 33-4 shows how the electric field $\vec{E}$ and the magnetic field $\vec{B}$ change with time as one wavelength of the wave sweeps past the distant point $P$ of Fig. 33-3; in each part of Fig. 33-4, the wave is traveling directly out of the page. (We choose a distant point so that the curvature of the waves suggested in Fig. 33-3 is small enough to neglect. At such points, the wave is said to be a *plane wave*, and discussion of the wave is much simplified.) Note several key features in Fig. 33-4; they are present regardless of how the wave is created:

1. The electric and magnetic fields $\vec{E}$ and $\vec{B}$ are always perpendicular to the direction in which the wave is traveling. Thus, the wave is a *transverse wave,* as discussed in Chapter 16.

2. The electric field is always perpendicular to the magnetic field.

3. The cross product $\vec{E} \times \vec{B}$ always gives the direction in which the wave travels.

4. The fields always vary sinusoidally, just like the transverse waves discussed in Chapter 16. Moreover, the fields vary with the same frequency and *in phase* (in step) with each other.

In keeping with these features, we can assume that the electromagnetic wave is traveling toward $P$ in the positive direction of an $x$ axis, that the electric field in Fig. 33-4 is oscillating parallel to the $y$ axis, and that the magnetic field is then oscillating parallel to the $z$ axis (using a right-handed coordinate system, of course). Then we can write the electric and magnetic fields as sinusoidal functions of position $x$ (along the path of the wave) and time $t$:

$$E = E_m \sin(kx - \omega t), \tag{33-1}$$

$$B = B_m \sin(kx - \omega t), \tag{33-2}$$

in which $E_m$ and $B_m$ are the amplitudes of the fields and, as in Chapter 16, $\omega$ and $k$ are the angular frequency and angular wave number of the wave, respectively. From these equations, we note that not only do the two fields form the electromagnetic wave but each also forms its own wave. Equation 33-1 gives the *electric wave component* of the electromagnetic wave, and Eq. 33-2 gives the *magnetic wave component*. As we shall discuss below, these two wave components cannot exist independently.

From Eq. 16-13, we know that the speed of the wave is $\omega/k$. However, because this is an electromagnetic wave, its speed (in vacuum) is given the symbol $c$ rather than $v$. In the next section you will see that $c$ has the value

$$c = \frac{1}{\sqrt{\mu_0 \varepsilon_0}} \quad \text{(wave speed)}, \tag{33-3}$$

which is about $3.0 \times 10^8$ m/s. In other words,

All electromagnetic waves, including visible light, have the same speed $c$ in vacuum.

You will also see that the wave speed $c$ and the amplitudes of the electric and magnetic fields are related by

$$\frac{E_m}{B_m} = c \qquad \text{(amplitude ratio)}. \qquad (33\text{-}4)$$

If we divide Eq. 33-1 by Eq. 33-2 and then substitute with Eq. 33-4, we find that the magnitudes of the fields at every instant and at any point are related by

$$\frac{E}{B} = c \qquad \text{(magnitude ratio)}. \qquad (33\text{-}5)$$

We can represent the electromagnetic wave as in Fig. 33-5a, with a *ray* (a directed line showing the wave's direction of travel) or with *wavefronts* (imaginary surfaces over which the wave has the same magnitude of electric field), or both. The two wavefronts shown in Fig. 33-5a are separated by one wavelength $\lambda \ (= 2\pi/k)$ of the wave. (Waves traveling in approximately the same direction form a *beam*, such as a laser beam, which can also be represented with a ray.)

We can also represent the wave as in Fig. 33-5b, which shows the electric and magnetic field vectors in a "snapshot" of the wave at a certain instant. The curves through the tips of the vectors represent the sinusoidal oscillations given by Eqs. 33-1 and 33-2; the wave components $\vec{E}$ and $\vec{B}$ are in phase, perpendicular to each other, and perpendicular to the wave's direction of travel.

Interpretation of Fig. 33-5b requires some care. The similar drawings for a transverse wave on a taut string that we discussed in Chapter 16 represented the up and down displacement of sections of the string as the wave passed (*something actually moved*). Figure 33-5b is more abstract. At the instant shown, the electric and magnetic fields each have a certain magnitude and direction (but always perpendicular to the $x$ axis) at each point along the $x$ axis. We choose to represent these vector quantities with a pair of arrows for each point, and so we must draw arrows of different lengths for different points, all directed away from the $x$ axis, like thorns on a rose stem. However, the arrows represent field values only at points that are on the $x$ axis. Neither the arrows nor the sinusoidal curves represent a sideways motion of anything, nor do the arrows connect points on the $x$ axis with points off the axis.

Drawings like Fig. 33-5 help us visualize what is actually a very complicated situation. First consider the magnetic field. Because it varies sinusoidally, it induces (via Faraday's law of induction) a perpendicular electric field that also varies sinusoidally. However, because that electric field is varying sinusoidally, it induces (via Maxwell's law of induction) a perpendicular magnetic field that also

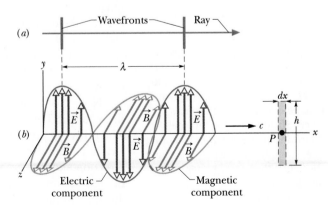

**FIG. 33-5** (a) An electromagnetic wave represented with a ray and two wavefronts; the wavefronts are separated by one wavelength $\lambda$. (b) The same wave represented in a "snapshot" of its electric field $\vec{E}$ and magnetic field $\vec{B}$ at points on the $x$ axis, along which the wave travels at speed $c$. As it travels past point $P$, the fields vary as shown in Fig. 33-4. The electric component of the wave consists of only the electric fields; the magnetic component consists of only the magnetic fields. The dashed rectangle at $P$ is used in Fig. 33-6.

varies sinusoidally. And so on. The two fields continuously create each other via induction, and the resulting sinusoidal variations in the fields travel as a wave — the electromagnetic wave. Without this amazing result, we could not see; indeed, because we need electromagnetic waves from the Sun to maintain Earth's temperature, without this result we could not even exist.

### A Most Curious Wave

The waves we discussed in Chapters 16 and 17 require a *medium* (some material) through which or along which to travel. We had waves traveling along a string, through Earth, and through the air. However, an electromagnetic wave (let's use the term *light wave* or *light*) is curiously different in that it requires no medium for its travel. It can, indeed, travel through a medium such as air or glass, but it can also travel through the vacuum of space between a star and us.

Once the special theory of relativity became accepted, long after Einstein published it in 1905, the speed of light waves was realized to be special. One reason is that light has the same speed regardless of the frame of reference from which it is measured. If you send a beam of light along an axis and ask several observers to measure its speed while they move at different speeds along that axis, either in the direction of the light or opposite it, they will all measure the *same speed* for the light. This result is an amazing one and quite different from what would have been found if those observers had measured the speed of any other type of wave; for other waves, the speed of the observers relative to the wave would have affected their measurements.

The meter has now been defined so that the speed of light (any electromagnetic wave) in vacuum has the exact value

$$c = 299\ 792\ 458\ \text{m/s},$$

which can be used as a standard. In fact, if you now measure the travel time of a pulse of light from one point to another, you are not really measuring the speed of the light but rather the distance between those two points.

## 33-4 | The Traveling Electromagnetic Wave, Quantitatively

We shall now derive Eqs. 33-3 and 33-4 and, even more important, explore the dual induction of electric and magnetic fields that gives us light.

### Equation 33-4 and the Induced Electric Field

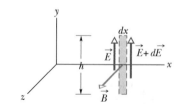

**FIG. 33-6**   As the electromagnetic wave travels rightward past point $P$ in Fig. 33-5$b$, the sinusoidal variation of the magnetic field $\vec{B}$ through a rectangle centered at $P$ induces electric fields along the rectangle. At the instant shown, $\vec{B}$ is decreasing in magnitude and the induced electric field is therefore greater in magnitude on the right side of the rectangle than on the left.

The dashed rectangle of dimensions $dx$ and $h$ in Fig. 33-6 is fixed at point $P$ on the $x$ axis and in the $xy$ plane (it is shown on the right in Fig. 33-5$b$). As the electromagnetic wave moves rightward past the rectangle, the magnetic flux $\Phi_B$ through the rectangle changes and—according to Faraday's law of induction—induced electric fields appear throughout the region of the rectangle. We take $\vec{E}$ and $\vec{E} + d\vec{E}$ to be the induced fields along the two long sides of the rectangle. These induced electric fields are, in fact, the electrical component of the electromagnetic wave.

Note the small red portion of the magnetic field component curve far from the $y$ axis in Fig. 33-5$b$. Let's consider the induced electric fields at the instant when this red portion of the magnetic component is passing through the rectangle. Just then, the magnetic field through the rectangle points in the positive $z$ direction and is decreasing in magnitude (the magnitude was greater just before the red section arrived). Because the magnetic field is decreasing, the magnetic flux $\Phi_B$ through the rectangle is also decreasing. According to Faraday's law, this change in flux is opposed by induced electric fields, which produce a magnetic field $\vec{B}$ in the positive $z$ direction.

According to Lenz's law, this in turn means that if we imagine the boundary of the rectangle to be a conducting loop, a counterclockwise induced current would have to appear in it. There is, of course, no conducting loop; but this analysis shows that the induced electric field vectors $\vec{E}$ and $\vec{E} + d\vec{E}$ are indeed oriented as shown in Fig. 33-6, with the magnitude of $\vec{E} + d\vec{E}$ greater than that of $\vec{E}$. Otherwise, the net induced electric field would not act counterclockwise around the rectangle.

Let us now apply Faraday's law of induction,

$$\oint \vec{E} \cdot d\vec{s} = -\frac{d\Phi_B}{dt}, \tag{33-6}$$

counterclockwise around the rectangle of Fig. 33-6. There is no contribution to the integral from the top or bottom of the rectangle because $\vec{E}$ and $d\vec{s}$ are perpendicular to each other there. The integral then has the value

$$\oint \vec{E} \cdot d\vec{s} = (E + dE)h - Eh = h\,dE. \tag{33-7}$$

The flux $\Phi_B$ through this rectangle is

$$\Phi_B = (B)(h\,dx), \tag{33-8}$$

where $B$ is the average magnitude of $\vec{B}$ within the rectangle and $h\,dx$ is the area of the rectangle. Differentiating Eq. 33-8 with respect to $t$ gives

$$\frac{d\Phi_B}{dt} = h\,dx\,\frac{dB}{dt}. \tag{33-9}$$

If we substitute Eqs. 33-7 and 33-9 into Eq. 33-6, we find

$$h\,dE = -h\,dx\,\frac{dB}{dt}$$

or

$$\frac{dE}{dx} = -\frac{dB}{dt}. \tag{33-10}$$

Actually, both $B$ and $E$ are functions of *two* variables, $x$ and $t$, as Eqs. 33-1 and 33-2 show. However, in evaluating $dE/dx$, we must assume that $t$ is constant because Fig. 33-6 is an "instantaneous snapshot." Also, in evaluating $dB/dt$ we must assume that $x$ is constant because we are dealing with the time rate of change of $B$ at a particular place, the point $P$ in Fig. 33-5$b$. The derivatives under these circumstances are *partial derivatives*, and Eq. 33-10 must be written

$$\frac{\partial E}{\partial x} = -\frac{\partial B}{\partial t}. \tag{33-11}$$

The minus sign in this equation is appropriate and necessary because, although $E$ is increasing with $x$ at the site of the rectangle in Fig. 33-6, $B$ is decreasing with $t$.

From Eq. 33-1 we have

$$\frac{\partial E}{\partial x} = kE_m \cos(kx - \omega t)$$

and from Eq. 33-2

$$\frac{\partial B}{\partial t} = -\omega B_m \cos(kx - \omega t).$$

Then Eq. 33-11 reduces to

$$kE_m \cos(kx - \omega t) = \omega B_m \cos(kx - \omega t). \tag{33-12}$$

The ratio $\omega/k$ for a traveling wave is its speed, which we are calling $c$. Equation

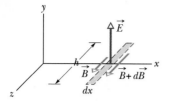

**FIG. 33-7** The sinusoidal variation of the electric field through this rectangle, located (but not shown) at point $P$ in Fig. 33-5$b$, induces magnetic fields along the rectangle. The instant shown is that of Fig. 33-6: $\vec{E}$ is decreasing in magnitude, and the magnitude of the induced magnetic field is greater on the right side of the rectangle than on the left.

33-12 then becomes

$$\frac{E_m}{B_m} = c \qquad \text{(amplitude ratio)}, \qquad (33\text{-}13)$$

which is just Eq. 33-4.

## Equation 33-3 and the Induced Magnetic Field

Figure 33-7 shows another dashed rectangle at point $P$ of Fig. 33-5$b$; this one is in the $xz$ plane. As the electromagnetic wave moves rightward past this new rectangle, the electric flux $\Phi_E$ through the rectangle changes and—according to Maxwell's law of induction—induced magnetic fields appear throughout the region of the rectangle. These induced magnetic fields are, in fact, the magnetic component of the electromagnetic wave.

We see from Fig. 33-5$b$ that at the instant chosen for the magnetic field represented in Fig. 33-6, marked in red on the magnetic component curve, the electric field through the rectangle of Fig. 33-7 is directed as shown. Recall that at the chosen instant, the magnetic field in Fig. 33-6 is decreasing. Because the two fields are in phase, the electric field in Fig. 33-7 must also be decreasing, and so must the electric flux $\Phi_E$ through the rectangle. By applying the same reasoning we applied to Fig. 33-6, we see that the changing flux $\Phi_E$ will induce a magnetic field with vectors $\vec{B}$ and $\vec{B} + d\vec{B}$ oriented as shown in Fig. 33-7, where field $\vec{B} + d\vec{B}$ is greater than field $\vec{B}$.

Let us apply Maxwell's law of induction,

$$\oint \vec{B} \cdot d\vec{s} = \mu_0 \varepsilon_0 \frac{d\Phi_E}{dt}, \qquad (33\text{-}14)$$

by proceeding counterclockwise around the dashed rectangle of Fig. 33-7. Only the long sides of the rectangle contribute to the integral, whose value is

$$\oint \vec{B} \cdot d\vec{s} = -(B + dB)h + Bh = -h\,dB. \qquad (33\text{-}15)$$

The flux $\Phi_E$ through the rectangle is

$$\Phi_E = (E)(h\,dx), \qquad (33\text{-}16)$$

where $E$ is the average magnitude of $\vec{E}$ within the rectangle. Differentiating Eq. 33-16 with respect to $t$ gives

$$\frac{d\Phi_E}{dt} = h\,dx\,\frac{dE}{dt}.$$

If we substitute this and Eq. 33-15 into Eq. 33-14, we find

$$-h\,dB = \mu_0 \varepsilon_0 \left( h\,dx\,\frac{dE}{dt} \right)$$

or, changing to partial-derivative notation as we did for Eq. 33-11,

$$-\frac{\partial B}{\partial x} = \mu_0 \varepsilon_0 \frac{\partial E}{\partial t}. \qquad (33\text{-}17)$$

Again, the minus sign in this equation is necessary because, although $B$ is increasing with $x$ at point $P$ in the rectangle in Fig. 33-7, $E$ is decreasing with $t$.

Evaluating Eq. 33-17 by using Eqs. 33-1 and 33-2 leads to

$$-kB_m \cos(kx - \omega t) = -\mu_0 \varepsilon_0 \omega E_m \cos(kx - \omega t),$$

which we can write as

$$\frac{E_m}{B_m} = \frac{1}{\mu_0 \varepsilon_0 (\omega/k)} = \frac{1}{\mu_0 \varepsilon_0 c}.$$

Combining this with Eq. 33-13 leads at once to

$$c = \frac{1}{\sqrt{\mu_0 \varepsilon_0}} \qquad \text{(wave speed)}, \qquad (33\text{-}18)$$

which is exactly Eq. 33-3.

✓**CHECKPOINT 1**
The magnetic field $\vec{B}$ through the rectangle of Fig. 33-6 is shown at a different instant in part 1 of the figure here; $\vec{B}$ is directed in the $xz$ plane, parallel to the $z$ axis, and its magnitude is increasing. (a) Complete part 1 by drawing the induced electric fields, indicating both directions and relative magnitudes (as in Fig. 33-6). (b) For the same instant, complete part 2 of the figure by drawing the electric field of the electromagnetic wave. Also draw the induced magnetic fields, indicating both directions and relative magnitudes (as in Fig. 33-7).

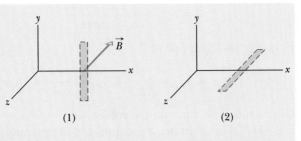

## 33-5 | Energy Transport and the Poynting Vector

All sunbathers know that an electromagnetic wave can transport energy and deliver it to a body on which the wave falls. The rate of energy transport per unit area in such a wave is described by a vector $\vec{S}$, called the **Poynting vector** after physicist John Henry Poynting (1852–1914), who first discussed its properties. This vector is defined as

$$\vec{S} = \frac{1}{\mu_0}\vec{E} \times \vec{B} \qquad \text{(Poynting vector)}. \qquad (33\text{-}19)$$

Its magnitude $S$ is related to the rate at which energy is transported by a wave across a unit area at any instant (inst):

$$S = \left(\frac{\text{energy/time}}{\text{area}}\right)_{\text{inst}} = \left(\frac{\text{power}}{\text{area}}\right)_{\text{inst}}. \qquad (33\text{-}20)$$

From this we can see that the SI unit for $\vec{S}$ is the watt per square meter (W/m²).

👉 The direction of the Poynting vector $\vec{S}$ of an electromagnetic wave at any point gives the wave's direction of travel and the direction of energy transport at that point.

Because $\vec{E}$ and $\vec{B}$ are perpendicular to each other in an electromagnetic wave, the magnitude of $\vec{E} \times \vec{B}$ is $EB$. Then the magnitude of $\vec{S}$ is

$$S = \frac{1}{\mu_0}EB, \qquad (33\text{-}21)$$

in which $S$, $E$, and $B$ are instantaneous values. The magnitudes $E$ and $B$ are so closely coupled to each other that we need to deal with only one of them; we choose $E$, largely because most instruments for detecting electromagnetic waves deal with the electric component of the wave rather than the magnetic component. Using $B = E/c$ from Eq. 33-5, we can rewrite Eq. 33-21 as

$$S = \frac{1}{c\mu_0}E^2 \qquad \text{(instantaneous energy flow rate)}. \qquad (33\text{-}22)$$

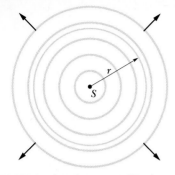

**FIG. 33-8** A point source $S$ emits electromagnetic waves uniformly in all directions. The spherical wavefronts pass through an imaginary sphere of radius $r$ that is centered on $S$.

By substituting $E = E_m \sin(kx - \omega t)$ into Eq. 33-22, we could obtain an equation for the energy transport rate as a function of time. More useful in practice, however, is the average energy transported over time; for that, we need to find the time-averaged value of $S$, written $S_{avg}$ and also called the **intensity** $I$ of the wave. Thus from Eq. 33-20, the intensity $I$ is

$$I = S_{avg} = \left( \frac{\text{energy/time}}{\text{area}} \right)_{avg} = \left( \frac{\text{power}}{\text{area}} \right)_{avg}. \qquad (33\text{-}23)$$

From Eq. 33-22, we find

$$I = S_{avg} = \frac{1}{c\mu_0} [E^2]_{avg} = \frac{1}{c\mu_0} [E_m^2 \sin^2(kx - \omega t)]_{avg}. \qquad (33\text{-}24)$$

Over a full cycle, the average value of $\sin^2 \theta$, for any angular variable $\theta$, is $\frac{1}{2}$ (see Fig. 31-17). In addition, we define a new quantity $E_{rms}$, the *root-mean-square* value of the electric field, as

$$E_{rms} = \frac{E_m}{\sqrt{2}}. \qquad (33\text{-}25)$$

We can then rewrite Eq. 33-24 as

$$I = \frac{1}{c\mu_0} E_{rms}^2. \qquad (33\text{-}26)$$

Because $E = cB$ and $c$ is such a very large number, you might conclude that the energy associated with the electric field is much greater than that associated with the magnetic field. That conclusion is incorrect; the two energies are exactly equal. To show this, we start with Eq. 25-25, which gives the energy density $u \ (= \frac{1}{2}\varepsilon_0 E^2)$ within an electric field, and substitute $cB$ for $E$; then we can write

$$u_E = \tfrac{1}{2}\varepsilon_0 E^2 = \tfrac{1}{2}\varepsilon_0 (cB)^2.$$

If we now substitute for $c$ with Eq. 33-3, we get

$$u_E = \tfrac{1}{2}\varepsilon_0 \frac{1}{\mu_0 \varepsilon_0} B^2 = \frac{B^2}{2\mu_0}.$$

However, Eq. 30-54 tells us that $B^2/2\mu_0$ is the energy density $u_B$ of a magnetic field $\vec{B}$; so we see that $u_E = u_B$ everywhere along an electromagnetic wave.

## Variation of Intensity with Distance

How intensity varies with distance from a real source of electromagnetic radiation is often complex—especially when the source (like a searchlight at a movie premier) beams the radiation in a particular direction. However, in some situations we can assume that the source is a *point source* that emits the light *isotropically*—that is, with equal intensity in all directions. The spherical wavefronts spreading from such an isotropic point source $S$ at a particular instant are shown in cross section in Fig. 33-8.

Let us assume that the energy of the waves is conserved as they spread from this source. Let us also center an imaginary sphere of radius $r$ on the source, as shown in Fig. 33-8. All the energy emitted by the source must pass through the sphere. Thus, the rate at which energy passes through the sphere via the radiation must equal the rate at which energy is emitted by the source—that is, the source power $P_s$. The intensity $I$ at the sphere must then be, from Eq. 33-23,

$$I = \frac{\text{power}}{\text{area}} = \frac{P_s}{4\pi r^2}, \qquad (33\text{-}27)$$

where $4\pi r^2$ is the area of the sphere. Equation 33-27 tells us that the intensity of the electromagnetic radiation from an isotropic point source decreases with the square of the distance $r$ from the source.

✓**CHECKPOINT 2**    The figure here gives the electric field of an electromagnetic wave at a certain point and a certain instant. The wave is transporting energy in the negative $z$ direction. What is the direction of the magnetic field of the wave at that point and instant?

---

**Sample Problem** | **33-1**

When you look at the North Star (Polaris), you intercept light from a star at a distance of 431 ly and emitting energy at a rate of $2.2 \times 10^3$ times that of our Sun ($P_{sun} = 3.90 \times 10^{26}$ W). Neglecting any atmospheric absorption, find the rms values of the electric and magnetic fields when the starlight reaches you.

**KEY IDEAS**

1. The rms value $E_{rms}$ of the electric field in light is related to the intensity $I$ of the light via Eq. 33-26 ($I = E_{rms}^2/c\mu_0$).

2. Because the source is so far away and emits light with equal intensity in all directions, the intensity $I$ at any distance $r$ from the source is related to the source's power $P_s$ via Eq. 33-27 ($I = P_s/4\pi r^2$).

3. The magnitudes of the electric field and magnetic field of an electromagnetic wave at any instant and at any point in the wave are related by the speed of light $c$ according to Eq. 33-5 ($E/B = c$). Thus, the rms values of those fields are also related by Eq. 33-5.

**Electric field:** Putting the first two ideas together gives us

$$I = \frac{P_s}{4\pi r^2} = \frac{E_{rms}^2}{c\mu_0}$$

and

$$E_{rms} = \sqrt{\frac{P_s c \mu_0}{4\pi r^2}}.$$

Substituting $P_s = (2.2 \times 10^3)(3.90 \times 10^{26}$ W$)$, $r = 431$ ly $= 4.08 \times 10^{18}$ m, and values for the constants, we find

$$E_{rms} = 1.24 \times 10^{-3} \text{ V/m} \approx 1.2 \text{ mV/m}. \quad \text{(Answer)}$$

**Magnetic field:** From Eq. 33-5, we write

$$B_{rms} = \frac{E_{rms}}{c} = \frac{1.24 \times 10^{-3} \text{ V/m}}{3.00 \times 10^8 \text{ m/s}}$$

$$= 4.1 \times 10^{-12} \text{ T} = 4.1 \text{ pT}.$$

**Cannot compare the fields:** Note that $E_{rms}$ ($= 1.2$ mV/m) is small as judged by ordinary laboratory standards, but $B_{rms}$ ($= 4.1$ pT) is quite small. This difference helps to explain why most instruments used for the detection and measurement of electromagnetic waves are designed to respond to the electric component of the wave. It is wrong, however, to say that the electric component of an electromagnetic wave is "stronger" than the magnetic component. You cannot compare quantities that are measured in different units. As we have seen, the electric and magnetic components are on an equal basis as far as the propagation of the wave is concerned because their average energies, which *can* be compared, are equal.

---

# 33-6 | Radiation Pressure

Electromagnetic waves have linear momentum as well as energy. This means that we can exert a pressure—a **radiation pressure**—on an object by shining light on it. However, the pressure must be very small because, for example, you do not feel a camera flash when it is used to take your photograph (which is good because otherwise every photographic flash could be like a punch).

To find an expression for the pressure, let us shine a beam of electromagnetic radiation—light, for example—on an object for a time interval $\Delta t$. Further, let us assume that the object is free to move and that the radiation is entirely **absorbed** (taken up) by the object. This means that during the interval $\Delta t$, the object gains an energy $\Delta U$ from the radiation. Maxwell showed that the object also gains linear momentum. The magnitude $\Delta p$ of the momentum change of the object is related to the energy change $\Delta U$ by

$$\Delta p = \frac{\Delta U}{c} \quad \text{(total absorption)}, \quad (33\text{-}28)$$

where $c$ is the speed of light. The direction of the momentum change of the object is the direction of the *incident* (incoming) beam that the object absorbs.

Instead of being absorbed, the radiation can be **reflected** by the object; that is, the radiation can be sent off in a new direction as if it bounced off the object. If the radiation is entirely reflected back along its original path, the magnitude of the momentum change of the object is twice that given above, or

$$\Delta p = \frac{2\,\Delta U}{c} \qquad \text{(total reflection back along path)}. \qquad (33\text{-}29)$$

In the same way, an object undergoes twice as much momentum change when a perfectly elastic tennis ball is bounced from it as when it is struck by a perfectly inelastic ball (a lump of wet putty, say) of the same mass and velocity. If the incident radiation is partly absorbed and partly reflected, the momentum change of the object is between $\Delta U/c$ and $2\,\Delta U/c$.

From Newton's second law in its linear momentum form (Section 9-4), we know that a change in momentum is related to a force by

$$F = \frac{\Delta p}{\Delta t}. \qquad (33\text{-}30)$$

To find expressions for the force exerted by radiation in terms of the intensity $I$ of the radiation, we first note that intensity is

$$I = \frac{\text{power}}{\text{area}} = \frac{\text{energy/time}}{\text{area}}.$$

Next, suppose that a flat surface of area $A$, perpendicular to the path of the radiation, intercepts the radiation. In time interval $\Delta t$, the energy intercepted by area $A$ is

$$\Delta U = IA\,\Delta t. \qquad (33\text{-}31)$$

If the energy is completely absorbed, then Eq. 33-28 tells us that $\Delta p = IA\,\Delta t/c$, and, from Eq. 33-30, the magnitude of the force on the area $A$ is

$$F = \frac{IA}{c} \qquad \text{(total absorption)}. \qquad (33\text{-}32)$$

Similarly, if the radiation is totally reflected back along its original path, Eq. 33-29 tells us that $\Delta p = 2IA\,\Delta t/c$ and, from Eq. 33-30,

$$F = \frac{2IA}{c} \qquad \text{(total reflection back along path)}. \qquad (33\text{-}33)$$

If the radiation is partly absorbed and partly reflected, the magnitude of the force on area $A$ is between the values of $IA/c$ and $2IA/c$.

The force per unit area on an object due to radiation is the radiation pressure $p_r$. We can find it for the situations of Eqs. 33-32 and 33-33 by dividing both sides of each equation by $A$. We obtain

$$p_r = \frac{I}{c} \qquad \text{(total absorption)} \qquad (33\text{-}34)$$

and

$$p_r = \frac{2I}{c} \qquad \text{(total reflection back along path)}. \qquad (33\text{-}35)$$

Be careful not to confuse the symbol $p_r$ for radiation pressure with the symbol $p$ for momentum. Just as with fluid pressure in Chapter 14, the SI unit of radiation pressure is the newton per square meter ($N/m^2$), which is called the pascal (Pa).

The development of laser technology has permitted researchers to achieve radiation pressures much greater than, say, that due to a camera flashlamp. This comes about because a beam of laser light—unlike a beam of light from a small lamp filament—can be focused to a tiny spot. This permits the delivery of great amounts of energy to small objects placed at that spot.

✓ **CHECKPOINT 3**    Light of uniform intensity shines perpendicularly on a totally absorbing surface, fully illuminating the surface. If the area of the surface is decreased, do (a) the radiation pressure and (b) the radiation force on the surface increase, decrease, or stay the same?

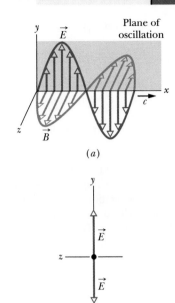

FIG. 33-9    (a) The plane of oscillation of a polarized electromagnetic wave. (b) To represent the polarization, we view the plane of oscillation head-on and indicate the directions of the oscillating electric field with a double arrow.

## 33-7 | Polarization

VHF (very high frequency) television antennas in England are oriented vertically, but those in North America are horizontal. The difference is due to the direction of oscillation of the electromagnetic waves carrying the TV signal. In England, the transmitting equipment is designed to produce waves that are **polarized** vertically; that is, their electric field oscillates vertically. Thus, for the electric field of the incident television waves to drive a current along an antenna (and provide a signal to a television set), the antenna must be vertical. In North America, the waves are polarized horizontally.

Figure 33-9a shows an electromagnetic wave with its electric field oscillating parallel to the vertical $y$ axis. The plane containing the $\vec{E}$ vectors is called the **plane of oscillation** of the wave (hence, the wave is said to be *plane-polarized* in the $y$ direction). We can represent the wave's *polarization* (state of being polarized) by showing the directions of the electric field oscillations in a head-on view of the plane of oscillation, as in Fig. 33-9b. The vertical double arrow in that figure indicates that as the wave travels past us, its electric field oscillates vertically—it continuously changes between being directed up and down the $y$ axis.

### Polarized Light

The electromagnetic waves emitted by a television station all have the same polarization, but the electromagnetic waves emitted by any common source of light (such as the Sun or a bulb) are **polarized randomly**, or **unpolarized** (the two terms mean the same thing). That is, the electric field at any given point is always perpendicular to the direction of travel of the waves but changes directions randomly. Thus, if we try to represent a head-on view of the oscillations over some time period, we do not have a simple drawing with a single double arrow like that of Fig. 33-9b; instead we have a mess of double arrows like that in Fig. 33-10a.

In principle, we can simplify the mess by resolving each electric field of Fig. 33-10a into $y$ and $z$ components. Then as the wave travels past us, the net $y$ component oscillates parallel to the $y$ axis and the net $z$ component oscillates parallel to the $z$ axis. We can then represent the unpolarized light with a pair of

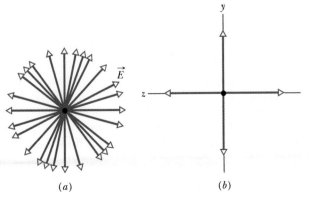

FIG. 33-10    (a) Unpolarized light consists of waves with randomly directed electric fields. Here the waves are all traveling along the same axis, directly out of the page, and all have the same amplitude $E$. (b) A second way of representing unpolarized light—the light is the superposition of two polarized waves whose planes of oscillation are perpendicular to each other.

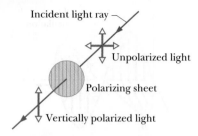

**FIG. 33-11** Unpolarized light becomes polarized when it is sent through a polarizing sheet. Its direction of polarization is then parallel to the polarizing direction of the sheet, which is represented here by the vertical lines drawn in the sheet.

double arrows as shown in Fig. 33-10$b$. The double arrow along the $y$ axis represents the oscillations of the net $y$ component of the electric field. The double arrow along the $z$ axis represents the oscillations of the net $z$ component of the electric field. In doing all this, we effectively change unpolarized light into the superposition of two polarized waves whose planes of oscillation are perpendicular to each other—one plane contains the $y$ axis and the other contains the $z$ axis. One reason to make this change is that drawing Fig. 33-10$b$ is a lot easier than drawing Fig. 33-10$a$.

We can draw similar figures to represent light that is **partially polarized** (its field oscillations are not completely random as in Fig. 33-10$a$, nor are they parallel to a single axis as in Fig. 33-9$b$). For this situation, we draw one of the double arrows in a perpendicular pair of double arrows longer than the other one.

We can transform unpolarized visible light into polarized light by sending it through a *polarizing sheet,* as is shown in Fig. 33-11. Such sheets, commercially known as Polaroids or Polaroid filters, were invented in 1932 by Edwin Land while he was an undergraduate student. A polarizing sheet consists of certain long molecules embedded in plastic. When the sheet is manufactured, it is stretched to align the molecules in parallel rows, like rows in a plowed field. When light is then sent through the sheet, electric field components along one direction pass through the sheet, while components perpendicular to that direction are absorbed by the molecules and disappear.

We shall not dwell on the molecules but, instead, shall assign to the sheet a *polarizing direction,* along which electric field components are passed:

> An electric field component parallel to the polarizing direction is passed (*transmitted*) by a polarizing sheet; a component perpendicular to it is absorbed.

Thus, the electric field of the light emerging from the sheet consists of only the components that are parallel to the polarizing direction of the sheet; hence the light is polarized in that direction. In Fig. 33-11, the vertical electric field components are transmitted by the sheet; the horizontal components are absorbed. The transmitted waves are then vertically polarized.

### Intensity of Transmitted Polarized Light

We now consider the intensity of light transmitted by a polarizing sheet. We start with unpolarized light, whose electric field oscillations we can resolve into $y$ and $z$ components as represented in Fig. 33-10$b$. Further, we can arrange for the $y$ axis to be parallel to the polarizing direction of the sheet. Then only the $y$ components of the light's electric field are passed by the sheet; the $z$ components are absorbed. As suggested by Fig. 33-10$b$, if the original waves are randomly oriented, the sum of the $y$ components and the sum of the $z$ components are equal. When the $z$ components are absorbed, half the intensity $I_0$ of the original light is lost. The intensity $I$ of the emerging polarized light is then

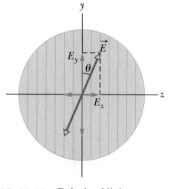

**FIG. 33-12** Polarized light approaching a polarizing sheet. The electric field $\vec{E}$ of the light can be resolved into components $E_y$ (parallel to the polarizing direction of the sheet) and $E_z$ (perpendicular to that direction). Component $E_y$ will be transmitted by the sheet; component $E_z$ will be absorbed.

$$I = \tfrac{1}{2}I_0. \qquad (33\text{-}36)$$

Let us call this the *one-half rule;* we can use it *only* when the light reaching a polarizing sheet is unpolarized.

Suppose now that the light reaching a polarizing sheet is already polarized. Figure 33-12 shows a polarizing sheet in the plane of the page and the electric field $\vec{E}$ of such a polarized light wave traveling toward the sheet (and thus prior to any absorption). We can resolve $\vec{E}$ into two components relative to the polarizing direction of the sheet: parallel component $E_y$ is transmitted by the sheet, and perpendicular component $E_z$ is absorbed. Since $\theta$ is the angle between $\vec{E}$ and the polarizing direction of the sheet, the transmitted parallel component is

$$E_y = E\cos\theta. \qquad (33\text{-}37)$$

Recall that the intensity of an electromagnetic wave (such as our light wave) is proportional to the square of the electric field's magnitude (Eq. 33-26, $I = E_{rms}^2/c\mu_0$). In our present case then, the intensity $I$ of the emerging wave is proportional to $E_y^2$ and the intensity $I_0$ of the original wave is proportional to $E^2$. Hence, from Eq. 33-37 we can write $I/I_0 = \cos^2 \theta$, or

$$I = I_0 \cos^2 \theta. \tag{33-38}$$

Let us call this the *cosine-squared rule;* we can use it *only* when the light reaching a polarizing sheet is already polarized. Then the transmitted intensity $I$ is a maximum and is equal to the original intensity $I_0$ when the original wave is polarized parallel to the polarizing direction of the sheet (when $\theta$ in Eq. 33-38 is 0° or 180°). The transmitted intensity is zero when the original wave is polarized perpendicular to the polarizing direction of the sheet (when $\theta$ is 90°).

Figure 33-13 shows an arrangement in which initially unpolarized light is sent through two polarizing sheets $P_1$ and $P_2$. (Often, the first sheet is called the *polarizer,* and the second the *analyzer.*) Because the polarizing direction of $P_1$ is vertical, the light transmitted by $P_1$ to $P_2$ is polarized vertically. If the polarizing direction of $P_2$ is also vertical, then all the light transmitted by $P_1$ is transmitted by $P_2$. If the polarizing direction of $P_2$ is horizontal, none of the light transmitted by $P_1$ is transmitted by $P_2$. We reach the same conclusions by considering only the *relative* orientations of the two sheets: If their polarizing directions are parallel, all the light passed by the first sheet is passed by the second sheet (Fig. 33-14a). If those directions are perpendicular (the sheets are said to be *crossed*), no light is passed by the second sheet (Fig. 33-14b). Finally, if the two polarizing directions of Fig. 33-13 make an angle between 0° and 90°, some of the light transmitted by $P_1$ will be transmitted by $P_2$, as set by Eq. 33-38.

Light can be polarized by means other than polarizing sheets, such as by reflection (discussed in Section 33-10) and by scattering from atoms or molecules. In *scattering,* light that is intercepted by an object, such as a molecule, is sent off in many, perhaps random, directions. An example is the scattering of sunlight by molecules in the atmosphere, which gives the sky its general glow.

Although direct sunlight is unpolarized, light from much of the sky is at least partially polarized by such scattering. Bees use the polarization of sky light in navigating to and from their hives. Similarly, the Vikings used it to navigate across the North Sea when the daytime Sun was below the horizon (because of the high latitude of the North Sea). These early seafarers had discovered certain crystals (now called cordierite) that changed color when rotated in polarized light. By looking at the sky through such a crystal while rotating it about their line of sight, they could locate the hidden Sun and thus determine which way was south.

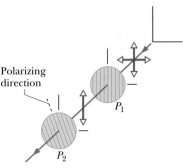

**FIG. 33-13** The light transmitted by polarizing sheet $P_1$ is vertically polarized, as represented by the vertical double arrow. The amount of that light that is then transmitted by polarizing sheet $P_2$ depends on the angle between the polarization direction of that light and the polarizing direction of $P_2$ (indicated by the lines drawn in the sheet and by the dashed line).

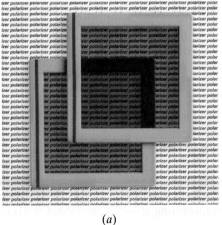
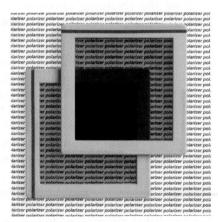

**FIG. 33-14** (a) Overlapping polarizing sheets transmit light fairly well when their polarizing directions have the same orientation, but (b) they block most of the light when they are crossed. (*Richard Megna/Fundamental Photographs.*)

(a)        (b)

✓ **CHECKPOINT 4**    The figure shows four pairs of polarizing sheets, seen face-on. Each pair is mounted in the path of initially unpolarized light. The polarizing direction of each sheet (indicated by the dashed line) is referenced to either a horizontal $x$ axis or a vertical $y$ axis. Rank the pairs according to the fraction of the initial intensity that they pass, greatest first.

---

**Sample Problem** | **33-2** | **Build your skill**

Figure 33-15$a$, drawn in perspective, shows a system of three polarizing sheets in the path of initially unpolarized light. The polarizing direction of the first sheet is parallel to the $y$ axis, that of the second sheet is at an angle of 60° counterclockwise from the $y$ axis, and that of the third sheet is parallel to the $x$ axis. What fraction of the initial intensity $I_0$ of the light emerges from the three-sheet system, and in which direction is that emerging light polarized?

**KEY IDEAS**

1. We work through the system sheet by sheet, from the first one encountered by the light to the last one.

2. To find the intensity transmitted by any sheet, we apply either the one-half rule or the cosine-squared rule, depending on whether the light reaching the sheet is unpolarized or already polarized.

3. The light that is transmitted by a polarizing sheet is always polarized parallel to the polarizing direction of the sheet.

**First sheet:** The original light wave is represented in Fig. 33-15$b$, using the head-on, double-arrow representation of Fig. 33-10$b$. Because the light is initially unpolarized, the intensity $I_1$ of the light transmitted by the first sheet is given by the one-half rule (Eq. 33-36):

$$I_1 = \tfrac{1}{2}I_0.$$

Because the polarizing direction of the first sheet is parallel to the $y$ axis, the polarization of the light transmitted by it is also, as shown in the head-on view of Fig. 33-15$c$.

**Second sheet:** Because the light reaching the second sheet is polarized, the intensity $I_2$ of the light transmitted by that sheet is given by the cosine-squared rule (Eq. 33-38). The angle $\theta$ in the rule is the angle between the polarization direction of the entering light (parallel to the $y$ axis) and the polarizing direction of the second sheet (60° counterclockwise from the $y$ axis), and so $\theta$ is 60°. Then

$$I_2 = I_1 \cos^2 60°.$$

The polarization of this transmitted light is parallel to the polarizing direction of the sheet transmitting it—

that is, 60° counterclockwise from the $y$ axis, as shown in the head-on view of Fig. 33-15$d$.

**Third sheet:** Because the light reaching the third sheet is polarized, the intensity $I_3$ of the light transmitted by that sheet is given by the cosine-squared rule. The angle $\theta$ is now the angle between the polarization direction of the entering light (Fig. 33-15$d$) and the polarizing direction of the third sheet (parallel to the $x$ axis), and so $\theta = 30°$. Thus,

$$I_3 = I_2 \cos^2 30°.$$

This final transmitted light is polarized parallel to the $x$ axis (Fig. 33-15$e$). We find its intensity by substituting first for $I_2$ and then for $I_1$ in the equation above:

$$I_3 = I_2 \cos^2 30° = (I_1 \cos^2 60°) \cos^2 30°$$
$$= (\tfrac{1}{2}I_0) \cos^2 60° \cos^2 30° = 0.094 I_0.$$

Thus,
$$\frac{I_3}{I_0} = 0.094. \qquad \text{(Answer)}$$

That is to say, 9.4% of the initial intensity emerges from the three-sheet system. (If we now remove the second sheet, what fraction of the initial intensity emerges from the system?)

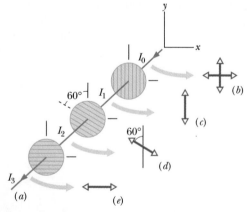

**FIG. 33-15**    ($a$) Initially unpolarized light of intensity $I_0$ is sent into a system of three polarizing sheets. The intensities $I_1$, $I_2$, and $I_3$ of the light transmitted by the sheets are labeled. Shown also are the polarizations, from head-on views, of ($b$) the initial light and the light transmitted by ($c$) the first sheet, ($d$) the second sheet, and ($e$) the third sheet.

## 33-8 | Reflection and Refraction

Although a light wave spreads as it moves away from its source, we can often approximate its travel as being in a straight line; we did so for the light wave in Fig. 33-5a. The study of the properties of light waves under that approximation is called *geometrical optics*. For the rest of this chapter and all of Chapter 34, we shall discuss the geometrical optics of visible light.

The photograph in Fig. 33-16a shows an example of light waves traveling in approximately straight lines. A narrow beam of light (the *incident* beam), angled downward from the left and traveling through air, encounters a *plane* (flat) water surface. Part of the light is **reflected** by the surface, forming a beam directed upward toward the right, traveling as if the original beam had bounced from the surface. The rest of the light travels through the surface and into the water, forming a beam directed downward to the right. Because light can travel through it, the water is said to be *transparent;* that is, we can see through it. (In this chapter we shall consider only transparent materials and not opaque materials, through which light cannot travel.)

The travel of light through a surface (or *interface*) that separates two media is called **refraction**, and the light is said to be *refracted*. Unless an incident beam of light is perpendicular to the surface, refraction changes the light's direction of travel. For this reason, the beam is said to be "bent" by the refraction. Note in Fig. 33-16a that the bending occurs only at the surface; within the water, the light travels in a straight line.

In Figure 33-16b, the beams of light in the photograph are represented with an *incident ray,* a *reflected ray,* and a *refracted ray* (and wavefronts). Each ray is oriented with respect to a line, called the *normal,* that is perpendicular to the surface at the point of reflection and refraction. In Fig. 33-16b, the **angle of incidence** is $\theta_1$, the **angle of reflection** is $\theta_1'$, and the **angle of refraction** is $\theta_2$, all measured *relative to the normal.* The plane containing the incident ray and the normal is the *plane of incidence,* which is in the plane of the page in Fig. 33-16b.

Experiment shows that reflection and refraction are governed by two laws:

**Law of reflection:** A reflected ray lies in the plane of incidence and has an angle of reflection equal to the angle of incidence. In Fig. 33-16b, this means that

$$\theta_1' = \theta_1 \quad \text{(reflection).} \tag{33-39}$$

(We shall now usually drop the prime on the angle of reflection.)

**Law of refraction:** A refracted ray lies in the plane of incidence and has an angle of refraction $\theta_2$ that is related to the angle of incidence $\theta_1$ by

$$n_2 \sin \theta_2 = n_1 \sin \theta_1 \quad \text{(refraction).} \tag{33-40}$$

Here each of the symbols $n_1$ and $n_2$ is a dimensionless constant, called the **index of refraction,** that is associated with a medium involved in the refraction. We derive this equation, called **Snell's law**, in Chapter 35. As we shall discuss there, the index of refraction of a medium is equal to $c/v$, where $v$ is the speed of light in that medium and $c$ is its speed in vacuum.

Table 33-1 gives the indexes of refraction of vacuum and some common substances. For vacuum, $n$ is defined to be exactly 1; for air, $n$ is very close to 1.0 (an approximation we shall often make). Nothing has an index of refraction below 1.

We can rearrange Eq. 33-40 as

$$\sin \theta_2 = \frac{n_1}{n_2} \sin \theta_1 \tag{33-41}$$

to compare the angle of refraction $\theta_2$ with the angle of incidence $\theta_1$. We can

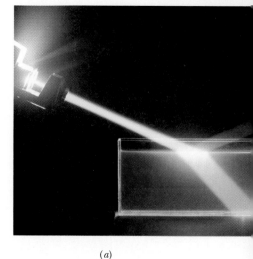

(a)

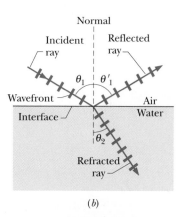

(b)

**FIG. 33-16** (a) A photograph showing an incident beam of light reflected and refracted by a horizontal water surface. (b) A ray representation of (a). The angles of incidence ($\theta_1$), reflection ($\theta_1'$), and refraction ($\theta_2$) are marked. (©1974 FP/Fundamentals Photography)

**TABLE 33-1**

**Some Indexes of Refraction**[a]

| Medium | Index | Medium | Index |
|---|---|---|---|
| Vacuum | Exactly 1 | Typical crown glass | 1.52 |
| Air (STP)[b] | 1.00029 | Sodium chloride | 1.54 |
| Water (20°C) | 1.33 | Polystyrene | 1.55 |
| Acetone | 1.36 | Carbon disulfide | 1.63 |
| Ethyl alcohol | 1.36 | Heavy flint glass | 1.65 |
| Sugar solution (30%) | 1.38 | Sapphire | 1.77 |
| Fused quartz | 1.46 | Heaviest flint glass | 1.89 |
| Sugar solution (80%) | 1.49 | Diamond | 2.42 |

[a]For a wavelength of 589 nm (yellow sodium light).
[b]STP means "standard temperature (0°C) and pressure (1 atm)."

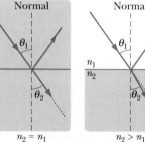
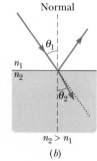

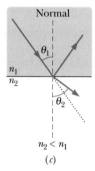

**FIG. 33-17** Refraction of light traveling from a medium with an index of refraction $n_1$ into a medium with an index of refraction $n_2$. (a) The beam does not bend when $n_2 = n_1$; the refracted light then travels in the *undeflected direction* (the dotted line), which is the same as the direction of the incident beam. The beam bends (b) toward the normal when $n_2 > n_1$ and (c) away from the normal when $n_2 < n_1$.

then see that the relative value of $\theta_2$ depends on the relative values of $n_2$ and $n_1$. In fact, we can have three basic results:

1. If $n_2$ is equal to $n_1$, then $\theta_2$ is equal to $\theta_1$ and refraction does not bend the light beam, which continues in the *undeflected direction*, as in Fig. 33-17a.

2. If $n_2$ is greater than $n_1$, then $\theta_2$ is less than $\theta_1$. In this case, refraction bends the light beam away from the undeflected direction and toward the normal, as in Fig. 33-17b.

3. If $n_2$ is less than $n_1$, then $\theta_2$ is greater than $\theta_1$. In this case, refraction bends the light beam away from the undeflected direction and away from the normal, as in Fig. 33-17c.

Refraction *cannot* bend a beam so much that the refracted ray is on the same side of the normal as the incident ray.

## Chromatic Dispersion

The index of refraction $n$ encountered by light in any medium except vacuum depends on the wavelength of the light. The dependence of $n$ on wavelength implies that when a light beam consists of rays of different wavelengths, the rays will be refracted at different angles by a surface; that is, the light will be spread out by the refraction. This spreading of light is called **chromatic dispersion,** in which "chromatic" refers to the colors associated with the individual wavelengths and "dispersion" refers to the spreading of the light according to its wavelengths or colors. The refractions of Figs. 33-16 and 33-17 do not show chromatic dispersion because the beams are *monochromatic* (of a single wavelength or color).

Generally, the index of refraction of a given medium is *greater* for a shorter wavelength (corresponding to, say, blue light) than for a longer wavelength (say, red light). As an example, Fig. 33-18 shows how the index of refraction of fused quartz depends on the wavelength of light. Such dependence means that when a beam made up of waves of both blue and red light is refracted through a surface, such as from air into quartz or vice versa, the blue *component* (the ray corresponding to the wave of blue light) bends more than the red component.

A beam of *white light* consists of components of all (or nearly all) the colors in the visible spectrum with approximately uniform intensities. When you see such a beam, you perceive white rather than the individual colors. In Fig. 33-19a, a beam of white light in air is incident on a glass surface. (Because the pages of this book are white, a beam of white light is represented with a gray ray here.

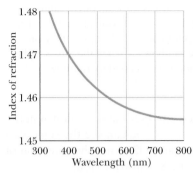

**FIG. 33-18** The index of refraction as a function of wavelength for fused quartz. The graph indicates that a beam of short-wavelength light, for which the index of refraction is higher, is bent more upon entering or leaving quartz than a beam of long-wavelength light.

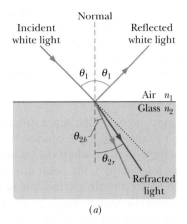
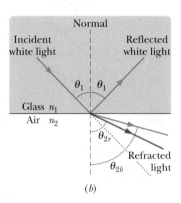

(a)    (b)

**FIG. 33-19** Chromatic dispersion of white light. The blue component is bent more than the red component. (a) Passing from air to glass, the blue component ends up with the smaller angle of refraction. (b) Passing from glass to air, the blue component ends up with the greater angle of refraction. Each dotted line represents the direction in which the light would continue to travel if it were not bent by the refraction.

Also, a beam of monochromatic light is generally represented with a red ray.) Of the refracted light in Fig. 33-19a, only the red and blue components are shown. Because the blue component is bent more than the red component, the angle of refraction $\theta_{2b}$ for the blue component is *smaller* than the angle of refraction $\theta_{2r}$ for the red component. (Remember, angles are measured relative to the normal.) In Fig. 33-19b, a ray of white light in glass is incident on a glass–air interface. Again, the blue component is bent more than the red component, but now $\theta_{2b}$ is greater than $\theta_{2r}$.

To increase the color separation, we can use a solid glass prism with a triangular cross section, as in Fig. 33-20a. The dispersion at the first surface (on the left in Figs. 33-20a, b) is then enhanced by the dispersion at the second surface.

## Rainbows

The most charming example of chromatic dispersion is a rainbow. When sunlight (which consists of all visible colors) is intercepted by a falling raindrop, some of the light refracts into the drop, reflects once from the drop's inner surface, and then refracts out of the drop. Figure 33-21a shows the situation when the Sun is on the horizon at the left (and thus when the rays of sunlight are horizontal). The first refraction separates the sunlight into its component colors, and the second

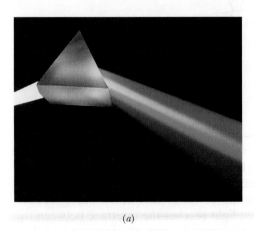

(a)

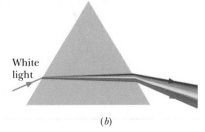

(b)

**FIG. 33-20** (a) A triangular prism separating white light into its component colors. (b) Chromatic dispersion occurs at the first surface and is increased at the second surface.
*(Courtesy Bausch & Lomb)*

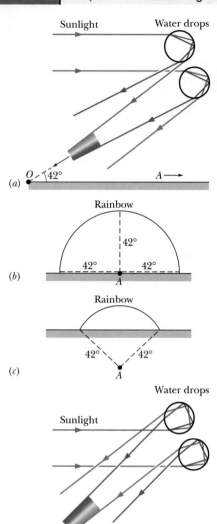

Sunlight    Water drops

(a)

Rainbow

(b)

Rainbow

(c)

Water drops

Sunlight

(d)

**FIG. 33-21** (a) The separation of colors when sunlight refracts into and out of falling rain-drops leads to a primary rainbow. The antisolar point A is on the horizon at the right. The rainbow colors appear at an angle of 42° from the direction of A. (b) Drops at 42° from A in any direction can contribute to the rainbow. (c) The rainbow arc when the Sun is higher (and thus A is lower). (d) The separation of colors leading to a secondary rainbow.

refraction increases the separation. (Only the red and blue rays are shown in the figure.) If many falling drops are brightly illuminated, you can see the separated colors they produce when the drops are at an angle of 42° from the direction of the *antisolar point A*, the point directly opposite the Sun in your view.

To locate the drops, face away from the Sun and point both arms directly away from the Sun, toward the shadow of your head. Then move your right arm directly up, directly rightward, or in any intermediate direction until the angle between your arms is 42°. If illuminated drops happen to be in the direction of your right arm, you see color in that direction.

Because any drop at an angle of 42° in any direction from A can contribute to the rainbow, the rainbow is always a 42° circular arc around A (Fig. 33-21b) and the top of a rainbow is never more than 42° above the horizon. When the Sun is above the horizon, the direction of A is below the horizon, and only a shorter, lower rainbow arc is possible (Fig. 33-21c).

Because rainbows formed in this way involve one reflection of light inside each drop, they are often called *primary rainbows*. A *secondary rainbow* involves two reflections inside a drop, as shown in Fig. 33-21d. Colors appear in the secondary rainbow at an angle of 52° from the direction of A. A secondary rainbow is wider and dimmer than a primary rainbow and thus is more difficult to see. Also, the order of colors in a secondary rainbow is reversed from the order in a primary rainbow, as you can see by comparing parts a and d of Fig. 33-21.

Rainbows involving three or four reflections occur in the direction of the Sun and cannot be seen against the glare of sunshine in that part of the sky. Rainbows involving even more reflections inside the drops can occur in other parts of the sky but are always too dim to see.

**✓CHECKPOINT 5**    Which of the three drawings here (if any) show physically possible refraction?

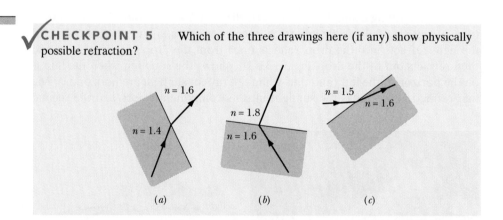

**Sample Problem | 33-3**

(a) In Fig. 33-22a, a beam of monochromatic light reflects and refracts at point A on the interface between material 1 with index of refraction $n_1 = 1.33$ and material 2 with index of refraction $n_2 = 1.77$. The incident beam makes an angle of 50° with the interface. What is the angle of reflection at point A? What is the angle of refraction there?

**KEY IDEAS** (1) The angle of reflection is equal to the angle of incidence, and both angles are measured relative to the normal to the surface at the point of reflection. (2) When light reaches the interface between two materials with different indexes of refraction (call them $n_1$ and $n_2$), part of the light can be refracted by the interface according to Snell's law, Eq. 33-40:

$$n_2 \sin \theta_2 = n_1 \sin \theta_1, \qquad (33\text{-}42)$$

where both angles are measured relative to the normal at the point of refraction.

*Calculations:* In Fig. 33-22a, the normal at point $A$ is drawn as a dashed line through the point. Note that the angle of incidence $\theta_1$ is not the given 50° but rather is $90° - 50° = 40°$. Thus, the angle of reflection is

$$\theta_1' = \theta_1 = 40°. \qquad \text{(Answer)}$$

The light that passes from material 1 into material 2 undergoes refraction at point $A$ on the interface between the two materials. Again we measure angles between light rays and a normal, here at the point of refraction. Thus, in Fig. 33-22a, the angle of refraction is the angle marked $\theta_2$. Solving Eq. 33-42 for $\theta_2$ gives us

$$\theta_2 = \sin^{-1}\left(\frac{n_1}{n_2}\sin\theta_1\right) = \sin^{-1}\left(\frac{1.33}{1.77}\sin 40°\right)$$

$$= 28.88° \approx 29°. \qquad \text{(Answer)}$$

This result means that the beam swings toward the normal (it was at 40° to the normal and is now at 29°). The reason is that when the light travels across the interface, it moves into a material with a greater index of refraction.

(b) The light that enters material 2 at point $A$ then reaches point $B$ on the interface between material 2 and material 3, which is air, as shown in Fig. 33-22b. The interface through $B$ is parallel to that through $A$. At $B$, some of the light reflects and the rest enters the air. What is the angle of reflection? What is the angle of refraction into the air?

*Calculations:* We first need to relate one of the angles at point $B$ with a known angle at point $A$. Because the interface through point $B$ is parallel to that through point $A$, the incident angle at $B$ must be equal to the

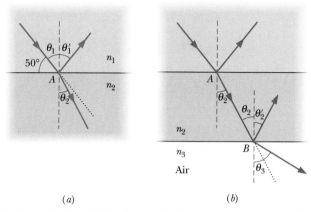

**FIG. 33-22** (*a*) Light reflects and refracts at point $A$ on the interface between materials 1 and 2. (*b*) The light that passes through material 2 reflects and refracts at point $B$ on the interface between materials 2 and 3 (air). Each dashed line is a normal. Each dotted line gives the incident direction of travel.

angle of refraction $\theta_2$, as shown in Fig. 33-22b. Then for reflection, we again use the law of reflection. Thus, the angle of reflection at $B$ is

$$\theta_2' = \theta_2 = 28.88° \approx 29°. \qquad \text{(Answer)}$$

Next, the light that passes from material 2 into the air undergoes refraction at point $B$, with refraction angle $\theta_3$. Thus, we again apply Snell's law of refraction, but this time we write Eq. 33-40 as

$$n_3 \sin \theta_3 = n_2 \sin \theta_2.$$

Solving for $\theta_3$ then leads to

$$\theta_3 = \sin^{-1}\left(\frac{n_2}{n_3}\sin\theta_2\right) = \sin^{-1}\left(\frac{1.77}{1.00}\sin 28.88°\right)$$

$$= 58.75° \approx 59°. \qquad \text{(Answer)}$$

This result means that the beam swings away from the normal (it was at 29° to the normal and is now at 59°). The reason is that when the light travels across the interface, it moves into a material (air) with a lower index of refraction.

---

**Sample Problem**  **33-4**

Sun dogs occur when sunlight passes through atmospheric ice crystals (Fig. 33-23a) that have their hexagonal cross sections horizontal. Such a crystal can turn a ray (Fig. 33-23b) from its original direction of travel by an angle that depends on the crystal's orientation. However, there is a minimum turning angle $\theta_{\text{dog}}$ at which rays bunch up to form an especially bright spot in the sky — a sun dog. The situation is similar to how white wine in a wine glass can bunch up candlelight to form bright lines on a tablecloth. A sunray undergoing such minimum turning follows a path through the crystal that is parallel to one of the hexagonal sides, and its

entrance and exit paths are symmetric. The minimum turning angle $\theta_{\text{dog}}$ is the angle to the left or right of the Sun at which you can see a sun dog. The index of refraction of ice is 1.31. Find $\theta_{\text{dog}}$.

---

**KEY IDEA**  When light reaches the interface between two materials with different indexes of refraction (call them $n_1$ and $n_2$), part of the light can be refracted by the interface according to Snell's law, Eq. 33-40:

$$n_2 \sin \theta_2 = n_1 \sin \theta_1. \qquad (33\text{-}43)$$

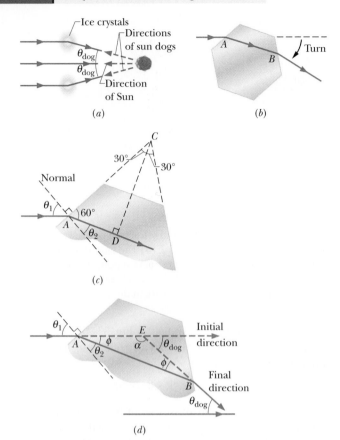

**FIG. 33-23** (a) Overhead view of sunrays redirected by atmospheric ice crystals. (b) Overhead view of hexagonal cross section of a crystal. Path of ray (c) into and (d) out of the cross section.

***Calculations:*** To get started, we mentally complete the corner of one side of the hexagon as in Fig. 33-23c and label the corner as having an angle of 60°. We are interested in the right triangle $ADC$ because the light path forms one leg of the triangle. We can see that one of the acute angles in the triangle is 30°, so the other one must be 60°. In Fig. 33-23c, we then see that the angle of refraction $\theta_2$ is the complement of 60°, so $\theta_2 = 30°$. We now use Snell's law to get the incident angle $\theta_1$ for the ray that ends up in the sun dog. Solving Eq. 33-43 for $\theta_1$, we obtain

$$\theta_1 = \sin^{-1}\left(\frac{n_2}{n_1}\sin\theta_2\right) = \sin^{-1}\left(\frac{1.31}{1.00}\sin 30°\right)$$
$$= 40.92°.$$

Next, note in Fig. 33-23d that $\theta_1$ on the left side of point $A$ is equal to the sum of $\theta_2$ and $\phi$ on the right side:

$$\theta_1 = \theta_2 + \phi,$$

so $\qquad \phi = 40.92° - 30° = 10.92°.$

Next, note that angle $\theta_{dog}$ at point $E$ is an *exterior angle* of triangle $ABE$ and thus (from Appendix E) we can write

$$\theta_{dog} = \phi + \phi = 2(10.92°)$$
$$= 21.8°.$$

This means that the sunray reaching you from the sun dog is turned by 21.8° from its initial direction of travel, and thus you see the sun dog at 21.8° from the direction of the Sun.

## 33-9 | Total Internal Reflection

Figure 33-24a shows rays of monochromatic light from a point source $S$ in glass incident on the interface between the glass and air. For ray $a$, which is perpendicular to the interface, part of the light reflects at the interface and the rest travels through it with no change in direction.

For rays $b$ through $e$, which have progressively larger angles of incidence at the interface, there are also both reflection and refraction at the interface. As the

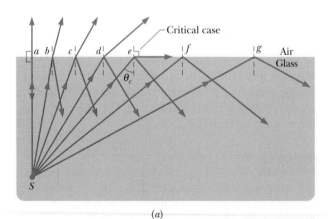

(a)

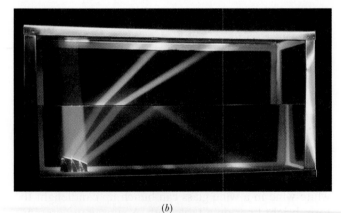

(b)

**FIG. 33-24** (a) Total internal reflection of light from a point source $S$ in glass occurs for all angles of incidence greater than the critical angle $\theta_c$. At the critical angle, the refracted ray points along the air–glass interface. (b) A source in a tank of water. *(Ken Kay/Fundamental Photographs)*

angle of incidence increases, the angle of refraction increases; for ray $e$ it is 90°, which means that the refracted ray points directly along the interface. The angle of incidence giving this situation is called the **critical angle** $\theta_c$. For angles of incidence larger than $\theta_c$, such as for rays $f$ and $g$, there is no refracted ray and *all* the light is reflected; this effect is called **total internal reflection.**

To find $\theta_c$, we use Eq. 33-40; we arbitrarily associate subscript 1 with the glass and subscript 2 with the air, and then we substitute $\theta_c$ for $\theta_1$ and 90° for $\theta_2$, finding

$$n_1 \sin \theta_c = n_2 \sin 90°,$$

which gives us

$$\theta_c = \sin^{-1} \frac{n_2}{n_1} \qquad \text{(critical angle).} \qquad (33\text{-}44)$$

Because the sine of an angle cannot exceed unity, $n_2$ cannot exceed $n_1$ in this equation. This restriction tells us that total internal reflection cannot occur when the incident light is in the medium of lower index of refraction. If source $S$ were in the air in Fig. 33-24a, all its rays that are incident on the air–glass interface (including $f$ and $g$) would be both reflected *and* refracted at the interface.

Total internal reflection has found many applications in medical technology. For example, a physician can view the interior of an artery of a patient by running two thin bundles of *optical fibers* through the chest wall and into an artery (Fig. 33-25). Light introduced at the outer end of one bundle undergoes repeated total internal reflection within the fibers so that, even though the bundle provides a curved path, most of the light ends up exiting the other end and illuminating the interior of the artery. Some of the light reflected from the interior then comes back up the second bundle in a similar way, to be detected and converted to an image on a monitor's screen for the physician to view.

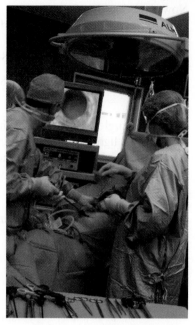

**FIG. 33-25** An endoscope used to inspect an artery. *(©Laurent/Phototake)*

---

**Sample Problem    33-5**

The purpose of a diamond on a ring is, of course, to sparkle. Part of the art of cutting a diamond is to ensure that all the light entering through the top face or side facets leaves through those surfaces, to participate in the sparkle. Figure 33-26 shows part of a cross-sectional slice through a brilliant-cut diamond, with a ray entering at point $A$ on the top face. In this type of cut, the top and bottom surfaces have normal lines that intersect at the indicated 48.84°. At point $B$, at least part of the light reflects and leaves the diamond properly, but part could refract and thus leak out of the diamond. Consider a light ray incident at angle $\theta_1 = 40°$ at $A$. Does light leak at $B$ if air ($n_4 = 1.00$) lies next to the bottom surface? Does light leak if greasy grime ($n_4 = 1.63$) coats the surface? The index of refraction of diamond is $n_{\text{dia}} = 2.419$.

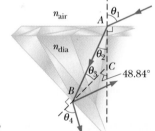

**FIG. 33-26** Light entering the top of a brilliant-cut diamond. Can the light leak from the bottom surface at point $B$?

**KEY IDEAS** When light reaches the interface between two materials with different indexes of refraction (call them $n_1$ and $n_2$), part or all of the light is reflected. The reflection is total if (1) the incident light is in the material with the *higher* index of refraction ($n_1 > n_2$) and (2) the incident angle exceeds a critical value given by Eq. 33-44:

$$\theta_c = \sin^{-1} \frac{n_2}{n_1}. \qquad (33\text{-}45)$$

If those two conditions are not met, part of the light is refracted across the interface according to Snell's law, Eq. 33-40:

$$n_2 \sin \theta_2 = n_1 \sin \theta_1. \qquad (33\text{-}46)$$

*Clean diamond:* We need to follow the light from point $A$ to point $B$, to see if it can leak at point $B$. The light incident at point $A$ is in the material (air, $n_1 = 1.00$) with a *lower* index of refraction than the material on the other side of the interface (diamond, $n_{\text{dia}} = 2.419$). Thus, some of the light is refracted across the interface (the reflected portion is not shown in Fig. 33-26), and we find the angle of refraction $\theta_2$ from Snell's law:

$$\theta_2 = \sin^{-1}\left(\frac{n_1}{n_2}\sin\theta_1\right) = \sin^{-1}\left(\frac{1.00}{2.419}\sin 40°\right)$$
$$= 15.41°.$$

Now note that the given angle of 48.84° at point $C$ is an *exterior angle* of triangle $ABC$ and thus (from Appendix E) we can write

$$\theta_2 + \theta_3 = 48.84°,$$

or $\quad \theta_3 = 48.84° - \theta_2 = 48.84° - 15.41°$

$$= 33.43°.$$

This is the angle of incidence of the ray at point $B$. Now the incident light is in the material (diamond) with the *greater* index of refraction than the material (air) on the other side of the interface. However, if we apply Snell's law at point $B$, we find

$$\theta_4 = \sin^{-1}\left(\frac{2.419}{1.00}\sin 33.43°\right) \qquad (33\text{-}47)$$

$$= \text{no answer.}$$

The reason is that the light is incident at an angle greater than the critical value of

$$\theta_c = \sin^{-1}\frac{n_4}{n_{\text{dia}}} = \sin^{-1}\frac{1.00}{2.419}$$

$$= 24.4°.$$

Thus, all the light reaching $B$ reflects and none leaks out in the air.

***Grimy diamond:*** We again follow the light from $A$ to $B$, with the only difference lying in the last two calculations. Now at $B$, the material on the other side of the interface is grime with an index $n_4 = 1.63$. The incident light is again in the material with the greater index, but now the critical angle is

$$\theta_c = \sin^{-1}\frac{n_4}{n_{\text{dia}}} = \sin^{-1}\frac{1.63}{2.419}$$

$$= 42.4°.$$

So, the incident angle $\theta_3 = 33.43°$ is *less* than the critical angle and light leaks through the bottom of the diamond. From Eq. 33-47, the angle of refraction is

$$\theta_4 = \sin^{-1}\left(\frac{2.419}{1.63}\sin 33.43°\right)$$

$$= 54.8°.$$

Thus, to keep a diamond sparkling, clean both the top *and* bottom surfaces.

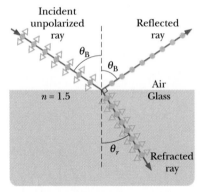

• Component perpendicular to page
◁–▷ Component parallel to page

**FIG. 33-27** A ray of unpolarized light in air is incident on a glass surface at the Brewster angle $\theta_B$. The electric fields along that ray have been resolved into components perpendicular to the page (the plane of incidence, reflection, and refraction) and components parallel to the page. The reflected light consists only of components perpendicular to the page and is thus polarized in that direction. The refracted light consists of the original components parallel to the page and weaker components perpendicular to the page; this light is partially polarized.

## 33-10 | Polarization by Reflection

You can vary the glare you see in sunlight that has been reflected from, say, water by looking through a polarizing sheet (such as a polarizing sunglass lens) and then rotating the sheet's polarizing axis around your line of sight. You can do so because any light that is reflected from a surface is either fully or partially polarized by the reflection.

Figure 33-27 shows a ray of unpolarized light incident on a glass surface. Let us resolve the electric field vectors of the light into two components. The *perpendicular components* are perpendicular to the plane of incidence and thus also to the page in Fig. 33-27; these components are represented with dots (as if we see the tips of the vectors). The *parallel components* are parallel to the plane of incidence and the page; they are represented with double-headed arrows. Because the light is unpolarized, these two components are of equal magnitude.

In general, the reflected light also has both components but with unequal magnitudes. This means that the reflected light is partially polarized—the electric fields oscillating along one direction have greater amplitudes than those oscillating along other directions. However, when the light is incident at a particular incident angle, called the *Brewster angle* $\theta_B$, the reflected light has only perpendicular components, as shown in Fig. 33-27. The reflected light is then fully polarized perpendicular to the plane of incidence. The parallel components of the incident light do not disappear but (along with perpendicular components) refract into the glass.

Glass, water, and the other dielectric materials discussed in Section 25-7 can partially and fully polarize light by reflection. When you intercept sunlight reflected from such a surface, you see a bright spot (the glare) on the surface where the reflection takes place. If the surface is horizontal as in Fig. 33-27, the reflected light is partially or fully polarized horizontally. To eliminate such glare from horizontal surfaces, the lenses in polarizing sunglasses are mounted with their polarizing direction vertical.

## Brewster's Law

For light incident at the Brewster angle $\theta_B$, we find experimentally that the reflected and refracted rays are perpendicular to each other. Because the reflected ray is reflected at the angle $\theta_B$ in Fig. 33-27 and the refracted ray is at an angle $\theta_r$, we have

$$\theta_B + \theta_r = 90°. \qquad (33\text{-}48)$$

These two angles can also be related with Eq. 33-40. Arbitrarily assigning subscript 1 in Eq. 33-40 to the material through which the incident and reflected rays travel, we have, from that equation,

$$n_1 \sin \theta_B = n_2 \sin \theta_r.$$

Combining these equations leads to

$$n_1 \sin \theta_B = n_2 \sin(90° - \theta_B) = n_2 \cos \theta_B,$$

which gives us

$$\theta_B = \tan^{-1} \frac{n_2}{n_1} \qquad \text{(Brewster angle)}. \qquad (33\text{-}49)$$

(Note carefully that the subscripts in Eq. 33-49 are *not* arbitrary because of our decision as to their meanings.) If the incident and reflected rays travel *in air,* we can approximate $n_1$ as unity and let $n$ represent $n_2$ in order to write Eq. 33-49 as

$$\theta_B = \tan^{-1} n \qquad \text{(Brewster's law)}. \qquad (33\text{-}50)$$

This simplified version of Eq. 33-49 is known as **Brewster's law.** Like $\theta_B$, it is named after Sir David Brewster, who found both experimentally in 1812.

## REVIEW & SUMMARY

**Electromagnetic Waves**  An electromagnetic wave consists of oscillating electric and magnetic fields. The various possible frequencies of electromagnetic waves form a *spectrum*, a small part of which is visible light. An electromagnetic wave traveling along an $x$ axis has an electric field $\vec{E}$ and a magnetic field $\vec{B}$ with magnitudes that depend on $x$ and $t$:

$$E = E_m \sin(kx - \omega t)$$

and

$$B = B_m \sin(kx - \omega t), \qquad (33\text{-}1, 33\text{-}2)$$

where $E_m$ and $B_m$ are the amplitudes of $\vec{E}$ and $\vec{B}$. The electric field induces the magnetic field and vice versa. The speed of any electromagnetic wave in vacuum is $c$, which can be written as

$$c = \frac{E}{B} = \frac{1}{\sqrt{\mu_0 \varepsilon_0}}, \qquad (33\text{-}5, 33\text{-}3)$$

where $E$ and $B$ are the simultaneous magnitudes of the fields.

**Energy Flow**  The rate per unit area at which energy is transported via an electromagnetic wave is given by the Poynting vector $\vec{S}$:

$$\vec{S} = \frac{1}{\mu_0} \vec{E} \times \vec{B}. \qquad (33\text{-}19)$$

The direction of $\vec{S}$ (and thus of the wave's travel and the energy transport) is perpendicular to the directions of both $\vec{E}$ and $\vec{B}$. The time-averaged rate per unit area at which energy is transported is $S_{avg}$, which is called the *intensity I* of the wave:

$$I = \frac{1}{c\mu_0} E_{rms}^2, \qquad (33\text{-}26)$$

in which $E_{rms} = E_m/\sqrt{2}$. A *point source* of electromagnetic waves emits the waves *isotropically*—that is, with equal intensity in all directions. The intensity of the waves at distance $r$ from a point source of power $P_s$ is

$$I = \frac{P_s}{4\pi r^2}. \qquad (33\text{-}27)$$

**Radiation Pressure**  When a surface intercepts electromagnetic radiation, a force and a pressure are exerted on the surface. If the radiation is totally absorbed by the surface, the force is

$$F = \frac{IA}{c} \qquad \text{(total absorption)}, \qquad (33\text{-}32)$$

in which $I$ is the intensity of the radiation and $A$ is the area of the surface perpendicular to the path of the radiation. If the radiation is totally reflected back along its original path, the force is

$$F = \frac{2IA}{c} \qquad \text{(total reflection back along path)}. \qquad (33\text{-}33)$$

The radiation pressure $p_r$ is the force per unit area:

$$p_r = \frac{I}{c} \qquad \text{(total absorption)} \qquad (33\text{-}34)$$

and $\qquad p_r = \dfrac{2I}{c}$ $\qquad$ (total reflection back along path). $\quad$ (33-35)

**Polarization** Electromagnetic waves are **polarized** if their electric field vectors are all in a single plane, called the *plane of oscillation*. Light waves from common sources are not polarized; that is, they are **unpolarized**, or **polarized randomly**.

**Polarizing Sheets** When a polarizing sheet is placed in the path of light, only electric field components of the light parallel to the sheet's **polarizing direction** are *transmitted* by the sheet; components perpendicular to the polarizing direction are absorbed. The light that emerges from a polarizing sheet is polarized parallel to the polarizing direction of the sheet.

If the original light is initially unpolarized, the transmitted intensity $I$ is half the original intensity $I_0$:

$$I = \tfrac{1}{2}I_0. \qquad (33\text{-}36)$$

If the original light is initially polarized, the transmitted intensity depends on the angle $\theta$ between the polarization direction of the original light and the polarizing direction of the sheet:

$$I = I_0 \cos^2 \theta. \qquad (33\text{-}38)$$

**Geometrical Optics** *Geometrical optics* is an approximate treatment of light in which light waves are represented as straight-line rays.

**Reflection and Refraction** When a light ray encounters a boundary between two transparent media, a **reflected** ray and a **refracted** ray generally appear. Both rays remain in the plane of incidence. The **angle of reflection** is equal to the angle of incidence, and the **angle of refraction** is related to the angle of incidence by Snell's law,

$$n_2 \sin \theta_2 = n_1 \sin \theta_1 \quad \text{(refraction)}, \qquad (33\text{-}40)$$

where $n_1$ and $n_2$ are the indexes of refraction of the media in which the incident and refracted rays travel.

**Total Internal Reflection** A wave encountering a boundary across which the index of refraction decreases will experience **total internal reflection** if the angle of incidence exceeds a **critical angle** $\theta_c$, where

$$\theta_c = \sin^{-1} \frac{n_2}{n_1} \quad \text{(critical angle)}. \qquad (33\text{-}44)$$

**Polarization by Reflection** A reflected wave will be fully **polarized**, with its $\vec{E}$ vectors perpendicular to the plane of incidence, if it strikes a boundary at the **Brewster angle** $\theta_B$, where

$$\theta_B = \tan^{-1} \frac{n_2}{n_1} \quad \text{(Brewster angle)}. \qquad (33\text{-}49)$$

## QUESTIONS

**1** If the magnetic field of a light wave oscillates parallel to a $y$ axis and is given by $B_y = B_m \sin(kz - \omega t)$, (a) in what direction does the wave travel and (b) parallel to which axis does the associated electric field oscillate?

**2** Figure 33-28 shows the electric and magnetic fields of an electromagnetic wave at a certain instant. Is the wave traveling into the page or out of it?

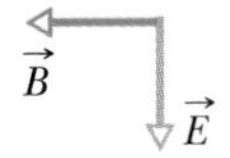

$\vec{B}$ $\quad$ $\vec{E}$

**FIG. 33-28** Question 2.

**3** In Fig. 33-15a, start with light that is initially polarized parallel to the $x$ axis, and write the ratio of its final intensity $I_3$ to its initial intensity $I_0$ as $I_3/I_0 = A \cos^n \theta$. What are $A$, $n$, and $\theta$ if we rotate the polarizing direction of the first sheet (a) 60° counterclockwise and (b) 90° clockwise from what is shown?

**4** Suppose we rotate the second sheet in Fig. 33-15a, starting with the polarization direction aligned with the $y$ axis ($\theta = 0$) and ending with it aligned with the $x$ axis ($\theta = 90°$). Which of the four curves in Fig. 33-29 best shows the intensity of the light through the three-sheet system during this 90° rotation?

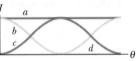

**FIG. 33-29** Question 4.

**5** (a) Figure 33-30 shows light reaching a polarizing sheet whose polarizing direction is parallel to a $y$ axis. We shall rotate the sheet 40° clockwise about the light's indicated line of travel. During this rotation, does the fraction of the initial light intensity passed by the sheet in-

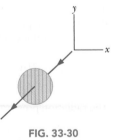

**FIG. 33-30** Question 5.

crease, decrease, or remain the same if the light is (a) initially unpolarized, (b) initially polarized parallel to the $x$ axis, and (c) initially polarized parallel to the $y$ axis?

**6** In Fig. 33-31, unpolarized light is sent into a system of five polarizing sheets. Their polarizing directions, measured counterclockwise from the positive direction of the $y$ axis, are the following: sheet 1, 35°; sheet 2, 0°; sheet 3, 0°; sheet 4, 110°; sheet 5, 45°. Sheet 3 is then rotated 180° counterclockwise about the light ray. During that rotation, at what angles (measured counterclockwise from the $y$ axis) is the transmission of light through the system eliminated?

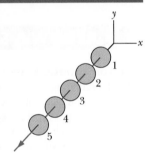

**FIG. 33-31** Question 6.

**7** Figure 33-32 shows rays of monochromatic light passing through three materials $a$, $b$, and $c$. Rank the materials according to index of refraction, greatest first.

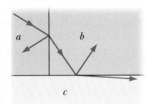

**FIG. 33-32** Question 7.

**8** Figure 33-33 shows the multiple reflections of a light ray along a glass corridor where the

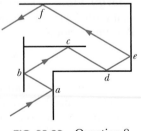

**FIG. 33-33** Question 8.

walls are either parallel or perpendicular to one another. If the angle of incidence at point *a* is 30°, what are the angles of reflection of the light ray at points *b*, *c*, *d*, *e*, and *f*?

**9** Each part of Fig. 33-34 shows light that refracts through an interface between two materials. The incident ray (shown gray in the figure) consists of red and blue light. The approximate index of refraction for visible light is indicated for each material. Which of the three parts show physically possible refraction?

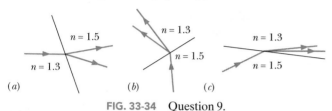

**FIG. 33-34** Question 9.

**10** In Fig. 33-35, light travels from material *a*, through three layers of other materials with surfaces parallel to one another, and then back into another layer of material *a*. The refractions (but not the associated reflections) at the surfaces are shown. Rank the materials according to index of refraction, greatest first.

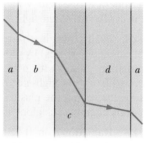

**FIG. 33-35** Question 10.

**11** Figure 33-36 shows four long horizontal layers *A–D* of different materials, with air above and below them. The index of refraction of each material is given. Rays of light are sent into the left end of each layer as shown. In which layer is there the possibility of totally trapping the light in that layer so that, after many reflections, all the light reaches the right end of the layer?

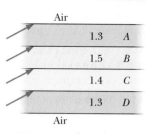

**FIG. 33-36** Question 11.

**12** The leftmost block in Fig. 33-37 depicts total internal reflection for light inside a material with an index of refraction $n_1$ when air is outside the material. A light ray reaching point *A* from anywhere within the shaded region at the left (such as the ray shown) fully reflects at that point and ends up in the shaded region at the right. The other blocks show similar situations for two other materials. Rank the indexes of refraction of the three materials, greatest first.

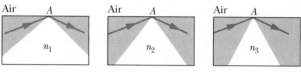

**FIG. 33-37** Question 12.

## PROBLEMS

 Tutoring problem available (at instructor's discretion) in *WileyPLUS* and WebAssign

**SSM** Worked-out solution available in Student Solutions Manual   **WWW** Worked-out solution is at

• – ••• Number of dots indicates level of problem difficulty   **ILW** Interactive solution is at ── http://www.wiley.com/college/halliday

 Additional information available in *The Flying Circus of Physics* and at flyingcircusofphysics.com

### sec. 33-2 Maxwell's Rainbow

•1 From Fig. 33-2, approximate the (a) smaller and (b) larger wavelength at which the eye of a standard observer has half the eye's maximum sensitivity. What are the (c) wavelength, (d) frequency, and (e) period of the light at which the eye is the most sensitive?

•2 About how far apart must you hold your hands for them to be separated by 1.0 nano-light-second (the distance light travels in 1.0 ns)?

•3 A certain helium–neon laser emits red light in a narrow band of wavelengths centered at 632.8 nm and with a "wavelength width" (such as on the scale of Fig. 33-1) of 0.0100 nm. What is the corresponding "frequency width" for the emission?

•4 Project Seafarer was an ambitious program to construct an enormous antenna, buried underground on a site about 10 000 km² in area. Its purpose was to transmit signals to submarines while they were deeply submerged. If the effective wavelength were $1.0 \times 10^4$ Earth radii, what would be the (a) frequency and (b) period of the radiations emitted? Ordinarily, electromagnetic radiations do not penetrate very far into conductors such as seawater.

### sec. 33-3 The Traveling Electromagnetic Wave, Qualitatively

•5 What inductance must be connected to a 17 pF capacitor in an oscillator capable of generating 550 nm (i.e., visible) electromagnetic waves? Comment on your answer.  **SSM**

•6 What is the wavelength of the electromagnetic wave emitted by the oscillator–antenna system of Fig. 33-3 if $L = 0.253\ \mu$H and $C = 25.0$ pF?

### sec. 33-5 Energy Transport and the Poynting Vector

•7 Some neodymium–glass lasers can provide 100 TW of power in 1.0 ns pulses at a wavelength of 0.26 $\mu$m. How much energy is contained in a single pulse?  **ILW**

•8 A plane electromagnetic wave has a maximum electric field magnitude of $3.20 \times 10^{-4}$ V/m. Find the magnetic field amplitude.

•9 A plane electromagnetic wave traveling in the positive direction of an *x* axis in vacuum has components $E_x = E_y = 0$ and $E_z = (2.0$ V/m$) \cos[(\pi \times 10^{15}$ s$^{-1})(t - x/c)]$. (a) What is the amplitude of the magnetic field component? (b) Parallel to which axis does the magnetic field oscillate? (c) When the electric field component is in the positive direction of the *z* axis at a certain point $P$, what is the direction of the magnetic field component there? **ILW**

•10 In a plane radio wave the maximum value of the electric field component is 5.00 V/m. Calculate (a) the maximum value of the magnetic field component and (b) the wave intensity.

•11 What is the intensity of a traveling plane electromagnetic wave if $B_m$ is $1.0 \times 10^{-4}$ T?

•12 Assume (unrealistically) that a TV station acts as a point source broadcasting isotropically at 1.0 MW. What is the intensity of the transmitted signal reaching Proxima Centauri, the star nearest our solar system, 4.3 ly away? (An alien civilization at that distance might be able to watch *X Files.*) A light-year (ly) is the distance light travels in one year.

••13 The maximum electric field 10 m from an isotropic point source of light is 2.0 V/m. What are (a) the maximum value of the magnetic field and (b) the average intensity of the light there? (c) What is the power of the source?

••14 Frank D. Drake, an investigator in the SETI (Search for Extra-Terrestrial Intelligence) program, once said that the large radio telescope in Arecibo, Puerto Rico (Fig. 33-38), "can detect a signal which lays down on the entire surface of the earth a power of only one picowatt." (a) What is the power that would be received by the Arecibo antenna for such a signal? The antenna diameter is 300 m. (b) What would be the power of an isotropic source at the center of our galaxy that could provide such a signal? The galactic center is $2.2 \times 10^4$ ly away. A light-year is the distance light travels in one year.

**FIG. 33-38** Problem 14. Radio telescope at Arecibo. (*Courtesy Cornell University*)

••15 Sunlight just outside Earth's atmosphere has an intensity of 1.40 kW/m². Calculate (a) $E_m$ and (b) $B_m$ for sunlight there, assuming it to be a plane wave.

••16 An electromagnetic wave with frequency $4.00 \times 10^{14}$ Hz travels through vacuum in the positive direction of an $x$ axis. The wave has its electric field directed parallel to the $y$ axis, with amplitude $E_m$. At time $t = 0$, the electric field at point $P$ on the $x$ axis has a value of $+E_m/4$ and is decreasing with time. What is the distance along the $x$ axis from point $P$ to the first point with $E = 0$ if we search in (a) the negative direction and (b) the positive direction of the $x$ axis?

••17 An airplane flying at a distance of 10 km from a radio transmitter receives a signal of intensity 10 $\mu$W/m². What is the

amplitude of the (a) electric and (b) magnetic component of the signal at the airplane? (c) If the transmitter radiates uniformly over a hemisphere, what is the transmission power?

••18 An isotropic point source emits light at wavelength 500 nm, at the rate of 200 W. A light detector is positioned 400 m from the source. What is the maximum rate $\partial B/\partial t$ at which the magnetic component of the light changes with time at the detector's location? **GO**

### sec. 33-6 Radiation Pressure

•19 What is the radiation pressure 1.5 m away from a 500 W lightbulb? Assume that the surface on which the pressure is exerted faces the bulb and is perfectly absorbing and that the bulb radiates uniformly in all directions. **ILW**

•20 A black, totally absorbing piece of cardboard of area $A = 2.0$ cm² intercepts light with an intensity of 10 W/m² from a camera strobe light. What radiation pressure is produced on the cardboard by the light?

•21 High-power lasers are used to compress a plasma (a gas of charged particles) by radiation pressure. A laser generating radiation pulses with peak power $1.5 \times 10^3$ MW is focused onto 1.0 mm² of high-electron-density plasma. Find the pressure exerted on the plasma if the plasma reflects all the light beams directly back along their paths. **SSM**

•22 Radiation from the Sun reaching Earth (just outside the atmosphere) has an intensity of 1.4 kW/m². (a) Assuming that Earth (and its atmosphere) behaves like a flat disk perpendicular to the Sun's rays and that all the incident energy is absorbed, calculate the force on Earth due to radiation pressure. (b) For comparison, calculate the force due to the Sun's gravitational attraction.

••23 Prove, for a plane electromagnetic wave that is normally incident on a flat surface, that the radiation pressure on the surface is equal to the energy density in the incident beam. (This relation between pressure and energy density holds no matter what fraction of the incident energy is reflected.) **SSM**

••24 A small laser emits light at power 5.00 mW and wavelength 633 nm. The laser beam is focused (narrowed) until its diameter matches the 1266 nm diameter of a sphere placed in its path. The sphere is perfectly absorbing and has density $5.00 \times 10^3$ kg/m³. What are (a) the beam intensity at the sphere's location, (b) the radiation pressure on the sphere, (c) the magnitude of the corresponding force, and (d) the magnitude of the acceleration that force alone would give the sphere?

••25 A plane electromagnetic wave, with wavelength 3.0 m, travels in vacuum in the positive direction of an $x$ axis. The electric field, of amplitude 300 V/m, oscillates parallel to the $y$ axis. What are the (a) frequency, (b) angular frequency, and (c) angular wave number of the wave? (d) What is the amplitude of the magnetic field component? (e) Parallel to which axis does the magnetic field oscillate? (f) What is the time-averaged rate of energy flow in watts per square meter associated with this wave? The wave uniformly illuminates a surface of area 2.0 m². If the surface totally absorbs the wave, what are (g) the rate at which momentum is transferred to the surface and (h) the radiation pressure on the surface? **SSM WWW**

••26 It has been proposed that a spaceship might be propelled in the solar system by radiation pressure, using a large sail made of foil. How large must the surface area of the sail be if the radiation force is to be equal in magnitude to the Sun's gravitational attraction? Assume that the mass of the ship + sail is 1500 kg, that the sail is perfectly reflecting, and that the sail is oriented perpendicular to the Sun's rays. See Appendix C for needed data. (With a larger sail, the ship is continuously driven away from the Sun.) **GO**

••27 A small spaceship whose mass is $1.5 \times 10^3$ kg (including an astronaut) is drifting in outer space with negligible gravitational forces acting on it. If the astronaut turns on a 10 kW laser beam, what speed will the ship attain in 1.0 day because of the momentum carried away by the beam? **SSM**

••28 In Fig. 33-39, a laser beam of power 4.60 W and diameter $D = 2.60$ mm is directed upward at one circular face (of diameter $d < 2.60$ mm) of a perfectly reflecting cylinder. The cylinder is levitated because the upward radiation force matches the downward gravitational force. If the cylinder's density is 1.20 g/cm³, what is its height $H$?

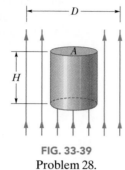

**FIG. 33-39**
Problem 28.

••29 Someone plans to float a small, totally absorbing sphere 0.500 m above an isotropic point source of light, so that the upward radiation force from the light matches the downward gravitational force on the sphere. The sphere's density is 19.0 g/cm³, and its radius is 2.00 mm. (a) What power would be required of the light source? (b) Even if such a source were made, why would the support of the sphere be unstable?

••30 The intensity $I$ of light from an isotropic point source is determined as a function of distance $r$ from the source. Figure 33-40 gives intensity $I$ versus the inverse square $r^{-2}$ of that distance. The vertical axis scale is set by $I_s = 200$ W/m², and the horizontal axis scale is set by $r_s^{-2} = 8.0$ m⁻². What is the power of the source?

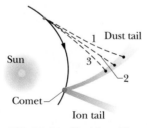

**FIG. 33-40** Problem 30.

•••31 As a comet swings around the Sun, ice on the comet's surface vaporizes, releasing trapped dust particles and ions. The ions, because they are

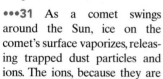

**FIG. 33-41** Problem 31.

electrically charged, are forced by the electrically charged *solar wind* into a straight *ion tail* that points radially away from the Sun (Fig. 33-41). The (electrically neutral) dust particles are pushed radially outward from the Sun by the radiation force on them from sunlight. Assume that the dust particles are spherical, have density $3.5 \times 10^3$ kg/m³, and are totally absorbing. (a) What radius must a particle have in order to follow a straight path, like path 2 in the figure? (b) If its radius is larger, does its path curve away from the Sun (like path 1) or toward the Sun (like path 3)?

### sec. 33-7 Polarization

•32 In Fig. 33-42, a beam of unpolarized light, with intensity 43 W/m², is sent into a system of two polarizing sheets with polarizing directions at angles $\theta_1 = 70°$ and $\theta_2 = 90°$ to the $y$ axis. What is the intensity of the light transmitted by the system? **GO**

•33 In Fig. 33-42, a beam of light, with intensity 43 W/m² and polarization parallel to a $y$ axis, is sent into a system of two polarizing sheets with polarizing directions at angles of $\theta_1 = 70°$ and $\theta_2 = 90°$ to the $y$ axis. What is the intensity of the light transmitted by the two-sheet system? **ILW**

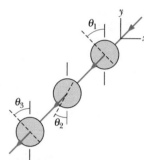

**FIG. 33-42**
Problems 32, 33,
and 36.

•34 In Fig. 33-43, initially unpolarized light is sent into a system of three polarizing sheets whose polarizing directions make angles of $\theta_1 = \theta_2 = \theta_3 = 50°$ with the direction of the $y$ axis. What percentage of the initial intensity is transmitted by the system? (*Hint:* Be careful with the angles.)

•35 In Fig. 33-43, initially unpolarized light is sent into a system of three polarizing sheets whose polarizing directions make angles of $\theta_1 = 40°$, $\theta_2 = 20°$, and $\theta_3 = 40°$ with the direction of the $y$ axis. What percentage of the light's initial intensity is transmitted by the system? (*Hint:* Be careful with the angles.) **SSM**

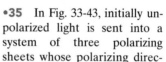

**FIG. 33-43** Problems 34 and 35.

••36 In Fig. 33-42, unpolarized light is sent into a system of two polarizing sheets. The angles $\theta_1$ and $\theta_2$ of the polarizing directions of the sheets are measured counterclockwise from the positive direction of the $y$ axis (they are not drawn to scale in the figure). Angle $\theta_1$ is fixed but angle $\theta_2$ can be varied. Figure 33-44 gives the intensity of the light emerging from sheet 2 as a function of $\theta_2$. (The scale of the intensity axis is not indicated.) What percentage of the light's initial intensity is transmitted by the two-sheet system when $\theta_2 = 90°$?

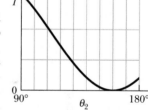

**FIG. 33-44** Problem 36.

••37 Unpolarized light of intensity 10 mW/m² is sent into a polarizing sheet as in Fig. 33-11. What are (a) the amplitude of the electric field component of the transmitted light and (b) the radiation pressure on the sheet due to its absorbing some of the light?

••38 At a beach the light is generally partially polarized due to reflections off sand and water. At a particular beach on a particular day near sundown, the horizontal component of the electric field vector is 2.3 times the vertical component. A standing sunbather puts on polarizing sunglasses; the glasses eliminate the horizontal field component. (a) What fraction of the light intensity received before the glasses were put on now reaches the sunbather's eyes? (b) The sunbather, still wearing the glasses, lies on his side. What fraction of the light intensity received before the glasses were put on now reaches his eyes?

••39 A beam of polarized light is sent into a system of

two polarizing sheets. Relative to the polarization direction of that incident light, the polarizing directions of the sheets are at angles $\theta$ for the first sheet and 90° for the second sheet. If 0.10 of the incident intensity is transmitted by the two sheets, what is $\theta$?

••**40** In Fig. 33-45, unpolarized light is sent into a system of three polarizing sheets. The angles $\theta_1$, $\theta_2$, and $\theta_3$ of the polarizing directions are measured counterclockwise from the positive direction of the $y$ axis (they are not drawn to scale). Angles $\theta_1$ and $\theta_3$ are fixed, but angle $\theta_2$ can be varied. Figure 33-46 gives the intensity of the light emerging from sheet 3 as a function of $\theta_2$. (The scale of the intensity axis is not indicated.) What percentage of the light's initial intensity is transmitted by the system when $\theta_2 = 30°$?

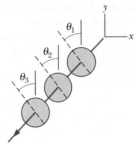

**FIG. 33-45** Problems 40, 42, and 44.

••**41** A beam of partially polarized light can be considered to be a mixture of polarized and unpolarized light. Suppose we send such a beam through a polarizing filter and then rotate the filter through 360° while keeping it perpendicular to the beam. If the transmitted intensity varies by a factor of 5.0 during the rotation, what fraction of the intensity of the original beam is associated with the beam's polarized light?

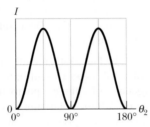

**FIG. 33-46** Problem 40.

••**42** In Fig. 33-45, unpolarized light is sent into a system of three polarizing sheets, which transmits 0.0500 of the initial light intensity. The polarizing directions of the first and third sheets are at angles $\theta_1 = 0°$ and $\theta_3 = 90°$. What are the (a) smaller and (b) larger possible values of angle $\theta_2$ ($< 90°$) for the polarizing direction of sheet 2?

••**43** We want to rotate the direction of polarization of a beam of polarized light through 90° by sending the beam through one or more polarizing sheets. (a) What is the minimum number of sheets required? (b) What is the minimum number of sheets required if the transmitted intensity is to be more than 60% of the original intensity? SSM WWW

••**44** In Fig. 33-45, unpolarized light is sent into a system of three polarizing sheets. The angles $\theta_1$, $\theta_2$, and $\theta_3$ of the polarizing directions are measured counterclockwise from the positive direction of the $y$ axis (they are not drawn to scale). Angles $\theta_1$ and $\theta_3$ are fixed, but angle $\theta_2$ can be varied. Figure 33-47 gives the intensity of the light emerging from sheet 3 as a function of $\theta_2$. (The scale of the intensity axis is not indicated.) What percentage of the light's initial intensity is transmitted by the three-sheet system when $\theta_2 = 90°$?

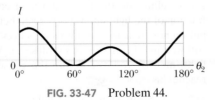

**FIG. 33-47** Problem 44.

### sec. 33-8 Reflection and Refraction

•**45** Light in vacuum is incident on the surface of a glass slab. In the vacuum the beam makes an angle of 32.0° with the normal to the surface, while in the glass it makes an angle of 21.0° with the normal. What is the index of refraction of the glass?

•**46** In Fig. 33-48a, a light ray in water is incident at angle $\theta_1$ on a boundary with an underlying material, into which some of the light refracts. There are two choices of underlying material. For each, the angle of refraction $\theta_2$ versus the incident angle $\theta_1$ is given in Fig. 33-48b. The vertical axis scale is set by $\theta_{2s} = 90°$. Without calculation, determine whether the index of refraction of (a) material 1 and (b) material 2 is greater or less than the index of water ($n = 1.33$). What is the index of refraction of (c) material 1 and (d) material 2?

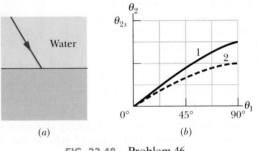

**FIG. 33-48** Problem 46.

•**47** Figure 33-49 shows light reflecting from two perpendicular reflecting surfaces $A$ and $B$. Find the angle between the incoming ray $i$ and the outgoing ray $r'$.

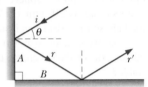

**FIG. 33-49** Problem 47.

•**48** In Fig. 33-50a, a light ray in an underlying material is incident at angle $\theta_1$ on a boundary with water, and some of the light refracts into the water. There are two choices of underlying material. For each, the angle of refraction $\theta_2$ versus the incident angle $\theta_1$ is given in Fig. 33-50b. The horizontal axis scale is set by $\theta_{1s} = 90°$. Without calculation, determine whether the index of refraction of (a) material 1 and (b) material 2 is greater or less than the index of water ($n = 1.33$). What is the index of refraction of (c) material 1 and (d) material 2?

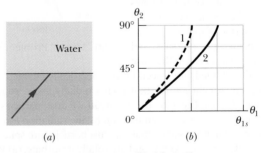

**FIG. 33-50** Problem 48.

•**49** When the rectangular metal tank in Fig. 33-51 is filled to the top with an unknown liquid, observer $O$, with eyes level with the top of the tank, can just see corner $E$. A ray that refracts toward $O$ at the top surface of the liquid is shown.

If $D = 85.0$ cm and $L = 1.10$ m, what is the index of refraction of the liquid?

**••50** In Fig. 33-52a, a beam of light in material 1 is incident on a boundary at an angle of $\theta_1 = 30°$. The extent of refraction of the light into material 2 depends, in part, on the index of refraction $n_2$ of material 2. Figure 33-52b gives the angle of refraction $\theta_2$ versus $n_2$ for a range of possible $n_2$ values. The vertical axis scale is set by $\theta_{2a} = 20.0°$ and $\theta_{2b} = 40.0°$. (a) What is the index of refraction of material 1? (b) If the incident angle is changed to 60° and material 2 has $n_2 = 2.4$, then what is angle $\theta_2$?

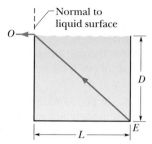

FIG. 33-51 Problem 49.

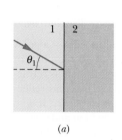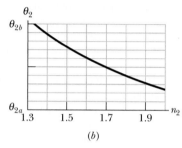

(a) (b)

FIG. 33-52 Problem 50.

**••51** In Fig. 33-53, a 2.00-m-long vertical pole extends from the bottom of a swimming pool to a point 50.0 cm above the water. Sunlight is incident at angle $\theta = 55.0°$. What is the length of the shadow of the pole on the level bottom of the pool? SSM

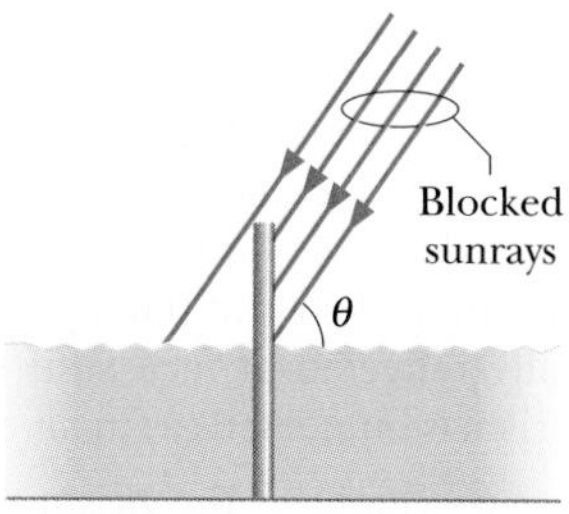

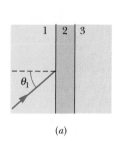

FIG. 33-53 Problem 51.

**••52** *Dispersion in a window pane.* In Fig. 33-54, a beam of white light is incident at angle $\theta = 50°$ on a common window pane (shown in cross section). For the pane's type of glass, the index of refraction for visible light ranges from 1.524 at the blue end of the spectrum to 1.509 at the red end. The two sides of the pane are parallel. What is the angular spread of the colors in the beam (a) when the light enters the pane and (b) when it emerges from the opposite side? (*Hint:* When you look at an object through a window pane, are the colors in the light from the object dispersed as shown in, say, Fig. 33-20?*)

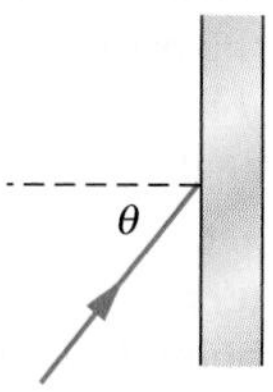

FIG. 33-54 Problem 52.

**••53** In Fig. 33-55, light is incident at angle $\theta_1 = 40.1°$ on a boundary between two transparent materials. Some of the light travels down through the next three layers of transparent materials, while some of it reflects upward and then escapes

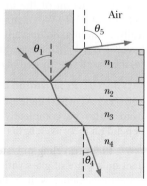

FIG. 33-55 Problem 53.

into the air. If $n_1 = 1.30$, $n_2 = 1.40$, $n_3 = 1.32$, and $n_4 = 1.45$, what is the value of (a) $\theta_5$ and (b) $\theta_4$?

**••54** In Fig. 33-56a, a beam of light in material 1 is incident on a boundary at an angle $\theta_1 = 40°$. Some of the light travels through material 2, and then some of it emerges into material 3. The two boundaries between the three materials are parallel. The final direction of the beam depends, in part, on the index of refraction $n_3$ of the third material. Figure 33-56b gives the angle of refraction $\theta_3$ in that material versus $n_3$ for a range of possible $n_3$ values. The vertical axis scale is set by $\theta_{3a} = 30.0°$ and $\theta_{3b} = 50.0°$. (a) What is the index of refraction of material 1, or is the index impossible to calculate without more information? (b) What is the index of refraction of material 2, or is the index impossible to calculate without more information? (c) If $\theta_1$ is changed to 70° and the index of refraction of material 3 is 2.4, what is $\theta_3$?

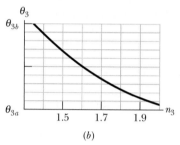

(a) (b)

FIG. 33-56 Problem 54.

**••55** In Fig. 33-57, a ray is incident on one face of a triangular glass prism in air. The angle of incidence $\theta$ is chosen so that the emerging ray also makes the same angle $\theta$ with the normal to the other face. Show that the index of refraction $n$ of the glass prism is given by

$$n = \frac{\sin\frac{1}{2}(\psi + \phi)}{\sin\frac{1}{2}\phi},$$

where $\phi$ is the vertex angle of the prism and $\psi$ is the *deviation angle,* the total angle through which the beam is turned in passing through the prism. (Under these conditions the deviation angle $\psi$ has the smallest possible value, which is called the *angle of minimum deviation.*) SSM ILW WWW

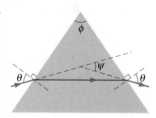

FIG. 33-57 Problems 55 and 66.

**••56** *Rainbows from square drops.* Suppose that, on some surreal world, raindrops had a square cross section and always fell with one face horizontal. Figure 33-58 shows such a falling drop, with a white beam of sunlight incident at $\theta = 70.0°$ at point $P$. The part of the light that enters the drop then travels to point $A$, where some of it refracts out into the air and the rest reflects. That reflected light then travels to point $B$, where again some of the light refracts out into the air and the rest reflects. What is the difference in the angles of the red light ($n = 1.331$) and the blue light ($n = 1.343$) that emerge at (a) point $A$ and

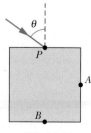

FIG. 33-58 Problem 56.

(b) point *B*? (This angular difference in the light emerging at, say, point *A* would be the angular width of the rainbow you would see were you to intercept the light emerging there.) ✈

### sec. 33-9 Total Internal Reflection

•**57** A point source of light is 80.0 cm below the surface of a body of water. Find the diameter of the circle at the surface through which light emerges from the water.

•**58** The index of refraction of benzene is 1.8. What is the critical angle for a light ray traveling in benzene toward a flat layer of air above the benzene?

••**59** In Fig. 33-59, light initially in material 1 refracts into material 2, crosses that material, and is then incident at the critical angle on the interface between materials 2 and 3. The indexes of refraction are $n_1$ = 1.60, $n_2$ = 1.40, and $n_3$ = 1.20. (a) What is angle $\theta$? (b) If $\theta$ is increased, is there refraction of light into material 3? GO

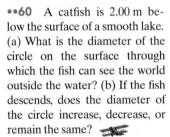

**FIG. 33-59** Problem 59.

••**60** A catfish is 2.00 m below the surface of a smooth lake. (a) What is the diameter of the circle on the surface through which the fish can see the world outside the water? (b) If the fish descends, does the diameter of the circle increase, decrease, or remain the same? ✈

••**61** In the ray diagram of Fig. 33-60, where the angles are not drawn to scale, the ray is incident at the critical angle on the interface between materials 2 and 3. Angle $\phi$ = 60.0°, and two of the indexes of refraction are $n_1$ = 1.70 and $n_2$ = 1.60. Find (a) index of refraction $n_3$ and (b) angle $\theta$. (c) If $\theta$ is decreased, does light refract into material 3?

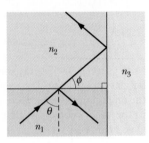

**FIG. 33-60** Problem 61.

••**62** In Fig. 33-61, light from ray *A* refracts from material 1 ($n_1$ = 1.60) into a thin layer of material 2 ($n_2$ = 1.80), crosses that layer, and is then incident at the critical angle on the interface between materials 2 and 3 ($n_3$ = 1.30). (a) What is the value of incident angle $\theta_A$? (b) If $\theta_A$ is decreased, does part of the light refract into material 3?

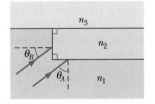

**FIG. 33-61** Problem 62.

Light from ray *B* refracts from material 1 into the thin layer, crosses that layer, and is then incident at the critical angle on the interface between materials 2 and 3. (c) What is the value of incident angle $\theta_B$? (d) If $\theta_B$ is decreased, does part of the light refract into material 3?

••**63** Figure 33-62 depicts a simplistic optical fiber: a plastic core ($n_1$ = 1.58) is surrounded by a plastic sheath ($n_2$ = 1.53). A light ray is incident on one end of the fiber at angle $\theta$. The ray is to undergo total internal reflection at point *A*, where it encounters the core–sheath boundary. (Thus there is no loss of light through that boundary.) What is the maximum value of $\theta$ that allows total internal reflection at *A*? GO

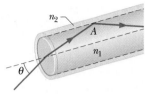

**FIG. 33-62** Problem 63.

••**64** In Fig. 33-63, a light ray in air is incident at angle $\theta_1$ on a block of transparent plastic with an index of refraction of 1.56. The dimensions indicated are *H* = 2.00 cm and *W* = 3.00 cm. The light passes through the block to one of its sides and there undergoes reflection (inside the block) and possibly refraction (out into the air). This is the point of *first reflection*. The reflected light then passes through the block to another of its sides—a point of *second reflection*. If $\theta_1$ = 40°, on which side is the point of (a) first reflection and (b) second reflection? If there is refraction at the point of (c) first reflection and (d) second reflection, give the angle of refraction; if not, answer "none." If $\theta_1$ = 70°, on which side is the point of (e) first reflection and (f) second reflection? If there is refraction at the point of (g) first reflection and (h) second reflection, give the angle of refraction; if not, answer "none."

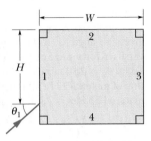

**FIG. 33-63** Problem 64.

••**65** In Fig. 33-64, a ray of light is perpendicular to the face *ab* of a glass prism (*n* = 1.52). Find the largest value for the angle $\phi$ so that the ray is totally reflected at face *ac* if the prism is immersed (a) in air and (b) in water. SSM ILW

**FIG. 33-64** Problem 65.

••**66** Suppose the prism of Fig. 33-57 has apex angle $\phi$ = 60.0° and index of refraction *n* = 1.60. (a) What is the smallest angle of incidence $\theta$ for which a ray can enter the left face of the prism and exit the right face? (b) What angle of incidence $\theta$ is required for the ray to exit the prism with an identical angle $\theta$ for its refraction, as it does in Fig. 33-57?

••**67** In Fig. 33-65, light enters a 90° triangular prism at point *P* with incident angle $\theta$, and then some of it refracts at point *Q* with an angle of refraction of 90°. (a) What is the index of refraction of the prism in terms of $\theta$? (b) What, numerically, is the maximum value that the index of refraction can have? Does light emerge at *Q* if the incident angle at *P* is (c) increased slightly and (d) decreased slightly?

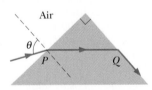

**FIG. 33-65** Problem 67.

### sec. 33-10 Polarization by Reflection

•**68** (a) At what angle of incidence will the light reflected from water be completely polarized? (b) Does this angle depend on the wavelength of the light?

•**69** Light traveling in water of refractive index 1.33 is incident on a plate of glass with index of refraction 1.53. At what angle of incidence is the reflected light fully polarized? SSM

**••70** In Fig. 33-66, a light ray in air is incident on a flat layer of material 2 that has an index of refraction $n_2 = 1.5$. Beneath material 2 is material 3 with an index of refraction $n_3$. The ray is incident on the air–material 2 interface at the Brewster angle for that interface. The ray of

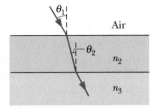

FIG. 33-66   Problem 70.

light refracted into material 3 happens to be incident on the material 2–material 3 interface at the Brewster angle for that interface. What is the value of $n_3$?

**Additional Problems**

**71**   *Rainbow.* Figure 33-67 shows a light ray entering and then leaving a falling, spherical raindrop after one internal reflection (see Fig. 33-21a). The final direction of travel is deviated (turned) from the initial direction of travel by angular deviation $\theta_{\text{dev}}$. (a) Show that $\theta_{\text{dev}}$ is

$$\theta_{\text{dev}} = 180° + 2\theta_i - 4\theta_r,$$

where $\theta_i$ is the angle of incidence of the ray on the drop and $\theta_r$ is the angle of refraction of the ray within the drop. (b) Using Snell's law, substitute for $\theta_r$ in terms of $\theta_i$ and the index of refraction $n$ of the water. Then, on a graphing calculator or with a computer graphing package, graph $\theta_{\text{dev}}$ versus $\theta_i$ for the

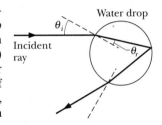

FIG. 33-67   Problem 71.

range of possible $\theta_i$ values and for $n = 1.331$ for red light and $n = 1.333$ for blue light.

The red-light curve and the blue-light curve have different minima, which means that there is a different *angle of minimum deviation* for each color. The light of any given color that leaves the drop at that color's angle of minimum deviation is especially bright because rays bunch up at that angle. Thus, the bright red light leaves the drop at one angle and the bright blue light leaves it at another angle.

Determine the angle of minimum deviation from the $\theta_{\text{dev}}$ curve for (c) red light and (d) blue light. (e) If these colors form the inner and outer edges of a rainbow (Fig. 33-21a), what is the angular width of the rainbow?

**72**   The *primary rainbow* described in Problem 71 is the type commonly seen in regions where rainbows appear. It is produced by light reflecting once inside the drops. Rarer is the *secondary rainbow* described in Section 33-8, produced by light reflecting twice inside the drops (Fig. 33-68a). (a) Show that the angular deviation of light entering and then leaving a

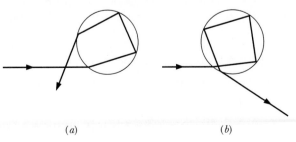

FIG. 33-68   Problem 72.

spherical water drop is

$$\theta_{\text{dev}} = (180°)k + 2\theta_i - 2(k + 1)\theta_r,$$

where $k$ is the number of internal reflections. Using the procedure of Problem 71, find the angle of minimum deviation for (b) red light and (c) blue light in a secondary rainbow. (d) What is the angular width of that rainbow (Fig. 33-21d)?

The *tertiary rainbow* depends on three internal reflections (Fig. 33-68b). It probably occurs but, as noted in Section 33-8, cannot be seen because it is very faint and lies in the bright sky surrounding the Sun. What is the angle of minimum deviation for (e) the red light and (f) the blue light in this rainbow? (g) What is the rainbow's angular width?

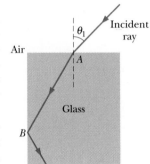

FIG. 33-69   Problem 73.

**73**   In Fig. 33-69, a light ray enters a glass slab at point $A$ at incident angle $\theta_1 = 45.0°$ and then undergoes total internal reflection at point $B$. What minimum value for the index of refraction of the glass can be inferred from this information? **SSM**

**74**   In Fig. 33-70, unpolarized light with an intensity of 25 W/m² is sent into a system of four polarizing sheets with polarizing directions at angles $\theta_1 = 40°$, $\theta_2 = 20°$, $\theta_3 = 20°$, and $\theta_4 = 30°$. What is the intensity of the light that emerges from the system? **GO**

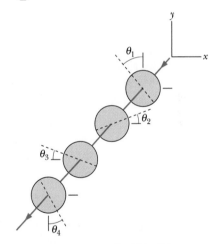

FIG. 33-70   Problem 74.

**75**   (a) Prove that a ray of light incident on the surface of a sheet of plate glass of thickness $t$ emerges from the opposite face parallel to its initial direction but displaced sideways, as in Fig. 33-71. (b) Show that, for small angles of incidence $\theta$, this displacement is given by

$$x = t\theta \frac{n - 1}{n},$$

where $n$ is the index of refraction of the glass and $\theta$ is measured in radians. **SSM**

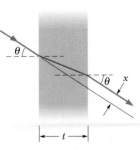

FIG. 33-71   Problem 75.

**76** In Fig. 33-72, two light rays pass from air through five layers of transparent plastic and then back into air. The layers have parallel interfaces and unknown thicknesses; their indexes of refraction are $n_1 = 1.7$, $n_2 = 1.6$, $n_3 = 1.5$, $n_4 = 1.4$, and $n_5 = 1.6$. Ray $b$ is incident at angle $\theta_b = 20°$. Relative to a normal at the last interface, at what angle do (a) ray $a$ and (b) ray $b$ emerge? (*Hint:* Solving the problem algebraically can save time.) If the air at the left and right sides in the figure were, instead, glass with index of refraction 1.5, at what angle would (c) ray $a$ and (d) ray $b$ emerge?

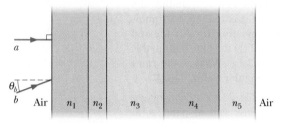

**FIG. 33-72** Problem 76.

**77** (a) How long does it take a radio signal to travel 150 km from a transmitter to a receiving antenna? (b) We see a full Moon by reflected sunlight. How much earlier did the light that enters our eye leave the Sun? The Earth–Moon and Earth–Sun distances are $3.8 \times 10^5$ km and $1.5 \times 10^8$ km, respectively. (c) What is the round-trip travel time for light between Earth and a spaceship orbiting Saturn, $1.3 \times 10^9$ km distant? (d) The Crab nebula, which is about 6500 light-years (ly) distant, is thought to be the result of a supernova explosion recorded by Chinese astronomers in A.D. 1054. In approximately what year did the explosion actually occur? **SSM**

**78** In about A.D. 150, Claudius Ptolemy gave the following measured values for the angle of incidence $\theta_1$ and the angle of refraction $\theta_2$ for a light beam passing from air to water:

| $\theta_1$ | $\theta_2$ | $\theta_1$ | $\theta_2$ |
|------------|------------|------------|------------|
| 10° | 8° | 50° | 35° |
| 20° | 15°30′ | 60° | 40°30′ |
| 30° | 22°30′ | 70° | 45°30′ |
| 40° | 29° | 80° | 50° |

Assuming these data are consistent with the law of refraction, use them to find the index of refraction of water. These data are interesting as perhaps the oldest recorded physical measurements.

**79** The electric component of a beam of polarized light is

$$E_y = (5.00 \text{ V/m}) \sin[(1.00 \times 10^6 \text{ m}^{-1})z + \omega t].$$

(a) Write an expression for the magnetic field component of the wave, including a value for $\omega$. What are the (b) wavelength, (c) period, and (d) intensity of this light? (e) Parallel to which axis does the magnetic field oscillate? (f) In which region of the electromagnetic spectrum is this wave? **SSM**

**80** The magnetic component of an electromagnetic wave in vacuum has an amplitude of 85.8 nT and an angular wave number of 4.00 m$^{-1}$. What are (a) the frequency of the wave,

(b) the rms value of the electric component, and (c) the intensity of the light?

**81** A helium–neon laser, radiating at 632.8 nm, has a power output of 3.0 mW. The beam diverges (spreads) at angle $\theta = 0.17$ mrad (Fig. 33-73). (a) What is the intensity of the beam 40 m from the laser? (b) What is the power of a point source providing that intensity at that distance?

**FIG. 33-73** Problem 81.

**82** The average intensity of the solar radiation that strikes normally on a surface just outside Earth's atmosphere is 1.4 kW/m². (a) What radiation pressure $p_r$ is exerted on this surface, assuming complete absorption? (b) For comparison, find the ratio of $p_r$ to Earth's sea-level atmospheric pressure, which is $1.0 \times 10^5$ Pa.

**83** During a test, a NATO surveillance radar system, operating at 12 GHz at 180 kW of power, attempts to detect an incoming stealth aircraft at 90 km. Assume that the radar beam is emitted uniformly over a hemisphere. (a) What is the intensity of the beam when the beam reaches the aircraft's location? The aircraft reflects radar waves as though it has a cross-sectional area of only 0.22 m². (b) What is the power of the aircraft's reflection? Assume that the beam is reflected uniformly over a hemisphere. Back at the radar site, what are (c) the intensity, (d) the maximum value of the electric field vector, and (e) the rms value of the magnetic field of the reflected radar beam? **SSM**

**84** An unpolarized beam of light is sent into a stack of four polarizing sheets, oriented so that the angle between the polarizing directions of adjacent sheets is 30°. What fraction of the incident intensity is transmitted by the system?

**85** A beam of initially unpolarized light is sent through two polarizing sheets placed one on top of the other. What must be the angle between the polarizing directions of the sheets if the intensity of the transmitted light is to be one-third the incident intensity?

**86** An electromagnetic wave is traveling in the negative direction of a $y$ axis. At a particular position and time, the electric field is directed along the positive direction of the $z$ axis and has a magnitude of 100 V/m. What are the (a) magnitude and (b) direction of the corresponding magnetic field?

**87** Calculate the (a) upper and (b) lower limit of the Brewster angle for white light incident on fused quartz. Assume that the wavelength limits of the light are 400 and 700 nm.

**88** A particle in the solar system is under the combined influence of the Sun's gravitational attraction and the radiation force due to the Sun's rays. Assume that the particle is a sphere of density $1.0 \times 10^3$ kg/m³ and that all the incident light is absorbed. (a) Show that, if its radius is less than some critical radius $R$, the particle will be blown out of the solar system. (b) Calculate the critical radius.

**89** The magnetic component of a polarized wave of light is

$$B_x = (4.0 \times 10^{-6} \text{ T}) \sin[(1.57 \times 10^7 \text{ m}^{-1})y + \omega t].$$

(a) Parallel to which axis is the light polarized? What are the (b) frequency and (c) intensity of the light?

**90** Three polarizing sheets are stacked. The first and third are crossed; the one between has its polarizing direction at 45.0° to the polarizing directions of the other two. What fraction of the intensity of an originally unpolarized beam is transmitted by the stack?

**91** A ray of white light traveling through fused quartz is incident at a quartz–air interface at angle $\theta_1$. Assume that the index of refraction of quartz is $n = 1.456$ at the red end of the visible range and $n = 1.470$ at the blue end. If $\theta_1$ is (a) 42.00°, (b) 43.10°, and (c) 44.00°, is the refracted light white, white dominated by the red end of the visible range, or white dominated by the blue end of the visible range, or is there no refracted light? SSM

**92** In Fig. 33-74, unpolarized light is sent into the system of three polarizing sheets, where the polarizing directions of the first and third sheets are at angles $\theta_1 = 30°$ (counterclockwise) and $\theta_3 = 30°$ (clockwise). What fraction of the initial light intensity emerges from the system?

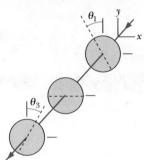

**93** In a region of space where gravitational forces can be neglected, a sphere is accelerated by a uniform light beam of intensity 6.0 mW/m². The sphere is totally absorbing and has a radius of 2.0 $\mu$m and a uniform density of $5.0 \times 10^3$ kg/m³. What is the magnitude of the sphere's acceleration due to the light?

**FIG. 33-74** Problem 92.

**94** In Fig. 33-75, unpolarized light is sent into a system of three polarizing sheets, where the polarizing directions of the first and second sheets are at angles $\theta_1 = 20°$ and $\theta_2 = 40°$. What fraction of the initial light intensity emerges from the system?

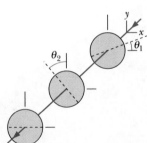

**95** In Fig. 33-76, unpolarized light is sent into a system of three polarizing sheets with polarizing directions at angles $\theta_1 = 20°$, $\theta_2 = 60°$, and $\theta_3 = 40°$. What fraction of the initial light intensity emerges from the system?

**FIG. 33-75** Problem 94.

**96** A square, perfectly reflecting surface is oriented in space to be perpendicular to the light rays from the Sun. The surface has an edge length of 2.0 m and is located $3.0 \times 10^{11}$ m from the Sun's center. What is the radiation force on the surface from the light rays?

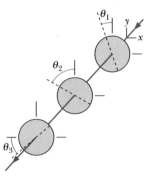

**FIG. 33-76** Problem 95.

**97** The rms value of the electric field in a certain light wave is 0.200 V/m. What is the amplitude of the associated magnetic field?

**98** In Fig. 33-77, an albatross glides at a constant 15 m/s horizontally above level ground, moving in a vertical plane that contains the Sun. It glides toward a wall of height $h = 2.0$ m, which it will just barely clear. At that time of day, the angle of the Sun relative to the ground is $\theta =$ 30°. At what speed does the shadow of the albatross move (a) across the level ground and then (b) up the wall? Suppose that later a hawk happens to glide along the same path, also at 15 m/s. You see that when its shadow reaches the wall, the speed of the shadow noticeably increases. (c) Is the Sun now higher or lower in the sky than when the albatross flew by earlier? (d) If the speed of the hawk's shadow on the wall is 45 m/s, what is the angle $\theta$ of the Sun just then?

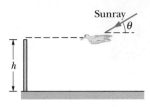

**FIG. 33-77** Problem 98.

**99** The magnetic component of a polarized wave of light is given by $B_x = (4.00 \ \mu T) \sin[ky + (2.00 \times 10^{15} \ s^{-1})t]$. (a) In which direction does the wave travel, (b) parallel to which axis is it polarized, and (c) what is its intensity? (d) Write an expression for the electric field of the wave, including a value for the angular wave number. (e) What is the wavelength? (f) In which region of the electromagnetic spectrum is this electromagnetic wave?

**100** In Fig. 33-78, where $n_1 = 1.70$, $n_2 = 1.50$, and $n_3 = 1.30$, light refracts from material 1 into material 2. If it is incident at point $A$ at the critical angle for the interface between materials 2 and 3, what are (a) the angle of refraction at point $B$ and (b) the initial angle $\theta$? If, instead, light is incident at $B$ at the critical angle for the interface between materials 2 and 3, what are (c) the angle of refraction at point $A$ and (d) the initial angle

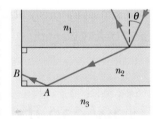

**FIG. 33-78** Problem 100.

$\theta$? If, instead of all that, light is incident at point $A$ at Brewster's angle for the interface between materials 2 and 3, what are (e) the angle of refraction at point $B$ and (f) the initial angle $\theta$?

**101** When red light in vacuum is incident at the Brewster angle on a certain glass slab, the angle of refraction is 32.0°. What are (a) the index of refraction of the glass and (b) the Brewster angle?

**102** Start from Eqs. 33-11 and 33-17 and show that $E(x, t)$ and $B(x, t)$, the electric and magnetic field components of a plane traveling electromagnetic wave, must satisfy the "wave equations"

$$\frac{\partial^2 E}{\partial t^2} = c^2 \frac{\partial^2 E}{\partial x^2} \quad \text{and} \quad \frac{\partial^2 B}{\partial t^2} = c^2 \frac{\partial^2 B}{\partial x^2}.$$

**103** (a) Show that Eqs. 33-1 and 33-2 satisfy the wave equations displayed in Problem 102. (b) Show that any expressions of the form $E = E_m f(kx \pm \omega t)$ and $B = B_m f(kx \pm \omega t)$, where $f(kx \pm \omega t)$ denotes an arbitrary function, also satisfy these wave equations. SSM

**104** A point source of light emits isotropically with a power of 200 W. What is the force due to the light on a totally absorbing sphere of radius 2.0 cm at a distance of 20 m from the source?

# 34 | Images

Underwater vision is usually difficult even if you have perfect vision above water. The reason has to do with how water affects the refraction of light entering the eye. The refraction may be appropriate in the air but quite wrong in the water. However, the peculiar fish **Anableps anableps** *swims with its eyes partially extending above the water surface so that it can see simultaneously above and below water.*

## How can its eyes function in both air and water?

The answer is in this chapter.

Paddy Ryan/Ryan Photographic

## 34-1 | WHAT IS PHYSICS?

One goal of physics is to discover the basic laws governing light, such as the law of refraction. A broader goal is to put those laws to use, and perhaps the most important use is the production of images. The first photographic images, made in 1824, were only novelties, but our world now thrives on images. Huge industries are based on the production of images on television, computer, and theater screens. Images from satellites guide military strategists during times of conflict and environmental strategists during times of blight. Camera surveillance can make a subway system more secure, but it can also invade the privacy of unsuspecting citizens. Physiologists and medical engineers are still puzzled by how images are produced by the human eye and the visual cortex of the brain, but they have managed to create mental images in some sightless people by electrical stimulation of the visual cortex.

Our first step in this chapter is to define and classify images. Then we examine several basic ways in which they can be produced.

## 34-2 | Two Types of Image

For you to see, say, a penguin, your eye must intercept some of the light rays spreading from the penguin and then redirect them onto the retina at the rear of the eye. Your visual system, starting with the retina and ending with the visual cortex at the rear of your brain, automatically and subconsciously processes the information provided by the light. That system identifies edges, orientations, textures, shapes, and colors and then rapidly brings to your consciousness an **image** (a reproduction derived from light) of the penguin; you perceive and recognize the penguin as being in the direction from which the light rays came and at the proper distance.

Your visual system goes through this processing and recognition even if the light rays do not come directly from the penguin, but instead reflect toward you from a mirror or refract through the lenses in a pair of binoculars. However, you now see the penguin in the direction from which the light rays came after they reflected or refracted, and the distance you perceive may be quite different from the penguin's true distance.

For example, if the light rays have been reflected toward you from a standard flat mirror, the penguin appears to be behind the mirror because the rays you intercept come from that direction. Of course, the penguin is not back there. This type of image, which is called a **virtual image,** truly exists only within the brain but nevertheless is *said* to exist at the perceived location.

A **real image** differs in that it can be formed on a surface, such as a card or a movie screen. You can see a real image (otherwise movie theaters would be empty), but the existence of the image does not depend on your seeing it and it is present even if you are not.

In this chapter we explore several ways in which virtual and real images are formed by reflection (as with mirrors) and refraction (as with lenses). We also distinguish between the two types of image more clearly, but here first is an example of a natural virtual image.

### A Common Mirage

A common example of a virtual image is a pool of water that appears to lie on the road some distance ahead of you on a sunny day, but that you can never reach. The pool is a *mirage* (a type of illusion), formed by light rays coming from the low section of the sky in front of you (Fig. 34-1a). As the rays approach the road, they travel through progressively warmer air that has been heated by

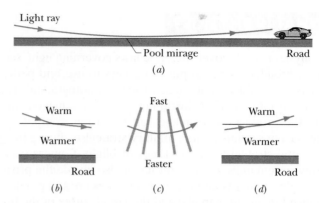

FIG. 34-1 (a) A ray from a low section of the sky refracts through air that is heated by a road (without reaching the road). An observer who intercepts the light perceives it to be from a pool of water on the road. (b) Bending (exaggerated) of a light ray descending across an imaginary boundary from warm air to warmer air. (c) Shifting of wavefronts and associated bending of a ray, which occur because the lower ends of wavefronts move faster in warmer air. (d) Bending of a ray ascending across an imaginary boundary to warm air from warmer air.

the road, which is usually relatively warm. With an increase in air temperature, the density of the air—and hence the index of refraction of the air—decreases slightly. Thus, as the rays descend, encountering progressively smaller indexes of refraction, they continuously bend toward the horizontal (Fig. 34-1b).

Once a ray is horizontal, somewhat above the road's surface, it still bends because the lower portion of each associated wavefront is in slightly warmer air and is moving slightly faster than the upper portion of the wavefront (Fig. 34-1c). This nonuniform motion of the wavefronts bends the ray upward. As the ray then ascends, it continues to bend upward through progressively greater indexes of refraction (Fig. 34-1d).

If you intercept some of this light, your visual system automatically infers that it originated along a backward extension of the rays you have intercepted and, to make sense of the light, assumes that it came from the road surface. If the light happens to be bluish from blue sky, the mirage appears bluish, like water. Because the air is probably turbulent due to the heating, the mirage shimmies, as if water waves were present. The bluish coloring and the shimmy enhance the illusion of a pool of water, but you are actually seeing a virtual image of a low section of the sky.

## 34-3 | Plane Mirrors

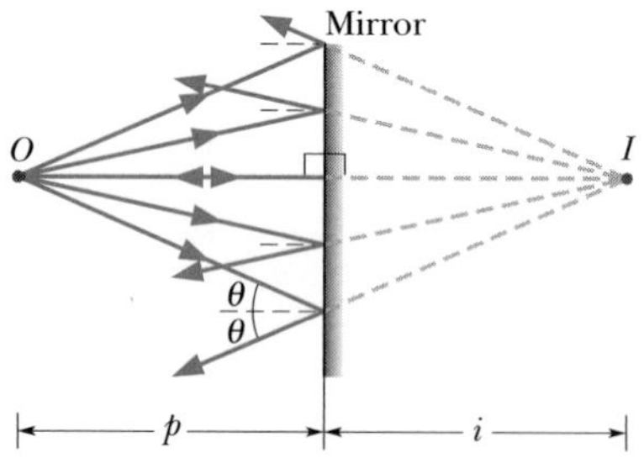

A **mirror** is a surface that can reflect a beam of light in one direction instead of either scattering it widely in many directions or absorbing it. A shiny metal surface acts as a mirror; a concrete wall does not. In this section we examine the images that a **plane mirror** (a flat reflecting surface) can produce.

Figure 34-2 shows a point source of light O, which we shall call the *object*, at a perpendicular distance p in front of a plane mirror. The light that is incident on the mirror is represented with rays spreading from O. The reflection of that light is represented with reflected rays spreading from the mirror. If we extend the reflected rays backward (behind the mirror), we find that the extensions intersect at a point that is a perpendicular distance i behind the mirror.

If you look into the mirror of Fig. 34-2, your eyes intercept some of the reflected light. To make sense of what you see, you perceive a point source of light located at the point of intersection of the extensions. This point source is the image I of object O. It is called a *point image* because it is a point, and it is a vir-

FIG. 34-2 A point source of light O, called the *object*, is a perpendicular distance p in front of a plane mirror. Light rays reaching the mirror from O reflect from the mirror. If your eye intercepts some of the reflected rays, you perceive a point source of light I to be behind the mirror, at a perpendicular distance i. The perceived source I is a virtual image of object O.

tual image because the rays do not actually pass through it. (As you will see, rays *do* pass through a point of intersection for a real image.)

Figure 34-3 shows two rays selected from the many rays in Fig. 34-2. One reaches the mirror at point *b*, perpendicularly. The other reaches it at an arbitrary point *a*, with an angle of incidence $\theta$. The extensions of the two reflected rays are also shown. The right triangles *aOba* and *aIba* have a common side and three equal angles and are thus congruent (equal in size); so their horizontal sides have the same length. That is,

$$Ib = Ob, \tag{34-1}$$

where *Ib* and *Ob* are the distances from the mirror to the image and the object, respectively. Equation 34-1 tells us that the image is as far behind the mirror as the object is in front of it. By convention (that is, to get our equations to work out), *object distances p* are taken to be positive quantities and *image distances i* for virtual images (as here) are taken to be negative quantities. Thus, Eq. 34-1 can be written as $|i| = p$ or as

$$i = -p \qquad \text{(plane mirror).} \tag{34-2}$$

Only rays that are fairly close together can enter the eye after reflection at a mirror. For the eye position shown in Fig. 34-4, only a small portion of the mirror near point *a* (a portion smaller than the pupil of the eye) is useful in forming the image. To find this portion, close one eye and look at the mirror image of a small object such as the tip of a pencil. Then move your fingertip over the mirror surface until you cannot see the image. Only that small portion of the mirror under your fingertip produced the image.

## Extended Objects

In Fig. 34-5, an extended object *O*, represented by an upright arrow, is at perpendicular distance *p* in front of a plane mirror. Each small portion of the object that faces the mirror acts like the point source *O* of Figs. 34-2 and 34-3. If you intercept the light reflected by the mirror, you perceive a virtual image *I* that is a composite of the virtual point images of all those portions of the object. This virtual image seems to be at (negative) distance *i* behind the mirror, with *i* and *p* related by Eq. 34-2.

We can also locate the image of an extended object as we did for a point object in Fig. 34-2: we draw some of the rays that reach the mirror from the top of the object, draw the corresponding reflected rays, and then extend those reflected rays behind the mirror until they intersect to form an image of the top of the object. We then do the same for rays from the bottom of the object. As shown in Fig. 34-5, we find that virtual image *I* has the same orientation and *height* (measured parallel to the mirror) as object *O*.

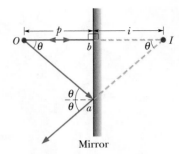

**FIG. 34-3** Two rays from Fig. 34-2. Ray *Oa* makes an arbitrary angle $\theta$ with the normal to the mirror surface. Ray *Ob* is perpendicular to the mirror.

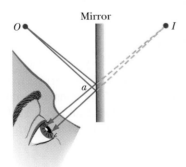

**FIG. 34-4** A "pencil" of rays from *O* enters the eye after reflection at the mirror. Only a small portion of the mirror near *a* is involved in this reflection. The light appears to originate at point *I* behind the mirror.

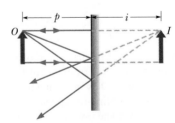

**FIG. 34-5** An extended object *O* and its virtual image *I* in a plane mirror.

**FIG. 34-6** A maze of mirrors. *(Courtesy Adrian Fisher. www.mazemaker.com)*

### Mirror Maze

In a mirror maze (Fig. 34-6), each wall is covered, floor to ceiling, with a mirror. Walk through such a maze and what you see in most directions is a confusing montage of reflections. In some directions, however, you see a hallway that seems to offer a path through the maze. Take these hallways, though, and you soon learn, after smacking into mirror after mirror, that the hallways are largely an illusion.

Figure 34-7a is an overhead view of a simple mirror maze in which differently painted floor sections form equilateral triangles (60° angles) and walls are covered with vertical mirrors. You look into the maze while standing at point *O* at the middle of the maze entrance. In most directions, you see a confusing jumble of images. However, you see something curious in the direction of the ray shown in Fig. 34-7a. That ray leaves the middle of mirror *B* and reflects to you at the middle of mirror *A*. (The reflection obeys the law of reflection, with the angle of incidence and the angle of reflection both equal to 30°.)

To make sense of the origin of the ray reaching you, your brain automatically extends the ray backward. It appears to originate at a point lying *behind* mirror *A*. That is, you perceive a virtual image of *B* behind *A*, at a distance equal to the actual distance between *A* and *B* (Fig. 34-7b). Thus, when you face into the maze in this direction, you see *B* along an apparent straight hallway consisting of four triangular floor sections.

This story is incomplete, however, because the ray reaching you does not *originate* at mirror *B*—it only reflects there. To find the origin, we continue to apply the law of reflection as we work backwards, reflection by reflection on the mirrors. Working through the four reflections shown in Fig. 34-7c, we finally come to the origin of the ray: you! What you see when you look along the apparent hallway is a virtual image of yourself, at a distance of nine triangular floor sections from you (Fig. 34-7d). (There is a second apparent hallway extending away from point *O*. Which way must you face to look along it?)

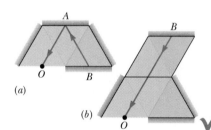

(a)

(b)

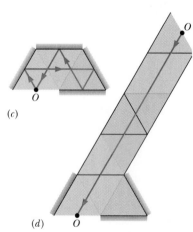

(c)

(d)

**FIG. 34-7** (a) Overhead view of a mirror maze. A ray from mirror *B* reaches you at *O* by reflecting from mirror *A*. (b) Mirror *B* appears to be behind *A*. (c) The ray reaching you comes from you. (d) You see a virtual image of yourself at the end of an apparent hallway.

✔ **CHECKPOINT 1** In the figure you are in a system of two vertical parallel mirrors *A* and *B* separated by distance *d*. A grinning gargoyle is perched at point *O*, a distance 0.2*d* from mirror *A*. Each mirror produces a *first* (least deep) image of the gargoyle. Then each mirror produces a *second* image with the object being the first image in the opposite mirror. Then each mirror produces a *third* image with the object being the second image in the opposite mirror, and so on—you might see hundreds of grinning gargoyle images. How deep behind mirror *A* are the first, second, and third images in mirror *A*?

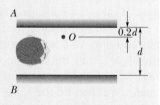

## 34-4 | Spherical Mirrors

We turn now from images produced by plane mirrors to images produced by mirrors with curved surfaces. In particular, we consider spherical mirrors, which are simply mirrors in the shape of a small section of the surface of a sphere. A plane mirror is in fact a spherical mirror with an infinitely large *radius of curvature* and thus an approximately flat surface.

### Making a Spherical Mirror

We start with the plane mirror of Fig. 34-8a, which faces leftward toward an object *O* that is shown and an observer that is not shown. We make a **concave**

**mirror** by curving the mirror's surface so it is *concave* ("caved in") as in Fig. 34-8*b*. Curving the surface in this way changes several characteristics of the mirror and the image it produces of the object:

1. The *center of curvature C* (the center of the sphere of which the mirror's surface is part) was infinitely far from the plane mirror; it is now closer but still in front of the concave mirror.

2. The *field of view*—the extent of the scene that is reflected to the observer—was wide; it is now smaller.

3. The image of the object was as far behind the plane mirror as the object was in front; the image is farther behind the concave mirror; that is, |*i*| is greater.

4. The height of the image was equal to the height of the object; the height of the image is now greater. This feature is why many makeup mirrors and shaving mirrors are concave—they produce a larger image of a face.

We can make a **convex mirror** by curving a plane mirror so its surface is *convex* ("flexed out") as in Fig. 34-8*c*. Curving the surface in this way (1) moves the center of curvature *C* to *behind* the mirror and (2) *increases* the field of view. It also (3) moves the image of the object *closer* to the mirror and (4) *shrinks* it. Store surveillance mirrors are usually convex to take advantage of the increase in the field of view—more of the store can then be seen with a single mirror.

## Focal Points of Spherical Mirrors

For a plane mirror, the magnitude of the image distance *i* is always equal to the object distance *p*. Before we can determine how these two distances are related for a spherical mirror, we must consider the reflection of light from an object *O* located an effectively infinite distance in front of a spherical mirror, on the mirror's *central axis*. That axis extends through the center of curvature *C* and the center *c* of the mirror. Because of the great distance between the object and the mirror, the light waves spreading from the object are plane waves when they reach the mirror along the central axis. This means that the rays representing the light waves are all parallel to the central axis when they reach the mirror.

When these parallel rays reach a concave mirror like that of Fig. 34-9*a*, those near the central axis are reflected through a common point *F*; two of these reflected rays are shown in the figure. If we placed a (small) card at *F*, a point image of the infinitely distant object *O* would appear on the card. (This would occur for any infinitely distant object.) Point *F* is called the **focal point** (or **focus**) of the mirror, and its distance from the center of the mirror *c* is the **focal length** *f* of the mirror.

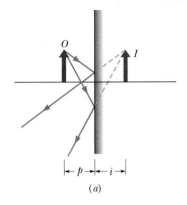
(a)

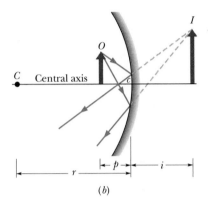
(b)

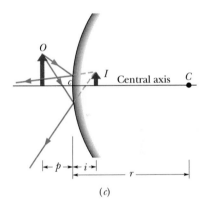
(c)

FIG. 34-8 (*a*) An object *O* forms a virtual image *I* in a plane mirror. (*b*) If the mirror is bent so that it becomes *concave*, the image moves farther away and becomes larger. (*c*) If the plane mirror is bent so that it becomes *convex*, the image moves closer and becomes smaller.

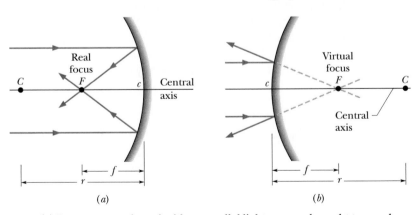
(a) (b)

FIG. 34-9 (*a*) In a concave mirror, incident parallel light rays are brought to a real focus at *F*, on the same side of the mirror as the incident light rays. (*b*) In a convex mirror, incident parallel light rays seem to diverge from a virtual focus at *F*, on the side of the mirror opposite the light rays.

If we now substitute a convex mirror for the concave mirror, we find that the parallel rays are no longer reflected through a common point. Instead, they diverge as shown in Fig. 34-9b. However, if your eye intercepts some of the reflected light, you perceive the light as originating from a point source behind the mirror. This perceived source is located where extensions of the reflected rays pass through a common point (*F* in Fig. 34-9b). That point is the focal point (or focus) *F* of the convex mirror, and its distance from the mirror surface is the focal length *f* of the mirror. If we placed a card at this focal point, an image of object *O* would *not* appear on the card; so this focal point is not like that of a concave mirror.

To distinguish the actual focal point of a concave mirror from the perceived focal point of a convex mirror, the former is said to be a *real focal point* and the latter is said to be a *virtual focal point*. Moreover, the focal length *f* of a concave mirror is taken to be a positive quantity, and that of a convex mirror a negative quantity. For mirrors of both types, the focal length *f* is related to the radius of curvature *r* of the mirror by

$$f = \tfrac{1}{2}r \quad \text{(spherical mirror)}, \tag{34-3}$$

where, consistent with the signs for the focal length, *r* is a positive quantity for a concave mirror and a negative quantity for a convex mirror.

## 34-5 | Images from Spherical Mirrors

With the focal point of a spherical mirror defined, we can find the relation between image distance *i* and object distance *p* for concave and convex spherical mirrors. We begin by placing the object *O* *inside the focal point* of the concave mirror—that is, between the mirror and its focal point *F* (Fig. 34-10a). An observer can then see a virtual image of *O* in the mirror: The image appears to be behind the mirror, and it has the same orientation as the object.

If we now move the object away from the mirror until it is at the focal point, the image moves farther and farther back from the mirror until, when the object is at the focal point, the image is at infinity (Fig. 34-10b). The image is then ambiguous and imperceptible because neither the rays reflected by the mirror nor the ray extensions behind the mirror cross to form an image of *O*.

If we next move the object *outside the focal point*—that is, farther away from the mirror than the focal point—the rays reflected by the mirror converge to form an *inverted* image of object *O* (Fig. 34-10c) in front of the mirror. That image moves in from infinity as we move the object farther outside *F*. If you were to hold a card at the position of the image, the image would show up on the card—the image is said to be *focused* on the card by the mirror. (The verb "focus," which in this context means to produce an image, differs from the noun "focus," which is another name for the focal point.) Because this image can actually appear on a surface, it is a real image—the rays actually intersect to create the image,

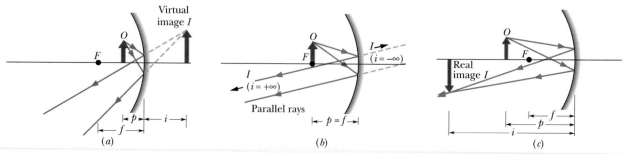

FIG. 34-10 (a) An object *O* inside the focal point of a concave mirror, and its virtual image *I*. (b) The object at the focal point *F*. (c) The object outside the focal point, and its real image *I*.

regardless of whether an observer is present. The image distance $i$ of a real image is a positive quantity, in contrast to that for a virtual image. We also see that

> Real images form on the side of a mirror where the object is, and virtual images form on the opposite side.

As we shall prove in Section 34-9, when light rays from an object make only small angles with the central axis of a spherical mirror, a simple equation relates the object distance $p$, the image distance $i$, and the focal length $f$:

$$\frac{1}{p} + \frac{1}{i} = \frac{1}{f} \qquad \text{(spherical mirror).} \qquad (34\text{-}4)$$

We assume such small angles in figures such as Fig. 34-10, but for clarity the rays are drawn with exaggerated angles. With that assumption, Eq. 34-4 applies to any concave, convex, or plane mirror. For a convex or plane mirror, only a virtual image can be formed, regardless of the object's location on the central axis. As shown in the example of a convex mirror in Fig. 34-8c, the image is always on the opposite side of the mirror from the object and has the same orientation as the object.

The size of an object or image, as measured *perpendicular* to the mirror's central axis, is called the object or image *height*. Let $h$ represent the height of the object, and $h'$ the height of the image. Then the ratio $h'/h$ is called the **lateral magnification** $m$ produced by the mirror. However, by convention, the lateral magnification always includes a plus sign when the image orientation is that of the object and a minus sign when the image orientation is opposite that of the object. For this reason, we write the formula for $m$ as

$$|m| = \frac{h'}{h} \qquad \text{(lateral magnification).} \qquad (34\text{-}5)$$

We shall soon prove that the lateral magnification can also be written as

$$m = -\frac{i}{p} \qquad \text{(lateral magnification).} \qquad (34\text{-}6)$$

For a plane mirror, for which $i = -p$, we have $m = +1$. The magnification of 1 means that the image is the same size as the object. The plus sign means that the image and the object have the same orientation. For the concave mirror of Fig. 34-10c, $m \approx -1.5$.

Equations 34-3 through 34-6 hold for all plane mirrors, concave spherical mirrors, and convex spherical mirrors. In addition to those equations, you have been asked to absorb a lot of information about these mirrors, and you should organize it for yourself by filling in Table 34-1. Under Image Location, note whether the image is on the *same* side of the mirror as the object or on the *oppo-*

**TABLE 34-1**

**Your Organizing Table for Mirrors**

| Mirror Type | Object Location | Image | | | Sign | | |
| --- | --- | --- | --- | --- | --- | --- | --- |
| | | Location | Type | Orientation | of $f$ | of $r$ | of $m$ |
| Plane | Anywhere | | | | | | |
| Concave | Inside $F$ | | | | | | |
| | Outside $F$ | | | | | | |
| Convex | Anywhere | | | | | | |

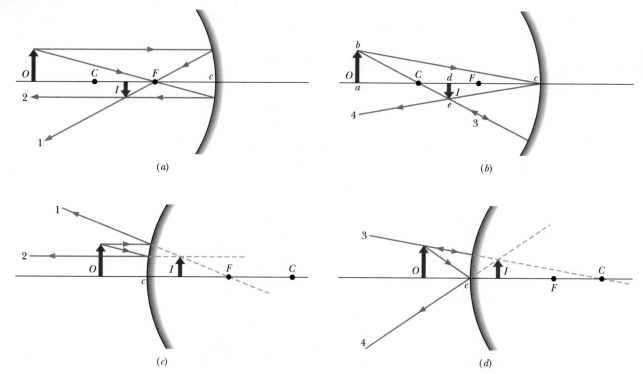

(a)

(b)

(c)

(d)

**FIG. 34-11** (a, b) Four rays that may be drawn to find the image formed by a concave mirror. For the object position shown, the image is real, inverted, and smaller than the object. (c, d) Four similar rays for the case of a convex mirror. For a convex mirror, the image is always virtual, oriented like the object, and smaller than the object. [In (c), ray 2 is initially directed toward focal point F. In (d), ray 3 is initially directed toward center of curvature C.]

site side. Under Image Type, note whether the image is *real* or *virtual*. Under Image Orientation, note whether the image has the *same* orientation as the object or is *inverted*. Under Sign, give the sign of the quantity or fill in ± if the sign is ambiguous. You will need this organization to tackle homework or a test.

### Locating Images by Drawing Rays

Figures 34-11a and b show an object O in front of a concave mirror. We can graphically locate the image of any off-axis point of the object by drawing a *ray diagram* with any two of four special rays through the point:

1. A ray that is initially parallel to the central axis reflects through the focal point F (ray 1 in Fig. 34-11a).

2. A ray that reflects from the mirror after passing through the focal point emerges parallel to the central axis (ray 2 in Fig. 34-11a).

3. A ray that reflects from the mirror after passing through the center of curvature C returns along itself (ray 3 in Fig. 34-11b).

4. A ray that reflects from the mirror at point c is reflected symmetrically about that axis (ray 4 in Fig. 34-11b).

The image of the point is at the intersection of the two special rays you choose. The image of the object can then be found by locating the images of two or more of its off-axis points. You need to modify the descriptions of the rays slightly to apply them to convex mirrors, as in Figs. 34-11c and d.

### Proof of Equation 34-6

We are now in a position to derive Eq. 34-6 ($m = -i/p$), the equation for the lateral magnification of an object reflected in a mirror. Consider ray 4 in Fig. 34-11b. It is reflected at point c so that the incident and reflected rays make equal angles with the axis of the mirror at that point.

The two right triangles $abc$ and $dec$ in the figure are similar (have the same set of angles); so we can write

$$\frac{de}{ab} = \frac{cd}{ca}.$$

The quantity on the left (apart from the question of sign) is the lateral magnification $m$ produced by the mirror. Because we indicate an inverted image as a *negative* magnification, we symbolize this as $-m$. However, $cd = i$ and $ca = p$; so we have

$$m = -\frac{i}{p} \qquad \text{(magnification)}, \qquad\qquad (34\text{-}7)$$

which is the relation we set out to prove.

✓ **CHECKPOINT 2**    A Central American vampire bat, dozing on the central axis of a spherical mirror, is magnified by $m = -4$. Is its image (a) real or virtual, (b) inverted or of the same orientation as the bat, and (c) on the same side of the mirror as the bat or on the opposite side?

## Sample Problem    34-1

A tarantula of height $h$ sits cautiously before a spherical mirror whose focal length has absolute value $|f| = 40$ cm. The image of the tarantula produced by the mirror has the same orientation as the tarantula and has height $h' = 0.20h$.

(a) Is the image real or virtual, and is it on the same side of the mirror as the tarantula or the opposite side?

**Reasoning:** Because the image has the same orientation as the tarantula (the object), it must be virtual and on the opposite side of the mirror. (You can easily see this result if you have filled out Table 34-1.)

(b) Is the mirror concave or convex, and what is its focal length $f$, sign included?

**KEY IDEA**    We *cannot* tell the type of mirror from the type of image because both types of mirror can produce virtual images. Similarly, we cannot tell the type of mirror from the sign of the focal length $f$, as obtained from Eq. 34-3 or Eq. 34-4, because we lack enough information to use either equation. However, we can make use of the magnification information.

**Calculations:** We know that the ratio of image height $h'$ to object height $h$ is 0.20. Thus, from Eq. 34-5 we have

$$|m| = \frac{h'}{h} = 0.20.$$

Because the object and image have the same orientation, we know that $m$ must be positive: $m = +0.20$. Substituting this into Eq. 34-6 and solving for, say, $i$ gives us

$$i = -0.20p,$$

which does not appear to be of help in finding $f$. However, it is helpful if we substitute it into Eq. 34-4. That equation gives us

$$\frac{1}{f} = \frac{1}{i} + \frac{1}{p} = \frac{1}{-0.20p} + \frac{1}{p} = \frac{1}{p}(-5 + 1),$$

from which we find

$$f = -p/4.$$

Now we have it: Because $p$ is positive, $f$ must be negative, which means that the mirror is convex with

$$f = -40 \text{ cm}. \qquad\qquad \text{(Answer)}$$

## 34-6 I Spherical Refracting Surfaces

We now turn from images formed by reflection to images formed by refraction through surfaces of transparent materials, such as glass. We shall consider only spherical surfaces, with radius of curvature $r$ and center of curvature $C$. The light will be emitted by a point object $O$ in a medium with index of refraction $n_1$; it will refract through a spherical surface into a medium of index of refraction $n_2$.

Our concern is whether the light rays, after refracting through the surface, form a real image (no observer necessary) or a virtual image (assuming that an observer intercepts the rays). The answer depends on the relative values of $n_1$ and $n_2$ and on the geometry of the situation.

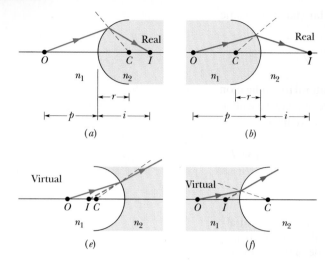

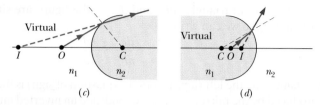

**FIG. 34-12** Six possible ways in which an image can be formed by refraction through a spherical surface of radius $r$ and center of curvature $C$. The surface separates a medium with index of refraction $n_1$ from a medium with index of refraction $n_2$. The point object $O$ is always in the medium with $n_1$, to the left of the surface. The material with the lesser index of refraction is unshaded (think of it as being air, and the other material as being glass). Real images are formed in (a) and (b); virtual images are formed in the other four situations.

Six possible results are shown in Fig. 34-12. In each part of the figure, the medium with the greater index of refraction is shaded, and object $O$ is always in the medium with index of refraction $n_1$, to the left of the refracting surface. In each part, a representative ray is shown refracting through the surface. (That ray and a ray along the central axis suffice to determine the position of the image in each case.)

At the point of refraction of each ray, the normal to the refracting surface is a radial line through the center of curvature $C$. Because of the refraction, the ray bends toward the normal if it is entering a medium of greater index of refraction and away from the normal if it is entering a medium of lesser index of refraction. If the bending sends the ray toward the central axis, that ray and others (undrawn) form a real image on that axis. If the bending sends the ray away from the central axis, the ray cannot form a real image; however, backward extensions of it and other refracted rays can form a virtual image, provided (as with mirrors) some of those rays are intercepted by an observer.

Real images $I$ are formed (at image distance $i$) in parts $a$ and $b$ of Fig. 34-12, where the refraction directs the ray *toward* the central axis. Virtual images are formed in parts $c$ and $d$, where the refraction directs the ray *away* from the central axis. Note, in these four parts, that real images are formed when the object is relatively far from the refracting surface and virtual images are formed when the object is nearer the refracting surface. In the final situations (Figs. 34-12$e$ and $f$), refraction always directs the ray away from the central axis and virtual images are always formed, regardless of the object distance.

Note the following major difference from reflected images:

> Real images form on the side of a refracting surface that is opposite the object, and virtual images form on the same side as the object.

In Section 34-9, we shall show that (for light rays making only small angles with the central axis)

$$\frac{n_1}{p} + \frac{n_2}{i} = \frac{n_2 - n_1}{r}. \tag{34-8}$$

This insect has been entombed in amber for about 25 million years. Because we view the insect through a curved refracting surface, the location of the image we see does not coincide with the location of the insect (see Fig. 34-12$d$). *(Dr. Paul A. Zahl/Photo Researchers)*

Just as with mirrors, the object distance $p$ is positive, and the image distance $i$ is positive for a real image and negative for a virtual image. However, to keep all the signs correct in Eq. 34-8, we must use the following rule for the sign of the radius of curvature $r$:

> When the object faces a convex refracting surface, the radius of curvature $r$ is positive. When it faces a concave surface, $r$ is negative.

Be careful: This is just the reverse of the sign convention we have for mirrors.

✓**CHECKPOINT 3**    A bee is hovering in front of the concave spherical refracting surface of a glass sculpture. (a) Which of the general situations of Fig. 34-11 is like this situation? (b) Is the image produced by the surface real or virtual, and is it on the same side as the bee or the opposite side?

## Sample Problem | 34-2

A Jurassic mosquito is discovered embedded in a chunk of amber, which has index of refraction 1.6. One surface of the amber is spherically convex with radius of curvature 3.0 mm (Fig. 34-13). The mosquito's head happens to be on the central axis of that surface and, when viewed along the axis, appears to be buried 5.0 mm into the amber. How deep is it really?

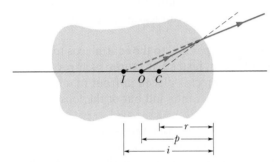

**FIG. 34-13**   A piece of amber with a mosquito from the Jurassic period, with the head buried at point $O$. The spherical refracting surface at the right end, with center of curvature $C$, provides an image $I$ to an observer intercepting rays from the object at $O$.

**KEY IDEAS**   The head appears to be 5.0 mm into the amber only because the light rays that the observer intercepts are bent by refraction at the convex amber surface. The image distance $i$ differs from the object distance $p$ according to Eq. 34-8. To use that equation to find the object distance, we first note:

1. Because the object (the head) and its image are on the same side of the refracting surface, the image must be virtual and so $i = -5.0$ mm.

2. Because the object is always taken to be in the medium of index of refraction $n_1$, we must have $n_1 = 1.6$ and $n_2 = 1.0$.

3. Because the *object* faces a concave refracting surface, the radius of curvature $r$ is negative, and so $r = -3.0$ mm.

*Calculations:* Making these substitutions in Eq. 34-8,

$$\frac{n_1}{p} + \frac{n_2}{i} = \frac{n_2 - n_1}{r},$$

yields

$$\frac{1.6}{p} + \frac{1.0}{-5.0 \text{ mm}} = \frac{1.0 - 1.6}{-3.0 \text{ mm}}$$

and

$$p = 4.0 \text{ mm}. \qquad \text{(Answer)}$$

## Sample Problem | 34-3

Figure 34-14 shows a vertical cross section through an eye of the *Anableps anableps* fish that swims with each eye half in and half out of the water, with a pigment band separating the two halves at the water surface. The front of the eye (the cornea) is a spherically convex refracting surface of radius $r = 1.95$ mm and index of refraction $n_2 = 1.335$. The refraction at the cornea is the first step in the eye's focusing of a *real* image onto the back of the eye (the retina), where visual processing begins. If the cornea faces an insect (lunch) at object distance $p = 0.20$ m, what is the image distance $i$ of that refraction for the cornea in air ($n_1 = 1.000$) and in water ($n_1 = 1.333$)?

**KEY IDEAS**   (1) Because the object and its image are on opposite sides of the refracting surface with its convex side facing the object, the situation is like either Fig. 34-12a (for a real image) or Fig. 34-12c (for a virtual image). (2) Image distance and object distance are related by Eq. 34-8.

*Calculations:* Solving Eq. 34-8 for $i$ gives us

$$i = \frac{n_2}{\dfrac{n_2 - n_1}{r} - \dfrac{n_1}{p}}.$$

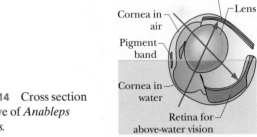

**FIG. 34-14**   Cross section of the eye of *Anableps anableps.*

Substituting the given values and using the index of refraction $n_1 = 1.000$ for air, we find

$$i = \frac{1.335}{\dfrac{1.335 - 1.000}{0.00195} - \dfrac{1.000}{0.20}}$$

$$= 8.00 \text{ mm}. \qquad \text{(Answer)}$$

Repeating the calculation but using the index of refraction $n_1 = 1.333$ for water, we find

$$i = \frac{1.335}{\dfrac{1.335 - 1.333}{0.00195} - \dfrac{1.333}{0.20}}$$

$$= -0.237 \text{ m}. \qquad \text{(Answer)}$$

After light is refracted by the cornea, it is further refracted by a lens to give a final real image on the retina. (The function of a lens is discussed in the next section.) Our first answer is positive (indicating a real image, as needed) and about twice the diameter of the eye. Thus the role of the lens in focusing the light onto the retina is moderate because so much refraction occurs at the cornea. Our second answer is quite different: It is negative (indicating a virtual image) and much larger. So, far more focusing is required by the lens to put a real image onto the retina. To provide moderate focusing of light from the air and much stronger focusing of light from the water, the eye lens in *Anableps anableps* is egg shaped, with much greater curvature in the bottom half than in the top.

## 34-7 | Thin Lenses

A **lens** is a transparent object with two refracting surfaces whose central axes coincide. The common central axis is the central axis of the lens. When a lens is surrounded by air, light refracts from the air into the lens, crosses through the lens, and then refracts back into the air. Each refraction can change the direction of travel of the light.

A lens that causes light rays initially parallel to the central axis to converge is (reasonably) called a **converging lens.** If, instead, it causes such rays to diverge, the lens is a **diverging lens.** When an object is placed in front of a lens of either type, light rays from the object that refract into and out of the lens can produce an image of the object.

We shall consider only the special case of a **thin lens**—that is, a lens in which the thickest part is thin relative to the object distance $p$, the image distance $i$, and the radii of curvature $r_1$ and $r_2$ of the two surfaces of the lens. We shall also consider only light rays that make small angles with the central axis (they are exaggerated in the figures here). In Section 34-9 we shall prove that for such rays, a thin lens has a focal length $f$. Moreover, $i$ and $p$ are related to each other by

$$\frac{1}{f} = \frac{1}{p} + \frac{1}{i} \quad \text{(thin lens),} \tag{34-9}$$

which is the same as we had for mirrors. We shall also prove that when a thin lens with index of refraction $n$ is surrounded by air, this focal length $f$ is given by

$$\frac{1}{f} = (n-1)\left(\frac{1}{r_1} - \frac{1}{r_2}\right) \quad \text{(thin lens in air),} \tag{34-10}$$

which is often called the *lens maker's equation.* Here $r_1$ is the radius of curvature of the lens surface nearer the object and $r_2$ is that of the other surface. The signs of these radii are found with the rules in Section 34-6 for the radii of spherical refracting surfaces. If the lens is surrounded by some medium other than air (say, corn oil) with index of refraction $n_{\text{medium}}$, we replace $n$ in Eq. 34-10 with $n/n_{\text{medium}}$. Keep in mind the basis of Eqs. 34-9 and 34-10:

> A lens can produce an image of an object only because the lens can bend light rays, but it can bend light rays only if its index of refraction differs from that of the surrounding medium.

A fire is being started by focusing sunlight onto newspaper by means of a converging lens made of clear ice. The lens was made by melting both sides of a flat piece of ice into a convex shape in the shallow vessel (which has a curved bottom). *(Courtesy Matthew G. Wheeler)*

Figure 34-15*a* shows a thin lens with convex refracting surfaces, or *sides.* When rays that are parallel to the central axis of the lens are sent through the lens, they refract twice, as is shown enlarged in Fig. 34-15*b*. This double refraction causes the rays to converge and pass through a common point $F_2$ at a distance $f$ from the center of the lens. Hence, this lens is a converging lens; further, a *real* focal point (or focus) exists at $F_2$ (because the rays really do pass through it), and

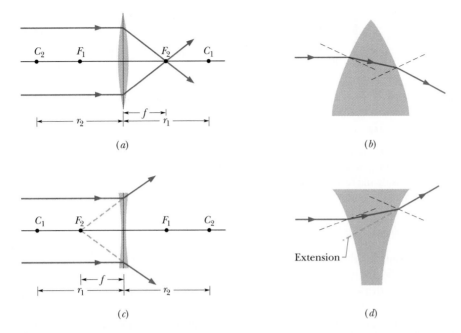

FIG. 34-15  (a) Rays initially parallel to the central axis of a converging lens are made to converge to a real focal point $F_2$ by the lens. The lens is thinner than drawn, with a width like that of the vertical line through it. We shall consider all the bending of rays as occurring at this central line. (b) An enlargement of the top part of the lens of (a); normals to the surfaces are shown dashed. Note that both refractions bend the ray downward, toward the central axis. (c) The same initially parallel rays are made to diverge by a diverging lens. Extensions of the diverging rays pass through a virtual focal point $F_2$. (d) An enlargement of the top part of the lens of (c). Note that both refractions bend the ray upward, away from the central axis.

the associated focal length is $f$. When rays parallel to the central axis are sent in the opposite direction through the lens, we find another real focal point at $F_1$ on the other side of the lens. For a thin lens, these two focal points are equidistant from the lens.

Because the focal points of a converging lens are real, we take the associated focal lengths $f$ to be positive, just as we do with a real focus of a concave mirror. However, signs in optics can be tricky; so we had better check this in Eq. 34-10. The left side of that equation is positive if $f$ is positive; how about the right side? We examine it term by term. Because the index of refraction $n$ of glass or any other material is greater than 1, the term $(n - 1)$ must be positive. Because the source of the light (which is the object) is at the left and faces the convex left side of the lens, the radius of curvature $r_1$ of that side must be positive according to the sign rule for refracting surfaces. Similarly, because the object faces a concave right side of the lens, the radius of curvature $r_2$ of that side must be negative according to that rule. Thus, the term $(1/r_1 - 1/r_2)$ is positive, the whole right side of Eq. 34-10 is positive, and all the signs are consistent.

Figure 34-15$c$ shows a thin lens with concave sides. When rays that are parallel to the central axis of the lens are sent through this lens, they refract twice, as is shown enlarged in Fig. 34-15$d$; these rays *diverge,* never passing through any common point, and so this lens is a diverging lens. However, extensions of the rays do pass through a common point $F_2$ at a distance $f$ from the center of the lens. Hence, the lens has a *virtual* focal point at $F_2$. (If your eye intercepts some of the diverging rays, you perceive a bright spot to be at $F_2$, as if it is the source of the light.) Another virtual focus exists on the opposite side of the lens at $F_1$, symmetrically placed if the lens is thin. Because the focal points of a diverging lens are virtual, we take the focal length $f$ to be negative.

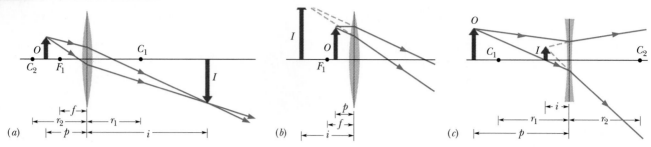

**FIG. 34-16** (*a*) A real, inverted image *I* is formed by a converging lens when the object *O* is outside the focal point $F_1$. (*b*) The image *I* is virtual and has the same orientation as *O* when *O* is inside the focal point. (*c*) A diverging lens forms a virtual image *I*, with the same orientation as the object *O*, whether *O* is inside or outside the focal point of the lens.

### Images from Thin Lenses

We now consider the types of image formed by converging and diverging lenses. Figure 34-16*a* shows an object *O* outside the focal point $F_1$ of a converging lens. The two rays drawn in the figure show that the lens forms a real, inverted image *I* of the object on the side of the lens opposite the object.

When the object is placed inside the focal point $F_1$, as in Fig. 34-16*b*, the lens forms a virtual image *I* on the same side of the lens as the object and with the same orientation. Hence, a converging lens can form either a real image or a virtual image, depending on whether the object is outside or inside the focal point, respectively.

Figure 34-16*c* shows an object *O* in front of a diverging lens. Regardless of the object distance (regardless of whether *O* is inside or outside the virtual focal point), this lens produces a virtual image that is on the same side of the lens as the object and has the same orientation.

As with mirrors, we take the image distance *i* to be positive when the image is real and negative when the image is virtual. However, the locations of real and virtual images from lenses are the reverse of those from mirrors:

☞ Real images form on the side of a lens that is opposite the object, and virtual images form on the side where the object is.

The lateral magnification *m* produced by converging and diverging lenses is given by Eqs. 34-5 and 34-6, the same as for mirrors.

You have been asked to absorb a lot of information in this section, and you should organize it for yourself by filling in Table 34-2 for thin *symmetric lenses* (both sides are convex or both sides are concave). Under Image Location note whether the image is on the *same* side of the lens as the object or on the *opposite* side. Under Image Type note whether the image is *real* or *virtual*. Under Image Orientation note whether the image has the *same* orientation as the object or is *inverted*.

**TABLE 34-2**

**Your Organizing Table for Thin Lenses**

| Lens Type | Object Location | Image | | | Sign | | |
|---|---|---|---|---|---|---|---|
| | | Location | Type | Orientation | of *f* | of *r* | of *m* |
| Converging | Inside *F* | | | | | | |
| | Outside *F* | | | | | | |
| Diverging | Anywhere | | | | | | |

TACTIC 1: *Signs of Trouble with Mirrors and Lenses*   Be careful: A mirror with a convex surface has a negative focal length $f$, just the opposite of a lens with convex surfaces. A mirror with a concave surface has a positive focal length $f$, just the opposite of a lens with concave surfaces. Confusing lens properties with mirror properties is a common mistake.

## Locating Images of Extended Objects by Drawing Rays

Figure 34-17a shows an object $O$ outside focal point $F_1$ of a converging lens. We can graphically locate the image of any off-axis point on such an object (such as the tip of the arrow in Fig. 34-17a) by drawing a ray diagram with any two of three special rays through the point. These special rays, chosen from all those that pass through the lens to form the image, are the following:

1. A ray that is initially parallel to the central axis of the lens will pass through focal point $F_2$ (ray 1 in Fig. 34-17a).

2. A ray that initially passes through focal point $F_1$ will emerge from the lens parallel to the central axis (ray 2 in Fig. 34-17a).

3. A ray that is initially directed toward the center of the lens will emerge from the lens with no change in its direction (ray 3 in Fig. 34-17a) because the ray encounters the two sides of the lens where they are almost parallel.

The image of the point is located where the rays intersect on the far side of the lens. The image of the object is found by locating the images of two or more of its points.

Figure 34-17b shows how the extensions of the three special rays can be used to locate the image of an object placed inside focal point $F_1$ of a converging lens. Note that the description of ray 2 requires modification (it is now a ray whose backward extension passes through $F_1$).

You need to modify the descriptions of rays 1 and 2 to use them to locate an image placed (anywhere) in front of a diverging lens. In Fig. 34-17c, for example, we find the point where ray 3 intersects the backward extensions of rays 1 and 2.

## Two-Lens Systems

When an object $O$ is placed in front of a system of two lenses whose central axes coincide, we can locate the final image of the system (that is, the image produced by the lens farther from the object) by working in steps. Let lens 1 be the nearer lens and lens 2 the farther lens.

**Step 1**   We let $p_1$ represent the distance of object $O$ from lens 1. We then find the distance $i_1$ of the image produced by lens 1, either by use of Eq. 34-9 or by drawing rays.

**Step 2**   Now, ignoring the presence of lens 1, we treat the image found in step 1

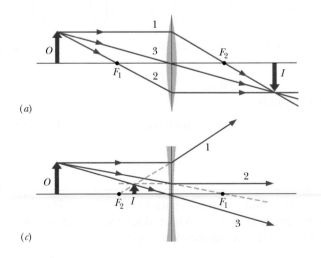

(a)

(c)

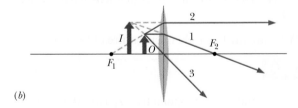

(b)

**FIG. 34-17**   Three special rays allow us to locate an image formed by a thin lens whether the object $O$ is (a) outside or (b) inside the focal point of a converging lens, or (c) anywhere in front of a diverging lens.

as the object for lens 2. If this new object is located beyond lens 2, the object distance $p_2$ for lens 2 is taken to be negative. (Note this exception to the rule that says the object distance is positive; the exception occurs because the object here is on the side opposite the source of light.) Otherwise, $p_2$ is taken to be positive as usual. We then find the distance $i_2$ of the (final) image produced by lens 2 by use of Eq. 34-9 or by drawing rays.

A similar step-by-step solution can be used for any number of lenses or if a mirror is substituted for lens 2.

The *overall*, or *net*, lateral magnification $M$ produced by a system of two lenses is the product of the lateral magnifications $m_1$ and $m_2$ produced by the two lenses:

$$M = m_1 m_2. \qquad (34\text{-}11)$$

✓ **CHECKPOINT 4**    A thin symmetric lens provides an image of a fingerprint with a magnification of $+0.2$ when the fingerprint is 1.0 cm farther from the lens than the focal point of the lens. What are the type and orientation of the image, and what is the type of lens?

---

## Sample Problem    34-4

A praying mantis preys along the central axis of a thin symmetric lens, 20 cm from the lens. The lateral magnification of the mantis provided by the lens is $m = -0.25$, and the index of refraction of the lens material is 1.65.

**(a)** Determine the type of image produced by the lens, the type of lens, whether the object (mantis) is inside or outside the focal point, on which side of the lens the image appears, and whether the image is inverted.

**Reasoning:** We can tell a lot about the lens and the image from the given value of $m$. From it and Eq. 34-6 ($m = -i/p$), we see that

$$i = -mp = 0.25p.$$

Even without finishing the calculation, we can answer the questions. Because $p$ is positive, $i$ here must be positive. That means we have a real image, which means we have a converging lens (the only lens that can produce a real image). The object must be outside the focal point (the only way a real image can be produced). Also, the image is inverted and on the side of the lens opposite the object. (That is how a converging lens makes a real image.)

**(b)** What are the two radii of curvature of the lens?

### KEY IDEAS

1. Because the lens is symmetric, $r_1$ (for the surface nearer the object) and $r_2$ have the same magnitude $r$.

2. Because the lens is a converging lens, the object faces a convex surface on the nearer side and so $r_1 = +r$. Similarly, it faces a concave surface on the farther side; so $r_2 = -r$.

3. We can relate these radii of curvature to the focal length $f$ via the lens maker's equation, Eq. 34-10 (our only equation involving the radii of curvature of a lens).

4. We can relate $f$ to the object distance $p$ and image distance $i$ via Eq. 34-9.

**Calculations:** We know $p$, but we do not know $i$. Thus, our starting point is to finish the calculation for $i$ in part (a); we obtain

$$i = (0.25)(20 \text{ cm}) = 5.0 \text{ cm}.$$

Now Eq. 34-9 gives us

$$\frac{1}{f} = \frac{1}{p} + \frac{1}{i} = \frac{1}{20 \text{ cm}} + \frac{1}{5.0 \text{ cm}},$$

from which we find $f = 4.0$ cm.

Equation 34-10 then gives us

$$\frac{1}{f} = (n - 1)\left(\frac{1}{r_1} - \frac{1}{r_2}\right) = (n - 1)\left(\frac{1}{+r} - \frac{1}{-r}\right)$$

or, with known values inserted,

$$\frac{1}{4.0 \text{ cm}} = (1.65 - 1)\frac{2}{r},$$

which yields

$$r = (0.65)(2)(4.0 \text{ cm}) = 5.2 \text{ cm}. \quad \text{(Answer)}$$

---

## Sample Problem    34-5    Build your skill

Figure 34-18a shows a jalapeño seed $O_1$ that is placed in front of two thin symmetrical coaxial lenses 1 and 2, with focal lengths $f_1 = +24$ cm and $f_2 = +9.0$ cm, re-

spectively, and with lens separation $L = 10$ cm. The seed is 6.0 cm from lens 1. Where does the system of two lenses produce an image of the seed?

**KEY IDEA** We could locate the image produced by the system of lenses by tracing light rays from the seed through the two lenses. However, we can, instead, calculate the location of that image by working through the system in steps, lens by lens. We begin with the lens closer to the seed. The image we seek is the final one—that is, image $I_2$ produced by lens 2.

**Lens 1:** Ignoring lens 2, we locate the image $I_1$ produced by lens 1 by applying Eq. 34-9 to lens 1 alone:

$$\frac{1}{p_1} + \frac{1}{i_1} = \frac{1}{f_1}.$$

The object $O_1$ for lens 1 is the seed, which is 6.0 cm from the lens; thus, we substitute $p_1 = +6.0$ cm. Also substituting the given value of $f_1$, we then have

$$\frac{1}{+6.0\text{ cm}} + \frac{1}{i_1} = \frac{1}{+24\text{ cm}},$$

which yields $i_1 = -8.0$ cm.

This tells us that image $I_1$ is 8.0 cm from lens 1 and virtual. (We could have guessed that it is virtual by noting that the seed is inside the focal point of lens 1, that is, between the lens and its focal point.) Because $I_1$ is virtual, it is on the same side of the lens as object $O_1$ and has the same orientation as the seed, as shown in Fig. 34-18b.

**Lens 2:** In the second step of our solution, we treat image $I_1$ as an object $O_2$ for the second lens and now ignore lens 1. We first note that this object $O_2$ is outside the focal point of lens 2. So the image $I_2$ produced by lens 2 must be real, inverted, and on the side of the lens opposite $O_2$. Let us see.

The distance $p_2$ between this object $O_2$ and lens 2 is, from Fig. 34-18c,

$$p_2 = L + |i_1| = 10\text{ cm} + 8.0\text{ cm} = 18\text{ cm}.$$

Then Eq. 34-9, now written for lens 2, yields

$$\frac{1}{+18\text{ cm}} + \frac{1}{i_2} = \frac{1}{+9.0\text{ cm}}.$$

Hence, $\qquad\qquad i_2 = +18$ cm. (Answer)

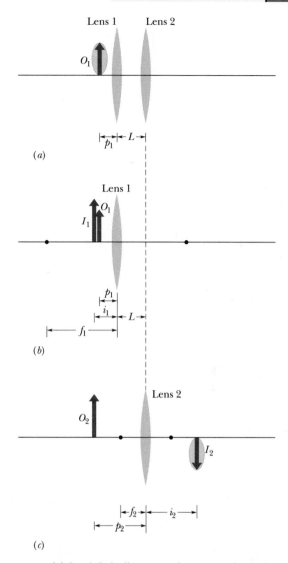

**FIG. 34-18** (a) Seed $O_1$ is distance $p_1$ from a two-lens system with lens separation $L$. We use the arrow to orient the seed. (b) The image $I_1$ produced by lens 1 alone. (c) Image $I_1$ acts as object $O_2$ for lens 2 alone, which produces the final image $I_2$.

The plus sign confirms our guess: Image $I_2$ produced by lens 2 is real, inverted, and on the side of lens 2 opposite $O_2$, as shown in Fig. 34-18c.

## 34-8 | Optical Instruments

The human eye is a remarkably effective organ, but its range can be extended in many ways by optical instruments such as eyeglasses, microscopes, and telescopes. Many such devices extend the scope of our vision beyond the visible range; satellite-borne infrared cameras and x-ray microscopes are just two examples.

The mirror and thin-lens formulas can be applied only as approximations to most sophisticated optical instruments. The lenses in typical laboratory microscopes are by no means "thin." In most optical instruments the lenses are compound lenses; that is, they are made of several components, the interfaces rarely being exactly spherical. Now we discuss three optical instruments, assuming, for simplicity, that the thin-lens formulas apply.

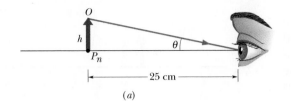

**FIG. 34-19** (a) An object O of height h placed at the near point of a human eye occupies angle θ in the eye's view. (b) The object is moved closer to increase the angle, but now the observer cannot bring the object into focus. (c) A converging lens is placed between the object and the eye, with the object just inside the focal point $F_1$ of the lens. The image produced by the lens is then far enough away to be focused by the eye, and the image occupies a larger angle θ' than object O does in (a).

## Simple Magnifying Lens

The normal human eye can focus a sharp image of an object on the retina (at the rear of the eye) if the object is located anywhere from infinity to a certain point called the *near point* $P_n$. If you move the object closer to the eye than the near point, the perceived retinal image becomes fuzzy. The location of the near point normally varies with age. We have all heard about people who claim not to need glasses but read their newspapers at arm's length; their near points are receding. To find your own near point, remove your glasses or contacts if you wear any, close one eye, and then bring this page closer to your open eye until it becomes indistinct. In what follows, we take the near point to be 25 cm from the eye, a bit more than the typical value for 20-year-olds.

Figure 34-19a shows an object O placed at the near point $P_n$ of an eye. The size of the image of the object produced on the retina depends on the angle θ that the object occupies in the field of view from that eye. By moving the object closer to the eye, as in Fig. 34-19b, you can increase the angle and, hence, the possibility of distinguishing details of the object. However, because the object is then closer than the near point, it is no longer *in focus;* that is, the image is no longer clear.

You can restore the clarity by looking at O through a converging lens, placed so that O is just inside the focal point $F_1$ of the lens, which is at focal length f (Fig. 34-19c). What you then see is the virtual image of O produced by the lens. That image is farther away than the near point; thus, the eye can see it clearly.

Moreover, the angle θ' occupied by the virtual image is larger than the largest angle θ that the object alone can occupy and still be seen clearly. The *angular magnification* $m_\theta$ (not to be confused with lateral magnification m) of what is seen is

$$m_\theta = \theta'/\theta.$$

In words, the angular magnification of a simple magnifying lens is a comparison of the angle occupied by the image the lens produces with the angle occupied by the object when the object is moved to the near point of the viewer.

From Fig. 34-19, assuming that O is at the focal point of the lens, and approximating tan θ as θ and tan θ' as θ' for small angles, we have

$$\theta \approx h/25 \text{ cm} \quad \text{and} \quad \theta' \approx h/f.$$

We then find that

$$m_\theta \approx \frac{25 \text{ cm}}{f} \quad \text{(simple magnifier)}. \tag{34-12}$$

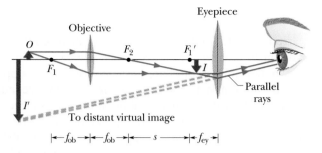

**FIG. 34-20** A thin-lens representation of a compound microscope (not to scale). The objective produces a real image $I$ of object $O$ just inside the focal point $F_1'$ of the eyepiece. Image $I$ then acts as an object for the eyepiece, which produces a virtual final image $I'$ that is seen by the observer. The objective has focal length $f_{ob}$; the eyepiece has focal length $f_{ey}$; and $s$ is the tube length.

## Compound Microscope

Figure 34-20 shows a thin-lens version of a compound microscope. The instrument consists of an *objective* (the front lens) of focal length $f_{ob}$ and an *eyepiece* (the lens near the eye) of focal length $f_{ey}$. It is used for viewing small objects that are very close to the objective.

The object $O$ to be viewed is placed just outside the first focal point $F_1$ of the objective, close enough to $F_1$ that we can approximate its distance $p$ from the lens as being $f_{ob}$. The separation between the lenses is then adjusted so that the enlarged, inverted, real image $I$ produced by the objective is located just inside the first focal point $F_1'$ of the eyepiece. The *tube length s* shown in Fig. 34-20 is actually large relative to $f_{ob}$, and therefore we can approximate the distance $i$ between the objective and the image $I$ as being length $s$.

From Eq. 34-6, and using our approximations for $p$ and $i$, we can write the lateral magnification produced by the objective as

$$m = -\frac{i}{p} = -\frac{s}{f_{ob}}. \qquad (34\text{-}13)$$

Because the image $I$ is located just inside the focal point $F_1'$ of the eyepiece, the eyepiece acts as a simple magnifying lens, and an observer sees a final (virtual, inverted) image $I'$ through it. The overall magnification of the instrument is the product of the lateral magnification $m$ produced by the objective, given by Eq. 34-13, and the angular magnification $m_\theta$ produced by the eyepiece, given by Eq. 34-12; that is,

$$M = mm_\theta = -\frac{s}{f_{ob}}\frac{25\ \text{cm}}{f_{ey}} \qquad \text{(microscope)}. \qquad (34\text{-}14)$$

## Refracting Telescope

Telescopes come in a variety of forms. The form we describe here is the simple refracting telescope that consists of an objective and an eyepiece; both are represented in Fig. 34-21 with simple lenses, although in practice, as is also true for most microscopes, each lens is actually a compound lens system.

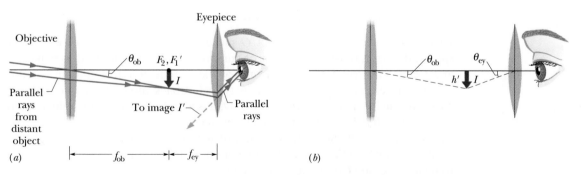

**FIG. 34-21** (*a*) A thin-lens representation of a refracting telescope. From rays that are approximately parallel when they reach the objective, the objective produces a real image $I$ of a distant source of light (the object). (One end of the object is assumed to lie on the central axis.) Image $I$, formed at the common focal points $F_2$ and $F_1'$, acts as an object for the eyepiece, which produces a virtual final image $I'$ at a great distance from the observer. The objective has focal length $f_{ob}$; the eyepiece has focal length $f_{ey}$. (*b*) Image $I$ has height $h'$ and takes up angle $\theta_{ob}$ measured from the objective and angle $\theta_{ey}$ measured from the eyepiece.

The lens arrangements for telescopes and for microscopes are similar, but telescopes are designed to view large objects, such as galaxies, stars, and planets, at large distances, whereas microscopes are designed for just the opposite purpose. This difference requires that in the telescope of Fig. 34-21 the second focal point of the objective $F_2$ coincide with the first focal point of the eyepiece $F_1'$, whereas in the microscope of Fig. 34-20 these points are separated by the tube length $s$.

In Fig. 34-21a, parallel rays from a distant object strike the objective, making an angle $\theta_{ob}$ with the telescope axis and forming a real, inverted image $I$ at the common focal point $F_2, F_1'$. This image $I$ acts as an object for the eyepiece, through which an observer sees a distant (still inverted) virtual image $I'$. The rays defining the image make an angle $\theta_{ey}$ with the telescope axis.

The angular magnification $m_\theta$ of the telescope is $\theta_{ey}/\theta_{ob}$. From Fig. 34-21b, for rays close to the central axis, we can write $\theta_{ob} = h'/f_{ob}$ and $\theta_{ey} \approx h'/f_{ey}$, which gives us

$$m_\theta = -\frac{f_{ob}}{f_{ey}} \quad \text{(telescope),} \tag{34-15}$$

where the minus sign indicates that $I'$ is inverted. In words, the angular magnification of a telescope is a comparison of the angle occupied by the image the telescope produces with the angle occupied by the distant object as seen without the telescope.

Magnification is only one of the design factors for an astronomical telescope and is indeed easily achieved. A good telescope needs *light-gathering power,* which determines how bright the image is. This is important for viewing faint objects such as distant galaxies and is accomplished by making the objective diameter as large as possible. A telescope also needs *resolving power,* which is the ability to distinguish between two distant objects (stars, say) whose angular separation is small. *Field of view* is another important design parameter. A telescope designed to look at galaxies (which occupy a tiny field of view) is much different from one designed to track meteors (which move over a wide field of view).

The telescope designer must also take into account the difference between real lenses and the ideal thin lenses we have discussed. A real lens with spherical surfaces does not form sharp images, a flaw called *spherical aberration.* Also, because refraction by the two surfaces of a real lens depends on wavelength, a real lens does not focus light of different wavelengths to the same point, a flaw called *chromatic aberration.*

This brief discussion by no means exhausts the design parameters of astronomical telescopes—many others are involved. We could make a similar listing for any other high-performance optical instrument.

## 34-9 | Three Proofs

### The Spherical Mirror Formula (Eq. 34-4)

Figure 34-22 shows a point object $O$ placed on the central axis of a concave spherical mirror, outside its center of curvature $C$. A ray from $O$ that makes an angle $\alpha$ with the axis intersects the axis at $I$ after reflection from the mirror at $a$. A ray that leaves $O$ along the axis is reflected back along itself at $c$ and also passes through $I$. Thus, $I$ is the image of $O$; it is a *real* image because light actually passes through it. Let us find the image distance $i$.

A trigonometry theorem that is useful here tells us that an exterior angle of a triangle is equal to the sum of the two opposite interior angles. Applying this to triangles $OaC$ and $OaI$ in Fig. 34-22 yields

$$\beta = \alpha + \theta \quad \text{and} \quad \gamma = \alpha + 2\theta.$$

If we eliminate $\theta$ between these two equations, we find

$$\alpha + \gamma = 2\beta. \tag{34-16}$$

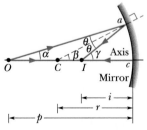

**FIG. 34-22** A concave spherical mirror forms a real point image $I$ by reflecting light rays from a point object $O$.

We can write angles $\alpha$, $\beta$, and $\gamma$, in radian measure, as

$$\alpha \approx \frac{\widehat{ac}}{cO} = \frac{\widehat{ac}}{p}, \qquad \beta = \frac{\widehat{ac}}{cC} = \frac{\widehat{ac}}{r},$$

and

$$\gamma \approx \frac{\widehat{ac}}{cI} = \frac{\widehat{ac}}{i}, \qquad (34\text{-}17)$$

where the overhead symbol means "arc." Only the equation for $\beta$ is exact, because the center of curvature of $\widehat{ac}$ is at $C$. However, the equations for $\alpha$ and $\gamma$ are approximately correct if these angles are small enough (that is, for rays close to the central axis). Substituting Eqs. 34-17 into Eq. 34-16, using Eq. 34-3 to replace $r$ with $2f$, and canceling $\widehat{ac}$ lead exactly to Eq. 34-4, the relation that we set out to prove.

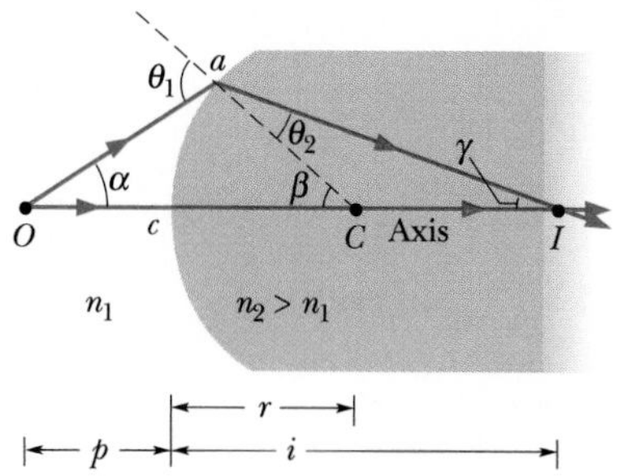

**FIG. 34-23** A real point image $I$ of a point object $O$ is formed by refraction at a spherical convex surface between two media.

### The Refracting Surface Formula (Eq. 34-8)

The incident ray from point object $O$ in Fig. 34-23 that falls on point $a$ of a spherical refracting surface is refracted there according to Eq. 33-40,

$$n_1 \sin \theta_1 = n_2 \sin \theta_2.$$

If $\alpha$ is small, $\theta_1$ and $\theta_2$ will also be small and we can replace the sines of these angles with the angles themselves. Thus, the equation above becomes

$$n_1 \theta_1 \approx n_2 \theta_2. \qquad (34\text{-}18)$$

We again use the fact that an exterior angle of a triangle is equal to the sum of the two opposite interior angles. Applying this to triangles $COa$ and $ICa$ yields

$$\theta_1 = \alpha + \beta \quad \text{and} \quad \beta = \theta_2 + \gamma. \qquad (34\text{-}19)$$

If we use Eqs. 34-19 to eliminate $\theta_1$ and $\theta_2$ from Eq. 34-18, we find

$$n_1 \alpha + n_2 \gamma = (n_2 - n_1)\beta. \qquad (34\text{-}20)$$

In radian measure the angles $\alpha$, $\beta$, and $\gamma$ are

$$\alpha \approx \frac{\widehat{ac}}{p}; \qquad \beta = \frac{\widehat{ac}}{r}; \qquad \gamma \approx \frac{\widehat{ac}}{i}. \qquad (34\text{-}21)$$

Only the second of these equations is exact. The other two are approximate because $I$ and $O$ are not the centers of circles of which $\widehat{ac}$ is a part. However, for $\alpha$ small enough (for rays close to the axis), the inaccuracies in Eqs. 34-21 are small. Substituting Eqs. 34-21 into Eq. 34-20 leads directly to Eq. 34-8, as we wanted.

### The Thin-Lens Formulas (Eqs. 34-9 and 34-10)

Our plan is to consider each lens surface as a separate refracting surface, and to use the image formed by the first surface as the object for the second.

We start with the thick glass "lens" of length $L$ in Fig. 34-24$a$ whose left and right refracting surfaces are ground to radii $r'$ and $r''$. A point object $O'$ is placed near the left surface as shown. A ray leaving $O'$ along the central axis is not deflected on entering or leaving the lens.

A second ray leaving $O'$ at an angle $\alpha$ with the central axis intersects the left surface at point $a'$, is refracted, and intersects the second (right) surface at point $a''$. The ray is again refracted and crosses the axis at $I''$, which, being the intersection of two rays from $O'$, is the image of point $O'$, formed after refraction at two surfaces.

Figure 34-24$b$ shows that the first (left) surface also forms a virtual image of $O'$ at $I'$. To locate $I'$, we use Eq. 34-8,

$$\frac{n_1}{p} + \frac{n_2}{i} = \frac{n_2 - n_1}{r}.$$

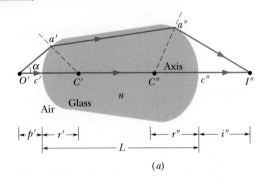

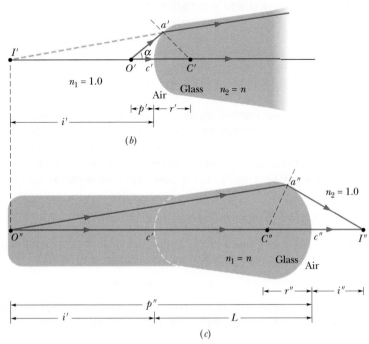

**FIG. 34-24**    (a) Two rays from point object $O'$ form a real image $I''$ after refracting through two spherical surfaces of a lens. The object faces a convex surface at the left side of the lens and a concave surface at the right side. The ray traveling through points $a'$ and $a''$ is actually close to the central axis through the lens. (b) The left side and (c) the right side of the lens in (a), shown separately.

Putting $n_1 = 1$ for air and $n_2 = n$ for lens glass and bearing in mind that the image distance is negative (that is, $i = -i'$ in Fig. 34-24b), we obtain

$$\frac{1}{p'} - \frac{n}{i'} = \frac{n-1}{r'}.$$
(34-22)

In this equation $i'$ will be a positive number because we have already introduced the minus sign appropriate to a virtual image.

Figure 34-24c shows the second surface again. Unless an observer at point $a''$ were aware of the existence of the first surface, the observer would think that the light striking that point originated at point $I'$ in Fig. 34-24b and that the region to the left of the surface was filled with glass as indicated. Thus, the (virtual) image $I'$ formed by the first surface serves as a real object $O''$ for the second surface. The distance of this object from the second surface is

$$p'' = i' + L.$$
(34-23)

To apply Eq. 34-8 to the second surface, we must insert $n_1 = n$ and $n_2 = 1$ because the object now is effectively imbedded in glass. If we substitute with Eq. 34-23, then Eq. 34-8 becomes

$$\frac{n}{i' + L} + \frac{1}{i''} = \frac{1-n}{r''}.$$
(34-24)

Let us now assume that the thickness $L$ of the "lens" in Fig. 34-24a is so small that we can neglect it in comparison with our other linear quantities (such as $p'$, $i'$, $p''$, $i''$, $r'$, and $r''$). In all that follows we make this *thin-lens approximation*. Putting $L = 0$ in Eq. 34-24 and rearranging the right side lead to

$$\frac{n}{i'} + \frac{1}{i''} = -\frac{n-1}{r''}.$$
(34-25)

Adding Eqs. 34-22 and 34-25 leads to

$$\frac{1}{p'} + \frac{1}{i''} = (n-1)\left(\frac{1}{r'} - \frac{1}{r''}\right).$$

Finally, calling the original object distance simply $p$ and the final image distance simply $i$ leads to

$$\frac{1}{p} + \frac{1}{i} = (n - 1)\left(\frac{1}{r'} - \frac{1}{r''}\right), \qquad (34\text{-}26)$$

which, with a small change in notation, is Eqs. 34-9 and 34-10, the relations we set out to prove.

## REVIEW & SUMMARY

**Real and Virtual Images** An *image* is a reproduction of an object via light. If the image can form on a surface, it is a *real image* and can exist even if no observer is present. If the image requires the visual system of an observer, it is a *virtual image*.

**Image Formation** *Spherical mirrors, spherical refracting surfaces,* and *thin lenses* can form images of a source of light — the object — by redirecting rays emerging from the source. The image occurs where the redirected rays cross (forming a real image) or where backward extensions of those rays cross (forming a virtual image). If the rays are sufficiently close to the *central axis* through the spherical mirror, refracting surface, or thin lens, we have the following relations between the *object distance p* (which is positive) and the *image distance i* (which is positive for real images and negative for virtual images):

**1. Spherical Mirror:**

$$\frac{1}{p} + \frac{1}{i} = \frac{1}{f} = \frac{2}{r}, \qquad (34\text{-}4, 34\text{-}3)$$

where $f$ is the mirror's focal length and $r$ is its radius of curvature. A *plane mirror* is a special case for which $r \rightarrow \infty$, so that $p = -i$. Real images form on the side of a mirror where the object is located, and virtual images form on the opposite side.

**2. Spherical Refracting Surface:**

$$\frac{n_1}{p} + \frac{n_2}{i} = \frac{n_2 - n_1}{r} \qquad \text{(single surface),} \qquad (34\text{-}8)$$

where $n_1$ is the index of refraction of the material where the object is located, $n_2$ is the index of refraction of the material on the other side of the refracting surface, and $r$ is the radius of curvature of the surface. When the object faces a convex refracting surface, the radius $r$ is positive. When it faces a concave surface, $r$ is negative. Real images form on the side of a refracting surface that is opposite the object, and virtual images form on the same side as the object.

**3. Thin Lens:**

$$\frac{1}{p} + \frac{1}{i} = \frac{1}{f} = (n - 1)\left(\frac{1}{r_1} - \frac{1}{r_2}\right), \qquad (34\text{-}9, 34\text{-}10)$$

where $f$ is the lens's focal length, $n$ is the index of refraction of

the lens material, and $r_1$ and $r_2$ are the radii of curvature of the two sides of the lens, which are spherical surfaces. A convex lens surface that faces the object has a positive radius of curvature; a concave lens surface that faces the object has a negative radius of curvature. Real images form on the side of a lens that is opposite the object, and virtual images form on the same side as the object.

**Lateral Magnification** The *lateral magnification m* produced by a spherical mirror or a thin lens is

$$m = -\frac{i}{p}. \qquad (34\text{-}6)$$

The magnitude of $m$ is given by

$$|m| = \frac{h'}{h}, \qquad (34\text{-}5)$$

where $h$ and $h'$ are the heights (measured perpendicular to the central axis) of the object and image, respectively.

**Optical Instruments** Three optical instruments that extend human vision are:

**1.** The *simple magnifying lens,* which produces an *angular magnification $m_\theta$* given by

$$m_\theta = \frac{25 \text{ cm}}{f}, \qquad (34\text{-}12)$$

where $f$ is the focal length of the magnifying lens.

**2.** The *compound microscope,* which produces an *overall magnification M* given by

$$M = mm_\theta = -\frac{s}{f_{ob}}\frac{25 \text{ cm}}{f_{ey}}, \qquad (34\text{-}14)$$

where $m$ is the lateral magnification produced by the objective, $m_\theta$ is the angular magnification produced by the eyepiece, $s$ is the tube length, and $f_{ob}$ and $f_{ey}$ are the focal lengths of the objective and eyepiece, respectively.

**3.** The *refracting telescope,* which produces an *angular magnification $m_\theta$* given by

$$m_\theta = -\frac{f_{ob}}{f_{ey}}. \qquad (34\text{-}15)$$

## QUESTIONS

**1** Figure 34-25 is an overhead view of a mirror maze based on floor sections that are equilateral triangles. Every wall within the maze is mirrored. If you stand at entrance $x$, (a) which of the maze monsters $a$, $b$, and $c$ hiding in the maze can you see along the virtual hallways extending from entrance $x$; (b) how many times does each visible monster appear in a hallway; and (c) what is at the far end of a hallway?

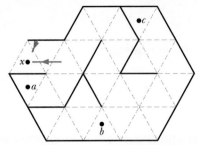

FIG. 34-25   Question 1.

**2** A penguin waddles along the central axis of a concave mirror, from the focal point to an effectively infinite distance. (a) How does its image move? (b) Does the height of its image increase continuously, decrease continuously, or change in some more complicated manner?

**3** Figure 34-26 shows a fish and a fish stalker in water. (a) Does the stalker see the fish in the general region of point $a$ or point $b$? (b) Does the fish see the (wild) eyes of the stalker in the general region of point $c$ or point $d$?

FIG. 34-26   Question 3.

**4** In Fig. 34-27, stick figure $O$ stands in front of a spherical mirror that is mounted within the boxed region; the central axis through the mirror is shown. The four stick figures $I_1$ to $I_4$ suggest general locations and orientations for the images that might be produced by the mirror. (The figures are only sketched in; neither their heights nor their distances from the mirror are drawn to scale.) (a) Which of the stick figures could not possibly represent images? Of the possible images, (b) which would be due to a concave mirror, (c) which would be due to a convex mirror, (d) which would be virtual, and (e) which would involve negative magnification?

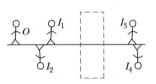

FIG. 34-27
Questions 4 and 8.

**5** Figure 34-28 shows four thin lenses, all of the same material, with sides that either are flat or have a radius of curvature of magnitude 10 cm. Without written calculation, rank the lenses according to the magnitude of the focal length, greatest first.

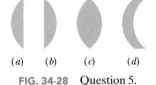

FIG. 34-28   Question 5.

**6** An object is placed against the center of a concave mirror and then moved along the central axis until it is 5.0 m from the mirror. During the motion, the distance $|i|$ between the mirror and the image it produces is measured. The procedure is then repeated with a convex mirror and a plane mirror. Figure 34-29 gives the results versus object distance $p$. Which curve corresponds to which mirror? (Curve 1 has two segments.)

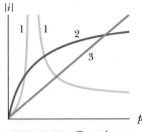

FIG. 34-29   Questions 6 and 10.

**7** When a *T. rex* pursues a jeep in the movie *Jurassic Park,* we see a reflected image of the *T. rex* via a side-view mirror, on which is printed the (then darkly humorous) warning: "Objects in mirror are closer than they appear." Is the mirror flat, convex, or concave?

**8** In Fig. 34-27, stick figure $O$ stands in front of a thin, symmetric lens that is mounted within the boxed region; the central axis through the lens is shown. The four stick figures $I_1$ to $I_4$ suggest general locations and orientations for the images that might be produced by the lens. (The figures are only sketched in; neither their height nor their distance from the lens is drawn to scale.) (a) Which of the stick figures could not possibly represent images? Of the possible images, (b) which would be due to a converging lens, (c) which would be due to a diverging lens, (d) which would be virtual, and (e) which would involve negative magnification?

**9** The table details six variations of the basic arrangement of two thin lenses represented in Fig. 34-30. (The points labeled $F_1$ and $F_2$ are the focal points of lenses 1 and 2.) An object is distance $p_1$ to the left of lens 1, as in Fig. 34-18. (a) For which variations can we tell, *without calculation,* whether the final image (that due to lens 2) is to the left or right of lens 2 and whether it has the same orientation as the object? (b) For those "easy" variations, give the image location as "left" or "right" and the orientation as "same" or "inverted."

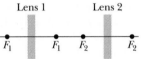

FIG. 34-30   Question 9.

| Variation | Lens 1 | Lens 2 | |
|---|---|---|---|
| 1 | Converging | Converging | $p_1 < \lvert f_1 \rvert$ |
| 2 | Converging | Converging | $p_1 > \lvert f_1 \rvert$ |
| 3 | Diverging | Converging | $p_1 < \lvert f_1 \rvert$ |
| 4 | Diverging | Converging | $p_1 > \lvert f_1 \rvert$ |
| 5 | Diverging | Diverging | $p_1 < \lvert f_1 \rvert$ |
| 6 | Diverging | Diverging | $p_1 > \lvert f_1 \rvert$ |

**10** An object is placed against the center of a converging lens and then moved along the central axis until it is 5.0 m from the lens. During the motion, the distance $|i|$ between the lens and the image it produces is measured. The procedure is then repeated with a diverging lens. Which of the curves in Fig. 34-29 best gives $|i|$ versus the object distance $p$ for these lenses? (Curve 1 consists of two segments. Curve 3 is straight.)

# PROBLEMS

GO  Tutoring problem available (at instructor's discretion) in *WileyPLUS* and WebAssign
SSM  Worked-out solution available in Student Solutions Manual  WWW  Worked-out solution is at
• – •••  Number of dots indicates level of problem difficulty  ILW  Interactive solution is at  http://www.wiley.com/college/halliday
✈  Additional information available in *The Flying Circus of Physics* and at flyingcircusofphysics.com

## sec. 34-3 Plane Mirrors

•1 A moth at about eye level is 10 cm in front of a plane mirror; you are behind the moth, 30 cm from the mirror. What is the distance between your eyes and the apparent position of the moth's image in the mirror? ILW

•2 You look through a camera toward an image of a hummingbird in a plane mirror. The camera is 4.30 m in front of the mirror. The bird is at camera level, 5.00 m to your right and 3.30 m from the mirror. What is the distance between the camera and the apparent position of the bird's image in the mirror?

••3 Figure 34-31 shows an overhead view of a corridor with a plane mirror $M$ mounted at one end. A burglar $B$ sneaks along the corridor directly toward the center of the mirror. If $d = 3.0$ m, how far from the mirror will she be when the security guard $S$ can first see her in the mirror?

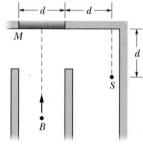

FIG. 34-31 Problem 3.

••4 In Fig. 34-32, an isotropic point source of light $S$ is positioned at distance $d$ from a viewing screen $A$ and the light intensity $I_P$ at point $P$ (level with $S$) is measured. Then a plane mirror $M$ is placed behind $S$ at distance $d$. By how much is $I_P$ multiplied by the presence of the mirror?

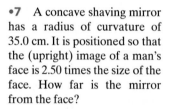

FIG. 34-32 Problem 4.

•••5 Figure 34-33 shows a small lightbulb suspended at distance $d_1 = 250$ cm above the surface of the water in a swimming pool where the water depth is $d_2 = 200$ cm. The bottom of the pool is a large mirror. How far below the mirror surface is the image of the bulb? (*Hint:* Construct a diagram of two rays like that of Fig. 34-3, but take into account the bending of light rays by refraction at the water surface. Assume that the rays are close to a vertical axis through the bulb, and use the small-angle approximation in which $\sin \theta \approx \tan \theta \approx \theta$.) SSM WWW

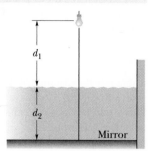

FIG. 34-33 Problem 5.

## sec. 34-5 Images from Spherical Mirrors

•6 An object is placed against the center of a spherical mirror and then moved 70 cm from it along the central axis as the image distance $i$ is measured. Figure 34-34 gives $i$ versus object distance $p$ out to $p_s = 40$ cm. What is the image distance when the object is on the central axis and 70 cm from the mirror?

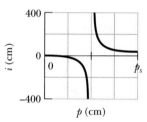

FIG. 34-34 Problem 6.

•7 A concave shaving mirror has a radius of curvature of 35.0 cm. It is positioned so that the (upright) image of a man's face is 2.50 times the size of the face. How far is the mirror from the face?

•8 An object is moved along the central axis of a spherical mirror while the lateral magnification $m$ of it is measured. Figure 34-35 gives $m$ versus object distance $p$ for the range $p_a = 2.0$ cm to $p_b = 8.0$ cm. What is the magnification of the object when the object is 14.0 cm from the mirror?

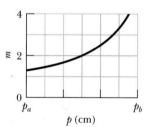

FIG. 34-35 Problem 8.

••9 through 16 *Spherical mirrors.* Object $O$ stands on the central axis of a spherical mirror. For this situation, each problem in Table 34-3 gives object distance $p_s$ (centimeters), the

### TABLE 34-3

**Problems 9 through 16: Spherical Mirrors. See the setup for these problems.**

|    | $p$  | Mirror       | (a) $r$ | (b) $i$ | (c) $m$ | (d) R/V | (e) I/NI | (f) Side |
|----|------|--------------|---------|---------|---------|---------|----------|----------|
| 9  | +12  | Concave, 18  |         |         |         |         |          |          |
| 10 | +24  | Concave, 36  |         |         |         |         |          |          |
| 11 | +18  | Concave, 12  |         |         |         |         |          |          |
| 12 | +15  | Concave, 10  |         |         |         |         |          |          |
| 13 | +10  | Convex, 8.0  |         |         |         |         |          |          |
| 14 | +17  | Convex, 14   |         |         |         |         |          |          |
| 15 | +8.0 | Convex, 10   |         |         |         |         |          |          |
| 16 | +22  | Convex, 35   |         |         |         |         |          |          |

type of mirror, and then the distance (centimeters, without proper sign) between the focal point and the mirror. Find (a) the radius of curvature $r$ (including sign), (b) the image distance $i$, and (c) the lateral magnification $m$. Also, determine whether the image is (d) real (R) or virtual (V), (e) inverted (I) from object $O$ or noninverted (NI), and (f) on the *same* side of the mirror as $O$ or on the *opposite* side. SSM 9, 11, 15 GO 10

**••17** (a) A luminous point is moving at speed $v_O$ toward a spherical mirror with radius of curvature $r$, along the central axis of the mirror. Show that the image of this point is moving at speed

$$v_I = -\left(\frac{r}{2p-r}\right)^2 v_O,$$

where $p$ is the distance of the luminous point from the mirror at any given time. Now assume the mirror is concave, with $r = 15$ cm, and let $v_O = 5.0$ cm/s. Find $v_I$ when (b) $p = 30$ cm (far outside the focal point), (c) $p = 8.0$ cm (just outside the focal point), and (d) $p = 10$ mm (very near the mirror).

**••18** Figure 34-36 gives the lateral magnification $m$ of an object versus the object distance $p$ from a spherical mirror as the object is moved along the mirror's central axis through a range of values for $p$. The horizontal scale is set by $p_s = 10.0$ cm. What is the magnification of the object when the object is 21 cm from the mirror?

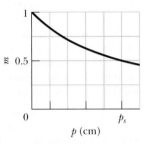

FIG. 34-36   Problem 18.

**••19 through 31** *More mirrors.* Object $O$ stands on the central axis of a spherical or plane mirror. For this situation, each problem in Table 34-4 refers to (a) the type of mirror, (b) the focal distance $f$, (c) the radius of curvature $r$, (d) the object distance $p$, (e) the image distance $i$, and (f) the lateral magnification $m$. (All distances are in centimeters.) It also refers to whether (g) the image is real (R) or virtual (V), (h)

inverted (I) or noninverted (NI) from $O$, and (i) on the *same* side of the mirror as object $O$ or on the *opposite* side. Fill in the missing information. Where only a sign is missing, answer with the sign. SSM 27, 29 GO 26

### sec. 34-6   Spherical Refracting Surfaces

**•32** A glass sphere has radius $R = 5.0$ cm and index of refraction 1.6. A paperweight is constructed by slicing through the sphere along a plane that is 2.0 cm from the center of the sphere, leaving height $h = 3.0$ cm. The paperweight is placed on a table and viewed from directly above by an observer who is distance $d = 8.0$ cm from the tabletop (Fig. 34-37). When viewed through the paperweight, how far away does the tabletop appear to be to the observer?

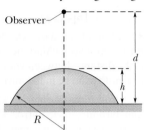

FIG. 34-37   Problem 32.

**•33** In Fig. 34-38, a beam of parallel light rays from a laser is incident on a solid transparent sphere of index of refraction $n$. (a) If a point image is produced at the back of the sphere, what is the index of refraction of the sphere? (b) What index of refraction, if any, will produce a point image at the center of the sphere?

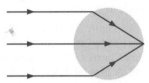

FIG. 34-38   Problem 33.

**••34 through 40** *Spherical refracting surfaces.* An object $O$ stands on the central axis of a spherical refracting surface. For this situation, each problem in Table 34-5 refers to the index of refraction $n_1$ where the object is located, (a) the index of refraction $n_2$ on the other side of the refracting surface, (b) the object distance $p$, (c) the radius of curvature $r$ of the surface, and (d) the image distance $i$. (All distances are in centimeters.) Fill in the missing information, including whether the image is

**TABLE 34-4**

**Problems 19 through 31: More Mirrors. See the setup for these problems.**

|    | (a) Type | (b) $f$ | (c) $r$ | (d) $p$ | (e) $i$ | (f) $m$ | (g) R/V | (h) I/NI | (i) Side |
|----|----------|---------|---------|---------|---------|---------|---------|----------|----------|
| 19 | Concave  | 20      |         | +10     |         |         |         |          |          |
| 20 |          |         |         | +60     |         | −0.50   |         |          |          |
| 21 |          |         | −40     |         | −10     |         |         |          |          |
| 22 |          |         |         | +10     |         | +1.0    |         |          |          |
| 23 |          | +20     |         | +30     |         |         |         |          |          |
| 24 |          |         |         | +24     |         | 0.50    |         | I        |          |
| 25 |          | −30     |         |         | −15     |         |         |          |          |
| 26 |          | 20      |         |         |         | +0.10   |         |          |          |
| 27 | Convex   |         | 40      |         | 4.0     |         |         |          |          |
| 28 |          |         |         | +40     |         | −0.70   |         |          |          |
| 29 |          | 30      |         |         |         | +0.20   |         |          |          |
| 30 |          | 20      |         | +60     |         |         |         |          | Same     |
| 31 |          |         |         | +30     |         | 0.40    |         | I        |          |

**TABLE 34-5**

Problems 34 through 40: Spherical Refracting Surfaces. See the setup for these problems.

| | | (a) | (b) | (c) | (d) | (e) | (f) |
|---|---|---|---|---|---|---|---|
| | $n_1$ | $n_2$ | $p$ | $r$ | $i$ | R/V | Side |
| **34** | 1.0 | 1.5 | | +30 | +600 | | |
| **35** | 1.0 | 1.5 | +10 | | −13 | | |
| **36** | 1.0 | 1.5 | +10 | +30 | | | |
| **37** | 1.5 | 1.0 | +70 | +30 | | | |
| **38** | 1.5 | | +100 | −30 | +600 | | |
| **39** | 1.5 | 1.0 | +10 | | −6.0 | | |
| **40** | 1.5 | 1.0 | | −30 | −7.5 | | |

(e) real (R) or virtual (V) and (f) on the *same* side of the surface as object O or on the *opposite* side. **SSM** 35,37 **GO** 39

### sec. 34-7 Thin Lenses

•**41** A double-convex lens is to be made of glass with an index of refraction of 1.5. One surface is to have twice the radius of curvature of the other and the focal length is to be 60 mm. What is the (a) smaller and (b) larger radius? **SSM WWW**

•**42** An object is placed against the center of a thin lens and then moved away from it along the central axis as the image distance $i$ is measured. Figure 34-39 gives $i$ versus object distance $p$ out to $p_s$ = 60 cm. What is the image distance when $p$ = 100 cm?

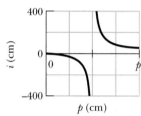

FIG. 34-39 Problem 42.

•**43** You produce an image of the Sun on a screen, using a thin lens whose focal length is 20.0 cm. What is the diameter of the image? (See Appendix C for needed data on the Sun.)

•**44** An object is placed against the center of a thin lens and then moved 70 cm from it along the central axis as the image distance $i$ is measured. Figure 34-40 gives $i$ versus object distance $p$ out to $p_s$ = 40 cm. What is the image distance when $p$ = 70 cm?

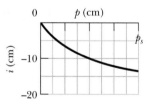

FIG. 34-40 Problem 44.

•**45** A movie camera with a (single) lens of focal length 75 mm takes a picture of a person standing 27 m away. If the person is 180 cm tall, what is the height of the image on the film?

•**46** An object is moved along the central axis of a thin lens while the lateral magnification $m$ is measured. Figure

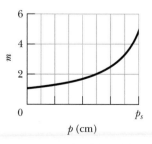

FIG. 34-41 Problem 46.

34-41 gives $m$ versus object distance $p$ out to $p_s$ = 8.0 cm. What is the magnification of the object when the object is 14.0 cm from the lens?

•**47** An illuminated slide is held 44 cm from a screen. How far from the slide must a lens of focal length 11 cm be placed (between the slide and the screen) to form an image of the slide's picture on the screen? **SSM**

•**48** Figure 34-42 gives the lateral magnification $m$ of an object versus the object distance $p$ from a lens as the object is moved along the central axis of the lens through a range of values for $p$ out to $p_s$ = 20.0 cm. What is the magnification of the object when the object is 35 cm from the lens?

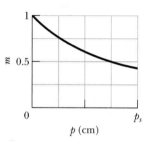

FIG. 34-42 Problem 48.

•**49** A lens is made of glass having an index of refraction of 1.5. One side of the lens is flat, and the other is convex with a radius of curvature of 20 cm. (a) Find the focal length of the lens. (b) If an object is placed 40 cm in front of the lens, where will the image be located?

••**50 through 57** *Thin lenses.* Object O stands on the central axis of a thin symmetric lens. For this situation, each problem in Table 34-6 gives object distance $p$ (centimeters), the type of lens (C stands for converging and D for diverging), and then the distance (centimeters, without proper sign) between a focal point and the lens. Find (a) the image distance $i$ and (b) the lateral magnification $m$ of the object, including signs. Also, determine whether the image is (c) real (R) or virtual (V), (d) inverted (I) from object O or noninverted (NI), and (e) on the *same* side of the lens as object O or on the *opposite* side. **SSM** 51,55 **GO** 57

••**58** In Fig. 34-43, a real inverted image I of an object O is formed by a certain lens (not shown); the object–image separation is $d$ = 40.0 cm, measured along the central axis of the lens. The image is just half the

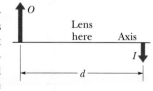

FIG. 34-43 Problem 58.

**TABLE 34-6**

**Problems 50 through 57: Thin Lenses. See the setup for these problems.**

| | $p$ | Lens | (a)<br>$i$ | (b)<br>$m$ | (c)<br>R/V | (d)<br>I/NI | (e)<br>Side |
|---|---|---|---|---|---|---|---|
| 50 | +10 | D, 6.0 | | | | | |
| 51 | +8.0 | D, 12 | | | | | |
| 52 | +16 | C, 4.0 | | | | | |
| 53 | +12 | C, 16 | | | | | |
| 54 | +12 | D, 31 | | | | | |
| 55 | +45 | C, 20 | | | | | |
| 56 | +25 | C, 35 | | | | | |
| 57 | +22 | D, 14 | | | | | |

size of the object. (a) What kind of lens must be used to produce this image? (b) How far from the object must the lens be placed? (c) What is the focal length of the lens?

**••59 through 68** *Lenses with given radii.* Object $O$ stands in front of a thin lens, on the central axis. For this situation, each problem in Table 34-7 gives object distance $p$, index of refraction $n$ of the lens, radius $r_1$ of the nearer lens surface, and radius $r_2$ of the farther lens surface. (All distances are in centimeters.) Find (a) the image distance $i$ and (b) the lateral magnification $m$ of the object, including signs. Also, determine whether the image is (c) real (R) or virtual (V), (d) inverted (I) from object $O$ or noninverted (NI), and (e) on the *same* side of the lens as object $O$ or on the *opposite* side. **SSM** 61 **GO** 62

**••69 through 79** *More lenses.* Object $O$ stands on the central axis of a thin symmetric lens. For this situation, each problem in Table 34-8 refers to (a) the lens type, converging (C) or diverging (D), (b) the focal distance $f$, (c) the object distance $p$, (d) the image distance $i$, and (e) the lateral magnification $m$. (All distances are in centimeters.) It also refers to whether (f) the image is real (R) or virtual (V), (g) inverted (I) or noninverted (NI) from $O$, and (h) on the *same* side of the lens as $O$ or on the *opposite* side. Fill in the missing information, including

the value of $m$ when only an inequality is given. Where only a sign is missing, answer with the sign. **SSM** 75, 79 **GO** 78

**••80 through 87** *Two-lens systems.* In Fig. 34-44, stick figure $O$ (the object) stands on the common central axis of two thin, symmetric lenses, which are mounted in the boxed regions. Lens 1 is mounted within the boxed region closer to $O$, which is at object distance $p_1$. Lens 2 is mounted within the farther boxed region, at distance $d$. Each problem in Table 34-9 refers to a different combination of lenses and different values for distances, which are given in centimeters. The type of lens is indicated by C for

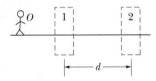

**FIG. 34-44**   Problems 80 through 87.

converging and D for diverging; the number after C or D is the distance between a lens and either of its focal points (the proper sign of the focal distance is not indicated).

Find (a) the image distance $i_2$ for the image produced by lens 2 (the final image produced by the system) and (b) the overall lateral magnification $M$ for the system, including signs. Also, determine whether the final image is (c) real (R) or

**TABLE 34-7**

**Problems 59 through 68: Lenses with Given Radii. See the setup for these problems.**

| | $p$ | $n$ | $r_1$ | $r_2$ | (a)<br>$i$ | (b)<br>$m$ | (c)<br>R/V | (d)<br>I/NI | (e)<br>Side |
|---|---|---|---|---|---|---|---|---|---|
| 59 | +35 | 1.70 | +42 | +33 | | | | | |
| 60 | +29 | 1.65 | +35 | ∞ | | | | | |
| 61 | +75 | 1.55 | +30 | −42 | | | | | |
| 62 | +18 | 1.60 | −27 | +24 | | | | | |
| 63 | +60 | 1.50 | +35 | −35 | | | | | |
| 64 | +6.0 | 1.70 | +10 | −12 | | | | | |
| 65 | +24 | 1.50 | −15 | −25 | | | | | |
| 66 | +10 | 1.50 | −30 | −60 | | | | | |
| 67 | +10 | 1.50 | −30 | +30 | | | | | |
| 68 | +10 | 1.50 | +30 | −30 | | | | | |

**TABLE 34-8**

Problems 69 through 79: More Lenses. See the setup for these problems.

| | (a) Type | (b) $f$ | (c) $p$ | (d) $i$ | (e) $m$ | (f) R/V | (g) I/NI | (h) Side |
|---|---|---|---|---|---|---|---|---|
| **69** | | | +16 | | +0.25 | | | |
| **70** | | | +16 | | −0.25 | | | |
| **71** | | 20 | +8.0 | | >1.0 | | | |
| **72** | | 20 | +8.0 | | <1.0 | | NI | |
| **73** | | +10 | +5.0 | | | | | |
| **74** | | 10 | +5.0 | | >1.0 | | | |
| **75** | | | +16 | | +1.25 | | | |
| **76** | C | 10 | +20 | | | | | |
| **77** | | | +10 | | −0.50 | | | |
| **78** | | | +10 | | 0.50 | | NI | |
| **79** | | 10 | +5.0 | | <1.0 | | | Same |

virtual (V), (d) inverted (I) from object $O$ or noninverted (NI), and (e) on the *same* side of lens 2 as object $O$ or on the *opposite* side. SSM WWW  81 GO  87

### sec. 34-8  Optical Instruments

•**88**  If the angular magnification of an astronomical telescope is 36 and the diameter of the objective is 75 mm, what is the minimum diameter of the eyepiece required to collect all the light entering the objective from a distant point source on the telescope axis?

•**89**  In a microscope of the type shown in Fig. 34-20, the focal length of the objective is 4.00 cm, and that of the eyepiece is 8.00 cm. The distance between the lenses is 25.0 cm. (a) What is the tube length $s$? (b) If image $I$ in Fig. 34-20 is to be just inside focal point $F'_1$, how far from the objective should the object be? What then are (c) the lateral magnification $m$ of the objective, (d) the angular magnification $m_\theta$ of the eyepiece, and (e) the overall magnification $M$ of the microscope?  SSM

••**90**  An object is 10.0 mm from the objective of a certain compound microscope. The lenses are 300 mm apart, and the intermediate image is 50.0 mm from the eyepiece. What overall magnification is produced by the instrument?

••**91**  Someone with a near point $P_n$ of 25 cm views a thimble through a simple magnifying lens of focal length 10 cm by placing the lens near his eye. What is the angular magnification of the thimble if it is positioned so that its image appears at (a) $P_n$ and (b) infinity?

••**92**  Figure 34-45a shows the basic structure of a camera. A lens can be moved forward or back to produce an image on film at the back of the camera. For a certain camera, with the distance $i$ between the lens and the film set at $f = 5.0$ cm, parallel light rays from a very distant object $O$ converge to a point image on the film, as shown. The object is now brought closer, to a distance of $p$ = 100 cm, and the lens–film dis-

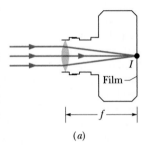

(a)

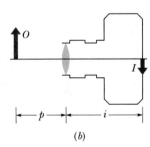

(b)

**FIG. 34-45**  Problem 92.

**TABLE 34-9**

Problems 80 through 87: Two-Lens Systems. See the setup for these problems.

| | $p_1$ | Lens 1 | $d$ | Lens 2 | (a) $i_2$ | (b) $M$ | (c) R/V | (d) I/NI | (e) Side |
|---|---|---|---|---|---|---|---|---|---|
| **80** | +8.0 | D, 6.0 | 12 | C, 6.0 | | | | | |
| **81** | +20 | C, 9.0 | 8.0 | C, 5.0 | | | | | |
| **82** | +12 | C, 8.0 | 30 | D, 8.0 | | | | | |
| **83** | +12 | C, 8.0 | 32 | C, 6.0 | | | | | |
| **84** | +10 | C, 15 | 10 | C, 8.0 | | | | | |
| **85** | +20 | D, 12 | 10 | D, 8.0 | | | | | |
| **86** | +15 | C, 12 | 67 | C, 10 | | | | | |
| **87** | +4.0 | C, 6.0 | 8.0 | D, 6.0 | | | | | |

tance is adjusted so that an inverted real image forms on the film (Fig. 34-45*b*). (a) What is the lens–film distance *i* now? (b) By how much was distance *i* changed?

••**93** Figure 34-46*a* shows the basic structure of a human eye. Light refracts into the eye through the cornea and is then further redirected by a lens whose shape (and thus ability to focus the light) is controlled by muscles. We can treat the cornea and eye lens as a single effective thin lens (Fig. 34-46*b*). A "normal" eye can focus parallel light rays from a distant object *O* to a point on the retina at the back of the eye, where processing of the visual information begins. As an object is brought close to the eye, however, the muscles must change the shape of the lens so that rays form an inverted real image on the retina (Fig. 34-46*c*). (a) Suppose that for the parallel rays of Figs. 34-46*a* and *b*, the focal length *f* of the effective thin lens of the eye is 2.50 cm. For an object at distance *p* =

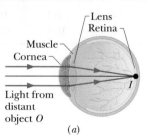

*(a)*

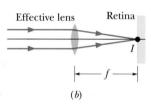

*(b)*

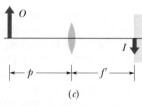

*(c)*

**FIG. 34-46** Problem 93.

40.0 cm, what focal length *f'* of the effective lens is required for the object to be seen clearly? (b) Must the eye muscles increase or decrease the radii of curvature of the eye lens to give focal length *f'*? SSM

**Additional Problems**

**94** An object is placed against the center of a spherical mirror and then moved 70 cm from it along the central axis as the image distance *i* is measured. Figure 34-47 gives *i* versus object distance *p* out to $p_s$ = 40 cm. What is the image distance when the object is 70 cm from the mirror?

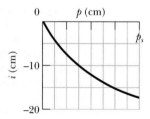

**FIG. 34-47** Problem 94.

**95** In Fig. 34-48, a box is somewhere at the left, on the central axis of the thin converging lens. The image $I_m$ of the box produced by the plane mirror is 4.00 cm "inside" the mirror. The lens–mirror separation is 10.0 cm, and the focal length of the lens is 2.00 cm. (a) What is the distance between the box and the lens? Light reflected by the mirror travels back through the lens, which produces a final image of the box. (b) What is the distance between the lens and that final image?

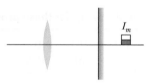

**FIG. 34-48** Problem 95.

**96** Two plane mirrors are placed parallel to each other and 40 cm apart. An object is placed 10 cm from one mirror.

Determine the (a) smallest, (b) second smallest, (c) third smallest (occurs twice), and (d) fourth smallest distance between the object and image of the object.

**97** Figure 34-49 shows a *beam expander* made with two coaxial converging lenses of focal lengths $f_1$ and $f_2$ and separation $d = f_1 + f_2$. The device can expand a laser beam while keeping the light rays in the beam parallel to the central axis through the lenses. Suppose a uniform laser beam of width $W_i$ = 2.5 mm and intensity $I_i$ = 9.0 kW/m² enters a beam expander for which $f_1$ = 12.5 cm and $f_2$ = 30.0 cm. What are (a) $W_f$ and (b) $I_f$ of the beam leaving the expander? (c) What value of *d* is needed for the beam expander if lens 1 is replaced with a diverging lens of focal length $f_1$ = −26.0 cm?

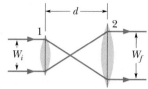

**FIG. 34-49** Problem 97.

**98** In Fig. 34-50, an object is placed in front of a converging lens at a distance equal to twice the focal length $f_1$ of the lens. On the other side of the lens is a concave mirror of focal length $f_2$ separated from the lens by a distance $2(f_1 + f_2)$. Light from the object passes rightward through the lens, reflects from the mirror, passes leftward through the lens, and forms a final image of the object. What are (a) the distance between the lens and that final image and (b) the overall lateral magnification *M* of the object? Is the image (c) real or virtual (if it is virtual, it requires someone looking through the lens toward the mirror), (d) to the left or right of the lens, and (e) inverted

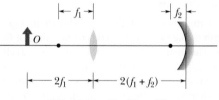

**FIG. 34-50** Problem 98.

or noninverted relative to the object?

**99** In Fig. 34-51, a fish watcher at point *P* watches a fish through a glass wall of a fish tank. The watcher is level with the fish; the index of refraction of the glass is 8/5, and that of the water is 4/3. The distances are $d_1$ = 8.0 cm, $d_2$ = 3.0 cm, and $d_3$ = 6.8 cm. (a) To the fish, how far away does the watcher appear to be? (*Hint:* The watcher is the object. Light from that object passes through the wall's outside surface, which acts as a refracting surface. Find the image produced by that surface. Then treat that image as an object whose light passes through the wall's inside surface, which acts as another refracting surface. Find the image produced by that surface, and there is the answer.) (b) To the watcher, how far away does the fish appear to be?

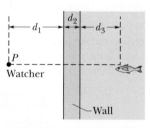

**FIG. 34-51** Problem 99.

**100** Figure 34-52*a* is an overhead view of two vertical plane mirrors with an object *O* placed between them. If you look into the mirrors, you see multiple images of $O$. You can find them by drawing the reflection in each mirror of the angular region between the mirrors, as is done in Fig. 34-52b for the left-hand mirror. Then draw the reflection of the reflection. Continue this on the left and on the right until the reflections meet or overlap at the rear of the mirrors. Then you can count the number of images of $O$. How many images of $O$ would you see if $\theta$ is (a) 90°, (b) 45°, and (c) 60°? If $\theta = 120°$, determine the (d) smallest and (e) largest number of images that can be seen, depending on your perspective and the location of $O$. (f) In each situation, draw the image locations and orientations as in Fig. 34-52b.

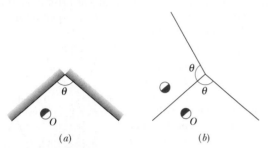

(a)         (b)

FIG. 34-52    Problem 100.

**101** A point object is 10 cm away from a plane mirror, and the eye of an observer (with pupil diameter 5.0 mm) is 20 cm away. Assuming the eye and the object to be on the same line perpendicular to the mirror surface, find the area of the mirror used in observing the reflection of the point. (*Hint:* Adapt Fig. 34-4.)

**102** You grind the lenses shown in Fig. 34-53 from flat glass disks ($n = 1.5$) using a machine that can grind a radius of curvature of either 40 cm or 60 cm. In a lens where either radius is appropriate, you select the 40 cm radius. Then you hold each lens in sunshine to form an image of the Sun. What are the (a) focal length $f$ and (b) image type (real or virtual) for (*bi-convex*) lens 1, (c) $f$ and (d) image type for (*plane-convex*) lens 2, (e) $f$ and (f) image type for (*meniscus convex*) lens 3, (g) $f$ and (h) image type for (*bi-concave*) lens 4, (i) $f$ and (j) image type for (*plane-concave*) lens 5, and (k) $f$ and (l) image type for (*meniscus concave*) lens 6?

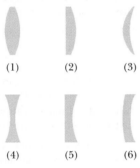

(1)    (2)    (3)

(4)    (5)    (6)

FIG. 34-53    Problem 102.

**103** The formula $1/p + 1/i = 1/f$ is called the *Gaussian* form of the thin-lens formula. Another form of this formula, the *Newtonian* form, is obtained by considering the distance $x$ from the object to the first focal point and the distance $x'$ from the second focal point to the image. Show that $xx' = f^2$ is the Newtonian form of the thin-lens formula. **SSM**

**104** Show that the distance between an object and its real image formed by a thin converging lens is always greater than or equal to four times the focal length of the lens.

**105** Two thin lenses of focal lengths $f_1$ and $f_2$ are in contact. Show that they are equivalent to a single thin lens for which the focal length is $f = f_1 f_2/(f_1 + f_2)$. **SSM**

**106** A luminous object and a screen are a fixed distance $D$ apart. (a) Show that a converging lens of focal length $f$, placed between object and screen, will form a real image on the screen for two lens positions that are separated by a distance $d = \sqrt{D(D - 4f)}$. (b) Show that

$$\left(\frac{D - d}{D + d}\right)^2$$

gives the ratio of the two image sizes for these two positions of the lens.

**107** A fruit fly of height $H$ sits in front of lens 1 on the central axis through the lens. The lens forms an image of the fly at a distance $d = 20$ cm from the fly; the image has the fly's orientation and height $H_I = 2.0H$. What are (a) the focal length $f_1$ of the lens and (b) the object distance $p_1$ of the fly? The fly then leaves lens 1 and sits in front of lens 2, which also forms an image at $d = 20$ cm that has the same orientation as the fly, but now $H_I = 0.50H$. What are (c) $f_2$ and (d) $p_2$? **SSM**

**108** A goldfish in a spherical fish bowl of radius $R$ is at the level of the center $C$ of the bowl and at distance $R/2$ from the glass (Fig. 34-54). What magnification of the fish is produced by the water in the bowl for a viewer looking along a line that includes the fish and the center, with the fish on the near side of the center? The index of refraction of the water is 1.33. Neglect the glass wall of the bowl. Assume the viewer looks with one eye. (*Hint:* Equation 34-5 holds, but Eq. 34-6 does not. You need to work with a ray diagram of the situation and assume that the rays are close to the observer's line of sight — that is, they deviate from that line by only small angles.)

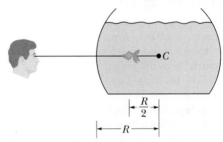

FIG. 34-54    Problem 108.

**109** An eraser of height 1.0 cm is placed 10.0 cm in front of a two-lens system. Lens 1 (nearer the eraser) has focal length $f_1 = -15$ cm, lens 2 has $f_2 = 12$ cm, and the lens separation is $d = 12$ cm. For the image produced by lens 2, what are (a) the image distance $i_2$ (including sign), (b) the image height, (c) the image type (real or virtual), and (d) the image orientation (inverted relative to the eraser or not inverted)?

**110** A peanut is placed 40 cm in front of a two-lens system: lens 1 (nearer the peanut) has focal length $f_1 = +20$ cm, lens 2 has $f_2 = -15$ cm, and the lens separation is $d = 10$ cm. For the image produced by lens 2, what are (a) the image distance $i_2$ (including sign), (b) the image orientation (inverted relative to the peanut or not inverted), and (c) the image type (real or virtual)? (d) What is the net lateral magnification?

**111** A coin is placed 20 cm in front of a two-lens system. Lens 1 (nearer the coin) has focal length $f_1 = +10$ cm, lens 2 has $f_2 = +12.5$ cm, and the lens separation is $d = 30$ cm. For the image produced by lens 2, what are (a) the image distance

$i_2$ (including sign), (b) the overall lateral magnification, (c) the image type (real or virtual), and (d) the image orientation (inverted relative to the coin or not inverted)?

**112** An object is 20 cm to the left of a thin diverging lens that has a 30 cm focal length. (a) What is the image distance $i$? (b) Draw a ray diagram showing the image position.

**113 through 118** *Three-lens systems.* In Fig. 34-55, stick figure $O$ (the object) stands on the common central axis of three thin, symmetric lenses, which are mounted in the boxed regions. Lens 1 is mounted within the boxed region closest

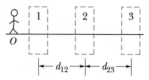

**FIG. 34-55** Problems 113 through 118.

to $O$, which is at object distance $p_1$. Lens 2 is mounted within the middle boxed region, at distance $d_{12}$ from lens 1. Lens 3 is mounted in the farthest boxed region, at distance $d_{23}$ from lens 2. Each problem in Table 34-10 refers to a different combination of lenses and different values for distances, which are given in centimeters. The type of lens is indicated by C for converging and D for diverging; the number after C or D is the distance between a lens and either of the focal points (the proper sign of the focal distance is not indicated).

Find (a) the image distance $i_3$ for the (final) image produced by lens 3 (the final image produced by the system) and (b) the overall lateral magnification $M$ for the system, including signs. Also, determine whether the final image is (c) real (R) or virtual (V), (d) inverted (I) from object $O$ or noninverted (NI), and (e) on the same side of lens 3 as object $O$ or on the opposite side. **GO** 113

**119** In Fig. 34-56, a pinecone is at distance $p_1 = 1.0$ m in front of a lens of focal length $f_1 = 0.50$ m; a flat mirror is at distance $d = 2.0$ m behind the lens. Light from the pinecone passes rightward through the lens, reflects

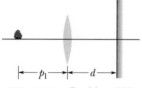

**FIG. 34-56** Problem 119.

from the mirror, passes leftward through the lens, and forms a final image of the pinecone. What are (a) the distance between the lens and that image and (b) the overall lateral magnification of the pinecone? Is the image (c) real or virtual (if it is virtual, it requires someone looking through the lens toward the mirror), (d) to the left or right of the lens, and (e) inverted relative to the pinecone or not inverted?

**120** One end of a long glass rod ($n = 1.5$) is a convex sur-

face of radius 6.0 cm. An object is located in air along the axis of the rod, at a distance of 10 cm from the convex end. (a) How far apart are the object and the image formed by the glass rod? (b) Within what range of distances from the end of the rod must the object be located in order to produce a virtual image?

**121** A short straight object of length $L$ lies along the central axis of a spherical mirror, a distance $p$ from the mirror. (a) Show that its image in the mirror has a length $L'$, where

$$L' = L\left(\frac{f}{p-f}\right)^2.$$

(*Hint:* Locate the two ends of the object.) (b) Show that the *longitudinal magnification* $m'$ ($= L'/L$) is equal to $m^2$, where $m$ is the lateral magnification.

**122** You look down at a coin that lies at the bottom of a pool of liquid of depth $d$ and index of refraction $n$ (Fig. 34-57). Because you view with two eyes, which intercept different rays of light from the coin, you perceive the coin to be where extensions of the intercepted rays cross, at depth $d_a$ instead of $d$. Assuming that the intercepted rays in Fig. 34-57 are close to a vertical axis through the coin, show that $d_a = d/n$. (*Hint:* Use the small-angle approximation $\sin\theta \approx \tan\theta \approx \theta$.)

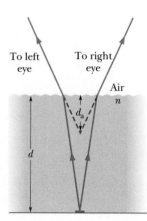

**FIG. 34-57** Problem 122.

**123** Prove that if a plane mirror is rotated through an angle $\alpha$, the reflected beam is rotated through an angle $2\alpha$. Show that this result is reasonable for $\alpha = 45°$.

**124** An object is 30.0 cm from a spherical mirror, along the mirror's central axis. The mirror produces an inverted image with a lateral magnification of absolute value 0.500. What is the focal length of the mirror?

**125** A concave mirror has a radius of curvature of 24 cm. How far is an object from the mirror if the image formed is (a) virtual and 3.0 times the size of the object, (b) real and 3.0 times the size of the object, and (c) real and 1/3 the size of the object?

**126** A pepper seed is placed in front of a lens. The lateral magnification of the seed is +0.300. The absolute value of the lens's focal length is 40.0 cm. How far from the lens is the image?

**TABLE 34-10**

**Problems 113 through 118: Three-Lens Systems. See the setup for these problems.**

| | $p_1$ | Lens 1 | $d_{12}$ | Lens 2 | $d_{23}$ | Lens 3 | (a) $i_3$ | (b) $M$ | (c) R/V | (d) I/NI | (e) Side |
|---|---|---|---|---|---|---|---|---|---|---|---|
| **113** | +12 | C, 8.0 | 28 | C, 6.0 | 8.0 | C, 6.0 | | | | | |
| **114** | +4.0 | C, 6.0 | 8.0 | D, 4.0 | 5.7 | D, 12 | | | | | |
| **115** | +18 | C, 6.0 | 15 | C, 3.0 | 11 | C, 3.0 | | | | | |
| **116** | +4.0 | D, 6.0 | 9.6 | C, 6.0 | 14 | C, 4.0 | | | | | |
| **117** | +8.0 | D, 8.0 | 8.0 | D, 16 | 5.1 | C, 8.0 | | | | | |
| **118** | +2.0 | C, 6.0 | 15 | C, 6.0 | 19 | C, 5.0 | | | | | |

**127** The equation $1/p + 1/i = 2/r$ for spherical mirrors is an approximation that is valid if the image is formed by rays that make only small angles with the central axis. In reality, many of the angles are large, which smears the image a little. You can determine how much. Refer to Fig. 34-22 and consider a ray that leaves a point source (the object) on the central axis and that makes an angle $\alpha$ with that axis.

First, find the point of intersection of the ray with the mirror. If the coordinates of this intersection point are $x$ and $y$ and the origin is placed at the center of curvature, then $y = (x + p − r) \tan \alpha$ and $x^2 + y^2 = r^2$, where $p$ is the object distance and $r$ is the mirror's radius of curvature. Next, use $\tan \beta = y/x$ to find the angle $\beta$ at the point of intersection, and then use $\alpha + \gamma = 2\beta$ to find the value of $\gamma$. Finally, use the relation $\tan \gamma = y/(x + i − r)$ to find the distance $i$ of the image.

(a) Suppose $r = 12$ cm and $p = 20$ cm. For each of the following values of $\alpha$, find the position of the image—that is, the position of the point where the reflected ray crosses the central axis: 0.500, 0.100, 0.0100 rad. Compare the results with those obtained with the equation $1/p + 1/i = 2/r$. (b) Repeat the calculations for $p = 4.00$ cm.

**128** A small cup of green tea is positioned on the central axis of a spherical mirror. The lateral magnification of the cup is +0.250, and the distance between the mirror and its focal point is 2.00 cm. (a) What is the distance between the mirror and the image it produces? (b) Is the focal length positive or negative? (c) Is the image real or virtual?

**129** A 20-mm-thick layer of water ($n = 1.33$) floats on a 40-mm-thick layer of carbon tetrachloride ($n = 1.46$) in a tank. A coin lies at the bottom of the tank. At what depth below the top water surface do you perceive the coin? (*Hint:* Use the result and assumptions of Problem 122 and work with a ray diagram of the situation.)

**130** A millipede sits 1.0 m in front of the nearest part of the surface of a shiny sphere of diameter 0.70 m. (a) How far from the surface does the millipede's image appear? (b) If the millipede's height is 2.0 mm, what is the image height? (c) Is the image inverted?

**131** (a) Show that if the object $O$ in Fig. 34-19c is moved from focal point $F_1$ toward the observer's eye, the image moves in from infinity and the angle $\theta'$ (and thus the angular magnification $m_\theta$) increases. (b) If you continue this process, where is the image when $m_\theta$ has its maximum usable value? (You can then still increase $m_\theta$, but the image will no longer be clear.) (c) Show that the maximum usable value of $m_\theta$ is $1 + (25 \text{ cm})/f$. (d) Show that in this situation the angular magnification is equal to the lateral magnification.

**132** Isaac Newton, having convinced himself (erroneously as it turned out) that chromatic aberration is an inherent property of refracting telescopes, invented the reflecting telescope, shown schematically in Fig. 34-58. He presented his second

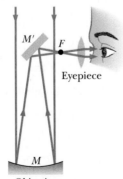

FIG. 34-58
Problem 132.

model of this telescope, with a magnifying power of 38, to the Royal Society (of London), which still has it. In Fig. 34-58 incident light falls, closely parallel to the telescope axis, on the objective mirror $M$. After reflection from small mirror $M'$ (the figure is not to scale), the rays form a real, inverted image in the *focal plane* (the plane perpendicular to the line of sight, at focal point $F$). This image is then viewed through an eyepiece. (a) Show that the angular magnification $m_\theta$ for the device is given by Eq. 34-15:

$$m_\theta = -f_{ob}/f_{ey},$$

where $f_{ob}$ is the focal length of the objective mirror and $f_{ey}$ is that of the eyepiece. (b) The 200 in. mirror in the reflecting telescope at Mt. Palomar in California has a focal length of 16.8 m. Estimate the size of the image formed by this mirror when the object is a meter stick 2.0 km away. Assume parallel incident rays. (c) The mirror of a different reflecting astronomical telescope has an effective radius of curvature of 10 m ("effective" because such mirrors are ground to a parabolic rather than a spherical shape, to eliminate spherical aberration defects). To give an angular magnification of 200, what must be the focal length of the eyepiece?

**133** A narrow beam of parallel light rays is incident on a glass sphere from the left, directed toward the center of the sphere. (The sphere is a lens but certainly not a *thin* lens.) Approximate the angle of incidence of the rays as 0°, and assume that the index of refraction of the glass is $n < 2.0$. (a) In terms of $n$ and the sphere radius $r$, what is the distance between the image produced by the sphere and the right side of the sphere? (b) Is the image to the left or right of that side? (*Hint:* Apply Eq. 34-8 to locate the image that is produced by refraction at the left side of the sphere; then use that image as the object for refraction at the right side of the sphere to locate the final image. In the second refraction, is the object distance $p$ positive or negative?)

**134** A *corner reflector,* much used in optical, microwave, and other applications, consists of three plane mirrors fastened together to form the corner of a cube. Show that after three reflections, an incident ray is returned with its direction exactly reversed.

**135** A cheese enchilada is 4.00 cm in front of a converging lens. The magnification of the enchilada is −2.00. What is the focal length of the lens?

**136** A grasshopper hops to a point on the central axis of a spherical mirror. The absolute magnitude of the mirror's focal length is 40.0 cm, and the lateral magnification of the image produced by the mirror is +0.200. (a) Is the mirror convex or concave? (b) How far from the mirror is the grasshopper?

**137** In Fig. 34-59, a sand grain is 3.00 cm from thin lens 1, on the central axis through the two symmetric lenses. The distance between focal point and lens is 4.00 cm for both lenses; the lenses are separated by 8.00 cm. (a) What is the distance between lens 2 and the image it produces of the sand grain? Is that image (b) to the left or right of lens 2, (c) real or virtual, and (d) inverted relative to the sand grain or not inverted?

FIG. 34-59  Problem 137.

# 35 | Interference

Amanda Hall/Robert Harding World Imagery/Getty Images, Inc.

*Governments worldwide scurry to stay ahead of counterfeiters who are quick to use the latest technology to duplicate paper currencies. Some of the security measures now used to thwart counterfeiters are security threads and special watermarks (both of which can be seen if the currency is held up against a light) and microprinting (which consists of dots too small to be picked up by a scanner). The feature that is probably the most difficult for a counterfeiter to duplicate is the variable tint that results from color-shifting inks. For example, the "20" in the lower right of the front face of a U.S. $20 bill contains color-shifting ink. If you look directly down on the number, it is red or red-yellow. If you then tilt the bill and look at it obliquely, the color shifts to green. A copy machine can duplicate color from only one perspective and therefore cannot duplicate this shift in color you see when you change your perspective.*

## How do color-shifting inks shift colors?

The answer is in this chapter.

FIG. 35-1   The blue of the top surface of a *Morpho* butterfly wing is due to optical interference and shifts in color as your viewing perspective changes. *(Philippe Colombi/PhotoDisc//Getty Images)*

## 35-1 | WHAT IS PHYSICS?

One of the major goals of physics is to understand the nature of light. This goal has been difficult to achieve (and has not yet fully been achieved) because light is complicated. However, this complication means that light offers many opportunities for applications, and some of the richest opportunities involve the interference of light waves — **optical interference.**

Nature has long used optical interference for coloring. For example, the wings of a *Morpho* butterfly are a dull, uninspiring brown, as can be seen on the bottom wing surface, but the brown is hidden on the top surface by an arresting blue due to the interference of light reflecting from that surface (Fig. 35-1). Moreover, the top surface is color-shifting; if you change your perspective or if the wing moves, the tint of the color changes. Similar color shifting is used in the inks on many currencies to thwart counterfeiters, whose copy machines can duplicate color from only one perspective and therefore cannot duplicate any shift in color caused by a change in perspective.

Before we get to color shifting in butterfly wings and currency inks, we first must discuss the basic physics of optical interference. That means we must largely abandon the simplicity of geometrical optics (in which we describe light as rays) and return to the wave nature of light.

## 35-2 | Light as a Wave

The first person to advance a convincing wave theory for light was Dutch physicist Christian Huygens, in 1678. Although much less comprehensive than the later electromagnetic theory of Maxwell, Huygens' theory was simpler mathematically and remains useful today. Its great advantages are that it accounts for the laws of reflection and refraction in terms of waves and gives physical meaning to the index of refraction.

Huygens' wave theory is based on a geometrical construction that allows us to tell where a given wavefront will be at any time in the future if we know its present position. This construction is based on **Huygens' principle,** which is:

> All points on a wavefront serve as point sources of spherical secondary wavelets. After a time *t*, the new position of the wavefront will be that of a surface tangent to these secondary wavelets.

Here is a simple example. At the left in Fig. 35-2, the present location of a wavefront of a plane wave traveling to the right in vacuum is represented by plane *ab*, perpendicular to the page. Where will the wavefront be at time $\Delta t$ later? We let several points on plane *ab* (the dots) serve as sources of spherical secondary wavelets that are emitted at $t = 0$. At time $\Delta t$, the radius of all these spherical wavelets will have grown to $c\,\Delta t$, where *c* is the speed of light in vacuum. We draw plane *de* tangent to these wavelets at time $\Delta t$. This plane represents the wavefront of the plane wave at time $\Delta t$; it is parallel to plane *ab* and a perpendicular distance $c\,\Delta t$ from it.

### The Law of Refraction

We now use Huygens' principle to derive the law of refraction, Eq. 33-40 (Snell's law). Figure 35-3 shows three stages in the refraction of several wavefronts at a flat interface between air (medium 1) and glass (medium 2). We arbitrarily choose the wavefronts in the incident light beam to be separated by $\lambda_1$, the wavelength in medium 1. Let the speed of light in air be $v_1$ and that in glass be $v_2$. We assume that $v_2 < v_1$, which happens to be true.

Angle $\theta_1$ in Fig. 35-3*a* is the angle between the wavefront and the interface; it has the same value as the angle between the *normal* to the wavefront (that is, the incident ray) and the *normal* to the interface. Thus, $\theta_1$ is the angle of incidence.

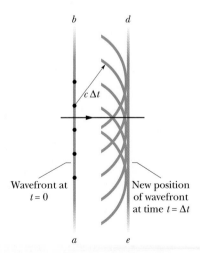

Wavefront at
$t = 0$

New position
of wavefront
at time $t = \Delta t$

FIG. 35-2   The propagation of a plane wave in vacuum, as portrayed by Huygens' principle.

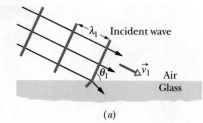

λ₁ Incident wave

$\theta_1$ $\vec{v}_1$ Air
Glass

(a)

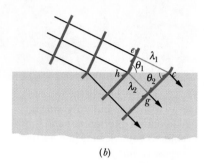

(b)

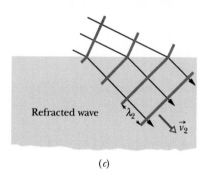

Refracted wave λ₂ $\vec{v}_2$

(c)

**FIG. 35-3** The refraction of a plane wave at an air–glass interface, as portrayed by Huygens' principle. The wavelength in glass is smaller than that in air. For simplicity, the reflected wave is not shown. Parts (a) through (c) represent three successive stages of the refraction.

As the wave moves into the glass, a Huygens wavelet at point $e$ in Fig. 35-3b will expand to pass through point $c$, at a distance of $\lambda_1$ from point $e$. The time interval required for this expansion is that distance divided by the speed of the wavelet, or $\lambda_1/v_1$. Now note that in this same time interval, a Huygens wavelet at point $h$ will expand to pass through point $g$, at the reduced speed $v_2$ and with wavelength $\lambda_2$. Thus, this time interval must also be equal to $\lambda_2/v_2$. By equating these times of travel, we obtain the relation

$$\frac{\lambda_1}{\lambda_2} = \frac{v_1}{v_2}, \tag{35-1}$$

which shows that the wavelengths of light in two media are proportional to the speeds of light in those media.

By Huygens' principle, the refracted wavefront must be tangent to an arc of radius $\lambda_2$ centered on $h$, say at point $g$. The refracted wavefront must also be tangent to an arc of radius $\lambda_1$ centered on $e$, say at $c$. Then the refracted wavefront must be oriented as shown. Note that $\theta_2$, the angle between the refracted wavefront and the interface, is actually the angle of refraction.

For the right triangles $hce$ and $hcg$ in Fig. 35-3b we may write

$$\sin \theta_1 = \frac{\lambda_1}{hc} \qquad \text{(for triangle } hce\text{)}$$

and

$$\sin \theta_2 = \frac{\lambda_2}{hc} \qquad \text{(for triangle } hcg\text{)}.$$

Dividing the first of these two equations by the second and using Eq. 35-1, we find

$$\frac{\sin \theta_1}{\sin \theta_2} = \frac{\lambda_1}{\lambda_2} = \frac{v_1}{v_2}. \tag{35-2}$$

We can define the **index of refraction** $n$ for each medium as the ratio of the speed of light in vacuum to the speed of light $v$ in the medium. Thus,

$$n = \frac{c}{v} \qquad \text{(index of refraction)}. \tag{35-3}$$

In particular, for our two media, we have

$$n_1 = \frac{c}{v_1} \quad \text{and} \quad n_2 = \frac{c}{v_2}. \tag{35-4}$$

If we combine Eqs. 35-2 and 35-4, we find

$$\frac{\sin \theta_1}{\sin \theta_2} = \frac{c/n_1}{c/n_2} = \frac{n_2}{n_1} \tag{35-5}$$

or

$$n_1 \sin \theta_1 = n_2 \sin \theta_2 \qquad \text{(law of refraction)}, \tag{35-6}$$

as introduced in Chapter 33.

✓ **CHECKPOINT 1** The figure shows a monochromatic ray of light traveling across parallel interfaces, from an original material $a$, through layers of materials $b$ and $c$, and then back into material $a$. Rank the materials according to the speed of light in them, greatest first.

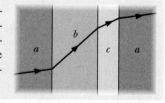

## Wavelength and Index of Refraction

We have now seen that the wavelength of light changes when the speed of the light changes, as happens when light crosses an interface from one medium into another. Further, the speed of light in any medium depends on the index of refraction of the medium, according to Eq. 35-3. Thus, the wavelength of light in any medium depends on the index of refraction of the medium. Let a certain monochromatic light have wavelength $\lambda$ and speed $c$ in vacuum and wavelength $\lambda_n$ and speed $v$ in a medium with an index of refraction $n$. Now we can rewrite Eq. 35-1 as

$$\lambda_n = \lambda \frac{v}{c}. \tag{35-7}$$

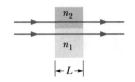

FIG. 35-4  Two light rays travel through two media having different indexes of refraction.

Using Eq. 35-3 to substitute $1/n$ for $v/c$ then yields

$$\lambda_n = \frac{\lambda}{n}. \tag{35-8}$$

This equation relates the wavelength of light in any medium to its wavelength in vacuum. It tells us that the greater the index of refraction of a medium, the smaller the wavelength of light in that medium.

What about the frequency of the light? Let $f_n$ represent the frequency of the light in a medium with index of refraction $n$. Then from the general relation of Eq. 16-13 ($v = \lambda f$), we can write

$$f_n = \frac{v}{\lambda_n}.$$

Substituting Eqs. 35-3 and 35-8 then gives us

$$f_n = \frac{c/n}{\lambda/n} = \frac{c}{\lambda} = f,$$

where $f$ is the frequency of the light in vacuum. Thus, although the speed and wavelength of light in the medium are different from what they are in vacuum, *the frequency of the light in the medium is the same as it is in vacuum.*

The fact that the wavelength of light depends on the index of refraction via Eq. 35-8 is important in certain situations involving the interference of light waves. For example, in Fig. 35-4, the *waves of the rays* (that is, the waves represented by the rays) have identical wavelengths $\lambda$ and are initially in phase in air ($n \approx 1$). One of the waves travels through medium 1 of index of refraction $n_1$ and length $L$. The other travels through medium 2 of index of refraction $n_2$ and the same length $L$. When the waves leave the two media, they will have the same wavelength—their wavelength $\lambda$ in air. However, because their wavelengths differed in the two media, the two waves may no longer be in phase.

☞ The phase difference between two light waves can change if the waves travel through different materials having different indexes of refraction.

As we shall discuss soon, this change in the phase difference can determine how the light waves will interfere if they reach some common point.

To find their new phase difference in terms of wavelengths, we first count the number $N_1$ of wavelengths there are in the length $L$ of medium 1. From Eq. 35-8, the wavelength in medium 1 is $\lambda_{n1} = \lambda/n_1$; so

$$N_1 = \frac{L}{\lambda_{n1}} = \frac{Ln_1}{\lambda}. \tag{35-9}$$

Similarly, we count the number $N_2$ of wavelengths there are in the length $L$ of medium 2, where the wavelength is $\lambda_{n2} = \lambda/n_2$:

$$N_2 = \frac{L}{\lambda_{n2}} = \frac{Ln_2}{\lambda}. \tag{35-10}$$

To find the new phase difference between the waves, we subtract the smaller of $N_1$ and $N_2$ from the larger. Assuming $n_2 > n_1$, we obtain

$$N_2 - N_1 = \frac{Ln_2}{\lambda} - \frac{Ln_1}{\lambda} = \frac{L}{\lambda}(n_2 - n_1). \tag{35-11}$$

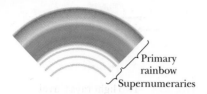

FIG. 35-5 A primary rainbow and the faint supernumeraries below it are due to optical interference.

Suppose Eq. 35-11 tells us that the waves now have a phase difference of 45.6 wavelengths. That is equivalent to taking the initially in-phase waves and shifting one of them by 45.6 wavelengths. However, a shift of an integer number of wavelengths (such as 45) would put the waves back in phase; so it is only the decimal fraction (here, 0.6) that is important. A phase difference of 45.6 wavelengths is equivalent to an *effective phase difference* of 0.6 wavelength.

A phase difference of 0.5 wavelength puts two waves exactly out of phase. If the waves had equal amplitudes and were to reach some common point, they would then undergo fully destructive interference, producing darkness at that point. With a phase difference of 0.0 or 1.0 wavelength, they would, instead, undergo fully constructive interference, resulting in brightness at the common point. Our phase difference of 0.6 wavelength is an intermediate situation but closer to fully destructive interference, and the waves would produce a dimly illuminated common point.

We can also express phase difference in terms of radians and degrees, as we have done already. A phase difference of one wavelength is equivalent to phase differences of $2\pi$ rad and 360°.

> ✓ **CHECKPOINT 2** The light waves of the rays in Fig. 35-4 have the same wavelength and amplitude and are initially in phase. (a) If 7.60 wavelengths fit within the length of the top material and 5.50 wavelengths fit within that of the bottom material, which material has the greater index of refraction? (b) If the rays are angled slightly so that they meet at the same point on a distant screen, will the interference there result in the brightest possible illumination, bright intermediate illumination, dark intermediate illumination, or darkness?

### Rainbows and Optical Interference

In Section 33-8, we discussed how the colors of sunlight are separated into a rainbow when sunlight travels through falling raindrops. We dealt with a simplified situation in which a single ray of white light entered a drop. Actually, light waves pass into a drop along the entire side that faces the Sun. Here we cannot discuss the details of how these waves travel through the drop and then emerge, but we can see that different parts of an incoming wave will travel different paths within the drop. That means waves will emerge from the drop with different phases. Thus, we can see that at some angles the emerging light will be in phase and give constructive interference. The rainbow is the result of such constructive interference. For example, the red of the rainbow appears because waves of red light emerge in phase from each raindrop in the direction in which you see that part of the rainbow.

If you are lucky and look carefully below a primary rainbow, you can see dimmer colored arcs called *supernumeraries* (Fig. 35-5). Like the main arcs of the rainbow, the supernumeraries are due to waves that emerge from each drop approximately in phase with one another to give constructive interference. If you are very lucky and look very carefully above a secondary rainbow, you might see even more (but even dimmer) supernumeraries. Keep in mind that both types of rainbows and both sets of supernumeraries are naturally occurring examples of optical interference and naturally occurring evidence that light consists of waves.

### Sample Problem 35-1

In Fig. 35-4, the two light waves that are represented by the rays have wavelength 550.0 nm before entering media 1 and 2. They also have equal amplitudes and are in phase. Medium 1 is now just air, and medium 2 is a transparent plastic layer of index of refraction 1.600 and thickness 2.600 $\mu$m.

(a) What is the phase difference of the emerging waves in wavelengths, radians, and degrees? What is their effective phase difference (in wavelengths)?

**KEY IDEA** The phase difference of two light waves can change if they travel through different media, with different indexes of refraction. The reason is that their wavelengths are different in the different media. We can calculate the change in phase difference by counting the number of wavelengths that fits into each medium and then subtracting those numbers.

*Calculations:* When the path lengths of the waves in the

two media are identical, Eq. 35-11 gives the result of the subtraction. Here we have $n_1 = 1.000$ (for the air), $n_2 = 1.600$, $L = 2.600 \ \mu m$, and $\lambda = 550.0$ nm. Thus, Eq. 35-11 yields

$$N_2 - N_1 = \frac{L}{\lambda}(n_2 - n_1)$$

$$= \frac{2.600 \times 10^{-6} \ m}{5.500 \times 10^{-7} \ m}(1.600 - 1.000)$$

$$= 2.84. \qquad \text{(Answer)}$$

Thus, the phase difference of the emerging waves is 2.84 wavelengths. Because 1.0 wavelength is equivalent to $2\pi$ rad and 360°, you can show that this phase difference is equivalent to

$$\text{phase difference} = 17.8 \ \text{rad} \approx 1020°. \quad \text{(Answer)}$$

The effective phase difference is the decimal part of the actual phase difference *expressed in wavelengths*. Thus, we have

effective phase difference = 0.84 wavelength. (Answer)

You can show that this is equivalent to 5.3 rad and about 300°. *Caution:* We do *not* find the effective phase difference by taking the decimal part of the actual phase difference as expressed in radians or degrees. For example, we do *not* take 0.8 rad from the actual phase difference of 17.8 rad.

(b) If the waves reached the same point on a distant screen, what type of interference would they produce?

*Reasoning:* We need to compare the effective phase difference of the waves with the phase differences that give the extreme types of interference. Here the effective phase difference of 0.84 wavelength is between 0.5 wavelength (for fully destructive interference, or the darkest possible result) and 1.0 wavelength (for fully constructive interference, or the brightest possible result), but closer to 1.0 wavelength. Thus, the waves would produce intermediate interference that is closer to fully constructive interference—they would produce a relatively bright spot.

## 35-3 | Diffraction

In the next section we shall discuss the experiment that first proved that light is a wave. To prepare for that discussion, we must introduce the idea of **diffraction** of waves, a phenomenon that we explore much more fully in Chapter 36. Its essence is this: If a wave encounters a barrier that has an opening of dimensions similar to the wavelength, the part of the wave that passes through the opening will flare (spread) out—will *diffract*—into the region beyond the barrier. The flaring is consistent with the spreading of wavelets in the Huygens construction of Fig. 35-2. Diffraction occurs for waves of all types, not just light waves; Fig. 35-6 shows the diffraction of water waves traveling across the surface of water in a shallow tank.

Figure 35-7a shows the situation schematically for an incident plane wave of wavelength $\lambda$ encountering a slit that has width $a = 6.0\lambda$ and extends into and out of the page. The part of the wave that passes through the slit flares out on the far side. Figures 35-7b (with $a = 3.0\lambda$) and 35-7c ($a = 1.5\lambda$) illustrate the main feature of diffraction: the narrower the slit, the greater the diffraction.

Diffraction limits geometrical optics, in which we represent an electromagnetic wave with a ray. If we actually try to form a ray by sending light through

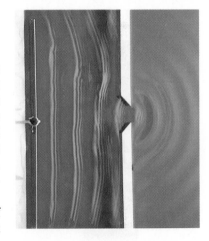

FIG. 35-6 Waves produced by an oscillating paddle at the left flare out through an opening in a barrier along the water surface. *(Runk Schoenberger/Grant Heilman Photography)*

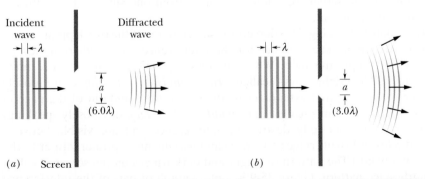
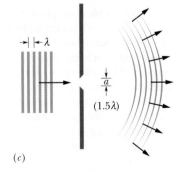

FIG. 35-7 Diffraction represented schematically. For a given wavelength $\lambda$, the diffraction is more pronounced the smaller the slit width $a$. The figures show the cases for (a) slit width $a = 6.0\lambda$, (b) slit width $a = 3.0\lambda$, and (c) slit width $a = 1.5\lambda$. In all three cases, the screen and the length of the slit extend well into and out of the page, perpendicular to it.

**FIG. 35-8** In Young's interference experiment, incident monochromatic light is diffracted by slit $S_0$, which then acts as a point source of light that emits semicircular wavefronts. As that light reaches screen $B$, it is diffracted by slits $S_1$ and $S_2$, which then act as two point sources of light. The light waves traveling from slits $S_1$ and $S_2$ overlap and undergo interference, forming an interference pattern of maxima and minima on viewing screen $C$. This figure is a cross section; the screens, slits, and interference pattern extend into and out of the page. Between screens $B$ and $C$, the semicircular wavefronts centered on $S_2$ depict the waves that would be there if only $S_2$ were open. Similarly, those centered on $S_1$ depict waves that would be there if only $S_1$ were open.

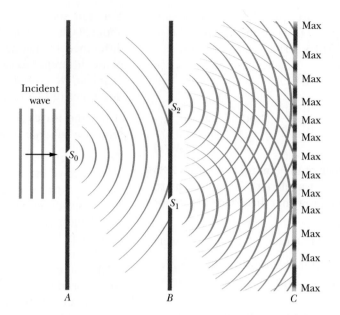

**FIG. 35-9** A photograph of the interference pattern produced by the arrangement shown in Fig. 35-8. (The photograph is a front view of part of screen $C$.) The alternating maxima and minima are called *interference fringes* (because they resemble the decorative fringe sometimes used on clothing and rugs). *(From Michel Cagnet, Maurice Franzon, and Jean Claude Thierr, Atlas of Optical Phenomena. Springer-Verlag, New York, 1962. Reproduced with permission.)*

a narrow slit, or through a series of narrow slits, diffraction will always defeat our effort because it always causes the light to spread. Indeed, the narrower we make the slits (in the hope of producing a narrower beam), the greater the spreading is. Thus, geometrical optics holds only when slits or other apertures that might be located in the path of light do not have dimensions comparable to or smaller than the wavelength of the light.

## 35-4 | Young's Interference Experiment

In 1801, Thomas Young experimentally proved that light is a wave, contrary to what most other scientists then thought. He did so by demonstrating that light undergoes interference, as do water waves, sound waves, and waves of all other types. In addition, he was able to measure the average wavelength of sunlight; his value, 570 nm, is impressively close to the modern accepted value of 555 nm. We shall here examine Young's experiment as an example of the interference of light waves.

Figure 35-8 gives the basic arrangement of Young's experiment. Light from a distant monochromatic source illuminates slit $S_0$ in screen $A$. The emerging light then spreads via diffraction to illuminate two slits $S_1$ and $S_2$ in screen $B$. Diffraction of the light by these two slits sends overlapping circular waves into the region beyond screen $B$, where the waves from one slit interfere with the waves from the other slit.

The "snapshot" of Fig. 35-8 depicts the interference of the overlapping waves. However, we cannot see evidence for the interference except where a viewing screen $C$ intercepts the light. Where it does so, points of interference maxima form visible bright rows—called *bright bands, bright fringes,* or (loosely speaking) *maxima*—that extend across the screen (into and out of the page in Fig. 35-8). Dark regions—called *dark bands, dark fringes,* or (loosely speaking) *minima*—result from fully destructive interference and are visible between adjacent pairs of bright fringes. (*Maxima* and *minima* more properly refer to the center of a band.) The pattern of bright and dark fringes on the screen is called an **interference pattern.** Figure 35-9 is a photograph of part of the interference pattern that would be seen by an observer standing to the left of screen $C$ in the arrangement of Fig. 35-8.

## Locating the Fringes

Light waves produce fringes in a *Young's double-slit interference experiment,* as it is called, but what exactly determines the locations of the fringes? To answer, we shall use the arrangement in Fig. 35-10a. There, a plane wave of monochromatic light is incident on two slits $S_1$ and $S_2$ in screen $B$; the light diffracts through the slits and produces an interference pattern on screen $C$. We draw a central axis from the point halfway between the slits to screen $C$ as a reference. We then pick, for discussion, an arbitrary point $P$ on the screen, at angle $\theta$ to the central axis. This point intercepts the wave of ray $r_1$ from the bottom slit and the wave of ray $r_2$ from the top slit.

These waves are in phase when they pass through the two slits because there they are just portions of the same incident wave. However, once they have passed the slits, the two waves must travel different distances to reach $P$. We saw a similar situation in Section 17-5 with sound waves and concluded that

> ☞ The phase difference between two waves can change if the waves travel paths of different lengths.

The change in phase difference is due to the *path length difference* $\Delta L$ in the paths taken by the waves. Consider two waves initially exactly in phase, traveling along paths with a path length difference $\Delta L$, and then passing through some common point. When $\Delta L$ is zero or an integer number of wavelengths, the waves arrive at the common point exactly in phase and they interfere fully constructively there. If that is true for the waves of rays $r_1$ and $r_2$ in Fig. 35-10, then point $P$ is part of a bright fringe. When, instead, $\Delta L$ is an odd multiple of half a wavelength, the waves arrive at the common point exactly out of phase and they interfere fully destructively there. If that is true for the waves of rays $r_1$ and $r_2$, then point $P$ is part of a dark fringe. (And, of course, we can have intermediate situations of interference and thus intermediate illumination at $P$.) Thus,

> ☞ What appears at each point on the viewing screen in a Young's double-slit interference experiment is determined by the path length difference $\Delta L$ of the rays reaching that point.

We can specify where each bright fringe and each dark fringe is located on the screen by giving the angle $\theta$ from the central axis to that fringe. To find $\theta$, we must relate it to $\Delta L$. We start with Fig. 35-10a by finding a point $b$ along ray $r_1$ such that the path length from $b$ to $P$ equals the path length from $S_2$ to $P$. Then the path length difference $\Delta L$ between the two rays is the distance from $S_1$ to $b$.

The relation between this $S_1$-to-$b$ distance and $\theta$ is complicated, but we can simplify it considerably if we arrange for the distance $D$ from the slits to the screen to be much greater than the slit separation $d$. Then we can approximate rays $r_1$ and $r_2$ as being parallel to each other and at angle $\theta$ to the central axis (Fig. 35-10b). We can also approximate the triangle formed by $S_1$, $S_2$, and $b$ as being a right triangle, and approximate the angle inside that triangle at $S_2$ as being $\theta$. Then, for that triangle, $\sin \theta = \Delta L/d$ and thus

$$\Delta L = d \sin \theta \qquad \text{(path length difference).} \qquad (35\text{-}12)$$

For a bright fringe, we saw that $\Delta L$ must be either zero or an integer number of wavelengths. Using Eq. 35-12, we can write this requirement as

$$\Delta L = d \sin \theta = (\text{integer})(\lambda), \qquad (35\text{-}13)$$

or as

$$d \sin \theta = m\lambda, \qquad \text{for } m = 0, 1, 2, \ldots \qquad \text{(maxima—bright fringes).} \qquad (35\text{-}14)$$

For a dark fringe, $\Delta L$ must be an odd multiple of half a wavelength. Again using

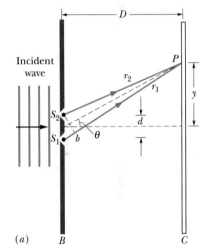

(a)

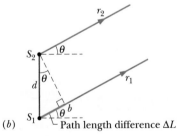

(b)

**FIG. 35-10** (a) Waves from slits $S_1$ and $S_2$ (which extend into and out of the page) combine at $P$, an arbitrary point on screen $C$ at distance $y$ from the central axis. The angle $\theta$ serves as a convenient locator for $P$. (b) For $D \gg d$, we can approximate rays $r_1$ and $r_2$ as being parallel, at angle $\theta$ to the central axis.

Eq. 35-12, we can write this requirement as

$$\Delta L = d \sin \theta = (\text{odd number})(\tfrac{1}{2}\lambda), \tag{35-15}$$

or as

$$d \sin \theta = (m + \tfrac{1}{2})\lambda, \qquad \text{for } m = 0, 1, 2, \dots \qquad \text{(minima—dark fringes).} \tag{35-16}$$

With Eqs. 35-14 and 35-16, we can find the angle $\theta$ to any fringe and thus locate that fringe; further, we can use the values of $m$ to label the fringes. For the value and label $m = 0$, Eq. 35-14 tells us that a bright fringe is at $\theta = 0$ and thus on the central axis. This *central maximum* is the point at which waves arriving from the two slits have a path length difference $\Delta L = 0$, hence zero phase difference.

For, say, $m = 2$, Eq. 35-14 tells us that *bright* fringes are at the angle

$$\theta = \sin^{-1}\left(\frac{2\lambda}{d}\right)$$

above and below the central axis. Waves from the two slits arrive at these two fringes with $\Delta L = 2\lambda$ and with a phase difference of two wavelengths. These fringes are said to be the *second-order bright fringes* (meaning $m = 2$) or the *second side maxima* (the second maxima to the side of the central maximum), or they are described as being the second bright fringes from the central maximum.

For $m = 1$, Eq. 35-16 tells us that *dark* fringes are at the angle

$$\theta = \sin^{-1}\left(\frac{1.5\lambda}{d}\right)$$

above and below the central axis. Waves from the two slits arrive at these two fringes with $\Delta L = 1.5\lambda$ and with a phase difference, in wavelengths, of 1.5. These fringes are called the *second-order dark fringes* or *second minima* because they are the second dark fringes to the side of the central axis. (The first dark fringes, or first minima, are at locations for which $m = 0$ in Eq. 35-16.)

We derived Eqs. 35-14 and 35-16 for the situation $D \gg d$. However, they also apply if we place a converging lens between the slits and the viewing screen and then move the viewing screen closer to the slits, to the focal point of the lens. (The screen is then said to be in the *focal plane* of the lens; that is, it is in the plane perpendicular to the central axis at the focal point.) One property of a converging lens is that it focuses all rays that are parallel to one another to the same point on its focal plane. Thus, the rays that now arrive at any point on the screen (in the focal plane) were exactly parallel (rather than approximately) when they left the slits. They are like the initially parallel rays in Fig. 34-15a that are directed to a point (the focal point) by a lens.

 **CHECKPOINT 3**  In Fig. 35-10a, what are $\Delta L$ (as a multiple of the wavelength) and the phase difference (in wavelengths) for the two rays if point $P$ is (a) a third side maximum and (b) a third minimum?

---

**Sample Problem** | **35-2**

What is the distance on screen $C$ in Fig. 35-10a between adjacent maxima near the center of the interference pattern? The wavelength $\lambda$ of the light is 546 nm, the slit separation $d$ is 0.12 mm, and the slit–screen separation $D$ is 55 cm. Assume that $\theta$ in Fig. 35-10 is small enough to permit use of the approximations $\sin \theta \approx \tan \theta \approx \theta$, in which $\theta$ is expressed in radian measure.

**KEY IDEAS**  (1) First, let us pick a maximum with a low value of $m$ to ensure that it is near the center of the pat-

tern. Then, from the geometry of Fig. 35-10a, the maximum's vertical distance $y_m$ from the center of the pattern is related to its angle $\theta$ from the central axis by

$$\tan \theta \approx \theta = \frac{y_m}{D}.$$

(2) From Eq. 35-14, this angle $\theta$ for the $m$th maximum is given by

$$\sin \theta \approx \theta = \frac{m\lambda}{d}.$$

*Calculations:* If we equate our two expressions for angle $\theta$ and then solve for $y_m$, we find

$$y_m = \frac{m\lambda D}{d}. \tag{35-17}$$

For the next maximum as we move away from the pattern's center, we have

$$y_{m+1} = \frac{(m+1)\lambda D}{d}. \tag{35-18}$$

We find the distance between these adjacent maxima by subtracting Eq. 35-17 from Eq. 35-18:

$$\Delta y = y_{m+1} - y_m = \frac{\lambda D}{d}$$

$$= \frac{(546 \times 10^{-9}\,\text{m})(55 \times 10^{-2}\,\text{m})}{0.12 \times 10^{-3}\,\text{m}}$$

$$= 2.50 \times 10^{-3}\,\text{m} \approx 2.5\,\text{mm}. \qquad \text{(Answer)}$$

As long as $d$ and $\theta$ in Fig. 35-10a are small, the separation of the interference fringes is independent of $m$; that is, the fringes are evenly spaced.

---

**Sample Problem** | **35-3** | **Build your skill**

A double-slit interference pattern is produced on a screen, as in Fig. 35-10; the light is monochromatic at a wavelength of 600 nm. A strip of transparent plastic with index of refraction $n = 1.50$ is to be placed over one of the slits. Its presence changes the interference between light waves from the two slits, causing the interference pattern to be shifted across the screen from the original pattern. Figure 35-11a shows the original locations of the central bright fringe ($m = 0$) and the first bright fringes ($m = 1$) above and below the central fringe. The purpose of the plastic is to shift the pattern upward so that the lower $m = 1$ bright fringe is shifted to the center of the pattern. Should the plastic be placed over the top slit (as arbitrarily drawn in Fig. 35-11b) or the bottom slit, and what thickness $L$ should it have?

**KEY IDEA** The interference at a point on the screen depends on the phase difference of the light rays arriving from the two slits. The light rays are in phase at the slits, but their relative phase can shift on the way to the screen due to (1) a difference in the length of the paths they follow and (2) a difference in the number of their internal wavelengths $\lambda_n$ in the materials through which they pass. Let's check both possibilities.

**Path-length difference:** Figure 35-11a shows rays $r_1$ and $r_2$ along which waves from the two slits travel to reach the lower $m = 1$ bright fringe. Those waves start in phase at the slits but arrive at the fringe with a phase difference of exactly 1 wavelength. To remind ourselves of this main characteristic of the fringe, let us call it the $1\lambda$ fringe. The one-wavelength phase difference is due to the one-wavelength path length difference between the rays reaching the fringe; that is, there is exactly one more wavelength along ray $r_2$ than along $r_1$.

Figure 35-11b shows the $1\lambda$ fringe shifted up to the center of the pattern with the plastic strip over the top slit (we still do not know whether the plastic should be there or over the bottom slit). The figure also shows the new orientations of rays $r_1$ and $r_2$ to reach that fringe. There still must be one more wavelength along $r_2$ than along $r_1$ (because they still produce the $1\lambda$ fringe), but now the path length difference between those rays is

zero, as we can tell from the geometry of Fig. 35-11b. However, $r_2$ now passes through the plastic.

***Internal wavelength:*** The wavelength $\lambda_n$ of light in a material with index of refraction $n$ is smaller than the wavelength in vacuum, as given by Eq. 35-8 ($\lambda_n = \lambda/n$). Here, this means that the wavelength of the light is smaller in the plastic than in the air. Thus, the ray that passes through the plastic will have more wavelengths along it than the ray that passes through only air—so we do get the one extra wavelength we need along ray $r_2$ by placing the plastic over the top slit, as drawn in Fig. 35-11b.

***Thickness:*** To determine the required thickness $L$ of the plastic, we use the procedure in Sample Problem 35-1a. Here, as there, waves that are initially in phase travel equal distances $L$ through different materials (plastic and air). However, here we know the phase difference and require $L$; so we again use Eq. 35-11,

$$N_2 - N_1 = \frac{L}{\lambda}(n_2 - n_1). \tag{35-19}$$

We know that $N_2 - N_1$ is 1 for a phase difference of one wavelength, $n_2$ is 1.50 for the plastic in front of the top slit, $n_1$ is 1.00 for the air in front of the bottom slit, and $\lambda$ is $600 \times 10^{-9}$ m. Then Eq. 35-19 yields

$$L = \frac{\lambda(N_2 - N_1)}{n_2 - n_1} = \frac{(600 \times 10^{-9}\,\text{m})(1)}{1.50 - 1.00}$$

$$= 1.2 \times 10^{-6}\,\text{m}. \qquad \text{(Answer)}$$

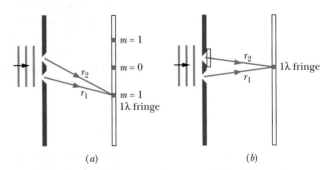

(a)                                    (b)

**FIG. 35-11** (a) Arrangement for two-slit interference (not to scale). The locations of three bright fringes (or maxima) are indicated. (b) A strip of plastic covers the top slit. We want the $1\lambda$ fringe to be at the center of the pattern.

## 35-5 | Coherence

For the interference pattern to appear on viewing screen $C$ in Fig. 35-8, the light waves reaching any point $P$ on the screen must have a phase difference that does not vary in time. That is the case in Fig. 35-8 because the waves passing through slits $S_1$ and $S_2$ are portions of the single light wave that illuminates the slits. Because the phase difference remains constant, the light from slits $S_1$ and $S_2$ is said to be completely **coherent.**

Direct sunlight is partially coherent; that is, sunlight waves intercepted at two points have a constant phase difference only if the points are very close. If you look closely at your fingernail in bright sunlight, you can see a faint interference pattern called *speckle* that causes the nail to appear to be covered with specks. You see this effect because light waves scattering from very close points on the nail are sufficiently coherent to interfere with one another at your eye. The slits in a double-slit experiment, however, are not close enough, and in direct sunlight, the light at the slits would be **incoherent.** To get coherent light, we would have to send the sunlight through a single slit as in Fig. 35-8; because that single slit is small, light that passes through it is coherent. In addition, the smallness of the slit causes the coherent light to spread via diffraction to illuminate both slits in the double-slit experiment.

If we replace the double slits with two similar but independent monochromatic light sources, such as two fine incandescent wires, the phase difference between the waves emitted by the sources varies rapidly and randomly. (This occurs because the light is emitted by vast numbers of atoms in the wires, acting randomly and independently for extremely short times—of the order of nanoseconds.) As a result, at any given point on the viewing screen, the interference between the waves from the two sources varies rapidly and randomly between fully constructive and fully destructive. The eye (and most common optical detectors) cannot follow such changes, and no interference pattern can be seen. The fringes disappear, and the screen is seen as being uniformly illuminated.

A *laser* differs from common light sources in that its atoms emit light in a cooperative manner, thereby making the light coherent. Moreover, the light is almost monochromatic, is emitted in a thin beam with little spreading, and can be focused to a width that almost matches the wavelength of the light.

## 35-6 | Intensity in Double-Slit Interference

Equations 35-14 and 35-16 tell us how to locate the maxima and minima of the double-slit interference pattern on screen $C$ of Fig. 35-10 as a function of the angle $\theta$ in that figure. Here we wish to derive an expression for the intensity $I$ of the fringes as a function of $\theta$.

The light leaving the slits is in phase. However, let us assume that the light waves from the two slits are not in phase when they arrive at point $P$. Instead, the electric field components of those waves at point $P$ are not in phase and vary with time as

$$E_1 = E_0 \sin \omega t \tag{35-20}$$

and
$$E_2 = E_0 \sin(\omega t + \phi), \tag{35-21}$$

where $\omega$ is the angular frequency of the waves and $\phi$ is the phase constant of wave $E_2$. Note that the two waves have the same amplitude $E_0$ and a phase difference of $\phi$. Because that phase difference does not vary, the waves are coherent. We shall show that these two waves will combine at $P$ to produce an intensity $I$

given by

$$I = 4I_0 \cos^2 \tfrac{1}{2}\phi, \qquad (35\text{-}22)$$

and that

$$\phi = \frac{2\pi d}{\lambda} \sin\theta. \qquad (35\text{-}23)$$

In Eq. 35-22, $I_0$ is the intensity of the light that arrives on the screen from one slit when the other slit is temporarily covered. We assume that the slits are so narrow in comparison to the wavelength that this single-slit intensity is essentially uniform over the region of the screen in which we wish to examine the fringes.

Equations 35-22 and 35-23, which together tell us how the intensity $I$ of the fringe pattern varies with the angle $\theta$ in Fig. 35-10, necessarily contain information about the location of the maxima and minima. Let us see if we can extract that information to find equations about those locations.

Study of Eq. 35-22 shows that intensity maxima will occur when

$$\tfrac{1}{2}\phi = m\pi, \qquad \text{for } m = 0, 1, 2, \ldots. \qquad (35\text{-}24)$$

If we put this result into Eq. 35-23, we find

$$2m\pi = \frac{2\pi d}{\lambda} \sin\theta, \qquad \text{for } m = 0, 1, 2, \ldots$$

or
$$d \sin\theta = m\lambda, \qquad \text{for } m = 0, 1, 2, \ldots \quad \text{(maxima)}, \qquad (35\text{-}25)$$

which is exactly Eq. 35-14, the expression that we derived earlier for the locations of the maxima.

The minima in the fringe pattern occur when

$$\tfrac{1}{2}\phi = (m + \tfrac{1}{2})\pi, \qquad \text{for } m = 0, 1, 2, \ldots. \qquad (35\text{-}26)$$

If we combine this relation with Eq. 35-23, we are led at once to

$$d \sin\theta = (m + \tfrac{1}{2})\lambda, \qquad \text{for } m = 0, 1, 2, \ldots \quad \text{(minima)}, \qquad (35\text{-}27)$$

which is just Eq. 35-16, the expression we derived earlier for the locations of the fringe minima.

Figure 35-12, which is a plot of Eq. 35-22, shows the intensity of double-slit interference patterns as a function of the phase difference $\phi$ between the waves at the screen. The horizontal solid line is $I_0$, the (uniform) intensity on the screen when one of the slits is covered up. Note in Eq. 35-22 and the graph that the intensity $I$ varies from zero at the fringe minima to $4I_0$ at the fringe maxima.

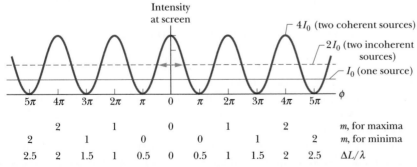

FIG. 35-12 A plot of Eq. 35-22, showing the intensity of a double-slit interference pattern as a function of the phase difference between the waves when they arrive from the two slits. $I_0$ is the (uniform) intensity that would appear on the screen if one slit were covered. The average intensity of the fringe pattern is $2I_0$, and the *maximum* intensity (for coherent light) is $4I_0$.

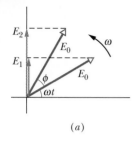

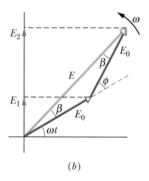

(a)

(b)

**FIG. 35-13** (a) Phasors representing, at time $t$, the electric field components given by Eqs. 35-20 and 35-21. Both phasors have magnitude $E_0$ and rotate with angular speed $\omega$. Their phase difference is $\phi$. (b) Vector addition of the two phasors gives the phasor representing the resultant wave, with amplitude $E$ and phase constant $\beta$.

If the waves from the two sources (slits) were *incoherent,* so that no enduring phase relation existed between them, there would be no fringe pattern and the intensity would have the uniform value $2I_0$ for all points on the screen; the horizontal dashed line in Fig. 35-12 shows this uniform value.

Interference cannot create or destroy energy but merely redistributes it over the screen. Thus, the *average* intensity on the screen must be the same $2I_0$ regardless of whether the sources are coherent. This follows at once from Eq. 35-22; if we substitute $\frac{1}{2}$, the average value of the cosine-squared function, this equation reduces to $I_{avg} = 2I_0$.

### Proof of Eqs. 35-22 and 35-23

We shall combine the electric field components $E_1$ and $E_2$, given by Eqs. 35-20 and 35-21, respectively, by the method of phasors as is discussed in Section 16-11. In Fig. 35-13a, the waves with components $E_1$ and $E_2$ are represented by phasors of magnitude $E_0$ that rotate around the origin at angular speed $\omega$. The values of $E_1$ and $E_2$ at any time are the projections of the corresponding phasors on the vertical axis. Figure 35-13a shows the phasors and their projections at an arbitrary time $t$. Consistent with Eqs. 35-20 and 35-21, the phasor for $E_1$ has a rotation angle $\omega t$ and the phasor for $E_2$ has a rotation angle $\omega t + \phi$.

To combine the field components $E_1$ and $E_2$ at any point $P$ in Fig. 35-10, we add their phasors vectorially, as shown in Fig. 35-13b. The magnitude of the vector sum is the amplitude $E$ of the resultant wave at point $P$, and that wave has a certain phase constant $\beta$. To find the amplitude $E$ in Fig. 35-13b, we first note that the two angles marked $\beta$ are equal because they are opposite equal-length sides of a triangle. From the theorem (for triangles) that an exterior angle (here $\phi$, as shown in Fig. 35-13b) is equal to the sum of the two opposite interior angles (here that sum is $\beta + \beta$), we see that $\beta = \frac{1}{2}\phi$. Thus, we have

$$E = 2(E_0 \cos \beta)$$
$$= 2E_0 \cos \tfrac{1}{2}\phi. \tag{35-28}$$

If we square each side of this relation, we obtain

$$E^2 = 4E_0^2 \cos^2 \tfrac{1}{2}\phi. \tag{35-29}$$

Now, from Eq. 33-24, we know that the intensity of an electromagnetic wave is proportional to the square of its amplitude. Therefore, the waves we are combining in Fig. 35-13b, whose amplitudes are $E_0$, each has an intensity $I_0$ that is proportional to $E_0^2$, and the resultant wave, with amplitude $E$, has an intensity $I$ that is proportional to $E^2$. Thus,

$$\frac{I}{I_0} = \frac{E^2}{E_0^2}.$$

Substituting Eq. 35-29 into this equation and rearranging then yield

$$I = 4I_0 \cos^2 \tfrac{1}{2}\phi,$$

which is Eq. 35-22, which we set out to prove.

We still must prove Eq. 35-23, which relates the phase difference $\phi$ between the waves arriving at any point $P$ on the screen of Fig. 35-10 to the angle $\theta$ that serves as a locator of that point.

The phase difference $\phi$ in Eq. 35-21 is associated with the path length difference $S_1b$ in Fig. 35-10b. If $S_1b$ is $\frac{1}{2}\lambda$, then $\phi$ is $\pi$; if $S_1b$ is $\lambda$, then $\phi$ is $2\pi$, and so on. This suggests

$$\left(\begin{array}{c}\text{phase}\\\text{difference}\end{array}\right) = \frac{2\pi}{\lambda}\left(\begin{array}{c}\text{path length}\\\text{difference}\end{array}\right). \tag{35-30}$$

The path length difference $S_1b$ in Fig. 35-10b is $d \sin \theta$; so Eq. 35-30 for the phase difference between the two waves arriving at point $P$ on the screen becomes

$$\phi = \frac{2\pi d}{\lambda}\sin\theta,$$

which is Eq. 35-23, the other equation that we set out to prove to relate $\phi$ to the angle $\theta$ that locates $P$.

## Combining More Than Two Waves

In a more general case, we might want to find the resultant of more than two sinusoidally varying waves at a point. Whatever the number of waves is, our general procedure is this:

1. Construct a series of phasors representing the waves to be combined. Draw them end to end, maintaining the proper phase relations between adjacent phasors.

2. Construct the vector sum of this array. The length of this vector sum gives the amplitude of the resultant phasor. The angle between the vector sum and the first phasor is the phase of the resultant with respect to this first phasor. The projection of this vector-sum phasor on the vertical axis gives the time variation of the resultant wave.

✓ **CHECKPOINT 4**    Each of four pairs of light waves arrives at a certain point on a screen. The waves have the same wavelength. At the arrival point, their amplitudes and phase differences are (a) $2E_0, 6E_0$, and $\pi$ rad; (b) $3E_0, 5E_0$, and $\pi$ rad; (c) $9E_0, 7E_0$, and $3\pi$ rad; (d) $2E_0, 2E_0$, and 0 rad. Rank the four pairs according to the intensity of the light at the arrival point, greatest first. (*Hint:* Draw phasors.)

---

**Sample Problem**   **35-4**

Three light waves combine at a certain point where their electric field components are

$$E_1 = E_0\sin\omega t,$$
$$E_2 = E_0\sin(\omega t + 60°),$$
$$E_3 = E_0\sin(\omega t - 30°).$$

Find their resultant component $E(t)$ at that point.

---

**KEY IDEA**    The resultant wave is

$$E(t) = E_1(t) + E_2(t) + E_3(t).$$

We can use the method of phasors to find this sum, and we are free to evaluate the phasors at any time $t$.

**Calculations:** To simplify the solution, we choose $t = 0$, for which the phasors representing the three waves are shown in Fig. 35-14. We can add these three phasors either directly on a vector-capable calculator or by components. For the component approach, we first write the sum of their horizontal components as

$$\sum E_h = E_0\cos 0 + E_0\cos 60° + E_0\cos(-30°) = 2.37E_0.$$

The sum of their vertical components, which is the value of $E$ at $t = 0$, is

$$\sum E_v = E_0\sin 0 + E_0\sin 60° + E_0\sin(-30°) = 0.366E_0.$$

The resultant wave $E(t)$ thus has an amplitude $E_R$ of

$$E_R = \sqrt{(2.37E_0)^2 + (0.366E_0)^2} = 2.4E_0,$$

and a phase angle $\beta$ relative to the phasor representing $E_1$ of

$$\beta = \tan^{-1}\left(\frac{0.366E_0}{2.37E_0}\right) = 8.8°.$$

We can now write, for the resultant wave $E(t)$,

$$E = E_R\sin(\omega t + \beta)$$
$$= 2.4E_0\sin(\omega t + 8.8°).\qquad\text{(Answer)}$$

Be careful to interpret the angle $\beta$ correctly in Fig. 35-14: It is the constant angle between $E_R$ and the phasor representing $E_1$ as the four phasors rotate as a single unit around the origin. The angle between $E_R$ and the horizontal axis in Fig. 35-14 does not remain equal to $\beta$.

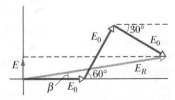

**FIG. 35-14**   Three phasors, representing waves with equal amplitudes $E_0$ and with phase constants $0°, 60°$, and $-30°$, shown at time $t = 0$. The phasors combine to give a resultant phasor with magnitude $E_R$, at angle $\beta$.

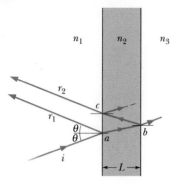

**FIG. 35-15** Light waves, represented with ray $i$, are incident on a thin film of thickness $L$ and index of refraction $n_2$. Rays $r_1$ and $r_2$ represent light waves that have been reflected by the front and back surfaces of the film, respectively. (All three rays are actually nearly perpendicular to the film.) The interference of the waves of $r_1$ and $r_2$ with each other depends on their phase difference. The index of refraction $n_1$ of the medium at the left can differ from the index of refraction $n_3$ of the medium at the right, but for now we assume that both media are air, with $n_1 = n_3 = 1.0$, which is less than $n_2$.

## 35-7 | Interference from Thin Films

The colors we see when sunlight illuminates a soap bubble or an oil slick are caused by the interference of light waves reflected from the front and back surfaces of a thin transparent film. The thickness of the soap or oil film is typically of the order of magnitude of the wavelength of the (visible) light involved. (Greater thicknesses spoil the coherence of the light needed to produce the colors due to interference.)

Figure 35-15 shows a thin transparent film of uniform thickness $L$ and index of refraction $n_2$, illuminated by bright light of wavelength $\lambda$ from a distant point source. For now, we assume that air lies on both sides of the film and thus that $n_1 = n_3$ in Fig. 35-15. For simplicity, we also assume that the light rays are almost perpendicular to the film ($\theta \approx 0$). We are interested in whether the film is bright or dark to an observer viewing it almost perpendicularly. (Since the film is brightly illuminated, how could it possibly be dark? You will see.)

The incident light, represented by ray $i$, intercepts the front (left) surface of the film at point $a$ and undergoes both reflection and refraction there. The reflected ray $r_1$ is intercepted by the observer's eye. The refracted light crosses the film to point $b$ on the back surface, where it undergoes both reflection and refraction. The light reflected at $b$ crosses back through the film to point $c$, where it undergoes both reflection and refraction. The light refracted at $c$, represented by ray $r_2$, is intercepted by the observer's eye.

If the light waves of rays $r_1$ and $r_2$ are exactly in phase at the eye, they produce an interference maximum and region $ac$ on the film is bright to the observer. If they are exactly out of phase, they produce an interference minimum and region $ac$ is dark to the observer, *even though it is illuminated*. If there is some intermediate phase difference, there are intermediate interference and brightness.

Thus, the key to what the observer sees is the phase difference between the waves of rays $r_1$ and $r_2$. Both rays are derived from the same ray $i$, but the path involved in producing $r_2$ involves light traveling twice across the film ($a$ to $b$, and then $b$ to $c$), whereas the path involved in producing $r_1$ involves no travel through the film. Because $\theta$ is about zero, we approximate the path length difference between the waves of $r_1$ and $r_2$ as $2L$. However, to find the phase difference between the waves, we cannot just find the number of wavelengths $\lambda$ that is equivalent to a path length difference of $2L$. This simple approach is impossible for two reasons: (1) the path length difference occurs in a medium other than air, and (2) reflections are involved, which can change the phase.

  The phase difference between two waves can change if one or both are reflected.

Before we continue our discussion of interference from thin films, we must discuss changes in phase that are caused by reflections.

### Reflection Phase Shifts

Refraction at an interface never causes a phase change—but reflection can, depending on the indexes of refraction on the two sides of the interface. Figure 35-16 shows what happens when reflection causes a phase change, using as an example pulses on a denser string (along which pulse travel is relatively slow) and a lighter string (along which pulse travel is relatively fast).

When a pulse traveling relatively slowly along the denser string in Fig. 35-16a reaches the interface with the lighter string, the pulse is partially transmitted and partially reflected, with no change in orientation. For light, this situation corresponds to the incident wave traveling in the medium of greater index of refraction $n$ (recall that greater $n$ means slower speed). In that case, the wave that is reflected at the interface does not undergo a change in phase; that is, its *reflection phase shift* is zero.

When a pulse traveling more quickly along the lighter string in Fig. 35-16b reaches the interface with the denser string, the pulse is again partially transmit-

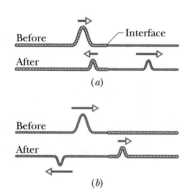

**FIG. 35-16** Phase changes when a pulse is reflected at the interface between two stretched strings of different linear densities. The wave speed is greater in the lighter string. (a) The incident pulse is in the denser string. (b) The incident pulse is in the lighter string. Only here is there a phase change, and only in the reflected wave.

ted and partially reflected. The transmitted pulse again has the same orientation as the incident pulse, but now the reflected pulse is inverted. For a sinusoidal wave, such an inversion involves a phase change of $\pi$ rad, or half a wavelength. For light, this situation corresponds to the incident wave traveling in the medium of lesser index of refraction (with greater speed). In that case, the wave that is reflected at the interface undergoes a phase shift of $\pi$ rad, or half a wavelength.

We can summarize these results for light in terms of the index of refraction of the medium off which (or from which) the light reflects:

| Reflection | Reflection phase shift |
|---|---|
| Off lower index | 0 |
| Off higher index | 0.5 wavelength |

This might be remembered as "higher means half."

### Equations for Thin-Film Interference

In this chapter we have now seen three ways in which the phase difference between two waves can change:

**1.** by reflection

**2.** by the waves traveling along paths of different lengths

**3.** by the waves traveling through media of different indexes of refraction

When light reflects from a thin film, producing the waves of rays $r_1$ and $r_2$ shown in Fig. 35-15, all three ways are involved. Let us consider them one by one.

We first reexamine the two reflections in Fig. 35-15. At point $a$ on the front interface, the incident wave (in air) reflects from the medium having the higher of the two indexes of refraction; so the wave of reflected ray $r_1$ has its phase shifted by 0.5 wavelength. At point $b$ on the back interface, the incident wave reflects from the medium (air) having the lower of the two indexes of refraction; so the wave reflected there is not shifted in phase by the reflection, and thus neither is the portion of it that exits the film as ray $r_2$. We can organize this information with the first line in Table 35-1, which refers to the simplified drawing in Fig. 35-17 for a thin film in air. So far, as a result of the reflection phase shifts, the waves of $r_1$ and $r_2$ have a phase difference of 0.5 wavelength and thus are exactly out of phase.

Now we must consider the path length difference $2L$ that occurs because the wave of ray $r_2$ crosses the film twice. (This difference $2L$ is shown on the second line in Table 35-1.) If the waves of $r_1$ and $r_2$ are to be exactly in phase so that they produce fully constructive interference, the path length $2L$ must cause an additional phase difference of $0.5, 1.5, 2.5, \ldots$ wavelengths. Only then will the net phase difference be an integer number of wavelengths. Thus, for a bright film, we must have

$$2L = \frac{\text{odd number}}{2} \times \text{wavelength} \quad \text{(in-phase waves).} \quad (35\text{-}31)$$

The wavelength we need here is the wavelength $\lambda_{n2}$ of the light in the medium containing path length $2L$—that is, in the medium with index of refraction $n_2$. Thus, we can rewrite Eq. 35-31 as

$$2L = \frac{\text{odd number}}{2} \times \lambda_{n2} \quad \text{(in-phase waves).} \quad (35\text{-}32)$$

If, instead, the waves are to be exactly out of phase so that there is fully destructive interference, the path length $2L$ must cause either no additional phase difference or a phase difference of $1, 2, 3, \ldots$ wavelengths. Only then will the net phase difference be an odd number of half-wavelengths. For a dark film,

**TABLE 35-1**

**An Organizing Table for Thin-Film Interference in Air (Fig. 35-17)[a]**

| Reflection phase shifts | $r_1$ | $r_2$ |
|---|---|---|
| | 0.5 wavelength | 0 |
| Path length difference | 2L | |
| Index in which path length difference occurs | $n_2$ | |

| In phase[a]: | $2L = \dfrac{\text{odd number}}{2} \times \dfrac{\lambda}{n_2}$ |
|---|---|
| Out of phase[a]: | $2L = \text{integer} \times \dfrac{\lambda}{n_2}$ |

[a]Valid for $n_2 > n_1$ and $n_2 > n_3$.

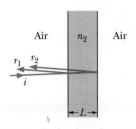

**FIG. 35-17** Reflections from a thin film in air.

we must have

$$2L = \text{integer} \times \text{wavelength}, \tag{35-33}$$

where, again, the wavelength is the wavelength $\lambda_{n2}$ in the medium containing $2L$. Thus, this time we have

$$2L = \text{integer} \times \lambda_{n2} \quad \text{(out-of-phase waves).} \tag{35-34}$$

Now we can use Eq. 35-8 ($\lambda_n = \lambda/n$) to write the wavelength of the wave of ray $r_2$ inside the film as

$$\lambda_{n2} = \frac{\lambda}{n_2}, \tag{35-35}$$

where $\lambda$ is the wavelength of the incident light in vacuum (and approximately also in air). Substituting Eq. 35-35 into Eq. 35-32 and replacing "odd number/2" with $(m + \frac{1}{2})$ give us

$$2L = (m + \tfrac{1}{2})\frac{\lambda}{n_2}, \quad \text{for } m = 0, 1, 2, \ldots \quad \text{(maxima—bright film in air).} \tag{35-36}$$

Similarly, with $m$ replacing "integer," Eq. 35-34 yields

$$2L = m\frac{\lambda}{n_2}, \quad \text{for } m = 0, 1, 2, \ldots \quad \text{(minima—dark film in air).} \tag{35-37}$$

For a given film thickness $L$, Eqs. 35-36 and 35-37 tell us the wavelengths of light for which the film appears bright and dark, respectively, one wavelength for each value of $m$. Intermediate wavelengths give intermediate brightnesses. For a given wavelength $\lambda$, Eqs. 35-36 and 35-37 tell us the thicknesses of the films that appear bright and dark in that light, respectively, one thickness for each value of $m$. Intermediate thicknesses give intermediate brightnesses.

### Film Thickness Much Less Than $\lambda$

A special situation arises when a film is so thin that $L$ is much less than $\lambda$, say, $L < 0.1\lambda$. Then the path length difference $2L$ can be neglected, and the phase difference between $r_1$ and $r_2$ is due *only* to reflection phase shifts. If the film of Fig. 35-17, where the reflections cause a phase difference of 0.5 wavelength, has thickness $L < 0.1\lambda$, then $r_1$ and $r_2$ are exactly out of phase, and thus the film is dark, regardless of the wavelength and intensity of the light. This special situation corresponds to $m = 0$ in Eq. 35-37. We shall count *any* thickness $L < 0.1\lambda$ as being the least thickness specified by Eq. 35-37 to make the film of Fig. 35-17 dark. (Every such thickness will correspond to $m = 0$.) The next greater thickness that will make the film dark is that corresponding to $m = 1$.

Figure 35-18 shows a vertical soap film whose thickness increases from top to bottom because gravitation has caused the film to slump. Bright white light illuminates the film. However, the top portion is so thin that it is dark. In the (somewhat thicker) middle we see fringes, or bands, whose color depends primarily on the wavelength at which reflected light undergoes fully constructive interference for a particular thickness. Toward the (thickest) bottom the fringes become progressively narrower and the colors begin to overlap and fade.

### Color Shifting by Butterflies, Inks, and Paints

A surface that displays colors due to thin-film interference is said to be *iridescent* because the tints of the colors change as you change your view of the surface. The iridescence of the top surface of a *Morpho* butterfly wing is due to thin-film interference of light reflected by thin terraces of transparent cuticle-like material

FIG. 35-18 The reflection of light from a soapy water film spanning a vertical loop. The top portion is so thin that the light reflected there undergoes destructive interference, making that portion dark. Colored interference fringes, or bands, decorate the rest of the film but are marred by circulation of liquid within the film as the liquid is gradually pulled downward by gravitation. *(Richard Megna/Fundamental Photographs)*

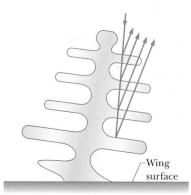

FIG. 35-19 Reflecting structure extending up from a Morpho butterfly wing. Reflections from the top surfaces of the transparent "terraces" give an interference color to the wing.

on the wing (Fig. 35-19). These terraces are arranged like wide, flat branches on a tree-like structure that extends perpendicular to the wing.

Suppose you look directly down on these terraces as white light shines directly down on the wing. Then the light reflected back up to you from the terraces undergoes fully constructive interference in the blue-green region of the visible spectrum. Light in the yellow and red regions, at the opposite end of the spectrum, is weaker because it undergoes only intermediate interference. Thus, the top surface of the wing looks blue-green to you.

If you intercept light that reflects from the wing in some other direction, the light has traveled along a slanted path through the terraces. Then the wavelength at which there is fully constructive interference is somewhat different from that for light reflected directly upward. Thus, if the wing moves in your view so that the angle at which you view it changes, the color at which the wing is brightest changes somewhat, producing the iridescence of the wing.

The color-shifting inks and paints used on paper currencies, cars, guitars, and other objects function in almost the same way as the color-shifting wing on a *Morpho* butterfly. Figure 35-20a shows a cross section of the ink layer used on some currencies. The color shifting is due to thin multilayered flakes suspended in regular ink. Figure 35-20b shows a cross section through one of the flakes. Light penetrating the regular ink above the flake travels through thin layers of chromium (Cr), magnesium fluoride (MgF$_2$), and aluminum (Al). The Cr layers function as weak mirrors, the Al layer functions as a better mirror, and the MgF$_2$ layers function like soap films. The result is that light reflected upward from each boundary between layers passes back through the regular ink and then undergoes interference at an observer's eye. Which color undergoes fully constructive interference depends on the thickness $L$ of the MgF$_2$ layers. In U.S. currency printed with color-shifting inks, the value of $L$ is designed to give fully constructive interference for red or red-yellow light when the observer looks directly down on the currency. When the observer tilts the currency and thus each flake, the light reaching the observer from the flakes undergoes constructive interference for green light. Thus, by changing the angle of view, the observer can shift the color. Other countries use other designs of thin-film flakes to achieve different shifts in the colors on their currencies.

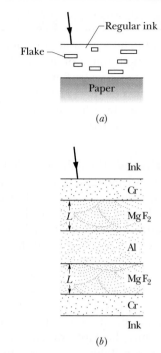

**FIG. 35-20** (a) Color-shifting ink on a paper currency consists of multilayered thin-film flakes suspended in regular ink. (b) Cross section of one of the flakes. Light penetrates the five layers, reflecting from each boundary. The color that results from the interference of these reflected light waves is determined by the thickness $L$ of the magnesium fluoride layers.

---

**PROBLEM-SOLVING TACTICS**

*Tactic 1: Thin-Film Equations* Some students believe that Eq. 35-36 gives the maxima and Eq. 35-37 gives the minima for *all* thin-film situations. This is not true. These relations were derived only for the situation in which $n_2 > n_1$ and $n_2 > n_3$ in Fig. 35-15 and for the situation of Fig. 35-17, in which a thin film lies in air.

The appropriate equations for other relative values of the indexes of refraction can be derived by following the reasoning of this section and constructing new versions of Table 35-1. In each case you will end up with Eqs. 35-36 and 35-37, but sometimes Eq. 35-36 will give the minima and Eq. 35-37 will give the maxima—the opposite of what we found here. Which equation gives which depends on whether the reflections at the two interfaces give the same reflection phase shift.

---

✓ **CHECKPOINT 5** The figure shows four situations in which light reflects perpendicularly from a thin film of thickness $L$, with indexes of refraction as given. (a) For which situations does reflection at the film interfaces cause a zero phase difference for the two reflected rays? (b) For which situations will the film be dark if the path length difference $2L$ causes a phase difference of 0.5 wavelength?

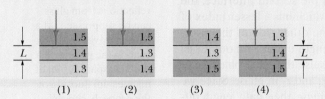

## Sample Problem 35-5

White light, with a uniform intensity across the visible wavelength range of 400 to 690 nm, is perpendicularly incident on a water film, of index of refraction $n_2 = 1.33$ and thickness $L = 320$ nm, that is suspended in air. At what wavelength $\lambda$ is the light reflected by the film brightest to an observer?

**KEY IDEA** The reflected light from the film is brightest at the wavelengths $\lambda$ for which the reflected rays are in phase with one another. The equation relating these wavelengths $\lambda$ to the given film thickness $L$ and film index of refraction $n_2$ is either Eq. 35-36 or Eq. 35-37, depending on the reflection phase shifts for this particular film.

*Calculations:* To determine which equation is needed, we should fill out an organizing table like Table 35-1. However, because there is air on both sides of the water film, the situation here is exactly like that in Fig. 35-17, and thus the table would be exactly like Table 35-1. Then from Table 35-1, we see that the reflected rays are

in phase (and thus the film is brightest) when

$$2L = \frac{\text{odd number}}{2} \times \frac{\lambda}{n_2},$$

which leads to Eq. 35-36:

$$2L = (m + \tfrac{1}{2})\frac{\lambda}{n_2}.$$

Solving for $\lambda$ and substituting for $L$ and $n_2$, we find

$$\lambda = \frac{2n_2 L}{m + \tfrac{1}{2}} = \frac{(2)(1.33)(320 \text{ nm})}{m + \tfrac{1}{2}} = \frac{851 \text{ nm}}{m + \tfrac{1}{2}}.$$

For $m = 0$, this gives us $\lambda = 1700$ nm, which is in the infrared region. For $m = 1$, we find $\lambda = 567$ nm, which is yellow-green light, near the middle of the visible spectrum. For $m = 2$, $\lambda = 340$ nm, which is in the ultraviolet region. Thus, the wavelength at which the light seen by the observer is brightest is

$$\lambda = 567 \text{ nm.} \qquad \text{(Answer)}$$

## Sample Problem 35-6

In Fig. 35-21, a glass lens is coated on one side with a thin film of magnesium fluoride ($MgF_2$) to reduce reflection from the lens surface. The index of refraction of $MgF_2$ is 1.38; that of the glass is 1.50. What is the least coating thickness that eliminates (via interference) the reflections at the middle of the visible spectrum ($\lambda = 550$ nm)? Assume that the light is approximately perpendicular to the lens surface.

**KEY IDEA** Reflection is eliminated if the film thickness $L$ is such that light waves reflected from the two film interfaces are exactly out of phase. The equation relating $L$ to the given wavelength $\lambda$ and the index of refraction $n_2$ of the thin film is either Eq. 35-36 or Eq. 35-37, depending on the reflection phase shifts at the interfaces.

*Calculations:* To determine which equation is needed, we fill out an organizing table like Table 35-1. At the first interface, the incident light is in air, which has a lesser index of refraction than the $MgF_2$ (the thin film). Thus, we fill in 0.5 wavelength under $r_1$ in our organizing table (meaning that the waves of ray $r_1$ are shifted by $0.5\lambda$ at the first interface). At the second interface, the incident light is in the $MgF_2$, which has a lesser index of refraction than the glass on the other side of the interface. Thus, we fill in 0.5 wavelength under $r_2$ in our table.

Because both reflections cause the same phase shift, they tend to put the waves of $r_1$ and $r_2$ in phase. Since we want those waves to be *out of phase*, their path length difference $2L$ must be an odd number of half-wavelengths:

$$2L = \frac{\text{odd number}}{2} \times \frac{\lambda}{n_2}.$$

This leads to Eq. 35-36. Solving that equation for $L$ then gives us the film thicknesses that will eliminate reflection from the lens and coating:

$$L = (m + \tfrac{1}{2})\frac{\lambda}{2n_2}, \qquad \text{for } m = 0, 1, 2, \dots . \quad (35\text{-}38)$$

We want the least thickness for the coating—that is, the smallest value of $L$. Thus, we choose $m = 0$, the smallest possible value of $m$. Substituting it and the given data in Eq. 35-38, we obtain

$$L = \frac{\lambda}{4n_2} = \frac{550 \text{ nm}}{(4)(1.38)} = 99.6 \text{ nm.} \quad \text{(Answer)}$$

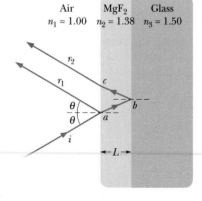

**FIG. 35-21** Unwanted reflections from glass can be suppressed (at a chosen wavelength) by coating the glass with a thin transparent film of magnesium fluoride of the properly chosen thickness.

Air
$n_1 = 1.00$

$MgF_2$
$n_2 = 1.38$

Glass
$n_3 = 1.50$

**Sample Problem** | **35-7** | **Build your skill**

Figure 35-22a shows a transparent plastic block with a thin wedge of air at the right. (The wedge thickness is exaggerated in the figure.) A broad beam of red light, with wavelength $\lambda = 632.8$ nm, is directed downward through the top of the block (at an incidence angle of 0°). Some of the light that passes into the plastic is reflected back up from the top and bottom surfaces of the wedge, which acts as a thin film (of air) with a thickness that varies uniformly and gradually from $L_L$ at the left-hand end to $L_R$ at the right-hand end. (The plastic layers above and below the wedge of air are too thick to act as thin films.) An observer looking down on the block sees an interference pattern consisting of six dark fringes and five bright red fringes along the wedge. What is the change in thickness $\Delta L$ ($= L_R - L_L$) along the wedge?

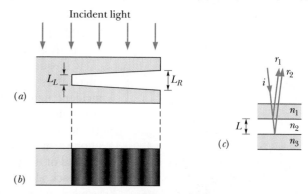

FIG. 35-22 (a) Red light is incident on a thin, air-filled wedge in the side of a transparent plastic block. The thickness of the wedge is $L_L$ at the left end and $L_R$ at the right end. (b) The view from above the block: an interference pattern of six dark fringes and five bright red fringes lies over the region of the wedge. (c) A representation of the incident ray $i$, reflected rays $r_1$ and $r_2$, and thickness $L$ of the wedge anywhere along the length of the wedge.

**KEY IDEAS** (1) The brightness at any point along the left–right length of the air wedge is due to the interference of the waves reflected at the top and bottom interfaces of the wedge. (2) The variation of brightness in the pattern of bright and dark fringes is due to the variation in the thickness of the wedge. In some regions, the thickness puts the reflected waves in phase and thus produces a bright reflection (a bright red fringe). In other regions, the thickness puts the reflected waves out of phase and thus produces no reflection (a dark fringe).

**Organizing the reflections:** Because the observer sees more dark fringes than bright fringes, we can assume that a dark fringe is produced at both the left and right ends of the wedge. Thus, the interference pattern is that shown in Fig. 35-22b.

We can represent the reflection of light at the top and bottom interfaces of the wedge, at any point along its length, with Fig. 35-22c, in which $L$ is the wedge thickness at that point. Let us apply this figure to the left end of the wedge, where the reflections give a dark fringe.

We know that, for a dark fringe, the waves of rays $r_1$ and $r_2$ in Fig. 35-22c must be out of phase. We also know that the equation relating the film thickness $L$ to the light's wavelength $\lambda$ and the film's index of refraction $n_2$ is either Eq. 35-36 or Eq. 35-37, depending on the reflection phase shifts. To determine which equation gives a dark fringe at the left end of the wedge, we should fill out an organizing table like Table 35-1.

At the top interface of the wedge, the incident light is in the plastic, which has a greater $n$ than the air beneath that interface. So, we fill in 0 under $r_1$ in our organizing table. At the bottom interface of the wedge, the incident light is in air, which has a lesser $n$ than the plastic beneath that interface. So we fill in 0.5 wavelength under $r_2$. Thus the reflections alone tend to put the waves of $r_1$ and $r_2$ out of phase.

**Reflections at left end:** Because the waves are out of phase at the left end of the air wedge, the path length difference $2L$ at that end of the wedge must be given by

$$2L = \text{integer} \times \frac{\lambda}{n_2},$$

which leads to Eq. 35-37:

$$2L = m\,\frac{\lambda}{n_2}, \qquad \text{for } m = 0, 1, 2, \dots . \quad (35\text{-}39)$$

**Reflections at right end:** Equation 35-39 holds not only for the left end of the wedge but also for any point along the wedge where a dark fringe is observed, including the right end, with a different integer value of $m$ for each fringe. The least value of $m$ is associated with the least thickness of the wedge where a dark fringe is observed. Progressively greater values of $m$ are associated with progressively greater thicknesses of the wedge where a dark fringe is observed. Let $m_L$ be the value at the left end. Then the value at the right end must be $m_L + 5$ because, from Fig. 35-22b, the right end is located at the fifth dark fringe from the left end.

**Thickness difference:** To find $\Delta L$, we first solve Eq. 35-39 twice—once for the thickness $L_L$ at the left end and once for the thickness $L_R$ at the right end:

$$L_L = (m_L)\,\frac{\lambda}{2n_2}, \qquad L_R = (m_L + 5)\,\frac{\lambda}{2n_2}. \quad (35\text{-}40)$$

We can now subtract $L_L$ from $L_R$ and substitute $n_2 = 1.00$ for the air within the wedge and $\lambda = 632.8 \times 10^{-9}$ m:

$$\Delta L = L_R - L_L = \frac{(m_L + 5)\lambda}{2n_2} - \frac{m_L\lambda}{2n_2} = \frac{5}{2}\,\frac{\lambda}{n_2}$$

$$= 1.58 \times 10^{-6}\ \text{m}. \qquad \text{(Answer)}$$

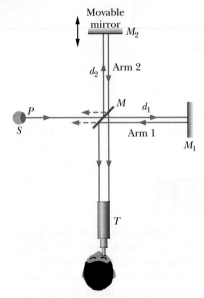

FIG. 35-23 Michelson's interferometer, showing the path of light originating at point $P$ of an extended source $S$. Mirror $M$ splits the light into two beams, which reflect from mirrors $M_1$ and $M_2$ back to $M$ and then to telescope $T$. In the telescope an observer sees a pattern of interference fringes.

## 35-8 | Michelson's Interferometer

An **interferometer** is a device that can be used to measure lengths or changes in length with great accuracy by means of interference fringes. We describe the form originally devised and built by A. A. Michelson in 1881.

Consider light that leaves point $P$ on extended source $S$ in Fig. 35-23 and encounters *beam splitter M*. A beam splitter is a mirror that transmits half the incident light and reflects the other half. In the figure we have assumed, for convenience, that this mirror possesses negligible thickness. At $M$ the light thus divides into two waves. One proceeds by transmission toward mirror $M_1$ at the end of one arm of the instrument; the other proceeds by reflection toward mirror $M_2$ at the end of the other arm. The waves are entirely reflected at these mirrors and are sent back along their directions of incidence, each wave eventually entering telescope $T$. What the observer sees is a pattern of curved or approximately straight interference fringes; in the latter case the fringes resemble the stripes on a zebra.

The path length difference for the two waves when they recombine at the telescope is $2d_2 - 2d_1$, and anything that changes this path length difference will cause a change in the phase difference between these two waves at the eye. As an example, if mirror $M_2$ is moved by a distance $\frac{1}{2}\lambda$, the path length difference is changed by $\lambda$ and the fringe pattern is shifted by one fringe (as if each dark stripe on a zebra had moved to where the adjacent dark stripe had been). Similarly, moving mirror $M_2$ by $\frac{1}{4}\lambda$ causes a shift by half a fringe (each dark zebra stripe shifts to where the adjacent white stripe was).

A shift in the fringe pattern can also be caused by the insertion of a thin transparent material into the optical path of one of the mirrors—say, $M_1$. If the material has thickness $L$ and index of refraction $n$, then the number of wavelengths along the light's to-and-fro path through the material is, from Eq. 35-9,

$$N_m = \frac{2L}{\lambda_n} = \frac{2Ln}{\lambda}. \quad (35\text{-}41)$$

The number of wavelengths in the same thickness $2L$ of air before the insertion of the material is

$$N_a = \frac{2L}{\lambda}. \quad (35\text{-}42)$$

When the material is inserted, the light returned from mirror $M_1$ undergoes a phase change (in terms of wavelengths) of

$$N_m - N_a = \frac{2Ln}{\lambda} - \frac{2L}{\lambda} = \frac{2L}{\lambda}(n - 1). \quad (35\text{-}43)$$

For each phase change of one wavelength, the fringe pattern is shifted by one fringe. Thus, by counting the number of fringes through which the material causes the pattern to shift, and substituting that number for $N_m - N_a$ in Eq. 35-43, you can determine the thickness $L$ of the material in terms of $\lambda$.

By such techniques the lengths of objects can be expressed in terms of the wavelengths of light. In Michelson's day, the standard of length—the meter—was chosen by international agreement to be the distance between two fine scratches on a certain metal bar preserved at Sèvres, near Paris. Michelson was able to show, using his interferometer, that the standard meter was equivalent to 1 553 163.5 wavelengths of a certain monochromatic red light emitted from a light source containing cadmium. For this careful measurement, Michelson received the 1907 Nobel Prize in physics. His work laid the foundation for the eventual abandonment (in 1961) of the meter bar as a standard of length and for the redefinition of the meter in terms of the wavelength of light. By 1983, even this wavelength standard was not precise enough to meet the growing requirements of science and technology, and it was replaced with a new standard based on a defined value for the speed of light.

# REVIEW & SUMMARY

**Huygens' Principle** The three-dimensional transmission of waves, including light, may often be predicted by *Huygens' principle,* which states that all points on a wavefront serve as point sources of spherical secondary wavelets. After a time $t$, the new position of the wavefront will be that of a surface tangent to these secondary wavelets.

The law of refraction can be derived from Huygens' principle by assuming that the index of refraction of any medium is $n = c/v$, in which $v$ is the speed of light in the medium and $c$ is the speed of light in vacuum.

**Wavelength and Index of Refraction** The wavelength $\lambda_n$ of light in a medium depends on the index of refraction $n$ of the medium:

$$\lambda_n = \frac{\lambda}{n}, \tag{35-8}$$

in which $\lambda$ is the wavelength of the light in vacuum. Because of this dependency, the phase difference between two waves can change if they pass through different materials with different indexes of refraction.

**Young's Experiment** In **Young's interference experiment,** light passing through a single slit falls on two slits in a screen. The light leaving these slits flares out (by diffraction), and interference occurs in the region beyond the screen. A fringe pattern, due to the interference, forms on a viewing screen.

The light intensity at any point on the viewing screen depends in part on the difference in the path lengths from the slits to that point. If this difference is an integer number of wavelengths, the waves interfere constructively and an intensity maximum results. If it is an odd number of half-wavelengths, there is destructive interference and an intensity minimum occurs. The conditions for maximum and minimum intensity are

$$d \sin \theta = m\lambda, \quad \text{for } m = 0, 1, 2, \ldots$$
$$\text{(maxima—bright fringes),} \tag{35-14}$$

$$d \sin \theta = (m + \tfrac{1}{2})\lambda, \quad \text{for } m = 0, 1, 2, \ldots$$
$$\text{(minima—dark fringes),} \tag{35-16}$$

where $\theta$ is the angle the light path makes with a central axis and $d$ is the slit separation.

**Coherence** If two light waves that meet at a point are to interfere perceptibly, the phase difference between them must remain constant with time; that is, the waves must be **coherent.** When two coherent waves meet, the resulting intensity may be found by using phasors.

**Intensity in Two-Slit Interference** In Young's interference experiment, two waves, each with intensity $I_0$, yield a resultant wave of intensity $I$ at the viewing screen, with

$$I = 4I_0 \cos^2 \tfrac{1}{2}\phi, \quad \text{where } \phi = \frac{2\pi d}{\lambda} \sin \theta. \tag{35-22, 35-23}$$

Equations 35-14 and 35-16, which identify the positions of the fringe maxima and minima, are contained within this relation.

**Thin-Film Interference** When light is incident on a thin transparent film, the light waves reflected from the front and back surfaces interfere. For near-normal incidence the wavelength conditions for maximum and minimum intensity of the light reflected from a *film in air* are

$$2L = (m + \tfrac{1}{2}) \frac{\lambda}{n_2}, \quad \text{for } m = 0, 1, 2, \ldots$$
$$\text{(maxima—bright film in air),} \tag{35-36}$$

$$2L = m \frac{\lambda}{n_2}, \quad \text{for } m = 0, 1, 2, \ldots$$
$$\text{(minima—dark film in air),} \tag{35-37}$$

where $n_2$ is the index of refraction of the film, $L$ is its thickness, and $\lambda$ is the wavelength of the light in air.

If the light incident at an interface between media with different indexes of refraction is in the medium with the smaller index of refraction, the reflection causes a phase change of $\pi$ rad, or half a wavelength, in the reflected wave. Otherwise, there is no phase change due to the reflection. Refraction at an interface does not cause a phase shift.

**The Michelson Interferometer** In *Michelson's interferometer* a light wave is split into two beams that, after traversing paths of different lengths, are recombined so they interfere and form a fringe pattern. Varying the path length of one of the beams allows distances to be accurately expressed in terms of wavelengths of light, by counting the number of fringes through which the pattern shifts because of the change.

# QUESTIONS

**1** Figure 35-24 shows two light rays that are initially exactly in phase and that reflect from several glass surfaces. Neglect the slight slant in the path of the light in the second arrangement. (a) What is the path length difference of the rays? In wavelengths $\lambda$, (b) what should that path length difference equal if the rays are to be exactly out of phase when they emerge, and (c) what is the smallest value of $d$ that will allow that final phase difference?

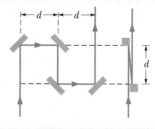

**FIG. 35-24** Question 1.

**2** In Fig. 35-25, three pulses of light—$a$, $b$, and $c$—of the same wavelength are sent through layers of plastic having the given indexes of refraction and along the paths indicated. Rank the pulses according to their travel time through the plastic layers, greatest first.

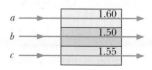

**FIG. 35-25** Question 2.

**3** Light travels along the length of a 1500-nm-long nano-structure. When a peak of the wave is at one end of the nanostructure, is there a peak or a valley at the other end if the wavelength is (a) 500 nm and (b) 1000 nm?

**4** (a) If you move from one bright fringe in a two-slit interference pattern to the next one farther out, (b) does the path length difference $\Delta L$ increase or decrease and (c) by how much does it change, in wavelengths $\lambda$?

**5** Does the spacing between fringes in a two-slit interference pattern increase, decrease, or stay the same if (a) the slit separation is increased, (b) the color of the light is switched from red to blue, and (c) the whole apparatus is submerged in cooking sherry? (d) If the slits are illuminated with white light, then at any side maximum, does the blue component or the red component peak closer to the central maximum?

**6** Figure 35-26 shows two rays of light, of wavelength 600 nm, that reflect from glass surfaces separated by 150 nm. The rays are initially in phase. (a) What is the path length difference of the rays? (b) When they have cleared the reflection region, are the rays exactly in phase, exactly out of phase, or in some intermediate state?

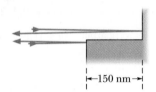

FIG. 35-26   Question 6.

**7** Is there an interference maximum, a minimum, an intermediate state closer to a maximum, or an intermediate state closer to a minimum at point $P$ in Fig. 35-10 if the path length difference of the two rays is (a) $2.2\lambda$, (b) $3.5\lambda$, (c) $1.8\lambda$, and (d) $1.0\lambda$? For each situation, give the value of $m$ associated with the maximum or minimum involved.

**8** Figure 35-27$a$ gives intensity $I$ versus position $x$ on the viewing screen for the central portion of a two-slit interference pattern. The other parts of the figure give phasor diagrams for the electric field components of the waves arriving at the screen from the two slits (as in Fig. 35-13$a$). Which numbered points on the screen best correspond to which phasor diagram?

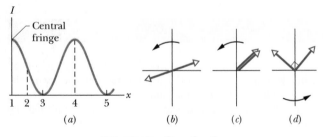

FIG. 35-27   Question 8.

**9** Figure 35-28 shows two sources $S_1$ and $S_2$ that emit radio waves of wavelength $\lambda$ in all directions. The sources are exactly in phase and are separated by a distance equal to $1.5\lambda$.

The vertical broken line is the perpendicular bisector of the distance between the sources. (a) If we start at the indicated start point and travel along path 1, does the interference produce a maximum all along the path, a minimum all along the path, or alternating maxima and minima? Repeat for (b) path 2 and (c) path 3.

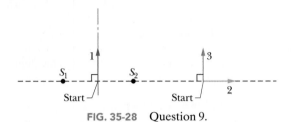

FIG. 35-28   Question 9.

**10** Figure 35-29 shows the transmission of light through a thin film in air by a perpendicular beam (tilted in the figure for clarity). (a) Did ray $r_3$ undergo a phase shift due to reflection? (b) In wavelengths, what is the reflection phase shift for ray $r_4$? (c) If the film thickness is $L$, what is the path length difference between rays $r_3$ and $r_4$?

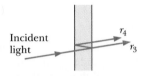

FIG. 35-29   Question 10.

**11** Figure 35-30 shows four situations in which light reflects perpendicularly from a thin film of thickness $L$ sandwiched between much thicker materials. The indexes of refraction are given. In which situations does Eq. 35-36 correspond to the reflections yielding maxima (that is, a bright film)?

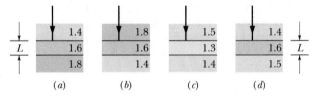

FIG. 35-30   Question 11.

**12** Figure 35-31$a$ shows the cross section of a vertical thin film whose width increases downward because gravitation causes slumping. Figure 35-31$b$ is a face-on view of the film, showing four bright (red) interference fringes that result when the film is illuminated with a perpendicular beam of red light. Points in the cross section corresponding to the bright fringes are labeled. In terms of the wavelength of the light inside the film, what is the difference in film thickness between (a) points $a$ and $b$ and (b) points $b$ and $d$?

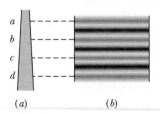

FIG. 35-31   Question 12.

# PROBLEMS

**GO** Tutoring problem available (at instructor's discretion) in *WileyPLUS* and WebAssign

**SSM** Worked-out solution available in Student Solutions Manual     **WWW** Worked-out solution is at

• – ••• Number of dots indicates level of problem difficulty     **ILW** Interactive solution is at ⸻ http://www.wiley.com/college/halliday

Additional information available in *The Flying Circus of Physics* and at flyingcircusofphysics.com

## sec. 35-2   Light as a Wave

**•1** The speed of yellow light (from a sodium lamp) in a certain liquid is measured to be $1.92 \times 10^8$ m/s. What is the index of refraction of this liquid for the light?

**•2** In Fig. 35-32a, a beam of light in material 1 is incident on a boundary at an angle of 30°. The extent to which the light is bent due to refraction depends, in part, on the index of refraction $n_2$ of material 2. Figure 35-32b gives the angle of refraction $\theta_2$ versus $n_2$ for a range of possible $n_2$ values, from $n_a = 1.30$ to $n_b = 1.90$. What is the speed of light in material 1?

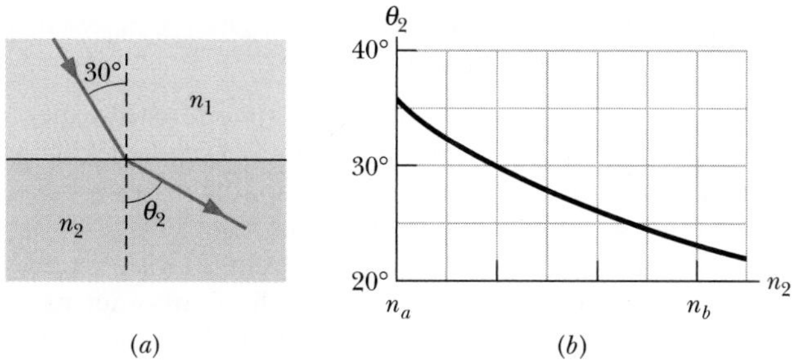

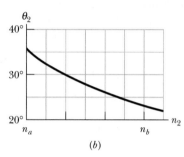

(a)             (b)

**FIG. 35-32**   Problem 2.

**•3** How much faster, in meters per second, does light travel in sapphire than in diamond? See Table 33-1.

**•4** The wavelength of yellow sodium light in air is 589 nm. (a) What is its frequency? (b) What is its wavelength in glass whose index of refraction is 1.52? (c) From the results of (a) and (b), find its speed in this glass.

**•5** In Fig. 35-4, assume that two waves of light in air, of wavelength 400 nm, are initially in phase. One travels through a glass layer of index of refraction $n_1 = 1.60$ and thickness $L$. The other travels through an equally thick plastic layer of index of refraction $n_2 = 1.50$. (a) What is the smallest value $L$ should have if the waves are to end up with a phase difference of 5.65 rad? (b) If the waves arrive at some common point with the same amplitude, is their interference fully constructive, fully destructive, intermediate but closer to fully constructive, or intermediate but closer to fully destructive? **SSM**

**•6** In Fig. 35-33, a light wave along ray $r_1$ reflects once from a mirror and a light wave along ray $r_2$ reflects twice from that same mirror and once from a tiny mirror at distance $L$ from the bigger mirror. (Neglect the slight tilt of the rays.) The waves have wavelength $\lambda$ and are initially exactly out of phase. What are the (a) smallest, (b) second smallest, and (c) third smallest values of $L/\lambda$ that result in the final waves being exactly in phase?

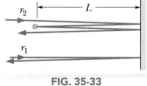

**FIG. 35-33**
Problems 6 and 7.

**•7** In Fig. 35-33, a light wave along ray $r_1$ reflects once from a mirror and a light wave along ray $r_2$ reflects twice from that same mirror and once from a tiny mirror at distance $L$ from the bigger mirror. (Neglect the slight tilt of the rays.) The waves have wavelength 620 nm and are initially in phase. (a) What is the smallest value of $L$ that puts the final light waves exactly out of phase? (b) With the tiny mirror initially at that value of $L$, how far must it be moved away from the bigger mirror to again put the final waves out of phase?

**•8** In Fig. 35-34, two light pulses are sent through layers of plastic with thicknesses of either $L$ or $2L$ as shown and indexes of refraction $n_1 = 1.55$, $n_2 = 1.70$, $n_3 = 1.60$, $n_4 = 1.45$, $n_5 = 1.59$, $n_6 = 1.65$, and $n_7 = 1.50$. (a) Which pulse travels through the plastic in less time? (b) What multiple of $L/c$ gives the difference in the traversal times of the pulses?

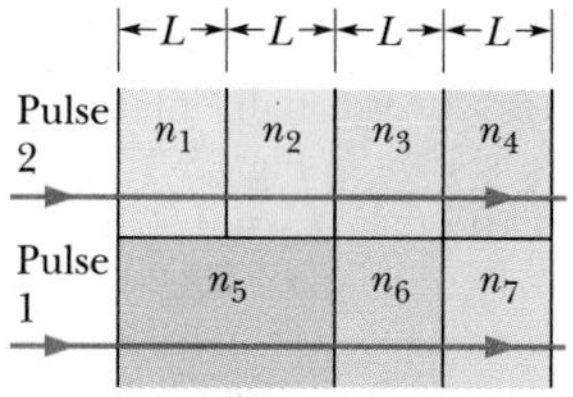

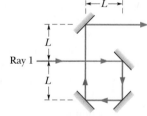

**FIG. 35-34**   Problem 8.

**••9** Suppose that the two waves in Fig. 35-4 have wavelength $\lambda = 500$ nm in air. What multiple of $\lambda$ gives their phase difference when they emerge if (a) $n_1 = 1.50$, $n_2 = 1.60$, and $L = 8.50$ $\mu$m; (b) $n_1 = 1.62$, $n_2 = 1.72$, and $L = 8.50$ $\mu$m; and (c) $n_1 = 1.59$, $n_2 = 1.79$, and $L = 3.25$ $\mu$m? (d) Suppose that in each of these three situations the waves arrive at a common point (with the same amplitude) after emerging. Rank the situations according to the brightness the waves produce at the common point.

**••10** In Fig. 35-35, two light rays go through different paths by reflecting from the various flat surfaces shown. The light waves have a wavelength of 420.0 nm and are initially in phase. What are the (a) smallest and (b) second smallest value of distance $L$ that will put the waves exactly out of phase as they emerge from the region?

**••11** In Fig. 35-4, assume that the two light waves, of wavelength 620 nm in air, are initially out of phase by $\pi$ rad. The indexes of refraction of the media are $n_1 = 1.45$ and $n_2 = 1.65$. What are the (a) smallest and (b) second smallest value of $L$ that will put the waves exactly in phase once they pass through the two media?

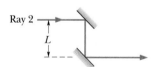

**FIG. 35-35**   Problems 10 and 94.

**••12** In Fig. 35-36, a light ray is incident at angle $\theta_1 = 50°$ on a series of five transparent layers with parallel boundaries. For layers 1 and 3, $L_1 = 20$ $\mu$m, $L_3 = 25$ $\mu$m, $n_1 = 1.6$, and $n_3 = 1.45$. (a) At what angle does the light emerge back into air at the right? (b) How much time does the light take to travel through layer 3?

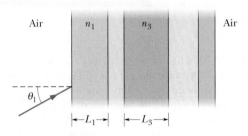

**FIG. 35-36** Problem 12.

**••13** Two waves of light in air, of wavelength $\lambda = 600.0$ nm, are initially in phase. They then travel through plastic layers as shown in Fig. 35-37, with $L_1 = 4.00$ $\mu$m, $L_2 = 3.50$ $\mu$m, $n_1 = 1.40$, and $n_2 = 1.60$. (a) What multiple of $\lambda$ gives their phase difference after they both have emerged from the layers? (b) If the waves later arrive at some common point with the same amplitude, is their interference fully constructive, fully destructive, intermediate but closer to fully constructive, or intermediate but closer to fully destructive? **ILW**

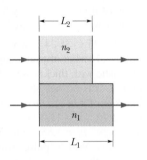

**FIG. 35-37** Problem 13.

### sec. 35-4 **Young's Interference Experiment**

**•14** Monochromatic green light, of wavelength 550 nm, illuminates two parallel narrow slits 7.70 $\mu$m apart. Calculate the angular deviation ($\theta$ in Fig. 35-10) of the third-order (for $m = 3$) bright fringe (a) in radians and (b) in degrees.

**•15** In Fig. 35-38, two radio-frequency point sources $S_1$ and $S_2$, separated by distance $d = 2.0$ m, are radiating in phase with $\lambda = 0.50$ m. A detector moves in a large circular path around the two sources in a plane containing them. How many maxima does it detect? **SSM**

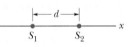

**FIG. 35-38** Problems 15 and 24.

**•16** In a double-slit arrangement the slits are separated by a distance equal to 100 times the wavelength of the light passing through the slits. (a) What is the angular separation in radians between the central maximum and an adjacent maximum? (b) What is the distance between these maxima on a screen 50.0 cm from the slits?

**•17** A double-slit arrangement produces interference fringes for sodium light ($\lambda = 589$ nm) that have an angular separation of $3.50 \times 10^{-3}$ rad. For what wavelength would the angular separation be 10.0% greater? **SSM**

**•18** A double-slit arrangement produces interference fringes for sodium light ($\lambda = 589$ nm) that are 0.20° apart. What is the angular fringe separation if the entire arrangement is immersed in water ($n = 1.33$)?

**•19** Suppose that Young's experiment is performed with blue-green light of wavelength 500 nm. The slits are 1.20 mm apart, and the viewing screen is 5.40 m from the slits. How far apart are the bright fringes near the center of the interference pattern? **SSM ILW**

**•20** In the two-slit experiment of Fig. 35-10, let angle $\theta$ be 20.0°, the slit separation be 4.24 $\mu$m, and the wavelength be $\lambda = 500$ nm. (a) What multiple of $\lambda$ gives the phase difference between the waves of rays $r_1$ and $r_2$ when they arrive at point $P$ on the distant screen? (b) What is the phase difference in radians? (c) Determine where in the interference pattern point $P$ lies by giving the maximum or minimum on which it lies, or the maximum and minimum between which it lies.

**••21** In Fig. 35-39, sources $A$ and $B$ emit long-range radio waves of wavelength 400 m, with the phase of the emission from $A$ ahead of that from source $B$ by 90°. The distance $r_A$ from $A$ to detector $D$ is

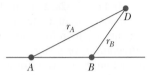

**FIG. 35-39** Problem 21.

greater than the corresponding distance $r_B$ by 100 m. What is the phase difference of the waves at $D$?

**••22** Sunlight is used in a double-slit interference experiment. The fourth-order maximum for a wavelength of 450 nm occurs at an angle of $\theta = 90°$. Thus, it is on the verge of being eliminated from the pattern because $\theta$ cannot exceed 90° in Eq. 35-14. (a) What range of wavelengths in the visible range (400 nm to 700 nm) are not present in the third-order maxima? To eliminate all of the visible light in the fourth-order maximum, (b) should the slit separation be increased or decreased and (c) what least change in separation is needed?

**••23** In Fig. 35-40, two isotropic point sources of light ($S_1$ and $S_2$) are separated by distance 2.70 $\mu$m along a $y$ axis and emit in phase at wavelength 900 nm and at the same amplitude. A light detector is located at point $P$ at coordinate $x_P$ on the $x$ axis. What is the greatest value of $x_P$ at which the detected light is minimum due to destructive interference? **GO**

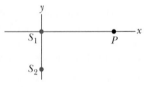

**FIG. 35-40** Problems 23, 28, and 122.

**••24** In Fig. 35-38, two isotropic point sources $S_1$ and $S_2$ emit identical light waves in phase at wavelength $\lambda$. The sources lie at separation $d$ on an $x$ axis, and a light detector is moved in a circle of large radius around the midpoint between them. It detects 30 points of zero intensity, including two on the $x$ axis, one of them to the left of the sources and the other to the right of the sources. What is the value of $d/\lambda$?

**••25** In a double-slit experiment, the distance between slits is 5.0 mm and the slits are 1.0 m from the screen. Two interference patterns can be seen on the screen: one due to light of wavelength 480 nm, and the other due to light of wavelength 600 nm. What is the separation on the screen between the third-order ($m = 3$) bright fringes of the two interference patterns?

**••26** In Fig. 35-41, two isotropic point sources $S_1$ and $S_2$ emit light in phase at wavelength $\lambda$ and at the same amplitude. The sources are separated by distance $2d = 6.00\lambda$. They lie on an axis that is parallel to an x axis, which runs along a viewing screen at distance $D = 20.0\lambda$. The origin lies on the perpendicular bisector between the sources. The figure shows two rays reaching point P on the screen, at position $x_P$. (a) At what value of $x_P$ do the rays have the minimum possible phase difference? (b) What multiple of $\lambda$ gives that minimum

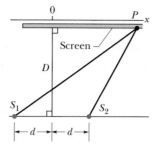

FIG. 35-41    Problem 26.

phase difference? (c) At what value of $x_P$ do the rays have the maximum possible phase difference? What multiple of $\lambda$ gives (d) that maximum phase difference and (e) the phase difference when $x_P = 6.00\lambda$? (f) When $x_P = 6.00\lambda$, is the resulting intensity at point P maximum, minimum, intermediate but closer to maximum, or intermediate but closer to minimum?

•••27    A thin flake of mica ($n = 1.58$) is used to cover one slit of a double-slit interference arrangement. The central point on the viewing screen is now occupied by what had been the seventh bright side fringe ($m = 7$). If $\lambda = 550$ nm, what is the thickness of the mica?

•••28    Figure 35-40 shows two isotropic point sources of light ($S_1$ and $S_2$) that emit in phase at wavelength 400 nm and at the same amplitude. A detection point P is shown on an x axis that extends through source $S_1$. The phase difference $\phi$ between the light arriving at point P from the two sources is to be measured as P is moved along the x axis from $x = 0$ out to $x = +\infty$. The results out to $x_s = 10 \times 10^{-7}$ m are given in Fig. 35-42. On the way out to $+\infty$, what is the greatest value of x at which the light arriving at P from $S_1$ is exactly out of phase with the light arriving at P from $S_2$?

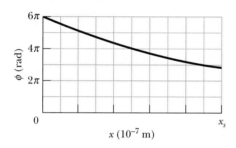

FIG. 35-42    Problem 28.

### sec. 35-6    Intensity in Double-Slit Interference

•29    Two waves of the same frequency have amplitudes 1.00 and 2.00. They interfere at a point where their phase difference is 60.0°. What is the resultant amplitude?    SSM

•30    Find the sum y of the following quantities:

$$y_1 = 10 \sin \omega t \quad \text{and} \quad y_2 = 8.0 \sin(\omega t + 30°).$$

••31    Three electromagnetic waves travel through a certain point P along an x axis. They are polarized parallel to a y axis, with the following variations in their amplitudes. Find their resultant at P.    GO

$$E_1 = (10.0 \ \mu\text{V/m}) \sin[(2.0 \times 10^{14} \ \text{rad/s})t]$$
$$E_2 = (5.00 \ \mu\text{V/m}) \sin[(2.0 \times 10^{14} \ \text{rad/s})t + 45.0°]$$
$$E_3 = (5.00 \ \mu\text{V/m}) \sin[(2.0 \times 10^{14} \ \text{rad/s})t - 45.0°]$$

••32    In the double-slit experiment of Fig. 35-10, the electric fields of the waves arriving at point P are given by

$$E_1 = (2.00 \ \mu\text{V/m}) \sin[(1.26 \times 10^{15})t]$$
$$E_2 = (2.00 \ \mu\text{V/m}) \sin[(1.26 \times 10^{15})t + 39.6 \ \text{rad}],$$

where time t is in seconds. (a) What is the amplitude of the resultant electric field at point P? (b) What is the ratio of the intensity $I_P$ at point P to the intensity $I_{cen}$ at the center of the interference pattern? (c) Describe where point P is in the interference pattern by giving the maximum or minimum on which it lies, or the maximum and minimum between which it lies. In a phasor diagram of the electric fields, (d) at what rate would the phasors rotate around the origin and (e) what is the angle between the phasors?

••33    Add the quantities $y_1 = 10 \sin \omega t$, $y_2 = 15 \sin(\omega t + 30°)$, and $y_3 = 5.0 \sin(\omega t - 45°)$ using the phasor method.    ILW

••34    In the double-slit experiment of Fig. 35-10, the viewing screen is at distance $D = 4.00$ m, point P lies at distance $y = 20.5$ cm from the center of the pattern, the slit separation d is 4.50 $\mu$m, and the wavelength $\lambda$ is 580 nm. (a) Determine where point P is in the interference pattern by giving the maximum or minimum on which it lies, or the maximum and minimum between which it lies. (b) What is the ratio of the intensity $I_P$ at point P to the intensity $I_{cen}$ at the center of the pattern?

### sec. 35-7    Interference from Thin Films

•35    The rhinestones in costume jewelry are glass with index of refraction 1.50. To make them more reflective, they are often coated with a layer of silicon monoxide of index of refraction 2.00. What is the minimum coating thickness needed to ensure that light of wavelength 560 nm and of perpendicular incidence will be reflected from the two surfaces of the coating with fully constructive interference?

•36    White light is sent downward onto a horizontal thin film that is sandwiched between two materials. The indexes of refraction are 1.80 for the top material, 1.70 for the thin film, and 1.50 for the bottom material. The film thickness is $5.00 \times 10^{-7}$ m. Of the visible wavelengths (400 to 700 nm) that result in fully constructive interference at an observer above the film, which is the (a) longer and (b) shorter wavelength? The materials and film are then heated so that the film thickness increases. (c) Does the light resulting in fully constructive interference shift toward longer or shorter wavelengths?

•37    Light of wavelength 624 nm is incident perpendicularly on a soap film ($n = 1.33$) suspended in air. What are the (a) least and (b) second least thicknesses of the film for which the reflections from the film undergo fully constructive interference?    ILW

•38    A 600-nm-thick soap film ($n = 1.40$) in air is illuminated with white light in a direction perpendicular to the film. For how many different wavelengths in the 300 to 700 nm range is there (a) fully constructive interference and (b) fully destructive interference in the reflected light?

•39    We wish to coat flat glass ($n = 1.50$) with a transparent material ($n = 1.25$) so that reflection of light at wavelength 600 nm is eliminated by interference. What minimum thickness can the coating have to do this?    SSM

**••40** A thin film of acetone ($n = 1.25$) coats a thick glass plate ($n = 1.50$). White light is incident normal to the film. In the reflections, fully destructive interference occurs at 600 nm and fully constructive interference at 700 nm. Calculate the thickness of the acetone film.

**••41 through 52** *Reflection by thin layers.* In Fig. 35-43, light is incident perpendicularly on a thin layer of material 2 that lies between (thicker) materials 1 and 3. (The rays are tilted only for clarity.) The waves of rays $r_1$ and $r_2$ interfere, and here we consider the type of interference to be either maximum (max) or minimum (min). For this situation, each problem in Table 35-2 refers to the indexes of refraction $n_1$, $n_2$, and $n_3$, the type of interference, the thin-layer thickness $L$ in nanometers, and the wavelength $\lambda$ in nanometers of the light as measured in air. Where $\lambda$ is missing, give the wavelength that is in the visible range. Where $L$ is missing, give the second least thickness or the third least thickness as indicated. **SSM** 41, 47 **GO** 43, 51

**••53** A disabled tanker leaks kerosene ($n = 1.20$) into the Persian Gulf, creating a large slick on top of the water ($n = 1.30$). (a) If you are looking straight down from an airplane, while the Sun is overhead, at a region of the slick where its thickness is 460 nm, for which wavelength(s) of visible light is the reflection brightest because of constructive interference? (b) If you are scuba diving directly under this same region of the slick, for which wavelength(s) of visible light is the transmitted intensity strongest?

SSM WWW

**••54** A thin film, with a thickness of 272.7 nm and with air on both sides, is illuminated with a beam of white light. The beam is perpendicular to the film and consists of the full range of wavelengths for the visible spectrum. In the light reflected by the film, light with a wavelength of 600.0 nm undergoes fully constructive interference. At what wavelength does the reflected light undergo fully destructive interference? (*Hint:* You must make a reasonable assumption about the index of refraction.)

**••55** The reflection of perpendicularly incident white light by a soap film in air has an interference maximum at 600 nm and a minimum at 450 nm, with no minimum in be-

**TABLE 35-2**

**Problems 41 through 52: Reflection by Thin Layers. See the setup for these problems.**

| | $n_1$ | $n_2$ | $n_3$ | Type | $L$ | $\lambda$ |
|---|---|---|---|---|---|---|
| **41** | 1.50 | 1.34 | 1.42 | min | 380 | |
| **42** | 1.32 | 1.75 | 1.39 | max | 325 | |
| **43** | 1.55 | 1.60 | 1.33 | max | 3rd | 612 |
| **44** | 1.55 | 1.60 | 1.33 | max | 285 | |
| **45** | 1.60 | 1.40 | 1.80 | min | 200 | |
| **46** | 1.40 | 1.46 | 1.75 | min | 2nd | 482 |
| **47** | 1.40 | 1.46 | 1.75 | min | 210 | |
| **48** | 1.50 | 1.34 | 1.42 | max | 2nd | 587 |
| **49** | 1.68 | 1.59 | 1.50 | min | 2nd | 342 |
| **50** | 1.68 | 1.59 | 1.50 | min | 415 | |
| **51** | 1.32 | 1.75 | 1.39 | max | 3rd | 382 |
| **52** | 1.60 | 1.40 | 1.80 | max | 2nd | 632 |

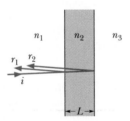

FIG. 35-43 Problems 41 through 52.

tween. If $n = 1.33$ for the film, what is the film thickness, assumed uniform?

**••56** A plane wave of monochromatic light is incident normally on a uniform thin film of oil that covers a glass plate. The wavelength of the source can be varied continuously. Fully destructive interference of the reflected light is observed for wavelengths of 500 and 700 nm and for no wavelengths in between. If the index of refraction of the oil is 1.30 and that of the glass is 1.50, find the thickness of the oil film.

**••57 through 68** *Transmission through thin layers.* In Fig. 35-44, light is incident perpendicularly on a thin layer of material 2 that lies between (thicker) materials 1 and 3. (The rays are tilted only for clarity.) Part of the light ends up in material 3 as ray $r_3$

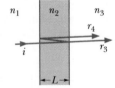

FIG. 35-44 Problems 57 through 68.

**TABLE 35-3**

**Problems 57 through 68: Transmission Through Thin Layers. See the setup for these problems.**

| | $n_1$ | $n_2$ | $n_3$ | Type | $L$ | $\lambda$ |
|---|---|---|---|---|---|---|
| **57** | 1.50 | 1.34 | 1.42 | min | 2nd | 587 |
| **58** | 1.68 | 1.59 | 1.50 | max | 2nd | 342 |
| **59** | 1.40 | 1.46 | 1.75 | max | 2nd | 482 |
| **60** | 1.32 | 1.75 | 1.39 | min | 3rd | 382 |
| **61** | 1.60 | 1.40 | 1.80 | min | 2nd | 632 |
| **62** | 1.55 | 1.60 | 1.33 | min | 3rd | 612 |
| **63** | 1.68 | 1.59 | 1.50 | max | 415 | |
| **64** | 1.50 | 1.34 | 1.42 | max | 380 | |
| **65** | 1.32 | 1.75 | 1.39 | min | 325 | |
| **66** | 1.40 | 1.46 | 1.75 | max | 210 | |
| **67** | 1.55 | 1.60 | 1.33 | min | 285 | |
| **68** | 1.60 | 1.40 | 1.80 | max | 200 | |

(the light does not reflect inside material 2) and $r_4$ (the light reflects twice inside material 2). The waves of $r_3$ and $r_4$ interfere, and here we consider the type of interference to be either maximum (max) or minimum (min). For this situation, each problem in Table 35-3 refers to the indexes of refraction $n_1$, $n_2$, and $n_3$, the type of interference, the thin-layer thickness $L$ in nanometers, and the wavelength $\lambda$ in nanometers of the light as measured in air. Where $\lambda$ is missing, give the wavelength that is in the visible range. Where $L$ is missing, give the second least thickness or the third least thickness as indicated. **SSM** 63 **GO** 61

**••69** In Fig. 35-45, a broad beam of light of wavelength 630 nm is incident at 90° on a thin, wedge-shaped film with index of refraction 1.50. An observer intercepting the light transmitted by the film sees 10 bright and 9 dark fringes along the length of the film. By how much does the film thickness change over this length? **GO**

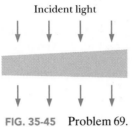

Incident light

**FIG. 35-45**    Problem 69.

**••70** Two rectangular glass plates ($n = 1.60$) are in contact along one edge and are separated along the opposite edge (Fig. 35-46). Light with a wavelength of 600 nm is incident perpendicularly onto the top plate. The air between the plates acts as a thin film. Nine dark fringes and eight bright fringes are observed from above the top plate. If the distance between the two plates along the separated edges is increased by 600 nm, how many dark fringes will there then be across the top plate?

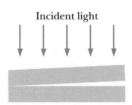

Incident light

**FIG. 35-46**    Problems 70, 71, 72–74, 106, 107, and 113.

**••71** In Fig. 35-46, a broad beam of light of wavelength 683 nm is sent directly downward through the top plate of a pair of glass plates. The plates are 120 mm long, touch at the left end, and are separated by 48.0 $\mu$m at the right end. The air between the plates acts as a thin film. How many bright fringes will be seen by an observer looking down through the top plate? **SSM**

**••72** In Fig. 35-46, a broad beam of light of wavelength 620 nm is sent directly downward through the top plate of a pair of glass plates touching at the left end. The air between the plates acts as a thin film, and an interference pattern can be seen from above the plates. Initially, a dark fringe lies at the left end, a bright fringe lies at the right end, and nine dark fringes lie between those two end fringes. The plates are then very gradually squeezed together at a constant rate to decrease the angle between them. As a result, the fringe at the right side changes between being bright to being dark every 15.0 s. (a) At what rate is the spacing between the plates at the right end being changed? (b) By how much has the spacing there changed when both left and right ends have a dark fringe and there are five dark fringes between them?

**••73** In Fig. 35-46, two microscope slides touch at one end and are separated at the other end. When light of wavelength 500 nm shines vertically down on the slides, an overhead observer sees an interference pattern on the slides with the dark fringes separated by 1.2 mm. What is the angle between the slides?

**••74** In Fig. 35-46, a broad beam of monochromatic light is directed perpendicularly through two glass plates that are held together at one end to create a wedge of air between them. An observer intercepting light reflected from the wedge of air, which acts as a thin film, sees 4001 dark fringes along the length of the wedge. When the air between the plates is evacuated, only 4000 dark fringes are seen. Calculate to six significant figures the index of refraction of air from these data.

**••75** Figure 35-47a shows a lens with radius of curvature $R$ lying on a flat glass plate and illuminated from above by light with wavelength $\lambda$. Figure 35-47b (a photograph taken from above the lens) shows that circular interference fringes (called *Newton's rings*) appear, associated with the variable thickness $d$ of the air film between the lens and the plate. Find the radii $r$ of the interference maxima assuming $r/R \ll 1$. **SSM ILW**

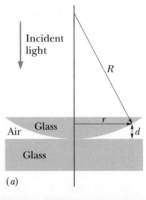

(a)

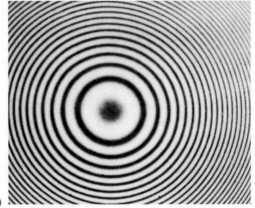

(b)

**FIG. 35-47**    Problems 75 and 77. *(Courtesy Bausch & Lomb)*

**••76** In a Newton's rings experiment (see Problem 75), the radius of curvature $R$ of the lens is 5.0 m and the lens diameter is 20 mm. (a) How many bright rings are produced? Assume that $\lambda = 589$ nm. (b) How many bright rings would be produced if the arrangement were immersed in water ($n = 1.33$)?

**••77** A Newton's rings apparatus is to be used to determine the radius of curvature of a lens (see Fig. 35-47 and Problem 75). The radii of the $n$th and $(n + 20)$th bright rings are measured and found to be 0.162 and 0.368 cm, respectively, in light of wavelength 546 nm. Calculate the radius of curvature of the lower surface of the lens.

**•••78** A thin film of liquid is held in a horizontal circular ring, with air on both sides of the film. A beam of light at wavelength 550 nm is directed perpendicularly onto the film, and the intensity $I$ of its reflection is monitored. Figure 35-48 gives intensity $I$ as a function of time $t$; the horizontal scale is set by $t_s = 20.0$ s. The intensity changes because of evaporation from the two sides of the film. Assume that the film is flat and has parallel sides, a radius of 1.80 cm, and an index of refraction of 1.40. Also assume that the film's volume decreases at a constant rate. Find that rate.

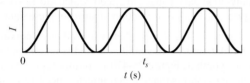

**FIG. 35-48** Problem 78.

## sec. 35-8 Michelson's Interferometer

**•79** If mirror $M_2$ in a Michelson interferometer (Fig. 35-23) is moved through 0.233 mm, a shift of 792 bright fringes occurs. What is the wavelength of the light producing the fringe pattern?

**•80** A thin film with index of refraction $n = 1.40$ is placed in one arm of a Michelson interferometer, perpendicular to the optical path. If this causes a shift of 7.0 bright fringes of the pattern produced by light of wavelength 589 nm, what is the film thickness?

**••81** In Fig. 35-49, an airtight chamber of length $d = 5.0$ cm is placed in one of the arms of a Michelson interferometer. (The glass window on each end of the chamber has negligible thickness.) Light of wavelength $\lambda = 500$ nm is used. Evacuating the air from the chamber causes a shift of 60 bright fringes. From these data and to six significant figures, find the index of refraction of air at atmospheric pressure. **SSM WWW**

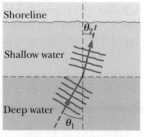

**FIG. 35-49** Problem 81.

**••82** The element sodium can emit light at two wavelengths, $\lambda_1 = 589.10$ nm and $\lambda_2 = 589.59$ nm. Light from sodium is being used in a Michelson interferometer (Fig. 35-23). Through what distance must mirror $M_2$ be moved if the shift in the fringe pattern for one wavelength is to be 1.00 fringe more than the shift in the fringe pattern for the other wavelength?

## Additional Problems

**83** Ocean waves moving at a speed of 4.0 m/s are approaching a beach at angle $\theta_1 = 30°$ to the normal, as shown from above in Fig. 35-50. Suppose the water depth changes abruptly at a certain distance from the beach and the wave speed there drops to 3.0 m/s. (a) Close to the beach, what is the angle $\theta_2$ between the direc-

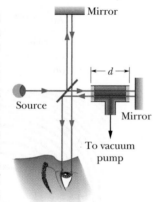

**FIG. 35-50** Problem 83.

tion of wave motion and the normal? (Assume the same law of refraction as for light.) (b) Explain why most waves come in normal to a shore even though at large distances they approach at a variety of angles.

**84** Figure 35-51a shows two light rays that are initially in phase as they travel upward through a block of plastic, with wavelength 400 nm as measured in air. Light ray $r_1$ exits directly into air. However, before light ray $r_2$ exits into air, it travels through a liquid in a hollow cylinder within the plastic. Initially the height $L_{liq}$ of the liquid is 40.0 $\mu$m, but then the liquid begins to evaporate. Let $\phi$ be the phase difference between rays $r_1$ and $r_2$ once they both exit into the air. Figure 35-51b shows $\phi$ versus the liquid's height $L_{liq}$ until the liquid disappears, with $\phi$ given in terms of wavelength and the horizontal scale set by $L_s = 40.00$ $\mu$m. What are (a) the index of refraction of the plastic and (b) the index of refraction of the liquid?

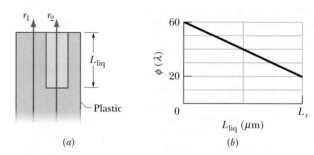

**FIG. 35-51** Problem 84.

**85** Two light rays, initially in phase and with a wavelength of 500 nm, go through different paths by reflecting from the various mirrors shown in Fig. 35-52. (Such a reflection does not itself produce a phase shift.) (a) What least value of distance $d$ will put the rays exactly out of phase when they emerge from the region? (Ignore the slight tilt of the path for ray 2.) (b) Repeat the question assuming that the entire apparatus is immersed in a protein solution with an index of refraction of 1.38. **GO**

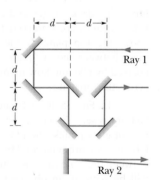

**FIG. 35-52** Problem 85.

**86** In Figure 35-53, two isotropic point sources $S_1$ and $S_2$ emit light in phase at wavelength $\lambda$ and at the same amplitude. The sources are separated by distance $d = 6.00\lambda$ on an $x$ axis. A viewing screen is at distance $D = 20.0\lambda$ from $S_2$ and parallel to the $y$ axis. The figure shows two rays reaching point $P$ on the screen, at height $y_P$. (a) At what value of $y_P$ do the rays have the minimum possible phase difference? (b) What multiple of $\lambda$ gives that minimum phase difference? (c) At what value of $y_P$ do the rays have the maximum possible phase difference? What multiple of $\lambda$ gives

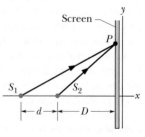

**FIG. 35-53** Problem 86.

(d) that maximum phase difference and (e) the phase difference when $y_P = d$? (f) When $y_P = d$, is the resulting intensity at point $P$ maximum, minimum, intermediate but closer to maximum, or intermediate but closer to minimum?

**87** In Fig. 35-54, a microwave transmitter at height $a$ above the water level of a wide lake transmits microwaves of wavelength $\lambda$ toward a receiver on the opposite shore, a distance $x$ above the water level. The microwaves reflecting from the water interfere with the microwaves arriving directly from the transmitter. Assuming that the lake width $D$ is much greater than $a$ and $x$, and that $\lambda \geq a$, find an expression that gives the values of $x$ for which the signal at the receiver is maximum. (*Hint:* Does the reflection cause a phase change?) SSM

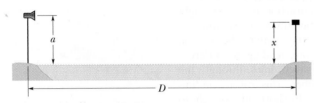

FIG. 35-54  Problem 87.

**88** In Fig. 35-55, two isotropic point sources $S_1$ and $S_2$ emit light at wavelength $\lambda = 400$ nm. Source $S_1$ is located at $y = 640$ nm; source $S_2$ is located at $y = -640$ nm. At point $P_1$ (at $x = 720$ nm), the wave from $S_2$ arrives ahead of the wave from $S_1$ by a phase difference of $0.600\pi$ rad. (a) What multiple of $\lambda$ gives the phase difference between the waves from the two sources as the waves arrive at point $P_2$, which is located at $y = 720$ nm. (The figure is not drawn to scale.) (b) If the waves arrive at $P_2$ with equal amplitudes, is the interference there fully constructive, fully destructive, intermediate but closer to fully constructive, or intermediate but closer to fully destructive?

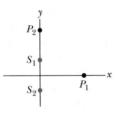

FIG. 35-55
Problem 88.

**89** A double-slit arrangement produces bright interference fringes for sodium light ($\lambda = 589$ nm) that are angularly separated by 0.30° near the center of the pattern. What is the angular fringe separation if the entire arrangement is immersed in water, which has an index of refraction of 1.33? SSM

**90** In Fig. 35-56a, the waves along rays 1 and 2 are initially in phase, with the same wavelength $\lambda$ in air. Ray 2 goes through a material with length $L$ and index of refraction $n$. The rays are then reflected by mirrors to a common point $P$ on a screen. Suppose that we can vary $n$ from $n = 1.0$ to $n = 2.5$. Suppose also that, from $n = 1.0$ to $n = n_s = 1.5$, the intensity $I$ of the

light at point $P$ varies with $n$ as given in Fig. 35-56b. At what values of $n$ greater than 1.4 is intensity $I$ (a) maximum and (b) zero? (c) What multiple of $\lambda$ gives the phase difference between the rays at point $P$ when $n = 2.0$?

**91** In a phasor diagram for the waves at any point on the viewing screen for the two-slit experiment in Fig. 35-10, the phasor of the resultant wave rotates 60.0° in $2.50 \times 10^{-16}$ s. What is the wavelength of the light?

**92** Light of wavelength 700.0 nm is sent along a route of length 2000 nm. The route is then filled with a medium having an index of refraction of 1.400. In degrees, by how much does the medium phase-shift the light? Give (a) the full shift and (b) the equivalent shift that has a value less than 360°.

**93** Two parallel slits are illuminated with monochromatic light of wavelength 500 nm. An interference pattern is formed on a screen some distance from the slits, and the fourth dark band is located 1.68 cm from the central bright band on the screen. (a) What is the path length difference corresponding to the fourth dark band? (b) What is the distance on the screen between the central bright band and the first bright band on either side of the central band? (*Hint:* The angle to the fourth dark band and the angle to the first bright band are small enough that $\tan \theta \approx \sin \theta$.) SSM

**94** In two experiments, light is to be sent along the two paths shown in Fig. 35-35 by reflecting it from the various flat surfaces shown. In the first experiment, rays 1 and 2 are initially in phase and have a wavelength of 620.0 nm. In the second experiment, rays 1 and 2 are initially in phase and have a wavelength of 496.0 nm. What least value of distance $L$ is required such that the 620.0 nm waves emerge from the region exactly in phase but the 496.0 nm waves emerge exactly out of phase?

**95** Find the slit separation of a double-slit arrangement that will produce interference fringes 0.018 rad apart on a distant screen when the light has wavelength $\lambda = 589$ nm.

**96** A thin film suspended in air is 0.410 $\mu$m thick and is illuminated with white light incident perpendicularly on its surface. The index of refraction of the film is 1.50. At what wavelength will visible light that is reflected from the two surfaces of the film undergo fully constructive interference?

**97** In Fig. 35-56a, the waves along rays 1 and 2 are initially in phase, with the same wavelength $\lambda$ in air. Ray 2 goes through a material with length $L$ and index of refraction $n$. The rays are then reflected by mirrors to a common point $P$ on a screen. Suppose that we can vary $L$ from 0 to 2400 nm. Suppose also that, from $L = 0$ to $L_s = 900$ nm, the intensity $I$

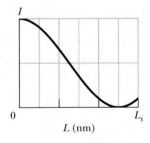

FIG. 35-57  Problem 97.

of the light at point $P$ varies with $L$ as given in Fig. 35-57. At what values of $L$ greater than $L_s$ is intensity $I$ (a) maximum and (b) zero? (c) What multiple of $\lambda$ gives the phase difference between ray 1 and ray 2 at common point $P$ when $L = 1200$ nm? SSM

**98** A camera lens with index of refraction greater than 1.30 is coated with a thin transparent film of index of refraction 1.25 to eliminate by interference the reflection of light at

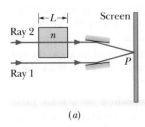

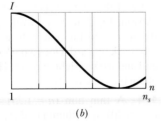

$(a)$  $(b)$

FIG. 35-56  Problems 90 and 97.

wavelength $\lambda$ that is incident perpendicularly on the lens. What multiple of $\lambda$ gives the minimum film thickness needed?

**99** If the distance between the first and tenth minima of a double-slit pattern is 18.0 mm and the slits are separated by 0.150 mm with the screen 50.0 cm from the slits, what is the wavelength of the light used? **SSM**

**100** What is the speed in fused quartz of light of wavelength 550 nm? (See Fig. 33-18.)

**101** In Sample Problem 35-6, assume that the coating eliminates the reflection of light of wavelength 550 nm at normal incidence. By what percentages is reflection diminished by the coating at (a) 450 nm and (b) 650 nm?

**102** Laser light of wavelength 632.8 nm passes through a double-slit arrangement at the front of a lecture room, reflects off a mirror 20.0 m away at the back of the room, and then produces an interference pattern on a screen at the front of the room. The distance between adjacent bright fringes is 10.0 cm. (a) What is the slit separation? (b) What is at the center of the pattern when the lecturer places a thin cellophane sheet over one slit, thereby increasing by 2.50 the number of wavelengths along the path that includes the cellophane?

**103** Light of wavelength $\lambda$ is used in a Michelson interferometer. Let $x$ be the position of the movable mirror, with $x = 0$ when the arms have equal lengths $d_2 = d_1$. Write an expression for the intensity of the observed light as a function of $x$, letting $I_m$ be the maximum intensity. **SSM**

**104** A sheet of glass having an index of refraction of 1.40 is to be coated with a film of material having an index of refraction of 1.55 in order that green light with a wavelength of 525 nm (in air) will be preferentially transmitted via constructive interference. (a) What is the minimum thickness of the film that will achieve the result? (b) Why are other parts of the visible spectrum not also preferentially transmitted? (c) Will the transmission of any colors be sharply reduced? If so, which colors?

**105** One slit of a double-slit arrangement is covered by a thin glass plate with index of refraction 1.4, and the other by a thin glass plate with index of refraction 1.7. The point on the screen at which the central maximum fell before the glass plates were inserted is now occupied by what had been the $m = 5$ bright fringe. Assuming that $\lambda = 480$ nm and that the plates have the same thickness $t$, find $t$.

**106** In Fig. 35-46, two glass plates are held together at one end to form a wedge of air that acts as a thin film. A broad beam of light of wavelength 480 nm is directed through the plates, perpendicular to the first plate. An observer intercepting light reflected from the plates sees on the plates an interference pattern that is due to the wedge of air. How much thicker is the wedge at the sixteenth bright fringe than it is at the sixth bright fringe, counting from where the plates touch?

**107** A broad beam of light of wavelength 600 nm is sent directly downward through the glass plate ($n = 1.50$) in Fig. 35-46. That plate and a plastic plate ($n = 1.20$) form a thin wedge of air that acts as a thin film. An observer looking

**FIG. 35-58** Problem 107.

down through the top plate sees the fringe pattern shown in Fig. 35-58, with dark fringes centered on the ends. (a) What is the thickness of the wedge at the right end? (b) How many dark fringes will the observer see if the air between the plates is replaced with water ($n = 1.33$)?

**108** Sodium light ($\lambda = 589$ nm) illuminates two slits separated by $d = 2.0$ mm. The slit–screen distance $D$ is 40 mm. What percentage error is made by using Eq. 35-14 to locate the $m = 10$ bright fringe on the screen rather than using the exact path length difference?

**109** Figure 35-59 shows an optical fiber in which a central plastic core of index of refraction $n_1 = 1.58$ is surrounded by a plastic sheath of index of refraction $n_2 = 1.53$. Light can travel along different paths within the central core, leading to different travel times through the fiber. This causes an initially short pulse of light to spread as it travels along

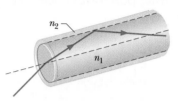
**FIG. 35-59** Problem 109.

the fiber, resulting in information loss. Consider light that travels directly along the central axis of the fiber and light that is repeatedly reflected at the critical angle along the core–sheath interface, reflecting from side to side as it travels down the central core. If the fiber length is 300 m, what is the difference in the travel times along these two routes?

**110** When an electron moves through a medium at a speed exceeding the speed of light in that medium, the electron radiates electromagnetic energy (the *Cerenkov effect*). What minimum speed must an electron have in a liquid with index of refraction 1.54 in order to radiate?

**111** Point sources $S_1$ and $S_2$ radiate in phase at wavelength 400 nm and at the same amplitude. The sources are located on an $x$ axis at $x = 6.5$ $\mu$m and $x = -6.0$ $\mu$m, respectively. (a) Determine the phase difference (in radians) at the origin between the radiation from $S_1$ and the radiation from $S_2$. (b) Suppose a slab of transparent material with thickness 1.5 $\mu$m and index of refraction $n = 1.5$ is placed between $x = 0$ and $x = 1.5$ $\mu$m. What then is the phase difference (in radians) at the origin between the radiation from $S_1$ and the radiation from $S_2$?

**112** The second dark band in a double-slit interference pattern is 1.2 cm from the central maximum of the pattern. The separation between the two slits is equal to 800 wavelengths of the monochromatic light incident (perpendicularly) on the slits. What is the distance between the plane of the slits and the viewing screen?

**113** In Fig. 35-46, two glass plates ($n = 1.60$) form a wedge, and a fluid ($n = 1.50$) fills the wedge-shaped space. At the left end the plates touch; at the right, they are separated by 580 nm. Light with a wavelength (in air) of 580 nm shines downward on the assembly, and an observer intercepts light sent back upward. Does a dark or bright band lie at (a) the left end and (b) the right end? (c) How many dark bands are there along the plates?

**114** A thin film ($n = 1.25$) is deposited on a glass plate ($n = 1.40$) and illuminated with light of wavelength 550 nm (measured in air). The light beam is perpendicular to the plate. What is the minimum (nonzero) thickness for the film that will (a) maximally transmit and (b) maximally reflect the light?

**115** A light beam with a wavelength of 600 nm in air passes through film 1 ($n_1 = 1.2$) of thickness 1.0 $\mu$m, then through film 2 (air) of thickness 1.5 $\mu$m, and finally through film 3 ($n_3 = 1.8$) of thickness 1.0 $\mu$m. The beam is perpendicular to the film surfaces, which are parallel to one another. (a) Which film does the light cross in the least time, and (b) what is that least time? (c) What is the total number of wavelengths (at any instant) across all three films together?

**116** Two light rays, initially in phase and having wavelength $\lambda = 6.00 \times 10^{-7}$ m, pass through different plastic layers of the same thickness, $7.00 \times 10^{-6}$ m. The indexes of refraction are 1.65 for one layer and 1.49 for the other. (a) What *least* multiple of $\lambda$ gives the phase difference between the rays when they emerge? (b) If those two rays then reach a common point with the same amplitude, does the interference result in complete darkness, maximum brightness, intermediate illumination but closer to complete darkness, or intermediate illumination but closer to maximum brightness? If the two rays are, instead, initially exactly out of phase, what are the answers to (c) part (a) and (d) part (b)?

**117** In a double-slit interference experiment, the slit separation is 2.00 $\mu$m, the light wavelength is 500 nm, and the separation between the slits and the screen is 4.00 m. (a) What is the angle between the center and the third side bright fringe? If we decrease the light frequency to 90.0% of its initial value, (b) does the third side bright fringe move along the screen toward or away from the pattern's center and (c) how far does it move?

**118** A plane monochromatic light wave in air is perpendicularly incident on a thin film of oil that covers a glass plate. The wavelength of the source may be varied continuously. Fully destructive interference of the reflected light occurs for wavelengths 500 and 700 nm and for no wavelength in between. The index of refraction of the glass is 1.50. Show that the index of refraction of the oil must be less than 1.50.

**119** Figure 35-60 shows the design of a Texas arcade game. Four laser pistols are pointed toward the center of an array of plastic layers where a clay armadillo is the target. The indexes of refraction of the layers are $n_1 = 1.55$, $n_2 = 1.70$, $n_3 = 1.45$, $n_4 = 1.60$, $n_5 = 1.45$, $n_6 = 1.61$, $n_7 = 1.59$, $n_8 = 1.70$, and $n_9 = 1.60$. The layer thicknesses are either 2.00 mm or 4.00 mm, as drawn. What is the travel time through the layers for the laser burst from (a) pistol 1, (b) pistol 2, (c) pistol 3, and (d) pistol 4? (e) If the pistols are fired simultaneously, which laser burst hits the target first?

**120** In Fig. 35-10, let the angle $\theta$ of the two rays be 20.0°, slit separation $d$ be 58.00 $\mu$m, and wavelength $\lambda$ be 500.9 nm. (a) What multiple of $\lambda$ gives the phase difference of the two rays when they reach a common point on a distant screen? (b) Does their interference result in complete darkness, maximum brightness, intermediate illumination but closer to complete darkness, or intermediate illumination but closer to maximum brightness?

**121** Slits of unequal widths are used in a double-slit arrangement to produce an interference pattern on a distant screen. If only the narrower slit 1 is illuminated (the wider slit 2 is covered), the light reaching the center of the pattern has amplitude $E_0$ and intensity $I_0$. If only slit 2 is illuminated, the amplitude of the light reaching the pattern center is $2E_0$. When both slits are illuminated and a two-slit interference pattern is on the screen, what is the intensity $I(\theta)$ in the pattern as a function of angle $\theta$? Answer in the style of Eqs. 35-22 and 35-23.

**122** Figure 35-40 shows two point sources $S_1$ and $S_2$ that emit light at wavelength $\lambda = 500$ nm and with the same amplitude. The emissions are isotropic and in phase, and the separation between the sources is $d = 2.00$ $\mu$m. At any point $P$ on the $x$ axis, the wave from $S_1$ and the wave from $S_2$ interfere. When $P$ is very far away ($x \approx \infty$), what are (a) the phase difference between the waves arriving from $S_1$ and $S_2$ and (b) the type of interference they produce (approximately fully constructive or fully destructive)? (c) As we then move $P$ along the $x$ axis toward $S_1$, does the phase difference between the waves from $S_1$ and $S_2$ increase or decrease? (d)–(o) Fill in Table 35-4 for the given phase differences by determining the type of interference and the $x$ coordinate at which the interference occurs.

**TABLE 35-4**

**Problem 122: Parts (d) through (o)**

| Phase Difference | Type | Position $x$ |
|---|---|---|
| 0 | (d) | (e) |
| 0.500$\lambda$ | (f) | (g) |
| 1.00$\lambda$ | (h) | (i) |
| 1.50$\lambda$ | (j) | (k) |
| 2.00$\lambda$ | (l) | (m) |
| 2.50$\lambda$ | (n) | (o) |

**123** (a) Use the result of Problem 75 to show that, in a Newton's rings experiment, the difference in radius between adjacent bright rings (maxima) is given by

$$\Delta r = r_{m+1} - r_m \approx \tfrac{1}{2}\sqrt{\lambda R/m},$$

assuming $m \gg 1$. (b) Now show that the *area* between adjacent bright rings is given by $A = \pi\lambda R$, assuming $m \gg 1$. Note that this area is independent of $m$.

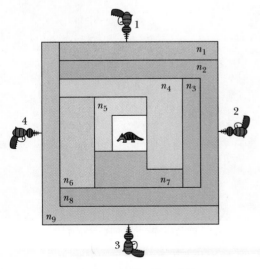

**FIG. 35-60** Problem 119.

# Diffraction

Many surfaces (such as shirts and pants) have color because dye molecules absorb visible light of certain wavelengths, allowing the rest of the light to reflect. Some surfaces (such as iridescent butterfly wings) have color because thin-film interference removes certain wavelengths and allows the rest of the light to be reflected. However, the arresting blue skin of this mandrill baboon is due to neither dye molecules nor thin films.

## What makes the face of a mandrill baboon blue?

The answer is in this chapter.

Timothy Laman/National Geographic/Getty Images, Inc.

## 36-1 | WHAT IS PHYSICS?

One focus of physics in the study of light is to understand and put to use the diffraction of light as it passes through a narrow slit or (as we shall discuss) past either a narrow obstacle or an edge. We touched on this phenomenon in Chapter 35 when we looked at how light flared—diffracted—through the slits in Young's experiment. Diffraction through a given slit is more complicated than simple flaring, however, because the light also interferes with itself and produces an interference pattern. It is because of such complications that light is rich with application opportunities. Even though the diffraction of light as it passes through a slit or past an obstacle seems awfully academic, countless engineers and scientists make their living using this physics, and the total worth of diffraction applications worldwide is probably incalculable.

Before we can discuss some of these applications, we first must discuss why diffraction is due to the wave nature of light.

## 36-2 | Diffraction and the Wave Theory of Light

In Chapter 35 we defined diffraction rather loosely as the flaring of light as it emerges from a narrow slit. More than just flaring occurs, however, because the light produces an interference pattern called a **diffraction pattern.** For example, when monochromatic light from a distant source (or a laser) passes through a narrow slit and is then intercepted by a viewing screen, the light produces on the screen a diffraction pattern like that in Fig. 36-1. This pattern consists of a broad and intense (very bright) central maximum plus a number of narrower and less intense maxima (called **secondary** or **side** maxima) to both sides. In between the maxima are minima.

Such a pattern would be totally unexpected in geometrical optics: If light traveled in straight lines as rays, then the slit would allow some of those rays through to form a sharp rendition of the slit on the viewing screen. As in Chapter 35, we must conclude that geometrical optics is only an approximation.

Diffraction is not limited to situations when light passes through a narrow opening (such as a slit or pinhole). It also occurs when light passes an edge, such as the edges of the razor blade whose diffraction pattern is shown in Fig. 36-2. Note the lines of maxima and minima that run approximately parallel to the edges, at both the inside edges of the blade and the outside edges. As the light passes, say, the vertical edge at the left, it flares left and right and undergoes interference, producing the pattern along the left edge. The rightmost portion of that pattern actually lies behind the blade, within what would be the blade's shadow if geometrical optics prevailed.

You encounter a common example of diffraction when you look at a clear blue sky and see tiny specks and hairlike structures floating in your view. These *floaters,* as they are called, are produced when light passes the edges of tiny deposits in the vitreous humor, the transparent material filling most of the eyeball. What you are seeing when a floater is in your field of vision is the diffraction pattern produced on the retina by one of these deposits. If you sight through a pinhole in a piece of cardboard so as to make the light entering your eye approximately a plane wave, you can distinguish individual maxima and minima in the patterns.

Diffraction is a wave effect. That is, it occurs because light is a wave and it occurs with other types of waves as well. For example, you have probably seen diffraction in action at football games. When a cheerleader near the playing field yells up at several thousand noisy fans, the yell can hardly be heard because the sound waves diffract when they pass through the narrow opening of the

FIG. 36-1   This diffraction pattern appeared on a viewing screen when light that had passed through a narrow vertical slit reached the screen. Diffraction caused the light to flare out perpendicular to the long sides of the slit. That flaring produced an interference pattern consisting of a broad central maximum plus less intense and narrower secondary (or side) maxima, with minima between them. *(Ken Kay/Fundamental Photographs)*

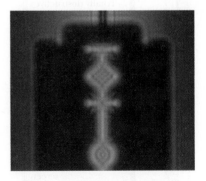

FIG. 36-2   The diffraction pattern produced by a razor blade in monochromatic light. Note the lines of alternating maximum and minimum intensity. *(Ken Kay/Fundamental Photographs)*

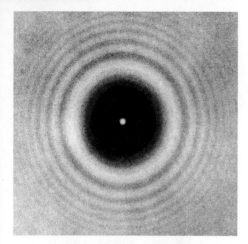

**FIG. 36-3** A photograph of the diffraction pattern of a disk. Note the concentric diffraction rings and the Fresnel bright spot at the center of the pattern. This experiment is essentially identical to that arranged by the committee testing Fresnel's theories, because both the sphere they used and the disk used here have a cross section with a circular edge. *(From Michel Cagnet, Maurice Franzon, and Jean Thierr, Atlas of Optical Phenomena. Springer-Verlag, New York, 1962. Reproduced with permission.)*

cheerleader's mouth. This flaring leaves little of the waves traveling toward the fans in front of the cheerleader. To offset the diffraction, the cheerleader can yell through a megaphone. The sound waves then emerge from the much wider opening at the end of the megaphone. The flaring is thus reduced, and much more of the sound reaches the fans in front of the cheerleader.

### The Fresnel Bright Spot

Diffraction finds a ready explanation in the wave theory of light. However, this theory, originally advanced in the late 1600s by Huygens and used 123 years later by Young to explain double-slit interference, was very slow in being adopted, largely because it ran counter to Newton's theory that light was a stream of particles.

Newton's view was the prevailing view in French scientific circles of the early 19th century, when Augustin Fresnel was a young military engineer. Fresnel, who believed in the wave theory of light, submitted a paper to the French Academy of Sciences describing his experiments with light and his wave-theory explanations of them.

In 1819, the Academy, dominated by supporters of Newton and thinking to challenge the wave point of view, organized a prize competition for an essay on the subject of diffraction. Fresnel won. The Newtonians, however, were neither converted nor silenced. One of them, S. D. Poisson, pointed out the "strange result" that if Fresnel's theories were correct, then light waves should flare into the shadow region of a sphere as they pass the edge of the sphere, producing a bright spot at the center of the shadow. The prize committee arranged a test of Poisson's prediction and discovered that the predicted *Fresnel bright spot*, as we call it today, was indeed there (Fig. 36-3). Nothing builds confidence in a theory so much as having one of its unexpected and counterintuitive predictions verified by experiment.

## 36-3 | Diffraction by a Single Slit: Locating the Minima

Let us now examine the diffraction pattern of plane waves of light of wavelength $\lambda$ that are diffracted by a single long, narrow slit of width $a$ in an otherwise opaque screen $B$, as shown in cross section in Fig. 36-4. (In that figure, the slit's length extends into and out of the page, and the incoming wavefronts are parallel to screen $B$.) When the diffracted light reaches viewing screen $C$, waves from different points within the slit undergo interference and produce a diffraction pattern of bright and dark fringes (interference maxima and minima) on the screen. To locate the fringes, we shall use a procedure somewhat similar to the one we used to locate the fringes in a two-slit interference pattern. However, diffraction is more mathematically challenging, and here we shall be able to find equations for only the dark fringes.

Before we do that, however, we can justify the central bright fringe seen in Fig. 36-1 by noting that the Huygens wavelets from all points in the slit travel about the same distance to reach the center of the pattern and thus are in phase there. As for the other bright fringes, we can say only that they are approximately halfway between adjacent dark fringes.

To find the dark fringes, we shall use a clever (and simplifying) strategy that involves pairing up all the rays coming through the slit and then finding what conditions cause the wavelets of the rays in each pair to cancel each other. We apply this strategy in Fig. 36-4 to locate the first dark fringe, at point $P_1$. First, we mentally divide the slit into two *zones* of equal widths $a/2$. Then we extend to $P_1$ a light ray $r_1$ from the top point of the top zone and a light ray $r_2$ from the

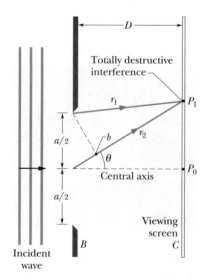

**FIG. 36-4** Waves from the top points of two zones of width $a/2$ undergo fully destructive interference at point $P_1$ on viewing screen $C$.

top point of the bottom zone. A central axis is drawn from the center of the slit to screen $C$, and $P_1$ is located at an angle $\theta$ to that axis.

The wavelets of the pair of rays $r_1$ and $r_2$ are in phase within the slit because they originate from the same wavefront passing through the slit, along the width of the slit. However, to produce the first dark fringe they must be out of phase by $\lambda/2$ when they reach $P_1$; this phase difference is due to their path length difference, with the path traveled by the wavelet of $r_2$ to reach $P_1$ being longer than the path traveled by the wavelet of $r_1$. To display this path length difference, we find a point $b$ on ray $r_2$ such that the path length from $b$ to $P_1$ matches the path length of ray $r_1$. Then the path length difference between the two rays is the distance from the center of the slit to $b$.

When viewing screen $C$ is near screen $B$, as in Fig. 36-4, the diffraction pattern on $C$ is difficult to describe mathematically. However, we can simplify the mathematics considerably if we arrange for the screen separation $D$ to be much larger than the slit width $a$. Then we can approximate rays $r_1$ and $r_2$ as being parallel, at angle $\theta$ to the central axis (Fig. 36-5). We can also approximate the triangle formed by point $b$, the top point of the slit, and the center point of the slit as being a right triangle, and one of the angles inside that triangle as being $\theta$. The path length difference between rays $r_1$ and $r_2$ (which is still the distance from center of the slit to point $b$) is then equal to $(a/2) \sin \theta$.

We can repeat this analysis for any other pair of rays originating at corresponding points in the two zones (say, at the midpoints of the zones) and extending to point $P_1$. Each such pair of rays has the same path length difference $(a/2) \sin \theta$. Setting this common path length difference equal to $\lambda/2$ (our condition for the first dark fringe), we have

$$\frac{a}{2} \sin \theta = \frac{\lambda}{2},$$

which gives us

$$a \sin \theta = \lambda \quad \text{(first minimum).} \quad (36\text{-}1)$$

Given slit width $a$ and wavelength $\lambda$, Eq. 36-1 tells us the angle $\theta$ of the first dark fringe above and (by symmetry) below the central axis.

Note that if we begin with $a > \lambda$ and then narrow the slit while holding the wavelength constant, we increase the angle at which the first dark fringes appear; that is, the extent of the diffraction (the extent of the flaring and the width of the pattern) is *greater* for a *narrower* slit. When we have reduced the slit width to the wavelength (that is, $a = \lambda$), the angle of the first dark fringes is $90°$. Since the first dark fringes mark the two edges of the central bright fringe, that bright fringe must then cover the entire viewing screen.

We find the second dark fringes above and below the central axis as we found the first dark fringes, except that we now divide the slit into *four* zones of equal widths $a/4$, as shown in Fig. 36-6a. We then extend rays $r_1, r_2, r_3,$ and $r_4$ from the top points of the zones to point $P_2$, the location of the second dark fringe above the central axis. To produce that fringe, the path length difference between $r_1$ and $r_2$, that between $r_2$ and $r_3$, and that between $r_3$ and $r_4$ must all be equal to $\lambda/2$.

For $D \gg a$, we can approximate these four rays as being parallel, at angle $\theta$ to the central axis. To display their path length differences, we extend a perpendicular line through each adjacent pair of rays, as shown in Fig. 36-6b, to form a series of right triangles, each of which has a path length difference as one side. We see from the top triangle that the path length difference between $r_1$ and $r_2$ is $(a/4) \sin \theta$. Similarly, from the bottom triangle, the path length difference between $r_3$ and $r_4$ is also $(a/4) \sin \theta$. In fact, the path length difference for any two rays that originate at corresponding points in two adjacent zones is $(a/4) \sin \theta$. Since in each such case the path length difference is equal to $\lambda/2$, we have

$$\frac{a}{4} \sin \theta = \frac{\lambda}{2},$$

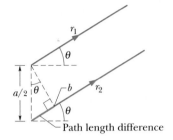

**FIG. 36-5** For $D \gg a$, we can approximate rays $r_1$ and $r_2$ as being parallel, at angle $\theta$ to the central axis.

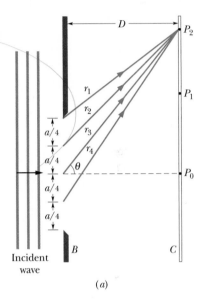

(a)

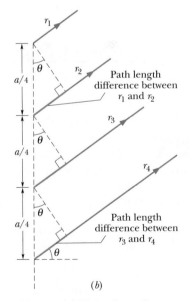

(b)

**FIG. 36-6** (a) Waves from the top points of four zones of width $a/4$ undergo fully destructive interference at point $P_2$. (b) For $D \gg a$, we can approximate rays $r_1, r_2, r_3,$ and $r_4$ as being parallel, at angle $\theta$ to the central axis.

which gives us

$$a \sin \theta = 2\lambda \qquad \text{(second minimum).} \qquad (36\text{-}2)$$

We could now continue to locate dark fringes in the diffraction pattern by splitting up the slit into more zones of equal width. We would always choose an even number of zones so that the zones (and their waves) could be paired as we have been doing. We would find that the dark fringes above and below the central axis can be located with the general equation

$$a \sin \theta = m\lambda, \qquad \text{for } m = 1, 2, 3, \ldots \qquad \text{(minima—dark fringes).} \qquad (36\text{-}3)$$

You can remember this result in the following way. Draw a triangle like the one in Fig. 36-5, but for the full slit width $a$, and note that the path length difference between the top and bottom rays equals $a \sin \theta$. Thus, Eq. 36-3 says:

> In a single-slit diffraction experiment, dark fringes are produced where the path length differences ($a \sin \theta$) between the top and bottom rays are equal to $\lambda, 2\lambda, 3\lambda, \ldots$.

This may seem to be wrong because the waves of those two particular rays will be exactly in phase with each other when their path length difference is an integer number of wavelengths. However, they each will still be part of a pair of waves that are exactly out of phase with each other; thus, *each* wave will be canceled by some other wave, resulting in darkness.

Equations 36-1, 36-2, and 36-3 are derived for the case of $D \gg a$. However, they also apply if we place a converging lens between the slit and the viewing screen and then move the screen in so that it coincides with the focal plane of the lens. The lens ensures that rays which now reach any point on the screen are *exactly* parallel (rather than approximately) back at the slit. They are like the initially parallel rays of Fig. 34-15a that are directed to the focal point by a converging lens.

✓ **CHECKPOINT 1** We produce a diffraction pattern on a viewing screen by means of a long narrow slit illuminated by blue light. Does the pattern expand away from the bright center (the maxima and minima shift away from the center) or contract toward it if we (a) switch to yellow light or (b) decrease the slit width?

---

## Sample Problem | 36-1

A slit of width $a$ is illuminated by white light.

(a) For what value of $a$ will the first minimum for red light of wavelength $\lambda = 650$ nm appear at $\theta = 15°$?

**KEY IDEA** Diffraction occurs separately for each wavelength in the range of wavelengths passing through the slit, with the locations of the minima for each wavelength given by Eq. 36-3 ($a \sin \theta = m\lambda$).

**Calculation:** When we set $m = 1$ (for the first minimum) and substitute the given values of $\theta$ and $\lambda$, Eq. 36-3 yields

$$a = \frac{m\lambda}{\sin \theta} = \frac{(1)(650 \text{ nm})}{\sin 15°}$$

$$= 2511 \text{ nm} \approx 2.5 \ \mu\text{m}. \qquad \text{(Answer)}$$

For the incident light to flare out that much ($\pm 15°$ to the first minima) the slit has to be very fine indeed—in this case, a mere four times the wavelength. For comparison, note that a fine human hair may be about 100 $\mu$m in diameter.

(b) What is the wavelength $\lambda'$ of the light whose first side diffraction maximum is at 15°, thus coinciding with the first minimum for the red light?

**KEY IDEA** The first side maximum for any wavelength is about halfway between the first and second minima for that wavelength.

**Calculations:** Those first and second minima can be located with Eq. 36-3 by setting $m = 1$ and $m = 2$, respectively. Thus, the first side maximum can be located *approximately* by setting $m = 1.5$. Then Eq. 36-3 becomes

$$a \sin \theta = 1.5\lambda'.$$

Solving for $\lambda'$ and substituting known data yield

$$\lambda' = \frac{a \sin \theta}{1.5} = \frac{(2511 \text{ nm})(\sin 15°)}{1.5}$$

$$= 430 \text{ nm}. \qquad \text{(Answer)}$$

Light of this wavelength is violet. From the two equations we used, can you see that the first side maximum for light of wavelength 430 nm will always coincide with the first minimum for light of wavelength 650 nm, no matter what the slit width is? However, the angle $\theta$ at which this overlap occurs does depend on slit width. If the slit is relatively narrow, the angle will be relatively large, and conversely.

## 36-4 | Intensity in Single-Slit Diffraction, Qualitatively

In Section 36-3 we saw how to find the positions of the minima and the maxima in a single-slit diffraction pattern. Now we turn to a more general problem: find an expression for the intensity $I$ of the pattern as a function of $\theta$, the angular position of a point on a viewing screen.

To do this, we divide the slit of Fig. 36-4 into $N$ zones of equal widths $\Delta x$ small enough that we can assume each zone acts as a source of Huygens wavelets. We wish to superimpose the wavelets arriving at an arbitrary point $P$ on the viewing screen, at angle $\theta$ to the central axis, so that we can determine the amplitude $E_\theta$ of the electric component of the resultant wave at $P$. The intensity of the light at $P$ is then proportional to the square of that amplitude.

To find $E_\theta$, we need the phase relationships among the arriving wavelets. The phase difference between wavelets from adjacent zones is given by

$$\left(\begin{array}{c}\text{phase}\\\text{difference}\end{array}\right) = \left(\frac{2\pi}{\lambda}\right)\left(\begin{array}{c}\text{path length}\\\text{difference}\end{array}\right).$$

For point $P$ at angle $\theta$, the path length difference between wavelets from adjacent zones is $\Delta x \sin \theta$; so the phase difference $\Delta\phi$ between wavelets from adjacent zones is

$$\Delta\phi = \left(\frac{2\pi}{\lambda}\right)(\Delta x \sin \theta). \tag{36-4}$$

We assume that the wavelets arriving at $P$ all have the same amplitude $\Delta E$. To find the amplitude $E_\theta$ of the resultant wave at $P$, we add the amplitude $\Delta E$ via phasors. To do this, we construct a diagram of $N$ phasors, one corresponding to the wavelet from each zone in the slit.

For point $P_0$ at $\theta = 0$ on the central axis of Fig. 36-4, Eq. 36-4 tells us that the phase difference $\Delta\phi$ between the wavelets is zero; that is, the wavelets all arrive in phase. Figure 36-7a is the corresponding phasor diagram; adjacent phasors represent wavelets from adjacent zones and are arranged head to tail. Because there is zero phase difference between the wavelets, there is zero angle between each pair of adjacent phasors. The amplitude $E_\theta$ of the net wave at $P_0$ is the vector sum of these phasors. This arrangement of the phasors turns out to be the one that gives the greatest value for the amplitude $E_\theta$. We call this value $E_m$; that is, $E_m$ is the value of $E_\theta$ for $\theta = 0$.

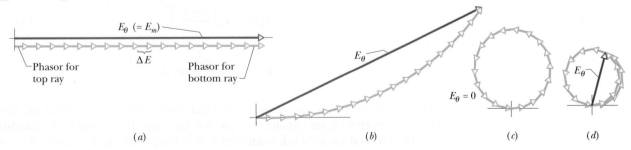

**FIG. 36-7** Phasor diagrams for $N = 18$ phasors, corresponding to the division of a single slit into 18 zones. Resultant amplitudes $E_\theta$ are shown for (a) the central maximum at $\theta = 0$, (b) a point on the screen lying at a small angle $\theta$ to the central axis, (c) the first minimum, and (d) the first side maximum.

We next consider a point $P$ that is at a small angle $\theta$ to the central axis. Equation 36-4 now tells us that the phase difference $\Delta\phi$ between wavelets from adjacent zones is no longer zero. Figure 36-7b shows the corresponding phasor diagram; as before, the phasors are arranged head to tail, but now there is an angle $\Delta\phi$ between adjacent phasors. The amplitude $E_\theta$ at this new point is still the vector sum of the phasors, but it is smaller than that in Fig. 36-7a, which means that the intensity of the light is less at this new point $P$ than at $P_0$.

If we continue to increase $\theta$, the angle $\Delta\phi$ between adjacent phasors increases, and eventually the chain of phasors curls completely around so that the head of the last phasor just reaches the tail of the first phasor (Fig. 36-7c). The amplitude $E_\theta$ is now zero, which means that the intensity of the light is also zero. We have reached the first minimum, or dark fringe, in the diffraction pattern. The first and last phasors now have a phase difference of $2\pi$ rad, which means that the path length difference between the top and bottom rays through the slit equals one wavelength. Recall that this is the condition we determined for the first diffraction minimum.

As we continue to increase $\theta$, the angle $\Delta\phi$ between adjacent phasors continues to increase, the chain of phasors begins to wrap back on itself, and the resulting coil begins to shrink. Amplitude $E_\theta$ now increases until it reaches a maximum value in the arrangement shown in Fig. 36-7d. This arrangement corresponds to the first side maximum in the diffraction pattern.

If we increase $\theta$ a bit more, the resulting shrinkage of the coil decreases $E_\theta$, which means that the intensity also decreases. When $\theta$ is increased enough, the head of the last phasor again meets the tail of the first phasor. We have then reached the second minimum.

We could continue this qualitative method of determining the maxima and minima of the diffraction pattern but, instead, we shall now turn to a quantitative method.

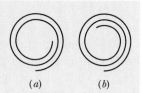

**CHECKPOINT 2** The figures represent, in smoother form (with more phasors) than Fig. 36-7, the phasor diagrams for two points of a diffraction pattern that are on opposite sides of a certain diffraction maximum. (a) Which maximum is it? (b) What is the approximate value of $m$ (in Eq. 36-3) that corresponds to this maximum?

(a)　　(b)

## 36-5 | Intensity in Single-Slit Diffraction, Quantitatively

Equation 36-3 tells us how to locate the minima of the single-slit diffraction pattern on screen $C$ of Fig. 36-4 as a function of the angle $\theta$ in that figure. Here we wish to derive an expression for the intensity $I(\theta)$ of the pattern as a function of $\theta$. We state, and shall prove below, that the intensity is given by

$$I(\theta) = I_m\left(\frac{\sin\alpha}{\alpha}\right)^2,\qquad(36\text{-}5)$$

where

$$\alpha = \tfrac{1}{2}\phi = \frac{\pi a}{\lambda}\sin\theta.\qquad(36\text{-}6)$$

The symbol $\alpha$ is just a convenient connection between the angle $\theta$ that locates a point on the viewing screen and the light intensity $I(\theta)$ at that point. The intensity $I_m$ is the greatest value of the intensities $I(\theta)$ in the pattern and occurs at the central maximum (where $\theta = 0$), and $\phi$ is the phase difference (in radians) between the top and bottom rays from the slit of width $a$.

Study of Eq. 36-5 shows that intensity minima will occur where

$$\alpha = m\pi,\qquad \text{for } m = 1, 2, 3, \ldots.\qquad(36\text{-}7)$$

If we put this result into Eq. 36-6, we find

$$m\pi = \frac{\pi a}{\lambda} \sin\theta, \qquad \text{for } m = 1, 2, 3, \ldots,$$

or    $a \sin\theta = m\lambda$,    for $m = 1, 2, 3, \ldots$   (minima—dark fringes),   (36-8)

which is exactly Eq. 36-3, the expression that we derived earlier for the location of the minima.

Figure 36-8 shows plots of the intensity of a single-slit diffraction pattern, calculated with Eqs. 36-5 and 36-6 for three slit widths: $a = \lambda$, $a = 5\lambda$, and $a = 10\lambda$. Note that as the slit width increases (relative to the wavelength), the width of the *central diffraction maximum* (the central hill-like region of the graphs) decreases; that is, the light undergoes less flaring by the slit. The secondary maxima also decrease in width (and become weaker). In the limit of slit width $a$ being much greater than wavelength $\lambda$, the secondary maxima due to the slit disappear; we then no longer have single-slit diffraction (but we still have diffraction due to the edges of the wide slit, like that produced by the edges of the razor blade in Fig. 36-2).

### Proof of Eqs. 36-5 and 36-6

The arc of phasors in Fig. 36-9 represents the wavelets that reach an arbitrary point $P$ on the viewing screen of Fig. 36-4, corresponding to a particular small angle $\theta$. The amplitude $E_\theta$ of the resultant wave at $P$ is the vector sum of these phasors. If we divide the slit of Fig. 36-4 into infinitesimal zones of width $\Delta x$, the arc of phasors in Fig. 36-9 approaches the arc of a circle; we call its radius $R$ as indicated in that figure. The length of the arc must be $E_m$, the amplitude at the center of the diffraction pattern, because if we straightened out the arc we would have the phasor arrangement of Fig. 36-7a (shown lightly in Fig. 36-9).

The angle $\phi$ in the lower part of Fig. 36-9 is the difference in phase between the infinitesimal vectors at the left and right ends of arc $E_m$. From the geometry, $\phi$ is also the angle between the two radii marked $R$ in Fig. 36-9. The dashed line in that figure, which bisects $\phi$, then forms two congruent right triangles. From either triangle we can write

$$\sin\tfrac{1}{2}\phi = \frac{E_\theta}{2R}. \qquad (36\text{-}9)$$

In radian measure, $\phi$ is (with $E_m$ considered to be a circular arc)

$$\phi = \frac{E_m}{R}.$$

Solving this equation for $R$ and substituting in Eq. 36-9 lead to

$$E_\theta = \frac{E_m}{\tfrac{1}{2}\phi} \sin\tfrac{1}{2}\phi. \qquad (36\text{-}10)$$

In Section 33-5 we saw that the intensity of an electromagnetic wave is proportional to the square of the amplitude of its electric field. Here, this means that the maximum intensity $I_m$ (which occurs at the center of the diffraction pattern) is proportional to $E_m^2$ and the intensity $I(\theta)$ at angle $\theta$ is proportional to $E_\theta^2$. Thus, we may write

$$\frac{I(\theta)}{I_m} = \frac{E_\theta^2}{E_m^2}. \qquad (36\text{-}11)$$

Substituting for $E_\theta$ with Eq. 36-10 and then substituting $\alpha = \tfrac{1}{2}\phi$, we are led to the following expression for the intensity as a function of $\theta$:

$$I(\theta) = I_m \left(\frac{\sin\alpha}{\alpha}\right)^2.$$

This is exactly Eq. 36-5, one of the two equations we set out to prove.

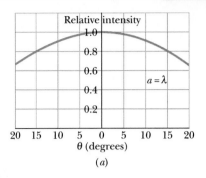

(a)

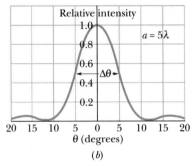

(b)

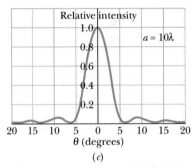

(c)

**FIG. 36-8**  The relative intensity in single-slit diffraction for three values of the ratio $a/\lambda$. The wider the slit is, the narrower is the central diffraction maximum.

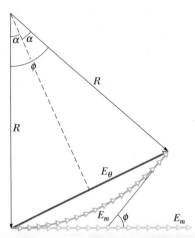

**FIG. 36-9**  A construction used to calculate the intensity in single-slit diffraction. The situation shown corresponds to that of Fig. 36-7b.

The second equation we wish to prove relates $\alpha$ to $\theta$. The phase difference $\phi$ between the rays from the top and bottom of the entire slit may be related to a path length difference with Eq. 36-4; it tells us that

$$\phi = \left(\frac{2\pi}{\lambda}\right)(a \sin \theta),$$

where $a$ is the sum of the widths $\Delta x$ of the infinitesimal zones. However, $\phi = 2\alpha$, so this equation reduces to Eq. 36-6.

✓ **CHECKPOINT 3** Two wavelengths, 650 and 430 nm, are used separately in a single-slit diffraction experiment. The figure shows the results as graphs of intensity $I$ versus angle $\theta$ for the two diffraction patterns. If both wavelengths are then used simultaneously, what color will be seen in the combined diffraction pattern at (a) angle $A$ and (b) angle $B$?

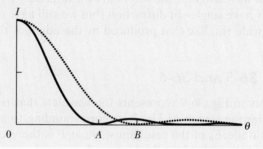

## Sample Problem 36-2

Find the intensities of the first three secondary maxima (side maxima) in the single-slit diffraction pattern of Fig. 36-1, measured as a percentage of the intensity of the central maximum.

**KEY IDEAS** The secondary maxima lie approximately halfway between the minima, whose angular locations are given by Eq. 36-7 ($\alpha = m\pi$). The locations of the secondary maxima are then given (approximately) by

$$a = (m + \tfrac{1}{2})\pi, \qquad \text{for } m = 1, 2, 3, \ldots,$$

with $\alpha$ in radian measure. We can relate the intensity $I$ at any point in the diffraction pattern to the intensity $I_m$ of the central maximum via Eq. 36-5.

**Calculations:** Substituting the approximate values of $\alpha$ for the secondary maxima into Eq. 36-5 to obtain the relative intensities at those maxima, we get

$$\frac{I}{I_m} = \left(\frac{\sin \alpha}{\alpha}\right)^2 = \left(\frac{\sin(m + \tfrac{1}{2})\pi}{(m + \tfrac{1}{2})\pi}\right)^2, \quad \text{for } m = 1, 2, 3, \ldots.$$

The first of the secondary maxima occurs for $m = 1$, and its relative intensity is

$$\frac{I_1}{I_m} = \left(\frac{\sin(1 + \tfrac{1}{2})\pi}{(1 + \tfrac{1}{2})\pi}\right)^2 = \left(\frac{\sin 1.5\pi}{1.5\pi}\right)^2$$

$$= 4.50 \times 10^{-2} \approx 4.5\%. \qquad \text{(Answer)}$$

For $m = 2$ and $m = 3$ we find that

$$\frac{I_2}{I_m} = 1.6\% \quad \text{and} \quad \frac{I_3}{I_m} = 0.83\%. \quad \text{(Answer)}$$

Successive secondary maxima decrease rapidly in intensity. Figure 36-1 was deliberately overexposed to reveal them.

## 36-6 | Diffraction by a Circular Aperture

Here we consider diffraction by a circular aperture—that is, a circular opening, such as a circular lens, through which light can pass. Figure 36-10 shows the image of a distant point source of light (a star, for instance) formed on photographic film placed in the focal plane of a converging lens. This image is not a point, as geometrical optics would suggest, but a circular disk surrounded by several progressively fainter secondary rings. Comparison with Fig. 36-1 leaves little doubt that we are dealing with a diffraction phenomenon. Here, however, the aperture is a circle of diameter $d$ rather than a rectangular slit.

The (complex) analysis of such patterns shows that the first minimum for the diffraction pattern of a circular aperture of diameter $d$ is located by

$$\sin \theta = 1.22 \frac{\lambda}{d} \quad \text{(first minimum—circular aperture).} \quad (36\text{-}12)$$

The angle $\theta$ here is the angle from the central axis to any point on that (circular) minimum. Compare this with Eq. 36-1,

$$\sin \theta = \frac{\lambda}{a} \quad \text{(first minimum—single slit),} \quad (36\text{-}13)$$

which locates the first minimum for a long narrow slit of width $a$. The main difference is the factor 1.22, which enters because of the circular shape of the aperture.

### Resolvability

The fact that lens images are diffraction patterns is important when we wish to *resolve* (distinguish) two distant point objects whose angular separation is small. Figure 36-11 shows, in three different cases, the visual appearance and corresponding intensity pattern for two distant point objects (stars, say) with small angular separation. In Figure 36-11a, the objects are not resolved because of diffraction; that is, their diffraction patterns (mainly their central maxima) overlap so much that the two objects cannot be distinguished from a single point object. In Fig. 36-11b the objects are barely resolved, and in Fig. 36-11c they are fully resolved.

In Fig. 36-11b the angular separation of the two point sources is such that the central maximum of the diffraction pattern of one source is centered on the first minimum of the diffraction pattern of the other, a condition called **Rayleigh's criterion** for resolvability. From Eq. 36-12, two objects that are barely resolvable by this criterion must have an angular separation $\theta_R$ of

$$\theta_R = \sin^{-1} \frac{1.22\lambda}{d}.$$

Since the angles are small, we can replace $\sin \theta_R$ with $\theta_R$ expressed in radians:

$$\theta_R = 1.22 \frac{\lambda}{d} \quad \text{(Rayleigh's criterion).} \quad (36\text{-}14)$$

Applying Rayleigh's criterion for resolvability to human vision is only an approximation because visual resolvability depends on many factors, such as the relative brightness of the sources and their surroundings, turbulence in the air

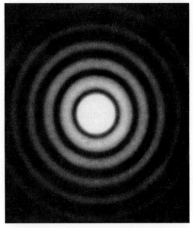

**FIG. 36-10** The diffraction pattern of a circular aperture. Note the central maximum and the circular secondary maxima. The figure has been overexposed to bring out these secondary maxima, which are much less intense than the central maximum. (*From Michel Cagnet, Maurice Franzon, and Jean Thierr, Atlas of Optical Phenomena. Springer-Verlag, New York, 1962. Reproduced with permission.*)

**FIG. 36-11** At the top, the images of two point sources (stars) formed by a converging lens. At the bottom, representations of the image intensities. In (*a*) the angular separation of the sources is too small for them to be distinguished, in (*b*) they can be marginally distinguished, and in (*c*) they are clearly distinguished. Rayleigh's criterion is satisfied in (*b*), with the central maximum of one diffraction pattern coinciding with the first minimum of the other.

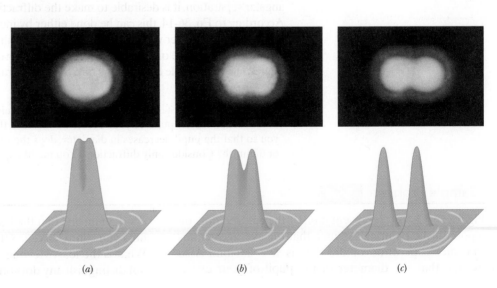

(a)       (b)       (c)

**FIG. 36-12** The pointillistic painting *The Seine at Herblay* by Maximilien Luce consists of thousands of colored dots. With the viewer very close to the canvas, the dots and their true colors are visible. At normal viewing distances, the dots are irresolvable and thus blend. *(Maxilien Luce, The Seine at Herblay, 1890. Musee d'Orsay, Paris, France. Photo by Erich Lessing/Art Resource)*

between the sources and the observer, and the functioning of the observer's visual system. Experimental results show that the least angular separation that can actually be resolved by a person is generally somewhat greater than the value given by Eq. 36-14. However, for calculations here, we shall take Eq. 36-14 as being a precise criterion: If the angular separation $\theta$ between the sources is greater than $\theta_R$, we can visually resolve the sources; if it is less, we cannot.

Rayleigh's criterion can explain the arresting illusions of color in the style of painting known as pointillism (Fig. 36-12). In this style, a painting is made not with brush strokes in the usual sense but rather with a myriad of small colored dots. One fascinating aspect of a pointillistic painting is that when you change your distance from it, the colors shift in subtle, almost subconscious ways. This color shifting has to do with whether you can resolve the colored dots. When you stand close enough to the painting, the angular separations $\theta$ of adjacent dots are greater than $\theta_R$ and thus the dots can be seen individually. Their colors are the true colors of the paints used. However, when you stand far enough from the painting, the angular separations $\theta$ are less than $\theta_R$ and the dots cannot be seen individually. The resulting blend of colors coming into your eye from any group of dots can then cause your brain to "make up" a color for that group—a color that may not actually exist in the group. In this way, a pointillistic painter uses your visual system to create the colors of the art.

When we wish to use a lens instead of our visual system to resolve objects of small angular separation, it is desirable to make the diffraction pattern as small as possible. According to Eq. 36-14, this can be done either by increasing the lens diameter or by using light of a shorter wavelength. For this reason ultraviolet light is often used with microscopes; because of its shorter wavelength, it permits finer detail to be examined than would be possible for the same microscope operated with visible light.

✓ **CHECKPOINT 4** Suppose that you can barely resolve two red dots because of diffraction by the pupil of your eye. If we increase the general illumination around you so that the pupil decreases in diameter, does the resolvability of the dots improve or diminish? Consider only diffraction. (You might experiment to check your answer.)

**Sample Problem** 36-3

Figure 36-13a is a representation of the colored dots on a pointillistic painting. Assume that the average center-to-center separation of the dots is $D = 2.0$ mm. Also assume that the diameter of the pupil of your eye is $d = 1.5$ mm and that the least angular separation between dots you can resolve is set only by Rayleigh's criterion. What is the least viewing distance from which you cannot distinguish any dots on the painting?

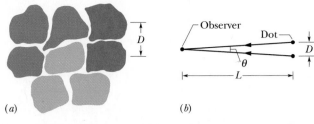

(a)                    (b)

**FIG. 36-13** (a) Representation of some dots on a pointillistic painting, showing an average center-to-center separation D. (b) The arrangement of separation D between two dots, their angular separation θ, and the viewing distance L.

$D/L$ is small, angle $\theta$ is also small and we can make the approximation

$$\theta = \frac{D}{L}. \qquad (36\text{-}16)$$

Setting $\theta$ of Eq. 36-16 equal to $\theta_R$ of Eq. 36-15 and solving for $L$, we then have

$$L = \frac{Dd}{1.22\lambda}. \qquad (36\text{-}17)$$

Equation 36-17 tells us that $L$ is larger for smaller $\lambda$. Thus, as you move away from the painting, adjacent red dots (long wavelengths) become indistinguishable before adjacent blue dots do. To find the least distance $L$ at which *no* colored dots are distinguishable, we substitute $\lambda = 400$ nm (blue or violet light) into Eq. 36-17:

---

**KEY IDEA** Consider any two adjacent dots that you can distinguish when you are close to the painting. As you move away, you continue to distinguish the dots until their angular separation $\theta$ (in your view) has decreased to the angle given by Rayleigh's criterion:

$$\theta_R = 1.22\,\frac{\lambda}{d}. \qquad (36\text{-}15)$$

**Calculations:** Figure 36-13b shows, from the side, the angular separation $\theta$ of the dots, their center-to-center separation $D$, and your distance $L$ from them. Because

$$L = \frac{(2.0 \times 10^{-3}\,\text{m})(1.5 \times 10^{-3}\,\text{m})}{(1.22)(400 \times 10^{-9}\,\text{m})} = 6.1\,\text{m}.$$
(Answer)

At this or a greater distance, the color you perceive at any given spot on the painting is a blended color that may not actually exist there.

---

## Sample Problem | 36-4

A circular converging lens, with diameter $d = 32$ mm and focal length $f = 24$ cm, forms images of distant point objects in the focal plane of the lens. The wavelength is $\lambda = 550$ nm.

(a) Considering diffraction by the lens, what angular separation must two distant point objects have to satisfy Rayleigh's criterion?

---

**KEY IDEA** Figure 36-14 shows two distant point objects $P_1$ and $P_2$, the lens, and a viewing screen in the focal plane of the lens. It also shows, on the right, plots of light intensity $I$ versus position on the screen for the central maxima of the images formed by the lens. Note that the angular separation $\theta_o$ of the objects equals the angular separation $\theta_i$ of the images. Thus, if the images are to satisfy Rayleigh's criterion for resolvability, the angular separations on both sides of the lens must be given by Eq. 36-14 (assuming small angles).

**Calculations:** From Eq. 36-14, we obtain

$$\theta_o = \theta_i = \theta_R = 1.22\,\frac{\lambda}{d}$$

$$= \frac{(1.22)(550 \times 10^{-9}\,\text{m})}{32 \times 10^{-3}\,\text{m}} = 2.1 \times 10^{-5}\,\text{rad}.$$
(Answer)

At this angular separation, each central maximum in the two intensity curves of Fig. 36-14 is centered on the first minimum of the other curve.

(b) What is the separation $\Delta x$ of the centers of the *images* in the focal plane? (That is, what is the separation of the *central* peaks in the two intensity-versus-position curves?)

**Calculations:** From either triangle between the lens and the screen in Fig. 36-14, we see that $\tan \theta_i/2 = \Delta x/2f$. Rearranging this equation and making the approximation $\tan \theta \approx \theta$, we find

$$\Delta x = f\theta_i, \qquad (36\text{-}18)$$

where $\theta_i$ is in radian measure. Substituting known data then yields

$$\Delta x = (0.24\,\text{m})(2.1 \times 10^{-5}\,\text{rad}) = 5.0\,\mu\text{m}. \quad \text{(Answer)}$$

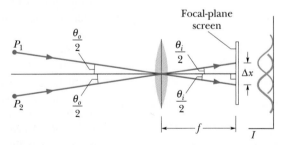

**FIG. 36-14** Light from two distant point objects $P_1$ and $P_2$ passes through a converging lens and forms images on a viewing screen in the focal plane of the lens. Only one representative ray from each object is shown. The images are not points but diffraction patterns, with intensities approximately as plotted at the right. The angular separation of the objects is $\theta_o$ and that of the images is $\theta_i$; the central maxima of the images have a separation $\Delta x$.

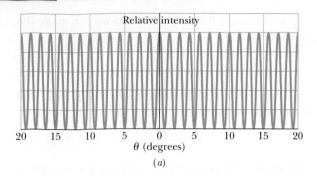

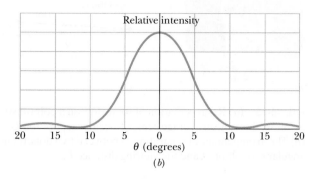

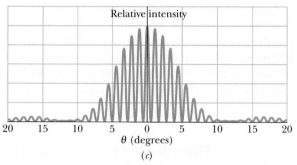

FIG. 36-15 (a) The intensity plot to be expected in a double-slit interference experiment with vanishingly narrow slits. (b) The intensity plot for diffraction by a typical slit of width a (not vanishingly narrow). (c) The intensity plot to be expected for two slits of width a. The curve of (b) acts as an envelope, limiting the intensity of the double-slit fringes in (a). Note that the first minima of the diffraction pattern of (b) eliminate the double-slit fringes that would occur near 12° in (c).

## 36-7 | Diffraction by a Double Slit

In the double-slit experiments of Chapter 35, we implicitly assumed that the slits were much narrower than the wavelength of the light illuminating them; that is, $a \ll \lambda$. For such narrow slits, the central maximum of the diffraction pattern of either slit covers the entire viewing screen. Moreover, the interference of light from the two slits produces bright fringes with approximately the same intensity (Fig. 35-12).

In practice with visible light, however, the condition $a \ll \lambda$ is often not met. For relatively wide slits, the interference of light from two slits produces bright fringes that do not all have the same intensity. That is, the intensities of the fringes produced by double-slit interference (as discussed in Chapter 35) are modified by diffraction of the light passing through each slit (as discussed in this chapter).

As an example, the intensity plot of Fig. 36-15a suggests the double-slit interference pattern that would occur if the slits were infinitely narrow (and thus $a \ll \lambda$); all the bright interference fringes would have the same intensity. The intensity plot of Fig. 36-15b is that for diffraction by a single actual slit; the diffraction pattern has a broad central maximum and weaker secondary maxima at ±17°. The plot of Fig. 36-15c suggests the interference pattern for two actual slits. That plot was constructed by using the curve of Fig. 36-15b as an *envelope* on the intensity plot in Fig. 36-15a. The positions of the fringes are not changed; only the intensities are affected.

Figure 36-16a shows an actual pattern in which both double-slit interference and diffraction are evident. If one slit is covered, the single-slit diffraction pattern of Fig. 36-16b results. Note the correspondence between Figs. 36-16a and 36-15c, and between Figs. 36-16b and 36-15b. In comparing these figures, bear in mind that Fig. 36-16 has been deliberately overexposed to bring out the faint secondary maxima and that two secondary maxima (rather than one) are shown.

With diffraction effects taken into account, the intensity of a double-slit interference pattern is given by

$$I(\theta) = I_m(\cos^2 \beta) \left( \frac{\sin \alpha}{\alpha} \right)^2 \qquad \text{(double slit)}, \qquad (36\text{-}19)$$

FIG. 36-16   (a) Interference fringes for an actual double-slit system; compare with Fig. 36-15c. (b) The diffraction pattern of a single slit; compare with Fig. 36-15b. (*From Michel Cagnet, Maurice Franzon, and Jean Thierr, Atlas of Optical Phenomena. Springer-Verlag, New York, 1962. Reproduced with permission.*)

(a)

(b)

in which
$$\beta = \frac{\pi d}{\lambda} \sin \theta \qquad (36\text{-}20)$$

and
$$\alpha = \frac{\pi a}{\lambda} \sin \theta. \qquad (36\text{-}21)$$

Here $d$ is the distance between the centers of the slits and $a$ is the slit width. Note carefully that the right side of Eq. 36-19 is the product of $I_m$ and two factors. (1) The *interference factor* $\cos^2 \beta$ is due to the interference between two slits with slit separation $d$ (as given by Eqs. 35-22 and 35-23). (2) The *diffraction factor* $[(\sin \alpha)/\alpha]^2$ is due to diffraction by a single slit of width $a$ (as given by Eqs. 36-5 and 36-6).

Let us check these factors. If we let $a \to 0$ in Eq. 36-21, for example, then $\alpha \to 0$ and $(\sin \alpha)/\alpha \to 1$. Equation 36-19 then reduces, as it must, to an equation describing the interference pattern for a pair of vanishingly narrow slits with slit separation $d$. Similarly, putting $d = 0$ in Eq. 36-20 is equivalent physically to causing the two slits to merge into a single slit of width $a$. Then Eq. 36-20 yields $\beta = 0$ and $\cos^2 \beta = 1$. In this case Eq. 36-19 reduces, as it must, to an equation describing the diffraction pattern for a single slit of width $a$.

The double-slit pattern described by Eq. 36-19 and displayed in Fig. 36-16a combines interference and diffraction in an intimate way. Both are superposition effects, in that they result from the combining of waves with different phases at a given point. If the combining waves originate from a small number of elementary coherent sources—as in a double-slit experiment with $a \ll \lambda$—we call the process *interference*. If the combining waves originate in a single wavefront—as in a single-slit experiment—we call the process *diffraction*. This distinction between interference and diffraction (which is somewhat arbitrary and not always adhered to) is a convenient one, but we should not forget that both are superposition effects and usually both are present simultaneously (as in Fig. 36-16a).

**Sample Problem    36-5    Build your skill**

In a double-slit experiment, the wavelength $\lambda$ of the light source is 405 nm, the slit separation $d$ is 19.44 $\mu$m, and the slit width $a$ is 4.050 $\mu$m. Consider the interference of the light from the two slits and also the diffraction of the light through each slit.

(a) How many bright interference fringes are within the central peak of the diffraction envelope?

**KEY IDEAS**   We first analyze the two basic mechanisms responsible for the optical pattern produced in the experiment:

1. *Single-slit diffraction:*   The limits of the central peak are the first minima in the diffraction pattern due to

either slit individually. (See Fig. 36-15.) The angular locations of those minima are given by Eq. 36-3 ($a \sin \theta = m\lambda$). Here let us rewrite this equation as $a \sin \theta = m_1 \lambda$, with the subscript 1 referring to the one-slit diffraction. For the first minima in the diffraction pattern, we substitute $m_1 = 1$, obtaining
$$a \sin \theta = \lambda. \qquad (36\text{-}22)$$

2. *Double-slit interference:* The angular locations of the bright fringes of the double-slit interference pattern are given by Eq. 35-14, which we can write as
$$d \sin \theta = m_2 \lambda, \qquad \text{for } m_2 = 0, 1, 2, \ldots . \quad (36\text{-}23)$$

Here the subscript 2 refers to the double-slit interference.

***Calculations:*** We can locate the first diffraction minimum within the double-slit fringe pattern by dividing
Eq. 36-23 by Eq. 36-22 and solving for $m_2$. By doing so
and then substituting the given data, we obtain

$$m_2 = \frac{d}{a} = \frac{19.44 \ \mu m}{4.050 \ \mu m} = 4.8.$$

This tells us that the bright interference fringe for $m_2 =$
4 fits into the central peak of the one-slit diffraction pattern, but the fringe for $m_2 = 5$ does not fit. Within the
central diffraction peak we have the central bright
fringe ($m_2 = 0$), and four bright fringes (up to $m_2 = 4$)
on each side of it. Thus, a total of nine bright fringes of the
double-slit interference pattern are within the central
peak of the diffraction envelope. The bright fringes to one
side of the central bright fringe are shown in Fig. 36-17.

**(b)** How many bright fringes are within either of the
first side peaks of the diffraction envelope?

**KEY IDEA** The outer limits of the first side diffraction
peaks are the second diffraction minima, each of which
is at the angle $\theta$ given by $a \sin \theta = m_1 \lambda$ with $m_1 = 2$:

$$a \sin \theta = 2\lambda. \qquad (36\text{-}24)$$

***Calculation:*** Dividing Eq. 36-23 by Eq. 36-24, we find

$$m_2 = \frac{2d}{a} = \frac{(2)(19.44 \ \mu m)}{4.050 \ \mu m} = 9.6.$$

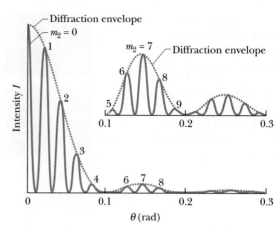

**FIG. 36-17** One side of the intensity plot for a two-slit interference experiment. The inset shows (vertically expanded) the
plot within the first and second side peaks of the diffraction
envelope.

This tells us that the second diffraction minimum occurs
just before the bright interference fringe for $m_2 = 10$ in
Eq. 36-23. Within either first side diffraction peak we
have the fringes from $m_2 = 5$ to $m_2 = 9$, for a total of
five bright fringes of the double-slit interference pattern
(shown in the inset of Fig. 36-17). However, if the $m_2 =$
5 bright fringe, which is almost eliminated by the first
diffraction minimum, is considered too dim to count,
then only four bright fringes are in the first side diffraction peak.

## 36-8 | Diffraction Gratings

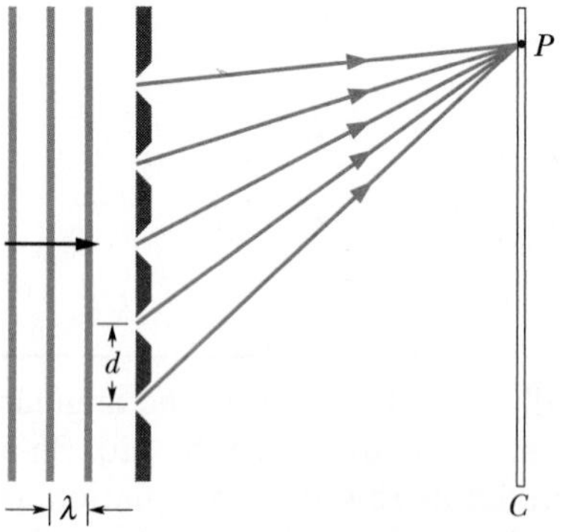

One of the most useful tools in the study of light and of objects that emit and
absorb light is the **diffraction grating.** This device is somewhat like the double-slit
arrangement of Fig. 35-10 but has a much greater number $N$ of slits, often called
*rulings,* perhaps as many as several thousand per millimeter. An idealized grating
consisting of only five slits is represented in Fig. 36-18. When monochromatic
light is sent through the slits, it forms narrow interference fringes that can be
analyzed to determine the wavelength of the light. (Diffraction gratings can also
be opaque surfaces with narrow parallel grooves arranged like the slits in Fig.
36-18. Light then scatters back from the grooves to form interference fringes
rather than being transmitted through open slits.)

With monochromatic light incident on a diffraction grating, if we gradually
increase the number of slits from two to a large number $N$, the intensity plot
changes from the typical double-slit plot of Fig. 36-15c to a much more complicated one and then eventually to a simple graph like that shown in Fig. 36-19a. The
pattern you would see on a viewing screen using monochromatic red light from,
say, a helium–neon laser is shown in Fig. 36-19b. The maxima are now very narrow
(and so are called *lines*); they are separated by relatively wide dark regions.

We use a familiar procedure to find the locations of the bright lines on the
viewing screen. We first assume that the screen is far enough from the grating so
that the rays reaching a particular point $P$ on the screen are approximately parallel when they leave the grating (Fig. 36-20). Then we apply to each pair of
adjacent rulings the same reasoning we used for double-slit interference. The separation $d$ between rulings is called the *grating spacing.* (If $N$ rulings occupy a total

**FIG. 36-18** An idealized diffraction
grating, consisting of only five rulings,
that produces an interference pattern
on a distant viewing screen $C$.

width $w$, then $d = w/N$.) The path length difference between adjacent rays is again $d \sin \theta$ (Fig. 36-20), where $\theta$ is the angle from the central axis of the grating (and of the diffraction pattern) to point $P$. A line will be located at $P$ if the path length difference between adjacent rays is an integer number of wavelengths—that is, if

$$d \sin \theta = m\lambda, \quad \text{for } m = 0, 1, 2, \ldots \quad \text{(maxima—lines)}, \quad (36\text{-}25)$$

where $\lambda$ is the wavelength of the light. Each integer $m$ represents a different line; hence these integers can be used to label the lines, as in Fig. 36-19. The integers are then called the *order numbers*, and the lines are called the zeroth-order line (the central line, with $m = 0$), the first-order line ($m = 1$), the second-order line ($m = 2$), and so on.

If we rewrite Eq. 36-25 as $\theta = \sin^{-1}(m\lambda/d)$, we see that, for a given diffraction grating, the angle from the central axis to any line (say, the third-order line) depends on the wavelength of the light being used. Thus, when light of an unknown wavelength is sent through a diffraction grating, measurements of the angles to the higher-order lines can be used in Eq. 36-25 to determine the wavelength. Even light of several unknown wavelengths can be distinguished and identified in this way. We cannot do that with the double-slit arrangement of Section 35-4, even though the same equation and wavelength dependence apply there. In double-slit interference, the bright fringes due to different wavelengths overlap too much to be distinguished.

## Width of the Lines

A grating's ability to resolve (separate) lines of different wavelengths depends on the width of the lines. We shall here derive an expression for the *half-width* of the central line (the line for which $m = 0$) and then state an expression for the half-widths of the higher-order lines. We define the **half-width** of the central line as being the angle $\Delta\theta_{hw}$ from the center of the line at $\theta = 0$ outward to where the line effectively ends and darkness effectively begins with the first minimum (Fig. 36-21). At such a minimum, the $N$ rays from the $N$ slits of the grating cancel one another. (The actual width of the central line is, of course, $2(\Delta\theta_{hw})$, but line widths are usually compared via half-widths.)

In Section 36-3 we were also concerned with the cancellation of a great many rays, there due to diffraction through a single slit. We obtained Eq. 36-3, which, because of the similarity of the two situations, we can use to find the first minimum here. It tells us that the first minimum occurs where the path length difference between the top and bottom rays equals $\lambda$. For single-slit diffraction, this difference is $a \sin \theta$. For a grating of $N$ rulings, each separated from the next by distance $d$, the distance between the top and bottom rulings is $Nd$ (Fig. 36-22),

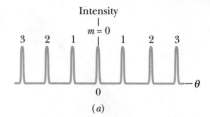
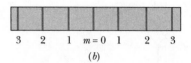

**FIG. 36-19** (*a*) The intensity plot produced by a diffraction grating with a great many rulings consists of narrow peaks, here labeled with their order numbers $m$. (*b*) The corresponding bright fringes seen on the screen are called lines and are here also labeled with order numbers $m$.

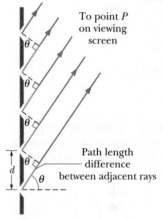

**FIG. 36-20** The rays from the rulings in a diffraction grating to a distant point $P$ are approximately parallel. The path length difference between each two adjacent rays is $d \sin \theta$, where $\theta$ is measured as shown. (The rulings extend into and out of the page.)

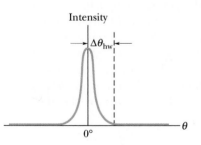

**FIG. 36-21** The half-width $\Delta\theta_{hw}$ of the central line is measured from the center of that line to the adjacent minimum on a plot of $I$ versus $\theta$ like Fig. 36-19a.

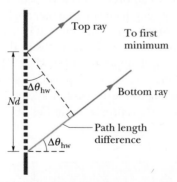

**FIG. 36-22** The top and bottom rulings of a diffraction grating of $N$ rulings are separated by $Nd$. The top and bottom rays passing through these rulings have a path length difference of $Nd \sin \Delta\theta_{hw}$, where $\Delta\theta_{hw}$ is the angle to the first minimum. (The angle is here greatly exaggerated for clarity.)

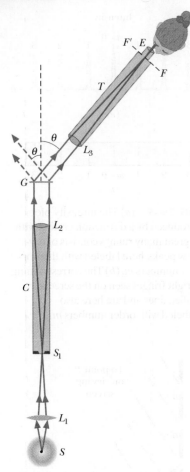

**FIG. 36-23** A simple type of grating spectroscope used to analyze the wavelengths of light emitted by source $S$.

and so the path length difference between the top and bottom rays here is $Nd \sin \Delta\theta_{hw}$. Thus, the first minimum occurs where

$$Nd \sin \Delta\theta_{hw} = \lambda. \qquad (36\text{-}26)$$

Because $\Delta\theta_{hw}$ is small, $\sin \Delta\theta_{hw} = \Delta\theta_{hw}$ (in radian measure). Substituting this in Eq. 36-26 gives the half-width of the central line as

$$\Delta\theta_{hw} = \frac{\lambda}{Nd} \qquad \text{(half-width of central line).} \qquad (36\text{-}27)$$

We state without proof that the half-width of any other line depends on its location relative to the central axis and is

$$\Delta\theta_{hw} = \frac{\lambda}{Nd \cos \theta} \qquad \text{(half-width of line at } \theta\text{).} \qquad (36\text{-}28)$$

Note that for light of a given wavelength $\lambda$ and a given ruling separation $d$, the widths of the lines decrease with an increase in the number $N$ of rulings. Thus, of two diffraction gratings, the grating with the larger value of $N$ is better able to distinguish between wavelengths because its diffraction lines are narrower and so produce less overlap.

## Grating Spectroscope

Diffraction gratings are widely used to determine the wavelengths that are emitted by sources of light ranging from lamps to stars. Figure 36-23 shows a simple *grating spectroscope* in which a grating is used for this purpose. Light from source $S$ is focused by lens $L_1$ on a vertical slit $S_1$ placed in the focal plane of lens $L_2$. The light emerging from tube $C$ (called a *collimator*) is a plane wave and is incident perpendicularly on grating $G$, where it is diffracted into a diffraction pattern, with the $m = 0$ order diffracted at angle $\theta = 0$ along the central axis of the grating.

We can view the diffraction pattern that would appear on a viewing screen at any angle $\theta$ simply by orienting telescope $T$ in Fig. 36-23 to that angle. Lens $L_3$ of the telescope then focuses the light diffracted at angle $\theta$ (and at slightly smaller and larger angles) onto a focal plane $FF'$ within the telescope. When we look through eyepiece $E$, we see a magnified view of this focused image.

By changing the angle $\theta$ of the telescope, we can examine the entire diffraction pattern. For any order number other than $m = 0$, the original light is spread out according to wavelength (or color) so that we can determine, with Eq. 36-25, just what wavelengths are being emitted by the source. If the source emits discrete wavelengths, what we see as we rotate the telescope horizontally through the angles corresponding to an order $m$ is a vertical line of color for each wavelength, with the shorter-wavelength line at a smaller angle $\theta$ than the longer-wavelength line.

For example, the light emitted by a hydrogen lamp, which contains hydrogen gas, has four discrete wavelengths in the visible range. If our eyes intercept this light directly, it appears to be white. If, instead, we view it through a grating spectroscope, we can distinguish, in several orders, the lines of the four colors corresponding to these visible wavelengths. (Such lines are called *emission lines.*) Four orders are represented in Fig. 36-24. In the central order ($m = 0$), the lines

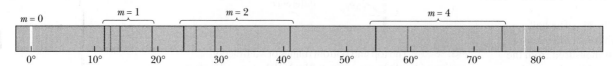

**FIG. 36-24** The zeroth, first, second, and fourth orders of the visible emission lines from hydrogen. Note that the lines are farther apart at greater angles. (They are also dimmer and wider, although that is not shown here.)

FIG. 36-25 The visible emission lines of cadmium, as seen through a grating spectroscope. *(Department of Physics, Imperial College/Science Photo Library/Photo Researchers)*

corresponding to all four wavelengths are superimposed, giving a single white line at $\theta = 0$. The colors are separated in the higher orders.

The third order is not shown in Fig. 36-24 for the sake of clarity; it actually overlaps the second and fourth orders. The fourth-order red line is missing because it is not formed by the grating used here. That is, when we attempt to solve Eq. 36-25 for the angle $\theta$ for the red wavelength when $m = 4$, we find that $\sin \theta$ is greater than unity, which is not possible. The fourth order is then said to be *incomplete* for this grating; it might not be incomplete for a grating with greater spacing $d$, which will spread the lines less than in Fig. 36-24. Figure 36-25 is a photograph of the visible emission lines produced by cadmium.

## Optically Variable Graphics

Holograms are made by having laser light scatter from an object onto an emulsion. Once the hologram is developed, an image of the object can be created by illuminating the hologram with the same type of laser light. The image is arresting because, unlike common photographs, it has depth, and you can change your perspective of the object by changing the angle at which you view the hologram.

Holograms were thought to be an ideal anticounterfeiting measure for credit cards and other types of personal cards. However, they have several disadvantages. (1) A holographic image can be sharp when viewed in laser light, which is coherent and incident from a single direction. However, the image is murky ("milky") when viewed in the normal light of a store. (Such *diffuse light* is incoherent and incident from many directions.) Thus a store clerk is unlikely to examine a display on a credit card closely enough to see whether it is a legitimate hologram. (2) A hologram can be easily counterfeited because it is a photograph of an actual object. A counterfeiter merely makes a model of that object, then makes a hologram of it, and then attaches the hologram to a counterfeit credit card.

Most credit cards and many identification cards now carry *optically variable graphics* (OVG), which produce an image via the diffraction of diffuse light by gratings embedded in the device (Fig. 36-26). The gratings send out hundreds or even thousands of different orders. Someone viewing the card intercepts some of these orders, and the combined light creates a virtual image that is part of, say, a credit-card logo. For example, in Fig. 36-27a, gratings at point $a$ produce a certain image when the viewer is at orientation $A$, and in Fig. 36-27b, gratings at point $b$ produce a different image when the viewer is at orientation $B$. These images are bright and sharp because the gratings have been designed to be viewed in diffuse light.

An OVG is very difficult to design because optical engineers must work backwards from a graphic, such as a given logo. The engineers must determine the grating properties across the OVG if a certain image is to be seen from one set of viewing angles and a different image is to be seen from a different set of viewing angles. Such work requires sophisticated programming on computers. Once designed, the OVG structure is so complicated that counterfeiting it is extremely difficult.

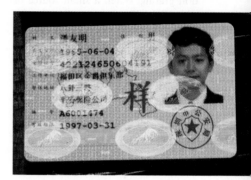

FIG. 36-26 An identification card with OVG. *(Pascal Goetgheluck/Photo Researchers)*

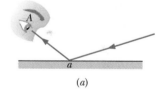

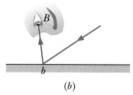

FIG. 36-27 (a) Gratings at point $a$ on the surface of an OVG device send light to a viewer at orientation $A$, creating a certain virtual image. (b) Gratings at point $b$ send light to the viewer at orientation $B$, creating a different virtual image.

✓ **CHECKPOINT 5** The figure shows lines of different orders produced by a diffraction grating in monochromatic red light. (a) Is the center of the pattern to the left or right? (b) In monochromatic green light, are the half-widths of the lines produced in the same orders greater than, less than, or the same as the half-widths of the lines shown?

The fine rulings, each 0.5 $\mu$m wide, on a compact disc function as a diffraction grating. When a small source of white light illuminates a disc, the diffracted light forms colored "lanes" that are the composite of the diffraction patterns from the rulings.
*(Kristen Brochmann/Fundamental Photographs)*

# 36-9 | Gratings: Dispersion and Resolving Power

## Dispersion

To be useful in distinguishing wavelengths that are close to each other (as in a grating spectroscope), a grating must spread apart the diffraction lines associated with the various wavelengths. This spreading, called **dispersion,** is defined as

$$D = \frac{\Delta\theta}{\Delta\lambda} \qquad \text{(dispersion defined).} \tag{36-29}$$

Here $\Delta\theta$ is the angular separation of two lines whose wavelengths differ by $\Delta\lambda$. The greater $D$ is, the greater is the distance between two emission lines whose wavelengths differ by $\Delta\lambda$. We show below that the dispersion of a grating at angle $\theta$ is given by

$$D = \frac{m}{d\cos\theta} \qquad \text{(dispersion of a grating).} \tag{36-30}$$

Thus, to achieve higher dispersion we must use a grating of smaller grating spacing $d$ and work in a higher-order $m$. Note that the dispersion does not depend on the number of rulings $N$ in the grating. The SI unit for $D$ is the degree per meter or the radian per meter.

## Resolving Power

To *resolve* lines whose wavelengths are close together (that is, to make the lines distinguishable), the line should also be as narrow as possible. Expressed otherwise, the grating should have a high **resolving power** $R$, defined as

$$R = \frac{\lambda_{\text{avg}}}{\Delta\lambda} \qquad \text{(resolving power defined).} \tag{36-31}$$

Here $\lambda_{\text{avg}}$ is the mean wavelength of two emission lines that can barely be recognized as separate, and $\Delta\lambda$ is the wavelength difference between them. The greater $R$ is, the closer two emission lines can be and still be resolved. We shall show below that the resolving power of a grating is given by the simple expression

$$R = Nm \qquad \text{(resolving power of a grating).} \tag{36-32}$$

To achieve high resolving power, we must use many rulings (large $N$).

## Proof of Eq. 36-30

Let us start with Eq. 36-25, the expression for the locations of the lines in the diffraction pattern of a grating:

$$d\sin\theta = m\lambda.$$

Let us regard $\theta$ and $\lambda$ as variables and take differentials of this equation. We find

$$d(\cos\theta)\,d\theta = m\,d\lambda.$$

For small enough angles, we can write these differentials as small differences, obtaining

$$d(\cos\theta)\,\Delta\theta = m\,\Delta\lambda \tag{36-33}$$

or

$$\frac{\Delta\theta}{\Delta\lambda} = \frac{m}{d\cos\theta}.$$

The ratio on the left is simply $D$ (see Eq. 36-29), and so we have indeed derived Eq. 36-30.

**TABLE 36-1**

**Three Gratings**[a]

| Grating | $N$ | $d$ (nm) | $\theta$ | $D$ (°/$\mu$m) | $R$ |
|---|---|---|---|---|---|
| A | 10 000 | 2540 | 13.4° | 23.2 | 10 000 |
| B | 20 000 | 2540 | 13.4° | 23.2 | 20 000 |
| C | 10 000 | 1360 | 25.5° | 46.3 | 10 000 |

[a]Data are for $\lambda = 589$ nm and $m = 1$.

## Proof of Eq. 36-32

We start with Eq. 36-33, which was derived from Eq. 36-25, the expression for the locations of the lines in the diffraction pattern formed by a grating. Here $\Delta\lambda$ is the small wavelength difference between two waves that are diffracted by the grating, and $\Delta\theta$ is the angular separation between them in the diffraction pattern. If $\Delta\theta$ is to be the smallest angle that will permit the two lines to be resolved, it must (by Rayleigh's criterion) be equal to the half-width of each line, which is given by Eq. 36-28:

$$\Delta\theta_{hw} = \frac{\lambda}{Nd\cos\theta}.$$

If we substitute $\Delta\theta_{hw}$ as given here for $\Delta\theta$ in Eq. 36-33, we find that

$$\frac{\lambda}{N} = m\,\Delta\lambda,$$

from which it readily follows that

$$R = \frac{\lambda}{\Delta\lambda} = Nm.$$

This is Eq. 36-32, which we set out to derive.

## Dispersion and Resolving Power Compared

The resolving power of a grating must not be confused with its dispersion. Table 36-1 shows the characteristics of three gratings, all illuminated with light of wavelength $\lambda = 589$ nm, whose diffracted light is viewed in the first order ($m = 1$ in Eq. 36-25). You should verify that the values of $D$ and $R$ as given in the table can be calculated with Eqs. 36-30 and 36-32, respectively. (In the calculations for $D$, you will need to convert radians per meter to degrees per micrometer.)

For the conditions noted in Table 36-1, gratings $A$ and $B$ have the same *dispersion D* and $A$ and $C$ have the same *resolving power R*.

Figure 36-28 shows the intensity patterns (also called *line shapes*) that would be produced by these gratings for two lines of wavelengths $\lambda_1$ and $\lambda_2$, in the vicinity of $\lambda = 589$ nm. Grating $B$, with the higher resolving power, produces narrower lines and thus is capable of distinguishing lines that are much closer together in wavelength than those in the figure. Grating $C$, with the higher dispersion, produces the greater angular separation between the lines.

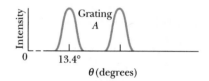
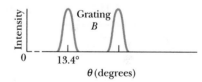
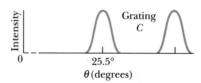

**FIG. 36-28** The intensity patterns for light of two wavelengths sent through the gratings of Table 36-1. Grating $B$ has the highest resolving power, and grating $C$ the highest dispersion.

---

**Sample Problem** | **36-6**

A diffraction grating has $1.26 \times 10^4$ rulings uniformly spaced over width $w = 25.4$ mm. It is illuminated at normal incidence by yellow light from a sodium vapor lamp. This light contains two closely spaced emission lines (known as the sodium doublet) of wavelengths 589.00 nm and 589.59 nm.

(a) At what angle does the first-order maximum occur (on either side of the center of the diffraction pattern) for the wavelength of 589.00 nm?

**KEY IDEA** The maxima produced by the diffraction grating can be determined with Eq. 36-25 ($d\sin\theta = m\lambda$).

*Calculations:* The grating spacing $d$ is

$$d = \frac{w}{N} = \frac{25.4 \times 10^{-3}\,\text{m}}{1.26 \times 10^4}$$

$$= 2.016 \times 10^{-6}\,\text{m} = 2016\,\text{nm}.$$

The first-order maximum corresponds to $m = 1$. Substituting these values for $d$ and $m$ into Eq. 36-25 leads to

$$\theta = \sin^{-1}\frac{m\lambda}{d} = \sin^{-1}\frac{(1)(589.00\,\text{nm})}{2016\,\text{nm}}$$

$$= 16.99° \approx 17.0°. \qquad \text{(Answer)}$$

(b) Using the dispersion of the grating, calculate the angular separation between the two lines in the first order.

**KEY IDEAS** (1) The angular separation $\Delta\theta$ between the two lines in the first order depends on their wavelength difference $\Delta\lambda$ and the dispersion $D$ of the grating, according to Eq. 36-29 ($D = \Delta\theta/\Delta\lambda$). (2) The dispersion $D$ depends on the angle $\theta$ at which it is to be evaluated.

*Calculations:* We can assume that, in the first order, the two sodium lines occur close enough to each other for us to evaluate $D$ at the angle $\theta = 16.99°$ we found in part (a) for one of those lines. Then Eq. 36-30 gives the dispersion as

$$D = \frac{m}{d\cos\theta} = \frac{1}{(2016\,\text{nm})(\cos 16.99°)}$$

$$= 5.187 \times 10^{-4}\,\text{rad/nm}.$$

From Eq. 36-29 and with $\Delta\lambda$ in nanometers, we then have

$$\Delta\theta = D\,\Delta\lambda = (5.187 \times 10^{-4}\,\text{rad/nm})(589.59 - 589.00)$$

$$= 3.06 \times 10^{-4}\,\text{rad} = 0.0175°. \qquad \text{(Answer)}$$

You can show that this result depends on the grating spacing $d$ but not on the number of rulings there are in the grating.

(c) What is the least number of rulings a grating can have and still be able to resolve the sodium doublet in the first order?

**KEY IDEAS** (1) The resolving power of a grating in any order $m$ is physically set by the number of rulings $N$ in the grating according to Eq. 36-32 ($R = Nm$). (2) The smallest wavelength difference $\Delta\lambda$ that can be resolved depends on the average wavelength involved and on the resolving power $R$ of the grating, according to Eq. 36-31 ($R = \lambda_{\text{avg}}/\Delta\lambda$).

*Calculation:* For the sodium doublet to be barely resolved, $\Delta\lambda$ must be their wavelength separation of 0.59 nm, and $\lambda_{\text{avg}}$ must be their average wavelength of 589.30 nm. Thus, we find that the smallest number of rulings for a grating to resolve the sodium doublet is

$$N = \frac{R}{m} = \frac{\lambda_{\text{avg}}}{m\,\Delta\lambda}$$

$$= \frac{589.30\,\text{nm}}{(1)(0.59\,\text{nm})} = 999\,\text{rulings}. \qquad \text{(Answer)}$$

## 36-10 | Diffraction by Organized Layers

Here we examine two examples of optical interference due to the diffraction of light by organized layers. First, we discuss x-ray diffraction by organized layers of atoms in crystals, an effect known for about 100 years. Then we discuss visible-light diffraction by organized layers in organic materials, including the skin of mandrill baboons, which is a current subject of research.

### X-Ray Diffraction

X rays are electromagnetic radiation whose wavelengths are of the order of 1 Å ($= 10^{-10}$ m). Compare this with a wavelength of 550 nm ($= 5.5 \times 10^{-7}$ m) at the center of the visible spectrum. Figure 36-29 shows that x rays are produced when electrons escaping from a heated filament $F$ are accelerated by a potential difference $V$ and strike a metal target $T$.

A standard optical diffraction grating cannot be used to discriminate between different wavelengths in the x-ray wavelength range. For $\lambda = 1$ Å ($= 0.1$ nm) and $d = 3000$ nm, for example, Eq. 36-25 shows that the first-order maximum occurs at

$$\theta = \sin^{-1}\frac{m\lambda}{d} = \sin^{-1}\frac{(1)(0.1\,\text{nm})}{3000\,\text{nm}} = 0.0019°.$$

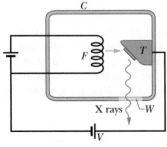

**FIG. 36-29** X rays are generated when electrons leaving heated filament $F$ are accelerated through a potential difference $V$ and strike a metal target $T$. The "window" $W$ in the evacuated chamber $C$ is transparent to x rays.

This is too close to the central maximum to be practical. A grating with $d \approx \lambda$ is desirable, but, because x-ray wavelengths are about equal to atomic diameters, such gratings cannot be constructed mechanically.

In 1912, it occurred to German physicist Max von Laue that a crystalline solid, which consists of a regular array of atoms, might form a natural three-dimensional "diffraction grating" for x-rays. The idea is that, in a crystal such as sodium chloride (NaCl), a basic unit of atoms (called the *unit cell*) repeats itself throughout the array. Figure 36-30a represents a section through a crystal of NaCl and identifies this basic unit. The unit cell is a cube measuring $a_0$ on each side.

When an x-ray beam enters a crystal such as NaCl, x rays are *scattered*—that is, redirected—in all directions by the crystal structure. In some directions the scattered waves undergo destructive interference, resulting in intensity minima; in other directions the interference is constructive, resulting in intensity maxima. This process of scattering and interference is a form of diffraction, although it is unlike the diffraction of light traveling through a slit or past an edge as we discussed earlier.

Although the process of diffraction of x rays by a crystal is complicated, the maxima turn out to be in directions *as if* the x rays were reflected by a family of parallel *reflecting planes* (or *crystal planes*) that extend through the atoms within the crystal and that contain regular arrays of the atoms. (The x rays are not actually reflected; we use these fictional planes only to simplify the analysis of the actual diffraction process.)

Figure 36-30b shows three reflecting planes (part of a family containing many parallel planes) with *interplanar spacing d,* from which the incident rays shown are said to reflect. Rays 1, 2, and 3 reflect from the first, second, and third planes, respectively. At each reflection the angle of incidence and the angle of reflection are represented with $\theta$. Contrary to the custom in optics, these angles are defined relative to the *surface* of the reflecting plane rather than a normal to that surface. For the situation of Fig. 36-30b, the interplanar spacing happens to be equal to the unit cell dimension $a_0$.

Figure 36-30c shows an edge-on view of reflection from an adjacent pair of planes. The waves of rays 1 and 2 arrive at the crystal in phase. After they are reflected, they must again be in phase because the reflections and the reflecting planes have been defined solely to explain the intensity maxima in the diffraction of x rays by a crystal. Unlike light rays, the x rays do not refract upon entering the crystal; moreover, we do not define an index of refraction for this situation. Thus, the relative phase between the waves of rays 1 and 2 as they leave the crystal is set solely by their path length difference. For these rays to be in phase, the path length difference must be equal to an integer multiple of the wavelength $\lambda$ of the x rays.

By drawing the dashed perpendiculars in Fig. 36-30c, we find that the path length difference is $2d \sin \theta$. In fact, this is true for any pair of adjacent planes in the family of planes represented in Fig. 36-30b. Thus, we have, as the criterion for intensity maxima for x-ray diffraction,

$$2d \sin \theta = m\lambda, \qquad \text{for } m = 1, 2, 3, \ldots \quad \text{(Bragg's law)}, \qquad (36\text{-}34)$$

where $m$ is the order number of an intensity maximum. Equation 36-34 is called **Bragg's law** after British physicist W. L. Bragg, who first derived it. (He and his father shared the 1915 Nobel Prize in physics for their use of x rays to study the structures of crystals.) The angle of incidence and reflection in Eq. 36-34 is called a *Bragg angle.*

Regardless of the angle at which x rays enter a crystal, there is always a family of planes from which they can be said to reflect so that we can apply Bragg's law. In Fig. 36-30d, notice that the crystal structure has the same orientation as it does in Fig. 36-30a, but the angle at which the beam enters the structure differs from that shown in Fig. 36-30b. This new angle requires a new family of reflecting

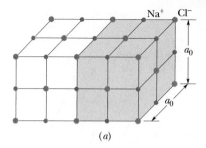

(a)

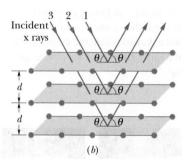

(b)

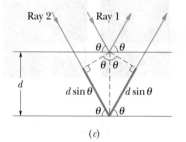

(c)

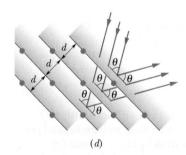

(d)

**FIG. 36-30** (a) The cubic structure of NaCl, showing the sodium and chlorine ions and a unit cell (shaded). (b) Incident x rays undergo diffraction by the structure of (a). The x rays are diffracted as if they were reflected by a family of parallel planes, with the angle of reflection equal to the angle of incidence, both angles measured relative to the planes (not relative to a normal as in optics). (c) The path length difference between waves effectively reflected by two adjacent planes is $2d \sin \theta$. (d) A different orientation of the incident x rays relative to the structure. A different family of parallel planes now effectively reflects the x rays.

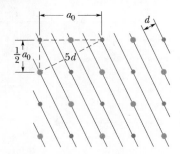

**FIG. 36-31** A family of planes through the structure of Fig. 36-30$a$, and a way to relate the edge length $a_0$ of a unit cell to the interplanar spacing $d$.

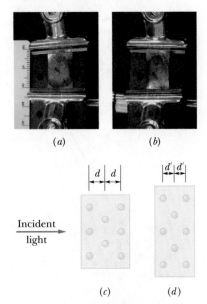

(a)        (b)

(c)        (d)

**FIG. 36-32** Color-shifting by a rubber sheet with embedded polystyrene spheres. ($a$) and ($c$) Unstretched, with the spheres forming planes of separation $d$. ($b$) and ($d$) Stretched, with the spheres forming planes of smaller separation $d'$. (*Courtesy Hiroshi Fudouzi*)

planes, with a different interplanar spacing $d$ and different Bragg angle $\theta$, in order to explain the x-ray diffraction via Bragg's law.

Figure 36-31 shows how the interplanar spacing $d$ can be related to the unit cell dimension $a_0$. For the particular family of planes shown there, the Pythagorean theorem gives

$$5d = \sqrt{\tfrac{5}{4}a_0^2},$$

or

$$d = \frac{a_0}{\sqrt{20}} = 0.2236a_0. \qquad (36\text{-}35)$$

Figure 36-31 suggests how the dimensions of the unit cell can be found once the interplanar spacing has been measured by means of x-ray diffraction.

X-ray diffraction is a powerful tool for studying both x-ray spectra and the arrangement of atoms in crystals. To study spectra, a particular set of crystal planes, having a known spacing $d$, is chosen. These planes effectively reflect different wavelengths at different angles. A detector that can discriminate one angle from another can then be used to determine the wavelength of radiation reaching it. The crystal itself can be studied with a monochromatic x-ray beam, to determine not only the spacing of various crystal planes but also the structure of the unit cell.

### Structural Coloring by Diffraction

Figure 36-32$a$ shows a transparent unstretched rubber sheet in which polystyrene spheres have been embedded fairly uniformly, forming planes with separation $d$ (Fig. 36-32$c$). When visible light penetrates the sheet, it is diffracted by the spheres and a portion emerges toward us. However, because there is a path length difference between the light that reaches us from any two planes, the light undergoes interference. With the sheet unstretched, constructive interference occurs at wavelengths around 590 nm, giving us red light (Fig. 36-32$a$). When the sheet is stretched, it thins, which causes the separation between planes to narrow to $d'$ (Fig. 36-32$d$) and thus the wavelengths for constructive interference to decrease. In Fig. 36-32$b$ the stretched sheet gives us green light at wavelengths around 563 nm. This coloration is said to be an example of *structural coloring* because it is due to the interference of light caused by the structural arrangement within a material.

Similar structural coloring occurs in the facial skin of mandrill baboons (and the skin on some other animals) because of diffraction from parallel collagen fibers beneath the skin surface. Light penetrates the skin, diffracts from the fibers, and then emerges from the skin. The fiber separation results in constructive interference for wavelengths around 460 nm, so the skin is blue. It is not iridescent (like the wings on some types of butterflies) because the parallel organization of collagen fibers is uniform only in microscopic regions, an arrangement said to be *quasi-ordered.*

## REVIEW & SUMMARY

**Diffraction** When waves encounter an edge, an obstacle, or an aperture the size of which is comparable to the wavelength of the waves, those waves spread out as they travel and, as a result, undergo interference. This is called **diffraction.**

**Single-Slit Diffraction** Waves passing through a long narrow slit of width $a$ produce, on a viewing screen, a **single-slit diffraction pattern** that includes a central maximum and other maxima, separated by minima located at angles $\theta$ to the central axis that satisfy

$$a \sin \theta = m\lambda, \quad \text{for } m = 1, 2, 3, \ldots \quad \text{(minima)}. \quad (36\text{-}3)$$

The intensity of the diffraction pattern at any given angle $\theta$ is

$$I(\theta) = I_m \left( \frac{\sin \alpha}{\alpha} \right)^2, \quad \text{where } \alpha = \frac{\pi a}{\lambda} \sin \theta \quad (36\text{-}5, 36\text{-}6)$$

and $I_m$ is the intensity at the center of the pattern.

**Circular-Aperture Diffraction** Diffraction by a circular aperture or a lens with diameter $d$ produces a central maximum and concentric maxima and minima, with the first minimum at an angle $\theta$ given by

$$\sin \theta = 1.22 \frac{\lambda}{d} \quad \text{(first minimum—circular aperture)}. \quad (36\text{-}12)$$

**Rayleigh's Criterion** *Rayleigh's criterion* suggests that two objects are on the verge of resolvability if the central diffraction maximum of one is at the first minimum of the other. Their angular separation must then be at least

$$\theta_R = 1.22 \frac{\lambda}{d} \qquad \text{(Rayleigh's criterion)}, \qquad (36\text{-}14)$$

in which $d$ is the diameter of the aperture through which the light passes.

**Double-Slit Diffraction** Waves passing through two slits, each of width $a$, whose centers are a distance $d$ apart, display diffraction patterns whose intensity $I$ at angle $\theta$ is

$$I(\theta) = I_m(\cos^2 \beta) \left( \frac{\sin \alpha}{\alpha} \right)^2 \qquad \text{(double slit)}, \qquad (36\text{-}19)$$

with $\beta = (\pi d/\lambda) \sin \theta$ and $\alpha$ as for single-slit diffraction.

**Diffraction Gratings** A *diffraction grating* is a series of "slits" used to separate an incident wave into its component wavelengths by separating and displaying their diffraction maxima. Diffraction by $N$ (multiple) slits results in maxima (lines) at angles $\theta$ such that

$$d \sin \theta = m\lambda, \qquad \text{for } m = 0, 1, 2, \dots \qquad \text{(maxima)}, \qquad (36\text{-}25)$$

with the **half-widths** of the lines given by

$$\Delta\theta_{hw} = \frac{\lambda}{Nd \cos \theta} \qquad \text{(half-widths)}. \qquad (36\text{-}28)$$

The dispersion $D$ and resolving power $R$ are given by

$$D = \frac{\Delta\theta}{\Delta\lambda} = \frac{m}{d \cos \theta} \qquad (36\text{-}29, 36\text{-}30)$$

$$R = \frac{\lambda_{avg}}{\Delta\lambda} = Nm. \qquad (36\text{-}31, 36\text{-}32)$$

**X-Ray Diffraction** The regular array of atoms in a crystal is a three-dimensional diffraction grating for short-wavelength waves such as x rays. For analysis purposes, the atoms can be visualized as being arranged in planes with characteristic interplanar spacing $d$. Diffraction maxima (due to constructive interference) occur if the incident direction of the wave, measured from the surfaces of these planes, and the wavelength $\lambda$ of the radiation satisfy **Bragg's law:**

$$2d \sin \theta = m\lambda, \quad \text{for } m = 1, 2, 3, \dots \qquad \text{(Bragg's law)}. \qquad (36\text{-}34)$$

# QUESTIONS

**1** For three experiments, Fig. 36-33 gives the parameter $\beta$ of Eq. 36-20 versus angle $\theta$ for two-slit interference using light of wavelength 500 nm. The slit separations in the three experiments differ. Rank the experiments according to (a) the slit separations and (b) the total number of two-slit interference maxima in the pattern, greatest first.

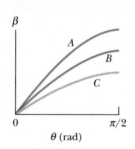

**FIG. 36-33** Question 1.

**2** For three experiments, Fig. 36-34 gives $\alpha$ versus angle $\theta$ in one-slit diffraction using light of wavelength 500 nm. Rank the experiments according to (a) the slit widths and (b) the total number of diffraction minima in the pattern, greatest first.

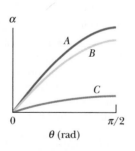

**FIG. 36-34** Question 2.

**3** Figure 36-35 shows four choices for the rectangular opening of a source of either sound waves or light waves. The sides have lengths of either $L$ or $2L$, with $L$ being 3.0 times the wavelength of the waves. Rank the openings according to the extent of (a) left–right spreading and (b) up–down spreading of the waves due to diffraction, greatest first.

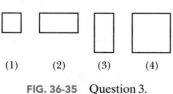

**FIG. 36-35** Question 3.

**4** Light of frequency $f$ illuminating a long narrow slit produces a diffraction pattern. (a) If we switch to light of frequency 1.3$f$, does the pattern expand away from the center or contract toward the center? (b) Does the pattern expand or contract if, instead, we submerge the equipment in clear corn syrup?

**5** You are conducting a single-slit diffraction experiment with light of wavelength $\lambda$. What appears, on a distant viewing screen, at a point at which the top and bottom rays through the slit have a path length difference equal to (a) $5\lambda$ and (b) $4.5\lambda$?

**6** In a single-slit diffraction experiment, the top and bottom rays through the slit arrive at a certain point on the viewing screen with a path length difference of 4.0 wavelengths. In a phasor representation like those in Fig 36-7, how many overlapping circles does the chain of phasors make?

**7** Figure 36-36 shows a red line and a green line of the same order in the pattern produced by a diffraction grating. If we increased the number of rulings in the grating—say, by removing tape that had covered the outer half of the rulings— would (a) the half-widths of the lines and (b) the separation of the lines increase, decrease, or remain the same? (c) Would the lines shift to the right, shift to the left, or remain in place?

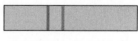

**FIG. 36-36** Questions 7 and 8.

**8** For the situation of Question 7 and Fig. 36-36, if instead we increased the grating spacing, would (a) the half-widths of the lines and (b) the separation of the lines increase, decrease, or remain the same? (c) Would the lines shift to the right, shift to the left, or remain in place?

**9** At night many people see rings (called *entoptic halos*) surrounding bright outdoor lamps in otherwise dark surroundings. The rings are the first of the side maxima in diffraction patterns produced by structures that are thought to be within the cornea (or possibly the lens) of the observer's eye. (The central maxima of such patterns overlap the lamp.) (a) Would a particular ring become smaller or larger if the lamp were switched from blue to red light? (b) If a lamp emits white light, is blue or red on the outside edge of the ring?

**10** (a) For a given diffraction grating, does the smallest difference $\Delta\lambda$ in two wavelengths that can be resolved increase, decrease, or remain the same as the wavelength increases? (b) For a given wavelength region (say, around 500 nm), is $\Delta\lambda$ greater in the first order or in the third order?

**11** (a) Figure 36-37a shows the lines produced by diffraction gratings A and B using light of the same wavelength; the lines

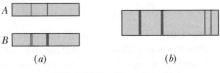

(a)

(b)

**FIG. 36-37** Question 11.

are of the same order and appear at the same angles $\theta$. Which grating has the greater number of rulings? (b) Figure 36-37b shows lines of two orders produced by a single diffraction grating using light of two wavelengths, both in the red region of the spectrum. Which lines, the left pair or right pair, are in the order with greater $m$? Is the center of the diffraction pattern located to the left or to the right in (c) Fig. 36-37a and (d) Fig. 36-37b?

**12** Figure 36-38 shows the bright fringes that lie within the central diffraction envelope in two double-slit diffraction experiments using the same wavelength of light. Are (a) the slit width $a$, (b) the slit separation $d$, and (c) the ratio $d/a$ in experiment B greater than, less than, or the same as those quantities in experiment A?

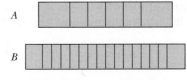

**FIG. 36-38** Question 12.

---

# PROBLEMS

(GO) Tutoring problem available (at instructor's discretion) in *WileyPLUS* and WebAssign

SSM Worked-out solution available in Student Solutions Manual  WWW Worked-out solution is at —

• – ••• Number of dots indicates level of problem difficulty  ILW Interactive solution is at — http://www.wiley.com/college/halliday

Additional information available in *The Flying Circus of Physics* and at flyingcircusofphysics.com

---

### sec. 36-3 Diffraction by a Single Slit: Locating the Minima

**•1** A single slit is illuminated by light of wavelengths $\lambda_a$ and $\lambda_b$, chosen so that the first diffraction minimum of the $\lambda_a$ component coincides with the second minimum of the $\lambda_b$ component. (a) If $\lambda_b = 350$ nm, what is $\lambda_a$? For what order number $m_b$ (if any) does a minimum of the $\lambda_b$ component coincide with the minimum of the $\lambda_a$ component in the order number (b) $m_a = 2$ and (c) $m_a = 3$?

**•2** Monochromatic light of wavelength 441 nm is incident on a narrow slit. On a screen 2.00 m away, the distance between the second diffraction minimum and the central maximum is 1.50 cm. (a) Calculate the angle of diffraction $\theta$ of the second minimum. (b) Find the width of the slit.

**•3** Light of wavelength 633 nm is incident on a narrow slit. The angle between the first diffraction minimum on one side of the central maximum and the first minimum on the other side is 1.20°. What is the width of the slit?

**•4** What must be the ratio of the slit width to the wavelength for a single slit to have the first diffraction minimum at $\theta = 45.0$°?

**•5** A plane wave of wavelength 590 nm is incident on a slit with a width of $a = 0.40$ mm. A thin converging lens of focal length +70 cm is placed between the slit and a viewing screen and focuses the light on the screen. (a) How far is the screen from the lens? (b) What is the distance on the screen from the center of the diffraction pattern to the first minimum?

**•6** In conventional television, signals are broadcast from towers to home receivers. Even when a receiver is not in direct view of a tower because of a hill or building, it can still intercept a signal if the signal diffracts enough around the obstacle, into the obstacle's "shadow region." Previously, television signals had a wavelength of about 50 cm, but digital

television signals that are transmitted from towers have a wavelength of about 10 mm. (a) Did this change in wavelength increase or decrease the diffraction of the signals into the shadow regions of obstacles? Assume that a signal passes through an opening of 5.0 m width between two adjacent buildings. What is the angular spread of the central diffraction maximum (out to the first minima) for wavelengths of (b) 50 cm and (c) 10 mm?

**•7** The distance between the first and fifth minima of a single-slit diffraction pattern is 0.35 mm with the screen 40 cm away from the slit, when light of wavelength 550 nm is used. (a) Find the slit width. (b) Calculate the angle $\theta$ of the first diffraction minimum. (GO)

**••8** Manufacturers of wire (and other objects of small dimension) sometimes use a laser to continually monitor the thickness of the product. The wire intercepts the laser beam, producing a diffraction pattern like that of a single slit of the same width as the wire diameter (Fig. 36-39). Suppose a helium–neon laser, of wavelength 632.8 nm, illuminates a

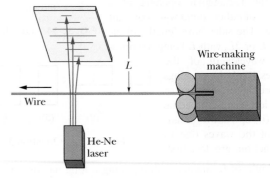

**FIG. 36-39** Problem 8.

wire, and the diffraction pattern appears on a screen at distance $L = 2.60$ m. If the desired wire diameter is 1.37 mm, what is the observed distance between the two tenth-order minima (one on each side of the central maximum)?

••9   A slit 1.00 mm wide is illuminated by light of wavelength 589 nm. We see a diffraction pattern on a screen 3.00 m away. What is the distance between the first two diffraction minima on the same side of the central diffraction maximum? SSM ILW

••10   Sound waves with frequency 3000 Hz and speed 343 m/s diffract through the rectangular opening of a speaker cabinet and into a large auditorium of length $d = 100$ m. The opening, which has a horizontal width of 30.0 cm,

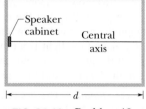

FIG. 36-40   Problem 10.

faces a wall 100 m away (Fig. 36-40). Along that wall, how far from the central axis will a listener be at the first diffraction minimum and thus have difficulty hearing the sound? (Neglect reflections.)

### sec. 36-5   Intensity in Single-Slit Diffraction, Quantitatively

•11   Monochromatic light with wavelength 538 nm is incident on a slit with width 0.025 mm. The distance from the slit to a screen is 3.5 m. Consider a point on the screen 1.1 cm from the central maximum. Calculate (a) $\theta$ for that point, (b) $\alpha$, and (c) the ratio of the intensity at that point to the intensity at the central maximum.

•12   In the single-slit diffraction experiment of Fig. 36-4, let the wavelength of the light be 500 nm, the slit width be 6.00 $\mu$m, and the viewing screen be at distance $D = 3.00$ m. Let a $y$ axis extend upward along the viewing screen, with its origin at the center of the diffraction pattern. Also let $I_P$ represent the intensity of the diffracted light at point $P$ at $y = 15.0$ cm. (a) What is the ratio of $I_P$ to the intensity $I_m$ at the center of the pattern? (b) Determine where point $P$ is in the diffraction pattern by giving the maximum and minimum between which it lies, or the two minima between which it lies.

•13   A 0.10-mm-wide slit is illuminated by light of wavelength 589 nm. Consider a point $P$ on a viewing screen on which the diffraction pattern of the slit is viewed; the point is at 30° from the central axis of the slit. What is the phase difference between the Huygens wavelets arriving at point $P$ from the top and midpoint of the slit? (*Hint:* See Eq. 36-4.)

•14   Figure 36-41 gives $\alpha$ versus the sine of the angle $\theta$ in a single-slit diffraction experiment using light of wavelength 610 nm. The vertical axis scale is set by $\alpha_s = 12$ rad. What are (a) the slit width, (b) the total number of diffraction minima in the pattern (count them on

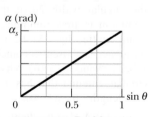

FIG. 36-41   Problem 14.

both sides of the center of the diffraction pattern), (c) the least angle for a minimum, and (d) the greatest angle for a minimum?

••15   (a) Show that the values of $\alpha$ at which intensity maxima for single-slit diffraction occur can be found exactly by differentiating Eq. 36-5 with respect to $\alpha$ and equating the result to zero, obtaining the condition $\tan \alpha = \alpha$. To find values

of $\alpha$ satisfying this relation, plot the curve $y = \tan \alpha$ and the straight line $y = \alpha$ and then find their intersections, or use a calculator to find an appropriate value of $\alpha$ by trial and error. Next, from $\alpha = (m + \frac{1}{2})\pi$, determine the values of $m$ associated with the maxima in the single-slit pattern. (These $m$ values are *not* integers because secondary maxima do not lie exactly halfway between minima.) What are the (b) smallest $\alpha$ and (c) associated $m$, the (d) second smallest $\alpha$ and (e) associated $m$, and the (f) third smallest $\alpha$ and (g) associated $m$?

••16   *Babinet's principle.* A monochromatic beam of parallel light is incident on a "collimating" hole of diameter $x \gg \lambda$. Point $P$ lies in the geometrical shadow region on a *distant* screen (Fig. 36-42a). Two diffracting objects, shown in Fig. 36-42b, are placed in turn over the collimating hole. Object $A$ is an opaque circle with a hole in it, and $B$ is the "photographic negative" of $A$. Using superposition concepts, show that the intensity at $P$ is identical for the two diffracting objects $A$ and $B$.

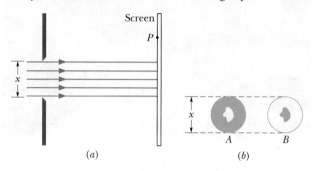

FIG. 36-42   Problem 16.

••17   The full width at half-maximum (FWHM) of a central diffraction maximum is defined as the angle between the two points in the pattern where the intensity is one-half that at the center of the pattern. (See Fig. 36-8b.) (a) Show that the intensity drops to one-half the maximum value when $\sin^2 \alpha = \alpha^2/2$. (b) Verify that $\alpha = 1.39$ rad (about 80°) is a solution to the transcendental equation of (a). (c) Show that the FWHM is $\Delta\theta = 2 \sin^{-1}(0.443\lambda/a)$, where $a$ is the slit width. Calculate the FWHM of the central maximum for slit width (d) $1.00\lambda$, (e) $5.00\lambda$, and (f) $10.0\lambda$.   SSM WWW

### sec. 36-6   Diffraction by a Circular Aperture

•18   The telescopes on some commercial surveillance satellites can resolve objects on the ground as small as 85 cm across (see Google Earth), and the telescopes on military surveillance satellites reportedly can resolve objects as small as 10 cm across. Assume first that object resolution is determined entirely by Rayleigh's criterion and is not degraded by turbulence in the atmosphere. Also assume that the satellites are at a typical altitude of 400 km and that the wavelength of visible light is 550 nm. What would be the required diameter of the telescope aperture for (a) 85 cm resolution and (b) 10 cm resolution? (c) Now, considering that turbulence is certain to degrade resolution and that the aperture diameter of the Hubble Space Telescope is 2.4 m, what can you say about the answer to (b) and about how the military surveillance resolutions are accomplished?

•19   If Superman really had x-ray vision at 0.10 nm wavelength and a 4.0 mm pupil diameter, at what maximum altitude could he distinguish villains from heroes, assuming that he needs to resolve points separated by 5.0 cm to do this?

•**20** Assume that Rayleigh's criterion gives the limit of resolution of an astronaut's eye looking down on Earth's surface from a typical space shuttle altitude of 400 km. (a) Under that idealized assumption, estimate the smallest linear width on Earth's surface that the astronaut can resolve. Take the astronaut's pupil diameter to be 5 mm and the wavelength of visible light to be 550 nm. (b) Can the astronaut resolve the Great Wall of China (Fig. 36-43), which is more than 3000 km long, 5 to 10 m thick at its base, 4 m thick at its top, and 8 m in height? (c) Would the astronaut be able to resolve any unmistakable sign of intelligent life on Earth's surface?

**FIG. 36-43** Problem 20. The Great Wall of China. *(AP/Wide World Photos)*

•**21** The two headlights of an approaching automobile are 1.4 m apart. At what (a) angular separation and (b) maximum distance will the eye resolve them? Assume that the pupil diameter is 5.0 mm, and use a wavelength of 550 nm for the light. Also assume that diffraction effects alone limit the resolution so that Rayleigh's criterion can be applied. SSM

•**22** *Entoptic halos.* If someone looks at a bright outdoor lamp in otherwise dark surroundings, the lamp appears to be surrounded by bright and dark rings (hence *halos*) that are actually a circular diffraction pattern as in Fig. 36-9, with the central maximum overlapping the direct light from the lamp. The diffraction is produced by structures within the cornea or lens of the eye (hence *entoptic*). If the lamp is monochromatic at wavelength 550 nm and the first dark ring subtends angular diameter 2.5° in the observer's view, what is the (linear) diameter of the structure producing the diffraction?

•**23** Find the separation of two points on the Moon's surface that can just be resolved by the 200 in. (= 5.1 m) telescope at Mount Palomar, assuming that this separation is determined by diffraction effects. The distance from Earth to the Moon is 3.8 × 10⁵ km. Assume a wavelength of 550 nm for the light. ILW

•**24** The radar system of a navy cruiser transmits at a wavelength of 1.6 cm, from a circular antenna with a diameter of 2.3 m. At a range of 6.2 km, what is the smallest distance that two speedboats can be from each other and still be resolved as two separate objects by the radar system?

•**25** Estimate the linear separation of two objects on Mars that can just be resolved under ideal conditions by an observer on Earth (a) using the naked eye and (b) using the 200 in. (= 5.1 m) Mount Palomar telescope. Use the following

data: distance to Mars = 8.0 × 10⁷ km, diameter of pupil = 5.0 mm, wavelength of light = 550 nm. SSM WWW

•**26** The wall of a large room is covered with acoustic tile in which small holes are drilled 5.0 mm from center to center. How far can a person be from such a tile and still distinguish the individual holes, assuming ideal conditions, the pupil diameter of the observer's eye to be 4.0 mm, and the wavelength of the room light to be 550 nm?

•**27** (a) How far from grains of red sand must you be to position yourself just at the limit of resolving the grains if your pupil diameter is 1.5 mm, the grains are spherical with radius 50 μm, and the light from the grains has wavelength 650 nm? (b) If the grains were blue and the light from them had wavelength 400 nm, would the answer to (a) be larger or smaller?

••**28** *Floaters.* The floaters you see when viewing a bright, featureless background are diffraction patterns of defects in the vitreous humor that fills most of your eye. Sighting through a pinhole sharpens the diffraction pattern. If you also view a small circular dot, you can approximate the defect's size. Assume that the defect diffracts light as a circular aperture does. Adjust the dot's distance $L$ from your eye (or eye lens) until the dot and the circle of the first minimum in the diffraction pattern appear to have the same size in your view. That is, until they have the same diameter $D'$ on the retina at distance $L' = 2.0$ cm from the front of the eye, as suggested in Fig. 36-44a, where the angles on the two sides of the eye lens are equal. Assume that the wavelength of visible light is $\lambda = 550$ nm. If the dot has diameter $D = 2.0$ mm and is distance $L = 45.0$ cm from the eye and the defect is $x = 6.0$ mm in front of the retina (Fig. 36-44b), what is the diameter of the defect?

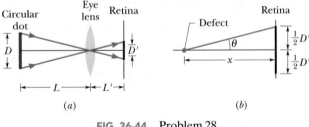

**FIG. 36-44** Problem 28.

••**29** Millimeter-wave radar generates a narrower beam than conventional microwave radar, making it less vulnerable to antiradar missiles than conventional radar. (a) Calculate the angular width $2\theta$ of the central maximum, from first minimum to first minimum, produced by a 220 GHz radar beam emitted by a 55.0-cm-diameter circular antenna. (The frequency is chosen to coincide with a low-absorption atmospheric "window.") (b) What is $2\theta$ for a more conventional circular antenna that has a diameter of 2.3 m and emits at wavelength 1.6 cm? SSM

••**30** (a) A circular diaphragm 60 cm in diameter oscillates at a frequency of 25 kHz as an underwater source of sound used for submarine detection. Far from the source, the sound intensity is distributed as the diffraction pattern of a circular hole whose diameter equals that of the diaphragm. Take the speed of sound in water to be 1450 m/s and find the angle between the normal to the diaphragm and a line from the diaphragm to the first minimum. (b) Is there such a minimum for a source having an (audible) frequency of 1.0 kHz?

••31 Nuclear-pumped x-ray lasers are seen as a possible weapon to destroy ICBM booster rockets at ranges up to 2000 km. One limitation on such a device is the spreading of the beam due to diffraction, with resulting dilution of beam intensity. Consider such a laser operating at a wavelength of 1.40 nm. The element that emits light is the end of a wire with diameter 0.200 mm. (a) Calculate the diameter of the central beam at a target 2000 km away from the beam source. (b) By what factor is the beam intensity reduced in transit to the target? (The laser is fired from space, so that atmospheric absorption can be ignored.) (GO)

••32 The wings of tiger beetles (Fig. 36-45) are colored by interference due to thin cuticle-like layers. In addition, these layers are arranged in patches that are 60 μm across and produce different colors. The color you see is a pointillistic mixture of thin-film interference colors that varies with perspective. Approximately what viewing distance from a wing puts you at the limit of resolving the different colored patches according to Rayleigh's criterion? Use 550 nm as the wavelength of light and 3.00 mm as the diameter of your pupil.

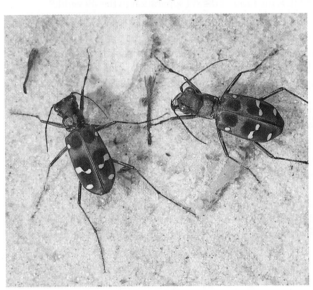

FIG. 36-45 Problem 32. Tiger beetles are colored by pointillistic mixtures of thin-film interference colors. *(Kjell B. Sandved/Bruce Coleman, Inc.)*

••33 (a) What is the angular separation of two stars if their images are barely resolved by the Thaw refracting telescope at the Allegheny Observatory in Pittsburgh? The lens diameter is 76 cm and its focal length is 14 m. Assume λ = 550 nm. (b) Find the distance between these barely resolved stars if each of them is 10 light-years distant from Earth. (c) For the image of a single star in this telescope, find the diameter of the first dark ring in the diffraction pattern, as measured on a photographic plate placed at the focal plane of the telescope lens. Assume that the structure of the image is associated entirely with diffraction at the lens aperture and not with lens "errors."

•••34 A circular obstacle produces the same diffraction pattern as a circular hole of the same diameter (except very near θ = 0). Airborne water drops are examples of such obstacles. When you see the Moon through suspended water drops, such as in a fog, you intercept the diffraction pattern from many drops. The composite of the central diffraction maxima

of those drops forms a white region that surrounds the Moon and may obscure it. Figure 36-46 is a photograph in which the Moon is obscured. There are two faint, colored rings around the Moon (the larger one may be too faint to be seen in your copy of the photograph). The smaller ring is on the outer edge of the central maxima from the drops; the somewhat larger ring is on the outer edge of the smallest of the secondary maxima from the drops (see Fig. 36-10). The color is visible because the rings are adjacent to the diffraction minima (dark rings) in the patterns. (Colors in other parts of the pattern overlap too much to be visible.)

(a) What is the color of these rings on the outer edges of the diffraction maxima? (b) The colored ring around the central maxima in Fig. 36-46 has an angular diameter that is 1.35 times the angular diameter of the Moon, which is 0.50°. Assume that the drops all have about the same diameter. Approximately what is that diameter? (GO)

FIG. 36-46 Problem 34. The corona around the Moon is a composite of the diffraction patterns of airborne water drops. *(Pekka Parvianen/Photo Researchers)*

sec. 36-7 **Diffraction by a Double Slit**

•35 In a double-slit experiment, the slit separation $d$ is 2.00 times the slit width $w$. How many bright interference fringes are in the central diffraction envelope?

•36 A beam of light of a single wavelength is incident perpendicularly on a double-slit arrangement, as in Fig. 35-10. The slit widths are each 46 μm and the slit separation is 0.30 mm. How many complete bright fringes appear between the two first-order minima of the diffraction pattern?

•37 Suppose that the central diffraction envelope of a double-slit diffraction pattern contains 11 bright fringes and the first diffraction minima eliminate (are coincident with) bright fringes. How many bright fringes lie between the first and second minima of the diffraction envelope?

••38 Two slits of width $a$ and separation $d$ are illuminated by a coherent beam of light of wavelength $\lambda$. What is the linear separation of the bright interference fringes observed on a screen that is at a distance $D$ away?

••39 (a) How many bright fringes appear between the

first diffraction-envelope minima to either side of the central maximum in a double-slit pattern if $\lambda = 550$ nm, $d = 0.150$ mm, and $a = 30.0$ $\mu$m? (b) What is the ratio of the intensity of the third bright fringe to the intensity of the central fringe? **SSM WWW**

**••40** (a) In a double-slit experiment, what ratio of $d$ to $a$ causes diffraction to eliminate the fourth bright side fringe? (b) What other bright fringes are also eliminated?

**••41** Light of wavelength 440 nm passes through a double slit, yielding a diffraction pattern whose graph of intensity $I$ versus angular position $\theta$ is shown in Fig. 36-47. Calculate (a) the slit width and (b) the slit separation. (c) Verify the displayed intensities of the $m = 1$ and $m = 2$ interference fringes.

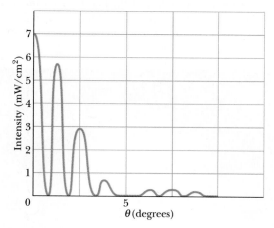

FIG. 36-47    Problem 41.

**••42** Figure 36-48 gives the parameter $\beta$ of Eq. 36-20 versus the sine of the angle $\theta$ in a two-slit interference experiment using light of wavelength 435 nm. The vertical axis scale is set by $\beta_s = 80.0$ rad. What are (a) the slit separation, (b) the total number of interference maxima (count them on both sides of the pattern's center), (c) the smallest angle for a maxima, and (d) the greatest angle for a minimum? Assume that none of the interference maxima are completely eliminated by a diffraction minimum.

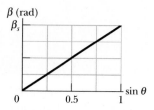

FIG. 36-48    Problem 42.

**••43** In the two-slit interference experiment of Fig. 35-10, the slit widths are each 12.0 $\mu$m, their separation is 24.0 $\mu$m, the wavelength is 600 nm, and the viewing screen is at a distance of 4.00 m. Let $I_P$ represent the intensity at point $P$ on the screen, at height $y = 70.0$ cm. (a) What is the ratio of $I_P$ to the intensity $I_m$ at the center of the pattern? (b) Determine where $P$ is in the two-slit interference pattern by giving the maximum or minimum on which it lies or the maximum and minimum between which it lies. (c) Next, for the diffraction that occurs, determine where point $P$ is in the diffraction pattern by giving the minimum on which it lies or the two minima between which it lies.

### sec. 36-8  Diffraction Gratings

**•44** Visible light is incident perpendicularly on a grating with 315 rulings/mm. What is the longest wavelength that can be seen in the fifth-order diffraction?

**•45** A grating has 400 lines/mm. How many orders of the entire visible spectrum (400–700 nm) can it produce in a diffraction experiment, in addition to the $m = 0$ order? **SSM ILW**

**•46** Perhaps to confuse a predator, some tropical gyrinid beetles (whirligig beetles) are colored by optical interference that is due to scales whose alignment forms a diffraction grating (which scatters light instead of transmiting it). When the incident light rays are perpendicular to the grating, the angle between the first-order maxima (on opposite sides of the zeroth-order maximum) is about 26° in light with a wavelength of 550 nm. What is the grating spacing of the beetle?

**•47** A diffraction grating 20.0 mm wide has 6000 rulings. Light of wavelength 589 nm is incident perpendicularly on the grating. What are the (a) largest, (b) second largest, and (c) third largest values of $\theta$ at which maxima appear on a distant viewing screen?

**•48** In a certain two-slit interference pattern, 10 bright fringes lie within the second side peak of the diffraction envelope and diffraction minima coincide with two-slit interference maxima. What is the ratio of the slit separation to the slit width?

**••49** A diffraction grating having 180 lines/mm is illuminated with a light signal containing only two wavelengths, $\lambda_1 = 400$ nm and $\lambda_2 = 500$ nm. The signal is incident perpendicularly on the grating. (a) What is the angular separation between the second-order maxima of these two wavelengths? (b) What is the smallest angle at which two of the resulting maxima are superimposed? (c) What is the highest order for which maxima for both wavelengths are present in the diffraction pattern?

**••50** A diffraction grating is made up of slits of width 300 nm with separation 900 nm. The grating is illuminated by monochromatic plane waves of wavelength $\lambda = 600$ nm at normal incidence. (a) How many maxima are there in the full diffraction pattern? (b) What is the angular width of a spectral line observed in the first order if the grating has 1000 slits?

**••51** Light of wavelength 600 nm is incident normally on a diffraction grating. Two adjacent maxima occur at angles given by $\sin \theta = 0.2$ and $\sin \theta = 0.3$. The fourth-order maxima are missing. (a) What is the separation between adjacent slits? (b) What is the smallest slit width this grating can have? For that slit width, what are the (c) largest, (d) second largest, and (e) third largest values of the order number $m$ of the maxima produced by the grating? **SSM WWW**

**••52** With light from a gaseous discharge tube incident normally on a grating with slit separation 1.73 $\mu$m, sharp maxima of green light are experimentally found at angles $\theta = \pm17.6°$, 37.3°, $-37.1°$, 65.2°, and $-65.0°$. Compute the wavelength of the green light that best fits these data.

**••53** Assume that the limits of the visible spectrum are arbitrarily chosen as 430 and 680 nm. Calculate the number of rulings per millimeter of a grating that will spread the first-order spectrum through an angle of 20.0°.

**••54** A beam of light consisting of wavelengths from 460.0 nm to 640.0 nm is directed perpendicularly onto a diffraction grating with 160 lines/mm. (a) What is the lowest order that is overlapped by another order? (b) What is the highest order for which the complete wavelength range of the beam is present? In that highest order, at what angle does the light at wavelength (c) 460.0 nm and (d) 640.0 nm appear? (e) What is the greatest angle at which the light at wavelength 460.0 nm appears? **GO**

**••55** A grating has 350 rulings/mm and is illuminated at normal incidence by white light. A spectrum is formed on a screen 30.0 cm from the grating. If a hole 10.0 mm square is cut in the screen, its inner edge being 50.0 mm from the central maximum and parallel to it, what are the (a) shortest and (b) longest wavelengths of the light that passes through the hole?

**•••56** Derive this expression for the intensity pattern for a three-slit "grating":

$$I = \tfrac{1}{9}I_m(1 + 4\cos\phi + 4\cos^2\phi),$$

where $\phi = (2\pi d \sin\theta)/\lambda$ and $a \ll \lambda$.

### sec. 36-9 Gratings: Dispersion and Resolving Power

**•57** A diffraction grating with a width of 2.0 cm contains 1000 lines/cm across that width. For an incident wavelength of 600 nm, what is the smallest wavelength difference this grating can resolve in the second order?

**•58** The $D$ line in the spectrum of sodium is a doublet with wavelengths 589.0 and 589.6 nm. Calculate the minimum number of lines needed in a grating that will resolve this doublet in the second-order spectrum. See Sample Problem 36-6.

**•59** Light at wavelength 589 nm from a sodium lamp is incident perpendicularly on a grating with 40 000 rulings over width 76 nm. What are the first-order (a) dispersion $D$ and (b) resolving power $R$, the second-order (c) $D$ and (d) $R$, and the third-order (e) $D$ and (f) $R$?

**•60** (a) How many rulings must a 4.00-cm-wide diffraction grating have to resolve the wavelengths 415.496 and 415.487 nm in the second order? (b) At what angle are the second-order maxima found?

**•61** A source containing a mixture of hydrogen and deuterium atoms emits red light at two wavelengths whose mean is 656.3 nm and whose separation is 0.180 nm. Find the minimum number of lines needed in a diffraction grating that can resolve these lines in the first order. **SSM ILW**

**•62** A grating has 600 rulings/mm and is 5.0 mm wide. (a) What is the smallest wavelength interval it can resolve in the third order at $\lambda = 500$ nm? (b) How many higher orders of maxima can be seen?

**•63** With a particular grating the sodium doublet (see Sample Problem 36-6) is viewed in the third order at 10° to the normal and is barely resolved. Find (a) the grating spacing and (b) the total width of the rulings.

**••64** A diffraction grating illuminated by monochromatic light normal to the grating produces a certain line at angle $\theta$. (a) What is the product of that line's half-width and the grating's resolving power? (b) Evaluate that product for the first order of a grating of slit separation 900 nm in light of wavelength 600 nm.

### sec. 36-10 Diffraction by Organized Layers

**•65** Figure 36-49 is a graph of intensity versus angular position $\theta$ for the diffraction of an x-ray beam by a crystal. The horizontal scale is set by $\theta_s = 2.00°$. The beam consists of two

wavelengths, and the spacing between the reflecting planes is 0.94 nm. What are the (a) shorter and (b) longer wavelengths in the beam?

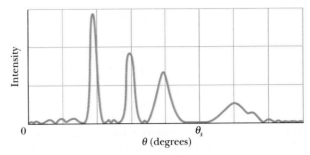

FIG. 36-49  Problem 65.

**•66** If first-order reflection occurs in a crystal at Bragg angle 3.4°, at what Bragg angle does second-order reflection occur from the same family of reflecting planes?

**•67** X rays of wavelength 0.12 nm are found to undergo second-order reflection at a Bragg angle of 28° from a lithium fluoride crystal. What is the interplanar spacing of the reflecting planes in the crystal?

**•68** What is the smallest Bragg angle for x rays of wavelength 30 pm to reflect from reflecting planes spaced 0.30 nm apart in a calcite crystal?

**•69** An x-ray beam of wavelength $A$ undergoes first-order reflection from a crystal when its angle of incidence to a crystal face is 23°, and an x-ray beam of wavelength 97 pm undergoes third-order reflection when its angle of incidence to that face is 60°. Assuming that the two beams reflect from the same family of reflecting planes, find (a) the interplanar spacing and (b) the wavelength $A$.

**•70** An x-ray beam of a certain wavelength is incident on a NaCl crystal, at 30.0° to a certain family of reflecting planes of spacing 39.8 pm. If the reflection from those planes is of the first order, what is the wavelength of the x rays?

**••71** Consider a two-dimensional square crystal structure, such as one side of the structure shown in Fig. 36-30a. The largest interplanar spacing of reflecting planes is the unit cell size $a_0$. Calculate and sketch the (a) second largest, (b) third largest, (c) fourth largest, (d) fifth largest, and (e) sixth largest interplanar spacing. (f) Show that your results in (a) through (e) are consistent with the general formula

$$d = \frac{a_0}{\sqrt{h^2 + k^2}},$$

where $h$ and $k$ are relatively prime integers (they have no common factor other than unity).

**••72** In Fig. 36-50, first-order reflection from the reflection planes shown occurs when an x-ray beam of wavelength 0.260 nm makes an angle $\theta = 63.8°$ with the top face of the crystal. What is the unit cell size $a_0$? **GO**

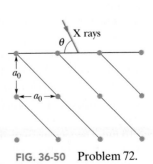

FIG. 36-50  Problem 72.

••73 In Fig. 36-51, let a beam of x rays of wavelength 0.125 nm be incident on an NaCl crystal at angle $\theta = 45.0°$ to the top face of the crystal and a family of reflecting planes. Let the reflecting planes have separation $d = 0.252$ nm. The crystal is turned through angle $\phi$ around an axis perpendicular to the plane of the page until these reflecting planes give diffraction maxima. What are the (a) smaller and (b) larger value of $\phi$ if the crystal is turned clockwise and the (c) smaller and (d) larger value of $\phi$ if it is turned counterclockwise? **SSM**

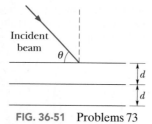

**FIG. 36-51** Problems 73 and 74.

••74 In Fig. 36-51, an x-ray beam of wavelengths from 95.0 to 140 pm is incident at $\theta = 45.0°$ to a family of reflecting planes with spacing $d = 275$ pm. What are the (a) longest wavelength $\lambda$ and (b) associated order number $m$ and the (c) shortest $\lambda$ and (d) associated $m$ of the intensity maxima in the diffraction of the beam?

### Additional Problems

75 In June 1985, a laser beam was sent out from the Air Force Optical Station on Maui, Hawaii, and reflected back from the shuttle *Discovery* as it sped by 354 km overhead. The diameter of the central maximum of the beam at the shuttle position was said to be 9.1 m, and the beam wavelength was 500 nm. What is the effective diameter of the laser aperture at the Maui ground station? (*Hint:* A laser beam spreads only because of diffraction; assume a circular exit aperture.)

76 An astronaut in a space shuttle claims she can just barely resolve two point sources on Earth's surface, 160 km below. Calculate their (a) angular and (b) linear separation, assuming ideal conditions. Take $\lambda = 540$ nm and the pupil diameter of the astronaut's eye to be 5.0 mm.

77 Visible light is incident perpendicularly on a diffraction grating of 200 rulings/mm. What are the (a) longest, (b) second longest, and (c) third longest wavelengths that can be associated with an intensity maximum at $\theta = 30.0°$? **SSM**

78 When monochromatic light is incident on a slit 22.0 $\mu$m wide, the first diffraction minimum lies at 1.80° from the direction of the incident light. What is the wavelength?

79 In a single-slit diffraction experiment, there is a minimum of intensity for orange light ($\lambda = 600$ nm) and a minimum of intensity for blue-green light ($\lambda = 500$ nm) at the same angle of 1.00 mrad. For what minimum slit width is this possible? **SSM**

80 In a two-slit interference pattern, what is the ratio of slit separation to slit width if there are 17 bright fringes within the central diffraction envelope and the diffraction minima coincide with two-slit interference maxima?

81 In two-slit interference, if the slit separation is 14 $\mu$m and the slit widths are each 2.0 $\mu$m, (a) how many two-slit maxima are in the central peak of the diffraction envelope and (b) how many are in either of the first side peak of the diffraction envelope? **SSM**

82 A single-slit diffraction experiment is set up with light of wavelength 420 nm, incident perpendicularly on a slit of width 5.10 $\mu$m. The viewing screen is 3.20 m distant. On the screen, what is the distance between the center of the diffraction pattern and the second diffraction minimum?

83 A beam of light with a narrow wavelength range centered on 450 nm is incident perpendicularly on a diffraction grating with a width of 1.80 cm and a line density of 1400 lines/cm across that width. For this light, what is the smallest wavelength difference this grating can resolve in the third order?

84 If you look at something 40 m from you, what is the smallest length (perpendicular to your line of sight) that you can resolve, according to Rayleigh's criterion? Assume the pupil of your eye has a diameter of 4.00 mm, and use 500 nm as the wavelength of the light reaching you.

85 Two yellow flowers are separated by 60 cm along a line perpendicular to your line of sight to the flowers. How far are you from the flowers when they are at the limit of resolution according to the Rayleigh criterion? Assume the light from the flowers has a single wavelength of 550 nm and that your pupil has a diameter of 5.5 mm.

86 In a single-slit diffraction experiment, what must be the ratio of the slit width to the wavelength if the second diffraction minima are to occur at an angle of 37.0° from the center of the diffraction pattern on a viewing screen?

87 A diffraction grating 3.00 cm wide produces the second order at 33.0° with light of wavelength 600 nm. What is the total number of lines on the grating?

88 A beam of light consists of two wavelengths, 590.159 nm and 590.220 nm, that are to be resolved with a diffraction grating. If the grating has lines across a width of 3.80 cm, what is the minimum number of lines required for the two wavelengths to be resolved in the second order?

89 A spy satellite orbiting at 160 km above Earth's surface has a lens with a focal length of 3.6 m and can resolve objects on the ground as small as 30 cm. For example, it can easily measure the size of an aircraft's air intake port. What is the effective diameter of the lens as determined by diffraction consideration alone? Assume $\lambda = 550$ nm.

90 The pupil of a person's eye has a diameter of 5.00 mm. According to Rayleigh's criterion, what distance apart must two small objects be if their images are just barely resolved when they are 250 mm from the eye? Assume they are illuminated with light of wavelength 500 nm.

91 Light is incident on a grating at an angle $\psi$ as shown in Fig. 36-52. Show that bright fringes occur at angles $\theta$ that satisfy the equation

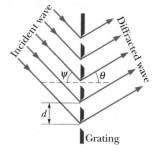

**FIG. 36-52** Problem 91.

$$d(\sin \psi + \sin \theta) = m\lambda, \quad \text{for } m = 0, 1, 2, \ldots .$$

(Compare this equation with Eq. 36-25.) Only the special case $\psi = 0$ has been treated in this chapter.

92 A grating with $d = 1.50$ $\mu$m is illuminated at various angles of incidence by light of wavelength 600 nm. Plot, as a function of the angle of incidence (0 to 90°), the angular deviation of the first-order maximum from the incident direction. (See Problem 91.)

93 If you double the width of a single slit, the intensity of the

central maximum of the diffraction pattern increases by a factor of 4, even though the energy passing through the slit only doubles. Explain this quantitatively. **SSM**

**94** In an experiment to monitor the Moon's surface with a light beam, pulsed radiation from a ruby laser ($\lambda = 0.69\ \mu m$) was directed to the Moon through a reflecting telescope with a mirror radius of 1.3 m. A reflector on the Moon behaved like a circular flat mirror with radius 10 cm, reflecting the light directly back toward the telescope on Earth. The reflected light was then detected after being brought to a focus by this telescope. Approximately what fraction of the original light energy was picked up by the detector? Assume that for each direction of travel all the energy is in the central diffraction peak.

**95** A diffraction grating has resolving power $R = \lambda_{avg}/\Delta\lambda = Nm$. (a) Show that the corresponding frequency range $\Delta f$ that can just be resolved is given by $\Delta f = c/Nm\lambda$. (b) From Fig. 36-22, show that the times required for light to travel along the ray at the bottom of the figure and the ray at the top differ by $\Delta t = (Nd/c)\ \sin\theta$. (c) Show that $(\Delta f)(\Delta t) = 1$, this relation being independent of the various grating parameters. Assume $N \gg 1$. **SSM**

**96** A double-slit system with individual slit widths of 0.030 mm and a slit separation of 0.18 mm is illuminated with 500 nm light directed perpendicular to the plane of the slits. What is the total number of complete bright fringes appearing between the two first-order minima of the diffraction pattern? (Do not count the fringes that coincide with the minima of the diffraction pattern.) **GO**

**97** A diffraction grating has 8900 slits across 1.20 cm. If light with a wavelength of 500 nm is sent through it, how many orders (maxima) lie to one side of the central maximum?

**98** A diffraction grating 1.00 cm wide has 10 000 parallel slits. Monochromatic light that is incident normally is diffracted through 30° in the first order. What is the wavelength of the light?

**99** A diffraction grating has 200 lines/mm. Light consisting of a continuous range of wavelengths between 550 nm and 700 nm is incident perpendicularly on the grating. (a) What is the lowest order that is overlapped by another order? (b) What is the highest order for which the complete spectrum is present?

**100** Suppose that two points are separated by 2.0 cm. If they are viewed by an eye with a pupil opening of 5.0 mm, what distance from the viewer puts them at the Rayleigh limit of resolution? Assume a light wavelength of 500 nm.

**101** Show that the dispersion of a grating is $D = (\tan\theta)/\lambda$. **SSM**

**102** Monochromatic light (wavelength = 450 nm) is incident perpendicularly on a single slit (width = 0.40 mm). A screen is placed parallel to the slit plane, and on it the distance between the two minima on either side of the central maximum is 1.8 mm. (a) What is the distance from the slit to the screen? (*Hint:* The angle to either minimum is small enough that $\sin\theta \approx \tan\theta$.) (b) What is the distance on the screen between the first minimum and the third minimum on the same side of the central maximum?

**103** Light containing a mixture of two wavelengths, 500 and 600 nm, is incident normally on a diffraction grating. It is desired (1) that the first and second maxima for each wavelength appear at $\theta \le 30°$, (2) that the dispersion be as high as possible, and (3) that the third order for the 600 nm light be a missing order. (a) What should be the slit separation?

(b) What is the smallest individual slit width that can be used? (c) For the values calculated in (a) and (b) and the light of wavelength 600 nm, what is the largest order of maxima produced by the grating?

**104** A beam of x rays with wavelengths ranging from 0.120 nm to 0.0700 nm scatters from a family of reflecting planes in a crystal. The plane separation is 0.250 nm. It is observed that scattered beams are produced for 0.100 nm and 0.0750 nm. What is the angle between the incident and scattered beams?

**105** Show that a grating made up of alternately transparent and opaque strips of equal width eliminates all the even orders of maxima (except $m = 0$).

**106** Light of wavelength 500 nm diffracts through a slit of width 2.00 $\mu m$ and onto a screen that is 2.00 m away. On the screen, what is the distance between the center of the diffraction pattern and the third diffraction minimum?

**107** If, in a two-slit interference pattern, there are 8 bright fringes within the first side peak of the diffraction envelope and diffraction minima coincide with two-slit interference maxima, then what is the ratio of slit separation to slit width?

**108** White light (consisting of wavelengths from 400 nm to 700 nm) is normally incident on a grating. Show that, no matter what the value of the grating spacing $d$, the second order and third order overlap.

**109** If we make $d = a$ in Fig. 36-53, the two slits coalesce into a single slit of width $2a$. Show that Eq. 36-19 reduces to give the diffraction pattern for such a slit.

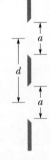

**FIG. 36-53** Problem 109.

**110** Derive Eq. 36-28, the expression for the half-width of lines in a grating's diffraction pattern.

**111** Prove that it is not possible to determine both wavelength of incident radiation and spacing of reflecting planes in a crystal by measuring the Bragg angles for several orders.

**112** How many orders of the entire visible spectrum (400–700 nm) can be produced by a grating of 500 lines/mm?

**113** An acoustic double-slit system (of slit separation $d$ and slit width $a$) is driven by two loudspeakers as shown in Fig. 36-54. By use of a variable delay line, the phase of one of the speakers may be varied relative to the other speaker. Describe in detail what changes occur in the double-slit diffraction pattern at large distances as the phase difference between the speakers is varied from zero to $2\pi$. Take both interference and diffraction effects into account.

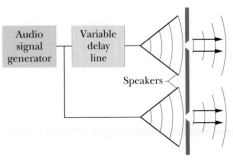

**FIG. 36-54** Problem 113.

# 37 Relativity

Courtesy J. A. Biretta, et al., Hubble Heritage Team (STScI/AURA), NASA

*A jet 5000 light-years long streams from the center of galaxy M87 (the bright spot at the upper left end of the jet), which is $5 \times 10^7$ light-years from us. The jet is formed by electrons moving at nearly the speed of light. Something very strange must be occurring at the center of M87 in order for the electrons to be ejected from it at such high speed. Is there some type of monster there? Unfortunately, M87 is too distant for us to see any object at its center.*

## How can we tell what monster lies at the center of M87?

The answer is in this chapter.

## 37-1 | WHAT IS PHYSICS?

One principal subject of physics is **relativity,** the field of study that measures events (things that happen): where and when they happen, and by how much any two events are separated in space and in time. In addition, relativity has to do with transforming such measurements between reference frames that move relative to each other. (Hence the name *relativity.*)

Transformations and moving reference frames, such as those we discussed in Sections 4-8 and 4-9, were well understood and quite routine to physicists in 1905. Then Albert Einstein (Fig. 37-1) published his **special theory of relativity.** The adjective *special* means that the theory deals only with **inertial reference frames,** which are frames in which Newton's laws are valid. (Einstein's *general theory of relativity* treats the more challenging situation in which reference frames can undergo gravitational acceleration; in this chapter the term *relativity* implies only inertial reference frames.)

Starting with two deceivingly simple postulates, Einstein stunned the scientific world by showing that the old ideas about relativity were wrong, even though everyone was so accustomed to them that they seemed to be unquestionable common sense. This supposed common sense, however, was derived only from experience with things that move rather slowly. Einstein's relativity, which turns out to be correct for all possible speeds, predicted many effects that were, at first study, bizarre because no one had ever experienced them.

In particular, Einstein demonstrated that space and time are entangled; that is, the time between two events depends on how far apart they occur, and vice versa. Also, the entanglement is different for observers who move relative to each other. One result is that time does not pass at a fixed rate, as if it were ticked off with mechanical regularity on some master grandfather clock that controls the universe. Rather, that rate is adjustable: Relative motion can change the rate at which time passes. Prior to 1905, no one but a few daydreamers would have thought that. Now, engineers and scientists take it for granted because their experience with special relativity has reshaped their common sense. For example, any engineer involved with the Global Positioning System of the NAVSTAR satellites must routinely use relativity to determine the rate at which time passes on the satellites because that rate differs from the rate on Earth's surface.

Special relativity has the reputation of being difficult. It is not difficult mathematically, at least not here. However, it is difficult in that we must be very careful about *who* measures *what* about an event and just *how* that measurement is made—and it can be difficult because it can contradict routine experience.

**FIG. 37-1** Einstein posing for a photograph as fame began to accumulate. *(Corbis Images)*

## 37-2 | The Postulates

We now examine the two postulates of relativity, on which Einstein's theory is based:

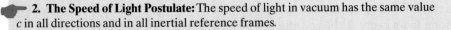
**1. The Relativity Postulate:** The laws of physics are the same for observers in all inertial reference frames. No one frame is preferred over any other.

Galileo assumed that the laws of *mechanics* were the same in all inertial reference frames. Einstein extended that idea to include *all* the laws of physics, especially those of electromagnetism and optics. This postulate does *not* say that the measured values of all physical quantities are the same for all inertial observers; most are not the same. It is the *laws of physics,* which relate these measurements to one another, that are the same.

**2. The Speed of Light Postulate:** The speed of light in vacuum has the same value *c* in all directions and in all inertial reference frames.

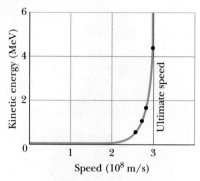

**FIG. 37-2** The dots show measured values of the kinetic energy of an electron plotted against its measured speed. No matter how much energy is given to an electron (or to any other particle having mass), its speed can never equal or exceed the ultimate limiting speed $c$. (The plotted curve through the dots shows the predictions of Einstein's special theory of relativity.)

We can also phrase this postulate to say that there is in nature an *ultimate speed c*, the same in all directions and in all inertial reference frames. Light happens to travel at this ultimate speed. However, no entity that carries energy or information can exceed this limit. Moreover, no particle that has mass can actually reach speed $c$, no matter how much or for how long that particle is accelerated. (Alas, the faster-than-light warp drive used in many science fiction stories appears to be impossible.)

Both postulates have been exhaustively tested, and no exceptions have ever been found.

### The Ultimate Speed

The existence of a limit to the speed of accelerated electrons was shown in a 1964 experiment by W. Bertozzi, who accelerated electrons to various measured speeds and—by an independent method—measured their kinetic energies. He found that as the force on a very fast electron is increased, the electron's measured kinetic energy increases toward very large values but its speed does not increase appreciably (Fig. 37-2). Electrons have been accelerated in laboratories to at least 0.999 999 999 95 times the speed of light but—close though it may be—that speed is still less than the ultimate speed $c$.

This ultimate speed has been defined to be exactly

$$c = 299\ 792\ 458 \text{ m/s.} \qquad (37\text{-}1)$$

*Caution:* So far in this book we have (appropriately) approximated $c$ as $3.0 \times 10^8$ m/s, but in this chapter we shall often use the exact value. You might want to store the exact value in your calculator's memory (if it is not there already), to be called up when needed.

### Testing the Speed of Light Postulate

If the speed of light is the same in all inertial reference frames, then the speed of light emitted by a source relative to, say, a laboratory should be the same as the speed of light that is emitted by a source at rest in the laboratory. This claim has been tested directly, in an experiment of high precision. The "light source" was the *neutral pion* (symbol $\pi^0$), an unstable, short-lived particle that can be produced by collisions in a particle accelerator. It decays (transforms) into two gamma rays by the process

$$\pi^0 \rightarrow \gamma + \gamma. \qquad (37\text{-}2)$$

Gamma rays are part of the electromagnetic spectrum (at very high frequencies) and so obey the speed of light postulate, just as visible light does. (In this chapter we shall use the term light for any type of electromagnetic wave, visible or not.)

In 1964, physicists at CERN, the European particle-physics laboratory near Geneva, generated a beam of pions moving at a speed of 0.999 75$c$ with respect to the laboratory. The experimenters then measured the speed of the gamma rays emitted from these very rapidly moving sources. They found that the speed of the light emitted by the pions was the same as it would be if the pions were at rest in the laboratory, namely $c$.

## 37-3 | Measuring an Event

An **event** is something that happens, and every event can be assigned three space coordinates and one time coordinate. Among many possible events are (1) the turning on or off of a tiny lightbulb, (2) the collision of two particles, (3) the passage of a pulse of light through a specified point, (4) an explosion, and (5) the sweeping of the hand of a clock past a marker on the rim of the clock. A certain observer, fixed in a certain inertial reference frame, might, for example, assign to an event $A$ the coordinates given in Table 37-1. Because space and time are entangled with each other in relativity, we can describe these coordinates collectively as *spacetime* coordinates. The coordinate system itself is part of the reference frame of the observer.

A given event may be recorded by any number of observers, each in a different inertial reference frame. In general, different observers will assign different spacetime coordinates to the same event. Note that an event does not "belong" to any particular inertial reference frame. An event is just something that happens, and anyone in any reference frame may detect it and assign spacetime coordinates to it.

Making such an assignment can be complicated by a practical problem. For example, suppose a balloon bursts 1 km to your right while a firecracker pops 2 km to your left, both at 9:00 A.M. However, you do not detect either event precisely at 9:00 A.M. because at that instant light from the events has not yet reached you. Because light from the firecracker pop has farther to go, it arrives at your eyes later than does light from the balloon burst, and thus the pop will seem to have occurred later than the burst. To sort out the actual times and to assign 9:00 A.M. as the happening time for both events, you must calculate the travel times of the light and then subtract these times from the arrival times.

This procedure can be very messy in more challenging situations, and we need an easier procedure that automatically eliminates any concern about the travel time from an event to an observer. To set up such a procedure, we shall construct an imaginary array of measuring rods and clocks throughout the observer's inertial frame (the array moves rigidly with the observer). This construction may seem contrived, but it spares us much confusion and calculation and allows us to find the space coordinates, the time coordinate, and the spacetime coordinates, as follows.

1. **The Space Coordinates.** We imagine the observer's coordinate system fitted with a close-packed, three-dimensional array of measuring rods, one set of rods parallel to each of the three coordinate axes. These rods provide a way to determine coordinates along the axes. Thus, if the event is, say, the turning on of a small lightbulb, the observer, in order to locate the position of the event, need only read the three space coordinates at the bulb's location.

2. **The Time Coordinate.** For the time coordinate, we imagine that every point of intersection in the array of measuring rods includes a tiny clock, which the observer can read because the clock is illuminated by the light generated by the event. Figure 37-3 suggests one plane in the "jungle gym" of clocks and measuring rods we have described.

   The array of clocks must be synchronized properly. It is not enough to assemble a set of identical clocks, set them all to the same time, and then move them to their assigned positions. We do not know, for example, whether moving the clocks will change their rates. (Actually, it will.) We must put the clocks in place and *then* synchronize them.

   If we had a method of transmitting signals at infinite speed, synchronization would be a simple matter. However, no known signal has this property. We therefore choose light (any part of the electromagnetic spectrum) to send out our synchronizing signals because, in vacuum, light travels at the greatest possible speed, the limiting speed $c$.

   Here is one of many ways in which an observer might synchronize an array of clocks using light signals: The observer enlists the help of a great number of temporary helpers, one for each clock. The observer then stands at a point selected as the origin and sends out a pulse of light when the origin clock reads $t = 0$. When the light pulse reaches the location of a helper, that helper sets the clock there to read $t = r/c$, where $r$ is the distance between the helper and the origin. The clocks are then synchronized.

3. **The Spacetime Coordinates.** The observer can now assign spacetime coordinates to an event by simply recording the time on the clock nearest the event

**TABLE 37-1**

**Record of Event A**

| Coordinate | Value |
|------------|-------|
| $x$ | 3.58 m |
| $y$ | 1.29 m |
| $z$ | 0 m |
| $t$ | 34.5 s |

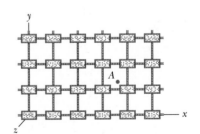

**FIG. 37-3** One section of a three-dimensional array of clocks and measuring rods by which an observer can assign spacetime coordinates to an event, such as a flash of light at point $A$. The event's space coordinates are approximately $x = 3.6$ rod lengths, $y = 1.3$ rod lengths, and $z = 0$. The time coordinate is whatever time appears on the clock closest to $A$ at the instant of the flash.

and the position as measured on the nearest measuring rods. If there are two events, the observer computes their separation in time as the difference in the times on clocks near each and their separation in space from the differences in coordinates on rods near each. We thus avoid the practical problem of calculating the travel times of the signals to the observer from the events.

## 37-4 | The Relativity of Simultaneity

Suppose that one observer (Sam) notes that two independent events (event Red and event Blue) occur at the same time. Suppose also that another observer (Sally), who is moving at a constant velocity $\vec{v}$ with respect to Sam, also records these same two events. Will Sally also find that they occur at the same time?

The answer is that in general she will not:

> If two observers are in relative motion, they will not, in general, agree as to whether two events are simultaneous. If one observer finds them to be simultaneous, the other generally will not.

We cannot say that one observer is right and the other wrong. Their observations are equally valid, and there is no reason to favor one over the other.

The realization that two contradictory statements about the same natural event can be correct is a seemingly strange outcome of Einstein's theory. However, in Chapter 17 we saw another way in which motion can affect measurement without balking at the contradictory results: In the Doppler effect, the frequency an observer measures for a sound wave depends on the relative motion of observer and source. Thus, two observers moving relative to each other can measure different frequencies for the same wave, and both measurements are correct.

We conclude the following:

> Simultaneity is not an absolute concept but rather a relative one, depending on the motion of the observer.

If the relative speed of the observers is very much less than the speed of light, then measured departures from simultaneity are so small that they are not noticeable. Such is the case for all our experiences of daily living; that is why the relativity of simultaneity is unfamiliar.

### A Closer Look at Simultaneity

Let us clarify the relativity of simultaneity with an example based on the postulates of relativity, no clocks or measuring rods being directly involved. Figure 37-4 shows two long spaceships (the SS *Sally* and the SS *Sam*), which can serve as inertial reference frames for observers Sally and Sam. The two observers are stationed at the midpoints of their ships. The ships are separating along a common *x* axis, the relative velocity of *Sally* with respect to *Sam* being $\vec{v}$. Figure 37-4*a* shows the ships with the two observer stations momentarily aligned opposite each other.

Two large meteorites strike the ships, one setting off a red flare (event Red) and the other a blue flare (event Blue), not necessarily simultaneously. Each event leaves a permanent mark on each ship, at positions $RR'$ and $BB'$.

Let us suppose that the expanding wavefronts from the two events happen to reach Sam at the same time, as Fig. 37-4*b* shows. Let us further suppose that, after the episode, Sam finds, by measuring the marks on his spaceship, that he was indeed stationed exactly halfway between the markers *B* and *R* on his ship when the two events occurred. He will say:

Sam    Light from event Red and light from event Blue reached me at the same time. From the marks on my spaceship, I find that I was standing halfway between the two sources. Therefore, event Red and event Blue were simultaneous events.

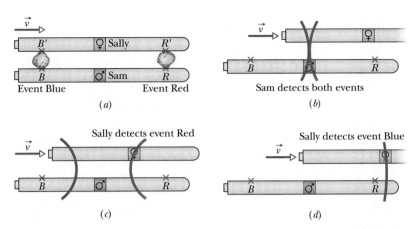

**FIG. 37-4** The spaceships of Sally and Sam and the occurrences of events from Sam's view. Sally's ship moves rightward with velocity $\vec{v}$. (a) Event Red occurs at positions $RR'$ and event Blue occurs at positions $BB'$; each event sends out a wave of light. (b) Sam simultaneously detects the waves from event Red and event Blue. (c) Sally detects the wave from event Red. (d) Sally detects the wave from event Blue.

As study of Fig. 37-4 shows, Sally and the expanding wavefront from event Red are moving *toward* each other, while she and the expanding wavefront from event Blue are moving in the *same direction*. Thus, the wavefront from event Red will reach Sally *before* the wavefront from event Blue does. She will say:

Sally   Light from event Red reached me before light from event Blue did. From the marks on my spaceship, I found that I too was standing halfway between the two sources. Therefore, the events were not simultaneous; event Red occurred first, followed by event Blue.

These reports do not agree. Nevertheless, *both* observers are correct.

Note carefully that there is only one wavefront expanding from the site of each event and that *this wavefront travels with the same speed c in both reference frames,* exactly as the speed of light postulate requires.

It *might* have happened that the meteorites struck the ships in such a way that the two hits appeared to Sally to be simultaneous. If that had been the case, then Sam would have declared them not to be simultaneous.

## 37-5 I The Relativity of Time

If observers who move relative to each other measure the time interval (or *temporal separation*) between two events, they generally will find different results. Why? Because the spatial separation of the events can affect the time intervals measured by the observers.

> The time interval between two events depends on how far apart they occur in both space and time; that is, their spatial and temporal separations are entangled.

In this section we discuss this entanglement by means of an example; however, the example is restricted in a crucial way: *To one of two observers, the two events occur at the same location.* We shall not get to more general examples until Section 37-7.

Figure 37-5a shows the basics of an experiment Sally conducts while she and her equipment—a light source, a mirror, and a clock—ride in a train moving with constant velocity $\vec{v}$ relative to a station. A pulse of light leaves the light source $B$ (event 1), travels vertically upward, is reflected vertically downward by the mirror, and then is detected back at the source (event 2). Sally measures a certain time interval $\Delta t_0$ between the two events, related to the distance $D$ from source to mirror by

$$\Delta t_0 = \frac{2D}{c} \quad \text{(Sally).} \tag{37-3}$$

**FIG. 37-5** (*a*) Sally, on the train, measures the time interval $\Delta t_0$ between events 1 and 2 using a single clock $C$ on the train. That clock is shown twice: first for event 1 and then for event 2. (*b*) Sam, watching from the station as the events occur, requires two synchronized clocks, $C_1$ at event 1 and $C_2$ at event 2, to measure the time interval between the two events; his measured time interval is $\Delta t$.

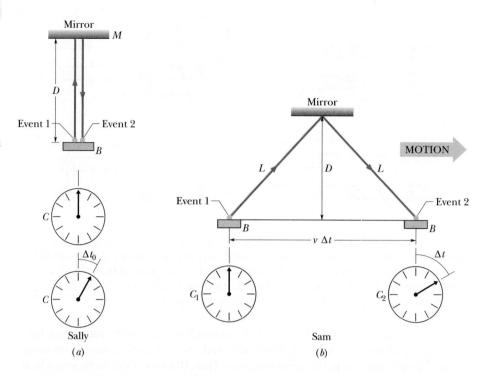

The two events occur at the same location in Sally's reference frame, and she needs only one clock $C$ at that location to measure the time interval. Clock $C$ is shown twice in Fig. 37-5*a*, at the beginning and end of the interval.

Consider now how these same two events are measured by Sam, who is standing on the station platform as the train passes. Because the equipment moves with the train during the travel time of the light, Sam sees the path of the light as shown in Fig. 37-5*b*. For him, the two events occur at different places in his reference frame, and so to measure the time interval between events, Sam must use *two* synchronized clocks, $C_1$ and $C_2$, one at each event. According to Einstein's speed of light postulate, the light travels at the same speed $c$ for Sam as for Sally. Now, however, the light travels distance $2L$ between events 1 and 2. The time interval measured by Sam between the two events is

$$\Delta t = \frac{2L}{c} \quad \text{(Sam)}, \tag{37-4}$$

in which

$$L = \sqrt{(\tfrac{1}{2}v\,\Delta t)^2 + D^2}. \tag{37-5}$$

From Eq. 37-3, we can write this as

$$L = \sqrt{(\tfrac{1}{2}v\,\Delta t)^2 + (\tfrac{1}{2}c\,\Delta t_0)^2}. \tag{37-6}$$

If we eliminate $L$ between Eqs. 37-4 and 37-6 and solve for $\Delta t$, we find

$$\Delta t = \frac{\Delta t_0}{\sqrt{1 - (v/c)^2}}. \tag{37-7}$$

Equation 37-7 tells us how Sam's measured interval $\Delta t$ between the events compares with Sally's interval $\Delta t_0$. Because $v$ must be less than $c$, the denominator in Eq. 37-7 must be less than unity. Thus, $\Delta t$ must be greater than $\Delta t_0$: Sam measures a *greater* time interval between the two events than does Sally. Sam and Sally have measured the time interval between the *same* two events, but the relative motion between Sam and Sally made their measurements *different*. We conclude that relative motion can change the *rate* at which time passes between two events; the key to this effect is the fact that the speed of light is the same for both observers.

We distinguish between the measurements of Sam and Sally with the following terminology:

When two events occur at the same location in an inertial reference frame, the time interval between them, measured in that frame, is called the **proper time interval** or the **proper time.** Measurements of the same time interval from any other inertial reference frame are always greater.

Thus, Sally measures a proper time interval, and Sam measures a greater time interval. (The term *proper* is unfortunate in that it implies that any other measurement is improper or nonreal. That is just not so.) The amount by which a measured time interval is greater than the corresponding proper time interval is called **time dilation.** (To dilate is to expand or stretch; here the time interval is expanded or stretched.)

Often the dimensionless ratio $v/c$ in Eq. 37-7 is replaced with $\beta$, called the **speed parameter,** and the dimensionless inverse square root in Eq. 37-7 is often replaced with $\gamma$, called the **Lorentz factor:**

$$\gamma = \frac{1}{\sqrt{1 - \beta^2}} = \frac{1}{\sqrt{1 - (v/c)^2}}. \tag{37-8}$$

With these replacements, we can rewrite Eq. 37-7 as

$$\Delta t = \gamma \, \Delta t_0 \quad \text{(time dilation)}. \tag{37-9}$$

The speed parameter $\beta$ is always less than unity, and, provided $v$ is not zero, $\gamma$ is always greater than unity. However, the difference between $\gamma$ and 1 is not significant unless $v > 0.1c$. Thus, in general, "old relativity" works well enough for $v < 0.1c$, but we must use special relativity for greater values of $v$. As shown in Fig. 37-6, $\gamma$ increases rapidly in magnitude as $\beta$ approaches 1 (as $v$ approaches $c$). Therefore, the greater the relative speed between Sally and Sam is, the greater will be the time interval measured by Sam, until at a great enough speed, the interval takes "forever."

You might wonder what Sally says about Sam's having measured a greater time interval than she did. His measurement comes as no surprise to her, because to her, he failed to synchronize his clocks $C_1$ and $C_2$ in spite of his insistence that he did. Recall that observers in relative motion generally do not agree about simultaneity. Here, Sam insists that his two clocks simultaneously read the same time when event 1 occurred. To Sally, however, Sam's clock $C_2$ was erroneously set ahead during the synchronization process. Thus, when Sam read the time of event 2 on it, to Sally he was reading off a time that was too large, and that is why the time interval he measured between the two events was greater than the interval she measured.

## Two Tests of Time Dilation

1. **Microscopic Clocks.** Subatomic particles called *muons* are unstable; that is, when a muon is produced, it lasts for only a short time before it *decays* (transforms into particles of other types). The *lifetime* of a muon is the time interval between its production (event 1) and its decay (event 2). When muons are stationary and their lifetimes are measured with stationary clocks (say, in a laboratory), their average lifetime is 2.200 $\mu$s. This is a proper time interval because, for each muon, events 1 and 2 occur at the same location in the reference frame of the muon—namely, at the muon itself. We can represent this proper time interval with $\Delta t_0$; moreover, we can call the reference frame in which it is measured the *rest frame* of the muon.

   If, instead, the muons are moving, say, through a laboratory, then measurements of their lifetimes made with the laboratory clocks should yield a greater average lifetime (a dilated average lifetime). To check this conclusion, measurements were made of the average lifetime of muons moving with a

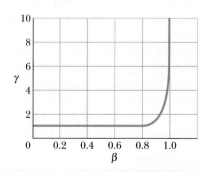

**FIG. 37-6** A plot of the Lorentz factor $\gamma$ as a function of the speed parameter $\beta$ ($= v/c$).

speed of 0.9994c relative to laboratory clocks. From Eq. 37-8, with $\beta = 0.9994$, the Lorentz factor for this speed is

$$\gamma = \frac{1}{\sqrt{1 - \beta^2}} = \frac{1}{\sqrt{1 - (0.9994)^2}} = 28.87.$$

Equation 37-9 then yields, for the average dilated lifetime,

$$\Delta t = \gamma \, \Delta t_0 = (28.87)(2.200 \ \mu s) = 63.51 \ \mu s.$$

The actual measured value matched this result within experimental error.

2. **Macroscopic Clocks.** In October 1977, Joseph Hafele and Richard Keating carried out what must have been a grueling experiment. They flew four portable atomic clocks twice around the world on commercial airlines, once in each direction. Their purpose was "to test Einstein's theory of relativity with macroscopic clocks." As we have just seen, the time dilation predictions of Einstein's theory have been confirmed on a microscopic scale, but there is great comfort in seeing a confirmation made with an actual clock. Such macroscopic measurements became possible only because of the very high precision of modern atomic clocks. Hafele and Keating verified the predictions of the theory to within 10%. (Einstein's *general* theory of relativity, which predicts that the rate at which time passes on a clock is influenced by the gravitational force on the clock, also plays a role in this experiment.)

A few years later, physicists at the University of Maryland carried out a similar experiment with improved precision. They flew an atomic clock round and round over Chesapeake Bay for flights lasting 15 h and succeeded in checking the time dilation prediction to better than 1%. Today, when atomic clocks are transported from one place to another for calibration or other purposes, the time dilation caused by their motion is always taken into account.

✓ **CHECKPOINT 1**     Standing beside railroad tracks, we are suddenly startled by a relativistic boxcar traveling past us as shown in the figure. Inside, a well-equipped hobo fires a laser pulse from the front of the boxcar to its rear. (a) Is our measurement of the speed of the pulse greater than, less than, or the same as that measured by the hobo? (b) Is his measurement of the flight time of the pulse a proper time? (c) Are his measurement and our measurement of the flight time related by Eq. 37-9?

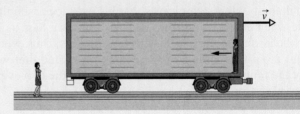

---

**Sample Problem** | **37-1**

Your starship passes Earth with a relative speed of 0.9990c. After traveling 10.0 y (your time), you stop at lookout post LP13, turn, and then travel back to Earth with the same relative speed. The trip back takes another 10.0 y (your time). How long does the round trip take according to measurements made on Earth? (Neglect any effects due to the accelerations involved with stopping, turning, and getting back up to speed.)

**KEY IDEAS**     We begin by analyzing the outward trip:

1. This problem involves measurements made from two (inertial) reference frames, one attached to

Earth and the other (your reference frame) attached to your ship.

2. The outward trip involves two events: the start of the trip at Earth and the end of the trip at LP13.

3. Your measurement of 10.0 y for the outward trip is the proper time $\Delta t_0$ between those two events, because the events occur at the same location in your reference frame—namely, on your ship.

4. The Earth-frame measurement of the time interval $\Delta t$ for the outward trip must be greater than $\Delta t_0$, according to Eq. 37-9 ($\Delta t = \gamma \, \Delta t_0$) for time dilation.

*Calculations:* Using Eq. 37-8 to substitute for $\gamma$ in Eq.

37-9, we find

$$\Delta t = \frac{\Delta t_0}{\sqrt{1 - (v/c)^2}}$$

$$= \frac{10.0 \text{ y}}{\sqrt{1 - (0.9990c/c)^2}} = (22.37)(10.0 \text{ y}) = 224 \text{ y}.$$

On the return trip, we have the same situation and the

same data. Thus, the round trip requires 20 y of your time but

$$\Delta t_{total} = (2)(224 \text{ y}) = 448 \text{ y} \qquad \text{(Answer)}$$

of Earth time. In other words, you have aged 20 y while the Earth has aged 448 y. Although you cannot travel into the past (as far as we know), you can travel into the future of, say, Earth, by using high-speed relative motion to adjust the rate at which time passes.

---

**Sample Problem** | **37-2** | **Build your skill**

The elementary particle known as the *positive kaon* ($K^+$) has, on average, a lifetime of $0.1237$ $\mu$s when stationary—that is, when the lifetime is measured in the rest frame of the kaon. If a positive kaon has a speed of $0.990c$ relative to a laboratory reference frame when the kaon is produced, how far can it travel in that frame during its lifetime according to *classical physics* (which is a reasonable approximation for speeds much less than $c$) and according to special relativity (which is correct for all physically possible speeds)?

**KEY IDEAS**

1. This problem involves measurements made from two (inertial) reference frames, one attached to the kaon and the other attached to the laboratory.

2. This problem also involves two events: the start of the kaon's travel (when the kaon is produced) and the end of that travel (at the end of the kaon's lifetime).

3. The distance traveled by the kaon between those two events is related to its speed $v$ and the time interval for the travel by

$$v = \frac{\text{distance}}{\text{time interval}}. \qquad (37\text{-}10)$$

With these ideas in mind, let us solve for the distance first with classical physics and then with special relativity.

**Classical physics:** In classical physics we would find the same distance and time interval (in Eq. 37-10) whether we measured them from the kaon frame or from the laboratory frame. Thus, we need not be careful about the frame in which the measurements are made. To find the kaon's travel distance $d_{cp}$ according to classical physics, we first rewrite Eq. 37-10 as

$$d_{cp} = v \Delta t, \qquad (37\text{-}11)$$

where $\Delta t$ is the time interval between the two events in either frame. Then, substituting $0.990c$ for $v$ and $0.1237$ $\mu$s for $\Delta t$ in Eq. 37-11, we find

$$d_{cp} = (0.990c) \Delta t$$
$$= (0.990)(299 \ 792 \ 458 \text{ m/s})(0.1237 \times 10^{-6} \text{ s})$$
$$= 36.7 \text{ m}. \qquad \text{(Answer)}$$

This is how far the kaon would travel if classical physics were correct at speeds close to $c$.

**Special relativity:** In special relativity we must be very careful that both the distance and the time interval in Eq. 37-10 are measured in the *same* reference frame—especially when the speed is close to $c$, as here. Thus, to find the actual travel distance $d_{sr}$ of the kaon *as measured from the laboratory frame* and according to special relativity, we rewrite Eq. 37-10 as

$$d_{sr} = v \Delta t, \qquad (37\text{-}12)$$

where $\Delta t$ is the time interval between the two events *as measured from the laboratory frame.*

Before we can evaluate $d_{sr}$ in Eq. 37-12, we must find $\Delta t$. The $0.1237$ $\mu$s time interval is a proper time because the two events occur at the same location in the kaon frame—namely, at the kaon itself. Therefore, let $\Delta t_0$ represent this proper time interval. Then we can use Eq. 37-9 ($\Delta t = \gamma \Delta t_0$) for time dilation to find the time interval $\Delta t$ as measured from the laboratory frame. Using Eq. 37-8 to substitute for $\gamma$ in Eq. 37-9 leads to

$$\Delta t = \frac{\Delta t_0}{\sqrt{1 - (v/c)^2}} = \frac{0.1237 \times 10^{-6} \text{ s}}{\sqrt{1 - (0.990c/c)^2}} = 8.769 \times 10^{-7} \text{ s}.$$

This is about seven times longer than the kaon's proper lifetime. That is, the kaon's lifetime is about seven times longer in the laboratory frame than in its own frame—the kaon's lifetime is dilated. We can now evaluate Eq. 37-12 for the travel distance $d_{sr}$ in the laboratory frame as

$$d_{sr} = v \Delta t = (0.990c) \Delta t$$
$$= (0.990)(299 \ 792 \ 458 \text{ m/s})(8.769 \times 10^{-7} \text{ s})$$
$$= 260 \text{ m}. \qquad \text{(Answer)}$$

This is about seven times $d_{cp}$. Experiments like the one outlined here, which verify special relativity, became routine in physics laboratories decades ago. The engineering design and the construction of any scientific or medical facility that employs high-speed particles must take relativity into account.

## 37-6 | The Relativity of Length

If you want to measure the length of a rod that is at rest with respect to you, you can—at your leisure—note the positions of its end points on a long stationary scale and subtract one reading from the other. If the rod is moving, however, you must note the positions of the end points *simultaneously* (in your reference frame) or your measurement cannot be called a length. Figure 37-7 suggests the difficulty of trying to measure the length of a moving penguin by locating its front and back at different times. Because simultaneity is relative and it enters into length measurements, length should also be a relative quantity. It is.

Let $L_0$ be the length of a rod that you measure when the rod is stationary (meaning you and it are in the same reference frame, the rod's rest frame). If, instead, there is relative motion at speed $v$ between you and the rod *along the length of the rod,* then with simultaneous measurements you obtain a length $L$ given by

$$L = L_0\sqrt{1 - \beta^2} = \frac{L_0}{\gamma} \qquad \text{(length contraction)}. \qquad (37\text{-}13)$$

Because the Lorentz factor $\gamma$ is always greater than unity if there is relative motion, $L$ is less than $L_0$. The relative motion causes a *length contraction,* and $L$ is called a *contracted length.* Because $\gamma$ increases with speed $v$, the length contraction also increases with $v$.

☞ The length $L_0$ of an object measured in the rest frame of the object is its **proper length** or **rest length.** Measurements of the length from any reference frame that is in relative motion parallel to that length are always less than the proper length.

Be careful: Length contraction occurs only along the direction of relative motion. Also, the length that is measured does not have to be that of an object like a rod or a circle. Instead, it can be the length (or distance) between two objects in the same rest frame—for example, the Sun and a nearby star (which are, at least approximately, at rest relative to each other).

Does a moving object *really* shrink? Reality is based on observations and measurements; if the results are always consistent and if no error can be determined, then what is observed and measured is real. In that sense, the object really does shrink. However, a more precise statement is that the object *is really measured* to shrink—motion affects that measurement and thus reality.

When you measure a contracted length for, say, a rod, what does an observer moving with the rod say of your measurement? To that observer, you did not locate the two ends of the rod simultaneously. (Recall that observers in motion relative to each other do not agree about simultaneity.) To the observer, you first located the rod's front end and then, slightly later, its rear end, and that is why you measured a length that is less than the proper length.

### Proof of Eq. 37-13

Length contraction is a direct consequence of time dilation. Consider once more our two observers. This time, both Sally, seated on a train moving through a station, and Sam, again on the station platform, want to measure the length of the platform. Sam, using a tape measure, finds the length to be $L_0$, a proper length because the platform is at rest with respect to him. Sam also notes that Sally, on the train, moves through this length in a time $\Delta t = L_0/v$, where $v$ is the speed of the train; that is,

$$L_0 = v\,\Delta t \qquad \text{(Sam)}. \qquad (37\text{-}14)$$

This time interval $\Delta t$ is not a proper time interval because the two events that

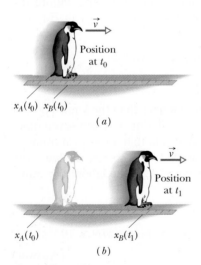

$x_A(t_0)$ $x_B(t_0)$

(a)

$x_A(t_0)$ $\qquad\qquad$ $x_B(t_1)$

(b)

**FIG. 37-7** If you want to measure the front-to-back length of a penguin while it is moving, you must mark the positions of its front and back simultaneously (in your reference frame), as in (a), rather than at different times, as in (b).

define it (Sally passes the back of the platform and Sally passes the front of the platform) occur at two different places, and therefore Sam must use two synchronized clocks to measure the time interval $\Delta t$.

For Sally, however, the platform is moving past her. She finds that the two events measured by Sam occur *at the same place* in her reference frame. She can time them with a single stationary clock, and so the interval $\Delta t_0$ that she measures is a proper time interval. To her, the length $L$ of the platform is given by

$$L = v\,\Delta t_0 \quad \text{(Sally).} \tag{37-15}$$

If we divide Eq. 37-15 by Eq. 37-14 and apply Eq. 37-9, the time dilation equation, we have

$$\frac{L}{L_0} = \frac{v\,\Delta t_0}{v\,\Delta t} = \frac{1}{\gamma},$$

or

$$L = \frac{L_0}{\gamma}, \tag{37-16}$$

which is Eq. 37-13, the length contraction equation.

## Sample Problem   37-3

In Fig. 37-8, Sally (at point $A$) and Sam's spaceship (of proper length $L_0 = 230$ m) pass each other with constant relative speed $v$. Sally measures a time interval of 3.57 $\mu$s for the ship to pass her (from the passage of point $B$ to the passage of point $C$). In terms of $c$, what is the relative speed $v$ between Sally and the ship?

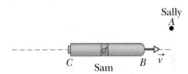

**FIG. 37-8**   Sally, at point $A$, measures the time a spaceship takes to pass her.

**KEY IDEAS**   Let's assume that speed $v$ is near $c$. Then,

1. This problem involves measurements made from two (inertial) reference frames, one attached to Sally and the other attached to Sam and his spaceship.

2. This problem also involves two events: the first is the passage of point $B$ past Sally and the second is the passage of point $C$ past her.

3. From either reference frame, the other reference frame passes at speed $v$ and moves a certain distance in the time interval between the two events:

$$v = \frac{\text{distance}}{\text{time interval}}. \tag{37-17}$$

Because speed $v$ is assumed to be near the speed of light, we must be careful that the distance and the time interval in Eq. 37-17 are measured in the *same* reference frame.

**Calculations:** We are free to use either frame for the measurements. Because we know that the time interval $\Delta t$ between the two events measured from Sally's frame is 3.57 $\mu$s, let us also use the distance $L$ between the two events measured from her frame. Equation 37-17 then becomes

$$v = \frac{L}{\Delta t}. \tag{37-18}$$

We do not know $L$, but we can relate it to the given

$L_0$: The distance between the two events as measured from Sam's frame is the ship's proper length $L_0$. Thus, the distance $L$ measured from Sally's frame must be less than $L_0$, as given by Eq. 37-13 ($L = L_0/\gamma$) for length contraction. Substituting $L_0/\gamma$ for $L$ in Eq. 37-18 and then substituting Eq. 37-8 for $\gamma$, we find

$$v = \frac{L_0/\gamma}{\Delta t} = \frac{L_0\sqrt{1 - (v/c)^2}}{\Delta t}.$$

Solving this equation for $v$ (notice that it is on the left and also buried in the Lorentz factor) leads us to

$$v = \frac{L_0 c}{\sqrt{(c\,\Delta t)^2 + L_0^2}}$$

$$= \frac{(230\text{ m})c}{\sqrt{(299\,792\,458\text{ m/s})^2(3.57 \times 10^{-6}\text{ s})^2 + (230\text{ m})^2}}$$

$$= 0.210c. \qquad\qquad \text{(Answer)}$$

Thus, the relative speed between Sally and the ship is 21% of the speed of light. Note that only the relative motion of Sally and Sam matters here; whether either is stationary relative to, say, a space station is irrelevant. In Fig. 37-8 we took Sally to be stationary, but we could instead have taken the ship to be stationary, with Sally moving to the left past it. Nothing would have changed in our result.

Caught by surprise near a supernova, you race away from the explosion in your spaceship, hoping to outrun the high-speed material ejected toward you. Your Lorentz factor $\gamma$ relative to the inertial reference frame of the local stars is 22.4.

(a) To reach a safe distance, you figure you need to cover $9.00 \times 10^{16}$ m as measured in the reference frame of the local stars. How long will the flight take, as measured in that frame?

**KEY IDEAS** Just as we did in Chapter 2, we can find the time required to travel a certain distance at constant speed with the definition of speed $v$:

$$\text{speed} = \frac{\text{distance}}{\text{time interval}}. \qquad (37\text{-}19)$$

From Fig. 37-6, we see that because your Lorentz factor $\gamma$ relative to the stars is 22.4 (large), your relative speed $v$ is almost $c$—so close that we can approximate it as $c$. Then for speed $v \approx c$, we must be careful that the distance and the time interval in Eq. 37-19 are measured in the *same* reference frame.

**Calculations:** The given distance $(9.00 \times 10^{16}$ m) for the length of your travel path is measured in the reference frame of the stars, and the requested time interval $\Delta t$ is to be measured in that same frame. Thus, we can write

$$\left(\begin{array}{c}\text{time interval} \\ \text{relative to stars}\end{array}\right) = \frac{\text{distance relative to stars}}{c}.$$

Then substituting the given distance, we find that

$$\left(\begin{array}{c}\text{time interval} \\ \text{relative to stars}\end{array}\right) = \frac{9.00 \times 10^{16}\ \text{m}}{299\ 792\ 458\ \text{m/s}}$$

$$= 3.00 \times 10^{8}\ \text{s} = 9.51\ \text{y}. \quad \text{(Answer)}$$

(b) How long does that trip take according to you (in your reference frame)?

**KEY IDEAS**

1. We now want the time interval measured in a different reference frame—namely, yours. Thus, we need to transform the data given in the reference frame of the stars to your frame.

2. The given path length of $9.00 \times 10^{16}$ m, measured in the reference frame of the stars, is a proper length $L_0$, because the two ends of the path are at rest in that frame. As observed from your reference frame, the stars' reference frame and those two ends of the path race past you at a relative speed of $v \approx c$.

3. You measure a contracted length $L_0/\gamma$ for the path, not the proper length $L_0$.

**Calculations:** We can now rewrite Eq. 37-19 as

$$\left(\begin{array}{c}\text{time interval} \\ \text{relative to you}\end{array}\right) = \frac{\text{distance relative to you}}{c} = \frac{L_0/\gamma}{c}.$$

Substituting known data, we find

$$\left(\begin{array}{c}\text{time interval} \\ \text{relative to you}\end{array}\right) = \frac{(9.00 \times 10^{16}\ \text{m})/22.4}{299\ 792\ 458\ \text{m/s}}$$

$$= 1.340 \times 10^{7}\ \text{s} = 0.425\ \text{y}. \quad \text{(Answer)}$$

In part (a) we found that the flight takes 9.51 y in the reference frame of the stars. However, here we find that it takes only 0.425 y in your frame, due to the relative motion and the resulting contracted length of the path.

## 37-7 | The Lorentz Transformation

Figure 37-9 shows inertial reference frame $S'$ moving with speed $v$ relative to frame $S$, in the common positive direction of their horizontal axes (marked $x$ and $x'$). An observer in $S$ reports spacetime coordinates $x, y, z, t$ for an event, and an observer in $S'$ reports $x', y', z', t'$ for the same event. How are these sets of numbers related?

We claim at once (although it requires proof) that the $y$ and $z$ coordinates, which are perpendicular to the motion, are not affected by the motion; that is, $y = y'$ and $z = z'$. Our interest then reduces to the relation between $x$ and $x'$ and that between $t$ and $t'$.

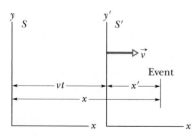

FIG. 37-9 Two inertial reference frames: frame $S'$ has velocity $\vec{v}$ relative to frame $S$.

### The Galilean Transformation Equations

Prior to Einstein's publication of his special theory of relativity, the four coordinates of interest were assumed to be related by the *Galilean transformation equations:*

$$x' = x - vt$$    (Galilean transformation equations;
$$t' = t$$    approximately valid at low speeds).    (37-20)

(These equations are written with the assumption that $t = t' = 0$ when the origins of $S$ and $S'$ coincide.) You can verify the first equation with Fig. 37-9. The second equation effectively claims that time passes at the same rate for observers in both reference frames. That would have been so obviously true to a scientist prior to Einstein that it would not even have been mentioned. When speed $v$ is small compared to $c$, Eqs. 37-20 generally work well.

## The Lorentz Transformation Equations

We state without proof that the correct transformation equations, which remain valid for all speeds up to the speed of light, can be derived from the postulates of relativity. The results, called the **Lorentz transformation equations*** or sometimes (more loosely) just the Lorentz transformations, are

$$x' = \gamma(x - vt),$$
$$y' = y,$$    (Lorentz transformation equations;
$$z' = z,$$    valid at all physically possible speeds).    (37-21)
$$t' = \gamma(t - vx/c^2)$$

(The equations are written with the assumption that $t = t' = 0$ when the origins of $S$ and $S'$ coincide.) Note that the spatial values $x$ and the temporal values $t$ are bound together in the first and last equations. This entanglement of space and time was a prime message of Einstein's theory, a message that was long rejected by many of his contemporaries.

It is a formal requirement of relativistic equations that they should reduce to familiar classical equations if we let $c$ approach infinity. That is, if the speed of light were infinitely great, *all* finite speeds would be "low" and classical equations would never fail. If we let $c \to \infty$ in Eqs. 37-21, $\gamma \to 1$ and these equations reduce—as we expect—to the Galilean equations (Eqs. 37-20). You should check this.

Equations 37-21 are written in a form that is useful if we are given $x$ and $t$ and wish to find $x'$ and $t'$. We may wish to go the other way, however. In that case we simply solve Eqs. 37-21 for $x$ and $t$, obtaining

$$x = \gamma(x' + vt') \quad \text{and} \quad t = \gamma(t' + vx'/c^2).$$    (37-22)

Comparison shows that, starting from either Eqs. 37-21 or Eqs. 37-22, you can find the other set by interchanging primed and unprimed quantities and reversing the sign of the relative velocity $v$.

Equations 37-21 and 37-22 relate the coordinates of a single event as seen by two observers. Sometimes we want to know not the coordinates of a single event but the differences between coordinates for a pair of events. That is, if we label our events 1 and 2, we may want to relate

$$\Delta x = x_2 - x_1 \quad \text{and} \quad \Delta t = t_2 - t_1,$$

as measured by an observer in $S$, and

$$\Delta x' = x'_2 - x'_1 \quad \text{and} \quad \Delta t' = t'_2 - t'_1,$$

as measured by an observer in $S'$.

* You may wonder why we do not call these the *Einstein transformation equations* (and why not the *Einstein factor* for $\gamma$). H. A. Lorentz actually derived these equations before Einstein did, but as the great Dutch physicist graciously conceded, he did not take the further bold step of interpreting these equations as describing the true nature of space and time. It is this interpretation, first made by Einstein, that is at the heart of relativity.

**TABLE 37-2**

**The Lorentz Transformation Equations for Pairs of Events**

| | |
|---|---|
| 1. $\Delta x = \gamma(\Delta x' + v\,\Delta t')$ | 1'. $\Delta x' = \gamma(\Delta x - v\,\Delta t)$ |
| 2. $\Delta t = \gamma(\Delta t' + v\,\Delta x'/c^2)$ | 2'. $\Delta t' = \gamma(\Delta t - v\,\Delta x/c^2)$ |

$$\gamma = \frac{1}{\sqrt{1 - (v/c)^2}} = \frac{1}{\sqrt{1 - \beta^2}}$$

Frame $S'$ moves at velocity $v$ relative to frame $S$.

Table 37-2 displays the Lorentz equations in difference form, suitable for analyzing pairs of events. The equations in the table were derived by simply substituting differences (such as $\Delta x$ and $\Delta x'$) for the four variables in Eqs. 37-21 and 37-22.

Be careful: When substituting values for these differences, you must be consistent and not mix the values for the first event with those for the second event. Also, if, say, $\Delta x$ is a negative quantity, you must be certain to include the minus sign in a substitution.

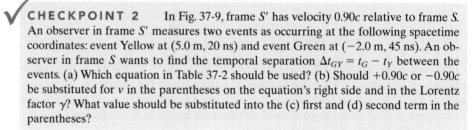

**✓ CHECKPOINT 2**    In Fig. 37-9, frame $S'$ has velocity $0.90c$ relative to frame $S$. An observer in frame $S'$ measures two events as occurring at the following spacetime coordinates: event Yellow at (5.0 m, 20 ns) and event Green at (−2.0 m, 45 ns). An observer in frame $S$ wants to find the temporal separation $\Delta t_{GY} = t_G - t_Y$ between the events. (a) Which equation in Table 37-2 should be used? (b) Should $+0.90c$ or $-0.90c$ be substituted for $v$ in the parentheses on the equation's right side and in the Lorentz factor $\gamma$? What value should be substituted into the (c) first and (d) second term in the parentheses?

## 37-8 | Some Consequences of the Lorentz Equations

Here we use the equations of Table 37-2 to affirm some of the conclusions that we reached earlier by arguments based directly on the postulates.

### Simultaneity

Consider Eq. 2 of Table 37-2,

$$\Delta t = \gamma\left(\Delta t' + \frac{v\,\Delta x'}{c^2}\right). \tag{37-23}$$

If two events occur at different places in reference frame $S'$ of Fig. 37-9, then $\Delta x'$ in this equation is not zero. It follows that even if the events are simultaneous in $S'$ (thus $\Delta t' = 0$), they will not be simultaneous in frame $S$. (This is in accord with our conclusion in Section 37-4.) The time interval between the events in $S$ will be

$$\Delta t = \gamma \frac{v\,\Delta x'}{c^2} \qquad \text{(simultaneous events in } S'\text{)}.$$

### Time Dilation

Suppose now that two events occur at the same place in $S'$ (thus $\Delta x' = 0$) but at different times (thus $\Delta t' \neq 0$). Equation 37-23 then reduces to

$$\Delta t = \gamma\,\Delta t' \qquad \text{(events in same place in } S'\text{)}. \tag{37-24}$$

This confirms time dilation between frames $S$ and $S'$. Moreover, because the two events occur at the same place in $S'$, the time interval $\Delta t'$ between them can be measured with a single clock, located at that place. Under these conditions, the measured interval is a proper time interval, and we can label it $\Delta t_0$ as we have previously labeled proper times. Thus, with that label Eq. 37-24 becomes

$$\Delta t = \gamma\, \Delta t_0 \quad \text{(time dilation)},$$

which is exactly Eq. 37-9, the time dilation equation.

## Length Contraction

Consider Eq. 1′ of Table 37-2,

$$\Delta x' = \gamma(\Delta x - v\, \Delta t). \tag{37-25}$$

If a rod lies parallel to the $x$ and $x'$ axes of Fig. 37-9 and is at rest in reference frame $S'$, an observer in $S'$ can measure its length at leisure. One way to do so is by subtracting the coordinates of the end points of the rod. The value of $\Delta x'$ that is obtained will be the proper length $L_0$ of the rod because the measurements are made in a frame where the rod is at rest.

Suppose the rod is moving in frame $S$. This means that $\Delta x$ can be identified as the length $L$ of the rod in frame $S$ only if the coordinates of the rod's end points are measured *simultaneously*—that is, if $\Delta t = 0$. If we put $\Delta x' = L_0$, $\Delta x = L$, and $\Delta t = 0$ in Eq. 37-25, we find

$$L = \frac{L_0}{\gamma} \quad \text{(length contraction)}, \tag{37-26}$$

which is exactly Eq. 37-13, the length contraction equation.

## Sample Problem  37-5

An Earth starship has been sent to check an Earth outpost on the planet P1407, whose moon houses a battle group of the often hostile Reptulians. As the ship follows a straight-line course first past the planet and then past the moon, it detects a high-energy microwave burst at the Reptulian moon base and then, 1.10 s later, an explosion at the Earth outpost, which is $4.00 \times 10^8$ m from the Reptulian base as measured from the ship's reference frame. The Reptulians have obviously attacked the Earth outpost, and so the starship begins to prepare for a confrontation with them.

(a) The speed of the ship relative to the planet and its moon is $0.980c$. What are the distance and time interval between the burst and the explosion as measured in the planet–moon inertial frame (and thus according to the occupants of the stations)?

**KEY IDEAS**

1. This problem involves measurements made from two reference frames, the planet–moon frame and the starship frame.

2. This problem involves two events: the burst and the explosion.

3. We need to transform the given data about the time and distance between the two events as measured in the starship frame to the corresponding data as measured in the planet–moon frame.

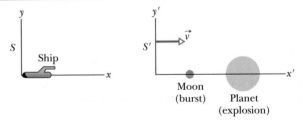

**FIG. 37-10** A planet and its moon in reference frame $S'$ move rightward with speed $v$ relative to a starship in reference frame $S$.

*Starship frame:* Before we get to the transformation, we need to carefully choose our notation. We begin with a sketch of the situation as shown in Fig. 37-10. There, we have chosen the ship's frame $S$ to be stationary and the planet–moon frame $S'$ to be moving with positive velocity (rightward). (This is an arbitrary choice; we could, instead, have chosen the planet–moon frame to be stationary. Then we would redraw $\vec{v}$ in Fig. 37-10 as being attached to the $S$ frame and indicating leftward motion; $v$ would then be a negative quantity. The results would be the same.) Let subscripts e and b represent the explosion and burst, respectively. Then the given data, all in the unprimed (starship) reference frame, are

$$\Delta x = x_e - x_b = +4.00 \times 10^8 \text{ m}$$

and $$\Delta t = t_e - t_b = +1.10 \text{ s}.$$

Here, $\Delta x$ is a positive quantity because in Fig. 37-10, the coordinate $x_e$ for the explosion is greater than the coordinate $x_b$ for the burst; $\Delta t$ is also a positive quantity because the time $t_e$ of the explosion is greater (later) than the time $t_b$ of the burst.

**Planet–moon frame:** We seek $\Delta x'$ and $\Delta t'$, which we shall get by transforming the given $S$-frame data to the planet–moon frame $S'$. Because we are considering a pair of events, we choose transformation equations from Table 37-2—namely, Eqs. 1' and 2':

$$\Delta x' = \gamma(\Delta x - v\,\Delta t) \qquad (37\text{-}27)$$

and $$\Delta t' = \gamma\left(\Delta t - \frac{v\,\Delta x}{c^2}\right). \qquad (37\text{-}28)$$

Here, $v = +0.980c$ and the Lorentz factor is

$$\gamma = \frac{1}{\sqrt{1 - (v/c)^2}} = \frac{1}{\sqrt{1 - (+0.980c/c)^2}} = 5.0252.$$

Equation 37-27 then becomes

$$\Delta x' = (5.0252)[4.00 \times 10^8\text{ m} - (+0.980\,c)(1.10\text{ s})]$$
$$= 3.86 \times 10^8\text{ m}, \qquad \text{(Answer)}$$

and Eq. 37-28 becomes

$$\Delta t' = (5.0252)\left[(1.10\,\text{s}) - \frac{(+0.980c)(4.00 \times 10^8\text{m})}{c^2}\right]$$
$$= -1.04\text{ s}. \qquad \text{(Answer)}$$

(b) What is the meaning of the minus sign in the value for $\Delta t'$?

**Reasoning:** We must be consistent with the notation we

set up in part (a). Recall how we originally defined the time interval between burst and explosion: $\Delta t = t_e - t_b = +1.10$ s. To be consistent with that choice of notation, our definition of $\Delta t'$ must be $t'_e - t'_b$; thus, we have found that

$$\Delta t' = t'_e - t'_b = -1.04\text{ s}.$$

The minus sign here tells us that $t'_b > t'_e$; that is, in the planet–moon reference frame, the burst occurred 1.04 s *after* the explosion, not 1.10 s *before* the explosion as detected in the ship frame.

(c) Did the burst cause the explosion, or vice versa?

**KEY IDEA** The sequence of events measured in the planet–moon reference frame is the reverse of that measured in the ship frame. In either situation, if there is a causal relationship between the two events, information must travel from the location of one event to the location of the other to cause it.

**Checking the speed:** Let us check the required speed of the information. In the ship frame, this speed is

$$v_{info} = \frac{\Delta x}{\Delta t} = \frac{4.00 \times 10^8\text{ m}}{1.10\text{ s}} = 3.64 \times 10^8\text{ m/s},$$

but that speed is impossible because it exceeds $c$. In the planet–moon frame, the speed comes out to be $3.70 \times 10^8$ m/s, also impossible. Therefore, neither event could possibly have caused the other event; that is, they are *unrelated* events. Thus, the starship should not confront the Reptulians.

## 37-9 | The Relativity of Velocities

Here we wish to use the Lorentz transformation equations to compare the velocities that two observers in different inertial reference frames $S$ and $S'$ would measure for the same moving particle. Let $S'$ move with velocity $v$ relative to $S$.

Suppose that the particle, moving with constant velocity parallel to the $x$ and $x'$ axes in Fig. 37-11, sends out two signals as it moves. Each observer measures the space interval and the time interval between these two events. These four measurements are related by Eqs. 1 and 2 of Table 37-2,

$$\Delta x = \gamma(\Delta x' + v\,\Delta t')$$

and $$\Delta t = \gamma\left(\Delta t' + \frac{v\,\Delta x'}{c^2}\right).$$

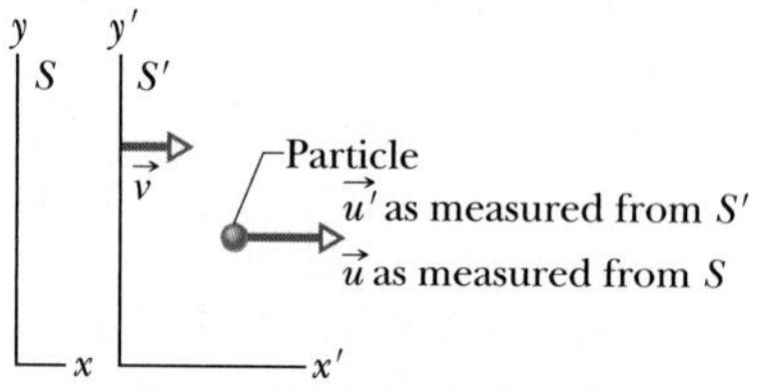

If we divide the first of these equations by the second, we find

$$\frac{\Delta x}{\Delta t} = \frac{\Delta x' + v\,\Delta t'}{\Delta t' + v\,\Delta x'/c^2}.$$

Dividing the numerator and denominator of the right side by $\Delta t'$, we find

$$\frac{\Delta x}{\Delta t} = \frac{\Delta x'/\Delta t' + v}{1 + v(\Delta x'/\Delta t')/c^2}.$$

**FIG. 37-11** Reference frame $S'$ moves with velocity $\vec{v}$ relative to frame $S$. A particle has velocity $\vec{u}'$ relative to reference frame $S'$ and velocity $\vec{u}$ relative to reference frame $S$.

However, in the differential limit, $\Delta x/\Delta t$ is $u$, the velocity of the particle as measured in $S$, and $\Delta x'/\Delta t'$ is $u'$, the velocity of the particle as measured in $S'$. Then we have, finally,

$$u = \frac{u' + v}{1 + u'v/c^2} \qquad \text{(relativistic velocity transformation)} \qquad (37\text{-}29)$$

as the relativistic velocity transformation equation. This equation reduces to the classical, or Galilean, velocity transformation equation,

$$u = u' + v \qquad \text{(classical velocity transformation)}, \qquad (37\text{-}30)$$

when we apply the formal test of letting $c \rightarrow \infty$. In other words, Eq. 37-29 is correct for all physically possible speeds, but Eq. 37-30 is approximately correct for speeds much less than $c$.

## 37-10 | Doppler Effect for Light

In Section 17-9 we discussed the Doppler effect (a shift in detected frequency) for sound waves traveling in air. For such waves, the Doppler effect depends on two velocities—namely, the velocities of the source and detector with respect to the air. Air is the medium that transmits the waves.

That is not the situation with light waves, for they (and other electromagnetic waves) require no medium, being able to travel even through vacuum. The Doppler effect for light waves depends on only one velocity, the relative velocity $\vec{v}$ between source and detector, as measured from the reference frame of either. Let $f_0$ represent the **proper frequency** of the source—that is, the frequency that is measured by an observer in the rest frame of the source. Let $f$ represent the frequency detected by an observer moving with velocity $\vec{v}$ relative to that rest frame. Then, when the direction of $\vec{v}$ is directly away from the source,

$$f = f_0 \sqrt{\frac{1 - \beta}{1 + \beta}} \qquad \text{(source and detector separating)}, \qquad (37\text{-}31)$$

where $\beta = v/c$. When the direction of $\vec{v}$ is directly toward the source, we must change the signs in front of both $\beta$ symbols in Eq. 37-31.

### Low-Speed Doppler Effect

For low speeds ($\beta \ll 1$), Eq. 37-31 can be expanded in a power series in $\beta$ and approximated as

$$f = f_0(1 - \beta + \tfrac{1}{2}\beta^2) \qquad \text{(source and detector separating, } \beta \ll 1). \qquad (37\text{-}32)$$

The corresponding low-speed equation for the Doppler effect with sound waves (or any waves except light waves) has the same first two terms but a different coefficient in the third term. Thus, the relativistic effect for low-speed light sources and detectors shows up only with the $\beta^2$ term.

A police radar unit employs the Doppler effect with microwaves to measure the speed $v$ of a car. A source in the radar unit emits a microwave beam at a certain (proper) frequency $f_0$ along the road. A car that is moving toward the unit intercepts that beam but at a frequency that is shifted upward by the Doppler effect due to the car's motion toward the radar unit. The car reflects the beam back toward the radar unit. Because the car is moving toward the radar unit, the detector in the unit intercepts a reflected beam that is further shifted up in frequency. The unit compares that detected frequency with $f_0$ and computes the speed $v$ of the car.

### Astronomical Doppler Effect

In astronomical observations of stars, galaxies, and other sources of light, we can determine how fast the sources are moving, either directly away from us or directly toward us, by measuring the *Doppler shift* of the light that reaches us. If a certain star were at rest relative to us, we would detect light from it with a certain proper frequency $f_0$. However, if the star is moving either directly away from us or directly toward us, the light we detect has a frequency $f$ that is shifted from $f_0$ by the Doppler effect. This Doppler shift is due only to the *radial* motion of the star (its motion directly toward us or away from us), and the speed we can determine by measuring this Doppler shift is only the *radial speed* $v$ of the star—that is, only the radial component of the star's velocity relative to us.

Suppose a star (or any other light source) moves away from us with a radial speed $v$ that is low enough ($\beta$ is small enough) for us to neglect the $\beta^2$ term in Eq. 37-32. Then we have

$$f = f_0(1 - \beta).  \tag{37-33}$$

Because astronomical measurements involving light are usually done in wavelengths rather than frequencies, let us replace $f$ with $c/\lambda$ and $f_0$ with $c/\lambda_0$, where $\lambda$ is the measured wavelength and $\lambda_0$ is the **proper wavelength** (the wavelength associated with $f_0$). We then have

$$\frac{c}{\lambda} = \frac{c}{\lambda_0}(1 - \beta),$$

or
$$\lambda = \lambda_0(1 - \beta)^{-1}.  \tag{37-34}$$

Because we assume $\beta$ is small, we can expand $(1 - \beta)^{-1}$ in a power series. Doing so and retaining only the first power of $\beta$, we have

$$\lambda = \lambda_0(1 + \beta),$$

or
$$\beta = \frac{\lambda - \lambda_0}{\lambda_0}.  \tag{37-35}$$

Replacing $\beta$ with $v/c$ and $\lambda - \lambda_0$ with $|\Delta\lambda|$ leads to

$$v = \frac{|\Delta\lambda|}{\lambda_0}c \qquad \text{(radial speed of light source, } v \ll c\text{).}  \tag{37-36}$$

The difference $\Delta\lambda$ is the *wavelength Doppler shift* of the light source. We enclose it with an absolute sign so that we always have a magnitude of the shift.

Equation 37-36 is an approximation that can be applied only when $v \ll c$. Under that condition, Eq. 37-36 can be applied whether the light source is moving toward or away from us. If it is moving away from us, then $\lambda$ is longer than $\lambda_0$, $\Delta\lambda$ is positive, and the Doppler shift is called a *red shift*. (The term *red* does not mean the detected light is red or even visible. It merely serves as a memory device because red is at the *long* wavelength end of the visible spectrum. Thus $\lambda$ is longer than $\lambda_0$.) If the light source is moving toward us, then $\lambda$ is shorter than $\lambda_0$, $\Delta\lambda$ is negative, and the Doppler shift is called a *blue shift*.

✓ CHECKPOINT 3    The figure shows a source that emits light of proper frequency $f_0$ while moving directly toward the right with speed $c/4$ as measured from reference frame $S$. The figure also shows a light detector, which measures a frequency $f > f_0$ for the emitted light. (a) Is the detector moving toward the left or the right? (b) Is the speed of the detector as measured from reference frame $S$ more than $c/4$, less than $c/4$, or equal to $c/4$?

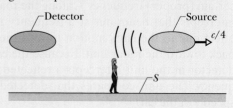

## Transverse Doppler Effect

So far, we have discussed the Doppler effect, here and in Chapter 17, only for situations in which the source and the detector move either directly toward or directly away from each other. Figure 37-12 shows a different arrangement, in which a source $S$ moves past a detector $D$. When $S$ reaches point $P$, the velocity of $S$ is perpendicular to the line joining $P$ and $D$, and at that instant $S$ is moving neither toward nor away from $D$. If the source is emitting sound waves of frequency $f_0$, $D$ detects that frequency (with no Doppler effect) when it intercepts the waves that were emitted at point $P$. However, if the source is emitting light waves, there is still a Doppler effect, called the **transverse Doppler effect.** In this situation, the detected frequency of the light emitted when the source is at point $P$ is

$$f = f_0 \sqrt{1 - \beta^2} \quad \text{(transverse Doppler effect).} \quad (37\text{-}37)$$

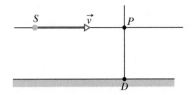

FIG. 37-12   A light source $S$ travels with velocity $\vec{v}$ past a detector at $D$. The special theory of relativity predicts a transverse Doppler effect as the source passes through point $P$, where the direction of travel is perpendicular to the line extending through $D$. Classical theory predicts no such effect.

For low speeds ($\beta \ll 1$), Eq. 37-37 can be expanded in a power series in $\beta$ and approximated as

$$f = f_0(1 - \tfrac{1}{2}\beta^2) \quad \text{(low speeds).} \quad (37\text{-}38)$$

Here the first term is what we would expect for sound waves, and again the relativistic effect for low-speed light sources and detectors appears with the $\beta^2$ term.

In principle, a police radar unit can determine the speed of a car even when the path of the radar beam is perpendicular (transverse) to the path of the car. However, Eq. 37-38 tells us that because $\beta$ is small even for a fast car, the relativistic term $\beta^2/2$ in the transverse Doppler effect is extremely small. Thus, $f \approx f_0$ and the radar unit computes a speed of zero. For this reason, police officers always try to direct the radar beam along the car's path to get a Doppler shift that gives the car's actual speed. Any deviation from that alignment works in favor of the motorist because it reduces the measured speed.

The transverse Doppler effect is really another test of time dilation. If we rewrite Eq. 37-37 in terms of the period $T$ of oscillation of the emitted light wave instead of the frequency, we have, because $T = 1/f$,

$$T = \frac{T_0}{\sqrt{1 - \beta^2}} = \gamma T_0, \quad (37\text{-}39)$$

in which $T_0 \ (= 1/f_0)$ is the **proper period** of the source. As comparison with Eq. 37-9 shows, Eq. 37-39 is simply the time dilation formula because a period is a time interval.

## The NAVSTAR Navigation System

Each NAVSTAR satellite in the Global Positioning System (GPS) continuously signals its location by broadcasting radio signals at a set frequency controlled by precise atomic clocks. When the signal is sensed by a GPS detector on, say, a commercial aircraft, the frequency has been Doppler shifted because of the relative motion between the satellite and the detector. By detecting the signals from several NAVSTAR satellites simultaneously, the detector can determine the direction to any one of them and the direction of the satellite velocity. From the Doppler shift of the signal, the detector then determines the speed of the aircraft.

Although the relativistic contribution to the Doppler shift is extremely slight (roughly $4.5 \times 10^{-10}$ of the shift is due to the relativity of this chapter), engineers must take the contribution into account to keep the NAVSTAR system accurate. They must also take into account the effects of Einstein's general theory of relativity because the gravitational acceleration at the altitude of a satellite is different from that at the altitude of a ground-based detector. Einstein's special and general theories of relativity were considered bizarre for a long time after they were published. Now they are indispensable in global positioning and long-range navigation.

**Sample Problem 37-6**

Figure 37-13a shows curves of intensity versus wavelength for light reaching us from interstellar gas on two opposite sides of galaxy M87 (Fig. 37-13b). One curve peaks at 499.8 nm, the other at 501.6 nm. The gas orbits the core of the galaxy at a radius $r = 100$ light-years, moving toward us on one side of the core and moving away from us on the opposite side.

(a) Which curve corresponds to the gas moving toward us? What is the speed of the gas relative to us (and relative to the galaxy's core)?

**KEY IDEAS**

1. If the gas were not moving, the light from it would be detected at a certain wavelength.

2. The motion of the gas changes the detected wavelength via the Doppler effect, increasing the wavelength for the gas moving away from us and decreasing it for the gas moving toward us.

Thus, the curve peaking at 501.6 nm corresponds to motion away from us, and that peaking at 499.8 nm corresponds to motion toward us.

**Calculations:** Let us assume that the increase and the decrease in wavelength due to the motion of the gas are equal in magnitude. Then the unshifted wavelength, which we shall take as the proper wavelength $\lambda_0$, must be the average of the two shifted wavelengths:

$$\lambda_0 = \frac{501.6 \text{ nm} + 499.8 \text{ nm}}{2} = 500.7 \text{ nm}.$$

The Doppler shift $\Delta\lambda$ of the light from the gas moving away from us is then

$$\Delta\lambda = |\lambda - \lambda_0| = 501.6 \text{ nm} - 500.7 \text{ nm}$$
$$= 0.90 \text{ nm}.$$

Substituting this and $\lambda_0 = 500.7$ nm into Eq. 37-36, we find that the speed of the gas is

$$v = \frac{\Delta\lambda}{\lambda_0} c = \frac{0.90 \text{ nm}}{500.7 \text{ nm}} 299 \ 792 \ 458 \text{ m/s}$$
$$= 5.39 \times 10^5 \text{ m/s}. \qquad \text{(Answer)}$$

(b) The gas orbits the core of the galaxy because the gas experiences a gravitational force due to the mass $M$ of the core. What is that mass in multiples of the Sun's mass $M_S$ ($= 1.99 \times 10^{30}$ kg)?

**KEY IDEAS**

1. From Eq. 13-1, the magnitude $F$ of the gravitational force on an orbiting gas element of mass $m$ at orbital radius $r$ is

$$F = \frac{GMm}{r^2}.$$

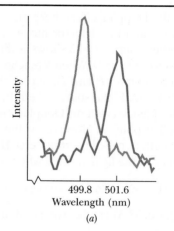
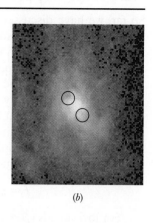

FIG. 37-13 (a) Plots of intensity versus wavelength for light emitted by gas on opposite sides of galaxy M87 and detected on Earth. (b) The central region of M87. The circles indicate the locations of the gas whose intensity is given in (a). The core of M87 is halfway between the circles. *(Courtesy NASA)*

2. If the gas element orbits the galaxy core in a circle, then the element must have a centripetal acceleration of magnitude $a = v^2/r$, directed toward the core.

3. Newton's second law, written for a radial axis extending from the core to the gas element, tells us that $F = ma$.

**Calculations:** Putting these three ideas together, we have

$$\frac{GMm}{r^2} = m \frac{v^2}{r}.$$

Solving this for $M$ and substituting known data, we find

$$M = \frac{v^2 r}{G}$$

$$= \frac{(5.39 \times 10^5 \text{ m/s})^2 (100 \text{ ly})(9.46 \times 10^{15} \text{ m/ly})}{6.67 \times 10^{-11} \text{ N} \cdot \text{m}^2/\text{kg}^2}$$

$$= 4.12 \times 10^{39} \text{ kg} = (2.1 \times 10^9) M_S. \qquad \text{(Answer)}$$

This result tells us that a mass equivalent to two billion suns has been compacted into the core of the galaxy, strongly suggesting that a *supermassive black hole* occupies the core.

***The monster of M87:*** The M87 black hole is about 1000 times larger than the supermassive black hole lying at the center of our Milky Way galaxy (Sample Problem 13-7) and thus is indeed a monster. It rotates about an axis, and as material (stars and dust) fall into it, the material generates strong electromagnetic forces that shoot electrons away from the black hole, outward along the rotation axis and out of M87. The ejection is so severe that the electrons move at nearly the speed of light. The image that opens this chapter shows the jet that is angled somewhat in our direction; we cannot see the jet in the opposite direction.

## 37-11 | A New Look at Momentum

Suppose that a number of observers, each in a different inertial reference frame, watch an isolated collision between two particles. In classical mechanics, we have seen that—even though the observers measure different velocities for the colliding particles—they all find that the law of conservation of momentum holds. That is, they find that the total momentum of the system of particles after the collision is the same as it was before the collision.

How is this situation affected by relativity? We find that if we continue to define the momentum $\vec{p}$ of a particle as $m\vec{v}$, the product of its mass and its velocity, total momentum is *not* conserved for the observers in different inertial frames. We have two choices: (1) Give up the law of conservation of momentum or (2) see whether we can refine our definition of momentum in some new way so that the law of conservation of momentum still holds. The correct choice is the second one.

Consider a particle moving with constant speed $v$ in the positive direction of an $x$ axis. Classically, its momentum has magnitude

$$p = mv = m\frac{\Delta x}{\Delta t} \qquad \text{(classical momentum)}, \qquad (37\text{-}40)$$

in which $\Delta x$ is the distance it travels in time $\Delta t$. To find a relativistic expression for momentum, we start with the new definition

$$p = m\frac{\Delta x}{\Delta t_0}.$$

Here, as before, $\Delta x$ is the distance traveled by a moving particle as viewed by an observer watching that particle. However, $\Delta t_0$ is the time required to travel that distance, measured not by the observer watching the moving particle but by an observer moving with the particle. The particle is at rest with respect to this second observer; thus that measured time is a proper time.

Using the time dilation formula, $\Delta t = \gamma \Delta t_0$ (Eq. 37-9), we can then write

$$p = m\frac{\Delta x}{\Delta t_0} = m\frac{\Delta x}{\Delta t}\frac{\Delta t}{\Delta t_0} = m\frac{\Delta x}{\Delta t}\gamma.$$

However, since $\Delta x/\Delta t$ is just the particle velocity $v$, we have

$$p = \gamma mv \qquad \text{(momentum)}. \qquad (37\text{-}41)$$

Note that this differs from the classical definition of Eq. 37-40 only by the Lorentz factor $\gamma$. However, that difference is important: Unlike classical momentum, relativistic momentum approaches an infinite value as $v$ approaches $c$.

We can generalize the definition of Eq. 37-41 to vector form as

$$\vec{p} = \gamma m\vec{v} \qquad \text{(momentum)}. \qquad (37\text{-}42)$$

This equation gives the correct definition of momentum for all physically possible speeds. For a speed much less than $c$, it reduces to the classical definition of momentum ($\vec{p} = m\vec{v}$).

## 37-12 | A New Look at Energy

### Mass Energy

The science of chemistry was initially developed with the assumption that in chemical reactions, energy and mass are conserved separately. In 1905, Einstein showed that as a consequence of his theory of special relativity, mass can be considered to be another form of energy. Thus, the law of conservation of energy is really the law of conservation of mass–energy.

In a *chemical reaction* (a process in which atoms or molecules interact), the amount of mass that is transferred into other forms of energy (or vice versa) is such a tiny fraction of the total mass involved that there is no hope of measuring the mass change with even the best laboratory balances. Mass and energy truly *seem* to be separately conserved. However, in a *nuclear reaction* (in which nuclei or fundamental particles interact), the energy released is often about a million times greater than in a chemical reaction, and the change in mass can easily be measured. Taking mass–energy transfers into account in nuclear reactions became routine long ago.

An object's mass $m$ and the equivalent energy $E_0$ are related by

$$E_0 = mc^2, \tag{37-43}$$

which, without the subscript 0, is the best-known science equation of all time. This energy that is associated with the mass of an object is called **mass energy** or **rest energy.** The second name suggests that $E_0$ is an energy that the object has even when it is at rest, simply because it has mass. (If you continue your study of physics beyond this book, you will see more refined discussions of the relation between mass and energy. You might even encounter disagreements about just what that relation is and means.)

Table 37-3 shows the (approximate) mass energy, or rest energy, of a few objects. The mass energy of, say, a U.S. penny is enormous; the equivalent amount of electrical energy would cost well over a million dollars. On the other hand, the entire annual U.S. electrical energy production corresponds to a mass of only a few hundred kilograms of matter (stones, burritos, or anything else).

In practice, SI units are rarely used with Eq. 37-43 because they are too large to be convenient. Masses are usually measured in atomic mass units, where

$$1\ \text{u} = 1.660\,538\,86 \times 10^{-27}\ \text{kg}, \tag{37-44}$$

and energies are usually measured in electron-volts or multiples of it, where

$$1\ \text{eV} = 1.602\,176\,462 \times 10^{-19}\ \text{J}. \tag{37-45}$$

In the units of Eqs. 37-44 and 37-45, the multiplying constant $c^2$ has the values

$$c^2 = 9.314\,940\,13 \times 10^8\ \text{eV/u} = 9.314\,940\,13 \times 10^5\ \text{keV/u}$$
$$= 931.494\,013\ \text{MeV/u}. \tag{37-46}$$

## Total Energy

Equation 37-43 gives, for any object, the mass energy $E_0$ that is associated with the object's mass $m$, regardless of whether the object is at rest or moving. If the object is moving, it has additional energy in the form of kinetic energy $K$. If we assume that the object's potential energy is zero, then its total energy $E$ is the sum of its mass energy and its kinetic energy:

$$E = E_0 + K = mc^2 + K. \tag{37-47}$$

**TABLE 37-3**

**The Energy Equivalents of a Few Objects**

| Object | Mass (kg) | Energy Equivalent | |
|---|---|---|---|
| Electron | $\approx 9.11 \times 10^{-31}$ | $\approx 8.19 \times 10^{-14}$ J | ($\approx 511$ keV) |
| Proton | $\approx 1.67 \times 10^{-27}$ | $\approx 1.50 \times 10^{-10}$ J | ($\approx 938$ MeV) |
| Uranium atom | $\approx 3.95 \times 10^{-25}$ | $\approx 3.55 \times 10^{-8}$ J | ($\approx 225$ GeV) |
| Dust particle | $\approx 1 \times 10^{-13}$ | $\approx 1 \times 10^4$ J | ($\approx 2$ kcal) |
| U.S. penny | $\approx 3.1 \times 10^{-3}$ | $\approx 2.8 \times 10^{14}$ J | ($\approx 78$ GW·h) |

Although we shall not prove it, the total energy $E$ can also be written as

$$E = \gamma mc^2, \tag{37-48}$$

where $\gamma$ is the Lorentz factor for the object's motion.

Since Chapter 7, we have discussed many examples involving changes in the total energy of a particle or a system of particles. However, we did not include mass energy in the discussions because the changes in mass energy were either zero or small enough to be neglected. The law of conservation of total energy still applies when changes in mass energy are significant. Thus, regardless of what happens to the mass energy, the following statement from Section 8-8 is still true:

 The total energy $E$ of an *isolated system* cannot change.

For example, if the total mass energy of two interacting particles in an isolated system decreases, some other type of energy in the system must increase because the total energy cannot change.

In a system undergoing a chemical or nuclear reaction, a change in the total mass energy of the system due to the reaction is often given as a $Q$ value. The $Q$ value for a reaction is obtained from the relation

$$\begin{pmatrix} \text{system's initial} \\ \text{total mass energy} \end{pmatrix} = \begin{pmatrix} \text{system's final} \\ \text{total mass energy} \end{pmatrix} + Q$$

or

$$E_{0i} = E_{0f} + Q. \tag{37-49}$$

Using Eq. 37-43 ($E_0 = mc^2$), we can rewrite this in terms of the initial *total* mass $M_i$ and the final *total* mass $M_f$ as

$$M_i c^2 = M_f c^2 + Q$$

or

$$Q = M_i c^2 - M_f c^2 = -\Delta M c^2, \tag{37-50}$$

where the change in mass due to the reaction is $\Delta M = M_f - M_i$.

If a reaction results in the transfer of energy from mass energy to, say, kinetic energy of the reaction products, the system's total mass energy $E_0$ (and total mass $M$) decreases and $Q$ is positive. If, instead, a reaction requires that energy be transferred to mass energy, the system's total mass energy $E_0$ (and its total mass $M$) increases and $Q$ is negative.

For example, suppose two hydrogen nuclei undergo a *fusion reaction* in which they join together to form a single nucleus and release two particles in the process. The total mass energy (and total mass) of the resultant single nucleus and two released particles is less than the total mass energy (and total mass) of the initial hydrogen nuclei. Thus, the $Q$ of the fusion reaction is positive, and energy is said to be *released* (transferred from mass energy) by the reaction. This release is important to you because the fusion of hydrogen nuclei in the Sun is one part of the process that results in sunshine on Earth and makes life here possible.

## Kinetic Energy

In Chapter 7 we defined the kinetic energy $K$ of an object of mass $m$ moving at speed $v$ well below $c$ to be

$$K = \tfrac{1}{2}mv^2. \tag{37-51}$$

However, this classical equation is only an approximation that is good enough when the speed is well below the speed of light.

Let us now find an expression for kinetic energy that is correct for *all* physically possible speeds, including speeds close to $c$. Solving Eq. 37-47 for $K$ and then

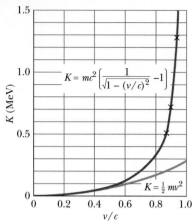

**FIG. 37-14** The relativistic (Eq. 37-52) and classical (Eq. 37-51) equations for the kinetic energy of an electron, plotted as a function of $v/c$, where $v$ is the speed of the electron and $c$ is the speed of light. Note that the two curves blend together at low speeds and diverge widely at high speeds. Experimental data (at the × marks) show that at high speeds the relativistic curve agrees with experiment but the classical curve does not.

substituting for $E$ from Eq. 37-48 lead to

$$K = E - mc^2 = \gamma mc^2 - mc^2$$
$$= mc^2(\gamma - 1) \quad \text{(kinetic energy)}, \tag{37-52}$$

where $\gamma \, (= 1/\sqrt{1 - (v/c)^2})$ is the Lorentz factor for the object's motion.

Figure 37-14 shows plots of the kinetic energy of an electron as calculated with the correct definition (Eq. 37-52) and the classical approximation (Eq. 37-51), both as functions of $v/c$. Note that on the left side of the graph the two plots coincide; this is the part of the graph—at lower speeds—where we have calculated kinetic energies so far in this book. This part of the graph tells us that we have been justified in calculating kinetic energy with the classical expression of Eq. 37-51. However, on the right side of the graph—at speeds near $c$—the two plots differ significantly. As $v/c$ approaches 1.0, the plot for the classical definition of kinetic energy increases only moderately while the plot for the correct definition of kinetic energy increases dramatically, approaching an infinite value as $v/c$ approaches 1.0. Thus, when an object's speed $v$ is near $c$, we *must* use Eq. 37-52 to calculate its kinetic energy.

Figure 37-14 also tells us something about the work we must do on an object to increase its speed by, say, 1%. The required work $W$ is equal to the resulting change $\Delta K$ in the object's kinetic energy. If the change is to occur on the low-speed, left side of Fig. 37-14, the required work might be modest. However, if the change is to occur on the high-speed, right side of Fig. 37-14, the required work could be enormous because the kinetic energy $K$ increases so rapidly there with an increase in speed $v$. To increase an object's speed to $c$ would require, in principle, an infinite amount of energy; thus, doing so is impossible.

The kinetic energies of electrons, protons, and other particles are often stated with the unit electron-volt or one of its multiples used as an adjective. For example, an electron with a kinetic energy of 20 MeV may be described as a 20 MeV electron.

### Momentum and Kinetic Energy

In classical mechanics, the momentum $p$ of a particle is $mv$ and its kinetic energy $K$ is $\frac{1}{2}mv^2$. If we eliminate $v$ between these two expressions, we find a direct relation between momentum and kinetic energy:

$$p^2 = 2Km \quad \text{(classical).} \tag{37-53}$$

We can find a similar connection in relativity by eliminating $v$ between the relativistic definition of momentum (Eq. 37-41) and the relativistic definition of kinetic energy (Eq. 37-52). Doing so leads, after some algebra, to

$$(pc)^2 = K^2 + 2Kmc^2. \tag{37-54}$$

With the aid of Eq. 37-47, we can transform Eq. 37-54 into a relation between the momentum $p$ and the total energy $E$ of a particle:

$$E^2 = (pc)^2 + (mc^2)^2. \tag{37-55}$$

The right triangle of Fig. 37-15 can help you keep these useful relations in mind. You can also show that, in that triangle,

$$\sin \theta = \beta \quad \text{and} \quad \cos \theta = 1/\gamma. \tag{37-56}$$

With Eq. 37-55 we can see that the product $pc$ must have the same unit as energy $E$; thus, we can express the unit of momentum $p$ as an energy unit divided by $c$, usually as MeV/$c$ or GeV/$c$ in fundamental particle physics.

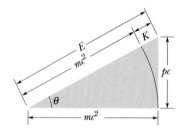

**FIG. 37-15** A useful memory diagram for the relativistic relations among the total energy $E$, the rest energy or mass energy $mc^2$, the kinetic energy $K$, and the momentum magnitude $p$.

✓**CHECKPOINT 4** Are (a) the kinetic energy and (b) the total energy of a 1 GeV electron more than, less than, or equal to those of a 1 GeV proton?

## Sample Problem    37-7

**(a)** What is the total energy $E$ of a 2.53 MeV electron?

**KEY IDEA**  From Eq. 37-47, the total energy $E$ is the sum of the electron's mass energy (or rest energy) $mc^2$ and its kinetic energy:

$$E = mc^2 + K. \qquad (37\text{-}57)$$

*Calculations:* The adjective "2.53 MeV" in the problem statement means that the electron's kinetic energy is 2.53 MeV. To evaluate the electron's mass energy $mc^2$, we substitute the electron's mass $m$ from Appendix B, obtaining

$$mc^2 = (9.109 \times 10^{-31}\ \text{kg})(299\ 792\ 458\ \text{m/s})^2$$
$$= 8.187 \times 10^{-14}\ \text{J}.$$

Then dividing this result by $1.602 \times 10^{-13}$ J/MeV gives us 0.511 MeV as the electron's mass energy (confirming the value in Table 37-3). Equation 37-57 then yields

$$E = 0.511\ \text{MeV} + 2.53\ \text{MeV} = 3.04\ \text{MeV}. \quad \text{(Answer)}$$

**(b)** What is the magnitude $p$ of the electron's momentum, in the unit MeV/$c$?

**KEY IDEA**  We can find $p$ from the total energy $E$ and the mass energy $mc^2$ via Eq. 37-55,

$$E^2 = (pc)^2 + (mc^2)^2.$$

*Calculations:* Solving for $pc$ gives us

$$pc = \sqrt{E^2 - (mc^2)^2}$$
$$= \sqrt{(3.04\ \text{MeV})^2 - (0.511\ \text{MeV})^2} = 3.00\ \text{MeV}.$$

Finally, dividing both sides by $c$ we find

$$p = 3.00\ \text{MeV}/c. \qquad \text{(Answer)}$$

## Sample Problem    37-8

The most energetic proton ever detected in the cosmic rays coming to Earth from space had an astounding kinetic energy of $3.0 \times 10^{20}$ eV (enough energy to warm a teaspoon of water by a few degrees).

**(a)** What were the proton's Lorentz factor $\gamma$ and speed $v$ (both relative to the ground-based detector)?

**KEY IDEAS**  (1) The proton's Lorentz factor $\gamma$ relates its total energy $E$ to its mass energy $mc^2$ via Eq. 37-48 ($E = \gamma mc^2$). (2) The proton's total energy is the sum of its mass energy $mc^2$ and its (given) kinetic energy $K$.

*Calculations:* Putting these ideas together we have

$$\gamma = \frac{E}{mc^2} = \frac{mc^2 + K}{mc^2} = 1 + \frac{K}{mc^2}. \quad (37\text{-}58)$$

We can calculate the proton's mass energy $mc^2$ from its mass given in Appendix B, as we did for the electron in Sample Problem 37-7a. We find that $mc^2$ is 938 MeV (as listed in Table 37-3). Substituting this and the given kinetic energy into Eq. 37-58, we obtain

$$\gamma = 1 + \frac{3.0 \times 10^{20}\ \text{eV}}{938 \times 10^6\ \text{eV}}$$
$$= 3.198 \times 10^{11} \approx 3.2 \times 10^{11}. \quad \text{(Answer)}$$

This computed value for $\gamma$ is so large that we cannot use the definition of $\gamma$ (Eq. 37-8) to find $v$. Try it; your calculator will tell you that $\beta$ is effectively equal to 1 and thus that $v$ is effectively equal to $c$. Actually, $v$ is almost $c$, but we want a more accurate answer, which we can obtain by first solving Eq. 37-8 for $1 - \beta$. To begin

we write

$$\gamma = \frac{1}{\sqrt{1 - \beta^2}} = \frac{1}{\sqrt{(1 - \beta)(1 + \beta)}} \approx \frac{1}{\sqrt{2(1 - \beta)}},$$

where we have used the fact that $\beta$ is so close to unity that $1 + \beta$ is very close to 2. The velocity we seek is contained in the $1 - \beta$ term. Solving for $1 - \beta$ then yields

$$1 - \beta = \frac{1}{2\gamma^2} = \frac{1}{(2)(3.198 \times 10^{11})^2}$$
$$= 4.9 \times 10^{-24} \approx 5 \times 10^{-24}.$$

Thus,    $\beta = 1 - 5 \times 10^{-24}$

and, since $v = \beta c$,

$$v \approx 0.999\ 999\ 999\ 999\ 999\ 999\ 995c. \quad \text{(Answer)}$$

**(b)** Suppose that the proton travels along a diameter ($9.8 \times 10^4$ ly) of the Milky Way galaxy. Approximately how long does the proton take to travel that diameter as measured from the common reference frame of Earth and the Galaxy?

*Reasoning:* We just saw that this *ultrarelativistic* proton is traveling at a speed barely less than $c$. By the definition of light-year, light takes 1 y to travel a distance of 1 ly, and so light should take $9.8 \times 10^4$ y to travel $9.8 \times 10^4$ ly, and this proton should take almost the same time. Thus, from our Earth–Milky Way reference frame, the proton's trip takes

$$\Delta t = 9.8 \times 10^4\ \text{y}. \qquad \text{(Answer)}$$

**(c)** How long does the trip take as measured in the reference frame of the proton?

**KEY IDEAS**

1. This problem involves measurements made from two (inertial) reference frames: one is the Earth–Milky Way frame and the other is attached to the proton.

2. This problem also involves two events: the first is when the proton passes one end of the diameter along the Galaxy, and the second is when it passes the opposite end.

3. The time interval between those two events as measured in the proton's reference frame is the proper time interval $\Delta t_0$ because the events occur at the same location in that frame—namely, at the proton itself.

4. We can find the proper time interval $\Delta t_0$ from the time interval $\Delta t$ measured in the Earth–Milky Way frame by using Eq. 37-9 ($\Delta t = \gamma \Delta t_0$) for time dilation.

*Calculation:* Solving Eq. 37-9 for $\Delta t_0$ and substituting $\gamma$ from (a) and $\Delta t$ from (b), we find

$$\Delta t_0 = \frac{\Delta t}{\gamma} = \frac{9.8 \times 10^4 \text{ y}}{3.198 \times 10^{11}}$$
$$= 3.06 \times 10^{-7} \text{ y} = 9.7 \text{ s.} \qquad \text{(Answer)}$$

In our frame, the trip takes 98 000 y. In the proton's frame, it takes 9.7 s! As promised at the start of this chapter, relative motion can alter the rate at which time passes, and we have here an extreme example.

## REVIEW & SUMMARY

**The Postulates** Einstein's **special theory of relativity** is based on two postulates:

1. The laws of physics are the same for observers in all inertial reference frames. No one frame is preferred over any other.

2. The speed of light in vacuum has the same value $c$ in all directions and in all inertial reference frames.

The speed of light $c$ in vacuum is an ultimate speed that cannot be exceeded by any entity carrying energy or information.

**Coordinates of an Event** Three space coordinates and one time coordinate specify an **event**. One task of special relativity is to relate these coordinates as assigned by two observers who are in uniform motion with respect to each other.

**Simultaneous Events** If two observers are in relative motion, they will not, in general, agree as to whether two events are simultaneous. If one of the observers finds two events at different locations to be simultaneous, the other will not, and conversely. Simultaneity is *not* an absolute concept but a relative one, depending on the motion of the observer. The relativity of simultaneity is a direct consequence of the finite ultimate speed $c$.

**Time Dilation** If two successive events occur at the same place in an inertial reference frame, the time interval $\Delta t_0$ between them, measured on a single clock where they occur, is the **proper time** between the events. *Observers in frames moving relative to that frame will measure a larger value for this interval.* For an observer moving with relative speed $v$, the measured time interval is

$$\Delta t = \frac{\Delta t_0}{\sqrt{1 - (v/c)^2}} = \frac{\Delta t_0}{\sqrt{1 - \beta^2}}$$
$$= \gamma \, \Delta t_0 \qquad \text{(time dilation).} \qquad \text{(37-7 to 37-9)}$$

Here $\beta = v/c$ is the **speed parameter** and $\gamma = 1/\sqrt{1 - \beta^2}$ is the **Lorentz factor.** An important result of time dilation is that moving clocks run slow as measured by an observer at rest.

**Length Contraction** The length $L_0$ of an object measured by an observer in an inertial reference frame in which the object is at rest is called its **proper length.** *Observers in frames moving relative to that frame and parallel to that length will measure a shorter length.* For an observer moving with relative speed $v$, the measured length is

$$L = L_0\sqrt{1 - \beta^2} = \frac{L_0}{\gamma} \qquad \text{(length contraction).} \qquad (37\text{-}13)$$

**The Lorentz Transformation** The *Lorentz transformation* equations relate the spacetime coordinates of a single event as seen by observers in two inertial frames, $S$ and $S'$, where $S'$ is moving relative to $S$ with velocity $v$ in the positive $x$ and $x'$ direction. The four coordinates are related by

$$\begin{aligned} x' &= \gamma(x - vt), \\ y' &= y, \\ z' &= z, \\ t' &= \gamma(t - vx/c^2) \end{aligned} \qquad (37\text{-}21)$$

**Relativity of Velocities** When a particle is moving with speed $u'$ in the positive $x'$ direction in an inertial reference frame $S'$ that itself is moving with speed $v$ parallel to the $x$ direction of a second inertial frame $S$, the speed $u$ of the particle as measured in $S$ is

$$u = \frac{u' + v}{1 + u'v/c^2} \qquad \text{(relativistic velocity).} \qquad (37\text{-}29)$$

**Relativistic Doppler Effect** If a source emitting light waves of frequency $f_0$ moves directly away from a detector with relative radial speed $v$ (and speed parameter $\beta = v/c$), the frequency $f$ measured by the detector is

$$f = f_0\sqrt{\frac{1 - \beta}{1 + \beta}}. \qquad (37\text{-}31)$$

If the source moves directly toward the detector, the signs in Eq. 37-31 are reversed.

For astronomical observations, the Doppler effect is measured in wavelengths. For speeds much less than $c$, Eq. 37-31 leads to

$$v = \frac{|\Delta\lambda|}{\lambda_0} c, \qquad (37\text{-}36)$$

where $\Delta\lambda$ $(= \lambda - \lambda_0)$ is the *Doppler shift* in wavelength due to the motion.

**Transverse Doppler Effect** If the relative motion of the light source is perpendicular to a line joining the source and detector, the Doppler frequency formula is

$$f = f_0 \sqrt{1 - \beta^2}. \qquad (37\text{-}37)$$

This **transverse Doppler effect** is due to time dilation.

**Momentum and Energy** The following definitions of linear momentum $\vec{p}$, kinetic energy $K$, and total energy $E$ for a particle of mass $m$ are valid at any physically possible speed:

$$\vec{p} = \gamma m \vec{v} \qquad \text{(momentum)}, \qquad (37\text{-}42)$$

$$E = mc^2 + K = \gamma mc^2 \qquad \text{(total energy)}, \qquad (37\text{-}47, 37\text{-}48)$$

$$K = mc^2(\gamma - 1) \qquad \text{(kinetic energy)}. \qquad (37\text{-}52)$$

Here $\gamma$ is the Lorentz factor for the particle's motion, and $mc^2$ is the *mass energy,* or *rest energy,* associated with the mass of the particle. These equations lead to the relationships

$$(pc)^2 = K^2 + 2Kmc^2, \qquad (37\text{-}54)$$

and

$$E^2 = (pc)^2 + (mc^2)^2. \qquad (37\text{-}55)$$

When a system of particles undergoes a chemical or nuclear reaction, the $Q$ of the reaction is the negative of the change in the system's total mass energy:

$$Q = M_i c^2 - M_f c^2 = -\Delta M c^2, \qquad (37\text{-}50)$$

where $M_i$ is the system's total mass before the reaction and $M_f$ is its total mass after the reaction.

## QUESTIONS

**1** Figure 37-16 shows two clocks in stationary frame $S$ (they are synchronized in that frame) and one clock in moving frame $S'$. Clocks $C_1$ and $C_1'$ read zero when they pass each other. When clocks $C_1'$ and $C_2$ pass each other, (a) which clock has the smaller reading and (b) which clock measures a proper time?

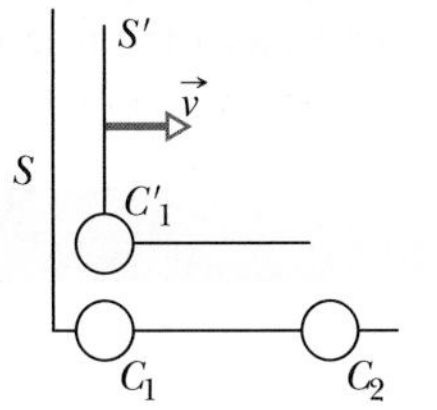

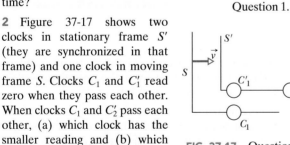

FIG. 37-16
Question 1.

**2** Figure 37-17 shows two clocks in stationary frame $S'$ (they are synchronized in that frame) and one clock in moving frame $S$. Clocks $C_1$ and $C_1'$ read zero when they pass each other. When clocks $C_1$ and $C_2'$ pass each other, (a) which clock has the smaller reading and (b) which clock measures a proper time?

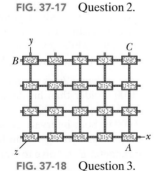

FIG. 37-17　Question 2.

**3** The plane of clocks and measuring rods in Fig. 37-18 is like that in Fig. 37-3. The clocks along the $x$ axis are separated (center to center) by 1 light-second, as are the clocks along the $y$ axis, and all the clocks are synchronized via the procedure described in Section 37-3. When the initial synchronizing signal of $t = 0$ from the origin reaches (a) clock $A$, (b) clock $B$,

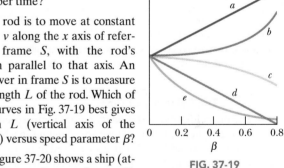

FIG. 37-18　Question 3.

and (c) clock $C$, what initial time is then set on those clocks? An event occurs at clock $A$ when it reads 10 s. (d) How long does the signal of that event take to travel to an observer stationed at the origin? (e) What time does that observer assign to the event?

**4** Sam leaves Venus in a spaceship headed to Mars and passes Sally, who is on Earth, with a relative speed of $0.5c$. (a) Each measures the Venus–Mars voyage time. Who measures a proper time: Sam, Sally, or neither? (b) On the way, Sam sends a pulse of light to Mars. Each measures the travel time of the pulse. Who measures a proper time?

**5** A rod is to move at constant speed $v$ along the $x$ axis of reference frame $S$, with the rod's length parallel to that axis. An observer in frame $S$ is to measure the length $L$ of the rod. Which of the curves in Fig. 37-19 best gives length $L$ (vertical axis of the graph) versus speed parameter $\beta$?

**6** Figure 37-20 shows a ship (attached to reference frame $S'$) passing us (standing in reference frame $S$). A proton is fired at nearly the speed of light along the length of the ship, from the front to the rear. (a) Is the spatial separation $\Delta x'$ between the point at which the proton is fired and the point at which it hits the ship's rear wall a positive or negative quantity? (b) Is the temporal separation $\Delta t'$ between those events a positive or negative quantity?

FIG. 37-19
Questions 5 and 7.

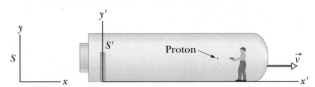

FIG. 37-20　Question 6 and Problem 64.

**7** Reference frame $S'$ is to pass reference frame $S$ at speed $v$ along the common direction of the $x'$ and $x$ axes, as in Fig. 37-9. An observer who rides along with frame $S'$ is to count off 25 s on his wristwatch. The corresponding time interval $\Delta t$ is to be measured by an observer in frame $S$. Which of the curves in Fig. 37-19 best gives $\Delta t$ (vertical axis of the graph) versus speed parameter $\beta$?

**8** While on board a starship, you intercept signals from four shuttle craft that are moving either directly toward or directly away from you. The signals have the same proper frequency $f_0$. The speed and direction (both relative to you) of the shuttle craft are (a) $0.3c$ toward, (b) $0.6c$ toward, (c) $0.3c$ away, and (d) $0.6c$ away. Rank the shuttle craft according to the frequency you receive, greatest first.

**9** Figure 37-21 shows one of four star cruisers that are in a race. As each cruiser passes the starting line, a shuttle craft leaves the cruiser and races toward the finish line. You, judging the race, are stationary relative to the starting and finish lines. The speeds $v_c$ of the cruisers relative to you and the speeds $v_s$ of the shuttle craft relative to their respective starships are, in that order, (1) $0.70c$, $0.40c$; (2) $0.40c$, $0.70c$;

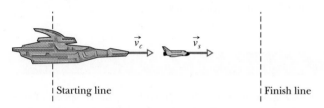

**FIG. 37-21** Question 9.

(3) $0.20c$, $0.90c$; (4) $0.50c$, $0.60c$. (a) Without written calculation, rank the shuttle craft according to their speeds relative to you, greatest first. (b) Still without written calculation, rank the shuttle craft according to the distances their pilots measure from the starting line to the finish line, greatest first. (c) Each starship sends a signal to its shuttle craft at a certain frequency $f_0$ as measured on board the starship. Again without written calculation, rank the shuttle craft according to the frequencies they detect, greatest first.

**10** The rest energy and total energy, respectively, of three particles, expressed in terms of a basic amount $A$ are (1) $A$, $2A$; (2) $A$, $3A$; (3) $3A$, $4A$. Without written calculation, rank the particles according to their (a) mass, (b) kinetic energy, (c) Lorentz factor, and (d) speed, greatest first.

**11** Figure 37-22 shows the triangle of Fig. 37-15 for six particles; the slanted lines 2 and 4 have the same length. Rank the particles according to (a) mass, (b) momentum magnitude, and (c) Lorentz factor, greatest first. (d) Identify which two particles have the same total energy. (e) Rank the three lowest-mass particles according to kinetic energy, greatest first.

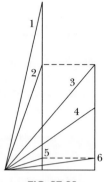

**FIG. 37-22** Question 11.

---

## PROBLEMS

**GO** Tutoring problem available (at instructor's discretion) in *WileyPLUS* and WebAssign
**SSM** Worked-out solution available in Student Solutions Manual     **WWW** Worked-out solution is at
**• – •••** Number of dots indicates level of problem difficulty     **ILW** Interactive solution is at ——— http://www.wiley.com/college/halliday
Additional information available in *The Flying Circus of Physics* and at flyingcircusofphysics.com

### sec. 37-5 The Relativity of Time

**•1** The mean lifetime of stationary muons is measured to be 2.2000 $\mu$s. The mean lifetime of high-speed muons in a burst of cosmic rays observed from Earth is measured to be 16.000 $\mu$s. To five significant figures, what is the speed parameter $\beta$ of these cosmic-ray muons relative to Earth?

**•2** To eight significant figures, what is the speed parameter $\beta$ if the Lorentz factor $\gamma$ is (a) 1.010 000 0, (b) 10.000 000, (c) 100.000 00, and (d) 1000.000 0?

**••3** An unstable high-energy particle enters a detector and leaves a track of length 1.05 mm before it decays. Its speed relative to the detector was $0.992c$. What is its proper lifetime? That is, how long would the particle have lasted before decay had it been at rest with respect to the detector? **ILW**

**••4** Reference frame $S'$ is to pass reference frame $S$ at speed $v$ along the common direction of the $x'$ and $x$ axes, as in Fig. 37-9. An observer who rides along with frame $S'$ is to count off a certain time interval on his wristwatch. The corresponding time interval $\Delta t$ is to be measured by

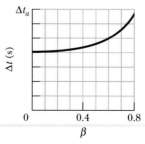

**FIG. 37-23** Problem 4.

an observer in frame $S$. Figure 37-23 gives $\Delta t$ versus speed parameter $\beta$ for a range of values for $\beta$. The vertical axis scale is set by $\Delta t_a = 14.0$ s. What is interval $\Delta t$ if $v = 0.98c$?

**••5** The premise of the *Planet of the Apes* movies and book is that hibernating astronauts travel far into Earth's future, to a time when human civilization has been replaced by an ape civilization. Considering only special relativity, determine how far into Earth's future the astronauts would travel if they slept for 120 y while traveling relative to Earth with a speed of $0.9990c$, first outward from Earth and then back again.

**••6** *(Come) back to the future.* Suppose that a father is 20.00 y older than his daughter. He wants to travel outward from Earth for 2.000 y and then back to Earth for another 2.000 y (both intervals as he measures them) such that he is then 20.00 y *younger* than his daughter. What constant speed parameter $\beta$ (relative to Earth) is required for the trip?

**••7** You wish to make a round trip from Earth in a spaceship, traveling at constant speed in a straight line for exactly 6 months (as you measure the time interval) and then returning at the same constant speed. You wish further, on your return, to find Earth as it will be exactly 1000 years in the future. (a) To eight significant figures, at what speed parameter $\beta$ must you travel? (b) Does it matter whether you travel in a straight line on your journey?

## sec. 37-6 The Relativity of Length

•8 A meter stick in frame $S'$ makes an angle of 30° with the $x'$ axis. If that frame moves parallel to the $x$ axis of frame $S$ with speed $0.90c$ relative to frame $S$, what is the length of the stick as measured from $S$?

•9 A rod lies parallel to the $x$ axis of reference frame $S$, moving along this axis at a speed of $0.630c$. Its rest length is 1.70 m. What will be its measured length in frame $S$?

•10 An electron of $\beta = 0.999\ 987$ moves along the axis of an evacuated tube that has a length of 3.00 m as measured by a laboratory observer $S$ at rest relative to the tube. An observer $S'$ at rest relative to the electron, however, would see this tube moving with speed $v\ (= \beta c)$. What length would observer $S'$ measure for the tube?

•11 A spaceship of rest length 130 m races past a timing station at a speed of $0.740c$. (a) What is the length of the spaceship as measured by the timing station? (b) What time interval will the station clock record between the passage of the front and back ends of the ship? SSM

••12 A rod is to move at constant speed $v$ along the $x$ axis of reference frame $S$, with the rod's length parallel to that axis. An observer in frame $S$ is to measure the length $L$ of the rod. Figure 37-24 gives length $L$ versus speed parameter $\beta$ for a range of values for $\beta$. The vertical axis scale is set by $L_a$ = 1.00 m. What is $L$ if $v = 0.95c$?

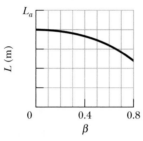

FIG. 37-24 Problem 12.

••13 The center of our Milky Way galaxy is about 23 000 ly away. (a) To eight significant figures, at what constant speed parameter would you need to travel exactly 23 000 ly (measured in the Galaxy frame) in exactly 30 y (measured in your frame)? (b) Measured in your frame and in light-years, what length of the Galaxy would pass by you during the trip? GO

••14 The length of a spaceship is measured to be exactly half its rest length. (a) To three significant figures, what is the speed parameter $\beta$ of the spaceship relative to the observer's frame? (b) By what factor do the spaceship's clocks run slow relative to clocks in the observer's frame?

••15 A space traveler takes off from Earth and moves at speed $0.9900c$ toward the star Vega, which is 26.00 ly distant. How much time will have elapsed by Earth clocks (a) when the traveler reaches Vega and (b) when Earth observers receive word from the traveler that she has arrived? (c) How much older will Earth observers calculate the traveler to be (measured from her frame) when she reaches Vega than she was when she started the trip? GO

## sec. 37-8 Some Consequences of the Lorentz Equations

•16 Inertial frame $S'$ moves at a speed of $0.60c$ with respect to frame $S$ (Fig. 37-9). Further, $x = x' = 0$ at $t = t' = 0$. Two events are recorded. In frame $S$, event 1 occurs at the origin at $t = 0$ and event 2 occurs on the $x$ axis at $x = 3.0$ km at $t = 4.0$ $\mu$s. According to observer $S'$, what is the time of (a) event 1 and (b) event 2? (c) Do the two observers see the two events in the same sequence or the reverse sequence?

•17 An experimenter arranges to trigger two flashbulbs simultaneously, producing a big flash located at the origin of his reference frame and a small flash at $x = 30.0$ km. An observer moving at a speed of $0.250c$ in the positive direction of $x$ also views the flashes. (a) What is the time interval between them according to her? (b) Which flash does she say occurs first?

•18 Observer $S$ reports that an event occurred on the $x$ axis of his reference frame at $x = 3.00 \times 10^8$ m at time $t = 2.50$ s. Observer $S'$ and her frame are moving in the positive direction of the $x$ axis at a speed of $0.400c$. Further, $x = x' = 0$ at $t = t' = 0$. What are the (a) spatial and (b) temporal coordinate of the event according to $S'$? If $S'$ were, instead, moving in the *negative* direction of the $x$ axis, what would be the (c) spatial and (d) temporal coordinate of the event according to $S'$?

•19 In Fig. 37-9, the origins of the two frames coincide at $t = t' = 0$ and the relative speed is $0.950c$. Two micrometeorites collide at coordinates $x = 100$ km and $t = 200$ $\mu$s according to an observer in frame $S$. What are the (a) spatial and (b) temporal coordinate of the collision according to an observer in frame $S'$? SSM WWW

••20 Bullwinkle in reference frame $S'$ passes you in reference frame $S$ along the common direction of the $x'$ and $x$ axes, as in Fig. 37-9. He carries three meter sticks: meter stick 1 is parallel to the $x'$ axis, meter stick 2 is parallel to the $y'$ axis, and meter stick 3 is parallel to the $z'$ axis. On his wristwatch he counts off 15.0 s, which takes 30.0 s according to you. Two events occur during his passage. According to you, event 1 occurs at $x_1 = 33.0$ m and $t_1 = 22.0$ ns, and event 2 occurs at $x_2 = 53.0$ m and $t_2 = 62.0$ ns. According to your measurements, what is the length of (a) meter stick 1, (b) meter stick 2, and (c) meter stick 3? According to Bullwinkle, what are (d) the spatial separation and (e) the temporal separation between events 1 and 2, and (f) which event occurs first?

••21 A clock moves along an $x$ axis at a speed of $0.600c$ and reads zero as it passes the origin. (a) Calculate the clock's Lorentz factor. (b) What time does the clock read as it passes $x = 180$ m? ILW

••22 As in Fig. 37-9, reference frame $S'$ passes reference frame $S$ with a certain velocity. Events 1 and 2 are to have a certain temporal separation $\Delta t'$ according to the $S'$ observer. However, their spatial separation $\Delta x'$ according to that observer has not been set yet. Figure 37-25 gives their temporal separation $\Delta t$ according to the $S$ observer as a function of $\Delta x'$ for a range of $\Delta x'$ values. The vertical axis scale is set by $\Delta t_a = 6.00$ $\mu$s. What is $\Delta t'$?

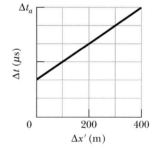

FIG. 37-25 Problem 22.

••23 In Fig. 37-9, observer $S$ detects two flashes of light. A big flash occurs at $x_1 = 1200$ m and, 5.00 $\mu$s later, a small flash occurs at $x_2 = 480$ m. As detected by observer $S'$, the two flashes occur at a single coordinate $x'$. (a) What is the speed parameter of $S'$, and (b) is $S'$ moving in the positive or negative direction of the $x$ axis? To $S'$, (c) which flash occurs first and (d) what is the time interval between the flashes?

••24 In Fig. 37-9, observer $S$ detects two flashes of light. A big flash occurs at $x_1 = 1200$ m and, slightly later, a small flash occurs at $x_2 = 480$ m. The time interval between the flashes is $\Delta t = t_2 - t_1$. What is the smallest value of $\Delta t$ for which observer $S'$ will determine that the two flashes occur at the same $x'$ coordinate?

••25 *Relativistic reversal of events.* Figures 37-26a and b show the (usual) situation in which a primed reference frame passes an unprimed reference frame, in the common positive direction of the $x$ and $x'$ axes, at a constant relative velocity of magnitude $v$. We are at rest in the unprimed frame; Bullwinkle, an astute student of relativity in spite of his cartoon upbringing, is at rest in the primed frame. The figures also indicate events $A$ and $B$ that occur at the following spacetime coordinates as measured in our unprimed frame and in Bullwinkle's primed frame:

| Event | Unprimed | Primed |
|-------|----------|--------|
| $A$ | $(x_A, t_A)$ | $(x'_A, t'_A)$ |
| $B$ | $(x_B, t_B)$ | $(x'_B, t'_B)$ |

In our frame, event $A$ occurs before event $B$, with temporal separation $\Delta t = t_B - t_A = 1.00$ μs and spatial separation $\Delta x = x_B - x_A = 400$ m. Let $\Delta t'$ be the temporal separation of the events according to Bullwinkle. (a) Find an expression for $\Delta t'$ in terms of the speed parameter $\beta\ (= v/c)$ and the given data. Graph $\Delta t'$ versus $\beta$ for the following two ranges of $\beta$:

    (b) 0 to 0.01    ($v$ is low, from 0 to 0.01$c$)
    (c) 0.1 to 1    ($v$ is high, from 0.1$c$ to the limit $c$)

(d) At what value of $\beta$ is $\Delta t' = 0$? For what range of $\beta$ is the sequence of events $A$ and $B$ according to Bullwinkle (e) the same as ours and (f) the reverse of ours? (g) Can event $A$ cause event $B$, or vice versa? Explain.

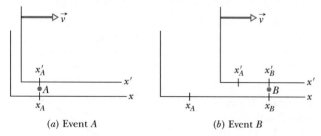

(a) Event $A$          (b) Event $B$

**FIG. 37-26** Problems 25, 26, 62, and 63.

••26 For the passing reference frames in Fig. 37-26, events $A$ and $B$ occur at the following spacetime coordinates: according to the unprimed frame, $(x_A, t_A)$ and $(x_B, t_B)$; according to the primed frame, $(x'_A, t'_A)$ and $(x'_B, t'_B)$. In the unprimed frame, $\Delta t = t_B - t_A = 1.00$ μs and $\Delta x = x_B - x_A = 400$ m. (a) Find an expression for $\Delta x'$ in terms of the speed parameter $\beta$ and the given data. Graph $\Delta x'$ versus $\beta$ for two ranges of $\beta$: (b) 0 to 0.01 and (c) 0.1 to 1. (d) At what value of $\beta$ is $\Delta x'$ minimum, and (e) what is that minimum?

### sec. 37-9 The Relativity of Velocities

•27 Galaxy A is reported to be receding from us with a speed of 0.35$c$. Galaxy B, located in precisely the opposite direction, is also found to be receding from us at this same speed. What multiple of $c$ gives the recessional speed an observer on Galaxy A would find for (a) our galaxy and (b) Galaxy B?

•28 Stellar system $Q_1$ moves away from us at a speed of 0.800$c$. Stellar system $Q_2$, which lies in the same direction in space but is closer to us, moves away from us at speed 0.400$c$. What multiple of $c$ gives the speed of $Q_2$ as measured by an observer in the reference frame of $Q_1$?

•29 A particle moves along the $x'$ axis of frame $S'$ with velocity 0.40$c$. Frame $S'$ moves with velocity 0.60$c$ with respect to frame $S$. What is the velocity of the particle with respect to frame $S$? **SSM**

•30 In Fig. 37-11, frame $S'$ moves relative to frame $S$ with velocity $0.62c\hat{i}$ while a particle moves parallel to the common $x$ and $x'$ axes. An observer attached to frame $S'$ measures the particle's velocity to be $0.47c\hat{i}$. In terms of $c$, what is the particle's velocity as measured by an observer attached to frame $S$ according to the (a) relativistic and (b) classical velocity transformation? Suppose, instead, that the $S'$ measure of the particle's velocity is $-0.47c\hat{i}$. What velocity does the observer in $S$ now measure according to the (c) relativistic and (d) classical velocity transformation?

••31 An armada of spaceships that is 1.00 ly long (in its rest frame) moves with speed 0.800$c$ relative to a ground station in frame $S$. A messenger travels from the rear of the armada to the front with a speed of 0.950$c$ relative to $S$. How long does the trip take as measured (a) in the messenger's rest frame, (b) in the armada's rest frame, and (c) by an observer in frame $S$? **GO**

••32 In Fig. 37-27a, particle $P$ is to move parallel to the $x$ and $x'$ axes of reference frames $S$ and $S'$, at a certain velocity relative to frame $S$. Frame $S'$ is to move parallel to the $x$ axis of frame $S$ at velocity $v$. Figure 37-27b gives the velocity $u'$ of the particle relative to frame $S'$ for a range of values for $v$. The vertical axis scale is set by $u'_a = 0.800c$. What value will $u'$ have if (a) $v = 0.90c$ and (b) $v \to c$?

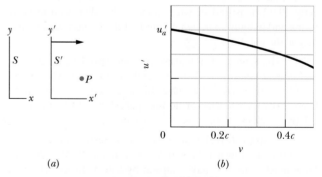

(a)             (b)

**FIG. 37-27** Problem 32.

••33 A spaceship whose rest length is 350 m has a speed of 0.82$c$ with respect to a certain reference frame. A micrometeorite, also with a speed of 0.82$c$ in this frame, passes the spaceship on an antiparallel track. How long does it take this object to pass the ship as measured on the ship? **SSM ILW WWW**

### sec. 37-10 Doppler Effect for Light

•34 Certain wavelengths in the light from a galaxy in the constellation Virgo are observed to be 0.4% longer than the corresponding light from Earth sources. (a) What is the radial speed of this galaxy with respect to Earth? (b) Is the galaxy approaching or receding from Earth?

•35 Assuming that Eq. 37-36 holds, find how fast you would

have to go through a red light to have it appear green. Take 620 nm as the wavelength of red light and 540 nm as the wavelength of green light.

•36 Figure 37-28 is a graph of intensity versus wavelength for light reaching Earth from galaxy NGC 7319, which is about $3 \times 10^8$ light-years away. The most intense light is emitted by the oxygen in NGC 7319. In a laboratory that emission is at wavelength $\lambda = 513$ nm, but in the light from NGC 7319 it has been shifted to 525 nm due to the Doppler effect (all the emissions from NGC 7319 have been shifted). (a) What is the radial speed of NGC 7319 relative to Earth? (b) Is the relative motion toward or away from our planet?

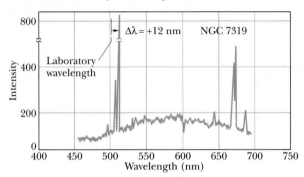

FIG. 37-28    Problem 36.

•37 A spaceship, moving away from Earth at a speed of $0.900c$, reports back by transmitting at a frequency (measured in the spaceship frame) of 100 MHz. To what frequency must Earth receivers be tuned to receive the report? SSM

•38 A sodium light source moves in a horizontal circle at a constant speed of $0.100c$ while emitting light at the proper wavelength of $\lambda_0 = 589.00$ nm. Wavelength $\lambda$ is measured for that light by a detector fixed at the center of the circle. What is the wavelength shift $\lambda - \lambda_0$?

••39 A spaceship is moving away from Earth at speed $0.20c$. A source on the rear of the ship emits light at wavelength 450 nm according to someone on the ship. What (a) wavelength and (b) color (blue, green, yellow, or red) are detected by someone on Earth watching the ship? SSM

### sec. 37-12    A New Look at Energy
•40 What is the minimum energy that is required to break a nucleus of $^{12}$C (of mass 11.996 71 u) into three nuclei of $^4$He (of mass 4.001 51 u each)?

•41 How much work must be done to increase the speed of an electron (a) from $0.18c$ to $0.19c$ and (b) from $0.98c$ to $0.99c$? Note that the speed increase is $0.01c$ in both cases.

•42 In the reaction $p + {}^{19}F \rightarrow \alpha + {}^{16}O$, the masses are

$m(p) = 1.007825$ u,      $m(\alpha) = 4.002603$ u,

$m(F) = 18.998405$ u,      $m(O) = 15.994915$ u.

Calculate the $Q$ of the reaction from these data.

•43 The mass of an electron is $9.109\ 381\ 88 \times 10^{-31}$ kg. To six significant figures, find (a) $\gamma$ and (b) $\beta$ for an electron with kinetic energy $K = 100.000$ MeV. SSM WWW

•44 How much work must be done to increase the speed of an electron from rest to (a) $0.500c$, (b) $0.990c$, and (c) $0.9990c$?

••45 What must be the momentum of a particle with mass $m$ so that the total energy of the particle is 3.00 times its rest energy? ILW

••46 The mass of an electron is $9.109\ 381\ 88 \times 10^{-31}$ kg. To eight significant figures, find the following for the given electron kinetic energy: (a) $\gamma$ and (b) $\beta$ for $K = 1.000\ 000\ 0$ keV, (c) $\gamma$ and (d) $\beta$ for $K = 1.000\ 000\ 0$ MeV, and (e) $\gamma$ and (f) $\beta$ for $K = 1.000\ 000\ 0$ GeV.

••47 As you read this page (on paper or monitor screen), a cosmic ray proton passes along the left–right width of the page with relative speed $v$ and a total energy of 14.24 nJ. According to your measurements, that left–right width is 21.0 cm. (a) What is the width according to the proton's reference frame? How much time did the passage take according to (b) your frame and (c) the proton's frame? GO

••48 (a) The energy released in the explosion of 1.00 mol of TNT is 3.40 MJ. The molar mass of TNT is 0.227 kg/mol. What weight of TNT is needed for an explosive release of $1.80 \times 10^{14}$ J? (b) Can you carry that weight in a backpack, or is a truck or train required? (c) Suppose that in an explosion of a fission bomb, 0.080% of the fissionable mass is converted to released energy. What weight of fissionable material is needed for an explosive release of $1.80 \times 10^{14}$ J? (d) Can you carry that weight in a backpack, or is a truck or train required?

••49 A certain particle of mass $m$ has momentum of magnitude $mc$. What are (a) $\beta$, (b) $\gamma$, and (c) the ratio $K/E_0$?

••50 What is $\beta$ for a particle with (a) $K = 2.00E_0$ and (b) $E = 2.00E_0$?

••51 Quasars are thought to be the nuclei of active galaxies in the early stages of their formation. A typical quasar radiates energy at the rate of $10^{41}$ W. At what rate is the mass of this quasar being reduced to supply this energy? Express your answer in solar mass units per year, where one solar mass unit (1 smu $= 2.0 \times 10^{30}$ kg) is the mass of our Sun.

••52 (a) If $m$ is a particle's mass, $p$ is its momentum magnitude, and $K$ is its kinetic energy, show that

$$m = \frac{(pc)^2 - K^2}{2Kc^2}.$$

(b) For low particle speeds, show that the right side of the equation reduces to $m$. (c) If a particle has $K = 55.0$ MeV when $p = 121$ MeV/c, what is the ratio $m/m_e$ of its mass to the electron mass?

••53 A 5.00-grain aspirin tablet has a mass of 320 mg. For how many kilometers would the energy equivalent of this mass power an automobile? Assume 12.75 km/L and a heat of combustion of $3.65 \times 10^7$ J/L for the gasoline used in the automobile. SSM

••54 To four significant figures, find the following when the kinetic energy is 10.00 MeV: (a) $\gamma$ and (b) $\beta$ for an electron ($E_0 = 0.510\ 998$ MeV), (c) $\gamma$ and (d) $\beta$ for a proton ($E_0 = 938.272$ MeV), and (e) $\gamma$ and (f) $\beta$ for an $\alpha$ particle ($E_0 = 3727.40$ MeV).

••55 In Section 28-6, we showed that a particle of charge $q$ and mass $m$ will move in a circle of radius $r = mv/|q|B$ when its velocity $\vec{v}$ is perpendicular to a uniform magnetic field $\vec{B}$. We also found that the period $T$ of the motion is independent of speed $v$. These two results are approximately correct if $v \ll c$. For relativistic speeds, we must use the correct equation for the radius:

$$r = \frac{p}{|q|B} = \frac{\gamma mv}{|q|B}.$$

(a) Using this equation and the definition of period ($T = 2\pi r/v$), find the correct expression for the period. (b) Is $T$ independent of $v$? If a 10.0 MeV electron moves in a circular path in a uniform magnetic field of magnitude 2.20 T, what are (c) the radius according to Chapter 28, (d) the correct radius, (e) the period according to Chapter 28, and (f) the correct period?

••**56** The mass of a muon is 207 times the electron mass; the average lifetime of muons at rest is 2.20 μs. In a certain experiment, muons moving through a laboratory are measured to have an average lifetime of 6.90 μs. For the moving muons, what are (a) $\beta$, (b) $K$, and (c) $p$ (in MeV/c)? **GO**

••**57** In a high-energy collision between a cosmic-ray particle and a particle near the top of Earth's atmosphere, 120 km above sea level, a pion is created. The pion has a total energy $E$ of $1.35 \times 10^5$ MeV and is traveling vertically downward. In the pion's rest frame, the pion decays 35.0 ns after its creation. At what altitude above sea level, as measured from Earth's reference frame, does the decay occur? The rest energy of a pion is 139.6 MeV.

••**58** Apply the binomial theorem (Appendix E) to the last part of Eq. 37-52 for the kinetic energy of a particle. (a) Retain the first two terms of the expansion to show the kinetic energy in the form

$$K = \text{(first term)} + \text{(second term)}.$$

The first term is the classical expression for kinetic energy. The second term is the first-order correction to the classical expression. Assume the particle is an electron. If its speed $v$ is $c/20$, what is the value of (b) the classical expression and (c) the first-order correction? If the electron's speed is $0.80c$, what is the value of (d) the classical expression and (e) the first-order correction? (f) At what speed parameter $\beta$ does the first-order correction become 10% or greater of the classical expression?

•••**59** An alpha particle with kinetic energy 7.70 MeV collides with an $^{14}$N nucleus at rest, and the two transform into an $^{17}$O nucleus and a proton. The proton is emitted at 90° to the direction of the incident alpha particle and has a kinetic energy of 4.44 MeV. The masses of the various particles are alpha particle, 4.00260 u; $^{14}$N, 14.00307 u; proton, 1.007825 u; and $^{17}$O, 16.99914 u. In MeV, what are (a) the kinetic energy of the oxygen nucleus and (b) the $Q$ of the reaction? (*Hint:* The speeds of the particles are much less than $c$.)

**Additional Problems**

**60** In Fig. 37-29a, particle $P$ is to move parallel to the $x$ and $x'$ axes of reference frames $S$ and $S'$, at a certain velocity relative to frame $S$. Frame $S'$ is to move parallel to the $x$ axis of frame $S$ at velocity $v$. Figure 37-29b gives the velocity $u'$ of the particle relative to frame $S'$ for a range of values for $v$.

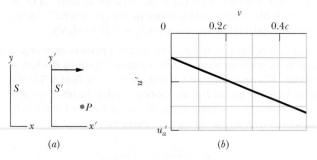

(a)     (b)

**FIG. 37-29** Problem 60.

The vertical axis scale is set by $u'_a = -0.800c$. What value will $u'$ have if (a) $v = 0.80c$ and (b) $v \to c$?

**61** *Superluminal jets.* Figure 37-30a shows the path taken by a knot in a jet of ionized gas that has been expelled from a galaxy. The knot travels at constant velocity $\vec{v}$ at angle $\theta$ from the direction of Earth. The knot occasionally emits a burst of light, which is eventually detected on Earth. Two bursts are indicated in Fig. 37-30a, separated by time $t$ as measured in a stationary frame near the bursts. The bursts are shown in Fig. 37-30b as if they were photographed on the same piece of film, first when light from burst 1 arrived on Earth and then later when light from burst 2 arrived. The apparent distance $D_{app}$ traveled by the knot between the two bursts is the distance across an Earth-observer's view of the knot's path. The apparent time $T_{app}$ between the bursts is the difference in the arrival times of the light from them. The apparent speed of the knot is then $V_{app} = D_{app}/T_{app}$. In terms of $v$, $t$, and $\theta$, what are (a) $D_{app}$ and (b) $T_{app}$? (c) Evaluate $V_{app}$ for $v = 0.980c$ and $\theta = 30.0°$. When superluminal (faster than light) jets were first observed, they seemed to defy special relativity—at least until the correct geometry (Fig. 37-30a) was understood. **GO**

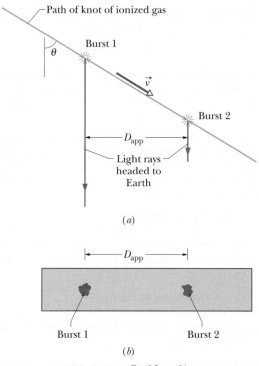

Path of knot of ionized gas

$\theta$      Burst 1

$\vec{v}$

Burst 2

$D_{app}$

Light rays
headed to
Earth

(a)

$D_{app}$

Burst 1                    Burst 2

(b)

**FIG. 37-30** Problem 61.

**62** *Temporal separation between two events.* Events $A$ and $B$ occur with the following spacetime coordinates in the reference frames of Fig. 37-26: according to the unprimed frame, ($x_A$, $t_A$) and ($x_B$, $t_B$); according to the primed frame, ($x'_A$, $t'_A$) and ($x'_B$, $t'_B$). In the unprimed frame, $\Delta t = t_B - t_A = 1.00$ μs and $\Delta x = x_B - x_A = 240$ m. (a) Find an expression for $\Delta t'$ in terms of the speed parameter $\beta$ and the given data. Graph $\Delta t'$ versus $\beta$ for the following two ranges of $\beta$: (b) 0 to 0.01 and (c) 0.1 to 1. (d) At what value of $\beta$ is $\Delta t'$ minimum and (e) what is that minimum? (f) Can one of these events cause the other? Explain.

**63** *Spatial separation between two events.* For the passing reference frames of Fig. 37-26, events $A$ and $B$ occur with the following spacetime coordinates: according to the unprimed frame, $(x_A, t_A)$ and $(x_B, t_B)$; according to the primed frame, $(x'_A, t'_A)$ and $(x'_B, t'_B)$. In the unprimed frame, $\Delta t = t_B - t_A = 1.00$ μs and $\Delta x = x_B - x_A = 240$ m. (a) Find an expression for $\Delta x'$ in terms of the speed parameter $\beta$ and the given data. Graph $\Delta x'$ versus $\beta$ for two ranges of $\beta$: (b) 0 to 0.01 and (c) 0.1 to 1. (d) At what value of $\beta$ is $\Delta x' = 0$?

**64** Figure 37-20 shows a ship (attached to reference frame $S'$) passing us (standing in reference frame $S$) with velocity $\vec{v} = 0.950c\hat{i}$. A proton is fired at speed $0.980c$ relative to the ship from the front of the ship to the rear. The proper length of the ship is 760 m. What is the temporal separation between the time the proton is fired and the time it hits the rear wall of the ship according to (a) a passenger in the ship and (b) us? Suppose that, instead, the proton is fired from the rear to the front. What then is the temporal separation between the time it is fired and the time it hits the front wall according to (c) the passenger and (d) us?

**65** *The car-in-the-garage problem.* Carman has just purchased the world's longest stretch limo, which has a proper length of $L_c = 30.5$ m. In Fig. 37-31*a*, it is shown parked in front of a garage with a proper length of $L_g = 6.00$ m. The garage has a front door (shown open) and a back door (shown closed). The limo is obviously longer than the garage. Still, Garageman, who owns the garage and knows something about relativistic length contraction, makes a bet with Carman that the limo can fit in the garage with both doors closed. Carman, who dropped his physics course before reaching special relativity, says such a thing, even in principle, is impossible.

To analyze Garageman's scheme, an $x_c$ axis is attached to the limo, with $x_c = 0$ at the rear bumper, and an $x_g$ axis is attached to the garage, with $x_g = 0$ at the (now open) front door. Then Carman is to drive the limo directly toward the front door at a velocity of $0.9980c$ (which is, of course, both technically and financially impossible). Carman is stationary in the $x_c$ reference frame; Garageman is stationary in the $x_g$ reference frame.

There are two events to consider. *Event 1:* When the rear bumper clears the front door, the front door is closed. Let the time of this event be zero to both Carman and Garageman: $t_{g1} = t_{c1} = 0$. The event occurs at $x_c = x_g = 0$. Figure 37-31*b* shows event 1 according to the $x_g$ reference frame. *Event 2:* When the front bumper reaches the back door, that door opens. Figure 37-31*c* shows event 2 according to the $x_g$ reference frame.

According to Garageman, (a) what is the length of the limo, and what are the spacetime coordinates (b) $x_{g2}$ and (c) $t_{g2}$ of event 2? (d) For how long is the limo temporarily "trapped"

inside the garage with both doors shut? Now consider the situation from the $x_c$ reference frame, in which the garage comes racing past the limo at a velocity of $-0.9980c$. According to Carman, (e) what is the length of the passing garage, what are the spacetime coordinates (f) $x_{c2}$ and (g) $t_{c2}$ of event 2, (h) is the limo ever in the garage with both doors shut, and (i) which event occurs first? (j) Sketch events 1 and 2 as seen by Carman. (k) Are the events causally related; that is, does one of them cause the other? (l) Finally, who wins the bet?

**66** Reference frame $S'$ passes reference frame $S$ with a certain velocity as in Fig. 37-9. Events 1 and 2 are to have a certain spatial separation $\Delta x'$ according to the $S'$ observer. However, their temporal separation $\Delta t'$ according to that observer has not been set yet. Figure 37-32 gives their spatial separation $\Delta x$ according to the $S$ observer as a function of $\Delta t'$ for a range of $\Delta t'$ values. The vertical axis scale is set by $\Delta x_a = 10.0$ m. What is $\Delta x'$?

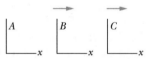

FIG. 37-32    Problem 66.

**67** *Another approach to velocity transformations.* In Fig. 37-33, reference frames $B$ and $C$ move past reference frame $A$ in the common direction of their $x$ axes. Represent the $x$ components of the velocities of one frame relative to another with a two-letter subscript. For example, $v_{AB}$ is the $x$ component of the velocity of $A$ relative to $B$.

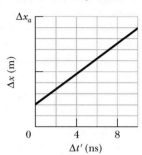

FIG. 37-33    Problems 67, 68, and 69.

Similarly, represent the corresponding speed parameters with two-letter subscripts. For example, $\beta_{AB} \ (= v_{AB}/c)$ is the speed parameter corresponding to $v_{AB}$. (a) Show that

$$\beta_{AC} = \frac{\beta_{AB} + \beta_{BC}}{1 + \beta_{AB}\beta_{BC}}.$$

Let $M_{AB}$ represent the ratio $(1 - \beta_{AB})/(1 + \beta_{AB})$, and let $M_{BC}$ and $M_{AC}$ represent similar ratios. (b) Show that the relation

$$M_{AC} = M_{AB}M_{BC}$$

is true by deriving the equation of part (a) from it.

**68** *Continuation of Problem 67.* Use the result of part (b) in Problem 67 for the motion along a single axis in the following situation. Frame $A$ in Fig. 37-33 is attached to a particle that moves with velocity $+0.500c$ past frame $B$, which moves past frame $C$ with a velocity of $+0.500c$. What are (a) $M_{AC}$, (b) $\beta_{AC}$, and (c) the velocity of the particle relative to frame $C$?

**69** *Continuation of Problem 67.* Let reference frame $C$ in Fig. 37-33 move past reference frame $D$ (not shown). (a) Show that

$$M_{AD} = M_{AB}M_{BC}M_{CD}.$$

(b) Now put this general result to work: Three particles move parallel to a single axis on which an observer is stationed. Let plus and minus signs indicate the directions of motion along that axis. Particle $A$ moves past particle $B$ at $\beta_{AB} = +0.20$. Particle $B$ moves past particle $C$ at $\beta_{BC} = -0.40$. Particle $C$ moves past observer $D$ at $\beta_{CD} = +0.60$. What is the velocity of particle $A$ relative to observer $D$? (The solution technique here is *much* faster than using Eq. 37-29.)

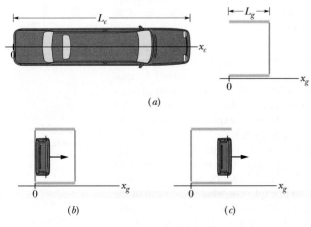

FIG. 37-31    Problem 65.

**70** The total energy of a proton passing through a laboratory apparatus is 10.611 nJ. What is its speed parameter $\beta$? Use the proton mass given in Appendix B under "Best Value," not the commonly remembered rounded number.

**71** If we intercept an electron having total energy 1533 MeV that came from Vega, which is 26 ly from us, how far in light-years was the trip in the rest frame of the electron? **SSM**

**72** A pion is created in the higher reaches of Earth's atmosphere when an incoming high-energy cosmic-ray particle collides with an atomic nucleus. A pion so formed descends toward Earth with a speed of $0.99c$. In a reference frame in which they are at rest, pions decay with an average life of 26 ns. As measured in a frame fixed with respect to Earth, how far (on the average) will such a pion move through the atmosphere before it decays?

**73** What is the momentum in MeV/$c$ of an electron with a kinetic energy of 2.00 MeV? **SSM**

**74** Find the speed parameter of a particle that takes 2.0 y longer than light to travel a distance of 6.0 ly.

**75** How much work is needed to accelerate a proton from a speed of $0.9850c$ to a speed of $0.9860c$? **SSM**

**76** An airplane whose rest length is 40.0 m is moving at uniform velocity with respect to Earth, at a speed of 630 m/s. (a) By what fraction of its rest length is it shortened to an observer on Earth? (b) How long would it take, according to Earth clocks, for the airplane's clock to fall behind by 1.00 $\mu$s? (Use special relativity in your calculations.)

**77** To circle Earth in low orbit, a satellite must have a speed of about $2.7 \times 10^4$ km/h. Suppose that two such satellites orbit Earth in opposite directions. (a) What is their relative speed as they pass, according to the classical Galilean velocity transformation equation? (b) What fractional error do you make in (a) by not using the (correct) relativistic transformation equation? **SSM**

**78** A radar transmitter $T$ is fixed to a reference frame $S'$ that is moving to the right with speed $v$ relative to reference frame $S$ (Fig. 37-34). A mechanical timer (essentially a clock) in frame $S'$, having a period $\tau_0$ (measured in $S'$), causes transmitter $T$ to emit timed radar pulses, which travel at the speed of light and are received by $R$, a receiver fixed in frame $S$. (a) What is the period $\tau$ of the timer as detected by observer $A$, who is fixed in frame $S$? (b) Show that at receiver $R$ the time interval between pulses arriving from $T$ is not $\tau$ or $\tau_0$, but

$$\tau_R = \tau_0 \sqrt{\frac{c+v}{c-v}}.$$

(c) Explain why receiver $R$ and observer $A$, who are in the same reference frame, measure a different period for the transmitter. (*Hint:* A clock and a radar pulse are not the same thing.)

**79** A particle with mass $m$ has speed $c/2$ relative to inertial frame $S$. The particle collides with an identical particle at rest relative to frame $S$. Relative to $S$, what is the speed of a frame $S'$ in which the total momentum of these particles is zero? This frame is called the *center of momentum frame.*

**80** An elementary particle produced in a laboratory experiment travels 0.230 mm through the lab at a relative speed of $0.960c$ before it decays (becomes another particle). (a) What is the proper lifetime of the particle? (b) What is the distance the particle travels as measured from its rest frame?

**81** What are (a) $K$, (b) $E$, and (c) $p$ (in GeV/$c$) for a proton moving at speed $0.990c$? What are (d) $K$, (e) $E$, and (f) $p$ (in MeV/$c$) for an electron moving at speed $0.990c$?

**82** In the red shift of radiation from a distant galaxy, a certain radiation, known to have a wavelength of 434 nm when observed in the laboratory, has a wavelength of 462 nm. (a) What is the radial speed of the galaxy relative to Earth? (b) Is the galaxy approaching or receding from Earth?

**83** (a) What potential difference would accelerate an electron to speed $c$ according to classical physics? (b) With this potential difference, what speed would the electron actually attain?

**84** The radius of Earth is 6370 km, and its orbital speed about the Sun is 30 km/s. Suppose Earth moves past an observer at this speed. To the observer, by how much does Earth's diameter contract along the direction of motion?

**85** A spaceship at rest in a certain reference frame $S$ is given a speed increment of $0.50c$. Relative to its new rest frame, it is then given a further $0.50c$ increment. This process is continued until its speed with respect to its original frame $S$ exceeds $0.999c$. How many increments does this process require?

**86** A Foron cruiser moving directly toward a Reptulian scout ship fires a decoy toward the scout ship. Relative to the scout ship, the speed of the decoy is $0.980c$ and the speed of the Foron cruiser is $0.900c$. What is the speed of the decoy relative to the cruiser?

**87** One cosmic-ray particle approaches Earth along Earth's north–south axis with a speed of $0.80c$ toward the geographic north pole, and another approaches with a speed of $0.60c$ toward the geographic south pole (Fig. 37-35). What is the relative speed of approach of one particle with respect to the other?

**88** (a) How much energy is released in the explosion of a fission bomb containing 3.0 kg of fissionable material? Assume that 0.10% of the mass is converted to released energy. (b) What mass of TNT would have to explode to provide the same energy release? Assume that each mole of TNT liberates 3.4 MJ of energy on exploding. The molecular mass of TNT is 0.227 kg/mol. (c) For the same mass of explosive, what is the ratio of the energy released in a nuclear explosion to that released in a TNT explosion?

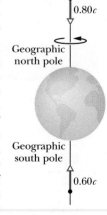

**FIG. 37-35** Problem 87.

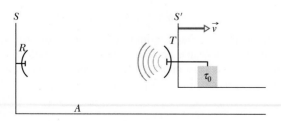

**FIG. 37-34** Problem 78.

# The International A
# System of Units (SI)*

**TABLE 1**

**The SI Base Units**

| Quantity | Name | Symbol | Definition |
|---|---|---|---|
| length | meter | m | ". . . the length of the path traveled by light in vacuum in 1/299,792,458 of a second." (1983) |
| mass | kilogram | kg | ". . . this prototype [a certain platinum–iridium cylinder] shall henceforth be considered to be the unit of mass." (1889) |
| time | second | s | ". . . the duration of 9,192,631,770 periods of the radiation corresponding to the transition between the two hyperfine levels of the ground state of the cesium-133 atom." (1967) |
| electric current | ampere | A | ". . . that constant current which, if maintained in two straight parallel conductors of infinite length, of negligible circular cross section, and placed 1 meter apart in vacuum, would produce between these conductors a force equal to $2 \times 10^{-7}$ newton per meter of length." (1946) |
| thermodynamic temperature | kelvin | K | ". . . the fraction 1/273.16 of the thermodynamic temperature of the triple point of water." (1967) |
| amount of substance | mole | mol | ". . . the amount of substance of a system which contains as many elementary entities as there are atoms in 0.012 kilogram of carbon-12." (1971) |
| luminous intensity | candela | cd | ". . . the luminous intensity, in a given direction, of a source that emits monochromatic radiation of frequency $540 \times 10^{12}$ hertz and that has a radiant intensity in that direction of 1/683 watt per steradian." (1979) |

*Adapted from "The International System of Units (SI)," National Bureau of Standards Special Publication 330, 1972 edition. The definitions above were adopted by the General Conference of Weights and Measures, an international body, on the dates shown. In this book we do not use the candela.

**TABLE 2**

**Some SI Derived Units**

| Quantity | Name of Unit | Symbol | |
|---|---|---|---|
| area | square meter | $m^2$ | |
| volume | cubic meter | $m^3$ | |
| frequency | hertz | Hz | $s^{-1}$ |
| mass density (density) | kilogram per cubic meter | $kg/m^3$ | |
| speed, velocity | meter per second | m/s | |
| angular velocity | radian per second | rad/s | |
| acceleration | meter per second per second | $m/s^2$ | |
| angular acceleration | radian per second per second | $rad/s^2$ | |
| force | newton | N | $kg \cdot m/s^2$ |
| pressure | pascal | Pa | $N/m^2$ |
| work, energy, quantity of heat | joule | J | $N \cdot m$ |
| power | watt | W | J/s |
| quantity of electric charge | coulomb | C | $A \cdot s$ |
| potential difference, electromotive force | volt | V | W/A |
| electric field strength | volt per meter (or newton per coulomb) | V/m | N/C |
| electric resistance | ohm | $\Omega$ | V/A |
| capacitance | farad | F | $A \cdot s/V$ |
| magnetic flux | weber | Wb | $V \cdot s$ |
| inductance | henry | H | $V \cdot s/A$ |
| magnetic flux density | tesla | T | $Wb/m^2$ |
| magnetic field strength | ampere per meter | A/m | |
| entropy | joule per kelvin | J/K | |
| specific heat | joule per kilogram kelvin | $J/(kg \cdot K)$ | |
| thermal conductivity | watt per meter kelvin | $W/(m \cdot K)$ | |
| radiant intensity | watt per steradian | W/sr | |

**TABLE 3**

**The SI Supplementary Units**

| Quantity | Name of Unit | Symbol |
|---|---|---|
| plane angle | radian | rad |
| solid angle | steradian | sr |

# Some Fundamental Constants of Physics* | B

| Constant | Symbol | Computational Value | Best (1998) Value | |
|---|---|---|---|---|
| | | | Value[a] | Uncertainty[b] |
| Speed of light in a vacuum | $c$ | $3.00 \times 10^8$ m/s | 2.997 924 58 | exact |
| Elementary charge | $e$ | $1.60 \times 10^{-19}$ C | 1.602 176 462 | 0.039 |
| Gravitational constant | $G$ | $6.67 \times 10^{-11}$ m³/s² · kg | 6.673 | 1500 |
| Universal gas constant | $R$ | 8.31 J/mol · K | 8.314 472 | 1.7 |
| Avogadro constant | $N_A$ | $6.02 \times 10^{23}$ mol⁻¹ | 6.022 141 99 | 0.079 |
| Boltzmann constant | $k$ | $1.38 \times 10^{-23}$ J/K | 1.380 650 3 | 1.7 |
| Stefan–Boltzmann constant | $\sigma$ | $5.67 \times 10^{-8}$ W/m² · K⁴ | 5.670 400 | 7.0 |
| Molar volume of ideal gas at STP[d] | $V_m$ | $2.27 \times 10^{-2}$ m³/mol | 2.271 098 1 | 1.7 |
| Permittivity constant | $\epsilon_0$ | $8.85 \times 10^{-12}$ F/m | 8.854 187 817 62 | exact |
| Permeability constant | $\mu_0$ | $1.26 \times 10^{-6}$ H/m | 1.256 637 061 43 | exact |
| Planck constant | $h$ | $6.63 \times 10^{-34}$ J · s | 6.626 068 76 | 0.078 |
| Electron mass[c] | $m_e$ | $9.11 \times 10^{-31}$ kg | 9.109 381 88 | 0.079 |
| | | $5.49 \times 10^{-4}$ u | 5.485 799 110 | 0.0021 |
| Proton mass[c] | $m_p$ | $1.67 \times 10^{-27}$ kg | 1.672 621 58 | 0.079 |
| | | 1.0073 u | 1.007 276 466 88 | $1.3 \times 10^{-4}$ |
| Ratio of proton mass to electron mass | $m_p/m_e$ | 1840 | 1836.152 667 5 | 0.0021 |
| Electron charge-to-mass ratio | $e/m_e$ | $1.76 \times 10^{11}$ C/kg | 1.758 820 174 | 0.040 |
| Neutron mass[c] | $m_n$ | $1.68 \times 10^{-27}$ kg | 1.674 927 16 | 0.079 |
| | | 1.0087 u | 1.008 664 915 78 | $5.4 \times 10^{-4}$ |
| Hydrogen atom mass[c] | $m_{1_H}$ | 1.0078 u | 1.007 825 031 6 | 0.0005 |
| Deuterium atom mass[c] | $m_{2_H}$ | 2.0141 u | 2.014 101 777 9 | 0.0005 |
| Helium atom mass[c] | $m_{4_{He}}$ | 4.0026 u | 4.002 603 2 | 0.067 |
| Muon mass | $m_\mu$ | $1.88 \times 10^{-28}$ kg | 1.883 531 09 | 0.084 |
| Electron magnetic moment | $\mu_e$ | $9.28 \times 10^{-24}$ J/T | 9.284 763 62 | 0.040 |
| Proton magnetic moment | $\mu_p$ | $1.41 \times 10^{-26}$ J/T | 1.410 606 663 | 0.041 |
| Bohr magneton | $\mu_B$ | $9.27 \times 10^{-24}$ J/T | 9.274 008 99 | 0.040 |
| Nuclear magneton | $\mu_N$ | $5.05 \times 10^{-27}$ J/T | 5.050 783 17 | 0.040 |
| Bohr radius | $a$ | $5.29 \times 10^{-11}$ m | 5.291 772 083 | 0.0037 |
| Rydberg constant | $R$ | $1.10 \times 10^7$ m⁻¹ | 1.097 373 156 854 8 | $7.6 \times 10^{-6}$ |
| Electron Compton wavelength | $\lambda_C$ | $2.43 \times 10^{-12}$ m | 2.426 310 215 | 0.0073 |

[a]Values given in this column should be given the same unit and power of 10 as the computational value.

[b]Parts per million.

[c]Masses given in u are in unified atomic mass units, where 1 u = 1.660 538 86 $\times 10^{-27}$ kg.

[d]STP means standard temperature and pressure: 0°C and 1.0 atm (0.1 MPa).

*The values in this table were selected from the 1998 CODATA recommended values (www.physics.nist.gov).

## Some Distances from Earth

| | | | |
|---|---|---|---|
| To the Moon* | $3.82 \times 10^8$ m | To the center of our galaxy | $2.2 \times 10^{20}$ m |
| To the Sun* | $1.50 \times 10^{11}$ m | To the Andromeda Galaxy | $2.1 \times 10^{22}$ m |
| To the nearest star (Proxima Centauri) | $4.04 \times 10^{16}$ m | To the edge of the observable universe | $\sim 10^{26}$ m |

*Mean distance.

## The Sun, Earth, and the Moon

| Property | Unit | Sun | Earth | Moon |
|---|---|---|---|---|
| Mass | kg | $1.99 \times 10^{30}$ | $5.98 \times 10^{24}$ | $7.36 \times 10^{22}$ |
| Mean radius | m | $6.96 \times 10^8$ | $6.37 \times 10^6$ | $1.74 \times 10^6$ |
| Mean density | kg/m$^3$ | 1410 | 5520 | 3340 |
| Free-fall acceleration at the surface | m/s$^2$ | 274 | 9.81 | 1.67 |
| Escape velocity | km/s | 618 | 11.2 | 2.38 |
| Period of rotation[a] | — | 37 d at poles[b]    26 d at equator[b] | 23 h 56 min | 27.3 d |
| Radiation power[c] | W | $3.90 \times 10^{26}$ | | |

[a]Measured with respect to the distant stars.

[b]The Sun, a ball of gas, does not rotate as a rigid body.

[c]Just outside Earth's atmosphere solar energy is received, assuming normal incidence, at the rate of 1340 W/m$^2$.

## Some Properties of the Planets

| | Mercury | Venus | Earth | Mars | Jupiter | Saturn | Uranus | Neptune | Pluto |
|---|---|---|---|---|---|---|---|---|---|
| Mean distance from Sun, $10^6$ km | 57.9 | 108 | 150 | 228 | 778 | 1430 | 2870 | 4500 | 5900 |
| Period of revolution, y | 0.241 | 0.615 | 1.00 | 1.88 | 11.9 | 29.5 | 84.0 | 165 | 248 |
| Period of rotation,[a] d | 58.7 | −243[b] | 0.997 | 1.03 | 0.409 | 0.426 | −0.451[b] | 0.658 | 6.39 |
| Orbital speed, km/s | 47.9 | 35.0 | 29.8 | 24.1 | 13.1 | 9.64 | 6.81 | 5.43 | 4.74 |
| Inclination of axis to orbit | <28° | ≈3° | 23.4° | 25.0° | 3.08° | 26.7° | 97.9° | 29.6° | 57.5° |
| Inclination of orbit to Earth's orbit | 7.00° | 3.39° | | 1.85° | 1.30° | 2.49° | 0.77° | 1.77° | 17.2° |
| Eccentricity of orbit | 0.206 | 0.0068 | 0.0167 | 0.0934 | 0.0485 | 0.0556 | 0.0472 | 0.0086 | 0.250 |
| Equatorial diameter, km | 4880 | 12 100 | 12 800 | 6790 | 143 000 | 120 000 | 51 800 | 49 500 | 2300 |
| Mass (Earth = 1) | 0.0558 | 0.815 | 1.000 | 0.107 | 318 | 95.1 | 14.5 | 17.2 | 0.002 |
| Density (water = 1) | 5.60 | 5.20 | 5.52 | 3.95 | 1.31 | 0.704 | 1.21 | 1.67 | 2.03 |
| Surface value of $g$,[c] m/s$^2$ | 3.78 | 8.60 | 9.78 | 3.72 | 22.9 | 9.05 | 7.77 | 11.0 | 0.5 |
| Escape velocity,[c] km/s | 4.3 | 10.3 | 11.2 | 5.0 | 59.5 | 35.6 | 21.2 | 23.6 | 1.1 |
| Known satellites | 0 | 0 | 1 | 2 | 60 + ring | 31 + rings | 21 + rings | 11 + rings | 1 |

[a]Measured with respect to the distant stars.

[b]Venus and Uranus rotate opposite their orbital motion.

[c]Gravitational acceleration measured at the planet's equator.

# Conversion Factors | D

Conversion factors may be read directly from these tables. For example, 1 degree = 2.778 × $10^{-3}$ revolutions, so 16.7° = 16.7 × 2.778 × $10^{-3}$ rev. The SI units are fully capitalized. Adapted in part from G. Shortley and D. Williams, *Elements of Physics,* 1971, Prentice-Hall, Englewood Cliffs, NJ.

## Plane Angle

| | ° | ′ | ″ | RADIAN | rev |
|---|---|---|---|---|---|
| 1 degree = | 1 | 60 | 3600 | $1.745 \times 10^{-2}$ | $2.778 \times 10^{-3}$ |
| 1 minute = | $1.667 \times 10^{-2}$ | 1 | 60 | $2.909 \times 10^{-4}$ | $4.630 \times 10^{-5}$ |
| 1 second = | $2.778 \times 10^{-4}$ | $1.667 \times 10^{-2}$ | 1 | $4.848 \times 10^{-6}$ | $7.716 \times 10^{-7}$ |
| 1 RADIAN = | 57.30 | 3438 | $2.063 \times 10^{5}$ | 1 | 0.1592 |
| 1 revolution = | 360 | $2.16 \times 10^{4}$ | $1.296 \times 10^{6}$ | 6.283 | 1 |

## Solid Angle

1 sphere = $4\pi$ steradians = 12.57 steradians

## Length

| | cm | METER | km | in. | ft | mi |
|---|---|---|---|---|---|---|
| 1 centimeter = | 1 | $10^{-2}$ | $10^{-5}$ | 0.3937 | $3.281 \times 10^{-2}$ | $6.214 \times 10^{-6}$ |
| 1 METER = | 100 | 1 | $10^{-3}$ | 39.37 | 3.281 | $6.214 \times 10^{-4}$ |
| 1 kilometer = | $10^{5}$ | 1000 | 1 | $3.937 \times 10^{4}$ | 3281 | 0.6214 |
| 1 inch = | 2.540 | $2.540 \times 10^{-2}$ | $2.540 \times 10^{-5}$ | 1 | $8.333 \times 10^{-2}$ | $1.578 \times 10^{-5}$ |
| 1 foot = | 30.48 | 0.3048 | $3.048 \times 10^{-4}$ | 12 | 1 | $1.894 \times 10^{-4}$ |
| 1 mile = | $1.609 \times 10^{5}$ | 1609 | 1.609 | $6.336 \times 10^{4}$ | 5280 | 1 |

1 angström = $10^{-10}$ m
1 nautical mile = 1852 m
  = 1.151 miles = 6076 ft

1 fermi = $10^{-15}$ m
1 light-year = $9.461 \times 10^{12}$ km
1 parsec = $3.084 \times 10^{13}$ km

1 fathom = 6 ft
1 Bohr radius = $5.292 \times 10^{-11}$ m
1 yard = 3 ft

1 rod = 16.5 ft
1 mil = $10^{-3}$ in.
1 nm = $10^{-9}$ m

## Area

| | METER$^2$ | cm$^2$ | ft$^2$ | in.$^2$ |
|---|---|---|---|---|
| 1 SQUARE METER = | 1 | $10^{4}$ | 10.76 | 1550 |
| 1 square centimeter = | $10^{-4}$ | 1 | $1.076 \times 10^{-3}$ | 0.1550 |
| 1 square foot = | $9.290 \times 10^{-2}$ | 929.0 | 1 | 144 |
| 1 square inch = | $6.452 \times 10^{-4}$ | 6.452 | $6.944 \times 10^{-3}$ | 1 |

1 square mile = $2.788 \times 10^{7}$ ft$^2$ = 640 acres
1 barn = $10^{-28}$ m$^2$

1 acre = 43 560 ft$^2$
1 hectare = $10^{4}$ m$^2$ = 2.471 acres

## Volume

| | METER$^3$ | cm$^3$ | L | ft$^3$ | in.$^3$ |
|---|---|---|---|---|---|
| 1 CUBIC METER = 1 | $10^6$ | 1000 | 35.31 | $6.102 \times 10^4$ |
| 1 cubic centimeter = $10^{-6}$ | 1 | $1.000 \times 10^{-3}$ | $3.531 \times 10^{-5}$ | $6.102 \times 10^{-2}$ |
| 1 liter = $1.000 \times 10^{-3}$ | 1000 | 1 | $3.531 \times 10^{-2}$ | 61.02 |
| 1 cubic foot = $2.832 \times 10^{-2}$ | $2.832 \times 10^4$ | 28.32 | 1 | 1728 |
| 1 cubic inch = $1.639 \times 10^{-5}$ | 16.39 | $1.639 \times 10^{-2}$ | $5.787 \times 10^{-4}$ | 1 |

1 U.S. fluid gallon = 4 U.S. fluid quarts = 8 U.S. pints = 128 U.S. fluid ounces = 231 in.$^3$
1 British imperial gallon = 277.4 in.$^3$ = 1.201 U.S. fluid gallons

## Mass

Quantities in the colored areas are not mass units but are often used as such. For example, when we write 1 kg "=" 2.205 lb, this means that a kilogram is a *mass* that *weighs* 2.205 pounds at a location where $g$ has the standard value of 9.80665 m/s$^2$.

| | g | KILOGRAM | slug | u | oz | lb | ton |
|---|---|---|---|---|---|---|---|
| 1 gram = 1 | 0.001 | $6.852 \times 10^{-5}$ | $6.022 \times 10^{23}$ | $3.527 \times 10^{-2}$ | $2.205 \times 10^{-3}$ | $1.102 \times 10^{-6}$ |
| 1 KILOGRAM = 1000 | 1 | $6.852 \times 10^{-2}$ | $6.022 \times 10^{26}$ | 35.27 | 2.205 | $1.102 \times 10^{-3}$ |
| 1 slug = $1.459 \times 10^4$ | 14.59 | 1 | $8.786 \times 10^{27}$ | 514.8 | 32.17 | $1.609 \times 10^{-2}$ |
| 1 atomic mass unit = $1.661 \times 10^{-24}$ | $1.661 \times 10^{-27}$ | $1.138 \times 10^{-28}$ | 1 | $5.857 \times 10^{-26}$ | $3.662 \times 10^{-27}$ | $1.830 \times 10^{-30}$ |
| 1 ounce = 28.35 | $2.835 \times 10^{-2}$ | $1.943 \times 10^{-3}$ | $1.718 \times 10^{25}$ | 1 | $6.250 \times 10^{-2}$ | $3.125 \times 10^{-5}$ |
| 1 pound = 453.6 | 0.4536 | $3.108 \times 10^{-2}$ | $2.732 \times 10^{26}$ | 16 | 1 | 0.0005 |
| 1 ton = $9.072 \times 10^5$ | 907.2 | 62.16 | $5.463 \times 10^{29}$ | $3.2 \times 10^4$ | 2000 | 1 |

1 metric ton = 1000 kg

## Density

Quantities in the colored areas are weight densities and, as such, are dimensionally different from mass densities. See the note for the mass table.

| | slug/ft$^3$ | KILOGRAM/ METER$^3$ | g/cm$^3$ | lb/ft$^3$ | lb/in.$^3$ |
|---|---|---|---|---|---|
| 1 slug per foot$^3$ = 1 | | 515.4 | 0.5154 | 32.17 | $1.862 \times 10^{-2}$ |
| 1 KILOGRAM per METER$^3$ = $1.940 \times 10^{-3}$ | | 1 | 0.001 | $6.243 \times 10^{-2}$ | $3.613 \times 10^{-5}$ |
| 1 gram per centimeter$^3$ = 1.940 | | 1000 | 1 | 62.43 | $3.613 \times 10^{-2}$ |
| 1 pound per foot$^3$ = $3.108 \times 10^{-2}$ | | 16.02 | $16.02 \times 10^{-2}$ | 1 | $5.787 \times 10^{-4}$ |
| 1 pound per inch$^3$ = 53.71 | | $2.768 \times 10^4$ | 27.68 | 1728 | 1 |

## Time

| | y | d | h | min | SECOND |
|---|---|---|---|---|---|
| 1 year = 1 | 365.25 | $8.766 \times 10^3$ | $5.259 \times 10^5$ | $3.156 \times 10^7$ |
| 1 day = $2.738 \times 10^{-3}$ | 1 | 24 | 1440 | $8.640 \times 10^4$ |
| 1 hour = $1.141 \times 10^{-4}$ | $4.167 \times 10^{-2}$ | 1 | 60 | 3600 |
| 1 minute = $1.901 \times 10^{-6}$ | $6.944 \times 10^{-4}$ | $1.667 \times 10^{-2}$ | 1 | 60 |
| 1 SECOND = $3.169 \times 10^{-8}$ | $1.157 \times 10^{-5}$ | $2.778 \times 10^{-4}$ | $1.667 \times 10^{-2}$ | 1 |

## Speed

|  | ft/s | km/h | METER/SECOND | mi/h | cm/s |
|---|---|---|---|---|---|
| 1 foot per second = 1 | 1 | 1.097 | 0.3048 | 0.6818 | 30.48 |
| 1 kilometer per hour = 0.9113 | | 1 | 0.2778 | 0.6214 | 27.78 |
| 1 METER per SECOND = 3.281 | | 3.6 | 1 | 2.237 | 100 |
| 1 mile per hour = 1.467 | | 1.609 | 0.4470 | 1 | 44.70 |
| 1 centimeter per second = $3.281 \times 10^{-2}$ | | $3.6 \times 10^{-2}$ | 0.01 | $2.237 \times 10^{-2}$ | 1 |

1 knot = 1 nautical mi/h = 1.688 ft/s      1 mi/min = 88.00 ft/s = 60.00 mi/h

## Force

Force units in the colored areas are now little used. To clarify: 1 gram-force (= 1 gf) is the force of gravity that would act on an object whose mass is 1 gram at a location where $g$ has the standard value of 9.80665 m/s$^2$.

|  | dyne | NEWTON | lb | pdl | gf | kgf |
|---|---|---|---|---|---|---|
| 1 dyne = 1 | | $10^{-5}$ | $2.248 \times 10^{-6}$ | $7.233 \times 10^{-5}$ | $1.020 \times 10^{-3}$ | $1.020 \times 10^{-6}$ |
| 1 NEWTON = $10^5$ | | 1 | 0.2248 | 7.233 | 102.0 | 0.1020 |
| 1 pound = $4.448 \times 10^5$ | | 4.448 | 1 | 32.17 | 453.6 | 0.4536 |
| 1 poundal = $1.383 \times 10^4$ | | 0.1383 | $3.108 \times 10^{-2}$ | 1 | 14.10 | $1.410 \times 10^2$ |
| 1 gram-force = 980.7 | | $9.807 \times 10^{-3}$ | $2.205 \times 10^{-3}$ | $7.093 \times 10^{-2}$ | 1 | 0.001 |
| 1 kilogram-force = $9.807 \times 10^5$ | | 9.807 | 2.205 | 70.93 | 1000 | 1 |

1 ton = 2000 lb

## Pressure

|  | atm | dyne/cm$^2$ | inch of water | cm Hg | PASCAL | lb/in.$^2$ | lb/ft$^2$ |
|---|---|---|---|---|---|---|---|
| 1 atmosphere = 1 | | $1.013 \times 10^6$ | 406.8 | 76 | $1.013 \times 10^5$ | 14.70 | 2116 |
| 1 dyne per centimeter$^2$ = $9.869 \times 10^{-7}$ | | 1 | $4.015 \times 10^{-4}$ | $7.501 \times 10^{-5}$ | 0.1 | $1.405 \times 10^{-5}$ | $2.089 \times 10^{-3}$ |
| 1 inch of water[a] at 4°C = $2.458 \times 10^{-3}$ | | 2491 | 1 | 0.1868 | 249.1 | $3.613 \times 10^{-2}$ | 5.202 |
| 1 centimeter of mercury[a] at 0°C = $1.316 \times 10^{-2}$ | | $1.333 \times 10^4$ | 5.353 | 1 | 1333 | 0.1934 | 27.85 |
| 1 PASCAL = $9.869 \times 10^{-6}$ | | 10 | $4.015 \times 10^{-3}$ | $7.501 \times 10^{-4}$ | 1 | $1.450 \times 10^{-4}$ | $2.089 \times 10^{-2}$ |
| 1 pound per inch$^2$ = $6.805 \times 10^{-2}$ | | $6.895 \times 10^4$ | 27.68 | 5.171 | $6.895 \times 10^3$ | 1 | 144 |
| 1 pound per foot$^2$ = $4.725 \times 10^{-4}$ | | 478.8 | 0.1922 | $3.591 \times 10^{-2}$ | 47.88 | $6.944 \times 10^{-3}$ | 1 |

[a]Where the acceleration of gravity has the standard value of 9.80665 m/s$^2$.

1 bar = $10^6$ dyne/cm$^2$ = 0.1 MPa          1 millibar = $10^3$ dyne/cm$^2$ = $10^2$ Pa          1 torr = 1 mm Hg

### Energy, Work, Heat

Quantities in the colored areas are not energy units but are included for convenience. They arise from the relativistic mass–energy equivalence formula $E = mc^2$ and represent the energy released if a kilogram or unified atomic mass unit (u) is completely converted to energy (bottom two rows) or the mass that would be completely converted to one unit of energy (rightmost two columns).

| | Btu | erg | ft·lb | hp·h | JOULE | cal | kW·h | eV | MeV | kg | u |
|---|---|---|---|---|---|---|---|---|---|---|---|
| 1 British thermal unit = | 1 | $1.055 \times 10^{10}$ | 777.9 | $3.929 \times 10^{-4}$ | 1055 | 252.0 | $2.930 \times 10^{-4}$ | $6.585 \times 10^{21}$ | $6.585 \times 10^{15}$ | $1.174 \times 10^{-14}$ | $7.070 \times 10^{12}$ |
| 1 erg = | $9.481 \times 10^{-11}$ | 1 | $7.376 \times 10^{-8}$ | $3.725 \times 10^{-14}$ | $10^{-7}$ | $2.389 \times 10^{-8}$ | $2.778 \times 10^{-14}$ | $6.242 \times 10^{11}$ | $6.242 \times 10^{5}$ | $1.113 \times 10^{-24}$ | 670.2 |
| 1 foot-pound = | $1.285 \times 10^{-3}$ | $1.356 \times 10^{7}$ | 1 | $5.051 \times 10^{-7}$ | 1.356 | 0.3238 | $3.766 \times 10^{-7}$ | $8.464 \times 10^{18}$ | $8.464 \times 10^{12}$ | $1.509 \times 10^{-17}$ | $9.037 \times 10^{9}$ |
| 1 horsepower-hour = | 2545 | $2.685 \times 10^{13}$ | $1.980 \times 10^{6}$ | 1 | $2.685 \times 10^{6}$ | $6.413 \times 10^{5}$ | 0.7457 | $1.676 \times 10^{25}$ | $1.676 \times 10^{19}$ | $2.988 \times 10^{-11}$ | $1.799 \times 10^{16}$ |
| 1 JOULE = | $9.481 \times 10^{-4}$ | $10^{7}$ | 0.7376 | $3.725 \times 10^{-7}$ | 1 | 0.2389 | $2.778 \times 10^{-7}$ | $6.242 \times 10^{18}$ | $6.242 \times 10^{12}$ | $1.113 \times 10^{-17}$ | $6.702 \times 10^{9}$ |
| 1 calorie = | $3.968 \times 10^{-3}$ | $4.1868 \times 10^{7}$ | 3.088 | $1.560 \times 10^{-6}$ | 4.1868 | 1 | $1.163 \times 10^{-6}$ | $2.613 \times 10^{19}$ | $2.613 \times 10^{13}$ | $4.660 \times 10^{-17}$ | $2.806 \times 10^{10}$ |
| 1 kilowatt-hour = | 3413 | $3.600 \times 10^{13}$ | $2.655 \times 10^{6}$ | 1.341 | $3.600 \times 10^{6}$ | $8.600 \times 10^{5}$ | 1 | $2.247 \times 10^{25}$ | $2.247 \times 10^{19}$ | $4.007 \times 10^{-11}$ | $2.413 \times 10^{16}$ |
| 1 electron-volt = | $1.519 \times 10^{-22}$ | $1.602 \times 10^{-12}$ | $1.182 \times 10^{-19}$ | $5.967 \times 10^{-26}$ | $1.602 \times 10^{-19}$ | $3.827 \times 10^{-20}$ | $4.450 \times 10^{-26}$ | 1 | $10^{-6}$ | $1.783 \times 10^{-36}$ | $1.074 \times 10^{-9}$ |
| 1 million electron-volts = | $1.519 \times 10^{-16}$ | $1.602 \times 10^{-6}$ | $1.182 \times 10^{-13}$ | $5.967 \times 10^{-20}$ | $1.602 \times 10^{-13}$ | $3.827 \times 10^{-14}$ | $4.450 \times 10^{-20}$ | $10^{6}$ | 1 | $1.783 \times 10^{-30}$ | $1.074 \times 10^{-3}$ |
| 1 kilogram = | $8.521 \times 10^{13}$ | $8.987 \times 10^{23}$ | $6.629 \times 10^{16}$ | $3.348 \times 10^{10}$ | $8.987 \times 10^{16}$ | $2.146 \times 10^{16}$ | $2.497 \times 10^{10}$ | $5.610 \times 10^{35}$ | $5.610 \times 10^{29}$ | 1 | $6.022 \times 10^{26}$ |
| 1 unified atomic mass unit = | $1.415 \times 10^{-13}$ | $1.492 \times 10^{-3}$ | $1.101 \times 10^{-10}$ | $5.559 \times 10^{-17}$ | $1.492 \times 10^{-10}$ | $3.564 \times 10^{-11}$ | $4.146 \times 10^{-17}$ | $9.320 \times 10^{8}$ | 932.0 | $1.661 \times 10^{-27}$ | 1 |

### Power

| | Btu/h | ft·lb/s | hp | cal/s | kW | WATT |
|---|---|---|---|---|---|---|
| 1 British thermal unit per hour = | 1 | 0.2161 | $3.929 \times 10^{-4}$ | $6.998 \times 10^{-2}$ | $2.930 \times 10^{-4}$ | 0.2930 |
| 1 foot-pound per second = | 4.628 | 1 | $1.818 \times 10^{-3}$ | 0.3239 | $1.356 \times 10^{-3}$ | 1.356 |
| 1 horsepower = | 2545 | 550 | 1 | 178.1 | 0.7457 | 745.7 |
| 1 calorie per second = | 14.29 | 3.088 | $5.615 \times 10^{-3}$ | 1 | $4.186 \times 10^{-3}$ | 4.186 |
| 1 kilowatt = | 3413 | 737.6 | 1.341 | 238.9 | 1 | 1000 |
| 1 WATT = | 3.413 | 0.7376 | $1.341 \times 10^{-3}$ | 0.2389 | 0.001 | 1 |

### Magnetic Field

| | gauss | TESLA | milligauss |
|---|---|---|---|
| 1 gauss = | 1 | $10^{-4}$ | 1000 |
| 1 TESLA = | $10^{4}$ | 1 | $10^{7}$ |
| 1 milligauss = | 0.001 | $10^{-7}$ | 1 |

### Magnetic Flux

| | maxwell | WEBER |
|---|---|---|
| 1 maxwell = | 1 | $10^{-8}$ |
| 1 WEBER = | $10^{8}$ | 1 |

1 tesla = 1 weber/meter$^2$

# Mathematical Formulas E

## Geometry

Circle of radius $r$: circumference $= 2\pi r$; area $= \pi r^2$.

Sphere of radius $r$: area $= 4\pi r^2$; volume $= \frac{4}{3}\pi r^3$.

Right circular cylinder of radius $r$ and height $h$:
area $= 2\pi r^2 + 2\pi rh$; volume $= \pi r^2 h$.

Triangle of base $a$ and altitude $h$: area $= \frac{1}{2}ah$.

## Quadratic Formula

If $ax^2 + bx + c = 0$, then $x = \dfrac{-b \pm \sqrt{b^2 - 4ac}}{2a}$.

## Trigonometric Functions of Angle $\theta$

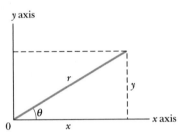

$$\sin\theta = \frac{y}{r} \qquad \cos\theta = \frac{x}{r}$$

$$\tan\theta = \frac{y}{x} \qquad \cot\theta = \frac{x}{y}$$

$$\sec\theta = \frac{r}{x} \qquad \csc\theta = \frac{r}{y}$$

## Pythagorean Theorem

In this right triangle,
$$a^2 + b^2 = c^2$$

## Triangles

Angles are $A, B, C$

Opposite sides are $a, b, c$

Angles $A + B + C = 180°$

$$\frac{\sin A}{a} = \frac{\sin B}{b} = \frac{\sin C}{c}$$

$$c^2 = a^2 + b^2 - 2ab\cos C$$

Exterior angle $D = A + C$

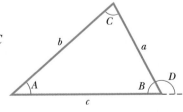

## Mathematical Signs and Symbols

$=$ equals

$\approx$ equals approximately

$\sim$ is the order of magnitude of

$\neq$ is not equal to

$\equiv$ is identical to, is defined as

$>$ is greater than ($\gg$ is much greater than)

$<$ is less than ($\ll$ is much less than)

$\geq$ is greater than or equal to (or, is no less than)

$\leq$ is less than or equal to (or, is no more than)

$\pm$ plus or minus

$\propto$ is proportional to

$\Sigma$ the sum of

$x_{\text{avg}}$ the average value of $x$

## Trigonometric Identities

$\sin(90° - \theta) = \cos\theta$

$\cos(90° - \theta) = \sin\theta$

$\sin\theta/\cos\theta = \tan\theta$

$\sin^2\theta + \cos^2\theta = 1$

$\sec^2\theta - \tan^2\theta = 1$

$\csc^2\theta - \cot^2\theta = 1$

$\sin 2\theta = 2\sin\theta\cos\theta$

$\cos 2\theta = \cos^2\theta - \sin^2\theta = 2\cos^2\theta - 1 = 1 - 2\sin^2\theta$

$\sin(\alpha \pm \beta) = \sin\alpha\cos\beta \pm \cos\alpha\sin\beta$

$\cos(\alpha \pm \beta) = \cos\alpha\cos\beta \mp \sin\alpha\sin\beta$

$$\tan(\alpha \pm \beta) = \frac{\tan\alpha \pm \tan\beta}{1 \mp \tan\alpha\tan\beta}$$

$\sin\alpha \pm \sin\beta = 2\sin\frac{1}{2}(\alpha \pm \beta)\cos\frac{1}{2}(\alpha \mp \beta)$

$\cos\alpha + \cos\beta = 2\cos\frac{1}{2}(\alpha + \beta)\cos\frac{1}{2}(\alpha - \beta)$

$\cos\alpha - \cos\beta = -2\sin\frac{1}{2}(\alpha + \beta)\sin\frac{1}{2}(\alpha - \beta)$

## Binomial Theorem

$$(1 + x)^n = 1 + \frac{nx}{1!} + \frac{n(n-1)x^2}{2!} + \cdots \qquad (x^2 < 1)$$

## Exponential Expansion

$$e^x = 1 + x + \frac{x^2}{2!} + \frac{x^3}{3!} + \cdots$$

## Logarithmic Expansion

$$\ln(1 + x) = x - \tfrac{1}{2}x^2 + \tfrac{1}{3}x^3 - \cdots \qquad (|x| < 1)$$

## Trigonometric Expansions
## ($\theta$ in radians)

$$\sin \theta = \theta - \frac{\theta^3}{3!} + \frac{\theta^5}{5!} - \cdots$$

$$\cos \theta = 1 - \frac{\theta^2}{2!} + \frac{\theta^4}{4!} - \cdots$$

$$\tan \theta = \theta + \frac{\theta^3}{3} + \frac{2\theta^5}{15} + \cdots$$

## Cramer's Rule

Two simultaneous equations in unknowns $x$ and $y$,

$$a_1 x + b_1 y = c_1 \quad \text{and} \quad a_2 x + b_2 y = c_2,$$

have the solutions

$$x = \frac{\begin{vmatrix} c_1 & b_1 \\ c_2 & b_2 \end{vmatrix}}{\begin{vmatrix} a_1 & b_1 \\ a_2 & b_2 \end{vmatrix}} = \frac{c_1 b_2 - c_2 b_1}{a_1 b_2 - a_2 b_1}$$

and

$$y = \frac{\begin{vmatrix} a_1 & c_1 \\ a_2 & c_2 \end{vmatrix}}{\begin{vmatrix} a_1 & b_1 \\ a_2 & b_2 \end{vmatrix}} = \frac{a_1 c_2 - a_2 c_1}{a_1 b_2 - a_2 b_1}.$$

## Products of Vectors

Let $\hat{i}$, $\hat{j}$, and $\hat{k}$ be unit vectors in the $x$, $y$, and $z$ directions. Then

$$\hat{i} \cdot \hat{i} = \hat{j} \cdot \hat{j} = \hat{k} \cdot \hat{k} = 1, \quad \hat{i} \cdot \hat{j} = \hat{j} \cdot \hat{k} = \hat{k} \cdot \hat{i} = 0,$$

$$\hat{i} \times \hat{i} = \hat{j} \times \hat{j} = \hat{k} \times \hat{k} = 0,$$

$$\hat{i} \times \hat{j} = \hat{k}, \quad \hat{j} \times \hat{k} = \hat{i}, \quad \hat{k} \times \hat{i} = \hat{j}$$

Any vector $\vec{a}$ with components $a_x$, $a_y$, and $a_z$ along the $x$, $y$, and $z$ axes can be written as

$$\vec{a} = a_x \hat{i} + a_y \hat{j} + a_z \hat{k}.$$

Let $\vec{a}$, $\vec{b}$, and $\vec{c}$ be arbitrary vectors with magnitudes $a$, $b$, and $c$. Then

$$\vec{a} \times (\vec{b} + \vec{c}) = (\vec{a} \times \vec{b}) + (\vec{a} \times \vec{c})$$

$$(s\vec{a}) \times \vec{b} = \vec{a} \times (s\vec{b}) = s(\vec{a} \times \vec{b}) \qquad (s = \text{a scalar}).$$

Let $\theta$ be the smaller of the two angles between $\vec{a}$ and $\vec{b}$. Then

$$\vec{a} \cdot \vec{b} = \vec{b} \cdot \vec{a} = a_x b_x + a_y b_y + a_z b_z = ab \cos \theta$$

$$\vec{a} \times \vec{b} = -\vec{b} \times \vec{a} = \begin{vmatrix} \hat{i} & \hat{j} & \hat{k} \\ a_x & a_y & a_z \\ b_x & b_y & b_z \end{vmatrix}$$

$$= \hat{i} \begin{vmatrix} a_y & a_z \\ b_y & b_z \end{vmatrix} - \hat{j} \begin{vmatrix} a_x & a_z \\ b_x & b_z \end{vmatrix} + \hat{k} \begin{vmatrix} a_x & a_y \\ b_x & b_y \end{vmatrix}$$

$$= (a_y b_z - b_y a_z)\hat{i} + (a_z b_x - b_z a_x)\hat{j}$$

$$\quad + (a_x b_y - b_x a_y)\hat{k}$$

$$|\vec{a} \times \vec{b}| = ab \sin \theta$$

$$\vec{a} \cdot (\vec{b} \times \vec{c}) = \vec{b} \cdot (\vec{c} \times \vec{a}) = \vec{c} \cdot (\vec{a} \times \vec{b})$$

$$\vec{a} \times (\vec{b} \times \vec{c}) = (\vec{a} \cdot \vec{c})\vec{b} - (\vec{a} \cdot \vec{b})\vec{c}$$

## Derivatives and Integrals

In what follows, the letters $u$ and $v$ stand for any functions of $x$, and $a$ and $m$ are constants. To each of the indefinite integrals should be added an arbitrary constant of integration. The *Handbook of Chemistry and Physics* (CRC Press Inc.) gives a more extensive tabulation.

1. $\dfrac{dx}{dx} = 1$

2. $\dfrac{d}{dx}(au) = a\dfrac{du}{dx}$

3. $\dfrac{d}{dx}(u + v) = \dfrac{du}{dx} + \dfrac{dv}{dx}$

4. $\dfrac{d}{dx}x^m = mx^{m-1}$

5. $\dfrac{d}{dx}\ln x = \dfrac{1}{x}$

6. $\dfrac{d}{dx}(uv) = u\dfrac{dv}{dx} + v\dfrac{du}{dx}$

7. $\dfrac{d}{dx}e^x = e^x$

8. $\dfrac{d}{dx}\sin x = \cos x$

9. $\dfrac{d}{dx}\cos x = -\sin x$

10. $\dfrac{d}{dx}\tan x = \sec^2 x$

11. $\dfrac{d}{dx}\cot x = -\csc^2 x$

12. $\dfrac{d}{dx}\sec x = \tan x \sec x$

13. $\dfrac{d}{dx}\csc x = -\cot x \csc x$

14. $\dfrac{d}{dx}e^u = e^u\dfrac{du}{dx}$

15. $\dfrac{d}{dx}\sin u = \cos u\dfrac{du}{dx}$

16. $\dfrac{d}{dx}\cos u = -\sin u\dfrac{du}{dx}$

---

1. $\displaystyle\int dx = x$

2. $\displaystyle\int au\,dx = a\int u\,dx$

3. $\displaystyle\int (u + v)\,dx = \int u\,dx + \int v\,dx$

4. $\displaystyle\int x^m\,dx = \dfrac{x^{m+1}}{m+1} \quad (m \neq -1)$

5. $\displaystyle\int \dfrac{dx}{x} = \ln |x|$

6. $\displaystyle\int u\dfrac{dv}{dx}\,dx = uv - \int v\dfrac{du}{dx}\,dx$

7. $\displaystyle\int e^x\,dx = e^x$

8. $\displaystyle\int \sin x\,dx = -\cos x$

9. $\displaystyle\int \cos x\,dx = \sin x$

10. $\displaystyle\int \tan x\,dx = \ln |\sec x|$

11. $\displaystyle\int \sin^2 x\,dx = \tfrac{1}{2}x - \tfrac{1}{4}\sin 2x$

12. $\displaystyle\int e^{-ax}\,dx = -\dfrac{1}{a}e^{-ax}$

13. $\displaystyle\int xe^{-ax}\,dx = -\dfrac{1}{a^2}(ax + 1)e^{-ax}$

14. $\displaystyle\int x^2 e^{-ax}\,dx = -\dfrac{1}{a^3}(a^2x^2 + 2ax + 2)e^{-ax}$

15. $\displaystyle\int_0^\infty x^n e^{-ax}\,dx = \dfrac{n!}{a^{n+1}}$

16. $\displaystyle\int_0^\infty x^{2n} e^{-ax^2}\,dx = \dfrac{1 \cdot 3 \cdot 5 \cdots (2n - 1)}{2^{n+1}a^n}\sqrt{\dfrac{\pi}{a}}$

17. $\displaystyle\int \dfrac{dx}{\sqrt{x^2 + a^2}} = \ln(x + \sqrt{x^2 + a^2})$

18. $\displaystyle\int \dfrac{x\,dx}{(x^2 + a^2)^{3/2}} = -\dfrac{1}{(x^2 + a^2)^{1/2}}$

19. $\displaystyle\int \dfrac{dx}{(x^2 + a^2)^{3/2}} = \dfrac{x}{a^2(x^2 + a^2)^{1/2}}$

20. $\displaystyle\int_0^\infty x^{2n+1} e^{-ax^2}\,dx = \dfrac{n!}{2a^{n+1}} \quad (a > 0)$

21. $\displaystyle\int \dfrac{x\,dx}{x + d} = x - d\ln(x + d)$

# Properties of the Elements

All physical properties are for a pressure of 1 atm unless otherwise specified.

| Element | Symbol | Atomic Number Z | Molar Mass, g/mol | Density, g/cm³ at 20°C | Melting Point, °C | Boiling Point, °C | Specific Heat, J/(g · °C) at 25°C |
|---|---|---|---|---|---|---|---|
| Actinium | Ac | 89 | (227) | 10.06 | 1323 | (3473) | 0.092 |
| Aluminum | Al | 13 | 26.9815 | 2.699 | 660 | 2450 | 0.900 |
| Americium | Am | 95 | (243) | 13.67 | 1541 | — | — |
| Antimony | Sb | 51 | 121.75 | 6.691 | 630.5 | 1380 | 0.205 |
| Argon | Ar | 18 | 39.948 | $1.6626 \times 10^{-3}$ | −189.4 | −185.8 | 0.523 |
| Arsenic | As | 33 | 74.9216 | 5.78 | 817 (28 atm) | 613 | 0.331 |
| Astatine | At | 85 | (210) | — | (302) | — | — |
| Barium | Ba | 56 | 137.34 | 3.594 | 729 | 1640 | 0.205 |
| Berkelium | Bk | 97 | (247) | 14.79 | — | — | — |
| Beryllium | Be | 4 | 9.0122 | 1.848 | 1287 | 2770 | 1.83 |
| Bismuth | Bi | 83 | 208.980 | 9.747 | 271.37 | 1560 | 0.122 |
| Bohrium | Bh | 107 | 262.12 | — | — | — | — |
| Boron | B | 5 | 10.811 | 2.34 | 2030 | — | 1.11 |
| Bromine | Br | 35 | 79.909 | 3.12 (liquid) | −7.2 | 58 | 0.293 |
| Cadmium | Cd | 48 | 112.40 | 8.65 | 321.03 | 765 | 0.226 |
| Calcium | Ca | 20 | 40.08 | 1.55 | 838 | 1440 | 0.624 |
| Californium | Cf | 98 | (251) | — | — | — | — |
| Carbon | C | 6 | 12.01115 | 2.26 | 3727 | 4830 | 0.691 |
| Cerium | Ce | 58 | 140.12 | 6.768 | 804 | 3470 | 0.188 |
| Cesium | Cs | 55 | 132.905 | 1.873 | 28.40 | 690 | 0.243 |
| Chlorine | Cl | 17 | 35.453 | $3.214 \times 10^{-3}$ (0°C) | −101 | −34.7 | 0.486 |
| Chromium | Cr | 24 | 51.996 | 7.19 | 1857 | 2665 | 0.448 |
| Cobalt | Co | 27 | 58.9332 | 8.85 | 1495 | 2900 | 0.423 |
| Copper | Cu | 29 | 63.54 | 8.96 | 1083.40 | 2595 | 0.385 |
| Curium | Cm | 96 | (247) | 13.3 | — | — | — |
| Darmstadtium | Ds | 110 | (271) | — | — | — | — |
| Dubnium | Db | 105 | 262.114 | — | — | — | — |
| Dysprosium | Dy | 66 | 162.50 | 8.55 | 1409 | 2330 | 0.172 |
| Einsteinium | Es | 99 | (254) | — | — | — | — |
| Erbium | Er | 68 | 167.26 | 9.15 | 1522 | 2630 | 0.167 |
| Europium | Eu | 63 | 151.96 | 5.243 | 817 | 1490 | 0.163 |
| Fermium | Fm | 100 | (237) | — | — | — | — |
| Fluorine | F | 9 | 18.9984 | $1.696 \times 10^{-3}$ (0°C) | −219.6 | −188.2 | 0.753 |
| Francium | Fr | 87 | (223) | — | (27) | — | — |
| Gadolinium | Gd | 64 | 157.25 | 7.90 | 1312 | 2730 | 0.234 |
| Gallium | Ga | 31 | 69.72 | 5.907 | 29.75 | 2237 | 0.377 |
| Germanium | Ge | 32 | 72.59 | 5.323 | 937.25 | 2830 | 0.322 |
| Gold | Au | 79 | 196.967 | 19.32 | 1064.43 | 2970 | 0.131 |

| Element | Symbol | Atomic Number Z | Molar Mass, g/mol | Density, g/cm³ at 20°C | Melting Point, °C | Boiling Point, °C | Specific Heat, J/(g·°C) at 25°C |
|---|---|---|---|---|---|---|---|
| Hafnium | Hf | 72 | 178.49 | 13.31 | 2227 | 5400 | 0.144 |
| Hassium | Hs | 108 | (265) | — | — | — | — |
| Helium | He | 2 | 4.0026 | $0.1664 \times 10^{-3}$ | −269.7 | −268.9 | 5.23 |
| Holmium | Ho | 67 | 164.930 | 8.79 | 1470 | 2330 | 0.165 |
| Hydrogen | H | 1 | 1.00797 | $0.08375 \times 10^{-3}$ | −259.19 | −252.7 | 14.4 |
| Indium | In | 49 | 114.82 | 7.31 | 156.634 | 2000 | 0.233 |
| Iodine | I | 53 | 126.9044 | 4.93 | 113.7 | 183 | 0.218 |
| Iridium | Ir | 77 | 192.2 | 22.5 | 2447 | (5300) | 0.130 |
| Iron | Fe | 26 | 55.847 | 7.874 | 1536.5 | 3000 | 0.447 |
| Krypton | Kr | 36 | 83.80 | $3.488 \times 10^{-3}$ | −157.37 | −152 | 0.247 |
| Lanthanum | La | 57 | 138.91 | 6.189 | 920 | 3470 | 0.195 |
| Lawrencium | Lr | 103 | (257) | — | — | — | — |
| Lead | Pb | 82 | 207.19 | 11.35 | 327.45 | 1725 | 0.129 |
| Lithium | Li | 3 | 6.939 | 0.534 | 180.55 | 1300 | 3.58 |
| Lutetium | Lu | 71 | 174.97 | 9.849 | 1663 | 1930 | 0.155 |
| Magnesium | Mg | 12 | 24.312 | 1.738 | 650 | 1107 | 1.03 |
| Manganese | Mn | 25 | 54.9380 | 7.44 | 1244 | 2150 | 0.481 |
| Meitnerium | Mt | 109 | (266) | — | — | — | — |
| Mendelevium | Md | 101 | (256) | — | — | — | — |
| Mercury | Hg | 80 | 200.59 | 13.55 | −38.87 | 357 | 0.138 |
| Molybdenum | Mo | 42 | 95.94 | 10.22 | 2617 | 5560 | 0.251 |
| Neodymium | Nd | 60 | 144.24 | 7.007 | 1016 | 3180 | 0.188 |
| Neon | Ne | 10 | 20.183 | $0.8387 \times 10^{-3}$ | −248.597 | −246.0 | 1.03 |
| Neptunium | Np | 93 | (237) | 20.25 | 637 | — | 1.26 |
| Nickel | Ni | 28 | 58.71 | 8.902 | 1453 | 2730 | 0.444 |
| Niobium | Nb | 41 | 92.906 | 8.57 | 2468 | 4927 | 0.264 |
| Nitrogen | N | 7 | 14.0067 | $1.1649 \times 10^{-3}$ | −210 | −195.8 | 1.03 |
| Nobelium | No | 102 | (255) | — | — | — | — |
| Osmium | Os | 76 | 190.2 | 22.59 | 3027 | 5500 | 0.130 |
| Oxygen | O | 8 | 15.9994 | $1.3318 \times 10^{-3}$ | −218.80 | −183.0 | 0.913 |
| Palladium | Pd | 46 | 106.4 | 12.02 | 1552 | 3980 | 0.243 |
| Phosphorus | P | 15 | 30.9738 | 1.83 | 44.25 | 280 | 0.741 |
| Platinum | Pt | 78 | 195.09 | 21.45 | 1769 | 4530 | 0.134 |
| Plutonium | Pu | 94 | (244) | 19.8 | 640 | 3235 | 0.130 |
| Polonium | Po | 84 | (210) | 9.32 | 254 | — | — |
| Potassium | K | 19 | 39.102 | 0.862 | 63.20 | 760 | 0.758 |
| Praseodymium | Pr | 59 | 140.907 | 6.773 | 931 | 3020 | 0.197 |
| Promethium | Pm | 61 | (145) | 7.22 | (1027) | — | — |
| Protactinium | Pa | 91 | (231) | 15.37 (estimated) | (1230) | — | — |
| Radium | Ra | 88 | (226) | 5.0 | 700 | — | — |
| Radon | Rn | 86 | (222) | $9.96 \times 10^{-3}$ (0°C) | (−71) | −61.8 | 0.092 |
| Rhenium | Re | 75 | 186.2 | 21.02 | 3180 | 5900 | 0.134 |
| Rhodium | Rh | 45 | 102.905 | 12.41 | 1963 | 4500 | 0.243 |
| Rubidium | Rb | 37 | 85.47 | 1.532 | 39.49 | 688 | 0.364 |
| Ruthenium | Ru | 44 | 101.107 | 12.37 | 2250 | 4900 | 0.239 |
| Rutherfordium | Rf | 104 | 261.11 | — | — | — | — |

| Element | Symbol | Atomic Number Z | Molar Mass, g/mol | Density, g/cm³ at 20°C | Melting Point, °C | Boiling Point, °C | Specific Heat, J/(g·°C) at 25°C |
|---|---|---|---|---|---|---|---|
| Samarium | Sm | 62 | 150.35 | 7.52 | 1072 | 1630 | 0.197 |
| Scandium | Sc | 21 | 44.956 | 2.99 | 1539 | 2730 | 0.569 |
| Seaborgium | Sg | 106 | 263.118 | — | — | — | — |
| Selenium | Se | 34 | 78.96 | 4.79 | 221 | 685 | 0.318 |
| Silicon | Si | 14 | 28.086 | 2.33 | 1412 | 2680 | 0.712 |
| Silver | Ag | 47 | 107.870 | 10.49 | 960.8 | 2210 | 0.234 |
| Sodium | Na | 11 | 22.9898 | 0.9712 | 97.85 | 892 | 1.23 |
| Strontium | Sr | 38 | 87.62 | 2.54 | 768 | 1380 | 0.737 |
| Sulfur | S | 16 | 32.064 | 2.07 | 119.0 | 444.6 | 0.707 |
| Tantalum | Ta | 73 | 180.948 | 16.6 | 3014 | 5425 | 0.138 |
| Technetium | Tc | 43 | (99) | 11.46 | 2200 | — | 0.209 |
| Tellurium | Te | 52 | 127.60 | 6.24 | 449.5 | 990 | 0.201 |
| Terbium | Tb | 65 | 158.924 | 8.229 | 1357 | 2530 | 0.180 |
| Thallium | Tl | 81 | 204.37 | 11.85 | 304 | 1457 | 0.130 |
| Thorium | Th | 90 | (232) | 11.72 | 1755 | (3850) | 0.117 |
| Thulium | Tm | 69 | 168.934 | 9.32 | 1545 | 1720 | 0.159 |
| Tin | Sn | 50 | 118.69 | 7.2984 | 231.868 | 2270 | 0.226 |
| Titanium | Ti | 22 | 47.90 | 4.54 | 1670 | 3260 | 0.523 |
| Tungsten | W | 74 | 183.85 | 19.3 | 3380 | 5930 | 0.134 |
| Unnamed | Uuu | 111 | (272) | — | — | — | — |
| Unnamed | Uub | 112 | (285) | — | — | — | — |
| Unnamed | Uut | 113 | — | — | — | — | — |
| Unnamed | Unq | 114 | (289) | — | — | — | — |
| Unnamed | Uup | 115 | — | — | — | — | — |
| Unnamed | Uuh | 116 | — | — | — | — | — |
| Unnamed | Uus | 117 | — | — | — | — | — |
| Unnamed | Uuo | 118 | (293) | — | — | — | — |
| Uranium | U | 92 | (238) | 18.95 | 1132 | 3818 | 0.117 |
| Vanadium | V | 23 | 50.942 | 6.11 | 1902 | 3400 | 0.490 |
| Xenon | Xe | 54 | 131.30 | $5.495 \times 10^{-3}$ | -111.79 | -108 | 0.159 |
| Ytterbium | Yb | 70 | 173.04 | 6.965 | 824 | 1530 | 0.155 |
| Yttrium | Y | 39 | 88.905 | 4.469 | 1526 | 3030 | 0.297 |
| Zinc | Zn | 30 | 65.37 | 7.133 | 419.58 | 906 | 0.389 |
| Zirconium | Zr | 40 | 91.22 | 6.506 | 1852 | 3580 | 0.276 |

The values in parentheses in the column of molar masses are the mass numbers of the longest-lived isotopes of those elements that are radioactive. Melting points and boiling points in parentheses are uncertain.

The data for gases are valid only when these are in their usual molecular state, such as $H_2$, He, $O_2$, Ne, etc. The specific heats of the gases are the values at constant pressure.

*Source:* Adapted from J. Emsley, *The Elements,* 3rd ed., 1998, Clarendon Press, Oxford. See also www.webelements.com for the latest values and newest elements.

# Periodic Table G
## of the Elements

**Legend:**
- Metals
- Metalloids
- Nonmetals
- Noble gases 0

THE HORIZONTAL PERIODS

| Period | IA | IIA | IIIB | IVB | VB | VIB | VIIB | VIIIB | | | IB | IIB | IIIA | IVA | VA | VIA | VIIA | 0 |
|---|---|---|---|---|---|---|---|---|---|---|---|---|---|---|---|---|---|---|
| 1 | 1 H | | | | | | | | | | | | | | | | | 2 He |
| 2 | 3 Li | 4 Be | | | | | | | | | | | 5 B | 6 C | 7 N | 8 O | 9 F | 10 Ne |
| 3 | 11 Na | 12 Mg | | | | | | | | | | | 13 Al | 14 Si | 15 P | 16 S | 17 Cl | 18 Ar |
| 4 | 19 K | 20 Ca | 21 Sc | 22 Ti | 23 V | 24 Cr | 25 Mn | 26 Fe | 27 Co | 28 Ni | 29 Cu | 30 Zn | 31 Ga | 32 Ge | 33 As | 34 Se | 35 Br | 36 Kr |
| 5 | 37 Rb | 38 Sr | 39 Y | 40 Zr | 41 Nb | 42 Mo | 43 Tc | 44 Ru | 45 Rh | 46 Pd | 47 Ag | 48 Cd | 49 In | 50 Sn | 51 Sb | 52 Te | 53 I | 54 Xe |
| 6 | 55 Cs | 56 Ba | 57-71 * | 72 Hf | 73 Ta | 74 W | 75 Re | 76 Os | 77 Ir | 78 Pt | 79 Au | 80 Hg | 81 Tl | 82 Pb | 83 Bi | 84 Po | 85 At | 86 Rn |
| 7 | 87 Fr | 88 Ra | 89-103 † | 104 Rf | 105 Db | 106 Sg | 107 Bh | 108 Hs | 109 Mt | 110 Ds | 111 | 112 | 113 | 114 | 115 | 116 | 117 | 118 |

Alkali metals IA — Transition metals — Noble gases 0

**Inner transition metals**

Lanthanide series *

| 57 La | 58 Ce | 59 Pr | 60 Nd | 61 Pm | 62 Sm | 63 Eu | 64 Gd | 65 Tb | 66 Dy | 67 Ho | 68 Er | 69 Tm | 70 Yb | 71 Lu |
|---|---|---|---|---|---|---|---|---|---|---|---|---|---|---|

Actinide series †

| 89 Ac | 90 Th | 91 Pa | 92 U | 93 Np | 94 Pu | 95 Am | 96 Cm | 97 Bk | 98 Cf | 99 Es | 100 Fm | 101 Md | 102 No | 103 Lr |
|---|---|---|---|---|---|---|---|---|---|---|---|---|---|---|

Elements 111, 112, 114, and 116 have been discovered but, as of 2003, have not yet been named. Evidence for the discovery of elements 113 and 115 has been reported. See www.webelements.com for the latest information and newest elements.

# Answers

## Chapter 1

**P** **1.** (a) $10^9$ μm; (b) $10^{-4}$; (c) $9.1 \times 10^5$ μm  **3.** (a) 160 rods; (b) 40 chains  **5.** (a) $4.00 \times 10^4$ km; (b) $5.10 \times 10^8$ km²; (c) $1.08 \times 10^{12}$ km³  **7.** $1.9 \times 10^{22}$ cm³  **9.** $1.1 \times 10^3$ acre-feet  **11.** $1.21 \times 10^{12}$ μs  **13.** (a) 1.43; (b) 0.864  **15.** (a) 495 s; (b) 141 s; (c) 198 s; (d) $-245$ s  **17.** C, D, A, B, E; the important criterion is the consistency of the daily variation, not its magnitude  **19.** $5.2 \times 10^6$ m  **21.** (a) $1 \times 10^3$ kg; (b) 158 kg/s  **23.** $9.0 \times 10^{49}$ atoms  **25.** (a) $1.18 \times 10^{-29}$ m³; (b) 0.282 nm  **27.** 1750 kg  **29.** $1.9 \times 10^5$ kg  **31.** 1.43 kg/min  **33.** (a) 22 pecks; (b) 5.5 Imperial bushels; (c) 200 L  **35.** (a) 18.8 gallons; (b) 22.5 gallons  **37.** (a) 11.3 m²/L; (b) $1.13 \times 10^4$ m⁻¹; (c) $2.17 \times 10^{-3}$ gal/ft²; (d) number of gallons to cover a square foot  **39.** 0.3 cord  **41.** (a) 293 U.S. bushels; (b) $3.81 \times 10^3$ U.S. bushels  **43.** $8 \times 10^2$ km  **45.** 0.12 AU/min  **47.** 3.8 mg/s  **49.** 10.7 habaneros  **51.** (a) yes; (b) 8.6 universe seconds  **53.** (a) 3.88; (b) 7.65; (c) 156 ken³; (d) $1.19 \times 10^3$ m³  **55.** 1.2 m  **57.** (a) $4.9 \times 10^{-6}$ pc; (b) $1.6 \times 10^{-5}$ ly  **59.** (a) 3.9 m, 4.8 m; (b) $3.9 \times 10^3$ mm, $4.8 \times 10^3$ mm; (c) 2.2 m³, 4.2 m³

## Chapter 2

**CP** **1.** b and c  **2.** (check the derivative $dx/dt$) (a) 1 and 4; (b) 2 and 3  **3.** (a) plus; (b) minus; (c) minus; (d) plus  **4.** 1 and 4 ($a = d^2x/dt^2$ must be constant)  **5.** (a) plus (upward displacement on $y$ axis); (b) minus (downward displacement on $y$ axis); (c) $a = -g = -9.8$ m/s²  **Q** **1.** (a) all tie; (b) 4, tie of 1 and 2, then 3  **3.** (a) negative; (b) positive; (c) yes; (d) positive; (e) constant  **5.** (a) positive direction; (b) negative direction; (c) 3 and 5; (d) 2 and 6 tie, then 3 and 5 tie, then 1 and 4 tie (zero)  **7.** (a) 3, 2, 1; (b) 1, 2, 3; (c) all tie; (d) 1, 2, 3  **9.** (a) $D$; (b) $E$  **P** **1.** (a) $+40$ km/h; (b) 40 km/h  **3.** 13 m  **5.** (a) 0; (b) $-2$ m; (c) 0; (d) 12 m; (e) $+12$ m; (f) $+7$ m/s  **7.** 1.4 m  **9.** 128 km/h  **11.** 60 km  **13.** (a) 73 km/h; (b) 68 km/h; (c) 70 km/h; (d) 0  **15.** (a) $-6$ m/s; (b) $-x$ direction; (c) 6 m/s; (d) decreasing; (e) 2 s; (f) no  **17.** (a) 28.5 cm/s; (b) 18.0 cm/s; (c) 40.5 cm/s; (d) 28.1 cm/s; (e) 30.3 cm/s  **19.** $-20$ m/s²  **21.** (a) m/s²; (b) m/s³; (c) 1.0 s; (d) 82 m; (e) $-80$ m; (f) 0; (g) $-12$ m/s; (h) $-36$ m/s; (i) $-72$ m/s; (j) $-6$ m/s²; (k) $-18$ m/s²; (l) $-30$ m/s²; (m) $-42$ m/s²  **23.** (a) $+1.6$ m/s; (b) $+18$ m/s  **25.** (a) $3.1 \times 10^6$ s; (b) $4.6 \times 10^{13}$ m  **27.** $1.62 \times 10^{15}$ m/s²  **29.** (a) 30 s; (b) 300 m  **31.** (a) 10.6 m; (b) 41.5 s  **33.** (a) 3.56 m/s²; (b) 8.43 m/s  **35.** (a) 4.0 m/s²; (b) $+x$  **37.** (a) $-2.5$ m/s²; (b) 1; (d) 0; (e) 2  **39.** 40 m  **41.** 0.90 m/s²  **43.** (a) 15.0 m; (b) 94 km/h  **45.** (a) 29.4 m; (b) 2.45 s  **47.** (a) 31 m/s; (b) 6.4 s  **49.** (a) 5.4 s; (b) 41 m/s  **51.** 4.0 m/s  **53.** (a) 20 m; (b) 59 m  **55.** (a) 857 m/s²; (b) up  **57.** (a) $1.26 \times 10^3$ m/s²; (b) up  **59.** (a) 89 cm; (b) 22 cm  **61.** 2.34 m  **63.** 20.4 m  **65.** (a) 2.25 m/s; (b) 3.90 m/s  **67.** 100 m  **69.** 0.56 m/s  **71.** (a) 82 m; (b) 19 m/s  **73.** (a) 2.00 s; (b) 12 cm; (c) $-9.00$ cm/s²; (d) right; (e) left; (f) 3.46 s  **75.** (a) 48.5 m/s; (b) 4.95 s; (c) 34.3 m/s; (d) 3.50 s  **77.** 414 ms  **79.** 90 m  **81.** (a) 3.0 s; (b) 9.0 m  **83.** 2.78 m/s²  **85.** (a) 0.74 s; (b) 6.2 m/s²  **87.** 17 m/s  **89.** $+47$ m/s  **91.** (a) 3.1 m/s²; (b) 45 m; (c) 13 s

**93.** (a) 1.23 cm; (b) 4 times; (c) 9 times; (d) 16 times; (e) 25 times  **95.** 25 km/h  **97.** 1.2 h  **99.** $4H$  **101.** (a) 3.2 s; (b) 1.3 s  **103.** (a) 10.2 s; (b) 10.0 m  **105.** (a) 8.85 m/s; (b) 1.00 m  **107.** (a) 2.0 m/s²; (b) 12 m/s; (c) 45 m  **109.** 3.75 ms  **111.** (a) 5.44 s; (b) 53.3 m/s; (c) 5.80 m  **113.** (a) 9.08 m/s²; (b) 0.926$g$; (c) 6.12 s; (d) 15.3$T_r$; (e) braking; (f) 5.56 m

## Chapter 3

**CP** **1.** (a) 7 m ($\vec{a}$ and $\vec{b}$ are in same direction); (b) 1 m ($\vec{a}$ and $\vec{b}$ are in opposite directions)  **2.** $c, d, f$ (components must be head-to-tail; $\vec{a}$ must extend from tail of one component to head of the other)  **3.** (a) $+, +$; (b) $+, -$; (c) $+, +$ (draw vector from tail of $\vec{d}_1$ to head of $\vec{d}_2$)  **4.** (a) 90°; (b) 0° (vectors are parallel—same direction); (c) 180° (vectors are antiparallel—opposite directions)  **5.** (a) 0° or 180°; (b) 90°  **Q** **1.** Either the sequence $\vec{d}_2, \vec{d}_1$ or the sequence $\vec{d}_2, \vec{d}_2, \vec{d}_3$  **3.** yes, when the vectors are in same direction  **5.** (a) yes; (b) yes; (c) no  **7.** all but ($e$)  **9.** (a) $+x$ for (1), $+z$ for (2), $+z$ for (3); (b) $-x$ for (1), $-z$ for (2), $-z$ for (3)  **P** **1.** (a) 47.2 m; (b) 122°  **3.** (a) $-2.5$ m; (b) $-6.9$ m  **5.** (a) 156 km; (b) 39.8° west of due north  **7.** (a) 6.42 m; (b) no; (c) yes; (d) yes; (e) a possible answer: $(4.30 \text{ m})\hat{i} + (3.70 \text{ m})\hat{j} + (3.00 \text{ m})\hat{k}$; (f) 7.96 m  **9.** (a) $(-9.0 \text{ m})\hat{i} + (10 \text{ m})\hat{j}$; (b) 13 m; (c) 132°  **11.** 4.74 km  **13.** (a) $(3.0 \text{ m})\hat{i} - (2.0 \text{ m})\hat{j} + (5.0 \text{ m})\hat{k}$; (b) $(5.0 \text{ m})\hat{i} - (4.0 \text{ m})\hat{j} - (3.0 \text{ m})\hat{k}$; (c) $(-5.0 \text{ m})\hat{i} + (4.0 \text{ m})\hat{j} + (3.0 \text{ m})\hat{k}$  **15.** (a) $-70.0$ cm; (b) 80.0 cm; (c) 141 cm; (d) $-172°$  **17.** (a) 1.59 m; (b) 12.1 m; (c) 12.2 m; (d) 82.5°  **19.** (a) 38 m; (b) $-37.5°$; (c) 130 m; (d) 1.2°; (e) 62 m; (f) 130°  **21.** 5.39 m at 21.8° left of forward  **23.** 2.6 km  **25.** 3.2  **27.** (a) 7.5 cm; (b) 90°; (c) 8.6 cm; (d) 48°  **29.** (a) $8\hat{i} + 16\hat{j}$; (b) $2\hat{i} + 4\hat{j}$  **31.** (a) $a\hat{i} + a\hat{j} + a\hat{k}$; (b) $-a\hat{i} + a\hat{j} + a\hat{k}$; (c) $a\hat{i} - a\hat{j} + a\hat{k}$; (d) $-a\hat{i} - a\hat{j} + a\hat{k}$; (e) 54.7°; (f) $3^{0.5}a$  **33.** (a) $-18.8$ units; (b) 26.9 units, $+z$ direction  **35.** (a) $-21$; (b) $-9$; (c) $5\hat{i} - 11\hat{j} - 9\hat{k}$  **37.** (a) 12; (b) $+z$; (c) 12; (d) $-z$; (e) 12; (f) $+z$  **39.** 22°  **41.** 70.5°  **43.** (a) 3.00 m; (b) 0; (c) 3.46 m; (d) 2.00 m; (e) $-5.00$ m; (f) 8.66 m; (g) $-6.67$; (h) 4.33  **45.** (a) 27.8 m; (b) 13.4 m  **47.** (a) 30; (b) 52  **49.** (a) $-2.83$ m; (b) $-2.83$ m; (c) 5.00 m; (d) 0; (e) 3.00 m; (f) 5.20 m; (g) 5.17 m; (h) 2.37 m; (i) 5.69 m; (j) 25° north of due east; (k) 5.69 m; (l) 25° south of due west  **51.** (a) 103 km; (b) 60.9° north of due west  **53.** (a) 140°; (b) 90.0°; (c) 99.1°  **55.** (a) $-83.4$; (b) $(1.14 \times 10^3)\hat{k}$; (c) $1.14 \times 10^3$, $\theta$ not defined, $\phi = 0°$; (d) 90.0°; (e) $-5.14\hat{i} + 6.13\hat{j} + 3.00\hat{k}$; (f) 8.54, $\theta = 130°$, $\phi = 69.4°$  **57.** (a) 3.0 m²; (b) 52 m³; (c) $(11 \text{ m}^2)\hat{i} + (9.0 \text{ m}^2)\hat{j} + (3.0 \text{ m}^2)\hat{k}$  **59.** (a) $+y$; (b) $-y$; (c) 0; (d) 0; (e) $+z$; (f) $-z$; (g) $ab$; (h) $ab$; (i) $ab/d$; (j) $+1$  **61.** (a) 0; (b) 0; (c) $-1$; (d) west; (e) up; (f) west  **63.** Walpole (where the state prison is located)  **65.** (a) $(9.19 \text{ m})\hat{i}' + (7.71 \text{ m})\hat{j}'$; (b) $(14.0 \text{ m})\hat{i}' + (3.41 \text{ m})\hat{j}'$  **67.** (a) $11\hat{i} + 5.0\hat{j} - 7.0\hat{k}$; (b) 120°; (c) $-4.9$; (d) 7.3  **69.** (a) $(-40\hat{i} - 20\hat{j} + 25\hat{k})$ m; (b) 45 m  **71.** 4.1

## Chapter 4

**CP** **1.** (draw $\vec{v}$ tangent to path, tail on path) (a) first; (b) third  **2.** (take second derivative with respect to time) (1) and (3) $a_x$

and $a_y$ are both constant and thus $\vec{a}$ is constant; (2) and (4) $a_y$ is constant but $a_x$ is not, thus $\vec{a}$ is not   **3.** no   **4.** (a) $v_x$ constant; (b) $v_y$ initially positive, decreases to zero, and then becomes progressively more negative; (c) $a_x = 0$ throughout; (d) $a_y = -g$ throughout   **5.** (a) $-(4\text{ m/s})\hat{\imath}$; (b) $-(8\text{ m/s}^2)\hat{\jmath}$
**Q** **1.** (a) $(7\text{ m})\hat{\imath} + (1\text{ m})\hat{\jmath} + (-2\text{ m})\hat{k}$; (b) $(5\text{ m})\hat{\imath} + (-3\text{ m})\hat{\jmath} + (1\text{ m})\hat{k}$; (c) $(-2\text{ m})\hat{\imath}$   **3.** (a) all tie; (b) 1 and 2 tie (the rocket is shot upward), then 3 and 4 tie (it is shot into the ground!)   **5.** decreases   **7.** (a) all tie; (b) all tie; (c) 3, 2, 1; (d) 3, 2, 1   **9.** (a) 0; (b) 350 km/h; (c) 350 km/h; (d) same (nothing changed about the vertical motion)   **11.** (a) 90° and 270°; (b) 0° and 180°; (c) 90° and 270°   **13.** 2, then 1 and 4 tie, then 3
**P** **1.** $(-2.0\text{ m})\hat{\imath} + (6.0\text{ m})\hat{\jmath} - (10\text{ m})\hat{k}$   **3.** (a) 6.2 m   **5.** $(-0.70\text{ m/s})\hat{\imath} + (1.4\text{ m/s})\hat{\jmath} - (0.40\text{ m/s})\hat{k}$   **7.** (a) 7.59 km/h; (b) 22.5° east of due north   **9.** (a) 0.83 cm/s; (b) 0°; (c) 0.11 m/s; (d) $-63°$   **11.** (a) $(8\text{ m/s}^2)t\hat{\jmath} + (1\text{ m/s})\hat{k}$; (b) $(8\text{ m/s}^2)\hat{\jmath}$   **13.** (a) $(6.00\text{ m})\hat{\imath} - (106\text{ m})\hat{\jmath}$; (b) $(19.0\text{ m/s})\hat{\imath} - (224\text{ m/s})\hat{\jmath}$; (c) $(24.0\text{ m/s}^2)\hat{\imath} - (336\text{ m/s}^2)\hat{\jmath}$; (d) $-85.2°$   **15.** $(32\text{ m/s})\hat{\imath}$   **17.** (a) $(-1.50\text{ m/s})\hat{\jmath}$; (b) $(4.50\text{ m})\hat{\imath} - (2.25\text{ m})\hat{\jmath}$   **19.** (a) $(72.0\text{ m})\hat{\imath} + (90.7\text{ m})\hat{\jmath}$; (b) 49.5°   **21.** (a) 3.03 s; (b) 758 m; (c) 29.7 m/s   **23.** 43.1 m/s (155 km/h)   **25.** (a) 18 cm; (b) 1.9 m   **27.** (a) 10.0 s; (b) 897 m   **29.** (a) 1.60 m; (b) 6.86 m; (c) 2.86 m   **31.** (a) 202 m/s; (b) 806 m; (c) 161 m/s; (d) $-171$ m/s   **33.** 3.35 m   **35.** 78.5°   **37.** (a) 11 m; (b) 23 m; (c) 17 m/s; (d) 63°   **39.** 4.84 cm   **41.** (a) 32.3 m; (b) 21.9 m/s; (c) 40.4°; (d) below   **43.** (a) ramp; (b) 5.82 m; (c) 31.0°   **45.** 64.8°   **47.** (a) yes; (b) 2.56 m   **49.** (a) 2.3°; (b) 1.4 m; (c) 18°   **51.** (a) 31°; (b) 63°   **53.** the third   **55.** (a) 75.0 m; (b) 31.9 m/s; (c) 66.9°; (d) 25.5 m   **57.** (a) 12 s; (b) 4.1 m/s²; (c) down; (d) 4.1 m/s²; (e) up   **59.** (a) $1.3 \times 10^5$ m/s; (b) $7.9 \times 10^5$ m/s²; (c) increase   **61.** (a) 7.32 m; (b) west; (c) north   **63.** $(3.00\text{ m/s}^2)\hat{\imath} + (6.00\text{ m/s}^2)\hat{\jmath}$   **65.** 2.92 m   **67.** 160 m/s²   **69.** (a) 13 m/s²; (b) eastward; (c) 13 m/s²; (d) eastward   **71.** 1.67   **73.** (a) 38 knots; (b) 1.5° east of due north; (c) 4.2 h; (d) 1.5° west of due south   **75.** 60°   **77.** 32 m/s   **79.** (a) $(80\text{ km/h})\hat{\imath} - (60\text{ km/h})\hat{\jmath}$; (b) 0°; (c) answers do not change   **81.** (a) $(-32\text{ km/h})\hat{\imath} - (46\text{ km/h})\hat{\jmath}$; (b) $[(2.5\text{ km}) - (32\text{ km/h})t]\hat{\imath} + [(4.0\text{ km}) - (46\text{ km/h})t]\hat{\jmath}$; (c) 0.084 h; (d) $2 \times 10^2$ m   **83.** (a) 2.7 km; (b) 76° clockwise   **85.** 2.64 m   **87.** (a) 2.5 m; (b) 0.82 m; (c) 9.8 m/s²; (d) 9.8 m/s²   **89.** (a) $-30°$; (b) 69 min; (c) 80 min; (d) 80 min; (e) 0°; (f) 60 min   **91.** (a) 62 ms; (b) $4.8 \times 10^2$ m/s   **93.** (a) $6.7 \times 10^6$ m/s; (b) $1.4 \times 10^{-7}$ s   **95.** (a) 4.2 m, 45°; (b) 5.5 m, 68°; (c) 6.0 m, 90°; (d) 4.2 m, 135°; (e) 0.85 m/s, 135°; (f) 0.94 m/s, 90°; (g) 0.94 m/s, 180°; (h) 0.30 m/s², 180°; (i) 0.30 m/s², 270°   **97.** (a) 6.79 km/h; (b) 6.96°   **99.** (a) 16 m/s; (b) 23°; (c) above; (d) 27 m/s; (e) 57°; (f) below   **101.** (a) 24 m/s; (b) 65°   **103.** (a) 1.5; (b) (36 m, 54 m)   **105.** (a) 0.034 m/s²; (b) 84 min   **107.** (a) 44 m; (b) 13 m; (c) 8.9 m   **109.** (a) $2.6 \times 10^2$ m/s; (b) 45 s; (c) increase   **111.** (a) 45 m; (b) 22 m/s   **113.** (a) 2.00 ns; (b) 2.00 mm; (c) $1.00 \times 10^7$ m/s; (d) $2.00 \times 10^6$ m/s   **115.** (a) $4.6 \times 10^{12}$ m; (b) $2.4 \times 10^5$ s   **117.** 93° from the car's direction of motion   **119.** (a) 8.43 m; (b) $-129°$   **121.** (a) 63 km; (b) 18° south of due east; (c) 0.70 km/h; (d) 18° south of due east; (e) 1.6 km/h; (f) 1.2 km/h; (g) 33° north of due east   **123.** $3 \times 10^1$ m   **125.** (a) 14 m/s; (b) 14 m/s; (c) $-10$ m; (d) $-4.9$ m; (e) $+10$ m; (f) $-4.9$ m   **127.** 67 km/h   **129.** (a) from 75° east of due south; (b) 30° east of due north. For a second set of solutions, substitute west for east in both answers.   **131.** (a) 11 m; (b) 45 m/s

## Chapter 5

**CP** **1.** $c$, $d$, and $e$ ($\vec{F}_1$ and $\vec{F}_2$ must be head-to-tail, $\vec{F}_{\text{net}}$ must be from tail of one of them to head of the other)   **2.** (a) and (b) 2 N, leftward (acceleration is zero in each situation)   **3.** (a) equal; (b) greater (acceleration is upward, thus net force on body must be upward)   **4.** (a) equal; (b) greater; (c) less   **5.** (a) increase; (b) yes; (c) same; (d) yes   **Q** **1.** increase   **3.** (a) 2 and 4; (b) 2 and 4   **5.** (a) 2, 3, 4; (b) 1, 3, 4; (c) 1, $+y$; 2, $+x$; 3, fourth quadrant; 4, third quadrant   **7.** (a) 20 kg; (b) 18 kg; (c) 10 kg; (d) all tie; (e) 3, 2, 1   **9.** (a) increases from initial value $mg$; (b) decreases from $mg$ to zero (after which the block moves up away from the floor)   **11.** (a) $M$; (b) $M$; (c) $M$; (d) $2M$; (e) $3M$   **P** **1.** (a) 1.88 N; (b) 0.684 N; (c) $(1.88\text{ N})\hat{\imath} + (0.684\text{ N})\hat{\jmath}$   **3.** 2.9 m/s²   **5.** (a) $(-32.0\text{ N})\hat{\imath} - (20.8\text{ N})\hat{\jmath}$; (b) 38.2 N; (c) $-147°$   **7.** (a) $(0.86\text{ m/s}^2)\hat{\imath} - (0.16\text{ m/s}^2)\hat{\jmath}$; (b) 0.88 m/s²; (c) $-11°$   **9.** 9.0 m/s   **11.** (a) 8.37 N; (b) $-133°$; (c) $-125°$   **13.** (a) 108 N; (b) 108 N; (c) 108 N   **15.** (a) 4.0 kg; (b) 1.0 kg; (c) 4.0 kg; (d) 1.0 kg   **17.** (a) $-9.80\hat{\jmath}$ m/s²; (b) $2.35\hat{\jmath}$ m/s²; (c) 1.37 s; (d) $(-5.56 \times 10^{-3}\text{ N})\hat{\jmath}$; (e) $(1.333 \times 10^{-3}\text{ N})\hat{\jmath}$   **19.** (a) 42 N; (b) 72 N; (c) 4.9 m/s²   **21.** (a) 11.7 N; (b) $-59.0°$   **23.** (a) 0.022 m/s²; (b) $8.3 \times 10^4$ km; (c) $1.9 \times 10^3$ m/s   **25.** $1.2 \times 10^5$ N   **27.** (a) 494 N; (b) up; (c) 494 N; (d) down   **29.** 1.5 mm   **31.** (a) 46.7°; (b) 28.0°   **33.** (a) 0.62 m/s²; (b) 0.13 m/s²; (c) 2.6 m   **35.** (a) 1.18 m; (b) 0.674 s; (c) 3.50 m/s   **37.** (a) $2.2 \times 10^{-3}$ N; (b) $3.7 \times 10^{-3}$ N   **39.** $1.8 \times 10^4$ N   **41.** (a) 31.3 kN; (b) 24.3 kN   **43.** (a) 1.4 m/s²; (b) 4.1 m/s   **45.** (a) 1.23 N; (b) 2.46 N; (c) 3.69 N; (d) 4.92 N; (e) 6.15 N; (f) 0.250 N   **47.** (a) 2.18 m/s²; (b) 116 N; (c) 21.0 m/s²   **49.** $6.4 \times 10^3$ N   **51.** (a) 0.970 m/s²; (b) 11.6 N; (c) 34.9 N   **53.** (a) 1.1 N   **55.** (a) 3.6 m/s²; (b) 17 N   **57.** (a) 4.9 m/s²; (b) 2.0 m/s²; (c) up; (d) 120 N   **59.** (a) 0.735 m/s²; (b) down; (c) 20.8 N   **61.** $2Ma/(a + g)$   **63.** (a) 0.653 m/s³; (b) 0.896 m/s³; (c) 6.50 s   **65.** 81.7 N   **67.** (a) 8.0 m/s; (b) $+x$   **69.** (a) 13 597 kg; (b) 4917 L; (c) 6172 kg; (d) 20 075 L;  (e) 45%   **71.** (a) 0; (b) 0.83 m/s²; (c) 0   **73.** (a) 0.74 m/s²; (b) 7.3 m/s²   **75.** (a) rope breaks; (b) 1.6 m/s²   **77.** 2.4 N   **79.** (a) 4.6 m/s²; (b) 2.6 m/s²   **81.** (a) 65 N; (b) 49 N   **83.** (a) 11 N; (b) 2.2 kg; (c) 0; (d) 2.2 kg   **85.** (a) $4.6 \times 10^3$ N; (b) $5.8 \times 10^3$ N   **87.** (a) 4 kg; (b) 6.5 m/s²; (c) 13 N   **89.** 195 N   **91.** (a) 44 N; (b) 78 N; (c) 54 N; (d) 152 N   **93.** 16 N   **95.** (a) $1.8 \times 10^2$ N; (b) $6.4 \times 10^2$ N   **97.** (a) $(5.0\text{ m/s})\hat{\imath} + (4.3\text{ m/s})\hat{\jmath}$; (b) $(15\text{ m})\hat{\imath} + (6.4\text{ m})\hat{\jmath}$   **99.** 16 N   **101.** (a) 2.6 N; (b) 17°   **103.** (a) 4.1 m/s²; (b) 836 N

## Chapter 6

**CP** **1.** (a) zero (because there is no attempt at sliding); (b) 5 N; (c) no; (d) yes; (e) 8 N   **2.** ($\vec{a}$ is directed toward center of circular path) (a) $\vec{a}$ downward, $\vec{F}_N$ upward; (b) $\vec{a}$ and $\vec{F}_N$ upward   **Q** **1.** (a) same; (b) increases; (c) increases; (d) no   **3.** (a) decrease; (b) decrease; (c) increase; (d) increase; (e) increase   **5.** (a) upward; (b) horizontal, toward you; (c) no change; (d) increases; (e) increases   **7.** At first, $\vec{f}_s$ is directed up the ramp and its magnitude increases from $mg \sin\theta$ until it reaches $f_{s,\max}$. Thereafter the force is kinetic friction directed up the ramp, with magnitude $f_k$ (a constant value smaller than $f_{s,\max}$).   **9.** (a) all tie; (b) all tie; (c) 2, 3, 1   **11.** 4, 3, then 1, 2, and 5 tie   **P** **1.** (a) $2.0 \times 10^2$ N; (b) $1.2 \times 10^2$ N   **3.** (a) $1.9 \times 10^2$ N; (b) 0.56 m/s²   **5.** 36 m   **7.** (a) 11 N; (b) 0.14 m/s²   **9.** (a) 6.0 N; (b) 3.6 N; (c) 3.1 N   **11.** (a) $1.3 \times 10^2$ N; (b) no; (c) $1.1 \times 10^2$ N; (d) 46 N; (e) 17 N   **13.** (a) $3.0 \times 10^2$ N; (b) 1.3 m/s²   **15.** 2°   **17.** (a) no;

---

(b) $(-12\text{ N})\hat{i} + (5.0\text{ N})\hat{j}$  **19.** (a) 19°; (b) 3.3 kN  **21.** (a) (17 N)$\hat{i}$; (b) (20 N)$\hat{i}$; (c) (15 N)$\hat{i}$  **23.** $1.0 \times 10^2$ N  **25.** 0.37  **27.** (a) 3.5 m/s²; (b) 0.21 N  **29.** (a) 0; (b) $(-3.9\text{ m/s}^2)\hat{i}$; (c) $(-1.0\text{ m/s}^2)\hat{i}$  **31.** (a) 66 N; (b) 2.3 m/s²  **33.** $4.9 \times 10^2$ N  **35.** 9.9 s  **37.** 2.3  **39.** (a) $3.2 \times 10^2$ km/h; (b) $6.5 \times 10^2$ km/h; (c) no  **41.** 21 m  **43.** 0.60  **45.** (a) 10 s; (b) $4.9 \times 10^2$ N; (c) $1.1 \times 10^3$ N  **47.** $1.37 \times 10^3$ N  **49.** (a) light; (b) 778 N; (c) 223 N; (d) 1.11 kN  **51.** 12°  **53.** 2.2 km  **55.** 1.81 m/s  **57.** $2.6 \times 10^3$ N  **59.** (a) 8.74 N; (b) 37.9 N; (c) 6.45 m/s; (d) radially inward  **61.** (a) 69 km/h; (b) 139 km/h; (c) yes  **63.** (a) 7.5 m/s²; (b) down; (c) 9.5 m/s²; (d) down  **65.** (a) 27 N; (b) 3.0 m/s²  **67.** (a) 35.3 N; (b) 39.7 N; (c) 320 N  **69.** $g(\sin\theta - 2^{0.5}\mu_k\cos\theta)$  **71.** (a) $3.0 \times 10^5$ N; (b) 1.2°  **73.** 147 m/s  **75.** (a) 56 N; (b) 59 N; (c) $1.1 \times 10^3$ N  **77.** (a) 275 N; (b) 877 N  **79.** (b) 240 N; (c) 0.60  **81.** (a) 13 N; (b) 1.6 m/s²  **83.** 0.76  **85.** (a) $3.21 \times 10^3$ N; (b) yes  **87.** 3.4 m/s²  **89.** (a) 84.2 N; (b) 52.8 N; (c) 1.87 m/s²  **91.** (a) 222 N; (b) 334 N; (c) 311 N; (d) 311 N; (e) c, d  **93.** (a) 6.80 s; (b) 6.76 s  **95.** 3.4%  **97.** (a) $\mu_k mg/(\sin\theta - \mu_k\cos\theta)$; (b) $\theta_0 = \tan^{-1}\mu_s$  **99.** (a) $v_0^2/(4g\sin\theta)$; (b) no  **101.** (a) 30 cm/s; (b) 180 cm/s²; (c) inward; (d) $3.6 \times 10^{-3}$ N; (e) inward; (f) 0.37  **103.** (a) 0.34; (b) 0.24  **105.** 0.18  **107.** 0.56  **109.** (a) 2.1 m/s²; (b) down the plane; (c) 3.9 m; (d) at rest

## Chapter 7

**CP** **1.** (a) decrease; (b) same; (c) negative, zero  **2.** (a) positive; (b) negative; (c) zero  **3.** zero  **Q** **1.** (a) positive; (b) negative; (c) negative  **3.** all tie  **5.** all tie  **7.** $b$ (positive work), $a$ (zero work), $c$ (negative work), $d$ (more negative work)  **9.** (a) $A$; (b) $B$  **P** **1.** (a) $5 \times 10^{14}$ J; (b) 0.1 megaton TNT; (c) 8 bombs  **3.** (a) $2.9 \times 10^7$ m/s; (b) $2.1 \times 10^{-13}$ J  **5.** (a) 2.4 m/s; (b) 4.8 m/s  **7.** 20 J  **9.** 0.96 J  **11.** (a) $1.7 \times 10^2$ N; (b) $3.4 \times 10^2$ m; (c) $-5.8 \times 10^4$ J; (d) $3.4 \times 10^2$ N; (e) $1.7 \times 10^2$ m; (f) $-5.8 \times 10^4$ J  **13.** (a) 1.50 J; (b) increases  **15.** (a) 62.3°; (b) 118°  **17.** (a) 12 kJ; (b) $-11$ kJ; (c) 1.1 kJ; (d) 5.4 m/s  **19.** (a) $-3Mgd/4$; (b) $Mgd$; (c) $Mgd/4$; (d) $(gd/2)^{0.5}$  **21.** 4.41 J  **23.** 25 J  **25.** (a) 25.9 kJ; (b) 2.45 N  **27.** (a) 7.2 J; (b) 7.2 J; (c) 0; (d) $-25$ J  **29.** (a) 6.6 m/s; (b) 4.7 m  **31.** (a) 0.90 J; (b) 2.1 J; (c) 0  **33.** (a) 0.12 m; (b) 0.36 J; (c) $-0.36$ J; (d) 0.060 m; (e) 0.090 J  **35.** (a) 0; (b) 0  **37.** $5.3 \times 10^2$ J  **39.** (a) 42 J; (b) 30 J; (c) 12 J; (d) 6.5 m/s, $+x$ axis; (e) 5.5 m/s, $+x$ axis; (f) 3.5 m/s, $+x$ axis  **41.** 4.00 N/m  **43.** $4.9 \times 10^2$ W  **45.** (a) 0.83 J; (b) 2.5 J; (c) 4.2 J; (d) 5.0 W  **47.** $7.4 \times 10^2$ W  **49.** (a) $1.0 \times 10^2$ J; (b) 8.4 W  **51.** (a) 32.0 J; (b) 8.00 W; (c) 78.2°  **53.** (a) $1 \times 10^5$ megatons TNT; (b) $1 \times 10^7$ bombs  **55.** $-6$ J  **57.** (a) 98 N; (b) 4.0 cm; (c) 3.9 J; (d) $-3.9$ J  **59.** $-37$ J  **61.** 165 kW  **63.** (a) $1.8 \times 10^5$ ft·lb; (b) 0.55 hp  **65.** (a) 797 N; (b) 0; (c) $-1.55$ kJ; (d) 0; (e) 1.55 kJ; (f) $F$ varies during displacement  **67.** (a) 1.20 J; (b) 1.10 m/s  **69.** (a) 314 J; (b) $-155$ J; (c) 0; (d) 158 J  **71.** (a) 23 mm; (b) 45 N  **73.** 235 kW  **75.** (a) 13 J; (b) 13 J  **77.** (a) 0.6 J; (b) 0; (c) $-0.6$ J  **79.** (a) 6 J; (b) 6.0 J

## Chapter 8

**CP** **1.** no (consider round trip on the small loop)  **2.** 3, 1, 2 (see Eq. 8-6)  **3.** (a) all tie; (b) all tie  **4.** (a) $CD, AB, BC$ (0) (check slope magnitudes); (b) positive direction of $x$  **5.** all tie  **Q** **1.** (a) 12 J; (b) $-2$ J  **3.** (a) 3, 2, 1; (b) 1, 2, 3  **5.** 2, 1, 3  **7.** $+30$ J  **9.** (a) increasing; (b) decreasing; (c) decreasing; (d) constant in $AB$ and $BC$, decreasing in $CD$

**P** **1.** (a) 167 J; (b) $-167$ J; (c) 196 J; (d) 29 J; (e) 167 J; (f) $-167$ J; (g) 296 J; (h) 129 J  **3.** (a) 4.31 mJ; (b) $-4.31$ mJ; (c) 4.31 mJ; (d) $-4.31$ mJ; (e) all increase  **5.** 89 N/cm  **7.** (a) 13.1 J; (b) $-13.1$ J; (c) 13.1 J; (d) all increase  **9.** (a) $2.6 \times 10^2$ m; (b) same; (c) decrease  **11.** (a) 2.08 m/s; (b) 2.08 m/s; (c) increase  **13.** (a) 17.0 m/s; (b) 26.5 m/s; (c) 33.4 m/s; (d) 56.7 m; (e) all the same  **15.** (a) 0.98 J; (b) $-0.98$ J; (c) 3.1 N/cm  **17.** (a) 8.35 m/s; (b) 4.33 m/s; (c) 7.45 m/s; (d) both decrease  **19.** (a) 2.5 N; (b) 0.31 N; (c) 30 cm  **21.** (a) 4.85 m/s; (b) 2.42 m/s  **23.** $-3.2 \times 10^2$ J  **25.** (a) no; (b) $9.3 \times 10^2$ N  **27.** (a) 784 N/m; (b) 62.7 J; (c) 62.7 J; (d) 80.0 cm  **29.** (a) 39.2 J; (b) 39.2 J; (c) 4.00 m  **31.** (a) 35 cm; (b) 1.7 m/s  **33.** (a) 2.40 m/s; (b) 4.19 m/s  **35.** $-18$ mJ  **37.** (a) 39.6 cm; (b) 3.64 cm  **39.** (a) 2.1 m/s; (b) 10 N; (c) $+x$ direction; (d) 5.7 m; (e) 30 N; (f) $-x$ direction  **41.** (a) $-3.7$ J; (c) 1.3 m; (d) 9.1 m; (e) 2.2 J; (f) 4.0 m; (g) $(4 - x)e^{-x/4}$; (h) 4.0 m  **43.** (a) 5.6 J; (b) 3.5 J  **45.** (a) 30.1 J; (b) 30.1 J; (c) 0.225  **47.** (a) $-2.9$ kJ; (b) $3.9 \times 10^2$ J; (c) $2.1 \times 10^2$ N  **49.** 0.53 J  **51.** (a) 1.5 MJ; (b) 0.51 MJ; (c) 1.0 MJ; (d) 63 m/s  **53.** 1.2 m  **55.** (a) 67 J; (b) 67 J; (c) 46 cm  **57.** (a) $1.5 \times 10^{-2}$ N; (b) $(3.8 \times 10^2)g$  **59.** (a) $-0.90$ J; (b) 0.46 J; (c) 1.0 m/s  **61.** (a) 19.4 m; (b) 19.0 m/s  **63.** 20 cm  **65.** (a) 7.4 m/s; (b) 90 cm; (c) 2.8 m; (d) 15 m  **67.** (a) 10 m; (b) 49 N; (c) 4.1 m; (d) $1.2 \times 10^2$ N  **69.** 4.33 m/s  **71.** (a) 5.5 m/s; (b) 5.4 m; (c) same  **73.** (a) 109 J; (b) 60.3 J; (c) 68.2 J; (d) 41.0 J  **75.** 3.7 J  **77.** 15 J  **79.** (a) 2.7 J; (b) 1.8 J; (c) 0.39 m  **81.** 80 mJ  **83.** (a) 7.0 J; (b) 22 J  **85.** (a) $7.4 \times 10^2$ J; (b) $2.4 \times 10^2$ J  **87.** 25 J  **89.** 24 W  **91.** $-12$ J  **93.** (a) 8.8 m/s; (b) 2.6 kJ; (c) 1.6 kW  **95.** (a) 300 J; (b) 93.8 J; (c) 6.38 m  **97.** 738 m  **99.** (a) $-0.80$ J; (b) $-0.80$ J; (c) $+1.1$ J  **101.** (a) $2.35 \times 10^3$ J; (b) 352 J  **103.** (a) $-3.8$ kJ; (b) 31 kN  **105.** (a) $2.1 \times 10^6$ kg; (b) $(100 + 1.5t)^{0.5}$ m/s; (c) $(1.5 \times 10^6)/(100 + 1.5t)^{0.5}$ N; (d) 6.7 km  **107.** (a) 5.6 J; (b) 12 J; (c) 13 J  **109.** (a) 4.9 m/s; (b) 4.5 N; (c) 71°; (d) same  **111.** (a) 1.2 J; (b) 11 m/s; (c) no; (d) no  **113.** 54%  **115.** (a) $2.7 \times 10^9$ J; (b) $2.7 \times 10^9$ W; (c) \$2.4 \times 10^8$  **117.** (a) 5.00 J; (b) 9.00 J; (c) 11.0 J; (d) 3.00 J; (e) 12.0 J; (f) 2.00 J; (g) 13.0 J; (h) 1.00 J; (i) 13.0 J; (j) 1.00 J; (l) 11.0 J; (m) 10.8 m; (n) It returns to $x = 0$ and stops.  **119.** (a) 3.7 J; (b) 4.3 J; (c) 4.3 J  **121.** (a) 4.8 N; (b) $+x$ direction; (c) 1.5 m; (d) 13.5 m; (e) 3.5 m/s  **123.** (a) 24 kJ; (b) $4.7 \times 10^2$ N  **125.** (a) 3.0 mm; (b) 1.1 J; (d) yes; (e) $\approx 40$ J; (f) no  **127.** (a) 6.0 kJ; (b) $6.0 \times 10^2$ W; (c) $3.0 \times 10^2$ W; (d) $9.0 \times 10^2$ W  **129.** $3.1 \times 10^{11}$ W  **131.** 880 MW  **133.** (a) $v_0 = (2gL)^{0.5}$; (b) $5mg$; (c) $-mgL$; (d) $-2mgL$  **135.** because your force on the cabbage (as you lower it) does work

## Chapter 9

**CP** **1.** (a) origin; (b) fourth quadrant; (c) on $y$ axis below origin; (d) origin; (e) third quadrant; (f) origin  **2.** (a)–(c) at the center of mass, still at the origin (their forces are internal to the system and cannot move the center of mass)  **3.** (Consider slopes and Eq. 9-23.) (a) 1, 3, and then 2 and 4 tie (zero force); (b) 3  **4.** (a) unchanged; (b) unchanged (see Eq. 9-32); (c) decrease (Eq. 9-35)  **5.** (a) zero; (b) positive (initial $p_y$ down $y$; final $p_y$ up $y$); (c) positive direction of $y$  **6.** (No net external force; $\vec{P}$ conserved.) (a) 0; (b) no; (c) $-x$  **7.** (a) 10 kg·m/s; (b) 14 kg·m/s; (c) 6 kg·m/s  **8.** (a) 4 kg·m/s; (b) 8 kg·m/s; (c) 3 J  **9.** (a) 2 kg·m/s (conserve momentum along $x$); (b) 3 kg·m/s (conserve momentum along $y$)

**Q 1.** (a) 2 N, rightward; (b) 2 N, rightward; (c) greater than 2 N, rightward **3.** (a) $x$ yes, $y$ no; (b) $x$ yes, $y$ no; (c) $x$ no, $y$ yes **5.** b, c, a **7.** (a) one was stationary; (b) 2; (c) 5; (d) equal (pool player's result) **9.** (a) $C$; (b) $B$; (c) 3 **11.** (a) $c$, kinetic energy cannot be negative; $d$, total kinetic energy cannot increase; (b) $a$; (c) $b$ **P 1.** (a) $-1.50$ m; (b) $-1.43$ m **3.** (a) $-0.45$ cm; (b) $-2.0$ cm **5.** (a) 0; (b) $3.13 \times 10^{-11}$ m **7.** (a) $-6.5$ cm; (b) 8.3 cm; (c) 1.4 cm **9.** $(-4.0 \text{ m})\hat{i} + (4.0 \text{ m})\hat{j}$ **11.** (a) 28 cm; (b) 2.3 m/s **13.** (a) $(2.35\hat{i} - 1.57\hat{j})$ m/s$^2$; (b) $(2.35\hat{i} - 1.57\hat{j})t$ m/s, with $t$ in seconds; (d) straight, at downward angle $34°$ **15.** 53 m **17.** 4.2 m **19.** (a) $7.5 \times 10^4$ J; (b) $3.8 \times 10^4$ kg·m/s; (c) $39°$ south of due east **21.** (a) 5.0 kg·m/s; (b) 10 kg·m/s **23.** (a) 67 m/s; (b) $-x$; (c) 1.2 kN; (d) $-x$ **25.** $1.0 \times 10^3$ to $1.2 \times 10^3$ kg·m/s **27.** (a) 42 N·s; (b) 2.1 kN **29.** 5 N **31.** (a) 5.86 kg·m/s; (b) $59.8°$; (c) 2.93 kN; (d) $59.8°$ **33.** (a) $2.39 \times 10^3$ N·s; (b) $4.78 \times 10^5$ N; (c) $1.76 \times 10^3$ N·s; (d) $3.52 \times 10^5$ N **35.** (a) 9.0 kg·m/s; (b) 3.0 kN; (c) 4.5 kN; (d) 20 m/s **37.** $9.9 \times 10^2$ N **39.** 3.0 mm/s **41.** 55 cm **43.** (a) $-(0.15 \text{ m/s})\hat{i}$; (b) 0.18 m **45.** (a) 14 m/s; (b) $45°$ **47.** (a) $(1.00\hat{i} - 0.167\hat{j})$ km/s; (b) 3.23 MJ **49.** $3.1 \times 10^2$ m/s **51.** (a) 33%; (b) 23%; (c) decreases **53.** (a) 721 m/s; (b) 937 m/s **55.** (a) 4.4 m/s; (b) 0.80 **57.** (a) $+2.0$ m/s; (b) $-1.3$ J; (c) $+40$ J; (d) system got energy from some source, such as a small explosion **59.** 25 cm **61.** (a) 99 g; (b) 1.9 m/s; (c) 0.93 m/s **63.** (a) 1.2 kg; (b) 2.5 m/s **65.** $-28$ cm **67.** (a) 3.00 m/s; (b) 6.00 m/s **69.** (a) 0.21 kg; (b) 7.2 m **71.** (a) 433 m/s; (b) 250 m/s **73.** (a) $4.15 \times 10^5$ m/s; (b) $4.84 \times 10^5$ m/s **75.** $120°$ **77.** (a) $1.57 \times 10^6$ N; (b) $1.35 \times 10^5$ kg; (c) 2.08 km/s **79.** (a) 46 N; (b) none **81.** (a) 1.78 m/s; (b) less; (c) less; (d) greater **83.** (a) 1.92 m; (b) 0.640 m **85.** 28.8 N **87.** 1.10 m/s **89.** (a) 7290 m/s; (b) 8200 m/s; (c) $1.271 \times 10^{10}$ J; (d) $1.275 \times 10^{10}$ J **91.** (a) 1.0 kg·m/s; (b) $2.5 \times 10^2$ J; (c) 10 N; (d) 1.7 kN; (e) answer for (c) includes time between pellet collisions **93.** (a) $(7.4 \times 10^3 \text{ N·s})\hat{i} - (7.4 \times 10^3 \text{ N·s})\hat{j}$; (b) $(-7.4 \times 10^3 \text{ N·s})\hat{i}$; (c) $2.3 \times 10^3$ N; (d) $2.1 \times 10^4$ N; (e) $-45°$ **95.** (a) 3.7 m/s; (b) 1.3 N·s; (c) $1.8 \times 10^2$ N **97.** $1.18 \times 10^4$ kg **99.** $+4.4$ m/s **101.** (a) 1.9 m/s; (b) $-30°$; (c) elastic **103.** (a) 6.9 m/s; (b) $30°$; (c) 6.9 m/s; (d) $-30°$; (e) 2.0 m/s; (f) $-180°$ **105.** (a) 25 mm; (b) 26 mm; (c) down; (d) $1.6 \times 10^{-2}$ m/s$^2$ **107.** (a) 0.745 mm; (b) $153°$; (c) 1.67 mJ **109.** (a) $(2.67 \text{ m/s})\hat{i} + (-3.00 \text{ m/s})\hat{j}$; (b) 4.01 m/s; (c) $48.4°$ **111.** 0.22% **113.** 190 m/s **115.** (a) $4.6 \times 10^3$ km; (b) 73% **117.** (a) 50 kg/s; (b) $1.6 \times 10^2$ kg/s **119.** (a) $-0.50$ m; (b) $-1.8$ cm; (c) 0.50 m **121.** (a) 0.800 kg·m/s; (b) 0.400 kg·m/s **123.** 29 J **125.** $5.0 \times 10^6$ N **127.** (a) 1; (b) $1.83 \times 10^3$; (c) $1.83 \times 10^3$; (d) all the same **129.** 5.0 kg **131.** 2.2 kg **133.** (a) 11.4 m/s; (b) $95.1°$ **135.** (a) 0; (b) 0; (c) 0

## Chapter 10

**CP 1.** b and c **2.** (a) and (d) ($\alpha = d^2\theta/dt^2$ must be a constant) **3.** (a) yes; (b) no; (c) yes; (d) yes **4.** all tie **5.** 1, 2, 4, 3 (see Eq. 10-36) **6.** (see Eq. 10-40) 1 and 3 tie, 4, then 2 and 5 tie (zero) **7.** (a) downward in the figure ($\tau_{net} = 0$); (b) less (consider moment arms) **Q 1.** (a) $c$, $a$, then $b$ and $d$ tie; (b) $b$, then $a$ and $c$ tie, then $d$ **3.** $c$, $a$, $b$ **5.** larger **7.** (a) decrease; (b) clockwise; (c) counterclockwise **9.** all tie **P 1.** 14 rev **3.** 11 rad/s **5.** (a) 4.0 rad/s; (b) 11.9 rad/s **7.** (a) 4.0 m/s; (b) no **9.** (a) 30 s; (b) $1.8 \times 10^3$ rad **11.** (a) 3.00 s; (b) 18.9 rad **13.** 8.0 s **15.** (a) 44 rad; (b) 5.5 s; (c) 32 s; (d) $-2.1$ s; (e) 40 s **17.** (a) $3.4 \times 10^2$ s; (b) $-4.5 \times 10^{-3}$ rad/s$^2$; (c) 98 s **19.** $6.9 \times 10^{-13}$ rad/s **21.** (a) 20.9 rad/s; (b) 12.5 m/s; (c) 800 rev/min$^2$; (d) 600 rev **23.** (a) $2.50 \times 10^{-3}$ rad/s;

(b) 20.2 m/s$^2$; (c) 0 **25.** (a) 40 s; (b) 2.0 rad/s$^2$ **27.** (a) $3.8 \times 10^3$ rad/s; (b) $1.9 \times 10^2$ m/s **29.** (a) $7.3 \times 10^{-5}$ rad/s; (b) $3.5 \times 10^2$ m/s; (c) $7.3 \times 10^{-5}$ rad/s; (d) $4.6 \times 10^2$ m/s **31.** (a) 73 cm/s$^2$; (b) 0.075; (c) 0.11 **33.** 12.3 kg·m$^2$ **35.** 0.097 kg·m$^2$ **37.** (a) 1.1 kJ; (b) 9.7 kJ **39.** (a) 0.023 kg·m$^2$; (b) 11 mJ **41.** $4.7 \times 10^{-4}$ kg·m$^2$ **43.** (a) 49 MJ; (b) $1.0 \times 10^2$ min **45.** 4.6 N·m **47.** $-3.85$ N·m **49.** (a) 28.2 rad/s$^2$; (b) 338 N·m **51.** 0.140 N **53.** $2.51 \times 10^{-4}$ kg·m$^2$ **55.** (a) 6.00 cm/s$^2$; (b) 4.87 N; (c) 4.54 N; (d) 1.20 rad/s$^2$; (e) 0.0138 kg·m$^2$ **57.** (a) $4.2 \times 10^2$ rad/s$^2$; (b) $5.0 \times 10^2$ rad/s **59.** (a) 19.8 kJ; (b) 1.32 kW **61.** 396 N·m **63.** 5.42 m/s **65.** 9.82 rad/s **67.** (a) 5.32 m/s$^2$; (b) 8.43 m/s$^2$; (c) $41.8°$ **69.** (a) 314 rad/s$^2$; (b) 7.54 m/s$^2$; (c) 14.0 N; (d) 4.36 N **71.** $6.16 \times 10^{-5}$ kg·m$^2$ **73.** (a) 1.57 m/s$^2$; (b) 4.55 N; (c) 4.94 N **75.** (a) $4.81 \times 10^5$ N; (b) $1.12 \times 10^4$ N·m; (c) $1.25 \times 10^6$ J **77.** 30 rev **79.** 3.1 rad/s **81.** (a) 0.791 kg·m$^2$; (b) $1.79 \times 10^{-2}$ N·m **83.** (a) 2.3 rad/s$^2$; (b) 1.4 rad/s$^2$ **85.** $1.4 \times 10^2$ N·m **87.** 4.6 rad/s$^2$ **89.** (a) $-67$ rev/min$^2$; (b) 8.3 rev **93.** 0.054 kg·m$^2$ **95.** (a) $5.92 \times 10^4$ m/s$^2$; (b) $4.39 \times 10^4$ s$^{-2}$ **97.** 2.6 J **99.** (a) 0.32 rad/s; (b) $1.0 \times 10^2$ km/h **101.** (a) 7.0 kg·m$^2$; (b) 7.2 m/s; (c) $71°$ **103.** (a) $1.4 \times 10^2$ rad; (b) 14 s **105.** (a) 221 kg·m$^2$; (b) $1.10 \times 10^4$ J **107.** 0.13 rad/s **109.** $6.75 \times 10^{12}$ rad/s **111.** (a) $1.5 \times 10^2$ cm/s; (b) 15 rad/s; (c) 15 rad/s; (d) 75 cm/s; (e) 3.0 rad/s **113.** 18 rad **115.** (a) 10 J; (b) 0.27 m

## Chapter 11

**CP 1.** (a) same; (b) less **2.** less (consider the transfer of energy from rotational kinetic energy to gravitational potential energy) **3.** (draw the vectors, use right-hand rule) (a) $\pm z$; (b) $\pm y$; (c) $-x$ **4.** (see Eq. 11-21) (a) 1 and 3 tie; then 2 and 4 tie, then 5 (zero); (b) 2 and 3 **5.** (see Eqs. 11-23 and 11-16) (a) 3, 1; then 2 and 4 tie (zero); (b) 3 **6.** (a) all tie (same $\tau$, same $t$, thus same $\Delta L$); (b) sphere, disk, hoop (reverse order of $I$) **7.** (a) decreases; (b) same ($\tau_{net} = 0$, so $L$ is conserved); (c) increases **Q 1.** (a) 1, 2, 3 (zero); (b) 1 and 2 tie, then 3; (c) 1 and 3 tie, then 2 **3.** (a) spins in place; (b) rolls toward you; (c) rolls away from you **5.** $a$, then $b$ and $c$ tie, then $e$, $d$ (zero) **7.** $D$, $B$, then $A$ and $C$ tie **9.** (a) same; (b) increase; (c) decrease; (d) same, decrease, increase **P 1.** (a) 0; (b) $(22 \text{ m/s})\hat{i}$; (c) $(-22 \text{ m/s})\hat{i}$; (d) 0; (e) $1.5 \times 10^3$ m/s$^2$; (f) $1.5 \times 10^3$ m/s$^2$; (g) $(22 \text{ m/s})\hat{i}$; (h) $(44 \text{ m/s})\hat{i}$; (i) 0; (j) 0; (k) $1.5 \times 10^3$ m/s$^2$; (l) $1.5 \times 10^3$ m/s$^2$ **3.** 0.020 **5.** $-3.15$ J **7.** (a) $(-4.0 \text{ N})\hat{i}$; (b) 0.60 kg·m$^2$ **9.** (a) 63 rad/s; (b) 4.0 m **11.** 4.8 m **13.** (a) $-(0.11 \text{ m})\omega$; (b) $-2.1$ m/s$^2$; (c) $-47$ rad/s$^2$; (d) 1.2 s; (e) 8.6 m; (f) 6.1 m/s **15.** 0.50 **17.** (a) 13 cm/s$^2$; (b) 4.4 s; (c) 55 cm/s; (d) 18 mJ; (e) 1.4 J; (f) 27 rev/s **19.** (a) $(6.0 \text{ N·m})\hat{j} + (8.0 \text{ N·m})\hat{k}$; (b) $(-22 \text{ N·m})\hat{i}$ **21.** $(-2.0 \text{ N·m})\hat{i}$ **23.** (a) $(50 \text{ N·m})\hat{k}$; (b) $90°$ **25.** (a) $(-1.5 \text{ N·m})\hat{i} - (4.0 \text{ N·m})\hat{j} - (1.0 \text{ N·m})\hat{k}$; (b) $(-1.5 \text{ N·m})\hat{i} - (4.0 \text{ N·m})\hat{j} - (1.0 \text{ N·m})\hat{k}$ **27.** (a) 9.8 kg·m$^2$/s; (b) $+z$ direction **29.** (a) 0; (b) $(8.0 \text{ N·m})\hat{i} + (8.0 \text{ N·m})\hat{k}$ **31.** (a) 0; (b) $-22.6$ kg·m$^2$/s; (c) $-7.84$ N·m; (d) $-7.84$ N·m **33.** (a) $(-1.7 \times 10^2 \text{ kg·m}^2/\text{s})\hat{k}$; (b) $(+56 \text{ N·m})\hat{k}$; (c) $(+56 \text{ kg·m}^2/\text{s}^2)\hat{k}$ **35.** (a) $48t\hat{k}$ N·m; (b) increasing **37.** (a) 1.47 N·m; (b) 20.4 rad; (c) $-29.9$ J; (d) 19.9 W **39.** (a) $4.6 \times 10^{-3}$ kg·m$^2$; (b) $1.1 \times 10^{-3}$ kg·m$^2$/s; (c) $3.9 \times 10^{-3}$ kg·m$^2$/s **41.** (a) 1.6 kg·m$^2$; (b) 4.0 kg·m$^2$/s **43.** (a) 3.6 rev/s; (b) 3.0; (c) forces on the bricks from the man transferred energy from the man's internal energy to kinetic energy **45.** (a) 267 rev/min; (b) 0.667 **47.** (a) 750 rev/min; (b) 450 rev/min; (c) clockwise **49.** 0.17 rad/s **51.** (a) 1.5 m; (b) 0.93 rad/s; (c) 98 J;

(d) 8.4 rad/s; (e) $8.8 \times 10^2$ J; (f) internal energy of the skaters
**53.** 3.4 rad/s **55.** $1.3 \times 10^3$ m/s **57.** 11.0 m/s **59.** (a) 18 rad/s;
(b) 0.92 **61.** 1.5 rad/s **63.** (a) 0.180 m; (b) clockwise
**65.** 0.070 rad/s **67.** (a) 0.148 rad/s; (b) 0.0123; (c) 181°
**69.** 0.041 rad/s **71.** 39.1 J **73.** (a) $6.65 \times 10^{-5}$ kg·m$^2$/s; (b) no;
(c) 0; (d) yes **75.** (a) 0.333; (b) 0.111 **77.** (a) 58.8 J; (b) 39.2 J
**79.** (a) 0.81 mJ; (b) 0.29; (c) $1.3 \times 10^{-2}$ N **81.** (a) $mR^2/2$;
(b) a solid circular cylinder **83.** rotational speed would
decrease; day would be about 0.8 s longer **85.** (a) 149 kg·m$^2$;
(b) 158 kg·m$^2$/s; (c) 0.744 rad/s **87.** (a) 0; (b) 0;
(c) $-30t^3\hat{k}$ kg·m$^2$/s; (d) $-90t^2\hat{k}$ N·m; (e) $30t^3\hat{k}$ kg·m$^2$/s;
(f) $90t^2\hat{k}$ N·m **89.** (a) 61.7 J; (b) 3.43 m; (c) no
**91.** (a) 12.7 rad/s; (b) clockwise **93.** (a) $mvR/(I + MR^2)$;
(b) $mvR^2/(I + MR^2)$ **95.** (a) 1.6 m/s$^2$; (b) 16 rad/s$^2$; (c) $(4.0$ N$)\hat{i}$
**97.** 0.47 kg·m$^2$/s

## Chapter 12

**CP** **1.** $c, e, f$ **2.** (a) no; (b) at site of $\vec{F}_1$, perpendicular to
plane of figure; (c) 45 N **3.** $d$ **Q** **1.** $a$ and $c$ (forces and
torques balance) **3.** (a) 12 kg; (b) 3 kg; (c) 1 kg **5.** (a) 1 and
3 tie, then 2; (b) all tie; (c) 1 and 3 tie, then 2 (zero) **7.** increase **9.** (a) at $C$ (to eliminate forces there from a torque
equation); (b) plus; (c) minus; (d) equal **P** **1.** (a) 1.00 m;
(b) 2.00 m; (c) 0.987 m; (d) 1.97 m **3.** 7.92 kN **5.** (a) 9.4 N;
(b) 4.4 N **7.** (a) 1.2 kN; (b) down; (c) 1.7 kN; (d) up; (e) left;
(f) right **9.** (a) $2.8 \times 10^2$ N; (b) $8.8 \times 10^2$ N; (c) 71° **11.** 74.4 g
**13.** (a) 5.0 N; (b) 30 N; (c) 1.3 m **15.** 8.7 N **17.** (a) 2.7 kN;
(b) up; (c) 3.6 kN; (d) down **19.** (a) 0.64 m; (b) increased
**21.** 13.6 N **23.** (a) 1.9 kN; (b) up; (c) 2.1 kN; (d) down
**25.** (a) 192 N; (b) 96.1 N; (c) 55.5 N **27.** (a) 6.63 kN;
(b) 5.74 kN; (c) 5.96 kN **29.** 2.20 m **31.** (a) $(-80$ N$)\hat{i}) +$
$(1.3 \times 10^2$ N$)\hat{j}$; (b) $(80$ N$)\hat{i} + (1.3 \times 10^2$ N$)\hat{j}$ **33.** (a) 445 N;
(b) 0.50; (c) 315 N **35.** (a) 60.0°; (b) 300 N **37.** 0.34
**39.** (a) slides; (b) 31°; (c) tips; (d) 34° **41.** (a) 211 N; (b) 534 N;
(c) 320 N **43.** (a) $6.5 \times 10^6$ N/m$^2$; (b) $1.1 \times 10^{-5}$ m
**45.** (a) 866 N; (b) 143 N; (c) 0.165 **47.** (a) 0.80; (b) 0.20; (c)
0.25 **49.** (a) $1.4 \times 10^9$ N; (b) 75 **51.** (a) $1.2 \times 10^2$ N; (b) 68 N
**53.** 76 N **55.** (a) 8.01 kN; (b) 3.65 kN; (c) 5.66 kN **57.** 71.7 N
**59.** (a) $L/2$; (b) $L/4$; (c) $L/6$; (d) $L/8$; (e) $25L/24$ **61.** (a) $1.8 \times$
$10^7$ N; (b) $1.4 \times 10^7$ N; (c) 16 **63.** 0.29 **65.** 60° **67.** (a) 270
N; (b) 72 N; (c) 19° **69.** (a) 106 N; (b) 64.0° **71.** $2.4 \times$
$10^9$ N/m$^2$ **73.** (a) 88 N; (b) $(30\hat{i} + 97\hat{j})$ N **75.** (a) $a_1 = L/2$,
$a_2 = 5L/8, h = 9L/8$; (b) $b_1 = 2L/3, b_2 = L/2, h = 7L/6$
**77.** (a) $BC, CD, DA$; (b) 535 N; (c) 757 N **79.** (a) 1.38 kN;
(b) 180 N **81.** (a) $\mu < 0.57$; (b) $\mu > 0.57$ **83.** $L/4$
**85.** (a) $(35\hat{i} + 200\hat{j})$ N; (b) $(-45\hat{i} + 200\hat{j})$ N; (c) $1.9 \times 10^2$ N

## Chapter 13

**CP** **1.** all tie **2.** (a) 1, tie of 2 and 4, then 3; (b) line $d$
**3.** (a) increase; (b) negative **4.** (a) 2; (b) 1 **5.** (a) path 1
(decreased $E$ (more negative) gives decreased $a$); (b) less
(decreased $a$ gives decreased $T$) **Q** **1.** $Gm^2/r^2$, upward
**3.** $b$ and $c$ tie, then $a$ (zero) **5.** $3GM^2/d^2$, leftward **7.** (a) positive $y$; (b) yes, rotates counterclockwise until it points toward
particle $B$ **9.** 1, tie of 2 and 4, then 3 **11.** $b, d,$ and $f$ all tie,
then $e, c, a$ **P** **1.** 19 m **3.** $\frac{1}{2}$ **5.** $-5.00d$ **7.** $2.60 \times 10^5$ km
**9.** 0.8 m **11.** (a) $M = m$; (b) 0 **13.** $8.31 \times 10^{-9}$ N **15.** (a)
$-1.88d$; (b) $-3.90d$; (c) $0.489d$ **17.** $2.6 \times 10^6$ m **19.** (a) 17 N;
(b) 2.4 **21.** (a) 7.6 m/s$^2$; (b) 4.2 m/s$^2$ **23.** $5 \times 10^{24}$ kg
**25.** (a) 9.83 m/s$^2$; (b) 9.84 m/s$^2$; (c) 9.79 m/s$^2$ **27.** (a) $(3.0 \times$
$10^{-7}$ N/kg$)m$; (b) $(3.3 \times 10^{-7}$ N/kg$)m$; (c) $(6.7 \times 10^{-7}$ N/kg·m$)mr$
**29.** (a) 0.74; (b) 3.8 m/s$^2$; (c) 5.0 km/s **31.** (a) 0.0451; (b) 28.5

**33.** $5.0 \times 10^9$ J **35.** (a) 0.50 pJ; (b) $-0.50$ pJ **37.** (a) 1.7 km/s;
(b) $2.5 \times 10^5$ m; (c) 1.4 km/s **39.** (a) 82 km/s; (b) $1.8 \times 10^4$ km/s
**41.** $-4.82 \times 10^{-13}$ J **43.** $6.5 \times 10^{23}$ kg **45.** $5 \times 10^{10}$ stars
**47.** (a) $6.64 \times 10^3$ km; (b) 0.0136 **49.** (a) 7.82 km/s; (b) 87.5 min
**51.** (a) $1.9 \times 10^{13}$ m; (b) $3.6R_P$ **55.** 0.71 y **57.** $5.8 \times 10^6$ m
**59.** $(GM/L)^{0.5}$ **61.** (a) 2.8 y; (b) $1.0 \times 10^{-4}$ **63.** (a) $3.19 \times$
$10^3$ km; (b) lifting **65.** (a) $r^{1.5}$; (b) $r^{-1}$; (c) $r^{0.5}$; (d) $r^{-0.5}$
**67.** (a) 7.5 km/s; (b) 97 min; (c) $4.1 \times 10^2$ km; (d) 7.7 km/s;
(e) 93 min; (f) $3.2 \times 10^{-3}$ N; (g) no; (h) yes **69.** 1.1 s
**71.** (a) $1.0 \times 10^3$ kg; (b) 1.5 km/s **73.** $-0.044\hat{j}$ μN **75.** (a) $2.15 \times$
$10^4$ s; (b) 12.3 km/s; (c) 12.0 km/s; (d) $2.17 \times 10^{11}$ J; (e) $-4.53 \times$
$10^{11}$ J; (f) $-2.35 \times 10^{11}$ J; (g) $4.04 \times 10^7$ m; (h) $1.22 \times 10^3$ s;
(i) elliptical **77.** $0.37\hat{j}$ μN **79.** 29 pN **81.** $2.5 \times 10^4$ km
**83.** (a) $2.2 \times 10^{-7}$ rad/s; (b) 89 km/s **85.** $3.2 \times 10^{-7}$ N
**87.** (a) 0; (b) $1.8 \times 10^{32}$ J; (c) $1.8 \times 10^{32}$ J; (d) 0.99 km/s
**89.** $GM_Em/12R_E$ **91.** (a) $1.4 \times 10^6$ m/s; (b) $3 \times 10^6$ m/s$^2$
**93.** $2\pi r^{1.5}G^{-0.5}(M + m/4)^{-0.5}$ **95.** $2.4 \times 10^4$ m/s **97.** $-1.87$ GJ
**99.** (a) $GMmx(x^2 + R^2)^{-3/2}$; (b) $[2GM(R^{-1} - (R^2 + x^2)^{-1/2})]^{1/2}$
**101.** (a) $Gm^2/R_i$; (b) $Gm^2/2R_i$; (c) $(Gm/R_i)^{0.5}$; (d) $2(Gm/R_i)^{0.5}$;
(e) $Gm^2/R_i$; (f) $(2Gm/R_i)^{0.5}$; (g) The center-of-mass frame is an
inertial frame, and in it the principle of conservation of energy
may be written as in Chapter 8; the reference frame attached
to body $A$ is noninertial, and the principle cannot be written as
in Chapter 8. Answer (d) is correct. **103.** (a) $1.9 \times 10^{11}$ m;
(b) $4.6 \times 10^4$ m/s

## Chapter 14

**CP** **1.** all tie **2.** (a) all tie (the gravitational force on the
penguin is the same); (b) $0.95\rho_0, \rho_0, 1.1\rho_0$ **3.** 13 cm$^3$/s, outward
**4.** (a) all tie; (b) 1, then 2 and 3 tie, 4 (wider means slower);
(c) 4, 3, 2, 1 (wider and lower mean more pressure)
**Q** **1.** $b$, then $a$ and $d$ tie (zero), then $c$ **3.** (a) moves downward; (b) moves downward **5.** (a) downward; (b) downward;
(c) same **7.** $B, C, A$ **9.** (a) 1 and 4; (b) 2; (c) 3
**P** **1.** $1.1 \times 10^5$ Pa **3.** $2.9 \times 10^4$ N **5.** 0.074 **7.** (b) 26 kN
**9.** $1.08 \times 10^3$ atm **11.** $7.2 \times 10^5$ N$^-$ **13.** $-2.6 \times 10^4$ Pa
**15.** (a) 94 torr; (b) $4.1 \times 10^2$ torr; (c) $3.1 \times 10^2$ torr **17.** (a) 1.0
$\times 10^3$ torr; (b) $1.7 \times 10^3$ torr **19.** 0.635 J **21.** 44 km
**23.** $4.69 \times 10^5$ N **25.** 739.26 torr **27.** (a) 7.9 km; (b) 16 km
**29.** 8.50 kg **31.** (a) $2.04 \times 10^{-2}$ m$^3$; (b) 1.57 kN **33.** five
**35.** (a) $6.7 \times 10^2$ kg/m$^3$; (b) $7.4 \times 10^2$ kg/m$^3$ **37.** (a) 1.2 kg;
(b) $1.3 \times 10^3$ kg/m$^3$ **39.** (a) 0.10; (b) 0.083 **41.** 57.3 cm
**43.** 0.126 m$^3$ **45.** (a) 1.80 m$^3$; (b) 4.75 m$^3$ **47.** (a) 637.8 cm$^3$;
(b) 5.102 m$^3$; (c) $5.102 \times 10^3$ kg **49.** 8.1 m/s **51.** (a) 3.0 m/s;
(b) 2.8 m/s **53.** 66 W **55.** (a) 2.5 m/s; (b) $2.6 \times 10^5$ Pa
**57.** (a) 3.9 m/s; (b) 88 kPa **59.** (a) $1.6 \times 10^{-3}$ m$^3$/s; (b) 0.90 m
**61.** $1.4 \times 10^5$ J **63.** (a) 74 N; (b) $1.5 \times 10^2$ m$^3$ **65.** (a) 35 cm;
(b) 30 cm; (c) 20 cm **67.** (b) $2.0 \times 10^{-2}$ m$^3$/s **69.** (a) 0.0776 m$^3$/s;
(b) 69.8 kg/s **71.** $1.1 \times 10^2$ m/s **73.** 44.2 g **75.** 45.3 cm$^3$
**77.** (a) 3.2 m/s; (b) $9.2 \times 10^4$ Pa; (c) 10.3 m **79.** $5.11 \times 10^{-7}$ kg
**81.** $1.07 \times 10^3$ g **83.** $6.0 \times 10^2$ kg/m$^3$ **85.** 1.5 g/cm$^3$

## Chapter 15

**CP** **1.** (sketch $x$ versus $t$) (a) $-x_m$; (b) $+x_m$; (c) 0
**2.** a ($F$ must have the form of Eq. 15-10) **3.** (a) 5 J; (b) 2 J;
(c) 5 J **4.** all tie (in Eq. 15-29, $m$ is included in $I$) **5.** 1, 2, 3
(the ratio $m/b$ matters; $k$ does not) **Q** **1.** (a) 2; (b) positive; (c) between 0 and $+x_m$ **3.** a and b **5.** (a) all tie; (b) 3,
then 1 and 2 tie; (c) 1, 2, 3 (zero); (d) 1, 2, 3 (zero); (e) 1, 3, 2
**7.** (a) between $D$ and $E$; (b) between $3\pi/2$ rad and $2\pi$ rad
**9.** (a) greater; (b) same; (c) same; (d) greater; (e) greater
**11.** $b$ (infinite period, does not oscillate), $c, a$

**P** **1.** 37.8 m/s$^2$ **3.** (a) 1.0 mm; (b) 0.75 m/s; (c) 5.7 × 10$^2$ m/s$^2$
**5.** (a) 0.50 s; (b) 2.0 Hz; (c) 18 cm **7.** (a) 0.500 s; (b) 2.00 Hz;
(c) 12.6 rad/s; (d) 79.0 N/m; (e) 4.40 m/s; (f) 27.6 N
**9.** (a) 498 Hz; (b) greater **11.** (a) 3.0 m; (b) −49 m/s;
(c) −2.7 × 10$^2$ m/s$^2$; (d) 20 rad; (e) 1.5 Hz; (f) 0.67 s **13.** 39.6 Hz
**15.** (a) 5.58 Hz; (b) 0.325 kg; (c) 0.400 m **17.** 3.1 cm
**19.** (a) 0.18$A$; (b) same direction **21.** (a) 25 cm; (b) 2.2 Hz
**23.** 54 Hz **25.** (a) 0.525 m; (b) 0.686 s **27.** 37 mJ
**29.** (a) 0.75; (b) 0.25; (c) $2^{-0.5}x_m$ **31.** (a) 2.25 Hz; (b) 125 J;
(c) 250 J; (d) 86.6 cm **33.** (a) 3.1 ms; (b) 4.0 m/s; (c) 0.080 J; (d)
80 N; (e) 40 N **35.** (a) 1.1 m/s; (b) 3.3 cm **37.** (a) 2.2 Hz;
(b) 56 cm/s; (c) 0.10 kg; (d) 20.0 cm **39.** (a) 39.5 rad/s;
(b) 34.2 rad/s; (c) 124 rad/s$^2$ **41.** (a) 1.64 s; (b) equal
**43.** (a) 0.205 kg·m$^2$; (b) 47.7 cm; (c) 1.50 s **45.** 0.366 s
**47.** 8.77 s **49.** (a) 0.53 m; (b) 2.1 s **51.** 0.0653 s
**53.** (a) 0.845 rad; (b) 0.0602 rad **55.** (a) 2.26 s; (b) increases;
(c) same **57.** (a) 14.3 s; (b) 5.27 **59.** 6.0% **61.** (a) $F_m/b\omega$;
(b) $F_m/b$ **63.** 5.0 cm **65.** (a) 1.2 J; (b) 50 **67.** 1.53 m
**69.** (a) 16.6 cm; (b) 1.23% **71.** (a) 2.8 × 10$^3$ rad/s; (b) 2.1 m/s;
(c) 5.7 km/s$^2$ **73.** (a) 0.735 kg · m$^2$; (b) 0.0240 N · m;
(c) 0.181 rad/s **75.** (a) 0.35 Hz; (b) 0.39 Hz; (c) 0 (no oscillation)
**77.** (a) 7.90 N/m; (b) 1.19 cm; (c) 2.00 Hz **79.** 1.6 kg **81.** (a) 3.5
m; (b) 0.75 s **83.** 7.2 m/s **85.** (a) 1.23 kN/m; (b) 76.0 N
**87.** (a) 1.1 Hz; (b) 5.0 cm **89.** (a) 1.3 × 10$^2$ N/m; (b) 0.62 s;
(c) 1.6 Hz; (d) 5.0 cm; (e) 0.51 m/s **91.** (a) 3.2 m; (b) 0.26 m;
(c) $x = (0.26 \text{ m}) \cos(20t - \pi/2)$, with $t$ in seconds
**93.** 0.079 kg·m$^2$ **95.** (a) 0.44 s; (b) 0.18 m **97.** (a) 245 N/m;
(b) 0.284 s **99.** 50 cm **101.** (a) 8.11 × 10$^{-5}$ kg · m$^2$; (b) 3.14 rad/s
**103.** 14.0° **105.** (a) 0.30 m; (b) 0.28 s; (c) 1.5 × 10$^2$ m/s$^2$;
(d) 11 J **107.** (a) 0.45 s; (b) 0.10 m above and 0.20 m below;
(c) 0.15 m; (d) 2.3 J **109.** 7 × 10$^2$ N/m **111.** (a) $F/m$;
(b) $2F/mL$; (c) 0

## Chapter 16
**CP** **1.** a, 2; b, 3; c, 1 (compare with phase in Eq. 16-2, then see
Eq. 16-5) **2.** (a) 2, 3, 1 (see Eq. 16-12); (b) 3, then 1 and 2 tie
(find amplitude of $dy/dt$) **3.** (a) same (independent of $f$);
(b) decrease ($\lambda = v/f$); (c) increase; (d) increase **4.** 0.20 and
0.80 tie, then 0.60, 0.45 **5.** (a) 1; (b) 3; (c) 2 **6.** (a) 75 Hz; (b) 525
Hz **Q** **1.** a, upward; b, upward; c, downward; d, down-
ward; e, downward; f, downward; g, upward; h, upward
**3.** (a) 1, 4, 2, 3; (b) 1, 4, 2, 3 **5.** (a) 0, 0.2 wavelength, 0.5 wavelength (zero); (b) $4P_{\text{avg,1}}$ **7.** intermediate (closer to fully
destructive) **9.** c, a, b **11.** d **P** **1.** (a) 3.49 m$^{-1}$;
(b) 31.5 m/s **3.** (a) 0.680 s; (b) 1.47 Hz; (c) 2.06 m/s **5.** 1.1 ms
**7.** (a) 11.7 cm; (b) $\pi$ rad **9.** (a) 64 Hz; (b) 1.3 m; (c) 4.0 cm;
(d) 5.0 m$^{-1}$; (e) 4.0 × 10$^2$ s$^{-1}$; (f) $\pi/2$ rad; (g) minus
**11.** (a) 3.0 mm; (b) 16 m$^{-1}$; (c) 2.4 × 10$^2$ s$^{-1}$; (d) minus
**13.** (a) negative; (b) 4.0 cm; (c) 0.31 cm$^{-1}$; (d) 0.63 s$^{-1}$;
(e) $\pi$ rad; (f) minus; (g) 2.0 cm/s; (h) −2.5 cm/s **15.** 129 m/s
**17.** (a) 0.12 mm; (b) 141 m$^{-1}$; (c) 628 s$^{-1}$; (d) plus
**19.** (a) 15 m/s; (b) 0.036 N **21.** (a) 5.0 cm; (b) 40 cm; (c) 12 m/s;
(d) 0.033 s; (e) 9.4 m/s; (f) 16 m$^{-1}$; (g) 1.9 × 10$^2$ s$^{-1}$; (h) 0.93 rad;
(i) plus **23.** 2.63 m **27.** 3.2 mm **29.** 0.20 m/s **31.** 1.41$y_m$
**33.** (a) 9.0 mm; (b) 16 m$^{-1}$; (c) 1.1 × 10$^3$ s$^{-1}$; (d) 2.7 rad;
(e) plus **35.** 5.0 cm **37.** 84° **39.** (a) 3.29 mm; (b) 1.55 rad;
(c) 1.55 rad **41.** (a) 7.91 Hz; (b) 15.8 Hz; (c) 23.7 Hz
**43.** (a) 82.0 m/s; (b) 16.8 m; (c) 4.88 Hz **45.** (a) 144 m/s;
(b) 60.0 cm; (c) 241 Hz **47.** (a) 105 Hz; (b) 158 m/s
**49.** 260 Hz **51.** (a) 0.25 cm; (b) 1.2 × 10$^2$ cm/s; (c) 3.0 cm;
(d) 0 **53.** (a) 0.50 cm; (b) 3.1 m$^{-1}$; (c) 3.1 × 10$^2$ s$^{-1}$; (d) minus

**55.** (a) 2.00 Hz; (b) 2.00 m; (c) 4.00 m/s; (d) 50.0 cm; (e) 150 cm;
(f) 250 cm; (g) 0; (h) 100 cm; (i) 200 cm **57.** 0.25 m **59.** (a) 324
Hz; (b) eight **61.** (a) 0.83$y_1$; (b) 37° **63.** (a) 0.31 m; (b) 1.64
rad; (c) 2.2 mm **65.** 1.2 rad **67.** (a) 3.77 m/s; (b) 12.3 N; (c) 0;
(d) 46.4 W; (e) 0; (f) 0; (g) ± 0.50 cm **69.** (a) $2\pi y_m/\lambda$; (b) no
**71.** (a) 1.00 cm; (b) 3.46 × 10$^3$ s$^{-1}$; (c) 10.5 m$^{-1}$; (d) plus
**73.** (a) 75 Hz; (b) 13 ms **75.** (a) 240 cm; (b) 120 cm; (c) 80 cm
**77.** (a) 144 m/s; (b) 3.00 m; (c) 1.50 m; (d) 48.0 Hz; (e) 96.0 Hz
**79.** (a) 2.0 mm; (b) 95 Hz; (c) + 30 m/s; (d) 31 cm; (e) 1.2 m/s
**81.** 36 N **83.** (a) 300 m/s; (b) no **85.** (a) 1.33 m/s; (b) 1.88
m/s; (c) 16.7 m/s$^2$; (d) 23.7 m/s$^2$ **87.** (a) 0.16 m; (b) 2.4 × 10$^2$
N; (c) $y(x,t) = (0.16 \text{ m}) \sin[(1.57 \text{ m}^{-1})x] \sin[(31.4 \text{ s}^{-1})t]$
**89.** (a) $[k\,\Delta\ell\,(\ell + \Delta\ell)/m]^{0.5}$ **91.** (a) 0.52 m; (b) 40 m/s;
(c) 0.40 m **93.** (c) 2.0 m/s; (d) −$x$

## Chapter 17
**CP** **1.** beginning to decrease (example: mentally move the
curves of Fig. 17-7 rightward past the point at $x = 42$ m)
**2.** (a) 1 and 2 tie, then 3 (see Eq. 17-28); (b) 3, then 1 and 2 tie
(see Eq. 17-26) **3.** second (see Eqs. 17-39 and 17-41)
**4.** a, greater; b, less; c, can't tell; d, can't tell; e, greater; f, less
**Q** **1.** C, then A and B tie **3.** (a) 0, 0.2 wavelength, 0.5 wavelength (zero); (b) $4P_{\text{avg,1}}$ **5.** 150 Hz and 450 Hz **7.** E, A, D, C,
B **9.** 1, 4, 3, 2 **P** **1.** (a) 2.6 km; (b) 2.0 × 10$^2$ **3.** (a) 79 m;
(b) 41 m; (c) 89 m **5.** 40.7 m **7.** 1.9 × 10$^3$ km **9.** (a) 76.2
$\mu$m; (b) 0.333 mm **11.** 0.23 ms **13.** (a) 2.3 × 10$^2$ Hz;
(b) higher **15.** 960 Hz **17.** (a) 14; (b) 14 **19.** (a) 343 Hz;
(b) 3; (c) 5; (d) 686 Hz; (e) 2; (f) 3 **21.** (a) 143 Hz; (b) 3; (c) 5;
(d) 286 Hz; (e) 2; (f) 3 **23.** (a) 0; (b) fully constructive;
(c) increase; (d) 128 m; (e) 63.0 m; (f) 41.2 m **25.** 15.0 mW
**27.** 36.8 nm **29.** (a) 1.0 × 10$^3$; (b) 32 **31.** 0.76 $\mu$m **33.** 2 $\mu$W
**35.** (a) 5.97 × 10$^{-5}$ W/m$^2$; (b) 4.48 nW **37.** (a) 0.34 nW;
(b) 0.68 nW; (c) 1.4 nW; (d) 0.88 nW; (e) 0 **39.** (a) 833 Hz;
(b) 0.418 m **41.** (a) 2; (b) 1 **43.** (a) 405 m/s; (b) 596 N;
(c) 44.0 cm; (d) 37.3 cm **45.** (a) 3; (b) 1129 Hz; (c) 1506 Hz
**47.** 45.3 N **49.** 12.4 m **51.** 2.25 ms **53.** 0.020 **55.** 0
**57.** (a) 526 Hz; (b) 555 Hz **59.** 155 Hz **61.** (a) 1.022 kHz;
(b) 1.045 kHz **63.** 41 kHz **65.** (a) 485.8 Hz; (b) 500.0 Hz;
(c) 486.2 Hz; (d) 500.0 Hz **67.** (a) 2.0 kHz; (b) 2.0 kHz
**69.** (a) 42°; (b) 11 s **71.** (a) 2.10 m; (b) 1.47 m **73.** (a) 21 nm;
(b) 35 cm; (c) 24 nm; (d) 35 cm **75.** 0.25 **77.** (a) 9.7 × 10$^2$
Hz; (b) 1.0 kHz; (c) 60 Hz, no **79.** (a) 39.7 $\mu$W/m$^2$; (b) 171
nm; (c) 0.893 Pa **81.** (a) 10 W; (b) 0.032 W/m$^2$; (c) 99 dB
**83.** (a) 7.70 Hz; (b) 7.70 Hz **85.** (a) 59.7; (b) 2.81 × 10$^{-4}$
**87.** (a) 5.2 kHz; (b) 2 **89.** 2.1 m **91.** 1 cm **93.** (a) 3.6 × 10$^2$
m/s; (b) 150 Hz **95.** (a) 0; (b) 0.572 m; (c) 1.14 m **97.** 171 m
**99.** (a) 11 ms; (b) 3.8 m **101.** (a) rightward; (b) 0.90 m/s; (c)
less **103.** (a) 5.5 × 10$^2$ m/s; (b) 1.1 × 10$^3$ m/s; (c) 1
**105.** 400 Hz **107.** (a) 14; (b) 12 **109.** (b) 0.8 to 1.6 $\mu$s
**111.** 4.8 × 10$^2$ Hz

## Chapter 18
**CP** **1.** (a) all tie; (b) 50°X, 50°Y, 50°W **2.** (a) 2 and 3 tie,
then 1, then 4; (b) 3, 2, then 1 and 4 tie (from Eqs. 18-9 and 18-
10, assume that change in area is proportional to initial area)
**3.** A (see Eq. 18-14) **4.** c and e (maximize area enclosed by a
clockwise cycle) **5.** (a) all tie ($\Delta E_{\text{int}}$ depends on $i$ and $f$, not
on path); (b) 4, 3, 2, 1 (compare areas under curves); (c) 4, 3, 2, 1
(see Eq. 18-26) **6.** (a) zero (closed cycle); (b) negative ($W_{\text{net}}$ is
negative; see Eq. 18-26) **7.** b and d tie, then a, c ($P_{\text{cond}}$ identical;
see Eq. 18-32) **Q** **1.** B, then A and C tie **3.** c, then the

rest tie  **5.** (a) both clockwise; (b) both clockwise
**7.** $c, b, a$  **9.** (a) $f$, because ice temperature will not rise to
freezing point and then drop; (b) $b$ and $c$ at freezing point, $d$
above, $e$ below; (c) in $b$ liquid partly freezes and no ice melts;
in $c$ no liquid freezes and no ice melts; in $d$ no liquid freezes
and ice fully melts; in $e$ liquid fully freezes and no ice melts
**11.** (a) greater; (b) 1, 2, 3; (c) 1, 3, 2; (d) 1, 2, 3; (e) 2, 3, 1
**P**  **1.** 348 K  **3.** 1.366  **5.** (a) 320°F; (b) $-12.3$°F  **7.** $-92.1$°X
**9.** 29 cm$^3$  **11.** 2.731 cm  **13.** 49.87 cm$^3$  **15.** 0.26 cm$^3$
**17.** 360°C  **19.** 0.13 mm  **21.** 7.5 cm  **23.** 94.6 L  **25.** 42.7 kJ
**27.** 160 s  **29.** 33 g  **31.** 3.0 min  **33.** 33 m$^2$  **35.** 13.5 C°
**37.** 742 kJ  **39.** (a) 5.3°C; (b) 0; (c) 0°C; (d) 60 g  **41.** (a) 0°C;
(b) 2.5°C  **43.** $-30$ J  **45.** (a) $1.2 \times 10^2$ J; (b) 75 J; (c) 30 J
**47.** 60 J  **49.** (a) 6.0 cal; (b) $-43$ cal; (c) 40 cal; (d) 18 cal; (e) 18
cal  **51.** 1.66 kJ/s  **53.** (a) 16 J/s; (b) 0.048 g/s  **55.** (a) 1.23
kW; (b) 2.28 kW; (c) 1.05 kW  **57.** 0.50 min  **59.** (a) $1.7 \times 10^4$
W/m$^2$; (b) 18 W/m$^2$  **61.** $-4.2$°C  **63.** 1.1 m  **65.** 0.40 cm/h
**67.** 10%  **69.** $4.5 \times 10^2$ J/kg·K  **71.** 0.432 cm$^3$
**73.** (a) $11 p_1 V_1$; (b) $6 p_1 V_1$  **75.** $4.83 \times 10^{-2}$ cm$^3$  **77.** 23 J
**79.** $3.1 \times 10^2$ J  **81.** 10.5°C  **83.** 79.5°C  **85.** 8.6 J  **87.** 333 J
**89.** (a) 90 W; (b) $2.3 \times 10^2$ W; (c) $3.3 \times 10^2$ W  **91.** (a) $1.87 \times 10^4$; (b) 10.4 h  **93.** (a) $-45$ J; (b) $+45$ J  **95.** (a) 80 J; (b) 80 J
**97.** $-6.1$ nW  **99.** 1.17 C°

## Chapter 19

**CP**  **1.** all but $c$  **2.** (a) all tie; (b) 3, 2, 1  **3.** gas $A$  **4.** 5
(greatest change in $T$), then tie of 1, 2, 3, and 4  **5.** 1, 2, 3 ($Q_3 = 0$, $Q_2$ goes into work $W_2$, but $Q_1$ goes into greater work $W_1$ and
increases gas temperature)  **Q**  **1.** 20 J  **3.** $d$, then $a$ and $b$
tie, then $c$  **5.** (a) 3; (b) 1; (c) 4; (d) 2; (e) yes  **7.** constant-volume process  **9.** (a) 1, 2, 3, 4; (b) 1, 2, 3  **P**  **1.** (a) 0.0127
mol; (b) $7.64 \times 10^{21}$ atoms  **3.** 25 molecules/cm$^3$  **5.** 186 kPa
**7.** (a) 0.0388 mol; (b) 220°C  **9.** (a) $3.14 \times 10^3$ J; (b) from
**11.** 360 K  **13.** 5.60 kJ  **15.** (a) 1.5 mol; (b) $1.8 \times 10^3$ K;
(c) $6.0 \times 10^2$ K; (d) 5.0 kJ  **17.** $2.0 \times 10^5$ Pa  **19.** $1.8 \times 10^2$ m/s
**21.** (a) 511 m/s; (b) $-200$°C; (c) 899°C  **23.** 1.9 kPa
**25.** (a) $5.65 \times 10^{-21}$ J; (b) $7.72 \times 10^{-21}$ J; (c) 3.40 kJ; (d) 4.65 kJ
**27.** (a) $6.76 \times 10^{-20}$ J; (b) 10.7  **29.** (a) $6 \times 10^9$ km
**31.** (a) $3.27 \times 10^{10}$ molecules/cm$^3$; (b) 172 m  **33.** (a) 420 m/s;
(b) 458 m/s; (c) yes  **35.** (a) 6.5 km/s; (b) 7.1 km/s  **37.** (a) $1.0 \times 10^4$ K; (b) $1.6 \times 10^5$ K; (c) $4.4 \times 10^2$ K; (d) $7.0 \times 10^3$ K; (e)
no; (f) yes  **39.** (a) 7.0 km/s; (b) $2.0 \times 10^{-8}$ cm; (c) $3.5 \times 10^{10}$
collisions/s  **41.** (a) 0.67; (b) 1.2; (c) 1.3; (d) 0.33  **43.** (a) 0;
(b) $+374$ J; (c) $+374$ J; (d) $+3.11 \times 10^{-22}$ J  **45.** 15.8 J/mol·K
**47.** (a) $6.6 \times 10^{-26}$ kg; (b) 40 g/mol  **49.** (a) 3.49 kJ; (b) 2.49
kJ; (c) 997 J; (d) 1.00 kJ  **51.** 8.0 kJ  **53.** (a) 6.98 kJ; (b) 4.99
kJ; (c) 1.99 kJ; (d) 2.99 kJ  **55.** (a) 14 atm; (b) $6.2 \times 10^2$ K
**57.** $-15$ J  **59.** $-20$ J  **61.** (a) diatomic; (b) 446 K; (c) 8.10 mol
**63.** (a) 3.74 kJ; (b) 3.74 kJ; (c) 0; (d) 0; (e) $-1.81$ kJ; (f) 1.81 kJ;
(g) $-3.22$ kJ; (h) $-1.93$ kJ; (i) $-1.29$ kJ; (j) 520 J; (k) 0; (l) 520
J; (m) 0.0246 m$^3$; (n) 2.00 atm; (o) 0.0373 m$^3$; (p) 1.00 atm
**65.** (a) 900 cal; (b) 0; (c) 900 cal; (d) 450 cal; (e) 1200 cal; (f) 300
cal; (g) 900 cal; (h) 450 cal; (i) 0; (j) $-900$ cal; (k) 900 cal; (l) 450 cal
**67.** 349 K  **69.** (a) $-374$ J; (b) 0; (c) $+374$ J; (d) $+3.11 \times 10^{-22}$ J
**71.** $7.03 \times 10^9$ s$^{-1}$  **73.** (a) 2.00 atm; 333 J; (c) 0.961 atm;
(d) 236 J  **75.** (a) monatomic; (b) $2.7 \times 10^4$ K; (c) $4.5 \times 10^4$ mol;
(d) 3.4 kJ; (e) $3.4 \times 10^2$ kJ; (f) 0.010  **77.** (a) 8.0 atm; (b) 300
K; (c) 4.4 kJ; (d) 3.2 atm; (e) 120 K; (f) 2.9 kJ; (g) 4.6 atm;
(h) 170 K; (i) 3.4 kJ  **79.** (a) 38 L; (b) 71 g  **81.** (a) $3/v_0^3$;
(b) $0.750 v_0$; (c) $0.775 v_0$  **83.** (a) $-2.37$ kJ; (b) 2.37 kJ
**85.** $-3.0$ J  **87.** (b) 125 J; (c) to

## Chapter 20

**CP**  **1.** a, b, c  **2.** smaller ($Q$ is smaller)  **3.** c, b, a  **4.** a, d, c, b
**5.** b  **Q**  **1.** $a$ and $c$ tie, then $b$ and $d$ tie  **3.** $b, a, c, d$
**5.** unchanged  **7.** A, first; B, first and second; C, second; D, neither  **9.** (a) same; (b) increase; (c) decrease  **P**  **1.** 14.4 J/K
**3.** (a) $5.79 \times 10^4$ J; (b) 173 J/K  **5.** (a) 9.22 kJ; (b) 23.1 J/K;
(c) 0  **7.** (a) 57.0°C; (b) $-22.1$ J/K; (c) $+24.9$ J/K; (d) $+2.8$ J/K
**9.** (a) $-710$ mJ/K; (b) $+710$ mJ/K; (c) $+723$ mJ/K; (d) $-723$
mJ/K; (e) $+13$ mJ/K; (f) 0  **11.** (a) 320 K; (b) 0; (c) $+1.72$ J/K
**13.** (a) 0.333; (b) 0.215; (c) 0.644; (d) 1.10; (e) 1.10; (f) 0;
(g) 1.10; (h) 0; (i) $-0.889$; (j) $-0.889$; (k) $-1.10$; (l) $-0.889$;
(m) 0; (n) 0.889; (o) 0  **15.** $+0.76$ J/K  **17.** (a) $-943$ J/K;
(b) $+943$ J/K; (c) yes  **19.** $-1.18$ J/K  **21.** (a) 0.693; (b) 4.50;
(c) 0.693; (d) 0; (e) 4.50; (f) 23.0 J/K; (g) $-0.693$; (h) 7.50;
(i) $-0.693$; (j) 3.00; (k) 4.50; (l) 23.0 J/K  **23.** (a) 266 K;
(b) 341 K  **25.** (a) 23.6%; (b) $1.49 \times 10^4$ J  **27.** 97 K
**29.** (a) 1.47 kJ; (b) 554 J; (c) 918 J; (d) 62.4%  **31.** (a) 2.27 kJ;
(b) 14.8 kJ; (c) 15.4%; (d) 75.0%; (e) greater
**33.** (a) 33 kJ; (b) 25 kJ; (c) 26 kJ; (d) 18 kJ  **35.** (a) 3.00; (b)
1.98; (c) 0.660; (d) 0.495; (e) 0.165; (f) 34.0%  **37.** 20 J
**39.** 440 W  **41.** 2.03  **43.** 0.25 hp  **47.** (a) $W = N!/(n_1! \, n_2! \, n_3!)$;
(b) $[(N/2)! \, (N/2)!]/[(N/3)! \, (N/3)! \, (N/3)!]$; (c) $4.2 \times 10^{16}$
**49.** (a) 87 m/s; (b) $1.2 \times 10^2$ m/s; (c) 22 J/K  **51.** (a) 78%;
(b) 82 kg/s  **53.** (a) 40.9°C; (b) $-27.1$ J/K; (c) 30.5 J/K;
(d) 3.4 J/K  **55.** $1.18 \times 10^3$ J/K  **57.** (a) 0; (b) 0; (c) $-23.0$ J/K;
(d) 23.0 J/K  **59.** (a) 25.5 kJ; (b) 4.73 kJ; (c) 18.5%
**61.** 0.141 J/K·s  **63.** (a) 42.6 kJ; (b) 7.61 kJ  **65.** (a) 4.45 J/K;
(b) no  **67.** (a) 1; (b) 1; (c) 3; (d) 10; (e) $1.5 \times 10^{-23}$ J/K;
(f) $3.2 \times 10^{-23}$ J/K  **69.** $+3.59$ J/K  **71.** (a) 1.95 J/K;
(b) 0.650 J/K; (c) 0.217 J/K; (d) 0.072 J/K; (e) decrease
**73.** (a) $1.26 \times 10^{14}$; (b) $4.71 \times 10^{13}$; (c) 0.37; (d) $1.01 \times 10^{29}$;
(e) $1.37 \times 10^{28}$; (f) 0.14; (g) $9.05 \times 10^{58}$; (h) $1.64 \times 10^{57}$; (i) 0.018;
(j) decrease

## Chapter 21

**CP**  **1.** $C$ and $D$ attract; $B$ and $D$ attract  **2.** (a) leftward;
(b) leftward; (c) leftward  **3.** (a) $a, c, b$; (b) less than
**4.** $-15e$ (net charge of $-30e$ is equally shared)
**Q**  **1.** $a$ and $b$  **3.** 3, 1, 2, 4 (zero)  **5.** $b$ and $c$ tie, then $a$
(zero)  **7.** $2kq^2/r^2$, up the page  **9.** (a) same; (b) less than;
(c) cancel; (d) add; (e) adding components; (f) positive
direction of $y$; (g) negative direction of $y$; (h) positive
direction of $x$; (i) negative direction of $x$  **P**  **1.** 1.39 m
**3.** 2.81 N  **5.** 0.500  **7.** (a) $-1.00$ μC; (b) 3.00 μC
**9.** (a) 0.17 N; (b) $-0.046$ N  **11.** $-4.00$  **13.** (a) 1.60 N;
(b) 2.77 N  **15.** (a) $-14$ cm; (b) 0  **17.** (a) 35 N; (b) $-10$°;
(c) $-8.4$ cm; (d) $+2.7$ cm  **19.** (a) 3.00 cm; (b) 0; (c) $-0.444$
**21.** (a) 0; (b) 12 cm; (c) 0; (d) $4.9 \times 10^{-26}$ N  **23.** $3.8 \times 10^{-8}$ C
**25.** (a) $3.2 \times 10^{-19}$ C; (b) 2  **27.** $6.3 \times 10^{11}$  **29.** 122 mA
**31.** $1.3 \times 10^7$ C  **33.** (a) $-6.05$ cm; (b) 6.05 cm
**35.** (a) 0; (b) $1.9 \times 10^{-9}$ N  **37.** (a) $^9$B; (b) $^{13}$N; (c) $^{12}$C
**39.** $1.31 \times 10^{-22}$ N  **41.** (a) $2.00 \times 10^{10}$ electrons;
(b) $1.33 \times 10^{10}$ electrons  **43.** 0.19 MC  **45.** 3.8 N
**47.** (a) $8.99 \times 10^9$ N; (b) 8.99 kN  **49.** $1.7 \times 10^8$ N
**51.** (a) 0.5; (b) 0.15; (c) 0.85  **53.** (a) $5.7 \times 10^{13}$ C;
(b) cancels out; (c) $6.0 \times 10^5$ kg  **55.** (b) 3.1 cm
**57.** $-1.32 \times 10^{13}$ C  **59.** (a) $(0.829 \text{ N})\hat{i}$; (b) $(-0.621 \text{ N})\hat{j}$
**61.** $2.2 \times 10^{-6}$ kg  **63.** $-45$ μC  **65.** (a) $5.1 \times 10^2$ N;
(b) $7.7 \times 10^{28}$ m/s$^2$  **67.** $4.68 \times 10^{-19}$ N  **69.** (a) $1.72L$; (b) 0

## Chapter 22

**CP** **1.** (a) rightward; (b) leftward; (c) leftward; (d) rightward (p and e have same charge magnitude, and p is farther) **2.** (a) toward positive $y$; (b) toward positive $x$; (c) toward negative $y$ **3.** (a) leftward; (b) leftward; (c) decrease **4.** (a) all tie; (b) 1 and 3 tie, then 2 and 4 tie **Q** **1.** $a, b, c$ **3.** (a) to their left; (b) no **5.** (a) yes; (b) toward; (c) no (the field vectors are not along the same line); (d) cancel; (e) add; (f) adding components; (g) toward negative $y$ **7.** $e, b$, then $a$ and $c$ tie, then $d$ (zero) **9.** (a) 4, 3, 1, 2; (b) 3, then 1 and 4 tie, then 2 **11.** $a, b, c$ **P** **1.** (a) $6.4 \times 10^{-18}$ N; (b) 20 N/C **3.** 56 pC **5.** (a) $3.07 \times 10^{21}$ N/C; (b) outward **7.** $-30$ cm **9.** $(1.02 \times 10^5$ N/C$)\hat{j}$ **11.** (a) $1.38 \times 10^{-10}$ N/C; (b) $180°$ **13.** (a) 160 N/C; (b) $45°$ **15.** (a) $3.60 \times 10^{-6}$ N/C; (b) $2.55 \times 10^{-6}$ N/C; (c) $3.60 \times 10^{-4}$ N/C; (d) $7.09 \times 10^{-7}$ N/C; (e) As the proton nears the disk, the forces on it from electrons $e_s$ more nearly cancel. **17.** (a) $-90°$; (b) $+2.0$ $\mu$C; (c) $-1.6$ $\mu$C **19.** (a) $qd/4\pi\varepsilon_0 r^3$; (b) $-90°$ **23.** 0.506 **25.** (a) 23.8 N/C; (b) $-90°$ **27.** (a) $-5.19 \times 10^{-14}$ C/m; (b) $1.57 \times 10^{-3}$ N/C; (c) $-180°$; (d) $1.52 \times 10^{-8}$ N/C; (e) $1.52 \times 10^{-8}$ N/C **29.** (a) $1.62 \times 10^6$ N/C; (b) $-45°$ **31.** 1.57 **35.** 0.346 m **37.** 28% **39.** $3.51 \times 10^{15}$ m/s² **41.** $6.6 \times 10^{-15}$ N **43.** (a) $1.5 \times 10^3$ N/C; (b) $2.4 \times 10^{-16}$ N; (c) up; (d) $1.6 \times 10^{-26}$ N; (e) $1.5 \times 10^{10}$ **45.** (a) $1.92 \times 10^{12}$ m/s²; (b) $1.96 \times 10^5$ m/s **47.** $-5e$ **49.** (a) $2.7 \times 10^6$ m/s; (b) 1.0 kN/C **51.** 27 $\mu$m **53.** (a) 0.245 N; (b) $-11.3°$; (c) 108 m; (d) $-21.6$ m **55.** (a) $2.6 \times 10^{-10}$ N; (b) $3.1 \times 10^{-8}$ N; (c) moves to stigma **57.** (a) $9.30 \times 10^{-15}$ C·m; (b) $2.05 \times 10^{-11}$ J **59.** $(1/2\pi)(pE/I)^{0.5}$ **61.** $1.22 \times 10^{-23}$ J **63.** $217°$ **65.** (a) 47 N/C; (b) 27 N/C **67.** (a) 6.0 mm; (b) $180°$ **69.** $+1.00$ $\mu$C **71.** (a) $8.87 \times 10^{-15}$ N; (b) 120 **73.** 38 N/C **75.** 9:30 **77.** (a) $-0.029$ C; (b) repulsive forces would explode the sphere **79.** (a) $-1.0$ cm; (b) 0; (c) 10 pC **81.** (a) $-1.49 \times 10^{-26}$ J; (b) $(-1.98 \times 10^{-26}$ N·m$)\hat{k}$; (c) $3.47 \times 10^{-26}$ J **83.** 61 N/C **85.** (a) $(-1.80$ N/C$)\hat{i}$; (b) $(43.2$ N/C$)\hat{i}$; (c) $(-6.29$ N/C$)\hat{i}$ **87.** (a) top row: 4, 8, 12; middle row: 5, 10, 14; bottom row: 7, 11, 16; (b) $1.63 \times 10^{-19}$ C

## Chapter 23

**CP** **1.** (a) $+EA$; (b) $-EA$; (c) 0; (d) 0 **2.** (a) 2; (b) 3; (c) 1 **3.** (a) equal; (b) equal; (c) equal **4.** 3 and 4 tie, then 2, 1 **Q** **1.** all tie **3.** (a) 8 N·m²/C; (b) 0 **5.** $a, c$, then $b$ and $d$ tie (zero) **7.** (a) all tie ($E = 0$); (b) all tie **9.** all tie **P** **1.** $-0.015$ N·m²/C **3.** (a) 0; (b) $-3.92$ N·m²/C; (c) 0; (d) 0 **5.** $2.0 \times 10^5$ N·m²/C **7.** 3.01 nN·m²/C **9.** 3.54 $\mu$C **11.** (a) 8.23 N·m²/C; (b) 72.9 pC; (c) 8.23 N·m²/C; (d) 72.9 pC **13.** (a) 0; (b) 0.0417 **15.** $-1.70$ nC **17.** (a) $4.5 \times 10^{-7}$ C/m²; (b) $5.1 \times 10^4$ N/C **19.** (a) 37 $\mu$C; (b) $4.1 \times 10^6$ N·m²/C **21.** (a) $-3.0 \times 10^{-6}$ C; (b) $+1.3 \times 10^{-5}$ C **23.** 5.0 $\mu$C/m **25.** (a) 0.32 $\mu$C; (b) 0.14 $\mu$C **27.** (a) 0.214 N/C; (b) inward; (c) 0.855 N/C; (d) outward; (e) $-3.40 \times 10^{-12}$ C; (f) $-3.40 \times 10^{-12}$ C **29.** (a) $2.3 \times 10^6$ N/C; (b) outward; (c) $4.5 \times 10^5$ N/C; (d) inward **31.** $3.8 \times 10^{-8}$ C/m² **33.** $-1.5$ **35.** (a) $5.3 \times 10^7$ N/C; (b) 60 N/C **37.** (a) 0; (b) 0; (c) $(-7.91 \times 10^{-11}$ N/C$)\hat{i}$ **39.** 0.44 mm **41.** 5.0 nC/m² **43.** (a) 0; (b) 1.31 $\mu$N/C; (c) 3.08 $\mu$N/C; (d) 3.08 $\mu$N/C **45.** $-7.5$ nC **47.** (a) $2.50 \times 10^4$ N/C; (b) $1.35 \times 10^4$ N/C **49.** $1.79 \times 10^{-11}$ C/m² **51.** (a) 0; (b) 56.2 mN/C; (c) 112 mN/C; (d) 49.9 mN/C; (e) 0; (f) 0; (g) $-5.00$ fC; (h) 0 **53.** $6K\varepsilon_0 r^3$ **55.** (a) 7.78 fC; (b) 0; (c) 5.58 mN/C; (d) 22.3 mN/C **57.** (a) 0.125; (b) 0.500 **59.** (a) $+2.0$ nC; (b) $-1.2$ nC; (c) $+1.2$ nC; (d) $+0.80$ nC **61.** (a) 5.4 N/C; (b) 6.8 N/C **63.** (a) 0; (b) $2.88 \times 10^4$ N/C; (c) 200 N/C **65.** $(5.65 \times 10^4$ N/C$)\hat{j}$ **67.** (a) $-2.53 \times 10^{-2}$ N·m²/C; (b) $+2.53 \times 10^{-2}$ N·m²/C **69.** (a) 0; (b) $q_a/4\pi\varepsilon_0 r^2$; (c) $(q_a + q_b)/4\pi\varepsilon_0 r^2$ **71.** $-1.04$ nC **73.** 3.6 nC

**75.** (a) 693 kg/s; (b) 693 kg/s; (c) 347 kg/s; (d) 347 kg/s; (e) 575 kg/s **79.** (a) $+4.0$ $\mu$C; (b) $-4.0$ $\mu$C **81.** (a) $0.25R$; (b) $2.0R$

## Chapter 24

**CP** **1.** (a) negative; (b) increase **2.** (a) positive; (b) higher **3.** (a) rightward; (b) 1, 2, 3, 5: positive; 4, negative; (c) 3, then 1, 2, and 5 tie, then 4 **4.** all tie **5.** $a, c$ (zero), $b$ **6.** (a) 2, then 1 and 3 tie; (b) 3; (c) accelerate leftward **Q** **1.** (a) 1 and 2; (b) none; (c) no; (d) 1 and 2, yes; 3 and 4, no **3.** $-4q/4\pi\varepsilon_0 d$ **5.** (a) higher; (b) positive; (c) negative; (d) all tie **7.** (a) 3 and 4 tie, then 1 and 2 tie; (b) 1 and 2, increase; 3 and 4, decrease **9.** (a) 0; (b) 0; (c) 0; (d) all three quantities still 0 **P** **1.** $2.8 \times 10^5$ **3.** (a) $3.0 \times 10^5$ C; (b) $3.6 \times 10^6$ J **5.** 8.8 mm **7.** (a) $1.87 \times 10^{-21}$ J; (b) $-11.7$ mV **9.** $-32.0$ V **11.** (a) $-0.268$ mV; (b) $-0.681$ mV **13.** (a) 3.3 nC; (b) 12 nC/m² **15.** 0.562 mV **17.** (a) 6.0 cm; (b) $-12.0$ cm **19.** (a) 0.54 mm; (b) 790 V **21.** 16.3 $\mu$V **23.** (a) $-2.30$ V; (b) $-1.78$ V **25.** (a) 24.3 mV; (b) 0 **27.** 32.4 mV **29.** 47.1 $\mu$V **31.** 13 kV **33.** 18.6 mV **35.** $(-12$ V/m$)\hat{i} + (12$ V/m$)\hat{j}$ **37.** $(-4.0 \times 10^{-16}$ N$)\hat{i} + (1.6 \times 10^{-16}$ N$)\hat{j}$ **39.** 150 N/C **41.** $-0.192$ pJ **43.** (a) 0.90 J; (b) 4.5 J **45.** (a) $+6.0 \times 10^4$ V; (b) $-7.8 \times 10^5$ V; (c) 2.5 J; (d) increase; (e) same; (f) same **47.** 2.5 km/s **49.** 22 km/s **51.** (a) 0.225 J; (b) $A$ 45.0 m/s², $B$ 22.5 m/s²; (c) $A$ 7.75 m/s, $B$ 3.87 m/s **53.** 0.32 km/s **55.** (a) proton; (b) 65.3 km/s **57.** $1.6 \times 10^{-9}$ m **59.** (a) 3.0 J; (b) $-8.5$ m **61.** (a) 12; (b) 2 **63.** $2.5 \times 10^{-8}$ C **65.** (a) $-1.8 \times 10^2$ V; (b) 2.9 kV; (c) $-8.9$ kV **67.** (a) 12 kN/C; (b) 1.8 kV; (c) 5.8 cm **69.** $7.0 \times 10^5$ m/s **71.** (a) 1.8 cm; (b) $8.4 \times 10^5$ m/s; (c) $2.1 \times 10^{-17}$ N; (d) positive; (e) $1.6 \times 10^{-17}$ N; (f) negative **73.** (a) $+7.19 \times 10^{-10}$ V; (b) $+2.30 \times 10^{-28}$ J; (c) $+2.43 \times 10^{-29}$ J **75.** 2.1 days **77.** (a) 64 N/C; (b) 2.9 V; (c) 0 **79.** $2.30 \times 10^{-28}$ J **81.** $2.30 \times 10^{-22}$ J **83.** (a) $3.6 \times 10^5$ V; (b) no **85.** $-1.92$ MV **87.** $1.48 \times 10^7$ m/s **89.** $6.4 \times 10^8$ V **93.** (a) $Q/4\pi\varepsilon_0 r$; (b) $(\rho/3\varepsilon_0)(1.5r_2^2 - 0.50r^2 - r_1^3 r^{-1})$, $\rho = Q/[(4\pi/3)(r_2^3 - r_1^3)]$; (c) $(\rho/2\varepsilon_0)(r_2^2 - r_1^2)$, with $\rho$ as in (b); (d) yes **95.** $p/2\pi\varepsilon_0 r^3$ **97.** 2.90 kV **99.** (a) 0.484 MeV; (b) 0 **103.** (a) 38 s; (b) 280 days **105.** $-1.7$ **107.** 1 **109.** (a) 1.48 nC; (b) 795 V **111.** $-187$ V **115.** (c) 4.2 V

## Chapter 25

**CP** **1.** (a) same; (b) same **2.** (a) decreases; (b) increases; (c) decreases **3.** (a) $V, q/2$; (b) $V/2; q$ **Q** **1.** $a, 2; b, 1; c, 3$ **3.** $a$, series; $b$, parallel; $c$, parallel **5.** (a) no; (b) yes; (c) all tie **7.** (a) same; (b) same; (c) more; (d) more **9.** parallel, $C_1$ alone, $C_2$ alone, series **11.** (a) increase; (b) same; (c) increase; (d) increase; (e) increase; (f) increase **P** **1.** (a) 3.5 pF; (b) 3.5 pF; (c) 57 V **3.** $6.79 \times 10^{-4}$ F/m² **5.** (a) 144 pF; (b) 17.3 nC **7.** 0.280 pF **9.** 3.16 $\mu$F **11.** 315 mC **13.** (a) 789 $\mu$C; (b) 78.9 V **15.** 43 pF **17.** (a) 3.00 $\mu$F; (b) 60.0 $\mu$C; (c) 10.0 V; (d) 30.0 $\mu$C; (e) 10.0 V; (f) 20.0 $\mu$C; (g) 5.00 V; (h) 20.0 $\mu$C **19.** (a) 50 V; (b) $5.0 \times 10^{-5}$ C; (c) $1.5 \times 10^{-4}$ C **21.** 3.6 pC **23.** (a) 4.0 $\mu$F; (b) 2.0 $\mu$F **25.** (a) $4.5 \times 10^{14}$; (b) $1.5 \times 10^{14}$; (c) $3.0 \times 10^{14}$; (d) $4.5 \times 10^{14}$; (e) up; (f) up **27.** (a) 9.00 $\mu$C; (b) 16.0 $\mu$C; (c) 9.00 $\mu$C; (d) 16.0 $\mu$C; (e) 8.40 $\mu$C; (f) 16.8 $\mu$C; (g) 10.8 $\mu$C; (h) 14.4 $\mu$C **29.** 0.27 J **31.** 72 F **33.** (a) $9.16 \times 10^{-18}$ J/m³; (b) $9.16 \times 10^{-6}$ J/m³; (c) $9.16 \times 10^6$ J/m³; (d) $9.16 \times 10^{18}$ J/m³; (e) $\infty$ **35.** (a) 16.0 V; (b) 45.1 pJ; (c) 120 pJ; (d) 75.2 pJ **37.** (a) 190 V; (b) 95 mJ **39.** 0.11 J/m³ **41.** Pyrex **43.** 81 pF/m **45.** 0.63 m² **47.** 66 $\mu$J **49.** 17.3 pF **51.** (a) 10 kV/m; (b) 5.0 nC; (c) 4.1 nC **53.** (a) 0.107 nF; (b) 7.79 nC; (c) 7.45 nC **55.** (a) 89 pF; (b) 0.12 nF; (c) 11 nC; (d) 11 nC; (e) 10 kV/m; (f) 2.1 kV/m; (g) 88 V; (h) $-0.17$ $\mu$J **57.** (a) 7.20 $\mu$C; (b) 18.0 $\mu$C; (c) Battery supplies charges only to plates to which it is connected; charges on other plates are due to electron transfers between plates, in accord with new distribution of voltages across the capacitors. So battery does not directly supply charge on capacitor 4. **59.** (a) 10 $\mu$C; (b) 20 $\mu$C **61.** 45 $\mu$C **63.** 16 $\mu$C **65.** (a) 2.40 $\mu$F; (b) 0.480 mC; (c) 80 V; (d) 0.480 mC; (e) 120 V **67.** 40 $\mu$F **69.** (a) 200 kV/m; (b) 200 kV/m; (c) 1.77 $\mu$C/m$^2$; (d) 4.60 $\mu$C/m$^2$; (e) $-2.83$ $\mu$C/m$^2$ **71.** 4.9% **73.** 1.06 nC **75.** (a) 0.708 pF; (b) 0.600; (c) 1.02 $\times$ 10$^{-9}$ J; (d) sucked in **77.** 5.3 V

## Chapter 26

**CP** **1.** 8 A, rightward **2.** (a)–(c) rightward **3.** $a$ and $c$ tie, then $b$ **4.** device 2 **5.** (a) and (b) tie, then (d), then (c) **Q** **1.** $a$, $b$, and $c$ all tie, then $d$ **3.** tie of $A$, $B$, and $C$, then tie of $A + B$ and $B + C$, then $A + B + C$ **5.** (a) top-bottom, front-back, left-right; (b) top-bottom, front-back, left-right; (c) top-bottom, front-back, left-right; (d) top-bottom, front-back, left-right **7.** (a) $C$, $B$, $A$; (b) all tie; (c) $A$, $B$, $C$; (d) all tie **9.** (a) $B$, $A$, $C$; (b) $B$, $A$, $C$ **P** **1.** (a) 1.2 kC; (b) 7.5 $\times$ 10$^{21}$ **3.** 6.7 $\mu$C/m$^2$ **5.** 0.38 mm **7.** (a) 6.4 A/m$^2$; (b) north; (c) cross-sectional area **9.** 13 min **11.** 18.1 $\mu$A **13.** (a) 1.33 A; (b) 0.666 A; (c) $J_a$ **15.** 2.0 $\times$ 10$^6$ ($\Omega\cdot$m)$^{-1}$ **17.** 2.0 $\times$ 10$^{-8}$ $\Omega\cdot$m **19.** 2.4 $\Omega$ **21.** 54 $\Omega$ **23.** 3.0 **25.** (1.8 $\times$ 10$^3$)$^\circ$C **27.** 3.35 $\times$ 10$^{-7}$ C **29.** 8.2 $\times$ 10$^{-4}$ $\Omega\cdot$m **31.** (a) 38.3 mA; (b) 109 A/m$^2$; (c) 1.28 cm/s; (d) 227 V/m **33.** (a) 6.00 mA; (b) 1.59 $\times$ 10$^{-8}$ V; (c) 21.2 n$\Omega$ **35.** 981 k$\Omega$ **39.** (a) 1.0 kW; (b) US$0.25 **41.** 0.135 W **43.** (a) 10.9 A; (b) 10.6 $\Omega$; (c) 4.50 MJ **45.** 150 s **47.** (a) 28.8 $\Omega$; (b) 2.60 $\times$ 10$^{19}$ s$^{-1}$ **49.** (a) 5.85 m; (b) 10.4 m **51.** (a) US$4.46; (b) 144 $\Omega$; (c) 0.833 A **53.** (a) 5.1 V; (b) 10 V; (c) 10 W; (d) 20 W **55.** (a) 2.3 $\times$ 10$^{12}$; (b) 5.0 $\times$ 10$^3$; (c) 10 MV **57.** (a) $-8.6\%$; (b) smaller **59.** 660 W **61.** (a) silver; (b) 51.6 n$\Omega$ **63.** (a) 1.37; (b) 0.730 **65.** 28.8 kC **67.** 146 kJ **69.** 3.0 $\times$ 10$^6$ J/kg **71.** 2.4 kW **73.** 560 W **75.** (a) 250°C; (b) yes

## Chapter 27

**CP** **1.** (a) rightward; (b) all tie; (c) $b$, then $a$ and $c$ tie; (d) $b$, then $a$ and $c$ tie **2.** (a) all tie; (b) $R_1$, $R_2$, $R_3$ **3.** (a) less; (b) greater; (c) equal **4.** (a) $V/2$, $i$; (b) $V$, $i/2$ **5.** (a) 1, 2, 4, 3; (b) 4, tie of 1 and 2; then 3 **Q** **1.** (a) series; (b) parallel; (c) parallel **3.** (a) equal; (b) more **5.** parallel, $R_2$, $R_1$, series **7.** (a) same; (b) same; (c) less; (d) more **9.** (a) less; (b) less; (c) more **11.** (a) all tie; (b) 1, 3, 2 **P** **1.** (a) 80 J; (b) 67 J; (c) 13 J **3.** 11 kJ **5.** (a) 14 V; (b) 1.0 $\times$ 10$^2$ W; (c) 6.0 $\times$ 10$^2$ W; (d) 10 V; (e) 1.0 $\times$ 10$^2$ W **7.** (a) 0.50 A; (b) 1.0 W; (c) 2.0 W; (d) 6.0 W; (e) 3.0 W; (f) supplied; (g) absorbed **9.** (a) 4.00 $\Omega$; (b) parallel **11.** 3.6 $\times$ 10$^3$ A **13.** (a) 50 V; (b) 48 V; (c) negative **15.** (a) 6.9 km; (b) 20 $\Omega$ **17.** (a) 0.333 A; (b) right; (c) 720 J **19.** 8.0 $\Omega$ **21.** (a) 0.004 $\Omega$; (b) 1 **23.** 4.50 $\Omega$ **25.** 5.56 A **27.** (a) 50 mA; (b) 60 mA; (c) 9.0 V **29.** 3$d$ **31.** 48.3 V **33.** (a) $-11$ V; (b) $-9.0$ V **35.** 1.43 $\Omega$ **37.** (a) 0.67 A; (b) down; (c) 0.33 A; (d) up; (e) 0.33 A; (f) up; (g) 3.3 V **39.** (a) 5.25 V; (b) 1.50 V; (c) 5.25 V; (d) 6.75 V **41.** (a) 0.150 $\Omega$; (b) 240 W **43.** (a) 0.709 W; (b) 0.050 W; (c) 0.346 W; (d) 1.26 W; (e) $-0.158$ W **45.** 9 **47.** (a) 1.11 A; (b) 0.893 A; (c) 126 m **49.** $-3.0\%$ **51.** (a) 0.45 A **53.** (a) 55.2 mA; (b) 4.86 V; (c) 88.0 $\Omega$; (d) decrease **57.** 4.61 **59.** (a) 2.41 $\mu$s; (b) 161 pF **61.** 0.208 ms **63.** 0.72 M$\Omega$ **65.** (a) 1.1 mA; (b) 0.55 mA; (c) 0.55 mA; (d) 0.82 mA; (e) 0.82 mA; (f) 0; (g) 4.0 $\times$ 10$^2$ V; (h) 6.0 $\times$ 10$^2$ V **67.** 411 $\mu$A **69.** (a) 0.955 $\mu$C/s; (b) 1.08 $\mu$W; (c) 2.74 $\mu$W; (d) 3.82 $\mu$W **71.** (a) 3.0 kV; (b) 10 s; (c) 11 G$\Omega$ **73.** (a) 24.8 $\Omega$; (b) 14.9 k$\Omega$ **75.** (a) 1.32 $\times$ 10$^7$ A/m$^2$; (b) 8.90 V; (c) copper; (d) 1.32 $\times$ 10$^7$ A/m$^2$; (e) 51.1 V; (f) iron **77.** the cable

**79.** (a) 3.00 A; (b) 3.75 A; (c) 3.94 A **81.** 20 $\Omega$ **83.** (a) 3.00 A; (b) down; (c) 1.60 A; (d) down; (e) supply; (f) 55.2 W; (g) supply; (h) 6.40 W **85.** (a) 85.0 $\Omega$; (b) 915 $\Omega$ **87.** (a) 1.0 V; (b) 50 m$\Omega$ **89.** $-13$ $\mu$C **91.** 4.0 V **93.** 3 **97.** (a) 1.5 mA; (b) 0; (c) 1.0 mA **99.** (a) 0; (b) 14.4 W **103.** (a) 60.0 mA; (b) down; (c) 180 mA; (d) left; (e) 240 mA; (f) up **105.** (a) 4.0 A; (b) up; (c) 0.50 A; (d) down; (e) 64 W; (f) 16 W; (g) supplied; (h) absorbed **107.** (a) 1.00 A; (b) 24.0 W **109.** 1.00 $\times$ 10$^{-6}$ **111.** (b) yes

## Chapter 28

**CP** **1.** $a$, $+z$; $b$, $-x$, $c$, $\vec{F}_B = 0$ **2.** (a) 2, then tie of 1 and 3 (zero); (b) 4 **3.** (a) electron; (b) clockwise **4.** $-y$ **5.** (a) all tie; (b) 1 and 4 tie, then 2 and 3 tie **Q** **1.** (a) $\vec{F}_E$; (b) $\vec{F}_B$ **3.** (a) no, because $\vec{v}$ and $\vec{F}_B$ must be perpendicular; (b) yes; (c) no, because $\vec{B}$ and $\vec{F}_B$ must be perpendicular **5.** (a) $+z$ and $-z$ tie, then $+y$ and $-y$ tie, then $+x$ and $-x$ tie (zero); (b) $+y$ **7.** (a) negative; (b) equal; (c) equal; (d) half-circle **9.** (a) $\vec{B}_1$; (b) $\vec{B}_1$ into page, $\vec{B}_2$ out of page; (c) less **11.** (a) positive; (b) 2 $\rightarrow$ 1 and 2 $\rightarrow$ 4 tie, then 2 $\rightarrow$ 3 (which is zero) **P** **1.** (a) (6.2 $\times$ 10$^{-14}$ N)$\hat{k}$; (b) ($-6.2 \times 10^{-14}$ N)$\hat{k}$ **3.** (a) 400 km/s; (b) 835 eV **5.** $-2.0$ T **7.** $-(0.267$ mT)$\hat{k}$ **9.** ($-11.4$ V/m)$\hat{i}$ $-$ (6.00 V/m)$\hat{j}$ + (4.80 V/m)$\hat{k}$ **11.** 0.68 MV/m **13.** 7.4 $\mu$V **15.** (a) ($-600$ mV/m)$\hat{k}$; (b) 1.20 V **17.** (a) 2.05 $\times$ 10$^7$ m/s; (b) 467 $\mu$T; (c) 13.1 MHz; (d) 76.3 ns **19.** 21.1 $\mu$T **21.** (a) 0.978 MHz; (b) 96.4 cm **23.** 1.2 $\times$ 10$^{-9}$ kg/C **25.** (a) 2.60 $\times$ 10$^6$ m/s; (b) 0.109 $\mu$s; (c) 0.140 MeV; (d) 70.0 kV **27.** 65.3 km/s **29.** (a) 0.358 ns; (b) 0.166 mm; (c) 1.51 mm **31.** (a) 495 mT; (b) 22.7 mA; (c) 8.17 MJ **33.** 5.07 ns **35.** 2.4 $\times$ 10$^2$ m **37.** (a) 200 eV; (b) 20.0 keV; (c) 0.499% **39.** (a) 467 mA; (b) right **41.** (a) 28.2 N; (b) horizontally west **43.** ($-2.50$ mN)$\hat{j}$ + (0.750 mN)$\hat{k}$ **45.** (a) 0.10 T; (b) 31° **47.** ($-4.3 \times 10^{-3}$ N$\cdot$m) $\hat{j}$ **49.** 0.60 $\mu$N **51.** (a) 542 $\Omega$; (b) series; (c) 2.52 $\Omega$; (d) parallel **53.** 2.45 A **55.** (a) 12.7 A; (b) 0.0805 N$\cdot$m **57.** (a) 0.30 A$\cdot$m$^2$; (b) 0.024 N$\cdot$m **59.** (a) 2.86 A$\cdot$m$^2$; (b) 1.10 A$\cdot$m$^2$ **61.** (a) $-(9.7 \times 10^{-4}$ N$\cdot$m)$\hat{i}$ $-$ (7.2 $\times$ 10$^{-4}$ N$\cdot$m)$\hat{j}$ + (8.0 $\times$ 10$^{-4}$ N$\cdot$m)$\hat{k}$; (b) $-6.0 \times 10^{-4}$ J **63.** (a) 90°; (b) 1; (c) 1.28 $\times$ 10$^{-7}$ N$\cdot$m **65.** (a) $-72.0$ $\mu$J; (b) (96.0$\hat{i}$ + 48.0$\hat{k}$) $\mu$N$\cdot$m **67.** 127 u **69.** (a) 20 min; (b) 5.9 $\times$ 10$^{-2}$ N$\cdot$m **71.** 8.2 mm **73.** ($-500$ V/m)$\hat{j}$ **75.** (a) 0.50; (b) 0.50; (c) 14 cm; (d) 14 cm **77.** $-40$ mC **79.** (a) (12.8$\hat{i}$ + 6.41$\hat{j}$) $\times$ 10$^{-22}$ N; (b) 90°; (c) 173° **81.** (a) 6.3 $\times$ 10$^{14}$ m/s$^2$; (b) 3.0 mm **83.** (a) 1.4; (b) 1.0 **85.** (0.80$\hat{j}$ $-$ 1.1$\hat{k}$) mN

## Chapter 29

**CP** **1.** $b$, $c$, $a$ **2.** $d$, tie of $a$ and $c$, then $b$ **3.** $d$, $a$, tie of $b$ and $c$ (zero) **Q** **1.** $c$, $d$, then $a$ and $b$ tie (zero) **3.** $a$, $c$, $b$ **5.** $c$, $a$, $b$ **7.** (a) 1, 3, 2; (b) less **9.** $c$ and $d$ tie, then $b$, $a$ **11.** $b$, $a$, $d$, $c$ (zero) **P** **1.** (a) 16 A; (b) east **3.** (a) 3.3 $\mu$T; (b) yes **5.** (a) 0.102 $\mu$T; (b) out **7.** (a) opposite; (b) 30 A **9.** (a) 4.3 A; (b) out **11.** (a) 1.0 mT; (b) out; (c) 0.80 mT; (d) out **13.** ($-7.75 \times 10^{-23}$ N)$\hat{i}$ **15.** 2.00 rad **17.** 50.3 nT **19.** (80 $\mu$T)$\hat{j}$ **21.** 132 nT **23.** (a) 1.7 $\mu$T; (b) into; (c) 6.7 $\mu$T; (d) into **25.** 61.3 mA **27.** 5.0 $\mu$T **29.** 256 nT **31.** (22.3 pT)$\hat{j}$ **33.** (a) 20 $\mu$T; (b) into **35.** 88.4 pN/m **37.** 800 nN/m **39.** ($-125$ $\mu$N/m)$\hat{i}$ + (41.7 $\mu$N/m)$\hat{j}$ **41.** (3.20 mN)$\hat{j}$ **43.** (a) $-2.5$ $\mu$T$\cdot$m; (b) 0 **45.** (a) 0; (b) 0.850 mT; (c) 1.70 mT; (d) 0.850 mT **47.** (a) 0; (b) 0.10 $\mu$T; (c) 0.40 $\mu$T **49.** 0.30 mT **51.** (a) 533 $\mu$T; (b) 400 $\mu$T **53.** (a) 4.77 cm; (b) 35.5 $\mu$T **55.** 0.272 A **57.** 0.47 A$\cdot$m$^2$ **59.** (a) 2.4 A$\cdot$m$^2$; (b) 46 cm **61.** (a) (0.060 A$\cdot$m$^2$)$\hat{j}$; (b) (96 pT)$\hat{j}$ **63.** (a) 79 $\mu$T; (b) 1.1 $\times$ 10$^{-6}$ N$\cdot$m **65.** (a) 15.3 $\mu$T **67.** (a) 15 A; (b) $-z$ **69.** 7.7 mT **71.** (a) (0.24$\hat{i}$) nT; (b) 0; (c) ($-43\hat{k}$) pT; (d) (0.14$\hat{k}$) nT **73.** 1.28 mm

**77.** $(-0.20\,\text{mT})\hat{\text{k}}$  **85.** (a) 4.8 mT; (b) 0.93 mT; (c) 0
**93.** (a) $\mu_0 ir/2\pi c^2$; (b) $\mu_0 i/2\pi r$; (c) $\mu_0 i(a^2 - r^2)/2\pi(a^2 - b^2)r$; (d) 0

## Chapter 30

**CP** **1.** $b$, then $d$ and $e$ tie, and then $a$ and $c$ tie (zero)  **2.** $a$ and $b$ tie, then $c$ (zero)  **3.** $c$ and $d$ tie, then $a$ and $b$ tie  **4.** $b$, out; $c$, out; $d$, into; $e$, into  **5.** d and e  **6.** (a) 2, 3, 1 (zero); (b) 2, 3, 1  **7.** $a$ and $b$ tie, then $c$   **Q** **1.** (a) all tie (zero); (b) 2, then 1 and 3 tie (zero)  **3.** out  **5.** $d$ and $c$ tie, then $b$, $a$  **7.** (a) all tie (zero); (b) 1 and 2 tie, then 3; (c) all tie (zero)  **9.** (a) more; (b) same; (c) same; (d) same (zero)   **P** **1.** (a) 31 mV; (b) left  **3.** 0.198 mV  **5.** 30 mA  **7.** 0  **9.** 0  **11.** (a) 21.7 V; (b) counterclockwise  **13.** (b) 0.796 $\text{m}^2$  **15.** (a) 40 Hz; (b) 3.2 mV  **17.** 5.50 kV  **19.** 29.5 mC  **21.** (a) $\mu_0 iR^2\pi r^2/2x^3$; (b) $3\mu_0 i\pi R^2 r^2 v/2x^4$; (c) counterclockwise  **23.** (a) $1.26\times10^{-4}$ T; (b) 0; (c) $1.26\times10^{-4}$ T; (d) yes; (e) $5.04\times10^{-8}$ V  **25.** (a) 80 $\mu$V; (b) clockwise  **27.** (a) 13 $\mu$Wb/m; (b) 17%; (c) 0  **29.** 3.68 $\mu$W  **31.** (a) 48.1 mV; (b) 2.67 mA; (c) 0.129 mW  **33.** (a) 0.60 V; (b) up; (c) 1.5 A; (d) clockwise; (e) 0.90 W; (f) 0.18 N; (g) 0.90 W  **35.** (a) 240 $\mu$V; (b) 0.600 mA; (c) 0.144 $\mu$W; (d) 2.87 $\times 10^{-8}$ N; (e) 0.144 $\mu$W  **37.** (a) 71.5 $\mu$V/m; (b) 143 $\mu$V/m  **39.** 0.15 V/m  **41.** (a) 2.45 mWb; (b) 0.645 mH  **43.** 1.81 $\mu$H/m  **45.** (a) decreasing; (b) 0.68 mH  **47.** (b) $L_{\text{eq}} = \Sigma L_j$, sum from $j = 1$ to $j = N$  **49.** 59.3 mH  **51.** 6.91  **53.** 46 $\Omega$  **55.** (a) 8.45 ns; (b) 7.37 mA  **57.** (a) 1.5 s  **59.** (a) $i[1 - \exp(-Rt/L)]$; (b) $(L/R)\ln 2$  **61.** 25.6 ms  **63.** (a) 97.9 H; (b) 0.196 mJ  **65.** (a) 18.7 J; (b) 5.10 J; (c) 13.6 J  **67.** $1.5\times10^8$ V/m  **69.** (a) 34.2 $\text{J/m}^3$; (b) 49.4 mJ  **71.** (a) 1.0 $\text{J/m}^3$; (b) $4.8\times 10^{-15}\,\text{J/m}^3$  **73.** (a) 1.67 mH; (b) 6.00 mWb  **75.** (b) have the turns of the two solenoids wrapped in opposite directions  **77.** 13 $\mu$H  **79.** (a) $(4.4\times10^7\,\text{m/s}^2)\hat{\text{i}}$; (b) 0; (c) $(-4.4\times10^7\,\text{m/s}^2)\hat{\text{i}}$  **81.** (a) 0.40 V; (b) 20 A  **83.** 1.15 W  **85.** (a) 2.0 A; (b) 0; (c) 2.0 A; (d) 0; (e) 10 V; (f) 2.0 A/s; (g) 2.0 A/s; (h) 1.0 A; (i) 3.0 A; (j) 10 V; (k) 0; (l) 0  **87.** 0.520 ms  **89.** 12 A/s  **91.** (a) 10 A; (b) $1.0\times10^2$ J  **93.** (a) 20 A/s; (b) 0.75 A  **95.** (a) 0; (b) $8.0\times10^2$ A/s; (c) 1.8 mA; (d) $4.4\times10^2$ A/s; (e) 4.0 mA; (f) 0  **97.** (a) 10 $\mu$T; (b) out; (c) 3.3 $\mu$T; (d) out  **99.** (a) 3.28 ms; (b) 6.45 ms; (c) infinite time; (d) 0; (e) 3.00 ms  **101.** (a) 400 A/s; (b) 200 A/s; (c) 0.600 A  **103.** 21 mA

## Chapter 31

**CP** **1.** (a) $T/2$; (b) $T$; (c) $T/2$; (d) $T/4$  **2.** (a) 5 V; (b) 150 $\mu$J  **3.** (a) remains the same; (b) remains the same  **4.** (a) $C, B, A$; (b) 1, $A$; 2, $B$; 3, $S$; 4, $C$; (c) $A$  **5.** (a) remains the same; (b) increases  **6.** (a) 1, lags; 2, leads; 3, in phase; (b) 3 ($\omega_d = \omega$ when $X_L = X_C$)  **7.** (a) increase (circuit is mainly capacitive; increase $C$ to decrease $X_C$ to be closer to resonance for maximum $P_{\text{avg}}$); (b) closer  **8.** (a) greater; (b) step-up   **Q** **1.** (a) $T/4$; (b) $T/4$; (c) $T/2$; (d) $T/2$  **3.** $b, a, c$  **5.** $c, b, a$  **7.** $a$ inductor; $b$ resistor; $c$ capacitor  **9.** (a) rightward, increase ($X_L$ increases, closer to resonance); (b) rightward, increase ($X_C$ decreases, closer to resonance); (c) rightward, increase ($\omega_d/\omega$ increases, closer to resonance)  **11.** (a) positive; (b) decreased (to decrease $X_L$ and get closer to resonance); (c) decreased (to increase $X_C$ and get closer to resonance)   **P** **1.** (a) 6.00 $\mu$s; (b) 167 kHz; (c) 3.00 $\mu$s  **3.** 45.2 mA  **5.** (a) 1.17 $\mu$J; (b) 5.58 mA  **7.** (a) 1.25 kg; (b) 372 N/m; (c) $1.75\times10^{-4}$ m; (d) 3.02 mm/s  **9.** $7.0\times10^{-4}$ s  **11.** (a) 275 Hz; (b) 365 mA  **13.** (a) 3.0 nC; (b) 1.7 mA; (c) 4.5 nJ  **15.** (a) 6.0; (b) 36 pF; (c) 0.22 mH  **17.** (a) 356 $\mu$s; (b) 2.50 mH; (c) 3.20 mJ  **19.** (a) 1.98 $\mu$J; (b) 5.56 $\mu$C; (c) 12.6 mA; (d) $-46.9°$; (e) $+46.9°$  **21.** (a) 0.180 mC; (b) 70.7 $\mu$s; (c) 66.7 W  **25.** 8.66 m$\Omega$

**29.** (a) 0.65 kHz; (b) 24 $\Omega$  **31.** (a) 95.5 mA; (b) 11.9 mA  **33.** (a) 6.73 ms; (b) 11.2 ms; (c) inductor; (d) 138 mH  **35.** (a) 267 $\Omega$; (b) $-41.5°$; (c) 135 mA  **37.** (a) 218 $\Omega$; (b) 23.4°; (c) 165 mA  **39.** (a) 206 $\Omega$; (b) 13.7°; (c) 175 mA  **41.** 7.61 A  **43.** 89 $\Omega$  **45.** (a) 224 rad/s; (b) 6.00 A; (c) 219 rad/s; (d) 228 rad/s; (e) 0.040  **47.** (a) yes; (b) 1.0 kV  **51.** (a) 796 Hz; (b) no change; (c) decreased; (d) increased  **53.** 1.84 A  **55.** (a) 12.1 $\Omega$; (b) 1.19 kW  **57.** (a) 0.743; (b) lead; (c) capacitive; (d) no; (e) yes; (f) no; (g) yes; (h) 33.4 W  **59.** (a) 117 $\mu$F; (b) 0; (c) 90.0 W; (d) 0°; (e) 1; (f) 0; (g) $-90°$; (h) 0  **61.** (a) 2.59 A; (b) 38.8 V; (c) 159 V; (d) 224 V; (e) 64.2 V; (f) 75.0 V; (g) 100 W; (h) 0; (i) 0  **63.** (a) 2.4 V; (b) 3.2 mA; (c) 0.16 A  **65.** (a) 1.9 V; (b) 5.9 W; (c) 19 V; (d) $5.9\times10^2$ W; (e) 0.19 kV; (f) 59 kW  **69.** (a) 39.1 $\Omega$; (b) 21.7 $\Omega$; (c) capacitive  **71.** (a) 45.0°; (b) 70.7 $\Omega$  **73.** (a) 0.689 $\mu$H; (b) 17.9 pJ; (c) 0.110 $\mu$C  **75.** (a) 2.41 $\mu$H; (b) 21.4 pJ; (c) 82.2 nC  **77.** (a) 64.0 $\Omega$; (b) 50.9 $\Omega$; (c) capacitive  **79.** (a) $-0.405$ rad; (b) 2.76 A; (c) capacitive  **81.** 1.84 kHz  **83.** (a) $0.577Q$; (b) 0.152  **87.** (a) 6.73 ms; (b) 2.24 ms; (c) capacitor; (d) 59.0 $\mu$F  **91.** (a) 165 $\Omega$; (b) 313 mH; (c) 14.9 $\mu$F  **93.** (a) 168 $\Omega$; (b) decrease; (c) decrease; (d) decrease  **95.** 7.08 mH  **97.** (a) 4.00 $\mu$F, 5.00 $\mu$F, 5.00 $\mu$F, 5.00 $\mu$F; (b) 1.78 kHz, 1.59 kHz, 1.59 kHz, 1.59 kHz; (c) 12.0 $\Omega$, 12.0 $\Omega$, 6.00 $\Omega$, 4.00 $\Omega$; (d) 19.8 $\Omega$, 22.4 $\Omega$, 19.9 $\Omega$, 19.4 $\Omega$; (e) 0.605 A, 0.535 A, 0.603 A, 0.619 A

## Chapter 32

**CP** **1.** $d, b, c, a$ (zero)  **2.** $a, c, b, d$ (zero)  **3.** tie of $b$, $c$, and $d$, then $a$  **4.** (a) 2; (b) 1  **5.** (a) away; (b) away; (c) less  **6.** (a) toward; (b) toward; (c) less   **Q** **1.** a, decreasing; b, decreasing  **3.** 1 $a$, 2 $b$, 3 $c$ and $d$  **5.** (a) $a$ and $b$ tie, then $c, d$; (b) none (because plate lacks circular symmetry, $\vec{B}$ not tangent to any circular loop); (c) none  **7.** (a) 1 up, 2 up, 3 down; (b) 1 down, 2 up, 3 zero  **9.** supplied  **11.** (a) 1, 3, 2; (b) 2   **P** **1.** $+3$ Wb  **3.** (a) 47.4 $\mu$Wb; (b) inward  **5.** $2.4\times10^{13}$ V/m·s  **7.** (a) 1.9 pT  **9.** (a) $1.18\times10^{-19}$ T; (b) $1.06\times10^{-19}$ T  **11.** (a) $5.01\times10^{-22}$ T; (b) $4.51\times10^{-22}$ T  **15.** $7.5\times10^5$ V/s  **17.** (a) 0.63 $\mu$T; (b) $2.3\times10^{12}$ V/m·s  **19.** (a) 0.71 A; (b) 0; (c) 2.8 A  **21.** (a) 2.0 A; (b) $2.3\times10^{11}$ V/m·s; (c) 0.50 A; (d) 0.63 $\mu$T·m  **23.** (a) 0.324 V/m; (b) $2.87\times10^{-16}$ A; (c) $2.87\times10^{-18}$  **25.** (a) 75.4 nT; (b) 67.9 nT  **27.** (a) 27.9 nT; (b) 15.1 nT  **29.** (a) 7.60 $\mu$A; (b) 859 kV·m/s; (c) 3.39 mm; (d) 5.16 pT  **31.** 55 $\mu$T  **33.** (a) $-9.3\times10^{-24}$ J/T; (b) $1.9\times10^{-23}$ J/T  **35.** (a) 0; (b) 0; (c) 0; (d) $\pm3.2\times10^{-25}$ J; (e) $-3.2\times10^{-34}$ J·s; (f) $2.8\times10^{-23}$ J/T; (g) $-9.7\times10^{-25}$ J; (h) $\pm3.2\times10^{-25}$ J  **37.** (b) $+x$; (c) clockwise; (d) $+x$  **39.** 20.8 mJ/T  **41.** yes  **43.** (b) $K_i/B$; (c) $-z$; (d) 0.31 kA/m  **47.** (a) 3.0 $\mu$T; (b) $5.6\times10^{-10}$ eV  **49.** $5.15\times10^{-24}$ A·$\text{m}^2$  **51.** (a) 0.14 A; (b) 79 $\mu$C  **53.** (a) $1.8\times10^2$ km; (b) $2.3\times10^{-5}$  **57.** (a) $6.3\times10^8$ A; (b) yes; (c) no  **59.** (a) 27.5 mm; (b) 110 mm  **61.** (a) 7; (b) 7; (c) $3h/2\pi$; (d) $3eh/4\pi m$; (e) $3.5h/2\pi$; (f) 8  **63.** (a) 9; (b) $3.71\times10^{-23}$ J/T; (c) $+9.27\times10^{-24}$ J; (d) $-9.27\times10^{-24}$ J  **65.** 0.84 kJ/T  **67.** (a) $-8.8\times10^{15}$ V/m·s; (b) $5.9\times10^{-7}$ T·m  **69.** (a) $(1.2\times10^{-13}$ T) exp$[-t/(0.012$ s)]; (b) $5.9\times10^{-15}$ T  **71.** 8.0 A  **73.** (b) $-x$; (c) counterclockwise; (d) $-x$  **75.** (b) sign is minus; (c) no, because there is compensating positive flux through open end nearer to magnet

## Chapter 33

**CP** **1.** (a) (Use Fig. 33-5.) On right side of rectangle, $\vec{E}$ is in negative $y$ direction; on left side, $\vec{E} + d\vec{E}$ is greater and in same direction; (b) $\vec{E}$ is downward. On right side, $\vec{B}$ is in negative $z$ direction; on left side, $\vec{B} + d\vec{B}$ is greater and in same direction. **2.** positive direction of $x$ **3.** (a) same; (b) decrease **4.** $a, d, b, c$ (zero) **5.** $a$ **Q 1.** (a) positive direction of $z$; (b) $x$ **3.** (a) and (b) $A = 1, n = 4, \theta = 30°$ **5.** (a) same; (b) increase; (c) decrease **7.** $a, b, c$ **9.** none **11.** $B$ **P 1.** (a) 515 nm; (b) 610 nm; (c) 555 nm; (d) $5.41 \times 10^{14}$ Hz; (e) $1.85 \times 10^{-15}$ s **3.** 7.49 GHz **5.** $5.0 \times 10^{-21}$ H **7.** 0.10 MJ **9.** (a) 6.7 nT; (b) $y$; (c) negative direction of $y$ **11.** 1.2 MW/m$^2$ **13.** (a) 6.7 nT; (b) 5.3 mW/m$^2$; (c) 6.7 W **15.** (a) 1.03 kV/m; (b) 3.43 $\mu$T **17.** (a) 87 mV/m; (b) 0.29 nT; (c) 6.3 kW **19.** $5.9 \times 10^{-8}$ Pa **21.** $1.0 \times 10^7$ Pa **25.** (a) $1.0 \times 10^8$ Hz; (b) $6.3 \times 10^8$ rad/s; (c) 2.1 m$^{-1}$; (d) 1.0 $\mu$T; (e) $z$; (f) $1.2 \times 10^2$ W/m$^2$; (g) $8.0 \times 10^{-7}$ N; (h) $4.0 \times 10^{-7}$ Pa **27.** 1.9 mm/s **29.** (a) $4.68 \times 10^{11}$ W; (b) any chance disturbance could move sphere from directly above source—the two force vectors no longer along the same axis **31.** (a) 0.17 $\mu$m; (b) toward the Sun **33.** 4.4 W/m$^2$ **35.** 3.1% **37.** (a) 1.9 V/m; (b) $1.7 \times 10^{-11}$ Pa **39.** 20° or 70° **41.** 0.67 **43.** (a) 2 sheets; (b) 5 sheets **45.** 1.48 **47.** 180° **49.** 1.26 **51.** 1.07 m **53.** (a) 56.9°; (b) 35.3° **57.** 182 cm **59.** (a) 26.8°; (b) yes **61.** (a) 1.39; (b) 28.1°; (c) no **63.** 23.2° **65.** (a) 48.9°; (b) 29.0° **67.** (a) $(1 + \sin^2 \theta)^{0.5}$; (b) $2^{0.5}$; (c) yes; (d) no **69.** 49.0° **71.** (c) 137.6°; (d) 139.4°; (e) 1.7° **73.** 1.22 **77.** (a) 0.50 ms; (b) 8.4 min; (c) 2.4 h; (d) 5446 B.C. **79.** (a) (16.7 nT) $\sin[(1.00 \times 10^6$ m$^{-1})z + (3.00 \times 10^{14}$ s$^{-1})t]$; (b) 6.28 $\mu$m; (c) 20.9 fs; (d) 33.2 mW/m$^2$; (e) $x$; (f) infrared **81.** (a) 83 W/m$^2$; (b) 1.7 MW **83.** (a) 3.5 $\mu$W/m$^2$; (b) 0.78 $\mu$W; (c) $1.5 \times 10^{-17}$ W/m$^2$; (d) $1.1 \times 10^{-7}$ V/m; (e) 0.25 fT **85.** 35° **87.** (a) 55.8°; (b) 55.5° **89.** (a) $z$ axis; (b) $7.5 \times 10^{14}$ Hz; (c) 1.9 kW/m$^2$ **91.** (a) white; (b) white dominated by red end; (c) no refracted light **93.** $1.5 \times 10^{-9}$ m/s$^2$ **95.** 0.034 **97.** $9.43 \times 10^{-10}$ T **99.** (a) $-y$; (b) $z$; (c) 1.91 kW/m$^2$; (d) $E_z = (1.20$ kV/m$) \sin[(6.67 \times 10^6$ m$^{-1})y + (2.00 \times 10^{15}$ s$^{-1})$ $t]$; (e) 942 nm; (f) infrared **101.** (a) 1.60; (b) 58.0°

**Chapter 34**

**CP 1.** $0.2d, 1.8d, 2.2d$ **2.** (a) real; (b) inverted; (c) same **3.** (a) $e$; (b) virtual, same **4.** virtual, same as object, diverging **Q 1.** (a) $a$ and $c$; (b) three times; (c) you **3.** (a) $a$; (b) $c$ **5.** $d$ (infinite), tie of $a$ and $b$, then $c$ **7.** convex **9.** (a) all but variation 2; (b) 1, 3, 4: right, inverted; 5, 6: left, same **P 1.** 40 cm **3.** 1.5 m **5.** 351 cm **7.** 10.5 cm **9.** (a) +36 cm; (b) −36 cm; (c) +3.0; (d) V; (e) NI; (f) opposite **11.** (a) +24 cm; (b) +36 cm; (c) −2.0; (d) R; (e) I; (f) same **13.** (a) −16 cm; (b) −4.4 cm; (c) +0.44; (d) V; (e) NI; (f) opposite **15.** (a) −20 cm; (b) −4.4 cm; (c) +0.56; (d) V; (e) NI; (f) opposite **17.** (b) 0.56 cm/s; (c) 11 m/s; (d) 6.7 cm/s **19.** (b) plus; (c) +40 cm; (e) −20 cm; (f) +2.0; (g) V; (h) NI; (i) opposite **21.** (a) convex; (b) −20 cm; (d) +20 cm; (f) +0.50; (g) V; (h) NI; (i) opposite **23.** (a) concave; (c) +40 cm; (e) +60 cm; (f) −2.0; (g) R; (h) I; (i) same **25.** (a) convex; (c) −60 cm; (d) +30 cm; (f) +0.50; (g) V; (h) NI; (i) opposite **27.** (b) −20 cm; (c) minus; (d) +5.0 cm; (e) minus; (f) +0.80; (g) V; (h) NI; (i) opposite **29.** (a) convex; (b) minus; (c) −60 cm; (d) +1.2 m; (e) −24 cm; (g) V; (h) NI; (i) opposite **31.** (a) concave; (b) +8.6 cm; (c) +17 cm; (e) +12 cm; (f) minus; (g) R; (i) same **33.** (a) 2.00; (b) none **35.** (c) −33 cm; (e) V; (f) same **37.** (d) −26 cm; (e) V; (f) same **39.** (c) +30 cm; (e) V; (f) same **41.** (a) 45 mm; (b) 90 mm **43.** 1.86 mm **45.** 5.0 mm **47.** 22 cm **49.** (a) +40 cm; (b) ∞ **51.** (a) −4.8 cm; (b) +0.60; (c) V; (d) NI; (e) same **53.** (a) −48 cm; (b) +4.0; (c) V; (d) NI; (e) same **55.** (a) +36 cm; (b) −0.80; (c) R; (d) I; (e) opposite **57.** (a) −8.6 cm; (b) +0.39;

(c) V; (d) NI; (e) same **59.** (a) −30 cm; (b) +0.86; (c) V; (d) NI; (e) same **61.** (a) +55 cm; (b) −0.74; (c) R; (d) I; (e) opposite **63.** (a) +84 cm; (b) −1.4; (c) R; (d) I; (e) opposite **65.** (a) −18 cm; (b) +0.76; (c) V; (d) NI; (e) same **67.** (a) −7.5 cm; (b) +0.75; (c) V; (d) NI; (e) same **69.** (a) D; (b) −5.3 cm; (d) −4.0 cm; (f) V; (g) NI; (h) same **71.** (a) C; (b) plus; (d) −13 cm; (e) +1.7; (f) V; (g) NI; (h) same **73.** (a) C; (d) −10 cm; (e) +2.0; (f) V; (g) NI; (h) same **75.** (a) C; (b) +80 cm; (d) −20 cm; (f) V; (g) NI; (h) same **77.** (a) C; (b) +3.3 cm; (d) +5.0 cm; (f) R; (g) I; (h) opposite **79.** (a) D; (b) minus; (d) −3.3 cm; (e) +0.67; (f) V; (g) NI **81.** (a) +3.1 cm; (b) −0.31; (c) R; (d) I; (e) opposite **83.** (a) +24 cm; (b) +6.0; (c) R; (d) NI; (e) opposite **85.** (a) −5.5 cm; (b) +0.12; (c) V; (d) NI; (e) same **87.** (a) −4.6 cm; (b) +0.69; (c) V; (d) NI; (e) same **89.** (a) 13.0 cm; (b) 5.23 cm; (c) −3.25; (d) 3.13; (e) −10.2 **91.** (a) 3.5; (b) 2.5 **93.** (a) 2.35 cm; (b) decrease **95.** (a) 3.00 cm; (b) 2.33 cm **97.** (a) 6.0 mm; (b) 1.6 kW/m$^2$; (c) 4.0 cm **99.** (a) 20 cm; (b) 15 cm **101.** 2.2 mm$^2$ **107.** (a) 40 cm; (b) 20 cm; (c) −40 cm; (d) 40 cm **109.** (a) +36 cm; (b) 1.2 cm; (c) real; (d) inverted **111.** (a) −50 cm; (b) 5.0; (c) virtual; (d) inverted **113.** (a) +8.6 cm; (b) +2.6; (c) R; (d) NI; (e) opposite **115.** (a) +7.5 cm; (b) −0.75; (c) R; (d) I; (e) opposite **117.** (a) +24 cm; (b) −0.58; (c) R; (d) I; (e) opposite **119.** (a) 0.60 m; (b) +0.20; (c) real; (d) left; (e) not inverted **125.** (a) 8.0 cm; (b) 16 cm; (c) 48 cm **127.** (a) $\alpha = 0.500$ rad: 7.799 cm; $\alpha = 0.100$ rad: 8.544 cm; $\alpha = 0.0100$ rad: 8.571 cm; mirror equation: 8.571 cm; (b) $\alpha = 0.500$ rad: −13.56 cm; $\alpha = 0.100$ rad: −12.05 cm; $\alpha = 0.0100$ rad: −12.00 cm; mirror equation: −12.00 cm **129.** 42 mm **131.** (b) $P_n$ **133.** (a) $(0.5)(2 - n)r/(n - 1)$; (b) right **135.** 2.67 cm **137.** (a) 3.33 cm; (b) left; (c) virtual; (d) not inverted

**Chapter 35**

**CP 1.** $b$ (least $n$), $c, a$ **2.** (a) top; (b) bright intermediate illumination (phase difference is 2.1 wavelengths) **3.** (a) $3\lambda$, 3; (b) $2.5\lambda, 2.5$ **4.** a and d tie (amplitude of resultant wave is $4E_0$), then b and c tie (amplitude of resultant wave is $2E_0$) **5.** (a) 1 and 4; (b) 1 and 4 **Q 1.** (a) $2d$; (b) (odd number)$\lambda/2$; (c) $\lambda/4$ **3.** (a) peak; (b) valley **5.** (a) decrease; (b) decrease; (c) decrease; (d) blue **7.** (a) intermediate closer to maximum, $m = 2$; (b) minimum, $m = 3$; (c) intermediate closer to maximum, $m = 2$; (d) maximum, $m = 1$ **9.** (a) maximum; (b) minimum; (c) alternates **11.** $c, d$ **P 1.** 1.56 **3.** $4.55 \times 10^7$ m/s **5.** (a) 3.60 $\mu$m; (b) intermediate closer to fully constructive **7.** (a) 155 nm; (b) 310 nm **9.** (a) 1.70; (b) 1.70; (c) 1.30; (d) all tie **11.** (a) 1.55 $\mu$m; (b) 4.65 $\mu$m **13.** (a) 0.833; (b) intermediate closer to fully constructive **15.** 16 **17.** 648 nm **19.** 2.25 mm **21.** 0 **23.** 7.88 $\mu$m **25.** 72 $\mu$m **27.** 6.64 $\mu$m **29.** 2.65 **31.** (17.1 $\mu$V/m) $\sin[(2.0 \times 10^{14}$ rad/s$)t]$ **33.** 27 $\sin(\omega t + 8.5°)$ **35.** 70.0 nm **37.** (a) 0.117 $\mu$m; (b) 0.352 $\mu$m **39.** 120 nm **41.** 509 nm **43.** 478 nm **45.** 560 nm **47.** 409 nm **49.** 161 nm **51.** 273 nm **53.** (a) 552 nm; (b) 442 nm **55.** 338 nm **57.** 329 nm **59.** 248 nm **61.** 339 nm **63.** 528 nm **65.** 455 nm **67.** 608 nm **69.** 1.89 $\mu$m **71.** 140 **73.** 0.012° **75.** $[(m + \frac{1}{2})\lambda R]^{0.5}$, for $m = 0, 1, 2, \ldots$ **77.** 1.00 m **79.** 588 nm **81.** 1.00030 **83.** (a) 22°; (b) refraction reduces $\theta$ **85.** (a) 50.0 nm; (b) 36.2 nm **87.** $x = (D/2a)(m + 0.5)\lambda$, for $m = 0, 1, 2, \ldots$ **89.** 0.23° **91.** 450 nm **93.** (a) 1.75 $\mu$m; (b) 4.8 mm **95.** 33 $\mu$m **97.** (a) 1500 nm; (b) 2250 nm; (c) 0.80 **99.** 600 nm **101.** (a) 88%; (b) 94% **103.** $I_m \cos^2(2\pi x/\lambda)$ **105.** 8.0 $\mu$m

**107.** (a) 1.80 $\mu$m; (b) 9 **109.** 51.6 ns **111.** (a) 1.6 rad; (b) 0.79 rad **113.** (a) dark; (b) dark; (c) 4 **115.** (a) 1; (b) 4.0 fs; (c) 7.5 **117.** (a) 48.6°; (b) away; (c) 1.49 m **119.** (a) 42.0 ps; (b) 42.3 ps; (c) 43.2 ps; (d) 41.8 ps; (e) 4 **121.** $I_0[1 + 8\cos^2(\phi/2)]$, with $\phi = (2\pi d/\lambda)\sin\theta$

## Chapter 36

**CP** **1.** (a) expand; (b) expand **2.** (a) second side maximum; (b) 2.5 **3.** (a) red; (b) violet **4.** diminish **5.** (a) left; (b) less **Q** **1.** (a) $A, B, C$; (b) $A, B, C$ **3.** (a) 1 and 3 tie, then 2 and 4 tie; (b) 1 and 2 tie, then 3 and 4 tie **5.** (a) $m = 5$ minimum; (b) (approximately) maximum between the $m = 4$ and $m = 5$ minima **7.** (a) decrease; (b) same; (c) remain in place **9.** (a) larger; (b) red **11.** (a) $A$; (b) left; (c) left; (d) right **P** **1.** (a) 700 nm; (b) 4; (c) 6 **3.** 60.4 $\mu$m **5.** (a) 70 cm; (b) 1.0 mm **7.** (a) 2.5 mm; (b) $2.2 \times 10^{-4}$ rad **9.** 1.77 mm **11.** (a) 0.18°; (b) 0.46 rad; (c) 0.93 **13.** 160° **15.** (b) 0; (c) $-0.500$; (d) 4.493 rad; (e) 0.930; (f) 7.725 rad; (g) 1.96 **17.** (d) 52.5°; (e) 10.1°; (f) 5.06° **19.** $1.6 \times 10^3$ km **21.** (a) $1.3 \times 10^{-4}$ rad; (b) 10 km **23.** 50 m **25.** (a) $1.1 \times 10^4$ km; (b) 11 km **27.** (a) 19 cm; (b) larger **29.** (a) 0.346°; (b) 0.97° **31.** (a) 17.1 m; (b) $1.37 \times 10^{-10}$ **33.** (a) $8.8 \times 10^{-7}$ rad; (b) $8.4 \times 10^7$ km; (c) 0.025 mm **35.** 3 **37.** 5 **39.** (a) 9; (b) 0.255 **41.** (a) 5.0 $\mu$m; (b) 20 $\mu$m **43.** (a) $7.43 \times 10^{-3}$; (b) between the $m = 6$ minimum (the seventh one) and the $m = 7$ maximum (the seventh side maximum); (c) between the $m = 3$ minimum (the third one) and the $m = 4$ minimum (the fourth one) **45.** 3 **47.** (a) 62.1°; (b) 45.0°; (c) 32.0° **49.** (a) 2.1°; (b) 21°; (c) 11 **51.** (a) 6.0 $\mu$m; (b) 1.5 $\mu$m; (c) 9; (d) 7; (e) 6 **53.** $1.09 \times 10^3$ rulings/mm **55.** (a) 470 nm; (b) 560 nm **57.** 0.15 nm **59.** (a) 0.032°/nm; (b) $4.0 \times 10^4$; (c) 0.076°/nm; (d) $8.0 \times 10^4$; (e) 0.24°/nm; (f) $1.2 \times 10^5$ **61.** $3.65 \times 10^3$ **63.** (a) 10 $\mu$m; (b) 3.3 mm **65.** (a) 25 pm; (b) 38 pm **67.** 0.26 nm **69.** (a) 0.17 nm; (b) 0.13 nm **71.** (a) $0.7071a_0$; (b) $0.4472a_0$; (c) $0.3162a_0$; (d) $0.2774a_0$; (e) $0.2425a_0$ **73.** (a) 15.3°; (b) 30.6°; (c) 3.1°; (d) 37.8° **75.** 4.7 cm **77.** (a) 625 nm; (b) 500 nm; (c) 416 nm **79.** 3.0 mm **81.** (a) 13; (b) 6 **83.** 59.5 pm **85.** 4.9 km **87.** $1.36 \times 10^4$ **89.** 36 cm **97.** 2 **99.** (a) fourth; (b) seventh **103.** (a) 2.4 $\mu$m; (b) 0.80 $\mu$m; (c) 2 **107.** 9

## Chapter 37

**CP** **1.** (a) same (speed of light postulate); (b) no (the start and end of the flight are spatially separated); (c) no (because his measurement is not a proper time) **2.** (a) Eq. 2; (b) $+0.90c$; (c) 25 ns; (d) $-7.0$ m **3.** (a) right; (b) more **4.** (a) equal; (b) less **Q** **1.** (a) $C_1'$; (b) $C_1'$ **3.** (a) 4 s; (b) 3 s; (c) 5 s; (d) 4 s; (e) 10 s **5.** $c$ **7.** $b$ **9.** (a) 3, tie of 1 and 2, then 4; (b) 4, tie of 1 and 2, then 3; (c) 1, 4, 2, 3 **11.** (a) a tie of 3, 4, and 6, then a tie of 1, 2, and 5; (b) 1, then a tie of 2 and 3, then 4, then a tie of 5 and 6; (c) 1, 2, 3, 4, 5, 6; (d) 2 and 4; (e) 1, 2, 5 **P** **1.** 0.990 50 **3.** 0.446 ps **5.** $2.68 \times 10^3$ y **7.** (a) 0.999 999 50 **9.** 1.32 m **11.** (a) 87.4 m; (b) 394 ns **13.** (a) 0.999 999 15; (b) 30 ly **15.** (a) 26.26 y; (b) 52.26 y; (c) 3.705 y **17.** (a) 25.8 $\mu$s; (b) small flash **19.** (a) 138 km; (b) $-374$ $\mu$s **21.** (a) 1.25; (b) 0.800 $\mu$s **23.** (a) 0.480; (b) negative; (c) big flash; (d) 4.39 $\mu$s **25.** (a) $\gamma[1.00\ \mu s - \beta(400\ m)/(2.998 \times 10^8\ m/s)]$; (d) 0.750; (e) $0 < \beta < 0.750$; (f) $0.750 < \beta < 1$; (g) no **27.** (a) 0.35; (b) 0.62 **29.** $0.81c$ **31.** (a) 1.25 y; (b) 1.60 y; (c) 4.00 y **33.** 1.2 $\mu$s **35.** $0.13c$ **37.** 22.9 MHz **39.** (a) 550 nm; (b) yellow **41.** (a) 1.0 keV; (b) 1.1 MeV **43.** (a) 196.695; (b) 0.999 987 **45.** $2.83mc$ **47.** (a) 0.222 cm; (b) 701 ps; (c) 7.40 ps **49.** (a) 0.707; (b) 1.41; (c) 0.414 **51.** 18 smu/y **53.** $1.01 \times 10^7$ km **55.** (a) $\gamma(2\pi m/|q|B)$; (b) no; (c) 4.85 mm; (d) 15.9 mm; (e) 16.3 ps; (f) 0.334 ns **57.** 110 km **59.** (a) 2.08 MeV; (b) $-1.21$ MeV **61.** (a) $vt \sin\theta$; (b) $t[1 - (v/c)\cos\theta]$; (c) $3.24c$ **63.** (d) 0.801 **65.** (a) 1.93 m; (b) 6.00 m; (c) 13.6 ns; (d) 13.6 ns; (e) 0.379 m; (f) 30.5 m; (g) $-101$ ns; (h) no; (i) 2; (k) no; (l) both **69.** (b) $+0.44c$ **71.** $8.7 \times 10^{-3}$ ly **73.** 2.46 MeV/$c$ **75.** 189 MeV **77.** (a) $5.4 \times 10^4$ km/h; (b) $6.3 \times 10^{-10}$ **79.** $0.27c$ **81.** (a) 5.71 GeV; (b) 6.65 GeV; (c) 6.58 GeV/$c$; (d) 3.11 MeV; (e) 3.62 MeV; (f) 3.59 MeV/$c$ **83.** (a) 256 kV; (b) $0.745c$ **85.** 7 **87.** $0.95c$

Figures are noted by page numbers in *italics*, tables are indicated by t following the page number.